Encyclopaedia of

GENETICS

Volume 1

Encyclopaedia of GENETICS

Volume 1

Editor

Eric C.R. Reeve

Assistant Editor

Isobel Black

PENTAGON PRESS

Encyclopedia of GENETICS / Eric C.R. Reeve (Editor)

ISBN 978-81-8274-241-2 (Set)

2010

Published with arrangement FITZROY DEARBORN PUBLISHERS, USA.

Published by

PENTAGON PRESS
206, Peacock Lane, Shahpur Jat, New Delhi-110049
Phones: 011-64706243, 26491568
Telefax: 011-26490600
email: rajan@pentagonpress.in
website: www.pentagonpress.in

Printed at
New Elegant Printers, New Delhi-110064

Dedicated to:

My Wife EDITH SIMON (Mrs Edith Reeve)

Our Children ANTONIA, SIMON and JESSICA REEVE

Our Granddaughter INDIGO ANDERSON REEVE

And the GENES WE SHARE

CONTENTS

CONTRIBUTORS

Abbott, Cathy. Lecturer, Medical Genetics Section, Department of Medical Genetics, University of Edinburgh, Western General Hospital, UK. Essay: Trinucleotide repeat instability as a cause of human genetic disease

Alston, Frank H. Emeritus Fellow, Plant Breeding and Biotechnology department, Horticulture Research International – East Malling, Kent, UK. Essays: Apples: the genetics of resistance to diseases and insect pests, Apples: the genetics of fruit quality

Amyes, Sebastian G.B. Head of Department, Department of Medical Microbiology, The Medical School, University of Edinburgh, UK. Essay: Antibiotic resistanc: theory into practice

Arkhipova, Irina. Staff Scientist, Department of Molecular and Cellular Biology, Harvard University, Cambridge, Massachusetts, USA. Essay: Retrotransposons

Arnott, David E. Senior Lecturer, Institute of Cell, Animal and Population Biology, University of Edinburgh, UK. Essay: Molecular biology of malaria parasites

Austin, Jeremy J. Research Fellow, Department of Palaeontology and Department of Zoology, The Natural History Museum, London, UK. Essay: DNA from museum specimens

Barrett, John. Department of Genetics, University of Cambridge, UK. Essay: Environmental risk assessment and transgenic plants

Barritt, Suzanne M. Armed Forces DNA Identification Laboratory, Armed Forces Institute of Pathology, Rockville, USA. Essay: Forensic DNA analysis of the last Russian royal family

Barton, James C. Medical Director, Southern Iron Disorders Center, Birmingham, Alabama, USA. Essay: Haemochromatosis

Beale, Geoffrey H. Retired research worker, Institute of Cell, Animal and Population Biology, University of Edinburgh, UK. Essay: Genetics of malaria parasites

Beggs, Jean. Professor, Wellcome Trust Centre for Cell Biology, University of Edinburgh, UK. Essay: Introns

Bond, Jeff. Lecturer, Institute of Cell and Molecular Biology, University of Edinburgh, UK. Essays: Aneuploidy, Fungal genetics, Genetics for newcomers

Bopp, Daniel. Lecturer, Zoological Institute, University of Zürich, Switzerland. Essay: Sex determination in Drosophila

Borghi, Basilio. Director, Research Centre, Department of Agriculture in San Michele All' Adige (TN), Italy. Essay: Pinpointing einkorn wheat domestication by DNA fingerprinting

Bowers, John E. Plant Gene Mapping Laboratory, University of Georgia, USA. Essay: Winegrape origins revealed by DNA marker analysis

Bownes, Mary. Professor, Institute of Cell and Molecular Biology, University of Edinburgh, UK. Essay: Development of the egg and early embryo of Drosophila

Brem, Gottfried. Institute for Animal Breeding and Genetics, Veterinary University Vienna, Austria. Essay: Large transgenic animals – their making and their use

Brown, Susan C. Lecturer in Muscle Pathology, Dubowitz Neuromuscular Centre, Imperial College, School of Medicine, UK. Essay: Genetics of Duchenne muscular dystrophy

Brookfield, John F.Y. Reader, Institute of Genetics, University of Nottingham, UK. Essays: DNA fingerprinting, Mitochondrial "Eve"

Buis, Reinoud C. Associate Professor, Department of Animal Sciences, Animal Breeding and Genetics Group, Wageningen University, The Netherlands. Essay: Body weight limits in mice

Bünger, Lutz. Honorary Research Fellow, Institute of Cell, Animal and Population Biology, University of Edinburgh, UK. Essay: Body weight limits in mice

Buratovich, Michael. Assistant Professor of Biology, Spring Arbor College, Michigan, USA. Essay: Genetic dissection of wing patterning in *Drosophila melanogaster*

Cahill, Mary R. Consultant Haematologist, Mid-Western Regional Hospital, Limerick and University College Cork, Republic of Ireland. Essay: Haemophilia: patients, genes and therapy

Carlson, Karin. Professor, Department of Cell and Molecular Biology, University of Uppsala, Sweden. Essay: Bacteriophages

Charlesworth, Deborah. Research Fellow, Institute of Cell, Animal and Population Biology, University of Edinburgh, Edinburgh, UK. Essay: Plant self-incompatibility, Separation of the sexes and sex chromosomes in plants

Clarke, Alan R. Professor, Cardiff School of Biosciences, University of Cardiff, UK. Essay: Gene targeting

Clayton-Smith, Jill. Honorary Senior Lecturer, University Department of Medical Genetics, St.Mary's Hospital, Manchester, UK. Essays: Genetics of mental retardation

Colvin, Brian T. Senior Lecturer and Honorary Consultant Haematologist, Royal Hospitals Trust and St Bartholemews and the Royal London Scool of Medicine and Dentistry, London, UK. Essay: Haemophilia: patients, genes and therapy

Cook, Kevin R. Assistant Scientist, Department of Biology, Indiana University, USA. Essay: Genetic information online: a brief introduction to FlyBase and other organismal databases

Couble, Pierre. Centre of Molecular and Cellular Genetics, CNRS, Villeurbanne, France. Essay: The silkworm, *Bombyx mori*, a model genetic system

Cox, Brian S. Visiting Professor, School of Biological Sciences, University of Kent at Canterbury, UK. Essay: Fungal prions

Crow, James F. Professor Emeritus, Genetics Department, University of Wisconsin-Madison, USA. Essay: Motoo Kimura and the neutral theory

Crusio, Wim E. Professor, Brudnick Neuropsychiatric Research Institute, Department of Psychiatry, University of Massachusetts Medical School, Worcester, Massachusetts USA. Essay: Behavioural and neural genetics of the mouse

Davies, Julian. Professor Emeritus, University of British Columbia, Vancouver, B.C., Canada. Essay: Antibiotic resistance in bacteria

Davis, Allan P. Mouse Genome Informatics, The Jackson Laboratory, Maine, USA. Essay: Ethylnitrosourea (ENU) as a genetic tool in mouse mutagenesis

Dickson, George. Professor and Chair of Molecular Cell Biology, School of Biological Sciences, Royal Holloway College – University of London, UK. Essay: Genetics of Duchenne muscular dystrophy

Dix, Ian. Senior Bioinformatics Specialist, AstraZeneca, Macclesfield, UK. Essay: Introns

Donachie, William D. Professor, Institute of Cell and Molecular Biology, University of Edinburgh, UK. Essay: *Escherichia coli*: genes and the cell cycle

Drew, D.R. Department of Microbiology and Immunology, University of Melbourne, Australia. Essay: DNA vaccination

Dübendorfer, Andreas. Professor, Institute of Zoology, University of Zürich, Switzerland. Essay: Genetic control of sex determination in the housefly

Dudley, John. Professor, Plant Genetics and Associate Head, Department of Crop Sciences, University of Illinois at Urbana-Champaign, USA. Essay: Long term selection for oil and protein in maize

Edwards, A.W.F. Reader in Biometry, University of Cambridge, UK. Essay: Darwin and Mendel united

Ehrman, Lee. Distinguished Professor of Biology, State University of New York, USA. Essays: Handedness: a look at laterality, Reproductive isolating mechanisms

Elder, George H. Professor of Medical Biochemistry, University of Wales College of Medicine, Cardiff, UK. Essay: Porphyrias

Elgar, Greg. Group Leader, UK HGMP Resource Centre, Wellcome Trust Genome Campus, Cambridgeshire, UK. Essay: The pufferfish as a model organism for comparative vertebrate genome analysis

Eley, Thalia C. Lecturer and MRC Fellow, Social, Genetic and Developmental Psychiatry Research Centre, Institute of Psychiatry, King's College London, UK. Essay: Genetics of human behaviour

Ellis, Noel. John Innes Centre, Norwich Research Park, UK. Essay: Pea genetics

Eppig, J.T. Staff Scientist, The Jackson Laboratory, Bar Harbor, Maine, USA. Essay: Comparative mapping: tracking gene homologues among mammals

Evans, K.M. Research Scientist, Plant Breeding and Biotechnology department, Horticulture Research International – East Malling, Kent, UK. Essay: Apples: the genetics of resistance to diseases and insect pests, Apples: the genetics of fruit quality

Ewan, Kenneth B.R. Postdoctoral Fellow, Life Sciences Division, Lawrence Berkeley National Laboratory, Berkeley, USA. Essay: The role of *Pax-6* homologues in the development of the eye and other tissues during the embryogenesis of vertebrates and invertebrates

Fincham, John S. Retired. Honorary Fellow, Institute of Cell and Molecular Biology, University of Edinburgh, and Emeritus Professor, University of Cambridge, UK. Essays: Barbara McClintock and transposable genetic sequences in maize (*Zea mays*), Charles Darwin and the theory of evolution

Foster, Eugene. Professor Emeritus, Department of Pathology, School of Medicine, Tufts University, Boston, USA. Essay: Y-chromosomal DNA analysis and the Jefferson–Hemings controversy

Frankham, Richard. Professor, Key Centre for Biodiversity and Bioresources, Department of Biological Sciences, Macquarie University, Australia. Essay: Conservation genetics

French, Vernon. Institute of Cell, Animal and Population Biology, University of Edinburgh, UK. Essay: Development of the egg and early embryo of Drosophila

Galinat, Walton C. Professor Emeritus, Department of Plant and Soil Sciences, University of Massachusetts, Amherst, USA. Essay: Origin of maize from teosinte by applied genetics

Gillham, Nicholas W. Professor, Developmental, Cell and Molecular Biology Group, Departments of Botany and Zoology, Duke University, North Carolina, USA. Essays: Chloroplasts and mitochondria, Francis Galton

Gregersen, Peter K. Professor of Medicine and Pathology, New York University School of Medicine, and Chief, Division of Biology and Human Genetics, North Shore University Hospital, Manhasset, USA. Essay: Absolute pitch

Gruss, Peter. Department of Molecular Cell Biology, Max Planck Institute of Biophysical Chemistry, Göttingen, Germany. Essay: The role of *Pax-6* homologues in the development of the eye and other tissues during the embryogenesis of vertebrates and invertebrates

Hall, Judith. Professor, Departments of Pediatrics and Medical Genetics, University of British Columbia, Canada. Essay: Twinning

Hardison, Ross. Professor, Department of Biochemistry and Molecular Biology, Pennsylvania State University, USA. Essay: Evolution of the haemoglobins

Harris, S.A. Druce Curator of the Oxford University Herbaria and Xylarium, Department of Plant Sciences, University of Oxford, UK. Essay: Origin of the apple (*Malus domestica* Borkh.)

Heun, Manfred. Professor, Department of Biology and Nature Conservation, Agricultural University of Norway (NLH), Norway. Essay: Pinpointing einkorn wheat domestication by DNA fingerprinting

Hill, Anne. Postdoctoral Scientist, Moredun Research Institute, International Research Centre, Pentlands Science Park, Midlothian, UK. Essays: *Escherichia coli* O157:H7, Influenza, Meningitis

Hodgins, G.W.L. Research Laboratory for Archaeology and the History of Art, University of Oxford, UK. Essay: Neanderthals

Hunter, Nora. Head of Scrapie Sheep Genetics Group, Institute for Animal Health, Edinburgh, UK. Essay: Genetics of animal and human prion diseases

Hurst, Laurence. Department of Biology and Biochemistry, University of Bath, UK. Essay: Selfish genes

Jackson, Ian J. Senior Scientist, MRC Human Genetics Unit, Western General Hospital, Edinburgh, UK. Essay: Coat colour genetics of mouse, man and other animals

Jones, D.T. Professor of Bioinformatics, Department of Biological Sciences, Brunel University, Middlesex, UK. Essay: The role of bioinformatics in the postsequencing phase of genomics

Juniper, B.E. Reader Emeritus in Plant Sciences, University of Oxford, UK. Essay: Origin of the apple (*Malus domestica* Borkh.)

Justice, Prof. Monica J. Department of Molecular and Human Genetics, Baylor College of Medicine, Houston, Texas, USA. Essay: Ethylnitrosourea (ENU) as a genetic tool in mouse mutagenesis

Kalatzis, Vasiliki. Research Scientist, Unité de Génétique des Déficits Sensoriels, Institut Pasteur, France. Essay: Hereditary hearing loss

Kanowski, Peter. Professor of Forestry and Head of Department of Forestry, Australian National University. Essay: Trees: still challenging breeders at the millennium

Khush, Gurdev S. Head, Department of Plant Breeding, Genetics and Biochemistry, International Rice Research Institute, Philippines. Essay: Super rice for increasing the genetic yield potential

Kidwell, Margaret G. Regents' Professor, Department of Ecology and Evolutionary Biology, University of Arizona, USA. Essay: Hybrid dysgenesis determinants and other useful transposable elements in Drosophila

Kipling, David. Senior Lecturer, Department of Pathology, University of Wales College of Medicine, Cardiff, Wales. Essay: Telomeres

Klimenko, V. Institute of Animal Science, Kharkov, Ukraine. Essay: The silkworm, *Bombyx mori*, a model genetic system

Kövér, Szilvia. Lecturer, Department of Plant Taxonomy and Ecology, Eötvös University, Budapest, Hungary. Essay: Evolution of sex

Lansdell, Katherine A. Research Scientist, Department of Pharmacology, Quintiles Scotland Ltd, Edinburgh, UK. Essay: Cystic fibrosis: therapy based on a molecular understanding of CFTR

Laurie, Graeme T. Lecturer in Law, Faculty of Law, University of Edinburgh, UK. Essay: Law, ethics and genetics

Leach, David R.F. Reader, Institute of Cell and Molecular Biology, University of Edinburgh, UK. Essay: Genetic recombination

Lindsey, Gillian. Fitzroy Dearborn Publishers, London, UK. Essay: Human Genome Project

Lithgow, Gordon J. Biological Gerontology Group, School of Biological Sciences, University of Manchester, UK. Essay: Genetics of ageing

Lovejoy, Elizabeth A. Department of Veterinary Pathology, University of Edinburgh, UK. Essay: Gene targeting

Lyon, Mary F. Scientist, MRC Mammalian Genetics Unit, Harwell, Didcot, UK. Essay: X-chromosome inactivation

Masters, Millicent. Reader, Institute of Cell and Molecular Biology, University of Edinburgh, Edinburgh, UK. Essay: *Escherichia coli*: Genome, genetic exchange, genetic analysis

Matthews, Kathleen A. Associate Scientist, Department of Biology, Indiana University, USA. Essay: Genetic information online: a brief introduction to FlyBase and other organismal databases

Maule, John. Senior Scientific Officer, Comparative and Developmental Genetics Section, MRC Human Genetics Unit, University of Edinburgh, UK. Essay: Techniques in molecular genetics

McClearn, Gerald E. Director, Centre for Developmental and Health Genetics, Biobehavioral Health Department, Pennsylvania State University, USA. Essay: The mouse as model for the study of genetic aspects of alcoholism

McGuffin, Peter. Professor and Director of Psychiatric Genetics, MRC Social, Genetic and Developmental Psychiatry Research Centre, Institute of Psychiatry, King's College London, UK. Essay: Genetics of psychiatric disorders

McGuigan, Fiona E.A. Research Assistant, Department of Medicine and Therapeutics, University of Aberdeen, UK. Essay: Genetics of osteoporosis

Meredith, Carole P. Professor, Department of Viticulture and Enology, University of California at Davis, USA. Essay: Winegrape origins revealed by DNA marker analysis

Mitchison, J. Murdoch. Professor, Institute of Cell, Animal and Population Biology, University of Edinburgh, Edinburgh, UK. Essay: Cell cycle genes

Moore, Adrian W. Research Fellow, Howard Hughes Medical Institute, University of California at San Francisco, California, USA. Essay: The Wilms' tumour 1 gene *WT1*: a tool to examine the links between cancer and development

Moore, Graham. Cereals Research Department, John Innes Centre, Colney, Norwich, UK. Essay: Cereal chromosome evolution and pairing

Müller, Mathias. Professor, Institute for Animal Breeding and Genetics, Veterinary University Vienna, Austria. Essay: Large transgenic animals – their making and their use

Nagaraju, J.W. Laboratory of Molecular Genetics, Centre for DNA Fingerprinting and Diagnostics, Nacharacham, Hyderabad, India. Essay: The silkworm, *Bombyx mori*, a model genetic system

Nasir, Jamal. MRC Research Fellow, Division of Genomic Medicine, University of Sheffield, Royal Hallamshire Hospital, Sheffield, UK. Essay: Genetics of Huntington's disease: when more is less

Neel, James V. Formerly, Professor, Department of Human Genetics, The University of Michigan Medical School, Ann Arbor, Michigan, USA. Essay: Genetic effects of the atomic bombs

Nelson, Oliver E. Professor of Genetics, Emeritus, Laboratory of Genetics, University of Wisconsin, USA. Essay: Maize: the long trail to QPM

Novelli, Giuseppe. Professor, Department of Biopathology and Diagnostic Imaging, Tor Vergata University, Rome, Italy. Essay: Genetics of baldness

Old, John. Reader in Haematology, Director, The National Haemoglobinopathy Reference Laboratory, The Institute of Molecular Medicine, University of Oxford John Radcliffe Hospital, UK. Essay: Haemoglobinopathies

Orias, Eduardo. Department of Molecular, Cellular and Developmental Biology, University of California at Santa Barbara, USA. Essay: Genetics of *Tetrahymena thermophila*

Ostedgaard, Lynda S. Associate Research Scientist, Department of Internal Medicine, College of Medicine, University of Iowa, USA. Essay: Cystic fibrosis: therapy based on a molecular understanding of CFTR

Owen, Michael J. Professor and Head of Department, Department of Psychological Medicine, University of Wales College of Medicine, UK. Essay: Genetics of psychiatric disorders

Parsons, Thomas J. Chief Scientist, Armed Forces DNA Identification Laboratory, Armed Forces Institute of Pathology, Rockville, USA. Essay: Forensic DNA analysis of the last Russian royal family

Perelle, Ira B. Professor, Departments of Psychology, and Business and Economics, Mercy College, Dobbs Ferry, New York, USA. Essay: Handedness: a look at laterality

Petit, Christine. Head of Department, Unité de Génétique des Déficits Sensoriels, Institut Pasteur, France. Essay: Hereditary hearing loss

Pettitt, Paul B. Research Laboratory for Archaeology and the History of Art, University of Oxford, UK. Essay: Neanderthals

Pickard, Ben. Post-doctoral scientist, Medical Genetics Section, Molecular Medicine Centre, University of Edinburgh, UK. Essay: Genomic imprinting

Plomin, Robert. Professor and Deputy Director, Social, Genetic and Developmental Psychiatry Research Centre, Institute of Psychiatry, King's College London. Essay: Genetics of human behaviour

Preer, John R. Jr. Distinguished Professor Emeritus, Department of Biology, Indiana University, USA. Essay: Genetics of Paramecium

Promislow, Daniel E.L. Associate Professor, Department of Genetics, University of Georgia, USA. Essay: Genetics of ageing

Raeburn, Sandy. Professor, Centre for Medical Genetics, University of Nottingham, UK. Essay: Genetic aspects of bioethics

Ralston, Stuart H. Professor of Medicine, Department of Medicine and Therapeutics, University of Aberdeen, UK. Essay: Genetics of osteoporosis

Reeve, Eric C.R. Honorary Senior Lecturer, Institute of Cell and Molecular Biology, University of Edinburgh, UK. Essays: From the great apes to Gregor Mendel, Genetics of quantitative characters, Gregor Mendel

Reik, Wolf. Head of Programme, Laboratory of Developmental Genetics and Imprinting, Babraham Institute, Cambridge, UK. Essay: Genomic imprinting

Renne, Ulla. Scientist, Research Institute for the Biology of Farm Animals, Department of Genetics and Biometry, Dummerstorf, Germany. Essay: Body Weight Limits in Mice

Riley, Monica. Senior Scientist, Marine Biological Laboratory, Woods Hole, USA. Essay: Microbial genomics

Robinson, J. Department of Plant Sciences, University of Oxford, UK. Essay: The Origin of the Apple (*Malus domestica* Borkh.)

Ryder, Michael L. Southampton, UK. Essays: Genetics of wool production, Textile fibres from goats

Salamini, Francesco. Professor and Director, Department Plant Breeding and Yield Physiology, Max-Planck-Institut (MPI) Züchtungsforschung, Köln, Germany. Essay: Pinpointing einkorn wheat domestication by DNA fingerprinting

Sang, James H. Professor, Department of Biological Sciences, University of Sussex. Essay: *Drosophila melanogaster*: the fruit fly

Schimenti, John C. Staff Scientist, The Jackson Laboratory, Bar Harbor, Maine, USA. Essay: The mouse *t* complex

Scholl, Randall. Associate Professor, Department of Plant Biology, Ohio State University, USA. Essay: Arabidopsis genetics and genome analysis

Serreze, David V. Staff Scientist, The Jackson Laboratory, Maine, USA. Essay: Genetic control of immunological responses

Sheppard, David N. Lecturer, Department of Physiology, School of Medical Sciences, University of Bristol, UK. Essay: Cystic fibrosis: therapy based on a molecular understanding of CFTR

Short, Roger V. Professorial Fellow, Department of Obstetrics and Gynaecology, University of Melbourne, Australia. Essay: The testis: the witness of the mating system, the site of mutation, and the engine of desire

Strugnell, R.A. Department of Microbiology and Immunology, University of Melbourne, Australia. Essay: DNA vaccination

Sweet, J.B. Head of Environmental Research, National Institute of Agricultural Botany, Cambridge. Essay: Determining the impact and consequences of genetically modified crops

Szathmáry, Eörs. Professor and Head of Department, Department of Plant Taxonomy and Ecology, Eötvös University, Budapest, Hungary. Essay: Evolution of sex

Sumner, Adrian. North Berwick, East Lothian, UK. Essays: Chromosome banding and the longitudinal differentiation of chromosomes, Human cytogenetics

Sunkel, Claudio E. Associate Professor, Instituto de Biologia Molecular e Celular, Universidade do Porto, Portugal. Essay: Centromeres and kinetochores

Tarantino, Lisa. Postdoctoral Associate, Center for Developmental and Health Genetics, Pennsylvania State University, USA. Current address: The Genomics Institute of the Novartis Research Foundation, USA. Essay: The mouse as model for the study of genetic aspects of alcoholism

Thapar, Anita. Professor of Child and Adolescent Psychiatry, Department of Psychological Medicine, University of Wales College of Medicine, Cardiff, UK. Essay: Genetics of mental retardation

Thomas, Christopher M. Professor of Molecular Genetics, School of Biosciences, University of Birmingham, UK. Essay: Acquisition and spread of antibiotic resistance genes by promiscuous plasmids

Tilley, Michael. Graduate Research Associate, Department of Molecular Genetics, Ohio State University, USA. Essay: Arabidopsis genetics and genome analysis

Tweedie, Susan. Research Fellow, Institute of Cell and Molecular Biology, University of Edinburgh, UK. Essay: DNA methylation in animals and plants

Tyler-Smith, Chris. University Research Lecturer, Department of Biochemistry, University of Oxford, UK. Essay: Y-chromosomal DNA analysis and the Jefferson–Hemings controversy

Viney, Mark. School of Biological Sciences, University of Bristol, UK. Essay: *Caenorhabditis elegans*

Wadhams, Mark J. Armed Forces DNA Identification Laboratory, Armed Forces Institute of Pathology, Rockville, USA. Essay: Forensic DNA analysis of the last Russian royal family

Walker, Catherine A. PhD student, Medical Genetics Section, Department of Medical Genetics, University of Edinburgh, Western General Hospital, UK. Essay: Trinucleotide repeat instability as a cause of human genetic disease

Watkins, R. Axminster, Devon, UK. Essay: Origin of the apple (*Malus domestica Borkh.*)

West, John D. Senior Lecturer, University of Edinburgh, UK. Essay: Genetic studies with mouse chimaeras

Whitelaw, Bruce. Principal Investigator, Department of Gene Expression and Development, Roslin Institute, Edinburgh, UK. Essay: Pharmaceutical proteins from milk of transgenic animals

Whittle, J.R.S. Reader in Genetics, School of Biological Sciences, University of Sussex, UK. Essay: Genetic dissection of wing patterning in *Drosophila melanogaster*

Wright, A.F. Professor, Cell and Molecular Genetics Section, MRC Human Genetics Unit, Western General Hospital, Edinburgh, UK. Essay: Genetic eye disorders

Wurtzel, Eleanore. Professor, Department of Biological Sciences, Lehman College and the Graduate Center of the City University of New York, USA. Essay: Rice genetics: engineering vitamin A

Young, H.-K. Senior Lecturer, Department of Biological Sciences, University of Dundee, UK. Essay: Antibiotic resistance – theory into practice

Zallen, DorisTeichler. Professor, Centre for Interdisciplinary tudies, Virginia Polytechnic Institute and State University, Blacksburg, Virginia, USA. Essay: US gene therapy in crisis

Zeng, Zhao-Bang. Professor, Program in Statistical Genetics, North Carolina State University, USA. Essay: Quantitative trait loci: statistical methods for mapping their positions

PREFACE

The first encyclopedia, I believe, to have been created in Scotland was the *Encyclopaedia Britannica*, founded by a "Society of Gentlemen" in 1764 and issued in weekly numbers at sixpence each from 1768 to 1771 when the whole work was completed. Two Edinburgh printers, Colin MacFarquhar and William Smellie, and the leading Scottish engraver Andrew Bell wrote, illustrated and edited it, and an interesting point is that the subjects of history and biography were omitted as "beneath the dignity of Encyclopaedias". New editions without such omissions followed, published and printed first in Edinburgh, then in Cambridge, UK, and more recently in the US. The *Encyclopaedia Britannica* is now in its 15th edition.

So here was Edinburgh, capital of Scotland, future home of the proposed Scottish Parliament, and a uniquely beautiful city, which I and my family had long been enamoured of, which had launched surely the greatest encyclopedia of all time during the period of the Scottish Enlightenment and the founding of Edinburgh's "New Town".

Then along came Fitzroy Dearborn of London and Chicago, publishers of encyclopedias and other compendia, who wished to commission an Encyclopedia of Genetics. When they approached me I almost replied that "Such a task would be quite impossible!" but then I thought to myself "Where else but in Edinburgh, where the *Encyclopedia Britannica* was launched, should the creation of an encyclopedia of genetics be attempted?" Encouraged by my experience with *Genetical Research*, the journal I founded and edited for many years, I agreed to take on this much more formidable editing task.

The science of genetics essentially began in 1900 with the rediscovery of Mendel's laws, and the short period of 1900 to 2000 has witnessed an enormous growth in our genetic knowledge, so this seems an appropriate time for an encyclopedia of genetics to be launched. There were, however, problems to be faced and if possible, overcome.

First was the question of the size (and therefore the price) of the encyclopedia. Many large multi-author volumes have been devoted to a single organism of especial genetic importance, for example two on the very small nematode worm *Caenorhabditis elegans* (1988, 1997), two on the very much smaller bacteriophage *Lambda* (1971, 1983) and one on a small herb of the Crucifer family *Arabidopsis thaliana* (1994), all published by Cold Spring Harbor Laboratory Press. In addition to even larger volumes on the fruit fly Drosophila and the laboratory mouse, these form only a small part of the genetics literature of the last 30 years, so clearly we cannot get all genetics into one, or even several volumes.

The alternative is to be selective in the content of the encyclopedia; and here we have the problem that every genetic topic under serious study has several or many teams of geneticists applying the latest techniques to its further analysis. There are now very many such topics, with new ones coming forward, and one has to rely on the good will of those involved in this exciting and absorbing research to find time to write, and to write articles suitable for the encyclopedia, by which I mean articles that should not be so simple as to be of no interest to professional geneticists working in other branches of the science, and preferably not too difficult for the majority of our likely readers to fathom.

So for the last four or five years I have corresponded with the international community of geneticists; inviting them to contribute to the encyclopedia. Meanwhile, of course, new research of great interest is continually published, and I have obtained articles from some of the authors working on these newer topics. Bearing in mind the intense competition among those at the forefront of current genetic research, I have had a pleasingly positive response from contributors.

This encyclopedia is a collection of articles of varying length on a great variety of genetic topics, supported by plenty of references that show how each topic is built on many strands of research. The References and Further Reading give access to the primary literature of genetics, and an Appendix of useful web addresses on page 925 provides some suggestions for further exploration.

Every branch and even "twig" of modern genetics has acquired its own special terms, which help the reader who knows them but baffle non-specialists, so we have attached a glossary of special terms to many of the articles. A Glossary describing terms more generally used is also included on page 919 before the Index. The Introducing Genetics section includes a long article on "Genetics for the newcomer" (page 3) to provide those readers who need it with some additional useful background information. This is followed by an article giving details of techniques in molecular genetics.

The remaining articles are grouped in sections of broadly similar topics, progressing from The Origins of Genetics through the genetics of microorganisms, animals, humans and plants, to the more general topics of Genetics of Cell Organelles, Structures and Function and DNA-based Genetic Analysis and Biotechnology. Model Organisms provide a section in their own right, and the encyclopedia concludes with Population Genetics and Evolutionary Studies. Some of these sections have an introduction with relevant or late-breaking information that could not be included in the individual articles, which reflects the fact that, more than any other science, genetics is like a rapidly moving staircase. An example of this is that much of the science is being almost taken over by DNA sequencing, leading to genomics, bioinformatics and the emergence of proteomics. Enormous computers are being designed to help this transformation and large numbers of new specialists have now joined this important field, but the newer a specialism the more hesitant are its *aficionados* to write explanatory articles for an encyclopedia. If it is too early to be able to put the reader fully in the frame on these new developments, I hope the readers will forgive me.

Acknowledgements

I am particularly indebted to the many authors who wrote the articles in this Encyclopedia, thereby making it possible. Nick Barton of the Institute of Cell, Animal and Population Biology, Edinburgh, Mary Bownes of the Institute of Cell and Molecular Biology, Edinburgh and John C. Schimenti of the Jackson Laboratory, Bar Harbor USA suggested new topics and authors, whom they encouraged to write. Isobel Black, who became my assistant editor, has been of immense help in dealing with all the problems that most trouble editors, and Anne Hill, having just completed her PhD at the Institute of Cell and Molecular Biology, took over converting figures in both colour and black and white into electronic form, wrote several articles, checked or created glossaries and looked after some problems to which my computer was prone. Lesley Henderson and more recently Gillian Lindsey, new commissioning editor at our Publishers Fitzroy Dearborn, have been very helpful through my editing period, and it has been a pleasure to work with them. My family, who all live nearby, are a constant source of pleasure even when I don't have the time to tell them so. These acknowledgements explain why editing this encyclopedia has been much less of a burden that I had expected.

Eric C.R. Reeve

A INTRODUCING GENETICS

GENETICS FOR NEWCOMERS

Jeff Bond
Institute of Cell and Molecular Biology, University of Edinburgh, Edinburgh, UK

1. Introduction

The intention of this chapter is to introduce the subject of genetics to those with little previous experience of the subject. The plan is to survey progress in genetics in a semi-historical way. The article will examine how our understanding has developed over the decades and how the subject itself has changed so dramatically in recent times. While doing so, some key genetic terms will be introduced and explained because one of the barriers to understanding genetics is the technical vocabulary of the subject. Technical terms are not used simply to make things difficult, they are usually very convenient shorthand to say something succinctly when the alternative is a very lengthy explanation. Their very convenience is seductive to the writer and their use is unavoidable when there are practical constraints placed on the length of an article. The main glossary also provides a useful reference for technical terms used in this Encyclopedia.

2. Genetics in the classical period

Nearly everyone knows that the science of genetics had its origin in experiments carried out by Gregor Mendel in the 1860s. Working with garden peas, he established the basic rules underlying the inheritance of biological characteristics. He was the first to realize that by counting and classifying the numbers of different types of plant he could work out how the differences between the plants (such as round *vs.* wrinkled peas) were inherited. He worked with seven pairs of characters, namely:

Round *vs.* wrinkled pea seeds
Yellow *vs.* green endosperm
Coloured *vs.* white flowers (associated with a difference in seed coat)
Inflated *vs.* constricted seedpods
Green *vs.* yellow seedpods
Axial *vs.* terminal flower position.
Long *vs.* short stems.

He deduced (Mendel, 1866) that the differences were each determined by a pair of factors (which we now know as genes) that were passed on from one generation to the next. He discovered that genetic differences segregated during the formation of pollen and ovules. The genetic differences were passed on unchanged from one generation to the next in predictable ratios. His work remained unappreciated for several decades until its significance was realized at the turn of the century (see article on "Gregor Mendel"). It was this rediscovery that heralded the classical genetics era.

The rediscovery coincided with an appreciation that, during cell division, chromosomes in a cell are passed to the two daughter cells exactly like Mendel's factors were in his breeding experiments. In 1902 and 1903, Theodor Boveri and Walter Sutton published this idea separately and almost at the same time (Boveri, 1902; Sutton, 1903). With this discovery, the chromosome theory was born; it was realized that the chromosomes in a cell carried the genetic information and that from one cell generation to the next the chromosomes were duplicated so that each daughter cell inherited a copy of the chromosomes and, hence, the genes. Some extraordinarily perceptive statements were made around this time. For example, in 1896, the cytogeneticist E.B. Wilson wrote:

> These facts . . . support [the idea] of the nucleus as the bearer of hereditary qualities. The chromosomal substance, the *chromatin*, is to be regarded as the physical basis of inheritance . . . and thus we reach the remarkable conclusion that inheritance may be, perhaps, effected by the physical transmission of a particular chemical compound from parent to offspring.
>
> (Wilson, 1896)

Remember that this was written before the work of Mendel had been appreciated and before the formulation of the Sutton–Boveri chromosome hypothesis. The speculation was based on the analysis of the chemical content of gametes.

Following the rediscovery of Mendel's work, the first few decades of the 20th century were characterized by rapid progress in understanding the laws of inheritance and the many different ways in which genes can interact within an organism. At this time, investigations were confined to higher organisms, the eukaryotes, in which the genetic material is organized into chromosomes located in a nucleus. Only later were genetic studies carried out on bacteria and viruses, the prokaryotes. The most fundamental conclusion that emerged at this time was that the laws applied universally. Any minor differences that were shown to exist between eukaryotes did not detract from this generalization. This meant that any conclusions derived from work on one organism could be applied to others. This, in turn, led to the development of model organisms, notably the fruit fly *Drosophila melanogaster* and maize, in which increasingly sophisticated genetic experiments were carried out and the results used to generalize about inheritance in other organisms, especially man. The experiments in Drosophila that led to the confirmation of the chromosome theory resulted in the award of the Nobel Prize to Thomas Hunt Morgan in 1933.

An important fact arose directly from Mendel's work: the genetic constitution of an organism cannot be worked out from its appearance. Mendel had discovered that some mutations (white pea flowers, for example) are recessive and can be hidden for one or more generations. In 1909, the Danish geneticist Wilhelm Johannsen coined two important genetic terms to acknowledge this; the "phenotype" is the appearance of an organism and has to be distinguished from the "genotype" which is its genetic constitution as revealed by breeding experiments (Johanssen, 1909). The realization that some genetic characteristics are inherited in a recessive manner meant that variation in the genes possessed by any one individual could only be worked out by careful breeding experiments (or crosses). It is the controlled breeding and analysis of the resulting progeny that forms the cornerstone of classical genetics. In the beginning, some workers, notably William Bateson, believed that when a gene was changed (or mutated) it was absent from the organism. Although it seems obvious that evolution cannot proceed through the progressive loss of genes, this idea was quite difficult to disprove. We now know that mutated genes are altered rather than missing and that any gene can be altered in many different ways. These different forms of a gene are called alleles. An individual may contain two different alleles of a particular gene, in which case that individual is said to be heterozygous. Conversely, when the two alleles in an individual are the same as each other, the individual is said to be homozygous. It was Bateson and Saunders (1902) who invented these terms.

(i) Chromosomes and linkage groups

When the chromosomes of any organism are examined by looking at them down a microscope, there are normally a relatively small number of them. On the other hand, the number of genes is large; even a superficial investigation into the genetics of an organism soon reveals the existence of a large number of genes controlling diverse characteristics. In 1927, the American geneticist Hermann Muller discovered that exposure to X-rays greatly increased the frequency of mutations (Muller, 1927). He was awarded the Nobel Prize for this discovery in 1946. Chemicals were also discovered that had the same effect. By exposing the popular experimental organisms of the period to these mutagenic agents, a large number of mutations were obtained. It follows from the fact that the number of genes is much greater than the number of chromosomes that any one chromosome contains many genes. Genes are arranged in linear array along the chromosomes and an early breakthrough in the classical genetics era was the realization that the position of different genes along the chromosome could be determined. When two genes are located close together on the same chromosome, they will probably be passed on to the next generation together. Genes transmitted this way are said to be linked. Carl Correns made the first report of linkage of two characteristics in 1900 (Correns, 1900), but he could have been dealing with two effects of a single gene. Bateson and Reginald Punnett later described linkage in the sweet pea in which this explanation could not hold: the two characters they were studying (purple *vs.* red flowers and long *vs.* round pollen grains) were obviously caused by two separate genes (Bateson & Punnett, 1905–8). **Figure 1** illustrates the idea of linkage in terms of the chromosome theory.

When two genes are located on the same chromosome, they are not always inherited together. This is because, at the specialized cell division (called meiosis) which generates the gametes, the chromosomes are broken and rejoined so that the genetic material which is passed on to the next generation is a mixture of the two homologous chromosomes. In 1909, Frans Janssens was the first to put forward this idea following his work with salamanders (Janssens, 1909). He showed that paired chromosomes at meiosis are split longitudinally into chromatids and that an exchange takes place between two of the four chromatids. This process is called crossing-over and results in the recombination of the genes (see **Figure 2**).

Crossing-over is brought about by chiasmata (singular = chiasma) which can be observed in the early stages of meiosis as cross-shaped structures – hence the name. These are the places at which the shuffling of the chromosomes and recombination of genes takes place. When two genes are close together, the chances are small that a chiasma will form between them. If two other genes are further apart then the chance of recombination is greater. Alfred Sturtevant was the first geneticist to realize that this could be used as a basis for locating, or mapping, genes. In 1913, he suggested using

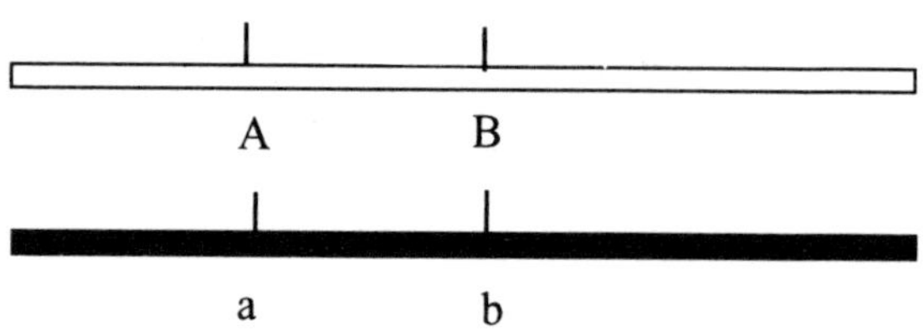

The vertical lines mark the positions of two genes A and B. One chromosome of the pair has a mutant form of gene A (called a) and also a mutant form of B (called b). The two chromosomes are shaded differently simply to represent this genetic difference

When the products of meiosis are formed A and B will be transmitted together and likewise a and b will be passed on as a pair.

Figure 1 The chromosomal basis of linkage. The inheritance of genes A and B is followed in **Figure 2**.

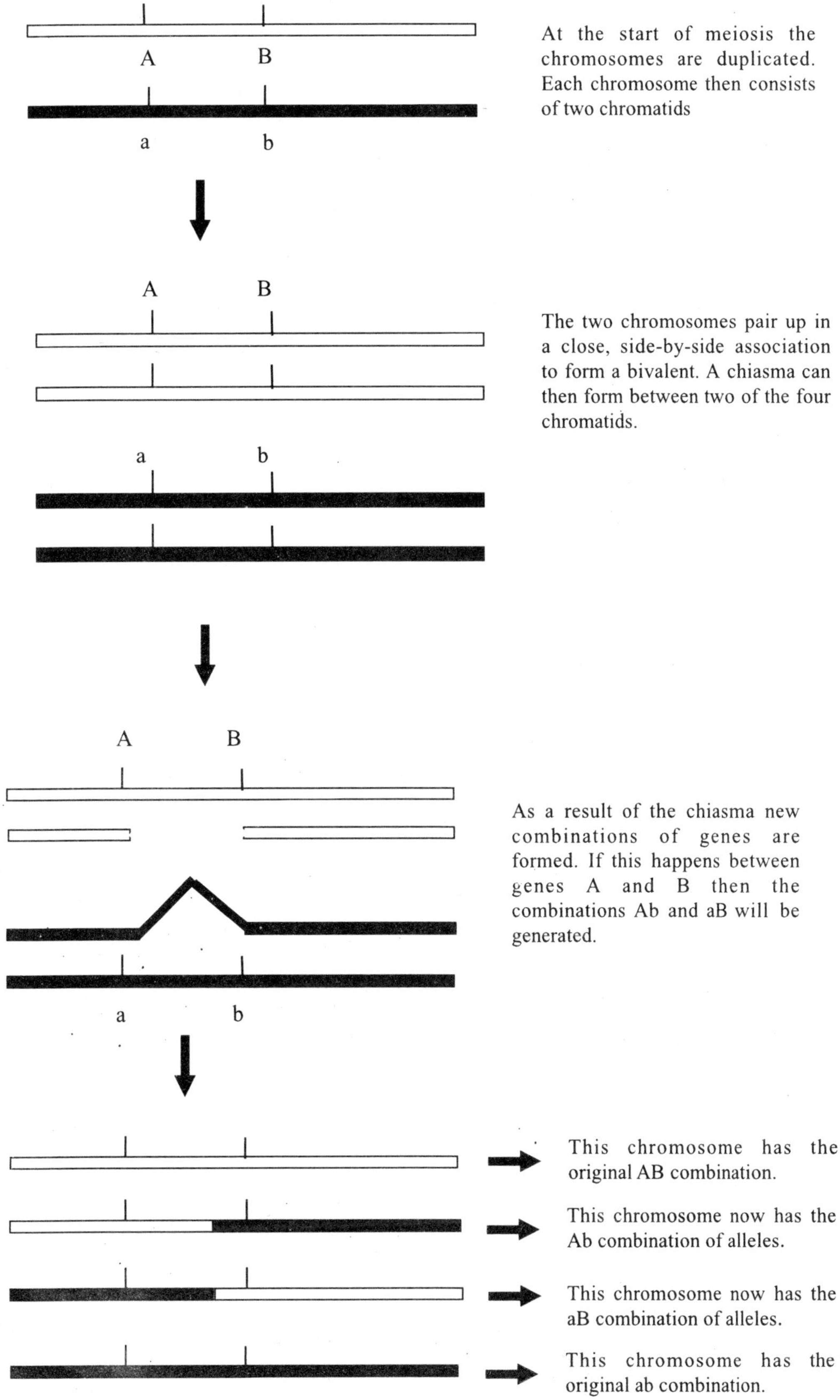

Figure 2 The genetic consequences of chiasma formation. Recombination by crossing over results in a new combination of genetic material. How often this occurs will depend on the physical distance separating the genes A and B.

the frequency of recombination as a measure of the distance separating two genes and this forms the basis on which linkage maps are drawn (Sturtevant, 1913). A linkage map represents the positions of a set of genes along a chromosome; in well-studied organisms, there will be one linkage map for each chromosome. **Figure 3** shows the linkage maps for the fruit fly *D. melanogaster* that has four pairs of chromosomes and, therefore, four separate linkage groups of genes.

(ii) Cytoplasmic inheritance

In the early experiments to establish the rules underlying the transmission of genes from one generation to the next, an important conclusion was that, for the most part, reciprocal crosses gave identical results; the phenotypes of the male and female parents were immaterial to the result. There were, however, some important exceptions. Not all genes are located on chromosomes within the nucleus; cell organelles, notably the mitochondria and (in plants) chloroplasts, also contain genes and these are not inherited equally from each parent. The female parent normally contributes the cytoplasm to the developing embryo and any genetic characteristic determined by genes in these organelles shows extranuclear or cytoplasmic inheritance. An example from humans is the disease myoclonic epilepsy associated with ragged-red muscle fibres. This is a disease of the central nervous system and of skeletal muscles, and it is caused by a mutation in the mitochondrial DNA. If all the mitochondria are mutant, the disease is lethal, but the inheritance, exclusively from the mother, is normally of a mixture of

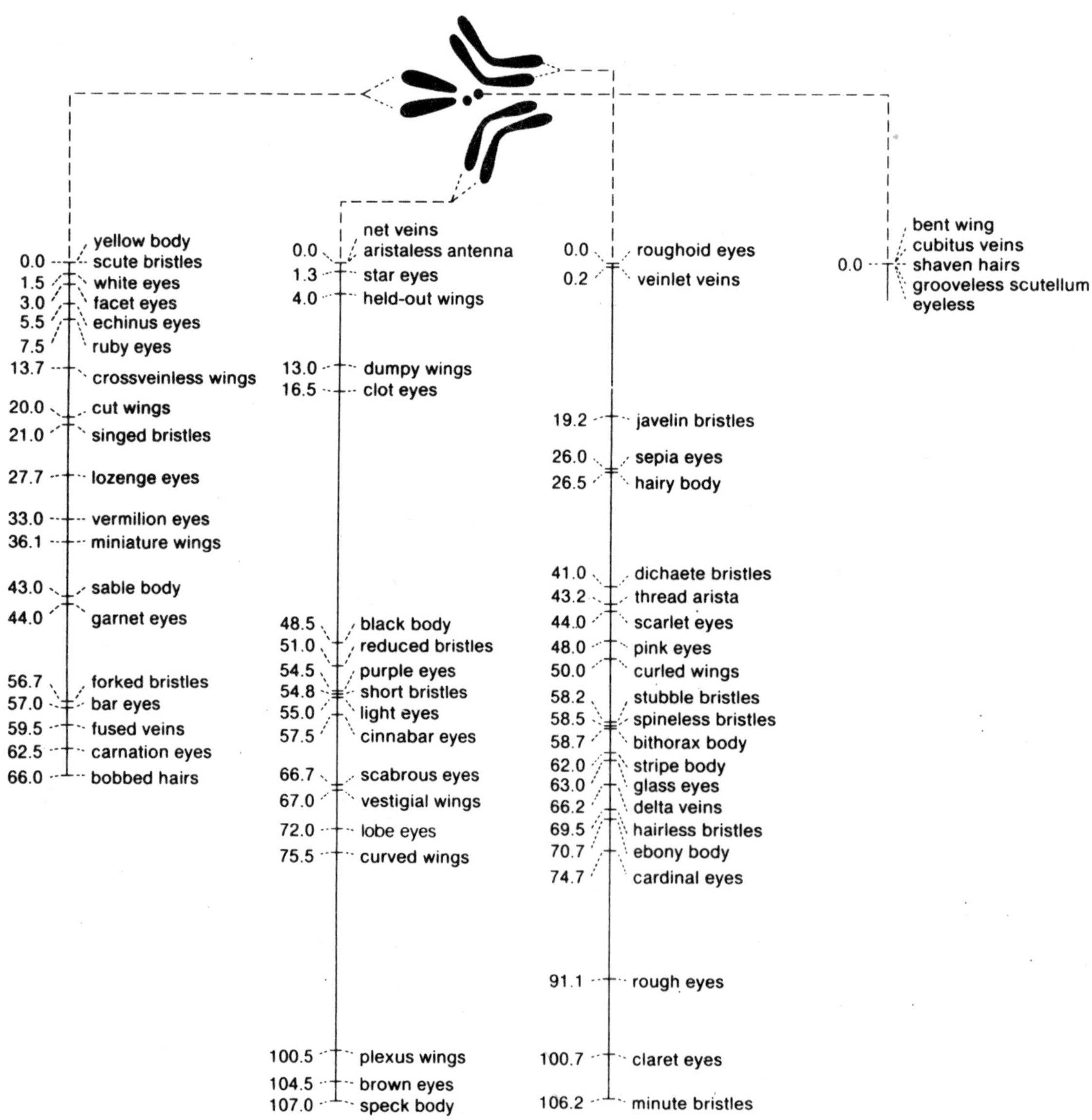

Figure 3 Genetic linkage maps of the fruit fly *D. melanogaster.* The diagram shows how each linkage group corresponds to a chromosome. The numerical values along each group are units of genetic distance known as map units. Reproduced with permission from the McGraw–Hill Companies, from Sinnot, E.W. *et al.*, *Principles of Genetics*, 6th edition, 1962.

mutant and non-mutant mitochondria. Consequently, the disease symptoms vary due to the proportions of normal and mutant mitochondria in the cells of affected people.

(iii) Sex-linked inheritance

Another example of inheritance in which reciprocality is not necessarily observed does involve chromosomally transmitted genes. In 1891, Hermann Henking discovered that the males of a species of Hemiptera (*Pyrrhocoris apterus*) possessed 11 pairs of chromosomes and one additional element which he called an "X body" (Henking, 1891). The X body segregated at meiosis so that half the sperm received it and half did not. Clarence McClung was the first to suggest (McClung, 1902) that this segregation was connected to sex determination. Later, Edmund B. Wilson noticed that the males of another Hemiptera (*Protenor belfragei*) had five chromosome pairs and an additional sixth chromosome (the X chromosome) which was without a partner. In females of *P. belfragei*, there are six chromosome pairs, the extra pair consisting of two X chromosomes. Wilson suggested that sperm that contained an X chromosome will give rise to females and those without one will give rise to males (Wilson, 1905). At about the same time, Nettie Stevens (Stevens, 1905) discovered that, in the beetle Tenebrio, males and females possessed the same number of chromosomes, but in males two of the chromosomes are of a different size and shape to each other. These heteromorphic chromosomes pair with each other to make a bivalent comprising one X and one Y chromosome. In this case, sperm which contain a Y chromosome following meiosis will give rise to a male. It should be noted that not all species have sex determined by a genetic mechanism. In some reptiles, for example, the temperature at which the eggs are incubated determines whether the hatchling will be a male or a female. The mechanism of sex determination varies between different groups of organisms even when it is determined genetically (see below), but there are consequences for the inheritance of genes when they happen to be on the sex chromosomes.

Many of the genes on the X chromosome control characteristics other than sex. When a gene is located on the X chromosome, it exhibits a characteristic pattern of inheritance that distinguishes it from genes which are carried on the other chromosomes (the autosomes). In organisms in which the male has an X and a Y chromosome, males determine the sex of their offspring by passing their X chromosome to female offspring and the Y chromosome to males. Any mutant genes that happen to be on the X chromosome are expressed in males when they inherit the X chromosome (with its mutant gene) from the mother because they do not have a second X chromosome to mask the mutation. (Males are said to be hemizygous.) It is this feature (a high frequency of affected males) which is the most noticeable characteristic of these genes. Genes on the X chromosome are said to be sex linked. Sex linkage was first discovered by L. Doncaster and G.H. Raynor (1906) working with the Magpie moth *Abraxus grossulariata*, but the equivalent work of T. H. Morgan (1910) using *D. melanogaster* is both more extensive and more familiar. **Figure 4** illustrates the "criss-cross" pattern of inheritance characteristic of sex-linked conditions (diagonal arrows trace the inheritance of a mutant X chromosome). The key features to note from **Figure 4** are that mothers transmit the mutant chromosome to both sons and daughters, but, if the condition is recessive, only the sons will be affected. Affected males, on the other hand, pass their X chromosome to all daughters and all are, therefore, unaffected carriers of the condition. Males never pass their X chromosome to their sons and, therefore, father-to-son transmission of a sex-linked disease is not possible.

It is interesting to note that this pattern was recognized long before the discovery of Mendelian inheritance. Haemophilia, which is a disorder of the blood resulting in a failure to clot following injury to the blood vessels, can be caused by mutation of a gene on the X chromosome. The Talmud contains quite accurate instructions specifying which boys were not to be circumcised when this blood clotting disorder was present in a family. Charles Darwin wrote a very accurate description of the characteristic pattern of inheritance without appreciating its significance.

> Generally with [haemophilia], and often with colour-blindness, and in some other cases, the sons never inherit the peculiarity directly from their fathers, but the daughters alone transmit the latent tendency, so that the sons of daughters alone exhibit it. Thus, the father, grandson and the great-great-grandson will exhibit the peculiarity, the grandmother, daughter and great-granddaughter having transmitted it in latent state.
>
> (Darwin, 1882, p. 49)

(iv) Sex determination

The mechanism of sex determination varies between different groups of organisms. Calvin Bridges first gained insight into sex determination of *D. melanogaster* when he investigated an abnormal strain of the fly. It had two X chromosomes and a Y chromosome and was female. Bridges (1914, 1916) used the pattern of inheritance obtained from breeding this fly to establish the chromosome theory of inheritance, but flies with an abnormal number of sex chromosomes also helped establish how sex is determined in Drosophila. Sexual development is regulated by the ratio of X chromosomes to autosomes. The Y chromosome plays no role in determining maleness, although it does play a part in sperm development. Flies which have one X chromosome and no Y (XO flies) develop as males.

The situation is different in humans. Here, too, individuals with an abnormal number of sex chromosomes provide important clues into the role of the X and Y chromosomes. In contrast to Drosophila, XO individuals are female and XXY and XYYs are male. XO females have Turner's syndrome, named after Henry Turner who first described the condition in three women in 1938 (Turner, 1938). The chromosomal basis of the condition was described much later, by Charles Ford and colleagues in 1959 (Ford *et al.*, 1959). XXY males also develop abnormally and have Klinefelter's syndrome, named after Harry Klinefelter who first described the condition in 1942 (Klinefelter *et al.*, 1942). The chromosomal basis of this syndrome

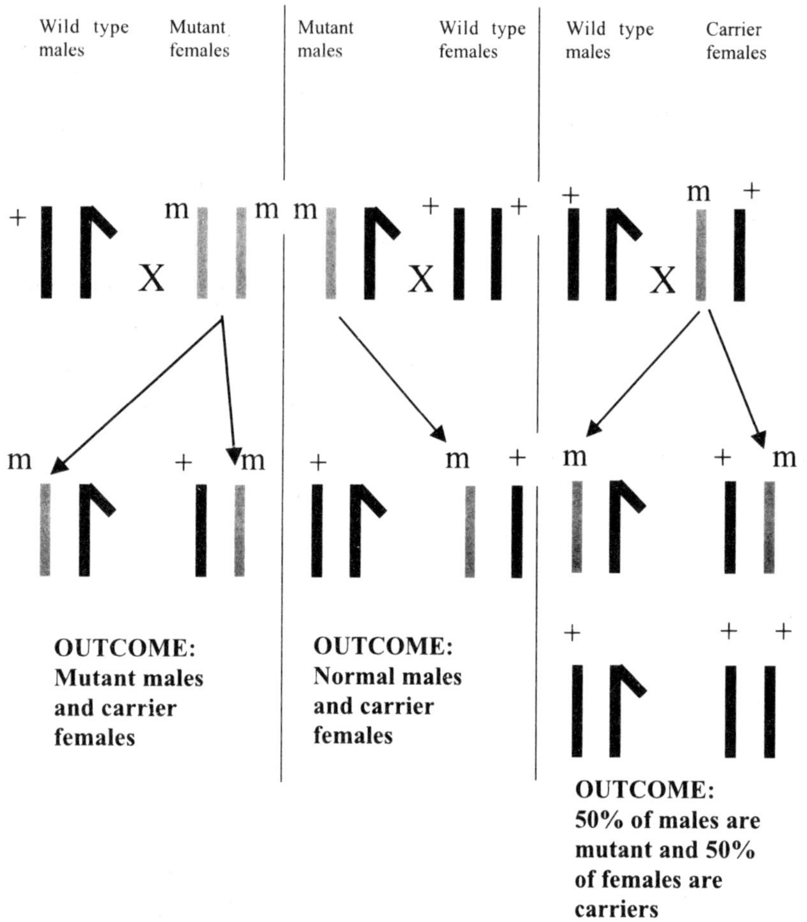

Figure 4 The chromosomal basis of sex-linked inheritance. An X chromosome with a recessive mutant allele (grey shading) affects males more often than females. The arrows follow the inheritance of the chromosome with the mutant allele from one generation to the next.

was reported by Patricia Jacobs and John Strong in 1959 (Jacobs & Strong, 1959). From these observations, it follows that the Y chromosome plays a pivotal role in determining maleness – people with a Y chromosome develop as males, those without as females.

(v) Genes and development

Sexual development is just one example of how genetics influences the development of an individual. It may be the most obvious case, but it is not the only one. In multicellular organisms such as Drosophila, the mouse or ourselves, the individual cells that make up the adult can be very different from each other. In each case, the adult organism starts as a single-celled zygote, but the cells, during the course of growth and development, undergo differentiation so that, for example, the cells of the kidney are quite different from those in the eye. How does this differentiation come about? Morgan (1934) wrote an excellent account of early experiments investigating the relationship between genetics and embryological development. Detailed understanding of the process came about only after the nature of the genetic material was discovered and the techniques for manipulating genes in a test tube were developed.

Nuclear and tissue transplantation experiments gave some early insights into differentiation. Differences in the developing embryo could be detected from a very early stage and evidence was soon obtained that differentiation started long before any obvious morphological changes became apparent. Weiss (1939) and Barth (1949) contain comprehensive accounts of these experiments. For example, Hörstadius (1939) carried out experiments with single-cell embryos of the echinoderm *Paracentrotus lividus* that showed that even at this early stage the cell is not a uniform structure. Fertilized eggs of a sea urchin possess a band of red pigment granules that stay in place as the cell divides. These granules allow the orientation of the first few cell divisions to be followed. Experiments in which the eggs at the single-cell stage were cut in two showed that the cytoplasm of the fertilized egg was already non-uniform; cells in which either the bottom or the top portion of cytoplasm was removed failed to develop properly, but in characteristically different ways. Transplanting cells of a developing amphibian embryo from one region to another and following their subsequent fate led to an important discovery. The differentiation of the transplanted cells could be influenced by their new position. Sometimes, the transplanted cells changed the fate of neighbouring, non-transplanted, cells too. The precise effects depended on the stage to which the embryo had developed before the cells were transplanted, but there was clear evidence that cells interacted with each other. These first experiments, which probed differences between different parts of the developing organism, were the forerunners of later experiments using precise molecular probes to address the same question.

Early experiments also allowed two quite different ideas to be distinguished. One idea suggested that differentiation is accompanied by the loss of genes from particular cells, so that differentiated cell types each contained a different constellation of genetic material. An alternative idea was that all cells contained the same set of genes, but that differentiation was accompanied by changes in gene activity so that different genes were active in different types of cell. A classic experiment by John Gurdon in 1968 (Gurdon, 1968) showed that differentiated cells from the intestine of tadpoles of the amphibian *Xenopus laevis* were capable of developing into a complete adult frog. In the experiment, he injected nuclei from tadpole intestine into unfertilized eggs which had been irradiated to remove the nucleus normally present. The eggs with transplanted nuclei developed further and, following further nuclear transplants from these, eventually some adult frogs were obtained. In a much-publicized recent development, Ian Wilmut and colleagues (Wilmut *et al.*, 1997) introduced nuclei from adult tissue into an egg. The birth of Dolly the cloned sheep resulted from this experiment. The conclusion to be drawn is that, although it is not easy to produce an adult animal from a differentiated cell, it can be done with at least some differentiated cells and these must contain all the necessary genetic instructions for development of an adult.

In the early decades of classical Drosophila genetics, many mutations that affected development were obtained. A particularly important set of mutations in this respect consists of the homeotic mutations that alter the identity of particular segments of the adult fly. The term "homeosis" was originally introduced into the genetics literature by Bateson for mutant organisms in which one body part was transformed into another. Edward Lewis carried out a classical series of experiments on a cluster of genes called the *bithorax* complex (for an elegant review see Lewis, 1978). Many of these mutations are lethal, but some allow the development of adults with abnormal adult structures. For example, in one particular mutant combination, an adult fly develops with an extra pair of wings on one of its thoracic segments. These mutations, and others like them, provided a starting point for the genetic dissection of the control of development. They were later to play an important part in understanding the molecular details of how the adult Drosophila develops.

(vi) Genes in populations

In the early years of the 20th century, a major barrier to the general acceptance of evolution through natural selection was the lack of an adequate theory of heredity. Following the rediscovery of Mendel's work it was, at first, not at all obvious that Mendelian genetics helped explain Darwinism and some early Mendelians were opposed to the theory of natural selection. It is now accepted that genetics is a discipline that provides key elements in the understanding of evolution. This change in thinking can be attributed mainly to the work of Ronald Fisher, J.B.S. Haldane and Sewall Wright who specified mathematically the major features of evolution by natural selection (Fisher, 1918, 1930; Wright, 1931; Haldane, 1932). The synthesis of Darwin's theory of evolution and Mendelian genetics is known as neo-Darwinism or the modern synthesis (Huxley, 1942; see article on "Darwin and Mendel united: the contributions of Fisher, Haldane and Wright up to 1932"). Two major ideas are accommodated in the synthesis. The first was an insight into what happens to genes in populations rather than individuals; this is the domain of population genetics. The second was the development of quantitative genetics and the realization that Mendelian inheritance could provide the basis for understanding the inheritance of small differences (such as height or weight) which are a characteristic of much of the variation that is commonly seen in populations. Douglas Falconer's book *An Introduction to Quantitative Genetics* is a classic introduction to this area of genetics (Falconer & Mackay, 1996).

In population genetics, an early development was the formulation of the Hardy–Weinberg principle of genetic equilibrium. In 1908, working independently, Godfrey Hardy, a mathematician at Cambridge University, and Wilhelm Weinberg, a physician from Stuttgart, pointed out, in clear mathematical terms, that the frequency of alleles of any gene in a population will remain constant from one generation to the next unless factors, such as natural selection, disturb the equilibrium (Hardy, 1908; Weinberg, 1908). To this day, the Hardy–Weinberg law remains the core around which population genetics revolves. Theodosius Dobzhansky's work on evolution in populations of fruit flies provided experimental support for the theoretical ideas (Dobzhansky, 1937).

In quantitative genetics, statistical and mathematical techniques were developed to study the inheritance of characteristics that showed continuous variation in the population. There are two principal reasons why a trait or character can show continuous variation in the population. The trait in question may be influenced by a large number of loci, each having a small effect on it. As the number of loci influencing the trait increases, the number of possible combinations of genes becomes very large. Any one combination of genes may differ only slightly from another in its consequences for the phenotype and so variation appears to be continuous. The second reason a trait may show continuous variation is that it may also be influenced by the environment. By analysing resemblance between relatives, estimates can be made of the extent to which variation in continuously varying characters is determined genetically and how much of the variation is attributable to environmental effects. Heritability measures the extent to which variation in a population is determined by variation among genes. If, in a particular population, the heritability of a character is estimated to be 0.6 (or 60%) then this means that 60% of the observed variation can be attributed to variation in the genes.

In selecting suitable strains of plants and animals for agricultural purposes, early farmers had carried out artificial selection without any knowledge of the underlying genetics of the organisms (see article on "Origin and evolution of modern maize"). In a classic experiment using the broad bean, Wilhelm Johannsen showed that success in artificial selection experiments (selecting for increased and decreased weight of the seeds in this particular case) depended on there being genetic variation present in the population under selection and that selection could not extend the variation already present. For a period, this was used as evidence against natural selection as a force in evolution until it was realized that mutation provides a source of variation and selection acts on this variation, eliminating some, but not all, of it.

3. The dawn of the modern era

The modern era of genetics can be considered to start when clear insight was gained into the molecular nature of inheritance. Two questions spring to mind when considering this matter: what do genes do, and what is the chemical nature of the genetic material? Answers to these questions were sought very early on in the history of genetics, but clear insight has been gained only in the last 50 years or so.

(i) What do genes do?

Speculation occurred very early on in genetics about the role of genes. Early studies on humans who suffered from inherited conditions and had abnormal metabolism by Archibald Garrod (1909) and studies into pigment production in plants by Haldane (1932) are just two examples of early investigations into the relationship between the genetics and the biochemistry of an organism. The relationship was put onto a formal basis in 1941 by George Beadle and Edward Tatum, working on the fungus *Neurospora crassa*, who proposed the "one gene–one enzyme" hypothesis (Beadle & Tatum, 1941). Put simply, they suggested that the genes carried the information that allowed a cell to make an enzyme. Mutation to a particular gene destroyed the cell's ability to make a specific enzyme. The formation of other proteins that are not enzymes, for example the muscle protein actin, is also under the control of genes, so the original hypothesis has become modified with the passage of time. Also, some genes do not directly encode proteins and some can instead be considered as switches, determining whether or not a particular gene is active within a given cell.

The increasingly detailed understanding of what genes do can perhaps best be summarized as follows. Genes determine the biochemistry of an organism by containing the instructions for making its complex molecules. Not all the instructions are acted on in all cells. In any one cell type, some genes are switched off and others are switched on; in a different cell type, a different constellation of genes will be active. In this context, switched on means that the instructions are acted on. The cell makes the relevant protein and the gene is said to be "on". This differential gene activity is the means by which, in multicellular organisms, different types of cell can be derived from a single cell zygote. The development of this idea will be considered later, in section (iii).

(ii) What is the chemical nature of the genetic material?

As we have seen from the Wilson quotation in section 1, speculation about the chemical nature of the genetic material was made very early on in the development of genetics. At first, it was thought that genes would be made up of proteins because these molecules were known to be both diverse and complex. It was considered likely that genes would have to be very complicated because they controlled so many features of an organism and proteins were a strong candidate to be the genetic material. However, experiments by Oswald Avery, Colin MacLeod and Maclyn McCarty (Avery *et al.*, 1944) identified the chemical substance deoxyribonucleic acid (DNA) as the hereditary determinant. They exposed bacteria of one particular type to DNA extracted from a second, different, strain and were able to transform one type into another. At the time of these experiments, the structure of the DNA was unknown and an obvious objective was to work out the detailed structure of DNA.

The story of how the details about DNA were worked out is well known and has been the subject of several documentary films and popular books. In 1953, James Watson and Francis Crick, working with data obtained in London by Rosalind Franklin and Maurice Wilkins, proposed a model for the structure of DNA that was both elegant and insightful (Watson & Crick, 1953). Genes, as identified by classical crossing experiments, have three essential properties: they must be able to carry information from one generation to the next; they must be capable of replication; and they must be able to mutate so that the information they carry is altered in some way. When mutated, it is this altered property that is inherited. Watson and Crick's model for DNA, put forward in 1953, provided an explanation for each of these properties. It is the sequence of the nucleotide bases that provides the basis of the information for the cell. The DNA molecule can replicate to generate two identical copies from one starting molecule. The cell achieves this by using each of the two helices of the DNA as a template to synthesize a new duplex. The experimental proof that DNA replicates in this semi-conservative way was obtained by Matthew Meselson and Frank Stahl in 1958 (Meselson & Stahl, 1958). Mutations in the DNA were simply explained as alterations to the sequence of bases, which resulted in disruption of a particular piece of information. Once altered, the changed sequence of bases was itself faithfully replicated and passed to daughter cells.

Much of the work spanning the next few decades was aimed at confirming ideas about how the information carried in genes was acted on by the cell. The result of this activity was a detailed description of how the genetic information in the DNA is used by the cell. The information present in the genes is carried in the form of a genetic code. The code is almost universal; with some relatively minor exceptions, all organisms use the same set of instructions. The sequence of the bases form the instructions that determine which genes will be active and, hence, which proteins a cell will make. More specifically, the instructions are in the form of a triplet code. An early experimental indication on the nature of the genetic code came from a brilliant series of genetic experiments by Crick *et al.* in 1961 using *rII* mutants of the T4 virus. If a sequence of bases in the DNA is the instruction for making a protein then, within that sequence, any one set of three bases is an instruction either to start making that protein, to insert one particular amino acid building block into the protein or to terminate the protein synthesis. Each of the 20 amino acids commonly found in proteins is encoded by at least one triplet of bases. Most amino acids are, in fact, encoded by more than one triplet and the code is said to be redundant.

The cell uses the information in the genes by decoding the information in the genes. The sequence of bases in the DNA is not decoded directly into an amino acid sequence and hence a protein. There are several steps to the process. First, the DNA sequence is transcribed into a sequence of bases in another type of nucleic acid molecule – messenger RNA (mRNA). One difference between RNA and DNA is that, in the former, the base uracil is used instead of thymine. The mRNA molecules formed after transcription may undergo complex processing (outlined below) before the information in the RNA is translated into a sequence of amino acids which make up a particular protein. In higher organisms that contain a well-defined nucleus within the cell, transcription is carried out in the nucleus, but the mRNA molecules are then transported into the cytoplasm where translation takes place. The translation process takes place using ribosomes and several different species of transfer RNA (tRNA) molecules (small RNA molecules), each species becoming chemically attached to one specific amino acid. The tRNA + amino acid complex base pairs with a particular triplet of bases (the codon) in the mRNA. This takes place in ribosomes, which are the workbenches of the cell where the protein synthesis is carried out. A polypeptide chain is synthesized by the addition of one amino acid at a time; each amino acid residue is brought to the ribosome by a tRNA molecule and a new chemical bond is formed between the existing peptide and the newly imported amino acid. The sequence of the amino acids in the growing peptide is determined by the sequence of the bases in the mRNA. This information, in turn, has been encoded in the DNA. Thus, the DNA sequence determines the sequence of amino acids in the protein. **Table 1** shows the meaning that each triplet of bases has when translated.

The information in **Table 1** was gleaned from biochemical experiments in which mRNA molecules were synthesized in the laboratory. These were used to see which amino acids were incorporated into polypeptides made under instruction

Table 1 The codon assignments in the genetic code. Each triplet of bases has a specific meaning, usually for the insertion of a particular amino acid during translation [but see footnotes [a b c]]. Some minor differences in the genetic code are known for some organisms; these are not included in this table.

		Second position								
First position		U		C		A		G		Third position
	U	UUU	⇒ Phe	UCU	⇒ Ser	UAU	⇒ Tyr	UGU	⇒ Cys	U
		UUC		UCC		UAC		UGC		C
		UUA	⇒ Leu	UCA		UAA [c]	STOP	UGA [c]	STOP	A
		UUG		UCG		UAG [c]	STOP	UGG	⇒ Trp	G
	C	CUU	⇒ Leu	CCU	⇒ Pro	CAU	⇒ His	CGU	⇒ Arg	U
		CUC		CCC		CAC		CGC		C
		CUA		CCA		CAA	⇒ Gln	CGA		A
		CUG		CCG		CAG		CGG		G
	A	AUU	⇒ Ile	ACU	⇒ Thr	AAU	⇒ Asn	AGU	⇒ Ser	U
		AUC		ACC		AAC		AGC		C
		AUA		ACA		AAA	⇒Lys	AGA	⇒ Arg	A
		AUG [a]	⇒ Met	ACG		AAG		AGG		G
	G	GUU	⇒ Val	GCU	⇒ Ala	GAU	⇒ Asp	GGU	⇒ Gly	U
		GUC		GCC		GAC		GGC		C
		GUA		GCA		GAA	⇒ Glu	GGA		A
		GUG [b]		GCG		GAG		GGG		G

Ala, alanine; Arg, arginine; Asn, asparagine; Asp, aspartic acid; Cys, cysteine; Gln, glutamine; Glu, glutamic acid; Gly, glycine; His, histidine; Ile, isoleucine; Leu, leucine; Lys, lysine; Met, methionine; Phe, phenylalanine; Pro, proline; Ser, serine; Thr, threonine; Trp, tryptophan; Tyr, tyrosine; Val, valine.

[a] Used to specify the initiation of translation. The START codon.

[b] GUG is an ambiguous codon. It usually specifies valine but can code for methionine to initiate translation from the mRNA.

[c] UAA, UAG and UGA do not specify an amino acid. They signal the termination of translation and are known as STOP or "nonsense" codons.

In some circumstance, some codons have exceptional meanings. For example, in the mitochondria of many organisms UGA results in the insertion of Trp into the growing polypeptide chain. In these cases it does not mean STOP.

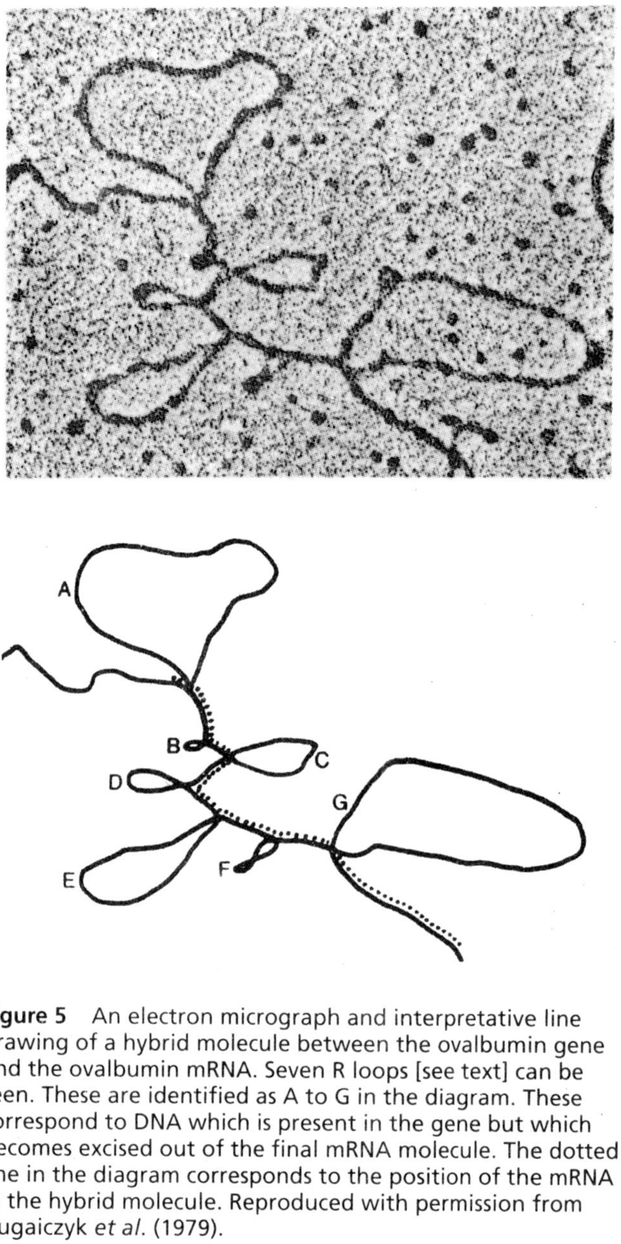

Figure 5 An electron micrograph and interpretative line drawing of a hybrid molecule between the ovalbumin gene and the ovalbumin mRNA. Seven R loops [see text] can be seen. These are identified as A to G in the diagram. These correspond to DNA which is present in the gene but which becomes excised out of the final mRNA molecule. The dotted line in the diagram corresponds to the position of the mRNA in the hybrid molecule. Reproduced with permission from Dugaiczyk *et al.* (1979).

from the artificial messages. In other experiments, a triplet-binding assay was developed that allowed tRNA molecules, carrying a specific amino acid, to bind to very small mRNA molecules consisting of only three bases. The binding was specific. Any one triplet of RNA bases bound only one amino acid. This work led to the elucidation of the genetic code and the award of the Nobel Prize to Marshall Nirenberg and Gobind Khorana in 1968.

Towards the end of 1977, two groups of workers (Jeffreys and Flavell, working with β-globin, and Breatnach *et al.* working on ovalbumin in chickens) reported results which gave a new insight into how genes are organized in the chromosomes. RNA molecules that encode information for the synthesis of proteins can act as templates for the synthesis of DNA using an enzyme called reverse transcriptase. The DNA so formed is called cDNA and is an exact copy of the messenger. When these cDNA molecules were compared with the genomic DNA from which they were transcribed, a surprising result was obtained. The β-globin gene in the genomic DNA contained more nucleotide bases than the cDNA. The inference that was drawn was that some DNA present in the genes in the nucleus did not end up in the mRNA. Direct evidence for this conclusion was obtained by Tilghman *et al.* (1978) and Dugaiczyk *et al.* (1979) for the β-globin and ovalbumin genes, respectively. Both groups used a method in which they annealed mRNA and DNA molecules and examined the results using an electron microscope. The mRNA molecule is complementary to one of the DNA chains of the double helix from which it has been transcribed. The mRNA can, therefore, be annealed to the DNA to form a DNA–RNA duplex. This process is called hybridization. When double-stranded DNA is denatured and then mixed with mRNA, the DNA–RNA duplex prevents the two DNA chains from re-annealing. The result is a displaced chain of DNA called an R loop and this can be visualized directly using an electron microscope. The double-stranded DNA will be a different thickness from the single-stranded R loop. When this was done for both the β-globin and ovalbumin genes, several R loops were seen, each separated by regions of double-stranded DNA. These results can be seen in **Figure 5** together with an interpretive diagram.

To explain these observations, it was suggested that the unpaired regions in the DNA did not end up in the mRNA. This idea turned out to be correct – both the β-globin gene and the ovalbumin genes contain intervening sequences or introns. These turn out to be a common feature of genes of higher organisms. The initial RNA product of transcription, the primary transcript, is a molecule that contains the introns, but these are removed in a complex series of chemical reactions known as mRNA processing or splicing. This removes the introns and joins the exons into a continuous mRNA molecule that can then be translated. In the nucleus, the primary RNA transcript becomes associated with nuclear proteins and other nuclear RNAs in giant heterogeneous nuclear particles. These probably have several different functions in the processing of the mRNA and its transport out of the nucleus. In 1993, Philip Sharp and Richard Roberts were awarded the Nobel Prize for their work on RNA processing of split genes.

(iii) Control of gene activity

Early genetic experiments tackled the problem of understanding the basis of variation between individuals. A different problem involves understanding variation observed within an individual, the most obvious being the development of a multicellular organism from a single-cell zygote. Many different cell types are formed during the growth of the adult organism, but, if successive mitotic divisions provide each cell with an identical set of chromosomes and genes, how can the genes form the basis of this differentiation? It now seems obvious that the molecular basis of this differentiation will lie in differential gene activity, with some genes being active in one tissue and a different set of genes being active in another. The precise origin of this idea is, however, quite difficult to trace. In the early stages of genetics, many different mutations were isolated in many organisms that affected their development and morphology. These were a clear indication that genes were involved. The two ideas to be distinguished are, in some ways, quite subtly different. Mutations in genes might affect development because the mutation changes the structure of gene's protein product. This altered product might then affect development by being less or more active, or less or more stable, than the non-mutant product. Alternatively, gene products might be structurally unaffected by a mutation, but the time and place at which the gene is expressed might be altered by mutation. This idea seems obvious now, but, before the exact nature of the gene was discovered, the idea was more abstract. Direct evidence for the existence of genes that control gene activity was obtained by François Jacob and Jacques Monod working with the bacterium *Escherichia coli* (see Jacob & Monod, 1961). The work, which was to result in the award (together with Andre Lwoff) of the Nobel Prize for Medicine in 1965, was a watershed in understanding the control of gene activity. They discovered genetic elements that were involved in the utilization of the sugar lactose. The *i* gene encodes a protein molecule which switches off (or represses) genes used in the utilization of lactose, so that the enzymes needed to use the sugar are not produced when there is no lactose present. The repressor protein recognizes a genetic element, the operator, and binds to it (Gilbert & Müller-Hill, 1965). This prevents the transcription of the genes needed for the bacterium to use lactose. The complex of genes and genetic elements responsible for lactose utilization is called an operon. When lactose is present in the growth medium, and the bacterium needs to synthesize the enzymes needed to utilize this sugar as an energy source, lactose (or allolactose, a close metabolic relative) binds to the repressor and this binding results in a conformational change to the repressor. The change of shape results in a loss of affinity for the operator and transcription is initiated. When the repressor is not bound to the operator, RNA polymerase initiates transcription by binding to the DNA at a specific region, 20–200 bases in length, called the promoter (Burgess *et al.*, 1969). There is substantial sequence variation between different promoter regions, but certain consensus sequences or patterns are found when promoters are compared (Pribnow, 1975; Hawley & McClure 1983). In *E. coli*, the consensus sequences are TTGACA, about 35 bases upstream (i.e. preceding) from the start of transcription and TATAAT, the TATA box, which is 10 bases upstream. Experiments in which DNA signal sequences were removed from their normal position and reinserted elsewhere (using the DNA manipulation techniques described briefly in the next section) demonstrated their importance directly in many cases. The experiments also led directly to the development of useful molecular tools that have played an important part in the cloning revolution described in the next section.

Regulation of gene activity in higher organisms differs from the system in bacteria and is more complex. In particular, genes are not organized into operons in higher organisms. Transcription and translation are uncoupled, with the former taking place in the nucleus and the latter in the cytoplasm after the RNA molecules have been processed and exported. Additional elements are also involved in the control of gene expression, but the central idea that DNA sequences outside the coding sequence are important elements in switching genes on and off still holds. mRNA molecules are synthesized by RNA polymerase II. Although there is considerable variation in the details of the control from one gene to the next, a generalization is possible. *Cis*-acting DNA sequences and *trans*-acting proteins are the two major components in the control of transcription.

In most cases, the TATA box directs RNA polymerase II to begin transcription about 30 base pairs (bp) downstream in mammals and about 100 bp downstream in yeast. Other consensus sequences, located within 100–200 bp from the start of transcription, increase the efficiency of the TATA box. The CCAAT consensus is one of these and there is often a GC-rich sequence too. Enhancers (Khoury & Gruss, 1983; Schaffer *et al.*, 1985) are more distantly located *cis*-acting elements (they can be up to 50 kilobases away from the promoter elements) that greatly increase transcription from promoters on the same DNA molecule.

Apart from RNA polymerase II itself, transcription factors are foremost among the *trans*-acting molecules. A large number of these have now been identified. Some bind to enhancer sequences, others to the promoter. They help to initiate transcription by forming a preinitiation complex with RNA polymerase II. A number of proteins interact with the consensus sequence elements and, in combination, a large number of possibilities exist. It is this combinatorial feature which underlies the variation in gene activity which exists from one tissue to the next.

mRNA levels after transcription are determined not only by the initial level of transcription but also by the stability of the mRNA molecules. The half-life of mRNA is influenced by specific sequences at its untranslated 3′ end. It has been shown that repeats of AUUUA affect the half-life of mRNA molecules by exchanging these sequences between different species of mRNA. A poly-A tail is added to the mRNA by a poly(A) polymerase that can add up to 200 residues at the 3′ end of the mRNA molecules. These residues stabilize the molecules and protect them from degradation.

4. The cloning revolution

The last two or three decades of the 20th century witnessed the most remarkable change in the science of genetics. The change, known as the cloning revolution, has been in the way in which geneticists work. The repercussions are widespread and are still being felt. The cloning revolution has had a direct effect on many different aspects of biology and its application to society.

The objectives in this section are to describe these changes and to consider why they have had such a profound impact. The revolution grew directly out of the detailed knowledge about the genetic material; this knowledge has allowed genetic experiments to proceed in ways that were almost unimaginable just a few years ago. The genetic make-up of organisms, including ourselves, can be examined in ever-increasing detail because very sensitive techniques for isolating and analysing the genetic material have been developed; these techniques exploit an attribute that genetic material must have, namely the ability to replicate itself.

Once the nature of the genetic material and the universality of the genetic code had been worked out, detailed studies into the working of individual genes was possible. Bacteria and bacterial viruses (called bacteriophages) played a prominent part in the exploration of the genetic material because they had an extremely short generation time and were easily grown in the laboratory. It was research with these organisms that led to the discovery of some of the molecular tools necessary for manipulating genes in the test tube.

When bacteriophages are grown in the laboratory, a very convenient way to assess how many viral particles are present is to carry out a plaque assay. A dilute suspension of viruses is mixed with a fairly concentrated suspension of bacteria and the mixture spread onto a Petri dish of nutrient agar. After incubating for a few hours, the bacteria grow to produce a lawn covering the surface of the agar. In places, the viruses have attacked and killed the bacterial cells. This killing zone can be seen as a clearing in the lawn and is called a plaque. Each plaque represents one instance of infection of a bacterium by a virus followed by reinfection and killing of neighbouring bacteria, so that the number of plaques corresponds to the number of virus particles originally present. Several early workers in this area of study noticed that sometimes the ability of the bacteriophage to grow on a bacterial host was restricted in some way. Werner Arber discovered in the early 1960s that bacteria (*E. coli* in this particular case) resisted virus attack by enzymatically cutting up the DNA of the invading virus into small pieces (Arber, 1974). The enzymes that do this are called restriction enzymes and they are now one of the basic tools of genetic analysis. The most useful enzymes are those which recognize particular sequences in the DNA and cut it there to produce staggered breaks. These are called type II restriction enzymes. Hamilton Smith reported the purification of one type II restriction enzyme in 1970 and, in 1975, Daniel Nathans published a paper which showed how the restriction sites could be used for chromosome mapping (Nathans & Smith, 1975). Arber, Smith and Nathans were awarded the 1978 Nobel Prize for Medicine for this work. The diagram below illustrates this for a DNA sequence that has a site recognized by one particular restriction enzyme known as *Bam*HI.

```
          ↓
ATGGGGATCCTTAGCG      ATGGG          GATCCTTAGCG
TACCCCTAGGAATCGC  →   YACCCCTAG   +      GAATCGC
          ↑
```

The *Bam*HI enzyme recognizes the sequence GGATCC in the DNA. The vertical arrows identify the place in this sequence where the *Bam*HI enzyme cuts the DNA. The result of the enzyme action is to generate pieces of DNA, each piece ending with a single-stranded end protruding from the DNA duplex. This is an important feature of the DNA following enzymatic digestion with some type II restriction enzymes; the single-stranded ends are complementary and can be joined up again by simply using the base-pairing properties of the nucleotide bases. Another enzyme, DNA ligase, is used to complete the rejoining of the DNA molecules by reforming the chemical bonds of the DNA backbone. The important point to realize about the use of these enzymes in the laboratory is that they allow DNA molecules to be constructed from widely different sources. For example, human DNA cut with the *Bam*HI enzyme can be mixed with the DNA of, say, the bacterium *E. coli* and novel DNA molecules can be generated because both the human and *E. coli* DNA have the same single-stranded compatible ends. DNA ligase can be used to complete the joining process. Since the discovery of the first restriction enzymes, many different enzymes have been isolated so that a wide range of DNA sequences can be targeted.

Another important technical development takes advantage of a universal property of the genetic material, namely the ability to replicate. If the purification of a protein is carried out in the laboratory, each step in the process reduces the amount of material available. Rigorous purification often results in only tiny amounts of the protein being available for experiments. The reverse is true of experiments involving DNA thanks to the development of a technique that can produce a large amount of DNA starting from minute quantities of starting material. The technique, called the polymerase chain reaction (or PCR for short), uses cycles of DNA synthesis followed by denaturing of the DNA into single strands so that the synthesis can be restarted (Saiki *et al.*, 1985). The result is an exponential increase in the amount of DNA formed. The technique is illustrated in **Figure 6**.

The property of replication of DNA also finds application in DNA sequencing technologies that enable sequencing of extensive tracts of DNA. The Human Genome Project has, as its ultimate goal, the objective of sequencing the DNA of all the human chromosomes (see article on "Human Genome Project").

The rationale of DNA sequencing was initially developed by Frederick Sanger using dideoxynucleotides (**Figure 7**) (Sanger *et al.*, 1977). DNA synthesis is set up in the laboratory by including all the essential components for the reaction to occur. However, dideoxynucleotides (ddNTPs) are also included at a low concentration. These molecules differ only slightly from the deoxynucleotides that are the normal building blocks of the DNA and are incorporated into the growing DNA chain. However, whenever a dideoxynucleotide is incorporated instead of the normal nucleotide, the difference (a lack of hydroxyl groups on the 2′ and 3′ carbon atoms of the sugar) is sufficient to prevent the addition of another nucleotide to a growing DNA chain. Polymerization of each growing chain therefore stops once a ddNTP has been built into the chain. Four reaction mixtures are set up, one with each of a different ddNTP. The reactions that, for example, contain the dideoxycytosine triphosphate will terminate with a cytosine residue. Once the reactions are completed, the DNA strands are separated on a polyacrylamide gel. When sifted through the gel, the distance the

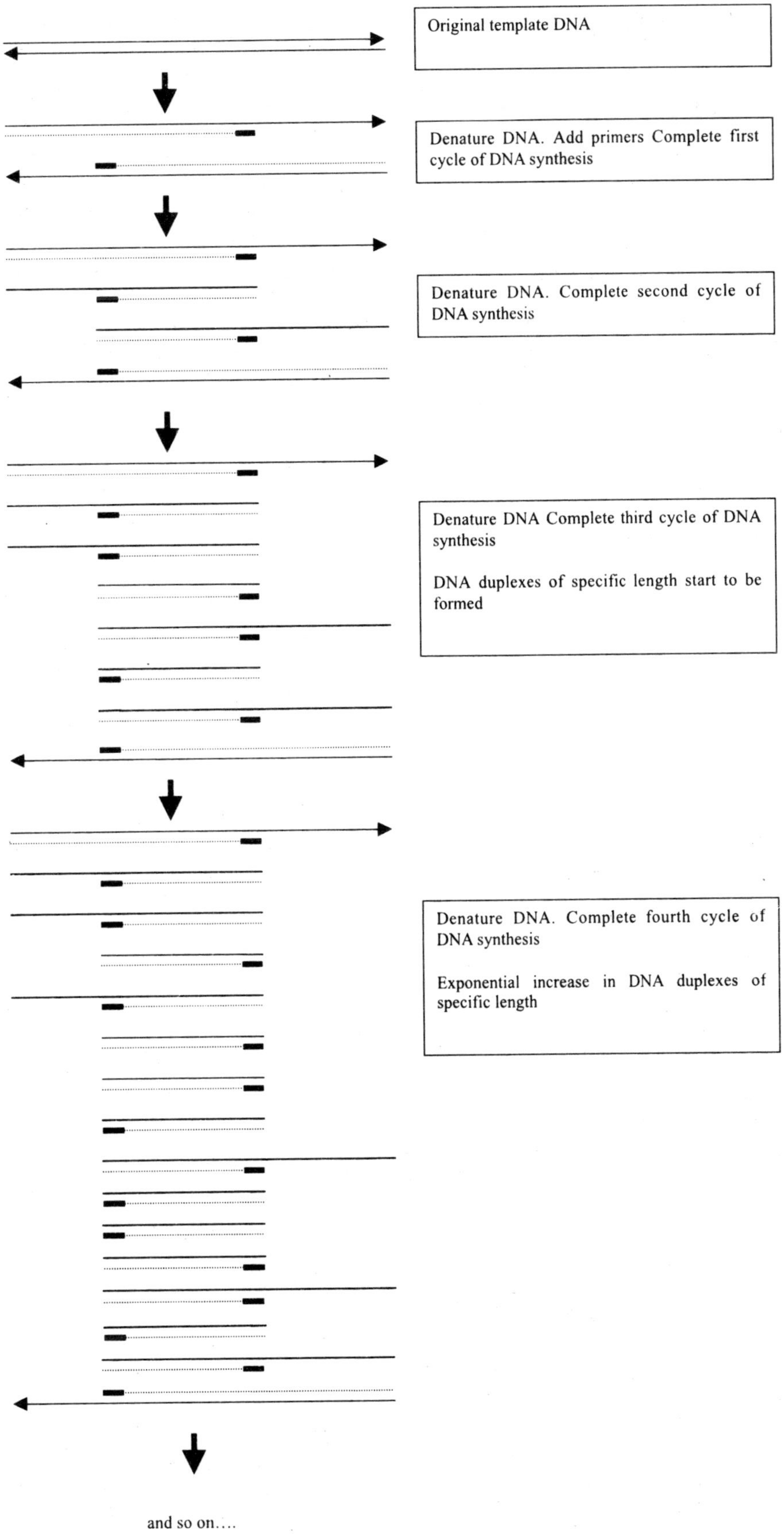

Figure 6 The polymerase chain reaction. Alternate cycles of DNA synthesis and denaturation result in the exponential build-up of DNA fragments of one particular size.

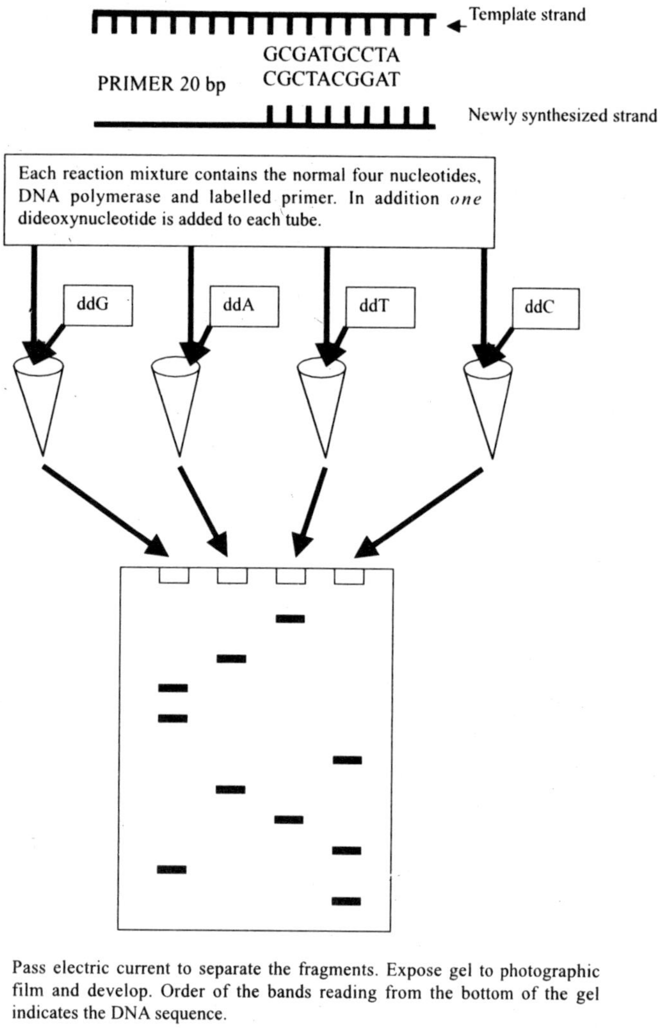

Figure 7 The Sanger method of determining the sequence of bases in a DNA molecule. See text for an explanation.

DNA molecules migrate depends on their length. Radioactivity in the primer molecules enables the position of each band to be determined after exposure of the gel to photographic film. The sequence of bases can then be read off the exposed and developed film.

Automated DNA sequencing uses a modification of the Sanger method. Instead of using radioactive primers to label the reaction products, each ddNTP is labelled with a different fluorescent dye. These cause the molecules to fluoresce differently and each nucleotide can be distinguished. The four ddNTPs can be included in one reaction tube. The molecules are separated, as in manual sequencing, by electrophoresis through a polyacrylamide gel. As the molecules come off the end of the gel, a laser beam is used to excite fluorescence and the machine measures this.

(i) Cloning and molecular evolution

One of the immediate consequences of detailed analysis of individuals at the molecular level was the realization that populations of any sexually reproducing species are highly polymorphic. This was first appreciated when proteins were studied at the molecular level. A change in the DNA encoding a protein can result in a substitution of one amino acid for another. For example, if the DNA sequence encoding a protein reads GGT instead of GAT at a particular site, then glycine, instead of aspartic acid, will be inserted when the protein is made. This change may not affect the function of the protein to any measurable extent, but it may well result in a change of electric charge and this can be detected using gel electrophoresis. This technique can be used to separate protein molecules on the basis of either their charge or their molecular weight. A very large number of species have been analysed using this method and it turns out that differences can be detected in approximately one-third of protein-encoding loci.

When direct analysis of DNA became possible, first by using restriction enzymes and later by direct sequencing, it became very clear that there was a great deal of genetic variation between individuals in a population. The idea that individuals are genetically unique is now a familiar one through the publicity associated with genetic fingerprinting and its use in forensic science. With the exception of identical twins, DNA fingerprinting allows any two individuals to be distinguished by characterizing the DNA at highly variable regions.

Measurements of the extent of genetic polymorphism led directly to controversy over the explanation for it. The controversy started in the 1960s and 1970s and was concerned with the relative importance of natural selection in accounting for the variation. In the late 1960s, Motoo Kimura (1968) and Jack King and Tom Jukes (1969) suggested that most of the variation seen at the molecular level is driven by random drift and did not result from natural selection. According to the neutral theory (Kimura, 1983; see article on "Motoo Kimura and the neutral theory"), the molecular variation seen in a population is too great to be explained by natural selection; according to this view, the majority of variants are selectively neutral and confer no significant advantage to the individual possessing them. The neutral theory suggests that any disadvantageous mutation is rapidly eliminated from the population and so the observed variation is due to selectively neutral mutations. Selectionists take a different view: there are enough selectively advantageous mutations to account for the observed polymorphism in populations. The crucial difference between the two theories lies in the relative frequencies of neutral and selectively advantageous mutations (Ridley, 1993). **Figure 8** attempts to summarize the difference between the two points of view.

The lively debate between selectionists and neutralists made an important contribution to the comparison of gene sequences between species. One exciting claim made when comparisons were first made was that gene sequences evolve at a constant rate. This gives rise to a molecular clock and enables an estimate to be made of the time of species' divergence from a simple comparison of the DNA sequences of two species. It seems likely that both natural selection and genetic drift determine the fate of DNA sequence changes. Changes in non-coding DNA probably have no measurable

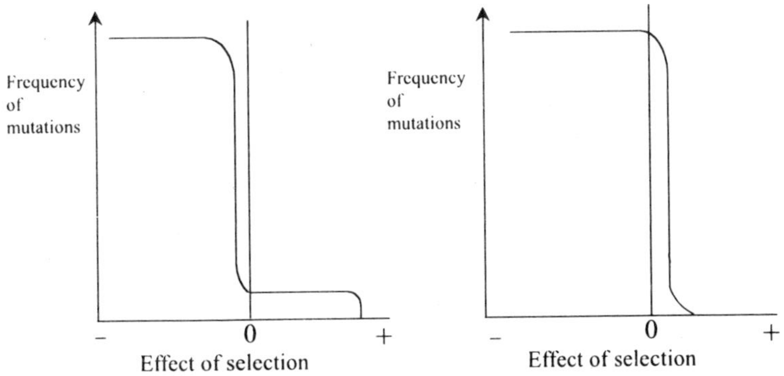

(a) The selectionist view – neutral alleles (effect of selection = 0) are rare.

(b) The neutralist view – neutral alleles (effect of selection = 0) are very frequent.

Figure 8 Illustrating the different views of "selectionists" and "neutralists". The key difference is summed up in (a) and (b) below the graphs. After Ridley (1993).

effect on fitness, but it is likely that changes to the coding regions will do so.

One of the consequences of sequencing DNA has been the development of techniques to infer molecular phylogenies. The theory of evolution suggests that all of life has a common ancestry. The discovery that all of life shares the same genetic code supports this idea and the analysis of DNA sequences is just one (very powerful) method of inferring this ancestry. Phylogenetic tree building methods using DNA sequences allow the evolutionary past to be reconstructed. Roderic Page and Edward Holmes (1998) provide a good introduction to these methods.

(ii) Cloning and developmental genetics

There is no doubt that developmental biology is one of the great growth areas of modern biology. Great advances have been made in understanding the details of the control of development by utilizing molecular techniques, such as *in situ* hybridization or reporter genes. The idea underlying the development of reporter genes is that the *cis*-acting DNA sequences responsible for determining where and when a gene is active are analysed by manipulating them so that they are attached to a gene, the activity of which can be monitored directly. Green fluorescent protein and β-galactosidase are two such proteins; these allow the spatial and temporal expression of genes to be monitored and demonstrate very elegantly that specific gene products are localized to particular areas of the developing organism. Only a flavour of recent findings can be given in this short article. The control of Drosophila development provides a very good example and is recounted in Peter Lawrence's book *The Making of a Fly – The Genetics of Animal Design* (Lawrence, 1992). Comprehensive collections of lethal mutations that are defective at different stages of early development are the starting point of the analysis. Some disturb the development of the anterior–posterior axis of the fly; some disturb the dorsal–ventral axis. Others, the homeotic mutants noted above, transform one set of body structures into another. Gradients of gene expression are established across the developing embryo from maternal effect genes. The developing zygote inherits a set of proteins and mRNA molecules transcribed from the diploid maternal genome. An example is the *bicoid* gene discovered by Frohnhöfer and Nüsslein-Volhard (1986). If this gene is mutant in the mother, the developing embryos are all grossly abnormal and lack all anterior structures. Both the *bicoid* mRNA and the *bicoid*-encoded protein are present as a steep anterior-to-posterior gradient in the developing embryo. The *bicoid* protein, in turn, regulates other genes, notably a gene called *hunchback*. This gene is a recessive lethal; mutants, like *bicoid*, lack anterior structures. Other genes are involved in establishing the anterior–posterior pattern of the Drosophila embryo. These encode proteins that have DNA binding motifs. These genes are expressed in the zygote and are called *gap* genes because the mutations, when homozygous, create large gaps in the sequence of structures normally seen in the anterior–posterior axis.

(iii) The application of reverse genetics

Techniques that are used to manipulate the genetic material in the test tube form the essential basis of reverse genetics. In the classical genetics phase, genetics was studied using a simple universal methodology. Two individuals with a different phenotype were mated to establish that the difference between them was due to pre-existing mutation. By analysing the results of the mating, inferences were drawn about the organism's genetic constitution (i.e. the genotype). The classical genetic route to understanding can, therefore, be symbolized as:

$$\text{PHENOTYPE} \Rightarrow \text{MUTATION} \Rightarrow \text{GENOTYPE}$$

Reverse genetics became possible with the development of methods to reintroduce DNA into the cell where it could become reintegrated into the chromosomes. Following this development, it is possible to take any fragment of DNA, the role of which in the life of the organism is unknown, and mutate it in the test tube. After transformation back into the cell, the consequence, if any, on the phenotype of the organism can be seen. The reverse genetics route can be symbolized as:

$$\text{GENOTYPE} \Rightarrow \text{MUTATION} \Rightarrow \text{PHENOTYPE}$$

Why should a change in the way genetics is studied have such a profound impact? If the current revolution is compared with previous scientific ones, then you can see an obvious difference. For example, Darwin put forward the evidence for evolution and changed the way in which biologists thought about their subject – he brought about a revolution in biological *thought*. The cloning revolution, however, has changed the way in which geneticists *work*. Perhaps the most obvious reason for the impact of the cloning revolution lies in the fact that it allows scientists to address questions in a completely new way and to obtain insights which were previously unattainable. An example is the investigations that have taken place into the genetic basis of memory. When genetics was concerned with the inheritance of characteristics from one generation to the next, it was obvious that memory has no genetic basis because we do not inherit any memories of the past from our parents. It is possible, however, to investigate this subject by asking whether differences in the ability to remember things have any genetic foundation. This has been done in mice, for example, by taking a piece of DNA that might have some involvement in establishing and maintaining nerve connections in the brain, and mutating it. Following the reintroduction of this altered DNA into a mouse embryo, a novel mouse can be produced with a specific alteration to one gene. This mouse can then be tested for impaired memory and the role of the gene in sustaining memory verified. Mark Mayford and Eric Kandel (1999) have reviewed this fast-moving subject.

Perhaps the biggest impact of the cloning revolution has been on human genetics. Genetic analysis is no longer dependent on designed mating experiments; in principle, any piece of human DNA can be cloned and characterized by DNA sequencing and the sequence information used to identify whether or not the piece of DNA corresponds to a gene. An ambitious project is underway to sequence the DNA of all the human chromosomes. This is known as the Human Genome Project where the term genome refers to the total amount of genetic material in an organism's chromosome set. For humans, this corresponds to 3×10^9 bp of DNA containing between 60 000 and 100 000 genes. It comes as a surprise to many that we can only guess at the number of genes we possess. Only a small minority of human genes has been characterized by their effect on us when they are mutated; the majority have still to be identified and their role in the cell determined. As infectious diseases become treatable, so genetic disease becomes relatively more common as a cause of illness and the importance of fully characterizing the human genome increases. Some of that determination will involve the detection of genes which influence behaviour and other, essentially human, traits, and society as a whole will need to accommodate and deal with the consequences of this increased knowledge (see article on "Human Genome Project"). **Figure 9** attempts to illustrate just some of the applied aspects of the cloning revolution. It is not meant to be an exhaustive statement on the impact of the new technologies, but instead tries to cover a range of diverse applications.

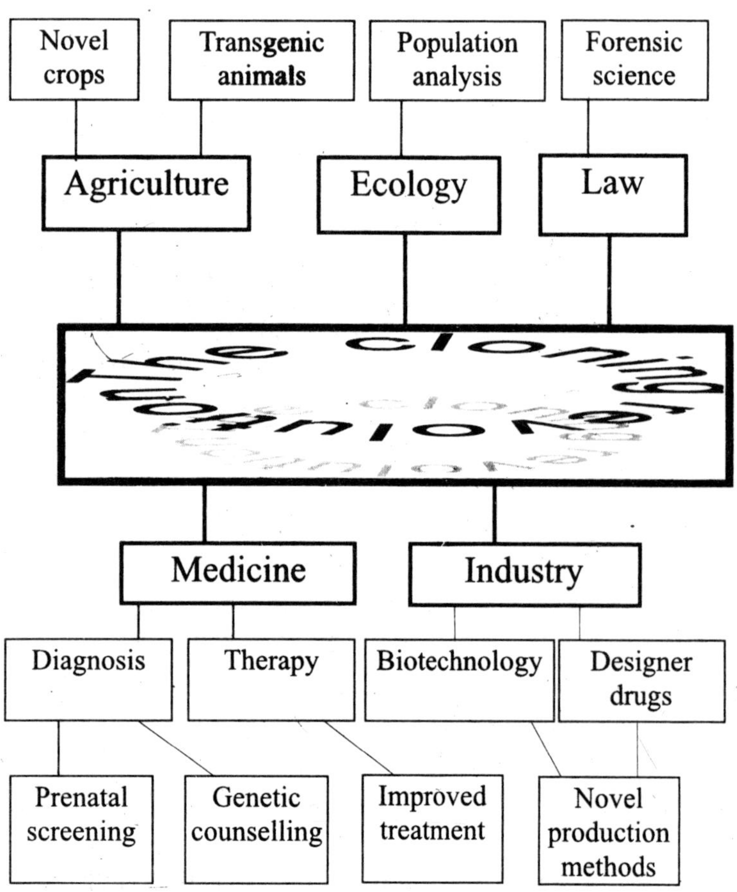

Figure 9 Some practical applications and consequences of the cloning revolution.

The impact of these developments on society is only just being felt. There are many ethical issues that need to be confronted. There is no doubt that legislation to regulate the impact of the cloning revolution will be needed and increased public awareness is necessary for this to occur – the controversy over genetically manipulated food is one example. There are two quite separate issues involved in this particular issue and the popular press often makes little effort to distinguish these. In the first place, there is the question of food safety. Are genetically modified crops safe to eat? Despite the controversy surrounding this question, this is, in principle, not a difficult question to address. Each new genetically manipulated food has to be tested for its safety in the same way as the safety of any new drug is evaluated. If a particular crop proves to be unsafe, its use is discontinued and that is the end of the matter (see article on "Environmental risk and transgenic plants"). Then, there is the separate question of any likely environmental impact if manipulated crops are to be grown extensively (see article on "Determining the impact and consequences of genetically modified crops"). This is a much more difficult problem to assess if only because assessment of the ecological impact requires growing the plants in the open and it is difficult to be sure that any adverse environmental consequences are not irreversible.

The potential for genetic discrimination is another distinct area of concern. There is an increasing ability to analyse the genetic make-up of individuals, especially for characteristics, such as predisposition to heart attacks, which are not clear cut but which result from both genetic and environmental factors. This detailed genetic characterization has serious implications in both employment and insurance. There are many concerns here. As genetic testing becomes more widespread, insurance companies will be concerned if individuals have information about their genetic make-up that is not available to them when they assess life expectancy. On the other hand, there is something intensely personal and private about the genes we inherit from our parents and the extent to which we should have to share this information with others. There is no question that, in this and other areas, society needs to balance the conflicting needs of confidentiality and privacy of the individual on the one hand and the potential abuses of the information on the other.

REFERENCES

Arber, W. (1974) DNA modification and restriction. *Progress in Nucleic Acids Research* 14: 1–37

Avery, O.T., MacLeod, C.M. & McCarty, M. (1944) Studies on the chemical nature of the substance inducing transformation of pneumococcal types. *Journal of Experimental Medicine* 79: 137–58

Barth, L.G. (1949) *Embryology*, New York: Dryden Press

Bateson, W. & Punnett, R.C. (1905–8) Experimental studies in the physiology of heredity. *Reports to the Evolution Committee Royal Society* Reports 2, 3 and 4

Bateson, W. & Saunders, E.R. (1902) Experimental studies in the physiology of heredity. *Reports to the Evolution Committee Royal Society* 1: 1–160

Beadle, G.W. & Tatum E.L. (1941) Genetic control of biochemical reactions in *Neurospora*. *Proceedings of the National Academy of Sciences USA* 27: 499–506

Boveri, T. (1902) Über mehrpolige Mitosen als Mittel zur Analyse des Zellkerns. *Verhandlungen Physikalisch-Medizinische Gesellschaft in Würzburg* N.F. 35: 67–90

Breatnach, R., Mandel, J.L. & Chambon, P. (1977) Ovalbumin gene is split in chicken DNA. *Nature* 270: 314–19

Bridges, C.B. (1914) Direct proof through non-disjunction that the sex-linked genes of *Drosophila* are borne by the X-chromosome. *Science* 40: 107–09

Bridges, C.B. (1916) Non-disjunction as proof of the chromosome theory of heredity. *Genetics* 1: 1–52; 107–63

Burgess, R., Travers, A.A., Dunn, J.J. & Bautz, E.K.F. (1969) Factor stimulating transcription by RNA polymerase. *Nature* 222: 537–40

Correns, C. (1900) G. Mendel's Regel über das Verhalten der Nachkommenschaft der Rassenbastarde. *Berichte der deutschen botanischen Gesellschaft* 18; as G. Mendel's law concerning the behavior of progeny of varietal hybrids in *Genetics* 35 (Suppl., 1950): 33–41

Crick, F.H.C., Barnett, L., Brenner, S. & Watts-Tobin, R.J. (1961) General nature of the genetic code for proteins. *Nature* 192: 1227–32

Darwin, C. (1882) *The Variation of Animals and Plants under Domestication*, vol. 2, 2nd edition, London: John Murray

Dobzhansky, T. (1937) *Genetics and the Origin of Species*, New York: Columbia University Press; many subsequent editions

Doncaster, L. & Raynor, G.H. (1906) On breeding experiments with Lepidoptera. *Proceedings of the Zoological Society London* 1: 125–33

Dugaiczyk, A., Woo, S.L.C., Colbert, D.A. *et al.* (1979) The ovalbumin gene: cloning and molecular organization of the entire natural gene. *Proceedings of the National Academy of Sciences USA* 76(5): 2253–57

Falconer, D.S. & Mackay, T.F.C. (1996) *An Introduction to Quantitative Genetics*, Edinburgh: Oliver and Boyd, and New York: Ronald Press

Fisher, R.A. (1918) The correlation between relatives with the supposition of Mendelian inheritance. *Transactions of the Royal Society of Edinburgh* 52: 399–433

Fisher, R.A. (1930) *The Genetical Theory of Natural Selection*, Oxford: Clarendon Press

Ford, C.E., Miller, O.J., Polani, P.E., de Almeida, J.C. & Briggs J.H. (1959) A sex chromosome anomaly in a case of gonadal dysgenesis (Turner's syndrome). *Lancet* i: 711–13

Frohnhöfer, H.G. & Nüsslein-Volhard, C. (1986) Organization of anterior pattern in the *Drosophila* embryo by the maternal gene *bicoid*. *Nature* 324: 120–25

Garrod, A.E. (1909) *Inborn Errors of Metabolism*, London: Frowde and Hodder and Stoughton

Gilbert, W. & Müller-Hill, B. (1965) The *lac* operator is DNA. *Proceedings of the National Academy of Sciences USA* 58: 2415–21

Gurdon, J.B. (1968) Transplanted nuclei and cell differentiation. *Scientific American* 219: 24–35

Haldane, J.B.S. (1932) *The Causes of Evolution*, London: Longmans Green; many subsequent reprints

Hardy, G.H. (1908) Mendelian proportions in a mixed population. *Science* 28: 49–50

Hawley, D.K. & McClure, W.R. (1983) Compilation and analysis of *Escherichia coli* promoter DNA sequences. *Nucleic Acids Research* 11: 2237–55

Henking, H. von (1891) Untersuchungen über die ersten Entwickslungvorgänge in den Eiern der Insekten. II. Über Spermatogenese und deren Beziehung zur Eientwicklung bei *Pyrrhocoris apterus L. Zeitschrift für Wissenschaftliche Zoologie* 51: 685–736

Hörstadius, S. (1939) The mechanisms of sea urchin development studied by operative methods. *Biological Reviews* 14: 132–79

Huxley, J. (1942) *Evolution: The Modern Synthesis*, London: Allen and Unwin

Jacob, F. & Monod, J. (1961) Genetic regulatory mechanisms in the synthesis of proteins. *Journal of Molecular Biology* 3: 318

Jacobs, P.A. & Strong, J.A. (1959) A case of human intersexuality having a possible XXY sex-determining mechanism. *Nature* 183: 302–03

Janssens, F.A. (1909) Spermatogénèse dans les Batraciens. V. La théorie de la chiasmatypie. Novelles interprétation des cinèsesde maturation. *Cellule* 25: 387–411

Jeffreys, A.J. & Flavell, R.A. (1977) The rabbit beta-globin gene contains a large insert in the coding region. *Cell* 12: 1097–08

Johannsen, W. (1909) *Elemente der Exakten Erblichkeitslehre*, Jena: Gustav Fischer

Khoury, G. & Gruss, P. (1983) Enhancer elements. *Cell* 33: 313–14

Kimura, M. (1968) Evolutionary rate at the molecular level. *Nature* 217: 624–26

Kimura, M. (1983) *The Neutral Theory of Molecular Evolution*, Cambridge and New York: Cambridge University Press

King, J.L. & Jukes, T.H. (1969) Non-Darwinian evolution. *Science* 164: 788–89

Klinefelter, H.F. Jr, Reifenstein, E.C. Jr & Albright, F. (1942) Syndrome characterized by gynaecomastia, aspermatogenesis without α-Leydigism, and increased excretion of follicle stimulating hormone. *Journal of Clinical Endocrinology* 2: 615–27

Lawrence, P.A. (1992) *The Making of a Fly: The Genetics of Animal Design*, Oxford, and Cambridge, Massachusetts: Blackwell Scientific

Lewis, E.B. (1978) A gene complex controlling segmentation in *Drosophila. Nature* 276: 565–70

Mayford, M. & Kandel, E.R. (1999) Genetic approaches to memory storage. *Trends in Genetics* 15: 463–70

McClung, C.E. (1902) The accessory chromosome – sex determinant? *Biological Bulletin Marine Biology Laboratory, Woods Hole* 3: 43–84

Mendel, G. (1866) Versuche über Pflanzenhybriden [Experiments on Plant Hybridization]. *Verhandlungen. Naturforschender verein in Brünn* 4: 3–44. English translation in *Journal of the Royal Horticultural Society* (1901) 26: 1–32; also with commentary in *Experiments in Plant Hybridisation*, edited by J.H. Bennett, Edinburgh: Oliver and Boyd, 1965

Meselson, M. & Stahl, F.W. (1958) The replication of DNA in *Escherichia coli. Proceedings of the National Academy of Sciences USA* 44: 671–82

Morgan, T.H. (1910) Sex limited inheritance in *Drosophila. Science* 32: 120–22

Morgan, T.H. (1934) *Embryology and Genetics*, New York: Columbia University Press

Muller, H.J. (1927) Artificial transmutation of the gene. *Science* 66: 84–87

Nathans, D. & Smith, H.O. (1975) Restriction endonucleases in the analysis and restructuring of DNA molecules. *Annual Review of Biochemistry* 44: 273–93

Page, R.D.M. & Holmes, E.C. (1998) *Molecular Evolution: A Phylogenetic Approach*, Oxford: Blackwell Science

Pribnow, D. (1975) Nucleotide sequence of an RNA polymerase binding site at an early T7 promoter. *Proceedings of the National Academy of Sciences USA* 72: 784–88

Ridley, M. (1993) *Evolution*, Oxford, and Cambridge, Massachusetts: Blackwell Science; 2nd edition, 1997

Saiki, R.K., Scharf, S., Falcona, F. *et al.* (1985) Enzymatic amplification of β-globin genomic sequences and restriction site analysis for diagnosis of sickle cell anaemia. *Science* 230: 1350–54

Sanger, F., Nicklen, S. & Coulson, A.R. (1977) DNA sequencing with chain terminating inhibitors. *Proceedings of the National Academy of Sciences USA* 74: 5463–67

Schaffer, W.E., Serfling, E. & Jasin, M. (1985) Enhancers and eukaryotic gene expression. *Trends in Genetics* 1: 224–30

Stevens, N.M. (1905) Studies in spermatogenesis with especial reference to the "accessory chromosome". *Publication of the Carnegie Institute* 36: 1–32

Sturtevant, A.H. (1913) The linear arrangement of six sex-linked factors in *Drosophila*, as shown by their mode of association. *Journal of Experimental Zoology* 14: 43–59

Sutton, W.S. (1903) The chromosomes in heredity. *Biological Bulletin Marine Biology Laboratory, Woods Hole* 4: 231–48

Tilghman, S.M., Curtis, P.J., Tiemeier, D.C. *et al.* (1978) The intervening sequence of a mouse beta-globin gene is transcribed within the beta-globin mRNA precursor. *Proceedings of the National Academy of Sciences USA* 75: 1309–13

Turner, H.H. (1938) A syndrome of infantilism, congenital webbed neck and cubitus valgus. *Endocrinology* 23: 73–97

Watson, J.D. & Crick, F.H.C. (1953) A structure for deoxyribose nucleic acid. *Nature* 171: 737–38

Weinberg, W. (1908) Über deb Nachweis der Vererbung beim Menschen. *Jahresheft Verein fuer Vaterlaendische Naturekunde in Württemberg* 64: 369–82

Weiss, P. (1939) *Principles of Development: A Text in Experimental Embryology*, New York: Henry Holt

Wilmut, I., Scneike, A.E., McWhir, J., Kind, A.J. & Campbell, K.H.S. (1997) Viable offspring derived from fetal and adult mammalian cells. *Nature* 385: 810–13

Wilson, E.B. (1896) *The Cell in Development and Inheritance*, New York: Macmillan; many later editions and reprints

Wilson, E.B. (1905) The chromosomes in relation to the determination of sex in insects. *Science* 22: 500–02

Wright, S. (1931) Evolution in Mendelian Populations. *Genetics* 16: 97–159

FURTHER READING

Brooker, R.J. (1999) *Genetics: Analysis and Principles*, Menlo Park, California: Addison Wesley

Fairbanks, D.J. & Andersen, W.R. (1999) *Genetics: The Continuity of Life*, Pacific Grove, California: Brooks Cole

Griffiths, A.J.F., Gelbart, W.M., Miller, J.H. & Lewontin, R.C. (1999) *Modern Genetic Analysis*, New York: Freeman

Hartl, D.L. & Jones E.W. (1998) *Genetics: Principles and Analysis*, Sudbury, Massachusetts: Jones and Bartlett

Peters, J.A. (1959) *Classic Papers in Genetics*, Englewood Cliffs, New Jersey: Prentice Hall

See also **Gregor Mendel** (p.62); **Darwin and Mendel united** (p.77); **Human Genome Project** (p.606); **Determining the impact and consequences of genetically modified crops** (p.692); **Environmental risk assessment and transgenic plants** (p.695); **Motoo Kimura and the neutral theory** (p.871)

TECHNIQUES IN MOLECULAR GENETICS

John Maule
MRC Human Genetics Unit, Western General Hospital, Edinburgh, UK

CONTENTS

1. Bacterial artificial chromosomes (BACs)

BACs provide a cloning system capable of accommodating inserts of about 100 kilobases (kb), although recombinants containing at least 300 kb have been isolated (Shizuya *et al.*, 1992). A vector (pBAC108L) based on the *Escherichia coli* F factor has been developed to allow the creation of BACs (see **Figure 1**). The vector is 6.7 kb in size and carries the gene for chloramphenicol resistance (Cm). It also retains the F factor regulatory genes *oriS*, *repE*, *parA* and *parB*. The genes *oriS* and *repE* control plasmid replication whereas *parA* and *parB* control plasmid copy number, maintaining just one or two copies per cell. The unique *cosN* site is cleaved by bacteriophage λ terminase.

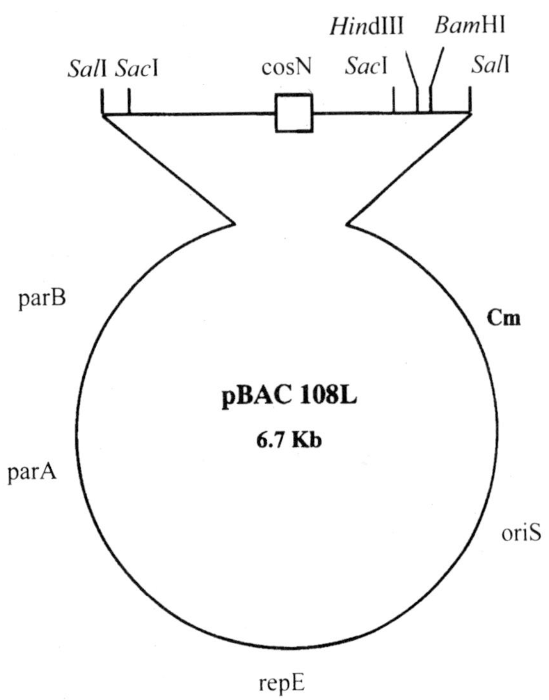

Figure 1 Map of the cloning vector pBAC108L.

The region of the plasmid containing the cloning sites and flanked by the *Sac*I and *Sal*I sites is shown in more detail in **Figure 2**. The unique *lox P* site is cut by the P1 Cre protein. Both unique sites provide useful features for restriction mapping by partial digestion. The plasmid provides *Hin*dIII and *Bam*HI cloning sites so that large insert recombinants can be created by partial digestion of genomic DNA using either of these enzymes and then cloning into linearized pBAC108L vector. Several rare cutter restriction sites (*Xma*I, *Sma*I, *Not*I, *Bgl*I and *Sfi*I) have been created and are useful in long-range mapping. The two *Not*I sites flanking the cloning site allow the recombinant insert to be sized by pulsed field gel electrophoresis. The cloning site is also flanked by sp6 and T7 promoters which can be used for creating riboprobes and for sequencing using sp6 and T7 primers.

Advantages and disadvantages, and some examples

BACs provide several advantages over yeast artificial chromosomes (YACs). The low copy number of the vector reduces the possibility of recombination between plasmids in the same cell and the generation of chimeric molecules. BAC recombinants seem to be very stable and have the advantage that they exist as supercoiled molecules and are thus easy to separate away from the host chromosome. Recombinants can also be easily identified by ethidium bromide-stained pulsed field gels after cleavage with *Not*I. Finally, BACs exhibit a much higher transformation frequency than YACs. There are, however, two distinct disadvantages of the BAC cloning system. First, the low copy number means that DNA preparations yield small amounts

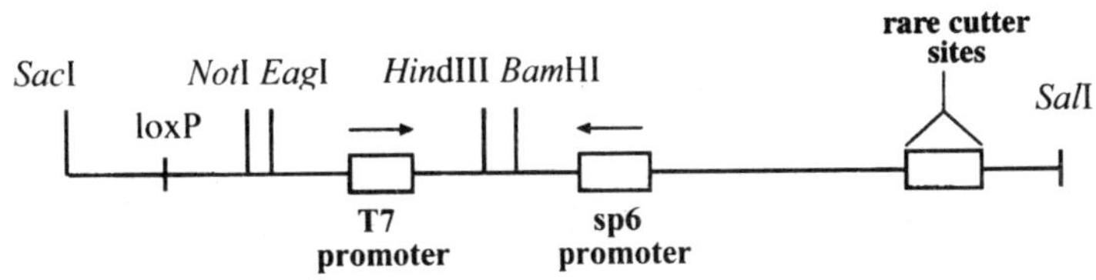

Figure 2 Detailed map of the cloning site of pBAC108L, showing the region between the *Sac*I and *Sal*I sites.

of product. Secondly, there is no positive selection for recombinants.

BACs have been successfully introduced into mammalian cells, where they are able to complement mutations and are stably inherited (Hejna *et al.*, 1998). In one experiment, a BAC carrying the human adenosine phosphoribosyltransferase (*aprt*) gene was electroporated into a mouse fibroblast cell line deleted for *aprt*. The BAC was shown to complement this defect and fluorescent *in situ* hybridization (FISH) analysis demonstrated that it had integrated into a host cell chromosome and was stably inherited. In another experiment, a BAC carrying the hygromycin gene was introduced into a human fibroblast cell line and recombinants were shown to be resistant to hygromycin. Once again, the BAC had integrated into a host cell chromosome.

REFERENCES

Hejna, J.A., Johnstone, P.L., Kohler, S.L. *et al.* (1998) Functional complementation by electroporation of human BACs into mammalian fibroblast cells. *Nucleic Acids Research* 26: 1124–25

Shizuya, H., Birren, B., Kim, U.-J. *et al.* (1992) Cloning and stable maintenance of 300-kilobase-pair fragments of human DNA in *Escherichia coli* using an F-factor-based vector. *Proceedings of the National Academy of Sciences USA* 89: 8794–97

2. DNA sequencing

Methods

The many different approaches to sequencing DNA now rely almost exclusively on the dideoxynucleotide terminator chemistry of Sanger *et al.* (1977). This method involves annealing a primer to the template to be sequenced, and extending out from this primer in a $5' \rightarrow 3'$ direction using DNA polymerase to incorporate the four deoxynucleotides A, C, G and T. If double-stranded DNA is the starting material, then a denaturation step is required to create a single-stranded template. This incorporation of bases will continue, but, if a dideoxynucleotide is included in the reaction mix, then as soon as it is incorporated in the growing chain, the reaction will terminate at this point. So, by including a dideoxynucleotide in the reaction, in the correct ratio, a series of specifically terminated fragments will be generated, all of which share a common 5′ end. Thus, four separate reactions are required for each template, with each reaction mixture containing one of the four dideoxynucleotides ddATP, ddGTP, ddCTP or ddTTP. The reaction products are labelled by either end-labelling the primer, or using a labelled deoxynucleotide or dideoxynucleotide. The labelled reaction products are then loaded onto four consecutive tracks on a denaturing polyacrylamide gel. The labelled fragments are separated and appear as ladders from which the order of bases can be determined (see **Figure 3A**). Originally, ^{32}P was used to label the products, but more recently ^{35}S and ^{33}P have been used (Evans & Read, 1992). By using labelled dideoxynucleotides in the reaction, only correctly terminated sequences will be labelled and prematurely terminated primer extensions, which often occur in regions of secondary structure, will not be labelled. There are also several non-radioactive methods available for terminator sequencing, including biotin and digoxigenin labelling.

The drive to generate high throughput sequencing strategies has led to the development of automated methods for DNA sequencing. These involve labelling the sequencing products with fluorophores, which can be excited by lasers as they approach the bottom of the gel. Computer analysis of the fluorescent output generates a base order in the form of peaks (see **Figure 3B**). Two systems are in common use, one of which makes use of a single fluorophore, such as fluoroscein, and, in this case, four separate reactions – each

Figure 3A Partial sequence of the cystic fibrosis transmembrane conductance regulator (CFTR) gene obtained using ^{33}P-labelled dideoxynucleotide chain terminators.

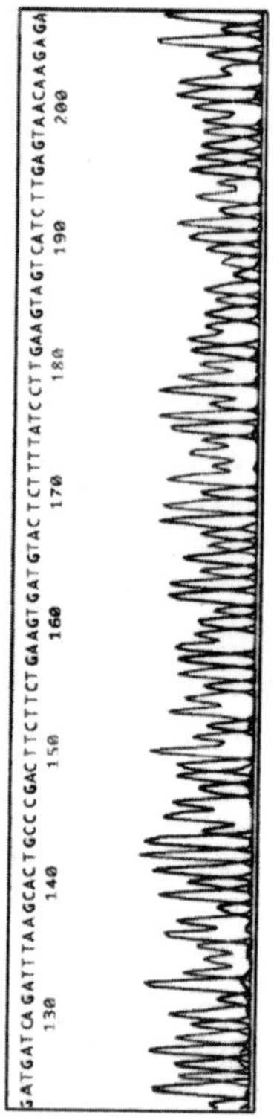

Figure 3B Partial sequence of a PAC derived from human chromosome 4p16, using the ABI Prism™ dRhodamine terminator cycle sequencing method.

using one specific dideoxynucleotide – are run on four consecutive lanes on a sequencing gel. Products are labelled by using either a fluorescently labelled primer or deoxynucleotide. In contrast, the dye-terminator system uses labelled dideoxynucleotides, each of which carries a different fluorophore (Prober *et al.*, 1987). Each fluorophore exhibits a specific emission spectrum so that the four different dideoxynucleotide terminated products can be distinguished. The advantage of this system is that the products of a single reaction, which includes the substrate, a primer, all four deoxynucleotides and all four differentially labelled dideoxynucleotides, can be run on a single sequencing gel track. A variant of this technique, called the dye-primer system, involves labelling the primer with four different fluorophores and carrying out four separate reactions, each with a different dideoxynucleotide.

Cycle sequencing involves performing PCR in the presence of labelled primers, deoxynucleotides or dideoxynucleotides and separating the products on a denaturing polyacrylamide gel (Murray, 1989). Taq DNA polymerase is used and mutant forms of the enzyme have been developed which give more even peak heights. Unlike most PCR reactions, only a single primer is used. This approach has several advantages over other sequencing techniques. First, the amount of starting template can be small (typically a few hundred nanograms). This can be important where the amount of material is limited and it also minimizes the level of contaminants introduced into the reaction. Secondly, since cycle sequencing typically involves many cycles of denaturation at 96°C, annealing at 50°C and primer extension at 60°C, both single- and double-stranded templates can be used without the need for a separate denaturation step. Since the reaction is carried out at an elevated temperature, problems associated with secondary structure are also minimized. As with other sequencing methods, primer, deoxynucleotides or dideoxynucleotides can be radioactively or non-radioactively labelled and methods using different fluorophores allow the process to be automated.

DNA from many different sources can be sequenced by any of the standard methods available. Thus, plasmids, cosmids, M13, lambda and PCR products can be used as templates for sequencing. There are many methods for purifying DNA prior to sequencing, including column methods, magnetic separations and purification from gel bands. DNA samples should be checked for purity, prior to sequencing, by running out an aliquot on an agarose gel. PCR products are frequently used for sequencing and their purification is particularly important. By the conclusion of a PCR reaction, there are still unused single-strand primers and deoxynucleotides as well as uncompleted partial single-strand products. All these reagents must be removed, for they will interfere with the sequencing reaction. A very effective method for accomplishing this is by treatment with exonuclease 1 and shrimp alkaline phosphatase prior to sequencing. The exonuclease 1 removes all single-stranded products and unused primers, and the alkaline phosphatase digests remaining deoxynucleotides. Both enzymes are active in the PCR reaction buffer and can be heat-killed by a short incubation at 80°C. This is a very simple and effective method and can be easily automated. A PCR reaction may produce a poor yield of DNA or reveal multiple bands when examined on an agarose gel. In this situation, the quality of the sequencing reaction can be improved by using a nested primer, i.e. a primer internal to one of the ones used for PCR. Alternatively, the desired band can be excised from an agarose gel and purified before sequencing.

Summary

Current sequencing technology allows the process to be automated and maximizes the length of DNA that can be read. Future developments are aimed at increasing the signal-to-noise ratio and producing cleaner results over longer lengths of DNA.

REFERENCES

Evans, M.R. & Read, C.A. (1992) ^{32}P, ^{33}P and ^{35}S: selecting a label for nucleic acid analysis. *Nature* 358: 520–21

Murray, V. (1989) Improved double-stranded DNA sequencing using the linear polymerase chain reaction. *Nucleic Acids Research* 17: 8889

Prober, J.M., Trainor, G.L., Dam, R.J. *et al.* (1987) A system for rapid DNA sequencing with fluorescent chain-terminating dideoxynucleotides. *Science* 238: 336–41

Sanger, F., Nicklen, S. & Coulson, A.R. (1977) DNA sequencing with chain-terminating inhibitors. *Proceedings of the National Academy of Sciences USA* 74: 5463–67

3. Gene delivery by particle bombardment

Methods

Particle bombardment can be used as a method for introducing DNA into cells by using DNA-coated metallic microparticles. Several systems have been developed to facilitate this and DNA has been successfully introduced into many cell types in a variety of organisms (Franks & Birch, 1991).

Microparticles (or microcarriers) can be made from tungsten or gold, since these two metals are relatively inert and yet sufficiently dense to achieve the necessary momentum to penetrate cell walls. Particles range in size between 0.6 and 3 μm in diameter, the size employed being mainly determined by the size and type of the target cell. Gold is the preferred material since it has been shown that this metal has less toxicity than tungsten when introduced into a variety of living cells. There are several methods for coating the microparticles. DNA is bound to the particles by either $CaCl_2$/spermidine coprecipitation or ethanol precipitation followed by drying before use. A combination of these two methods is sometimes employed. Coated microparticles are then introduced into a chamber and accelerated at high speed towards the target cells.

Originally, gunpowder was used as a propellant, but this had the disadvantage of damaging the target cells and creating unreproducible terminal particle velocities. Two principal methods are now currently in common use, one relying on compressed gas and the other on a high-voltage discharge. The compressed gas method (Sanford *et al.*, 1991), available commercially as the PDS-1000/He system, uses helium as the propellant since this gas is safer than hydrogen and, for a given pressure, projects material faster than other gases. When the supply of gas to the device is connected, the rupture disc tears at a predetermined pressure and accelerates the macrocarrier towards the stopping screen (see **Figure 4A**).

Rupture discs are available in nine thicknesses and designed to burst at between 450 and 2200 pounds per square inch (psi) pressure. Prior to firing, the macrocarrier is coated with DNA-bearing microparticles, which are dried onto the surface of the carrier that faces the target. During operation, the accelerating macrocarrier is abruptly halted by the perforated stopping screen, which allows the microparticles to pass through at high velocity and continue on towards the target cells. The rupture disc, macrocarrier and stopping screen have to be replaced between successive firing events. To reduce the atmospheric drag on the particles to a minimum, a vacuum is created in the chamber between the stopping screen and the target.

There are several variables that can be adjusted to achieve maximum particle penetration (Sanford *et al.*, 1993). Apart from the applied pressure, which is controlled by the thickness of the rupture disc, the size of the microparticles can be varied, as can the distance between the stopping screen and the target and between the rupture disk and the macrocarrier. Gold microparticles are commonly available in diameters of 0.6, 1.0 and 1.6 μm. The strength of the vacuum can also be regulated.

One possible disadvantage associated with this apparatus is the maximum size limitation of the target, which has to be small enough to fit within the bombardment chamber. An alternative helium-powered device in the form of a hand-held gun allows ballistic particles to be delivered to any size of target and opens up the ability to introduce DNA or RNA under *in vivo* conditions (Williams *et al.*, 1991).

The commercially available Helios Gene Gun operates in the relatively low pressure range of 100–600 psi and microparticles are accelerated through a flight chamber maintained at atmospheric pressure. In this particular device, up to 12 preformed cartridges can be loaded into the gun and fired sequentially. Each cartridge consists of a 0.5 inch length of tubing coated with DNA- or RNA-covered gold microparticles. When the gun is operated, helium under pressure is released into the cartridge and sweeps the gold particles from the tubing wall and accelerates them down the barrel towards the target (see **Figure 4B**). A vented spacer at the end of the barrel allows the propellant gas to be released, thus minimizing the physical damage to the target cells. Several other hand-held gun designs have been published, some employing microparticle-coated macrocarriers and stopping screens, similar to the PDS-1000/He device.

While the ballistic devices so far described use compressed gas to accelerate microparticles, a different approach, which relies on a high-voltage discharge, is commercially available as the Accell® gene gun (McCabe & Christou, 1993). Electrical power for this apparatus is provided by a 25 kilovolt (kV) DC supply which discharges through a 25 kV, 2 μF capacitor. A 10 μl drop of water is placed between arc points, 0.5 mm apart, prior to the discharge, which then generates a shock wave and accelerates a particle-coated carrier sheet towards a mesh stopping screen (see **Figure 4C**). The DNA-containing gold particles are released when the sheet is brought to a sudden halt by the stopping screen and they accelerate and strike the target cells. The apparatus is partially evacuated to minimize atmospheric drag on the gold particles.

Since this method does not rely on the release of high pressure gas, which can cause damage to cells, the target can

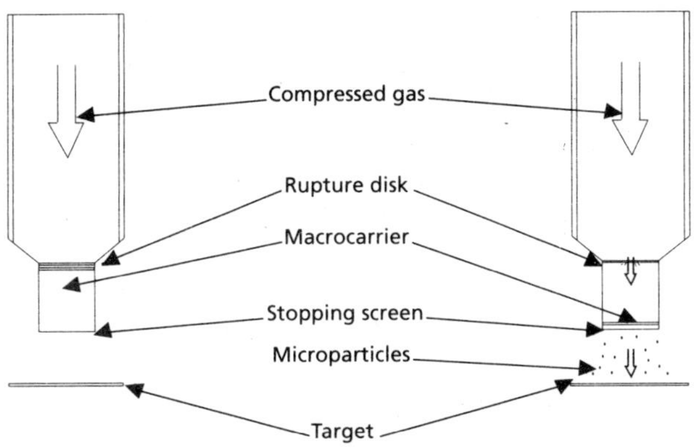

Figure 4A PDS-1000/He system. Left: prior to the tearing of the rupture disk. Right: immediately after the rupture disk has torn and the accelerating macrocarrier has been halted by the stopping screen.

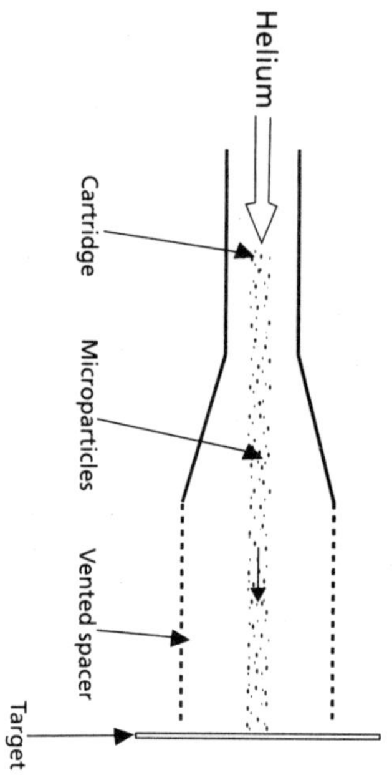

Figure 4B Principle of the Helios Gene Gun. Immediately after firing, microparticles are swept from the walls of the cartridge and accelerated towards the target.

be placed at a very short distance from the stopping screen. The two important variables associated with this device are applied voltage and degree of evacuation. If the vacuum is too high, cells can be dehydrated. Since the speed and penetration of the particles is directly proportional to the applied voltage, care must be taken in optimizing this variable. Too high a voltage can lead to extensive damage to the cells as they become saturated with particles.

There is also a hand-held version of the Accell® device and this operates at atmospheric pressure. It can be used for bombardment of tissues and organs *in situ*. Under these conditions, an applied voltage of 15–22 kV using 1–3 µm particles gives optimum transformation results for a variety of tissues.

Examples of gene delivery

Hand-held guns can be used in particle bombardment of whole animals and plants. *In vivo* transformation of leaves and meristems has been accomplished by particle bombardment of plants *in situ*. Specific tissues or organs can be targeted in animals. Internal organs such as liver, kidney and heart are exposed by surgical incision and the hand-held gun positioned right up against the target. Skin cells can be bombarded after removing any hair and layers of dead skin. Many mammalian species have been subjected to transformation in this way and successful transient gene expression demonstrated. This method has also been used to achieve genetic immunization by DNA introduced into skin cells (Tang *et al.*, 1992). Expression plasmids containing human growth hormone (hGH) and human alpha-1-antitrypsin (hAAT) have been introduced via mouse skin and antibodies against these proteins detected in the serum.

Whichever ballistic system is used, it is important not only to determine the number of particles that penetrate the target cells, but also to be able to measure DNA uptake and transient gene expression. DNA sequences to be introduced into cells are generally cloned into high copy number plasmids that also carry reporter or marker genes under the control of an appropriately strong promoter. Two reporter systems have been used extensively in ballistic studies. The first employs the GUS (β-glucuronidase) reporter (Jefferson *et al.*, 1987). Expression of the *gus* (uidA) gene in target cells is measured 2 days after bombardment by incubating with the histochemical substrate 5-bromo-4-chloro-3-indoyl-β-D-glucuronic acid. Expressing cells stain blue, but this treatment is toxic and the cells die.

A second, non-lethal, reporter system uses the firefly luciferase (*luc*) gene (Howell *et al.*, 1991). When incubated with luciferin, expressing cells emit light that can be detected and quantified by a photon-counting camera. Chloramphenicol acetyl transferase (CAT) has also been used as a reporter gene. Since only a small percentage of transiently expressing cells will lead to stable transformants, it is important to optimize the delivery system to produce a high number of expressing cells. Care must be taken in establishing that exogenous DNA really has been stably integrated into the host genome. *In situ* hybridization and Southern blot analysis provide good evidence for integration as does transmission of the integrated sequences to progeny.

Conclusions

Particle bombardment has been used to create transgenic progeny in many species of plants including tobacco, maize, Arabidopsis and rice as well as woody plants such as spruce and larch (reviewed in Klein *et al.*, 1992). This technique has also been used to deliver DNA into cellular organelles.

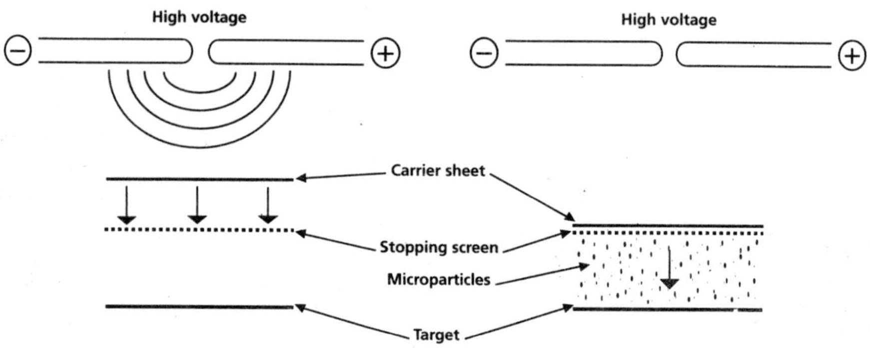

Figure 4C Accell® gene gun. Left: high voltage discharged across an electrode gap creates a shock wave, which displaces the carrier sheet. Right: the accelerated carrier sheet is brought to an abrupt halt by the stopping screen, releasing its load of microparticles, which travel onwards and collide with the target.

Mitochondria and chloroplasts, which are difficult to target by other methods, have been successfully penetrated by ballistic bombardment. Genes involved in photosynthesis or electron transport have been successfully integrated. The yeast *Saccharomyces cerevisiae*, which can be difficult to transform by traditional methods, has been successfully transformed. Thus, in a typical experiment, 1 μg of plasmid DNA containing the selectable marker uracil will transform several thousand uracil auxotrophs to prototrophic progeny using selection on a uracil-deficient medium. Ballistic gene delivery has certain advantages in facilitating gene therapy. It may be biologically safer than viral delivery and experiments have already shown that some of the particles penetrate the nucleus. Since many gene delivery systems introduce DNA into the cytoplasm, where a high percentage is degraded, then any method which achieves nuclear targeting will increase the chances for incorporating exogenous DNA into the host genome.

REFERENCES

Franks, T. & Birch, R.G. (1991) Microprojectile techniques for direct gene transfer into intact plant cells. In *Advanced Methods in Plant Breeding and Biotechnology*, edited by D. Murray, Wallingford: CAB International

Howell, S.H., Ow, D.W. & Schneider, M. (1991) Use of firefly luciferase gene as a reporter of gene expression in plants. In *Plant Molecular Biology Manual*, edited by S.B. Gelvin, R.A. Schilperoot & D.P.S. Verman, Dordrecht: Kluwer

Jefferson, J.G., Kavanagh, T.A. & Bevan. M.W. (1987) GUS fusions: β-glucuronidase as a sensitive and versatile gene fusion marker in higher plants. *EMBO Journal* 6: 3901–07

Klein, T.M., Arentzen, R., Lewis, P.A. & Fitzpatrick-McElligot, S. (1992) Transformation of microbes, plants and animals by particle bombardment. *Bio/Technology* 10: 286–91

McCabe, D. & Christou, P. (1993) Direct DNA transfer using electric discharge particle acceleration (Accell® technology). *Plant Cell, Tissue and Organ Culture* 33: 227–36

Sanford, J.C., DeVit, M.J., Russell, J.A. *et al.* (1991) An improved helium-driven biolistic device. *Technique* 3: 3–16

Sanford, J.C., Smith, F.D. & Russell, J.A. (1993) Optimizing the biolistic process for different biological applications. *Methods in Enzymology* 217: 483–509

Tang, D.C., DeVit, M. & Johnston, S.A. (1992) Genetic immunization is a simple method for eliciting an immune response. *Nature* 356: 152–54

Williams, R.S., Johnston, S.A., Riedy, M. *et al.* (1991) Introduction of foreign genes into tissues of living mice by DNA-coated microprojectiles. *Proceedings of the National Academy of Sciences USA* 88: 2726–30

4. *In situ* hybridization (ISH)

Methods

ISH is a method for detecting nucleic acid sequences within the cell (Leitch *et al.*, 1994). DNA or RNA probes are labelled, denatured and then hybridized to cellular targets that have themselves been denatured to create single-stranded sequences. Hybridization is detected by several methods depending on how the probe is labelled. Early ISH procedures involved labelling with radioisotopes such as tritium, ^{35}S or ^{32}P. While the high energy β particle-emitter ^{32}P produces a hybridization signal which can be detected within a few days by autoradiography, the resolution is poor. A lower energy source such as tritium produces more discrete signals, but the exposure time for detection can be several weeks or months.

More recently, radioisotopes have been largely replaced by non-radioactive labels such as fluorochromes, biotin or digoxigenin. Modified nucleotides chemically linked to such fluorochromes as rhodamine, FITC or AMCA are incorporated into DNA or RNA and the hybridization signal detected by microscopy and excitation of the target by appropriate wavelength light. Probes labelled in this way are not subject to decay and can be stored for long periods of time before use.

There are several methods for incorporating label into probes:

1. RNA probes can be labelled by cloning an appropriate sequence into a vector containing an RNA polymerase promoter. The labelling reaction takes place in the presence of RNA polymerase and a mixture of labelled and unlabelled nucleotides.
2. DNA probes can be labelled by nick translation, random prime labelling, PCR or end-labelling.
   - *Nick translation* involves the treatment of double-stranded DNA with DNAase I, which creates single-stranded nicks at regular intervals, and then the nicks are repaired by DNA polymerase I, which catalyses the incorporation of complementary labelled and unlabelled nucleotides.
   - *Random prime labelling* involves incubating denatured DNA with a mixture of all possible oligonucleotides, which anneal to the single-stranded sequence at regular intervals of about 100 bases apart. The oligonucleotides serve as primers for the Klenow fragment of *Escherichia coli* DNA polymerase which incorporates labelled and unlabelled nucleotides during strand synthesis. This method is particularly valuable when starting with small DNA fragments of low concentration.
   - *PCR* allows both efficient labelling and amplification from small quantities of starting material. Each of many cycles of amplification involves a denaturation step followed by primer annealing and finally DNA synthesis in the presence of labelled and unlabelled nucleotides and a thermostable DNA polymerase. There are several more specialized forms of PCR labelling useful for preparing probes for ISH. Interspersed repetitive sequence PCR (IRS-PCR) utilizes primers based on repetitive sequences and allows synthesis between, for example, Alu repeats which are on average 3–4 kilobases (kb) apart in the human genome. This technique is particularly appropriate for generating probes from yeast artificial chromosomes (YACs), since only the human component is amplifiable. An alternative approach is catch-linker PCR (Shibasaki *et al.*, 1995) in which oligonucleotides are ligated onto the ends of short YAC restriction fragments and used as

primers for PCR. Degenerate oligonucleotide-primed PCR (DOP-PCR) and sequence-independent amplification (SIA) use partially degenerate primers allowing amplification from dispersed sites which are species independent.

- *End-labelling* of DNA involves the addition of labelled and unlabelled nucleotides to the end of a DNA fragment and is catalysed by either terminal deoxynucleotidyl transferase or polynucleotide kinase. Oligonucleotides which cannot be labelled by other methods can be used in this procedure.

Primed *in situ* labelling (PRINS) (Gosden, 1997) provides another method for simultaneously incorporating labelled nucleotides and hybridizing them to chromosomes. Chromosomal material fixed to microscope slides is denatured and then annealed to complementary oligonucleotide primers, which act as templates for the incorporation of exogenous nucleotides in a PCR reaction. A labelled nucleotide is supplied for the reaction, which can be performed on a thermocycler modified to accommodate microscope slides.

Once hybridization of the probe to the target is complete (this generally involves an overnight incubation), the position at which the probe binds can be revealed by a variety of detection systems. One method involves antibodies raised against the label and conjugated to a signal-generating system, which allows the label to be visualized. Thus, digoxigenin(DIG)- and biotin-labelled probes can be detected by antiDIG or antibiotin, respectively. Biotin can also be detected by a signal-generating system conjugated to avidin or streptavidin, both of which bind strongly to biotin. Signal-generating systems can be fluorochromes, enzymatic procedures or metals.

- *Fluorochromes* generate signals by excitation with light of an appropriate wavelength which is detected following passage through optical filters. Fluoroscein isothiocyanate (FITC), rhodamine and Texas red are three commonly used fluorochromes which have different excitation and emission wavelengths. FITC produces a green fluorescence whereas rhodamine and Texas red produce red. Different probes can be detected simultaneously if differentially labelled and detected with different fluorochromes having characteristic colours.
- *Enzymatic detection* involves conjugating an enzyme to an antibody raised against the hybridized label. Horseradish peroxidase is a commonly used enzyme detector and utilizes diaminobenzidine (DAR) as a substrate, producing a brown precipitate localized at the point of hybridization. Alkaline phosphatase is another enzyme, acting on 5-bromo-4-chloro-3-indoylphosphate/nitroblue tetrazolium (BCIP/NBT) as substrate and producing a blue deposit. These detection methods have the advantage of yielding permanent signals that can be detected easily by simple microscopy.
- One of the most sensitive *metal-detecting* systems uses colloidal gold conjugated to antibodies which can be visualized by light or electron microscopy and has the advantage of allowing a semiquantitative estimation of signal strength.

The microscopic method chosen to visualize an ISH signal depends very much on the signal-generating system employed. Transmission light microscopy is useful for detecting radioactive label and enzymatically generated signals. By using appropriate filters and counterstaining with cytological stains, the signal can be localized accurately to the target. Epifluorescent microscopy involves illuminating the specimen with filtered light from a mercury vapour lamp. This incident light causes excitation of the fluorochrome which then emits fluorescence at a longer wavelength. The output signal is selected and filtered ready for observation. More recently, laser scanning confocal microscopy has been used to observe fluorescent signals. The specimen is excited by a spot of laser light that scans across the slide and the fluorescent signal has to pass through a confocal aperture which only allows the passage of light from within the focal plane. Signals from above and below the focal plane are excluded from the aperture. The signal ultimately passes through a photomultiplier and is then displayed on a monitor screen. Since the images are digital, they can be stored, processed and quantified by a computer. Where sensitivity is a major consideration, then the cooled charge-coupled device (CCD) camera allows the efficient collection of photons from very weak signals and displays the image on a monitor.

Uses of fluorescent in situ *hybridization*

Since the advent of fluorescent *in situ* hybridization (FISH) in the late 1970s, this method has gradually become the method of choice for ISH (Espinosa & Le Beau, 1997). Labelled probes can be hybridized to chromosomes or sectioned tissue, immobilized on a microscope slide. Probes can be generated from plasmids, lambda phage, cosmids, bacterial artificial chromosomes (BACs), P1-derived artificial chromosomes (PACs) or yeast artificial chromosomes (YACs). Clones derived from genomic material will probably contain repeated sequences and the detection of single-copy genomic targets is dependent on the suppression of repeats by hybridization in the presence of sonicated total genomic DNA or the *Cot*1 DNA fraction, which is enriched for repeated sequences.

Gene mapping by FISH generally involves simultaneous hybridization with more than one probe, each differentially labelled. Fluorochrome-linked nucleotides incorporated in a probe can generate different colours under fluorescent microscopy. Thus, rhodamine gives a red signal, fluorescein green, coumarin blue and Cy3 orange. Likewise, probes containing biotin and DIG and detected by different antibody conjugates can be distinguished by emitting different colour signals. Two probes differentially labelled and juxtaposed can emit a characteristic fluorescence, as in the case of red and green signals which, when combined, give yellow. The hybridization targets can be aligned to the chromosome structure by counterstaining with propidium iodide or DAPI (4,6-diamidino-2-phenylindole), the latter producing a G-band staining pattern.

Recently, differentially labelled probes specific for individual human chromosomes have produced a karyotype picture with each chromosome having a distinct colour

(Heng *et al.*, 1997). Collections of probes for each chromosome when simultaneously hybridized and detected by fluorescence yield an image in which the whole chromosome fluoresces. This technique, known as chromosome painting, can yield important information on chromosome structure and rearrangements. Translocations, insertions, deletions and duplications can be detected by painting chromosomes with different colours so that changes in chromosome architecture can be traced visually. FISH has been used successfully to detect aneuploidy and, in some situations, has been used as a test in clinical diagnosis. Reverse chromosome painting can be used to map chromosomal aberrations to normal chromosomes. In this approach, aberrant chromosomes are isolated by flow-sorting or microdissection, PCR amplified, labelled and used as probes onto normal human chromosomes.

Comparative genomic hybridization (CGH) allows a complete genomic analysis by FISH and has been used successfully to visualize chromosome changes associated with tumour cell development. Normal and abnormal genomic DNAs are differentially labelled and hybridized to normal metaphase chromosomes and the ratio of one colour to the other becomes an indicator of the extent of chromosome changes in the abnormal sample.

FISH has been used to correlate the distribution of the short interspersed repeat sequences (SINEs; Alu) and long interspersed repeat sequences (LINE; L1) with human chromosome banding. R bands have been found to be Alu-rich and G/Q bands abundant in L1 repeats. Similarly, FISH has provided important insights into the distribution of C + G-rich sequences and the association of CpG islands with coding regions of the human genome. High resolution FISH has also been used to map trinucleotide repeat expansions, which are often associated with neurological diseases in humans.

Evolutionary studies demonstrating the origin and relationships between mammalian chromosomes have been revolutionized by FISH, which has provided an alternative to the laborious study of chromosome banding. Likewise, the repeat sequence $(TTAGGG)^n$ has been shown by FISH to be conserved across vertebrate chromosome telomeres.

High resolution FISH

While the bulk of studies utilizing FISH have been performed on metaphase chromosomes, which are sufficiently condensed to be clearly visible under the light microscope, high resolution FISH, used principally in gene mapping, has demanded the exploitation of more extended chromosomal material. Several alternative approaches are available (see **Table 1**).

Table 1 Resolution provided by the various FISH techniques.

FISH technique	Resolution
Metaphase spreads	2–3 Mb
Prometaphase spreads	1 Mb
Mechanically stretched chromosomes	200–3000 kb
Interphase nuclei	50–1000 kb
Fiber FISH	1–300 kb

Mb, megabase = 1 000 000 base pairs

While capturing chromosomes in the early stages of metaphase (prometaphase), during which they are not fully condensed, can increase the resolution to 1 Megabase, a 20-fold increase in resolution can be achieved by using interphase chromosomes which are present to some extent in all metaphase chromosome preparations and particularly abundant if nuclei are arrested in G_1 phase.

A further improvement in resolution can be achieved by using extended chromosome fibres. Chromatin is released from nuclei by chemical treatments such as alkali or detergents and using a variety of techniques including HaloFISH or DIRVISH. Resolution down to 1 kb is achievable and many published images clearly show fluorescent probes hybridized to extended chromatin like beads on a string. One disadvantage of this method is the absence of familiar structural landmarks such as the centromere and telomere.

An alternative approach which spans the resolution gap between fibre FISH and metaphase chromosomes involves mechanically stretching chromosomes (Laan *et al.*, 1995). This can be achieved by cytocentrifugation, which deposits chromosomes onto a glass slide. While banding studies are not possible using stretched chromosomes, the centromere and telomeres are still discernible.

REFERENCES

Espinosa, R. III & Le Beau, M.M. (1997) Gene mapping by FISH. In *Gene Isolation and Mapping Protocols*, edited by J. Boultwood, Totowa, New Jersey: Humana Press

Gosden, J.R. (1997) *PRINS and In Situ PCR Protocols*, Totowa, New Jersey: Humana Press

Heng, H.H.Q., Spyropoulos, B. & Moens, P.B. (1997) Fish technology in chromosome and genome research. *BioEssays* 19: 75–84

Laan, M., Kallioniemi, O.-P., Hellsten, E. *et al.* (1995) Mechanically stretched chromosomes as targets for high-resolution FISH mapping. *Genome Research* 5: 13–20

Leitch, A.R., Schwarzacher, T., Jackson, D. & Leitch, I.J. (1994) *In Situ Hybridisation*, Oxford: BIOS Scientific

Shibasaki, Y., Maule, J.C., Devon, R.S. *et al.* (1995) Catchlinker+PCR labelling: a simple method to generate fluorescence *in situ* hybridization probes from yeast artificial chromosomes. *PCR Methods and Applications* 4: 209–11

5. Mammalian artificial chromosomes (MACs)

MACs allow us to study the minimal chromosomal requirements for accurate replication, segregation and chromosomal function. The large cloning capacity of MACs also makes them ideal vectors for introducing and maintaining exogenous DNA in mammalian cells. Their key feature is that they are autonomously replicating units that do not integrate into host chromosomes and disrupt their function (Monaco & Larin, 1994).

There are three essential components in a MAC – telomeres, a centromere and origins of DNA replication (**Figure 5**). Of these elements, only telomeres are well characterized. They are essentially DNA–protein complexes that maintain the integrity of chromosomes by binding at the ends and preventing DNA loss during replication, and they also prevent fusion of chromosomal ends. The telomeric DNA content

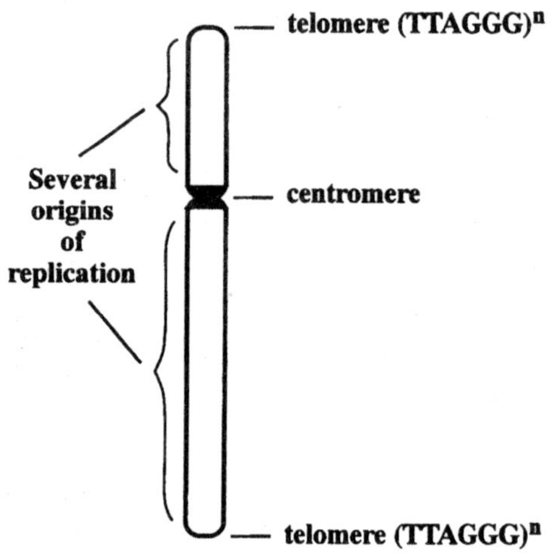

Figure 5 A typical mammalian chromosome, showing the key components: telomeres, a centromere and origins of replication.

consists of long arrays of TTAGGG repeats extending 5–25 kilobases (kb) in length. The telomere length is controlled jointly by telomerase (a DNA polymerase) and the repeat-binding protein TRF1.

Centromeres are the constricted regions of chromosomes and their kinetochore component attaches to spindle fibres during meiosis and mitosis, thus ensuring the correct segregation of chromosomes during cell division. Centromeres are more difficult to isolate and characterize, but consist of alphoid satellite sequences, with a monomer size of 170 base pairs (bp), arranged as repeated sequences in 230 kb–5 Megabase (Mb) arrays, although lengths as small as 140 kb seem to be capable of possessing some of the properties of a fully functional centromere in human cells. Non-alphoid repeated sequences are also present in native centromeres and the higher order centromeric repeat organization confers locus specificity. Several centromere proteins (CENPs), which bind to the centromere/kinetochore complex, have been identified. CENP-B is an alpha satellite-binding protein found associated with both functional and non-functional centromeres. CENP-C and CENP-E are present only at functional centromeres.

Mammalian chromosomal origins of replication are difficult to define and the identification of sites of initiation of DNA replication suggests that there are many origins of replication occurring in human chromosomes at 50–350 kb intervals. The exact sequence of an origin of replication is unknown.

The construction of MACs has followed two different approaches. In a bottom-up approach, the essential components of a chromosome, cloned into yeast artificial chromosomes (YACs), are assembled by homologous recombination in *Saccharomyces cerevisiae*, in much the same way that YACs are created. The ill-defined nature of centromeres and origins of replication, however, has created problems in constructing MACs by this method. Furthermore, the randomly repeated sequences present in centromeres are unstable in yeast. There are also difficulties in transferring the large MACs, constructed in yeast, back into mammalian cells by the conventional methods of lipofection, yeast spheroplast fusion or microinjection. There may also be limitations in assembling even a minimum-sized MAC as a YAC construct, although the technology now exists to fuse individual YACs (Larin *et al.*, 1996).

A less well-defined approach has succeeded in creating a series of human artificial chromosomes (HACs) (Harrington *et al.*, 1997), in which telomeric DNA, 1 Mb of concatemerized alpha satellite sequence from chromosome 17 or Y (each containing a selectable marker) and human genomic DNA are mixed with lipofectin and transfected into a human tumour cell line. The genomic DNA may perform an essential stuffer function and/or contain essential origin(s) of DNA replication.

Analysis of the products from this cloning experiment demonstrated that several different structures had been produced, some involving a mixture of exogenous and endogenous chromosomal material. Crucially, one clone seemed to contain just exogenous DNA and was mitotically stable. These minichromosomes were sized at between 6 and 10 Mb and appeared to be linear.

A more fruitful approach involves top-down methodology in which an existing functional mammalian chromosome is whittled down to the very minimum required for a fully functional unit (Brown *et al.*, 1996). This is achieved by the telomere-fragmentation technique in which cloned human telomeres are introduced into mammalian cells, where they induce chromosome breakage and seed a new telomere at this location. The products of this process can be acentric (no centromere) or centric chromosomes. Although this technique is inefficient, the proportion of fragmented chromosomes can be increased by introducing positive selection for the retention of one chromosome arm and negative selection for the loss of the other. In this way, minichromosomes derived from human X and Y chromosomes have been produced.

In one study (Farr *et al.*, 1995), an X chromosome was broken in the centromeric region producing an acrocentric minichromosome retaining just the short arm (Xp). A second round of telomeric fragmentation produced another breakage in the proximal region of Xp, resulting in a stably inherited minichromosome of 7 Mb, which could be resolved by pulsed field gel electrophoresis. Even smaller minichromosomes have been produced from Y chromosomes by a similar approach, resulting in stable products as small as 4 Mb and with one minichromosome of variable copy number sized at 2.5 Mb (Brown *et al.*, 1996). While the 4 Mb minichromosome was stably inherited in the Chinese hamster ovary (CHO) host cells, when it was transferred to mouse embryonal stem (ES) cells it became unstable and was lost in the absence of positive selection. One derivative of this minichromosome was stable in mouse ES cells, in the absence of selection, and analysis showed that it had acquired mouse centromeric sequences (Shen *et al.*, 1997). A variation of this technique involves incorporating defined homologous sequences along with cloned telomeres, so that the fragmentation efficiency is increased and targeted at specific locations by homologous recombination.

MACs have the potential to act as vectors for the introduction of exogenous genetic material into mammalian cells and may be powerful reagents in future attempts at gene therapy. Clearly, the development of efficient MAC vectors will involve defining the minimum requirements for chromosome function and thereby retaining as little superfluous genomic DNA as possible in the vector.

REFERENCES

Brown, W., Heller, R., Loupart, M.-L., Shen, M.-H. & Chand, A. (1996) Mammalian artificial chromosomes. *Current Opinion in Genetics and Development* 6: 281–88

Farr, C.J., Bayne, R.A.L., Kipling, D. *et al.* (1995) Generation of a human X-derived minichromosome using telomere associated chromosome fragmentation. *EMBO Journal* 14: 5444–54

Harrington, J.J., Bokkelen, G.V., Mays, R.W., Gustashaw, K. & Willard, H.F. (1997) Formation of *de novo* centromeres and construction of first-generation human artificial microchromosomes. *Nature Genetics* 15: 345–55

Larin, Z., Taylor, S.S. & Tyler-Smith, C. (1996) A method for linking yeast artificial chromosomes. *Nucleic Acids Research* 24: 4192–96

Monaco, A.P. & Larin, Z. (1994) YACs, BACs, PACs and MACs: artificial chromosomes as research tools. *Trends in Biotechnology* 12: 280–86

Shen, M.H., Yang, J., Loupart, M.-L., Smith, A. & Brown, W. (1997) Human mini-chromosomes in mouse embryonal stem cells. *Human Molecular Genetics* 6: 1375–82

6. Mutation detection

Methods

There are many methods for detecting mutations and, with the current interest in mapping and characterizing inherited diseases, this area of molecular genetics has assumed major importance (Mashal & Sklar, 1996). Mutations can result from gross changes in DNA structure – such as large deletions, insertions, inversions or duplications – or smaller changes resulting from minor versions of the above, together with point mutations. Major changes in DNA can be detected cytogenetically, by *in situ* hybridization, pulsed field gel electrophoresis, Southern blot hybridization or restriction enzyme analysis. Point mutations can be more difficult to identify and map because, in many cases, only a single nucleotide change is implicated. Nevertheless, many techniques have been developed to address this problem and several new approaches are emerging.

The success of any method depends not only on the efficiency with which it is able to detect mutations but also whether it can simultaneously map the DNA change and how much effort is required to achieve a result. Techniques can be divided into those appropriate for searching for mutations of unknown location and those designed to detect mutations at specific loci. Methods that exploit DNA conformational changes as a result of mispairing are commonly used to search for random mutations.

Conformational change methods

Single strand conformation polymorphism analysis (SSCP)

This is currently one of the most commonly used techniques as it is rapid and requires no specialized apparatus. DNA samples are prepared by PCR which can include the incorporation of a labelled nucleotide. Fragment lengths of less than 300 base pairs (bp) are preferred, so primers should be spaced no more than this distance apart. Should this prove impracticable, then longer amplified products can be cleaved to the correct length by appropriate restriction enzymes. Prior to loading onto the gel, the sample is heated in the presence of formamide and sodium hydroxide to produce single-stranded fragments and then run on a thin non-denaturing polyacrylamide gel.

Mutations are detected by changes in mobility as a result of alterations in DNA tertiary structure when compared with wild-type controls. Detection rates of about 80% have been reported for fragments less than 300 bp long, but the sensitivity of the technique decreases with increased fragment length. Various gel running conditions can be used including varying the amount of cross-linker, temperature of the run or including glycerol in the gel. For maximum detection efficiency, it is recommended to run samples under at least two different run conditions. Alternatively, acrylamide can be replaced by a vinyl polymer called MDE (mutation detection enhancement gel) which can increase the efficiency of detection. After the run, the gel is stained and, if radioactively labelled, then autoradiographed. The most sensitive stains for SSCP gels are either silver or some of the newer products like SYBR™ green I. Bands exhibiting altered mobility can be isolated and reamplified using the same original primers or can be sequenced.

Heteroduplex analysis (HA)

The technique of HA also involves running samples on SSCP gels, but, in this case, heteroduplexes are formed, under carefully controlled conditions, between mutant and wild-type DNA. The mismatched region corresponding to the site of the mutation leads to a change in mobility compared with homoduplex controls. Fragments of less than 500 bp give the best results and, if prepared by PCR, then the amplification conditions must be carefully controlled to minimize any unwanted products that would interfere with heteroduplex formation.

Restriction endonuclease fingerprinting (REF)

PCR-amplified DNA is cut with, for example, five or six restriction enzymes, the fragments mixed, end-labelled, denatured and then run on an SSCP gel. In this situation, the mutation will be present in 10 or 12 different restriction fragments which will increase the likelihood of generating a mobility difference. The mutation may, of course, also produce a restriction fragment length polymorphism (RFLP).

Dideoxy fingerprinting (ddF)

This approach involves sequencing a PCR product or a restriction fragment by the Sanger dideoxy method, using a single dideoxynucleotide. Products are separated on an

SSCP gel, and changes in band mobility compared with wild-type may be observed.

Electrophoretic techniques

There are also electrophoretic techniques which exploit the melting behaviour of double-stranded DNA and the detection of mutations by changes in this property:

- *Denaturing gradient gel electrophoresis (DGGE)* uses a polyacrylamide gel containing a linear gradient of urea/formamide as denaturant. Wild-type and mutant homo- or heteroduplexes can be distinguished because double-stranded DNA melts or separates into single-stranded molecules at specific denaturant concentrations and, in this condition, the DNA exhibits decreased mobility. There are two types of DGGE gel – parallel and perpendicular. In the former, the denaturant concentration increases in the same direction as the electric field and several different samples can be run on a single gel. In the perpendicular gel, the denaturant gradient increases across the width of the gel, at right angles to the electric field. Only one sample at a time can be loaded on this type of gel.

  In both types of gel, mutant and wild-type molecules occupy different positions in the gel depending on their melting characteristics. A DNA fragment of less than 200 bp will probably have several different melting domains. Mutations are best detected if contained within a low melting domain and this can sometimes be achieved by using appropriately positioned PCR primers. Optimum detection occurs with the mutation in a low melting domain, flanked by a high melting domain. This situation can be created by the attachment of a GC clamp to one of the primers. This is a 30–40 base GC-rich oligonucleotide which exhibits a high melting domain and can influence the melting profile of the rest of the molecule. Once optimum conditions for mutation detection have been established by DGGE, then specific melting domains can be analysed by constant denaturant gel electrophoresis (CDGE), which employs a single denaturant concentration and which can provide enhanced conditions for mutation detection. Computer programs are available which predict the denaturant concentration necessary to achieve melting for a given sequence (Lerman & Silverstein, 1987).
- *Capillary electrophoresis* provides a very rapid method for DNA separation and can be used to detect mutations. Constant denaturant capillary electrophoresis (CDCE) is analogous to CDGE in that it employs a uniform concentration of polyacrylamide and denaturant. Separation of mutant and wild-type sequences is achieved by heating the portion of the capillary just prior to the detector to a temperature necessary to create partial melting of the DNA. Fluorescein or rhodamine are used to label the DNA, which is detected by laser-induced fluorescence.
- *Temperature gradient gel electrophoresis (TGGE)* relies on the same principles as DGGE, but it maintains a constant concentration of denaturant and instead varies the temperature across the gel. Mutations are detected by differences in their thermal stabilities compared with the wild-type. In a typical situation, samples are loaded onto a polyacrylamide gel containing 8 M urea and a temperature gradient established by maintaining the gel in contact with a copper plate, the temperature of which varies from one end to the other. In a situation analogous to DGGE, two types of gel can be created utilizing parallel and perpendicular temperature gradients. In the former case, the gradient is established in the same direction as the electric field, so samples gradually experience an increase in temperature as they move down the gel. Several different samples can be run on a single gel. In contrast, perpendicular gels can accommodate just one sample that is loaded into a long slot across the top of the gel and a temperature gradient is established across the width of the gel, at right angles to the direction of the electric field. Samples run on this type of gel exhibit a characteristic profile with a retardation transition at the top of the gel, where, at the lower temperature, the double strands are beginning to open up at the ends. Further down the gel, samples exhibit characteristic transition curves depending on their thermal stability and at the bottom of the gel there is a region of dissociation transition, at the higher temperature, in which molecules are fully dissociated (Tee *et al.*, 1992).
- *Temporal temperature gradient electrophoresis (TTGE) or temperature sweep gel electrophoresis (TSGE)*, like TGGE, uses polyacrylamide gels containing a single concentration of urea, but, in this case, the temperature of the gel is gradually and uniformly increased during the electrophoretic run. The result is equivalent to parallel TGGE. Like DGGE (and TGGE), an amplified sample with a GC clamp gives the best results and software is available to design appropriate primers and position the GC clamp (Yoshino *et al.*, 1991).

Non-conformational change methods

Several other methods that do not rely on conformation changes can be used to detect mutations.

Ribonuclease A protection method

The ribonuclease A protection method has been used with some success (Meyers *et al.*, 1985). Ribonuclease A cleaves single-stranded RNA after a pyrimidine base so that heteroduplexes formed between wild-type and mutant RNAs or RNA–DNA hybrids will be cleaved at the region of mismatch. In practice, hybrids are usually formed between genomic DNA and RNA derived from cDNA. Thus, exons are examined for mutations which are detected at the mismatch. This method has the advantage of not only detecting mutations but also revealing their positions. There are 12 possible mismatches between DNA and RNA and six of these involve pyrimidines in RNA. Four of these are cleaved efficiently by RNAase A: C-A, C-C, C-T and U-T. Thus, eight (66%) of 12 mismatches are detectable if sense and antisense DNA are used in the hybridization. The remaining pyrimidines U-C and U-G are cleaved less efficiently, but can boost the detection rate beyond 66%.

Chemical cleavage method (CCM)

This method cleaves mismatched pyrimidines by using hydroxylamine or osmium tetroxide to destroy the 5,6 double bond allowing subsequent cleavage by alkali or piperidine (Cotton *et al.*, 1988). Although the method specifically cleaves unmatched pyrimidines, even if the mismatch involves purines, then neighbouring perfectly matched pyrimidines up to 6 bases away will be susceptible to cleavage due to the destabilizing effect of the mismatch. In practical terms, a heteroduplex is formed between labelled wild-type and mutant strands. These can be generated by PCR and the wild-type end-labelled. Fragments up to 1.5 kilobases (kb) can be analysed. Controls are homoduplexes between labelled and unlabelled wild-type strands. Once the chemical cleavage reaction has been completed, the products are separated on a denaturing polyacrylamide gel and fragments are detected by autoradiography, or silver staining if label has not been used. Sizing the products allows the location of the mutation to be determined. This method is very sensitive and detects mutations with an efficiency of 90–95%. It suffers from the drawback of employing dangerous reagents, although hydroxylamine and osmium tetroxide can be replaced by the water-soluble carbodiimide.

Fluorescence-assisted mismatch analysis (FAMA)

This is a modification of the chemical cleavage method and utilizes PCR primers labelled with different fluorochromes, so that wild-type and mutant strands can be distinguished (Verpy *et al.*, 1994). The heteroduplex is subjected to chemical cleavage and the products separated on a denaturing polyacrylamide sequencing gel. Product bands are detected by laser excitation and the data analysed by specialist software. This forms the basis of the commercially available ABI 672 Genescan methodology.

Enzyme mismatch cleavage (EMC)

Mismatches in heteroduplexes between wild-type and mutant DNA strands are cleaved by the resolvases T4 endonuclease VII (T4E7) or T7 endonuclease I (T7EI) near the site of the mismatch. These enzymes have slightly different preferences depending on the sequence context, but, when used in combination, about 95% of mismatches are detectable. Fragment lengths of about 1.5 kb, generated by PCR, are suitable substrates and the location of the mismatch can be deduced from the size of the cleavage product, since the enzymes cleave within 6 bases 3′ of a mismatch. The products are separated on a 8% polyacrylamide urea gel and since either wild-type or mutant DNA is end-labelled, then the bands can be located by autoradiography (Youil *et al.*, 1995).

Mismatch repair enzyme cleavage (MREC)

This method makes use of bacterial proteins that recognize misincorporated bases and lead to their replacement. Several proteins have been described and they include MutL, MutS and MutY. The former two merely bind to the mismatch region whereas MutY not only binds but cleaves the DNA too. MutL does not recognize all mismatches whereas MutS binds to all eight mismatch combinations. Binding by these proteins protects the mismatch region against DNase activity, which – together with gel shift assays – form the basis for mutation detection. When MutS protein is bound to a nitrocellulose filter, then its affinity for mismatched DNA is greatly increased. *Escherichia coli* MutY protein recognizes A/G and A/C mismatches, removes the adenine base and cleaves the phosphodiester bond 3′ to the mismatch. Topoisomerase I also recognizes all eight mismatch combinations. Experimentally, DNA from wild-type and mutant cells is PCR amplified, labelled, denatured and then annealed to form heteroduplexes. Oligonucleotides containing the A/G mismatch can be used as a positive control and C/G oligonucleotides as a negative. After incubation with MutY, samples are run on a 14% polyacrylamide sequencing gel, which is then autoradiographed to detect cleaved and uncleaved samples. MutY activity can also be detected by assaying for binding of the protein to a mismatched target. After treatment with MutY, samples are run on a 8% non-denaturing polyacrylamide gel. Autoradiography will reveal that MutY-bound DNA migrates slower than unbound DNA (Ellis *et al.*, 1994).

Low stringency single specific primer-PCR (LSSP-PCR)

This method employs a single primer that anneals to one of the strands at the periphery of the region under analysis (Pena *et al.*, 1994). By employing low stringency conditions that include a low annealing temperature and a high concentration of primer and Taq polymerase, the primer anneals at other non-specific locations as well as to the PCR products as they are synthesized. The result is a complex fingerprint characteristic of the genomic sequence and mutational changes in the substrate will change this fingerprint. The starting material is a DNA fragment of between 250 and 1000 bp that is itself the product of a high stringency PCR reaction, preferably employing a proofreading thermostable DNA polymerase. Following low stringency PCR employing a typical annealing temperature of 30°C, the products are separated on a 6% polyacrylamide gel and the DNA detected by silver staining. This method requires no special equipment and so is cheap and rapid, although the outcome only detects mutations without revealing their location.

Allele-specific PCR amplification (ASPCR)

This method is useful when the location of the suspected mutation is known. One of the PCR primers precisely matches one of the allelic forms but is unmatched to the other. If the mismatch occurs towards the 3′ end of the primer, then only the matched allele will be efficiently amplified. This technique is also referred to as amplification refractory mutation system (ARMS) and allele-specific amplification (ASA).

Sequence-specific oligonucleotide hybridization (SSO)

An oligonucleotide is designed to be exactly complementary to one of the DNA strands in the mutation region. The substrate DNA will be a PCR product and either the substrate or the oligonucleotide can be bound to a membrane with the other component in solution. The probe can be labelled radioactively or fluorescently and the correct design of the oligonucleotide will allow a mismatch to be determined by a poor or negative hybridization result.

Protein truncation test (PTT)

This method detects mutations that interrupt gene reading frames, leading to premature translation termination (Roest *et al.*, 1993). In the first stage of the procedure, PCR templates are generated by the isolation of genomic DNA or the production of cDNA from isolated mRNA that has been amplified by reverse transcription PCR (RT-PCR). These templates are then amplified by PCR using forward and reverse primers. In addition to the target sequence, the forward primer incorporates a number of extra features. At the 5′ end, a T7 (or sp6 or T3) promoter sequence is incorporated followed by a short spacer sequence and then the eukaryotic translation initiation sequence, with the ATG codon in frame with the target gene sequence. Often a restriction site is incorporated into the 5′ end of the primer to facilitate cloning of the amplified product. The PCR product is then transcribed and translated using the rabbit reticulocyte system and the appropriate phage RNA polymerase. A radioactively labelled amino acid such as leucine or lysine is included in the reaction mixture and protein products are separated by SDS-PAGE (sodium dodecyl sulphate-polyacrylamide gel electrophoresis), together with the products of control reactions. Finally, labelled proteins are detected by autoradiography or fluorography. More recently, non-radioactive reagents have been used. For example, biotin-labelled lysine can be incorporated into proteins and, after Western blotting, these can be detected by a streptavidin conjugate followed by visualization using a chemiluminescent reagent. The size of the truncated protein product can be used to locate the position of the mutation.

The PTT method has some limitations. Thus, it will not detect in-frame small deletions, insertions or missense mutations. Furthermore, if the primer binding site is mutated then amplification will not occur. Similarly, gross rearrangements may preclude amplification.

Ligase chain reaction (LCR) or ligation amplification reaction (LAR)

This approach is appropriate in situations where the position of the mutation is known. Amplification is template dependent and utilizes two pairs of oligonucleotides, one pair complementary to the sense strand and one complementary to the antisense strand. Following denaturation of the template, the strand-specific oligonucleotides hybridize juxtaposed such that the 5′ phosphate of one oligonucleotide can be ligated to the neighbouring 3′ hydroxyl of the other member of the pair. The ligation reaction is template dependent, however, so that if there is a mismatch at the junction between the oligonucleotides then ligation is inhibited. If ligation is successful, the product can itself serve as a template for the next round of hybridization. Thus, by using the heat-stable Taq DNA ligase, the reaction can undergo a number of amplification cycles, each involving a denaturation step at 94°C followed by a longer ligation step at 62°C. If the phosphorylated oligonucleotides are end-labelled with γ-^{32}P ATP, then the ligated products can be detected using PAGE and autoradiography. LCR can be used to distinguish between two alleles, so oligonucleotide pairs specific for each allele can be used and positive ligation of a particular oligonucleotide pair will indicate which allele is present (Wu & Wallace, 1989).

Repeat expansion detection (RED)

This technique has been used successfully to detect simple sequence repeats of 100 nucleotides or more in genomic DNA. It is particularly useful in detecting the trinucleotide repeat expansions often associated with human neuropsychiatric diseases (Schalling *et al.*, 1993). Many of these conditions exhibit a phenomenon called anticipation, where successive generations present the disease at an earlier age of onset or with increased severity. The technique involves denaturing genomic DNA by heating at 94°C and then annealing phosphorylated oligonucleotides at a temperature close to the melting temperature of the oligonucleotide used. These oligonucleotides are composed of several tandem repeats, consisting of multiples of three nucleotides and typically ranging in size from 21 to 51 residues. Thus, a 51-mer $(CTG)_{17}$ would hybridize to a CAG expansion. A thermostable ligase such as Taq DNA ligase is included in the reaction and serves to join adjacently annealed oligonucleotides. Many cycles (typically 500) of denaturation, annealing and ligation are performed, using a thermocycler, and the oligonucleotide products are separated on a denaturing polyacrylamide gel. Under appropriate conditions, the size of the largest ligated oligonucleotide product will be equal to the size of the longest repeat expansion in the genomic target. In fact, a range of product sizes is usually seen, representing the end-products from successive reaction cycles. Following electrophoresis and Southern blotting, products are detected by hybridizing with an end-labelled complementary oligonucleotide, followed by autoradiography. Alternatively, by using fluorescein-labelled oligonucleotides in the reaction cycles, the products can be directly detected using a DNA sequencer.

REFERENCES

Cotton, R.G., Rodrigues, N.R. & Campbell, R.D. (1988) Reactivity of cytosine and thymine in single base-pair mismatches with hydroxylamine and osmium tetroxide and its application to the study of mutations. *Proceedings of the National Academy of Sciences USA* 85: 4397–401

Ellis, L.A., Taylor, G.R., Banks, R. & Baumberg, S. (1994) MutS binding protects heteroduplex DNA from exonuclease digestion *in vitro*: a simple method for detecting mutations. *Nucleic Acids Research* 22: 2710–11

Lerman, L.S. & Silverstein, K. (1987) Computational simulation of DNA melting and its application to denaturing gradient gel electrophoresis. *Methods in Enzymology* 155: 482–501

Mashal, R.D. & Sklar, J. (1996) Practical methods of mutation detection. *Current Opinion in Genetics and Development* 6: 275–80

Meyers, R.M., Larin, Z. & Maniatis, T. (1985) Detection of single base substitutions by ribonuclease cleavage at mismatches in RNA:DNA duplexes. *Science* 230: 1242–46

Pena, S.D.J., Barreto, G., Vago, A.R. *et al.* (1994) Sequence-specific "gene signatures" can be obtained by PCR with single specific primers at low stringency. *Proceedings of the National Academy of Sciences USA* 91: 1946–49

Roest, P.A.M., Roberts, R.G., Sugino, S., van Ommen, G.J.B. & den Dunnen, J.T. (1993) Protein truncation test (PTT) for rapid detection of translation-terminating mutations. *Human Molecular Genetics* 2: 1719–21

Schalling, M., Hudson, T., Buetow, K. & Housman, D. (1993) Direct detection of novel expanded trinucleotide repeats in the human genome. *Nature Genetics* 4: 135–39

Tee, M.K., Moran, C. & Nicholas, F.W. (1992) Temperature gradient gel electrophoresis: detection of a single base substitution in the cattle β-lactoglobulin gene. *Animal Genetics* 23: 431–35

Verpy, E., Biasotto, M., Meo, T. & Tosi, M. (1994) Efficient detection of point mutations on color-coded strands of target DNA. *Proceedings of the National Academy of Sciences USA* 91: 1873–77

Wu., D.Y. & Wallace, R.B. (1989) The ligation amplification reaction (LAR) – amplification of specific DNA sequences using sequential rounds of template-dependent ligation. *Genomics* 4: 560–69

Yoshino, K., Nishigaki, K. & Husimi, Y. (1991) Temperature sweep gel electrophoresis: a simple method to detect point mutations. *Nucleic Acids Research* 19: 3153

Youil, R., Kemper, B.W. & Cotton, R.G.H. (1995) Screening for mutations by enzyme mismatch cleavage with T4 endonuclease VII. *Proceedings of the National Academy of Sciences USA* 92: 87–91

7. P1-derived artificial chromosomes (PACs)

Methods

PACs are used for cloning genomic DNA and the vector exhibits features of both the *Escherichia coli* F factor and the P1 phage cloning system (Ioannou *et al.*, 1994).

PAC cloning vectors

PACs offer several advantages over yeast artificial chromosomes (YACs) including an improved cloning efficiency and ease of propagation in *E. coli.* Crucially, they exhibit a reduced capacity for rearrangement due to the vector's single-copy replication property. While cloning into P1 phage imposes an upper limit of about 95 kilobases (kb), due to the constraints of packaging into the phage head, PACs can accommodate inserts of up to 300 kb with an average insert size of 120 kb.

The PAC cloning vector pCYPAC-2 (see **Figure 6A**) has a number of important features designed to maximize efficient cloning and subsequent manipulation and analysis. The vector carries a kanamycin (*kan*) resistance gene that allows positive vector selection. The P1 replicon maintains the single-copy replication facility so that insert-containing transformants can be propagated in single-copy mode. In contrast, the P1 lytic replicon allows multicopy replication, which is a useful feature for DNA amplification. The lytic replicon is regulated by the lactose operon promoter and is turned off if the host cell carries a lacI^q repressor. When multicopy regulation is required, the cells are grown in the presence of the lac inducer isopropyl β-D-thiogalactoside (IPTG). The loxP sequence is used for loxP–Cre homologous recombination. The cloning site is sandwiched between the *E. coli* promoter and the SacB11 gene. In the absence of a cloned insert, the *E. coli* promoter drives the SacB11 gene that converts exogenous sucrose to levan, which is toxic to *E. coli.* This feature provides a positive selection for cloned inserts.

The cloning site (see **Figure 6B**) carries a pUC19 insert which incorporates the gene for ampicillin resistance. For vector-only propagation, the cells are grown in the presence of both ampicillin and kanamycin with the pUC19 insert reducing the effect of the *E. coli* promoter on the transcription of the SacB11 gene.

A modified version of pCYPAC-2 has been constructed in Pieter de Jong's laboratory and is called pPAC4 (http://bacpac.med.buffalo.edu). It carries various additional features that facilitate the generation, analysis and expression of PAC recombinants. These extra features include a site for the yeast intron-encoded enzyme I *–SceI.* This enzyme recognizes an 18 base pair sequence and does not cut the human genome at all. Thus, PAC recombinants carrying human inserts can be linearized with this enzyme. The vector also contains the Epstein–Barr virus replicon (EBV oriP), which permits plasmid replication in human cells. pPAC4 carries the gene for blasticidin-S-methylase, which is under the control of a SV40 promoter, and confers

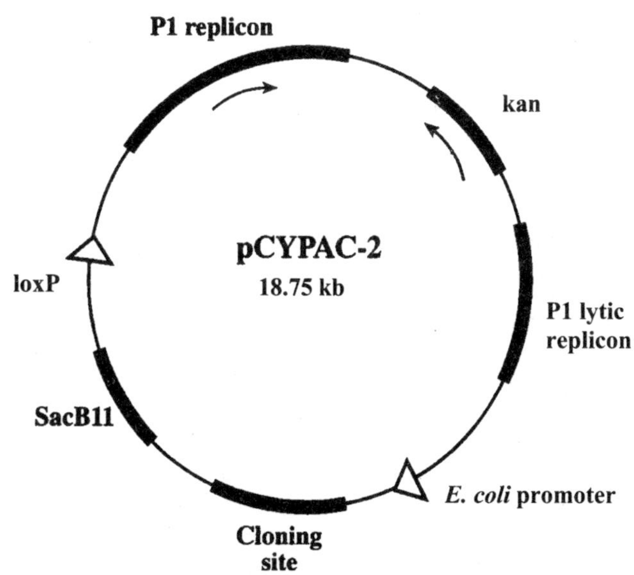

Figure 6A Map of the PAC cloning vector pCYPAC-2.

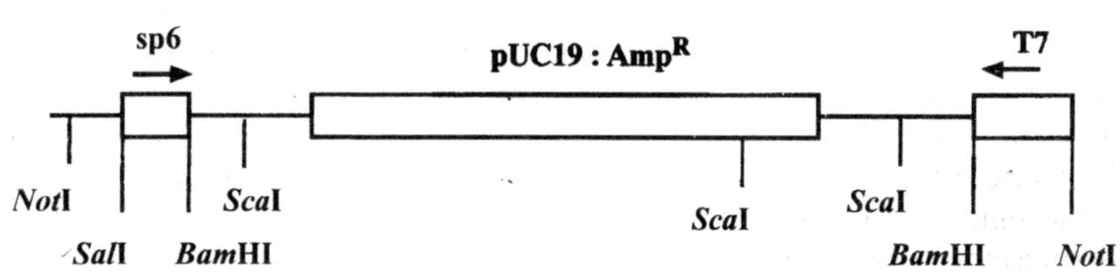

Figure 6B Detailed map of the cloning site of pCYPAC-2 showing the unique *Bam*HI sites used for cloning genomic DNA.

resistance to the antibiotic blasticidin which is used in mammalian cell culture.

Cloning inserts into the pCYPAC-2 vector begins with cleavage by *Sca*I, which destroys the pUC19 insert. Digestion with *Bam*HI generates the unique cloning site and, following phosphatase treatment (to reduce vector self-ligation), the vector preparation is subjected to spin dialysis to remove small *Sca*I/*Bam*HI fragments. Genomic DNA is partially digested with *Sau*3A or *Mbo*I, size selected by pulsed field gel electrophoresis and cloned into the vector. Transformants are generated by electroporating the ligated DNA into electrocompetent cells. Inserts can be subsequently isolated and sized by digestion with *Not*I. The cloning site is flanked by sp6 and T7 promoters that can be used for making RNA probes by *in vitro* transcription. Furthermore, sp6 and T7 primers can be used to sequence the insert ends.

Summary

PACs are proving crucial in many genome mapping programmes and several libraries are now in the public domain. These include human (de Jong, http://bacpac.med.buffalo.edu/), rat (de Jong, http://bacpac.med.buffalo.edu/), mouse (de Jong, http://bacpac.med.buffalo.edu/), Cryptosporidium (UK Human Genome Mapping Project Resource Centre, http://www.hgmp.mrc.ac.uk) and many invertebrate species (Amemiya *et al.*, 1996).

REFERENCES

Amemiya, C.T., Ota, T. & Litman, G.W. (1996) Construction of P1 artificial chromosome (PAC) libraries from lower vertebrates. In *Nonmammalian Genomic Analysis: A Practical Guide*, edited by B.W. Birren, New York: Academic Press

Ioannou, P.A., Amemiya, C.T., Garnes, J. *et al.* (1994) A new bacteriophage P1-derived vector for the propagation of large human DNA fragments. *Nature Genetics* 6: 84–89

8. Polymerase chain reaction (PCR)

Methods

PCR is a method used to synthesize DNA starting from minute quantities of material. Picogram quantities of DNA can be amplified to produce sufficient product to visualize by gel electrophoresis. The technique involves many consecutive cycles in which DNA is gradually synthesized by a thermostable DNA polymerase (Saiki *et al.*, 1988). Starting from small quantities of DNA (the template), short oligonucleotides (primers) are used to initiate the synthesis of DNA by the incorporation of exogenous nucleotide bases.

The cycle of events begins by denaturation in which the DNA template is heated to about 96°C, which separates the strands of the double helix and eliminates any secondary structure. After a few minutes, the temperature is dropped to between 37°C and 65°C and maintained for 2–3 minutes. During this stage, the oligonucleotide primers base pair (anneal) to the denatured single-stranded template. The third stage of the cycle involves incubating at 72°C, which is the optimal temperature for DNA synthesis using the thermostable DNA polymerase (Taq polymerase) isolated from *Thermophilus aquaticus*. During this stage, the enzyme sequentially incorporates the four exogenous deoxynucleotides adenine, guanosine, cytosine and thymine, which are supplied in excess. These reagents base pair to the single-stranded template using the bound oligonucleotide as a primer and extending the DNA in the 5′→3′ direction. Each cycle consists of sequential denaturation, annealing and extension; the number of cycles employed can be as many as 35 (see **Figure 7**).

There are several variables which have to be considered in order to achieve a successful amplification reaction. Sufficient template sequence information needs to be available to design and synthesize oligonucleotide primers. Computer programs are available to design primers of the optimal length and sequence (Rychlik & Rhoads, 1989). Care must also be taken to avoid selecting primers that will spontaneously dimerize or form secondary structures. Under these conditions, the primers will fail to anneal efficiently to the template. The annealing temperature employed during the reaction is related to the length of the oligonucleotide and its base composition. Increasing the annealing temperature when using a particular primer tends to increase the specificity of the reaction.

While a relatively long initial denaturation stage is required to achieve maximum single-strand status, subsequent denaturation stages should be as short as possible in order to maintain DNA polymerase activity. The length of the extension stage depends on the template, but Taq polymerase generally synthesizes 2–4 kilobases (kb) of DNA per minute. The final extension stage is usually prolonged (from 5–15 minutes) and allows the completion of DNA synthesis and the generation of the maximum number of complete products. The efficiency of synthesis can be determined by running an aliquot of the product out on an agarose gel.

Since the introduction of PCR and the use of Taq polymerase as an enzyme capable of maintaining activity even when subjected to intermittent sequential incubation at high temperature, several other DNA polymerases have been isolated from thermophilic bacteria and their activity in PCR demonstrated. Some of these enzymes have similar properties to Taq polymerase, but others possess additional features such as enhanced stability at high temperatures, proofreading functions or the ability to perform additional enzymatic functions.

Taq polymerase exhibits highly processive 5′→3′ activity, but no proofreading function. It has a half-life of less than 5 minutes at 100°C, but about 40 minutes at 95°C. It is capable of incorporating modified nucleotides tagged with, for example, biotin or digoxigenin. Tfl DNA polymerase is isolated from *Thermus flavis* and has similar properties to Taq polymerase. Tli polymerase, from *Thermococcus litoralis*, however, has increased thermal stability and possesses proofreading activity in the form of a 3′→5′ exonuclease that removes single-stranded unpaired bases. These arise during primer extension as a result of the incorporation of incorrect bases by the polymerase enzyme. Taq polymerase exhibits 2×10^{-4} errors per base.

There are situations which require the increased fidelity offered by a polymerase with a proofreading activity. These

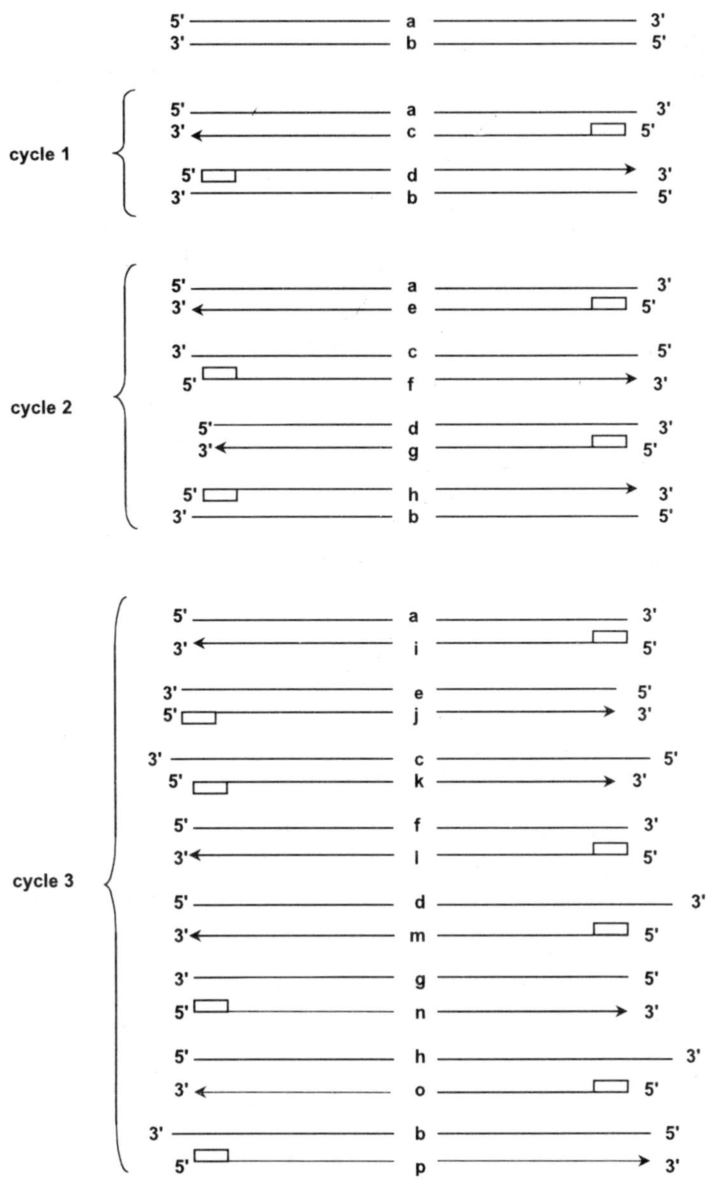

Figure 7 The first three cycles of a PCR reaction, starting from a double-stranded DNA template, the two strands labelled "a" and "b". During cycle 1, DNA strands "c" and "d" are synthesized in the 5'→3' direction starting from two different primers that are complementary to denatured strands "a" and "b". In cycle 2, all four strands from the previous cycle act as templates, producing four newly synthesized strands: "e", "f", "g" and "h". Note that the lengths of strands "f" and "g" are determined by the position of the primer at the 5' end and the 5' termini of the substrates "c" and "d". During cycle 3, eight new strands are synthesized from the eight substrates produced in cycle 2.

include procedures that result in the cloning of PCR products, the investigation of rare mutations or the study of allelic variations between DNA samples. Pwo (from *Pyrococcus woesei*) and Pfu (from *Pyrococcus furiosus*) are two more thermophilic DNA polymerases exhibiting not only highly processive 5'→3' activity but also having the proofreading 3'→5' exonuclease which results in a ten times more accurate amplification of template compared with Taq polymerase.

One disadvantage of the proofreading activity is that the polymerase can digest primers from the 3' end, so it is important to start the PCR reaction as soon as the enzyme is added and possibly use longer primers with the complementarity concentrated at the 5' end. Both these enzymes (Pwo and Pfu) have increased stability at high temperatures. Tth polymerase (isolated from *Thermus thermophilus*) is unusual in not only possessing a 5'→3' function but also having reverse transcription activity at high temperature in the presence of Mn^{2+}. This enzyme can be used to synthesize cDNA from RNA and then to amplify the cDNA if Mg^{2+} is substituted for Mn^{2+}. Tth is particularly valuable in situations where RNA secondary structure is a problem, for the synthesis of cDNA at an elevated temperature removes these structural constraints.

Cloning from PCR products

PCR can be used as an efficient and precise method for producing products for cloning, and several methods have been developed to optimize this process (Costa *et al.*, 1994). The ends of PCR products need to be compatible with appropriate restriction enzyme sites in the cloning vector of choice and this can only be achieved by incorporating suitable restriction sites into the 5' ends of the PCR primers. The amplification primers are incorporated into the 5' ends of PCR products, so long primers can be designed in which their 3' ends are compatible with the target sequences, but the 5′ ends are not, containing instead the recognition sequence for a restriction enzyme. Care must be taken in the primer design, since many restriction enzymes do not cut efficiently if the recognition sequence is at the end of a substrate molecule, so sufficient sequence must be incorporated between the site and the end of the primer. Following PCR, the product is cleaved by a restriction enzyme and can be cloned into a compatible site in a cloning vector (see **Figure 8A**). One disadvantage of this approach is the problem caused by the presence of internal restriction sites, which will create undesired fragments. The use of infrequent cutting enzymes can help to solve this problem.

Another cloning approach relies on the fact that Taq polymerase exhibits terminal transferase activity and adds an additional nucleotide (usually dATP) onto the 3′ end of PCR products (Marchuk *et al.*, 1991). So-called TA cloning vectors carry a single 3′ overhanging thymidine and allow base pairing to the single adenosine residue. This cloning method, however, cannot be used with thermostable DNA polymerases that lack terminal transferase activity. The proofreading polymerases generate PCR products with blunt ends and these must be cloned by the relatively inefficient method of blunt-end ligation. Furthermore, PCR products normally have 5′-OH ends and so must first be phosphorylated with T4 polynucleotide kinase. Blunt-end cloning of PCR products derived from Taq polymerase must be preceded by the removal of the 3′ overhanging nucleotide by 3′→5′ exonuclease treatment, using, for example, the Klenow fragment of DNA polymerase I.

A method that claims to be more efficient than ligase cloning involves the use of the enzyme topoisomerase I. This enzyme recognizes the sequence 5′-yCCTT-3′ (where y is C or T), binds to the phosphate attached to the 3′ thymine and then cleaves just one strand of the DNA substrate. This

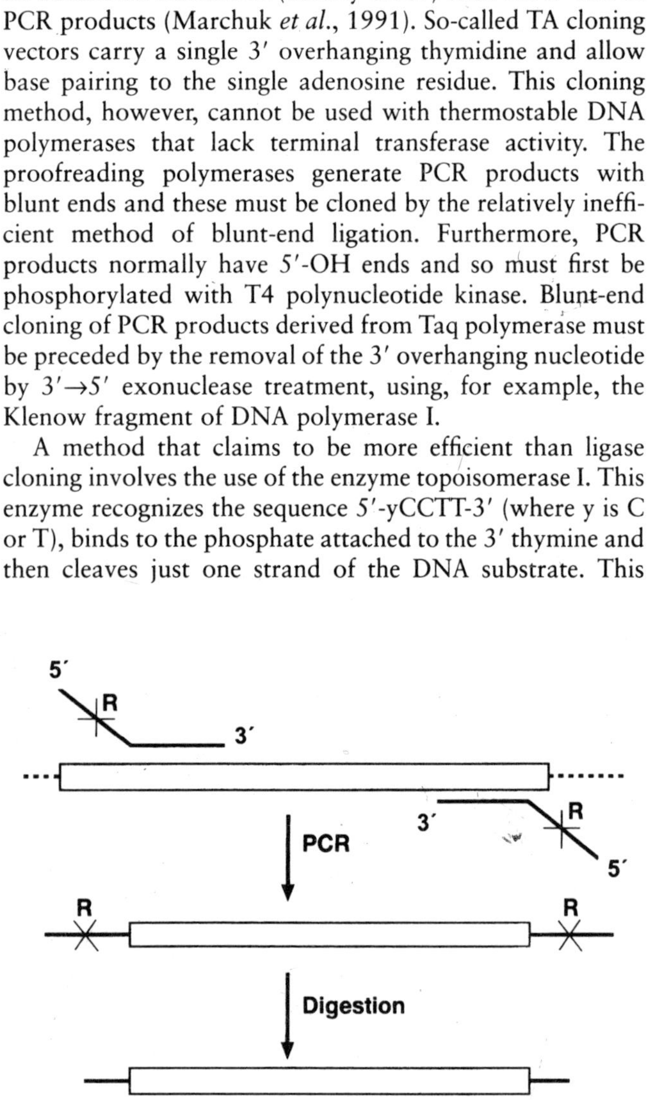

Figure 8A PCR, utilizing primers carrying a restriction enzyme site, "R", in the 5′ region. Note that only the 3′ region of the primer is complementary to the substrate DNA. During PCR, the primers are incorporated into the 5′ ends of the products, which can then be digested with the appropriate restriction enzyme to yield suitable ends for cloning.

enzyme is implicated in DNA replication and allows the double helix to unwind and then religates the cleaved strand. Cloning vectors can be created which, when linearized, carry the yCCTT sequence at their 3′ end. When topoisomerase is added, it binds to the 3′ phosphate and will ligate a fragment carrying a compatible end to the vector:

```
5′ CCCTT 3′   5′  NNNNNNNNN 3′
3′ GGGA 5′    3′ ANNNNNNNNN 5′
```

In this example, a PCR product (NNNNNNN . . .), created by a Taq polymerase catalysed reaction, possesses a 3′ overhanging adenosine residue and will pair with the overhanging thymine present on the vector in the presence of topoisomerase I.

Reverse transcription polymerase chain reaction (RT-PCR)

PCR can be used not only for amplifying DNA but, when used in combination with reverse transcriptase, will also amplify RNA. This process, called RT-PCR, involves producing a cDNA copy of the RNA with the enzyme reverse transcriptase and then amplifying this product by PCR (Hamoui *et al.*, 1994). RT-PCR can be most profitably employed in looking at differential expression of genes during tissue development, cloning cDNA from rare messages or in the analysis of genetic diseases. RT-PCR can be performed with either of the two common reverse transcriptases, from avian myoblastosis virus (AMV) or moloney murine leukaemia virus (M-MuLV), to synthesize the first cDNA strand, followed by a thermostable DNA polymerase for the synthesis of the second cDNA strand and, thereafter, amplification of the double-stranded product. Mutant AMV enzymes have been produced which can be used at higher temperatures (60°C), which helps to overcome the problems of RNA secondary structure and improves primer specificity. These enzymes also tend to exhibit lower levels of RNaseH activity, which leads to RNA degradation. Residual RNase activity can be reduced by deploying a RNase inhibitor such as RNasin®. Both reverse transcriptases can synthesize up to 5–10 kb of cDNA. Recent developments have led to the introduction of buffers compatible with both the AMV enzyme and a proofreading DNA polymerase so that RT-PCR can be performed as a single reaction.

An alternative strategy employs Tth DNA polymerase which exhibits reverse transcriptase activity when Mn^{2+} ions are substituted for Mg^{2+} ions in the PCR reaction. In this situation, however, only products of 1–2 kb are synthesized because Mn^{2+} leads to increases in the PCR error rate.

High-quality RNA template is necessary for successful RT-PCR and can be performed on total or poly(A)$^+$ RNA. Contaminating DNA must be absent. In general, poly(A)$^+$ RNA requires less template for a successful reaction. RT-PCR can be performed using sequence-specific or random primers, or with oligo dT primers which are complementary to poly(A)$^+$ template. Finally, long RT-PCR products can be synthesized by combining AMV enzyme with a long RT-PCR combination of DNA polymerases.

Amplification of long fragments

Taq DNA polymerase can amplify up to 15 kb of λ DNA and up to 3 kb of human genomic DNA. Presumably, this upper limit is determined by the rate of exogenous nucleotide misincorporation, since products containing incorrect bases will not act as perfect templates for subsequent reactions. While proofreading thermostable DNA polymerases can reduce this error rate 10-fold, they will still not amplify targets larger than Taq polymerase. This may be due to primer degradation associated with their 3′→5′ exonuclease activity. Several critical factors have been identified in the quest to amplify longer fragments, one of the most crucial being the use of a combination of DNA polymerases. Thus, amplification of up to 40 kb of λ DNA and 25 kb of human genomic DNA can be achieved by using a combination of Taq DNA polymerase and a smaller concentration of a proofreading DNA polymerase such as Pwo or Pfu (Barnes, 1994; Cheng *et al.*, 1994). There are other factors that affect the length of template that can be amplified. While initial denaturation conditions are typically 2 minutes at 92–94°C, subsequent denaturation steps should last no longer than 10 seconds within this temperature range. The inclusion of various cosolvents (e.g. dimethyl sulphoxide [DMSO], glycerol and Tween®20) helps to lower the strand separation and melting temperatures, and the use of a high pH buffer (typically Tris pH 9.2) provides an environment in which depurination and nicking is minimized. Using reduced potassium concentrations and the inclusion of NH_4^+ can further enhance the reaction. Correctly designed primers are crucial for these reactions in which optimum specificity is essential. Primers tend to be longer than in standard PCR reactions and lengths of up to 34 mers are often employed with melting temperatures of up to 68°C. The use of high annealing temperatures helps to improve specificity. Elongation times tend to be long (10–20 minutes) at 68°C to increase the yield of completely synthesized products. Finally, optimized template preparation is important, with lengths in excess of 50 kb, prepared by agarose gel electrophoresis. The ability to perform long PCR has important applications in physical mapping, cloning and subcloning.

Degenerate oligonucleotide primer polymerase chain reaction (DOP-PCR)

Amplification using a partially degenerate oligonucleotide primer is referred to as DOP-PCR (Wu *et al.*, 1996). It is particularly useful for amplifying limited amounts of DNA or isolating lengths of DNA from unknown sequences. 6-MW is a typical DOP primer and has the sequence: 5′-CCGACTCGAGNNNNNNATGTGG-3′, where the Ns refer to A, C, T and G in roughly equal amounts. DOP primers can be used in pairs to amplify random DNA sequences or in combination with a defined primer to amplify a specific sequence. Thus, the junction fragment spanning the cloning site in a bacterial artificial chromosome (BAC), P1-derived artificial chromosome (PAC) or a yeast artificial chromosome (YAC) can be amplified by using a DOP primer, which anneals to several sites within the insert, and a vector-specific primer designed from the vector sequence. Amplification from both ends of an artificial clone can be achieved by selecting a specific primer from either side of the cloning site. In YACs, left and right arm primers can be designed and, in PACs, oligonucleotides from the sp6 or T7 sequences. DOP primer binding occurs at any sequence that is complementary to 5 or 6 of the 3′ primer nucleotides, so there are usually several binding sites. In amplifying a junction fragment, the DOP-PCR reaction is begun by a single cycle at a low stringency annealing temperature, such as 37°C, in the presence of just the degenerate primer and using a modified T7 DNA polymerase (e.g. Sequenase version 2.0®). Primer annealing and extension take place from several sites. Next, amplification takes place more selectively by using both the degenerate primer and the specific primer in the presence of Taq polymerase and using high stringency PCR conditions – for example, 35 cycles, each consisting of 95°C for 1 minute, 58°C for 2 minutes and 72°C for 2 minutes. Alternatively, the low stringency reactions can involve five cycles with just the degenerate primer in the presence of Taq polymerase, followed by 35 high stringency cycles with both primers present (see **Figure 8B**). The success of the DOP-PCR method depends very much on the production of degenerate primers and correct cycling conditions.

Rapid amplification of cDNA ends (RACE)

RACE is a method used to amplify specific sequences from the termini of cDNA. Many different approaches have been developed (Schaefer, 1995), but all begin with the synthesis of single-stranded cDNA from RNA using the enzyme reverse transcriptase. The exact method employed depends very much on whether 5′ or 3′ ends are to be amplified.

5′ RACE

Amplifying the 5′ terminus of poly $(A)^+$ mRNA involves using a sequence-specific primer to initiate the reverse transcriptase reaction, which synthesizes the complementary single-stranded cDNA strand in a 5′→3′ direction. Next, the RNA is digested away with RNase, so that just the cDNA remains. At this stage, several different approaches

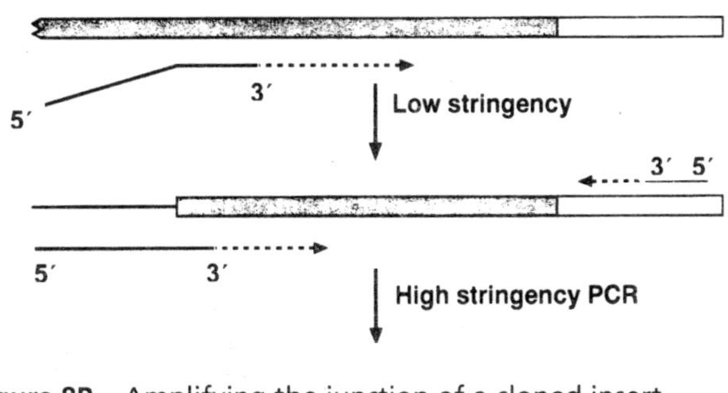

Figure 8B Amplifying the junction of a cloned insert (shaded region) and the vector. PCR begins with low stringency cycles during which the 3′ end of the DOP primer anneals to sites within the insert. There follow many high stringency PCR cycles during which the products of the low stringency cycles are amplified with the DOP primer and a primer specific to the vector. Note that the degenerate primer can now anneal perfectly with the substrates that have incorporated the DOP primer during the low stringency cycles.

are possible, two of which are described below (see **Figure 8C**):

(i) The cDNA can be tailed at the 3′ end with a single deoxynucleotide, using the enzyme terminal deoxynucleotidyl transferase. Assuming cytosine is chosen as the deoxynucleotide, then PCR can be initiated by using a primer containing a complementary poly(G)$^+$ tract as part of its sequence.

(ii) An anchor oligonucleotide can be ligated onto the 3′ end of the single-stranded cDNA using RNA ligase. PCR can commence between a primer complementary to the ligated oligonucleotide and a sequence-specific primer.

3′ RACE

3′ end amplification does not require the addition of any linkers or adapters prior to PCR. The poly(A)$^+$ mRNA is reverse transcribed using an anchor primer containing a poly(T) tract, complementary to the RNA poly(A)$^+$, with the remainder of the primer sequence composed of G, C and A residues. Following digestion of the RNA with RNase, PCR can begin using a sequence-specific primer and the poly(T) anchor primer (see **Figure 8D**).

Once the ends of a cDNA are amplified, the products can be sequenced or cloned and can be used to design further primers which, when used in long-distance PCR, can ultimately provide the complete cDNA sequence.

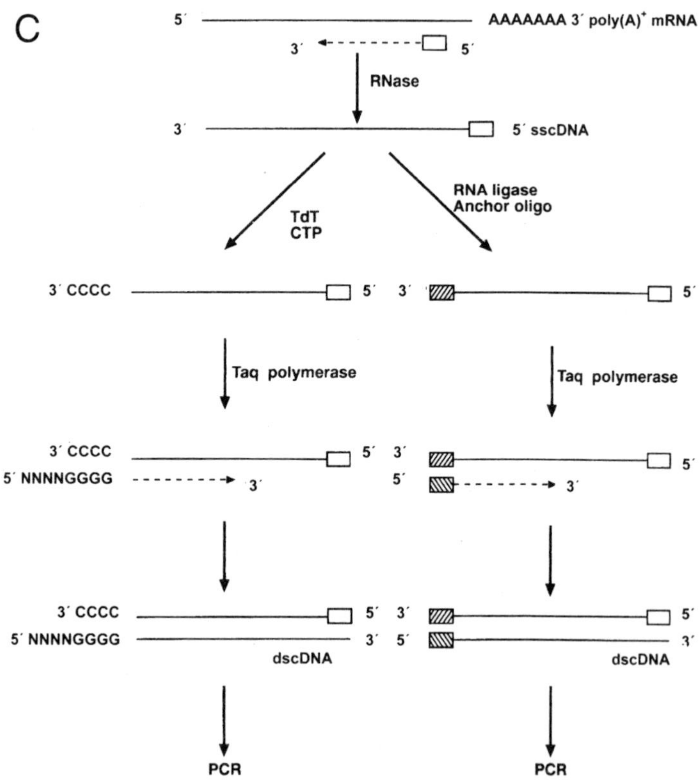

Figure 8C Amplifying the 5′ end of mRNA begins with the synthesis of cDNA by the enzyme reverse transcriptase, which synthesizes single-stranded cDNA (sscDNA) from a sequence-specific primer using the mRNA as substrate. Following RNase digestion, the sscDNA can be treated in two ways. Terminal deoxynucleotidyl transferase can be used to tail the 3′ end of the sscDNA with a single nucleotide, in this case cytosine, then a primer containing a poly(G) tract at the 3′ end can be used to synthesize a complementary strand, thus creating a double-stranded cDNA (dscDNA). (Note: "N" refers to any of the four nucleotides.) PCR can then continue using the sequence-specific primer and the poly(G) primer. Alternatively, an anchor oligonucleotide can be added to the 3′ end of the sscDNA by RNA ligase. Then, a primer complementary to the anchor oligonucleotide can be used in the synthesis of a second strand, creating a dscDNA. This molecule can be amplified by PCR using the sequence specific primer and the complementary anchor primer.

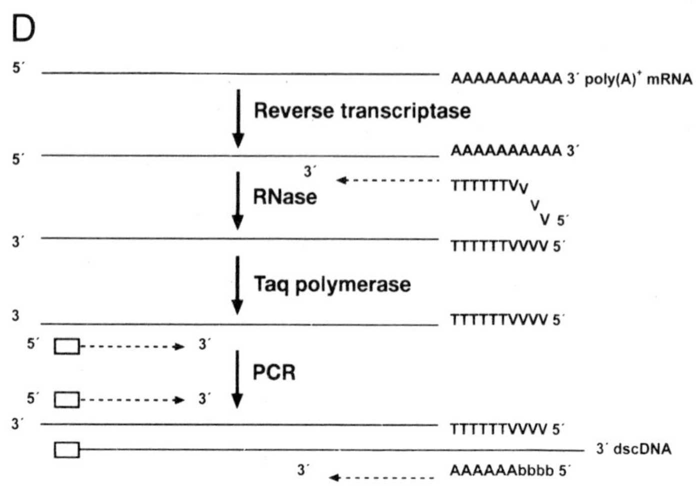

Figure 8D The 3′ end of mRNA is amplified by reverse transcribing the poly(A) mRNA using a primer composed of a poly(T) tract at the 3′ end and a combination of the other three nucleotides in the 5′ region ("V" denotes nucleotides A, C or G). The 5′ region is degenerate and does not anneal to the mRNA. Following RNase digestion to remove mRNA, a double-stranded cDNA (dscDNA) is synthesized using a sequence-specific primer and Taq polymerase. Amplification of this molecule can occur by PCR, using the sequence-specific primer and a primer complementary to the poly(T) primer. (Note: "b" denotes nucleotides C, G or T.)

REFERENCES

Barnes, W.M. (1994) PCR amplification of up to 35-kb DNA with high fidelity and high yield from λ bacteriophage templates. *Proceedings of the National Academy of Sciences USA* 91: 2216–20

Cheng, S., Fockler, C., Barnes, W.M. & Higuchi, R. (1994) Effective amplification of long targets from cloned inserts and human genomic DNA. *Proceedings of the National Academy of Sciences USA* 91: 5695–99

Costa, G.L., Grafsky, A. & Weiner, M.P. (1994) Cloning and analysis of PCR-generated DNA fragments. *PCR Methods and Applications* 3: 338–45

Hamoui, S., Benedetto, J.P., Garret, M. & Bonnet, J. (1994) Quantitation of mRNA species by RT-PCR on total mRNA populations. *PCR Methods and Applications* 4: 160–66

Marchuk, D.M., Drumm, M., Saulino, A. & Collins, F.S. (1991) Construction of T-vectors, a rapid and general system for direct cloning of unmodified PCR products. *Nucleic Acids Research* 19: 1154

Rychlik, W. & Rhoads, R.E. (1989) A computer program for choosing optimal oligonucleotides for filter hybridisation, sequencing and for in vitro amplification of DNA. *Nucleic Acids Research* 17: 8543

Saiki, R.K., Gelfand, D.H., Stoffel, S. *et al.* (1988) Primer-directed enzymatic amplification of DNA with a thermostable DNA polymerase. *Science* 239: 487–91

Schaefer, B.C. (1995) Revolutions in rapid amplification of cDNA ends: new strategies for polymerase chain reaction cloning of full-length cDNA ends. *Analytical Biochemistry* 27: 255–73

Wu, C., Zhu, S., Simpson, S. & de Jong, P.J. (1996) DOP-vector PCR: a method for rapid isolation and sequencing of insert termini from PAC clones. *Nucleic Acids Research* 24: 2614–15

9. Pulsed field gel electrophoresis (PFGE)

Methods

PFGE is a technique particularly suited for separating DNA larger than about 20 kilobases (kb). During conventional agarose gel electrophoresis, DNA larger than this size becomes trapped in the agarose matrix so that molecules of differing size fail to resolve. The upper size limit for resolution can be increased by using low percentage gels and weak electric fields, but running times become excessively long and gels are too fragile to handle easily. In contrast, PFGE has resolved DNA molecules as large as 12 Megabases (Mb). The technique employs a DC electric field of varying direction and/or intensity to the gel, allowing DNA to zigzag through the agarose in response to the changing field conditions. The single most important variable during this process is the pulse time, which is the length of time the field is maintained in any one direction. Short pulse times favour the separation of small molecules and, as the pulse time is increased, progressively larger molecules are resolved.

Various pieces of specialized equipment have been developed to facilitate this technique, but probably the most popular system is the contour-clamped homogeneous electric field (CHEF). This features a hexagonal array of electrodes enclosing the gel. A typical apparatus may contain 24 individual electrodes and each one is "clamped" or fixed at a particular voltage relative to the other electrodes. Thus, in **Figure 9**, if the apparatus is being run at 200 volts (V), then the electrodes in the A^- side of the hexagon will be at zero V and the electrodes along the A^+ side will be maintained at 200 V. Electrodes in the remaining four sides will be fixed at intermediate voltages which gradually extend from 0 to 200 V in the direction $A^- \rightarrow A^+$. The sizes of the "+" symbols in **Figure 9(i)** and **(ii)** are intended to indicate the magnitude of the voltage. The direction of the field in **Figure 9(i)** will be as indicated and DNA will migrate in this direction as long as the field is maintained in that orientation. At the conclusion of this pulse time, the field will switch through 120° with the electrodes in side B^- now at zero V and side B^+ at 200 V. Once again, the intervening electrodes will be at intermediate voltages as depicted in **Figure 9(ii)**, with the electric field being in the direction indicated. At the conclusion of the pulse time, the field will revert to the **Figure 9(i)** orientation. The net effect of the field changes is that as long as the pulse times are of equal duration then DNA will migrate in a straight line down the gel (see **Figure 9(iii)**). Since the CHEF apparatus employs electrodes clamped at specific voltages, the electric field across the gel is totally uniform or homogeneous and samples migrate in a straight line regardless of which well they occupy. Other apparatus employ gels that rotate in a fixed electric field or static gels

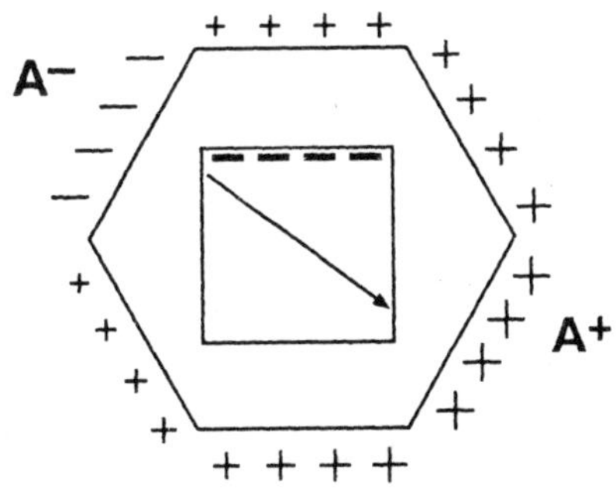

Figure 9(i) A CHEF apparatus depicting all the electrodes on side A^- at 0 V and all electrodes on side A^+ at high voltage. The electric field is in the direction $A^- \rightarrow A^+$.

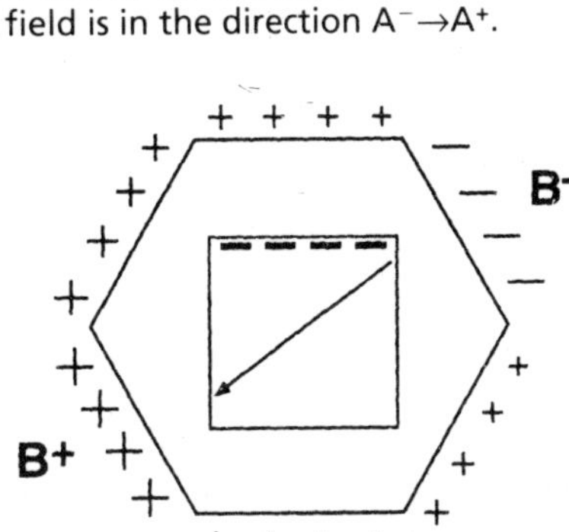

Figure 9(ii) The CHEF apparatus shown in Figure 9(i), in which the electric field has rotated through 120°. The electrodes on side B^- are at 0 V and the electrodes on side B^+ are at high voltage. The electric field is orientated in the direction $B^- \rightarrow B^+$.

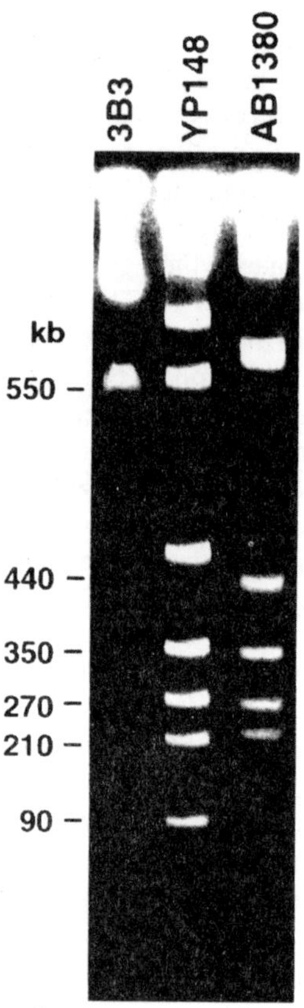

Figure 9(iii)

with rotating electrodes or static gels in which the field switches through 180° and is greater in the forward than in the reverse direction (reviewed in Maule, 1998).

A typical CHEF apparatus consists of a tank containing the gel and electrode array and a pump for circulating the buffer through a cooling device. Pulsed field gels are typically run at 12–15°C and this temperature can only be maintained by cooling the buffer during the run. In addition, a DC power supply to deliver the required voltage and an electronic switching device to change the electric field are required. As well as the pulse time, there are several other variables that determine the size range of molecules separated. The strength of the electric field determines the speed of separation up to molecules as large as 1 Mb. Beyond this size, a reduced voltage must be employed to achieve effective separation. The strength and composition of the buffer can influence the degree of separation, as can the buffer temperature and agarose type and concentration (Birren *et al.*, 1988). Finally, the angle through which the electric field changes direction can be varied by a variation of the CHEF. This apparatus is called PACE (programmable autonomously controlled electrodes) and, while similar in electrode configuration to the CHEF, it also allows the operator to alter the relative voltages applied to each electrode and, in so doing, change the 120° angle characteristic of the CHEF to any angle from 90° to 180°. Careful use of this feature allows resolution and run time to be optimized (Clark *et al.*, 1988).

Since PFGE is a method used for the analysis of large DNA molecules, then the preparation of samples must maintain the starting material intact as high molecular weight entities. Conventional methods for DNA preparation result in the shearing of large molecules by the various procedures involved. Intact DNA can be prepared by embedding the molecules in a protective agarose environment in the form of plugs or microbeads. Both methods of preparation lead to samples that can be directly loaded into the wells of a pulsed field gel. In both cases, cells at an appropriate concentration are embedded in molten low melting temperature agarose. When this is set, then the DNA can be released from the cells by various enzymatic procedures. With organisms containing a cell wall, prior incubation of the agarose inserts with a cell wall-digesting enzyme is required. Thereafter, treatment with a detergent and proteinase K results in the production of high molecular weight intact DNA suitable for PFGE analysis. Agarose plugs are formed by dispensing a small volume (50–100 µl) of a homogeneous mixture of molten agarose and cells into moulds and allowing the agarose to set. Large numbers of plugs can be quickly produced using an adapted microtitre plate mould and a multichannel pipette (Porteous & Maule, 1990). The alternative approach involves similar methodology but forms microbeads by violently agitating the molten agarose/cell mixture with immiscible liquid paraffin producing an emulsion which (when cooled) yields tiny droplets of agarose (Koob & Szybalski, 1992).

DNA encapsulated in either plugs or microbeads can be successfully digested with restriction enzymes, once the residual proteinase K has been inactivated. Larger concentrations of restriction enzymes and longer incubation times are required than when digesting naked DNA and the agarose inserts can then be directly loaded onto a pulsed field gel. A variety of molecular weight markers is available to size fragments or whole molecules on pulsed field gels. Concatemers of λ (monomer size 48.5 kb) or cloning vectors like pBR328 (4.4 kb) provide size markers up to several hundred kb. Beyond this, advantage can be taken of whole chromosomes from several different types of yeast. *Saccharomyces cerevisiae* has 16 chromosomes ranging in size from 200 to 2200 kb and all are resolvable by PFGE. The heterothallic yeast *Hansenula wingei* possesses seven chromosomes over the size range 1–3.3 Mb and the fission yeast *Schizosaccharomyces pombe* has three chromosomes of 3.5, 4.6 and 5.7 Mb, although some strains also possess minichromosomes of less than 1 Mb in size (see **Table 2**). At the conclusion of a gel run, DNA can be visualized by staining in ethidium bromide solution and photographed under ultraviolet (UV) light. Southern blotting is accomplished following fragmentation of the DNA with hydrochloric acid (depurination) or by nicking with UV light. Alternatively, gels can be dried under vacuum and can be hybridized directly using probes that have been fragmented with hydrochloric acid (Stoye *et al.*, 1991).

Uses of PFGE

Since the inception of PFGE in 1982, this technique has been used in many ways to visualize and analyse large DNA molecules. PFGE was first used to provide an electrophoretic karyotype of several microbial genomes and has been

Table 2 Chromosome sizes (kb).

Saccharomyces cerevisiae **AB1380**	*Saccharomyces cerevisiae* **YP148**	*Hansenula wingei*	*Schizosaccharomyces pombe* **3B3**
2200	2200	3300	5700
1640	1640	2900	4600
1130	1125	2600	3500
1120	1030	1800	550
955	1000	1500	
930	920	1250	
830	830	1030	
790	790		
750	750		
690	700		
585	600		
585	550		
420	440		
350	350		
285	270		
240	210		
	90		

particularly useful in analysing chromosomes that cannot be observed by any other method. Chromosomes from yeast, fungi and protozoa have been visualized and their sizes established. For particularly large chromosomes, restriction fragments (created by cleavage with rare cutting enzymes like *Not*I and *Sfi*I) can be sized more accurately and, if all the fragments can be resolved, then their sizes can be summed to establish an overall size for the complete chromosome. This approach has been particularly successful in the analysis of bacterial chromosomes, which can be as large as 9.5 Mb (Smith & Condemine, 1990). Small circular chromosomes can be sized by first linearizing them enzymatically or with gamma irradiation and then running the intact molecule on a pulsed field gel. Larger structures must be fragmented by restriction enzyme digestion. The choice of enzymes to give a limited number of resolvable fragments is determined by the G + C content of the genome or the prevalence of particular sequence motifs. Thus, bacterial DNA having a high G + C content will be cleaved infrequently by enzymes like *Dra*I, *Ssp*I and *Ase*I which have only As and Ts in their recognition sequences. For low G + C genomes, *Sac*II, *Sma*I, *Nae*I and *Not*I are enzymes that recognize only Gs and Cs and so will cut infrequently and produce a limited number of discrete fragments.

PFGE has provided the means to establish long-range restriction maps of genomic DNA and to clone and analyse very large DNA fragments. The production of long-range physical maps relies on the use of rare cutting restriction enzymes. In mammalian DNA, the CpG dinucleotide is under-represented and is also frequently methylated. Conveniently, unmethylated CpGs tend to be clustered at the 5′ ends of genes (so-called CpG islands). There are several restriction enzymes that recognize only unmethylated Cs and Gs and, if these comprise a lengthy recognition sequence such as seen with *Not*I (GCGGCCGC), these enzymes will not only cut infrequently, but also will preferentially cleave at CpG islands. **Table 3** shows a comprehensive list of rare cutter restriction enzymes. Long-range physical maps can be constructed by observing which markers map to a common restriction fragment. By using several different enzymes and performing double digests, markers can be ordered and their physical separation established. Bacterial artificial chromosomes (BACs), P1-derived artificial chromosomes (PACs) and yeast artificial chromosomes (YACs) provide vehicles for the cloning of large genomic fragments and these can be sized by PFGE and markers mapped by using rare cutter enzymes.

Table 3 Rare cutter restriction enzymes.

Enzyme	Recognition Sequence[a]
*Not*I	GC/GGCCGC
*Asc*I	GG/CGCGCC
*Fse*I	GGCCGG/CC
*Srf*I	GCCC/GGGC
*Bss*HII	G/CGCGC
*Xma*III, *Eag*I, *Ecl*XI	C/GGCCG
*Sst*II, *Sac*II, *Ksp*I	CCGC/GG
*Nae*I	GCC/GGC
*Nar*I	GG/CGCC
*Sma*I	CCC/GGG
*Sse*83871	CCTGCA/GG
*Rsr*II	CG/GwCCG
*Sgr*AI	Cr/CCGGyG
*Mlu*I	A/CGCGT
*Pvu*I	CGAT/CG
*Nru*I	TCG/CGA
*Aat*II	GACGT/C
*Sal*I	G/TCGAC
*BsiW*I, *Spl*I	C/GTACG
*Sna*BI	TAC/GTA
*Sfi*I	GGCCNNNN/NGGCC
*Xho*I	C/TCGAG
*Bse*AI, *Acc*III, *Mro*I	T/CCGGA
*Cla*I	AT/CGAT
*Sfu*I, *Asu*II	TT/CGAA

[a] N = A, C, G or T; r = A or G; w = A or T; y = C or T

It should be remembered that mammalian genomic DNA cloned into YACs and cultured in *S. cerevisiae* will present a different methylation status compared with the original starting material. Cloned DNA fragments can also be arranged into contiguous arrays (or contigs) by looking for common restriction fragments between clones when hybridized with the same probe. Since clones are terminated by vector sequences, then end clones can be isolated and mapped by screening for vector sequences following the digestion of the clones. An end clone sequence can be used as a probe to determine which other clones hybridize and thereby build up a contig of overlapping clones. End clones also provide a means to order markers and restriction sites relative to the ends. Partial digests can be used to map sites some distance from the end. PFGE can also be used to detect chromosomal rearrangements such as deletions, duplications and translocations, since these changes often lead to a redistribution of restriction sites, which can be detected as band shifts on a gel.

PFGE has been used to study the replication of bacterial genomes, the vast majority of which are circular chromosomes. Replication is first synchronized by introducing a metabolic block. Bacterial cells are then incubated with a radiolabelled nucleotide, following the removal of the block, and the incorporation stopped at various time points (Dingwall *et al.*, 1990). If the genomic DNA is cleaved by a rare cutter restriction enzyme, then the progress of the replication fork can be followed by running fragments out on a pulsed field gel. As the replication fork progresses around the chromosome, then gradually additional labelled fragments appear. The rate of replication can be calculated by a time course experiment.

PFGE can be used to study the location and appearance of double-strand breaks (DSBs) in chromosomal DNA. The generation of DSBs in chromosomal DNA leads to the appearance of faster migrating, low molecular weight fragments that can be detected and sized by PFGE. DSBs can be artificially generated by the irradiation of cells with γ- or X-rays. In one study (Wong *et al.*, 1995), radiolabelled Chinese hamster ovary cells were irradiated, the products fractionated by PFGE and detected and quantified by liquid scintillation counting. The authors demonstrated that heat treatment increased the number of DSBs because the cells were unable to repair the damage. DSBs are also generated *in vivo* during meiosis when recombination leads to the exchange of genetic material between chromosomes. Thus, the appearance of DSBs has been studied in yeast (Game, 1992) and their locations, during meiosis, mapped to specific chromosomal locations.

Mapping of bacterial genomes has also been accomplished using 2-dimensional PFGE. For example, two approaches have been adopted in preparing a physical map of *Pseudomonas aeruginosa* PAO (Romling & Timmler, 1991). In the first, genomic DNA was partially digested with a restriction enzyme, the products separated by PFGE and the gel track excised and cut to completion with the same enzyme. The gel slice was laid across the top of another gel and the products separated by PFGE. All the fragments detected in the second dimension were derived from specific partial products separated by the first dimensional gel. The second approach involved performing complete digests with two different enzymes and hybridizing the second dimensional gel with labelled genomic clones. Reciprocal digests can be performed and used to order fragments.

REFERENCES

Birren, B.W., Lai, E., Clark, S.M., Hood, L. & Simon, M.I. (1988) Optimized conditions for pulsed field gel electrophoretic separation of DNA. *Nucleic Acids Research* 16: 7563–82

Clark, S.M., Lai, E., Birren, B.W. & Hood, L. (1988) A novel instrument for separating large DNA molecules with pulsed homogeneous electric fields. *Science* 241: 1203–05

Dingwall, A., Shapiro, L. & Ely, B. (1990) Analysis of bacterial genome organisation and replication using pulsed-field gel electrophoresis. *Methods* 1: 160–68

Game, J.C. (1992) Pulsed-field gel analysis of the pattern of DNA double-strand breaks in the *Saccharomyces* genome during meiosis. *Developmental Genetics* 13: 485–97

Koob, M. & Szybalski, W. (1992) Preparing and using agarose microbeads. *Methods in Enzymology* 216: 13–20

Maule, J.C. (1998) Pulsed field gel electrophoresis. *Molecular Biotechnology* 9: 107–26

Porteous, D.J. & Maule, J.C. (1990) Casting multiple aliquots of agarose embedded cells for PFGE analysis. *Trends in Genetics* 6: 346

Romling, U. & Timmler, B. (1991) The impact of two-dimensional pulsed field gel electrophoresis techniques for the consistent and complete mapping of bacterial genomes; refined physical map of *Pseudomonas aeruginosa* PAO. *Nucleic Acids Research* 19: 3199–206

Smith, C.L. & Condemine, G. (1990) New approaches for physical mapping of small genomes. *Journal of Bacteriology* 172: 1167–72

Stoye, J.P., Frankel, W.N. & Coffin, J.M. (1991) DNA hybridization in dried gels with fragmented probes: an improvement over blotting techniques. *Technique* 3: 123–28

Wong, R.S.L., Dynlacht, J.R., Cedervall, B. & Dewey, W.C. (1995) Analysis by pulsed-field gel electrophoresis of DNA double-strand breaks induced by heat and/or X-irradiation in bulk and replicating DNA of CHO cells. *International Journal of Radiation Biology* 68: 141–52

10. Yeast artificial chromosomes (YACs)

Methods

YACs consist of various yeast chromosomal elements necessary for replication and segregation into which exogenous DNA can be cloned. A typical YAC vector has a number of essential features (see **Figure 10A**). It contains a yeast centromere (CEN) which controls accurate chromosome segregation during cell division. A yeast autonomous replicating sequence (ARS) carries all the information required for the origin of chromosome replication and the telomere (TEL) sequences maintain the ends of the linear chromosomes and allow accurate replication. The YAC vector replicates as a circular molecule in yeast and bacteria.

The *ampicillin* (*Amp*) gene allows positive selection by ampicillin resistance in bacteria and the *Ura* and *Trp* genes provide selection when the vector is maintained in a yeast

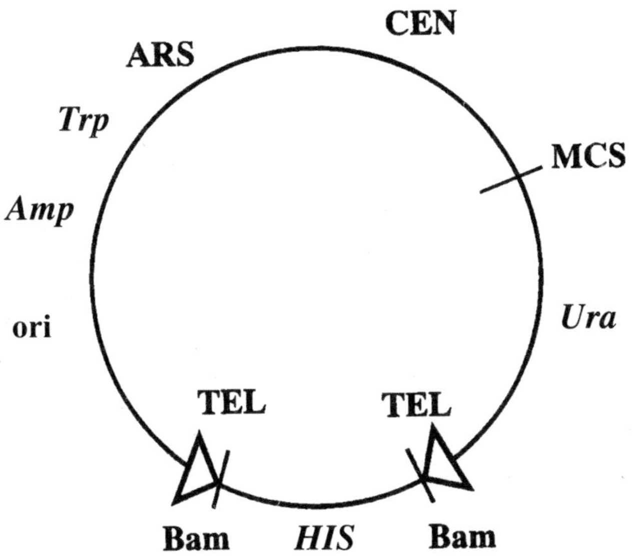

Figure 10A A typical YAC cloning vector, showing the intact stuffer fragment between the telomere (TEL) sequences.

host auxotrophic for these two markers. A bacterial site (ori) allows the replication of the vector in *Escherichia coli*. Prior to cloning into the YAC vector, the molecule is digested with *Bam*HI, which releases the stuffer fragment containing the *HIS* gene and generates a linear molecule terminating in the telomere sequences. Exogenous DNA can be cloned into the multiple cloning site (MCS), which contains one or more unique restriction enzyme sites, depending on the vector.

Thus, pYAC4 (Burke *et al.*, 1987) contains a unique *Eco*RI cloning site whereas pYAC-RC (Marchuk & Collins, 1988) contains an MCS with sites which can be cut by a variety of rare cutter restriction enzymes. The cloning site is within the *sup4* gene, which is disrupted by the introduction of exogenous DNA. The *sup4* gene encodes an ochre suppressor for the *ade2–1* ochre mutation. Vectors carrying inserts, in which the *sup4* gene is disrupted, produce red colonies whereas non-recombinants yield white colonies.

YACs provide a cloning system capable of accommodating exogenous DNA inserts up to 40 times larger than the capacity of cosmids. Whole genes, together with their controlling elements, can be cloned as one syntenic unit. YACs, when retrofitted with a suitable selectable marker, can be introduced into mammalian cells and used to complement mutations or in functional studies. Several methods are available which facilitate the transfer of YACs to mammalian cells. These include fusing yeast spheroplasts with target cells in the presence of polyethylene glycol or isolating the intact YAC DNA (Maule *et al.*, 1994) and introducing it into cells by lipofection or microinjection. Transgenic mice have been created by injecting YAC DNA into the nucleus (Schedl *et al.*, 1993) or by its introduction into embryonic stem cells which are subsequently injected into blastocysts. YACs can be manipulated in yeast, which has a powerful and well-characterized homologous recombination system. Yeast can also be transformed easily by spheroplast formation, treatment with lithium acetate or by electroporation.

YACs have been used as vectors to clone large DNA fragments and create libraries from a variety of organisms including bacteria, plants, Drosophila, *Caenorhabditis elegans*, mouse, human, sheep and pig. Inserts as big as 2 Megabases (Mb) in size have been cloned by using a size-selection procedure. The most common approach involves embedding DNA from cell lines, blood, somatic cell hybrids or flow-sorted chromosomes in low melting point agarose plugs (to prevent shearing) and partially digesting with a restriction enzyme to produce large DNA fragments. Alternatively, the DNA can be completely digested with a rare cutter enzyme. The products are fractionated by pulsed field gel electrophoresis (PFGE) and large fragments cut out of the gel, purified and ligated to phosphatased YAC vector (Arveiler, 1994). Once again, the products of this reaction are fractionated by PFGE to remove unligated vector. Finally, yeast is transformed with the ligated products and cells plated on selective media.

YAC libraries are created by picking the appropriate transformants into 96- or 384-well microtitre dishes, either manually or using a robot device. These dishes can be used for long-term storage at −70°C and also as a resource for preparing filters for library screening. Filters can be produced by inoculating from microtitre dishes using a 96- or 384-pin robotic tool, as appropriate. Using the 96-pin tool, each of the 96 locations on the filter can be inoculated from 16 96-well microtitre plates by arranging the inocula in a 4 × 4 grid:

o o o o
o o o o
o o o o
o o o o

Thus, one such filter will contain 96 × 16 different inocula, giving a total of 1536. If the average size of the YAC is 300 kilobases (kb) and the human genome is 3 × 10^6 kb, then 10 000 YACs would be required to cover the genome once. Since there is no guarantee that the genome would be fully represented under these circumstances, it is normal practice to cover the genome several times over, so that in this example 5-fold redundancy would require 50 000 YACs which could be accommodated on 32 filters. Inoculated filters are grown by laying on selective agar plates and incubating overnight. DNA can be prepared *in situ* by lysing the cells with the cell wall-digesting enzyme zymolyase and then denaturing the DNA with alkali, neutralizing and finally treating the filter with proteinase K. Filters can be hybridized with a labelled probe and positive YACs identified, traced back to a particular well in a microtitre dish and then picked for further analysis.

Alternatively, a YAC library can be screened by PCR, using primers specific for a particular locus. Clearly, it is unrealistic to prepare DNA from each individual YAC, so DNA is prepared from pools of YACs for screening by PCR. YACs can be replicated from microtitre dishes onto filters as already described, grown and DNA prepared after washing off all the growth from one or more filters. If any positives are identified after PCR screening, then DNA can be prepared from the individual constituents of the pool and eventually the specific YAC(s) identified.

Individual YAC clones can be characterized by several techniques. The physical size is established by separating the YAC on a pulsed field gel. In cases where a YAC co-migrates with a yeast chromosome, the identity of the YAC can be established by hybridization with a genomic probe. Fine analysis involves creating a restriction map initially by digesting with rare cutter enzymes and using available markers and end clones to order the fragments (Maule, 1997). The complexity of the genomic insert can be established by fingerprinting, but, because the yeast genome is always present along with the YAC, the traditional approach used for cosmid and phage clones cannot be used. Instead the recombinant is digested and the bands well resolved on a gel, which is then blotted and probed with a repeat sequence. In the case of human YACs, probes containing Alu or L1 repeats can be used and B1 repeats used for mouse YACs.

Alternatively, fingerprinting can be accomplished by Alu-PCR, which selectively amplifies just the human component of a YAC. The most efficient method involves using two primers directed outwards from the left and right ends of the consensus Alu sequence. To amplify the interAlu sequence, the repeats must be in an inverted orientation and must be close enough for PCR to be successful. The most informative fingerprinting technique involves an initial PCR reaction with both primers, an aliquot of this reaction can be analysed on an agarose gel to check that amplification has been successful and the remainder diluted and used in a second PCR reaction employing primers end-labelled with [γ-^{32}P]dATP (Coffey *et al.*, 1996).The products of this reaction are resolved on a denaturing polyacrylamide gel which is dried down and autoradiographed or phosphoimaged.

Fingerprinting involves comparing the band profiles from different YACs and looking for similarities in terms of size and clustering. In this way, overlaps between YACs can be identified and a contig assembled. The unusual appearance of additional bands when comparing similar YACs can be an early indication of chimerism. Alternatively, the labelled products can be hybridized against the Alu-PCR products from somatic cell hybrids or radiation-reduced hybrids that have been immobilized on a nylon membrane. A collection of monochromosome hybrids covering the whole genome can be displayed as a dot blot so that the hybridization of a YAC to more than one dot will be an indication of chimerism.

The chromosomal origin of a YAC insert can be established by preparing a probe from the YAC and using it in fluorescence *in situ* hybridization (FISH) with an appropriate chromosome preparation. YAC probes can be generated by a variety of methods including labelling whole yeast DNA or PCR-based approaches such as Alu-PCR. One of the most efficient methods for preparing probes is by catch-linker + PCR (Shibasaki *et al.*, 1995), which is particularly valuable where the amount of material is limited. YAC DNA is isolated by preparative PFGE, cut with *Sau*3A1 and catch-linkers, consisting of an oligonucleotide duplex with a compatible end, ligated onto the restriction fragment. One of the oligonucleotides can be synthesized with a 5′ biotin moiety if the probe is to be used for FISH. Oligonucleotides homologous to the catch-linkers can now be used to amplify the YAC fragment by PCR and if a biotin-containing oligonucleotide is employed then the product can be used as a probe for FISH. The catch-linkers also contain recognition sites for several restriction enzymes so that amplified products can be subcloned into appropriate vectors for further analysis.

YAC contigs are assembled by isolating end clones and using them to probe neighbouring YACs. There are several methods for isolating YAC end fragments. This can be accomplished by plasmid rescue, as envisaged in the original paper describing YAC vectors (Burke *et al.*, 1987). The left YAC arm contains the CEN, ARS and *Trp* sequences, but also carries the *Amp* gene and ori. Thus, by choosing a restriction enzyme which cuts between ori and the telomere and at some unspecified location in the genomic insert, the fragment generated can be self-ligated and propagated in *E. coli* as a plasmid using the ori function for replication and the *Amp* gene for selection (Bates, 1996). Useful enzymes for creating these fragments in the left arm are *Xho*I, *Sal*I and *Nde*I. The YAC vector right arm does not carry an ori function so cannot be propagated directly as a bacterial plasmid. Once again though, if an enzyme which cleaves between *Ura* and the telomere (and at some unspecified location in the insert) is employed, the resulting fragment can be cloned into a plasmid vector and transformed into an *E. coli* strain carrying the *PyrF* mutation. The *Ura* gene will complement this mutation.

An alternative method for isolating end clones involves inverse PCR (Arveiler & Porteous, 1991) (see **Figure 10B**). In this approach, primer pairs are designed which are complementary to YAC vector sequences adjacent to the *Eco*RI cloning site. Each pair of primers is specific to the left or right vector arm and is complementary to sequences in either the sense or antisense strand. Restriction enzymes are chosen that cut in specific vector positions between *sup4* and CEN (in the left arm) and *sup4* and *Ura* (in the right arm). The enzymes will cleave the genomic insert at unspecified positions. Examples of suitable enzymes are *Eco*RV, *Hae*III or *Taq*I for left arm isolation and *Hha*I, *Hinc*II or *Sal*I for the right arm. Following restriction enzyme digestion, the products are self-ligated to create monomer circles, some of

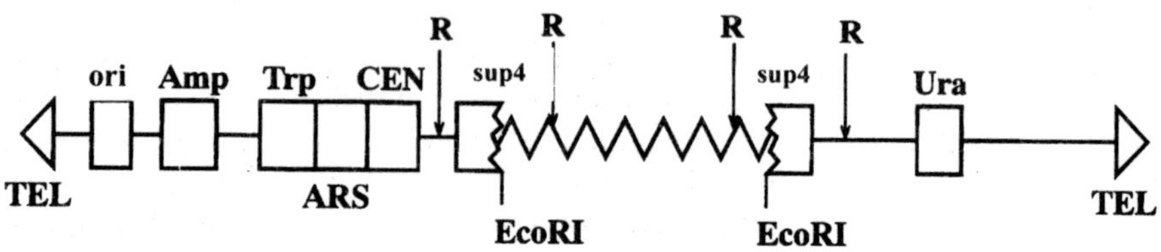

Figure 10B A linear YAC, with genomic DNA depicted as a sawtooth cloned into the *Eco*RI site. Restriction enzyme sites (R) used for end cloning are shown in the vector and genomic insert.

which will span the vector-insert boundary. The ligation products are subjected to inverse PCR, using the oligonucleotide pairs which prime outwards from the vector and amplify into the genomic insert sequences (see **Figure 10C**). The products of this reaction can be sequenced or used as probes in hybridization with other YACs to establish overlaps and build contigs. Nearly the whole human genome is now covered by YAC contigs (Chumakov *et al.*, 1995). YAC libraries can be screened with end clones and further overlapping YACs isolated. Finally, YACs can be artificially reduced in size by a variety of targeted fragmentation techniques. Vectors used to fragment human YACs carry a consensus Alu sequence which is homologous to the Alu repeat sequences which occur on average every 4–5 kb in the human genome. The vectors also carry a yeast telomere and a selectable marker such as ADE2 (Heus *et al.*, 1997). In addition, if YACs are to be fragmented from the YAC vector left arm end, the fragmentation vector will also carry a cloned centromere. The fragmentation process involves the vector recombining with Alu sequences in the genomic insert and thereby simultaneously introducing a new telomere (from the vector) onto the truncated YAC. Recombinants are selected by the presence of the yeast selectable marker ADE2, since most YAC libraries are constructed in the host strain AB1380, which carries the *ade2* mutation. Thus, YACs truncated from the right arm end will be Ade^+ Ura^- and from the left arm Ade^+ Trp^-. By using this method, a series of YACs truncated from either end can be created and this approach is particularly valuable in physical mapping studies. Alternatively, bigger YACs can be created by recombining neighbouring YACs.

Advantages and disadvantages

Although YACs have been widely used as mapping and cloning tools, they do suffer from a number of disadvantages. From a mapping point of view, chimerism presents a serious drawback. This is a not uncommon feature, particularly among larger YACs, and arises during the cloning procedure when two or more fragments from non-syntenic locations are simultaneously cloned into a single vector. Alternatively, two DNA molecules can recombine if introduced together into the same yeast cell. FISH is frequently used to demonstrate chimerism by showing the hybridization of a YAC probe to more than one chromosomal location. YACs can also exhibit instability in which internal deletions and rearrangements occur, compromising their usefulness in mapping studies. This is particularly a problem for small YACs of less than 100 kb. Other limitations include the difficulties in isolating YACs away from the host yeast genome, their low copy number, slow growth rate and poor transformation efficiency.

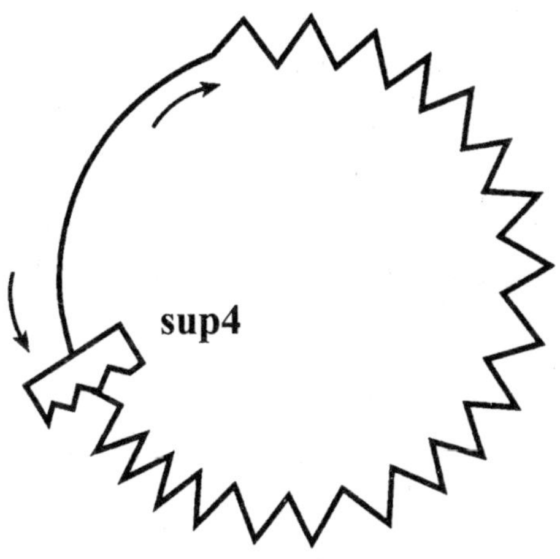

Figure 10C A YAC end clone spanning the vector-insert junction. Sense and antisense primer pairs are shown as arrows, priming out into the genomic insert (depicted as a sawtooth).

Summary

There are several YAC libraries readily available to the scientific community. The Centre d'Etude du Polymorphisme Humain (CEPH) megaYAC library (Albertson *et al.*, 1990) contains the largest human inserts of 1–2 Mb in size. Other human YAC libraries have been produced by the Imperial Chemical Industries plc (ICI) (Anand *et al.*, 1990), Imperial Cancer Research Fund (ICRF) (Larin *et al.*, 1991) and Washington University (Brownstein *et al.*, 1989). There are many libraries from non-human eukaryotes including mouse (Burke *et al.*, 1991; Larin *et al.*, 1991), Arabidopsis (Guzman & Ecker, 1988) and *Schizosaccharomyces pombe* (Maier *et al.*, 1992).

REFERENCES

Albertson, H.M., Abderrahim, H., Cann, H.M. *et al.* (1990) Construction and characterisation of a yeast artificial chromosome library containing seven haploid genome equivalents. *Proceedings of the National Academy of Sciences USA* 87: 4256–60

Anand, R., Riley, J.H., Butler, R., Smith, J.C. & Markham, A.F. (1990) A 3.5 genome equivalent multi-access YAC library: construction, characterisation, screening and storage. *Nucleic Acids Research* 18: 1951–56

Arveiler, B. (1994) Construction of chromosome-specific libraries of yeast artificial chromosome recombinants from somatic cell lines. In *Chromosome Analysis Protocols*, edited by J.R. Gosden, Totowa, New Jersey: Humana Press

Arveiler, B. & Porteous, D.J. (1991) Amplification of end fragments of YAC recombinants by inverse-polymerase chain reaction. *Technique* 3: 24–28

Bates, G. (1996) Isolation of YAC ends by plasmid rescue. In *YAC Protocols*, edited by D. Markie, Totowa, New Jersey: Humana Press

Brownstein, B.H., Silverman, G.A., Little, R.D. *et al.* (1989) Isolation of single-copy human genes from a library of yeast artificial chromosome clones. *Science* 244: 1348–51

Burke, D.T., Carle, G.F. & Olson, M.V. (1987) Cloning of large DNA segments of exogenous DNA into yeast by means of artificial chromosome vectors. *Science* 236: 806–12

Burke, D.J., Rossi, J.M., Leung, J., Koos, D.S. & Tilghman, S.M. (1991) A mouse genome library of yeast artificial chromosome clones. *Mammalian Genome* 1: 65

Chumakov, I.M., Rigault, P., Le Gall, I. *et al.* (1995) A YAC contig map of the human genome. *Nature* 377: 175–298

Coffey, A., Gregory, S. & Cole, C.G. (1996) *Alu*-PCR fingerprinting of YACs. In *YAC Protocols*, edited by D. Markie, Totowa, New Jersey: Humana Press

Guzman, P. & Ecker, J. (1988) Development of large DNA methods for plants: molecular cloning of large segments of Arabidopsis and carrot DNA into yeast. *Nucleic Acids Research* 16: 11091–105

Heus, J.J., de Winther, M.P.J., van de Vosse, E., van Ommen, G.-J.B. & den Dunnen, J.T. (1997) Centromeric and non-centromeric ADE2-selectable fragmentation vectors for yeast artificial chromosomes in AB1380. *Genome Research* 7: 657–60

Larin, Z., Monaco, A.P. & Lehrach, H. (1991) Yeast artificial chromosome libraries containing large inserts from mouse and human DNA. *Proceedings of the National Academy of Sciences USA* 88: 4123–27

Maier, E., Howeisel, J., McCarthy, L. *et al.* (1992) Complete coverage of the *Schizosaccharomyces pombe* genome in yeast artificial chromosomes. *Nature Genetics* 1: 273–97

Marchuk, D. & Collins, F.S. (1988) pYAC-RC, a yeast artificial chromosome vector for cloning DNA cut with infrequently cutting restriction endonucleases. *Nucleic Acids Research* 16: 7743

Maule, J.C. (1997) Physical mapping by pulsed field gel electrophoresis. In *Gene Isolation and Mapping Protocols*, edited by J. Boultwood, Totowa, New Jersey: Humana Press

Maule, J.C., Porteous, D.J. & Brookes, A.J. (1994) An improved method for recovering intact pulsed field gel purified DNA, of at least 1.6 megabases. *Nucleic Acids Research* 22: 3245–46

Schedl, A., Montoliu, L., Kelsey, G. & Schutz, G. (1993) A yeast artificial chromosome covering the tyrosinase gene confers copy number-dependent expression in transgenic mice. *Nature* 362: 258–60

Shibasaki, Y., Maule, J.C., Devon, R.S. *et al.* (1995) Catch-linker+PCR labeling: a simple method to generate fluorescence in situ hybridisation probes from yeast artificial chromosomes. *PCR Methods and Applications* 4: 209–11

11. Keeping up to date

Every laboratory working in molecular genetics has its own technical preferences, and new methods or improvements on current techniques are frequently developed. For those wanting to apply these techniques, some recent manuals, which cover a range of techniques, are listed below. Each of these publications has contributions from experts in the field and detailed protocols, which allow even the novice to attempt the most demanding procedure.

REFERENCES

Ausubel, F.M., Brent, R., Kingston, R.E. *et al.* (eds) (1998) *Current Protocols in Molecular Biology*, New York: Wiley

Birren, B., Green, E.D., Klapholz, S., Myers, R.M. & Roskams, J. (eds) *Genome Analysis: A Laboratory Manual Series*: Volume 1: *Analyzing DNA* (1997), Volume 2: *Detecting Genes* (1998), Volume 3: *Cloning Systems* (1999), Volume 4: *Mapping Genomes* (1999). Cold Spring Harbor, New York: Cold Spring Harbor Laboratory Press

Rapley, R. (2000) *The Nucleic Acid Protocols Handbook*, Totowa, New Jersey: Humana Press

Sambrook, J. & Russell, D. (2000) *Molecular Cloning: A Laboratory Manual*, 3rd edition, Cold Spring Harbor, New York: Cold Spring Harbor Laboratory Press

GLOSSARY

agarose polysaccharide made from agar-agar (which is extracted from red algae) used to manufacture gels for electrophoresis

alkaline phosphatase enzyme that catalyses cleavage of inorganic phosphate

antisense DNA the complementary strand of a coding sequence of DNA or mRNA

blunt end end of a double-stranded DNA molecule, cut at the same site on both strands without producing sticky ends

chromosome painting the labelling of specific whole chromosomes by *in situ* hybridization. Hybridization probes are usually made by selecting a particular chromosome by flow sorting, and using the DNA from such chromosomes. Chromosome painting can be used to identify specific chromosomes not only in normal metaphases, but also in interphase nuclei, in chromosomal rerrangements, and to study chromosomal changes during evolution

copy number number of molecules of a specific type on or in a cell

CpG island a region rich in CpG dinucleotides. It is usually located at the 5′ end of an expressed housekeeping gene where it is found in the unmethylated state

cross-linker an agent that creates bonds between polymer chains, thereby increasing their strength

cytocentrifugation a centrifugal method used to deposit a thin layer of cells on to microscope slides

deoxynucleotide a purine or pyrimidine covalently bonded to a phosphorylated deoxyribose moiety

deoxyribonuclease (DNase) enzyme that cleaves DNA into shorter oligonucleotides or degrades it totally into its constituent deoxyribonucleotides

dideoxynucleotide a deoxyribonucleotide lacking a 3′ hydroxyl on the deoxyribose moiety.

digestion the process of fragmenting a substrate by enzymatic activity

DNA ligase enzyme involved in joining two completed DNA segments

DNA polymerase removes the RNA primers and replaces them with DNA until it reaches the start of the previously made DNA segment

electrocompetent cell cell prepared by extensive washing in low ionic strength medium and used in electrotransformation

endonuclease enzyme that splits the nucleic acid chain at internal sites (*see also* exonuclease)

ethidium bromide reagent demonstrating orange fluorescence on binding to double-stranded DNA and when viewed under ultraviolet light

exonuclease enzyme that digests the ends of a piece of DNA

F factor in bacterial cells, a plasmid that confers the ability to conjugate (unite)

fluorescence *in situ* hybridization (FISH) technique of directly mapping the position of a gene within a genome (*see also* chromosome painting)
fluorochromes fluorescent dyes, which have the advantage, in general, when compared with dyes that merely absorb light, of much greater sensitivity. Valuable for *in situ* hybridization and for immunocytochemistry, where very small amounts of material have to be detected. Also very important for staining chromosomes: fluorochromes such as acridine orange, ethidium bromide, and propidium iodide bind to DNA without any base specificity, produce uniformly stained chromosomes, and are useful as counterstains (e.g. for immunocytochemical labelling of chromosomes). Others show base-specific binding to DNA, or base-specific enhancement of fluorescence, and produce banding patterns on chromosomes
fluorophore a moiety responsible for giving a molecule fluorescent properties
G_1 phase (gap phase) in the eukaryotic cell cycle, phase between the end of cell division and the commencement of DNA synthesis
Klenow fragment fragment of DNA polymerase I from *E. coli* containing both the polymerase and 3′→5′ exonuclease, but lacking 5′→3′ exonuclease activity
lambda [phage] bacterial DNA virus, first isolated from *E. coli*
lipofection transfer of material into a cell by enveloping it in liposomes, which then fuse with the cell membrane
nick translation method used to label DNA with radioactive or non-radioactive moieties.
operon groups of bacterial genes with a common promoter, clustered together and transcribed as a unit into a polycistronic mRNA
origin of DNA replication regions of DNA necessary for its replication to commence
PAGE (polyacrylamide gel electrophoresis) technique in which molecules (especially proteins) are separated by their different electrophoretic mobilities on a hydrated gel
primer short RNA or DNA that must be present on a DNA template before DNA polymerase can commence elongation of a new DNA chain
repressor in bacteria, a protein that prevents transcription of an operon by binding to the operator region on DNA
resolvase enzyme that mediates site-specific recombination in prokaryotes
restriction map map of DNA showing the location of sites recognized and cut by a variety of restriction endonucleases
riboprobe an RNA probe used for gene isolation and identification
RNase (ribonuclease) enzyme that cleaves ribonucleic acid (RNA)
RNase H activity action of a ribonuclease that specifically cleaves an RNA base-paired to a complementary DNA strand
scintillation counter apparatus for measuring the quantity of a radioactive isotope in a sample
Southern blot technique in which DNA molecules, separated by electrophoresis in an agarose gel, are transferred by "blotting" to a nitrocellulose filter, then hybridized with labelled nucleic acid probes for identification and isolation of sequences of interest
spheroplast bacterial or yeast cell from which the wall has been removed, leaving a naked protoplast
telomerase (telomere terminal transferase) enzyme involved in forming the telomeres
terminal transferase enzyme catalysing the addition of deoxynucleotides to the 3′-OH end of DNA molecules
terminase an enzyme responsible for the introduction of staggered nicks at the *cos* site of bacteriophage lambda
topoisomerase action change in the degree of supercoiling in DNA by enzymatic cutting of one or both strands; type I topoisomerase cuts only one strand; type II topoisomerase cuts both strands
transformation any change in the properties of a cell stably inherited by its progeny
zymolase an enzyme isolated from *Arthrobacter luteus* and used to lyse cell walls by hydrolysing glucose polymers with 1→3 linkages

See also **Gene targeting** (p.791)

B THE ORIGINS OF GENETICS

Introduction to the Origins of Genetics

While it is now common knowledge that DNA is the hereditary material (see p.10), the first article in this section describes how beliefs about heredity are recorded from the Old Testament and the writings of Hippocrates, to 17th- and 18th-century writers. Subsequent articles in this section focus on the scientific theories of Mendel, Darwin, Galton and finally Fisher, Haldane and Wright, the leaders of the modern "Neo-Darwinian" revolution. Problems in modern evolutionary theory are highlighted in ideas such as punctuated equilibrium and selfish genes (see also Selfish genes in section N).

As genetics concerns the study of differences between species, it is interesting to consider here the origin of humans themselves. Although there are some gaps in our knowledge, it seems that Neanderthals were not in the direct line of human ancestry.

FROM THE GREAT APES TO GREGOR MENDEL

Eric C.R. Reeve
Institute of Cell, Animal and Population Biology, University of Edinburgh, Edinburgh, UK

1. Introduction

Recorded opinions about the nature of heredity only go back a few millennia, while archaeological evidence showing human intervention to make certain plants more suitable as foods, and to domesticate various animals, can be found at least 10 000 years ago, and these processes may have begun much earlier. So let us start with the view of human evolution which is emerging from many years of intensive study by palaeoanthropologists, although there are still controversies and some gaps in our knowledge which may never be filled.

2. Human origins

Our closest living relative is now generally accepted to be the chimpanzee, from whose ancestral line the human lineage descended in East Africa, probably 5–6 million years ago when change to a drier climate caused replacement of the dense forests by savannah and patches of more open woodland. The resulting pressure on the tree-living apes for food and space would have forced some of them, including our hominid ancestors, to move down to the savannah and to learn how to avoid its predators. This must have been a major factor in their further evolution.

Comparison of the human and great ape genomes shows that we and the chimpanzee are more closely related to each other genetically than either is to the gorilla or orang-utan, and we are very much closer genetically to the chimpanzee than our respective appearances and abilities would lead us to assume. The human and chimpanzee genomes both have about 70 000 genes in 3 billion base pairs of nuclear DNA, and a high proportion of this DNA is non-coding, and so presumably cannot contribute to the differences between human and ape. Roger Lewin (1999) stated the current view that: "Humans are known to share 98.3% identity in nuclear non-coding DNA sequence and more than 99.5% identity in nuclear coding DNA, or genes, with chimpanzees." The apes also have 24 pairs of chromosomes instead of the 23 pairs in the human genome, because human chromosome 2 was created by fusion of two chromosomes in the ape ancestry, although this does not appear to have had a significant genetic effect (Strachan & Read, 1996).

Among the changes of consequence on the way from ape to human were the move to full bipedal locomotion, the marked increase in brain size and complexity, and the increased hand (i.e. forelimb) skills which this made possible, and, finally, development of the ability to make a wide range of sounds, convert them into meaningful speech and thence into complex language with an ever-increasing vocabulary – a progress which would inevitably have led to the creation of a written language. It is not certain when these various abilities developed nor their order of succession, but there are suggestive clues.

3. From forelimbs to two legs

Laetoli in the eastern branch of the African Great Rift Valley contains two parallel trails of fossilized footprints, dated at 3.4–3.8 million years ago and believed to have been made by bipedal hominids, so bipedalism seems to have evolved within about 2 million years of the ape–human divergence (Agnew & Demas, 1998). The complex skeletal changes leading from knuckle walking by the apes as they do today, to full bipedalism with an upright stance, are described by Lewin (Lewin, 1999). However, in a recent study of fossil skeletons, Richmond and Strait produce compelling evidence that the early hominids, *Australopithecus anamensis* and *A. afarensis*, presumed to be human ancestors, were knuckle walkers like the apes (Richmond & Strait, 2000). Commenting on this analysis, Collard and Aiello conclude that it makes the way hominids converted to bipedalism even more difficult to explain (Collard & Aiello, 2000).

4. The development of language and use of stone tools

(i) Development of language

Apes are very limited in their sound-making ability, which has obviously prevented the growth of an elaborate ape language (Savage-Rumbaugh & Lewin, 1994; De Waal & Lanting, 1997), and a recent suggestion is that upright bipedal motion and posture are essential to enable vocal cords, speech and language to evolve. So, the striking difference between apes and humans in their abilities to communicate may depend on fewer gene differences than is generally assumed.

Lewin in Unit 32 of his book (Lewin, 1999) gives an excellent discussion of recent theories about the evolution of language, which I summarize here. Endocasts of fossil skulls show the surface features of their brains, but with insufficient detail for palaeoneurologists Dean Falk and Ralph Holloway to agree on their interpretation. Comparison of the vocal tracts of chimpanzees and humans is more revealing, and shows that in the chimpanzee, as in all mammals, the larynx is high in the neck, enabling simultaneous breathing and swallowing. In mature humans, on the other hand, the larynx is low in the neck, making simultaneous breathing and swallowing impossible since swallowing now requires temporary closure of the air passage. This change, however, increases the size of the pharynx and the scope of vocal production. It is of particular interest that newborn humans maintain the basic mammalian pattern until aged 1.5–2 years; then the larynx migrates lower in the neck during the next 12 years to produce the adult human pattern. The australopithecine skull resembles that of the chimpanzee, but many later fossil skulls cannot be checked for this character, leaving the time sequence of speech development uncertain.

(ii) Use of stone tools

Stone tool making seems to have originated at about the same time as significant brain expansion, roughly 2.6 million years ago, and was probably developed by early Homo species (*Homo habilis*, *H. rudolfensis*), suggesting that they already possessed a good deal of manual skill. Early Homo gave rise in Africa to a larger bodied and larger brained species about 2 million years ago, which has been named *H. erectus*, *Sinanthropus pekinensis* or *H. ergaster* by different researchers, depending on where the fossils were found, and is now generally referred to as *H. ergaster*.

5. Emergence of *H. sapiens*

There is more than one theory about the evolution and spread of the modern human species, *H. sapiens*, but the "Out of Africa" theory seems the most likely. This postulates that *H. ergaster* had spread widely from Africa into Asia by 1.8 million years ago, and gave rise to *H. erectus*, which spread more widely into Eurasia and back into Africa. A mutation or other speciation event then led to modern humans (*H. sapiens*) about 150 000 years ago, which spread into the Old World, replacing the more primitive human species and later spreading into Australia and the Americas (see Lewin, 1999). This 150 000 years, equivalent to less than 10 000 generations, seems a very short time for our present-day civilization (with all its complex known history and enormous number of languages) to have evolved. However, the oldest stone tools date from about 2.5 million years ago, so modern civilization may have begun to develop before *H. sapiens* was detected in fossil sites.

6. Cave paintings

The discovery of European cave paintings of breathtaking skill and complexity, particularly those of Chauvet (which are about 35 000 years old) and the equally or even more remarkable Lascaux (about 17 000 years old), has led to the conclusion that there was a great flowering of art in Europe from at least 40 000 years ago, which is considered to be the time when *H. sapiens* moved on a large scale from Africa into Eurasia (Bahn, 1995; Chauvet *et al.*, 1996; Lewin, 1999; see also articles in *Cambridge Encyclopedia of Human Evolution* (Jones, Martin & Pilbeam, 1992) and *Science* vol. 282, 1998). These cave paintings were created where there was no natural light and, whatever their motivation was, they suggest that the inhabitants at the time had developed considerable technical abilities, not limited to painting.

7. Settled farming and agricultural progress

An important stage in human progress was the development of settled farming communities, based on the regular planting and harvesting of crops to feed sizeable populations. This began at least 10 000 years ago, as judged by the earliest Neolithic villages discovered in the Near East. However, the much earlier cave paintings are of such skill and sophistication that their creators, living up to 40 000 years ago, could have discovered how to apply selection to improve the plants and animals that they were beginning to cultivate.

Early agriculture was turning the wild grass ancestors of our modern crops into food plants able to sustain their families, and a study by Zohary (1999) on the crops that founded Neolithic agriculture in the Near East concludes that emmer wheat (*Triticum turgidum* subspecies *dicoccum*), einkorn wheat (*T. monococcum*), pea (*Pisum sativum*) and lentil (*Lens culinaris*) were very probably taken into cultivation only once, while the same seems to apply to chickpea (*Cicer arietinum*), bitter vetch (*Vicia ervilia*) and flax (*Linum usitatissimum*), but with less certainty, and only barley (*Hordeum vulgare*) was almost certainly domesticated more than once. These plants differed from their wild progenitors in characters such as non-brittle ears and broad kernels, which indicated that they were intentionally sown and harvested. Domestication of sheep, goats, cattle and pigs, and no doubt of dogs, was also taking place over the same period, if not earlier (Zohary & Hopf, 1999). (See article on "Pinpointing einkorn wheat domestication by DNA fingerprinting".)

8. Early records of beliefs

The first record of writing comes from about 6000 years ago, but the earliest record we have of actual theories or beliefs about heredity is the trick Jacob played on Laban, who was not only his uncle but also his father-in-law and employer (Book of Genesis, chapter 30), in order to extract his unpaid wages. Laban's flock (cattle, sheep and goats?) contained a few "ring-straked, speckled and spotted" animals (let us call them striped), and Jacob offered to take only striped animals as his wages (a very small wage, to Laban's delight). Jacob then placed striped sticks where the flocks were watered and the pregnant animals duly gave birth to mainly striped progeny, so that Jacob took over a high proportion of Laban's flock.

This is an apparent example of "maternal impression" (the influence of local environment at a time of stress on heredity), which has been accepted by many people over the centuries as a cause of physical abnormalities at birth. (Thus, my grandmother assured me that a pregnant woman gave birth to a child with a lobster claw instead of a hand after seeing a lobster landed at Great Yarmouth.) Jacob and Laban lived before the Exodus from Egypt, perhaps 4–5000 years ago, and the truth of Jacob's successful application of "maternal impression" either has not been put to the test since or led to failure and was not reported. However, we can assume that, whenever this passage from the Book of Genesis was written, the theory of "maternal impression" was widely believed.

9. Greek theories

The next recorded genetic theory is due to Hippocrates, the Greek physician and philosopher, known as the father of medicine and author of the Hippocratic oath. He was probably born on the Greek island of Cos, lived from *c.*460–370 BC and practised and taught medicine, based on objective observation and deductive reasoning. He thought that each part of the body produces something (later called gemmules by Darwin) which is released into the "semen" at the time of copulation and transferred to the embryo. The gemmules form the characters of the offspring.

This theory, later named pangenesis by Darwin, could explain the inheritance of acquired characters (i.e. character changes caused by the environment), since the gemmules would carry the current states of the parts of the body to the egg. The theory was criticized by Aristotle as too simple – he pointed out that the effects of mutilations or loss of parts in plants and animals are often not inherited and that characters not yet present in an individual may nevertheless be inherited. Aristotle lived from 384–322 BC, and was a naturalist as well as a philosopher and teacher. He described many kinds of animals and knew the hybrid nature of the mule. He thought that many other animals were species hybrids; for example, the giraffe was a hybrid between the camel and the leopard. He said that in Libya there were few places where water is available and that many animals congregate in these water holes and may mate with each other – this is the basis for the saying that "something new is always coming from Libya".

10. Ideas in the 17th–19th centuries

Aristotle believed in the inheritance of acquired characters, like Darwin and other scientists of the 18th and 19th centuries. Over this period, there was a great interest in nature and the collecting of plants and animals, but recording them presented difficulties until the advent of Linnaeus.

(i) Linnaeus

Carl Linnaeus (1707–1778) was the son of a Swedish pastor who at an early age showed a consuming interest in botany, with disastrous results for his school work. Since he had no chance of going to the seminary, his father decided to apprentice him to a cobbler, but was dissuaded by the family physician, who thought the boy would yet distinguish himself in medicine and natural history, and taught him physiology, which helped him enter the University of Lund. From there he moved to Uppsala, where he became Professor of Medicine in 1741 and soon exchanged this position for Professor of Botany. He travelled widely in northern Europe on collecting and recording expeditions and astonished many scientists by his materials and knowledge. In 1735, he travelled to Holland and obtained his MD with a thesis on the cause of intermittent fever. He showed the manuscript of his most important work, the *Systema Naturae*, to the Professor at Leiden, who was so greatly astonished by it that he arranged for its publication at his own expense.

This work for the first time provided literally a system of nature, a logical way to order the creatures of the world, which was received with tremendous enthusiasm by the numerous amateur collectors who were accumulating specimens. The method of Linnaeus was to examine individual organisms, note their observed characteristics and group them into categories that reflected ordered relationships, giving them species and genus names. Genera were grouped into families and so on, and all individuals that could mate together successfully were put into a single species. Each species was represented by a type specimen, which carried the genus and species name, e.g. *H. sapiens* or *Arabidopsis thaliana*, and the species was officially represented by this specimen. It should be emphasized, however, that Linnaeus believed that all organisms were created by God and fixed, so that his system was based on his urgent need to create an ordered system and had nothing to do with any suspicion about evolution.

(ii) Peloria

A student of Linnaeus discovered specimens of the flowering plant *Linaria vulgaris* having five spurs instead of one. Linnaeus was startled by this variant, which he named peloria (Greek for monster), and wrote of it

> Nothing can be more fantastic than what has here happened: a highly deviating progeny of a plant that has always given irregular (i.e. zygomorphic) flowers has produced regular ones. By this character it deviates from its entire ancestry and also from the entire class. This is certainly no less fantastic than if a cow had given birth to a calf with the head of a wolf.

Pelorism was soon found in a number of plant species and Darwin discussed how it might appear and disappear. He presented data on segregation of peloric types in crosses of Antirrhinum, which fitted a 3:1 ratio, but did not understand its genetic significance (the classic 1911 edition of *The Encyclopedia Britannica* gives a useful biography of Linnaeus and his ideas on peloria, see also Darwin, 1868 pp. 90–92).

11. Another view of Linnaeus

A new biography of Carl Linnaeus entitled *Linnaeus, Nature and Nation* by the Swedish author Lisbet Koerner (1999), contains much new information and has been reviewed by Londa Schiebinger in *Science* (10 March 2000, p. 1761). Schiebinger notes that Linnaeus attempted to defraud the Uppsala Science Society by doubling his expenses for field work in Lapland, anonymously wrote glowing reviews of his books for Stockholm newspapers, and too easily took credit due to others who contributed to his theories. These do not seem to me to severely affect his reputation, and the reader should study the new biography and Linda Schiebinger's review of it before making up his or her mind.

12. Inheritance of acquired characters

The most prominent supporter of the theory of the inheritance of acquired characters was the French naturalist, Chevalier de Lamarck (1744–1829), whose early life would make excellent material for a Hollywood film. In brief, he was the eleventh child of the lord of the manor in a Picardy village and was destined for the Church, but left the Jesuit College on his father's death. Not yet 17, he set off to join his three elder brothers in the local war at Bergen-op-Zoom on an old horse with a village boy in attendance. He posted himself in front of a body of grenadiers and in the ensuing battle only fourteen grenadiers without any officers were left, along with the young Lamarck. He was rapidly made an officer and, after the war, went to Paris to study medicine, supporting himself by working in a banker's office. He soon changed over to botany and became the leading naturalist of his time.

Lamarck's only defect as a scientist seems to have been his excessive love of speculation. Believing in the popular theory of the inheritance of acquired characters, he argued that the giraffe acquired a long neck from stretching up towards the foliage above and passed this valuable character on to its progeny. The same hypothesis could also explain why the sons of blacksmiths developed strong arm muscles. No-one thought of putting this ingenious hypothesis to the test, until August Weissmann (1834–1913) cut off the tails of 22 generations of a pure-bred line of mice and found that identical tails were grown by mice of generation 23. However, this experiment had little impact on those who believed in the theory (a good biography of Lamarck can be found in the classic 1911 *Encyclopedia Britannica*).

13. Telegony

Another theory long held by breeders of both domestic pets and animals was that of telegony – that a pure-bred mare or bitch can be "infected" by mating to a male of another breed, so that she will pass some of his characters on to her later progeny from matings to her own breed. This theory was also very widely believed during the 19th century and was supported by a report of Lord Morton to the Royal Society of London in 1820, stating that a chestnut mare, after having a hybrid foal by a quagga (a South African striped wild ass, *Equus quagga*), was mated to a black Arabian horse and produced three foals showing a number of stripes, one of them more than the quagga hybrid. Even Darwin accepted this theory and wrote in *The Variation of Animals and Plants under Domestication* (vol. 1, p. 436) in 1868: "It is worth notice that farmers in south Brazil . . . are convinced that mares which have once borne mules, when subsequently put to horses frequently produce colts striped like a mule."

Later, critics were not convinced, since stripes are often seen on high-caste Arab horses and, in a much larger experiment, the quagga having become extinct, a number of mares were put to a richly striped Birchell's zebra and then to Arab thoroughbred horses, producing 17 zebra hybrids and, subsequently, 20 pure-bred foals, while a control cross without zebra intervention produced ten pure-bred foals (see Ewart, 1899). All the zebra hybrids were richly striped, some with far more stripes than their zebra parent, but the 20 pure-bred foals from these "infected" mares were as Arab horse like as the foals from the control mares, and no infection by zebra coat characters appeared in further breeding to Arab horses. It should be noted that Ewart's experiment (the first test of a genetic theory carried out in Scotland) had little effect on the agricultural community. Cossar Ewart was Professor of Natural History at the University of Edinburgh.

14. Early ideas on human heredity

Medical reports in the 18th and early 19th centuries, long before Gregor Mendel's work, showed some understanding of inherited human disease. Thus, Maupertuis in 1752 described a family of four generations with polydactyly (presence of extra fingers or toes) and demonstrated that this characteristic could be transmitted equally by the father or mother, and also that chance alone could not explain the frequency of this trait in the family. More remarkably, a British doctor, Joseph Adams, in 1814 published *A Treatise on the Supposed Hereditary Properties of Diseases*, which showed that he had a very good understanding of the bases of human hereditary diseases. He distinguished "familial" (i.e. recessive) and "hereditary" (i.e. dominant) diseases, and knew that the (healthy) parents were often near relatives in familial diseases. Also, familial diseases were more frequent in isolated populations, which could be due to inbreeding. He noted that clinically identical diseases may have different hereditary causes and that many patients with hereditary diseases produce few or no children, so that these diseases would eventually disappear if they did not turn up from time to time afresh among children of healthy parents (Emery, 1989).

C.F. Nasse in 1820 recognized the X-linked (i.e. sex-linked) recessive mode of inheritance of haemophilia; he reported that bleeders were always males and that daughters always transmitted this character to their male children even when they were married to non-afflicted males. These women never had the character themselves and sometimes produced male children who were not bleeders. Other medical authors were reporting on various inherited diseases and suggesting rules for the influence of heredity on disease, long before the rediscovery of Mendel's paper in 1900 (see article on "Gregor Mendel" for further details of Mendel's ideas on heredity).

In 1865, Francis Galton, a cousin of Charles Darwin, published two papers entitled *Hereditary Talent and Character* (Galton, 1865a,b). He wrote: "It is commonly asserted that the children of eminent men are stupid, that, where great power of intellect seems to have been inherited, it has descended through the mother's side, and that one son commonly runs away with the talent of the whole family", describing these beliefs as a remarkable misapprehension. From his studies of outstanding men, he concluded that eminent achievement and high talent were strongly influenced by heredity. The article "Francis Galton" contains a more detailed review of Galton's remarkable and varied achievements.

REFERENCES

Adams, J. (1814) *A Treatise on the Supposed Hereditary Properties of Diseases*, London: Callow (see Emery, 1989)

Agnew, N. & Demas, M. (1998) Preserving the Laetoli footprints. *Scientific American* 279: 44–55

Bahn, P.G. (1995) Foreword to *Chauvet Cave: The Discovery of the World's Oldest Paintings*, J.-M. Chauvet, E.B. Deschamps & C. Hillaire

Chauvet, J.-M., Deschamps, E.B. & Hillaire, C. (1996) *Chauvet Cave: The Discovery of the World's Oldest Paintings*, London: Thames and Hudson; as *Dawn of Art: The Chauvet Cave: The Oldest Known Paintings in the World*, New York: Abrams

Collard, M. & Aiello, L.C. (2000) From forelimbs to two legs. *Nature* 404: 339–40

Darwin, C. (1868) *The Variation of Animals and Plants under Domestication*, London: John Murray and New York: Judd; reprinted Baltimore: Johns Hopkins University Press, 1998

De Waal, F. & Lanting, F. (1997) *Bonobo, the Forgotten Ape*, Berkeley: University of California Press

Emery, A.E.H. (1989) Portraits in medical genetics: Joseph Adams 1756–1818. *Journal of Medical Genetics* 26: 116–18

Ewart, C. (1899) *The Penicuik Experiments*, London: Adam and Charles Black

Galton, F. (1865a) Hereditary talent and character. *Macmillan's Magazine* 12: 157–66

Galton, F. (1865b) Hereditary talent and character. *Macmillan's Magazine* 12: 318–27

Koerner, L. (1999) *Linnaeus, Nature and Nation*, Cambridge, Massachusetts: Harvard University Press

Lewin, R. (1999) *Human Evolution: An Illustrated Introduction*, 4th edition, Oxford and Malden, Massachusetts: Blackwell Science

Richmond, B.G. & Strait, D.S. (2000) Evidence that humans evolved from a knuckle-walking ancestor. *Nature* 404: 382–85

Savage-Rumbaugh, S. & Lewin, R. (1994) *Kanzi: The Ape at the Brink of the Human Mind*, New York: Wiley and London: Doubleday

Schiebinger, L. (2000) *Science* Book Reviews: review of *Linnaeus, Nature and Nation*. *Science* 287: 1761

Strachan, T. & Read, A.P. (1996) *Human Molecular Genetics 2*, 2nd edition, Oxford: Bios Scientific and New York: Wiley

The Encyclopedia Britannica, 11th edition (1911) volume 16, Cambridge: Cambridge University Press

Zohary, D. (1999) Monophyletic vs. polyphyletic origin of the crops on which agriculture was founded in the Near East. *Genetic Resources and Crop Evolution* 46: 133–42

Zohary, D. & Hopf, M. (1999) *Domestication of Plants in the Old World: The Origin and Spread of Cultivated Plants in West Asia, Europe, and the Nile Valley*, 3rd edition, Oxford: Clarendon Press and New York: Oxford University Press

FURTHER READING

Bahn, P.G. & Vertut, J. (1988) *Images of the Ice Age*, London, Windward and New York: Facts on File

Cavalli-Sforza, L.L. & Cavalli-Sforza, F. (1995) *The Great Human Diasporas: A History of Diversity and Evolution*, Reading, Massachusetts: Addison Wesley

Jones, S., Martin, R. & Pilbeam, D. (eds) (1992) *Cambridge Encyclopedia of Human Evolution*, Cambridge and New York: Cambridge University Press

Leakey, R. & Lewin, R. (1992) *Origins Reconsidered: In Search of What Makes us Human*, London: Abacus and New York: Doubleday

Manning, A. & Serpell, J. (eds) (1994) *Animals and Human Society: Changing Perspectives*, London and New York: Routledge

McGrew, W.C., Marchant, L.F. & Toshisada, N. (eds) (1996) *Great Ape Societies*, Cambridge and New York: Cambridge University Press

Ruspoli, M. (1987) *Cave of Lascaux: A Final Photographic Record*, London: Thames and Hudson and New York: Abrams

Science (1998) 282: 1441–58

See also **Neanderthals** (p.58); **Gregor Mendel** (p.62); **Charles Darwin and the theory of evolution** (p.68); **Francis Galton** (p.84); **Pinpointing einkorn wheat domestication by DNA fingerprinting** (p.634); **Mitochondrial "Eve"** (p.891)

NEANDERTHALS

P.B. Pettitt and G.W.L. Hodgins
Research Laboratory for Archaeology and the History of Art, University of Oxford, UK

1. Introduction

The Neanderthals (*Homo neanderthalensis*; King, 1864) were a Later Pleistocene archaic taxonomic human group in Eurasia, distributed from Iberia to Uzbekistan, the UK to the Near East. It is likely that Neanderthals evolved in Eurasia from robust Middle Pleistocene archaic humans that are usually referred to as *H. heidelbergensis*, probably as a biological response to the cold, dry environments that higher latitudes experienced over much of the Pleistocene.

2. Anatomical traits

The partial remains of over 400 individuals are known for classic Neanderthals. Anatomical traits which together are characteristic of the Neanderthals begin to appear *c.*450 000 years ago, with the Mauer (Heidelberg) mandible for example, and are well under way by *c.*300 000–200 000 years ago (e.g. in the >32 individuals represented in the Sima de los Huesos, Sierra de Atapuerca, Spain, and on the occipital [cranial] fragment from Swanscombe, England). The classic Neanderthal traits, i.e. those found only in this species (autapomorphies), include a well-developed supraorbital torus (brow ridge), relatively rounded occipital (rear) cranial morphology with a distinct protrusion (bunning), a large brain (larger than modern humans on average) and reduced midfacial prognathion (projection). Furthermore, in relation to previous archaic humans, Neanderthals had a large nasal aperture, a gap between the third molar and ascending mandibular ramus, lack of a pronounced chin, robust (thick-walled) postcranial bones and hyperarctic body proportions (short limbed, minimizing body surface area). All of these characteristics coalesced at least by *c.*75 000 years ago, e.g. in Oxygen Isotope Stage 4 of the Upper Pleistocene, where they are present throughout their range and particularly in the Near East, in Israel (Tabun, Amud, Kebara) and Iraq (Shanidar), and persist as late as 28 000 years (^{14}C dating, see glossary) ago in certain regions (e.g. Vindija Cave, Croatia, two fossils from which have been directly dated by AMS radiocarbon to *c.*28–29 000 years (^{14}C dating) ago).

Enough fossil specimens are known to give information on the mortality patterns of the Neanderthals: they generally lived to their early 30s, rarely beyond 40, with most deaths occurring either during infancy or in adolescence/early adulthood. The ubiquity of healed fractures, particularly on the bones of the upper body, indicates that the Neanderthal lifestyle required constant physical exertion; the nearest modern parallel to their injuries are rodeo riders. There is no anatomical evidence for disease in any surviving Neanderthal skeletal material, although healed breaks, trauma-related degenerative joint disease, lumbar abnormalities and lesions possibly relating to cancers are well known.

Historically, the Neanderthals have been depicted as the archetypal "shambling cavemen" since their discovery. This arose largely through a part-faulty reconstruction of their gait by the French anatomist Marcellin Boule in the early years of the 20th century, and the name itself can be employed in a derogatory sense. This is clearly an exaggeration.

3. Neanderthal behaviour

A good deal is known about Neanderthal behaviour. Neanderthals used caves and rock shelters where they existed, but, in all probability, constructed shelters in open environments. They probably organized themselves in small groups, and ranged over relatively small and fairly circumscribed territories. Pointed tools of bone indicate that they probably wore furs. They were habitual and successful hunters of medium (e.g. reindeer, bison) and large (e.g. mammoth) sized herbivores; their injuries may be explained by such hunting. Their stone tool technology was sophisticated, despite the fact that it was not as advanced as that of the earliest modern humans in Europe.

They could predetermine the shape and size of flakes from stone cores, but there is no convincing evidence that Neanderthals produced art, although they did – at least on occasion – bury their dead. Reconstructions of the voice box from points on the basicranium, in addition to a preserved hyoid bone from Kebara, Israel (Arensberg 1989; Arensberg *et al.*, 1990), indicate that Neanderthals could have coped physically with most consonants and vowels in the Western vocabulary, and there is general agreement with the theory that Neanderthals possessed at least some capability for language.

4. Neanderthal fate

Historically, questions as to the fate of the Neanderthals have usually centred on their relationship – or not – to our own species, *H. sapiens sapiens*. Their sudden disappearance from the fossil record has fuelled such questions, with early interpretations favouring epidemics, climatic change, outcompetition by or genetic incorporation into modern humans as they colonized the Old World. Among the most widely debated hypotheses of Neanderthal fate, three stand out and are described below.

(i) Hypothesis 1: Neanderthal phase of man

The first hypothesis, termed the "Neanderthal phase of man", saw a gradual evolution from *H. heidelbergensis* through *H. (sapiens) neanderthalensis* to *H. sapiens sapiens*. This has been rejected since the early 1980s, given that Neanderthals clearly persisted in some regions long after modern humans had appeared there.

(ii) Hypothesis 2: Presapiens

The second, "Presapiens", hypothesis held that a European pre-*H. sapiens sapiens* lineage existed alongside the Neanderthals, eventually giving rise to our own species independent of the Neanderthals. Recent research has called this into doubt as Africa appears to have been the cradle of our own species and as early "presapient" fossils such as Swanscombe appear more comfortably to fit a model which sees them as ancestral to Neanderthals.

(iii) Hypothesis 3: Preneanderthal

The third, "Preneanderthal", hypothesis sees the Neanderthals arising from Middle Pleistocene archaic humans in relatively northern latitudes through a process of evolutionary adaptation to cold environments, leaving them as a side branch of Late Pleistocene humans. Thus, while an ancestral evolutionary link to modern humans can be ruled out based on anatomical and geological evidence, the issue as to whether the two populations met and whether gene flow occurred between them is hotly debated. At present, the issue revolves most clearly around two specimens, both burials, from Germany and Portugal; crucial genetic evidence has been recovered from the former.

5. Neanderthal remains

The original find, Neanderthal 1, was discovered in 1856 during limestone quarrying in the Neander Valley (Thal), through which the Duessel River flows 13 km east of Duesseldorf. The finds came to light during the quarrying, by dynamite, of the Feldhoffer Grotte (now quarried away), and were delivered into the hands of a local schoolteacher and palaeontology enthusiast Johann Fuhrott.

The finds were publicized in several newspapers towards the end of 1856 and subsequently published by Hermann Schaafhausen, a Professor of Anatomy at Bonn (1857). They comprise a calvarium (skullcap), the right clavicle and scapula, both humeri, five rib fragments, both ulnae, the right radius, the left hipbone and two femora, all belonging to one individual – an adult male who was in all probability deliberately interred in the cave. A recent reinvestigation of the cave and surrounding sediments was begun in 1991 and has, so far, yielded further human material in the form of a molar, fragments of a right humerus, a cervical vertebra and fragments of a right tibia, right ischium, right humerus, right femur and ribs. All of these pieces complement and complete the original finds, with the exception of a midshaft fragment of a right humerus which is duplicated and, therefore, indicates the existence of a second individual, the sequencing of which is to be undertaken. One of the femoral fragments refitted perfectly to the original right femur confirming that the new finds belong to Neanderthal 1, which is now dated to 40 360 ± 760 years (^{14}C) before present on the basis of an AMS radiocarbon date on the right humerus (Schmitz, 1997; Schmitz & Thissen, 1998).

6. DNA evidence

The publication of a 360 nucleotide-long sequence from the Neanderthal 1 mitochondrial genome by Krings *et al.* (1997) provided the first genetic sequence from an extinct form of human. The DNA preserved in Neanderthal 1 provided evidence that the Neanderthal and modern human mitochondrial lineages are separate, and that they diverged significantly earlier than the mitochondrial lineages found within the modern human population.

The Neanderthal mitochondrial DNA sequence from hypervariable region I was assembled from short overlapping clones produced from PCR products amplified from bone extracts. Initially, the primer sequences were based on the modern human reference sequence (Anderson *et al.*, 1981); however, some of the subsequent amplifications included Neanderthal-specific primers.

The extraction and amplification of ancient DNA has been fraught with difficulties. Krings *et al.* (1997) addressed these difficulties through establishing, by a number of criteria, that:

- the Neanderthal bone should contain amplifiable DNA;
- the amplified DNA was likely to have been derived from endogenous DNA templates; and
- the observed sequence differences were neither a consequence of amplification from damaged DNA in early rounds of the PCR reaction nor of amplification of nuclear insertions of mitochondrial DNA sequences.

(i) Comparison of Neanderthal and 20th century gene pools

The relationship between the 360 nucleotide Neanderthal sequence and the modern gene pool was examined in several ways. The 360 nucleotide sequence was compared with mitochondrial lineages in modern populations from Europe, Africa, Asia, America (Native Americans) and Australia/Oceania. The human sequences differed among themselves by roughly 8.0 ± 3.1 substitutions, whereas the Neanderthal sequence differed from the human sequences by 27.2 ± 2.2 substitutions (see **Figure 1**). In spite of the archaeological evidence of an extensive period of coexistence in certain regions of Europe, the Neanderthal 1 sequence was not more closely related to the European lineage.

(ii) Phylogenetic trees

Phylogenetic analyses were also performed. Neighbour-joining trees (Saitou & Nei, 1987), rooted in chimpanzee mitochondrial DNA sequences, consistently showed the Neanderthal sequence branching off before the divergences seen within the human mitochondrial DNA lineages. An estimate of the divergence date, based on the estimated 4–5 million year divergence date for the human and chimpanzee lineages put forth by Takahata *et al.* (1995), suggested that the Neanderthal and human lineages split between 550 000 and 690 000 years ago. Using the same method, the age of the common human mitochondrial ancestor was found to be 120 000–150 000 years, consistent with current estimates (Vigilant *et al.*, 1991). Clearly, these estimates of divergence time are influenced by the accuracy of the assumed chimpanzee–human split; however, the relative divergence times of the Neanderthal–human *vs.* human–human mitochondrial DNA lineages will be unaffected. The mitochondrial sequence does not define a species and so the estimates

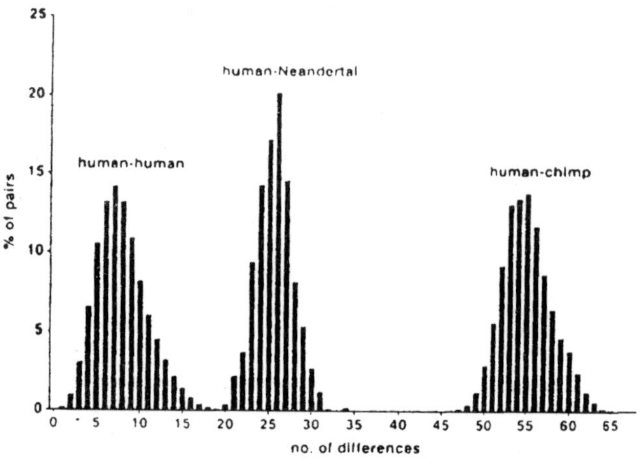

Figure 1 Distributions of the pairwise sequence differences among humans, humans–Neanderthal and humans–chimpanzees. X axis, the number of sequence differences; Y axis, pairwise comparisons. (Reproduced from Krings *et al.* (1997) with permission.)

should not be interpreted as estimates of the date of speciation (see article on "Mitochondrial 'Eve'").

A second mitochondrial DNA sequence has now been obtained from a Neanderthal from the Mezmaiskaya Cave in the northern Caucasus, over 2500 km from the Neander Valley (Ovchinnikov *et al.*, 2000). The analysis was performed in two independent laboratories – Glasgow and Stockholm – on a rib fragment from a partial skeleton of a Neanderthal infant dating to *c.*30 000 years before present. A total of 345 base pairs of HVRI was determined from two overlapping PCR fragments with lengths of 232 and 256 bp. Comparison of these with the Neanderthal sequence revealed 12 differences, with the two Neanderthals sharing 19 substitutions relative to a reference sequence. The length of pairwise difference between the two was comparable to a random sample of 300 Africans (8.36 ± 3.2) where 37% of pairs differed at 12 or more positions. Given that the mtDNA of the two individuals is closely related and displays only a moderate level of diversity compared with other primates the data suggest very low gene flow between Neanderthals and early modern humans.

7. Future investigations

Although the possibility of interbreeding between Neanderthals and early modern humans appears diminished, it cannot be ruled out based on a single specimen. Indeed, the recent recovery of a burial of a *c.*4-year-old boy from the Abrigo do Lagar Velho, Portugal (Lagar Velho 1), has crystallized the issue. The boy, dating to *c.*24 000–25 000 (^{14}C) years ago, displays a mosaic of modern human and Neanderthal anatomical traits, despite the fact that he lived and died some 5000–6000 years after the Neanderthals became extinct in the region. As these traits persisted over several millennia, they are taken to indicate that significant admixture occurred between the two populations in the region. The question of whether this case is the exception or the rule will remain open until further DNA sequences are available. A cautious interpretation of the evidence is that the process of Neanderthal extinction differed from region to region.

The "hybrid" from Lagar Velho and other Neanderthal specimens, e.g. Vindija, are being similarly studied with genetic techniques. Undoubtedly, this will shed light on the diversity among the known "Neanderthal" specimens, and begin to provide a picture of the size and distribution of past populations.

REFERENCES

Anderson, S., Bankier, A.T., Barrell, B.G. *et al.* (1981) Sequence and organisation of the human mitochondrial genome. *Nature* 290: 457–74

Arensberg, B. (1989) The hyoid bone from the Kebara 2 hominid. In *Investigations in South Levantine Prehistory*, edited by O. Bar Yosef & B. Vandermeersch, Oxford: British Archaeological Reports

Arensberg, B., Schepartz, L.A., Tillier, A.-M., Vandermeersch, B. & Rak, J. (1990) A reappraisal of the anatomical basis for speech in Middle Palaeolithic hominids. *American Journal of Physical Anthropology* 83: 137–46

King, N. (1864) The reputed fossil man of the Neanderthal. *Quarterly Journal of Science* 1: 96

Krings, M., Stone, A., Schmitz, R.W. *et al.* (1997) Neanderthal DNA sequences and the origin of modern humans. *Cell* 90: 19–30

Ovchinnikov, I.V., Gotterstrom, A., Romanova, G.P., Kharitonov, V.M., Liden, K. & Goodwin, W. (2000) Molecular analysis of Neanderthal DNA from the northern Caucasus. *Nature* 404: 490–93

Saitou, N. & Nei, M. (1987) The neighbor joining method: a new method for reconstructing phylogenetic trees. *Molecular Biology and Evolution* 4: 406–25

Schmitz, R.W. (1997) Neandertal (Feldhoffer Grotte). In *History of Physical Anthropology: an Encyclopaedia*, 2 vols, edited by F. Spencer, New York and London: Garland

Schmitz, R.W. & Thissen, J. (1998) Archaologie im Neandertal – nicht nur auf den Spuren des Neandertalers. *Archaologie im Rheinland* 1997: 20–21

Takahata, N., Satta, Y. & Klein, J. (1995) A genetic perspective on the origin and history of humans. *Annual Review of Ecology and Systematics* 26: 343–72

Vigilant, L., Stoneking, M., Harpending, H., Hawkes, K. & Wilson, A.C. (1991) African populations and the evolution of mitochondrial DNA. *Science* 253: 1503, 1507

GLOSSARY

AMS radiocarbon dating Radiocarbon is formed in the upper atmosphere by the action of cosmic rays upon ^{14}N, which then becomes incorporated into plants through photosynthesis, and then up the trophic pyramid through consumption. It is therefore present in all living organisms. Radiocarbon dating is based on the decay of ^{14}C back to ^{14}N by beta emission, and given its half life of 5730 years is applicable back to *c.*50 000 years before present. Although ^{14}C constantly decays at the rate of about 1% loss of existing atoms per 83 years, it is constantly replaced during an organism's life; this replacement stops with death and this forms the basis for its use in dating. In relation to other chronometric dating methods it is relatively precise and is therefore of great use for later Neanderthals and the earliest modern humans in Eurasia. Since the 1980s radiocarbon dating by accelerator mass spectrometry (AMS) has enabled the dating of very small samples, up to one thousandth the size of those required for conventional dating. This has facilitated the direct dating of fossil human remains such as the Neanderthals as the size of samples required are in the region of 100–500 mg. Carbon is isolated from amino acids in bone and is combusted, the resulting gas dated in an accelerator mass spectrometer. In a given carbon molecule the ratio of the two stable isotopes of carbon – ^{12}C and ^{13}C – are known relative to the unstable isotope ^{14}C, and from this the original amount of ^{14}C the datable sample had in life can be predicted. By AMS the carbon atoms can be counted and the amount of ^{14}C remaining can be ascertained at a given level of confidence. As the half life of ^{14}C is well know an age can therefore be computed and is expressed with an error at one standard deviation, for example 40360 ± 760 for the Neanderthal 1 specimen (*see text*).

See also **Mitochondrial "Eve"** (p.891)

GREGOR MENDEL

Eric C.R. Reeve
Institute of Cell, Animal and Population Biology, University of Edinburgh, Edinburgh, UK

1. Mendel's life

Gregor Johann Mendel, the founder of the science of genetics, was born on 22 July 1822 in the village of Heinzendorf bei Odrau in Austrian Silesia, the only son of a small peasant proprietor who took a special interest in fruit culture and who initiated his son at an early age into the methods of grafting. Mendel impressed his teacher at the village school and was sent on to Leipnik at the age of 11 and then to the gymnasium at Troppau, where he continued to do well. However, the cost of his education severely taxed the resources of his parents, and he was only able to complete his studies with help from a younger sister, Theresa, who contributed part of her dowry.

In 1843, 13 candidates were put forward for admission as novices to the Augustinian Monastery of St. Thomas in Brünn, four were short-listed and Mendel was the only one accepted. He took the name Gregor and became a priest in 1847. Teaching science, philosophy and theology was an important function of the monks, and required the teacher to hold a teaching certificate. However, Gregor failed in the natural science section of his examination in 1850.

One of his exam questions, set by the zoology professor, R. Kner, consisted of an essay to be written within 8 weeks, and the question and Mendel's answer of 23 foolscap pages have been rediscovered and translated into English by Orel *et al.* (1983). Kner, well known for his publications in palaeontology and ichthyology, set Gregor the following:

> The chief differences between rocks formed by water and those formed by fire, detailing the main varieties of the Neptunian strata in serial order according to their age and giving a short characterization thereof, and, in conclusion, giving in like manner a review of the igneous rocks, both Plutonian and Vulcanian.

Kner stated that Mendel in his essay had used outdated references and had given too much attention to describing minerals and rocks, and this criticism evidently prevented him from receiving the teaching certificate. Mendel's second essay was to explain the mechanical and chemical properties of air and the origin of wind, which appears to have been Mendel's first contact with meteorology. It was well received by the examiner and caused Mendel to become very interested in meteorology when he returned to the monastery.

Abbott Napp, head of the monastery, then wrote to the Bishop of Brünn to explain that Father Gregor was not suitable to be a parish priest, but had shown exceptional intellectual capacity and industry in studying the natural sciences, and Napp arranged for Mendel to spend 2 years at the University of Vienna, at the monastery's expense, to prepare for the reexamination and study mathematics, physics and natural sciences under teachers who included leading botanical scientists.

Returning to the monastery at Brünn in autumn 1853, Mendel became a teacher, especially of physics, and was reported to be extraordinarily successful, both in his teaching skill and in the care of his pupils. He continued teaching, along with his experimental studies, until 1868, when he was elected Abbott; from then onwards, administrative duties and the handling of a controversy with the government on the taxation of monastery property prevented him from continuing his productive scientific work. He died on 6 January 1884.

2. Mendel's experiments

The only details of Mendel's research surviving are those given in his scientific paper "Versuche über Pflanzenhybryden", published in the *Verhandlungen des Naturforschenden Vereins in Brünn* (Mendel, 1866). An English translation by the Royal Horticultural Society of London was published in volume 26 of that Society's Journal in 1901, and a modified version of this translation was included by William Bateson in his book *Mendel's Principles of Heredity* (1909). This version was later checked against the German original by Miss Denise Ryan and Dr. J.H. Bennett and was included in the book entitled *Experiments in Plant Hybridization*, with commentary and assessment by the late Sir Ronald Fisher and the biographical notice of Mendel from W. Bateson's 1909 book (Bennett, 1965). This book forms the main basis for this article, with additional information from Edwards (1986) and Orel (1996).

(i) Experimental materials

Mendel's experiments, carried out in the monastery garden and greenhouse, began in 1854 shortly after his return from Vienna to Brünn. Having decided on the garden pea (*Pisum sativum*) as experimental material, he obtained 34 strains from a few seedsmen and, during the following 2 years, tested them for constancy of form and ability to hybridize successfully with each other. The 22 strains that bred true for one or more of seven pairs of alternative characters that could be easily distinguished from each other were selected for the experiments and were bred throughout the period of the experiments, during which time they remained constant without exception.

(ii) Choice of characters and methods of breeding

The characters, in Mendel's terminology as translated from the German, were:

1. round *vs.* wrinkled seed
2. yellow *vs.* green endosperm
3. white seed-coat and white flowers *vs.* grey or brown seed-coat with violet-purple flowers

4. ripe pods simply inflated *vs.* pods deeply constricted between the seeds and more or less wrinkled
5. unripe pods light to dark green or vividly yellow
6. flowers axial (i.e. arranged along the main stem) or terminal (bunched at the top of the stem)
7. stem long (6–7 feet) or dwarf (¾ to 1½ feet).

P. sativum is normally self-pollinating, the seed and pollen both becoming ripe while completely enclosed in the keel surrounding them, so that the seed cannot be fertilized by foreign pollen. Cross-fertilization is, however, easy to achieve by opening the keel before seed and pollen are ripe, removing the pollen tubes and shaking pollen from another plant onto the seed. This was Mendel's method of cross-fertilization. The plants could also be easily cultivated in the monastery garden or greenhouse, so they made excellent experimental material.

(iii) Crosses of parental lines

Mendel first crossed a number of plants differing in one pair of characters, making reciprocal crosses to test the contributions of each parent. Using modern terminology (which differs slightly from that of Mendel) to distinguish the generations, the parental generation is called P (or P1 and P2, when the two parents need to be distinguished) and the hybrids form the first filial generation, i.e. F_1. Mendel found that the hybrids always closely resembled one of the two parents, regardless of whether this parent was the donor of seeds or pollen. This fact immediately demonstrated that seed and pollen made equal contributions to their progeny. Mendel named the character that was manifest in the hybrid progeny "dominant" and the other character "recessive", two terms that have become standard.

The dominant characters in his experiments were: *round seeds, yellow seed endosperm, coloured seed coat and flowers, ripe pods simply inflated, unripe pods green, flowers axial* and *long stem.*

(iv) Self-pollination of hybrid F_1 plants

Mendel next planted seeds from each hybrid and allowed them to self-pollinate to produce the F_2 (second filial) generation. His great innovation was to plant large numbers of seeds from each cross, examine all the resulting plants carefully and count the numbers showing each alternative character. He found that in every case the recessive character, invisible and apparently lost in the hybrids, reappeared in their progeny, but that the dominant character always occurred approximately three times as often as the recessive character. This is most clearly demonstrated by the seed characters (numbers 1 and 2 as listed above), since a large number of seeds can be scored on a small number of plants.

As examples, in experiment 1, on form of seed: 253 plants hybrid for form of seed were grown, all of course from round seeds, and these plants produced 7324 seeds, of which 5474 were round and 1850 wrinkled, a ratio of 2.96:1.

In experiment 2, on colour of seed albumen: 258 plants hybrid for colour of seed albumen were planted, all from yellow seeds, and gave 8023 seeds, of which 6022 were yellow and 2001 were green, a ratio of 3.01:1.

In each experiment, individual plants always produced both types of seed and, to illustrate the variation shown by individual plants, Mendel lists the numbers of dominant and recessive seeds obtained from the first ten plants of each experiment (Table 1).

Experiments 3–7 give Mendel's results with F_1 hybrids for characters 3–7, where plants rather than seeds had to be counted, so that the numbers are much smaller, and these are summarized in Table 2.

Mendel points out that: "If now the results of the whole of the experiments be brought together, there is found, as between the number of forms with the dominant and recessive characters, an average ratio of 2.98:1, or 3:1."

Mendel then makes another important point:

> The dominant character can have here a *double signification* – viz. that of a parental character, or a hybrid character. In which of the two significations it appears in each separate case can only be determined by the following generation. As a parental character it must pass over unchanged to the whole of the offspring; as a hybrid-character, on the other hand, it must maintain the same behaviour as in the first generation.

(v) Tests on the second generation bred from the hybrids

Mendel next planted seeds of the self-pollinated F_2 generation, the first generation in which the recessive character had reappeared. He found that seeds showing the recessive character (wrinkled and green, respectively, for the two seed characters) gave plants which always bred true for the recessive character. However, seeds showing the dominant character gave two types of plant: one-third of these bred

Table 1 Form of seed and colour of albumen from the first 10 plants in Experiments 1 and 2.

	Experiment 1 *Form of seed*		Experiment 2 *Colour of albumen*	
	Round	**Angular**	**Yellow**	**Green**
Plants				
1	45	12	25	11
2	27	8	32	7
3	24	7	14	5
4	19	10	70	27
5	32	11	24	13
6	26	6	20	6
7	88	24	32	13
8	22	10	44	9
9	28	6	50	14
10	25	7	44	18

Table 2 Counts from F_1 hybrid plants: Experiments 3–7.

Experiment	Seeds planted	Dominants	Recessives	Ratio
3	929	705	224	3.15 to 1
4	1181	883	229	2.95 to 1
5	580	428	152	2.82 to 1
6	858	651	207	3.14 to 1
7	1064	787	277	28.4 to 1

Average ratio of Dominants to Recessives: 2.98 to 1

pure for the dominant character while two-thirds behaved like the F_1 plants, i.e. each plant produced dominant to recessive type seeds in the ratio of 3:1.

Thus, for the first seed character (round *vs.* wrinkled seed), Mendel grew 565 plants from round seeds and, of these, 193 plants gave entirely round seeds (the dominant character), while 372 plants produced round and wrinkled seeds in the ratio of 3:1. The ratio of 372:193 is 1.93:1, which is almost exactly 2:1.

For the other seed character, Mendel raised 519 plants from yellow seed and obtained 166 plants which bred true for yellow seed and 353 plants which gave yellow and green seeds in the ratio of 3:1. The ratio of 353:166 is 2.13:1, which is again very close to 2:1.

Mendel made the same tests with the other five pairs of characters, but on a much smaller scale, since he had to count plants instead of seeds. He found essentially the same result: plants always showing the recessive character bred true in subsequent generations, while dominants appeared approximately in the ratio of 3:1 and, of these, two-thirds gave a 3:1 ratio in the next generation, while one-third bred pure for the dominant character.

Mendel realized that the 3:1 and 2:1 ratios fitted the same hypothesis. He wrote:

> Since the members of the first generation (F2) spring directly from the seeds of the hybrids (F1), it is clear that the hybrids form seeds having one or other of the two differentiating characters, and of these one-half develop again the hybrid form, while the other half yield plants which remain constant and receive the dominant or the recessive characters (respectively) in equal numbers.

He used the symbols *A*, *Aa* and *a* to denote the dominant, the hybrid and the recessive form, and stated that "*A* + 2*Aa* + *a* shows the terms in the series for the progeny of the hybrids of two differentiating characters". Nowadays, we would write this formula as "*AA* + 2*Aa* + *aa*", to show that each plant or fertilized seed carries two copies of the gene in question, *A* and *a* being the dominant and recessive forms of the gene, respectively.

(vi) Effect of repeated self-pollination of hybrid

Mendel goes on to calculate what happens if a hybrid such as those in his experiments is allowed to self-pollinate for a number of generations, provided there is equal fertility of all plants in all generations. The successive ratios of the three types *A*, *Aa* and *a* will be: (1:2:1); (3:2:3); (7:2:7); and so on, with the proportion of the hybrid type *Aa* halving each generation (i.e. 2/4, 2/8, 2/16, etc.).

(vii) Hybrids with several differentiating characters

Mendel then gives the rules governing cases where several differentiating characters are associated in the same hybrid and illustrates them for the combination of his two seed characters, round/wrinkled (symbolized *A/a*) and yellow/green (symbolized *B/b*). He crossed *AB* seed parents with *ab* pollen parents to obtain hybrid plants whose seeds were all round and yellow. Fifteen of these plants gave a total of 556 seeds (of the F_2 generation) of which:

315 were round and yellow,
101 were wrinkled and yellow,
108 were round and green, and
32 were wrinkled and green.

This shows that the two pairs of characters, round/wrinkled and yellow/green, were passed on to the progeny independently, each in the ratio 3 dominants:1 recessive, the overall ratios being close to 9:3:3:1, as expected from the formula (3 + 1)(3 + 1). Mendel gave the results of a second generation of self-pollination, which produced nine types, and also data and expectations for selfing a hybrid containing three pairs of characters. These results and Mendel's interpretation show that he was very well aware of the way that recognizable characters in hybrids were distributed among their progeny, and that the elements determining each behaved as factors which retained their individual characteristics, even if a recessive was hidden when present with the dominant.

3. Publication and reception of his results

Mendel's experiments on the garden pea were continued over 8 years, and his results and conclusions were presented to the Brünn Natural History Society verbally on 8 February and 8 March 1865, and his celebrated paper on "Inheritance in the Garden Pea" (*Versuche über Pflanzenhybryden)* was published in the following year in the proceedings of the Society for 1865. Mendel received 40 reprints of his paper, which were doubtless sent to the leading European biologists studying plant breeding and hybridization. In addition, the *Proceedings of the Brünn Natural History Society* was widely distributed and was known to have reached the Royal Society and the Linnaean Society in London. Nevertheless, Mendel's paper was completely ignored for 34 years, until it was discovered by the plant biologists de Vries in Holland, Correns in Germany and Tschermak in Austria at almost the same time in 1900.

Almost instantly, it was recognized as a work of blazing importance, explaining the phenomenon of heredity, bafflingly complex as this had seemed, in terms of the transmission unchanged from generation to generation of permanent units, for which Johannsen gave the name "genes". It was also quickly appreciated that Mendel's discovery was applicable not only to plants, but also to animals, including man, since suitable human pedigrees already existed. Appropriate terms for discussing the new subject of "genetics" soon came into common use. These were "allelomorph" (later shortened to allele) for genetic units such as *A*, *a*, *B* and *b*; "homozygote" for *AA*, *BB*; "heterozygote" for *Aa* etc.; and "homozygous/heterozygous" for the corresponding states.

Some problems remained of course. Mendel's seven pairs of characters all showed dominance and it was suggested that the recessive allele was actually the loss of the gene. However, recessive alleles were soon discovered which showed dominance over another allele, indicating that they must retain genetic activity, and many pairs of alleles have since been discovered for which the heterozygote has an intermediate "phenotype" or shows "heterosis" (the phenomenon in which the heterozygote exceeds the values of both homozygotes).

Mendel did not claim that hereditary factors were always either dominant or recessive, but he selected his seven pairs of "differentiating characters" as dominant *vs.* recessive because they were all clearly distinguishable. Of these, only the two seed characters (his numbers 1 and 2) could be tested together on a large enough scale to show that they segregated independently. Knowing nothing about chromosomes and linkage, Mendel assumed that his other characters would also behave independently. There are in fact seven pairs of chromosomes in the garden pea and Mendel's seven characters were not each located on a separate chromosome. However, no two of them were located closely together on the same chromosome so that linkage, in the form of unexpected segregation frequencies, would be obvious from his data.

Charles Darwin's *Origin of Species* was published in 1859 and his *The Effects of Cross and Self Fertilisation in the Vegetable Kingdom* in 1876. Darwin's major problem was how characters were inherited and, in particular, how new genetic variation on which natural selection could act was created, since he believed that inheritance is blending. So he would have found Mendel's paper of absorbing interest, and clearly never received a copy (or never read it). Another remarkable fact is that, although Mendel was a member of the Brünn Natural History Society which had leading botanical scientists among its members, none of them can have appreciated the significance of his research or indeed taken any notice of it.

4. Rediscovery of Mendel's paper

Enormous enthusiasm as well as arguments about his work followed the rediscovery of Mendel's paper. William Bateson, Professor of Zoology at Cambridge, was so enthusiastic about Mendel's paper in his lectures and writing that it was very widely believed that it was he – and not the three botanists – who had actually rediscovered it. In the sweet pea, Bateson and Punnett crossed two pure breeding plants with white flowers and obtained coloured flowers in the F_1. Selfing the F_1 gave a ratio of approximately 9 coloured:7 whites; tests on individual F_2 plants showed that two independent dominant pairs of alleles were segregating and that both dominant alleles were needed to give coloured flowers (if the two pairs are *A*/*a* and *B*/*b*, white is obtained if either *A* or *B* is not present). The Andalusian fowl gave a similar puzzle. *Rose comb* was dominant to *single*, and *pea* was also dominant to *single*, but *rose* crossed to *pea* gave a new type called "walnut", which did not look structurally like a combination of rose and pea. The F_2 cross of walnut × walnut gave ratios of 9 walnut:3 rose:3 pea:1 single, showing that two independent pairs of alleles were segregating and that these were both dominant alleles which affected the comb and combined to give a new phenotype.

A quarrel interfered with the general acceptance of Mendel's theories in Britain for some years after their rediscovery. Karl Pearson (1857–1936), a mathematical physicist, and W.F.R. Weldon (1860–1906), Professor of Zoology at Oxford, had become interested in Francis Galton's attempts at a mathematical analysis of parent–offspring resemblances. The personal quarrel developed between Weldon and Bateson in 1890 over the origin of cultivated races of Cineraria, each accusing the other of misrepresenting their published statements. This seems to have pushed Weldon and Pearson into disputing that the operation of Mendelian principles could explain the correlations between relatives in a random mating population for a quantitative character such as human height. Pearson supported this view in 1904 with mathematical arguments, which were not disposed of until R.A. Fisher's 1918 paper published in the *Transactions of the Royal Society of Edinburgh*.

This personal quarrel seems to have led to particularly marked enthusiasm of Bateson for Mendel's theory and for his claim that Mendel was not sympathetic to Darwin's theory about the origin of species by natural selection. Bateson wrote in his biographical notice on Mendel (see Bennett, 1965): "With the views of Darwin which at that time were coming into prominence Mendel did not find himself in full agreement, and he embarked on his experiments with peas." The implication here is that Mendel started his experiments to find fault with Darwin's view, but in fact Mendel could not have known about Darwin's theories, which were not published until after he started his Pisum experiments, and there is no evidence that Mendel then or later disagreed with Darwin.

5. Some mysteries remain about Mendel and his work

(i) The origin of Mendel's theory about inheritance

Mendel's paper gives the strong impression that he knew precisely what results to expect from his experiments and that they were designed as a demonstration of his novel theory rather than as an investigation of a new phenomenon. He also showed a remarkably modern scientific approach in carefully choosing the organism, the series of strains, and the characters and tests that would best suit his purpose. The results were most successful and his one disappointment could only have been that none of the large number of expert and enthusiastic students of plant hybridization could understand, or believe, his very straightforward presentation – which should have created as much excitement then as it did when rediscovered 35 years later.

Introducing his paper, Mendel wrote:

> Experience of artificial fertilization, such as is effected with ornamental plants in order to obtain new variations in colour, has led to the experiments which will here be discussed. The striking regularity with which the same hybrid forms always reappeared whenever fertilization took place between the same species induced further experiments to be undertaken, the object of which was to follow up the developments of the hybrids in their progeny.

(ii) Mendel's notebooks and other documents

Mendel's papers would certainly be of extraordinary interest. As a result, when Mendel's name became famous almost overnight in 1900, British and European biologists visited his monastery in search of all the information available there, but none of the notebooks he would have kept nor other books and articles he must certainly have accumulated during his long period as monk and abbot could be found. Thus, Bateson reports having visited the monastery at Brünn in 1905 and searching unsuccessfully for such materials with the help of the Abbot; while Iltis who wrote a biography of Mendel published in German in 1924 and in English translation in 1932:

> ... had several opportunities of talking about Mendel with an old man named *Joseph Maresch*, the Monastery gardener who died only a few years ago. Unfortunately Maresch was of bibulous inclinations ... He told me he had some of Mendel's notebooks, but none of them could be found. (Iltis, 1932)

(iii) Were Mendel's numerical results "cooked"?

The most interesting mystery over Mendel's work was raised by R.A. Fisher in his introduction and commentary for an English translation of Mendel's paper, which he wrote in 1936. This project fell through until after Fisher's death, when it was edited by J.H. Bennett and published in 1965 with added material, including a biography of Mendel by Bateson (Bennett, 1965). Fisher analysed Mendel's data as far as they were published, and reached two remarkable conclusions:

1. Statistical tests applied by Fisher show that the very close agreement between Mendel's data and the results he expected is most unlikely to have arisen by chance, suggesting that the data have been "doctored" to fit his theory. Modern statistical methods developed largely by Fisher himself can test whether experimental results and theory agree too closely, just as they can test whether the difference is too great for the theory to be acceptable.
2. There is a particular discrepancy in the results for Mendel's characters 3–7, where the first generation hybrids (dominants of constitution *Aa*) were self-pollinated to produce dominants to recessives in the ratio 3:1 and the plants showing dominance were then selfed to test the proportion of *AA* to *Aa* types. Expecting this proportion to be one *AA* (giving entirely dominant progeny) to two *Aa* (giving dominant and recessive plants in the ratio 3:1), Mendel had to plant a number of seeds from each plant to establish its constitution. Altogether, he tested 600 progenies from F_2 plants showing dominance and, of these, 201 were recorded as non-segregating, which is extremely close to the 1:2 ratio he expected.

However, as Fisher points out, each F_2 plant was characterized from growing 10 progeny plants and the 1:2 expectation is not exact for such small samples. This is because there is a small probability that 10 seeds chosen at random from a selfed parent plant of constitution *Aa* will contain no recessives, so that the parent will be classified as *AA* and not *Aa*. This probability is $(3/4)^{10}$, or 5.631%, which means that, instead of expecting one non-segregating to two segregating progenies, he should have expected 1.1126:1.8874, giving a total expectation of 223 non-segregating progenies out of the 600 he tested. As noted above, he recorded 201, which is exactly one-third of the number tested and substantially less than 223.

Fisher concludes:

> There is no easy way out of the difficulty. The same discrepancy must have occurred with the 473 test progenies needed to complete the trifactorial experiment; and an examination of the general level of agreement between Mendel's expectations and his reported results shows that it is closer than would be expected in the best of several thousand repetitions. The data have evidently been sophisticated systematically, and after examining various possibilities, I have no doubt that Mendel was deceived by a gardening assistant, who knew too well what his principal expected from each trial made. (Fisher, 1936)

(iv) Further statistical analysis by A.W.F. Edwards (1986)

Fisher's conclusion that Mendel's data were in general too close to the expected values of his theory to be accepted at face value, with the implication (which Fisher strongly rejected) that Mendel had "cooked" his data, led to a number of papers generally motivated by the belief that Mendel could not have manipulated his data and suggesting ways in which the data might have been created without any scientific fraud. Edwards (1986) surveys these suggestions, which are of interest for their ingenuity and variety, but which fail to provide any adequate explanation, so I merely refer the curious reader to the paper by Edwards.

Edwards also points out that Bateson and Killby (1905) and Darbishire (1908, 1909) made large-scale tests of segregation in the pea, to look for breeding behaviour which could explain the deviations in Mendel's data, and these were analysed by Weiling (1971, 1985). Edwards comments that these data and analyses show that the pea is a good randomizer, so that abnormal segregation was not a likely explanation of Mendel's results.

Edwards then presents a more detailed statistical analysis of the whole of Mendel's data than Fisher or others had published, which involves individual consideration of the 84 independent segregations reported by Mendel. This analysis is too long and complex to be presented here, and needs reference to Edwards (1972) to help the reader understand his discussion of the analysis based on χ, which he prefers to χ^2. Edwards concludes from his analysis:

> We now see that Mendel's data exhibit two independent peculiarities both of which point to the same conclusion, namely, that in general the segregations agree more closely with what Mendel expected than chance would dictate. Where Mendel expected a 2:1 ratio even though the true expectation must have been different it was the former ratio that was achieved, and throughout the rest of his results there is a persistent lack of extreme segregations. (Edwards, 1986)

It is possible, if unlikely, that further searches will lead to the discovery of notebooks throwing light on Mendel's attitude to his analytical results. But meanwhile we have to accept Fisher's verdict, strengthened by the further analysis of Edwards, that only "systematic sophistication" of Mendel's results (probably by an assistant gardener) can have given such excellent agreement between the results and theory. This agreement can only have strengthened Mendel's belief that he was correct, which would have done no harm.

REFERENCES

Bateson, W. (1909) *Mendel's Principles of Heredity*, Cambridge: Cambridge University Press

Bateson, W. & Killby, H. (1905) Experimental studies in the physiology of heredity: peas (*Pisum sativum*). *Royal Society Reports to the Evolution Committee* 2: 55–80

Bennett, J.H. (ed.) (1965) *Experiments in Plant Hybridisation* (Mendel's original paper in English translation with commentary and assessment by the late Sir Ronald Aylmer Fisher, together with a reprint of W. Bateson's biographical notice of Mendel), Edinburgh: Oliver and Boyd

Darbishire, A.D. (1908) On the result of crossing round with wrinkled peas, with especial reference to their starch-grains. *Proceedings of the Royal Society* B 80: 122–35

Darbishire, A.D. (1909) An experimental estimation of the theory of ancestral contributions in heredity. *Proceedings of the Royal Society* B 81: 61–79

Darwin, C. (1859) *On the Origin of Species by Means of Natural Selection, or the Preservation of Favoured Races in the Struggle for Life*, London: John Murray; many reprints

Darwin, C. (1876) *The Effects of Cross and Self Fertilisation in the Vegetable Kingdom*, London: John Murray; revised 1878, 1891

Edwards, A.W.F. (1972) *Likelihood: An Account of the Statistical Concept of Likelihood and its Application to Scientific Inference*, Cambridge: Cambridge University Press; revised edition, Baltimore: Johns Hopkins University Press, 1992

Edwards, A.W.F. (1986) Are Mendel's results really too close? *Biological Reviews* 61: 295–312

Fisher, R.A. (1918) The correlation between relatives on the supposition of Mendelian inheritance. *Transactions of the Royal Society of Edinburgh* LII: 399–433

Fisher, R.A. (1936) Has Mendel's work been rediscovered? *Annals of Science* 1: 115–37; reprinted in Bennett, 1965, pp.59–85

Iltis, H. (1932) *Life of Mendel*, London: Allen and Unwin (German edition, 1924)

Mendel, G. (1866) Versuche über Pflanzenhybryden. *Verhandlungen des Naturforschenden Vereins in Brünn* 4: 1

Orel, V. (1996) *Gregor Mendel: The First Geneticist*, Oxford: Oxford University Press

Orel, V., Czihak, G. & Wieseneder, H. (1983) Mendel's examination paper on the geological formation of the earth of 1850. *Folio Mendeliana* 18: 227–72

Weiling, F. (1971) Mendel's 'too good' data in *Pisum* experiments. *Folio Mendeliana* 6: 75–77

Weiling, F. (1985) What about R.A. Fisher's statement of the 'too good' data of J.G. Mendel's *Pisum*-paper? 45th Session of the International Statistical Institute, Amsterdam

See also **From the great apes to Gregor Mendel** (p.53); **Darwin and Mendel united** (p.77)

CHARLES DARWIN AND THE THEORY OF EVOLUTION

John R.S. Fincham
Institute of Cell and Molecular Biology, University of Edinburgh, Edinburgh, UK

1. Early life and influences

Charles Darwin is commonly credited with the theory of evolution, but the idea that species were subject to change was not original with him. It was the idea of evolution by natural selection that was his special contribution.

Charles Darwin was born in Shrewsbury in 1809, the son of Robert Darwin, a successful and much-respected physician. Robert's father, Erasmus Darwin (1731–1802), had also been a physician, and a leading one in his day, but is better remembered as a poet, a versatile inventor of mechanical devices, a founding member of the West Midland club of intellectuals and pioneering technologists known as the Lunaticks (King-Hele, 1998) and, most notably in the present context, for his vision of the evolutionary development of life forms from submicroscopic beginnings:

> First, forms minute, unseen by spheric glass,
> Move on the mud, or pierce the watery mass,
> These, as successive generations bloom
> New powers acquire and larger limbs assume.
> (From Erasmus Darwin, *Zoonomia*, 1794)

Charles had influential connections on both sides of his family, for his mother, Susannah, was the daughter of Josiah Wedgwood of the great pottery dynasty. She died in 1817 when Charles was eight years old, and he was left largely in the care of his elder sisters Marianne, Caroline and Susan, who were respectively eleven, nine and six years his senior. He also had an elder brother, Erasmus (born 1803), and a younger sister, Emily (born 1810).

From his own account, Charles Darwin does not appear to have gained much from his schooling in Shrewsbury, consisting as it did mainly of Latin and ancient Greek. However, he was able to augment his education in other ways. He was taught Euclidian geometry by a private tutor and took pleasure in the elegance of the proofs. He learned something about chemistry by assisting his brother, who conducted experiments in the garden shed, and he showed an interest in the natural world, collecting minerals and insects and (under the influence of Gilbert White's *The Natural History and Antiquities of Selbourne*) observing the habits of birds. He also read English poetry and became fascinated by the plays of Shakespeare.

2. Edinburgh

In 1829, Robert Darwin, seeing that Charles was making little progress in his regular school subjects, took him away from school and sent him, at the early age of 16, to Edinburgh to study medicine. It was a natural assumption that Charles would follow in his father's footsteps. Robert, who was reckoned to be an outstandingly good judge of character, considered that his son, who had already given him some assistance in his medical practice, had the makings of a good, confidence-inspiring physician. As it turned out, however, Charles, although he did not neglect his medical studies, took no pleasure in them. He found the lectures very dull and he had no stomach for the surgical operations that he was expected to witness – before the use of chloroform, they must indeed have been harrowing. However, he was able to cultivate some new interests in Edinburgh. He attended lectures on zoology and geology which, although he thought them very badly delivered, may have done something for his future career. He was a regular attendee at, and occasional contributor to, the Plinian Society, a group which met to discuss scientific topics, including natural history. He became interested in marine life and befriended fishermen. On one occasion, he presented to the Society a paper with some genuinely novel observations on the larvae of Flustra, a polyzoan of tidal pools. His natural history interests were combined with sporting and he got most absorbed in developing his skill with the shotgun, which he used for shooting birds, taking lessons from a local taxidermist so that he could preserve his specimens.

3. Cambridge

After Charles had spent a little over two years in Edinburgh, his father saw that he was not, after all, going to succeed in medicine and decided, as a second best, that he should enter the Church. As Darwin later recalled in his autobiography (Barlow, 1958), his father had been "very properly vehement against my becoming an idle sporting man". From his son's account, Robert does not appear to have been a religious believer himself, but he saw the Church as a beneficial institution (especially for women) and a good second-best career option.

To become a minister, Charles needed a degree from an English university, and he went up to Cambridge early in 1828. Half a century later he wrote in his autobiography (Barlow, 1958) that his three years spent at that university were "sadly wasted . . . and worse than wasted". From his passion for shooting and hunting he had got into "a sporting set, including some dissipated and low-minded young men". They drank too much and frittered away their evenings in jolly singing and card games. However, Darwin had to admit, "as some of my friends were very pleasant and we were all in the highest spirits, I cannot help looking back on those times with much pleasure".

In fact, Cambridge was far from an intellectual desert so far as Darwin was concerned. He made the important acquaintance of two professors: John Henslow, whose lectures on botany he attended and much enjoyed, and, through Henslow, Adam Sedgewick, who fostered his interest in geology and whom he accompanied on a geological and fossil-hunting tour of north Wales in 1831. He also pursued his own interests in natural history, especially of beetles, of which he became an avid collector. He read, and was much impressed by the works of William Paley, *A View of the Evidences of Christianity*, *The Principles of Moral and Political Philosophy* and *Natural Theology*, the first two of which were then obligatory reading for Cambridge students. From Paley, Darwin learned much about the wonders of biological adaptation, which Paley took as evidence for divine craftsmanship, but which Darwin later came to interpret as due to natural selection.

Almost incidentally, it seems, Darwin obtained his degree, securing a creditable tenth place among that "crowd of men who do not go in for honours". He managed to resuscitate his almost forgotten school Latin and Greek, and scored with a good paper in Euclid. In later years, he regretted not having taken his mathematics further, but his attempt to do so at Cambridge foundered on the algebra.

4. The voyage of the Beagle

At the end of his time in Cambridge, Darwin was still bound for the Church. He rather fancied the life of a country parson and was still a believer in biblical authority. However, in 1831, the course of his career was permanently diverted. On his return from his Welsh excursion with Sedgewick he received a letter from Henslow informing him of an offer from Captain Robert Fitzroy, of the Royal Navy, of a place in his own cabin to any young man prepared to travel as an unpaid naturalist on the forthcoming survey voyage of H.M.S. Beagle. Charles would have volunteered immediately, but his father, whom he greatly respected and on whose financial support he was still dependent, had strong objections; he considered the project a distraction and further evidence of his son's frivolity. Yet, he added a saving clause in his letter: "If you can find any man of common sense who advises you to go, I will give my consent." Fortunately, Charles soon had the opportunity of consulting his uncle Josiah Wedgwood, whose estate he was visiting for the opening of the shooting season. On being told of the situation, Uncle Josiah was rather in favour of the proposed voyage and wrote to Robert Darwin accordingly. Among other cogent arguments, he pointed out that Charles was not doing anything so important that it could not be interrupted, and also that seafaring men often settled down to quiet and stable domestic lives once their voyages were completed. So Robert, who had always regarded Josiah as the epitome of common sense, finally consented. The naval authorities duly received, and accepted, Charles's application.

Fitzroy's Beagle, a small vessel only 90 feet long, left Plymouth with Darwin on board at the end of December 1831, and did not return until October 1836. The first two years were spent surveying the eastern coast of South America, with special emphasis on Patagonia. The initial ports of call were Bahia (Salvador), Rio de Janeiro and Montevideo, and then the voyage followed a complicated itinerary back and forth between Montevideo and Cape Horn. In early summer 1834, Fitzroy rounded the Cape (an earlier attempt had been driven back by heavy seas) and then another year was spent sailing up and down the west coast. In September and October 1835, the expedition visited the Galapagos Islands, which were to assume such importance in Darwin's evolutionary thinking. The journey round the world was completed in one more year via Tahiti, New Zealand, Australia, the Cocos Keeling Islands, Mauritius, Cape Town, Ascension Islands, Salvador (again) and the Azores.

After an initial period of severe sea-sickness, Darwin adapted well to life on board ship. On the whole he had good relations with the captain, who was, in fact, only four years his senior. Fitzroy's tendency to take grave offence when contradicted (as when Darwin argued with him against slavery) was offset by a readiness to make peace after he had calmed down. As well as being a good sea captain, he was a hydrologist and meteorologist, and interested in scientific matters generally; he was also an inflexible believer in the truth of the Bible and in later years was to come to regret the unwitting part he had played in the genesis of evolutionary theory.

During the long days at sea, Darwin had the opportunity to educate himself in preparation for his forthcoming explorations. Probably the most important of the books that he read was Lyell's *Principles of Geology* (1830–1833), which Darwin came to regard as the definitive text. It may have prepared his mind for his later gradualist view of evolution, for it laid emphasis on gradual geological change, due to long-continued processes of the kinds visible in the current world, rather than on catastrophes of kinds outside our experience. Lyell was ahead of his time in his deduction that the earth must be a great deal older than was then generally believed.

From Beagle's ports of call, Darwin made many excursions on foot and horseback, some of them very extensive. One of his trips, into the interior of Argentina from Montevideo, took nearly four months. For that tour, and for the rest of the voyage, Fitzroy provided him with an assistant, Syms Covington, who had been previously employed as poop fiddler and cabin boy. Covington became very knowledgeable, and remained with Darwin as an amanuensis for some years after their return to England. Wherever he went, Darwin collected specimens, and was as interested in the rocks, and the fossils that they contained, as he was in the animals and plants. He displayed an energy and general hardiness quite at variance with the semi-invalid condition that was to overtake him in later life. It was said that he could not see a mountain without trying to climb it. In his travel through the wild country, he suffered a number of injuries and infections, but had the luck and fortitude to overcome them.

Darwin made his first scientific reputation in geology. From time to time during his voyage he was able to send reports and specimens back to Henslow, who showed some of them to Sedgewick, who was greatly impressed. In 1836, on the last leg of the voyage, Darwin was highly delighted to learn, from a letter from his sisters, that Sedgewick, during a visit to their father, had predicted a great scientific future for him. What probably did most for his growing reputation was his development of an explanation for the origin of coral atolls, which stands essentially uncorrected to this day. Lyell, when he heard of this theory after Darwin's return to England, was instantly and enthusiastically convinced by it. This was the beginning of a lifelong friendship between Darwin and Lyell. There is little doubt that it was the great impression that Darwin made on senior geologists that led to his election in 1839 to the Royal Society.

Darwin's biological observations, though massive, took a little longer to bear fruit. The most significant conclusion to which he found himself driven was that species could change, or transmute as the term then was. Several kinds of evidence confirmed him in this view. First, he found some very impressive fossils, for example a massive armadillo-like creature, which, though extinct, was evidently related to other species still alive in the same area. Secondly, as he traversed the South American continent, he found that some species disappeared to be replaced by other related species. Perhaps most impressively, different islands in the Galapagos had different, though obviously closely related birds (Darwin's finches etc.), all distinct from, but most similar to, those on the nearest mainland. All these kinds of evidence suggested the origin of new species by divergence from common ancestors.

In fact, the idea of transmutation of species, though by no means generally accepted, was already rather familiar. Darwin's grandfather, Erasmus, had clearly believed in it. Darwin had, of course, read Erasmus's work and had initially been very impressed by it, though he was later to discount its influence on him – it had too much poetic imagery and not enough hard reasoning. Also, in 1809, Lamarck had published his own theory of evolution, in which the driving force for evolutionary change came from an innate tendency towards perfection of function. So Darwin never claimed to have invented the idea of evolution. His Beagle observations convinced him that evolution had happened, but he still did not see how.

5. The origin of the theory of natural selection

When writing his autobiography, Darwin looked back on the two years following his return to England in October 1836 as the most active of his life. All thoughts of entering the Church behind him, he took lodgings, first in Cambridge, but soon in London, and "went a little into society". He became one of the secretaries of the Geological Society, but far more important was the writing that he had to do. He produced a Catalogue of Observations from the voyage and (at the request of Lyell) a geological review of the coast of Chile; he also put the journal that he had kept on the voyage into publishable form. The *Journal of Researches* was first published in 1839 as part of Fitzroy's overall report and Darwin edited it for more popular reading in 1845 (Darwin, 1839, 1845). As Darwin's *The Voyage of the Beagle* it still ranks among the world's greatest travel books.

In July 1837, Darwin also began a notebook of facts that he thought relevant to the origin of species. In October 1838, he happened to come across Malthus's *Essay on the Principle of Population* (1826), which was a revelation to him. Given the idea that natural populations would continue to expand exponentially unless checked by predation or limitation of food or space, he came readily to the conclusion that there must be a "struggle for existence" and, therefore, "favourable variations would tend to be preserved and unfavourable ones to be destroyed". Darwin was very familiar with the breeders' belief in heritability – "like begets like" – and it needed only the assumption that variations in fitness were heritable for natural selection to follow inevitably. Curiously, as it seems now, it took him several more years to grasp the full power of natural selection to diversify a single ancestral species to fit multiple niches in the environment – the process we would now call adaptive radiation. When writing his autobiography, Darwin was able to recall the exact spot on the road where, when riding in his carriage, he was first struck by this sudden insight.

The year 1838 also set the course of Darwin's domestic life. After carefully weighing the pros and cons (the notes that he made on the question survived), he decided that he should marry. It is not clear from his notes whether or not he had already identified his lady of choice. At all events, he proposed to and was accepted by his cousin, Emma Wedgwood, and they were married on 29 January 1839. In 1842, they moved permanently to Down House, near the Weald in Kent. They remained happily married until his death 43 years later and had ten children, seven of whom lived well into the 20th century (Leonard until 1943, when he was 93 years old) and two died in infancy. Looking back on his married life, Darwin recalled only one time of real unhappiness; the death at the age of ten of his daughter Annie.

The year 1839 was also notable, among other things, for the beginning of his lifelong friendship with Joseph Hooker, who was to become one of the leading botanists of his day and one of Darwin's strongest allies.

During his first 15 years at Down House, Darwin, although increasingly handicapped by ill health, was busy in several directions. In the earlier years, he was still completing books based on his voyage. Apart from *The Voyage of the Beagle*, these included *The Structure and Distribution of Coral Reefs* (1842), *Geological Observations on the Volcanic Islands* (1844) and *Geological Observations on South America* (1846). He worked for eight years on the morphology and classification of barnacles, and his detailed systematization of this difficult group was published in 1851 and 1856 in four heavy volumes, two of them on fossil forms. He also became keenly interested in pigeons and their manifold varieties, joining two pigeon fanciers' clubs.

Natural selection was by no means forgotten, but he left it in the background. He confided his theory to his friends Lyell and Hooker, who were impressed if not wholly convinced. In 1842, he had already written a sketch of his theory and, in 1844, he expanded it into an essay of 230 pages (see Darwin & Wallace, 1958). He was urged by his brother Erasmus, as well as by Lyell, to publish the essay before someone else had the same idea, but for one reason or another he held back. One explanation, advanced by Gavin de Beer, is that he did not want to get into a public discussion on evolution at that time, because he felt that the waters had been muddied by the popular success of *Vestiges of Creation*, a book published anonymously in 1844. This was a rather ill-informed amalgam of Paley's argument from design, Lamarckian evolution and scriptural orthodoxy, but it was evidently what many of the public wanted. It is now known to have been authored by Robert Chambers, a Scottish public figure.

At last, in 1856, Darwin started work on a book to be called *Natural Selection*. It was planned on a large scale, but, before he had half finished it, he was forced into more urgent action. On 18 June 1856 he received a manuscript from the naturalist and explorer Alfred Russel Wallace, entitled *On the tendency of varieties to depart infinitely from the original type*, and setting out in compact form what was essentially the very same theory. It was accompanied by a request that Darwin forward it to Lyell for an opinion. Darwin did as he was asked, lamenting, in his covering letter, that "all my originality, whatever it may amount to, has been smashed". Acutely disappointed though he was, he was anxious to be fair to Wallace and not to try to rush into publication ahead of him. The dilemma was resolved by Lyell and Hooker, under whose influence an extract from Darwin's Essay on Evolution was published together with Wallace's article in the *Proceedings of the Linnean Society* of 20 August (see Darwin & Wallace, 1958). It is to the credit of both men that they appeared to harbour no mutual resentment and remained friends for the rest of Darwin's life.

Now that natural selection was in the public domain, Darwin felt impelled to elaborate the theory in more extended form and with the least delay. The result was *The Origin of Species*, which cost him 13 months of intensive work and which was published in 1859. It was written for the educated general public and, at a little under 400 pages, was only, he thought, a quarter or a fifth of what would have been necessary had he fully used the information in his notebooks. The first edition of 1250 copies sold out almost immediately, and a second edition appeared in January 1860. In all, it ran to six editions, the last in 1872, with many subsequent reprints.

(i) The Origin of Species

Darwin's first two chapters dealt with biological variability, the first with domesticated and the second with wild organisms. He had no theory about how variation originated, but he thought it must arise more freely under conditions of domestication than in the wild. Consistent with this view, he noted that domestic plants and animals sometimes showed conspicuous and abrupt variations ("sports"), but thought that conspicuous individual variants in the wild were "extremely rare". He thought that both natural and artificial selection had their effects through the accumulation of small differences.

In fact, Darwin was unable to cite examples in wild species of natural selection caught in the act, as it were. His evidence that it had happened was rather the established presence within species of more or less variant populations – varieties grading into subspecies. In the case of domesticated plants and animals, on the other hand, he was able to cite the success of practical breeders, especially animal breeders:

> The key is man's power of accumulative selection: nature gives successive variations; man adds them up in certain directions useful to him. (Chapter 1)

He noted, but did not give great importance to, the tendency of variation to follow hybridization. He commented that first-generation hybrids could be relatively uniform, but the second and further generations much more variable, yet he was unable to make anything of this early glimpse of Mendelian heredity.

Chapter 3, *The Struggle for Existence*, was essentially an explanation of the Malthusian argument. Darwin argued that the most intense struggle was with other organisms, especially members of the same species.

Chapter 4, *Natural Selection*, was perhaps the most strongly argued in the book. Its message is contained in the following passages, presenting selection in both its positive and its negative aspects: "Under nature, the slightest difference in structure or constitution may well turn the nicely-balanced scale in the struggle for life and so be preserved." And: ". . . natural selection will ensure that modifications . . . shall not be in the least injurious; for if they became so they would ensure the extinction of the species."

As examples of the results of selection, Darwin cited camouflage, through colour or pattern, and species-specific patterns or displays apparently designed to attract mates (sexual selection). His observations had led him to conclude that outcrossing was on the whole good for organisms and continued inbreeding debilitating, and this provided him with a selectionist explanation of the many kinds of device seemingly designed to promote outbreeding. To illustrate his concept of adaptive radiation, he presented a hypothetical phylogenetic tree, with branches, sub-branches and clusters of twigs, which, though imaginary at the time, looks much like many supposedly real trees that have been constructed in our own day.

In Darwin's discussion of selection, he was clear that, in order to work, it must benefit the individual selected; so modern proponents of group selection cannot count him as a supporter.

The title of Chapter 5, *Laws of Variation*, was somewhat misleading, as Darwin was not in a position to define any real laws in this area. He did, however, make some relevant generalizations and speculations. He doubted whether environmentally induced variation was heritable, but was inclined to think that natural selection could be reinforced by heritability of the effects of use and disuse, such as might occur when muscles were exercised or not. He wrote that the question of whether acclimatization of species to new climates was due to "mere habit" (presumably not heritable) or to selection of "innate constitution" was a very obscure one, but he thought that both explanations might be true. Anticipating later evolutionary writings, he pointed out that the alteration, under natural selection, of one biological function or structure part might well have an effect, not necessarily beneficial, on another feature of the organism (the "Law of Correlation").

Chapter 6, *Difficulties of Theory*, was an attempt to meet anticipated objections. Probably the most serious problem was the difficulty, then as now, of explaining how complex structures, such as the eye, could have evolved step-by-step with a selective advantage at every stage. Darwin's statement on this question is much the same as we would make today:

> ... if we know of a long series of gradations, each good for its possessor, then, under changing conditions of life, there is no logical impossibility in the acquirement of any conceivable degree of perfection through natural selection. In the cases in which we know of no intermediate or transitional states, we should be very cautious in concluding that none could have existed.

In support of this view of the great adaptability of organisms, Darwin went on to cite examples in which already evolved structures appeared to have been modified and put to new uses. He did not, however, claim that the result was always perfection.

In Chapter 7, *On Instinct*, Darwin argued that behavioural traits should be as much subject to natural selection as anything else.

Chapter 8 dealt with hybridism, and hence with the species concept, which is still something of a problem. Darwin's position on hybrid sterility was that, although it was a common consequence of interspecific crosses, it was not a firm criterion for defining species. He thought that, as species diverged, hybrid sterility developed as a secondary consequence of their divergence and increasing physiological incompatibility. However, remarks both here and elsewhere in the book show that he recognized that some degree of reproductive isolation must be necessary for divergence in the first place.

Chapter 9, obviously related to Chapter 6, dealt with imperfections in the geological record and why gaps were to be expected. Chapter 10 was on the geological succession and, here, Darwin interpreted the fossil record as supporting his consistently gradualist view of evolution. Chapter 11, *Geographical Distribution*, presented the argument that closely related species, supposedly with a comparatively recent common ancestor, tended to reside in mutually accessible territories; where they did not, their present separation could be explained as a result of kinds of disturbance well known to geologists, such as advance or retreat of the ice caps, or elevation or depression of the land surface. Finally, Chapter 12 dealt with biological classification. Darwin's position was, in brief, that the taxonomic system should reflect evolutionary relationships.

Darwin concluded his book with a Recapitulation and Conclusion that included a strong attack on the special creationists: "... do they really believe that at innumerable periods in the earth's history certain elemental atoms have been commanded to flash into living tissues?" He came down boldly in favour of a single origin of life:

> ... all living things have much in common, in their chemical composition, their germinal vesicles, their cellular structure and their laws of growth and reproduction. ... I should infer from analogy that all the organic beings that have ever lived on earth have descended from a single primordial form into which life was first breathed.

He was not here necessarily discounting the possibility of God as the primary originator, though did later propose a chemical origin of life, at least in correspondence (see below).

(ii) After publication of The Origin of Species

The book made an immediate impact. The first edition of 1250 copies sold out almost at once and a second edition followed in a little over a month. Two of Darwin's closest scientific friends, Hooker and Thomas Huxley (who had befriended Darwin after returning from his own navy-sponsored voyage in 1850), were enthusiastic converts from the start. Huxley, who had not noticed the earlier synchronized papers by Darwin and Wallace, is said to have exclaimed: "How stupid of me not to have thought of that!" Lyell, too, was also full of praise for the book.

There was also much opposition, notably from Sedgewick, Darwin's geological mentor and earlier supporter. Sedgewick wrote: "I have read your book with more pain than pleasure ... You have deserted ... the true method of induction ...". Sedgewick belonged to the empirical school, claiming descent from Francis Bacon, which held that observation (supplemented by scripture) was everything, with theory best avoided and hypothesis virtually unmentionable. In fact, Darwin also considered himself a follower of Bacon, but was a more liberal empiricist. He believed that although theory should be banished from the mind when one was actually making observations, it could be very helpful beforehand and afterwards.

The controversy generated by *The Origin of Species* reached its high point at the meeting of the British Association in 1860, where a strong attack on the book was mounted by Samuel Wilberforce, Bishop of Oxford, who is said to have been coached for the purpose by Richard Owen, a professor at the Royal College of Surgeons. Darwin himself was not present – he had neither the strength nor the inclination for controversy – but his defence was taken up by Huxley ("Darwin's bulldog"). The result was a remarkable contest of wit and venom which Huxley is generally considered to have won. Amid the excitement, Robert Fitzroy, now a senior figure in the Meteorological Office, was seen stalking about the hall brandishing the Bible and exclaiming "The book, the book ... !" John Henslow, who had the misfortune to be in the chair, must have been torn between his long friendship with Darwin and his own religious orthodoxy.

The reception of the *Origin* in the press was mixed. A particularly long and polemical review, by Richard Owen, appeared in the *Edinburgh Review*. Another review, in the *Gardeners' Chronicle*, not hostile in itself, prompted an unexpected protest from Patrick Matthew, a Scottish gentleman farmer and fruit grower, who pointed out that he had published the idea of natural selection himself, 28 years previously. Darwin had to acknowledge Matthew's priority, but excused his own failure to notice it on the grounds of the obscurity of the place of publication – an appendix to a book entitled *Naval Timber and Arboriculture* (1831). Though received with cynicism by some, Darwin's explanation here was surely entirely reasonable.

Of much more serious concern to him was an 1867 article in the *North British Review* by Fleeming Jenkin, a professor of engineering in Edinburgh University. Jenkin asked how Darwin's evolutionary model could be reconciled with the prevailing idea, which Darwin accepted, that the characteristics of parents were blended, that is to say, averaged, in their progeny. Jenkin pointed out that, in that case, the small differences on which Darwin relied for gradual change should be quickly blended out of existence. Darwin took the criticism very seriously and, in later editions of the *Origin*, referred to Jenkin's "able and valuable" critique. In an interesting exchange of letters, recently summarized by Gayon (1998), Wallace suggested that Darwin might now be inclined to give more importance to "sports" and less to nearly imperceptible variations, but Darwin's view was quite otherwise. He thought that Jenkin's criticism weighed against a role for any kind of single variant, small or large, and looked instead for some kind of populational property to drive variation. In his fifth edition of 1869, he suggested that the propensity to vary in a certain direction might itself be inherited; in modern terms this would be bringing mutation pressure to the aid of natural selection.

What Darwin needed was a genetic theory in which heredity was carried by non-blending particles of some kind. Such a theory, of course, became generally available with the rediscovery of Mendel's Laws in 1900, but Mendel's work remained virtually unknown during Darwin's lifetime. Darwin himself proposed a kind of particulate theory in his *Variation of Animals and Plants under Domestication*, published in 1868. According to this, every part of a plant or animal, as it developed, shed minute particles ("gemmules" or "pangenes") which were inherited through the germ cells, providing for both heredity and development. This scheme was compatible both with blending inheritance and, because of the particulate nature of the pangenes, with the occasional reappearance of ancestral characters, though Darwin did not explicitly propose it as an answer to Jenkin. Pangenesis attracted little support, and, in his autobiography, Darwin referred to it as "my well-abused hypothesis" and added, as if to apologize for his gross departure from empiricism, "An unverified hypothesis is of little or no value".

During the 23 years between the publication of the *Origin* and Darwin's death, the idea of natural selection became very widely accepted. His own late thoughts on the success of his main project seem very fair:

> Some of my critics have said, 'Oh, he is a good observer, but he has no power of reasoning'. I do not think that this can be true, for *The Origin of Species* is one long argument from beginning to end, and it has convinced not a few able men.
>
> (from *The Autobiography*; Barlow, 1958, p. 140)

Darwin's own participation in public controversy, to which he was in any case strongly disinclined, was severely limited by his ill health. Almost from the time of his return from the voyage of the Beagle he had been subject to spells of nausea and fatigue, and these became progressively worse. Although he remained active up to, and a few years beyond, the time of his marriage, he became, from the mid-1840s, increasingly reluctant to move outside the immediate neighbourhood of his own house and garden. His degree of incapacity can be overstated; for example, he was, over the period 1854–1856, able to travel to London for meetings of the Council of the Royal Society. For a number of years, he continued to receive visitors at Down House, but eventually began to find even visits from friends too fatiguing. He became increasingly unable to pay attention to anything outside his own family and his own work. In his autobiography, written for his family in 1876, he deplored the loss, not only of his physical vigour, but also of his aesthetic sense. He found himself no longer moved by poetry or music or the beauties of landscape. Shakespeare had become intolerably boring and though he still took pleasure in novels, which his family sometimes read to him, they had to have happy endings and, preferably, young and attractive heroines.

Nevertheless, with the support of his devoted wife and family, Darwin managed to maintain an intense scholarly activity, interrupted only when he felt particularly unwell. He produced books over a broad range of biological subjects: *The Various Contrivances by which Orchids are Fertilised by Insects* (1862), *Variation of Animals and Plants under Domestication* (1868), *The Descent of Man, and Selection in Relation to Sex* (1871), *The Expression of the Emotions in Man and Animals* (1872), *Insectivorous Plants* and *The Movements and Habits of Climbing Plants* (1875a,b), *The Effects of Cross and Self Fertilisation in the Vegetable Kingdom* (1876), *The Various Contrivances by which Orchids are Fertilised by Insects* (2nd edition) and *The Different Forms of Flowers on Plants of the Same Species* (1877), and *The Formation of Vegetable Mould, through the Action of Worms* (1881). He may be said to have worked until he dropped.

The nature of Darwin's illness has always been a matter of doubt and some controversy. Those who have hated his ideas have often depicted him as a pathetic hypochondriac. More reasonably, it has been suggested that he was a victim of Chagas disease, a condition due to a trypanosomal parasite transmitted by a species of blood-sucking bug, to which (as explained by de Beer, 1963) Darwin is known to have been exposed in South America. More recently, attention has been drawn by D.A.B. Young to the similarity between Darwin's symptoms and those of lupus erythematosus, an autoimmune condition (Young, 1997). In Darwin's time, of course, the means for diagnosis of these diseases did not exist.

Darwin's religious belief, which had initially been quite conventional, withered away over the years – painlessly, he wrote, as the process had been so gradual. On board the Beagle he had still believed in the Bible, but later he found himself unable to accept parts of it, particularly the doctrine of the damnation of the ungodly. He could not accept that his father and brother and most of his best friends were destined for everlasting torment. Later, he began to think as a theist – one who could believe in a God who set life going, but then left it to its own devices. Later, he thought that life could have originated chemically. A letter that he wrote in 1871 anticipated much more recent thinking:

> It is often said that all the conditions for the first production of life are now present, which could ever have been present. But if (and oh! what a big if) we could convince [that] in some warm little pond, with all sorts of ammonia and phosphoric salts, light, heat, electricity, etc., present, that a protein compound

> was chemically formed ready to undergo still more complex changes; at the present day such matter would be instantly devoured or absorbed, which would not have been the case until living creatures were formed.
>
> (Quoted in de Beer, 1963, p. 471)

However, in his autobiography, Darwin reserved his position: "The mystery of the beginnings of all things is insoluble to us; and I for one must be content to remain an Agnostic" (Barlow, 1958).

6. Darwinism after Darwin

The rise of Mendelian genetics after 1900 should have been seen as a rescue of Darwinism. It showed how heritable variation in populations, essential for effective selection, could be maintained from generation to generation without being diluted away. This was, indeed, how the matter was regarded later in the century. Darwinian selection and Mendelian segregation were both essential parts of the evolutionary theory expounded in Julian Huxley's landmark book *Evolution: The Modern Synthesis*, published in 1942.

Perversely, as it seems now, the immediate effect of the rediscovery of Mendel was to put Darwinism into eclipse. The earliest Mendelians, particularly William Bateson, the most vociferous of them, thought that Mendelian heredity was a property only of striking and clear-cut differences such as Mendel had used in his experiments. Bateson, indeed, does not appear to have believed that small differences were heritable at all. Consequently, he thought that natural selection, to the extent that it was significant, must bring about sharp changes and not the gradual modifications on which Darwin had placed so much emphasis. The possibility that Mendelian allelic differences might have small effects as well as large was lost on Bateson, though it was noted at the time by less committed observers. Bateson's obduracy in this matter can be seen as a consequence of his bitter feud with the group of biometricians, headed by the statistician Carl Pearson and the zoologist Walter Weldon, who wanted to analyse heredity statistically. To put the dispute in over-simple terms, the biometricians disbelieved Mendelism because it appeared to ignore measurement, while the Mendelians thought that statistics was irrelevant.

The controversy between the Mendelians and the biometricians had more or less burnt itself out by the end of the second decade of the 20th century. Observations of various kinds, including the demonstration by the Drosophila school in the US of allelic differences of small effect, and the successful experimental selection of quantitative traits in rodents, showed the extreme position taken by the Bateson school to be untenable. It became apparent, on the one hand, that the statistical treatment of heredity had to be based on the Mendelian rules and, on the other hand, that small differences could indeed be heritable. The statistician and geneticist R.A. Fisher contributed much to the new thinking, showing by mathematics that selection based on gene differences of small effect could indeed be effective on a feasible time-scale, and arguing, with the aid of a geometrical diagram, that small variations were more likely to be viable and, therefore, selectable than large ones. Gradualism became, once again, respectable.

So, by the mid-1920s, the foundations of the modern synthesis were laid. At the same time, however, a new disruptive influence appeared in the person of the American geneticist Sewall Wright, who pointed out that selection was not the only way in which genetic change could occur in populations. If small numbers of organisms became isolated from each other, their genetic compositions could well be very different, setting different conditions for further change, purely because of the accidents of sampling. The need to recognize the role of random sampling, or drift, is now universally accepted, although its importance in comparison with that of selection was for years a matter of dispute between Wright and Fisher and their respective followers.

Drift acquired a new importance from the end of the 1960s, as large numbers of molecular variants, first in proteins and then in DNA, and all inherited in Mendelian fashion, began to be discovered in organisms of all kinds. Most of this diversity, so far as could be seen, was without effect on fitness. This led Motoo Kimura, in 1983, to propose his *Neutral Theory of Molecular Evolution*, in which evolutionary changes in macromolecular structure could occur entirely by drift without selection. One can argue here about what one actually means by evolution and whether or not changes in protein or DNA sequence can properly count as such if they have no consequences for the life of the organism. More recently, Kimura's followers have turned their attention to "nearly neutral" mutations, which are those, no doubt very numerous, whose fate is largely determined by drift, but to some extent by selection too. It also has to be remembered, in considering the importance of drift, that a mutation that is virtually neutral when it arises may become functionally important when conditions change.

In recent years, Darwinism in the gradualist sense has been vigorously challenged by the American palaeontologists Stephen Gould and Niles Eldredge. Citing fossil evidence, they claim that evolution is not in general slow and steady, but consists rather of periods of relatively rapid change, when new species originate, and much longer periods of stasis in which very little happens. This theory, called "punctuated equilibrium" (Gould & Eldredge, 1993), gives a major role to contingency, that is to say, chance. Organisms are considered to stick to their tried and established ways of life until confronted with some major change in circumstances – perhaps a drastic change in climate, the arrival or extinction of a competitor, or even the impact of a meteor – which breaks the mould and leads either to extinction or the establishment of a new system. The punctuationists tend to view speciation as an abrupt process, involving some radical restructuring of the living system, rather than gradual selection-driven divergence of more-or-less isolated sub-populations, which is the more usual neo-Darwinian view.

It is questionable whether the quite fierce controversy that has developed about punctuated equilibrium has enough substance to justify the heat. The punctuationists claimed to have a radical new insight that others have missed. On the other side, other evolutionary theorists (for example Maynard Smith, 1988, chapter 16), dubbed ultradarwinians by Eldredge, replied that they had long understood that evolution proceeds at different rates in different organisms, and pointed out that evolutionary changes that, on the geological time scale, seemed to have been comparatively rapid, could still have taken hundreds of thousands of years. They also cited examples of gradual change within species, and so maintained that stasis was not as general a rule as the punctuationists believed. A point for the neo-Darwinians is that, leaving aside allopolyploidy, a well-understood mechanism almost confined to plants, there is no clear explanation of how abrupt speciation could occur. In any case, there can be no disagreement about the importance of natural selection, since any evolutionary departure, whether moderate or radical, must have been able to pass the test of survivability.

A new and quite unanticipated aspect of evolution has come to light as a result of molecular analysis of the genome. It turns out that virtually all organisms have, embedded in their chromosomes, DNA segments that are transposable from one chromosome locus to another and may have the ability to increase in number. These transposable elements ("transposons") generally contain no genes with functions for the organisms that they inhabit, and may be regarded as molecular parasites, or "selfish DNA" (see article on "Selfish genes"). The transposons first studied by Barbara McClintock in maize (see article on "Barbara McClintock and transposable genetic sequences in maize") move by excision and reinsertion, with an increase in their number per genome only when they "jump" from a locus already replicated to another still to be replicated. A much greater potential for proliferation is possessed by "retrotransposons", which, without excision, are transcribed into an indefinite number of RNA copies, which can then (in the presence of the enzyme reverse transcriptase) be reverse-transcribed back into DNA and reinserted into the chromosome at new loci without having vacated their previous ones. In the fruit fly Drosophila and in various mammalian species, retrotransposons have, over evolutionary time, come to comprise a substantial fraction of the total genome – about 15% in the case of the human L1 element. Retrotransposons commonly include a gene encoding reverse transcriptase, and in this respect are akin to retroviruses, though without the capacity to make infective protein-coated virus particles. The existence of selfish DNA, though of course it could not have been imagined by Darwin, does not contravene his principle of natural selection, but rather suggests new levels at which it may act. The movable elements themselves must, at least during their periods of great expansion (usually in the distant past, fortunately), have been subject to selection for efficient proliferation. Their host organisms must have been subject to selection for toleration of the changes within its own DNA as well as for adaptation to the external environment. It has been suggested that L1 and other successful transposons have been important agents of evolution, bringing about structural changes in the genome that may on rare occasions have been of selective advantage to the whole organism (Kazazian, 2000).

The philosopher Karl Popper criticized Darwinism on the grounds that "the survival of the fittest" (actually a phrase coined not by Darwin but by his supporter Herbert Spencer) was a tautology. What, after all, did fitness mean if not survivability? However, this criticism missed the key point, which is that the variation, which affects capacity to survive and leave viable progeny, is in some degree heritable. If that is accepted, and it can scarcely be denied, then at least some effect of natural selection cannot logically be avoided. It cannot, of course, be demonstrated that natural selection was responsible for all the varied, elaborate and evidently adaptive features that we find in living organisms – they all arose far too long ago – but it is the only non-miraculous explanation available. Darwinism's continued strength depends not only on its plausibility, but also on our disinclination to believe in miracles.

REFERENCES

Barlow, N. (ed.) (1958) *The Autobiography of Charles Darwin*, and other notes, London: Collins and New York: Harcourt Brace

Chambers, R. (1844) *Vestiges of The Natural History of Creation*, London: Churchill; reprinted Chicago: University of Chicago Press, 1994

Darwin, C. (1839) *Journal of Researches into the Geology and Natural History of the Various Counties visited by H.M.S. Beagle*, London: Colburn; reprinted New York and London: Hafner, 1952

Darwin, C. (1842) *The Structure and Distribution of Coral Reefs*, London: Smith Elder; reprinted Berkeley: University of California Press, 1976

Darwin, C. (1844) *Geological Observations on the Volcanic Islands Visited during the Voyage of H.M.S. 'Beagle'*, London: Smith Elder

Darwin, C. (1845) *Journal of Researches into the Natural History and Geology of the Countries Visited During the Voyage of H.M.S. Beagle Round the World: Under the Command of Capt. Fitz Roy*. London, John Murray; many reprints as *The Voyage of the Beagle*

Darwin, C. (1846) *Geological Observations on South America*, London: Smith Elder

Darwin, C. (1851–54) *A Monograph on the Sub-Class Cirripedia*, 2 vols, London: Ray Society

Darwin, C. (1851) *A Monograph on the Fossil Lepadidae, or, Pedunculated Cirripedes of Great Britain*, London: Palaeontographical Society

Darwin, C. (1854) *A Monograph on the Fossil Balanidae and Verrucidae of Great Britain*, London: Palaeontographical Society

Darwin, C. (1859) *On the Origin of Species by Means of Natural Selection, or the Preservation of Favoured Races in the Struggle for Life*, London: John Murray; many reprints

Darwin, C. (1862) *The Various Contrivances by which Orchids are Fertilised by Insects*, London: John Murray; 2nd edition 1877

Darwin, C. (1868) *The Variation of Animals and Plants under Domestication*, London: John Murray and New York: Judd; 2nd edition 1875; reprinted Baltimore: Johns Hopkins University Press, 1998

Darwin, C. (1871) *The Descent of Man, and Selection in Relation to Sex*, 2 vols, London: John Murray; 2nd edition 1874

Darwin, C. (1872) *The Expression of the Emotions in Man and Animals*, London: John Murray

Darwin, C. (1875a) *Insectivorous Plants*, London: John Murray; 2nd edition 1888

Darwin, C. (1875b) *The Movements and Habits of Climbing Plants*, London: John Murray

Darwin, C. (1876) *The Effects of Cross and Self Fertilisation in the Vegetable Kingdom*, London: John Murray, revised 1878, 1891

Darwin, C. (1877) *The Different Forms of Flowers on Plants of the Same Species*, London: John Murray, revised 1880

Darwin, C. (1881) *The Formation of Vegetable Mould, through the Action of Worms, with Observations on Their Habits*, London: John Murray

Darwin, C. & Wallace, A.R. (1958) *Evolution by Natural Selection*, Cambridge: Cambridge University Press. (Essays by Darwin and Wallace, including Darwin's Sketch of 1842, his Essay of 1844 and the Darwin–Wallace Papers of 1856. Introduction by Sir Francis Darwin and foreword by Sir Gavin de Beer)

Darwin, E. (1794–96) *Zoonomia, or the Laws of Organic Life*, 2 vols, London: Johnson

de Beer, G. (1963) *Charles Darwin: Evolution by Natural Selection*, London: Nelson; New York: Doubleday, 1964

Gayon, J. (1998) *Darwinism's Struggle for Survival: Heredity and the Hypothesis of Natural Selection*, Cambridge and New York: Cambridge University Press

Gould, S.J. & Eldredge, N. (1993) Punctuated equilibrium comes of age. *Nature* 334: 19–22

Huxley, J. (1942) *Evolution: The Modern Synthesis*, London: Allen and Unwin, and New York: Harper

Jenkin, H.C.F. (1867) The Origin of Species. *North British Review* 46: 277–302

Kazazian, H.H. Jr (2000) L1 retrotransposons shape the mammalian genome. *Science* 289: 1152–53

Kimura, M. (1983) *The Neutral Theory of Molecular Evolution*, Cambridge and New York: Cambridge University Press

King-Hele, D. (1998) Erasmus Darwin, the Lunaticks and Evolution. *Notes and Records of the Royal Society* 52: 153–80

Lamarck, J.B. (1809) *Philosophie Zoologique, ou exposition des consideration relative a l'histoire naturelle des animaux*, 2 vols, Paris: Dentu et l'Auteur

Lyell, C. (1830–33) *Principles of Geology: Being an Attempt to Explain the Former Changes of the Earth's Surface, by Reference to Causes Now in Operation*, 3 vols, London: John Murray

Malthus, T.R. (1826) *An Essay on the Principle of Population; or, a View of Its Past and Present Effects on Human Happiness, with an Inquiry into our Prospects Respecting the Future Removal or Mitigation of the Evils which it Occasions*, 2 vols, 6th edition, London: John Murray

Matthew, P. (1831) *On Naval Timber and Arboriculture*, Edinburgh: Black

Maynard Smith, J. (1988) *Did Darwin Get it Right? Essays on Games, Sex and Evolution*, New York: Chapman and Hall

Paley, W. (1785) *The Principles of Moral and Political Philosophy*, London: Faulder; 17th edition, 1810

Paley, W. (1795) *A View of the Evidences of Christianity*, 4th edition, London: Faulder

Paley, W. (1802) *Natural Theology: or, Evidences of the Existence and Attributes of the Deity, Collected from the Appearances of Nature*, London: Faulder; 10th edition, 1805

White, G. (1789) *The Natural History and Antiquities of Selbourne*, London: Archer; many reprints

Young, D.A.B. (1997) Darwin's illness and systemic lupus erythematosus. *Notes and Records of the Royal Society* 51: 77–86

FURTHER READING

Darwin, C. (1998) *The Expression of the Emotions in Man and Animals*, edited by P. Ekman, 3rd edition, Oxford and New York: Oxford University Press (originally published 1873)

Desmond, A. & Moore, J. (1991) *Darwin*, New York: Warner and London: Joseph

Eldridge, N. (1995) *Reinventing Darwinism: The Great Evolutionary Debate*, London: Weidenfeld and Nicholson

See also **Darwin and Mendel united** (p.77); **Barbara McClintock and transposable genetic sequences in maize** (p.640); **Motoo Kimura and the neutral theory** (p.871); **Selfish genes** (p.875)

DARWIN AND MENDEL UNITED: THE CONTRIBUTIONS OF FISHER, HALDANE AND WRIGHT UP TO 1932

A.W.F. Edwards
Gonville and Caius College, Cambridge University, Cambridge, UK

Since the 1930s, it has been widely recognized and oft repeated that R.A. Fisher (1890–1962), J.B.S. Haldane (1892–1964) and S. Wright (1889–1988) successfully developed evolutionary Mendelian genetics and applied it to Darwin's theory of natural selection, thereby creating "neo-Darwinism" by supplying the mechanism which that theory had perforce lacked. Their achievement was all the more important because, before the rediscovery of Mendel's work in 1900, and for some years afterwards, Darwinism had been in the doldrums and many biologists had continued to doubt the possibility of accounting for evolution in terms of natural selection or of natural selection alone. In showing that the Mendelian mechanism could indeed account for the main features of evolution, Fisher, Haldane and Wright thus revivified Darwin's theory and ensured it a place not only in biology but also in the development of human thought as a whole. Modern evolutionary biology rests on their achievements, but their influence is also felt as far afield as game theory, sociology, economics and in specialist areas such as genetic algorithms in computer science.

R.A. Fisher (Sir Ronald Fisher) was born in London and educated at Cambridge, where he studied mathematics and mathematical physics. He was successively statistician to Rothamsted Experimental Station, Galton Professor of Eugenics at University College London and Arthur Balfour Professor of Genetics at Cambridge. He died in Adelaide, South Australia. J.B.S. Haldane was born and educated in Oxford, also studying mathematics, but then transferring to "Greats", that is, classics and philosophy. Initially a don at Oxford, Haldane became Reader in Biochemistry at Cambridge and subsequently a professor at University College London, finally holding the Weldon Professorship of Biometry. S. Wright, who normally published under his full name Sewall Wright, was born in Massachusetts and educated in Illinois, undertaking graduate study in biology at the University of Illinois followed by 3 years at Harvard. He worked at the Animal Husbandry Division of the US Department of Agriculture in Washington, DC, and subsequently at the University of Chicago, and enjoyed a productive retirement at the University of Wisconsin, Madison. Each man also made contributions outside evolutionary biology: Fisher notably creating much of modern statistical theory, Haldane ranging widely over most of genetics as well as being a gifted popular writer and Wright contributing the inbreeding coefficient, path analysis and many studies in physiological genetics. Each has been the subject of a full biography (Clark, 1968; Box, 1978; Provine, 1986).

The first of the three to show an interest in the application of Mendelism was Fisher who, as an undergraduate in 1911, gave a lecture (a copy of which has survived) entitled "Mendelism and biometry" which foreshadows his subsequent work in many respects. Fisher's early reading was an ideal preparation for the synthesis of Darwinism and Mendelism, for he had chosen the 13-volume edition of the collected works of Charles Darwin as a school prize in 1906 and had bought William Bateson's *Mendel's Principles of Heredity*, with its translation of Mendel's paper, in his first undergraduate year (1909). Studying for the mathematical tripos while relaxing with Darwin and Mendel could not have been a better preparation for the work that was to follow. Fisher published his first paper in 1912 and by 1916 had completed the paper which was to appear 2 years later as *The correlation between relatives on the supposition of Mendelian inheritance*, which established his early reputation.

In a 1914 paper (originally read in 1912), Fisher pointed out how the genes of a childless man could nevertheless prosper if he had enough nephews, thus formulating the idea of inclusive fitness or "kin selection". Another early paper of evolutionary significance was *The evolution of sexual preference* published in 1915 in which, starting from Darwin's theory of sexual selection, he took the fundamental further step of "showing how mate choice itself would evolve as a consequence of the very process of sexual selection it produced" (O'Donald, 1990). His next work of evolutionary significance was the seminal 1922 paper which we consider below.

Haldane similarly showed an early interest in genetics, helping his sister Naomi to breed guinea pigs in 1908. In 1915, they and a fellow-student, A.D. Sprunt, were able to publish a paper reporting linkage from experiments with mice. Haldane's first paper, like Fisher's, had appeared in 1912. In 1919, he published two papers on the statistical estimation of linkage, but not until 1924 did the first paper in his series *A mathematical theory of natural and artificial selection* appear. In 1927, he wrote in one of his popular scientific essays about how the concept of inclusive fitness would explain the evolution of altruism.

Wright also published his first paper in 1912 and, a few months before the Haldane–Sprunt paper announced a linkage in the mouse, Wright and W.E. Castle published a similar observation in the rat, the first linkage to be observed in mammals. Throughout the 1920s, Wright worked on the statistical analysis of pedigrees, inventing path coefficients, the inbreeding coefficient and the concept of effective population size in the process. An important result in this work was the calculation, in 1921, of the rate of decrease in heterozygosis in regular systems of mating (without selection).

In 1929, Wright's first papers on evolution appeared, although it is known that in 1925 he had prepared a long paper on the subject, which was to be the basis of his influential 1931 paper and to which we will return.

As early as 1931, Lancelot Hogben (Hogben, 1931) linked together the names of Fisher, Haldane and Wright as the leaders of the neo-Darwinian revolution and, in 1932, Haldane, with characteristic immodesty, repeated the claim. The key period requiring detailed study is the decade 1922–32, starting with Fisher's 1922 paper and ending with the *Proceedings of the Sixth International Congress of Genetics* and Haldane's book *The Causes of Evolution*, both of which appeared in 1932. The decade embraces the publication of Fisher's *The Genetical Theory of Natural Selection* (1930) and Wright's influential 1931 and 1932 papers. We shall explore it chronologically.

In his 1922 paper, Fisher introduced stochastic models into evolutionary genetics. He was concerned "to discuss the distribution of the frequency ratio of the allelomorphs of dimorphic factors, and the conditions under which the variance of the population may be maintained". The second point is a kind of lodestar in Fisher's work of this period, for he had been the first to perceive clearly that one major distinction between Mendel's theory and pre-Mendelian theories of blending inheritance was that, under the former, the rate of loss of genetic variability would be infinitely less and indeed due only to finite population size and the consequential stochastic losses. How would the rate of loss compare with the gains through mutation under differing assumptions about selection? Fisher's concentration on the variance (a word he had coined in his 1918 paper) and its analysis ran right through to his enunciation of the "fundamental theorem of natural selection" in 1930.

First Fisher breaks new ground by analysing the simple model of selection at a single diallelic locus with constant fitnesses, pointing out that "if the selection favours the heterozygote, there is a condition of stable equilibrium". Given that there had been calculations on the effects of selection on Mendelian models over the previous 7 years by workers such as H.T.J. Norton and H.C. Warren, it is remarkable that this theory of the balanced polymorphism had to wait until 1922. Even when Haldane inaugurated his long series of papers in 1924, he did not notice Fisher's result (all the more remarkable given that the address on Haldane's papers was Trinity College and on Fisher's was Gonville and Caius College, neighbouring colleges of the University of Cambridge).

Next in the paper, Fisher introduces stochastic considerations, first treating "the survival of individual genes" by means of a branching process analysed by functional iteration and then setting up the "chain-binomial" model and analysing it by a diffusion approximation. The latter model is the keystone of stochastic population genetics (and, incidentally, stochastic diffusion theory itself), yet it often goes by the name of "the Fisher–Wright model" or even the "Wright–Fisher model". Not only is it due to Fisher alone, but Wright's main paper on it did not appear until a further 9 years had elapsed. We consider it below. Fisher goes on to consider the distributions of the gene frequency (actually in the form of the logarithm of the frequency ratios) for various models of selection, and for no selection, finding that in the latter case a population of n individuals would have its variance reduced by a factor e in $4n$ generations (the "relaxation time"). He concludes: "The rate of mutation necessary to maintain the variance of the species may be calculated from these distributions."

Wright first met Fisher in 1924 in Washington, DC, at which time Wright had not yet seen Fisher's paper, but Fisher promised him a copy and sent it on his return to England. "Until reading Fisher's 1922 paper, it had not occurred to Wright to extend his own quantitative analysis to the statistical distribution of genes in populations" (Provine, 1986). Fisher's 1922 paper may indeed be said to have inaugurated neo-Darwinism.

Between 1924 and 1934, Haldane published a series of ten papers under the general title *A mathematical theory of natural and artificial selection* in which he worked out the consequences of selection for a large variety of Mendelian models, frequently solving the resulting difference equations by deriving and solving the corresponding differential equation in continuous time. For the most part, these papers report the necessary, but somewhat ponderous, mathematical analysis of particular models with infinite populations and discrete generations, and they have naturally tended to be neglected because of the existence of Haldane's 1932 book *The Causes of Evolution*, whose appendix *Outline of the mathematical theory on natural selection* summarizes them. However, certain historical points may be noted, as follows.

In Part I (1924), Haldane makes the first attempt to estimate the "intensity of natural selection", from data on the spread of melanism in the peppered moth in Manchester. In Part IV (1927) he gives the first account of the theory of natural selection in an age-structured population, making a generous acknowledgement of H.T.J. Norton's priority in unpublished work that Haldane had seen in 1922 and which eventually appeared in 1928. In Part V (also 1927), Haldane takes a rare excursion into the stochastic theory, developing Fisher's 1922 branching-process treatment to make the first calculation of the probability of fixation of an advantageous mutant gene; in the Appendix to *The Causes of Evolution*, Haldane wrote with uncharacteristic modesty "The investigation of the case where the total population is finite has been wholly due to Fisher (1930) and Wright (1931)". In Part VIII (1931), Haldane inaugurates the mathematical study of two linked loci in a paper whose single diagram may have had some influence on Wright because it depicted population trajectories in the two-dimensional space of gene frequencies.

Continuing chronologically, we should next notice in passing two 1928 papers of Fisher's on the evolution of dominance, because these drew a comment from Wright the following year and inaugurated a topic which eventually involved many people, including Haldane. Important though the ensuing controversy is for the light it throws on the differing emphases of Fisher and Wright on the mechanisms of evolution, it is too much of a side issue to be considered in detail here. Also to be recorded is the 1929 abstract of a paper in which Wright gave the first published intimation of his "shifting balance" theory of evolution.

In 1929, Wright sent Fisher a draft of his forthcoming 1931 paper *Evolution in Mendelian populations* in which, as we shall see, he used a different method of determining the "relaxation time" required for a population to have its variance reduced by a factor e and found it to be $2n$ generations rather than Fisher's $4n$. Fisher acknowledged his mistake in 1930 in a paper in which he also gave "a more rigorous and comprehensive treatment of the whole subject" of his 1922 paper. Of this paper Motoo Kimura later wrote "In my opinion this is one of the most beautiful papers ever written on the mathematical theory of population genetics". However, the principal event of 1930 was the publication of Fisher's *magnum opus* in this field, *The Genetical Theory of Natural Selection.*

The depth, difficulty and originality of *The Genetical Theory* ensured that its initial impact was limited, just as Leonard Darwin, Charles's fourth son, to whom the book is dedicated, had predicted: "it will be slowly recognized as a very important contribution to the subject. But I am afraid it will be slow, because so few will really grasp all that it means" (Bennett, 1983). Haldane and Wright were perhaps the only two people really capable of appreciating it at the time. In the introduction to *The Causes of Evolution*, Haldane said that he was "much indebted" to Fisher's "brilliant book" and, in a 1931 essay (Haldane, 1931), he remarked:

> The book before us is of the latter class [books which, though often inaccurate in detail, state a new point of view, and lay the foundations of new branches of science] . . . No serious future discussion either of evolution or eugenics can possibly ignore it.

and

> during the next generation any discussions of the problem of gradual evolution which are likely to be of permanent value will take the form of a development, discussion, and perhaps in some cases a refutation, of the arguments stated in the book before us.

Wright reviewed *The Genetical Theory* at length for the *Journal of Heredity*, stating: "It is a book which is certain to take rank as one of the major contributions to the theory of evolution".

Only a brief description of *The Genetical Theory* is possible here. Chapter 1 contrasts the conservation of variance under Mendelian inheritance with its decay under the blending theory of inheritance assumed by Darwin. In Chapter 2, the Malthusian parameter of population increase is introduced (which Fisher had previously acknowledged to be equivalent to A.J. Lotka's "intrinsic rate of natural increase"), along with the novel concept of "reproductive value" for each age of life. The genic element in the overall variance is then discussed, leading to the "fundamental theorem of natural selection", which relates the partial rate of increase in the mean fitness of a population ascribable to natural selection acting through changes in gene frequencies to this genic variance in fitness. Long misunderstood, this theorem is now leading to new insights in population genetics and evolutionary biology. Chapter 3 is devoted to the evolution of dominance. Chapters 4 and 5 *Variation as determined by mutation and selection* contain Fisher's mature treatment of the topics he first examined in 1922. Once again, his concentration on the variance of a population is evident. Some section headings are: *The chances of survival of an individual gene*; *Low mutation rates of beneficial mutations*; *Single origins not improbable*; *Distribution of gene ratios in factors contributing to the variance*; and *The observed connection between variability and abundance*. Then, in *Stable gene ratios*, Fisher reproduces his 1922 result about an equilibrium under heterozygotic advantage, but this time demonstrating convergence to the equilibrium, and, in *Equilibrium involving two factors*, he opens the debate on selection at two linked loci (soon to be discussed by Haldane) and, in *Simple metrical characters*, the debate on selection for continuous characters. Chapter 6 contrasts sexual and asexual reproduction and the "runaway" process of sexual selection which Fisher had discussed in 1915, and ends with the highly influential discussion of *Natural selection and the sex-ratio* which, though not original to Fisher, has had ramifications in many areas of evolutionary biology. In Chapter 7, *Mimicry*, kin selection, already mentioned in 1914, is invoked to explain the evolution of distastefulness in insects. The last five chapters are on the implications of the entire theory for man and society, and include an emphasis on the influential concept of parental expenditure, an idea which Fisher attributes to Leonard Darwin.

The Genetical Theory of Natural Selection was a watershed in the development of neo-Darwinism because within the compass of a single book it provided, for the first time, a comprehensive account of the synthesis of Mendelian genetics and Darwinian evolutionary theory. As Fisher later remarked, one of the main purposes of his book was to demonstrate how little basis there was for the opinion that "the discovery of Mendel's laws of inheritance was unfavourable, or even fatal, to the theory of natural selection". However, it had the added importance that its argument was based on selection acting on the individual rather than on the group, and it was consequently to be highly influential in disposing not only of objections to Darwinism, but also of the group selection version of Darwinism then in the ascendant. Moreover, where Haldane's papers treated mainly the immediate effects of selection acting in various models, Fisher's book already discussed the second-order effects, such as the evolutionary modification of dominance, the tightening of linkage between loci, natural selection and the sex ratio, and selection for fertility, which occupy such an important place in modern evolutionary biology.

Wright's sole publication of any kind in 1930 was his review of *The Genetical Theory*, but this in itself was to have lasting influence in an unexpected way. In it, Wright commented on the importance Fisher evidently attached to his fundamental theorem of natural selection, which in turn prompted a comment from Fisher in the letter to Wright in which he expressed his delight at the review. Wright's letter in reply, dated 3 February 1931, included the following statement:

> Some aspects of the ideas which I tried to express in pages 353 to 355 of my review might be visualised as follows: Think of the field of visible joint frequencies of all genes as spread out in a multidimensional space. Add another dimension measuring degree of fitness. The field would be very humpy in relation to the latter because of epistatic relations, groups of mutations which were deleterious individually producing a harmonious result in combination.

The letter contains a drawing of a continuous line with three humps of different height (fitness) followed by a description of how a species might be seen to evolve in this representation. It would "tend to move steadily" up the slope towards the nearest maximum "under the influence of selection", but, if this maximum was not the absolute maximum, the species would only be able to escape from it and make further progress in fitness through "something other than the steady pressure of selection". Wright suggested four factors: (1) a changing environment would continually change the system of humps, (2) new mutations would add further dimensions to the field, allowing new paths of advance, and (3) random genetic drift in a small species would allow stochastic jumps from one hump to another, as would (4) a subdivided large species. Wright's image was that of a potential function expressing the selective effect in a multidimensional space, accompanied by stochastic variation. It was an image that Fisher was quick to dismiss, but one which was to achieve great popularity and influence.

In January 1931, Haldane delivered lectures at the National University of Wales, Aberystwyth, which the following year were to become his book *The Causes of Evolution*, but before describing the book the sequence of events first requires an examination of Wright's major 1931 paper *Evolution in Mendelian populations*, the final form of the manuscript which he had been polishing since 1925.

The paper occupies 63 pages of the journal *Genetics*. After a brief introduction *Theories of evolution*, it continues with a discussion of simple Mendelian selection models in *Variation of gene frequency*, with mentions of mutation, migration and multiple alleles, and a section *Random variation of gene frequency*. This opens with a brief discussion of Wright's own 1921 result on the rate of decrease of heterozygosis under inbreeding and continues: "Another phase of this question was opened by Fisher (1922) who attempted to discover the distribution of gene frequencies ultimately reached in a population as a result of the above process." The remainder of Wright's long paper is largely devoted to the elucidation of such distributions under various conditions, in *The distribution of gene frequencies and its immediate consequences*, and to the conclusions to be drawn from them, in the final section *The evolution of Mendelian systems*. Notable among these conclusions is his "shifting balance" theory of evolution according to which selection will be most effective in populations of intermediate size (defined, of course, in relation to the mutation rate) or in large populations which are subdivided into "partially isolated local races of small size".

As we noted earlier, this famous paper of Wright's was based on a typescript which he wrote in 1925 (which has not survived), the year after he had received from Fisher the offprint of the latter's 1922 paper. By the time it was published in 1931, *The Genetical Theory of Natural Selection* had already appeared and was indeed the object of four of his footnotes. Just as Wright had reviewed *The Genetical Theory*, Fisher published a notice of *Evolution in Mendelian populations*. Short and friendly, it is of interest for including Fisher's first criticism of the "shifting balance" theory of evolution. It is important to note that Wright's 1931 paper did not include any reference to the "adaptive surface" metaphor, whose origin is thus almost certainly no earlier than the February letter to Fisher. However, by August 1932, when he spoke at the 6th International Congress of Genetics at Cornell, Ithaca, US, he had seen how to modify this adaptive surface metaphor so as to give pictorial representation to his shifting balance theory, by replacing the space of *gene frequencies*, in which each point represents a *population* as in Haldane's 1931 diagram for two linked loci, by a space in which each point represents a *genotype*, and a population is thus represented by a sort of bounded cloud of points depicting its individual members. Fitness was now an attribute of the genotype, not of the population. Fisher and Haldane both heard Wright's paper at the Congress, which his biographer, W.B. Provine, called "probably the most influential paper he ever published", for it contained the first drawings of this "adaptive (or fitness) surface" diagram which had had its origin in Wright's contemplation of Fisher's fundamental theorem. More importantly for the future popularity of the adaptive surface, the lecture was heard (and seen) by Th. Dobzhansky, who was enthralled by it.

The metaphor captivated many others besides Dobzhansky and dominated evolutionary biology for a generation. Its appearance in 1932 essentially completed the theoretical basis of the neo-Darwinian revolution, not so much because it contributed to the mathematical theory itself, but because it used a graphical metaphor which, whether or not it was an accurate reflection of nature, was simple enough to persuade evolutionary biologists that the mathematical geneticists really had succeeded in understanding the Mendelian mechanism of evolutionary change.

Fisher's and Haldane's papers at the Congress did not have the same impact as Wright's. Haldane's was a general discussion entitled *Can evolution be explained in terms of known genetical facts?* (to which question he answered "probably") while in *The evolutionary modification of genetic phenomena* Fisher stood Haldane's "analytic and deductive" approach on its head by adopting the "inductive and statistical" approach involved in enquiring "Can genetical phenomena be explained in terms of known evolutionary causes?" under which Fisher then discussed the evolution of dominance.

1932 also saw the publication of Haldane's *The Causes of Evolution*, which is now summarized. It opens in Chapter 1 with a discussion of "what is meant by evolution" and of Darwinism and its critics. Chapter 2 discusses variation within a species and Chapter 3 the genetical analysis of interspecific differences, both chapters with many examples taken from nature. Chapter 4 describes natural selection at work and in Chapter 5 the question *What is fitness?* is answered with emphasis on "the fallacy . . . that natural selection will always make an organism fitter in its struggle with the environment". Here Haldane is echoing *The Genetical Theory* in stressing the importance of selection within populations as opposed to group selection, and there is a further similarity as Haldane turns to man and, like Fisher, invokes the example of social insects as a simile for the evolution of social behaviour. However, Haldane is more specific in discussing altruism: "For in so far as it makes for the survival of one's descendants and near relations, altruistic behaviour is a kind of Darwinian fitness, and may be expected to spread as the result of natural selection." Chapter 6 is a wide-ranging *Conclusion*.

The main interest of the book, however, is the Appendix, *Outline of the mathematical theory on natural selection*, which was not part of the lectures. Naturally, it summarizes many of the author's results from *A mathematical theory of natural and artificial selection* and also from Fisher's *Genetical Theory*. In a brief section, *Wright's theory*, Haldane mentions that Wright's 1931 paper was only published after his book was written. The most original section of the Appendix is *Socially valuable but individually disadvantageous characters* in which Haldane initiates the mathematical discussion of the selective effects of altruism.

Fisher, in an unpublished review of *The Causes of Evolution* (Bennett, 1983), reciprocated Haldane's epithet of 'brilliant' for *The Genetical Theory*, but complained about the loose and egocentric argument which the lecture format had encouraged. Privately, in a letter to Leonard Darwin, Fisher wrote that the book was "an utter disappointment" to him (Bennett, 1983). Wright, in an assessment of *Haldane's contribution to population and evolutionary genetics* written in 1969, mentioned the series of papers *A mathematical theory of natural and artificial selection*, but not *The Causes of Evolution*.

The publication of *The Genetical Theory of Natural Selection* in 1930, of *Evolution in Mendelian populations* in 1931, of *The Causes of Evolution* and the *Proceedings of the 6th International Congress of Genetics* in 1932, brings to a close the decade which opened with Fisher's 1922 paper. The theory of neo-Darwinism had been established, but its propagation through the biological community was yet to come. As with all scientific revolutions, its long-term impact is likely to be rather different from its immediate reception and this article therefore closes with an attempt at a prediction of the former rather than a description of the latter. For the history of the spread of the ideas in the period 1930 to 1950, the reader is referred to W.B. Provine's article *The role of mathematical population geneticists in the evolutionary synthesis of the 1930s and 1940s* (Provine, 1978).

The achievement of the decade 1922–32 was a revolution in the understanding of the role of natural selection in evolution. Mendelian inheritance was shown to be an adequate mechanism for the operation of Darwin's theory, so that the theory, shorn of its Lamarckian tendencies, could itself be accepted as the explanation of evolutionary change. Dismissive references to the theory as being incapable of accounting for evolution, such as Sir John Herschel's description of it in 1861 as "the law of the higgledy-piggledy", were replaced by the realization that "natural selection is a mechanism for generating an exceedingly high degree of improbability" (reported by Julian Huxley (Huxley *et al.*, 1954) to be one of Fisher's favourite remarks). The mathematical theory developed by Fisher, Haldane and Wright contained within it sufficient flexibility to accommodate the enormous variety of observations from nature and from selection experiments.

In the years following 1932, the different emphases of Fisher and Wright, Fisher thinking that "evolution would proceed most surely and rapidly in a large, panmictic, population" and Wright arguing that "the most favourable condition for evolution lies in a structured population" (the descriptions are those of Crow, 1992), were the subject of much controversy, but nature is so varied that there is surely a need for both points of view, and the same can be said in connection with the evolution of dominance. The one contribution which, though it was extremely influential, has not stood the test of time, is Wright's "adaptive surface" metaphor which, as we have seen, started life as a corollary to Fisher's fundamental theorem with gene frequency as the variate, and then underwent a metamorphosis into a picture in which there was no variate as such but the fitness of individual genotypes was "mapped". In 1935, the picture reverted to a surface in which the mean fitness of the population was once again graphed against gene frequencies and, in this form, was used as the basis for a great deal of theorizing. Unfortunately, the loose identification of this adaptive surface with the potential functions of physics then led to misapprehensions about the role of Fisher's fundamental theorem, and the net result was a model of the operation of natural selection which was untenable in detail and historically extremely confusing, Fisher being supposed to have championed a "potential function" view about which his resolute denials were of no avail. It is unlikely that this controversy will be long remembered.

The notable features of the neo-Darwinian theory are, first, that the Mendelian mechanism of inheritance (in conjunction, of course, with other known influences) accounts satisfactorily for the similarities between relatives and for the changes observed in selection experiments and inbred populations. Second, although changes under natural selection are difficult to observe, the Mendelian mechanism coupled with reasonable values for such parameters as mutation rates and effective population sizes (the latter determined by population structure as well as size) accounts satisfactorily for the kind of intrapopulation selection which appears to be the raw material of evolution. Third, detailed mathematical analysis of the stochastic properties of populations demonstrates that the survival probabilities of genes of varying degrees of advantage or disadvantage are compatible with the requirements of the evolutionary model. Fourth, the theory has been successfully applied to the evolution of genetic systems themselves, such as linkage, dominance, altruism and the sex ratio.

So successful has the neo-Darwinian revolution been in incorporating Mendelian genetics into the theory of evolution that the modern tendency in evolutionary biology is to take the genetics for granted by formulating models in terms of game theory and other similar concepts. This has its own dangers, both conceptual and technical. On the one hand, it introduces inappropriate words with a teleological flavour such as "strategy" and, on the other hand, it presumes that the underlying Mendelian mechanism will always be able to deliver the predicted result even though it has been ignored in the formulation of the problem. A study of the original papers and books by Fisher, Haldane and Wright is a sensible antidote to this tendency.

R.A. Fisher, J.B.S. Haldane and Sewall Wright jointly engineered a revolution. Though they often disagreed over its details, Fisher's remark to Wright in 1931 could be held to apply to all three equally: "However, there is a substantial body of theory on which I think we do agree and that after all is of infinitely more interest to the world at large than the very obscure points still in dispute."

Primary literature

All the papers of R.A. Fisher referred to in this article will be found in *Collected Papers of R.A. Fisher*, vols 1–5, edited by J.H. Bennett, University of Adelaide (1971–1974). A bibliography appears in vol. 1, and also in the biography of Fisher (Box, 1978). Fisher's 1911 lecture *Mendelism and biometry* is reprinted in *Natural Selection, Heredity, and Eugenics: Including Selected Correspondence of R.A. Fisher with Leonard Darwin and Others*, edited by J.H. Bennett, Oxford: Clarendon Press and New York: Oxford University Press (1983), which also includes correspondence with Haldane and Wright and a very valuable *Introduction* by Bennett which covers the period of the present article in detail. Fisher's 1922 paper is: On the dominance ratio, *Proceedings of the Royal Society of Edinburgh* 42: 321–41. *The Genetical Theory of Natural Selection* was published by Clarendon Press, Oxford, in 1930, and reprinted in a revised edition by Dover, New York, in 1958. A variorum edition prepared by J.H. Bennett was published in 1999 by Oxford University Press.

There is no collected edition of J.B.S. Haldane's papers, but a bibliography appears in his biography (Clark, 1968), and shortened references to the series of papers *A mathematical theory of natural and artificial selection* are given in his book *The Causes of Evolution*, originally published in London and New York by Longmans Green in 1932. There is a facsimile reprint of *The Causes of Evolution* published by Cornell University Press in 1966 and a reprint with a new *Introduction* and *Afterword* by E.G. Leigh Jr was published by Princeton University Press in 1990.

Most of the papers of Sewall Wright referred to in this article will be found in *Evolution: Selected Papers*, edited by W.B. Provine, Chicago University Press (1986), which also includes a complete bibliography, as does Provine's biography of Wright (see below). Wright's 1931 paper is: Evolution in Mendelian populations, *Genetics* 16: 97–159, and his 1932 paper is: The roles of mutation, inbreeding, crossbreeding and selection in evolution, *Proceedings of the Sixth International Congress of Genetics* 1: 356–66. Many years after the period described in this article Wright published *Evolution and the Genetics of Populations* in four volumes, University of Chicago Press (1968–78).

The important papers of Fisher, Haldane and Wright have been frequently reprinted. Two valuable collections are in the *Benchmark Papers in Genetics* series published by Dowden, Hutchinson and Ross, Stroudsburg, Pennsylvania: no. 7, *Stochastic Models in Population Genetics*, edited by W-H. Li (1977) and no. 8, *Evolutionary Genetics*, edited by D.L. Jameson (1977).

Secondary literature referred to in the text

Bennett, J.H. (1983) *Natural Selection, Heredity, and Eugenics. Including Selected Correspondence of R.A. Fisher with Leonard Darwin and Others*, Oxford: Clarendon Press, and New York: Oxford University Press

Box, J.F. (1978) *R.A. Fisher: The Life of a Scientist*, New York: Wiley

Clark, R. (1968) *J.B.S: The Life and Work of J.B.S. Haldane*, London: Hodder and Stoughton

Crow, J.F. (1992) Haldane, Fisher and Wright. In *J.B.S. Haldane: A Tribute*, Calcutta: Indian Statistical Institute

Haldane, J.B.S. (1931) Mathematical Darwinism. *Eugenics Review* 23: 115–17

Hogben, L. (1931) *Genetic Principles in Medicine and Social Science*, London: Williams and Norgate

Huxley, J.S., Hardy, A.C. & Ford, E.B. (eds) (1954) *Evolution as a Process*, London: Allen and Unwin; 2nd edition, 1958

O'Donald, P. (1990) Fisher's contributions to the theory of sexual selection as the basis of recent research. *Theoretical Population Biology* 38: 285–300

Provine, W.B. (1978) The role of mathematical population geneticists in the evolutionary synthesis of the 1930s and 1940s. *Studies in History of Biology* 2: 167–92

Provine, W.B. (1986) *Sewall Wright and Evolutionary Biology*, Chicago: Chicago University Press

Other secondary literature of interest

The secondary literature about Fisher, Haldane and Wright is now quite extensive. Many important contributions are to be found in the editorial material of the sources listed under **Primary literature** and in obituaries published by learned societies. The occasion of the centenary of Fisher's birth resulted in a number of publications in 1990 which are listed in Edwards (1992) and the publication of the variorum edition of *The Genetical Theory of Natural Selection* in 1999 prompted a number of articles including Edwards (2000).

Edwards, A.W.F. (1992) Celebration of the centenary of the birth of Sir Ronald Aylmer Fisher, 1990. *Historia Mathematica* 19: 81–82

Edwards, A.W.F. (2000) The genetical theory of natural selection. In *Perspectives*, edited by J.F. Crow & W.F. Dove, *Genetics* 154: 1419–26

See also:

Provine, W.B. (1971) *The Origins of Theoretical Population Genetics*, Chicago: University of Chicago Press

Sarkar, S. (ed.) (1992) *The Founders of Evolutionary Genetics*, Dordrecht: Kluwer

and articles in the series *Oxford Surveys in Evolutionary Biology* published by Oxford University Press.

GLOSSARY

allelomorph former term for allele

diallelic locus specific site on a chromosome where two different alleles are located

game theory mathematical theory to determine the most favourable strategy in situations of competition or conflict

heterozygosis percentage of heterozygotes for a given locus in a population

inbreeding coefficient of an individual, the probability that the pair of alleles carried by the male and female gametes that produced it is identical by descent from a common ancestor, as a result of inbreeding

path coefficient a standardized regression coefficient measuring the influence of one of a number of variables on another variable

See also **Gregor Mendel** (p.62); **Charles Darwin and the Theory of Evolution** (p.68)

FRANCIS GALTON

Nicholas W. Gillham
DCMB Group, Duke University, Durham, USA

1. Introduction

Sir Francis Galton (1822–1911) (**Figure 1**) was a man of diverse interests and achievements. To those interested in the history of Africa, he was a 19th-century explorer and geographer. He was also a well-known travel writer, the author of an immensely useful guidebook for novice and expert alike who ventured into the bush over 100 years ago. To meteorologists, he is remembered as the discoverer of the anticyclone. Those who plumb the history of statistics will find Francis Galton's name associated with regression analysis, the coefficient of correlation and the founding of biometrics. Psychologists, especially those interested in mental imagery, claim him as one of their own. Forensic experts will recognize Francis Galton as the man who played a central role in establishing the use of fingerprints as evidence on a firm scientific footing. Last, but certainly not least, Francis Galton's name will always be associated with eugenics, the founding of human genetics and the use of pedigree analysis.

Figure 1 Francis Galton photographed at age 80. (Reprinted from K. Pearson, *The Life, Labours and Letters of Francis Galton*, vol. 3, 1930 with permission from Cambridge University Press.)

2. Galton's life and early interests

Since he had a seemingly endless array of interests, Francis Galton is sometimes called a dilettante. This reveals both a misunderstanding of the man's achievements and the nature of Victorian science. Galton's research and published work revolve around two distinct sets of problems. During the first part of his career, he was interested in exploration and geography. Travel writing related to these interests, as did meteorology, for the explorer is forever having to take into account the vicissitudes of the weather. The second part of Galton's career opened when he read the *Origin of Species* by his cousin Charles Darwin. He became interested in human nature, human heredity and the possibility of human improvement through selection of the fittest which led him to coin the word eugenics. In order to investigate the heritability of what Galton referred to as "talent and character", in the age before the rediscovery of Mendel's principles and the development of IQ tests, Galton made use of pedigrees, twin studies and anthropometric measurements. However, while he believed that favourable physical characteristics correlated with superior mental qualities, he had no way to measure the latter directly. Consequently, he became interested in psychology and personal identification, eventually lighting on fingerprinting which appeared to provide a foolproof way to distinguish different individuals. To analyse the masses of data he accumulated he had to develop new statistical techniques which included regression and correlation. Galton believed that the data he had obtained, particularly from pedigree studies, demonstrated that heredity was the predominant influence in determining what we are, with environment playing a secondary role, and so his ultimate goal was to see the implementation of a rational programme of "national eugenics" to improve the human stock in Britain.

The diversity of Galton's interests was not atypical for a Victorian scientist. His grandfather, Erasmus Darwin (1731–1802), was a highly successful physician, a serious student of botany and zoology, an inventor and also a talented poet. William Whewell (1794–1866), the Master of Trinity College, Cambridge, while Galton was a student there, had studied and written about philosophy, mathematics, mechanics, theology and moral philosophy. He proposed and wrote a book about his theory of Gothic design, taught mineralogy and authored a treatise on the classification of minerals. Galton, like Charles Darwin, was independently wealthy and could spend all of his time on whatever interested him, whereas most scientists were not so fortunate. T.H. Huxley, for example, had to work hard to support himself as a scientist and teacher as he had no fortune to fall back on. Many other scientists combined their investigations with an independent business that put bread on the table. Galton's friend, the mathematician William Spottiswood, was printer to the Queen, and Charles Booth, whose 17-volume work *Life and Labour of the People of London* (1891–1903) is a classic in early sociology, founded and chaired a successful steamship company with his brother Alfred.

Francis Galton was born on 16 February 1822, the youngest of nine children, seven of whom, four girls and three boys, survived infancy, but none of whom would achieve his eminence. His mother, Violetta Darwin Galton, was the daughter of Erasmus Darwin by his second marriage to Elizabeth Collier Sacheverall-Pole. Charles Darwin was Erasmus Darwin's grandson by his first marriage to Mary (Polly) Howard. Galton's father, Samuel Tertius Galton, was the scion of an old and wealthy Quaker family who converted to the established church of England. After a brief fling at medical studies, Francis Galton went up to Trinity College, Cambridge, where he read mathematics for the Tripos in that subject. A nervous breakdown forced him to pull out as an honours student and he graduated from Cambridge with an ordinary pass, or poll, degree – just as Charles Darwin had. Following Cambridge, it was expected that Galton would return to his medical studies, but, with the sudden death of his father, he was free to do as he chose, so he made a trip up the Nile to Khartoum with a couple of friends, following which he visited Lebanon, Israel and Syria.

After returning from the Middle East, Galton had spent a couple of years aimlessly hunting with friends in England and visiting the Shetlands, when he read of an expedition recently mounted by David Livingstone (1813–1873) to Lake Ngami in the northern part of what is now the Republic of South Africa. In 1850, Galton determined to head an expedition of his own to the same lake. To facilitate this goal, he obtained membership in the Royal Geographical Society (RGS) through the agency of his cousins Douglas Galton and Charles Darwin. The RGS lent its name to the expedition, but Galton provided the financial support for the undertaking as well as purchasing all of the necessary equipment and supplies and eventually hiring all of the support staff, including his right hand man, the naturalist Charles Andersson. Galton was unable to proceed from the Cape Colony to Lake Ngami via the route followed by Livingstone because of difficulties with rebellious Afrikaners in the Orange River Territory, so he was diverted instead to what is now Namibia, the plan being to approach Lake Ngami from the west. Although Galton never reached the lake, he trekked into northern Namibia, a region never before visited by Europeans and discovered a previously unknown people, the Ovampo.

Galton returned home in the spring of 1852 and, by the following spring, had completed an account of his expedition, *Tropical South Africa* (Galton, 1853), which was well received. He also won the Founders Medal, one of the two gold medals awarded annually by the RGS for particularly notable exploratory work. That same year, he met and married Louisa Butler, daughter of the Dean of Peterborough Cathedral, George Butler. He now entered a phase of his life in which travel writing and service to the RGS figured prominently. He was soon at work on a paper for a special edition of the RGS Journal devoted to the needs of explorers called *Hints to Travellers* (Galton, 1854). This was to become the most successful publication ever to emanate from the RGS and Galton helped to edit the second and third editions of the work and was the sole editor of the fourth edition published in 1883. He was also at work on the *Art of Travel* (Galton, 1855), which was without doubt the most popular book he ever wrote. It went through eight editions between 1855 and 1893. This is a sort of "how to" book designed for the amateur traveller and seasoned explorer alike. Galton always had a strong quantitative bent and this is evident in his travel and geographical writing, where he is concerned with subjects like the load a pack animal can carry over a specific distance, boiling point thermometer readings and careful measurements of longitude and latitude.

Galton loved to tinker and was always inventing new gadgets that might be useful in connection with a specific problem he was working on. For example, he devised a hand-held heliostat that enabled the operator to catch a flash from the sun and direct it to a distant point. A version of this instrument was manufactured commercially under the name Galton's Sun Signal and was used in late 19th-century surveys by the British to enable shore parties to make their exact whereabouts visible to those on shipboard. From the mid-1850s until well into the 1870s, Galton was very active in the RGS, usually serving on its Council. He was especially interested in the daring explorations of this period that led to the discovery of the Great Lakes of Africa and the source of the White Nile. He knew all of the *dramatis personae* and helped to raise funds for their expeditions, provided written instructions for them and asked pointed questions about specific geographical observations made by them. He was also involved in the formation of yet another literary venture, *The Reader*, whose purpose was to cover literature, science and art. Published from 1863 to 1866, *The Reader* had a very distinguished group of supporters from science, literature, economics, classics, etc. However, it was not a commercial success, but metamorphosed several years after its founding into one of the greatest scientific journals of all time, *Nature*.

Francis Galton's foray into meteorology began as an outgrowth of his interest in collecting data, in this case about climates, and in map making. In 1861, he hit on a method for combining these two objectives and began printing retrospective maps of the weather of Britain for specific dates. He soon realized that the key to making proper weather maps of Britain and Europe was to obtain meteorological observations synchronously from a wide variety of sources. Then data like barometric pressure, wind speed and direction, clouds and precipitation could be plotted precisely for a given time on a given day for many geographically separated locations. To gather this information, he sent circulars to meteorologists throughout the British Isles and Europe. Although Galton was by no means the first person to construct synoptic weather charts, his careful analysis of European weather patterns for the month of December 1861, published in his book *Meteorographica* (Galton, 1863), resulted in his discovery of the anticyclone, the clockwise circulation of air associated with a high barometer and a clear sky in the northern hemisphere, whose direction is reversed in the southern hemisphere.

3. Interest in heredity and eugenics

Charles Darwin (1809–1882) sent his cousin a copy of the *Origin of Species*, shortly after its publication in 1859, and this marked a major turning point in Galton's career. Although he continued to be active in geography and meteorology, he now became enamoured with the notion that natural selection must also apply to human beings and also that most human traits, whether physical or mental, were determined by heredity rather than environment. In the early 1860s, he developed a method of pedigree analysis which he would use throughout the rest of his life. The basic methodology and assumptions are quite simple. He collected dictionaries, lists, etc., of eminent men and asked the frequency with which they had eminent male relatives. His first attempts at this sort of analysis were published in two articles in *Macmillan's Magazine* in 1865 (Galton, 1865) entitled "Hereditary talent and character". One of the most remarkable concepts elaborated in these papers is the notion that embryos of the next generation spring forth from embryos of the preceding generation. This anticipates August Weismann's experimentally supported theory of the continuity of the germline by almost 20 years. Over 40 years later Galton wrote that on "re-reading these articles, I must say that, considering the novel conditions under which they were composed, and notwithstanding some crudeness here and there, I am surprised by their justness and comprehensiveness" (Galton, 1908, p.289).

The *Macmillan's* articles were followed by a major work taking the same tack, *Hereditary Genius* (Galton, 1869). The major conclusion reached in the two articles and the book was that the closer a relative to an eminent man, the higher the probability that the man in question would also be talented. Thus, fathers, sons and brothers were more likely to be eminent than grandfathers, grandsons and uncles. For Galton, this meant that there must be a strong hereditary component to "talent and character". Publication of *Hereditary Genius* also made apparent a new interest of Galton's. He had become fascinated by the normal distribution and how it might be applied to the quantification and analysis of different human traits. However, the kinds of data that Galton had collected for *Hereditary Genius* were not amenable to statistical analysis – that would have to wait until later when he obtained large amounts of data on human physical parameters by canvassing public school students about their heights and weights and, even more comprehensively, from the thousands of individuals who flowed through his Anthropometric Laboratory.

Meanwhile, Galton was beginning to think about the hereditary mechanisms that might underlie human traits. Darwin too was concerned about heredity, as heredity must provide the variations on which natural selection could act. In his lengthy work, *The Variation of Animals and Plants under Domestication* (1868), Darwin laid out his "Provisional Hypothesis of Pangenesis". His central assumption was that the cells of the body threw off granules (gemmules) which were gathered from all parts of the organism "to constitute the sexual elements, and their development in the next generation forms a new being" (Darwin, 1868, p. 321). It was also essential that Darwin's hypothesis explain the origin of the variations essential for the workings of natural selection and evolution. To create new variations, he posited two mechanisms. The first resulted when injury to the reproductive organs occurred. This might lead to improper aggregation of gemmules so that some were in excess and others in deficit which, in turn, resulted in variation and modification. There is a certain resemblance to the modern theory of mutation here, except that Darwin is talking about quantitative variations in numbers rather than qualitative alterations of individual elements. Darwin's second mechanism for creating variation caused Galton concern. Here Darwin proposed a role for modified (mutant) gemmules, but he imagined that these modifications were the result of the direct action of changed environmental conditions causing certain parts of the body to be affected specifically. These modified gemmules would then be transmitted to the offspring. Galton's problem with this notion was that it not only implied that an organism could acquire new characteristics in response to environmental change, but that gemmules from the somatic tissue contributed directly to what would now be called the germline. Nevertheless, Galton was enchanted with the quantitative possibilities that gemmules provided and rushed to include a chapter on pangenesis in *Hereditary Genius* which was probably nearly ready to be set in proof.

Galton was anxious to test his cousin's hypothesis and wrote to Darwin in December 1869 suggesting that they enter into collaborative experiments to do so. The experiments were designed to establish whether gemmules from various parts of the body were transmitted via the bloodstream to the sex organs. Galton chose a pure-breeding line of silver grey rabbits to act as recipients and rabbits with different inherited characteristics such as colour to be used as blood donors. The progeny of the transfused silver greys were then examined and no heritable changes that could be attributed to the blood donor were ever found. The level of sophistication of the experiments reached such a point that the donor and recipient were joined at their carotid arteries to ensure adequate blood exchange. Galton wrote up these negative results, apparently without consulting Darwin, who seems to have been quite annoyed. Darwin wrote a letter in refutation to *Nature* (Darwin, 1871), saying that in his provisional hypothesis of pangenesis he had "not said one word about blood, or about any fluid proper to any circulating system". His point was that his hypothesis was also meant to apply to plants, which lack true blood, and to protozoa which lack blood or vessels of any kind. Galton, who greatly admired Darwin, sent a letter of apology to *Nature* (Galton, 1871) which ended "Vive Pangenesis!"

Galton was simply being gallant, however, for he now set

about constructing his own theory of heredity which did, nevertheless, derive an important notion from Darwin. This involved the problem of reversion. Darwin had pondered over the problem of how to account for the reappearance of a characteristic present in an earlier generation. Today, we know the explanation could be the emergence of an individual expressing an autosomal or sex-linked recessive mutation, but for Darwin such observations presented a puzzle. He had to assume that, unlike the majority, a few gemmules were latent or dormant and remained hidden for several generations. Multiplication of such gemmules would result in their visible manifestation in the sense of reversion. In an almost incomprehensible paper, "On blood-relationship", published in 1872, Galton, like Darwin, recognized two classes of hereditary elements which he called latent and patent. Both differentiated from a common group of structureless elements in the ovum and converged once again to dedifferentiate into structureless elements to be transmitted to the next generation. However, Galton argued that the pool of latent elements must be much more greatly varied than the patent elements actually expressed would suggest, because of the phenomenon of reversion – and this led him to an important departure from Darwin's theory. Galton assumed that only a subset of the patent elements, those which differentiated into adult elements, could be transmitted to the progeny. The same was true of the latent elements, but for a different reason. Since no two parents are exactly alike, the number of latent elements would increase in each generation unless some limit were imposed on this increase. Hence, a subset, a family, must be selected at each generation. Although Galton never explicitly made the point, he seems to recognize the problem that a geometric increase in the number of elements is going to pose at each generation and solves it by invoking a mechanism that picks a subset. This predicts that a reduction in elements must occur at each generation, just as we know it does when the chromosome number is halved at meiosis.

Galton expanded this notion in a paper published in 1875 in the *Contemporary Review*, a journal that catered to the literate layman (Galton, 1875a). Here he proposed that within the ovum is the sum total of what he called germs or gemmules and this he termed the *stirp* which derives from the Latin, stirpes, a root. The stirp was all the fertilized ovum received from the parents. What Galton is enunciating here, much more clearly than he had in his *Macmillan's* papers of 1865, is a variant of the germline theory and Weismann acknowledged this in a letter to Galton in 1889 (Pearson, 1930), when he wrote that Galton's paper was brought to his attention after he had published his own papers on the germline theory: "I regret not to have known it before, as you have exposed in your paper an idea which is one essential point nearly allied to the main idea contained in my theory of the continuity of the germ plasm."

Galton's theory was based on four postulates: (1) Each of the enormous number of "quasi-independent units the body consists of, has a separate origin, or germ". The modern equivalent of this statement would be that each gene specifies a different protein. (2) The germs in the stirp are much greater in number and variety than the structural units derived from them. Galton needed this postulate to account for variation and reversion. (3) The undifferentiated germs propagate themselves in the latent state and contribute to the stirp of the offspring. This postulate is, presumably, the one to which Weismann referred for it clearly implies the existence of what he would have called the germline. (4) The final structure, organization and appearance of the adult organism depends on the mutual affinities and repulsions of separate germs within the stirp and during development. In modern terms, this might be defined by the interaction of gene products and gene regulation.

Galton's theory also predicts the existence of what we now refer to as the deleterious recessive mutation. He reasons that unisexual systems are rare because they tend to die out, since "a deficiency of some of the structural elements gradually sets in, and the race ultimately perishes". In bisexual systems, in contrast, "the chance deficiency in the contribution from either of them, of any particular species of germ, will be supplied by the other". Yet, Galton realized that if the fertilized ovum contains the entire stirp from each parent, the progeny will contain double the stirp of the two parents. So, while "the stirp whence the child sprang can be only half the size of the combined stirps of the two parents, it follows that one half of his possible heritage must have been suppressed". On theoretical grounds, Galton is again predicting that a process like meiosis must exist.

But Galton's real interest was not in developing and proving a theory of inheritance – what he really wanted to do was gather lots of data relating to human heredity and develop methods for their analysis. In 1874, he published a book called *English Men of Science: Their Nature and Nurture* based on a lengthy questionnaire he had sent out to 180 members of the Royal Society who had eminence beyond membership in the Society itself which he referred to as a "pass examination". In this book, Galton gave his rationale for using the famous words: "The phrase 'nature and nurture' is a convenient jingle of words, for it separates under two distinct heads the innumerable elements of which personality is composed." However, what Galton had actually done was to copy a line from Shakespeare's play, *The Tempest*, where, in Act IV, Scene 1, Prospero, the Duke of Milan, says, in a conversation with the spirit Ariel, in reference to that strange creature Caliban: "A devil, a born devil, on whose nature nurture can never stick".

Meanwhile, Galton had discovered another line of inquiry into human heredity, twin studies. In 1875, he published an article in the popular monthly *Fraser's Magazine* called "The history of twins, as a criterion of the relative powers of nature and nurture" (Galton, 1875b). Galton recognized that there were two classes of twins, those that were "closely alike in boyhood and youth" and those "who were exceedingly unlike in childhood". He also recognized that the first class, identical twins, derived from the same ovum while the second class, non-identical twins, must derive from separate ova. Again, Galton made use of questionnaires to collect masses of data. He eventually obtained data on 35 sets of identical twins and 20 sets of non-identical twins and he did exactly what psychologists studying identical twins have done ever since. He collected a series of anecdotes about uniquely similar behavioural characteristics found among certain of identical twins and recounted these. In contrast, Galton's non-identical twins, despite similar nurture, were, as he recounted for one pair, "otherwise as dissimilar as two boys could be, physically and mentally, and in their emotional nature".

However, Galton's pedigree and twin studies were really not very useful for quantitative analysis. He wanted data on human beings which he could analyse statistically using the properties of the normal distribution. However, he could not do proper experiments with people whose generation times were lengthy anyway. To speed things up in a system where he could exert control of breeding conditions, Galton chose, in 1874, to carry out some experiments on sweet peas. He cited three reasons for choosing sweet peas: they had little tendency to cross-fertilize as far as he knew; they were hardy and prolific; and the seeds within a pod were uniform in size. He sent off packets of seeds that had been carefully graded by size to various friends and acquaintances throughout Britain. These individuals planted the seeds, grew up the progeny and returned their seeds to Galton.

What Galton discovered was something very interesting. Seed sizes were distributed normally among progeny as they had been among the parents, but he also found that the mean size of the progeny seeds obtained, for example, from parental seeds selected for large size had reverted (or regressed) towards the original mean seen for the entire parental distribution. He was able to illustrate this point using an ingenious device he had invented called the quincunx (**Figure 2**). This was a contraption Galton had originally hammered together to illustrate the normal distribution. Lead shot was poured into a funnel which, after passing through its narrow neck, cascaded through a series of rows of pins into individual bins at the bottom. Each row consisted of a series of arrays of five pins arranged as a quincunx – that is, with a pin at each of the four corners of a rectangle and the fifth in the middle. The uppermost row consisted of a single quincunx with the rows successively expanding in length so that the final assembly of pins resembled an equilateral triangle. When the lead shot had bounced back and forth through the quincuncial array and into the bins at the bottom, they were distributed normally. By using a modification of this design, Galton was also able to illustrate regression to the mean. Galton also found that if he plotted the average diameters of parental seeds on the X axis against the average diameter of the progeny seeds they produced on the Y axis, he could connect the points with a straight line (**Figure 3**). This was the first regression line ever plotted and Galton referred to its slope as the "coefficient of reversion", but, on finding that this coefficient was not a hereditary property, and the result of his own statistical manipulations instead, he changed the name to coefficient of regression.

By the early 1870s, Galton was thinking seriously about improving the human race through selective breeding and he set out his agenda for public consumption in *Fraser's Magazine* in an article published in 1873 entitled "Hereditary improvement". His aim was to summarize the current human condition as he understood it and "to propose a scheme for its improvement whose seeds would be planted almost without knowing it, and would slowly but steadily grow, until it had transformed the nation". Galton made use of the normal distribution to illustrate for his readers how men could be graded according to "natural ability". Highest natural ability was to be found at one end of the distribution and lowest at the other. Galton hoped that "the average standard of a civilised race might be raised to the average standard of the pick of them" and what he obviously had in mind was selective breeding. However, before such a scheme could be implemented, Galton needed to collect a lot more anthropometric data and here he made a crucial assumption. He believed that good physical attributes were a prerequisite for an active mind. Since he had no measure equivalent to IQ, this meant that he could use the measurement of physical characteristics as a proxy for measuring intelligence.

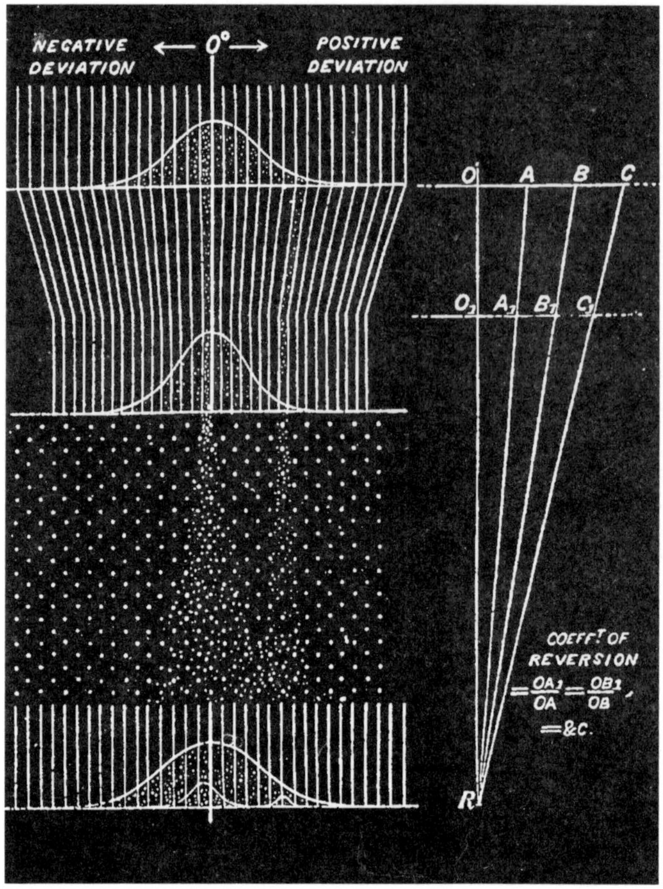

Figure 2 Galton's quincunx illustrating the nature of regression. (Reprinted from K. Pearson, *The Life, Labours and Letters of Francis Galton*, vol. 3, 1930 with permission from Cambridge University Press.)

To begin to collect the necessary data he needed, Galton in 1874, with the sponsorship of the Anthropological Institute, proposed to obtain information on the heights and weights of students at selected English public schools. He argued that this would enable the collection of data sets on homogeneous groups of boys and girls. Galton was soon

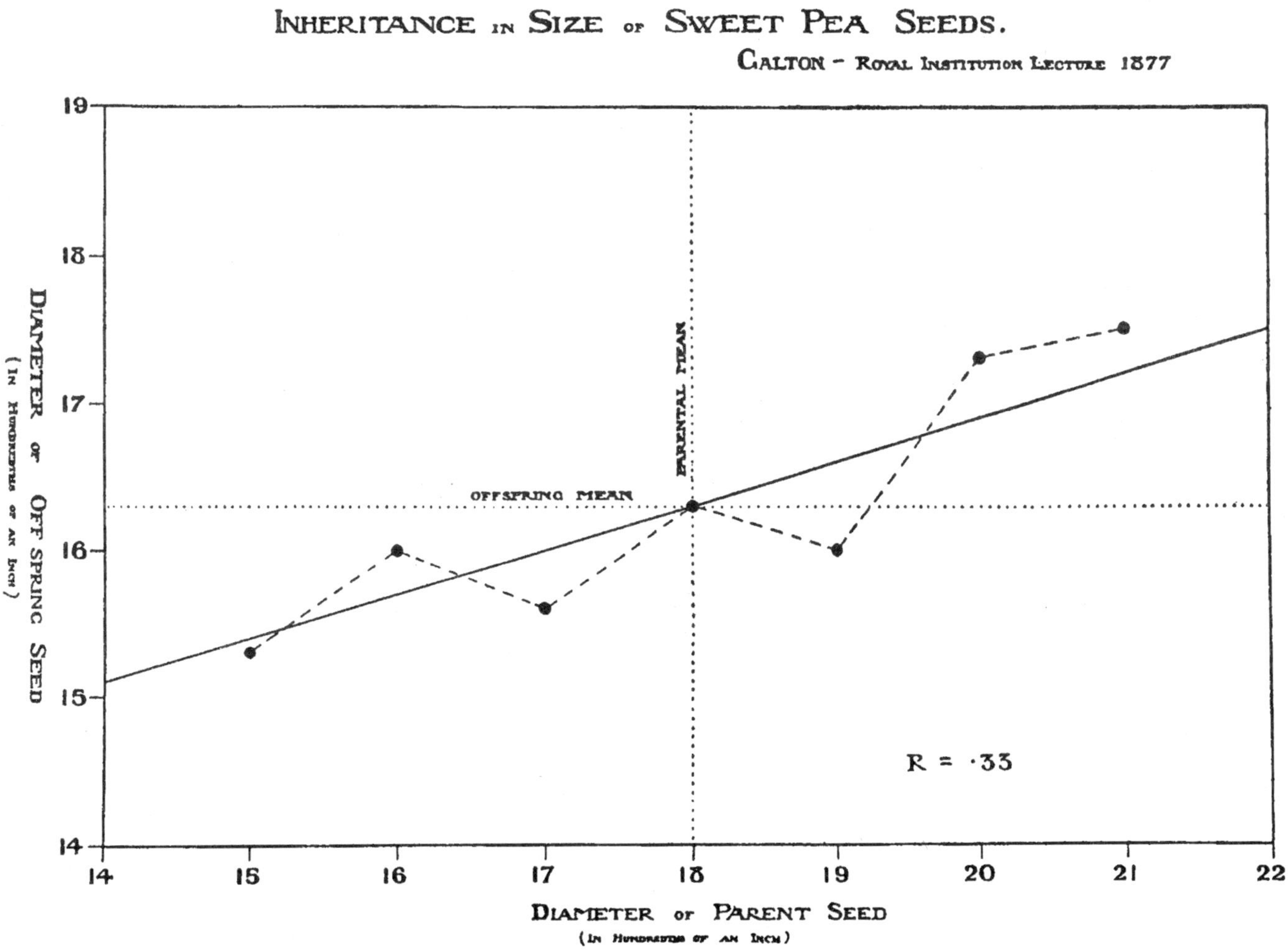

Figure 3 The first regression line. (Reprinted from K. Pearson, *The Life, Labours and Letters of Francis Galton*, vol. 3, 1930 with permission from Cambridge University Press.)

publishing his findings and, as a result, Charles Roberts, a doctor at St George's Hospital in London, contacted Galton concerning a study in which he was involved that had yielded measurements on some 10 000 children of both sexes employed in textile factories or similar establishments. Yet, even these data were not enough to satiate Galton's appetite for quantifiable observations that could be analysed statistically and he proposed the formation of anthropometric laboratories to gather the requisite information in a *Fortnightly Review* article published in 1882. Soon the opportunity came for Galton to form his own Anthropometric Laboratory and he did so in connection with the International Health Exhibition which opened in London in 1884. He equipped and maintained this laboratory at his own expense: first in connection with the Exhibition and, after it had closed, in South Kensington, where he collected enormous amounts of data not only on height and weight, but on reaction time, strength of pull and squeeze, colour sense, etc. In 1883, Galton published a book, *Inquiries into Human Faculty and its Development*, where he pulled together the results of his twin studies, his thoughts on anthropometrics and statistics, and also touched on topics like psychometrics, psychology, race and population. It was also in this book that he coined the term eugenics.

In the midst of his inquiries into heredity, anthropometrics and statistics, Francis Galton took a detour. From 1877 until 1885, he became preoccupied largely with psychological studies, but what he really wanted was some way of measuring mental ability, but the IQ test was still far in the future. His first approach was one rooted in physiognomy. If a certain group of individuals shared a particular mental trait, and this was somehow reflected in their physiognomy, the common features might be extracted by superimposing photographs of their faces on one another. This should factor out the unique features and emphasize shared attributes creating, as it were, a photographic mean or average. With the aid of Sir Edmund Du Cane (1830–1903), the director-general of prisons, Galton was able to examine many thousands of photographs of thieves, murderers, etc., hoping that composite photography would reveal features that typified different groups of criminals. These, and other studies using composite photography in race and pedigree analysis, did not provide Galton with the key to the mind he sought, so his interest turned to other psychological approaches. The study of mental imagery, the nature of the evanescent impressions that pass through the mind, was the area that interested him most. Once again, Galton wanted to apply quantification to this psychological process and used questionnaires addressed to public school boys and entreaties to friends in various learned societies to help him gain the information he sought. Galton's pioneering work on this problem is remembered by psychologists even today.

In searching for methods of personal identification, Galton came across the classification system of the French criminologist Alphonse Bertillon (1853–1914) who used a combination of photography (full face and profile), together with precise measurements (height, limb length, head width, etc.), to characterize each individual felon. Galton became intensely interested in Bertillonage, as it was called, as a means of personal identification and actually visited Bertillon himself to learn more about the system. However, he was also on the threshold of developing the concept of correlation and realized that the set of individual measurements Bertillon was making on each criminal did not necessarily represent independent variables. In the meantime, he had also become absorbed with another method of personal identification that seemed to have great promise, fingerprinting. He became aware of the power of this method from a letter sent by a young Scottish doctor, Henry Faulds, to Charles Darwin which Darwin had forwarded to him in the spring of 1880. This was followed by two papers published in *Nature* in the same year (Faulds, 1880; Herschel, 1880). The first was from Faulds, who was attached to the Tsukiji Hospital in Tokyo. His interest in fingerprints was piqued when he observed the intricate structures of fingermarks found on prehistoric pottery discovered on the beaches of Tokyo Bay. He then expanded his study to the fingerprints of monkeys, noting their similarity to those of humans, and finally to Japanese citizens. He actually developed a method for fingerprinting and suggested that it might be used to classify people. The second letter was from Sir William Herschel who wrote that he had been using fingerprints for over 20 years, while serving in the Bengal Civil Service, to identify Indian pensioners in registration offices under his administration and also for criminal identification. Herschel also made the crucial observation that fingerprints did not change over time.

The potential of fingerprinting became obvious to Galton and he threw himself into the study of fingerprints with characteristic zeal. He was to publish several major papers and books (e.g. *Finger Prints*, 1892; *Finger Print Directories*, 1895) on the subject. He was able to use his Anthropometric Laboratory to gather thousands of fingerprints and, based on careful scrutiny, Galton developed a highly detailed system for classifying fingerprints. His painstaking comparison of the structure of fingerprints enabled him to confirm with precision Herschel's original claim that the structure of individual fingerprints does not change over time. He also estimated that the probability of any two individuals having the same print on a single finger was vanishingly small. Galton's strong advocacy of fingerprints helped to bring them into use by Scotland Yard, but his classification system was cumbrous and difficult to apply, so it fell to Sir Edward Henry (1850–1931) to develop a workable system of fingerprint classification that would be generally useful in criminal identification.

In the years 1888–89, Francis Galton brought to completion what were in many ways his two most important works. Both were based largely on the great mounds of data he had collected in his Anthropometric Laboratory. The first, a paper published in the *Proceedings of the Royal Society* in 1888, described the correlations he had found in parameters like the length of the arm and leg in a specific individual. Thus, he observed, a person with long arms usually has long legs. In this same paper, Galton determined his first set of correlation coefficients. Galton's second major work, *Natural Inheritance* (1889), is the book that really launched biometrics. The text of *Natural Inheritance* follows a logical progression from Galton's view of heredity, to an essay on statistics (frequency distribution and normal variation), to his anthropometric data and, finally, to statistical analysis of the data. This was the book that inspired Francis Galton's two most devoted disciples, the great statistician Karl Pearson (1857–1936) and the marine biologist Walter F.R. Weldon (1860–1906). Pearson would take the statistical tools Galton had formed clumsily and laboriously and, almost effortlessly, transform and expand them to define many of the elements present today in the modern armamentarium of biometrics. Weldon, whose life was brought to a close prematurely by pneumonia, would take the methods that Galton and Pearson had developed and apply them successfully to data he had gathered on shrimps and crabs. Together, this triumvirate of mathematically inclined scientists would launch biometrics as a science and *Biometrika* as its flagship journal.

Galton also had another disciple, at least for a while, and that was William Bateson (1861–1926). Like Pearson and Weldon, Bateson read, and was impressed by, *Natural Inheritance*, but he also became deeply concerned with a very different problem, the nature of variation. Whereas biometrics concerned itself with continuously varying traits like height and length, Bateson had been mightily impressed by discontinuously varying traits like white and red flower colour. He collected hundreds of examples of discontinuously varying traits in his lengthy monograph *Materials for the Study of Variation* (Bateson, 1894). For Bateson, discontinuous variation seemed to pose a difficulty for Darwin's theory of evolution. It was not clear to Bateson how natural selection could act progressively on selected small alterations in a species when discontinuous variation seemed so much more likely a source of diversity on which the process could act. Galton was worried about the same problem, but for a very different reason. He had discovered, from his sweet pea and anthropometric data, the phenomenon of regression towards the mean. Thus, the progeny of tall people tended to be closer to the mean of the population as a whole. If this were the case, it seemed to Galton that evolution could never progress. Instead, he visualized evolution as proceeding by a sudden change in the equilibrium of a race – "a leap from one position of organic stability to another". His model was a polygon. If he nudged it slightly, it would return to the same position (regression to the mean), but if he pushed it slightly harder, it would land on another face, a new position of organic stability. Hence, Galton was lavish in his praise for Bateson's book for they both believed that evolution must proceed via some sort of salutatory process.

In 1897, Galton promulgated a new law of heredity which would be dubbed by Pearson "Galton's Law of Ancestral Inheritance" (Galton, 1897). The basic concept was quite simple. Galton contemplated a continuous series in which parents contributed one-half (0.5) of the total heritage of the offspring; the four grandparents one-quarter $(0.5)^2$; the eight great-grandparents one-eighth $(0.5)^3$ and so forth. By adding these contributions $(0.5) + (0.5)^2 + (0.5)^3$ . . . the whole series sums to 1. Galton also tried to apply his law in a quantitative sense, taking into account the mean and deviation of offspring from parents, parents from grandparents, etc., an approach which was greatly refined mathematically later on by Pearson. While Galton's law described the total, average genomic contribution of ancestors to progeny, it did not describe the fate of individual, discontinuously varying traits, which was one of the great beauties of Mendel's theory. These disparate theories of heredity led to a falling out between Bateson, one of Mendel's great defenders, and the biometricians which would not be resolved until R.A. Fisher, in a paper published in 1918, showed that continuously varying traits could be reconciled with Mendelian inheritance.

During the last decade of his life, Francis Galton was largely engaged in promoting eugenics through presentations and papers. The idea of eugenic improvement of the human race through selective breeding fitted well with the prevailing attitude favouring social Darwinism at the turn of the century. Consequently, eugenic concepts took root not only in England, but in Europe, particularly in France and Germany, and in the US which led the way in eugenic legislation until Hitler rose to power in 1932. Galton himself always thought of eugenics in a positive sense as a way of improving the human stock by concentrating on the breeding of the best and brightest. These were the ones to be encouraged to marry early and have plenty of children and, to do this, Galton felt that they, rather than the poor, should receive financial assistance. Galton rarely dealt with negative eugenics, the attempt to limit the propagation of the eugenically unfit, leaving this problem to others, while acknowledging that practising negative eugenics would probably abet the cause of his utopian notion of human improvement. There was one problem which Galton could never come to grips with successfully and that was how to classify eugenically fit women. All of his life he had concentrated on men, although he clearly understood, as his Ancestral Law shows, that men and women contribute equally in a genetic sense to their children. He simply did not know how to set up suitable criteria, except the obvious physical ones, for identifying eugenically fit women. Yet, he was, after all, a Victorian who lived his life largely in a man's world.

Galton was recognized by his peers throughout his career as a scientist of stature, but won unprecedented kudos in the last decade of his life when he was awarded both the Darwin and Copley Medals of the Royal Society, was knighted and was asked to write his autobiography, *Memories of My Life* (1908), a charming and readable account of an eminent Victorian scientist. There are two published biographies of Francis Galton. One is an enormous work by Karl Pearson, whose first volume was published in 1914 with the fourth and last appearing in 1930. The other is a much slimmer and more readable biography published by D.W. Forrest in 1974.

REFERENCES

Bateson, W. (1894) *Materials for the Study of Variation Treated with Especial Regard to Discontinuity in the Origin of Species*, London and New York: Macmillan; reprinted Baltimore: Johns Hopkins University Press, 1992

Darwin, C. (1868) *The Variation of Animals and Plants under Domestication*, London: John Murray, and New York: Judd; reprinted Baltimore: Johns Hopkins University Press, 1998

Darwin, C. (1871) Pangenesis. *Nature* 3: 502–3

Faulds, H. (1880) On skin furrows in the hands. *Nature* 22: 605

Fisher, R.A. (1918) The correlation between relatives on the supposition of Mendelian inheritance. *Transactions of the Royal Society (Edinburgh)* 52: 399–433

Forrest, D.W. (1974) *Francis Galton: The Life and Work of a Victorian Genius*, New York: Taplinger, and London: Elek

Galton, F. (1853) *The Narrative of an Explorer in Tropical South Africa*, London: John Murray

Galton, F. (1854) List of astronomical instruments, etc. In "Hints to Travellers." *Journal of the Royal Geographical Society* 24: 1–13

Galton, F. (1855) *The Art of Travel; or Shifts and Contrivances Available in Wild Countries*, London: John Murray

Galton, F. (1863) *Meteorographica, or Methods of Mapping the Weather*, London and Cambridge: Macmillan

Galton, F. (1865) Hereditary talent and character. *Macmillan's Magazine* 12: 157–66, 318–27

Galton, F. (1869) *Hereditary Genius: An Inquiry into its Laws and Consequences*, London: Macmillan, and New York: Appleton, 1870; reprinted London: Friedmann and New York: St Martin's Press, 1978

Galton, F. (1871) Pangenesis. *Nature* 4: 5–6

Galton, F. (1872) On blood-relationship. *Proceedings of the Royal Society* 20: 394–402

Galton, F. (1873) Hereditary improvement. *Fraser's Magazine* 7: 116–30

Galton, F. (1874) *English Men of Science: Their Nature and Nurture*, London: Macmillan, and New York: Appleton, 1875; reprinted London: Cass, 1970

Galton, F. (1875a) A theory of heredity. *Contemporary Review* 27: 80–95

Galton, F. (1875b) The history of twins, as a criterion of the relative powers of nature and nurture. *Fraser's Magazine* 12: 566–76

Galton, F. (1882) The anthropometric laboratory. *Fortnightly Review* 183: 332–38

Galton, F. (1883) *Inquiries into Human Faculty and Its Development*, London: Macmillan; reprinted London: Dent, and New York: Dutton, 1973

Galton, F. (1888) Co-relations and their measurement, chiefly from anthropometric data. *Proceedings of the Royal Society* 45: 135–45

Galton, F. (1889) *Natural Inheritance*, London and New York: Macmillan

Galton, F. (1892) *Finger Prints*, London and New York: Macmillan; reprinted New York: Da Capo Press, 1965

Galton, F. (1895) *Finger Print Directories*, London: Macmillan

Galton, F. (1897) The average contribution of each several ancestor to the total heritage of the offspring. *Proceedings of the Royal Society* 61: 401–13

Galton, F. (1908) *Memories of My Life*, London: Methuen and New York: Dutton, 1909

Herschel, W. (1880) Skin furrows of the hand. *Nature* 23: 76

Pearson, K. (1914–1930) *The Life, Labours and Letters of Francis Galton*, Cambridge: Cambridge University Press, 3 vols

C GENETICS OF BACTERIA AND VIRUSES

Introduction to Genetics of Bacteria and Viruses

Ever since the discovery of the microbial world in the 17th century, scientists have been fascinated by the ability of these tiny cells to grow and reproduce. It is only relatively recently, however, that the mechanisms by which bacterial cells replicate have been fully elucidated. It was not until 1942 that Salvador Luria and Max Delbrück first provided statistical evidence that inheritance in bacteria follows Darwinian principles, through studies on bacterial mutants. And the hereditary material of the cell was only demonstrated to be deoxyribonucleic acid (DNA), and not protein, by Oswald Avery, Colin MacLeod and Maclyn McCarty in 1944. Further significant occasions in the history of bacterial genetics include the discovery of bacteriophage by Fred Twort in 1915, and the discovery, by Joseph Lederberg during the 1950s, that phage are able to transfer genetic material from one bacterium to another.

In 1929, Alexander Fleming first described the effects of penicillin on Gram positive bacterial cells. The large-scale production of penicillin began in the 1940s, heralding the beginning of the antibiotic era. The development of antibiotic resistance has subsequently led to increased studies of bacterial genetics to identify the mechanisms by which antibiotic resistance is spread. Many answers have been provided through DNA sequencing, and it is hoped that the publication of bacterial genome sequences will help further in the design of future antibiotics.

The majority of information concerning bacterial genetics has been gained through studies of *Escherichia coli*. The complete sequence of an *E. coli* strain was first published in 1997. The approaches developed during sequencing of this genome, and the variety of other genome projects started since, have proved to be key in the progress towards sequencing of "higher" organisms, including humans.

While the *E. coli* strain sequenced in 1997 is harmless, related strains of *E. coli* are responsible for several human diseases; the genome of *E. coli* O157: H7, a toxic strain, has recently been sequenced (Perna *et al.*, 2001). Comparing it with that of a harmless strain will help suggest why some forms of this bacterium cause disease. We have learned many lessons from the study of bacterial genetics and our studies will continue to make significant contributions in the future.

REFERENCES

Perna, N.T., Plunkett, G., Burland, V., Mau, B., Glasner, J.D. *et al.* (2001) Genome sequence of enterohaemorrhagic *Escherichia coli* O157:H7. *Nature* 409: 529–33

BACTERIOPHAGES

Karin Carlson
Department of Cell and Molecular Biology, University of Uppsala, Uppsala, Sweden

1. Introduction

A thorough treatise and illustrations for the material covered in the first two sections of this review, as well as some background for the remaining sections, can be found in Snyder and Champness (1997) or other university textbooks in microbiology or molecular genetics. Calendar (1988) gives a comprehensive treatment of many different bacteriophages.

A bacteriophage (or phage) is a bacterial virus, an obligatory parasite capable of either killing its bacterial host while producing numerous copies of itself (lytic growth) or establishing a mutually beneficial symbiosis with its host (lysogeny). Phages contain genomes that may be either RNA or DNA, in either single-stranded (ss) or double-stranded (ds) form, linear or circular. The nucleic acid is surrounded by a protein capsid, formed as a cylinder (in filamentous phage, such as coliphage M13) or a regular or elongated icosahedron; some phages (such as PRD1) contain a lipid layer inside or outside the protein shell. Some phage particles (virions) are made up from just an icosahedral capsid surrounding the genome (such as coliphage ΦX174), while others carry a tail attached to the capsid which may be short (as for coliphage T7), long and contractile (as for the coliphage T4), or long and non-contractile (as for coliphages P1 and λ). Sometimes, appendages of various kinds are attached to the tail, and some virions also contain host-derived structures. Properties of some commonly studied coliphages (phages infecting *Escherichia coli*) are shown in **Table 1**.

The infection process starts by an interaction between virion adhesins (some capsid, tail or appendage component) and receptor(s) on the host cell surface, mediating attachment of the phage. Phages may utilize a variety of cell surface components as receptors, as exemplified in Table 1. This interaction usually determines the host range; intracellular development often can proceed in a larger spectrum of hosts than just those displaying the proper receptor(s) on their surface, as evidenced by successful phage development in spheroplasts (wall-less cells) of bacteria lacking receptors for that particular phage strain. After attachment, phage nucleic acids enter the host cell, through the tail of tailed phage. In most cases the capsid remains on the outside, but some phages (e.g. M13) one or more of the capsid components also enter(s) the cell. Linear phage molecules may circularize once inside the cell (e.g. λ) or replicate as linear molecules (e.g. T4, T7); no RNA genomes are known to circularize.

Table 1 Some properties of commonly used coliphages.

Phage	capsid[a]	life style[b]	nucleic acid[c]	genome size[d] kb	genome size[d] no. genes	host receptor[e]
MS2	I	L	ssRNA	**4**	**4**	side of F-pili[f]
M13	F	L[g]	ssDNA, circular	**6**	**10**	tip of F-pili[f]
ϕX174	I	L	ssDNA, circular	**5**	**11**	LPS core
PRD1	I[h]	L	dsDNA, linear	**15**	**22**[i]	transfer apparatus encoded by plasmids of P,N or W compatibility[f]
Mu	T	T	dsDNA, linear	**37**	**53**	terminal glucose of LPS core, 1,2-glycosidic linkage
T7	T	L	dsDNA, linear	**40**	**60**	LPS core
λ	T	T	dsDNA, linear	**49**	**67**	LamB
P1	T	T	dsDNA, linear	100	100[j]	terminal glucose of LPS core
T4	T	L	dsDNA, linear	**169**	**270**[k]	diglucosyl LPS core residues or OmpC

[a] I, icosahedral; F, filamentous (cylindrical); T, tailed (icosahedron + cylindrical appendage)
[b] L, lytic; T, temperate
[c] dsRNA is found for instance in *Pseudomonas phaseolicola* phage ϕ6, which has a genome consisting of three linear dsRNA molecules (and a lipid membrane surrounding the protein capsid), and circular dsDNA (and lipids) in phage PM2 infecting *Alteromonas espejiano*
[d] bold-faced numbers are from complete genome sequences (http://www.ncbi.nlm.nih.gov/PMGifs/Genomes/phg.html)
[e] the outermost layer of *Escherichia coli* cell walls is a membrane ("outer membrane") containing a lipopolysaccharide (LPS) facing outwards, and inserted proteins, such as porins permitting membrane transport. OmpC, outer membrane porin C; LamB, maltoporin (maltose transport protein)
[f] many large plasmids encode pili, appendages of bacterial cells that permit plasmid transfer from cell to cell. Plasmid compatibility is determined by the mechanism of regulation of DNA replication; plasmids belonging to the same incompatibility group (e.g. F, P, N or W) control their replication the same way and mutually exclude each other
[g] phage released by budding through cell wall, no lysis
[h] lipid layer inside capsid
[i] open reading frames; nine genes characterized
[j] number based on expected coding capacity; less than half defined by mutations
[k] open reading frames; about 130 genes have been characterized

During lytic growth of phage, early expression immediately after infection typically includes catalytic proteins required for phage macromolecule synthesis and proteins mediating host shut off and killing. After this follows replication of the phage genome and late expression (dependent on some early expression) resulting in production and assembly of virion components, filling these with genomes in an assembly-line process and finally lysing the cell to liberate the progeny. Some 20–1000 new virions (different numbers depending on what phage infected what host under what external conditions) are produced and the host cell is destroyed, often in less than an hour, but sometimes with considerable delay.

Upon lysogenization, the phage genome is largely shut off – often integrated in the host chromosome (e.g. λ, Mu) but sometimes forming a plasmid (e.g. P1) – and is called a prophage. A few prophage genes are active, notably one producing a repressor shutting off expression from most other phage genes; other active prophage genes may produce toxins, such as diphtheria and botulinum toxins, and other bacterial virulence factors. As a result of prophage expression, the lysogenic cell is immune to infection by similar phages; sometimes, other phages are excluded as well. A prophage may become induced by conditions adversely affecting the survival of its host, such as irradiation, and initiate lytic growth. Phages capable of both lytic growth and lysogeny are termed temperate, while lytic, or virulent, phages are only capable of lytic growth. After infection with a temperate phage, the choice of lytic growth or lysogeny is the result of an elaborate balance between pathways leading to the expression of lytic genes and pathways leading to their shut off.

The tight and elaborate regulatory systems made phage interesting to molecular geneticists aiming to solve the riddle of how genes work. These studies were facilitated by the small size of phage genomes, enabling more detailed studies than were possible with cellular organisms.

2. Phage and the development of molecular genetics

Starting in the 1940s, primarily the temperate phage λ and the lytic phage T4, both dsDNA phages infecting *E. coli*, were used to analyse in detail how gene expression is regulated, how DNA molecules replicate and recombine, and how proteins are folded and assembled into large structures (see Cairns *et al.*, 1966). Genes, their RNA messages and many features of the genetic code were defined using phage systems, as well as much of the work determining colinearity between genes and their protein products, leading to our current view of a gene as the sequence encoding a peptide chain, and the mechanisms of recombination as involving breakage and exchange as well as copying and degradation.

Lysogeny of phage λ is established and maintained through finely tuned molecular interactions between phage and host functions (see Ptashne, 1992). Immediately after infection only the two lytic promoters, p_L and p_R, are available for transcription. One of the proteins produced from the p_R-initiated transcript is CII, which is needed to start transcription of the repressor gene, *cI*. Once repressor is made, it binds at operator sites O_L and O_R that overlap p_L and p_R, and thereby prevents further transcription from these promoters. Binding of repressor at O_R activates another *cI* promoter that now ensures the stable production of repressor and maintenance of lysogeny. CII is a very unstable protein, subject to proteolysis by host-encoded proteases whose activities, in turn, are tuned by the metabolic state of the host. If a cell is infected with several phages at the same time, as is the case when the phage has been growing lytically on a bacterial population for a while so that few cells remain, relatively more copies of CII will be present in the cell as compared with singly infected cells, favouring lysogeny. Thus, establishment of lysogeny is keyed to both host metabolism and phage–host ratio, steering the phage towards lysogeny when the prospects of propagating successfully by lytic growth are dim, either because the host is starving or because there are too few uninfected cells around for the progeny to infect. Lytic growth of λ ensues when too little CII is available to turn on efficient repressor synthesis, through another cascade of RNA molecules and proteins interacting with each other and DNA sequences.

Although the repressor–operator concept had been established earlier in a bacterial system – the *lac* operon encoding three proteins required for lactose catabolism – the λ system enabled further refinement of this concept and an understanding of how molecular interactions among DNA and RNA sequences and proteins can regulate gene expression. Molecular details of protein–DNA interactions were worked out in this system, leading to the first general model of such interactions, the helix–turn–helix model. The λ repressor was shown to contain two λ-helical segments, separated by a short random coil, that bind to DNA. One of these helical segments binds in the major groove of the DNA, contacting the bases and providing the sequence specificity of the interaction. The other helical segment contacts mainly phosphates of the sugar–phosphate backbone of the DNA, adding stability but little specificity to the interaction. Many other proteins interact with DNA in the same way, including not only bacterial repressors but also proteins regulating eukaryotic gene expression. By now, many different interfaces between proteins and DNA have been identified, positioning both α-helical segments and β-sheets in contact with DNA bases.

Studies of lytic λ development led to the elucidation of factor-mediated transcription antitermination. As mentioned above, lytic development is a cascade of events that necessitates very elaborate regulation of gene expression. The *E. coli* transcription termination factor ρ was identified through its effect on λ transcription. In the presence of ρ, λ transcripts initiated at the lytic promoter p_L terminate shortly after the *N* gene, which limits expression to promoter-proximal genes. The product of the *N* gene interacts with the RNA polymerase at a particular RNA sequence, permitting transcription to proceed further. The N protein also antiterminates p_R-initiated transcription, permitting expression of a gene whose product enables expression of the late genes through yet another antitermination event.

In comparison with this, the regulatory repertoire of the lytic coliphage T4 (reviewed in Karam *et al.*, 1994) is more complex, with several overlapping regulatory circuits. Several products of early T4 genes contribute to switching the expression pattern, by covalently modifying and unmodifying subunits of the RNA polymerase, by binding to RNA polymerase subunits to alter their activity and affinity for other components of the transcriptional apparatus, or binding to DNA and thereby altering the interaction of RNA polymerase with promoters. Later modifications, including a new σ factor (the promoter-recognizing RNA polymerase subunit), are essential for the transcription of late genes. On top of this, post-transcriptional regulation plays important roles in fine tuning of the differential gene expression during growth. These mechanisms include mRNA splicing, translational repression, selective mRNA processing and degradation and post-translational protein modification.

All ssRNA phages contain (+)-strand RNA – that is, the phage genome serves as mRNA immediately after infection. Here, the temporal sequence of gene expression is regulated through RNA secondary structure. Work on these phages has yielded important understanding of translational regulation.

Molecular chaperones, proteins whose job it is to fold nascent peptide chains into functional proteins, refold damaged proteins and target badly damaged proteins for degradation, were defined through studies of assembly of λ capsids and, later, also T4 capsids. Studies of the assembly of these virions provided insights into the very complex, largely autocatalytic pathways of assembly of large protein complexes.

Around the middle of the 1970s, eukaryotic systems, including eukaryotic viruses, had been developed to the point where they could be used to address many of the questions previously possible to address only with phage. However, the unique properties of phage maintain them as interesting genetic systems and important genetic tools. Over the past 10 years, increasing insights into phage biology has opened up entirely new avenues of phage research.

3. Phage as genetic tools

From the 1970s, the potential of phage as tools in recombinant DNA work and analyses of gene function in other systems was apparent. Restriction endonucleases, the basic tools of the recombinant DNA era, were discovered and elucidated in phage systems. Perhaps because of the selection for very efficient pathways in lytic growth, phage enzymes such as RNA and DNA ligases and polynucleotide kinase are quite efficient and often the choice for recombinant DNA work. Some applications capitalizing on the very nature of phage are summarized below.

A phage is an excellent tool for the transport of DNA between bacteria since its life cycle involves being produced in one bacterium and then introducing its genome into another one. Some temperate phages may bring with them pieces of bacterial DNA adjacent to their chromosomal integration sites, through faulty excision on induction of the lysogen (specialized transduction, e.g. λ); other phages (both lytic and temperate) may simply pack bacterial DNA instead of viral DNA (generalized transduction, e.g. T4 and P1). The temperate phage Mu, a transposon in the clothes of a phage, repeatedly inserts and excises itself in the bacterial chromosome during its lytic growth in the same way that transposons integrate and excise. Variants of Mu lacking the ability to replicate autonomously (so-called Mud elements) can be used to ferry DNA segments into any location of the bacterial chromosome.

Cosmids are the vehicles of choice (together with artificial chromosomes) for cloning large DNA segments, such as bacterial pathogenicity islands, and for construction of genomic libraries. Cosmids are plasmids that contain those sequences (*cos*) of a temperate phage that are recognized by the maturase that packs DNA into phage heads, together with a replicative origin permitting independent replication. DNA cloned between two *cos* sites can then be packed *in vitro* into infective particles that can be used to deliver the construct into a bacterial cell where it can be propagated as a plasmid. The amount of DNA that can be cloned depends on the size of the phage head; cosmids based on phage λ take about 50 kilobases (kb) of cloned DNA while those based on phage P1 may take as much as 85 kb.

The ssDNA of coliphage ΦX174 was the first complete genome sequenced (Sanger *et al.*, 1977), and coliphage M13 subsequently led the way into the DNA sequencing era. This filamentous phage packs circular ssDNA, but replicates intracellularly as a double-stranded plasmid. Thus, DNA fragments can be cloned into M13 DNA using the same techniques as for plasmid cloning and then sequenced from the viral ssDNA. Although this system is less frequently used now, many preferring instead to sequence from PCR fragments, many vectors designed for easy sequencing of inserts still use the M13 primers developed for the original phage system.

Since the evolution of phage undoubtedly has involved selection for efficient production, many phage promoters are quite strong. At the same time, the need to control tightly the switch between lytic and lysogenic development of temperate phage has selected very precise repression mechanisms. In expression vectors utilizing the lytic p_L promoter of phage λ, phage expression can be modulated by the λ repressor. Some small DNA phage (such as coliphage T7) encode their own RNA polymerases, recognizing promoters very different from those of cells. This has been utilized in the generation of a very successful series of expression vectors placing the gene to be expressed after a promoter from phage T7 which is engineered so that it is regulated by the Lac repressor from *E. coli*. A gene encoding T7 RNA polymerase is also placed under Lac repressor control, so that induction (lifting Lac repression) results in expression from the T7 promoter (see Studier *et al.*, 1990).

A very useful application of the T7 regulatory system is in SELEX (systematic evolution of ligands by exponential enrichment), which was originally developed to study the affinity of coliphage T4 DNA polymerase for its translational operator (Tuerk & Gold, 1990). This method involves selecting from a heterogeneous population of nucleic acids those species that bind a particular protein. The nucleic acid population is chemically synthesized with a variable region, flanked by two regions of invariant sequence, one of which contains a T7 promoter. Those variant nucleic acid species that bind to the protein are separated from free nucleic acids and amplified by *in vitro* transcription with T7 RNA polymerase. This selection is repeated until no further increase in protein affinity to the target is seen or the variable region becomes non-random.

Phage display utilizes the unique composition of phage capsids, which contain only proteins. As originally described by Smith (1985), a DNA sequence is cloned in the middle of the gene encoding a minor capsid protein of a filamentous DNA phage, so as to generate a translational fusion. The fusion protein will then become incorporated into the virion, which retains infectivity and displays the foreign peptide on its surface in an immunologically accessible form. The usefulness of this concept for combinatorial chemistry and the identification of receptor–ligand interactions recently made it the topic of an entire volume of *Methods in Enzymology* (Abelson, 1996). Today, other phage are also used successfully for phage display analyses.

4. Modern phage genetics

To give a glimpse of ongoing phage genetics research, a number of areas will be highlighted where phage today provide important insights into fundamental problems. These will deal predominantly with coliphages which are still the main subject of such research.

(i) DNA maintenance and proliferation

Evolution has selected very efficient replicative machineries in phage, and thereby provided researchers with very efficient tools to dissect the different mechanisms used for replication initiation as well as for producing clean, packable DNA out of branched recombinants. Coliphage T4 provided the first system where the role of recombination in DNA replication could be studied in detail. Later, recombination has been found to be important for chromosome segregation in eukaryotes and to be involved in the initiation of DNA replication in bacteria, mitochondria, chloroplasts and yeast. However, phage T4 continues to reveal the multitude of different pathways and mechanisms that entwine replication, recombination and DNA repair molecules, shaping a view that these processes are just different aspects of general DNA maintenance (reviewed by Mosig, 1999).

Soon after infection, T4 DNA replication is initiated at one or more of several independent replicative origins (the choice of initiation site(s) is dependent on growth conditions), dependent on host RNA polymerase and phage primase. Phage-induced modifications of the RNA polymerase soon disable this function and several different pathways, involving different functions and operating at different times after infection, enable recombinational initiation of T4 DNA replication and, at the same time, mediate genetic exchange and repair of DNA damage. There are also several pathways for resolution of the resulting branched intermediates to produce linear molecules that can be packed into new virions. As a result, recombination is extremely efficient in this phage, contributing to the intensive horizontal exchange evident from genome analysis.

The self-splicing group I introns of coliphage T4 were the first such elements discovered in prokaryotic systems; later, such introns were also found in phage infecting several Gram-positive bacteria. Since the introns are unevenly distributed among T4-like phage, it is most likely that they have entered the genomes through horizontal transfer fairly recently. Intron-encoded homing endonucleases spread the intron through a double-strand break–repair mechanism that has been analysed in considerable molecular detail (e.g. Derbyshire *et al.*, 1997). The best-studied phage homing endonuclease, I-*Tev*I from the T4 *td* intron, belongs to a large family of endonucleases involved in recombination and repair that also includes, for instance, the bacterial UvrC excisionase.

(ii) Molecular machines

One strong focus of current research is macromolecular interactions, from structural as well as functional viewpoints. As a result of the efficiency and variety of phage systems, and the ease of their genetic dissection, they are much used as model systems in this area.

Bacillus subtilis phage Φ29 and its relatives, e.g. coliphage PRD1, pioneered the studies of protein priming mechanisms that still contribute important knowledge on how DNA polymerases interact with their substrates (see de Vega *et al.*, 1999). The replication apparatus of coliphage T4 is probably the simplest replication system including all the separate assemblies present in cellular systems, and continues to serve as a useful paradigm in elucidating the mechanics of DNA replication. Recent determinations of the structures of some of the components (e.g. Moarefi *et al.*, 2000) increase the usefulness of this approach even more. Analyses of T4 recombination mechanisms are aided by the known structure of the Holliday resolvase, endonuclease VII and the structural and genetic dissection of the packing machine where it participates.

The transcription machinery in T4-infected cells is subjected to extensive mutational dissection, with the surprising result that none of the early events is essential for the switch in promoter recognition that takes place soon after infection (e.g. Pene & Uzan, 2000). Instead, multiple pathways lead to the switches observed, as discussed above for recombination mechanisms. This bewildering array of seemingly redundant pathways has led to the concept of interdigitated three-dimensional regulatory networks rather than two-dimensional cascades, finely tuned to permit maximal efficiency under what must be varying natural conditions, and thereby has led the way to dissecting eukaryotic developmental regulation.

Other phage proteins recently subjected to structural analyses, through NMR or X-ray crystallography are T4 lysozyme, Mu repressor, λ Cro repressor and many structural proteins of different phage.

Molecular machines involve not only proteins and DNA, but also RNA participating directly in various regulatory and catalytic events. The λ-related temperate coliphage HK022 provides an interesting system where regulatory RNA can be studied in molecular detail, since this phage does not use a protein to antiterminate its early transcription (see description of the λ system above), but instead uses an RNA molecule (review by Weisberg *et al.*, 1999).

5. Phage genomics

Several thousand different phage infecting different bacteria have been described. As of August 2000, 71 complete bacteriophage genomes, 40 of which are non-*coli* phages, and, in addition, two genomes of archaeal viruses, have been sequenced and deposited in public databases (http://www.ncbi.nlm.nih.gov/PMGifs/Genomes/phg.html). A number of unknown prophage genomes likely dwell in the 33 (as of August 31, 2000) completely sequenced microbial genomes. Studies of these genomes enforce the concept of mosaic evolution of phage: when comparing any two phage, regions of obvious sequence similarity are interspersed with segments that are apparently unrelated. Even phage lacking discernible sequence similarity at either the nucleic acid or protein level may show similarities in genetic or protein structure (e.g. Casjens *et al.*, 1992; Hendrix *et al.*, 1999). Such similarities most likely reflect common ancestry, since genetic homogenization through recombination (which would produce a similar result) cannot be very frequent among genomes that lack sequence similarity. Structural similarities extend to animal viruses (e.g. Belnap & Steven, 2000), suggesting phylogenetic relatedness among all viruses and an origin of viruses coincident with or predating the origin of the three domains of life. This, and the occurrence of phage-like RNA polymerases in cellular organelles, themselves early bacterial endosymbionts, raises interesting questions concerning which reproductive concept arose first: the viral assembly-line process or cellular binary fission.

Being a genetic parasite, the natural sexual partner for a phage is its host, not other phage (except those coincidentally infecting the same host). Yet, as mentioned above, analyses of phage genomes reveal closely similar sequences in phage infecting very distantly related hosts not believed to exchange genetic material naturally, interspersed by unrelated sequences. The only plausible explanation for this genetic mosaicism is ongoing extensive horizontal exchange. This can be viewed as a sort of random walk through phylogenetic space, where sequences travel through phage with overlapping host ranges; favouring this interpretation, sequence similarities between phage infecting phylogenetically distal hosts are at a modest level of identity. On the other hand, even sequences originating in eukaryotic organisms (such as fungi and green algae) appear to have made their way into phage, suggesting exchange mechanisms not based on coinfection. One possibility is random uptake into bacteria of DNA liberated from dying organisms of different kinds, DNA that, once having entered a bacterial cell, can spread to its viruses.

The efficiency of horizontal exchange among phage extends also to other DNA elements. Bacterial virulence factors, such as pathogenicity islands and genes encoding toxins, commonly reside in temperate phage genomes and are expressed from repressed prophages. They may spread to different bacteria through the random walk procedure described above, move to plasmids permitting spreading through different mechanisms or take up permanent residency in bacterial chromosomes either by loss of lytic phage genes that would enable escape from the host or through recombination. As an example, the virulence cassette including the gene for cholera toxin resides in the genome of a temperate *Vibrio* phage (Waldor & Mekalanos, 1996), while homologous genes are located on plasmids in enterobacteria.

6. The biology of phage

Little of the work discussed above has any relevance for how phage live in nature. Until the late 1970s, the common contention was that phage were of low abundance and played insignificant roles in nature. At the current time, however, new methods (electron microscopy and, somewhat later, various PCR-based methods) led to the surprising conclusion that phage are about one order of magnitude more abundant than bacteria in natural ecosystems, and that phage and other viruses, therefore, are probably the most abundant group on Earth. Phages have a significant role in controlling bacterial populations; on the average, 10–30% of all bacterial cells in a natural aquatic population of bacteria are infected by phage at any one time (see review by Fuhrman, 1999). Since phages frequently show quite narrow host range, blooms by a single type of bacteria are easily controlled by phage predating on this particular strain; this has also been demonstrated in natural systems. This phage predation is self-limiting, in that at low bacterial concentrations the likelihood of a phage finding a new host is reduced. This way, phage predation may actually be an evolutionary advantage to bacterial communities, allowing other bacteria to feed on released bacterial macromolecules rather than having them removed from the bacterial community through grazing by protozoa.

The potential use for phage predation in controlling human pathogens was realized early after phage were discovered in the early 1900s. At this time, little was known about phage biology, and results were unpredictable and irreproducible. Phage therapy was abandoned in the Western world soon after the discovery of antibiotics in the 1930s, but was used continuously and eventually with considerable success in eastern Europe and the Soviet Union (review by Alisky *et al.*, 1998). Due to increasing problems with antibiotic resistance in bacteria, people are again looking at phage therapy in the West, now with considerably better knowledge of phage biology and with the long experience in the East as a solid foundation. Some pilot experiments with animal models have been carried out, generally with good results and improvements of key factors such as targeting virulence determinants of bacteria and developing new phage variants capable of longer survival or broader host specificity (e.g. Merril *et al.*, 1996), but much work remains to be done before phage–host interactions are understood to the point that phage therapy can become routine.

FURTHER READING

Space constraints prevent full acknowledgement of all sources of information and inspiration for this review; the choices here are to some extent arbitrary in that many other papers could have been quoted for the same criteria of interest and pertinence.

Original literature and reviews

Alisky, J., Iczkowski, K., Rapoport, A. & Troitsky, N. (1998) Bacteriophages show promise as antimicrobial agents. *Journal of Infection* 36: 5–15

Belnap, D.M. & Steven, A.C. (2000) "*Déjà vu* all over again": the similar structures of bacteriophage PRD1 and adenovirus. *Trends in Microbiology* 8: 91–93

Casjens, S., Hatfull, G. & Hendrix, R. (1992) Evolution of dsDNA tailed-bacteriophage genomes. *Seminars in Virology* 3: 383–97

de Vega, M., Blanco, L. & Salas, M. (1999) Processive proofreading and the spatial relationship between polymerase and exonuclease active sites of bacteriophage Φ29 DNA polymerase. *Journal of Molecular Biology* 292: 39–51

Derbyshire, V., Kowalski, J.C., Dansereau, J.T., Hauer, C.R. & Belfort, M. (1997) Two-domain structure of the *td* intron-encoded endonuclease I-*Tev*I correlates with the two-domain configuration of the homing site. *Journal of Molecular Biology* 265: 494–506

Fuhrman, J.A. (1999) Marine viruses and their biogeochemical and ecological effects. *Science* 399: 541–48

Hendrix, R.W., Smith, M.C.M., Burns, R.N., Ford, M.E. & Hatfull, G. (1999) Evolutionary relationships among diverse bacteriophages and prophages: all the world's a phage. *Proceedings of the National Academy of Sciences USA* 96: 2192–97

Merril, C.R., Biswas, B., Carlton, R. *et al.* (1996) Long-circulating bacteriophage as antibacterial agents. *Proceedings of the National Academy of Sciences USA* 93: 3188–92

Moarefi, I., Jeruzalmi, D., Turner, J., O'Donnell, M. & Kuriyan, J. (2000) Crystal structure of the DNA polymerase processivity factor of T4 bacteriophage. *Journal of Molecular Biology* 296: 1215–23

Mosig, G. (1999) Recombination and recombination-dependent DNA replication in bacteriophage T4. *Annual Reviews of Genetics* 32: 379–413

Oaks, E.V., Wingfield, M.E. & Formal, S.B. (1985) Plaque formation by virulent *Shigella flexneri*. *Infection and Immunity* 48: 124

Pene, C. & Uzan, M. (2000) The bacteriophage T4 anti-sigma factor AsiA is not necessary for the inhibition of early promoters *in vivo*. *Molecular Microbiology* 35: 1180–91

Sanger, F., Air, G.M., Barrell, B.G. *et al.* (1977) Nucleotide sequence of bacteriophage ΦX174 DNA. *Nature* 265: 687–95

Smith, G.P. (1985) Filamentous fusion phage: novel expression vectors that display cloned antigens on the virion surface. *Science* 228: 1315–17

Studier, F.W., Rosenberg, A.H., Dunn, J.J. & Dubendorff, J.W. (1990) Use of T7 RNA polymerase to direct expression of cloned genes. *Methods in Enzymology* 185: 60–89

Tuerk, C. & Gold, L. (1990) Systematic evolution of ligands by exponential enrichment: RNA ligands to bacteriophage T4 DNA polymerase. *Science* 249: 505–10

Waldor, M.K. & Mekalanos, J.J. (1996) Lysogenic conversion by a filamentous phage encoding cholera toxin. *Science* 272: 1910–14

Weisberg, R.A., Gottesmann, M.E., Hendrix, R.W. & Little, J.W. (1999) Family values in the age of genomics: comparative analysis of temperate bacteriophage HK022. *Annual Review of Genetics* 33: 565–602

Books

Abelson, J.N. (1996) *Combinatorial Chemistry*, San Diego: Academic Press

Cairns, J., Stent, G.S. & Watson, J.D. (eds) (1966) *Phage and the Origin of Molecular Biology*, Cold Spring Harbor, New York: Cold Spring Harbor Laboratory Press; expanded edition 1992

Calendar, R. (1988) *The Bacteriophages*, 2 volumes, New York and London: Plenum Press

Karam, J.D., Drake, J.W. & Kreuzer, K.N. *et al.* (eds) (1994) *Molecular Biology of Bacteriophage T4*, Washington, DC: American Society for Microbiology Press

Ptashne, M. (1992) *A Genetic Switch*, 2nd edition, Cambridge, Massachusetts: Blackwell Scientific and Cell Press

Snyder, L. & Champness, W. (1997) *Molecular Genetics of Bacteria*, Washington, DC: American Society for Microbiology Press

GLOSSARY

capsid the protein shell of a virus

cosmid cloning vector that consists of a bacterial plasmid into which the *cos* sequences of phage have been inserted, permitting packing of vector plus insert into a phage capsid

lysogeny the presence in a host bacterium of a phage genome (prophage) that does not undergo lytic growth

lytic growth replication cycle resulting in the release of multiple new phages, usually by lysis (breaking open the host cell wall)

pathogenicity island DNA segment containing genes that confers on a bacterium a variety of virulence traits

prophage phage chromosome residing in a bacterial cell without undergoing lytic growth

repressor protein that suppresses the expression of a gene or group of genes

temperate phage phage that on infection can establish either lysogeny (forming a prophage) or lytic growth

transduction transfer of genes from one bacterium to another using a phage as the vector

virion a virus particle

virulent (or lytic) phage phage that on infection always grows lytically

INFLUENZA

Anne E. Hill
Moredun Research Institute, Pentlands Science Park, Edinburgh, UK

1. Introduction

Twenty million people died during the great influenza pandemic of 1918. Nearly a century later, an estimated 20 000 people in the US still die each year because of influenza or influenza-related pneumonia, and influenza and pneumonia together are the sixth most common cause of death in the US. In this article, we will examine the molecular biology of the parasite responsible for these horrifying statistics and the steps that have been taken to control its impact. First, however, we will look at the events that have marked influenza out as such a formidable pathogen.

2. History and origins

The first recorded epidemic of influenza occurred in 1173. Since then, over 300 epidemics have been described, a rate of approximately one every 2½ years. Epidemics by definition are confined in space and time, and mortality rates vary between 5–30 000 per outbreak. More seriously, global epidemics, or pandemics, kill millions of people. Over the past 2½ centuries between 10 and 20 influenza pandemics have swept the globe, the most dramatic of which was "Spanish 'flu" in 1918. The 1918 outbreak was the biggest and most lethal pandemic in history, causing more than 20 million deaths and affecting more than 200 million people.

In the period from 1920 onward, a further three pandemics have been recorded: Asian 'flu in 1957, Hong Kong 'flu in 1968 and Russian 'flu in 1977. Mortalities in 1957 and 1968 were nearly 6 million, although fewer people were affected in 1977.

According to phylogenetic data, the influenza virus which caused the 1918 pandemic originated between 1905 and 1914 from an avian source (Taubenberger *et al.*, 1997; Oxford *et al.*, 1999). It is thought to have then passed the species barrier to pigs and then to humans. The general agreement for the cause of the rapid global spread is that the American Expeditionary Force carried the virus to France in 1918 during World War I. The subsequent movement of troops during and after the war contributed to the spread of influenza. There had also been widespread scattered episodes during the preceding year in China and Hungary (Reid *et al.*, 1999). The 1957 Asian pandemic had its origin in the province of Kweichow, China. The 1968 and 1977 pandemics are similarly thought to have originated in China.

The origin of viruses in general is poorly understood, but it is presumed that each virus began "life" by copying a few useful genes from their host's genome. These stolen genes are most often genes involved in the host cells' reproduction, energy uptake or communication. Over generations, viruses then evolve to maximize their own survival rate.

A number of questions are generated from looking at the history and origin of influenza viruses. For instance, what are influenza viruses and what is it that makes them able repeatedly to cause such devastating outbreaks? To answer these questions, we first need to look at the genetic make-up and then at the structure of influenza viruses.

3. Genetic make-up of influenza viruses

Viruses are minimalist in design, consisting of nucleic acid (either DNA or RNA) contained within a lipoprotein coat of approximately 20–300 nm in diameter. As a result of their simplicity, viruses are obligate intracellular parasites that take over the cells they infect by substituting their own genes for those of their host. This causes the host machinery to produce viral rather than host gene products.

Influenza viruses belong to the Orthomyxovirus class of viruses that contain RNA as their genetic material. RNA viruses contain an enzyme called reverse transcriptase that permits the usual sequence of DNA to RNA to be reversed so that the virus can make a DNA version of itself. Influenza viruses are divided into three types – A, B and C – based on various proteins found within the virus: type A causes the worldwide epidemics of influenza and can infect other mammals and birds; type B infects only humans and causes local outbreaks; and type C causes only a mild infection in humans.

The influenza genome is unusual in that the genes are distributed over a number of segments of antimessenger RNA. In type A and type B viruses, there are ten genes encoded on eight segments of RNA, whereas in type C viruses there are nine genes encoded on seven segments (Lamb, 1989). The genes encoded in the influenza genome are as follows.

(i) Segments 1–3: proteins PB1, PB2 and PA

Each of the RNA segments 1–3 codes for a single gene. Segments 1 and 2 are 2341 nucleotides in length and code for proteins of 759 (PA) and 757 (PB1) amino acids, respectively. Segment 3 is 2233 nucleotides in length and codes for a protein, designated PB2, of 716 amino acids. All three proteins are involved in RNA transcription (Hiromoto *et al.*, 2000): protein PB1 possesses RNA polymerase catalytic activity and PB2 is involved in the binding, cleavage and recruitment of cellular mRNA cap-1 structures essential for viral mRNA synthesis. The function of PA, on the other hand, is not yet fully elucidated, but it is possibly involved in switching from mRNA synthesis to cRNA synthesis during replication.

(ii) Segment 4: haemagglutinin

The protein coded for on segment 4 is one of the most studied integral membrane proteins in virology and cell biology. It was first identified because of its ability to cause red blood cells to agglutinate and was subsequently called "haemagglutinin" (HA). The segment encoding HA is 562–566 amino acids in length, producing a protein with a molecular weight of 77 000 Daltons.

HA is a glycoprotein consisting of two polypeptides, HA_1 and HA_2, held together by disulfide bonds (Perdue & Suarez, 2000). The HA_1 part of the molecule contains at least four variable antigenic sites, a feature, which we will see later, that plays an important role in the virulence of type A influenza viruses. A total of 15 different HAs have been identified in avian species; five (H1, H2, H3, H5 and H9) have been recovered from humans, but only three of these (H1–3) are known to have caused pandemics in humans.

HA has three major functions. First, it binds to sialic acid-containing receptors found on the surface of host cells, helping to establish the attachment of the infectious virus particle to the plasma membrane of the host cell. Secondly, HA initiates the infection process by mediating the fusion of the endocytosed virus particle with the endosomal membrane, thus liberating viral cores into the cytoplasm. Lastly, HA is the major antigen of the virus against which neutralizing antibodies are produced. As we shall see later, recurrent influenza epidemics are associated with changes in its antigenic character.

(iii) Segment 5: the nucleocapsid protein

The nucleocapsid protein is a major structural protein that interacts with a viral RNA segment to form a ribonucleoprotein particle (RNP). The RNP structures contain the individual genome RNA segments bound on the outside with the 3′ and 5′ ends base paired. Associated with each RNP are the three transcriptase-associated polypeptides, PA, PB1 and PB2. Each RNP contains a promoter for transcriptase activity and each of the RNPs has transcriptase activity. The nucleocapsid protein itself is 498 amino acids in length, encoded on an RNA segment of 1565 nucleotides (Lamb, 1989).

(iv) Segment 6: neuraminidase

Segment 6 encodes the glycoprotein neuraminidase (NA). NA is a major "spike" protein on the surface of the influenza virion. It is made up of four monomers of approximately 50 000 Daltons. Each of the monomers is linked by disulfide bonds to make up a mushroom structure that projects out of the viral envelope. NA catalyses the cleavage of the α-ketosialic linkage between a terminal sialic acid and an adjacent D-galactose/D-galactosamine (Colman, 1989). This activity is thought to free the virus from sialic acid-containing structures found on host cells. The function of the NA is thought to be to permit the subsequent transport of the virus through mucin in the respiratory tract, enabling it to reach target epithelial cells and also to destroy HA receptors in the host cell, thereby aiding the elution of progeny viruses. NA is a member of a group of internal membrane proteins that have an uncleaved N-terminal extended signal domain that both targets the protein to the membrane during synthesis and anchors the protein in the lipoprotein bilayer.

A total of nine different NA sequences have been identified in avian strains of influenza, but only three of these have been identified in humans. As with variation in HA, this plays a major role in the virulence of type A influenza viruses. Influenza C viruses do not contain an RNA segment that encodes NA activity. The functions performed by NA in type A and type B viruses are assumed to be taken over by HA, which exhibits esterase activity.

(v) Segment 7: M_1 and M_2 proteins

Segment 7 encodes two known proteins: M_1, which is the most abundant polypeptide in the virion, and M_2, which is expressed at the cell surface. Three RNA transcripts are derived from segment 7: a collinear transcript encoding M_1, a spliced mRNA encoding M_2 and an alternatively spliced mRNA that has the potential to encode M_3 – a small peptide for which there is no evidence of existence *in vitro*.

M_1 is 252 amino acids in length with a molecular weight of 27 801 Daltons. It is a matrix protein responsible for maintaining the structural integrity of the virion. M_1 protein is found underneath the lipid bilayer, adding rigidity to the whole structure. M_1 plays a crucial role in recruiting viral RNPs to the plasma membrane as a prerequisite for virus assembly.

A second open reading frame (ORF) which encodes the second matrix protein, M_2, overlaps the M_1 ORF by 68 nucleotides. The initiation codon and subsequent eight amino-terminal amino acids of M_1 and M_2 are predicted to be common as they are encoded before the 5′ splice junction of the M_2 mRNA. The remaining 88 amino acids of the M_2 protein are predicted to be translated from the second ORF. M_2 is a cation ion channel. A secondary function is thought to be to assist in the assembly of the virion by chaperoning the M_1 protein either in transport to the cell surface or in the formation of the virus particle.

(vi) Segment 8: non-structural proteins NS_1 and NS_2

The last segment encodes two proteins, NS_1 and NS_2, that are found only in infected cells. NS_1 is 202–237 amino acids in length encoded by a collinear mRNA transcript. NS_1 is thought to be involved in the shut off of the host cell's protein synthesis or in the synthesis of viral RNA. NS_2 is encoded on an interrupted RNA strand. The function of NS_2 is unknown at present and results concerning its cellular location are conflicting.

4. Influenza virus structure

Influenza viruses are simple in structure. The viral envelope contains two major integral membrane glycoproteins, HA and NA. These protrude from the lipid bilayer-based envelope as 5–600 projections. Approximately 80% of the spikes are HA, the other 20% being NA. When negatively stained, the HA and NA spikes form a characteristic halo of projections around the viral envelope. Contained within the envelope are the viral matrix proteins M_1 and M_2 and the seven or eight viral RNA segments found as helical RNPs.

5. Viral infection

Influenza is acquired from an infected person by the respiratory route. After an incubation period of 1–3 days, headache, fever and muscle aches commence, followed in most cases by a cough and sore throat. Most people who get the 'flu recover completely in 1–2 weeks, but some people develop complications, such as pneumonia, that are potentially life threatening.

Influenza virus infection is divided into two distinct phases. First, the virus lodges in the upper respiratory tract as a result of inhalation of infectious particles. Viral NA then rapidly lowers the viscosity of the surface mucus, laying bare cell surface receptors and allowing the virus to attach to the outside of the host cell. Then, penetration of the virus into vacuoles, or intake of RNA from particles disrupted at the surface, initiates viral reproduction. Secondly, once inside the host cell, the protein envelope is shed and the viral genes are transcribed and translated by the cell's enzymes and ribosomes. In this way, the virus takes over the cell's productivity. The cell then produces hundreds of new virus particles that are eventually released from the cell and drift off. Release of newly synthesized virus is facilitated by NA activity and a new cycle of adsorption, penetration, synthesis and release is begun.

6. Virulence

Containing so few genes, how do influenza viruses actually cause so many fatalities? A number of studies have attempted to correlate virulence to individual gene products, but the overall findings are that both host- and virus-determined properties play a role in pathogenesis. Virulence is generally a polygenic phenomenon; in other words, many gene products play a role in determining the pathogenicity of influenza viruses.

How then do influenza viruses continue to cause disease year after year? The answers lie in variation: the RNA genome of influenza viruses is capable of changing in a number of ways.

(i) Antigenic shift

There are currently 15 known antigenic types of HA and nine known types of NA. Of these, only three HA and two NA types have been found to cause pandemics in humans. Pandemics usually result from the appearance of a new influenza strain containing a novel antigenic type of HA and/or NA, because the population is not immunologically prepared to tackle the new antigen (Webster, 1999). Influenza viruses are unique among the 120 viruses that cause respiratory disease in humans in that they undergo antigenic shift by a process known as genetic reassortment. Due to the segmented nature of the influenza genome, if a cell is simultaneously infected by two different strains, the offspring virus may contain mixtures of the parents' genes as a result of recombination (Smith & Palese, 1989). All eight segments in the influenza genome may take part in genetic reassortment, creating genomes that are very different from their parents and probably not seen before by the immune system. However, antigenic variation primarily involves the HA and NA glycoproteins which may differ by up to 80% between strains. It is the dramatic changes in these surface antigens that cause global epidemics.

No antigenic shift is observed in influenza B HA, making the strain less of a public health problem than influenza A strains.

(ii) Antigenic drift

More commonly than the abrupt changes in antigenic shift, HA and NA proteins undergo minor changes or mutations that cause a gradual evolution of the virus (Claas, 2000). Influenza viruses have a high error rate during the transcription of their genomes because of the low RNA polymerase fidelity; HA and NA proteins are subjected to the highest rates of change. The resulting accumulation of mutations is known as antigenic drift. This constant changing enables the virus to evade the immune system of the host so that people are susceptible to 'flu virus throughout life; immunity to one strain does not necessarily confer immunity to a new strain which has undergone antigenic drift. Antigenic drift is responsible for many of the localized outbreaks and it occurs gradually without changes in gene associations.

Most drift in the HA protein occurs in the HA_1 part of the molecule that contains the major antigenic sites. As the mutations accumulate, the shape of the antigen changes until eventually it drifts so much that the original host antibody can no longer bind to it.

(iii) Animal viruses become virulent for humans

The influenza virus has a remarkably broad host range. The natural hosts are wild waterfowl, shorebirds and gulls, but aberrant hosts include chickens, turkeys, swine, mink, horses and humans. Sequence analysis of influenza strains from a variety of sources has shown that the strains that caused pandemics in the 20th century were all derived from avian sources. Experimentally, however, viruses from one species of animal do not grow efficiently in another species. How then did avian strains cause pandemics in humans? The answer is thought to lie with pigs.

Two different receptors are found in the cells lining the pig trachea: one contains α-2,3-galactose–sialic acid linkages, the other contains α-2,6-galactose–sialic acid linkages. Avian species contain only receptors with α-2,3-galactose linkages and humans contain only receptors with α-2,6-galactose linkages (Colman, 1989). Pigs have a relatively weak species barrier against infection by either avian or human influenza strains and, therefore, represent an ideal candidate for the role of intermediate host (Ludwig *et al.*, 1995). If both an avian and a human strain infect swine at the same time, antigenic shift may lead to the generation of a new strain which may be transmitted to other species, including humans. It is also possible that, in order to form a stable lineage, an avian strain must first be transferred through pigs so that it might be adapted to mammalian species. The close proximity of humans, birds and swine in farming communities in southern China enhances the chances of mixing events and these geographical regions are subsequently thought to be the breeding ground of strains capable of causing pandemics.

New viruses may also appear in humans when an animal virus acquires mutations that confer infectivity in the human population. Evidence of direct transmission of an avian strain to humans was provided during 1997 when six people in Hong Kong died following infection with a chicken strain of influenza (H5N1; Shortridge *et al.*, 2000). Fortunately, such strains tend to be unable to transfer efficiently from human to human.

(iv) Subtype recycling

One of the main defences of humans against influenza infection is the possession of neutralizing antibodies to the influenza surface antigens. Once exposed to a particular influenza strain, any antibodies raised to the surface antigens are "remembered" so that effectively humans are self-vaccinated. If we look at the last five major epidemics, the surface antigens HA and NA varied as follows: 1889: H2N2; 1900: H3N8; 1918: H1N1; 1957: H2N2; and 1968: H3N2. These results suggest that there may be recycling of the HA on the viral surface. Influenza may take advantage of the disappearance of antibodies to these surface antigens in the population as generations of people previously exposed to the viral strain die or their immune system "forgets".

7. Controlling influenza viruses

Prevention of the nationwide spread of novel influenza strains has never been achieved. We have seen how influenza is capable of repeatedly evading our immune system, but what can we do to prevent infection or to control any spread? The first step is to keep track of circulating strains.

(i) Surveillance

In order to monitor influenza activity and to detect antigenic changes in the circulating strains, the American Centers for Disease Control (CDC) and the World Heath Organization (WHO) have entered a collaborative programme. Four reference centres in London, Atlanta, Melbourne and Tokyo collate information on influenza activity from 200 laboratories in 79 countries. The information is used to assess the potential of new influenza strains to cause epidemics. In addition, this information includes evaluation of how well current vaccines or existing antibodies in the population would work to protect against any new viral strain. The results of the collaborative effort are vital in planning appropriate preventative measures – in particular, vaccination.

(ii) Vaccines

Vaccination against influenza remains the most effective method of reducing the impact of the virus. Inactivated (killed) influenza virus preparations are currently the only influenza vaccines licensed in the US. These multisubunit vaccines are designed on the basis of information supplied by the WHO and are usually administered to people over 65 years of age or under 5 years and to a variety of immunocompromized patients. The 2000 vaccine consisted of two type A strains, H3N2 and H1N1, along with a type B strain. Two further influenza A strains, H5N1 and H9N2, are also currently in circulation, but there is no evidence of human-to-human transmission and these strains were not included in the 2000 vaccine (Claas, 2000; Shortridge *et al.*, 2000).

In addition to killed vaccines, a number of live strains have been very effective against current influenza A and B strains. One such vaccine is cold adapted to elicit an immune response in the upper respiratory tract where there is the highest concentration of receptors for influenza viruses. A second, more recent, vaccine contains an altered NS_1 protein. This strain has attenuated virulence, but is capable of generating good protection against both A- and B-type influenza strains (Talon *et al.*, 2000).

A number of problems are found with both live and killed vaccines. First, it takes over 7 months to prepare vaccine stocks once the strain information has been produced by the WHO. This delay in vaccination may be too long to abort a fast-spreading epidemic. Secondly, because vaccines are designed on current circulating strains, which are capable of antigenic shift or drift, they may remain effective for, at best, 1 year. Immunity also declines throughout the year, particularly in the elderly. This means that vaccination must be performed each year. An alternative is to develop a universal vaccine, but technical difficulties associated with production have stalled investigation. Finally, no vaccine is 100% effective. Effectiveness is dependent on how well the strains used to make the vaccine "match" the circulating strains (Riberdy *et al.*, 1999).

More recently, DNA vaccination has been investigated (Fomsgaard *et al.*, 2000; Olsen, 2000). DNA vaccines are as effective as live or killed vaccines, but many of the problems remain in that the vaccines are also strain specific.

(iii) Antiviral agents

Vaccination is the primary method of preventing influenza infection. However, a number of antiviral agents are available to both treat infection and afford protection. Amantadine and rimantidine are chemically related drugs that interfere with the viral replication cycle to both reduce and shorten the duration of illness if given within 48 hours of illness onset. In addition to treating influenza, they are also 70–90% effective in preventing infection. These drugs are effective only against influenza A viruses and not against type B or C viruses.

One problem with amantidine and rimantidine, apart from cost, is the evolution of viruses that are resistant to the drugs. Although there is no evidence that resistant strains are more virulent or more infectious than sensitive strains, it is vitally important to reduce the generation of resistant strains so that the drugs remain effective.

In 1999, two new anti-influenza drugs were released that are effective for the treatment of both influenza A and B viruses. Zanamivir and oseltamivir are NA blockers (Jack, 1996; Leneva *et al.*, 2000) and work by selectively blocking NA activity late in the viral replication cycle. The residues in NA involved in binding host receptors are conserved across all influenza strains enabling the NA blocker to retain efficacy irrespective of shift or drift in viral antigens. Studies are currently ongoing to determine whether zanamivir and oseltamivir can be used in the prevention of influenza A and B infection.

8. Summary

Despite over 70 years of active research, influenza still presents us with many puzzles. What is evident is that, due to the avian reservoir, we cannot eradicate influenza viruses. We can, however, substantially reduce the impact of the virus through continual monitoring and improvement of vaccination protocols and antiviral drugs.

REFERENCES

Claas, E.C.J. (2000) Pandemic influenza is a zoonosis, as it requires introduction of avian-like gene segments in the human population. *Veterinary Microbiology* 74: 133–39

Colman, P.M. (1989) Neuraminidase: enzyme and antigen. In *The Influenza Viruses*, edited by R.M. Krug, New York: Plenum Press

Fomsgaard, A., Nielsen, H.V., Kirkby, N. *et al.* (2000) Induction of cytotoxic T-cell responses by gene gun DNA vaccination with minigenes encoding influenza A virus HA and NP CTL-epitopes. *Vaccine* 18: 681–91

Hiromoto, Y., Saito, T., Lindstrom, S.E. *et al.* (2000) Phylogenetic analysis of the three polymerase genes (PB1, PB2 and PA) of influenza B virus. *Journal of General Virology* 81: 929–37

Jack, D.B. (1996) Getting ready for the next influenza pandemic. *Lancet* 347: 1252

Lamb, R.A. (1989) Genes and proteins of the influenza viruses. In *The Influenza Viruses*, edited by R.M. Krug, New York: Plenum Press

Leneva, I.A., Roberts, N. & Webster, R.G. (2000) The neuraminidase inhibitor oseltamivir is efficacious against A/Hong Kong/156/97 (H5N1) and H9N2 influenza viruses. *Antiviral Research* 46(1): 60

Ludwig, S., Stitz, L., Planz, O. *et al.* (1995) European swine virus as a possible source for the next influenza pandemic? *Virology* 212: 555–61

Olsen, C.W. (2000) DNA vaccination against influenza viruses: a review with emphasis on equine and swine influenza. *Veterinary Microbiology* 74: 149–64

Oxford, J.S., Sefton, A., Jackson, R., Johnson, N.P.A.S. & Daniels, R.S. (1999) Who's that lady? *Nature Medicine* 5(12): 1351–52

Perdue, M.L. & Suarez, D.L. (2000) Structural features of the avian influenza virus hemagglutinin that influence virulence. *Veterinary Microbiology* 74: 77–86

Reid, A., Fanning, T.G., Hultin, J.V. & Taubenberger, J.K. (1999) Origin and evolution of the 1918 "Spanish" influenza virus haemagglutinin gene. *Proceedings of the National Academy of Sciences USA* 96: 1651–56

Riberdy, J.M., Flynn, K.J., Stech, J. *et al.* (1999) Protection against a lethal avian influenza A virus in a mammalian system. *Journal of Virology* 73(2): 1452–59

Shortridge, K.F., Gao, P., Guan, Y. *et al.* (2000) Interspecies transmission of influenza viruses: H5N1 virus and a Hong Kong SAR perspective. *Veterinary Microbiology* 74: 141–47

Smith, F.I. & Palese, P. (1989) Variation in influenza genes: epidemiological, pathogenic, and evolutionary consequences. In *The Influenza Viruses*, edited by R.M. Krug, New York: Plenum Press

Talon, J., Salvatore, M., O'Neill, R.E. *et al.* (2000) Influenza A and B viruses expressing altered NS1 proteins: a vaccine approach. *Proceedings of the National Academy of Sciences USA* 97(8): 4309–14

Taubenberger, J.K., Reid, A.H., Krafft, A.E., Bijwaard, K.E. & Fanning, T.G. (1997) Initial characterisation of the 1918 "Spanish" influenza virus. *Science* 275: 1793–96

Webster, R.G. (1999) Antigenic variation in influenza viruses. In *Origin and Evolution of Viruses*, edited by E. Domingo, R. Webster & J. Holland, San Diego: Academic Press

USEFUL WEB SITES

Centers for Disease Control and Prevention: http://www.cdc.gov/
National Foundation for Infectious Diseases: http://www.nfid.org/
World Health Organisation: http://www.oms2.b3e.jussieu.fr/flunet and http://www.who.int/emc/diseases/flu/index.html

GLOSSARY

alternative splicing splicing different strands of messenger RNA to form different mRNA transcripts

antigenic site a molecule (such as a glycoprotein) on the surface of a microbe or macromolecule that triggers an immune response

antimessenger RNA RNA that serves as a template for protein synthesis using reverse transcriptase

cRNA/copy RNA RNA synthesized from an antimessenger RNA template using reverse transcriptase

endosome a large membrane-bound structure that is the result after several coated vesicles fuse together following endocytosis. The structure contains the combined contents of all of the former vesicles

glycoprotein a protein linked to a sugar or polysaccharide which are components of receptor molecules on the outer surface of cells

See also **DNA vaccination** (p.521)

MENINGITIS

Anne E. Hill
Moredun Research Institute, Pentlands Science Park, Edinburgh, UK

1. Introduction

Meningitis is the inflammation of the meninges, the lining surrounding the brain and spinal cord. This potentially life-threatening condition may be caused by a wide spectrum of different bacteria, viruses, fungi and protozoa. There are reportedly over 500 000 cases of meningitis annually in the US alone, a cost to the tax payer of over US$2 billion per year.

In this article, we take a look at the major causes of meningitis, the tactics they employ to create the condition, the symptoms and treatment, and finally the recent advances in preventative methods.

2. Bacterial meningitis

Meningitis caused by bacteria tends to be severe, frequently causing brain damage, hearing loss and learning disability. It is fatal in approximately one in ten cases, and one in seven survivors are left with serious disability. The major players in bacterial meningitis are *Neisseria meningitidis*, *Streptococcus pneumoniae* and *Haemophilus influenzae*, although cases have been attributed to *Mycobacterium tuberculosis*, *Staphylococcus aureus*, *Streptococcus agalactiae*, *Streptococcus suis*, *Klebsiella pneumoniae*, *Listeria monocytogenes*, *Flavobacterium meningosepticum*, *Pseudomonas aeruginosa* and *Escherichia coli* (Cabellos *et al.*, 1999; Walia & Hoskyns, 2000). Many of these organisms are common commensals found in the nasopharynx of healthy individuals. It is only occasionally that they are able to break down the host's natural defences and cause meningitis or septicaemia.

The majority of information regarding bacterial meningitis has been obtained through studies of *N. meningitidis*, *S. pneumoniae* and *H. influenzae*. In the next few sections, we will take a closer look at these pathogens and the genetic attributes that enable them to cause meningitis.

(i) N. meningitidis

N. meningitidis is an obligate intracellular parasite, unable to survive in the environment. This Gram-negative diplococcus is an opportunist pathogen that colonizes the nasopharynges of asymptomatic carriers, but which occasionally manages to gain access to the blood and cerebrospinal fluid (CSF) to cause septicaemia and meningitis. There are five pathogenic serogroups, A, B, C, Y and W135, based on capsular polysaccharide typing. Outbreaks of type A meningitis tend to be epidemic or pandemic in nature; type B outbreaks tend to be hyperendemic. Over the past few years, there has been a significant rise in the number of type C cases of meningitis mainly due to the availability of vaccines for type A (see later). Types Y and W135 are now rarely isolated from meningitis cases.

The complete genome sequence has been published for both serogroup A (strain Z2491; Parkhill *et al.*, 2000) and serogroup B types of *N. meningitidis* (strain MC58; Tettelin *et al.*, 2000). Examination of their 2.2 Megabase genomes has shown that the majority of coding sequences are devoted to attributes necessary for colonization and survival. However, a number of potential virulence factors, which enable infection, have also been identified (**Table 1**) (Stojiljkovic *et al.*, 1995; Feavers *et al.*, 1999; Ferreiros *et al.*, 1999; Moe *et al.*, 1999; Nassif, 1999; Perrin *et al.*, 1999; Sawaya *et al.*, 1999; Zhu *et al.*, 1999).

Virulence factors are expressed at strategic points of the pathogenesis process. For *N. meningitidis*, the proposed infection process is as follows:

Table 1 Virulence factors of *Neisseria meningitidis*.

Virulence factor	Gene(s)	Function
Type IV Pili	*pilC2, pilD, pilE, pilS, pilG, pilF, pilT*	Promote adhesion to epithelial and endothelial cells; interact with host cells and trigger a signal transduction response in them
Adhesins	*hsf, hap*	Enable attachment to host cells
Lipopolysaccharide	*lgtA, lgtB, lgtE, lgaG, lic1A, lic2A, lic3A*	Lipopolysaccharide on the bacterial surface impedes clearance and killing of meningococci by host phagocytes
Capsule	*siaB, siaC, siaD, synX, ctrA, ctrB, ctrC, ctrD, lipA, lipB*	Provokes immune response; impedes clearance and killing of meningococci by host phagocytes
Porins (PorA, PorB)	*porA, porB*	Pathogenic Neisseria produce two porins which promote bacterial invasiveness
Membrane protein	*nspA*	Acts as a protective antibody
Ferric binding protein	*fbpA–C*	Necessary for obtaining iron during infection
Transferrin binding protein	*tbpA*	Necessary for obtaining iron during infection
Haemoglobin receptor	*hmbR*	Necessary for obtaining iron from haemoglobin during infection

1. *N. meningitidis* bacteria adhere to the epithelium of the nasopharynx and may then be transferred into the bloodstream. The virulence factors involved include pili, adhesins, lipopolysaccharide and capsule (Nassif *et al.*, 1999; Zhang & Tuomanen, 1999; Hardy *et al.*, 2000).
2. The bacteria adhere to the endothelial cells and cross the blood–brain barrier into the CSF. Here, the virulence factors involved include the Opa and Opc proteins and porins.
3. Replication of bacteria in the CSF induces meningitis.

Within this infection process, adhesion to both epithelial and endothelial cells is crucial and is usually viewed as a two-step process for *N. meningitidis*: (1) initial adhesion which is pilus mediated and (2) intimate adhesion which involves other bacterial and cellular structures.

A further feature of pathogenic Neisseria is the ability to incorporate foreign DNA into the genome (Kroll *et al.*, 1998). Genes encoded on foreign DNA may include novel virulence factors and multidrug resistance genes (Maiden, 1998).

(ii) S. pneumoniae

Approximately 10% of bacterial meningitis cases are the result of *S. pneumoniae*, or pneumococcal, infection. Over 80 serotypes of this Gram-positive coccus have been identified. The majority are harmless commensals, but a number are also capable of causing severe bacteraemic infections including pneumonia, endocarditis, sinusitis, otitis, bronchitis, laryngitis and meningitis.

The genome sequence of a serotype 4 strain of *S. pneumoniae*, isolated from a child with meningitis, is complete and, as with *N. meningitidis*, a number of virulence factors have been identified (**Table 2**) (Paton, 1998; Polissi *et al.*, 1998; Rossjohn *et al.*, 1998; Rubins & Janoff, 1998; Tu *et al.*, 1999; Tuomanen, 1999; Whatmore & Dowson, 1999; Whatmore *et al.*, 1999; Serino & Virji, 2000).

Many of the virulence factors shown in **Table 2** are common to Neisseria – in particular, the adhesins and outer membrane proteins necessary for initial attachment to host cells. *S. pneumoniae* have additionally been found to possess a natural ability to incorporate non-self genetic material into the genome. As with Neisseria, this "natural competence" enables *S. pneumoniae* to evolve rapidly, acquiring new virulence factors and, more worryingly, multidrug resistance (see later).

The pathogenesis of *S. pneumoniae* infection is identical to that described for *N. meningitidis* (Molinari & Chhatwal, 1999):

1. Pneumococci colonize the nasopharynges, a process mediated by surface adhesins which bind to epithelial cell receptors.
2. The bacteria cross the blood–brain barrier into the CSF.
3. Replication of pneumococci in the CSF induces meningitis.

S. pneumoniae meningitis is renowned as being particularly serious. The possession of potent bacterial toxins combined with the ability of the pneumococcal cell wall to induce a dramatic inflammatory response frequently leads to lethal increases in intracranial pressure on the meninges.

(iii) H. influenzae

H. influenzae, like *N. meningitidis*, is a strict parasite of humans found principally in the upper respiratory tract. Six antigenically distinct serotypes, a–f, have been identified, along with a number of non-typable (no capsule present) strains. The non-encapsulated forms are normal commensals of the upper respiratory tract whereas *H. influenzae* type b causes over 10 000 cases of meningitis annually in the US. *H. influenzae* also causes a number of other invasive infections such as septic arthritis, cellulitis and epiglottitis, in addition to pneumonia and lower respiratory disease.

In 1995, *H. influenzae* became the first free-living organism to have its genome completely sequenced (Fleischmann *et al.*, 1995). Since then, a large number of groups have been involved in identifying and characterizing the virulence factors expressed by this organism (**Table 3**) (Moxon & Kroll, 1988; Murley *et al.*, 1998; Hosking *et al.*, 1999; White *et al.*, 1999; Serino & Virji, 2000).

In addition to the polysaccharide capsule and pili, natural genetic exchange is another feature shared with pathogenic *N. meningitidis* and *S. pneumoniae* (Kroll *et al.*, 1998; Kubiet *et al.*, 2000). As with the infection processes described above, the mechanism by which *H. influenzae* causes meningitis involves initial colonization of the nasopharynges, followed by spread into the CSF and replication.

3. Viral meningitis

The majority of meningitis cases are caused by viruses. Fortunately viral, or aseptic, meningitis is generally less serious than bacterial or fungal meningitis and is rarely fatal. There are three major viruses that are responsible for meningitis

Table 2 Virulence factors of *Streptococcus pneumoniae*.

Virulence factor	Gene(s)	Function
Adhesin	*cbpA*	Participate in binding of pneumococci to activated human cells and in nasopharyngeal colonization
Permease	*psaA*	Pneumococcal surface adhesin
Surface proteins	*pspA, spsA*	Protective antigens; presence is required for full virulence. Reduce clearance by host
Neuraminidase	*nanA, nanB*	Hydrolysis of *N*-acetylneuraminic acid
Hyaluronidase	*hyl*	Catalyses breakdown of host hyaluronic acid
Immunoglobin A1 protease	*iga*	Cleavage of human IgA1

Table 3 Virulence factors of *Haemophilus influenzae*.

Virulence factor	Gene(s)	Function
Pili	*hifA–E*	Putative role in colonization/adherence
Capsule type b	*cap*	Produces sustained bactaeremia; prevents host clearing
Outer membrane proteins	*omp1, omp2, omp4–6*	P2 has porin activity
Phosphorylcholine	*choP*	Surface protein involved in immune avoidance
Haemocin	*hmc*	A bacteriocin involved in invasion
Transferrin receptor	*hmbR, hitA*	Required for obtaining iron from haemoglobin during infection
IgA protease	*iga*	Direct proof not found; mutants have reduced virulence

outbreaks: enteroviruses, herpesviruses and paramyxoviruses, although lymphocytic choriomeningitis virus is still looked for routinely (Barton & Hyndman, 2000).

(i) Enteroviruses

Enteroviruses, such as Coxsackie and Echovirus, cause 80–90% of viral meningitis cases. As with many of the bacterial causes of meningitis, these viruses are found as harmless commensals in the back of the throat and neck, and it is only occasionally that they break the host's defences and multiply in the CSF (Muir & van Loon, 1997).

The enteroviruses are small icosahedral viruses with a plus-strand RNA chromosome which serves as messenger RNA during replication. They are unique in that the viral chromosome is translated into a single polypeptide which is then proteolytically cleaved into six essential viral proteins. Four of these viral proteins are structural, involved in formation of the protein capsid.

All viruses are simple in design and are obligate parasites; they need to exploit the host cell's metabolism in order to multiply. The factors that enable the viruses to cause disease are therefore polygenic – that is, both host and viral factors are responsible for pathogenesis. Studies of the haemagglutinin–neuraminidase (HN) gene of Echovirus 30 have, however, revealed that avirulent strains differ in only one nucleotide to virulent strains in this gene. It is also thought that genetic drift of the HN gene may contribute to new antigenic variants of echovirus in a similar way to the H and N genes of influenza viruses.

(ii) Herpesviruses

The herpesvirus family includes herpes simplex virus (HSV), varicella zoster and Epstein–Barr virus, more commonly associated with cold sores, chickenpox and glandular fever, respectively (Tenser, 1984; Portegies & Corssmit, 2000). Herpesviruses are much larger than the enteroviruses. They are enveloped viruses containing a double-stranded DNA genome.

HSV is responsible for 10–30% of adult cases of meningitis, whereas varicella zoster and Epstein–Barr are rarely isolated. As with the enteroviruses, virulence is associated with both host and viral factors. Meningitis tends to arise in patients with specific weaknesses in their immune system.

(iii) Paramyxoviruses

Measles, mumps and Rubella are all caused by paramyxoviruses. These are small minus-strand RNA viruses more commonly associated with childhood diseases. Before the introduction of the measles, mumps and rubella (MMR) vaccine in 1996, meningitis was frequently found as a complication of mumps. However, vaccination has not proved to be the end of meningitis caused by these paramyxoviruses; a number of vaccine-associated meningitis cases have been documented (Brown *et al.*, 1996; Dourado *et al.*, 2000) and it has subsequently been demonstrated that one of the viruses included in the vaccine, Urabe AM9, is able to revert from the avirulent vaccine virus to the virulent wild-type virus. Further, it is now known that the difference between the avirulent and virulent forms is the result of a single nucleotide change in the HN gene, similar to that of Echovirus 30 described above. Although virulence is determined by both viral and host genes, the importance of the HN gene product in virulence is beyond doubt.

4. Fungal meningitis

Fungal meningitis is found predominantly in immunosuppressed patients such as those suffering from AIDS or transplant patients. Two different fungi, *Cryptococcus neoformans* and *Candida albicans*, have been identified as the most common causes of fungal meningitis, although cases have been attributed to *Clavispora lusitaniae*, *Coccidiodes immitis* and *Rhodotorula rubra* (Krcmery *et al.*, 1999; Kleinschmidt-DeMasters *et al.*, 2000; Krcmery *et al.*, 2000; Odds, 2000).

(i) C. neoformans

Cryptococcal meningitis is caused by the yeast-like fungus *C. neoformans*. Cases of cryptococcal meningitis are increasing in incidence as a result of HIV infection, chemotherapy, and the use of high doses of steroids and immunosuppressive drugs (Bogaerts *et al.*, 1999; Marra, 1999). A staggering 10% of AIDS patients develop cryptococcal meningitis which is fatal if untreated. As a result, research into the virulence factors and pathogenesis of cryptococcal meningitis has intensified.

C. neoformans is dimorphic: it may grow either as a budding yeast or as filamentous hyphal forms. Both prevalence and virulence of the fungus are associated with mating type, of which there are two, MATa and MATα, in *C. neoformans*. MATα-type *C. neoformans* are significantly more prevalent and significantly more virulent than MATa type. For this reason, the search for virulence factors has focused on the MATα locus.

Virulence is associated with the production of melanin and a polysaccharide capsule, and the ability of the fungus to grow at body temperature. Studies of the MATα locus have identified a small peptide pheromone homologue and an STE12 transcription factor homologue that stimulate the production of melanin. In addition, gene disruption has recently been successfully used to identify a number of other putative virulence factors (**Table 4**) (Alspaugh *et al.*, 1997; Odom *et al.*, 1997; Kwon-Chung, 1998; Wang & Heitman, 1999).

(ii) C. albicans

Candida species are within the top four most common microorganisms isolated from blood cultures in the US. The number of cases of meningitis associated with Candida species is few, but the severity of the infection has warranted investigation of virulence factors. As with *C. neoformans*, gene disruption techniques have recently been developed and used successfully to identify a number of virulence factors in *C. albicans* (**Table 5**) (Sanchez-Portocarrero *et al.*, 2000).

Pathogenesis of the disease is suspected, but has not demonstrated, to involve initial adhesion to host cells and subsequent invasion of the CSF.

5. Symptoms

Meningitis is characterized by the rapid onset of disease. Independent of cause, the symptoms of meningitis develop over hours to days, and are often confused with those of influenza. The first symptoms are high fever, nausea, vomiting and severe headache as the meninges become inflamed and the CSF thickens. Fatigue, confusion and memory loss are also common, along with a dislike of light and a stiff neck. Stiffness of the neck is due to irritation of the spinal nerves supplying the neck muscles.

In babies, it is often more difficult to spot symptoms early, but they may include fever with hands and feet feeling cold, high-pitched moaning cry, neck retraction with arching back and a pale, blotchy complexion.

In patients suffering septicaemia in addition to meningitis, a rash may appear which does not fade when a glass is pressed against it. This blotchy skin rash is caused by bleeding under the skin.

With viral meningitis, symptoms usually last for 7–10 days and the person recovers completely. However, in chronic cases of bacterial meningitis, symptoms may persist for over 4 weeks.

6. Diagnosis

Rapid diagnosis of meningitis is essential for determining the most effective disease management strategy. The most efficient method is through characterization of CSF obtained by lumbar puncture, which is, unfortunately, an extremely painful procedure.

CSF may be examined directly under the microscope; CSF cell type can provide clues to the underlying disease. Gram-staining of the CSF also enables the identification of bacteria.

Instrumental in the rapid diagnosis has been the development of PCR techniques. A multiplex PCR assay may be used for clinical differentiation of septic (bacterial) or aseptic meningitis. Primers are available for the major bacterial causes, enteroviruses, Epstein–Barr virus and *C. neoformans* (Greisen *et al.*, 1994; Chen *et al.*, 1998; Rappelli *et al.*, 1998; Glustein *et al.*, 1999; Ramers *et al.*, 2000). Combinations of hybridization using specific oligoprobes and PCR of 16S RNA sequences may be used to identify specific bacterial serotypes (Dicuonzo *et al.*, 1999).

PCR diagnosis is usually confirmed by culture of the infectious agent from CSF. The majority of bacteria may be cultured on blood agar plates, whereas shell vial culture is required for viruses. Fungi often require complex media for growth and may take several days to grow.

Additional diagnostic methods include detection of bacterial capsular antigens in the CSF using immunoelectrophoresis, latex agglutination or enzyme-linked immunosorbent assay.

Recently, an assay for procalcitonin has been developed. Increased blood levels of procalcitonin are thought to be an early indicator of systemic infection.

7. Treatment

Following diagnosis, it is essential that treatment is instituted rapidly in order to reduce the complications of meningitis.

Table 4 Virulence factors of *Cryptococcus neoformans.*

Virulence factor	Gene(s)	Function
Capsule	*cap*	Mutants are avirulent
Laccase	*cnlac1*	Laccase catalyses melanin formation
Calcineurin	*cna1*	Ca^{2+}/calmodulin-dependent protein phosphatase involved in regulating expression of virulence factors
MAPK cascade proteins	*gpa1, ste12α*	Regulates mating, capsule formation and laccase production; ste12α is found within the MATα locus

Table 5 Virulence factors of *Candida albicans.*

Virulence factor	Gene(s)	Function
Adhesin	*cph1*	Involved in adhesion to host cells
Hyphae	*efg1, sap*	Hyphal formation and virulence
Integrin	*int*	Mutant has decreased filamentous growth and adhesion to epithelial cells

For all causes of meningitis, treatment usually involves large doses of intravenous drugs.

The cephalosporin group of antibiotics, such as cefotaxime, have proved to be key in the treatment of bacterial meningitis, although chloramphenicol is the most effective drug to date against *H. influenzae*, *N. meningitidis* and pneumococci (Mulholland, 1999). Combinations of drugs such as cefotaxime and penicillin may also be employed. With bacterial meningitis, very large doses of antibiotic are necessary and problems with drug resistance are particularly worrying. Bacterial strains resistant to chloramphenicol, ampicillin, benzyl penicillin and vancomycin have already been isolated, leading to increased research into alternative compounds (Novak *et al.*, 1999).

With viral meningitis, patients usually make a full recovery without treatment. Indeed, many of those infected do not even realize they have contracted the disease. Bed rest, plenty of fluids and painkillers for fever and headache are usually the only recommendations. For more serious cases, however, a number of antiviral drugs are available such as acyclovir and pleconaril.

Long-term treatment is often needed in cases of fungal meningitis, mainly to prevent any relapses. With cryptococcal meningitis, amphotericin B is the first drug of choice, followed by fluconazole, intraconazole and flucytosine (Davis, 1999). Amphotericin B is also used in cases of candidiasis, as well as 5-fluorocytosine (Sanchez-Portocarrero *et al.*, 2000). A number of novel antifungal drugs have recently been described by DiDomenico (1999).

8. Prevention

Vaccination has proved to be a formidable weapon in preventing bacterial meningitis. The first vaccination programme began over 25 years ago with the production of a polysaccharide vaccine against type b *H. influenzae* (Campbell & Carter, 1993; Moxon *et al.*, 1999; Peltola, 2000). Since then, a number of improved vaccines have been developed along with vaccines against *N. meningitidis* (types A and C) and *S. pneumoniae* (Fedson & Scott, 1999; Mulholland, 1999; Rubin, 2000). Vaccination against *H. influenzae* type b has been part of routine vaccination of babies in the UK since 1992. For all three of these bacteria, the polysaccharide capsule has been the base on which vaccines have been designed. In each case, the polysaccharide has been linked to a protein which is immunogenic, since the polysaccharide itself does not elicit a very effective immune response. Such vaccines are known as conjugate vaccines. The most recent conjugal vaccine to be produced was released in the UK in November 1999 for type C *S. pneumoniae*; it is estimated that the vaccine will be more than 95% effective. A bivalent vaccine against both type C and type B *S. pneumoniae* has also recently been trialled (Fukasawa *et al.*, 1999). No exclusive type B vaccine has yet been produced (Moxon *et al.*, 1999).

A number of virulence factors from *S. pneumoniae*, such as CbpA, PsaA and PspA (**Table 2**), have recently been investigated as protein-based vaccines. The plasticity of bacterial genomes means that polysaccharide vaccines may not remain effective, and protein-based vaccines may also be able to protect against a wider range of serotypes; only 23 of the 90 known serotypes of *S. pneumoniae* are covered by existing vaccines.

Currently, no vaccine is available for protection against enteroviral infection, the most common cause of viral meningitis. DNA vaccines have been trialled, but immune responses have proven to be short lived. As described above, the MMR vaccine has significantly reduced the number of mumps-associated meningitis cases. A vaccine which prevents persistent Epstein–Barr infection is also currently available (von Herrath *et al.*, 2000).

Viral meningitis is spread through direct contact with respiratory secretions (e.g. saliva, nasal mucus), but, typically, less than one in every 1000 persons infected actually develops meningitis. The most effective prevention is to wash your hands thoroughly and often. Dilute bleach on worktops etc. can also inactivate viruses.

Fluconazole, a drug used in treatment of cryptococcal meningitis, may also be used to prevent infection for severely immunosuppressed patients. Concern about drug resistance does, however, mean that drug treatment is rarely used as a preventative measure (Mondon *et al.*, 1999).

9. Summary

A common theme links all causes of meningitis; analysis of virulence factors (**Tables 1–5**) and pathogenesis reveals the importance of adhesion to host cells and subsequent infiltration of the CSF. Any organism that is capable of overcoming host defences and replicating in the CSF is able to cause meningitis, making the task of preventing the condition seem virtually impossible. However, as we move further into the "postgenome" era, our increasing knowledge of bacterial, viral and fungal genetics will facilitate the production of improved diagnostic techniques and, more importantly, will lead to improved vaccines for all of the major causes of meningitis.

REFERENCES

Alspaugh, J.A., Perfect, J.R. & Neitman, J. (1997) *Cryptococcus neoformans* mating and virulence are regulated by the G-protein alpha subunit GPA1 and cAMP. *Genes and Development* 11: 3206–217

Barton, L.L. & Hyndman, N.J. (2000) Lymphocytic choriomeningitis virus: re-emerging central nervous system pathogen. *Pediatrics* 105: E35

Bogaerts, J., Rouvroy, D., Taelman, H. *et al.* (1999) AIDS-associated cryptococcal meningitis in Rwanda (1983–1992): epidemiologic and diagnostic features. *Journal of Infection* 39: 32–37

Brown, E.G., Dimock, K. & Wright, K.E. (1996) The Urabe AM9 mumps vaccine is a mixture of viruses differing at amino acid 335 of the hemagglutinin-neuraminidase gene with one form associated with disease. *Journal of Infectious Diseases* 174: 619–22

Cabellos, C., Viladrich, P.F., Corredoira, J. *et al.* (1999) Streptococcal meningitis in adult patients: current epidemiology and clinical spectrum. *Clinical Infectious Diseases* 28: 1104–08

Campbell, H. & Carter, H. (1993) Rational use of *Haemophilus influenzae* type b vaccine. *Drugs* 46: 378–83

Chen, Z., Dong, Y. & Cui, W. (1998) Detection and identification of enterovirus RNA by using polymerase chain reaction. *Journal of Tongji Medical University* 18: 156–60

Davis, L.E. (1999) Fungal infections and the central nervous system. *Neurologic Clinics* 17: 761–81

DiDomenico, B. (1999) Novel antifungal drugs. *Current Opinion in Microbiology* 2: 509–15

Dicuonzo, G., Lorino, G., Lilli., D., Rivanera, D. & Guarina, F. (1999) Use of oligoprobes on amplified DNA in the diagnosis of bacterial meningitis. *European Journal of Clinical Microbiology and Infectious Diseases* 18: 352–57

Dourado, I., Cunha, S., Teixeira, M.G. *et al.* (2000) Outbreak of aseptic meningitis associated with mass vaccination with a urabe-containing measles-mumps-rubella vaccine: implications for immunisation programs. *American Journal of Epidemiology* 151: 524–30

Feavers, I.M., Gray, S.G., Urwin, R. *et al.* (1999) Multilocus sequence typing and antigen gene sequencing in the investigation of a meningococcal disease outbreak. *Journal of Clinical Microbiology* 37: 3883–87

Fedson, D.S. & Scott, J.A.G. (1999) The burden of pneumococcal disease among adults in developing countries: what is and is not known. *Vaccine* 17: S11–S18

Ferreiros, C., Criado, M.T. & Gomez, J.A. (1999) The Neisserial 37 kDa ferric binding protein (FbpA). *Comparative Biochemistry and Physiology, Biochemistry and Molecular Biology* 123: 1–7

Fleischmann, R.D., Adams, M.D., White, O. *et al.* (1995) Whole-genome random sequencing and assembly of *Haemophilus influenzae* Rd. *Science* 269: 496–512

Fukasawa, L.O., Gorlas, M.C.O., Schenkman, R.P.F. *et al.* (1999) *Neisseria meningitidis* serogroup C polysaccharide and serogroup B outer membrane vesicle conjugate as a bivalent meningococcus vaccine candidate. *Vaccine* 17: 2951–58

Glustein, J.Z., Zhang, Y.Z., Wadowsky, R.M. & Ehrlich, G.D. (1999) Development of a simple polymerase chain reaction-based assay for the detection of *Neisseria meningitidis*. *Molecular Diagnosis* 4: 233–39

Greisen, K., Loeffelholz, M., Purohit, A. & Leong, D. (1994) PCR primers and probes for the 16S rRNA gene of most species of pathogenic bacteria, including bacteria found in the cerebrospinal fluid. *Journal of Clinical Microbiology* 32: 335–51

Hardy, S.J., Christoulides, M., Weller, R.O. & Heckels, J.E. (2000) Interactions of *Neisseria meningitidis* with cells of the human meninges. *Molecular Microbiology* 36: 817–29

Hosking, S.L., Craig, J.E. & High, N.J. (1999) Phase variation of *lic*1A, *lic*2A and *lic*3A in colonisation of the nasopharynx, bloodstream and cerebrospinal fluid by *Haemophilus influenzae* type b. *Microbiology* 145: 3005–11

Kleinschmidt-DeMasters, B.K., Mazowiecki, M., Bonds, L.A., Cohn, D.L. & Wilson, M.L. (2000) Coccidiomycosis meningitis with massive dural and cerebral venous thrombosis and tissue arthroconidia. *Archives of Pathological Laboratory Medicine* 124: 310–14

Krcmery, V., Mateicka, F., Grasova, S., Kunova, A. & Hanzen, J. (1999) Invasive infections due to *Clavispora lusitaniae*. *FEMS Immunology and Medical Microbiology* 23: 75–78

Krcmery, V., Paradisi, F. & the Pediatric Nosocomial Meningitis Study Group (2000) Nosocomial bacterial and fungal meningitis in children; an eight year national survey reporting 101 cases. *International Journal of Antimicrobial Agents* 15: 143–47

Kroll, J.S., Wilks, K.E., Farrant, J.L. & Langford, P.R. (1998) Natural genetic exchange between Haemophilus and Neisseria: intergeneric transfer of chromosomal genes between major human pathogens. *Proceedings of the National Academy of Sciences USA* 95: 12381–85

Kubiet, M., Ramphal, R., Weber, A. & Smith, A. (2000) Pilus-mediated adherence of *Haemophilus influenzae* to human respiratory mucins. *Infection and Immunity* 68: 3362–67

Kwon-Chung, K.J. (1998) Gene disruption to evaluate the role of fungal candidate virulence genes. *Current Opinion in Microbiology* 1: 381–89

Maiden, M.C.J. (1998) Horizontal genetic exchange, evolution, and spread of antibiotic resistance in bacteria. *Clinical Infectious Diseases* 27: S12–S20

Marra, C.M. (1999) Bacterial and fungal brain infections in AIDS. *Seminars in Neurology* 19: 177–84

Moe, G.R., Tan, S.Q. & Granoff, D.M. (1999) Molecular mimetics of polysaccharide epitopes as vaccine candidates for prevention of *Neisseria meningitidis* serogroup B disease. *FEMS Immunology and Medical Microbiology* 26: 209–26

Molinari, G. & Chhatwal, G.S. (1999) Streptococcal invasion. *Current Opinion in Microbiology* 2: 56–61

Mondon, P., Petter, R., Amalfitano, G. *et al.* (1999) Heteroresistance to fluconazole and voriconazole in *Cryptococcus neoformans*. *Antimicrobial Agents and Chemotherapy* 43: 1856–61

Moxon, E.R. & Kroll, J.S. (1988) Type b capsular polysaccharide as a virulence factor of *Haemophilus influenzae*. *Vaccine* 6: 113–15

Moxon, E.R., Heath, P.T., Booy, R. *et al.* (1999) The impact of Hib conjugate vaccines in preventing invasive *H. influenzae* diseases in the UK. *Vaccine* 17: S11–S13

Muir, P. & van Loon, A.M. (1997) Enterovirus infections of the central nervous system. *Intervirology* 40: 153–66

Mulholland, K. (1999) Strategies for the control of pneumococcal diseases. *Vaccine* 17: S79–S84

Murley, Y.M., Edlind, T.D., Plett, P.A. & LiPuma, J.J. (1998) Cloning of the haemocin locus of *Haemophilus influenzae* type b and assessment of the role of haemocin in virulence. *Microbiology* 144: 2531–38

Nassif, X. (1999) Interaction mechanisms of encapsulated meningococci with eucaryotic cells: what does this tell us about the crossing of the blood–brain barrier by *Neisseria meningitidis*. *Current Opinion in Microbiology* 2: 71–77

Nassif, X., Pujol, C., Morand, P. & Eugene, E. (1999) Interactions of pathogenic Neisseria with host cells. Is it possible to assemble the puzzle? *Molecular Microbiology* 32: 1124–32

Novak, R., Henriques, B., Charpentier, E., Normark, S. & Tuomanen, E. (1999) Emergence of vancomycin tolerance in *Streptococcus pneumoniae*. *Nature* 339: 590–93

Odds, F.C. (2000) Pathogenic fungi in the 21st century. *Trends in Microbiology* 8: 200–02

Odom, A., Muir, S., Lim, E. *et al.* (1997) Calcineurin is required for virulence of *Cryptococcus neoformans*. *EMBO Journal* 16: 2576–89

Parkhill, J., Achtman, M., James, K.D. *et al.* (2000) Complete DNA sequence of a serogroup A strain of *Neisseria meningitidis* Z2491. *Nature* 404: 502–06

Paton, J.C. (1998) Novel pneumococcal surface proteins: role in virulence and vaccine potential. *Trends in Microbiology* 6: 85–87

Peltola, H. (2000) Worldwide *Haemophilus influenzae* type b disease at the beginning of the 21st century: global analysis of the disease burden 25 years after the use of the polysaccharide vaccine and a decade after the advent of conjugates. *Clinical Microbiology Reviews* 13: 302–17

Perrin, A., Nassif, X. & Tinsley, C. (1999) Identification of regions of the chromosome of *Neisseria meningitidis* and *Neisseria gonorrhoeae* which are specific to the pathogenic Neisseria species. *Infection and Immunity* 67: 6119–29

Polissi, A., Pontiggia, A., Feger, G. *et al.* (1998) Large-scale identification of virulence genes from *Streptococcus pneumoniae. Infection and Immunity* 66(12): 5620–29

Portegies, P. & Corssmit, N. (2000) Epstein-Barr virus and the nervous system. *Current Opinion in Neurology* 13: 301–04

Ramers, C., Billman, G., Hartin, M., Ho, S. & Sawyer, M.H. (2000) Impact of a diagnostic cerebrospinal fluid enterovirus polymerase chain reaction test on patient management. *Journal of the American Medical Association* 283: 2680–85

Rappelli, P., Are, R., Casu, G. *et al.* (1998) Development of a nested PCR for detection of Cryptococcus neoformans in cerebrospinal fluid. *Journal of Clinical Microbiology* 36: 3438–40

Rossjohn, J., Gilbert, R.J., Crane, D. *et al.* (1998) The molecular mechanism of pneumolysin, a virulence factor from *Streptococcus pneumoniae. Journal of Molecular Biology* 284: 449–61

Rubin, L.G. (2000) Pneumococcal vaccine. *Pediatric Clinics of North America* 47: 269–85

Rubins, J.B. & Janoff, E.N. (1998) Pneumolysin: a multifunctional pneumococcal virulence factor. *Journal of Laboratory Clinical Medicine* 131: 21–27

Sanchez-Portocarrero, J., Perez-Cecilia, E., Corral, O., Romero-Vivas, J. & Picazo, J.J. (2000) The central nervous system and infection by Candida species. *Diagnostic Microbiology of Infectious Diseases* 37: 169–79

Sawaya, R., Arhin, F.F., Moreau, F., Coulton, J.W. & Mills, E.L. (1999) Mutational analysis of the promoter region of the *por*A gene of *Neisseria meningitidis. Gene* 233: 49–57

Serino, L. & Virji, M. (2000) Phosphorylcholine decoration of lipopolysaccharide differentiates commensal Neisseriae from pathogenic strains: identification of *lic*A-type genes in commensal Neisseriae. *Molecular Microbiology* 35: 1550–59

Stojiljkovic, I., Hwa, V., deSaint Martin, L. *et al.* (1995) The *Neisseria meningitidis* haemoglobin receptor: its role in iron utilisation and virulence. *Molecular Microbiology* 15: 531–41

Tenser, R.B. (1984) Herpes simplex and herpes zoster: nervous system involvement. *Neurological Clinics* 2: 215–40

Tettelin, H., Saunders, N.J., Heidelberg, J. *et al.* (2000) Complete genome sequence of *Neisseria meningitidis* serogroup B strain MC58. *Science* 287: 1809–15

Tu, A.H., Fulgram, R.L., McCrory, A., Briles, D.E. & Szalai, A.J. (1999) Pneumococcal surface protein A inhibits complement activation by *Streptococcus pneumoniae. Infection and Immunity* 67: 4720–24

Tuomanen, E. (1999) Molecular and cellular biology of pneumococcal infection. *Current Opinion in Microbiology* 2: 35–39

von Herrath, M.G., Berger, D.P., Homann, D. *et al.* (2000) Vaccination to treat persistent viral infection. *Virology* 268: 411–19

Walia, R. & Hoskyns, W. (2000) Tuberculosis meningitis in children: problem to be addressed effectively with thorough contact tracing. *European Journal of Pediatrics* 159: 535–38

Wang, P. & Heitman, J. (1999) Signal transduction cascades regulating mating, filamentation, and virulence in *Cryptococcus neoformans. Current Opinion in Microbiology* 2: 358–62

Whatmore, A.M. & Dowson, C.G. (1999) The autolysin-encoding gene (*lyt*A) of *Streptococcus pneumoniae* displays restricted allelic variation despite localised recombination events with genes of pneumococcal bacteriophage encoding cell wall-lytic enzymes. *Infection and Immunity* 67: 4551–56

Whatmore, A.M., King, S.J., Doherty, N.C. *et al.* (1999) Molecular characterisation of Equine isolates of *Streptococcus pneumoniae*: natural disruption of genes encoding the virulence factors pneumolysin and autolysin. *Infection and Immunity* 67: 2776–82

White, K.A., Lin, S., Cotter, R.J. & Raetz, C.R. (1999) A *Haemophilus influenzae* gene that encodes a membrane bound 3-deoxy-D-manno-octulosonic acid (Kdo) kinase. Possible involvement of kdo phosphorylation in bacterial virulence. *Journal of Biological Chemistry* 274: 31391–400

Zhang, J.R. & Tuomanen, E. (1999) Molecular and cellular mechanisms for microbial entry into the CNS. *Journal of Neurovirology* 5: 591–603

Zhu, P.X., Morelli, G. & Achtman, M. (1999) The *opc*A and psi *opc*B regions in Neisseria: genes, pseudogenes, deletions, insertion elements and DNA islands. *Molecular Microbiology* 33: 635–50

Useful web sites

Meningitis Research Foundation: http://www.meningitis.org
Meningitis Foundation of America: http://www.musa.org
Centers for Disease Control and Prevention: http://www.cdc.gov

GLOSSARY

16S RNA one of the structural RNA components of the ribosome. Prokaryotes have 5S and 23S species in the large subunit and a 16S species in the small subunit. Eukaryotes have a 5S, 5.8S and 28S species in the large subunit and an 18S species in the small subunit

dsDNA (double-stranded DNA) two base-pairing strands of DNA in the double-helix form

ELISA (enzyme-linked immunosorbent assay) sensitive technique for the detection of small amounts of protein or other antigenic substances

endothelial cell lining cell of a body cavity

epithelial cell any of the surface layer of cells covering the cutaneous, mucous and serous surfaces of the body

immunoelectrophoresis technique for the analysis of antigens by separation of the antigen components by electrophoresis prior to testing against antibodies

oligoprobe a peptide of a small number of component amino acids that has been labelled (either radioactively, with biotin, digoxygenin or fluorescein) and will hybridize to complementary sequences that contain the gene of interest

phagocyte a cell that is capable of phagocytosis, which is a five-step process by which particular cells engulf and destroy microorganisms and cell debris: (1) invagination;

(2) engulfment; (3) formation of phagocyte vacuole; (4) digestion of phagocytosed material; and (5) release of digested microbial products

proteolytic cleavage separation of proteins and peptides into their basic amino acids by hydrolysis of the peptide bonds

septicaemia systemic infection by pathogens circulating in the bloodstream, having spread from infection in any part of the body

serotype/serogroup subdivision of a species, especially bacteria or viruses, characterized by its antigenic character

virulence factor the extent to which a microorganism can produce disease

See also **Influenza** (p.101)

ANTIBIOTIC RESISTANCE IN BACTERIA

Julian Davies
Department of Microbiology and Immunology, University of British Columbia, Canada

1. Introduction

The introduction of antibiotics for the treatment of infectious diseases in the late 1940s drastically disturbed human and animal microbial ecology: the response of microbes (especially pathogens) to the threat of extinction has been to develop resistance to every antimicrobial agent used, with the result that there is a large pool of resistance determinants in the environment that can, by a variety of genetic mechanisms, become available to the microbial population in general. The origins, evolution and dissemination of these resistance genes is the subject of this review.

2. Biochemistry

The first antimicrobial agents used extensively in human therapy were the chemically synthesized sulfonamides, introduced in the late 1930s and still employed to this day. Streptococci and gonococci resistant to these agents were isolated soon after the first clinical use of the drugs, and resistance is now common in all bacterial genera, both Gram-positive and Gram-negative (Huovinen *et al.*, 1995). The early isolates have not been studied in any detail, but it is likely that the resistance mechanism involved mutations that led either to amplified expression of the gene for the target of the drug, the enzyme dihydropteroate synthase, or to the production of an altered enzyme that was less susceptible to inhibition. Plasmid-determined resistance to the sulfonamides may well have played a part in the early period of drug use, but was not identified definitively until the mid-1950s; this form of resistance is now the most common.

The introduction of penicillin, the first natural product antibiotic, was clearly the landmark in antimicrobial therapy, and the β-lactams and their many classes (cephalosporins, penicillins, etc.) have been the most widely used of all the antibiotics; in 1996, they accounted for the major share of the antibiotic market. Interestingly, microbial resistance to the β-lactams was presaged in 1940 when Abraham and Chain identified a penicillinase (β-lactamase) in a strain of *Escherichia coli*, several years before penicillin was introduced for general therapeutic use. The array of β-lactamase genes currently distributed in the bacterial population is the major impediment to successful β-lactam therapy (Bush *et al.*, 1995). The introduction of each novel member of this class of antibiotic has been followed by the development of resistance, in some cases in as little as 2–3 years. In some penicillin-resistant pneumococcal infections, no significant problem due to resistant bacteria appeared for some considerable time, but penicillin-resistant *Streptococcus pneumoniae* are now an increasing problem, especially within the general population (Appelbaum, 1992).

This phenomenon of development of resistance during therapy has been noted for all the antibiotic classes. Some ten different biochemical processes have been identified and the majority of antibiotics are affected by at least two different mechanisms of resistance. Tables 1–3 list the commonly used antibiotics with their modes of action and associated resistance mechanisms.

3. Genetics of resistance

When antimicrobial agents were first introduced for general clinical use, resistance probably arose by mutation of the gene for the target of the antibiotic, leading to overexpression and "out-titration" of the drug or reduced binding of the drug to the target, as happened with the sulfonamides; another example is the case of streptomycin, which has been used since the late 1940s for the treatment of tuberculosis. Streptomycin is an inhibitor of protein synthesis targeting the ribosome, and resistance developed in *Mycobacterium tuberculosis* as a result of mutations in structural genes for 16S ribosomal RNA (Finken *et al.*, 1993) or ribosomal protein S12 (Honore & Cole, 1994); these mutations essentially eliminated the receptor site for the drug on the ribosome. It should be noted that, in the case of *M. tuberculosis*, resistance to the commonly used therapeutic agents (streptomycin, rifampicin, isonicotinic hydrazide, etc.) in clinical situations occurs only as a result of mutations (Zhang & Young, 1994). As we shall see, another, unsuspected genetic determinant was responsible for antibiotic resistance in other pathogens.

In the early 1950s, bacterial dysentery occurred on an epidemic scale in Japan and, for the first time, antibiotics were dispensed on a large scale to try to control the disease. In 1955, antibiotic-resistant *Shigella dysenteriae* were reported and before long strains with resistance to up to four different antibiotics were identified (Davies, 1995). The genetic basis of this unexpected multiple antibiotic resistance was revealed to be the presence of transmissible resistance (R) factors that promoted the rapid spread of resistance determinants throughout the bacterial population. Other reports of this phenomenon of extrachromosomally determined antibiotic resistance were soon identified in Europe and North America, and it became obvious that R-plasmids could be the most common genetic element responsible for bacterial resistance to antibiotics in any clinical situation (Falkow, 1975).

The discovery of antibiotic resistance plasmids was unexpected and controversial. The prevailing thinking in bacterial genetics anticipated that microbes might undergo spontaneous mutation generating resistance to a single antibiotic; the concept of resistance developing simultaneously to two or more different classes of antibiotic (non-identical targets) could not be explained on the basis of mutation. However, the isolation of multiple-drug-resistant bacterial pathogens worldwide soon provided convincing proof of the validity of the findings; extrachromosomal determinants for resistance

Table 1 Biochemical mechanisms of resistance.

Mechanism	Class of antibiotic
1. Reduced uptake into cell	Chloramphenicol β-lactams
2. Active efflux from cell	Tetracycline Quinolones Sulfonamides Trimethoprim
3. Modification of target to eliminate or reduce binding of antibiotic	β-lactams Erythromycin Rifampin Aminoglycosides
4. Detoxification of antibiotic by enzymic modification:	
i) hydrolysis	β-lactams Erythromycin
ii) derivatization	Aminoglycosides Chloramphenicol Fosfomycin Lincomycin
5. Sequestration of antibiotic by protein binding	β-lactams Fusidic acid
6. Metabolic bypass of inhibited reaction	Sulfonamides Trimethoprim
7. Binding of specific immunity protein to antibiotic	Bleomycin
8. Overproduction of antibiotic target (titration)	Sulfonamides Trimethoprim
9. Failure to activate prodrug	Isoniazid

Table 2 Commonly used antibiotics and their mechanisms of resistance.

Class of antibiotic	Examples	**Resistance mechanism** (see Table 1)
β-lactam	penicillin, methicillin, cephamycin	1, 3, 4(i), 5
Aminoglycoside	kanamycin, gentamicin, tobramycin	3, 4(ii)
Tetracycline	tetracycline, minocycline	1, 3
Macrolide	erythromycin, clarithromycin	3, 4(i)
Sulfonamide combined with Trimethoprim	co-trimoxazole	2, 6, 8
Quinolone	nalidixic acid	2, 3
Ansamycin	rifampin	3
Nicotinic hydrazide	isoniazid	9
Glycopeptide	vancomycin	3

to antibiotics, toxic metals, detergents and even bacterial virulence factors were identified (Timmis *et al.*, 1986). Plasmids come in a variety of shapes, sizes and forms, and it is not possible to describe all their characteristics in this review. Perhaps the two most important properties of bacterial plasmids (apart from their ability to encode antibiotic resistance) are the traits of self-transmission and bacterial host range. The majority of R-plasmids harbour genes that promote transmission by bacterial conjugation (that is, the plasmids are bacterial sex factors); this permits both vertical and horizontal transmission of the plasmid-encoded determinants and a given population rapidly acquires antibiotic resistance. The R-plasmids present in the first resistant isolates identified in Japan had limited host ranges, being propagated only within the Enterobacteriaceae. Subsequently, resistance plasmids with the capacity to be transmitted and maintained in a much broader range of bacterial genera were identified. These "promiscuous" plasmids possessed conjugation systems that permitted transfer among a wide spectrum of unrelated bacterial genera which could include Gram-negative, Gram-positive and even anaerobic hosts (Thomas, 1989). Some classes of plasmid which were non-conjugative could be promoted to transfer in the presence of broad–host range fertility factors resident in the same bacterial host (mobilization). An important characteristic of broad–host range plasmids is the property of replication and stable maintenance in many different bacterial genera. Obviously the ability to "parasitize" the replication systems of unrelated bacterial hosts is an important factor in broad–host range transfer and this property of the R-plasmids is not well understood.

Table 3 Antibiotics and their modes of action.

Antibiotic	Use	Mode of action
Aminoglycosides	potent broad-spectrum inhibitors	block protein synthesis in susceptible hosts by binding to ribosomes
Ansamycins	primarily used in the treatment of tuberculosis	block transcription (RNA synthesis) by interfering with the function of RNA polymerase
β-lactams	most used antibiotic with broad-spectrum of activity	interfere with synthesis of cell wall peptidoglycan
Macrolides	primarily used for Gram-positive bacterial infections	block protein synthesis by binding to ribosomes
Quinolones	broad-spectrum agents	prevent DNA synthesis by interfering with the function of topoisomerase enzymes
Sulfa-trimethoprim	primarily used in the treatment of urinary tract infections caused by sensitive organisms	blocks the biosynthesis of folic acid, thus interfering with the formation of precursors of DNA
Tetracyclines	broad-spectrum inhibitors	block protein synthesis by preventing ribosome function

R-plasmids are not necessarily autonomous in their bacterial hosts. Many possess short regions of nucleotide sequence homology with the bacterial chromosome which resemble those associated with temperate bacteriophage, and the R-plasmid may insert into the host chromosome by site-specific integration promoted by the appropriate recombination enzyme (gd, etc.). Other plasmids associate with the host chromosome by participating in homologous, recombinative integration across identical insertion sequences (IS) on the plasmid and chromosome. Many different IS sequences are found on R-plasmids and they have been shown to play important roles in the distribution of resistance genes among related bacterial species. With the advent of molecular techniques, studies of R-plasmid structure revealed that antibiotic resistance genes are often clustered within discrete elements on the plasmid genome and, furthermore, that the same element is often found associated with a variety of different plasmid types or with the chromosomes of different bacterial genera, implying an independent transmission mechanism known as transposition (Craig, 1996). These transposable resistance elements are composed of one or more resistance genes flanked by directly or indirectly repeated IS elements. The process of transposition is an illegitimate (non-homologous) recombination event mediated by transposases that are encoded in the transposon sequence flanked by the IS elements or in one of the IS elements itself. Two different types of transposition mechanism have been characterized: (1) excision of the transposon as a discrete structure from its host replicon and direct insertion into another, and (2) formation of a cointegrate structure between the donor and recipient replicons with subsequent resolution of the structure to generate a copy of the transposable element in the recipient. The antibiotic resistance genes found on transposons are often organized in the form of an integron-related cluster (see below), which may include genes encoding resistance to heavy metals (mercury) and disinfectants (quaternary ammonium salts).

Many of the transposons found in Gram-positive bacteria (enterococci, Streptococci, Staphylococci and Bacteroides spp.) have been shown to encode functions that promote their own conjugative transfer, permitting inter- or intraspecies exchange (Clewell & Flannagan, 1993). The conjugative transposons often have a very broad host range and are likely to have been responsible for the widespread distribution of tetracycline resistance in many bacterial genera. This transfer is actually enhanced by the use of the antibiotic; it has been demonstrated recently that the expression of the genes necessary for conjugative transposition is actually induced by the presence of low concentrations of tetracycline (Salyers *et al.*, 1995). In other words, use of the antibiotic provides the selective pressure for both the appearance of resistance and its dissemination within the bacterial population.

While these two processes (mutation or plasmid inheritance) usually contribute independently to the aetiology of antibiotic resistance, there is increasing evidence of the further evolution of plasmid-encoded functions by point mutation. The best example is that of the β-lactamases in Gram-negative bacteria (Bush *et al.*, 1995). In attempts to combat increasing resistance to β-lactam antibiotics in the 1960s, semisynthetic compounds were designed to have activity against β-lactamase-containing strains, provoking one of the most striking demonstrations of natural genetic engineering: the effective substrate range of the β-lactamases was broadened by mutation through a series of single base changes in the genes that introduced successive amino acid substitutions in the enzymes to inactivate the "new" antibiotics. The pattern of evolution by mutation has continued with the subsequent development and use of a large family of antibiotics, the third and fourth generation cephalosporins: each introduction of a new and apparently more robust antibiotic (in terms of resistance) provoked the appearance of a novel β-lactamase. More than 100 different β-lactamases have now been identified among the Enterobacteriaceae and this plethora of plasmid-encoded, extended-spectrum β-lactamases can hydrolyse all the available β-lactam antibiotics (penicillins and cephalosporins). There are no other reported cases wherein an acquired antibiotic resistance gene associated with an R-plasmid has evolved through mutation in such a dramatic fashion. Thus, while genes recruited from the environment on plasmids remain the most common resistance determinants in clinical isolates, evolution by subsequent mutational steps in response to the selection pressure of a "new" antibiotic or derivative is a significant threat to antibiotic therapy (Davies, 1994).

4. Evolution (acquisition) of resistance

The above example of the β-lactamase enzymes emphasizes the rapidity with which resistance genes may evolve in response to the selective pressures of antibiotic usage. However, the origins and evolution of the genetic determinants of antibiotic resistance are not well understood. The rapidity with which bacterial pathogens become resistant to the antibiotics introduced to control them, usually within a few years after the introduction of the antibiotic, lends support to the notion that resistance genes (or their close relatives) are already present in the bacterial gene pool and are accessible to all microbes in the environment (Davies, 1997).

Microorganisms that produce molecules with cytotoxic or antibiotic activity (Actinomycetes are the most significant group in this respect) must, by necessity, possess the wherewithal to be resistant to the toxic agents they produce, otherwise they would commit suicide when the antibiotic was made. This fact gave rise to the proposition that antibiotic-producing organisms are likely to be the source of most antibiotic resistance genes (Benveniste & Davies, 1973). The nucleic acid sequences of clusters of antibiotic resistance genes found in a variety of bacterial hosts often differ in percentage G + C composition relative to the flanking sequences of the resident plasmid or chromosome, which supports the concept of gene recruitment: the resistance genes form "islands" of distinct nucleic acid sequence. There is substantial evidence consistent with the notion of the origin of antibiotic resistance genes in producing organisms (biochemical similarities, nucleic acid sequence relationships), but absolute proof as evidenced by direct genetic transfer is lacking. The resistance genes could well have been derived from other sources, such as cryptic or latent genes in bacterial chromosomes. For example, the aminoglycoside resistance gene (aminoglycoside 2′-O-acetyltransferase) is present in all strains of *Providencia stuartii* as a chromosomal gene and its expression is regulated by other chromosomal genes. The acetyltransferase may be involved in some cellular housekeeping function, and it appears that the enzyme plays a role in the maintenance of cell-membrane peptidoglycan structure (Payie *et al.*, 1995). A number of bacterial genera encode chromosomal β-lactamases, and it is assumed that they also play a role in cell-wall function, in addition to their roles as β-lactam hydrolases.

The fact that resistance genes are acquired so readily within the bacterial population begs the question of how such heterologous gene "pickup" occurred. In the case of the Enterobacteriaceae, an answer to this question comes from the studies of Hall and Stokes and colleagues (Recchia & Hall, 1995) who have analysed the structures of clusters of antibiotic resistance genes found on plasmids or transposons in these bacteria. Their analysis led to the proposal that DNA fragments encoding different antibiotic resistance genes, once inside the cell, can insert by non-homologous recombination, in a tandem fashion, into a functional genetic expression element known as an integron. Detailed molecular analyses of integron resistance clusters and their formation show that resistance genes, in the form of circular cassettes, are integrated by the mediation of a recombinase at a specific site in a linear array. The resistance genes are subsequently transcribed from a strong promoter that is an intrinsic component of the integron structure. This mechanism has been confirmed by *in vitro* demonstrations of the integration of circular DNA cassettes encoding resistance genes into the attachment site adjacent to the recombinase gene.

There is no evidence that structures similar to integrons exist outside the Enterobacteriaceae, and the mechanisms of acquisition and integration of resistance genes into R-plasmids of other microbes, such as Gram-positive bacteria, remain to be elucidated. This information is crucial to an understanding of the evolution of antibiotic resistance, since resistance determinants appear to have been transferred initially from Gram-positive to Gram-negative bacteria in clinical situations (Courvalin, 1994). It should be noted that the process of gene acquisition is random, and it is not uncommon to find bacterial strains that have more than one resistance determinant for the same class of antibiotic.

5. Conclusion

The use of antibiotics in the treatment of infectious diseases has been and is seriously threatened by the development of antibiotic-resistant organisms. Since there is a direct correlation between antibiotic use and the appearance of resistant strains, the controlled (prudent) utilization of antibiotics would seem logical. Responsible scientists and physicians have been advocating this for many years. Yet, in North America, up to 40% of all hospital patients receive antibiotics and evidence indicates that half of this use is inappropriate (Gaynes & Monnet, 1997). Moreover, 50% of all antibiotics produced in the world are employed for non-human applications, such as growth promotion in farm animals, prophylaxis in fish farming and protective spraying of fruit trees. Not surprisingly, this has resulted in antibiotic resistance genes accessible to all bacteria. Regrettably, the use (or misuse) of antibiotics for both medical and non-medical applications is expanding. The efficacy of some of the most valuable human therapeutics, such as ciprofloxacin and vancomycin, is being threatened by extensive use in agriculture. Vancomycin is a glycopeptide antibiotic that is reserved for hospital treatment of serious Gram-positive infections; unhappily, related antibiotics have been used widely in animal husbandry with a corresponding increase in vancomycin-resistant enterococcal strains in the environment (Klare *et al.*, 1995). Enterococcal strains that are resistant to all of the agents currently approved for their control have been isolated. Of even greater concern is the possibility that transfer of the resistance genes will lead to vancomycin-resistant Staphylococci; this would render a significant pathogen untreatable by antibiotics, thus effectively creating a pre-antibiotic era in infectious disease. The knowledge that there are very few "replacement" antibiotics undergoing pharmaceutical development underlines the seriousness of the problem of antibiotic resistance.

REFERENCES

Abraham, E.P. & Chain, E. (1940) An enzyme from bacteria able to destroy penicillin. *Nature* 146: 837

Appelbaum, P.C. (1992) Antimicrobial resistance in *Streptococcus pneumoniae*: an overview. *Clinical Infectious Diseases* 15: 77–83

Benveniste, R. & Davies, J. (1973) Aminoglycoside antibiotic-inactivating enzymes in actinomycetes similar to those present in clinical isolates of antibiotic-resistant bacteria. *Proceedings of the National Academy of Sciences USA* 70: 2276–80

Bush, K., Jacoby, G.A. & Medeiros, A.A. (1995) A functional classification scheme for beta-lactamases and its correlation with molecular structure. *Antimicrobial Agents and Chemotherapy* 39: 1211–33

Clewell, D.B. & Flannagan, S.E. (1993) The conjugative transposons of Gram-positive bacteria. In *Bacterial Conjugation*, edited by D.B. Clewell, New York and London: Plenum Press

Courvalin, P. (1994) Transfer of antibiotic resistance genes between gram-positive and gram-negative bacteria. *Antimicrobial Agents and Chemotherapy* 38: 1447–51

Craig, N.L. (1996) Transposition. In *Escherichia coli and Salmonella typhimurium: Cellular and Molecular Biology*, 2nd edition, edited by F.C. Neidhardt *et al.*, Washington, DC: American Society for Microbiology Press

Davies, J. (1994) Inactivation of antibiotics and the dissemination of resistance genes. *Science* 264: 375–82

Davies, J. (1995) Vicious circles: looking back on resistance plasmids. *Genetics* 139: 1465–68

Davies, J. (1997) Origins, acquisitions and dissemination of antibiotic resistance determinants. In *Antibiotic Resistance: Origins, Evolution, Selection and Spread*, Ciba Foundation Symposium No. 207, Chichester and New York: Wiley

Falkow, S. (1975) *Infectious Multiple Drug Resistance*, London: Pion

Finken, M., Kirschner, P., Meier, A., Wrede, A. & Bottger, E.C. (1993) Molecular basis of streptomycin resistance in *Mycobacterium tuberculosis*: alterations of the ribosomal protein S12 gene and point mutations within a functional 16S ribosomal RNA pseudoknot. *Molecular Microbiology* 9: 1239–46

Gaynes, R. & Monnet, D. (1997) The contribution of antibiotic use on the frequency of antibiotic resistance in hospitals. In *Antibiotic Resistance: Origins, Evolution, Selection and Spread*, Ciba Foundation Symposium No. 207, Chichester and New York: Wiley

Honore, N. & Cole, S.T. (1994) Streptomycin resistance in mycobacteria. *Antimicrobial Agents and Chemotherapy* 38: 238–42

Huovinen, P., Sundstrom, L., Swedberg, G. & Skold. O. (1995) Trimethoprim and sulfonamide resistance. *Antimicrobial Agents and Chemotherapy* 39: 279–89

Klare, I., Heier, H., Claus, H. *et al.* (1995) *Enterococcus faecium* strains with vanA-mediated high-level glycopeptide resistance isolated from animal foodstuffs and fecal samples of humans in the community. *Microbial Drug Resistance* 1: 265–72

Payie, K.G., Rather, P.N. & Clarke, A.J. (1995) Contribution of gentamicin 2′-N-acetyltransferase to the O acetylation of peptidoglycan in *Providencia stuartii*. *Journal of Bacteriology* 177: 4303–10

Recchia, G.D. & Hall, R.M. (1995) Gene cassettes: a new class of mobile element. *Microbiology* 141: 3015–27

Salyers, A.A., Shoemaker, N.B. & Li, L.-Y. (1995) In the driver's seat: the Bacteroides conjugative transposons and the elements they mobilize. *Journal of Bacteriology* 177: 5727–31

Thomas, C.M. (1989) *Promiscuous Plasmids of Gram-negative Bacteria*, London and San Diego: Academic Press

Timmis, K.N., Gonzalez-Carrero, M.I., Sekizaki, T. & Rojo, F. (1986) Biological activities specified by antibiotic resistance plasmids. *Journal of Antimicrobial Chemotherapy* 18 (Suppl. C): 1–12

Zhang, Y. & Young, D. (1994) Molecular genetics of drug resistance in Mycobacterium tuberculosis. *Journal of Antimicrobial Chemotherapy* 34: 313–19

FURTHER READING

Ciba Foundation Symposium No. 207 (1997) *Antibiotic Resistance: Origins, Evolution, Selection and Spread*, Chichester and New York: Wiley

Davies, J. & Webb, V. (1998) Antibiotic resistance in bacteria. In *Emerging Infections*, edited by R.M. Krause, New York: Academic Press

Levy, S.B. (1992) *The Antibiotic Paradox: How Miracle Drugs are Destroying the Miracle*, New York and London: Plenum Press

Mazel, D. & Davies, J. (1999) Antibiotic resistance in microbes. *Cellular and Molecular Life Sciences* 56: 742–54

Nikaido, H. (1994) Prevention of drug access to bacterial targets: permeability barriers and active efflux. *Science* 264: 382–87

Spratt, B.G. (1994) Resistance to antibiotics mediated by target alterations. *Science* 264: 388–93

GLOSSARY

anaerobic host an organism that has a very low tolerance for oxygen

bacterial dysentery (*Shigella dysenteriae*) inflammation of the intestine, particularly of the colon

integron class of DNA element composed of a DNA integrase gene adjacent to a recombination site at which one or more genes can be found inserted

insertion sequence (IS) elements mobile nucleotide sequences that occur naturally in the genomes of bacterial populations

replicons a replication unit of the genome. It contains an origin of transfer. Most plasmids are independent replicons

See also **Acquisition and spread of antibiotic resistance genes by promiscuous plasmids** (p.119); **Antibiotic resistance: theory into practice** (p.125); **Genetic recombination** (p.701)

ACQUISITION AND SPREAD OF ANTIBIOTIC RESISTANCE GENES BY PROMISCUOUS PLASMIDS

Christopher M. Thomas
School of Biosciences, University of Birmingham, Birmingham, UK

1. Introduction

Resistance to antibiotics can arise in bacteria in a number of ways, including: mutation in the gene encoding the target for the antibiotic so that the antibiotic no longer binds to and inhibits the vital molecule in the cell; active export of the antibiotic out of the cell, so that the intracellular concentration is maintained below an inhibitory level; and, finally, chemical alteration of the antibiotic so that it can no longer bind to its target. For each of these mechanisms a single resistance gene is often found which, if transported into a sensitive cell, would render it partly or completely resistant. The appearance of resistance genes by new mutation or recruitment from bacteria in which they have served a function for many years (as, for example, bacteria that need to be resistant because they produce the antibiotic) and their spread to bacteria of clinical and veterinary importance has raised the spectre of a time when antibiotics will no longer be an effective way of treating bacterial infection (Davies, 1994; Bennett, 1995).

2. Gene transfer mechanisms

The carriage of resistance genes from one cell to another can occur by a number of mechanisms: transformation (uptake of naked DNA), transduction (carriage of the DNA between cells packaged in a virus particle) or conjugative transfer (DNA transfer through a pore at a junction where two cells fuse temporarily) (Mazodier & Davies, 1991). If the resistance gene is chromosomally encoded then its establishment in the recipient bacteria will normally involve homologous recombination, in which related sequences line up and undergo breaking/joining to replace resident DNA by incoming DNA. The sequences need not match perfectly for this to happen, but must be ≥70% identical, depending on the species that is undergoing recombination. However, if the gene becomes part of a plasmid (a small piece of DNA that can multiply autonomously in the bacterial cytoplasm), then plasmid transfer can allow the spread of the resistance gene independently of homologous recombination so long as the plasmid is able to replicate in the new strain into which its DNA has been transferred. The spread of the resistance gene will, therefore, be most rapid between bacteria which can participate in conjugative transfer and in which the plasmid replicates efficiently. If the gene happens also to be part of a defined DNA segment, termed a transposon, within the plasmid then it has an additional means of establishment: transposition from the plasmid location into the chromosome of the new host (this is, once again, independent of homologous recombination).

3. Acquisition of resistance genes by plasmids and transposable elements

Evidence for the relatively recent spread of antibiotic resistance genes from antibiotic-producing bacteria, that may have possessed them for many centuries as a means of protection against the antibiotics they produce, is based on comparison of DNA sequences which reveal closely related genes in diverse organisms (Davies, 1992). The polymerase chain reaction (PCR) has been used to demonstrate that crude antibiotic preparations of a grade which can be used as a food supplement to promote growth in animal husbandry contain DNA of the antibiotic-producing organism, including resistance genes (Webb & Davies, 1993). This DNA could be taken up by bacteria in the resident flora of an animal and integrated into chromosomal or plasmid DNA. A possible route by which such foreign DNA may be integrated into the genome has been indicated by the discovery of a group of transposable elements encoding an integrase enzyme that is associated with the ability to recruit a variety of gene cassettes via a specific sequence termed the 59 base pair (bp) element because of the length of the first and dominant member of this family (**Figure 1**) (Hall & Collis, 1995). The first step, conversion to a cassette, may involve illegitimate recombination or reverse transcription with the 59 bp element possibly acting as a primer to generate a DNA fragment with a promoter-less protein coding sequence, the 59 bp element and a short motif including GTT which defines the later integration point. After this "mutational event", rapid evolution can occur to produce complex combinations of selectable genes. Members of a ubiquitous family of integrase-associated transposons carry mercury resistance. Since low levels of mercury are released from amalgam tooth fillings during chewing, it has been proposed that we constantly select mercury-resistant gut flora and, therefore, favour those bacteria with transposons that are most likely to recruit new resistance genes (Lorscheider *et al.*, 1995). Indeed, mercury resistance is one of the commonest traits found on plasmids. Plasmids with a broad host range could help to recruit and transport genes between many groups of bacteria.

4. Broad host range replicons

The reasons why some plasmids are able to multiply in many different hosts while others are confined to one or a group of closely related species is not understood (Kues & Stahl, 1989; Del Solar *et al.*, 1996). DNA sequencing reveals that related replication systems are found in many different species. In some cases, it is likely that a plasmid has evolved

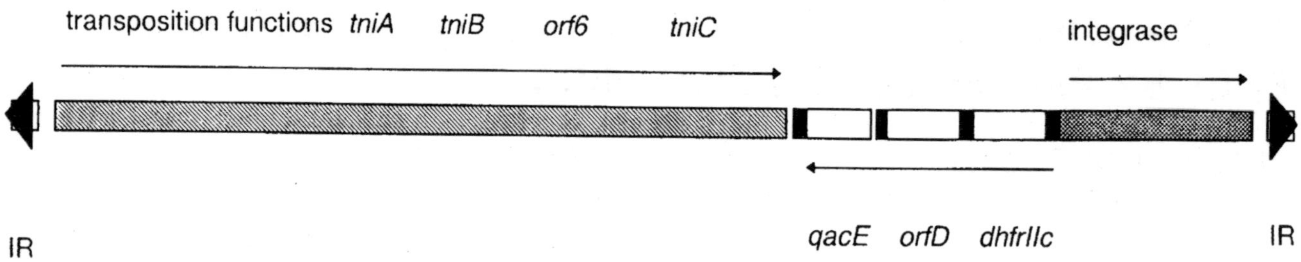

Figure 1 Organization of the typical transposon Tn*402*/Tn*5090* (Radstrom *et al.*, 1994). Transposition proteins are encoded by *tniA, tniB* and *tniC*. These proteins act on the inverted repeats (IR) at each end of the transposon. They cut the DNA symmetrically on either end of the element and splice these ends to random nick sites in the target DNA molecule. *tniR* encodes a DNA-binding protein that acts both as a repressor and a site-specific recombinase which promotes resolution of transposition intermediates containing two copies of the transposon. The result is one copy of the transposon left at its original location and one copy inserted at the new site. Not all transposons move in this replicative fashion. Some effectively splice themselves out and insert themselves at a new location in a "conservative" (non-replicative) fashion. The integrase is another site-specific recombinase which acts on the 59 bp elements (■) to promote integration of cassettes containing the same 59 bp sequence. How these cassettes arise is not known (perhaps by reverse transcription or by a one-ended rolling circle copying process), but once formed these cassettes can accumulate downstream of the anti-*int* promoter which directs their expression. The *dhfrII* cassette encodes a mutant dihydrofolate reductase that gives resistance to the antibiotic trimethoprim. *qacE* encodes an export system which gives resistance to quaternary ammonium compounds used as disinfectants. No phenotype for *orfD* is known. The element is thus an efficient machine for acquisition of resistance genes and their random movement from one genome to another. Although variants with new genes arise rarely, if they confer a selective advantage they will spread rapidly through a population.

in a single host while, in other cases, nearly identical plasmids in different hosts indicate clearly that horizontal transfer has taken place. The most widespread family of replication systems initiates the replication process by introducing a nick into one strand at the replication origin (**Figure 2A**) (Gruss & Ehrlich, 1989; Novick, 1989). The 3′ end generated acts as the primer for rolling circle replication, displacing the other strand which is then converted back to double-stranded form by replication from a second origin. This sort of replicon is found in both Gram-negative and Gram-positive bacteria. It is common for a single plasmid to be able to replicate in many different Gram-positive species and at least one such plasmid, originating in Streptococcus, will also replicate in Gram-negative bacteria.

Another replication strategy is the activation of the replication origin by the binding of a plasmid protein, Rep, to a repeated sequence, which thus recruits host replication proteins to unwind the DNA and allows the copying process to begin (**Figure 2B**) (Bramhill & Kornberg, 1988). Some plasmids which use this strategy have a narrow host range while others have a broad host range. In the plasmid pPS10 from *Pseudomonas savastanoi*, a single point mutation in the plasmid Rep protein can change the host range of the replication system. The IncQ plasmids code for extra proteins (their own helicase and primase) which allows them to be relatively independent of the host replication system. These plasmids can replicate in at least one Gram-positive species (Mycobacteria) as well as in every Gram-negative species tested. Plasmids of the IncP group have a simple replicon, but their Rep protein is made in two forms which may give it a flexibility in the ways that it can interact with its host's replication machinery. Thus, there does not seem to be a simple set of rules defining the properties necessary to possess a broad host range.

5. The IncP plasmids

The IncP plasmids are the best studied broad host range plasmids (Thomas & Smith, 1987). Originally isolated as the agents that caused an outbreak of carbenicillin-resistant *Pseudomonas aeruginosa* infections in the Burns Unit of the Birmingham Accident Hospital in Britain in 1969, related plasmids have subsequently been isolated from bacteria all over the world. The plasmids are found in many different species and confer a variety of different phenotypes. There are two main groups of IncP plasmids, designated α and β. The complete sequence of the archetypal IncPα plasmids, RK2 and RP4, as well as the best understood IncPβ plasmids, have been compiled (Pansegrau *et al.*, 1994).

In combination with detailed molecular genetic studies, a number of properties of these plasmids have been determined (**Figure 3**). First, it is possible to identify a backbone of plasmid replication and transfer functions that is common to both groups that must predate the acquisition of the different resistance genes which are now carried by different plasmids (Smith & Thomas, 1987). Secondly, the expression of replication and transfer genes is coordinated by a set of global regulators encoded in a central control operon (Jagura-Burdzy & Thomas, 1994), two of whose genes are related to partitioning functions found on many other plasmids and in the chromosomes of many bacteria (Motallebi-Veshareh *et al.*, 1990). Thirdly, there are primarily two places in the plasmid backbone where it is possible to insert foreign blocks of genes carrying antibiotic resistance and other genes which may confer a selective advantage on the bacterial host. Within the IncPβ plasmids, the two regions which appear to be hotspots for acquisition of foreign DNA show a striking sequence motif which may be related to the tendency of these regions to attract these insertions (Smith *et al.*, 1993). Molecular tools for hybridization and PCR

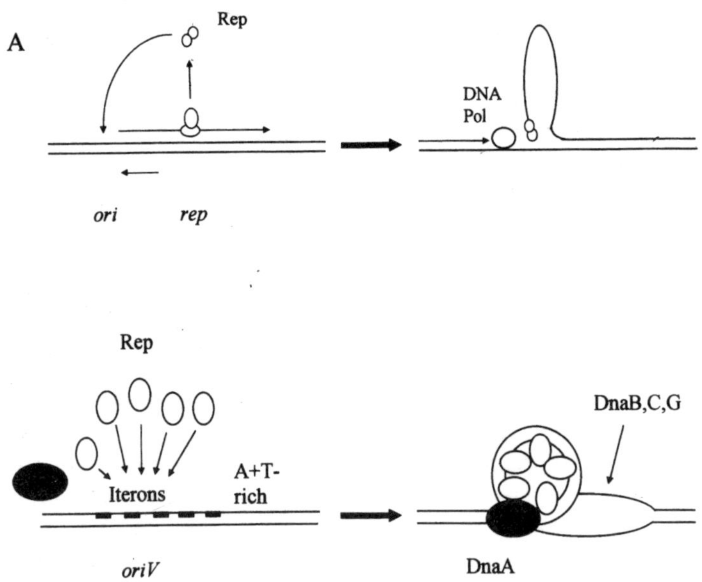

Figure 2 Broad host range strategies for initiation of replication. In both (A) and (B), the double lines represent just the replication origin region of circular plasmids prior to initiation. (A) Rolling-circle replication as found in plasmids of Gram-positive species such as *Bacillus subtilis* and *Staphylococcus aureus*. The replication protein Rep is encoded in the origin region. Transcription and translation of the *rep* mRNA is controlled by a short antisense RNA which is made constitutively and which shuts off expression when copy number rises too high. The Rep protein attacks the origin region at a specific site and nicks one strand, becoming covalently attached to the 5′ phosphate side of the gap. The 3′ OH acts as a primer for DNA polymerase which displaces the parental strand as it copies the unnicked strand. The attached Rep protein attacks the reformed origin when copying has completed the circle and creates a single-stranded circle from the displaced strand. (B) Iteron-activated theta replication as found in many low- to medium-copy number plasmids from Gram-negative bacteria. A series of directly repeated sequences in the origin binds monomers of the plasmid-encoded Rep protein. In many cases, the host protein DnaA also binds close by, often being helped to do so by the Rep protein. A nucleoprotein complex is formed which distorts the DNA and facilitates strand separation. DnaA recruits DnaBC helicase activity which causes further separation and the DnaG primase enters to initiate leading strand synthesis. A primosome (not shown) will then assemble to initiate replication on the lagging strand.

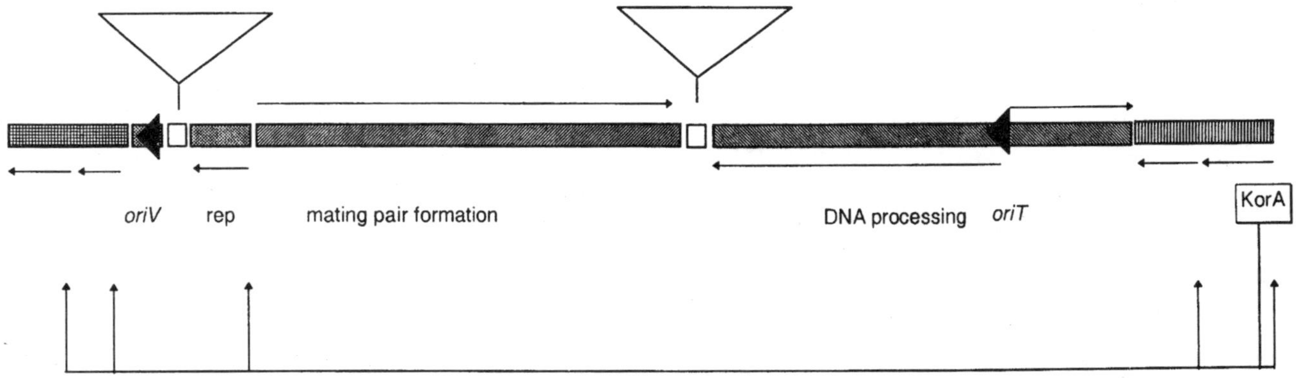

Figure 3 Organization of a typical plasmid that can transfer itself from one bacterium to another, modelled on the IncP-1 group plasmids that can transfer between and maintain themselves in all Gram-negative bacteria and can also transfer to Gram-positive bacteria and to yeast. Only two small parts of the plasmid are needed for replication, *rep* encoding the Rep protein which acts on *oriV* where replication is initiated (see **Figure 2**). Transfer between bacteria depends on genes that promote local fusion between bacteria (see **Figure 4**), so-called mating pair formation genes of which there are approximately ten. The genes for DNA processing (about three in total) act on *oriT*, the transfer origin, where a nick is introduced into one DNA strand which allows rolling circle replication to be initiated. The displaced strand transfers across to the recipient (see **Figure 4**). Thus, more than two-thirds of the plasmid is occupied by transfer genes. Low copy number plasmids need to be distributed evenly in the cell prior to cell division so that cells without plasmid are not produced. Much of the rest of the plasmid backbone is taken up with these functions. Finally, an increasing number of plasmids are recognized as having mechanisms which ensure that expression of replication transfer and stable inheritance is coordinated. In IncP-1 plasmids, KorA is one protein that performs this function, binding to the plasmid DNA as indicated by the arrows to control transcription. There are a few places where foreign DNA (often as part of transposable elements) can insert into this backbone. These places are indicated as white triangles. DNA inserted at such sites can add up to the same amount again as the "backbone" functions shown.

amplification to detect and follow specific plasmids in the environment are now available (Gotz *et al.*, 1996; Pukall *et al.*, 1996).

6. Promiscuous transfer systems

One of the most remarkable features of the IncP plasmids is their conjugative system which can promote very efficient transfer between all Gram-negative bacteria. It can also promote transfer from Gram-negative bacteria to Gram-positive bacteria and even to the yeasts *Saccharomyces cerevisiae* and *Schizosaccharomyces pombe* (Mazodier & Davies, 1991; Pansegrau *et al.*, 1994). The conjugative process depends on two distinct steps (**Figure 4**). The first step is mating pair formation which depends on the pilus that is characteristic of self-transmissible plasmids. In contrast to the first self-transmissible plasmid identified, F, the mating pair apparatus of IncP plasmids does not create a strong bridge between the bacteria when the bacteria are free in suspension. Efficient mating occurs only in dense layers of bacteria on a solid surface. This fits with the notion that efficient transfer in nature is probably confined to biofilms. Mating pair formation is thought to involve attachment of the pilus to the surface of the recipient, retraction to bring the bacterial surfaces together and localized membrane fusion with concomitant formation of a pore which may involve other transfer proteins in addition to, or instead of, the pilus subunit protein, pilin. DNA sequence analysis has shown that the *mpf* genes of IncP plasmids are remarkably similar in DNA sequence and organization, not only to the *mpf* genes of other self-transmissible plasmids, but also to the virulence system that is needed for Ti plasmid to transfer T-DNA from *Agrobacterium tumefaciens* to plant cells (Lessl *et al.*, 1992). Much of the system may be needed to export the pilus protein(s) to the outside of the bacteria as indicated by its additional similarity to the toxin export system of *Bordetella pertussis*. Given that this apparatus uses proteins which seem very flexible in the cell envelope-associated function they have evolved to perform, it makes sense that these plasmids are apparently able to interact with any type of surface to generate a pore through which the DNA can transfer. It is possible that the low efficiency mating that the IncP plasmids promote with non-Gram-negative surfaces has nothing to do with the normal *mpf* system. However, mutations in the *mpf* genes which abolish transfer between Gram negatives also abolish the low frequency mating with other Gram-positive bacteria and with yeast.

The second part of the transfer involves the duplication of the plasmid molecule by rolling circle replication with concomitant transfer (Lanka & Wilkins, 1995). This is initiated by the nicking of one strand at the transfer origin. The nick site shows sequence motifs characteristic of many other nick sequences, including not only other self-transmissible plasmids of Gram-negative bacteria, but also the rolling circle replication plasmids of Gram-positive bacteria. A number of the proteins which are needed for this DNA processing have relatives in many other analogous systems. In addition, the plasmid carries a DNA primase which appears to be transported from donor to recipient bacteria and which helps to promote conversion of the plasmid back from single-stranded to double-stranded form. In at least some species, the primase provides the plasmid with an advantage after transfer and this may be accentuated at low temperature (20°C–25°C).

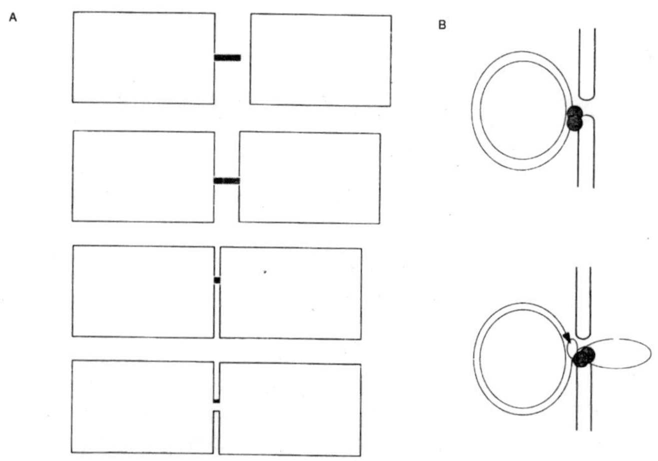

Figure 4 The two major steps in transfer of self-transmissible plasmids of Gram-negative bacteria. (A) A bacterium carrying a plasmid synthesizes a protein hair called a pilus. Pilus structure and synthesis is encoded by the mating pair formation genes of the plasmid (see **Figure 3**). The pilus tip binds to another bacterium and depolymerizes from its base, thus facilitating the close contact between the bacteria. In an as yet poorly understood process, a local membrane fusion and cell wall breakdown create a cytoplasmic bridge. (B) Once a mating bridge has formed this apparently triggers the activation of a protein–DNA complex called a relaxasome, located at *oriT* on the plasmid (see **Figure 3**). The relaxasome is so called because, when activated, it introduces a nick in one DNA strand thus "relaxing" the normally supercoiled molecule and allowing rolling circle replication to be initiated (see **Figure 2**) with the generation of a single-stranded loop which is extruded into the recipient bacterium, where it forms a circle, is converted back to double-stranded form and proceeds to establish itself at normal copy number in the new bacterium.

Conjugative plasmids and conjugative transposons of Gram-positive bacteria are also known to have a broad host range, but transfer is less well understood, despite involving many fewer genes (Scott, 1993; Clewell *et al.*, 1995). Pheromones can provide attraction between plasmid positive and negative bacteria. A pilus does not appear to be needed for mating pair formation, which may involve simple cell fusion followed by DNA exchange or replicative transfer from a rolling circle origin.

7. Gene expression and regulation

One factor that limits the host range of a plasmid is expression of the genes that the plasmid needs for replication and transfer. In this respect, the IncP plasmids are interesting because nearly all of the identified promoters are strong and match closely the consensus sequences for Eσ^{70}-type core RNA polymerase which occurs in most eubacteria. These promoters are all repressed by one of a number of directly or indirectly autogenous control circuits. Thus, the genes are prevented from overexpression, but can compensate for variations between hosts in the efficiency with which they are recognized.

8. Conclusions

The steps by which a resistance gene arises and spreads are beginning to be understood at a molecular level, but the factors that favour or disfavour these processes in mixed bacterial communities in soil, water or clinical specimens are poorly defined (Thomas, 2000). The references cited below may provide more information with which to approach the various aspects of the subject.

REFERENCES

Bennett, P.M. (1995) The spread of drug resistance. *SGM Symposium* 52: 317–44

Bramhill, D. & Kornberg, A. (1988) A model for initiation at origins of DNA replication. *Cell* 54: 915–18

Clewell, D.B., Flannagan, S.E. & Jaworski, D.D. (1995) Unconstrained bacterial promiscuity: the Tn*916*-Tn*1545* family of conjugative transposons. *Trends in Microbiology* 3: 229–36

Davies, J. (1992) Another look at antibiotic resistance. *Journal of General Microbiology* 138: 1553–59

Davies, J. (1994) Inactivation of antibiotics and the dissemination of resistance genes. *Science* 264: 375–82

Del Solar, G., Alonso, J.C., Espinosa, M. & Diaz-Orejas, R. (1996) Broad host range plasmid replication: an open question. *Molecular Microbiology* 21: 661–66

Gotz, A., Pukall, R., Smit, E. *et al.* (1996) Detection and characterisation of broad-host-range plasmids in environment bacteria by PCR. *Applied and Environmental Microbiology* 62: 2621–28

Gruss, A. & Ehrlich, S.D. (1989) The family of highly interrelated single-stranded deoxyribonucleic acid plasmids. *Microbiological Reviews* 53: 231–41

Hall, R.M. & Collis, C.M. (1995) Mobile gene cassettes and integrons: capture and spread of genes by site-specific recombination. *Molecular Microbiology* 15: 593–600

Jagura-Burdzy, G. & Thomas, C.M. (1994) KorA protein of promiscuous plasmid RK2 controls a transcriptional switch between divergent operons for plasmid replication and conjugative transfer. *Proceedings of the National Academy of Sciences USA* 91: 10571–75

Kues, U. & Stahl, U. (1989) Replication of plasmids in Gram negative bacteria. *Microbiological Reviews* 53: 491–516

Lanka, E. & Wilkins, B.M. (1995) DNA processing reactions in bacterial conjugation. *Annual Review of Biochemistry* 64: 141–69

Lessl M., Balzer, D., Pansegrau, W. & Lanka, E. (1992) Sequence similarities between the RP4 Tra2 and Ti VirB region strongly support the conjugative model for T-DNA-transfer. *Journal of Biological Chemistry* 267: 20471–80

Lorscheider, F.L., Vimy, M.J., Summers, A.O. & Zwiers, H. (1995) The dental amalgam controversy – inorganic mercury and the CNS – genetic linkage of mercury and antibiotic resistance in intestinal bacteria. *Toxicology* 97: 19–22

Mazodier, P. & Davies, J. (1991) Gene transfer between distantly related bacteria. *Annual Review of Genetics* 25: 147–71

Motallebi-Veshareh, M., Rouch, D.A. & Thomas, C.M. (1990) A family of ATPases involved in active partitioning of diverse bacterial plasmids. *Molecular Microbiology* 4: 1455–63

Novick, R.P. (1989) Staphylococcal plasmids and their replication. *Annual Review of Microbiology* 43: 537–65

Pansegrau, W., Lanka, E., Barth, P.T. *et al.* (1994) Complete nucleotide sequence of Birmingham IncP′ plasmids. Compilation and comparative analysis. *Journal of Molecular Biology* 239: 623–63

Pukall, R., Tschape, H. & Smalla, K. (1996) Monitoring the spread of broad and narrow host range plasmids in soil microcosms. *FEMS Microbiology Letters* 20: 53–66

Radstrom, P., Skold, O., Swedberg, G. *et al.* (1994) Transposon Tn*5090* of plasmid R751, which carries an integron, is related to Tn7, Mu and the retroelements. *Journal of Bacteriology* 176: 3257–68

Scott, J. (1993) Sex and the single circle: conjugative transposition. *Journal of Bacteriology* 174: 6004–10

Smith, C.A. & Thomas, C.M. (1987) Comparison of the organisation of the genomes of phenotypically diverse plasmids of incompatibility group P: members of the IncPβ subgroup are closely related. *Molecular and General Genetics* 206: 419–27

Smith, C.A., Pinkney, M., Guiney, D.G. & Thomas, C.M. (1993) The ancestral IncP replication system consisted of contiguous *oriV* and *trfA* segments as deduced from a comparison of the nucleotide sequences of diverse IncP plasmids. *Journal of General Microbiology* 139: 1761–66

Thomas, C.M. & Smith, C.A. (1987) Incompatibility group P plasmids: genetics, evolution and use in genetic manipulation. *Annual Review of Microbiology* 41: 77–101

Thomas, C.M. (2000) (ed) *The Horizontal Gene Pool: Bacterial Plasmids and Gene Spread*, Amsterdam: Harwood Academic

Webb, V. & Davies, J. (1993) Antibiotic preparations contain DNA – a source of drug resistance genes? *Antimicrobial Agents and Chemotherapy* 37: 2379–84

GLOSSARY

conjugation the temporary union of two bacteria that results in genetic exchange

DNA primase an enzyme that makes complementary

RNA strands hydrogen-bonded to each of the single DNA strands

helicase (unwindase) enzyme that can unwind the DNA helix at the replication fork

integrase enzymes of family defined by that of bacteriophage lambda that catalyses the integration of phage DNA into the host DNA

mating pair formation the process that leads to localized fusion of the surfaces of two bacteria to provide a bridge through which DNA can be transferred

pilus hair-like projection from the surface of some bacteria

primer short RNA or DNA that must be present on a DNA template before DNA polymerase can commence elongation of a new DNA chain

replicon genetic region that is capable of directing replication of the DNA to which it is joined: at its simplest, a replication origin and those elements needed to activate it

RNA polymerase an enzyme that is capable of synthesizing complementary RNA from a DNA template: normally producing messenger RNA from genes coding for proteins, or ribosomal and transfer RNA from their respective genes

rolling circle replication method of DNA replication in some circular DNAs, where replication of only one strand is initiated at the origin, with the newly synthesized strand displacing the other parental strand

See also **Antibiotic resistance in bacteria** (p.114); **Antibiotic resistance: theory into practice** (p.125)

ANTIBIOTIC RESISTANCE: THEORY INTO PRACTICE

S.G.B. Amyes[1] and H.-K. Young[2]
[1]Department of Medical Microbiology, University of Edinburgh, Edinburgh, UK
[2]Department of Biological Sciences, University of Dundee, Dundee, UK

1. Introduction

Ever since the Swann Report in 1969, there have been many scientists who have been alarmed at the prospect that antibiotic resistance might restrict the capability of physicians to treat bacterial infections. It has to be stated that this concern was not shared by the medical profession as a whole. Indeed, in 1979, the Surgeon General of the US stated that we could "close the book on infectious diseases". Unfortunately, this rash prediction did not take account of two factors: first, that modern medical procedures were radically altering the use of antibiotics in hospitals and, secondly, the general failure of pharmaceutical companies to identify and develop new classes of antibiotics.

2. Traditional usage of antibiotics

In 1935, when Domagk identified the potential of the first clinically successful antibacterial drugs – the sulfonamides – the medical problems that were paramount were community infections, sepsis in hospitals and infections in wounds caused in battle. The latter two at least were manifested as severe, acute and often life-threatening infections usually caused by Gram-positive bacteria. Severe community infections were mainly in the chest, the most drastic of which was caused by *Mycobacterium tuberculosis*. The early antibiotics were very effective against these infections, but they were often associated with toxic side-effects. At the time, this was not considered to be important as the infections being treated were often life threatening and the alternative to treatment was unthinkable.

The "Golden Age" of antibiotics was in the 1950s when almost all the chemical structures that we currently use for antibiotics were discovered. The 1960s were spent improving the structure of the natural antibiotics, increasing their spectrum of activity and attempting to improve their safety profile. During this period, early problems with antibiotic resistance were experienced in hospitals. Following the extensive use of penicillin, strains of the ubiquitous pathogen *Staphylococcus aureus* which produced β-lactamases (penicillinases), which render this drug ineffective, emerged. Semisynthetic modification of penicillin produced methicillin. Although a far less potent antibiotic, methicillin was successful in overcoming these β-lactamases. Nevertheless, by the end of the 1960s, many isolates of clinical *S. aureus* were reported to be resistant to methicillin. This caused a surprising lack of concern at the time, perhaps because the bacteria still remained sensitive to the far more toxic aminoglycosides, particularly the recently introduced gentamicin. Use of gentamicin, a hospital antibiotic, which has to be used carefully because of its toxic properties, considerably reduced the level of methicillin resistance and allowed almost complete control of staphylococcal infections throughout the 1970s.

3. Modern medical procedures

The 1970s also saw a rapid expansion in transplantation and the introduction of chemotherapy regimens to treat cancer and leukaemia. These life-saving procedures require suppression of the immune system, the body's normal defence against infection. Thus, aggressive antibiotic therapies had to be given to these patients to defend them against the threat of bacterial infection. This heralded a new era for antibiotics. Up until this time, it was largely considered irrelevant whether an antibiotic killed the pathogen (bactericidal) or merely prevented its growth (bacteriostatic). The success of antibiotic treatments had been based on the premise that the antibiotic was only required to prevent the pathogen from growing and dividing, and that the residue of the infection would be cleared by the body's own immune system. Indeed, it was believed that unless the patient could clear the residue of the infection, antibiotic therapy would ultimately fail.

When antibiotics are administered to the immunosuppressed, the issue of the antibiotic's capability to deal successfully with the pathogen became crucial. The antibiotic is now required not only to prevent the pathogen from dividing, but also to clear up the residue of the infection, fulfilling a role previously performed by the patient. The use of bactericidal antibiotics, therefore, increased both in volume and intensity in certain parts of the hospital. Faced with this bombardment of antibiotics, bacteria that had never, in the long term, been particularly effective in overcoming the patient's immune system soon proved to be much more adept at overcoming antibiotics. Ultimately, some strains of bacteria emerged that were capable of resisting almost all of the antibiotics used to control them.

(i) Methicillin-resistant S. aureus

The first of these problematic bacteria was *S. aureus*. By the end of the 1970s, methicillin resistance was again found in hospital isolates; however, these bacteria were not the same as those found at the beginning of the decade. In addition to methicillin resistance, they had also acquired resistance to gentamicin. Indeed, careful examination revealed that these bacteria had acquired resistance to all available antibiotics except the glycopeptide vancomycin. The acquisition of methicillin resistance gave these bacteria the name methicillin-resistant *S. aureus* (MRSA), but in reality these were multidrug-resistant *S. aureus* (Wadsworth *et al.*, 1992). They became rampant in hospitals especially in Australia and England (Turnidge *et al.*, 1989; Casewell, 1995), but are now found universally in almost all hospitals (Brunbuisson, 1998). They were also robust and were spread from patient to patient, particularly in hospitals of newer design where the traditional "Nightingale" wards and pavilions had been removed.

Treatment of MRSA was almost exclusively with vancomycin, although a second glycopeptide, teicoplanin, was employed in some hospitals. Vancomycin was a comparatively old antibiotic which, until the emergence of MRSA, had been sidelined because it was considered too toxic for routine use. The threat of large numbers of severely infected patients who were otherwise untreatable allayed any reservations about the reintroduction of vancomycin into hospital use, and thus vancomycin became the drug of choice for the treatment of MRSA for 15 years with a remarkable lack of treatment failures.

Problems were first recognized in 1995 in Japan when Hiramatsu found an infant whose MRSA infection did not respond to vancomycin treatment. Examination of the causative organism indicated that it was an MRSA with a thickened cell wall (Hiramatsu *et al.*, 1997a). Subsequent analysis revealed that the organism had not only activated cell-wall synthesis, but also exhibited an increased rate of cell-wall turnover which was concluded to be a prerequisite for the expression of vancomycin resistance in this species (Hanaki *et al.*, 1998). This thickened cell wall enabled the MRSA strain to overcome vancomycin up to 8 mg/l, which is of clinical significance. Hiramatsu then initiated a search in Japan to establish how widespread these vancomycin-resistant MRSA might be. He found there was a small but finite number of them (Hiramatsu *et al.*, 1997b), but his study provoked others to look elsewhere in the world and similar bacteria were found in almost every hospital environment.

It is currently difficult to state categorically whether these multidrug-resistant *S. aureus* originated as sensitive bacteria which have successively acquired resistance genes to each antibiotic as it was introduced. Certainly, multidrug-resistant *S. aureus* have all the pathogenic characteristics of their sensitive counterparts. However, perhaps significantly, the methicillin resistance gene *mecA* in MRSA strains is found within a large chromosomal region which is absent in the methicillin-susceptible *S. aureus* chromosome. This region, designated *mec* DNA, is 52 kilobases (kb) and, in addition to the *mecA* gene complex, carries a complete copy of the resistance transposon Tn*554* (encoding resistance to spectinomycin and erythromycin) and an integrated copy of the staphylococcal resistance plasmid pUB110 (encoding resistance to tobramycin, kanamycin and bleomycin; Ito *et al.*, 1999). Two further open reading frames encoding proteins homologous to site-specific recombinases have been identified in the *mec* DNA and the entire 52 kb region has recently been shown by PCR analysis to excise spontaneously from the chromosome at a low frequency (less than 1 in 10^4 cells). Although excision is precise, the element appears to be unlike other large mobile genetic elements such as bacteriophage, conjugative transposons or pathogenicity islands. While these types of genetic data support the view that MRSA strains may well have evolved through acquisition of additional DNA elements by their sensitive counterparts (Fey *et al.*, 1998), we shall not be able to conclude unequivocally that this is the origin of multidrug-resistant *S. aureus* until microbial genome sequencing projects, in which a number of resistant and sensitive *S. aureus* are examined, have been completed. The spread of MRSA, however, appears mainly to be clonal, passing from person to person by direct cross-infection and although these bacteria may possess plasmids, these elements have not played a major role in the spread of resistance from hospital strain to hospital strain. It begs the question as to whether multidrug-resistant bacteria, such as MRSA, are subpopulations that have emerged because of our antibiotic regimens.

The acquisition of vancomycin resistance can make MRSA untreatable. Bearing in mind the universal distribution of MRSA in modern hospitals, if the ability to treat it is reduced or lost, we face a major medical crisis.

(ii) Vancomycin-resistant enterococci

We are already facing such crises, albeit on a reduced scale. *Enterococcus faecalis* and *E. faecium* are species normally found in human gut flora and, apart from their rare occurrence as the cause of urinary tract infections, they were not usually considered as pathogenic. However, when patients were given immunosuppressive therapy, enterococci were increasingly found as the causative organism of infection, particularly in patients in intensive therapy units. More worrying was the realization that these bacteria were becoming increasingly resistant to antibiotics. Like MRSA, the final defence against these Gram-positive bacteria was considered to be vancomycin, but, by the end of the 1980s, they had acquired the ability to resist high levels of vancomycin (Woodford *et al.*, 1993)

The mechanism of vancomycin resistance is particularly sophisticated and, although there are different variations of it, they all involve the reversal of the bacterium's normal manufacture of the cell wall. This is accompanied by the synthesis of a modified cell wall which includes D-alanyl–D-lactate or D-alanyl–D-serine terminal dipeptides, which are less capable of binding vancomycin than the normal bacterial dipeptide D-alanyl–D-alanine (Arthur *et al.*, 1992, 1996). Many of the vancomycin-resistance operons in enterococci are known to be on transferable elements and this may be important in the establishment of the genes in clinical bacteria. Nevertheless, outbreaks of van^R enterococci appear to result primarily as a result of clonal spread through a hospital.

Interestingly, if we look at the causes and prevalence of infections caused by the genus Enterococcus, the species *faecalis* is far more common than *faecium* in antibiotic-sensitive strains; however, when vancomycin-resistant strains are examined, the species *faecium* is much more prevalent (Woodford *et al.*, 1993), which suggests that the vancomycin-resistant variants may derive from a resistant subpopulation. As with MRSA, vancomycin resistance is really a synonym for multidrug resistance, but, in this case, it means that there is no ready therapy to treat these bacteria. In an outbreak in Edinburgh in 1995, which was responsible for a number of deaths, the only control measure that could be applied was to restrict the availability of antibiotics, particularly vancomycin, to infected patients and other patients in infected hospital units (Brown *et al.*, 1998). Totally resistant Enterococcus species are now a reality in many hospitals, particularly in the US (Emori & Gaynes, 1993). Although these bacteria cause severe infections, mainly in the seriously ill (e.g. the immunosuppressed patients of intensive therapy units), the presence of these untreatable bacteria should not cause immediate alarm to healthy members of the community. Nevertheless, their widespread dissemination will require risk assessments to be made in some hospitals to establish whether the potential benefits of immunosuppressing a patient outweigh the catastrophic risk of acquiring an infection caused by an untreatable bacterium during a period when the patient is unable to protect himself.

(iii) Carbapenem-resistant Acinetobacter baumannii

S. aureus and enterococci are Gram-positive bacteria and it has been argued that resistant variants have emerged because development of most antibiotic chemical nuclei has concentrated on improvement of their antiGram-negative capabilities since Gram-negative bacteria have traditionally been perceived as the major problematic hospital-acquired infections. When antibiotics were first targeted at Gram-negative hospital infections, *Pseudomonas aeruginosa* was considered to be the most difficult organism to treat. Small porins in the outer membrane ensured that this species had an inbuilt capability to resist many antibiotic challenges. As antibiotics were modified to increase activity against this organism, *P. aeruginosa* infections largely became controllable. Even when *P. aeruginosa* started to develop resistance mechanisms to the latest generation antibiotics, the apparent importance of this bacterium as a hospital-acquired pathogen declined. Although we certainly see *P. aeruginosa* that are resistant to almost all antibiotics, they are often not responsible for life-threatening infections. The same cannot be said for another non-fermenting Gram-negative pathogen, *Acinetobacter baumannii.*

A. baumannii is a soil bacterium that was virtually unheard of in hospitals 30 years ago. As each new class of hospital antibiotic was introduced, the prevalence of *A. baumannii* increased. Its particular attribute is that it appears to acquire resistance at high speed (Bergogne-Berezin *et al.*, 1971). Within a matter of a few years, a population, which was totally sensitive to an antibiotic, can become totally resistant. This is a rare capability for a bacterium and is not even found with *S. aureus* or with enterococci. The two antibiotic classes that appear to have promoted the prevalence of *A. baumannii* are the second-generation cephalosporins and the fluoroquinolones. Most hospitals now have problems with *A. baumannii* and it is a most severe Gram-negative problem in intensive therapy units, where it is a significant cause of hospital-acquired pneumonia and regularly causes bacteraemia. Although it can infect the immune competent, it causes maximum damage in the immune-suppressed patient. In most major medical centres, *A. baumannii* is resistant to all antiGram-negative antibiotics except the carbapenems, imipenem and meropenem. Surveys in the late 1980s suggested that some clinical strains were able to resist even this class (Vallee *et al.*, 1990). A strain from a Scottish hospital isolated in 1985 showed clinical resistance in imipenem. This resistance was manifested by a novel carbapenem-hydrolyzing β-lactamase, ARI-1, which was transferable from one strain to another (Scaife *et al.*, 1995). In the early 1990s, a second β-lactamase, ARI-2, emerged in *A. baumannii* strains responsible for hospital-acquired infections in Buenos Aires, Argentina (Brown *et al.*, 1998). This β-lactamase has been subsequently found to be much more successful and has been found in *A. baumannii* worldwide, and its presence usually elevates the organism to the category of hospital-acquired infection that is untreatable with antibiotics.

4. Are there any other problems?

S. aureus, *E. faecium* and *A. baumannii* have shown us that the return to a pre-antibiotic era when infectious diseases were untreatable is again possible, even probable, for some species. We now have to be very imaginative in the treatment of infections caused by totally resistant variants of the species and outcomes are by no means certain. There are other hospital-acquired bacteria that are rapidly becoming resistant to most available antibiotics and these include *Klebsiella pneumoniae* and Enterobacter species. However, perhaps even more worrying is the rapid acquisition of multidrug resistance in community-acquired infections, as this has the potential to affect us all. Respiratory infections spread rapidly through the population, particularly in areas where we congregate the very young (day-care centres, schools) and the elderly (residential homes). Indeed, multidrug resistance has become a problem in the severest form of community-acquired pneumonia, *Streptococcus pneumoniae*, and conventional therapies against this pathogen are being rendered far less effective. This has been particularly evident in some European countries, South Africa and the US (Linares *et al.*, 1992; Friedland & McCracken, 1994; Doern *et al.*, 1998). In recent years, there has also been a resurgence in tuberculosis, particularly among the underprivileged in North America, and of grave concern is the fact that the bacteria responsible are resistant to virtually all the available first-line antituberculosis therapies (Cole & Telenti, 1995).

5. New antibiotics

The Chief Medical Officer of Scotland predicted that we will have run out of antibiotics by the year 2020 (Kendell, 1994). He came to that view because the problems of antibiotic resistance have been overcome in the past by the introduction of new variants of the currently available antibiotics; the chemists were able to modify the main antibiotic structures to overcome the bacterial resistance mechanisms. However, bacteria have continued to acquire new or improved resistance mechanisms and we are now largely incapable of modifying currently used structures to overcome resistance mechanisms. While we can still modify structures to improve the spectrum of activity of an antibiotic, if an individual bacterium has started to develop a resistance mechanism, current and future modifications of existing chemical structures look impotent. The alternative is to devise new chemical structures, but we have been spectacularly unsuccessful at this over the past 35 years, with only one new structure, the fluoroquinolones, being forthcoming. However, some pharmaceutical companies are now recognizing the potential of the new genomic and gene-chip technologies and are investing in the analysis of microbial genome sequences, to identify potential targets of inhibition, and in screening microarrays of new compounds for their inhibitory properties. This is a vast undertaking and will mean the screening of literally millions of potential new compounds.

6. Conclusion

These drug-hunting programmes are unlikely to provide new antibiotics in the next 5 years and it may even be 10 years before clinically useful drugs are available. What is certain, however, is that antibiotic resistance will become even more severe during that time. For this reason, we need to understand how antibiotic resistance spreads and what factors promote its dissemination, not just from a biological perspective, but also with respect to the epidemiology of resistant bacteria within the hospital. New antibiotics alone will not solve the problem – an acceptance that we all have a responsibility to reduce the impact of resistance, particularly those who are responsible for prescribing antibiotics, will also be needed.

REFERENCES

Arthur, M., Molinas, C. & Courvalin, P. (1992) The VanS–VanR two-component regulatory system controls synthesis of depsipeptide peptidoglycan precursors in *Enterococcus faecium* BM4147. *Journal of Bacteriology* 174: 2582–591

Arthur, M., Reynolds, P.E., Depardieu, F. *et al.* (1996) Mechanisms of glycopeptide resistance in enterococci. *Journal of Infection* 32: 11–16

Bergogne-Berezin, E., Zechchovsky, N., Piechaud, M., Vieu, J.F. & Bordini, A. (1971) Antibiotic sensitivity of 240 strains of oxidase-negative Moraxella (Acinetobacter) isolated from hospital infections. Evolution during 3 years. [Article in French] *Pathologie Biologie* (Paris) 19: 981–90

Brown, A.R., Amyes, S.G.B., Paton, R. *et al.* (1998) Epidemiology and control of vancomycin-resistant enterococci (VRE) in a renal unit. *Journal of Hospital Infection* 40: 115–24

Brown, S., Bantar, C., Young, H.-K. & Amyes, S.G.B. (1998) Limitation of *Acinetobacter baumannii* treatment by plasmid-mediated carbapenemase ARI-2. *Lancet* 351: 186–87

Brunbuisson, C. (1998) Methicillin-resistant *Staphylococcus aureus*. Trends, epidemiology, clinical impact, and prevention. *Pathologie Biologie* 46: 227–34

Casewell, M.W. (1995) New threats to the control of methicillin-resistant *Staphylococcus aureus*. *Journal of Hospital Infection* 30 (Suppl.): 465–71

Cole, S.T. & Telenti, A. (1995) Drug resistance in *Mycobacterium tuberculosis*. *European Respiratory Journal* 8: S701–13

Doern, G.V., Pfaller, M.A., Kugler, K., Freeman, J. & Jones, R.N. (1998) Prevalence of antimicrobial resistance among respiratory tract isolates of *Streptococcus pneumoniae* in North America: 1997 results from the SENTRY antimicrobial surveillance program. *Clinical Infectious Diseases* 27: 764–70

Domagk G. (1935) Ein Beitrag zur chemotherapie der bakteriellen infectionen. *Deutsch Medizinishe Wochenschrift* 61: 253–56

Emori, T.G. & Gaynes, R.P. (1993) An overview of nosocomial infections, including the role of the microbiology laboratory. *Clinical Microbiology Reviews* 6: 428–42

Fey, P.D., Climo, M.W. & Archer, G.L. (1998) Determination of the chromosomal relationship between *mecA* and *gyrA* in methicillin-resistant coagulase negative staphylococci. *Antimicrobial Agents and Chemotherapy* 42: 306–12

Friedland, I.R. & McCracken, G.H. (1994) Drug therapy – management of infections caused by antibiotic-resistant *Streptococcus pneumoniae*. *New England Journal of Medicine* 331: 377–82

Hanaki, H., KuwaharaArai, K., BoyleVavra, S. *et al.* (1998) Activated cell-wall synthesis is associated with vancomycin resistance in methicillin-resistant *Staphylococcus aureus* clinical strains Mu3 and Mu50. *Journal of Antimicrobial Chemotherapy* 42: 199–209

Hiramatsu, K., Hanaki, H., Ino, T. *et al.* (1997a) Methicillin-resistant *Staphylococcus aureus* clinical strain with reduced vancomycin susceptibility. *Journal of Antimicrobial Chemotherapy* 40: 135–46

Hiramatsu, K., Aritaka, N., Hanaki, H. *et al.* (1997b) Dissemination in Japanese hospitals of strains of *Staphylococcus aureus* heterogeneously resistant to vancomycin. *Lancet* 350: 1670–673

Ito, T., Katayama, Y. & Hiramatsu, K. (1999) Cloning and nucleotide sequence determination of the entire mec DNA of pre-methicillin-resistant *Staphylococcus aureus* N315. *Antimicrobial Agents and Chemotherapy* 43: 1449–458

Kendell, R. (1994) From the Chief Medical Officer. *Health Bulletin* 52: 311–12

Linares, J., Alonso, T., Perez, J.L. *et al.* (1992) Decreased susceptibility of penicillin-resistant pneumococci to twenty-four β-lactam antibiotics. *Journal of Antimicrobial Chemotherapy* 30: 279–88

Scaife, W., Young, H.-K., Paton, R.H. & Amyes, S.G.B. (1995) Transferable imipenem-resistance in *Acinetobacter* species from a clinical source. *Journal of Antimicrobial Chemotherapy* 36: 585–86

Swann, M.M. (1969) The Joint Committee on the use of Antibiotics in Animal Husbandry and Veterinary Medicine (Cmnd 4190). HMSO, London

Turnidge, J., Lawson, P., Munro, R. & Benn, R. (1989) A national survey of antimicrobial resistance in *Staphylococcus aureus* in Australian teaching hospitals. *Medical Journal of Australia* 150: 65–69

Vallee, E., Joly-Guillou, M.L. & Bergogne-Berezin, E. (1990) Comparative activity of imipenem, ceftazidime and cefotaxime against *Acinetobacter calcoaceticus*. *Presse Medicale* 19: 588–91

Wadsworth, S.J., Kim, K.H., Satishchandran, V. *et al.* (1992) Development of new antibiotic resistance in methicillin-resistant but not methicillin-susceptible *Staphylococcus aureus*. *Journal of Antimicrobial Chemotherapy* 30: 821–26

Woodford, N., Morrison, D., Johnson, A.P. & George, R.C. (1993) Antimicrobial resistance amongst enterococci isolated in the United Kingdom: a reference laboratory perspective. *Journal of Antimicrobial Chemotherapy* 32: 344–46

GLOSSARY

bacteriophage a virus which infects bacteria

conjugative transposon a circular DNA element that can transfer itself but has to integrate into a replicon to divide

microarray an automated microsystem for measuring the expression of genes

pathogenicity island inserted DNA segments within the chromosome that confer upon the host bacterium a variety of virulence traits

See also **Antibiotic resistance in bacteria** (p.114); **Acquisition and spread of antibiotic resistance genes by promiscuous plasmids** (p.119)

ESCHERICHIA COLI: GENES AND THE CELL CYCLE

William D. Donachie
Institute of Cell and Molecular Biology, University of Edinburgh, Edinburgh, UK

1. Why study *Escherichia coli*?

Bacteria are minute single cells of simple organization that have existed on this planet for at least 3700 million years and which, by their activities, have largely shaped the Earth and its atmosphere, and continue to maintain the Earth's ability to support more complex life forms, such as ourselves. They are present in scarcely conceivable numbers in every possible place where life can exist – including places where no eukaryotic cell could survive (Knoll & Bauld, 1989) and even kilometres down in the solid rock of the Earth's crust (Frederickson & Onstott, 1996). A few kinds are responsible for deadly or debilitating diseases of plants and animals, again including ourselves. Very recently, bacterial cells have been employed as "factories" for the products of genetic engineering. The importance of this invisible empire can, therefore, not be overestimated although, because of our metaphorical myopia and literal "macroscopia" (inability to see anything less than a few fractions of a millimetre across), its very existence is largely ignored by most of our species – almost as much as is ours by the bacteria.

The ecological and economic importance of bacteria is not the only reason why we study them so intensively, for, although by now adapted and modifed to suit many strange environments, the essential organization of the bacterial cell still reflects the earliest successful organization of the processes of life within self-replicating and self-modifying closed units. If we hope to understand the way in which cells survive and multiply, then we should certainly study these simple, but immensely successful, cells.

Since there is an unguessably large number of bacterial species, each with its own peculiarities and abilities, we cannot hope to study more than a few kinds in any detail. For historical reasons, the main "model" bacterium for study has come to be the common enteric bacterium, *Escherichia coli*. This species normally lives in the guts of vertebrates, including ourselves, and is adapted to a warm environment of occasional plenty and frequent dearth, as well as to survival for prolonged periods outside its favourite niche. It can grow with or without oxygen, utilize a wide variety of molecules as its food supply and repair the damage caused by numerous kinds of injurious agents. Close relatives (such as Salmonella species) can survive and thrive inside cells of the host organism (some adapted to animal hosts and some to plants), causing disease and even death, and some strains of *E. coli* itself are dangerous pathogens (e.g. strain O157 that can contaminate meat and cause serious food-poisoning epidemics).

Almost every aspect of *E. coli* has been studied in immense detail (Neidhardt, 1996), as also has the process of cell growth and self-replication, the cell cycle. Many more people are now studying the cell cycle of eukaryotes than have ever studied the cell cycle of prokaryotes, but, partly because of the early start to the study of *E. coli* and partly because it is a simpler system to investigate, it remains true that the cell cycle of this bacterium is better understood than that of any other organism. Nevertheless, ignorance about the bacterial cell cycle is widespread, even among biologists, mostly because of the recent emphasis on the study of our own species and those organisms perceived to be most like us in fundamental organization (such as yeasts). Perhaps this short account of the replicative cycle of *E. coli* may, therefore, provide a useful overview of this arcane knowledge.

2. Overview of the cell cycle (Donachie, 1993)

E. coli cells will grow in a synthetic ("minimal") medium that contains certain inorganic salts (providing sources of nitrogen, phosphorus, sulfur, etc.) and a carbon and energy source, such as glucose. At 37°C in well-aerated minimal medium with glucose, cells of a typical laboratory strain will double every 40 minutes. This growth rate can easily be altered by changing the carbon source to some other organic compound (usually giving a lower growth rate than glucose) or by supplementing the "minimal + glucose" medium with various ready-made organic compounds, such as amino acids or nucleotides, causing an increase in the rate of growth. In a "fully supplemented" medium, such as a yeast- or meat-extract "broth", the generation time can be as short as 20 minutes.

The cell cycle looks extremely simple; rod-shaped cells grow by elongation and, when they reach a certain length, divide into two equal-length rods (**Figure 1**). Nevertheless, this regular cycle of growth and division needs complex control systems, because it requires that every biosynthetic process be coordinated so that each cellular component will double in amount in the same period of time (a period set by the culture medium) and then be distributed equally between two new sister cells. This process may be simple for those molecules that are present in large numbers, or which can readily be made again in a new cell, but the replication and segregation of the bacterial chromosome and certain plasmids present special problems. Certain events, such as the replication of the genome, must be coordinated with cell growth and also with the timing and location of cell division.

The cell cycle of *E. coli* growing at different growth rates at 37°C can be simply summarized (**Figure 1**). Cells grow until they reach a critical cell mass or volume ("initiation mass"; Donachie, 1968) when they start to replicate their single circular "chromosome". Replication begins at a specific locus, *oriC*, and two complexes of DNA-synthesizing proteins ("replisomes") proceed in opposite directions away from *oriC*, replicating the two complementary DNA strands at constant rate until they meet on the opposite side of the circle in the "terminus" region. The formation of a "division septum" then begins. Some of the details of chromosomal DNA replication and of septum formation are described below, but first it is necessary to describe a problem that is implicit in this form of cell cycle, and then show how *E. coli* has solved it.

E. coli replicates DNA very efficiently, at a rate of about 1000 base pairs (bp) per second per replisome at 37°C, but the *E. coli* chromosome is also very large, consisting of about 4.7 million bp. Consequently, it requires about 40 minutes for a single pair of replisomes, starting at *oriC*, to completely replicate the chromosome (**Figure 1**). It also requires about 20 minutes more for a septum to be synthesized and cell division to be completed (**Figure 1**). Therefore, the period between reaching initiation mass and first cell division is about 60 minutes. Despite this limitation, an *E. coli* cell can duplicate itself in as little as 20 minutes (in rich media). *E. coli*'s solution to this apparent paradox is simple; every time it doubles its initiation mass, it initiates chromosome replication from every copy of *oriC*. Even if the previous pair of replisomes have not yet completed their transit of the chromosome, the new pairs nevertheless start to replicate from each of the (duplicated) copies of *oriC*. So, if the cell is doubling its mass every 20 minutes, new rounds of chromosome replication are also initiated every 20 minutes and a pair of new chromosomes is completed every 20 minutes. R.H. Pritchard (pers. comm.) has likened this process to lorries travelling at constant speed along a fixed length of road: the frequency at which lorries exit the road (cell division) is the same as the frequency at which lorries enter it (initiate rounds of chromosome replication). Thus, in fast-growing cells the chromosomal road is "busy", with many replisomes following one another along the road at the same time. In slow-growing cells, there may be only a single pair of replisomes on the chromosome and, if the generation time of the cells is longer than 40 minutes, the "road" will be empty some of the time.

There is another consequence of this mode of replication – cells get bigger when they grow faster. The easiest way to understand this is to think of a resting cell starting into growth; first it must reach initiation mass, after which it will first divide 60 minutes later. During this initial period of growth, the cell will have initiated new rounds of DNA replication every time it doubled its initiation mass (M_i), but its first division will not occur until 60 minutes after the first initiation. If the cell is doubling in mass, say, every 20 minutes, then its mass at first (and every subsequent) division will be $M_i \times 2^3$. A very slowly growing cell may scarcely increase in mass after initiation and so the average size of *E. coli* cells in culture can vary over an approximately 8-fold range, depending on their growth rate.

Eukaryotic cells, which synthesize DNA at slower rates per replisome, but have much more DNA to replicate, solve this problem, partly by having large numbers of sites on each chromosome at which replication can be initiated simultaneously, partly by growing much more slowly than *E. coli* and partly by having an elaborate system of "checkpoints" that prevent mitosis and cell division, or the reinitiation of DNA synthesis, until every chromosome has been completely replicated. (*E. coli* also has a system which prevents cell division if chromosome replication has been interrupted or delayed: see below.)

3. Cell cycle genes

E. coli cells make about 1200 different kinds of protein, while growing under laboratory conditions. Most of these proteins are used in general biosynthetic activities, allowing the cell to grow in size, swim, etc., but only a small number are required solely to carry out the periodic events that are necessary for cell replication. These periodic events include the initiation of chromosome replication (when the cell reaches its initiation mass), chromosome replication itself, separation of completed sister chromosomes and cell division (**Figure 1**).

4. Chromosome replication and segregation

The process of DNA replication in *E. coli* is well understood (Kornberg, 1980, 1982).

(i) Initiation of chromosome replication

Chromosomal replication begins when a cell reaches the initiation mass (Donachie, 1968). At that time, DnaA proteins (activated by having ATP bound to them) bind to four, nearly identical, 9 bp sequences located at *oriC*. The binding of the DnaA–ATP molecules is cooperative and continues until there are about 20–50 bound at *oriC*. This binding, together with bending by associated DNA-binding proteins, distorts the DNA double helix at *oriC*, causing the complementary DNA strands to separate in an adjacent region of weak hydrogen bonding (rich in adenine–thymine base pairs). The local separation into two single strands allows a second complex of proteins ($[\text{DnaB–DnaC–ATP}]_6$ + single-stranded binding protein) to bind to the separated strands. DNA primase, an enzyme that makes complementary RNA strands hydrogen-bonded to each of the single DNA strands, then joins this complex, followed by DNA polymerase III holoenzyme, which actually makes the new complementary DNA strands, and DNA gyrase, which unwinds the double helix for replication. The formation of this replisome completes the steps needed to initiate DNA replication.

(ii) DNA synthesis

The short RNAs synthesized by DNA primase are then used by DNA polymerase III as primers for the synthesis of new DNA strands. The different components of the replisome ensure that the double helix of DNA is progressively unwound and the two strands separated as the polymerase moves forward along the chromosome. In fact, because DNA polymerases can add new bases on to one end only (the 3′-hydroxyl end) of a pre-existing RNA or DNA chain, and because the two complementary strands of the double helix are running in opposite directions to one another (are antiparallel), DNA synthesis proceeds continuously along only one of the two separated DNA strands. In order for new complementary DNA to be copied from the other strand, DNA primase must first make a short RNA primer on this strand, which can then be used by the

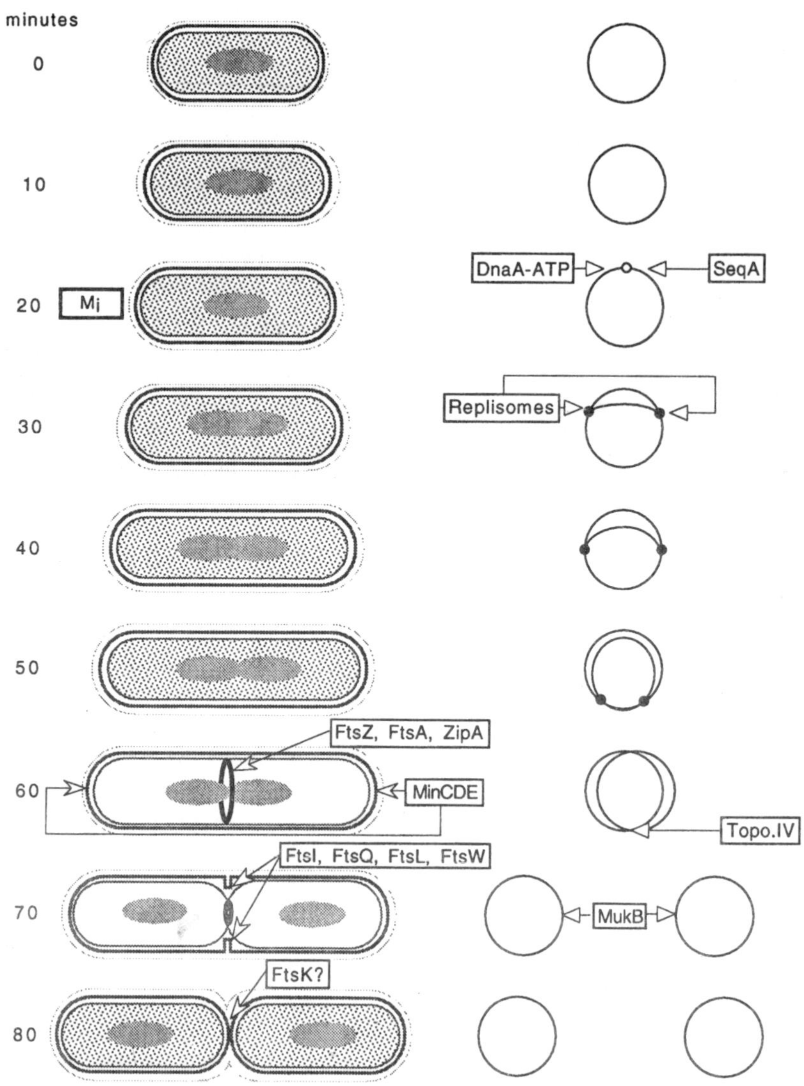

replisome to make a short strand of DNA in the "backwards" direction. Another enzyme, DNA polymerase I, removes the RNA primers and replaces them with DNA until it reaches the start of the previously made DNA segment. The two completed segments are then joined together by another enzyme, DNA ligase. Synthesis of new DNA, therefore, proceeds differently on the two complementary strands of the double helix, with one strand extending continuously (the "leading" strand), but the other (the "lagging" strand) being made in the opposite direction in short segments that are subsequently joined together by DNA ligase.

(iii) Termination of chromosome replication

Since the bacterial chromosome is circular, the replisomes eventually meet in a region (the terminus) opposite to the point (*oriC*) at which they started. Any asynchrony in arrival times is corrected by the presence of special *ter* sequences that, in association with the Tus protein, prevent early arriving replisomes from proceeding right through the terminus region.

(iv) Resolution

Replication of a circular DNA molecule has intrinsic advantages, in that DNA polymerases that can copy each DNA strand in only one direction can nevertheless succeed in replicating the whole molecule. (This presents a problem for linear DNA molecules, such as eukaryotic chromosomes and some virus genomes, which require special mechanisms to accomplish replication of their ends.) However, the replication of covalently closed DNA circles also has its own special problem because, although the end product of replication is two complete double-helical circles of DNA, these two circles are interlinked like conjurer's rings. In order to separate the two circles, it is, therefore, necessary to break and rejoin them. This process requires a specific DNA-unlinking enzyme, topoisomerase IV.

(v) Prevention of premature reinitiation

We have seen how chromosome replication begins when an initiation complex forms at *oriC*, but why does this complex not reform immediately after the *oriC* site has been replicated? Obviously, if this were to happen, then DNA synthesis would become continuous and would not be linked to cell growth and the cell would fill up with DNA. Premature reinitiation at *oriC* is prevented because each newly replicated double-stranded copy of *oriC* DNA consists of one old strand and one newly synthesized complementary strand which the cell is able to distinguish from one another. This happens because there is an enzyme, DNA-adenine methylase (DAM), that adds a methyl group to every adenine in a specific sequence of bases (GATC). Adenines in newly synthesized GATC sequences are therefore unmethylated, whereas the old complementary (CTAG) strand has its adenine methylated. Such "hemimethylated" DNA cannot be replicated (for as yet unknown reasons). Methylation of new adenines is accomplished within a few minutes of the synthesis of the strand in most parts of the chromosome, but the newly replicated *oriC* region remains hemimethylated for about one-third of a cell cycle. How this comes about is currently unknown, but it seems that the hemimethylated *oriC* DNA is somehow sequestered in the cell membrane for this period, making it impossible to reinitiate DNA replication prematurely. This event requires yet another protein, SeqA.

(vi) Partition

Replicating the chromosome is not enough. The two sister chromosomes must somehow be localized in the cell in such a way that the cell division septum, when it forms, will lie between them. This process, by which chromosomes are segregated into the two halves of the cell before division, is generally referred to as partition, and is the prokaryotic equivalent of mitosis in eukaryotes. However, unlike mitosis, almost nothing is known about this process and there is even debate about when it occurs, i.e. whether it takes place suddenly after the completion of replication or whether chromosomal DNA, perhaps attached to the cell membrane, is somehow continually pulled apart during cell growth. What is known is that separation of sister chromosomes will not take place unless there is a short period of new protein synthesis after the completion of replication. The approach that has been so successful in dissecting the mechanics of the rest of the cell cycle, the study of mutants, has not yet been very helpful in understanding partition. Most mutants that are unable to separate their sister chromosomes have been found to be defective in either the resolution or monomerization steps described above. The only exception so far are cells with mutations in a gene, *mukB*, which have defective chromosome partition.

5. Cell division

Soon after chromosome partition is completed, the cell becomes committed to divide, i.e. division will take place even if all further DNA, RNA and protein synthesis is stopped. Preformed proteins are therefore used in cell division. The division process requires the concerted action of a number of proteins which appear to have no other function. Absence of any of these proteins blocks division, but does not prevent any other aspect of growth, so that cells grow into long, undivided, multichromosomal "filaments".

Figure 1 (see opposite) The cell cycle of *E. coli*. This diagram shows the main events in the replication of a cell. The cell is assumed to be growing relatively slowly (with a generation time of 80 minutes at 37°C), as it would in certain artificial "minimal" media. The state of the cell is shown every 10 minutes through the duplication process. Left: The "newborn" cell, at 0 minutes, is a short cylinder (about 1.7 mm long) with rounded ends. The cytoplasm is surrounded by a cell membrane, outside of which is a rigid "cell wall" (in *E. coli* this is actually a single net-like molecule of peptidoglycan, a polymer unique to bacteria) which gives the cell its shape. In Gram-negative bacteria like *E. coli* there is a further "outer membrane" that encloses the cell wall. The cell has a single "chromosome", a covalently closed circle of DNA that is folded up in the cell to form a "nucleoid" (shown diagrammatically as a grey ellipsoid). Various cell division proteins are present in the cytoplasm (and others in the cell membrane) and these are represented by dots in the diagram. The cell grows by elongation at a rate (in this particular growth medium) which doubles its length in 80 minutes. At this growth rate, the cell attains the critical "initiation mass" (M_i) after 20 minutes, and DNA replication is then initiated. Right: Diagram of the chromosome as it would appear if it were spread out into a circle (except that the actual diameter of this circle would be about 1400 μm, much larger than the dimensions of the cell itself, and so it is shown at a reduced magnification). When initiation mass is reached at 20 minutes, the DnaA–ATP protein binds to *oriC*, a unique sequence of bases that serves as the origin of chromosome replication, and this binding causes the two strands of the double helix to become separated locally. A complex of DNA-synthesizing proteins (the "replisome", including DNA polymerase III, the *dnaE* gene product) is then able to bind to the single-stranded sections and start replication of the DNA. (Further binding of DnaA–ATP to the newly replicated *oriC* regions, which would cause premature reinitiation of replication, is prevented by the action of the SeqA protein.) Once initiated, DNA replication, carried out by two replisomes, proceeds from both directions away from *oriC*. This replication takes 40 minutes to complete, when the two replisomes meet on the far side of the circle (i.e. after 60 minutes' growth of the cell). The two completed sister circles of DNA are interlocked at this stage and require to be unlinked by the action of the enzyme topoisomerase IV. Immediately after separation, the two sister chromosomes are somehow moved apart into the two cell halves, a process that involves the MukB protein. At about this same time, the cytoplasmic cell division proteins, FtsZ and FtsA, become associated with a membrane protein, ZipA, to form a ring around the inside of the cell membrane in the centre of the cell. (The ring of proteins needs to be prevented from forming, inappropriately, at either end of the cell by the action of three proteins, MinC, MinD and MinE.) The "cytokinetic ring" of proteins then contracts, probably pulling the cell membrane with it. At the same time a set of membrane proteins (FtsQ, FtsL, FtsW), together with the peptidoglycan-synthesizing enzyme FtsI, cause new cell wall to be laid down in a double layer that follows the closing cytokinetic ring across the cell centre. When this septum is complete, the two layers are split apart and pushed out by the internal pressure of the cell to form new cell ends and complete the process of cell division. The completion of the septum is thought to require the action of a final protein, FtsK. After closure of the cytokinetic ring, the FtsZ and FtsA proteins probably disassemble and go back into the cytoplasm.

The genes controlling division are therefore called "*fts*" genes, and the proteins, "Fts" proteins.

(i) Initiation of division

Cell division begins when an abundant cytoplasmic protein, FtsZ, assembles in the centre of the cell in a ring around the inside of the cell membrane (Bi & Lutkenhaus, 1991). FtsZ is thought to attach to the inside of the cell membrane and to form a contractile ring that physically pulls in the cell membrane until it is constricted completely and reseals as two separate membranes bounding two sister cell compartments.

FtsZ is found in all bacteria and Archaebacteria and it has recently been shown that it is probably required for the division of chloroplasts in eukaryote plant cells (Ousteryoung & Vierling, 1995). This fits well with the accepted idea that chloroplasts had their origins as prokaryotic symbionts of early eukaryotic cells. It is possibly the only protein required for cell division in wall-less prokaryotes, such as Mycoplasmas (which have a homologue of FtsZ, but not of any other known cell division proteins).

FtsZ protein binds GTP and appears to be (de Boer *et al.*, 1992; RayChaudhuri & Park, 1992; Wang & Lutkenhaus, 1993) an archaic form of the tubulin proteins of eukaryotic cells, perhaps close to the ancestral form.

FtsZ is not the only protein in the preseptal ring around the cell: a second cytoplasmic protein, the actin-like FtsA, is associated with FtsZ (Addinall & Lutkenhaus, 1996), as is a transmembrane protein, ZipA, which may anchor the ring to the cell membrane (Hale & de Boer, 1997).

(ii) Formation of the septum

FtsZ is absolutely required for cell division, but, in bacteria with cell walls, it is not sufficient by itself. A number of other proteins are also required for the formation of the crosswall or septum. These include FtsI (PBP3), which is a septum-synthesizing peptidoglycan (cell wall) enzyme, located in the periplasm (where the cell wall is assembled), but which is anchored to the cell membrane. (FtsI is the most sensitive target for the action of β-lactam antibiotics, such as penicillins, which kill cells by inhibiting septum formation and, thus, allowing localized cell lysis at the cell-division site. This lysis is caused by the localized action of peptidoglycan hydrolases that are presumed to be active in remodelling peptidoglycan during septum formation.) A number of other proteins are also required for division, including FtsA in the cytoplasm, and the membrane proteins FtsK, FtsL, FtsN, FtsQ and FtsW. The individual activities of these septation proteins are still largely unknown.

(iii) The "topological transition"

There is an interesting topological problem in cell division, which is how a single peptidoglycan surface bounding the incipient sister cells is converted into two surfaces. This "topological transition" must take place when the ingrowing septum is almost complete. At this stage, the septum is like an almost-closed annulus composed of an infolded double layer of peptidoglycan (**Figure 1**). In order to complete the crosswall and separate the two sister cells, the folded edge of the annulus must be cut, to separate the two layers, and then somehow the cut edges must be sewn together to close the holes in the ends of the cells. The transient existence of such a hole need not be lethal because, up until a late stage, the two apposed layers of the septum are linked together by covalent bonds. After the "cut and darn" stage, the two septal layers must be separated to form the new cell poles.

The *ftsK* gene was identified in a search for mutants that are blocked at this theoretical terminal stage in cell division. Cells with mutated FtsK protein can almost complete division, but sister cells remain connected by a narrow constriction (Begg *et al.*, 1995).

(iv) Cell separation

Mutants in three other genes (*envA*, *cha* and *rlpB*) are defective in the final separation of sister cells, perhaps because they are defective in the final breaking of bonds between the septal layers or perhaps in an even later stage, such as the ingrowth and completion of the outer membrane layer that surrounds Gram-negative cells.

(v) Choice of the division site

Normal bacterial cells always divide exactly in the middle, but the choice of this site for septum formation is governed by the action of a set of three proteins, MinC, MinD and MinE. In the absence of these proteins, *E. coli* cells still divide at the same frequency, once for each doubling of cell mass or volume, but septa are formed, with apparently equal probability, in the cell centre or near one or other of the cell poles. Division near a cell pole produces only a minute "minicell" that contains no chromosomal DNA, plus an almost full-sized "mother" cell with two chromosomes. The mother cell, however, cannot divide again (because it has somehow used up its "division potential" in making a septum) until it has once again doubled in size and has four chromosomes. At that stage, it can make two new septa, but, once again, these will be laid down with equal probability between any two chromosomes, or between a chromosome and a cell pole. The result of this "lottery" system of choice of division site is that about one-half of all divisions produce only minicells, and the average size of cells with chromosomes is doubled. A population of *min* mutant cells is, therefore, very heterogeneous, consisting of normal-sized cells mixed with minicells and numerous elongated cells (Teather *et al.*, 1974).

The way in which the Min proteins prevent septa from forming at the poles is partly understood (de Boer *et al.*, 1989). The MinC protein (after activation by MinD protein) is a general inhibitor of cell division. Its target appears to be the primary cell division protein, FtsZ. Expression of MinC and MinD by themselves, therefore, blocks all cell division, but expression of the small third protein, MinE, somehow prevents the MinC protein from acting at the cell centre, which thus becomes the only location at which septa can be laid down. In the absence of this selective system, septa often form near the ends of the cell instead of in the centre.

In cells in which chromosome positioning is disturbed (e.g. by mutations in proteins required for decatenation after replication), septa form at any location in the cell at which there is no DNA (e.g. the *parCE* mutant cell shown in **Figure 2**).

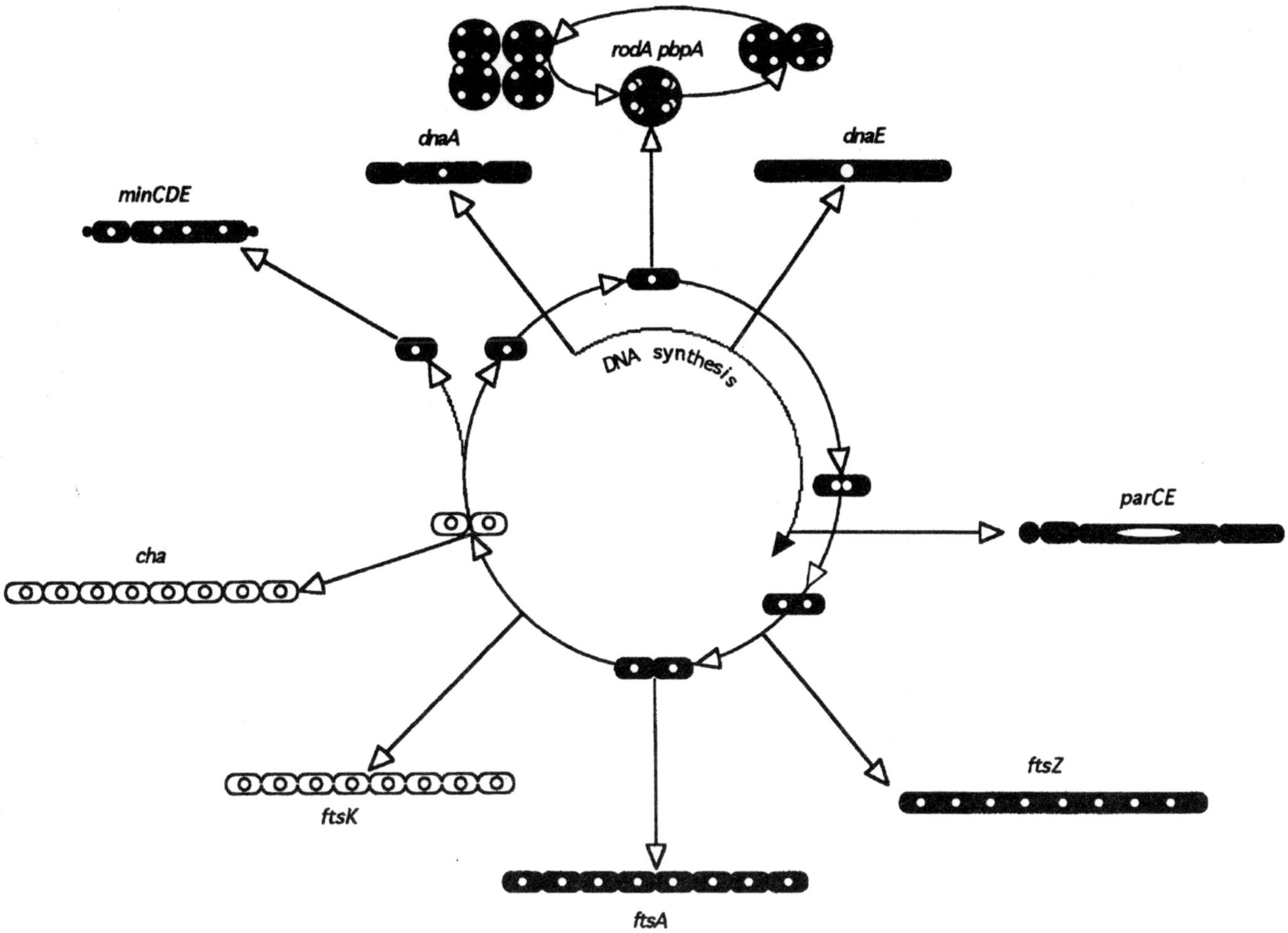

Figure 2 The effect of mutations on the cell cycle. The circle shows the sequence of events in the normal cell cycle: a newborn cell elongates until it reaches initiation mass, when DNA replication is initiated; DNA synthesis and cell elongation continue until chromosome replication is complete (inner arc); sister chromosomes then separate and a septum forms across the cell centre; after completion of the septum, it is split apart to form two new cell ends and the two new cells separate. So-called "conditional" mutations in a number of genes affect various steps in this cycle; such mutations allow cells to go through their cycles normally at low temperature (usually 30°C), but cause a block in some step at high temperature (usually 42°C). The appearance of such mutant cells is shown as they would appear after two mass doublings at 42°C. Thus, a mutation in the *dnaA* gene would prevent initiation of DNA replication, so that the cell would elongate without increasing its number of chromosomes (such cells go on to divide, producing DNA-less daughters). A mutation in *dnaE*, encoding DNA polymerase III, would stop DNA synthesis after the DnaA-dependent initiation step. These cells would also continue to grow without further DNA synthesis, but in this case the SOS system would be induced and cell division would be blocked. *ParCE* mutants would be able to replicate their chromosomes but would not be able to separate the linked circles, so that all the DNA would accumulate in the cell centre and DNA-less daughter cells would be produced. Mutations in various cell division genes (e.g. *ftsZ, ftsA, ftsK, cha*) would cause blocks at various different stages in division (producing filaments or chains of cells). Mutations in the *minC, minD* and *minE* genes allow cell division to take place near the cell poles, producing tiny DNA-less "minicells" at the expense of normal divisions. Finally, mutations in the *rodA* or *pbpA* genes cause cells to lose their rod-shape and grow as large, multichromosomal cocci.

6. Regulation of the cycle

As has been shown above, the mechanics of bacterial cell duplication are beginning to be quite well understood, at least in the sense that we seem to know most of the specific genes and proteins that are required to replicate the chromosome and make the cell divide. However, knowledge of how these processes are regulated is much more limited. We know that certain general controls exist, because cells only initiate chromosome replication when they reach initiation mass, and that they will form a septum only if there is a sufficiently large DNA-free zone in the cell. What we do not know is how the activities of the cycle-specific proteins (DnaA, FtsZ, etc.) are regulated so that they act only at the correct times and in the correct locations. A few things, however, are known.

(i) Growth rate regulation of ftsQ, A and Z transcription

Most of the cell division genes (i.e. *ftsL, I, W, Q, A* and *Z*) are located in a single large gene cluster that also contains many of the genes encoding enzymes of cell wall (peptidoglycan) synthesis. Transcription of these genes into messenger RNAs takes place from a number of start sites (promoters) that are scattered throughout the cluster. *FtsQ*, *A* and *Z* are together at one end of this cluster and are transcribed together from promoters "upstream". In addition, *ftsZ* is transcribed from promoters that lie within *ftsA* itself. Those that result in transcription of *ftsZ* are regulated in such a way that they produce about the same amount of message per cell at all growth rates. This may reflect a requirement for a fixed amount of FtsZ (and other division proteins) for every division (independent of the size of the cell, which increases markedly with increasing growth rate, as we have explained above).

Recently, it has become known that one of the promoters that transcribes the *ftsQAZ* genes is turned on by a factor (σ^S) that is produced when cells run out of nutrients (Miguel Vicente, personal communication and Ballesteros *et al.*, 1998). Under these conditions, cells divide for a while more rapidly than they grow in size, so that cells get progressively smaller, eventually ending up as small "resting" cells, each with a single chromosome copy. Turning on the essential division genes under these conditions may therefore be required to ensure this reduction to the minimum size.

(ii) Antisense regulation

Another level of regulation is post-transcriptional. The *ftsZ* gene is transcribed from one strand of DNA. On the complementary strand, however, there appears to be a small "gene", called *stfZ* because it runs counter to *ftsZ* (Dewar & Donachie, 1993), that is transcribed to produce a short RNA which, because it is complementary to the actual *ftsZ* messenger RNA, can bind to it and prevent its translation into FtsZ protein. The role of *stfZ* in the normal cell cycle is still unknown, but its overexpression blocks all cell division.

(iii) Regulation in response to DNA damage and inhibition of DNA replication

Exposure to ultraviolet light, or to certain chemicals, can cause adjacent pyrimidines in the same DNA strand to become crosslinked ("pyrimidine dimers"). Such dimers distort the DNA backbone and prevent replication. The cell has a number of enzymes that can recognize and repair such damage (by reversing the dimerization, or by cutting out the damaged section and copying the undamaged complementary strand to fill in the gap, or, sometimes, by copying the damaged section as a random set of bases, causing a mutation) and the synthesis of many of these enzymes is specifically induced by the damage itself (the "SOS" response). During the repair of damaged DNA, replication is stalled, but the cell continues to grow. If nothing else happened, the cell would grow long enough for a septum to form, even though its chromosome had not yet been duplicated, and a DNA-free cell would be divided off. This does not normally happen because a part of the SOS response is the production of a protein (SulA) that binds to FtsZ and prevents cell division from beginning. The SulA protein is very unstable so that, as soon as the damaged DNA has been repaired and the SOS response turned off, the SulA protein rapidly decays and the FtsZ protein can act once more.

The SOS response in *E. coli* may be compared to what are called "checkpoints" in the eukaryotic cell cycle, which ensure that mitosis does not begin before chromosome replication is complete. However, the *E. coli* cell does not actually check directly whether or not it has two chromosomes: it only recognizes blocked replication. Consequently, if DNA replication is prevented from initiating (e.g. in a *dnaA* mutant) the cells will continue to grow and then divide to give pairs of cells, one with DNA and one without (**Figure 2**).

(iv) Regulation by viruses

When a virus invades a bacterial cell, its "purpose" often appears to be to make as many new virus particles as possible. One event that could reduce the number of progeny virus by half would be if the infected cell managed to divide before the virus had started replicating. Perhaps this is the reason why several bacterial viruses have genes that block the division of the host cell. The mechanism of action of one set of such virus genes has been worked out. In some strains of *E. coli* there exists, integrated into the chromosome, a set of "cryptic" genes that appear to have been derived in the past from a virus that integrated part of its genome into the host chromosome (Faubladier & Bouché, 1994). These genes have no function in the absence of the other viral genes and are passively replicated as part of the host chromosome. However, these genes are potentially lethal. This is because the products of at least two of the genes are able to block cell division. This does not happen normally because a third gene (*dicA*) prevents transcription of the others. If, however, the *dicA* gene is inactivated, for example by mutation, then the remaining genes are activated and produce their lethal products. One of these (DicB protein) is able to activate the host's MinC protein (see above) which, in turn, inactivates FtsZ. Another gene (*dicF*) is transcribed into an "antisense" RNA that is complementary to *ftsZ*-messenger RNA and is able to prevent its translation into FtsZ protein. It is intriguing that the virus inhibits host cell division by mimicking two of the processes that the host cell itself uses to regulate division during the normal cell cycle: activation of MinC by MinD and antisense regulation of FtsZ translation by *stfZ*. This is true "molecular mimicry" and not the recruitment of host genes by the virus, since the DicB protein is quite different in sequence from MinD and the *dicF*-RNA is produced in a quite different way (by post-transcriptional excision from a longer RNA that also encodes other viral proteins) from *stfZ*-RNA.

One thing is very striking; all regulation of cell division – whether by the normal cell cycle regulators (e.g. *stfZ*-RNA, MinC, MinD, MinE), by emergency regulators (SulA) or by invading viruses (DicB, *dicF*-RNA) – seems to have as its sole target the key protein, the universal initiator of cell division (in walled bacteria, wall-less mycoplasmas, Archaebacteria and chloroplasts), FtsZ.

REFERENCES

Addinall, S.G. & Lutkenhaus, J. (1996) FtsA is localized to the septum in an FtsZ-dependent manner. *Journal of Bacteriology* 178: 7167–72

Ballesteros, M., Kusano, S., Ishihama, A. & Vicente, M. (1998) The *ftsQ*1p Gearbox promoter of *Escherichia coli* is a major sigma S-dependent promoter in the *ddlB-fts* region. *Molecular Microbiology* 30: 419–30

Begg, K.J., Dewar, S.J. & Donachie, W.D. (1995) A new *E. coli* cell division gene, *ftsK*. *Journal of Bacteriology* 177: 6211–27

Bi, E. & Lutkenhaus, J. (1991) FtsZ structure associated with cell division in *Escherichia coli*. *Nature* 354: 161–64

de Boer, P.A.J., Crossley, R.E. & Rothfield, L.I. (1989) A division inhibitor and a topological specificity factor coded for by the minicell locus determine the proper placement of the division site in *Escerichia coli*. *Cell* 56: 641–49

de Boer, P.A.J., Crossley, R.E. & Rothfield, L.I. (1992) The essential bacterial cell division protein FtsZ is a GTPase. *Nature* 359: 254–56

Dewar, S.J. & Donachie, W.D. (1993) Antisense transcription of the *ftsA–ftsZ* gene junction inhibits cell division in *Escherichia coli*. *Journal of Bacteriology* 175: 7097–101

Donachie, W.D. (1968) Relationship between cell size and time of initiation of DNA replication. *Nature* 219: 1077–079

Donachie, W.D. (1993) The cell cycle of *Escherichia coli*. *Annual Review of Microbiology* 47: 199–230

Faubladier, M. & Bouché, J.-P. (1994) Division inhibition gene *dicF* of *Escherichia coli* reveals a widespread group of prophage sequences in bacterial genomes. *Journal of Bacteriology* 176: 1150–56

Frederickson, J.K. & Onstott, T.C. (1996) Microbes deep inside the Earth. *Scientific American* 275: 42–47

Hale, C.A. & de Boer, P.A.J. (1997) Direct binding of FtsZ to ZipA, an essential component of the septal ring structure that mediates cell division in *E. coli*. *Cell* 88: 175–85

Knoll, A.H. & Bauld, J. (1989) The evolution of ecological tolerance in prokaryotes. *Transactions of the Royal Society of Edinburgh: Earth Sciences* 80: 209–23

Kornberg, A. (1980) *DNA Replication*, San Francisco: Freeman; 2nd edition 1992

Kornberg, A. (1982) *Supplement to DNA Replication*, San Francisco: Freeman

Neidhardt, F.C. (1996) Escherichia coli *and Salmonella: Cellular and Molecular Biology*, 2nd edition, Washington, DC: American Society for Microbiology Press

Ousteryoung, K.W. & Vierling, E. (1995) Conserved cell and organelle division. *Nature* 376: 473–74

RayChaudhuri, D. & Park, J.T. (1992) *Escherichia coli* cell division gene *ftsZ* encodes a novel GTP-binding protein. *Nature* 359: 254–56

Teather, R.M., Collins, J.F. & Donachie, W.D. (1974) Quantal behaviour of a diffusible factor which initiates septum formation at potential division sites in *E. coli*. *Journal of Bacteriology* 118: 407–13

Wang, X. & Lutkenhaus, J. (1993) The FtsZ protein of *Bacillus subtilus* is localized to the division site and has a GTPase activity that is dependent on the FTsZ concentration. *Molecular Microbiology* 9: 435–44

GLOSSARY

division septum peptidoglycan partition separating a cell into two parts

DNA gyrase unwinds the DNA double helix during replication

DnaA *E. coli* protein that binds to *oriC* and, by separating the two complementary strands there, allows DNA replication to begin

DNA-ligase enzyme that joins together two completed DNA segments

DNA-polymerase I removes the RNA primers and replaces them with DNA until it reaches the start of the previously made DNA segment. These are then joined together by DNA ligase

DNA-polymerase III holoenzyme forms new complementary DNA strands

DNA-primase an enzyme that makes complementary RNA strands hydrogen-bonded to each of the single DNA strands. These provide the primers for the synthesis of complementary DNA strands by the replisome

initiation mass the critical cell mass or volume at which initiation of chromosomal DNA replication by DnaA at *oriC* takes place

oriC the locus on the bacterial chromosome at which replication is initiated

partition process by which chromosomes are segregated into the two halves of a cell before division

peptidoglycan cell wall polymer, unique to bacteria, in which long polysaccharide chains are cross-linked by short peptides to form a stress-bearing layer around the cell

peptidoglycan hydrolases enzymes which break preexisting peptidoglycan chains to allow the insertion of new strands during cell wall growth

replisome multiprotein assembly (including DNA polymerase III) that replicates DNA

single-stranded binding protein protein that binds single-stranded DNA, preventing reformation of the duplex, thus permitting the formation of new complementary strands during replication

terminus the region of the *E. coli* circular chromosome diametrically opposite *oriC* at which pairs of replisomes meet and replication is completed

topoisomerase IV DNA unlinking enzyme that breaks and rejoins the DNA backbone after completion of replication at the terminus in such a way as to separate the two circular sister chromosomes

Tus protein protein that binds to special sites (*ter* sites) that surround the terminus and which unidirectionally blocks replication from continuing past the terminus

See also **Cell cycle genes** (p.736)

ESCHERICHIA COLI: GENOME, GENETIC EXCHANGE, GENETIC ANALYSIS

Millicent Masters
Institute of Cell and Molecular Biology, University of Edinburgh, Edinburgh, UK

1. Introduction

When, towards the middle of the 20th century, the discipline of molecular genetics was invented, the genes and genetic systems of *Escherichia coli* (and its associated parasites and symbionts, phages and plasmids) were its principal subject matter. *E. coli* is a rod-shaped Gram-negative bacterium 0.5 × 3–5 μm in size, abundant in the large intestines of mammals; most strains are non-pathogenic. Although the occasional practitioner worked with the Ascomycete fungus Neurospora, or with one of a small number of other bacterial species, or persevered with fruit flies or maize, the momentum of the time was toward the analysis of events in this chosen model organism.

As a consequence, we now know an immense amount about the K-12 strain of *E. coli*, the favoured research tool of that and later periods. An overview of that knowledge can be found in an encyclopedic, 155-chapter, two-volume tome *Escherichia coli and Salmonella typhimurium: Cellular and Molecular Biology* (Neidhardt *et al*., 1996). In an epilogue, the editors point out that, despite the size of this work, the last chapter has not yet been written – although more is known about *E. coli* than about any other organism, we remain a long way from completely understanding the many systems that comprise it. *E. coli* has provided many of the necessary signposts that have steered the way toward the study of other ("higher") organisms. While these have been and are being studied by many, those of us who have spent our research careers using molecular genetics to understand *E. coli* hope that the full characterization of this most nearly understood of organisms will not be abandoned incomplete.

The K-12 strain of *E. coli*, isolated from a human patient in the 1920s, was found, in the 1940s and 1950s, to be capable of transferring genetic markers from certain donor to recipient lines. This, and the fact that it carries no pathogenic characters, led to its adoption as the strain of preference for subsequent genetic studies. It is interesting to note that *E. coli* K-12, isolated well before the discovery of antibiotics, did not carry infectious antibiotic resistance.

2. The genome of *E. coli*

(i) The chromosome

The genomes of eukaryotic cells comprise multiple chromosomes (consisting of protein–DNA structures, normally linear) segregated from the cytoplasm within a nuclear membrane. The genomes of prokaryotes, such as *E. coli*, are different. *E. coli* has a single circular chromosome, without a structural protein scaffold or nuclear membrane. In actively growing cells, the chromosome penetrates to all parts of the cytoplasmic compartment, leaving only the periplasm (the space between inner and outer cell membranes) free of DNA. This chromosome contains 4.7 million base pairs (bp) of DNA and is densely packed with genes. It is replicated once per cell cycle from a unique site (the replication origin, or *oriC*) bidirectionally to a terminus opposite. Replication, at 37°C, begins at a fixed cell mass and requires about 40 minutes; this can be longer than a cell-division cycle (which, at its shortest, takes only 20 minutes), necessitating overlapping rounds of replication in fast-growing cells (see article on "*Escherichia coli*: genes and the cell cycle"). When replication is completed, daughter chromosomes are intertwined and must be separated by topoisomerase action. The mechanism by which partition between daughter cells is achieved remains unclear.

The primary genome of most bacteria studied, Gram positive or Gram negative, is, like that of *E. coli*, a single circle, although length may vary; the sequenced chromosome of *Mycoplasma genitalium* is only 0.58 kilobases (kb) in size. Recently, however, several bacterial chromosomes have been demonstrated to be linear; such chromosomes have been found, among other bacteria, in both the spirochete Borrelia (the cause of Lyme disease) and in Streptomyces (a genus of Gram-positive spore-forming bacteria important as the source of many antibiotics). Recently, it has been demonstrated that some species of bacteria have more than one chromosome (using the working definition that a chromosome must carry essential genes). It is to be anticipated that, as studies are extended beyond the few model organisms so far considered, other exceptions to the "singular circular chromosome" generalization will come to light.

(ii) Plasmids

If a genome can be said to consist of that DNA which is stably inherited, then *E. coli* and other bacteria can possess a variety of secondary genetic elements in addition to the primary chromosome. Most worthy of mention are *plasmids*. These are DNA circles (although linear types are being discovered, mainly in species which also have linear chromosomes) with their own, specific, origins of replication (Hardy, 1993, Summers, 1996). Each also has a replication control system that, by controlling the number of replications permitted per cell cycle, ensures that a fixed copy number per cell of that kind of plasmid (copy numbers vary from one to hundreds per cell) is maintained. Replication control systems rely on positive or negative regulatory loops, varying with plasmid type. Plasmids that share the elements of a control system are classified as belonging to the same "incompatibility" group; separately identifiable members of such a group cannot be stably maintained in a single cell line. Plasmids mostly vary from about five to several hundred kb in length and are thus generally too small to be confused with the primary chromosome. Plasmids, by definition, do not encode products necessary for the functioning of the basic cell machinery; rather, they depend on their host cells to supply most of the functions needed for their own replication and transcription.

What are plasmids for? It may be easiest to consider them as separate life forms whose niche is to exist as bacterial symbionts, commensals or parasites. They do not rely on linear transmission from mother to daughter as might a bona fide bacterial chromosome, however small; instead, plasmids are equipped with systems that allow horizontal transmission. Larger plasmids specify transfer systems (to be considered further below) that permit direct transfer of the plasmid genome between cells; these systems are complex and comprise many gene products. Smaller plasmids feature "mobilization" regions that co-opt the transfer systems of larger cohabiting plasmids to promote their own transfer. Although many plasmids have, in the course of their promiscuous travels between species, acquired genes that might be of use to new bacterial hosts (such as drug or toxic metal resistance, the ability to metabolize complex organic compounds or pathogenicity determinants), they also possess determinants more appropriate to parasites. These bring about the death of plasmid-free cells that may arise in, or be encountered by, populations of plasmid-bearing cells. Some of these systems promote the synthesis of secreted toxic compounds to which plasmid-containing cells are immune; others have so-called postsegregational killing systems that cause the death of the occasional segregants that fail to inherit plasmid copies at division. These systems are composed of two components: a stable and an unstable component. The action of the stable toxic component is normally prevented by the inhibitory action of the unstable component. When the latter ceases to be made, because the encoding DNA has failed to be inherited, the segregants succumb.

(iii) Temperate bacteriophage

Bacteriophages are viruses parasitic on bacteria. While virulent phages have a "grab and smash" life style – they infect cells, subvert them to make virus particles and then lyse them – temperate phages have developed a strategy of co-existence in which they become part of the genome of the bacterial host (in a process called lysogenization). Some, such as the lambdoid phages, integrate themselves into the continuity of the bacterial chromosome and are replicated passively, while others take up residence as plasmids. Once ensconced, these genomes tend to remain a permanent feature until adverse conditions cause them to "jump ship", reproducing themselves and destroying the host cell in the process. However, there is an alternative fate for lysogenized DNA; there is now considerable evidence that the *E. coli* chromosome contains remnants of lysogenized phage genomes which, although now too incomplete to promote phage production, can and have been co-opted in part for purely cellular purposes.

3. A closer look at the chromosome

(i) Gene arrangement

Two large-scale systematic genome sequencing projects (one in the US, the other in Japan) for different substrains of *E. coli* K-12, have now been completed (Blattner *et al.*, 1997; see http://mol.genes.nig.ac.jp/ecoli/ for a full description of the W3110 sequence). Between these two projects, the entire genome has now been sequenced. Analysis of this sequence has, obviously, revealed a great deal about the arrangement of the ~4300 genes identified on the chromosome. The genes are densely packed – about 85% of the DNA appears to encode proteins and another 4% is transcribed to yield non-messenger RNAs. Much of the remaining intergenic sequence is in short stretches and is probably concerned with the control of transcription. Although bacterial genes are commonly thought to be arranged in operons, only about two-thirds of *E. coli* genes appear to belong to multigene transcriptional units. Despite numerous exceptions, cotranscribed genes tend to be functionally related. About 40% of genes identified by sequencing can be assigned definite functions. A further 20–25% have sufficient similarity to genes of known function to be classified functionally (preliminary functions hypothesized in this way need, of course, to be verified). However, a substantial minority, perhaps as much as 40%, have similarity only to other sequenced genes of unknown function or appear, so far, to encode proteins that are unique. Thus, there remain, even in this best understood of organisms, many mysterious genes.

Pairwise comparisons of over 2000 *E. coli* protein sequences show that about half have some sequence similarity over an extended region. They can, on the basis of these similarities, be assigned to families that are thought to be evolutionary groupings whose members have diverged from a single ancestral protein (Riley & Lebedon, 1997).

Since replication of the *E. coli* chromosome requires a long time relative to the cell division cycle, genes close to the origin of replication have a greater average concentration than those near to the terminus, especially in rapidly growing cells. The locations of certain genes, whose products need to vary with growth rate, reflect this. Thus, many genes whose products are concerned with transcription and translation are origin proximal – this helps to maximize the concentrations of their products in rapidly growing cells (Lin & Lynch, 1996). There are seven copies of the genes encoding ribosomal RNA, all near to the origin.

(ii) Repeated sequences

Several per cent of the genome consists of short, interspersed repeated sequences. These belong to several different classes. Some are thought to be native to *E. coli* and to have functional significance, either defined or suspected. Others appear to have originated as a result of horizontal transmission.

(iii) Native repeated sequences

Chi is an 8 bp sequence that promotes homologous recombination in its vicinity. *Chi*-stimulated recombination is mediated by the major *E. coli* recombinational enzyme, the RecBCD complex, and *Chi* is thought to be the recognition site for the RecD subunit of this complex. Consistent with its possessing a functional role, *Chi* occurs 20 times more frequently than would be predicted by chance. Remarkably, *Chi* appears on one strand only, the leading strand of newly replicated DNA (Blattner *et al.*, 1997). This suggests that it is acted on while at the replication fork, and that recombination principally occurs during replication.

REP is a 30–40 bp palindromic sequence that can form a stable stem–loop structure. *REP* sequences often occur in clusters and, although each is short, they are so numerous that they compose, in aggregate, 0.54% of the chromosome. They are not confined to *E. coli*, but can be found in quite diverse bacterial species. Genome locations are not well conserved, even between closely related enterobacteria. The function of *REP* sequences has not yet been defined, although there are numerous clues and suggestions. *REP* sequences are located between genes, such that they are always transcribed, but never translated. As parts of messages, they can stabilize upstream RNA or act as transcription terminators. *REP* DNA can bind both DNA gyrase and DNA polymerase I and it has therefore been proposed that *REP* sequences have a role in chromosome organization. Further research is awaited.

ter is a repeated sequence with an 11 bp conserved core that directionally blocks replication forks approaching it. There are eight *ter* sites, mostly near the terminus, and all oriented so as to prevent replication from proceeding past the terminus back toward the origin.

(iv) Sequences acquired by horizontal transmission

Insertion sequences (IS) are transposable elements of roughly 0.8–1.5 kb in length. There are many kinds, but all are similarly organized, with repeated sequences flanking a transposase gene (these encode enzymes that can move the IS element from one location to another). In the chromosome of the completely sequenced strain MG1655, there are ten different species of IS elements, which have a range in number of copies of 1–12. They move or are lost at frequencies well above those at which point mutations occur. Thus, strains vary greatly in the numbers and positions of particular elements, and spontaneous mutations are often the result of IS insertion. Transposons are related to IS elements and are probably derived from them by the acquisition of a selectable transposable trait. Although transposons are absent from wild-type K-12 strains, they occur frequently on plasmids (and include genes specifying drug resistance, among other phenotypes) and are transmitted by the plasmids in natural populations. Transposons have become a major tool in molecular genetics (see below), as well as a major scourge of medicine.

Rhs elements are a family of long (7 kb) complex sequences that structurally resemble eukaryotic transposons. Since their base composition is different from that of *E. coli*, and because they are present in only some strains, it has been concluded that they are of heterospecific origin. In total, *Rhs* elements account for close to 1% of the *E. coli* chromosome. *E. coli* K-12 DNA also contains several defective prophages (the partial genomes of ancient lysogenized phages), some with sequence similarity to lambda. MG1655 contains at least eight such elements, comprising several per cent of the genome. Finally, there are several other classes of repeated sequence of unknown function.

4. Routes of genetic exchange

Eukaryotes are characterized by having dedicated, evolved systems to ensure genetic exchange. In higher eukaryotes, two parents are generally required to produce offspring. In lower eukaryotes, both sexual and asexual systems may exist side by side and be used in different circumstances. Bacteria, on the other hand, do not exhibit sexual reproduction in the classic sense.

Reproduction is always asexual; each new cell is the daughter of one other. Superimposed on this asexual system of reproduction, however, are several mechanisms by which chromosomal genetic material can be introduced into a cell from outside and substituted for, or added to, existing DNA. Except for transformation in some organisms, the processes by which bacterial DNA transfer occurs appear to be accidental concomitants of systems that have evolved to ensure transfer of parasite DNA between cells. The gene transfer that does occur, although it clearly can provide selective advantages, and probably has contributed in a major way to evolution, is never an integral part of reproduction.

So how does DNA move between cells? There are three types of mechanism by which this occurs: transduction, conjugation and transformation.

(i) Transduction

Transduction occurs when chromosomal DNA is incorporated, by error, into a bacteriophage coat during lytic infection of a donor cell. Such a phage then acts as a vector to deliver host genetic material to a recipient cell by whatever mechanism (e.g. injection) that the phage ordinarily uses to deliver its own infective DNA. There are two types of transduction, termed specialized and generalized.

Specialized transduction is carried out by temperate phages (see 2(iii) above) that lysogenize by integrating, usually at a specific site, into the continuity of the genome. The occasional aberrant excision of phage DNA from a lysogen, followed by multiplication and lysis, can create a phage in which part of the phage DNA is replaced by neighbouring chromosomal DNA. If this phage lysogenizes a recipient cell, it will bring with it the incorporated fragment of donor DNA which will then be added to the recipient genome along with the lysogenizing phage. If there is an allelic difference between donor and recipient in the host genes incorporated into the phage, the recipient lysogen will be heterozygous for that marker. This type of transduction is termed "specialized" because only the DNA adjacent to the site of lysogenization is normally transduced.

Generalized transducing particles are formed during the lytic infection of a donor by phages which do not degrade host DNA completely during infection and which, principally, package their genomes by "biting off" headfuls from a longer length of DNA. Occasionally, such a DNA packaging system will accidentally select host DNA rather than phage DNA as a substrate, thus creating phage particles which contain only host DNA (of the same length as the phage genome). This DNA enters the recipient cell as a linear double-stranded molecule and is reliant on recipient recombination/repair mechanisms and a double crossover for integration into the recipient chromosome in place of the resident homologue. Since the major vector for generalized transduction in *E. coli* is bacteriophage P1, with a genome of about 100 kb, the maximum fraction of the bacterial DNA that can be replaced in such an event is about 2% of the total. Formation of generalized transducing phages is rare and successful integration of the transduced fragment rarer, so that only about 1 in 10^5 infected cells will be successfully transduced for a specific selected marker. Fortunately, the power of bacterial genetics is such that, provided you can think of a way to select for a particular phenotype, finding a cell that exhibits that phenotype (if it is present in the population) is no trouble at all.

(ii) Conjugation

Plasmids, as described above, encode complex transfer systems that allow them to enter and colonize neighbouring cells. This is achieved by direct contact between donor and recipient cells. Conjugation is a widespread phenomenon, which can occur in both Gram-positive and Gram-negative species; however, because of their differently structured cell envelopes, the mechanism probably differs between these two groups. Conjugation can even occur between bacterial and eukaryotic species and is a part of the normal mechanism by which the Gram-negative, rod-shaped, flagellated bacterium, *Agrobacterium tumefaciens*, induces crown gall tumour disease in a wide range of dicotyledonous plants. The bacterium only enters dead, broken plant cells and then may transmit a tumour-inducing (Ti) plasmid into adjacent living plant cells. This system can be used as a natural form of genetic engineering, since the Ti plasmid can be used to transfer foreign genes of choice into plant cells from which whole new plants can be derived after tissue culture.

In Gram-negative bacterial species, conjugational DNA transfer is mediated by so-called sex-pili – hair-like, plasmid-encoded structures on the surfaces of donor cells. These attach to recipient cells and are probably then retracted to promote cell–cell contact of mating pairs. The plasmid DNA circle is nicked and a single strand of donor DNA transferred; this is replaced in the donor and a complement synthesized in the recipient, creating two plasmid molecules, each in a separate cell, in place of the original. A large proportion of plasmid gene products, perhaps 30–40 in all, are involved in the conjugal process.

These mechanisms are very effective for transferring plasmids, but how do they promote chromosomal DNA transfer between cells? The answer is, by chance. This can happen in one of two ways: either the plasmid can become part of the chromosome so that the chromosome (or, more usually, because of its large size, a part of it) is transferred passively whenever plasmid transfer is initiated. Alternatively, part of the chromosome may become incorporated into a plasmid, replicated passively as a part of it and then transferred along with it during mating.

The first plasmid discovered, the F plasmid, was so named because it conferred fertility; F plasmids containing chromosomal DNA are called F primes (F′s). *E. coli* strains in which the F plasmid has become a part of the chromosome are called high frequency transfer (Hfr) strains, because they were originally discovered through their enhanced ability to transfer chromosomal markers during conjugation. Hfr strains are thought to arise by integration of F into the chromosome through a single crossover in a region of homology. Homology is most often provided by an IS present on both plasmid and chromosome. Since F contains at least three different types of IS, two of which are present in multiple copies on the chromosome, a moderately varied number of possible positions of integration are provided by this mechanism. It is also possible, although apparently less frequently, for the IS to insert into sites of non-homology by a transposition event.

When transfer is initiated in an Hfr strain, the proximal part of the integrated F is transferred, followed by the adjacent chromosomal DNA; transfer is usually interrupted by accidental breakage of the junction between the cells so that the distal portion of F is not transferred. The transferred DNA cannot circularize and, if not degraded, can be recombined into the recipient chromosome, replacing homologous DNA. Because the recipients in such a mating rarely receive a complete F plasmid they remain F^-. When the Hfr strain is mixed with a recipient population and pair formation allowed only briefly, transfer is initiated more or less synchronously. DNA transfer can be interrupted in aliquots of the mixture by rapid agitation and the genes transferred measured for each sample if a multiply marked recipient is used. Variants of this procedure have been used to show that a particular Hfr will transfer its chromosome from a particular point (later established to be the F insertion site) at a fixed rate and that different Hfrs transfer from different points on the chromosome or in opposite directions (see Hayes, 1968). The combined data from a variety of donors provided the first evidence that the *E. coli* chromosome is a circle. Although, in many situations, other mapping methods have overtaken Hfr mapping, it is still used to map newly isolated mutations.

F′ strains usually arise secondarily from Hfr strains by a recombinational event, legitimate or illegitimate, at sequences different from those at which F was inserted. A new plasmid is created, which can be transferred in its entirety. Unless a transient Hfr is formed, only markers included on the F′ will be transferred when an F′ strain is used as a donor. Progeny of F′ crosses will be diploid for the chromosomal markers included on the F′. Plasmids other than F, such as R factors encoding antibiotic resistance determinants, can also achieve "prime" status and mobilize chromosomes.

Plasmids can also gain genes as a result of a transposition event, without having been integrated into a chromosome. For instance, a cohabiting plasmid can acquire a transposon carried by an incoming plasmid. The spread of infectious drug resistance is, unfortunately, favoured by such events.

(iii) Transformation

The third method by which DNA can be transferred between bacterial cells, and the only one that does not require the participation of a non-chromosomal replicon, is called transformation. In transformation, naked DNA, either prepared in the laboratory or produced by a natural event, such as cell lysis, is taken up by a "competent" recipient cell (i.e. one in a state able to take up DNA) and is incorporated into its genome. This may be either by replacement, in the case of homologous chromosomal DNA, or by addition (plasmid establishment).

There is no evidence to suggest that transformation occurs naturally in *E. coli*, but its cells can be made competent by chemical or electrical treatment that alters the cell membrane so that DNA can pass through it. Linear DNA introduced into *E. coli* in this way is normally degraded by exonucleases and so has no genetic impact. Circular plasmid DNA, however, is not so readily degraded and can become established with good efficiency in a competent population of recipient cells. The ability to achieve this uptake of naked DNA by *E. coli* is one of the cornerstones of the molecular genetics revolution, for it allows DNA manipulated *in vitro* to be introduced into a cell and propagated there. Thus, although we are dealing with an artificial process, its value and consequences, for both technology and studies at the molecular level, cannot be overestimated.

If we digress for a moment from *E. coli*, we can find parts of the microbial world where transformation appears to be a natural and evolved process which enables gene transfer. Among the bacterial species for which transformation has been documented and its mechanism elucidated are the Gram-positive species *Streptococcus pneumoniae* and *Bacillus subtilis*, the former a pathogen which requires complex media to grow in the laboratory, the latter a free-living soil organism with no exotic nutritional requirements. There is considerable similarity in the way in which transformation occurs in these two species. Both can take up DNA from any source and both degrade one of the two strands in the course of uptake. The remaining strand can then be recombined with the recipient chromosome to effect a genetic exchange. Both species become competent for DNA uptake as they approach the stationary phase of growth, a time at which *S. pneumoniae* also excretes a competence factor, which can render cells from other phases of growth competent, and at which *B. subtilis* undergoes considerable cell lysis – thereby liberating DNA which the competent, intact, members of the population can take up. The third species in which transformation has been thoroughly studied is the Gram-negative *Haemophilus influenzae*. The process here is different. Only homologous DNA is taken up (there is a specific base sequence, common in *H. influenzae*, but rare elsewhere, which is required) and the DNA enters the cell and recombines while in the double-stranded form. In each of these species, there are many genes and an entire pathway devoted to the transformation process. It is difficult to imagine any purpose for such pathways, other than to ensure genetic exchange.

Using suitably marked donors and recipients, each of the methods described above can be used, and of course has been, to deliver the DNA needed either to map gene locations roughly on the chromosome (time of entry in Hfr crosses, linkage in transductional crosses) or to perform more detailed crosses for the ordering of closely linked genes.

5. *E. coli* genetics today

What is the role of genetics and genetic methods in the study of an organism once its DNA sequence and the chromosomal locations of putative genes are known? The answer is, a very important one: not until each gene has been functionally characterized will the genetic story be complete. Even then, genetic techniques will continue to be used for manipulation of the organism if, as is the case for *E. coli*, it is a tool in its own right.

(i) Large-scale genetic questions that are still to be addressed

First, close to one-half of the *E. coli* genes predicted from their sequences have either hypothetical or completely unknown functions. Although 4200 open reading frames have been identified, the current genetic map (Berlyn, 1998) lists only 2200 genes which have been individually studied and shown to be expressed (and even for these mapped genes, a detailed knowledge of function does not necessarily exist). Hypothetical functions thus need to be verified and studies need to be undertaken to characterize functionally those genes for which functional predictions have not been made. Directed mutations will play a large part in such studies.

Secondly, genetic analysis in *E. coli* is far more than positional analysis – it includes the assignment of genes to transcriptional units (operons) and the identification of the controlling factors and conditions that turn particular transcriptional units on and off. Sequence analysis has not provided many definitive answers here.

Thirdly, there are few molecular geneticists who start with the question, what does this gene do? Most begin with a process that interests them (cell division, chemotaxis, etc.) and aim to identify the genes involved, and the exact roles of those genes, in order to better understand the process. Initial identification of the relevant genes still requires mutation, selection and analysis of the mutants, although modern methods and tools can make this far easier than it once was.

(ii) Examples of current molecular genetic methods and tools

Deletions. Initial characterization of genes of unknown function (provided they are not essential for survival) is considerably facilitated if deletion mutants are available for study. Deletions of target genes can be achieved in *E. coli* by employing "crossover PCR", followed by use of a conditionally replicating vector to facilitate gene exchange (Link *et al.*, 1997). In brief, DNA flanking the gene to be deleted, with appropriate restriction endonuclease cutting sites incorporated for later cloning, is amplified by PCR. A further PCR reaction joins the two flanking fragments (now equivalent to a fragment deleted for the target gene) which is then cloned into a plasmid that is temperature sensitive for replication functions. An antibiotic resistance marker may also be included within the cloned DNA to allow selection. This plasmid is introduced into a host cell and forced to integrate into the chromosome (by temperature and antibiotic selection). Later resolution will give rise to progeny in which allele exchange has occurred. Selection against the parent plasmid will eliminate the plasmid, now carrying the wild-type allele, from the cell.

Genetically engineered transposons. A multitude of "minitransposons" which are delivered to recipient cells by suicide vectors (plasmids or phages that cannot replicate) have been developed in order to facilitate mutant hunts. Minitransposons are engineered so that the DNA encoding the enzymes required for transposition is not itself transposed. Therefore, the insertions obtained do not encode transposition functions, so that the transposition products obtained are quite stable. Minitransposons have been used to deliver promoters, reporter genes and selectable markers to the chromosome of *E. coli* and many other bacteria (this is particularly useful where other genetic systems are lacking) and also to a variety of plasmids.

Minitransposons are particularly useful when insertions at many, or unknown, locations are sought. For example, by using reporter genes which allow colony colour or fluorescence screening, it is possible to identify genes turned on by particular treatments or active in particular environments (e.g. heat-shock, cold-shock, high-temperature, response to ultraviolet light damage, etc.). Simple insertions, selected by drug resistance, can provide a collection of mutants unable to grow under screening conditions. PCR technology (in combination with the known chromosomal DNA sequence) allows the site of insertion to be determined by one of several methods (these include cloning and sequencing, inverse PCR and other PCR techniques (see article "Techniques in molecular genetics" for more detail on these techniques).

By selecting for transposition to a plasmid, plasmid function can be analysed or cloned genes targeted. (The desired products are found by transfer of the plasmid to a new host.) Use of minitransposons designed to yield protein fusions with a "reporter" can provide information on cellular location and topology of the protein encoded by the target gene. This is possible because certain enzymes used as reporters are active only in a particular cellular compartment. Thus, an alkaline phosphatase domain of a fusion protein is enzymatically active only if the topology of the fusion places it in the periplasm, while β-galactosidase functions only in the cytoplasm. More recently fusions with fluorescing reporters (green fluorescent protein) have been used to localize proteins within cells.

Mutations in essential genes. Deletion or insertion mutants of essential genes obviously cannot be successful in haploid cells. Mapping of temperature-sensitive mutations in essential genes can be facilitated by finding closely linked transposon insertions and determining the location of the insertion. If the function of a defined essential gene is to be investigated, conditional expression of the native protein can be achieved by cloning the gene under the control of an inducible promoter. The promoter most commonly used for this purpose in *E. coli* at present is that of the arabinose operon, since it can be effectively switched off to allow observation of cells as the essential protein is depleted.

Thus, we have seen that *E. coli*, despite a long history as a subject of investigation, still has a lot to tell us. More and more sophisticated tools have continued to be developed for its analysis. It can be anticipated that, in the next few decades, the function and the way in which expression is controlled will be described for each of its genes. We will then be in a good position to understand how at least one organism can be put together from its component parts.

REFERENCES

Berlyn, M.K.B. (1998) Linkage map of *Escherichia coli* K-12, edition 10: the traditional map (Review). *Microbiology and Molecular Biology Review* 62: 814–984

Blattner, F.R., Plunkett, G. III, Bloch, C.A. *et al.* (1997) The complete genome sequence of *Escherichia coli* K-12. *Science* 277: 1453–474

Hardy, K.G. (ed.) (1993) *Plasmids: A Practical Approach*, 2nd edition, Oxford: Oxford University Press

Hayes, W. (1968) *The Genetics of Bacteria and their Viruses*, 2nd edition, Oxford: Blackwell Scientific

Lin, E.C.C. & Lynch, A. Simon (1996) *Regulation of Gene Expression in E. coli*, London: Chapman and Hall

Link, A.J., Phillips, D. & Church, G.M. (1997) Methods for generating precise deletions and insertions in the genome of wild-type *Escherichia coli*: application to open reading frame characterization. *Journal of Bacteriology* 179: 6228–237

Neidhardt, F.C. *et al.* (eds) (1996) *Escherichia coli and Salmonella typhimurium: Cellular and Molecular Biology*, 2nd edition, Washington, DC: American Society for Microbiology Press. This huge work can be used as a reference source for most *E. coli* work prior to its publication date.

Riley, M. & Lebedon, B. (1997) Protein evolution viewed through *E. coli* protein sequences: introducing the notion of a structural segment of homology, the module. *Journal of Structural Biology* 268: 857–68

Summers, D.K. (1996) *The Biology of Plasmids*, Oxford: Blackwell Science

GLOSSARY

cytoplasm (bacterial) contents of the innermost compartment of the cell; contains DNA, protein synthesizing machinery and most proteins

DNA gyrase a Type II topoisomerase (see below)

DNA polymerase I removes the RNA primers and replaces them with DNA until it reaches the start of the previously made DNA segment

lysogen a bacterial cell in which the DNA of a bacteriophage has become part of the genome, either by being integrated into the chromosome, or by establishment as a self-replicating plasmid

lysogenization the process of forming a lysogen

messenger RNA (mRNA) single-stranded RNA molecules which specify the amino acid sequence of one or more polypeptide chains and act as a template for their synthesis

non-messenger RNA RNA that does not specify an amino acid sequence for part of a polypeptide chain (*see* messenger RNA)

operon groups of bacterial genes with a common promoter, clustered together and transcribed as a unit into a polycistronic mRNA

topoisomerase an enzyme which changes the degree of supercoiling of a DNA molecule by cutting, twisting or untwisting and resealing the cut single-stranded end (Type I topoisomerase) or cut double-stranded ends (Type II topoisomerase)

See also **Techniques in molecular genetics** (p.22); ***Escherichia coli*: genes and the cell cycle** (p.130)

ESCHERICHIA COLI O157:H7

Anne E. Hill
Moredun Research Institute, Pentlands Science Park, Edinburgh, UK

1. Introduction

Escherichia coli O157:H7 was first identified as a human pathogen in 1982. A rare but dangerous type of *E. coli*, the strain O157:H7 is now a major public health problem worldwide. The Center for Disease Control and Prevention (CDC) currently estimates that O157:H7 causes 20–70 000 cases of illness per year, with 100 deaths each year in the US alone. In this article, we take a look at *E. coli* O157:H7, the illness it causes and methods of prevention. Firstly, however, we will look at how this strain emerged as such an effective pathogen.

2. History and origins

The intestinal tract of most warm-blooded animals is colonized by *E. coli* within a few hours or days after birth. These Gram-negative facultative anaerobes normally serve a useful function in the body by suppressing the growth of harmful bacterial species and by synthesizing appreciable amounts of vitamins. *E. coli* is also widely used as an experimental subject in bacterial genetics and as a host bacterium in recombinant DNA work. Indeed, discoveries from *E. coli* have facilitated the new era in genetic engineering. We therefore usually view *E. coli* as microbes on which we depend. As a consequence, how *E. coli* O157:H7 evolved is a subject of interest not only to health workers and medical researchers, but also to molecular geneticists and evolutionary biologists.

The main method of identifying pathogenic strains of *E. coli* has, for many years, been serotyping based on specific markers found on the bacterial cell surface, namely the O, H and K antigens. The "O" antigen of *E. coli* O157:H7 was the 157th cell surface lipopolysaccharide, or somatic antigen, to be identified; the "H" antigen was the seventh flagellar antigen identified. Over 700 antigenic types, or serotypes, of *E. coli* have now been recognized and serotyping, as a method of identification, remains important in distinguishing the small number of strains that actually cause disease.

However, although serotyping reveals relatedness of *E. coli* strains and some clues as to the origin of *E. coli* O157:H7, the majority of the information concerning its evolution has come from sequencing results.

Reid *et al.* (1999) sequenced flagellin (*fli*C) alleles from 15 pathogenic strains including O157:H7. H7 sequences from strains with varying O antigens were found to be remarkably similar, indicating the exchange of an entire *fli*C allele between distant clonal lineages. Phylogenetic analysis demonstrates that the *fli*C sequences of O157:H7 and O55:H7 serotypes are nearly identical and highly divergent from those of *E. coli* strains expressing H6 and H2 flagellar antigens.

Several other studies have highlighted the similarities between O157:H7 and O55:H7. Perna *et al.* (1998) sequenced a number of genes from O157:H7 and O55:H7 and found that most show greater than 95% identity. For example, only a single point mutation is found in the β-glucuronidase gene of O157:H7 and its non-motile relatives (O157:H$^-$) on comparison with the O55:H7 gene (Feng *et al.*, 1998). In addition, several DNA segments have been identified in laboratory and wild-type *E. coli* strains that are absent from O157:H7 and O55:H7 (Stumpfle *et al.*, 1999).

(i) Production of toxins

One of the main characteristics of *E. coli* O157:H7 is the production of toxins that are closely related to those produced by *Shigella dysenteriae*. Several segments of the *E. coli* O157:H7 genome may be traced directly to *S. dysenteriae*, and sequencing studies have, as with O55:H7, highlighted remarkable similarities between genes found in O157:H7 and *S. dysenteriae*. Identity of 99.5% is found between *shu*A and *chu*A, genes for haem iron transport from *S. dysenteriae* and *E. coli*, respectively (Mills & Payne, 1995). Most significantly, genes encoding virulence factors identified in Shigella and O157:H7 tend to be encoded on mobile elements, such as plasmids and transposons, which offer a mechanism for genetic exchange between these two organisms (Stumpfle *et al.*, 1999).

(ii) Evolution of E. coli *O157*

The current consensus for the evolution of *E. coli* O157 is that DNA exchange occurred relatively recently between an *E. coli* O55:H7 ancestor and *S. dysenteriae* (or a *S. dysenteriae*-like organism). *E. coli* O157:H7 built on the backbone of O55:H7, an organism associated with infantile diarrhoea and, therefore, already adapted for causing diarrhoeal disease, by acquiring a plethora of virulence traits on mobile genetic elements.

As with many Gram-negative pathogens, the *E. coli* O157:H7 genome displays plasticity. Extrachromosomal elements such as plasmids and bacteriophages continually import factors that improve the disease potential of the organism. In *E. coli* O157:H7, the three main virulence traits – toxin, intimin and haemolysin – are encoded on phage, transposon and plasmid, respectively (see below). This horizontal gene transfer represents a key mechanism in the evolution of pathogens.

The imminent completion of the *E. coli* O157:H7 genome (http://www.genome.wisc.edu/) will undoubtedly answer many questions that remain about its origin, but it is important to note that the rate of evolution cannot be followed by genome sequencing alone. *E. coli* O157:H7 is an extremely versatile organism. As we have found, it is continually evolving and adapting. But how does this pathogen actually cause disease? What are the clinical manifestations of the illness? How can we treat the disease symptoms and how can we prevent infection?

3. Virulence factors

The capacity of *E. coli* O157:H7 to cause disease is a direct result of its possession of a number of genetic elements for virulence factors. Due to the severity of O157:H7 infection, research into the virulence factors has been intense. Common virulence determinants are given in Table 1, but here the most significant *E. coli* O157:H7 factors are discussed. It is important to note that a single determinant does not cause disease in itself: several virulence factors act in concert to produce the disease.

(i) Toxins

One of the most important virulence factors of *E. coli* is the production of large quantities of one or more Shigella-like (or Shiga-like) toxins (SLTs). Two distinct types of SLT are produced by *E. coli* O157:H7.

Shiga-like toxin 1 (SLT1). SLT1 is structurally very similar to Shiga toxin, the toxin produced by *S. dysenteriae*. SLT1 consists of six subunits, one of which is an A subunit of approximately 33 kiloDaltons (kDa) and five are B subunits of about 7.5 kDa each. Following proteolysis and reduction, the A subunit produces a large N-terminal A1 fragment, in which the enzymatic activity resides, and a small C-terminal A2 fragment. The five B subunits form a ring with a central pore. The B subunits bind to globotriaosylceramide (Gb_3) present on human platelets, renal epithelial cells and erythrocytes.

Shiga-like toxin 2 (SLT2). SLT2 shares approximately 56% homology with SLT1. The overall structure remains identical to SLT1, but both A and B subunits are larger, with molecular weights of 35 kDa and 10.7 kDa each, respectively (Plunkett *et al.*, 1999).

Two further toxins from *E. coli* O157:H7 have been described, one of which is similar to SLT2. The second has a molecular weight of 64 kDa and has an unknown function.

The *E. coli* O157:H7 toxins are all encoded on temperate bacteriophage inserted into the O157 chromosome (pathogenicity islands, PAIs); in SLT2, the PAI is phage 933W which is integrated into the genome. Integration into the chromosome leads to the continuous production of the toxin(s) at a low level. The mechanism by which toxin levels are increased is unknown.

Table 1 Common Gram-negative bacterial virulence characteristics.

Virulence trait	Example
Adhesins	Intimin
Invasins	Haemolysins, siderophores
Motility/chemotaxis	Flagella
Toxins	Shiga-like toxins
Antiphagocytic surface properties	Capsules, lipopolysaccharides, K antigens
Defence against serum bactericidal reaction	Lipopolysaccharides, K antigens
Defence against immune responses	Capsules, K antigens, lipopolysaccharides, antigenic variation
Genetic attributes	Genetic exchange by transformation and conjugation, transmissible plasmids, r-factors, drug resistance

(ii) Adhesins

A key aspect of infection by *E. coli* O157:H7 is colonization of the intestinal tract mediated by specific adherence factors. The main adhesion molecule in O157:H7 is intimin. Intimin, an outer membrane protein, is encoded on the *eae*A gene which includes part of a 43 359 base pair chromosomal PAI, termed the locus of enterocyte effacement (LEE; Jerse *et al.*, 1991; Perna *et al.*, 1998). The LEE consists of three functional domains: the *eae* and *tir* genes in the central region, a type III secretion system, and genes for other secreted proteins (*esp* loci; Karch *et al.*, 1999). This locus is required for the distinctive attaching and effacing lesions produced by *E. coli* O157:H7.

The *eae*A gene is necessary for intimate attachment to epithelial cells *in vitro*. A mutant in *eae*A from *E. coli* O157:H7 was constructed by Donnenberg *et al.* (1993) and shown to be deficient in its ability to attach intimately to colonic epithelial cells in a piglet model.

An additional adherence factor unlinked to LEE has recently been identified (Nicholls *et al.*, 2000). Mutants in *efa*1 (for EHEC *f*actor for *a*dherence) were found to be markedly deficient in adherence to epithelial cells *in vitro*, suggesting that *efa*1 contributes to the adhesive capacity of *E. coli* O157:H7.

(iii) Invasins

A further key virulence factor is enterohaemolysin. *E. coli* O157:H7 haemolysin belongs to the repeats in toxin (RTX) family of exoproteins. It is a 60 kDa outer membrane protein that serves as a pore-forming cytolysin late in the growth phase. Lipid bilayer experiments by Schmidt *et al.* (1996) showed that EHEC haemolysin caused the formation of transient ion-permeable channels by integration. These channels are cation selective at neutral pH. *E. coli* O157:H7 haemolysin is encoded on a 60 MDa/90 kilobase virulence plasmid known as pO157 or pEHEC. Two open reading frames (ORFs) are found on this virulence plasmid which show approximately 60% homology to the *hly*C and *hly*A genes of the chromosomally encoded *E. coli* alpha-haemolysin operon. *E. coli* O157:H7 *hly* genes are similar, but not identical to alpha-haemolysin.

Several reports correlate the presence of pO157 with virulence of *E. coli* O157:H7 (Karch *et al.*, 1987; Burland *et al.*, 1998; Nicholls *et al.*, 2000). The complete sequence was published in 1998 (Burland *et al.*, 1998), when 100 ORFs were identified, including 19 previously identified virulence genes. A large ORF has similarity to the N-terminal domain of large Clostridial toxins (LCT) that include ToxA and ToxB.

(iv) Iron acquisition

One nutrient essential for all pathogenic bacteria is iron. Access to this essential element is severely limited in the host and enteric pathogens have developed several mechanisms to acquire it. One method is the production of siderophores that scavenge free iron. Specific receptors and uptake mechanisms for ferric–siderophore complexes have evolved to optimize acquisition.

Most *E. coli* produce the catechol siderophore, enterobactin, but an additional siderophore, aerobactin, has been identified in some pathogenic *E. coli* isolates. Aerobactin is a conjugate of 6-(N-acetyl-N-hydroxylamine)–2-aminohexanoic acid and citric acid, which forms an octahedral complex with ferric iron. This siderophore can be recycled and, unlike enterobactin, aerobactin is not bound by serum albumin. Loss of the ability to produce aerobactin, but not enterobactin, results in reduced virulence.

Alternatively, pathogenic bacteria may obtain iron directly from host iron sources. The ability to use haem or haemoglobin as an iron source has been implicated as a determinant of virulence in *E. coli*. In *E. coli* O157:H7, an iron-regulated haem-binding outer membrane protein has been identified. Using ShuA, the haem receptor in Shigella, Torres and Payne (1997) found a homologous DNA sequence in O157:H7. The product, named ChuA, is a 69 kDa, TonB-dependent protein that is required for the utilization of haem and haemoglobin as iron sources by *E. coli*.

(v) Acid tolerance

The low infectious dose of *E. coli* O157:H7 may be due, in part, to its exceptional tolerance of acid pH. The factors that influence the induction and extent of acid tolerance of O157 have not yet been fully elucidated. Several studies have examined the survival of heat-shocked *E. coli* O157:H7 cells at low pH and show that the survival of "stressed" cells is about 10–100 times greater at low pH than untreated cells. A number of heat-shock proteins are induced in the outer membrane that may contribute to acid tolerance, but these have yet to be investigated.

4. Pathogenesis

Charged with the variety of virulence factors described above, how does *E. coli* O157:H7 actually cause illness? The answer lies in the expression of the various virulence factors at defined points as the disease progresses through the body.

The pathogenesis of *E. coli* O157:H7 involves two steps: intestinal colonization, followed by production of diarrhoeagenic enterotoxin(s). Based on current knowledge, the likely sequence of events is as follows:

(a) *E. coli* bacteria are ingested and pass through the stomach and small intestine, then adhere to the intestinal mucosa to produce the characteristic "attaching and effacing" lesions.
(b) SLT(s) then stimulate intestinal guanylate cyclase, the enzyme that converts guanosine 5′-triphosphate (GTP) to cyclic guanosine 5′-monophosphate (cGMP). Increased intracellular cGMP inhibits intestinal fluid uptake, resulting in net fluid secretion. The B subunits of toxin bind it to target cells via a specific receptor, Gb_3.
(c) The A subunit is then activated by cleavage of a peptide bond and internalized where it catalyses the ADP-ribosylation (transfer of ADP-ribose from NAD) of a regulatory subunit of membrane-bound adenylate cyclase, the enzyme that converts ATP to cAMP. This activates the adenylate cyclase, which produces extracellular cAMP, leading to hypersecretion of water and electrolytes into the bowel lumen.

This invasion of the intestinal cells causes damage to the lining of the large intestine. Further, as the SLT are produced and enter the bloodstream, they cause damage to the cells lining the blood vessels. In extreme cases, toxin production leads to damage of the kidneys.

5. Symptoms

Infection with *E. coli* O157:H7 produces many symptoms common to other food-borne diseases. Typically, stomach cramps and diarrhoea develop within 1–3 days after exposure. Occasionally, sickness occurs; fever is usually low grade or absent. At first, diarrhoea is watery, but it may become bloody (haemorrhagic colitis; HC) after 2–3 days. Illness is usually self-limited and lasts for an average of 8 days.

Approximately 10% of patients go from HC to develop haemolytic uraemic syndrome (HUS). HUS is a life-threatening complication of *E. coli* O157:H7 infection that is characterized by acute renal failure, haemolytic anaemia and thrombocytopaenia. HUS is particularly serious for young children and those over 65 years of age. In the US, HUS is the principal cause of acute kidney failure in children. On average, 2–7% of patients with HUS die, but in some outbreaks the mortality rate may be as high as 50%.

Some patients with *E. coli* O157:H7 infection, particularly the elderly, develop thrombotic thrombocytopenic purpura (TTP) in which the clinical features of HUS are seen together with neurological complications.

6. Diagnosis

Traditionally, *E. coli* O157:H7 infections are diagnosed by isolating the organism from diarrhoeal stools. This process involves enrichment cultures and plating in sorbitol MacConkey (SMAC) agar. *E. coli* O157:H7 is unlike more than 90% of *E. coli* strains in that it ferments sorbitol and will give rise to colourless colonies on SMAC plates. Further biochemical tests and latex agglutination antibodies for O157 and H7 antigen are then used to identify the strain. Stools may also be tested for the presence of toxins by cell culture and neutralizing antibodies. No other reliable phenotypic markers are known for O157:H7. Culturing and typing tests such as these usually take from days to weeks to perform, and a number of faster alternative diagnostic tools have been investigated.

A selection of DNA probes is available to detect the presence of toxin and intimin genes in food and clinical samples (Samadpour *et al.*, 1994). However, probing for such virulence factors may prove problematic; Shiga-like toxins, for example, are not obligatory in O157:H7 infections and, therefore, misdiagnosis may result.

One way to circumvent problems with varying virulence genes in *E. coli* O157:H7 is to use probes for O157 and H7 antigens. Li *et al.* (1999) developed a multiplex polymerase chain reaction (PCR) approach where primer sets are designed on genes for the biosynthesis of O157 and H7 antigens as well as for Shiga toxin and intimin. The combination of PCR reactions allows simultaneous detection of serotype O157, H7 and a variety of possible virulence traits. Such techniques will undoubtedly aid rapid diagnosis of O157:H7 infection.

7. Treatment

Most persons with O157:H7 infection recover without antibiotics or other specific treatments within 5–10 days. However, those patients with HUS or TTP require meticulous attention to fluid and electrolyte balance, nutritional support, treatment of severe anaemia and control of hypertension, seizures and azotaemia. Dialysis is necessary in approximately 50% of cases. With intensive care, the death rate of HUS is 3–5%.

The use of antimicrobials in treatment of O157:H7 infections is controversial. Although the strain is sensitive to a wide variety of antibiotics, including cefixime and potassium tellurite, their use may in fact increase the risk of HUS. Antidiarrhoeal agents such as loperamide (Imodium) should also be avoided.

8. Prevention

Most human *E. coli* O157:H7 infections are caused by consumption of contaminated food and/or water, and cattle are generally considered the major reservoir for this organism. Most outbreaks have been associated with the consumption of foods of bovine origin such as ground beef and milk. However, the pathogen has also been isolated from acidic foods such as apple cider, fermented sausage, blueberries, yoghurt and mayonnaise-containing food (Ansay & Kaspar, 1997; Bolton *et al.*, 1999; Hu *et al.*, 1999; Sout, 1999). Reports of *E. coli* O157:H7 in melon, sliced raw tuna, salad, hydroponically grown radishes and even mineral water have also been published (Fukushima *et al.*, 1999; Kerr *et al.*, 1999).

The Pennington report (1997; http://www.scotland.gov.uk/library/documents-w4/pgr-00.htm) detailed measures of preventing the spread of *E. coli* O157:H7 and highlighted the importance of good hygiene and thorough cooking of foods. The incidence of *E. coli* O157:H7 has been clearly linked to hygiene, and control should be implemented at all levels of the food chain with strict attention to sanitary practices during cattle slaughter. Rutala *et al.* (2000) have shown various home disinfectants and antibacterial kitchen cleaner to be very effective in eliminating *E. coli*.

A number of reports have also demonstrated the importance of person-to-person contact in the spread of *E. coli* O157:H7 (Pennings *et al.*, 1994; Ludwig *et al.*, 1997). The infective dose of *E. coli* O157:H7 is estimated at ten organisms. This small dose may even be spread on coinage.

Research into protective methods for *E. coli* O157:H7 infection is only just beginning. Natural infection with O157:H7 does not confer immunity and no human vaccine is currently available, although Shiga toxoid vaccines have been shown to be effective in preventing related disease in animals (Konadu *et al.*, 1998). Lactoferrin and its peptides, as naturally occurring antimicrobial agents, may contribute to protection against infection (Shin *et al.*, 1998).

Biocontrol of *E. coli* O157:H7 using virulent O157 antigen-specific phages is currently being investigated (Kudva *et al.*, 1999). *E. coli* O157:H7 may be lysed by a number of bacteriocidal bacteriophages isolated from the environment. The main concern with using bacteriophage is the development of phage-resistant strains. However, with our increasing knowledge of phage genetics it may be possible to circumvent any problems. Biocontrol using bacteriophage may prove to be a very efficient method of preventing the spread of *E. coli* O157:H7 in the future.

9. Putting *E. coli* O157:H7 in perspective

It is recognized that *E. coli* O157:H7 can produce severe, potentially life-threatening illness, but the absolute numbers of infections are low in comparison with other enteric bacteria such as Salmonella and Campylobacter. For every 1000 *E. coli* infections, Salmonella causes 35 000 and Campylobacter is held responsible for 47 600 cases.

Members of other *E. coli* serotypes have also been the cause of serious illness and death including O26, O104, O111 and O145 (Fukushima *et al.*, 1999; Todd *et al.*, 1999). As described above, strains are continually evolving, but the implementation of simple preventative methods can help significantly in reducing the risk of infection.

REFERENCES

Ansay, S.E. & Kaspar, C.W. (1997) Survey of retail cheeses, dairy processing environments and raw milk for *Escherichia coli* O157:H7. *Letters in Applied Microbiology* 25: 131–34

Bolton, D.J., Byrne, C.M., Sheridan, J.J., McDowell, D.A. & Blair, I.S. (1999) The survival characteristics of a non-toxigenic strain of *Escherichia coli* O157:H7. *Journal of Applied Microbiology* 86: 407–11

Burland, V., Saho, Y., Perna, N.T. *et al.* (1998) The complete DNA sequence and analysis of the large virulence plasmid of *Escherichia coli* O157:H7. *Nucleic Acids Research* 26: 4196–204

Donnenberg, M.S., Tzipori, S., McKee, M.L. *et al.* (1993) The role of the *eae* gene of enterohemorrhagic *Escherichia coli* in intimate attachment *in-vitro* and in a porcine model. *Journal of Clinical Investigation* 92: 1418–24

Feng, P., Lampel, K.A., Karch, H. & Whittam, T.S. (1998) Genotypic and phenotypic changes in the emergence of *Escherichia coli* O157:H7. *Journal of Infectious Diseases* 177: 1750–53

Fukushima, H., Hoshina, K. & Gomyoda, M. (1999) Long-term survival of Shiga toxin-producing *Escherichia coli* O26, O113, and O157 in bovine feces. *Applied and Environmental Microbiology* 65: 5177–81

Hu, Y., Zhang, Q. & Meitzler, J.C. (1999) Rapid and sensitive detection of *Escherichia coli* O157:H7 in bovine faeces by a multiplex PCR. *Journal of Applied Microbiology* 87: 867–76

Jerse, A.E., Gicquelais, K.G. & Kaper, J.B. (1991) Plasmid and chromosomal elements involved in the pathogenesis of attaching and effacing *Escherichia-coli*. *Infection and Immunity* 59: 3869–75

Karch, H., Heesemann, J., Laufs, R. *et al.* (1987) A plasmid of enterohemorrhagic *Escherichia-coli* O157-H7 is required for expression of a new fimbrial antigen and for adhesion to epithelial cells. *Infection and Immunity* 55: 455–61

Karch, H., Schubert, S., Zhang, D. *et al.* (1999) A genomic island, termed high-pathogenicity island, is present in certain non-O157 shiga toxin-producing *Escherichia coli* clonal lineages. *Infection and Immunity* 67: 5994–6001

Kerr, M., Fitzgerald, M., Sheridan, J.J., McDowell, D.A. & Blair, I.S. (1999) Survival of *Escherichia coli* O157:H7 in bottled natural mineral water. *Journal of Applied Microbiology* 87: 833–41

Konadu, E.Y., Parke, J.C., Tran, H.T. *et al.* (1998) Investigational vaccine for *Escherichia coli* O157: phase I study of O157 O-specific polysaccharide *Pseudomonas aeruginosa* recombinant exoprotein A conjugates in adults. *Journal of Infectious Diseases* 177: 383–87

Kudva, I.T., Jelacic, S., Tarr, P.I. *et al.* (1999) Biocontrol of *Escherichia coli* O157 with O157-specific bacteriophages. *Applied and Environmental Microbiology* 65/9: 3767–73

Li, Z., Elliott, E., Payne, J. *et al.* (1999) Shiga toxin-producing *Escherichia coli* can impair T84 cell structure and function without inducing attaching/effacing lesions. *Infection and Immunity* 67: 5938–945

Ludwig, K., Ruder, H., Bitzan, M., Zimmermann, S. & Karch, H. (1997) Outbreak of *Escherichia coli* O157:H7 infection in a large family. *European Journal of Clinical Microbiology and Infectious Diseases* 16: 238–41

Mills, M. & Payne, S.M. (1995) Genetics and regulation of heme iron transport in *Shigella dysenteriae* and detection of an analogous system in *Escherichia coli* O157:H7. *Journal of Bacteriology* 177: 3004–09

Nicholls, L., Grant, T.H. & Robins-Browne, R.M. (2000) Identification of a novel genetic locus that is required for in vitro adhesion of a clinical isolate of enterohaemorrhagic *Escherichia coli* to epithelial cells. *Molecular Microbiology* 35: 275–88

Pennings, C.M., Seitz, R.C., Karch, H. & Lenard, H.G. (1994) Hemolytic-anemia in association with *Escherichia-coli*-O157 infection in 2 sisters. *European Journal of Pediatrics* 153: 656–58

Perna, N.T., Mayhew, G.F., Posfai, G. *et al.* (1998) Molecular evolution of a pathogenicity island from enterohemorrhagic *Escherichia coli* O157:H7. *Infection and Immunity* 66: 3810–17

Plunkett, G., Rose, D.J., Durfee, T.J. & Blattner, F.R. (1999) Sequence of Shiga toxin 2 phage 933W from *Escherichia coli* O157:H7: shiga toxin as a phage late-gene product. *Journal of Bacteriology* 181: 1767–78

Reid, S.D., Selander, R.K. & Whittam, T.S. (1999) Sequence diversity of flagellin (*fli*C) alleles in pathogenic *Escherichia coli*. *Journal of Bacteriology* 181: 153–60

Rutala, W.A., Barbee, S.L., Aguiar, N.C., Sobsey, M.D. & Weber, D.J. (2000) Antimicrobial activity of home disinfectants and natural products against potential human pathogens. *Infection Control and Hospital Epidemiology* 21: 33–38

Samadpour, M., Ongerth, J.E., Liston, J. *et al.* (1994) Occurrence of shiga-like toxin-producing *Escherichia-coli* in retail fresh seafood, beef, lamb, and poultry from grocery stores in Seattle, Washington. *Applied and Environmental Microbiology* 60: 1038–40

Schmidt, H., Maier, E., Karch, H. & Benz, R. (1996) Pore-forming properties of the plasmid-encoded hemolysin of enterohemorrhagic *Escherichia coli* O157:H7. *European Journal of Biochemistry* 241: 594–601

Shin, K., Yamauchi, K., Teraguchi, S. *et al.* (1998) Antibacterial activity of bovine lactoferrin and its peptides against enterohaemorrhagic *Escherichia coli* O157:H7. *Letters in Applied Microbiology* 26: 407–11

Sout, D. (1999) Study puts US food poisoning 76 million yearly. *New York Times* 17 September 1999

Stumpfle, P., Broll, H. & Beutin, L. (1999) Absence of DNA sequence homology with genes of the *Escherichia coli hem*B locus in Shiga-toxin producing *E-coli* (STEC) O157 strains. *FEMS Microbiology Letters* 174: 97–103

Todd, E.C.D., Szabo, R.A., MacKenzie, J.M. *et al.* (1999) Application of a DNA hybridization-hydrophobic-grid membrane filter method for detection and isolation of verotoxigenic *Escherichia coli*. *Applied and Environmental Microbiology* 65: 4775–80

Torres, A.G. & Payne, S.M. (1997) Haem iron-transport system in enterohaemorrhagic *Escherichia coli* O157:H7. *Molecular Microbiology* 23: 825–33

USEFUL WEB ADDRESSES

E. coli reference centre home page: http://www.ecoli.cas.psu edu/ecoli/index.htm

E. coli O157:H7 genome sequencing group: http://www.genome wisc.edu/

Pennington group report (1997): http://www.scotland.gov.uk. library/documents-w4/pgr-00.htm

US Department of Agriculture Food Safety and Inspection Service: http://www.fsis.usda.gov/

GLOSSARY

adenylate cyclase an enzyme occurring in cell membranes that transforms adenosine triphosphate (ATP) into adenosine 3′,5′-cyclic phosphate (cAMP) and inorganic pyrophosphate. cAMP serves as a metabolic regulator

ADP-ribosylation a reaction in which ADP-ribose is covalently attached to another compound

azotaemia a higher than normal blood level of nitrogen-containing compounds in the blood, for example urea

cytolysin a toxin which lyses (usually eukaryotic) cells

enterocytes cells of the intestinal epithelium

pathogen any disease-causing organism

pathogenicity island inserted DNA segments within the chromosome that confer upon the host bacterium a variety of virulence traits

shiga toxin bacterial toxin from *Shigella dysenteriae* that blocks eukaryotic protein synthesis

shiga-like toxin a group of structurally related toxins that block eukaryotic protein synthesis by cleaving the 28S rRNA subunit of ribosomes

siderophores low molecular weight ferric iron-chelating compounds which are synthesized and exported by most microorganisms for the sequestration and uptake of iron

thrombocytopenia a decrease in the number of platelets in the blood, resulting in the potential for increased bleeding and decreased ability for clotting

TonB a periplasmic protein in *E. coli* that may act as a transmitter of ferric-iron-chelate complexes, vitamin B_{12} and metabolic energy

MICROBIAL GENOMICS

Monica Riley
The Marine Biological Laboratory, Woods Hole, USA

1. Introduction

Improvements in the efficiency and accuracy of methods for determining the sequence of bases in DNA in recent years have made it possible, in fact reasonable, for scientists to undertake to sequence fully the entire genetic complement of an organism. At the time of writing, there are 18 completely sequenced microbial genomes, some available to the public and some proprietary, and another 20 or so in the works (**Table 1**; for updated lists, see web sites such as http://genomes.rockefeller.edu/magpie/magpie.html and http://www.tigr.org). For the completely sequenced organisms, the sequence of all of the nucleotides of the DNA of the chromosome(s) has been determined and, in many cases, the sequences of extrachromosomal plasmids and plastids are known as well. From the point of view of genetics, one can ask what information or understanding the new approach – termed "genomics" – offers us that had not come to us earlier from (a) classical genetic studies on these organisms or (b) work utilizing the powerful combination of microbial genetics and molecular biology. Where do the data that emerge from the completed nucleotide sequences of genomes take us? They give us the translated amino acid sequences of all the proteins the cell can make, the nucleotide sequences of all the RNA molecules and the sequences of the many regulatory signals embedded in the DNA. This information brings within reach an understanding of all the genes and gene products, and their interactions, that constitute a living, reproducing cell. Genomic data also open doors to understanding the processes of molecular evolution in the context of the origin of life. In addition, the information has practical, applied uses, such as the development of pharmaceutical drugs that act on specific targets of disease-causing microorganisms. Other practical applications are in the sphere of agriculture, such as improving nitrogen fixation processes in symbiotic microorganisms.

2. Full sequences of microorganisms

Due to their small size, the first genomes to be completely sequenced were those of microorganisms. Some of the microorganisms are members of the domains of Bacteria and Archaea, while one, *Saccharomyces cerevisiae*, is a eukaryote. In most cases, the so-called "genomic sequence" equals the chromosomal DNA sequence of the organism. However, in some cases, full genomes have been sequenced, including not only the chromosomal DNA but also plasmid elements, as in *Borrelia burgdorferi* and the pathogenic strain of *Escherichia coli* (H7:O157), and the mitochondrial DNA of *Schizosaccharomyces cerevisiae*.

What does the information of genomic studies bring us? One of the first implications of having the full sequence of all the genes of any organism is the prospect of knowing exactly what comprises the total information needed to make a free-living cell that has the ability to generate energy and cell substance from simple foods, and the ability to reproduce itself indefinitely. Organisms such as *E. coli* and *S. cerevisiae* have been the subject of biological and genetic experimentation for many decades and have the best possibility of being completely understood in terms of their total genetic information. To report on progress to date in this article, *E. coli* will be used as an example.

3. How far are we from understanding the biology of the full genetic complement of *E. coli*?

The *E. coli* chromosome is larger than those of many bacteria, but not as large as some (**Table 1**). As it stands today, 1781 of the 4405 genes in the fully sequenced chromosome of *E. coli* are known and understood in the sense that their gene products have been identified experimentally and the physiological roles of their proteins or RNAs have been experimentally defined (Blattner *et al.*, 1997; Table 2; Genbank Accession no. U00096; http://genprotec.mbl.edu/dbase). For an additional 97 genes, the genetic location and mutant phenotype is known, but the biology of the gene product is only partially understood. The rest of the sequences have been analysed for open reading frames (ORFs; see article on "The role of bioinformatics in the post-sequencing analysis of genomic data") and, of these, 1122 bear similarity to at least one other sequence in the databases of all sequenced genes and can be assigned a putative function (**Table 2**). If sequences in a match are closely similar, the function of the biologically known gene product

Table 1 Fully sequenced microorganisms.

Species	Megabases DNA
Mycoplasma genitalium	0.58
Mycoplasma pneumoniae	0.81
Chlamydia trachomatis	1.05
Rickettsia prowazeki	1.10
Treponema pallidum	1.14
Borrelia burgdorferi	1.44
Aquiflex aeolicus	1.50
Methanococcus jannaschii	1.66
Helicobacter pylori	1.66
Methanobacterium thermoautotrophicum	1.75
Pyrococcus horikoshii	1.80
Haemophilus influenzae Rd	1.83
Archeoglobus fulgidus	2.18
Synechocystis sp.	3.57
Bacillus subtilis	4.20
Mycobacterium tuberculosis	4.40
Escherichia coli	4.62
Saccharomyces cerevisiae	13.0

Table 2 Types of gene product in *Escherichia coli*.

No. of genes	Type of gene product
950	Enzymes
481	Putative enzymes*
219	Transporters
275	Putative transporters
188	Regulators
165	Putative regulators
93	Structural components
53	Putative structural components
75	Factors
47	Putative factors
45	Membrane proteins
73	Putative membrane proteins
19	Carriers
11	Putative carriers
12	Leaders
64	Foreign proteins (prophages, transposons)
14	Apparent foreign proteins
116	RNAs
4	Putative RNAs
97	Known only as a mutant phenotype
1405	ORFs, no homologs found

* Putative functions are attributed as functions of sequence-similar proteins from either *E. coli* or other species.

is imputed to the unknown. Similarity matches for *E. coli* ORFs range from convincingly close at one extreme to borderline similarity at the other. Accordingly, some ORFs can be confidently labelled with a "putative" function based on close similarity, whereas others can only be assigned to a general class. After this type of identification, 1405 of the ORFs in *E. coli* remain that have no match in the databases of today.

How will we learn what the functions of the unknown ORFs are? Several approaches can be used.

(i) The in vivo *approach*

By an *in vivo* biological approach, the respective genes can be mutated by directed mutagenesis and the phenotype of the deficient bacteria can be observed. This should give information on genes whose absence confers a phenotype that we can see under specific conditions of culture. To exhibit a phenotype, the gene product must be unique, in that there are no similar cell components that can supply adequate function in the absence of the gene in question.

One of the problems of this approach is that not all *E. coli* genes give an observable phenotype when mutated. At present, we know at least 80 examples of multiple enzymes such that, when one is lost, another gene product can compensate for a missing gene product, resulting in leaky mutants or mutants with no phenotype (Riley & Labedan, 1996). Multiple transporters are also common, so that mutation of only one gene may not give physiological information about the gene product.

(ii) The in vitro *approach*

In vitro approaches have the promise of being far more efficient. One method uses the power of silicon chips with attached arrays of oligo- or polynucleotides of known sequence, several for any one gene, and uses them to identify by hybridization what mRNAs are found in the cell under specified conditions of culture.

(iii) Other approaches

Another method is to spot cloned nuclear DNA (or cDNA) on nitrocellulose filters and then hybridize the cDNA to RNA. By these, and other means (such as two-dimensional protein gels), the gene products of all *E. coli* genes will ultimately be identified and their functions known. Of the *E. coli* genes of presently known function, the largest class of gene products consists of enzymes; the next largest class consists of transporters of molecules in and out of the cell, followed by a close third, the regulators (**Table 2**). Distant followers include classes consisting of cell structure components, factors, carriers and RNAs. What additional information do we have from 50 years of bench experimentation with *E. coli* that will help us characterize all its genes? There are 96 biochemically characterized enzymes known to be present in *E. coli* that have never been associated with any gene, and these will account for some of the ORFs when the connection with the gene is made. Also, there are 252 mapped *E. coli* genes with recorded mutant phenotypes that are not yet associated with any DNA sequence (Bachmann, 1990). Both *in vivo* and *in vitro* studies will help us to assign these to specific ORFs.

4. Protein sequences provide detail for understanding the molecular architecture of proteins

Some proteins are multimodular in the sense that large-sized pieces (100–350 amino acids) seem to have become shuffled and re-ordered over evolutionary time (Riley & Labedan, 1997). **Figure 1** shows an example: a response regulator protein, Che. This large protein contains both an enzyme function, a methylase, and a regulatory function, a protein kinase. These two modules are found in other orders and combinations in other proteins. Evidently, such modules have become rearranged and fused with different partners during evolution to make many large multifunctional proteins. Terminology for these units is not uniform. According to some authors, the module is a functional component within a gene that corresponds to a polypeptide chain of independent function. Multimodular genes have undergone fusion in the course of evolution to generate multifunctional proteins (Guigo *et al.*, 1996). A module can be as large as an independently functioning protein or as small as a structural domain, a unit of protein structure. Some refer to such a subdivision of a gene as a domain (Henikoff *et al.*, 1997). It might be better to reserve the word "domain" for protein structures. A structural domain has been defined as a part of a protein which can fold independently into a stable tertiary structure, often with an associated function (Chothia, 1984). By contrast, a motif is a short sequence composed of some specified amino acids, such as the 22 amino acid helix–turn–helix motif of DNA-binding proteins. The connections between motifs in database sequences have been explored by Tatusov and colleagues (Tatusov *et al.*, 1997). Both modules and domains usually contain more than one motif. In the literature, the uses of these terms (module, domain and motif) vary, and it will be important to have

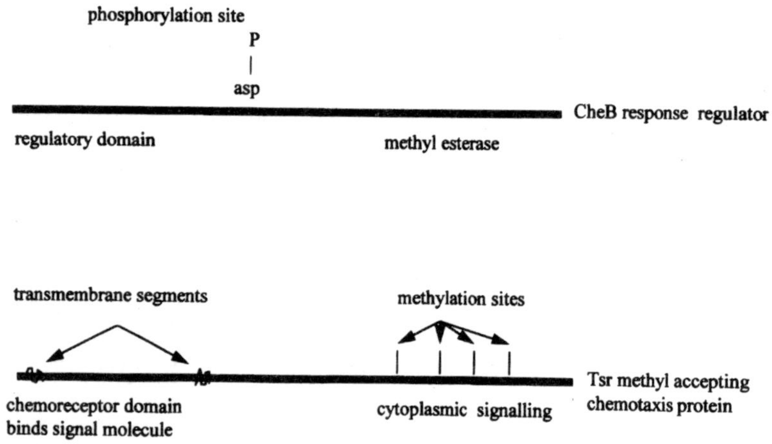

Figure 1 Illustration of multiple functions of two modular proteins, CheB and Tsr. Multiple functions of proteins and different homologies for the N- and C-terminal parts probably reflect fusions of genes during evolution.

formal nomenclature for these different elements of gene and protein sequences. Nomenclature aside, gene and protein sequences are being used to illuminate evolutionary history.

5. Molecular evolution: phylogeny of organisms

The evolutionary tree of all organisms showing the order of emergence of new species over time is referred to as the tree of life. Its structure has been elucidated by comparative analysis of nucleotide sequences of highly conserved ribosomal RNA (rRNA) molecules (Woese *et al.*, 1990). As for proteins, some well-conserved proteins, such as triose phosphate dehydrogenase or some of the cytochrome proteins, have also been used to construct evolutionary trees of organisms. Now, with complete genome sequence data for some organisms, many comparable protein sequences are known in multiple species. However, evolutionary trees made with amino acid sequence analysis do not always agree with rRNA trees and are not always consistent from one protein to another. The role of horizontal transfer across species lines probably accounts for some of this inconsistency. Also, descendants of one protein may adopt different biological roles in different species. This can introduce problems of different selection pressures on proteins of similar sequence, creating problems of knowing which genes and proteins of different species are evolutionarily comparable. With more data, the complex history of proteins concurrent with speciation events may become decipherable.

Extensive data for the construction of protein-engendered trees is accumulating from the efforts of many groups around the world now engaged in sequencing higher organisms. At present, the species range for completed genomes is restricted to microorganisms, but major eukaryote genomic sequences are well under way.

6. Molecular evolution: phylogeny of proteins

Genomic data are useful in the context of the molecular evolution of proteins (to be distinguished from the molecular evolution of organisms). The "last universal common ancestor" had to be a complex cell with many cellular functions already in place before the generation of the three kingdoms of the tree of life: the Archaea, the Bacteria and the Eukaryotes. The proteins that existed at this stage of evolution and performed functions that were passed on to all three branches of life can be called ancestral proteins. Molecular events occurring since the last universal common ancestor are pertinent to reconstructing organismal evolutionary trees from contemporary protein sequences from a variety of organisms (see previous section).

However, another important aspect of evolution can be examined through the relationships of protein sequences in any one organism. This aspect is the process of evolution of the ancestral proteins themselves. In early evolutionary stages, preceding the last common universal ancestor, there existed the earliest functional ancient proteins that ultimately evolved into the ancestral proteins in the last common universal ancestor (Ycas, 1974). It is possible to learn about the processes of evolution of proteins from the sequences of contemporary proteins in a single organism and to use this sequence information to make trees of the evolution of proteins.

Using the genes of one single organism to reconstruct the ancient events of protein phylogeny, one delineates sequence-related groups of proteins that might have descended from a single parental protein. Analysing the protein sequences of *E. coli*, sets of proteins are found, each composed of a set of similar sequences with similar functions (Riley & Labedan, 1997). The sizes of such groups range from two to well over 100 for the largest group. The large groups are typically the transporters and regulators; the smaller groups are typically enzymes. Evidently, the transporter and regulator proteins have been more conserved over time than the enzymes, which have diverged much more as they achieved narrow specificities. Similar analysis of completed genome sequences of other organisms will tell us how similar the groupings are to those of *E. coli*. Universal groups will be those that date back to earliest times. These groups of proteins, present in most modern cells (many being metabolic enzymes, with some types involved in transport and others in regulation), will provide a picture of the early stages of making, from a few ancient proteins, the larger number of ancestral proteins of differentiated function that we believe were in the last universal common ancestor.

7. Practical applications of genomics

Besides increasing our knowledge of details of microbial cell physiology and molecular evolution, the revolution in genomics brings information of practical value. Systems are being developed in the area of "functional genomics" to automate the attribution of function to genes that are known only by their sequence. Attribution of function depends on the fact that proteins of closely similar sequence have closely similar functions. The parallel weakens with evolutionary distance. As the distance between the proteins increases, the confidence in similarity of function decreases. When using sequence similarity as an instrument to attribute function to an unknown gene, it goes without saying that it is important that the level of confidence in any attribution be explicitly expressed to avoid misattribution and introduction of errors in the genomic databases.

8. Conclusion

Used with appropriate caution, the methods of functional genomics have the capability to expand rapidly the body of our information on microorganisms that have not to date been the subject of very much experimental investigation. Knowledge acquired by analogy through sequence similarity accelerates understanding of the physiology of bacterial pathogens and opens up possibilities of identifying "targets" in disease-causing microorganisms that may be susceptible to pharmaceutical assault, while not affecting the human host. Likewise, in the field of agriculture, more detailed information on the genomics of nitrogen-fixing organisms and symbiosis with legumes, as well as the metabolism of plant pathogens, will rapidly augment knowledge already in hand through decades of genetic and biochemical investigations.

Ultimately, our knowledge about all microbes that are either useful or harmful to man will be augmented by genomic approaches, with the development of genetic methods that may be able to enhance the desirable activities of microorganisms or eliminate undesirable ones.

REFERENCES

Bachmann, B.J. (1990) Linkage map of *Escherichia coli* K-12. *Microbiological Reviews* 54: 130–97

Blattner, F.R., Plunkett, III, G., Bloch, C.A. *et al.* (1997) The complete genome sequence of *Escherichia coli* K-12. *Science* 277: 1453–74

Chothia, C. (1984) Principles that determine the structure of proteins. *Annual Reviews of Biochemistry* 53: 537–72

Guigo, R., Muchnik, I. & Smith, T.F. (1996) Reconstruction of ancient molecular phylogeny. *Molecular Phylogenetics and Evolution* 6: 189–213

Henikoff, S., Greene, E.A., Pietrokovski, S. *et al.* (1997) Gene families: the taxonomy of protein paralogs and chimeras. *Science* 278: 609–14

Riley, M. & Labedan, B. (1996) *E. coli* gene products: physiological functions and common ancestries. In *Escherichia coli and Salmonella, Cellular and Molecular Biology*, 2nd edition, edited by F.C. Neidhardt *et al.*, Washington, DC: American Society for Microbiology Press

Riley, M. & Labedan, B. (1997) Protein evolution viewed through *Escherichia coli* protein sequences: introducing the notion of a structural segment of homology, the module. *Journal of Molecular Biology* 268: 857–68

Tatusov, R.L., Koonin, E.V. & Lipman, D.J. (1997) A genomic perspective on protein families. *Science* 278: 631–37

Woese, C.R., Kandler, O. & Wheelis, M.L. (1990) Towards a natural system of organisms: proposal for the domains Archaea, Bacteria, and Eucarya. *Proceedings of the National Academy of Sciences USA* 87: 4576–79

Ycas, M. (1974) On earlier states of the biochemical system. *Journal of Theoretical Biology* 44: 145–60

GLOSSARY

cytochrome enzyme pigmented as a result of its haem prosthetic groups

horizontal transfer transfer of genes from one evolutionarily unrelated organism to another

methylase enzyme that attaches a methyl group to a molecule

open reading frame (ORF) stretch of DNA containing a signal for the start of translation, followed by an adequate number of amino acid-encoding triplets in the correct register to form a protein, followed by a signal for termination of translation

See also **The role of bioinformatics in the postsequencing analysis of genomic data** (p.833)

D GENETICS OF DROSOPHILA AND OTHER INSECTS

Introduction to Genetics of Drosophila and Other Insects

The sequencing of the fruit fly *Drosophila melanogaster* (Adams *et al.*, 2000) marks almost a century of research on one of the most intensively studied organisms in biology. As described in the opening articles of this section (and also described in the article in section N on Genetics of Quantitative Characters), the model organism *D. melanogaster* has contributed much to the investigation of many developmental and cellular processes common to higher eukaryotes, including humans. The implications of the Drosophila genome for future research in biology and medicine are described in a series of articles in a special issue of *Science* in which the Adams *et al.* article appears.

The 120 million base pairs (Mb) of sequence of the *D. melanogaster* genome could soon be joined by the sequence of a second member of the genus, *D. pseudoobscura*. The mapping of the silkworm too, *Bombyx mori*, should enable geneticists to prepare comparative maps of insect gene function, just as the mouse and other vertebrate genomes are expected to aid annotation of the human sequence.

REFERENCES

Adams, M.D., Celniker, S.E., Holt, R.A., Evans, C.A., Gocayne, J.D. *et al.* (2000) The genome sequence of *Drosophila melanogaster*. *Science* 287: 2185–95

DROSOPHILA MELANOGASTER: THE FRUIT FLY

James H. Sang
University of Sussex, Brighton, UK

1. Introduction

There was a great expansion of biological experimentation at the turn of the 20th century, sometimes simply triggered by the growth of academic institutions but also by the pervading interest in Darwinism and Mendelism. Fieldwork was carried out, following metrical changes in organisms, in time and space, to see if they did evolve, but more especially work was done in laboratories selecting variants of many domesticated plants, and of mice, rabbits, chickens and a range of easily kept insects. The most productive work employed the fruit fly, *Drosophila melanogaster*, first cultured in the laboratory in 1901.

Most people have seen this cosmopolitan fly without recognizing it. It is quite literally a commensal, for on a warm evening it will be found hovering over a wine glass, attracted by the scent of alcohol, or by any fermenting fruit which carries its food – yeasts. It is a dipteran like the blowfly or the housefly, only smaller, and it weighs about 1 mg. The smaller male has a black abdomen (hence *melanogaster)* and the sexes are therefore easily distinguished. The female is very fertile and can lay up to 100 eggs per day at peak, and perhaps 2000 eggs during her 60–70 day lifetime. The flies can be easily cultivated on any source of yeasts. Originally it was cultivated on banana in the equivalent of half pint milk bottles (and if you place a rotting banana by your window and then keep it in a closed jar for a few days, you will almost certainly find Drosophila larvae growing there). Nowadays a maize meal – molasses mix firmed with agar which supports the tunnelling larvae for their three feeding days prior to pupation – is generally used as a standard. The egg–adult life cycle takes about 10 days at the optimum temperature of 25°C, so one can reasonably expect to grow up to 30 generations per year. It was this short operation time, minimal space requirement and low handling costs which gave Drosophila its advantage over other animals (e.g. mice) being used for genetic experiments. However, it also brought other benefits.

2. Drosophila as an experimental animal

Drosophila soon became an experimental animal, used for testing responses to heat, light, gravity and various chemicals; it was a cheap, easily replaceable teaching material in the days when biology students were expected to handle whole living organisms. The exploitation that made Drosophila the most important organism for genetical research was its selection by the embryologist Thomas Hunt Morgan for his studies of mutation and his establishment of what became known as the Fly Room at Columbia University, New York. A detailed account of this exciting time in the development of Drosophila genetics can be found in the book *Lords of the Fly* (Kohler, 1994).

By 1910 Morgan had isolated a mutation affecting a gene (most mutations are called "genes" since that is what they identify), changing the brick-red colour of the fly's eye to *white*, symbol *w*; wild-type symbol always plus (+), implying that *w* and other mutations are by comparison minus (−), or defective. By luck, *w* was a sex-linked gene on the X chromosome; it is inherited by sons from their mother, and therefore easily identified (**Figure 1**). This started many studies by Morgan and his assistant C.B. Bridges on the important subject of sex determination, and many breeding experiments which, in their turn, produced a multitude of new mutations affecting wings, limbs, bristles, etc. of the, by now standard, fly. Genes were for many years named according to the effect that the mutant allele had on the appearance (phenotype) of the fly; for example, the *white* gene because of the white-eyed phenotype caused by the first mutation. Many genes are now identified from molecular rather than mutational analysis and this has led to some unusual names that are not related to any phenotype associated with the genes concerned. Information on all genes that have been identified can be found on FlyBase (http://flybase.bio.indiana.edu/).

3. Linkage group analysis

During this early period, most of the mutations found were recessive; that is, they were not expressed in the presence of a partner (or homologous) wild-type gene. The collection of genes affecting, say, the eye were not inherited together so classifying genes according to the organs they affected therefore had to be abandoned. It had been known for some time from work on the sweet pea that some genes were inherited together (linked) and Morgan showed that the genes for white eye, yellow body and miniature wing were so bound on the X chromosome. New mutations were thereafter classified by their linkage group and hence it was presumed by the chromosome that carried them. Drosophila has four pairs of chromosomes (compared with the human 23 pairs), thus serendipitously simplifying the breeder's problems to four linkage groups. Some genes, such as *w*, were found to give mutations with different phenotypes to that of the

		Male gametes X	Male gametes Y
Female gametes	X^w	X^wX	X^wY
	X^w	X^wX	X^wY

Figure 1 Diagram showing gametic combinations in the progeny of a female homozygous for an X-linked mutation such as *w* (X^wX^w) after mating to a wild-type male (XY).

mutation by which the gene had been originally identified. This implied a gene locus that could be altered in different ways by mutation. But before this problem – the nature of a locus – could be explored, a simpler question had to be answered: how were loci organized in relation to one another on the chromosome?

(i) Test crosses in Drosophila mutants

Data from test crosses of multiply marked (mutant) chromosomes had already provided information that resolved this problem. First, new arrangements of the marker genes were found which could be accounted for by recombination or crossing-over between chromosomes. Second, crossing-over was confined to females although this restriction is not generally true. Third, by crossing an F_1 heterozygous multiply marked female by a parental male carrying the same marked chromosome (a backcross), the recessive, unexpressed genes in the female can be identified by the appearances of the progeny (i.e. their phenotypes as opposed to their genotypes).

For example, if we take a female heterozygous for recessive mutations in three genes A, B and C on the first (or X) chromosome ($X^{ABC}X^{abc}$) (i.e. its phenotype will be wild type) and mate it to a male (X^{abc}) carrying the same mutations (all expressed in its phenotype since it has no homologous X chromosome) then a simple linkage map can be constructed from the numbers of progeny of each phenotypic class. For example, the numbers in the different classes might show that for about 80% of the progeny there is no recombination, for about 9% crossing-over occurs in the a–b region and for just over 12% in the b–c region. Sturtevant and Bridges reckoned that the further apart two loci were on a chromosome the more likely a crossover was to occur, so that the percentage crossing over (called a centiMorgan (cM) after their boss) measured the distance apart of two loci. That is, a–9 cM–b–12 cM–c provides a primitive map, the accuracy of which depends on the numbers scored, the ease of identification of the mutations and the assumption of proportionality to distance, which proved to be wrong! In larger experiments infrequent double crossovers were found when crossing occurred in a–b and b–c regions simultaneously. This double exchange could occur only around the central locus (here b) thus giving the gene order (which we had assumed). By extending the map progressively the whole ordering of genes could be established. This started an "industry" of mapping to which most of the last generation of geneticists contributed, and the technique was taken over for other organisms, including man. The location (its site on the chromosome) gave the best available definition of a gene.

The three-point test cross described above was soon elaborated by increasing the number of marker (mutant) genes on a chromosome; with up to eight markers as devised by Bridges, and with 12 markers created by Muller, whose X-irradiation experiments were producing a never-ending stream of new genetic material. The investment of labour and technique in these new tester strains was considerable, but despite this they were made freely available to Drosophilists worldwide: the open exchange of ideas and materials was maintained under a considerable pressure to monopolize. Muller's work showed that most mutations (~75%) were recessive lethals: dominant lethals eliminate themselves. Such mutations blocked or otherwise affected embryonic or larval development and therefore merited saving. For this Muller devised the ingenious balanced lethal system, which consisted of flies carrying two different lethal genes, one on each member of a pair of homologous chromosomes (**Figure 2**) where half the progeny perpetuate the parents and one-quarter create homozygotes of each of the two lethals and die. The stock is self-perpetuating.

4. Chromosome mutations: deficiencies, translocations, inversions

Mutation studies also threw up the phenomenon of chromosome breakage or chromosomal mutations, of which there are three types: deficiencies, translocations and inversions. A deficiency occurs when a chromosome segment is lost. In a translocation the segment moves its position within the genome either intrachromosomally or interchromosomally, or there may be a reciprocal translocation when there is an exact exchange of segments between non-homologous chromosomes. An inversion is when a piece of chromosome is reversed

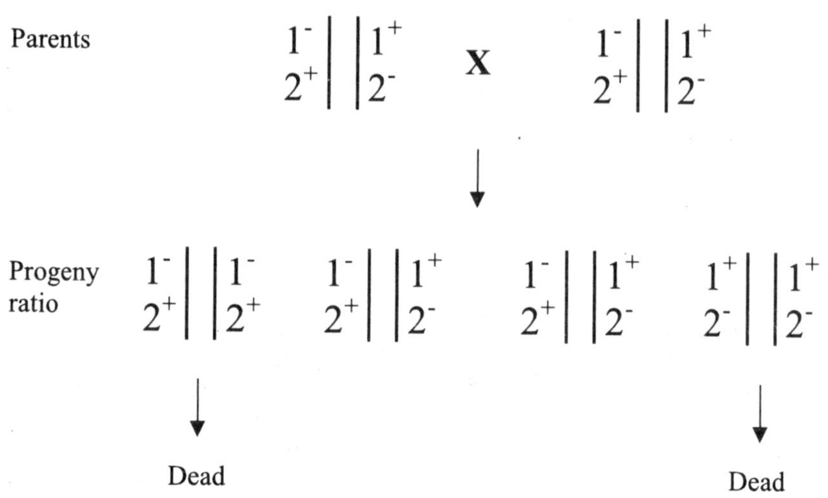

Figure 2 The balanced lethal system in Drosophila: 1^- and 2^- are different recessive lethals on the same chromosome, so homozygotes for 1^- or 2^- die. There is no recombination in the production of male gametes, so homozygous + chromosome pairs cannot appear.

in position (inverted); the sequence *abcde* then becomes *acbde*, for example (see **Figure 3A**). If the inverted segment involves the centromere it is called a pericentric inversion; if it does not involve the centromere it is termed a paracentric inversion (see **Figure 3B**). Since there is no recombination in males most of these rearrangements have only minor effects on population survival, and they could be used to supplement mapping data, as seems obvious since a gene moved must be in the company of the others in the segment.

Inversions were discovered to block recombination within their sequence and in adjacent regions of the chromosome and, indeed, it was this failure to recombine that led to their discovery. Inversions have been used to "stabilize" chromosomes in various situations, the most important being for mutation studies. Muller synthesized an X chromosome with a crossover suppressor (C), comprising an inversion covering more than half the chromosome, a lethal (l) and a dominant marker (Bar eye). A Bar-eyed female carrying a ClB X chromosome must have a wild-type homologue (from her father) to survive. She would thus have two classes of male offspring – ClB/Y and +/Y – and the former would all die. The latter class would survive unless it, too, carried a new lethal mutation, either spontaneous or induced. It was easy, and required no personal qualitative judgement, to scan cultures for the absence of males and so to determine the frequency of induced X-linked lethal mutations. The ClB chromosome became the primary instrument

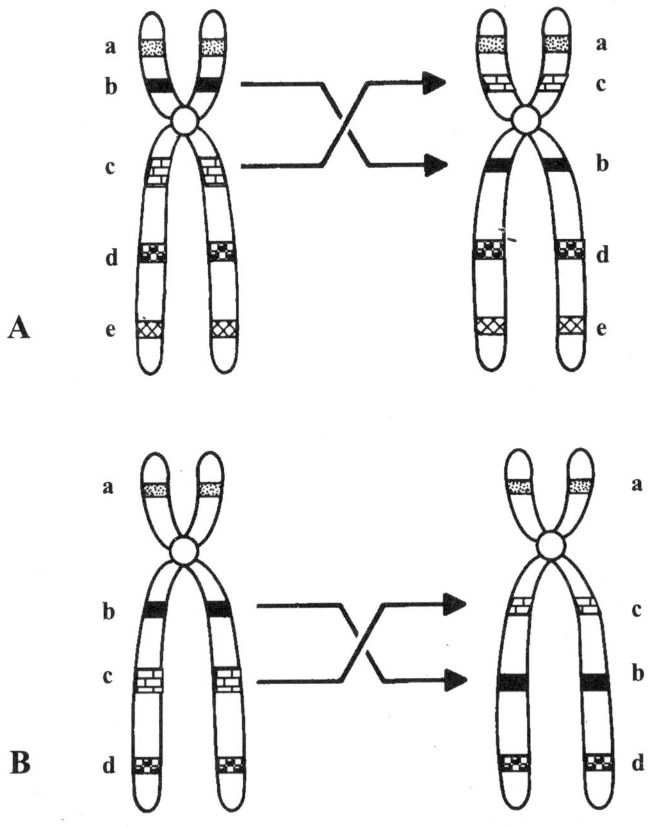

Figure 3 Diagrams of (A) pericentric inversion; (B) paracentric inversion. (Adapted from Emery & Muller, 1992.)

for mutation studies of the effects of radiation (Muller, 1928) and then, later, of chemical mutagens by Auerbach (1949).

5. Development of cytogenetic maps

Drosophila had one considerable disadvantage: its chromosomes are very small and it was difficult to identify changes from the standard, even of deletions. The story so far was therefore exclusively an intellectual one, of deduction and hypothesis. But in 1933 Painter found that the large cells of the larval salivary glands contained polytene chromosomes. That is, although the cells grew the number of nuclei did not and their chromosomes multiplied in number with this growth to produce many-stranded giant chromosomes (**Figure 4**). Since the chromatin is hypercoiled in regions and the chromatids are in register, a banding pattern is produced at right angles to the long axis of the chromosome. Painter guessed that this constant and distinct patterning would lead "to the lair of the gene". And, indeed, Bridges' last contribution to Drosophila genetics was to relate his crossover maps to the banding patterns of the salivary gland chromosomes (Bridges, 1935). These cytogenetic maps have since been changed in only minor aspects – of course they have also been added to – and are still currently used by researchers who have elaborated and refined them. They mark the end of the first major period of Drosophila research.

6. Evolution and development

Originally Morgan's interest was in the role of genes in evolution and in how they might be used to elucidate the then intractable problems of development. A number of individuals, inside and outside the main American laboratory, attempted to approach these issues, and we shall look at them in that order. The theoretical studies of R.A. Fisher, J.B.S. Haldane and S. Wright provided a stimulating framework for practical researches into wild populations.

D. melanogaster's advantage was that it was adapted to a great variety of environments including laboratories, and it soon became obvious that this flexible physiology ruled it out from serious studies of adaptation and evolution. There was, for instance, very little variation of chromosome banding among flies sampled from different parts of the temperate landmasses. True, it was shown that a population cage started with a mixed wild-type and mutant population eventually resulted in the elimination of the mutant gene; a sort of model of natural selection but not very inspiring – would another mutation behave in the same way? During the period 1924–1934 a Russian school of evolutionists under Chetverikov was studying wild populations of Drosophila – frequencies of lethals and the like (Adams, 1968) – but it was only when Sturtevant and Dobzhansky found that populations of *D. pseudoobscura* sampled in the wild carried genetic variations that differed from location to location in their natural wooded environment, that Drosophila became the object of detailed evolutionary studies (Dobzhansky & Sturtevant, 1938). These variations were chromosomal mutations, particularly inversions, easily identified from preparations of their salivary glands, but

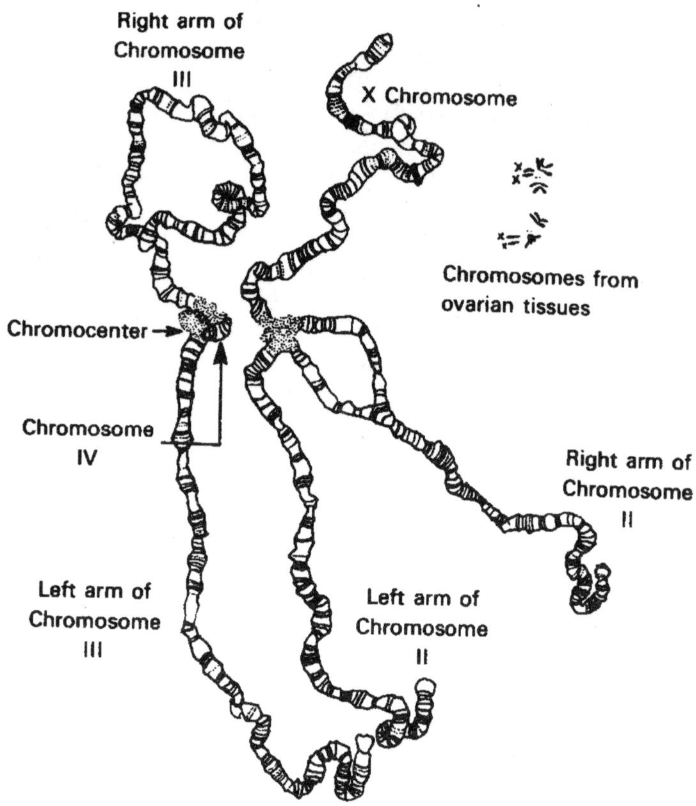

Figure 4 Giant (polytene) chromosomes of Drosophila. (Reproduced from *Heredity* (1934) 25: 464–76 with permission from Blackwell Science Ltd.)

large samples had to be captured at different locations to provide valid population measures, involving much detailed and often hurried work. Sturtevant and Dobzhansky found that the varied and isolated ecosystems of western North America preserved an exceptional richness of diverse genotypes, and that the putative phylogenetic relationships between them could be inferred from their overlapping chromosomal inversion patterns. This was a very exciting finding, implying as it did an insight into the origin of speciation, and in his enthusiasm Dobzhansky wrote, "This is the first time in my life that I believe in constructing phylogenies" (Dobzhansky, 1937).

Part of this optimism derived from the belief that these results supported the theory implying that speciation depended on environmental selection of better adapted mutations, or on their chance of separation by geographic isolation (genetic drift); cage experiments again showed that in time mixed populations from two locations moved in the direction indicated by the phylogenetic map. The data were too crude to support such generalizations, but they greatly stimulated interest in linking Mendelism to evolutionary studies, later summarized in Dobzhansky's *Genetics of the Evolutionary Process* (1937) and in other texts.

The most dramatic change of view came when Hubby & Lewontin (1966) used the new technique of protein separation (electrophoresis) to look at the levels of heterozygosis among a number of enzymes (known to be directly coded by the relevant genes) in *D. pseudoobscura* and found this to be unexpectedly great. Also in 1966 Harris found the same to be true from a survey of human enzymes. The "standard fly" (and human!) concealed a great measure of unanticipated genetic variability. And the more this issue has been studied the more certain it has become that an immense array of genetic variation exists in populations of sexually reproducing organisms. The relationship between theory and the growing multiplicity of facts is now being reconsidered: no doubt data from Drosophila populations will continue to contribute to the solution of this problem.

7. Gene function

One issue this raised is the functioning of genes during the complex life history of the fly and this sorting out of the activity of genes during development was one of the problems Morgan had on his agenda when he chose to work on Drosophila. It initially proved very difficult to approach this topic, partly because the developmental details of egg, larva and fly were then poorly understood. Of course some mutants were found with striking developmental effects, particularly the class known as homeotic mutations where one normal organ was found to replace another, e.g. where the antenna was replaced by a limb *(antennapedia)* or the haltere by a second wing *(bithorax)* to give a four-winged dipteran! (**Figure 5**). This obviously implied that there were general pattern switches, but it was not until Lewis surveyed these mutations as a group in 1978 that the meaning of these pattern controls became evident. Again, numerous treatments (temperature shock, chemicals, etc.) were found to cause non-heritable copies (phenocopies) of mutations already discovered; these were usually like most mutations of defective development such as the *eyeless* mutant. However, so little was understood about the biochemistry of these processes that most observations ended as "interesting", but

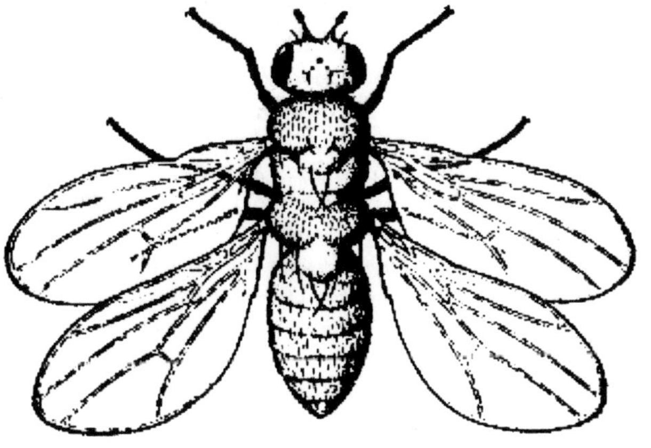

Figure 5 The four-winged fly, produced by combining *bithorax and postbithorax* mutations.

suggested no road forward to the solution of the major developmental problem. The right questions were being asked but the technology for answering them had to await the arrival of molecular biology. At the time of Dobzhansky's studies in the 1930s it looked as if George Beadle had found how to resolve this issue technically. But, in fact, his work led to the temporary demise of Drosophila as the major organism in genetical research.

8. Cell structure and differentiation

The structures which are to form the adult Drosophila, which was the focus of all studies up to that time, are groups of cells which grow in compact assemblies (imaginal discs) but differentiate only during pupation when the separate discs essentially come together to form the adult. Antennal discs, eye discs, wing discs, leg discs and genital discs can all be readily identified in the larva. Of these, the eye discs are most easily separated since they lie on either side of the larval brain. With the French embryologist Ephrussi, Beadle worked on the transplantation of mutant eye-colour eye discs into wild-type larvae; a two-man job – one preparing the transplant and the other injecting it into the prepared larva (Beadle & Ephrussi, 1936). Twenty six eye colour mutants had been identified and when these were all tested only two proved to be non-autonomous: vermilion (*v*) and cinnabar (*cn*); a third, claret (*ca*), was doubtful. Culturing these mutant eye discs in wild-type larvae/pupae corrected their defective phenotype, implying that the body fluids (haemolymph) of the wild type provided a pigment precursor that the mutants could not synthesize. Supposing that the *v* and *cn* substances were different, injecting the disc of one type into a host of the other should normally correct the defect. But this did not work out: a *v* disc in a *cn* host did make a wild-type eye, but a *cn* disc in a *v* host gave a *cn* eye. The explanation of this apparent anomaly was that *v* and *cn* were intermediates in a chain of chemical reactions $v^+ \rightarrow cn^+ \rightarrow$ wild-type pigment. Drosophila had suddenly become a potential vehicle not for the biochemistry of development but for the elucidation of pigment synthesis reactions.

The carefully accumulated understanding of the genetics of eye pigments was brought together by Tatum (1939) to make flies with only the lightest of pigmentation so that they could be used as sensitive testers for intermediates in pigment synthesis, at first by injecting haemolymph and then from nutrition. Only the amino acid tryptophan proved effective and it was shown to be the precursor of kynutenine. Drosophila was now being used to analyse the then unknown biochemistry of a synthetic pathway, not as a genetic animal any more.

9. Prototrophs versus auxotrophs

Before this change of direction was properly appreciated attempts were made to use the Tatum-type technique to study the development of other imaginal discs. But they all failed. It was soon appreciated that eye-colour synthesis was a trivial, superficial instance of development. On the other hand, Beadle saw that with the right organism, mutation could be used to analyse synthetic pathways but that an auxotroph like Drosophila (i.e. an organism requiring preformed essential nutrients) could not serve this purpose. He and Tatum therefore switched their interest to the mould Neurospora (already established in the laboratory) since it was a prototroph requiring only the vitamin biotin as a preformed element of its diet. Neurospora was followed by the even more convenient *Escherichia coli* and Drosophila took a secondary place in genetical research during the 1950–60s, only to recover its role when attention was paid to the lethal mutations blocking embryogenesis and larval development.

10. Conclusion

Of course, the many Drosophila laboratories established worldwide continued to work with the fly. In the UK in particular, it was used as a model for selection experiments (quantitative genetics) relevant to agricultural practice, where it exposed the limitations of simple selection techniques. In Russia much work was done on natural population variability and in the US Dobzhansky's evolutionary experiments were continued and elaborated, as indeed was the technique of mapping (fine structure maps). Much of this work systematized our understanding of the Drosophila genome, in ways that could be exploited when it became the organism of choice for development studies. The 25 years following World War II was a period of consolidation which provided the base for the second great phase of Drosophila work which was triggered by the cloning of the first Drosophila DNA in 1974 (Wensink *et al.*, 1974) and the systematic isolation and analysis of mutations affecting development in 1980 (Nusslein-Volhard & Wieschaus, 1980).

REFERENCES

Adams, M.B. (1968) The founding of population genetics: contributions of the Chetverikov school, 1924–1934. *Journal of the History of Biology* 1: 23–39

Auerbach, C. (1949) Chemical mutagenesis. *Biological Reviews* 24: 355–91

Beadle, G.W. & Ephrussi, B. (1936) The differentiation of eye pigments in Drosophila as studied by transplantation. *Genetics* 21: 225–47

Bridges, C.B. (1935) Salivary chromosome maps with a key to

the banding of the chromosomes of *Drosophila melanogaster*. *Journal of Heredity* 26: 60–64

Dobzhansky, T. (1937) *Genetics of the Evolutionary Process*, New York: Columbia University Press; many subsequent editions

Dobzhansky, T. & Sturtevant, A.H. (1938) Inversions in the chromosomes of *Drosophila pseudoobscura*. *Genetics* 23: 28–64

Emery, A.E.H. & Mueller, R.F. (1992) *Elements of Medical Genetics*, 8th edition Edinburgh: Churchill Livingstone

Harris, H. (1966) Enzyme polymorphisms in man. *Proceedings of the Royal Society, Series B* 164: 298–316

Hubby, J.L. & Lewontin, R.C. (1966) A molecular approach to the study of genic heterozygosity in natural populations. *Genetics* 54: 577–94

Kohler, R.E. (1994) *Lords of the Fly: Drosophila Genetics and the Experimental Life*, Chicago: University of Chicago Press

Lewis, E.B. (1978) A gene complex controlling segmentation in Drosophila. *Nature* 276: 565–70

Muller, H.J. (1928) The effects of X-radiation on genes and chromosomes. *Science* LXVII: 82

Nusslain-Volhard, C. & Wieschaus, E. (1980) Mutations affecting segment number and polarity in Drosophila. *Nature* 287: 795–801

Painter, T.S. (1933) A new method for the study of chromosome rearrangements and the plotting of chromosome maps. *Science* 78: 585–86

Tatum, E.L. (1939) Development of eye colours in Drosophila: bacterial synthesis of v+ hormone. *Proceedings of the National Academy of Sciences USA* 25: 486–90

Wensink, P.C., Finnegan, D.J., Donaldson, J.E. & Hogness, D.S. (1974) A system for mapping DNA sequences in the chromosomes of *Drosophila melanogaster*. *Cell* 3: 315–25

FURTHER READING

Ashburner, M. (1989) *Drosophila: A Laboratory Handbook*, Cold Spring Harbor, New York: Cold Spring Harbor Laboratory Press

Dobzhansky, T. (1951) *Genetics and the Origin of Species*, 3rd edition, New York: Columbia University Press

Lawrence, P.A. (1992) *The Making of a Fly*, Oxford: Blackwell Scientific

Muller, H.J. (1962) *Studies in Genetics*, Bloomington, Indiana: Indiana University Press

GLOSSARY

cytogenetic map map of the microscopic structure of chromosomes

dipteran relating to insects with one pair of wings

homogametic sex that produces gametes of only one type, e.g. females producing only X-bearing ova

phenocopy phenotype resulting from environmental factors which simulates a genetically determined change

polytene chromosomes giant chromosomes produced by repeated replication of chromosomes without chromosome separation or nuclear division (endoreduplication).

prototroph wild-type strain of bacterium or fungus with no special nutritional requirements

See also **Darwin and Mendel united** (p.77); **Development of the egg and early embryo of Drosophila** (p.163); **Genetic control of sex determination in the housefly** (p.190)

DEVELOPMENT OF THE EGG AND EARLY EMBRYO OF DROSOPHILA

Mary Bownes[1] and Vernon French[2]
[1] Institute of Cell and Molecular Biology, University of Edinburgh, Edinburgh, UK
[2] Institute of Cell, Animal and Population Biology, University of Edinburgh, Edinburgh, UK

1. Introduction

In Drosophila, as in other insects, the egg hatches into a larva that is clearly segmented and highly differentiated along its anterior–posterior and dorsal–ventral body axes. This morphological pattern appears gradually during the 24 hours of embryonic development, but we now know that spatial organization originates early in oogenesis, as the egg forms in the mother's ovary. In this article, we will describe oogenesis and the early development of the embryo, and then we will outline the processes by which the body plan is established and discuss some of the genes that are involved.

2. Oogenesis and early embryogenesis

The maturing oocyte passes down the ovary as part of an egg chamber, together with 15 nurse cells and a layer of follicle cells (**Figure 1A**). RNA and protein are synthesized in the nurse cells and pass through cytoplasmic bridges into the oocyte. The surrounding layer of follicle cells pass yolk proteins into the oocyte, before secreting the egg coverings (the vitelline membrane and chorion). Finally, nurse and follicle cells die, the oocyte is fertilized and then the egg is laid (King, 1970). The egg has a thin layer of cytoplasm surrounding the central core of yolk, and the embryo is curious in that it remains syncitial throughout the initial rounds of rapid nuclear division (**Figure 1B**). Nuclei move to the periphery and then, at the posterior end, the egg membrane pinches off the first cells, which will form the germ cells of the embryo. By about 3 hours after the egg was laid, a complete cell layer, the blastoderm, has formed and the most ventral cells then gastrulate, folding in to form the

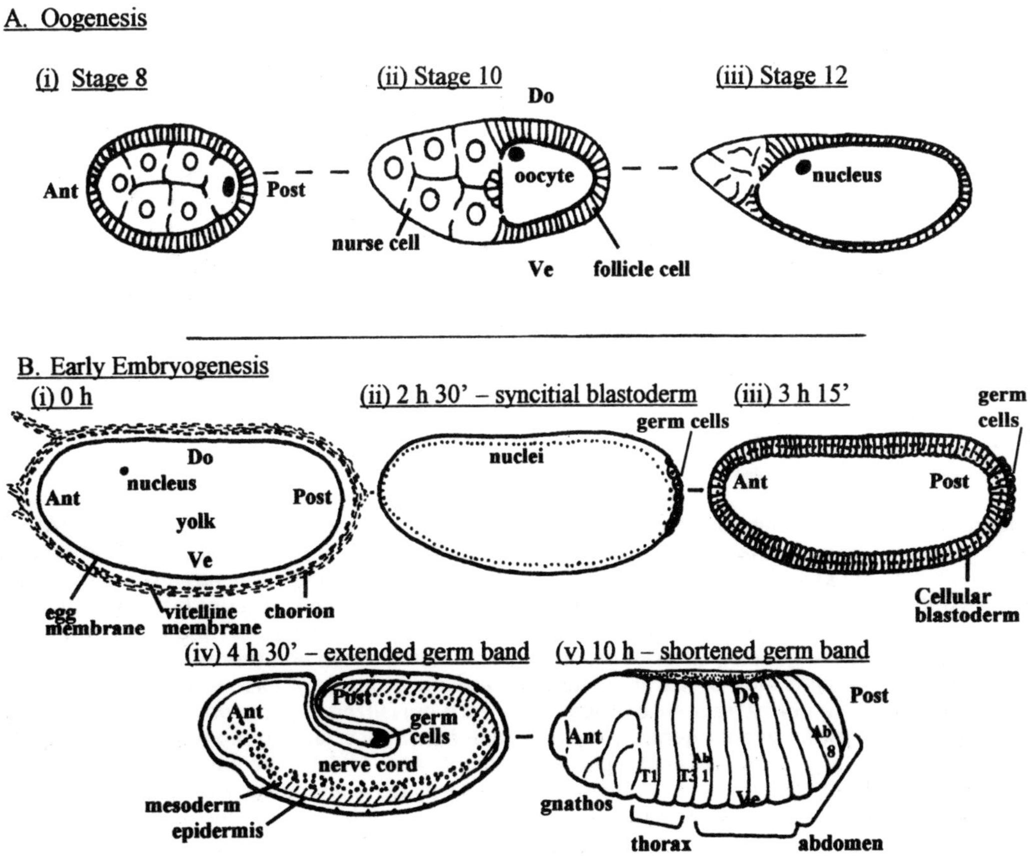

Figure 1 Development of (A) the oocyte and (B) the early embryo of Drosophila. The egg chamber (comprising the oocyte plus the accessory nurse and follicle cells) and the embryo are shown in schematic section (or in surface view, Bv). From early stages, there is clear morphological anterior–posteror (Ant, Post) and dorsal–ventral (Do, Ve) polarity in the egg chamber (e.g. Aii). The egg coverings are shown only on the newly laid egg (Bi). After cellularization of the embryo (Biv), invagination of the most ventral cells forms the mesoderm, some ventrolateral cells invaginate to form the nerve cord and the posterior tip of the embryo extends forward along the dorsal side of the egg. By 10 hours (Bv), the embryo has shortened again and shows clearly the segments of the posterior head (gnathos), the thorax (T1–T3) and abdomen (Ab1–Ab8).

mesoderm. The embryo then elongates around the posterior tip of the egg, and shallow grooves, the first morphological sign of segmentation, appear along its anterior–posterior axis (**Figure 1B**) (Campos-Ortega & Hartenstein, 1985).

3. The role of the maternal genome in establishing the embryonic axes

The egg is a large structure which is provisioned with protein and mRNA and which has a spatial organization that is derived from the expression of many genes in the maternal nurse and follicle cells. These genes are crucial for the later development of the embryo and mutation in any of them affects the embryo developing from an egg laid by the homozygous mutant female, regardless of the genotype of the fertilizing sperm. The lethal phenotypes of these maternally acting genes fall into discrete classes, lacking specific parts of the embryonic pattern (**Figure 2A,B**). This suggests, surprisingly, that the genes function in largely independent processes that establish the anterior–posterior axis, the ends and the dorsal–ventral axis of the embryo. Once the genes were cloned, it became possible to understand their roles in specifying embryonic pattern (St Johnston & Nusslein-Volhard, 1992).

(i) Maternal genes and the anterior–posterior axis

About 13 maternal genes are involved specifically in formation of the anterior and posterior regions of the embryo. The key gene for anterior embryonic development is *bicoid.* Mothers lacking *bicoid* function lay eggs that develop into embryos that have no head or thorax (**Figure 2B**). These structures will form, however, at a site of injection of cytoplasm taken from the anterior of a wild-type egg (Frohnhofer & Nusslein-Volhard, 1986). *bicoid* is transcribed

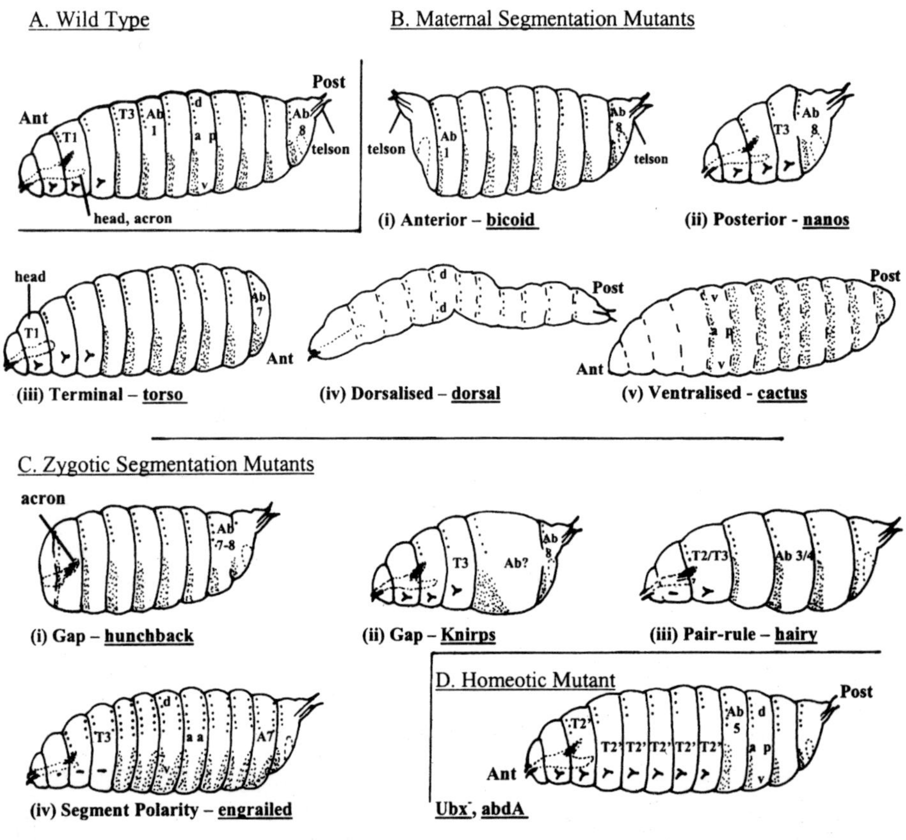

Figure 2 Genes controlling segment number, polarity and identity. Schematic drawings of (A) the wild-type hatchling larva, and of mature embryos that are (B) derived from mothers mutant for maternal segmentation genes, or (C) mutant for zygotic segmentation genes or (D) mutant for homeotic genes. (A) The normal larva has distinct thoracic (T1–T3) and abdominal (Ab1–Ab8) segments, each bearing characteristic dorsal (d), ventral (v), anterior (a) and posterior (p) cuticular structures. The head is invaginated and there are unsegmented terminal regions – the acron and telson. (B) Phenotypes resulting from loss-of-function of maternal genes: i) anterior group (e.g. *bicoid* lacks head and thorax; acron is transformed to telson); ii) posterior group (e.g. *nanos* lacks most of the abdomen); iii) terminal group (e.g. *torso* lacks both the acron and telson); and (iv, v) the dorsal group with a normal sequence of segments, each dorsalized (iv), e.g. *dorsal*, or ventralized (v), e.g. *cactus.* (C) Phenotypes mutant for zygotic genes: i) the gap gene *hunchback*, lacking segments of the head and thorax; ii) the gap gene *knirps*, with one large segment (Ab?) replacing segments Ab1–Ab7; iii) the pair-rule gene *hairy*, with pairwise fusion of segments (T2/T3, Ab1/2, etc.); and (iv) the segment polarity gene *engrailed*, with duplication of the anterior (a) and deletion of the posterior of each segment. (D) Phenotype of larva with a deletion of homeotic genes *Ultrabithorax* and *abdominal-A* (*Ubx⁻, abdA*) showing normal segment number and polarity, but the transformation of segments T3–Ab4 into copies of the mesothorax (T2′).

only during oogenesis, in the nurse cells, and the mRNA is transported to the oocyte, where it becomes localized at the anterior tip (**Figure 3A**). Bicoid protein is produced shortly after the egg is laid and it diffuses away, producing a concentration gradient along the syncitial embryo (Driever & Nusslein-Volhard, 1988). The protein has a homeodomain through which it binds to the regulatory regions of specific genes, controlling their transcription in the embryonic nuclei (see below).

Many maternal genes are needed for the development of the posterior segments of the embryo and the related process of germ cell determination. Mothers mutant for any of these genes (e.g. *oskar*, *nanos*) lay eggs that fail to develop an abdomen (**Figure 2B**). As with the anterior mutants, these eggs can be rescued by injection of cytoplasm, but, in this case, it must come from the posterior end of a wild-type egg. These genes are also transcribed in the nurse cells, but their products accumulate sequentially at the posterior pole of the oocyte, some as proteins and others as mRNAs (**Figure 3A**). Later, at the posterior of the early embryo, the Oskar protein is required for germ cell determination, while *nanos* RNA is translated and the resulting protein gradient indirectly regulates gene expression (Ephrussi & Lehmann, 1991; Wang & Lehmann, 1991).

(ii) Maternal genes and the ends of the embryo

Mothers that are mutant for a "terminal" gene produce embryos that lack the unsegmented terminal regions, the acron and telson (**Figure 2B**). By transplantation of germ cells between embryos of different genotypes, it was shown that activity of some genes (e.g. *torso*) is needed in the germline nurse cells, whereas others (e.g. *torso-like*) are required in the follicle cells. *torso* transcript moves into the oocyte and is translated in the early embryo to produce a receptor that becomes uniformly distributed in the egg membrane. During oogenesis, groups of polar (anterior and posterior) follicle cells express *torso-like* and it is probable that the resulting protein is secreted onto the vitelline membrane (Martin *et al.*, 1994), to be released later and provide a localized ligand that activates the Torso receptor, initiating signal transduction which ultimately controls transcription at both ends of the embryo (**Figure 3A**).

(iii) Maternal genes and dorsal–ventral polarity

The dorsal–ventral axis is set up by a complex pathway that includes interactions between the oocyte and surrounding follicle cells and which eventually results in the cells around the circumference of the embryo becoming committed to different fates (Anderson, 1987; Deng & Bownes, 1998).

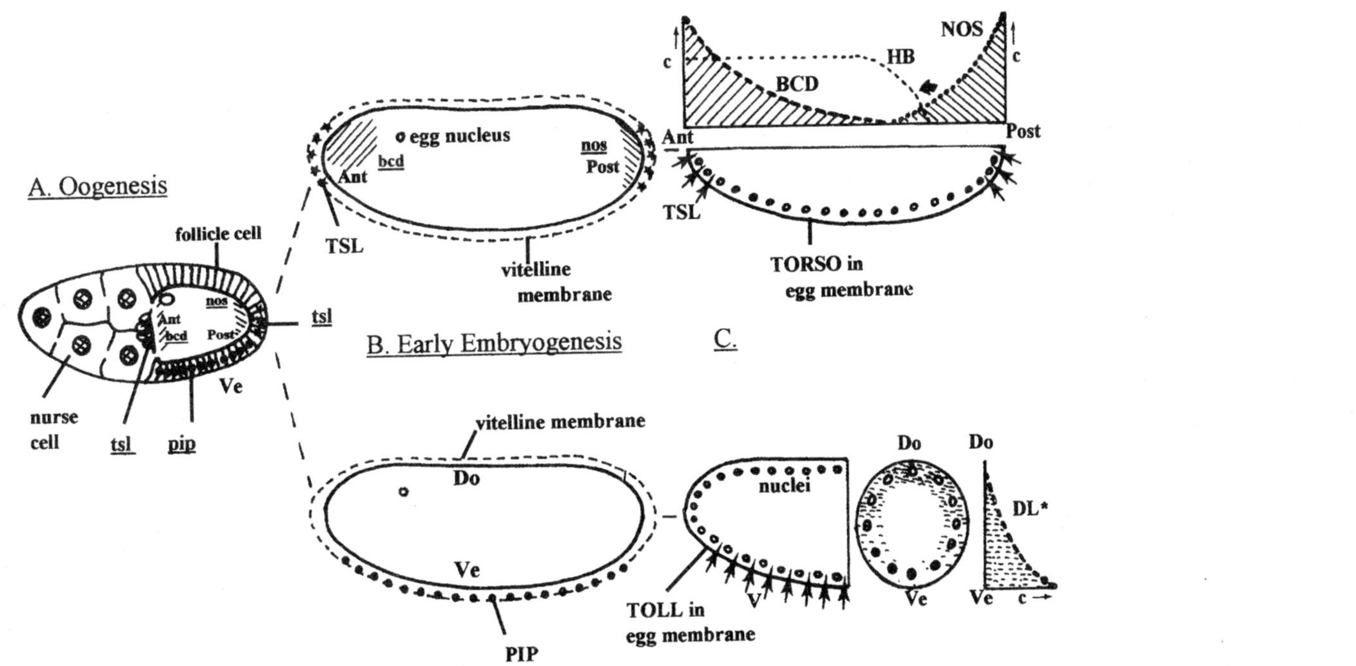

Figure 3 Expression and function of maternal segmentation genes. (Note: conventional abbreviations for genes are used in Figures 3 and 5; the genes and transcripts are in italics and the proteins are in capitals.) (A) Egg chamber, showing transcripts of *bicoid* (*bcd*) and *nanos* (*nos*) produced in nurse cell nuclei (cross hatched) and localized at anterior (Ant) and posterior (Post) ends of the oocyte; and transcripts of *torso-like* (*tsl*) and *pipe* (*pip*) in polar and ventral (Ve) follicle cells, respectively. (B) Newly laid eggs, showing separately the anterior and posterior cues (*bcd* and *nos* transcripts and the TSL protein anchored to terminal regions of the vitelline membrane), and the ventral cue (PIP) localized on the vitelline membrane. (C) Generation of graded spatial information in the early syncitial blastoderm. The upper composite diagram shows concentration gradients (c) of BCD and NOS protein (the latter leads to local degradation (curved arrow) of maternal *hunchback* transcript, affecting distribution of the protein [HB]), and also the terminal signal (arrows) resulting from activation of the egg membrane receptor (TORSO) by localized external TSL. The lower diagram shows the ventral signal (V), an external ligand modified by localized PIP, causing a change in location of DORSAL protein (stipple) from cytoplasm to nucleus, generating a concentration gradient in nuclear DORSAL (DL*).

Inactivation of one of a group of maternal genes (e.g. *Toll*, *dorsal*, *spatzle*, *pipe*) results in a "dorsalized" embryo lacking ventral structures (**Figure 2B**), while some gain-of-function alleles produce ventralized phenotypes. In contrast, loss-of-function mutants of *cactus* ventralize the embryo. Germ cell transplantation experiments indicate that most of the maternal genes are active in the nurse cells, whereas a few (e.g. *pipe*) are required in the follicle cells. Most of these genes have now been cloned and many aspects of their function are understood.

One gene, *Toll*, encodes a membrane receptor protein that is activated ventrally by an external ligand which appears to depend on the expression of *pipe* in a ventral strip of follicle cells (Sen *et al.*, 1998). The ligand, however, is likely to be a different protein (probably Spatzle) that is produced within the oocyte or early embryo, and then secreted and modified externally through interaction with Pipe (or another Pipe-dependent protein) that is localized on the ventral vitelline membrane (**Figure 3A**). The modified ligand then binds back, activating the Toll receptor in the egg membrane. This signal is transduced through several other maternal gene products, including Dorsal and Cactus, finally to determine the axis of the embryo.

dorsal and *cactus* transcripts pass from the nurse cells into the oocyte and the resulting proteins form a complex which becomes distributed evenly in the cytoplasm of the early embryo. On the ventral side of the blastoderm, however, the activation of Toll receptor leads to the release of Dorsal, which then enters nearby nuclei where it functions as a transcription factor. Thus, this very complex pathway results finally in a gradient of nuclear Dorsal, with maximum levels ventrally, that controls gene expression in nuclei around the circumference of the embryo (Roth *et al.*, 1989).

(iv) Polarization of the egg chamber

We have seen that embryonic development depends on the localization of specific maternal gene products in and around the oocyte (Lane & Kalderon, 1994; Pokrywka, 1995). As oogenesis proceeds, both the oocyte and the overlying follicle cell layer become polarized. These processes are not completely understood, although many necessary genes have been identified through mutations that disrupt polarity and the roles of some of these genes have now been demonstrated (Gonzalez-Reyes & St Johnston, 1994).

The single oocyte forms at the posterior of the cluster of related germline cells and the remaining 15 nurse cells start to transfer gene products along connecting microtubules that are orientated initially with their "minus" ends in the oocyte (**Figure 4**) (Theurkauf, 1994). A critical transcript is that of *gurken*, which is translated in the oocyte, generating a membrane protein which signals to the adjacent polar follicle cells. This initiates a kinase transduction pathway that determines their "posterior" fate. These cells signal back to the oocyte by an as yet unknown mechanism, causing the microtubules to reorientate, "plus" ends to the posterior. The oocyte nucleus now moves along to the minus end, as do *bicoid* transcripts that enter from the adjacent nurse cells, whereas other maternal gene products are conveyed down to the posterior plus end of the enlarging oocyte (Gonzalez-Reyes *et al.*, 1995; Roth *et al.*, 1995). These localizations define the anterior–posterior axis of the future embryo.

Dorsal–ventral polarity in the egg chamber requires a further interaction mediated by Gurken protein, which continues to be produced around the oocyte nucleus and enters the nearby membrane. This provides a signal to the overlying non-polar follicle cells which then become determined as "dorsal" (Price *et al.*, 1989). Somehow, this leads to the formation of "ventral" follicle cells on the opposite side and, as described above, those cells eventually provide the return signal that finally polarizes the dorsal–ventral axis of the embryo (Hsu *et al.*, 1996).

Details of the genes referred to in sections 2 and 3 are given in Table 1.

4. Gene expression and pattern formation in the embryo

When the egg is laid, it contains maternal gene products, some of which are localized in the anterior or the posterior cytoplasm and others probably on the vitelline membrane, and over the end or along the ventral side of the vitelline membrane. This positional framework controls gene expression in the nuclei of the early embryo, leading to the spatial patterns of cell differentiation that construct the segmented body of the larva and, ultimately, of the adult fly. As in the case of oogenesis, the mechanisms of embryonic pattern formation have been dissected by the identification and study of the genes involved.

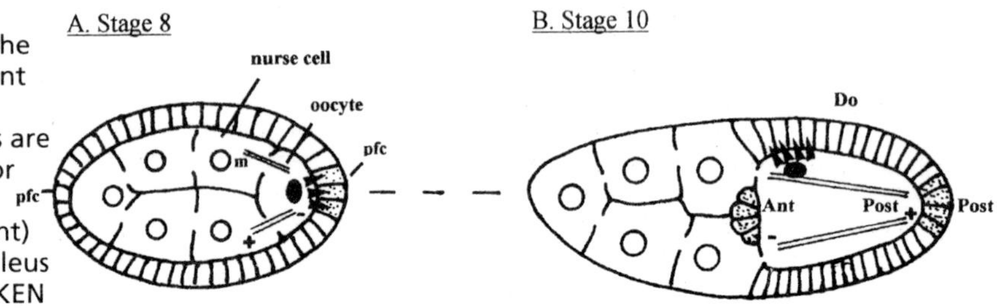

Figure 4 Polarization of the egg chamber. (A) At early stages, microtubules (m) extend from the oocyte to the nurse cells. GURKEN protein is incorporated into the oocyte membrane over the nucleus and signals (arrowheads) to the adjacent polar follicle cells (pfc), specifying them as posterior (Post) – the other group of polar cells are still at the anterior of the chamber. (B) Posterior follicle cells then signal back, the microtubules become reorientated, defining the anterior (Ant) and posterior (Post) of the oocyte, and the nucleus moves to an anterior peripheral position. GURKEN again signals, now to adjacent non-polar follicle cells, specifying them as dorsal (Do).

Table 1 Oogenesis and early embryogenesis – function and expression of genes discussed in the text.

Gene	Protein*	Function	Expression**
1. The anterior–posterior axis and segmentation			
gurken	TGF-α family – S	specifies posterior follicle cells in egg chamber	tr – localized around oocyte nucleus pr – oocyte; secreted from the region near the nucleus
bicoid	homeodomain – TF	activates *hunchback* and other anterior gap genes	tr – nurse cells; localized in anterior oocyte pr – early embryo; forms anterior gradient
oskar	RNA/protein binding factor	binds *nanos* RNA, specifies germ cells	tr – nurse cells pr – becomes localized in posterior oocyte
nanos	translational repressor	inactivates maternal *hunchback* transcript	tr – nurse cells; localized in posterior oocyte pr – early embryo; forms posterior gradient
torso	receptor tyrosine kinase – MR	activation specifies terminal regions (leads to *tailless* expression)	tr – nurse cells; passed to oocyte pr – early embryo; uniformly distributed in membrane
torso-like	S	ligand for Torso	tr – polar follicle cells pr – incorporated into terminal vitelline membrane
hunchback	zinc finger – TF	primary gap gene; controls expression of gap, pair-rule and homeotic genes	tr – anterior of early embryo (also maternal transcription) pr – forms anterior gradient
tailless	zinc finger – TF	primary gap gene; controls expression of gap, pair-rule and homeotic genes	tr – at both ends of early embryo pr – forms terminal gradients
Kruppel	zinc finger – TF	gap gene; controls expression of pair-rule and homeotic genes	tr – centrally in early embryo pr – forms central gradient
knirps	zinc finger – TF	gap gene; controls expression of pair-rule and homeotic genes	tr – centrally in early embryo pr – forms central gradient
hairy	helix–loop–helix – TF	primary pair-rule gene; controls the expression of pair-rule and segment polarity genes	tr & pr – seven stripes in syncitial blastoderm embryo
even-skipped	homeodomain – TF	primary pair-rule gene; controls the expression of pair-rule and segment polarity genes	tr & pr – seven stripes in syncitial blastoderm embryo
engrailed	homeodomain – TF	defines posterior compartment; controls other segment polarity genes	tr & pr – posterior of each segment from cellular blastoderm stage
patched	membrane protein – MR	receptor required for segmental pattern formation	tr & pr – segmental stripes from cellular blastoderm stage
wingless	Wnt family – S	positional signal in anterior–posterior axis of segment	tr – posterior of each segment from cellular blastoderm stage pr – secreted to form segmental gradients
2. Segment identity			
Antennapedia	homeodomain – TF	homeotic gene; specifies segment identity	tr & pr – thoracic segments, from cellular blastoderm stage
Ultrabithorax	homeodomain – TF	homeotic gene; specifies segment identity	tr & pr – thoracic and anterior abdominal segments
abdominal-A	homeodomain – TF	homeotic gene; specifies segment identity	tr and pr – abdominal segments from cellular blastoderm stage
3. The dorsal–ventral axis			
gurken	TGF-β family – S	specifies dorsal follicle cells in egg chamber	tr – localized around oocyte nucleus pr – oocyte; secreted from region near the nucleus
Toll	MR	activation specifies ventral regions (leads to dissociation of Cactus and Dorsal)	tr – nurse cells; passed to oocyte pr – early embryo; uniform in membrane
spatzle	S	ligand for Toll, when cleaved	tr – nurse cells; passed to oocyte pr – early embryo; secreted into perivitelline space
pipe	protease	localized protease, acting on Spatzle	tr – ventral follicle cell pr – probably localized on ventral vitelline membrane
cactus	binding protein	forms cytoplasmic complex with Dorsal	tr – nurse cells pr – early embryo

Table 1 Continued.

Gene	Protein*	Function	Expression**
Dorsal	TF	controls zygotic gene expression in dorsal–ventral axis	tr – nurse cells pr – early embryo; forms ventral gradient in nuclei of syncitial blastoderm
twist	helix–loop–helix – TF	activated by high level of Dorsal; specifies mesoderm	tr & pr – ventral blastoderm cells
snail	zinc finger – TF	activated by high level of Dorsal; specifies mesoderm	tr & pr – ventral blastoderm cells
rhomboid	membrane protein	activated by medium level of Dorsal	tr & pr – lateral blastoderm cells
short gastrulation	secreted binding protein	activated by medium level of Dorsal; inactivates Dpp	tr – lateral blastoderm cells pr – secreted, forming gradient
Decapentaplegic (dpp)	TGF-β family – S	positional signal in dorsal ventral axis of segment	tr – dorsal blastoderm cells pr – secreted, forming dorsal gradient of active protein

* abbreviations for nature of encoded proteins: TF, transcription factor; MR, membrane receptor; S, secreted signal protein
** abbreviations for expression data: tr, site of transcription and location of transcript; pr, site of translation and location of protein product

(i) Segmentation along the anterior–posterior axis

By large-scale mutagenesis, many zygotic (embryonically acting) "segmentation" genes have been identified on the basis of mutant phenotypes that are altered in segment number or structure (Nusslein-Volhard & Wieschaus, 1980). They may be grouped into three discrete classes (see **Figure 2C**):

1. embryos mutant for a "gap" gene (e.g. *hunchback*, *Kruppel*, *knirps*) lack a block of segments (and *tailless* lacks both anterior and posterior ends);
2. mutants for a "pair-rule" gene (e.g. *hairy*, *even-skipped*) have a regular two-into-one fusion of segments; and
3. mutants for a "segment polarity" gene (e.g. *wingless*, *engrailed*) have a deletion/duplication defect in every segment of the embryo (**Figure 3A**).

These phenotypes suggested that segmentation proceeds in stages, gradually defining the repeating segmental unit, and this model has been supported by subsequent studies of the expression and function of the genes.

Gap genes. The gap genes are the first to be transcribed, in bands roughly corresponding to the embryonic regions missing in the mutants. *hunchback* is expressed in nuclei in the anterior half of the early embryo, activated directly by above-threshold levels of Bicoid protein. This was clearly demonstrated by manipulating both the Bicoid gradient and the DNA sites to which Bicoid binds upstream of the *hunchback* promoter (Driever & Nusslein-Volhard, 1989). Curiously, *hunchback* was also transcribed in oogenesis and its maternal transcript also becomes confined to the anterior of the early embryo as, posteriorly, it binds to Nanos and is degraded (Struhl, 1989). At the ends of the embryo, *tailless* is transcribed, in response to the activation of membrane Torso. After translation, diffusion generates anterior (Hunchback) and terminal (Tailless) protein gradients, and the other gap genes (e.g. *Kruppel*, *knirps*) are then activated or repressed by specific concentration ranges of these proteins, plus Bicoid (Hulskamp & Tautz, 1991; Rivera-Pomar & Jackle, 1996). This results in bands of transcription at particular locations along the axis and, again, the protein products diffuse in the syncitial cytoplasm, giving a set of overlapping gradients (**Figure 5A**). The gap proteins, all transcription factors, control the next phase – the expression of the pair-rule genes.

Pair-rule genes. The pair-rule genes are transcribed in periodic patterns of seven broad stripes, roughly corresponding in width and spacing to future segments. For at least some of these genes (e.g. *hairy*, *even-skipped*), the periodic pattern is controlled directly, but in a complex way, by gap gene expression (Pankratz & Jackle, 1990). Each of the stripes reflects activation of the pair-rule gene, through an independent enhancer, by a specific combination of gap protein concentrations. Thus, the expression pattern is a montage of differently controlled elements at different locations. Pair-rule genes mostly encode transcription factors which modulate each other's expression and which also control expression of the primary segment polarity genes (e.g. *engrailed*, *wingless*). The stripe patterns of the different pair-rule genes are out of register and, by a mechanism not yet completely understood (Ingham & Martinez Arias, 1992), the overlap activates the segment polarity genes in narrow stripes within every segment at around 3 hours, as the blastoderm becomes cellular (**Figure 5A**).

Segment polarity genes. The expression of the maternal, gap and pair-rule genes is transient and each serves to activate the next stage of the segmentation process, but the segmental expression of the segment polarity genes is more permanent. These genes mostly encode secreted signals (e.g. *wingless*), receptors (e.g. *patched*), transcription factors (e.g. *engrailed*) and other components of signalling pathways that mediate cell interaction. These interactions establish the segment borders and control anterior–posterior fate within each segment, throughout embryonic and subsequent larval development (DiNardo *et al.*, 1994).

(ii) The dorsal–ventral axis

We have seen that maternal gene expression results eventually in graded nuclear concentration of the transcription factor, Dorsal, around the syncitial blastoderm. This gradient establishes a pattern of zygotic gene expression which leads the ventral cells to develop as mesoderm, cells in the ventrolateral position to form ventral epidermis and

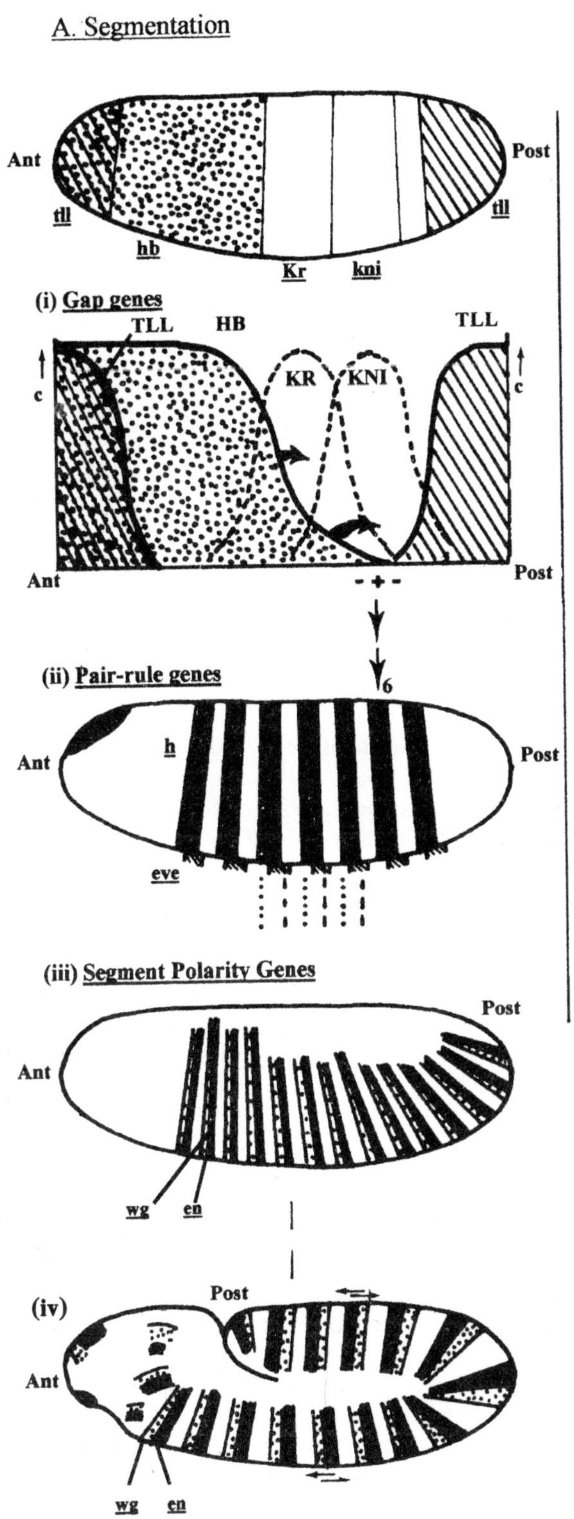

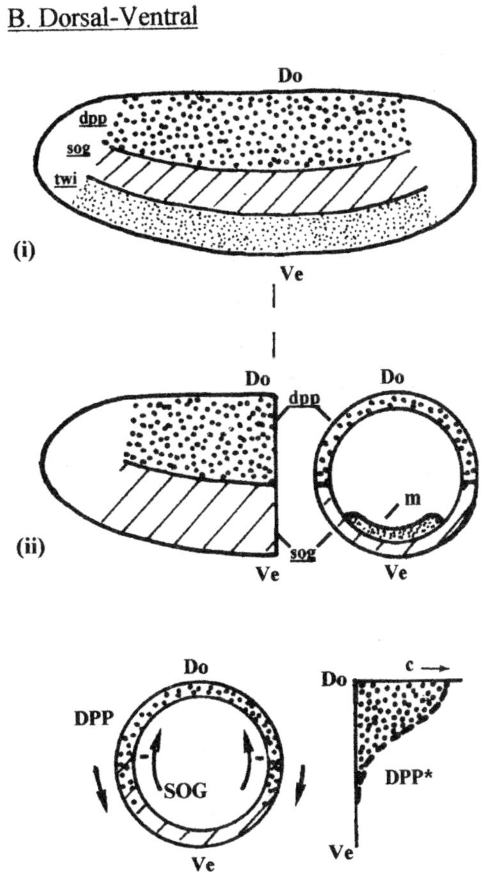

Figure 5 Zygotic genes and pattern formation in the early embryo. (A) Sequential expression of zygotic segmentation genes. (i) Gap genes: bands of transcription (upper) and the resulting gradients of protein concentration (lower). Transcription of *tailless* (*tll*) follows activation of maternal TORSO and *hunchback* (*hb*) is activated by BICOID (see Figure 3B). *Kruppel* (*Kr*) and *knirps* (*kni*) are activated mainly by medium and low levels of HB, respectively (arrows). (ii) Expression of pair-rule gene *hairy* (*h*) in seven stripes, each activated in response to a unique combination of gap protein concentrations. Diagram shows stripe 6 activated (+) by KNI and limited by the flanking inhibition exerted by KR and TLL. Also indicated ventrally is the offset seven-stripe expression pattern of the pair-rule gene *even-skipped* (*eve*). (iii) Expression in the late cellular blastoderm of segment polarity genes *wingless* (*wg*) and *engrailed* (*en*) in 14 adjacent pairs of segmental stripes, in response to repeating combinations of pair-rule proteins (indicated by dashed and dotted lines). (iv) Later expression of *wg* and *en*, with the 14 stripes of the gnathal, thoracic and abdominal segments supplemented by localized expression in the anterior head and posterior abdomen. At this stage, these genes are required for cell interactions (indicated by arrows). (B) The dorsal–ventral axis. (i) At early syncitial stages, the gradient in DORSAL (see Figure 3B) results in the activation of *twist* (*twi*) and *short gastrulation* (*sog*) ventrally and laterally, respectively, and the inhibition of *decapentaplegic* (*dpp*), which is thus restricted to the dorsal nuclei. After cellularization and gastrulation (ii), the *twi*-expressing mesoderm forms an inner cell layer (m). The composite diagram shows (above) lateral and transverse views of gene expression and (below) the diffusion of the secreted proteins (arrows) and the inactivation of DPP by SOG (–), resulting in a dorsal-to-ventral (Do–Ve) gradient of DPP activity (DPP*) which controls further gene expression in this axis.

neuroblasts, dorsolateral cells to form the dorsal epidermis, and the most dorsal cells to form the extra-embryonic serosa (Roth *et al.*, 1989). This process of pattern formation occurs in two stages (**Figure 5B**).

A number of genes are directly regulated by Dorsal in a concentration-dependent way (Rusch & Levine, 1996). *twist* and *snail*, which encode transcription factors essential for mesodermal development, have promoter regions with low-affinity Dorsal binding sites and, hence, they are activated only by the high concentrations found in the most ventral nuclei. The promoter of *rhomboid*, however, has high-affinity sites that are able to bind even the low levels of Dorsal found laterally, but it is inhibited by Snail and, hence, expression is repressed in the ventral cells. *short gastrulation* is also activated in lateral stripes and is likely to be controlled in a similar way. Dorsal acts, however, as a repressor for a number of other genes (e.g. *decapentaplegic*), even at low concentrations, and thereby limits their expression to the dorsal side of the embryo.

Response to the Dorsal gradient distinguishes broad ventral, lateral and dorsal regions. After gastrulation, further patterning of the embryo depends critically on *decapentaplegic*, which is transcribed uniformly in the dorsal half of the embryo and encodes a secreted protein homologous to the vertebrate Transforming Growth Factor β (TGF-β) family (Wharton *et al.*, 1993). Complete loss of function results in embryos with no dorsal structures and an expansion of the ventral neurectoderm, but some alleles cause loss of only the extreme dorsal serosa. This, plus the results of injecting different amounts of *decapentaplegic* transcript, suggests that activity is graded, with high levels of the protein (Dpp) specifying the dorsal serosa and the level declining laterally to specify the dorsal and lateral epidermis. *short gastrulation*, which is expressed laterally, also encodes a secreted protein (Sog) and there is genetic evidence that Sog inactivates Dpp, suggesting that the graded Dpp activity may result indirectly from a diffusion gradient of Sog (Ferguson & Anderson, 1992). Within the dorsal part of each segment, the level of active Dpp directs cell fate, both in the ectoderm and in the mesoderm which has now spread out beneath it (Frasch, 1995). The mechanism is not yet fully understood, but there are several different membrane receptors which may be activated at different Dpp concentrations and which may have different effects on gene expression.

(iii) Homeotic genes and segment identity

The segment polarity genes and the genes controlling dorsal–ventral pattern are expressed in the same way in each body segment, yet these go on to develop differently. Some of the genes involved in controlling segmental differences were identified long ago through mutations that cause spectacular homeotic transformations (Lewis, 1978), such as the "bithorax" or four-winged fly with sets of mesothoracic structures developing on both the second and third thoracic segments. As these genes were analysed, cloned and studied, it became clear that segmental identity in the gnathos, thorax and abdomen is specified by the expression of members of a cluster of eight homeotic (or "Hox") genes (Morata, 1993). In response to the transient cues provided at cellular blastoderm stage by the distribution of the gap and pair-rule proteins (**Figure 5A**), *Antennapedia* is expressed in the future thorax, *Ultrabithorax* in the posterior thorax and anterior abdomen, *abdominal-A* in the mid abdomen, and the other homeotic genes in the more anterior or more posterior regions of the embryo. There are also regulatory interactions between the genes so that, for example, the absence of both *Ultrabithorax* and *abdominal-A* function results in the expression of *Antennapedia* spreading posteriorly and five further segments developing as "mesothorax" (**Figure 2D**). Homeotic genes encode transcription factors and we are now beginning to understand how their expression controls many other genes and leads to the formation of characteristic and different structures on the segments of the larva and, eventually, of the fly (Akam, 1998).

5. Conclusions: Drosophila and other early embryos

Through the application of genetics (traditional and then molecular), we now understand more about the mechanisms of early development in Drosophila than in any other organism. It remains to be seen to what extent the detailed mechanisms are conserved across the insects where, for example, the more ancient orders differ from Drosophila by lacking nurse cells during oogenesis and by forming the embryonic segments sequentially, during the posterior growth of the embryo (Patel, 1994). Also, it seems that the very distantly related vertebrates differ considerably, in that they specify very little pattern during oogenesis and probably have very different mechanisms of segmentation. There are, however, some spectacular similarities between Drosophila and the vertebrates – in the structure and function of the homeotic genes (Carroll, 1995) and in at least some aspects of dorsal–ventral patterning (Holley *et al.*, 1995) – suggesting that development of the main embryonic axes involves some mechanisms that are very ancient and that have been retained throughout animal evolution. Thus, the detailed information obtained from Drosophila may be of wide relevance and a comparable molecular genetic approach will certainly be vital in exploring the early development of other animals.

REFERENCES

Akam, M. (1998) Hox genes: from master genes to micromanagers. *Current Biology* 8: R676–78

Anderson K.V. (1987) Dorsal–ventral embryonic pattern genes of *Drosophila*. *Trends in Genetics* 3: 91–97

Campos-Ortega, J.A. & Hartenstein, V. (1985) *The Embryonic Development of* Drosophila melanogaster, Berlin: Springer; 2nd edition 1997

Carroll, S.B. (1995) Homeotic genes and the evolution of arthropods and chordates. *Nature* 376: 479–85

Deng, W.-M. & Bownes, M. (1998) Patterning and morphogenesis of the follicle cell epithelium during *Drosophila* oogenesis. *International Journal of Developmental Biology* 42: 541–42

DiNardo, S., Heemskerk, J., Dougan, S. & O'Farrell, P.H. (1994) The making of a maggot: patterning the *Drosophila* embryonic epidermis. *Current Opinion in Genetic Development* 4: 529–34

Driever, W. & Nusslein-Volhard, C. (1988) A gradient of *bicoid* protein in *Drosophila* embryos. *Cell* 54: 83–93

Driever, W. & Nusslein-Volhard, C. (1989) The *bicoid* protein is a positive regulator of *hunchback* transcription in the early *Drosophila* embryo. *Nature* 337: 138–43

Ephrussi, A. & Lehmann, R. (1991) Induction of germ cell formation by *oskar*. *Nature* 358: 387–92

Ferguson, E. & Anderson, K.V. (1992) Localised enhancement and repression of the activity of the TGF-β family member, *decapentaplegic*, is necessary for dorsal–ventral pattern formation in the *Drosophila* embryo. *Development* 114: 583–97

Frasch, M. (1995) Induction of visceral and cardiac mesoderm by ectodermal Dpp in the early *Drosophila* embryo. *Nature* 374: 454–67

Frohnhofer, H.G. & Nusslein-Volhard, C. (1986) Organisation of anterior pattern in the *Drosophila* embryo by the maternal gene *bicoid*. *Nature* 324: 120–25

Gonzalez-Reyes, A. & St Johnston, D. (1994) Role of oocyte position in establishment of anterior–posterior polarity in *Drosophila*. *Science* 266: 639–42

Gonzalez-Reyes, A., Elliott, H. & St Johnston, D. (1995) Polarization of both major body axes in *Drosophila* by gurken–torpedo signaling. *Nature* 375: 654–58

Holley, S.A., Jackson, P.D., Sasai, Y. *et al.* (1995) A conserved system for dorsal–ventral patterning in insects and vertebrates involving *sog* and *chordin*. *Nature* 376: 249–53

Hsu, T., Bagni, C.J., Sutherland, D. & Kafatos, F.C. (1996) The transcriptional factor CF2 is a mediator of EGF-R-activated dorsoventral patterning in *Drosophila* oogenesis. *Genes and Development* 10: 1411–21

Hulskamp, M. & Tautz, D. (1991) Gap genes and gradients – the logic behind the gaps. *BioEssays* 13: 261–68

Ingham, P. & Martinez Arias, A. (1992) Boundaries and fields in early embryos. *Cell* 68: 221–35

King, R.C. (1970) *Ovarian Development in Drosophila melanogaster*, New York and London: Academic Press

Lane, M.E. & Kalderon, D. (1994) RNA localization along the anteroposterior axis of the *Drosophila* oocyte requires PKA-mediated signal-transduction to direct normal microtubule organization. *Genes and Development* 8: 2986–95

Lewis, E.B. (1978) A gene complex controlling segmentation in *Drosophila*. *Nature* 276: 565–70

Martin J.-R., Raibaud, A. & Ollo, R. (1994) Terminal pattern elements in *Drosophila* embryo induced by the *torso-like* protein. *Nature* 367: 741–45

Morata, G. (1993) Homeotic genes of *Drosophila*. *Current Opinion in Genetic Development* 3: 606–14

Nusslein-Volhard, C. & Wieschaus, E. (1980) Mutations affecting segment number and polarity in *Drosophila*. *Nature* 287: 795–801

Pankratz, M. & Jackle, H. (1990) Making stripes in the *Drosophila* embryo. *Trends in Genetics* 6: 287–92

Patel, N. (1994) Developmental evolution: insights from studies of insect segmentation. *Science* 266: 581–90

Pokrywka, N.J. (1995) RNA localization and the cytoskeleton in *Drosophila* oocytes. *Current Topics in Developmental Biology* 31: 139–66

Price, J.V., Clifford, R.J. & Schüpbach, T. (1989) The maternal ventralizing locus torpedo is allelic to faint-little-ball, an embryonic lethal, and encodes the *Drosophila* EGF receptor homolog. *Cell* 56: 1085–92

Rivera-Pomar, R. & Jackle, H. (1996) From gradients to stripes in *Drosophila* embryogenesis: filling in the gaps. *Trends in Genetics* 12: 478–83

Roth, S., Neuman-Silberberg, F.S., Barcelo, G. & Schüpbach, T. (1995) *Cornichon* and the EGF receptor signalling process are necessary for both anterior–posterior and dorsal–ventral pattern formation in *Drosophila*. *Cell* 81: 967–78

Roth, S., Stein, D. & Nusslein-Volhard, C. (1989) A gradient of nuclear localization of the *dorsal* protein determines dorsoventral pattern in the *Drosophila* embryo. *Cell* 59: 1189–202

Rusch, J. & Levine, M. (1996) Threshold responses to the dorsal regulatory gradient and the subdivision of primary tissue territories in the *Drosophila* embryo. *Current Opinion in Genetic Development* 6: 416–23

Sen, J., Goltz, J.S., Stevens, L. & Stein, D. (1998) Spatially restricted expression of *pipe* in the *Drosophila* egg chamber defines embryonic dorsal–ventral polarity. *Cell* 95: 471–81

St Johnston, D. & Nusslein-Volhard, C. (1992) The origin of pattern and polarity in the *Drosophila* embryo. *Cell* 68: 201–19

Struhl, G. (1989) Differing strategies for organising anterior and posterior body pattern in *Drosophila* embryos. *Nature* 338: 741–44

Theurkauf, W.E. (1994) Microtubules and cytoplasm organization during *Drosophila* oogenesis. *Developmental Biology* 165: 352–60

Wang, C. & Lehmann, R. (1991) *Nanos* is the localised posterior determinant in *Drosophila*. *Cell* 66: 637–47

Wharton, K.A., Ray, R.P. & Gelbart, W.M. (1993) An activity gradient of *decapentaplegic* is necessary for the specification of dorsal pattern elements in the *Drosophila* embryo. *Development* 117: 807–22

GLOSSARY

blastoderm layer of cells surrounding the central yolk in the early stages of embryonic development

gnathos the three mouth-part segments of the posterior head (mandible, maxilla and labium)

oogenesis formation, development and maturation of the female gamete or egg

serosa temporary region of extra-embryonic cells (limited to a thin dorsal strip in Drosophila)

syncitial describes a tissue where there is cytoplasmic continuity between the cells of which it is composed

GENETIC DISSECTION OF WING PATTERNING IN DROSOPHILA

J.R.S. Whittle and M. Buratovich

School of Biological Sciences, University of Sussex, Brighton, UK

1. Introduction

Unravelling the cellular machinery underlying the organization of the developing insect wing has become a paradigm for the genetic analysis of a complex developmental event. Since there is very little variation in wing structure between different individuals reared under the same conditions, the developmental mechanism is expected to be robust. The cells of a simple sac-like epithelium, known as the wing imaginal disc, produce the structures of the insect wing and adjacent body wall (**Figure 1**). Success in understanding this patterning process has been built on a series of genetic technologies and insights. This article illustrates how these approaches have led to the current picture.

2. The cell biology of imaginal discs

A late-stage Drosophila larva contains a fixed number of imaginal discs, each composed of a folded diploid epithelium, and having a characteristic shape, size and location (**Figure 2**). Each disc is committed to a particular developmental fate and is named after the appendage it forms (e.g. wing disc, leg disc). After metamorphosis, in which the discs turn inside out, each one produces a fixed sector of the body surface and secretes structures that together form the cuticle, i.e. the exoskeleton.

Most imaginal discs arise as an infolding of a small number of "progenitor" cells on the lateral flank of the embryo. The wing and haltere discs bud off the second and third leg discs, respectively. At its inception, the wing disc contains around 50 cells and is larger than the other disc primordia. The wing disc increases in size to 50 000 cells through cell proliferation, the pattern of which is genetically programmed. Mutations in Drosophila tumour-suppressor genes will cause hyperplastic growth (*giant discs, discs overgrown, expanded, hyperplastic discs, fat*) or neoplastic overgrowth (*discs large, giant larvae*) of some or all of the imaginal discs. The cells of each primordium derive from segmental stripes of cells expressing several genes that have been induced in the lateral epithelium of the embryo (see **Plate 1A–D** in the colour plate section).

At its initiation, the wing disc contains two adjacent small groups ("compartments") of cells; a dichotomous cell-fate decision made during embryogenesis has separated the cells into two non-identical populations and has made a border between these two compartments. The two compartments of cells are heritably different and separate, reflecting the continuous expression in the posterior (P) compartment of a gene called *engrailed* (*en*), which encodes a homeodomain-containing transcription factor. The two compartments of the wing disc primordium also express unique combinations of genes, encoding transcription factors containing homeodomains, which determine in a cell-heritable way the developmental fate of each compartment. As an illustration, if a fragment of a wing disc is implanted in either a young

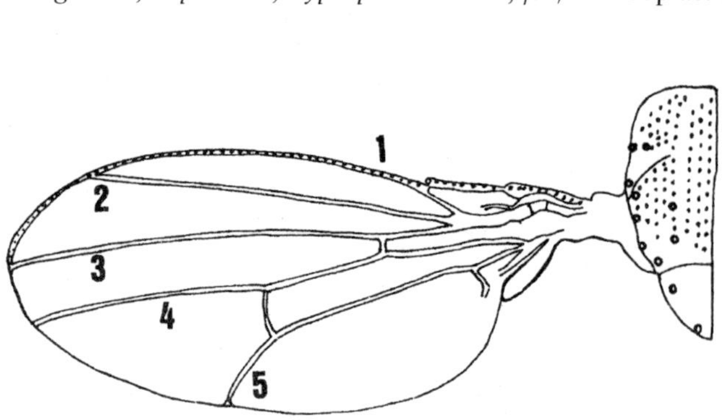

Figure 1 Structures formed by the wing imaginal disc. Large and small circles indicate the sites of mechanosensory bristles on the dorsal thoracic body wall. The leading (anterior) edge of the wing blade carries mechano- and chemosensory bristles. The pattern of veins (numbered) is also invariant.

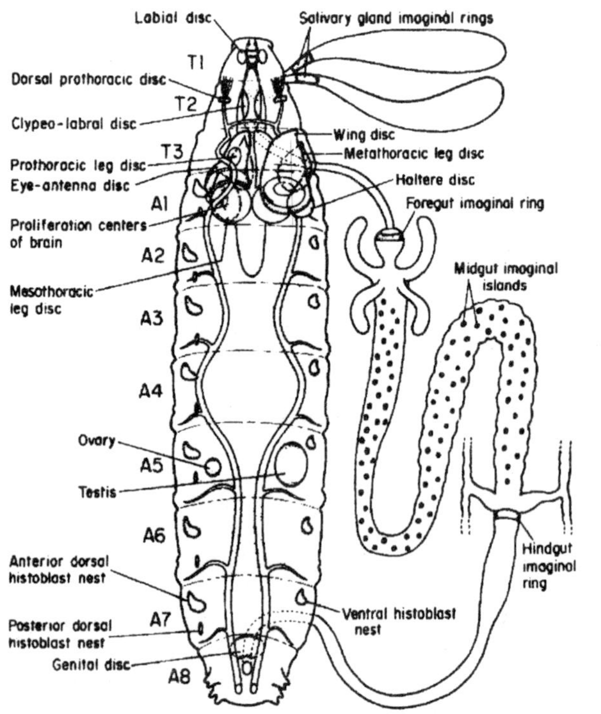

Figure 2 Location of imaginal primordia within the larva. The location of the wing disc and other imaginal discs is shown. For clarity, each disc is drawn only on one side; with the exception of the midline genital disc, these are all paired structures. (Reprinted from Bryant & Levinson, *Developmental Biology* 107: 355–63 © 1985, with permission from Academic Press.)

larva or the abdomen of an adult female, it will undergo mitotic proliferation. A metamorphosis test made after many extra cell divisions shows that the commitment of the cells to form wing rather than tissue of any other disc type has persisted in the descendent cells, dependent on the stable maintenance of expression of the genes.

These genes dictate the segmental type of the structure formed, but the detailed elaboration and spatial siting of characteristic structures of the insect wing rather than, for instance, the leg, depend on an interaction between this "developmental address label" and cell machinery that sets up local differences (or "positional signals") between cells in each tissue. The characteristic differences between a wing and a leg must derive from changes unique to each disc, which alter the expression of components of the cellular machinery that generate local fate differences between neighbouring cells.

The basement membrane on which the tissue is built is important in the patterning and morphogenesis because, at the end of disc development, ecdysterone hormones prompt each disc to evert so that the surface previously in the lumen of the sac is now presented to the outside world. This morphogenetic change depends on the structural integrity of the tissue.

3. Patterning along the antero-posterior (A–P) axis of the wing

The "border" between the A and P compartments of a disc plays a crucial role in coordinating pattern and is itself the result of complex interactions between different genes. The activity of *en* and a very similar adjacent gene called *invected* (*inv*) in all P cells, but not in A cells, causes P cells to form a minimum-contact border with A cells. Mitotic clones completely lacking the activities of both *en* and *inv*, however, can still respect compartment boundaries. These two genes are, therefore, not the only contributors to the formation of the boundary. The specification of posterior structures is due to *inv* and not *en*, since overexpression of *en* in P cells downregulates *inv* and causes posterior to anterior transformations. Engrailed acts as a regulatory protein that is able to control the expression of several posterior- and anterior-specific genes. The expression of *en* in P cells prevents the expression in them of two A cell-specific genes: *patched* (*ptc*), which encodes a multiple-pass transmembrane protein with structural similarity to those found in channels and transporters, and *cubitus interruptus* (*ci*), encoding a zinc finger-containing transcription factor while activating the manufacture and secretion of the product of the gene *hedgehog* (*hh*) in P cells (see **Plate 2A–C**).

The following picture of the organizing role of the dorsoventral (A–P) border domain has emerged from examining the expression of genes in mitotic clones and in discs where other genes have been misexpressed. Hh acts as a ligand (a short-range inducer moving only a few cell diameters from its site of secretion) and is recognized by the Patched (Ptc) protein, which is only present on surfaces of A cells (**Plate 2A**). The binding of Hh to Ptc inactivates Ptc and activates a serpentine transmembrane protein, the product of the *smoothened* (*smo*) gene. Activated Smoothened (Smo) transduces the Hh signal within those A cells near the P border and prompts them to express the gene *decapentaplegic* (*dpp*) (**Plate 2B,C**) and export its product, Dpp (a transforming growth factor-β homologue), to neighbouring cells. Induction of the synthesis of Dpp protein in A cells at the A–P border by Hh is specific to A cells, because *engrailed* represses *dpp* expression in P cells. In fact, mitotic clones lacking activity of *en* in P cells cause the ectopic expression of *ci*, *dpp* and *ptc* in P cells. A normal (wild-type) concentration of Dpp at the A–P boundary is important for wing growth and is crucial for maintenance of the boundary because mitotic clones induced in *dpp* hypomorphic backgrounds do cross the normally strictly maintained compartment boundary.

If the cells immediately anterior to the A–P border have no functional *dpp* gene at all, then the wing fails to develop. On the other hand, if the *dpp* gene is activated at the wrong location within the A compartment, distant from the place where Dpp would normally be secreted, then the resultant wing is grossly distorted in the patterning of its anterior structures. To keep Dpp synthesis localized to the A–P boundary, the secretion of Hh also induces elevated levels of the Ptc protein. Ptc suppresses Dpp expression in A cells. It is only the high concentration of secreted Hh at the compartment boundary that ensures Dpp expression. Ptc prevents the inappropriate expression of *en* in A cells. Ptc is aided in this venture by a basic helix–loop–helix protein, encoded by the *groucho* (*gro*) gene, which acts to repress *en* at the dorso-ventral margin. The inactivation of Ptc by Hh also causes elevated levels of the Cubitus Interruptus transcription factor at the A–P border, and the increased expression of the *ptc*, *dpp* and *ci* genes at the A–P border is probably due to Cubitus Interruptus. In summary, the juxtaposition of the non-identical A and P cells has created a cellular discontinuity which, in turn, becomes the cornerstone to organize this biological pattern (Blair, 1995).

The Dpp protein, which has its peak concentration at the compartment border, behaves as what developmental biologists call a "morphogen"; that is, it can elicit one of several alternative outcomes in cells that receive it, depending on the concentration reaching those particular cells. This has put to rest a long-standing debate in developmental biology. The alternative mechanism considered was that Dpp might activate in immediately adjacent cells the first of a series of signalling molecules and that the spatial patterning might have been the outcome of a signalling cascade of nearest-neighbour interactions (formally rather like "numbering from the left"), as it was propagated across the epithelial sheet of cells.

The receptor for Dpp is heteromeric; two genes, *thickveins* and *punt*, encode two of its essential components. These genes were modified by recombinant DNA technology so that they were constitutively active (active without being dependent on the arrival of a signal) and yet had lost their extracellular Dpp-binding domain. The modified versions of the *thickveins* and *punt* genes were reintroduced to flies and were deliberately expressed in clones of cells within the A compartment of the wing disc (**Plate 3A–C**). The cells of the clone responded by expressing *optomotor blind* (*omb*)

(**Plate 3B**), one of two genes that Dpp normally activates in the wing disc, although no cell neighbouring the clone changed its behaviour. This made it clear that activating the Dpp receptor can only trigger a response limited to the cell in which it is activated and, therefore, could not be part of any cascade of cell-to-cell signalling. In a wild-type wing disc, a broad stripe of cells centred on the Dpp-secreting cells at the A–P border expresses *omb* and a narrower stripe expresses a gene called *spalt* (**Plate 3A**). When *dpp* itself was expressed in clones distant from its normal position, larger haloes of *omb* and smaller haloes of *spalt*-expressing cells surrounded the clones, strongly suggesting that the level of Dpp protein determines whether only *omb* or both *omb* and *spalt* are activated in neighbouring cells.

This aspect of the patterning machinery can be summarized by saying that, from an initial starting point of two non-mixing heritably different cell groups (which we have abbreviated as A and P cells), a relatively straight-line border is created, from which a diffusible molecule, Dpp, triggers different responses in target cells through its receptors, depending on its concentration. In other words, the response depends on the distance of the target cells from the source of Dpp. The genes *omb* and *spalt* are transcription factors, so the detail of the final pattern will, in turn, require finding out the target genes of the Omb and Spalt proteins and may involve further local signalling events to increase the precision of specification of cell fate. The graded distribution of Dpp from its peak at the A–P border necessarily leads to a spatial order in the response of cells to it. In addition, the orientation or "polarity" of individual cellular structures formed in the wing may also reflect the slope of this gradient.

4. Mutational dissection: the recognition of the genes involved in a process through their functional disruption by mutation

For geneticists, it has always been an accessible strategy to attempt to find and analyse the genes necessary for a biological process (e.g. the formation of the wing) by searching for new mutations that upset the process. The simplicity of the strategy is deceptive, as shown by the fact that three geneticists (Edward B. Lewis, Christiane Nüsslein-Volhard, Eric F. Wieschaus) working with Drosophila were jointly awarded the Nobel prize in 1995 for their achievements using this approach to study embryogenesis, after the research field had been active for 80 years. Several caveats are worth noting in the context of wing patterning.

- When the complete loss of a gene results in a fly with altered or no wings, this is strong circumstantial evidence to implicate the gene in the process. Determining its precise contribution to the process is the subsequent challenge, because the relationship between gene products and developmental effects is certainly not a "jigsaw" of separate contributions, but the product of successive interactions and ramifications.
- Developmental processes are self-regulating, so that the phenotype following the removal of one gene has superimposed on it the attempts of the system to minimize or compensate for the disturbance.
- The situation can be further complicated by the possibility that a mutational event may have upset the timing or the context in which a particular gene is expressed, so that it wreaks havoc through its "maverick" activity during a process in which it is normally not a participant. Such a mutant may offer a false trail for the investigator.

Many genes necessary for wing formation are also essential for earlier cell patterning events including embryogenesis. Mutations in these genes commonly arrest development during embryogenesis, making it impossible to examine their impact on wing patterning directly. However, genetic mosaic analysis can be used to "expose" these recessive mutations for investigation in the wing disc after embryogenesis has been completed. Once a putative candidate "patterning" gene has been cloned and its expression pattern defined during imaginal disc growth, its role can be investigated directly.

5. D–V patterning and the wing border

The wing disc is divided into compartments that will give rise to the wing blade or the thorax. Early in the second instar, the expression of *wingless* (*wg*), the prototypical member of the Wnt (named for the original family members, *int-1* in mice and *wg* in flies) family of secreted glycoproteins, which is ubiquitous within vertebrates and invertebrates, is necessary to establish the wing field. Mutations that disrupt *wg* expression at this early stage convert the wing field into thorax tissue, and the resultant animals show mirror-image duplicated thoraces, but have no wings.

The wing blade is made from two sheets of cells – an upper (dorsal, D) and a lower (ventral, V) layer – and at the margin between these layers a complex array of sense organs is constructed. The generation of this pattern is mechanistically similar to that of the A–P axis, but involves different gene products (**Plate 4**). A LIM-family transcription factor encoded by the *apterous* (*ap*) gene becomes expressed in what eventually will become dorsal cells (D cells), establishing them as a separate compartment from the remainder (the V cells). At this time, V cells express Wg and the *Delta* (*Dl*) gene, which encodes a ligand for the Notch receptor. Thus, their identity or developmental fate becomes heritably fixed, but, in addition, Ap protein triggers the genes *fringe* (*fng*) and *Serrate* (*Ser*) in D cells.

The genes *fng* and *Ser* both encode secreted glycoproteins that are presumably transported to just within the dorsal boundary. The induction of either *ap* or *fng* mutant clones in the dorsal compartment will produce an ectopic wing margin. Thus, it is the boundary between *ap*-ON and *ap*-OFF or *fng*-ON and *fng*-OFF cells where a wing margin can be made. The ectopic expression of *fng* in the ventral compartment induces ectopic Wg, Ser and *vestigial* (*vg*), a gene necessary for wing growth, and induces the formation of an extra wing margin from ventral tissue. This demonstrates that Fng and Ser are necessary for dorso-ventral signalling, but are not necessary for the formation of dorsal structures. Another gene under the control of Ap called *Dorsal Wing* is probably responsible for the production of dorsal wing structures. Ectopic expression of Ser will induce ectopic Wg and an extra margin without the activation of ectopic Fng. Thus, Fng seems to act "upstream" of Ser. At the D–V boundary, Ser, which is a ligand for the Notch receptor, binds to Notch protein and induces the expression of Wg at the presumptive margin, and activates the Suppressor of Hairless (*Su(H)*) and Enhancer of split (*E(spl)*) proteins, which mediate Notch function.

These interactions are further complicated by the presence of another ligand of Notch, the product of the *Delta* (*Dl*) gene. *Dl* expression is elevated at the D–V boundary and in the ventral compartment during the second larval instar, when this boundary is formed. Ectopic Dl (Dl expressed at an atypical location) causes the induction of *wg* and *vg*, and causes outgrowths in the dorsal compartment. These outgrowths can be suppressed by mutations in *Notch*, thus demonstrating the role of the Notch receptor in mediating *Dl* function. *cut*, a gene that is necessary for the formation of a particular subset of marginal bristles, is also induced by ectopic *Dl*, but only in the dorsal compartment (**Plate 5**). On the other hand, ectopic Ser induces *cut* in the ventral compartment only. Since both of these genes encode ligands for the Notch receptor, these results strongly suggest that Dl and Ser act as compartment-specific signals to activate Notch differentially and bring about compartment-specific gene expression (Doherty *et al.*, 1996).

Vestigial protein is required for proliferation of the wing blade. In fact, ectopic expression of *vg* will cause wing-like outgrowths in a variety of tissues. *vg* expression is activated in two parts: first at the D–V boundary and then in the ventral and dorsal areas that surround, but do not include, the D–V boundary. The activation of Vg at the boundary requires the Su(H) protein. The early presence of Wg in the ventral compartment, and the requirement for Wg, might mean that Ser and Wg interact to activate Su(H). Su(H) turns on *vg* expression at the D–V boundary by specifically binding to an enhancer located in the second intron of the *vg* gene. Due to the sustained requirement throughout the entire wing pouch for Vg, another enhancer in the fourth intron of the *vg* gene activates Vg expression in the remainder of the wing pouch, but not at the D–V boundary. This enhancer requires a functional D–V boundary and the expression of the *dpp* gene for its activation, as any perturbation of either prevents expression from the second enhancer. Together, these two activation systems provide the wing pouch with enough Vg for its sustained proliferation. Expression of *wingless* at the wing margin is required for sustained Vg expression throughout the wing pouch, which implicates Wg in the long-range activation of Vg.

These exchanges involve the expression of genes *wg* and *Dl* across the presumptive margin, which establishes a new unique group of cells along the borderline that act as a signalling centre and create the complex cellular differentiation pattern at the border itself. It has long been known that mutations in many genes could lead to upsets in the integrity of the D–V margin structures, but this is now understandable because any failure in the cascade of molecular events onwards from the initial correct activation of *apterous* will exert "knock-on" consequences for the subsequent precision of the patterning process.

6. The siting of individual sense organs in the wing disc

The mechanism underlying the spatial arrangement of the larger bristles on the dorsal thorax of Drosophila has become a paradigm for the analysis of patterning cell differentiation in an epithelium. Each of the bristles, which are mechanosensory organs, arises from four cells which are the products of division of a single "sense organ precursor" (SOP). Operationally, the process can be divided into two stages, the first of which is the emergence of clusters of cells related by their relative position in the disc rather than their exact cell lineage ancestry. The second stage is the resolution of the site of origin of a single SOP more precisely within the proneural cluster in a "one-from-many" selection process. A plethora of mutations affecting the number and location of the larger bristles has long implied that the process is complex. One technique in particular, the use of genetic mosaics for these mutations, has been paramount in understanding these developmental events, but this has only been successful in combination with the ability to document gene expression within genetic mosaics and around their borders.

During the middle to late third instar, the expression of the Wg glycoprotein resolves into an inner and outer ring of expression bordering the wing pouch and a stripe 3–6 cells wide at the presumptive wing margin. At the margin, individual bristles, as defined by expression of the *achaete* gene, arise in fixed relation to the distribution of Wg protein. When this secretion is interrupted using temperature-sensitive mutations of *wingless*, a set of the bristles fails to appear, suggesting that the Wg protein creates a discontinuity at which an SOP forms. The binding of Wg to a receptor encoded by the *frizzled-2* gene activates an intracellular signal transduction pathway. This pathway requires the participation of the products of the *dishevelled* (*dsh*) and *armadillo* (*arm*) genes. Dsh is responsible for receiving the Wg signal from the cell membrane, as it is a cytoplasmic protein, and Arm is a β-catenin homologue that is enriched at adherens junctions and which associates with cadherins. One of the downstream consequences is the inactivation of a specific kinase, from the *zeste white3/shaggy* gene. The Arm protein is a target of the Shaggy kinase, and phosphorylation of Arm causes it Arm to associate much more readily with the plasma membrane. Reception of Wg antagonizes the effect of the Shaggy kinase and causes Arm to localize in the cytoplasm, forming a tertiary complex with two other proteins, APC and Axin, which interact with the product of the gene *pangolin*. At the margin, reception of the Wg signal causes the transcriptional activation of the "proneural" genes *achaete* (*ac*) and *scute* (*sc*), which encode basic helix–loop–helix transcriptional activators, in cells that border the Wg-secreting cells. In the centre of the Wg stripe, the *cut* gene is expressed, which seems to prevent the expression of *ac* inside the Wg stripe (**Plate 6**).

Proximal structures of the wing include the hinge, which attaches the wing to the body wall of the adult and facilitates flight movement. The proximal rings of Wg expression are necessary for hinge formation as shown by a regulatory

Table 1 Genes and the role they play in wing patterning.

Gene	Gene product	Role in wing development
abrupt	zinc finger protein	vein formation
achaete	helix–loop–helix transcription factor	expressed in sense organ precursors
APC	associates with Arm	wingless signal transduction
apterous	homeodomain and LIM domain transcription factor	compartment identity
armadillo	β-katenin homologue	wingless signal transduction
axin	axin/conductin	wingless signal transduction
blister	serum response factor homologue	vein formation
cubitus interruptus	zinc-finger-containing transcription factor	Hh signal transduction
cut	homeodomain-containing transcription factor	wing margin patterning
decapentaplegic	transforming growth factor-β	spatial patterning signal
Delta	ligand for the receptor Notch	cell–cell local signalling
Dichaete	high mobility group-domain protein	
discs large	membrane-associated protein	septate junction structure
discs overgrown		apico-basal cell polarity and proliferation control
dishevelled	PDZ-containing domain	wingless signal transduction
Dorsal Wing		regulated by apterous
engrailed	homeodomain-containing transcription factor	determines the posterior compartment, regulates *ptc* and *ci*
Enhancer of split	basic helix–loop–helix transcription factor	mediating Notch function
epidermal growth factor	ligand for epidermal growth factor receptor	spatial patterning
fat discs	cadherin	cell–cell adhesion
fringe	secreted protein, probably modifying action of Notch	establishing wing margin
frizzled-2	7-pass serpentine membrane protein	wingless receptor
giant discs		proliferation control
giant larvae	plasma membrane protein	chromosome condensation
groucho	basic helix–loop–helix protein with WD40 domain	part of repressor complex
hedgehog	transforming growth factor-β secreted protein superfamily member	signalling ligand for Ptc receptor
hyperplastic discs	ubiquitin	
invected	homeodomain-containing transcription factor	determines the posterior compartment
knot (synonym: *collier*)	DNA-binding protein	vein formation
net		vein formation
Notch	transmembrane receptor	binds the ligands Delta and Serrate
nubbin	POU-domain containing protein	activation of *wingless*
optomotor blind	transcription factor, brachyury T homologue	response element downstream of Dpp signalling
pangolin	HMG-containing transcription factor	wingless signal transduction
patched	multiple pass transmembrane domain protein	receptor for Hedgehog
plexus		vein formation
punt	receptor protein	Dpp receptor type II
radius incompletus		vein formation
ras signalling	signal transduction element	spatial patterning
scute	helix–loop–helix transcription factor	expressed in sense organ precursors
Serrate	transmembrane transforming growth factor homologue, ligand for the receptor Notch	cell–cell local signalling
smoothened	serpentine class membrane protein	transduction of Hh signal
spalt	transcription factor	response element downstream of Dpp signalling
Suppressor of Hairless	transcription factor with integrase domain	transducer of Notch function
thickveins	receptor protein	Dpp receptor

Table 1 Continued.

Gene	Gene product	Role in wing development
veinlet	transmembrane protein interacting with epidermal growth factor receptor	vein formation
ventral veins lacking	Cf1a transcription factor	vein formation
vestigial	transcription factor?	essential for growth of the wing blade
wingless	secreted glycoprotein	signalling ligand
zeste white3/shaggy	GSK3a kinase homologue	wingless signal transduction

mutation of *wg* called *spade*, which causes loss of the inner (distal) ring of Wg expression, as well as decreasing Wg levels at the presumptive wing margin, thus causing occasional loss of the margin. The distal ring of *wg* expression also requires the product of the *nubbin* gene, a POU-domain protein. Dominant *Dichaete* mutations show defects in the hinge and repression of the distal ring of Wg expression, and are due to misexpression of the high mobility group-domain protein Dichaete in the hinge primordium. Once again, the availability of reagents for detecting transcription and the appearance of these gene products allows monitoring of the behaviour of individual genes in the presence or absence of the activity of other genes implicated in the same pathway. The distinction can then be made as to whether or not the action of one gene depends on another particular gene, and whether the two elements are simultaneously required in the same cell for the developmental step to proceed (De Celis, 1998).

7. Formation of the wing veins

On pupariation, the wing disc becomes exposed to a flood of ecdysterone hormones, which cause the disc to undergo a complex set of cell rearrangements called disc eversion. The wing disc begins eversion with the extension of the D–V boundary. This extension brings the dorsal and ventral disc primordia into close contact with each other, which promotes signalling events between the two epithelial surfaces. During pupariation, the wing surfaces detach and then rejoin. Those cells that express the *veinlet* (*ve*) gene never appose, and also fail to appose during pupariation, but go on to form hollow tubes, through which neurones, haemocytes and tracheae pass. These tubes are the wing veins and adult Drosophila wings have five longitudinal veins and two cross veins (**Figure 1**). Expression of *ve* begins in the late third instar, and cells that express *ve* invariably go on to form vein tissue (**Plate 7A–B**). *ve* has been found to be a part of the *epidermal growth factor/ras signalling* pathway, and mutations in or overexpression of any member of this pathway will affect vein formation. Veinlet protein is thought to increase the epidermal growth factor signal. Several genes modify *ve* expression and, therefore, modify vein formation. Mutations in the genes *net* or *plexus*, for example, cause ectopic expression of *ve* and, subsequently, extra vein tissue. Other mutations cause loss of *ve* expression, particularly in vein primordia and with the subsequent loss of specific veins. *radius incompletus* (*ri*) mutations cause loss of vein 2 and concomitant loss of *ve* expression in the vein primordium of vein 2. *abrupt* (*ab*) mutations cause loss of vein 5 and loss of *ve* in vein primordia 5. Other mutations, like *knot* (*kn*), upset the positioning of veins. Wing imaginal discs from *kn* mutants also cause *ve* expression in vein primordia 3 and 4 to be closer together and sometimes to fuse. Loss-of-function mutations in Notch, or the Dpp receptor encoded by the *thickveins* gene, cause expansion of the areas of *ve* expression and thickened veins, which suggests a signalling system that prevents the spread of vein tissue. One potential downstream target of Ve is a Cf1a transcription factor, encoded by the *ventral veins lacking* gene.

Attachment between the D–V surfaces of the wing blade is crucial for proper vein and intervein formation. Integrins, which play an important role in the adhesion of tissue culture cells to extracellular matrix, and laminin A are important in D–V surface apposition and show differential regulation in vein and intervein regions during pupariation. Mutations that prevent proper apposition of the dorsal and ventral surfaces also disturb vein formation. The *blister* (*bs*) gene, encoding a Drosophila homologue of the serum response factor protein, is required for basal attachment between dorsal and ventral wing blades. *bs* mutations produce extensive ectopic veins, and ectopic *ve* expression, which further confirms the role of *ve* in wing vein formation.

8. Conclusion

The robust and self-regulating nature of wing disc growth and development, and the precision of the pattern formed, can both, therefore, be understood in terms of the dynamic nature of the cellular "machinery", which has been revealed by methods of genetic dissection (see "The Interactive Fly" in the References). The completion of the genome project for Drosophila will now permit a complete definition of the genes involved and a full characterization of the roles of their products (see "FlyBase" in the References). The genes discovered so far, and discussed in this article (Table 1), are therefore a subset of the complete inventory. Even so, our understanding of the developmental genetics of the wing disc has become an excellent paradigm for analysing pattern formation in other biological systems.

REFERENCES

Blair, S.S. (1995) Compartments and appendage development in *Drosophila*. *BioEssays* 17: 299–309

Bryant, P.J. & Levinson, P. (1985) Intrinsic growth control in the imaginal primordia of Drosophila, and the autonomous action of a lethal mutation causing overgrowth. *Developmental Biology* 107: 355–63

Chen, Y. & Struhl, G. (1996) Dual roles for patched in sequestering and transducing hedgehog. *Cell* 87: 553–63

Couso, J.-P., Bate, M. & Martinez-Arias, A. (1993) A wingless-dependent polar coordinate system in Drosophila imaginal disks. *Science* 259: 484–89

Couso, J.-P., Bishop, S.A. & Martinez-Arias, A. (1994) The wingless signalling pathway and the patterning of the wing margin. *Development* 120: 621–36

Couso, J.-P., Knust, E. & Martinez-Arias, A. (1995) Serrate and wingless co-operate to induce vestigial gene expression and wing formation on Drosophila. *Current Biology* 5: 1437–48

De Celis, J. (1998) Positioning and differentiation of veins in the *Drosophila* wing. *International Journal of Developmental Biology* 42: 335ff.

Doherty, D., Feger, G., Younger-Shepherd, S., Jan, L.Y. & Jan Y.N. (1996) Delta is a ventral to dorsal signal, complementary to Serrate, another Notch ligand in *Drosophila* wing formation. *Genes and Development* 10: 421–34

FlyBase* The FlyBase Database of the Drosophila Genome Projects and Community Literature. Available from http://flybase.bio.indiana.edu/

Sturtevant, M.A., Roarck, M. & Bier, E. (1993) The Drosophila rhomboid gene mediates the localized formation of wing veins and interacts genetically with components of the EGF-R signaling pathway. *Genes and Development* 7: 961–73

The Interactive Fly* http://sdb.bio.purdue.edu/fly/aimain/1aahome.htm

* These references are to continuously updated databases.

GLOSSARY

dichotomy/dichotomous branching resulting from division of growing points into two equal parts

ecdysterone hormone steroid hormone that stimulates growth and moulting

epithelium sheet of tightly bound cells covering internal and external organs of the body

heteromeric an association of two or more sub-units that are of different, rather than identical, composition

homeodomain DNA-binding protein domain encoded by the homeobox sequence

hyperplastic growth the increase in size of a tissue resulting from excessive cell division that does not perturb normal cell associations

hypomorphic background - an altered genotype in which the genes under consideration have a reduced effect

LIM domain/family found in proteins required for developmental decisions. Contain 60-residue conserved, cysteine-rich repeats. Named after first three genes in group: Lin-11 (*Caenorhabditis elegans*, required for asymmetric division of blast cells), Isl-1 (mammalian insulin gene-binding enhancer protein), mec-3 (*C. elegans*, required for differentiation of a set of sensory neurons)

neoplastic overgrowth - excessive cell division that is accompanied by loss of normal cell associations or cell adhesion, and where the cells can often separate from their tissue of origin and invade other tissues

PDZ domain domains found in various intracellular signalling proteins associated with the plasma membrane; named for the postsynaptic density, Discs-large, ZO-1 proteins in which they were first described

POU domain a conserved protein domain of around 150 amino acids, composed of a 20 amino acid homeobox domain and a larger POU-specific domain, and therefore is the target of some transcription factors. Named POU (Pit-Oct-Unc) after three such proteins: Pit-1 that regulated expression of certain pituitary genes; Oct-1 and 2, that bind an octamer sequence in the promoters of histone H2A and some immunoglobulin genes; and Unc-86, involved in nematode sensory neuron development

See also **Development of the egg and early embyro of Drosophila** (p.163)

SEX DETERMINATION IN DROSOPHILA

Daniel Bopp
Institute of Zoology, University of Zürich, Switzerland

1. Introduction

In the course of ontogenesis, cells are instructed to adopt different fates to form a functional multicellular entity. One of the most ambitious areas in biological research during the last decades has been the study of the molecular mechanisms that control specification of cellular identities. The fruit fly, *Drosophila melanogaster*, has been instrumental in identifying the genetic elements that are involved in these processes. Largely because of the fly's amenability to genetic analysis and manipulation, rapid progress has been made in understanding some of the principles of developmental control. In particular, sexual development of Drosophila has become an attractive system in which to study the genetic control of developmental fates. How does the zygote select between two alternative programmes, and what are the underlying mechanisms that stably propagate and execute this decision at the single-cell level? The purpose of this review is to present our current knowledge about the mechanisms involved, with a focus on the first determinative steps in the pathway.

2. Sexual dimorphic development

Unlike the complexity of body patterning which requires the definition of a multitude of different cellular fates, the process of sex determination can be reduced to that of a simple binary decision. Fertilized eggs select between two fates, namely to develop either into a female or a male individual. Early in ontogenesis, the embryo receives instructions to follow one of the two pathways, but the consequences thereof become noticeable not until much later in development when sexually dimorphic traits are manifested at the level of morphology, physiology and behaviour. The first clear sign of dimorphism is seen in the gonads of the larva. The male gonad develops far ahead of the female and contains a substantially larger number of germ-cell progenitors (Kerkis, 1931). After metamorphosis, adult males and females can be easily distinguished by appearance. The internal and the external genital organs are composed of different structures (reviewed in Garcia-Bellido, 1983). In addition, males have a distinctive dark pigmentation of the distal regions of the abdomen and carry large specialized bristles on the forelegs (see **Figure 1**). Females, on the other hand, have a larger body size.

Besides differences in appearance, males and females also have specific internal structures and differ in some aspects of their physiology. One example is the presence of a male-specific muscle in the fifth abdominal segment (Lawrence & Johnston, 1986). It has been suggested that this muscle might allow the male to curl his abdomen for proper positioning for mating. A well-documented example in physiology is the female-specific synthesis of yolk proteins in fat-body cells (reviewed in Bownes, 1994). These cells are present in both sexes, but only those of the female produce yolk proteins which are secreted into the haemolymph and deposited into the developing eggs of the ovary.

Another major target for sex-specific development is the nervous system. For example, females have a 10% higher number of Kenyon fibres in the mushroom body, a region of the brain that processes primarily olfactory information and that may be important for the mating process (Heisenberg, 1980). Similarly, sex-specific differences have also been seen in the relative numbers of certain sense organs in the antennae. These differences might be related to the sensing of pheromones by the sexes. The mating process itself, of course, relies on a distinctive behaviour of the male and female fly. The male follows a precisely defined ritual when approaching a virgin female (reviewed in Greenspan, 1995). The various elements of this behaviour are specified in parts of the brain, e.g. the mushroom body, and in the thoracic ganglion of the male (Hotta & Benzer, 1976; Hall, 1979; Ferveur *et al.*, 1995; O'Dell & Kaiser, 1997). The mated

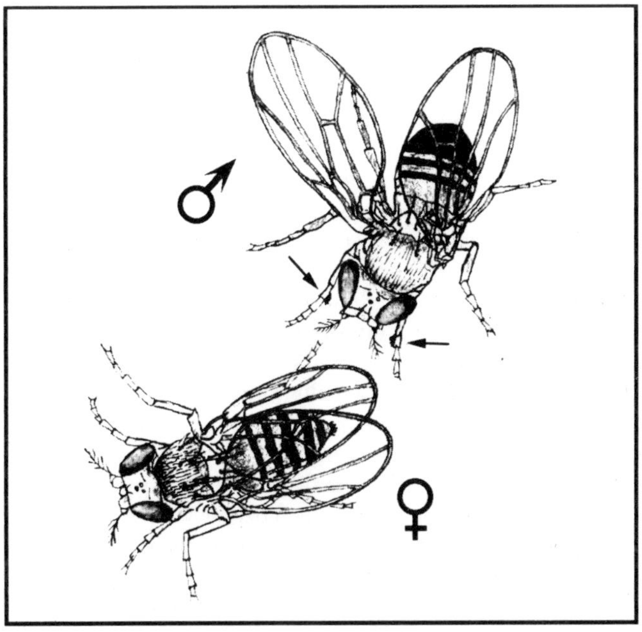

Figure 1 Sex-specific traits in Drosophila. The male fly has darker pigmentation in the posterior part of the abdomen than the female and carries a row of specialized bristles, the sex combs, on its forelegs (arrows). In addition, the male depicted here displays a typical element of courtship when approaching a female. It extends one of its wings perpendicularly to its body axis and vibrates it at a fast rate, a part of the courtship ritual called the "love song".

female, in turn, displays a rejective conduct after copulation to prevent successive mating attempts by other males (Hall, 1985). This rejection behaviour is induced by chemical compounds which are transferred with the male ejaculate (Chen, 1984; Kubli, 1992).

To achieve this high order of physiological and neural complexity, the system implements a cascade of regulatory activities which act early in the zygote to govern every aspect of sex-specific development. The next section will introduce the key components of the pathway and how they interact to transmit the genetic instructions received early in development down to the level of the genes that express the corresponding dimorphic traits.

3. The sex determination cascade in Drosophila

The first studies on sex determination in Drosophila were initiated more than 70 years ago when Bridges discovered the nature of the primary sex-determining signal (Bridges, 1922, 1925). Unlike many other systems that use dominant determiners as the primary signal, sex determination in Drosophila relies on a polygenic signal which was at that time described as a balance between feminizing tendencies on the X chromosome and masculinizing tendencies on the autosomes. Based on experimental tests, the signal was formally defined as the ratio of X chromosomes to sets of autosomes (X:A). In animals with a diploid set of autosomes, the presence of two X chromosomes will implement female development, whereas one X chromosome will determine male development. The X:A signal is assessed in each cell of the zygote independently. Thus, unlike the mammalian system, where primary events in sex determination are under hormonal control, each cell of the Drosophila zygote determines its sex autonomously. When the signal is ambiguous, as for instance in XX:AAA animals, sexual mosaicism can occur which produces intersexes that differentiate male and female tissue next to each other.

The execution of the corresponding developmental programme is achieved through a cascade of regulatory genes which are under the direct control of the primary signal. This cascade, in turn, controls all genes responsible for the expression of sexually dimorphic traits. While much is known about the genes involved in transducing the primary signal, only a few targets of the cascade have so far been identified. Nevertheless, sexually dimorphic development is expected to employ a large number of differentiation genes performing different tasks in males and females. They are collectively referred to as male differentiation genes and female differentiation genes. The current view is that all of these genes are controlled by two antagonistic states of the cascade. In XX animals, the female mode will repress male differentiation genes and activate the female differentiation genes. Conversely, in XY animals, the cascade represses the female differentiation genes, while it permits the male differentiation genes to be expressed.

Rapid progress has been made in identifying the key players in transducing the sex-determining signal. These genes were initially discovered by recessive loss-of-function mutations and dominant gain-of-function mutations that revert the sexual fate of affected animals (reviewed in Baker & Belote, 1983; Steinmann-Zwicky *et al.*, 1990; Cline & Meyer, 1996). Studying interactions between these genes made it possible to determine at what level a given gene acts in the cascade. In recent years, many of the relevant genes in the pathway have been successively isolated and characterized at the molecular level. This information has permitted us to reconstruct the course of events from the signal input at the top down to the bottom level where the signal is translated into sex-specific expression of differentiation genes.

The cascade essentially consists of three hierarchically arranged levels, of which the first two act as ON/OFF switches, whereas the bottom-most level expresses two regulatory activities with opposite effects (**Figure 2**). The gene *Sex-lethal* (*Sxl*) occupies a central position in the cascade and is directly controlled by the primary signal. In the XX zygote, where the X:A ratio is 1, *Sxl* is ON, while in X(Y) animals, where the X:A ratio is 0.5, it remains functionally OFF. *Sxl*, in turn, will regulate *transformer* (*tra*), the next gene in the cascade. An active *Sxl* will launch the female-specific activity of *tra*. In contrast, absence of active *Sxl* gene products in X(Y) animals will leave *tra* in the OFF state. The next and last level in the sexual differentiation pathway is occupied by the gene *doublesex* (*dsx*). The presence of active *tra* gene products will direct the female mode of *dsx* expression. In the absence of a functional *tra* gene, *dsx* will be expressed in the male-specific mode. Unlike the genes upstream in the cascade, this gene is bifunctional, producing either a male-specific or a female-specific protein. The male-specific function of *dsx*, DSX_M, is needed to repress the activity of female-specific differentiation genes and to stimulate expression of male differentiation genes. The female-specific function, DSX_F, on the other hand, has the opposite effect and represses male-specific differentiation genes and assists expression of female differentiation genes. In concordance with this view, the only known targets of *dsx* so far, the yolk protein genes, are active when *dsx* is set in the female mode, while their expression is repressed when it is in the male mode.

Apart from the genes described above, a number of other genes that are not sex-specifically regulated participate in the cascade. As they are expressed in males and females, they

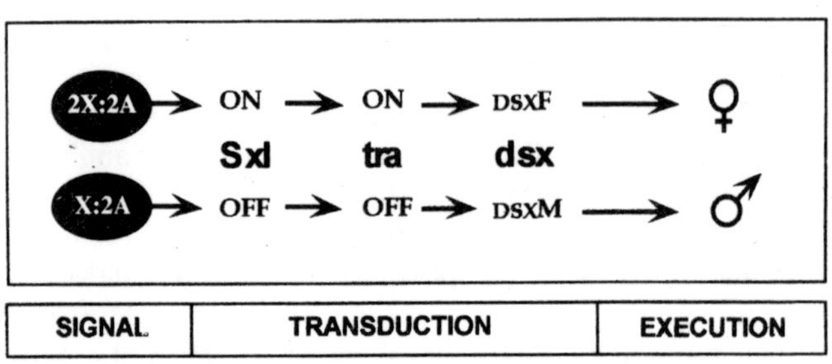

Figure 2 The sex determination pathway in *D. melanogaster* is composed of genetic switches transducing the signal through three hierarchically organized levels. The bottom-most switch is bifunctional, expressing two activities with opposite effects on the terminal differentiation genes.

have no discriminatory properties, but they act as auxiliary factors essential for the correct transduction of the primary signal in the cascade. For instance, *Sxl* requires the presence of wild-type activity of *fl(2)d*, *sans-fille* (*snf*) and *virilizer* (*vir*) to perform its function properly (Granadino *et al.*, 1990; Flickinger & Salz, 1994; Hilfiker *et al.*, 1995). On the other hand, *tra* needs the activity of *transformer2* (*tra2*) to impose the female mode of *dsx* expression (Fujihara *et al.*, 1978; Boggs *et al.*, 1987; Amrein *et al.*, 1988). The female activity of this gene, in turn, depends on the presence of *intersex* (*ix*) and *hermaphrodite* (*her*) to incite female differentiation (Pultz & Baker, 1994; Case & Baker, 1995). These few examples are likely to represent only a fraction of all the factors needed to propagate the signal. It is conceivable that a number of other essential cofactors have escaped detection so far, as they perform additional vital tasks in development besides sex determination.

In Figure 2, the flow of sex-determining instructions appears as a linear course of decisions. However, there is mounting evidence that the cascade is, in fact, more complex and branches at various levels in the hierarchy. For instance, many sex-specific features of the nervous system, such as courtship behaviour, are under the control of *tra*, but not of *dsx*. There is now substantial evidence for the presence of at least one *dsx*-independent branch below *tra* which specifies the sexual fate of neuronal tissues. A likely target in this branch has been found in the gene *fruitless*. Mutations in this gene affect sexual behaviour of males as well as the differentiation of some male-specific neurones (Gailey & Hall, 1989; Gailey *et al.*, 1991; Ryner *et al.*, 1996). Another important branch in the cascade is found downstream of the gene *Sxl*. Besides controlling sexual differentiation, this gene performs a vital task in the developing fly which is described in the next section.

4. Dosage compensation in Drosophila

Animals with heteromorphic sex chromosomes are confronted with the problem that genes linked to these chromosomes are present in different doses in the two sexes. In Drosophila, females carry twice the dose of X-linked genes than males. As the X chromosome constitutes about one-fifth of the total genome, a 2-fold difference would have drastic effects on global gene activity. To prevent an imbalance of gene product expression from the X relative to autosomes, a mechanism exists, referred to as dosage compensation, that equalizes the level of X-linked gene activity in animals with one (XY) or two X chromosomes (XX) (see article on "X-chromosome inactivation"). In Drosophila, this compensation is achieved by an increased rate of transcription of genes on the single X chromosome in males (reviewed in Baker *et al.*, 1994). By doubling this rate, males express the same amount of X-linked gene products as females with two X chromosomes. The higher transcription rate in males is based on differences in chromatin structure. The male X contains a specifically modified isoform of the H4 histone (a major component of the nucleosome) not found in the autosomal chromosomes or in the female X chromosomes (Turner *et al.*, 1992). This isoform of H4 is monoacetylated at lysine 16 and, because of this modification, the male X chromosome is less condensed than the female X chromosomes. It is assumed that this partially decondensed state makes the male X more accessible to components of the general transcription machinery, thus permitting a higher rate of transcription.

The question arises as to why only the male X becomes modified and not the female X chromosomes or any of the autosomal chromosomes. A genetic dissection of this problem has led to the identification of at least five genes involved in this process. These loci were originally found as recessive mutations that cause male-specific lethality and, therefore, were collectively referred to as *male-specific lethals* (*msls*). They have no effect on sexual differentiation, but are confined to the pathway that controls dosage compensation (reviewed in Lucchesi, 1996). Mutations in these genes reduce transcription rates to the level of that seen for a single X in female cells. The current model is that the MSLS proteins form a multimeric complex which specifically associates with many sites along the single X chromosome. This complex sequesters an acetylase to the X, and increased histone acetylation will intensify the action of transcriptional regulators. Absence of any of these *msls* will prevent the formation of a stable complex and histone 4–lysine 16 acetylation of the X chromosome, which results in a female-specific low rate of transcription (**Figure 3**).

While association of the MSLS is essential for survival of XY flies, it must be repressed in animals with two X chromosomes. Hypertranscription in XX animals will otherwise lead to an overexpression of X-linked gene products, a situation which is incompatible with viability, the affected females dying at the embryonic stage. Therefore, the rate of X-linked transcription, like sexual differentiation, must be differentially controlled in the two sexes. Drosophila has adopted an elegant solution by coupling both processes under the same control (Cline & Meyer, 1996). Their control converges at the level of *Sxl* which acts as the ON/OFF switch common to both pathways. Active *Sxl* is thus not only needed to impose female differentiation through the activation of *tra*, but also to repress hypertranscription in XX cells. In XY cells, on the other hand, *Sxl* must be OFF to permit male sexual development and a 2-fold higher rate of X-linked transcription.

The name *Sex-lethal* already implies that this gene plays a vital role in development. In fact, the lethal effects resulting from inappropriate activity of *Sxl*, e.g. off in XX cells or on in XY cells, have initially obscured its participation in the control of sexual differentiation. Its dual function in sexual differentiation and dosage compensation became first known by a carefully performed genetic dissection of the gene (Cline, 1979). Recently, one of the *msls* genes, *msl-2*, has been identified as the direct target of *Sxl* regulation in dosage compensation (Bashaw & Baker, 1995; Kelley *et al.*, 1995; Zhou *et al.*, 1995). In XX cells, active *Sxl* prevents *msl-2* expression and thereby prevents the formation of functional MSLS complexes. The absence of H4 modification in these cells causes X-linked genes to be transcribed at a basal level.

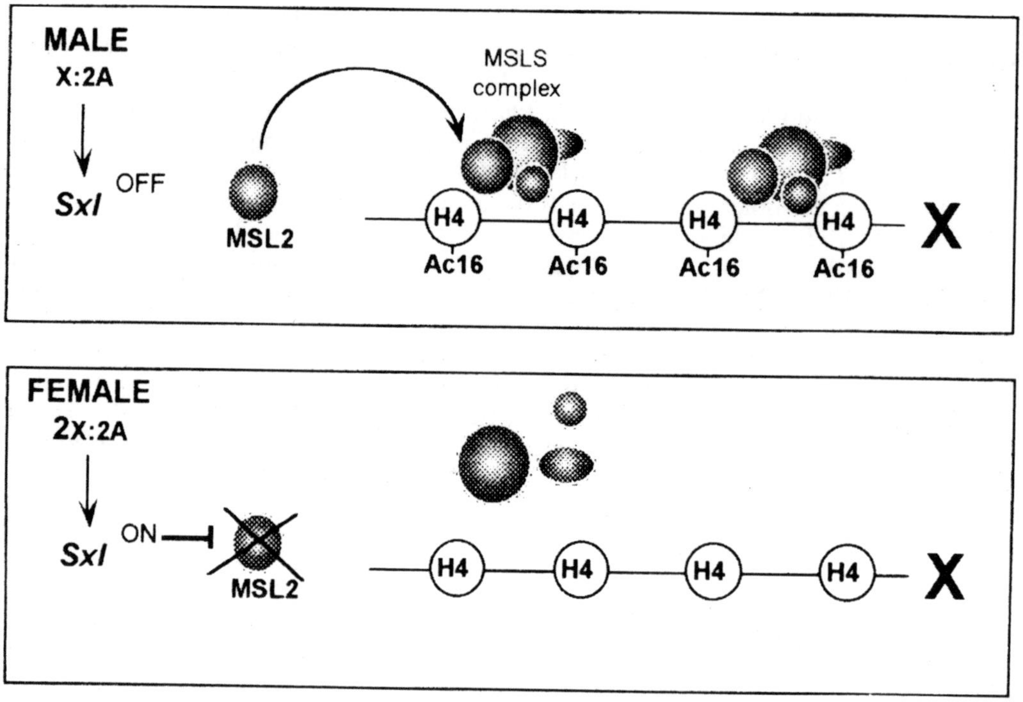

Figure 3 Dosage compensation in Drosophila. When all of the MSLS products are present in the cell (upper panel), they form a complex and assemble to the X chromosome and acetylate H4 histones at lysine 16 (Ac16). This modification will open up the chromatin structure permitting a higher rate of transcription. In the presence of active Sxl (lower panel), one of the MSL components, MSL-2, is not expressed, the complex cannot form and H4 histones remain unmodified – this chromosome is transcribed at a basal level.

5. The molecular basis of sex determination

This section will discuss the molecular mechanisms by which the sex-determining signal is interpreted and routed to the sex differentiation genes. The focus will be on the early steps of the pathway. *Sxl* has a pivotal role in this part of the pathway. It is not only important in assessing the signal and establishing the sexual fate of a particular cell, but it also serves as a cellular memory for the selected state throughout the rest of the fly's life cycle. This dual function of the gene, namely establishment and maintenance of a given sexual identity, involves different regulatory mechanisms.

(i) Establishment of sexual fate

Since the discovery in 1922 of the primary signal, the X:A ratio, by Bridges, many biologists have speculated about the nature of the underlying counting mechanism. For instance, a long-time favoured theory was based on titration models in which the X chromosome was depicted as a sink for a limiting number of autosomal factors. It was not until the 1980s that the first experimental steps were made towards the elucidation of the counting mechanism. Genetic mosaics revealed that the X:A signal is only transiently active in the early zygote around the blastoderm stage and that its action is mediated through the regulation of *Sxl* (Sanchez & Nöthiger, 1982, 1983) Also, it was demonstrated that the counting mechanism relies on a small number of discrete elements on the X chromosome, referred to as numerator elements. The genetic properties of two of these genes, *sisterless-A* (*sisA*) and *sisterless-B* (*sisB*), were concordant with the expected behaviour of counting elements (Cline, 1988). Elevating the dose of *sisterless* activity promotes female development by activating *Sxl*, while lowering the dose has the opposite effect; *Sxl* remains inactive and the male mode of development follows.

Despite a significant leap forward in understanding the nature of the X:A signal, the mechanistic principle still remained elusive, until a molecular analysis of the participating elements became possible. The first important piece of evidence that transcriptional regulators are involved in the counting process was obtained in studies with transgenic flies. Precocious expression of the segmentation gene *hairy* (*h*) prevented activation of *Sxl* in XX embryos. The gene *h* encodes a protein which belongs to the family of transcriptional repressors containing a DNA-binding domain (Parkhurst *et al.*, 1990; Parkhurst & Meneeley, 1994). It is normally involved in the process of segmental patterning of the early embryo where it takes part in complex regulatory circuitry to specify cellular fates along the main body axis. Although *h* itself is not involved in sex determination, those transgenic studies suggested that assessment of the X:A ratio is achieved by transcriptional control of *Sxl*. In concordance with this notion, the counting element *sis-b* encodes a transcription factor of the bHLH class which, when elevating its levels in XY embryos by ectopic expression, will activate *Sxl*.

If *Sxl* is transcriptionally controlled, we would expect that the gene contains a promoter that must respond in a dose-dependent manner to the X:A signal. Indeed, in blastoderm embryos, a specific set of early *Sxl* transcripts is transiently expressed only in XX, but not in XY, animals. The emergence of these *Sxl* transcripts coincides with the time when the X:A signal is read. These transcripts derive from an internal promoter P_E (establishment) located in the first intron of the gene (Keyes *et al.*, 1992; Estes *et al.*, 1995). Activation of P_E was demonstrated to depend directly on the dose of *sisterless* activity expressed in the blastoderm embryo. Remarkably, this promoter can respond in an ON/OFF mode to a 2-fold difference in *sisterless* activity. It was already concluded from genetic studies that an unambiguous interpretation of a 2-fold difference in gene dose can only be achieved by synergistic interactions between the transcription factors with each other and the promoter sequences. Currently, at least four numerator genes are known to be located on the X chromosome; in addition to *sisA* and *sisB*, these are *runt* and *sisterless-C* (reviewed in Cline, 1993) (**Figure 4**).

The identification of the autosomal counterpart of the X:A signal, the denominator elements, proved to be more difficult than anticipated. Only one denominator element has been identified so far. Similar to *h*, this gene, *deadpan* (*dpn*), encodes a transcriptional repressor of the bHLH type (Younger-Shepherd *et al.*, 1992; Barbash & Cline, 1995). As expected, it antagonizes the activities of numerator elements. Changing the dose of *dpn* in the early embryo will elicit the opposite effects on *Sxl* transcription to that of the *sisterless* genes. Elevating the levels of *dpn* activity will obstruct transcriptional activation of *Sxl* in XX animals, while decreasing its dose will facilitate the activation of *Sxl* in XY animals.

To describe the mechanism of how these elements act on early *Sxl* transcription, two models can be envisioned (Cline, 1993). Both models assume a threshold requirement for the activation of P_E depending on the absolute levels of the transcriptional activators inside the nucleus. One model describes a competition between the products of numerator and denominator genes at the level of protein complex formation (**Figure 5**). For instance, *dpn* could actively sequester numerator components of the activation complex by directly binding to these and rendering them in a non-functional complex. This titration effect reduces the level of functional numerator complexes to a level below the threshold required for activating *Sxl*. On the other hand, it is possible that competition occurs directly at the level of DNA binding to the promoter. In this model, denominators and numerators compete for binding to the same or overlapping DNA sites. It is also conceivable that both modes of control, namely competition at the level of complex formation and DNA binding, are used.

Though some details of the molecular interactions still remain to be experimentally addressed, we do understand the basic principle of the X:A signalling. There are a number of other genes that participate in the counting process as important cofactors and, since they are maternally deposited into the zygote, they are present in XX and XY zygotes in the same amount and hence have no discriminatory function. For instance, *daughterless* (*da*) and *hermaphrodite* (*her*) are essential maternal cofactors for activating *Sxl*, while *groucho* (*go*) and *extramacrochaetae* (*emc*) are necessary maternal cofactors for the denominator function of the X:A signal (Cline, 1978; Younger-Shepherd *et al.*, 1992; Paroush *et al.*, 1994; Pultz & Baker, 1994). In accordance with their participation in transcriptional control of *Sxl*, these genes also encode transcription factors.

In summary, the X:A signal is composed of a distinct set of X-linked transcriptional activators and autosomal transcriptional repressors which, by synergistic and antagonistic interactions, regulate a promoter of *Sxl*, P_E, in a dose-dependent manner. In cells with two X chromosomes, the dose of activators is sufficiently high to launch early *Sxl* transcription, while in cells with only one X the level of activators is below the threshold for activation of P_E. All somatic cells at the blastoderm stage receive and assess the chromosomal signal independently and set the activity of *Sxl* accordingly. Hence, differential gene activity of *Sxl* becomes the first manifestation of sexually dimorphic development.

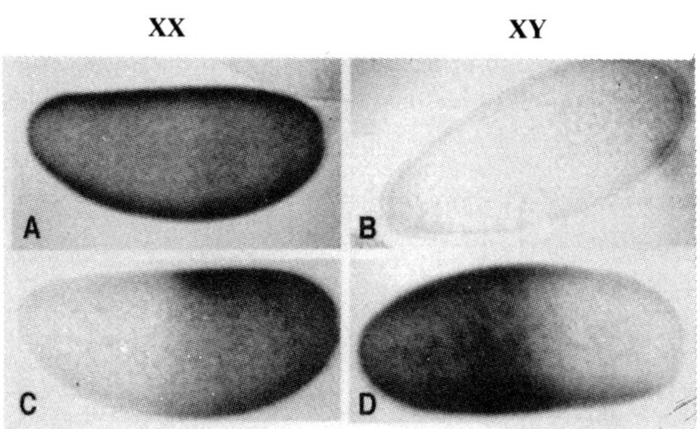

Figure 4 Initial expression of *Sxl* is controlled by helix–loop–helix (HLH) transcription factors. Dark staining marks the presence of Sxl protein. (A) A wild-type XX embryo displays ubiquitous staining, whereas the wild-type XY embryo in (B) is completely devoid of staining. (C) Enhanced expression of *hairy* gene products in the head region of XX animals prevents *Sxl* expression in this domain. (D) Increased levels of *sisB* activity in the head region has the opposite effect and leads to activation of *Sxl* in XY embryos (anterior is to the left).

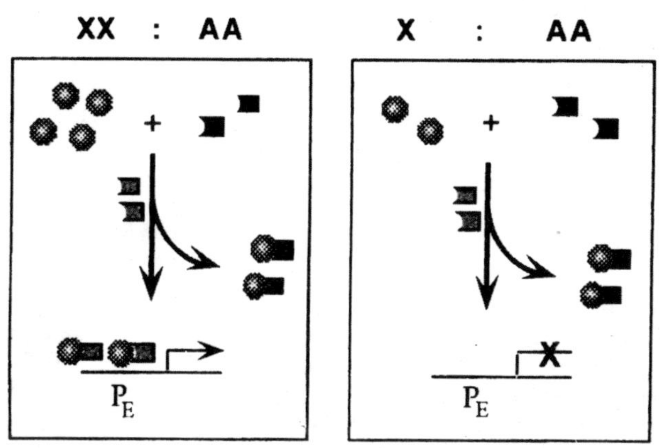

Figure 5 A model for the selective activation of the early promoter (P_E) of *Sxl*. In XX cells (left panel), the number of numerator products (spheres) exceeds that of denominator products (black boxes). The non-sequestered numerators can form complexes with other positively acting zygotic and maternal factors (shaded boxes). These complexes bind to the promoter region (P_E) and activate transcription of *Sxl*. In XY cells (right panel), all numerator elements become sequestered by denominators in non-functional complexes. As a result, *Sxl* is not activated.

(ii) Maintenance of sexual fate

The important prerequisite for a discriminatory X:A signal is that the presence of one or two X-chromosomes leads to a quantitative difference of numerator products in the cell. Implicit to this regulation is the absence of dosage compensation at the time when the X:A signal is produced and read. By activating *Sxl*, however, it is precisely this initial difference of X-linked gene products that becomes equalized and hence can no longer serve as a discriminating signal. XX cells will assume the same level of X-linked transcription as XY cells and the level of numerator activity will consequently drop below the threshold for activating P_E. This poses the question of how the activity of *Sxl* can be maintained. Evidently, a different mechanism must be operational to ascertain a continuous expression of *Sxl* in XX animals. Different mechanisms for establishing and maintaining the sexual fate were already proposed by T.W. Cline. His postulate was based on studies of two complementing classes of mutations in *Sxl*, one of which affects the early function of *Sxl*, while the other class defined a late function (Cline, 1984). He could show that the maintenance of *Sxl* activity is not regulated by the X:A signal, but depends on an autoregulatory function of the gene. Once activated, *Sxl* will maintain its active state through a self-sustaining feedback loop.

In accordance with these predicted features of *Sxl*, molecular analysis of the *Sxl* gene disclosed a different level of regulation following initial transcriptional activation. The first evidence came from RNA analysis of *Sxl* in XX and XY animals. About 2 hours after activation of P_E, the gene becomes transcribed from a more distally located promoter, P_M (maintenance). This promoter is active in both sexes and is constitutively transcribed throughout development until the adult stage. The important finding was that transcripts in females and males are different in structure, implying that the gene is subject to sex-specific RNA processing (Bell *et al.*, 1988). Male-specific transcripts differ from female-specific transcripts in that they contain an additional internal exon of 200 base pairs. This male-specific exon serves as a translation-terminating device which prematurely truncates the long open reading frame (ORF) by introducing in-frame translational stop signals. As a result, mature male transcripts can only produce small non-functional peptides. In contrast, this exon is removed in the mature female mRNA to preserve a long ORF and the potential to encode a functional polypeptide. Thus, the mode of ON/OFF regulation has changed now from the transcriptional level to the splicing level. This discovery became one of the first examples of the use of differential splicing to control the activity of a developmental regulator (Baker, 1989).

The model of how ON/OFF regulation of *Sxl* is maintained is depicted in **Figure 6**. Presence of functional Sxl protein in the cell achieves the exclusion of the translation-terminating exon, thereby allowing a continuous production of full-length protein. *Sxl* gene products which result from the activation of P_E serve as a first source of functional protein to activate this self-sustaining feedback loop. In male cells, no such early Sxl protein (P_E derived) is produced and, hence, the loop cannot be activated. By default, *Sxl* mRNAs include the translation-terminating exon and the gene remains functionally OFF for the rest of the fly's life. Other splicing factors have been implicated to play an essential role in regulation of *Sxl*. For instance, the gene *sans-fille* (*snf*) encodes a homologue of the mammalian B^R snRNP (Flickinger & Salz, 1994). It is not only required for regulated splicing of *Sxl*, but it is also an essential splicing component in both sexes.

(iii) Coordinate control of sexual differentiation and dosage compensation by Sxl

Besides sustaining its active mode of expression, *Sxl* must execute the selected programme by controlling the next downstream level of the cascade. The gene *tra* heads the branch of the hierarchy that specifically controls somatic sexual differentiation, while *msl-2* heads a branch of the pathway that controls dosage compensation. Both genes are under the direct control of *Sxl*. They are regulated at the level of RNA processing, expressing different sets of transcripts in males and in females. The gene *tra* is transcribed in both sexes, but only the female-specific messages encode

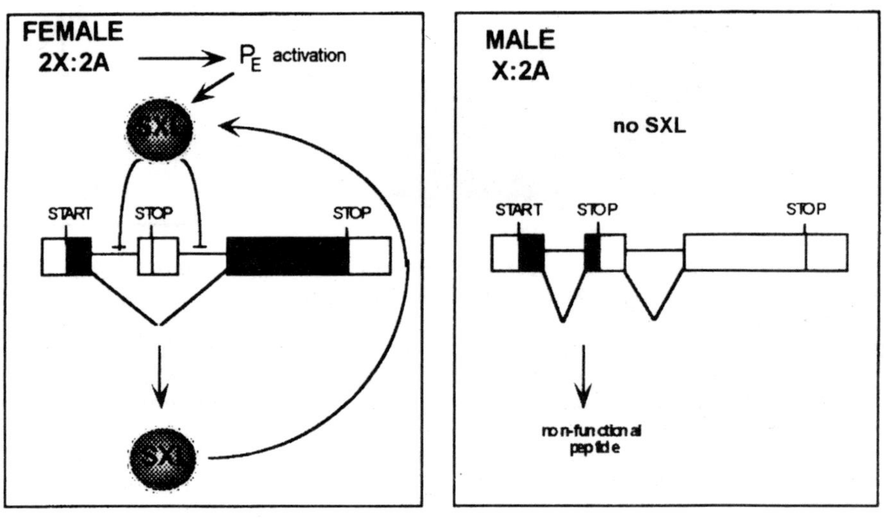

Figure 6 ON/OFF regulation of *Sxl* by differential splicing. In the absence of Sxl (right panel), male splicing occurs incorporating an additional exon. The translation-terminating signals (STOP) in this exon shorten the open reading frame (filled boxes) to a small non-functional peptide of about 42 amino acids. In the presence of Sxl (left panel), incorporation of this exon is suppressed by binding of Sxl to sites flanking the exon. The female splice reconstitutes a long open reading frame (filled boxes) which encodes a functional polypeptide of about 360 amino acids.

a functional product (Boggs *et al.*, 1987; Belote *et al.*, 1989). The mechanism behind this regulation is based on the differential utilization of alternative splice acceptor sites in the *tra* nascent RNA (**Figure 7**). In absence of Sxl protein, the first acceptor site is used, thereby incorporating sequences into the mRNA that prematurely truncate the long ORF. Thus, this mRNA in XY animals expresses a small non-functional peptide. The presence of Sxl protein in XX animals will block this acceptor site enforcing the use of the next site located downstream (Sosnowski *et al.*, 1989; Inoue *et al.*, 1990). The long ORF is preserved in the resulting mRNA, and this message encodes a functional Tra product.

The gene *msl-2* has been identified as the target gene of *Sxl* in the dosage compensation pathway. Sxl proteins bind to target sites in the untranslated leader and trailer sequences (**Figure 7**). This association prevents *msl-2* RNA in XX animals from being translated. In XY animals, when Sxl is absent, *msl-2* is translated and produces a functional component of the MSLS complex. Thus, *Sxl* affects gene expression not only by altering the splicing pattern, but also by acting as a translational repressor (Bashaw & Baker, 1997; Kelley *et al.*, 1997).

Sxl belongs to the family of RNA-binding proteins. It contains two RNA recognition motifs (RRMs) which bind to RNA with high affinity for long stretches of uridyl nucleotides in a row (Sakashita & Sakamoto, 1994; Samuels *et al.*, 1994). Such binding sites were found near the regulated splice sites in *Sxl*, *tra* and *msl-2* pre-mRNA, suggesting that binding may have a direct effect on splicing. Indeed, it has been shown that binding of Sxl protein to these (U)-rich sites will prevent nearby splice sites from being recognized by general components of the splicing machinery. In the case of *tra* splicing, Sxl acts as an antagonist of the essential splice factor U2AF; it competes for binding and prevents U2AF from recognizing this acceptor site (Valcarcel *et al.*, 1993). Instead, the next available downstream splice acceptor site will be bound by U2AF and used for the assembly of a functional splice complex. For regulation of *Sxl* splicing, a slightly different mechanism seems to be operational. Multiple stretches of (U)-rich sequences are located not only upstream, but also downstream of the male-specific exon. *In vivo* studies by Horabin and Schedl revealed that *Sxl* regulation depends to a higher degree on the binding of Sxl to the downstream sites than to the upstream binding sites (Horabin & Schedl, 1993a,b). This observation suggests that interference with the use of the male-specific donor site may have a more important role in preventing the inclusion of the male-specific exon than blocking the male-specific acceptor site. Finally, in the regulation of *msl-2*, *Sxl* again appears to work on a different level. Its binding affects *msl-2* gene expression not by truncating the coding sequence, but by preventing translation (Bashaw & Baker, 1997; Kelley *et al.*, 1997). In all cases, the effect of Sxl binding is to block the male splice, although the mechanism of that blocking and its specific molecular consequences differ.

(iv) Control of the bifunctional gene dsx

The last step in the sex-determining cascade is also controlled at the post-transcriptional level. The bifunctional gene *dsx* is transcribed in both sexes, but its pre-mRNA is differentially processed (Nagoshi & Baker, 1990). In XX

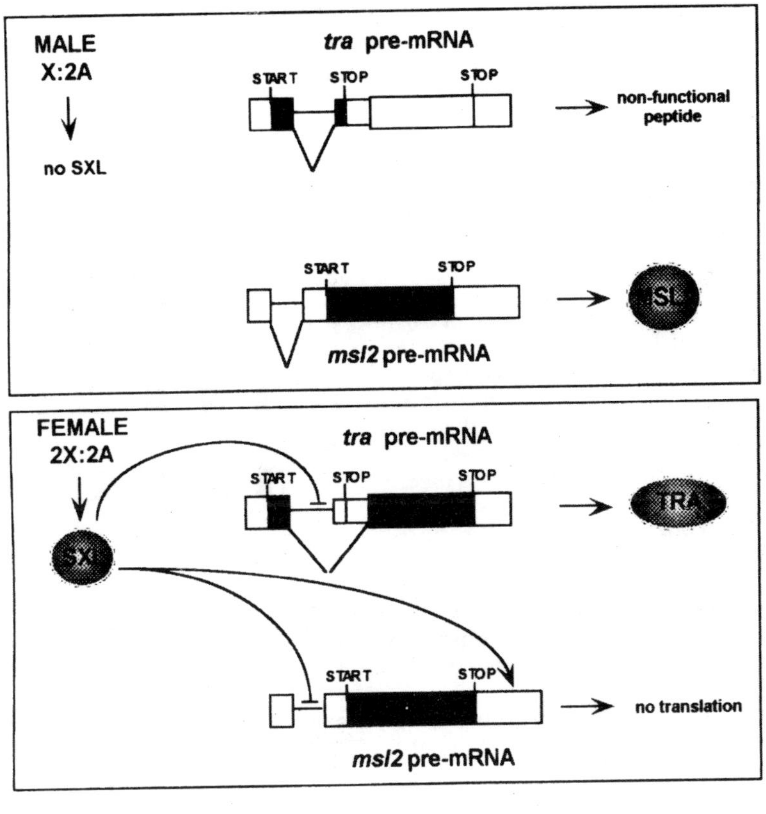

Figure 7 Regulation of *tra* and *msl-2* by *Sxl*. In the absence of Sxl protein (upper panel), male splicing uses the first available acceptor site. The resulting mRNA contains a short open reading frame (filled box) encoding a small non-functional peptide. Binding of Sxl protein upstream of this acceptor site in female cells (lower panel) forces the use of the next acceptor site downstream and the corresponding mRNA encodes a functional Tra protein. In the case of *msl-2*, Sxl binds to sequences in the 5′ and 3′ untranslated regions of *msl* pre-mRNA. Binding to the 5′ UTR (untranslated region) will prevent splicing of a short sequence in the leader. Unlike *tra*, this association prevents the *msl-2* message from being translated.

animals, functional Tra product, in association with Tra2, assembles at a repeat element (dsxRE) in the *dsx* pre-mRNA to promote the use of a cryptic acceptor site which produces a female-specific mRNA (Hedley & Maniatis, 1991; Ryner & Baker, 1991). This mRNA encodes the Dsx_F protein (panel A in **Figure 8**). In the absence of Tra protein, as in XY animals, an alternative splice site is used, and the mRNA derived from this splice mode encodes a slightly different protein isoform of *dsx*. This variant, the male-specific Dsx_M protein, only differs at the carboxyl end (Burtis & Baker, 1989). Tra and Tra2 proteins contain a domain rich in arginine and serine, characteristic of splicing factors belonging to the SR family. Tra2, in addition, has a conserved RRM domain which allows the protein to bind directly to the *dsxRE* sites on the *dsx* pre-mRNA. It has been proposed that binding of Tra2 recruits a complex that includes Tra and other SR proteins to activate the female-specific acceptor site upstream of the *dsxRE* binding sites (Tian & Maniatis, 1993; Lynch & Maniatis, 1995).

The *dsx* gene encodes a transcription factor containing an atypical zinc-finger DNA-binding domain (Erdman & Burtis, 1993). This domain is encoded by the part of the mRNA that is common to the male- and female-specific mRNA. Therefore, Dsx_M and Dsx_F can bind to the same promoter sequences, but their binding will elicit opposite effects. For instance, binding of Dsx_M to the enhancer region of the yolk protein genes represses transcription, whereas binding of Dsx_F to the same site enhances it (Coschigano & Wensink, 1993). These protein variants work both by preventing gene expression that is only appropriate for differentiation of the opposite sex and by inducing or enhancing gene expression appropriate to the same sex. The molecular basis for the antagonistic behaviour of Dsx_M and Dsx_F is not understood. It has been speculated that differences in structure at the carboxyl terminus may attract different types of cofactors which will interact with the protein. These factors may play a major part in determining whether the complex acts as a repressor or enhancer of transcription.

It is rather peculiar that, in contrast to the bulk of information about the regulatory cascade, we know little about its final targets. Currently, the three yolk protein genes present in the Drosophila genome are the only known targets of *dsx*. There are two general ways in which the cascade controls the expression of sexually dimorphic traits. In most cases, there are defined critical periods during development at which the sex-determining cascade exerts control and after which the sexual phenotype is irreversibly fixed. In other cases, like that of the yolk protein genes, continual regulation is required to maintain sexual phenotype. Despite the profound difference between an irreversible control involving a critical period of sex determination gene action and continual control by the cascade, these modes of control could involve analogous mechanisms. In the former case, *dsx* may be targeting other regulatory genes that initiate sex-specific development of a particular tissue. In the case of continual control, the direct target of *dsx* would be the gene for the final differentiation product rather than a tissue-specifying gene.

Identification of terminal differentiation genes in the sex-determining pathway presents a major challenge for future studies. We can predict that at this level many developmental pathways must intersect to instruct the final fate of a cell. For example, the formation of the male sex comb depends on instructions coming from three distinct regulation cascades. Segmentation genes define the body segment in which the sex comb will appear, proximal–distal patterning genes define the precise location on the leg and, finally, the sex of these cells determines whether or not male bristles will be formed. The isolation and analysis of target genes may help to understand how the information of multiple hierarchies is integrated at the molecular level.

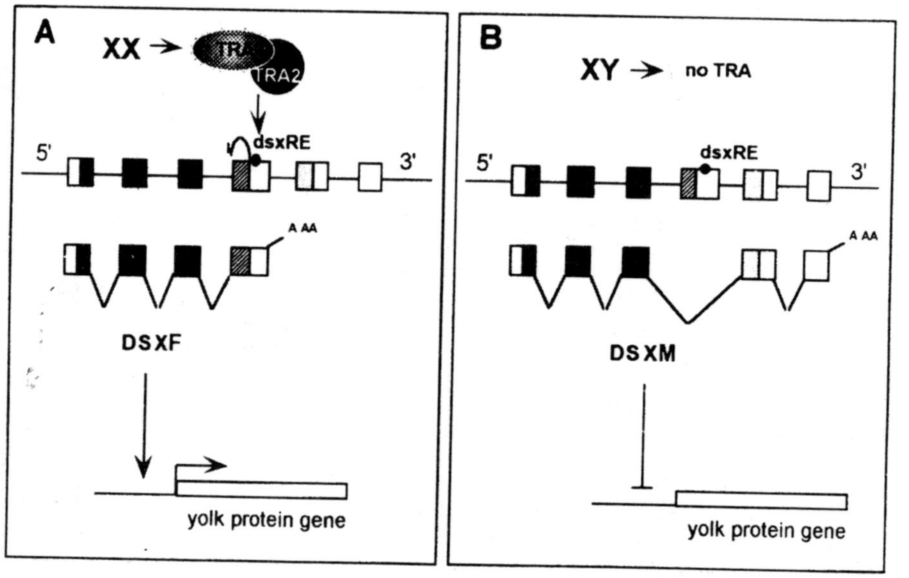

Figure 8 The gene *dsx* is sex-specifically processed. In XX flies (A), binding of Tra/Tra2 to dsxRE in the *dsx* pre-RNA activates an upstream splice acceptor site in the fourth exon. In XY flies (B), where Tra is absent, this site is not used; instead a conventional site further downstream (exon 5) is used. As a result, the carboxyl ends of Dsx_F (shaded box) and Dsx_M (spotted box) differ in sequence, but most of the upstream coding sequence (filled box) is shared between both variants. The two proteins exert opposite effects on transcription of target genes. For instance, Dsx_F will stimulate transcription of the yolk protein genes, whereas Dsx_M will repress the use of this promoter.

Acknowledgement

I wish to express my sincere gratitude to Dr Rolf Nöthiger for his continuous support, for critical reading and many helpful suggestions. Financial support from the Swiss National Science Foundation (grants 31-56795-99 and 31-47180-96).

REFERENCES

Amrein, H., Gorman, M. & Nöthiger, R. (1988) The sex-determining gene tra-2 of Drosophila encodes a putative RNA binding protein. *Cell* 55: 1025–35

Baker, B.S. (1989) Sex in flies: the splice of life. *Nature* 340: 521–24

Baker, B.S. & Belote, J.M. (1983) Sex determination and dosage compensation in *Drosophila*. *Annual Review of Genetics* 17: 345–79

Baker, B.S., Gorman, M. & Marin, I. (1994) Dosage compensation in *Drosophila*. *Annual Review of Genetics* 28: 491–521

Barbash, D.W. & Cline, T.W. (1995) Genetic and molecular analysis of the autosomal component of the primary sex determination signal of *Drosophila melanogaster*. *Genetics* 141: 1451–71

Bashaw, G.J. & Baker, B.S. (1995) The *msl-2* dosage compensation gene of *Drosophila* encodes a putative DNA-binding protein whose expression is sex-specifically regulated by *Sex-lethal*. *Development* 121: 3245–58

Bashaw, G.J. & Baker, B.S. (1997) The regulation of the *Drosophila msl-2* gene reveals a function for Sex-lethal in translational control. *Cell* 89: 789–98

Bell, L.R., Maine, E.M., Schedl, P. & Cline, T.W. (1988) *Sex-lethal*, a *Drosophila* sex determination switch gene, exhibits sex-specific RNA splicing and sequence similarity to RNA binding proteins. *Cell* 55: 1037–46

Belote, J.M., McKeown, M., Boggs, R.T., Ohkawa, R. & Sosnowski, B.A. (1989) Molecular genetics of *transformer*, a genetic switch controlling sexual differentiation in *Drosophila*. *Developmental Genetics* 10: 143–54

Boggs, R.T., Gregor, P., Idriss, S., Belote, J.M. & McKeown, M. (1987) Regulation of sexual differentiation in *D. melanogaster* via alternative splicing of RNA from the *transformer* gene. *Cell* 50: 739–47

Bownes, M. (1994) The regulation of the yolk protein genes, a family of sex differentiation genes in *Drosophila melanogaster*. *BioEssays* 16: 745–52

Bridges, C.B. (1922) The origin of variations in sexual and sex limited characters. *American Naturalist* 56: 51–63

Bridges, C.B. (1925) Sex in relation to chromosomes. *American Naturalist* 59: 127–37

Burtis, K.C. & Baker, B.S. (1989) Drosophila *doublesex* gene controls somatic sexual differentiation by producing alternatively spliced mRNAs encoding related sex-specific polypeptides. *Cell* 56: 997–1010

Case, B.A. & Baker, B.S. (1995) A genetic analysis of intersex, a gene regulating sexual differentiation in *Drosophila melanogaster* females. *Genetics* 139: 1649–61

Chen, P.S. (1984) The functional morphology and biochemistry of insect male accessory glands and their secretions. *Annual Review of Entomology* 29: 233–55

Cline, T.W. (1978) Two closely linked mutations in *Drosophila melanogaster* that are lethal to opposite sexes and interact with *daughterless*. *Genetics* 90: 683–98

Cline, T.W. (1979) A male-specific lethal mutation in *Drosophila melanogaster* that transforms sex. *Developmental Biology* 72: 266–75

Cline, T.W. (1984) Autoregulation functioning of a *Drosophila* gene product that establishes and maintains the sexually determined state. *Genetics* 107: 231–77

Cline, T.W. (1988) Evidence that *sisterless-a* and *sisterless-b* are two of several discrete "numerator elements" of the X/A sex determination signal in *Drosophila* that switch *Sxl* between two alternative stable expression states. *Genetics* 119: 829–62

Cline, T.W. (1993) The *Drosophila* sex determination signal: how do flies count to two? *Trends in Genetics* 9: 385–90

Cline, T.W. & Meyer, B.J. (1996) Vive la différence: males vs females in flies vs worms. *Annual Review of Genetics* 30: 637–702

Coschigano, T. & Wensink, P.C. (1993) Sex-specific transcriptional regulation by the male and female *doublesex* proteins of *Drosophila*. *Genes and Development* 7: 42–54

Erdman, S.E. & Burtis, K.C. (1993) The *Drosophila doublesex* proteins share novel zinc finger related DNA binding domain. *EMBO Journal* 12: 527–35

Estes, P.A., Keyes, L.N. & Schedl, P. (1995) Multiple response elements in the *Sex-lethal* early promoter ensure its female-specific expression pattern. *Molecular and Cellular Biology* 15: 904–17

Ferveur, J.F., Stortkuhl, K.F., Stocker, R.F. & Greenspan, R.J. (1995) Genetic feminization of brain structures and changed sexual orientation in male *Drosophila*. *Science* 267: 902–05

Flickinger, T.W. & Salz, H.K. (1994) The *Drosophila* sex determination gene *snf* encodes a nuclear protein with sequence and functional similarity to the mammalian U1A snRNP. *Genes and Development* 8: 914–25

Fujihara, T., Kawabe, M. & Oishi, K. (1978) A sex-transformation gene in *Drosophila melanogaster*. *Journal of Heredity* 69: 229–36

Gailey, D.A. & Hall, J.C. (1989) Behavior and cytogenetics of *fruitless* in *Drosophila melanogaster*: different courtship defects caused by separate, closely linked lesions. *Genetics* 121: 773–85

Gailey, D.A., Taylor, B.J. & Hall, J.C. (1991) Elements of the *fruitless* locus regulate development of the muscle of Lawrence, a male-specific structure in the abdomen of *Drosophila melanogaster* adults. *Development* 113: 879–90

Garcia-Bellido, A. (1983) Comparative anatomy of cuticular patterns in the genus Drosophila. In *Development and Evolution*, edited by B.C. Goodwin, N. Holder & C.C. Wylie, Cambridge: Cambridge University Press

Granadino, B., Campuzano, S. & Sanchez, L. (1990) The *Drosophila melanogaster* fl(2)d gene is needed for the female-specific splicing of *Sex-lethal* RNA. *EMBO Journal* 9: 2597–602

Greenspan, R.J. (1995) Understanding the genetic construction of behavior. *Scientific American* 272: 72–78

Hall, J. (1979) Control of male reproductive behavior by the central nervous system of *Drosophila*; dissection of a courtship pathway by genetic mosaics. *Genetics* 92: 437–57

Hall, J.C. (1985) Genetic analysis of behavior in insects. In *Comprehensive Insect Physiology, Biochemistry and Pharmacology*, edited by G.A. Kerkut & L.I. Gilbert, Oxford and New York: Pergamon Press

Hedley, M.L. & Maniatis, T. (1991) Sex-specific splicing and polyadenylation of *dsx* pre-mRNA requires a sequence that binds specifically to *tra-2* protein in vivo. *Cell* 65: 579–86

Heisenberg, M. (1980) Mutants of brain structure and function: what is the significance of the mushroom bodies for behavior? *Basic Life Sciences* 16: 373–90

Hilfiker, A., Amrein, H., Dübendorfer, A., Schneiter, R. & Nöthiger, R. (1995) The gene *virilizer* is required for female-specific splicing controlled by *Sxl*, the master gene for sexual development. *Development* 121: 4017–26

Horabin, J.I. & Schedl, P. (1993a) Regulated splicing of the *Drosophila Sex-lethal* male exon involves a blockage mechanism. *Molecular and Cellular Biology* 13: 1408–14

Horabin, J.I. & Schedl, P. (1993b) *Sex-lethal* autoregulation requires multiple cis-acting elements upstream and downstream of the male exon and appears to depend largely on controlling the use of the male exon 5′ splice site. *Molecular and Cellular Biology* 13: 7734–46

Hotta, Y. & Benzer, S. (1976) Courtship in *Drosophila* mosaics: sex-specific foci for sequential action patterns. *Proceedings of the National Academy of Sciences USA* 73: 4154–58

Inoue, K., Hoshijima, K., Sakamoto, H. & Shimura, Y. (1990) Binding of the *Drosophila Sex-lethal* gene product to the alternative splice site of *transformer* primary transcript. *Nature* 344: 461–63

Kelley, R.L., Solovyeva, I., Lyman, L.M. *et al.* (1995) Expression of *Msl-2* causes assembly of dosage compensation regulators on the X-chromosome and female lethality in *Drosophila. Cell* 81: 867–77

Kelley, R.L., Wang, J., Bell, L. & Kuroda, M.I. (1997) *Sex-lethal* controls dosage compensation in *Drosophila* by a non-splicing mechanism. *Nature* 387: 195–99

Kerkis, J. (1931) The growth of the gonads in *Drosophila melanogaster. Genetics* 16: 212–44

Keyes, L.N., Cline, T.W. & Schedl, P. (1992) The primary sex determination signal of *Drosophila* acts at the level of transcription. *Cell* 68: 933–43

Kubli, E. (1992) The sex-peptide. *BioEssays* 14: 779–84

Lawrence, P.A. & Johnston, P. (1986) The muscle pattern of a segment of *Drosophila* may be determined by neurons and not by contributing myoblasts. *Cell* 45: 501–13

Lucchesi, J.C. (1996) Dosage compensation in *Drosophila* and the "complex" world of transcriptional regulation. *BioEssays* 18: 541–47

Lynch, K. & Maniatis, T. (1995) Synergistic interactions between two distinct elements of a regulated splicing enhancer. *Genes and Development* 9: 284–93

Nagoshi, R.N. & Baker, B.S. (1990) Regulation of sex-specific RNA splicing at the *Drosophila doublesex* gene: cis-acting mutations in exon sequences alter sex-specific RNA splicing patterns. *Genes and Development* 4: 89–97

O'Dell, K.M.C. & Kaiser, K. (1997) Sexual behaviour: secrets and flies. *Current Biology* 7: 345–47

Parkhurst, S.M. & Meneeley, P.M. (1994) Sex determination and dosage compensation: lessons from flies and worms. *Science* 264: 924–32

Parkhurst, S.M., Bopp, D. & Ish-Horowicz, D. (1990) X:A ratio, the primary sex-determining signal in *Drosophila*, is transduced by helix–loop–helix proteins. *Cell* 63: 1179–91

Paroush, Z., Finley, R.L. Jr, Kidd, T. *et al.* (1994) *Groucho* is required for *Drosophila* neurogenesis, segmentation, and sex determination and interacts directly with *Hairy*-related bHLH proteins. *Cell* 79: 805–15

Pultz, M.A. & Baker, B.S. (1994) The dual role of *hermaphrodite* in the *Drosophila* sex determination regulatory hierarchy. *Development* 121: 99–111

Ryner, L.C. & Baker, B.S. (1991) Regulation of *doublesex* pre-mRNA processing occurs by 3′ splice site activation. *Genes and Development* 5: 2071–85

Ryner, L.C., Goodwin, S.F., Castrillon, D.H. *et al.* (1996) Control of male sexual behavior and sexual orientation in *Drosophila* by the *fruitless* gene. *Cell* 87: 1079–89

Sakashita, E. & Sakamoto, H. (1994) Characterization of RNA binding specificity of the *Drosophila* Sex-lethal protein by *in vitro* ligand selection. *Nucleic Acids Research* 22: 4082–86

Samuels, M.E., Bopp, D., Colvin, R.A. *et al.* (1994) RNA binding by *Sxl* proteins in vitro and vivo. *Molecular and Cellular Biology* 14: 4975–90

Sanchez, L. & Nöthiger, R. (1982) Clonal analysis of *Sex-lethal*, a gene needed for female sexual development in *Drosophila melanogaster. Roux's Archives of Developmental Biology* 191: 211–14

Sanchez, L. & Nöthiger, R. (1983) Sex determination and dosage compensation in *Drosophila melanogaster*: production of male clones in XX females. *EMBO Journal* 2: 485–91

Sosnowski, B.A., Belote, J.M. & McKeown, M. (1989) Sex-specific alternative splicing of RNA from the *transformer* gene results from sequence-dependent splice site blockage. *Cell* 58: 449–59

Steinmann-Zwicky, M., Amrein, H. & Nöthiger, R. (1990) Genetic control of sex determination in *Drosophila. Advances in Genetics* 27: 189–237

Tian, M. & Maniatis, T. (1993) A splicing enhancer complex controls alternative splicing of *doublesex* pre-mRNA. *Cell* 74: 105–14

Turner, B.M., Birley, A.J. & Lavender, J. (1992) Histone H4 isoforms acetylated at specific lysine residues define individual chromosomes and chromatin domains in *Drosophila* polytene nuclei. *Cell* 69: 375–84

Valcarcel, J., Singh, R., Zamore, P.D. & Green M.R. (1993) The protein *Sex-lethal* antagonizes the splicing factor U2AF to regulate alternative splicing of *transformer* pre-mRNA. *Nature* 362: 171–73

Younger-Shepherd, S., Vaessin, H., Bier, R., Jan, L.Y. & Jan, Y.N. (1992) *deadpan*, an essential pan-neural gene encoding an HLH protein, acts as a denominator in *Drosophila* sex determination. *Cell* 70: 911–22

Zhou, S., Yang, Y., Scott, M.J. *et al.* (1995) *Male-specific lethal-2*, a dosage compensation gene of *Drosophila* undergoes sex-specific regulation and encodes a protein with a RING finger and a metallothionein-like cystein cluster. *EMBO Journal* 14: 2884–95

GLOSSARY

blastoderm layer of cells forming the blastula wall in the early stages of embryonic development of Drosophila

differential splicing utilization of alternative splice sites, which produces mRNAs with differences in structure

haemolymph circulating body fluid of some invertebrates, e.g. insects, that have sinuses and spaces between organs rather than a closed circulatory system; regarded as equivalent to blood and lymph of more complex organisms

hypertranscription augmented rate (twofold) of transcription of the single X chromosome in Drosophila males to compensate for the presence of two X chromosomes in females

isoform form of a protein with slightly different amino acid sequences, which may differ in activity, function, distribution, etc

ontogenesis chronicle of development and growth of an individual

See also **X-chromosome inactivation** (p.780)

GENETIC CONTROL OF SEX DETERMINATION IN THE HOUSEFLY

Andreas Dübendorfer
Institute of Zoology, University of Zürich, Zürich, Switzerland

1. Why housefly genetics?

The problem of insecticide resistance and the need to design strategies for the suppression of field populations first promoted genetic analyses of the housefly, *Musca domestica* L. In this context, the mechanism of sex determination was obviously of foremost interest, and researchers, much to their surprise, found a variety of different genetic mechanisms in geographically separate populations (for reviews see Milani, 1967, 1975; Franco *et al.*, 1982; Rubini *et al.*, 1989). These mechanisms may all have evolved from one standard system by mutations in the genes that control sex determination. Experimenting with these mutations, the geneticist can now reproduce in the laboratory the changes from one type of sex determination to another type. This not only helps us to understand the genetic control mechanisms in Musca (Dübendorfer *et al.*, 1992), but may reflect, at least in part, the evolution of the diversity of sex-determining mechanisms in dipteran insects (Nöthiger & Steinmann-Zwicky, 1985).

Before looking at the sex-determining signals in detail, we should note that the two main cell populations of an early housefly embryo, namely the somatic cells and those of the germline, follow different principles of sex determination. Whereas somatic cells use the genetic signals contributed by the egg and the sperm, the progenitor cells of the germline, called pole cells because they first form at the posterior pole of the embryo, do not use these signals. When transplanted from a male embryo to a female embryo (**Figure 1**), such cells integrate into the female gonads and produce eggs despite their male genotype. Likewise, genetically female pole cells can integrate into male gonads and non-autonomously differentiate into functional sperm (Hilfiker-Kleiner *et al.*, 1994). This means that pole cells are sexually undetermined and that, in contrast to the situation in *Drosophila melanogaster* (Steinmann-Zwicky *et al.*, 1989), their sexual development depends entirely on the surrounding gonadal soma. With this simple mechanism, the organism avoids mismatch problems between the soma and germline. It should be noted that this non-autonomous sex determination of the germline has great experimental potential, permitting the production of eggs and sperm from transplanted pole cells of any genotype.

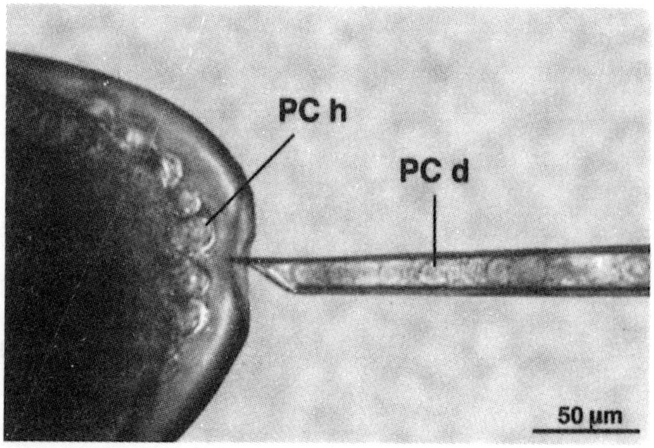

Figure 1 Posterior end of a 3½ hour-old embryo with clearly visible pole cells, the progenitors of the germline. When pole cells from an embryo of the opposite sex are implanted, they become integrated into the gonads of the host and non-autonomously form gametes of the host's sex. PC h, pole cells of the host embryo; PC d, pole cells from a donor embryo.

2. The sex-determining mechanisms of Musca

Somatic sex determination in *M. domestica* is under strict genetic control. A key gene of the zygotic nucleus determines the sexual development of the animal (**Figure 2**). This key gene is regulated by a primary genetic signal that is also present in the zygote nucleus, along with a maternal signal that originates from the maternal germline. These signals can be analysed by comparing naturally occurring variants of the system.

The chromosome complement (**Figure 3**) comprises five pairs of autosomes (I–V) and one pair of heterosomes (X and Y); in the so-called standard strains, females are X/X and males X/Y (Perje, 1948). The Y chromosome carries a male-determining factor (M^Y) which constitutes the primary

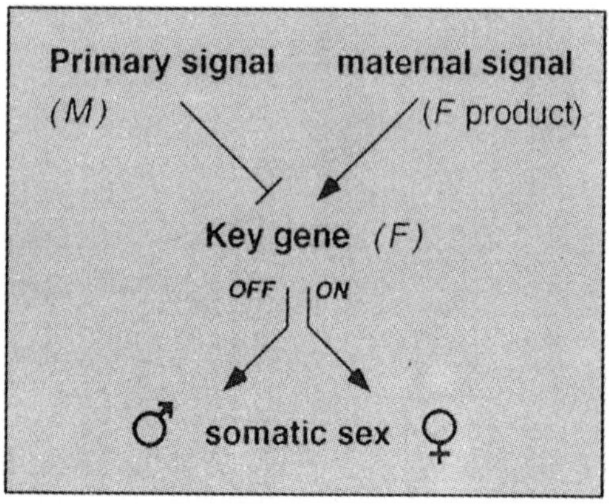

Figure 2 In a fertilized egg of the housefly, the activity of a key gene decides on the somatic sex of the developing embryo. The key gene is regulated by a genetic primary signal and a signal from the maternal germline.

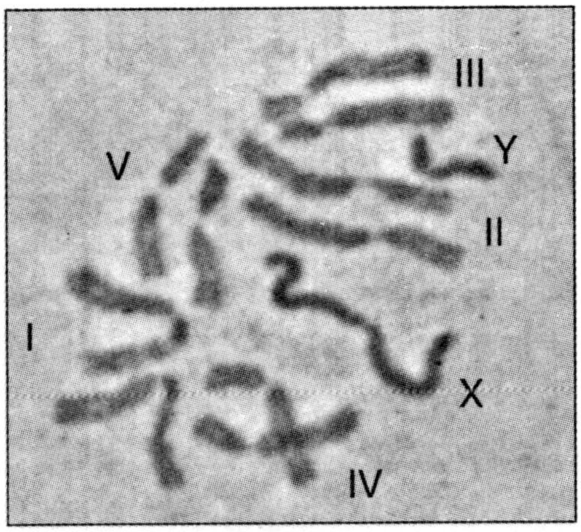

Figure 3 Chromosomal complement ($2n = 12$) of a standard wild-type male of *Musca domestica*. Homologous autosomes (I–V) are loosely paired, but the heterosomes (X, Y) do not pair. These orcein-stained chromosomes are from a spermatogonial cell. (Photograph courtesy of M. Hediger.)

signal for somatic sex determination; the number of X chromosomes is irrelevant (Milani *et al.*, 1967). In other populations, the *M* factor is located on one of the autosomes, which is evident from the genetic linkage of male sex with markers of that linkage group (Hiroyoshi, 1964; Rubini & Palenzona, 1967; Wagoner, 1969a). All autosomes, and even the X chromosome, can be carriers of *M*, which suggests that *M* is a mobile element. Strains with autosomal *M* may lose the Y chromosome altogether, such that all animals are X/X. In any case, irrespective of whether *M* is located on the Y or on an autosome, its presence in a fertilized egg leads to male development and its absence to female development (**Table 1**).

In a third type of strain, males and females are both homozygous for an autosomal *M* factor. This can easily be demonstrated by crossing these males to females of a standard strain; all offspring will be of male sex (heterozygous for *M*). Yet, how can a strain with *M/M* in both sexes produce females? This is achieved by a dominant allele of a gene on chromosome IV, F^D ($F^{Dominant}$), that is epistatic to *M* and determines femaleness, in the presence of one or several *M* factors (Rubini *et al.*, 1972; McDonald *et al.*, 1978; Malacrida *et al.*, 1982; Dübendorfer *et al.*, 1992). Hence, in these strains, the females represent the heterogametic sex, passing on F^D to 50% of their offspring which then develop as females. This finding prompted Nöthiger and Steinmann-Zwicky (1985) to conclude that F^D was a constitutive allele of a sex-determining gene *F*, necessary for female development and, as a wild-type allele (F^+), negatively controlled by *M* factors. *M* can thus be seen as the primary sex-determining signal which controls the activity of the key gene *F*; an active *F* promotes female, and an inactive *F* male, development.

3. The maternal signal

Laboratory strains have revealed that sex is not only determined by the interplay of zygotic *M* and *F*, but that female development also requires a maternal product. The first evidence for this maternal activity came from a dominant mutation, *Arrhenogenic* (*Ag*), which has no morphological effect on female carriers, but which causes *Ag* mothers to produce male progeny with female genotype (Vanossi Este, 1971; Rubini *et al.*, 1972; Vanossi Este & Rovati, 1982). *Ag* is considered to be allelic to M^I, as both map to the same locus on autosome I (Vanossi Este *et al.*, 1974; Rovati *et al.*, 1983), but it differs from M^I in that it is not a zygotic masculinizer, but rather blocks a maternal activity needed for female development of the offspring.

Two recessive mutations, *transformer* (*tra*; Inoue & Hiroyoshi, 1981, 1986) and *masculinizer* (*man*; Schmidt *et al.*, 1997a), both transform XX zygotes into fertile males when homozygous. Heterozygosity in females causes a male-determining maternal effect similar to that of *Ag*. The mutations *tra* and *man* map to the same position on autosome IV, and there is evidence that they are both allelic to F^D. Here, they will, therefore, be called F^{tra} and F^{man}, respectively.

How *Ag*, F^{tra} and F^{man} work is shown by experiments that mimic their effects (**Figure 4**). When the pole cells of a male embryo (*M*/+) are transplanted to a female embryo, the adult fly can produce eggs from these genetically male cells. Half of the resulting eggs contain the *M* factor and accordingly develop as males, in a similar way to eggs that are fertilized by *M* sperm. Surprisingly, however, even those zygotes that do not receive any *M* factor, either from the maternal germline or from the fertilizing sperm, invariably develop as

Table 1 Five types of genetic sex determination in Musca.

Type	Female genotype a	b	c	Male genotype a	b	c
Standard strains: Dominant male determiner on the Y chromosome (M^Y)	X/X;	+/+;	+/+	X/Y;	+/+;	+/+
Male autosomal heterogamety (autosomal *M*) ([d])	X/X;	+/+;	+/+	X/X;	M/+;	+/+
Female autosomal heterogamety (F^D)	X/X;	M/M;	F^D/+	X/X;	M/M;	+/+
Recessive autosomal male determiner (F^{man})	X/X;	+/+;	F^{man}/+	X/X;	+/+;	F^{man}/F^{man}
Male-determining maternal effect (*Ag*) ([e])	X/X;	Ag/+; ([f]) or +/+; ([g])	+/+	X/X;	Ag/+; ([h]) or +/+; ([i])	+/+

[a] Heterosomes; [b] autosome pair I, II, III or V; [c] autosome pair IV; [d] the table does not include those cases where *M* is on autosome IV or on the X; [e] *Ag* is homozygous lethal; [f] these females are arrhenogenic (produce sons); [g] these females are thelygenic (produce daughters); [h] 50% of the daughters of these males are arrhenogenic; [i] all daughters of these males are thelygenic.

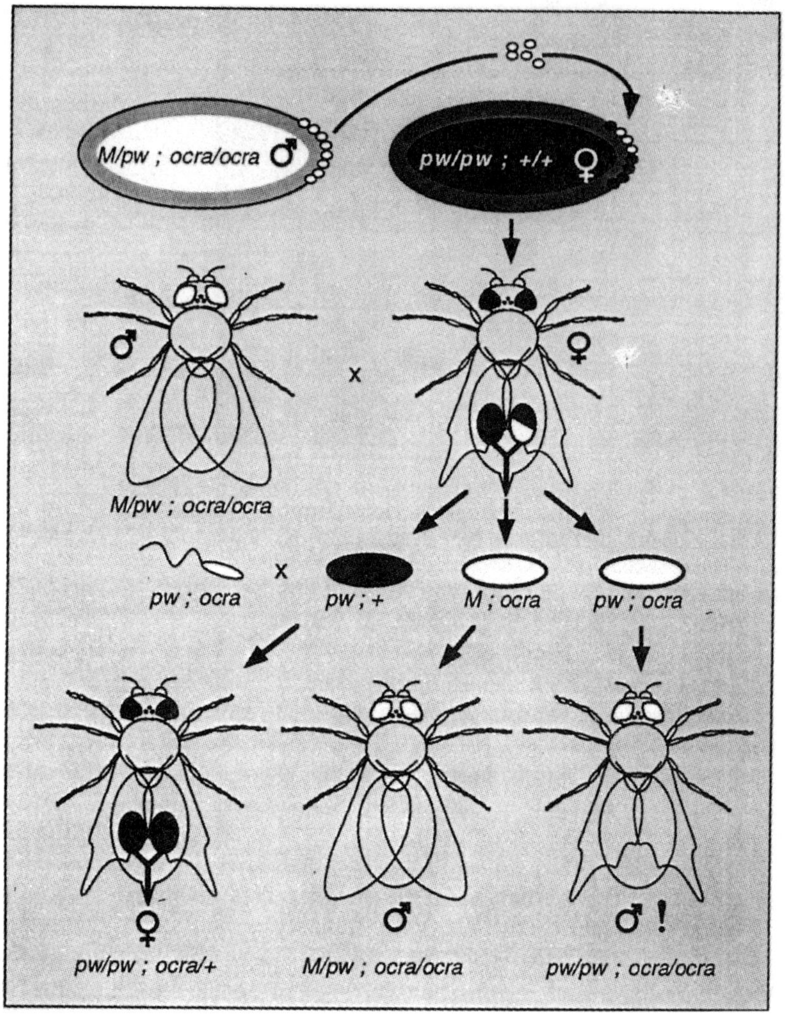

Figure 4 Transplantation of genetically male pole cells (white) into an embryo of female genotype (black). The male donor is marked with the mutation *ocra* (whitish ochre eyes) and has *M* on chromosome III. The homologous chromosome III is marked with *pw* (*pointed wings*), such that flies without *M* can be recognized by their pointed wing shape. The experiment shows, first, that genetically male pole cells can form functional eggs and, secondly, that genetically female zygotes (no *M* present) develop as fertile males if they originate from a maternal germline heterozygous for *M* (bottom right).

males despite their female genotype. This experiment was reproduced with a variety of different *M* factors (M^Y, M^I, M^{II}, M^{III}, M^V; Hilfiker-Kleiner *et al.*, 1994; Schmidt *et al.*, 1997b), always with the same result; *M* in the female germline exerts a maternal effect that prevents future female differentiation and forces the zygotes into male development. The same result is obtained when the female germline is made mutant for *Ag* or F^{man} (F^{tra} could not be tested since this mutant is no longer available). These male-determining maternal effects are all suppressed if F^D is also present in the germline (Rovati & Vanossi-Este, 1978; Hilfiker-Kleiner *et al.*, 1994; Schmidt *et al.*, 1997a) or if it is provided to the zygote by the sperm (Dübendorfer & Hediger, 1998). This shows, first, that the known male-determining maternal effects all come about by the inactivation of *F* in the maternal germline, and second, that the activity of zygotic *F*, however achieved, is the sole requirement for female development. From this, the conclusion can be drawn that, in the standard wild-type, the germline function of *F* is required to activate *F* in the zygote. Although the product of *F* is not yet biochemically characterized, these discoveries identify it as the maternal signal necessary in standard-type zygotes that are to develop as females.

The described experiments also reveal an interesting difference between the soma and the germline; the somatic activity of *F* is responsible for female differentiation of the fly, including ovarian development and the production of the somatic signal that dictates the female development of the germ cells. *M* in the female germline is active and prevents *F* activity just as it does in the soma, but oogenesis nevertheless proceeds normally. This indicates that the somatic signal for determination of the female sex in the germline does not affect *F* activity in the germline, but rather that it affects a gene, or genes, downstream of *F*.

4. The current model for sex determination

The standard mechanism for sex determination in the housefly can now be visualized by the following model (**Figure 5**). When an egg matures in the ovary, the *F* gene is active in the cells of the germline (the nurse cells and perhaps also the oocyte) and deposits its product in the egg. If the fertilizing sperm does not carry *M*, the maternal *F* product activates the zygotic *F* gene which results in the continuous expression of *F* and, thus, in female development (Hilfiker-Kleiner *et al.*, 1993). If the sperm contributes *M*, this factor prevents the activation of *F* in the zygote, which results in male development (default pathway).

This model also explains all non-standard cases of sex

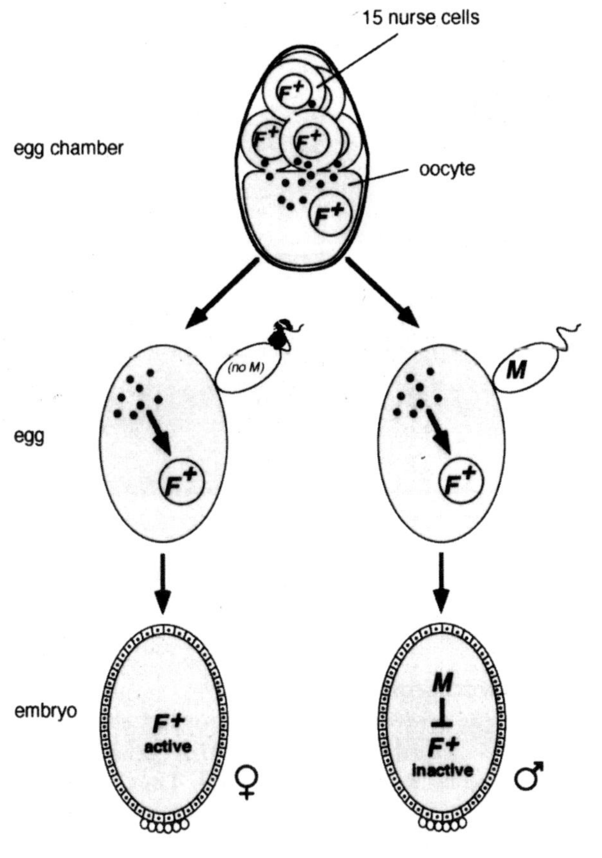

Figure 5 Current model of sex determination in the soma of *Musca domestica*. Cells of the maternal germline express the gene *F* whose product (black dots) is transferred to the egg and later is required to activate zygotic *F*. If no *M* is introduced by the sperm, *F* in the zygote becomes active and determines female sex. Presence of *M* prevents *F* activity, and male development ensues.

determination in Musca. The key gene for female development, *F*, is constitutively expressed from the allele F^D, and hence zygotes that carry F^D always develop as females, even in the absence of maternal *F* (Dübendorfer & Hediger, 1998) or in the presence of several *M* factors (Rubini *et al.*, 1972). The known male-determining maternal effects result from the inactivity of *F* in the female germline, either caused by repressing factors (*Ag*, *M*) or by loss-of-function mutations of *F* (F^{tra}, F^{man}).

5. Sex ratios

Animals with the described non-standard modes of genetic sex determination were originally isolated from wild populations and then kept as true breeding stocks. Such stocks do not reflect the situation in natural populations; these are normally mixed, which affects the sex ratio of the offspring. **Table 2** presents a choice of crosses that produce offspring with different sex ratios, varying from all female to all male. Intermediate sex ratios may be observed in wild populations depending on the allele frequencies of the sex-determining genes involved (Denholm *et al.*, 1986; Rubini *et al.*, 1989).

Table 2 Sex ratios in the progenies of sample crosses, arranged by falling proportion of females.

Female parent	Male parent	Sex ratio of progeny females:males
X/X; +/+	X/X; F^{man}/F^{man}	1:0
X/X; $M^{III}/+$; $F^D/+$	X/X; +/+; F^{man}/F^{man}	3:1
X/X; $M^{III}/+$; $F^D/+$	X/X; $M^{III}/+$; +/+	5:3
X/X; $M^{III}/+$; $F^D/+$; +/+	X/X; $M^{III}/+$; +/+; $M^V/+$	9:7
X/X	X/Y	1: 1
X/X; +/+	X/Y; $M^{III}/+$	1:3
X/X; $Ag/+$	X/Y; +/+	>0:<1 [a]
X/X; +/+	X/X; M^{III}/M^{III}	0:1

[a] Depending on the genetic background, the male-determining maternal effect of *Ag* can vary and allow the development of some intersexes and even rare females.

6. Intersexuality and gynandromorphism

All reports on intersexuality in Musca (reviews by Milani, 1967, 1975) describe these animals as mosaics composed of typically female and male structures (**Figure 6**). Animals with intermediate cuticular patterns, recognizable neither as typically female nor as clearly male, have not yet been observed. Such intermediate differentiation is known from *D. melanogaster*, caused by mutations in *doublesex* (*dsx*), the last member of the cascade of sex-determining genes (Hildreth, 1965; Nöthiger *et al.*, 1987). Despite recent evidence that the *dsx* gene may also be functional in *M. domestica* (Bopp *et al.*, 1998), such mutations have not yet been found.

Sexual mosaicism can be recognized in the cuticular patterns of the head and the postabdomen as well as in the internal genitalia (gonads and their ducts) and the organs producing yolk proteins. Such mosaicism can be purely phenotypic when it occurs in genetically homogeneous animals with an ambiguous genetic signal for sex determination. These mosaics are called intersexes. Gynandromorphs, on the other hand, are genetic mosaics, consisting of genetically male and female cells that differentiate autonomously according to their genetic sex. Since the causes for the occurrence of these two types of sexual mosaicism are so distinctly different, they will be dealt with separately.

(i) Intersexes – origin and significance

Intersexes arise as a consequence of an ambiguous sex-determining signal. Some cells interpret this signal as male, others as female, which then results in sexual mosaicism. A strong masculinizing signal prevents *F* activity more rigorously, and results in more male cells, whereas a weak signal allows more *F* activity and thus more cells to differentiate in the female mode. Wild-type Y chromosomes and the known autosomal *M* factors of chromosomes II, III and V are strong male determiners with a rigorous masculinizing effect in the zygote as well as in the maternal germline, when introduced there by pole cell transplantation (cf. **Figure 4**) (Schmidt *et al.*, 1997b). Defective masculinizers, however, such as M^I or truncated Y chromosomes, are weaker and produce a varying proportion of intersexual flies.

Let us look at some examples of masculinizing factors of different strength, as recently studied by Schmidt *et al.*

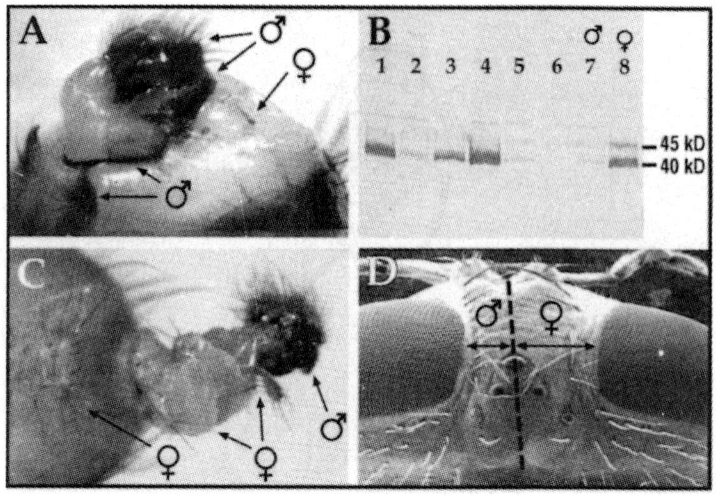

Figure 6 Intersexuality in flies with truncated Y chromosomes. (A) Fertile male carrier of the ring-Y chromosome (R[YS]1) which consists of the short arm and the centromere of the Y. The postabdomen is perfectly male, but there are also some small elements typical for the female ovipositor. (B) A Western blot shows yolk proteins in the haemolymph of ring-Y intersexes (lanes 1–6) which otherwise were fertile males. Lanes 7 and 8 are control samples of a standard XY male and XX female. (C) Weak masculinizing effect of the long arm of the Y chromosome of the translocation stock *T(Y; II)2,ar*$^+$. The postabdomen forms an ovipositor with male genital elements. (D) Head of a ring-Y intersex with a wide frons and eye shape typical for the female on the right, and a typically male narrow frons and male eye shape on the left. (Figure modified from Hediger *et al.*, 1998a.)

(1997b) and Hediger *et al.* (1998a). The isolated long arm of the Y chromosome, tested in a translocation that involves the Y and autosome II, is a particularly weak masculinizer. Animals with this fragment of the Y develop as intersexes of mostly female appearance, often with yolk proteins in the haemolymph (71%), but some 20% show patches of male cuticular structures in the head or the postabdomen (**Figure 6C**). The short arm of the Y, present in a ring Y chromosome, is stronger; such flies develop as males with minute female cuticular patches (**Figure 6A,D**), although 43% of them produce yolk proteins (**Figure 6B**), a feature that is strictly female specific.

Another *M* factor, M^I of autosome I, causes its carriers to develop as externally normal, fertile males, but more than half of these produce yolk proteins. *Ag*, also on chromosome I, has no masculinizing effect in the soma, but is strong in the female germline. The maternal effects of both, M^I and *Ag*, are incomplete, allowing for some intersexual flies in the progeny, and they are probably the consequence of maternal *F* activity being reduced to near a threshold level. The cells of the developing embryo might then not all react in the same way to this low level of *F* product, some activating their *F* gene and developing into female structures, and others leaving *F* inactive and differentiating male elements. The same effect is seen in females that, due to heterozygosity for the loss-of-function mutation F^{man}, have only one functional *F* allele. This allows normal female development in the soma and germline, but it is not sufficient for an unambiguous maternal signal; about half of the progeny of such females develop as intersexes (Schmidt *et al.*, 1997a).

Cytological studies have provided a possible explanation of why *M* factors on autosome I are weaker than those on other autosomes. Chromosome I is the only autosome with prominent stretches of heterochromatin (Hediger *et al.*, 1998b), and this may exert a position effect on *M*, reducing its activity. Should this be the case, how then can *M* be active on the Y chromosome, of which the entire long arm and parts of the short arm are heterochromatic? Different fragments of the Y chromosome strongly suggest that the normal Y harbours more than one *M* factor, a stronger one on the more euchromatic short arm and a weaker one on the heterochromatic long arm. Together, these *M* factors guarantee strong masculinizing activity.

(ii) Gynandromorphs

The first clear evidence of sexual mosaics that were not intersexes as described above, but true genetic mosaics, was presented by Rubini *et al.* (1980). They worked with the so-called "Orlando" strain in which the males are M^{III} bwb^+/bwb and the females *bwb*/*bwb* (*brown body*). Since crossing-over in males is very rare, this stock is quite stable, constantly producing *brown body* females and phenotypically wild-type males. All sexual mosaics that occur in this strain are also mosaic for the body colour marker, with brown female parts and wild-type male parts. Based on an earlier report on mitotic recombination in Musca (Nöthiger & Dübendorfer, 1971), Rubini *et al.* (1980) interpreted their gynandromorphs as a result of mitotic recombination in the early embryonic nuclear divisions. Such an event, in a nucleus of a male embryo, would generate daughter nuclei of the genotypes *M*/*M* and *bwb*/*bwb* and would thus lead to the observed gynandromorphic phenotype with congruent mosaicism of cuticle colour and sex.

The hypothesis of Rubini *et al.* (1980) is compatible with the observed gynandromorphs, but it does not explain other cases where, in the same animal, the boundaries between male and female regions are not only coincident with markers of the chromosome carrying *M*, but also with those of other linkage groups (Milani, 1967, 1975). In the author's laboratory, 27 gynandromorphic flies have been collected where several linkage groups were involved (S. Roth, M. Hediger & A. Dübendorfer, unpublished data). These mosaics show that double fertilization is a major, though perhaps not the only, cause of gynandromorphic development (for an example see **Figure 7**). This conclusion receives support from LaChance and Leopold (1969) who recorded frequent polyspermy in *M. domestica*; two or more sperm heads were observed in over 50% of the fertilized eggs. The genetic indication of double fertilization suggests that fertilization in Musca is gonomeric, as in Drosophila, i.e. the

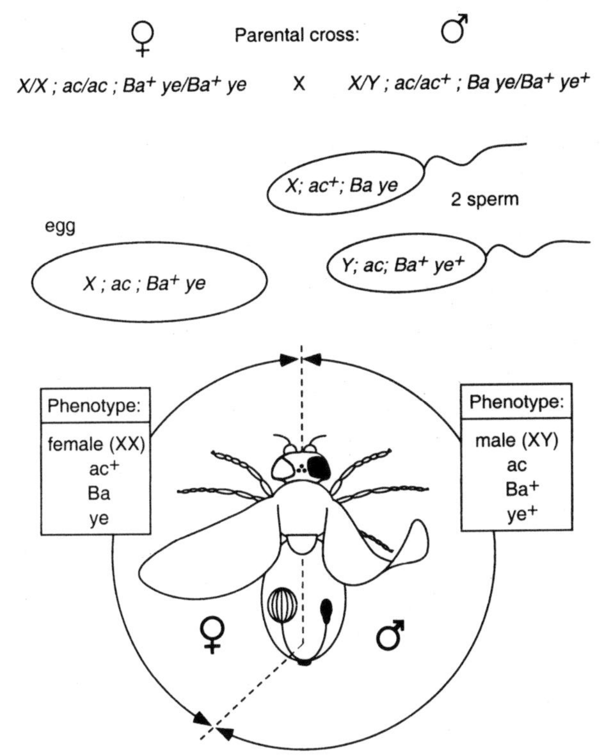

Figure 7 Schematic representation of a gynandromorph resulting from double fertilization, as indicated by the coincidence of the genetic markers of three chromosomes (X/Y; I; IV). Right: male tissue (head with narrow frons, male eye shape, testis, male external genitalia), wild-type eye (ye^+), curved wing (ac) and wild-type bristle pattern (Ba^+). Left: female tissue (head with wide frons, female eye shape, ovary), yellow eye (ye), wild-type wing (ac^+), and reduced bristles on head, thorax, and abdomen (Ba). The distribution of the markers *ye* (*yellow*) and *Ba* (*Bald abdomen*) shows that the mosaic cannot have arisen by mitotic recombination; the paternal *Ba ye* chromosome is required for the Ba phenotype, and the Ba^+ ye^+ chromosome for the ye^+ phenotype.

female and the male pronuclei both replicate prior to synkaryon formation, and each daughter nucleus of the oocyte joins one daughter nucleus of the sperm, simultaneously forming two diploid zygote nuclei (King, 1970). If one egg is fertilized by two sperm, the two daughter nuclei of the oocyte may fuse with nuclei originating from different sperm which results in genetic mosaicism of the embryo.

7. Dosage compensation

Aneuploidy for autosomes is extremely rare in Musca (Rubini *et al.*, 1972), probably because it is lethal. In contrast, aneuploidy for heterosomes presents no problem. Up to six heterosomes were found in the chromosome complement of individual flies, with any combination of 0–5 X chromosomes and 0–4 Y chromosomes, without effect on viability, sexual development or fertility. Only one heterosome, either an X or a Y chromosome, is required in the complement for survival and fertility (Boyes, 1967; Milani, 1967; Rubini & Palenzona, 1967; Denholm *et al.*, 1983). Apart from the *M* factor, X and Y chromosomes have no discriminatory function in sex determination; *YY*; F^D/F^+ animals are normal females (Rubini *et al.*, 1972) and *XX*; +/+ flies can be perfect males if maleness is determined by an autosomal *M*, or even without *M*, by maternal effect (see Section 3, The maternal signal).

Both X and Y chromosomes are mostly heterochromatic, as seen from their condensed appearance in Figure 2, and, apart from the *M* factor on the Y, none of the more than 180 described visible mutations has been located to one of them (Milani, 1967; Wagoner, 1969b; Hiroyoshi, 1977). The requirement for either an X or a Y chromosome in the set, however, suggests that there are vital genes on the heterosomes and that, with respect to these genes, X and Y are homologous and equivalent. The tolerance of any number and combination of X and Y chromosomes could mean that the housefly utilizes a perfect dosage compensation mechanism that is uncoupled from sex determination. It could, however, also mean that the low content of genetically active chromatin in both heterosomes renders dosage differences insignificant and, therefore, tolerable to the animal.

8. Conclusions and outlook

The variety of naturally occurring mechanisms of sex determination in the housefly, and the ease of generating new systems in the laboratory, give us the impression of evolutionary processes in action. Several variations of a ground state are tried out and, if sex ratios get distorted (e.g. by accumulation of *M* factors in the genome), the system adapts with new mutations (F^D). Such variants are useful for the researcher to understand the ground state, and they have revealed that the key gene for female development is regulated by its own maternal product (Dübendorfer & Hediger, 1998). Only similar studies on other insects can show whether or not this is a general and phylogenetically ancient mechanism.

In contrast to the housefly, *D. melanogaster* uses an extremely specialized sex-determining mechanism. By genetically coupling three processes – sex determination in the soma, sex determination in the germline and dosage compensation – and placing them all under the control of the same gene, *Sex-lethal* (*Sxl*), the system became rigid, so that most mutations that occur in the sex-determining genes result in lethality or sterility because of mismatch of the three pathways. Drosophila, having arrived at this level of specialization, became dead-locked with respect to sex determination and thus has probably lost much of its plasticity for further evolutionary changes in the sex-determining system.

Still, it is possible that, despite the obvious differences in the sex-determining mechanisms, the main genes involved in this process are conserved and used in different insect species or even in different animal phyla. The first of these genes isolated in Musca was *Sxl*, the key gene for sex determination in Drosophila. *Sxl* is well conserved in Musca (Meise *et al.*, 1998), but it is apparently not involved in sex determination. The same is true for the *Sxl* gene of the blowfly *Chrysomya rufifacies*, the phorid fly *Megaselia scalaris* and the Mediterranean fruit fly *Ceratitis capitata* (Müller-Holtkamp, 1995; Sievert *et al.*, 1997; Saccone *et al.*, 1998), which suggests that its use for sex determination is part of the specialization of *D. melanogaster*. However, the bottom-most gene of the sex-determining cascade, *doublesex* (*dsx*), is not only conserved in its sequence, but also in its sex-determining function, as predicted by Wilkins (1995). First molecularily identified and characterized in *D. melanogaster* (Nagoshi & Baker, 1990), *dsx* has now also been found in the dipterans *Megaselia scalaris* (Sievert *et al.*, 1997) and *Bactrocera tryoni* (Shearman & Frommer, 1998), in the nematode *Caenorhabditis elegans* and in man (Raymond *et al.*, 1998), and it is also present and differentially spliced in *M. domestica* (Bopp *et al.*, 1998). All evidence suggests that *dsx* is used in the sex-determination cascade of these very distantly related species whose mechanisms of sex determination apparently have not diverged by variations of this switch gene, but rather by mutating the genes that regulate it.

Acknowledgements

I wish to thank my colleagues Drs Rolf Nöthiger, Monika Hediger and Mary Bownes for helpful comments, and the Swiss National Science Foundation for financial support (grant 31-56795.99).

REFERENCES

Bopp, D., Burghardt, G., Hilfiker-Kleiner, D., Dübendorfer, A. & Nöthiger, R. (1998) Evolution of sex-determining mechanisms in dipteran insects. VI International Congress of Entomology, Budejovice (Abstract)

Boyes, J.W. (1967) The cytology of muscoid flies. In *Genetics of Insect Vectors of Disease*, edited by J.W. Wright & R. Pal, Amsterdam and New York: Elsevier

Denholm, I., Franco, M.G., Rubini, P.G. & Vecchi, M. (1983) Identification of a male determinant on the X chromosome of housefly (*Musca domestica* L.) populations in south-east England. *Genetical Research* 42: 311–22

Denholm, I., Franco, M.G., Rubini, P.G. & Vecchi, M. (1986) Geographical variation in house-fly (*Musca domestica* L.) sex determinants within the British Isles. *Genetical Research* 47: 19–27

Dübendorfer, A. & Hediger, M. (1998) The female-determining gene *F* of the housefly, *Musca domestica*, acts maternally to regulate its own zygotic activity. *Genetics* 150: 221–26

Dübendorfer, A., Hilfiker-Kleiner, D. & Nöthiger, R. (1992) Sex determination mechanisms in dipteran insects: the case of *Musca domestica*. *Seminars in Developmental Biology* 3: 349–56

Franco, M.G., Rubini, P.G. & Vecchi, M. (1982) Sex-determinants and their distribution in various populations of *Musca domestica* L. of western Europe. *Genetical Research* 40: 279–93

Hediger, M., Minet, A.D., Niessen, M. *et al.* (1998a) The male-determining activity on the Y chromosome of the housefly (*Musca domestica* L.) consists of separable elements. *Genetics* 150: 651–61

Hediger, M., Niessen, M., Müller-Navia, J., Nöthiger, R. & Dübendorfer, A. (1998b) Distribution of different types of heterochromatin on the mitotic chromosomes of the housefly *Musca domestica* L. *Chromosoma* 107: 267–71

Hildreth, P.E. (1965) *Doublesex*, a recessive gene that transforms both males and females of *Drosophila* into intersexes. *Genetics* 51: 659–78

Hilfiker-Kleiner, D., Dübendorfer, A., Hilfiker, A. & Nöthiger, R. (1993) Developmental analysis of two sex-determining genes, *M* and *F*, in the housefly, *Musca domestica*. *Genetics* 134: 1189–94

Hilfiker-Kleiner, D., Dübendorfer, A., Hilfiker, A. & Nöthiger, R. (1994) Genetic control of sex determination in the germline and soma of the housefly, *Musca domestica*. *Development* 120: 2531–38

Hiroyoshi, T. (1964) Sex-limited inheritance and abnormal sex ratios in strains of the housefly. *Genetics* 50: 373–85

Hiroyoshi, T. (1977) Some new mutants and revised linkage maps of the housefly, *Musca domestica* L. *Japanese Journal of Genetics* 52: 275–88

Inoue, H. & Hiroyoshi, T. (1981) A maternal effect mutant of the housefly that transforms sex. *Japanese Journal of Genetics* 56: 604–05

Inoue, H. & Hiroyoshi, T. (1986) A maternal-effect sex-transformation mutant of the housefly, *Musca domestica* L. *Genetics* 112: 469–82

King, R.C. (1970) *Ovarian Development in Drosophila melanogaster*, New York: Academic Press

LaChance, L.E. & Leopold, R.A. (1969) Cytogenetic effect of chemosterilants in house fly sperm: incidence of polyspermy and expression of dominant lethal mutations in early cleavage divisions. *Canadian Journal of Genetics* 11: 648–59

Malacrida, A., Gasperi, G., Biscaldi, G.F., Milani, R. & Franco, G.M. (1982) Evidence for the existence of a female determining factor *F* on the 4th linkage group of *Musca domestica* L. *Atti associazione Genetica Italiana* 28: 243–44

McDonald, I.C., Evenson, P., Nickel, C.A. & Johnson, O.A. (1978) House fly genetics: isolation of a female determining factor on chromosome 4. *Annals of the Entomological Society of America* 71: 692–94

Meise, M., Hilfiker-Kleiner, D., Dübendorfer, A. *et al.* (1998) *Sex-lethal*, the master sex-determining gene in *Drosophila*, is not sex-specifically regulated in *Musca domestica*. *Development* 125: 1487–94

Milani, R. (1967) The genetics of *Musca domestica* and of other muscoid flies. In *Genetics of Insect Vectors of Disease*, edited by J.W. Wright & R. Pal, Amsterdam and New York: Elsevier

Milani, R. (1975) The house fly, *Musca domestica*. In *Handbook of Genetics*, edited by R.C. King, New York: Plenum Press

Milani, R., Rubini, P.G. & Franco, M.G. (1967) Sex determination in the housefly. *Genetica agraria* 21: 385–411

Müller-Holtkamp, F. (1995) The *Sex-lethal* gene homologue in *Chrysomya rufifacies* is highly conserved in sequence and exon–intron organization. *Journal of Molecular Evolution* 41: 467–77

Nagoshi, R.N. & Baker, B.S. (1990) Regulation of sex-specific RNA splicing at the *Drosophila doublesex* gene: cis-acting mutations in exon sequences alter sex-specific RNA splicing patterns. *Genes and Development* 4: 89–97

Nöthiger, R. & Dübendorfer, A. (1971) Somatic crossing-over in the housefly. *Molecular and General Genetics* 112: 9–13

Nöthiger, R. & Steinmann-Zwicky, M. (1985) A single principle for sex determination in insects. *Cold Spring Harbor Symposia* 50: 615–21

Nöthiger, R., Leuthold, M., Andersen, N. *et al.* (1987) Genetic and developmental analysis of the sex-determining gene 'double sex' (*dsx*) of *Drosophila melanogaster. Genetical Research* 50: 113–23

Perje, A.-M. (1948) Studies on the spermatogenesis in *Musca domestica. Hereditas* 34: 209–32

Raymond, C.S., Shamu, C.E., Shen, M.M. *et al.* (1998) Evidence for evolutionary conservation of sex-determining genes. *Nature* 391: 691–95

Rovati, C. & Vanossi-Este, S. (1978) Determinazione del sesso in *Musca domestica* L. Soppressione dell' effeto materno del fattore *Ag* (arrenogeno) ad opera del fattore di femminilità *F. Bollettino di Zoologia* 45: 240

Rovati, C., Vanossi Este, S., Cima, L. & Milani, R. (1983) Recombination rates of the loci *M*1, *Ag*, *ac*, and *Mdh* (1st CHR.) of *Musca domestica* L. *Atti del XIII Congresso Nazionale Italiano di Entomologia*, Torino: Sestriere

Rubini, P.G. & Palenzona, D. (1967) Response to selection for high number of heterochromosomes in *Musca domestica* L. *Genetica agraria* 21: 101–10

Rubini, P.G., Franco, M.G. & Vanossi Este, S. (1972) Polymorphisms for heterochromosomes and autosomal sex-determinants in *Musca domestica* L. *Atti del IX Congresso Nazionale Italiano di Entomologia* 341–52

Rubini, P.G., Vecchi, M. & Franco, M.G. (1980) Mitotic recombination in *Musca domestica* L. and its influence on mosaicism, gynandromorphism and recombination in males. *Genetical Research* 35: 121–30

Rubini, P.G., Franco, M.G., Rovati, C. & Vecchi, M. (1989) Genetic variability of the sex balance in European populations of *Musca domestica* L. In *Regulation of Insect Reproduction IV*, edited by M. Tonner, T. Soldan & B. Bennettova, Prague: Czech Academy of Science

Saccone, G., Peluso, I., Artiaco, D. *et al.* (1998) The *Ceratitis capitata* homologue of the *Drosophila* sex-determining gene *Sex-lethal* is structurally conserved, but not sex-specifically regulated. *Development* 125: 1495–500

Schmidt, R., Hediger, M., Nöthiger, R. & Dübendorfer, A. (1997a) The mutation *masculinizer* (*man*) defines a sex-determining gene with maternal and zygotic functions in *Musca domestica* L. *Genetics* 145: 173–83

Schmidt, R., Hediger, M., Roth, S., Nöthiger, R. & Dübendorfer, A. (1997b) The Y-chromosomal and autosomal male-determining *M* factors of *Musca domestica* are equivalent. *Genetics* 147: 271–80

Shearman, D.C.A. & Frommer, M. (1998) The *Bactrocera tryoni* homologue of the *Drosophila melanogaster* sex-determination gene *doublesex. Insect Molecular Biology* 7: 1–12

Sievert, V., Kuhn, S. & Traut, W. (1997) Expression of the sex determining cascade genes *Sex-lethal* and *doublesex* in the phorid fly *Megaselia scalaris. Genome* 40: 211–14

Steinmann-Zwicky, M., Schmid, H. & Nöthiger, R. (1989) Cell-autonomous and inductive signals can determine the sex of the germ line of *Drosophila* by regulating the gene *Sxl. Cell* 57: 157–66

Vanossi Este, S. (1971) Nuovi equilibri nella determinazione del sesso in *Musca domestica* L. *Bollettino di Zoologia* 38: 566

Vanossi Este, S. & Rovati, C. (1982) Inheritance of the arrhenogenic factor *Ag* of *Musca domestica* L. *Bollettino di Zoologia* 49: 269–78

Vanossi Este, S., Rovati, C., Franco, M.G. & Rubini, P. G. (1974) Localizzazione genetica del fattore arrenogeno *Ag* di *Musca domestica* L. *Bollettino di Zoologia* 41: 532–33

Wagoner, D.E. (1969a) Linkage group–karyotype correlation in the house fly, *Musca domestica* L., confirmed by cytological analysis of X-ray induced Y-autosomal translocations. *Genetics* 62: 115–21

Wagoner, D.E. (1969b) Presence of male determing factors found on three autosomes in the house fly, *Musca domestica. Nature* 223: 187–88

Wilkins, A.C. (1995) Moving up the hierarchy: a hypothesis on the evolution of a genetic sex determination pathway. *BioEssays* 17: 71–77

GLOSSARY

cytological studies studies of cells and especially their chromosomes, usually employing light or electron microscopy

heterogametic sex the sex that possesses a pair of non-homologous sex chromosomes and therefore produces two different types of gametes (e.g. X-sperms and Y-sperms)

heterosome the sex chromosome (e.g. Y-chromosome in man) that renders the heterogametic sex chromosomally different from the homogametic sex

oogenesis formation, development and maturation of the female gamete or egg

soma animal body as a whole with the exception of the germline (from which the gametes originate)

synkaryon zygote nucleus resulting from fusion of gametic nuclei

See also **Sex determination in Drosophila** (p.179)

HYBRID DYSGENESIS DETERMINANTS AND OTHER USEFUL TRANSPOSABLE ELEMENTS IN DROSOPHILA

Margaret G. Kidwell
Department of Ecology and Evolutionary Biology, University of Arizona, USA

1. Introduction

The term "hybrid dysgenesis" refers to a genetically caused syndrome of abnormal traits, such as high mutation rates and sterility, observed in hybrids resulting from interstrain crosses. Usually, hybrid dysgenesis is observed in only one direction when reciprocal crosses between two parental strains are made. Several years after its first description in the mid-1970s (Kidwell *et al.*, 1977), the cause of hybrid dysgenesis was traced to several new families of transposable genetic elements, including the *P*, *I* and *hobo* elements in *Drosophila melanogaster* and several elements in *D. virilis*. More recently, some of these elements have been developed for use in genetically manipulating genes, both within and between species, as described below. The discovery and description of hybrid dysgenesis is an excellent example of seemingly esoteric basic research that eventually leads to broad and useful applications – in this case, the development of sophisticated tools for genetic engineering.

2. Phenomenology of hybrid dysgenesis

Early phenomenological observations in France, Australia and the US were explained in terms of two different systems of hybrid dysgenesis in *D. melanogaster*, namely the P–M and I–R systems. For simplicity, the description here will largely refer to the P–M system. Most strains of *D. melanogaster* can be classified into two categories: P, for paternal, and M, for maternal. Typically, strains of *D. melanogaster* that have been established for a long time in a laboratory are of the M type and strains collected from natural populations in the wild are of the P type. When females from an M strain are crossed with males from a P strain, the F_1 hybrids often have reduced fertility, particularly when they are raised at high temperature. Hybrids that are fertile may show evidence of other dysgenic traits, including high frequencies of offspring inviability, male recombination (which does not normally occur in Drosophila), mutations, chromosomal aberrations and transmission ratio distortion (Kidwell *et al.*, 1977). These traits result from instability of the germline of M female × P male F_1 hybrids. The DNA in the soma (the body cells) is usually not affected. Hybrids resulting from the P female × M male reciprocal cross are mostly normal, as are the progeny from P × P and M × M matings (**Figure 1**). In the I–R system, the two types of interacting strain are called I (inducer) and R (reactive). Similar, but not identical, traits are again produced non-reciprocally in the germlines of F_1 hybrids.

3. Regulation of hybrid dysgenesis

The expression of P–M hybrid dysgenesis is regulated at several different levels (Engels, 1989). Its occurrence in germ cells, but not somatic ones, provides evidence for a tissue-specific regulatory function. Evidence for another level of regulation came from the lack of dysgenesis in hybrids produced by the reciprocal P female × M male cross and in the P strains themselves. Genetic analysis showed that multiple factors present on all the chromosomes of P strains were responsible for regulation. Dysgenesis is suppressed by a state called the P cytotype that is established only in a P strain genome. The alternative state, called the M cytotype, permits the expression of hybrid dysgenesis. The inheritance of cytotype was shown to be dependent on both chromosomal and maternal transmission, and temperature and maternal age affect its expression. Similar regulatory mechanisms probably exist in other dysgenesis systems (Bregliano & Kidwell, 1983), but the details have not yet been worked out.

4. Hybrid dysgenesis is caused by the activation of transposable elements

Following the first description of hybrid dysgenesis, it took several years for transposable elements (TEs) to be established as the causal mechanism. Although TEs had been discovered in maize by Barbara McClintock about 30 years earlier, very little was known in the early 1970s about the variety, distribution and properties of TEs in other organisms. Following the cloning of the X-linked *white* gene in

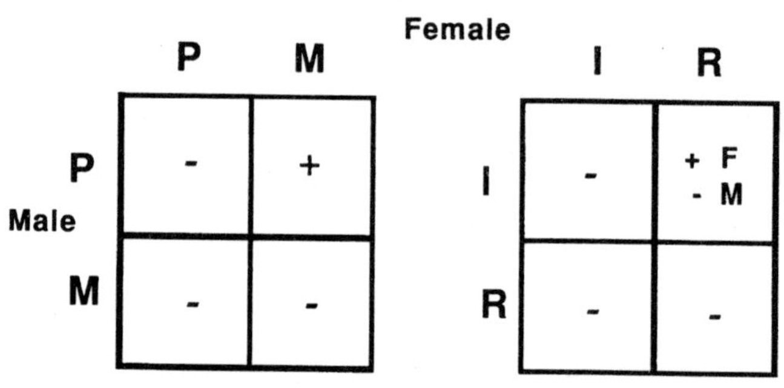

Figure 1 A schematic diagram showing the strain combinations that result in expression of hybrid dysgenesis in the P–M system (left) and I–R system (right). I, inducer strain; M, maternal contributing strain; P, paternal contributing strain; R, reactive strain; minus sign, no hybrid dysgenesis in the F_1 generation; plus sign, hybrid dysgenesis in the F_1 generation. (Within right diagram: F, F_1 female; M, F_1 male.)

D. melanogaster, a strategy was adopted to identify the genetic basis of P–M hybrid dysgenesis. Dysgenic crosses were made using knowledge about the instability of hybrids (Kidwell *et al.*, 1977). Following the screening of about 70 000 F_1 hybrid offspring, six independent white-eyed mutations were observed and found to be caused by the insertion of TE DNA into the *white* gene. A new TE, called the *P* element, was thus discovered and identified as the cause of the P–M hybrid dysgenesis syndrome (Bingham *et al.*, 1982). The TEs responsible for the other systems of hybrid dysgenesis were later identified in similar ways (Berg & Howe, 1989).

5. Hybrid dysgenesis in non-Drosophilid insects

In addition to Drosophila, several potential examples of hybrid dysgenesis have been reported in other insects, including the fly Chironomus, the flour beetle, Tribolium, and the Mediterranean fruit fly, *Ceratis capitata*, but few of these have been clearly linked to the activation of TEs. In addition to hybrid dysgenesis, it is well known that hybrid inviability and sterility can be caused by a diverse group of phenomena including X-linked genes and their suppressors associated with speciation, microorganism-mediated cytoplasmic incompatibility and selfish Mendelian genes. It is not clear why hybrid dysgenesis has so far been observed almost exclusively in Drosophila.

6. Classification of TEs

TEs are classified into two main groups (Finnegan, 1992) depending on their structure and mode of transposition (**Figure 2**). Class I elements transpose by means of an RNA intermediate, and include the retrotransposons, while Class II elements use a DNA intermediate. Hybrid dysgenesis determinants are found in both classes. For example, the *I* element responsible for the I–R system of hybrid dysgenesis (Bucheton *et al.*, 1984) is a Class I element, and *P* and *hobo* are Class II elements. There is considerable evidence that the hybrid dysgenesis syndromes caused by the *P*, *I* and *hobo* elements are independent of one another. In contrast, at least six unrelated TE families are simultaneously activated in the same dysgenic crosses in *D. virilis* (Petrov *et al.*, 1995; Andrianov *et al.*, 1999), including the Class I elements *Penelope*, *Ulysses*, *Helena*, *Telemac* and *Tv1* and the Class II element *Paris*. The mechanism(s) involved in this simultaneous activation is currently unknown. See article on "Retrotransposons" for the structures of Class I elements.

The structure of the *P* element (**Figure 3**) illustrates the main features of a typical Class II element. *P* elements are characterized by having 31 base pair (bp) perfect, inverted, terminal repeats and they generate an 8 bp duplication of host DNA at the site of insertion. Autonomous *P* elements are 2.9 kilobases (kb) in length and have four open reading

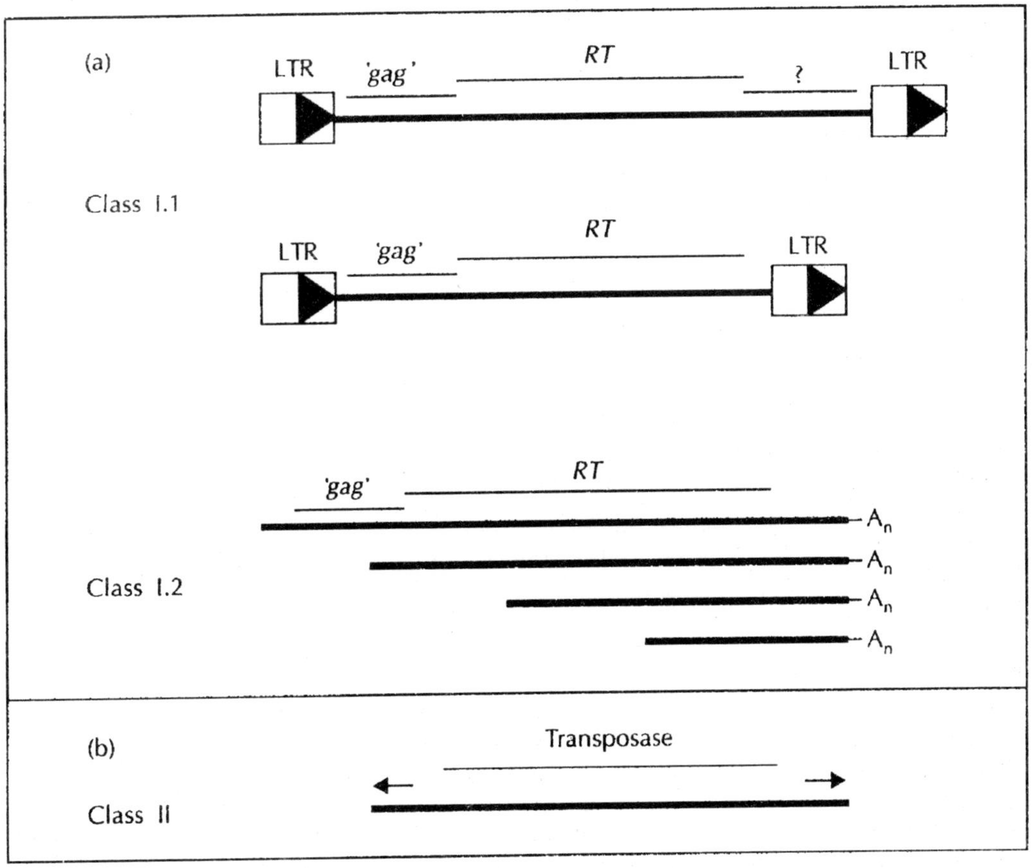

Figure 2 Schematic representation of the structures of Class I and Class II eukaryotic transposable elements (redrawn from Finnegan, 1992). A_n, A-rich sequences; 'gag', gag-like open reading frame; LTR, long terminal repeat; RT, reverse transcriptase.

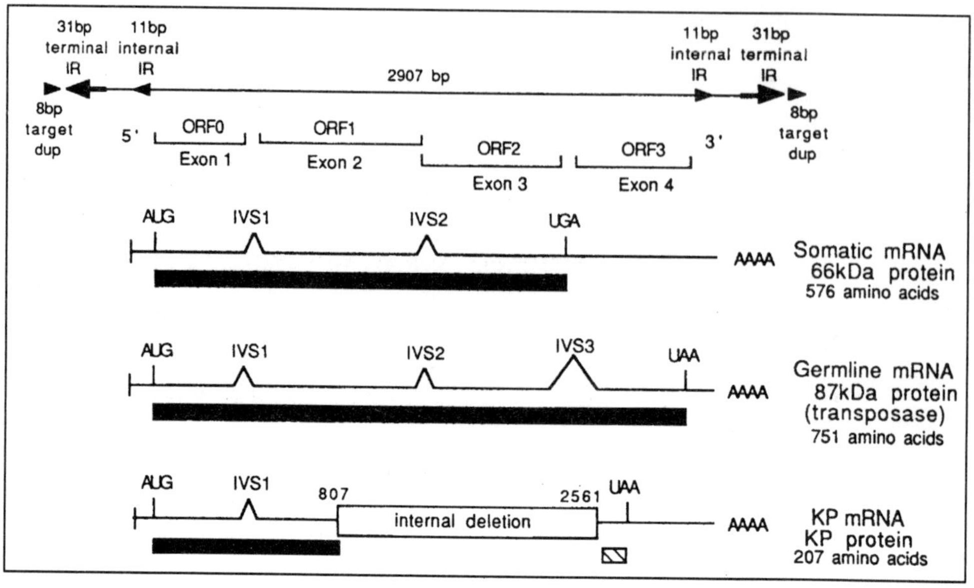

Figure 3 Diagram of the structures of an autonomous *P* element in *D. melanogaster* and the proteins that it encodes (modified from Rio, 1991). IR, inverted repeat; IVS, intervening sequence; KP, KP element; ORF, open reading frame.

frames, all of which are required to encode a functional transposase enzyme. They can catalyse their own transposition and that of defective, non-autonomous *P* elements present in the same genome. Defective *P* elements are generally smaller and variable in size and are derived from complete elements by internal deletions (**Figure 3**). In most *D. melanogaster* natural populations, the genomic copy number of *D. melanogaster P* elements varies from 0 to about 60 per genome. Usually, a minority of these are autonomous elements, while the majority are internally deleted, non-autonomous ones (see Engels (1989) for more information on *P* elements).

7. Evolution of TEs

There is now considerable evidence that many TEs have an ancient origin. When the sequences of transposable elements from different species are compared (for example, Clark & Kidwell, 1997), phylogenetic trees can be reconstructed and evolutionary histories traced. With the growing number of TEs that have been identified and sequenced from diverse organisms, relationships are being established on the basis of sequence similarities that go increasingly deep into the past. For example, it has been shown that Class I TEs are related to retroviruses. However, phylogenetic trees based on different retroelement genes may not always be completely congruent with one another due to domain reshuffling during evolution (McClure, 1999). The evolution histories of three Class II families of Drosophila TEs that are actually or potentially useful in genetic engineering are briefly described below.

(i) The P *element family*

Studies of the population biology of *P* elements indicate that this element has only become established in the cosmopolitan species *D. melanogaster* during the last century. It appears that the *P* element was transmitted horizontally, like a virus, to *D. melanogaster*, from the distantly related *D. willistoni*. The evolutionary history of *P* elements in Drosophila over the last 50 million years is now being studied in considerable detail. *P* element phylogenies provide evidence for additional, more ancient, *P* element horizontal transfers. A clear evolutionary link has not yet been made between the *P* family and other TE families. So far, *P* elements have been found only in Drosophila and closely related families of insects. Thus, they have a relatively narrow host range and are not active when introduced into more distantly related individuals.

(ii) The hAT *super family of TEs*

In contrast to the *P* element, the hybrid dysgenesis determinant *hobo*, from *D. melanogaster*, is a member of a large family of TEs that is broadly distributed in insects and includes the *Hermes* element from the house fly, *Musca domestica*. In turn, this family has been shown to be part of the *hAT* (*hoboAc Tam*) superfamily which includes *hobo*, the *Ac* element from *Zea maize* and the *Tam* element from *Antirrhinum majus* (Warren *et al.*, 1994). Both the *hobo* and *Hermes* elements have recently been shown to have potential as insect transformation vectors.

(iii) The mariner-Tc1 *superfamily of TEs*

The *mariner* element is a Class II element that was originally identified in *D. mauritiana*, but which does not show hybrid dysgenesis. It belongs to the *mariner-Tc1* family of elements which is taxonomically widespread (Robertson, 1993). This family also includes the *Tc1* elements originally identified in the nematode *Caenorhabditis elegans*. In turn, it has been suggested that, on the basis of a common sequence motif, the *mariner-Tc1* family belongs to a superfamily which includes bacterial insertion sequences and retrons (Class I) sequences. Due to their relatively broad host range, some members of this family have promise for use as general insect transformation vectors (see below). Two of these are the *S* element of *D. melanogaster* (Merriman *et al.*, 1995) and the *Minos* element of *D. hydei* (Loukeris *et al.*, 1995).

8. Are TEs strictly selfish genes or molecular parasites?

TEs are often considered as essentially selfish DNA, molecular parasites that are integrated into the chromosomes of their host organisms (e.g. Orgel & Crick, 1980). According to this view, these elements can survive and multiply over long periods of time essentially because of their ability to replicate faster than the DNA of their host organisms by means of transposition. They may spread rapidly in a population despite a negative effect on the fitness of their hosts. There is, however, another school of thought (e.g. McDonald, 1995) that does not deny that there are selfish aspects to the behaviour of these elements, yet also maintains that they may confer important advantages to their hosts over evolutionary time. These advantages may be conferred either directly, due to the advantages which the presence of some TEs may confer on their hosts, or indirectly, through the increase in host genetic variability due to TE activity. A lively debate on this issue is likely to continue for some time. As with many contentious issues, the resolution of this one may lie somewhere in the middle ground between the two extreme positions.

9. Using TEs in genetic engineering

The discovery that *P* TEs were responsible for the P–M system of hybrid dysgenesis allowed the development of a new generation of tools for the genetic manipulation and engineering of Drosophila. These tools have revolutionized Drosophila research through the introduction of powerful methodologies, such as germline transformation, the analysis of *in vitro* mutagenized genes, gene cloning by transposon tagging and enhancer trapping (Engels, 1995; Kaiser *et al.*, 1995).

In general, Class II elements are more suitable than Class I elements as genetic engineering tools because many Class I elements do not transpose at a sufficiently high frequency to be useful. In addition, a number of other technical difficulties are associated with the use of Class I elements. For example, although the *I* element transposes at a relatively high rate in I–R hybrid dysgenesis, further genetic manipulation is problematic because all *D. melanogaster* strains carry deleted *I* elements.

10. *P* element germline transformation

One of the most important uses of *P* elements is as transformation vectors (Spradling & Rubin, 1982) which allow the production of transgenic flies through the introduction and manipulation of Drosophila germline DNA. A gene of interest is first "loaded" into a defective *P* element within a bacterial plasmid by placing it between the inverted repeats by standard ligation procedures. Germline transformation is then achieved by injecting this artificial DNA "construct" into early Drosophila embryos along with a source of *P* element transposase. Successful transformation is dependent on the *P* element construct transposing from the injected plasmid into a random chromosomal site in a germline cell of the developing embryo (transposition to a somatic cell will only result, at most, in transient expression in progeny of the next generation). If a successfully transformed embryo completes development and survives to reproduce successfully, some of its germline cells will carry the introduced sequences. All the cells of individuals inheriting this lineage in the following generation will then be expected to carry the *P* element construct. Transformants are usually identified at the phenotypic level by the expression of a dominant marker gene that is incorporated into the construct for that purpose (see Spradling (1986) for more details).

11. Using *P* elements as mutagens

The ability of *P* elements to act as powerful mutagenic agents has been utilized in increasingly more sophisticated ways to produce mutations in the absence of chemicals or radiation (Kaiser *et al.*, 1995). *P* elements are now widely used to obtain the sequences of genes responsible for mutations that were previously identified only by phenotype. This is achieved by obtaining a *P* element insertion mutation that has the phenotype of interest. The inserted *P* element then serves as a molecular identification tag that allows the cloning and sequencing of the gene. *P* elements are also used to solve the reverse problem of obtaining mutations in DNA sequences that lack any clearly defined phenotype. In this case, mutations lacking any phenotypic effect can be identified at the molecular level.

Strains constructed to carry single *P* elements provide a versatile genetic method for identifying and cloning *D. melanogaster* genes. Mutagenesis is initiated by crossing two strains, each of which contains a specially designed *P* element. This method has been used to construct a library of single *P* element insertion stocks that have been screened for recessive mutations and have many uses in Drosophila molecular genetics. In a sophisticated extension of the method of *P* element mutagenesis, a *P* element has been modified to incorporate binding regions isolated from the *gypsy* element, which is a Class I TE. This composite transposon combines the mutagenic efficacy of the *gypsy* element with the controllable transposition of *P* elements. It causes an expanded repertoire of mutations, produces alleles that are suppressed by mutations at another suppressor locus and is useful for studying the structural and functional properties of heterochromatin (Roseman *et al.*, 1995).

12. Use of *P* elements for targeted gene replacement

Although the transformation procedure has been a key technique, it has one major limitation: it provides no control over the location in the genome where the DNA construct will be inserted. Therefore, it does not allow genes, or parts of genes, to be replaced in their original genomic locations. In order to overcome this obstacle, *P* elements have been used in a different way to provide a method of targeted gene replacement based on the discovery that they replicate by a cut-and-paste mechanism that involves gap repair (Engels, 1995). A gene conversion-like process is used to repair such gaps by copying and inserting information from a homologous template sequence. The method is designed to provide a target site, an ectopic template and a transposase source, all in the same individual.

13. Enhancer traps in molecular and developmental genetics

Another important way that *P* elements have been used is in the development of enhancer trap elements for identifying and studying genes in Drosophila development. An enhancer trap element is a modified *P* element containing a reporter gene with a very low level of intrinsic expression due to being transcribed by a weak promoter. Single copies of the transposon can be introduced into flies at many different genomic locations. Screening of the progeny allows the identification of genes on the basis of the expression patterns of the mutations produced. First generation enhancer trap elements contain the bacterial reporter gene *lacZ*, encoding the enzyme β-galactosidase. A second generation enhancer trap element includes the yeast transcription factor GAL4, instead of *lacZ*. This can be used to target expression of any desired gene product to the marked cells (Kaiser *et al.*, 1995).

14. TEs as vectors for gene transfer in other insects

There is an increasing need to develop transformation vectors that can be used in species other than Drosophila for many purposes, including the development of methods of biological control of insect pest species. Since the narrow host range of the *P* element severely limits its use for this purpose, candidate TEs have been sought among other DNA transposons. Knowledge of the population and evolutionary biology of these elements provides useful information for assessing their potential as generalized insect vectors for transformation. The prospect of using members of the *mariner-Tc1* family as transformation vectors is particularly attractive because of their widespread taxonomic distribution. These elements have high potential for horizontal gene transfer across species boundaries because they are not dependent on host factors for transposition.

The *D. hydei* element *Minos* belongs to the *mariner* family and has been used successfully for germline transfer of the Mediterranean fruit fly, *Ceratis capitata*, a major pest of fruit crops (Loukeris *et al.*, 1995). Both *mariner* and *hobo* elements have been similarly used for transformation of the yellow fever mosquito, *Aedes aegypti*. Species that are taxonomically distant from insects, such as the zebrafish, *Danio rerio*, have also been successfully transformed by the *mariner* element (Weinberg, 1998).

Anopheline mosquitoes are obligatory vectors for human malaria and their transformation presents considerable technical challenges. However a transposon based on the *Minos* element has now been used for stable germline transformation of *Anopheles stephensi* (Catteruccia *et al.*, 2000). In addition to *mariner*, the Lepidopteran transposable element *piggyBac* is being recognized as a useful vector for genetic engineering in a variety of insect species. The *piggyBac* transposon can mediate transformation of the Mediterranean fruit fly. It can also potentially serve as a versatile vector for transformation of a wide variety of insect species, such as *Drosophila melanogaster*, the yellow fever mosquito, *Aedes aegypti*, and the silkworm, *Bombyx mori* (Tamura *et al.*, 2000).

REFERENCES

Andrianov, B.V., Zakharyev, V.M., Reznik, N.L., Gorelova, T.V. & Evgen'ev, M.B. (1999) *Gypsy* group retrotransposon *Tv1* from *Drosophila virilus*. *Gene* 239: 193–99

Berg, D.E. & Howe, M.M. (eds) (1989) *Mobile DNA*, Washington, DC: American Society of Microbiology

Bingham, P.M., Kidwell, M.G. & Rubin, G.M. (1982) The molecular basis of P–M hybrid dysgenesis: the role of the P element, a P strain-specific transposon family. *Cell* 29: 995–1004

Bregliano, J.C. & Kidwell, M.G. (1983) Hybrid dysgenesis determinants. In *Mobile Genetic Elements*, edited by J.A. Shapiro, New York: Academic Press

Bucheton, A., Paro, R., Sang, H.M., Pelisson, A. & Finnegan, D.J. (1984) The molecular basis of I–R hybrid dysgenesis: identification, cloning and properties of the *I* factor. *Cell* 38: 153–63

Catteruccia, F., Nolan, T., Loukeris, T.G. *et al.* (2000) Stable germline transformation of the malaria mosquito *Anopheles stephensi*. *Nature* 405: 959–62

Clark, J.B. & Kidwell, M.G. (1997) A phylogenetic perspective on *P* transposable element evolution in *Drosophila*. *Proceedings of the National Academy of Sciences USA* 94: 11428–33

Engels, W.R. (1989) P elements in Drosophila. In *Mobile DNA*, edited by D.E. Berg & M.M. Howe

Engels, W.R. (1995) P elements in Drosophila. In *Transposable Elements*, edited by H. Saedler & A. Gierl, Berlin: Springer

Finnegan, D.J. (1992) Transposable elements. *Current Opinion in Genetical Development* 2: 861–67

Kaiser, K., Sentry, J.W. & Finnegan, D.J. (1995) Eukaryotic transposable elements to study gene structure and function. In *Mobile Genetic Elements*, edited by D.J. Sheratt, Oxford: IRL Press

Kidwell, M.G., Kidwell, J.F. & Sved, J.A. (1977) Hybrid dysgenesis in *Drosophila melanogaster*: a syndrome of aberrant traits including mutation, sterility and male recombination. *Genetics* 36: 813–33

Loukeris, T.G., Livadaras, I., Arca, B., Zabalou, S. & Savakis, C. (1995) Gene transfer into the medfly, *Ceratitis capitata*, with a *Drosophila hydei* transposable element. *Science* 270: 2002–05

McClure, M.A. (1999) The retroid agents: disease, function, and evolution. In *Origin and Evolution of Viruses*, edited by E. Domingo, R. Webster & J. Holland, San Diego and London: Academic Press

McDonald, J.F. (1995) Transposable elements: possible catalysts of organic evolution. *Trends in Ecology and Evolution* 10: 123–26

Merriman, P.J., Grimes, C.D., Ambroziak, J. *et al.* (1995) S elements: a family of Tc1-like transposons in the genome of Drosophila melanogaster. *Genetics* 141: 1425–38

Orgel, L.E. & Crick, F.H.C. (1980) Selfish DNA: the ultimate parasite. *Nature* 284: 604–07

Petrov, D.A., Schutzman, J.L., Hartl, D.L. & Lozovskaya, E.R. (1995) Diverse transposable elements are mobilized in hybrid dysgenesis in *Drosophila virilis*. *Proceedings of the National Academy of Sciences USA* 92: 8050–54

Rio, D.C. (1991) Regulation of Drosophila P element transposition. *Trends in Genetics* 7: 282–87

Robertson, H.M. (1993) The mariner transposable element is widespread in insects. *Nature* 362: 241–45

Roseman, R.R., Johnson, E.A., Rodesch, C.K. *et al.* (1995) A *P* element containing suppressor of hairy-wing binding regions has novel properties for mutagenesis in *Drosophila melanogaster*. *Genetics* 141: 1061–74

Spradling, A.C. (1986) P element-mediated transformation. In *Drosophila: A Practical Approach*, edited by D.B. Roberts, Oxford and Washington, DC: IRL Press

Spradling, A.C. & Rubin, G.M. (1982) Transposition of cloned P elements into *Drosophila* germ line chromosomes. *Science* 218: 341–47

Tamura, T., Thibert, C., Royer, C. *et al.* (2000) Germline transformation of the silkworm *Bombyx mori* L. using a piggyBac transposon-derived vector. *Nature Biotechnology* 18: 81–84

Warren, W.D., Atkinson, P.W. & O'Brochta, D.A. (1994) The *Hermes* transposable element from the housefly, *Musca domestica*, is a short inverted repeat element of the *hobo*, *Ac* and *Tam 3* (*hAT*) element family. *Genetical Research* 64: 87–97

Weinberg, E.S. (1998) Zebrafish genetics: harnessing horizontal gene transfer. *Current Biology* 8: R244–R247

GLOSSARY

β-galactosidase/LacZ enzyme that hydrolyses lactose to galactose and glucose

cytotype an inherited cellular environment, described in *D. melanogaster*, that is dependent on chromosomal and cytoplasmic factors, and that determines whether or not the *P* transposable element is active (M cytotype) or quiescent (P cytotype)

enhancer trapping a method to identify genes based on their pattern of expression

ligation end-to-end joining of two nucleic acid molecules to form a continuous DNA molecule

retron a simple type of retroelement found in some bacteria that encodes a reverse transcriptase, but is unable to transpose

transposon tagging a method for cloning genes after they have been "tagged" by having a transposon inserted into them

See also **Retrotransposons** (p.204); **The silkworm *Bombyx mori*** (p.219); **Selfish genes** (p.875)

RETROTRANSPOSONS

Irina Arkhipova
Department of Molecular and Cellular Biology, Harvard University, Cambridge, Massachusetts, USA

1. Introduction

Retrotransposons are one of the two classes of transposable genetic elements. In contrast to the other class (DNA transposons, which transpose directly from DNA to DNA), retrotransposons are capable of spreading throughout the genome by means of reverse transcription or RNA-dependent DNA synthesis. For this purpose, retrotransposons use an enzyme called reverse transcriptase (RT) to make DNA copies of their RNAs, with subsequent integration of these newly synthesized DNA copies into other genomic locations (DNA–RNA–DNA pathway). Integration is usually accompanied by a staggered cut in the host DNA and a small segment of DNA at the target site is duplicated on insertion of the transposon. The original DNA copy of a retrotransposon is not excised, resulting in an increase of retrotransposon copy number (replicative transposition).

First discovered in Drosophila, retrotransposons were later shown to be integral components of all the eukaryotic genomes which were examined for them. They occur in families of repeated elements and members of each family are similar in sequence and are dispersed throughout the genome in multiple copies, the number of which may vary from less than a dozen to hundreds of thousands. In mammalian species, retrotransposon copy number and location are mostly conserved, while in invertebrates there are often differences between strains of the same species. In Drosophila, certain crosses between strains containing and lacking particular transposon families (including retrotransposons) lead to a dramatic increase in their mobility, or transposition rate, which can in turn lead to a set of abnormal traits (sterility, high mutability, chromosome non-disjunction, etc.) known collectively as hybrid dysgenesis. In principle, transposition events can occur at any developmental stage, but only germline transposition would lead to heritable changes. Normally, transposition occurs at very low frequency, but this may be increased under certain circumstances. Repression of transposition may be achieved by different mechanisms for individual retrotransposon families, since the control may be exerted at any level (transcription, splicing, RNA stability, translation, post-translational modifications, genetic background, etc.). Mutations in a number of cellular genes may suppress or enhance the effect of retrotransposon-induced mutations by disrupting interactions between retrotransposons and the products of these so-called allele-specific modifier (suppressor or enhancer) genes.

2. Molecular structure

Retrotransposons possess a distinct molecular structure that allows one to distinguish them from DNA transposons and which reflects the usage of reverse transcription in their life cycle. Knowledge of their structural organization has been essential for understanding the mechanisms of retrotransposition and its control. A necessary condition for retrotransposon transposition is its ability to be transcribed into a full-length RNA containing all the genetic information to be reverse transcribed into DNA and, at the same time, to be used as a template for the synthesis of proteins necessary for transposition. Transcription and translation are carried out by cellular systems, and retrotransposons usually contain within themselves the *cis*-acting regulatory sequences (promoters, enhancers, RNA processing signals, etc.) which allow their expression by the host machinery and provide the targets for its stage- and tissue-specific control.

3. Autonomous and non-autonomous retrotransposons

Retrotransposons can be divided into autonomous and non-autonomous types. The former encode the enzymatic apparatus for their reverse transcription and integration into the host genome, while the latter are not capable of transposing on their own and presumably use the enzymes produced by the autonomous elements. Non-autonomous retrotransposons, also known as SINEs (short interspersed repeat sequences), are several hundred base pairs (bp) in length, usually bear similarity to tRNA or other small RNA genes, are transcribed by RNA polymerase III from an internal promoter and do not code for proteins. The human Alu elements, up to 10^6 copies of which are present in the genome, are the best-known representatives of this type. Autonomous retrotransposons are several thousand bp in length, are transcribed by RNA polymerase II and code for RT and several other proteins. They fall into two major groups: (1) retrovirus-like or long terminal repeat (LTR)-containing retrotransposons (**Figure 1A,B**); and (2) non-LTR retrotransposons, which are also called retroposons or long interspersed repeat sequence (LINE)-like elements (**Figure 1C**).

(i) LTR-containing retrotransposons

Members of the first group bear striking structural and functional similarities to vertebrate retroviruses, which can cause cancers, immunodeficiencies and other diseases. They contain LTRs (usually 200–600 bp in length) at both ends of their DNA copies and open reading frames (ORFs), which are usually separated by a translational frameshift or by stop codons, but which sometimes may be fused into a single ORF. These ORFs are very similar to the retroviral *gag* and *pol* genes and encode nucleic acid (NA)-binding proteins, aspartic protease (PR), RT with an RNase H (RH) activity and integrase (IN) (**Figure 1A,B**). RT performs the key reaction of copying RNA into DNA and then synthesizes the

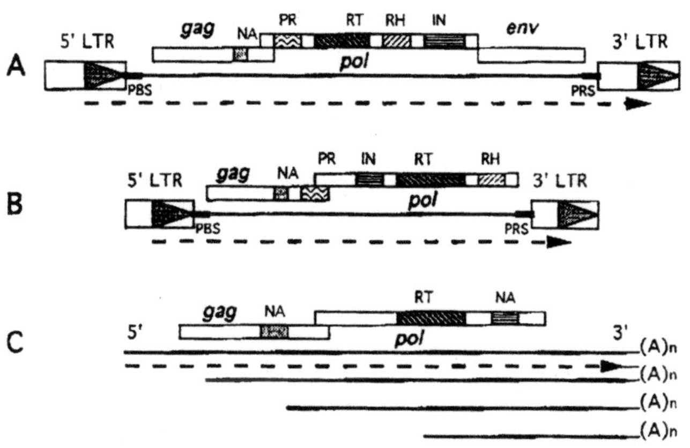

Figure 1 Structure of long-terminal repeat (LTR)-containing (A) *gypsy*/Ty3-like group, (B) *copia*/Ty1-like group and (C) non-LTR retrotransposons. The groups are named after *gypsy* and *copia* retrotransposons of Drosophila and *Ty* retrotransposons of yeast. LTRs, arrowed boxes; open reading frames (ORFs), open boxes; domains within ORFs, filled boxes; RNA transcripts, dashed lines; A-rich terminal sequences, A_n. Other abbreviations are explained in the text. The drawing is not to scale.

second DNA strand, using the RH activity to remove RNA from the hybrid. The IN activity is responsible for the endonucleolytic cleavage of the target site in the host DNA and performs integration of double-stranded DNA into the chromosome. PR is used for cleavage of the primary translation product into separate structural and enzymatic components. NA-binding proteins play a structural role and are responsible for the formation of the viral nucleoprotein core. Products of the structural *gag* gene are required in larger amounts than those of the *pol* gene, which is expressed at lower levels by mechanisms involving frameshift, reinitiation of translation or alternative splicing.

Discrimination between retrotransposons and retroviruses is usually achieved through the existence of extracellular infectious virus particles in the latter, formed with the aid of envelope proteins essential for penetration through the cellular membrane during infection and encoded by ORF3 (*env*). A subgroup of LTR-containing retrotransposons, traditionally regarded as retrotransposons, do contain a third *env*-like ORF, however (**Figure 1A**). One element of this type, the *gypsy* element from Drosophila, is capable of forming such particles, thus representing an insect retrovirus. Elements with a third ORF are widespread and have also been found in yeast and plants. Members of the other subgroup do not contain an *env*-like sequence and are only capable of forming non-infectious intracellular virus-like particles. This subgroup is more distant from retroviruses in having a different order of *pol* gene domains, with the integrase domain preceding the RT domain (**Figure 1B**). Elements of an intermediate type have also been found. It is not known if these can form infectious particles.

LTR-containing retrotransposons use an elaborate mechanism for reverse transcription of their RNAs, which takes place within virus-like particles and requires participation of specific cellular tRNAs as primers for RT. One of the LTRs is copied from the other during reverse transcription, and the identity of both LTRs provides evidence that the element has recently transposed. The LTRs and the regions adjacent to them (the tRNA primer binding site [PBS] and the purine-rich sequence [PRS]; **Figure 1A,B**) play an important role in initiation and termination of transcription and in priming of the first and second DNA strand synthesis during reverse transcription. The terminal regions are essential for integration of the resulting double-stranded linear DNA into the host genome. All these regions need to be preserved in retrovirus-based vectors for gene transfer.

(ii) Non-LTR retrotransposons

Members of the second group (non-LTR retrotransposons) are characterized by the absence of terminal repeats, either direct or inverted. Most of them also possess an oligo(A) stretch at the 3′ end, which apparently takes its origin from the poly(A) tail of the corresponding reverse-transcribed RNAs. LINE or L1 family of elements, which make up several per cent of mammalian genomes, are the best-known representatives of this group. Non-LTR retrotransposons are more structurally divergent than retrovirus-like retrotransposons. The only common enzymatic activity encoded by them is RT, with some type of NA-binding protein usually also being present (**Figure 1C**). In a few families, the RH and endonuclease domains can also be identified. No LTRs are generated to restore the upstream promoter elements after retrotransposition, and these retrotransposons may be transcribed either from internal RNA polymerase II promoters or from adjacent cellular promoters. The most widely accepted model for retrotransposition of retrotransposons belonging to this group involves RT-mediated direct association of the 3′ end of the RNA template with a break in genomic DNA, which is used for priming of RNA-directed DNA synthesis *in situ*. This model explains frequent 5′-end truncations of these retrotransposons (**Figure 1**) by incomplete reverse transcription, which is always initiated at the conserved 3′ end of the element.

4. The role of retrotransposons

There is significant discussion of the roles that retrotransposons may play in the organization and function of eukaryotic genomes. No organism is known to be devoid of such elements and their ability to multiply and spread within, and even between, species and to insert into new chromosomal locations can have numerous consequences for their hosts, both harmful and advantageous. On the one hand, there are examples of retrotransposon-induced deleterious chromosome rearrangements (duplications, deficiencies, inversions) which can result if recombination occurs between two different members of the same family. In addition, gene expression may be severely disrupted by retrotransposon insertion into the gene itself or its regulatory regions, and the majority of spontaneous mutations in Drosophila may result from transposon insertion. On the other hand, some genes are known to have acquired novel stage- or tissue-specific expression patterns associated with insertion of retrotransposons with novel regulatory elements.

Retrotransposons constitute a substantial fraction of genomic repetitive DNA, making up to 10–15% of the entire genome (sometimes called its "fluid component"). They are often associated with the centric and terminal regions of chromosomes and may play an important structural role there. Indeed, in Drosophila, two particular non-LTR retrotransposon families are closely involved in formation and maintenance of chromosome ends by retrotransposing specifically to chromosomal termini and preventing them from degradation. This example contrasts with a widespread view of retrotransposons as "genomic parasites" and "selfish DNA" existing only because of the ability to self-replicate and to exploit the host cell environment for propagation. In this scenario, the tendency of retrotransposons to amplify indefinitely is counterbalanced by negative selection acting on deleterious mutations and chromosome rearrangements caused by retrotransposon insertion, leading to an equilibrium in copy number.

In reality, the influence of retrotransposons on host genomes appears to be rather complex, depending on individual features of each retrotransposon family, location of insertion sites and other factors. Action of retrotransposon RT on cellular mRNAs can occasionally lead to the formation of pseudogenes – silent cDNA copies reinserted into chromosomes that have lost their introns and upstream promoter regions. Since retrotransposons are able to acquire various regulatory elements, their properties and functions may be quite diverse and not necessarily conserved. However, they do possess a potent apparatus for genome restructuring and reprogramming and are apparently able to play a prominent role in evolutionary processes.

FURTHER READING

Arkhipova, I.R., Lyubomirskaya, N.V. & Ilyin, Y.V. (1995) *Drosophila Retrotransposons*, Austin, Texas: Landes Bioscience

Berg, D.E. & Howe, M.M. (1989) *Mobile DNA*, Washington, DC: American Society for Microbiology

Capy, P., Bazin, C., Higuet, D. & Langin, T. (1998) *Dynamics and Evolution of Transposable Elements*, Austin, Texas: Landes Bioscience

Charlesworth, B., Sniegowski, P. & Stephan, W. (1994) The evolutionary dynamics of repetitive DNA in eukaryotes. *Nature* 371: 215–20

Eickbush, T.H. (1994) Origin and evolutionary relationships of retroelements. In *The Evolutionary Biology of Viruses*, edited by S.S. Morse, New York: Raven Press

Gabriel, A. & Boeke, J.D. (1993) Retrotransposon reverse transcription. In *Reverse Transcriptase*, edited by A.M. Skalka & S.P. Goff, Cold Spring Harbor, New York: Cold Spring Harbor Laboratory Press

Labrador, M. & Corces, V.G. (1997) Transposable element–host interactions: regulation of insertion and excision. *Annual Review of Genetics* 31: 381–404

McDonald, J.F. (1993) *Transposable Elements and Evolution*, Dordrecht: Kluwer

Sherratt, D.J. (ed.) (1995) *Mobile Genetic Elements*, Oxford: IRL Press

GLOSSARY

hybrid dysgenesis production of sterile offspring which display chromosomal abnormalities, mutations, etc

RNase (ribonuclease) enzyme that cleaves ribonucleic acid (RNA)

See also **Barbara McClintock and transposable genetic sequences in maize** (p.640)

THE ROLE OF *PAX-6* HOMOLOGUES IN THE DEVELOPMENT OF THE EYE AND OTHER TISSUES DURING THE EMBRYOGENESIS OF VERTEBRATES AND INVERTEBRATES

Kenneth B.R. Ewan[1] and Peter Gruss[2]
[1]*Life Sciences Division, Lawrence Berkeley National Laboratory, Berkeley, USA*
[2]*Department of Molecular Cell Biology, Max Planck Institute, Göttingen, Germany*

1. Introduction: the *Pax* gene family

Pax proteins are transcription factors that regulate the expression of genes coding for proteins as functionally varied as other transcription factors, crystallins, receptors, peptide hormones and cell adhesion molecules (reviewed in Gruss & Walther, 1992; Noll, 1993; Mansouri *et al.*, 1996a), which in turn modulate embryonic development. Pax proteins regulate gene transcription by binding to specific sequences in the enhancer and promoter regions upstream of the coding regions of genes. These sequences are bound by domains that have highly conserved amino acid sequences. All Pax proteins contain the paired domain, a 136 amino acid domain that has DNA-binding activity (reviewed in Gruss & Walther, 1992; Noll, 1993; Mansouri *et al.*, 1996a). Changes to amino acid sequences in this domain have been demonstrated to inhibit strongly the binding of Pax proteins to DNA (Chalepakis *et al.*, 1991; Vogan *et al.*, 1993) and particular amino acid residues of this domain determine the specificity of the DNA sequence recognized (Czerny & Busslinger, 1995). As well as the paired domain, Pax proteins also carry one or both of two other conserved domains: the paired-type homeobox and the octapeptide sequence. The paired-type homeodomain is crucial for the functions of the Pax proteins that carry it; mutations in this domain, for example, severely reduce protein activity in *Pax-3* (Epstein *et al.*, 1991). The presence or absence of the octapeptide and the paired-type homeodomain is used to classify the Pax proteins, which are shown in **Table 1**.

Pax transcription factors have been found in many animal phyla. In vertebrates, Pax factors help to regulate the morphogenesis of many tissues during embryonic development, including processes such as cell lineage commitment, the maintenance of boundaries between regions in segmented structures such as the central nervous system, the establishment of competence within structures to respond to external signals and regulation of extracellular matrix composition. Thus, Pax factors are required for the correct development of such tissues as the eye, olfactory structures, the forebrain, the region separating the midbrain and hindbrain, the B-cell lineage, the follicular cells of the thyroid, the vertebral column, limb muscle, kidney and neural crest-derived structures such as the dorsal root and autonomic ganglia (see **Table 1** for gene functions). In Drosophila, some *Pax* genes have long been known. These include *paired*, which codes for a segmentation regulator, and which was the first of the *Pax* factors to be isolated; other factors include *gooseberry* and *gooseberry neuro*, which regulate the development of the segmented cuticle and neural structures, respectively (reviewed in Noll, 1993). A recently characterized factor is *sparkling*, related to vertebrate *Pax-2*, *Pax-5* and *Pax-8*, which is instrumental in the specification of the non-photoreceptor cells in the Drosophila eye (Fu & Noll, 1997). *Pax* genes have also been isolated from ascidians, *Caenorhabditis elegans* and squid.

By far the most studied Pax factor is *Pax-6*, which was first found in vertebrates. Highly conserved homologues have been isolated in non-vertebrate chordates, Drosophila, squid, *C. elegans*, ribbonworm and the sea urchin. In many of these organisms, *Pax-6* expression is associated with the development of eyes, chemosensory structures and anterior neural tissue. *Pax-6* has multiple roles in vertebrate embryogenesis, specifically in the eye, nasal and forebrain structures, the endocrine pancreas, hindbrain and spinal cord, and a demonstrated early function in Drosophila eye morphogenesis. Thus, a discussion of *Pax-6* not only details the activity of the most remarkable Pax protein, but also serves to illustrate the mechanisms of action of other Pax family members.

2. Conservation of sequence, structure and biological activity of *Pax-6* across phyla

Pax-6 is a member of the Pax transcription factor family whose distinguishing feature is the paired DNA-binding domain (reviewed in Gruss & Walther, 1992; Noll, 1993; Mansouri *et al.*, 1996a). In the case of *Pax-6*, there is another DNA-binding domain (the paired-type homeodomain), a small N-terminus region, a linker region between the paired and homeodomains, and a C terminal region (see **Figure 1**). Within the vertebrates, the sequence of the whole protein is extremely conserved (95–100%; see **Table 2**). Indeed, the proteins encoded by the human, rat and mouse genes are identical (Ton *et al.*, 1991; Walther & Gruss, 1991; Matsuo *et al.*, 1993). The two DNA-binding domains – the paired domain and the paired class homeodomain – retain 90% amino acid sequence identity across phyla, but the other domains are not conserved across phyla (see **Table 2**). For example, the C-terminal domain of the protein encoded by *eyeless*, the vertebrate *Pax-6* homologue in Drosophila, has only 8% identity with Xenopus *Pax-6* – while mouse and zebrafish have 98% and 93% identity, respectively (Hirsch & Harris, 1997) – and varies in length by as much as 100%.

Table 1 Function and presence of DNA-binding domains of the *Pax* genes of vertebrates and Drosophila.

Gene	Species	Octapeptide	Paired-like homeodomain	Function or expression	Reference
Pax-1	Vertebrate	Yes	No	Roles in vertebral formation, thymus epithelium development	Chalepakis et al., 1991; Wallin et al., 1994, 1996
Pax-2	Vertebrate	Yes	Truncated	Roles in formation of kidney, optic nerve and chiasma, midbrain/hindbrain boundary maintenance	Sanyanusin *et al.*, 1995; Schimmenti *et al.*, 1995 Torres *et al.*, 1995, 1996; Macdonald *et al.*, 1997; Schwarz *et al.*, 1997; Urbanek *et al.*, 1997
Pax-3	Vertebrate	Yes	Yes	Roles in the migration/proliferation of neural crest cells and limb muscle precursor cells	Franz & Kothary, 1990, 1993; Epstein *et al.*, 1991; Steel & Smith, 1992; Tassabehji *et al.*, 1992; Vogan *et al.*, 1993; Daston *et al.*, 1996; Yang *et al.*, 1996; Conway *et al.*, 1997
Pax-4	Vertebrate	No	Yes	Specifies insulin producing endocrine cell lineage in pancreas	Sosa-Pineda *et al.*, 1997
Pax-5	Vertebrate	Yes	Truncated	B-cell lineage specification; midbrain/hindbrain boundary maintenance	Schwarz *et al.*, 1997; Urbanek *et al.*, 1994, 1997
Pax-6	Vertebrate	No	Yes	Direct lens development; roles in retina development, forebrain boundary maintenance; specification of glucagon producing endocrine cell lineage in the pancreas and of certain motorneuron and interneuron cell types in the hindbrain and spinal cord	Hogan *et al.*, 1986; Hill *et al.*, 1991; Ton *et al.*, 1991; Walther & Gruss, 1991; Jordan *et al.*, 1992; Plueschel *et al.*, 1992; Matsuo *et al.*, 1993; Epstein *et al.*, 1994; Fujiwara *et al.*, 1994; Glaser *et al.*, 1994; Hanson *et al.*, 1994; Li *et al.*, 1994; Czerny & Busslinger, 1995; Grindley *et al.*, 1995, 1997; Cvekl & Piatigorsky, 1996; Quinn *et al.*, 1996; Schedl *et al.*, 1996; Stoykova *et al.*, 1996, 1997; Altman *et al.*, 1997; Ericson *et al.*, 1997; Hirsch & Harris, 1997; Macdonald & Wilson, 1997; Mastick *et al.*, 1997; Osumi *et al.*, 1997; St-Onge *et al.*, 1997; Warren & Price, 1997
Pax-7	Vertebrate	Yes	Yes	Role in the migration of anterior neural crest cells; expression in dermomyotome and myotome of somite but function here unknown	Mansouri *et al.*, 1996b
Pax-8	Vertebrate	Yes	Truncated	Specification and/or proliferation of thyroid follicular cells	Mansouri & Gruss, 1998
Pax-9	Vertebrate	Yes	Yes	Expressed in vertebral and tooth primordia, function unknown	Neubuser *et al.*, 1995, 1997
eyeless (Pax-6 homologue)	Vertebrate	No	Yes	Required to initiate eye morphogenesis, possible role in photoreceptor specification; expressed in developing neural tissue, function unknown	Quiring *et al.*, 1994; Halder *et al.*, 1995a, 1998; Sheng *et al.*, 1997

Gene	Species	Octapeptide	Paired-like homeodomain	Function or expression	Reference
sparkling[1]	Drosophila	Yes	Truncated	Specification of non-neural ommatidal cells	Fu & Noll, 1997
paired	Drosophila	No	Yes	Primary segmentation of the Drosophila embryo	Nusselein-Volhard & Wieschaus, 1980; Kilcherr *et al.*, 1986; Gutjahr *et al.*, 1993a
gooseberry	Drosophila	No	Yes	Patterning of cuticle; induces *gooseberry neuro*	Baumgartner *et al.*, 1987; Patel *et al.*, 1989; Gutjahr *et al.*, 1993b
gooseberry neuro	Drosophila	No	Yes	Patterning of central nervous system, specification of posterior commisures	Baumgartner *et al.*, 1987; Patel *et al.*, 1989; Gutjahr *et al.*, 1993b
pox meso	Drosophila	Yes	No	Expressed in somatic mesoderm, function unknown	Bopp *et al.*, 1989
pox neuro	Drosophila	Yes	No	Specification of chemosensory neurons	Bopp *et al.*, 1989; Dambly-Chaudiere *et al.*, 1992; Nottebohm *et al.*, 1992

[1] Drosophila homologue to the *Pax-2*, *Pax-5* and *Pax-8* genes found in vertebrates

Table 2 Level of conservation of amino acid sequence of the different domains of *Pax-6*.

Species	Percentage identity to mouse domain (AA)				Reference
	Paired domain	Linker	Homeodomain	C-terminal	
Human	100	100	100	100	Chalepakis *et al.*, 1991
Rat	100	100	100	100	Noll, 1993
Xenopus	98	88	100	97	Czerny & Busslinger, 1995
Zebrafish	99	94	100	92	Fu & Noll, 1997
Ascidian (*P. mammillata*)	87	3	95	10	Callaerts *et al.*, 1997
Drosophila eyeless	95	6	90	1	Tomarev *et al.*, 1997
Squid (*L. opalescens*)	96	26	92	23	Hill *et al.*, 1991
Ribbonworm	93	19	93	13	Hara-Nashimura *et al.*, 1993
Nematode (*C. elegans*)	80	18	93	7	Stoykova *et al.*, 1996, 1997
Sea urchin	89	40	85	47	Matsuo *et al.*, 1993

However, the C-terminal domain has been shown to be required for transactivation in *in vitro* studies and some natural truncations of the C-terminus (Sey^{1Neu}, *rSey* and in human *Pax-6*) display phenotypes similar, if less severe, to mutations in the DNA-binding domains (in both paired and paired-type homeobox) (Hill *et al.*, 1991; Matsuo *et al.*, 1993; Glaser *et al.*, 1994; Czerny & Busslinger, 1995).

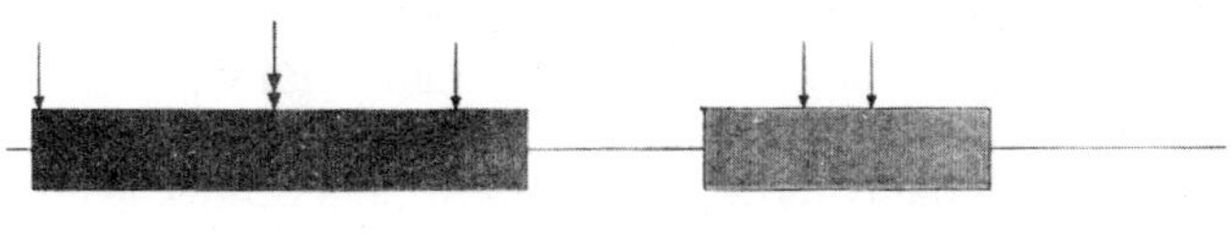

Figure 1 Structure of *Pax-6*. The single arrows show the approximate locations of splicing sites that are conserved across the phyla. The double arrow shows the insertion site for the extra 46 amino acids in the paired domains of vertebrate splice variants.

The genomic organization of the *Pax-6* homologues also displays features of conservation. Two splice sites in the paired domain and two in the homeodomain are found in all the vertebrate genes and in the genes of all invertebrates studied except *C. elegans* (Callaerts *et al.*, 1997; see also **Figure 1**). Outside of these domains, the splice sites are not conserved across the phyla (Callaerts *et al.*, 1997; Hirsch & Harris, 1997). Vertebrates also have an extra splice form in which an extra 42 base pairs (bp; equivalent to 14 amino acids) is spliced into the paired homeobox (Ton *et al.*, 1991; Walther & Gruss, 1991; Plueschel *et al.*, 1992; Hirsch & Harris, 1997). *Pax-6*'s DNA interactions can involve either or both of the two DNA-binding domains. Conserved binding sites for the paired domain have been found in vertebrates (Epstein *et al.*, 1994; Czerny & Busslinger, 1995) and homeodomain sites have been found in the *cis*-regulatory regions of Drosophila and vertebrate rhodopsin genes (Walter & Gruss, 1991; Hirsch & Harris, 1997).

In Drosophila, the biological activity of Pax-6 homologue proteins has been examined by assaying their ability to induce ectopic eyes in antennae, legs and haltere imaginal discs, using a Gal4 ectopic expression system that had previously been used with the *eyeless* gene (Halder *et al.*, 1995a). When mouse (Halder *et al.*, 1995a), squid (Tomarev *et al.*, 1997) or ascidian *Pax-6* (Glardon *et al.*, 1997) were expressed in Drosophila imaginal discs using this system, ectopic eyes were induced. Although a proper rescue of the eye-specific *eyeless* mutants (ey^2 and ey^R)was not attempted, this finding does demonstrate that there is at least some conservation of biological activity between vertebrates and invertebrates.

Although the vertebrate gene may be able to replace the Drosophila gene, the reverse does not necessarily happen. An example of this is the Drosophila gene *orthodenticle* and its mouse homologue *Otx-1*. Targeted expression of *Otx-1* in the *orthodenticle* mutant (Leuzinger *et al.*, 1998) results in a full rescue of the phenotype, but some defects were not rescued at all in the reverse experiment of expressing *orthodenticle* in the *Otx-1* null mutant mouse (Acampora *et al.*, 1998). This result may be attributable to the non-conserved domains in the proteins.

In the case of *Pax-6*, the non-conserved C-terminal domains of mouse and sea urchin proteins have been shown to transactivate genes (Glaser *et al.*, 1994; Czerny & Busslinger, 1995). Thus, the expression of *eyeless* in *Pax-6* mutant mice may discriminate between murine *Pax-6* functions that depend only on the paired and paired-type homeobox domains from other functions that additionally require the non-conserved C-terminal and linker domains.

3. Conserved *Pax-6* interactions

Pax-6 homologues have been implicated in the regulation of two classes of genes that are conserved across phyla. The first group includes the genes for the transcriptional regulators *eyes absent* (*eya*) in Drosophila and their vertebrate homologues *Eya-1* and *-2*, *sine oculis* and *dachshund*. The second class consists of the opsin genes that code for the protein components of photosensitive pigments. These two groups will be considered in turn.

After the characterization of the *eyeless* locus in Drosophila, more transcriptional regulators that induce eye formation were found in three other loci associated with eye agenesis. These are *dachshund* or *dac* (Shen & Mardon, 1997), *eya* (Bonini *et al.*, 1993) and *sine oculis* or *so* (Cheyette *et al.*, 1994). Initial experiments suggested that these factors had a very similar role to that of *eyeless* as defined by Gehring and co-workers (Cheyette *et al.*, 1994; Quiring *et al.*, 1994), because of their eye induction capability. Later studies showed, however, that expression of all of these factors can be induced by *eyeless* and that their expression fails to occur in *eyeless* mutants (Halder *et al.*, 1995a; Shen & Mardon, 1997). Null mutants *of sine oculis*, *dachshund* and *eya* fail to develop eyes (Bonini *et al.*, 1993; Cheyette *et al.*, 1994; Mardon *et al.*, 1994). Moreover, ectopic eye formation in imaginal discs forced to express *eyeless* fails to occur in null mutants for these genes (Bonini *et al.*, 1997; Shen & Mardon, 1997; Halder *et al.*, 1998).

The Eya protein interacts with So or with Dac to form complexes that synergistically induce eye development (Chen *et al.*, 1997; Pignoni *et al.*, 1997). Complexes of *eya* with *sine oculis* or with *dachshund* induce *eyeless* expression (Bonini *et al.*, 1997; Chen *et al.*, 1997; Pignoni *et al.*, 1997; Halder *et al.*, 1998). Thus, *eyeless* could be part of a regulatory loop (Desplan, 1997). Vertebrate homologues of these Drosophila genes have been shown to be coexpressed with *Pax-6* in the mouse and zebrafish eye. These include the *sine oculis* homologues *Six-3*, *Six-4*, *Six-5* and *Six-6* (Oliver *et al.*, 1995; Kawakami *et al.*, 1996; Seo *et al.*, 1998), and the *eyes absent* homologues *Eya-1* and *Eya-2* (Xu *et al.*, 1997). A vertebrate homologue of *dachshund* has been cloned (Shen & Mardon, 1997), but has not yet been described. The vertebrate *Eya* and *Six* genes show conservation at the amino acid level with Eya and So in the domains that have been shown to interact in Drosophila (Oliver *et al.*, 1997; Pignoni *et al.*, 1997; Xu *et al.*, 1997). Overexpression of *Pax-6* and *Six-3* in Xenopus and medaka fish, respectively, results in the induction of crystallin-expressing lens-like elements in the ectoderm in both cases (Altman *et al.*, 1997; Oliver *et al.*, 1997). These findings suggest that the role of the *Pax-6* and *Six-3* genes in the presumptive lens ectoderm could be analogous to that of *eyeless* and *sine oculis* in the eye imaginal disc of Drosophila. Furthermore, *Eya-2* can rescue the phenotype in Drosophila *eya* mutants, demonstrating a conservation of biological activity between the two protein products (Bonini *et al.*, 1997). The roles of the *Eya* genes and *Six-4*, *-5* and *-6* in vertebrate eye development are presently unknown.

Pax-6 has been implicated in relationships with the *opsin* genes that are conserved across phyla. *Opsin* genes encode a class of proteins, including rhodopsin, which form photosensitive pigments (Nathans, 1992; Maden, 1995). The *Pax-6* P3 homeodomain DNA-binding site has been found in the *cis*-regulatory elements of the Drosophila and vertebrate *rhodopsin* genes (Wilson *et al.*, 1995). Additionally, *eyeless* can transactivate the P3 site in Drosophila *rhodopsin-1* upstream sequences in a dimerized form (Sheng *et al.*, 1997). Squid opsins have also been cloned, but the role of squid *Pax-6* in their regulation is presently unknown (Hara-Nashimura *et al.*, 1993; Maden, 1995). In Drosophila, *eyeless* is expressed in ommatidia just prior to and during the terminal stages of photoreceptor differentiation and in larval photoreceptor (Bollwig) cells (Sheng *et al.*, 1997) and is also able to transactivate the rhodopsin promoter (Sheng *et al.*, 1997). The significance of the *Pax-6* binding sites in the upstream sequences of *rhodopsin* genes (Wilson *et al.*, 1995) is still unclear, because of the lack of expression of rhodopsin in vertebrate photoreceptors (Walther & Gruss, 1991; Hirsch & Harris, 1997; Macdonald & Wilson, 1997). Overdosage of mice with copies of human *PAX-6* genomic loci has been shown to inhibit photoreceptor differentiation partially or totally (Schedl *et al.*, 1996), but ectopic expression of *Pax-6* in Xenopus has shown no effect on photoreceptor differentiation (Hirsch & Harris, 1997).

4. *Pax-6* and eye development

Pax-6 homologues have been associated with eye development in all model organisms in a number of phyla, including the vertebrates (Hogan *et al.*, 1986; Hill *et al.*, 1991; Ton *et al.*, 1991; Walther & Gruss, 1991; Plueschel *et al.*, 1992; Matsuo *et al.*, 1993; Li *et al.*, 1994; Grindley *et al.*, 1995; Schedl *et al.*, 1996; Hirsch & Harris, 1997), Drosophila (Quiring *et al.*, 1994; Halder *et al.*, 1995b; Sheng *et al.*, 1997), squid (Tomarev *et al.*, 1997), ribbonworm (Loosli *et al.*, 1996) and the ascidian (Glardon *et al.*, 1997). The visual structures in different phyla are very dissimilar (reviewed in Land & Fernald, 1992; Fernald, 1997) and comparison of the structure and morphogenesis of eyes across the phyla has led to the widely held view that eye structures evolved independently and separately (Salvini-Plawen & Mayr, 1977), rather than from simple patches of opsin-expressing photosensitive cells. Even the most fundamental cell of the eye, the photoreceptor, varies in morphology and in its physiological response to light stimulus (reviewed in Salvini-Plawen & Mayr, 1977; Goldsmith, 1990; Land & Fernald, 1992; Fernald, 1997). The permutations in eye morphology range from patches of photoreceptors and photosensitive pits to compound eyes and eyes with lenses, irises and corneas (see **Figure 2** for examples). Although some molecules in eyes, such as *Pax-6* and opsins (see below), are homologous, this in no way infers that the structures are anatomically or evolutionarily homologous (reviewed in Dickinson, 1995; Fernald, 1997). Indeed, other molecules in eyes are not conserved across phyla. Crystallins, for example, are found in both vertebrate and invertebrate species, but the invertebrate crystallins are not conserved across phyla and have no sequence similarity with vertebrate proteins (reviewed in Tomarev & Piatigorsky, 1996). This suggests *Pax-6* has functional roles in the morphogenesis of analogous structures, not homologous structures.

The Drosophila compound eye has around 750 facets or ommatidia. Each ommatidium consists of photoreceptors, cone cells, pigment cells, accessory cells and a secreted lens (reviewed in Wolff & Ready, 1991). Patterning of the eye portion of the eye/antenna imaginal disc occurs when a groove known as the morphogenetic furrow progresses anteriorly from the posterior edge of the imaginal disc in a process requiring the secreted factors Decapentaplegic and Hedgehog (Heberlein *et al.*, 1993; Ma *et al.*, 1993). This process patterns the ommatidia of the eye. Expression of *eyeless* occurs in the unpatterned region of the eye disc anterior to the morphogenetic furrow and is attenuated by the arrival of the furrow (Quiring *et al.*, 1994; see **Figure 2**). Eye-specific ablation of *eyeless* expression in the ey^2 and ey^R mutants results in partial or total lack of eye formation (Quiring *et al.*, 1994), while expression of *eyeless* in leg, wing or antenna imaginal discs results in the generation of ectopic eyes. This has led to the suggestion that *eyeless* is a "master regulator of eye morphogenesis" (Quiring *et al.*, 1994; Halder *et al.*, 1995a,b). Later findings that the expression of *dachshund, eya* and *sine oculis* in imaginal discs can also generate ectopic eyes (Bonini *et al.*, 1993; Cheyette *et al.*, 1994; Shen & Mardon, 1997) prompted the suggestion that there was not just one "master regulator", but a group of genes acting as a regulatory loop (Desplan, 1997). The Dachshund, So and Eya proteins directly control the progression of the morphogenetic furrow and the expression of *hedgehog* and *decapentaplegic* (Bonini *et al.*, 1993; Cheyette *et al.*, 1994; Mardon *et al.*, 1994; Bonini *et al.*, 1997; Shen & Mardon, 1997). Both So and Dachshund have been found to promote these processes synergistically with the Eya protein, possibly as components of a transactivation complex (Chen *et al.*, 1997; Pignoni *et al.*, 1997). Thus, the role of eyeless is to initiate and maintain the expression of these proteins, which then direct the morphogenetic furrow in patterning the eye. Later expression of *eyeless* is seen in the ommatidia before and during receptor differentiation (Sheng *et al.*, 1997) and the ability of the eyeless protein to transactivate the *rhodopsin-1* gene (Sheng *et al.*, 1997) shows that *eyeless* could have a later role in specifying photoreceptors. Thus, *eyeless* has a fundamental role in early eye morphogenesis and a more specific role in later development in Drosophila.

The vertebrate Pax-6 protein is fundamental to lens formation, a process which is needed for the correct development of the optic cup into the retina. The earliest phases of vertebrate eye morphogenesis, the formation of the optic vesicle and the secretion of a lens-inducing factor by the optic vesicle are not directed by *Pax-6*. The paired-type homeodomain factor, *Rx*, is required to develop optic vesicles (Mathers *et al.*, 1997) and the transcription factor, Lhx-2, is needed to develop lens placodes (Porter *et al.*, 1997). The optic vesicles developed by homozygotes for *Pax-6* mutations have abnormal morphology (Hogan *et al.*, 1986; Matsuo *et al.*, 1993; Grindley *et al.*, 1995; Oliver *et al.*, 1997), but can still induce lens placodes expressing lens-specific markers in wild-type ectoderm in recombination assays (Fujiwara *et al.*, 1994). However, *Pax-6* is of crucial importance in conferring lens competence in the cranial ectoderm. In mice or rats homozygous for the *Pax-6* mutations *Sey*, Sey^{1Neu} and *rSey* (Hogan *et al.*, 1986; Matsuo *et al.*, 1993; Grindley *et al.*, 1995; Oliver *et al.*, 1997), lens development fails to occur. Expression of *Pax-6* in the cranial ectoderm and in the induced lens placode (Hill *et al.*, 1991; Walther & Gruss, 1991; Li *et al.*, 1994; Grindley *et al.*, 1995; Hirsch & Harris, 1997) suggested a cell autonomous function for *Pax-6* in lens placode specification. A number of other findings also suggest cell autonomy. In explant recombination experiments, the homozygous *rSey* ectoderm cannot be induced to form lens tissue by optic vesicle tissue (Fujiwara *et al.*, 1994). Additionally, *sey* homozygote cells are excluded from the lens placode in chimeras consisting of wild-type and *sey* homozygote embryonic cells (Quinn *et al.*, 1996). This has been complemented by a Xenopus overexpression study in which crystallin-expressing lens-like structures were induced in the ectoderm (Altman *et al.*, 1997). *Pax-6* has been shown to bind to the promoter/enhancer sequences of several crystallin genes (reviewed in Cvekl & Piatigorsky, 1996), implicating a role for *Pax-6* in later lens development.

Pax-6 is less prominent in retinal development because of the critical role of the lens in optic cup and retinal formation. Removal of the lens placode or developing lens vesicle severely impairs development of the neural retina (Pittack *et al.*, 1997). This can at least partially be compensated for by the addition of exogenous fibroblast growth factor-1 or -2 (Pittack *et al.*, 1997). However, a study of chimeric embryos

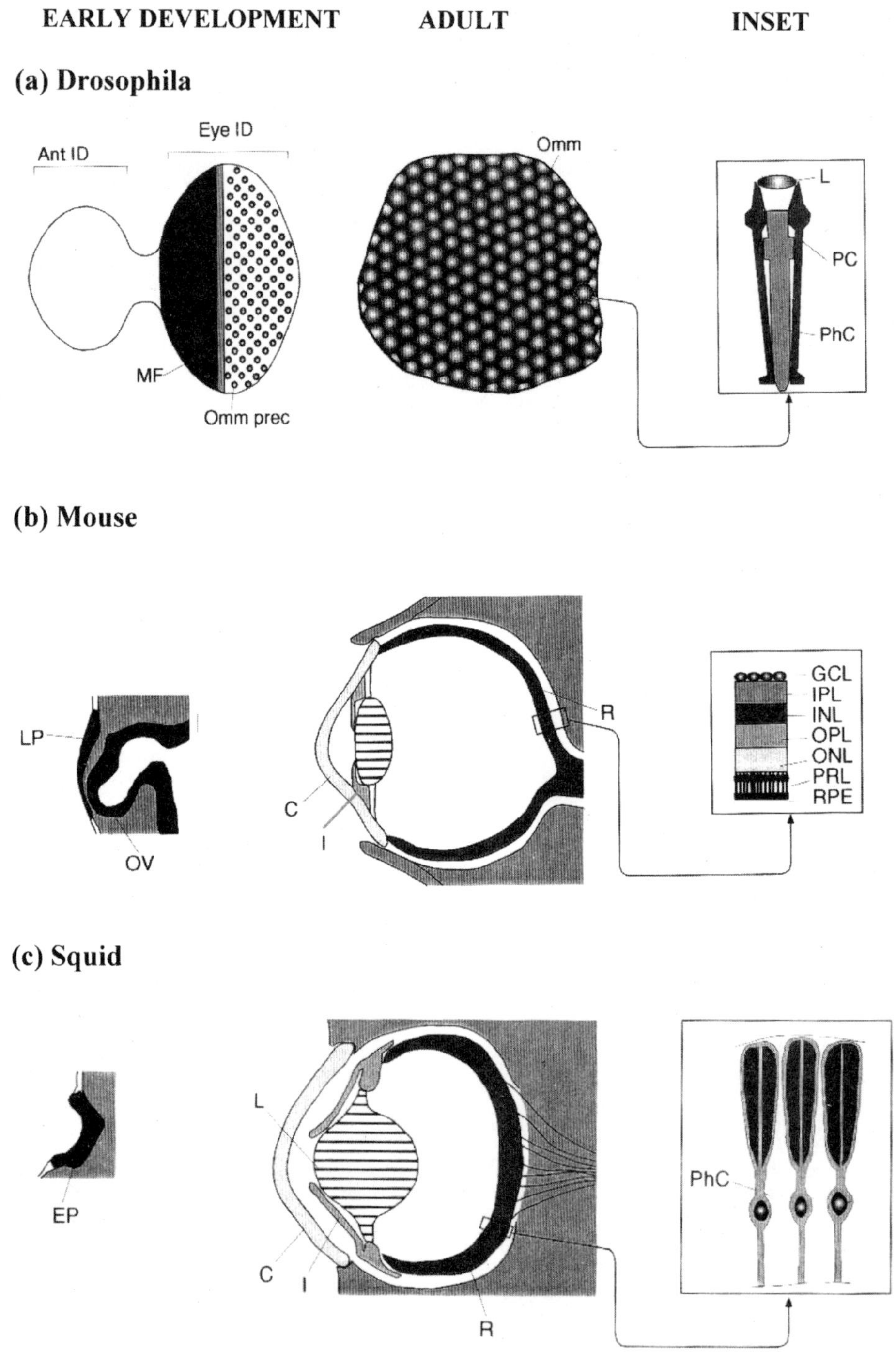

Figure 2 *Pax-6* expression during the development of analogous visual structures. Expression in the early developing structures is shown in black. Ant ID, antenna imaginal disc; C, cornea; EP, eye placode; Eye ID, eye imaginal disc; GCL, ganglionic cell layer; I, iris; INL, inner nuclear layer; IPL, inner plexiform layer; L, lens; LP, lens placode; MF, morphogenetic furrow; Omm prec, ommatidia precursors; Omm,-? ommatida; ONL, outer nuclear layer; OPL, outer plexiform layer; OV, optic vesicle; PC, pigmented cells; PhC, photoreceptor cells; PRL, photoreceptor layer; R, retina; and RPE, retinal pigmented epithelium.

composed of *Pax-6* null mutant cells (*sey/sey*) and wild-type cells showed that the *sey/sey* cells were at a selective disadvantage in the presumptive neural retina layer of the optic cup of eyes with relatively well-developed lenses and were excluded from the pigmented retinal epithelium altogether (Quinn *et al.*, 1996). Thus, the impairment of the further development of the optic cup is due to both the lack of lens tissue and the cell autonomous effects of *Pax-6*. In later stages, *Pax-6* expression is found in the inner nuclear layer and the retinal ganglion cell layer, but is downregulated in the photoreceptor cell layer (Walther & Gruss, 1991; Hirsch & Harris, 1997; Macdonald & Wilson, 1997). This suggests that *Pax-6* has a role in the final differentiation of some retinal cell types.

Analyses of *Pax-6* genes in squid, ribbonworm and ascidian have found an association of *Pax-6* homologue expression with developing or regenerating eye structures in these organisms. The squid (*Loligo opalescens*) eye has an overall similarity to the vertebrate eye in that it has a retina, lens, cornea and iris (see **Figure 2**). However, the squid retina is not multilayered like the vertebrate retina and the light-sensitive regions of squid photoreceptor cells face the incident light instead of away as in vertebrate photoreceptors. The greatest difference between squid eye morphogenesis and that of the vertebrate eye is that the entire squid eye develops from one structure, the eye placode, rather than from two as in vertebrates (see **Figure 2**). This is evidence for the hypothesis that squid and vertebrate eyes are products of convergent evolution rather than being descended from the same ancestral structure (reviewed in Salvini-Plawen & Mayr, 1977; Land & Fernald, 1992; Fernald, 1997).

The squid *Pax-6* gene is expressed in the invaginating placode and is later found in the developing cornea (Hill *et al.*, 1991). In ribbonworms, expression of a *Pax-6* homologue is found in developing and regenerating pit eyes consisting of photoreceptors and pigmented cells (Hara-Nashimura *et al.*, 1993). Another *Pax-6* homologue is expressed in the ocellus (primitive eye) of the ascidian (Glardon *et al.*, 1997). Thus, *Pax-6* is associated with developing and regenerating eyes in certain species in invertebrate phyla and in the non-vertebrate chordates.

In Drosophila, *eyeless* is fundamental in determining imaginal disc fate along with *eyes absent*, *sine oculis* and *dachshund*. Like *eyeless* in the Drosophila imaginal disc, *Pax-6* plays a critical role in the development of a region of the vertebrate cranial ectoderm into the lens along with the *sine oculis* homologue *Six-3*. However, optic vesicle development does not require the action of Pax-6 protein (Hogan *et al.*, 1986; Hill *et al.*, 1991; Grindley *et al.*, 1995), but does require the action of another factor, *Rx* (Mathers *et al.*, 1997). *Pax-6* has a function in optic cup development (Quinn *et al.*, 1996), but there are clearly other influences such as the fibroblast growth factors secreted from the lens vesicle (Pittack *et al.*, 1997). Both vertebrate *Pax-6* and Drosophila *eyeless* have later roles transactivating genes for functional proteins such as rhodopsin-1 in Drosophila and the crystallins in vertebrates.

5. Other *Pax-6* functions

Pax-6 homologues are also involved in the morphogenesis of non-visual tissues, including the central nervous system, some chemosensory structures and the vertebrate endocrine pancreas. *Pax-6* is involved both in the early phase of regulating the morphogenesis of primordial structures and in the later phase of cell type specification, which is similar to the action of *Pax-6* in the eye.

Pax-6 has functions in patterning the central nervous system of vertebrates and in specifying neuronal cells in *C. elegans*. It is also expressed in developing the neural tissues of squid and Drosophila. In the vertebrate forebrain, *Pax-6* has been shown to establish correctly the boundaries between the structures within the forebrain and between the forebrain and the midbrain. In *Sey* homozygous embryos, the caudal forebrain expresses markers that are characteristic of the rostral midbrain (Grindley *et al.*, 1997; Mastick *et al.*, 1997; Warren & Price, 1997). Moreover, there is distortion of the boundary between the cortical and striatal region (Stoykova *et al.*, 1997) and of that between the hypothalamus and telencephalon (Stoykova *et al.*, 1996). *Pax* genes have also been implicated in establishing regional identity in the hindbrain and the spinal cord. *Pax* gene expression domains are regulated by sonic hedgehog signalling from the notochord and floorplate and by transforming growth factor-β (TGF-β superfamily members signalling from the overlying ectoderm and roofplate (reviewed in Tanabe & Jessell, 1996; Pituello, 1997). This results in a *Pax-6* expression domain in the middle region of the hindbrain and spinal cord, which is excluded from the dorsal and ventral extremities (Ericson *et al.*, 1997). The level of Pax-6 protein has been postulated to help in specifying the type and target tissue of neurones (Ericson *et al.*, 1997). In *Sey* homozygote mutants, motor neurones supplying the hypoglossal nerve from the hindbrain are absent and the number of ventral interneurones is reduced (Ericson *et al.*, 1997; Osumi *et al.*, 1997). This neuronal specification function of *Pax-6* in vertebrates is similar to that found in *C. elegans*, where head neuronal fates are transformed in null mutants (Chisholm & Horvitz, 1995). In Drosophila, *eyeless* gene expression is seen in neuroblasts of the ventral nervecord that give rise to the optic lobes and in some ganglion mother cells. Although there were a number of natural *eyeless* mutants including embryonic lethals, all but the ey^2 and ey^R eye phenotype mutants have now died out. Thus, the role of *eyeless* in the Drosophila brain is not yet known. The squid *Pax-6* gene is expressed in the developing brain (Tomarev *et al.*, 1997), but, again, its functional role has yet to be determined.

Expression of *Pax-6* homologues occurs in some chemosensory tissues: the nasal structures of vertebrates, the olfactory organ and the chemosensory cells of the arms of the squid, and also the ray cells of *C. elegans*. In homozygote *Sey* mice, the olfactory bulbs and the nasal placode fail to form and later the nasal processes suffer massive cell death (Hogan *et al.*, 1986; Hill *et al.*, 1991; Grindley *et al.*, 1995). In chimeric embryos composed of *sey/sey* and wild-type cells, the *sey/sey* cells are excluded from the nasal placode (which is similar to what happens in the lens placode), demonstrating a cell autonomous effect (Quinn *et al.*, 1996). In *C. elegans*, a *Pax-6* splice variant lacking the paired domain is required for the specification of some of the sensory cells known as rays (Zhang & Emmons, 1995). *Pax-6* gene expression is also seen in the olfactory organ of the squid and in the epithelium of its suckers, which contain chemosensory cells (Tomarev *et al.*, 1997).

The endocrine cell populations of the vertebrate pancreas are found in the islets of Langerhans (reviewed in Slack, 1995). α, β, γ and PP cell lineages synthesize glucagon, insulin, somatostatin and pancreatic polypeptide, respectively. *Pax-6* null mutants lack glucagon-expressing cells (St-Onge *et al.*, 1997) and fail to form islets of Langerhans. Moreover, *Pax-6* binds to glucagon, insulin and somatostatin gene regulatory elements and can transactivate both the glucagon promoter and the insulin promoter (Sander *et al.*, 1997). Inactivation of the *Pax-4* gene has been shown to result in the failure of insulin- and somatostatin-producing cell populations to differentiate and in increased numbers of glucagon-producing cells (Sosa-Pineda *et al.*, 1997). In embryos lacking both *Pax-4* and *Pax-6*, endocrine pancreas cells fail to develop altogether (St-Onge *et al.*, 1997). Hence, the actions of *Pax-6*, together with *Pax-4*, are required to generate the full complement of endocrine pancreas cell types from a single endocrine progenitor.

6. Conclusions and perspectives

Pax-6 is a protein containing conserved DNA-binding domains that has been found in chordates, arthropods and molluscs, with identity ranging upwards from 80% in the paired domain and from 85% upwards in the homeodomain (Table 1). The conservation of these domains is sufficient to preserve biological activity in the context of inducing ectopic eyes in the imaginal discs of Drosophila pupae (Hill *et al.*, 1991; Halder *et al.*, 1995a; Glardon *et al.*, 1997). Although the C-terminal region of *Pax-6* has been found to be required for transactivation on the basis of *in vitro* experiments and analysis of naturally occurring human and rodent mutations, the imaginal disc expression assay has showed that changes in the sequence of this region have little effect on *Pax-6*'s interactions with downstream genes. Although the sequence of this region is not conserved, the regions all share the characteristic of being rich in proline, serine and threonine residues.

Another aspect of conserved *Pax-6* function is the relationship between the Pax-6 protein and the regulatory elements of conserved genes such as *eya* and *rhodopsin*.

Eye morphogenesis varies from phylum to phylum. The Drosophila eye develops from a single structure, the eye imaginal disc, whose structure is, in part, determined by *eyeless*. Later in eye development, eyeless protein is expressed in photoreceptors and may regulate their differentiation.

The vertebrate eye is formed from two structures, the optic vesicle and the lens placode. *Pax-6* facilitates the formation of the lens placode, but not the optic vesicle. Early events in the eye imaginal disc of Drosophila and in the developing lens placode of vertebrates are in some ways analogous. The homologues of *eyes absent* and *sine oculis*, *Eya-1* and -2, and *Six-3*, respectively, are expressed in the lens placode on its formation. In both Drosophila and vertebrates, *Pax-6* homologues in the developing eye act first in the determination of early eye structures and then later in the differentiation of cell types.

One hypothesis of *Pax-6*'s action is that it is a "master regulator of morphogenesis and evolution" of the eye (Halder *et al.*, 1995b, 1998). This hypothesis has arisen from three findings: ectopic expression of the Drosophila homologue, *eyeless*, induces eye development in the leg, wing and haltere imaginal discs (Quiring *et al.*, 1994; Halder *et al.*, 1995b); mouse *Pax-6* has important roles in eye morphogenesis (Hogan *et al.*, 1986; Hill *et al.*, 1991; Grindley *et al.*, 1995); and expression of mouse, squid and ascidian *Pax-6* genes can also induce ectopic eyes in these structures (Halder *et al.*, 1995a,b; Glardon *et al.*, 1997). This theory suffers from a number of drawbacks.

One argument against this hypothesis is that many researchers consider that eyes of different phyla have evolved independently (Salvini-Plawen & Mayr, 1977; Goldsmith, 1990; Land & Fernald, 1992; Fernald, 1997) and, thus, that the mechanisms regulating the development of these structures should have evolved independently. Drosophila eyes develop from one primordial structure (the imaginal disc), while vertebrate eyes develop from two (the optic vesicle and the lens placode). Of these two structures, only one – the lens placode – requires *Pax-6* for development (Hogan *et al.*, 1986; Hill *et al.*, 1991; Grindley *et al.*, 1995). Another problem arises from the placing of *Pax-6* in the regulatory mechanisms directing eye development in both phyla. In the fly, eyeless acts directly on the *sine oculis*, *eyes absent* and *dachshund* genes (Mardon *et al.*, 1994; Bonini *et al.*, 1997; Halder *et al.*, 1998). However, other genes in the morphogenetic furrow, whose proteins help pattern the ommatidia, cannot be directly activated by eyeless protein since eyeless protein is not expressed there (Quiring *et al.*, 1994). In the development of the mouse eye, the action of *Pax-6* is preceded by that of two other transcriptional regulators, Rx and Lhx-2 (Mathers *et al.*, 1997; Porter *et al.*, 1997). Thus, other transcriptional regulators are similarly critical to eye development in both Drosophila and the mouse.

The final argument against the "master regulator" hypothesis is that the expression and action of *Pax-6* homologues is not restricted to the eye. In vertebrates, *Pax-6* is expressed and has definitive instructive roles in the development of the central nervous system, nasal epithelium and the endocrine pancreas. Similarly, expression of *Pax-6* homologues is found in the developing nervous system and chemosensory structures of invertebrates. Thus, *Pax-6* can only have an instructive role in certain tissues, such as in the imaginal discs of Drosophila.

Comparison of the activity of *Pax-6* homologues in the morphogenesis of both visual and non-visual tissues has shown that *Pax-6* acts in two ways. The first is to regulate the genesis of early structures, such as the development of the Drosophila eye imaginal disc and the induction of the lens and nasal placodes (Hogan *et al.*, 1986; Hill *et al.*, 1991; Grindley *et al.*, 1995). In these contexts, *Pax-6* acts in the commitment of these structures to following certain developmental pathways. The second type of interaction is the regulation of genes expressed in specific cell types such as the *crystallins*, *rhodopsin-1*, *insulin* and *glucagon* genes (Wilson *et al.*, 1995; Cvekl & Piatigorsky, 1996; Sander *et al.*, 1997; Sheng *et al.*, 1997). Thus, *Pax-6* helps to specify cell types as well as patterning early tissues. The action of *Pax-6* homologues in maturing vertebrate lens fibres and Drosophila photoreceptors is implied by assays showing transactivation of *crystallin* (reviewed in Cvekl & Piatigorsky, 1996) and *rhodopsin-1* (Sheng *et al.*, 1997) genes, respectively, and by *Pax-6* expression in the differentiating cells. These modes of action are shared with other Pax proteins (reviewed in Gruss & Walther, 1992; Noll, 1993; Mansouri *et al.*, 1996a). In conclusion, then, *Pax-6* regulates both early and late events in vertebrate morphogenesis.

REFERENCES

Acampora, D., Avantaggio, T., Tuorto, F. *et al.* (1998) A murine *Otx-1* and *Drosophila otd* genes share conserved genetic functions required in invertebrate and vertebrate brain development. *Development* 125: 1691–702

Altman, C.R., Chow, R.L., Lang, R.A. & Hemmati-Brivanlou, A. (1997) Lens induction by *Pax-6* in *Xenopus laevis*. *Developmental Biology* 185: 119–23

Baumgartner, S., Bopp, D., Burri, M. & Noll, M. (1987) Structure of two genes at the *gooseberry* locus related to the *paired* gene and their spatial expression during *Drosophila* embryogenesis. *Genes and Development* 1: 1247–67

Bonini, N.M., Leiserson, W.M. & Benzer, S. (1993) The *eyes absent* gene: genetic control of cell survival and differentiation in the developing *Drosophila* eye. *Cell* 72: 379–95

Bonini, N.M., Bui, Q.T., Gray-Board, G.L. & Warrick, J.M. (1997) The *Drosophila eyes absent* gene directs ectopic eye formation in a pathway conserved between flies and vertebrates. *Development* 124: 4819–26

Bopp, D., Jamet, E., Baumgartner, S., Burri, M. & Noll, M. (1989) Isolation of two tissue-specific *Drosophila* paired box genes, *pox meso* and *pox neuro*. *EMBO Journal* 8: 3447–57

Callaerts, P., Halder, G. & Gehring, W.J. (1997) Pax-6 in development and evolution. *Annual Review of Neuroscience* 20: 483–532

Chalepakis, G., Fritsch, R., Fickenscher, H. *et al.* (1991) The molecular basis of the *undulated/Pax-1* mutation. *Cell* 66: 873–84

Chen, R., Amoui, M., Zhang, Z. & Mardon, G. (1997) Dachshund and eyes absent proteins form a complex and function synergistically to induce ectopic eye development in *Drosophila*. *Cell* 91: 893–903

Cheyette, B.N., Green, P.J., Martin, K. *et al.* (1994) The *Drosophila sine oculis* gene locus encodes a homeodomain protein-containing protein required for the development of the entire visual system. *Neuron* 12: 977–96

Chisholm, A.D. & Horvitz, H.R. (1995) Patterning of the *Caenorhabditis elegans* head region by the *Pax-6* family member *vab-3*. *Nature* 377: 52–55

Conway, S.J., Henderson, D.J. & Copp, A.J. (1997) Pax-3 is required for cardiac neural crest migration in the mouse: evidence from the *Splotch* (*Sp2H*) mutant. *Development* 124: 505–14

Cvekl, A. & Piatigorsky, J. (1996) Lens development and *crystallin* gene expression: many roles for Pax-6. *BioEssays* 18: 621–30

Czerny, T. & Busslinger, M. (1995) DNA binding and transactivation properties of Pax-6 — three amino acids in the paired domain are responsible for the different sequence recognition of Pax-6 and BSAP (Pax-5). *Molecular and Cellular Biology* 15: 2858–71

Dambly-Chaudiere, C., Jamet, E., Burri, M. *et al.* (1992) The paired box gene *pox neuro*: a determinant of poly-innervated sense organs in *Drosophila*. *Cell* 69: 159–72

Daston, G., Lamar, E., Olivier, M. & Goulding, M. (1996) Pax-3 is necessary for migration but not differentiation of limb muscle precursors in the mouse. *Development* 122: 1017–027

Desplan, C. (1997) Eye development governed by a dictator or a junta? *Cell* 91: 861–64

Dickinson, W.J. (1995) Molecules and morphology: where's the homology? *Trends in Genetics* 11: 119–21

Epstein, D.J., Verkemans, M. & Gros, P. (1991) Splotch (Sp2H), a mutation affecting development of the mouse neural tube, shows a deletion within the paired homeodomain of Pax-3. *Cell* 67: 767–74

Epstein, J.A., Cai, J., Glaser, T., Jepeal, L. & Maas, R. (1994) Identification of a Pax paired domain recognition sequence and evidence for DNA-dependent conformational changes. *Journal of Biological Chemistry* 269: 8355–61

Ericson, J., Rashbass, P., Schedl, A. *et al.* (1997) Pax-6 controls progenitor cell identity and neuronal fate in response to graded Shh signalling. *Cell* 90: 169–80

Fernald, R.D. (1997) The evolution of eyes. *Brain, Behaviour and Evolution* 50: 253–59

Franz, T. & Kothary, R. (1990) The *Splotch* mutation interferes with muscle development in the limbs. *Anatomy and Embryology* 138: 246–53

Franz, T. & Kothary, R. (1993) Characterization of the neural crest defect in *Splotch* (*Sp1H*) mutant mice using a lacZ transgene. *Developmental Brain Research* 72: 99–105

Fu, W. & Noll, M. (1997) The Pax2 homolog *sparkling* is required for development of cone and pigment cells in the Drosophila eye. *Genes and Development* 11: 2066–078

Fujiwara, M., Uchida, T., Osumi-Yamashita, N. & Eto, K. (1994) Uchida rat (*rSey*): a new mutant with craniofacial abnormalities resembling those of the mouse *Sey* mutant. *Differentiation* 57: 31–38

Glardon, S., Callaerts, P., Halder, G. & Gehring, W.J. (1997) Conservation of Pax-6 in a lower chordate, the ascidian *Phallusia mammillata*. *Development* 124: 817–25

Glaser, T., Jepeal, I., Edwards, J.G. *et al.* (1994) PAX6 gene dosage in a family with congenital cataracts, aniridia, anophthalmia and central nervous system defects. *Nature Genetics* 7: 463–71

Goldsmith, T.H. (1990) Optimization, constraint and history in the evolution of eyes. *Quarterly Review of Biology* 65: 281–322

Grindley, J.C., Davidson. D.R. & Hill, R.E. (1995) The role of Pax-6 in eye and nasal development. *Development* 121: 1433–42

Grindley, J.C., Hargett, L.K., Hill, R.E. *et al.* (1997) Disruption of Pax-6 by the Sey1Neu mutation produces abnormalities in the early development and regionalization of the diencephalon. *Mechanisms of Development* 64: 111–26

Gruss, P. & Walther, C. (1992) Pax in development. *Cell* 69: 719–22

Gutjahr, T., Frei, E. & Noll, M. (1993a) Complex regulation of early paired expression: initial activation by gap genes and pattern modulation by pair-rule genes. *Development* 117: 609–23

Gutjahr, T., Patel, N.H., Li, X., Goodman, C.S. & Noll, M. (1993b) Analysis of the gooseberry locus in *Drosophila* embryos: *gooseberry* defines cuticular pattern and activates *gooseberry neuro*. *Development* 118: 21–31

Halder, G., Callaerts, P. & Gehring, W.J. (1995a) Induction of ectopic eyes by targeted expression of the *eyeless* gene in Drosophila. *Science* 267: 1788–92

Halder, G., Callaerts, P. & Gehring, W.J. (1995b) New perspectives on eye evolution. *Current Opinion in Genes and Development* 5: 602–09

Halder, G., Callaerts, P., Fister, S., Kloster, U. & Gehring, W.J. (1998) *Eyeless* initiates the expression of both *sine oculis* and *eyes absent* during *Drosophila* compound eye development. *Development* 125: 2181–91

Hanson, I.M., Fletcher, J.M., Jordan, T. *et al.* (1994) Mutations at the PAX-6 locus are found in heterogeneous anterior segment malformations including Peter's anomaly. *Nature Genetics* 6: 168–73

Hara-Nashimura, I., Kondo, M., Nishimura, M., Hara, R. & Hara, T. (1993) Cloning and nucleotide sequence of cDNA for rhodopsin of the squid *Todores pacificus*. *FEBS Letters* 232: 69–72

Heberlein, U., Wolff, T. & Rubin, G.M. (1993) The TGF beta homolog dpp and the segment polarity gene hedgehog are required for the propagation of a morphogenetic wave in the Drosophila retina. *Cell* 75: 913–26

Hill, R.E., Favor, J., Hogan, B.L.M. *et al.* (1991) Mouse *small eye* results from mutations in a paired-like homeobox containing gene. *Nature* 354: 522–25

Hirsch, N. & Harris, W.A. (1997) *Xenopus Pax-6* and retinal development. *Journal of Neurobiology* 32: 45–61

Hogan, B.L.M., Horsburgh, G., Cohen, J. *et al.* (1986) *Small eyes* (*Sey*): a homozygous lethal mutation on chromosome 2 which affects the differentiation of both lens and nasal placodes in the mouse. *Journal of Embryology and Experimental Morphology* 97: 95–110

Jordan, T., Hanson, I.M., Zaletayev, D. *et al.* (1992) The human PAX6 gene is mutated in two patients with aniridia. *Nature Genetics* 1: 328–32

Kawakami, K., Ohto, H., Takizawa, T. & Saito, T. (1996) Identification and expression of *Six* family genes in mouse retina. *FEBS Letters* 393: 259–63

Kilcherr, F., Baumgartner, S., Bopp, D., Frei, E. & Noll, M. (1986) Isolation of the *paired* gene of *Drosophila*. *Nature* 321: 493–99

Land, M.F. & Fernald, R.D. (1992) The evolution of eyes. *Annual Review of Neuroscience* 15: 1–29

Leuzinger, S., Hirth, F., Gerlich, D. *et al.* (1998) Equivalence of the fly *orthodenticle* gene and the human *Otx* genes in the embryonic brain development of *Drosophila*. *Development* 125: 1703–10

Li, H.S., Yang, J.M., Jacobson, R.D., Pasko, D. & Sundin, O. (1994) *Pax-6* is first expressed in a region of ectoderm anterior to the early neural plate: implications for stepwise determination of the lens. *Developmental Biology* 162: 181–94

Loosli, F., Kmita-Cunisse, M. & Gehring, W.J. (1996) Isolation of a *Pax-6* homolog from the ribbonworm *Lineus sanguineus*. *Proceedings of the National Academy of Sciences USA* 93: 2658–63

Ma, C., Zhou, Y., Beachy, P.A. & Moses, K. (1993) The segment polarity gene hedgehog is required for the progression of the morphogenetic furrow in the developing Drosophila eye. *Genes and Development* 5: 583–93

Macdonald, R. & Wilson, S.W. (1997) Distribution of Pax 6 protein during eye development suggests discrete roles in proliferative and differentiated visual cells. *Developments in Genes and Evolution* 206: 363–69

Macdonald, R., Scholes, J., Strahle, U. *et al.* (1997) The Pax protein Noi is required for commissural axon pathway formation in the rostral forebrain. *Development* 124: 2397–408

Maden, B.E.H. (1995) *Opsin* genes. *Essays in Biochemistry* 29: 87–111

Mansouri, A. & Gruss, P. (1998) Follicular cells of the thyroid gland require Pax-8 functions. *Nature Genetics* 19: 87–90

Mansouri, A., Hallonet, M. & Gruss, P. (1996a) Pax genes and their roles in cell differentiation and development. *Current Opinion in Cell Biology* 8: 851–57

Mansouri, A., Stoykova, A., Torres, M. & Gruss, P. (1996b) Dysgenesis of cephalic neural crest derivatives in Pax-7 -/- mutant mice. *Development* 122: 831–38

Mardon, G., Solomon, N.M. & Rubin, G.M. (1994) *dachshund* encodes a nuclear protein required for normal eye and leg development in *Drosophila*. *Development* 120: 3473–86

Mastick, G.S., Davis, N.M., Andrew, G.L. & Easter, S.S. (1997) Pax-6 functions in boundary formation and axon guidance in the embryonic mouse forebrain. *Development* 124: 1985–97

Mathers, P.H., Grinberg, A., Mahon, K.A. & Jamrich, M. (1997) The *Rx* homeobox gene is essential for vertebrate eye development. *Nature* 387: 603–07

Matsuo, T., Osumi-Yamashita, N., Noji, S. *et al.* (1993) A mutation in the *Pax-6* gene in rat *small eye* is associated with impaired migration of midbrain crest cells. *Nature Genetics* 3: 299–304

Nathans, J. (1992) Rhodopsin: structure, function and genetics. *Biochemistry* 31: 4923–31

Neubuser, A., Koseki, H. & Balling, R. (1995) Characterization and developmental expression of Pax-9, a paired-box containing gene related to Pax-1. *Developmental Biology* 170: 701–16

Neubuser, A., Peters, H., Balling, R. & Martin, G.R. (1997) Antagonistic interactions between FGF and BMP signalling pathways: a mechanism for positioning the sites of tooth formation. *Cell* 90: 247–55

Noll, M. (1993) Evolution and role of Pax genes. *Current Opinion in Genes and Development* 3: 595–605

Nottebohm, E., Dambly-Chaudiere, C. & Ghysen, A. (1992) Connectivity of chemosensory neurons is controlled by the gene *poxn* in *Drosophila*. *Nature* 359: 829–32

Nusselein-Volhard, C. & Wieschaus, E. (1980) Mutations affecting segment number and polarity in *Drosophila*. *Nature* 287: 795–801

Oliver, G., Mailhos, A., Wehr, R. *et al.* (1995) *Six-3*, a murine homologue of the *sine oculis* gene, demarcates the most anterior border of the developing neural plate and is expressed during eye development. *Development* 121: 4045–55

Oliver, G., Loosli, F., Koster, R., Wittbrodt, J. & Gruss, P. (1997) Ectopic lens induction in fish in response to the murine homeobox gene *Six-3*. *Mechanisms of Development* 60: 233–39

Osumi, N., Hirota, A., Ohuchi, H. *et al.* (1997) Pax-6 is involved in the specification of hindbrain motor neuron subtype. *Development* 124: 2961–72

Patel, N.H., Schafer, C., Goodman, C.S. & Holmgren, R. (1989)

The role of segment polarity genes during *Drosophila* neurogenesis. *Genes and Development* 3: 890–904

Pignoni, F., Hu, B., Zavitz, K.N. *et al.* (1997) The eye-specification proteins So and Eya form a complex and regulate multiple steps in *Drosophila* eye development. *Cell* 91: 881–91

Pittack, C., Grunwald, G.B. & Reh, T.A. (1997) Fibroblast growth factors are necessary for neural retina but not pigmented epithelium differentiation in chick embryos. *Development* 124: 805–16

Pituello, F. (1997) Neuronal specification: generating diversity in the spinal cord. *Current Biology* 7: R701–04

Plueschel, A.W., Gruss, P. & Westerfield, M. (1992) Sequence and expression pattern of *Pax-6* are highly conserved between zebrafish and mice. *Development* 114: 643–51

Porter, F.D., Drago, J., Xu, Y. *et al.* (1997) *Lhx-2*, a LIM homeobox gene, is required for eye, forebrain, and definitive erythrocyte development. *Development* 124: 2935–44

Quinn, J.C., West, J.D. & Hill, R.E. (1996) Multiple role functions for *Pax-6* in mouse eye and nasal development. *Genes and Development* 10: 435–46

Quiring, R., Walldorf, U., Kloster, U. & Gehring, W.J. (1994) Homology of the eyeless gene of *Drosophila* to the small eye gene in mice and Aniridia in humans. *Science* 265: 785–89

Salvini-Plawen, L.V. & Mayr, E. (1977) On the evolution of photoreceptors and eyes. *Evolutionary Biology* 10: 207–63

Sander, M., Neubuser, A., Kalamaras, J. *et al.* (1997) Genetic analysis reveals that PAX-6 is required for normal transcription of pancreatic hormone genes and islet development. *Genes and Development* 11: 1662–73

Sanyanusin, P., Schimmenti, L.A., McNoe, L.A., *et al.* (1995) Mutation of the PAX2 gene in a family with optic nerve colobomas, renal anomalies and vesicoureteral reflux. *Nature Genetics* 9: 358–63

Schedl, A., Ross, A., Lee, M. *et al.* (1996) Influence of *PAX-6* gene dosage on development: overexpression causes severe eye abnormalities. *Cell* 86: 71–82

Schimmenti, L.A., Pierpont, M.E., Carpenter, B.L. *et al.* (1995) Autosomal dominant optic nerve colobomas, vesicoureteral reflux, and renal anomalies. *American Journal of Medical Genetics* 59: 204–08

Schwarz, M., Alvarez-Bolado, G., Urbanek, P., Busslinger, M. & Gruss, P. (1997) Conserved biological function between Pax-2 and Pax-S in midbrain and cerebellum development: evidence from targeted mutations. *Proceedings of the National Academy of Sciences USA* 94: 14518–23

Seo, H.C., Drivenes, O., Ellingsen, S. & Fjose, A. (1998) Expression of two zebrafish homologues of the murine *Six-3* gene demarcates the initial eye primordia. *Mechanisms of Development* 73: 45–57

Shen, W. & Mardon, G. (1997) Ectopic eye development in *Drosophila* induced by directed *dachshund* expression. *Development* 124: 45–52

Sheng, G., Thouvenot, E., Schmucker, D., Wilson, D.S. & Desplan, C. (1997) Direct regulation of *rhodopsin 1* by *Pax-6/eyeless* in Drosophila: evidence for a conserved function in photoreceptors. *Genes and Development* 11: 1122–31

Slack, J.M.W. (1995) Developmental biology of the pancreas. *Development* 121: 1569–80

Sosa-Pineda, B., Chowdhury, K., Torres, M., Oliver, G. & Gruss, P. (1997) The *Pax-4* gene is essential for differentiation of insulin producing beta cells in the mammalian pancreas. *Nature* 386: 399–402

Steel, K.P. & Smith, R.J. (1992) Normal hearing in Splotch (Sp/+), the mouse homologue of Waardenburg syndrome type 1. *Nature Genetics* 2: 75–79

St-Onge, L., Sosa-Pineda, B., Chowdhury, K., Mansouri, A. & Gruss, P. (1997) *Pax-6* is required for differentiation of glucagon-producing alpha-cells in mouse pancreas. *Nature* 387: 406–09

Stoykova, A., Fritsch, R., Walther, C. & Gruss, P. (1996) Forebrain patterning defects in Small eye mice. *Development* 122: 3453–65

Stoykova, A., Gotz, M., Gruss, P. & Price, J. (1997) Pax-6 dependent regulation of adhesive patterning, R-cadherin expression and boundary formation in the developing forebrain. *Development* 124: 3765–77

Tanabe, Y. & Jessell, T.M. (1996) Diversity and patterning in the developing spinal cord. *Science* 274: 1115–23

Tassabehji, M., Read, A.P., Newton, V.E. *et al.* (1992) Waárdenburg's syndrome patients have mutations in the human homologue of the Pax-3 paired box gene. *Nature* 355: 635–36

Tomarev, S.I. & Piatigorsky, J. (1996) Lens crystallins of invertebrates. *European Journal of Biochemistry* 235: 449–65

Tomarev, S., Callaerts, P., Kos, L. *et al.* (1997) Squid Pax-6 and eye development. *Proceedings of the National Academy of Sciences USA* 94: 2421–26

Ton, C.C., Hirvonen, H., Miwa, H. *et al.* (1991) Positional cloning and characterization of a paired box- and homeobox-containing gene from the *Aniridia* region. *Cell* 67: 1059–74

Torres, M., Gomez-Pardo, E., Dressler, G.R. & Gruss, P. (1995) Pax-2 controls multiple steps of urogenital development. *Development* 121: 4057–65

Torres, M., Gomez-Pardo, E. & Gruss, P. (1996) Pax-2 contributes to inner ear patterning and optic nerve trajectory. *Development* 122: 3381–91

Urbanek, P., Wang, Z.Q., Fetka, I., Wagner, E.F. & Busslinger, M. (1994) Complete block of early B-cell differentiation and altered patterning of the posterior midbrain in mice lacking Pax-5. *Cell* 79: 901–12

Urbanek, P., Fetka, I., Meisler, M.H. & Busslinger, M. (1997) Cooperation of Pax-2 and Pax-S in midbrain and cerebellum development. *Proceedings of the National Academy of Sciences USA* 94: 5703–08

Vogan, K.J., Epstein, D.J., Trasler, D.G. & Gros, P. (1993) The splotch-delayed (SpD) mouse mutant carries a point mutation within the paired-box of the Pax-3 gene. *Genomics* 17: 364–69

Wallin, J., Wilting, J., Koseki, H. *et al.* (1994) The role of Pax-1 in axial development. *Development* 120: 1109–21

Wallin, J., Eibel, H., Neubuser, A. *et al.* (1996) Pax-1 is expressed during development of the thymus epithelium and is required for normal T cell maturation. *Development* 122: 23–30

Walther, C. & Gruss, P. (1991) *Pax-6*, a murine multi-gene family member is expressed in the developing CNS. *Development* 113: 1435–49

Warren, N. & Price, D.J. (1997) Roles of Pax-6 in murine diencephalon development. *Development* 124: 1573–82

Wilson, D.S., Guenther, B., Desplan, C. & Kuriyan, J. (1995) High resolution crystal structure of a paired (Pax) class cooperative homeodomain dimer on DNA. *Cell* 82: 709–19

Wolff, T. & Ready, D.F. (1991) Pattern formation in the *Drosophila* retina. In *The Development of Drosophila melanogaster*, edited by M. Bate & A.M. Arias, Cold Spring Harbor, New York: Cold Spring Harbor Laboratory Press

Xu, P.X., Woo, I., Her, H., Beier, D. & Maas, R. (1997) Mouse Eya homologues of the *Drosophila eyes absent* gene require Pax-6 for expression in cranial placodes. *Development* 124: 219–31

Yang , X.M., Vogan, K., Gros, P. & Park, M. (1996) Expression of the *c-met* receptor tyrosine kinase in muscle progenitor cells in somites and limbs is absent in *Splotch* mice. *Development* 122: 2163–71

Zhang, Y. & Emmons, S.W. (1995) Specification of a sense-organ identity by a *Caenorhabditis elegans Pax-6* homologue. *Nature* 377: 55–59

GLOSSARY

ascidian a non-vertebrate chordate

crystallins major protein components of vertebrate and invertebrate lenses

dachshund transcriptional regulator of eye development in Drosophila

decapentaplegic factor secreted in the morphogenetic furrow that regulates eye patterning

Eya-1, Eya-2 vertebrate homologues of eyes absent; expressed in the lens placode

ey^2, ey^R mutations of eyeless gene that ablate eye expression of eyeless, causing eye phenotype

eyeless Drosophila homologue of *Pax-6*

eyes absent transcriptional regulator of eye development in Drosophila

hedgehog factor secreted in the morphogenetic furrow that regulates eye patterning

imaginal disc progenitor structure for adult tissues found in the embryo

lens placode local thickening of ectoderm that develops into the lens

Lhx-2 Lim class transcription factor required to make optic vesicle induce lens placode development

ommatidia (sing. ommatidium) single units of an insect eye, containing photoreceptors, pigmented cells, cone cells and a secreted lens

optic vesicle outpocket of forebrain that develops into the retina and optic nerve

orthodenticle Drosophila transcription factor that regulates anterior tissue development

Otx-1 vertebrate homologue of orthodenticle

paired first Pax factor isolated; found in Dresophila.

Pax factors category of transcription factors that have the paired DNA-binding domain

Pax-6 Pax transcription factor that has major role in eye and CNS development

rhodopsin light-sensitive protein found in photoreceptor cells; member of opsin protein family

Rx paired-class homeodomain transcription factor required for optic vesicle development

Sey, Sey^{1Neu}, rSey natural mutations of *Pax-6* gene in mice (*Sey, Sey^{1Neu}*) and rat (*rSey*)

sine oculis transcriptional regulator of eye development in Drosophila

Six-3 homologue of *sine oculis*; expressed in the lens placode

See also **Genetic dissection of wing patterning in Drosophila** (p.172)

THE SILKWORM, *BOMBYX MORI*: A MODEL GENETIC SYSTEM

Jaware Gowda Nagaraju[1], Viacheslav Klimenko[2] and Pierre Couble[3]
[1]*Laboratory of Molecular Genetics, Centre for DNA Fingerprinting and Diagnostics, Hyderbad, India*
[2]*Institute of Animal Science, Kharkov, Ukraine*
[3]*Centre of Molecular and Cellular Genetics, CNRS, Villeurbanne, France*

1. Historical background

Archaeological and bibliographical evidence shows that sericulture was practised in China about 2500BC. The earliest known silk textile is almost 5000 years old (Kuhn, 1988), which conforms well with the observation that the domesticated silkworm, *Bombyx mori*, originated from its wild relative *B. mandarina* almost 4600 years ago (Hirobe, 1968). Sericulture probably began some 45 centuries ago in North China along the bank of the Hwang Ho River. In the 12th century BC, it expanded outside China, as mulberry seeds and silkworm eggs were smuggled out. Thereafter, the secrets of silkworm rearing and silk production spread to neighbouring countries and to the rest of the world through the "silk road", leading to the establishment of a sericulture industry in many countries.

The silkworm was introduced from China into Japan more than 1000 years ago. By 1970, China and Japan together accounted for 70% of world silk production. Together with the technological revolutions that improved silkworm rearing and breeding, the application of genetic principles to silkworm race amelioration in the 20th century brought a sharp rise in silk productivity, both qualitatively and quantitatively. Sericulture has enjoyed glorious years and has also suffered setbacks. During the 1920s and 1930s, the world silk yield exceeded 60–70 thousand tons, while in the 1950s and 1960s it decreased to 30 thousand tons. After 1970, sericulture showed rapid development and has become the vocation of small agricultural families in many populous developing countries like China, India, Vietnam and Thailand. Current annual world production is about 496 thousand tons. Due to an enhanced consumer demand for silk, it appears that the upward trend in silk production will continue in the future.

2. The silkworm and its relatives

Most silkworm species belong to either the family Bombycidae or the family Saturniidae, in the superfamily Bombycoidea. Phylogenetic studies based on interspecific hybridization (Aratake & Kayamura, 1973), chromosome-pairing behaviour (Murakami & Imai, 1974) and gene structure comparisons (Kusuda *et al.*, 1986; Maekawa *et al.*, 1988; Shimada *et al.*, 1995) suggest that the present-day domesticated silkworm, *B. mori*, is closely related to *B. mandarina*, the wild silkworm. *B. mandarina* collected from Japan has 27 chromosomes whereas that from Manchuria and central China has 28, the same as *B. mori*. Cytological investigations of interspecific hybrids of Japanese *B. mandarina* and *B. mori* have revealed the formation of a trivalent with one chromosome of the former and two chromosomes of the latter (Murakami & Imai, 1974), suggesting that one mandarina chromosome could have resulted from Robertsonian translocation (fusion of two chromosomes at the centromere, with loss of the two short arms) between two chromosomes of the continental form.

As in most Lepidoptera, the chromosomes of the silkworm are holocentric, i.e. they possess centromeres throughout the chromosome body. This was shown by the persistence of chromosomal fragments after irradiation during cell division (Murakami & Imai, 1974), the dispersion of microtubule attachments (Friedlander & Wahrman, 1970), the absence of recombination nodules in females (Rasmussen & Holm, 1982), the presence of supernumerary chromosomes in close relatives of the silkworm (Puttaraju & Nagaraju, 1985) and the chromosome-pairing behaviour and fertility of interspecific hybrids of two saturniid species with diverse chromosome numbers (Nagaraju & Jolly, 1985). *B. mori* chromosomes are usually highly condensed and appear dot-shaped at meiotic and mitotic metaphase stages, their diffused centromeres and the lack of special features making them difficult to identify individually. This has resulted in only limited applications of modern cytogenetic tools for genetic and molecular genetic studies.

A general consensus phylogenetic tree of saturniid and bombycoid moths based on nucleotide and amino acid sequences of the arylophorin gene (Shimada *et al.*, 1995) illustrates the phylogenetic status of *B. mori* in relation to other silkmoths (**Figure 1**).

3. Chromosomes and sex determination

The sex chromosome constitution in the silkworm is ZZ in males and ZW in females. Tanaka (1916) first discovered sex-linked inheritance in the silkworm and Hasimoto (1933) ascertained that the W chromosome is the determinant of femaleness through the finding that ZZW and ZZZW individuals of triploids and tetraploids are normal females. More recent observations, however, have brought some contradictory views to this question (Shshegelskaya & Klimenko, 1987). Implantation of sheathless testes into the male or the female haemocoel at the 2nd larval instar (see **Figure 2** for life cycle diagram) results in the differentiation of egg chambers with one oocyte and 15 nurse cells as in the normal situation. However, the occurrence of egg chamber-like structures is 30–300 per testis implanted in females and 10–60 in grafted males, whereas no such event is observed in undamaged implanted testes. This suggests that the external testicular sheath acts as a barrier against feminizing factors that would exist in both sexes, irrespective of the presence or absence of the W chromosome. This sex-reversal

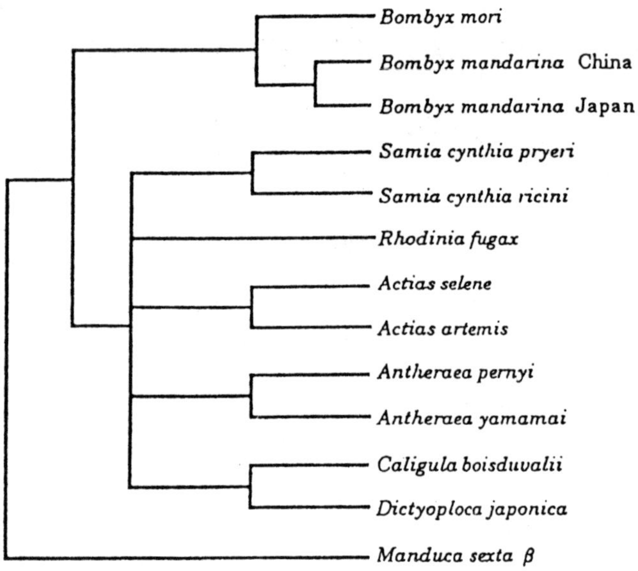

Figure 1 A general consensus phylogenetic tree of silkmoths based on the arylophorin nucleotide and amino acid sequences.

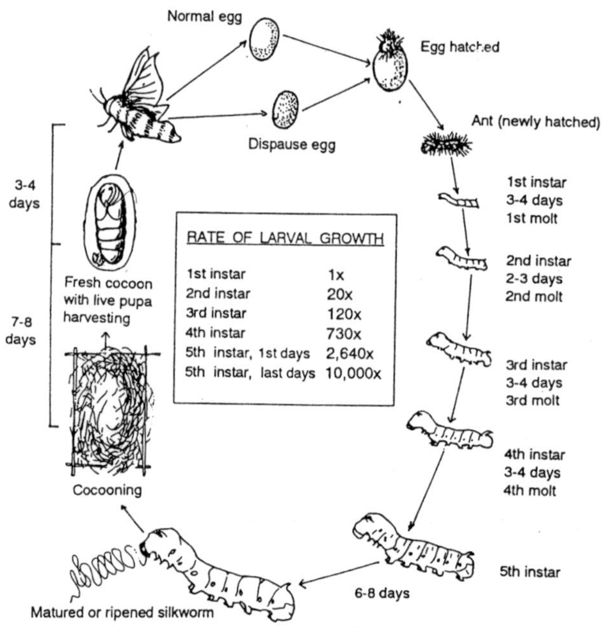

Figure 2 Lifecycle of the silkworm.

phenomenon is an interesting situation for further investigation of sex determination in the Lepidoptera.

No crossing-over occurs in the heterogametic sex. A synaptonemal complex is observed, but chiasmata, the visible expressions of crossing-over, are absent in females (Rasmussen & Holm, 1982). During female diakinesis, the bivalents can be seen lying side by side without formation of any chiasma (**Figure 3**) (Murakami & Imai, 1974).

4. Experimental sex control

Interesting peculiarities of reproduction in the silkworm *B. mori* arise spontaneously or can be induced by artificial

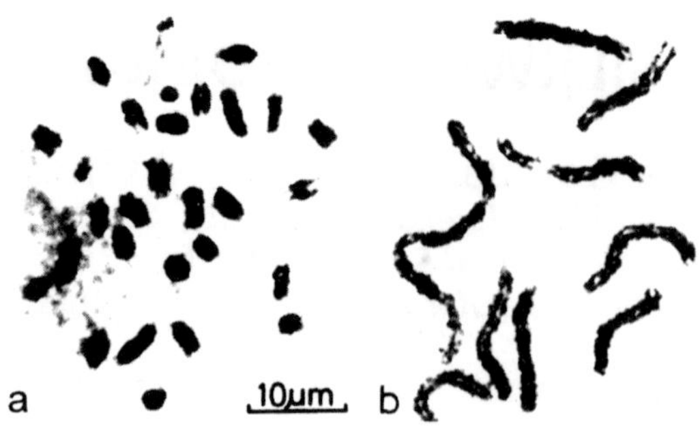

Figure 3 Silkworm chromosomes in oogenesis: (A) achiasmatic bivalents at diakinesis and (B) diplotene.

manipulation (**Table 1**). These may result in parthenogenesis, androgenesis, gynogenesis or polyploidogenesis (Astaurov, 1967). The resulting alterations in chromosome number have been exploited in fixed heterozygotic clones for many generations and for further investigation of the chromosomal basis of sex determination.

(i) Parthenogenesis

Spontaneous parthenogenesis is very rare in *B. mori* (one case in 10^{5-6}) and seems to be of the meiotic type since it occurs by fusion of the two haploid derivatives following meiosis (Kawaguchi, 1934). No natural ameiotic parthenogenesis has yet been demonstrated. In contrast, high yields of ameiotic parthenogenesis can be achieved artificially (Astaurov, 1967). Parthenogenetic cloning thus became a basic tool for silkworm developmental cytogenetics and different protocols were developed, leading to either ameiotic or meiotic parthenogenetic animals, at a very high level of efficiency.

Ameiotic parthenogenesis. A controlled heat shock applied to non-fertilized eggs can disrupt the metaphase I spindle and prevent the reductional division, giving rise to diploid female pronuclei identical to the mother's genotype since no crossing-over occurs in females (Klimenko, 1982, 1990). Thus, parthenogenetic daughters are genotypically identical, and stable clones can be derived from a given moth. Parthenoclones propagated in this way can be maintained for many years without visible changes as efficiently as normal bisexual stocks. The degree of individual heterozygosity positively correlates with the ability to complete parthenogenetic development; hybrid vigour is thus a good prerequisite for successful cloning (Altukhov & Klimenko, 1978). This leads to the advantage of being able to maintain individual stocks with a large number of useful genetic markers in a fixed genetic background, without the problems of inbreeding encountered by maintaining homozygous stocks.

Meiotic parthenogenesis. Another specific temperature regime (**Table 1**) applied to parthenoclones leads to very high proportions of meiotic parthenogenetic development. In this case, the absence of crossing-over in females (ZW) and the restoration of diploidy through fusion (or non-separation) of the two genetically identical mitotic derivatives of the haploid female pronucleus (Z or W) ends in rare male individuals of ZZ genotype, homozygous at all loci. The eggs in such experiments show very low hatchability.

Table 1 Reproductive modes in the silkworm *Bombyx mori*, in natural and experimentally-induced conditions.

Development	FF	MF	Artificial treatments	Sex produced	References
Normal	normal	normal	none	males + females	Tazima, 1964
Parthenogenesis (spontaneous)	normal	absent	subnormal temperature, high humidity	males, rare females	Kawaguchi, 1934
Parthenogenesis (artificial)	moths or eggs	absent	+46°C −11°C	males females	Astaurov, 1967 Terskaya & Strunnikov, 1974
1. meiotic 2. ameiotic			3–5°C, 30 min 15–20°C, 120 min	males + rare females females + rare males	Klimenko, 1990 Astaurov, 1940
Androgenesis	egg nucleus inactivation by X-irradiation of moth	normal	inseminated eggs treated at 40°C for 60 min or CO_2 for 30–240 min, or gamma irradiation	males	Astaurov, 1967; Xu *et al.*, 1997
Gynogenesis	normal	sperm nucleus inactivation	subnormal temperature, high humidity		Tazima & Onuma, 1967
Polyploidy	normal	normal	inseminated eggs heated at +46°C for 18 min	females of increased ploidy	Klimenko, 1990

FF, female factors; MF, male factors.

No female of WW constitution arises since this genotype is not viable (Terskaya & Strunnikov, 1974).

Parthenoclones are also characterized by the occurrence (at 0.1–1%) of tetraploid oocytes (ZZWW) among normal ones. Completion of meiosis in these tetraploid eggs leads to reversion to maternal ploidy ($2n$) giving rise to recombinant females (ZW) and males (ZZ). These should not be confused with the above-mentioned homozygotic males (Klimenko & Spiridonova, 1982). Interestingly, such tetraploid oocytes can develop by themselves or after egg insemination. In this case, the progeny is triploid with [$2n$ (ZW or ZZ) + $1n$ (Z)] constitution, being either females (ZZW) or males (ZZZ). By this method, Astaurov (1967) obtained clones of *B. mori* females with increasing ploidy. Applied to hybrids between Bombyx and mandarina, this led him also to create the first fertile tetraploid bisexual in animals.

(ii) Androgenesis and gynogenesis

Androgenesis occurs when the oocyte nucleus is excluded from development of the egg, which then relies exclusively on paternal nuclear material. X-ray irradiation of moths of *B. mori* and subsequent heating or cooling of the eggs has also been described as a reliable method for the production of androgenetic silkworms (Astaurov, 1967; Strunnikov, 1995). Recent observations confirm and extend this method by carrying out X-ray inactivation at the pupal stage (Xu *et al.*, 1997).

Sperm activation of the egg without contribution of the sperm pronucleus triggers gynogenesis. Such cytogenetic variants were obtained by carbon dioxide treatment of the eggs after insemination. The yield is, however, very poor, and high proportions of mosaic individuals are obtained in such experiments.

(iii) Polyploidogenesis

Silkworm eggs are arrested in meiotic metaphase I and complete meiosis after insemination, which occurs as the eggs are laid using sperm kept in spermatophores stored in the female bursa copulatrix. If freshly inseminated eggs (3–5 minutes after laying) are heat treated, the ameiotic pronucleus ZW can fuse with the sperm pronucleus giving rise to a triploid female, ZZW. The incidence of triploids is then >90%. The procedure can be repeated at the following generations, leading to females whose ploidy increases each time by one haploid set of chromosomes (Klimenko & Spiridinova, 1982). Hepta- (ZZZZZZW) and decaploid (ZZZZZZZZZW) females have been obtained by this method (Klimenko, unpublished data), leaving open the possibility that bisexual polyploid races can be developed. Theoretically, this would be feasible for genotypes of even ploidy for which paired conjugation of homologues could occur, but this has not yet been realized.

5. Genetic studies in the silkworm

Inspired by the re-discovery of Mendel's laws in the 1900s, genetical research on the silkworm rapidly promoted this organism as a valuable model system for fundamental studies.

In 1906, Toyama reported for the first time the utilization of the principles of hybrid vigour to enhance viability and yield by crossing different silkworm genotypes. As noted earlier, the phenomenon of sex linkage in *B. mori* was first reported by Tanaka (1916), shortly after its initial report in *Drosophila melanogaster*. Many genetic phenomena such as maternal inheritance, parthenogenesis and polyploidogenesis, radiation-induced translocations and complex loci were reported in the early 1930s (Tazima, 1964).

Together with the many inherent merits of the silkworm, the growing collection of world genetic stocks, which encompass around 3000 genotypes, places *B. mori* second only to Drosophila as an insect model for genetic studies. The formulation of artificial diets that freed rearing from the use of fresh mulberry leaves helped to expand the study of Bombyx to academic laboratories. The accumulation of information on the nutritional requirements of the silkworm led to the formulation of artificial diet as early as 1965. Since then, the composition of artificial diet has gradually improved and almost no differences are observed in the growth rate of larvae reared on mulberry leaves and on artificial diets. The composition of artificial diets used for practical rearing of silkworm is given in **Table 2**.

(i) Basic biological processes

Many of the research findings have already provided important insights into the understanding of basic biological processes (see Goldsmith & Wilkins, 1995). *B. mori* was the first insect shown to have an interspersed pattern of repetitive and non-repetitive sequences typical of mammalian genomes (Gage, 1974; Gage *et al.*, 1976). Molecular characterization of the repetitive elements from *B. mori* has revealed the presence of transposable elements typical of the Drosophila genome as well as retroposons typical of mammalian genomes (see review by Eickbush, 1995). The *B. mori* genome harbours abundant short interspersed nucleotide elements (SINEs) like *Bm1* and *Bm2* which account for 5–10% of the total genome, similar to the Alu and Alu-like elements found in the mammalian genome (total haploid genome content of the silkworm is 5.3×10^8 bp). Interestingly, the fruit fly is devoid of such SINEs. The Bombyx genome also contains several types of long interspersed nucleotide elements (LINEs), some with long terminal repeats (LTRs) like *Pao* and *Mag*, or devoid of LTRs like *R1Bm*, *R2Bm* and *BMC1*. Unlike the *D. melanogaster* genome, the *B. mori* genome harbours DNA-mediated mobile elements like mariner (Ueda *et al.*, 1986; Robertson & Asplund, 1996).

Telomeric DNA motifs have been characterized from the silkworm (Okazaki *et al.*, 1993) and are made of a pentanucleotide repeat (TTAGG), characteristic of many insects, instead of the hexanucleotides (TTNGGG) found in many vertebrates.

Table 2 Composition of artificial diets used for silkworm rearing.

Substance	Diet for 1–4 instars (g)	Diet for 5th instar (g)
Mulberry leaf powder	25.0	25.0
Soybean oil	1.5	3.0
Defatted soybean meal	36.0	45.0
Cholesterol	0.2	0.2
Citric acid	4.0	4.0
Ascorbic acid	2.0	2.0
Sorbic acid	0.2	0.2
Agar	7.5	5.0
Salt mixture	3.0	3.0
Glucose	8.0	10.0
Potato starch	7.5	15.0
Cellulose powder	20.8	–
Vitamin B mixture	Added	Added
Antiseptic	Added	Added
Total	115.7	112.4
Water	300 ml	220 ml

(ii) Quantitative adaptation

The quantitative adaptation of tRNA species to the population of mRNA codons to be translated was elucidated in the silk gland before it appeared to be a general phenomenon in living organisms (Garel, 1976; see review by Sprague, 1995). The quantitative adaptation of tRNA populations was first described in the posterior silk gland where the fibroin heavy chain is produced. This protein, the synthesis of which mobilizes most of the cellular proteosynthetic activity, is 4700 amino acids long and glycine residues account for 48% of the molecule. The translation of its mRNA occurs in cells which contain an unbalanced representation of glycyl tRNAs that quantitatively satisfies the demand for glycine amino acids to be incorporated into the nascent chains. This adaptation is also observed for the glycyl iso-tRNA species whose relative representation in the cell fits the frequencies of the four glycine isocodons in the H-fibroin mRNA. A similar adaptation between populations of tRNAs and mRNA codons was also detected in the middle silk gland that synthesizes the serine-rich sericins.

(iii) Protein function

Many sequenced *B. mori* genes are now available that encode proteins with diverse functions (**Table 3**). In this regard, the silk gland (see section 6 below) and the chorion provided effective model systems to help unravel the molecular basis of gene expression. Functional assays for regulatory elements of chorion genes in follicular cells have been elegantly demonstrated using the heterologous transgenic system of Drosophila (Mitsialis & Kafatos, 1985), and follicular cell regulatory factors and their genes are now well characterized (Drevet *et al.*, 1994). Drosophila homologous homeobox genes, *Bombyx ultrabithorax* (BmUbx) and *Bombyx abdominal-A* (Bmabd-A) and *-B* (Bmabd-B), which constitute part of the E-pseudoallelic complex which specifies the identity of body segments (Hasimoto, 1941), are now cloned and characterized (Ueno *et al.*, 1992, 1996). From the initial molecular analysis of the homeodomain, it was shown that the E-complex is analogous to the bithorax complex (BX-C) of Drosophila. However, some mutations of the E-complex reveal aspects that are not found in Drosophila *bithorax* mutants. E-mutant phenotypes, such as E^{kp}/E^{kp} and $E^{D}/+$, cause shifts in one direction on the dorsal side and in the opposite direction on the ventral side. Such an independent determination of the dorsal and ventral sides is not observed in Drosophila. Another homeotic complex, the Nc locus, has been characterized. It carries the homeotic genes *Antennapedia* required for thoracic segments (Bmantp) and *sex comb reduced* (BmScr) which specifies the identity of maxillae, labial and prothoracic segments (Ueno *et al.*, 1996). The Nc locus corresponds to the ANT-C complex of Drosophila. The Nc locus lies 1 centiMorgan (cM) from the E-complex on chromosome 6, similar to the way the ANT-C and the BX-C loci are positioned close together in the Drosophila genome.

Table 3 Selected genes cloned from the silkworm, *Bombyx mori*.

Cloned genes	Encoded products	References
Enzymes		
sorbitol dehydrogenase	sorbitol dehydrogenase	Niimi *et al*., 1996
trehalase (cDNA)	trehalase	Su *et al*., 1993
glycyl-tRNA synthetase	glycyl-tRNA synthetase	Nada *et al*., 1993
cdc2 and cdc2-related (cDNA)	cell cycle kinases	Iwasaki *et al*., 1997
Homeoproteins		
Bmftz-F1	Fujitaratzu-related	Sun *et al*., 1994
Bmantp	Antennapedia	Ueno *et al*., 1996
BmUbx	Ultrabithorax	Ueno *et al*., 1992
Bmcdl	Caudal	Xu, X. *et al*., 1994
Bmabd-A, Bmabd-B	Abdominal	Ueno *et al*., 1992
Bmen	Engrailed	Hui *et al*., 1992
Bmin	Invected	Hui *et al*., 1992
BmScr	Sex Comb Reduced	Kokubo *et al*., 1997
Bmdef	Deformed	Kokubo *et al*., 1997
Hormones and receptors		
PTTH	Prothoracicotropic hormone	Kawakami *et al*.,1990
DH/P-BAN	Diapause hormone/Pheromone biosynthesis activating neuropeptide	Sato *et al*., 1993; Kawano *et al*., 1992
Bombyxin A, B, C and D	Insulin related peptides	Kawakami *et al*., 1989; Kondo *et al*., 1996
EcR	Ecdysone receptor	Swevers *et al*., 1995
BmEcR-A and B1 (cDNA)	Ecdysone receptor isoforms	Kamimura *et al*., 1997
Immune proteins		
BmLeb	Lebocin	Furukawa *et al*., 1997
GNBP (cDNA)	Gram-negative bacteria-binding protein	Lee *et al*., 1996
FPI-F (cDNA)	Serine protease inhibitor	Pham *et al*., 1996
BmAtt	Attacin	Taniai *et al*., 1996
MORICIN (cDNA)	Moricin	Hara & Yamakawa, 1996
CecB1 and 2	Cecropin	Taniai *et al*., 1995
BmLyz	Lysozyme	Lee & Brey, 1995
Pro-PO (cDNA)	Prophenol oxidase	Kawabata *et al*., 1995
Regulatory proteins		
SCF (cDNA)	Supercoiling factor	Ohta *et al*., 1995
SGF-1	Silk gland cell regulatory protein	Mach *et al*., 1995
POU-M1/SGF3	Sericin gene regulatory protein	Xu, P.X. *et al*., 1994
BmGATAα and β	GATA-like regulatory proteins	Drevet *et al*., 1994
RNAs		
rDNA	Ribosomal DNA	Lecanidou *et al*., 1984
tRNA^{gly}	Silk gland glycine tRNA	Fournier *et al*., 1993
tRNA^{Ala}	Silk gland alanine tRNA	Young *et al*., 1986; Sullivan *et al*., 1994
tRNA^{Ala}	Constitutive alanine tRNA	Hagenbuchle *et al*., 1979
Structural proteins		
BmColl (cDNA)	Collagen-like peptide	Chareyre *et al*., 1996
Bmβtub (cDNA)	β-tubulin	Mita *et al*., 1995
A3	Cytoplasmic actin	Mounier & Prudhomme, 1986
A4	Cytoplasmic actin	Mangé *et al*., 1996
A1 and 2	Muscle actins	Mounier *et al*., 1987
ErA and ErB	Early chorion proteins	Hibner *et al*., 1991
A and B	Middle chorion proteins	Spoerel *et al*., 1986
HcA and HcB	High cysteine late chorion proteins	Iatrou *et al*., 1984; Rodakis *et al*., 1984
PCP-2	Pupal cuticle protein	Nakato *et al*., 1992
H-Fib	Heavy fibroin chain (silk protein)	Suzuki *et al*, 1972; Mita *et al*., 1994
L-Fib	Light fibroin chain (silk protein)	Yamaguchi *et al*., 1989
P25	Fibroin-associated peptide	Couble *et al*., 1985
Ser-1	Sericins (silk gum protein)	Michaille *et al*., 1986
Ser-2	Sericins (silk gum protein)	Michaille *et al*., 1990
ESP	Egg storage protein	Sato and Yamashita, 1991
SP1, SP2	Larval haemolymph storage proteins	Sakurai *et al*.,1988; Fujii *et al*.,1989

(iv) Immune response

The immediate immune response to microbial infection, first well-characterized in the wild silkworm *Hyalophora cecropia* (Boman, 1995), has also been extensively studied in Bombyx. A large spectrum of antibacterial and antifungal proteins has been identified, some of which are known only in the silkworm and await identification in other species (Hara & Yamakawa, 1996; Furukawa *et al.*, 1997).

6. The silk gland as a model system

The silk gland has long been used as a model for studying cell differentiation (see reviews by Bello *et al.*, 1994; Hui & Suzuki, 1995). An attractive feature of this tissue is its simple construction: a one-cell layered glandular epithelium with a succession of territories comprising a fixed number of cells, each expressing a specific set of genes encoding silk proteins (**Figure 4**). Silk gland organogenesis mobilizes the homeoprotein SCR, the homologue of the Drosophila sex comb-reduced protein (Kokubo *et al.*, 1997), and is completed at late embryogenesis, following a tubulogenesis-driven morphogenesis along the antero-posterior axis. The subsequent growth of the organ during larval life occurs without mitosis and single cells accumulate up to 400 000–800 000 haploid genomes by endomitosis at the onset of metamorphosis, the highest known degree of polyploidy ever reported. This impressive amplification raises the question of the nature of the mechanisms that perpetuate the chromatin structure at each round of endomitosis during the course of larval development.

Silk is made of fibroin and sericins, two distinct families of proteins. Fibroin is a complex of three associated polypeptides encoded by single genes: the heavy and light chains that are linked by disulfide bonds, and P25 or fibrohexamerin (S. Mizuno, personal communication) loosely attached to six dimers of the two others. Sericins are the products of two distinct genes whose precursor mRNAs are the target of differential splicing leading to 4- and 2-sericin isoforms from the genes *Ser-1* and *Ser-2*, respectively. Fibroin is secreted in the 500 silk gland posterior cells and makes the core of the silk thread that confers on the cocoon its strong mechanical resistance. Sericins are synthesized in three median and contiguous territories of 115, 260 and 150 cells and serve to glue the silk threads together and shape the cocoon.

The architecture of the silk thread prior to spinning results from an accurate spatial distribution of cells expressing the fibroin, P25 and sericin genes, and of the spatially and temporally regulated splicing of the *Ser-1* pre-mRNA (**Figure 4**) (Couble *et al.*, 1987). The spatial restriction of expression of the different silk proteins is controlled by selective activation of their genes (see reviews by Bello *et al.*, 1994; Hui & Suzuki, 1995). A series of transactivators involved in transcriptional activation has been identified by their role in promoting transcription *in vitro* or *in vivo* in biolistic-transfected silk gland cells (Horard *et al.*, 1997). Posterior cell-specific expression is driven by at least two DNA-binding factors: one named SGFB (or OBF1), a silk gland-specific regulatory protein, and the other PSGF, a factor that links SGFB to the protein complex that binds to the TATA box. *Ser-1* expression in the middle silk gland is under the control of a POU-type transacting factor called POU-M1 (Fukuta *et al.*, 1993). The transcript of this factor is restricted to the median cells where *Ser-1* transcription occurs, strongly suggesting that it is a true territory-specific regulatory protein (Xu, X. *et al.*, 1994).

Activity of silk protein gene promoters has been reported in transgenic Drosophila (Bello & Couble, 1990; Bello *et al.*, 1994). *P25* and *H-Fib* promoter activities are restricted to

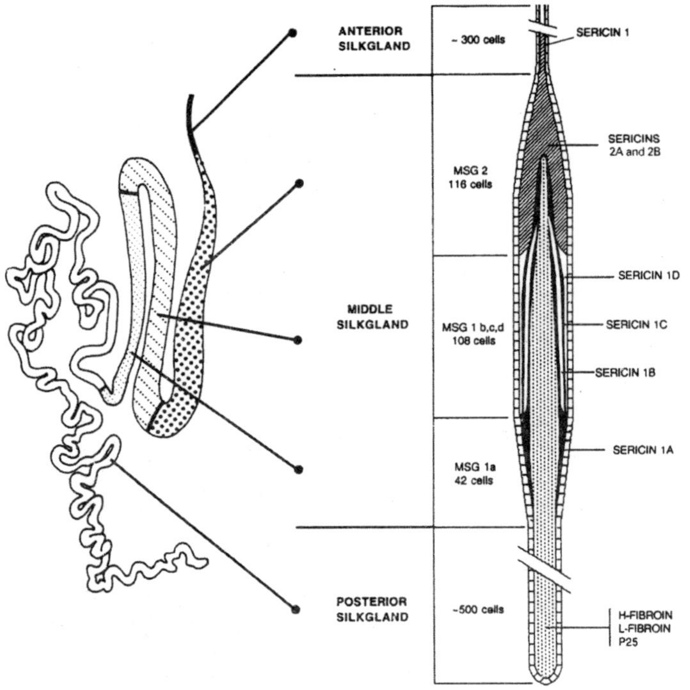

Figure 4 Functional organization of the silk gland of *B. mori*. The different territories of the anterior, middle and posterior silk gland are shown on the left and the topological characteristics of silk protein gene expression on the right. The posterior silk gland cells secrete fibroin, a mixture of the heavy and light fibroin chains and of the chaperone P25. MSG1a, b, c and d refer to middle silk gland cells expressing the sericin gene, *Ser-1*. MSG1a cells perform the splicing modality that leads to the expression of sericin 1A, whereas the MSG1b, c and d cells realize three other spicing pathways sequentially, leading to the successive deposition of sericin 1B, 1C and 1D onto fibroin. The MSG2 cells coexpress sericin 2A and 2B encoded by the single gene, *Ser-2*. (Data from Couble *et al.*, 1987.)

the anterior salivary gland of the fruit fly, the organ orthologous to the silk gland of the silkmoth larva. Both the silk gland and the salivary gland secrete proteins that are exported from the organism (silk proteins used for making the cocoon and glue proteins for attaching the pupae on the substratum). Silk and glue proteins are different, albeit displaying a repeated structure. This shows that, despite a 240-million-year divergence, the transcriptional machinery in the two organs has been conserved while the coding sequences of the most highly transcribed genes, those of the glue proteins in the salivary gland and of the silk proteins in the silkworm, have largely diverged. These experiments show also the genetical potential of using Drosophila for unravelling the mechanisms of transcription that operate in the silk gland itself (Nony *et al.*, 1995).

7. Application of genetic principles for improving silk production

(i) Hybrid vigour

The silkworm is one of the earliest organisms to which genetic principles were applied for enhancing its economic potential, namely, silk production. Cross-breeding strategies have been extensively used as a means of harnessing heterosis (Harada *et al.*, 1961; Nagaraju *et al.*, 1996). Indeed, the two big leaps involving the use of hybrid vigour were taken with the silkworm and maize, the former coming slightly earlier than the latter. The credit for introducing F_1 hybrids with a clear demonstration of their superiority over parental genotypes goes to Toyama (1906) in Japan. The hybrids of Chinese and Japanese parents became so popular in Japan that, by 1919, over 90% of the eggs produced were of hybrid origin, reaching 100% by 1928 (Yokoyama, 1979). The average weight of the cocoon shell, brought by slow selection from around 180 mg in 1804 to about 200 mg in 1910, was increased thereafter to more than 400 mg in less than 20 years, when the use of hybrids took over silk production in Japan (**Figure 5**). The phenomenal contribution of hybrid vigour to silkworm improvement has continued to the present, with the use of breeding systems for commercial egg production that include not only simple F_1 hybrids (usually between Japanese and Chinese races, see below) but also tetraparental hybrids involving the crossing of two F_1 hybrids, each produced by two different combinations of Japanese or Chinese stocks with fixed and selected economic traits.

(ii) Selection

In addition to the use of controlled hybridization, the systematic improvement of silkworm varieties for silk production and associated traits was carried out by artificial selection. This resulted in an increase in silk production and enabled geneticists to understand the correlation and heritability for various attributes associated with silk yield. For example, in a classical experiment by Miyahara (1978), the silk filament length was increased from 1900 m to a remarkable length of 2700 m at the 50th generation of selection. A decrease in size of the silk fibre and an increase in overall weight of the cocoon accompanied the increase in filament length.

(iii) Sexing animals

Since the discovery of the sex-determining mechanism by Tanaka (1916), many genes responsible for morphological traits have been mapped on the Z chromosome, but no gene of morphological significance has been found on the W chromosome. In an irradiated batch of eggs, Tazima (1951) discovered a translocation between the W chromosome and chromosome 2 carrying a dominant gene for larval markings ($+^P$). When the translocated dominant gene was present, the larvae carrying the markings were invariably females, while all the non-marked larvae were males (**Figure 6**). This discovery not only confirmed the female-determining nature of the W chromosome but also led to the discovery of so-called sex-limited strains of practical utility to screen males and females at the larval stage (see Nagaraju, 1996), a discrimination otherwise possible at the pupal stage, but which requires cutting every cocoon, rendering them useless for silk reeling. Since most of the silkworms reared for commercial silk production are hybrids, the separation of sexes within the parental varieties helps to prevent "selfing", an important task in the preparation of the hybrids.

After this initial discovery, strains with sex-limited cocoon or egg colour were generated by translocating the yellow cocoon colour gene from chromosome 2 and the black egg colour gene from chromosome 10 to the W

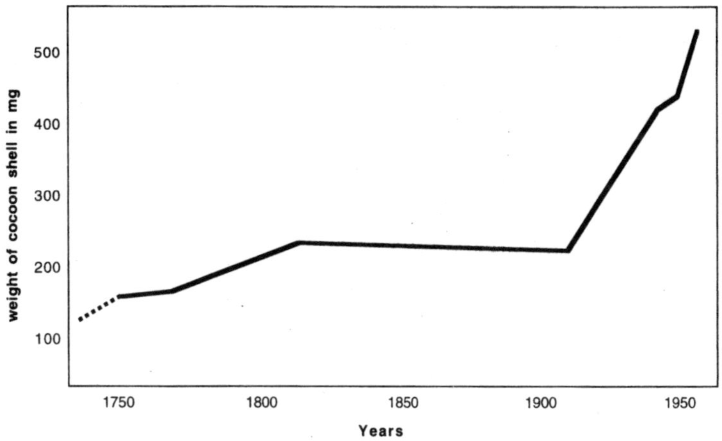

Figure 5 Increase in cocoon shell weight after the introduction of hybrid silkworm rearing in Japan.

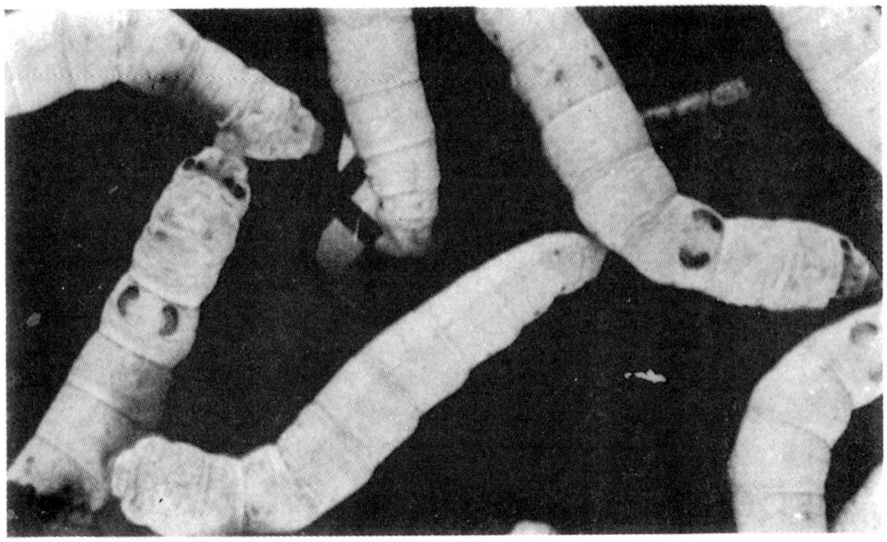

Figure 6 Sex-limited strain that carries larval markings (+^{P}) only in females. The W chromosome in this strain carries a translocated part of chromosome 2 that harbours the dominant allele of +^{P} (W+^{P}).

chromosome. Sex-limited breeds became a reality and heralded a new era in the application of fundamental genetic principles to the commercial use in sericulture.

(iv) Virus resistance

Controlling virus propagation is crucial in sericulture and recurrent losses due to virus infection are registered. Resistance to cytoplasmic polyhedrosis virus (CPV), nuclear polyhedrosis virus (NPV) and infectious flacherie virus (IFV) has been found to be under the control of multiple genes (Aratake 1973a,b), except for one instance of a strain harbouring a major gene for CPV resistance (Watanabe, 1965). In contrast, the genetic control of resistance to densonucleosis virus (DNV) is simpler. A single recessive gene (*nsd-1*) controlling non-susceptibility (refractoriness) to DNV-1 has been reported (Watanabe & Maeda, 1981) and mapped on chromosome 21 (Eguchi *et al.*, 1991). Similarly, resistance to DNV-2 has also been reported to be under the control of a major recessive gene (*nsd-2*) (Seki, 1984) located on a different chromosome from that of the *nsd-1* gene (Abe *et al.*, 1987). Recently, molecular tags closely linked to the DNV-1 resistance gene were discovered and converted into codominant markers (Abe *et al.*, 1995).

(v) Towards transgenesis

Germline transformation may also be viewed as a way towards improving silk production. Such a strategy depends on the availability of a protocol to transform the germline of the silkmoth, as has been done in *D. melanogaster* for more than 15 years (Spradling & Rubin, 1982). The construction of gene vectors to catalyse the stable chromosomal insertion of foreign sequences is the current bottleneck for transgenesis (Nagaraju *et al.*, 1996). Fortunately, the recent development of vectors derived from mammalian retroviruses and capable of infecting invertebrate cells (Burns *et al.*, 1993), and the discovery of lepidopteran mobile elements such as piggyBac (Fraser *et al.*, 1996), allow the prediction of rapid progress towards reaching the goal.

Improving silk production by controlled transformation also implies that candidate genes for introduction are identified. These can fall into various categories such as viral resistance genes, supernumerary silk protein-encoding genes from either Bombyx or other silk-making organisms, or genes whose products would sustain quantitative or qualitative control of silk production. Although some of these genes are already available, many remain to be identified and cloned after more is known about their mapping and the organization of the genome. This is the project of a consortium of several laboratories worldwide.

8. Molecular genetic maps of silkworm

(i) Silkworm genetic stocks

The silkworm genetic stocks maintained around the world comprise geographical races, inbred lines and mutants that carry numerous morphological, developmental, behavioural and biochemical features. It is estimated that more than 3000 silkworm genotypes are being maintained in Asia and Europe. The genetic diversity found in the present-day inbred lines and elite stocks is mostly a result of cross-breeding of geographical races which display a variety of qualitative, quantitative and biochemical traits. Broadly, four sets of geographical races have been identified – Japanese, Chinese, European and Tropical (**Table 4**) – which, in addition to many visible characters, differ widely in qualitative and quantitative characters that affect silk yield. The genotypes of temperate origin produce high quantities of rated silk, while the tropical races are poor producers, but hardy and able to survive under adverse climatic conditions.

The genetic stocks of the silkworm comprise more than 500 mutants for a wide variety of characters (serosal colours; larval and adult integument colours, skin markings and body shapes; cocoon colours and shapes; physiological traits such as diapause, number of larval moults and timing of larval maturity; food habits and biochemical features such as digestive amylase, blood amylase, blood and egg esterases, larval integument esterase, alkaline and acid phosphatases, haemolymph proteins; silk production and fibroin secretion; homeoproteins and body plan determination, etc.) (**Table 5**; see Chikushi, 1972).

Table 4 Characteristics of geographical races of silkworms.

	Japanese	Chinese	European	Tropical
Egg	Greyish-purple serosa, translucent chorion, occasionally coloured non-diapausing eggs, lethality after body pigmentation stage	Pale greyish-purple serosa, light yellow chorion, scarcely white rot egg	Large egg, hatching asynchronous	Light yellow colour egg, weak in wintering-over
Larva	Wild-type larval marking, many red-ripening and yellowish moulters, rather slow growth, susceptible to viruses, N and F, insensitive to mulberry leaf quality	Plain larval marking, many yellow-ripening silkworms and whitish moulters, fast growing, susceptible to muscardine, insensitive to temperature	Slender body shape, many red-ripening and yellowish moulters, very slow growing, susceptible to pebrine* and virus C, sensitive to unfavourable conditions	Plain marking, small and slender body, very rapid growth, susceptible to muscardine, very strong against diseases
Cocoon	Ball shape, white, occasionally double cocoon, fibre thick and short	Elliptical or spherical shape, white or yellow, fibre thin in diameter and long in length	Long elliptical shape, white or flesh color, abundant in sericins, heavy cocoon shell weight, fibre thick in diameter	Spindle shape, white or light green, large quantities of floss, light cocoon shell weight, fibre thin in diameter, no double cocoon
Voltinism	Univoltine (one generation per year) Bivoltine (two generations per year)	Univoltine; bivoltine	Univoltine	Polyvoltine

* pebrine, silkworm disease provoked by the sporozoa, Nozema
C, cytoplasmic polyhedrosis; F, flacherie; N, nuclear polyhedrosis

B. mori, the only lepidopteran with such a well-developed genetics, proves an ideal organism for applying the emerging molecular technologies to construct a high density map. Such a map would serve multiple applications: positional or map-based cloning, localization of complex traits, comparative studies on genome evolution, marker-assisted breeding and selection, and development of tools for fingerprinting. Furthermore, the information generated in the silkworm significantly benefits many species of wild silkmoths that belong to the family of Saturniidae whose genetics is hardly studied.

(ii) Experimental strategies

The haploid silkworm genome contains 530 million base pairs (0.5 pg DNA) (Gage, 1974). This is approximately 3.5 times the size of the *D. melanogaster* genome (140 million bp) (Rasch *et al.*, 1971) and one-sixth the size of the human genome. Given the total length of the genome as approximately 1000 cM and 500 million bp, 1 cM roughly corresponds to 500 kilobases (kb). On this scale, about 4000 molecular markers would be required to place one marker within every 200 kb, an average insert size which could be packaged in yeast artificial chromosomes (YACs) or bacterial artificial chromosomes (BACs) if one plans to use the genetic map for positional cloning with YAC or BAC libraries. In order to pursue this scale of molecular map, one requires high volume marker technology.

The recent explosion of knowledge based on the use of molecular techniques has led to high density mapping of various genomes, map-based cloning of genes of interest for which no prior information was available, identification of quantitative trait loci (QTL) for economic characters and utilization of molecular markers for "marker-assisted selection" and estimation of genetic diversity. Such an integrated approach would form the basis for the analysis and manipulation of the silkworm genome, and provide important raw material for comparative genomics and germline transformation.

(iii) Molecular markers

Molecular markers have several advantages over the traditional phenotypic markers available in the silkworm genetic stocks: they are unaffected by environment, detectable at all stages of development and cover the entire genome. A variety of techniques have been employed to reveal the genome-wide polymorphism, i.e. restriction fragment length polymorphism (RFLP), random amplified polymorphic DNA (RAPD), simple sequence repeat polymorphism (SSR), sequence characterized amplified regions (SCARs) and sequence tagged sites (STS).

Among the various molecular markers developed, RFLPs were the first to be used in human genome mapping and, later, this technique found extensive use in plant and animal genome studies. The polymorphisms detected by RFLPs are highly reliable as they are codominant and can be accurately scored. The method involves the digestion of genomic DNA with a restriction enzyme, agarose gel electrophoretic separation of the digested DNA, transfer of the digested DNA to a membrane and hybridization with labelled probes. Developing RFLP markers involves the cloning of single-copy

Table 5 List of selected silkworm mutants.

Serosal colour

The serosa, lying beneath the chorion, consists of cells that derive from cleavage nuclei and has the embryo's genotype. Wild-type serosal colour is dark brown; mutants are white, pink, red or brown. The pathway involved in the formation of brown pigments is: tryptophan → formylkynurenine → kynurenine → 3-hydroxykynurenine → xanthommatin.

Character	Gene	Locus	Phenotype
Pink-eyed white egg	*pe*	5–0.0	Light orange serosal cell pigment; pink compound eyes.
Red egg	*re*	5–31.7	Red serosal cell pigment; red compound eyes; low hatchability.
White egg 1	*w-1*	10–12.7	Serosa contains no pigment; white compound eyes due to lack of enzyme that converts kynurenine → 3-hydroxykynurenine; maternal inheritance for egg colour, normal inheritance for eye colour.
White egg 2	*w-2*	10–16.1	Yellowish-white freshly laid eggs, gradually becoming red tinted; white compound eyes; 3-hydroxykynurenine accumulates in pupae and eggs due to absence of xanthommatin-producing enzyme; normal inheritance.
White egg 3	*w-3*	10–19.6	Light yellowish-brown serosa; black compound eyes.
Brown egg 1	*b-1*	6–8.0	Brown serosal cells; black compound eyes; inheritance shows typical maternal determination; transmission of 3-hydroxykynurenine from haemolymph to ovum is somehow prevented.
Brown egg 2	*b-2*	6–8.0	Light brown serosal cells; black compound eyes. Complex inheritance pattern: F_1 eggs of +/++ × *b-2*/*b-2* male are normal black, those of reciprocal cross are similar to recessive *b-2* mother. In F_2 generation, segregation occurs within a brood laid by a single female in the ratio 3 black (+/+):1 light brown (*b-2*/*b-2*). In BF_1 of cross +/*b-2*+ × *b-2*/*b-2* male, segregation occurs in eggs of a single batch in the ratio 1 black:1 light brown.
Brown egg 4	*b-4*	20–21.9	Reddish-brown serosal cells; black/reddish-tinted compound eyes; normal inheritance.

Egg size and shape

The silkworm egg is short, elliptic, slightly narrowed at the anterior end where the micropyle is situated. The egg shape is dependent on the shape of the chorion that is formed in the maternal body before fertilization.

Egg size

Character	Gene	Locus	Phenotype
Small egg	*Sm*	3–41.8	Oocytes degenerate; reduced yolk uptake; females sterile.
Small egg-2	*Sm-2*	13–0.0	Small eggs with reduced yolk uptake; females sterile.
Giant egg	*Ge*	1–14.0	40% size increase, yolk composition normal. Hatched larvae larger than normal; hatchability reduced.

Egg shape

Character	Gene	Locus	Phenotype
Kidney-shaped	*ki*	6–8.6	Kidney- or bean-shaped egg, lethal at embryonic stage.
Spindle-shaped	*sp*	23–22.9	Long narrow eggs with attenuated ends.
Ellipsoid	*elp*	18–16.1	Long elliptic eggs.

Chorion colours

In the normal egg, the chorion is colourless and transparent.

Character	Gene	Locus	Phenotype
Gray egg	*Gr*	2–6.9	Elliptical eggs; chorion transparent in homozygotes, opaque in heterozygotes. *Gr* causes abnormality in lamellar structure of mid-chorion layer.
Gray egg (collapsing)	Gr^{col}	2–6.9	Thin collapsing chorion due to reduced secretion of early and early/middle chorion proteins.
Gray egg (European[16] gray)	Gr^{16}	2–6.9	Opaque chorion; homozygote wrinkled due to altered post-translational modification of putative framework proteins.
Bird egg	Gr^{B}	2–6.9	Heterozygotes show thin collapsing chorion, transparent egg centre and opaque rim.
Mottled gray egg	*mgr*	6–8.9	Transparent and opaque areas in chorion.
White sided egg	*se*	15–16.9	Chorion opaque, wrinkled; homozygote sterile.

Larval characteristics

Newly hatched larva

The newly hatched larva is dark brown and covered with long black setal.

Character	Gene	Locus	Phenotype
Chocolate	*ch*	13–9.6	Reddish-brown newly hatched larva, gradually diminishing as larva grows to advanced instars.
Dominant chocolate	*l-a*	9–5.9	Phenotypically like chocolate; head is black.
Sex-linked chocolate	*Sch*	1–21.5	Reddish-brown newly hatched larva like the chocolate; fully grown larva is normal.

Larval markings

Larval colour patterns depend mainly on the nature and distribution of pigments in hypodermal cells and epicuticle. Markings are clearly visible after the third moult. Normal larval pattern is composed of eye spots on the second thoracic segment, crescents (lunules) on the second abdominal segment and star spots on the fifth abdominal segment.

P-alleles

The dominance relationship of major alleles of this group is $P^B > P^S \geq P^M + P > P$

Character	Gene	Locus	Phenotype
Plain	*P*	2–0.0	Devoid of any markings.
Moricaud	P^M	2–0.0	Many dark, greyish-brown lines and spots.
Striped	P^S	2–0.0	Solid black body surface; distinct white stripe on posterior margin of each segment.
Black	P^B	2–0.0	Black
Ventral striped	P^G	2–0.0	Ventral striped
Whitish shaped	P^{SW}	2–0.0	Whitish shaped
Sable	P^{Sa}	2–0.0	Sable

S-alleles

Character	Gene	Locus	Phenotype
New striped	*S*	2–6.1	Resembles striped group of P-alleles and cannot be discriminated phenotypically.
Dilute striped	S^d	2–6.1	May be due to duplication or rearrangement involving *p* locus.
Zebra	*Ze*	3–20.8	Narrow black band on anterior rim of each larval segment; dark brown cheek spots on both sides of head cuticle.
Speckled	*Spc*	4–33.1	Dark brown spots scattered over whole larval body surface. Female moths have difficulty in egg laying (i.e. lay few eggs).
Multilunar	*L*	4–0.0	Paired large brown or yellow spots appear on dorsal side of each segment; many modifying genes affect number of spots; temperature during embryonic development affects spot number.
Dirty	*Di*	14–43.2	Irregular black lines and dots cover larval body surface; homozygotes lack pigments in stars and crescents.
Ursa	*U*	14–40.5	Tegument covered with dark brown pigments except in region of dorsal blood vessel.
Quail	*q*	7–0.0	Body tinted reddish-purple; posterior outline of crescent does not have intense pigmentation.
Multistars	*ms*	12–15.5	Additional paired star spots on dorsal side of abdominal segments 3–7; temperature during embryonic development affects spot number.
Blind	*bl*	15–0.0	Solid black eye spots on second thoracic segments.
Knobbed	*k*	11–0.0	Dermal protrusions on sites of crescents and star spots and on spots manifested additionally in *L* and *ms*.
Homeotic mutants (or genes affecting segment)	*E alleles*	6–0.0	35 alleles known, affecting number, position of larval segments, markings, prolegs, reproductive organs and nervous system. Lethal in embryonic stages, homologues of *Drosophila melanogaster bithorax* complex.
No crescent	*Nc*	6–1.4	Missing crescents, narrow eye spots, highly pigmented embryonic lethal, homologue of *Antennapedia* complex.
No lunule	*Nl*	14–35.2	Missing crescents and stars, embryonic lethal due to chromosome deficiency.

Table 5 Continued

Larval skin mutants
The larval integument is normally opaque and contains urate crystals. Many mutants have a translucent skin. Degree of translucency depends on mutant strain and degree of accumulation of uric acid. Higher translucency means less uric acid storage.

Character	Gene	Locus	Phenotype
Distinct translucent	*od*	1–49.6	Highly translucent due to reduced uric acid uptake.
Sex-linked translucent	*os*	1–0.0	Weakly translucent due to reduced uric acid uptake.
Oal-mottle translucent	*oal*	2–26.7	Mottled skin with translucent and opaque areas.
Mutator of oal	*mu-oal*	13–40.4	Induces somatic mosaicism of mutable alleles of *oal*.
Chinese translucent	*oc*	5–40.8	Moderately translucent.
Giallo Ascoli translucent	*og*	9–7.4	Highly translucent; high pupal mortality; both sexes sterile; low xanthine dehydrogenase activity excretes xanthine and hypoxanthine.
Aojuku translucent	*oa*	14–42.2	Moderately translucent, oa^2 allele mottled.
Waxy translucent	*ow*	17–36.4	Moderately translucent.
Japanese translucent	*oj*	9–0.0	Moderately translucent, light egg colour.
Kinshiryu translucent	*ok*	5–4.7	Highly translucent, early larval mortality high.
Tanaka's mottled translucent	*otm*	5–15.2	Moderately translucent, many fine opaque dots.
Dominant obese translucent	*obs*	18–6.2	Third instar distinctly translucent, fifth less conspicuous; body short/stout; difficulty ecdysing at fourth and pupal moults; male poor copulation. Exhibits inhibitory epistatic interactions with genes affecting stars, crescents, e.g. P^s, *U*, and *K*.
Inhibitor of lemon	*i-lem*	2–29.5	Faint yellow in combination with *lem*.
Albino	*al*	5–37.9	Albino after first moult; cuticle porous, incomplete hardening, larva dehydrates, accumulates fluorescent sepiapterin.
Dilute black	*bd*	9–6.7	Greyish-black larval trunk; female moths completely sterile.
Sooty	*So*	26–2	Head dark coloured; thorax/abdomen sooty in larva and moth.

Blood colours
Normal haemolymph is colourless and turns black when exposed to air due to phenoloxidase activity.

Character	Gene	Locus	Phenotype
Yellow blood	*Y*	2–25.6	Deep yellow haemolymph due to presence of carotenoids; yellow-blooded larva usually spins yellow cocoon.
Yellow Inhibitor	*I*	9–0.0	This gene completely inhibits *Y*, thus producing white cocoons.
Red haemolymph	*rb*	21–0.0	Haemolymph of *rb*/*rb* silkworm shows faint reddish coloration after brief exposure to air; mutant lacks enzyme that converts 3-hydroxykynurenine into anthranilic acid in tryptophan metabolism, accumulating abundant 3-hydroxykynurenine in haemolymph.

Larval body shapes

Character	Gene	Locus	Phenotype
Elongated	*E*	1–36.4	First and second abdominal segments unusually elongated due to fully stretched intersegmental regions.
Stick	*sk*	4–25.8	Withered twig appearance; larval body less flexible, slender and firm.
Gooseneck	*gn*	9–5.8	Slender larval body, constricted between segments.
Narrow breast	*nb*	19–0.0	Spindle-shaped larval body, narrow thorax; stout abdomen flabby to touch.

Cocoon characters

Cocoon colours
Almost all silkworm varieties reared in Japan, China and Korea are white cocoon varieties; those reared in India and Thailand are yellow cocoon colour varieties. There are many colour mutants with golden yellow, pinkish and green cocoons. Yellow and pinkish colours are due to carotenoids, carotenes and xanthophylls derived from mulberry leaves; they are taken up in sericin proteins that cement the fibroin threads. Green colour is due to flavinoids, which, unlike carotenoids, are distributed in both sericins and fibroin. White colour ($+^C$) is considered to be wild type.

C-alleles

Character	Gene	Locus	Phenotype
Golden yellow	*C*	12–14.2	Cocoon colour golden yellow on outside, nearly white inside.
	C^I		Deep yellow inner cocoon layer; outer layer faintly pigmented.
	C^D		All layers pale yellow.

Character	Gene	Locus	Phenotype
Straw colour	C^{St}	12–14.2	Faint yellow outer and inner cocoon shell.
Yellow colour	*Y*	2–25.6	Yellow cocoon colour.
Flesh	*F*	6–13.6	In combination with *Y* and $+^{C}$, reddish-yellow colour produced on outer layer of cocoon.
Pink colour	*Pk*	?	Cocoon pink on outside, coexists with *Y* and $+^{C}$. *F* needed for manifestation of *Pk*.
Green colour	*Ga*	?	Cocoon light green when combined with *Gb*.
	Gb	7–7.0	Cocoon light green when combined with *Ga*.
	Gc	15–?	Cocoon light green

Silk production mutants

Character	Gene	Locus	Phenotype
Naked pupa	*Nd*	25–0.0	Cocoon contains only sericin; some homozygous lines lack posterior silk gland.
Sericin cocoon	*Nd-s*	14–19.2	Cocoon contains only sericin; posterior silk gland degenerates.
Flimsy cocoon	*fle*	3–49.0	Cocoon contains excess sericin relative to fibroin; posterior silk gland degenerates.

Pupal and adult characteristics

Character	Gene	Locus	Phenotype
Wingless	*fl*	10–13.0	Wings lacking in pupa/moth; legs on second and third thoracic segments poorly developed.
Vestigial	*vg*	1–38.7	Wings poorly developed, do not extend fully; lethal in homo- and hemizygous conditions.
Crayfish	*cf*	13–20.9	Wingbuds inflated with haemolymph; protruding, fragile, bleed easily.
Crayfish of Eguchi	*cf-e*	4–0.0	Pupal wings swollen, protrude laterally from body; wing buds swollen, fragile, bleed easily.
Minute wing	*mw*	22–25.2	Pupal/adult wings extremely small.
Micropterous	*mp*	11–51.8	Wings 80% of normal size.
Wrinkled wings	*wri*	14–0.0	Wings poorly developed, do not extend fully.

Body and wing colours

Character	Gene	Locus	Phenotype
Black pupa	*bp*	11–14.3	Pupal cuticle black; expression temperature-sensitive for several hours before pupation; involves prophenoloxidase cascade.
White banded black wing	*wb*	5–35.8	Outer margin/proximal region of wings dark brown, leaving wide white band, white body.
Black moth	*Bm*	17–0.0	Black scales on body and wings.
Wild wing spot	*ws*	17–14.7	Black spot on wing apex, less conspicuous in heterozygotes, especially females.
Sooty	*so*	26–0.0	Body smoky in larva and moth; pupal case black, especially at ventral tip of abdomen.
Lustrous	*lu*	16–0.0	Facets of compound eyes small and irregular.
Varnished	*ve*	6–11.1	Compound eyes extremely small, lustrous; facets sparsely distributed.

Moulting characteristics

There are at least five different types of moulting in the silkworm, namely di-, tri-, tetra-, penta- and hexamoulters. All present-day commercial varieties are tetramoulters. Tetramoulting character is considered wild type in silkworm genetics. The dominant relations among the M alleles are $M^3 > +M > M^5$. The expression of moulting character is also partly affected by the sex-linked maturity genes (Lm).

Character	Gene	Locus	Phenotype
M-alleles	M^3	6–3.0	Trimoulter
	+M		Tetramoulter
	M^5		Pentamoulter
Recessive trimoulting	*rt*	7–9.0	Trimoulters, but occasionally dimoulters are seen

(present only once in the haploid genome) DNA fragments or cDNA (DNA complementary to mRNA) and their use as probes to analyse the extent of polymorphism.

The first RFLP map was developed in silkworm using 52 progenies of an F_2 cross from a pair mating of two parental genotypes, C_{108} and P^{50}, a bivoltine (two generations reared per year) and a polyvoltine strain, respectively. C_{108} and P^{50} were selected since they show contrasting features for various characters – such as diapause, larval growth, larval size, larval span, fecundity, silk fibre length, pupal weight, quantity of sericin and fibroin proteins, biochemical traits such as digestive and haemolymph amylases, resistance to viral diseases, etc. – suggesting that there would be considerable polymorphism at their DNA level. The RFLP probes included 15 characterized single-copy sequences, 36 anonymous sequences derived from a follicular cDNA library and ten loci corresponding to a low copy number retrotransposon. The RFLP map covered a total recombination length of 413 cM dispersed over 15 linkage groups. Out of the 15 known single-copy sequences, ten had already been localized on the conventional linkage map using biochemical and morphological variants and could be correlated with the molecular map.

Although RFLP analysis offers several advantages, it requires large quantities of DNA and is labour intensive, time consuming and expensive. The availability of PCR technology increased the capacity of detecting DNA polymorphisms, and offered efficient alternatives to the RFLP hybridization methods. The PCR-based techniques meet the wide spectrum of requirements for large-scale genetic studies since they apply to very small quantities of DNA, avoid DNA blotting and probe making, and are amenable to automation. These features of PCR analysis resulted in its widespread use in silkworm genomic research.

Thus, PCR-aided RAPD marker technology was used to construct new genetic maps of silkworm (Promboon *et al.*, 1995). This method involves amplification of genomic DNA template primed onto a 10-oligonucleotide primer of random sequence, but with a minimum guanidine–cytosine content of 50%. A low annealing temperature (37–40°C) is maintained throughout all (35–40) cycles, which results in the amplification of several discrete fragments usually separated on agarose gels. These fragments vary in size from 200–2000 bp. They are amplified from regions of the genome that contain two DNA sites bearing homologies to the primer. Polymorphisms occur as the presence or absence of one or several specific fragments among individuals, which can be caused by point mutations at one or both annealing sites on DNA strands, insertions surrounding a site or insertions that separate annealing sites by a larger distance than can be sustained by amplification.

The silkworm RAPD linkage map was constructed using 102 F_2 individuals of the same parental strains (C_{108} and P^{50}) used for the RFLP map. The screening of 320 primers resulted in 243 clear polymorphic products between C_{108} and P^{50}, out of which 168 bands revealed Mendelian inheritance. The MAPMAKER program sorted these 168 bands into 29 linkage groups and ten unlinked loci covering the 900 cM map distance (Shi *et al.*, 1995).

The routine application of RAPD analysis for genetic mapping is not without limitations. First, the efficiency of amplification of fragments using decamers at low annealing temperatures is very sensitive to reaction conditions and may be inconsistently reproducible in different experiments or in different laboratories. This is particularly problematic if one is interested in integrating mapping data from different crosses by comparison with common reference markers. Second, because RAPD markers are inherited as dominant markers, the lack of allelism is a disadvantage as the heterozygous and homozygous loci are not distinguishable. In many cases, cloning and sequencing the RAPD fragments of interest and designing primers that are complementary to the ends of the original RAPD product can circumvent this. When such primers are used with the original template DNA, single loci (or SCARs) are specifically amplified. These SCARs retain the dominant segregation behaviour of the original RAPD fragments and can be easily evidenced. However, since the primer binding site is often the source of strain specificity, the sequence obtained between the primers may not be polymorphic.

The three criteria that define ideal polymorphic loci (easy to identify, highly polymorphic and amenable to automation) are met by the microsatellite-based genetic markers. These markers, which are distributed across most of the eukaryotic genomes, have proven an efficient means to complete genetic mapping. Microsatellites are short stretches of DNA which consist of an array of simple di-, tri- or tetranucleotide repeats like $(CA)_n$, $(GA)_n$, $(GTA)_n$, $(ATT)_n$, $(GATA)_n$, etc. The successful application of microsatellite DNA in genetic analysis requires the cloning of microsatellite repeats from the genome, the characterization of their flanking sequences for primer design, the PCR amplification of the template DNA, the electrophoretic separation of the PCR products and the detection of alleles of microsatellite loci.

Recent studies show that the silkworm genome is abundantly interspersed with CA/GT and GA/CT repeats (Reddy *et al.*, 1999). The $(GT)_n$ repeats occur on average every 49 kb, while $(CT)_n$ repeats occur every 104 kb (**Table 6**). These frequencies are similar to those estimated in the mouse and

Table 6 Distribution of microsatellite sequences in different animal species.

	Silkworm	Honey bee	Bumble bee	Salmon	Rat	Human
$(GT)_n$	49	34	500	12	15	30
$(CT)_n$	104	15	40	ND	50	113

For each microsatellite tested, the average distance between two loci was obtained by dividing the length of the DNA by the number of positive signals given by a particular microsatellite probe. In the silkworm, 3513 clones were screened, corresponding to 3513 × 0.8 kb (mean insert size of the library screened) = 2810 kbp of DNA. For each microsatellite, this value was divided by the number of signals observed.

human genomes. Initial studies show that the primers designed from the flanking microsatellite loci reveal allelic differences between silkworm strains and are inherited in a Mendelian fashion (**Figure 7**). These findings show that microsatellite markers can provide high volume marker technology for intensive mapping of the silkworm genome.

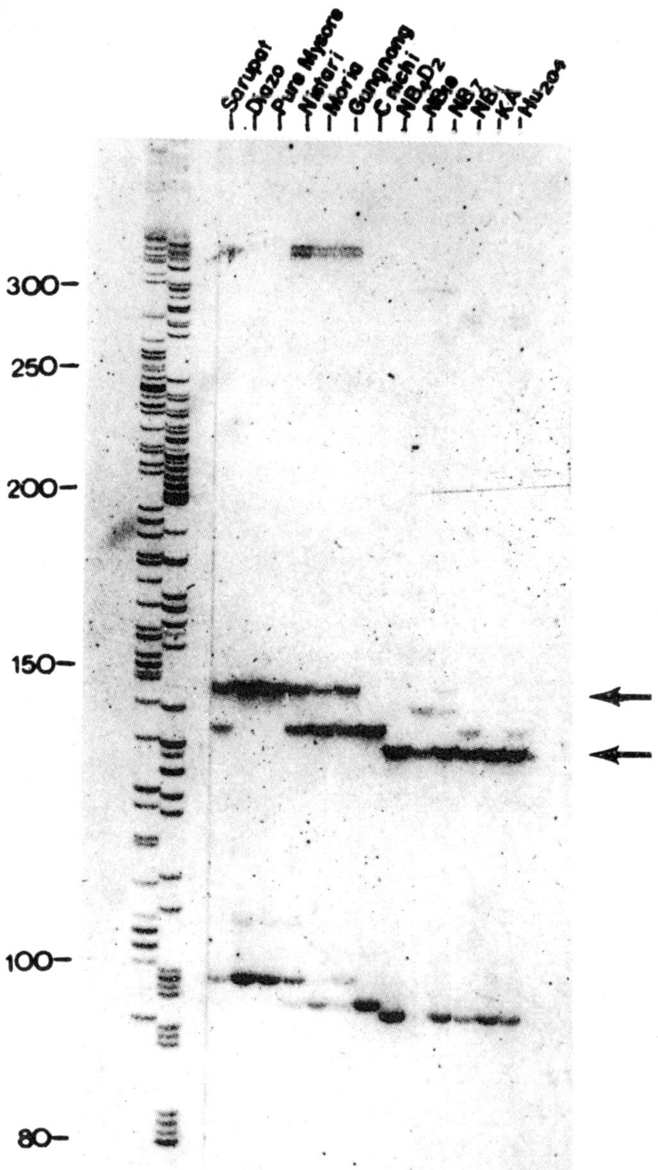

Figure 7 Polymorphism of microsatellite locus BmSat 3513 in 13 silkworm genotypes. Microsatellites were PCR amplified using unique flanking primer sequences and separated on a sequencing gel. The size in bases is indicated. Arrows indicate diapause and non-diapause genotype-specific products. Allelic length polymorphisms were detected by the microsatellite locus BmSat 3513. The forward (5′-CGCAATTCTGTATTAGATAA-3′) and reverse (5′-TAAAGGTATTATTCTTATTCG-3′) primers were designed for the flanking sequences of the BmSat 3513 which is a compound microsatellite repeat motif, $(GT)_5$, $(TGCG)_6(TA)_2$. The PCR reactions (20 μl) were performed with 20 ng genomic DNA. (Data from Reddy *et al.*, 1999)

Efforts are now underway to combine these markers with the RFLP and RAPD markers in a single coherent map. Some of the flanking sequences of the Bombyx microsatellite loci are conserved in the saturniid silkmoths *Antheraea mylitta*, *A. pernyi*, *A. polyphemus*, *A. yamamai* and *Philosamia cynthia ricini* and, hence, could be used for genetic analysis of these economically valuable insects (Nagaraju & Reddy, unpublished data).

More variations of PCR technology are emerging in genome analysis. One takes advantage of the ubiquitous distribution of $(CA)_n$ repeats in the silkworm chromosomes and uses primers in the repeated sequence to amplify the genomic regions between two microsatellites. Anchored primers of such repeats lead to discrete amplified products, with Mendelian inheritance. This method, a variant of the SSR-anchored PCR, is useful to fingerprint diverse silkworm genotypes (**Figure 8**) and to carry out genetic mapping without any prior cloning and characterization of microsatellite sequences (Zietkiewicz *et al.*, 1994).

The banded krait (an Indian snake) minor satellite DNA, Bkm-2(8), containing a 545 bp sequence consisting mainly of 22 GATA repeats, was also used as a probe to reveal polymorphisms in different silkworm races. Although the polymorphism level is lower than with SSR-anchored PCR, the Bkm-2(8) probe revealed very clear fingerprint profiles (**Figure 9**). Such DNA fingerprint-based information will be useful for estimating the genetic diversity of silkworm germplasm (Nagaraju *et al.*, 1995; Nagaraju & Singh, 1997).

Efforts are being made by different groups to construct a high density molecular map of the silkworm that would integrate RFLPs (M.R. Goldsmith, D. Heckel, US; W. Hara, Japan), RAPDs (J. Nagaraju, India; T. Shimada, Japan; Y. Yasukochi, Japan), microsatellites (J. Nagaraju, India) and expressed sequence tags (ESTs) (S. Maeda & K. Mita, Japan).

9. Conclusions and future prospects

The progress made in the construction of linkage maps and the high volume marker technology based on microsatellites, RAPDs and ESTs shows that a high density linkage map of the silkworm is not far from reality. These developments will make possible the positional cloning of genes and investigation of their functional and evolutionary aspects. The refinements that have taken place in recent years in mouse, human and Drosophila, such as chromosome-specific libraries, *in situ* hybridization and large insert libraries, could also hasten the analysis of the silkworm genome.

The wealth of silkworm genotypes which represent an array of differences for various quantitative and qualitative characters could serve as a genetic repertoire to identify and map QTLs. Such QTL mapping will provide knowledge about the number of loci influencing quantitative traits in the silkworm as well as the mode of action of the underlying genes (e.g. pleiotropic effects). Such an effort would require a collaborative venture by silkworm breeders, molecular geneticists, silkworm pathologists and physiologists, and biometricians, and promises to open up new genetic systems for research.

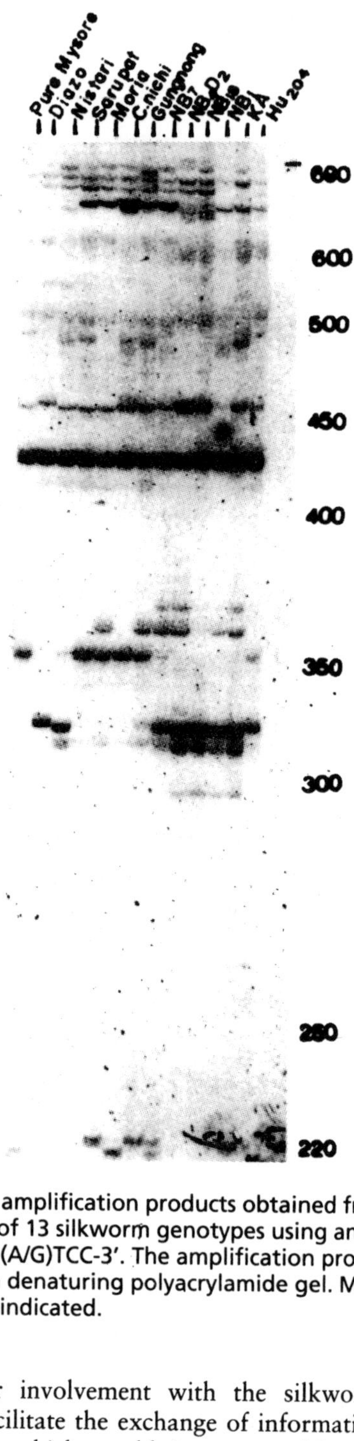

Figure 8 PCR amplification products obtained from genomic DNA of 13 silkworm genotypes using anchored primer 5′-(GT)$_8$(A/G)TCC-3′. The amplification products were separated on a denaturing polyacrylamide gel. Molecular size in bases is indicated.

The deeper involvement with the silkworm genome would also facilitate the exchange of information between related insects, which would, in turn, enable geneticists to prepare gross comparative maps. Such comparative maps will serve as blueprints for dissecting the evolutionary aspects of genes as well as the genome as a whole.

The silkworm mapping project has many practical applications, such as using molecular markers for marker-assisted selection (MAS) to hasten the breeding programme and to develop strains resistant to various viral and bacterial diseases. Overall, the mapping project could improve silkworm health and productivity, thus improving the economy of sericulture. The new molecular genetic methods should not be viewed as replacements for the existing methods, but rather should be used to complement the well-established tools such as biometrics and quantitative genetics to improve silkworm health and productivity.

10. A late-breaking result

Bombyx transgenesis has recently been achieved using a lepidopteran transposon (piggyBac) to catalyse the insertion of genes of interest into the chromosomes of the germline cells (Tamura *et al.*, 2000). The method consists in coinjecting a helper vector (providing a transient boost of expression of the piggyBac transposase) and a vector carrying the gene of interest placed between the inverted repeats of the element. The integration proceeds via a cut and paste mechanism whereby the transgene is excised from the vector and inserted into the chromosomal DNA. The efficiency of the method is similar to that observed in Drosophila.

Stable germline transformation of *B. mori* will be of considerable help in unravelling the fundamental mechanisms of gene regulation and in improving strains of sericultural interest. Also, the high proteosynthesis that occurs in the silk glands of the silkworm could be exploited to produce proteins of pharmaceutical or veterinarian interest that could be purified directly from the cocoon. The textile industry could also benefit from novel fibres made by silkworms transformed with heterologous genes encoding fibrous proteins.

REFERENCES

* *In Russian with English abstract*
** *In Japanese with English abstract*

Abe, H., Watanabe, H. & Eguchi, R. (1987) Genetical relationship between nonsusceptibilities of the silkworm, *Bombyx mori*, to two densonucleosis viruses. *Journal of Sericultural Science of Japan* 56: 443–44

Abe, H., Shimada, T., Yokoyama, T., Oshiki, T. & Kobayashi, M. (1995) Identification of random amplified polymorphic DNA linked to the densonucleosis virus type-1 susceptible gene of the silkworm, *Bombyx mori*. *Journal of Sericultural Science of Japan* 64: 19–22

*Altukhov, Y.P. & Klimenko, V.V. (1978) Positive correlation between the level of individual heterozygosity and a tendency towards complete parthenogenesis in the silkworm. *Doklady Akademii Nauk SSSR (USSR)* 239: 460–62

Aratake, Y. (1973a) Difference in the resistance to infectious flacherie virus between the strains of the silkworm, *Bombyx mori* L. *Sanshi-Kenkyu* 86: 48–57

Aratake, Y. (1973b) Strain differences of the silkworm, *Bombyx mori* L., in the resistance to a nuclear polyhedrosis virus. *Journal of Sericultural Science of Japan* 42: 230–38

Aratake, Y. & Kayamura, T. (1973) "Scattered" and "small sized" silkworm strains selected from the progeny of hybrid between *Bombyx mori* and *Theophila mandarina*. *Journal of Sericultural Science of Japan* 42: 331–39

*Astaurov, B.L. (1940) *Artificial Parthenogenesis in the Silkworm (Bombyx mori). An Experimental Study*, Moscow: USSR Academy Press

Astaurov, B.L. (1967) Experimental alterations of the developmental cytogenetic mechanisms in mulberry silkworms (artificial parthenogenesis, polyploidy, gynogenesis and androgenesis. *Advances in Morphogenesis* 6: 199–257

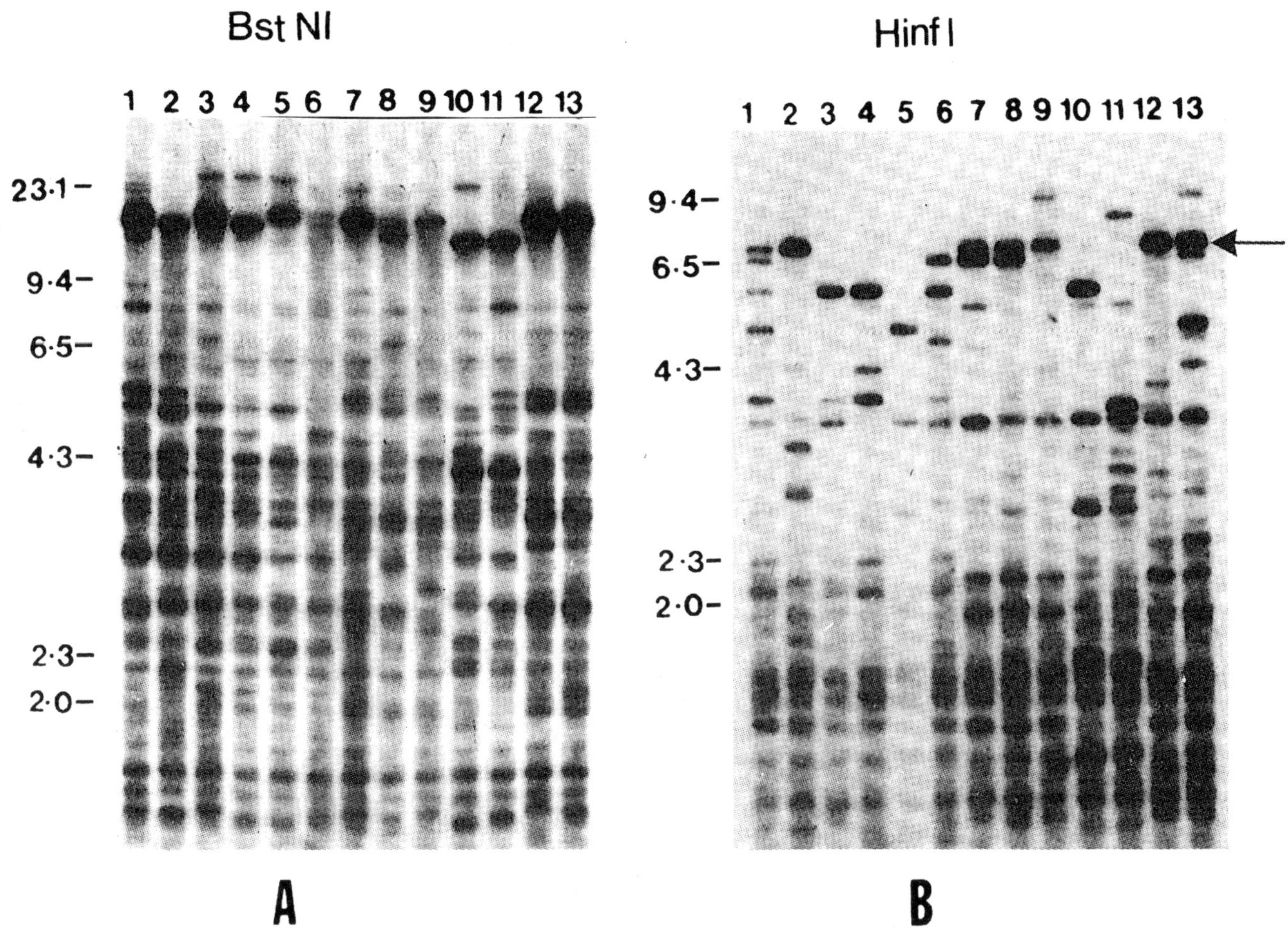

Figure 9 Bkm-2(8) hybridization pattern of (A) *Bst*NI and (B) *Hinf*I digested DNA from 13 silkworm genotypes. Lanes 1–13 are pooled DNA samples of Hu_{204}, C. nichi, gungnong, Moria, Diazo, Sarupat, KA, NB_1, NB_7, Nistari, Pure Mysore, NB_4D_2 and NB_{18}, respectively. Each lane contains 10–12 µg of DNA pooled from 12 individuals. *Hinf*I profiles (B) are unique to each genotype. The arrow indicates the band shared by all diapausing genotypes (lanes 1, 7, 8, 9, 12 and 13) and one non-diapausing genotype, C. nichi (lane 2). Note its absence in all the other non-diapausing genotypes (lanes 3, 4, 5 6, 10 and 11). Numbers on the left indicate DNA fragment sizes in kb.

Bello, B. & Couble, P. (1990) Specific expression of a silk encoding gene of *Bombyx* in the anterior salivary gland of *Drosophila. Nature* 346: 480–82

Bello, B., Horard, B. & Couble, P. (1994) The selective expression of silk protein encoding genes in *Bombyx mori* silk gland. *Bulletin d'Institut Pasteur* 92: 81–100

Boman, H.G. (1995) Peptide antibiotics and their role in innate immunity. *Annual Review of Immunology* 13: 33–42

Burns, J., Friedman, T., Driever, W., Burrascano, M. & Yee, J.K. (1993) Vesicular somatitis virus G glycoprotein pseudotyped retroviral vectors: concentration to very high titer and efficient gene transfer to mammalian and non mammalian cells. *Proceedings of the National Academy of Sciences USA* 90: 8033–37

Chareyre, P., Besson, M.T., Fourche, J. & Bosquet, G. (1996) Identification of a *Bombyx* collagenous protein with multiple short domains of Gly-Xaa-Yaa repeats: cDNA characterization and regulation of expression. *Insect Biochemistry and Molecular Biology* 26: 677–85

Chikushi, H. (1972) *Genes and Genetical Stocks of the Silkworm*, Tokyo: Keigaku Publishing

Couble, P., Chevillard, M., Moine, A., Ravel-Chapuis, P. & Prudhomme, J.C. (1985) Structural organization of the P25 gene of *Bombyx mori* and comparative analysis of its 5′ flanking DNA with that of the fibroin gene. *Nucleic Acids Research* 13: 1801–14

Couble, P., Michaille, J.J., Garel, A., Couble, M.L. & Prudhomme, J.C. (1987) Developmental switches of sericin mRNA splicing in individual cells of *Bombyx mori* silkgland. *Developmental Biology* 124: 431–40

Drevet, J.R., Seiky, Y.A. & Iatrou, K. (1994) GATA-type zinc finger motif-containing sequences and chorion gene transcription factors of the silkworm *Bombyx mori. Journal of Biological Chemistry* 269: 10660–67

**Eguchi, R., Ninaki, O. & Hara, W. (1991) Genetical analysis on the non-susceptibility to densonucleosis virus in the silkworm, *Bombyx mori. Journal of Sericultural Science of Japan* 60: 384–89

Eickbush, T.H. (1995) Mobile elements of lepidopteran genomes. In *Molecular Model Systems in the Lepidoptera*, edited by M.R. Goldsmith & A.S. Wilkins

Fournier, A., Taneja, R., Gopalkrishnan, R., Prudhomme, J.C. & Gopinathan, K.P. (1993) Differential transcription of multiple copies of a silkworm gene encoding tRNA (Gly1). *Gene* 134: 183–90

Fraser, M., Ciszczon, T., Elick, T. & Bauser, C. (1996) Precise excision of TTAA-specific lepidopteran transposons piggybac (IFP2) and tagalong (TFP3) from the baculovirus genome in cell lines from two species of *Lepidoptera*. *Insect Biochemistry and Molecular Biology* 5: 141–51

Friedlander, M. & Wahrman, J. (1970) The spindle as a basal body distributor. A study in the meiosis of the male silkworm moth, *Bombyx mori*. *Journal of Cell Science* 7: 65–89

Fujii, T., Sakurai, H., Izumi, S. & Tomino, S. (1989) Structure of the gene for the arylphorin-type storage protein SP_2 of *Bombyx mori*. *Journal of Biological Chemistry* 264: 11020–25

Fukuta, M., Matsuno, K., Hui, C.C. *et al.* (1993) Molecular cloning of a POU domain-containing factor involved in the regulation of the *Bombyx* sericin-1 gene. *Journal of Biological Chemistry* 268: 19471–75

Furukawa, S., Taniai, K., Ishibashi, J. *et al.* (1997) A novel member of lebocin gene family from the silkworm, *Bombyx mori*. *Biochemical and Biophysical Research Communications* 238: 769–74

Gage, L.P. (1974) The *Bombyx mori* genome: analysis by DNA association kinetics. *Chromosoma* 45: 27–42

Gage, L.P., Friedlander, E. & Manning, R.F. (1976) Interspersion of inverted and middle repeated sequences within the genome of silkworm, *Bombyx mori*. In *Molecular Mechanisms in the Control of Gene Expression*, edited by D.P. Nierlich, W.J. Rutter & C.F. Fox, New York: Academic Press

Garel, J.P. (1976) Quantitative adaptation of isoacceptor tRNAs to mRNA codons of alanine, glycine and serine. *Nature* 260: 805–06

Goldsmith, M.R. & Wilkins, A.S. (eds) (1995) *Molecular Model Systems in the Lepidoptera*, Cambridge and New York: Cambridge University Press

Hagenbuchle, O., Larson, D., Hall, G.I. & Sprague, K.U. (1979) The primary transcription product of a silkworm alanine tRNA gene: identification of *in vitro* sites of initiation, termination and processing. *Cell* 18: 1217–29

Hara, S. & Yamakawa, M. (1996) Production in *Escherichia* of moricin, a novel type of antibacterial peptide from the silkworm, *Bombyx mori*. *Biochemical and Biophysical Research Communications* 220: 664–69

**Harada, C., Kimura, K. & Aoki, H. (1961) On the effect of hybrid vigour on the quantitative characters concerned with reeling of cocoons. *Acta Sericologica* 37: 42–45

Hasimoto, H. (1933) Genetic studies on the tetraploid female in the silkworm. *Bulletin of the Sericultural Experiment Station Japan* 8: 359–81 (*In Japanese*)

**Hasimoto, H. (1941) Linkage studies in the silkworm. *Bulletin of the Sericultural Experiment Station Japan* 10: 328–63

Hibner, B.L., Burke, W.D. & Eickbush, T.H. (1991) Sequence identity in an early chorion multigene family as the result of localised gene conversion. *Genetics* 128: 595–606

Hirobe, T. (1968) Evolution, differentiation and breeding of the silkworms – the Silk Road, past and present. *Genetics in Asian Countries: XII International Congress of Genetics*

Horard, B., Julien, E., Nony, P., Garel, A. & Couble, P. (1997) Differential binding of the *Bombyx* silk gland specific factor SGFB onto its target DNA sequence drives posterior cell restricted expression. *Molecular and Cellular Biology* 17: 1572–79

Hui, C.C. & Suzuki, Y. (1995) Regulation of the silk protein genes and the homeobox genes in silk gland development. In *Molecular Model Systems in the Lepidoptera*, edited by M. Goldsmith & A.S. Wilkins

Hui, C.C., Matsuno, K., Ueno, K. & Suzuki, Y. (1992) Molecular characterization and silkgland expression of *Bombyx* engrailed and invected genes. *Proceedings of the National Academy of Sciences USA* 89: 167–71

Iatrou, K., Tsitilou, S.G. & Kafatos, F.C. (1984) DNA sequence transfer between two high-cysteine chorion gene families in the silkmoth, *Bombyx mori*. *Proceedings of the National Academy of Sciences USA* 81: 4452–56

Iwasaki, H., Takahashi, M., Niimi, T., Yamashita, O. & Yaginuma, T. (1997) Cloning of cDNAs encoding *Bombyx* homologues of cdc2 and cdc2-related kinase from eggs. *Insect Molecular Biology* 6: 131–41

Kamimura, M., Tomita, S., Kiuchi, M. & Fujiwara, H. (1997) Tissue-specific and stage-specific expression of two silkworm ecdysone receptor isoforms-ecdysone dependent transcription in cultured anterior silk glands. *European Journal of Biochemistry* 248: 786–93

Kawabata, T., Yasuhara, Y., Ochiai, M., Matsuura, S. & Ashida, M. (1995) Molecular cloning of insect prophenol oxidase: a copper-containing protein homologous to arthropod hemocyanin. *Proceedings of the National Academy of Sciences USA* 92: 7774–78

Kawaguchi, E. (1934) Genetical and cytological analysis of parthenogenetically raised silkworm. *Journal of Sericultural Science of Japan* 5: 1–20 (*In Japanese*)

Kawakami, A., Iwami, M., Nagasawa, H., Suzuki, A. & Ishizaki, H. (1989) Structure and organization of four clustered genes that encode bombyxin, an insulin-related brain secretory peptide of the silkmoth, *Bombyx mori*. *Proceedings of the National Academy of Sciences USA* 86: 6843–47

Kawakami, A., Kataoka, H., Oka, T. *et al.* (1990) Molecular cloning of the *Bombyx mori* prothoracicotropic hormone. *Science* 247: 1333–35

Kawano, T., Kataoka, H., Nagasawa, H., Isogai, A. & Suzuki, A. (1992) The DNA cloning and sequence determination of the pheromone biosynthesis activating neuropeptide of the silkmoth, *Bombyx mori*. *Biochemical and Biophysical Research Communications* 189: 221–26

Klimenko, V.V. (1982) Mechanism of artificial parthenogenesis. *Genetika* (USSR) 18: 64–72

Klimenko, V.V. (1990) The silkworm *Bombyx mori L.* In *Animal Species in Developmental Studies, Volume 1: Invertebrates*, edited by T.A. Detlaff & S.G. Vassetsky, New York: Consultants Bureau

*Klimenko, V.V. & Spiridonova, T.L. (1982) Polyploidy and parthenogenesis in the silkworm. *Seriya Biologhicheskikh i Khimicheskikh Nauk* 4: 32–36

Kokubo, H., Ueno, K., Amanai, K. & Suzuki, Y. (1997) Involvement of the *Bombyx* Scr gene in development of the embryonic silk gland. *Developmental Biology* 186: 46–57

Kondo, H., Ino, M., Suzuki, A., Ishizaki, H. & Iwami, M. (1996) Multiple gene copies for Bombyxin, an insulin-related peptide of the silkmoth *Bombyx mori*: structural signs for gene rearrangement and duplication responsible for generation of multiple molecular forms of bombyxin. *Journal of Molecular Biology* 259: 926–37

Kuhn, D. (1988) Textile technology: spinning and reeling. In *Science and Civilization in China*, Part XI, edited by J. Needham, Cambridge: Cambridge University Press

Kusuda, J., Tazima, Y., Onumaru, K., Ninaki, O. & Suzuki, Y. (1986) The sequence around the 5′ end of the fibroin gene from the wild silkworm, *Bombyx mandarina* and comparison with that of the domesticated species, *B. mori*. *Molecular and General Genetics* 203: 359–64

Lecanidou, R., Eickbush, T.H. & Kafatos, F.C. (1984) Ribosomal DNA genes of *Bombyx mori*: a minor fraction of the repeating units contain insertions. *Nucleic Acids Research* 12: 4703–13

Lee, W.J. & Brey, P.T. (1995) Isolation and characterization of the lysozyme encoding gene from the silkworm, *Bombyx mori*. *Gene* 161: 199–203

Lee, W.J., Lee, J.D., Kravchenko, V.V., Ulevitch, R.J. & Brey, P.T. (1996) Purification and molecular cloning of an inducible gram-negative bacteria binding protein from the silkworm, *Bombyx mori*. *Proceedings of the National Academy of Sciences USA* 93: 7888–93

Mach, V., Takiya, S., Ohno, K., Handa, H., Imai, T. & Suzuki, Y. (1995) Silk gland factor-1 involved in the regulation of *Bombyx* sericin-1 gene contains fork-head motif. *Journal of Biological Chemistry* 270: 9340–46

Maekawa, H., Takada, N., Mikitani, K. *et al.* (1988) Nucleolus organisers in the wild silkworm *Bombyx mandarina* and the domesticated silkworm *B. mori*. *Chromosoma* 96: 263–69

Mangé, A., Couble, P. & Prudhomme, J.C. (1996) Two alternative promoters drive the expression of the cytoplasmic actin A4 gene of *Bombyx mori*. *Gene* 183: 191–99

Michaille, J.J., Couble, P., Prudhomme, J.C. & Garel, A. (1986) A single gene produces multiple sericin messenger RNAs in the silk gland of *Bombyx mori*. *Biochimie* 68: 1165–73

Michaille, J.J., Garel, A. & Prudhomme, J.C. (1990) Cloning and characterization of the highly polymorphic *Ser2* gene of *Bombyx mori*. *Gene* 86: 177–84

Mita, K., Ichimura, S. & James, T.C. (1994) Highly repetitive structure and its organization of the silk fibroin gene. *Journal of Molecular Evolution* 38: 583–92

Mita, K., Nenoi, M., Morimyo, M. *et al.* (1995) Expression of the *Bombyx mori* beta-tubulin-encoding gene in testis. *Gene* 162: 329–30

Mitsialis, S.A. & Kafatos, F.C. (1985) Regulatory elements controlling chorion gene expression are conserved between flies and moths. *Nature* 317: 453–56

**Miyahara, T. (1978) Selection of long filament length basic variety, MK – effect of selection on the later generations. *Acta Sericologica* 106: 73–78

Mounier, N. & Prudhomme, J.C. (1986) Isolation of actin genes in *Bombyx mori*: the coding sequence of a cytoplasmic actin gene expressed in the silk gland interrupted by a single intron in an unusual position. *Biochimie* 68: 1053–61

Mounier, N., Gaillard, J. & Prudhomme, J.C. (1987) Nucleotide sequence of the coding region of two actin genes in *Bombyx mori*. *Nucleic Acids Research* 6: 2781

Murakami, A. & Imai, H. (1974) Cytological evidence for holocentric chromosomes of the silkworms, *Bombyx mori* and *B. mandarina* (*Bombycidae*, *Lepidoptera*). *Chromosoma* 47: 167–78

Nada, S., Chang, P.K. & Dignam, J.D. (1993) Primary structure of the gene for glycyl-tRNA synthetase from *Bombyx mori*. *Journal of Biological Chemistry* 268: 7660–67

Nagaraju, J. (1996) Sex determination and sex-limited traits in the silkworm, *Bombyx mori*: their applications in sericulture. *Indian Journal of Sericulture* 35: 83–89

Nagaraju, J. & Jolly, M.S. (1985) Interspecific hybrids of *Antheraea roylei* and *A. pernyi* – a cytogenetic reassessment. *Theoretical and Applied Genetics* 72: 269–73

Nagaraju, J. & Singh, L. (1997) Assessment of genetic diversity by DNA profiling and its significance in silkworm, *Bombyx mori*. *Electrophoresis* 18: 1676–81

Nagaraju, J., Sharma, A., Sethuraman, B.N., Rao, J.V. & Singh, L. (1995) DNA fingerprinting in silkworm *Bombyx mori* using banded krait minor satellite DNA-derived probe. *Electrophoresis* 16: 1639–42

Nagaraju, J., Urs, R. & Datta, R.K. (1996) Cross breeding and heterosis in the silkworm, *Bombyx mori*, a review. *Sericologia* 36: 1–20

Nakato, H., Izumi, S. & Tomino, S. (1992) Structure and expression of gene coding for a pupal cuticle protein of *Bombyx mori*. *Biochimica et Biophysica Acta* 1218: 64–74

Niimi, T., Yamashita, O. & Yaginuma, T. (1996) Structure of the *Bombyx* sorbitol dehydrogenase gene: a possible alternative use of the promoter. *Insect Molecular Biology* 5: 269–80

Nony, P., Prudhomme, J.C. & Couble, P. (1995) Regulation of the *P25* transcription in the silk gland of *Bombyx mori*. *Biochemistry and Cell Biology* 84: 43–52

Ohta, T., Kobayashi, M. & Hirose, S. (1995) Cloning of a cDNA for DNA supercoiling factor reveals a distinctive Ca(2+)-binding protein. *Journal of Biological Chemistry* 270: 15571–75

Okazaki, S., Tsuchida, K., Maekawa, H., Ishikawa, H. & Fujiwara, H. (1993) Identification of pentanucleotide telomeric sequences $(TTAGG)_n$ in the silkworm, *Bombyx mori* and other insects. *Molecular and Cellular Biology* 13: 1424–32

Pham, T.N., Hayashi, K., Takano, R. *et al.* (1996) Expression of *Bombyx* family fungal protease inhibitor F from *Bombyx mori* by baculovirus vector. *Journal of Biochemistry* 119: 1080–85

Promboon, A., Shimada, T., Fujiwara, F. & Kobayashi, M. (1995) Linkage map of random amplified polymorphic DNAs (RAPDs) in the silkworm, *Bombyx mori*. *Genetical Research* 66: 1–7

Puttaraju, H.P. & Nagaraju, J. (1985) Preliminary observations on the occurrence of B-chromosome in the silkworm, *Antheraea roylei* (*Lepidoptera*: *Saturniidae*). *Current Science* 54: 471–72

Rasch, E.M., Barr, H.J. & Rasch, R.W. (1971) The DNA content of sperm of *Drosophila melanogaster*. *Chromosoma* 33: 1–18

Rasmussen, S.W. & Holm, P.B. (1982) The meiotic prophase in *Bombyx mori*. In *Insect Ultrastructure*, edited by R.C. King & H. Akai, New York: Plenum Press

Reddy, K.D., Abraham, E.G. & Nagaraju, J. (1999) Microsatellites in the silkworm, *Bombyx mori*: Abundance, polymorphism, and strain characterization. *Genome* 42: 1057–65

Robertson, H.M. & Asplund, M.L. (1996) *Bmmar1*: a basal

lineage of *mariner* family of transposable elements in the silkworm moth, *Bombyx mori*. *Insect Biochemistry and Molecular Biology* 26: 945–54

Rodakis, G.C., Lecanidou, R. & Eickbush, T.H. (1984) Diversity in a chorion multigene family created by tandem duplications and a putative gene conversion event. *Journal of Molecular Biology* 20: 265–73

Sakurai, H., Fujii, T., Izumi, S. & Tomino, S. (1988) Structure and expression of a gene coding for sex-specific storage protein of *Bombyx mori*. *Journal of Biological Chemistry* 263: 7876–80

Sato, Y. & Yamashita, O. (1991) Structure and expression of a gene coding for egg-specific protein in the silkworm, *Bombyx mori*. *Insect Biochemistry and Molecular Biology* 21: 495–503

Sato, Y., Oguchi, M., Menjo, N. *et al.* (1993) Precursor polyprotein for multiple neuropeptides secreted from the suboesophageal ganglion of the silkworm, *Bombyx mori*: characterization of the DNA encoding the diapause hormone and presence and identification of additional peptides. *Proceedings of the National Academy of Sciences USA* 90: 3251–55

Seki, H. (1984) Mode of inheritance of the resistance to the infection with the densonucleosis virus (Yamanashi isolate) in the silkworm, *Bombyx mori*. *Journal of Sericultural Science of Japan* 53: 472–75

Shi, J., Heckel, D.G. & Goldsmith, M.R. (1995) A genetic linkage map for the domesticated silkworm, *Bombyx mori*, based on restriction fragment length polymorphisms. *Genetical Research* 66: 109–26

Shimada, T., Kurimoto, Y. & Kobayashi, M. (1995) Phylogenetic relationship of silkmoths inferred from sequence data of the Arylphorin gene. *Molecular Phylogeny and Evolution* 4: 223–34

Shshegelskaya, E.A. & Klimenko, V.V. (1987) Phenotypic reversal in cells of the silkworm germ line. *Tsitologiya i Genetika* 21: 87–90

Spoerel, N., Nguyen, H.T. & Kafatos, F.C. (1986) Gene regulation and evolution in the chorion locus of *Bombyx mori*. Structural and developmental characterization of four egg shell genes and their flanking regions. *Journal of Molecular Biology* 190: 23–35

Spradling, A. & Rubin, G. (1982) Transposition of cloned P elements into *Drosophila* germ line chromosomes. *Science* 218: 341–47

Sprague, K.V. (1995) Control of *Bombyx mori* RNA polymerase III. In *Molecular Model Systems in the Lepidoptera*, edited by M.R. Goldsmith & A.S. Wilkins

Strunnikov, V.A. (1995) *Control over Reproduction, Sex and Heterosis of the Silkworm*, Luxembourg: Harwood Academic

Su, Z.H., Sato, Y. & Yamashita, O. (1993) Purification, cDNA cloning and Northern blot analysis of trehalase of pupal midgut of the silkworm *Bombyx mori*. *Biochimica et Biophysica Acta* 1173: 217–24

Sullivan, H.S., Young, L.S., White, C.N. & Sprague, K.U. (1994) Silk gland-specific tRNA(Ala) genes interact more weakly than constitutive tRNA(Ala) genes with silkworm TFIIIB and polymerase III fractions. *Molecular and Cellular Biology* 14: 1806–14

Sun, G.C., Hirose, S. & Ueda, H. (1994) Intermittent expression of BmFTZ-F1, a member of the nuclear hormone receptor superfamily during development of the silkworm *Bombyx mori*. *Developmental Biology* 162: 426–37

Suzuki, Y., Gage, L.P. & Brown, D.D. (1972) The gene for silk fibroin in *Bombyx mori*. *Journal of Molecular Biology* 70: 637–49

Swevers, L., Drevet, J.R., Lunke, M.D. & Iatrou, K. (1995) The silkmoth homolog of the *Drosophila* ecdysone receptor (B1 isoform): cloning and analysis of expression during follicular cell differentiation. *Insect Biochemistry and Molecular Biology* 25: 857–66

Tamura, T., Thibert, C., Royer, C. *et al.* (2000) Germline transformation of the silkworm *Bombyx mori L.* using a piggyBac transposon-derived vector. *Nature Biotechnology* 18: 81–84

Tanaka, Y. (1916) Genetic studies in the silkworm. *Journal of the College of Agriculture, Sapporo* 7: 129–255

Taniai, K., Kodano-Okuda, K., Kato, Y. *et al.* (1995) Structure of two cecropin B-encoding genes and bacteria-inducible DNA-binding proteins which bind to the 5′-upstream regulatory region in the silkworm, *Bombyx mori*. *Gene* 163: 215–19

Taniai, K., Ishii, T., Sugiyama, M., Miyanoshita, A. & Yamakawa, M. (1996) Nucleotide sequence of the 5′-upstream region and expression of a silkworm gene encoding a new member of the attacin family. *Biochemical and Biophysical Research Communications* 220: 594–99

Tazima, Y. (1951) Separation of male and female silkworms in egg stage now becomes possible. *Silk Digest, Tokyo* 61: 1–3

Tazima, Y. (1964) *The Genetics of the Silkworm*, London: Logos Press, and Englewood Cliffs, New Jersey: Prentice Hall

**Tazima, Y. & Onuma, A. (1967) Experimental induction of androgenesis, gynogenesis and polyploidy in *Bombyx mori* by treatment with CO_2 gas. *Journal of Sericultural Science of Japan* 36: 286–92

*Terskaya, E.R. & Strunnikov, V.A. (1974) Activation of silkworm eggs towards meiotic parthenogenesis. *Doklady Akademii Nauk SSSR (USSR)* 219: 1238–41

Toyama, K. (1906) Breeding methods of silkworm. *Sangyo Shimpo* 158: 282–86 (*In Japanese*)

Ueda, H., Mizuno, S. & Shimura, K. (1986) Transposable genetic element found in the 5′ flanking region of the fibroin H-chain gene in a genomic clone from the silkworm *Bombyx mori*. *Journal of Molecular Biology* 190: 319–27

Ueno, K., Hui, C.C., Fukuta, M. & Suzuki, Y. (1992) Molecular analysis of the deletion mutants in the E homeotic complex of the silkworm *Bombyx mori*. *Development* 114: 555–63

Ueno, K., Nagata, K. & Suzuki, Y. (1996) Roles of homeotic genes in the *Bombyx* body plan. In *Molecular Model Systems in the Lepidoptera*, edited by M.R. Goldsmith & A.S. Wilkins

Watanabe, H. (1965) Resistance to peroral infection by the cytoplasmic-polyhedrosis virus in the silkworm, *Bombyx mori*. *Journal of Invertebrate Pathology* 7: 257–58

Watanabe, H. & Maeda, S. (1981) Genetically determined nonsusceptibility of the silkworm, *Bombyx mori*, to infection with a densonucleosis virus (Densovirus). *Journal of Invertebrate Pathology* 38: 370–73

Xu, A.Y., Li, W., Fang, A., Fei, M.H. & Huang, J.T. (1997) Isolation of a self-bred line of *Bombyx mori L.*, by means of dispermic androgenesis. *Sericologia* 37: 199–204

Xu, P.X., Fukuta, S., Takiya, S. *et al.* (1994) Promoter of the POU-M1/SGF-3 gene involved in the expression of *Bombyx* silk genes. *Journal of Biological Chemistry* 269: 2733–42

Xu, X., Xu, P.X. & Suzuki, Y. (1994) A maternal homeobox gene, *Bombyx caudal*, forms both mRNA and protein concentration gradients spanning antero-posterior axis during gastrulation. *Development* 120: 277–85

Yamaguchi, K., Kikuchi, Y., Takagai, T. *et al.* (1989) Primary structure of the silk fibroin light chain determined by cDNA sequencing and peptide analysis. *Journal of Molecular Biology* 210: 127–39

Yokoyama, T. (1979) Silkworm selection and hybridizations. In *Genetics in Relation to Insect Management*, edited by M.A. Hoy and J.J. McKelvey, New York: Rockefeller Foundation

Young, L.S., Takahashi, N. & Sprague, K.U. (1986) Upstream sequences confer distinctive transcriptional properties on genes encoding silkgland-specific $tRNA^{Ala}$. *Proceedings of the National Academy of Sciences USA* 83: 374–78

Zietkiewicz, E., Rafalski, A. & Labuda, D. (1994) Genome fingerprinting by simple sequence repeat (SSR)-anchored polymerase chain reaction amplification. *Genomics* 20: 176–83

GLOSSARY

arylophorin genes genes that encode storage proteins

biolistic a method of transfection by which DNA-coated metallic particles are launched by bombardment onto cells or tissues, from plants or animals

CA repeat the most abundant class of microsatellites found eukaryotic genomes

chaperones members of several structurally unrelated protein families that interact with non-native conformations of other proteins

codominance the situation in which an organism heterozygous for two alleles (A_1 and A_2 at the A locus) expresses both of the phenotypes observed in the corresponding homozygotes. Thus, the heterozygote (A_1/A_2) and both homozygotes (A_1/A_1 and A_2/A_2) are all distinguishable from each other and A_1 and A_2 would be considered "codominant". This term is used to describe alternate allelic forms of DNA markers such as different sized restriction fragments or PCR products

cross-breeding a cross between two strains that are genetically different from each other. For example, a F_1 hybrid obtained from a cross between two inbred strains with different genetic backgrounds

diakinesis a stage of meiotic division just before metaphase I in which the bivalents are shortened and thickened

haemocoel the body cavity in insects, where haemolymph circulation takes place

microsatellite a very short unit of DNA segment (2–4 bp) that is repeated multiple times in tandem. They are also called simple sequence repeats (SSRs) and have emerged as ideal genetic markers for linkage analysis

minisatellite a highly polymorphic locus containing tandemly repeated sequences having a unit length of 10–40 bp. The polymorphism can be assayed by RFLP or by PCR. They are also referred to as variable number of tandem repeats (VNTR) loci

parthenogenesis reproduction by development of an unfertilized egg, i.e. no genetic contribution from a male

POU-domain containing factors transcription factors that share similarities in their DNA binding region that consists of a 75–82 amino acid POU-specific domain and a 60 amino acid POU specific homeodomain. POU refers to the factors of this family that were originally described: the *P*it-1/growth hormone factor-1, the ubiquitous *O*ct-1 and the B cell-specific Oct2, and the *C. elegans* *U*nc 86 cell lineage control factor

sex comb reduced a homeotic gene isolated in Drosophila that participates in the determination of the primordia of the salivary gland in the labial ectoderm. The Bombyx *Src* also has the potential to invaginate the placode of the silk gland, the organ orthologue to the salivary gland of the fruit fly

synaptonemal complex a ribbon-like structure formed between synaposed homologues at the end of the first prophase, binding the chromatids along their length and facilitating chromatid exchange

TBP TATA box binding protein, a DNA binding factor that recruits other proteins via protein–protein interactions

TATA box a tetranucleotide sequence found in the proximal upstream region of most polymerase II-transcribed genes. TBP binding to this sequence element is required to initiate transcription

E GENETICS OF EUKARYOTIC MICRO-ORGANISMS AND THEIR ORGANELLES

Introduction to Genetics of Eukaryotic Micro-organisms and their Organelles

As described in the articles on Paramecium, Fungal genetics and Tetrahymena, study of the lower eukaryotes led to confirmation of Mendel's principles in the lower organisms and has contributed to an increased understanding of fundamental molecular and cellular mechanisms.

1. Yeast

Yeast is a model eukaryote, and the genome sequence of the yeast *Saccharomyces cerevisiae*, completed in 1996, was one of the early goals of the Human Genome Project. About 40% of yeast genes may be homologous in structure or even function to humans, while progress in sequencing technology and analysis helped to accelerate the human project.

Recently, it has been proposed that yeast prions can transmit inheritable phenotypic states with no underlying changes in nucleic acids, create new genetic diversity and provide an explanation for the sometimes puzzlingly rapid pace of evolution (True & Lindquist, 2000).

2. Sequencing the genome of the malaria parasite

The Malaria Genome Sequencing Consortium was founded in 1997 to sequence the genome of *Plasmodium falciparum*, the organism responsible for virtually all of the enormous number of deaths caused annually by malaria in tropical and subtropical areas of the world. The *P. falciparum* genome contains about 30 Mb of DNA, divided between 14 chromosomes ranging in size from 0.7 to 3.5 Mb, and sequencing the genome has presented particular difficulties because of its biased nucleotide composition, whose overall (A + T) content is estimated at 82%. Special techniques can deal with this problem: chromosome 2 was sequenced in 1998 by a team of American scientists based mainly at Rockville, Maryland (Gardner *et al.*) and chromosome 3 in 1999 by a team at the Pathogen Sequencing Unit, Sanger Centre, Hinxton near Cambridge, UK (Bowman *et al.*). Chromosome 2 contains 947 103 base pairs and encodes 210 predicted genes, of which 90 (43%) contain introns, and 87 (42%) have homologues in other species. Gardner *et al.* point out that chromosomes from different wild isolates exhibit extensive size polymorphism.

Chromosome 3 contains 1 060 106 base pairs with 215 protein-coding genes predicted, giving a gene density of 4.8 kb per gene compared with 4.5 for chromosome 2, while the number of genes predicted to have introns are 47.4% compared with 43.1% for chromosome 2. As the sequences of the other 12 chromosomes are added to these two, many further details of the genes and types of proteins in the complete genome will certainly be revealed. These should be of great importance in solving the health problems caused by *P. falciparum*.

REFERENCES

Bowman, S., Lawson, D., Basham, D., Brown, D. *et al.* (1999) The complete nucleotide sequence of chromosome 3 of *Plasmodium falciparum*. *Nature* 400: 532–38

Gardner, M.J., Tettelin, H., Carucci, D.J., Cummings, L.M. *et al.* (1998) Chromosome 2 sequence of the human malaria parasite *Plasmodium falciparum*. *Science* 282: 1126–32

True, H. & Lindquist, S.L. (2000) A yeast prion provides a mechanism for genetic variation and phenotypic diversity. *Nature* 407: 477–83

GENETICS OF PARAMECIUM

John R. Preer Jr
Department of Biology, Indiana University, Bloomington, Indiana, USA

1. Introduction

Most ciliated protozoans, like Paramecium, are distinguished by having two kinds of nuclei. Most genes in the small diploid macronuclei are not transcribed: they act only during sexual reproduction. Genes of the macronuclei, however, are transcribed. They reside on segments of DNA that are highly polyploid and carry out the functions of the somatic nuclei in higher organisms. The specialized functions of these two kinds of nuclei have led to numerous differences in the organization of their DNA. Moreover, in higher organisms, the traits that geneticists follow result from complex developmental processes involving many cells and many cell generations. In protozoa, however, developmental events are confined to the single cell. While the somatic cells of higher organisms cannot be crossed to study cellular inheritance, many protozoan cells can become competent to mate. Although DNA and RNA and the chromosome mechanics that lead to Mendelism are essentially the same in higher organisms and ciliates, we might expect, and indeed do find, that the genetics of ciliates exhibits many unique features. Studies on the base sequence of genes coding for 17S ribosomal RNA have shown that ciliated protozoans diverged evolutionarily from other organisms very early. Several exceptions to the "universal" genetic code are found in ciliates. UAA and UAG code for the amino acid glutamine in Paramecium, Stylonichia and Oxytricha, instead of acting as translational stop signals as they do in most other organisms (Preer *et al.*, 1985; Caron & Meyer, 1985a). In *Euplotes octocarinatus*, UGA, instead of acting as a stop signal, codes for cysteine (Grimm *et al.*, 1998).

Paramecium was the first protozoan to be studied genetically. Today, other single-celled eukaryotes, such as unicellular fungi and algae, trypanosomes, Plasmodium and other ciliates such as Tetrahymena and Euplotes, are also favourite objects of study. Studies on the genetics of Paramecium began in the late 1800s, but since conjugation could not be controlled, progress was slow. When Sonneborn discovered mating types in Paramecium (Sonneborn, 1937) and began the first controlled crosses between hereditarily diverse strains, these studies quickly led to a confirmation of Mendel's principles in the lower organisms, but also revealed a series of surprising and unique phenomena. Today, the techniques of molecular biology are revealing the structure of the macronucleus and giving us new insights into these unusual phenomena.

2. Species

There are several species of Paramecium that are easily distinguished morphologically by size, shape and by the number and structure of their micronuclei (Wichterman, 1986). Within each morphological species, there are many different mating types that define different sexually isolated groups (Sonneborn, 1975). These groups were first referred to as "varieties", then later as "syngens". In the species formerly known as *P. aurelia*, the 14 known syngens have been elevated to the rank of species, *P. primaurelia*, *P. biaurelia*, etc. In *P. primaurelia*, there are mating types I and II, which mate with each other, but not with the mating types found in other species. *P. biaurelia* consists of mating types III and IV, which mate only with each other, and so on. Perhaps the most extensively investigated species is *P. tetraurelia* with mating types VII and VIII. All species of Paramecium have one macronucleus, but the number of micronuclei varies from one species to another. For example, there is one in *P. caudatum*, two in the *P. aurelia* array of species and a variable number (but more than two) in *P. multimicronucleatum*.

3. Binary fission, conjugation and autogamy

Reproduction and genetics are summarized by Beale (1954) and Sonneborn (1975).

(i) Binary fission

Vegetative cells reproduce by binary fission. The diploid micronuclei divide mitotically and the polyploid macronuclei simply pinch in two amitotically.

(ii) Conjugation

Whenever competent cells of opposite mating types come into contact with each other, conjugation is initiated. In a species such as *P. tetraurelia*, each vegetative cell contains two diploid micronuclei and one polyploid macronucleus. The cytological events that occur in each member of the conjugating pair are identical. The macronucleus breaks into 20 or 30 fragments that are normally destined to be lost; each fragment contains many macronuclear chromosomes, each chromosome consisting of one DNA molecule. The two diploid micronuclei in each cell undergo two meiotic divisions to yield eight haploid micronuclei. All but one of these eight nuclei degenerate and the one remaining haploid nucleus undergoes one mitotic division to produce two haploid gametic nuclei of identical genotype. One of the two is the stationary gametic nucleus and the other is the migratory gametic nucleus. Now the two migratory gametic nuclei, one in each conjugant, move to the opposite cell and fuse there with the stationary nucleus. The resulting diploid syncaryon in each cell then divides mitotically twice to produce four diploid micronuclei. At this point, the cells separate to produce two exconjugants. In each exconjugant, two of the four diploid micronuclei develop into polyploid macronuclei and the other two remain as diploid micronuclei. At the next cell division, the two macronuclei in each exconjugant are simply segregated, one to each daughter cell, while the two micronuclei undergo mitosis. In the resulting daughter cells, the original condition of two diploid micronuclei and one polyploid macronucleus per cell is restored.

Ciliates usually will not mate again immediately after conjugation. This phase of the life cycle, called the immature period, can vary from zero fissions in a species such as *P. tetraurelia* up to hundreds of fissions in *P. bursaria*.

(iii) Autogamy

A related uniparental process called autogamy also occurs in many species of Paramecium. The cytological events that occur in autogamy are identical to those occurring in conjugation, except that the process occurs in single cells and, consequently, cross-fertilization does not take place. The two gametic nuclei formed in each cell simply fuse with each other to produce the diploid syncaryon.

In *P. tetraurelia*, autogamy takes place whenever cells starve and have undergone approximately 10–60 fissions since the last conjugation or autogamy. Autogamy can be suppressed by providing an excess of culture medium, but, if both autogamy and conjugation are suppressed, cell lines eventually age and die. Although the basis for senescence has been shown to lie in the macronucleus (Aufderheide, 1987), its mechanism is unknown.

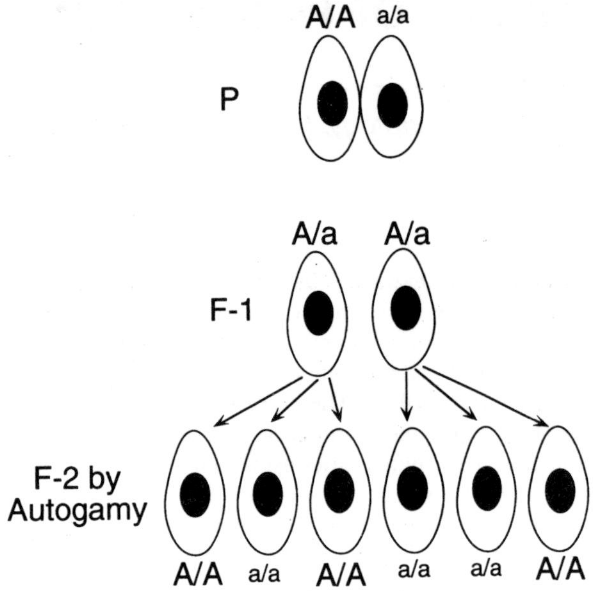

Figure 1 Simple Mendelian inheritance in *Paramecium tetraurelia*. The F_1 hybrid cells from each parent are alike and are all heterozygous in a single factor cross. When autogamy is induced, all cells become homozygous with either one of the two alleles being chosen at random. Hence an approximate 1:1 ratio is obtained from each exconjugant.

4. Genetic analysis

In conjugation, the two identical gametic nuclei originally produced from a product of meiosis in one cell are likely to have a different genotype from the corresponding nuclei in the partner conjugant. Therefore, after cross-fertilization, the syncarya will probably be heterozygous at many loci. Nevertheless, after fertilization, the two haploid nuclei producing the syncaryon in one cell have the same genotypes as the two haploid nuclei producing the syncaryon in the partner. Consequently, the genotypes of the two exconjugant lines derived from a single conjugating pair must be identical to each other. In autogamy, on the other hand, all newly forming micronuclei and macronuclei are derived from a diploid syncaryon which arose from the fusion of gametic nuclei, both of which arose from a single haploid nucleus. Therefore, all exautogamous cells become homozygous for all their genes. Autogamy provides a very useful way of obtaining homozygous individuals for genetic studies.

In a cross of *A*/*A* by *a*/*a*, all exconjugants are *A*/*a* (see **Figure 1**). If such cells are allowed to produce an F_2 by autogamy, there is an equal chance that the surviving meiotic product in each autogamous cell will be *A* or *a*. After the surviving nucleus divides, fusion takes place and autogamy is completed, the resultant cells being either *A*/*A* or *a*/*a* in equal frequency. In this way, a very convenient 1:1 genotypic and phenotypic Mendelian ratio of the two homozygotes is obtained. For this reason, crossing lines of diverse genotype and then inducing autogamy is usually carried out in genetic analysis.

Several additional "tricks" devised by Sonneborn are useful in genetic analysis. Sometimes, fragments of the old macronucleus are not lost (Sonneborn, 1975), but regenerate into new macronuclei and this regeneration makes it possible to obtain cells whose micronuclei and macronuclei have different genetic constitutions. Another technique (Sonneborn, 1975) makes it possible to measure the amount of cytoplasm exchanged between mates at conjugation.

Although many genes have been found in ciliates, recombination frequencies at meiosis appear to be so high that classical genetic linkage maps have not been produced.

5. Macronuclear structure

For many years, the structure of the macronucleus was an enigma. We now know that when a micronucleus in Paramecium develops into a macronucleus, chromosome breakage occurs, producing hundreds of fragments of DNA, each on the order of several thousand base pairs (see **Figure 2**). Each fragment then acquires a telomere and becomes a macronuclear segment. The size of the macronuclear segments varies greatly both within and among different species of ciliates. If one separates the macronuclear DNA on a gel into an array of different sizes, it is found that the array extends over a large size range, is constant and is characteristic for the species (but differs for different species). Moreover, individual genes are usually found to be located on different sized molecules. For some genes, the size of the segment on which a specific gene is located may vary (due to variations in DNA processing during macronuclear formation). In Tetrahymena, the base sequence at which the breaks occur, the chromosome breakage site, consists of 15 bases (AAAGAGGTTGGTTTA) (Yao *et al.*, 1990) and the role of this and surrounding sequences involved in the excision process is being determined.

Micronuclear genes in ciliates often contain non-coding

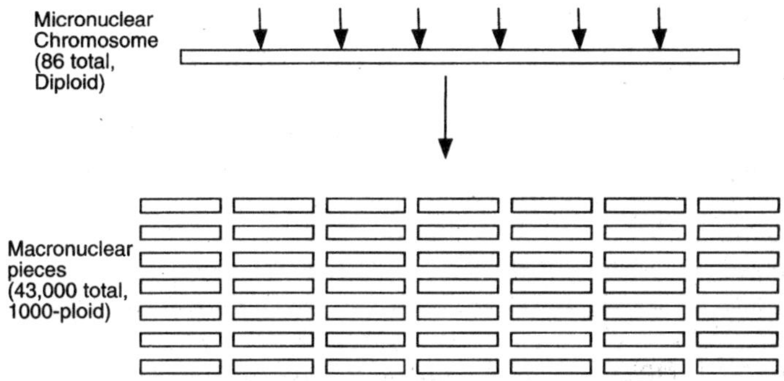

Figure 2 Chromosomal constitution of micronucleus and macronucleus in *Paramecium tetraurelia*. Micronuclear chromosmes are about 86 in number and are diploid. When a micronucleus makes a macronucleus, each chromosome is cut into about seven pieces, and then they are amplified about 1000-fold to *c*.43 000 total pieces.

sequences called internal eliminated sequences (IESs) that are precisely excised when the micronucleus produces a macronucleus (Preer *et al*., 1992). The number of IESs has been estimated at 6000–170 000 per genome in different species of ciliates. The immobilization antigen gene *A* in *P. tetraurelia* contains seven different IESs, while a gene in Oxytricha has 43 IESs (Prescott & Dubois, 1996). The characteristic base sequences in the DNA at the points of excision have been determined and replicative intermediates in the excision process are known in some ciliates. In Paramecium and Euplotes, the junction at which excision occurs is between the bases T and A. Adjacent bases vary from one IES to another, but there are similarities and a consensus sequence has been identified. Some ciliates have been found to contain families of repeated elements, which have sequence homology with known transposons from other organisms. Like IESs, these repeated elements are normally eliminated when micronuclei produce macronuclei and the break points are also at the sequence TA. It has been suggested that IESs are evolutionarily degraded forms of transposons that invaded ciliates early in their evolutionary history. These remnants of unwanted DNA have to be eliminated precisely in order to produce functional macronuclear genes with intact coding regions.

An intriguing group of IESs has been studied in Oxytricha (Prescott & Dubois, 1996). The order of the sections of a given gene found in the micronucleus is different from the order found in the macronucleus. Reordering occurs when the IESs are removed and it appears that pairs of repeated sequences found at the borders between the IESs and coding sequences undergo homologous pairing prior to excision. At this point, the molecules are highly convoluted, for the pairs of repeats are placed in such a way that the proper order of sections within the macronuclear gene results when breakage and rejoining occur within the paired regions. The molecular mechanisms of breakage and rejoining are unknown. Breakage does not occur at TA in Oxytricha, as it does in Paramecium and Euplotes.

All fragments acquire characteristic telomeric sequences (repeats of CCCCAA and CCCAAA in Paramecium) that are synthesized onto the broken ends (Forney & Blackburn, 1988). Moreover, the fragments are amplified to a copy number of 1000–2000. The problem of how genic balance is maintained during amitotic cell divisions is still unsolved, although it has been suggested that, like the bacterial plasmids, each DNA fragment can sense its own copy number and adjust it by differential replication (Preer & Preer, 1979; Brunk, 1986).

The finding that any DNA sequence can be injected into the macronucleus of Paramecium where it will obtain telomeres and replicate has proved to be an important genetic tool (Godiska *et al*., 1987). Transformation frequencies approach 100% and injections are easy to carry out in such large cells. Unlike Paramecium, Tetrahymena cannot add telomeres to any fragment of DNA. Consequently, a somewhat different transformation system has been devised for Tetrahymena. Due to the numerous small chromosomes and the resulting large numbers of chromosomal ends, ciliates are uniquely suited for the study of telomeres. In Tetrahymena, the enzyme telomerase has been characterized. It is a ribonucleoprotein, containing a complementary copy of the telomeric base sequence in its RNA component (Greider & Blackburn, 1985). It is also interesting to note that another enzyme, consisting entirely of RNA, has also been discovered in Tetrahymena (Been & Cech, 1988). It is one of the first non-protein enzymes ever to be discovered. It is interesting that most multicellular organisms have solved the problem of the need for many rDNA genes with tandem chromosomal duplications, making conventional genetic analysis impossible. In the ciliate Tetrahymena, there is only one micronuclear rDNA gene, but very numerous macronuclear copies of the gene (Orias *et al*., 1982). Paramecium has several micronuclear rDNA genes of importance, and they have been identified (Preer *et al*., 1999).

6. Caryonidal inheritance

The two macronuclei produced in each cell after autogamy and conjugation are segregated to different cells at the next cell division and produce sister clones that are called caryonides. Since the two macronuclei came by mitosis from a single diploid nucleus, one would expect sister caryonides to be identical. Surprisingly, however, they differ in some cases and the traits are said to be caryonidally inherited (Sonneborn, 1975). Caryonidal differences are now known to arise because of differences that occur during the processing of micronuclear DNA into macronuclear DNA. Once formed, however, the macronuclei remain constant. Mating type itself exhibits caryonidal inheritance in some species and other traits, as well as numerous differences in DNA sequences, are caryonidally inherited.

7. Macronuclear inheritance

Inheritance of mating type in *P. tetraurelia* is caryonidal, but caryonidal with a "twist". Crosses between the two mating types sometimes produce different caryonides, but usually give no change in the mating type of the two exconjugant clones. Since, as pointed out above, exconjugants are always alike in their micronuclear genes, but remain different in their cytoplasms, these results suggest cytoplasmic inheritance. A cross of mating type VII by mating type VIII without cytoplasmic exchange gives no subsequent change in type (**Figure 3**). Mating type follows the cytoplasm. If cytoplasm is exchanged, however, both exconjugants come out mating type VIII. So there is a VIII cytoplasmic factor. Sonneborn (1954) was able to show that the cytoplasm acts on the newly forming macronuclei, and then the mature macronuclei act to determine the cytoplasm. Thus, the genetic properties of the old macronucleus determine the genetic properties of the new, a phenomenon called macronuclear inheritance. Macronuclear inheritance, while caryonidal, exhibits a remarkable kind of hereditary transmission not seen in simple caryonidal inheritance. This remarkable situation has been considerably illuminated by a study of the *A* gene, which controls a surface protein in *P. tetraurelia*. In the d48 non-Mendelian mutation (Epstein & Forney, 1984), the *A* gene in the micronucleus fails to be included in the new macronucleus during autogamy and conjugation. However, normal *A* transmission can be completely and permanently restored by injecting isolated *A* gene, or specific portions of the *A* gene, into the macronucleus of d48 (Aufderheide, 1987; Koizumi & Kobayashi, 1989). The only real defect in d48 is that it lacks the *A* gene in its macronucleus. Thus, *A* genes are necessary for their own proper processing when micronuclei form macronuclei. *B* gene mutants similar to d48 have been constructed, and act specifically and independently of the *A* gene. A number of other cases are known in which the presence of specific sequences of DNA injected into the old macronucleus affect DNA processing at autogamy (Meyer, 1992). It is clear that the processing of specific regions of DNA is influenced by the presence of those same sequences in the old macronucleus. How this influence is exerted through the cytoplasm on the newly forming macronuclei is an intriguing and currently unsolved problem.

8. Endosymbionts

Intracellular bacterial symbionts abound in ciliates and represent many genera and species of bacteria (Preer *et al.*, 1972). They are site specific, with some restricted to the cytoplasm, some to the macronuclei and even some to the micronuclei. Fully half of the individuals of *P. biaurelia* from ponds and streams are infected. The symbionts do not appear to harm their hosts, and virtually none can be cultured outside of the cells in which they are found. Studies on the DNA genes coding for 17S rRNA show that some are related to the rickettsia. The symbiont that has been most investigated is called kappa (*Caedibacter taeniospiralis*), which grows in paramecia bearing the gene *K*. Its presence makes paramecia into killers, which kill sensitive paramecia lacking kappa. Kappa contains minute, compact, rolled protein structures called R-bodies. The presence of R-bodies is dependent on plasmids and phage-like elements found in kappa. Killing occurs when sensitives take R-body-containing kappas from the medium into their food vacuoles. There, the R-bodies unroll suddenly into long filamentous structures that contain the toxins that kill the sensitives.

9. Serotype inheritance

Each cell of *P. tetraurelia* is covered by one of a family of very large antigenic proteins (Caron & Meyer, 1985b; Preer, 1986). Exposure of paramecia to homologous antiserum results in immobilization of the cells. The proteins are called immobilization antigens and a clone of cells bearing one of the alternative antigens is called a serotype. Each strain of paramecia can produce about a dozen mutually exclusive

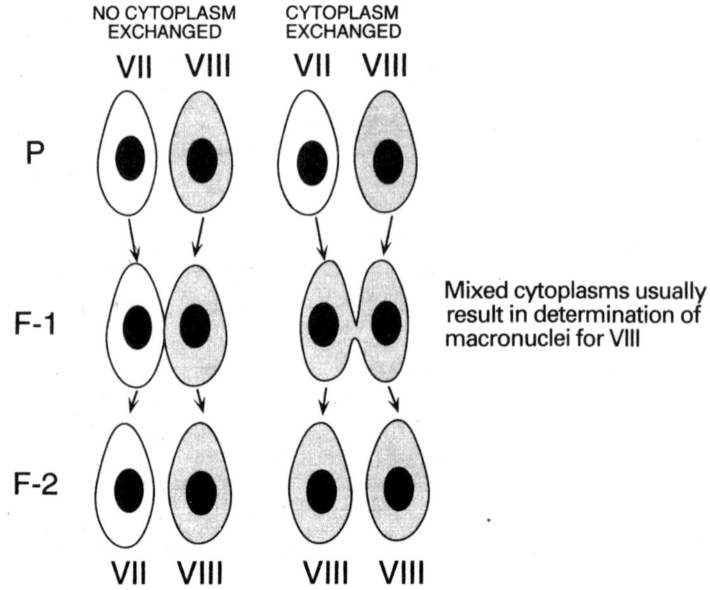

Figure 3 Mating type inheritance in *Paramecium tetraurelia*. A cross of VII x VIII usually leads to no change in phenotype, unless cytoplasms are exchanged, when the VII exconjugant becomes VIII.

antigens and types can be induced to switch by changing environmental conditions. A separate unlinked gene codes each antigen. The proteins have been isolated and many of the genes sequenced.

Serotype switching is seen as a turning off of a previously active gene and activation of a new one. Serotypes appear to be rather stable and many can be cultured under the same environmental conditions for hundreds of fissions. Crosses between different serotypes with the same genotype reveal cytoplasmic inheritance. Chimaeric genes have been constructed *in vitro* by combining parts of cloned genes which produce serotypes of different stabilities. These have then been injected into the macronuclei of cells that have had their normal genes deleted and the injected constructs acquire telomeres, replicate and are expressed. These studies (Thai & Forney, 1999) show that the DNA of the coding regions of the antigen genes themselves appear to be involved in the stability of their expression and suggest that a positive feedback mechanism accounts for the observed stability of expression.

10. Cortical inheritance

Cortical inheritance has been reviewed by Frankel (1989) and by Grimes & Aufderheide (1991). The cortex of Paramecium is highly complex, containing structures such as the gullet, mouth, contractile vacuole pores and anal pore. There are rows, called kineties, of repeating asymmetrical structural units. The units contain one or two cilia, a trichocyst insertion point, a parasomal sac and associated fibres and membranes. On occasion, modifications in these structures can be observed microscopically. In one of the best studied cases, one or more kineties have reversed symmetry and the kineties are said to be inverted. Crosses have revealed that determination is due neither to micronuclei/macronuclei nor to the fluid cytoplasm. Thus, pre-existing cortical structure determines new cortical structure by an as yet unknown mechanism. Sometimes, conjugants fail to separate and reproduce as double animals, again showing cortical inheritance.

The influence of pre-existing structure in determining new structure has been called cytotaxis. Cytotaxis has been studied extensively in several ciliates and it has been suggested that cytotaxis may be involved in determining the structure of certain membrane systems and other structural features in higher organisms.

11. Behavioural mutants

The behaviour of Paramecium has been studied for many years (Saimi & Kung, 1987; Kung *et al.*, 2000), and numerous mutants have been found. Two examples are "paranoiac", which reverses swimming direction at the slightest disturbance, and "pawn", which always moves forward. Most of the mutants derive from defects in calcium, sodium and potassium ion channels and are telling us much about membrane channels and membrane potentials and their effects on the behaviour of single cells.

REFERENCES

Aufderheide, K.J. (1987) Clonal aging in *Paramecium tetraurelia*. II. Evidence of functional changes in the macronucleus with age. *Mechanisms of Aging and Development* 37: 265–79

Beale, G.H. (1954) *The Genetics of Paramecium aurelia*, Cambridge: Cambridge University Press

Been, M.D. & Cech, T.R. (1988) RNA as an RNA polymerase: net elongation of an RNA primer catalysed by the Tetrahymena ribozyme. *Science* 239: 1412–16

Brunk, D.F. (1986) Genome reorganization in Tetrahymena. *International Review of Cytology* 99: 49–83

Caron, C.F. & Meyer, E. (1985a) Does *Paramecium primaurelia* use a different genetic code in its macronucleus? *Nature* 314: 185–88

Caron, C.F. & Meyer, E. (1985b) Molecular basis of surface antigen variation in paramecia. *Annual Review of Micrbiology* 43: 23–42

Epstein, L.M. & Forney, J.D. (1984) Mendelian and non-Mendelian mutations affecting surface antigen expression in *Paramecium tetraurelia. Molecular and Cellular Biology* 4: 1583–90

Forney, J.D. & Blackburn, E.H. (1988) Developmentally controlled telomere addition in wild-type and mutant paramecia. *Molecular and Cellular Biology* 8: 251–58

Frankel, J. (1989) *Pattern Formation: Ciliate Studies and Models*, New York: Oxford University Press

Godiska, R., Aufderheide, K.J., Gilley, D. *et al.* (1987) Transformation of *Paramecium* by microinjection of a cloned serotype gene. *Proceedings of the National Academy of Sciences USA* 84: 7590–94

Greider, C.W. & Blackburn, E.H. (1985) Identification of a specific telomere terminal transferase activity in Tetrahymena extracts. *Cell* 43: 405–13

Grimes, G.W. & Aufderheide, K.J. (1991) *Cellular Aspects of Pattern Formation: The Problem of Assembly*, Basel and New York: Karger

Grimm, M., Brünen-Niewler, C., Junker, V., Heckmann, K. & Beier, H. (1998) The hypotrichous ciliate *Euplotes octocarinatus* has only one type of $tRNA^{Cys}$ with GCA anticodon encoded on a single macronuclear DNA molecule. *Nucleic Acids Research* 26: 4557–65

Koizumi, S. & Kobayashi, S. (1989) Microinjection of plasmid DNA encoding the *A* surface antigen of *Paramecium tetraurelia* restores the ability to regenerate a wild-type macronucleus. *Molecular and Cellular Biology* 9: 4398–401

Kung, C., Saimi, Y., Haynes, W.J. *et al.* (2000) Recent advances in the molecular genetics of Paramecium. *Journal of Eukaryotic Microbiology* 47: 11–14

Meyer, E. (1992) Induction of specific macronuclear developmental mutations by microinjection of a cloned telomeric gene in *Paramecium primaurelia. Genes and Development* 6: 211–22

Orias, E., Pan, W.C., Orias, E., Flacks, M. & Blackburn, E.H. (1982) Allele-specific, selective amplification of a ribosomal RNA gene in *Tetrahymena thermophila. Cell* 28: 595–604

Preer, J.R. Jr (1986) Surface antigens of Paramecium. In *Molecular Biology of the Ciliated Protozoa*, edited by J. Gall, New York: Academic Press

Preer, J.R. Jr (2000) Epigenetic mechanisms affecting macronuclear development in Paramecium and Tetrahymena. *Journal of Eukaryotic Microbiology* 47(6): 515–24

Preer, J.R. Jr & Preer, L.B. (1979) The size of macronuclear DNA and its relationship to models for maintaining genetic balance. *Journal of Protozoology* 26: 14–18

Preer, L.B., Jurand, A., Preer, J.R. Jr *et al* (1972) The classes of kappa in *Paramecium aurelia. Journal of Cell Science* 11: 581–600

Preer, J.R. Jr, Preer, L.B., Rudman, B.M & Barnett, A.J. (1985) A deviation from the universal genetic code: the gene for surface protein 51A in Paramecium. *Nature* 314: 188–90

Preer, L.B., Hamilton, G. & Preer, J.R. Jr (1992) Micronuclear DNA from Paramecium tetraurelia: serotype 51 *A* gene has internally eliminated sequences. *Journal of Protozoology* 39: 678–82

Preer, L.B., Rudman, B., Pollack, S. & Preer, J.R. Jr (1999) Does ribosomal DNA get out of the micronuclear chromosome in *Paramecium tetraurelia* by means of a rolling circle? *Molecular and Cellular Biology* 19: 7792–800

Prescott, D.M. & Dubois, M.L. (1996) Internal eliminated segments (IESs) of Oxytrichidae. *Journal of Eukaryotic Microbiology* 43: 432–41

Saimi, Y. & Kung, C. (1987) Behavioral genetics of Paramecium. *Annual Review of Genetics* 21: 47–65

Sonneborn, T.M. (1937) Sex, sex inheritance and sex determination in *Paramecium aurelia. Proceedings of the National Academy of Sciences USA* 23: 378–85

Sonneborn, T.M. (1954) Patterns of nucleo-cytoplasmic integration in Paramecium. *Caryologia* 6 (Suppl.): 307–25

Sonneborn, T.M. (1975) *Paramecium aurelia*. In *Handbook of Genetics*, edited by R.C. King, New York: Plenum Press

Thaî, K.Y. & Forney, J.D. (1999) Evidence for transcriptional self-regulation of variable surface antigens in *Paramecium tetraurelia. Gene Expression* 8: 263–72

Wichterman, R. (1986) *The Biology of Paramecium*, 2nd edition, New York and London: Plenum Press

Yao, M.-C., Yao, C.-H. & Monks, B. (1990) The controlling sequence for site-specific chromosome breakage in Tetrahymena. *Cell* 63: 763–72

FURTHER READING

Kipling, D. (1995) *The Telomere*, Oxford and New York: Oxford University Press

Preer, J.R. Jr (1993) Nonconventional genetic systems. *Perspectives in Biological Medicine* 36: 395–419

Prescott, D.M. (1994) The DNA of ciliated protozoa. *Microbiological Reviews* 58: 233–67

Yao, M.-C. (1988) Site-specific chromosome breakage and DNA deletion in ciliates. In *Mobile DNA*, edited by D.E. Berg & M.M. Howe, Washington, DC: American Society of Microbiology Press

GLOSSARY

binary fission the principal mode of division in prokaryotic organisms, in which a cell divides into two equal daughter cells

ciliate having small, hair-like outgrowths on the outer surface of the cell

conjugation the temporary union of two cells that results in genetic exchange

transformation any change in the properties of a cell stably inherited by its progeny

vegetative cell a cell not involved in a sexual process

See also **Genetics of *Tetrahymena thermophila*** (p.258)

GENETICS OF MALARIA PARASITES

Geoffrey Beale
Institute of Cell, Animal and Population Biology, University of Edinburgh, Edinburgh, UK

1. Introduction

Malaria is the most widespread parasite-caused disease of man, leading to millions of cases of sickness and death annually in tropical countries. The disease arises from infection by protists belonging to the genus Plasmodium and is transmitted by mosquitoes. Travellers are advised to have a sample of blood examined under the microscope as soon as possible if they experience fever on returning from a visit to an endemic area. Treatment requires the administration of drugs such as quinine, chloroquine, pyrimethamine and others, although, in recent years, such drug treatment has become markedly less effective than it was formerly, because of the development of drug-resistant forms of the parasite. Hence, study of the genetics of drug resistance has become particularly important. Moreover, because of the failure of drug treatment, attempts have been made to develop an antimalarial vaccine – attempts which, so far, have met with little success. The genetics of antigen variation is, therefore, a second area of great importance for research.

Due to the parasitic nature of the causative organism, and difficulties in maintaining some of its life cycle stages in laboratory cultures, genetic study by classical Mendelian methods has been difficult. However, with the aid of molecular biological techniques, considerable advances have been made recently and it is likely that the complete DNA genome of some species of Plasmodium will soon be mapped (Gardner *et al.*, 1992; see also information on the Internet at http://www.sanger.ac.uk/projects/P_falciparum).

2. Life cycle

The life cycle of malaria parasites is very complex, involving growth in two hosts: mosquitoes and man (or another vertebrate). There are four species of Plasmodium that infect human beings, the most life-threatening being *P. falciparum*. The life cycle and principal parasite stages of this species are shown in **Figure 1**. Minor variations occur in the life cycles of other species.

In brief, the infection of human patients is initiated by bites of mosquitoes, whereby elongated (11 μm × 1 μm) sporozoites are injected into the bloodstream and, within a few minutes, pass into the liver, where they undergo numerous divisions and produce thousands of the oval (1.5 μm × 1 μm) merozoites. The latter are released into the blood system and invade the red blood cells (erythrocytes). About 48 hours later, multinuclear schizonts develop in the infected red cells, each of which is then ruptured and releases about 20 new merozoites, which can immediately start a new cycle of red cell invasion. In this way, huge numbers of parasites are quickly produced in the infected individual.

Parallel with this asexual (mitotic) replication process, other merozoites prepare for the sexual stage by differentiating into micro- and macrogametocytes, and these, when mature, pass into the midgut of a mosquito with a blood meal and there transform into micro- and macrogametes. Sexual fusion then occurs and the zygotes thus produced undergo meiosis almost immediately and give rise to the elongated (20 μm) ookinetes, which penetrate the wall of the mosquito's midgut to form the oocysts and, subsequently, new sporozoites. The latter migrate into the salivary glands, and a new human infection can then take place.

Malaria parasites are eukaryotic organisms – that is, they consist of cells basically like those of higher animals or plants, with chromosomes lying within membrane-bound nuclei and with cytoplasm containing mitochondria and other organelles. However, the parasite's chromosomes are very small, do not condense into typical rod-shaped stainable structures during mitosis and meiosis, and can only be studied by special techniques such as pulsed field gel electrophoresis (PFGE). During the entire life cycle, apart from the zygote and possibly the ookinete stages, all the cells of Plasmodium contain the haploid number (14) of chromosomes, which vary

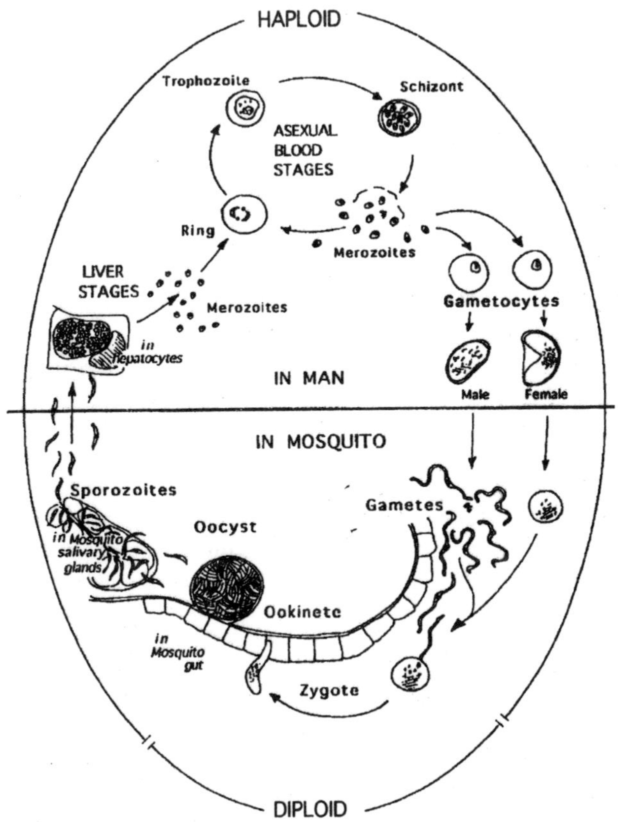

Figure 1 Life cycle of *P. falciparum* (from Creasey, 1996).

from ~800 kilobases (kb) to 3500 kb in length and are numbered 1–14 according to their size. There are, however, considerable variations in the size of the individually numbered chromosomes in different clones of parasite.

3. Genetic analysis

Different isolates of malaria parasites are genetically very diverse and the populations even within a single human patient often consist of a mixed assortment of genotypes (Thaithong *et al.*, 1984). Therefore, before starting a genetic analysis, it is important to make pure clones from any isolate. This can be done by a dilution method or, preferably, by micromanipulation. For both technical and ethical reasons, early genetic work was done with species of Plasmodium (e.g. *P. chabaudi*) that infect rodents rather than human beings. In one experiment, two clones of this species – which differed in the allelic forms of each of two different enzymes (one clone containing 6PGD-2 and LDH-3, and the other clone containing 6PGD-3 and LDH-2) – were mixed together and injected into mice, which were then caused to be bitten by mosquitoes. After oocysts and sporozoites had appeared in the latter, they were allowed to feed on and infect other mice. Clones of progeny parasites were then obtained and characterized for enzyme types. Some of these progeny were found to show recombination involving different enzyme-determining genes previously present in the two parental clones, proving that fertilization and meiosis had occurred in the mosquitoes and that the parasites had undergone a typical Mendelian process of inheritance. Both cross- and self-fertilization were found to have occurred. Progeny clones which had been produced (Beale *et al.*, 1978) as a result of cross-fertilization could be identified by the simultaneous presence within them of one or more characters previously present in the two different parent clones.

It is also worth mentioning that enzymes, being specific products of single genes, were of great value in the identification of different species and varieties of rodent Plasmodia. Apart from enzymes, other characters (e.g. antigens, virulence and drug susceptibility) (Beal & Walliker, 1988) were also studied in rodent malaria parasites. Such studies showed that there were separate genes for chloroquine and pyrimethamine resistance in these Plasmodium species.

Turning now to research on human Plasmodium species (e.g. *P. falciparum*), two sets of crosses between genetically diverse clones (HB3 × 3D7 and HB3 × Dd2) have been made. These experiments involved taking infected blood from different malaria patients, establishing parasite clones in artificial media, feeding mosquitoes on mixtures of two clones of parasites, allowing the mosquitoes to infect chimpanzees from the sporozoites thus produced and, finally, establishing new clones of progeny parasites in laboratory cultures derived from the chimpanzees' blood. The progeny clones were then analysed with regard to various genetic characters distinguishing the original parental clones.

From these two crosses, some very informative results were obtained (Walliker *et al.*, 1987), a few of which will be described here. The clones HB3 and 3D7 differed in a number of genetic characters, such as: (1) enzyme type (ADA/1 or ADA/2); (2) sensitivity or resistance to the drug pyrimethamine; and (3) and (4) different variants of the two proteins MSP/1 and MSP/2, respectively. Parasites derived from mixed infections of mosquitoes displayed a remarkable range of variants. In a relatively small sample of 14 descendant clones, recombinants involving all these four pairs of alternative characters were obtained. The second cross, between clones HB3 and Dd2, yielded segregants involving variations in response to the drug chloroquine, as well as in other characters. Thus, these crosses showed that passage through mosquitoes could result in a large increase in genetic variability affecting many characters, due to the reassortment of genes on different chromosomes or to crossing-over between two genes on a pair of homologous chromosomes, provided of course that the original infections involved more than a single clone of parasites.

These experiments also shed light on the genetics of the chromosomal size variations, which are an unusual feature of Plasmodium. Several of the 14 chromosomes of isolates HB3 and 3D7 of *P. falciparum* differ in size; for example, chromosomes 3 and 4 are longer in 3D7 than in HB3, and some of the progeny clones derived from the cross between HB3 and 3D7 contained chromosomes differing in size from those in either parent. Detailed study of one chromosome (chromosome 4) by restriction enzyme analysis showed that the size differences between the parental chromosomes involved mainly segments near their ends (the so-called "subtelomeric" regions). These regions contain repetitive sequences of nucleotides denoted *Rep 20*. One of the progeny clones obtained following hybridization of clones HB3 and 3D7 possessed a variant of chromosome 4 which was intermediate in size between those of the two parents, and this was shown to have been produced by crossing-over at a site in the central region of the chromosomes.

Further details of the inheritance process in *P. falciparum* were obtained by the analysis of 16 recombinant clones obtained from the cross HB3 × Dd2. By PFGE, each of the 14 chromosomes in these clones was separated and identified by means of specific gene probes. With the aid of restriction fragment length polymorphisms, small segments of each chromosome were marked and the approximate location of alleles from either of the two parents was established. By such analysis, it was shown that every one of the 16 progeny clones analysed contained unique combinations of segments of each of the 14 chromosomes inherited from the two parents. In most of these clones there were chromosomes containing some alleles from one parent and other alleles from the other parent. It was therefore clear that extensive recombination, due either to reassortment of genes on different chromosomes or to crossing-over between pairs of homologous chromosomes, had occurred in the mosquitoes from which the progeny parasites were derived. This confirmed that abundant recombination, and therefore an increase in genetic variation, occurs following the passage through mosquitoes of mixtures of genetically diverse parasites. Such analysis also made possible a rudimentary mapping of the parasite genome.

Recently the technique (Gardner *et al.*, 1992) of incorporating small groups of Plasmodium genes into yeast artificial chromosomes has been used for more detailed mapping of the genes on each of the 14 chromosomes. Provisional maps have already been published for chromosomes 2 and 4 and it is expected that the complete genome will be mapped within the next few years.

We will now consider a few details about some of the more important genetic characters.

4. Genetics of drug resistance

(i) General

In recent years, there has been a rise in the resistance of malaria parasites to all the frequently used antimalarial drugs, especially chloroquine, pyrimethamine, quinine and mefloquine, depending to some extent on the time and place of occurrence of the disease, intensity of drug use, etc. This resistance is due to spontaneous mutation in the parasites, followed by the selection of drug-resistant mutants in the presence of one or other of the drugs. The process occurred first with chloroquine, and then later with pyrimethamine and other drugs. A variety of different mechanisms are thought to operate to produce resistance to different drugs (Hyde, 1990).

(ii) Chloroquine

The biochemical action of chloroquine on malaria parasites is not well understood at present, although it is known that genetically resistant parasites have an increased ability to expel the drug from their cells compared with sensitive parasites. Some data on the genetics of chloroquine resistance have been obtained. Following a cross between clone HB3 of *P. falciparum*, which is sensitive, and clone Dd2, which is resistant, segregation of a gene for resistance located on chromosome 7 takes place, as shown by linkage studies. In other work (see Fidouk *et al.*, 2000) with the species *P. chabaudi*, an increase in chloroquine resistance was obtained by selection in the laboratory in the presence of the drug and, in that species, mutation was found to have occurred on its chromosome 11 (Carlton *et al.*, 1998). However, it seems likely that in both species chloroquine resistance is caused by mutations of a number of differently located genes.

(iii) Pyrimethamine

Resistance of *P. falciparum* to pyrimethamine has been found to arise rapidly in endemic areas where this drug has been used intensively, thus rendering it largely ineffective. At the present time, wild isolates collected from different places display a huge range in the concentrations of this drug which are necessary to inhibit growth of the parasites in culture, varying from 10^{-9} M with the most sensitive parasites to 10^{-4} M with the most resistant. In some cases, there is a mixed assortment of pyrimethamine-sensitive and pyrimethamine-resistant clones of parasites in a single patient.

Considerable information about the action of pyrimethamine on the parasites is available. The drug binds to and inhibits the action of dihydrofolate reductase (DHFR), an enzyme that is essential for the growth of the parasites and which is controlled by a gene on chromosome 4 of *P. falciparum*, and by a gene on chromosome 7 of *P. chabaudi*. In resistant parasites, the binding of drug to enzyme is reduced, rendering the drug less inhibitory to the enzyme and hence to growth of the parasite. The complete amino acid sequence of parasite DHFR is known, and a change in the amino acid residue from serine to asparagine at site 108 occurs in the molecule of DHFR in resistant *P. falciparum*. It has been stated that this mutation in the gene for DHFR is the main cause of the increase in pyrimethamine resistance. However, changes at other nucleotide positions in the DHFR-determining gene, and probably at other so far unknown positions elsewhere in the genome, are also likely to contribute to the increase in resistance of malaria parasites which has taken place in recent years.

5. Genetics of antigens of Plasmodium

Many different proteins of Plasmodium are antigenic and some have been considered as possible constituents of an antimalarial vaccine, a few of which are mentioned below. Unfortunately, in spite of much work, an effective vaccine has not yet been developed. No doubt this has been made difficult by the fact that the parasites themselves contain built-in mechanisms favouring their own survival, rather than that of their human hosts.

(i) MSP/1 (merozoite surface protein/1)

This protein (also known as MSA/1, P/190, P/195, PSA, PMMSA, etc.) has been shown to confer on injected monkeys a partial degree of protection from subsequent malaria infection. It has a molecular weight of 180–220 kilodaltons (kD), depending on the particular parasite clone from which it has been extracted, and it is governed by a gene on chromosome 9 of *P. falciparum*. Studies with monoclonal antibodies have revealed that there is a large amount of immunological diversity affecting this protein; for example, collections of parasites from The Gambia in West Africa have been grouped into more than 36 different types (Conway *et al.*, 1991).

Biochemical analysis (Miller *et al.*, 1993) of the MSP/1 protein has shown that it can be subdivided into some 17 "blocks" of amino acids, roughly classified as "conserved", "semiconserved" and "variable" in different isolates, depending on the percentage of similar amino acid residues present. There was also shown to be a curious "dimorphic" system of variation in blocks 5–16 inclusive, though with intercalated patches of conserved sections. In block 4, and to some extent in block 3, there were two types of variant that did not necessarily correspond with those in blocks 5–16 (inclusive). In block 2, the dimorphic system was not present at all, since there were three possible variant types, two of which consisted of tandem sets of triplet repeats involving particular amino acids. Taking together all these variations, it can be seen that many combinations of different amino acid sequences exist in the MSP/1 proteins of different isolates of *P. falciparum*. It has been suggested that different alleles, comprising different combinations of the blocks, may have arisen by some process of intra-allelic recombination (possibly by rare crossing-over or gene conversion) in hybrids between long-separated populations.

Although the MSP/1 antigens may be of little value in the production of a vaccine, they do provide many convenient markers for genetic experiments with different clones, where accidental contaminations or mistakes in labelling cultures may occur, which then need to be rectified during research.

(ii) MSP/2 (merozoite surface protein/2)

This is a second protein antigen situated on the merozoite surface. It has a molecular weight of about 45 kD and is governed by a gene on chromosome 2. Like MSP/1, it is highly polymorphic and can also act as an excellent genetic marker in research. Members of both MSP/1 and MSP/2 allelic series can be readily identified with the aid of the polymerase chain reaction (PCR).

(iii) CSP (circumsporozoite protein)

CSP was formerly considered to be the most promising candidate antigen for use in an antimalarial vaccine, being a protein covering the surface of the sporozoites, which (unlike the merozoites) are extracellular and, therefore, readily able to stimulate the immune system of the host. If the circumsporozoites could be killed or prevented from entering the hepatocytes, the parasite would be unable to infect the blood system and the disease would be prevented. The protein, which is governed by a gene on chromosome 3 of *P. falciparum*, contains a central region containing some 36–43 tandem repeats of sequences of the four amino acids asparagine–valine–asparagine–proline, interspersed with three to four repeats of the sequence asparagine–valine–aspartic acid–proline. Large amounts of these repeats have been synthesized by molecular biological methods and have been shown to render sporozoites inactive *in vitro*. However, when such material was injected into a patient, little, if any, protective effect against attacks of malaria was obtained. This may be due to the fact that the central repetitive section of the CSP molecule is flanked on both sides by regions containing non-repetitive peptides, and these contain certain T-cell epitopes that are important in initiating antibody production and for controlling the development of cell-mediated immunity. Unfortunately, although the central repeat section is largely conserved in different isolates of the parasite, the T-cell epitopes are highly polymorphic and this is thought to be the reason why vaccines based on the CSP protein have turned out to be unsuccessful.

6. Extranuclear genomes of Plasmodium

In addition to the 14 chromosomes in their nuclei, malaria parasites possess two kinds of extranuclear genetic system: a 6 kb mitochondrial genome and a 35 kb circular genome (Creasey *et al.*, 1994). The mitochondria are unlike those of all other organisms in being extremely small and coding for only three proteins: cytochrome b, subunits I and III of cytochrome oxidase, and a group of fragmentary ribosomal RNAs. In *P. falciparum*, there are approximately 20 copies of the 6 kb genome per cell, arranged mainly as linear repeats. Merozoites contain only a single mitochondrion, but more develop at later stages of the parasite.

The 35 kb circular genome encodes about 60 genes, including those for ribosomal RNAs, transfer RNAs, several ribosomal proteins and subunits of an RNA polymerase. The position in the cell of these organelles has not yet been observed, but there are one to three copies per cell in *P. falciparum*. They are presumed to be essential to the parasites and, surprisingly, have some homologies with the plastids of green plants. Both 6 kb and 35 kb extranuclear genomes are inherited solely through the macrogametes (Creasey *et al.*, 1994).

7. Conclusion

Compared with other organisms used in genetic research, malaria parasites are unusual in a number of respects. They are obligatory parasites living at different times in two different hosts – mosquitoes and vertebrates – and this imposes a whole series of different lifestyles and nutritional situations on the parasites. Since they are unicellular, all their metabolic activities must take place within the confines of a single cell and they develop a great diversity of different stages during their life cycles. They pass through sexual processes at rather frequent intervals and this may result in a frequent change of genotype, leading to opportunities for the selection of new phenotypes, some of which are highly relevant to human disease and its treatment. Such characteristics offer us a number of interesting opportunities for the study of the genetics of these organisms and provide hopes of increasing our control over an exceedingly important disease.

REFERENCES

Beale, G.H. & Walliker, D. (1988) Genetics of malaria parasites. In *Malaria: Principles and Practice of Malariology*, edited by W.H. Wernsdorfer & I. McGregor, Edinburgh and New York: Churchill Livingstone

Beale, G.H., Carter, R. & Walliker D. (1978) Genetics. In *Rodent Malaria*, edited by R. Killick-Kendrick & W. Peters, London and New York: Academic Press

Carlton, J., MacKinnon, M. & Walliker, D. (1998) A chloroquine resistance locus in the rodent malaria species *Plasmodium chabaudi. Molecular Biochemical Parasitology* 93: 57–72

Conway, D.J., Rosario, V., Oduola, A.M.J. *et al.* (1991) *Plasmodium falciparum*: intragenic recombination and nonrandom associations between polymorphic domains of the precursor to the major merozoite surface antigens. *Experimental Parasitology* 73: 469–80

Conway D.J., Roper, C., Oduola, A.M.J. *et al.* (1999) High recombination rate in natural populations in *Plasmodium falciparum. Proceedings of the National Academy of Sciences USA* 96: 4506–11

Creasey, A. (1996) The inheritance of extranuclear DNA in malaria parasites. Ph.D. thesis, University of Edinburgh

Creasey, A., Mendis, K., Carlton, J. *et al.* (1994) Maternal inheritance of extra-chromosomal DNA in malaria parasites. *Molecular Biochemistry and Parasitology* 65: 95–98

Fidouk, D.A. *et al.* (2000) Mutations in the *P. falciparum* digestive vacuole transmembrane protein PfCRT and evidence for this role in chloroquine resistance. *Molecular Cell* 6: 861–71

Gardner, M.J., Tettelin, H., Carucci, D.J. *et al.* (1992) Chromosome 2 sequence of the human malaria parasite *Plasmodium falciparum. Science* 282: 1126–32

Hyde, J.E. (1990) *Molecular Parasitology*, Milton Keynes,

Buckinghamshire: Open University Press and New York: Van Nostrand Reinhold

Miller, L.H., Roberts, T., Shahabuddin, M. & McCutchan, T.F. (1993) Analysis of sequence diversity in the *Plasmodium falciparum* merozoite surface protein (MSP/1). *Molecular Biochemistry and Parasitology* 59: 1–14

Thaithong, S., Beale, G.H., Fenton, B. *et al.* (1984) Clonal diversity in a single isolate of the malaria parasite *Plasmodium falciparum. Transactions of the Royal Society of Tropical Medicine and Hygiene* 78: 242–45

Triglia, T., Wellems, T.E. & Kemp, D.J. (1992) Towards a high resolution map of the *Plasmodium falciparum* genome. *Parasitology Today* 8: 225–29

Walliker, D., Quakyi, I.A., Wellems, T.E. *et al.* (1987) Genetic analysis of the human malaria parasite *Plasmodium falciparum. Science* 236: 1661–66

FURTHER READING

Much of the recent literature is mentioned in:

Sherman, I.W. (ed.) (1998) *Malaria: Parasite Biology, Biogenesis and Protection*, Washington, DC: ASM Press

http://www.sanger.ac.uk/projects/P-falciparum

GLOSSARY

allelic pertaining to alleles; where two or more mutations, non-complementary in the heterozygous state, map to the same area

epitope (antigenic determinant) exposed structure of an antigen that causes a specific reaction by an immunoglobulin

gametocyte (micro- and macro-) cell from which gamete is produced

hepatocyte liver cell

merozoite (schizozoite) small cell produced by multiple fission of a schizont

oocyst sporozoan developmental stage where the zygote develops an enclosing cyst wall

ookinete motile elongated zygote formed by fertilization of the macrogamete during the reproductive stage of the sporozoan life cycle

protist a unicellular organism

segregant variant parasite produced by segregation

schizont life cycle stage of some protozoans (esp. sporozoan parasites) following the trophozoite stage

sporozoite in malaria, spore released from salivary glands of the mosquito and transmitted to humans

telomeric relating to the ends of chromosomes

See also **Molecular biology of malaria parasites** (p.254)

MOLECULAR BIOLOGY OF MALARIA PARASITES

David E. Arnot
Institute of Cell, Animal and Population Biology, University of Edinburgh, Edinburgh, UK

1. Introduction

Despite the accumulation of a century of scientific knowledge of malaria, around 40% of the world's population remains at risk from infection and 1–2 million people, particularly children, die from the disease annually. There are no currently effective vaccines and their development has proven more difficult than was anticipated at the start of the biotechnology revolution, which was triggered by developments in genetic engineering and monoclonal antibody technology almost 20 years ago. In many parts of the world, *Plasmodium falciparum* has become resistant to most of the older generation of inexpensive drugs (e.g. chloroquine) and is rapidly evolving resistance to the "new" generation of more expensive drugs. Current knowledge of the molecular biology of the parasite is recognized to be inadequate for the rational development of drugs, vaccines and other possible interventions. Therefore, renewed attempts to gain control of the global malaria situation involve basic molecular and cellular biological study of the parasite as an integral part of efforts to develop novel and more effective ways to attack the parasite. Some of the main lines emerging in this research are outlined here.

2. The Plasmodium life cycle as a cell biological process

Human malaria parasites have a singularly complex life cycle, containing four major replication and cell division phases and at least five discrete phases of DNA synthesis. The four replication phases occur in two separate hosts (humans living at 37°C and Anopheline mosquitoes living around 18–25°C) and four different cellular environments. The mosquito-transmitted sporozoites invade and replicate within liver hepatocytes (a process referred to as exo-erythrocytic schizogony), ultimately lysing them to release merozoites which then invade, replicate within and ultimately lyse the host red blood cells within 48 hours (erythrocytic schizogony). Merozoites can bypass the asexual replicative and lytic cycles and develop into sexual macro- and microgametocytes which are the only forms capable of carrying out the insect phases of the life cycle: gametogenesis, zygote and oocyst formation, and sporogenesis. The molecular processes underlying this "transmission *vs.* replication" decision are not understood.

Although erythrocytic schizogony is rather different from the cell cycles of mammalian cells or yeasts, this is the most experimentally accessible Plasmodium life cycle phase, the only one beginning to be analysed in terms of signal transduction and the eukaryotic cell cycle and, therefore, the only one considered here. None of the discrete phases of DNA replication and cell division in the Plasmodium life cycle follows a yeast-type simple cell cycle model and the precise nature of the Plasmodium cell cycle is not entirely clear (Leete & Rubin, 1996, and see **Figure 1**). Intraerythrocytic schizogony is characterized morphologically by early ring forms, developing into trophozoites, then undergoing schizogony (essentially repeated rounds of nuclear division) and forming merozoites which are released on red cell rupture. The nuclear divisions of Plasmodium are quite different from those of yeasts or higher cells and have been described as a "cryptomitosis", where chromosomes do not condense and the nuclear membrane remains intact throughout genome replication. DNA synthesis in synchronous *P. falciparum* cultures starts 28–31 hours after merozoite invasion in small, early trophozoites and DNA content then

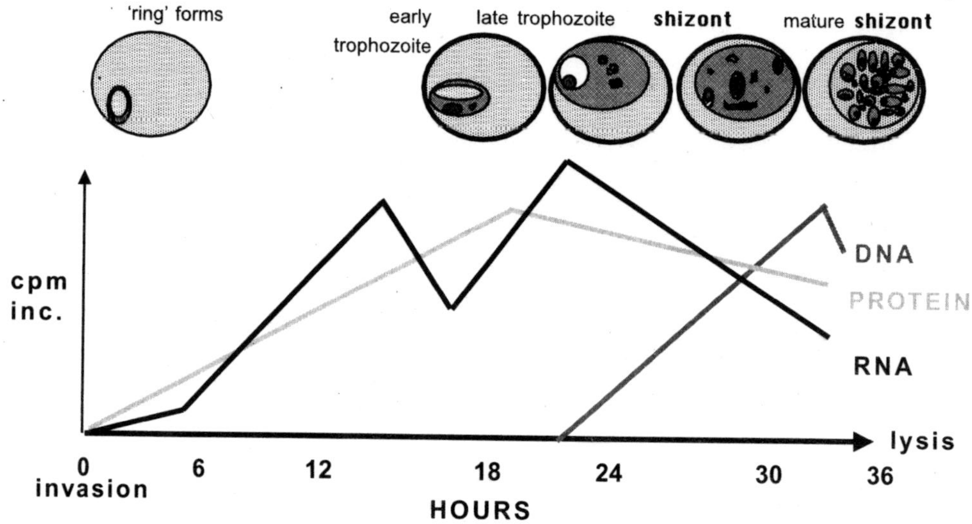

Figure 1 DNA, protein and RNA synthesis during the development of *P. falciparum* in synchonized cultures.

increases for 8–10 hours (Graeser *et al.*, 1996, and see **Figure 1**).

A current interpretation of the relationship between the intraerythrocytic Plasmodium cell cycle and the morphological development of the parasite is as follows. The merozoite, both free and in late segmentors, is in a G_0-like state with condensed chromatin. The trophozoite is formally similar to the gap G_1 stage of mammalian cells or yeast, and rings and early trophozoites are equivalent to G_{1A} and G_{1B}, respectively. Since DNA synthesis begins in relatively small trophozoites, but subdivision of the nuclear material (the morphologically defined onset of schizogony) does not appear until the trophozoite is much enlarged, growth and DNA synthesis in trophozoites must occur concurrently. Prior to DNA synthesis, trophozoites have pre-accumulated high levels of RNA. Schizogony is S phase and G_2, and then M (mitotic) phases occur during progression from the haploid (1C) DNA content to the final genome complement of the mature segmentor (~8–30C). Segmentation occurs as an independent stage at the end of the asexual erythrocytic cycle and merozoites are not formed after S phase. Since the transitions from one stage to the next are not clearly separable biochemically or morphologically, the relationship of schizogony to both the DNA synthetic cycle and the cell cycle is poorly defined, but it is likely that a brief G_2+M phase occurs after each genome doubling (6–8 hours for four to five rounds of mitosis, Arnot & Gull, 1998). Since haploid (1C) merozoites can differentiate sexually into micro- and macrogametocytes, the simplest cell cycle model for this process would be that gametocytes have been arrested in G_1 prior to the onset of S phase (although in *P. falciparum* a considerable excess of DNA over the 1C value has been measured after Feulgen staining and fluorometry; Janse *et al.*, 1988).

3. Gene expression

Plasmodium gene expression is not well understood and transcription has some unusual features, although the *trans*-spliced leader sequences found in several protozoan groups have not been found. Plasmodium RNA can be translated *in vitro* and is thus inferred to be capped. However, the mRNAs usually possess long >90% A+T rich, 5′ and 3′ untranslated regions whose function is unclear. Contrary to what was thought after the first few genes had been sequenced (Bowman *et al.*, 1999), introns are not uncommon in Plasmodium genes and probably occur in the majority of loci, although they are usually small (<150 base pairs) and tend to occur near the 5′ or 3′ ends of the genes, leaving the central transcribed region uninterrupted. Although the parasite rapidly progresses through a complex series of morphological and metabolic changes which bring about abrupt changes in gene expression, the molecular switches regulating the parasite's responses to extracellular signals are poorly understood and, so far, few of the components of the higher eukaryotic transcriptional regulation apparatus have been identified (Ji & Arnot, 1997).

4. Metabolic and signalling pathways

Again, the most studied phase of the complex life cycle is that spent replicating within the host erythrocyte. The host cell cytoplasmic constituents, particularly haemoglobin, are digested to provide some, but not all, of the parasite's essential nutrients and metabolic precursors. The process of subversion of the host cell to reorientate it towards the parasite's needs is extremely complex, requiring imports and exports across the parasite's own cell membrane, the parasitophorous vacuole membrane, the erythrocyte cytoplasm and the erythrocyte membrane (Elford *et al.*, 1995). Once again, Plasmodium molecular cell biology appears to have unusual features. Not all of the secreted proteins from parasitized erythrocytes appear to contain signal sequences (Lingelbach, 1993), and the secretion process itself is not as Brefeldin A sensitive as are higher eukaryotic cells to this Golgi apparatus-disrupting fungal metabolite (Ward *et al.*, 1997). Characterization of some parasite-encoded membrane transporters has been achieved, particularly of one which may be associated with chloroquine resistance, the ATPase-containing P-glycoprotein homologues (Rubio & Cowman, 1996). Detection in malaria parasites of the enzymology and mediators of signal transduction is in its early stages, but pathways dependent on cAMP, cGMP, MAP kinase and phosphatidylinositol have all been identified (Doerig, 1997).

5. Pathogenesis and antigenic variation

Human red blood cells infected with the malaria parasite *P. falciparum* adhere to the endothelial lining of venular capillaries. This process allows the developing parasite to evade elimination by the reticuloendothelial system of the host spleen and it is mediated through parasite-encoded erythrocyte membrane proteins of the PfEMP-1 family (Baruch *et al.*, 1995). These polymorphic proteins are encoded by the *var* genes, each of which consists of two exons, the first containing between two and four Duffy binding-like (DBL) domains. These cysteine-rich sub-regions of the protein probably serve as "anchoring" sites for host cell surface receptors and are homologous to motifs within the *P. vivax* Duffy antigen-binding protein and the *P. falciparum* EBA (erythrocyte-binding antigen) 175 protein. The second exon encodes an intraerythrocytic amino terminal segment (Su *et al.*, 1995). A gene family of around 50 *var* loci, each of which appears to be polymorphic in different isolates, is located on all 14 of the *P. falciparum* chromosomes. Each chromosome appears to possess two subtelomeric *var* genes and additional small clusters of *var* genes exist on several chromosomes. It has been proven that clonal antigenic variation via *var* gene switching occurs (Smith *et al.*, 1995), although the mechanism for the activation of *var* gene transcription is unknown and does not appear to be related to the transpositional activation mechanism frequently observed with trypanosome variant surface protein genes. The significance of this phenomenon, and its possible relationship to the sequestration-mediated severe pathological syndromes such as cerebral malaria, is being intensively studied.

6. The Plasmodium Genome Project approach

Integrating the research efforts of individual scientists and groups within a collaborative Genome Project offers a novel, rational and comprehensive framework for exploiting the advances in Plasmodium molecular biology in the treatment and control of malaria. Originally aimed at mapping and sequencing all the genes on the human chromosomes, the Human Genome Project has brought about advances in biotechnology and information technology which make it feasible to obtain rapidly genomic data from any organism. Microorganisms of medical importance are obvious candidates for complete genome analysis and several bacterial pathogens have now been sequenced (Fleischmann *et al.*, 1995). *P. falciparum* has a genome size of 25–30 Megabases of DNA contained on 14 chromosomes which can be separated by pulsed field gel electrophoresis in agarose. This property can be exploited to map genes onto separated chromosomes by Southern blotting and then the genes can be orientated relative to restriction fragment length polymorphism maps (Walker-Jonah *et al.*, 1992). Ordered arrays of overlapping clones have been assembled across each of the 14 chromosomes using yeast artificial chromosomes (YACs). Ordered "YAC contigs" are used to map the location of randomly sequenced cDNA and genomic clones in order to place each gene in its genomic location. A near-complete genome sequence and map of the genetic organization of the malaria parasite are planned to be available to the research community by the year 2002 (Dame *et al.*, 1996). Based on the 30–40 currently completed (non-viral) genome sequences (all bacteria or achaebacteria, except the yeast *Saccharomyces cerevisiae*), around 50–70% of the estimated 5-7000 *P. falciparum* genes will have a recognizable sequence homologue in the gene banks. Although sequence homology is an interesting clue rather than a proof of functional homology, this information will clearly be a powerful resource for molecular biologists working on the aspects of malaria parasites outlined here.

7. Can molecular biology cure malaria?

As new global efforts on malaria control (such as the World Health Organisation Roll back Malaria Programme) are being put into place as part of a post-millennial effort to attack the links between ill health and poverty, questions have been raised about whether high-tech laboratory science can ever deliver malaria control, let alone eradication (see *The Malaria Capers: More Tales of Parasites and People, Research and Reality*, Desowitz 1991 for an interesting critique). While it is reasonable to point out that scientific research and the improvements in living conditions brought about by technological advance are our only real weapons against any disease, critics such as Desowitz make a fair point that as resources allocated to laboratory science in the North have increased, resources allocated to health in the South have decreased. The solution is easily stated but difficult to put into practice: global cooperation to increase the efforts to produce new anti-malarial therapies and then apply them to produce real improvements in the malaria situation. New knowledge of the molecular biology of the parasite will unquestionably play a key role in the continuing struggle against malaria through identification of new drug and vaccine targets and much-needed improvements in ease of production and screening of such therapeutics. The search for the Achilles heel of malaria continues.

REFERENCES

Arnot, D.E. & Gull, K. (1998) The *Plasmodium* cell cycle: facts and questions. *Annals of Tropical Medicine and Parasitology.* 92: 361–65

Baruch, D.I., Pasloske, B.L., Singh, H.B. *et al.* (1995) Cloning the *P. falciparum* gene encoding PfEMP-1, a malarial variant antigen and adherence receptor on the surface of parasitized human erythrocytes. *Cell* 82: 77–87

Bowman, S., Lawson, D., Basham, D. *et al.* (1999) The complete nucleotide sequence of chromosome 3 of *Plasmodium falciparum. Nature* 400: 532–38

Dame, J.B., Arnot, D.E., Burke, P.F. *et al.* (1996) Current status of the *Plasmodium falciparum* Genome Project. *Molecular and Biochemical Parasitology* 79: 1–12

Doerig, C.D. (1997) Signal transduction in malaria parsites. *Parasitology Today* 13: 307–13

Elford, B.C., Cowan, G.M. & Fergusson, D.J.P. (1995) Parasite-regulated membrane transport processes and metabolic control in malaria-infected erythrocytes. *Biochemical Journal* 308: 361–74

Fleischmann, R.D., Adams, M.D., White, O. *et al.* (1995) Whole-genome random sequencing and assembly of *Haemophilus influenzae* Rd. *Science* 269: 496–512

Graeser, R., Wernli, B., Franklin, R.M. & Kappes, B. (1996) *Plasmodium falciparum* protein kinase 5 and the malarial nuclear division cycles. *Molecular and Biochemical Parasitology* 82: 37–49

Janse, C.J., Ponnudurai, T., Lensen, A.H. *et al.* (1988) DNA synthesis in gametocytes of *Plasmodium falciparum. Parasitology* 96: 1–7

Ji, D.-D. & Arnot, D.E. (1997) A *Plasmodium falciparum* homologue of the ATPase subunit of a multi-protein complex involved in chromatin remodelling for transcription. *Molecular and Biochemical Parasitology* 88: 151–62

Leete, T.H. & Rubin, H. (1996) Malaria and the cell cycle. *Parasitology Today* 12: 442–43

Lingelbach, K.R. (1993) *Plasmodium falciparum*: a molecular view of protein transport from the parasite into the host erythrocyte. *Experimental Parasitology* 76: 318–27

Rubio, J.P. & Cowman, A.F. (1996) The ATP binding cassette (ABC) gene family of *Plasmodium falciparum. Parasitology Today* 12: 135–40

Smith, J.D., Chitnis, C.E., Craig, A.G. *et al.* (1995) Switches in expression of *Plasmodium falciparum var* genes correlate with changes in antigenic and cytoadherant phenotypes of infected erythrocytes. *Cell* 82: 101–10

Su, X.Z., Heatwole, V.M., Wertheimer, S.P. *et al.* (1995) The large diverse gene family *var* encodes proteins involved in cytoadherence and antigenic variation of *Plasmodium falciparum*-infected erythrocytes. *Cell* 82: 89–100

Walker-Jonah, A., Dolan, S.A., Gwadz, R.W., Panton, L.J. & Wellems, T.E. (1992) An RFLP map of the *Plasmodium falciparum* genome, recombination rates and favored linkage groups in a genetic cross. *Molecular and Biochemical Parasitology* 51: 313–20

Ward, G.E.,Tilney, L.G. & Langsley, G. (1997) Rab GTPases and the unusual secretory pathway of *Plasmodium. Parasitology Today* 13: 57–61

GLOSSARY

A+T rich rich (replete) in the purine base adenine and the pyrimidine base thymine

agarose polysaccharide made from agar-agar (which is extracted from red algae) used to manufacture gels for electrophoresis

cAMP (cyclic 3′,5′ adenosine monophosphate) this universal "second messenger" is generated from adenosine triphosphate (ATP) by adenylate cyclase, has a hormonally controlled structure, and triggers reactions that produce a cellular response to the specific stimulus

cGMP (cyclic guanosine monophosphate) molecule formed from guanosine monophosphate by the enzyme guanylate cyclase; acts as a second messenger in some cellular reactions

contig large continuous DNA sequence formed from assembly of overlapping shorter sequences

cryptomitosis a form of nuclear division in protozoans where the mass of chromatin assembles in the equatorial region without apparent formation of chromosomes

erythrocytic schizogony the process whereby a single uninucleate merozoite undergoes DNA replication and nuclear division to produce ~8–30 daughter merozoites within a host red blood cell

exo-erythrocytic schizogony the process by which an invading uni-nucleate sporozoite undergoes DNA replication and nuclear division to produce 10–30 000 first generation merozoites within a host liver hepatocyte

G_0-like stage phase of the eukaryotic cell cycle in which the cycle is arrested

G_1 phase (gap phase) in the eukaryotic cell cycle, phase between the end of cell division and the commencement of DNA synthesis (*see also* G_0, G_2 and S phases)

G_2 phase phase of the eukaryotic cell cycle between DNA replication and mitosis

gametocyte (micro- and macro-) cell from which gamete is produced

gametogenesis formation of gametes (germ cells containing a haploid number of chromosomes)

hepatocyte liver cell

intraerythrocytic Plasmodium cell cycle the phases of malaria parasite development within host red blood cells

MAP (mitogen-activated protein) kinase serine–threonine kinase activated when quiescent cells are treated with mitogens. Lying at the end of a multicomponent signal transduction pathway it phosphorylates transcription factors, enabling them to stimulate gene expression

merozoite (schizozoite) small cell produced by multiple fission of a schizont

monoclonal antibody antibody produced by a single clone of B cells, thereby consisting of a population of identical antibody molecules specific for a single antigenic determinant

oocyst sporozoan developmental stage where the zygote develops an enclosing cyst wall

ring form form adopted by the immature trophozoite of the malaria parasite in red blood cells

S phase phase of the cell cycle when DNA replication occurs

Southern blot technique in which DNA molecules, separated by electrophoresis in an agarose gel, are transferred to "blotting" to a nitrocellulose filter, then hybridized with radioactively labelled nucleic acid probes for identification and isolation of sequences of interest

sporogenesis spore formation

telomeric relating to the ends of chromosomes

transcriptional regulation apparatus regulatory proteins that bind to DNA at specific control sites to initiate or prevent transcription

trophozoite adult stage of a sporozoan

See also **Genetics of malaria parasites** (p.249)

GENETICS OF *TETRAHYMENA THERMOPHILA*

Eduardo Orias
Department of Molecular, Cellular and Developmental Biology, University of California at Santa Barbara, California, USA

1. *Tetrahymena thermophila*

(i) The organism

T. thermophila (**Figure 1**) belongs to the ciliated Protozoa (or Ciliates), a major, successful and diversified evolutionary lineage of unicellular eukaryotes. It is a freshwater organism that commonly inhabits streams, lakes and ponds. The cells are large, reaching 40–50 μm along the anterior–posterior axis. Like other ciliates, Tetrahymena cells have a striking variety of highly complex and specialized cell structures. Additional views of Tetrahymena cells can be found in the published literature: by transmission electron microscopy (Hill, 1972, p.5) and optical microscopy of stained cells (Nanney, 1980, p.180). The Tetrahymena genome size (roughly 220 Megabase pairs) is of the same order of magnitude as that of Drosophila – and one order larger than yeast (Saccharomyces) and one smaller than human.

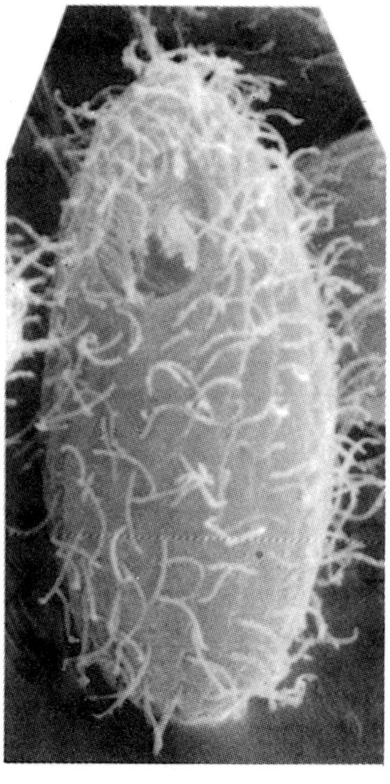

Figure 1 Scanning electron micrograph of a Tetrahymena cell. Note the somatic cilia, arranged in rows (meridians), and the oral apparatus, near the anterior end, with tightly spaced cilia that comprise the oral "membranelles". This micrograph, taken by E. Marlo Nelsen, is reproduced with the kind permission of Professor Joseph Frankel, University of Iowa, USA.

This microbial eukaryote has great value as an experimental system due to its rapid growth, its structural and functional differentiations, its accessibility to genetic and molecular approaches, its large evolutionary distance from other commonly used genetic model systems and its potential for biotechnological applications coupled with its biosafety. It has provided an excellent system for the discovery and investigation of fundamental molecular and cellular mechanisms (e.g. ribozymes and telomeres); the availability of dozens of related species has proven extremely useful for identifying evolutionarily conserved domains in macromolecules and, particularly, for the dissection of important secondary structure and functional domains in RNAs that have, or participate in, catalytic activity; and its nuclear dimorphism provides an additional and unique dimension of biological versatility and genetic manipulability.

The species now called *T. thermophila* was first considered to be *T. pyriformis*. Later, *T. pyriformis* was found to comprise many sibling (cryptic) species, sexually isolated from one another, but morphologically indistinguishable. As knowledge advanced, it was successively renamed variety 1 and syngen 1 of *T. pyriformis*, before acquiring its current name (Nanney & McCoy, 1976). Genetically, *thermophila* is the species of Tetrahymena which is by far the most extensively characterized.

For the purpose of brevity, this article does not review knowledge on the molecular basis of gene expression. (As a point of interest, Tetrahymena and certain other ciliates have a variant genetic code in which UAG and UAA – normally stop codons – are additional glutamine codons. For a useful recent review of this area see Prescott, 1994.) Whenever possible, useful reviews will be referred to, rather than primary research articles; Nanney's (1980) introduction to the experimental biology of ciliates, with a strong coverage of Tetrahymena, is particularly useful. A broad review of the biology of Tetrahymena was edited by Elliott (1973) and a comprehensive review of ciliate molecular biology was edited by Gall (1986), the latter including relevant chapters on conjugation, genetics and DNA organization, all with extensive coverage of *T. thermophila*. Prescott's (1994) comprehensive review of ciliate molecular genetics includes much about Tetrahymena, and a more detailed version of the present article, as well as a summary of micronucleus and macronucleus genetic maps of nearly 400 DNA polymorphisms, can be found on the Internet at: http://lifesci.ucsb.edu/~genome/Tetrahymena. Bleyman (1996) has recently reviewed the genetics of ciliates and the loci determining interesting mutant phenotypes, including many in Tetrahymena.

(ii) Nuclear dimorphism

As is typical of ciliates, the nuclear apparatus of Tetrahymena is composed of two structurally and functionally differentiated types of nuclei, a phenomenon known as nuclear dimorphism. The micronucleus (MIC), which is the germline, i.e. the store of genetic information for the sexual progeny, is diploid and contains five pairs of chromosomes. No known genes are expressed in the MIC. Amicronucleate Tetrahymena cells (i.e. cells lacking a MIC) are frequently collected in nature, but in laboratory strains of *T. thermophila*, the loss of the MIC leads to clone death and only one viable laboratory-obtained amicronucleate cell line has been described. At cell division, the MIC divides mitotically with the formation of kinetochores and an intranuclear mitotic spindle.

The macronucleus (MAC) is the somatic nucleus, i.e. the nucleus actively expressed during vegetative multiplication. No known MAC DNA is transmitted to the sexual progeny. The MAC contains 200–300 autonomously replicating DNA pieces derived from the five MIC chromosomes by site-specific fragmentation (further described below). The bulk of these DNA pieces is present at the average level of 45 copies per MAC. There are none of the visible structures expected for the mitotic distribution of MAC pieces, such as kinetochores or mitotic spindle. The MAC is thus said to divide by amitosis. Approximately (but seldom exactly) half of the MAC DNA is distributed to each daughter MAC at cell division and alternative allele copies of a locus segregate at random during MAC division (described later). Physical methods are available to separate preparatively and purify MICs and MACs from one another.

(iii) Life cycle

The life cycle consists of an alternation of haploid and diploid stages (haplophase and diplophase, respectively) with reference to the germline. Cell reproduction is exclusively by binary fission; it is exclusively asexual and occurs only in the diplophase. It is remarkable that such highly differentiated cells can divide by binary fission (e.g. see Nanney, 1980, Plate II, p. 32). Cell division is accompanied by a variety of morphogenetic events that result in the development of duplicate sets of cell structures, one for each daughter.

Conjugation is the sexual stage of the Tetrahymena life cycle: two cells pair, form a temporary junction, exchange gamete nuclei and generate and differentiate the nuclear apparatus of their sexual progeny. The nuclear events of conjugation (**Figure 2**) normally include meiosis, gamete nucleus formation, fertilization and nuclear differentiation. It is remarkable that, at the time of exconjugant separation (stage 7 in **Figure 2**), there are five nuclei experiencing four extremely diverse fates, all within a common cytoplasm. Conjugation includes the only – and very brief – haploid stage of the life cycle; it follows meiosis and quickly ends at fertilization. The haplophase is limited to a single nuclear division, without any cell division.

In order to conjugate, Tetrahymena cells must satisfy the following requirements:

- they must be starved for at least one required nutrient;

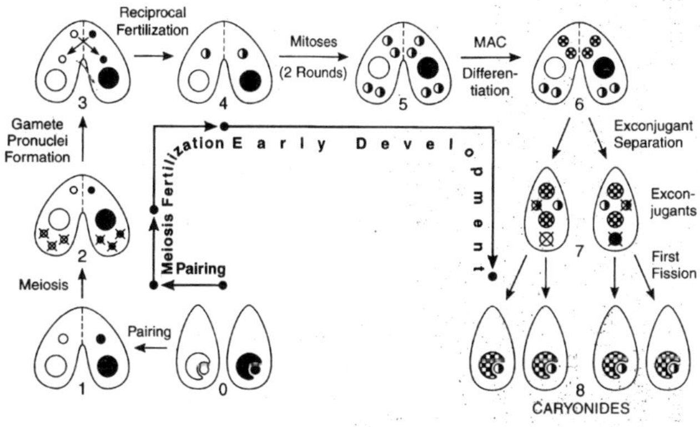

Figure 2 Nuclear events in Tetrahymena conjugation. (0) Vegetative cells homozygous for alternative alleles at one locus. The MIC (small circles) is shown nested in – but physically separate from – the MAC (large circles), the normal arrangement in non-dividing vegetative cells. (1) Paired cells. (2) MICs undergo meiosis and four haploid nuclei are produced. Only the anterior meiotic product remains functional, while the other three disintegrate. This is the stage at which meiotic crossing-over, used for making genetic maps of the MIC genome, occurs. (3) Mitotic division of functional meiotic product yields genetically identical migratory (anterior) and stationary (posterior) gamete pronuclei. (4) Migratory pronuclei are reciprocally exchanged and fuse with stationary pronuclei of the recipient cell, forming the zygote nucleus. (5) The zygote nucleus undergoes two mitotic divisions, giving rise to four genetically identical diploid nuclei. (6) Anterior products differentiate into macronuclei, while posterior products remain diploid micronuclei. This is the stage at which site-specific DNA rearrangements and mating-type determination (discussed in the text) occur in the MAC. (7) Exconjugants separate. The old macronucleus, and one of the two new micronuclei, are destroyed. (8) Exconjugants undergo the first postzygotic cell division, forming the four caryonide cells. Each caryonide receives an independently differentiated new macronucleus and a mitotic copy of the functional micronucleus. Caryonides then begin vegetative multiplication by binary fission.

- they must be of different mating type; and
- they must have reached a sufficient level of sexual maturity.

Seven mating types (I–VII) have been known since the earliest investigations of *T. thermophila*; the number has not increased after testing thousands of additional independent isolates from the wild. Sexual progeny normally are unable to mate again immediately after finishing conjugation. After 50–80 fissions, they reach adolescence, i.e. they can mate with mature, but not adolescent cells; cells reach full sexual maturity 20–25 fissions later (Rogers & Karrer, 1985). In the wild, the frequencies of the seven mating types tend to be equal and conjugation seems to occur frequently (Doerder *et al*., 1995).

It is virtually certain that *T. thermophila* is an outbreeder in nature, i.e. with little or no mating among close relatives. Nevertheless, inbred strains have been successfully developed in the laboratory. Unlike Paramecium, there is no programmed somatic (vegetative) senescence and death:

Tetrahymena cell lines are in practice immortal. However, the MIC of a cell line tends to lose chromosomes at a variable and unpredictable rate (germinal senescence). The lack of MIC gene expression prevents direct selection against such loss, just as in multicellular organisms possessing differentiated germline and soma. A useful (and still up-to-date) discussion of these topics is contained in Nanney, 1980 (Chapter 9).

Tetrahymena (and most ciliates) are remarkably advanced among unicellular eukaryotes: no cell reproduction in the haplophase; differentiation of somatic and germline nuclei; binary fission in the face of highly specialized compound cell structures; conjugation in diplophase; and internal fertilization at conjugation. Given the enormous biological versatility of ciliates, it is puzzling why this evolutionary group includes no known multicellular forms.

2. Micronuclear (germline) genetics

(i) Mendelian genetics

The nuclear events of conjugation (**Figure 2**) have two noteworthy genetic consequences:

1. they generate Mendelian processes, as the diploid MIC undergoes meiosis and the zygote nuclei (from which progeny MIC and MAC are derived) are formed by the fusion of two haploid gamete nuclei; and
2. the MIC and MAC of a caryonide start out genetically identical to one another – normal progeny are thus said to be homokaryons.

The differentiated MICs and MACs of the entire set of four caryonides derived from a given conjugating pair all have genetically identical diploid progenitors and, thus, the four caryonides are expected to have identical phenotypes, no matter how different the parental cells were. It follows that pairs – and not exconjugants or caryonides – are the units to be counted in determining phenotypic ratios among the progeny of a cross.

There are several useful and well-characterized variants of the normal conjugation pathway. Under some circumstances, conjugation is aborted and the differentiation of a new MAC does not occur, the old MAC being retained instead. *MAC* retention can be readily identified (and excluded from progeny phenotypic ratio calculations) because the exconjugants remain sexually mature and continue to express parental phenotypes, including mating type. MAC retention is one way to generate heterokaryons, i.e. cells with genetically different MIC and MAC. Heterokaryons are useful for:

- positively selecting true progeny of conjugation in mass culture (e.g. drug-resistant MIC and drug-sensitive MAC);
- perpetuating lethal genotypes in the MIC (e.g. nullisomy, explained below); and
- sorting out the contributions of old and new MACs toward the execution of early developmental events.

With low frequency, three cells can conjugate with one another in a mixture of cells of two mating types. Triplet conjugation usually generates either a set of three diploid exconjugants or a haploid–diploid–triploid set, depending on how well the three junctions are developed (see Nanney, 1980). The first well-characterized nullisomic strains, i.e. strains lacking one or more pairs of chromosomes in the MIC, were obtained after meiosis of a haploid MIC within cells derived from a conjugating triplet.

Other useful and well-characterized conjugation variants that occur spontaneously and can be induced at will are genomic exclusion, cytogamy (i.e. self-fertilization), uniparental cytogamy and pronuclear fusion failure (Bruns, 1986; Orias, 1986; Cole & Bruns, 1992). Some of these variant conjugation pathways generate cells that are homozygous for their entire genome, either in the MIC only (genomic exclusion) or in both MICs and MACs (cytogamy and fusion failure). Cytogamy generates homokaryons, while fusion failure generates both hetero- and homokaryons. Cytogamy is used to isolate laboratory-induced recessive mutants efficiently. Pronuclear fusion failure is used for the generation of strains that are homozygous yet simultaneously heterokaryons for several loci.

(ii) Mapping genetically the MIC genome

Genetic differences in Tetrahymena are mapped to chromosome arms by using nullisomic strains. These are strains lacking both copies of a chromosome or chromosome arm in their MIC., but which survive because they are heterokaryons with a normal MAC. When a diploid strain homozygous for a recessive mutation is crossed to a nullisomic strain, the resulting monosomic progeny express the recessive genotype if the chromosome carrying the locus in question is absent from the nullisomic parent (illustrated in Brickner *et al.* (1996) for mapping randomly amplified polymorphic DNA (RAPD) polymorphisms). By doing parallel crosses of the mutant to an appropriate set of nullisomics, the locus can be assigned to a chromosome arm. Bruns and Cassidy-Hanley (1993) and the Tetrahymena Genome webpage (Orias, 1996) list the chromosome arm assignments of many conventional loci, cloned DNA sequences and DNA polymorphisms. Parenthetically, the newly differentiated monosomic MAC starts out in a state of major gene dosage unbalance. While this condition would be lethal in most eukaryotes, Tetrahymena cells monosomic even for several chromosomes survive to be useful and reach normal growth rate because of mechanisms that regulate the copy number of MAC pieces during asexual multiplication.

As in other eukaryotes, genetic linkage maps of loci in individual chromosomes are constructed by exploiting the fact that meiotic crossing-over generates recombinant genotypes with a frequency which, in the linear portion of the range, is roughly proportional to physical distance (see Griffiths *et al.*, 1996, Chapter 5). Major progress has recently been made in constructing a solid framework for a genetic map of the Tetrahymena genome (Orias, 1996). This framework is currently based on the results of testing for linkage nearly 400 RAPD segments as well as some "classical" Tetrahymena genetic loci. A map of a segment of the left arm of chromosome 2 is shown in **Figure 6.**

3. Genetics of the macronucleus

(i) DNA rearrangements during macronuclear differentiation

Several types of developmentally programmed DNA rearrangement occur during MAC differentiation (**Figure 3**) (reviewed in Prescott, 1994). One type is the site-specific fragmentation of the five MIC chromosomes into 200–300 subchromosomal molecular DNA pieces, here called MAC ARPs (autonomously replicating pieces) – they have also been called minichromosomes. The average size of these pieces is roughly 700 kilobases (kb), and they range in size from a few hundred to a few thousand kb; thus, each ARP must contain many genes. The rDNA ARP (described later) is exceptionally small (21 kb) and contains only the gene for the rRNA 45S precursor as an inverted repeat.

The *Cbs* (chromosome breakage sequence) is a unique DNA 15-mer (5′ AAAGAGGTTGGTTTA3′ in one strand), necessary and sufficient for chromosome fragmentation during MAC differentiation. Telomeric repeats (GGGGTT at the 3′ end) are added by telomerase at the ends of these pieces, which are then amplified to the 45-ploid level; the rDNA, with 10 000 copies per MAC, is the only known exception. The Tetrahymena telomerase is a reverse transcriptase that includes the RNA template for telomeric sequence synthesis as an integral component of the enzyme. Telomeres are hundreds of base pairs (bp) long, and the number of GGGGTT/CCCCAA repeats varies from telomere to telomere. Tetrahymena cells with certain mutant telomere sequences senesce and die. The very high copy number of the rDNA telomere, and the intensive telomere synthesis occurring during MAC differentiation, made Tetrahymena a favourable system for the discovery of the molecular basis of eukaryotic telomeres, and of telomerase, and for their continued investigation. The regulation of telomere length has important implications for carcinogenesis in mammalian cells (see Blackburn & Greider (1995) for a comprehensive review of telomeres).

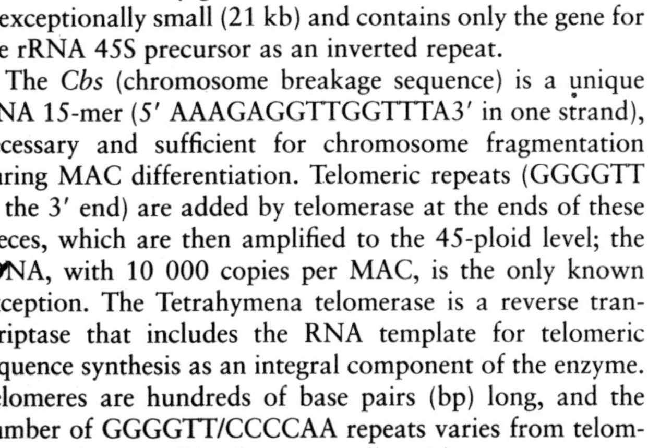

Figure 3 Types of developmentally programmed DNA rearrangements occurring during MAC differentiation. Top line: arbitrary segment of MIC DNA. Thick bar: MAC-destined DNA segments. Open segments: internally deleted (MIC-limited) DNA sequences. Thin connector: DNA sequences (including Cbs site – vertical arrow) lost during chromosome fragmentation. Second line: MAC ARPs derived from MIC chromosome segment. Wavy line: *de novo* added telomeric repeats. Dashed lines: boundaries of MIC-limited sequences. Bottom lines: result of amplification in newly differentiated MAC. MAC ARPs (averaging ~700 kb) are not drawn at the same scale as MIC-limited sequences (generally in the order of 1 kb) or telomeric repeats (hundreds of bp).

A seemingly unrelated type of site-specific rearrangement involves the deletion of internal DNA segments from the MAC during its differentiation. These are known as MIC-limited sequences or internally eliminated sequences (reviewed in Yao, 1996). There are roughly 6000 different deletion sites per haploid genome, at least a quarter of which are estimated to have alternative deletion ends. The alternative ends can be accurately specified or can loosely fall within domains of the order of 100 bp. The deletion size generally ranges from a few hundred to a few thousand bp. Some imprecision may be tolerated, because none of the deletion systems so far characterized in Tetrahymena has ends within protein-coding sequences. In total, roughly 15% of the MIC-derived DNA sequences are lost during MAC differentiation. MIC-limited sequences are AT-rich DNA. So far, they have no known function and **are proposed to be remnants of ancient invasions and dispersions of transposons [reference?].**

(ii) Phenotypic assortment

When cells with a MAC initially heterozygous at a given locus undergo asexual multiplication, subclones that irreversibly express phenotypes associated with either homozygote are generated. Since recessive phenotypes come to expression in this way, the phenomenon was termed phenotypic assortment. Starting with a heterozygous cell with a mixed MAC, the steady state rate at which subclones pure for either allele arise is 0.011/fission. Assortment is attributed to the random distribution of allelic copies in a compound MAC (**Figure 4**; see also Nanney, 1980). Mathematically, the steady state rate of assortment of pure MACs from mixed MACs approaches $\frac{1}{2}N$-1 per fission for a large N, where N is the number of copies just after MAC division (G_1 stage). The measured rate led to the first determination of 45 as the average G_1 ploidy of the MAC. This ploidy was subsequently confirmed by molecular measurements of MIC and MAC DNA amount and sequence complexity. With increased knowledge about the molecular structure of the MAC, phenotypic assortment is now attributed to the random distribution of the acentromeric ARP copies at MAC division (**Figure 4**). Phenotypic assortment has some similarities to the segregation of multicopy incompatible plasmids in bacterial clones.

Phenotypic assortment allows any recessive allele to come to full expression. It also allows even a single mutant allele, generated by mutation or transformation, to replace completely the 45 wild-type alleles in the MAC. Independent assortment, in combination with site-specific fragmentation, ensures extensive hereditary phenotypic diversification of the members of a vegetative clone, a situation commonly expected for Tetrahymena in the wild, given the multiple heterozygosis predicted in an outbreeding species. Phenotypic assortment thus provides a second, somatic, shorter-term level of natural selection and adaptation for the species.

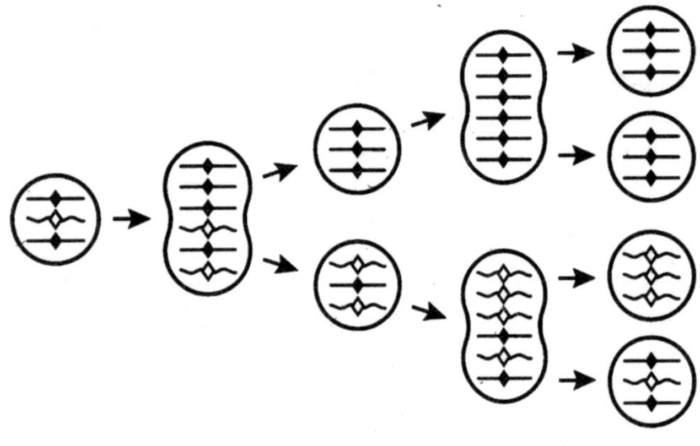

Figure 4 Phenotypic assortment. Only three of the 45 ARP copies are shown for simplicity. Circles, G_1 MACs; "peanut-shells", amitotically dividing MACs. Each ARP copy has been replicated and its two copies have equal probability of going to the same or to different daughter MAC. Straight lines and solid loci symbols (diamonds), DNA from one parental cell; wavy lines and open symbols, DNA from the other parental cell. The diagram illustrates how a cell with a mixed MAC generates, through random distribution at successive MAC divisions, vegetative descendants that are pure for one allele or the other (phenotypic assortment). If one allele is dominant (e.g. solid diamonds), the initial cell expresses the dominant phenotype; later assortants pure for open diamonds, which express the recessive phenotype, arise. Note: (i) At the first MAC division, the two wavy copies could, with equal probability, have been segregated to the two daughter MACs. Thus, the rate of assortment of pure MACs with three ARP copies is 0.25 per fission. With 45 ARP copies, the steady state rate assortment of pure MACs is merely slower – 0.011 per fission. (ii) While each ARP copy is shown here to double exactly at each S period and the daughter MACs are shown here to receive exactly the same number of copies, such precision is probably rare *in vivo* in Tetrahymena MACs with 45 copies at G_1.

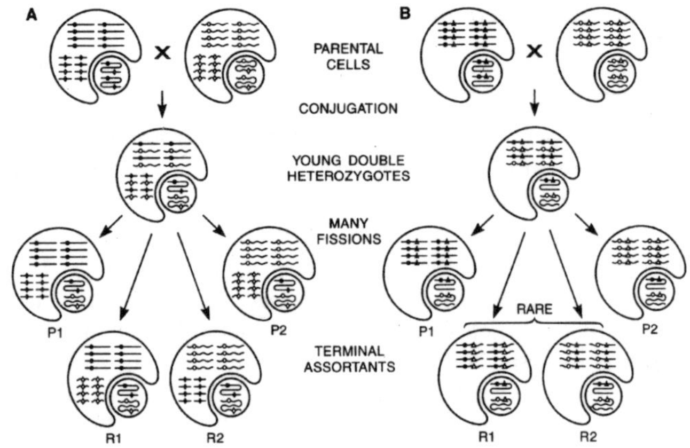

Figure 5 Macronuclear coassortment of two loci in the course of asexual multiplication. (A) *Independent assortment* of loci on two separate ARPs. (B) Coassortment of loci on same ARP. Circular figure, MIC; "helmet"-shaped figure, MAC. Two different MAC ARPs (long and short) and three different loci (circles, triangles and diamonds) are shown. Only eight of the 45 copies of each ARP are shown. The short ARP is omitted in panel B, as it is not needed for illustrating coassortment. Straight lines and solid loci symbols, DNA from one parental cell; wavy lines and open symbols, DNA from the other parental cell. Both parental cells are double homozygotes. P1 and P2, terminal assortants with parental genotypes; R1 and R2, terminal assortants with recombinant genotypes. Note that: (i) Phenotypic assortment generated terminal assortants with MACs that are pure at each locus. (ii) Coassortment between two loci is defined by a strong statistical excess of parental over recombinant types, not seen for independently assorting loci. (iii) Loci on different MIC chromosomes (not shown here) are expected always to assort independently. (iv) There is no assortment in the mitotically dividing MIC, which remains doubly heterozygous.

(iii) Coassortment

When doubly heterozygous cell lines are independently cultured asexually for hundreds of fissions, they exhibit virtually identical assortment patterns for certain pairwise combinations of neighbouring loci (Longcor *et al.*, 1996). In other words, starting with an *AB*/*ab* MAC, most of the vegetative descendants end up pure for either the *AB* or the *ab* (parental) combinations (**Figure 5**); less than 10% of the descendants become pure for either of the recombinant types (*Ab* or *aB*). This phenomenon is termed coassortment and allows the identification of coassortment groups, i.e. groups of loci such that its members all coassort with one another. A coassortment group is the MAC analogue of a MIC meiotic linkage group. ARPs are the physical basis of coassortment groups; i.e. loci carried on the same ARP coassort, while loci carried on different ARPs assort independently. Coassortment makes it possible to map loci to MAC ARPs and to determine ARP boundaries in the MAC purely by genetic means. Finding the physical ARP that carries a conventional locus can now be done indirectly by detecting its genetic coassortment with a physically mapped DNA polymorphism. These features may in the future facilitate the cloning of novel mutant genes in Tetrahymena that can only be cloned by complementation or chromosome walking.

(iv) Genetic and molecular mapping of the MAC genome

Genetic mapping of the MAC on the basis of coassortment is now underway. Coassortment groups currently identified are listed on the Tetrahymena Genome webpage (Orias, 1996). The most carefully mapped coassortment groups so far represent contiguous segments of MIC DNA. A few cloned DNA segments have been physically assigned to ARPs by hybridizing labelled probes to Southern blots of whole cell DNA separated on agarose gels by pulsed field electrophoresis (illustrated in Longcor *et al.*, 1996).

MIC-limited polymorphic DNA sequences are expected to show no assortment during vegetative multiplication because the signal originates in the MIC, which divides mitotically and remains heterozygous. This lack of assortment has been used to detect potential MIC-limited RAPD polymorphisms. Their status has been confirmed by PCR or Southern blot analysis: the signal disappears when whole cell DNA from a nullisomic strain missing the chromosome that carries the MIC-limited sequence is tested (Orias' laboratory, unpublished observations). A list

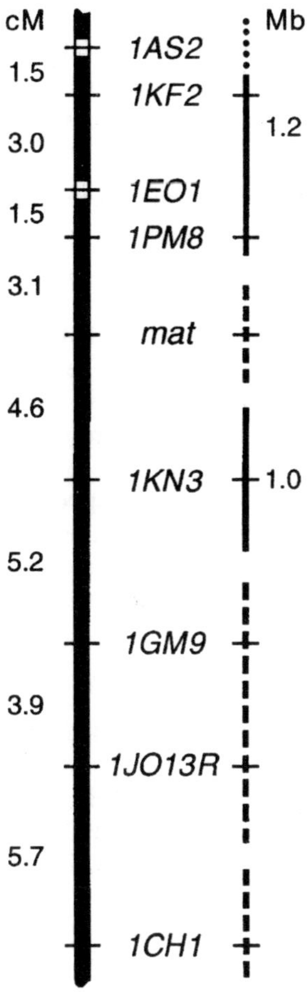

Figure 6 MIC/MAC maps of a segment of the left arm of chromosome 2. Thick bar and open segments, MIC chromosome segment and MIC-limited sequences, respectively; tick marks, genetic loci with distances shown in centiMorgans (cM). Solid thin bars, MAC coassortment groups physically related to ARPs, whose size is shown in Megabase (Mb) pairs; possible extension of an ARP is shown with dotted lines. Dashed thin bars, coassortment groups not yet related to physical ARPs. Two loci in the same MAC coassortment group coassort with one another, while two loci on different coassortment groups assort independently of one another.

of MIC-limited RAPDs and their MIC map location is given in Orias (1996). The map in **Figure 6** illustrates the relationship between MAC coassortment groups, MIC-limited sequences and MIC linkage groups in a segment of the left arm of chromosome 2L (Orias' laboratory, unpublished observations).

4. Special topics of MAC genetics

(i) Sexual maturity

Sexual maturation is a stable, somatically inherited differentiation of the MAC, remarkable in that it must occur in response to some endogenous fission counting programme, as these are free-living unicells. The differentiation of new MACs under conditions of physiological stress induces early maturity.

(ii) Mating-type differentiation

The MIC carries the potential for five to seven mating types, depending on the genotype at the mating type (*mat*) locus. Nevertheless, a cell generally expresses only one mating type, which is the result of a somatically inherited, irreversible, stochastic event that occurs in the differentiating new MAC. The frequencies with which various mating types arise are not necessarily equal and are affected by the environmental conditions prevailing at the time of MAC differentiation, such as temperature and starvation (see Orias, 1981).

Normally, about 50% of newly differentiated MACs are pure for determinants for a single mating type. The rest of the new MACs are mixed, but, as a result of phenotypic assortment, most MACs have become pure for a single mating-type determinant by the time sexual maturity is reached. Sexually mature cells with MACs that remain mixed give rise to clones within which pairing can occur – these are called selfers. Continued asexual multiplication of selfers generates descendants with pure MACs. The genetics of selfers was the unusual context within which phenotypic assortment was first described and analysed in Tetrahymena. A model of mating type determination based on an alternative deletion system has been proposed (Orias, 1981), which explains quantitatively the genetic phenomenology of mating-type determination and predicts the order of mating type genes within the *mat* locus. This model, however, has yet to be tested molecularly.

(iii) Genetics of the ribosomal RNA gene (rDNA)

Tetrahymena is an exceptional organism in having only a single copy (per haploid genome) of the gene for the 45S rRNA precursor, which is processed post-transcriptionally to the 18S and 28S rRNAs. During MAC differentiation, this copy is excised by *Cbs*-directed fragmentation, turned into a 21 kb inverted repeat (palindrome), supplied with telomeres and amplified to the level of approximately 10 000 copies per MAC (illustrated in Kapler, 1993). All but the central 29 bp of the rDNA is palindromic and the two halves of the palindrome are transcribed divergently. Each half consists of a central transcribed region, flanked by the 5′ and the 3′ non-transcribed spacers (NTS); the 5′-NTS contains the origin of rDNA replication. Due to its high copy number, the rDNA telomere represents roughly half of the telomeres in the MAC. Tetrahymena rDNA telomeres are built into the vector used to clone large DNA inserts in Saccharomyces as yeast artificial chromosomes (YACs). The 28S rRNA coding sequence includes an intron and understanding of its self-splicing nature led to the codiscovery of ribozymes (Kruger *et al.*, 1982). Paromomycin resistance is a useful selectable trait, determined by a mutation located near the 3′ end of the 16S rRNA coding region. Most current Tetrahymena transformation vectors incorporate the rDNA origin of replication, which is required for their autonomous maintenance in transformants.

5. Other non-Mendelian inheritance

(i) Mitochondrial genetics

Tetrahymena cells contain 600–800 mitochondria (see Hill, 1972, p. 75) whose DNA is a linear molecule, an uncommon occurrence among eukaryotic mitochondria (but also found in Paramecium). The mitocondria's telomeres consist of a variable number of repeats of a 53 bp unique sequence. Chloramphenicol resistance, determined by a mutation virtually certain to reside in the mitochondria, shows cytoplasmic inheritance (see Bleyman, 1996), i.e. the exconjugant clone derived from a resistant conjugant remains resistant, while the exconjugant clone derived from the sensitive parent remains sensitive. The genetic evidence suggests that, in Tetrahymena, the exchange of mitochondria across the conjugal junction occurs rarely, if ever.

(ii) 10 kb plasmid

A linear 10 kb plasmid (*pTtL10*), present in roughly 20 000 copies per cell, has been described in certain wild isolates of *T. thermophila* (Ortiz, 1993). Its ends are terminal inverted repeats of a 47 bp unique sequence. This organization – but not the repeat sequence itself – is reminiscent of the Tetrahymena mitochondrial DNA telomeres. Its 5′ (but not 3′) ends are resistant to exonuclease digestion. This plasmid is suspected of having protein-primed (rather than RNA-primed) DNA replication, such as found in adenovirus, for example. When plasmid-free cells conjugate with plasmid-containing cells, the plasmid is transferred to the plasmid-free conjugant in some but not all the pairs. No plasmid transfer has been detected in mixed vegetative culture. The cellular location of the plasmid DNA may well be mitochondrial. It has been proposed that, in the course of evolution, the Tetrahymena mitochondrial DNA became linear as a consequence of recombination with a *pTtL10*-type plasmid.

6. On the evolution of nuclear dimorphism

The nuclear dimorphism of the ciliates is a remarkable phenomenon. It staggers the imagination to consider how amitosis of an acentromeric, fragmented, somatic genome could have evolved from mitosis of a diploid nucleus through a series of functional intermediates. The following steps have been proposed for the evolution of ciliate MACs, based on observations of the life cycles, cell biology and molecular biology of extant ciliates and other unicellular eukaryotes (Orias, 1991a,b).

- A differentiated, division-inhibited somatic MAC could have evolved in a unicellular eukaryote in the context of a diplophase stage consisting exclusively of a division-inhibited trophic (feeding) stage, as observed in certain Foraminifera. MAC mitotic mechanisms in diplophase cells could have been lost, since there would be no selective pressure to maintain them.
- Cell division in the diplophase could later have been reacquired, through events akin to cancer-causing mutations in multicellular organisms. This would have generated division-competent cells with division-inhibited MACs, a unique feature of karyorelict ciliates. Ploidy increase and chromosome fragmentation could have evolved at this stage and the lack of centromeres on the fragments would carry no risk of chromosomal imbalance in division-inhibited MACs.
- Reacquisition of nuclear division in MACs with relatively high ploidy, as a superior alternative to MAC differentiation at every cell cycle, could have generated extant higher ciliates, e.g., Tetrahymena. The high ploidy of the MAC genome would have initially lessened the risks of chromosomal loss or imbalance associated with unequal distributions.

The validity of these evolutionary speculations remains to be challenged by future research.

Acknowledgements

The author is grateful to his many collaborators, who are the ones mainly responsible for the research contributions of his laboratory to the field of Tetrahymena genetics; to the National Institutes of Health (current grant RR09231), the National Science Foundation and the American Cancer Society, which over the years have supported the work of his laboratory; to Eileen Hamilton, Judy Orias and Steven Wickert, who critically read this manuscript; and to Dottie McLaren, who skilfully translated rough sketches into presentable figures.

REFERENCES

Blackburn, E.H. & Greider, C.W. (1995) *Telomeres*, Cold Spring Harbor, New York: Cold Spring Harbor Laboratory Press

Bleyman, L.K. (1996) Ciliate genetics. In *Ciliates: Cells as Organisms*, edited by K. Hausmann & P.C. Bradbury, Stuttgart: Gustav Fisher and Deerfield Beach, Florida: Chemie International

Brickner, J.H., Lynch, T.J., Zeilinger, D. & Orias, E. (1996) Identification, mapping and linkage analysis of randomly amplified DNA polymorphisms in *Tetrahymena thermophila*. *Genetics* 143: 811–21

Bruns, P.J. (1986) Genetic organization of *Tetrahymena*. In *Molecular Biology of the Ciliated Protozoa*, edited by J.G. Gall, New York: Academic Press

Bruns, P.J. & Cassidy-Hanley, D. (1993) Tetrahymena thermophila. In *Genetic Maps: Locus Maps of Complex Genomes*, 6th edition, edited by S.J. O'Brien, Cold Spring Harbor, New York: Cold Spring Harbor Laboratory Press

Cole, E.S. & Bruns, P.J. (1992) Uniparental cytogamy: a novel method for bringing micronuclear mutations of *Tetrahymena* into homozygous macronuclear expression with precocious sexual maturity. *Genetics* 132: 1017–031

Doerder, F.P.M., Gates, M.A., Eberhardt, F.P. & Arslanyolu, M. (1995) High frequency of sex and equal frequencies of mating types in natural populations of the ciliate *Tetrahymena thermophila. Proceedings of the National Academy of Sciences USA* 92: 8715–718

Elliott, A.M. (ed.) (1973) *Biology of Tetrahymena*, Stroudsburg, Philadelphia: Dowden, Hutchinson and Ross

Gall, J.G. (ed.) (1986) *Molecular Biology of the Ciliated Protozoa*, New York: Academic Press

Griffiths, A.J.F., Miller, J.H., Suzuki, D.T., Lewontin, R.C. & Gelbart, W.M. (1996) *An Introduction to Genetic Analysis*, 6th edition, New York: Freeman

Hill, D.L. (1972) *The Biochemistry and Physiology of Tetrahymena*, New York and London: Academic Press

Kapler, G.M. (1993) Developmentally regulated processing and replication of the *Tetrahymena* rDNA minichromosome. *Current Opinion in Genetics and Development* 3: 730–35

Kruger, K., Grabowski, P.J., Zaug, A.J. *et al.* (1982) Self-splicing RNA: autoexcision and autocyclization of the ribosomal RNA intervening sequence in *Tetrahymena. Cell* 31: 147–57

Longcor, M.A., Wickert, S.A., Chau, M.-F. & Orias, E. (1996) Coassortment of genetic loci during macronuclear division in *Tetrahymena thermophila. European Journal of Protistology* 32: 85–89

Nanney, D.L. (1980) *Experimental Ciliatology: An Introduction to Genetic and Developmental Analysis in Ciliates*, New York: Wiley

Nanney, D.L. & McCoy, J.W. (1976) Characterization of the species of the *Tetrahymena pyriformis* complex. *Transactions of the American Microscopical Society* 95: 664–82

Orias, E. (1981) Probable somatic DNA rearrangements in mating type determination in *Tetrahymena thermophila*: a review and a model. *Developmental Genetics* 2: 185–202

Orias, E. (1986) Ciliate conjugation. In *Molecular Biology of the Ciliated Protozoa*, edited by J.G. Gall, New York: Academic Press

Orias, E. (1991a) On the evolution of the karyorelict ciliate life cycle: heterophasic ciliates and the origin of ciliate binary fission. *BioSystems* 25: 67–73

Orias, E. (1991b) Evolution of amitosis of the ciliate macronucleus: gain of the capacity to divide. *Journal of Protozoology* 38: 217–21

Orias, E. (1996) Website: *Tetrahymena* Genome Project. http://lifesci.ucsb.edu/~genome/Tetrahymena

Ortiz, J.M. (1993) Characterization of an extrachromosomal linear DNA molecule discovered in *Tetrahymena*, Ph.D. thesis, Santa Barbara: University of California at Santa Barbara

Prescott, D.M. (1994) The DNA of ciliated protozoa. *Microbiology Reviews* 58: 233–67

Rogers, M.B. & Karrer, K.M. (1985) Adolescence in *T. thermophila. Proceedings of the National Academy of Sciences USA* 82: 436–39

Yao, M.-C. (1996) Programmed DNA deletions in *Tetrahymena*: mechanisms and implications. *Trends in Genetics* 12: 26–30

GLOSSARY

acentromeric lacking a centromere

amitosis rare type of nuclear division where the nucleus constricts without chromosome condensation or spindle formation

chromosome walking a technique that produces sets of overlapping DNA clones from one continuous segment of a chromosome that is too large to be cloned itself. The method begins at a known genetic marker and walks along the chromosome

complementation a genetic test to determine whether two mutations producing a similar phenotype are allelic

conjugation the temporary union of two bacteria that results in genetic exchange

cytogamy cell conjugation

diplophase stage in life history of an organism when nuclei are diploid

fission cleavage of cells

haplophase stage in the life history of an organism when nuclei are haploid

heterokaryon naturally formed cell containing two or more genetically different nuclei

homokaryon cell containing more than one genetically identical nucleus

kinetochore structure within the centromere of a chromosome to which spindle microtubules affix during meiosis or mitosis

macronucleus larger of the two types of nucleus found in cells of ciliate protozoans

micronucleus smaller of the two types of nucleus found in cells of ciliate protozoans

mitotically produced by mitosis

random amplified polymorphic DNA (RAPD) variant of the polymerase chain reaction (PCR) used to identify differentially expressed genes

ribozyme RNA molecule with catalytic capacity; made of nucleic acid, not protein

telomerase (telomere terminal transferase) enzyme involved in forming the telomeres

transformation any change in the properties of a cell stably inherited by its progeny

FUNGAL GENETICS

Jeff Bond
Institute of Cell and Molecular Biology, University of Edinburgh, Edinburgh, UK

1. Introduction

Fungi have made an important contribution to the understanding of many different aspects of heredity. Initially, their simple life cycle, short generation time and the fact that at least some species could be cultured on a simple, chemically defined growth medium made them attractive model organisms for studying genetics. These advantages led to important discoveries and helped to put in place foundation stones of genetics (a brief appreciation of some of these historical landmarks is given in section 2). In recent years, several new discoveries have ensured that this group of organisms has continued to contribute to increased understanding of basic genetics. Fungi are an important source of enzymes, metabolites and drugs, and the importance of this group of organisms is sure to grow. Ostergaard *et al.* (2000) have reviewed developments in the yeast *Saccharomyces cerevisiae*, while Gouka *et al.* (1977) have looked at the production of secreted proteins by the filamentous fungus Aspergillus.

In the early days of fungal genetics, the orange-coloured breadmould, *Neurospora crassa*, was used as a research tool to confirm and consolidate some of the basic tenets of genetics (for reviews see Perkins, 1992; Davis, 2000). Bernard Dodge – a key figure in the early development of Neurospora as a model organism – described the mating-type system (now known to be an important aspect of the life cycle of many fungi). He also discovered that the ascospores of the species were stimulated to germinate by exposure to high temperature and carried out the first tetrad analysis, a technique in which the products of individual meioses can be directly analysed (Shear & Dodge, 1927). Following the realization that the laws of inheritance applied to fungi just as to other organisms, other fungi were used to investigate a variety of basic genetic phenomena. Studies using ascomycete fungi, notably *Aspergillus nidulans*, the budding yeast *S. cerevisiae* and the fission yeast *Schizosaccharomyces pombe*, and using basidiomycetes, especially *Coprinus cinereus*, *Schizophylum commune* and *Ustilago maydis*, have made important contributions to understanding genetics. Some of the discoveries have been shown to apply almost universally while others are unique to fungi.

2. Biochemical genetics – the classical period and molecular analysis

Neurospora was used by Beadle and Tatum (1941) to investigate the relationship between genetics and biochemistry, their experiments leading to the idea that the information carried in the genes enabled organisms to make enzymes. This is known as the "one gene–one enzyme" hypothesis and led to the era of biochemical genetics in which the underlying metabolism of organisms was investigated by gene mutation. The idea that defects in the biochemistry and physiology of an organism can result from the mutation of genes is a very old one. At the turn of the 20th century, Garrod (1909) had studied the inherited disease alkaptonuria in humans and had concluded that the build-up of homogentisic acid in affected people came about through a block in the biochemistry of those affected. He wrote: "It is among the highly complex proteins that such specific differences are to be looked for". This conclusion was largely ignored and it was the systematic investigation of auxotrophic mutations in Neurospora and the bacterium *Escherichia coli* that led to the generalized rule and the award of the Nobel Prize to Beadle and Tatum in 1958.

The one gene–one enzyme hypothesis was not immediately accepted. There was some debate at the time about alternative explanations and whether or not the discovery of mutations resulting in specific enzyme deficiencies could be generalized to cover all (or even most) mutations. The demonstration that when auxotrophic mutants reverted to prototrophic growth the revertants were not always completely wild-type in their enzymatic properties led to the conclusion that the genes directly encoded the information for making the proteins. A systematic analysis of temperature-sensitive mutations in *N. crassa* (Horowitz & Leupold, 1951) showed that a high proportion of these conditional mutations had a simple growth requirement so that they would grow at the restrictive temperature following the addition of a chemical to the growth medium. This led to the conclusion that auxotrophic mutations were a representative sample of all mutations and that many genes did carry the information for encoding proteins. Some important qualifications were made to this general idea following the discovery of interactions between pathways. In Neurospora, for example, Davis (1967) and Reissig *et al.* (1967) showed that the phenotypic effect of mutation in the *arg-12* gene could be suppressed by a mutation in another gene (*pyr-3d*) and a mutation (*arg-12^S*) had a similar effect in suppressing the pyrimidine requirement conferred by *pyr-3a*. Similar studies led to the discovery of two genes for one enzyme and duplicate enzyme activities contributing to two pathways. As the knowledge of genetics increased, the position of the one gene–one enzyme hypothesis in genetics has been further modified. Transcribed genes need not be translated into proteins, and, even when they are, we now know that one gene can encode non-enzymatic proteins and also that one gene can carry the information for making more than one protein product (e.g. through the use of alternative splicing). The essence of the idea, however, remains intact.

Mutations that conferred specific growth requirements on fungi played an important part in working out the details of biochemical pathways, and knowledge of basic metabolic pathways increased enormously as a result. Perkins *et al.*

(1982) reviewed much of the work on specific pathways and the enzymes involved, while Fincham (1985) compiled an excellent historical overview of the development of fungal biochemical genetics from its origins through to the cloning era. The discovery by Fincham and Pateman (1957) that mutations within a single gene could sometimes complement each other (allelic complementation) led to advances in understanding about the structure of enzymes and the interaction of enzymatic subunits both *in vivo* and *in vitro*. It also became apparent, from the study of the ways in which some mutations interacted, that a metabolic structure existed within the fungal cell so that some enzymes were compartmentalized and that the products of the enzyme action were isolated within particular parts of the cell, such as the cell vacuole (Weiss, 1973). Unlinked genes which encode enzymes involved in the same pathway turned out to be coordinately controlled and this led to detailed studies into the control of gene expression and the identification of regulatory genes central to this control.

(i) Regulation of gene activity

Lang-Hinrichs (1995) and Jacobs and Stahl (1995) have, respectively, reviewed gene regulation in yeast and mycelial fungi. In the latter, there are differences in detail of the control mechanisms from both yeast and from other eukaryotes. Before the advent of cloning techniques, understanding was confined largely to deductions made from the study of mutants and their interactions. Repressor and loss of function mutations and epistatic interactions were used to work out the basic control mechanisms of several different pathways. The regulation of various aspects of nitrogen and carbon metabolism represents a very good example dating from this period (for review see Davis & Hynes, 1991). More recent detailed molecular analysis has resulted in the identification of the major structural elements of mRNA and of the mode of action of regulatory proteins associated with transcriptional control in both these and other pathways.

Analysis of fungal transcripts has revealed several general features. The 5′ non-coding region contains the main signals associated with gene regulation. Both *cis*-acting and enhancer elements have been reported for many fungal genes. Typically, they are relatively short, ranging from 6 to 20 nucleotides. The regulatory elements can themselves be present in more than one copy. These elements can be binding sites for regulatory proteins and transcription factors, some of which have been studied in detail. Genes associated with the same pathways may be clustered, as are, for example, the *qa* genes of *N. crassa* and the *spoC1* gene cluster of *A. nidulans*. Introns, when present, are typically small (50–100 base pairs [bp]). Their position is sometimes conserved between different species but can also differ; for example, analysis of glyceraldehyde-3-dehydrogenase in ascomycete and basidiomycete fungi has revealed that introns are conserved in position within the two groups but differ markedly between them.

Transcription factors can be classified into those that act generally on a wide range of genes and those that act specifically on a specific set. One of the best-studied examples is the transcriptional factor TFIID that binds to the TATA box and acts to activate transcription. Commonly, transcription factors have two domains: one involved in DNA binding and one involved in transcriptional activation. Some proteins control gene activity by acting negatively (repressing the genes) while others act positively (by activating the genes).

3. Tetrad analysis, chromosome rearrangements and the mechanism of recombination

In many fungi, the products of meiosis are found as tetrads (or octads) of spores, each being the product of a single meiotic division. Tetrad analysis remains the simplest, most direct demonstration that crossing-over of linked genes occurs at meiosis when the paired chromosomes in a bivalent consist of four chromatids. A single chiasma recombines two of the four chromatids. **Figure 1** presents an example of octads of spores from the fungus *Sordaria brevicollis*, showing asci in which recombination of two spore colour mutations has occurred to generate both wild-type and double mutant spores (see also **Plate 8** in the colour plate section). Importantly, the crossing-over has generated one pair of each recombinant, but there are also two pairs of non-recombinant spores. These spores contain the parental chromosomes that were not involved in the crossing-over event.

Tetrad analysis has also facilitated research into chromosome rearrangements. Meioses in which heterozygous chromosome rearrangements are segregating give patterns of spore abortion that allow detailed characterization of the various different rearrangements. In particular, an extensive set of rearrangements has been identified in Neurospora where the genetic analysis has often been supplemented by

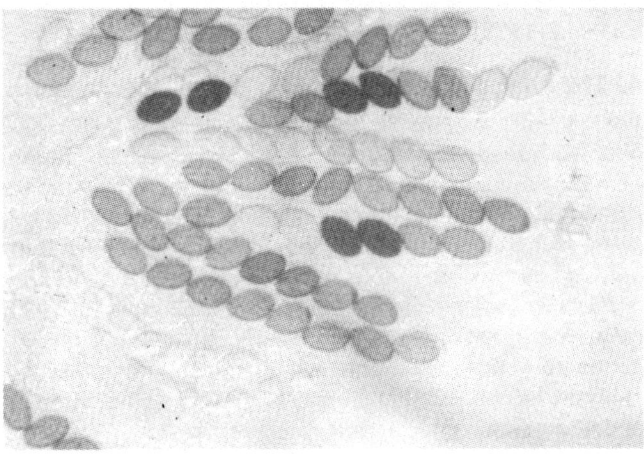

Figure 1 The photograph shows asci resulting from crossing buff and yellow spore colour mutant strains of the fungus *Sordaria brevicollis*. Some of the asci show the result of recombination between the genes; such an event results in a tetratype ascus. Three tetratypes are present, each containing a pair of black, white yellow and buff spores. In the other four asci no recombination has ocurred. These contain four buff and four yellow spores. See also **Plate 8**.

confirming cytogenetic analysis (for review see Perkins, 1997).

The utility of fungi in enabling the analysis of recombination, both in *N. crassa* and in other fungi, especially yeast, led to the discovery of gene conversion and postmeiotic segregation and detailed investigation into the mechanism of recombination. In a classic paper, Mitchell (1955) demonstrated gene conversion, resulting in aberrant non-Mendelian segregation, at a pyridoxine locus in *N. crassa*. Importantly, she showed, by using closely linked flanking markers that gave normal segregation, that the aberrant event was confined to a small chromosomal region. This eliminated other explanations for the aberrant segregation and convinced geneticists of the reality of gene conversion. When gene conversion occurs, information on one chromosome is transferred to the homologous chromosome in a non-reciprocal manner. Mitchell's experiments, and others like them, led to the development of models of recombination and a deeper understanding of the mechanics underlying the process by which chromosomes break and rejoin at meiosis. Following the initial proof of gene conversion in *N. crassa*, experiments on the mechanism of recombination were continued and extended in yeast, which is now the best-studied eukaryote for detailed understanding of both the mechanism and enzymology of recombination. Studies on gene conversion and the associated phenomenon of postmeiotic segregation led to the development of hybrid DNA models of recombination (Holliday, 1964) and, later, to the double-strand break model (Szostak *et al.*, 1983). Early reviews of the mechanism of gene conversion concentrated on the explanation for this phenomenon, but, as a more detailed understanding of the enzymes involved in recombination emerged, the emphasis became increasingly more molecularly slanted. An introductory article on "Genetic recombination" by Leach can be found in this volume; more comprehensive reviews of recombination have been written by Stahl (1979), Whitehouse (1982) and Leach (1996).

4. The analysis of the cell cycle

Fission yeast, *S. pombe*, and budding yeast, *S. cerevisiae*, have both made major contributions to our understanding of how the various component parts of the cell cycle are regulated. The different events that take place during the cell cycle, such as DNA synthesis, assembly of cell division spindles and mitosis itself, need to be coordinated in a carefully orchestrated time-scale. In both species, mutants in which these events are disturbed are readily isolated and investigation into how these interact, together with molecular analysis, has provided great insight into the genetic control of cell division.

(i) Fission yeast

Fission yeast is a unicellular organism in which the cells grow by extension at the apex. In doing so, the cells increase in length from 7 to 14 μm, then divide to become 7 μm long again. Mutations which disrupt the control of events in the cell cycle can result in cells that are of a different size (either longer or shorter than normal) or cells which are unable to divide at all. The latter can be analysed if the mutations result in a conditional change. Classically, temperature-sensitive mutations which are mutant only at a particular temperature have been used in this context. Mutant cells which show arrest of the cell cycle are known as cell division cycle (cdc) mutants. They are typically longer than normal because cell metabolism, including growth, goes on while cell division is arrested. In contrast, cells in which division is uninhibited will be shorter than normal. Control of the cell cycle in fission yeast has been reviewed by MacNeill and Fantes (1995) where three genes have a central role in the process, as discussed below.

cdc2⁺. This gene encodes a protein that puts phosphate groups onto other proteins; such a protein is known as a kinase. Activity of the cdc2-encoded protein kinase is required at two points in the cell cycle and this activity is controlled in two ways. The activity is dependent on interaction with another cellular component – cyclin B encoded by the gene *cdc13*⁺. The cdc2 protein is, therefore, known as a cyclin-dependent kinase. Interaction between the kinase and cyclin can only occur if the kinase is itself phosphorylated on a threonine residue at position 167. The activity of the kinase–cyclin complex (there are, in fact, more components to the complex than the two named above) is regulated by the phosphorylation and dephosphorylation of the cyclin-dependent kinase on the tyrosine residue at position 15. Dephosphorylation results in activation of the kinase–cyclin complex. The dephosphorylase that brings this about is encoded by another cdc gene – *cdc25*⁺.

cdc25⁺. The activity of the *cdc2* gene product is a balance between dephosphorylation of tyrosine 15 (under the control of *cdc25*) and phosphorylation of the same residue (which inactivates the gene product). This phosphorylation step is under the control of another kinase encoded by the gene *wee1*⁺.

wee1⁺. This gene gets its name through the effect of inactivating mutations on the size of the cell. When inactivated, the gene product no longer phosphorylates the tyrosine 15 residue of the cyclin-dependent kinase; the kinase is, therefore, active and the cell divides more often than it should. This results in a small, or "wee", cell. Mutations that result in overproduction of the *wee1* gene product, on the other hand, inhibit entry into cell division and the cells are consequently longer than normal.

(ii) Budding yeast

Budding yeast has, on first sight, a very different life cycle. Although both species are yeasts, they are, in evolutionary terms, remotely related, being as divergent from each other as they are from vertebrates. When comparisons are made between their cell cycles, however, there are distinctly related genes and gene products associated with the control of the cell cycle and this is a powerful indicator of their importance. The same gene products can also be identified in higher eukaryotes, where they perform similar functions. This emphasizes that the control of the cell cycle has been conserved in evolution over millions of years. The two yeasts are important model organisms in the study of the control of cell division because such findings have important applications in the understanding of cancer.

As in fission yeast, investigation of cell cycle control in budding yeast started with the isolation of mutant genes whose functions are required for the successful completion of cell division (for a review of cell cycle regulation in both yeasts see Forsburg & Nurse, 1991 and MacNeill, 1994).

(iii) Signal transduction pathways

The kinase signalling pathway by which yeast cells monitor and regulate their cell division is just one of several signal transduction pathways that have been studied in fungi. Response to a variety of environmental signals is transduced to the nucleus where it results in an appropriate change in gene activity. Mitogen-activated protein kinases (MAP kinases) play a central role in this process. Six different MAP kinases have been identified in *S. cerevisiae* and these have been implicated in responses to particular signals. One of these kinases plays a role in the mating of yeast cells where cells respond by mating when the appropriate mating pheromone is present nearby.

5. Mating-type systems in fungi

It is not surprising, given the nature of fungal life styles, that recognition systems (whereby members of a species recognize each other and exchange genetic material) play an important part in the life cycle. Fungi are non-photosynthetic and rely for energy on utilizing pre-existing organic material. Fungi invade and digest a variety of substrates and, in so doing, often encounter a variety of competitor species. Mechanisms have evolved so that an individual fungal colony can distinguish members of the same species from competitors for the same substrate. The principal and most obvious component of recognition at the organismal level is the mating-type system that allows compatible individuals to reproduce sexually. Some fungal species have an obvious mating-type system whereby mating occurs only between individuals of different mating types; such species are said to be heterothallic. In contrast, other species have no obvious mating type; single spore isolates are self-fertile and can produce fruiting bodies and spores via meiosis without mating with another fungus; such species are said to be homothallic.

The mating-type system in *N. crassa* was the first to be recognized. Shear and Dodge (1927) described the mating-type system in which two alternative mating types (now called *mat A* and *mat a*) exist. Sexual reproduction resulting in meiosis and the production of ascospores is only possible between two cultures of opposite (i.e. different) mating types. The existence of clear genetic segregation in heterothallic species permitted the analysis of the mating-type genes in these fungi, but, in most species, the underlying basis of homothallism remained mysterious until it was characterized by cloning. The cloned mating-type genes from heterothallic species could then be used to probe the molecular basis of homothallism. The mating-type system of the budding yeast *S. cerevisiae* was the first to be characterized in detail at the molecular level, but, in this species, classical genetic analysis had given some insight into the nature of homothallism before the genes were cloned. The cloning of the mating-type genes of other fungi has subsequently shown that considerable variation exists between different species in the details of the mating-type mechanism (for reviews see Glass & Lorimer, 1991 and Casselton & Olesnicky, 1998).

(i) The mating-type system in ascomycete fungi

(A) Saccharomyces cerevisiae

The life cycle of yeast is, in outline, a simple one. Haploid cells of opposite mating types signal their presence to each other through pheromones and pheromone receptors located on the cell surface. In the presence of pheromone produced by the opposite mating type, the morphology of the cells changes to become pear-shaped (called "schmoos" after a strip-cartoon character). These cells are arrested at late G_1 phase of the cell cycle and initiate transcription of genes required for cell fusion. The diploid cells formed after fusion of cells and nuclei can divide mitotically, but meiosis can be induced by suitable nutritional signals. Meiosis results in the production of four haploid ascospores, the germination of which completes the life cycle.

Classical genetic studies which preceded molecular cloning revealed unusual features of the mating-type system of *S. cerevisiae*. Some strains were heterothallic, possessing one of two mating types, *MAT*α or *MAT***a**, but others were homothallic and were self-fertile. Crosses between a homothallic and a heterothallic strain revealed that the difference segregated at meiosis in a 1:1 ratio. This led to the conclusion that the difference between homothallic and heterothallic strains was caused by the mutation of a gene to give two alternative alleles called *HO* (conferring homothallism) and *ho* (conferring heterothallism). An instructive result was obtained by Oeser (1962) when an *HO* strain was crossed with both an *ho MAT*α and an *ho MAT***a** strain. As expected, homothallic and heterothallic strains could be recovered, reflecting the segregation of *HO vs. ho*. However, both crosses gave rise to heterothallic strains of both mating types as summarized in the diagram below:

Cross		**Progeny**
HO × *ho MAT*α	⇒	*HO* (homothalic) and *ho MAT*α and *ho MAT***a** (heterothalic)
HO × *ho MAT***a**	⇒	*HO* (homothalic) and *ho MAT*α and *ho MAT***a** (heterothalic)

It seems clear that the *HO* strain contains both the *MAT*α and the *MAT***a** alleles. Pedigree analysis by Strathern and Herskowitz (1979) demonstrated that haploid *HO* cells do indeed possess both mating-type alleles and can switch from *MAT*α *to MAT***a** and vice versa.

The cloning and characterization of the mating-type genes of yeast at the molecular level revealed several unusual features of mating-type organization and also confirmed some of the ideas that emerged from the genetic analysis.

Mating-type switching and the cassette model

The switching of mating type in yeast has been very extensively studied and reviewed (Herskowitz *et al.*, 1992; Schmidt & Gutz, 1994). There are three loci containing mating-type information, all located on chromosome III but not very closely linked. Only the copy at the *MAT* locus is expressed; the two other copies, at *HML* and *HMR*, are silent. Homothallic strains of yeast have a functional *HO* gene that encodes a site-specific endonuclease that facilitates the movement of a copy of the mating-type gene from silent repositories or cassettes. The endonuclease makes a double-strand cut at the *MAT* locus and this ultimately results in the copying of information from one or other of the silent cassettes into the *MAT* locus via gene conversion. **Figure 2** illustrates the relationship between the three loci and also summarizes the movement of copies of the mating-type genes from one position to another. The silencing of the left and right cassettes depends both on the function of other genes and on regions of DNA flanking the silent cassettes.

Molecular organization and functional analysis of the mating-type genes

The *MAT*α and the *MAT*a loci contain sequences that are completely dissimilar to each other. Unlike alleles, which have evolved via mutation one from the other, these sequences are unlikely to be related to each other through a common ancestor. Metzenberg and Glass (1990) introduced the term "idiomorph" to denote unrelated sequences, such as *MAT*α and *MAT*a, which occupy the same locus on a chromosome. The *MAT*α idiomorph is composed of 747 bp; the *MAT*a idiomorph has 642 bp of unique sequence. Other sequences are found surrounding the idiomorphic sequences: X (704 bp) and Z1 (239 bp) flank both the *MAT* locus and the loci containing the silent cassettes. Two further sequences (called W and Z2) are found at both the *MAT* and *HML* loci, but not at the *HMR* locus. Details of the organization are shown in **Figure 3**.

The two mating types act as master switches, determining the activity of other genes involved in mating, meiosis and ascospore formation. Two divergent transcripts are produced by both the *MAT*α and the *MAT*a sequences. The region of the *MAT* locus that encodes these transcripts is

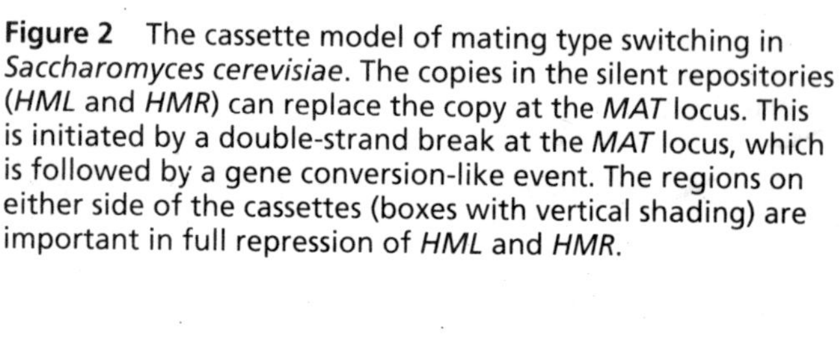

Figure 2 The cassette model of mating type switching in *Saccharomyces cerevisiae*. The copies in the silent repositories (*HML* and *HMR*) can replace the copy at the *MAT* locus. This is initiated by a double-strand break at the *MAT* locus, which is followed by a gene conversion-like event. The regions on either side of the cassettes (boxes with vertical shading) are important in full repression of *HML* and *HMR*.

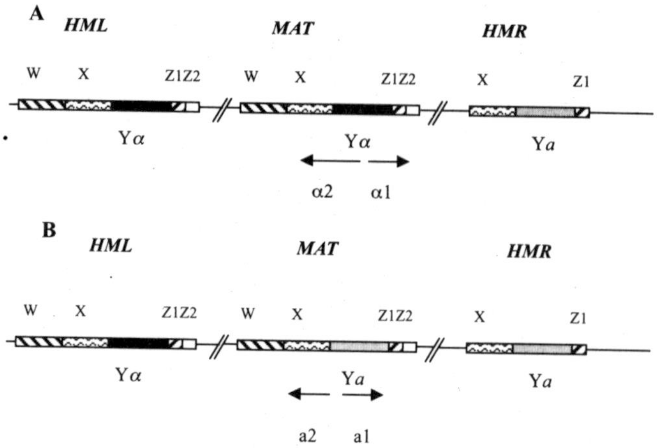

Figure 3 Organization of mating type in *Saccharomyces cerevisiae*. X and Z1 are regions of homology common to all three loci (*HML, MAT* and *HMR*). W and Z2 are homologies that are missing from *HMR*. (A) The structure of *MAT*α. Two divergently transcribed genes α*1* and α*2* are expressed from the *MAT*α locus. (B) The structure of *MAT***a**. Two divergently transcribed genes a*1* and a*2* are expressed from the *MAT***a** locus.

depicted in **Figure 3**; the presence of the two transcripts in the α mating type was inferred earlier. Strathern *et al.* (1981) put forward the α*1*–α*2* hypothesis following elegant genetic experiments which extended earlier studies of MacKay and Manney (1974a,b). Tatchell *et al.* (1981) obtained confirmation of this idea in experiments in which mutations were introduced *in vitro* into the mating-type transcripts. These mutations were then reintroduced into the yeast cell by transformation. The α*1* transcript encodes an activator of α-specific functions, while α2 encodes a repressor protein which acts to switch off *MAT*a-specific functions. A *MAT*a cell transcribes and translates its **a**-specific genes because there is no repressor present. The *MAT*a transcript acts in combination with the α2 gene product to repress a further set of genes. Foremost among these is a gene "repressor of meiosis" – *RME*. This gene is active in haploid cells and, consequently, meiosis is repressed; in diploid cells, *RME* is itself repressed through the activity of the a1–α2 repressor. **Figure 4** summarizes the action of the mating-type proteins on the various pathways.

(B) N. crassa and Podospora anserina

The life cycle of filamentous ascomycetes is a little more complicated than that of *S. cerevisiae*. Sexual reproduction involves the differentiation of complex multicellular structures and the mating types play a vital role in the latter stages of this process. When a spore of *N. crassa* germinates, the fungal hyphae that emerge from the spore colonize the available substrate and extract nutrients from it. They form a fungal colony, on the surface of which specialized hyphae intertwine and coalesce to form a female reproductive structure called a "protoperithecium". In a heterothallic fungus such as *N. crassa*, this structure develops further (into a perithecium) only following fertilization by a strain of the opposite mating type. When this happens, the nuclei

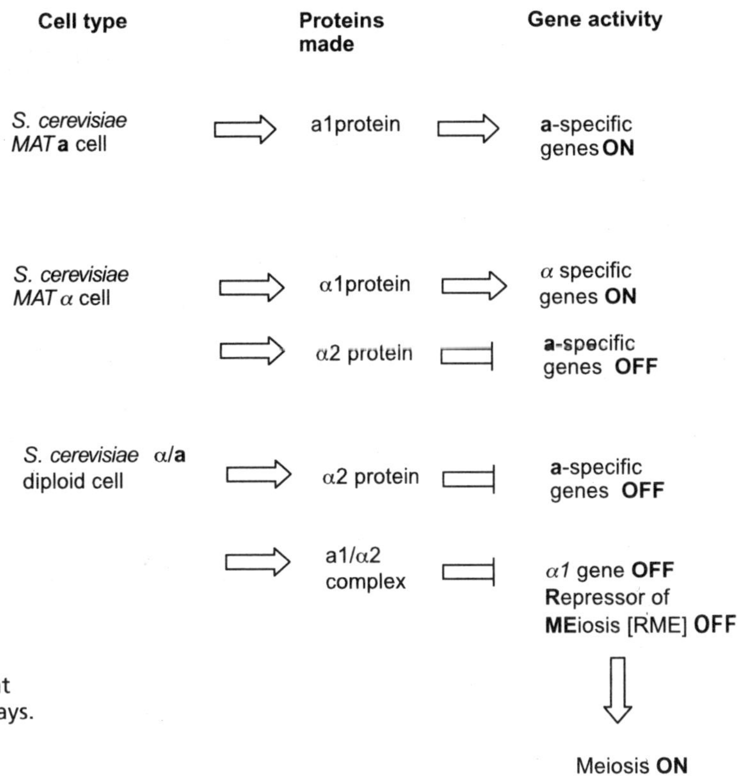

Figure 4 The action of mating type proteins on gene expression in *Saccharomyces cerevisiae*. Arrows represent active steps. Truncated arrows represent blocked pathways. The a2 protein has no known role in mating-type determination.

in the developing perithecium divide synchronously in pairs so that the fruiting body comes to contain many hundreds of paired nuclei. Eventually, the nuclei fuse to form many diploid nuclei, fusion always occurring between two nuclei of opposite mating type, i.e. between a nucleus containing the *mat A* locus and one with the *mat a* locus. The diploid phase exists only very briefly. Meiosis immediately follows the fusion to restore the haploid state and spore walls are laid down around the haploid nuclei to form highly resistant ascospores. The life cycle is completed when these spores are discharged from the perithecium and germinate. The morphological stages in a homothallic species are similar, but asci and ascospores develop inside the perithecium without an obvious fertilization step.

Idiomorphs. When the mating-type genes of *N. crassa* were first characterized at the molecular level, it was discovered that the *mat A* and *mat a* alleles were completely different from each other. They were called idiomorphs in recognition of this fact (Metzenberg & Glass, 1990). Normally, at equivalent points, two homologous chromosomes will possess DNA sequences that are obviously related to each other and differ at only a few bases, these differences having arisen by mutation. That is why such chromosomes are said to be homologous – they are similar without being identical. Chromosomes that contain the mating-type genes are similar to this except that, at the mating-type gene itself, the DNA sequences diverge dramatically and are not at all alike. It is not clear how the mating-type regions have evolved to their present-day form. However, it is clear that, unlike yeast, the difference does not come about through switching of mating type from copies held elsewhere in the genome. Southern blotting of genomic DNA shows that there are no "silent" copies of the other mating type, so transposition of mating-type sequences from one place to another cannot arise. **Figure 5** summarizes what is known about the structure of the two mating-type regions.

Both idiomorphs encode proteins which are transcription binding factors that control gene activity following mating. The *mat a* idiomorph contains one gene that encodes a protein that has been shown to bind to DNA (Philley & Staben, 1994). The HMG DNA-binding motif in the encoded protein is similar to that found in mating-type genes of other fungi. The HMG domain is found in *Cochliobolus heterostrophus* (*MAT-1*), *S. pombe* (M^c) and *P. anserina* (*FPR1*). Although the *mat a* region contains more DNA than that encoding the protein, substituting the DNA into another strain has shown that the part that encodes the polypeptide is all that is necessary to determine the mating type (Chang & Staben, 1994). Recently, however, Pöggeler and Kück (2000) have identified another transcript in this region, the function of which (if any) remains unknown.

The *mat A* region is more complicated. It too encodes a DNA binding protein at a gene called *mat A-1*. The DNA binding motif of the protein has also been recognized in previous studies and is related to the motif found in the *MATa*1 of *S. cerevisiae* and other fungi. In addition, the *mat A* locus has two other genes that play an important role in events following fertilization. A related fungus, *P. anserina*, has a very similar structure to *N. crassa* mating type (Debuchy *et al.*, 1993) and experiments in which these genes were deleted have shown that they are responsible for determining the correct behaviour of the nuclei following

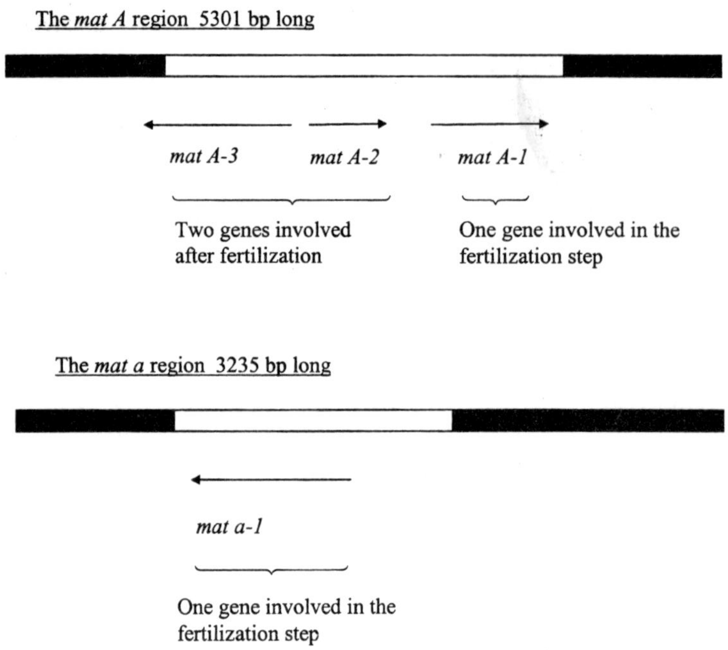

Figure 5 The structure of the *Neurospora crassa* mating-type region. Filled-in boxes represent regions of homology flanking the unique idiomorph region (empty box).

fertilization (Zickler *et al.*, 1995). If either of the genes (called *SMR1* and *SMR2* in Podospora) is mutated, the nuclei behave "selfishly" in the developing perithecia and proceed through meiosis without fusing with a nucleus of a different mating type. In a way that is still mysterious at the molecular level, the gene product of *SMR2* "labels" the nucleus in which it was encoded (Arnaise *et al.*, 1997).

Homothallism in species belonging to the same family as Neurospora has been investigated using the Neurospora mating-type genes. The conclusion to be drawn from these studies is that homothallism has evolved independently several times in the comparatively recent past, with several different types of homothallism being recognized. In Neurospora, most homothallic strains possess only the *mat A* idiomorph. In evolutionary terms, they have bypassed the functions of the *mat a* locus and complete the life cycle without it. However, one species, *N. terricola*, has both mating-type genes. Homothallism in this case has evolved by a rearrangement of the genetic material so that both unique idiomorphic sequences are present within one nucleus. In the very closely related Sordaria genus, the opposite situation exists. Most species contain both idiomorphs, but one species, *S. equina*, contains only *mat A*. In Gelasinospora (another genus in this family of fungi), at least some homothallic species contain both *mat A* and *mat a* sequences. There is a clear indication that in the different homothallic species separate evolutionary events brought the idiomorphs together in one nucleus. In *N. terricola*, the two sequences are linked but separated by at least 20 kilobases (Beatty *et al.*, 1994), in *S. macrospora* (Pöggeler *et al.*, 1997) the two sequences are very closely linked, while, in *Gelasinospora udagawae*, the indications are that *mat A* and *mat a* are widely separated.

Although *N. crassa* and *P. anserina* are the two best-studied ascomycete species, other species have been investigated and, as might be expected, some differences have been discovered. In *C. heterostrophus*, the two mating-type genes each encodes a single transcript but, for both, differential splicing occurs resulting in a heterogeneous population of transcripts.

Faretra *et al.* (1988) and Faretra and Pollastro (1996) have reported the analysis of mating type in *Botryotinia fuckeliana*. This species is primarily heterothallic, but a small percentage of ascospore progeny from field isolates were homothallic. Tetrad analysis showed that seven out of 105 asci gave pairs of homothallic spores that involved the unidirectional change of one of the mating-type alleles. This might be an example of mating-type switching in filamentous ascomycetes, although molecular analysis is required to confirm the details.

Cisar and Tebeest (1999) have shown that isolates of *Glomerella cingulata* obtained from the field contained more than two mating types. This is the first example of multiple allelism in an ascomycete, although mating types of this sort are well known in basidiomycete fungi.

(ii) Mating-type systems in basidiomycete fungi

Mating types have an essential role in the life cycle of many basidiomycetes, but detailed analysis of a few model species has shown that, generally speaking, they are more complex than those of the ascomycetes. The basidiomycetes are a large group, each sharing a common feature: meiosis occurs in a specialized cell called the basidium, with the resulting basidiospores being formed outside the cell. The mating-type organization of three species (*U. maydis*, *C. cinereus* and *S. commune*) will be considered briefly.

(A) U. maydis

The corn smut fungus *U. maydis* exists in two forms: one is unicellular, haploid and divides by budding; the other is dikaryotic, filamentous and is pathogenic on *Zea mays*. The dikaryon arises by fusion between two compatible haploids. The life cycle is summarized in **Figure 6**.

Compatibility of any two isolates is determined by the mating type at two loci. The *a* locus encodes a pheromone and a pheromone receptor; there are two alleles, *a1* and *a2*, and these are idiomorphic. The fact that the *a1* allele shares no homology with *a2* is possibly quite important; no recombination can occur within this region and, consequently, the pheromone and receptor are always transmitted as a pair. This ensures that any one cell is able both to transmit a pheromone signal and to receive the signal of the opposite mating type.

The *b* locus, on the other hand, is multiallelic. It has been estimated that at least 25 alleles exist in nature (Puhalla, 1970). Each allele is complex and encodes two divergently transcribed genes called *bE* and *bW* that are separated by a 260 bp spacer region. This spacer region is different for every *b* allele so that, once again, the two genes found in any one *b* locus are inseparable by recombination; they remain linked as one genetic unit. Both proteins are regulatory proteins with a well-characterized DNA binding motif called the homeodomain. *bE* alleles encode HD1 proteins and *bW* alleles encode HD2 proteins. When two compatible Ustilago strains fuse, four different *b* polypeptides are present in the cell. If these are *bE1*, *bE2*, *bW1* and *bW2*, it was shown by deletion experiments that only two of the four are necessary to bring about successful mating. The two homeodomain proteins, however, must originate from different isolates. Thus, either *bE1* and *bW2* or *bE2* and *bW1* must be present to establish a successful filamentous dikaryon. There are 25 pairs of *b* genes and, potentially, 600 possible successful combinations.

Yee and Kronstad (1993) carried out experiments to determine where the specificity resided in the *bE* alleles. By making chimaeric alleles *in vitro*, they were able to show that the region between amino acids 38 and 87 was responsible for specificity. A join anywhere within this region generated a new allele with the specificity of neither parent. Similarly, Dahl *et al.* (1991) showed that the region between amino acids 56 and 115 conferred specificity for the *bW* alleles.

(B) C. cinereus and S. commune

The life cycle of *C. cinereus* is depicted in **Figure 7**; *S. commune* has essentially the same control systems, with only the details of the morphogenesis differing. Both these species are filamentous basidiomycetes with complex mating-type systems.

The mating-type loci of both species were investigated using classical genetic analysis (Raper *et al.*, 1958; Day, 1960, 1963; for a review see Raper, 1966). The mating-type system of both is said to be tetrapolar because four mating

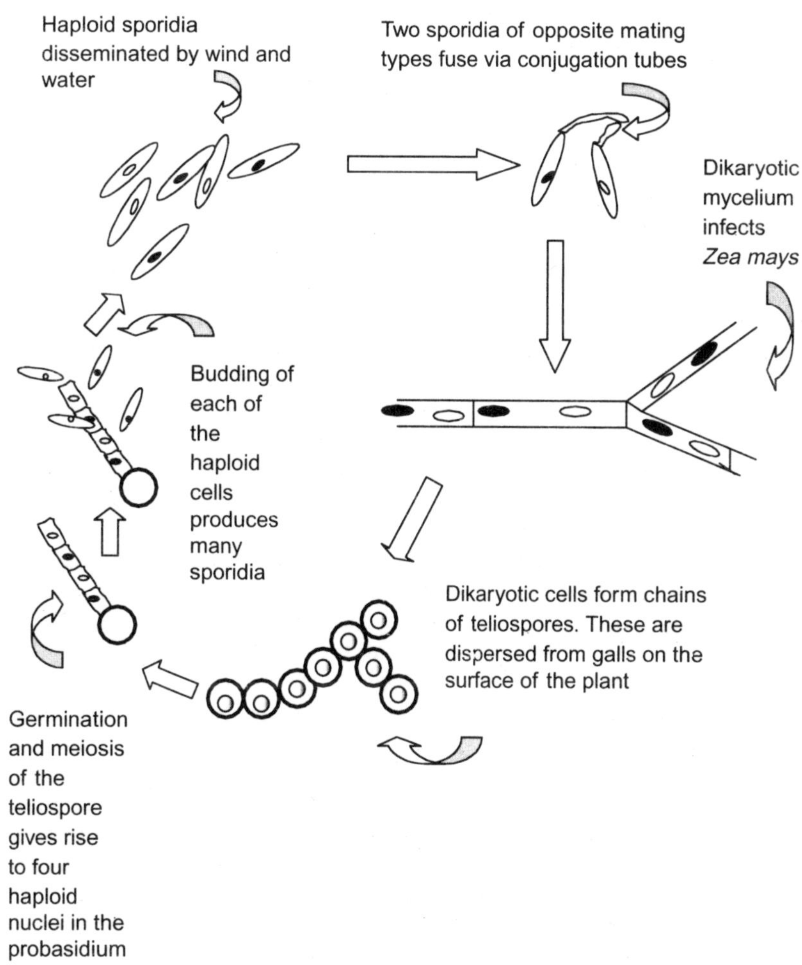

Figure 6 Life cycle of the smut fungus *Ustilago maydis*.

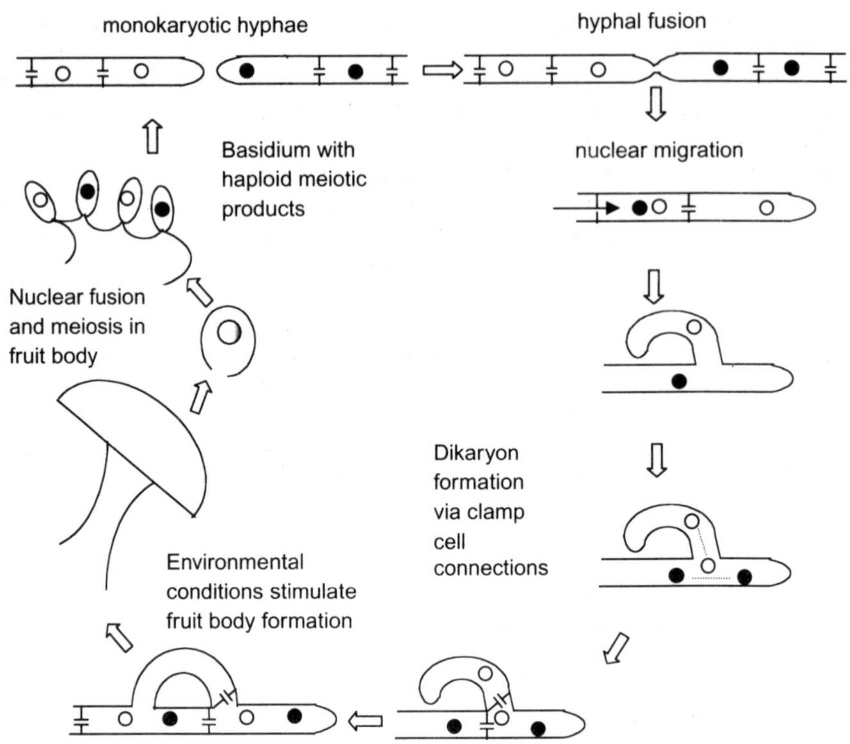

Figure 7 Life cycle of *Coprinus cinereus.*

types segregate from a single fruiting body. Two factors (A and B) determine whether or not two cultures are compatible. A and B are termed factors because each is complex and subdivided into an α and β locus. The A factor controls nuclear pairing and coordinates nuclear division and the formation of clamp cells, which are specialized structures ensuring the maintenance of the dikaryon. The B factor controls nuclear migration and dissolution of the septa following fusion of two monokaryons. Many different alleles exist at each of the two mating-type factors and the specificity of mating type is derived from the combination of α and β. Thus, if each α and β allele is distinguished by a number then Aα1β1 is an allelic combination which is cross-compatible with all other A factors other than itself. In *C. cinereus*, Aα and Aβ are only 0.07 map units apart whereas in *S. commune* both Aα and Aβ and Bα and Bβ are more loosely linked.

Molecular analysis of the A factor

The mating-type genes of both species have been extensively characterized at the molecular level. The A factor of *C. cinereus* was cloned by a chromosome walk started from the nearby *pab-1* locus. Mating-type activity was identified by transformation and, because Aα and Aβ are closely linked, the entire A factor from the A42 strain was isolated (Mutasa *et al.*, 1990). Although classical analysis had suggested the A factor was subdivided into two components, the A factor turned out to be much more complicated than this. The A factor encodes homeodomain proteins, HD1 and HD2, and there is a great deal of functional redundancy at the locus. Thus, any particular A allele may have lost one or more encoding regions through mutation. It will still retain mating-type function providing it retains a pair of HD1 and HD2 proteins. **Figure 8** illustrates details of the organization of the A complex, portraying the predicted archetypal A locus as revealed by the work of Casselton and her group (Kües *et al.*, 1994). Most alleles analysed from nature do not have all the genes intact, some mating types having lost several genes. The three pairs of A factor genes are functionally independent, only one compatible HD1/HD2 combination being required to stimulate sexual development. Figure 8 also depicts the interactions between the HD1 and HD2 proteins as revealed by transformation experiments, which showed that for a compatible reaction it was necessary to have one HD1 and one HD2 protein from any pair. Only a few of all the possible incompatible reactions are depicted.

Several A factor mutants have been isolated that are constitutive for the development of clamp cell connections. Two of these, analysed at the molecular level, have come about through fusion of an HD1 and an HD2 gene to generate a chimaeric gene (Kües *et al.*, 1994). Surprisingly, the fusion protein does not contain both DNA binding domains: only the HD2 domain is present in the fused gene, leading to the conclusion that the HD1 domain is not essential for function of the heterodimer.

Molecular analysis of the B factor

The B factor has been characterized in both *S. commune* and *C. cinereus*, appearing to have a similar function in both. Like the *U. maydis a* gene, the factors encode pheromones and pheromone receptors. Unlike the smut fungus, however, the B factor in these mushrooms is multifactorial and multiallelic, each B factor encoding more than one pair of pheromones and pheromone receptors (**Figure 9** summarizes the organization of the *C. cinereus* B factor). The two alleles

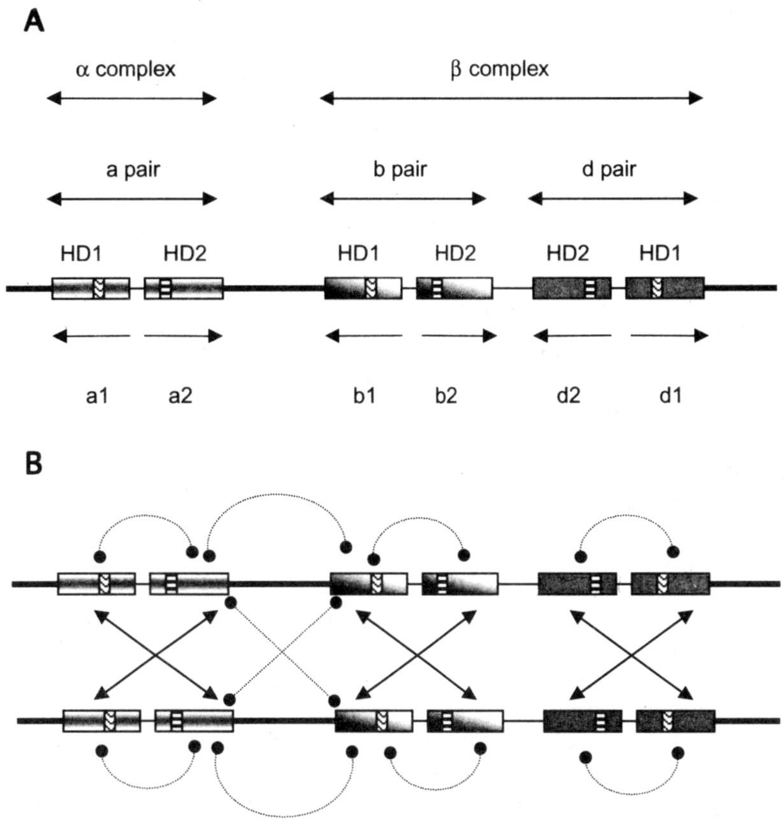

Figure 8 Organization of the mating-type A complex of *Coprinus cinereus*. (A) Three pairs of homeodomain-encoding genes are depicted. Not all alleles analysed from nature have all three loci with functional genes. For successful mating, a genetic difference between HD1 and HD2 of one pair is all that is necessary. (B) Successful interactions between HD1 and HD2 are represented by double-headed arrows. Some incompatible pairings are represented by dotted lines and arcs.

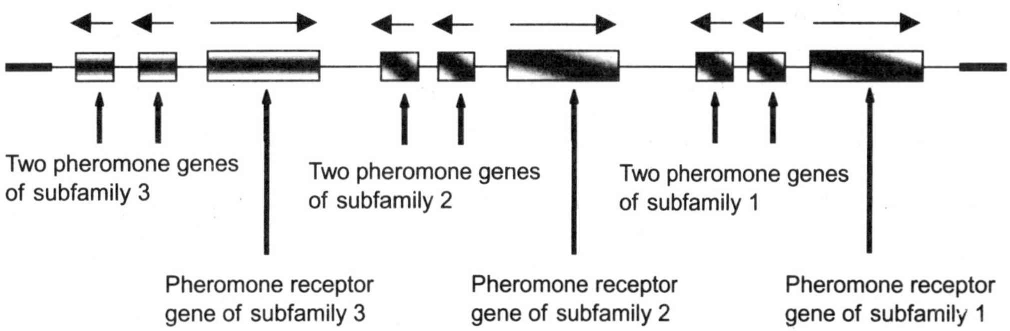

Figure 9 Structure of the B factor in *C. cinereus*. Each subfamily consists of a pheromone receptor gene and two pheromone genes. Genetic differences between two mating strains within only one subfamily are necessary for successful mating.

that were first investigated showed an organization into three subfamilies, each subfamily encoding two pheromone genes and one pheromone receptor. Once again, extensive tracts of non-homologous DNA are present which prevent recombination and, therefore, disruption of the pheromone/pheromone receptor pairs.

6. Molecular recognition systems

(i) Repeat-induced point mutation

DNA-mediated transformation is an indispensable tool in the armoury of the molecular geneticist that allows the transfer of specific genetic material from one organism to another without carrying out a conventional cross. When transformation was developed in *N. crassa* (Mishra *et al.*, 1973), the transformants obtained were transmitted with very poor efficiency through meiosis. Selker and co-workers (Cogoni *et al.*, 1996) showed that this inefficiency was the result of a process in which duplicated DNA sequences were mutated during the nuclear proliferation phase just prior to meiosis – the acronym RIP (repeat-induced point mutation) is used to describe this process. If, during transformation, a DNA sequence is integrated into a chromosome without replacing the resident sequence, then the RIP process introduces mutation into the duplicated DNA. RIP can alter one, both or neither sequence. If the duplicated sequences are tandem repeats then these are especially sensitive to the RIP process and are almost always mutated. Unlinked duplicated sequences are less prone to RIP, but even so mutations occur in about 50% of the progeny.

G:C base pairs are the target for RIP. About 10% of G:C pairs can be mutated to A:T pairs in a single cross. This process of mutation can continue in subsequent crosses providing the duplicated regions retain enough homology for the RIP process to recognize the duplication. It is thought that RIP may be a defence mechanism to prevent invasion of the Neurospora genome by foreign DNA and, consistent with this possibility, foreign DNA if introduced into Neurospora triggers RIP. The RIP process is closely associated with methylation of the duplicated DNA. Tandemly duplicated DNA seems to be a signal for DNA methylation and this becomes associated with G:C to A:T transitions in the RIP process. The RIP process extends into unique sequences that border any duplicated regions.

Fincham *et al.* (1989) used tetrad analysis to explore the RIP process when transformants contained either two or three copies of the *am* gene. The observations were consistent with the pairing of DNA segments during the RIP process. When two copies were present, either both or neither were mutated. When triplicated sequences of the *am* were crossed, the process tended to mutate two of the three copies in most asci. However, there were some examples where all three copies were mutated. These can be explained either by successive rounds of pairing during the RIP process or by occasional association of all three copies in a triplex.

The high frequency with which mutations are induced in duplicated DNA has proved to be a very useful tool in the analysis of gene function. Since mutations are induced in both copies of the DNA, crossing transformants with two copies of any piece of DNA results in mutation of the resident gene and the phenotype associated with the mutation can be investigated. For example, the effect of mutating the mating-type genes in the *MAT A* idiomorph was explored using RIP (for review see Selker, 1990).

(ii) Methylation induced premeiotically

DNA methylation is also a feature of gene inactivation in *Ascobolus immersus*. In this organism, there is no associated mutation of the DNA bases, but methylation does induce silencing epigenetically. Methylation is induced premeiotically (hence the acronym MIP) and transformed strains containing ectopically integrated DNA are unstable when crossed. Faugeron *et al.* (1988) and Goyon and Faugeron (1989) showed that extensive cytosine methylation was induced at some point during (or just before) the sexual cycle. Unlike RIP in *N. crassa*, the altered DNA was not stable and gene function was regained following several mitotic divisions as methylation was lost. Colot *et al.* (1996) showed that the methylation state induced by MIP could be transferred from one chromosome to another by a mechanism that appeared to be closely related to gene conversion.

(iii) Quelling or gene silencing

The introduction of extra DNA sequences by transformation can lead to inactivation during the vegetative phase of the life cycle. In *N. crassa*, the phenomenon has been called quelling; in other groups of organisms (e.g. plants), it has been called silencing. Quelling is distinguished from both RIP and MIP not only because it operates in the vegetative phase but also by the mechanism of action. It appears to operate post-transcriptionally and not via methylation of cytosine residues. Cogoni *et al.* (1996) used fragments of the *al-1* gene to silence the *albino-1* gene and showed that only part of the gene is required to bring about quelling; specific sequences from within the gene were not required. The quelled strains had greatly reduced levels of mRNA and this reduction was specific for the duplicated sequences. This phenomenon is likely to be of considerable importance because of its widespread nature. Similar silencing mechanisms seem to operate in plants and at least some animals.

7. Fungal genetics in the future

DNA manipulation techniques, which are readily applicable to fungi, the development of transformation methods (so that the sexual life cycle can be bypassed) and the development of expression vectors allowing the production of specific gene products all ensure that fungi will continue to make an important contribution in the future. Many fungal products are unique and some of these will prove to be of great utility in an applied sense. The biotechnology industry has still to tap the potential of many fungal metabolites, but clearly the production of new antibiotics, utilization of lignin and cellulose degradation pathways, and an increased understanding of fungal pathogenesis are just three applied areas in which progress can be expected.

REFERENCES

Arnaise, S., Debuchy, R. & Picard, M. (1997) What is a *bona fide* mating-type gene? Internuclear complementation of *mat* mutants in *Podospora anserina*. *Molecular and General Genetics* 256: 169–78

Beadle, G.W. & Tatum, E.L. (1941) Genetic control of biochemical reactions in Neurospora. *Proceedings of the National Academy of Sciences USA* 27: 499–506

Beatty, N.P., Smith, M.L. & Glass, N.L. (1994) Molecular characterization of mating-type loci in selected homothallic species of *Neurospora*, *Gelasinospora* and *Anixiella*. *Mycological Research* 98: 1309–16

Casselton, L.A. & Olesnicky, N.S. (1998) Molecular genetics of mating recognition in basidiomycete fungi. *Microbiology and Molecular Biology Reviews* 62: 55–70

Chang, S. & Staben, C (1994) Directed replacement of *mtA* by *mt-1* effects a mating type switch in *Neurospora crassa*. *Genetics* 138: 75–81

Cisar, C.R. & Tebeest, D.O. (1999) Mating system of the filamentous ascomycete, *Glomerella cingulata*. *Current Genetics* 35: 127–33

Cogoni, C., Irelan, J.T., Schumacher, M. *et al.* (1996) Transgene silencing of the *al-1* gene in vegetative cells of *Neurospora* is mediated by a cytoplasmic effector and does not depend on DNA–DNA interactions or DNA methylation. *EMBO Journal* 15: 3153–63

Colot, V., Maloisel, L. & Rossignol, J.-L. (1996) Interchromosomal transfer of epigenetic states in Ascolbolus: transfer of DNA methylation is mechanistically related to homologous recombination. *Cell* 86: 855–64

Dahl, M., Bölker, M., Gillisen, B. *et al.* (1991) The *b* locus of *Ustilago maydis*: molecular analysis of allele specificity. In *Advances in Molecular Genetics of Plant–Microbe Interactions*, edited by H. Hennecke & D.P.S. Verma, Dordrecht and Boston: Kluwer

Davis, M.A. & Hynes M.J. (1991) Regulatory circuits in *Aspergillus nidulans*. In *More Gene Manipulations in Fungi*, edited by J.W. Bennet & L.L. Lasure, San Diego: Academic Press

Davis, R.H. (1967) Channelling in Neurospora metabolism. In *Organizational Biosynthesis*, edited by H.J.Vogel, J.O. Lampen & V. Bryson, New York: Academic Press

Davis, R.H. (2000) *Neurospora: Contributors of a Model Organism*, Oxford and New York: Oxford University Press

Day, P.R. (1960) The structure of the *A* mating type locus in *Coprinus lagopus*. *Genetics* 45: 641–50

Day, P.R. (1963) The structure of the *A* mating type locus in *Coprinus lagopus*. Wild alleles. *Genetical Research* 4: 323–25

Debuchy, R., Arnaise, S. & Lecellier, G. (1993) The *mat⁻* allele of *Podospora anserina* contains three regulatory genes required for the development of fertilised female organs. *Molecular and General Genetics* 241: 667–73

Faretra, F. & Pollastro, S. (1996) Genetic studies of the phytopathogenic fungus *Botryotinia fuckeliana* (*Botrytis cinerea*) by analysis of ordered tetrads. *Mycological Research* 100: 620–24

Faretra, F., Antonacci, E. & Pollastro, S. (1988) Sexual behaviour and mating system of *Botryotinia fuckeliana*, teleomorph of *Botrytis cinerea*. *Journal of General Microbiology* 134: 2543–50

Faugeron, G., Goyon, C. & Gregoire, A. (1988) Stable allele replacement and stable and unstable non-homologous integration events during transformation of *Ascobolus immersus*. *Gene* 76: 109–19

Fincham, J.R.S. (1985) From auxotrophic mutants to DNA sequences. In *Gene Manipulations in Fungi*, edited by J.W. Bennett & L.L. Lasure, San Diego: Academic Press

Fincham, J.R.S. & Pateman, J.A. (1957) Formation of an enzyme through complementary action of mutant "alleles" in different nuclei of a heterocaryon. *Nature* 179: 741–42

Fincham, J.R.S., Connerton, I.F., Notarianni, E. & Harrington, K. (1989) Premeiotic disruption of duplicated and triplicated copies of the *Neurospora crassa am* (glutamate dehydrogenase) gene. *Current Genetics* 15: 327–34

Forsburg, S.L. & Nurse, P. (1991) Cell cycle regulation in the yeasts *Saccharomyces cerevisiae* and *Schizosaccharomyces pombe*. *Annual Review of Cell Biology* 7: 227–56

Garrod, A.E. (1909) *Inborn Errors of Metabolism*, London: Hodder and Stoughton; reprinted with a supplement, London and New York: Oxford University Press, 1963

Glass, N.L. & Lorimer, I.A.J. (1991) Ascomycete mating types. In *More Gene Manipulations in Fungi*, edited by J.W. Bennett & L.L. Lasure, San Diego: Academic Press

Gouka, R.J., Punt, P.J. & Van den Hondel, C.A.M.J.J. (1997) Efficient production of secreted proteins by Aspergillus: progress, limitations and prospects. *Applied Micobiology and Biotechnology* 47: 1–11

Goyon, C. & Faugeron, G. (1989) Targeted transformation of *Ascobolus immersus* and de novo methylation of the resulting duplicated DNA sequences. *Molecular and Cellular Biology* 9: 2818–27

Herskowitz, I., Rine, J. & Strathern, J. (1992) Mating type determination and mating-type interconversion in *Saccharomyces cerevisiae*. In *The Molecular and Cellular Biology of the Yeast Saccharomyces*, edited by J.R. Broach, J.R. Pringle & E.W. Jones, Cold Spring Harbor, New York: Cold Spring Harbor Laboratory Press

Holliday, R. (1964) A mechanism for gene conversion. *Genetical Research* 5: 282–304

Horowitz, N.H. & Leupold, U. (1951) Some recent studies bearing on the one gene–one enzyme hypothesis. *Cold Spring Harbor Symposium on Quantitative Biology* 16: 65–74

Jacobs, M. & Stahl, U. (1995) Gene regulation in mycelial fungi. In *The Mycota: A Comprehensive Treatise on Fungi as Experimental Systems for Basic and Applied Research*, volume 2, edited by K. Esser & P.A. Lemke, Berlin and New York: Springer

Kües, U., Göttgens, B., Stratmann, R. *et al.* (1994) A chimeric homeodomain protein causes self-compatibility and constitutive sexual development in the mushroom *Coprinus cinereus*. *EMBO Journal* 13: 4054–59

Lang-Hinrichs, C. (1995) Gene regulation in yeast. In *The Mycota: A Comprehensive Treatise on Fungi as Experimental Systems for Basic and Applied Research*, volume 2, edited by K. Esser & P.A. Lemke, Berlin and New York: Springer

Leach, D.R.F. (1996) *Genetic Recombination*, Oxford, and Cambridge, Massachusetts: Blackwell Science

MacKay, V. & Manney, T.R. (1974a) Mutations affecting sexual conjugation and related processes in *Saccharomyces cerevisiae*. I. Isolation and phenotypic characterization of nonmating mutants. *Genetics* 72: 255–71

MacKay, V. & Manney, T.R. (1974b) Mutations affecting sexual conjugation and related processes in *Saccharomyces cerevisiae*. II. Genetic analysis of nonmating mutants. *Genetics* 72: 273–88

MacNeill, S.A. (1994) Cell cycle control in yeasts. In *The Mycota: A Comprehensive Treatise on Fungi as Experimental Systems for Basic and Applied Research*, volume 1, edited by K. Esser & P.A. Lemke, Berlin and New York: Springer

MacNeill, S.A. & Fantes, P.A. (1995) Controlling entry into mitosis in fission yeast. In *Cell Cycle Control*, edited by C. Hutchinson & D. Glover, Oxford and New York: Oxford University Press

Metzenberg, R.L. & Glass, N.L. (1990) Mating type and mating strategies in *Neurospora*. *BioEssays* 12: 53–60

Mishra, N.C., Szabo, G. & Tatum, E.L. (1973) Nucleic acid induced genetic changes in Neurospora. In *The Role of RNA in Reproduction and Development*, edited by M.C. Niu & S.J. Segal, Amsterdam: North Holland, and New York: Elsevier

Mitchell, M.B. (1955) Aberrant recombination of pyridoxine mutants of *Neurospora crassa*. *Proceeding of the National Academy of Sciences USA* 41: 215–20

Mutasa, E.S., Tymon, A.M., Göttgens, B. *et al.* (1990) Molecular organisation of an A mating type factor of the basidiomycete fungus *Coprinus cinereus*. *Current Genetics* 18: 223–29

Oeser, H. (1962) Genetische Untersuchungen üdas Paarungstypverhalten bei *Saccharomyces* und die Maltose-Gene einiger untergäriger Bierhafen. *Archiv fuer Mikrobiologie* 44: 47–74

Ostergaard, S., Olsson, L. & Nielsen, J. (2000) Metabolic engineering of *Saccharomyces cerevisiae*. *Microbiological and Molecular Biology Reviews* 64: 34–50

Perkins, D.D. (1992) Neurospora: the organism behind the molecular revolution. *Genetics* 130: 687–701

Perkins, D.D. (1997) Chromosome rearrangements in Neurospora and other filamentous fungi. *Advances in Genetics* 36: 239–398

Perkins, D.D., Radford, A., Newmeyer, D. & Borkman, M. (1982) Chromosomal loci of *Neurospora crassa*. *Microbiological Reviews* 46: 426–570

Philley, M.L. & Staben, C. (1994) Functional analysis of the

Neurospora crassa Mta-1 mating type polypeptide. *Genetics* 137: 715–22

Pöggeler, S. & Kück, U. (2000) Comparative analysis of the mating-type loci from *Neurospora crassa* and *Sordaria macrospora*: identification of novel transcribed ORFs. *Molecular and General Genetics* 263: 292–301

Pöggeler, S., Risch, S., Kück, U. & Osiewacz, H.D. (1997) Mating type genes from the homothallic fungus *Sordaria macrospora* are functionally expressed in a heterothallic Ascomycete. *Genetics* 147: 567–80

Puhalla, J.E. (1970) Genetic studies of the *b* incompatibility locus of *Ustilago maydis*. *Genetical Research* 16: 229–306

Raper, J.R. (1966) *Genetics of Sexuality in Higher Fungi*, New York: Ronald Press

Raper, J.R., Krongelb, G.S. & Baxter, M.G. (1958) The number and distribution of incompatibility factors of *Schizophyllum commune*. *American Naturalist* 92: 221–32

Reissig, J.L., Issaly, A.S. & Issaly, I.M. (1967) Arginine–pyrimidine pathways in micro-organisms. *National Cancer Institute Monographs* 27: 259–71

Schmidt, H. & Gutz, H. (1994) The mating type switching in yeasts. In *The Mycota: A Comprehensive Treatise on Fungi as Experimental Systems for Basic and Applied Research*, volume 1, edited by K. Esser & P.A. Lemke, Berlin and New York: Springer

Selker, E.U. (1990) Premeiotic instability of repeated sequences in *Neurospora crassa*. *Annual Review of Genetics* 24: 579–613

Shear, C.L. & Dodge, B.O. (1927) Life histories and heterothallism of the red bread-mold fungi of the *Monila sitophila* group. *Journal of Agricultural Research* 34: 1019–42

Stahl, F.W. (1979) *Genetic Recombination: Thinking About It in Phage and Fungi*, San Francisco: Freeman

Strathern, J.N. & Herskowitz, I. (1979) Asymmetry and directionality in production of new cell types during clonal growth: the switching pattern of homothallic yeast. *Cell* 17: 371–81

Strathern, J., Hicks, J. & Herskowitz, I. (1981) Control of cell type in yeast by the mating type locus; the α1–α2 hypothesis. *Journal of Molecular Biology* 147: 357–72

Szostak, J.W., Orr-Weaver, T.L., Rothstein, R.J. & Stahl, F.W. (1983) The double-strand break repair model for conversion and crossing over. *Cell* 33: 25–35

Tatchell, K., Nasmyth, K.A., Hall, B.D., Astell, C. & Smith, M. (1981) *In vitro* mutation analysis of the mating type locus in yeast. *Cell* 27: 25–35

Weiss, R.L. (1973) Intracellular localisation of ornithine and arginine in Neurospora. *Journal of Biological Chemistry* 248: 5409–13

Whitehouse, H.L.K. (1982) *Genetic Recombination: Understanding the Mechanisms*, Chichester and New York: Wiley

Yee, A.R. & Kronstad, J.W. (1993) Construction of chimeric alleles with altered specificity at the *b* incompatibility locus of *Ustilago maydis*. *Proceedings of the National Academy of Sciences USA* 90: 664–68

Zickler, D., Arnaise, S., Coppin, E., Debuchy, R. & Picard, M. (1995) Altered mating-type identity in the fungus *Podospora anserina* leads to selfish nuclei, uniparental progeny and haploid meiosis. *Genetics* 140: 493–503

FURTHER READING

Bennett, J.W. & Lasure L.L. (eds) (1985) *Genetic Manipulations in Fungi*, San Diego: Academic Press

Bennett, J.W. & Lasure L.L (eds) (1991) *More Genetic Manipulations in Fungi*, San Diego: Academic Press

Esser, K. & Lemke, P.A. (eds) (1995) *The Mycota: A Comprehensive Treatise on Fungi as Experimental Systems for Basic and Applied Research*, volume 1: *Growth, Differentiation and Development*, edited by J.G.H. Wessels & F. Meinhardt, Berlin and New York: Springer

Esser, K. & Lemke, P.A. (eds) (1995) *The Mycota: A Comprehensive Treatise on Fungi as Experimental Systems for Basic and Applied Research*, volume 2: *Genetics and Biotechnology*, edited by U. Kück, Berlin and New York: Springer

GLOSSARY

allelic complementation ability of two mutant alleles to restore normal (or near normal) function to a cell. Allelic complementation can result from formation of a hybrid protein from two different mutant monomers or from cooperation between two mutants, each defective in a different functional domain

ascospore haploid spore formed by ascomycete fungi, contained within an ascus

auxotrophic any organism with a nutritional requirement for a specific substance

bivalents structures composed of four double-stranded DNA molecules generated in meiosis from two homologous chromosomes that have replicated and paired. Each bivalent is usually held together by chiasmata

compartmentalized the subdivision of a cell or organism so that functions are not evenly distributed. Sometimes the compartments may correspond to subcellular organelles, but on other occasions there may be no obvious explanation for the polarized organisation

heterothallic cells, thalli or mycelia of algae or fungi that can only reproduce with members of a physiologically different strain

homothallic cells, thalli or mycelia of algae or fungi that can reproduce with a similar strain or a branch of its own thallus or mycelium

idiomorph(s) the two (or more) alternative forms of mating type genes occupying equivalent loci on homologous chromosomes

octad an ascus containing eight, haploid, ascospores, produced in a species in which the tetrad normally undergoes a post-meiotic mitotic division. More generally, any group of eight objects

perithecium flask-shaped structure opening in a terminal hole, which contains the asci

protoperithecium primary haploid perithecium

tetrad four haploid products of a single meiosis. More generally, any group of four objects

See also **Genetic recombination** (p.701)

FUNGAL PRIONS

Brian Cox
Department of Biosciences, University of Kent at Canterbury, UK

1. Introduction

The word "prion" was coined by Prusiner in 1982 (Prusiner, 1982; Prusiner *et al.*, 1982) to describe the infectious principle of scrapie-like diseases in vertebrates. He proposed that infectious units were composed only of protein, with no nucleic acid component. The principle on which such proteins would operate was that a prion protein could exist in two forms, one of which was pathogenic. The pathogenic form had the property of converting the normal form to a likeness of itself. The conditions, then, for infectivity were the passage of sufficient pathogenic conformers into a cell or tissue for a conversion of the normal molecules to occur. In these diseases, transmissible spongiform encephalopathies (TSEs) indeed, a protein is found to accumulate in certain brain tissues in the form of plaques or fibrils and its conformation in these plaques differs from that found in normal tissue. The protein, which has been named PrP, normally folds so that approximately 50% is α-helix. In morbid tissue, its conformer in plaques is about 90% β-sheet and about two-thirds of the C-terminal length of the molecule is resistant to extended treatment with proteases (Bolton *et al.*, 1982; Pan *et al.*, 1993).

2. Prions in yeast

In 1994, Wickner wrote an article in *Science* in which he pointed out that certain traits in the yeast *Saccharomyces cerevisiae* which were inherited in a so-called non-Mendelian pattern and with which no nucleic acid had been associated might also be due to prion-like proteins. He presented evidence that this was the case for one of these traits, known as *[URE3]*. In this case, the prion-like properties of the protein would manifest themselves as an inherited trait, because, when two haploid cells which differ (e.g. *[URE3+]* and *[URE3−]*) are mated, the diploid zygote would become *[URE3+]* and all haploid spore products likewise *[URE3+]* as a result of the prion form, *[URE3+]*, converting the normal form, *[URE3−]*, to its own likeness. This gives rise to the observed non-Mendelian inheritance. Normally, when diploids sporulate following meiosis, any parental trait for which the diploid is heterozygous segregates in the haploid spore products as described by Gregor Mendel. In the case of *[URE3+]* × *[URE3−]* zygotes, the *[URE3−]* trait does not appear in any of the progeny.

The trait controlled by this factor is the ability to take up the uracil precursor ureidosuccinic acid (USA) in the presence of other nitrogen sources. This allows *ura2* mutants to grow in the presence of this substrate – *ura2* controls an enzyme for the synthesis of USA. This property, namely the uptake of USA in the presence of nitrogen sources, is also exhibited by mutants of the *URE2* gene which are inherited in a normal Mendelian fashion, the mutants (designated *ure2*) being recessive. Wickner proposed that the product of this gene had prion-like properties and then went on to demonstrate that *ure2* mutants were unable to maintain the *[URE3+]* condition and also that overexpression of the *URE2* gene on a plasmid enhanced the mutation of *[URE3−]* to *[URE3+]* cells about 1000-fold over normal levels. The phenotypes caused by these changes are explained by the requirement for a functional *URE2* product for regulating the uptake of nitrogenous compounds and that this product is removed either by mutation of the *URE2* gene or by conversion of its protein product into the prion form.

At this time, there were two other traits known, unrelated to *[URE3+/−]*, which were also inherited in a similar non-Mendelian pattern and for which no nucleic acid association had been found. These were Ψ psi (Cox, 1965) and *[eta]* (Liebman & All-Robyn, 1984). At the end of his 1994 article, Wickner suggested that these, too, might be prion-like phenomena.

The phenotype controlled by Ψ is nonsense suppression. Weak suppressors which fail to suppress certain nonsense mutations become strong suppressors in Ψ+ strains; strong suppressors which suppress in Ψ− strains are lethal in Ψ+. This phenotype, namely the increase of read-through at chain-termination codons is mimicked by mutations which are recessive and inherited in a Mendelian pattern – that is, are mutations of chromosomal genes (Cox *et al.*, 1988). Thus, as in the case of *[URE3]*, an inherited trait can be determined by either a dominant non-Mendelian determinant or by recessive genic mutations.

In 1994, Cox suggested that the product of the yeast *SUP35* gene might have prion-like properties, causing the Ψ phenomenon. The reasons for this were, first, that it had been shown by Ter-Avanesyan and his colleagues that one particular domain of this gene, the N-terminal portion, was necessary for maintainance of the Ψ+ state (Ter-Avanesyan *et al.*, 1994). Secondly, Chernoff and his colleagues had shown that overexpression of either the full-length gene or its N-domain alone could cause a change from Ψ− to Ψ+ at a high frequency (Chernoff *et al.*, 1993). Thirdly, it had been found by Doel *et al.* (1994) that a point mutation in the N-terminal domain of this gene caused cells to become and remain Ψ− even when mated with Ψ+ cells.

[ETA] is a non-Mendelian trait which causes some mutant alleles of the *SUP35* gene to become lethal. Derkatch and colleagues have shown that *[ETA]* is, in fact, a form of the Ψ factor in that the gene product involved is that of *SUP35* itself and, like Ψ, is dependent specifically on the N-terminal domain of *SUP35* (Zhou *et al.*, 1999). Liebman and her colleagues have also shown that Ψ+ can exist in various forms (Derkatch *et al.*, 1996). In other words, if this

is indeed a prion-based phenomenon, then different prion conformers of Sup35p can occur, having different properties and transmitting their individual characteristics to normal conformers.

Finally, a fourth possible prion has been described in yeast. This is *[PIN]*, discovered also by Liebman and her colleagues (Derkatch *et al.*, 1997). *[PIN]* controls the ability of Ψ− cells to revert to Ψ+, either spontaneously or through overexpression of the *SUP35* gene. *[PIN+]* is inherited in a non-Mendelian fashion, but occurs independently of *[PSI]*; that is, both *[PIN+]* and *[PIN−]* states can occur in either Ψ+ or Ψ− cells. *[PIN+]* does not depend on the N-terminal domain of *SUP35* and so, if it is a prion, the protein must either be the product of a different gene or of the C-terminal portion of *SUP35*. The properties of these four prion-like phenomena in yeast are summarized in **Table 1**.

Apart from being inherited in a non-Mendelian fashion, all four traits have another property in common, namely that their propagation in cells is sensitive to the presence of very low concentrations of guanidinium chloride (and one or two other guanidinium salts) (Tuite *et al.*, 1981). Elimination of the prion depends on continued growth of the cells and the pattern of elimination suggests that guanidinium inhibits the replication of prion conformers in the cell, without destroying them (Eaglestone *et al.*, 2000). In addition, in the case of Ψ, propagation has been found to depend on or be affected by a number of genes other than *SUP35*. One of these is *HSP104*, a heat-shock protein gene. This has the ambivalent property of being essential for the maintainance of Ψ+ and yet causing its elimination if it is overexpressed (Chernoff *et al.*, 1995). Another gene which is not essential for Ψ maintenance, but which clearly assists and whose product binds to both pHsp104 and to eRF3, is *SLA1*, which codes for an actin assembly protein (Bailleul *et al.*, 1999). *SUP35* is part of the chain termination complex and is found bound to SUP45p (eRF1) (Stansfield *et al.*, 1995). Overexpression of *SUP45* leads to loss of the ability of Ψ− cells to be induced to Ψ+, presumably due to dilution of the Sup35p available to aggregate into prion form (Paushkin *et al.*, 1997a). Other genes affecting Ψ maintenance have been identified genetically, but have not been characterized molecularly.

Table 1 Properties of prion-like phenomena in yeast.

Name	Phenotype	Mendelian gene	Protein function
[URE3]	uptake of USA*	*URE2*	negative regulator of gene for uptake of nitrogenous compounds
Ψ (PSI)	enhances nonsense suppression	*SUP35*	polypeptide chain release
η (eta)	lethal with some *SUP35* mutants	*SUP35*	polypeptide chain release
[PIN]	enhances induction of Ψ+ from Ψ−	?	?

*USA, ureidosuccinic acid

(i) Properties of the proteins

Both Ure2p and Sup35p (eRF3) have been studied with a view to establishing that they have the properties characteristic of the model prion, PrP, responsible for TSEs. These characteristics are the existence of two conformers which can be physically distinguished: the formation *in vivo* and *in vitro* of self-aggregates, and the ability *in vitro* for the prion form to convert the normal form into prion.

Structural homologies. The amino acid sequences of PrP, Ure3p and Sup35p have little or nothing in common. However, there are similarities in the overall organization of all three proteins (**Figure 1**). Each has distinct N- and C-terminal domains. In the yeast genes, the C-terminal domains are essential and sufficient for the normal function of the protein in the cell, with the N-terminal domains being dispensible. In the case of Sup35p, the cellular function is essential for viability. In both yeast genes, the N-terminal domain is sufficient and necessary for the propagation of the prion+ state *in vivo* and *in vitro* (Chernoff *et al.*, 1993; Ter-Avanesyan *et al.*, 1993; Doel *et al.*, 1994; Masison & Wickner, 1995; Patino *et al.*, 1996; Glover *et al.*, 1997; Masison *et al.*, 1997; DePace *et al.*, 1998). With PrP, the converse appears to be the case, since mice carrying coding sequences for only the 150 C-terminal residues are capable of developing scrapie (Fischer *et al.*, 1996). It is not known quite which, if any, particular part of the PrP gene is essential by itself for the gene's function in synaptic transmission.

Both yeast genes have, close to their N-terminals, tracts rich in the amino acids asparagine and glutamine. This they have in common with several proteins implicated in

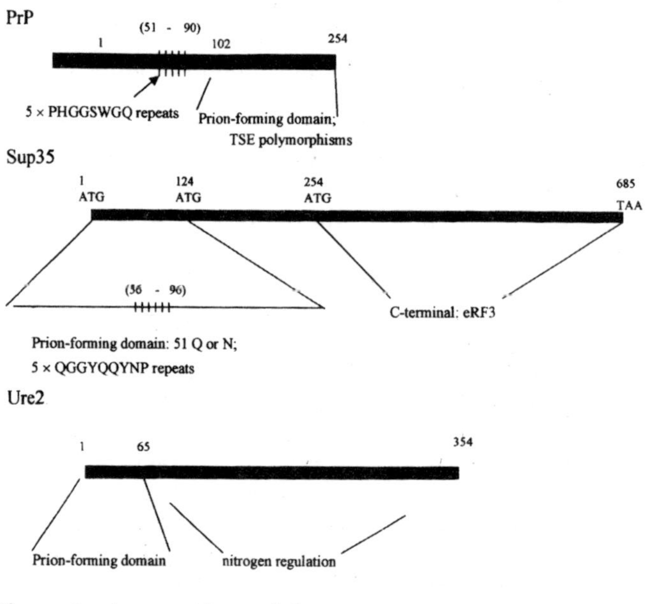

Figure 1 A comparison of the organization of the proteins said to behave as prions in infection in animals or in heredity in yeast.

neurodegenerative diseases, including huntingtin; these diseases are characterized by the formation of aggregates of the protein concerned. Familial Huntington's disease is associated with expansions of the glutamine-coding trinucleotide repeat tracts. In the case of *SUP35*, it has been shown by mutation studies that this glutamine- and asparagine-rich region is implicated in maintenance of the Ψ+ state *in vivo* and in the formation of aggregates from soluble protein *in vitro* (DePace *et al.*, 1998).

In addition, the Sup35p and PrP proteins have in common a set of octo- (PrP) or nona-peptide tandem repeats. The N-terminal-deleted PrP gene described above, which can support scrapie in mice, does not include these repeats; however, in several cases of familial Creutzfeldt–Jacob disease, the PrP gene is found to have expansions of this set of repeats (Poulter *et al.*, 1992; Prusiner, 1994). In yeast, it has been found that deletion of a part of the set renders the Sup35p unable to support Ψ+ (**Figure 1**). Furthermore, the point mutation described by Doel *et al.* (1994), which eliminates Ψ+ from Ψ− cells, occurs in a residue in the first of these repeats. The role of these repeats in promoting or propagating prion conformers remains ambiguous for PrP protein, but the evidence so far suggests that they are important in propagation of Ψ+.

Physicochemical properties. All the prion proteins described so far are characterized by the formation of insoluble aggregates. PrP itself was first purified and sequenced from the fibrillary tangles and plaques formed in the diseased brains of infected animals. Sup35p can be recovered from pellets after high-speed centrifugation (100 krpm) of whole-cell lysates of Ψ+ cells, but not of Ψ− cells (Paushkin *et al.*, 1997b).

In yeast, aggregates induced by the N-terminal prion-forming domains of either *URE2* or *SUP35* can be visualized *in vivo* by making gene fusions of these domains with green fluorescent protein (GFP). When the fusion protein is expressed in the appropriate genetic background, *[URE3+]* or Ψ+ respectively, GFP is observed to have a punctate or clumpy distribution in the cytoplasm of the cells, whereas it is diffuse in *[URE3−]* or Ψ− cells (Patino *et al.*, 1996; Wickner, R.B., personal communication).

The PrP recovered from plaques and fibrils in diseased brain tissue is mostly in the form of amyloid fibrils. This is a structure characteristic of a large number of neurodegenerative and other diseases classed as amyloidoses. The list includes, apart from the TSEs, Huntington's and Alzheimer's diseases. Amyloid is defined as a filamentous protein structure that stains with the dye Congo Red to produce green birefringence under polarized light. The amyloid fibre structure is formed *in vitro* by the purified prion-forming domains of *URE2* and *SUP35*, and in Ure2p is also induced in the full-length protein by these fragments (Kelly, 1996; Glover *et al.*, 1997; King *et al.*, 1997; Taylor *et al.*, 1999).

The prion conformer is distinguished from the normal one by the extreme resistance of the C-terminal part of the molecule to protease digestion and this property is used as a diagnostic (Bolton *et al.*, 1982). Protease resistance is characteristic also of the *[URE3]* and Ψ prions (King *et al.*, 1997; Taylor *et al.*, 1999).

Finally, both PrP and the *[URE3]* prion show a change in folding at the molecular level, in that the normal ratio of α-helix to β-sheet is altered. In the case of PrP, the normal β-sheet content is negligible and, in the fibrils, it goes up to about 80%, with α-helix dropping from 60% to <10% (Pan *et al.*, 1993). Similarly, in the N-terminal fragment of the URE2 protein, prion filaments are predominently β-sheet, whereas the native protein has both α-helix and β-sheet (Taylor *et al.*, 1999). β-sheet is a significant structure in the formation of amyloid fibrils; the fibrils are thought to be composed of stacks of interleaved β-sheets (Kelly, 1996). There is evidence that the prion domain of Sup35p may form β-sheet during prolonged incubation (King *et al.*, 1997). These properties of PrP, Ure2p and Sup35p are summarized in Table 2.

In vitro *propagation.* The prion theory makes the prediction that transmission of the altered protein is the consequence of the templated conversion of normal molecules into the abnormal form by the abnormal form and, naturally, this has prompted attempts to demonstrate this autocatalytic event *in vitro*. Successful conversion *in vitro* of protease-sensitive PrP^C to a protease-resistant form by the addition of excess PrP^{Sc} has been reported (Kocisko *et al.*, 1994). In the case of Ure2p, purified prion domain fragments denatured in 6M guanidine HCl spontaneously form amyloid fibres when diluted into buffer. When diluted into buffer containing excess full-length protein, amyloid cofilaments form. These cofilaments can be used to recruit native Ure2p into amyloid precipitates over a period of 2 weeks at 4°C, which demonstrates seeded conversion, if not autocatalysis. Cofilaments do not form with added Alzheimer's Aβ peptide or with the Ure2p non-prion domain fragment, demonstrating the specificity of the interaction (Taylor *et al.*, 1999).

With *SUP35*, there is an even more encouraging picture. As with Ure2p, prolonged incubation of purified prion domain protein results in the formation of aggregates, but this lag can be eliminated or considerably reduced by the

Table 2 Properties of PrP, Ure2p and Sup35p.

Protein (length of amino acids)	Prion domain	Protease resistance	Amyloid fibre	β-sheet in prion	*in vivo*
PrPSc (254)	121–231	high	yes	yes	plaques and fibrils
Sup35 (630)	1–100	low	Prion domain – amyloid like	yes	aggregates
			full length?	?	punctate GFP* from fusion protein
Ure2 (354)	1–65	high	yes	yes	punctate GFP from fusion protein

*GFP, green fluorescent protein

addition of preformed aggregate, either from cell-free extracts or from previous *in vitro* experiments. The reaction can be enormously speeded up by slow rotation of the test tubes. The rate of formation of new aggregate in the presence of seed of preformed aggregate is physiological. The formation of aggregate can be propagated through serial transfers such that effectively none of the original seed aggregate remains. It is affected by point mutations in the prion-forming domain that prevent Ψ+ propagation *in vivo* and is enhanced by the addition of Hsp104p (q.v.) (Glover *et al.*, 1997; Paushkin *et al.*, 1997b; DePace *et al.*, 1998).

(ii) The basis of the prion hypothesis

It is clear that workers in a diversity of laboratories have found in the protein products of PrP, *URE2* and *SUP35* properties which are expected of prions. The most fundamental of these is the existence of the protein in two different metastable folding configurations and, in these examples, this is signalled by a shift from a normal form which contains much α-helix to the pathological variant containing mostly β-sheet. It also happens that these three examples exhibit the property of aggregation into amyloid or amyloid-like fibres. In the case of PrP, these aggregates are found *in vivo* in most forms of spongiform encephalopathy. In yeast, intracellular aggregates have been visualized using gene fusions of prion-forming domains with GFP. It has also been found possible to use aggregated prion-form protein to promote the conversion of normal form to prion form *in vitro* and, in the case of *SUP35*, this can be achieved at physiologically relevant rates and stoichiometry (Paushkin *et al.*, 1997b; DePace *et al.*, 1998). What is missing from all these studies is the demonstration that the introduction of the prion form of the protein into cells (or into intercellular space in brain tissue) is sufficient to generate the effects described *in vivo*. Infectivity of nearly pure PrP^{Sc} has been demonstrated by Bolton and his colleagues (1982), but infectivity has not been demonstrated in yeast, although purely synthetic prion forms of the two proteins can be made.

The question also remains about what the process of conversion in living tissue consists of. It is clear in the case of *SUP35* that the process involves more than just the passive recruitment of soluble protein by aggregates. The involvement of the chaperone-like protein Hsp104p and the actin-binding Ssalp suggests a more active and complex process, and the sensitivity of propagation to very low concentrations of guanidine hydrochloride likewise suggests that that compound may be interfering with some important aspect of cell metabolism or cell organization. Likewise, it has been suggested that another factor, referred to as "protein X", is involved in PrP propagation *in vivo* and, indeed, that there may in each of these examples be an intermediate soluble form which acts as a "seed" for the recruitment of normal molecules to the prion aggregates.

3. Prions in *Podospora anserina*

Filamentous fungi exhibit many examples of extranuclear or non-Mendelian inheritance. Many have been characterized at the molecular level and have been found to correlate with nucleic acid modifications, often associated with mitochondrial DNA or self-propagating introns (Dujon & Belcour, 1989). Among the remaining uncharacterized phenomena, there are perhaps some that may be prion based and one such example has been described by Coustou *et al.* (1997). In Podospora, as in many filamentous fungi, hyphae from different mycelia often fuse and hyphal fusion is under genetic control. Mycelia which fail to fuse are said to be incompatible, and one of the ten or so loci which affect incompatibility is Het-S/Het-s. Mycelia which differ at this locus fail to fuse. However, Het-s comes in two variants, an "active" and an "inactive" type. The active type is called Het-s*. Het-s* is "neutral"; that is, it fuses with either Het-S or Het-s. In this respect, it behaves like a mycelium which carries a null allele of Het-s. This locus does not affect sexual fusion, so the Het-s/Het-s* difference can be analysed through meiotic segregation and is found to exhibit maternal inheritance. In other words, all the ascospores derived from a Het-s × Het-s* cross have the character of the maternal parent, either s or s*; the other character does not segregate.

When a Het-s mycelium fuses with a Het-s* mycelium, the latter rapidly becomes Het-s throughout. This happens without hyphal growth and is independent of any transfer of nuclei from one mycelium to another. Mycelia are coenocytic and there is considerable cytoplasmic mixing through the hyphae. The proposal that this is a prion-based phenomenon arises from the fact that conversion of Het-s* to Het-s by Het-s depends on the presence in the recipient mycelium of an intact *Het-s* gene: null alleles of *Het-s* do not support the conversion. Furthermore, the conversion can be achieved by the overexpression of the *Het-s* gene fused with a high-activity promoter. It has also been found that *Het-s* protein is more resistant to proteinase K than is Het-s* protein and that it forms multimeric structures *in vivo*.

The interesting difference compared with other prion phenomena is that it is the ostensible prion form of the protein, the invasive, dominant form Het-s, that has an "active" phenotype – of causing incompatibility with Het-S – rather than having the phenotype of the null mutant which is neutral in controlling hyphal fusion. Prions described so far in yeast, although dominant, have the non-functional phenotype of the recessive null mutation of the gene which codes for the prion.

REFERENCES

Bailleul, P.A., Newnam, G.P., Steenbergen, J.N. & Chernoff, Y.O. (1999) Genetic study of interactions between the cytoskeletal assembly protein Sla1 and prion-forming domain of the release factor Sup35 (eRF3) in *Saccharomyces cerevisiae*. *Genetics* 153: 81–94

Bolton, D.C., McKinley, M.C. & Prusiner, S.B. (1982) Identification of a protein that purifies with the scrapie prion. *Science* 218: 1309–11

Chernoff, Y.O., Derkach, I.L. & Inge-Vechtomov, S.G. (1993) Multicopy *SUP35* gene induces *de novo* appearance of *psi*-like factors in the yeast *Saccharomyces cerevisiae*. *Current Genetics* 24: 268–70

Chernoff, Y.O., Lindquist, S., Ono, B.-I., Inge-Vechtomov, S.G. & Liebman, S.W. (1995) Role of the chaperone protein

Hsp104 in propagation of the yeast prion-like factor [*psi*+]. *Science* 268: 880–84

Coustou, V., Deleu, C., Saupe, S. & Begueret, J. (1997) The protein product of the het-s heterokaryon incompatibility gene of the fungus *Podospora anserina* behaves as a prion analog. *Proceedings of the National Academy of Sciences USA* 94: 9773–78

Cox, B.S. (1965) Ψ, a cytoplasmic suppressor of super-suppressor in yeast. *Heredity* 20: 505–21

Cox, B.S. (1994) Prion-like factors in yeast. *Current Biology* 4: 744–48

Cox, B.S., Tuite, M.F. & McLaughlin, C.S. (1988) The Ψ factor of yeast: a problem in inheritance. *Yeast* 4: 159–78

DePace, A.H., Santoso, A., Hillner, P. & Weissman, J.S. (1998) A critical role for amino-terminal glutamine/asparagine repeats in the formation and propagation of a yeast prion. *Cell* 93: 1241–52

Derkatch, I.L., Chernoff, Y.O., Kushnirov, V.V., Inge-Vechtomov, S.G. & Liebman, S.W. (1996) Genesis and variability of [*PSI*] prion factors in *Saccharomyces cerevisiae. Genetics* 144: 1357–86

Derkatch, I.L., Bradley, M.E., Zhou, P., Chernoff, Y.O. & Liebman, S.W. (1997) Genetic and environmental factors affecting the *de novo* appearance of the [*PSI*+] prion in *Saccharomyces cerevisiae. Genetics* 147: 507–19

Doel, S.M., McCready, S.J., Nierras, C.R. & Cox, B.S. (1994) The dominant PNM2 mutation which eliminates the Ψ factor of *Saccharomyces cerevisiae* is the result of a missense mutation in the SUP35 gene. *Genetics* 137: 1–12

Dujon, B. & Belcour, L. (1989) Mitochondrial DNA instabilities and rearrangements in yeasts and fungi. In *Mobile DNA*, edited by D.E. Berg & M.M. Howe Washington, DC: American Society for Microbiology

Eaglestone, S.S., Ruddock, L.W., Cox, B.S. & Tuite, M.F. (2000) Guanidine hydrochloride blocks a critical step in the propagation of the prion-like determinant [*PSI*+] of *Saccharomyces cerevisiae. Proceedings of the National Academy of Sciences USA* 97: 240–44

Fischer, M., Rulicke, T., Raeber, A. *et al.* (1996) Prion protein (PrP) with amino-proximal deletions restoring susceptibility of PrP knockout mice to scrapie. *EMBO Journal* 15: 1255–64

Glover, J.R., Kowall, A.S., Schirmer, E.C. *et al.* (1997) Self-seeded fibres formed by Sup35, the protein determinant of [*PSI*+], a heritable prion-like factor of *S. cerevisiae. Cell* 89: 811–19

Kelly, J.W. (1996) Alternative conformations of amyloidogenic proteins govern their behavior. *Current Opinions in Structural Biology* 6: 11–17

King, C.-Y., Tittman, P., Gross, H. *et al.* (1997) Prion-inducing domain 2–114 of yeast Sup35 protein transforms *in vitro* into amyloid-like filaments. *Proceedings of the National Academy of Sciences USA* 94: 6618–22

Kocisko, D.A., Come, J.H., Priola, S.A. *et al.* (1994) Cell-free formation of protease-resistant prion protein. *Nature* 370: 471–74

Liebman, S.W. & All-Robyn, J.A. (1984) A non-Mendelian factor, [eta+], causes lethality of yeast omnipotent-suppressor strains. *Current Genetics* 8: 567–73

Masison, D.C. & Wickner, R.B. (1995) Prion-inducing domain of yeast Ure2p and protease resistance of Ure2p in prion-containing cells. *Science* 270: 93–95

Masison, D.C., Maddelein, M.-L. & Wickner, R.B. (1997) The prion model for [URE3] of yeast: spontaneous generation and requirements for propagation. *Proceedings of the National Academy of Sciences USA* 94: 12503–08

Pan, K.M., Baldwin, M., Nguyen, J. *et al.* (1993) Conversion of α-helices into β-sheets features in the formation of the scrapie prion proteins. *Proceedings of the National Academy of Sciences USA* 90: 10962–66

Patino, M.M., Liu, J.J., Glover, J.R. & Lindquist, S. (1996) Support for the prion hypothesis for inheritance of a phenotypic trait in yeast. *Science* 273: 622–26

Paushkin, S.V., Kushnirov, V.V., Smirnov, V.N. & Ter-Avanesyan, M.D. (1997a) Interaction between yeast Sup45p (eRF1) and Sup35p (eRF3) polypeptide chain release factors: implications for prion-dependent regulation. *Molecular and Cellular Biology* 17: 2798–805

Paushkin, S.V., Kushnirov, V.V., Smirnov, V.N. & Ter-Avanesyan, M.D. (1997b) *In vitro* propagation of the prion-like state of yeast Sup35 protein. *Science* 277: 381–83

Poulter, M., Baker, H.F., Frith, C.D. *et al.* (1992) Inherited prion disease with 144 base pair gene insertion. 1. Genealogical and molecular studies. *Brain* 115: 675–85

Prusiner, S.B. (1982) Novel proteinaceous particles cause scrapie. *Science* 216: 136–44

Prusiner, S.B. (1994) Prion diseases of humans and animals. *Journal of the Royal College of Physicians* 28: 1–30

Prusiner, S.B., Groth, D.F., McKinley, M.P. *et al.* (1982) Scrapie agent is a novel proteinaceous infectious particle. *Federation Proceedings* 41: 695

Stansfield, I., Jones, K.M., Kushnirov, V.V. *et al.* (1995) The products of the SUP45 (eRF1) and SUP35 genes interact to mediate translation termination in *Saccharomyces cerevisiae. EMBO Journal* 14: 4365–73

Taylor, K.L., Cheng, N., Williams, R.W., Steven, A.C. & Wickner, R.B. (1999) Prion domain initiates amyloid formation *in vitro* by native Ure2p. *Science* 283: 1339–43

Ter-Avanesyan, M.D., Kushnirov, V.V., Dagkesamanskaya, A.R., Didchenko, S.A. & Chernoff, Y.O. (1993) Deletion analysis of the SUP35 gene of the yeast *Saccharomyces cerevisiae* reveals two non-overlapping functional regions in the encoded protein. *Molecular Microbiology* 7: 683–92

Ter-Avanesyan, M.D., Dagkesamanskaya, A.R., Kushnirov, V.V. & Smirnov, V.N. (1994) The SUP35 omnipotent suppressor gene is involved in the maintenance of the non-Mendelian determinant [psi+] in the yeast *Saccharomyces cerevisiae. Genetics* 137: 671–76

Tuite, M.F., Mundy, C.R. & Cox, B.S. (1981) Agents that cause a high frequency of genetic change from [psi+] to [psi−] in *Saccharomyces cerevisiae. Genetics* 98: 691–711

Wickner, R.B. (1994) [URE3] as an altered URE2 protein: evidence for a prion analog in *Saccharomyces cerevisiae. Science* 264: 566–69

Zhou, P., Derkatch, I.L., Patino, M.M., Lindquist, S. & Liebman, S.W. (1999) The yeast non-Mendelian factor [*ETA*+] is a variant of [*PSI*+], a prion-like form of release factor eRF3. *Science* 18: 1182–91

GLOSSARY

α-helix secondary structure common in proteins where the polypeptide backbone is twisted in a right-handed spiral to form a rigid rod-like structure

β-sheet secondary structure common in proteins comprising a folded polypeptide chain where the strands that lie parallel to each other are bound together by hydrogen bonding

coenocytic where protoplasm of fungal or algal tissue is not divided by cell walls

C-terminus end of a polypeptide chain that contains a free COO^- group, the last part of the chain synthesized

N-terminus the other end of a polypeptide chain from the C-terminus. The N-terminal amino acid is the first to be incorporated when a polypeptide is synthesized

null allele mutant allele lacking a functional gene product

protease digestion degradation of proteins by splitting internal peptide bonds to create peptidesz

See also **Genetics of animal and human prion diseases (p.451)**

F MOUSE GENETICS

The Mouse *t* complex

Genetic studies with mouse chimaeras

Ethylnitrosourea (ENU) as a genetic tool in mouse mutagenesis

Coat colour genetics of mouse, man and other animals

Behavioural and neural genetics of the mouse

The mouse as a model for the study of genetic aspects of alcoholism

Body weight limits in mice

Introduction to Mouse Genetics

1. Origin of the house mouse and laboratory strains of mice

The house mouse is believed to have lived in close association with human populations since the end of the last ice age, about 12 000 years ago, when nomadic hunters and gatherers in the Fertile Crescent of the Middle East began to settle down and cultivate plants and livestock. Settled farming required dry food storage under cover to feed the family or local population and livestock, and provided an ideal food source also for small mice of the species *Mus musculus*. This commensal relation between mice and humans spread with human agricultural populations throughout Europe and Asia, and later to the rest of the world, and it was undoubtedly assisted by the speed at which mice can move and their ability to squeeze through holes as small as a centimetre in diameter.

Another important factor was the development of mice as "pets", particularly in Japan and China, encouraged by the fact that they became easy to handle and to breed, while the inbreeding this entailed led to the appearance of a variety of colour and other phenotypic variations, due to the fixing of recessive mutations.

2. Development of inbred strains for genetic study

After the rediscovery of Mendel's laws in 1900, geneticists began to purchase animals from the commercial breeders of "fancy" mice, from which they derived the classical inbred strains of laboratory mice that formed the material for future genetic studies. The main supplier in the US was Abbie Lathrop, from her farm in Massachusetts. Mouse genetics soon became an important subject, with laboratories starting at Harvard University in 1908, Cold Spring Harbor, Long Island in 1919 and the Jackson Laboratory at Bar Harbor, Maine started by Clarence Little in 1929 with money from Detroit industrialists. The first International Committee on Standard Nomenclature began in 1939, when there were only 31 known gene loci and seven linkage groups; the British Medical Research Council's Radiation Unit began to study mouse mutations in 1947 and the *Mouse News Letter* began 40 years of publication under that name in 1949, perhaps taking its lead from *Drosophila Information Service*, founded in the early 1930s. Among an increasing flow of new discoveries, the "obese" mouse (later shown to have a key mutation in the *leptin* gene) was reported at the Jackson Laboratory in 1950 and became the first animal model for human obesity.

The first edition of *Genetic Variants and Strains of the Laboratory Mouse*, edited by M.F. Lyon, appeared in 1982, and the second edition, edited by M.F. Lyon and A.G. Searle, both at the MRC Radiobiology Unit at Harwell, was published in 1989. These volumes cover all that was known about mouse gene mutants, polymorphic loci, recombination percentages and chromosomal assignments, linkage and synteny homologies in mouse and man, the wild mouse and its relatives and much more that is still of absorbing interest (though the third, two-volume edition appeared recently with much more information).

3. The expanding mouse economy

A News Focus section in *Science* for 14 April 2000 (288: 248–57, The Rise of the Mouse, Biomedicine's Model Mammal) describes recent growth in supplies, academic facilities and property claims over laboratory mouse strains. Some of the points made are that in the year 2000 probably more than 25 million mice were bred for biomedical and genetic research, more than 90% of all mammals used and double the number used a decade ago. "Once a modest regional business, the mouse trade is now a global enterprise that is being transformed by scientists' growing ability to fine-tune the genetic variation of these model animals." This rapidly growing trade has to meet the problems of continual demand for new genetic strains, supplying precisely what is required, avoiding all possible sources of infection or contamination and keeping their prices competitive. Transgenic mice are said to routinely cost $175 each, while some rare pairs will be worth $30 000. Legal problems also affect research due to the mass of patents being applied for.

4. Sequencing the mouse genome

The mouse genome is about the same size as the human genome, despite the mouse being much smaller, so the sequencing of its genome should need about the same time as the human genome. *Nature* (14 December 2000, pp.758–59) explains that the mouse genome has "already been sequenced once over", but it should be sequenced three times over and assembled before being released to the public. This should be achieved by April 2001.

THE MOUSE *T* COMPLEX

John Schimenti
The Jackson Laboratory, Bar Harbor, Maine, USA

1. Introduction

The mouse *t* haplotype occupies a prominent place in mouse genetics, beginning with its discovery reported in 1927 by Dobrovolskaia-Zavadskaia and Kobozieff. It was found quite by accident in crosses of a laboratory-derived mutation called *T*, or Brachyury, to wild-derived animals. *T* is a semidominant mutation that is lethal when homozygous and causes a short tail in heterozygotes (*T*/+). However, tailless animals were produced in some crosses of *T* mutants with the wild mice. The gene carried in these mice was presumed to be an allele of *T* and was named *t* (now called *tct*, for *t* complex tail interaction). This was just the tip of the iceberg.

The *t* "alleles" had an array of remarkable phenotypes. Aside from the interaction with *T* to cause taillessness, they caused homozygous lethality. Strangely, mice heterozygous for two different *t* alleles (t^x/t^y) had normal tails, but the males were invariably sterile. Furthermore, *t* alleles caused a strong recombination suppression in +/*t* heterozygotes across a 20 centiMorgan (cM) region of proximal chromosome 17. Finally, and most startling, was that male mice heterozygous for a *t* haplotype (such as +/*t* or *T*/*t*) exhibited preferential transmission of the *t* allele to nearly all offspring.

For decades, it was assumed that a single gene – *T* – was responsible for this array of diverse phenotypes. Eventually, genetic and molecular tools revealed that these phenotypes were caused by distinct genes and chromosomal rearrangements and arose during the evolution of a selfish genetic element we now call the *t* haplotype. The variety of unusual biological properties associated with the *t* haplotypes has attracted researchers from many disciplines. The combined interest in these phenomena has driven an extensive molecular analysis of the *t* complex, resulting in one of the most comprehensive genetic and physical maps in the mouse genome. The richness of mutations and molecular resources in the *t* complex continue to make this a fascinating experimental system for biological research. Our current understanding of this topic, while still far from complete, is summarized in this article.

2. The mouse *t* complex and *t* haplotypes in a nutshell

t haplotypes are naturally occurring variant forms of the *t* complex, a 20 cM stretch of DNA on the proximal end of mouse chromosome 17, representing approximately 1% of the mouse genome. Although most *t* haplotypes contain a recessive developmental lethal mutation, and males heterozygous for two complementing *t* haplotypes are sterile, these variant chromosomes propagate in wild mouse populations due to male transmission ratio distortion (TRD). This process causes male mice heterozygous for a *t* haplotype and a wild-type form of the *t* complex (*t*/+) to transmit the *t* chromosome to nearly all of their offspring. Recombination in +/*t* heterozygotes is greatly suppressed due to four relative inversions of genetic material between *t* and + chromosomes. This region of recombination suppression formally defines the *t* complex, and allows *t* haplotypes, with the collection of alleles contained therein, to propagate as intact genetic units in mouse populations.

3. Evolution of *t* haplotypes

The diverse mutations associated with *t* haplotypes appear to be intricately linked. Genetic evidence suggests that the genes responsible for homozygous male sterility (the *t* complex sterility loci or *tcs*), are identical to a class of *t* haplotype genes called *t* complex distorters (*Tcd*), which are essential for TRD (Lyon, 1986). If true, this indicates that TRD comes at a price: it carries the "side-effect" of homozygous sterility. As detailed below, all the distorters must be present for a high-level TRD to occur. The five distorter loci are distributed across the entire length of the *t* complex (**Figure 1**). Ordinarily, these alleles would be shuffled constantly via meiotic recombination; however, *t* haplotypes contain four chromosomal inversions relative to the wild-type, which genetically "lock" all the TRD genes together. Each of the TRD genes is associated with one of the inversions (Hammer *et al.*, 1989).

The evolution of these inversions is one of the more interesting stories concerning *t* haplotypes. Since the *t* chromosomes are viewed as the mutants, it was assumed that they independently acquired each of the four inversions. While this appears to be true for inversions *in(17)1*, *in(17)3* and *in(17)4*, genetic and molecular analyses led to the conclusion that *in(17)2* actually occurred in the lineage of present day "wild-type" chromosomes (Schimenti *et al.*, 1987; Hammer *et al.*, 1989). Hence, *t* haplotypes have preserved the primordial version of this 3 cM region. This finding indicates that *in(17)2* must have existed as a polymorphism in mouse populations, prior to the emergence of "complete" *t* haplotypes, a view which is supported by molecular analysis showing that the divergence between *t* and + is greatest in *in(17)2* (Morita *et al.*, 1992; Hammer & Silver, 1993). The other three inversions occurred within the last ~2 million years and became fixed, presumably because they acquired TRD loci that provided an advantage over the wild-type (Silver, 1993). Indeed, all modern day *t* haplotypes are derived from a recent single common ancestor, as evidenced by the extremely low degree of polymorphism between isolates trapped around the world (Morita *et al.*, 1992; Hammer & Silver, 1993). It has been suggested that the original inversion in the + lineage contained a gene(s) that conferred some advantage over non-inverted chromosomes. In

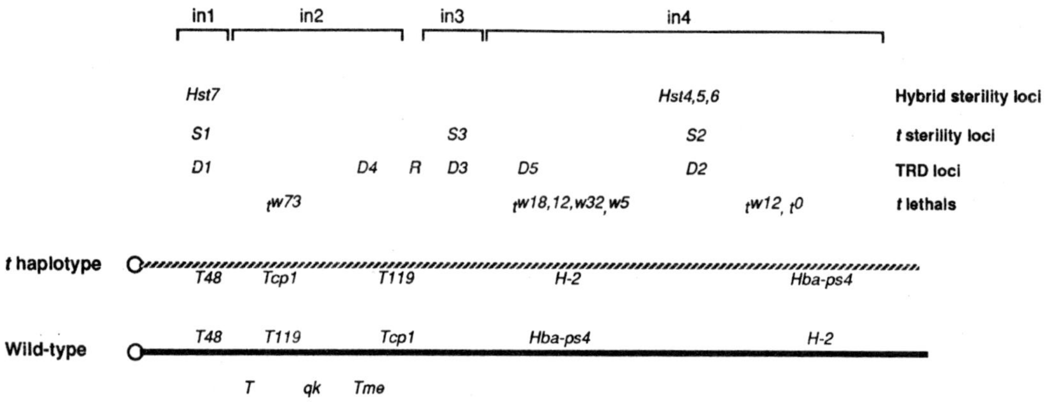

Figure 1 Maps of the mouse *t* complex. Structure of the *t* haplotype and wild-type forms of the *t* complex are shown. The locations of the inversions *in(1)17–in(4)17* are shown above, and abbreviated in1–in4. Loci of interest are shown, with the distorter genes *Tcd1* through *Tcd5* abbreviated as D1–D5. Between the *t* and + chromosomes are landmark loci that illustrate the two major inversions. Below the + chromosome are mutations found only in wild-type: *T* (Brachyury), *qk* (quaking) and *Tme* (T maternal effect). Mapping of *t* lethals was reported by Artzt (1984).

retaliation, a chromosome lacking *in(17)2* evolved its own series of distortion loci and inversions in order to survive (Silver, 1993). Our current understanding of the mechanism of TRD is discussed later.

The recessive developmental lethal mutations present in *t* haplotypes, of which at least 16 are known to exist (Committee for Chromosome 17, 1992), oddly appear to be essential for the successful spread of *t* haplotypes (Lyon, 1986). Since mice live in small groups called demes, in which a single male will mate with several females, a sterile dominant male would lead to extinction of that deme. It has been hypothesized that the recessive lethal mutations were selectively retained as a means to eliminate sterile (*t/t*) males from demes carrying a *t* haplotype (Lyon, 1986). The fact that so many independent *t*-associated lethals have been fixed in such a short time following the origin of *t* haplotypes (about 1–2 million years ago) argues in favour of the idea that the lethals were selectively maintained. In a sense, *t* haplotypes can be considered as a genome microcosm that has evolved a multi-tiered strategy for reproductive success, and has essentially begun to "speciate" away from wild-type chromosome 17.

4. Transmission ratio distortion and male sterility

Although +/*t* males produce equal numbers of + and *t* sperm, they transmit the *t* haplotype to nearly all of their offspring. Artificial insemination experiments have shown that *t* sperm are not superior to +, but that the + sperm from +/*t* heterozygotes are functionally inactivated (Olds-Clarke & Peitz, 1985). Beyond that, the nature of this inactivation is unclear. One set of experiments has shown that about half the sperm in a +/*t* male undergo a premature acrosome reaction, an event that normally occurs only upon egg contact. Although the genotypes of these unfortunate sperm could not be identified, the implication was that they contained the + chromosome (Brown *et al.*, 1989).

TRD is believed to occur through the action of at least five *trans*-acting *t* complex distorter (*Tcd*) loci on a *t* complex responder (*Tcr*) locus (Lyon, 1984; Silver & Remis, 1987). *Tcr* is the central locus in TRD; if it is absent, no distortion occurs. A male that carries all four *Tcd* loci and is heterozygous for *Tcr* can transmit the *Tcr*-containing chromosome 17 homologue to more than 95% of progeny. However, if a male is heterozygous for *Tcr* and no *Tcd* loci are present, the *Tcr*-containing homologue is transmitted to less than 15% of offspring. The distorters act in an additive fashion; as more are present, *Tcr* transmission increases (Lyon, 1984). These phenomena are summarized in **Figure 2**.

Tcr resides within a small region (<200 kilobases) in the centre of *t* haplotypes (the *D17Leh66B* locus), as defined by molecular and genetic analyses of recombinant chromosomes called partial *t* haplotypes (**Figures 1** and **2**) (Fox *et al.*, 1985; Schimenti *et al.*, 1988; Rosen *et al.*, 1990). Partial *t* haplotypes are the products of rare recombination events between a *t* haplotype and a wild-type form of the *t* complex. These recombinants contain only a portion of *t* haplotype DNA and often contain duplications or deletions because the crossovers occurred across inversions. In conjunction with genetic studies that determined whether particular partial *t* haplotypes contained a responder, it was found that only those partial *t* haplotypes with the *D17Leh66B* locus displayed *Tcr* activity (**Figure 2**) (Lyon, 1984; Fox *et al.*, 1985; Schimenti *et al.*, 1988). The search for *Tcr* in this interval uncovered a candidate called $Tcp10^{bt}$ (Schimenti *et al.*, 1988), but transgenic and gene targeting experiments later ruled this candidate out (Bullard *et al.*, 1992; Snyder & Silver, 1992; Ewulonu *et al.*, 1996). Recently, *Tcr* was identified just upstream of $Tcp10^{bt}$, and is a hybrid between a novel "Smok" kinase and ribosomal 56 kinase (Herrmann *et al.*, 1999). The function of this mutant, hybrid molecule remains to be elucidated. The male sterility associated with homozygosity of various *t* haplotype regions is thought to be directly linked to TRD. Specifically, each of the three major *t* complex distorter loci, *Tcd1–Tcd3*, co-maps with a *t* complex sterility (*tcs*) locus. An intriguing piece of data that lends insight into the distortion and sterility phenomena is that a deletion called T^{22H}, which lacks sequences corresponding to the proximal *in(17)4* region of the *t* complex, behaves like alleles of *Tcd1* and *tcs1* (Lyon,

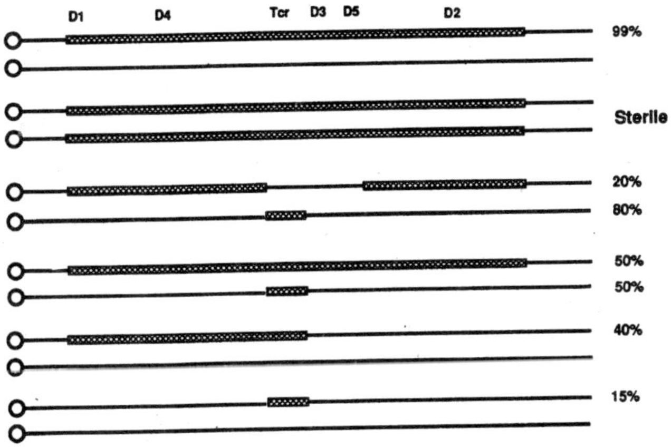

Figure 2 Genetics of transmission ratio distortion (TRD). A general summary of segregation values of males containing various combinations of partial and complete *t* haplotypes is shown. *T* haplotype chromatin is indicated by the hatched boxes, and + DNA by solid horizontal lines. Locations of TRD genes are shown at the top where D1–D5 stand for distorters *Tcd1*–*Tcd5*. *Tcr* is the *T* complex responder.

1992). This suggested that the distorter and sterility factors in *t* haplotypes are loss-of-function mutations. Furthermore, it strengthened the idea that the distorter and sterility genes are identical, or at least closely linked. A model based on distorter–responder interaction and gene dosage has been proposed to account for how a distorter gene can confer distortion when heterozygous and sterility when homozygous (Lyon, 1986).

A major obstacle to further investigations into TRD and eventual positional cloning of the responsible genes is the issue of recombination suppression, which precludes standard genetic mapping. Localization of these loci has, therefore, been achieved using the limited set of partial *t* haplotypes (Schimenti *et al.*, 1987). Hence, the regions in which distorter/sterility genes map are extremely large (some in excess of several cM). While several candidate genes for sterility/distortion factors have been identified on the basis of testis-specific expression patterns (Rappold *et al.*, 1987; Lader *et al.*, 1989; Mazarakis *et al.*, 1991), the absence of recombinational data renders it difficult to either include or exclude them as true candidates. The finding that the TRD genes may be null mutations opens the door for applying newly developed methods of deletion mapping to localize these genes more effectively (Ramirez *et al.*, 1995; You *et al.*, 1997). The several *t* haplotype genes causing TRD and sterility, and the genetic evidence to suggest that they interact, provide an attractive system for elucidating sperm development. An understanding of the TRD system will also reveal molecular strategies in meiotic drive, the nature of mutations that cause infertility and the rapid co-evolution of unrelated genes.

5. Hybrid sterility genes

Hybrid sterility is a mechanism that maintains the genetic integrity of species by preventing productive interspecific breeding. This phenomenon acts across the gamut of sexually reproducing life forms and, almost without exception, Haldane's rule applies – the heterogametic sex (males, in the case of mammals) is infertile in interspecific hybrids.

In addition to the *t* haplotype distortion/sterility alleles, it is remarkable that five of the six known (and verified) mouse hybrid sterility genes reside within the *t* complex. It is unclear whether the *t* complex indeed houses the bulk of these genes or if the special attention paid to this region is responsible for the skewing. Nevertheless, it is attractive to speculate, given the data summarized below, that the hybrid sterility genes are identical to some of the TRD genes. The behaviour, then, of the *t* haplotypes in house mice would be semi-analogous to the effects of hybrid sterility genes in interspecific hybrids.

Hst1 (*Hybrid sterility 1*) was the first hybrid sterility gene to be identified in mammals (Forejt & Ivanyi, 1974). It maps to a 1 Megabase region in the central *t* complex (**Figure 1**) (Gregorova *et al.*, 1996). This gene is responsible for causing infertility in F_1 hybrid males generated from matings of certain laboratory strains of mice with wild-derived *Mus musculus*.

Additional hybrid sterility genes, *Hst4*–*Hst7*, were identified more recently. They cause sterility in different genetic combinations of *M. spretus* and *M. musculus domesticus* (Pilder *et al.*, 1991, 1993). The *Hst6* gene has an interesting property that implies possible identity to the distorter, *Tcd2* – the $Hst6^S/Hst6^t$ combination (the *M. m. domesticus*, *t* haplotype and *M. spretus* alleles of these genes are denoted with the superscripts d, t and s, respectively) causes a "curlicue" flagellar motility phenotype indistinguishable from mice homozygous for the *t* allele of *Tcd2* (Pilder *et al.*, 1993). However, whereas males homozygous for $Hst6^s$ in a *M. m. domesticus* background are also sterile, their sperm do not display the flagellar phenotype exhibited by Hst^t/Hst^t sperm. Instead, they have morphologically abnormal "flagellae" that do not move at all. Additionally, Hst^t/Hst^d mice are fertile. These data suggest that the *M. m. domesticus* allele is dominant, and the spretus allele is a loss-of-function mutation in the context of the *M. m. domesticus* genome.

A final gene, *Hst7*, has been found in the proximal *t* complex, near *Tcd1* (Pilder, 1997). The *M. spretus* allele of this gene causes sterility against either a *t* or *d* allele. However, the sperm abnormalities are distinct. The accumulated data suggest that the *t* allele of *Hst7* might be identical to the *t* allele of the distorter, *Tcd1*.

The plethora of mutations involved in spermatogenesis raises the question of whether the *t* complex is a special "sperm development" chromosomal region. In other words, is the density of spermatogenesis genes in the *t* complex higher than the average chromosome or has the special scrutiny on the *t* complex been responsible for the discovery of more phenotypes? The answer to this is unknown, but investigators in the field nevertheless marvel at the striking array of sperm phenotypes harboured by *t* and + forms of the *t* complex, and the fact that the distortion and sterility genes interact in a genetically cooperative fashion.

6. Developmental mutations in the *t* complex

A major focus of the study of *t* haplotypes over the decades has been the series of lethal mutations. Most *t* haplotypes trapped around the world possess a recessive lethal mutation that acts early in development. Even though haplotypes all share a recent common ancestor, they have managed to acquire and maintain such mutations. As described above, it is thought that such mutations were selectively acquired to counteract the deleterious population effects of homozygous male sterility.

To reiterate, it was thought that all the "*T* locus" effects – TRD, sterility, recombination suppression, developmental abnormalities, tail length alterations – were due to alleles of a single gene, *T*. The Brachyury allele was dominant over the developmental lethal and sterility phenotypes carried by "*t*" alleles, and different *t* alleles affected processes at different times in embryonic development. Of course, we now know that: (1) *T* and *t* are not allelic; (2) the different *t* alleles affecting development are in fact distinct loci; (3) the recombination suppression is due to inversions; and (4) TRD is caused by another set of genes. Much of the terminology remains, however, which can confuse non-*t* aficionados.

Even after the realization that the *t* alleles represented different genes, they remained of high interest due to their interesting effects on a range of developmental timepoints and processes. Furthermore, because of the interaction of *tct* with Brachyury mutations to cause taillessness, it was simple for scientists to maintain the mutations in the days before DNA typing. For example, a cross of the type *T/t* × *T/t*, in which the parents are both tailless, only produces more tailless offspring. This is because *T/T* is lethal from severe notochord defects and *t/t*, which would have a normal tail, is not viable due to homozygosity of a t^{lethal} allele. In some cases, the tail effects could also allow a genotype determination of mutant embryos.

Unfortunately, despite decades of developmental characterization of the *t* mutations, none has been cloned to date (see below for a description of the non-*t* haplotype genes within the *t* complex that have been identified); the recombination suppression precludes high resolution genetic mapping in +/*t* animals. However, recombination does occur in *t/t* mice, which has been exploited to localize some of the lethal mutations (Artzt *et al.*, 1982; Artzt, 1984). Additionally, the partial *t* haplotypes have been useful in this regard. A brief description of the phenotypes of some of the better characterized *t* lethals is given below.

(i) t^{12}

Embryos homozygous for this allele progress normally to the morula stage, but then degenerate and do not progress to become blastocysts (Bennett, 1964). This is one of the earliest acting embryonic lethals known in mammals.

(ii) t^{w73}

Mutant embryos die within 5 days of implantation due to faulty trophectoderm development (Spiegelman *et al.*, 1976). These defects appear shortly after implantation and the mutants can be identified at 6 days of development, when normal litter-mates are double-layered, elongate egg cylinders. The mutants form egg cylinders, but are smaller, do not form close associations with the uterine wall and have a markedly absent or abnormal ectoplacental cone. While there are severe defects with extraembryonic cells and structures, the embryo proper develops relatively normally, but is retarded in growth.

(iii) t^{0}

Homozygous embryos implant and grow normally until the point when embryonic and extraembryonic ectoderm become differentiated from the former inner cell mass (ICM) (Gluecksohn & Hoenheimer, 1940). The ICM grows little and the differentiation never takes place. The defects first appear at day 5.5 postcoitum, followed by death and resorption within 2 days. Mesoderm formation never begins.

(iv) t^{w5}

These mutants progress normally through implantation, separation of embryonic and extra-embryonic ectoderm, and the beginnings of forming the elongate egg cylinder. At this point, the embryonic ectoderm becomes *pycnotic* and ceases growth. The extra-embryonic portion of the embryo remains essentially unaffected. The embryo, which then consists almost entirely of extra-embryonic cell types, is resorbed 2 or 3 days later.

(v) t^{w18}

This is a recombinant partial *t* haplotype whose defect comes from a deletion caused by recombination across *in(17)4* (Bucan *et al.*, 1987). This mutation affects the growth of the primitive streak cells. The effects are apparent at the late egg cylinder stage and are characterized by the formation of a much enlarged primitive streak that bulges into the surrounding ectoderm, often causing a duplication of the neural tube by this apparently mechanical distortion of the embryonic shape (Bennett & Dunn, 1960). The primitive streak cells do not complete the transition of mesoderm, due to a failure to migrate and form proper interactions typical of a gastrulating embryo (Spiegelman & Bennett, 1974).

(vi) t^{w1}

This is the latest acting of all the *t* lethals (Bennett, 1964). The effect is specifically directed towards the ventral portion of the neural tube and brain, and affected animals can be discerned as early as 9 days' gestation. Some, however, survive until birth, showing abnormal brains and spinal cord.

Several other very interesting mutations exist within the *t* complex, but in the non-*t* haplotype forms. The most famous is *T*, which was cloned several years ago (Herrmann *et al.*, 1990), and which is a key molecule in mesoderm formation. Another is *T* maternal effect (*Tme*), which was the first imprinted locus to be discovered in mice. Deletions of this gene cannot be transmitted by the mother, indicating that the paternal allele is inactive or insufficient during development. *Tme* appears to be the insulin-like growth factor receptor gene (*Igf2r*) (Barlow *et al.*, 1991). Another notable gene that has recently been cloned is quaking (*qk*), a neurological mutation that was historically a key genetic marker for mapping experiments in the *t* complex (Ebersole *et al.*, 1996). A gene involved in sex determination, *Tas* (*t*-associated sex reversal), was localized to the proximal *t* complex by virtue of deletions in the region (Washburn & Eicher, 1983, 1989; Washburn *et al.*, 1990). Finally, no listing of significant loci on chromosome 17 can be complete without mentioning that the entire major histocompatibility complex (*H-2*) resides in the distal *t* complex. Curiously, some *t* lethals appear to be embedded within the *H-2*.

7. Future prospects

At one time, the *t* complex provided one of the most compelling genetic systems in mouse genetics. However, in the last decade or two, interest has waned because of the discovery that the *T/t* "locus" was not a single gene, but a collection of genes spread over a large chromosomal region, and the realization that issues such as recombination suppression were going to make the identification of the interesting mutations very difficult. Furthermore, the revolution in mouse developmental genetics precipitated by the advent of transgenic and knockout technologies reduced the near monopoly the *t* complex once held with regards to developmental mutations. Consequently, the flow of new investigators into *t* complex research has seriously decreased and some scientists have stopped studying the *t* complex altogether.

Nevertheless, the *t* complex, with its extraordinary properties, still hold its own mystique and scientific novelty. The mouse genome project, which has led to high resolution genetic and physical maps, is providing the tools necessary to overcome the barriers to the structural characterization of *t* haplotypes. These advances, coupled with increasingly powerful transgenic technologies that enable the deletion or addition of extremely large chromosomal regions in mice, will enable scientists to home in on *t* haplotype genes by complementation or mutation.

In conclusion, after a period of decline, it would seem that the *t* complex is destined for a period of renaissance – the evolution of the *t* haplotype has been described as a "peculiar journey" (Silver, 1993), and so also has its history been in genetics research.

Acknowledgements

I would like to thank Lee Silver for introducing me to the *t* complex and a career in mouse genetics. This work was supported by grant HD24374 from the National Institutes of Health.

REFERENCES

Artzt, K. (1984) Gene mapping within the T/t complex of the mouse. III: *t*-Lethal genes are arranged in three clusters on chromosome 17. *Cell* 39: 565–72

Artzt, K., Shin, H.-S. & Bennett, D. (1982) Gene mapping within the T/t complex in the mouse: anomalous position of the H-2 complex in *t* haplotypes. *Cell* 28: 471–76

Barlow, D.P., Stoger, R., Herrmann, B.C., Saito, K. & Schweifer, N. (1991) The mouse insulin-like growth factor type-2 receptor is imprinted and closely linked to the Tme locus. *Nature* 349: 84–87

Bennett, D. (1964) Embryological effects of lethal alleles in the t-region. *Science* 144: 263–67

Bennett, D. & Dunn, L.C. (1960) A lethal mutant (t^{w18}) in the house mouse showing partial duplications. *Journal of Experimental Zoology* 143: 203–19

Brown, J., Cebra-Thomas, J.A., Bleil, J.D., Wassarman, P.M. & Silver, L.M. (1989) A premature acrosome reaction is programmed by mouse *t* haplotypes during sperm differentiation and could play a role in transmission ratio distortion. *Development* 106: 769–73

Bucan, M., Herrmann, B.C., Frischauf, A.M. *et al.* (1987) Deletion and duplication of DNA sequences is associated with the embryonic lethal-phenotype of the t9 complementation group of the mouse t complex. *Genes and Development* 1: 376–85

Bullard, D., Ticknor, C. & Schimenti, J. (1992) Functional analysis of a *t complex responder* locus transgene in mice. *Mammalian Genome* 3: 579–87

Committee for Chromosome 17 (1992) Mouse chromosome 17. *Mammalian Genome* 3: S241–60

Dobrovolskaia-Zavadskaia, N. & Kobozieff, N. (1927) Sur la reproduction des souris anoures. *Comptes Rendus des Seances de la Societe de Biologie et des Ses Filiales* 97: 116–19

Ebersole, T.A., Chen, Q., Justice, M.J. & Artzt, K. (1996) The quaking gene product necessary in embryogenesis and myelination combines features of RNA binding and signal transduction proteins. *Nature Genetics* 12: 260–65

Ewulonu, U.K., Schimenti, K.J., Kuemerle, B., Magnuson, T. & Schimenti, J. (1996) Targeted mutagenesis of a candidate *t* complex responder gene in mouse *t* haplotypes does not eliminate transmission ratio distortion. *Genetics* 144: 785–92

Forejt, J, & Ivanyi, P. (1974) Genetic studies on male sterility of hybrids between laboratory and wild mice (***Mus musculus*** L.). *Genetical Research* 24: 189–206

Fox, H., Martin, G., Lyon, M.F. *et al.* (1985) Molecular probes define different regions of the mouse *t* complex. *Cell* 49: 63–69

Gluecksohn, S. & Hoenheimer, S. (1940) The effect of an early lethal (t^0) in the house mouse. *Genetics* 25: 391–400

Gregorova, S., Mnukova-Fajdeloya, M., Trachulec, Z. *et al.* (1996) Sub-milliMorgan map of the proximal part of mouse chromosome 17 including the hybrid sterility 1 gene. *Mammalian Genome* 7: 107–13

Hammer, M.F. & Silver, L.M. (1993) Phylogenetic analysis of the alpha-globin pseudogene-4 (Hba-ps4) locus in the house mouse species complex reveals a stepwise evolution of t haplotypes. *Molecular and Biological Evolution* 10: 971–1001

Hammer, M.H., Schimenti, J.C. & Silver, L.M. (1989) On the origins of *t* complex inversions. *Proceedings of the National Academy of Sciences USA* 86: 3261–65

Herrmann, B.C., Labeit, S., Poustka, A., King, T.R. & Lehrach, H. (1990) Cloning of the T gene required in mesoderm formation in the mouse. *Nature* 343: 617–22

Herrmann, B.C., Koschorz, B., Wortz, K., McLaughlin, K. & Kisport, A. (1999) A protein kinase encoded by the t complex responder gene causes non-Mendelian inheritance. *Nature* 402: 141–46

Lader, E., Ha, H.-S., O'Neill, M., Artzt, K. & Bennett, D. (1989) *tctex-l*: a candidate gene for a mouse *t* complex sterility locus. *Cell* 58: 621–28

Lyon, M.F. (1984) Transmission ratio distortion in mouse *t* haplotypes is due to multiple distorter genes acting on a responder locus. *Cell* 37: 621–28

Lyon, M.F. (1986) Male sterility of the mouse *t*-complex is due to homozygosity of the distorter genes. *Cell* 44: 357–63

Lyon, M. (1992) Deletion of mouse *t*-complex distorter-1

produces an effect like that of that of the *t*-form of the distorter. *Genetical Research* 59: 27–33

Mazarakis, N.D., Nelki, D., Lyon, M.F. *et al.* (1991) Isolation and characterisation of a testis-expressed developmentally expressed gene from the distal inversion of the mouse *t*-complex. *Development* 111: 561–71

Morita, T., Kubota, H., Murata, K. *et al.* (1992) Evolution of the mouse *t* haplotype: recent and worldwide introgression to *Mus musculus*. *Proceedings of the National Academy of Sciences USA* 89: 6851–55

Olds-Clarke, P. & Peitz, B. (1985) Fertility of sperm from +/*t* mice: evidence that +-bearing sperm are dysfunctional. *Genetical Research* 47: 49–52

Pilder, S.H. (1997) Identification and linkage mapping of *Hst7*, a new *M. spretus*/*M. m. domesticus* chromosome 17 hybrid sterility locus. *Mammalian Genome* 8: 290–91

Pilder, S.H., Hammer, M.F. & Silver, L.M. (1991) A novel mouse chromosome 17 hybrid sterility locus: implications for the origin of *t* haplotypes. *Genetics* 129: 237–46

Pilder, S., Olds-Clarke, P., Phillips, D.L.S. (1993) *Hybrid sterility-6*: a mouse *t* complex locus controlling sperm flagellar assembly and movement. *Developmental Biology* 159: 631–42

Ramirez Solis, R., Liu, P. & Bradley, A. (1995) Chromosome engineering in mice. *Nature* 378: 720–24

Rappold, G., Stubbs, L., Labeit, S., Crkvenjakov, R. & Lehrach, H. (1987) Identification of a testis-specific gene from the mouse *t* complex next to a CpG-rich island. *EMBO Journal* 6: 1975–80

Rosen, L., Builard, D. & Schimenti, J. (1990) Molecular cloning of the *t* complex responder genetic locus. *Genomics* 8: 134–40

Schimenti, J., Vold, L., Socolow, D. & Silver, L.M. (1987) An unstable family of large DNA elements in the center of mouse t haplotypes. *Journal of Molecular Biology* 194: 583–94

Schimenti, J.C., Cebra-Thomas, J., Decker, C.L. *et al.* (1988) A candidate gene family for the mouse *t complex responder* (*Tcr*) locus responsible for haploid effects on sperm function. *Cell* 55: 71–78

Silver, L. (1993) The peculiar journey of a selfish chromosome: mouse *t* haplotypes and meiotic drive. *Trends in Genetics* 9: 250–54

Silver, L.M. & Remis, D. (1987) Five of the nine genetically defined regions of mouse *t* haplotypes are involved in transmission ratio distortion. *Genetical Research* 49: 51–69

Snyder, L.C. & Silver, L.M. (1992) Distortion of transmission ratio by a candidate *t* complex responder locus transgene. *Mammalian Genome* 3: 588–96

Spiegelman, M. & Bennett, D. (1974) Fine structural study of cell migration in the early mesoderm of normal and mutant mouse embryos (*T*-locus: t^9/t^9). *Journal of Embryology and Experimental Morphology* 32: 723–38

Spiegelman, M., Artzt, K. & Bennett, D. (1976) Embryological study of a *T*/*t* locus mutation (t^{w73}) affecting trophectoderm development. *Journal of Embryology and Experimental Morphology* 36: 373–81

Washburn, L.L. & Eicher, E.M. (1983) Sex reversal in XY mice caused by dominant mutation on chromosome 17. *Nature* 303: 338–40

Washburn, L.L. & Eicher, E.M. (1989) Normal testis determination in the mouse depends on genetic interaction of a locus on chromosome 17 and the Y chromosome. *Genetics* 123: 173–79

Washburn, L.L., Lee, B.K. & Eicher, E.M. (1990) Inheritance of T-associated sex reversal in mice. *Genetical Research* 56: 185–91

You, Y., Bergstrom, R., Klemm, M. *et al.* (1997) Chromosomal deletion complexes in mice by radiation of embryonic stem cells. *Nature Genetics* 15: 285–88

GLOSSARY

deme small, local, closely-related interbreeding population

ectoplacental cone a cluster of extraembryonic cells at the distal end of a late egg cylinder-stage embryo, destined to form the placenta

gastrulating embryo embryo in the gastrulation stage, i.e. where the body plan is set up.

gene dosage number of genes, or numbers of copies of a specific gene, in a nucleus, cell, etc

morula mammalian embryo immediately before the blastocyst stage (8–16 cells in the mouse)

notochord cells of mesodermal origin running along the back in the early chordate embryo, directing formation of the neural tube; in vertebrates is replaced by the spinal column

null mutation mutation lacking a functional gene product

pycnotic relating to cells that degenerate and form intensely staining clumps of chromosomes

recombination suppression a circumstance under which meiotic crossing over in a genetic interval is reduced to a level substantially lower than is typical

trophectoderm also known as trophoblast, the trophectoderm comprises the tissue layer that forms the blastocyst wall

GENETIC STUDIES WITH MOUSE CHIMAERAS

John D. West
Department of Obstetrics and Gynaecology, University of Edinburgh, Edinburgh, UK

Experimental mouse chimaeras were first produced in the early 1960s and were particularly enthusiastically used for studies in developmental biology during the 1970s. In recent years, chimaeras have mostly been relegated to providing a means for making "genetic knockout" mice via embryonic stem cell chimaeras. It is now clear, however, that chimaeras can provide powerful tools for the phenotypic analysis of abnormal genotypes (including those generated by genetic knockout technology) and this has rekindled a wider interest in mouse chimaeras. This article reviews many of the early contributions of chimaeras to mouse genetics which laid the foundations for current research. Also highlighted are more recent studies which demonstrate the power of chimaeras, when used with modern transgenic cell markers. This increased analytical power explains why the use of chimaeras is now undergoing a renaissance.

1. What are chimaeras and how are they made?

Chimaeras, like mosaics, are multicellular organisms composed of two or more genetically distinct cell populations (other than those normally present in the germline). In mosaic animals, the different cell populations arise from a single zygote, but, in a chimaeric animal, they are derived from more than one zygote. (In the plant kingdom, no such distinction is made and both types are considered to be chimaeras.) Most geneticists are familiar with mosaicism in Drosophila and mammals, where it is frequently encountered either as chromosome mosaicism or functional mosaicism, resulting from random X-chromosome inactivation (see section 4).

Mammalian chimaeras are usually subdivided into primary and secondary chimaeras. Secondary chimaeras are formed when tissues are combined by a variety of tissue grafting or transplantation techniques from two or more adults or post-implantation embryos. Chimaerism is usually restricted to one or a few tissues, but the results may be dramatic, as in the case of interspecies rat→mouse secondary chimaeras resulting in rat spermatogenesis in a mouse testis (Clouthier *et al.*, 1996). Primary chimaeras are formed at a very early stage of development, so that all body tissues may be involved. Rare cases of spontaneous human primary chimaeras have been documented and, for other mammalian species, primary chimaeras may be produced experimentally – either by aggregation of cleavage stage embryos (Tarkowski, 1961) or by microinjection of cells into the cavity of a blastocyst stage embryo (Gardner, 1968), where they become incorporated into the inner cell mass (**Figure 1**). Chimaeras were originally produced by combining two embryos, but it is also possible to incorporate pluripotent embryonic stem cells into chimaeras by injection, aggregation or co-culture techniques (Wood *et al.*, 1993).

Aggregation chimaeras are usually designated A↔B, where A and B represent the genotypes or strains of the two aggregated embryos or cells; in some early studies, they were referred to as allophenics or tetraparentals, rather than chimaeras. Aggregation of two embryos means that the chimaeric embryo is initially double the normal size, but size regulation occurs soon after implantation and the chimaeric pup is normal size at term. The A↔B convention may also be used to represent the genotype combination of injection chimaeras, but A→B is sometimes a more useful notation (where "A→B" implies the injection of A-type cells into a B-type blastocyst). Primary chimaeras can also be produced by recombining different tissues of two blastocyst stage embryos. Typically, an inner cell mass of one genotype will be combined with a trophectoderm of another genotype to form a blastocyst reconstitution chimaera.

Coat and eye pigment (**Figure 2a,b**) provide convenient genetic markers for visualizing patches of clonally related cells in a primary chimaera, but other genetic markers have to be included to identify the contribution of the two genetically distinct cell populations to non-pigmented tissues. Electrophoretic variants of enzymes have been extensively used as genetic markers in mouse chimaeras since the late 1960s (**Figure 2c**) and are still widely used for both qualitative and quantitative analysis. Cytogenetic and tissue-specific histochemical markers were also used in early experiments, but these have now been largely superseded by transgenic markers which can provide spatial information in many tissue types. Transgenic cells are typically identified in histological sections either by detection of reporter gene expression, such as β-galactosidase histochemical staining for *lacZ* expression (**Figure 2d**), or by DNA *in situ* hybridization (**Figure 2e**).

If chimaeras are required with a population of cells that are homozygous for a lethal gene or a gene that causes infertility, this may cause further complications. Such homozygous mutant embryos must be produced by intercrosses between two heterozygotes (*m*/+ × *m*/+). Usually, the required *m*/*m* embryos are not identified before the chimaeras are produced, so the different chimaeric genotypes (*m*/*m*↔+/+, *m*/+↔+/+ and +/+↔+/+) must be identified retrospectively. This may be difficult because both *m*/*m*↔+/+ and *m*/+↔+/+ chimaeras contain *m* and + alleles. Some genetic approaches to this problem have been discussed elsewhere (West, 1999), but it can also be avoided by producing chimaeras from previously genotyped *m*/*m* embryonic stem cell lines that are derived from *m*/+ × *m*/+ embryos (Varlet *et al.*, 1997; see sections 2 and 9).

Although primary chimaeras have been produced for several mammalian species, including interspecies combinations, mouse aggregation chimaeras and injection

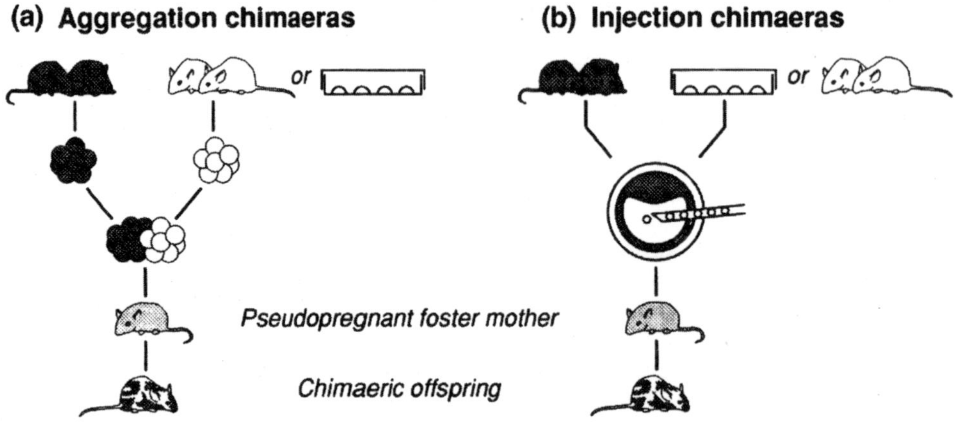

Figure 1 (a) Production of aggregation chimaeras (e.g. albino↔pigmented) (Tarkowski, 1961). Typically, two or more genetically distinct preimplantation stage embryos are recovered from the reproductive tract, their zonae pellucidae are removed (usually by exposure to acidic Tyrode's solution), they are aggregated together in a culture drop, cultured overnight to the blastocyst stage and surgically transferred to the uterus of a pseudopregnant female recipient (previously mated to a sterile male) for further development. (Sometimes adhesion of the embryos is facilitated by various means such as exposure of the embryos to phytohaemagglutinin, making indentations in the culture dish or by using a multiwell plate which either has small V-shaped wells or can be briefly centrifuged.) Preimplantation embryos may also be aggregated with cultured, pluripotent embryonic stem (ES) cells to produce chimaeras (Wood *et al.*, 1993). (b) Production of injection chimaeras (e.g. albino→pigmented) (Gardner, 1968). Pluripotent cells for injection may be isolated from a preimplantation stage embryo (e.g. inner cell mass cells) or maintained in an undifferentiated state in culture (e.g. ES cells). These cells are microinjected into the cavity of a genetically distinct blastocyst stage embryo with a micromanipulator. The injected blastocyst is cultured briefly and then surgically transferred to the uterus of a pseudopregnant female recipient for further development. The blastocyst consists of an outer layer of trophectoderm cells which envelopes the inner cell mass (epiblast plus primitive endoderm) and blastocyst cavity. (Reproduced with permission from West, 1999.)

chimaeras have been the most widely used. The uses of mouse chimaeras in developmental biology have been reviewed elsewhere (McLaren, 1976; Le Douarin & McLaren, 1984; Gardner, 1998; West, 1999) and this article will focus on their uses in genetics.

2. Production of transgenics with embryonic stem cell chimaeras

As noted above, primary chimaeras can be made by combining mouse embryos with pluripotent cells that have been cultured *in vitro*. The aim of this approach is to make genetic changes in tissue culture cells, select *in vitro* for the mutant or transgenic cells, use these to make germline chimaeras and transmit the altered genotype to mice. Such chimaeras were first produced with embryonal carcinoma (EC) cells, but the chimaeras often developed tumours and, more critically, the EC cells failed to colonize the germline. The production of embryonic stem (ES) cells overcame this problem. When injected into blastocysts or aggregated with morulae, ES cells produce a high rate of chimaerism in tissues derived from the epiblast (foetal lineage) including the germline, but colonize the extraembryonic primitive endoderm and trophectoderm extraembryonic lineages less frequently. The HPRT (hypoxanthine phosphoribosyltransferase)-deficient mouse was the first mutant produced by ES cell chimaera technology, because HPRT-deficient ES cells could be easily selected in culture. Other selection strategies were subsequently devised and mouse ES cell chimaeras are now widely produced as a means of making specific changes in the mouse genome (Hooper, 1992; Joyner, 1993).

3. Genetic effects on the composition of chimaeras

The composition of individual mouse chimaeras of the same strain combination usually varies widely, but chimaeras of some strain combinations are consistently unbalanced so that one strain predominates. This may reflect a generalized selective advantage of cells of one genotype or might occur if those cells are preferentially allocated to the foetal lineage early in development. Within an individual chimaera, the composition of different tissues is usually remarkably similar so that the compositions of most organs in an adult chimaera tend to be positively correlated with one another (Falconer *et al.*, 1981). Nevertheless, there are notable exceptions to this generalization and tissue-specific effects occur in some strain combinations which argue for genotype-dependent, tissue-specific selection pressures. Mintz (1970) introduced the concept of "SAM", the statistical allophenic mouse, to describe the most likely bias in the composition of different tissues (relative to the overall body composition) for a specific strain combination.

4. Validation of genetic mosaicism with chimaeras

The similarity between the variegated patterns seen in the coats of chimaeras (which are known to contain two genetically distinct cell populations) and X-linked heterozygotes provided important evidence in favour of the single active X hypothesis (Lyon hypothesis) rather than the alternative complemental-X hypothesis. (The latter proposed that both X chromosomes were active in each cell, but that gene expression was regulated so that total X-linked genetic activity was equal in XX and XY cells.) It is now

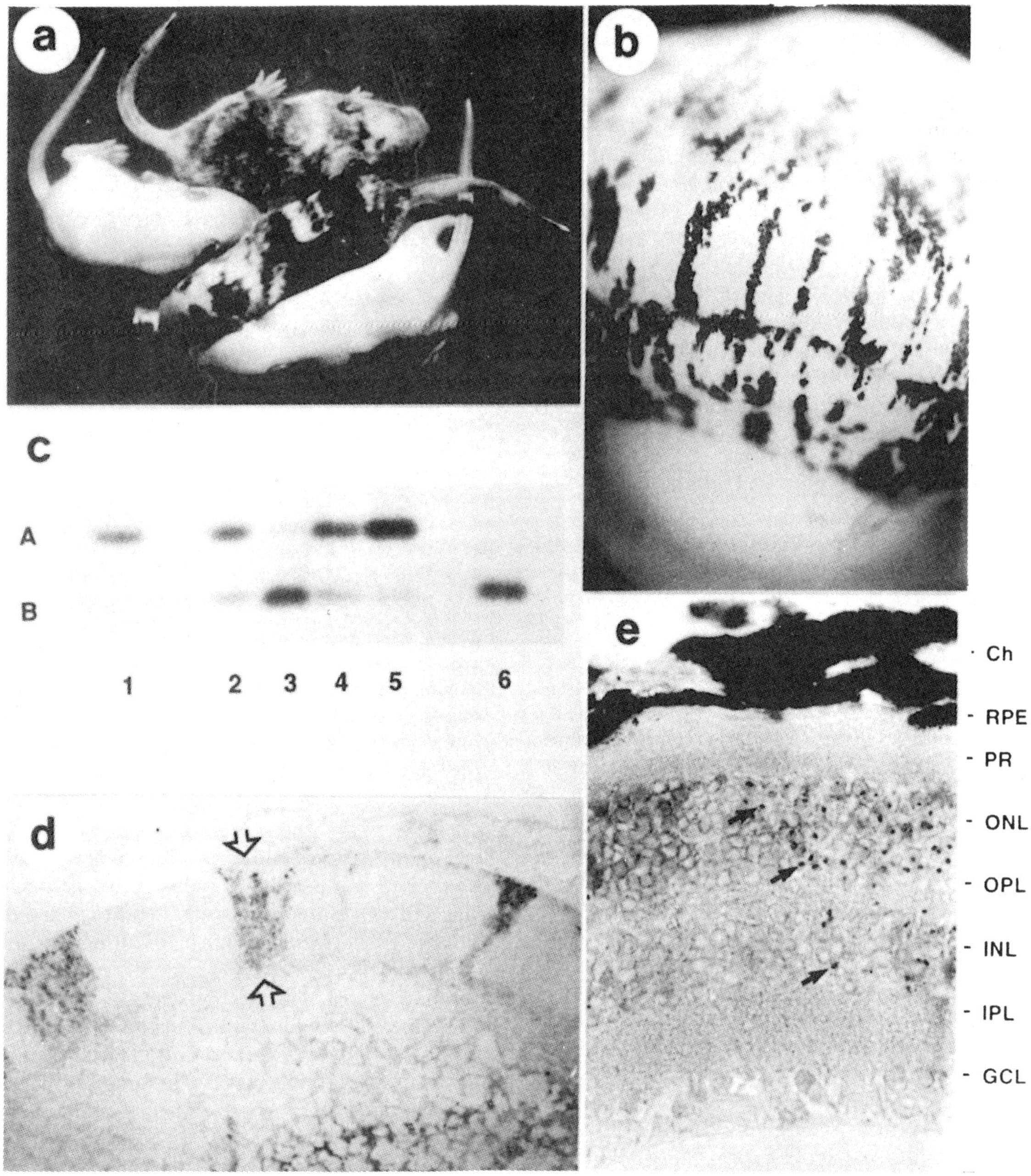

Figure 2 Some useful genetic markers for chimaera studies. (a) Two pigmented↔albino chimaeric mouse pups with their non-chimaeric, albino littermates. Coat pigmentation provides an excellent marker for identification of the chimaeras. Melanocyte clones migrate from the neural crest and often form visible stripes either side of the dorsal midline. (b) Part of an adult pigmented↔albino chimaeric eye, showing stripes of pigment in the chimaeric retinal pigment epithelium, RPE (top two-thirds of the photograph), near where it meets the iris. The RPE is visible in this eye because the overlying choroid is largely unpigmented. (c) Electrophoretic variants of glucose phosphate isomerase (GPI1). Samples 2–5 are from *Gpi1*^{a}/*Gpi1*^{a}↔*Gpi1*^{b}/*Gpi1*^{b} chimaeras and produced a mixture of GPI1A and GPI1B allozymes (labelled a and b), which were visualized by electrophoresis and histochemical staining for GPI1. (d) Transgenic marker TgR(ROSA26)26Sor, detectable by β-galactosidase histochemical staining for *lacZ* expression in a section of a hemizygous transgenic↔non-transgenic (*LacZ*+/−↔−/−) chimaeric mouse adrenal gland. Patches of stained (transgenic) and unstained (non-transgenic) cells form distinct stripes across the adrenal cortex (e.g. open arrows), as previously described for chimaeric rat adrenals (Iannaccone & Weinberg, 1987). (E) Transgenic marker TgN(Hbb-b1)83Clo, detectable by DNA *in situ* hybridization in a section of part of an eye from a pigmented, transgenic↔albino, non-transgenic mouse chimaera. Most of the hemizygous transgenic cells (with a single hybridization signal in their nucleus) are on the right of the photograph in the INL and ONL (e.g. arrows). This plane of focus shows the majority of the hybridization signals, but the cells are out of focus; the RPE is also out of focus. Abbreviations: Ch, choroid (outer dark pigmented layer); RPE, retinal pigment epithelium (inner pigmented layer); PR, photoreceptor cells; ONL, outer nuclear layer; OPL, outer plexiform layer; INL, inner nuclear layer; IPL, inner plexiform layer; GCL, ganglion cell layer. (Some photographs are reproduced from West, 1999.)

known that X-chromosome inactivation occurs early in development, so that X-linked heterozygotes are considered to be functional mosaics with each cell only expressing one of the two available X chromosomes. The pattern of variegation in these X-inactivation mosaics is essentially equivalent to that seen in chimaeras except that the composition of X-inactivation mosaics is less variable and is seldom grossly unbalanced. More recently, several cases of variegated patterns of gene expression in transgenic mice have been attributed to mosaicism, including that arising by position effect variegation (Dobie *et al.*, 1996). The variegated pattern of expression of a steroid 21-hydroxylase/ β-galactosidase transgene in the mouse adrenal gland (Morley *et al.*, 1996) is sufficiently similar to the variegated patterns seen in rat (Iannaccone & Weinberg, 1987) and mouse (**Figure 2d**) chimaeras to make mosaicism the most likely explanation.

5. Studies of quantitative genetic traits with chimaeras

Chimaeras have been used for studies of several inherited quantitative traits. The general approach is to make chimaeras from two stocks that differ for the quantitative trait, measure the trait in the chimaera and estimate the percentage contributions of the two parental stocks in a number of body organs. In three early studies of skeletal morphogenesis (Grüneberg & McLaren, 1972), body size (Falconer *et al.*, 1981) and behaviour (Nesbitt, 1984), the measured values in chimaeras were intermediate between those in the parent strains. This implies that these traits are controlled by a mixture of cells from both genotypes rather than a single clone of cells of like genotype. In another study (Dewey & Maxson, 1982), C57BL/6↔DBA chimaeras with balanced coat colour composition were very variable in their susceptibility to sound-induced seizures (DBA strain mice are susceptible). This suggests that this phenotype may be controlled by a small number of cells, whose composition does not correlate with the general level of chimaerism reflected by the coat colour. Attempts to determine whether body size (Falconer *et al.*, 1981) and behaviour (Nesbitt, 1984) are controlled by specific organs were made by testing whether the measured trait covaried with the composition of any particular organ(s). Unfortunately, this aspect of the analysis proved relatively unrewarding because the composition of most organs in a chimaera tend to be positively correlated with one another (Falconer *et al.*, 1981) and different tissues within each organ could not be analysed separately with the markers available at the time. Other studies of behaviour using mouse chimaeras have been reviewed elsewhere (Goldowitz, 1992) and the aggressive behaviour of parthenogenetic↔normal chimaeras is discussed briefly in section 8. There is also evidence for "vegetative heterosis" for body weight (Falconer *et al.*, 1981), litter size (Mikami & Onishi, 1985) and hippocampal anatomy (Crusio *et al.*, 1990) in chimaeras (i.e. the measured value exceeds the quantitative range defined by the two parental genotypes). Unlike heterosis seen in F_1 hybrids (hybrid vigour), "vegetative heterosis" must involve interactions between cells of different genotypes in chimaeric tissues.

6. Sex chromosomes in chimaeras

When chimaeras are made by randomly aggregating pairs of 8-cell stage embryos, 50% will be of mixed sex chromosome composition. The development of these XX↔XY chimaeras has been a source of interest to geneticists and reproductive biologists alike. Intuitively, one might predict that they would all develop as intersexes. Indeed, at foetal stages, many XX↔XY chimaeras have ovotestes (Bradbury, 1987), but the ovarian tissue usually regresses so that the postnatal gonad becomes a testis. Thus, in a balanced strain combination, the majority of XX↔XY chimaeras develop as phenotypic males and the sex ratio is close to 3 males:1 female. XX↔XY chimaeras appear only to develop as females (or occasionally as intersexes) if the proportion of XY cells is low.

The Sertoli cells of the testis and the follicle cells of the ovary are supporting cells derived from epithelial cells in the gonadal ridge. In XX↔XY female chimaeras, a significant proportion of follicle cells can be XY (Burgoyne *et al.*, 1988; Patek *et al.*, 1991). Although XX↔XY male chimaeras often have some XX Sertoli cells, the proportion of XX cells is usually much lower than for other testicular cells (such as Leydig cells) which more closely reflects the XX contribution to non-gonadal tissues (Palmer & Burgoyne, 1991; Patek *et al.*, 1991). One possibility is that *Sry* (the Y-linked, testis-determining gene) acts in XY pre-Sertoli cells, but that a few neighbouring XX prefollicle cells are somehow recruited into the Sertoli cell population even though they lack *Sry*. The presence of only a few XX Sertoli cells in XX↔XY male chimaeras does not itself distinguish between: (i) cell autonomous action of *Sry* (in XY cells), followed by recruitment of some XX prefollicle cells as Sertoli cells, and (ii) non-autonomous action of *Sry* to promote Sertoli cell differentiation in neighbouring XY pre-Sertoli or XX prefollicle cells. (If XX and XY cells were not finely intermixed, most cells neighbouring an XY cell would also be XY and this could account for the paucity of XX Sertoli cells.) Although *Sry* is a transcription factor that binds DNA and so almost certainly acts cell-autonomously, downstream genes activated by *Sry* could act either cell autonomously or non-autonomously.

In many chimaeric tissues, the two cell populations are finely intermixed, but the ovarian and testicular tissues in a foetal ovotestis are commonly spatially separated into testicular and ovarian domains (Bradbury, 1987). This may indicate that "sorting out" occurs during the aggregation of Sertoli cells to form testicular cords. Sex determination in XX↔XY chimaeras is unusual in that it includes a mechanism for converting an ovotestis into a testis. The current interpretation of sex determination in XX↔XY chimaeras (McLaren, 1991; Burgoyne & Palmer, 1992; Burgoyne, PS, personal communication) is that, if there are sufficient XY pre-Sertoli cells, some testicular cords are formed and, in these cords, the germ cells enter mitotic arrest as prospermatogonia. In areas where there are insufficient pre-Sertoli cells, testicular cords fail to form and the germ cells enter meiosis to become oocytes. These oocytes are subsequently eliminated and the ovarian component then regresses, thereby transforming the ovotestis into a testis. (Elimination of the meiotic oocytes is probably mediated by antiMullerian hormone, which is produced by the Sertoli cells in the testicular regions of the ovotestis.) Occasionally, when there are too few XY pre-Sertoli cells, testicular cords do not form. Consequently, little or no antiMullerian hormone is produced and the oocytes survive and direct the prefollicle cells (and a few pre-Sertoli cells) to form ovarian follicles.

7. Chimaeras with autosomal anomalies

Trisomic (Ts) mouse or human embryos mostly die during postimplantation development and monosomic (Ms) embryos usually die before implantation. In contrast, human trisomy/diploid (Ts/2n) mosaics are often viable and frequently occur as confined placental mosaicism in humans. Several trisomies have been incorporated into chimaeras in order to test whether trisomic cells can be rescued in different tissues of the foetus or adult (Epstein, 1985) or provide models for human mosaicism. Trisomic cells usually contribute to both foetal and extraembryonic tissues, but there is evidence for modest selection against them in certain tissues of the foetus or neonate. A much stronger selection was apparent against Ms19 cells in 9.5 day Ms19↔$2n$ mouse chimaeras (Magnuson *et al.*, 1982) and few Ms19 cells would be likely survive to term.

At least one tetraploid↔diploid ($4n$↔$2n$) mouse chimaera has survived postnatally with a tetraploid cell contribution (Lu & Markert, 1980). Usually, however, the tetraploid cells fail to contribute to the foetus or other derivatives of the epiblast lineage (amnion, visceral yolk sac mesoderm, allantois), but contribute well to the trophectoderm and primitive endoderm lineages (placenta, visceral yolk sac endoderm and parietal endoderm of Reichert's membrane) (Nagy *et al.*, 1990; James *et al.*, 1995). Unlike the Ts and Ms chimaeras so far studied, $4n$↔$2n$ chimaeras do provide an animal model for some types of human confined placental mosaicism. The poor ability of tetraploid cells to contribute to the foetus has also been exploited to maximize the contribution of ES cells to the foetuses of $4n$-embryo↔$2n$-ES cell chimaeras (Nagy *et al.*, 1990).

8. Studies of genomic imprinting with chimaeras

Elegant pronuclear transplantation studies have demonstrated that the maternal and paternal genomes are not equivalent, so implying the existence of genes that are differentially imprinted, depending on the parent of origin (McGrath & Solter, 1984; Surani *et al.*, 1984). Diploid parthenogenetic embryos (produced from a female gamete without participation of a male gamete), diploid gynogenetic embryos (male pronucleus replaced with a female pronucleus after fertilization) and diploid androgenetic embryos (female pronucleus replaced with a male pronucleus after fertilization) all fail to develop to term. Aggregation chimaeras, incorporating transgenic markers (**Figure 2d,e**), were produced to investigate the developmental potential of parthenogenetic and androgenetic cells beyond the time when the parthenogenetic and androgenetic embryos die.

Parthenogenetic cells in parthenogenetic↔normal and androgenetic cells in androgenetic↔normal aggregation chimaeras are initially present in all lineages at the blastocyst stage (Thomson & Solter, 1989), but later they show a more restricted distribution. Parthenogenetic cells are excluded from the primitive endoderm and trophectoderm lineages and, although they survive in the foetus and other epiblast derivatives, the parthenogenetic cells are gradually depleted in most tissues. They contribute well to the oocytes and forebrain, but not to the hypothalamus/midbrain or most other tissues (Nagy *et al.*, 1989; Allen *et al.*, 1995; Bender *et al.*, 1995). Parthenogenetic↔normal chimaeras have small bodies, but their brains are normal in size (or possibly enlarged) and male chimaeras with a high parthenogenetic contribution in their brains tend to be more aggressive than normal (Allen *et al.*, 1995). These studies suggest a role for imprinted genes in the development of the central nervous system and the control of behaviour.

Androgenetic cells usually become confined to the trophectoderm lineage in postimplantation stage androgenetic↔normal aggregation chimaeras, but they will contribute to the foetus if they are injected into normal blastocysts to make androgenetic→normal injection chimaeras. Some features of androgenetic→normal chimaeras are the reciprocal of parthenogenetic↔normal chimaeras. At the early foetal stage, these chimaeras are larger than normal and androgenetic cells tend to colonize those tissues that parthenogenetic cells fail to colonize (hypothalamus/midbrain and most proliferating tissues) (Fundele *et al.*, 1995; Keverne *et al.*, 1996). Striking embryonic growth enhancement also occurred when inner cell mass cells, with a paternal duplication of the distal region of chromosome 7 (PatDi7 cells), were injected into wild-type blastocysts to produce PatDi7→normal injection chimaeras (Ferguson-Smith *et al.*, 1991). The authors suggested that rapid growth may be caused by overexpression of an imprinted gene, such as the insulin-like growth factor gene (*Igf2*), which maps to the distal region of chromosome 7 and shows exclusive paternal expression.

9. Phenotypic analysis of mutant genes and genetic knockouts with chimaeras

The current vogue for producing genetic knockouts attests to the power of the genetic approach to developmental biology. However, if the genetic defect causes death of the embryo or abnormal development of a whole organ it will be unclear, from this gross phenotype alone, which tissues are primarily affected. Analysis of mouse embryos with an early-acting lethal null mutation will reveal the earliest critical time for normal gene expression, but later roles and the range of tissues affected will be obscured. Sometimes, these issues can be clarified by a chimaeric-rescue analysis where mutant and normal (wild-type) cells are combined in a mouse chimaera. Chimaeras are extremely useful for the phenotypic analysis of abnormal genotypes because the presence of the wild-type cells will often allow the mutant cells to survive beyond the time at which a fully mutant embryo would die. If the mutant cells are excluded from a specific tissue, this is likely to mean that normal gene function is required in that tissue.

The following examples illustrate how chimaeras can

reveal whether genes act in a cell-autonomous fashion and determine the developmental potential of mutant cells. The first two examples (dystrophia muscularis and retinal degeneration) illustrate early studies of mutant genes with chimaeras. Although they were necessarily limited by the lack of suitable cell markers that could be visualized *in situ*, they used chimaeras in elegant ways to address questions that are still relevant today. The remaining examples illustrate how improved cell markers (transgenes detectable either by DNA *in situ* hybridization or by *lacZ* histochemistry) have improved the resolution of this type of analysis. The last two examples (*Mash2* and nodal) also demonstrate how chimaeras, made with cells with a restricted tissue distribution, can be exploited to determine the site of gene action.

(i) A muscular dystrophy gene that acts outside the muscle

Mice homozygous for the dystrophia muscularis gene (*dy* or dy^{2J}) on chromosome 10 show a progressive weakness and paralysis beginning with the hindlimbs at about 3½ weeks and usually die by 6 months. Aggregation chimaeras were made by aggregating normal (+/+) embryos and homozygous *dy*/*dy* or dy^{2J}/dy^{2J} embryos (Peterson, 1974, 1979). The muscle phenotype (normal or dystrophic) was assessed histologically and the muscle genotype (+/+, dy^{2J}/dy^{2J} or mixed) was determined by enzyme electrophoresis, using variants of malic enzyme or glucose phosphate isomerase (see **Figure 2c**) that differed between the wild-type and dystrophic strains. The first experiment showed that some phenotypically normal muscles had a high proportion of genotypically dystrophic muscle fibres. The second experiment showed that many muscle fibres were phenotypically normal, but genotypically completely dystrophic (entirely from the *dy*/*dy* strain). This implies that the primary site of *dy* gene action is outside the muscle fibre.

(ii) Chimaeras reveal three different modes of action of retinal degeneration genes

Several genetic defects result in the degeneration of the photoreceptor cells, the nuclei of which lie in the outer nuclear layer (ONL) of the neural retina (see **Figure 2e**). **Figure 3** illustrates examples of three genotypes that produce a similar phenotype (degeneration of the photoreceptors/ONL but not the inner layers) that can be readily distinguished by chimaera experiments. **Figure 3** depicts the retinal pigment epithelium (RPE) and neural retina of five different types of adult mouse chimaera. They are all pigmented↔unpigmented, so that the RPE has patches of pigmented and unpigmented cells, but they differ for genes that cause degeneration of the ONL. **Figure 3a** represents a chimaeric eye where both of the cell populations in the chimaera are wild-type with respect to retinal degeneration (+/+↔+/+) and the neural retina is uniformly of full thickness. **Figure 3b** shows the phenotype if both cell populations are homozygous for the mouse retinal degeneration (*rd*) gene (*rd*/*rd*↔*rd*/*rd*). In this case, the neural retina is uniformly thin because the ONL has degenerated. In the pigmented↔albino eyes depicted in **Figure 3c–e**, the genotype of one cell population causes retinal degeneration whereas the other is wild-type. Homozygosity for each of the three mutant genotypes results in degeneration of the photoreceptors/ONL, similar to that shown in **Figure 3b** for *rd*, but the three genotypes are readily distinguishable when incorporated into chimaeras.

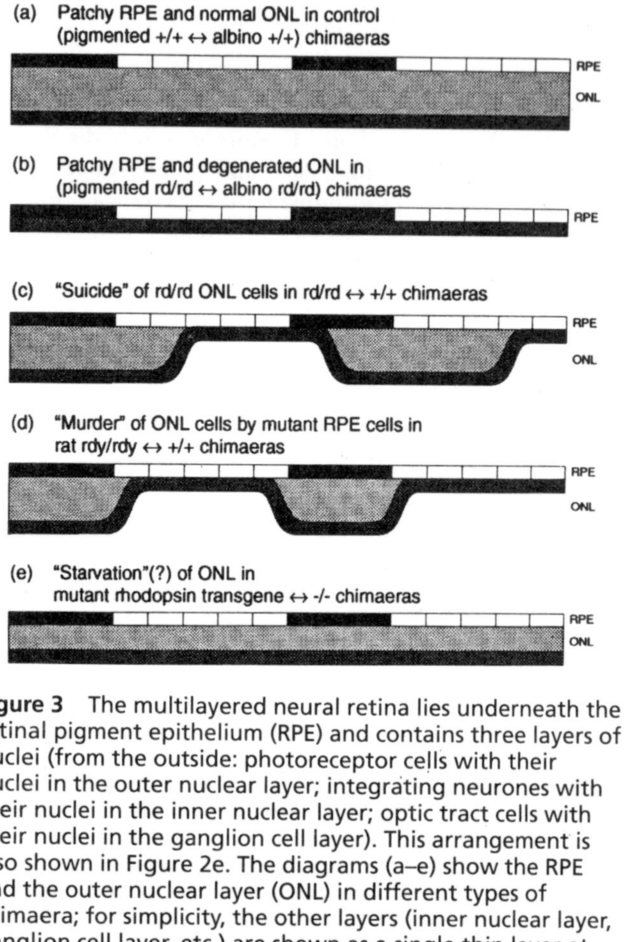

Figure 3 The multilayered neural retina lies underneath the retinal pigment epithelium (RPE) and contains three layers of nuclei (from the outside: photoreceptor cells with their nuclei in the outer nuclear layer; integrating neurones with their nuclei in the inner nuclear layer; optic tract cells with their nuclei in the ganglion cell layer). This arrangement is also shown in Figure 2e. The diagrams (a–e) show the RPE and the outer nuclear layer (ONL) in different types of chimaera; for simplicity, the other layers (inner nuclear layer, ganglion cell layer, etc.) are shown as a single thin layer at the bottom of each diagram. All of the chimaeras are pigmented↔unpigmented, so the RPE has patches of pigmented and unpigmented cells, but they differ for genes that cause degeneration of the photoreceptors/ONL. (a and b) represent control chimaeras and (c–e) depict chimaeras where one cell population carries a gene causing degeneration of the photoreceptors/ONL. (a) Both cell populations in the chimaera are wild-type with respect to retinal degeneration (+/+↔+/+) and the neural retina is uniformly of full thickness. (b) Both cell populations are homozygous for the mouse retinal degeneration, *rd*, gene (*rd*/*rd*↔*rd*/*rd*) so the neural retina is uniformly thin. (c) One cell population carries retinal degeneration (*rd*/*rd*↔+/+) and patches of normal and degenerate ONL are visible, but these show no spatial relationship with the patches in the RPE. (d) One cell population is homozygous for rat retinal dystrophy (*rdy*), which causes retinal degeneration (unpigmented *rdy*/*rdy*↔pigmented +/+); the ONL degenerates only in regions adjacent to unpigmented RPE. (e) Mouse chimaeras where one cell population carries a mutant pig rhodopsin transgene which causes retinal degeneration: uniform ONL of intermediate thickness. (Reproduced with permission from West, 1999.)

Figure 3c represents a mouse chimaera where one cell population carries retinal degeneration (*rd*/*rd*↔+/+). Several authors have reported patches of normal and degenerate photoreceptors/ONL in such chimaeras. Studies of both pigmented *rd*/*rd*↔albino +/+ and pigmented +/+↔albino *rd*/*rd* chimaeras showed that there was no spatial relationship between the patches of pigmented and albino cells in the RPE and the patches of degeneration in the ONL (LaVail & Mullen, 1976). The *rd* gene, therefore, acts in the neural retina and probably in the photoreceptors themselves.

Figure 3d illustrates a rat chimaera where one cell population is homozygous for retinal dystrophy (*rdy*), which causes retinal degeneration. Experiments with non-pigmented *rdy*/*rdy*↔pigmented +/+ rat chimaeras showed that the ONL degenerated only in regions that were adjacent to unpigmented RPE (Mullen & LaVail, 1976). This elegantly demonstrated that, in this case, the primary defect was in the overlying RPE and not in the neural retina itself. In effect, the non-pigmented *rdy*/*rdy* cells in the RPE "murdered" the underlying neural retina cells.

Figure 3e represents a mouse chimaera with one cell population that carries a pig rhodopsin transgene, that causes retinal degeneration (Huang *et al.*, 1993). *In situ* hybridization with a probe specific for pig rhodopsin RNA revealed the expected patchy distribution of transgenic and wild-type cells in the neural retina. However, instead of patches of degenerate and normal ONL, the chimaeras had a uniform ONL of intermediate thickness implying uniform degeneration of both wild-type and transgenic photoreceptor cells. As both wild-type and transgenic cells degenerate, the gene causing the degeneration must act non-autonomously and cell interactions are probably involved. One explanation, suggested by the researchers, is that each photoreceptor cell releases a trophic factor, but it must also take up this nutrient to survive (each cell contributes to and draws on a common pool present in the retina). If the transgenic photoreceptors released less of the factor, but required the normal amount, the pool would become depleted. Eventually, the pool of nutrients would be insufficient and both wild-type and transgenic photoreceptor cells would die.

(iii) *Mouse* Pax6 *acts cell-autonomously in the lens and nasal placodes*

The mouse *Pax6* gene encodes a transcription factor and mutations in this gene are responsible for the small eye mutant phenotype. Mouse embryos, homozygous for the small eye ($Pax6^{Sey}$) mutation, die soon after birth with severe facial abnormalities that result from the failure of the eyes and nasal cavities to develop. As a general disruption of eye and nasal development occurs in the homozygous $Pax6^{Sey}$ embryos, it is unclear which eye and nose tissues require functional *Pax6*. Foetal mouse chimaeras were made by aggregating wild-type embryos with embryos from matings between mice heterozygous for different small eye alleles ($Pax6^{Sey\text{-}Neu}/+ \times Pax6^{Sey}/+$) (Quinn *et al.*, 1996). The $+/+ \leftrightarrow Pax6^{Sey\text{-}Neu}/Pax6^{Sey}$ foetal chimaeras were identified by PCR and the contribution of mutant $Pax6^{Sey\text{-}Neu}/Pax6^{Sey}$ cells was analysed in tissue sections by *in situ* hybridization to a reiterated transgenic marker. The morphology of the optic cup was severely affected in these chimaeras. Mutant cells were excluded from the retinal pigment epithelium and did not intermix with wild-type cells in other regions of the optic cup, suggesting that *Pax6* affects the cell surface properties of these cells. Even more strikingly, mutant cells were excluded from both the lens and nasal epithelium and both tissues were smaller (sometimes absent) in chimaeras with high proportions of mutant cells. Since the wild-type cells were unable to rescue the mutant cells in these tissues, *Pax6* must act cell-autonomously. *Pax6* function is presumably required in the ectoderm from which the lens and nasal placodes are derived and mutant ectoderm cells are, therefore, unable to differentiate into lens or nasal epithelium.

(iv) *Chimaeras show that* Mash2 *acts in the placenta not in the foetus*

Mash2 (mammalian achaete-scute homologue 2) encodes a transcription factor that is expressed in the extraembryonic trophoblast lineage. $Mash2^{-/-}$ knockout mouse embryos die from placental failure at 10 days postcoitum (placental spongiotrophoblast cells and their precursors are absent and chorionic ectoderm is reduced). Tetraploid $Mash2^{+/+} \leftrightarrow$ diploid $Mash2^{-/-}$ chimaeras have been used to exploit the restricted colonization pattern of tetraploid cells and rescue $Mash2^{-/-}$ foetuses (Guillemot *et al.*, 1994). The chimaeras were produced by aggregating 4-cell tetraploid $Mash2^{+/+}$ embryos with morula stage embryos produced from ($Mash2^{+/-} \times Mash2^{+/-}$) intercross matings. It was predicted that, in the tetraploid $Mash2^{+/+} \leftrightarrow$diploid $Mash2^{-/-}$ chimaeras, the tetraploid (wild-type $Mash2^{+/+}$) cells would colonize the trophectoderm lineage (trophoblast) and extraembryonic endoderm lineage, but be excluded from the foetus itself (see section 7). This would produce a foetus composed entirely of mutant $Mash2^{-/-}$ (diploid) cells, but, in some cases, the foetus would be supported by a significant contribution of wild-type $Mash2^{+/+}$ (tetraploid) cells in the extraembryonic lineages. This expectation was fulfilled and about 15% of the viable chimaeric pups at term were uniformly mutant $Mash2^{-/-}$ (no $Mash2^{+}$ DNA identified). This neatly demonstrates that, normally, the lethality of $Mash2^{-/-}$ conceptuses is a consequence of the absence of *Mash2* in the trophectoderm (and/or the primitive endoderm) extraembryonic lineage. Together with other evidence, the viability of $Mash2^{-/-}$ foetuses implies that *Mash2* has no major role in the embryo itself, but plays a critical part in the development of the mammalian trophoblast lineage.

(v) *Anterior embryonic structures depend on gene expression in an adjacent extraembryonic tissue*

The mouse nodal gene encodes a member of the TGF-β family of secreted growth factors, which is expressed in both the epiblast (which produces the foetus) and the overlying primitive endoderm (which produces only extraembryonic endoderm). Homozygous nodal-deficient embryos ($nodal^{-/-}$) fail to initiate primitive streak formation and so arrest at gastrulation without forming any mesoderm. Wild-type or $nodal^{-/-}$ ES cells (carrying a *lacZ* lineage marker) were injected into wild-type or nodal$^{-/-}$ blastocysts and the embryonic phenotype and the contribution of the injected ES cells were examined to test whether wild-type cells could rescue the mutant phenotype (Varlet *et al.*, 1997).

Embryonic, +/+ ES cell→nodal$^{-/-}$ injection chimaeras were small, but gastrulation had occurred, indicating that the presence of wild-type cells had overcome the block to gastrulation. Use of the *lacZ* lineage marker indicated that chimaeras with 10–30% wild-type cells were almost fully rescued in the posterior region, but, despite good contributions of wild-type cells, the anterior structures were absent. Comparisons between +/+ ES cell→nodal$^{-/-}$ chimaeras and the reciprocal nodal$^{-/-}$ ES cell→+/+ chimaeras, showed that anteriorly truncated embryos were only produced when nodal$^{-/-}$ cells were derived from the host blastocyst, not from the injected ES cells, despite good contributions of nodal$^{-/-}$ cells in nodal$^{-/-}$ ES cell→+/+ chimaeric embryos.

The difference between these two groups of injection chimaeras is accounted for by the previous observation that ES cells contribute well to the epiblast (foetal lineage), but not to the overlying primitive endoderm (see section 2). The anterior truncation only occurred when nodal$^{-/-}$ was introduced from the host blastocyst and so contributed to the primitive endoderm as well as the epiblast. This implies that normal development of the anterior structures depends on nodal expression in the primitive endoderm rather than in the epiblast (foetal lineage) itself.

10. Concluding remarks

During the last 25 years, studies with experimentally produced primary chimaeras have made many contributions to mouse genetics. The introduction of transgenic lineage markers in the last decade has significantly improved the power of chimaeras for the analysis of the developmental potential of abnormally imprinted cells or cells with cytogenetic or genetic anomalies. The analysis of *Mash2* and nodal illustrate how the analytical power of chimaeras can be enhanced by the incorporation of cells with a restricted developmental potential (tetraploid and ES cells, respectively). Other cell types, such as parthenogenetic cells, which survive preferentially in the forebrain and oocytes, or *Pax6^{Sey}/Pax6^{Sey}* cells, which are excluded from the lens and nasal epithelium, could be used in a similar fashion. Chimaeric rescue analysis is likely to play an increasingly important role in the phenotypic evaluation of numerous new mutations that are being produced by genetic knockout techniques.

Acknowledgements

I thank Duncan Borthwick, Paul Burgoyne, Anne McLaren and Katrine West for helpful comments on the manuscript and Tom McFetters and Ted Pinner for help in preparing the figures. Most of all, I thank Anne McLaren for showing me how interesting mouse chimaeras are. I am also grateful to the Wellcome Trust (046359) and the Medical Research Council (G9630132MB) for financial support for my own current work with chimaeras. Some parts of this article are reproduced from West, 1999.

REFERENCES

Allen, N.D., Logan, K., Lally, G. *et al.* (1995) Distribution of parthenogenetic cells in the mouse brain and their influence on brain development and behavior. *Proceedings of the National Academy of Sciences USA* 92: 10782–86

Bender, R., Surani, M.A., Kothary, R. *et al.* (1995) Tissue-specific loss of proliferative capacity of parthenogenetic cells in fetal mouse chimeras. *Roux's Archives of Developmental Biology* 204: 436–43

Bradbury, M.W. (1987) Testes of XX↔XY chimeric mice develop from fetal ovotestes. *Developmental Genetics* 8: 207–18

Burgoyne, P.S. & Palmer, S.J. (1992) Cellular basis of sex determination and sex reversal in mammals. In *Gonadal Development and Function. Serono Symposium*, edited by S.G. Hillier, New York: Raven Press

Burgoyne, P.S., Buehr, M. & McLaren, A. (1988) XY follicle cells in ovaries of XX↔XY female mouse chimaeras. *Development* 104: 683–88

Clouthier, D.E., Avarbock, M.R., Maika, S.D., Hammer, R.E. & Brinster, R.L. (1996) Rat spermatogenesis in mouse testis. *Nature* 381: 418–21

Crusio, W.E., Bar, I.M., Schwegler, H. & Buselmaier, W. (1990) A multivariate morphometric analysis of hippocampal anatomical variation in C57BL/6↔BALB/c chimeric mice. *Brain Research* 535: 343–46

Dewey, M.J. & Maxson, S.C. (1982) Audiogenic seizure susceptibility of C57BL/6↔DBA/2 allophenic mice. *Brain Research* 246: 154–56

Dobie, K.W., Lee, M., Fantes, J.A. *et al.* (1996) Variegated transgene expression in mouse mammary-gland is determined by the transgene integration locus. *Proceedings of the National Academy of Sciences USA* 93: 6659–64

Epstein, C.J. (1985) Mouse monosomies and trisomies as experimental systems for studying mammalian aneuploidy. *Trends in Genetics* 1: 129–34

Falconer, D.S., Gauld, I.K., Roberts, R.C. & Williams, D.A. (1981) The control of body size in mouse chimaeras. *Genetical Research* 38: 25–46

Ferguson-Smith, A.C., Cattanach, B.M., Barton, S.C., Beechey, C.V. & Surani, M.A. (1991) Embryological and molecular investigations of parental imprinting on mouse chromosome-7. *Nature* 351: 667–70

Fundele, R., Barton, S.C., Christ, B., Krause, R. & Surani, M.A. (1995) Distribution of androgenetic cells in fetal mouse chimeras. *Roux's Archives of Developmental Biology* 204: 484–93

Gardner, R.L. (1968) Mouse chimaeras obtained by the injection of cells into the blastocyst. *Nature* 220: 596–97

Gardner, R.L. (1998) Contributions of blastocyst micromanipulation to the study of mammalian development. *BioEssays* 20: 168–80

Goldowitz, D. (1992) Mouse chimeras in the study of genetic and structural determinants of behavior. In *Techniques for the Genetic Analysis of Brain and Behavior*, edited by D. Goldowitz, D. Wahlstein & R.E. Wimer, New York: Elsevier

Grüneberg, H. & McLaren, A. (1972) The skeletal phenotype of some mouse chimaeras. *Proceedings of the Royal Society of London, Series B* 182: 9–23

Guillemot, F., Nagy, A., Auerbach, A., Rossant, J. & Joyner, A.L.

(1994) Essential role of *Mash-2* in extraembryonic development. *Nature* 371: 333–36
Hooper, M.L. (1992) *Embryonal Stem Cells*, Chur, Switzerland: Harwood Academic
Huang, P.C., Gaitan, A.E., Hao, Y., Petters, R.M. & Wong, F. (1993) Cellular interactions implicated in the mechanism of photoreceptor degeneration in transgenic mice expressing a mutant rhodopsin gene. *Proceedings of the National Academy of Sciences USA* 90: 8484–88
Iannaccone, P.M. & Weinberg, W.C. (1987) The histogenesis of the rat adrenal cortex – a study based on histologic analysis of mosaic pattern in chimeras. *Journal of Experimental Zoology* 243: 217–23
James, R.M., Klerkx, A., Keighren, M., Flockhart, J.H. & West, J.D. (1995) Restricted distribution of tetraploid cells in mouse tetraploid↔diploid chimaeras. *Developmental Biology* 167: 213–26
Joyner, A.L. (ed.) (1993) *Gene Targeting. A Practical Approach*, Oxford: IRL Press
Keverne, E.B., Fundele, R., Narasimha, M., Barton, S.C. & Surani, M.A. (1996) Genomic imprinting and the differential roles of parental genomes in brain development. *Developmental Brain Research* 92: 91–100
LaVail, M.M. & Mullen, R.J. (1976) Role of the pigment epithelium in inherited degeneration analysed with experimental mouse chimaeras. *Experimental Eye Research* 23: 227–45
Le Douarin, N. & McLaren, A. (eds) (1984) *Chimeras in Developmental Biology*, London: Academic Press
Lu, T.Y. & Markert, C.L. (1980) Manufacture of diploid/tetraploid chimeric mice. *Proceedings of the National Academy of Sciences USA* 77: 6012–16
Magnuson, T., Smith, S. & Epstein, C.J. (1982) The development of monosomy 19 mouse embryos. *Journal of Embryology and Experimental Morphology* 69: 223–36
McGrath, J. & Solter, D. (1984) Completion of mouse embryogenesis requires both the maternal and paternal genome. *Cell* 37: 179–83
McLaren, A. (1976) *Mammalian Chimaeras*, Cambridge: Cambridge University Press
McLaren, A. (1991) Development of the mammalian gonad – the fate of the supporting cell lineage. *BioEssays* 13: 151–56
Mikami, H. & Onishi, A. (1985) Heterosis in litter size of chimaeric mice. *Genetical Research* 46: 85–94
Mintz, B. (1970) Neoplasia and gene activity in allophenic mice. In *Genetic Concepts and Neoplasia. 23rd Annual Symposium on Fundamental Cancer Research*, Baltimore: Williams and Wilkins
Morley, S.D., Viard, I., Chung, B.C. *et al.* (1996) Variegated expression of a mouse steroid 21-hydroxylase/β-galactosidase transgene suggests centripetal migration of adrenocortical cells. *Molecular Endocrinology* 10: 585–98
Mullen, R.J. & LaVail, M.M. (1976) Inherited retinal dystrophy: primary defect in pigment epithelium determined with experimental rat chimaeras. *Science* 192: 799–801
Nagy, A., Sass, M. & Markkula, M. (1989) Systematic non-uniform distribution of parthenogenetic cells in adult mouse chimaeras. *Development* 106: 321–24
Nagy, A., Gocza, E., Merentes Diaz, E. *et al.* (1990) Embryonic stem cells alone are able to support fetal development in the mouse. *Development* 110: 815–21
Nesbitt, M.N. (1984) Mouse chimeras in the study of behavior. In *Chimeras in Developmental Biology*, edited by N. Le Douarin & A. McLaren, London: Academic Press
Palmer, S.J. & Burgoyne, P.S. (1991) In situ analysis of fetal, prepubertal and adult XX↔XY testes: Sertoli cells are predominantly, but not exclusively, XY. *Development* 112: 265–68
Patek, C.E., Kerr, J.B., Gosden, R.G. *et al.* (1991) Sex chimerism, fertility and sex determination in the mouse. *Development* 113: 311–25
Peterson, A.C. (1974) Chimaera mouse study shows absence of disease in genetically dystrophic muscle. *Nature* 248: 561–64
Peterson, A.C. (1979) Mosaic analysis of dystrophic↔normal chimeras: an approach to mapping the site of gene expression. *Annals of the New York Academy of Sciences* 317: 630–48
Quinn, J.C., West, J.D. & Hill, R.E. (1996) Multiple functions for *Pax6* in mouse eye and nasal development. *Genes and Development* 10: 435–46
Rossant, J. & Spence, A. (1998) Chimeras and mosaics in mouse mutant analysis. *Trends in Genetics* 14: 358–63
Surani, M.A., Barton, S.C. & Norris, M.L. (1984) Development of reconstituted mouse eggs suggests imprinting of the genome during gametogenesis. *Nature* 308: 548–50
Tarkowski, A.K. (1961) Mouse chimaeras developed from fused eggs. *Nature* 190: 857–60
Thomson, J.A. & Solter, D. (1989) Chimeras between parthenogenetic or androgenetic blastomeres and normal embryos: allocation to the inner cell mass and trophectoderm. *Developmental Biology* 131: 580–83
Varlet, I., Collignon, J. & Robertson, E.J. (1997) Nodal expression in the primitive endoderm is required for specification of the anterior axis during mouse gastrulation. *Development* 124: 1033–44
West, J.D. (1999) Insights into development and genetics from mouse chimaeras. *Current Topics in Developmental Biology* 44: 21–66
Wood, S.A., Pascoe, W.S., Schmidt, C. *et al.* (1993) Simple and efficient production of embryonic stem-cell embryo chimeras by co-culture. *Proceedings of the National Academy of Sciences USA* 90: 4582–85

Note Added in Proof:

Since this article was written the use of chimaeras for analysing mutant phenotypes has been reviewed by Rossant and Spence (1998).

GLOSSARY

allophenic (*see* chimaera)

cell autonomous gene action the phenotypic effects are only seen in the cells which express the gene. In chimaeras comprising a mixture of wild-type and mutant cells, only the mutant cells show the phenotypic effect if the gene acts cell autonomously (for example they may be absent, depleted in numbers or abnormally distributed). If the gene does not act cell autonomously the wild-type cells may rescue the mutant cells so that neither show an abnormal phenotype or both mutant and wild-type cells may be affected

chimaera (or chimera) an individual comprising two or more genetically distinct populations of cells derived from more than one zygote. Previously used terms "allophenic" and "tetraparental" are now redundant. (The definition for plant chimaeras differs in that they need not be derived from more than one zygote)

embryonal carcinoma (EC) **cells** pluripotent malignant stem cells of teratocarcinomas. Often have abnormal karyotypes and rarely (if ever) colonize the germline of chimaeras

morula mammalian embryo immediately before the blastocyst stage (8–16 cells in the mouse)

mosaic an individual comprising two or more genetically distinct populations of cells derived from one zygote

position effect variegation variegation (mosaic phenotype) caused by the inactivation of a gene in some cells resulting from its chromosomal position (usually proximity to heterochromatin)

tetraparental (*see* chimaera)

See also **The role of *Pax-6* homologues in the development of the eye and other tissues** (p.207); **X chromosome inactivation** (p.780)

ETHYLNITROSOUREA (ENU) AS A GENETIC TOOL IN MOUSE MUTAGENESIS

Allan Peter Davis[1] and Monica J. Justice[2]
[1]*The Jackson Laboratory, Bar Harbor, Maine, USA*
[2]*Department of Molecular and Human Genetics, Baylor College of Medicine, Houston, Texas, USA*

1. Introduction

Genetics is a science of variation. Aberrations of the normal phenotype are a curiosity, from "waltzing" mice to a multitude of coat colour mutations with descriptive names like ashen (*ash*), chocolate (*ctl*), gunmetal (*gm*) or cream (*Crm*), the last of which has the peculiar property of glowing under ultraviolet light. Historically, however, this extensive collection of mutants was more of happenstance than directed experimentation. As with most other animal genetics, there are few spontaneous mutations in the mouse. However, when something of noticeable interest or curiosity did arise, the trait was perpetuated in stocks, maintained initially by mouse fanciers who collected, showed and traded such pets with other enthusiasts. It was only in the early part of the 20th century, with the rediscovery of Mendel's Laws and the appreciation that the mouse resembles the human in physiology and reproduction, that a small collection of scientists started to use the animal as an experimental model organism, especially towards the understanding of cancer. From this humble beginning, mouse genetics is now recognized as a Rosetta stone for exploring human biology, from birth defects to cancer treatment to behaviour abnormalities.

In the last 20 years, however, mouse genetics has undergone a revolution. Up to that point, most mouse variants were spontaneous mutations (with the exception of deletion stocks induced by irradiation). These variants provided a wealth of insight into the understanding of developmental biology, immunology and mammalian genetics. Nonetheless, the nature of the mutation and the gene itself often remained unknown. Now, with the breakthroughs in molecular biology, we stand on the cusp of a new wave in mammalian genetics. By the year 2003, the mouse genome will be sequenced, but what will the nucleotide data tell us? A wealth of knowledge may pour forth from computer-generated sequence analyses with algorithms searching for functional motifs, but it is mutation analysis that reveals a gene's significance in the context of the whole animal. Today, scientists employ the powerful technique of gene targeting to disrupt ("knock out") their favourite gene *in vitro*, insert those changes into an embryonic stem cell and then implant the cell into an embryo ultimately to generate a mouse lacking that gene (Capecchi, 1989). The number of these so-called "knockout" mice has grown exponentially in the literature in the last decade and has more than doubled the known number of available mouse mutants.

2. The need for alleles

The power of genetic functional analysis, however, lies in collecting an allelic series for a gene (Davis & Justice, 1998b). Gene targeting, for the most part, has been used to make null mutations – a sledgehammer approach to genetics: bash the gene to oblivion and see what happens to a mouse lacking it. This technique has been invaluable because, when combined with the wild-type allele, it provides two ends of an all-or-nothing spectrum: the normal and null phenotypes. Two lessons were quickly learned from this approach.

1. Genes that were predicted to have important functions, surprisingly produced very normal looking mice when mutated. Currently, this situation is being re-examined with the realization that an investigator cannot hope to understand everything about a mouse when it is kept in an artificial laboratory setting. By analysing mice more precisely or under more natural conditions or simply by moving the mutation onto different genetic backgrounds, phenotypes do emerge (Doetschman, 1999).
2. It was realized that genes are used during different times in a mouse's life – for example, in embryogenesis and again in adulthood. Thus, many knockout mutations cause embryonic lethality, long before some predicted phenotype (e.g. mating behaviour) could ever be tested. This dilemma can also be circumvented by clever molecular tricks to create conditional mutations that allow the "null" mutation to be expressed in specific tissues or at specific times (Meyers *et al.*, 1998).

Between these two phenotypic extremes (wild and null), other types of mutation can exist: hypomorphs, hypermorphs, neomorphs and antimorphs. These mutations usually involve more subtle variations at the DNA level that do not "knock out" the gene's activity, but rather influence the function and perhaps produce more informative phenotypes. Therefore, when more alleles for a specific gene are analysed, a greater informational spectrum is acquired. By correlating phenotypic variations with molecular lesions, a fine structure–function analysis of the gene can be annotated. The only limitation is generating those alleles to form a series. For other genetic systems (such as phage, bacteria, yeast and Drosophila), this has not been a problem, because these organisms reproduce very quickly and are cheap to raise in large quantities. The mouse, however, lacks such temporal and financial luxury. Again, gene targeting can be employed to make these alleles or even recapitulate in the mouse known subtle mutations responsible for specific human diseases (Zeiher *et al.*, 1995). However, implementing such targeted mutagenesis for this purpose is only applicable if two conditions exist: (1) the gene is cloned and (2) the molecular lesion necessary to make a hypermorph (or any other subtle variant) is also known. So, if neither of these conditions can be met, what is a mouse geneticist to do if additional alleles are desired? One could turn to chemical mutagenesis: the process by which a compound is injected into a mouse to cause a heritable mutation in the germline, recoverable by breeding the animal. This philosophy uses a phenotype-driven approach instead of a genotype-driven one.

3. Chemical mutagenesis

Attempts at chemical mutagenesis in the mouse have early origins. Following on from the discovery that X-rays could induce mutations, it was thought that chemicals might also cause heritable variations (Falconer *et al.*, 1952). In addition, concerns were raised in the populace over the military use of chemicals in warfare, and the British geneticist J.B.S. Haldane pondered "the mutagenic effect of substances which are frequently added to human food as preservatives" (Haldane, 1956). Most of the compounds tested in the mouse, however, proved to be ineffective at inducing mutations in male spermatogonia (see Ehling (1978) for review). One exception was procarbazine, a drug used in the treatment of Hodgkin's disease; yet even this compound was not nearly as effective as X-rays (Ehling & Neuhauser, 1979). It was beginning to look as if mouse spermatogonia were protected against chemical insults. Even diethylnitrosamine (DEN), a compound known to be strongly mutagenic in Drosophila, was completely ineffective in mice (Russell & Kelly, 1979). To be mutagenic, however, DEN is enzymatically converted into an alkylating agent and it was possible this activation process did not occur in mammals. Subsequently, Bill Russell at the Oak Ridge National Laboratory in the US tried ethylnitrosourea (ENU), a chemical that forms the same alkylating species as DEN, but does not require metabolism. Russell's team quickly realized that, after treatment of male mice with ENU, mutations were recovered from premeiotic mutagenized spermatogonia at a rate 12 times higher than that of X-rays, 36 times better than procarbazine and over 200 times the spontaneous rate of mutation (Russell *et al.*, 1979; Hitotsumachi *et al.*, 1985). ENU acts as a point mutagen by transferring its ethyl group to oxygen or nitrogen radicals in DNA, resulting in mispairing and base pair substitution if not repaired. The highest mutation rates occur in premeiotic spermatogonial stem cells, with single locus mutation frequencies of 6×10^{-3}–1.5×10^{-3}, equivalent to isolating a mutation in a single locus in every 175–655 gametes screened (Hitotsumachi *et al.*, 1985; Shedlovsky *et al.*, 1993). Heralded as a "supermutagen", ENU was now being described as "the mutagen of choice for the production of any kind of desired new gene mutations in the mouse" (Russell *et al.*, 1979).

4. ENU as a genetic tool

Due to its power in isolating mutations in any gene or region of interest, ENU has been used in the mouse to: (1) obtain multiallelic series of single genes; (2) dissect biochemical or developmental pathways; (3) obtain new recessive mutations on a single chromosome or throughout the genome; and (4) saturate regions that are uncovered by deletions (Bode, 1984; Justice & Bode, 1986; Shedlovsky *et al.*, 1986, 1988; Rinchik *et al.*, 1995; Kasarkis *et al.*, 1998; Rinchik & Carpenter, 1999). The analysis of 62 sequenced germline mutations from 24 genes reveals that ENU predominantly modifies AT base pairs, with 43% AT–TA transversions, 39% AT–GC transitions, 8% GC–AT transitions, 3% GC–CG transversions, 5% AT–CG transitions and 2% GC–TA transitions (Table 1; the lesion was not found in two mutations with apparent splice defects). When translated into a protein product, these changes result in 63% missense mutations, 9% nonsense mutations and 27% splicing errors (**Table 1**).

The easiest genetic screen for recovering ENU-induced mutations is called the specific locus test (SLT), a non-complementation screen that rapidly collects recessive alleles of an already known mutation (Russell, 1951). In the SLT, the germ cells of the male are mutagenized and then the mouse is mated with a series of females that are already homozygous mutant for the gene of interest. New ENU-induced alleles from the male will fail to complement the female's genotype and the progeny will be recognizable as mutants. All that is needed is a quick and sensitive assay to detect the phenotype. For example, the developmental gene bone morphogenetic protein 5 (*Bmp5*), when mutated, causes mice to have short ears. ENU injected into wild-type males (+/+) will generate random point mutations throughout the entire genome, including perhaps some at *Bmp5*. When males are crossed to females homozygous mutant for *Bmp5* (*m*/*m*), almost all the progeny will be heterozygotes (+/*m*), appear normal and be discarded; however, those that were derived from sperm that carry a mutation in the *Bmp5* gene (*) will be (*/*m*) and recognized by their short-ear phenotype.

ENU mutagenesis is a simple, cost-effective genetic tool for any laboratory. To begin a mutagenesis screen, male mice 8–10 weeks of age are injected with an appropriate dose of ENU (Justice *et al.*, 1999). After the last injection, the males are allowed to recover, during which time they undergo a temporary period of sterility that ranges between 10 and 20 weeks (Davis *et al.*, 1999). ENU causes point mutations in the spermatogonial stem cells that repopulate the testis to generate clones of mutant spermatids. Unlike chemical mutagens that affect postmeiotic stages of germ cell development (such as chlorambucil), the ENU-generated mutations are recoverable so long as the male lives and breeds. To increase breeding capacity, each fertile male is rotated on a weekly basis to a new breeding cage with two females for 7 weeks. By that time, the first two females will have become pregnant, given birth and had their pups weaned, allowing the male to be rotated back into the first box with the original females. This system, called a 7-week rotation, is a

Table 1 Summary of sequenced ENU-induced mutations.

Locus	Number of sequenced ENU mutations	Mutation	Functional classification	References
Agouti (*A*)	1	AT–GC	Missense	Hustad *et al.*, 1995
Adenomatous polyposis coli (Apc^{Min})	1	AT–TA	Nonsense	Su *et al.*, 1992
Bone morphogenetic protein–5 (*Bmp5*)	11	4 AT–GC	6 Missense	Marker *et al.*, 1997
		4 AT–TA	2 Nonsense	
		2 GC–AT	2 Splicing with deletion	
		1 Unknown	1 Splicing with deletion	
Clock	1	AT–TA	Splicing with deletion	King *et al.*, 1997
γ-Crystallin (*Cryge*)	1	AT–GC	Missense	Klopp *et al.*, 1998
Dystrophin (*Dmd*)	4	3 AT–TA	3 Splicing with frameshift	
		1 GC–AT	1 Nonsense	Im *et al.*, 1996
Embryonic endoderm (*eed*)	2	AT–GC		
		AT–TA	2 Missense	Schumacher *et al.*, 1996; Rinchik & Carpenter, 1999
Connexin 50 (*Gja8*)	1	AT–CG	Missense	Steele *et al.*, 1998
Glucose phosphate isomerase (*Gpi1*)	4	3 AT–GC	3 Missense	Pearce *et al.*, 1995
		1 Unknown	1 Splicing with deletion	
Glucose 6 phosphate dehydrogenase (*G6pdx*)	1	AT–TA	Splicing, reduced protein	Sanders *et al.*, 1997
Haemoglobin alpha (*Hba*)	1	AT–TA	Missense	Popp *et al.*, 1983
Haemoglobin beta (*Hbb*)	2	AT–GC	2 Missense	Peters *et al.*, 1985;
		AT–TA		Jones & Peters, 1991
Kreisler (*Krm1*)	1	AT–GC	Missense	Cordes & Barsh, 1994
Lactate dehydrogenase (*Ldh1*)	5	4 AT–GC	4 Missense	Sandaluche *et al.*, 1994;
		1 AT–CG	1 Splicing with reduced protein	Pretch *et al.*, 1998
Mast cell growth factor (*Mgf*)	1	AT–TA	Splicing with frameshift	Brannan *et al.*, 1992
Microphthalmia (*Mitf*)	1	AT–TA	Missense	Steingrimsson *et al.*, 1998
Myosin VA (*Myo5a*)	10	2 GC–CG	8 Missense	Huang *et al.*, 1998a;
		5 AT–TA	2 Splicing, 1 protein	Huang *et al.*, 1998b;
		2 AT–GC	product from 1 of 3	Russell *et al.*, 1990
		1 GC–AT	isoforms	
Myosin VIIA (*Myo7a*)	5	3 AT–TA	1 Missense	Rinchik & Carpenter 1999;
		1 AT–GC	1 Splicing with deletion	Gibson *et al.*, 1995;
		1 GC–AT	1 Splicing with frameshift	Mburu *et al.*, 1997
			2 Nonsense	
Phenylalanine hydroxylase (*Pah*)	2	2 AT–GC	2 Missense	McDonald & Charlton 1997; Shedlovsky *et al.*, 1993
Paired box homeobox 6 (*Pax6*)	1	GC–TA	Splicing with frameshift	Hill *et al.*, 1991
Quaking (*qk*)	3	2 AT–GC	2 Missense	Ebersole *et al.*, 1996;
		1 AT–TA	1 Disruption of one isoform	Cox *et al.*, 1998
Triosephosphate isomerase (*Tpi*)	4	3 AT–TA	3 Missense	
		1 AT–CG	1 Stop to C	Zingg *et al.*, 1995
Tyrosinase related protein 1 (*Tyrp1*)	1	AT–GC	Missense	Zdarsky *et al.*, 1990
TOTAL	**64**	**27 AT–TA**	**40 Missense 63%**	
		24 AT–GC	**6 Nonsense 9%**	
		3 AT–CG	**17 Splicing 27%**	
		5 GC–AT	**1 Other 1%**	
		2 GC–CG		
		1 GC–TA		
		2 Unknown		

highly ordered and efficient procedure that can be run in any small mouse colony. With one rotation, a mutagenized male can breed with 14 different females. If each female were to become pregnant and give birth to an average litter of six pups, then 84 descendants of the mutagenized male could be screened. In two rotations, over 160 gametes have been screened per male. If a screen starts out with 15 mutagenized males, by the end of 14 weeks as many as 2500 gametes may have been examined for mutations in a gene of interest, all conducted on a few shelves of an average-sized mouse room. This approach can easily be applied to most viable mutations to yield an extensive allelic series.

Due to the randomness of ENU mutagenesis, a set of treated males can be mated to females carrying completely different test genes to increase dramatically productivity and results. For example, in one screen the males could breed to a female with a *Bmp5* mutation and another female with a myosin gene mutation (*Myo5a*), or, even more efficiently, both mutations could be in the same female. In fact, this is exactly how the original SLTs were performed. By being mated to females of a special test stock that harboured seven recessive homozygous mutations, each mutagenized male was simultaneously testing seven completely different loci. The result was a treasure trove of new alleles at non-agouti (*a*), brown (*Tryp1*), albino (*Tyr*), piebald (*Ednrb*), pink-eyed dilution (*p*), dilute (*Myo5a*) and short ear (*Bmp5*) (see Davis & Justice (1998a) for review).

The advantages of using ENU as a genetic tool in the mouse are numerous.

(i) No molecular knowledge of the gene is necessary

In chemical mutagenesis, the investigator does not need to worry about what type of an engineered point mutation might (or might not) generate a hypomorph *vs.* a hypermorph *vs.* a neomorph *vs.* an amorph. In fact, the gene itself does not even have to be cloned. Instead, the mice do all the work. Provided the assay is sensitive enough to detect a phenotype, any functional mutation will be collected. If a point mutation occurs in a certain codon, but makes no biological difference to the function/phenotype of the gene, then that mutation will go unnoticed; only the mutations that yield a detectable phenotype will be collected. Since the screens are phenotype driven, no *a priori* molecular knowledge of the gene is necessary. In fact, ENU mutagenesis has actually been used to confirm the identity of candidate genes (by producing sequence lesions in the affected gene) that were positionally cloned from complex genetic lesions, such as kreisler (*Krml*), quaking (*qk*) and the developmental gene *eed* (Cordes & Barsh, 1994; Ebersole *et al.*, 1996; Schumacher *et al.*, 1996).

(ii) Any type of mutation can be collected

ENU mutagenesis is presumed to allow for random mutational spectra to unfold, completely unencumbered by all preconceived and limited notions of the experimenter. One caveat is that ENU primarily alters AT base pairs in mouse spermatogonia, so its randomness is actually just an assumption. Furthermore, it is not inconceivable that chromatin structure and the state of gene expression during the time of mutagenesis might influence the process. Nonetheless, for the most part, ENU mutagenesis appears fairly random. Theoretically, this allows for mutations in any part of the gene to be collected, including regulatory, coding or intronic regions. In addition, targeted mutant lines can be used in sensitized screens, similar to work undertaken in Drosophila, to identify mutations in other genes that enhance or suppress the targeted phenotype. Non-allelic non-complementation is the phenomenon whereby an induced mutation at another locus (*) can interact with the test gene (*m*), thus failing to complement (*/+; +/*m*) but still yielding a phenotype (Harris & Juriloff, 1998). This has the potential to generate extensive functional pathways of genetic interactions between the test gene and other loci.

(iii) Point mutagens produce alleles similar to those responsible for human diseases

Many human diseases are not the result of null mutations, but of more subtle alleles. Early on, ENU was recognized as yielding phenotypes in between those of wild and null (Russell *et al.*, 1990). These more subtle alleles were assumed to be point mutations and this was confirmed when it was resolved that ENU had induced an A to T transversion in a histidine codon of a haemoglobin variant (Popp *et al.*, 1983). Since chemical mutagenesis screens are phenotype driven, mice can be mutagenized and screened for phenotypes resembling clinical disorders. Phenylketonuria, one of the first inborn errors of metabolism characterized in humans, was also one of the first diseases reproduced in the mouse using ENU (McDonald *et al.*, 1990; Shedlovsky *et al.*, 1993).

(iv) Dominant mutations can also be collected

One inherent consequence of an ENU screen is that non-related dominant mutations will be noticed as a side product in the screen, some of which may warrant further study. The most famous mutation captured in an ENU-generated dominant hunt is *Clock*, an antimorphic allele captured by assaying mice for abnormal wheel-running activity, resulting in the first cloned mouse mutation to disrupt circadian rhythm (King *et al.*, 1997).

(v) Complex genetic screens: using deletions

In addition to single locus non-complementation strategies, ENU can be used in large-scale screens to recover recessive mutations in many different loci. An under-represented class of mutations in the mouse are the lethal and detrimental mutations. It is estimated that these types of mutation may account for 30–50% of mutations in all mouse loci (Justice *et al.*, 1997). The isolation and management of these phenotypes requires the use of genetic reagents such as deletions and inversions, and more complicated two- and three-generation mating strategies.

ENU mutagenesis can be applied to define functional units in deleted segments of the genome. The first example of this employed a large deletion at the albino (*Tyr*) locus in an elegant screen (**Figure 1**) (Rinchik *et al.*, 1990). Male albino mice (*c/c*) were mutagenized with ENU and then mated to wild-type females (+/+) to produce F_1 heterozygotes, some of which carried an ENU-induced mutation linked to the coat colour gene (c*/++). These F_1s were then mated to a female

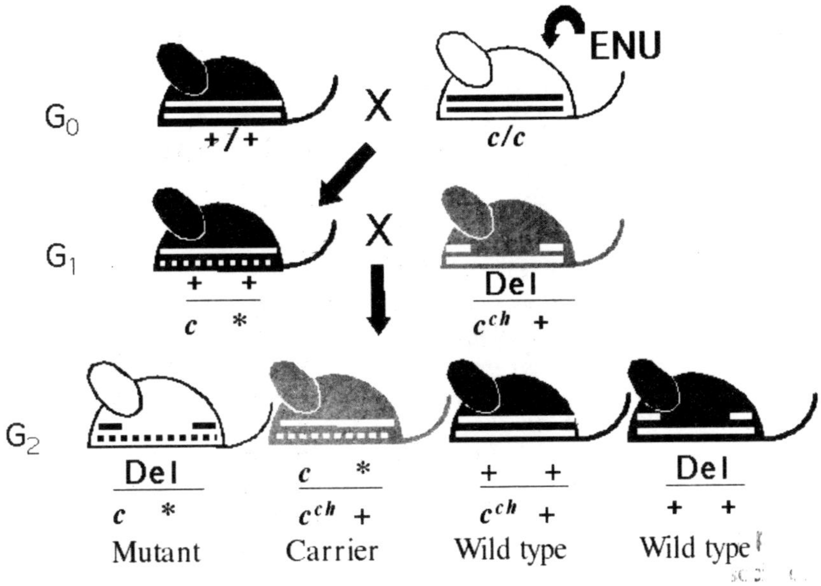

Figure 1 A genetic scheme used for fine structure–function analysis of the tyrosinase or albino (*Tyr*, designated *c*) region of mouse chromosome 7 (Rinchik *et al.*, 1990). Albino males (*c/c*; white) are mutagenized with ENU and mated with wild-type females (black). The dashed line in subsequent generations indicates the ENU-treated chromosome inherited from the male. G_1 females carrying a new mutation (*) induced near *c* were mated to males carrying a 6–11 cM deletion, Del$(c)^{26DVT}$, designated Del, carrying the chinchilla allele of albino (c^{ch}). If a new mutation mapped within the limits of the deletion, a new recessive mutation could be detected in the white G_2 animals as a hemizygous phenotype. Lethal and detrimental phenotypes would result in the absence of white animals among the G_2 offspring, but could be rescued as carriers with a light chinchilla coat (c*/c^{ch}+; grey).

heterozygous for the deleted region who also carried the chinchilla allele of albino (c^{ch}+/Df). Thus, if the F_1 male did indeed carry an ENU mutation linked to *c*, then four different types of progeny could arise, all distinguishable by their coat colours: (++/c^{ch}+) and (++/Df) which would appear as dark chinchillas and represent the wild-type class which could be discarded; (c*/c^{ch}+) which would appear as light chinchillas and act as heterozygous carriers of the newly induced ENU mutation; and (c*/Df) which would be albino and should yield a phenotype because the allele is haploid. If no albino mice are seen in a litter, the experimenter immediately knows that a new, ENU-induced recessive lethal mutation has been uncovered in the region of the deletion. This recessive lethal is easily retrievable from the light chinchilla heterozygous carrier. Using this elegant screen, the albino deletion was saturated with ENU-induced mutations to define the functional units in the region (Rinchik & Carpenter, 1999). This technique, however, is not limited to pre-existing deletions at known coat colour mutations.

Nowadays, through chromosomal engineering, any region of the mouse genome can simultaneously be deleted and marked with coat colour genes to mimic the above strategy (Justice *et al.*, 1997; Justice, 1999). This exciting approach affords the opportunity to define precisely the functional units at any segment of the genome.

(vi) Complex genetic screens: using balancer chromosomes

Although deletions are useful genetic tools, their use can be limited by their size and by haplo-insufficiency in certain regions. A straightforward approach to isolate viable recessive mutations is a simple three-generation pedigree (Justice, 1999). Alternatively, balancer chromosomes are useful tools to isolate and maintain lethal or detrimental mutations (**Figure 2**). An ideal balancer chromosome has three characteristics: first, it contains an inversion that suppresses recombination over a large interval, perhaps 20–60 centiMorgans (cM); secondly, it carries a dominant phenotypic marker for easy identification; and, thirdly, it is homozygous lethal, preventing its confusion with an inversion heterozygote. A few balancer chromosomes are already available, generated at random by radiation and chemical mutagens (see Rinchik (1999) for review). Additional inversions can be rapidly generated by *Cre/loxP* engineering techniques, and tagged with dominant markers affecting the coat pigment (Justice *et al.*, 1997; Justice, 1999; Zheng *et al.*, 1999) or fluorescence (Hadjantonakis *et al.*, 1998). In a genetic screen using balancer chromosomes, fertile ENU-treated males are mated to females carrying the balancer chromosome and a dominant phenotypic or molecular marker (In^{Ag}/D+). G_1 animals heterozygous for the balancer and a new mutation (*) are mated with animals heterozygous for the balancer chromosome. Three classes of offspring can be identified in the second generation, and the fourth class, which is homozygous for the balancer chromosome, dies. The useful G_2 animals (In^{Ag}/+*) are brother–sister mated. The G_3 offspring are easily classified as: (1) the mutant class, which, if missing, indicates the likelihood of a lethal mutation and (2) a carrier class used to rescue any lethal mutations, which carries the balanced point mutation, ideal for stock

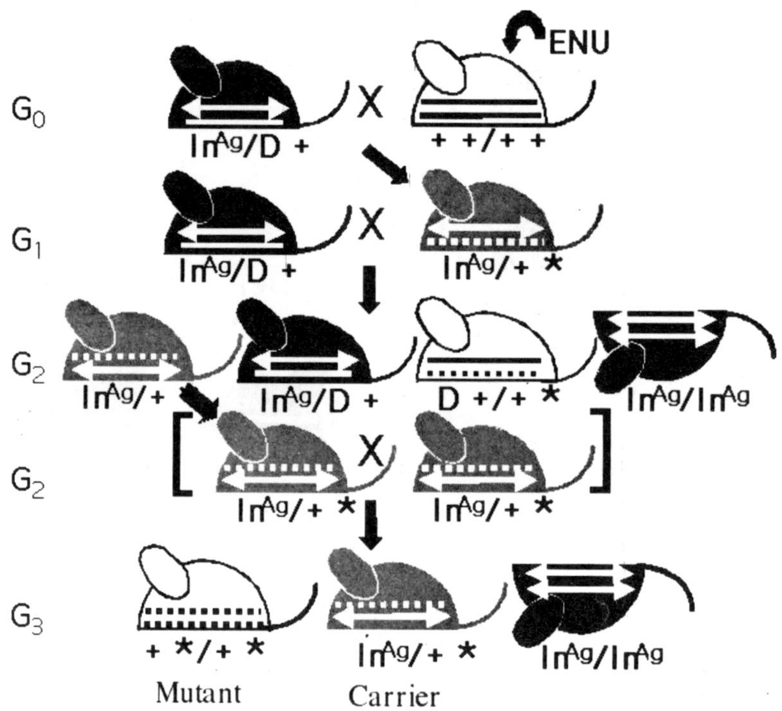

Figure 2 A genetic scheme for fine structure–function analysis of mouse chromosome 11. Wild-type males are mutagenized with ENU and mated with females carrying a balancer chromosome (In^{Ag}), which carries an inversion with one breakpoint in a gene that is homozygous lethal and is tagged with the K14-agouti transgene that confers a dominant yellow coat colour (Zheng *et al.*, 1999). The other chromosome is marked with a dominant mutation (D). Since In^{Ag} acts as a single locus covering both D and any mutant loci, it is given a single locus designation in the figure. For illustrative purposes, mice carrying both In^{Ag} and D are black; mice carrying In^{Ag} and a new ENU-induced mutation are grey; and mice that are wild-type or lack In^{Ag} are white in the cartoon – this does not, however, designate their actual coat colour. The ENU-treated chromosome is shown as a dashed line and the inversion is shown as a two-sided arrow. Each class of animal can be distinguished in the G_1 and G_2 generations. Mice from the G_2 generation carrying a new ENU-induced mutation that is balanced to suppress recombination will have a yellow coat and are brother–sister mated to generate G_3 offspring. Any new recessive mutations can be detected in the wild-type coat class of animal (white mice). A new lethal mutation would result in the absence of this class of animal, but the mutation can be rescued in the carrier class of animals, which would have a yellow coat (grey mice). Mice homozygous for the balancer chromosome will die (upside down).

maintenance. This approach has the advantage that a larger region of the genome can be screened for mutations.

5. Summary

ENU mutagenesis in the mouse is a quick and relatively easy process to generate new mutations and allelic series. The strategy is not intended to make gene targeting passé, but rather to complement it in order to add depth to our knowledge of mammalian gene function. Though first discovered in the late 1970s, ENU mutagenesis in the mouse quickly fell out of favour because it did not allow the induced mutation to be molecularly resolved. In the intervening time, advances were instead made in large-scale mapping and sequencing efforts. Today, we stand at an exciting crossroad. The mouse genome will be sequenced by 2003, and the ability to analyse any region will become an easy task. Thus, ENU mutagenesis is now undergoing a renaissance. The era of positional cloning is coming to an end and, instead, focus is turning to understanding the biology of the gene sequence. ENU contains the power to address that very question.

REFERENCES

Bode, V.C. (1984) Ethylnitrosourea mutagenesis and the isolation of mutant alleles for specific genes located in the *T* region of mouse chromosome 17. *Genetics* 108: 457–70

Brannan, C.I., Bedell, M.A., Resnick, J.K. *et al.* (1992) Developmental abnormalities in *Steel*[17H] mice result from a splicing defect in the steel factor cytoplasmic tail. *Genes and Development* 6: 1832–42

Capecchi, M.R. (1989) Altering the genome by homologous recombination. *Science* 244: 1288–92

Cordes, S.P. & Barsh, G.S. (1994) The mouse segmentation gene kr encodes a novel basic domain-leucine zipper transcription factor. *Cell* 79: 1025–34

Cox, R.D., Hugill, A., Shedlovsky, A. *et al.* (1998) Contrasting effects of ENU-induced embryonic lethal mutations of the *quaking* gene. *Genomics* 57: 333–41

Davis, A.P. & Justice, M.J. (1998a) An Oak Ridge legacy: the specific locus test and its role in mouse mutagenesis. *Genetics* 148: 7–12

Davis, A.P. & Justice, M.J. (1998b) Mouse alleles: if you've seen one, you haven't seen them all. *Trends in Genetics* 14: 438–41

Davis, A.P. Woychik, R.P. & Justice, M.J. (1999) Effective

chemical mutagenesis in FVB/N mice requires low doses of ethylnitrosourea. *Mammalian Genome* 10: 308–10

Doetschman, T. (1999) Interpretation of phenotype in genetically engineered mice. *Laboratory Animal Science* 49: 137–43

Ebersole, T.A., Chen, Q., Justice, M.J. & Artzt, K. (1996) The *quaking* gene product necessary in embryogenesis and myelination combines features of RNA binding and signal transduction proteins. *Nature Genetics* 12: 260–65

Ehling, U.H. (1978) Specific-locus mutations in mice. In *Chemical Mutagens: Principles and Methods for their Detection*, edited by A. Hollaender & F.J. de Serres, New York: Plenum Press

Ehling, U.H. & Neuhauser, A. (1979) Procarbazine induced specific-locus mutations in male mice. *Mutation Research* 59: 245–56

Falconer, D.S., Slizynski, B.M. & Auerbach, C. (1952) Genetical effects of nitrogen mustard in the house mouse. *Journal of Genetics* 51: 81–88

Gibson, F., Walsh, J., Mburu, P. *et al.* (1995) A type VII myosin encoded by the mouse deafness gene *shaker-1*. *Nature* 374: 62–64

Hadjantonakis, A.K., Gertsenstein, M., Ikawa, M., Okabe, M. & Nagy, A. (1998) Generating green fluorescent mice by germline transmission of green fluorescent ES cells. *Mechanism of Development* 76: 79–90

Haldane, J.B.S. (1956) The detection of autosomal lethals in mice induced by mutagenic agents. *Journal of Genetics* 54: 327–42

Harris, J.J. & Juriloff, D.M. (1998) Nonallelic noncomplementation models in mice: the first arch and lidgap-Gates mutations. *Genome* 41: 789–96

Hill, R.E., Favor, J., Hogan, B.L.M. *et al.* (1991) Mouse Small eye results from mutations in a paired-like homeobox-containing gene. *Nature* 354: 522–25

Hitotsumachi, S., Carpenter, D.A. & Russell, W.L. (1985) Dose-repetition increases the mutagenic effectiveness of N-ethyl-N-nitrosourea in mouse spermatogonia. *Proceedings of the National Academy of Sciences USA* 82: 6619–621

Huang, J.-D., Cope, M.J.T.V., Mermall, V. *et al.* (1998a) Molecular genetic dissection of mouse unconventional myosin-VA: head region mutations. *Genetics* 148: 1951–61

Huang, J.-D., Mermall, V., Strobel, M.C. *et al.* (1998b) Molecular genetic dissection of mouse unconventional myosin-VA: tail region mutations. *Genetics* 148: 1963–72

Hustad, C.M., Perry, W.L., Siracusa, L.D. *et al.* (1995) Molecular genetic characterization of six recessive viable alleles of the mouse *agouti* locus. *Genetics* 140: 255–65

Im, W.B., Phelps, S.F., Copen, E.H. *et al.* (1996) Differential expression of dystrophin isoforms in strains of mdx mice with different mutations. *Human Molecular Genetics* 5: 1149–53

Jones, J. & Peters, J. (1991) The molecular characterization of an A:T to G:C transition in the *Hbb-b1* gene of the murine homologue of hemoglobin Rainier. *Biochemical Genetics* 29: 617–26

Justice, M.J. (1999) Mutagenesis of the mouse germline. In *Mouse Genetics and Transgenics: A Practical Approach*, edited by I.J. Jackson & C.M. Abbott, Oxford and New York: Oxford University Press

Justice, M.J. & Bode, V.C. (1986) Induction of new mutations in a mouse t-haplotype using ethylnitrosourea mutagenesis. *Genetical Research* 47: 187–92

Justice, M.J., Zheng, B., Woychik, R.P. & Bradley, A. (1997) Using targeted large deletions and high-efficiency N-ethyl-N-nitrosourea mutagenesis for functional analyses of the mammalian genome. *Methods* 13: 423–36

Justice, M.J., Noveroske, J.N., Weber, J.S., Zheng, B. & Bradley, A. (1999) Mouse ENU mutagenesis. *Human Molecular Genetics* 8: 1955–65

Kasarkis, A., Manova, K. & Anderson, K.V. (1998) A phenotype-based screen for embryonic lethal mutations in the mouse. *Proceedings of the National Academy of Sciences USA* 95: 7485–90

King, D.P., Zhao, Y., Sangoram, A.M. *et al.* (1997) Positional cloning of the mouse circadian *Clock* gene. *Cell* 89: 641–53

Klopp, N., Favor, J., Loster, J. *et al.* (1998) Three murine cataract mutants (Cat2) are defective in different γ-crystallin genes. *Genomics* 52: 152–58

Marker, P.C., Seung, K., Bland, A.E., Russell, L.B. & Kingsley, D.M. (1997) Spectrum of Bmp5 mutations from germline mutagenesis experiments in mice. *Genetics* 145: 435–43

Mburu, P., Liu, X.Z., Walsh, J. *et al.* (1997) Mutation analysis of the mouse myosin VIIA deafness gene. *Genes and Function* 1: 191–203

McDonald, J.D. & Charlton, C.K. (1997) Characterization of mutations at the mouse phenylalanine hydroxylase locus. *Genomics* 39: 402–05

McDonald, J.D., Bode, V.C., Dove, W.F. & Shedlovsky, A. (1990) Pahhph3: a mouse mutant deficient in phenylalanine hydroxylase. *Proceedings of the National Academy of Sciences USA* 87: 1965–67

Meyers, E.N., Lewandoski, M. & Martin, G.R. (1998) An *Fgf8* mutant allelic series generated by Cre- and Flp-mediated recombination. *Nature Genetics* 18: 136–41

Pearce, S.R., Peters, J., Ball, S. *et al.* (1995) Sequence characterization of ENU-induced mutants of glucose phosphate isomerase in mouse. *Mammalian Genome* 6: 858–61

Peters, J., Andrews, S.J., Loutit, J.F. & Glegg, J.B. (1985) A mouse β-globin mutant that is an exact model of hemoglobin Rainier in man. *Genetics* 110: 709–21

Popp, R.A., Bailiff, E.G., Skow, L.C., Johnson, F.M. & Lewis, S.E. (1983) Analysis of a mouse α-globin gene mutation induced by ethylnitrosourea. *Genetics* 105: 157–67

Pretch, W., Chatterjee, B., Favor, J., Merkle, S. & Sandaluche, R. (1998) Molecular, genetic and biochemical characterization of lactate dehydrogenase-A enzyme activity mutations in *Mus musculus*. *Mammalian Genome* 9: 144–49

Rinchik, E.M. (2000) Developing genetic reagents to facilitate recovery, analysis, and maintenance of mouse mutations. *Mammalian Genome* Special Issue 11: 489–99

Rinchik, E.M. & Carpenter, D.A. (1999) N-ethyl-N-nitrosourea mutagenesis of a 6- to 11-cM subregion of the *Fah-Hbb* interval of mouse chromosome 7: completed testing of 4,557 gametes and deletion mapping and complementation analysis of 31 mutations. *Genetics* 152: 373–83

Rinchik, E.M., Carpenter, D.A. & Selby, P.B. (1990) A strategy for fine-structure functional analysis of a 6- to 11-centiMorgan region of mouse chromosome 7 by high-efficiency mutagenesis. *Proceedings of the National Academy of Sciences USA* 87: 896–900

Rinchik, E.M., Carpenter, D.A. & Handel, M.A. (1995) Pleiotropy in microdeletion syndromes: neurologic and spermatogenic abnormalities in mice homozygous for the p^{6H} deletion are likely due to dysfunction of a single gene. *Genetics* 92: 6394–98

Russell, L.B., Russell, W.L., Rinchik, E.M. & Hunsicker, P.R. (1990) Factors affecting the nature of induced mutations. In *Biology of Mammalian Germ Cell Mutagenesis*, Cold Spring Harbor, New York: Cold Spring Harbor Laboratory Press

Russell, W.L. (1951) X-ray induced mutations in mice. *Cold Spring Harbor Symposium on Quantitative Biology* 16: 327–35

Russell, W.L. & Kelly, E.M. (1979) Ineffectiveness of diethylnitrosamine in the induction of specific-locus mutations in mice. *Genetics* 91: S109–10

Russell, W.L., Kelly, E.M., Hunsicker, P.R. *et al.* (1979) Specific-locus test shows ethylnitrosourea to be the most potent mutagen in the mouse. *Proceedings of the National Academy of Sciences USA* 76: 5818–19

Sandaluche, R., Pretsch, W., Chatterjee, B. *et al.* (1994) Molecular analysis of four lactate dehydrogenase-A mutants in the mouse. *Mammalian Genome* 5: 777–80

Sanders, S., Smith, D.P., Thomas, G.A. & Williams, E.D. (1997) A glucose-6-phosphate dehydrogenase (G6PD) splice site consensus sequence mutation associated with G6PD enzyme deficiency. *Mutation Research* 374: 79–87

Schumacher, A., Faust, C. & Magnuson, T. (1996) Positional cloning of a global regulator of anterior–posterior patterning in mice. *Nature* 383: 250–53

Shedlovsky, A., Guenet, J.-L., Johnson, L.L. & Dove, W.F. (1986) Induction of recessive lethal mutations in the *T/t-H-2* region of the mouse genome by a point mutagen. *Genetical Research* 47: 135–42

Shedlovsky, A., King, T.R. & Dove, W.F. (1988) Saturation germ line mutagenesis of the murine *t* region including a lethal allele at the quaking locus. *Proceedings of the National Academy of Sciences USA* 85: 180–84

Shedlovsky, A., McDonald, J.D., Symula, D. & Dove, W.F. (1993) Mouse models of human phenylketonuria. *Genetics* 134: 1205–10

Steele, E.C., Lyon, M.F., Favor, J. *et al.* (1998) A mutation in the connexin 50 (*Cx50*) gene is a candidate for the *No2* mouse cataract. *Current Eye Research* 17: 883–89

Steingrimsson, E., Favor, J., Ferre-D'Amara, A.F., Copeland, N.G. & Jenkins, N.A. (1998) *Mitfmi-enu122* is a missense mutation in the HLH dimerization domain. *Mammalian Genome* 9: 250–52

Su, L.K., Kinzler, K.W., Vogelstein, B. *et al.* (1992) Multiple intestinal neoplasia caused by a mutation in the murine homolog of the APC gene. *Science* 256: 668–70

Zdarsky, E., Favor, J. & Jackson, I.J. (1990) The molecular basis of brown, an old mouse mutation, and of an induced revertant to wild type. *Genetics* 126: 443–49

Zeiher, B.G., Eichwald, E., Zabner, J. *et al.* (1995) A mouse model for the delta F508 allele of cystic fibrosis. *Journal of Clinical Investigation* 96: 2051–64

Zheng, B., Sage, M., Cai, W.W. *et al.* (1999) Engineering a balancer chromosome in the mouse. *Nature Genetics* 22: 375–78

Zingg, B.C., Pretsch, W. & Mohrenweiser, H.W. (1995) Molecular analysis of four ENU-induced triosephosphate isomerase null mutants in *Mus musculus*. *Mutation Research* 328: 163–73

GLOSSARY

alkylating agent any reagent that inserts an alkyl group in the place of a nucleophilic group in a molecule

antimorph mutant allele that has an opposite effect to the normal allele

circadian rhythm a biological rhythm with a period of about 24 hours

gene targeting replacement or mutation of a specific gene using recombinant DNA techniques

haplo-insufficiency haplo-insufficiency occurs if the loss of one allele of a gene is enough to give rise to a phenotypic effect. Haplo-insufficiency hence infers that the correct level of a gene product is essential for its normal function

hypermorph mutant allele that produces a more exaggerated version of the effect of the wild-type gene

hypomorph mutant allele that behaves in a similar fashion to the normal gene but has a weaker effect

neomorph mutant allele that produces changes in developmental processes, resulting in the appearance of a new character

stem cell undifferentiated cell capable of undergoing unlimited division; daughter cells can give rise to several different cell types

See also **Gene targeting** (p.791)

COAT COLOUR GENETICS OF MOUSE, MAN AND OTHER ANIMALS

Ian J. Jackson
MRC Human Genetics Unit, Western General Hospital, Edinburgh, UK

1. Introduction

One of the most striking features of any animal is their colour, and variation in colour between individuals within a species has been noted for centuries. In humans, there is considerable variation in skin pigmentation and hair colour. By contrast, variation in wild mammals is not common, probably because natural selection acts against animals who, for example, become more visible to their predators or to their prey, or who are no longer chosen as mates.

Once certain animals were domesticated, selection pressure was released or even imposed in a different direction, and variation accumulated. Agricultural animals such as horses, cattle, sheep and pigs demonstrate a range of pigmentation types. The variation is usually of colour intensity, of shade within the range of red to black or of patches of colour on a white background. In some agricultural species, the colour variation may have been co-selected along with characters of economic importance. On the other hand, companion animals and show or "fancy" animals, such as dogs, cats, mice and guinea pigs, often have been specifically selected for particular pigmentation varieties and exhibit a remarkable range of colour. The earliest geneticists used the fancy varieties to test Mendel's laws at the beginning of the 20th century. The mouse quickly came to be the animal of choice for much of mammalian genetics and new mutations have been found in the subsequent decades. There are now over 80 different genetic loci that affect pigmentation and at some loci there are many alleles. It is only in the last few years that another model animal has been developed which has a large number of identified pigmentation genes and that is the zebra fish (*Danio rario*), another animal initially kept as a pet but later studied in the laboratory. Following the zebra fish mutation screens, there are now over 100 zebra fish pigmentation mutations (Kelsh *et al.*, 1996).

2. Melanocyte biology

Melanin is the pigment that colours mammalian skin, hair and eyes and is produced by cells known as melanocytes. These are found principally in the basal layer of the epidermis of the skin and in the outer sheath of the hair follicle, where melanin is packaged into granules and secreted from the melanocytes into keratinocytes, which make up most of the epidermis of the skin, and into the cortical cells which form the hair shaft.

Within the mammalian inner ear, a layer of cells, the stria vascularis, contains pigmented melanocytes, whose function is enigmatic; their presence is essential for normal hearing function, but, curiously, pigment synthesis is not essential (Steel & Barkway, 1989). The skin, hair and inner ear melanocytes, and some of those in the eye, originate from an embryonic structure known as the neural crest. By contrast, the layer of melanin-producing cells that lies behind the retina, known as the retinal pigmented epithelium (RPE), has a different origin; these cells develop from a bilayered cup of neuroepithelium that differentiates to retina in the inner layer and RPE in the outer (**Figure 1**).

The neural crest is formed during mammalian embryogenesis when the neural plate folds to form a tube and cells migrate away from the crest on either side. These cells differentiate into a number of different types, such as facial structures, adrenal cells, several neuronal populations, including enteric ganglia that innervate the gut, and melanocytes. The melanocyte precursors (or melanoblasts) migrate from the trunk neural crest along pathways below the surface ectoderm from where they enter the basal layer of the epidermis and the hair follicles as they form. Melanocytes in the head migrate anteriorly from the trunk neural crest and enter the developing eye and inner ear, in addition to the skin and hair.

Mutations have been identified that may affect any one of the series of developmental and cellular processes leading from the cells of the neural crest to the secreted melanin in hair and skin. The nature of the products of many of the genes is known and a hierarchy (**Figure 2**) of interacting transcription factors, receptors, ligands, structural proteins and enzymes can be deduced (Jackson, 1994).

3. Genes involved in neural crest differentiation

An old dominant mouse mutation known as *splotch* has turned out to be a mutation in a transcription factor, a paired-box-containing gene called *Pax3*. The *splotch* phenotype consists of white patches on the belly and feet and is seen when one *Pax3* gene is mutant but the other is normal. It is due to haplo-insufficiency of the transcription factor, meaning that the one wild-type gene cannot produce enough protein for completely normal development, resulting in a deficiency of melanoblasts. As the belly is the last place the migrating cells reach, the deficiency is revealed by a lack of cells at this site. When both copies of the *Pax3* gene are defective (i.e. in homozygous *splotch* animals), the embryos die. They not only lack all melanocyte precursors, but they also have defects in other neural crest derivatives, such as spinal ganglia, and have neural overgrowth. The Pax3 protein seems to be a transcription factor required for the differentiation or migration of neural crest cells.

PAX3 is also defective in a human genetic condition named Waardenburg's syndrome, type I. Patients with this

MELANOCYTES and DEVELOPMENT

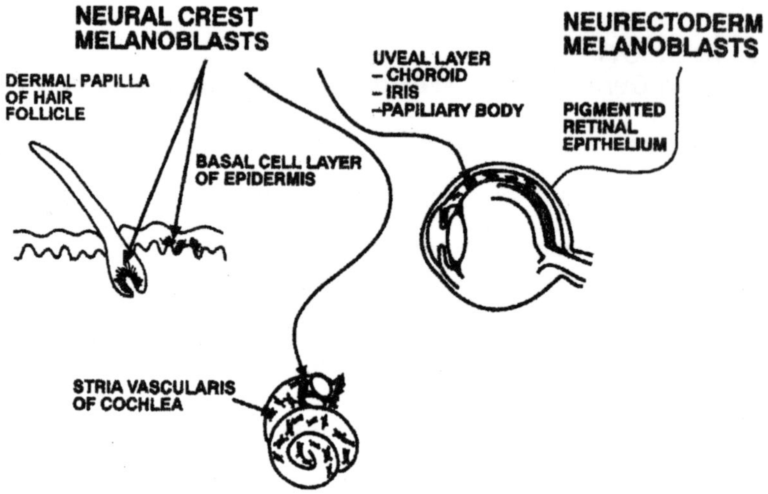

Figure 1 A schematic view of the origin of pigmented cells during mammalian development. Neural crest-derived melanocytes populate the epidermis, the hair follicle, the stria vascularis of the cochlea and several regions associated with the eye. Neuroectoderm cells behind the retina differentiate *in situ* to form the pigmented retinal epithelium.

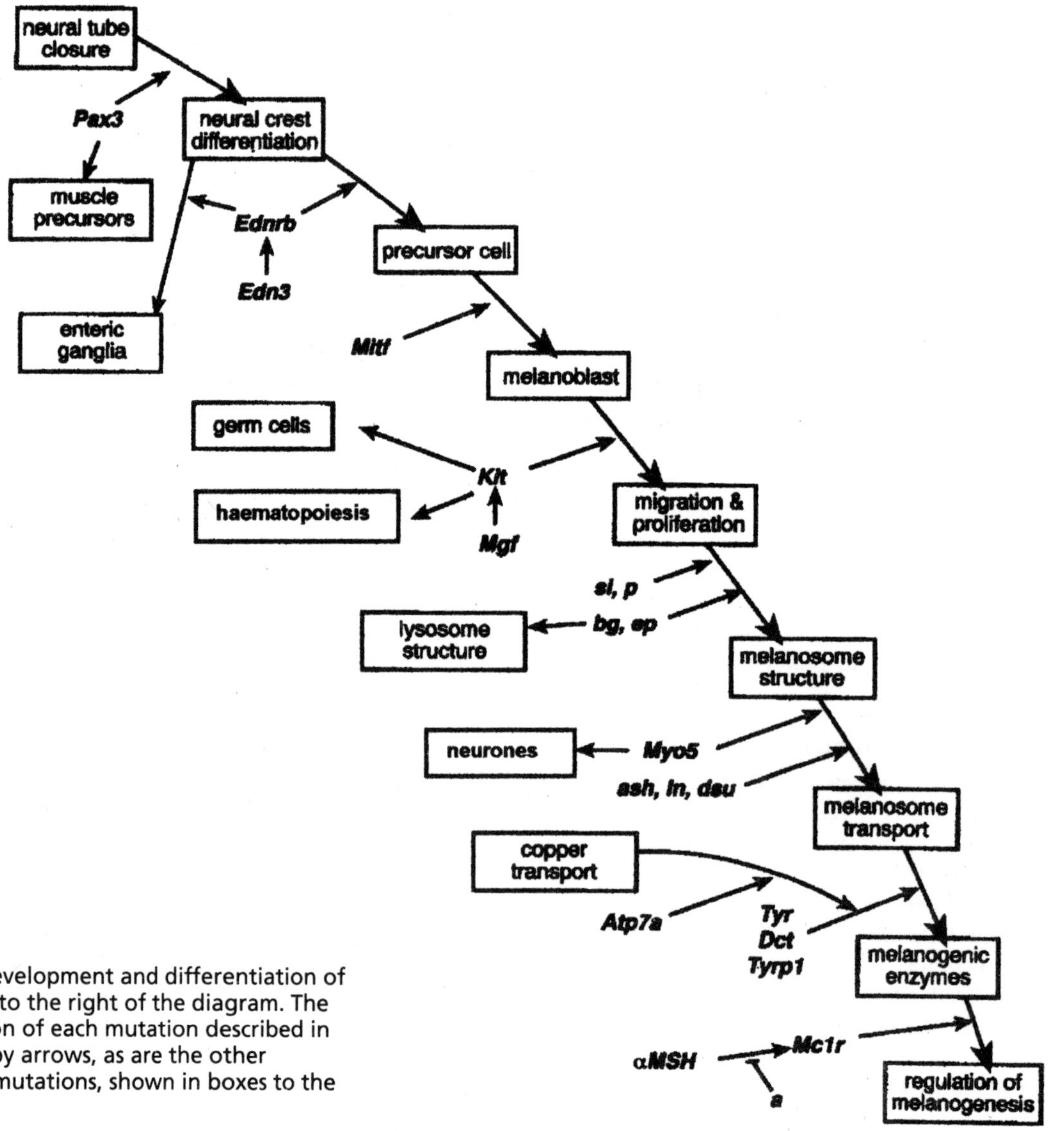

Figure 2 The stages of development and differentiation of melanocytes, in the boxes to the right of the diagram. The likely primary sites of action of each mutation described in this article are illustrated by arrows, as are the other processes affected by the mutations, shown in boxes to the left.

disease are characterized by different coloured irises (heterochromirides) and a white forelock. They also typically have a slightly abnormal facial morphology and are often deaf, due to the absence of melanocytes in the inner ear.

While *Pax3* is apparently needed for early neural crest development, two other genes have been identified that act slightly later, in the differentiation of melanoblasts and enteric ganglia. The genes encode a seven membrane-spanning receptor, the endothelin receptor B (*Ednrb*), and its peptide ligand, endothelin 3 (*Edn3*). Mouse mutations that cause loss of function of either gene (which are known as *piebald lethal* and *lethal spotting*, respectively) are recessive and, when homozygous, give rise to animals that lack virtually all coat colour (but have black eyes) and which eventually die of megacolon, caused by a functional blockage of the colon due to a lack of enteric ganglia (Baynash *et al.*, 1994; Hosoda *et al.*, 1994). The action of endothelin 3 on a cell that is a precursor of both melanoblasts and enteric ganglia is apparently necessary for the precursor either to survive or to differentiate into the melanoblast/enteric ganglia lineage.

There is a recessive rat mutation called *spotting lethal* that has a phenotype very similar to the mouse *piebald lethal* phenotype and which is caused by a deletion of the *Ednrb* gene.

A human genetic disorder, Hirschsprung disease (HD), is very similar to Waardenburg's syndrome. Patients have heterochromirides, white forelock, hypopigmentation and deafness, but, in addition, have gut paralysis, which results from the loss of ganglia in the intestine. Some patients have dominant mutations in the *RET* gene, a gene not involved in any classical mouse mutations. Other patients have mutations either in *EDNRB* or in *EDN3*, but the genetics is complex and the phenotypes are not completely penetrant (Chakravarti, 1996). Some HD patients also have another disorder, congenital central hypoventilation syndrome (CCVS) or "Ondine's Curse", which produces abnormal control of respiration, such that patients hypoventilate when asleep. Interestingly, one CCVS patient who has no pigmentation defects has, nevertheless, been found to have a loss-of-function mutation in *EDN3*.

4. Genes involved in melanoblast growth or survival

Developmentally downstream of *Ednrb* and *Edn3* is another receptor/ligand pair that is required for melanoblast survival and proliferation. Mutations called *dominant white spotting* or *W* mutations when heterozygous normally produce mice that have a coat that is somewhat paler than normal, but which also have white spots on the head and belly and have white feet. There are numerous alleles of the gene, which have a more or less severe phenotype. When homozygous, some of these mutations are lethal, but those that are not result in mice that have virtually no pigment derived from neural crest cells (but have black eyes). Both homozygotes and heterozygotes have deficiencies in haematopoiesis, and in mast and germ cells, and hence are anaemic and the homozygotes that do not die from severe anaemia are usually sterile or subfertile.

A second locus, which also has numerous alleles, results in so-called *steel* mutations. The phenotype of these mutations is remarkably similar to that of *W* mutations, affecting melanocytes, haematopoietic stem cells and germ cells. However, experiments in which skin was transplanted between mice with the two different mutations, and others in which the deficient haematopoiesis of mice with one mutation was reconstituted with the stem cells from the other, indicated that *W* mutations act cell-autonomously, but that *steel* mutations act through the environment in which the cells are located. These experiments suggested a receptor/ligand relationship between the two gene products.

Subsequently it was found that the gene, which when mutant, produces *W* mice, encodes a receptor tyrosine kinase known as Kit. Given the proposed relationships between the gene products, it was no surprise that the ligand that activates Kit, a protein now called mast cell growth factor (Mgf), was mutated in *steel* animals.

Simple loss-of-function mutations of both *Kit* and *Mgf* have a dominant phenotype as they are haplo-insufficient. A number of *Kit* mutations have a more severe dominant phenotype. The Kit receptor functions as a dimer and point mutations that reduce or eliminate the kinase function of the protein, while retaining its ability to dimerize, result in interference with the function of the normal protein. The effective level of functional Kit (i.e. normal dimers) is reduced to lower levels than in a heterozygote for a loss-of-function mutation, through a so-called "dominant negative", "dominant interfering" or "antimorph" action.

The skin colour of the Large White and several other pig breeds is caused by a dominant mutation resulting in a lack of melanocytes, which in turn is associated with abnormalities of the *KIT* gene. Human patients with piebaldism, characterized by a white patch on the ventral trunk on the forehead and the extremities, have been found to have point mutations in the *KIT* gene (Ezoe *et al.*, 1995). No human piebald patient has been found to be mutant in *MGF*; it may be that MGF is not haplo-insufficient in humans.

5. Gene encoding a melanocyte-specific transcription factor

Mutant alleles of the microphthalmia mutation have been known for some time. Recently, the gene was found to encode a transcription factor, Mitf, belonging to a family known as "basic helix–loop–helix leucine zipper" (bHLH-ZIP) proteins (Jackson & Raymond, 1994).

The phenotype of loss-of-function mutations of *Mitf* is complete absence of pigment cells both from skin and hair and from RPE. They are also deaf, because they are missing melanocytes in the inner ear, and have abnormal eye development, which is probably due to the abnormal RPE. It seems that this transcription factor is necessary for the normal expression of melanocyte genes and, indeed, for the survival of pigment cells. Studies on cultured neural crest cells suggest that *Kit* expression is not dependent on Mitf, but the response to Mgf binding to its receptor, Kit, and the response to endothelin 3 is absent if *Mitf* is mutant.

The bHLH-ZIP class of transcription factor proteins to which Mitf belongs contains several domains. The basic domain is the means by which the factor binds to DNA, which probably upregulates transcription through an activation domain. The factor will only bind to DNA as a dimer, either as a homodimer or as a heterodimer with another member of the same family. The requirement for dimerization means that mutations in the protein that affect DNA binding, but do not affect dimerization, can have a "dominant interfering" effect by sequestering the normal partner into inactive complexes. On the other hand, mutations that affect dimerization are recessive; in heterozygotes, there is sufficient normal protein to execute normal function and the mutant protein does not interfere.

If Mitf is a melanocyte-specific transcription factor, then what genes does it regulate? A DNA element, the M-box, has been found just upstream of the genes for the three melanogenic enzymes, tyrosinase, tyrosinase-related protein-1 (TRP-1) and TRP-2 (or DCT), that may be required for their transcription. The M-box is a very good binding site for Mitf, and melanocyte-specific transcription of these genes may well be mediated through this interaction. However, there are probably other targets for Mitf earlier in melanocyte development. Some work has suggested that Mitf might in fact be a "master regulator" of melanocyte development; if Mitf is made in certain cells that do not normally express it, a number of melanocyte-specific genes are activated and the cells adopt some of the characteristics of melanocytes (Tachibana *et al.*, 1996).

Human patients with one form of Waardenburg's syndrome (WS2) have the characteristic white forelock, heterochromirides and deafness, but not the facial dysmorphology of WS1 (*PAX3* mutant) patients. WS2 is due to a mutation in one copy of the MITF gene. Curiously, these mutations, unlike the dominant mouse mutations, seem to be a simple loss of function which, in the mouse, would be recessive. Apparently, human melanocytes show a haplo-insufficient phenotype with respect to MITF, while mouse melanoblasts do not.

6. Genes that affect the structure of the melanocyte

The earliest mutant mouse pigmentation locus to be cloned was the *dilute* gene. Some years later, the encoded protein was found to be an unconventional myosin, known as myosin type V (Myo5), that is expressed in melanocytes and neurones. Mice that are homozygous for the loss of Myo5 function have a diluted pigmentation, but also normally have opisthotony, a fatal neurological disorder characterized by upwards arching of the back and tail. However, the classical *dilute* allele does not result in opisthotony, but only affects pigmentation. The reason for this is that the mutation is due to the insertion of a retrovirus into an intron of *Myo5* that severely affects splicing of the mRNA in melanocytes, but which has no effect on the neuronal isoforms of the mRNA. An interesting consequence of the retroviral insertion is that this particular mutant allele reverts to a wild-type phenotype at relatively high frequency (3×10^{-6}) by intragenic recombination between the long terminal repeat (LTR) elements that flank the retrovirus, resulting in the *Myo5* gene containing only a single LTR, which no longer has an effect on melanocyte expression.

The myosin V molecule appears to be required in melanocytes to transport melanin granules out to the periphery. In *Myo* V^d mutant melanocytes, the granules cluster around the nucleus rather than move out into the dendrites and, as a consequence, pigment is transferred much less effectively to the growing hair.

The phenotypes of two other mouse mutant loci, *leaden* and *ashen*, are very similar to the dilute pigmentation phenotype of *Myo5*^{d}, although their product is unknown at present. The dilute pigmentation of all three mutations is suppressed in mice that are homozygous for a mutation, *dilute suppressor* (*dsu*). However, *dsu* has no effect on the neurological phenotype of *Myo5* mutations.

7. Genes involved in the structure and function of the melanosome

Melanin is synthesized in organelles known as melanosomes. These are bound by membrane, which encloses a protein framework onto which melanin is deposited. Once full of pigment, the organelle is transferred to neighbouring cells as a melanin granule.

A number of genes have been isolated whose mutant phenotypes are well characterized and which encode proteins that are part of the melanosomal structure or are inserted into the membrane, but whose function still eludes identification. These include the *pink-eyed dilution*, *silver*, *beige* and *pale ear* genes.

Beige (*bg*) and *pale ear* (*ep*) are members of a family of about a dozen mutant loci that, when homozygous, result in reduced pigmentation of eyes and hair, with abnormal melanosomes, but which also have a defect in platelet synthesis, so they have a prolonged bleeding time and have defective lysosomes in many tissues. Melanosomes and lysosomes have numerous proteins in common and perhaps share a common origin. Two human disorders, Chediak–Higashi syndrome (CHS) and Hermansky–Pudlak syndrome (HPS) affect the same three systems and, in fact, are the human homologues of *bg* and *ep*, respectively (Ramsey, 1996).

Beige encodes a protein that has some similarity to a yeast protein that is involved with the vacuole, the equivalent of the yeast lysosome. It is possibly associated with the cytoplasmic side of the melanosome and lysosome, and interacts with both cytoplasmic and other melanosomal or lysosomal proteins. CHS patients are mutant in the human homologue of *beige*.

The *HPS* gene was cloned from humans and found to encode a ubiquitously expressed protein with two transmembrane domains. It probably is integral to the membrane of the melanosome, but its function is as yet unknown. The homologous gene is inactivated in the *ep* mutation of mice.

Mice that have lost function of the pink-eyed dilution (*p*) gene have very little pigmentation in the coat or the eyes. Although the gene has been cloned, and it appears to be a protein with 12 membrane-spanning domains, its function is not yet known. It may be a melanosomal membrane protein, or it may be on the surface of the melanocyte or both. It is possibly a transport molecule, bringing the substrates of melanin synthesis into the cell or the organelle, but this is still only conjecture.

Oculocutaneous albinism (OCA) in humans is a genetic defect in which all pigment is absent from skin, hair and eyes (Spritz, 1994). There are two major forms of OCA, classified according to the presence of tyrosinase activity in the melanocytes. Tyrosinase negative OCA (type 1) is discussed below. The tyrosinase positive form (OCA2) is superficially indistinguishable from OCA1 and individuals with this disease have defects in the homologue of the *p* gene.

The protein matrix that occupies the lumen of the melanosomes is made up of several proteins. The gene for one of these, MMP115, was isolated from chickens and homologues have been found in humans, cattle and mice. The mouse gene is the product of the *silver* locus, mutations at which result in variably pigmented hair, probably as a result of melanocyte death. It is not known if these matrix proteins are simply structural or whether they have some enzymatic function in melanin synthesis.

8. Genes encoding the melanogenic enzymes

There are two forms of melanin synthesized by neural crest melanocytes: eumelanin, which is normally black or brown, and phaeomelanin which is yellow or red. The chemical synthesis of the two forms is regulated by the action of α-MSH (see below and illustrated in **Figure 3**). Both forms of melanin are synthesized from the amino acid tyrosine and both share two common initial reactions, the hydroxylation of tyrosine to 3,4-dihydroxyphenylalanine (DOPA) and the oxidation of DOPA to DOPAquinone, which are both catalysed by the same enzyme, tyrosinase. DOPAquinone will spontaneously form a dark pigment, but the actions of other melanogenic enzymes are needed to synthesize genuine eumelanin. Phaeomelanin is formed following the cysteinylation of DOPA, but the enzymatic requirement is unknown

The enzymes required for eumelanin synthesis are also illustrated in **Figure 3**. Two proteins were identified (initially by molecular cloning) as enzymes related to tyrosinase. While assigning an enzymic function to a protein is not straightforward, it appears that the one initially designated as TRP-2 converts DOPAchrome to 5,6-dihydroxyindole carboxylic acid (DHICA) and is now termed DOPAchrome tautomerase (DCT). The other related protein, TRP-1, catalyses the oxidation of DHICA to indole-5,6-quinone-carboxylic acid. Mutations have been found in the genes encoding these enzymes in coat colour mutant mice and humans.

Mice which have a loss of function of the tyrosinase gene (*Tyr*) have no melanin pigment at all; these animals have the classic albino phenotype of white fur and pink eyes. The original albino mutation was used in 1902 by Bateson and Cuenot independently to confirm the validity of Mendel's laws and is widespread among laboratory strains, which must all have come from a common mutant ancestor. Many other alleles have been identified, many of which have the same albino phenotype, but others have intermediate phenotypes. These intermediate mutations have a reduced, but not absent, tyrosinase activity, leading to a reduced quantity of melanin synthesis. Other *Tyr* alleles result in a patchy or mottled phenotype. This is a result of uneven expression of the tyrosinase gene, in which some melanocytes produce melanin, but others do not, and is usually due to chromosomal rearrangements that separate the gene from an important regulatory element. Another allele, *himalayan*, has a mottled phenotype because the encoded tyrosinase

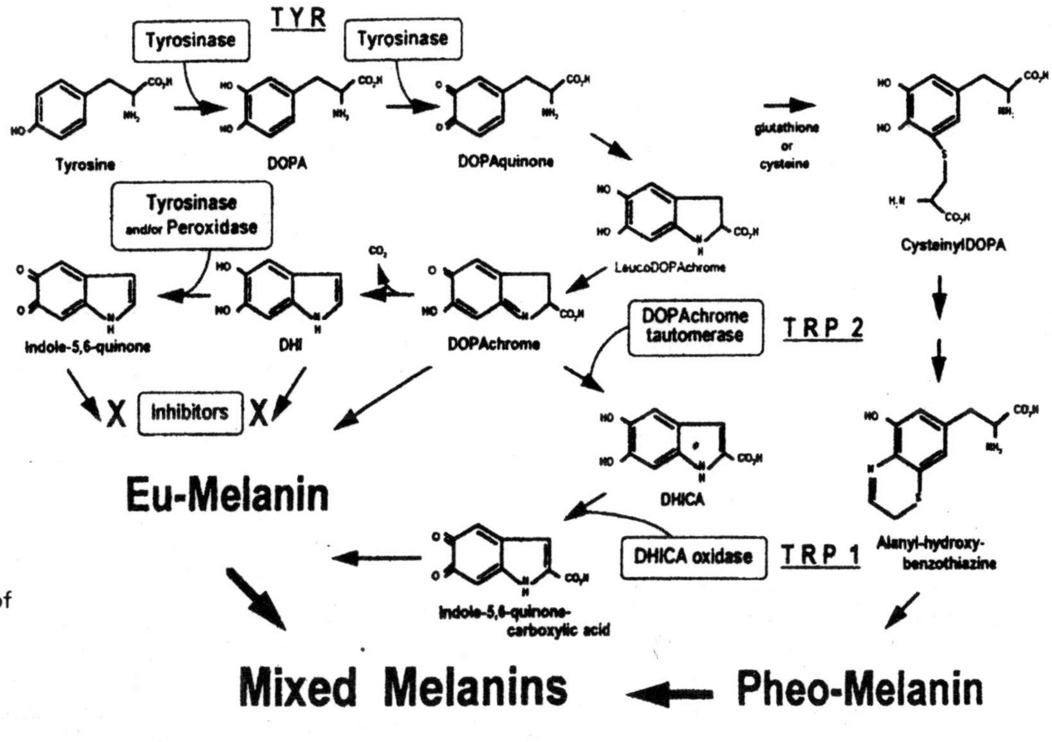

Figure 3 A summary of the chemical reactions leading to the formation of melanins. The enzymes known to be involved are indicated. (Reproduced from Winder *et al.*, 1994)

enzyme is temperature sensitive and so is inactive over the warmer trunk and head regions, but becomes active and so allows pigment synthesis on the cooler tail, ears and snout.

The tyrosinase-negative form of oculocutaneous albinism (OCA1) in humans is caused by mutations in the tyrosinase gene (Spritz, 1994). One characteristic feature of albino patients, in addition to their white skin and hair and pale eyes, is nystagmus, an involuntary flickering movement of the eyes caused by partial misrouting of the nerve connections between the retina and brain. This defect is not due directly to the absence of tyrosinase, but is a consequence of a lack of pigmentation, as numerous other human genetic defects characterized by reduced pigmentation (e.g. OCA2, CHS and HPS) also have nystagmus. Misrouting of neurones in the optic nerve does occur in albino mice, but they do not have the eye defect, probably because they do not have stereoscopic vision.

Other human patients with partial loss of tyrosinase function have "yellow" albinism, or OCA1B, and have residual tyrosinase activity that allows some pale pigment to be synthesized. One patient has been described who has a temperature-sensitive tyrosinase enzyme, equivalent to the *himalayan* mouse, and has pale head hair, but darker hair on the arms.

Surprisingly, although the albino phenotype is well known in many other animals, the molecular basis of only one other has been defined. The tyrosinase gene of albino medaka fish is inactivated by insertion of a transposable element. Among mammals, it has been established that albino rabbits have a defective tyrosinase enzyme as animals transgenic for the normal mouse *Tyr* gene are rescued to wild-type. The phenotype of the Siamese cat has the hallmark of a temperature-sensitive tyrosinase (as in the *himalayan* mouse), but this has not been proven.

The TRP-1 or DHICA oxidase gene, *Tyrp1*, is mutated in mice with the classical *brown* mutation. Mice with this or other loss-of-function alleles of *Tyrp1* produce brown rather than black eumelanin, and partial loss of function results in dark brown eumelanin. The *Tyrp1* gene in rats localizes to the *brown* mutation, but the sequence change involved is not yet known.

Human patients with mutations in *TYRP1* seem to be rare, although perhaps difficult to diagnose. A condition called brown albinism has been described which has only minimal depigmentation, but which still results in nystagmus. One individual who has been described in detail is one of a pair of (dizygotic) twins of Afro-American origin, which meant a comparison of pigmentation with his sib was possible. The individual has no detectable TRP-1 protein or mRNA, and this forms the basis of definition of a new form of albinism, OCA3 (Spritz, 1994).

The gene (*Dct*) encoding the third melanogenic enzyme, dopachrome tautomerase, is affected in the mouse *slaty* mutation. These mice, and those with a second mutant allele, may not be a complete loss of function and produce eumelanin that is dark brown. A third allele, *slaty-light*, has much paler eumelanin when homozygous, which, considering the enzyme's position in the synthetic pathway proximal to TRP-1, may be the loss-of-function phenotype.

These three related enzymes form a multienzyme, membrane-bound complex within the melanosome. All contain two metal-binding motifs and tyrosinase, at least, contains two copper atoms. This copper is essential for function and two mutations that affect pigmentation appear to act by decreasing the access of copper to the enzyme. The X-linked mouse mutation, *mottled*, produces depigmentation, curly hair and skeletal abnormalities. The homologous human phenotype, Menkes disease, is very similar and also includes neurological degeneration. In both cases, the disease is caused by mutations in the *Atp7a* gene, which encodes a membrane ATPase that transports copper into the cell. Another human disorder, cystathionine β-synthase deficiency, also includes hypopigmentation as part of the range of symptoms. This disorder results in increased synthesis of the amino acid homocyst(e)ine which acts directly on the copper in tyrosinase and inhibits the enzyme.

9. Genes involved in regulation of melanogenesis

The colour of wild-type mice (and many other mammals) is known as agouti and is not uniform, but is made up of alternating bands of yellow phaeomelanin and black eumelanin. A typical dorsal hair has a black tip and base separated by a single yellow band across the middle. This pattern is regulated indirectly by the action of α-melanocyte-stimulating hormone (α-MSH) on mature melanocytes. Melanocytes in culture respond to α-MSH by increasing the expression and activity of melanogenic enzymes and increasing melanin synthesis. *In vivo*, however, α-MSH appears to act on neural crest-derived melanocytes to switch synthesis from phaeomelanin (which may be the default synthesis) to eumelanin.

Circulating α-MSH is made as a part of preopiomelanocortin (POMC), a precursor molecule synthesized in the pituitary. However, α-MSH which acts on melanocytes is probably made locally around the hair follicle, because hypophysectomized mice, which have no circulating α-MSH, still synthesize eumelanin. There are at least four different α-MSH or melanocortin receptors, which are G-protein coupled receptors, consisting of seven membrane-spanning domains, whose stimulation by ligand binding results in the activation of adenylyl cyclase and an increase in intracellular cAMP. One of them, *Mc1r*, encodes the melanocyte-specific receptor.

Mc1r is mutant in mice with two opposite phenotypes. A loss-of-function mutation results in mice unable to respond to α-MSH and which are yellow because only phaeomelanin is made. Several other alleles are dominant and give rise to black mice, in which only eumelanin is made. These are all changes that give rise to receptors that have permanent, ligand-independent activation, or are hyper-responsive, and so constantly signal within the melanocyte to maintain the eumelanin pathway.

Other mammals with recessive phaeomelanic phenotypes also have mutations in the *Mc1r* gene. Red guinea pigs, red cattle, chestnut horses and wheaten (red/yellow) chickens are all recessive mutants for this receptor. Alaska foxes have a dominant, active form of the receptor (see below) and certain breeds of black cattle have a variant MC1R that may be constitutively active, as do black chickens. Red-haired humans also have variants of MC1R, but so also do some dark-haired individuals who, like redheads, burn in the sun rather than tanning (Valverde *et al.*, 1995). The genetics in humans may not be straightforward, because some of this group are heterozygous for a variant allele and a minority have a wild-type sequence at both alleles. The function of these variant receptors has yet to be tested.

The alternating black and yellow stripes of the mouse hair is due to alternating activation and inactivation of the melanocyte Mc1r. However, this is not in response to pulsatile expression or secretion of α-MSH; rather, it is caused by a pulsatile expression of an antagonist of Mc1r, the product of the *agouti* gene. Mice that have lost *agouti* gene function are completely black because there is no antagonism of the action of α-MSH, *Mc1r* is permanently stimulated and eumelanin is constantly made. On the other hand, mutations that result in constant expression of agouti protein result in mice that have an entirely yellow coat. These dominant *yellow* mutations are invariably caused by chromosomal rearrangements or insertions that juxtapose an active promoter next to the *agouti* gene. Dominant yellow mice have additional effects due to this widespread expression of agouti protein. They are obese and have a greater susceptibility to tumours, probably because the agouti protein is antagonizing other receptors. The other melanocortin receptors are obvious targets and it is notable that mice lacking the *Mc4r* gene are genetically obese, suggesting that it is this receptor that is the target (Huszar *et al.*, 1997).

The mechanism by which agouti protein antagonizes the action of α-MSH is not yet known. Direct antagonism with the hormone can be demonstrated on cultured cells that express Mc1r (Lu *et al.*, 1994), but there are effects of agouti protein alone. Whether all the actions of agouti are through the Mc1r, by direct antagonism or inverse agonism, or whether there is another agouti receptor, is at present unknown.

The *agouti* gene normally has two promoters: one promoter produces the pulsed agouti expression throughout the skin of wild-type mice, which thus have the striped agouti pattern, while the other promoter is active only in ventral skin and produces a constant expression of the gene which overrides the pulsed expression to give a yellow belly (the white-bellied agouti phenotype). Some so-called wild-type laboratory strains do not have the yellow ventrum, because the ventral promoter is inactive. Another mouse mutant, black and tan, has an inactive pulsatile promoter and a black back, but an active ventral promoter, and hence a yellow or tan belly.

Foxes have a very interesting interaction between the *agouti* and *Mc1r* genes (Vage *et al.*, 1997). The wild-type colouration is a red body, but with black areas on the ears, tail and feet. It appears that the balance between agouti and α-MSH in the fox is weighted towards agouti and antagonism of Mc1r. The Silver fox is homozygous for a recessive mutation of the *agouti* gene, caused by a deletion of a large part of the coding region. Heterozygotes for the deletion show a degree of haplo-insufficiency, and have a smoky red colour. The darkly pigmented Alaska fox has a dominant mutation of *Mc1r* that is constitutively active. However, animals heterozygous for the mutation still have some phaeomelanic regions, as long as there is at least one wild-type *agouti* allele, indicating that agouti protein can inhibit even a constitutively active receptor, either as a negative antagonist or by acting through a different receptor. Two copies of the dominant *Mc1r* mutation cannot be antagonized and produce a fully eumelanic coat indistinguishable from the recessive agouti homozygotes.

10. Summary

Coat colour genetics is as old as the study of genetics itself. The mutations have provided excellent models for genetic and developmental processes, and continue to produce insights into biological phenomena. The molecules identified through the study of coat colour are involved in fundamental aspects of cell biology, such as transcriptional regulation, organelle trafficking and cell signalling. Using a variety of genetic systems, in particular those of humans, mice and zebra fish, a great deal more of importance to both basic and applied biology and medicine can still be learned. Further details of many of the mutations in mouse can be found in Jackson (1994). There are two useful electronic databases; information about mouse genes can be found at http://www.informatics.jax.org/mgd.html, while details of human genetic disorders are at http://www3.ncbi.nlm.nih.gov/omim/.

REFERENCES

Baynash, A.G., Hosoda, K., Giaid, A. *et al.* (1994) Interaction of endothelin-3 with endothelin-B receptor is essential for development of epidermal melanocytes and enteric neurons. *Cell* 79: 1277–85

Chakravarti, A. (1996) Endothelin receptor-mediated signaling in Hirschsprung disease. *Human Molecular Genetics* 5: 303–07

Ezoe, K., Holmes, S.A., Ho, L. *et al.* (1995) Novel mutations and deletions of the KIT (steel factor receptor) gene in human piebaldism. *American Journal of Human Genetics* 56: 58–66

Hosoda, K., Hammer, R.E., Richardson, J.A. *et al.* (1994) Targeted and natural (piebald-lethal) mutations of endothelin-B receptor gene produce megacolon associated with spotted coat color in mice. *Cell* 79: 1267–76

Huszar, D., Lynch, C.A., Fairchild-Huntress, V. *et al.* (1997) Targeted disruption of melanocortin-4 receptor results in obesity in mice. *Cell* 88: 131–41

Jackson, I.J. & Raymond, S. (1994) Manifestations of microphthalmia. *Nature Genetics* 8: 209–10

Jackson, I.J. (1994) Molecular and developmental genetics of mouse coat color. *Annual Review of Genetics* 28: 189–217

Kelsh, R.N., Brand, M., Jiang, Y.J. *et al.* (1996) Zebra fish pigmentation mutations and the processes of neural crest development. *Development* 123: 369–89

Lu, D., Willard, D., Patel, I.R. *et al.* (1994) Agouti protein is an antagonist of the melanocyte-stimulating-hormone receptor. *Nature* 371: 799–802

Ramsey, M. (1996) Protein trafficking violations. *Nature Genetics* 14: 242–45

Spritz, R.A. (1994) Molecular genetics of oculocutaneous albinism. *Human Molecular Genetics* 3: 1469–75

Steel, K.P. & Barkway, C. (1989) Another role for melanocytes: their importance for normal stria vascularis development in the mammalian inner ear. *Development* 115: 1111–19

Tachibana, M., Takeda, K., Nobukuni, Y. *et al.* (1996) Ectopic expression of MITF, a gene for Waardenburg syndrome type 2, converts fibroblasts to cells with melanocyte characteristics. *Nature Genetics* 14: 50–54

Vage, D.I., Lu, D., Klungland, H. *et al.* (1997) A non-epistatic interaction of agouti and extension in the fox, Vulpes vulpes. *Nature Genetics* 15: 311–15

Valverde, P., Healy, E., Jackson, I.J., Rees, J.L. & Thody, A.J. (1995) Variants of the MSH receptor gene are associated with red hair in humans. *Nature Genetics* 11: 328–30

Winder, A., Kobayashi, T., Tsukamoto, K. *et al.* (1994) The tyrosinase gene family: interactions of melanogenic proteins to regulate melanogenesis. *Cellular and Molecular Biology Research* 40: 613–26

GLOSSARY

adenylyl cyclase membrane-associated enzyme that converts AMP into cyclic AMP (cAMP)

cAMP (cyclic 3′,5′ adenosine monophosphate) this universal "second messenger" is generated from adenosine triphosphate (ATP) by adenylate cyclase, has a hormonally controlled structure, and triggers reactions that produce a cellular response to the specific stimulus

cysteinylation the covalent addition of a molecule of cysteine to an organic molecule through a sulfhydryl bond

haematopoiesis normal formation and development of blood cells in the bone marrow

haematopoietic stem cell actively dividing cell that is the source of the blood cells

haplo-insufficiency haplo-insufficiency occurs if the loss of one allele of a gene is enough to give rise to a phenotypic effect. Haplo-insufficiency hence infers that the correct level of a gene product is essential for its normal function

hypophesectomized where the pituity gland has been removed.

keratinocyte epidermal cell that synthesizes keratin

leucine zipper structural motif found in many proteins concerned with gene regulation; though to be involved in protein–protein interactions

lysosome membrane-bound organelle containing various digestive enzymes that process material taken in by phagocytosis and damaged/redundant cellular components

mast cell cell with large basophilic granules containing heparin, serotonin, bradykinin and histamine, which are released from the cell in response to injury or infection

melanocyte cell capable of producing melanin

neural crest cells embryonic cells that separate from the neural plate during formation of the neural tube and migrate to give several different lines of adult cells

tautomerase an enzyme that catalyses the rearrangement of molecular structure without altering its chemical composition

BEHAVIOURAL AND NEURAL GENETICS OF THE MOUSE

Wim E. Crusio

Brudnick Neuropsychiatric Research Insitute, University of Massachusetts Medical School, Worcester, Massachusetts, USA

1. Introduction

Behaviour is an animal's way of interacting with its environment and it is, therefore, a prime target for natural selection. As behaviour is the output of an animal's nervous system, this indirectly leads to selection pressures on its neuronal structures. In consequence, each species' behaviour and nervous system have co-evolved in the context of that species' natural habitat and can be properly comprehended only when these interrelationships are taken into account.

This notion implies that, in order to arrive at a profound understanding of neural and behavioural traits, one will have to consider problems of causation. Van Abeelen (1979) distinguished between the phylogenetic and the phenogenetic aspects of causation; both concern the genetic correlates of neural and behavioural traits, the first in an evolutionary sense, the latter in a gene–physiological sense. Stated in another way, neurobehavioural geneticists attempt to provide an answer to the question of what exactly the adaptive value of a trait for an organism is and to uncover the physiological pathways underlying the expression of this trait.

2. The phylogenetic aspect of causation

Selection pressures mould the species' genetic make-up which consequently will show traces of this past selection. Therefore, information about the genetic architecture of neurobehavioural traits might permit us to make inferences about the evolutionary history of these traits (Broadhurst & Jinks, 1974). Information about the presence and nature of dominance is especially important in this respect (Crusio, 1992). Thus, stabilizing and directional selection have predictably different consequences for the genetic architecture of a trait: the first leading to situations where dominance is either absent or ambidirectional, the latter resulting in dominance in the same direction as the selection.

Traditionally, behavioural geneticists have used quantitative genetic methods to address problems related to the phylogenetic aspect of causation. Unfortunately, the quantitative genetic designs that may be applied to human populations have a very low power to detect genetic variation due to dominance and cannot distinguish between different types of dominance. For animal studies, the diallel cross has been most often the design of choice. As an illustration of this research strategy, the results of an elegant series of experiments concerning mouse locomotor behaviour will be presented.

Henderson (1986) argued that not only is the selection pressure exerted on a certain character a function of the ecological niche of the population, but also that the ecological niche may vary during the development of an organism. Hence, selection pressure, and the resulting genetic architecture, would vary too. To test this hypothesis, the locomotor activity of mouse pups was measured at different stages during development. It would be expected that the genetic architecture of this trait would change in a predictable manner.

Around the age of 4 days, mouse pups are not yet able to orient themselves in their environment and have poor locomotor coordination abilities. If such a pup is accidentally dragged out of the nest, it would probably be more adaptive for it to depend on retrieval by the dam than try to return by itself. In agreement with this hypothesis, Henderson (1978) observed directional dominance for low activity at this age. By contrast, at about 11 days old, pups start having a better locomotor coordination and some ability to orient themselves towards the nest, probably by means of olfaction. Indeed, dominance was found to be directional for higher activity at this age, although, when pups of the same age were placed in an open field, without the possibility to return to the nest, no such dominance was observed (Henderson, 1981).

3. The phenogenetic aspect of causation

(i) The correlational approach

To tackle problems connected with the phenogenetic aspect of causation – that is, the underlying physiological mechanisms of a behavioural trait – several possible approaches exist. The most classic is where one exploits naturally occurring individual differences as a tool for understanding brain function. No one brain is like another and every individual behaves differently. The assumption that there is a link between the variability of the brain and individual talents and propensities seems quite plausible. This approach differs in two important aspects from the usual one in neuropsychology, which consists of observing the behavioural effects of lesions to discrete brain structures. First, no subjects are studied that, by accident or by design, have damaged brains. Rather, all subjects fall within the range of normal, non-pathological variation (provided animals carrying deleterious neurological mutations are excluded). Secondly, instead of comparing a damaged group with normal controls, a whole range of subjects are studied and correlation in variation at the behavioural level with that at the neuronal level is sought. This strategy, especially when used in combination with methods permitting the estimation of genetic correlations (Crusio, 1992), yields a very powerful approach.

A weakness inherent in correlational studies is that a phenotypical correlation between characters does not necessarily reflect a functional relationship. On the other hand, if two independent processes, one causing a positive relationship, the other causing a negative relationship, act simultaneously on two characters, the effects may cancel each other out so that no detectable correlation can emerge. These problems can, to a large extent, be avoided by looking at the genetic correlations; that is, at correlations between the genetic effects that influence certain characters, which are the products of either genes with pleiotropic effects or of linkage disequilibrium. By using inbred strains that are only distantly related, the probability that a linkage disequilibrium occurs may be minimized so that a possible genetic correlation will most probably be caused by pleiotropy – that is, there exists a (set of) gene(s) influencing both characters simultaneously. Thus, for these characters, at least part of the physiological pathways leading from genotype to phenotype must be shared and a causal, perhaps also functional, relationship must exist. It is this special property that renders the genetic correlational approach so uniquely valuable.

As an example of this approach, some work has been undertaken concerning the involvement of the hippocampus in the regulation of the exploratory behaviour that a mouse exhibits when it enters a novel, unfamiliar environment. Both hippocampal neuroanatomy and exploratory behaviour differ widely among different individual mice and this variation is, to a large extent, heritable (Crusio & van Abeelen, 1986; Crusio *et al.*, 1986). If the hippocampus is indeed involved in the regulation of this type of behaviour, then we might expect that genes that influence hippocampal neuroanatomy would also affect exploratory behaviour. In other words, notable genetic correlations should exist. Although a number of different aspects of exploratory behaviour have been studied, we limit ourselves here to one component – rearing. By rearing up, an animal may survey its environment and the frequency of this behaviour is under genetic control, as is shown, for example, by the successful bidirectional selection of lines expressing high or low frequencies of this behaviour, carried out by van Abeelen (1970).

A strong additive genetic correlation ($r = 0.48$) was found between the size of the hippocampal intra- and infrapyramidal mossy fibre (IIPMF) terminal field and rearing. Due to the low environmental correlation ($r = -0.08$), which in addition had a sign opposite to that of the additive genetic correlation, only a low and non-significant phenotypic correlation ($r = 0.14$) was obtained (Crusio *et al.*, 1989b). Conventional correlational analysis would, therefore, not have detected any relationship between these neuroanatomical and behavioural variables. Yet, the sizeable, positive additive genetic correlation implies that there exist pleiotropic genes that influence these two phenotypes in the same direction. Therefore, selective pressures on rearing should provoke neuroanatomical changes in the hippocampus. This hypothesis was tested by examining the inbred selection lines SRH (selection for rearing: high) and SRL (selection for rearing: low) that have been developed by van Abeelen (1970). As expected, SRH mice possessed IIPMF terminal fields that were larger than those of SRL mice (Crusio *et al.*, 1989a).

Subsequently, the serendipitous appearance of a mutation in the C57BL/6J inbred strain permitted a further test of the relationship between the IIPMF and rearing. The C57BL/6J//Nmg subline displayed a marked drop in the frequency with which this behaviour is displayed in an open field when compared with the original C57BL/6J subline. Again, as expected on the basis of the positive genetic correlation between rearing and the IIPMF, the mutated subline was shown to have smaller IIPMF (Crusio *et al.*, 1991). Furthermore, it showed a poorer learning performance in a spatial radial-maze task (Jamot *et al.*, 1994a). A cross-breeding analysis indicated that the behavioural and neuroanatomical differences between these two sublines are most probably due to only one single genetic unit (Jamot *et al.*, 1994b), providing the opportunity to dissect this relationship between neuroanatomy and behaviour further by using neurochemical and molecular genetic methods.

(ii) The mutational approach

The study of the behavioural effects of spontaneous mutations in a single gene is one of the oldest approaches in behavioural genetics. Initially, these studies were carried out in Drosophila, but later also in mice and rats (see van Abeelen (1965) for an overview). Many of these studies focused on so-called neurological mutants that show deviations in their nervous systems entailing behavioural changes (e.g. the cerebellar mutants *staggerer*, *hot-foot* and *lurcher*; Guastavino *et al.*, 1990, 1991; Lalonde, 1994).

Of course, mutations can also be induced by applying a mutagenic treatment (radiation or chemical) to breeding animals. Subsequently, their descendants are then screened for phenotypic (behavioural) changes. This technique is known as the mutant screen and has been used with great success in *Drosophila melanogaster* and *Caenorhabditis elegans* (see, for example, Greenspan (1990) for a review). In recent years, Takahashi and his collaborators have applied this approach to the circadian rhythm that mice exhibit for wheel-running activity, a behaviour that can easily be measured in a great number of individuals simultaneously (Takahashi *et al.*, 1994). They succeeded in isolating a gene, since called *Clock*, that must be present in its wild-type form for the mouse to display its normal behaviour. It appears, however, that this method is less suitable for the analysis of behaviours that are more time-consuming to measure, such as learning behaviour, for example.

Finally, recent developments in molecular genetics have made it possible to insert a supplementary gene, sometimes even from another specific origin [species], into the mouse genome and to bring it to expression, or to target specific genes and disrupt their expression by means of homologous recombination. Although both methods have their problems (Gerlai, 1996), they may be used to increase our insight into the underlying physiological mechanisms regulating behaviour. A recent example of the use of these methods is the study by Cases *et al.* (1995) of a mouse deficient for monoamine oxidase A (MAOA) which exhibits increased aggressiveness and may be used as a model for a similar affliction in the human male (Brunner *et al.*, 1993). Very new developments include methods to limit the gene deletion to certain specified brain regions or cell types (Tsien *et al.*, 1996) or to regulate the expression of transgenes (Mayford *et al.*, 1996). These approaches clearly suffer from fewer drawbacks than more simple methods (Gerlai, 1996).

Studies employing some of the above methods generally follow one of two strategies. The first one is where existing physiological information leads one to suspect that a certain gene product, such as a neuroreceptor, is important for the expression of some behaviour, such as learning. The expression of the gene coding for this product may then be caused to increase by the insertion of a transgene or to decrease by inducing a null mutation by means of homologous recombination. The resulting animals are subsequently studied to verify whether the expected effects do indeed occur. The other strategy is more reminiscent of that generally employed with spontaneous mutants, which is to screen mutated animals on a number of behavioural tests to identify possible behavioural alterations, which may then lead to ideas of how the gene is involved in the regulation of the changed behaviour.

4. Conclusion

There exists a rather fundamental difference in the type of genetic variation being studied in the classical genetic analyses employing cross-breeding and inbred strains, on the one hand, and in those using modern DNA-manipulating techniques, on the other. This lies in the fact that in the former type of study naturally occurring variation is being investigated, whereas in the latter type of experiment genetic variation is generated artificially. Often, genes manipulated by these methods are not polymorphic in nature. Therefore, although this type of study is contributing considerably to our knowledge of the physiological mechanisms underlying the regulation of behaviour in general, they will almost never contribute substantially to our knowledge regarding naturally occurring differences between individuals. In other words, if we are interested in the question of *how* mice are able to learn a certain task, we should use mutagenesis or transgenic techniques. However, if we are interested in *why* some mice are able to learn a certain task better than others, we should use more classical methods.

REFERENCES

Broadhurst, P.L. & Jinks, J.L. (1974) What genetical architecture can tell us about the natural selection of behavioural traits. In *The Genetics of Behaviour*, edited by J.H.F. van Abeelen, Amsterdam: North Holland and New York: Elsevier

Brunner, H.G., Nelen, M., Breakefield, X.O., Ropers, H.H. & van Oost, B.A. (1993) Abnormal behavior associated with a point mutation in the structural gene for monoamine oxidase A. *Science* 262: 578–80

Cases, O., Seif, I., Grimsby, J. *et al.* (1995) Aggressive behavior and altered amounts of brain serotonin and norepinephrine in mice lacking MAOA. *Science* 268: 1763–66

Crusio, W.E. (1992) Quantitative genetics. In *Techniques for the Genetic Analysis of Brain and Behavior: Focus on the Mouse*, edited by D. Goldowitz, D. Wahlsten, & R.E. Wimer, Amsterdam and New York: Elsevier

Crusio, W.E. & van Abeelen, J.H.F. (1986) The genetic architecture of behavioural responses to novelty in mice. *Heredity* 56: 55–63

Crusio, W.E., Genthner-Grimm, G. & Schwegler, H. (1986) A quantitative-genetic analysis of hippocampal variation in the mouse. *Journal of Neurogenetics* 3: 203–14

Crusio, W.E., Schwegler, H., Brust, I. & van Abeelen, J.H.F. (1989a) Genetic selection for novelty-induced rearing behavior in mice produces changes in hippocampal mossy fiber distributions. *Journal of Neurogenetics* 5: 87–93

Crusio, W.E., Schwegler, H. & van Abeelen, J.H.F. (1989b) Behavioral responses to novelty and structural variation of the hippocampus in mice. II. Multivariate genetic analysis. *Behavoural Brain Research* 32: 81–88

Crusio, W.E., Schwegler, H. & van Abeelen, J.H.F. (1991) Behavioural and neuroanatomical divergence between two sublines of C57BL/6J inbred mice. *Behavoural Brain Research* 42: 93–97

Gerlai, R. (1996) Gene-targetting studies of mammalian behavior: is it the mutation or the background genotype? *Trends in Neurosciences* 19: 177–81

Greenspan, R.J. (1990) The emergence of neurogenetics. *Seminars in Neurosciences* 2: 145–57

Guastavino, J.-M., Sotelo, C. & Damez-Kinselle, I. (1990) Hotfoot murine mutation: behavioral effects and neuroanatomical alterations. *Brain Research* 523: 199–210

Guastavino, J.-M., Bertin, R. & Portet, R. (1991) Effects of the rearing temperature on the temporal feeding pattern of the staggerer mutant mouse. *Physiology and Behavior* 49: 405–09

Henderson, N.D. (1978) Genetic dominance for low activity in infant mice. *Journal of Comparative and Physiological Psychology* 92: 118–25

Henderson, N.D. (1981) Genetic influences on locomotor activity in 11-day-old house mice. *Behavior Genetics* 11: 209–25

Henderson, N.D. (1986) Predicting relationships between psychological constructs and genetic characters: an analysis of changing genetic influences on activity in mice. *Behavior Genetics* 16: 201–20

Jamot, L., Bertholet, J.-Y. & Crusio, W.E. (1994a) Hereditary neuroanatomical divergence between two substrains of C57BL/6J inbred mice entails differential radial-maze learning. *Brain Research* 644: 352–56

Jamot, L., Bertholet, J.-Y. & Crusio, W.E. (1994b) Genetic analysis of hippocampal mossy fibers and radial-maze learning in two substrains of C57BL/6J inbred mice. *Behavior Genetics* 24: 518

Lalonde, R. (1994) Motor learning in lurcher mutant mice. *Brain Research* 639: 351–53

Mayford, M., Bach, M.E., Huang, Y.-Y. *et al.* (1996) Control of memory formation through regulated expression of a CaMKII transgene. *Science* 274: 1678–83

Takahashi, J.S., Pinto, L.H. & Hotz Vitaterna, M. (1994) Forward and reverse genetic approaches to behavior in the mouse. *Science* 264: 1724–33

Tsien, J.Z., Chen, D.F., Gerber, D. *et al.* (1996) Subregion- and cell type-restricted gene knockout in mouse brain. *Cell* 87: 1317–26

van Abeelen, J.H.F. (1965) *An ethological investigation of single-gene differences in mice*. Ph.D. Thesis, University of Nijmegen, The Netherlands

van Abeelen, J.H.F. (1970) Genetics of rearing behavior in mice. *Behavior Genetics* 1: 71–76

van Abeelen, J.H.F. (1979) Ethology and the genetic foundations of animal behavior. In *Theoretical Advances in Behavior Genetics*, edited by J.R. Royce & L.P. Mos, Alphen aan den Rijn: Sijthoff and Noordhoff

GLOSSARY

homologous recombination genetic recombination involving exchange of homologous loci

linkage disequilibrium the alliance of two linked alleles more often than would be expected by chance

phenogenetic the investigation of the physiological substrates of phenotypes

THE MOUSE AS A MODEL FOR THE STUDY OF GENETIC ASPECTS OF ALCOHOLISM

Lisa M. Tarantino[1,2] and Gerald E. McClearn[1,3]
[1]Center for Developmental and Health Genetics, Pennsylvania State University, USA
[2]Intercollege Graduate Program in Genetics, Pennsylvania State University, USA
[3]Biobehavioral Health Department, Pennsylvania State University, USA

1. Introduction

Alcohol-related phenotypes have attracted researchers who bring to bear disciplinary perspectives from motivational psychology, personality theory, clinical medicine, pharmacology, epidemiology, biochemistry, neurobiology, sociology, social psychology and law, among other specialities. The diversity of interest in this domain testifies to the magnitude and variety of the problems deriving from human use and abuse of alcohol.

Many of the schemes put forward from these various disciplines to explain differences among individuals in avidity for alcohol and in sensitivity to its effects have emphasized factors located within the environment. An appreciation that alcohol abuse can "run in families" has long persisted, however, and a substantial research effort has explored the "hereditary factor", both in human beings and in animal models, as a source of some of the differential responses of humans to the ubiquitous drug, ethanol. A recent volume edited by Begleiter and Kissin (1995) reviews the literature relating to humans. This present article will concentrate on the animal model research domain, for which Crabbe and Harris (1991) have presented an excellent general review.

2. Inbred strains and derived generations

In the animal model research effort, obvious phenotypic targets for investigation have been behavioural phenomena associated with alcohol use by humans; particularly salient, has been voluntary alcohol ingestion. For example, a comparison of several fully inbred mouse strains in a two-bottle choice between water and a 10% ethanol solution (McClearn & Rodgers, 1959) identified mice of the C57BL/Crgl strain as high alcohol consumers. By contrast, C3H animals were low to intermediate in alcohol intake and A, BALB/c and DBA animals were alcohol avoiders. These initial observations on relatively small samples were soon verified in numerous laboratories, and various sublines of these strains came to be the animals of choice in subsequent research on a broad array of alcohol-related phenotypes. Phillips and Crabbe (1991) have summarized this work, and only a few illustrative results will be cited here.

Once ingested, alcohol can have obvious behavioural consequences, among which are motor disturbances. Pharmacological research has suggested that alcohol has a biphasic effect on activity, with a stimulating effect at low doses and an inhibiting effect at higher (but subsedative) doses. Crabbe (1983), who described wide strain variation in response to a standard dose level, showed that the magnitude of this effect is subject to genetic influence. In an extensive dose-response study, Dudek and Phillips (1990) found that all of the examined strains displayed activation at some dosage, except for the C57BL/6Abg strain, and further noted that some strains showed the classically expected biphasic response pattern while others did not.

Alcohol has effects on coordination as well as on activity level. Crabbe *et al.* (1982), for example, observed ethanol-induced ataxia in several strains, finding that BALB/c mice were less sensitive than C57BL/6N or DBA/2N mice, which did not differ from each other. By contrast, using a different test of ataxia, Belknap and Deutsch (1982) found DBA/2J mice to be less sensitive than C57BL/6J mice. There were differences both in substrain and apparatus between these studies. It is not clear whether either of these factors accounts for the different outcomes, but this example illustrates the important principle that the "model system" in any investigation involves all aspects of the measurement situation. In addition to the specific apparatus and procedure, and all of the associated variables such as time of day, type of diet, background illumination, etc., the substrain of animals must be considered in generalizing from the results of any particular study.

Other phenotypes for which inbred strain differences have been reported include alcohol-induced hypothermia, sensitivity to hypnotic doses of alcohol and acquisition of tolerance to administered alcohol, among others. The review of Phillips and Crabbe (1991) should be consulted for a review of this literature. In brief summary, differences among inbred strains are ubiquitous for a wide variety of phenomena representing partial models pertinent to human alcohol use and abuse. They provide reliable benchmarks in the exploration of the genetic basis for individuality in alcohol-related processes.

3. Selective breeding

Following the demonstration by Mardones *et al.* (1950) of the feasibility of selective breeding in alcohol-related studies, Eriksson (1968) undertook the first major sustained selection for alcohol preference in rats. Appreciation of the analytical power provided by this experimental mating procedure led to the development, by several investigators, of a number of bidirectionally selected lines. **Table 1** identifies the principal lines so generated.

Successful selective breeding provides unequivocal

Table 1 Selected lines for various ethanol-related behaviours.

Line designation	Phenotype selected	Reference
Mice		
LS/SS	Hypnotic dose sensitivity[a]	McClearn & Kakihana, 1981
WSP/WSR	Withdrawal severity[b]	Crabbe *et al*., 1985
SEW/MEW	Withdrawal severity[c]	Wilson *et al*., 1984
HOT/COLD	Hypothermia response[d]	Crabbe *et al*., 1987a
FAST/SLOW	Locomotor stimulation[e]	Crabbe *et al*., 1987b
Rats		
UCHA/UCHB	Alcohol consumption[f]	Mardones *et al*., 1953
AA/ANA	Alcohol consumption[g]	Eriksson, 1968
P/NP	Alcohol consumption[g]	Lumeng *et al*., 1977
HAD/LAD	Alcohol consumption[h]	Li *et al*., 1988
SP/SNP	Alcohol consumption[h]	Fadda *et al*., 1989
AT/ANT	Coordination impairment[i]	Eriksson & Rusi, 1981
MA/LA	Activity response[j]	Riley *et al*., 1976
HAS/LAS	Hypnotic dose sensitivity[k]	Spuhler *et al*., 1990

[a] duration of loss of righting response after intraperitoneal injection of 3.3 g/kg ethanol
[b] seizure activity after initial intraperitoneal injection of 1.5 g/kg and chronic inhalation of ethanol vapour for 72 hours
[c] multiple measures upon withdrawal from ethanol – seizures, body temperature, light/dark box and vertical screen
[d] sensitivity to acute hypothermia induced by injection of 3 g/kg ethanol
[e] locomotor activity in open field after intraperitoneal injection of 1.5 g/kg ethanol
[f] alcohol intake after 60 days on diet deficient in factor N_1 (B-complex vitamins)
[g] one-bottle forced ethanol consumption (10%) for 10 days, then two-bottle choice for four weeks
[h] two-bottle choice, 10% ethanol versus water for three weeks
[i] motor impairment on a motor plane following intraperitoneal injection of 2.0 g/kg ethanol[j] motor impairment after intraperitoneal injection of 1.5 g/kg ethanol
[k] duration of loss of righting response after intraperitoneal injection of 3.0 g/kg ethanol

evidence of the non-zero heritability of the phenotype, of course, and permits estimation of quantitative genetic parameters. In general, the heritabilities estimated from realized response to selection have been low to moderate – e.g. LS/SS: 0.18 (McClearn & Kakihana, 1981); WSP/WSR: 0.26 (Crabbe *et al*., 1985); HOT/COLD: 0.17 (Crabbe *et al*., 1987a); and HAS/LAS: 0.26 (Spuhler *et al*., 1990). Given the highly controlled environmental circumstances under which selected lines are maintained, the modest size of these heritability estimates speaks of the sensitivity of the systems involved to very small environmental variations. Nonetheless, sustained selection pressure has sufficed to generate lines that display enormous phenotypic mean differences and, in some cases, non-overlapping distributions. These generated lines constitute very powerful models for the elucidation of mechanisms. Indeed, characterization of correlated responses to selection by comparison of the "high" to the "low" selected lines on phenotypes other than the one selected has constituted a major research strategy in the genetics of alcohol actions. Some of these investigations have explored the relationship of other alcohol-related phenotypes to the one for which selection was successful; Long Sleep (LS) mice, for example, have been found, relative to Short Sleep (SS) mice, to be more sensitive to low-dose activity stimulation (Sanders, 1976), to be more sensitive to alcohol-induced ataxia (Church *et al*., 1976) and, among other attributes, to display a greater hypothermic response or drop in body temperature on exposure to ethanol (Kakihana, 1977).

Other investigations have sought to determine the commonality of phenotypes across drugs. For example, FAST and SLOW mice, selectively bred for the low-dose activation effect of ethanol, differ also with respect to responsivity to methanol, *t*-butanol and *d*-amphetamine. These lines were not differentially responsive to caffeine or pentobarbital, and their activity levels were depressed similarly by nicotine (Phillips *et al*., 1989).

Still other investigations have tested hypotheses about the mechanisms mediating the influence of the polygenic systems. Clearly, explorations into the correlated responses to selection provide potent data for the understanding of the determinant systems.

For the most part, until the early 1990s, the genes affecting alcohol-related phenotypes were represented by the variance component attributable to them. Some few single loci were proposed to have an influence on one or another phenotype (e.g. Goldman *et al*., 1987), but the overall picture was provided by anonymous genes of small effect size in the quantitative model. With the recent avalanche of information on structural genomics, it has become possible to investigate some of these hitherto anonymous loci in polygenic systems. The search for these "quantitative trait loci" (QTL), introduced to alcohol studies in 1991, has become a dominant strategy in recent alcohol–pharmacogenetics research. The remainder of this review will summarize the results.

4. Identifying the polygenes

In 1978, Klein advocated the use of a correlational approach in using recombinant inbred (RI) strains (formed by crossing two inbred strains, followed by 20 or more generations of brother × sister mating) to analyse gene effects of modest size. The paucity of available markers limited the power of this application, although Crabbe *et al.* (1983b) demonstrated the general utility of RI strains in their usual deployment in the search for major loci. By happy coincidence, the RI series with the largest number of strains available at the time, the BXD series, had as progenitor strains the C57BL/6 and DBA/2 strains, which had been shown to differ greatly in many alcohol-related phenotypes. The development of a relatively dense map of genetic markers in the mouse, and their subsequent mapping in the BXD RI mice, provided one of the major tools in the search for alcohol-related QTLs (Gora-Maslak *et al.*, 1991; McClearn *et al.*, 1991; Plomin *et al.*, 1991).

Other designs employed in the search for QTLs include the use of segregating populations such as F_2s, advanced intercrosses, backcrosses, short-term phenotypic selective breeding and genotypic selective breeding (Darvas, 1998; Belknap *et al.*, 1997b; McClearn *et al.*, 1997).

The explosive growth of this research area has inspired considerable methodological and statistical discussion. One prominent issue has been the proper way of dealing with the Type I errors (declaring the presence of a QTL when there is really nothing there) arising from multiple statistical comparisons. Lander and Kruglyak (1995) proposed the setting of stringent alpha levels, while acknowledging the risk of Type II (or false negative) errors. In an effort to limit both types of statistical error in QTL identification, many have adopted a multistage research plan. Rather than testing a single, large experimental group, a succession of studies is done and the results are pooled in order to determine the most likely regions containing QTLs. The merits of this general approach have been examined in several papers (Plomin *et al.*, 1991; Johnson *et al.*, 1992; Belknap *et al.*, 1997a).

A recent advance has been the use of permutation tests which are now offered in a variety of QTL mapping packages (Map Manager QT; QTL Cartographer; MQTL) allowing investigators to identify significance levels derived from their own data set. The use of permutation tests is outlined by Churchill and Doerge (1994). In this procedure, multiple random permutations of phenotype values are performed, eliminating any relationship between the phenotype and marker genotype. Interval mapping is performed on these permuted values and a distribution of statistic values is saved. A level of significance is then chosen that is only exceeded by 5% of all permutations, giving an overall level of significance for a genome scan based on the empirical data.

A mixture of these approaches has characterized the QTL research on alcohol-related phenotypes. The phenotypes emphasized have been consumption (alcohol preference or acceptance), withdrawal symptoms as a measure of dependence and various measures of sensitivity to the effects of administered ethanol.

5. Alcohol consumption

The search for QTLs for alcohol consumption in mice has utilized two different measures: alcohol preference and alcohol acceptance.

(i) Alcohol preference

Alcohol preference has been one of the most widely studied alcohol-related traits in mice. Phillips *et al.* (1994) published the first identification of a QTL for alcohol preference in BXD RIs. Among a variety of taste-related traits, Phillips tested 19 of the BXD RI strains (females only) for voluntary consumption of a 10% ethanol solution and identified QTLs on chromosomes 2, 3, 4, 7 and 9. A reanalysis of these data using a more recent marker set and a slightly modified phenotype identified additional QTLs on chromosomes 5, 13 and 15 (Belknap *et al.*, 1997b). Another study of alcohol preference by Rodriguez *et al.* (1995) using a larger sample of BXD RI strains ($n = 21$) and including males, identified QTLs for alcohol preference on chromosomes 1, 2, 6, 7, 10, 11, 12, 15, 16 and 17. These data were also reanalysed using the more recent marker set and a slightly different phenotypic index and additional QTLs on chromosomes 4, 5, 8 and 18 were identified (Tarantino *et al.*, 1998).

Due to the large marker set (>1500 markers) used for this type of analysis, and the number of statistical tests performed, a substantial fraction of putative QTLs identified in RIs will be false positives. In order to determine which QTLs are "true QTLs", a number of follow-up studies have been done. Belknap *et al.* (1997b) performed a short-term selective breeding experiment to test putative QTLs nominated by Phillips *et al.* (1994). In Phillips's study, (C57BL/6 × DBA/2) F_2 mice were used as progenitors in selective breeding for high and low alcohol preference. After four generations of breeding, the selected lines were genotyped and tested for allele frequencies at the markers identified in the original study. The rationale of this approach is that the frequency of the marker alleles should differ in the divergent lines (Belknap *et al.*, 1997b). Using short-term phenotypic selection, Belknap *et al.* confirmed the presence of QTLs for alcohol preference on chromosomes 2, 3, 7, 9 and 15.

In another study to confirm RI-nominated QTLs, Tarantino *et al.* (1998) performed a genome scan on a group ($n = 218$) of F_2 animals derived from C57BL/6 and DBA/2 progenitors. This study confirmed the presence of QTLs on chromosomes 1, 2, 4 and 10, and identified two QTLs not detected in the original RI study (Rodriguez *et al.*, 1995) on chromosomes 3 and 9.

Another search for alcohol preference QTLs (Melo *et al.*, 1996) identified "sex-specific" QTLs on chromosomes 2 and 11. This study was limited by the use of only a one-way backcross to the DBA/2 parent. Further, the stringent significance level used to report significant results might have limited the identification of true positive QTLs.

In one of the first studies for alcohol-related traits using knockout mice, Crabbe *et al.* (1996a) found elevated alcohol consumption in null mutant mice lacking 5-HT_{1B} serotonin receptors. Knockout mice for the gene encoding the serotonin receptors (*Htr1b*; chromosome 9) were tested in a standard two-bottle choice for increasing volumes of ethanol (3%, 6%, 10% and 20%) as compared with control animals of the wild-type 129 strain. Mice with the null mutation showed increased alcohol consumption, drinking at least twice as much as the wild-type mice. Intake of ethanol by the knockout mice was at or exceeded the maximum values reported for the high preference C57BL/6 strain (Belknap *et al.*, 1993a). These results suggest that the QTL identified on chromosome 9 by several research groups and in several different animal populations (Crabbe *et al.*, 1994a; Tarantino *et al.*, 1998) may be the serotonin receptor gene.

Thus far, the data seem to substantiate the existence of QTLs on chromosomes 2, 3 and 9 for alcohol preference (see figure 1). Other QTLs identified may be true QTLs as well, but further experimental evidence is needed to confirm them.

(ii) Alcohol acceptance

McClearn (1968) introduced the test of alcohol acceptance as an alternative measure to alcohol preference. The test assesses alcohol consumption from a single drinking tube over a 24-hour period after 24 hours of water deprivation. The measures have been shown to be correlated ($r = 0.27$, $P < 0.05$) in a heterogeneous population of mice (Anderson *et al.*, 1979), but this correlation is low, indicating that the two phenotypes are measuring different parts of the complex domain of alcohol consumption.

The study by Crabbe *et al.* (1983b) included alcohol acceptance, utilized 17 of the 23 available strains of BXD RI mice and was limited to male mice. The distribution of the RI strains for alcohol acceptance appeared to be bimodal, suggesting a major gene. The study identified two loci on chromosomes 1 and 4; the locus on chromosome 1 appeared to have a major effect, while the locus on chromosome 4 had only a minor effect (see **Figure 1**). Goldman *et al.* (1987) went on to test a putative locus on chromosome 1 (*Ltw-4*, currently *Aop2*; 83.6 centiMorgans [cM]) for its effect on ethanol acceptance in mice. Using two-dimensional electrophoresis, Goldman *et al.* determined that the protein encoded by *Aop2* had two variants, acidic and basic, and that animals possessing the basic allele showed significantly higher ethanol acceptance.

Gora-Maslak *et al.* (1991) reanalysed the data of Crabbe *et al.* (1983b) using correlational analysis and a much larger set of markers. This analysis identified the chromosome 1 locus, *Aop2*, and several additional loci. Another RI study by Rodriguez *et al.* (1995) utilized a greater number ($n = 21$) of the BXD RI strains and both sexes, and reported on QTLs across the genome. In 1997, the same laboratory (McClearn *et al.*, 1997) published a genotypic selection study which confirmed a QTL on chromosome 15 which had already been identified in both RIs and an F_2. The collective data thus far seem to substantiate QTLs for alcohol acceptance on chromosomes 1, 4 and 15.

6. Withdrawal/dependence

Physical dependence on alcohol is defined as the altered or adaptive physiological state produced in an individual with repeated administration (Hanson & Venturelli, 1995). This state is only revealed when alcohol use or administration is abruptly discontinued or the action of alcohol is diminished in some way. In mice, a standard procedure is to expose animals to ethanol vapour in a vapour chamber over a period of several days, with concurrent intraperitoneal administration of pyrazole hydrochloride. Pyrazole hydrochloride is used to elevate and stabilize blood ethanol concentrations because it interferes with alcohol metabolism by acting as a competitive inhibitor of alcohol dehydrogenase (ADH). Animals are removed from the chamber and withdrawn from the ethanol vapour. Withdrawal is then scored as the severity of handling induced convulsions (Goldstein & Pal, 1971).

This measure was used in the Crabbe *et al.* study (1983b) to characterize the BXD RI strains for ethanol withdrawal

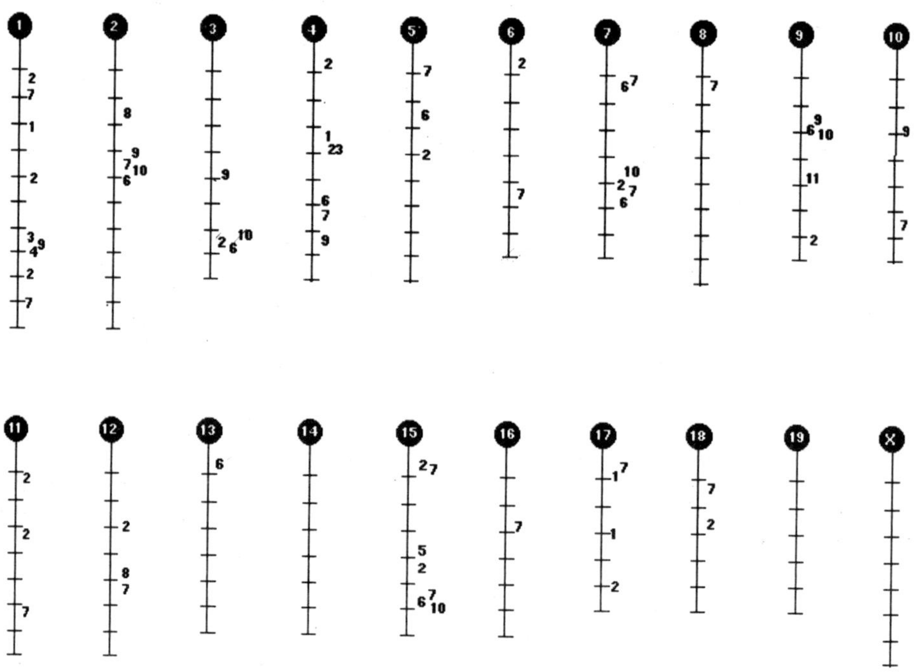

Figure 1 Alcohol consumption. Chromosomal localization of QTLs from multiple studies for alcohol acceptance (Nos 1–5) and alcohol preference (Nos 6–11). Numbers indicate location of QTLs and references as follows: 1, Gora-Maslak *et al.*, 1991; 2, Rodriguez *et al.*, 1995; 3, Crabbe *et al.*, 1983b; 4, Goldman *et al.*, 1987; 5, McClearn *et al.*, 1997; 6, Phillips *et al.*, 1994/Belknap *et al.*, 1997a (RI); 7, Rodriguez *et al.*, 1995/Tarantino *et al.* 1998 (RI); 8, Melo *et al.*, 1996; 9, Tarantino *et al.* 1998 (F_2); 10, Belknap *et al.*, 1997b (phenotypic selection); 11, Crabbe *et al.*, 1996a.

severity. Each of the RI strains was coded as either C57BL/6-like (B-like) or DBA/2-like (D-like) for withdrawal severity and the resulting strain distribution pattern (SDP) was compared with the then available markers in a linkage approach. The SDP for alcohol withdrawal in the BXD RIs resembled that for the gene *Car-2* on chromosome 3. The chromosome 3 locus appeared to be confirmed when the data were reanalysed by Gora-Maslak *et al.* (1991) using a larger marker set with a correlational approach. This reanalysis also identified QTLs on chromosomes 1, 9, 12, 13 and 18.

Additional QTLs for alcohol withdrawal in BXD RI mice were reported by Belknap *et al.* (1993b) using a slightly different measure. The discovery that the WSP and WSR selected lines of mice differ in withdrawal severity at least 10-fold following a single acute dose of ethanol led the authors to revise their technique. The new measure, acute withdrawal severity, is an index of handling induced convulsions (area under the curve) after a single intraperitoneal dose (4.0 g/kg) of ethanol. Using this measure, Belknap *et al.* (1993b) identified several QTLs on chromosomes 1, 2, 4 and 7.

Another study using acute ethanol withdrawal reported results on three different populations of mice; BXD RIs, F_2 and short-term selectively bred mice (Buck *et al.*, 1997). The BXD analysis yielded QTLs on chromosomes 1, 2, 4, 6, 8 and 11. Subsequent QTL analysis of the F_2 confirmed the QTLs on chromosomes 1, 2, 4 and 11. The short-term selection study produced significant results for chromosomes 1, 4 and 11. The chromosome 3 association (*Car-2*), which seemed so promising in early studies of chronic withdrawal, was not identified in any of the further QTL analyses. It is possible that it is either a false positive from the original study or that it is a QTL that does not influence both acute alcohol withdrawal and chronic withdrawal following vapour inhalation.

Based on the data thus far, the most promising areas containing QTLs for alcohol withdrawal are on chromosomes 1, 4 and 11 (see **Figure 2**). The results from both the acute withdrawal and the chronic withdrawal studies (Gora-Maslak *et al.*, 1991; Belknap *et al.*, 1993b; Buck *et al.*, 1997) indicate that the QTL on chromosome 1 influences both chronic and acute ethanol withdrawal, while the QTLs on chromosomes 4 and 11 are associated with acute ethanol withdrawal only.

7. Hypnotic dose sensitivity/sleep time/loss-of-righting response

The Long Sleep (LS) and Short Sleep (SS) mice are among the most widely studied groups of animals for alcohol-related traits. The LS/SS mice were bidirectionally selected by McClearn and Kakihana (1981) from a heterogeneous group of mice derived from an eight-way cross. The two lines were selected on sensitivity to the hypnotic action of ethanol after an intraperitoneal hypnotic dose of ethanol (3.3 g/kg body weight), measured as duration of loss-of-righting response (LORR) or "sleep-time". In early generations, the LS and SS mice did not differ in their rate of ethanol metabolism, but they did differ in the dose of ethanol required to cause a LORR and in blood ethanol level at the time of regaining the righting response (Heston *et al.*, 1974; Erwin *et al.*, 1976). Therefore, the LS and SS mice are models for central nervous system sensitivity to ethanol. In addition, inbred strains (ILS and ISS) were derived from the selected lines and these constituted progenitor strains for another RI strain series (LS × SS) of particular relevance to alcohol studies.

The earliest QTL work on the LS/SS mice was performed by DeFries *et al.* (1989) who, with a set of 27 LS × SS

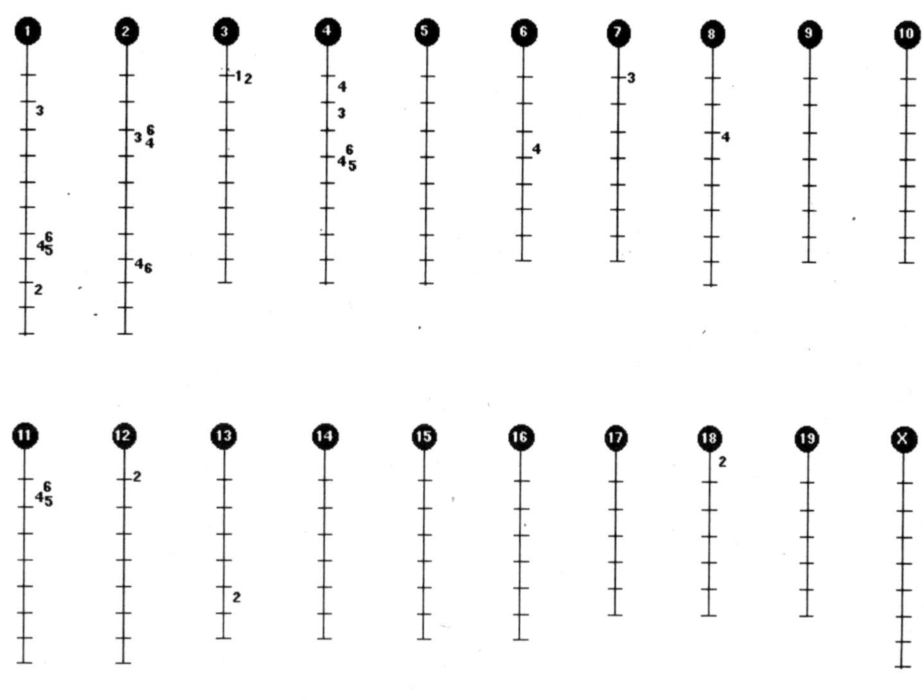

Figure 2 Alcohol withdrawal. Chromosomal localization of QTLs from multiple studies for both acute and chronic alcohol withdrawal. Numbers indicate location of QTLs and references as follows: 1, Crabbe *et al.*, 1983b (chronic withdrawal); 2, Gora-Maslak *et al.*, 1991 (chronic withdrawal); 3, Belknap *et al.*, 1993b (acute withdrawal); 4, Buck *et al.*, 1997 (RI) (acute withdrawal); 5, Buck *et al.*, 1997 (phenotypic selection) (acute withdrawal); 6, Buck *et al.*, 1997 (F2) (acute withdrawal).

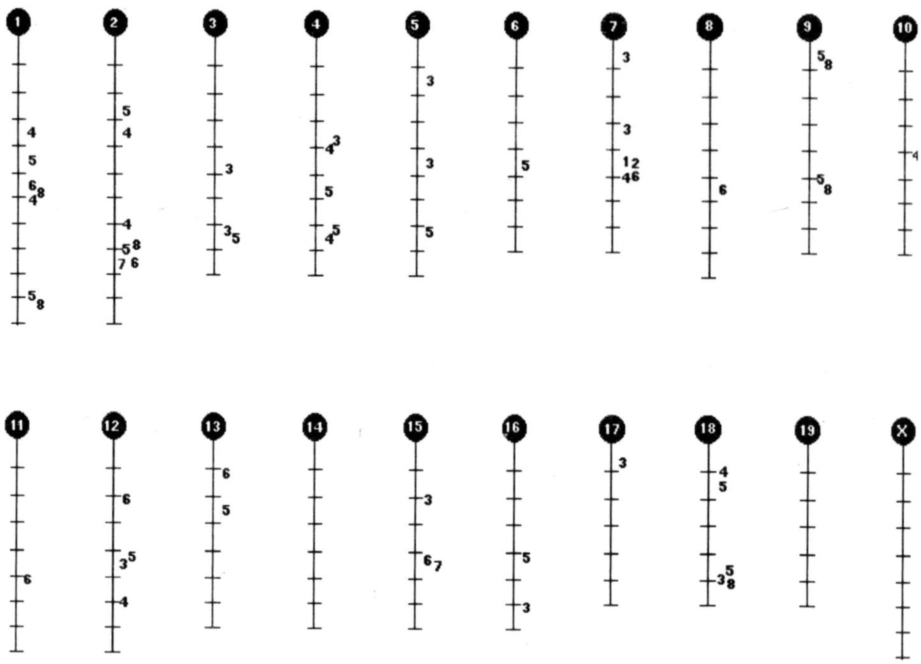

Figure 3 Hypnotic dose sensitivity. Chromosomal localization of QTLs from multiple studies for hypnotic dose sensitivity, also called loss-of-righting response or sleep time. Numbers indicate location of QTLs and references as follows: 1, DeFries *et al.*, 1989; 2, Johnson *et al.*, 1992; 3, Rodriguez *et al.*, 1995; 4, Christensen *et al.*, 1996; 5, Markel *et al.*, 1996; 6, Markel *et al.*, 1997; 7, Bennett *et al.*, 1997; 8, Erwin *et al.*, 1997a.

recombinant inbred lines, determined that there were approximately seven loci influencing ethanol-induced sleep time. This estimate is remarkably close to that of eight loci proposed by Dudek and Abbott (1984) based on their results of a classical cross between the LS and SS lines. It was also observed by DeFries *et al.* (1989) that LS mice are pigmented while SS mice are albino. An analysis of pigmented *vs.* non-pigmented RI strains showed that the average sleep time score for the pigmented strains was 94.0, while that of the albino strains was 75.6. The difference was only marginally significant ($t = 1.55$, $df = 24$, $P < 0.10$, one-tailed test), but accounted for 8.6% of the additive genetic variance in sleep time scores (DeFries *et al.*, 1989). These data suggested that a locus accounting for some of the variance in sleep time may be the albino locus (*Tyr*) on chromosome 7 or a locus which may be closely linked to it.

Johnson *et al.* (1992) proposed the use of a multistage strategy to locate the QTLs influencing ethanol-induced sleeping behaviour in the LS and SS strains. This strategy consisted of three steps: the detection of sequence variation between LS and SS strains, exploratory data analysis using the LS × SS RIs and confirmation of association by analysis of an F_2. The 1992 paper outlined results of a scan for restriction fragment length polymorphisms (RFLPs) on several chromosomes. A paper published by Markel *et al.* in 1996 reported on results from the second stage, using both point correlation analysis and interval mapping on the LS × SS RI data collected by DeFries *et al.* (1989) to nominate QTL regions. Point correlation identified QTLs for LORR on chromosomes 1, 2, 4, 5, 9, 13, 16 and 18, while interval mapping identified the same areas on chromosomes 1, 2, 4, 5, 13, 16 and 18 and identified other regions on chromosomes 3, 6, 7 and 12.

The third step in the multistage mapping project was published in 1997 by Markel *et al.* This paper reported on the use of 1072 F_2 mice generated from a cross between the inbred LS (ILS) and SS (ISS) progenitor strains and characterized for LORR using a repeated measures design (Markel *et al.*, 1995a,b). Scores for LORR were averaged individual scores from the two times of measure, resulting in a 39% reduction in environmental variance (Markel *et al.*, 1995a,b). This F_2 study tested the provisional QTLs already nominated by the RI analysis and confirmed QTLs on chromosomes 1, 2 and 18, and, possibly, 12 and 13. The genome scan of the F_2s also identified QTLs on chromosomes 8, 11 and 15 which had not been identified in either RI study (Christensen *et al.*, 1996; Markel *et al.*, 1996) and are subject to subsequent confirmation. An attempt at further confirmation was made using genotypic selection (Bennett *et al.*, 1997) for five of the seven QTLs identified in the RI study (Markel *et al.*, 1996) and/or the F_2 genome scan (Markel *et al.*, 1997). The genotypic selection confirmed only two of the seven QTLs, on chromosomes 2 and 15. The reason for the low rate of replication, however, may be the small sample sizes ($n = 2$–9) used in the study.

Another QTL analysis on sleep time induced by sedative–hypnotics in LS × SS RI strains was performed by Christensen *et al.* (1996) and included the DeFries data set as well as a data set for ethanol-induced sleep time collected by Wehner *et al.* (1992). This analysis identified QTLs on chromosomes 1, 2, 4, 7, 10, 12 and 18.

Erwin *et al.* (1997a) used the mean phenotypic values for the LS × SS RI strains published by DeFries *et al.* (1989) and Erwin *et al.* (1990) to perform a QTL analysis to determine common QTLs for hypnotic sensitivity to ethanol and central nervous system neurotensin measures. Using a one-way ANOVA, Erwin *et al.* (1997a) identified QTLs for LORR on chromosomes 1, 2, 9 and 18.

An analysis of LORR by Rodriguez *et al.* (1995) yielded provisional QTLs on chromosomes 2, 3, 4, 5, 7, 12, 15, 16,

17 and 18. The QTLs on chromosomes 2, 12, 15 and 18 are in the same locations as those identified in the previously reviewed studies and may be indicating the same genes.

8. Locomotor activation

Oliverio and Eleftheriou performed the earliest work in QTL mapping for ethanol-induced stimulation of locomotor activity in 1976. The testing paradigm used was a simple two-compartment toggle-floor box and mice were scored based on the number of crossings from one compartment to the other during a single test. Oliverio and Eleftheriou utilized the CXB RI strains of mice and, subsequently, two congenic strains of mice. The CXB RI strains are derived from the C57BL/6By inbred strain, which has a high basal activity level, and BALB/cBy, which has a low basal activity level. In the first part of the study, the seven CXB RI strains were tested for activity and were categorized as to which parent they resembled. They were assigned a SDP for activity based on this grouping. In the second part of the study, the authors found congenic lines that closely matched the SDPs obtained for activity in the RI strains. They found five lines and tested these for ethanol modification of basal activity. One of the lines (B6.C-H-16c) was identical to BALB/cBy and different from C57BL/6By in relation to alcohol-induced decrease in locomotor activity. This led the authors to conclude that a gene influencing decrease in basal activity on administration of ethanol was located at or near the gene *H(wl3K)* on chromosome 4. They named this locus *Eam* (ethanol activity modifier) because it lies on the same chromosome as *Exa* (exploratory activity), which was identified in a previous study (Oliverio *et al.*, 1973) as a major gene influencing basal activity levels.

Crabbe *et al.* (1983b) studied open field activity (an activity test measuring the number of boxes crossed in an open field) in C57BL/6J and DBA/2J progenitor strains and in the BXD RI series. The test paradigm differed from that of Oliverio *et al.* (1973) and consisted of a square box divided into a grid pattern. Mice were placed in the centre of the box and scored on how many lines they crossed within a 5-minute period. Mice were tested with saline injection on the first day and 10% ethanol (v/v) injection the second day. The measure was a difference score between baseline activity (under saline) and ethanol activity. The RI strains showed a range of values suggesting the influence of multiple loci. These data were reanalysed using the RI–QTL approach by Gora-Maslak *et al.* (1991) and QTLs for open field activity were detected on chromosomes 2, 4, 9 and 13. It is possible that the QTL on chromosome 4 is the same as *Eam* identified by Oliverio and Eleftheriou (1976).

Phillips *et al.* (1995) studied the effects of both acute and repeated ethanol exposures on locomotor activity in the BXD RI mice. Repeated exposure was studied to identify QTLs that may affect the magnitude of change in initial response that occurs with repeated treatments (e.g. sensitization). The study identified QTLs for acute ethanol effects on locomotor activity on chromosomes 3, 10, 11, 12, 17 and 18. QTLs for repeated ethanol exposure were identified on chromosomes 1, 8 and 10.

Another study published by Phillips *et al.* (1996) reported on QTLs for both acute and repeated ethanol exposures using a completely separate group of RI mice from the previous study. QTLs for acute ethanol effects on activity were reported on chromosomes 3, 4, 6, 7, 12, 13, 15, 17 and 18. QTLs for repeated exposure were identified on chromosomes 1, 7, 11 and 12. It is apparent that the same QTLs were not identified across the two studies and, in fact, there was poor agreement between strain means for the activity measures between the two studies (Phillips *et al.*, 1996). This fact was attributed to several variables, including the use of a different activity test apparatus and the demands of an ataxia test also administered to the same animals. The disparity between QTLs identified could be attributed to these

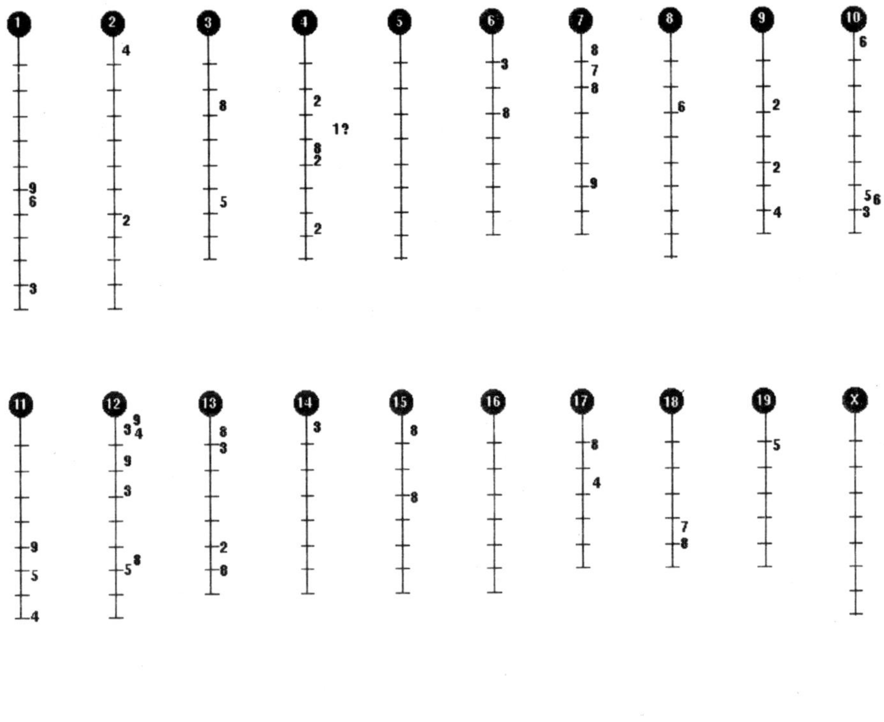

Figure 4 Locomotor activation. Chromosomal localization of QTLs from multiple studies for both acute locomotor activation and repeated sensitization. Numbers indicate location of QTLs and references as follows: 1, Oliverio & Eleftheriou, 1976; 2, Gora-Maslak *et al.*, 1991; 3, Erwin *et al.*, 1997b (activation slope); 4, Erwin *et al.*, 1997b (*Diff*); 5, Phillips *et al.*, 1995 (*acute*); 6, Phillips *et al.*, 1995 (*repeated*); 7, Cunningham, 1995; 8, Phillips *et al.*, 1996 (*acute*); 9, Phillips *et al.*, 1996 (*repeated*).

differences as well as to the use of a different, larger marker set.

A study of ethanol-induced locomotor activation using the LS × SS RI strains was published by Erwin *et al.* (1997b). Previous activation studies had used 2.0 g/kg as the standard dose and calculated a difference score using activity after an injection of saline minus activity after treatment with ethanol, in order to separate stress-induced activity from ethanol-induced activity. Erwin *et al.* noticed that the standard dose of ethanol poorly defined locomotor activity in the LS × SS RI strains, because the activity of most strains peaked at doses lower than 2.0 g/kg. Therefore, at 2.0 g/kg, most strains were already starting to experience some of the inhibitory effects of ethanol and identifying genes for only the stimulatory effects of ethanol would be difficult. In order to obtain an accurate measure for locomotor stimulation, the authors developed a new measure which they termed "ethanol activation slope" (EAS). This measure is the slope of the line resulting from a semilog plot of activity after three low doses of ethanol (1.0, 1.25 and 1.5 g/kg). Erwin *et al.* (1997b) felt that this measure represented a more accurate estimation of ethanol's stimulatory effects than a single dose.

A QTL analysis was performed on the measure of EAS and also on the standard measure of activation after administration of a single dose of ethanol. QTLs were identified on chromosomes 1, 6, 10, 12, 13 and 14 for EAS and on chromosomes 2, 9, 11, 12 and 17 for locomotor activity after a single dose.

The collective results of the aforementioned studies indicate concurrent QTLs at several locations throughout the genome. The *Eam* and *Exa* loci on chromosome 4 (Oliverio & Eleftheriou, 1976) were never localized to specific regions. However, the identification of QTLs on chromosome 4 in two subsequent studies (Gora-Maslak *et al.*, 1991; Phillips *et al.*, 1996) may indicate a replication of these early studies. Several studies have also identified coincident QTLs on chromosome 6 (Phillips *et al.*, 1995; Erwin *et al.*, 1997b) and chromosome 12 (Phillips *et al.*, 1995, 1996; Erwin *et al.*, 1997b).

9. Tolerance

Exposure to ethanol has ataxic, as well as stimulatory, effects on locomotor activity. Repeated exposure to ethanol can lead to sensitization to the stimulatory effects and has also been shown to decrease sensitivity to the ataxic or sedative effects of ethanol (tolerance). Studies conducted to find a common genetic relationship between these two effects of repeated exposure to ethanol have not been conclusive.

A study by Phillips *et al.* (1996) attempted to find a genetic link between these two effects under the assumption that strains of mice that are more susceptible to the development of locomotor sensitization should be more tolerant to ethanol's ataxic effects. Furthermore, Phillips *et al.* (1996) proposed that the same genetic elements might determine the tendency to develop both tolerance and sensitization; the results of this study, however, showed no significant genetic correlation between the development of tolerance and the development of locomotor sensitization. The study did report on the identification of QTLs for ataxia tolerance using a grid test. Since locomotor activity and ataxic tolerance were both tested using the same apparatus, the authors calculated a ratio of errors to activity counts as a mean of measuring tolerance. This measure takes into account the decrease in activity after ethanol treatment as a possible cause for a decrease in errors in a grid test. Using this measure, QTLs for tolerance were found on chromosomes 3, 4, 9, 13 and 18.

Another study tested for initial sensitivity, and rapid tolerance to ethanol-induced ataxia, using a rotarod test (Gallaher *et al.*, 1996). In the test for tolerance, mice were placed on a rotating dowel rod after being given an injection of ethanol. Treatment with ethanol caused the mouse to fall from the dowel rod. The mice were given repeated injections of ethanol until they could "pass" the dowel rod test. Blood ethanol concentrations were obtained and the measure of tolerance was reported as the fold increase between initial sensitivity or onset threshold and maximal threshold (the asymptote of the curve of blood ethanol concentration and time) or as a difference score (delta) between the maximal threshold and onset threshold. Initial sensitivity was measured using an assay for brain ethanol levels after a mouse initially fell from the dowel rod after an injection of ethanol.

QTLs were reported for all three measures: initial sensitivity, tolerance (delta) and tolerance (fold increase). QTLs for initial sensitivity were identified on chromosomes 1, 2, 3, 4, 5, 8, 9, 12, 14, 18 and X. QTLs for tolerance (delta) were identified on chromosomes 1, 2, 3, 4, 5, 7, 8, 9 and 18. Finally, QTLs for tolerance (fold increase) were identified on chromosomes 1, 2, 3, 4, 5, 8, 9, 12, 18 and X.

It is possible that the QTLs from both studies on chromosomes 3, 4 and 9 are indicating the same genes that influence tolerance. The QTLs on chromosome 18 identified in both studies are localized some distance apart on the chromosome. However, the Gallaher *et al.* (1996) study identified only the areas of highest correlation. Therefore, the actual confidence intervals containing the QTLs from both studies may overlap and indicate the same regions as well.

10. Conditioned place preference and conditioned taste aversion

Conditioned place preference is a Pavlovian differential conditioning procedure designed to study the hedonic value of ethanol while controlling for strain differences in the "preabsorptive" effects of ethanol, such as taste or odour, which can be present in the more prevalent test of ethanol drinking. This procedure is performed by pairing an environmental stimulus with ethanol and testing an animal's tendency to avoid or approach that environment. Ethanol can be administered by injection, therefore bypassing strain sensory differences for taste or odour.

A study by Cunningham (1995) employed this test using the BXD RI strains to identify QTLs for conditioned place preference. In the test, subjects from 20 BXD RI strains were grouped into either GRID+ or GRID− subgroups. The testing apparatus was a box in which one-half consisted of a grid

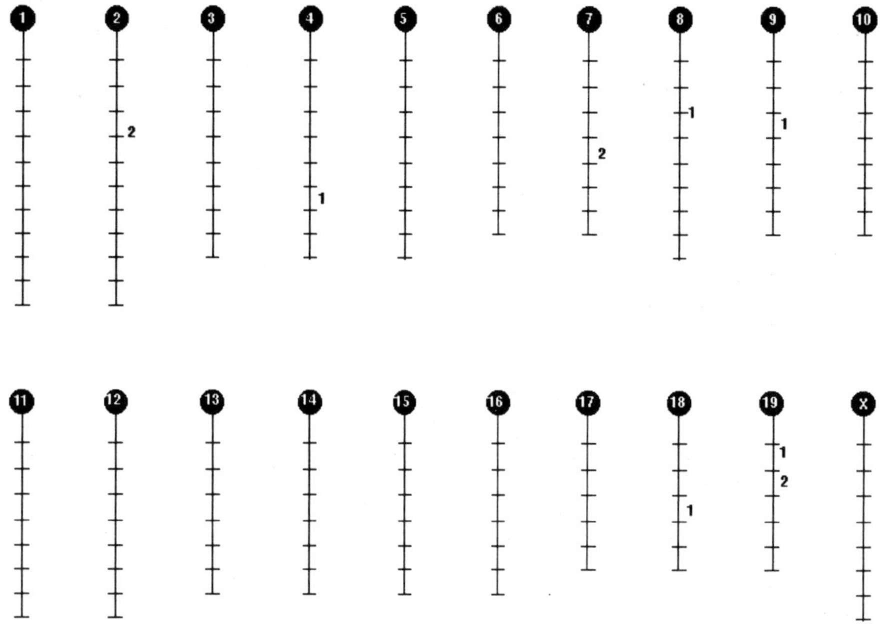

Figure 5 Conditioned place preference/conditioned taste aversion. Chromosomal localization of QTLs for both conditioned place preference and conditioned taste aversion. Numbers indicate location of QTLs and references as follows: 1, Cunningham, 1995 (conditioned place preference); Risinger and Cunningham, 1995 (conditioned taste aversion).

floor and the other half consisted of a perforated or hole floor. One-half of the animals in the GRID+ group were injected with ethanol prior to placement on a grid floor and half were injected with saline prior to placement on a hole floor. In the GRID− group, one-half of the animals were injected with ethanol prior to placement on the hole floor and the other half were injected with saline prior to placement on the grid floor. Control groups were injected with saline for all conditioning sessions. This conditioning procedure was conducted for four sessions within each group. After this conditioning phase, a place preference test was administered in which all animals were given an intraperitoneal injection of saline and then placed in the testing apparatus. The amount of time spent on the grid floor was recorded. The study reported on QTLs for conditioned place preference on chromosomes 1, 4, 5, 6, 7, 8, 9, 12, 13, 17, 18 and 19.

The testing procedure also permitted Cunningham to perform QTL analysis on measures of activity and sensitization, and tolerance. QTLs for activity were identified on chromosomes 3, 5, 13 and 17. QTLs for sensitization (measured as a change in activity response as a difference score between day 4 and day 1) were identified on chromosomes 1, 2, 3, 5, 7, 12, 15 and 19. QTLs for sensitization/tolerance (measured as a change in activity response to ethanol corrected for change in response to saline) were identified on chromosomes 1, 2, 12, 15 and 19.

Another Pavlovian conditioning procedure used for ethanol-related behaviours is conditioned taste aversion. This test is performed by giving animals access to a tastant such as saccharin and simultaneously giving them increasingly greater amounts of ethanol by injection. The result is ethanol-induced conditioned taste aversion in which consumption of saccharin decreases in a dose-dependent manner. A study by Risinger and Cunningham (1995) using BXD RI strains identified QTLs for ethanol-induced conditioned taste aversion on chromosome 2 after a single exposure to ethanol and on chromosomes 7 and 19 for longer-term ethanol exposure.

11. Hypothermia

Ethanol has a dose-dependent effect on thermoregulation in mice. This effect, measured by ethanol-induced hypothermia, has been used to study sensitivity and tolerance to the effects of ethanol. Various studies have shown that both initial response and development of tolerance to ethanol-induced hypothermia are under genetic control (Moore & Kakihana, 1978; Crabbe *et al.*, 1982) and that these two responses are genetically correlated (Crabbe *et al.*, 1982, 1989). Recently, studies have begun to identify the loci that may be associated with both initial response and tolerance to this alcohol-related trait. Two studies have utilized the BXD RI strains (Crabbe *et al.*, 1994b, 1996b). The C57BL/6 and DBA/2 inbred strains from which these RIs are derived have been shown to differ in both initial sensitivity and development of tolerance to ethanol-induced hypothermia. C57BL/6 mice have a substantial initial hypothermic response to various doses of ethanol and develop tolerance, while DBA/2 mice do not (Moore & Kakihana, 1978). The Crabbe *et al.* (1994b) study utilized 19 out of the 26 BXD RI strains and used three different doses of ethanol (2, 3 and 4 g/kg). The study identified QTLs for initial hypothermic response to 2 g/kg ethanol, 3 g/kg of ethanol and 4 g/kg ethanol. Crabbe *et al.* (1994b) also identified QTLs for tolerance after 3 days of treatment with 2, 3 and 4 g/kg ethanol.

Crabbe *et al.* (1994b) found several common QTLs for sensitivity and tolerance to hypothermia. For 2 g/kg treatment, common QTLs for hypothermic sensitivity and tolerance were co-localized on chromosomes 9 and 17. For 4 g/kg, common QTLs were co-localized on chromosome 16. Several QTLs were also identified which influence either sensitivity or tolerance at more than one dose of ethanol.

The data from the 1994 study were used again in a paper comparing QTLs for hypothermia with the ataxic effects of

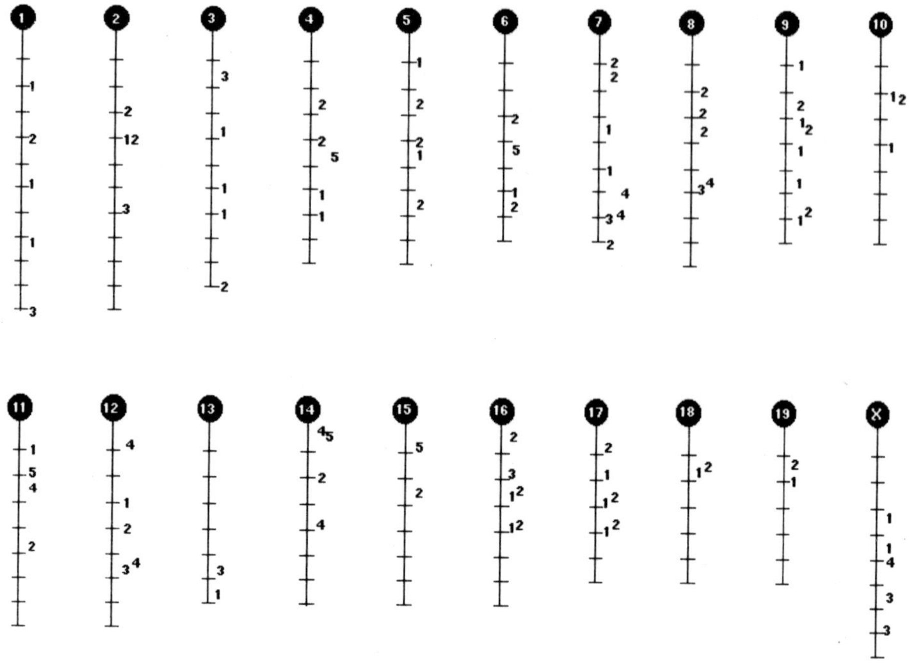

Figure 6 Hypothermia. Chromosomal localization of QTLs for hypothermic response to alcohol, measured as drop in body temperature on administration of ethanol, including both initial sensitivity and tolerance to ethanol-induced hypothermia. Numbers indicate location of QTLs and references as follows: 1, Crabbe *et al.*, 1994b (sensitivity); 2, Crabbe *et al.*, 1994b (tolerance); 3, Crabbe *et al.*, 1996b (sensitivity); 4, Crabbe *et al.*, 1996b (tolerance); 5, Erwin *et al.*, 1997a (sensitivity).

ethanol (Crabbe *et al.*, 1996b). The existing strain means from the first study were used, but four additional strains were added as well as additional polymorphic markers. Several additional QTL associations were identified and the location of several previously identified QTLs shifted due to reassignment of some marker locations.

A study using the LS × SS RI strains of mice and a slightly higher dose of ethanol (4.2 g/kg) identified QTLs for hypothermic sensitivity in several regions corresponding with those reported by Crabbe *et al.* (1994b, 1996b) on chromosomes 4, 6, 11 and 14 (Erwin *et al.*, 1997a). An additional QTL was also identified on chromosome 15.

12. Comments and discussion

(i) Overlapping QTLs – hotspots for alcohol-related behaviours

The figures indicating chromosomal regions containing QTLs for a variety of alcohol-related traits make it clear that several chromosomes appear to be "hotspots" for alcohol-related processes. An earlier review by Crabbe *et al.* (1994b) identified hotspots on chromosomes 1, 2, 4, 6, 8 and 9. It is apparent from **Figures 1–6** that a hotspot does exist on chromosome 1, with concurrent results from studies on preference, acceptance, locomotor activity, sleep time, hypothermia and withdrawal. Chromosome 2 has overlapping QTLs for several of the measures, including preference, hypothermia, conditioned taste aversion, LORR and withdrawal. There also appears to be a hotspot on chromosome 4, with evidence for QTLs for withdrawal, locomotor activation, sleep time, hypothermia and alcohol acceptance on proximal chromosome 4 and for conditioned place preference, locomotor activation, sleep time, hypothermia and preference on distal chromosome 4. Chromosome 9 also has some "coincidental" QTLs for conditioned place preference, locomotor activation, sleep time, hypothermia and alcohol preference, all localized near 30–40 cM.

(ii) Replication and fine mapping

Recent years have seen much progress in the field of QTL mapping for alcohol-related traits and an extensive body of literature is now available on QTLs for a variety of these traits, as outlined in previous reviews. The focus is now shifting towards the replication of results that have already been reported, and on identifying the genes underlying QTLs.

A variety of strategies are being used to replicate QTL studies which, until now, have primarily been performed on RIs. Recent studies have reported on replication results from an F_2 for alcohol preference (Tarantino *et al.*, 1998), from both F_2s and phenotypic selection for acute withdrawal severity (Buck *et al.*, 1997), and using genotypic selection to replicate a QTL for alcohol acceptance (McClearn *et al.*, 1997). These studies generally report a replication rate of ~25% from RI to F_2 when using an alpha level of $P < 0.05$ in RIs (Tarantino *et al.*, 1998), and 50% when using an alpha level of $P < 0.01$ in the preliminary RI scan (Buck *et al.*, 1997). These studies demonstrate the utility of the multistage approach for mapping and replicating QTL regions.

13. Future prospects

The strong convergent evidence for QTLs at many chromosomal locations for a variety of alcohol-related processes is providing the foundation for further dramatic advances. The most obvious next steps will be the identification and characterization of the actual genes located by the markers. A variety of approaches for mapping these genes has been reviewed including candidate gene identification and testing (Crabbe *et al.*, 1994a; Wehner & Bowers, 1995), generation of congenic strains and differential display analysis (Crabbe *et al.*, 1994a), positional cloning (Paterson, 1995) and random mutagenesis (Takahashi *et al.*, 1994).

The candidate gene approach has already met with some success in alcohol–QTL research, primarily through the use of knockout procedures. An example of this success is the observation that null mutant mice for the serotonin 1B receptor (5-HT_{1B}) consume more ethanol than their wild-type litter mates (Crabbe *et al.*, 1996a), indicating that a difference in the coding sequence or expression pattern of this gene may be involved in alcohol consumption. Null mutant mice lacking the gamma isoform of protein kinase C show reduced sensitivity to the effects of ethanol on LORR and hypothermia (Harris *et al.*, 1995). Another study showed that null mutant mice lacking the $GABA_A$ receptor alpha 6 subunit did not differ in tolerance and withdrawal to ethanol (Homanics *et al.*, 1998).

In addition to the candidate gene approach, which is dependent on genes that have already been identified, it is also going to be important to identify novel genes associated with alcohol-related traits. The ability to do this depends on the narrowing of the chromosomal regions containing QTLs to at least 0.5 cM or smaller in order to employ positional cloning techniques. This issue is being addressed by a variety of laboratories. A recent review by Darvasi (1998) outlines specific breeding schemes – advanced intercross lines and interval-specific congenic strains, for example – that can narrow the intervals containing QTLs from 25 cM to 1 cM.

Another method currently being explored is random mutagenesis screening. Random mutagenesis by chemical means (*N*-ethyl-*N*-nitrosourea or ENU; see article on "Ethylnitrosourea (ENU) as a genetic tool in mouse mutagenesis") results in point mutations at multiple locations throughout the genome. Screening mutagenized mice for a variety of behavioural assays may result in the identification of point mutations affecting behaviour traits, including alcohol-related traits. Positive results have already been seen using this method, with the identification of both the *Clock* gene (Takahashi *et al.*, 1994) and the *Wheels* gene (Pickard *et al.*, 1995) in mice, both of which influence circadian behaviour.

Another major research direction that will be enabled by the burgeoning information on QTLs will be the assembly of complex genetic systems element by element by various mating schemes, including most particularly that of genotypic selective breeding.

The identification of genes influencing alcohol-related behaviours in mice might aid in the understanding of alcoholism in humans. In fact, many of the QTL regions reported here contain candidate genes that have been and are being actively investigated in human samples. For example, the QTL for alcohol preference on chromosome 9 lies near the dopamine D2 receptor gene (*Drd2*) which has been associated with alcoholism in human studies (Blum *et al.*, 1990). However, the association between alcoholism and the human dopamine D2 receptor gene is still disputed. The gene for the serotonin 5-HT_{1B} receptor (*Htr1b*) also lies near the chromosome 9 QTL region and *Htr1b* knockout mice show elevated alcohol consumption (Crabbe *et al.*, 1996a). Recently, Lappalainen *et al.* (1998) have shown an association of the human 5-HT_{1B} receptor gene with antisocial alcoholism.

In conclusion, the utility of the mouse in genetic studies of alcohol behaviour has been validated with the identification of multiple chromosomal regions harbouring genes influencing a variety of alcohol phenotypes. The transfer of this information to human genetic studies of alcoholism will aid in concurrent mapping of genes influencing the intake and responses to alcohol in both species.

REFERENCES

Anderson, S.M., McClearn, G.E. & Erwin, V.G. (1979) Ethanol consumption and hepatic enzyme activity. *Pharmacology, Biochemistry and Behaviour* 11: 83–88

Begleiter, H. & Kissin, B. (eds) (1995) *Genetics of Alcoholism*, New York: Oxford University Press

Belknap, J.K. & Deutsch, C.K. (1982) Differential neurosensitivity to three alcohols and phenobarbital in C57BL/6J and DBA/2J mice. *Behavior Genetics* 12: 309–17

Belknap, J.K., Crabbe, J.C. & Young, E.R. (1993a) Voluntary consumption of ethanol in 15 inbred mouse strains. *Psychopharmacology* 112: 503–10

Belknap, J.K., Metten, P., Helms, M.L. *et al.* (1993b) Quantitative trait loci (QTL) applications to substances of abuse: physical dependence studies with nitrous oxide and ethanol in BXD mice. *Behavior Genetics* 23: 213–22

Belknap, J.K., Dubay, C., Crabbe, J.C. & Buck, K.J. (1997a) Mapping quantitative trait loci for behavioral traits in the mouse. In *Handbook of Psychiatric Genetics*, edited by K. Blum & E.P. Noble, Boca Raton, Florida: CRC Press

Belknap, J.K., Richards, S.P., O'Toole, L.A., Helms, M.L. & Phillips, T.J. (1997b) Short-term selective breeding as a tool for QTL mapping: ethanol preference drinking in mice. *Behavior Genetics* 27: 55–66

Bennett, B., Beeson, M., Gordon, L. & Johnson, T.E. (1997) Quick method for confirmation of quantitative trait loci. *Alcoholism, Clinical and Experimental Research* 21: 767–72

Blum, K., Noble, E.P., Sheridan, P.J. *et al.* (1990) Allelic association of human dopamine D2 receptor gene in alcoholism. *Journal of the American Medical Association* 263: 2055–060

Buck, K.J., Metten, P., Belknap, J.K. & Crabbe, J.C. (1997) Quantitative trait loci involved in genetic predisposition to acute alcohol withdrawal in mice. *Journal of Neuroscience* 17: 3946–955

Christensen, S.C., Johnson, T.E., Markel, P.D. *et al.* (1996) Quantitative trait locus analyses of sleep-times induced by sedative-hypnotics in LS × SS recombinant inbred strains of mice. *Alcoholism, Clinical and Experimental Research* 20: 543–50

Church, A.C., Fuller, J.L. & Dudek, B.C. (1976) Behavioral effects of salsolinol and ethanol on mice selected for sensitivity to alcohol. *Psychopharmacology* 47: 49–52

Churchill, D.A. & Doerge, R.W. (1994) Empirical threshold values for quantitative trait mapping. *Genetics* 138: 963–71

Crabbe, J.C. (1983) Sensitivity to ethanol in inbred mice: genotypic correlations among several behavioral responses. *Behavioral Neuroscience* 97: 280–89

Crabbe, J.C. & Harris, R.A. (eds) (1991) *The Genetic Basis of Alcohol and Drug Actions*, New York: Plenum Press

Crabbe, J.C., Janowsky, J.S., Young, E.R. *et al.* (1982) Tolerance to ethanol hypothermia in inbred mice: genotypic correlations with behavioral responses. *Alcoholism, Clinical and Experimental Research* 6: 446–58

Crabbe, J.C., Kosobud, A. & Young, E.R. (1983a) Genetic

selection for ethanol withdrawal severity: differences in replicate mouse lines. *Life Sciences* 33: 955–62

Crabbe, J.C., Kosobud, A., Young, E.R. & Janowsky, A. (1983b) Polygenic and single-gene determination of response to ethanol in BXD/Ty recombinant inbred mouse strains. *Neurobehavioral Toxicology and Teratology* 5: 181–87

Crabbe, J.C., Kosobud, A., Young, E.R., Tam, B.R. & McSwigan, J.D. (1985) Bidirectional selection for susceptibility to ethanol withdrawal seizures in *Mus musculus*. *Behavior Genetics* 15: 521–36

Crabbe, J.C., Kosobud, A., Tam, B.R., Young, E.R. & Deutsch, C.M. (1987a) Genetic selection of mouse lines sensitive (cold) and resistant (hot) to acute ethanol hypothermia. *Alcohol and Drug Research* 7: 163–74

Crabbe, J.C., Young, E.R., Deutsch, C.M., Tam, B.R. & Kosobud, A. (1987b) Mice genetically selected for differences in open-field activity after ethanol. *Pharmacology, Biochemistry and Behavior* 27: 577–81

Crabbe, J.C., Feller, D.J. & Dorow, J. (1989) Sensitivity and tolerance to ethanol-induced hypothermia in genetically selected mice. *Journal of Pharmacology and Experimental Therapeutics* 49: 456–61

Crabbe, J.C., Belknap, J.K. & Buck, K.J. (1994a) Genetic animal models of alcohol and drug abuse. *Science* 264: 1715–23

Crabbe, J.C., Belknap, J.K., Mitchell, S.R. & Crawshaw, L.I. (1994b) Quantitative trait loci mapping of genes that influence the sensitivity and tolerance to ethanol-induced hypothermia in BXD recombinant inbred mice. *Journal of Pharmacology and Experimental Therapeutics* 269: 184–92

Crabbe, J.C., Phillips, T.J., Feller, D.J. *et al.* (1996a) Elevated alcohol consumption in null mutant mice lacking 5-HT1B serotonin receptors. *Nature Genetics* 14: 98–101

Crabbe, J.C., Phillips, T.J., Gallaher, E.J., Crawshaw, L.I. & Mitchell, S.R. (1996b) Common genetic determinants of the ataxic and hypothermic effects of ethanol in BXD/Ty recombinant inbred mice: genetic correlations and quantitative trait loci. *Journal of Pharmacology and Experimental Therapeutics* 277: 624–32

Cunningham, C.L. (1995) Localization of genes influencing ethanol-induced conditioned place preference and locomotor activity in BXD recombinant inbred mice. *Psychopharmacology* 120: 28–41

Darvasi, A. (1998) Experimental strategies for the genetic dissection of complex traits in animal models. *Nature Genetics* 18: 19–23

DeFries, J.C., Wilson, J.R., Erwin, V.G. & Peterson, D.R. (1989) LS × SS recombinant inbred strains of mice: initial characterization. *Alcoholism, Clinical and Experimental Research* 13: 196–200

Dudek, B.C. & Abbott, M.E. (1984) A biometrical genetic analysis of ethanol response in selectively bred long-sleep and short-sleep mice. *Behavior Genetics* 14: 1–19

Dudek, B.C. & Phillips, T.J. (1990) Distinctions among sedative, disinhibitory and ataxic properties of ethanol in inbred and selectively bred mice. *Psychopharmacology* 101: 93–99

Eriksson, K. (1968) Genetic selection for voluntary alcohol consumption in the albino rat. *Science* 159: 739–41

Eriksson, K. & Rusi, M. (1981) Finnish selection studies on alcohol-related behaviors: general outline. In *Development of Animal Models As Pharmacogenetic Tools*, edited by G.E. McClearn, R.A. Deitrich & V.G. Erwin, Washington, DC: US Government Printing Office

Erwin, V.G., Heston, W.D.W., McClearn, G.E. & Deitrich, R.A. (1976) Effect of hypnotics on mice genetically selected for sensitivity to ethanol. *Pharmacology, Biochemistry and Behaviour* 4: 679–83

Erwin, V.G., Jones, B.C. & Radcliffe, R.A. (1990) Further characterization of LS × SS recombinant inbred strains of mice: activating and hypothermic effects of ethanol. *Alcoholism, Clinical and Experimental Research* 14: 200–04

Erwin, V.G., Markel, P.D., Johnson, T.E., Gehle, V.M. & Jones, B.C. (1997a) Common quantitative trait loci for alcohol-related behaviors and central nervous system neurotensin measures: hypnotic and hypothermic effects. *Journal of Pharmacology and Experimental Therapeutics* 280: 911–18

Erwin, V.G., Radcliffe, R.A., Gehle, V.M. & Jones, B.C. (1997b) Common quantitative trait loci for alcohol-related behaviors and central nervous system neurotensin measures: locomotor activation. *Journal of Pharmacology and Experimental Therapeutics* 280: 919–26

Fadda, F., Mosca, E., Colombo, G. & Gessa, G.L. (1989) Effect of spontaneous ingestion of ethanol on brain dopamine metabolism. *Life Sciences* 44: 281–87

Gallaher, E.J., Jones, G.E., Belknap, J.K. & Crabbe, J.C. (1996) Identification of genetic markers for initial sensitivity and rapid tolerance to ethanol-induced ataxia using quantitative trait locus analysis in BXD recombinant inbred mice. *Journal of Pharmacology and Experimental Therapeutics* 277: 604–12

Goldman, D., Lister, R.G. & Crabbe, J.C. (1987) Mapping of putative genetic locus determining ethanol intake in the mouse. *Brain Research* 420: 220–26

Goldstein, D.B. & Pal, N. (1971) Alcohol dependence produced in mice by inhalation of ethanol: grading the withdrawal reaction. *Science* 172: 288–90

Gora-Maslak, G., McClearn, G.E., Crabbe, J.C. *et al.* (1991) Use of recombinant inbred strains to identify quantitative trait loci in psychopharmacology. *Psychopharmacology* 104: 413–24

Hanson, G. & Venturelli, P.J. (1995) *Drugs and Society*, 4th edition, Boston: Jones and Bartlett

Harris, R.A., McQuilkin, S.J., Paylor, R. *et al.* (1995) Mutant mice lacking the gamma isoform of protein kinase C show decreased behavioral actions of ethanol and altered function of gamma-aminobutyrate type A receptors. *Proceedings of the National Academy of Sciences USA* 92: 3658–62

Heston, W.D.W., Erwin, V.G., Anderson, S.M. & Robbins, H. (1974) A comparison of the effects of alcohol in mice selectively bred for differences in ethanol sleep-time. *Life Sciences* 14: 365–70

Homanics, G.E., Le, N.Q., Kist, F. *et al.* (1998) Ethanol tolerance and withdrawal responses in $GABA_A$ receptor alpha 6 subunit null allele mice and in inbred C57BL/6J and strain 129/SvJ mice. *Alcoholism, Clinical and Experimental Research* 22: 259–65

Johnson, T.E., DeFries, J.C. & Markel, P.D. (1992) Mapping quantitative trait loci for behavioral traits in the mouse. *Behavior Genetics* 22: 635–53

Kakihana, R. (1977) Endocrine and autonomic studies in mice selectively bred for different sensitivity to ethanol. In *Alcohol*

Intoxication and Withdrawal III, edited by M.M. Gross, New York: Plenum Press

Klein, T.W. (1978) Analysis of major gene effects using recombinant inbred strains and related congenic lines. *Behavior Genetics* 8: 261–68

Lander, E.S. & Kruglyak, L. (1995) Genetic dissection of complex traits: guidelines for interpreting and reporting linkage results. *Nature Genetics* 11: 241–47

Lappalainen, J., Long, J.C., Eggert, M. *et al.* (1998) Linkage of antisocial alcoholism to the serotonin 5-HT1B receptor gene in 2 populations. *Archives of General Psychiatry* 55: 989–94

Li, T.-K., Lumeng, L., Doolittle, D.P. *et al.* (1988) Behavioral and neurochemical associations of alcohol-seeking behavior. In *Biomedical and Social Aspects of Alcohol and Alcoholism*, edited by K. Kuriyama, A. Takada & H. Ishii, Amsterdam and New York: Elsevier

Lumeng, L., Hawkins, T.D. & Li, T.-K. (1977) New strains of rats with alcohol preference and nonpreference. In *Alcohol and Aldehyde Metabolizing Systems*, Vol. III. Edited by R.G. Thurman, J.R. Williamson, H.E. Drott & B. Chance, New York: Academic Press

Mardones, R.J., Segovia, M.N. & Hederra, D.A. (1950) Herencio del alcoholismo in ratas. I. Comportamento de la primera generacion de ratas bebedoras. Colocadas en dieta carenciada en factor N. *Boletin de la Sociedad Biologia de Santiago* 7: 61–62

Mardones, R.J., Segovia, M.N. & Hederra, D.A. (1953) Heredity of experimental alcohol preference in rats: II. Coefficient of heredity. *Quarterly Journal of Studies on Alcohol* 14: 1–2

Markel, P.D., DeFries, J.C. & Johnson, T.E. (1995a) Ethanol-induced anesthesia in inbred strains of long-sleep and short-sleep mice: a genetic analysis using repeated measures. *Behavior Genetics* 25: 67–73

Markel, P.D., DeFries, J.C. & Johnson, T.E. (1995b) Use of repeated measures in an analysis of ethanol-induced righting reflex in inbred long-sleep and short-sleep mice. *Alcoholism, Clinical and Experimental Research* 19: 299–304

Markel, P.D., Fulker, D.W., Bennett, B. *et al.* (1996) Quantitative trait loci for ethanol sensitivity in the LS × SS recombinant inbred strains: interval mapping. *Behavior Genetics* 26: 447–58

Markel, P.D., Bennett, B., Beeson, M., Gordon, L. & Johnson, T.E. (1997) Confirmation of quantitative trait loci for ethanol sensitivity in long-sleep and short-sleep mice. *Genome Research* 7: 92–99

McClearn, G.E. (1968) The use of strain rank orders in assessing equivalence of techniques. *Behavior Research Methods and Instrumentation* 1: 49–51

McClearn, G.E. & Kakihana, R. (1981) Selective breeding for ethanol sensitivity: short-sleep and long-sleep mice. In *Development of Animal Models as Pharmacogenetic Tools*, edited by G.E. McClearn, R.A. Deitrich & V.G. Erwin, Washington, DC: US Government Printing Office

McClearn, G.E. & Rodgers, D. (1959) Differences in alcohol preference among inbred strains of mice. *Quarterly Journal of Studies on Alcohol* 20: 691–95

McClearn, G.E., Plomin, R., Gora-Maslak, G. & Crabbe, J.C. (1991) The gene chase in behavioral science. *Psychological Science* 2: 222–29

McClearn, G.E., Tarantino, L.M., Rodriguez, L.A. *et al.* (1997) Genotypic selection provides experimental confirmation for an alcohol consumption quantitative trait locus in mouse. *Molecular Psychiatry* 2: 486–89

Melo, J.A., Shendure, J., Pociask, K. & Silver, L.M. (1996) Identification of sex-specific quantitative trait loci controlling alcohol preference in C57BL/6 mice. *Nature Genetics* 13: 147–53

Moore, J.A. & Kakihana, R. (1978) Ethanol-induced hypothermia in mice. Influence of genotype on development of tolerance. *Life Sciences* 23: 2331–38

Oliverio, A. & Eleftheriou, B.E. (1976) Motor activity and alcohol: genetic analysis in the mouse. *Physiology and Behavior* 16: 577–81

Oliverio, A., Eleftheriou, B.E. & Bailey, D.W. (1973) Exploratory activity: a genetic analysis of its modifications by scopolamine and amphetamine. *Physiology and Behavior* 10: 893–99

Paterson, A.H. (1995) Molecular dissection of quantitative traits: progress and prospects. *Genome Research* 5: 321–33

Phillips, T.J. & Crabbe, J.C. (1991) Behavioral studies of genetic differences in alcohol action. In *The Genetic Basis of Alcohol and Drug Actions*, edited by J.C. Crabbe & R.A. Harris, New York: Plenum Press

Phillips, T.J., Feller, D.J. & Crabbe, J.C. (1989) Selected mouse lines, alcohol and behavior. *Experientia* 45: 805–27

Phillips, T.J., Crabbe, J.C., Metten, P. & Belknap, J. (1994) Localization of genes affecting alcohol drinking in mice. *Alcoholism, Clinical and Experimental Research* 18: 931–41

Phillips, T.J., Huson, M., Gwiazdon, C., Burkhart-Kasch, S. & Shen, E.H. (1995) Effects of acute and repeated ethanol exposures on the locomotor activity of BXD recombinant inbred mice. *Alcoholism, Clinical and Experimental Research* 19: 269–78

Phillips, T.J., Lessov, C.N., Harland, R.D. & Mitchell, S.R. (1996) Evaluation of potential genetic associations between ethanol tolerance and sensitization in BXD/Ty recombinant inbred mice. *Journal of Pharmacology and Experimental Therapeutics* 277: 613–23

Pickard, G.E., Sollars, P.J., Rinchik, E.M., Nolan, P.M. & Bucan, M. (1995) Mutagenesis and behavioral screening for altered circadian activity identifies the mouse mutant, *Wheels*. *Brain Research* 705: 255–66

Plomin, R., McClearn, G.E., Gora-Maslak, G. & Neiderhiser, J.M. (1991) Use of recombinant inbred strains to detect quantitative trait loci associated with behavior. *Behavior Genetics* 21: 99–116

Riley, E.P., Freed, E.X. & Lester, D. (1976) Selective breeding of rats for differences in reactivity to alcohol. An approach to an animal model of alcoholism. I. General procedures. *Quarterly Journal of Studies on Alcohol* 37: 1535–47

Risinger, F.O. & Cunningham, C.L. (1995) Genetic differences in ethanol-induced conditioned taste aversion after ethanol exposure. *Alcohol* 6: 535–39

Rodriguez, L.A., Plomin, R., Blizard, D.A., Jones, B.C. & McClearn, G.E. (1995) Alcohol acceptance, preference and sensitivity in mice. II. Quantitative trait loci mapping analysis using BXD recombinant inbred strains. *Alcoholism, Clinical and Experimental Research* 19: 367–73

Sanders, B. (1976) Sensitivity to low doses of ethanol and pentobarbital in mice selected for sensitivity to hypnotic doses of ethanol. *Journal of Comparative Physiological Psychology* 90: 394–98

Spuhler, K.P., Deitrich, R.A. & Baker, R.C. (1990) Selective breeding of rats differing in sensitivity to the hypnotic effects of acute ethanol administration. In *Initial Sensitivity to Alcohol*, edited by R.A. Deitrich & A.A. Pawlowski, Rockville, Maryland: National Institute on Alcohol Abuse and Alcoholism

Takahashi, J.S., Pinto, L.H. & Vitaterna, M.H. (1994) Forward and reverse genetic approaches to behavior in the mouse. *Science* 264: 1724–32

Tarantino, L.M., McClearn, G.E., Rodriguez, L.A. & Plomin, R. (1998) Confirmation of quantitative trait loci for alcohol preference in mice. *Alcoholism, Clinical and Experimental Research* 22: 1099–105

Wehner, J.M. & Bowers, B.J. (1995) Use of transgenics, null mutants and antisense approaches to study ethanol's actions. *Alcoholism, Clinical and Experimental Research* 19: 811–20

Wehner, J.M., Pounder, J.I., Parham, C. & Collins, A.C. (1992) A recombinant inbred strain analysis of sleep-time responses to several sedative-hypnotics. *Alcoholism, Clinical and Experimental Research* 16: 522–28

Wilson, J.R., Erwin, V.G., DeFries, J.C., Peterson, D.R. & Cole-Harding, S. (1984) Ethanol dependence in mice: direct and correlated responses to ten generations of selective breeding. *Behavior Genetics* 14: 235–56

GLOSSARY

advanced intercross lines (AILs) lines derived by random and sequential intercrossing of a population derived from two inbred strains for a number of generations (F_8–F_{10}). This provides increasing probability of recombination between any two loci and, consequently, the genetic length of the entire genome is increased, providing better mapping resolution for QTLs

ANOVA analysis of variance, a statistical test

ataxia the loss of full control of bodily movements

false negative (see type II error)

false positive (see type I error)

permutation in this context, a procedure in which quantitative trait values are randomly permuted, destroying any relationship between the trait values and the genotypes of the marker loci. A regression model is fitted for the permuted data at all positions in the genome and a maximum likelihood ratio statistic is recorded. This procedure is repeated hundreds or thousands of times, giving a distribution of likelihood ratio values expected if there were no QTL linked to any of the marker loci. Therefore, this test uses empirical data to establish thresholds for significance

recombinant inbred (RI) inbred strains generated by inbreeding from an F_2 population derived from two inbred progenitor strains. The genotype of each recombinant inbred strain constitutes a different combination of homozygous alleles for those loci for which the progenitor strains had different alleles

type I error (or false positive) rejection of a true null hypothesis. In the context of quantitative trait locus (QTL) analysis, a type I error or false positive occurs when a QTL is reported in a chromosomal region when, in fact, there is none. The chance for a false positive increases with the number of statistical tests being conducted

type II error (or false negative) acceptance of a false null hypothesis so that a genuine QTL is unrecognized. As statistical criteria are made more stringent to reduce type I errors, the probability of type II errors increases

See also **Ethylnitrosourea (ENU) as a genetic tool in mouse mutagenesis** (p.303); **Quantitative trait loci** (p.904)

BODY WEIGHT LIMITS IN MICE

Lutz Bünger[1], Ulla Renne[2], Reinoud C. Buis[3]
[1]Institute of Cell, Animal and Population Biology, University of Edinburgh, Edinburgh, UK
[2]Research Institute for the Biology of Farm Animals, Dummerstorf, Germany
[3]Animal Breeding and Genetics Group, Wageningen University, Wageningen, The Netherlands

1. Introduction

What is the weight of a mouse? How big or small can mice be made simply by breeding? How does this compare with the weights that can be achieved by transgenic procedures? Can selection infinitely increase or decrease body size? We will try to answer these questions and to explain why this is of importance for farm animals and for gene discovery.

The genetic improvement of livestock has always been a long and expensive process. It involves selection of probably large numbers of genes of mostly unknown effect (from which only very few are individually identified), long generation intervals, many generations of selection and large space requirement. This led H.D. Goodale, in 1930, to start long-term selection for increased body weight in a laboratory strain of mice, using them as a mammalian model organism with the aim of finding the maximum body size that could be achieved – would they, for example, become larger than rats?

His experimental animals were a random-bred albino strain, which he selected for maximum body weight at 60 days of age, continuing for more than 80 generations. Body weight increased from 24 to 42 g during the first 50 generations, with little further progress (Wilson *et al.*, 1971). This was a pioneering study of great interest, which demonstrated that a population of non-inbred animals, showing only small variations in a quantitative character such as body size, will respond continuously over many generations to selection and must contain numerous genes affecting that character.

Our main aim is to elucidate the total range of body weights in laboratory populations of mice, which can be achieved by long-term selection, using four experiments, starting with the classic Goodale experiment. We shall then discuss these results in relation to growth changes that have been achieved by manipulation of single genes using transgenic techniques, and by the use of known mutations affecting major metabolic pathways.

2. Statistical methods

The statistical methods are given here, since they will be needed throughout the article. See also the glossary of terms and abbreviations at the end of this article.

For fitting generation means against generation number or the cumulative selection differential a modified exponential model (1) (Bünger & Herrendörfer, 1994) was used:

$$y = A-(A-C)\, e^{\frac{-Bx}{(A-C)}} \qquad (1)$$

This modification was undertaken to have parameters with a biological meaning: A is the theoretical final value (i.e. theoretical selection limit), B is the maximal slope at $x = 0$ (i.e. selection response in the first generation), C is the y value at $x = 0$ (i.e. initial value) and x is an independent variable (i.e. generation or cumulative selection differential).

Based on this model the half life (HL, the time to get half way to the limit) can easily be obtained from:

$$HL = (\ln 0.5)\,(C-A)/B \qquad (2)$$

Model (1) was used for fitting generation means, selection response and the phenotypic variance and for estimating the "realized heritability function" from the slope of the fitted curve relating cumulative selection response (cumSR) and cumulative selection differential (cumSD), using the first derivative of model (1) (for details see Bünger & Herrendörfer, 1994). The slopes or ratio (b) yields the estimates for the realized heritability, which, depending on the selection method applied, has in some cases (e.g. sib selection in Experiment IV) to be converted to a mass selection basis (Falconer & Mackay, 1996, p.234).

In Section 7 (Comparison of growth between the experiments) a modified form of the logistic model described by Bünger *et al.* (1982) was used to describe growth curves. This modification was undertaken to use parameters with a biological meaning.

$$y = \frac{A}{1 + e^{\frac{4B(C-x)}{A}}} \qquad (3)$$

where A = asymptote, mature weight; B = maximum body weight gain; C = age at maximum gain (age at point of inflection); and x = age (days).

3. Experiment I: Goodale's selection for high body weight at 60 days

This was the first experiment ever reported in which laboratory mice were selected for high body weight. Started in 1930, it was continued for 84 generations, which took about 20 years even in mice. The initial population consisted of 5 males and 11 females of an albino outbred strain mated together, whose progeny Goodale selected for maximum body weight at 60 days after birth, using an undefined combination of individual, family and progeny-test selection. An average of 37 males and 100 females were recorded each generation from which about 43% were chosen as parents for the next generation ($N_e < 108$). No control line was maintained, and litter size was not restricted.

We have measured the response to selection in all the selection experiments described here using the modified exponential function explained in Section 2 above, which provides estimates of three parameters (A, B, C) with a biological meaning (**Figure 1**). As the selection approaches a plateau or limit asymptotically, the time and level of these statistics cannot be estimated exactly, but the formula fitted to the data provides an estimate of the half life (HL) from the three parameters.

The estimated initial male body weight at 60 days (BW60) was 22.2 g (parameter C), about 3 g less than the observed mean of the first three generations (**Figure 1**). During selection BW60 increased by 21.1 g (95%) to 43.4 g, equivalent to phenotypic and genetic standard deviations of about 8 s_p and 13 s_g (initial values: $s_p = 2.76$; $s_g = 1.66$). The estimated maximal selection response at the beginning of the experiment (parameter B) was 1.2 g per generation, which seems slightly overestimated because of the bad fit at the start. Thus the total selection response at the end of this experiment is at least 18 times the maximal response and the estimated selection limit is 43.6 g (parameter A), a value which is reached or even surpassed in generations 51–53, 64–72 and 82–83.

The independent variable "generation" is a suitable time scale for a descriptive presentation of results but in the genetical sense not appropriate if the selection intensity is not constant, as it does not take account of the amount of selection applied. The genetical situation is therefore better reflected by using the selection differential (SD) as the independent variable (Falconer, 1955). SD is basically the mean of the selected animals minus the mean of all animals; in other words a measure of the superiority of the selected animals or a measure of the amount of selection applied. If these differences are summed over generations we obtain the cumulative selection differential, cumSD. A selection plateau derived from a plot of selection response *vs.* generation could be misleading especially if there was actually no artificial selection at the end of the experiment (e.g. when the best animals did not reproduce). Therefore the selection response is plotted against the cumulative selection differential in **Figure 2.**

Probably because of variation in selection intensity and its rather small values in the first 10 generations (Wilson *et al.*, 1971), the fit of this relationship (**Figure 2**) was better than that in **Figure 1**. The total response estimated from **Figure 2** was 18 g (parameter A), smaller but more precise than that from **Figure 1**. Assuming an initial weight of 25 g the total selection response was about 72%, equivalent to 6.5 s_p or 11 s_g Using the parameters of **Figure 1**, HL is 12 generations, while HL derived from **Figure 2** is a little later (at 35 g of cumSD corresponding to 15–16 generations).

The coefficient of heritability (h^2) indicates how much of the observed variance in a population is of genetic origin; as variances change during the selection process the heritability will change. A realized heritability is derived from a particular experiment and estimates how much of the superiority of the selected animals is transmitted to their offspring. The relation between selection response and cumSD is the basis for estimating the "realized h^2 function", reflecting how the realized heritability will change during the experiment. This ratio is then transformed according to the method of selection applied, but not in this experiment as

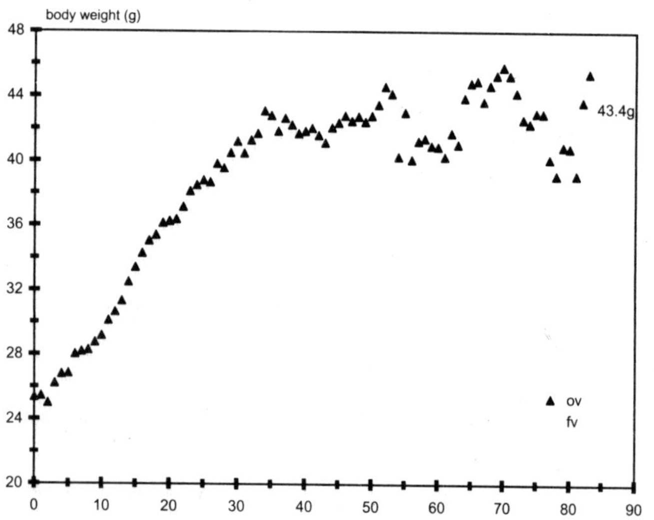

Figure 1 Direct selection response in body weight at 60 days (Expt I). Fitted exponential model [see section 2, model (1)]. Model parameters are: A: final value = theoretical selection limit, B: initial slope = maximum selection response, C: initial value = base population estimate, CD: coefficient of determination of quality (measuring how good the model describes the data; maximum is 1), HL: half life. A = 43.6 g, B = 1.2 g/generation, C = 22.3 g, CD = 0.923, HL = 12 generations. ov, fv: observed and fitted values. Figure modified from Bünger & Herrendörfer (1994).

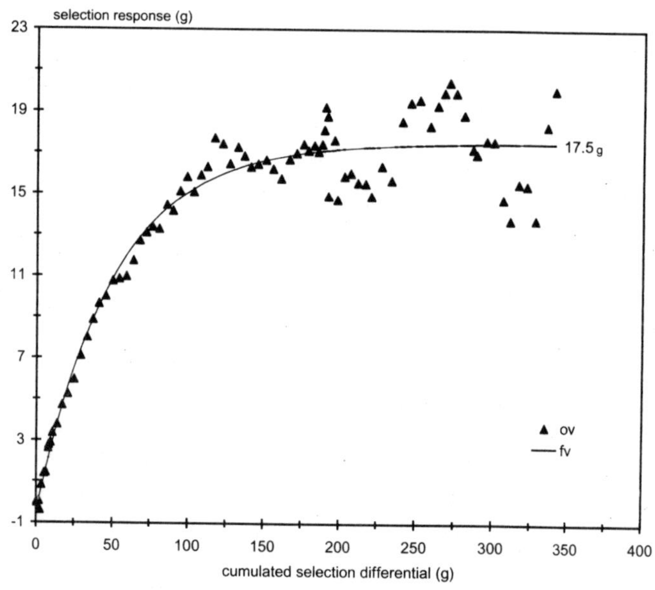

Figure 2 Direct selection response *vs.* cumulative selection differential (Expt I). A = 17.5 g, B = 0.36 g/generation, C = −0.56, CD = 0.943, HL = 35 g (15–16 generations). For further details see legend for Figure 1. Figure modified from Bünger & Herrendörfer (1994).

mass selection was applied. The h^2 function here is just the first derivative of model (1) used for the description of the selection response in proportion to the cumulated selection differential. The realized h^2 was initially 0.36, indicating that about 36% of the observed variance was genetic, and this proportion decreased to 0.0004 at the end of the experiment. The phenotypic variance increased from 7.6 to 39 g^2, showing that increased body weight is accompanied by higher variation. The coefficient of variation, measuring the variance in relation to the mean, averaged 11.5% and showed only a small but significant increase from 9.4 to 14% during the experiment (**Figure 3**). The genetic variance, calculated from the realized heritability and the fitted phenotypic variance, declined from 2.8 to 0.016 g^2.

4. Experiment II: Divergent selection on body weight at 56 days

Whereas selection in Expt I was only in the high direction, a selection for high and low body weight (divergent selection) was carried out in Wageningen, The Netherlands (Bakker, 1974). Starting from a Swiss random-bred laboratory mouse population (Cpb:SE) Bakker selected for high (BW56H) and low (BW56L) body weight for 27 generations (BW56L) and 44 generations (BW56H), respectively. An unselected control was designated BW56C. 16 males and 32 females were selected as parents per generation, resulting in a N_e of about 43. Mass selection was applied, taking not more than three mice per sex per litter. On average the best 37% and 49% in lines BW56H and BW56L, respectively, were selected. Litter size was not standardized.

Bakker (1974) reported the results up to generation 14. Means and phenotypic variances for later generations (Buis, unpublished) were available to present the analysis from generation 0 to 44 for the high line (BW56H) and from 0 to 27 for the low line (BW56L). The low line was terminated at that time because of fertility problems.

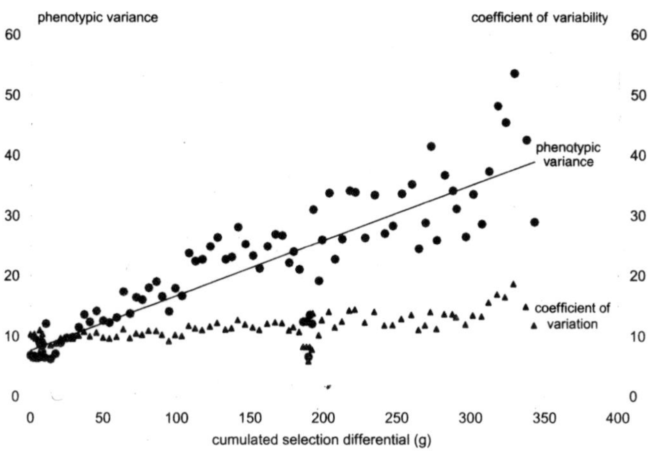

Figure 3 Change of phenotypic variance and coefficient of variation (Expt I). The phenotypic variance was fitted using model (1): A = 37625, B = 0.091, C = 7.62, CD = 0.74 and the coefficient of variation was fitted using a linear regression: $y = 9.44 + 0.0133x$. For further details see legend for Figure 1. Figure modified from Bünger & Herrendörfer (1994).

The change of body weight at 56 days (BW56) in both lines is presented in **Figure 4**. Starting from about 26 g in generation 0 the high line mice reached almost 52 g and the low line mice weighed only 12 g, a 4.3-fold difference between the two lines. The estimated initial body weight of line BW56H was 25.6 g (parameter C), which corresponds very well with the total mean for the control. At the end of the selection (generation 44) the BW56 was increased by 26 g (102%) to 51.6 g, about 10 times the initial phenotypic (s_p = 2.53) and 14 times the genetic standard deviation (s_g = 1.85 g). The highest response (parameter B) was 1.32 g/generation and thus the total selection response was about 19 times this value. At about 56.2 g a selection limit can be expected (parameter A) of which about 92% was actually reached in generation 44 and the half life was obtained after 16 generations.

The estimated initial BW56 in the low line (BW56L) was, at 27.0 g (parameter C), also close to the control line mean. At the end of selection (generation 27) the estimated BW56 was 12.0 g, a decrease of 15 g (56%), nearly 6 s_p and 9 s_g (estimated initial s_p and s_g were 2.62 and 1.60 g, respectively) or 9.6 times the maximal response (1.57 g, parameter C). The estimated theoretical selection limit (parameter A) is 10.8 g, of which 90% was reached at generation 27 and half of this response was obtained by generation 7. The plot of cumulative selection response (cumSR) *vs.* cumulative selection differential (cumSD) showed a very similar pattern (**Figure 5**).

The estimated total selection response for line BW56H is, at 23.2 g (= value in last generation, C), slightly lower than that from Figure 5 (26 g). The total selection response for

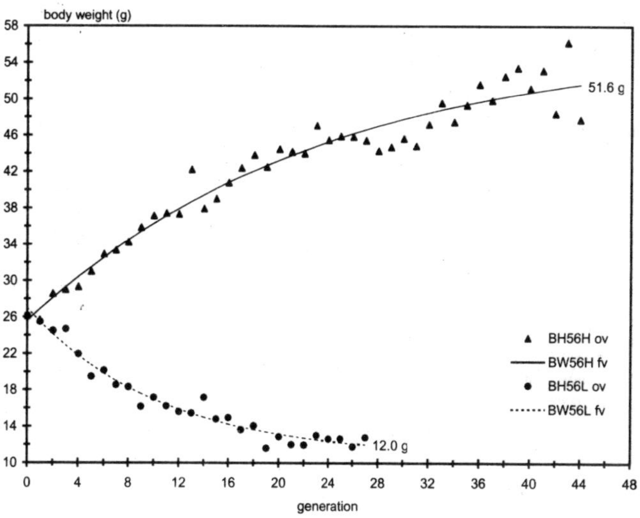

Figure 4 Direct selection response (Expt II). Generation means are weighted averages over sexes. The selection response in body weight at 56 days is expressed as difference to the control plus the total mean (over all generations) of the control ($\bar{X}$ = 26.2 g). Model (1) parameters for BH56H: A = 56.2 g, B = 1.32 g/generation, C = 25.6 g, CD = 0.95, HL = 16 generations. Model (1) parameters for BW56L: A = 10.8 g, B = −1.57 g/generation, C = 27.0 g, CD = 0.96, HL = 7 generations. For further details see legend for Figure 1.

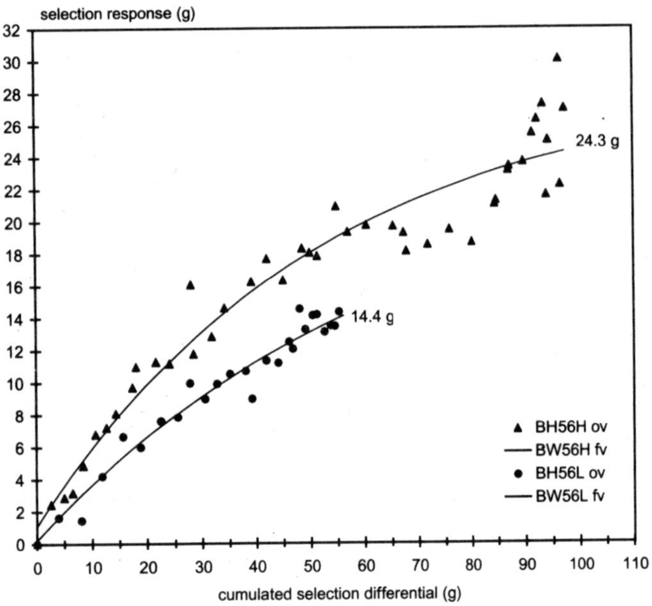

Figure 5 Direct selection response *vs.* cumulative selection differential (Expt II). Model (1) parameters for BH56H: A = 28.4 g, B = 0.53 g/generation, C = 1.08 g, CD = 0.94, HL = 36 g (= 16 generations). Model (1) parameters for BW56L: A = 23.8 g, B = 0.374 g/generation, C = 0.205 g, CD = 0.96, HL = 44 g (= 16 generations). For further details see legend for Figure 1.

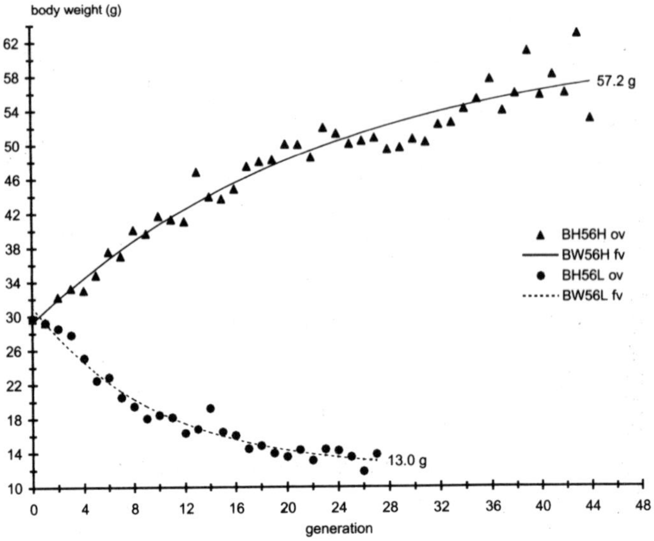

Figure 6 Change of male body weight at 56 days (Expt II). Body weights are presented as difference to the control, which was rescaled by adding the total mean (over all generations) of the control line ($\bar{X}$ = 29.7 g). Model (1) parameters A = 62.34 g, B = 1.40 g/generation, C = 29.27 g^2, CD = 0.95, HL = 16.4 generations. For further details see legend for Figure 1.

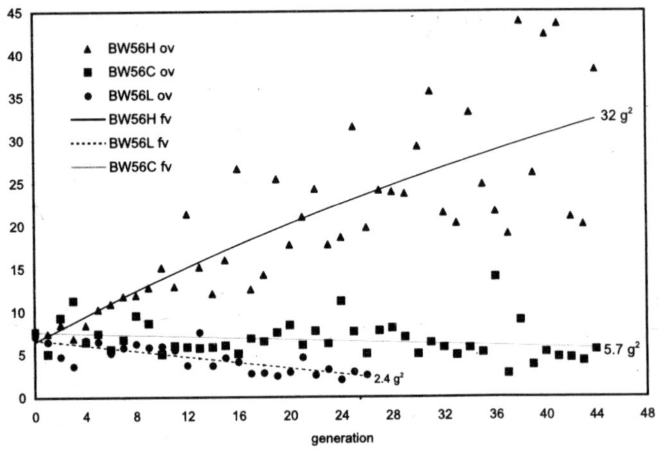

Figure 7 Change of phenotypic variance (Expt II). The phenotypic variances were calculated over sexes and corrected for the sex difference in body weight, using (2): $V_{p(m)} = V_{p(m,f)} - \frac{1}{4}(av_m - av_f)^2$ with $V_{p(m)}$ and $V_{p(m,f)}$ as the phenotypic variance in males and calculated over both sexes, respectively; av_m and av_f are the averages for males and females, respectively. The phenotypic variance of line BW56H was fitted using model (1): A = 62.86 g^2, B = 0.79 g^2/generation, C = 6.42 g^2 and of the other two lines using linear regression. BW56C: b_0 = 7.6, b_1 = −0.04 (−0.09, 0.003). BW56L: b_0 = 6.85, b_1 = −0.166 (−0.22, −0.11). For further details see legend for Figure 1.

line BW56L is 14.2 g, corresponding well with the estimate derived from **Figure 5** (15 g). This relation between selection response and cumSD is again used to describe the change of the realized heritability for BW56H and BW56L. They were initially 0.53 and 0.37, respectively, decreasing to 0.08 in both lines at the end of the experiment.

Since a comparison between different experiments is easier when body weights of one sex only are considered, male weights were chosen. The change of male body weights at 56 days (BW56) for both lines is presented in **Figure 6**. At the end of selection males from the BW56 high line weigh 57.2 g whereas low line males weigh 13.0 g. Thus they differ by about 44 g, a factor of 4.4.

Corresponding to Expt I, the phenotypic variance for the selection trait (BW56) in the high line BW56H increased significantly by 0.8 g^2 per generation from 6.4 to 32 g^2. In the low line it decreased significantly from a similar initial value (6.9 g^2) to 2.4 g^2 (**Figure 7**). There was no significant change in the phenotypic variance in the control line, averaging 6.6 g^2.

The coefficient of variation (**Figure 8**) increased significantly over the experiment in BW56L from 10.7 to 14.7%. There was also a significant but smaller increase in line BW56H (from 9.6 to 11.2) whereas CV% did not change in the control with an average of 9.7%.

5. Experiment III: Divergent selection on lean mass and body weight

Another divergent selection experiment was initiated in Edinburgh (UK) from an outbred base population, where the selection trait was not body weight but lean mass (Sharp *et al.*, 1984). Lines (P, or protein lines) were divergently selected for 20 generations for high (PH) and low (PL) lean mass, estimated from an index of body weight and gonadal fat pad weight in males. In subsequent generations body

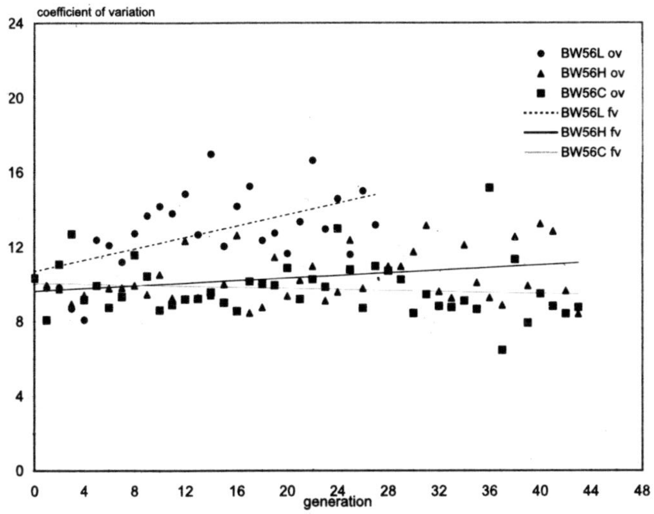

Figure 8 Change of coefficient of variation (Expt II). Model parameters of linear regression and the 95% confidence interval for b_0 were: BW56H: $b_0 = 9.61$, $b_1 = 0.036$ (0.005, 0.06). BW56C: $b_0 = 9.43$, $b_1 = -0.01$ (−0.05, 0.02). BW56L: $b_0 = 10.69$, $b_1 = 0.15$ (0.07, 0.24). ov, fv: observed and fitted values.

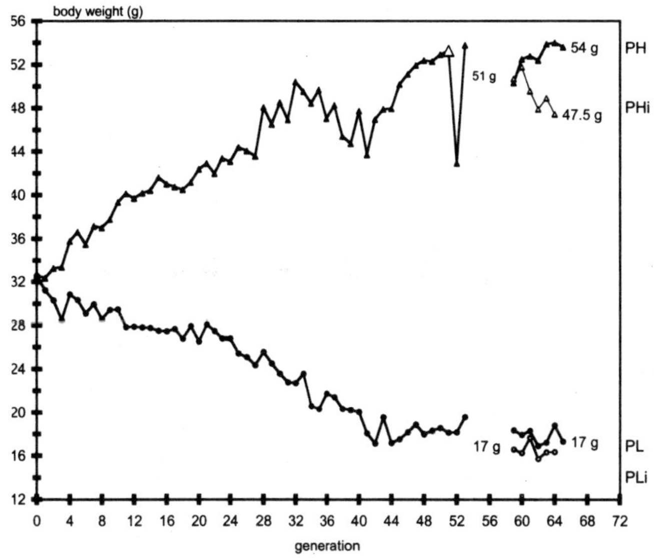

Figure 9 Change of male body weight at 10 weeks from the high (PH) and low (PL) body weight lines (Expt III). The means of the first 20 generations are pooled over the three replicates. Selection was suspended from generation 53 to generation 58 while all mouse stocks were transferred to a new mouse house by embryo transfer. Generation means before the transfer to the new animal unit are simple means and subsequently least square means fitting lines, generations, families nested within line, and interactions between line and generation. ANOVA was undertaken with GLM using the SAS System. Enlarged open symbols mark the beginning of derived inbred lines (PHi, PLi). Generation means up to generation 53 were used for fitting model (1). Model (1) parameters for PH: A = 56.28 g, B = 0.68 g/generation, C = 32.42 g, CD = 0.88, HL = 24 generations. Model (1) parameter for PL: A = −830 g, B: −0.28 g/generation, C = 31.8 g, CD = 0.94, HL = 2100. For further details see legend for Figure 1 and text. (Figure is adapted from Bünger & Hill, 1999a, with permission from Springer-Verlag.)

weight at 70 days of age was chosen as the selection criterion. For the first 20 generations three replicates were maintained per line, after which they were crossed to form one replicate. Selection was suspended from generation 53 to 58 (PH, PL) while all mouse stocks were transferred to a new mouse house by embryo transfer. The derivation of the inbred lines (PHi, PLi) from these selection lines was described by Bünger & Hill (1999a) in detail. During this experiment litter size was reduced when above 12.

The male body weight at 70 days, averaging over the three replicates, in the early stage increased in line PH from 32.4 g to 51 g (**Figure 9**). This is an increase of 18.6 g (57%) or 5.6 s_p and 7.7 s_g with initial values $s_p = 3.32$ and $s_g = 2.35$, using a coefficient of heritability of 0.5 (Beniwal *et al.*, 1992). The maximum response was 0.68 g/generation.

After the transfer of mouse lines to the new animal unit a similar level was reached with an increasing tendency, probably resulting from renewal of selection after the period of relaxed selection during the transfer. The estimate for the theoretical selection limit in line PH is 56 g and the HL was 24 generations.

The estimated initial weight (31.8 g) in PL was very similar to that of line PH (**Figure 9**) and the weight decreased until generation 53 to 17 g, by 14.8 g (47%) or 5 s_p and 7.1 s_g with initial values of $s_p = 2.94$ g^2 and $s_g = 2.1$ g^2, derived from **Figure 10**. Because of the linearity of selection response in the PL line, the estimate of the HL is above 2000 and thus meaningless. In generation 53 the PH line is 34 g or heavier by a factor of 3 than the PL, which corresponds to about 11 s_p or 15 s_g. The inbred lines (PHi, PLi) derived by full sib matings, in generation 51 and 46, respectively, have a similar divergence as the outbred lines, but their weights are lower (**Figure 9**). Such high divergence between inbred lines makes them a very good resource for QTL studies.

Regarding the change of the phenotypic variance, the results are similar to those of Expt II: an increase in PH (from 11 to 27 g^2) and a decrease in PL (from 8.7 to 2.6 g^2) (**Figure 10**). The coefficient of variation in both lines did not show any significant trend and averaged at about 10% in both lines (**Figure 11**).

6. Experiment IV: Selection for body weight at 42 days

The longest known selection experiment in mammals started in 1975 and was continued for 92 generations (almost 25 years) in Dummerstorf, Germany (Bünger *et al.*, 1983; Bünger, 1987) and is still in progress. This experiment used as a foundation population a very heterogeneous outbred strain (Fzt: DU), obtained in 1969–70 by systematic cross-breeding of four inbred and four outbred lines (Schüler, 1985). Sib selection (Falconer & Mackay, 1996) was applied, which means litters were selected on the sum of the performance of two "test males" per litter. 80 pair matings were made per generation, providing a $N_e = 60$, and

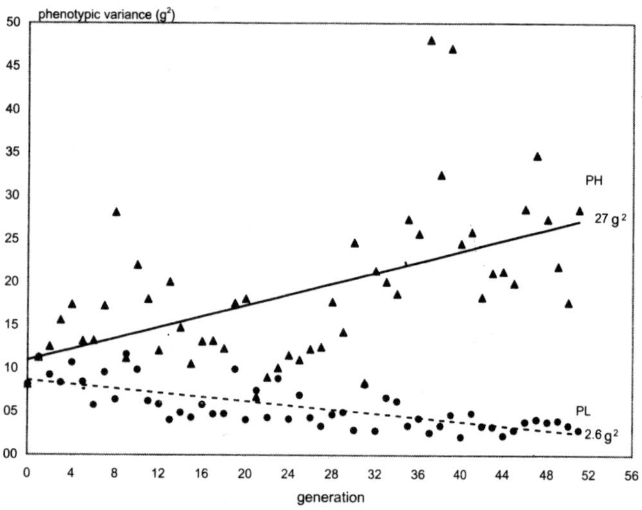

Figure 10 Change of phenotypic variance (Expt III). Phenotypic variances were fitted by linear regression using data up to generation 51. Model parameters of linear regression and the 95% confidence interval for b_0 were: PH: $b_0 = 11.0$ g², $b_1 = 0.32$ g²/generation (0.18, 0.45). PL: $b_0 = 8.66$ g², $b_1 = -0.12$ g²/generation (−0.15, −0.09). For further details see legend for Figure 1.

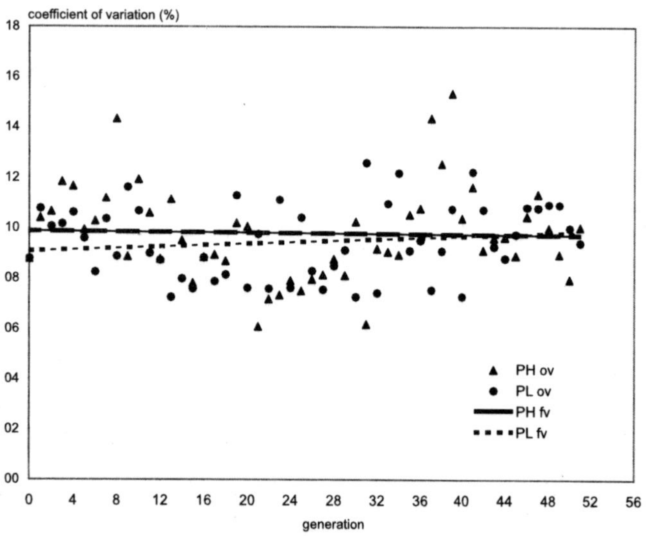

Figure 11 Change of coefficient of variation (Expt III). Phenotypic variances were fitted by linear regression using data up to generation 51. Model parameters of linear regression and the 95% confidence interval for b_0 were: PH: $b_0 = 9.87\%$, $b_1 = -0.003\%$/generation (−0.04, 0.03). PL: $b_0 = 9.08\%$, $b_1 = 0.014$ (−0.01, 0.04).

selection was on high body weight at 42 days of age in line BW. An unselected control line was also kept (CO). Litter size was standardized to eight in generations 0–15 and nine thereafter.

The male body weight at 42 days in line BW increased from 29.5 g in generation 0 (parameter C) to 63.5 g in generation 92 (**Figure 12**). This is an increase of 34 g (115%) or 16 s_p or 22 s_g with initial values $s_p = 2.15$ and $s_g = 1.57$. The

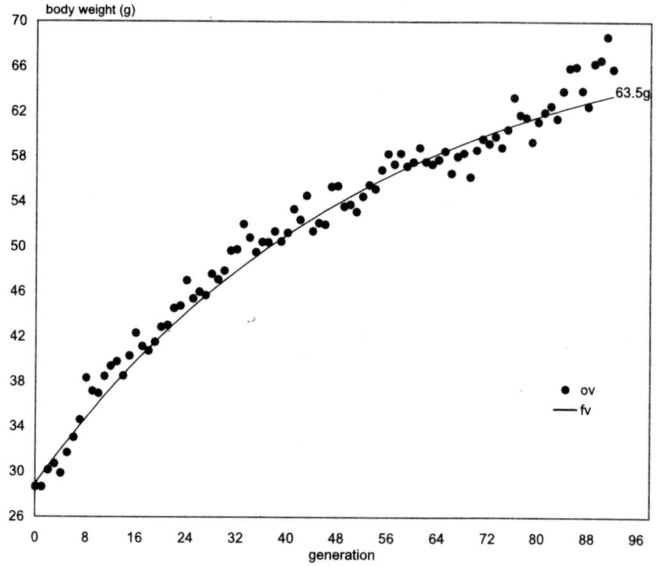

Figure 12 Direct selection response in male body weight at 42 days (Expt IV). Model (1) parameters: A = 71.6g, B = 0.788 g/generation, C = 29.5 g, CD = 0.977, HL = 31 generations. Presented is the difference to the control for male body weight at 42 days. The differences to the control are rescaled by adding the total mean of the control ($\bar{\bar{X}} = 28.9$ g) as a constant to every difference. Full sib sums were converted to an individual basis by dividing by 2. For further details see legend for Figure 1.

estimated maximum response was 0.79 g/generation and the estimate for the theoretical selection limit is nearly 72 g, corresponding to a selection response of 42 g, half of which was reached in generation 31.

The graph of cumulative selection response (cumSR) *vs.* cumulative selection differential (cumSD) showed a very similar pattern (**Figure 13**) and the estimated total selection response is 35 g, equal to that from **Figure 12.** Both figures show further selection response in the last few generations. The realized heritability for BW42 was initially 0.53, decreasing to 0.06 at the end of the experiment (**Figure 13**). The phenotypic variance in the control line showed no significant trend and averaged 23 g². There was a strong increase in line BW from initially 14 g² to 72 g² in generation 92 (**Figure 14**). The coefficient of variation in both lines showed no significant trend, averaging 6.9 and 8.3% in the BW and CO.

7. Comparison of growth between the experiments

Although selection traits differ between the experiments the selection responses can be compared when expressed in units of phenotypic or genetic standard deviations or as percentages but what about a comparison of the growth changes at other ages? Because selection in these four experiments was on body weights at different ages (42, 56, 60, 70 days) it would be advantageous to have weights in every experiment at all four selection ages or even better a growth curve, which would enable a comparison between the experiments and, moreover, with other studies, where weight records at single ages or growth curves are available. A comparison at these

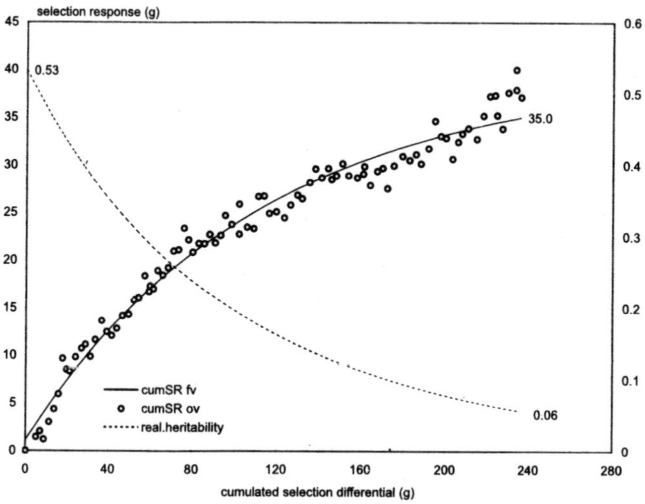

Figure 13 Selection response *vs.* cumulative selection differential and the change of the realized heritability (Expt IV). Model (1) parameters: A = 39.8 g, B = 0.342 g/generation, C = 1.11 g, CD = 0.974, HL = 78 g (≈ 34 generations). Selection response is the difference to the control, minus difference in generation 0. All values are transformed to an individual basis. The realized heritability estimates were converted to a mass selection basis, assuming a constant c^2; $h^2 = \frac{2b[1-(n-1)c^2]}{[n-(n-1)b]}$, which for n = 2 gives $h^2 = \frac{2b(1+c^2)}{(2-b)}$, with n as the number of full sibs. As estimate for $c^2 = 0.29$ was chosen (Bünger *et al.*, 1998a).

The cumSR in Expt IV is calculated as generation mean minus the mean in generation 0 and the cumulative selection differential cumSD is calculated as a realized SD, so SD is weighted according to number of offspring per litter used as parents and which have given birth to a litter in the next generation. Both are transformed to a single animal base.

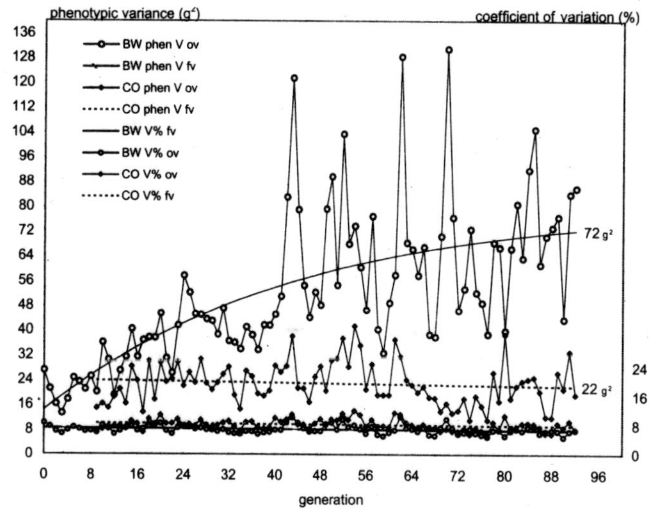

Figure 14 Change of phenotypic variance (phen V) and of coefficient of variation (CV%) in selection (BW) and control (CO) lines (Expt IV). Phenotypic variances in lines BW and CO were fitted by model (1) and linear regression, respectively. Model (1) parameters (for V_p in BW): A = 80.2 g^2, B = 1.52 g^2/generation, C = 14.34 g^2, CD = 0.45. Model parameters of linear regression and the 95% confidence interval for b_0 were: CO V_P: b_0 = 23.83 g^2, b_1 = −0.023 g^2/generation (−0.08, 0.037). BW CV%: b_0 = 7.24%, b_1 = −0.007%/generation (−0.015, 0.002). CO CV%: b_0 = 8.62%, b_1 = −0.009%/generation (−0.02, 0.004). The phenotypic variance of full sib sum ($\sigma^2_{P_{FS}}$) (cf. Table 4) was transformed to an individual basis ($\sigma^2_{P_i}$) as follows: $\sigma^2_{P_i} = \sigma^2_{P_{FS}}/(2\tau + 2)$ assuming $\tau = 0.55$. For further details see legend for Figure 1.

Table 1 Measured and predicted weights (g) for ages used as selection age in the four selection experiments and at maturity (parameter A).

Direction			Ages (days)				Mature weight (g)	Reference
			42	56	60	70	A	
High	Expt I		33.7	43.2	44.9	47.7	50.4	Fig. 15
	Expt II	BW56H (gen14)	33.4	45.0	47.1	50.3	53.1	Fig. 16
		BW56H (gen 44)	43.3	58.4	61.1	65.3	68.8	Fig. 16
	Expt III	PH	41.3	51.2	52.5	54.2	55.3	Fig. 17
	Expt IV	BW	64.3	79.2	81.3	84.1	85.9	Fig. 18
	"Expt I"	SM	16.0	17.5	17.7	17.9	18.0	Fig. 15
Low	Expt II	BW56L (gen 14)	16.6	17.9	18.0	18.2	18.4	Fig. 16
		BW56L (gen 27)	13.1	14.1	14.3	14.4	14.6	Fig. 16
	Expt III	PL	15.4	17.4	17.6	18.0	18.3	Fig. 17
	Expt III	LPL	7.29	8.24	8.46	8.92	10.1	Fig. 17

Responses in direct selection traits are underlined. Gen, generation; Expt, Experiment.

four different weights is given in **Table 1**, but first the growth curves in each of the four experiments will be considered.

(i) Experiment I

Selection in this line was on body weight at 60 days and its development over generations is presented in **Figure 1**. There is no growth curve for this line readily available; however Chai (1966) described the growth of an inbred line derived from these "Goodale mice" from birth to 14 weeks of age. According to the author this growth curve of the inbred line is typical for the selection line and there was no evident downward trend during inbreeding. This author

Table 2 Selection response and response in male body weight at 42 days of Experiments I–IV.

Experiment	Selection trait	N_e	Trait	generation	initial	end	abs.	%	s_p	s_g	h^2	HL
I	BW60+	<108	BW60 m	83	25.0	43.4	18.4	74	7	11	0.36	*c*.15
			BW42 m	after *c*.50	18.4	34.0	15.6					
II	BW56+	43	BW56	43	25.6	51.6	26.0	102	10	14	0.53	16
	BW56–	43	BW56	27	27.0	12.0	15.0	56	6	9	0.37	16
			BW56 1m	43	29.3	57.2	27.9	95				
			BW56 1m	27	31.1	13.0	18.1	58				
			BW42 1m	43	25.3	43.3	18.0	71				
			BW42 1m	27	25.3	13.1	12.2	48				
III	P70/BW70+	*c*.100/50	BW70 m	52	32.5	51.0	18.5	57	6	8	0.50	24
	P70/BW70–	*c*.100/50	BW70 m	52	31.8	17.0	14.8	47	5	7	0.50	
			BW42 m	65	26.5	41.3	14.8	56				
			BW42 m	65	26.5	15.4	11.2	42				
IV	BW42+	60	BW42 m	92	29.5	63.5	34.0	115	16	22	0.53	31

Table 3 Selection response or limits in some experiments on mice (adapted from Bünger *et al.*, 1998a).

					Selection response								
Base population	Selection trait	N_e	Trait	gen	"initial"	end	Abs.	%	s_P	s_A	$h^{2‡}$	HL	References
OB (6 × IB†)	BW60+	30	BW60 m	23	23.2	40.0	16.8	72	6.6	13	0.25		MacArthur, 1944a, 1949
			BW42 m	23	19.3	33.2	13.9						
	BW60–/BW42–	20	BW42	38	19.0	10.0	8.0	53	4.2			4	King, 1950; Roberts, 1966a
OB (4 × IB)	BW42+	15	BW42	52	21.6	28.0	6.4	30	3.4	7.1	0.35	8	Falconer, 1953, 1955
	BW42–	15	BW42	42	21.6	11.0	10.6	49	5.6	11.8	0.35	9	Roberts, 1966a
OB	G 21–42+/BW42+	17	BW42	53	24.5	35.0	10.5	43	4.6	8.1	0.33	7	Falconer, 1953, 1955, 1960
	G 21–42–/BW42–	19	BW42	53	24.5	14.0	10.5	43	4.6	8.1	0.33	10	Roberts, 1966b
OB (4 × S)	BW42+		BW42	18	32.0	41.0	9.0	28					Roberts, 1967
OB (3 × S)	BW42–		BW42	18	15.0	12.0	3.0	20					
OB	BW42+	32	BW42 m	23	24.5	37.0	12.5	51	5.1	8.3	0.40		Falconer, 1973
Q – strain	BW42–	32	BW42 m	23	24.5	15.0	9.5	39	6.1	6.3			
OB	G21–42+	41	G21–42	24/27	13.6	24.7	11.1	82	4.7	8.2	0.35	12	Eisen, 1975
ICR			BW42	–	27.0	44.0	17.0	63					Eisen *et al.*, 1988
OB (4 × IB)	G21–42+	33	G21–42 m	34/43	13.8	26.6	12.8	93	7.4	16.5	0.20	15	Barria & Bradford, 1981
			BW42 m	33	25.0	41.0	16.0	64					Eklund & Bradford, 1977
			BW70 f	31	22.8	41.9	19.1	84					Meyer & Bradford, 1974
OB (2 × IB)	BW42+	30	BW42	20	20.0	26.0	6	30	3	7	0.20		Heath et al., 1995
	BW42–		BW42		20.0	16.0	4	20	2	2.2			

†OB, IB: outbred, inbred; S: selection line
‡some h^2 values are given by the authors for within family-deviations (h), however $h^2 = h^2_w (1 - \tau)/(1 - r)$ (Falconer & Mackay, 1996) and r (coefficient of relationship) = 0.5 for full sibs and τ (intraclass-correlation of full sibs) ≈ 0.5 h^2 and h^2_w are almost identical
BW*xx* m or BW*xx* f: individual male or female body weight at *xx* days; G21–42: gain between 21 and 42 days of age
Approximations for BW42 from BW60 were obtained by multiplying by 0.83
If s_p was not given, CV% = 10% was assumed and s_p was calculated from the "initial" mean.

gave also growth data for a line which was selected on low body weight at 60 days (MacArthur, 1944a), a selection experiment which is not considered here in detail but used for comparison (**Table 3**, **Figure 15**).

Mice of the high and low lines have very different maximum gains: 0.98 g/day and 0.45 g/day, reached at about 33 and 21 days, respectively. Their mature weights were about 50 and 18 g. To compare weights in this line (Expt I) with those of the other three experiments at the four different selection ages used in these experiments, these weights are given in **Table 1**. Comparing the body weight at 60 days for the high line with those from **Figure 1** shows that the growth curves given by Chai (1966) and used here (**Figure 15**) seem to be representative for the plateaued line and therefore suitable for our purpose.

(ii) Experiment II

Divergent selection in this experiment was on BW56. No growth curve was available at the end of this experiment. However, in an earlier stage of the experiment (generation 14), weekly body weights from 1–8 weeks of age were available and a logistic model was used to fit these data (Bakker,

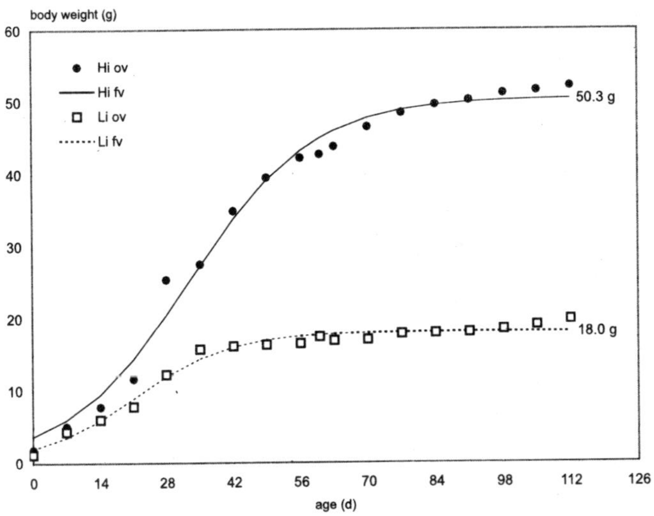

Figure 15 Growth of male mice (Expt I). Means were taken from graphs given by Chai (1966) on inbred lines derived from the line described in Expt I (here Hi) and from a selection line, which was selected on small body weights at 60 days (here low inbred = Li) (MacArthur, 1944a). Data were digitized from the original graph and may therefore deviate slightly from original values. A modified form of the logistic model, which is described in more detail in section 2, was used to fit the mean body weights. The three parameters, A (asymptote, mature weight), B (maximum body weight gain in g/day) and C = age at maximum gain (age at point of inflection) are always given. The parameters A, B and C and the coefficient of determination (CD) for the high (Hi) and low (Li) line were:

	Hi	Li
A (g)	50.43	18.04
B (g/day)	0.977	0.449
C (days)	32.96	21.29
CD	0.99	0.98

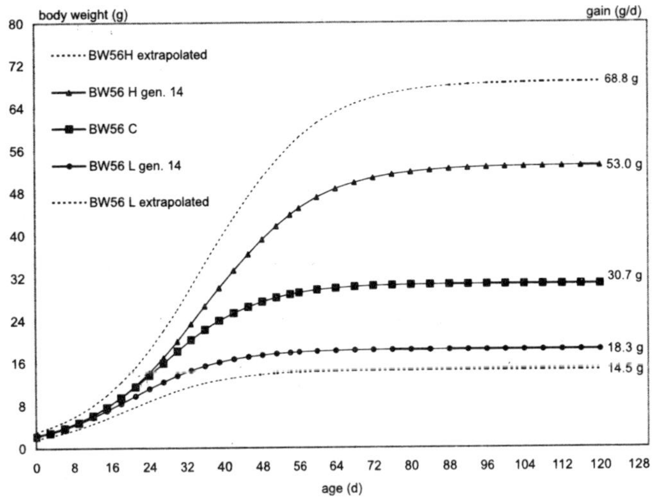

Figure 16 Growth of male mice (Expt II). Parameters for a logistic model were taken from Bakker (1974, p. 49), who fitted weekly measured male weights taken from 7 to 56 days of age in generation 14. Transforming these parameters into those of the modified version of the logistic model used herein results in parameter set **(a)**. From generation 14 to the end of both lines (high line generation 44 and low line generation 27) there was a further increase/decrease of BW56 in the high and low lines by a factor of 1.3 and 0.79, respectively (Figure 6). Assuming proportional changes of all body weights at the same ratio and no effects on age at maximum gain (parameter C), approximate growth curves for the final generations can be derived by multiplying A and B with these factors, which is the same as multiplying every weight by these factors. These extrapolated growth curves are given as dotted lines and the parameter are given in **(b)**:

a)

	gen 14	gen 14
	BW56H	BW56L
A (g)	53.1	18.4
B (g/day)	1.14	0.466
C (days)	35.9	19.8

b)

	gen 44	gen 27
	BW56H	BW56L
A (g)	68.8	14.5
B (g/day)	1.47	0.369
C (days)	35.9	19.8

1974). Using the parameters for the logistic function given by Bakker, the growth of males at this stage of the selection and strictly only up to 56 days of age can be described (**Figure 16**). However as there was still considerable selection response after generation 14 a further change of the growth curves has to be expected. Using the change of male body weights at 56 days from generations 14 to 44 in the high line and to generation 27 in the low line (**Figure 6**) a rough approximation for the growth curves at the end of the selection can be derived (see legend for **Figure 16**). Males from the high and low lines would possibly have reached an asymptotic weight of 69 g and 14.5 g, respectively. Weights at 42, 56, 60 and 70 days in the high and low lines would be 43.3, 58.4, 61.1, 65.3 g and 13.1, 14.1, 14.3, 14.4 g, respectively (**Table 1**).

(iii) Experiment III

Body weights at several ages were taken on males of generation 65 in this selection experiment. The growth curves of the lines PH and PL (**Figure 17**) are very different. Their maximum gains were 1.4 and 0.41 g/day, respectively, which were reached at about 31 and 23 days. The mature weights were about 55 and 18 g. The weights at the selection ages of the other lines are given in **Table 1**. Included in **Figure 17** are the growth curves for lines *l*PH and *l*PL, which present the growth curves of lines PH and PL, when the growth hormone system is "switched off" by the introgression of the *lit* mutation (Bünger & Hill, 1999b).

(iv) Experiment IV

Body weights of males from lines BW and CO were taken at different ages in generation 86 (**Figure 18**). They reach their maximum gains, which were 2.1 and 0.9 g/day, respectively, at 27 and 31 days. The mature weights were 35 and 86 g (**Figure 18**).

8. Discussion

(i) Total selection response

In terms of phenotypic standard deviations, upward and downward selection for body weight shifted the means by 3 to 16s_p and by 2 to 6s_p, respectively. Selection in all lines has

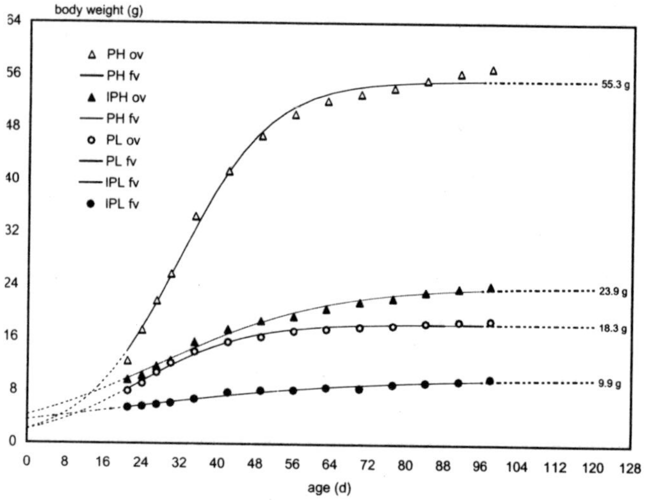

Figure 17 Growth of male mice (Expt III). (Figure modified from Bünger & Hill, 1999b.) Twenty male mice from generation 65 from lines PH, PL /PH and /PL were weighed on days 21, 24, 27, 30, 35, 42 and then weekly till 98 days of age. Weights were fitted using the logistic model (section 2). Lines /PH and /PL were produced by repeated backcrosses of the *lit* gene into the PH and PL lines, which knocks out the growth hormone production (for further explanation see section 8, Discussion). The parameters for the logistic model and the coefficient of determination (CD) were:

	PH	PL	/PH	/PL
A (g)	55.3	18.3	24.1	10.1
B (g/day)	1.42	0.41	0.34	0.097
C (days)	31.47	23.41	27.72	17.21
CD	0.995	0.992	0.991	0.977

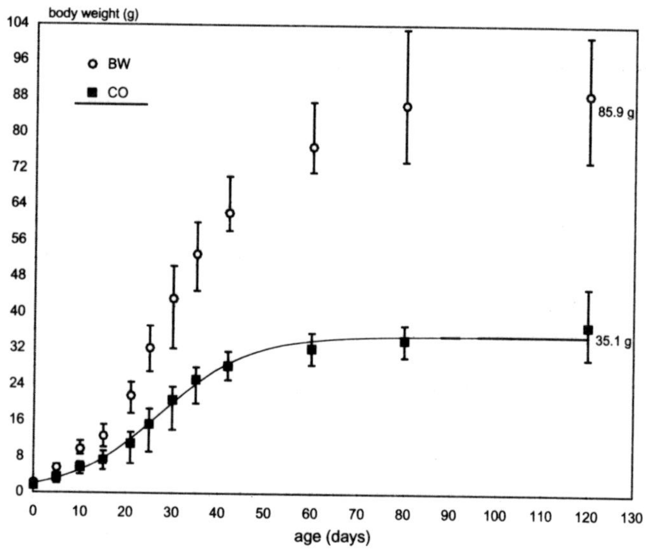

Figure 18 Fitted and observed growth curves for males (Expt IV). Males from line BW (n = 19) and CO (n = 18) in generation 86 were weighed at 0, 5, 10, 15, 21, 25, 30, 35, 42, 60, 80 and 120 days and the weights were fitted using the logistic model. Y error bars are given as minimum and maximum values, indicating the range for all observed values. The parameters for the logistic model and the coefficient of determination (CD) were:

	CO	BW
A (g)	34.9	85.9
B (g/day)	0.87	2.12
C (days)	27.74	30.98
CD	0.994	0.991

taken the mean far beyond the range of the initial variation in the base populations. This can be seen from the results of the four experiments considered here (**Table 2**) and from a comparison with results of some other long-term growth selection experiments (**Table 3**). In **Table 2** data or estimates for male body weights at 42 days (BW42 m) were included as a reference point because different selection criteria were used. The total selection response in relation to the initial mean seems to be very variable, but selection in the high direction can more than double the mean and selection in the low direction can reduce it by about 50%, which would mean a more or less symmetrical deviation from the initial level if data were measured on a log scale.

The weights reached in the different selection experiments can be used to find the heaviest and the lightest mouse line produced by selection. As body weight increases with age (up to an asymptotic value = parameter A in our logistic model) and sexes differ we will focus here on male body weights at 42 days as a reference point.

What then is the "total range" for those weights that can be produced just by breeding? The BW42 of nearly 64 g estimated by the non-linear model for line BW in generation 92 (Expt IV, **Figure 12**) could probably be considered as an approximate upper limit for this body weight in laboratory mice. Such weights were also reached by the BW line in the Edinburgh laboratory (**Figure 19A**, line designated DUH). However, it seems quite reasonable to take the estimated theoretical selection limit for male BW42 of 72 g (line BW; **Figure 12**) as an upper limit as the last four generation means were 67.4, 68.4, 71.1 and 68.4 g and there seems to be further selection response at the end of Expt IV (**Figures 12** and **13**).

Body weights of 38 widely used mouse lines (unselected, mostly inbred) over ages from 1 to usually over 100 days were reported (Poiley, 1972). The heaviest males were about 29 g at 42 days and the heaviest single mouse, found in the AKR/LwCr strain, weighed 46.2 g at 168 days and was probably at such an age relatively fat. The "high end" of these normal, unselected mouse lines is thus very much smaller than the weights reached by selection.

An estimate for a lower limit cannot be derived from Expt IV as there was no selection on low body weight. The lowest male BW42 in the four selection experiments considered here was about 13 g in line BW56L (Expt II). But results of Falconer and Roberts (**Table 3**) show that 10 to 12 g can possibly be reached. But even this is not "the end". After repeated backcrossing of the *lit* gene (a recessive mutation causing a growth hormone, GH, deficiency in homozygous *lit/lit* animals) into the low line (Expt III), the weight was reduced by about 50% (Bünger & Hill, 1999b) (**Figure 17**).

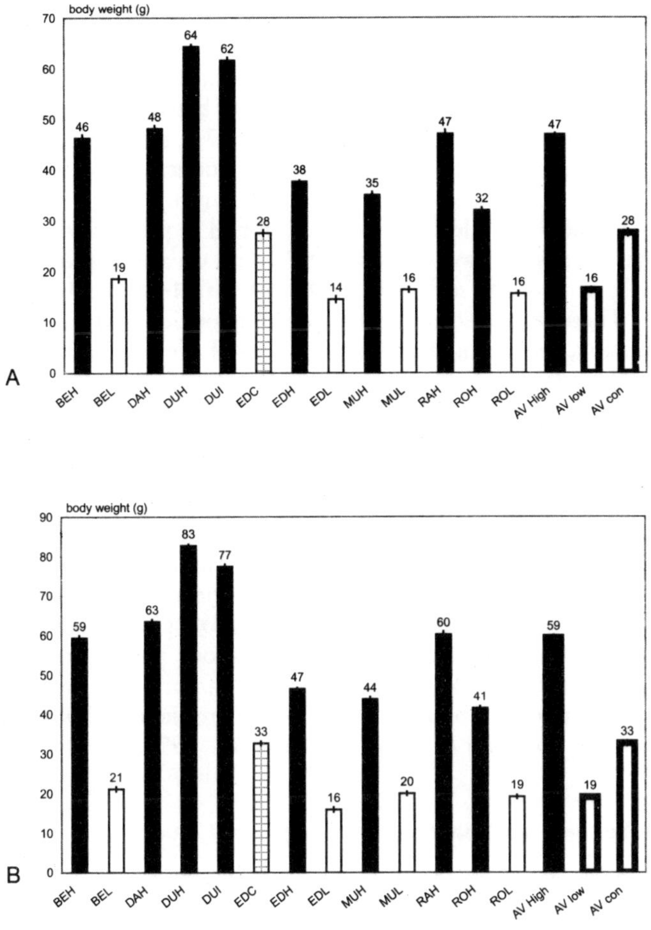

Figure 19 Male body weights at 42 (A) and 70 days (B) over generations 4 to 14 for different selection lines measured in the same laboratory (adapted from Bünger *et al.*, 2000). Least square means for BW42 and BW70 with their standard errors are given. For line description see Table 4 and for more details see Bünger *et al.* (2000). Lines are kept with about 24 matings per generation, litter size reduction to 12 and with small selection on BW70 until generation 8 and were afterwards inbred. Generation numbering started at time of immigration in all lines with 1. There were no measurements until generation 4 as all lines had been successfully transferred by embryo transfer to the laboratory in Edinburgh. Line EDC is a control line derived from the same base population as EDH and EDL lines.

Male mice of this PL line, homozygous for the *lit* mutation and producing no GH, weighed on average 7.3 g at 42 days. Despite their very small size, such mice are fertile and this mutation could therefore become fixed in a population assuming that it existed in the base population or arose spontaneously. From a hypothetical introgression of this mutation into the low lines of Falconer and Roberts (**Table 2**) one could probably expect male mice with a weight of 5–6 g at 42 days. Such expectation seems plausible because this proportional weight reduction by 50% due to the *lit* gene is almost independent of the genetic background (Bünger & Hill, 1999b). Mice with a body weight of 5–6 g at 42 days seem feasible. They would not fall below a biological or physiological limit for mammals, one might assume from thermoregulatory aspects, because the smallest known mammals, the Pygmy White toothed Shrew (*Suncus etruscus*) or the Bumblebee Bat (*Craseonycteris thonglongyai*) have body weights of 1.5–3 g and 2 g, respectively (Vogel, 1974; Sedlag, 1986; see also http://www.abdn.ac.uk/mammal/smallest.htm). The lowest body weights at 42 days reported by Poiley (Poiley, 1972) on 38 mouse strains was 14.4 g in inbred SM/JCr mice.

Taking the ranges for male body weight at 42 days as 10 to 64 g (without the *lit* mutation) or 7.3 to 64 g (for *lit/lit* mice) or 5 to 72 g (expected values, based on reasonable assumptions), the total range comes to about 54, 57 or 67 g, respectively. The ratio between the lower and the upper limits would be nearly 1:6, 1:9 or even 1:14. Similar but slightly lower ratios are known between high weight and low weight breeds of chickens and rabbits (Herre & Röhrs, 1973) but the ratio of heavy to light breeds of dogs is much bigger: 1:40 or 1:80 considering the Chihuahua (about 1–2 kg) and the St Bernard (about 80 kg) as limits for dogs (Kaiser, 1971; Sarowsky, 1986). Thus the total range for body weights in mice is impressive when reckoned in terms of variation present in the original populations, but is less spectacular when compared with the achievement of dog breeders. This higher range in dogs is probably due to the much higher population size involved, the time scale of domestication (probably about 100 000 years; Vila *et al.*, 1997), the different reasons for domestication (breeding goals: meat supply, sacrifice, guard, companion, hunting, racing, fighting, transportation) and the domestication from different subspecies of the wolf in different parts of the world, a nearly unrestricted exchange of wanted genes over continents, possibly allowing the use of very many mutations. In addition, episodes of admixture between wolves and dogs, allowing repeated genetic exchange between dog and wolf populations, may have been an important source of variation for artificial selection (Vila *et al.*, 1997).

The population sizes in the mice experiments were usually less than 100 and the longest experiment comprised 92 generations so far (Expt IV) – a long experiment (23 years) but extremely short on an evolutionary time scale. In some earlier experiments (Falconer & King, 1953; Roberts, 1967) new genetic variance was obtained by crossing previously growth-selected mouse lines. Subsequent selection using this new genetic variance increased the selection response significantly, but the number and initial divergence of the lines involved has been comparatively small, as has the population size, so that the total selection response did not exceed that of many other experiments (**Tables 2** and **3**).

Different laboratory environments, genotype–environment interactions and improvement in diets for laboratory animals over the years have doubtless contributed to the variation in selection response between experiments described in **Tables 2** and **3** and comparisons between recent and earlier selection experiments.

Measurement of mice from different selection lines in the same environment is only possible where the lines still exist. However, mice from different growth-selected lines from all over the world were recently brought into the Edinburgh laboratory by embryo transfer (**Table 4**). These mouse lines were selected from different base populations, selection was on different growth traits and the amount of selection applied and the selection history differ. Nevertheless these mice give a unique chance to look at their weights in a single environment regarding upper and lower limits for body weight achieved by breeding.

The body weights at 42 and 70 days pooled over generations 4 to 14 (counting started at transfer time regardless of generation number in the home laboratory) are shown in **Figure 19**. This corresponds to generations 11 to 21 and 15 to 25 in the inbred PHi and PLi lines derived from the P lines presented in Expt II (Bünger & Hill, 1999a), here called EDH and EDL for the sake of consistency.

The lines with highest male BW42 are clearly DUH (= BW line in Expt IV) and DUI (inbred line derived from DUH), which reach 64 g and 62 g, respectively. They are followed by lines DAH (48 g), RAH (47 g), BEH (46 g), EDH (38 g), MUH (35 g) and ROH (32 g). DUH is twice as heavy as the "lowest high line" (ROH). DUH mice perform very similarly in the Edinburgh and Dummerstorf laboratories (cf. **Figure 12**).

The low lines vary between 14.5 and 18.6 g. The "highest high line" (DUH) and the "lowest low lines" differ by about 50 g or by a factor of 4.4. A very similar picture emerges at 70 days (**Figure 19B**). DUH reaches weights of 83 g, followed at a distance of about 19 g or 6 s_p by the next high line, DAH. The lowest H lines (ROH and MUH) weigh 41 and 44 g respectively. The low lines vary between 16 and 21 g. DUH and EDL differ at 70 days by 67 g, a factor of 5.2.

Altogether this line comparison confirms the range derived in the discussion above and demonstrates that DUH is the heaviest selection line one can find at present, weighing 64 g at 42 days and 83 g at 70 days on average. At the other end of the scale is the EDL line with 14.5 and 16 g.

These lines also differ in body composition and conformation, suggesting that different loci and alleles contributed to the selection responses. A cross between the high lines or the low lines should provide much new genetic variance upon which continued selection could act. It is known that DUH does not have the most effective growth-promoting alleles on all loci. An earlier study looked at gene action underlying selection responses using reciprocal crosses between PH (= EDH) and PL (= EDL) lines. The difference in body weight at 10 weeks between these lines (Expt II, **Figure 9**) was largely (25%, equivalent to 5 g) explained by one QTL containing obviously at least one major gene on the X chromosome (Hastings & Veerkamp, 1993), which was mapped at about 23 cM from the proximal end with a peak LOD score of 24.4 (Rance *et al.*, 1997a,b). Recent

Table 4 Line description for contemporaneous line comparison in the same lab (adapted from Bünger *et al.*, 2000).

Place of origin line code	Original line code and alias	Original selection trait	Selection method	Base population	Key references
Berlin BEH/BEL	W+G+ and KW–	protein amount at 60 days/later BW60	sib selection/mass selection	not defined outbred (founders from different pet shops)	Weniger *et al.*, 1974; Barkemeyer *et al.*, 1989; Valle Zarate *et al.*, 1994
Davis DAH	subline of G, line 32, generation 31	G21–42	mass selection	cross of four inbred lines: C57BL/6J, AKR/J, C3H/J, DBA/2 J	Bradford, 1971; line without hg-mutation (Bradford & Famula, 1984)
Dummerstorf DUH	DU-6	BW42	sib selection	cross of four inbred (CBA/Bln, AB/Bln, C57BL/Bln, XVII/Bln) and four outbred (NMRI orig, Han:NMRI, Han:CFW, Han:CF1) strains	Experiment IV
Edinburgh EDH/EDL/EDC	PH, PL	lean mass until generation 20, later BW70 EDC = control	intrafamily selection	cross of two inbred (CBA, JU) and one outbred line (CFLP)	Experiment III
München MUH/MUL		BW56	intrafamily selection	cross of four inbred lines (C3H, C57Bl, BALBc, CBA)	Butler von & Pirchner, 1983; Butler von *et al.*, 1984
Raleigh RAH	M16	G21–42+	intrafamily selection	outbred population (ICR)	Eisen, 1975; Eisen *et al.*, 1988
Roslin ROH/ROL	X-lines	BW42+, BW42–	intrafamily selection	cross of two inbred lines: C57BL/6J × DBA/2J	Keightley & Bulfield, 1993; Heath *et al.*, 1995

B*x*, G*x*–*y*: body weight at *x* days or gain between *x* and *y* days of age reciprocal crosses between the DUH and EDH lines (Bünger & Hill, unpublished data) have shown that DUH probably has a low allele for this QTL. An introgression of this QTL into the DUH line is being made and is expected to increase its high growth limit. At the same time it would test for epistasis, measuring the effect of a QTL or a gene depending on the genetic background. Presumably there are more loci and alleles in which the selected lines differ, which will be investigated in due course.

(ii) Phenotypic variance

As selection is expected to decrease the genetic variance one could also expect lower phenotypic variance (V_P). However selection for high growth in all four experiments led to striking increases in phenotypic variance, by factors of 2.5 to 5 (**Figures 3, 7, 10** and **14**) whereas the low lines show a strong decrease in phenotypic variance (**Figures 7** and **10**). It is of note that the coefficient of variation remains relatively constant; although some of the changes are significant (**Figures 3, 8, 11** and **14**) they have been small compared with changes of V_p. This correlation between mean and variance, known as scale effect (Falconer & Mackay, 1996), suggests that a log scale would be more suitable for analysing body weights.

(iii) How many genes are involved in growth?

Having found lower and higher limits for body weights in mice, the total range can be used to estimate the number of genes influencing body weight. Such a calculation assumes an infinitesimal model and should be done on a logarithmic scale as stated above. That the range was worked out from different selection experiments will be ignored. How many genes are involved? The phenotypic and genetic standard deviation for BW42 can be assumed to be 2.8 g and 2 g ($h^2 = 0.5$) respectively, corresponding to 0.0433 and 0.0306 when log (base 10) transformed as described by Falconer & Mackay (1996, p. 293). The total range (5–72 g) on the log scale is 1.158, or about 27 s_p or 38 s_g. Using this range (R = 1.16), a genetic variance of 0.00094 (log scale) and the formula ($n = \frac{2R}{8V_A}$) given by Wright (1952; see Falconer & Mackay, 1996, p.226) an estimate of the number of loci for BW42 is n = 308, or about 300. Though such estimates are by their nature imprecise, they cover an area where little knowledge is available, especially for mammals (Roberts, 1966a; Eisen, 1975). This estimate for the "number of growth related loci" is much higher than values given by Roberts (1966a) for BW42 of about 8 to 20 or by Eisen (1975) for postweaning body weight gain of about 20 to 40. This is not surprising because: (i) the "total range" was derived here from selection responses in different experiments, and (ii) the upper limit is derived from Expt IV, where the selection response was very high and the lower limit assumed a fixation of the *lit* mutation in a low line. Using 10 and 64 g as "the limits" for BW42 in mice, the number of genes would be roughly 200, indicating that many genes affect growth.

The high response in line BW (Expt IV) seems to be primarily due to the comparatively high population size of about 60 in this experiment (Bünger *et al.*, 1998a), reflecting a more or less linear relation between $N_e \times i$ and the total selection response found in some selection experiments in mice (e.g. Eisen, 1975; Kownacki & Zuk, 1986) and in Drosophila (Jones *et al.*, 1968). The high heterogeneity of the base population, founded by a cross of four inbred and four outbred populations and the length of the experiment may also have contributed to this high response.

(iv) Response curve and selection limit

The classical selection response pattern in a long-term selection can be viewed as an initial linear change, followed by a gradual decrease of the response until a selection limit is reached. Without the creation of new variation by mutation the response cannot be expected to continue indefinitely, as genes segregating in the base population become fixed (or in equilibrium in the case of overdominance) by selection and the accompanying inbreeding. A progressive decline in the rate of response is inevitable and the population reaches a "selection limit". A limit may be reached well before the genetic variance is exhausted if the selection favours individuals that are heterozygous at some loci, or if natural selection opposes the direction of the artificial selection (for further details see, for example: Robertson, 1960; Bohren, 1975; Kress, 1975; Eisen, 1980; Hill, 1985; Falconer & Mackay, 1996, Hill & Mbaga, 1998). Though deviations from this general form are frequently encountered in practice (Falconer & Mackay, 1996) in most cases a stage is reached after which little or no further progress is made. The pattern of selection response in all four experiments fits very well to this general picture.

Such an approach is of course based on the assumption of no substantial contribution of mutation to the selection response, which might be not true, since other experiments have shown that spontaneous mutations contribute to the selection response especially in long-term experiments (Frankham, 1980; Bradford & Famula, 1984; Keightley & Hill, 1992; Hill *et al.*, 1994; Mackay *et al.*, 1994; Caballero *et al.*, 1995; Keightley *et al.*, 1998).

Caballero *et al.* (1995) found it impossible to give an unequivocal estimate of mutational heritability for body weight in the mouse, but it seemed likely that the total mutational heritability, including genes conferring significant fitness reduction, is at least 0.5% per generation, of which the component contributed by neutral or nearly neutral genes is about 0.1%, sufficient to contribute to long-term selection responses. But there are many selection experiments in which an obviously non-linear development or a plateau over a long time is shown (for examples see Expts I–IV and Bünger & Herrendörfer, 1994; Falconer & Mackay, 1996). Why then did mutations not contribute to an obvious extent to the selection response in those experiments? It could be that the fluctuations of observed means mask an underlying steady genetic change in some cases, or that many mutations may have negative pleiotropic effects on fitness (Keightley *et al.*, 1993).

To estimate and analyse theoretical selection limits does not necessarily mean to consider them as fixed and everlasting limits, as evolution shows that every limit is ultimately only a temporary one. The question is the time scale over which progress is measured, what population size is involved and the experimental environment. However, since in some populations limits are maintained over a very long time (e.g. 30 or 40 generations in Expt I) it seems worthwhile to describe them and to analyse their causes and factors affecting them.

Goodale's initial idea was that mice might be developed by selection which would weigh the same as rats and perhaps even more:

> Indeed, since the theory of evolution says that life began with a single-celled animal of no greater size than an amoeba, the very large animals such as mastodons, megatheria or titanotheres must have arisen from smaller ancestors. (Goodale, 1938)

From the selection experiments described above, we can devise a method to create the heaviest mice, but can breeding bring mice to a similar size as rats? We have discussed mice weights in detail, but how big are rats? In a recent study, body weights in male rats from the inbred F344/Shoe Wistar strains from birth to 249 days of age were measured (Gille *et al.*, 1996). The body weight development of these rats is presented in **Figure 20** together with the growth curve of male mice from mouse line DUH (Expt IV), which certainly can be considered the heaviest mice known, excluding ob/ob mice which are extremely obese (containing up to 50% of body fat), which might overtake the DUH in weight at higher ages.

It is clear that mice and rats are in "different leagues"; rats are still about five times as heavy as mice are. Since there is also much variation between different rat strains we should ask how representative these rat data are for rats at all. Using the parameter A of different growth models provides a rough estimate. The mature weight of this inbred line (A ≈ 450 g) would be in the upper third of a range derived from data on 16 different rat strains (Poiley, 1972) from estimates of A given by Gille *et al.* (1996). The heaviest rat line based on males (W/LCr) has an A value near 510 g whereas the low end seems to be marked by rat line AC19935/Cr or ACP9935/Cr with mature weights of about 305–320 g. Even compared with lines of small rats, the heaviest mice cannot even come close to rat weights. However as discussed above, on the scale of evolution, even the longest and biggest mouse experiment is extremely short and extremely small.

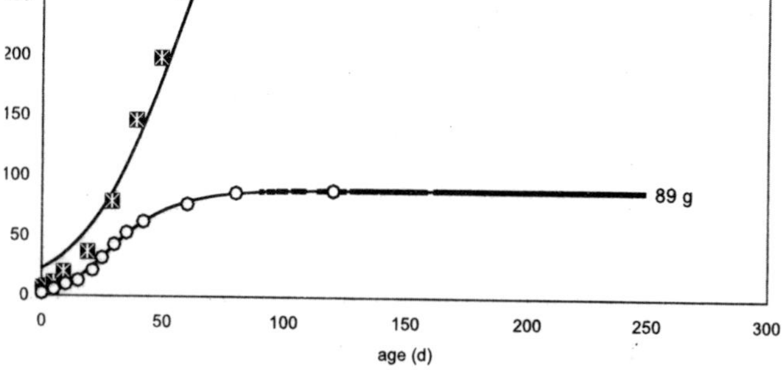

Figure 20 Growth of a "normal" rat (F344/Shoe Wistar) and of males from the heaviest mouse line (DUH, Expt IV). Rat data were taken from Gille *et al.* (1996). Parameters for the logistic function are:

	Rat	DUH
A (g)	449.2	85.9
B (g/day)	5.75	2.12
C (days)	57.3	31.0
CD	0.981	0.991

(v) *Effects of single genes on body weight*

Body weight is usually considered a typical polygenic trait, which means a very large number of genes is assumed with each having a small effect. However the number, location and effects of individual genes contributing to variation in those traits are mostly unknown. Therefore QTL mapping studies aim to identify the genes causing the variation, and some QTL with relatively high effects have been found (e.g. Winkelman & Hodgetts, 1992; Cheverud & Routman, 1993; Cheverud *et al.*, 1996; Rance *et al.*, 1997b; Brockmann *et al.*, 1998; Morris *et al.*, 1999). However, it is not possible to review them here.

In addition to these QTL a few single gene mutations on growth-related loci of major effects in mice are known, which alter the growth process in different ways and to different degrees (**Table 5**). Whereas for example the *hg* mutation can increase the weight gain (between 21 and 42 days) by 30–50% there are other mutations that decrease the body weight by up to 75%. A dependence on gene effects from the genetic background can be expected but is not well investigated yet.

Only a few experiments have analysed the effects of single gene mutations that depend on the genetic background of different selection lines (*dwarf*: Pidduck & Falconer, 1978; *lit*: Hastings *et al.*, 1993a; Bünger & Hill, 1999b). These experiments aimed to understand the basis of quantitative genetic variation and to identify the contributions of particular candidate loci or major hormonal pathways to the genetic change caused by selection. For example, the recessive *lit* mutation (for details see Doolittle *et al.*, 1996) was introduced into the high and low lines of Expt III (Bünger & Hill, 1999b) by repeated backcrosses. In brief, the *lit* mutation is known to cause a GH deprivation, due to a defect in the growth hormone releasing factor receptor gene (Lin *et al.*, 1993). Homozygotes fail to release significant levels of GH in response to the growth hormone releasing factor and are consequently dwarfs. Homozygotes are smaller than normal from about 2 weeks of age, with adult weights about 50 to 65% that of controls (Eicher & Beamer, 1976; Jansson *et al.*, 1986; Hastings *et al.*, 1993).

After six backcross generations the growth of male mice from these lines was investigated and compared with the growth of wild-type males from both wild-type selection

Table 5 Mutations affecting body weights in mice.

Code	Effect	Location	Reference
hg	Recessive Increase of weight gain by 30–50%	Chr 10	Bradford & Famula, 1984; Horvat & Medrano, 1996
lit	Recessive Homozygotes are smaller than *wt* from about 2 weeks of age. Decrease of adult weights by 35–50%	Chr 6	Eicher & Beamer, 1976; Phillips *et al*., 1982; Jansson *et al*., 1986; Hastings *et al*., 1993; Lin *et al*., 1993; Doolittle *et al*., 1996; Bünger & Hill, 1999a,b
dw and *dw*^{J}	Recessive Decrease in weight by 30% at 3 weeks and by 60–75% at 6 weeks of age	Chr 16	Snell, 1929; Pidduck & Falconer, 1978; Eicher & Beamer, 1980; Li *et al*., 1990; Doolittle *et al*., 1996
pg	Recessive Decrease of body weight in homozygotes by about 60–70%	Chr 10	MacArthur, 1944b; King, 1955; Nissly *et al*., 1980
cmpt	Intermediate dominance in males but almost fully recessive in females Hypermuscularity and an increase of carcass/body weight ratio Effects on body weight?	Chr 1	Weniger *et al*., 1974; Varga *et al*., 1997; Szabo *et al*., 1998

lines (PH, PL; Expt III) (**Figure 17**). Focusing here on the mature weights on the absolute scale there was obviously a much stronger decrease in body weight in the high line (55.3 − 23.9 g = 31.4 g) than in the low line (18.3 − 9.9 g = 8.4 g). Homozygous PH and PL animals reached 43 and 54%, respectively, of their wild-type controls (**Figure 17**). The GH metabolism, which might *a priori* have been expected to play a large part in the response of mouse lines selected for high or low body weight had similar proportionate effects on body weight in both lines of mice. These data show that changes in the GH system are not the only cause of line differences in growth resulting from long-term selection, but they are involved to a significant extent. The interactions between the mutation effect and the genetic background were, nevertheless, relatively small and therefore these support a "polygenic model" of the response to selection: that many genes with mostly small effects and with diverse metabolic roles all contribute to the divergence of phenotypes (for detailed discussion see Bünger & Hill, 1999b).

Six single gene mutations on fatness-related loci of major effects have been mapped in mice (*A* locus with two alleles A^{vy} and A^{y}, *Ad*, *Cpe*, *Tub*, Lep, Lepr; e.g. Doolittle *et al.* 1996), altering fatness in different ways. Resulting obesity differs in extent, ages of onset, progression to obesity and depends on genetic background. In some cases excessive deposition of adipose tissue can lead to a twofold increase of body weight (Falconer & Isaacson, 1959; Nobentrauth *et al*., 1996). This fact is important to bear in mind because body weight is a composite trait resulting from fat and non-fat "aggregation". Some genes might affect fat or lean whereas other genes might affect both. When the body composition is unknown it is difficult to disentangle the underlying genes and to estimate their effects.

The knowledge that can be expected from such studies is highly relevant for the study of human obesity and growth disorders, since conserved homology groups now span >90% of mice and human genomes. Isolating a mutant or major gene in one species may provide us with a candidate gene in a homologous region in the other species (Witherden *et al*., 1997).

(vi) Growth of transgenic mice

A discussion of growth limits in mice would be not complete without looking at the effects of transgenes, especially as the term "giant" accompanied many early transgenic studies. In a pioneer article in this field (Palmiter *et al*., 1982) the authors said: "On the other hand, the experiments described above show that gigantism can be created." How much growth can actually be stimulated by the overexpression of individual hormones, which are central for growth? And how does this compare to body weight changes caused by long-term selection, involving changes on many genes? Such a discussion ought to look at "dwarfism" creating transgenes as well.

The objective of this section is to elucidate the range of body weights that can be achieved by the introduction of foreign (usually single) genes into mouse embryos. This is another powerful approach for studying gene regulation and the genetic basis of growth and development. Several studies from the beginning of the 1980s have clearly established the feasibility of introducing foreign DNA into the mammalian genome by microinjection of DNA molecules of interest into pronuclei of fertilized eggs and the insertion of these eggs into the reproductive tracts of foster mothers. The transgenic animals developing from this procedure integrate the foreign DNA into one of the host chromosomes at an early stage of development and the foreign DNA is generally transmitted through the germline. By this procedure it was possible to obtain regulated expression of viral thymidine kinase by fusing the structural gene to the metallothionein-I (MT-I) gene promoter and inducing expression with heavy metals (Palmiter *et al*., 1982). These authors produced transgenic mice carrying a rat growth hormone gene by the microinjection of a fusion gene, comprised of the metallothionein-I (MT) promoter/regulator region joined to the rat growth hormone (rGH) structural gene. These MT–rGH mice have high levels of the fusion messenger RNA in several organs, elevated levels of foreign growth hormone and insulin-like growth factor I in serum, markedly increased growth rates and final adult size. In the studies of Palmiter and colleagues there was a wide range of response to the transgene; at 74 days of age growth ratios (body weight of transgenic animals/body weight of wild-type controls) of 1.0 to 1.9 were described. The body weights of some of these mice from about 40 to 100 days are given in **Figure 21**.

The heaviest transgenic male and female have been chosen for representation in comparison with the mean of non-transgenic male and female siblings. Both animals reach nearly double the weight of their non-transgenic siblings. At about 90 to 100 days these animals reach body weights of 49 g (♂) and 44 g (♀) whereas controls have weights of 28 and 24 g, respectively.Controls of both sexes have about 0.15 mg/ml GH and the range for the transgenic animals was from 0.31 to 112 mg/ml, or from 2 to over 700 times greater.

A cross-sectional study on the same transgenic mouse model in later generations was performed over a wider age range of 40 to 600 days (Shea *et al.*, 1990) (**Figure 22**). Body weights at the same ages vary greatly, partly due to not specifying sex, but are similar to those in **Figure 21**. Between 120 and 200 days body weights of the transgenic animals vary between about 42 to 62 g. On average the adult transgenic mice are almost twice as heavy as the adult controls. The highest weight (66 g) was observed in one mouse of about 600 days of age. Another study using the same transgenic mouse model reported 1.5-fold higher male body weights at 9 weeks, which confirms the above findings (Nagai *et al.*, 1993).

A porcine growth hormone (pGH) fusion gene under the transcriptional control of the human metallothionein promoter (Vize *et al.*, 1988) in mice caused 35-fold increases of circulating GH, a 1.5-fold increase in plasma IGF1 and an 1.3-fold increase in body weights at 9 weeks, where transgenic males weighed 39 g and the controls 30 g (Wilkinson & Luxford, 1990).

Transgenic mice carrying a sheep metallothionein 1a sheep growth hormone (bGH) (Shanahan *et al.*, 1989) were investigated from birth to 14 weeks (Searle *et al.*, 1992). A 1.5-fold increase in body weight was found, which was limited to females. At about 100 days transgenic females weighed 32 g and controls 21 g, and transgenic animals were leaner. Males from this transgenic mouse line were mated to F_1 females (C57Bl/6 × CBA). Body weights of transgenic and non-transgenic offspring were observed from 10 to

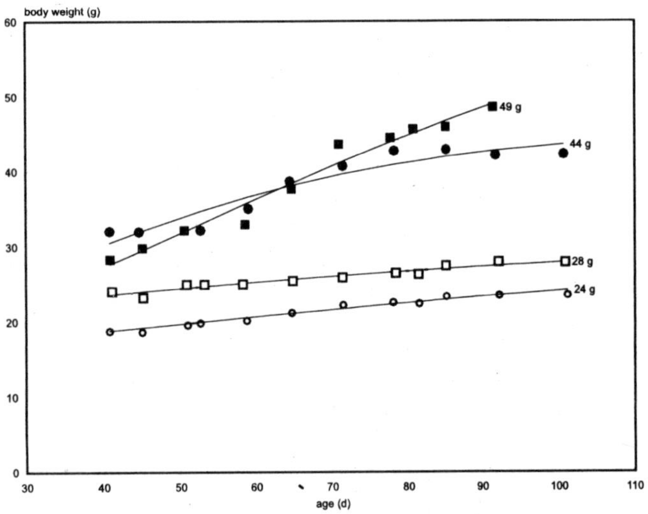

Figure 21 Growth of MT-rGH mice from about 6 to 14 weeks (adapted from Palmiter *et al.*, 1982). The weights of the heaviest transgenic male (tm, solid squares) and the heaviest transgenic female (tf, solid circles) are shown. The mean body weights of 14 siblings not containing MGH sequences (controls, c) are shown for comparison (open circles: females, cfs; open squares: males, cms). Data were digitized from the original graph (Figure 3 in Palmiter *et al.*, 1982) and may therefore deviate slightly from original values. Parameters for the logistic function are:

	cfs	cms	tm	tf
A (g)	32.60	42.65	65.13	45.86
B (g/day)	0.102	0.077	0.454	0.436
C (days)	16.14	8.81	51.72	22.68
CD	0.958	0.925	0.968	0.905

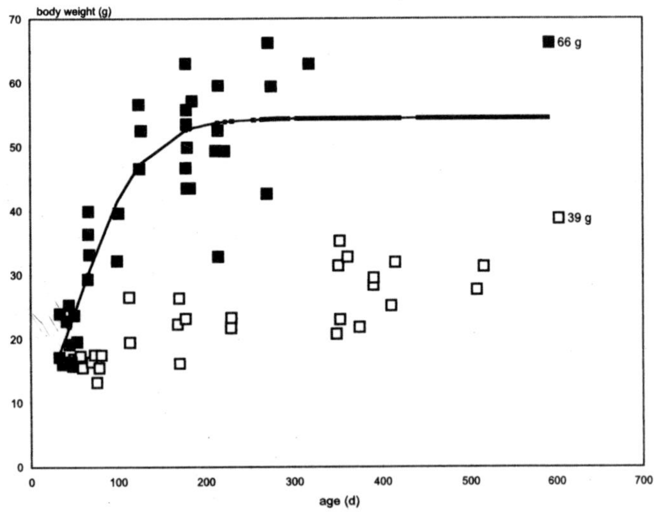

Figure 22 Growth of MT-rGH mice from about 6 to about 86 weeks (adapted from Shea *et al.*, 1990). Solid and open squares represent transgenic animals and their non-transgenic littermates, respectively. Data were digitized from the original graph (Figure 1 in Shea *et al.*, 1990) and may therefore deviate slightly from original values. Parameters for the logistic function are:

	non-transgenic	transgenic
A (g)	44.84	54.41
B (g/day)	0.035	0.381
C (days)	213.20	58.40
CD	0.674	0.793

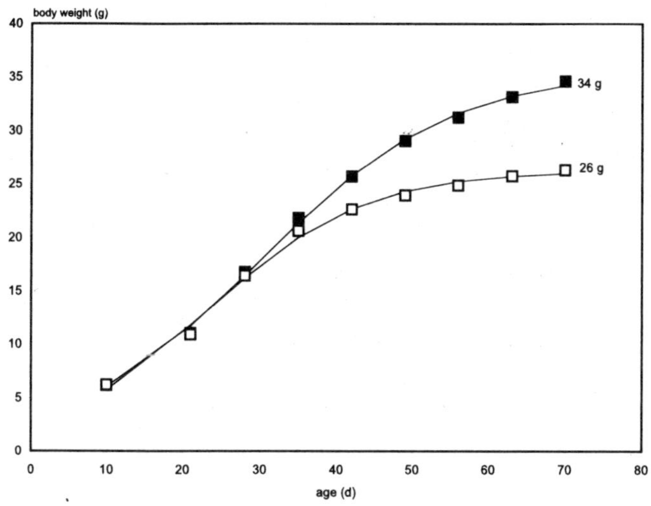

Figure 23 Growth of male MT-bGH transgenic mice (adapted from Pomp *et al.*, 1992, with permission from Elsevier Science). Open and solid symbols represent control (c) and transgenic males (t), respectively. The means were given by the authors in tabular form and here fitted using the logistic function yielding the following parameters:

	t	c
A (g)	35.60	26.24
B (g/day)	0.711	0.640
C (days)	29.93	23.09
CD	0.999	0.995

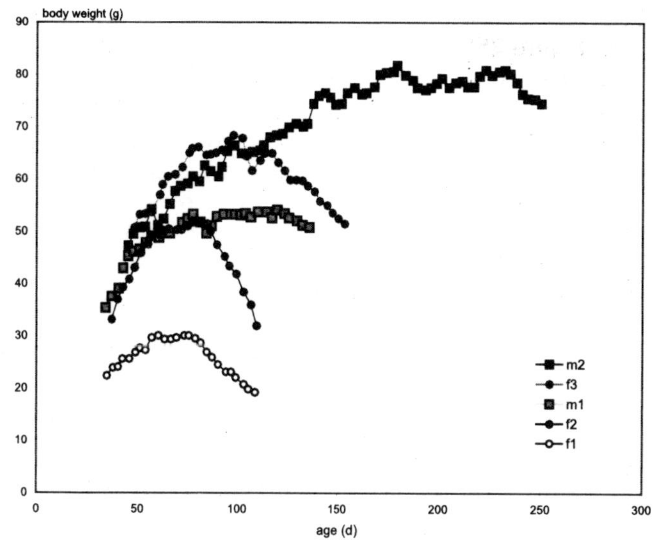

Figure 24 Growth of individual MT-hGH transgenic mice (adapted from Wolf *et al.*, 1991). Squares represent males and circles females. Data were digitized from the original graph and may differ slightly from the real values.

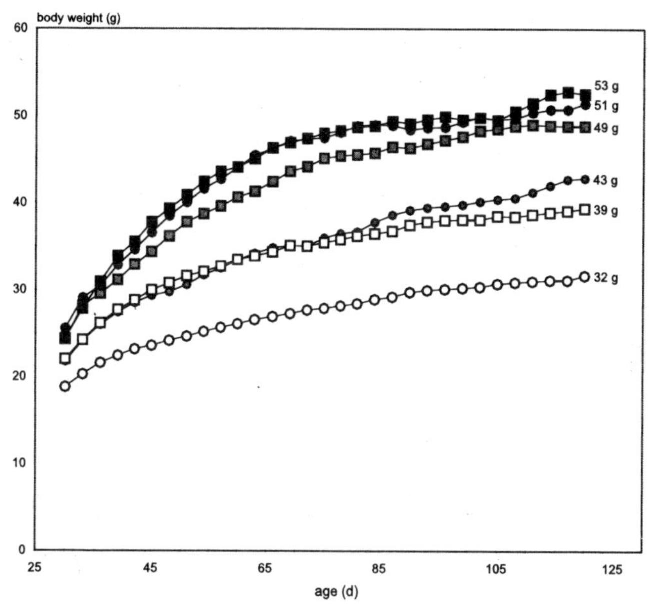

Figure 25 Growth of MT-hGH transgenic mice, of mice from a line selected for high body weight at 8 weeks and from a control line (adapted from Wolf *et al.*, 1991). Black squares (males) and circles represent the transgenic animals, grey symbols the selected animals and white symbols the controls. Data were digitized from the original graph and may differ slightly from the real values.

70 days of age (Pomp *et al.*, 1992). The growth of males is shown in **Figure 23**. The growth ratio between both male groups increased with age up to 1.3 at 70 days, when transgenic males reached 34 g and controls 26 g. In females this ratio was 1.5 resulting from body weights at 63 days of 29.7 g *vs.* 19.5 g. Again transgenic mice were leaner on a body weight basis.

Another study reported body weights of five (3 ♀, 2 ♂) mice (founder mice) with a human GH (MT–hGH) from about 30 days to approximately 250 days (Wolf *et al.*, 1991). There was an enormous inter-individual variation; for example, the maximum body weights in the founder population ranged from 30 g (at about 70 days) to 82 g (at about 200 days) (**Figure 24**), which is the highest body weight known for a transgenic mouse, however at a high age and for one individual only.

In extending the "usual" question about growth effects of the hGH transgene these authors (Wolf *et al.*, 1991) compared the body weights of MT–hGH transgenic mice with mice which were 28 generations selected for high 8-week body weight and an unselected control. All lines were derived from the same outbred population (NMRI) and body weights were taken from 30 to 120 days of age (**Figure 25**). Transgenic males and females reached similar weights: their maximum body weight was about 52 g and they differed only slightly from the selected males, which reached about 48 g. Transgenic males and females were up to about 1.4- and 1.7-fold, respectively, heavier than the controls.

It can probably be assumed that GH transgenic animals are in general leaner (for references see above). Body weights at higher ages in the selected and control lines might contain a substantial proportion of fat. A comparison on a fat-free body weight basis could result in even higher growth ratios (transgenic/non-transgenic). Interestingly there was almost no sexual dimorphism in the transgenic mice (**Figure 25**).

In a few experiments transgenes were used which increase growth and transgenes which decrease growth. Mice containing a wild-type bovine growth hormone (wt bGH) (Kopchick *et al.*, 1989) show accelerated growth rates and a mutated form (m bGH) resulted in decreased growth in mice. The body weight of mice (sex not specified) at 60 days was increased or decreased by a factor of about 1.6 and 0.7, respectively. The body weight at 60 days in the controls, wt bGH and m bGH were about 24, 39 and 17 g, respectively (Chen *et al.*, 1991). In another study (Chen *et al.*, 1990) mice carrying these transgenes were measured at 28, 56 and 84 days. The growth ratios of m bGH males and females vary between 0.72 and 0.85, respectively, with weights of 21 and 20 g at 84 days. Males and females with a wt bGH are 1.5- to 1.6-fold heavier than controls, reaching 45 and 39 g, respectively, at 84 days (**Figure 26**).

Another phenotypic analysis, using growth-decreasing transgenes, was performed on two alleles at the pygmy locus on chromosome 10, which arose by insertional mutagenesis in transgenic mice, analysing body weights from about 7 days of age up to 100 days (Benson & Chada, 1994). Despite normal GH level, homozygous mice of both transgenic lines had a growth ratio, compared with wild-type mice, of only 0.4 (**Figure 27**). Their body weights at *c.*100 days reached 9.5 g and 10 g, whereas the wild-type littermates weighed approximately 25 g. Interestingly heterozygous mice were lighter than the wild-type mice, suggesting that the pygmy gene is not a "true" recessive.

Having demonstrated that with an extreme overexpression of "just one" hormone, caused by the insertion of transgenes, mice can be created with nearly double the body weight of their wild-type controls and that long-term selection can produce mice more than double the weight of unselected controls, the question arises: what mice will result from a combined action? Is there any interaction between the genetic background and the transgenes? Could, for example, the weight of the DUH-mice (Expt IV) be further increased or even doubled again by the introgression of a

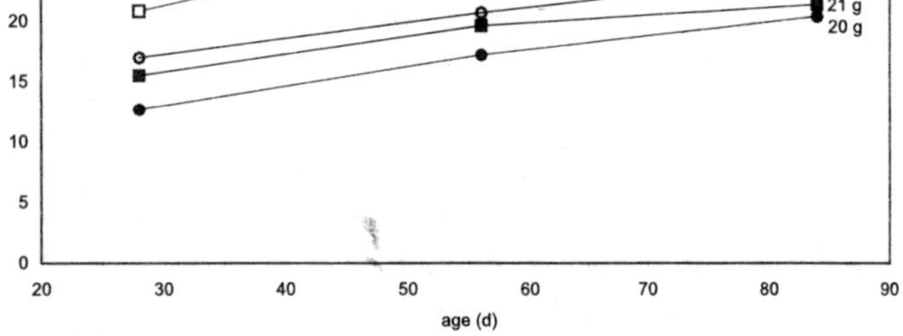

Figure 26 Growth of male transgenic mice with a wild-type bovine GH or a mutated bGH in comparison to non-transgenic mice (adapted from Chen *et al.*, 1990). Open symbols represent non-transgenic animals and black and grey symbols represent wt bGH and m bGH, respectively. Males are presented by squares and females by circles. Data were digitized from the original graph and may differ slightly from the real values.

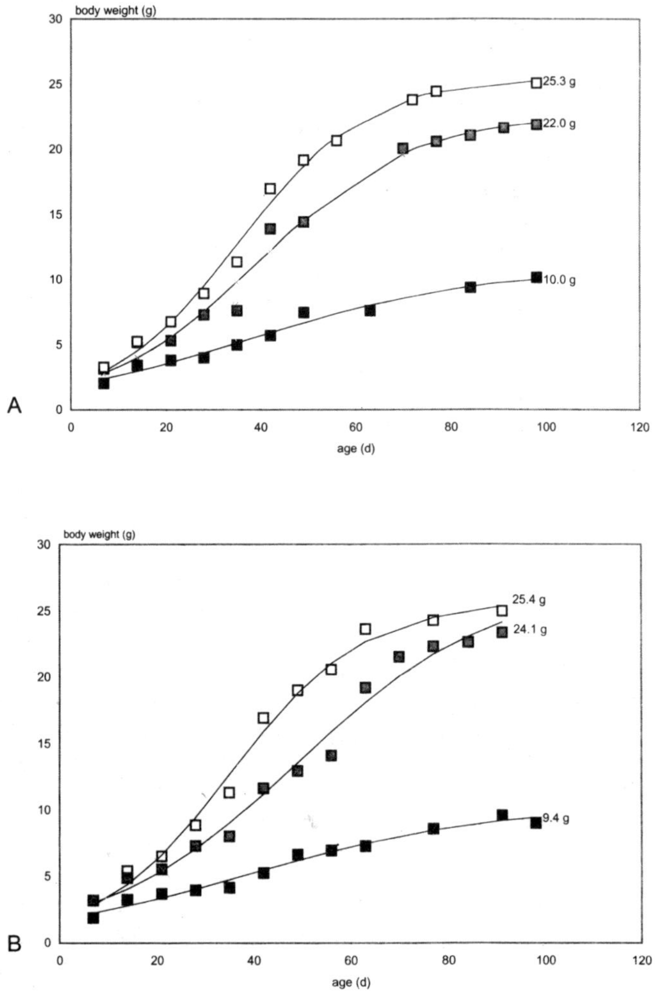

Figure 27 Growth of homozygous transgenic mice carrying an insertional mutation at the pygmy locus (adapted from Benson & Chada, 1994). White, grey and black symbols: wild-type (wt), heterozygous (he) and homozygous (ho) males. A: line $pg^{TGN40Acha}$; B: line $pg^{TgN49Bcha}$. ho, he, wt: homozygous, heterozygous and wild-type mice. Means were obtained from digitising original graphs and may show slight deviations from them. These means were fitted using the logistic model and the fitted curves are given. Parameters for the logistic model are:

	ho	he	wt	ho	he	wt
A	10.8	22.7	25.5	10.3	27.1	25.6
B	0.112	0.343	0.462	0.104	0.338	0.474
C	37.6	39.5	35.2	38.9	49.1	35.3
CD	0.98	0.98	0.99	0.98	0.98	0.99

suitable transgene? As this field is not yet well investigated, it is impossible to make predictions. Only a very few studies have so far focused on this problem. For example, a selection line (M16 [= RAH in Figure 19]), selected for high 3–6 week weight gain for 23 generations and maintained without selection for more than 60 further generations (Eisen, 1975), and animals from the base population (ICR) serving as control were mated with hemizygous mice from two different transgenic mouse lines (Eisen *et al.*, 1993). One line carried the wild-type bGH transgene and the other the m bGH, which acts as GH antagonist resulting in decreased growth (see Chen *et al.*, 1991). The resulting F_1 animals were classified as transgenic and non-transgenic using the PCR technique. Their weights were observed from 14 to 70 days and that for the males is given herein (**Figure 28A**). Transgenic males in both lines were substantially heavier by about 11 and 14 g in the selection lines and the control, respectively, corresponding to growth ratios of about 1.3 to 1.4, indicating a slightly lower reaction of the selection line. Final weights in the transgenic groups were approximately 52 and 49 g at 70 days; the non-transgenic selection line was nearly 20% heavier than the control. This ratio decreased to 1.06 when transgenic animals are compared.

A complication in this study could possibly arise from the use of a selection line that is known to have a substantially increased fat content (Eisen, 1987). In such a model a substantial part of the body weight is fat. Several of the studies looked at here have shown that a GH transgene might reduce fat. Therefore the net effect on body weight in a fat line remains uncertain.

The effects in the low direction were again dependent on the strain. The mutated bGH transgene caused a decrease in 70-day weight by 13 and 6 g in the selection line and the control, respectively, resulting in 70-day weights of 31 and 27 g, respectively (**Figure 28B**). The growth ratios were about 0.7 and 0.8 in selected and control males, respectively, indicating a stronger reaction of the selected line.

In such a one-step cross (F_1 cross) there is of course not only the transgene of importance but also the effect of all the other 50% of the genes coming on average from the genetic background of the line into which the transgene was originally incorporated. This disadvantage can be overcome by using a repeated backcross method, in which the genetic background of the lines of interest is step by step re-established, keeping mainly the gene of interest only. Using such an approach a growth-reducing transgene was introgressed into divergently selected mouse lines (lines from Expt III). This transgene caused a thyroid ablation, resulting in a deficiency in thyroid hormones and growth hormone (Bünger *et al.*, 1998b). The lines were selected over more than 50 generations and there were seven backcrosses so that the experimental animals had an expected proportion of 99.6% of their genes from the selected line. The body weights of transgenic (t) and non-transgenic (nt) mice were observed from 10 to about 100 days of age. Male weights are shown in **Figure 29**.

Non-transgenic males from the selection lines differ by a factor of up to 2.8, reaching final weights of about 54 and 19 g. In both lines there is a strong effect of the transgene. The growth ratio between t and nt PH males varied depending on age, starting at 0.89 (10 days) to 0.29 (35 days) and 0.64 (96 days). The same ratios for the low body weight line were 0.77, 0.36 and 0.58, indicating a slightly higher reaction of the PH line. The final weights at 96 days in the PH and PL lines were 34 and 12 g, respectively, showing the dramatic effects of genetical thyroid ablation on growth up to the adult state.

In general these results have shown that lines also differ in the absence of thyroid hormones and GH, indicating that

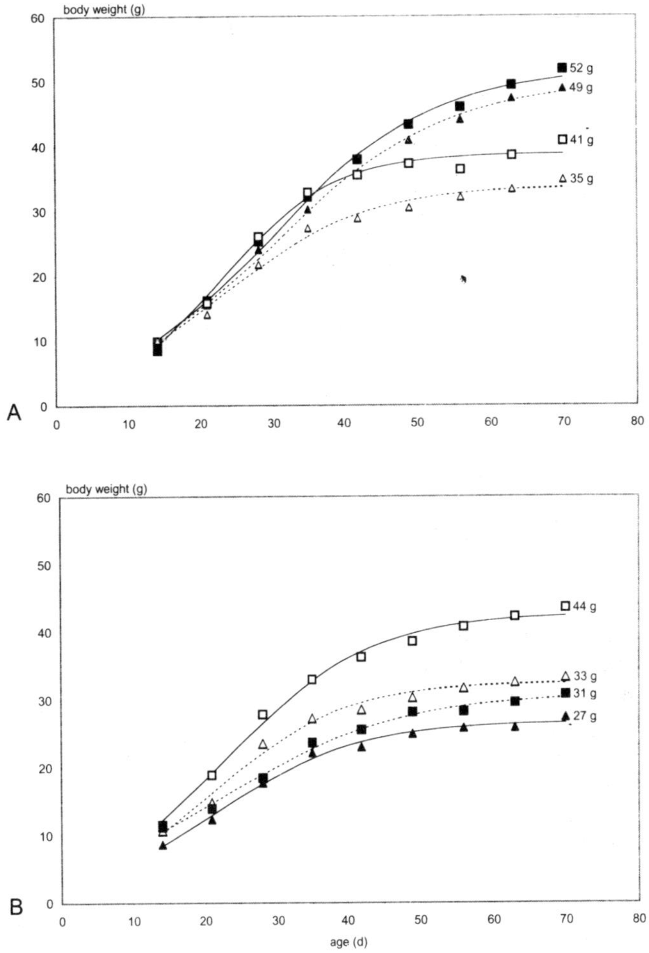

Figure 28 Growth of male F_1 progeny carrying the growth-promoting wild-type transgene (wt bGH) (A) and/or carrying the growth depressing mutated transgene (m bGH) (B) and their respective non-transgenic (nt) littermates (adapted from Eisen *et al.*, 1993, with permission from Springer-Verlag). S: selection line represented by squares; C: unselected control line represented by triangles; nt: non-transgenic (white symbols) and wt bGH or m bGH; transgenic (black symbols). Data were digitized from the original graph and may differ slightly from the real values. Parameters of the logistic function are:

	S nt	C nt	C wt bGH	S wt bGH	S nt	C nt	C m bGH	S m bGH
A (g)	38.8	33.8	50.0	52.0	42.7	32.6	26.7	31.0
B (g/day)	1.26	0.85	1.04	1.15	1.07	0.88	0.68	0.61
C (days)	23.0	23.0	30.3	30.0	23.1	21.1	21.6	22.3
CD	0.989	0.988	0.997	0.995	0.994	0.987	0.992	0.992

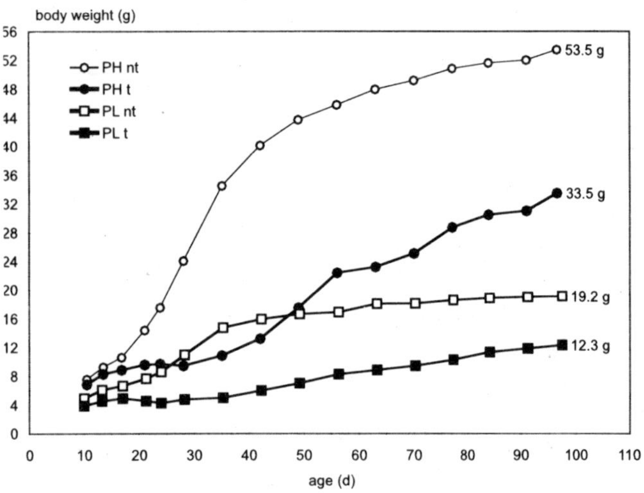

Figure 29 Growth of transgenic (t) and non-transgenic (nt) male mice from two selection lines: PH and PL (adapted from Bünger *et al.*, 1998b). PH and PL lines described in Expt II.

changes in these systems are not the only cause for line differences. The interaction between the transgene and the genetic background suggests that both systems are involved in the line divergence, but the interaction has been relatively small and therefore a polygenic model of selection response is supported (Bünger *et al.*, 1998b).

9. Summary

Long-term selection on high growth produces mice that can differ on average from their unselected control by a factor of up to 2.4 (**Figure 18**). For example, at 80 days these mice weigh about 86 g with a range of 74–103 g. In comparison, the corresponding unselected control mice weigh on average 34 g with a range of 30–37 g at 80 days. Selection for low body weights can result in a substantial body weight reduction to about 50% of the initial weight compared with an unselected control. Male body weights at 80 days were found to be between approximately 14 and 18 g (**Figures 16** and **17**), but may be approximately 2 g lower, when results of other selection experiments (**Table 3**) were considered. By introgression of the recessive *lit* mutation a further growth reduction in low lines by about 50% can be reached. Such mice have mature weights of about 10 g, are fertile and they reproduce. The realized heritability for body weights between 42 and 70 days was in the range of 0.36 to 0.53. Expressing the selection response in initial phenotypic and genetic standard deviations the selection response can produce deviations of up to 16 s_p and 22 s_g. A high correlation between the mean for body weight and the phenotypic variance (phenotypic variances changed in accordance with direction of selection) was found, accompanied by a relative constancy of the coefficient of variation, suggesting that body weights should be analysed on a log-transformed scale to account for scale effects (Falconer & Mackay, 1996). Approximate calculations of the number of effective genes for body weight resulted in estimates of about 200 to 300.

Effects of the knockout of major growth factors in selected lines were dependent on the genetic background (high lines reacted stronger than low lines) but these interactions were nevertheless relatively small and a polygenic model of selection response seems appropriate. However, in some selection lines single gene mutations and QTL with major effects have already been found and there are probably more to come as more selection lines are included. Interaction of these genes with the genetic background can be expected but are not yet well investigated. The availability of such divergent selection lines provides a unique opportunity for future mapping experiments and gene expression studies. In addition, further growth-promoting opportunities for the "highest high line", the DUH (considered as "reference line"), can be elucidated by marker-assisted introgression of discovered QTL, single gene mutations and of suitable transgenes.

Weight manipulations by the use of transgenes have shown impressive results, especially taking into consideration that the expression of only one hormone, although a major one with all its side effects, was manipulated. Body weights were nearly doubled in comparison to non-transgenic controls and using transgenic methods reductions of mature body weights to 9–12 g were reached, similar to the low limit after long term selection. The upper limits were still much lower than those produced by selection, where many genes are involved. On the other hand it seems unlikely that selection can result in such an overexpression, for example, of growth hormone by a factor of up to 700 as observed in transgenic animals, with all the known negative side effects (e.g. Wolf *et al.*, 1993; Wanke *et al.*, 1996). Growth-manipulated mice produced by long-term selection usually seem to have reduced fertility in addition to a shorter life expectancy but they appear more balanced than transgenic mice.

These results are of significant economical importance as they suggest that artificial selection, transgenic technology and environmental manipulation should be synergistic rather than antagonistic strategies. Research should take account of this by bringing these methods together on the same animal models.

Acknowledgement

The authors express their gratitude to William G. Hill, Eileen Wall and Charlotte Bruley for comments and suggestions on the manuscript. The recent work of LB on data analysis and on the manuscript was funded by the Biotechnology and Biological Science Research Council and by Cotswold Pig Development Company International.

For providing mice (line abbreviations as shown) from different selection lines (**Table 4**, **Figure 19**) we express our gratitude to:

- BE: Prof. W. Schlote, Humboldt-Universität Berlin, Landwirtschaftlich-Gärtnerische Fakultät, 14195 Berlin, BRD
- DAH: Prof. G.E. Bradford and Prof. J.F. Medrano, University of California, Department of Animal Science, Davis, CA 95616–8521, USA
- DUH: Dr Ulla Renne, Dr G. Dietl, Forschungsinstitut für

die Biologie landwirtschaftlicher Nutztiere, 18196 Dummerstorf, BRD

- MU: Prof. F. Pirchner, Technische Universität München, Lehrstuhl für Tierzucht, 85350 Freising-Weihenstephan, BRD
- RAH: Prof. E.J. Eisen, North Carolina State University, Department of Animal Science, Raleigh, NC 27695–7621, USA
- RO: Prof. G. Bulfield, Roslin Institute, Roslin, Midlothian EH25 9PS, UK

REFERENCES

Bakker, H. (1974) Effect of selection for relative growth rate and body weight of mice on rate, composition and efficiency of growth. *Mededelingen-Landbouwhogeschool-Wageningen* 74: 1–94

Barkemeyer, H., Horst, P. & Schlote, W. (1989) Antagonism between growth and fitness of mice as a consequence of long term selection for protein deposition (in German). *Zeitschrift für Tierzüchtung und Züchtungsbiologie* 106: 433–42

Barria, N. & Bradford, G.E. (1981) Long-term selection for rapid gain in mice. I. Genetic analysis at the limit of response. *Journal of Animal Science* 52: 729–38

Beniwal, B.K., Hastings, I.M., Thompson, R. & Hill, W.G. (1992) Estimation of changes in genetic parameters in selected lines of mice using REML with an animal model. 1. Lean mass. *Heredity* 69: 352–60

Benson, K.F. & Chada, K. (1994) Mini-mouse – phenotypic characterization of a transgenic insertional mutant allelic to pygmy. *Genetical Research* 64: 27–33

Bohren, B.B. (1975) Designing artificial selection experiments for specific objectives. *Genetics* 80: 205–20

Bradford, G.E. (1971) Growth and reproduction in mice selected for rapid body weight gain. *Genetics* 69: 499–512

Bradford, G.E. & Famula, T.R. (1984) Evidence for a major gene for rapid postweaning growth in mice. *Genetical Research* 44: 293–308

Brockmann, G.A., Haley, C.S., Renne, U., Knott, S.A. & Schwerin, M. (1998) Quantitative trait loci affecting body weight and fatness from a mouse line selected for extreme high growth. *Genetics* 150: 369–81

Bünger, L. (1987) Direct and correlated effects of long-term selection for different growth traits in laboratory mice (in German). *Genetische Probleme in der Tierzucht, FzT Dummerstorf-Rostock Heft* 13: 1–90

Bünger, L. & Herrendörfer, G. (1994) Analysis of a long-term selection experiment with an exponential model. *Zeitschrift für Tierzüchtung und Züchtungsbiologie* 111: 1–13

Bünger, L. & Hill, W.G. (1999a) Inbred lines derived from long-term divergent selection on fat content and body weight. *Mammalian Genome* 10: 645–48

Bünger, L. & Hill, W.G. (1999b) Role of growth hormone in the genetic change of mice divergently selected for body weight and fatness. *Genetical Research* 74: 351–60

Bünger, L., Schönfelder, E. & Schüler, L. (1982) Evaluation of stress tolerance in laboratory mice – ontogenesis of endurance fitness (in German). *Zoologische Jahrbücher-Abteilung Allgemeine Zoologie und Physiologie der Tiere* 86: 141–47

Bünger, L., Schüler, L., Kupatz, B. & Renne, U. (1983) Selection for growth in model animals (laboratory mice). 2. Direct selection response (in German). *Archiv für Tierzucht* 26: 281–93

Bünger, L., Renne, U., Dietl, G. & Kuhla, S. (1998a) Long-term selection for protein amount over 70 generations in mice. *Genetical Research* 72: 93–109

Bünger, L., Wallace, H., Bishop, J.O., Hastings, I.M. & Hill, W.G. (1998b) Effects of thyroid hormone deficiency on mice selected for increased and decreased body weight and fatness. *Genetical Research* 72: 39–53

Bünger, L., Hill, W.G., Laidlaw, A.H. *et al.* (2000) Inbred lines derived from long-term on growth selected mouse lines – unique resources for mapping growth related QTL. (*in preparation*)

Butler von, I. & Pirchner, F. (1983) Vergleich der Effizienz von Selektion innerhalb Familien und Massenselektion auf hohen Zuwachs in zwei verschiedenen Mäusepopulationen. *Züchtungskunde* 55: 241–46

Butler von, I., Willeke, H. & Pirchner, F. (1984) Two-way within-family and mass selection for 8-week body weight in different mouse populations. *Genetical Research* 43: 191–200

Caballero, A., Keightley, P.D. & Hill, W.G. (1995) Accumulation of mutations affecting body-weight in inbred mouse lines. *Genetical Research* 65: 145–49

Chai, C.K. (1966) Characteristics in inbred mouse populations plateaued by directional selection. *Genetics* 54: 743–53

Chen, W.Y., Wight, D.C., Wagner, T.E. & Kopchick, J.J. (1990) Expression of a mutated bovine growth-hormone gene suppresses growth of transgenic mice. *Proceedings of the National Academy of Sciences USA* 87: 5061–65

Chen, W.Y., White, M.E., Wagner, T.E. & Kopchick, J.J. (1991) Functional antagonism between endogenous mouse growth hormone (GH) and a GH analog results in dwarf transgenic mice. *Endocrinology* 129: 1402–08

Cheverud, J.M. & Routman, E. (1993) Quantitative trait loci – individual gene effects on quantitative characters – mini review. *Journal of Evolutionary Biology* 6: 463–80

Cheverud, J.M., Routman, E.J., Duarte, F.A.M. *et al.* (1996) Quantitative trait loci for murine growth. *Genetics* 142: 1305–19

Doolittle, D.P., Davisson, M.T., Guidi, J.N. & Green, M.C. (1996) Catalog of mutant genes and polymorphic loci. In *Genetic Variants and Strains of the Laboratory Mouse*, 3rd edition, edited by M.F. Lyon, S. Rastan & S.D.M. Brown, Oxford and New York: Oxford University Press

Eicher, E.M. & Beamer, W.G. (1976) Inherited ateliotic dwarfism in mice. *Journal of Heredity* 67: 87–91

Eicher, E.M. & Beamer, W.G. (1980) New mouse dw allele: genetic location and effects on lifespan and growth hormone levels. *Journal of Heredity* 71: 187–90

Eisen, E.J. (1975) Population size and selection intensity effects on long-term selection response in mice. *Genetics* 79: 305–23

Eisen, E.J. (1980) Conclusions from long-term selection experiments with mice. *Zeitschrift für Tierzüchtung und Züchtungsbiologie* 97: 305–19

Eisen, E.J. (1987) Effects of selection for rapid postweaning gain on maturing patterns of fat depots in mice. *Journal of Animal Science* 64: 133–47

Eisen, E.J., Croom, W.J. Jr & Helton, S.W. (1988) Differential response to the beta-adrenergic agonist cimaterol in mice selected for rapid gain and unselected controls. *Journal of Animal Science* 66: 361–71

Eisen, E.J., Fortman, M., Chen, W.Y. & Kopchick, J.J. (1993) Effect of genetic background on growth of mice hemizygotic for wild-type or dwarf mutated bovine growth-hormone transgenes. *Theoretical and Applied Genetics* 87: 161–69

Eklund, J. & Bradford, G.E. (1977) Genetic analysis of a strain of mice plateaued for litter size. *Genetics* 85: 529–42

Falconer, D.S. (1953) Selection for large and small size in mice. *Journal of Genetics* 51: 470–501

Falconer, D.S. (1955) Patterns of response in selection experiments with mice. *Cold Spring Harbor Symposia on Quantitative Biology* 20: 178–96

Falconer, D.S. (1960) Selection of mice for growth on high and low plans of nutrition. *Genetical Research* 1: 91–113

Falconer, D.S. (1973) Replicated selection for body weight in mice. *Genetical Research* 22: 291–321

Falconer, D.S. & Isaacson, J.H. (1959) Adipose, a new inherited obesity of the mouse. *Journal of Heredity* 50: 290–92

Falconer, D.S. & King, J.W.B. (1953) A study of selection limits in the mouse. *Genetics* 51: 561–71

Falconer, D.S. & Mackay, T.F.C. (1996) *Introduction to Quantitative Genetics*, 4th edition, Harlow: Longman

Frankham, R. (1980) Origin of genetic variation in selection lines [Drosophila]. In *Selection Experiments in Laboratory and Domestic Animals*, edited by A. Robertson, Slough: Commonwealth Agricultural Bureaux

Gille, U., Salomon, F.V., Rieck, O., Gericke, A. & Ludwig, B. (1996) Growth in rats (Rattus-norvegicus berkenhout). 1. Growth of body-mass – a comparison of different models. *Journal of Experimental Animal Science* 37: 190–99

Goodale, H.D. (1938) A study of the inheritance of body weight in the albino mouse by selection. *Journal of Heredity* 29: 101–12

Hastings, I.M. & Veerkamp, R.F. (1993) The genetic basis of response in mouse lines divergently selected for body weight or fat content. I. The relative contributions of autosomal and sex-linked genes. *Genetical Research* 62: 169–75

Hastings, I.M., Bootland, L.H. & Hill, W.G. (1993) The role of growth hormone in lines of mice divergently selected on body weight. *Genetical Research* 61: 101–06

Heath, S.C., Bulfield, G., Thompson, R. & Keightley, P.D. (1995) Rates of change of genetic-parameters of body-weight in selected mouse lines. *Genetical Research* 66: 19–25

Herre, W. & Röhrs, M. (1973) *Haustiere-zoologisch gesehen*, Jena: Gustav Fischer

Hill, W.G. (1985) Effects of population size on response to short and long-term selection. *Zeitschrift für Tierzüchtung und Züchtungsbiologie* 102: 161–73

Hill, W.G. & Mbaga, S.H. (1998) Mutation and conflicts between artificial and natural selection for quantitative traits. *Genetica* 103: 171–81

Hill, W.G., Caballero, A. & Keightley, P.D. (1994) Variation from spontaneous mutation for body size in the mouse. *Proceedings of the 5th WCGALP, Guelph* 19: 67–70

Horvat, S. & Medrano, J.F. (1996) The high growth (hg) locus maps to a deletion in mouse chromosome-10. *Genomics* 36: 546–49

Jansson, J.O., Downs, T.R., Beamer, W.G. & Frohman, L.A. (1986) Receptor-associated resistance to growth hormone-releasing factor in dwarf "little" mice. *Science* 232: 511–12

Jones, L.P., Frankham, R. & Barker, J.S.F. (1968) The effects of population size and selection intensity in selection for a quantitative character in Drosophila. II. Long-term response to selection. *Genetical Research* 12: 249–66

Kaiser, G. (1971) Die Reproduktionsleistung der Haushunde in ihrer Beziehung zur Körpergröße und zum Gewicht der Rassen. 1. Teil. *Zeitschrift für Tierzüchtung und Züchtungsbiologie* 88: 118–68

Keightley, P.D. & Bulfield, G. (1993) Detection of quantitative trait loci from frequency changes of marker alleles under selection. *Genetical Research* 62: 195–203

Keightley, P.D. & Hill, W.G. (1992) Quantitative genetic variation in body size of mice from new mutations. *Genetics* 131: 693–700

Keightley, P.D., Mackay, T.F.C. & Caballero, A. (1993) Accounting for bias in estimates of the rate of polygenic mutation. *Proceedings of the Royal Society of London, Series B* 253: 291–96

Keightley, P.D., Caballero, A. & Garciadorado, A. (1998) Population genetics: surviving under mutation pressure. *Current Biology* 8: R235

King, J.W.B. (1950) Pygmy, a dwarfing gene in the house mouse. *Journal of Heredity* 41: 249–52

King, J.W.B. (1955) Observations on the mutant "pygmy" in the house mouse. *Journal of Genetics* 53: 487–97

Kopchick, J.J., McAndrew, S.J., Shafer, A. *et al.* (1989) In vitro mutagenesis of the bovine growth hormone gene. *Journal of Reproduction and Fertility* (Suppl.) 41: 25–35

Kownacki, M. & Zuk, B. (1986) Populations size and response to selection for postweaning weight gain in mice. *Genetica Polonica* 27: 99–111

Kress, D.D. (1975) Results from long-term selection experiments relative to selection limits. *Genetics Lectures* 4: 253–71

Li, T., Crenshaw, E.B., Rawson, R. *et al.* (1990) Dwarf locus mutants lacking three pituitary cell types result from mutations in the POU-domain gene *pit-1*. *Nature* 347: 533

Lin, S.C., Lin, C.J.R., Gukovsky, I. *et al.* (1993) Molecular basis of the little mouse phenotype and implications for cell type-specific growth. *Nature* 364: 208–13

MacArthur, J.W. (1944a) Genetics of body size and related characters. *American Naturalist* 78: 224–37

MacArthur, J.W. (1944b) Genetics of body size and related characters. I. Selecting small and large races of the laboratory mouse. *American Naturalist* 78: 142–57

MacArthur, J.W. (1949) Selection for small and large body size in the house mouse. *Genetics* 34: 194–209

Mackay, T.F.C., Fry, J.D., Lyman, R.F. & Nuzhdin, S.V. (1994) Polygenic mutation in *Drosophila melanogaster* – estimates from response to selection of inbred strains. *Genetics* 136: 937–51

Meyer, H.H. & Bradford, G.E. (1974) Estrus, ovulation rate and body composition of selected strains of mice on ad libitum and restricted feed intake. *Journal of Animal Science* 38: 271–78

Morris, K.H., Ishikawa, A. & Keightley, P.D. (1999) Quantitative trait loci for growth traits in C57BL/6J × DBA/2J mice. *Mammalian Genome* 10: 225–28

Nagai, J., Lin, C.Y. & Sabour, P. (1993) Selection for increased adult body weight in mouse lines with and without the rat growth hormone transgene. *Zeitschrift für Tierzüchtung und Züchtungsbiologie* 110: 374–84

Nissly, S.P., Knazek, R.A. & Wolff, G.L. (1980) Somatomedin

activity in sera of genetically small mice. *Hormone and Metabolic Research* 12: 158–64

Nobentrauth, K., Naggert, J.K., North, M.A. & Nishina, P.M. (1996) A candidate gene for the mouse mutation tubby. *Nature* 380: 534–38

Palmiter, R.D., Brinster, R.L., Hammer, R.E. *et al.* (1982) Dramatic growth of mice that develop from eggs microinjected with metallothionein-growth hormone fusing genes. *Nature* 300: 611–15

Phillips, J.A., Beamer, W.G. & Bartke, A. (1982) Analysis of growth-hormone genes in mice with genetic-defects of growth-hormone expression. *Journal of Endocrinology* 92: 405

Pidduck, H.G. & Falconer, D.S. (1978) Growth hormone function in strains of mice selected for large and small size. *Genetical Research* 32: 195–206

Poiley, S.M. (1972) Growth tables for 66 strains and stocks of laboratory animals. *Laboratory Animal Science* 22: 759–79

Pomp, D., Nancarrow, C.D., Ward, K.A. & Murray, J.D. (1992) Growth, feed-efficiency and body-composition of transgenic mice expressing a sheep metallothionein 1a-sheep growth-hormone fusion gene. *Livestock Production Science* 31: 335–50

Rance, K.A., Heath, S.C. & Keightley, P.D. (1997a) Mapping quantitative trait loci for body weight on the X chromosome in mice. II. Analysis of congenic backcrosses. *Genetical Research* 70: 125–33

Rance, K.A., Hill, W.G. & Keightley, P.D. (1997b) Mapping quantitative trait loci for body weight on the X chromosome in mice. I. Analysis of a reciprocal F-2 population. *Genetical Research* 70: 117–24

Roberts, R.C. (1966a) The limits to artificial selection for body weight in the mouse. I. The limits attained in earlier experiments. *Genetical Research* 8: 347–60

Roberts, R.C. (1966b) The limits to artificial selection for body weight in the mouse. II. The genetic nature of limits. *Genetical Research* 8: 361–75

Roberts, R.C. (1967) The limits to artificial selection for body weight in the mouse. IV. Sources of new genetic variance – irradiation and out-crossing. *Genetical Research* 9: 87–98

Robertson, A. (1960) A theory of selection limits in artificial selection. *Proceedings of the Royal Society of London, Series B* 153: 234–49

Sarowsky, H.J. (1986) *BI-Lexikon-Hunderassen*, Leipzig: VEB Bibliographisches Institut

Schüler, L. (1985) The mice outbred strain Fzt: DU and its application in animal breeding research (in German). *Archiv für Tierzucht* 28: 357–63

Searle, T.W., Murray, J.D. & Baker, P.J. (1992) Effect of increased production of growth-hormone on body-composition in mice – transgenic versus control. *Journal of Endocrinology* 132: 285–91

Sedlag, U. (1986) Von der Größe der Tiere. *Wissenschaftliche Zeitschrift Ernst-Moritz-Arndt-Universitaet Greifswald, Mathematisch-Naturwissenschaftlich. Reihe* 35: 27–29

Shanahan, C.M., Rigby, N.W., Murray, J.D. *et al.* (1989) Regulation of expression of a sheep metallothionein 1a-sheep growth-hormone fusion gene in transgenic mice. *Molecular and Cellular Biology* 9: 5473–79

Sharp, G.L., Hill, W.G. & Robertson, A. (1984) Effects of selection on growth, body composition and food intake in mice. 1. Responses in selected traits. *Genetical Research* 43: 75–92

Shea, B.T., Hammer, R.E., Brinster, R.L. & Ravosa, M.R. (1990) Relative growth of the skull and postcranium in giant transgenic mice. *Genetical Research* 56: 21–34

Snell, G.D. (1929) Dwarf, a new Mendelian recessive character in the house mouse. *Proceedings of the National Academy of Sciences USA* 15: 733–34

Szabo, G., Dallmann, G., Müller, G. *et al.* (1998) A deletion in the myostatin gene causes the compact (Cmpt) hypermuscular mutation in mice. *Mammalian Genome* 9: 671–72

Valle Zarate, A., Horst, P. & Weniger, J.H. (1994) Antagonism between growth and productive adaptability in mice (in German). *Archiv für Tierzucht* 37: 185–98

Varga, L., Szabo, G., Darvasi, A. *et al.* (1997) Inheritance and mapping of compact (cmpt), a new mutation causing hypermuscularity in mice. *Genetics* 147: 755–64

Vila, C., Savolainen, P., Maldonado, J.E. *et al.* (1997) Multiple and ancient origins of the domestic dog. *Science* 276: 1687–89

Vize, P.D., Michalska, A.E., Ashman, R. *et al.* (1988) Introduction of a porcine growth-hormone fusion gene into transgenic pigs promotes growth. *Journal of Cell Science* 90: 295–300

Vogel, P. (1974) Kälteresistenz und versible Hypothermie der Etruskerspitzmaus. *Zeitschrift für Säugetierkunde* 39: 78–88

Wanke, R., Wolf, E., Brem, G. & Hermanns, W. (1996) Physiology and pathology of growth – studies in GH transgenic mice. *Zeitschrift für Tierzüchtung und Züchtungsbiologie* 113: 445–56

Weniger, J.H., Horst, P., Steinhauf, D. *et al.* (1974) Model experiments on selection for endurance and its relation to growth. Part I. Introduction, methods and preliminary investigations on the basic population. *Zeitschrift für Tierzüchtung und Züchtungsbiologie* 91: 265–70

Wilkinson, J.L. & Luxford, B.G. (1990) The effect of the transgene for porcine growth hormone on genetic and phenotypic parameters for growth and reproduction in mice. *Proceedings of the 4th WCGALP, Edinburgh* XIII: 333–36

Wilson, S.P., Goodale, H.D., Kyle, W.H. & Godfrey, E.F. (1971) Long term selection for body weight in mice. *Journal of Heredity* 62: 228–34

Winkelman, D.C. & Hodgetts, R.B. (1992) RFLPs for somatotropic genes identify quantitative trait loci for growth in mice. *Genetics* 131: 929–37

Witherden, A.S., Nicholson, S.J. & Fisher, E.M.C. (1997) The values of mouse mutants. *Mouse Genome* 95: 874

Wolf, E., Rapp, K., Wanke, R. *et al.* (1991) Growth characteristics of metallothionein-Human growth hormone transgenic mice as compared to mice selected for high eight-week body weight and unselected controls. II. Skeleton. *Growth, Development and Aging* 55: 237–48

Wolf, E., Kahnt, E., Ehrlein, J. *et al.* (1993) Effects of long-term elevated serum levels of growth-hormone on life expectancy of mice – lessons from transgenic animal-models. *Mechanisms of Ageing And Development* 68: 71–87

Wright, S. (1952) The genetics of quantitative variability. In *Quantitative Inheritance*, edited by E.C.R. Reeve & C.H. Waddington, London: Her Majesty's Stationery Office

Xiang, X., Benson, K.F. & Chada, K. (1990) Mini-mouse: disruption of the pygmy locus in a transgenic insertional mutant. *Science* 247: 967–69

GLOSSARY

BWx (body weight) body weight at x days of age

cumSD (cumulated SD) selection differential cumulated

over generations (measure of total amount of selection applied)

CV% (coefficient of variation) measures the variance in relation to the mean. CV% = mean/σ where σ is the standard deviation, or square root of the variance

HL (half life) time when the half of the maximum selection response (selection limit) is reached

h^2 heritability. Heritability is here used in the narrow sense, meaning the ratio between additive variance (V_A) and phenotypic variance (V_P): $h^2 = V_A/V_P$. It expresses the degree of resemblance between relatives and the extent to which phenotypes are determined by the genes transmitted from their parents

infinitesimal model genetic model used in animal breeding. It assumes that complex traits such as growth and carcass composition are controlled by a large number of unlinked genes each with only a small additive effect on the trait

N_e effective population size. N_e is the effective number of breeding animals or number of individuals that would give rise a calculated rate of inbreeding, if they bred as in an idealized population. $N_e = 1/2\Delta F$ with ΔF = the rate of inbreeding

ov, fv (observed and fitted values) fitted value obtained by fitting model (1) or the logistic model if used to fit a growth curve

realized heritability estimate derived from a selection experiment, expressing the selection response in relation to the amount of selection applied (SD)

SR (selection response) change of the population mean produced by selection

SD (selection differential) amount of selection applied, which measures the average superiority of the individuals selected as parents: SD = mean of selected individuals – mean of all individuals

Falconer's classic book (Falconer & Mackay, 1996) should be referred to for further explanation of statistical methods of analysis presented and of terms not included in this list.

G GENETICS OF OTHER MAMMALS

Introduction to Genetics of other Mammals

Although the human genome was the first vertebrate genome to be sequenced, no mammal has been studied as thoroughly as the mouse, as described in the articles in the previous section. Genomic projects in other mammals are less advanced, but should benefit from comparative mapping with the mouse and human genomes. For livestock animals there is the economic importance of improving disease resistance, wool, meat, milk and also understanding the genetics of prion diseases, for example scrapie in sheep and BSE in cattle.

1. Horse genetics

It is surprising that although stud books of several breeds of horse go back more than 200 years, work on horse genetics lags far behind that on other companion and production animals. There are obviously genetic variations affecting performance, coat colour and health (including genetic diseases) of different types of racehorse; temperament will also be affected by genes as well as by training. The horse has 31 pairs of autosomes and the usual large X and small Y chromosomes that determine sex: all of these are large enough to be individually recognizable, in contrast to the group of very small chromosomes in the dog genome. The International Equine Gene Mapping Workshop (IEGMW) was formed in 1995, and genetic maps (Raudsepp *et al.*, 1996; Marti & Binns, 1998) show that there is strong conservation of gene order between horse and human genomes. The review article by Marti & Binns illustrates the relationships between human and equine chromosomal regions, and gives references to studies on inherited diseases, coat colour and racing performances.

2. Sheep and other livestock

Dolly, a lamb cloned from the DNA of an adult sheep by the technique of nuclear cell transfer (Campbell *et al.*, 1996), became a media vehicle for debates about the ethics and potential of cloning and genetic modification, though Dolly was not herself genetically modified. Subsequent births of cloned lambs that had been genetically engineered to produce therapeutically useful proteins in their milk show the potential for transgenic animals as pharmacological sources, and could eventually lead to transgenic organs for xenotransplantation. Sheep genetic research at the Roslin Institute where these experiments were carried out has also provided the opportunity to test the telomere hypothesis of ageing (Sheils *et al.*, 1999).

The Roslin Institute is one of several institutes that are carrying out genome mapping of sheep, pigs, chickens and cattle, all with the aim of identifying QTLs or genetic markers for growth and other economically important production traits.

REFERENCES

Campbell, K.H.S., McWhir, J., Ritchie,W.A. & Wilmut, I. (1996) Sheep cloned by nuclear transfer from a cultured cell line. *Nature* 380: 64–66

Marti, E. & Binns, M. (1998) Horse genome mapping: a new era in horse genetics? *Equine Veterinary Journal* 30: 13–17

Raudsepp, T., Frönicke, L., Scherthan, H., Gustavsson, I. & Chowdhary, B.P. (1996) "Zoo-FISH delineates conserved chromosomal segments in horse and man. *Chromosome Research* 4: 1–8

Sheils, P.G. *et al.* (1999) Analysis of telomere length in cloned sheep. *Nature* 399: 316–17

LARGE TRANSGENIC ANIMALS – THEIR MAKING AND THEIR USE

Mathias Müller[1] and Gottfried Brem[2]
[1]*Institute for Animal Breeding and Genetics, UVW, Vienna, Austria*
[2]*Ludwig Boltzmann Institute for Immuno-, Cyto- and Molecular Genetics, Vienna, Austria*

1. Introduction

The term "transgenic" (Gordon & Ruddle, 1981; Palmiter & Brinster, 1985) refers to organisms which carry in their genome, and/or express, new genetic material. The extraneous DNA is usually referred to as the gene construct and encompasses the elements controlling gene expression (5′ promoter region, 3′ control regions) and the sequences (cDNA, genomic DNA) encoding the transgene product. For animal breeding, the three crucial aspects of transgenesis are integration, expression and transmission of the gene construct(s), i.e. gene transfer into the germline. Somatic gene transfer approaches result in (mostly transient) gene expression, with the longest duration being a life span. The main use of this technology is human gene therapy; in some cases, however, it might be beneficial in animal production (e.g. genetic vaccination; Beard & Mason, 1998).

Four main routes to transgenesis in mammals have been described:

1. integration of (retro)viral vectors into an early embryo;
2. microinjection of DNA into the pronucleus of a fertilized oocyte;
3. incorporation of genetically manipulated embryonic stem (ES) cells into an early embryo; and
4. transfer of genetically altered nuclei into enucleated oocytes.

The generation of transgenic farm animals was first reported in the mid-1980s (Brem *et al.*, 1985; Hammer *et al.*, 1985). The gene transfers were carried out by the microinjection of DNA constructs into the pronuclei of fertilized oocytes. This technique still represents the method of choice for generating large transgenic animals (for review, see Brem & Müller, 1994). Other methods such as using sperm cells or liposomes as a DNA vehicle have not yet been established for practical use (reviewed by Brem, 1993). Depending on the species used for gene transfer, the efficiency (transgenic newborns/microinjected zygotes) is routinely about 0.5–3%. Recent experiments using improved gene transfer protocols have given much higher total efficiencies. Approaches to increase the efficiency of transferring gene constructs into farm animal species mainly aim at the improvement of the collection and culture of early embryonic stage embryos, the microinjection procedure (including the microinjection solution) and the methods for embryo transfer. The main steps in reproductive and molecular biology required for the establishment of transgenic mammals are depicted in **Figure 1** and described in greater detail by Brem and Müller (1994).

In mice, the handling of pluripotential ES cell lines has become a routine method for altering the genome. The establishment of ES cell lines in farm animals is eagerly awaited, but cannot be guaranteed for most of these species in the near future. Pluripotent ES cells able to contribute to the germline have so far only been derived in mice. Recently, there have been descriptions of chimeric farm animals generated with ES-like cells and primordial germ cells, but, so far, no evidence of germline transmission has been published (Anderson, 1999). Cow foetuses obtained following the nuclear transfer of ES cell-derived nuclei were reported to give birth to living calves (Sims & First, 1994). In a recent breakthrough, three healthy lambs were born from embryos generated by the transfer of nuclei isolated from embryo-derived epithelial cells into enucleated oocytes (Campbell *et al.*, 1996). Moreover, the nuclear transfer technique was extended to differentiated cells resulting in the birth of the clone sheep "Dolly", the first report of successful cloning from adult cells (Wilmut *et al.*, 1997). Differentiated cells can be more easily maintained

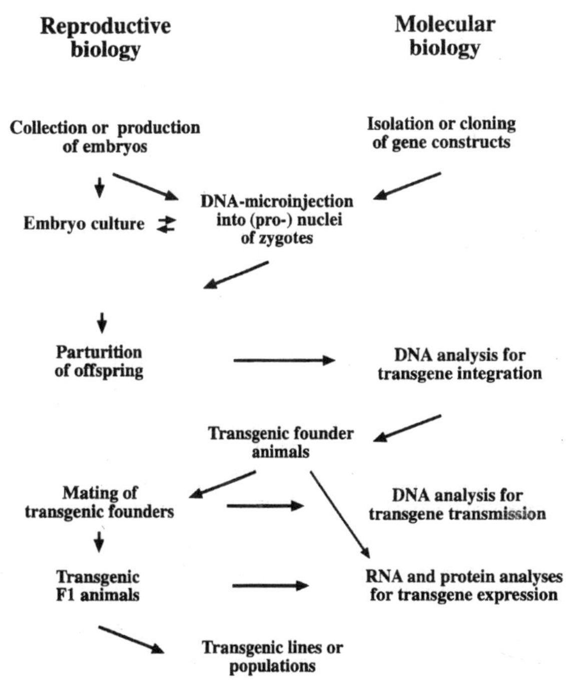

Figure 1 Steps and techniques in reproductive and molecular biology required for the production of large transgenic animals.

under *in vitro* culture conditions and can be altered by conventional transfection technologies. Hence, the first transgenic sheep and cattle using transformed foetal fibroblast as donor nuclei for nuclear transplantation were reported soon after the landmark experiments of the group headed by Ian Wilmut (Schnieke *et al.*, 1997; Cibelli *et al.*, 1998). One of the most important advantages of this technique is the possibility of insertion (additive gene transfer, gain of function), replacement (functional exchange, knockin) or removal (deletive gene transfer, loss of function, knockout) of the gene(s) of interest *in vitro*. In addition, the transgenesis by nuclear transfer ensures a more consistent and controllable expression pattern in transgenic animals and may overcome the obstacle of transgenesis by microinjection, i.e. low efficiency.

Gene transfer into farm animals focuses on: (a) the improvement of productivity traits; (b) production of heterologous proteins in animals ("gene pharming"); (c) the establishment of animal homologues for human diseases; and (d) the production of organs for xenotransplantation.

2. Improvement of productivity traits

Transgenic approaches to improve the efficiency and quality of animal production aim at: (a) the production performance (meat, milk, wool, etc.), (b) the reproduction performance, (c) health traits (disease resistance), (d) the quality of animal products and (e) the processing of animal products (reviewed by Pursel, 1998).

(i) Growth performance

Initially, gene transfer experiments concentrated on the change of growth performance and carcass composition in farm animals. Growth is a very complex process which is influenced by the interactions of hormones and autocrine/paracrine factors, nutritional conditions and environmental factors set against a discrete genetic background. Among the genetically determined factors, the genes encoding polypeptides of the growth hormone cascade are of particular interest. The positive-acting growth hormone-releasing hormone (somatoliberin) and its antagonist, somatotropin release-inhibiting factor (somatostatin), control the production of growth hormone (GH) and somatotropic hormone (somatotropin). The GH action is highly dependent on the metabolic situation of the organism; low blood glucose levels result in catabolic effects (lipolysis) and a positive energy balance causes anabolic effects which are mainly governed by insulin-like growth factor 1 (IGF-1, somatomedin C).

Gene constructs consisting of the genes of the GH cascade controlled by various regulatory elements (promoters) have been used to generate transgenic farm animals. The majority of experiments have been carried out in pigs, but GH-transgenic pigs did not *a priori* show increased growth performance. This was due to reduction in the food uptake combined with an increased utilization of nutrients in these animals. Only after the feeding of a special protein-enriched diet did GH-transgenic pigs attain a higher (15%) daily weight gain than control animals (reviewed by Pursel *et al.*, 1989). In terms of carcass composition, the transgenic pigs showed a massive reduction of the back fat thickness. It turned out that constitutive and/or high-level expression of GH in pigs caused a variety of pathological side-effects, including nephritis, cardiomegaly, synovitis, gastritis, dermatitis, pneumonia and reduced fertility. In additional experiments, promoter elements were used which provided lower constitutive or inducible production of GH. This resulted in the desired increase in carcass leanness and a reduction of the detrimental side-effects (reviewed by Müller & Brem, 1996a; Nottle *et al.*, 1999).

In order to overcome the problems of systemic overexpression of growth-promoting genes, Pursel *et al.* (1999) directed the expression of IGF-1 into skeletal muscle by using an appropriate tissue-specific promoter. No increased serum levels of GH or IGF-1 were observed; the expected stimulation of myofibre growth remains to be examined. Transgenic ruminants (cattle, sheep, goat) carrying growth-promoting genes have also been generated. However, no positive effects on growth performance or carcass composition have been reported (reviewed by Müller & Brem, 1996a).

An alternative approach to altering the growth performance involves the differentiation process of muscle cells themselves. For example, the chicken *c-ski* proto-oncogene was found to induce myogenic differentiation, and this muscle differentiation gene was introduced into pigs and cattle (Pursel *et al.*, 1992; Bowen *et al.*, 1994). However, as observed with the GH cascade genes, no or mainly deleterious effects were reported. Accepting the complexity of exogenous factors influencing the growth of an individual, and the fine interplay of endogenous growth-promoting and inhibiting factors, this is not completely surprising. Considering the difficulties experienced in transferring growth-related genes into farm animals and the level of growth performance and carcass leanness reached by conventional breeding programmes (at least for middle European livestock species), gene transfer approaches aimed at growth promotion in meat production do not have a high, if any, priority. There may be a possible application in countries and areas where the genetic potential of growth in animal production has not reached a plateau by breeding and where the import and husbandry of highly productive breeds is hampered by environmental or other factors.

(ii) Disease resistance

An "indirect" approach to enhance the growth performance is the improvement of disease resistance by gene transfer (Müller & Brem, 1991, 1996a,b, 1998). In contrast to the selection for production performance, attempts to select for improved disease resistance by conventional breeding programmes have not been successful. The reduction of disease susceptibility of livestock would be a benefit in terms of animal welfare and would also be of economic importance – improved health status in animal production results in improved production and reproduction performance. **Figure 2** shows the theoretically possible approaches to influence disease resistance by transferring DNA constructs.

The term "intracellular immunization" (Baltimore, 1988) was originally used for the overexpression in the host of an

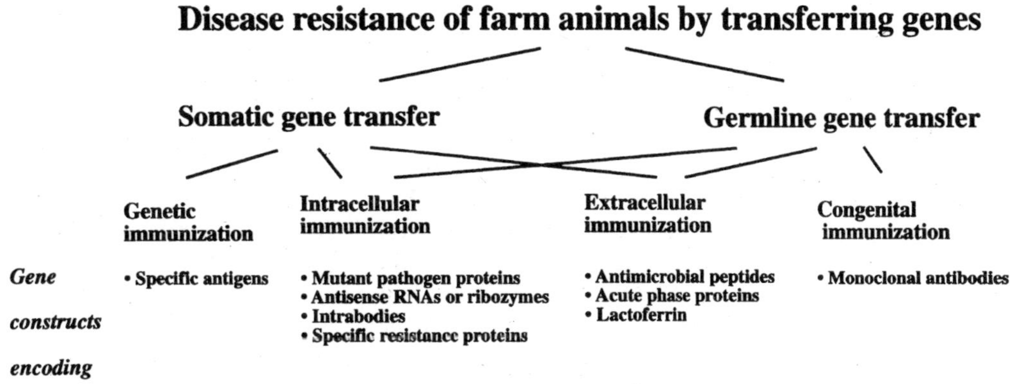

Figure 2 Gene transfer strategies aiming at the improvement of disease resistance. The possibility of disrupting genes known to confer specific susceptibility to diseases is not depicted since homologous recombination ("gene knockout") in totipotential embryonic cells of large mammals is currently not feasible.

aberrant form (dominant–negative mutant) of a viral protein that is able to interfere strongly with the replication of the wild-type virus. This definition was then extended to all approaches based on intracellular expression of transgene products which inhibit the replication of pathogens in host organisms (**Figure 2**) (see Müller & Brem (1998) and references therein). Both somatic and germline gene transfer can be applied; with respect to introducing new disease-resistance traits in farm animals, germline transmission is required. Initial studies in farm animals included the "classical" approach of overexpression of a viral protein in transgenic sheep, transgenic rabbits expressing antisense constructs complementary to adenovirus RNA and the transfer of the specific disease-resistance gene *Mx1* of mice, which confers influenza resistance, into swine (see Müller *et al.* (1992b) and Müller & Brem (1998) and references therein). The integration, stable germline transmission and, in some cases, the expression of the transferred genes could be demonstrated. However, the final proof of a successful "intracellular immunization" of farm animals (i.e. the challenge with an infectious pathogen) has not been reported. Nevertheless, the transfer of the mouse *Mx1* gene into swine is a challenging task since pigs are not only affected by swine influenza epidemics, but are also known as mixed vessels for the creation of new pandemic strains, since they are infected by human as well as avian influenza strains.

Mx1 gene constructs have been extensively studied in transgenic mice, where they conferred increased resistance to influenza virus, tick-borne influenza-like virus and vesicular stomatitis virus (Arnheiter *et al.*, 1996). Pigs have *Mx* genes of their own (Müller *et al.*, 1992a), but they are susceptible to influenza and provide a substantial reservoir for swine influenza viruses. Different gene constructs containing the mouse *Mx1* cDNA controlled by two constitutive promoters and the inducible murine *Mx1* promoter were transferred into swine. The results of the gene transfer experiments with the constitutive promoters indicated that permanent high-level synthesis of Mx1 might be toxic for the organism and not compatible with embryonic development (Müller *et al.*, 1992b). This was also observed for constitutive *Mx1* expression in transgenic mice (Arnheiter *et al.*, 1996). The use of the inducible *Mx1* gene construct resulted in two transgenic pig lines which expressed inducible mouse *Mx1* mRNA. However, extensive protein analysis did not detect mouse Mx1 protein in transgenic pigs. The underlying reasons may be transgene rearrangement or, more likely, the possibility that the mouse *Mx1* promoter is not functioning optimally in pig cells. Mx proteins are host-derived antiviral proteins synthesized in most vertebrates. Thus, the concept of intracellular immunization by the expression of Mx proteins might only lead to enhanced resistance if the susceptibility of an animal to infection is, in fact, due to a comparable weak Mx system. For future Mx gene transfers into pigs, this point has to be rigorously clarified. In addition, future gene constructs should guarantee a tightly controlled transgene expression only at required times and sites. This could be achieved, for example, by using one of the endogenous porcine *Mx* promoters.

"Congenital immunization" is defined as transgenic expression and germline transmission of a gene encoding an immunoglobulin specific for a pathogen and, therefore, providing congenital immunity without prior exposure to that pathogen. The approach was tested in farm animals by expressing the gene constructs encoding mouse monoclonal antibodies in transgenic rabbits, pigs and sheep (Lo *et al.*, 1991; Weidle *et al.*, 1991). Both experiments resulted in transgene expression, but also revealed some unexpected findings (e.g. aberrant sizes of the transgenic antibody or little antigen binding capacity). It remains to be investigated, however, whether or not the efforts required for optimizing the concept of congenital immunization are justified by its benefits in terms of increasing disease resistance in a certain species. Following this route, one has also to keep in mind that a given infectious pathogen will readily be able to escape the transgenic animal's immunity by changing its antigenic determinants. However, for the use of farm animals in gene pharming, the transfer of antibody-encoding genes allows the large scale production of therapeutic antibodies for human medicine (Logan, 1993; Brüggemann & Neuberger, 1996).

The concept of "genetic immunization" (Tang *et al.*, 1992) is based on somatic gene transfer and utilizes antigen-encoding gene constructs rather than antigens *per se* for vaccination. DNA constructs have been delivered to individuals via several routes, including intramuscular, intravenous, intradermal and needleless (jet) injections and microbombardment (Beard & Mason, 1998). DNA vaccines have the great advantage of avoiding the time-consuming steps for antigen purification. Therefore, it is possible to react more directly to antigenic variations. DNA constructs can be used for prophylactic vaccination against infections as well as for large-scale production of polyclonal antibodies (reviewed by Waine & Mcmanus, 1995).

"Extracellular immunization" refers to transgene products which exhibit their antipathogenic activity extracellularly. This strategy includes the expression of cytokines, antimicrobial peptides (magainins, defensins, melittins and cecropins) and other pathogen defence molecules, including lactoferrin and acute phase proteins (reviewed in Müller & Brem, 1998).

(iii) Wool production

Additional strategies in the genetic engineering of production traits concentrate on the modification of animal biochemistry by the introduction of novel or altered biosynthetic pathways (Pursel, 1998; Ward *et al.*, 1999). The most advanced of these approaches is targeting of gene expression to the wool follicle and the improvement of wool production in transgenic sheep (Damak *et al.*, 1996; Bawden *et al.*, 1998).

3. Gene pharming – production of proteins of high value in transgenic farm animals

Recombinant DNA technology has provided the possibility of producing specific proteins in high quantity at relatively low cost in bacteria or yeast expression systems. Using prokaryotic or primitive eukaryotic production systems has a variety of disadvantages. They frequently lack the ability to perform the required post-translational modifications of the recombinant protein, which in turn is necessary for biological activity, and intracellular inclusion bodies or protein degradation sometimes make the isolation and purification of the recombinant protein difficult. To avoid some of these problems, higher eukaryotic production systems were established by using genetically transformed human and animal cells. In addition, the possibility of establishing farm animals has opened up new ways for the production of recombinant proteins by a method called gene pharming (Logan, 1993; Echelard, 1996). The idea is to produce recombinant proteins in specified organs or body fluids of transgenic animals; the mammary gland is the most interesting organ for the production of recombinant proteins (Clark *et al.*, 1987). The potential of large transgenic animals, especially ruminants, to be bioreactors is discussed by Whitelaw in the article on "Pharmaceutical proteins from milk of transgenic animals".

The choice criteria for selecting the most suitable species for gene pharming are usually based on the quantity of protein needed per year. A simplified rule is: the production of a protein in tons should be carried out by cows, in hundreds of kg by sheep or goats, and in kg per year by rabbits. Pigs are the species of choice for expression of foreign proteins in blood or somatic cells.

Rabbit appears more and more to be an intermediate animal, well adapted for the production of limited amounts of protein. Rabbit husbandry can be done under specific pathogen-free conditions, and rabbits have a short generation time so that transgenic founders can be generated with a reasonable efficiency. The endogenous milk proteins are well characterized and milking can be performed semiautomatically resulting in a milk yield of 10 kg per rabbit per year (Brem *et al.*, 1994; Besenfelder *et al.*, 1998). Thus, considering both economical and hygienic aspects, the species is suitable for gene pharming. Mammary-gland-specific expression of recombinant proteins in rabbits has been achieved with gene constructs driven by various mammary-gland-specific promoters (reviewed by Besenfelder *et al.*, 1998).

Expression of recombinant proteins in the blood is only feasible for proteins of which the biological activity of abnormally high levels is not detrimental to the health of the animal (Logan, 1993). This approach is applicable to produce human antibodies (Brüggemann & Neuberger, 1996), human haemoglobin (Hb) or inactive fusion proteins to be cleaved to their biologically active forms *ex vivo*. The production of human Hb in the erythrocytes of transgenic swine was achieved by combining the locus control region (LCR) of the human β-globin locus and the genes encoding the human α- and β-chains of Hb (reviewed by Logan, 1993). The human Hb purified from porcine proteins (including porcine Hb) exhibited similar oxygen equilibrium curves to Hb derived from human serum. However, the level of expression was low, with recoverable human Hb constituting only 10% of the total Hb in the porcine erythrocytes. The use of an isologous porcine promoter instead of the human regulatory sequences has resulted in swine with an expression level of 24% recoverable human Hb (Sharma *et al.*, 1994). Although more difficult than collecting milk, blood can be collected aseptically and at an earlier stage and is appropriate with a large number of animals. Thus, transgenic swine might serve as an alternative to human erythrocytes as a source of human Hb.

4. Animal genetic homologues for human diseases

Transgenic animals have proven to be an invaluable tool in the studies of human genetic and infectious disorders and of cancer. Livestock species are often good animal models of human disease. The effects of a given transgene *in vivo* are studied in transgenic farm animals (Mullins & Mullins, 1996) in addition to transgenic laboratory animals – mice and rats – in order to evaluate the observed results for the application in other species. The appropriate species for a genetic homologue of human disease should be the species which best models the organ or system under consideration. Furthermore, farm animals have a more diverse genetic background than inbred and outbred mouse strains. This might be favourable when studying complex disease models or therapeutic applications, since it resembles more accurately the genetic background in humans. Due to limitations in efficiency and cost-effectiveness, transgene technology in biomedical research with farm animals mainly concentrates on rabbits (Besenfelder *et al.*, 1998).

5. Transgenic pigs as organ donors

Many groups are currently studying the transplantation of organs across species (xenografts) to eliminate the shortage of organs available for transplantation. For a variety of medical, practical, ethical and economic reasons, the animal of choice in these studies is the pig (reviewed by Cozzi & White, 1995; Platt, 1996). According to current understanding, the major barrier to successful xenotransplantation is hyperacute rejection, which is, in part, mediated by the deposition of high-titre preformed natural antibodies that activate the complement system. It has been established that these preformed, naturally occurring, antipig antibodies react with a single disaccharide, Galα-1,3-Gal (α-galactosyl epitope) (reviewed by Squinto, 1996). This epitope is not expressed in Old World primates and humans because the α-1,3-galactosyltransferase is inactive in these species. The identification of this major xenogenic epitope led to the working hypothesis that genetic modification of the donor xenograft by stably downregulating the expression of the Galα-1,3-Gal epitope would significantly inhibit natural antibody binding and thus the activation of the complement cascade resulting in the hyperacute rejection. The α-1,3-galactosyltransferase would be an ideal target for producing "knockout" pigs. Gene targeting in pigs now can be performed since both the availability of pluripotential cells and cloning by nuclear transfer has been reported (Müller *et al.*, 1999; Onishi *et al.*, 2000; Polejaeva *et al.*, 2000). An alternative method involves the introduction of a gene for another enzyme (α-1,2-fucosyltransferase) competing with α-1,3-galactosyltransferase for adding a terminal sugar onto a carbohydrate chain. By overexpression of this enzyme, the synthesis of the xenogenic disaccharide is expected to be suppressed. Recently, tissue culture experiments and preliminary experiments in transgenic mice and pigs demonstrated convincingly the feasibility of this approach (for reviews, see Cozzi & White, 1995; Platt, 1996; Squinto, 1996). The complement cascade is downregulated at specific points by regulators such as decay-accelerating factor (DAF, CD55), membrane cofactor protein (MCP, CD46) and/or CD59. These proteins are collectively referred to as regulators of complement activation (RCAs) (reviewed by Cozzi & White, 1995). Transgenic swine expressing a particular RCA or combinations of RCAs have been generated. Transgenic pig organs perfused by human blood or transplanted into primates indicated the potential of this approach to trick the human immune response into seeing swine organs as human (reviewed in Cozzi & White, 1995; Platt, 1996; Squinto, 1996). The results reported to date look promising, although, prior to routine xenotransplantation, many problems, not only in basic research but also in ethical aspects, remain to be solved.

6. Conclusions

The transfer of genetic material by recombinant DNA technologies has a plethora of applications in the fields of basic medical research. The most prominent areas include animal homologues of human diseases and studies of function and regulation of specific genes. Most gene transfer experiments in livestock have been performed with swine. This is due in part to the economic importance of pig production in many countries and in part to the reproductive nature of pigs. Transgenic ruminants and rabbits show great potential in the production of proteins of high value in the mammary gland. Current limitations of the transgene technology for livestock include the still unsatisfactory efficiency and the lack of knowledge in multigene interactions. Recently, the first transgenic farm animals have been reported carrying large DNA fragments cloned into a yeast artificial chromosome vector (Langford *et al.*, 1994; Brem *et al.*, 1996). The feasibility of transferring gene constructs comprising large regions of the regulatory sequences is an important step in guaranteeing the strict spatial–temporal expression of the transgene. In addition, new gene transfer techniques (i.e. retroviral vector-driven gene insertion into oocytes (Chan *et al.*, 1998), and especially the above-mentioned nuclear transfer methods) promise to overcome the limitations of generating transgenic farm animals.

The technique of cloning by nucleus transfer enabled the first gene homologous recombination experiments in a species other than mouse (McCreath *et al.*, 2000). Therefore, in the near future it will be possible to alter the genome of farm animals by targeted introduction of desired genes into predetermined chromosomal sites.

In contrast to the situation of genetic engineering in plant breeding, transgenic livestock cannot replace the entire population within a couple of years even if the transgenic founder animal generated is of high value. It is predictable that the use of genetic engineering in livestock production will focus on highly specialized products, which will not appear in the normal agricultural food production chain.

REFERENCES

Anderson, G.B. (1999) Embryonic stem cells in agricultural species. In *Transgenic Animals in Agriculture*, edited by J.D. Murray, G.B. Anderson, A.M. Oberbauer & M.M. McGloughlin, Wallingford and New York: CABI

Arnheiter, H., Frese, M., Kambadur, R., Meier, E. & Haller, O. (1996) Mx transgenic mice: animal models of health. In *Transgenic Models of Human Viral and Immunological Disease*, edited by F.V. Chisari & M.B.A. Oldstone, Berlin: Springer

Baltimore, D. (1988) Intracellular immunization. *Nature* 335: 395–96

Bawden, C.S., Powell, B.C., Walker, S.K. & Rogers, G.E. (1998) Expression of a wool intermediate filament keratin transgene in sheep fibre alters structure. *Transgenic Research* 7: 273–87

Beard, C.W. & Mason, P.W. (1998) Out on the farm with DNA vaccines. *Nature Biotechnology* 16: 1325–28

Besenfelder, U., Aigner, A., Müller, M. & Brem, G. (1998) Generation and application of transgenic rabbits. In *Microinjection and Transgenesis: Strategies and Protocols*, edited by A. Cid-Arregui & A. García-Carrancá, Berlin and New York: Springer

Bowen, R.A., Reed, M.L., Schnieke, A. *et al.* (1994) Transgenic cattle resulting from biopsied embryos: expression of c-ski in a transgenic calf. *Biology of Reproduction* 50: 664–68

Brem, G. (1993) Transgenic animals. In *Biotechnology: A Multi-volume Comprehensive Treatment*, 2nd edition, edited by H.J. Rehm, G. Reed, A. Pühler, & P. Stadler, Weinheim and New York: VCH

Brem, G. & Müller, M. (1994) Large transgenic animals. In *Animals with Novel Genes*, edited by N. Maclean, Cambridge and New York: Cambridge University Press

Brem, G., Brenig, B., Goodman, H.M. *et al.* (1985) Production of transgenic mice, rabbits and pigs by microinjection into pronuclei. *Zuchthygiene* 20: 251–52

Brem, G., Hartl, P., Besenfelder, U. *et al.* (1994) Expression of synthetic cDNA sequences encoding human insulin-like growth factor-1 (IGF-1) in the mammary gland of transgenic rabbits. *Gene* 149: 351–55

Brem, G., Besenfelder, U., Aigner, B. *et al.* (1996) YAC transgenesis in farm animals: rescue of albinism in rabbits. *Molecular Reproduction and Development* 44: 56–62

Brüggemann, M. & Neuberger, M.S. (1996) Strategies for expressing human antibody repertoires in transgenic mice. *Immunology Today* 17: 39–45

Campbell, K.H.S., McWhir, J., Ritchie, W.A. & Wilmut, I. (1996) Sheep cloned by nuclear transfer from a cultured cell line. *Nature* 380: 64–66

Chan, A.W.S., Homan, E.J., Ballou, L.U., Burns, J.C. & Bremel, R.D. (1998) Transgenic cattle produced by reverse-transcribed gene transfer in oocytes. *Proceedings of the National Academy of Sciences USA* 95: 14028–33

Cibelli, J.B., Stice, S.L., Golueke, P.J. *et al.* (1998) Cloned transgenic calves produced from nonquiescent fetal fibroblasts. *Science* 280: 1256–58

Clark, A.J., Simons, P., Wilmut, I. & Lathe, R. (1987) Pharmaceuticals from transgenic livestock. *Trends in Biotechnology* 5: 20–24

Cozzi, E. & White, D.J.G. (1995) The generation of transgenic pigs as potential donors for humans. *Nature Medicine* 1: 964–66

Damak, S., Jay, N.P., Barrell, G.K. & Bullock, D.W. (1996) Targeting gene expression to the wool follicle in transgenic sheep. *Bio/Technology* 14: 181–84

Echelard, Y. (1996) Recombinant protein production in transgenic animals. *Current Opinion in Biotechnology* 7: 536–40

Gordon, J.W. & Ruddle, F.H. (1981) Integration and stable germ line transmission of genes integrated into mouse pronuclei. *Science* 214: 1244–46

Hammer, R.E., Pursel, V.G., Rexroad, C.E. Jr *et al.* (1985) Production of transgenic rabbits, sheep and pigs by microinjection. *Nature* 315: 680–83

Langford, G.A., Yannoutsos, N., Cozzi, E. *et al.* (1994) Production of pigs transgenic for human decay accelerating factor. *Transplantation Proceedings* 26: 1400–01

Lo, D., Pursel, V., Linto, P.J. *et al.* (1991) Expression of mouse IgA by transgenic mice, pigs and sheep. *European Journal of Immunology* 21: 25–30

Logan, J.S. (1993) Transgenic animals: beyond 'funny milk'. *Current Opinion in Biotechnology* 4: 591–95

McCreath, K.J., Howcroft, J., Campbell, K.H.S. *et al.* (2000) Production of gene-targeted sheep by nuclear transfer from cultured somatic cells. *Nature* 405: 1066–69

Müller, M. & Brem, G. (1991) Disease resistance in farm animals. *Experientia* 47: 923–34

Müller, M. & Brem, G. (1996a) Approaches to influence growth promotion of farm animals by transgenic means. In *Proceedings of the Scientific Conference on Growth Promotion in Meat Production*, edited by the European Commission, Luxembourg: Office for Official Publications of the European Communities

Müller, M. & Brem, G. (1996b) Intracellular, genetic or congenital immunisation – transgenic approaches to increase disease resistance of farm animals. *Journal of Biotechnology* 44: 233–42

Müller, M., & Brem, G. (1998) Transgenic approaches to the increase of disease resistance in farm animals. *Revue Scientifique et Technique O.I.E.* 17: 365–78

Müller, M., Winnacker, E.-L. & Brem, G. (1992a) Molecular cloning of porcine Mx cDNAs: new members of a family of interferon-inducible proteins with homology to GTP binding proteins. *Journal of Interferon Research* 12: 119–29

Müller, M., Brenig, B., Winnacker, E.L. & Brem, G. (1992b) Transgenic pigs carrying cDNA copies encoding the murine Mx1 protein which confers resistance to influenza virus infection. *Gene* 121: 263–70

Müller, S., Prelle, K., Rieger, N. *et al.* (1999) Chimeric pigs following blastocyst injection of transgenic porcine primordial germ cells. *Molecular Reproduction and Development* 54: 244–54

Mullins, L. & Mullins, J. (1996) Transgenesis in the rat and larger mammals. *Journal of Clinical Investigation* 97: 1557–60

Nottle, M.B., Nagashima, H., Verma, P.J. *et al.* (1999) Production and analysis of transgenic pigs containing a methallothionein porcine growth hormone gene construct. In *Transgenic Animals in Agriculture*, edited by J.D. Murray, G.B. Anderson, A.M. Oberbauer & M.M. McGloughlin, Wallingford and New York: CABI

Onishi, A., Iwamoto, M., Akita, T. *et al.* (2000) Pig cloning by microinjection of fetal fibroblast nuclei. *Science* 289: 1188–90

Palmiter, R.D. & Brinster, R.L. (1985) Transgenic mice. *Cell* 41: 343–45

Platt, J.L. (1996) Xenotransplantation: recent progress and current perspectives. *Current Opinion in Immunology* 8: 721–28

Polejaeva, I.A., Chen, S.-H., Vaught, T.D. *et al.* (2000) Cloned pigs produced by nuclear transfer from adult somatic cells. *Nature* 407: 505–09

Pursel, V.G. (1998) Modification of production traits. In *Animal Breeding: Technology for the 21st Century*, edited by A.J. Clark, Amsterdam: Harwood Academic

Pursel, V.G., Pinkert, C.A., Miller, K.F. *et al.* (1989) Genetic engineering of livestock. *Science* 244: 1281–88

Pursel, V.G., Sutrave, P., Wall, R.J., Kelly, A.M. & Hughes, S.H. (1992) Transfer of c-ski gene into swine to enhance muscle development. *Theriogenology* 37: 278

Pursel, V.G., Wall, R.J., Mitchell, A.D. *et al.* (1999) Expression of insulin-like growth factor in skeletal muscle of transgenic swine. In *Transgenic Animals in Agriculture*, edited by J.D. Murray, G.B. Anderson, A.M. Oberbauer & M.M. McGloughlin, Wallingford and New York: CABI

Schnieke, A.E., Kind, A.J., Ritchie, W.A. *et al.* (1997) Human factor IX transgenic sheep produced by transfer of nuclei from transfected fetal fibroblasts. *Science* 278: 2130–33

Sharma, A., Martin, M.J., Okabe, J.F. *et al.* (1994) An isologous porcine promoter permits high level expression of human hemoglobin in transgenic swine. *Bio/Technology* 12: 55–59

Sims, M. & First, N. (1994) Production of calves by transfer of nuclei from cultured inner cell mass cells. *Proceedings of the National Academy of Sciences USA* 91: 6143–47

Squinto, S.P. (1996) Xenogenic organ transplantation. *Current Opinion in Biotechnology* 7: 641–45

Tang, D., DeVit, M. & Johnston, S.A. (1992) Genetic immunization is a simple method for eliciting an immune response. *Nature* 356: 152–54

Waine, G.J. & Mcmanus, D.P. (1995) Nucleic acids: vaccines for the future. *Parasitology Today* 11: 113–16

Ward, K.A., Leish, Z., Brownlee, A.G. *et al.* (1999) The utilization of bacterial genes to modify domestic animal biochemistry. In *Transgenic Animals in Agriculture*, edited by J.D. Murray, G.B. Anderson, A.M. Oberbauer & M.M. McGloughlin, Wallingford and New York: CABI

Weidle, U.H., Lenz, H. & Brem, G. (1991) Genes encoding a mouse monoclonal antibody are expressed in transgenic mice, rabbits and pigs. *Gene* 98: 185–91

Wilmut, I., Schnieke, A.E., McWhir, J., Kind, A.J. & Campbell, K.H.S. (1997) Viable offspring derived from fetal and adult mammalian cells. *Nature* 385: 810–13

GLOSSARY

antisense the protein-encoding information in a gene can only be obtained by transcription in one direction along the gene. This is described as the sense direction. Antisense transcription (reading the gene in the reverse direction) does not usually produce a translatable product but may act to regulate that gene's expression by forming an RNA–RNA duplex with the natural sense mRNA of the gene, thereby preventing its translation

autocrine secretion of a substance that stimulates the secretory cell itself

chimaera (or chimera) an individual comprising two or more genetically distinct populations of cells derived from more than one zygote

epitope (antigenic determinant) exposed structure of an antigen that causes a specific reaction by an immunoglobulin

fibroblast flat, elongated cell of connective tissue that secretes fibrillar procollagen, fibronectin and collagenase

liposome artificially formed spheres of lipid bilayer enclosing an aqueous compartment

monoclonal/polyclonal antibody antibody produced by a single clone/number of founder plasma cells

myogenic having origin in muscle cells

nuclear transfer the basis of all techniques currently used in cloning. A single diploid somatic cell is cloned and inserted into an ovum from which the nucleus has been removed. The developing embryo is transplanted into a surrogate mother

outbreeding mating of unrelated individuals

paracrine type of signalling where the target cell is close to the signal-releasing cell.

pluripotent [cells] cells that are able to develop into various different types of cell.

totipotential relates to cells capable of forming any cell type

See also **Pharmaceutical proteins from milk of transgenic animals** (p.370)

PHARMACEUTICAL PROTEINS FROM MILK OF TRANSGENIC ANIMALS

Bruce Whitelaw
Department of Gene Expression and Development, Roslin Institute (Edinburgh), Roslin, Midlothian, UK

1. Introduction

Although man has used animal proteins for food, clothing and medicine for centuries, the choice of products has been limited to those that occur naturally in the animal. Recent advances, however, in recombinant DNA technology and embryology now enable genes to be transferred between species, allowing this limitation to be overcome. One application of this biotechnology has generated a novel production system for pharmaceutical proteins. This project is now quite advanced, with the first products close to clinical trials. This is likely to be the first successful use of genetically modified livestock.

2. The need for a novel production system

Many human diseases are treated therapeutically by the administration of proteins with pharmacological effects. An obvious source of pharmaceuticals is man. Indeed, many proteins of biomedical importance are obtained from human tissue or blood (e.g. clotting factor VIII, growth hormone). However, the isolation processes involved are time consuming and expensive. Furthermore, products from this source carry the obvious risks of contamination with infectious agents: the haemophilia sufferers who developed AIDS after being treated with clotting factor VIII obtained from the blood of HIV-positive donors are a widely known example of this risk. In addition, the supply of human material is limited and, therefore, cannot, in many cases, satisfy fluctuations in demand. These limitations provided the impetus to develop recombinant DNA strategies to produce pharmaceuticals.

Genetically modified microorganisms are used for the production of some proteins, e.g. insulin. This approach, however, is restricted to certain proteins by the inability of microorganisms to perform the post-translational modifications that many mammalian proteins require for activity. By contrast, mammalian tissue culture cells do contain the enzymatic machinery required to modify proteins in the appropriate manner. Unfortunately, batch fermentation of animal cells is expensive and technically demanding. Thus, although used for certain proteins, the limitations of these production systems indicated that an alternative approach was required. This search has resulted in the use of transgenic animals as bioreactors for human pharmaceuticals.

3. Transgenic animals

In the early 1980s, it was speculated that transgenic animals could be utilized to produce human biomedically important proteins. It was perceived that transgenic animal bioreactors could overcome the limitations of the other recombinant production systems and provide appropriately post-translationally modified proteins in large amounts. In addition, after the initial development cost, the maintenance cost associated with animal bioreactors should be relatively small.

The techniques enabling gene transfer into the germline of higher animals were developed initially in the mouse. By adapting this technology, transgenic sheep, goats, pigs, cattle and rabbits have been produced, with the first transgenic livestock reported in 1985 (Hammer *et al.*). The transgene (the DNA fragment encoding the desired protein) is microinjected into one of the pronuclei within a fertilized egg. In a proportion of the injected eggs, the transgene integrates into one of the host chromosomes, becomes part of that animal's genetic repertoire and is usually inherited by subsequent generations in a Mendelian manner. In large animals, the usual success rate is about 1% – one transgenic animal per 100 injected eggs.

4. Choice of tissue for expression

A number of factors must be taken into consideration with regard to the target tissue: sites where high levels of production may be detrimental to the animal should be precluded, the protein should be easily purified, it should be present in large amounts and, preferably, be present at a high relative concentration. It would also be desirable if the source was renewable and could be harvested with the minimum of discomfort to the animal. Thus, the obvious medium is a fluid, especially since most proteins of biomedical importance are themselves secreted into body fluids.

Blood, saliva, urine and milk have all been proposed as potential targets in this regard. The use of urine or saliva has not been fully pursued and there are a number of disadvantages associated with using blood. Blood tends to coagulate and contains many biologically active peptides, thus complicating purification strategies. Furthermore, any biological activity of the transgene product would be passed systemically around the body. It is currently considered that, in general, these properties make blood unattractive for the production of pharmaceuticals, although this may not always be the case and future applications may well utilize this fluid for specific proteins (Sharma *et al.*, 1994).

The route now pursued with most vigour is to target expression of the transgene to milk. Milk is renewable, easily harvested and is not circulated around the body, but isolated in the mammary gland.

5. Targeting transgene expression to milk

A transgene product can be targeted to milk by linking the regulatory DNA elements from a gene normally expressed in the mammary gland to the protein-coding sequences of the gene of interest. Therefore, the first step in the generation of an hybrid transgene (**Figure 1**) capable of targeting expression to milk was to identify these mammary-specific DNA regulatory elements. This search has concentrated on the milk protein-encoding genes, which comprise the casein and whey proteins, and has relied extensively on the transgenic mouse model. The now extensive literature indicates that most, if not all, milk protein gene promoters will target expression of a heterologous gene to the mammary gland in transgenic animals. The actual expression cassette constructed is dependent on the source of the regulatory sequences and a variety of successful designs are available. Furthermore, there is no real technical limitation to the size of the transgene, so even large protein-coding sequences can

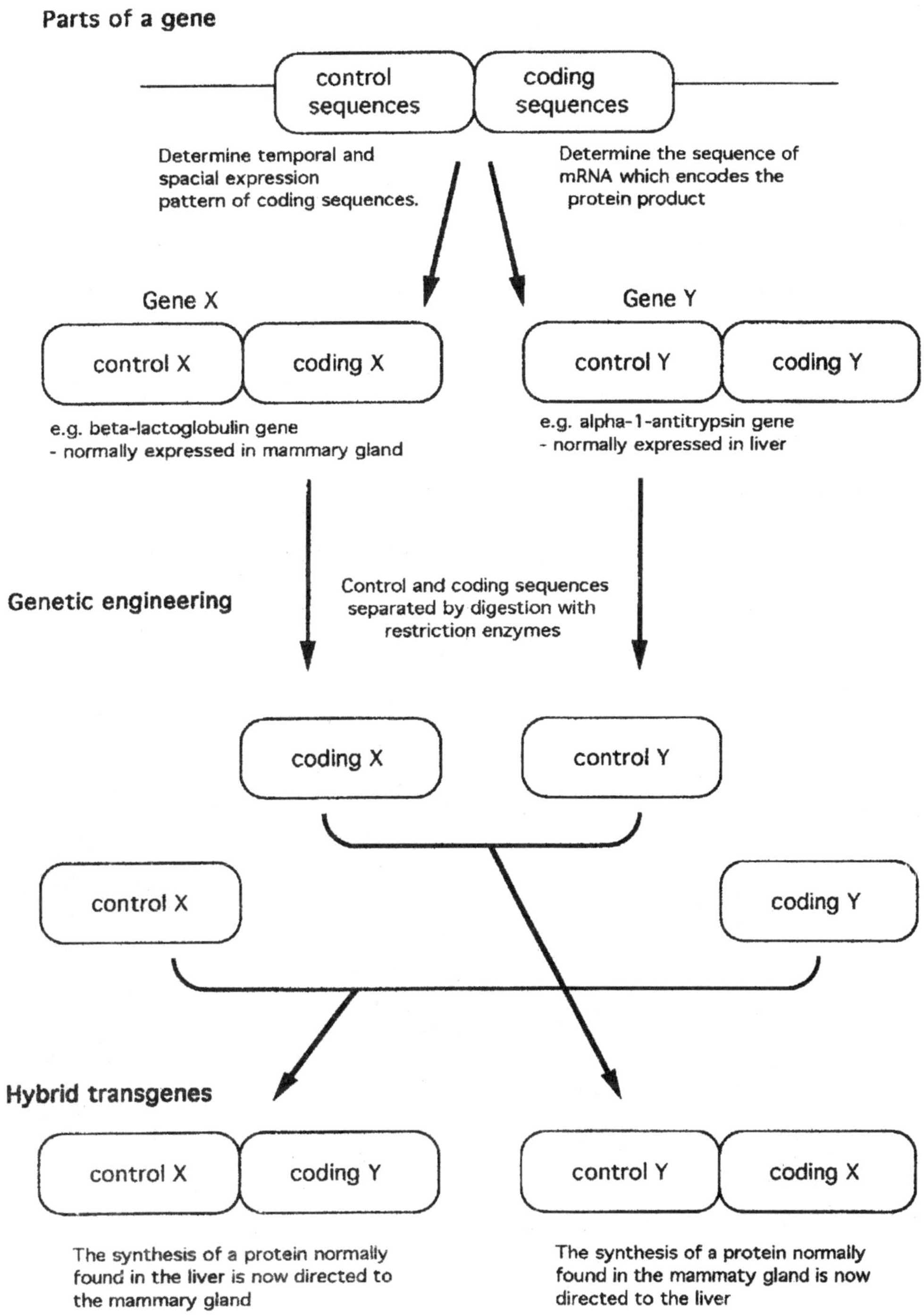

Figure 1 What is an hybrid transgene?

be used. In addition, there are strategies that increase the likelihood of expression, e.g. coinjecting a dominantly acting transgene can often rescue expression of a normally poorly expressed transgene. In the longer term, it may become possible to modify transgenes so that their expression can be regulated by the presence of specific components in the animal's feed.

The first demonstration of the feasibility of this approach in large animals was by Clark and colleagues in 1989. In their study, transgenic sheep were generated which expressed a transgene encoding human clotting factor IX. Unfortunately, in this early experiment, the level of product detected in the milk was very low (25 ng/l). The problem of low expression has now been solved for many proteins and this general approach has been applied to several species of animal, including rabbits, goats, sheep, pigs and cattle. In some cases, expression levels which are deemed to be economically viable have been obtained.

6. Pharmaceuticals in milk

A great variety of different proteins have been produced in the milk of transgenic animals (Wilmut & Whitelaw, 1994). As yet, most of these have been in the mouse model and range from relatively simple proteins, such as albumin, to highly complex proteins, such as monoclonal IgA antibodies and the hexameric fibrinogen. The successes in large animals include: tissue plasminogen activator in rabbits and goats, antithrombin III in goats, α-1-antitrypsin and clotting factor IX in sheep and protein C in pigs (**Figure 2**).

In one of these studies, by a private company especially established to exploit this biotechnology, very high-level production was obtained in transgenic sheep. This has earned Tracy the transgenic sheep considerable fame (**Figure 3**). Tracy produced human α-1-antitrypsin in her milk at 30 g/l, which approaches half that of the total protein content. This expression level was sustained in a flock derived from her (Carver *et al.*, 1992). Both this product and antithrombin III produced in the milk of goats are now in phase III clinical trials. If successful, larger trials will follow in the next few years. For the companies responsible for making these products, it has taken about 10 years, from their instigation, to reach this stage. Most of that time has been spent in establishing the technology and deriving expression vectors. Now that these hurdles have been overcome, further products can be expected in the near future.

Although transgenic cattle have been generated with transgenes which encode for human pharmaceutical proteins, as yet no expression data are available (Krimpenfort *et al.*, 1991). This current lack of data reflects the restricted number of experiments performed on cattle due to the high costs involved and the more lengthy gestation time for this large animal compared with the smaller ruminants. It is anticipated that, within the next few years, there will be transgenic cattle expressing pharmaceuticals.

It may be that certain species are more appropriate production animals for the synthesis of specific proteins. For example, rabbits have been suggested as appropriate for those products required in small quantities, while the recent generation of transgenic chickens using a DNA microinjection approach opens up the possibility of producing pharmaceuticals in hens' eggs (Love *et al.*, 1994). Alternatively, different tissues may have as yet unidentified advantages. For example, it is possible that specific post-translational modifications are only efficiently performed by certain cell types. Whatever the direction, it is very likely that in the near future many more proteins will be produced in a variety of species, as reflected in the growing number of patents associated with this technology.

Generally, the recombinant proteins purified from milk of transgenic animals are biologically active. There are some cases, however, where they are post-translationally modified in a different way to the natural human-derived material,

date	*human protein*	*amount in milk*	*animal*
1989	factor IX	25ng/l	sheep
1991	α-1-antitrypsin	30g/l	sheep
1991	tissue plasminogen activator	6mg/l	goat
1992	protein C	1g/l	pig

Figure 2 A list of the first transgenic livestock producing pharmaceuticals in their milk.

Figure 3 Tracy – PPL Therapeutics' transgenic ewe – produces 30 g/l of human α-1-antitrypsin in her milk.

e.g. with regard to glycosylation or the precision of proteolytic processing. These subtle differences could affect the antigenicity or stability of the protein. This aspect requires, and is now starting to attract, the experimental analysis it deserves. For example, in the mouse model, approaches that attempt to improve the synthetic ability of the mammary gland to convert a precursor protein into the mature protein are being evaluated.

7. Future prospects

Transgenic animal bioreactors producing pharmaceutical proteins are probably going to be the first application of gene transfer in large animals to be commercially successful. Pending the success of the clinical trials, and barring the emergence of any competing technology, the first milk-derived recombinant products will soon be available. Nevertheless, from a technical viewpoint, there are still several aspects which can be improved on; there is, for example, scope for improving the design of transgenes to increase the level of expression. It is, however, improvements to the method of gene transfer that are most sought after. The currently used microinjection approach is inefficient and only allows for the addition of a gene to the germline. In the mouse model, elegant strategies which enable precise genetic changes, including gene replacement and single base changes, are available. These strategies utilize mouse embryonic stem (ES) cells and the transgene is introduced by homologous recombination. Considerable effort has been made to generate ruminant ES cells, but to no avail; however, it is possible that the recently developed embryonic cell selection strategies will now overcome this limitation (McWhir *et al.*, 1996). Alternatively, it may be possible to use non-ES cells in combination with nuclear transfer technology (Campbell *et al.*, 1996) to generate transgenic animals. If either of these approaches is successful, it would quickly become the method of choice for generating large animal transgenics and would dramatically increase the potential applications of this biotechnology.

The technology which enabled the development of transgenic bioreactors was developed by scientists. It is now being commercialized by the business sector and the near future may see pharmaceuticals derived from the milk of transgenic animals being used by clinicians to enhance the quality of patient life. Thus, this technology is integral to the open debate regarding the ethics of using animals to enhance the welfare of humans. It will be through discussion between all of these interested parties that the future of transgenic animals will be decided.

REFERENCES

Campbell, K.H.S., McWhir, J., Ritchie, W.A. & Wilmut, I. (1996) Sheep cloned by nuclear transfer from a cultured cell line. *Nature* 380: 64–66

Carver, A.S., Dalrymple, M.A., Wright, G. *et al.* (1992) Transgenic livestock as bioreactors: stable expression of human alpha-1-antitrypsin by a flock of sheep. *Bio/Technology* 11: 1263–70

Clark, A.J., Bessos, H., Bishop, J.O. *et al.* (1989) Expression of human anti-hemophilic factor IX in the milk of transgenic sheep. *Bio/Technology* 7: 487–92

Hammer, R.E., Pursel, V.G., Rexroad, C.E. *et al.* (1985) Production of transgenic rabbits, sheep and pigs by microinjection. *Nature* 315: 680–83

Krimpenfort, K., Rademakers, A., Eyestone, W. *et al.* (1991) Generation of transgenic dairy cattle using *in vitro* embryo production. *Bio/Technology* 9: 844–47

Love, J., Gribbin, C., Mather, C. & Sang, H. (1994) Transgenic birds by DNA microinjection. *Bio/Technology* 12: 60–63

McWhir, J, Schnieke, A.E., Ansell, R. *et al.* (1996) Selective ablation of differentiated cells permits isolation of embryonic stem cell lines from murine embryos with a non-permissive genetic background. *Nature Genetics* 14: 223–26

Sharma, A., Martin, M.J., Okabe, J.F. *et al.* (1994) An isologous porcine promoter permits high level expression of human haemoglobin in transgenic swine. *Bio/Technology* 12: 55–59

Wilmut, I. & Whitelaw, C.B.A. (1994) Strategies for production of pharmaceutical proteins in milk. *Reproduction, Fertility and Development* 6: 625–30

GLOSSARY

nuclear transfer the basis of all techniques currently used in cloning. A single diploid somatic cell is cloned and inserted into an ovum from which the haploid nucleus has been removed. The developing embryo is transplanted into a surrogate mother

post-translational modification modification occurring after translation, e.g. glycosylation, cleavage of pre-proteins

pronuclei haploid nuclei resulting from meiosis. The female pronucleus in animals is the nucleus of the ovum prior to fusion with the male pronucleus; the male pronucleus is the sperm nucleus after it has entered the ovum but before fusion with the female pronucleus

See also **Large transgenic animals** (p.363); **Gene targeting** (p.791)

GENETICS OF WOOL PRODUCTION

Michael L. Ryder
Southampton, UK

1. Introduction

Wool is a kind of hair and is one of the most fascinating products of any living organism. Although originating as the underwool of wild sheep, the range of fleece types we see today is a product of human selective breeding. Sheep were domesticated *c.*9000BC where Turkey borders on Iraq and Iran. The main ancestor was the Mouflon wild sheep, which, like goats and deer, have a coloured hairy coat that moults annually in spring. After domestication, there was selective breeding for white, woolly fleeces that grow continuously. The value of wool as a clothing fibre gave sheep a unique role in human history. Wool was economically important in countries ranging from ancient Mesopotamia to medieval England, and still is in countries like modern Australia.

2. Different kinds of fleece

The wool fibres of modern sheep hang together in staples (**Figure 1**) and different fleece types have varying proportions of different fibre types (see below). The wild ancestor of domestic sheep has a brown double coat in which thick, bristly outer-hairs obscure a shorter and very fine, woolly undercoat (**Figure 2**). Human selective breeding towards a finer outer coat has been accompanied by a coarsening of the underwool (**Figure 3**). Thus, modern fleeces are derived from the outer coat as well as the underwool. By the Bronze Age, sheep had acquired a fleece, i.e. a coat that is largely composed of wool fibres, but it was still brown. During the Iron Age, there were black, grey and white sheep, in addition to brown, but there was still an annual moult.

The narrowing of the "outer" fibres continued towards modern times with a change from medium to moderately fine fibres, producing the Semi-fine, shortwool fleece. Further narrowing to produce truly fine fibres gave the Fine fleece type seen in the Merino breed. This originated in Spain and is now the main wool-producing sheep of the southern hemisphere. On another line, the fine wool became coarser to give a Medium fleece composed entirely of medium fibres, which is found in longwools. This development of modern fleece types included the elimination of colour and breeding for continuously growing wool that could be shorn when required. Since the genes that control the length of the wool growth cycle do not have a "product" (such as finer fibres or longer staples), it has not been possible to identify them. Since shearing is so costly, it would be useful today to be able to reintroduce a controlled moult and, because breeds that moult have hairy fleeces, it has not been possible to do this by crossbreeding without making the fleece unacceptably coarser.

The investigation of wool in archaeological remains has shown that until after the Middle Ages the predominant fleeces were of Hairy-medium and Generalized-medium type, types which survive in the modern Orkney and

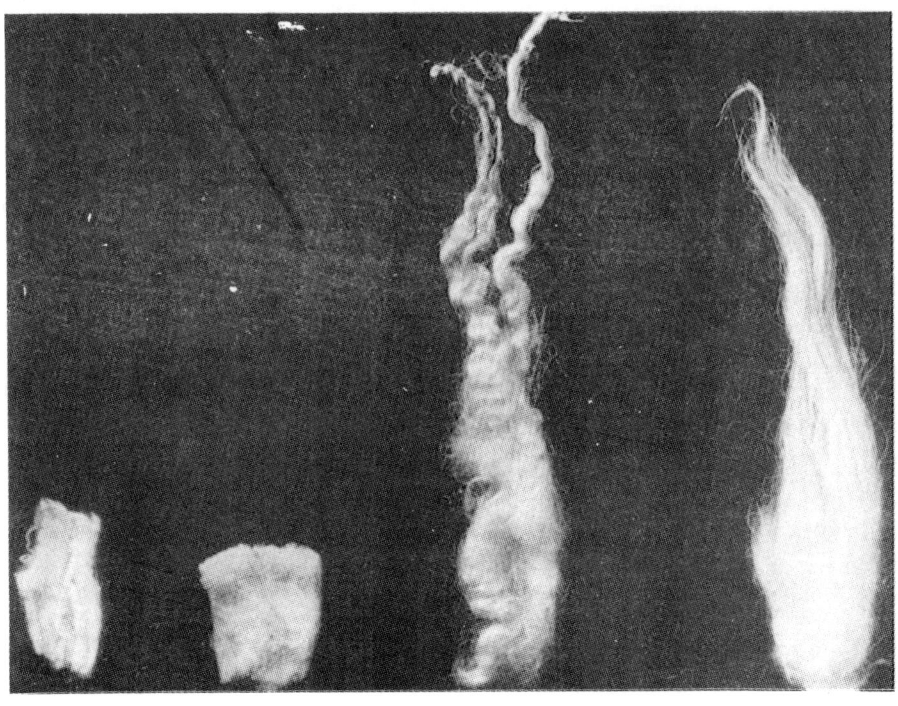

Figure 1 Staples from the main types of fleece. From left to right: Fine (unique to the Merino breed); Semi-fine, shortwool (English Down breed); Medium type (illustrated by Longwoolled breed); and Hairy (Scottish Blackface). (Reproduced with permission from Ryder, 1983.)

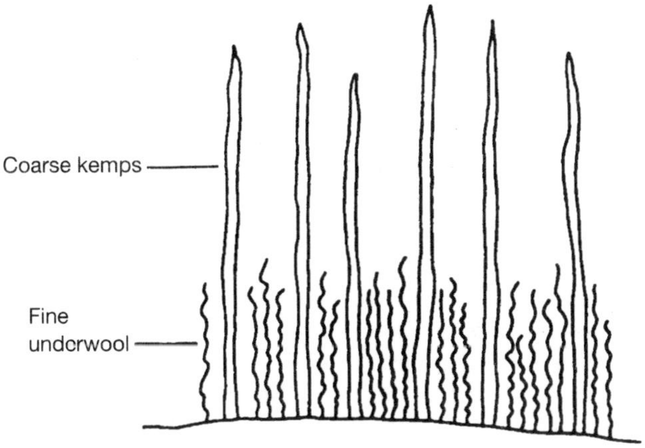

Figure 2 Double-coat structure of wild sheep in which very coarse, outer kemp-hairs obscure very fine underwool. (Reproduced with permission from Ryder, 1983.)

Shetland breeds (**Figure 3**). Breeds as we know them did not emerge until the 17th and 18th centuries. Descriptions *c.*1800AD indicate several breed types localized in different parts of Britain.

Scotland and the western parts of Britain had white- or tan-faced sheep in which only the rams were horned. Modern examples of these are the Cheviot and Welsh Mountain breeds. A black-faced horned and hairy type extended from East Anglia to the Pennines and the mountains of Scotland; examples of this are the Swaledale and the Scottish Blackface. Hornless sheep noted for fine wool were the Ryeland of Herefordshire and the Southdown of Sussex. The white-faced polled (hornless) Romney breed of Kent was already in existence, and represents a primitive longwool. The more typical lustre longwools first appeared in the midlands from the Cotswolds to Lincoln. These big, white-faced and polled sheep with a fat carcass became fashionable during the 18th century when improvements by Robert Bakewell (an English landowner who pioneered methods of livestock selection and breeding) created the New Leicester. The first breeds originated from local types bred in isolation. From the 18th century onwards, new breeds were created by crossbreeding. One such is the Border Leicester, which results from a cross between the Cheviot breed and the New (English) Leicester.

Britain's interest in fine wool declined after the Middle Ages and finally came to an end in the early 19th century

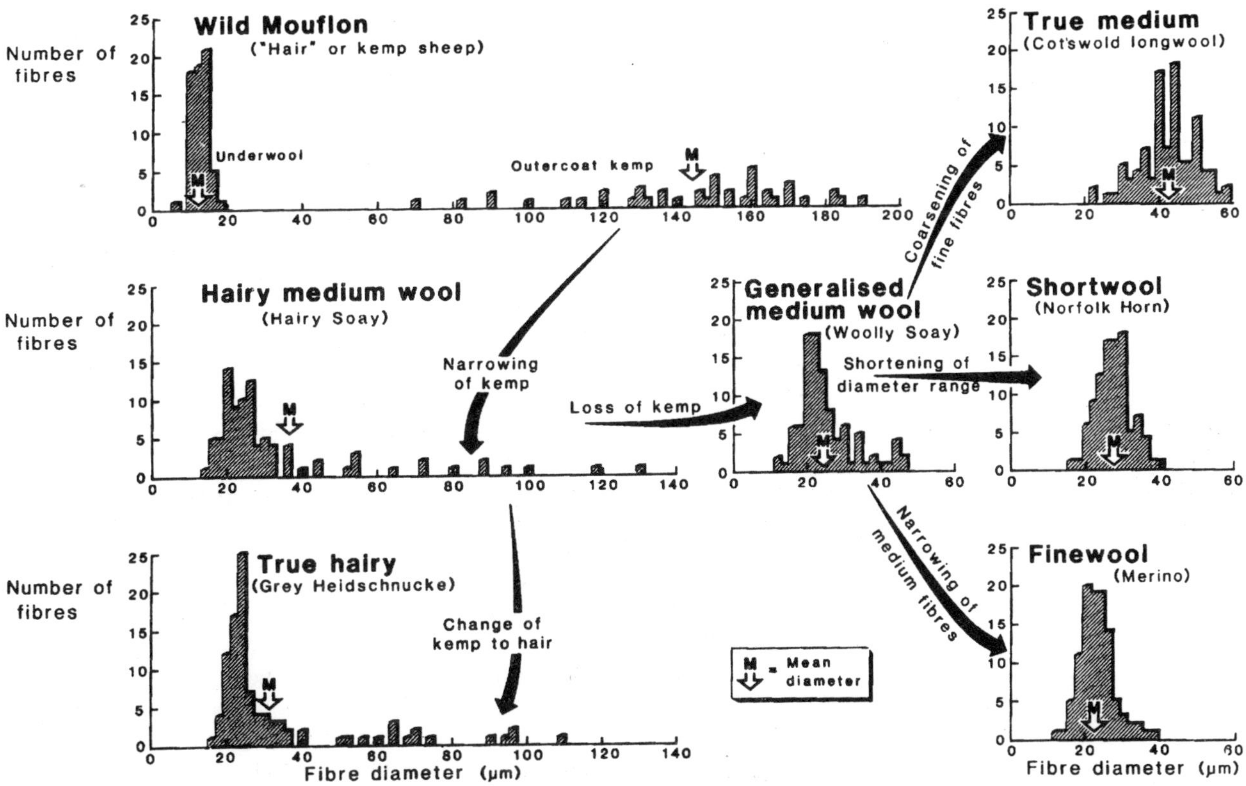

Figure 3 Changes in the coat during fleece evolution. Each diagram in the form of a histogram shows the distribution of fibre diameter which defines fleece type. The arrows indicate the changes that occurred when breeding caused one type of fleece to change into another. The main change has been a progressive narrowing of the outer kemp-hairs. These became finer as the coat of the first domestic sheep (top left) changed into the first (Hairy-medium) fleece. The remaining finer kemp-hairs then changed into medium fibres to give the Generalized-medium fleece (centre). Further narrowing changed them into the fine fibres of the Finewool (bottom right). Other changes resulted in the Semi-fine (shortwool) and the Medium wool (later seen as the longwool). The Hairy type (bottom left) developed when short kemp-hairs changed into long hairs. The breeds named are examples used to illustrate the different fleece types. (Reproduced with permission from Ryder, 1983.)

when it became easier to grow fine wool using Merino sheep in Australia. In Britain, the emphasis changed to early maturing lamb to feed the expanding human population. This was brought about by the development of the black-faced, polled Down breeds by crossing local types with the Southdown. The resulting Suffolk, Hampshire, Oxford and Dorset Down breeds with springy wool used in hosiery are the legacy from the 19th century. Early 19th-century breed lists contain only 15–20 breeds compared with double that number today, which now includes imported continental breeds.

3. Different fibre types

Although all types of fibre grown by sheep are collectively known as wool, they are divided into thicker, hairy fibres, which have a hollow core (the medulla), and finer, true-wool fibres, which mostly do not have an internal medulla. Hairy fibres are divided into short, coarse kemps akin to the outer hairs of wild sheep and long, less coarse hairs (**Figure 4**). Kemp fibres are chalky white; the hollow medulla occupies most of the width of the fibre and the lack of solid substance not only makes them brittle, but restricts the take-up of dyes so that they appear paler. These disadvantages make kemp a fault in most wool.

Hair fibres in sheep are usually not as thick as kemps, and the central medulla is not as wide. Hairs are longer than kemp fibres, being the longest fibres of the fleece. In winter, when kemp fibres stop growing, hairs merely thin down and continue to grow.

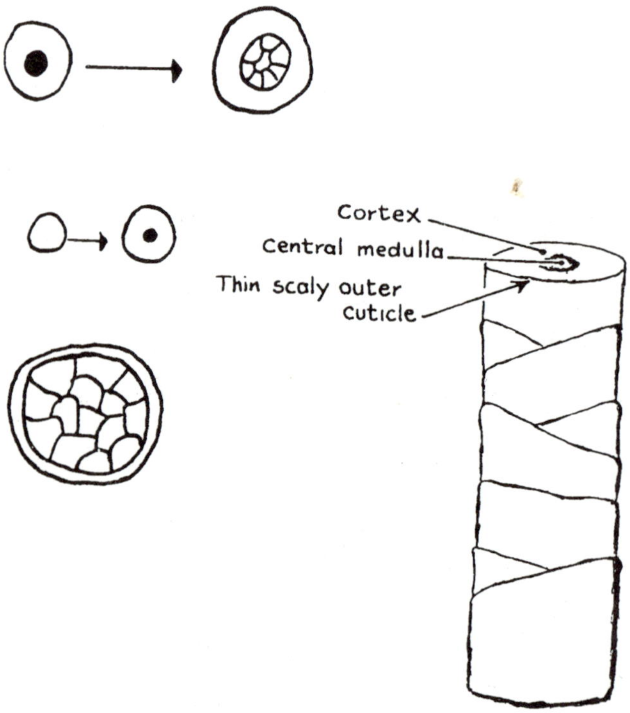

Figure 4 Left, different wool fibre types: bottom, kemp-hair with wide, latticed medulla; middle, fine fibre without a medulla, coarse fibre with narrow, non-latticed medulla; top, hairs showing how the medulla varies with fibre thickness. Hairs are never as thick as kemps, nor the medulla as wide as in kemps. Right, the main parts of a wool fibre. (Reprinted from M.L. Ryder & S.K. Stephenson, *Wool Growth*, © 1968, with permission from Academic Press.)

4. Increasing fleece weight and improving wool quality

Fleece weight is the most important factor for the wool grower, so how can it be increased without increasing fibre diameter, which is the most important characteristic for the wool user? Heredity contributes more to fleece variation than does the environment, which includes such factors as day length as well as nutrition. Most fleece characteristics are controlled by a number of genes, the effects of which will be additive, i.e. their inheritance is multifactorial. Genetic differences in fleeces can be exploited by selective breeding, which is a slow process, but it is advantageous that most fleece characteristics are highly heritable, the heritability of fleece weight being 0.3–0.6. In no breed other than the Merino, however, is it economical to apply selective breeding solely to wool characteristics.

In most other breeds, meat or milk provide such a high proportion (90–95%) of the income that wool cannot be put first in breeding programmes and so the rate of genetic progress is reduced. However, this does not mean that one cannot increase fleece weight and eliminate wool faults. There is no reason why a good meat ram (in any breed) should not also have a good fleece. Improved husbandry by reducing environmental variation leads to an immediate response in the fleece, although it is not usually economical to provide extra food solely to get more wool and, unlike genetic gains, any improvement is not permanent.

A simple formula worked out by Helen Newton Turner for Merino sheep in Australia helps one to understand what the different components of fleece weight are and how they are interrelated: fleece weight $W = L \times A \times N \times D \times S$. In this formula, L = the mean fibre length; A = the mean cross-sectional area of the fibres (measured as fibre diameter); N = the mean number of fibres per unit area of skin (fibre density); D = the density (specific gravity) of wool substance; and S = the total area of skin growing wool. To simplify, to get more wool you need larger sheep with longer and denser wool. Thicker fibres may be heavier, but they are undesirable since they reduce wool quality, which equates with fineness. The only component that cannot be changed is the specific gravity of wool, which has the constant value of 1.31.

Seasonal pasture variations mean that sheep rarely reach their genetic potential for wool production. Extra food in the form of long-term pasture improvement therefore stimulates wool growth and the greater fleece weight is brought about by increasing fibre length (L) and diameter (A). Fibre density (N) is unaffected by nutrition since most wool follicles are formed before birth. Better nutrition can also increase body size and, therefore, the skin area (S), but this has no effect on fleece weight since the result is merely a lower wool follicle density.

Breeding for increased fleece weight (W) could theoretically increase all the components except specific gravity. In some experiments with the Australian Merino, the selective breeding caused an increase in fibre length (L) and diameter (A), while in others the increased fleece weight was brought about by an increased fibre density, which was actually associated with a desirable decrease in mean fibre diameter of 1 μm. The best practical advice with Merino sheep is to select for finer wool at the same time as selecting for a greater fleece weight, i.e. to breed from animals that not only have heavier fleeces, but finer wool. In the Merino, there is no genetic correlation between fleece weight and body weight. There is, therefore, no danger of obtaining heavier sheep that would eat more and so reduce efficiency. In other breeds, fleece weight is correlated with body weight, but this is useful since any selection for increased body weight (for meat production) will incidentally increase wool production.

5. Fleece characteristics

Since fibre diameter is the most important textile characteristic, fleece weight should not be increased by making the fibres thicker, and certainly not in the Merino breed. Since poor nutrition reduces fibre diameter, its expected heritability of 0.5 can be less than 0.2. Crimpiness (waviness) has a heritability of 0.4–0.5, but the number of crimps per unit length (there are generally more in finer wool) should not be used in selection for fibre fineness because crimp number is poorly correlated with fibre diameter. Fleece weight can be raised by increasing staple length, provided that the increased length does not put the wool into an inferior class, which can happen in the Merino. Staple length is less affected by diet than is fibre diameter and has the higher heritability of 0.3–0.6.

Lack of uniformity in fibre length and diameter is a common complaint among wool users. In addition to the obvious genetic differences between breeds, there are genetic differences between flocks, between individual sheep, between different parts of the fleece and even within the wool staples. There is considerable scope in breeds other than the Merino for reducing this variability by selective breeding, e.g. the variation of individual fibre length within the staples and of staple length between sheep can be reduced in this way. The recent successful cloning (q.v.) of sheep in Edinburgh points the way to greater uniformity in wool.

6. Wool faults

Given that there might be restraints on actual improvement, at least one can aim at reducing wool faults. Common hereditary faults are colour and kemp. Natural black fibres and kemp are rare in the Merino, but can be a serious fault in crossbred and carpet wools. Pigment and kemp are entirely genetic in origin and so can be tackled by choosing rams lacking such fibres and, at the same time, culling badly affected ewes. Hairs, too, are basically genetic in origin, but they are subject to environmental influences in that the central medulla of hairs is only fully formed during summer; hairs become narrow and lose their medulla during winter, when they can appear like wool fibres. The heritability of hairiness is in a range of 0.3–0.7.

7. Carpet wool

Since hairy fleeces are commonly used in carpets, the impression is gained that coarse fibres with a wide central medulla are a desirable feature in carpet wools. In fact, hairy fleeces are used because they are cheaper; many such fleeces have too many kemps and coloured fibres for use in carpets, and so are used for mattress filling. The main requirement for carpets is relatively coarse, non-medullated wool. In modern carpet manufacture, this is achieved by blending wools of several types.

Here, there is the possibility of either crossing with an improved type or of selective breeding within existing breeds. Crossing of British hairy breeds with hairy strains of New Zealand Romney gave heavier fleece weights and lack of pigment, but the crossbred fleeces were kempy. In a 20-year selection experiment with the Scottish Blackface breed, 8-week-old lambs were selected on a medullation index and put into a Hairy or a Fine line. The Hairy line eventually acquired very coarse hair fibres (not kemps) in each primary (outer hair) follicle and the secondary (wool) fibres acquired a medulla, as well as becoming more numerous.

In the Fine line, some primary follicles grew wool fibres in place of hairs, but others retained kemps. The secondary fibres became finer and increased in number, which in genetic terms was a truly correlated response. Breeding together of the Hairy and Fine lines gave no evidence of a single-gene effect, but some evidence of heterosis in hairiness, such that crosses of Hairy with Fine sheep are likely to have more hairy fleeces than the intermediate value. In crosses with the longwoolled Wensleydale breed, offspring from the (kempy) Fine line lacked kemp, but offspring from the Hairy line showed much less reduction in medullation.

8. Crossbred clothing wool

British sheep farming is organized in a stratification system in which ewes of such breeds as the Scottish Blackface with hairy fleeces drafted from the hills are crossed with longwoolled rams of such breeds as the Border Leicester and Wensleydale. The resulting crossbred (Halfbred and Masham) ewes are crossed with a Down ram (e.g. of the Suffolk breed) to produce fat lamb. This means that crossbred sheep (the mothers of fat lamb) provide most of the medium and semi-fine wool, which, since crossbred, is not of clearly defined grade, and more semi-fine (Down cross) wool is obtained as skin wool removed after the slaughter of the lambs. This system means that improvement such as a Merino influence can be injected at the hill stage and again at the longwool-cross stage. The Merino breed is unique in combining a heavy fleece weight with a low fibre diameter, and Merino wool is always in steady demand because it has the widest range of uses.

9. Producing desired wool properties during fleece growth

Most textile research funds go into the modification of the fibre during processing, e.g. applying more crimp and creating shrink resistance. The philosophy of the wool biologist is to do this during growth, with any genetic modification being permanent. The current research on the genetic engineering of wool makes this more of a possibility.

FURTHER READING

Ryder, M.L. (1983) *Sheep and Man*, London: Duckworth

Ryder, M.L. (1987) Wool. In *New Techniques in Sheep Production*, edited by I.F.M. Marai & J.B. Owen, London and Boston: Butterworths

Ryder, M.L. & Stephenson, S.K. (1968) *Wool Growth*, London and New York: Academic Press

Turner, H.N. (1956) Measurement as an aid to selection in breeding sheep for wool production. *Animal Breeding Abstracts* 24: 87–118

Turner, H.N. (1958) Relationships among clean wool weight and its components. *Australian Journal of Agricultural Research* 9: 521–52

Turner, H.N. & Young, S.S.G. (1969) *Quantitative Genetics in Sheep Breeding*, Ithaca, New York: Cornell University Press

Wilmut, I., Campbell, K.H.S. & Young, L (1997) Modern reproduction technologies and transgenics. In *The Genetics of Sheep*, edited by L. Piper & A. Ruvinsky, Oxford and New York: CAB International. Other articles in this book are also of interest.

See also **Textile fibres from goats** (p.379)

TEXTILE FIBRES FROM GOATS

Michael L. Ryder
Southampton, UK

1. Introduction

The domestication of goats took place at about the same date and in the same place as the domestication of sheep (q.v.). The bezoar wild goat (*Capra aegagrus*) was the main ancestor of the domestic goat (*C. hircus*). Goats spread with the first farmers of the Neolithic period and reached China at this time. There is no evidence that the markhor wild goat (*C. falconeri*) of Afghanistan contributed to domestic goats, but it is possible that the ibex (*C. ibex*) contributed to the cashmere type since it appears to have more underwool than the other wild goats and its Asiatic range coincides with that of cashmere goats (Ryder, 1990).

The three basic kinds of goat fibre are ordinary goat hair, in which the outer hair rather than the underwool is used, cashmere, which is the underwool, and mohair, which lacks hair and is like the fleece of a sheep. A fourth, new fibre, cashgora, originated from the crossing of double-coated goats with the Angora in "grading-up" towards mohair.

2. Ordinary goat hair

Most domestic goats have a hairy outer coat, which obscures short, fine underwool (**Figure 1**). This structure has changed little from that of the wild ancestor. The thick "guard" hairs of the outer coat provide physical protection, while the underwool (down) gives thermal insulation. Since domestication, the development of a range of colours has taken place, along with an increase in coat length, so that primitive breeds tend to have long hairs. Although there is a long history of the use of ordinary goat hair in cloth and also in carpets, the hairy outer coat makes it on the whole too coarse for textiles. There is still a market for goat hair in Europe and it is used entire, with no separation of the underwool.

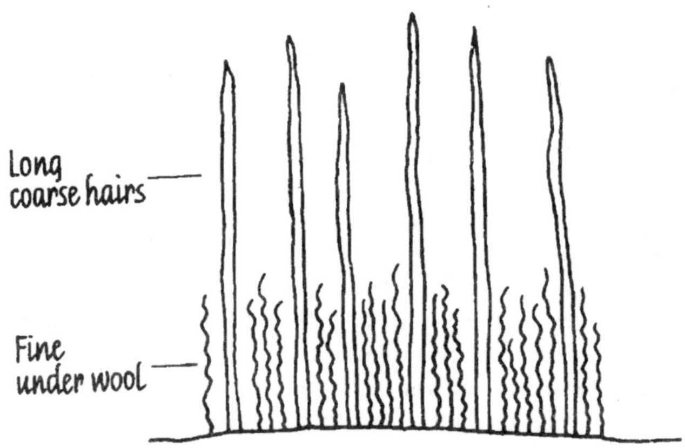

Figure 1 The double coat of "ordinary" (dairy) goats. The main change since domestication has been a lengthening of the outer hair and the breeding of a range of colours – black, white, and tan.

3. Cashmere

Cashmere fibre is the soft underwool of a double-coated type of domestic goat native to the mountainous region of central Asia. There is no specific breed, but cashmere goats are generally white with spiral horns (**Figure 2**). The fibre is obtained by combing during the spring moult and has a similar diameter to the underwool of "ordinary" goats. The mean diameter of commercial Chinese fibre, which is noted for fineness and which forms the bulk of world supplies, is 15.5 µm, yet that of British dairy goats is frequently less than 14 µm. Even the softness of Chinese cashmere is not unique, since the finer any animal fibre becomes, the softer it feels.

Chinese cashmere goats also have the distinction of growing more underwool than other double-coated goats. This probably became denser by natural selection in a cold environment before it was further increased by human selective breeding. Goats with a white coat have been selectively bred since white cashmere fetches the highest price. The outer hairs average 15 cm in length and can be coloured; manufacturers prefer the underwool to be at least 4 cm in length and any underwool pigmentation is paler than that of the hair. The amount of hair remaining in the down after harvesting comes third after fineness and colour in the grading of the raw material.

The Chinese believe that it is not possible to produce cashmere at altitudes lower than 3000 feet or south of latitude 36°. The Chinese national average of 250 g cashmere per goat "hides" a range from 25–50 g (similar to the weight produced by British dairy goats), in unimproved breeds, to 500 g grown by the Liaoning breed. This greater production, while maintaining fibre fineness, has been achieved by selective breeding and Liaoning males are being mated with females of local breeds in order to increase individual production.

Each area has a breeding centre where improved goats are being multiplied in superior herds and from which frozen semen as well as males are distributed to local peasant-owned herds. In this way, rapid and widespread progress is being made. In some herds, the first crosses were being crossed back to the Liaoning to give three-quarter Liaoning goats and, in others, the first-cross animals were being interbred (Ryder, 1990).

4. Mohair

Mohair comes from a single breed of goat, the Angora, which is named after a Turkish province. Like cashmere goats, Angora animals are mainly white with spiral horns, but they have a coat that is unique among goats in virtually

Figure 2 Cashmere goats in China. (Photo reproduced with permission from the late Li Jian-ping.)

lacking hair and in being comprised of long, lustrous and curly fibres (**Figure 3**). The single coat of Angora goats is comparable with the fleeces of sheep. The individual fibres are at least twice the thickness of cashmere and not only have the underwool fibres become longer and coarser, but they are also more numerous.

There appears to have been a mutation of the coat that produces a fleece like that of lustre-longwoolled sheep. How it occurred in sheep is no better understood than it is in goats, but this unique "underwool" is dominant in crosses. When ordinary goats are crossed with the Angora, the underwool assumes the character of mohair and grows much longer than the outer coat, completely obscuring the hair, which is only eliminated by further crossing. The tendency to moult has been lost and it is interesting to speculate whether or not the origin of the occasional gare hairs, which are akin to heterotype hairs in sheep (see article on "The genetics of wool production"), is associated with this change to continuous fibre growth, as it appears to be in sheep. Mohair is therefore harvested by shearing, commonly twice a year when the fleece is 12 cm long and weighs about 2.7 kg. Although mohair is much less valuable than cashmere, the greater weight grown means a greater annual income from each animal.

Figure 3 Turkish Angora goat, the coat of which is mohair. (Woodcut from *A General History of Quadrupeds* by Thomas Bewick, 1790.)

If the biological origin of mohair is a mystery, then so is the historical origin of the breed. One hypothesis holds that Angora goats evolved in Turkey and emerged after the Middle Ages. The second hypothesis states that the Angora goat was introduced by Turkish tribes who invaded Anatolia from central Asia during the 11th–13th centuries.

The first European record of the Angora was made in 1554 by Belon, and Tournefort in his *Levant Voyage*(1654) wrote that: "The finest goats in the world are bred in Angora. They are dazzlingly white, with hair as fine as silk, which falls in curly tresses 8–9 inches [21.5 cm] long." Raw mohair first reached Europe in 1820 and, by 1839, more raw fibre than yarn was leaving Turkey. The amount exported increased 10-fold by 1895. The increased demand was met by upgrading native goats, but the deterioration of the "true" Angora type described by Conolly in 1840 was deplored: "The long-framed goat peculiar to Angora is invariably white, and its coat is of one sort; it is silky and hangs in long curly locks. It is clipped annually in spring, progressively [in successive years] yielding from 1 lb to 4 lb [0.45–1.8 kg] of mohair" (Ryder, 1993).

South Africa was the first country to receive Angoras from Turkey in 1838 and, by 1893, there were 2.8 million Angoras there. Turkey and South Africa then provided most of the world production of 9 million kg mohair. Angoras were first imported into USA in 1848, but it had only 0.25 million goats by the turn of the century. Mohair production has fluctuated widely as a result of changes in fashion. From the mid-1950s to the mid-1980s, world production varied from 13–30 million kg. The current production is somewhat lower than the 25 million kg peak of 1987.

5. The growth of fibre outwith the main production areas

In the 19th century, there were small imports of Angora and cashmere goats into Australia and New Zealand. A nucleus of Angora goats was kept pure in Australia, but some of these escaped to join the feral population of ordinary goats. The cashmere goats were not maintained and it is possible that they all became feral, as did both types in New Zealand. However, re-examination of the Australian record shows that the cashmere animals introduced were "cachmere-Angora" crosses, what are now termed cashgora goats.

The Australian mohair industry began to expand about 1970. Import regulations meant that numbers could be increased only by crossing existing Angoras on to dairy goats and feral animals, followed by selective breeding to eliminate hairy and coloured fibres. By 1990, mohair production in Australia was over 1 million kg.

A similar development occurred in New Zealand from 1000 purebred Angora goats in 1979. A large population of feral goats allowed a rapid increase in numbers by cross-breeding, so that by 1987 there were 70 000 head and mohair production was 0.25 million kg. Other countries with Angora goats include Russia and Argentina. Angoras have been established in Spain using goats from Texas and they have been introduced into Britain. From an initial import of three males and 15 females from New Zealand in 1981, numbers have increased by breeding and further imports so that there are now over 5000 Angora goats in Britain plus 6000 crossbred animals.

The extension of cashmere production outside Asia presents a different picture. The main stimulus has been declining supplies of the fibre from China. It is not easy to obtain cashmere-growing animals from the areas of fibre production. However, most double-coated breeds grow cashmere down and the problem is to increase the amount produced up to that grown by Chinese goats.

Australia was the first non-Asiatic country to develop a cashmere industry, from 1975, and now has several hundred thousand goats with a production of over 250 g each. Feral goats were used as base stock and some of these have down as fine as 13.5 μm; the great individual variation in the weight of fibre grown has enabled rapid progress to be made by selective breeding.

Some breeders have increased individual production by crossing with the Angora breed, but this is associated with a coarsening of the fibres. More rapid progress was made towards cashmere production in New Zealand, where the feral animals had greater individual production and 150 000 cashmere goats were being farmed by 1985.

Cashmere production in Britain started with the use of goats to keep hill sheep grazings free from weeds. Although the dairy goats first used grew only 50 g of down, it was of cashmere fineness. Production was increased by mating the dairy goats with feral males. The offspring combine the greater percentage of down in the feral goat with the larger size of a modern breed. This, followed by selective breeding, raised the fibre production to 100 g. In order to achieve further major gains, semen was imported and crosses with Tasmanian goats have produced 250 g. Total British production increased from 54 kg in 1988 to 315 kg in 1989, but the fibre diameter increased from 14.5 to 17.2 μm.

6. Fibre growth in the animal

Knowledge of wool genetics can be applied to fibre growth in goats – and, more directly, to mohair. Only the Angora breed has developed a fleece – mohair is of medium diameter, long and lustrous like the wool of lustre-longwoolled sheep. Although unlike sheep where there is only one fleece type, Angora goats do have different hereditary staple forms like those of sheep – ringlet, flat (which is wavy) and straight ("sheepy"). The last has the most medullation, while the ringlet has the finest fibres, and this can be selectively bred.

Kemps in mohair are similar to those in sheep (and gare is the same as hair) and are a fault for the same reasons. Since these hairy fibres have the high heritability of 0.7, this fault can be tackled by selective breeding.

FURTHER READING

Li, J-P. (1988) Cashmere combing in Shaanxi Province. *FAO World Animal Review* 65: 57–59

Mason, I.L. (1981) Breeds. In *Goat Production*, edited by C. Gall, London and New York: Academic Press

Ryder, M.L. (1990) Goat fibre and its production. *Proceedings of the 8th International Wool Textile Research Conference, New Zealand* II: 241–66

Ryder, M.L. (1993) The use of goat hair – an introductory historical review. *Anthropozoologica* 17: 37–46

See also **Genetics of wool production (p.374)**

DOG GENETICS

Matthew Binns
Centre for Preventive Medicine, Animal Health Trust, Newmarket, UK

1. Introduction

It can be argued that the modern dog more than any other animal represents what can be achieved by genetic selection. Indeed, it has been stated that within the 300 or so breeds of pedigree dogs there exists a greater range of morphological types than is seen in any other mammalian species. In considering size and weight, for example, it is clear that there is at least a 30-fold difference between the Chihuahua and the St Bernard. It also surprises most people to learn that the vast majority of pedigree dog breeds have arisen over the last 200 years, although the origin of many of the breeds is controversial. The Kennel Club in the UK was established in 1873; before this time dogs were mainly bred for a range of functions, many of which are still evident in the names of present breeds such as retriever, deerhound and shepherd. Over the past hundred years the balance has shifted such that dogs are now bred largely for appearance.

2. Origin of the domestic dog

The dog family, Canidae, is thought to have diverged from other carnivore families 50–60 million years ago. The family, which now comprises 34 species, shows a wide range of chromosome morphologies, with the diploid chromosome number ranging from $2n = 36$ in the red fox (*Vulpes vulpes*), to $2n = 78$ in the grey wolf (*Canis lupus*). The chromosomal rearrangements observed in the different species have been used to deduce the phylogenetic history of the group. Mitochondrial sequences have also been used to examine the evolution of the Canidae and the origins of the domestic dog. These results (Wayne, 1993) demonstrate that the domestic dog is an extremely close relative of the grey wolf, with as little as 0.2% variation in mitochondrial DNA sequence between the two species. This contrasts with a 4% variation in mitochondrial sequences between grey wolves and their nearest wild relative, the coyote. In 1993 the domestic dog was reclassified as a subspecies of wolf: *C. lupus familiaris*, changed from *C. familiaris*..

The dog was first domesticated about 12 000 years ago and there are several historical sources which depict the type of dogs used by the ancient Greeks and Romans among others. It is clear that even several thousand years ago there were several basic "types" of dog, and it is likely that the dog was domesticated independently in several places followed by selection for particular functions. The ancient types of dog include examples that closely resemble the modern day Greyhound, Mastiff, Pekinese and Spitz breeds. Few studies have been undertaken investigating the genetic variation between different dog breeds. The evidence available suggests that relatively few genes control the diverse range of morphologies among breeds. The interbreeding of F_1 animals derived from cross-breeding will frequently result in the generation of offspring resembling the grandparental types. The stabilization of breeds in pure breeding lines within a short time span also suggests that a small number of genes control the large morphological differences. The application of microsatellite genotyping to different breeds is in progress and may reveal more about the origins of the various breeds. In one limited study, a particular allele was found only in miniature and giant Schnauzers, pointing to a common origin for these two breeds (Hershfield *et al.*, 1993).

3. Coat colour genetics

Coat colour genetics in dogs resembles that in other mammalian species, with a series of genes operating in a complex manner to generate the final coat colour. These series genes include A (Agouti), B/C (Albino), D (Dilution), E (Extension), G (Greying), M (Merle), S (Spotting) and T (Ticking). Further details of the series operating in different breeds can be found in Willis (1989) and Robinson (1990).

4. Cytogenetics

The domestic dog (*C. lupus familiaris*), with $2n = 78$, has a karyotype which has until recently defeated efforts at standardization, due to the large number of small acrocentric chromosomes present. The recent development of a set of chromosome paints containing every dog chromosome (Langford *et al.*, 1996) and the use of fluorescence *in situ* hybridization (FISH) using canine cosmid clones containing microsatellites (Fischer *et al.*, 1996) has facilitated a standardization of the dog karyotype (Breen *et al.*, 1999). These reagents should also be useful in studying the chromosomal evolution of the Canidae. The application of human chromosome paints to dog chromosomes should also prove valuable in identifying suitable candidate disease genes once markers linked to those diseases have been identified through genetic linkage studies.

5. Progress towards a canine linkage map

Dog genetic linkage maps are currently under construction by several groups; the published third generation linkage map (Werner *et al.* 1999) has 342 markers with an average interval distance of 9.0 cM, and integration with an expanded radiation-hybrid map has generated an integrated map of 724 unique markers (Mellersh *et al.*, 2000). The markers used in the construction of the maps are mainly microsatellites with about one third genes. Several hundred polymorphic dinucleotide microsatellites have been characterized (Ostrander *et al.*, 1995). In general tetranucleotide microsatellites are highly polymorphic in dogs (Francisco *et al.*, 1996), with GAAA repeats being particularly polymorphic. As tetranucleotide repeats have definite advantages over dinucleotide repeats in scoring alleles, their use is becoming increasingly popular. At present the reference families being used for the construction of the maps are not ideal. The Dogmap group is using a collection of about 140 animals of two breeds. Unfortunately, the amount of DNA available from these families is limited, although it should be sufficient for a map with 3–400 markers. Further details of the linkage studies can be at the Dogmap home page at http://www.dogmap.ch/index.html; the FHCRC Dog Genome project at http://www.fhcrc.org/science/dog_genome/dog.html; and the University of California Dog Genome project at http://mendel.berkeley.edu/dog.html. Many of the microsatellites derived from the domestic dog are polymorphic in other canids and indeed they have been used to look at wild canid populations. For example, microsatellites derived from the domestic dog were used to analyse hybridization between the Ethiopian Wolf (the world's most endangered canid) and domestic dogs. Results indicated that hybridization had already occurred in one population of wolves and also showed that the variability within and between populations was very low, suggesting that captive breeding may be necessary to preserve genetic variability (Gottelli *et al.*, 1994).

6. Inherited diseases in dogs

Much of the recent surge of interest in dog genetics has resulted from a desire on the part of veterinary scientists to tackle the problem of inherited diseases in pedigree dogs. At least 350 inherited disorders have been identified, although the mode of inheritance has not been established for more than a few. The majority of those that have been characterized are inherited in a simple recessive fashion. This situation reflects the high level of inbreeding that has been practised, as well as the small number of animals used in the foundation of many of the breeds. Examples also exist of diseases inherited in an autosomal dominant and X-linked manner and several disorders such as hip dysplasia are thought to be multifactorial. Many of the disorders are breed-specific, and even in conditions such as retinal dysplasia, which occurs in several breeds, cross-breeding experiments have revealed that at least three different genes (*rcd1*, *rcd2* and *erd*) exist in different breeds (Acland *et al.*, 1989). In contrast, for progressive rod-cone degeneration cross-breeding has revealed that Labrador Retriever, Miniature Toy Poodle and English and American Cocker Spaniels all harbour the same mutation (Aguirre & Acland, 1988).

Many of the inherited disorders in dogs are thought to be homologues of human inherited disorders, the gene *pra* representing a model for retinitis pigmentosa in man. Dog diseases thus have potential as models in which gene therapy experiments can be carried out. While dogs have many disadvantages as experimental animals compared with small animals such as mice, it is thought that the failure in humans of some therapies developed in mice may be due in part to differences in physical size. Dogs could provide suitable intermediate-sized models and their lifespan also allows longer-term studies to be undertaken than is possible with mice.

At present, very few of the inherited diseases in dogs have been characterized at the molecular level. The mutation for *pra* in Irish Setters has recently been identified within the β-subunit of the cGMP phosphodiesterase gene (Suber *et al.*, 1993), the same gene being mutated in the *rd* mouse (Pittler & Baehr, 1991) and in humans with retinitis pigmentosa (McLaughlin *et al.*, 1993). All affected Irish Setters tested so far possess the same mutation (a G to A transition at position 2420) that truncates the β-subunit by 49 amino acid residues (Ray *et al.*, 1994). Genetic screening tests are now being used by Irish Setter breeders to identify *pra* carriers and exclude them from breeding programmes. Single-stranded conformation polymorphism (SSCP) studies have indicated that the mutation in β cGMP phosphodiesterase is probably not responsible for *pra* in any of the other breeds associated with this disease. In addition, mutations in the opsin gene have been excluded from being the cause of generalized progressive atrophies in the Tibetan Terrier, the Miniature Schnauzer, the Irish Setter, the Miniature Poodle, the Labrador Retriever and English Cocker Spaniel breeds (Gould *et al.*, 1995).

Beddlington Terriers suffer from copper toxicosis, which has many similarities to Wilson's disease in man. A microsatellite marker linked to the disease has recently been identified which enables the identification of affected and carrier animals in pedigrees containing at least one member with confirmed copper toxicosis (Yuzbasiyan-Gurkan *et al.*, 1997). This screening test should result in improved health of the breed, and lead ultimately to the identification of the disease gene.

A genetic screening test has been developed for pyruvate kinase (PK) deficiency in Basenjis which is more reliable than previous tests, which estimated PK activity in erythrocytes. The test is based on PCR amplification of exon 5 of the PK gene and subsequent digestion with *Aci*I that cleaves the normal PK allele, while the single base pair deletion in the disease allele removes the *Aci*I site.

Linkage mapping and FISH studies have mapped the canine X-linked severe combined immunodeficiency disease locus to proximal Xq, suggesting that the mutation resides within the γ chain of the IL-2 receptor gene, as is the case in humans (Deschenes *et al.*, 1994). The mutation responsible for X-linked hereditary nephrosis (HN) in the family of Samoyed dogs has recently been identified within the a5 chain of collagen type IV and should provide an excellent model for HN in humans, where mutations in this gene are common (Zheng *et al.*, 1994).

7. Genetics of behaviour

Behaviour is an important characteristic of different dog breeds and a feature that has been subject to strong selection pressure since the domestication of the dog. The range of instinctive behaviours inherited from the dog's wild ancestors have been selected to varying degrees in different breeds such that certain behaviour patterns are now strongly associated with particular breeds. Thus the differing behaviour one would expect from a Border Collie, a Doberman or a Poodle is almost as strong a characteristic as their appearance. One study underway in the US involves the cross-breeding of Border Collies (well known for their herding instincts) with Newfoundlands (known for their love of water). Offspring are being scored for their behaviour and the results correlated with microsatellite genotypes in an attempt to identify markers linked to particular behaviour patterns, and eventually to locate the genes associated with the particular behaviour.

REFERENCES

Acland, G.M., Fletcher, R.T., Gentleman, S., Chader, G.J. & Aguirre, G.D. (1989) Non-allelism of three genes (*rcd1*, *rcd2* and *erd*) for early-onset hereditary retinal degeneration. *Experimental Eye Research* 49: 983–98

Aguirre, G.D. & Acland, G.M. (1988) Variation in retinal degeneration phenotype inherited at the PRCD locus. *Experimental Eye Research* 46: 663–87

Breen, M., Langford, C.F., Carter, N.P. *et al.* (1999) FISH mapping and identification of canine chromosomes. *Journal of Heredity* 90: 27–30

Deschenes, S.M., Puck, J.M. & Dutra, A.S. *et al.* (1994) Comparative mapping of canine and human proximal Xq and generic analysis of canine X-linked severe combined immunodeficiency. *Genomics* 23: 62–68

Fischer, P.E., Holmes, N.G., Dickens, H.F., Thomas, R., Binns, M.M. & Nacheva, E.P. (1996) The application of FISH techniques for physical mapping in the dog (*Canis familiaris*). *Mammalian Genome* 7: 37–41

Francisco, L.V., Langton, A.A., Mellersch, C.S., Neal, C.L. & Ostrander, E.A. (1996) A class of highly polymorphic tetranucleotide repeats for canine genetic mapping. *Mammalian Genome* 7: 359–62

Gottelli, D., Sillero-Zubiri, C. & Applebaum, G.D. *et al.* (1994) Molecular genetics of the most endangered canid: the Ethiopian wolf *Canis simensis*. *Molecular Ecology* 3: 301–12

Gould, D.J., Peterson-Jones, S.M. & Sargan, D.R. (1995) Investigation of the role of opsin gene polymorphism in generalized progressive retinal atrophies in dogs. *Animal Genetics* 26: 261–67

Hershfield, B., Chader, G. & Aguirre, G. (1993) Cloning of a polymorphic canine genetic marker which maps to human chromosome 9. *Animal Genetics* 24: 293–95

Langford, C.F., Fischer, P.E., Binns, M.M., Holmes, N.G. & Carter, N.P. (1996) Chromosome-specific paints from a high-resolution flow karyotype of the dog. *Chromosome Research* 4: 115–23

McLaughlin, M.E., Sandberg, M.A., Berson, E.L. & Dryja, T.P. (1993) Recessive mutations in the gene encoding the β-subunit of rod phosphodiesterase in patients with retinitis pigmentosa. *Nature Genetics* 4: 130–34

Mellersh, C.S., Hitte, C., Richman, M. *et al.* (2000) An integrated linkage-radiation hybrid map of the canine genome. *Mammalian Genome* 11: 120–30

Ostrander, E.A., Mapa, F.A., Yee, M. & Rine, J. (1995) One hundred new simple sequence repeat-based markers for the canine genome. *Mammalian Genome* 6: 192–95

Pittler, S.J. & Baehr, W. (1991) Identification of a nonsense mutation in the rod photoreceptor cGMP phosphodiesterase β-subunit gene of the *rd* mouse. *Proceedings of the National Academy of Sciences USA* 88: 8322–26

Ray, K., Baldwin, V.J., Acland, G.M., Blanton, S.H. & Aguirre, G.D. (1994) Cosegregation of codon 807 mutation of the canine rod cGMP phosphodiesterase β gene and *rccil*. *Investigative Ophthalmology and Visual Science* 35: 4291–99

Robinson, R. (1990) *Genetics for Dog Breeders*, 2nd edition, Oxford and New York: Pergamon Press

Suber, M.L., Pirtler, S.J. & Qin, N. *et al.* (1993) Irish setter dogs affected with rod/cone dysplasia contain a nonsense mutation in the rod cGMP phosphodiesterase β-subunit gene. *Proceedings of the National Academy of Sciences USA* 90: 3968–72

Wayne, R.K. (1993) Molecular evolution of the dog family. *Trends in Genetics* 9: 218–24

Werner, P., Mellersh, C.S. & Raducha, M.G. *et al.* (1999) Anchoring of canine linkage groups with chromosome-specific markers. *Mammalian Genome* 10: 814–23

Willis, M.B. (1989) *Genetics of the Dog*, London: Witherby and New York: Howell Book House

Yuzbasiyan-Gurkan, V., Blanron, S.H. & Cao, Y. *et al.* (1997) Linkage of a microsatellite marker to the canine copper toxicosis gene in the Beddlington terrier. *American Journal of Veterinary Research* 58: 23–27

Zheng, K., Thorner, P.S., Marrano, P., Baumal, R. & McInnes, R.R. (1994) Canine X chromosome-linked hereditary nephritis: a genetic model for human X-linked hereditary nephritis resulting from a single base mutation in the gene encoding the a5 chain of collagen type IV. *Proceedings of the National Academy of Sciences USA* 91: 3989–93

FURTHER READING

Padgett, G.A. (1998) *Control of Canine Genetic Diseases*, New York: Howell Book House

Scott, J.P. & Fuller, J.L. (1965) *Genetics and the Social Behavior of the Dog*, Chicago: University of Chicago Press

GLOSSARY

acrocentric chromosome a chromosome whose centromere lies very near one end

dysplasia abnormal development of tissues

fluorescence in situ hybridization (FISH) the labelling of specific whole chromosomes by *in situ* hybridization. Hybridization probes are usually made by selecting a particular chromosome by flow sorting, and using the DNA from such chromosomes, after removing repeated sequences that are common to other chromosomes. Also known as chromosome painting, this technique can be used to identify specific chromosomes not only in normal metaphases, but also in interphase nuclei, in chromosomal rearrangements, and to study chromosomal changes during evolution

radiation hybrid (RH) mapping one of several methods for ordering markers along a chromosome, and estimating the physical distances between them. Human cell lines are irradiated (which fragments the DNA) and fused with rodent cells to make hybrids. Assessing the frequency of marker sites remaining together in the same fragment can establish the order and distance between the markers

retinitis pigmentosa a genetic eye disease that affects night vision and peripheral vision

single-stranded conformation polymorphism (SSCP) analysis a method used to detect polymorphisms without sequencing the nucleotide affected

H HUMAN CLINICAL GENETICS

Introduction to Human Clinical Genetics

1. Genetic diseases

The genetic basis of both rare and common diseases provides the subject of most articles in this section, including improvements in predictive testing, and, in some cases, possible gene therapies. In future, a great deal of research into human clinical genetics will be based on the results of the complete human genome sequence. Causal genes for diseases such as cancer, diabetes, obesity, and genes with key roles, for example in immunological processes or addiction vulnerability, will be isolated, leading to the possible development of human gene therapies. While gene therapy remains controversial due to some widely reported failures, clinical trials continue, and successes in animals – e.g. delivering an insulin-encoding gene into diabetic rats and mice – show promise for future human treatment.

Pharmacogenomics – how genes influence a patient's response to drug – also offers the possibility of using information from individual patients' DNA to determine how they are likely to respond to a particular medicine.

Animal models of human disease, traditionally based on transgenic mice, could in future be developed in primates. Previous technical barriers to inserting human genes into primates have been overcome, and the first genetically modified non-human primate (Chan *et al.*, 2001) paves the way for genetically engineering a primate to have a human genetic mutation linked to disease – though this would surely be subject to very stringent ethical assessment. Any real advantages in using primates over mice or rats would also have to be sufficient to ameliorate the extra expense.

2. Cloning

Research on human stem cells could also revolutionize medicine, since stem cells could be used to grow specific transplant cells or tissue to treat a number of serious diseases or injuries. While there are several approaches now in human clinical trials that utilize stem cells isolated from adults, stem cells for all tissue or cell types have not yet been found in adults and the number of stem cells decreases with age, hence the interest in obtaining stem cells from human embryos, and in particular those derived from cloning. Cell nuclear replacement – the process of transferring a patient's DNA into an egg with its nucleus removed – results in a cloned human embryo. Cells or tissue derived from such an embryo are genetically identical to the patient (apart from mitochondrial DNA) and avoids the risk of immune rejection. Such "therapeutic" cloning has been called the "slippery slope" towards full cloning of a human being, though this has been widely condemned.

Regulations on embryonic stem cell research varies in various countries, and there is debate on the ethical issues of destroying embryos, whether these are derived from cloning or fertilized *in vitro* and donated for research. Britain was the first country in the world explicitly to permit the cloning of human embryos for research; in the US, researchers will not be allowed to work with human stem cells generated by cloning, or to use public funds to obtain stem cells by destroying donated embryos (though privately funded labs are permitted).

Stem cells are regarded as having considerable potential as a source of new tissue for therapeutic use, and embryo research may reveal how adult cells can be reprogrammed to make them behave with the full potential of embryonic stem cells. However, a lot of work still remains to be done on how to get the stem cells to form the required tissue – this work could be done with animal cells.

REFERENCES

Chan, A.W.S. *et al.* (2001) Transgenic monkeys produced by retroviral gene transfer into mature oocytes. *Science* 291: 309–12

Department of Health, *Stem Cell Research: Medical Progress with Responsibility*, 2000; available from http://www.doh.gov.uk/ cegc/stemcellreport.htm

THE *WILMS' TUMOUR 1* GENE *WT1*: A TOOL TO EXAMINE THE LINKS BETWEEN CANCER AND DEVELOPMENT

Adrian W. Moore
HHMI, University of California at San Francisco, California, USA

1. Introduction

Paediatric tumours can be regarded as cancers that arise as the result of normal development gone awry. Wilms' tumour of the kidney (nephroblastoma) is perhaps one of the best examples of such a tumour. Presenting as a solid tumour, it affects around one in 10 000 children, usually within the first 5 years of life, and can reach up to 1–2 kg in size. Wilms' tumours arise from a single kidney precursor cell (metanephric blastema cell) in which developmental control has been lost. This cell continues to divide long beyond the point where it would stop during normal kidney formation and its progeny attempt to differentiate into components of the kidney, but are unable to do so correctly.

The *Wilms' Tumour 1* gene (*WT1*) was isolated from 11p13, a chromosomal region disrupted or deleted in a large number of Wilms' tumours. The gene is expressed during the development of the urogenital system and loss of both copies of *WT1* leads to Wilms' tumour formation. In addition to sporadic Wilms' tumours, different mutations (alleles) of *WT1* give rise to three different human congenital syndromes – Wilms' tumour, aniridia (absence of the iris), genital abnormalities and mental retardation (WAGR) syndrome, Denys–Drash syndrome (DDS) and Frasier syndrome. Individuals with these syndromes show genital abnormalities and, in the case of DDS and Frasier syndrome, also nephropathy. Analysis of the phenotype of the different *WT1*-associated syndromes, and that of mice in which the *Wt1* gene has been deleted, identifies *WT1* functions at several different stages of gonad and kidney development. The finding that *WT1* is an important player in kidney organogenesis supports the premise that Wilms' tumours arise as a result of normal kidney developmental processes gone awry. It further allows a more detailed investigation of this method of tumorigenesis.

The protein encoded by the *WT1* gene appears to act as a transcription factor and may also play a role in mRNA processing. Two alternative splice sites, an alternative translation start site and RNA editing lead to the production of at least 16 different versions of the WT1 protein from the *WT1* gene and preliminary evidence suggests that different isoforms of the protein may perhaps be carrying out different molecular functions within the cell.

2. Isolation of the Wilms' tumour gene (*WT1*)

Isolation of a gene, disruption of which predisposes to Wilms' tumour, was aided by loss of heterozygosity (LOH) studies. Such studies aim to determine chromosomal regions that are regularly disrupted within tumours. In chromosomal regions showing LOH, DNA derived from a locus on one chromosome is lost such that the only sequences left are derived solely from the other. LOH can lead to tumour formation if a tumour suppressor gene is carried at the locus affected. The loss of both functional copies of a tumour suppressor gene leads to cancer formation; hence, if one copy of the tumour suppressor is already inactive, LOH may serve to inactivate the other. LOH can occur by a variety of mechanisms, such as mitotic recombination, chromosome loss followed by reduplication of the sister chromosome and deletion of a region of one of two sister chromosomes, as illustrated in **Figure 1**.

LOH studies on Wilms' tumour tissue have demonstrated the major chromosome which harbours a Wilms' tumour gene to be 11; a region of LOH was detected at chromosome 11p in around 30–40% of Wilms' tumours (Mannens *et al.*, 1988). In addition, chromosome 16q showed LOH in around 20% (Maw *et al.*, 1992), and other chromosomes in a smaller percentage, of Wilms' tumours.

LOH studies give only an indication of where a gene involved in tumorigenesis may lie. Identification of chromosome deletions associated with Wilms' tumour allowed investigators to narrow further the region in which a Wilms' tumour suppressor gene was situated. Wilms' tumour is one of an associated congenital set of abnormalities in WAGR syndrome – Wilms' tumour, aniridia, genital abnormalities and mental retardation. Deletions of 11p13 were identified in several WAGR cases (Lewis *et al.*, 1988) and, from the region spanned by these deletions, a potential predisposition gene (*WT1*) was isolated (Call *et al.*, 1990; Gessler *et al.*, 1990). Confirmation that *WT1* was indeed a Wilms' tumour predisposition gene came from the identification of mutations specifically inactivating it within Wilms' tumour tissue.

It is now clear that *WT1* is deleted or mutated in only approximately 15% of Wilms' tumours. LOH studies have indicated that a second Wilms' tumour predisposition gene (*WT2*) is located at 11p15 (**Figure 1**; Koufos *et al.*, 1989). The *WT2* gene may be imprinted – a process whereby the ability of a gene to be expressed depends on the parent from which it is inherited – and loss of imprinting (LOI) occurring in the 11p15 region may also lead to Wilms' tumour formation. *WT1* and *WT2* are both associated with sporadically occurring Wilms' tumours. In addition, a familial Wilms' tumour predisposition gene, *FWT1*, has been localized to the chromosomal region 17q12–q21 (Rahman *et al.*, 1996).

3. The *WT1* gene encodes a potential DNA/RNA-binding protein with several different isoforms

The *WT1* gene consists of ten exons spanning 50 kilobases of genomic sequence and at least 16 different versions

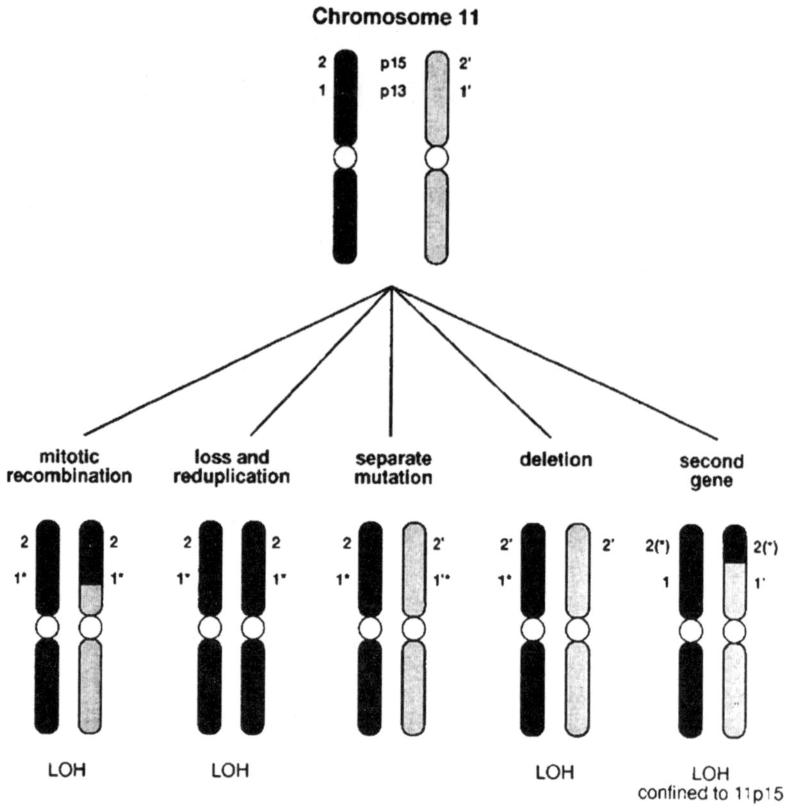

Figure 1 Mechanisms for the functional loss of both copies of putative Wilms' tumour suppressor genes at chromosome 11p. Following inactivation of the first copy of a putative Wilms' tumour suppressor gene, the second copy can be lost either by loss of heterozygosity (LOH) or a second mutation. While *WT1* lies at 11p13, LOH distal to this region or confined to 11p15 may also be involved in Wilms' tumorigenesis, implying that a second predisposition gene (*WT2*) is present in this region.

(isoforms) of the WT1 protein are produced from this gene. Alternative splicing of *WT1* exon 5 inserts 17 amino acids into the amino-terminal (N^+) domain of the WT1 protein (**Figure 2**). Utilization of an alternative splice donor site at the end of *WT1* exon 9 inserts an extra three amino acids, lysine (K), threonine (T) and serine (S), collectively termed KTS, at the carboxy-terminus (C^-). Hence, the WT1 protein can be ±exon 5 and ±KTS (Haber *et al.*, 1991). *WT1*, in addition, utilizes two different translation initiation sites and shows editing of its mRNA. In this editing process, $uracil^{839}$ of the *WT1* message may be converted to cytosine, leading to the replacement of $leucine^{280}$ of the protein by a proline (Sharma *et al.*, 1994).

The *WT1* gene has been isolated from a variety of different vertebrate organisms, but is not found outside this lineage. Both the ±KTS and ±exon 5 isoforms are present in all mammals investigated. Examples of birds (chicken), reptiles (alligator and turtle), amphibians (frog and axolotl) and fish (pufferfish and zebrafish) all still have the ±KTS, but not the ±exon 5, isoforms. The evolutionary conservation of these alternative WT1 protein isoforms, in particular the ± KTS versions, implies that each is of functional importance (Kent *et al.*, 1995).

The WT1 protein acts to regulate transcription within the cell. Its carboxy-terminal domain contains four zinc fingers. Zinc finger structures contain two cysteine (C) and two histidine (H) amino acid residues which chelate a zinc ion to create the 3-dimensional tertiary structure of the finger (**Figure 2**). They are found in over 200 proteins and bind DNA with a specificity dependent on the sequence of the target. The N-terminal domain of the WT1 protein is rich in the amino acids proline and glutamine, and such domains, along with zinc fingers, are indicative of transcription factors (proteins which regulate the transcription of genes). As with other transcription factors, WT1 protein is localized within the cell nucleus and is indeed able to regulate transcription of artificial target genes introduced into cultured cells.

WT1 may act not only at the level of transcription, but also in RNA splicing. In foetal kidneys, testis and *WT1*-expressing cell lines, the WT1 protein colocalizes within the nucleus along with, and can be purified complexed to, members of the RNA splicing machinery. WT1 is capable of binding RNA through its first zinc finger and also contains an RNA recognition motif which is common to many proteins interacting with RNA. This type of dual activity for a gene, first at the level of gene transcription and secondly at RNA processing, is not unprecedented – the TFIIIA transcription factor, for example, binds both DNA and RNA via its zinc fingers in order to regulate transcription and post-transcription processes (Theunissen *et al.*, 1992).

The separate functions of WT1 in transcription and splicing may be dependent on the different WT1(+KTS) and WT1(−KTS) protein isoforms. The three amino acids inserted in the WT1(+KTS) isoform lie between the third and fourth zinc fingers of the protein. This insertion changes the distance between the two zinc fingers; isoforms −KTS bind DNA with much higher affinity and a different sequence specificity from those containing the three amino acids. Hence, it is possible that these two different WT1

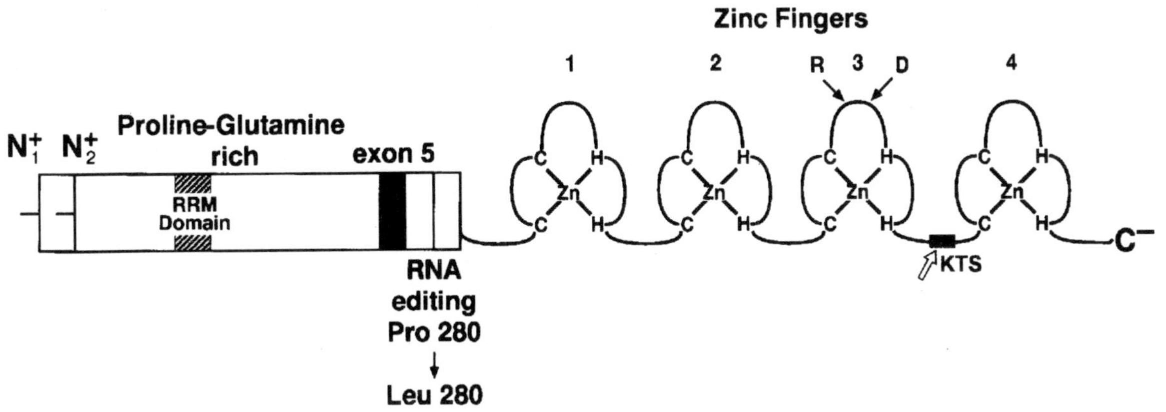

Figure 2 Structure of the 16 different isoforms of the mammalian WT1 protein. The mammalian WT1 protein has at least 16 different isoforms due to two alternative splice sites +/−exon 5 and +/−KTS (black boxes), RNA editing and two alternative translation start sites (dotted lines). The most common mutations in Denys–Drash syndrome (DDS) are illustrated: R, arginine[394]; D, aspartate[396]. Mutations in Frasier syndrome occur in the splice donor site at the end of exon 9 (open arrow).

protein isoforms regulate different sets of target genes. Alternatively, WT1(−KTS) may preferentially act as a transcription factor and the WT1(+KTS) isoform may be involved in RNA splicing. When artificially introduced into cultured cells, the WT1(+KTS) and the WT1(−KTS) isoforms localize to different compartments within the nucleus. WT1(−KTS) isoforms localize to areas containing transcription factors, while WT1(+KTS) isoforms localize to areas containing components of the RNA splicing machinery (Larsson *et al.*, 1995).

4. How do *WT1* mutations lead to Wilms' tumour formation?

Mutation of *WT1* causes around 15% of Wilms' tumours. We are only just beginning to perceive how the loss of two copies of the *WT1* gene leads to the formation of tumours. Wilms' tumours contain immature structures which are normally present during embryonic kidney formation, but lack mature structures such as glomeruli. The presence of embryonic structures within the tumours implies that they have derived from a single undifferentiated kidney precursor cell (metanephric blastema cell) which has exited the normal developmental pathway. Clusters of cells resembling metanephric blastema, termed "nephrogenic rests", are often found in kidneys of children with Wilms' tumour. It has been suggested that they could represent precursors to Wilms' tumours in which one copy of the *WT1* gene has been lost; the loss of the second *WT1* copy within the rest could then lead to tumour formation (Beckwith *et al.*, 1990). The relationship between nephrogenic rests and Wilms' tumours is, in practice, unclear; however, a nephrogenic rest and a neighbouring Wilms' tumour sharing the same *WT1* mutation have been reported.

One idea of how the loss of a second copy of *WT1* triggers tumour formation is that WT1 protein in normal kidney development could act to reduce the expression of growth factor genes, expression which often induces cells to proliferate. Loss of WT1 activity in kidney stem cells could, hence, hypothetically lead to increased growth factor activity which would, in turn, cause unchecked metanephric blastema proliferation. It has been demonstrated in artificial cell culture experiments that WT1 can reduce the expression of the *insulin-like growth factor 2* (*IGF2*) gene (Drummond *et al.*, 1992) and the *epidermal growth factor receptor* (*EGFR*) gene (Englert *et al.*, 1995). Both of these genes are expressed during kidney development and at high levels in some tumours and, thus, are potential targets for WT1 repression *in vivo*.

It is very likely that the *IGF2* gene is a target for WT1 activity. Indeed, *IGF2* itself may be the second Wilms' tumour predisposition gene *WT2*. *IGF2* lies at 11p15 and is imprinted, and this region shows LOH and LOI in a significant proportion of Wilms' tumours (**Figure 1**). In the great majority of these tumours, LOI occurs on the maternally inherited chromosome such that *IGF2* becomes overexpressed (Ogawa *et al.*, 1993). Wilms' tumours associated with LOH and LOI at 11p15 are often part of another congenital syndrome, Beckwith–Wiedemann syndrome (BWS).

A further set of targets for WT1 regulation may be the genes of the muscle differentiation pathway. In some Wilms' tumours, in addition to nephric structures, foci of muscle, cartilage and bone occur. In particular, it has been shown that a high level of muscle within a Wilms' tumour correlates with *WT1* mutation having led to the tumour formation (Miyagawa *et al.*, 1998).

5. *WT1* expression during embryonic development correlates with a role in the development of the genitourinary system

Following the isolation of the *WT1* gene, its expression during embryonic development was investigated. It quickly became clear that, in addition to its tumour suppressor function, *WT1* was involved in the development of the urogenital system. *WT1* is expressed in the kidneys and gonads and also the mesothelium, an epithelial sheet of cells which surrounds the internal organs of the body. Interestingly, the ratio of different WT1 protein isoforms remains fixed in all these structures throughout development, and

WT1 expression within the urogenital system and mesothelium is conserved throughout vertebrates including birds, reptiles and amphibians (Kent *et al.*, 1995). Is there any common theme running between these different areas of *WT1* expression? It has been suggested that these areas are sites where cells are undergoing a transition from a mesenchymal (irregular packing) to an epithelial (flat sheet-like) state and *WT1* could possibly be important for this process.

WT1 expression in the developing kidney has been carefully studied. The expression pattern of the gene during this process implies that it may have different roles at several stages of nephrogenesis and not simply in a mesenchymal to an epithelial cell-type transition (**Figure 3**). Development of the kidney begins when the ureteric bud, an outgrowing branch from the mesonephric duct, invades the metanephric blastema. The metanephric blastema cells lie within the body wall of the embryo; they represent the precursor cells from which the kidney derives and express *WT1* at low levels. Invasion of the ureteric bud induces these cells to condense at its tip and an upregulation of *WT1* levels then occurs within them. The spectra of nephric structures found within a Wilms' tumour imply that it is at this stage of development where loss of *WT1* function can lead to tumorigenesis. Following invasion, the ureteric bud proceeds to divide, forming many branches within the metanephric blastema and inducing the blastema to condense at each branch. This condensed mesenchyme undergoes a burst of proliferation and then differentiates into a hollow ball of epithelial cells known as the renal vesicle. *WT1* continues to be transcribed within the cells of the renal vesicle, which now undergoes a series of morphological changes. It first forms a comma-shaped then S-shaped body, with *WT1* expression localizing to cells at the proximal part of the S-shaped body. These cells then become flattened and form the glomerular podocyte cells. *WT1* expression continues in the podocytes after birth and into adult life, implying that the gene has a role in the maintenance of these terminally differentiated cells.

6. Several roles for *WT1* in the normal development of the genitourinary system are revealed by the study of three human congenital syndromes

WT1 gene expression patterns during embryonic development imply that the gene is involved in urogenital system formation. Confirmation that this was indeed the case came from the discovery that *WT1* mutation is responsible for developmental abnormalities of the gonad and kidney in three human congenital syndromes: WAGR, DDS and Frasier syndrome. Disruption of the same gene is able to cause these three different disorders because the *WT1* mutations associated with each of these disorders have a different effect at the molecular level. Nearly all mutations described in sporadic Wilms' tumours are nonsense or frameshift mutations, which are localized anywhere throughout the gene. However, WAGR appears to be caused by *WT1* haplo-insufficiency, DDS by dominant negative mutations and Frasier syndrome by disruption of the splice ratio between different WT1 protein isoforms (**Table 1**).

(i) *WAGR:* WT1 *haplo-insuffiency causes failure of gonad development and increases the probability of Wilms' tumour*

WAGR patients have a deletion of the region of chromosome 11p13 including the *WT1* gene. Male WAGR patients (i.e. with one X and one Y chromosome) develop genital abnormalities ranging from undescended testes and misplaced urethral opening on the penis to external female genitalia (pseudohermaphroditism); such abnormalities do not occur in females because being female is the default state for a developing mammalian embryo. WAGR-associated genital abnormalities correlate with the loss of one copy of the *WT1* gene and hence halving of the gene dose (haplo-insufficiency) (van Heyningen *et al.*, 1997). Aniridia in WAGR patients is similarly due to the loss of one copy of the *PAX6* gene which lies adjacent to *WT1* at 11p13.

(ii) *DDS: dominant negative* WT1 *mutations cause nephropathy, failure of gonad development and increase the probability of Wilms' tumours*

Wilms' tumours and genitourinary abnormalities along with nephropathy (diffuse mesangial sclerosis) also appear in DDS. Genital abnormalities in DDS patients are more severe than those in WAGR: male (XY) DDS patients often have pseudohermaphroditism, their gonads being a mixture of male (testis) and female (ovary) structures. In other patients, symptoms may be more severe still, with both male (XY) and female (XX) patients failing to develop any gonad structures at all. The nephropathy which occurs in DDS is also a product of *WT1* mutation and leads to renal failure soon after birth.

Table 1 *WT1* mutations and mechanism of action associated with different human pathologies.

Syndrome	Symptoms	*WT1* mutation	Mechanism of action
Sporadic WTs	WT	Nonsense or frameshift mutations	Both copies of *WT1* inactivated
WAGR	Predisposition to WT, GU abnormalities	Deletion of one allele	Haplo-insufficiency
Denys–Drash syndrome	Predisposition to WT, severe GU abnormalities, nephropathy	Truncation before, or disruption of, zinc fingers	Dominant negative
Frasier syndrome	Severe GU abnormalities, nephropathy	Disruption of exon 9 splice donor site	Reduction in WT1 +KTS:−KTS ratio

WAGR, Wilms' tumour, aniridia, genital abnormalities and mental retardation syndrome; WT, Wilms' tumour; GU, genitourinary system.

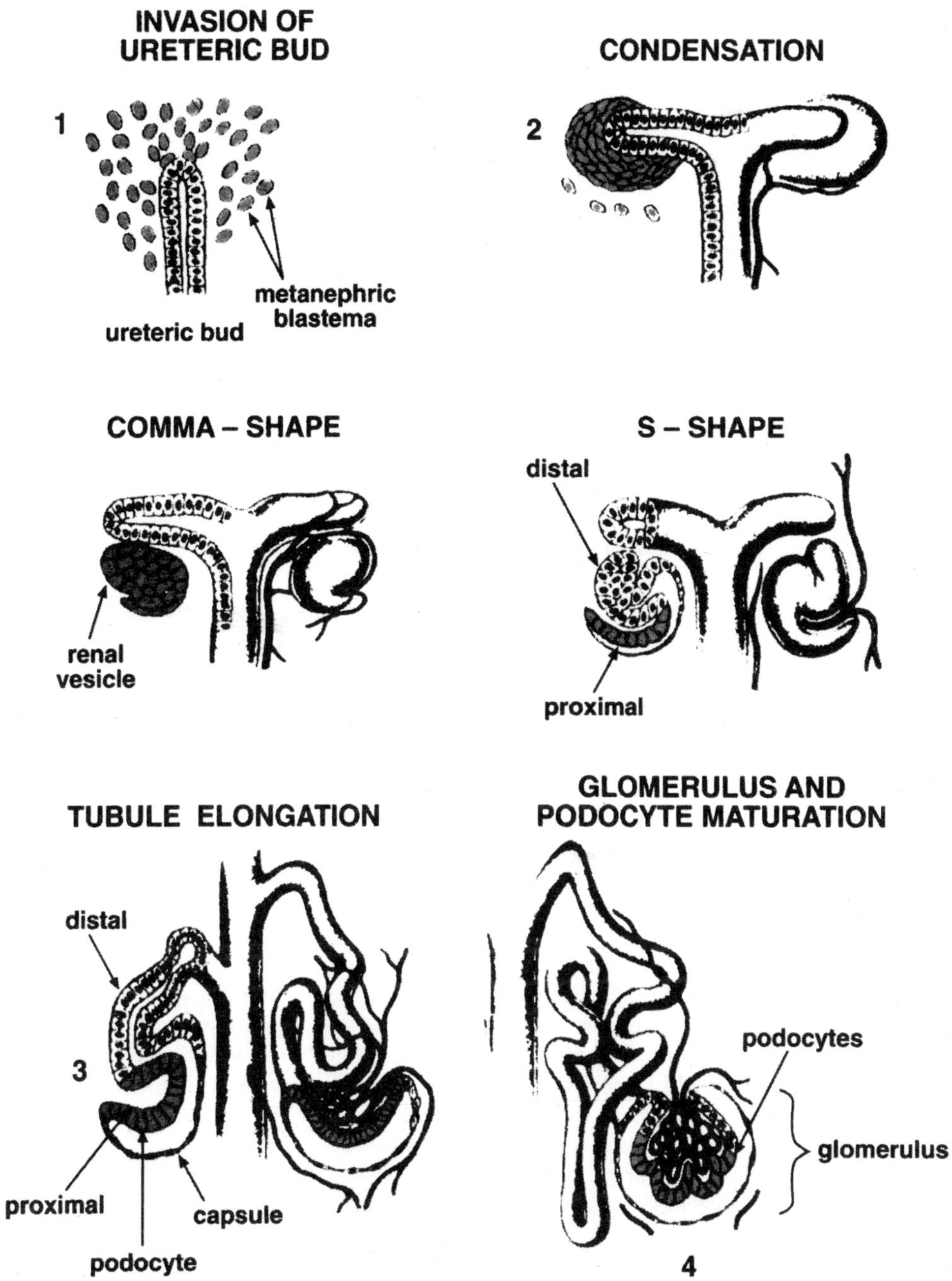

Figure 3 Schematic illustration of *WT1* expression and potential function during nephrogenesis. Light shading, low levels of *WT1* expression; dark shading, high level of *WT1* expression. *WT1* is expressed in the metanephric blastema prior to induction by the ureteric bud. It is progressively localized during nephrogenesis to the podocytes of the glomerulus where it remains expressed in adult life. *WT1* has several functions during nephrogenesis. (1) *WT1* is required for the initial outgrowth of the ureteric bud and the subsequent induction of the metanephric blastema. (2) Loss of *WT1* at the condensation stage may lead to Wilms' tumour formation. (3) *WT1* is required for the transition from S-shaped body to mature glomerulus. (4) *WT1* is required for the correct functioning of the mature glomerulus.

WT1 mutations which cause DDS either truncate the WT1 protein before the zinc fingers or occur within the fingers and disrupt their DNA-binding capacity (Pelletier *et al.*, 1991). The majority of DDS patients have missense mutations in WT1 zinc fingers 2 and 3 (**Figure 2**). These mutations specifically affect amino acids which either chelate a zinc ion or interact with the DNA target, with 60% of these mutations affecting either arginine^{394} (R) or aspartate^{396} (D) of zinc finger 3 (**Figure 2**). The crystal structure of DNA binding by the zinc finger protein early growth response 1, which has zinc fingers identical to WT1 zinc fingers 2–4, demonstrates that R^{394} binds a guanidine base in the target DNA molecule and D^{396} interacts with R^{394} to stabilize this interaction (Pavletich & Pabo, 1991).

Intriguingly, the loss of DNA-binding capacity in DDS-associated mutant WT1 proteins causes these proteins to localize within the same nuclear compartments as the WT1(+KTS) proteins isoform. Therefore, this subnuclear localization may be part of the mechanism by which DDS-associated mutant WT1 proteins exert their effect (Larsson *et al.*, 1995). As DDS mutant proteins are unable to bind DNA, it is unlikely that they act by having gained new functional targets (a neomorphic mutation). The WT1 protein is able to bind to itself and, hence, the protein may act within the cell, not as a single molecule but as a complex structure containing several WT1 molecules. When extracts of cell nuclei are separated (fractionated) according to size, WT1 can be detected in fractions containing structures much larger than free WT1 protein. Hence, it is more likely that mutations associated with DDS exert a dominant negative effect; that is, in WT1-containing complexes, the binding of a single mutant WT1 molecule can poison the functioning of the whole complex. In this way, just a small amount of mutant protein in a cell can severely inhibit gene activity and, in DDS patients, potentially half the WT1 protein within the cell is mutant.

(iii) Frasier syndrome: disruption of the normal ratio of WT1 *isoforms leads to nephropathy and failure of gonad development*

Frasier syndrome has characteristics similar to those of DDS. Patients with Frasier syndrome never have Wilms' tumours, but males have female external genitalia and failure of gonad development. They also have nephropathy (glomerular sclerosis) which leads to kidney failure in adolescence or early adulthood. Mutations in the *WT1* gene leading to Frasier syndrome occur in the splice donor site at the end of exon 9 (open arrow, **Figure 2**) and result in the loss of +KTS isoform translation from the mutated allele. These mutations occur in a heterozygous state alongside a normal *WT1* allele and disrupt the ratio of +KTS:−KTS WT1 isoforms within the cell (Barbaux *et al.*, 1997). This ratio alteration is enough to lead to the drastic symptoms of Frasier syndrome. Intriguingly, however, predisposition to Wilms' tumours is not part of Frasier syndrome; perhaps the presence of the full complement of WT1(−KTS) is enough to prevent tumour development.

The analysis of the spectra of *WT1* mutations and their associated phenotypes in humans allows us to propose a number of functions for the *WT1* gene. In addition to having a kidney tumour suppressor phenotype, it is required for the correct functioning of the mature glomerulus, as was implied by expression of the gene in the podocyte throughout adulthood. Furthermore, the gene is absolutely required for gonad development. In some DDS patients, no gonads develop at all. However, in other DDS, WAGR and Frasier syndrome, patients' gonads do form but, in males in particular, cannot follow a proper developmental pathway. These patients may develop external female structures and even internal female gonads, implying that *WT1* may in some way influence the sex determination pathway.

7. Is *WT1* involved in tumorigenesis in tissues other than the kidney?

(i) Gonad tumours

As *WT1* is expressed in the gonad it seems probable that *WT1* may have a tumour suppressor role in this organ in addition to the kidney. An extensive search has failed to connect *WT1* mutation to any sporadic gonadal tumour types. However, a small number of girls with DDS have developed (granulosa cell) tumours of the ovary in which the second allele of *WT1* has been lost.

(ii) Leukaemia

WT1 is involved in leukaemia formation, although, in this cancer, it may not act as a tumour suppressor. A small number of acute leukaemia patients have heterozygous mutations in the *WT1* gene leading to gene truncation; however, over 80% of patients show abnormally high levels of *WT1* expression within their tumours. The elevated level of expression of the gene in these acute leukaemias correlates with a poor outlook for cancer treatment and survival (Inoue *et al.*, 1994). The clinical importance of *WT1* gene function in leukaemia development is clear; however, what is much less clear is what role the gene is playing. *WT1* expression has been detected in a small subset of early haematopoietic (blood) precursor cells. Hence, perhaps the high expression levels of *WT1* in some leukaemias indicates that these tumours are derived from haematopoietic stem cells stuck at an early stage of differentiation.

(iii) Desmoplastic small round-cell tumour

Alongside the urogenital system, *WT1* is expressed in the mesothelium. No cases of tumours derived from this tissue are associated with mutations of *WT1*. However, desmoplastic small round-cell tumour (DSRCT), which arises mainly in tissues of mesothelial origin, is associated with the *WT1* gene. In DSRCT, the N-terminal domain of the *Ewing's sarcoma* gene and the C terminal end of *WT1* become fused by chromosomal rearrangements (Ladanyi & Gerald, 1994).

In general, therefore, *WT1* does not appear to play a tumour suppressor role in tissues other than the kidney. This is not uncommon; a number of tumour suppressor genes show suppressor activity in only a subset of the tissues in which they are expressed. For example, the *retinoblastoma* (*RB*) gene is expressed widely in the body, but loss of *RB* function only leads to cancer in the retina of the eye. In leukaemia and DSRCT, *WT1* does not necessarily exert a tumour suppressor effect; the mechanism of tumorigenesis may be quite different. In fact, *WT1* expression in some circumstances can actually cause cancer (oncogenesis): cells artificially expressing *WT1*, transplanted into mice, can cause tumour formation, for example.

8. Use of the mouse as a model organism to study *WT1* function

(i) Knockout of Wt1 *demonstrates its requirement for kidney and gonad induction and heart development*

For more direct experimental analysis of *WT1* function in development and tumorigenesis, a model organism is required. To this end, the *Wt1* gene in mice was disrupted to make a null (completely non-functional) allele. Mice heterozygous for one *Wt1* null allele are phenotypically normal and, surprisingly, never develop Wilms' tumours. Both the kidneys and gonads of homozygous null *Wt1* mice fail to develop at all, which reflects the importance for *WT1* in urogenital development illustrated by WAGR, DDS and Frasier syndrome patients. The effect of the *Wt1* null mutation is best characterized in the kidney. The metanephric blastema forms, but the ureteric bud does not branch from the mesonephric duct and, as no kidney induction occurs, these cells die. Isolated and cultured *Wt1* null metanephric blastema is unable to condense and begin to proliferate even if it is given proper induction (Kreidberg *et al.*, 1993).

Hence, *WT1* plays two roles at this early stage of kidney development: it is required for both the branching of the ureteric bud and the ability of the metanephric blastema to respond to this branching.

Mice homozygous for the *Wt1* null allele die *in utero* around halfway through gestation (day 13–15). The body cavities of these *Wt1* null embryos are swollen with excess fluid, which is symptomatic of heart failure. The heart of these animals is malformed although it itself does not express the *Wt1* gene. However, *Wt1* is expressed in the epicardium (visceral pericardium), a mesothelial tissue which surrounds the heart. This tissue fails to develop correctly in the *Wt1* null embryos and is required for the correct functioning of the heart itself.

(ii) Investigation of WT1 *functions in later nephrogenesis*

The continued expression of *WT1* throughout nephrogenesis, its expression in the podocytes and the nephropathy of DDS and Frasier syndrome patients implies that *WT1* has functions in nephrogenesis other than those identified in the *Wt1* null mice. We have used the mouse model to identify further functions of *WT1* at stages of nephrogenesis later than induction by introducing a yeast artificial chromosome (YAC) which carried the entire human *WT1* locus into mice as a transgene. Mice null for the endogenous *Wt1* gene, but carrying the *WT1* YAC, do not die *in utero* but have a normal heart and survive until birth. In many of these mice, no kidney or gonads were formed. In a subset of the *WT1* YAC-containing *Wt1* null animals, a kidney began to form but did not complete nephrogenesis; in these animals, kidney development arrested at the transition from S-shaped body to mature glomerulus, demonstrating that *WT1* is required for the final maturation of the glomerulus. The requirement for *WT1* in glomerulus maturation is especially interesting given that in Wilms' tumours, although elements representing all stages of the process of nephrogenesis can be present, no fully formed glomeruli are ever found (Moore *et al.*, 1999).

A further experiment replaced the endogenous mouse *Wt1* gene with a truncated version similar to that found in some DDS patients. Mice carrying this mutant gene now begin to develop nephropathy and, very rarely, Wilms' tumour (Patek *et al.*, 1999).

9. Summary

The *WT1* gene does not just act as a tumour suppressor, but has multiple roles in development. Some of these developmental functions of *WT1* are most clearly illustrated during nephrogenesis (summarized in **Figure 3**), although the gene probably plays similar roles in the developing gonad. At the first stage of nephrogenesis, *WT1* is active in the metanephric blastema, but not the ureteric bud. Activity of the gene is required within the metanephric blastema cells, first, to promote the branching of the ureteric bud from the mesonephric duct and, secondly, to allow the metanephric blastema, once induced, to begin to proliferate (**Figure 3.1**). Intriguingly, loss of *WT1* function at this stage leads to the death of the blastema cells, but once the cells have condensed and begun to proliferate, loss of *WT1* function leads to tumour formation (**Figure 3.2**). What is peculiar to a blastemal cell or the environment in which it lies at this potentially tumorigenic stage remains an unanswered question. During the later stages of nephrogenesis, *WT1* becomes localized to the presumptive podocytes of the glomerulus, and it is required for the final step in the maturation of these cells and the glomerulus as a whole (**Figure 3.3**). Finally, *WT1* is required for the correct functioning of the glomerulus once it has fully developed. In the absence of proper gene function, either due to a dominant DDS mutant WT1 protein or to a disruption in WT1 isoform ratios, nephropathy occurs (**Figure 3.4**).

The +KTS and −KTS isoforms of the WT1 protein have altered spacing between the third and fourth zinc fingers of the protein. They are conserved throughout vertebrate evolution and are localized to different compartments of the nucleus. Frasier syndrome-associated *WT1* mutations reduce the amount of WT1(+KTS) within the cell and disrupt the invariant WT1 isoform ratio. Intriguingly, Frasier syndrome patients do not get Wilms' tumours. Could, then, the tumour suppressor activity of *WT1* be solely due to the −KTS isoform? WT1(−KTS) binds DNA with strong affinity and localizes within the nucleus with other transcription factors. We could hypothesize that the WT1(−KTS) isoform acts as a tumour suppressor by repressing the activity of growth factor genes such as *IGF2* and *EGFR*. The WT1(+KTS) isoform may have an entirely different role – perhaps in RNA processing.

Study of the *WT1* gene has already given us new insights into the development of the urogenital system, but we have much yet to understand. What is it about the condensation stage of nephrogenesis which means that when a cell loses *WT1* function it proliferates and forms a tumour, while at other stages it dies? What is the significance of the different isoforms of WT1? We intend to use the mouse model system to disrupt the levels of all the different WT1 isoforms. Furthermore, WAGR, DDS and Frasier syndrome all demonstrate that *WT1* is probably required at several stages of gonad development; we will also use the mouse model to attempt to investigate this.

At the molecular level, we need to determine which DNA targets WT1 binds to within the cell. Do *IGF2* and *EGFR* represent true *in vivo* targets for WT1? Is the transcription factor activity of WT1 solely the preserve of the WT1(−KTS) isoform and WT1(+KTS) associated with the splicing machinery? If WT1 is associated with splicing, does it alter the activity of this machinery and with which mRNA molecules is it associated?

We are beginning to try and answer some of these new questions. Although the *WT1* gene was originally isolated as a Wilms' tumour suppressor gene, it is now clear that it has a multitude of functions beyond this. It still remains to be understood what causes a kidney stem cell to step outside its normal developmental pathway and go on to form a tumour. However, whatever this function is, it is peculiar to the effect of the *WT1* gene in one organ, the kidney, and at one stage of development, the condensation stage.

REFERENCES

Barbaux, S. Niaudet, P., Gubler, M.C. *et al.* (1997) Donor splice-site mutations in WT1 are responsible for Frasier syndrome. *Nature Genetics* 17(4): 467–70

Beckwith, J.B., Kiviat, N.B. & Bonadio, J.F. (1990) Nephrogenic rests, nephroblastomatosis, and the pathogenesis of Wilms' tumour. *Pediatric Pathology* 10(1–2): 1–36

Call, K.M., Glaser, T., Ito, C.Y. *et al.* (1990) Isolation and characterization of a zinc finger polypeptide gene at the human chromosome 11 Wilms' tumour locus. *Cell* 60(3): 509–20

Drummond, I.A., Madden, S.L., Rohwer-Nutter, P. *et al.* (1992) Repression of the insulin-like growth factor II gene by the Wilms tumour suppressor WT1. *Science* 257: 674–78

Englert, C., Hou, X., Maheswaran, S., Bennett, P. *et al.* (1995) WT1 suppresses synthesis of the epidermal growth factor receptor and induces apoptosis. *EMBO Journal* 14(19): 4662–75

Gessler, M., Poustka, A., Cavenee, W. *et al.* (1990) Homozygous deletion in Wilms' tumours of a zinc-finger gene identified by chromosome jumping. *Nature* 343: 774–78

Haber, D.A., Sohn, R.L., Buckler, A.J. *et al.* (1991) Alternative splicing and genomic structure of the Wilms' tumour gene WT1. *Proceedings of the National Academy of Sciences USA* 88(21): 9618–22

Inoue, K., Sugiyama, H., Ogawa, H. *et al.* (1994) WT1 as a new prognostic factor and a new marker for the detection of minimal residual disease in acute leukemia. *Blood* 84(9): 3071–79

Kent, J., Coriat, A.M., Sharpe, P.T., Hastie, N.D. & van Heyningen, V. (1995) The evolution of WT1 sequence and expression pattern in the vertebrates. *Oncogene* 11(9): 1781–92

Koufos, A., Grundy, P., Morgan, K., Aleck, K.A. *et al.* (1989) Familial Wiedemann–Beckwith syndrome and a second Wilms' tumour locus both map to 11p15. *American Journal of Human Genetics* 44(5): 711–19

Kreidberg, J.A., Sariola, H., Loring, J.M. *et al.* (1993) WT1 is required for early kidney development. *Cell* 74(4): 679–91

Ladanyi, M. & Gerald, W. (1994) Fusion of the EWS and WT1 genes in the desmoplastic small round cell tumour. *Cancer Research* 54(11): 2837–40

Larsson, S.H., Charlieu, J.P., Miyagawa, K. *et al.* (1995) Subnuclear localization of WT1 in splicing or transcription factor domains is regulated by alternative splicing. *Cell* 81(3): 391–401

Lewis, W.H., Yeger, H., Bonetta, L., Chan, H.S. *et al.* (1988) Homozygous deletion of a DNA marker from chromosome 11p13 in sporadic Wilms' tumour. *Genomics* 3(1): 25–31

Mannens, M., Slater, R.M., Heyting, C. *et al.* (1988) Molecular nature of genetic changes resulting in loss of heterozygosity of chromosome 11 in Wilms' tumours. *Human Genetics* 81(1): 41–48

Maw, M.A., Grundy, P.E., Millow, L.J. *et al.* (1992) A third Wilms' tumour locus on chromosome 16q. *Cancer Research* 52(11): 3094–98

Miyagawa, K., Kent, J., Moore, A. *et al.* (1998) Loss of WT1 function leads to ectopic myogenesis in Wilms' tumour. *Nature Genetics* 18(1): 15–17

Moore, A.W., McInnes, L., Kreidberg, J. *et al.* (1999) YAC complementation shows a requirement for WT1 in the development of epicardium, adrenal gland and throughout nephrogenesis. *Development* 126(9): 1845–57

Ogawa, O., Eccles, M.R., Szeto, J. *et al.* (1993) Relaxation of insulin-like growth factor II gene imprinting implicated in Wilms'. *Nature* 362: 749–51

Patek, C.E., Little, M.H., Fleming, S. *et al.* (1999) A zinc finger truncation of murine WT1 results in the characteristic urogenital abnormalities of Denys–Drash syndrome. *Proceedings of the National Academy of Sciences USA* 96(6): 2931–36

Pavletich, N.P. & Pabo, C.O. (1991) Zinc-finger DNA recognition: crystal structure of a Zif268-DNA complex at 2.1A. *Science* 252: 809–17

Pelletier, J., Bruening, W., Kashtan, C.E. *et al.* (1991) Germline mutations in the Wilms' tumor suppressor gene are associated with abnormal urogenital development in Denys–Drash syndrome. *Cell* 67(2): 437–47

Pritchard-Jones, K., Fleming, S. & Davidson, D. (1990) The candidate Wilms' tumour gene is involved in genitourinary development. *Nature* 346: 194–97

Rahman, N., Arbour, L., Tonin, P. *et al.* (1996) Evidence for a familial Wilms' tumour gene (FWT1) on chromosome 17q12-q21. *Nature Genetics* 13(4): 461–63

Sharma, P.M., Bowman, M., Madden, S.L. *et al.* (1994) RNA editing in the Wilms' tumour susceptibility gene, WT1. *Genes and Development* 8(6): 720–31

Theunissen, O., Rudt, F., Guddat, U. *et al.* (1992) RNA and DNA binding zinc fingers in Xenopus TFIIIA. *Cell* 71(4): 679–90

van Heyningen, V. (1997) Sugar and spice and all things splice? *Nature Genetics* 17: 367–68

FURTHER READING

Hastie, N.D. (1994) The genetics of Wilms' tumour – a case of disrupted development. *Annual Review of Genetics* 28: 523–58

Menke, A.L. (1998) The Wilms' tumour 1 gene: oncogene or tumour suppressor gene? *International Journal of Cytology* 181: 151–212

GLOSSARY

aniridia absence of the iris caused by mutation of one copy of the *PAX6* gene

congenital syndrome a collection of associated disorders which are present at or before birth

Denys–Drash syndrome (DDS) a congenital syndrome consisting of severe genitourinary abnormalities, glomerular nephropathy (diffuse mesangial sclerosis) and Wilms' tumour

desmoplastic small round-cell tumour (DSRCT) an aggressive cancer which is usually associated with the abdominal mesothelium. It primarily affects children and young adults

dominant negative a single mutated allele causes a dominant effect by disrupting the function of the other (wild-type) allele

***Ewing's Sarcoma* gene** (*EWS*) the *EWS* gene is mutated in some Ewing's sarcomas (a form of bone cancer). It is also involved in several tumour related chromosomal translocations which result in the fusion of *EWS* with other genes proposed to act as transcription factors

haplo-insufficiency haplo-insufficiency occurs if the loss of one allele of a gene is enough to give rise to a phenotypic effect. Haplo-insufficiency hence infers that the correct level of a gene product is essential for its normal function

insulin-like growth factor **2 gene** (*IGF2*) a growth factor gene involved in foetal development which is a putative target for regulation by WT1. *IGF2* is a causative factor in Beckwith–Wiedemann syndrome and may be involved in the genesis of Wilms' tumours showing LOH and LOI at 11p15

KTS a triplet of amino acids which consists of lysine (K), threonine (T) and serine (S). The KTS amino acid triplet may be inserted into the WT1 protein via alternative splicing at the end of exon 9

loss of heterozygosity (LOH) DNA derived from a region on one chromosome is lost. Hence the only DNA sequences remaining in this chromosomal region derive solely from the second chromosome

loss of imprinting (LOI) In certain regions of the genome the ability of an allele of a gene to be activated depends upon which parent it was inherited from (termed imprinting). Loss of imprinting occurs when this mechanism breaks down and both alleles are now able to be simultaneously activated or repressed

nephropathy kidney disease

pseudohermaphroditism a condition in which an individual chromosomally of one sex has the external genitals of the other

tumour suppressor gene the loss of function of a tumour suppressor gene leads to cancer formation

WAGR syndrome a congenital syndrome that consists of Wilms' tumour, aniridia, genital abnormalities and mental retardation

TRINUCLEOTIDE REPEAT INSTABILITY AS A CAUSE OF HUMAN GENETIC DISEASE

Catherine A. Walker and Catherine M. Abbott
University of Edinburgh Medical Genetics Section, Western General Hospital, Edinburgh, UK

1. Introduction

The phenomenon of anticipation, where a single gene disorder demonstrates a progressively earlier age of onset in successive generations, was first described as early as 1911, in relation to myotonic dystrophy (DM). Controversy as to whether this was a genuine observation or simply ascertainment bias was still occurring until the early 1990s when, for the first time, a molecular mechanism was discovered which could explain and validate the observation of anticipation. This molecular mechanism is the unstable expansion of trinucleotide (or triplet) (TR) repeats, and it has subsequently been found to be the mutational basis for more than a dozen human single-gene disorders. In each case, the TR concerned is polymorphic in terms of repeat number in normal individuals, but becomes pathogenic when it expands into an abnormal size range, the exact size of which depends on the disorder concerned. A key feature of this molecular mechanism is what has been termed "dynamic mutation"; that is, the longer the repeat, the greater its tendency to expand even further. It is this feature which explains so neatly the observation of anticipation. An additional feature of some, but not all, TR expansion disorders is the occurrence of somatic instability, where a variety of repeat lengths are seen even within a single tissue type of one individual.

TRs can be made up from any one of ten different sequence motifs, as shown in **Figure 1**. Of the ten possible motifs, only three (CCG, CAG and GAA) have so far been implicated in disease. The other classes of repeat motifs are found throughout the genome, but many are not associated with genes and are, therefore, unlikely to cause phenotypes; the remainder have not undergone expansion or have not yet been associated with a phenotype.

The group of disorders caused by TR expansion can be broken down into a number of subgroups, each of which has a particular key feature. These are summarized in **Table 1**, and several diseases are described in more detail below.

2. Fragile sites

There are three cases of disorders associated with rare, folate-sensitive fragile sites which are now known to be caused by TR expansions. Fragile sites are cytogenetically visible, non-staining gaps or breaks on metaphase chromosomes which can be induced in cultured cells under certain conditions. Folate-sensitive fragile sites can be detected in the chromosomes of cells cultured in media containing low levels of deoxythymidine triphosphate (dTTP) or deoxycytidine triphosphate (dCTP). These conditions can be achieved by omitting folic acid (hence, folate-sensitive sites) and thymidine, by the addition of folate metabolism inhibitors or thymidylate synthesis inhibitors, or by increasing the concentration of thymidine.

The most well studied of these folate-sensitive sites causes

AAT ATA TAA ATT TAT TTA	AAC ACA CAA GTT TGT TTG	ACC CAC CCA GGT GTG TGG	ACT TAC CTA AGT TAG GTA	ACG CGA GAC CGT GTC TCG
AAG ***AGA*** ***GAA*** ***TCT*** ***CTT*** ***TTC***	***AGC*** ***GCA*** ***CAG*** ***GCT*** ***CTG*** ***TGC***	AGG GAG GGA CCT CTC TCC	ATC TCA CAT GAT ATG TGA	***CCG*** ***CGC*** ***GCC*** ***CGG*** ***GCG*** ***GGC***

Figure 1 Trinucleotide sequence motifs, showing all possible units of trinucleotide repeats; for each class, there are six possible ways of writing the triplet. The classes shown italicized in bold are those which have been implicated in human diseases.

Table 1 Trinucleotide repeat (TR) expansion in humans. ORF = open reading frame.

Disease	Chromosome	Locus	Inheritance	Functional consequence of TR expansion	Normal allele repeat size	Mutant allele repeat size	Repeat location
Fragile X syndrome (A)	Xq27.3	FMR1 (FRAXA)	X-linked dominant	Fragile site Loss of function	$(CCG)_{6\text{–}52}$	$(CCG)_{60\text{–}200}$ pre $(CCG)_{230\text{–}1000}$ full	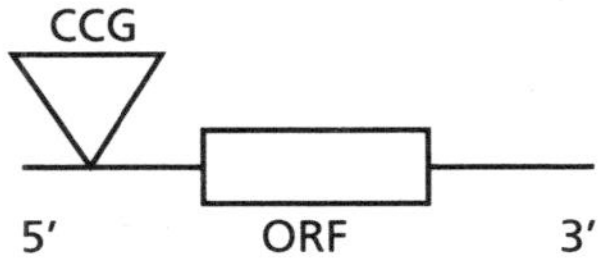
Fragile X (E) mental retardation	Xq28	FRAXE	X-linked	Fragile site Loss of function?	$(CCG)_{7\text{–}35}$	$(CCG)_{130\text{–}150}$ pre $(CCG)_{230\text{–}750}$ full	5′ untranslated region of the first exon of the gene
Jacobsen syndrome	11q23.3	FRA11B (CBL2)	Autosomal	Fragile site	$(CCG)_{11}$	$(CCG)_{80}$ pre $(CCG)_{100\text{–}1000}$ full	
None	Xq27–28	FRAXF	X-linked	Fragile site	$(CCG)_{6\text{–}29}$	$(CCG)_{300\text{–}1000}$	Not determined, but thought not to be in the vicinity of any genes
None	16p13	FRA16A	Autosomal	Fragile site	$(CCG)_{16\text{–}49}$	$(CCG)_{1000\text{–}1900}$	
Spinal + bulbar muscular atrophy/Kennedy disease	Xq11–12	AR-Androgen receptor	X-linked recessive	Gain and partial loss of function	$(CAG)_{11\text{–}33}$	$(CAG)_{38\text{–}66}$	
Huntington's disease	4p16.3	HD (IT15)	Autosomal dom.	Gain of function	$(CAG)_{10\text{–}35}$	$(CAG)_{36\text{–}121}$	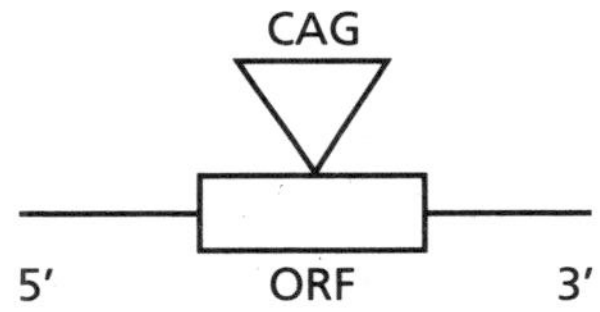
Spinocerebellar ataxia (1)	6p22–23	SCA1	Autosomal dom.	Gain of function	$(CAG)_{6\text{–}39}$	$(CAG)_{41\text{–}81}$	
Spinocerebellar ataxia (2)	12q24.1	SCA2	Autosomal dom.	Gain of function	$(CAG)_{15\text{–}24}$	$(CAG)_{35\text{–}59}$	
Spinocerebellar ataxia (3)/ Machado–Joseph disease	14p24.3–q31	SCA3 (MJD1)	Autosomal dom.	Gain of function	$(CAG)_{12\text{–}37}$	$(CAG)_{61\text{–}84}$	
Spinocerebellar ataxia (7)	3p12–13	SCA7	Autosomal dom.	Gain of function	$(CAG)_{7\text{–}17}$	$(CAG)_{38\text{–}130}$	Coding region of the gene
Dentatorubral + pallidoluysian atrophy/Haw River syndrome (phenotypic variant)	12p13.31	DRPLA (B37)	Autosomal dom.	Gain of function	$(CAG)_{7\text{–}25}$	$(CAG)_{49\text{–}75}$ $(CAG)_{63\text{–}68}$	
Myotonic dystrophy	19q13.3	DMPK	Autosomal dom.	Processing of DMPK message abnormal?	$(CAG)_{5\text{–}37}$	$(CAG)_{50\text{–}3000}$	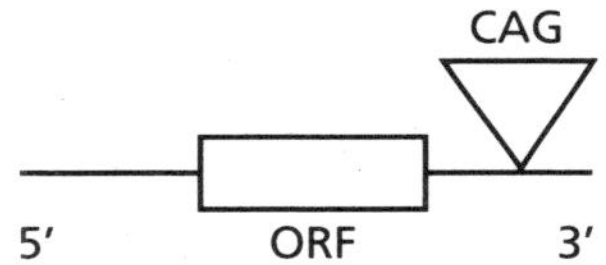 3′ untranslated region of the final exon of the DM protein kinase gene
Friedreich's ataxia	9q13	FRDA (X25)	Autosomal recessive	Loss of function	$(GAA)_{11\text{–}21}$	$(GAA)_{200\text{–}900}$	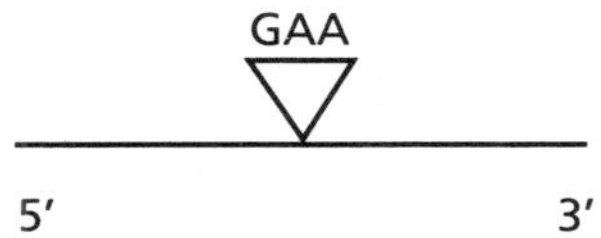 The first intron of the X25 gene

fragile X syndrome and was the first disorder in which TR expansion was implicated. Fragile X syndrome (FRAXA) is the most common cause of inherited mental retardation and is associated with a folate-sensitive fragile site on the X chromosome. The vast majority of cases are now known to be caused by the abnormal expansion of a (CCG)n TR in the 5′ untranslated region of the FMR-1 gene. The inheritance pattern in families affected by fragile X syndrome is unusual, since transmitting normal males can be identified. This had been known as the Sherman paradox, but was explained once the causative mutation had been identified. Transmitting normal males were found to carry a premutation, i.e. an allele in which the (CCG)n repeat had expanded to between 60 and 200 repeats (compared with a range of 6–50 in the normal population). Passage of this repeat through the female germline is required in order for the repeat to expand into the disease-causing range of 230–1000 repeats. It is now known that sperm, in fact, appear not to be able to tolerate repeats of this length in the FMR-1 gene, as even male individuals with a full expansion in their somatic tissues carry only premutation-length alleles in their sperm. The phenotype of FRAXA males seems to result from loss of function of the FMR-1 gene. Fully expanded (CCG)n repeats almost invariably become methylated and the gene is no longer transcribed.

Another fragile site which has recently been identified at the molecular level is FRA11B. This is also a rare, folate-sensitive fragile site, situated on chromosome 11q23.3. The fragile site occurs at an expanded (CCG)n repeat within the CBL2 proto-oncogene. An extremely interesting finding is that a number of individuals with Jacobsen syndrome have inherited a chromosome carrying FRA11B from their mother. Jacobsen syndrome, characterized by specific dysmorphic features and severe mental retardation, is the clinical presentation of a deletion in the terminal long arm of chromosome 11 (typically band 11q23–qter). It has thus been proposed that the deletion found in these Jacobsen patients is due to a breakage at FRA11B during an early stage of development. Prior to this study, it was generally assumed that fragile sites were a phenomenon seen only in cultured cells, but the finding of an association between a fragile site and a deletion syndrome suggests that chromosomal fragility may occur *in vivo*, with obvious implications for disease pathogenesis.

Two other folate-sensitive fragile sites (FRAXF and FRA16A), which are caused by the expansion of (CCG)n repeats, are situated outside the region of any known genes and, despite the considerable expansions that have been reported (>1000), no noticeable phenotype has been associated with them. Therefore, there may well be an obvious link between expanded TR location and phenotypic effects. The expansion of these repeats is also associated with the methylation of the repeat and, as there is no evidence of the methylation of such sites in normal individuals, this methylation is assumed to be a result of the repeat expansion rather than a cause of it.

3. Polyglutamine disorders

A number of neurodegenerative disorders are caused by the abnormal expansion of a (CAG)n TR within the coding region of certain genes. These TR expansions give rise to an enlarged polyglutamine tract in the resulting gene products. It is worth noting that the TRs located in untranslated regions of genes may undergo massive expansion, from normal repeat lengths of 5–52 to repeats of several thousand. However, repeats such as those found within the coding regions of genes tend to undergo a more modest expansion from normal repeat lengths of 6–39, to no more than an overall 3-fold increase in length.

Spinal and bulbar muscular atrophy (SBMA), also known as Kennedy disease, is caused by such a mutation within the androgen receptor (AR) gene. SBMA is a rare, X-linked recessive motor neuropathy with an adult onset of proximal and progressive muscle weakness and atrophy of bulbar muscles that affects one in 50 000 males. Normal individuals carry 11–33 repeats in the first exon of (the protein-coding region of) the AR gene, whereas affected individuals have 38–66 repeats. Both expansions and contractions of the AR repeat have been documented, although the bias of instability leans strongly towards expansion.

Huntington's disease (HD) is caused by a (CAG)n expansion in the IT15 gene, which maps to 4p16.3. The mode of inheritance is autosomal dominant and the disorder affects one in 10 000 Caucasians. Characteristic features of the disease include a progressive cognitive decline and chorea which usually manifest insidiously in mid-life, but can occur anywhere between 2 and 90 years of age. Normal individuals carry 10–35 CAG repeats at the IT15 locus, whereas those affected by the disorder may have anything in the range of 36–121 repeats. "Intermediate alleles" of 30–37 repeats (which are above normal but below the disease range) may not cause readily detectable disease but are liable to increase on transmission by a few copies and produce affected individuals. Both expansions and reductions of the repeat have been documented, but expansion is more common and occurs primarily through the paternal line.

The expanded (CAG)n repeats found in spinocerebellar ataxia type 1 (SCA1), SCA2, SCA3/Machado–Joseph disease (MJD), SCA7 and dentatorubral and pallidoluysian atrophy (DRPLA)/Haw River syndrome (HRS) all result in autosomal dominant, neurodegenerative disorders characterized by the progressive degeneration of the posterior spinal cord columns, with varying signs of central and peripheral nervous system impairment. The genes involved are located on different chromosomes (6p22–23, 12q24.1, 14q24.3–q31, 3p12–13 and 12p13.31, respectively) and are all of unknown function. In all cases, the phenotype is predicted to be a consequence of a deleterious gain of function.

DRPLA, caused by an expanded CAG repeat in the B37 gene, is extremely rare outside of Japan where it has an incidence of one in 1 million. DRPLA is characterized clinically by progressive dementia, epilepsy, ataxia and involuntary movement (chorea and myoclonus), and pathologically by combined dentatorubral and pallidoluysian degeneration. The gene for DRPLA encodes a 4.5 kilobase (kb) transcript expressed in all tissues, with the highest levels in brain, ovary, testis and prostate. The TR sizes are in a range of

7–25 in the normal population, with expansion to 49–75 repeats in affected individuals. HRS, which has affected five generations of an African–American family in North Carolina, is also caused by the expansion of a CAG repeat in the B37 gene. Individuals affected by HRS carry a similar number of TRs to those observed in DRPLA (63–68), but the clinical and neuropathological features of the two disorders are different. Individuals affected by HRS do not exhibit the myoclonic seizures characteristic of DRPLA, but additional pathologies are evident, including demyelination in the subcortical white matter, calcifications of the basal ganglia and neuroaxonal dystrophy. HRS does not appear to be an allelic variant of DRPLA, but may represent the influence of a modifying gene.

Several of the TRs described, including SCA1, SCA3 and FRAXA, are cryptic (the pure repeat tract is interrupted by triplet motifs from other classes) in the normal population; this phenomenon appears to "stabilize" the repeat tract. Once these cryptic repeats are lost, however, the repeat tracts of "pure" TRs appear to become more susceptible to expansion.

4. Myotonic dystrophy (DM)

DM is the most common cause of adult muscular dystrophy, affecting one in 8000 Caucasian individuals, and is caused by an expanded (CAG)n repeat in the 3′ untranslated region of the DM protein kinase gene (DMPK). The disorder is characterized by progressive muscle weakness, myotonia, cataracts, cardiac arrhythmia and diabetes, with frontal balding and testicular atrophy in affected males. The phenotype is highly variable, from balding or cataracts in the mildly affected, to severe symptoms and mental retardation in congenital DM. DMPK, located at 19q13.3, is expressed in muscle and codes for a cyclic AMP-dependent protein kinase. The DM gene comprises 15 exons spanning approximately 14 kb and is predominantly expressed in skeletal and cardiac muscle. The CAG motif is repeated 5–37 times in the normal population, but 50–3000 copies can occur in affected individuals. Reductions in the size of the repeat and "reverse" mutations (affected individuals who have produced phenotypically normal offspring with contracted repeats) have been described on the paternal transmission of this repeat. Myotonic dystrophy alleles with 50–80 (CAG) repeats are associated with non-penetrance or minimal expression termed "protomutation".

5. Friedreich's ataxia

Friedreich's ataxia (FRDA) is the most common cause of hereditary ataxia, affecting one in 50 000 individuals. As a TR disease, it is unique by virtue of being an autosomal recessive disorder, caused by the expansion of a (GAA)n repeat. FRDA is characterized by a progressive gait and limb ataxia, a lack of tendon reflexes in the legs, loss of position sense, dysarthria (slurred speech) and pyramidal weakness of the legs. Most affected individuals also exhibit hypertrophic cardiomyopathy, and some incidence of carbohydrate intolerance and diabetes mellitus has also been described. Manifestation of the disease usually occurs around adolescence and, unlike the other TR diseases, onset after the age of 25 is exceedingly rare. The gene for FRDA (X25) is located at 9q13 and encodes a 210 amino acid protein referred to as Frataxin. Fraxatin is a mitochondrial protein involved in iron homeostasis.

6. Animal models of TR disorders

To date, dynamic mutations have been identified only in the human species. The naturally occurring mouse homologues of human genes containing unstable repeats that have been identified exhibit shorter repeat lengths, repeat interruptions and a lower degree of polymorphism. It has been postulated that the apparent lack of TR instability in animal genomes (especially at homologous loci) may indicate that the process of dynamic mutation is unique to the human genome. However, it may be that similar types of late-onset diseases caused by trinucleotide expansion exist, but do not have sufficient time to manifest in animals with relatively short life spans. It is also possible that factors affecting repeat stability such as mismatch-repair deficiencies do not occur naturally in the mouse population.

To facilitate the study of expanded TR diseases transgenic murine FMR1 and HD gene homologue "knockouts" have been constructed. These models should serve as valuable tools in the elucidation of the physiological role of the gene products and the pathological mechanisms leading to the disease phenotypes. However, these models will not permit the study of the TRs, their mechanisms of expansion and instability, or their role in causing disease phenotypes. For these reasons, transgenic mice containing expanded TRs have been constructed as models for SBMA, HD, SCA1, SCA3 and DM. These models are more representative of the disease conditions observed in humans, and some of these artificially expanded repeats have exhibited instability on their transmission to subsequent generations. Such models will doubtless prove invaluable for the designing and testing of therapies.

FURTHER READING

Ashley, C.T. & Warren, S.T. (1995) Trinucleotide repeat expansion and human disease. *Annual Review of Genetics* 29: 703–28

Korneluk, R.J. & Narang, M.A. (1997) Anticipating anticipation. *Nature Genetics* 15: 119–20

McMurray, C.T. (1995) Mechanisms of DNA expansion. *Chromosoma* 104: 2–13

Reddy, P.S. & Housman, D.E. (1997) The complex pathology of trinucleotide repeats. *Current Opinion in Cell Biology* 9: 364–72

Richards, R.I. & Sutherland, G.R. (1992) Dynamic mutations: a new class of mutations causing human disease. *Cell* 70: 709–12

Sutherland, G.R. & Richards, R.I. (1995) Simple tandem DNA repeats and human genetic disease. *Proceedings of the National Academy of Sciences USA* 92: 3636–41

Warren, S.T. (1996) The expanding world of trinucleotide repeats. *Science* 271: 1374–75

GLOSSARY

anticipation a tendency for some diseases to manifest at an earlier age and to increase in severity with each succeeding generation

ascertainment tracing of individuals/families with a hereditary disorder

ataxia the loss of full control of bodily movements

cardiomyopathy any disease affecting the structure and function of the heart

chorea a nervous disorder characterized by involuntary twitching of muscles

dentatorubral and pallidoluysian atrophy (DRPLA) a neurodegenerative syndrome characterized by epilepsy, dementia and ataxia

diabetes mellitus disorder resulting from deficiency or complete lack of secretion of insulin by the pancreas or from a defect(s) in insulin receptors. Individuals with type I diabetes mellitus are insulin-dependent (IDDM); those with type II are non-insulin dependent (NIDDM)

dysarthria awkward, poorly articulated speech as a result of lack of control over the vocal muscles

fragile X syndrome (FRAXA) most common inherited cause of mental retardation; karyotype shows one tip of a nearly broken X chromosome "hanging by a thread"

Friedreich's ataxia (FRDA) characterized by muscular weakness and cavus deformity (clawfoot), resulting in an abnormal gait, this condition can be either dominantly or recessively inherited and commonly manifests between the ages of 5 and 20 years

Haw River syndrome a phenotype variant of DRPLA

Jacobsen syndrome a disorder characterized by specific dysmorphic features and severe mental retardation

Kennedy disease (*see* spinal and bulbar muscular atrophy)

Machado–Joseph disease a cerebellar ataxia of autosomal dominant inheritance, usually of late onset

myoclonus spasm of a muscle

myotonic dystrophy (DM) a severe form of muscular dystrophy characterized by eyelid droop (ptosis), facial weakness and dysarthria

Sherman paradox the phenomenon of normal transmitting males in an X-linked pedigree

spinal and bulbar muscular atrophy (Kennedy disease) a rare, slowly progressive, X-linked recessive neuroendocrine disorder of spinal muscular atrophy

spinocerebellar ataxia syndrome characterized by progressive degeneration of the spinal cord and cerebellum, as well as other parts of the nervous system. Of dominant and recessive inheritance, onset is usually in childhood/adolescence

---

See also **Genetics of mental retardation** (p.437); **Genetics of Huntingdon's disease** (p.459)

EVOLUTION OF THE HAEMOGLOBINS

Ross Hardison
Department of Biochemistry and Molecular Biology, Pennsylvania State University, Pennsylvania, USA

1. Introduction

Haemoglobins are found in almost all kingdoms of organisms, including eubacteria, unicellular eukaryotes, plants and animals. These homologous proteins serve diverse functions ranging from the transport of oxygen between tissues to catalysis of electron transfer reactions. Haemoglobins appear to be one branch of a larger family of haemoproteins whose ancestor predates the divergence of eubacteria and eukaryotes. The ancient and widespread family of haemoglobin genes is regulated by a variety of mechanisms that appear to be evolving rapidly.

2. Connections between light, oxygen and haem

Sometime around 2.0–2.3 billion years ago, a remarkable event transformed the earth and life on it – molecular oxygen (O_2) was generated by photosynthetic bacteria. Evidence of photosynthetic cyanobacteria (formerly called blue–green algae) can be found in fossilized stromatolites dating back much earlier, to 3.3–3.5 billion years ago (Cloud, 1983). The fossil record and phylogenetic reconstructions indicate that life evolved around 4 billion years ago, possibly within a few hundred million years after the earth was formed. However, these earliest organisms lived in a largely reducing environment, in an atmosphere containing water vapour, nitrogen, methane, ammonia and other components. The early cyanobacteria fossilized in the stromatolites appear to be similar to contemporary photosynthetic bacteria that harvest the energy in sunlight to remove hydrogen atoms (an electron and a proton) from a compound such as hydrogen sulfide (H_2S) to form elemental sulfur. In contemporary organisms, and likely also in these early cyanobacteria, the protons from the hydrogen atoms form electrochemical gradients across membranes that are used to generate the energy currency of the cell, ATP. This ATP plus the electrons from the hydrogen atoms are used to make complex organic compounds such as carbohydrates from carbon dioxide (CO_2).

The element oxygen (O) existed in the prebiotic earth and during early evolution of life, but in a form combined with other elements, not as free molecular oxygen (O_2). Perhaps most abundant was the reduced form, water (H_2O). When cyanobacteria acquired the ability use H_2Oinstead of H_2S as the reduced substrate for photosynthesis, they began a profound change in the biosphere (Chapman & Schopf, 1983). The oxidation of H_2O formed O_2, molecular oxygen. With the introduction of O_2 into the atmosphere, life could evolve around processes that yielded an abundance of energy.

Probably all contemporary organisms that utilize O_2, ranging from humans to aerobic bacteria, have respiratory chains in which electrons from reduced compounds are transferred eventually to O_2, generating H_2O and a large amount of the energy currency of the cell, ATP, in the process. Anaerobic organisms (growing in the absence of O_2) have similar systems to transfer electrons to a different acceptor. However, the electrochemical potential for the reduction of O_2 is higher than that for other electron acceptors found in biological systems and, thus, respiratory electron transfer systems (those that result in reduction of O_2) generate more energy per electron.

The more efficient energy production from respiration likely allowed more complex, multicellular organisms to evolve. This increased efficiency comes at a price, however. Oxygen is capable of damaging many molecules, and keeping oxygen under control while using it in energy production has been one of the great successful compromises of life on earth (Fridovich, 1998).

Porphyrin rings, including chlorophyll and haem, play a key role in contemporary light-harvesting and electron-transfer reactions now, and it is likely that they were doing so in ancient organisms, long before the appearance of molecular oxygen in the atmosphere. A porphyrin molecule is a planar group of four connected rings, with a nitrogen (N) in each, facing the centre of the cluster of rings (**Figure 1**). These four nitrogens, all in the same plane, provide an ideal environment in which metals can insert (forming coordinate covalent bonds between the metal and the porphyrin nitrogens). When the metal is magnesium (Mg) in the +2 oxidation state (and with some alterations to the side chains and one ring on the porphyrin ring), the resulting organometallic compound is called chlorophyll. This is the familiar green compound that, when bound to appropriate proteins, serves to harvest the electromagnetic energy of sunlight and transfer it to membrane-bound reaction centres for photosynthesis. A porphyrin ring containing a different metal, iron (Fe), in the +2 oxidation state (a ferrous ion) is the familiar red haem. Haem bound to globin polypeptides makes haemoglobin, which colours our blood red and carries O_2 from the lungs to other tissues. Haem can also be bound to other proteins that carry out many different functions. Some illustrative examples include cytochromes, which transfer electrons in respiratory chains, oxygenases, which catalyse the oxidation of a wide variety of compounds, and fungal ligninases, which are extracellular peroxidases that degrade the non-cellulose polymer lignin in decaying wood.

If metalloporphyrin compounds precede the evolution of photosynthesis and other early processes, then it is also possible that some of the genes encoding the proteins to which they bind could have equally deep evolutionary roots, with descendants of the ancestral gene being found in species in all major branches of life, including archaebacteria, eubacteria and eukaryotes (reviewed in Hardison, 1998). This is an example of divergent evolution among homologous

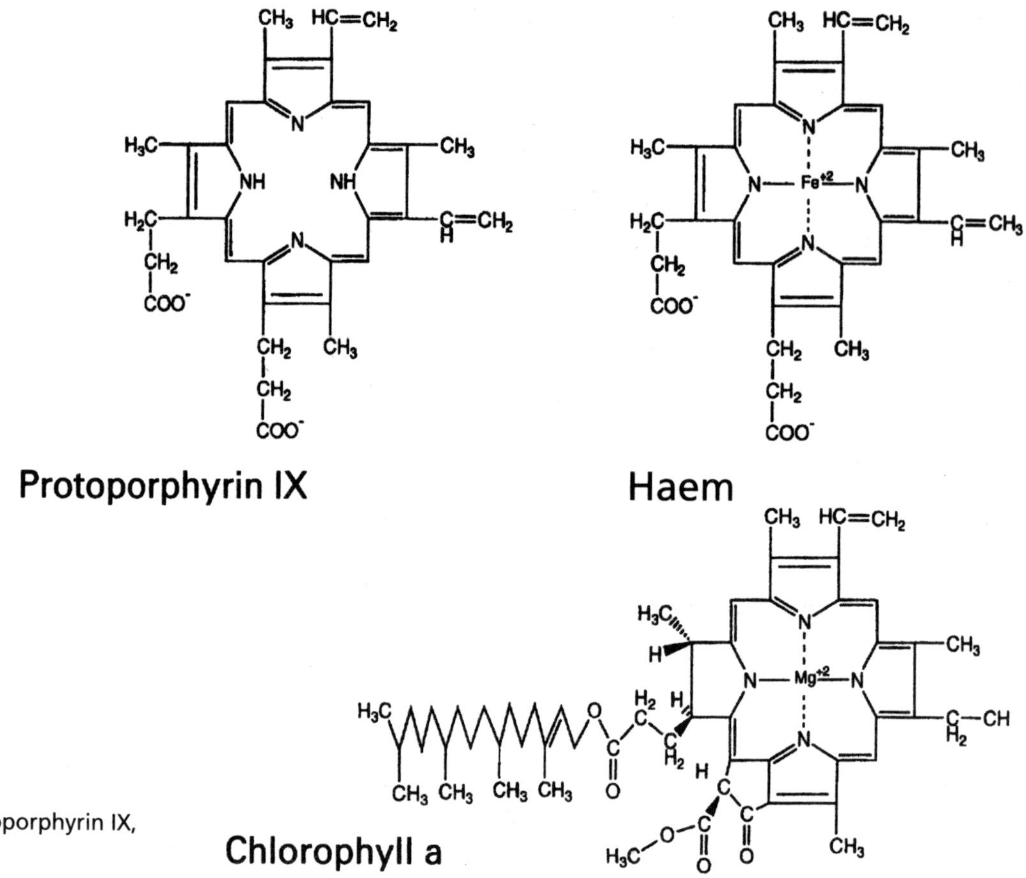

Figure 1 Structures of protoporphyrin IX, haem and chlorophyll a.

genes. In other cases, genes of different ancestry may have evolved independently to encode proteins that bind haem or other metalloporphyrins. These would be cases of convergent evolution to generate analogous genes.

How far back in evolution can the "family tree" be traced for proteins that bind metalloporphyrins? Keilin (1966) suggested that haemoglobins may have evolved from haem enzymes that utilize oxygen. One can extend this hypothesis in the following scenario. As mentioned earlier, some of the earliest known organisms appear to be cyanobacteria whose photosynthetic activity did not generate O_2. It is reasonable to expect that metal-bound porphyrin rings similar or identical to contemporary chlorophylls, along with appropriate binding proteins, were utilized in early photosynthetic processes to capture light energy. Haemoproteins, perhaps similar to contemporary cytochromes, could have served as electron-transfer agents even before the onset of photosynthesis. Once some organisms switched to photosynthesis that produced molecular oxygen, new roles would be opened for the use of metalloporphyrin proteins, in general, and haemoproteins, in particular. Given the capacity of oxygen to damage various cellular components, oxygen-binding haemoproteins may have functioned initially to protect cells from this highly reactive molecule. Later, once O_2 was available as an electron acceptor in respiratory chains, oxygen-bound haemoproteins may have served as the terminal electron acceptors. The concentration of O_2 in the atmosphere was probably much lower than now for hundreds of millions of years after oxygen-generating photosynthesis began (Cloud, 1983). Thus, once some organisms could benefit from oxygen-utilizing respiration, they would have needed to scavenge scarce oxygen to provide to the respiratory chain. At this stage, a capacity for intracellular transport of oxygen would have been advantageous, and that could have led to the evolution of proteins ancestral to contemporary myoglobins in animals and non-symbiotic haemoglobins in plants. In multicellular organisms, the oxygen-scavenging haemoglobins could have evolved into the abundant haemoglobins now used to transport oxygen. Further gene duplications and divergence would allow the encoded haemoproteins to evolve the capacity to catalyse redox and other reactions involving oxygen.

The preceding conjectures follow a parsimonious approach to evolutionary reconstructions, i.e. finding the path with the least number of steps. This scenario is one of divergent evolution; it assumes that, once porphyrin-binding proteins evolved, they acquired many different functions as the environment and needs of the organism changed. However, one can invoke convergent evolution at each step in the story as well, assuming that a new porphyrin-binding protein arose from some different ancestral protein to provide each new function. The next section will summarize molecular data that support divergent evolution of haemoglobins as well as some proteins that bind a relative of porphyrin rings in a light-harvesting apparatus. It is more difficult, at present, to distinguish between convergence and divergence for all haemoproteins.

3. Haemoglobins found in eubacteria, fungi, protists, plants and animals have functions ranging from O_2 transport to catalysis

Haemoglobin is the extremely abundant, red protein in vertebrate erythrocytes that binds O_2 in the lungs and delivers it to respiring tissues in the body. Haemoglobins were first found in blood simply because they are so abundant, with a concentration in normal human blood of 15 g/100 ml. The use of haemoglobins in O_2 transport is common to all known vertebrates and, until recently, this role was thought to be the dominant function of haemoglobins. The haemoglobins used in O_2 transport between tissues in vertebrates are heterotetramers of two α-globin and two β-globin polypeptides, with a haem tightly bound to a pocket in each globin monomer. Binding of O_2 to one subunit leads to allosteric changes in its conformation, and the consequent movements and changes in the interactions between the α- and β-globin subunits are the basis for cooperative binding of oxygen to this haemoglobin (Dickerson & Geis, 1983; Bunn & Forget, 1986). Cooperative binding of O_2 allows haemoglobin to pick up oxygen readily in the lungs and to unload it efficiently in the peripheral, respiring tissues. The O_2 can be stored intracellularly when bound to the monomeric haemoprotein called myoglobin. The amino acid sequences of the α- and β-globins and myoglobin are related to each other, indicating a common ancestor in early vertebrates approximately 500 million years ago (Goodman *et al.*, 1987). This is illustrated in the phylogenetic tree of haemoglobins (**Figure 2**). Myoglobin falls within the group of vertebrate globins, showing that it is more similar to the haemoglobins of vertebrates than non-vertebrates, supporting the conclusion of a common ancestor for myoglobin and both α- and β-globins early in the vertebrate lineage. Thus, vertebrate haemoglobins are used both for the delivery of O_2 between tissues (haemoglobins in erythrocytes) and for intracellular binding and delivery of O_2 (myoglobin) (Wittenberg & Wittenberg, 1987).

O_2-binding and transport proteins are found in other multicellular organisms, such as the extracellular haemoglobins in invertebrates (e.g. Riggs, 1991; Sherman *et al.*, 1992). Sequence data (amino acid sequences of the proteins and nucleotide sequences of the genes encoding them) have shown that these proteins are similar to the vertebrate haemoglobins and, in many cases, the proteins are gigantic (up to 4 million daltons) because they are multimers of the basic globin unit (Zhu *et al.*, 1996). A phylogenetic analysis (**Figure 2**) shows that the invertebrate haemoglobins from annelids, arthropods and nematodes form a clade distinct from the vertebrate haemoglobins (and other haemoglobins), indicating they are most similar among themselves and share a common ancestor, presumably early in the invertebrate lineage.

Plants can make their own O_2 from photosynthesis, and there is no obvious need for transport of O_2 from one tissue to another. However, it is now clear that haemoglobins are widespread in plants. The non-symbiotic haemoglobins have been found in all plants examined, and are present in all tissues of the plants (e.g. Andersson *et al.*, 1996). They may play a role similar to that of the vertebrate myoglobin, storing O_2 inside cells, or possibly scavenging O_2 when it is less abundant. The first haemoglobin discovered in plants, leghaemoglobin (reviewed in Appleby, 1984), is found in the symbiotic tissues of plants involved in N_2 fixation, such as the root nodules of legumes. Leghaemoglobins have been proposed to sequester O_2 away from the O_2-sensitive nitrogenase complex and to transport O_2 intracellularly to the electron transport chain in mitochondria of the nodule. Both classes of plant haemoglobins, the non-symbiotic and the symbiotic, form distinct branches within a monophyletic grouping on the tree of haemoglobins. This indicates that the contemporary plant haemoglobins are derived from a common globin gene in an ancestral plant.

This summary shows that haemoglobins are used in the

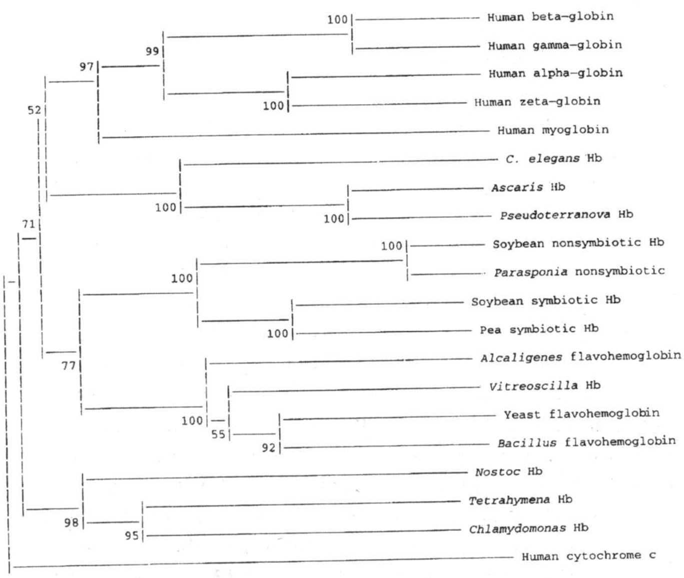

Figure 2 Phylogenetic tree of haemoglobins, using the neighbour-joining method to build the tree and fraction of amino acids that differ as the distance measure. The percentage of times a node is supported in 1000 bootstrap replicates is given to the left of the node.

major groups of multicellular organisms – plants, invertebrates and vertebrate animals – for the storage and transport of O_2. The ability to bind O_2 reversibly is essential to these functions, and reversible binding of O_2 occurs only when the iron in the haem is in the reduced (+2) oxidation state, i.e. when it is a ferrous ion. Thus, one would not have expected haemoglobin to participate in electron transfer reactions, where the iron would change reversibly between the +2 and +3 oxidation states (ferrous and ferric, respectively). Nor would one expect haemoglobins to catalyse oxidation–reduction reactions, since such reactions would require changes in the oxidation state of the haem-bound iron. Other haemoproteins catalyse these reactions, such as cytochromes transporting electrons or any number of oxygenases catalysing the addition of one or more oxygen atoms to a substrate.

The discovery of haemoglobins in unicellular organisms has been doubly surprising. First, it was commonly thought that simple diffusion was sufficient to provide adequate O_2 inside the cells of unicellular, free-living organisms. However, it is now known that haemoglobins are present in the fungus *Saccharomyces cerevisiae* (Zhu & Riggs, 1992) and other unicellular eukaryotes such as the alga Chlamydomonas (Couture *et al.*, 1994) and the protozoan Paramecium (Yamauchi *et al.*, 1995). In fact, haemoglobins have been characterized in eubacteria, such as *Alcaligenes eutrophus* (Cramm *et al.*, 1994), Vitreoscilla (Wakabayashi *et al.*, 1986), *Bacillus subtilis* (LaCelle *et al.*, 1996), *Escherichia coli* (Vasudevan *et al.*, 1991) and *Nostoc commune* (Potts *et al.*, 1992).

The second surprise is that many of these haemoglobins catalyse oxidation–reduction reactions. Biochemical analysis showed that haemoglobins from Chlamydomonas, Alcaligenes and Saccharomyces can participate in electron transfer reactions. The latter two haemoglobins are actually two-domain proteins, one binding haem and the other binding flavin cofactors, which usually play a role in oxidation and reduction. Thus, one can propose electrons being transferred in catalysis – for example, from a reduced flavin cofactor, via haem to a substrate. The haemoglobin from Vitreoscilla can serve as a terminal electron acceptor during respiration, illustrating a potential role in electron-transfer reactions (Dikshit *et al.*, 1992). The Alcaligenes flavohaemoglobin has been implicated in the metabolism of nitrous oxide (Cramm *et al.*, 1994), which raises the possibility that haemoglobins may have been involved in nitrogen metabolism long before their adaptation to be an O_2 carrier. Indeed, recent data show that haemoglobins participate in nitric oxide metabolism in a wide range of organisms, including bacteria and invertebrates. Biochemical and genetic analysis has shown that the flavohaemoglobin encoded by the *hmpA* gene of *E. coli* is a dioxygenase that catalyses the oxidation of toxic nitric oxide into nitrate (Gardner *et al.*, 1998). A haemoglobin with extraordinarily high affinity for oxygen from the parasitic worm *Ascaris lumbricoides* acts as a "deoxygenase", catalysing the consumption of oxygen by reaction with nitric oxide. This maintains a required hypoxic condition in the perienteric fluid of the worm (Minning *et al.*, 1999). These recent results will likely stimulate renewed investigations of the catalytic capacities of haemoglobins from many organisms, including the non-symbiotic haemoglobins of plants, with special attention to reactions involving nitric oxide.

4. Divergent evolution of haemoglobin genes from an ancient gene

These unexpected results for the haemoglobins from unicellular organisms raise the issue of whether these proteins are truly homologous to the haemoglobins from plants and animals, i.e. have they all descended from a common ancestral gene? One approach to answering this question is by phylogenetic reconstructions based on amino acid sequences. In the phylogenetic tree in Figure 2, the bacterial haemoglobins from a monophyletic group (which also includes the yeast flavohaemoglobin) and the haemoglobins from other unicellular eukaryotes form a different group (which includes the haemoglobin from the cyanobacterium Nostoc). If the haemoglobins from these unicellular organisms were encoded by a different gene family, then they would be shown as distinctive branches on a tree separate from that containing the plant and animal haemoglobins. This is not seen, but rather the amino acid sequences support the bacterial, protist, plant and animal haemoglobins all being in the same large family. Although the branching pattern at the early nodes in the tree (to the left of **Figure 2**) is not definitive, trees with a similar topology are obtained by using three different, independent methods of phylogenetic analysis, so the overall result appears to be quite robust. Also, one can see that a representative of a different family of haemoproteins, human cytochrome *c*, is treated as an outgroup, i.e. this cytochrome is significantly more different from all the haemoglobins than the haemoglobins are from themselves.

One can also investigate this issue within the context of all known protein sequences. The BLOCKS server (Pietrokovski *et al.*, 1998) searches all protein sequences in the public databases to find those with blocks of similar sequences and organizes the resulting groups in a large database. The searches and organization are done automatically and objectively, and thus should be free from biases introduced by an investigator's selection of sequences to include in the analysis. The haemoglobins discussed in this article are included in the large family of globins, thus supporting the hypothesized divergent evolution.

Independent and strong support for the conclusion that all haemoglobins share a common ancestor is provided by comparisons of 3-dimensional structures determined by X-ray crystallography. The seminal studies by Perutz and Kendrew to solve the 3-dimensional structures of mammalian myoglobin and haemoglobin established the field of protein crystallography. Both proteins have a series of α-helices that form the haem-binding pocket. This structure, the globin fold, is readily seen in myoglobin, α-globin and β-globin, and is characteristic of all members of the haemoglobin family of proteins, including leghaemoglobin (reviewed in Dickerson & Geis, 1983). The structures of the bacterial flavohaemoglobin from Alcaligenes and haemoglobin from Vitreoscilla have been determined (Ermler *et al.*, 1995; Tarricone *et al.*, 1997), and the haem-binding domain in both consists of the unmistakable globin fold first seen in myoglobin (**Figure 3**; see also **Plate 14** in the colour plate section).

Thus, both amino acid sequence comparisons and comparisons of 3-dimensional structure argue that the haemoglobins found in eubacteria, unicellular eukaryotes, plants and animals share a common ancestor very early in organismal evolution, and the several haemoglobins with widely different functions discussed here are homologous. This means that the gene encoding haemoglobin is truly ancient; that is, it appears to have been present in the ancestor to eubacteria and eukaryotes. To date, no haemoglobins have been reported in archaebacteria. Since the archaebacteria are thought to have diverged from eubacteria very early in evolution, between 3 and 4 billion years ago, whereas the divergence between archaebacteria and eukaryotes occurred approximately 2 billion years ago (Feng *et al.*, 1997), one may anticipate finding homologues to haemoglobins in archaebacteria as well. Alternatively, the haemoglobin genes may have been carried into eukaryotes by early eubacterial endosymbionts, in which case haemoglobins may be absent from archaebacteria. Such horizontal transfers have to be invoked to explain some aspects of the phylogenetic trees of haemoglobins. It is important to emphasize that the tree in Figure 2 is a "gene" tree (showing the relationships among individual proteins) and not a species tree. Although the flavohaemoglobin from yeast is included in the branch with bacterial haemoglobins, yeast is clearly not a bacterium. This similarity can be explained by a horizontal gene transfer between bacteria and yeast. Likewise, the inclusion of the haemoglobin from the cyanobacterium Nostoc with the protist haemoglobins may indicate another horizontal gene transfer.

The conclusion that haemoglobins share an ancestor prior to the eubacterial/eukaryote divergence has several

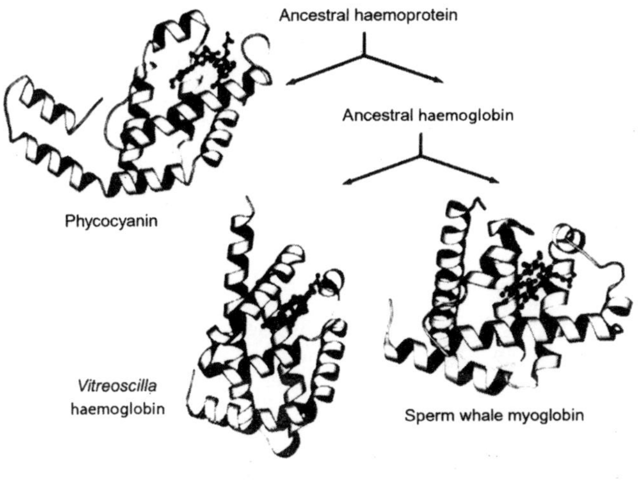

Figure 3 Ribbon diagrams of the 3-dimensional structures of sperm whale myoglobin, Vitreoscilla haemoglobin and C-phycocyanin from Mastigocladus, arranged in the deduced evolutionary relationships. The ribbons diagrams were generated by Dr Greg Farber using the atomic coordinates deposited in the Protein DataBank. (See also **Plate 14**)

ramifications, three of which will be explored briefly in the following sections. Are haemoglobins related to other haemoproteins? Can the family of haemoglobin genes inform us about the origins of introns? And, finally, given a common ancestor for proteins that now differ in function, how has the regulation of the genes encoding them changed to allow for those differences in function to be manifested?

(i) Three-dimensional structures indicate homology between haemoglobins and other porphyrin-binding proteins

Most studies that use amino acid or nucleotide sequences to explore the pattern and dates of deep divergences in the history of life restrict their analysis to very slowly changing macromolecules, such as ribosomal RNAs and enzymes with key catalytic functions in cellular metabolism. Thus, it is interesting that the amino acid sequences of haemoglobins are sufficient to trace ancestry back as far as the ancestor to eubacteria and eukaryotes, even though the proteins are involved in different functions. By contrast, amino acid sequences may not allow an exploration of earlier divergences, such as testing whether other families of haemoproteins share an ancestor with the haemoglobins. The percentage of amino acids that differs between pairs of proteins that are widely separated on the tree in Figure 2 is already quite high (e.g. 84% between human β-globin and the haemoglobin from Chlamydomonas), and it is unlikely that a simple distance measure based on pairwise similarity will be adequate to explore deeper roots (e.g. human cytochrome *c* and β-globin differ at 95% of the amino acids).

However, information in the 3-dimensional structure may be used to test hypothesized relationships between gene families. Currently, this structural information is available from a much smaller number of haemoproteins than is the primary sequence, and clear criteria for structural similarity have not yet been established, but this situation will doubtless improve in the future. Already, very intriguing similarities are known; for instance, the light-harvesting biliprotein, C-phycocyanin, from the cyanobacterium *Mastigocladus laminosus*, has a 3-dimensional structure very similar to that of a globin (Schirmer *et al.*, 1985). Although this is not a haem-binding protein *per se*, it does bind a linear tetrapyrrole pigment derived from haem. As illustrated in Figure 3 (see also Plate 14 in the colour plate section), a detailed comparison of the 3-dimensional structures of globins and phycocyanins supports the hypothesis of a distant evolutionary relationship between the two protein families (Pastore & Lesk, 1990). Since the phycocyanins are part of the light-harvesting apparatus of some cyanobacteria, their resemblance to globins supports the notion that at least some haemoproteins are ancestrally related to other, non-haem porphyrin-binding proteins, as speculated in the scenario in the first section of this article.

Just as in haemoglobins, haem binds between two α-helices, coordinated to histidine, in proteins as diverse as lignin peroxidase (Edwards *et al.*, 1993) and cytochrome b562 (Mathews *et al.*, 1979). However, the topology of these helices differs from the globin fold. Further studies of more haemoproteins will be needed to decide whether the differences arise from divergent or convergent evolution, thereby determining how far the superfamily of haemoglobin genes reaches.

(ii) Intron–exon structures help resolve some but not all issues about the evolution of haemoglobin genes

In many genes from eukaryotes, the segments that encode mRNA (and hence protein) are separated by segments of DNA that are not represented in the mature mRNA. The gene segments that encode mature mRNA are called exons, and those that do not are called introns (see article on "Introns"). The intron–exon structure of haemoglobin genes adds information that is independent of that from the coding regions. Only coding region information (leading to the amino acid sequences) is used in the alignments and analysis that led to the tree shown in Figure 2. Use of intron–exon structure allows some of the earlier branches (farther to the left in the tree) to be examined more thoroughly. Statistical tests of the reliability of these early branches show that they are not uniquely assigned by the tree-building algorithms, meaning that other branching patterns are about equally likely. Thus, for example, one may wish to examine whether the plant haemoglobin genes show more similarity in intron–exon structure to the bacterial (and yeast) haemoglobin genes than to other haemoglobin genes, consistent with the amino acid sequence-based tree in Figure 2. In fact, that is not the case, but rather the haemoglobin genes of plants have intron–exon structures rather similar to those of invertebrates, with four exons separated by three introns (**Figure 4**). Furthermore, the three exon–two intron structure of vertebrate haemoglobin and myoglobin genes appears to have arisen by a loss of the central intron from this ancestral globin gene. Thus, the intron–exon structure of the genes argues for a ancestral haemoglobin gene with three introns prior to the divergence of plants and animals approximately 1.5 billion years ago.

The haemoglobin genes from the protozoan Paramecium and the alga Chlamydomonas have introns as well, but these are *not* in positions clearly homologous to those found in haemoglobin genes from multicellular organisms (Figure 4). It has been difficult to resolve whether these introns arose by independent insertions (Stoltzfus *et al.*, 1994) or differential intron loss from an ancestor with many introns. The issue is further complicated by the hypothesis that introns may "slide" during evolution, caused by multiple mutations at the intron–exon junctions that allow formerly intronic sequences to become part of exons, and *vice versa* (reviewed in Goldberg, 1995). The bacterial and fungal (flavo)haemoglobin genes lack introns, but it is difficult to distinguish between intron loss *vs.* derivation from an intronless ancestral gene to account for this observation. Thus, at the current level of understanding, the intron–exon structures provide improved resolution to the relationships of the plant and animal globin genes, but they also raise important and difficult questions about the history of these genes in unicellular organisms.

(iii) The regulatory mechanisms for globin genes have changed more dramatically than have the protein structures

Given that the widespread haemoglobins are encoded by homologous genes, then their different functions illustrate the acquisition of new roles by a pre-existing structural gene. This requires changes not only in the protein-coding regions, but also in the DNA segments that regulate expression of the genes. Examination of the regulatory mechanisms for these genes indicates that, in general, the evolution of

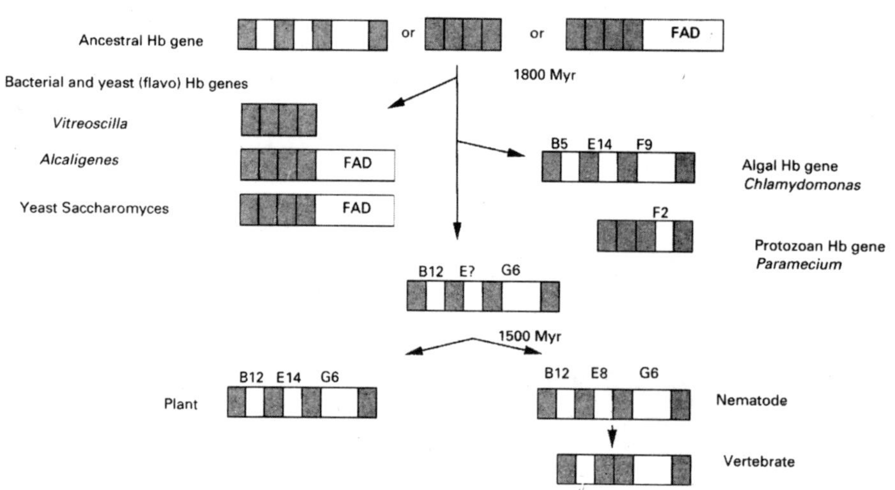

Figure 4 Number and positions of introns in some haemoglobin genes. Exons of the haemoglobin genes are shown as boxes with dark fill, introns are white boxes and the portions of the genes encoding flavin nucleotide-binding domains of flavohaemoglobins are boxes labelled "FAD". The letter and number above each intron indicates the position in the alpha-helical structure of globins occupied by the amino acid encoded by the codon interrupted by or adjacent to the intron. The intron–exon structure of the ancestral haemoglobin gene is difficult to determine. Introns occur in different places in haemoglobin genes of unicellular eukaryotes, suggestive of insertion of at least some introns in these lineages. The intron–exon structure indicates a common ancestor to the haemoglobin genes in plants and animals.

control elements appears to be very rapid, with regulatory regions changing much more dramatically than the protein-coding regions. In some instances, quite different solutions to similar problems have evolved. In many cases, the regulatory changes and evolutionary distance are so large that no remnant of the ancestral state is left to guide inferences from sequence alignments. Thus, at least for these loci, the use of structural information in the form of genomic DNA sequences requires a comparison of relatively close phyla.

Most of the genes encoding the haemoglobins discussed here are regulated by O_2 levels (**Figure 5**), as may be expected given the central role of haemoglobin in delivering, scavenging and sequestering O_2 in various organisms. Production of haemoglobins is often induced by *low* O_2 concentration (referred to as hypoxic or anaerobic conditions), thereby increasing the amount of O_2 delivered to cells for aerobic respiration and energy production. However, it appears that no mechanism for anaerobic induction has been conserved in all species. In several bacteria, the sensing of low O_2 can lead directly to an increased production of haemoglobin in the same cell. As illustrated in **Figure 6**, the common anaerobic regulator FNR (fumarate nitrate reduction) protein is used to increase expression of the haemoglobin gene of Vitreoscilla (Dikshit *et al.*, 1990; Joshi & Dikshit, 1994) and possibly Alcaligenes (Cramm *et al.*, 1994). FNR induces expression of a large number of genes when, as a result of low O_2 concentration, electron transport switches to alternative electron acceptors such as fumarate and nitrate. Under anaerobic conditions, the FNR protein is a transcriptional activator that acts directly to increase expression of the Vitreoscilla haemoglobin gene. Likewise, in *Bacillus subtilis*, anaerobic induction of the flavohaemoglobin gene *hmp* requires FNR as well as a two-component regulatory system (LaCelle *et al.*, 1996). However, in this case, the role of FNR appears to be to stimulate nitrite production, which then leads to induction of *hmp*. Thus, the same protein, FNR, has been implicated in anaerobic induction of homologous haemoglobin genes in two different genera of eubacteria, but it appears to be working directly on the target gene in one case and more indirectly in the other.

Haemoglobin production in vertebrates is also increased when the concentration of O_2 is low, but by the indirect mechanism of increasing the number of erythrocytes (**Figure 6**). Each erythrocyte contains an abundance of haemoglobin (about 280 million haemoglobin molecules per cell), so this boost in erythropoiesis effectively increases the amount of haemoglobin substantially. Hypoxia is actually sensed by cells in the kidney and liver, where it signals an increase in production of the polypeptide hormone erythropoietin, which then acts on the erythroid progenitor cells in the bone marrow to increase proliferation, stimulate further erythroid differentiation and block apoptosis (reviewed in Migliaccio *et al.*, 1996). Thus, the O_2 sensing system in vertebrates does not act directly on the haemoglobin genes, but rather acts on a hormone gene in a different tissue, eventually leading to an increase in number of cells carrying haemoglobin.

Expression of the erythropoietin gene is stimulated under conditions of low O_2 by the action the protein hypoxia induction factor 1 (HIF1) (Wang & Semenza, 1993). HIF is a heterodimer of an α and β subunit. Increased amounts of the unstable HIF1 α subunit are produced when the concentration of O_2 is low, generating the active form of HIF1, which then binds to the enhancer and stimulates transcription of the erythropoietin gene (Wang *et al.*, 1995). The two subunits of HIF are members of a large family of sensor proteins that contain PAS domains. Proteins with the PAS domain are found in eubacteria, archaebacteria, fungi, plants and animals and are involved in sensing a variety of stimuli, including oxygen, light and circadian rhythm (Zhulin *et al.*, 1997). Although none of the bacterial PAS-containing proteins has been implicated in the regulation of haemoglobin genes to date, several respond to changes in oxygen concentration. Thus, it will be of considerable interest to see if additional work demonstrates an involvement of PAS-containing proteins in addition to HIF in regulation, either directly or indirectly, of haemoglobin genes in other species.

Once erythroid cell proliferation and maturation is stimulated by erythropoietin in vertebrates, one might expect that the expression of globin genes to produce abundant haemoglobin would be regulated by rather similar mechanisms, but this is not the case. Contemporary α- and β-globin genes probably arose via duplication of a common ancestral gene in an early vertebrate. The two genes would have been iden-

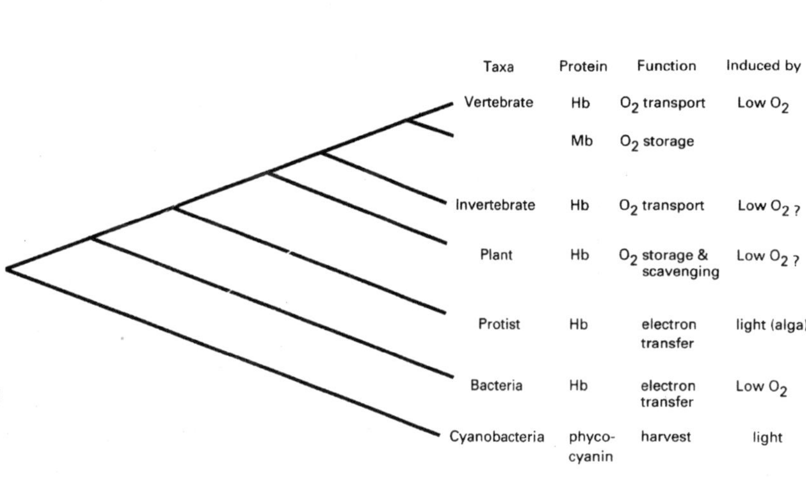

Figure 5 Generalized evolutionary relationships among haemoglobins, with their general functions and responses to O_2.

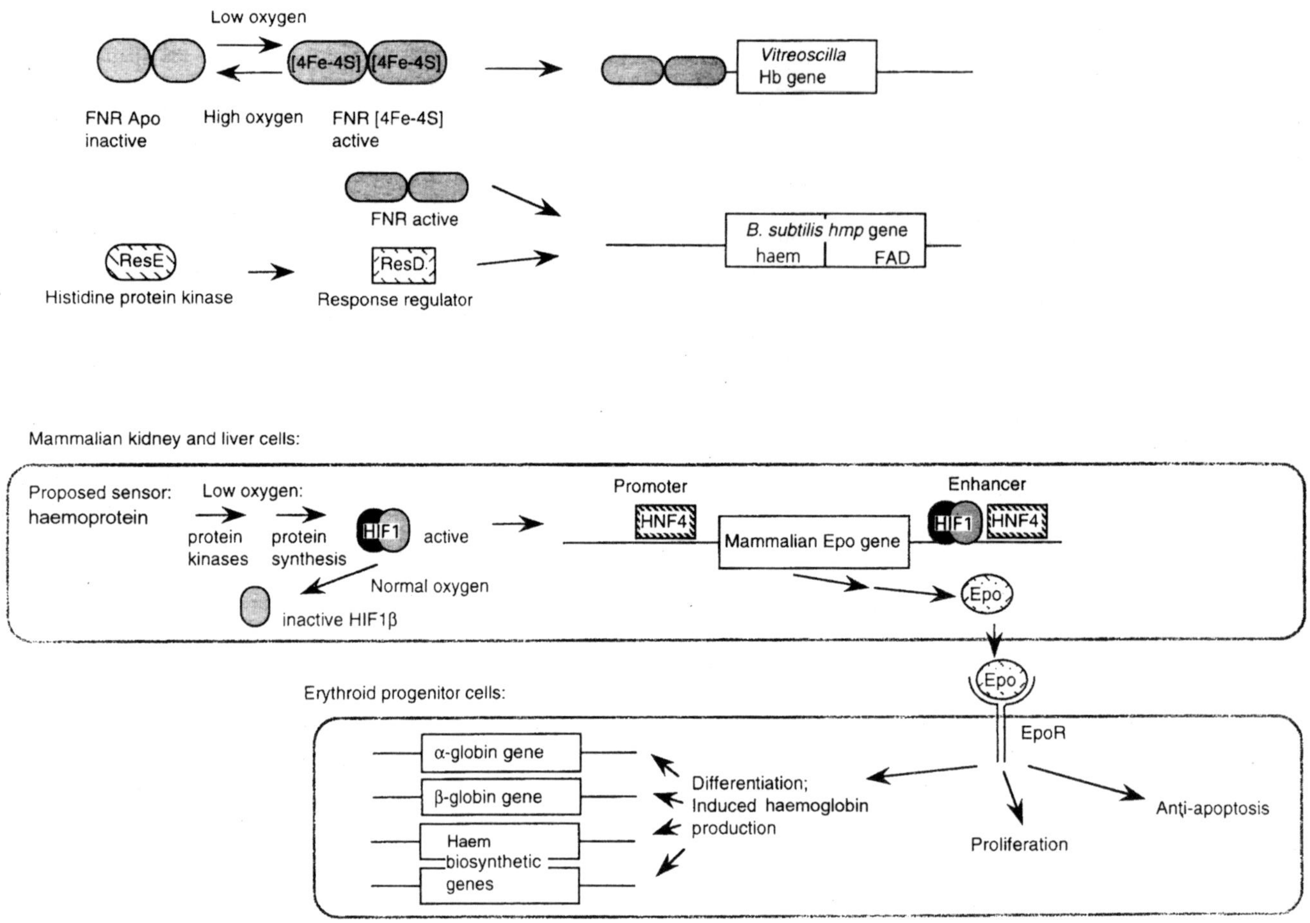

Figure 6 Mechanism of haemoglobin gene induction in response to O_2 in different species of bacteria and in mammals. The top panel illustrates the induction of the Vitreoscilla haemoglobin gene by the regulatory protein FNR. Under hypoxic conditions, an iron–sulfur (4Fe–4S) cluster forms in the protein, causing it to adopt a conformation that allows binding to specific sites on the DNA close to the haemoglobin gene and activation of transcription. The second panel illustrates two pathways of regulation of the *hmp* gene of *Bacillus subtilis*, which encodes a flavohaemoglobin. Active FNR is required, but indirectly – it appears to be regulating other genes whose products are needed for nitrite production. A separate two-component system consisting of the ResE and ResD proteins is more directly involved in regulating this *hmp* gene. The third panel illustrates the more complex, multicellular response to hypoxia in mammals. Liver and kidney cells detect low oxygen through an as-yet undefined sensor, which may be a haemoprotein. In response to hypoxia, a series of steps, including protein kinase reactions and new protein synthesis, is needed to make the active heterodimer, HIF1. This protein binds to an enhancer located 3′ to the gene encoding the glycoprotein hormone erythropoietin, stimulating its transcription and leading to production and release of the hormone. Erythropoietin binds to specific receptors on red cell progenitor cells, leading to blocks to apoptosis, proliferation and increased expression of the genes encoding globins and haem biosynthetic enzymes.

tical after the initial duplication, with identical regulatory elements and, since the genes are expressed coordinately in the same amounts, one would expect selection to keep the regulatory elements very similar. However, much has changed between the α-like and β-like globin gene clusters since their duplication. They are now on separate chromosomes in birds and mammals, and in the latter they are in radically different genomic contexts. The β-globin gene clusters have an A+T content comparable to the bulk of mammalian DNA, they have no CpG islands and the locus is in a DNase accessible, "open" chromatin conformation only in erythroid cells, where the genes are expressed (Groudine *et al.*, 1983). By contrast, the α-like globin gene clusters are highly G+C rich, they have a CpG island associated with each active gene and the locus is in a constitutively "open" chromatin conformation in all cells (Craddock *et al.*, 1995). Thus, although tissue-specific gene expression is frequently correlated with an increased accessibility of the chromatin in a locus only in expressing cells, this "opening" of a chromatin domain is a key step in the activation of β-like globin genes of mammals, but *not* the α-like globin genes.

Not only do the α-globin and β-globin genes differ in genomic context and chromatin structure, but the DNA sequences implicated in their regulation show many differences. The promoters for these genes have very few sequences in common. Some of the same proteins act at the distal, powerful regulatory elements for these genes, such as the enhancer for α-globin genes and the locus control region

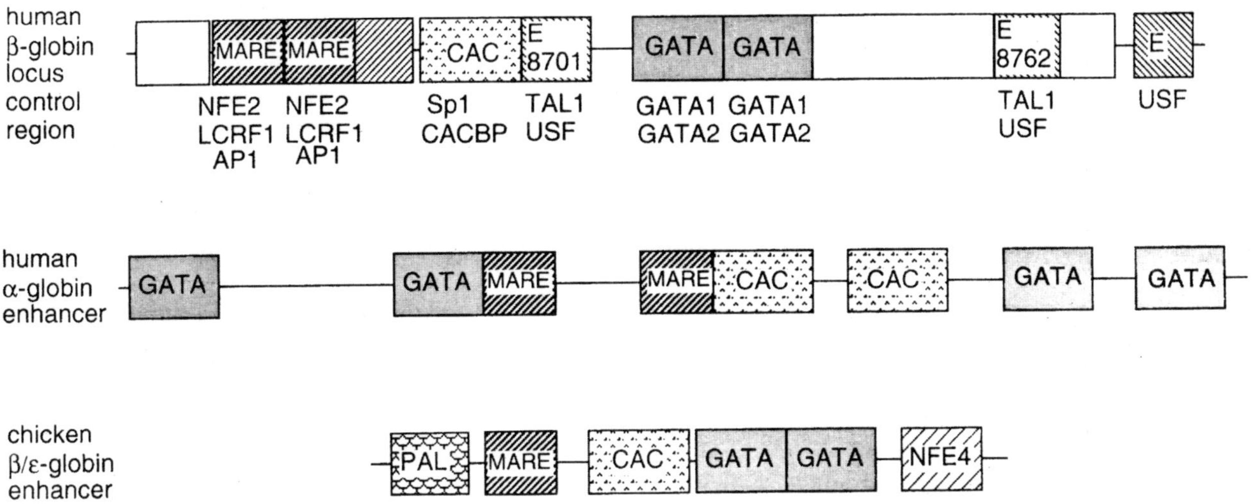

Figure 7 Summary of protein binding sites in distal control elements of vertebrate globin genes. The binding sites are shown as boxes with distinctive fill and are labelled with their names. Proteins implicated in binding to these sites are listed below the sites on the top line. A portion of the human β-globin locus control region called HS2 is shown. The DNA sequence motif MARE (GNTGASTCA, where N is any nucleotide and S is G or C) is a Maf-response element. Proteins of the basic zipper family such as AP1 (a heterodimer of Jun and Fos) bind to this site, and some that are heterodimers containing Maf proteins (such as NFE2 and LCRF1) have been specifically implicated in erythroid regulation. The CAC motif has this trinucleotide as an invariant part of the site. Proteins with Krüppel-like zinc fingers, such as Sp1, ELKF and others (abbreviated in this figure at CACBP, for CAC-binding proteins), bind to this motif. E boxes, such as E8701 and E8762, have the consensus sequence CANNTG and bind helix–loop–helix proteins. These two E boxes can bind heterodimeric proteins containing TAL1/SCL, which is required for haematopoiesis in mouse and causes leukaemias when aberrantly expressed in T cells or stem cells, the protein USF (upstream stimulatory factor) and other proteins. The GATA motif is WGATAR, where W is A or T, and R is A or G. A family of GATA proteins can bind to this site, and GATA1 and GATA2 play important roles in erythroid development. Similar motifs are found in HS2 of the β-globin locus control region, the α-globin enhancer and the chicken β/ε-globin enhancer. Additional motifs in the chicken enhancer include PAL, consisting of a palindromic sequence, and the site for binding NFE4 (nuclear factor erythroid 4).

of the β-globin genes. However, as illustrated in **Figure 7**, the number and positions of the binding sites for these proteins differ between the two (reviewed in Hardison, 1998).

Even when the analysis is done over a shorter phylogenetic distance, for example between the β-globin gene clusters of birds and mammals (Reitman *et al.*, 1993; Hardison, 1998), no statistically significant alignments are seen in the promoter regions or in the distal control elements. Only matches in parts of the protein-coding regions are seen. Thus, in the time since birds and mammals diverged, the DNA sequences of control regions of β-globin genes have diverged to the point where no matches can be detected in pairwise alignments. The control regions are still operational, as seen in functional tests and, in fact, as shown in Figure 7, homologous proteins have been implicated in the function of control regions of β-globin genes in birds and mammals (Evans *et al.*, 1990).

When non-coding DNA sequences of the β-globin gene clusters from a set of mammals are aligned, many blocks of conserved sequences are found in the gene regulatory regions (Hardison & Miller, 1993; Hardison *et al.*, 1997). Strongly conserved sequences identified in these studies are reliable guides to gene regulatory elements. Thus, when the comparisons are restricted to an appropriate phylogenetic distance, conserved regulatory elements are detectable in analysis of DNA sequences. It is curious that such conserved regulatory elements were not detected over a longer phylogenetic distance – in comparisons of β-globin gene clusters between birds and mammals, for instance, as discussed above. In this latter case, the functions of the encoded haemoglobins are the same in birds and mammals, and, to the best of our current understanding, the genes are regulated by rather similar mechanisms, including opening a chromatin domain in erythroid cells and using homologous regulatory proteins in both phyla. Perhaps some aspect of the regulation is different, but that aspect is not yet recognized. Alternatively, in this case, it could be that the regulatory sequences are in fact homologous, but have diverged so much that their similarity is no longer measurable above that of random sequences. This would be surprising given the level of conservation of regulatory elements in mammals.

5. Recent developments

Several more examples of haemoglobins catalysing redox reactions have been documented, with most involved in nitric oxide metabolism. The flavohaemoglobins from the enteric bacteria *Escherichia coli* and *Salmonella typhimurium* (Crawford & Goldberg, 1998; Gardner *et al.*, 1998; Hausladen *et al.*, 1998) and from yeast (Liu *et al.*, 2000) are nitric oxide dioxygenases, catalysing the conversion of nitric oxide to nitrate. These enzymes protect the microorganisms from the highly reactive free radical compound, nitric oxide. A haemoglobin found in the perienteric fluid of the parasitic worm *Ascaris lumbricoides* also catalyses reactions between oxygen and nitric oxide, producing nitrate (Minning *et al.*, 1999). However, the chemical mechanism is different from that of the microbial flavohaemoglobins, and Minning *et al.* (1999) propose that this haemoglobin functions to remove oxygen from the perienteric fluid via a series of reactions driven by nitric oxide.

In mammals, haemoglobins not only transport oxygen, but they may also help to regulate nitric oxide levels. Gow *et al.* (1999) show that at physiological concentrations, nitric oxide will react with a cysteine in oxyhaemoglobin to form S-nitrosohaemoglobin. Nitric oxide can subsequently be released from deoxyhaemoglobin (Stamler *et al.*, 1997). Since nitric oxide is a major regulator of blood pressure, these new findings lead to a proposal that haemoglobin is involved in the control of blood pressure as well as transport of oxygen, thereby facilitating efficient delivery of oxygen to tissues. Furthermore, the interplay between binding of oxygen and nitric oxide to haemoglobin and effects on vasodilation and constriction may have therapeutic applications (e.g., Bonaventura *et al.*, 1999; Gladwin *et al.*, 1999).

REFERENCES

Andersson, C.R., Jensen, E.O., Llewellyn, D.J., Dennis, E.S. & Peacock, W.J. (1996) A new hemoglobin gene from soybean: a role for hemoglobin in all plants. *Proceedings of the National Academy of Sciences USA* 93: 5682–87

Appleby, C.A. (1984) Leghemoglobin and Rhizobium respiration. *Annual Review of Plant Physiology* 35: 443–78

Bonaventura, C., Ferruzzi, G., Tesh, S. & Stevens, R.D. (1999) Effects of S-nitrosation on oxygen binding by normal and sickle cell hemoglobin. *Journal of Biological Chemistry* 274: 24742–48

Bunn, H.F. & Forget, B.G. (1986) *Hemoglobin: Molecular, Genetic and Clinical Aspects*, Philadelphia: Saunders

Chapman, D. & Schopf, J.W. (1983) Biological and biochemical effects of the development of an aerobic environment. In *Earth's Earliest Biosphere: Its Origin and Evolution*, edited by J.W. Schopf, Princeton, New Jersey: Princeton University Press

Cloud, P. (1983) Early biogeologic history: the emergence of a paradigm. In *Earth's Earliest Biosphere: Its Origin and Evolution*, edited by J.W. Schopf, Princeton, New Jersey: Princeton University Press

Couture, M., Chamberland, H., St.-Pierre, B., Lafontaine, J. & Guertin, M. (1994) Nuclear genes encoding chloroplast hemoglobins in the unicellular green alga *Chlamydomonas eugametos*. *Molecular and General Genetics* 243: 185–97

Craddock, C.F., Vyas, P., Sharpe, J.A. *et al.* (1995) Contrasting effects of alpha and beta globin regulatory elements on chromatin structure may be related to their different chromosomal environments. *EMBO Journal* 14: 1718–26

Cramm, R., Siddiqui, R.A. & Friedrich, B. (1994) Primary structure and evidence for a physiological function of the flavohemoprotein of *Alcaligenes eutrophus*. *Journal of Biological Chemistry* 269: 7349–54

Crawford, M.J. & Goldberg, D.E. (1998) Role for the Salmonella flavohemoglobin in protection from nitric oxide. *Journal of Biological Chemistry* 273: 12543–47

Dickerson, R.E. & Geis, I. (1983) *Hemoglobin: Structure, Function, Evolution and Pathology*, Menlo Park, California: Benjamin Cummings

Dikshit, K.L., Dikshit, R.P. & Webster, D.A. (1990) Study of *Vitreoscilla* globin (*vgb*) gene expression and promoter activity in *E. coli* through transcriptional fusion. *Nucleic Acids Research* 18: 4149–55

Dikshit, R.P., Dikshit, K.L., Liu, Y. & Webster, D.A. (1992) The bacterial hemoglobin from *Vitreoscilla* can support aerobic growth of *Escherichia coli* lacking terminal oxidases. *Archives of Biochemistry and Biophysics* 293: 241–45

Edwards, S.L., Raag, R., Wariishi, H., Gold, M.H. & Poulos, T.L. (1993) Crystal structure of lignin peroxidase. *Proceedings of the National Academy of Sciences USA* 90: 750–54

Ermler, U., Siddiqui, R.A., Cramm, R. & Friedrich, B. (1995) Crystal structure of the flavohemoglobin from *Alcaligenes eutrophus* at 1.75 Angstrom resolution. *EMBO Journal* 14: 6067–77

Evans, T., Felsenfeld, G. & Reitman, M. (1990) Control of globin gene transcription. *Annual Review of Cell Biology* 6: 95–124

Feng, D.F., Cho, G. & Doolittle, R.F. (1997) Determining divergence times with a protein clock: update and reevaluation. *Proceedings of the National Academy of Sciences USA* 94: 13028–33

Fridovich, I. (1998) Oxygen toxicity: a radical explanation. *Journal of Experimental Biology* 201: 1203–09

Gardner, P.R., Gardner, A.M., Martin, L.A. & Salzman, A.L. (1998) Nitric oxide dioxygenase: an enzymic function for flavohemoglobin. *Proceedings of the National Academy of Sciences USA* 95: 10378–83

Gladwin, M.T., Schechter, A.N., Shelhamer, J.H. *et al.* (1999) Inhaled nitric oxide augments nitric oxide transport on sickle cell hemoglobin without affecting oxygen affinity. *Journal of Clinical Investigation* 104: 937–45

Goldberg, D.E. (1995) The enigmatic oxygen-avid hemoglobin of *Ascaris*. *BioEssays* 17: 177–82

Goodman, M., Czelusniak, J., Koop, B., Tagle, D. & Slightom, J. (1987) Globins: a case study in molecular phylogeny. *Cold Spring Harbor Symposium in Quantitative Biology* 52: 875–90

Gow, A.J., Luchsinger, B.P., Pawloski, J.R., Singel, D.J. & Stamler, J.S. (1999) The oxyhemoglobin reaction of nitric oxide. *Proceedings of the National Academy of Sciences USA* 96: 9027–32

Groudine, J., Kohwi-Shigematsu, T., Gelinas, R., Stamatoyannopoylos, G. & Papyannopoulou, T. (1983) Human fetal to adult hemoglobin switching: changes in chromatin structure of the β-globin gene locus. *Proceedings of the National Academy of Sciences USA* 80: 7551–55

Hardison, R. (1998) Hemoglobins from bacteria to man: evolution of different patterns of gene expression. *Journal of Experimental Biology* 201: 1099–117

Hardison, R. & Miller, W. (1993) Use of long sequence alignments to study the evolution and regulation of mammalian globin gene clusters. *Molecular and Biological Evolution* 10: 73–102

Hardison, R., Slightom, J.L., Gumucio, D.L. *et al.* (1997) Locus control regions of mammalian β-globin gene clusters: combining phylogenetic analyses and experimental results to gain functional insights. *Gene* 205: 73–94

Hausladen, A., Gow, A.J. & Stamler, J.S. (1998) Nitrosative stress: metabolic pathway involving the flavohemoglobin. *Proceedings of the National Academy of Sciences USA* 95: 14100–05

Joshi, M. & Dikshit, K.L. (1994) Oxygen dependent regulation of *Vitreoscilla* globin gene: evidence for positive regulation by FNR. *Biochemical and Biophysical Research Communications* 202: 535–42

Keilin, D. (1966) *The History of Cell Respiration and Cytochrome*, Cambridge: Cambridge University Press

LaCelle, M., Kumano, M., Kurita, K. *et al.* (1996) Oxygen-controlled regulation of the flavohemoglobin gene in *Bacillus subtilis*. *Journal of Bacteriology* 178: 3803–08

Liu, L., Zeng, M., Hausladen, A., Heitman, J. & Stamler, J.S. (2000) Protection from nitrosative stress by yeast flavohemoglobin. *Proceedings of the National Academy of Sciences USA* 97: 4672–76

Mathews, F.S., Bethge, P.H. & Czerwinski, E.W. (1979) The structure of cytochrome b562 from *Escherichia coli* at 2.5 Å resolution. *Journal of Biological Chemistry* 254: 1699–706

Migliaccio, A.R., Vannucchi, A.M. & Migliaccio, G. (1996) Molecular control of erythroid differentiation. *International Journal of Hematology* 64: 1–29

Minning, D.M., Gow, A.J., Bonaventura, J. *et al.* (1999) Ascaris haemoglobin is a nitric oxide-activated 'deoxygenase'. *Nature* 401: 497–502

Pastore, A. & Lesk, A.M. (1990) Comparison of the structures of globins and phycocyanins: evidence for evolutionary relationship. *Proteins* 8: 133–55

Pietrokovski, S., Henikoff, J.G. & Henikoff, S. (1998) Exploring protein homology with the Blocks server. *Trends in Genetics* 14: 162–63

Potts, M., Angeloni, S.V., Ebel, R.E. & Bassam, D. (1992) Myoglobin in a cyanobacterium. *Science* 256: 1690–91

Reitman, M., Grasso, J.A., Blumentahl, R. & Lewit, P. (1993) Primary sequence, evolution and repetitive elements of the *G. gallus* (chicken) β-globin cluster. *Genomics* 18: 616–26

Riggs, A.F. (1991) Aspects of the origin and evolution of non-vertebrate hemoglobins. *American Zoologist* 31: 535–45

Schirmer, T., Bode, W., Huber, R., Sidler, W. & Zuber, H. (1985) X-ray crystallographic structure of the light harvesting biliprotein C-phycocyanin from the thermophilic cyanobacterium *Mastigocladus laminosus* and its resemblance to globin structures. *Journal of Molecular Biology* 184: 257–77

Sherman, D.R., Kloek, A.P., Krishnan, B.R., Guinn, B. & Goldberg, D.E. (1992) *Ascaris* hemoglobin gene: plant-like structure reflects the ancestral globin gene. *Proceedings of the National Academy of Sciences USA* 89: 11696–700

Stamler, J.S., Jia, L., Eu, J.P. *et al.* (1997) Blood flow regulation by S-nitrosohemoglobin in the physiological oxygen gradient. *Science* 276: 2034–37

Stoltzfus, A., Spencer, D.F., Zuker, M., Logsdon, J.M. Jr & Doolittle, W.F. (1994) Testing the exon theory of genes: the evidence from protein structure. *Science* 265: 202–07

Tarricone, C., Galizzi, A., Coda, A., Ascenzi, P. & Bolognesi, M. (1997) Unusual structure of the oxygen-binding site in the dimeric bacterial hemoglobin from *Vitreoscilla* sp. *Structure* 5: 497–507

Vasudevan, S.G., Armarego, W.L., Shaw, D.C. *et al.* (1991) Isolation and nucleotide sequence of the *hmp* gene that encodes a haemoglobin-like protein in *Escherichia coli* K-12. *Molecular and General Genetics* 226: 49–58

Wakabayashi, S., Matsubara, H. & Webster, D.A. (1986) Primary sequence of a dimeric bacterial haemoglobin from *Vitreoscilla*. *Nature* 322: 481–83

Wang, G.L. & Semenza, G.L. (1993) General involvement of hypoxia-inducible factor 1 in transcriptional response to hypoxia. *Proceedings of the National Academy of Sciences USA* 90: 4304–08

Wang, G.L., Jiang, B.H., Rue, E.A. & Semenza, G.L. (1995) Hypoxia-inducible factor 1 is a basic-helix–loop–helix–PAS heterodimer regulated by cellular O_2 tension. *Proceedings of the National Academy of Sciences USA* 92: 5510–14

Wittenberg, B.A. & Wittenberg, J.B. (1987) Myoglobin-mediated oxygen delivery to mitochondria of isolated cardiac myocytes. *Proceedings of the National Academy of Sciences USA* 84: 7503–07

Yamauchi, K., Tada, H. & Usuki, I. (1995) Structure and evolution of *Paramecium* hemoglobin genes. *Biochimica et Biophysica Acta* 1264: 53–62

Zhu, H. & Riggs, A.F. (1992) Yeast flavohemoglobin is an ancient protein related to globins and a reductase family. *Proceedings of the National Academy of Sciences USA* 89: 5015–19

Zhu, H., Ownby, D., Riggs, C., Nolasco, N., Stoops, J. & Riggs, A. (1996) Assembly of the gigantic hemoglobin of the earthworm *Lumbricus terrestris*. Roles of subunit equilibria, non-globin linker chains, and valence of the heme iron. *Journal of Biological Chemistry* 271: 30007–21

Zhulin, I.B., Taylor, B.L. & Dixon, R. (1997) PAS domain S-boxes in Archaea, bacteria and sensors for oxygen and redox. *Trends in Biochemical Sciences* 22: 331–33

GLOSSARY

allosteric relating to proteins in which the binding of a small molecule at one site brings about changes in properties at a separate, distant, site

apoptosis active process of cell death requiring metabolic activity by the dying cell

CpG island a region relatively rich in CpG dinucleotides. It is often located at the 5′ end of an expressed housekeeping gene where it is found in the unmethylated state

erythrocyte mature red blood cell

erythropoiesis/erythropoietic production of erythrocytes in the bone marrow in adult mammals

G+C rich rich in nucleotides containing the purine base guanine and the pyrimidine base cytosine

two-component regulatory system a member of a class of regulatory systems common in bacteria comprised of two proteins, a sensor of some environmental signal (e.g. concentration of nitrogen) and a regulator of gene expression that responds to that signal

See also **Haemoglobinopathies** (p.412); **Introns** (p.712)

HAEMOGLOBINOPATHIES

John Old
National Haemoglobinopathy Reference Laboratory, Institute of Molecular Medicine, John Radcliffe Hospital, Oxford, UK

1. Introduction

The haemoglobinopathies are a group of recessive genetic disorders affecting the genes which code for the globin chains making up haemoglobin (Hb). The disorders result from mutations that either alter the structure of a globin chain (the Hb variants such as Hb S, C and E) or those that drastically reduce the synthesis of one or more normal globin chains (the thalassaemias). The haemoglobin gene disorders are the most common single gene disorders in the world population. It has been estimated that there are over 200 million carriers for the haemoglobinopathies and 300 000 severely affected children are born each year (World Health Organization, 1983).

The haemoglobinopathies are most common in the Mediterranean region, Africa and Asia, but have now become a global problem through the migration of populations throughout Europe, the Americas and Australia (Flint *et al.*, 1993). The high carrier rates observed in many populations are thought to have resulted from a positive selection pressure due to falciparum malaria. Individuals with sickle cell trait suffer from malaria less frequently and less severely than normal persons and thus have a higher survival rate during the critical period of early childhood in areas where malaria is endemic. The mechanism by which thalassaemia trait provides protection against malaria is less clear, although, as with sickle cell anaemia, there is a strong geographical correlation of gene frequencies with the incidence of malaria.

The haemoglobinopathies are regionally specific with each local population having its own characteristic spectrum of thalassaemia mutations and Hb variants. The identification of all the mutant genes occurring in an at-risk population is the first step in the control of the clinically serious disorders of sickle cell disease and β-thalassaemia by an integrated programme of carrier screening, genetic counselling and prenatal diagnosis. The thalassaemia genes were the first genetic disorders to have their molecular defects characterized by gene-cloning techniques; prenatal diagnosis of the disorders by DNA analysis methods quickly followed, with the first prenatal diagnosis of α-thalassaemia reported in 1976, sickle cell anaemia in 1978 and β-thalassaemia in 1980 (Kazazian *et al.*, 1980). These early foetal diagnoses were performed on DNA from cultured amniotic fluid cells and then, in 1981, the analysis of DNA from chorionic villi biopsies in the first trimester of pregnancy was developed, allowing a safer, earlier termination of pregnancy if required (Old *et al.*, 1982). However, despite this major advance in foetal diagnosis, the uptake of prenatal diagnosis programmes for the haemoglobinopathies in developing countries was very slow due to the complexity of the Southern blotting method required for mutation identification until the discovery of the polymerase chain reaction (PCR) which amplifies specific gene sequences. The globin genes and their mutations served as a model genetic disease for the development of PCR and, in return, PCR has since revolutionized the molecular diagnosis of the haemoglobinopathies (Old, 1996). It has provided a variety of quick, cheap and sensitive methods of foetal diagnosis for thalassaemia and sickle cell disease, allowing first trimester foetal diagnosis to be established in India, Pakistan and many other developing countries (Petrou & Modell, 1995). Finally, PCR is permitting new approaches based on the analysis of DNA from a single cell to be developed for the future, such as preimplantation diagnosis and the determination of a foetal genotype from the analysis of foetal cells circulating in maternal blood (Cheung *et al.*, 1996).

2. The globin genes

Haemoglobin is a tetrameric protein normally made up of two identical α-globin-like chains (α or ζ) and two β-globin-like chains (β, δ, γ or ε). Each globin chain is associated with one iron-containing haem molecule which is involved in the binding and release of oxygen. The globin chains are synthesized independently from genes located in two clusters (**Figure 1**), the α-globin genes in a 20 kilobase (kb) region on chromosome 16 and the β-globin like genes in a 60 kb region on the short arm of chromosome 11.

The globin genes are very short compared with most other eukaryotic genes, being only 1700 base pairs (bp) in length with just three exons and two introns. The α-globin gene cluster contains three expressed genes (α1, α2 and ζ), each producing a protein of 141 amino acids. The β-globin gene cluster contains five expressed genes (β, δ, $^G\gamma$, $^A\gamma$ and ε), each producing a slightly larger protein of 146 amino acids. The globin genes are not expressed in equal amounts. The α1- and α2-genes have identical coding sequences, but are expressed in a ratio of approximately 1:3. The two foetal γ-globin genes also have very similar exons, differing only at codon 136, at which point the $^G\gamma$-gene codes for glycine and the $^A\gamma$-gene codes for alanine. The two γ-globin genes are also usually expressed in a ratio of 1:3 ($^A\gamma$:$^G\gamma$). There is a greater difference in expression between the δ-gene and the β-gene, two adult globin genes in the β-gene cluster. The δ-globin gene is expressed at approximately one-fortieth of the level of that of the β-globin gene, resulting in an average Hb A_2 level of 2.5% in normal adults. However, except in individuals with thalassaemia, the expression of the α-globin-like genes is always in balance with the expression of the β-globin-like genes (Weatherall & Clegg, 1981).

The globin gene clusters also contain four pseudogenes

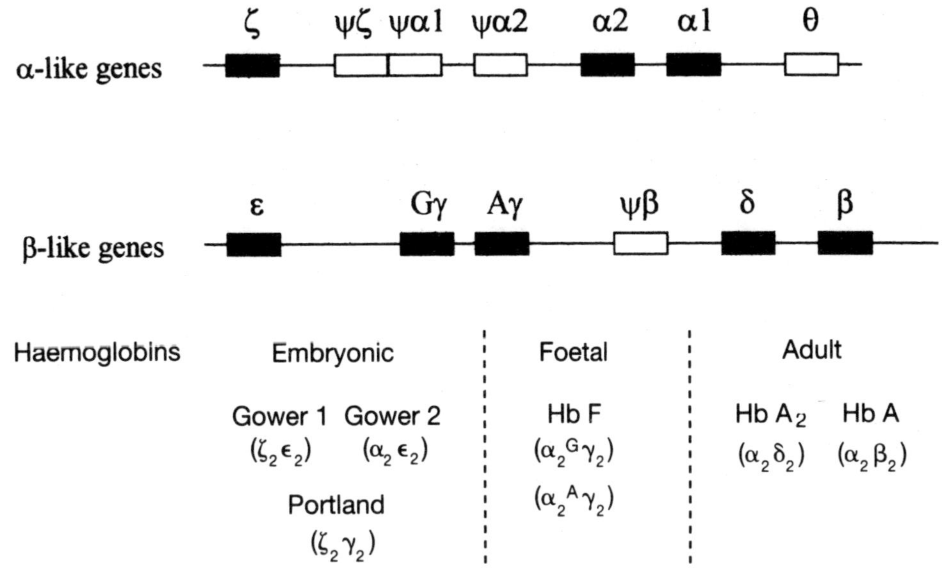

Figure 1 Diagram of the globin gene clusters and the various haemoglobins that are produced in embryonic, foetal and adult life.

($\Psi\beta$, $\Psi\zeta$, $\Psi\alpha_1$, $\Psi\alpha_2$) and a gene whose function is unknown (θ). The pseudogenes are genes homologous to various globin genes but which contain molecular defects that prevent their expression. These functionless genes are thought to be the remnants of once-active genes made redundant by evolution. The θ-gene is more puzzling as it is clearly a pseudogene in some species, but in humans the gene contains no obvious inactivating mutations and appears to be expressed at very low levels. However, no haemoglobin-containing θ-globin has yet been identified and it seems unlikely that it can produce a viable haemoglobin (Higgs, 1993).

The globin genes in both clusters are arranged in the order that they are expressed during development (**Figure 1**). During the first 8 weeks of intrauterine development, there is a coordinated change from embryonic haemoglobin to foetal haemoglobin, involving a switch in the expression of the embryonic ζ-gene to the foetal/adult α-gene in the α-like cluster, and a switch from the embryonic ε-gene to the foetal $^{G}\gamma$ and $^{A}\gamma$-genes in the β-like cluster. The β-like genes undergo an additional switch at around the time of birth when there is a change from foetal to adult haemoglobin and, by the end of the first year of life, the Hb F level has dropped to less than 1% in normal individuals. However, in some individuals, a significant quantity of Hb F persists into adult life and a higher than normal Hb F level in patients with sickle cell disease or β-thalassaemia is associated with milder disease. Thus, the molecular mechanism responsible for the silencing of the γ-globin genes has been the subject of intense research for many years in the hope that an understanding of the regulatory process will lead to the ability to induce Hb F production in affected patients and ameliorate their clinical symptoms. Several chemical agents such as hydroxyurea and phenylbutyrate have been shown to activate the expression of the γ-globin genes and, although the mechanisms of action of these agents are poorly understood, clinical trials involving the administration of these compounds in patients with β-thalassaemia or sickle cell disease are in progress, with promising early results in some patients with sickle cell disease and thalassaemia (Charache *et al.*, 1995).

3. Abnormal haemoglobins

Abnormal haemoglobins or haemoglobin variants arise from mutations in the globin genes which result in the synthesis of a globin chain with an amino acid sequence different from normal. The vast majority are caused by a point mutation in a globin gene exon resulting in a single amino acid substitution in the peptide chain. For example, Hb S differs from Hb A by the substitution of valine for glutamic acid in position 6 of the β-globin chain. Hb S was the first variant to be discovered when, in 1949, Linus Pauling demonstrated by electrophoresis that haemoglobin from patients with sickle cell disease had a different mobility when compared with normal Hb. He called the variant Hb S for sickle and as other variants were discovered by electrophoresis they were labelled with further letters of the alphabet until exhaustion, and then by the place names of their discovery. To date, 731 structurally different human haemoglobins have been discovered by electrophoresis techniques and the more recently developed technologies, such as isoelectric focusing, high-performance liquid chromatography, mass spectrometry and PCR-mediated DNA sequence analysis. The majority of the Hb variants are rare and are not associated with any clinical manifestations, having being identified through population surveys (Huisman *et al.*, 1996).

(i) Molecular aspects

The molecular defects underlying the abnormal haemoglobins are listed in **Table 1**. The vast majority have just one amino acid replacement resulting from a single base change in the globin gene coding sequence. The amino acid substitution leads to the synthesis of an unstable protein and/or one with a difference in charge allowing detection with an electrophoretic method. Neutral amino acid replacements

Table 1 Molecular defects underlying the abnormal haemoglobin.

Type of defect	Examples	α	β	δ	γ
Single point mutation (one amino acid replacement)	S, C, E	216	348	30	70
More than one point mutation (two amino acid replacements)	C-Harlem	1	24		
Deletion (shortened chain)	Gun Hill	4	12		
Insertion (elongated chain)	Grady	4	2		
Deletion/insertion	Montreal		3		
Extended polypeptide chain:					
Termination codon mutant	Constant Spring	5			
Frameshift	Wayne	1	3		
Extensions (retention of initiator methionine)	Thionville	1	3		
Fusion haemoglobins: (unequal crossing-over)					
δβ – 3	Lepore				
βδ – 5	Miyada				
δβδ – 1	Parchman				
γβ – 1	Kenya				

are possible, but not generally observed because mass spectrometry or DNA sequencing would have to be used for screening.

There are 25 Hb variants characterized by two amino acid replacements. These have arisen by either the occurrence of a new mutation in a globin gene that already carries a single base change or as a crossover between two genes, each carrying its own point mutation. All but one are β-chain variants and it is interesting to note that six are Hb S genes, each with a different additional base change, e.g. Hb C Harlem, which consists of the Hb S substitution of valine for glutamic acid at position β6, and the Hb Korle-Bu substitution of asparagine for aspartic acid at position β73. The second substitution at β73 alters its electrophoretic mobility so that it moves like Hb C in electrophoresis at alkaline pH, although its sickling properties are identical to Hb S.

A further group of haemoglobin variants is characterized by a deletion or insertion of a small number of amino acids in the polypeptide chain. The variants are caused by crossovers between misaligned chromosomes, resulting in the loss or addition of nucleotides in the codon sequence. Unequal crossovers between misaligned chromosomes result in the fusion haemoglobins Hb Lepore and Hb Kenya. Hb Lepore, which starts as a δ-chain sequence and changes to the β-chain sequence, is the most important because the gene acts like a β-thalassaemia gene since the hybrid gene is under the influence of the δ-gene promoter and synthesized in a much lower amount than β-globin. Three types of Hb Lepore have been identified, each with a different crossover point from the δ- to the β-gene DNA sequence. Hb Miyada is one of five abnormal variants discovered to have β-chain amino acids at the N-terminal and δ-chain at the C-terminal, and thus are products from the anti-Lepore gene created on the other chromosome at the time of the unequal crossover (**Figure 2**).

Finally, there is a group of variants with elongated α- or β-chains. The most important variants in this group are the five α-chain variants caused by a mutation in the termination codon of the α-globin gene. They all have an α-thalassaemia phenotype, the extended α-chain being unstable and occurring at a level of only 1–2% in heterozygous individuals. Apart from Hb Constant Spring, which is found in modest frequencies in many south-east Asian populations, the other variants in this group are all very rare. Four variants have been found with an extension at the C-terminus due to an internal frameshift mutation and four have an N-terminal extension due to mutations affecting the initiation codon, preventing the removal of the initiator methionine.

(ii) Clinical aspects

The majority of haemoglobin variants are not associated with disease; they are harmless and do not interfere with normal haemoglobin function. However, a few do interfere with function and carriers come to the attention of clinicians because the mutation affects haemoglobin solubility, oxygenation or synthesis (**Table 2**).

Unstable haemoglobins precipitate inside the red cell, forming aggregates (Heinz bodies) which damage the red cell membrane and result in haemolytic anaemia. Variants associated with a high oxygen affinity result in polycythaemia (an excess of the number of red cells in the blood) due to impaired oxygen delivery to the tissues. Less common are variants associated with a low affinity for oxygen. Many of these cause cyanosis, as do the M haemoglobins in which the haem iron is locked in the oxidized ferric state owing to an amino acid substitution in the haem pocket of the globin chain, therefore preventing it playing any part in oxygen transport.

4. Hb S

Clinically, the most important Hb variant is Hb S which, in the homozygous state, or in combination with either Hb C, Hb O-Arab, Hb D-Punjab or β-thalassaemia, causes sickle cell disease. Hb S is less soluble than normal haemoglobin during deoxygenation, crystallizing out into polymers in the form of long fibres which cause the classic sickle-shaped

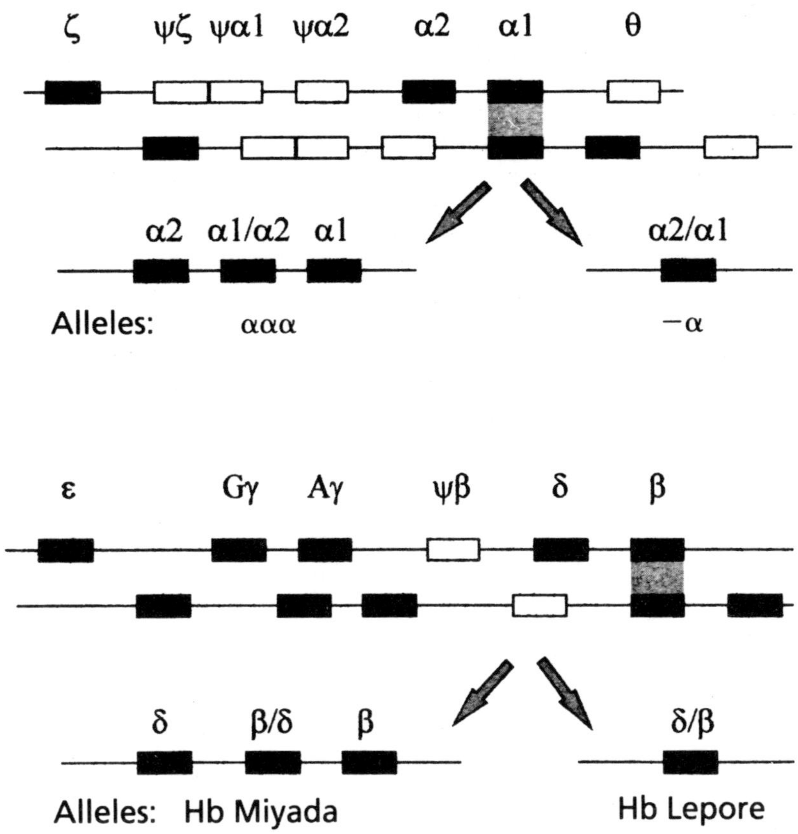

Figure 2 Diagram illustrating how unequal crossing-over between homologous sequences in the α1 and α2 genes (shaded area) creates a triplicated α-gene allele and an α^+-thalassaemia allele with a single α-gene and, similarly, between homologous sequences in the δ- and β-genes creates a triplicated gene allele with a normal δ-, normal β- and hybrid β/δ-gene (Hb Miyada) and a β-thalassaemic allele with a single hybrid δ/β-gene (Hb Lepore).

Table 2 Clinically important Hb variants.

Clinical features	Examples
Sickling disorders	Hb SS Hb SC Hb SD-Punjab Hb SO-Arab Hb S β-thalassaemia
Unstable Hb – haemolytic anaemia	Hb Kolne
Thalassaemia phenotype	Hb Constant Spring
Abnormal oxygen affinity	
High affinity – polycythaemia	Hb Luton
Low affinity – familial cyanosis	Hb Kansas
The M haemoglobins – familial cyanosis	Hb M (Boston)

deformation of the red cell. The sickle-shaped cells are more rigid than normal red cells and tend to block small arteries, resulting in an inadequate oxygen supply to the tissues and organs. In addition, the sickle cells have a shorter life span resulting in a lifelong haemolytic anaemia. The clinical expression of sickle cell disease is very variable, with some patients having an almost normal life, but others developing painful vascular–occlusive crises and an increased propensity to infection which may cause death in early childhood.

The sickle cell mutation at codon 6 (A→T) is thought to have arisen independently at least four times in Africa and once in Asia. The evidence for this comes from a study of the association of the sickle cell gene with β-globin gene cluster haplotypes. Four different haplotypes have been discovered in African individuals, named after the geographical area in which they are most frequently found: the Benin, Central African Republic (the CAR or also termed the Bantu haplotype), Senegal and Cameroon (Pagnier *et al.*, 1984). The sickle cell gene found in individuals of Mediterranean origin is the Benin type, providing evidence of the migration of the sickle cell gene from West Africa to the countries bordering the Mediterranean Sea. The fifth haplotype is found on sickle cell genes in the tribal peoples of central and southern India and Arab populations living on the eastern side of the Arabian Peninsula. The origin of this fifth mutation and the direction of the subsequent migration has yet to be resolved. The CAR haplotype is associated with very low levels of Hb F and the most clinically severe disease, while the Arab–India haplotype is associated with high levels of Hb F and the mildest course of the disease (Powars, 1991).

Sickle cell disease can also result from the combination of Hb S trait with Hb C (see below), Hb O-Arab, Hb D-Punjab and β-thalassaemia trait. The interactions with Hb O-Arab and Hb D-Punjab result in a severe disorder, indistinguishable from Hb SS disease. The interaction with β-thalassaemia is more variable, depending on the nature of the β-thalassaemia mutation. Sickle cell β-thalassaemia ranges from very severe in patients with a β^o-type mutation to very mild in patients with a mild β^+-type mutation, such as one of the promoter region mutations commonly found in Africans.

5. Hb C

The β-globin gene expressing Hb C contains a single nucleotide change in the same codon as Hb S, but is different in that lysine is substituted for glutamic acid at codon 6. The gene can reach frequencies of up to 0.15 in parts of West Africa where it coexists with that for Hb S. The Hb C mutation causes a decrease in solubility of both the oxygenated and the deoxyenated forms of the haemoglobin, resulting in the formation of crystals instead of long polymers. In individuals homozygous for Hb C, the red cells become dehydrated and rigid, causing a haemolytic anaemia, but such patients do not develop any sickling symptoms. The clinical importance of Hb C lies in its interaction with Hb S. The inheritance of Hb C trait with Hb S trait results in Hb SC disease, a milder version of sickle cell disease.

6. Hb E

The Hb E variant results from a point mutation (G→A) at codon 26 of the β-globin gene and is associated with a mild β-thalassaemia phenotype. The point mutation activates a cryptic splice site present between codons 24 and 27, causing two forms of β-globin mRNA to be produced. The abnormally spliced mRNA does not produce a functional β-globin chain. The normally spliced mRNA produces a globin chain containing the Hb E substitution of lysine for glycine, but the level of correctly spliced mRNA is lower than normal, leading to a deficient production of β^E-globin. Like Hb C, the heterozygous and homozygous state for Hb E is not associated with any clinical disability. Individuals homozygous for Hb E have an extremely mild disorder, with microcytic and hypochromic red cells, but no significant anaemia or haemolysis.

The clinical importance of Hb E lies in its interaction with β-thalassaemia, resulting in a variable clinical picture ranging from a mild form of thalassaemia intermedia to a disorder indistinguishable from transfusion-dependent β-thalassaemia major. In populations in eastern India and south-east Asia, where Hb E reaches very high frequencies, individuals with both Hb E trait and β-thalassaemia trait, i.e. Hb E thalassaemia, pose a major public health problem. As with Hb S, the interaction of Hb E with mild β^+ thalassaemia mutations generally results in the mildest form of Hb E thalassaemia and interaction with β^o thalassaemia mutations, the severest clinical disorder (Weatherall & Clegg, 1981).

7. Thalassaemia

The thalassaemia disorders are characterized by an absence or reduction in synthesis in one or more globin chains causing an imbalance in the ratio of the products of the two globin gene clusters which are normally expressed in equal amounts. They are a heterogeneous group of disorders, classified according to the particular globin chains that are synthesized in reduced amounts: α, β, δβ or γδβ-thalassaemia. Each type leads to an imbalanced globin chain synthesis with the precipitation of the globin chains that are produced in excess, resulting in microcytic and hypochromic red cells with a varying degree of anaemia.

8. α-Thalassaemia

(i) Molecular defects

α-Thalassaemia is characterized by a deficiency oı α-globin chain synthesis. Defective gene expression may occur in either one globin gene (called α-2 or α^+-thalassaemia) or in both (α-1 or α^o-thalassaemia). Most α-thalassaemia alleles result from the deletion of gene sequences in the α-globin gene cluster. The deletion of one of the two α-globin genes gives rise to the α^+-thalassaemia phenotype and, although five different deletions have been identified (Higgs, 1990), only two are commonly encountered in practice. These are the 3.7 kb deletion ($-\alpha^{3.7}$), which has reached high frequencies in the populations of Africa, the Mediterranean area, the Middle East, the Indian subcontinent and Melanesia, and the 4.2 kb deletion ($-\alpha^{4.2}$) which is commonly found in south-east Asian and Pacific populations. These deletions were created by unequal crossing-over between homologous sequences in the α-globin gene cluster, resulting in one chromosome with only one α-gene (−α) and the other chromosome with three α-genes (ααα). Further recombination events between the resulting chromosomes have given rise to quadruplicated α-genes (αααα).

Various non-deletion defects have also been found to cause α^+-thalassaemia and a total of 17 mutations have been described to date, mostly in populations from the Mediterranean area, Africa and south-east Asia. All of the mutations except one are located in the dominant α2-globin gene and are denoted $\alpha^T\alpha$ in order to signify the affected gene. As there appears to be no associated change in expression of the remaining functional α1-globin gene (as occurs with the −α alleles), they give rise to a more severe reduction in α-chain synthesis than in deletion α^+-thalassaemia. The interaction of these non-deletional mutations with α^o-thalassaemia results in severe Hb H disease or, very rarely, Hb Bart's hydrops fetalis syndrome in infants with a very low level of α-globin chain synthesis.

α^o-Thalassaemia results from deletions which involve both α-globin genes in the α-globin gene cluster. At least 14 different such deletions have been described (Higgs, 1990). The deletions which have attained the highest gene frequencies are found in individuals from south-east Asia and south China ($--^{SEA}$), the Philippine Islands ($--^{FIL}$), Thailand ($--^{THAI}$) and a few Mediterranean countries such as Greece and Cyprus ($--^{MED}$ and $-(\alpha)^{20.5}$). Although an α^o-thalassaemia mutation ($--^{SA}$) has been described in Asian Indians, it is extremely uncommon and no α^o-thalassaemia deletions have been reported in individuals from subSaharan Africa. In northern Europe, α-thalassaemia occurs sporadically because of the lack of natural selection, but several α^o-thalassaemia deletions have been reported in single British families and one particular molecular defect ($--^{BRIT}$) appears to be quite common in families living in Cheshire and Lancashire.

(ii) Clinical aspects

The most severe form of α-thalassaemia is the homozygous state for α^o-thalassaemia, known as Hb Bart's hydrops fetalis syndrome. This condition results from a deletion of all four globin genes and an affected foetus cannot synthesize any α-globin to make Hb F or Hb A. Foetal blood contains only the abnormal haemoglobin Bart's (γ_4) and a small amount of Hb Portland. The resulting severe foetal anaemia leads to asphyxia, hydrops fetalis and stillbirth or neonatal death; prenatal diagnosis is always indicated in order to avoid the severe toxaemic complications that occur frequently in pregnancy with hydropic foetuses.

Individuals with one functional α-gene have Hb H disease and are compound heterozygotes for α^+ and α^o-thalassaemia. They have a moderately severe hypochromic microcytic anaemia and produce large amounts of Hb H (β_4) as a result of the excess β-chains in the reticulocyte. Patients with deletional Hb H disease suffer from fatigue, general discomfort and splenomegaly, but they rarely require hospitalization and lead a relatively normal life. However, patients with non-deletional Hb H disease seem to exhibit more severe symptoms, with a possible requirement of recurrent blood transfusions and splenectomy. In some situations, couples at risk for the severe form of Hb H disease have opted for prenatal diagnosis and termination of an affected foetus.

Individuals with two functional α-globin genes are characterized by a normal Hb A_2 level and clearly reduced values for the red cell (mean corpuscular) haemoglobin (MCH) and mean cell volume (MCV). Such persons may be either homozygous for α^+-thalassaemia or heterozygous for α^o-thalassaemia. The two conditions have very similar phenotypes and can only be distinguished with certainty by DNA analysis. Finally, carriers of α^+-thalassaemia have three functional α-globin genes and exhibit slightly reduced red cell indices, but the overlap in these indices makes it impossible to distinguish carriers from normal individuals in many cases on the basis of MCV and MCH alone. The detection of Hb Bart's in neonates and the demonstration of occasional cells containing Hb H in adults also signify the presence of α-thalassaemia trait, but again the tests do not clearly distinguish the various α-genotypes and do not detect all cases of α-thalassaemia trait.

9. β-Thalassaemia

(i) Molecular defects

β-thalassaemia is caused by at least 170 different point mutations or small insertions/deletions of DNA sequence in and around the β-globin gene, together with a much smaller number of gene deletions ranging from 25 bp to 67 kb (Baysal, 1995). The mutations either reduce the expression of the β-globin gene (β^+-type) or result in the complete absence of β-globin (β^o-type). They affect globin gene transcription, RNA processing or translation, RNA cleavage and polyadenylation, or result in a highly unstable globin chain. Frameshift and nonsense codon mutations have been observed in all three exons and RNA processing mutations have been found in both introns and all four splice junctions. Examples of each type of mutation are shown in **Figure 3**.

Many of the mutations have attained high frequencies through a positive selection mechanism due to malaria, and Table 3 lists the most common β-thalassaemia mutations found in selected countries from the Mediterranean area,

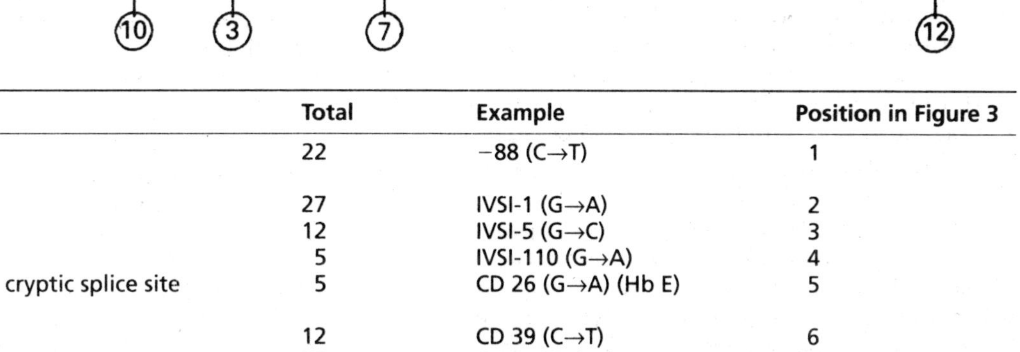

Figure 3 Diagram illustrating the β-globin gene and the position of 13 examples of the different types of β-thalassaemia mutations. Non-coding regions, light shaded areas; exons, dark shaded areas; introns or intervening sequences (IVS), blank areas.

Type of mutation	Total	Example	Position in Figure 3
Transcriptional mutations	22	−88 (C→T)	1
RNA processing mutations			
Splice junction changes	27	IVSI-1 (G→A)	2
Consensus site changes	12	IVSI-5 (G→C)	3
New internal splice site	5	IVSI-110 (G→A)	4
Activation of coding region cryptic splice site	5	CD 26 (G→A) (Hb E)	5
RNA translation mutations			
Nonsense mutations	12	CD 39 (C→T)	6
Frameshift mutations	55	CD 71/72 (+A)	7
Initiation codon mutations	6	ATG→GTG	8
Highly unstable β chain variants	30	CD 121 (G→T)	9
Cap site mutations	1	CAP+1 (A→C)	10
Polyadenylation site mutations	6	AATAAA→AACAAA	11
3′ Untranslated region mutations	3	3′ UTR+6 (C→G)	12
Gene deletions (25 bp–67 kb)	17	619 bp deletion	13

Table 3 The regionally specific distribution of the common β-thalassaemia mutations (expressed as percentage gene frequencies of the total number of thalassaemia chromosomes studied).

Mutation	**Mediterranean**			**Asian–Indian**		**Chinese**		**African**
	Italy	***Greece***	***Turkey***	***Pakistan***	***India***	***China***	***Thailand***	***African–American***
−88 (C→T)					0.8			21.4
−87 (C→G)	0.4	1.8	1.2					
−30 (T→A)			2.5					
−29 (A→G)						1.9		60.3
−28 (A→G)						11.6	4.9	
CAP+1 (A→C)					1.7			
CD 5 (−CT)		1.2	0.8					
CD 6 (−A)	0.4	2.9	0.6					
CD 8 (−AA)		0.6	7.4					
CD 8/9 (+G)				28.9	12.0			
CD 15 (G→A)				3.5	0.8			0.8
CD 16 (−C)				1.3	1.7			
CD 17 (A→T)						10.5	24.7	
CD 24 (T→A)								7.9
CD 39 (C→T)	40.1	17.4	3.5					
CD 41/42 (−TCTT)				7.9	13.7	38.6	46.4	
CD 71/72 (+A)						12.4	2.3	
IVSI-1 (G→A)	4.3	13.6	2.5					
IVSI-1 (G→T)				8.2	6.6			
IVSI-5 (G→C)				26.4	48.5	2.5	4.9	
IVSI-6 (T→C)	16.3	7.4	17.4					
IVSI-110 (G→A)	29.8	43.7	41.9					
IVSII-1 (G→A)	1.1	2.1	9.7					
IVSII-654 (C→T)						15.7	8.9	
IVSII-745 (C→G)	3.5	7.1	2.7					
619 bp deletion				23.3	13.3			
Others	4.1	2.2	9.7	0.5	0.9	6.8	7.9	10.6

CD, codon; IVS, intervening sequence.

Asian India, south-east Asia and Africa. For each region, there is a small number of mutations, accounting for more than 90% of those found, and the mutations are regionally specific, with each country within a region having its own characteristic spectrum of mutations.

A small number of β^+ mutations result in a milder phenotype than the majority of the β^+ and β^0-type mutations. These mutations include the transcription mutations located in the promoter region upstream of the β-globin gene, such as the mutations at nucleotides −88, −87, −30, −29 and −28 (listed in Table 3), the CAP+1 site mutation, mutations between codon 24 and 27 (such as the codon 24 mutation in Africans and the codon 26 mutation which creates Hb E) and the mutation at IVSI-6 found in Mediterranean countries. The latter mutation, together with CAP+1 found in Indians, results in a Hb A_2 level of 3.5–4.0%, a value only just raised above the normal range, and carriers of these mutations may be mistaken for having α-thalassaemia trait. However, these mutations do result in reduced MCV and MCH values, in contrast to three mutations observed in Italians: the promoter mutations −101 (C→T) and −92 (C→T), and the mutation IVSII-844 (G→C). Heterozygotes for these mutations exhibit completely normal MCV and MCH values in addition to a normal/borderline HbA_2 value and cannot be detected by haematological screening tests, thus having a truly silent type of β-thalassaemia.

The mild β^+ mutations have been characterized by their identification in patients with the milder clinical condition of thalassaemia intermedia, either in the homozygous state or in the compound heterozygous state with a severe β^+- or β^0-type mutation (see below). However, predicting the phenotype of thalassaemia intermedia on the basis of a known genotype remains a problem. Homozygosity for a mild mutation nearly always results in intermedia, but the combination of a mild and a severe mutation results in thalassaemia major in some cases. In addition, the inheritance of ameliorating genetic factors, such as α-thalassaemia, a high Hb F determinant or other hereditary factors for which the molecular basis remains unclear, can result in a patient with a thalassaemia major genotype developing thalassaemia intermedia. Other causes of thalassaemia intermedia include the compound heterozygosity for a β-thalassaemia mutation and a triple or quadruple α-gene locus, and the heterozygous state for one of the rare β-thalassaemia mutations in exon 3 which result in a highly unstable Hb variant and cause inclusion-body haemolytic anaemia, sometimes called dominantly inherited β-thalassaemia (Thein *et al.*, 1990) because the thalassaemia phenotype is observed in the carrier state.

(ii) Clinical aspects

β-thalassaemia is characterized by a deficiency of β-globin chain synthesis. Individuals with β-thalassaemia trait are essentially healthy although they may have a slightly reduced Hb level and have typically low MCH and MCV values. In contrast to β-thalassaemia, the Hb A_2 concentration is elevated above the normal range to a level of 3.5–7.0% in all but a small minority of cases. β-Thalassaemia trait with a normal Hb A_2 level is due to either the coinheritance of δ-thalassaemia (12 different mutations are known to restrict the output of the δ-globin gene) or to the inheritance of one of the six mild β-thalassaemia alleles known as silent β-thalassaemia mutations. β-Thalassaemia trait with an unusually high level of Hb A_2 (above 7.0%) results from one of the larger deletions affecting the β-globin gene.

The majority of individuals homozygous for β-thalassaemia have the transfusion-dependent condition called β-thalassaemia major. This condition results from the homozygous state for a severe β-thalassaemia mutation or, more commonly, the compound heterozygous state for two different mutations. At birth, β-thalassaemia homozygotes are asymptomatic because of the high production of Hb F, but, as this declines, affected infants present with severe anaemia during the first or second year of life. Treatment is by frequent blood transfusion to maintain a haemoglobin level above 10 g/dl, coupled with iron chelation therapy to control iron overload, otherwise death results in the second or third decade from cardiac failure. This treatment does not cure β-thalassaemia major, although some patients have now reached the age of 40 in good health and have married and produced children. With the prospects for gene therapy remaining as distant as ever, the only cure for β-thalassaemia for the foreseeable future is bone marrow transplantation. Although this form of treatment has proved successful when carried out in young children, it is limited by the requirement of a human leukocyte antigen-matched sibling or relative.

Some individuals homozygous for β-thalassaemia have a milder clinical condition called thalassaemia intermedia. Such patients present later in life relative to those with thalassaemia major and are capable of maintaining a haemoglobin level above 6 g/dl without transfusion. Thalassaemia intermedia is caused by a wide variety of genotypes and covers a broad clinical spectrum. Patients with a severe condition present between 2 and 6 years of age and, although they are capable of surviving with an Hb level of 5–7 g/dl, they will not develop normally and are treated with minimal blood transfusion. At the other end of the spectrum are patients who do not become symptomatic until they reach adult life and remain transfusion independent with Hb levels of 8–10 g/dl. However, even these more mildly affected patients tend to accumulate iron with age and many thalassaemia intermedia patients develop clinical problems relating to iron overload after the third decade.

10. δβ-Thalassaemia and HPFH

δβ-Thalassaemia and the deletion types of hereditary persistence of foetal haemoglobin (HPFH) are characterized by the complete absence of Hb A and Hb A_2 in homozygotes and an elevated level of Hb F in heterozygotes. Both conditions are caused by large DNA deletions involving the β-globin gene cluster affecting the β and δ genes, but leaving either one or both of the γ-globin genes intact. More than 50 different deletion mutations have been identified and they can be classified into the $(\delta\beta)^o$ and $(^{A}\gamma\delta\beta)^o$ thalassaemias, HPFH conditions and the fusion chain variants Hb Lepore and Hb Kenya. Included in this group are the rare $(\varepsilon\gamma\delta\beta)^o$ thalassaemia alleles (Weatherall *et al.*, 1989), in which all of the β-globin gene cluster genes are either deleted or not expressed.

The $(\delta\beta)^o$ thalassaemias are characterized by the Hb F consisting of both $^{G}\gamma$- and $^{A}\gamma$-globin chains, as both γ-globin genes remain intact in these conditions. Heterozygotes have normal levels of Hb A_2 and an Hb F level of 5–15% which, for the majority of mutations, is heterogeneously distributed in the red cells. There is a reduction of the non-α-globin chains compared with α-globin and the red cells are hypochromic and microcytic. Homozygotes for this condition have thalassaemia intermedia.

The $(^{A}\gamma\delta\beta)^o$ thalassaemias are characterized by the Hb F containing only $^{G}\gamma$-globin chains as the $^{A}\gamma$-globin gene has been deleted in these conditions. Apart from this distinction, the phenotypes of the heterozygous and homozygous states are identical to those for $(\delta\beta)^o$ thalassaemia.

The deletional HPFH conditions can be regarded as a type of δβ-thalassaemia in which the reduction in β-globin chain production is almost completely compensated for by the increased γ-globin chain production. Homozygous individuals have 100% Hb F comprised of both $^{A}\gamma$- and $^{G}\gamma$-globin chains, but, in contrast to $(\delta\beta)^o$ thalassaemia homozygotes, are clinically normal. Heterozygotes have an elevated Hb F level of 17–35%, higher than that found in δβ thalassaemia heterozygotes, and the Hb F is distributed uniformly (pancellular) in red cells with near normal MCH and MCV values.

Two deletions inside the β-globin gene cluster create an abnormal Hb chain as a result of unequal crossing-over between different globin genes. Hb Lepore is a hybrid globin chain comprising δ- and β-gene sequences (**Figure 2**) and Hb Kenya is a hybrid globin chain comprising of γ- and β-gene sequences. Hb Lepore homozygotes have a phenotype similar to thalassaemia major or severe thalassaemia intermedia. Hb Kenya has only been observed in the heterozygous state and is similar to heterozygous HPFH, with individuals having 5–10% Hb F, normal red cell morphology and balanced globin chain synthesis.

The $(\varepsilon\gamma\delta\beta)^o$ thalassaemias are rare conditions that result from several different long deletions that start upstream of the ε-gene and remove all of the globin genes in the β-globin gene cluster or, in two cases, the deletion ends between the δ- and β-genes thus sparing the β-globin gene, but causing no β-globin synthesis to occur. This is because the deletions remove the β-globin gene cluster locus control region, a regulatory region 50 kb upstream of the ε-gene which regulates the expression of the entire β-globin gene cluster. In adult life, heterozygotes for this condition have a similar haematological picture to β-thalassaemia trait with a normal Hb A_2 level. The homozygous condition is presumed to be incompatible with foetal survival.

Finally, there is a group of conditions called non-deletion HPFH in which heterozygous individuals have normal red cells and no clinical abnormalities, and an elevated Hb F level as a result of a point mutation in the promoter region of the $^{A}\gamma$- or $^{G}\gamma$-globin gene in most cases. The percentage Hb F is variable, in a range of 1–3% in the Swiss type to 10–20% in the Greek type. The only recorded homozygotes for non-deletion HPFH are for the British type described in a single family.

REFERENCES

Baysal, E. (1995) The β- and δ-thalassemia repository. *Hemoglobin* 19: 213–36

Charache, S., Terrin, M.L., Moore, R.D. *et al.* (1995) Effect of hydroxyurea on the frequency of painful crises in sickle cell anemia. *New England Journal of Medicine* 332: 1317–22

Cheung, M.-C., Goldberg, J.D. & Kan, Y.W. (1996) Prenatal diagnosis of sickle cell anemia and thalassemia by analysis of fetal cells in maternal blood. *Nature Genetics* 14: 264–68

Flint, J., Harding, R.M., Boyce, A.J. & Clegg, J.B. (1993) The population genetics of the haemoglobinopathies. In *The Haemoglobinopathies*, edited by D.R. Higgs & D.J. Weatherall, London: Baillière Tindall

Higgs, D.R. (1990) The molecular-genetics of the alpha-globin gene family. *European Journal of Clinical Investigation* 20: 340–47

Higgs, D.R. (1993) Alpha-thalassaemia. In *The Haemoglobinopathies*, edited by D.R. Higgs & D.J. Weatherall, London: Baillière Tindall

Huisman, T.H.J., Carver, M.H. & Efremov, G.D. (1996) *A Syllabus of Human Hemoglobin Variants*, Augusta, Georgia: Sickle Cell Anemia Foundation

Kazazian, H.H. Jr, Philips, J.A. III, Boehm, C.D. *et al.* (1980) Prenatal diagnosis of β-thalassaemia by amniocentesis: linkage analysis using multiple polymorphic restriction endonuclease sites. *Blood* 56: 926–30

Old, J. (1996) Haemoglobinopathies. *Prenatal Diagnosis* 16: 1181–86

Old, J.M., Ward, R.H.T., Petrou, M. *et al.* (1982) First-trimester fetal diagnosis for haemoglobinopathies: three cases. *Lancet* 2: 1413–16

Pagnier, J., Mears, J.G., Dunda-Belkodja, O. *et al.* (1984) Evidence for the multicentric origin of the sickle cell hemoglobin in Africa. *Proceedings of the National Academy of Sciences USA* 81: 1771–73

Petrou, M. & Modell, B. (1995) Prenatal screening for haemoglobin disorders. *Prenatal Diagnosis* 15: 1275–95

Powars, D.R. (1991) β^{s}-gene cluster haplotypes in sickle-cell-anemia – clinical and hematological features. *Hematology/Oncology Clinics of North America* 5: 475–93

Thein, S.L., Hesketh, C., Taylor, P. *et al.* (1990) Molecular basis for dominantly inherited inclusion body β-thalassemia. *Proceedings of the National Academy of Sciences USA* 87: 3924–28

Weatherall, D.J. & Clegg, J.B. (1981) *The Thalassemia Syndromes*, 3rd edition, Oxford: Blackwell

Weatherall, D.J., Clegg, J.B., Higgs, D.R. & Wood, W.G. (1989) The haemoglobinopathies. In *The Metabolic Basis of Inherited Disease*, 6th edition, volume 2, edited by R. Scriver, A.L. Beaudet, W.S. Sly & D. Valle, New York: McGraw-Hill

World Health Organization (1983) Community control of hereditary anaemias. Memorandum from a WHP meeting. *Bulletin of the World Health Organization* 61: 63

GLOSSARY

C-terminus the end of a polypeptide chain with a free carboxyl (COOH) group

frameshift mutation mutation that causes a change in the reading frame, consisting of an insertion or deletion of non-multiples of three consecutive nucleotides in a DNA sequence

Hb Bart's hydrops fetalis syndrome condition characterized by massive oedema in the foetus or newborn, typically accompanied by severe anaemia and effusions of the pericardial, pleural and peritoneal spaces. Usually leads to death despite immediate exchange transfusions post-delivery

high performance liquid chromatography (HPLC; alt. high pressure liquid chromatography) technique used to identify haemoglobin molecules by chromatography in a column subjected to a pressure of 4000 psi

hypochromic having less than normal colour

initiation codon first codon of the coding region in mRNA, the starting point of translation

isoelectric focusing electrophoresis in a stabilized pH gradient, in which molecules migrate to the pH matching their isoelectric point to form stationary bands

microcytic pertaining to a smaller than normal cell

N-terminus the end of a polypeptide chain with a free amino (NH_2) group

polyadenylation addition of a poly(A) tail to eukaryotic messenger RNA precursors in the nucleus

Southern blot technique in which DNA molecules, separated by electrophoresis in an agarose gel, are transferred by "blotting" to a nitrocellulose or nylon filter, then hybridized with radioactive or otherwise labelled nucleic acid probes for identification and location of sequences of interest

HAEMOCHROMATOSIS

James C. Barton
Southern Iron Disorders Center, Birmingham, Alabama, USA

1. Introduction

Haemochromatosis, the most common known hereditary disorder of Caucasians of western European descent, affects approximately 0.5% of these peoples. Iron absorption in homozygotes for this autosomal recessive abnormality is inappropriately high for body iron content. Many affected persons have progressive iron deposition that injures the liver, joints, pancreas, heart and other organs, and reduces longevity. The known complications of haemochromatosis are, however, prevented by early diagnosis and simple, effective treatment to remove excess iron and maintain low normal body iron stores (therapeutic phlebotomy). Thus while haemochromatosis is frequent, the simple diagnosis and effective iron depletion therapy makes haemochromatosis an important public health concern in a large fraction of the world's population (Witte *et al.*, 1996).

2. History

First described by European pathologists in the late 19th century from a few cases discovered at autopsy, haemochromatosis was initially believed to be very rare. Sheldon's monograph on haemochromatosis published in 1935 included observations on all reported cases. Haemochromatosis was regarded to be a very uncommon malady of men, most of whom had bronze discolouration of the skin, diabetes mellitus, hepatic cirrhosis and cardiac failure. During the next several decades, many additional cases were reported, including those of women, and the hereditary nature of the disorder became more apparent (Finch & Finch, 1955). Advances in the understanding of haemochromatosis included the introduction of routine serum iron parameter testing in clinical practice, the description of a peculiar form of degenerative arthritis associated with haemochromatosis (Schumacher, 1964) and recognition that the haemochromatosis gene is linked to the human leucocyte antigen (HLA) loci of the short arm of chromosome 6 (6p) (Simon *et al.*, 1975). Surveys of normal persons have confirmed the approximate frequency of haemochromatosis – and of its associated gene(s) – and emphasized the variation in clinical abnormalities that occur in probands diagnosed in routine medical care delivery, in population screening programmes and in relatives of probands (Edwards *et al.*, 1988). In 1996, two nonsense mutations in a major histocompatibility (MHC) class I gene on 6p (*HFE*) were identified that account for approximately 80% of haemochromatosis cases in most western Caucasian populations (Feder *et al.*, 1996).

3. Genetics

The common classical haemochromatosis-associated gene is linked to the HLA region on 6p, particularly with HLA-A3 (Simon *et al.*, 1975; Witte *et al.*, 1996). This abnormal gene is attributable to a coding region missense mutation in an MHC-like gene on 6p (exon 4: nucleotide 845G→A; C282Y; *HFE* gene) that occurs in 60–100% of unrelated haemochromatosis patients (Beutler *et al.*, 1996; Feder *et al.*, 1996; Barton *et al.*, 1997a). Approximately 14% of Caucasians are heterozygous for the C282Y mutation. A second common *HFE* coding region mutation (exon 2: cDNA nucleotide 187 C→G; H63D) occurs in approximately 20% of western Caucasians. In some studies, H63D has been linked to the HLA-A29 gene (Porto *et al.*, 1998), but it is uncommonly associated with a haemochromatosis clinical phenotype. Nevertheless, some compound heterozygotes (C282Y/H63D) have mild iron overload with little or no iron-associated organ injury (Mura *et al.*, 1997; Sham *et al.*, 1997; Adams & Chakrabarti, 1998; Porto *et al.*, 1998).

However, the C282Y and H63D mutations do not explain all haemochromatosis cases among Caucasians (Barton *et al.*, 1997a; Carella *et al.*, 1997) or the variability of iron overload among haemochromatosis patients. Some adults with haemochromatosis do not have mutations involving iron absorption or metabolism pathways that can be currently detected, although some have severe iron overload. *HFE* coding region mutations other than C282Y and H63D have been identified in persons with haemochromatosis and in members of general Caucasian populations (Bernard *et al.*, 1998; Douabin *et al.*, 1998). It seems likely that these and additional future discoveries will explain why haemochromatosis phenotypes sometimes appear among persons who are heterozygous (or negative) for C282Y or H63D. The frequency of haemochromatosis and haemochromatosis heterozygosity accounts for the frequent appearance of haemochromatosis in successive generations, mimicking dominant inheritance (Witte *et al.*, 1996; Barton *et al.*, 1999). An especially aggressive form of progressive iron overload that occurs in teenagers or young adults (juvenile haemochromatosis) appears to be unrelated to the common mutations associated with the classical form of the disorder.

Estimates of the age of the original C282Y mutation suggest that it occurred 69–113 generations ago (*c.*800BC–*c.*300AD) (Ajioka *et al.*, 1997; Malfroy *et al.*, 1997). Whether the mutation occurred originally in northern European Celts or in Vikings is unclear, although the age of the classical C282Y mutation and its geographic distribution in Europe indicate with certainty that dissemination of the C282Y mutation in northern Europe is largely attributable to the Vikings (Merryweather-Clarke *et al.*, 1997; Olsson *et al.*, 1998). The H63D mutation is probably much older and is more widely distributed in Caucasian and non-Caucasian populations (Porto *et al.*, 1998). H63D, even in a homozygous configuration, appears to have much less penetrance than does C282Y in causing a clinical phenotype of disturbed iron absorption and metabolism (Beutler *et al.*, 1996; Feder *et al.*, 1996; Barton *et al.*, 1997a; Porto *et al.*, 1998).

The frequency of haemochromatosis often varies significantly among regions of the same country in western Europe (Witte *et al.*, 1996; Merryweather-Clarke *et al.*, 1997). A few reports from several countries of middle Europe indicate that HLA-linked haemochromatosis exists there (and may be common). However, the eastern-most geographical extent of haemochromatosis is largely unknown. The haplotypes associated with haemochromatosis gene mutations vary from country to country. In derivative countries of western Europeans, e.g. Australia, Canada, Republic of South Africa and the US, the HLA phenotypes of persons with haemochromatosis may differ somewhat from those observed in western Europe. These differences are attributable to founder effects in some circumstances and to the creations of new populations in others. Hispanic persons in the US appear to have haemochromatosis with a frequency similar to that of other Caucasians. By contrast, haemochromatosis appears to be rare or absent among native persons in the Americas, in Ashkenazi Jews and in persons of African descent (Merryweather-Clarke *et al.*, 1997). A common iron overload disorder occurs in subSaharan African natives and in African–Americans, and this may also be heritable (Gordeuk *et al.*, 1992). Unlike haemochromatosis, this disorder is not HLA linked, is phenotypically distinguishable from Caucasian haemochromatosis and is caused by the interaction of genetic factors and dietary iron. According to the population examined, this disorder is designated African iron overload (Bantu siderosis) or African–American iron overload, respectively. Asians with unexplained systemic iron overload also appear to lack Caucasian-associated haemochromatosis mutations.

The frequency of the haemochromatosis gene(s) suggests that these mutations have conferred an evolutionary advantage. By analogy to other common mutations, haemochromatosis heterozygotes, not the clinically prominent homozygotes, may have been the beneficiaries of natural selection. The capability to absorb a greater fraction of iron from food and drink may permit greater vigour, resistance to chronic gastrointestinal haemorrhage caused by intestinal parasites and a lower likelihood of developing iron deficiency. Similarly, females heterozygous for haemochromatosis mutations may have an increased reproductive potential or be able to impart a greater quantity of iron to their developing foetuses and nursing infants. Alternatively, haemochromatosis mutations may increase resistance to one or more infections. The common haemochromatosis-associated mutations C282Y and H63D occur in an area of the genome devoted to diverse processes of immunity. Iron itself is important for defence against microbes. Likewise, mutations involving iron metabolic pathways may result in altered immunity. In mice, for example, mutations in a gene for a macrophage metal transporter protein (*Nramp1*) cause increased resistance to infection by species of the genera Mycobacterium, Leishmania and Salmonella. Among sub-Saharan African natives, mutations in the homologous gene (*NRAMP1*) are associated with resistance against tuberculosis and may be fundamentally related to African iron overload (Gordeuk *et al.*, 1992; Bellamy *et al.*, 1998). The infections against which the inheritance of haemochromatosis genes may have possibly provided greater resistance in early inhabitants of western Europe include tuberculosis, plague, hookworm infestation or malaria. However, the precise advantage(s) of inheriting haemochromatosis gene(s) remain unproven.

4. Absorption and metabolism of non-ferrous metals

In haemochromatosis, absorption of inorganic forms of some non-ferrous metals, including cobalt, manganese, zinc and lead, is increased (Barton *et al.*, 1994). It seems likely that the absorption of other metals is also increased. Even among haemochromatosis heterozygotes, there is supranormal absorption of certain non-ferrous metals. Inorganic cobalt absorbed in excess of normal is rapidly excreted. Manganese and zinc, however, are retained in the liver and other tissues, although their concentrations appear to be unrelated to the presence or absence of hepatic cirrhosis. Lead, toxic in small amounts, is retained in many tissues for prolonged periods (Barton *et al.*, 1994). The role of these non-ferrous metals in the pathogenesis of symptoms and tissue injury associated with haemochromatosis has not been reported. Likewise, the effects of occupational, avocational or environmental exposure to non-ferrous metals on persons who have inherited the haemochromatosis gene(s) remain unstudied.

5. Pathophysiology of iron overload

Normal adult males and postmenopausal females absorb approximately 1 mg of iron daily to replace unavoidable losses through desquamation, sweating and gastrointestinal and urinary excretion. Women must absorb approximately twice this amount of iron during their reproductive years. Many persons with haemochromatosis, especially males, absorb ~1 mg of elemental iron daily in excess of normal. This may be explained in part by the altered affinity of transferrin for the transferrin receptor in intestinal mucosal cells that express mutant *HFE* protein (Feder *et al.*, 1998). Since excretory mechanisms for iron are limited, iron overload of varying severity develops in most haemochromatosis homozygotes. In normal persons, small amounts of iron are stored in ferritin molecules synthesized by and accumulated in the cytoplasm of monocytes and macrophages. In persons with haemochromatosis, most excess iron is not found in these normal storage sites. Instead, iron accumulates free in the cytoplasm of many cell types. The presence of non-transferrin-bound iron in the serum in haemochromatosis may contribute significantly to the abnormal parenchymal cell deposition of excess iron. Approximately 90% of the total body excess iron in persons with haemochromatosis is retained by hepatocytes, the main functional cells of the liver. Much of the remaining iron is stored in the functional cells of the endocrine organs, e.g. pancreas (both exocrine epithelia and insulin-producing B-cells), anterior pituitary gland (especially gonadotrophin-producing cells), the joints and the heart. However, almost all organs and cell types contain excess iron in patients with advanced iron overload due to haemochromatosis, particularly in the liver (Witte *et al.*, 1996).

The severity of iron overload in haemochromatosis (the clinical phenotype) is related to the inheritance of haemochromatosis-associated mutations (the genotype). Typically, most persons with haemochromatosis who are homozygous for HLA-A3 and/or C282Y have developed iron overload that is often severe (Barton *et al.*, 1996; Beutler *et al.*, 1996; Feder *et al.*, 1996; Barton *et al.*, 1997a; Mura *et al.*, 1997; Sham *et al.*, 1997; Adams & Chakrabarti, 1998; Crawford *et al.*, 1998). Penetrance of the clinical phenotype among persons who are heterozygous for or who lack the C282Y mutation is much less than that of C282Y homozygotes, and iron overload is usually less severe in probands with these genotypes. Likewise, compound C282Y/H63D heterozygotes and H63D homozygotes develop a clinical phenotype infrequently; iron overload, if present, is typically mild. Caucasians with haemochromatosis who inherit neither of these genes may have iron overload of mild, moderate or extreme severity.

Age, sex, endocrine status, nutritional factors and occurrence of hereditary anaemia and coincidental liver disorders also influence the expression of the clinical phenotype, particularly when other disorders occur that also augment iron absorption (Witte *et al.*, 1996; Barton *et al.*, 1997b; Moirand *et al.*, 1997). These disorders include porphyria cutanea tarda, some haemolytic anaemias and, possibly, ingestion of excessive ethanol or iron (Witte *et al.*, 1996). Iron overload is also developed by 1–3% of haemochromatosis heterozygotes (Witte *et al.*, 1996). Whether these persons have additional undetected mutations that increase iron absorption is unclear. Many untreated men with haemochromatosis develop symptoms due to complications of iron overload by middle age; women typically develop iron overload after menopause or hysterectomy. Since the iron content of blood is high, haemochromatosis may elude detection among regular volunteer blood donors or among those who have gastrointestinal disease which otherwise inhibits iron absorption or causes chronic blood loss (Witte *et al.*, 1996).

6. Diagnosis

Transferrin, a 78 kilodalton protein that binds two moles of iron per mole, is essential for normal iron transport in body fluids, especially plasma. By its attachment to specific cell-surface transferrin receptors and internalization by the cell in clathrin-coated pits, transferrin conveys iron to (or removes iron from) cells for normal iron homeostasis. Approximately one-third of the iron-binding sites on plasma transferrin are saturated with iron in normal persons. In haemochromatosis, transferrin saturation with iron is almost invariably increased (Witte *et al.*, 1996). Further, elevated transferrin saturation values are often observed long before systemic iron overload from continued iron absorption and retention occur. Although the precise basis of this phenomenon has not been elucidated, it can nonetheless be used as a presumptive diagnostic test for haemochromatosis. Accordingly, an elevated transferrin saturation value (≥60% for males, ≥50% for females) on at least two occasions in the absence of other known causes strongly suggests the diagnosis of haemochromatosis (Witte *et al.*, 1996; Powell *et al.*, 1998).

Until recently, confirmation of the diagnosis of haemochromatosis depended primarily on the demonstration of hepatic or systemic iron overload and associated complications. Since affected individuals are assumed to be homozygotes or compound heterozygotes for the haemochromatosis gene(s), the demonstration of C282Y homozygosity in an individual with elevated serum transferrin saturation values also establishes the diagnosis. Persons who are heterozygous for the C282Y mutation, those who are compound heterozygotes for the C282Y and H63D mutations and those who are H63D homozygotes sometimes have elevated values of transferrin saturation. Iron overload in these persons, if any, is usually mild. In some persons who are suspected of having haemochromatosis (particularly those who have clinical evidence of hepatic abnormalities), biopsy of the liver may be indicated. This procedure is most helpful to evaluate persons for the occurrence of hepatic cirrhosis due to iron overload, thereby helping to define both diagnosis and prognosis (Powell *et al.*, 1998).

Widespread testing as a public health manoeuvre in Caucasian populations is expected to identify increasing numbers of persons with haemochromatosis, particularly young, asymptomatic individuals, whose only discernible abnormality is elevation of serum transferrin saturation. Currently, such screening efforts are limited to research-based demonstration projects, but molecular genetic analysis may become useful in confirming the diagnosis in these persons. However, the general applicability of genetic testing to haemochromatosis diagnosis in routine medical care delivery or in population studies has not been completely defined. HLA-matched siblings of a proband who has classical haemochromatosis are considered to have haemochromatosis, regardless of their transferrin saturation values. HLA typing is not otherwise useful as a diagnostic test (Powell *et al.*, 1998).

7. Evaluation of iron overload

The severity of iron overload is assessed most frequently using the serum ferritin concentration. Persons suspected of having haemochromatosis who have elevated serum concentrations of hepatic enzymes, hepatomegaly or other evidence of a liver disorder should undergo liver biopsy. These specimens should be evaluated for iron by histochemical (Perls' Prussian blue staining) and quantitative (atomic absorption spectrometric or biochemical) methods. The hepatic iron index helps to distinguish haemochromatosis homozygotes from heterozygotes and from persons with non-haemochromatosis hepatic disorders. However, hepatic biopsy is of limited usefulness to diagnose haemochromatosis before iron overload has occurred (Witte *et al.*, 1996; Powell *et al.*, 1998). The quantity of iron removed by phlebotomy is a retrospective indicator of the severity of iron overload, but cannot be used alone for diagnosis. Liver imaging techniques can visualize hepatic iron deposits, but are too insensitive to evaluate most young, asymptomatic persons. The sequelae of iron overload – hepatopathy, arthropathy, diabetes mellitus and other endocrine disorders, and cardiac abnormalities – may require additional evaluations (Witte *et al.*, 1996; Powell *et al.*, 1998). Some patients have coincidental abnormalities which augment iron absorption and increase iron overload, including excessive dietary iron supplementation or ethanol ingestion, porphyria cutanea tarda or haemolytic anaemia. Viral hepatitis with (and without) haemochromatosis is often associated with elevated serum iron parameters, elevated concentrations of hepatic enzymes and increased stainable iron in the liver.

The results of clinical studies reveal that longevity in persons with haemochromatosis who have hepatic cirrhosis or diabetes mellitus is reduced. The occurrence of these two complications of iron overload in the same patient may interact in an additive manner to reduce life expectancy further (Niederau *et al.*, 1996). The cardiac effects of haemochromatosis can also be life threatening, although these serious complications are uncommon and do not significantly affect the survival statistics in large case series. Significant morbidity in haemochromatosis also occurs due to arthropathy, complications of hepatic cirrhosis and diabetes mellitus, and endocrine dysfunction. In persons whose haemochromatosis is diagnosed before the development of iron overload, longevity is normal and haemochromatosis-associated morbidity is minimal or non-detectable when therapeutic phlebotomy is performed routinely to maintain body iron stores at a normal low level.

8. Therapeutic phlebotomy

Most persons with haemochromatosis benefit from phlebotomy when iron overload is present (Barton *et al.*, 1998). Typically, females of child-bearing years whose serum ferritin is >200 ng/ml and other adults with serum ferritin >300 ng/ml should be treated without delay. Therapeutic phlebotomy removes 200–250 mg of elemental iron per unit of blood (one unit = 450–500 ml). Children and adolescents with haemochromatosis sometimes have severe iron overload (often with cardiac and anterior pituitary failure) and need aggressive phlebotomy. Avoiding phlebotomy because of advanced age alone is not justifiable and delaying therapy until symptoms of iron overload develop is not recommended in any case. Patients with advanced hepatic cirrhosis, severe atherosclerosis or malignancy may not be candidates for phlebotomy treatment.

The depletion of body iron stores typically involves the removal of one unit of blood weekly (Barton *et al.*, 1998). Many females, smaller persons, the elderly and those with coexisting anaemia or cardiac or pulmonary problems can sustain the removal of only 0.5 units of blood weekly or 1 unit every 2 weeks. The haemoglobin concentration or haematocrit should be quantified before each treatment; subjects whose values are <11.0 g/dl or <33.0%, respectively, often experience undue fatigue and other consequences of hypovolaemia and anaemia. The serum ferritin concentration (in the absence of inflammation) and the hepatic iron concentration permit an estimation of the amount of phlebotomy needed. On average, men require twice as many units of phlebotomy as women. Iron depletion is complete when the serum ferritin concentration is 10–20 ng/ml. This indicates that mild iron deficiency has been induced and that pathogenic iron deposits have been removed. Iron depletion should be accomplished within 1 year of diagnosis (Niederau *et al.*, 1996; Barton *et al.*, 1998).

Lifelong phlebotomy should be performed to keep the serum ferritin at <50 ng/ml. This requires 3–4 units of phlebotomy per year in men and 1–2 units per year in women, on average. Some persons, particularly the elderly, appear to require no maintenance phlebotomy. Sustaining overt iron deficiency by phlebotomy is not desirable. Hyperferraemia and elevated transferrin saturation – essential attributes of haemochromatosis – recur quickly after iron depletion. Thus, the serum iron concentration and transferrin saturation are unsuitable as measures of body iron stores and the progress of phlebotomy; the serum ferritin concentration must be used for this purpose (Barton *et al.*, 1998).

9. Management of target organ complications

(i) Liver disorders

Obtaining a liver biopsy specimen is advisable in patients with hepatomegaly elevated serum concentrations of hepatic enzymes or serum ferritin >750 ng/ml (Witte *et al.*, 1996; Powell *et al.*, 1998). The presence or absence of hepatic cirrhosis is a major prognostic factor, and biopsy is the only means by which the occurrence of cirrhosis can be ascertained (Deugnier *et al.*, 1993; Niederau *et al.*, 1996; Witte *et al.*, 1996; Powell *et al.*, 1998). It is unnecessary to repeat the biopsy after iron depletion. Approximately 5% of patients need evaluation for Gilbert's syndrome, viral hepatitis, ethanol- and drug-associated liver injury, primary or secondary liver neoplasm and other abnormalities of the liver (Barton *et al.*, 1997b). Some patients with hepatic cirrhosis and ethanol-associated hepatic injury or viral hepatitis experience persistently elevated serum concentrations of hepatic enzymes or deterioration of hepatic function more rapidly than expected from iron overload alone. Hepatic failure and haemorrhage from oesophagogastric varices cause significant morbidity and require routine management.

(ii) Primary liver cancer

Primary liver cancer is the leading cause of death among haemochromatosis patients with hepatic cirrhosis, and causes 10–30% of all haemochromatosis-related deaths. With few exceptions, only patients with hepatic cirrhosis have an increased risk of developing primary liver cancer (~200 times more than normal) (Deugnier *et al.*, 1993; Witte *et al.*, 1996). The relative risks are increased in patients ≥55 years of age, with hepatitis B surface antigen seropositivity or with excess ethanol intake. One-third of patients with haemochromatosis and primary liver cancer have elevated serum concentrations of α-fetoprotein. Serial testing of patients with hepatic cirrhosis using serum α-fetoprotein concentrations and liver ultrasonography can detect primary liver cancers which can be treated successfully, but the efficacy and cost-effectiveness of this testing have not been clearly established (Barton *et al.*, 1998). At present, it is recommended that patients with hepatic cirrhosis, particularly males ≥50 years of age, undergo measurement of serum α-fetoprotein levels and hepatic imaging using ultrasonography every 6 months. Some patients who have primary, non-metastatic liver cancer undergo hepatic resection and have long-term, cancer-free survival; others may benefit from non-surgical treatments. Liver transplantation has been used successfully to treat primary liver cancer or severe cirrhosis (Kowdley *et al.*, 1995).

(iii) Complications in target organs other than liver

Arthropathy often progresses despite therapeutic phlebotomy; non-haemochromatosis joint disorders can also cause progressive discomfort and disability after iron depletion. Diabetes mellitus should be treated with diet, exercise, oral hypoglycaemic agents and insulin like that in non-haemochromatosis patients. Sexual impotence and the symptoms of premature menopause are treated with hormone replacement. With gonadotrophin therapy, successful pregnancy can occur. Severe cardiomyopathy and arrhythmias are typical of the massive iron overload in teenagers or young adults. Medical therapy to control congestive heart failure and minimize serious arrhythmias must be applied until vigorous therapeutic phlebotomy (possibly combined with iron chelation therapy) relieves the myocardial siderosis. In middle-aged or elderly subjects with haemochromatosis, cardiac dysfunction is occasionally due to iron overload, but coronary atherosclerosis or other forms of heart disease are more common (Witte *et al.*, 1996; Barton *et al.*, 1997b).

10. Dietary recommendations

Persons with haemochromatosis should avoid supplemental or medicinal iron, consume red meats and alcohol in moderation and limit supplemental vitamin C to 500 mg daily. Since the absorption and retention of many non-ferrous metals is increased in haemochromatosis, mineral supplements should be used only for demonstrated deficiencies. *Vibrio vulnificus*, a spiral bacterium, can cause infection when ingested in raw or improperly cooked shellfish or by entering open wounds of those who handle contaminated seafood or bathe in contaminated waters (including the Gulf of Mexico and other warm seas). Bacteraemia due to *V. vulnificus* is often fatal. Persons with haemochromatosis, chronic liver disease or immune deficiency states should consume only thoroughly cooked seafood items from these waters, and take other measures to prevent *V. vulnificus* infections. Since dietary regimens do not enhance iron excretion or prevent iron reaccumulation, all haemochromatosis patients must understand that there is no substitute for iron depletion therapy (Barton *et al.*, 1998).

11. Genetic counselling

The first-degree family members (parents, children and siblings) of each proband should be tested by measuring serum transferrin saturation and ferritin concentration (Witte *et al.*, 1996). All probands and their siblings should also undergo HLA-A and -B typing or HFE mutation analysis. When a proband has the classical form of haemochromatosis linked to chromosome 6p, siblings who are HLA identical to the proband are also presumed to have haemochromatosis. If a proband is homozygous for the C282Y mutation, testing for this mutation may replace HLA typing in family studies (Mercier *et al.*, 1998; Barton *et al.*, 1999). In addition, the possible genotypes of children can usually be deduced by analysis of their parents' *HFE* genotypes. In some kinships, this obviates the need to perform phenotypic testing (Adams, 1998). *HFE* genotyping can also identify asymptomatic siblings who have haemochromatosis, especially prepubertal children and premenopausal females whose serum iron parameters are sometimes normal (Barton *et al.*, 1999). Approximately 25% of siblings and approximately 7% of parents or children of probands have haemochromatosis, usually attributable to C282Y homozygosity. Affected family members should be evaluated and managed in the same manner as probands (Witte *et al.*, 1996; Barton *et al.*, 1998; Powell *et al.*, 1998; Barton *et al.*, 1999).

REFERENCES

Adams, P.C. (1998) Implications of genotyping of spouses to limit investigation of children in genetic hemochromatosis. *Clinical Genetics* 53: 176–78

Adams, P.C. & Chakrabarti, S. (1998) Genotype/phenotype correlations in genetic hemochromatosis: evolution of diagnostic criteria. *Gastroenterology* 114: 319–23

Ajioka, R.S., Jorde, L.B., Gruen, J.R. *et al.* (1997) Haplotype analysis of hemochromatosis: evaluation of different linkage-disequilibrium approaches and evolution of disease chromosomes. *American Journal of Human Genetics* 60: 1439–47

Barton, J.C., Patton, M.A., Edwards, C.Q. *et al.* (1994) Blood lead concentrations in hereditary hemochromatosis. *Journal of Laboratory and Clinical Medicine* 124: 193–98

Barton, J.C., Harmon, L., Rivers, C. & Acton R.T. (1996) Hemochromatosis: association of severity of iron overload with genetic markers. *Blood Cells, Molecules and Diseases* 22: 195–204

Barton, J.C., Shih, W.W.H., Sawada-Hirai, R. *et al.* (1997a) Genetic and clinical description of hemochromatosis probands and heterozygotes: evidence that multiple genes linked to the major histocompatibility complex are responsible for hemochromatosis. *Blood Cells, Molecules and Diseases* 23: 135–45

Barton, J.C., Barton, N.H. & Alford T.J. (1997b) Diagnosis of hemochromatosis in a community hospital. *American Journal of Medicine* 103: 498–503

Barton, J.C., McDonnell, S.E., Adams, P.C. *et al.* (1998) Management of hemochromatosis. *Annals of Internal Medicine* 129· 932–39

Barton, J.C., Rothenberg, B.E., Bertoli, L.F. & Acton, R.T. (1999) Diagnosis of hemochromatosis in family members of probands: a comparison of phenotyping and *HFE* genotyping. *Genetics in Medicine* 1(2): 1–5

Bellamy, R., Ruwende, C., Corrah, T. *et al.* (1998) Variation in the *NRAMP1* gene and susceptibility to tuberculosis in West Africans. *New England Journal of Medicine* 338: 640–44

Bernard, P.S., Ajioka, R.S., Kushner, J.P. & Wittwer, C.T. (1998) Homogeneous multiplex genotyping of hemochromatosis mutations with fluorescent hybridization probes. *American Journal of Pathology* 153: 1055–61

Beutler, E., Gelbart, T., West, C. *et al.* (1996) Mutation analysis in hereditary hemochromatosis. *Blood Cells, Molecules and Diseases* 22: 187–94

Carella, M., D'Ambrosio, L., Totaro, A. *et al.* (1997) Mutation analysis of the HLA-H gene in Italian hemochromatosis patients. *American Journal of Human Genetics* 60: 828–32

Crawford, D.H.G., Jazwinska, E.C., Cullen, L.M. & Powell, L.W. (1998) Expression of HLA-linked hemochromatosis in subjects homozygous or heterozygous for the C282Y mutation. *Gastroenterology* 114: 1003–08

Deugnier, Y.M., Guyader, D., Crantock, L. *et al.* (1993) Primary liver cancer in genetic hemochromatosis: a clinical, pathological, and pathogenetic study of 54 cases. *Gastroenterology* 104: 228–34

Douabin, V., Deugnier, Y., Jouanolle, A.M. *et al.* (1998) Polymorphisms in the haemochromatosis gene. *Proceedings of the International Symposium on Iron in Biology and Medicine*, Saint-Malo, France, June 1998

Edwards, C.Q., Griffen, L.M., Goldgar, D. *et al.* (1988) Prevalence of hemochromatosis among 11,065 presumably healthy blood donors. *New England Journal of Medicine* 318: 1355–62

Feder, J.N., Gnirke, A., Thomas, W. *et al.* (1996) A novel MHC class-I-like gene is mutated in patients with hereditary hemochromatosis. *Nature Genetics* 14: 399–408

Feder, J.N., Penny, D.M., Irrinki, A. *et al.* (1998) The hemochromatosis gene product complexes with the transferrin receptor and lowers its affinity for ligand binding. *Proceedings of the National Academy of Sciences USA* 95: 1472–77

Finch, S.C. & Finch, C.A. (1955) Idiopathic hemochromatosis, an iron storage disease. *Medicine* 34: 381–430

Gordeuk, V., Mukibi, J., Hasstedt, S.J. *et al.* (1992) Iron overload in Africa. Interaction between a gene and dietary iron content. *New England Journal of Medicine* 326: 95–100

Kowdley, K.V., Hassanein, T., Kaur, S. *et al.* (1995) Primary liver cancer and survival in patients undergoing liver transplantation for hemochromatosis. *Liver Transplant Surgery* 1: 237–41

Malfroy, L., Roth, M.P., Carrington, M. *et al.* (1997) Heterogeneity in rates of recombination in the 6-Mb region telomeric to the human major histocompatibility complex. *Genomics* 43: 226–31

Mercier, G., Burckel, A., Bathelier, C., Boillat, E. & Lucotte, G. (1998) Mutation analysis of the HLA-H gene in French hemochromatosis patients, and genetic counselling in families. *Genetic Counselling* 9: 181–86

Merryweather-Clarke, A.T., Pointon, J.J., Shearman, J.D. & Robson, K.J. (1997) Global prevalence of putative haemochromatosis mutations. *Journal of Medical Genetics* 34: 275–78

Mura, C., Nousbaum, J.B., Verger, P. *et al.* (1997) Phenotype–genotype correlation in haemochromatosis subjects. *Human Genetics* 101: 271–76

Niederau, C., Fischer, R., Purschel, A. *et al.* (1996) Long-term survival in patients with hereditary hemochromatosis. *Gastroenterology* 110: 1107–09

Olsson, K.S., Ritter, B., Sandberg, L. *et al.* (1998) The ancestral haplotype in patients with genetic hemochromatosis from central and western Sweden. *Proceedings of the International Symposium on Iron in Biology and Medicine*, Saint-Malo, France, June 1998

Porto, G., Alves, H., Rodrigues, P. *et al.* (1998) Major histocompatibility complex class I association in iron overload: evidence for a new link between the HFE H63D mutation, HLA-A29, and non-classical forms of hemochromatosis. *Immunogenetics* 47: 404–10

Powell, L.W., George, D.K., McDonnel, S.M. & Kowdley, K.V. (1998) Diagnosis of hemochromatosis. *Annals of Internal Medicine* 129: 925–31

Schumacher, H.R. Jr (1964) Hemochromatosis and arthritis. *Arthritis and Rheumatism* 7: 41–50

Sham, R.L., Ou, C.Y., Cappuccio, J. *et al.* (1997) Correlation between genotype and phenotype in hereditary hemochromatosis: analysis of 61 cases. *Blood Cells, Molecules and Diseases* 23: 314–20

Sheldon J.H. (1935) *Haemochromatosis*, London: Oxford University Press

Simon, M., Pawlotsky, Y., Bourel, M., Fauchet, R. & Genetet, B. (1975) Hémochromatose idiopathique: maladie associée a l'antigéne tissulaire HLA 3. *Nouvelle Presse Médicale* 19: 1432

Witte, D.L., Crosby, W.H., Edwards, C.Q., Fairbanks, V.F. & Mitros, F.A. (1996) Practice parameter for hereditary hemochromatosis. *Clinica Chimica Acta* 245: 139–200

GLOSSARY

hepatocyte liver cell

homeostasis relative constancy in the internal environment of the body

leukocyte a white blood cell

parenchymal resembling the functional tissues of an organ or gland

pathogenesis source/cause of an illness or abnormal condition

proband (index case) affected individual from whom a pattern of inheritance of a familial disorder is traced

HAEMOPHILIA: PATIENTS, GENES AND THERAPY

Mary R. Cahill[1] and Brian T. Colvin[2]
[1] *Department of Haematology, Mid-Western Regional Hospital, Limerick, Republic of Ireland*
[2] *Haemophilia Centre, Royal Hospitals Trust, London, UK*

1. What is haemophilia?

Haemophilia is a disabling, inherited bleeding tendency (Hoyer, 1994). Two types have been described and designated haemophilia A and haemophilia B, the latter also being known as Christmas disease. Individuals with haemophilia A suffer from a lack of a coagulation factor, factor VIII, which is essential for blood clotting, while in haemophilia B factor IX is deficient.

2. Classification

Haemophilia is classified according to the patient's factor VIII or IX levels. Severely affected individuals have <1 international unit (IU)/dl (<1% of normal) factor levels, moderately affected patients have 1–4 IU/dl (1–4%) and mild disease is associated with levels between 5–25 IU/dl (5–25%) and extending to the lower limit of the normal range, usually taken as 50 IU/dl (50%) (where dl = decalitres or 0.1 litres).

3. Bleeding manifestations and prognosis

Patients with severe haemophilia have spontaneous and sometimes severe bleeds 1–3 times per month. Those with moderate or mild disease typically have few spontaneous clinical manifestations, but may, nonetheless, suffer serious bleeding after trauma which will not resolve without effective treatment. Carriers of haemophilia usually have factor VIII or IX levels of approximately 50% of normal, but lower levels are occasionally seen, probably due to X-chromosome inactivation in favour of the abnormal chromosome (Lyon, 1993). As factor VIII levels tend to rise in pregnancy, haemorrhagic complications during childbirth are rare and therapeutic intervention is not often required.

The most frequent clinical manifestations of haemophilia are easy bruising, haemarthroses (bleedings into joints) and bleeding into muscles. Before the advent of appropriate clotting factor concentrate, boys with haemophilia suffered from repeated joint bleeding. This frequently led to the development of arthritis with joint deformity, and haemophilic arthropathy is associated with severe disability, even in children and teenagers.

Other forms of bleeding that can occur include haematuria (bleeding into urine) and bleeding after injury and surgery. Life-threatening bleeding most commonly takes the form of intracranial haemorrhage which used to be the most common cause of death in men with haemophilia. Data from the Swedish National Death Registry show how adequate treatment for bleeding episodes had restored life expectancy from a mere 11 years in the 1920s to 56 years by 1980 (Hoyer, 1994).

The remarkable improvements in morbidity and mortality attributable to the appropriate treatment of haemophilia with clotting concentrates (Nilsson *et al.*, 1992) have recently been reversed by the inadvertent transmission of the HIV virus to patients with haemophilia by clotting factor concentrates (Hoyer, 1994). The majority of patients with severe haemophilia A treated before 1986 have been infected with HIV, and many who were infected in the early 1980s have already died of AIDS-related illnesses. Between 1985 and 1992 AIDS accounted for 85% of deaths in seropositive patients and became the leading cause of death in haemophiliacs (Darby *et al.*, 1995).

4. The genetics of haemophilia

Both haemophilia A and B are X-linked disorders and, therefore, occur almost exclusively in males. The sons of haemophilia patients will be unaffected and all the daughters will be carriers. For carrier females with normal partners, half their sons will have haemophilia and half their daughters will be carriers (Hoyer, 1994).

The genes for factors VIII and IX have been mapped to the distal end of the long arm of the X chromosome, bands Xq28 and Xq27.1, respectively. The whole factor IX gene and the coding sequence of the factor VIII gene have been determined. The gene for factor IX was cloned in 1982 (Choo *et al.*, 1982; Kurachi & Davie, 1982) and that for factor VIII in 1984 (Gitschier *et al.*, 1984; Toole *et al.*, 1984; Wood *et al.*, 1984).

The factor VIII gene is very large and comprises 186 kilobases (kb) of DNA with 26 exons (for review see Antonarakis, 1995). Haemophilia A is due to a large number of different mutations and over 80 have been characterized (Tuddenham *et al.*, 1991). Favoured sites for mutations include CG dinucleotides (Youssoufian *et al.*, 1986), a pattern also found in haemophilia B. Mutations include deletions, insertions, point mutations and, most importantly, DNA inversions which have been shown to account for almost 50% of cases of severe haemophilia A (Naylor *et al.*, 1992; Lakich *et al.*, 1993; Naylor *et al.*, 1993; Goodeve *et al.*, 1994). These mutations occur in a region of intron 22 which is repeated twice upstream of and outside the factor VIII gene. Intron 22 can recombine with either one of the upstream homologous regions of the same chromosome or chromatid, causing a large DNA inversion and major functional disruption. Some mutations result from a C to T substitution at CGA sequences. The resulting TGA is a stop codon. A premature stop results in the production of a truncated factor VIII molecule and severe haemophilia. Deletion of DNA is found in about 5% of patients. Patients with large deletions have no detectable factor VIII.

Detailed knowledge of the genetic abnormalities which

cause haemophilia allows us to detect female carriers accurately. This is important because carriers can rarely be identified with certainty by the phenotypic analysis of clotting factor levels. Accurate carrier detection and antenatal diagnosis can be provided using either the direct detection of genetic abnormalities or, when the family history allows, by detection of associated polymorphisms within or close to the gene. From a gene therapy perspective, carrier detection is currently irrelevant. All the ongoing research in gene therapy for haemophilia is directed at the somatic DNA of the sufferer himself, although, in theory, alteration of germline DNA is possible and has been achieved in mice (Pasi, 1996).

The factor IX gene is 34 kb in length and the essential genetic information is present in eight exons (for review see Roberts, 1993). The following domains can be distinguished: signal peptide, propeptide, gamma-carboxyglutamic acid, epidermal growth factor-like domains (EGF1 and 2) and a serine protease domain. Over 400 distinct mutations have been documented in haemophilia B (Giannelli *et al.*, 1992), affecting all protein domains. Within the gene, over 30% of mutations are located at CG dinucleotides, frequently resulting in substitution at arginine residues and major molecular dysfunction (Monroe *et al.*, 1989). Mutations in the regulatory regions of the factor IX gene are described and, with only two exceptions so far, these result in haemophilia B at birth with a gradual improvement in factor IX levels until near normal levels are achieved around puberty. This is thought to be due to a beneficial effect of androgens on the binding of transcription factors to the mutated promoter. An example of this phenotype is haemophilia B Leiden (Crossley *et al.*, 1992; Reijnen *et al.*, 1992).

More than one-third of patients with haemophilia have no family history and arise from a new mutation (Gitschier *et al.*, 1991).

5. Current treatments

For a review of current treatments, see Cahill and Colvin (1997b).

(i) Non-blood product haemostatic treatment

In general, non-blood product haemostatic treatment is only useful in patients with mild or moderate haemophilia. The hormone DDAVP (desmopressin) is capable of raising the plasma level of factor VIII 2- or 3-fold (Mannucci *et al.*, 1977). It works by releasing stores of factor VIII which has already been synthesized and, therefore, release decreases to insignificant amounts after two or three doses until reaccumulation of stores can take place. Administration of DDAVP is associated with activation of fibrinolysis and the concomitant use of a fibrinolytic inhibitor, such as tranexamic acid or epsilon amino-caproic acid (EACA), is often recommended. Fibrinolytic inhibitors can also be usefully administered in conjunction with factor VIII, especially for mucosal bleeds, but antifibrinolytic therapy is contraindicated in the presence of haematuria and is not used with factor IX concentrate. These treatments will remain important after the successful implementation of gene therapy in patients with haemophilia. It is unlikely that patients with mild haemophilia will benefit from gene therapy as their endogenous secretion will exceed that which is currently agreed to be achievable from successful gene replacement.

(ii) Blood products

Treatment of haemophilia with blood products has had a stormy evolution (Pool *et al.*, 1964; Mannucci, 1993). Initially, fresh frozen plasma was the only product available for the treatment of bleeding episodes and huge volumes would have been required for successful treatment. After the production of small quantities of factor VIII concentrate in Oxford in the 1950s, the discovery that "antihaemophilic globulin" was concentrated in cryoprecipitate (Pool *et al.*, 1964) helped pave the way for the large-scale production of intermediate purity concentrates of factors VIII and IX. These concentrates were the mainstay of haemophilia care through the 1970s and 1980s and, despite the transmission of hepatitis and HIV viruses as a result, it is important to note that the patients who received them would otherwise have suffered serious disability or died of bleeding. Concentrates have been safe with respect to HIV and hepatitis C transmission since more effective donor selection and viral inactivation have become available.

There has been some concern that intermediate purity factor VIII concentrates are associated with immunosuppressive effects (Cuthbert *et al.*, 1992). These effects have been attributed to the non-factor VIII plasma proteins in the concentrates and have contributed to the move towards high-purity products which have an increased specific activity and are largely free from other plasma proteins although albumin is added after processing to stabilize them. High-purity (UK Regional Haemophilia Centre Directors Committee, 1992) or recombinant (UK Haemophilia Centre Directors Organisation, 1997) products have been recommended for the treatment of patients infected with HIV and are increasingly being used for all patients in the developed world.

For patients with factor IX deficiency, intermediate-purity factor IX is unacceptably thrombogenic, especially in the context of orthopaedic surgery. Thrombotic potential is probably due to contaminating activated-coagulation factors. High-purity factor IX is available and has replaced the intermediate-purity product in the UK. Inhibitors of fibrinolysis are not generally used in conjunction with factor IX therapy.

In 1997, the UK Haemophilia Centre Directors Organisation issued further recommendations for the treatment of patients with haemophilia which identified recombinant factor VIII as the treatment of choice for the management of haemophilia A. Recombinant Factor IX is increasingly used.

(iii) Administration of treatment

Treatment can be administered "on-demand" or prophylactically. For severe haemophilia, home treatment is now established as safe and effective and, in general, is acceptable to this well-motivated group of patients and families (Jones, 1992). For patients with moderate or mild haemophilia, on-demand treatment in hospital is sufficient and it should be remembered that even mild haemophilia may result in bleeding severe enough to necessitate vigorous treatment and hospital admission.

6. Problems relating to current treatments

There are a number of problems with replacement therapy for haemophilia which drive the search for gene therapy in this condition.

(i) Viruses and concentrates

Clotting concentrates have transmitted a number of viral infections (Kasper & Kipnis, 1972; Mannucci, 1992; Santagostino *et al.*, 1994; Telfer *et al.*, 1994; Darby *et al.*, 1995; Ludlam, 1997), but infection has been reduced dramatically since 1985 and has culminated in the production of a generation of products which are safe and free from lipid-coated viruses (Mannucci, 1993). Some products have undergone double viral inactivation steps – for example, dry heating or ultrafiltration may be combined with a chemical inactivation step such as the use of solvent detergent processes or sodium thiocyanate. Thanks to these measures, there have been no cases of seroconversion to HIV, hepatitis B or C recorded in the UK since 1986. Despite this significant achievement, caution is warranted, because documented transmission of hepatitis A and parvovirus B19 has occurred in Europe (Mannucci, 1992; Santagostino *et al.*, 1994) and this indicates that non-lipid-coated viruses can escape inactivation procedures. There is a theoretical possibility that variant Creutzfeldt–Jakob disease (vCJD) might be transmitted in plasma-derived products (Ludlam, 1997) and the advent of vCJD reminds us that new infectious agents, potentially transmitted by transfusion, may continue to be recognized in future years.

(ii) Inhibitors

Patients with haemophilia may develop inhibitory antibodies to exogenous factor VIII and factor IX. These inhibitors, if of high titre, are a most difficult complication in the clinical management of haemophilia and imply that haemostasis cannot be achieved by standard amounts of the concentrate against which the antibody is directed. The development of inhibitors is related to the basic genetic defect in a partly predictable way; patients whose haemophilia is due to deletions (especially in haemophilia B) or nonsense mutations (and who, therefore, have no native protein) are more likely to form inhibitors either to factor VIII (Tuddenham *et al.*, 1991) or to factor IX.

The treatment of inhibitors is difficult and costly and the potential seriousness of a high-titre inhibitor cannot be overestimated. The mainstay of treatment is the induction of immune tolerance using regimens in which clotting factor is administered daily; these regimens are most successful when administered as soon as inhibitors are detected, and are especially useful in children. Treatment strategies vary and fall outside the scope of this article, but they have been outlined elsewhere (UK Haemophilia Centre Directors Organisation, 2000).

(iii) Expense

Treatment is costly and may be an insurmountable difficulty for patients in many Third World countries.

(iv) Availability

Currently, the majority of factor VIII in use worldwide is derived from blood donors and availability depends on local production and the financial resources to import concentrate from other countries. Although many countries aim for self-sufficiency (Leikola, 1998), most countries, from time to time, have imported factor VIII. Following viral inactivation, factor VIII has a shelf life which allows potential shortfalls to be predicted and acted on. For many developing countries, plasma-derived factor VIII is unavailable and the move to manufactured (recombinant) factor VIII should ensure adequate supplies of therapeutic material worldwide, if the cost can be brought down and sufficient supplies can be made available. It is vital that the capacity to produce plasma-derived material be retained so that, in the event of manufacturing problems or unforeseen side-effects of recombinant material, alternatives remain available.

The incidence of haemophilia is constant worldwide and no racial group is spared. However, in developing countries, the problems of product availability are compounded by difficulties with storage conditions, adverse transport conditions and long distances between patients and medical centres. It is in such conditions that gene therapy has the most to offer and once established, could be cost-effective.

7. Recombinant factor VIII (rVIII) and IX

Factor VIII derived by recombinant technology is now available in the UK and should carry no viral risk, but, so far, most manufacturers have added small amounts of human albumin in order to stabilize their products. Albumin-free concentrates are now available for use. The Republic of Ireland in 1997 became the first country to opt to provide recombinant material for all its citizens with haemophilia A and other countries may follow suit. Most haemophilia centre directors will commence all new or previously untreated patients on recombinant material.

Recombinant factor IX is now licensed and is used in the treatment of haemophilia B.

8. Other types of care needed by patients with haemophilia – comprehensive care

Treating haemophilia involves far more than replacing the missing factor concentrate. The physical and psychological consequences of haemophilia combine to ensure that specialists in rheumatology, orthopaedics, HIV and hepatitis, dentistry, social work, physiotherapy and occupational health all have a role to play in the prevention and treatment of complications. Diagnosis and the counselling of carriers and patients with regard to genetics and antenatal diagnosis is also important. Such care will still be needed in the wake of the successful implementation of a gene therapy programme.

9. Gene therapy

(i) The origins

DNA was first successfully introduced into mammalian cells in culture in the early 1970s (Graham & Van der Eb, 1973). The next necessary step was the cloning of target DNA, the techniques for which have been increasing in sophistication in the last three decades. For haemophilia sufferers, these techniques have culminated in the breakthrough cloning of the factor IX gene in 1982 (Choo *et al.*, 1982; Kurachi & Davie, 1982) and then, in 1984, the much larger factor VIII gene (Toole *et al.*, 1984; Wood *et al.*, 1984), as described above.

(ii) The expectations from gene therapy in severe haemophilia

Achieving a life that is both free from the bleeding manifestations of haemophilia and relatively independent of regular treatment is the goal of every patient with haemophilia and their doctor (Felgner & Rhodes, 1991). For patients with haemophilia, compared with those with other chronic genetic diseases, the achievement of this goal is relatively advanced; the relevant genes and their defects have been identified and much is known about the structure and function of both factors VIII and IX. Only a small increase in the levels of factors VIII and IX is needed to improve the clinical manifestations of the disease, which contributes to the attractiveness of gene therapy in haemophilia. For patients with severe haemophilia (factor level <1% of normal), a rise to 5% of normal would transform their lives. This is about the level of increase that might be expected as a result of somatic "gene addition". Also, in the unlikely event of gene overexpression, this should have no adverse consequences.

(iii) Outline of problems

There are a number of obstacles to be overcome, however, before this goal is realized in patients with either type of haemophilia. Practical problems so far encountered include unreliable gene transfer into target cells, inadequate and waning levels of gene expression and concomitant inadequate protein synthesis (Eisensmith & Woo, 1997). There are also issues of vector safety in the human host. A number of ethical issues (such as which group of patients should be used for initial trials) remain to be addressed. So far, research has been directed towards modifying somatic cells. However, it is already possible to alter germline DNA in mice, raising further ethical issues for debate (Pasi, 1996).

(iv) The mechanisms

For a review of mechanisms, see Connelly and Kaleko (1997).

Outline. Appropriate cDNA is engineered to address the patient's genetic defect. This cDNA is introduced into cells using a vector and genetic manipulation of cells can take place *ex vivo* or *in vivo*. In the *ex vivo* approach, cells to undergo gene therapy are harvested and cultured in the laboratory where the DNA transfer is effected using the chosen vector system. For the *in vivo* method, direct intravenous injection of cDNA and vector can result in efficient transduction of hepatocytes and secretion of the relevant clotting factor. Successful gene function can be checked before the cells are reimplanted into the patient with haemophilia (*ex vivo*) or by measuring the rise in the level of the deficient protein in the patient afterwards (*in vivo* therapy). The *ex vivo* approach allows the use of allogeneic cells, which is attractive because a standard cell line could be developed expressing the appropriate gene for factor VIII or IX at adequate levels and used to treat many patients. The immunogenicity of the cells is, of course, a significant drawback, but one which could be addressed by cellular isolation in a physical capsule (Liu *et al.*, 1993) or, less attractively, by pharmacological immunosuppression of the host immune system. The advantages and disadvantages of *in vivo* or *ex vivo* therapy are summarized in **Table 1**.

The vector issue (for review see Thompson, 1995). Gene delivery into cells in experimental models of haemophilia is most commonly achieved by the use of retroviruses. These are organisms which have the unique capacity to integrate their DNA into the DNA of the host cell. Retrovirus-mediated gene therapy usually results in long-term constitutive transgene expression with low levels of transduction. Adenoviruses and parvoviruses have also been used and adenovirus-mediated gene therapy produces higher levels of gene expression. Their main disadvantage is the tendency to stimulate a significant immune response, resulting in transient genetic expression, although parvovirus has been associated with sustained expression over 36 months in one study (Xiao *et al.*, 1996). In response to these problems a novel adenoviral vector, "mini-ad", has been developed. This vector does not contain adenoviral coding sequences but has the capacity to hold more than 30 kb of DNA. These features lead to this viral vector being also termed "gutless", "helper-dependent" or "high-capacity" adenoviral vector. The development of this vector has been advanced by several groups and is summarized by Kochanek (1999). Lentiviruses have also been used and can integrate DNA into non-dividing cells (Naldini *et al.*, 1996). Lentiviruses are not a

Table 1 Advantages/disadvantages of *in vivo* and *ex vivo* approaches to gene therapy in haemophilia.

In vivo approach	*Ex vivo* approach
target cell *in situ*	target cell isolated
only autologous cell manipulation possible	immunoisolation of transfected cells possible – therefore, allogenic cells can be used
increased plasma levels of target protein is first evidence of integration	controlled culture system – possible to check for integration
more efficient	more cumbersome
less invasive	removal and reimplantation of cells necessary
cost-effective	costly
limited by disadvantages of chosen vector system	historically used retroviral vectors
less labour intensive	time-consuming

popular choice for gene therapy in haemophilia, perhaps because of the adverse experiences of haemophilia patients with the HIV lentivirus.

Non-viral vectors. Concerns about the safety of viral vectors in a community which has suffered disproportionately from viral illness are an acknowledged impetus to seek alternative methods of gene transfer. Non-viral vectors also have the advantages of assembly in a "cell-free" system and they are less immunogenic. One such non-viral vector involves receptor-mediated gene transfer. DNA is linked to polylysine/asialoglycoprotein conjugates which bind to hepatocyte-specific receptors allowing the entry of DNA (Wu & Wu, 1988). Initial success has been limited, but modification of the conjugate molecule has led to documented sustained expression (Perales *et al.*, 1994) and direct DNA injection into cells (Wolff *et al.*, 1990) is also possible. The features of an ideal vector are summarized as follows:

- safe;
- able to integrate into non-dividing cells;
- non-immunogenic;
- stable integration of target DNA into host genome;
- capable of target cell transfection after intravenous injection;
- active cell division of target cell unnecessary to achieve transfection;
- no expression of viral genes; and
- broad host cell range.

(v) Gene therapy for haemophilia B

Gene therapy for factor IX deficiency is more advanced than for factor VIII deficiency. These factors highlight a number of general principles with regard to the development of gene therapy and are summarized in **Table 2.**

Numerous studies have been carried out in rodent and canine models of haemophilia B. Sustained partial correction of the factor IX defect has been achieved in the Chapel Hill profoundly deficient dog (Kay *et al.*, 1993). Retroviral vectors carrying canine factor IX cDNA were expressed in hepatocytes for more than 2 years after therapy. Measurable expression of factor IX was only 0.1% of normal canine levels. However, the authors report a decrease in whole blood clotting times in the animals.

Using the same animal model, but an adenoviral vector, complete correction of the factor IX defect was achieved, but the effect lasted only days (Kay *et al.*, 1994).

(vi) Gene therapy for haemophilia A

For a review see Connelly & Kaleko (1997). Due to the larger size of the factor VIII molecule, integrating its DNA into host cells has provided more of a challenge than for factor IX. This problem has been partially overcome by creating a truncated version of cDNA for factor VIII, the 4.5 kb version of which compares favourably with the standard >7 kb of cDNA derived from the full 186 kb DNA gene and which is achieved by removing the B domain. This deletion has no effect on the function, structure or immunogenicity of the factor VIII molecule.

Complete but short-lived secretion of human factor VIII has been achieved in a canine model of haemophilia A using an adenoviral vector injected into a peripheral vein. Significant antifactor VIII antibodies were provoked and gene expression was short lived (Connelly *et al.*, 1996).

(vii) The current status of gene therapy for haemophilia

The advances in theory and practice described above make this goal possible. Problems with implementation and concerns about safety are being addressed for both types of haemophilia using animal models – dogs and rodents. Successful transfer of both genes to both types of animal has been effected and protein secretion achieved. The principal problems are sustaining the level of secretion and outstanding concerns about vector safety. Patient selection considerations are very important (Thompson, 2000) and should include international and institutional ethical debate.

Three phase one clinical studies have been underway for more than one year. One of them has published early results (Kay *et al.*, 2000). Another has been presented in abstract form (Hurst *et al.*, 2000). Early results show that vector expression of factor gene occurs and factor levels rise modestly, but no sustained rises have yet been demonstrated. Two more phase one studies have begun and further results are eagerly awaited.

Acknowledgements

We thank Professor F. Giannelli and Professor G.C. White for comments on the manuscript.

Table 2 Principles of gene therapy applied to haemophilia B.

General principle	Application to haemophilia B
availability of cDNA	human factor IX was cloned in 1982
gene size	factor IX gene is considerably smaller than factor VIII gene and this facilitates its insertion into vectors and cells
secretion of the protein from accessible cells	hepatocytes for factor IX
half-life of missing protein	long is best – 16 hours for factor IX
the ability of a small amount of protein to make a clinical difference	for severe haemophilia secretion of 1% factor IX could transform the clinical phenotype from severe to moderate
availability of an animal model	Chapel Hill factor IX deficient dog
regulation of expression required	no need for tight control of factor IX expression
stability of transfected genetic material	sustained partial expression of factor IX gene obtained in animal models

REFERENCES

Antonarakis, S.E. (1995) Molecular genetics of coagulation factor VIII gene and haemophilia A. *Thrombosis and Haemostasis* 74: 322–28

Cahill, M.R. & Colvin, B.T. (1997a) Haemophilia; current management and future directions. *Postgraduate Medical Journal*: 73: 201–06

Cahill, M.R. & Colvin, B.T. (1997b) Current practice in the treatment of haemophilia. *Haematology* 2: 351–58

Choo, K.H., Gould, K.G., Rees, D.J. & Brownlee, G.G. (1982) Molecular cloning of the gene for human anti-haemophilic factor IX. *Nature* 299: 178–80

Connelly, S. & Kaleko, M. (1997) Gene therapy for haemophilia A. *Thrombosis and Haemostasis* 78: 31–36

Connelly, S., Mount, J., Mauser, A. *et al.* (1996) Complete short term correction of canine hemophilia A by *in vivo* gene therapy. *Blood* 88: 3846–53

Crossley, M., Ludwig, M., Stowell, K.M. *et al.* (1992) Recovery from haemophilia B Leyden: an androgen-responsive element in the factor IX promoter. *Science* 257: 377–79

Cuthbert, R.J.G., Ludlam, C.A., Steel, C.M., Beatson, D. & Peutherer, J.F. (1992) Immunological studies in HIV seronegative haemophiliacs: relationship to blood product therapy. *British Journal of Haematology* 80: 364–69

Darby, S.C., Ewart, D.W., Giangrande, P.L.F. *et al.* (1995) Mortality before and after HIV infection in the complete UK population of haemophiliacs. *Nature* 377: 79–82

Department of Health (1993) Provision of haemophilia treatment and care: NHS Management Executive. *Health Service Guidelines* 93: 30

Eisensmith, R.C. & Woo, S.L.C. (1997) Viral vector-mediated gene therapy for hemophilia B. *Thrombosis and Haemostasis* 78: 24–30

Felgner, P.L. & Rhodes, G. (1991) Gene therapeutics. *Nature* 349: 351–52

Giannelli, F., Green, P.M., High, K.A. *et al.* (1992) Haemophilia B: database of point mutations and short additions and deletions, 3rd edition. *Nucleic Acids Research* 20 (Suppl.): 2027–63

Gitschier, J., Wood, W.I., Goralka, T.M. *et al.* (1984) Characterization of the human factor VIII gene. *Nature* 312: 326–30

Gitschier, J., Kogan, S., Diamond, C. & Levinson, B. (1991) Genetic basis of haemophilia A. *Thrombosis and Haemostasis* 66: 37–39

Goodeve, A., Preston, F.E. & Peake, I.R. (1994) Factor VIII gene rearrangements in patients with severe haemophilia A. *Lancet* 343: 329–30

Graham, F.L. & Van der Eb, A.J. (1973) A new technique for the assay of infectivity of human adenovirus 5 DNA. *Virology* 52: 456–67

Hoyer, L.W. (1994) Haemophilia A. *New England Journal of Medicine* 330: 38–47

Hurst, D., Cole, V., Libbrandt, M.E.I. *et al.* (2000) Phase I trial of factor VIII gene transfer using a retroviral vector administered by peripheral intravenous infusion: interim safety update. Presented at the XXIV International Congress of the World Federation of Haemophilia. Montreal, Canada, July 16–21, 2000

Jones, P. (1992) Haemophilia home therapy. *Haemostasis* 22: 247–50

Kasper, C.K. & Kipnis, S.A. (1972) Hepatitis and clotting-factor concentrates. *Journal of the American Medical Association* 221: 510

Kay, M.A., Rothenberg, S., Landen, S. *et al.* (1993) In vivo gene therapy of haemophilia B: sustained partial correction in factor IX deficient dogs. *Science* 262: 117–19

Kay, M.A., Landen, C.N., Rothenberg, S.R. *et al.* (1994) In vivo hepatic gene therapy; complete albeit transient correction of factor IX deficiency in haemophilia B dogs. *Proceedings of the National Academy of Sciences USA* 91: 2353–57

Kay, M.A., Manno, C.S., Ragni, M.V. *et al.* (2000) Evidence for gene transfer and expression of factor IX in haemophilia B patients treated with an AAV vector. *Nature Genetics* 24(3): 257–61

Kochanek, S. (1999) Development of high-capacity adenoviral vectors for gene therapy. *Thrombosis and Haemostasis* 82: 547–51

Kurachi, K. & Davie, E.W. (1982) Isolation and characterisation of a cDNA coding for human factor IX. *Proceedings of the National Academy of Sciences USA* 79: 6461–64

Lakich, D., Kazazian, H.H. Jr, Antonarakis, S.E. & Gitschier, J. (1993) Inversions disrupting the factor VIII gene are a common cause of severe haemophilia A. *Nature Genetics* 5: 236–41

Leikola, J. (1998) Achieving self sufficiency in blood across Europe. *British Medical Journal* 316: 489–90

Liu, H.W., Ofosu, F.A. & Chang, P.L. (1993) Expression of human factor IX by microencapsulated recombinant fibroblasts. *Human Gene Therapy* 4: 291–301

Ludlam, C.A. (1997) New-variant Creutzfeldt–Jakob disease and treatment of haemophilia. *Lancet* 350: 1704

Lyon, M.F. (1993) Epigenetic inheritance in mammals. *Trends in Genetics* 9: 123–28

Mannucci, P.M. (1992) Outbreak of hepatitis A among Italian patients with haemophilia. *Lancet* 339: 819

Mannucci, P.M. (1993) Modern treatment of haemophilia: from the shadows towards the light. *Thrombosis and Haemostasis* 70: 17–23

Mannucci, P.M., Ruggeri, Z.M., Pareti, F.I. & Capitanio, A. (1977) 1-Deamino-8-d-arginine vasopressin: a new pharmacological approach to the management of haemophilia and von Willebrand's diseases. *Lancet* i: 869–72

Monroe, D.M., McCord, D.M., Huang, M.N. *et al.* (1989) Functional consequences of an arginine 180 to glutamine mutation in factor IX Hilo. *Blood* 73(6): 1540–44

Naldini, L., Blomer, U., Gallay, P. *et al.* (1996) In vivo gene delivery and stable transduction of nondividing cells by a lentiviral vector. *Science* 272: 263–67

Naylor, J.A., Green, P.M., Rizza, C.R. & Giannelli, F. (1992) Factor VIII gene explains all cases of haemophilia A. *Lancet* 340: 1066–67

Naylor, J., Brinke, A., Hassock, S., Green, P.M. & Giannelli, F. (1993) Characteristic mRNA abnormality found in half the patients with severe haemophilia A is due to large DNA inversions. *Human Molecular Genetics* 2: 1773–78

Nilsson, I.M., Berntorp, E., Lofqvist, T. & Pettersson, H. (1992) Twenty-five years experience of prophylactic treatment in severe haemophilia A and B. *Journal of Internal Medicine* 232: 25–32

Pasi, K.J. (1996) Gene therapy for haemophilia. *Bailliere's Clinical Haematology* 9: 305–17

Perales, J.C., Ferkol, T., Beegen, H., Ratnoff, O.D. & Hanson, R.W. (1994) Gene transfer *in vivo*: sustained expression and regulation of genes introduced into the liver by receptor-targeted uptake. *Proceedings of the National Academy of Sciences USA* 91: 4086–90

Pool, J.G., Hershgold, E.J. & Pappenhagen, A.R. (1964) High-potency antihaemophilic factor concentrate prepared from cryoglobulin precipitate. *Nature* 203: 312

Reijnen, M.J., Sladek, F.M., Bertina, R.M. & Reitsma, P.H. (1992) Disruption of a binding site for hepatocyte nuclear factor 4 results in haemophilia B Leyden. *Proceedings of the National Academy of Sciences USA* 89: 6300–03

Roberts, H.R. (1993) Molecular biology of haemophilia B. *Thrombosis and Haemostasis* 70: 1–9

Santagostino, E., Mannucci, P.M., Gringeri, A., Azzi, A. & Morfini, M. (1994) Eliminating parvovirus B19 from blood products. *Lancet* 343: 798

Telfer, P., Sabin, C., Devereux, H., Scott, F., Dusheiko, G. & Lee, C. (1994) The progression of HCV-associated liver disease in a cohort of haemophilic patients. *British Journal of Haematology* 87: 555–61

Thompson, A.R. (1995) Progress towards gene therapy for the haemophilias. *Thrombosis and Haemostasis* 74: 45–51

Thompson, A.R. (2000) Gene therapy for the haemophilias. *Haemophilia* 6 (suppl. 1): 115–19

Toole, J.J., Knopf, J.L., Wozney, J.M. *et al.* (1984) Molecular cloning of a cDNA encoding human antihaemophilic factor. *Nature* 312: 342–47

Tuddenham, E.G.D., Cooper, D.N., Gitschier, J. *et al.* (1991) Haemophilia A: database of nucleotide substitutions, deletions, insertions and rearrangements of the factor VIII gene. *Nucleic Acids Research* 19: 4821–33

UK Regional Haemophilia Centre Directors Committee (1992) Recommendations on choice of therapeutic products for the treatment of patients with haemophilia A, haemophilia B and von Willebrand's disease. *Blood Coagulation and Fibrinolysis* 3: 205–14

UK Haemophilia Centre Directors Organisation (1997) Guidelines on therapeutic products to treat haemophilia and other hereditary coagulation disorders. *Haemophilia* 3: 63–77

UK Haemophilia Centre Directors Organisation (2000) The diagnosis and management of factor VIII and IX inhibitors: a guideline from the UK Haemophilia Centre Doctors' Organisation (UKHCDO). *British Journal of Haematology* 111: 78–90

Wolff, J.A., Malone, R.W., Williams, P. *et al.* (1990) Direct gene transfer into mouse muscle *in vivo*. *Science* 247: 1465–68

Wood, W.I., Capon, D.J., Simonsen, C.C. *et al.* (1984) Expression of active human factor VIII from recombinant DNA clones. *Nature* 312: 330–37

Wu, G.Y. & Wu, C.H. (1988) Receptor-mediated gene delivery and expression in vivo. *Journal of Biological Chemistry* 263: 14621–24

Xiao, X., Li, J. & Samulski, R.J. (1996) Efficient long term gene transfer into muscle tissue of immunocompetent mice by adeno associated virus vector. *Journal of Virology* 70: 8098–108

Youssoufian, H., Kazazian, H.H. Jr, Philips, D.G. *et al.* (1986) Recurrent mutations in haemophilia A give evidence of CpG mutation hot-spots. *Nature* 324: 380–82

PORPHYRIAS

George H. Elder
Department of Medical Biochemistry, University of Wales College of Medicine, Cardiff, UK

1. Introduction

The porphyrias are diseases of haem biosynthesis (Kappas *et al.*, 1995). Each of the seven main types is caused by partial deficiency of one of the enzymes that synthesizes haem from 5-aminolaevulinate (ALA); complete abolition of activity eliminates the vital process of haem formation and is incompatible with life. These deficiencies, apart from the liver-specific enzyme defect in sporadic (type I) porphyria cutanea tarda (PCT), are inherited in monogenic patterns. Porphyria has been called the "little imitator", because the variety of its symptoms rival those of syphilis and, likewise, are a boon to those interested in the historical art of retrospective diagnosis and a spur to molecular biographers of George III and his descendants (Röhl *et al.*, 1998).

Life-threatening acute attacks occur in four types of porphyria (**Table 1**) (Kappas *et al.*, 1995; Elder *et al.*, 1997). These attacks are commoner in women, often start between the ages of 15 and 40 years, are very rare before puberty and are frequently provoked by drugs (particularly those that induce hepatic cytochrome P-450s that require haem for their assembly), alcohol, calorie restriction and endocrine factors. Severe abdominal pain with vomiting is the usual dominant symptom; peripheral motor neuropathy which may progress to complete paralysis, mental confusion and psychotic behaviour, convulsions often related to hyponatraemia, hypertension and tachycardia are other features. The mechanism by which these attacks occur is unknown. All their clinical features are neurological in origin and induction of hepatic ALA synthase, the rate-limiting enzyme of haem biosynthesis, with overproduction of ALA is always present. A full explanation of the link between the biochemistry and the neurology has been elusive in spite of the recent development of a transgenic mouse model for acute intermittent porphyria (AIP) (Lindberg *et al.*, 1996). Replacement of haem by intravenous infusion during the early phase of an attack decreases its severity.

Skin lesions on sun-exposed skin occur in all but two types of porphyria (**Table 1**). Accumulated porphyrin in the skin catalyses the photoactivation (at wavelengths of around 400–500 nm) of oxygen to reactive species, mainly singlet oxygen, that cause tissue damage, the primary site of photodamage probably being the endothelial cells of small blood vessels. In erythropoietic protoporphyria (EPP), the clinical consequence is acute, painful photosensitivity. In all other cutaneous porphyrias, mechanical fragility of the skin, subepidermal bullae, increased hair growth and patchy pigmentation are the main lesions.

Table 1 shows that the clinical features of the porphyrias are not specific to a particular disorder. Accurate differentiation between them depends on the biochemical analysis of urine, faeces and blood, in order to identify the specific pattern of overproduction of haem precursors that indicates the presence of the enzyme deficiency that defines each main type (**Table 2**). The human genes for these enzymes have now been characterized (**Table 2**).

2. Autosomal dominant porphyrias

Allelic heterogeneity and low clinical penetrance are important genetic features of all five autosomal dominant porphyrias. Haplo-insufficiency with abolition or markedly decreased enzyme activity from the mutant allele produces a uniform biochemical pattern of half-normal enzyme activity in all who inherit the genes for these disorders. In contrast, at the DNA level, most mutations are restricted to

Table 1 The main types of porphyria.

		Clinical features			
Disorder	**MIM#**	**Acute attacks**	**Skin lesions**	**Defective enzyme**	**Inheritance**
ALA dehydratase porphyria (ALADP)	125270	+	–	ALA dehydratase	AR
Acute intermittent porphyria (AIP)	176000	+	–	PBG deaminase	AD
Congenital erythropoietic porphyria (CEP)	263700	–	+	Uro'gen III synthase	AR
Porphyria cutanea tarda (PCT)	176090; 176100	–	+	Uro'gen decarboxylase	Complex; 20% AD
Hereditary coproporphyria (HCP)	121300	+	+	Copro'gen oxidase	AD
Variegate porphyria (VP)	176200	+	+	Proto'gen oxidase	AD
Erythropoietic protoporphyria (EPP)	177000	–	+	Ferrochelatase	AD

AD, autosomal dominant; AR, autosomal recessive; copro'gen, coproporphyrinogen; MIM#, Mendelian Inheritance in Man number; PBG, porphobilogen; proto'gen, protoporphyrinogen; uro'gen, uroporphyrinogen.

Table 2 Human genes for the porphyrias.

Porphyria	Gene symbol	Chromosome	Size (kilobases)	No. of exons
ALADP	*ALAD*	9q34	13	13
AIP	*HMBS*	11q24.1–24.2	10	15
CEP	*UROS*	10q25.2–26.3	45	9
PCT	*UROD*	1p34	3	10
HCP	*CPX*	3q12	14	7
VP	*PPOX*	1q21–23	5	13
EPP	*FECH*	18q21.3	45	11

one or a few families. Over 130 different mutations have been identified in AIP and similar heterogeneity has been found in the other, less common disorders. Two founder effects provide striking exceptions. In South Africa, an estimated 10–20 000 individuals have inherited variegate porphyria (VP) from an immigrant Dutch couple who married at the Cape in 1688. All show the same mutation, 175C→T(R59W), in the *PPOX* gene (Jenkins, 1996). Similarly, the high prevalence of AIP in northern Sweden (one per 1000 population) is explained by inheritance of a mutation, 593G→A (W198X), in the *HMBS* (hydroxymethyl bilane synthesase) gene from a common ancestor.

Most of those who inherit an autosomal dominant porphyria are asymptomatic throughout life and many of these individuals show no evidence of overproduction of haem precursors. In the autosomal dominant acute porphyrias, detection of such individuals with latent porphyria is important because advice about avoiding drugs and other factors that provoke acute attacks decrease their risk of illness. Mutation analysis is fast becoming the method of choice for screening families, and estimates from family studies suggest that about 10–20% of affected individuals develop overt porphyria. Acute attacks of AIP occur in about one per 50–100 000 population in most countries, giving a prevalence for latent disease of about one per 5–10 000 population. However, screening blood donors for disease-specific *HBMS* mutations gives a far higher figure, e.g. 0.6% of the population in one study from France (Nordmann *et al.*, 1997). Although this discrepancy remains unexplained, the idea that mutations for the autosomal dominant porphyrias are relatively common in the general population is consistent with the sporadic presentation of about a quarter of cases, in spite of a low, probably less than 3%, *de novo* mutation rate and with the rare occurrence of homozygous variants.

EPP differs from other autosomal dominant porphyrias in that enzyme activity in all tissues is significantly less than half-normal in those with acute photosensitivity. Recent evidence suggests that inheritance of a *FECH* allele from the unaffected parent that is expressed at a lower level than normal is needed for the development of overt disease (Gouya *et al.*, 1999). Explanations for low penetrance in the other autosomal dominant porphyrias are less clear, with the interactions between environmental and genetic factors remaining to be defined.

3. Autosomal recessive porphyrias

Both the autosomal recessive porphyrias (**Table 1**) are rare; ALA dehydratase porphyria has been reported in only six patients worldwide (Kappas *et al.*, 1995). Congenital erythropoietic porphyria (CEP), on the other hand, is one of the best known of all porphyrias (Fritsch *et al.*, 1997). It was the first to be described (by Schultz in 1874), was one of two additional conditions included in the second (1923) edition of Garrod's *Inborn Errors of Metabolism* and acted as a source of natural porphyrins for Hans Fischer's Nobel prize-winning work on tetrapyrrole chemistry. Bullous skin lesions usually start in infancy and progress to severe photomutilation; the associated haemolytic anaemia may be severe and require frequent transfusion. Porphyrin is deposited in growing bones and stains the teeth brown (erythrodontia). Patients are either homo- or heteroallelic for mutations in the *UROS* gene. One mutation (C73R) occurs on about 40% of European CEP alleles and is associated with particularly severe disease. The disease can be diagnosed prenatally by porphyrin, enzyme or DNA analysis. The only curative treatment is allogeneic bone marrow transplantation, although gene transfer therapy offers hope for the future.

REFERENCES

Elder, G.H., Hift, R.J. & Meissner, P.N. (1997) The acute porphyrias. *Lancet* 349: 1613–17

Fritsch, C., Bolsen, K., Ruzicka, T. & Günter, G. (1997) Congenital erythropoietic porphyria. *Journal of the American Academy of Dermatology* 36: 594–610

Gouya, L., Puy, H., Lamoril, J. *et al.* (1999) Inheritance in erthropoietic porphyria: a common wild-type ferrochelatase allelic variant with low exposure accounts for clinical manifestation. *Blood* 93: 2105–10

Jenkins, T. (1996) The South African malady. *Nature Genetics* 13: 7–9

Kappas, A., Sassa, S., Galbraith, R.A. & Nordmann, Y. (1995) The porphyrias. In *The Molecular and Metabolic Basis of Inherited Disease*, 7th edition, edited by C.R. Scriver, A.L. Beaudet, W.S. Sly & D. Valle, New York: McGraw-Hill

Lindberg, R.L.P., Porcher, C., Grandchamp, B. *et al.* (1996) Porphobilinogen deaminase deficiency in mice causes a neuropathy resembling that of human hepatic porphyria. *Nature Genetics* 12: 195–99

Nordmann, Y., Puy, H., Da Silva, V. *et al.* (1997) Acute intermittent porphyria: prevalence of mutations in the porphobilinogen deaminase gene in blood donors in France. *Journal of Internal Medicine* 242: 213–17

Röhl, J.C.G., Warren, M. & Hunt, D. (1998) *Purple Secret: Genes, "Madness" and the Royal Houses of Europe*. London and New York: Bantam

GLOSSARY

bulla/bullous a rounded, bubble-like structure; when applied to skin or mucous membrane describes a blister containing serous fluid

cytochrome P450 specialized cytochrome (an enzyme pigmented as a result of its haem prosthetic groups) of the electron transport chain in adrenal mitochondria and liver microsomes, involved in hydroxylating reactions

endothelial cell lining cell of a body cavity

erythropoiesis/erythropoietic production of erythrocytes in the bone marrow in adult mammals

haem protoporphyrin IX with a ferrous iron ion (Fe^{2+}) in the centre. When part of haemoglobin, the ferrous iron ion in the haem will bind oxygen reversibly

haplo-insufficiency haplo-insufficiency occurs if the loss of one allele of a gene is enough to give rise to a phenotypic effect. Haplo-insufficiency hence infers that the correct level of a gene product is essential for its normal function

hyponatraemia lower than normal concentration of salt (sodium) in the blood

monogenic controlled by a single gene

photomutilation disfiguring changes produced by light in unusually photosensitized skin and underlying structures

tachycardia accelerated heart rate

GENETICS OF MENTAL RETARDATION

Anita Thapar[1] and Jill Clayton-Smith[2]
[1]Child and Adolescent Psychiatry Section, Department of Psychological Medicine, University of Wales College of Medicine, Cardiff, UK
[2]The University Department of Medical Genetics, St. Mary's Hospital, Manchester, UK

1. Introduction

Mental retardation (currently known as learning disability in the UK) is defined as early-onset, significantly below-average intellectual functioning accompanied by impairments in functioning. The most commonly used index of intellectual functioning is an intelligence quotient (IQ) score obtained using an IQ test which assesses different cognitive abilities. The average IQ in the general population was originally 100, with 95% of individuals scoring between 70 and 130. Mental retardation is defined primarily on the basis of having an IQ test score of less than 70. However, the term "mental retardation" refers to a heterogeneous group of individuals who differ widely in terms of abilities and social adaptation. For example, a substantial proportion of those who have mild mental retardation (MMR; IQ scores of between 50 and 70) require no special schooling and progress to employment in adult life. However, those with moderate to severe mental retardation (IQ scores <50) have special educational needs, need assistance or supervision with general living skills and may need institutionalization when severely affected. A "two group approach" to subdividing mental retardation on the basis of IQ scores into MMR and moderate–severe mental retardation has been influential for many years (Penrose, 1938). The two groups have often been considered as aetiologically distinct categories, with MMR having been viewed as representing the lower end of the normal IQ distribution and as being primarily influenced by psychosocial factors such as social disadvantage. By contrast, moderate–severe mental retardation has been considered as being qualitatively distinct from the general population and mainly accounted for by identifiable causes such as genetic disorders and birth trauma, infections and other environmental insults. There is some empirical support for this distinction into two groups, in that the recurrence risk to relatives of those with MMR is increased, whereas the IQ of relatives of those with moderate–severe mental retardation tends to be normal.

However, more recent work suggests that this aetiological disjunction is not quite as clear cut as previously thought. For example, 4–19% of those with MMR show chromosomal abnormalities (Thapar *et al.*, 1994). The wide range of reported frequencies of chromosomal aberrations may reflect differences between studies in sample selection and the methods used to detect abnormalities, particularly the number of cells analysed for mosaicism. This illustrates the point that MMR can be associated with recognizable medical conditions and is not necessarily "idiopathic". Similarly, a definite causation for moderate–severe mental retardation is not always found.

Recent advances in genetics are having a major impact on our understanding of the aetiology of mental retardation. Many single gene disorders and chromosomal defects are associated with mental retardation (Thapar *et al.*, 1994) and increasing numbers of genes for these disorders are being located. However, most mental retardation is not attributable to these genetic syndromes and, indeed, often no definite cause can be identified (idiopathic). Thus, we will begin by considering the genetics of this group with so-called idiopathic mental retardation before moving on to review the main types of specific genetic disorders.

2. Idiopathic mental retardation

Given that most individuals with idiopathic mental retardation are mildly affected and that MMR is sometimes considered as representing the lower end of the normal IQ distribution, we will begin with a brief overview of the genetics of intelligence.

It has consistently been shown that intelligence is a highly familial trait with family studies of IQ scores showing an average correlation of approximately 0.4 between first-degree relatives (Plomin *et al.*, 1997). Twin studies which allow us to disentangle the effects of genes and environmental factors shared by family members (e.g. social adversity) suggest that most of this familial clustering is accounted for by genetic factors. Genetically identical (monozygotic) twins who have been reared apart also show high correlations in IQ. In a report based on data from five reared-apart twin studies, the weighted average correlation for IQ between monozygotic twins was estimated as 0.75 (Bouchard, 1996). Similarly, findings from adoption studies also suggest the importance of genetic factors – by adulthood, unlike biological relatives, average correlations in IQ for relatives, by adoption, approach zero (Plomin *et al.*, 1997).

Thus, overall, there is now a wealth of evidence which shows that genetic factors influence normal intelligence. Most studies estimate the heritability of IQ to be around 50%, with environmental and other non-heritable effects accounting for the other 50% (Plomin *et al.*, 1997). Although the importance of environmental factors is well recognized, it appears that the types of environmental factors that influence IQ vary with age. In two French studies of adopted-away children (Capron & Duyme, 1989), it was found that the children's IQ was influenced by the social status of the adoptive home as well as by the occupational status of the biological parents. Environmental influences such as social status are shared by different members of the same family and are, thus, termed "shared environmental influences". Twin and adoption studies show that this type of shared environmental influence contributes to IQ in childhood. However, as children grow into adolescence and early adulthood, these shared environmental factors decrease in influence and genetic factors become more important. The heritability of IQ continues to remain substantial even into later life (McClearn *et al.*, 1997). Thus, by adult life, the total variance in IQ scores is explained by genetic factors and environmental influences of the non-shared type (Plomin *et al.*, 1997).

Given the strength of the evidence that genetic factors influence IQ, the next question is whether susceptibility genes for IQ can actually be identified. Attempts to map genes contributing to continuous or quantitative traits in humans including traits such as intelligence are indeed now being made. Quantitative trait loci (QTL) studies of IQ are currently in progress and preliminary findings are now emerging (Daniels *et al.*, 1996), although they await replication.

Unfortunately, there has been very much less interest in the genetics of MMR. Indeed, there have been no systematic family or twin studies of MMR. Moreover, results of family studies will clearly depend on how the proband (affected individual) is identified (e.g. from community or institutions or special schools), how MMR is defined and, finally, how the population is screened for recognizable syndromes. A few older studies suggest that the recurrence risk of MMR is between 13% and 33% for first-degree relatives (20–25% for siblings) and between 5% and 9% for second-degree relatives (Thapar *et al.*, 1994). These rates are clearly much higher than the expected rate of 20 per 1000 in the general population. However, as mentioned, there have as yet been no published systematic twin studies. Two older studies of twins showed higher concordance rates in mental retardation for monozygotic twins than dizygotic twins (Thapar *et al.*, 1994), which suggests a genetic influence. However, the sample sizes were small and the twins were non-systematically ascertained. Thus, overall, it appears that MMR is familial and, possibly, genetically influenced. Further systematic twin and family studies are needed to examine properly the relative effects of genetic and environmental influences.

It has generally been assumed that idiopathic MMR is transmitted in a multifactorial fashion, i.e. it is influenced by a large number of genes together with environmental influences. However, the term "idiopathic" is not necessarily accurate, even where gross chromosomal defects have been excluded. Technological advances now allow us to detect chromosomal defects which are not shown by standard cytology and recent work suggests that a proportion of those with mild, apparently idiopathic, mental retardation have submicroscopic chromosomal abnormalities which would have previously remained undetected. In a study where molecular analysis and FISH (fluorescence *in situ* hybridization) were used to search for telomeric abnormalities, deletions were found in three of 99 patients with so-called idiopathic mental retardation (Flint *et al.*, 1995). More recently, Knight and colleagues (1999) found that 7.4% of moderate–severe and 0.5% of mild unexplained mental retardation could be accounted for by these so-called cryptic chromosomal abnormalities. Thus, it appears that those with mild, so-called idiopathic mental retardation represent an aetiologically heterogeneous group consisting of some affected individuals who will be shown to have a specific genetic defect, others who will be affected because of specific environmental insults and a remaining proportion who are perhaps still best accounted for by a multifactorial model.

3. Single gene defects

Single gene defects account for only a small proportion of mental retardation overall and these disorders are more likely to be seen in individuals with moderate to severe mental retardation. In a review of the literature, Bundey (1996) estimated that single gene defects account for 7–14% of moderate to severe mental retardation and 5% of MMR, although this figure is likely to be an underestimate as it did not include individuals with multiple congenital anomaly syndromes, some of which are also the result of abnormalities in single genes. The group of single gene disorders can be divided into *autosomal disorders* (dominant and recessive) and *X-linked mental retardation.* A detailed family history may suggest the possibility and the mode of inheritance of a single gene disorder, and recurrence risks for this group of conditions may be high. In most cases, there are associated phenotypic features in addition to the mental retardation, such as physical malformations or skeletal abnormalities which can be visualized by radiography. Recognition of these associated features aids diagnosis and has allowed discrete phenotypes to be delineated.

One large group of single gene disorders causing mental retardation includes the *inborn errors of metabolism*, which have been well documented by Scriver *et al.* (1995). In these conditions, the underlying genetic defects are found within genes coding for enzymes or cofactors within metabolic pathways. These pathways usually involve several steps and, in some conditions, the biochemical defect may occur at different points along the pathway, giving rise to variation in clinical features (phenotypic variability). Phenotypic variability may be due to genetic heterogeneity. For example, the condition phenylketonuria is due to a biochemical defect in the metabolism of the amino acid phenylalanine, which is usually converted into tyrosine in the blood. If there is a block in this pathway, phenylalanine accumulates in the blood where it is converted into phenylpyruvic acid, and toxic effects result. Raised blood phenylalanine levels may arise as a result of an abnormality of the phenylalanine hydroxylase gene on chromosome 12q or of the gene encoding dihydropteridine reductase on chromosome 4p. In the latter case, the effects are much more severe. Variation in phenotype may also be due to allelic heterogeneity, where the different mutations within the same gene have differing effects on the clinical course of the condition. The frequency of different mutations varies from one population to another.

A very large number of single gene disorders have now been identified and many of the genes responsible have been localized. In some cases, the sequence and structure of the genes themselves are known (**Table 1**). Genes causing mental

Table 1 Autosomal single gene disorders causing mental retardation.

Disorder	Inheritance	Gene	Location (chromosome)	Population frequency	Degree of MR
Tuberous sclerosis	AD	TSC1	9q34	1 in 30 000	ranges from normal IQ to severe retardation
			TSC2	16p13.3	
Apert syndrome	AD	FGFR2	10q	1 in 150 000	mild to severe
Treacher Collins syndrome	AD	Treacle	5q31–33	1 in 40 000	mild to moderate
Noonan syndrome	AD	unknown	12	1 in 1500	mild
Bardet–Biedl syndrome	AR	BBS1,2,3, etc.	various	unknown	mild
Hurler syndrome	AR	IDUA	4p16.3	1 in 100 000	gradual deterioration after age of 2 years
Seckel syndrome	AR	unknown		1 in 40 000	moderate to severe

AD = autosomal dominant
AR = autosomal recessive
MR = mental retardation

retardation are usually expressed within the brain, but many different types of gene may be involved including tumour suppressor genes, cell adhesion molecules and fibroblast growth factor receptor genes.

(i) X-linked mental retardation

As long ago as 1938, Penrose recognized that mental retardation was commoner among males than females, although this was not initially attributed to genes on the X-chromosome, but rather to a greater susceptibility of males to environmental effects. Following his study, several large pedigrees with mental retardation which clearly followed an X-linked pattern of inheritance were published. On close scrutiny of these families, the mother was more likely to be the parent transmitting the mental retardation and there was a disproportionate number of families in which mental retardation affected only male children. Further studies demonstrated the significant contribution of X-linked genes to mental retardation and many of these genes have now been identified (**Table 2**) (Sutherland *et al.*, 1994). X-linked disorders give rise mostly to mild to moderate mental retardation and the prevalence of X-linked mental retardation (XLMR) has been estimated at 1.83 per 1000 (Glass, 1991), with a carrier frequency in females of 2.44 per 1000. Some X-linked genes cause only mental retardation (non-specific XLMR), whereas others may have additional features such as macrocephaly, hydrocephalus and biochemical or endocrine abnormalities. The commonest cause of XLMR is the fragile X syndrome (see below).

There are a large number of reports of non-specific XLMR segregating within individual families and it is likely that some of these are due to the same underlying genes, but show phenotypic variability. Manifesting females have been described in several of the XLMR syndromes, e.g. Coffin–Lowry syndrome. In other disorders, carrier females are phenotypically normal, but may occasionally manifest symptoms as a result of altered X-chromosome inactivation or the presence of an X-autosome translocation with the break in the X chromosome occurring near to, or through, the XLMR gene. Study of X-autosome translocations may facilitate the regional localization of the genes involved in XLMR. Finally, there are some forms of XLMR which occur almost exclusively in females. These are caused by X-linked dominant genes, which cause lethality in males. This group includes incontinentia pigmenti and oro-facio-digital syndrome Type I. One other disorder which deserves mention is Rett syndrome, a neurodevelopmental disorder affecting only females who typically develop normally for the first 6–18 months of life, then have a period of regression with loss of acquired skills, deceleration of head growth and development of stereotypic hand-wringing or patting movements and episodic hyperventilation. This condition is relatively common with an incidence of one in 10 000 females and most cases are sporadic. Mutations in MECP2 which encodes a protein thought to be involved in transcriptional silencing of other genes, have now been found (Amir *et al.*, 1999).

(ii) Fragile X syndrome

Fragile X syndrome is the single commonest known inherited cause of mental retardation. It affects approximately one in 2000 to one in 3000 people and accounts for approximately 7% of moderate retardation and 4% of MMR among males and approximately 2.5% of moderate retardation and 3% of MMR in females (Thapar *et al.*, 1994). Fragile X syndrome is characterized by the presence of a

Table 2 X-linked recessive disorders associated with mental retardation.

Disorder	Associated clinical features	Gene	Location
X-linked hydrocephalus	adducted thumbs, spasticity, aphasia	L1CAM	Xq28
Fragile X syndrome	macrocephaly, large ears, macro-orchidism	FMR1	Xq27–28
ATR-X syndrome	microcephaly, dysmorphic facies, genital abnormalities	XNP	Xq13
Coffin–Lowry syndrome	coarse face, hypertelorism, tapering fingers	CLS	Xp22
Simpson–Golabi–Behmel syndrome	tall stature, polydactyly, supernumerary nipples	GPC3	Xq25–27

fragile site on the X chromosome and, prior to the availability of molecular testing, fragile X was diagnosed by the presence of the fragile site or non-staining band at Xq27.3 which is only apparent with certain culture media. The criterion for diagnosis was that 4% or more of lymphocytes should show this fragile site.

The characteristic features of this syndrome include mental retardation, large ears and a long face, and macroorchidism (large testes). Affected individuals also show higher rates of speech and language problems, overactivity and attentional difficulties. Many also manifest autistic-type features such as gaze avoidance and hand flapping. However, it appears that the rate of fragile X among those with autism is no higher than that among those with mental retardation. One of the most interesting features of this syndrome is that it does not show typical X-linked recessive inheritance.

It is now known that an unstable repetitive sequence of trinucleotides (CGG) within the first exon of the FMR1 gene is the mutation associated with fragile X syndrome. It appears that when the repeat sequence reaches a critical length (about 200 copies), there is an absence of FMR1 mRNA expression (Pieretti *et al.*, 1991) and individuals then show the phenotypic features of fragile X syndrome. The discovery of heritable unstable DNA provides an explanation as to why the disorder becomes progressively more severe with subsequent generations and why phenotypic expression depends on whether the defect is transmitted by a male or by a female (the increase in size from one generation to the next only occurs when the genes are transmitted by a female and not a male). The inheritance of unstable repeat sequences within other genes is now known to be responsible for many other conditions such as Huntington's disease and myotonic dystrophy.

4. Chromosomal abnormalities

Chromosomal abnormalities are a major cause of mental retardation, accounting for approximately 35% of severe mental retardation and 10% of MMR. As mentioned earlier, the development of new techniques for chromosome analysis in recent years has enabled the detection of submicroscopic chromosome abnormalities in individuals where previous analyses have been normal, suggesting that the current prevalence figures are an underestimate. Almost all chromosome aneuploidies which involve an alteration in the amount of chromosome material are associated with mental retardation. The majority of chromosomal abnormalities occur *de novo*, with the parents having a normal karyotype. Some chromosome abnormalities tend to occur only in the mosaic form, with the affected individual possessing two different types of cell. One of these cell lines is chromosomally abnormal, whereas the other has an entirely normal karyotype; presumably the abnormality would be lethal if present in every cell. Such mosaic karyotypes are more likely to be identified within cultured fibroblasts than in blood cells. The commonest autosomal abnormalities are trisomies, particularly involving chromosomes 21, 13 and 18. These are all associated with increased maternal age.

(i) Down's syndrome

Trisomy 21 (Down's syndrome) is the commonest chromosomal cause of mental retardation and occurs in one in 750 pregnancies, with incidence figures ranging from one in 1500 at age 20 to one in 44 at age 45. In the majority of cases, the extra chromosome arises as a result of failure of separation (*non-disjunction*) of the two maternal chromosome 21s in meiosis during the formation of the oocyte and the affected individual has three separate copies of chromosome 21. In 5% of cases, there is a *translocation* between the extra copy of chromosome 21 and one of the other autosomes, most commonly chromosome 14. If one of the parents carries a balanced form of the same translocation then there will be a significant risk of recurrence of Down's syndrome during another pregnancy. The clinical features of Down's syndrome include neonatal hypotonia, brachycephaly, small, simply formed ears, protruding tongue, single palmar creases, incurving of the fifth fingers and wide gaps between the first and second toes. Frequent structural anomalies include congenital cardiac defects in 35% and a failure of development of part of the small intestine (duodenal atresia). The majority of patients have moderate to severe mental retardation. Some individuals who have a milder phenotype have been shown to have a mosaic karyotype. The genes responsible for the dysmorphic features seen in Down's syndrome appear to lie within a small region of chromosome 21 encompassing band 21q22. Individuals with Down's syndrome are predisposed to Hirschsprung's disease (a failure of development of the nervous system to the large bowel), hypothyroidism and leukaemia. Adults with Down's syndrome often develop early-onset dementia with Alzheimer-like changes within the brain and this is likely to be related to the fact that the gene for amyloid precursor protein lies on chromosome 21 and changes within this gene have been shown to be responsible for some cases of familial Alzheimer's disease (Goate *et al.*, 1991).

(ii) Microdeletion syndromes

The advent of improved cytogenetic techniques such as high resolution chromosome banding and FISH, together with molecular genetic techniques such as microsatellite analysis (Strachan & Read, 1996), has enabled the detection of small chromosomal deletions which had previously gone unnoticed. Several fairly common multiple congenital anomaly syndromes have been shown to be due to chromosomal microdeletions. These disorders are generally sporadic with a low recurrence risk, although chromosome rearrangements which predispose to the occurrence of microdeletions are present in some families.

(iii) DiGeorge syndrome

The commonest microdeletion appears to involve chromosome 22 at 22q11 and gives rise to features of the DiGeorge syndrome (Ryan *et al.*, 1997), also known as velo-cardio-facial syndrome or Schprintzen syndrome. The incidence of this microdeletion is estimated to be about one in 8000 and affected individuals may have congenital heart disease, cleft palate, a low blood calcium level, speech and swallowing difficulties, and characteristic facial features with a broad nasal bridge, prominent nose and small mouth and chin.

Most affected individuals have mild to moderate mental retardation. The features are variable from person to person, but the speech and swallowing difficulties are almost always present.

(iv) *Angelman syndrome and Prader–Willi syndrome*

Angelman syndrome and Prader–Willi syndrome may both be caused by a microdeletion of chromosome 15q11–13. In the case of Angelman syndrome, the deletion usually arises *de novo* on the chromosome 15 which was inherited from the mother and in Prader–Willi syndrome the deletion is paternally derived. The clinical features of Angelman syndrome are severe developmental delay, seizures, ataxia, a happy sociable affect, wide mouth and prominent chin. Prader–Willi syndrome is clinically quite different and causes neonatal hypotonia, mild to moderate developmental delay, short stature, excessive appetite leading to obesity, hypogonadism and behavioural problems. The hands and feet are often small and the facial features are characteristic, with narrowing of the face in the region of the temples, almond-shaped eyes and a high palate. Individuals with both conditions are more likely to have fair hair and blue eyes. This feature is likely to be related to the fact that the gene for Type II oculocutaneous albinism also lies within the region of chromosome 15. It is important to note that only 60% of cases of Angelman and Prader–Willi syndrome are caused by microdeletions, with the remaining cases arising due to other genetic mechanisms involving this same area of chromosome 15.

(v) *William's syndrome*

One further relatively common microdeletion syndrome is William's syndrome. This occurs with a frequency of one in 40 000 live births and gives rise to raised blood calcium levels in infancy, a characteristic narrowing of the aorta just above the aortic valve of the heart, moderate mental retardation and a characteristic face with saggy cheeks, thick lips, upturned nose, puffy appearance of the skin around the eyes and a lacy pattern to the iris. This condition is due to a microdeletion of chromosome 7q. The deleted region includes the elastin gene.

(vi) *Sex chromosome abnormalities*

Most sex chromosome anomalies are due to chromosomal non-disjunction. Although the risk of this increases with increasing maternal age, no definite cause for these anomalies has yet been identified.

Generally, extra X chromosomes are associated with a higher risk of mental retardation. Although it was originally thought that sex chromosome abnormalities were associated with higher rates of mental retardation and psychiatric disorder, it is recognized that many early studies included highly selected samples. It is now known that most affected individuals show no signs of either mental retardation or psychiatric disorder and remain unrecognized in the community.

(vii) *Klinefelter's syndrome*

Klinefelter's syndrome is characterized by the karyotype 47XXY. It affects approximately one in 1000 newborn males and can be attributed to maternal non-disjunction in approximately two-thirds of cases. There appears to be no increased recurrence risk in relatives. Earlier work suggested an association with criminality, mental retardation and psychiatric disorder (Thapar *et al.*, 1994). However, these studies were based on samples of institutionalized individuals and it is now recognized that most individuals with XXY are of normal intelligence with no increased rates of psychiatric disorder or criminality.

(viii) *Turner's syndrome*

Individuals with Turner's syndrome are phenotypically female, but all or part of one X chromosome is missing (45XO). The incidence of Turner's syndrome has been estimated as between one in 2000 and one in 5000 live female births, and the condition is characterized by short stature, failure to develop secondary sexual characteristics and abnormalities such as a webbed neck and increased angle of the elbow. It has recently been suggested that SHOX (short stature homeobox-containing gene), a novel gene which has recently been cloned and is located in the pseudoautosomal region of the sex chromosome, may be implicated in the growth failure associated with Turner's syndrome (Rao *et al.*, 1997). Although not associated with mental retardation, recent work on Turner's syndrome suggests the importance of the X chromosome for other sorts of cognitive abilities, namely social cognitive skills (Skuse *et al.*, 1997). These are skills that facilitate social interaction, such as the ability to respond to cues in the behaviour of others, to inhibit distractions and to develop strategies of action. In a recent report, Skuse *et al.* (1997) found that subjects with Turner's syndrome where the X chromosome was of maternal origin showed significantly poorer social cognitive skills than those with a paternally derived X chromosome. These observations might be explained by an imprinted X-linked locus influencing social cognition. Boys whose only X chromosome comes from their mother would not receive the imprinted locus. The authors suggested that this could provide some explanation as to why developmental disorders such as autism which affect social cognition are commoner in males.

(ix) *XYY syndrome*

Individuals with XYY syndrome are phenotypically male, show an above average height and a mean IQ below that of the general population. The incidence has been estimated at between one and two in 2000. There has been considerable interest in the possibility that the XYY karyotype is associated with excessive aggression. This was fuelled by findings from a study of inmates in a hospital for mentally abnormal offenders where 3% showed the XYY karyotype. Similar findings in other institutions also provided additional evidence of an association of XYY with criminality and psychiatric problems (Thapar *et al.*, 1994). However, one study (Witkin *et al.*, 1976) focused on an unselected sample of over 4000 non-institutionalized apparently normal males of height greater than 1.84 metres where 12 individuals had an XYY karyotype. Although five (42%) of these affected individuals had a criminal record, compared with 9% of normal males, the offences were mostly relatively minor and non-aggressive in nature. Thus, there does not appear to be a true association with serious violent criminality.

5. Conclusion

It is apparent that mental retardation is a term used to describe a clinically and aetiologically heterogeneous group. Although an increasing number of chromosomal abnormalities and single gene defects associated with mental retardation are being recognized and mapped, it is important to remember that the vast majority of mental retardation cannot be explained by these syndromes. Thus, the real challenge remains in understanding more about this remaining group. It is likely that new single gene disorders as well as an increasing number of submicroscopic defects will be recognized. However, it is clear that further twin and adoption studies of mental retardation are needed to examine the role of both genes and environment and to understand more about their interplay. It is also likely that a QTL approach will be fruitful in identifying genes of small effect that contribute to human traits such as intelligence. In addition, these genes may well be of importance for MMR. Moreover, identifying susceptibility loci should also help inform us about the aetiological distinction and continuities between normal intelligence and mental retardation, as well as providing greater understanding about the role of environmental risk and protective factors.

REFERENCES

Amir, R.E., Van den Veyver, I.B., Wan., M. *et al.* (1999) Rett syndrome is caused by mutations in X-linked MECP2, encoding methyl-CpG-binding protein 2. *Nature Genetics* 23: 185–88

Bouchard, T.J. Jr (1996) Behaviour genetic studies of intelligence, yesterday and today: the long journey from plausibility to proof. *Journal of Biosocial Science* 28: 527–55

Bundey, S. (1996) Abnormal mental development. In *Principles and Practice of Medical Genetics*, 3rd edition, 2 vols, edited by D.L. Rimoin, J.M. Connor & R.E. Pyeritz, Edinburgh and New York: Churchill Livingstone

Capron, C. & Duyme, M. (1989) Assessment of effects of socioeconomic status on I.Q. in a full cross-fostering study. *Nature* 340: 552–53

Daniels, J., McGuffin, P. & Owen, M. (1996) Molecular genetic research on IQ: can it be done? Should it be done? *Journal of Biosocial Science* 28: 491–507

Flint, J., Wilkie, A.O.M., Buckle, V.J. *et al.* (1995) The detection of subtelomeric chromosomal rearrangements in idiopathic mental retardation. *Nature Genetics* 9: 132–39

Glass, I.A. (1991) X linked mental retardation. *Journal of Medical Genetics* 28: 361–71

Goate, A., Chartier-Harlin, M-C , Mullan, M. *et al.* (1991) Segregation of a missense mutation in the amyloid precursor protein gene with familial Alzheimer's disease. *Nature* 349: 704–06

Knight, S.J.L., Regan, R., Nicod, A. *et al.* (1999) Subtle chromosomal rearrangements in children with unexplained mental retardation. *The Lancet* 354: 1676–81

McClearn, G.E., Johansson, B., Berg, S. *et al.* (1997) Substantial genetic influence on cognitive abilities in twins 80 or more years old. *Science* 276: 1560–63

Penrose, L.S. (1938) *A Clinical and Genetic Study of 1,280 Cases of Mental Defect*. London: HMSO

Pieretti, M., Zhang, F., Fu, Y-H. *et al.* (1991) Absence of expression of the FMR-1 gene in fragile X syndrome. *Cell* 66: 817–22

Plomin, R., DeFries, J.C., McClearn, G.E. & Rutter, M. (1997) *Behavioral Genetics*, 3rd edition, New York: Freeman

Rao, E., Weiss, B., Fukami, M. *et al.* (1997) Pseudoautosomal deletions encompassing a novel homeobox gene cause growth failure in idiopathic short stature and Turner syndrome. *Nature Genetics* 16: 54–63

Ryan, A.K., Goodship, J.A., Wilson, D. *et al.* (1997) Spectrum of clinical features associated with interstitial chromosome 22q11 deletions: a European collaborative study. *Journal of Medical Genetics* 34: 798–804

Scriver, C.R., Beaudet, A.L., Sly, W.S. & Valle, D. (eds) (1995) *Metabolic and Molecular Basis of Inherited Disease*, 7th edition, New York: McGraw Hill

Skuse, D., James, R.S., Bishop, D.V.M. *et al.* (1997) Evidence from Turner's syndrome of an imprinted X-linked locus affecting cognitive function. *Nature* 387: 705–08

Strachan, T. & Read, A.P. (eds) (1996) Genetic mapping. In *Human Molecular Genetics*, Oxford: Bios Scientific and New York: Wiley

Sutherland, G.R., Brown, W.T., Hagerman, R. *et al.* (1994) Conference report: sixth international workshop on the fragile X and X-linked mental retardation. *American Journal of Medical Genetics* 51: 281–93

Thapar, A., Gottesman, I.I., Owen, M.J., O'Donovan, M. & McGuffin, P. (1994) The genetics of mental retardation. *British Journal of Psychiatry* 164: 747–58

Witkin, H.A., Mednick, S.A., Schulsinger, F. (1976) Criminality in XYY and XXY men. *Science* 193: 547–55

GLOSSARY

autosomal disorder disorder resulting from a defect in an autosome (i.e. any chromosome other than a sex chromosome)

cytology the study of cells

dizygotic (DZ) developing from separately fertilized eggs, as in non-identical twins

fluorescence in situ hybridization (FISH) the labelling of specific whole chromosomes by *in situ* hybridization. Hybridization probes are usually made by selecting a particular chromosome by flow sorting, and using the DNA from such chromosomes, after removing repeated sequences that are common to other chromosomes. Also known as chromosome painting, this technique can be used to identify specific chromosomes not only in normal metaphases, but also in interphase nuclei, in chromosomal rearrangements, and to study chromosomal changes during evolution

monozygotic (MZ) developing from a single fertilized egg, as in identical twins

non-disjunction failure of separation of the two chromosomes of a pair during production of the egg or sperm at meiosis

proband (index case) affected individual from whom a pattern of inheritance of a familial disorder is traced

telomeric relating to the ends of chromosomes

See also **Trinucleotide repeat instability as a cause of human genetic disease** (p.396); **Aneuploidy** (p.531); **Genomic imprinting** (p.757); **X-chromosome inactivation** (p.780)

GENETICS OF PSYCHIATRIC DISORDERS

Peter McGuffin[1] and Michael J. Owen[2]
[1]*Institute of Psychiatry, King's College London, UK*
[2]*Department of Psychological Medicine, University of Wales College of Medicine, Cardiff, UK*

1. Introduction

The notion that mental disorders tend to run in families dates back to antiquity, but, as elsewhere in genetics, it was not until the beginning of the 20th century that the inheritance of psychiatric illness became the subject of systematic scientific enquiry. The rediscovery of Mendel's laws in 1900 coincided with the emergence of a workable system of classification of psychiatric disorders that was in large part attributable to the work of the German psychiatrist Emil Kraepelin. Current classifications, as embodied in the International Classification of Diseases, 10th edition (ICD 10) (World Health Organization, 1993) or the American Psychiatric Association's (1994) Diagnostic and Statistical Manual, 4th edition (DSM IV), are still recognizably Kraepelinian in their general outline. Most of the major categories of disorders are listed in **Table 1**, using a nomenclature based on ICD 10.

It is worth noting that, in contrast to diagnosis in other branches of medicine, psychiatric diagnosis is usually based entirely on clinical signs and symptoms and is rarely helped by specific laboratory or radiological tests. Indeed, the only circumstances in which such tests are of routine use in psychiatry are the detection of known physical causes of psychiatric symptoms (e.g. hyperthyroidism presenting as anxiety or a cerebral tumour presenting as behavioural change). Even in the dementias and other so-called organic disorders, diagnosis, in most cases, is a matter of clinical judgement. Postmortem examination of the brain does, of course, reveal a characteristic neuropathology in conditions such as Alzheimer's disease or vascular dementia which is in contrast with other conditions such as schizophrenia where gross brain changes, if any, are non-specific. Nevertheless, the inaccessibility of the brain during life has meant that understanding the biological basis of psychiatric disorders remains one of the most daunting challenges of modern medicine. Against this background, genetic approaches with the potential of leading to an understanding of the molecular biology of psychiatric disorders are, along with new techniques in brain imaging, widely seen as the best way forward in biological psychiatry.

Table 1 An abbreviated classification of psychiatric disorders based on ICD 10.

Organic disorders
Alzheimer's disease
Vascular dementia
Dementia in other diseases (e.g. Huntington's, Pick's, Creutzfeldt–Jakob)
Schizophrenia
Mood disorders
Bipolar affective disorder
Depressive disorder (unipolar disorder)
Neurotic disorders
Phobic anxiety disorder
Obsessive–compulsive disorder
Other anxiety disorders, including panic and generalized anxiety disorder
Behavioural syndromes associated with physiological disturbance
Anorexia nervosa
Bulimia nervosa
Disorders in adult personality
Dissocial personality disorder
Paranoid personality disorder
Dependent personality disorder
Disorders due to psychoactive substance use
Harmful use
Dependence
Disorders of psychological development
Childhood autism
Specific reading disorder
Behavioural and emotional disorders in childhood
Hyperkinetic disorder
Conduct disorder

2. Quantitative genetics and psychiatric disorders

The obvious starting point in the investigation of the genetics of any disorder is to find out whether it is more common in the relatives of index cases than in the population at large. However, given our earlier comments about diagnosis being based purely on clinical judgement, family studies of psychiatric disorders may be vulnerable to observer error or bias. The recognition of these difficulties has led to considerable elaboration of methodology so that in modern studies investigators will typically use a standardized diagnostic interview, such as the schedules of clinical assessment in neuropsychiatry (SCAN) (Wing *et al.*, 1990) or the diagnostic interview for genetic studies (DIGS) (Nurnberger *et al.*, 1994). These allow diagnosis to be made according to so-called operational diagnostic criteria that are both explicit and reliable. Also, observer bias can be reduced by using interviewers who examine family members "blindfolded" to the diagnosis of the index cases.

Even when all these precautions are taken, the finding of familial aggregation does not, of course, mean that a disorder necessarily has a genetic component. With rare exceptions, such as Huntington's disease and some early-onset forms of Alzheimer's disease, disorders that present with psychiatric symptoms tend not to have a straightforward Mendelizing pattern of segregation within families. Therefore, the question arises as to what extent familial clustering reflects shared family environment rather than shared genes. It would seem to be common sense to suppose that many aspects of behaviour that cluster in families, ranging from levels of alcohol consumption to political persuasion and to career choice, might be largely influenced by imitation or family culture. It is also important to be aware that even when traits appear to conform to Mendelian patterns, this does not necessarily imply straightforward single gene inheritance (Edwards, 1960). For example, a study of the educational history of the relatives of medical students attending the University of Wales College of Medicine showed that attending medical school fulfilled most of the statistical test criteria for an autosomal recessive trait (McGuffin & Huckle, 1990)!

Faced with these difficulties, psychiatric geneticists have made full use of two kinds of natural experiment: the biological phenomenon of twinning and the social phenomenon of adoption. The use of twins in distinguishing between the effects of genes (or nature) and family environment (or nurture) was first suggested by Francis Galton in 1876, but it was not until the 1920s that twins began to be used in a comprehensive way to study common behavioural traits and disease (Plomin *et al.*, 1994). The basic premise of the twin method is that identical or monozygotic twins share 100% of their genes plus a common environment. Fraternal or dizygotic twins on the other hand share just 50% of their genes as well as a common environment. If it is assumed that the environmental sharing is about the same for monozygotic and dizygotic twins, then any greater similarity in monozygotic compared with dizygotic pairs should reflect genetic factors. Although this "equal environment assumption" has often been criticized on the basis that monozygotic twins are likely to be treated more similarly by family and friends than dizygotic twins, a variety of methods is available to check its validity.

One such method is to compare the similarity in twin pairs when the parents and twins are mistaken about their true zygosity (i.e. where they have misclassified themselves as monozygotic when they are truly dizygotic or vice versa). Another method is to attempt to quantify environmental sharing and a third is to study those (inevitably rare) monozygotic twins who have been raised apart from an early age and compare their degree of similarity with monozygotic twins reared together. In practice, these approaches tend to support the equal environments assumption and, thus, the use of twin studies in separating out the effects of genes and shared environment. Nevertheless, many regard adoption studies as a more satisfactory approach.

In adoption studies, one can either start with a series of offspring of parents with a particular disorder who have been fostered or adopted away from their parents early in life and compare the rate of disorder with that in control adoptees, or one can compare the rate of the disorder in the biological and adopted relatives or adoptees who are affected by the condition. Adoption studies have been used most extensively in studying schizophrenia, but have also played a role in pointing to a genetic component in mood disorders, alcoholism and personality disorder.

The class of genetic models that has been most useful in conceptualizing the transmission of psychiatric disorders embodies the notions of liability and threshold (Falconer, 1965; Reich *et al.*, 1972). It is assumed that what is inherited is not so much the disease as a liability to disease and that that liability is continuously distributed in the population such that only those who at some point exceed a certain threshold manifest the disorder. In familial disorders, relatives of affected cases on average have an increased liability such that more of them lie beyond the threshold than are found in the population at large. Although it is usually assumed that liability is contributed to by multiple genes and multiple environmental factors such that it will tend to have a normal distribution, the concepts of liability and threshold can be applied to single gene traits or traits where there is a combination of a major gene and multiple minor genetic and environmental factors (Morton, 1982; McGuffin *et al.*, 1994). Quantitative genetic studies of psychiatric traits have applied liability-threshold models to estimate heritability or the proportion of variance in liability that can be explained by genes, but they have also gone beyond this. For example, some studies have explored definitions of disorders in order to see what combinations of symptoms are most heritable (Farmer *et al.*, 1987) or to use more sophisticated multivariate methods of analysis to explore the extent to which symptomatically overlapping disorders such as anxiety and depression are influenced by the same sets of genes (Kendler *et al.*, 1987).

3. Molecular genetic studies in psychiatry

Positional cloning has proved to be dramatically successful in the study of single gene diseases. Here, one starts off with a disorder such as Huntington's disease where little is known about the aetiology and pathogenesis other than that it is a central nervous system disease with an autosomal dominant mode of transmission. The cosegregation of genetic markers and the disorder is then investigated within families to detect linkage. This points to the locus at which the gene resides, which, in the case of Huntington's disease, is on chromosome 4p (Gusella *et al.*, 1983). The next step of moving from locus to gene may be complicated and time consuming – for example, for Huntington's disease, it took 10 years – but inevitably will succeed. It is then a matter of identifying the sequence and structure of the gene, defining the gene products and determining where these are expressed. Ultimately, once pathogenic mechanisms have been understood, there should be scope for devising specific targeted treatments.

Unfortunately, from the viewpoint of the molecular geneticist wishing to study psychiatric disorders, Mendelian disorders are comparatively rare and the transmission of common disorders such as schizophrenia, depression or dementia is likely to be more complicated. The question, therefore, is whether attempts to carry out positional cloning are still viable. One approach is simply to concentrate on multigeneration families containing multiple members affected by the condition. This has proven to be successful in studies of Alzheimer's disease where a minority of cases have an early onset, are highly familial and show a pattern of transmission compatible with autosomal dominance. Three different genes have now been identified that can cause Alzheimer's disease of this type. By contrast, linkage studies of schizophrenia and bipolar manic depression have so far yielded inconclusive and conflicting results, suggesting that Mendelian subforms are rare or non-existent and that the majority (perhaps all) cases will be explained by the combined effects of multiple genes. Consequently, much of the attention in attempting to locate and identify genes contributing to common psychiatric disorders has switched away from large multiply affected pedigrees to studies of affected sibling pairs. In this approach, no assumption about the mode of transmission of the disorder is required. In the absence of linkage, the probability of a pair of siblings sharing 0, 1 or 2 alleles at a particular locus is 1/4, 1/2 and 1/4, respectively. Therefore, when both siblings are affected by the same disease, a significant increase in allele sharing at a particular marker, above the expected rate, implies linkage between the marker and a disease susceptibility locus. So far, affected sibling pair methods have proved to be successful in identifying susceptibility loci contributing to other common diseases, such as insulin-dependent diabetes mellitus, that present many difficulties for genetic analysis in common with psychiatric disorders (Davies *et al.*, 1994).

A complementary approach, that has the attraction of being able to detect genes of very small effect, is to carry out allelic association studies comparing the frequency of marker alleles in a sample of patients with the disorder and healthy controls. So far, allelic association studies in psychiatric diseases have mainly focused on polymorphisms at candidate genes; that is, genes encoding for proteins that might plausibly be involved in a pathogenesis of the condition. These include, for example, the genes that encode for the receptors for serotonin or dopamine and the enzymes involved in the metabolism of these neurotransmitters.

4. An outline of current knowledge on the genetics of psychiatric disorders

(i) Organic disorders

The term "organic" disorders is a vexed one in psychiatry. One of the implications of finding a genetic contribution to a disorder is that it has a biological substrate and is, to that extent, organic. Nevertheless, the term is used traditionally to apply to those disorders where there is either a characteristic neuropathology or demonstrable gross metabolic disturbance. The dementias are numerically the most important of the organic psychiatric disorders, affecting between 5% and 10% of the over 65-year-olds in Western industrialized nations and up to 20% of those aged over 80. Dementia is characterized by a global and persistent deterioration in memory, intellect and personality. The commonest form of dementia is Alzheimer's disease, which is difficult to differentiate with certainty from other forms of dementia during life. However, at postmortem examination, there are typical brain changes that, in addition to cell death, include neurofibrillary tangles within cells, plaques consisting mainly of a protein called β-amyloid and deposition of amyloid around the blood vessels. A minority of cases of Alzheimer's disease, about 1% or less, show an early onset in their forties or fifties with familial aggregation and an autosomal dominant pattern of transmission. Three different genes have now been identified by positional cloning that together account for the majority of families of this type. The most common is the gene designated as presenilin 1 (PS1) on chromosome 14. The presenilin 2 gene on chromosome 1 shows a high degree of homology to PS1, but only accounts for a small proportion of known families with early-onset Alzheimer's disease. However, the first mutations causing Alzheimer's disease to be discovered were in neither of these genes, but in the amyloid precursor protein (APP) gene on chromosome 21q, now thought to account for 5–10% of autosomal dominant Alzheimer's disease.

Chromosome 21 had long been implicated in the aetiology of Alzheimer's disease by the fact that Alzheimer-type neuropathological findings are almost universal in patients with Down's syndrome dying in their forties and Down's syndrome, in turn, is usually a result of trisomy 21 or, less commonly, a translocation involving extra chromosome 21 material. The mapping of the APP gene to chromosome 21 seemed, therefore, too much of a coincidence for it not to be involved in Alzheimer's disease. However, the initial linkage study results in familial Alzheimer's disease using markers in the region of the APP gene provided conflicting results, with some showing evidence of linkage and others not. The problem, as it turned out, was an incorrect assumption that the same gene accounted for all cases. It was left to Goate *et al.* (1991) to demonstrate by sequencing that there was a mutation in exon 17 of the APP gene in a multiply affected British family that showed no recombinations between the APP gene and the disease. The sequencing showed a C to T transition which causes a valine to isoleucine substitution at amino acid 717. The same and other mutations in exon 17 were subsequently detected in American and Japanese families and a different mutation was discovered, this time in exon 16, in two Swedish families with early-onset Alzheimer's disease (Mullan *et al.*, 1992).

By contrast with the autosomal dominant early-onset forms, the vastly more common late-onset Alzheimer's disease is only weakly familial, with a risk in siblings of about 2–3 times that in the general population. Allelic association studies have demonstrated an increased risk associated with carrying allele e4 of the Apolipoprotein E gene. Apolipoprotein E is a cholesterol-carrying protein and the precise reason for an association with Alzheimer's disease is so far unknown. However, it binds avidly to the β-amyloid found in the plaques in Alzheimer's disease and it is thought that the e4 form may facilitate amyloid deposition.

Vascular or multi-infarct dementia is the next commonest form after Alzheimer's disease and, at postmortem examination, mixed forms showing both types of neuropathology are common. Vascular dementia shows some degree of familiality, but its genetics is truly multifactorial in that it overlaps with those aspects of arterial disease and hypertension that are under genetic influence. Some rare single gene forms exist including hereditary cerebral haemorrhage with amyloid (HCHWA) caused by a mutation in the APP gene that is distinct from the mutation causing early-onset Alzheimer's disease.

Other forms of hereditary dementia are rare by comparison with Alzheimer's disease and vascular dementia. They include Pick's disease, which is most commonly said to show an autosomal dominant pattern of transmission, and inherited forms of spongiform encephalopathies or prion diseases such as Creutzfeldt–Jakob disease and Gerstmann–Straussler syndrome (see article on "Genetics of human and animal TSEs or prion diseases").

Huntington's disease has a lifetime risk of about one in 10 000 in northern European populations. It is an autosomal dominant disease which usually presents its first symptoms in the forties and, although the characteristic feature is dementia with jerky abnormal movements called chorea, it can present with personality change or symptoms mimicking schizophrenia or mood disorders. It also shows the phenomenon of anticipation, whereby the age of onset may become lower from one generation to the next. Once thought to be a statistical artefact, this is now known to have a molecular basis. The Huntington's disease gene on chromosome 4p contains an expanded trinucleotide CAG repeat in the first exon in patients with the disease. Those affected have 42 or more repeats, while the normal gene contains 11–34 copies (Huntington's Disease Collaborative Group, 1993). There is a significant negative correlation between repeat length and age of onset (Snell *et al.*, 1993). Interestingly, Huntington's disease now appears to be but one of a group of disorders affecting the central nervous system and involving trinucleotide repeat expansion, a group that includes myotonic dystrophy and fragile X mental retardation (see articles on "Genetics of Huntingdon's disease: when more is less" and "Trinucleotide repeat stability as a cause of human genetic disease").

(ii) Schizophrenia

Schizophrenia affects about 1% of the population at some point in their lives with symptoms usually occurring for the first time when they are young adults. The most striking symptoms are auditory hallucinations and bizarre but unshakeable beliefs of, for example, a persecutory nature or of the patient's thoughts or actions being controlled. Even more disabling are so-called negative symptoms which result in social withdrawal, lack of self care, inability to maintain social rapport and a loss of normal "drive". The risk of the disorder in siblings or offspring of schizophrenics is about ten times the population risk and this increases further when multiple relatives are affected. For example, there is a lifetime risk of 16% for those who have both a parent and sibling already affected (Gottesman, 1991). Twin study evidence supports a genetic explanation of familiality in that the concordance for monozygotic twins using modern explicit diagnostic criteria is 48% compared with a dizygotic concordance of 10% or less (Farmer *et al.*, 1987; Onstad *et al.*, 1991; Cardno *et al.*, 1999). This is further supported by adoption studies which have been used extensively in schizophrenia, the classic study being that of Heston (1966) who studied 47 adoptees who had a schizophrenic biological mother and found that five had become schizophrenic by their mid-thirties compared with none of a sample of 50 control adoptees. The results were subsequently confirmed by a series of studies in Denmark and more recently in Finland that matched psychiatric with adoption registers to compare the rates of schizophrenia in biological relatives with those not biologically related to schizophrenics. The consistent finding was an excess of schizophrenia among biological relatives (McGuffin *et al.*, 1995).

As noted earlier, studies in schizophrenia that have attempted to locate genes in families containing multiply affected individuals have failed to produce evidence of Mendelian forms of the disorder. There does, however, appear to be support from more than one study favouring linkages on chromosomes 1q, 6p, 8p and 22q (Riley & McGuffin, 2000). The data so far suggest that none of these linkages are with genes that, on their own, are either sufficient or necessary to cause schizophrenia, but rather that they are susceptibility loci that contribute to the disease in concert with other factors.

It is unsurprising, because of the sites of action of the drugs that alleviate schizophrenic symptoms, that researchers have focused on dopaminergic and serotinergic pathways in the search for candidate genes. There is evidence from a large multicentre European study and a meta-analysis of studies elsewhere of involvement in the 5HT2A receptor gene in schizophrenia (Williams *et al.*, 1997). More controversially, it has been suggested that homozygosity at a dopamine D3 polymorphism also confers increased risk.

(iii) Mood disorders

Following the precedent of two influential family studies carried out independently in Switzerland and in Sweden in the 1960s (Angst, 1966; Perris, 1966), mood disorders are now categorized as unipolar disorder, consisting of recurrent episodes of depression alone, and bipolar disorder, consisting of episodes of both mania and depression. The relatives of unipolar index cases have increased risk of unipolar depression whereas the relatives of index cases with bipolar disorder have increased risk both of bipolar disorder and unipolar depression.

Depressive symptoms are extremely common and hence, in practice, there is probably a continuum between "normal" low mood and clinically significant depression. Nevertheless, the latter differs from ordinary low spirits in being more persistent, more profound and often associated with physiological disturbance including loss of appetite, weight loss and disturbed sleep. Suicidal thoughts are common and the long-term risk of death by suicide in severe cases is about 15%. Depression of a severity that requires medical treatment affects about one in ten of the population at some point in life and is commoner in women than in men. Bipolar disorder is less common, affecting less than 1% of the population per lifetime. Manic episodes are characterized by elation and an abundance of energy and ideas, but this is accompanied by a reckless disinhibition to a degree that alarms others and may be damaging to the patients themselves.

Twin studies support a genetic contribution by consistently showing higher monozygotic than dizygotic concordance. They also show the same pattern as is found in family studies, with high rates of depression alone in the co-twins of unipolar index cases and high rates of both unipolar and bipolar disorders in the co-twins of bipolar index cases. The classic twin study is that of Bertelsen and colleagues (1977) which suggests that there is a high degree of genetic overlap between unipolar and bipolar disorder, with bipolar disorder probably being more heritable. Nevertheless, a recent study of unipolar depression suggests that it is also substantially heritable (McGuffin *et al.*, 1996) and earlier suggestions of lower heritability might simply reflect low diagnostic reliability (Kendler *et al.*, 1993). Adoption studies also support a genetic contribution to both unipolar and bipolar mood disorders (Tsuang & Faraone, 1990).

As in schizophrenia, attempts at linkage analysis have been carried out under the assumption that a major gene or genes for affective disorder explains the transmission in multiply affected families. The first such study suggested X-linkage in two large pedigrees where mood disorder appeared to be coinherited with colour blindness (Reich *et al.*, 1969). Subsequently, the demonstration that many families show father–son transmission, together with the results of segregation analysis, suggest that X-linkage could at best account for a minority of cases. Studies using a variety of markers have produced conflicting results and these, together with the lack of formal evidence of X-linked inheritance, suggest that it is unlikely that a major gene for mood disorder is to be found on the X-chromosome (Hebebrand, 1992). More recently, several studies have suggested a locus contributing to mood disorders on chromosome 18 and there might be parent-of-origin effect, suggesting imprinting. Again, however, not all the data are consistent (Baron, 1997). The fact that support for even the most promising loci is not universal is, of course, to be expected for a complex disorder in which the susceptibility genes exert modest individual effect sizes and interact with both other genes and environmental factors. Recent candidate gene studies have attempted to detect allelic association with a polymorphism in the gene that encodes the serotonin transporter (Craddock & Owen, 1996; Ogilvie *et al.*, 1996). This is of interest because the serotonin transporter is the site of action of newer antidepressant drugs of the selective serotonin reuptake inhibitor (SSRI) type such as "Prozac" as well as of the older tricyclic antidepressants.

(iv) Other psychiatric disorders in adult life

Phobias or morbid fears associated with severe anxiety that may amount to panic and avoidance of a feared object or situation to a degree that interferes with life are the hallmarks of phobic anxiety disorder. Agoraphobia, usually manifesting as a fear of shops, supermarkets and other crowded public places, is the commonest form, but social phobias and specific fears, for example, of animals, blood or needles, are also fairly common. Phobias show a familial tendency and twin studies suggest a genetic component (McGuffin *et al.*, 1994).

Obsessive–compulsive disorder is characterized by recurrent intrusive and disturbing ideas that the patient tries in vain to resist. These can include contamination by dirt or germs and again tend to lead to the avoidance of contaminating objects and to compulsive rituals such as recurrent hand washing. The results of family studies are somewhat conflicting with some showing an increase in obsessive–compulsive disorder among relatives, but others showing a less specific increase in neurosis generally. Twin studies also suggest that what is inherited may be a combination of a tendency to obsessive–compulsive symptoms and a more general neuroticism (McGuffin *et al.*, 1994).

In generalized anxiety disorder, there is also a question of whether an inherited component is as much related to a personality trait of neuroticism as to the development of a disorder as such. There is also recent evidence from twin studies that the genetic overlap between depression and anxiety is substantial and that the form of the symptoms – that is, presentation mainly with anxiety rather than depression or vice versa – is largely attributable to environmental stressors (Kendler *et al.*, 1987). Consequently, there is much interest in the molecular genetic basis of neuroticism and, again, attention has focused on the serotonin transporter as a possible candidate gene. Neuroticism levels as measured by questionnaire are associated with a variation in the promoter region that has been found to have a functional significance in levels of gene expression (Lesch *et al.*, 1997).

Anorexia nervosa is a disorder predominantly of young adults and adolescents where affected women outnumber men 10:1. Restriction of calorie intake and attempts to lose weight through exercise result in amenorrhoea and loss of secondary sexual characteristics and can, in severe cases, when combined with self-induced vomiting and purgation, lead to metabolic disturbance and even to death. The disorder shows familiality that includes an overlap with depression and a twin study has shown a concordance of 59% in monozygotic compared with 8% in dizygotic twins (Holland *et al.*, 1984). There is also evidence from a study of normal female twins that attitudes to eating and dieting are moderately heritable (Rutherford *et al.*, 1993). Bulimia nervosa, characterized by bouts of over-eating followed by self-induced vomiting in young women (or less commonly young men) who are not markedly underweight, appears to be less influenced by genes (Holland *et al.*, 1984).

Personality disorders are characterized by ingrained maladaptive behaviour that is usually present from adolescence onwards and persists throughout adult life. In practice, this tends to be an imprecise and unsatisfactory group of diagnostic categories where the boundaries between normal and abnormal behaviour are ill defined. Nevertheless, the genetics of dissocial, or antisocial, behaviour has been the subject of much interest and there seems to be little doubt that genes do contribute to some aspects of antisocial traits. Inevitably, because antisocial behaviour may result in criminal convictions, criminality has often been used in genetic studies as a "marker" for antisocial personality. As such, it is far from perfect since crime is defined legally rather than medically or biologically and the definition differs between societies and even within a society over time. Studies of antisocial behaviour have, therefore, become one of the most controversial areas in genetics. There is, nevertheless, consistent evidence from twin studies of a genetic contribution to broadly defined criminal behaviour (Cloninger & Gottesman, 1987) and self-reported antisocial behaviour (Lyons *et al.*, 1995). A combined effect of genes and family environment is suggested by adoption studies which have shown that those adopted away from antisocial parents remain at high risk of behaving antisocially themselves, but that the risk is increased further when the adopting parents also show antisocial behaviour or where the environment in which the adoptees are raised is disturbed (Mednick *et al.*, 1984; Cadoret *et al.*, 1995).

Much interest has focused on the XYY karyotype and antisocial behaviour. Although XYY males do appear to have higher rates of criminality than normal XY males (Witkin *et al.*, 1976), this does not usually include serious criminality and accounts for only a very small proportion of crime overall. More recent studies have attempted to dissect out personality traits that may underlie antisocial behaviour such as aggression or impulsiveness. Hints about what genes may be involved have come from animal experiments including observation of the behaviours of "knockout" mice. In particular, mice lacking either nitric oxide synthase or serotonin 5HT1B receptors show over-aggressive behaviour. Increased aggression has also been shown in mice lacking monoamine oxidase A genes and there has been a report of a family in which X-linked mental retardation in humans is associated with aggression and a mutation in the monoamine oxidase A gene (Brunner *et al.*, 1993).

(v) Alcoholism

As shown in **Table 1**, modern terminology refers to drug addiction and alcoholism as "substance use", making the distinction between harmful use and actual dependence. In practice, few genetic researchers have observed this distinction and the great bulk of genetic research on substance use has focused on alcohol. Again, twin studies provide higher concordance in monozygotic than dizygotic twins, suggesting a genetic effect, but not all of the data are consistent and most studies suggest higher heritability in men than in women (McGue, 1993). Normal alcohol use also appears to be influenced by genes and affects such dimensions as frequency and level of consumption (Heath, 1993). Adoption studies support the existence of a modest genetic contribution to alcoholism that is more evident in men than women (Bohman *et al.*, 1981).

Selective breeding experiments on rats and mice have produced strains which consistently prefer dilute alcohol to water and also strains that differ in other ways such as behavioural response to alcohol intake or withdrawal (Crabb & Belknap, 1992). There is currently considerable effort underway to attempt to detect and map the quantitative trait loci (QTL) that influence alcohol preference or metabolism in rodents. In humans, a gene that influences alcohol metabolism and consumption is that coding for mitochondrial acetaldehyde dehydrogenase (ALDH2) on chromosome 12. A common form in ALDH2 in Asians results in about half of these populations having an inactive variant of the enzyme, resulting in a build-up of acetaldehyde after alcohol consumption. This is accompanied by unpleasant sensations including nausea and flushing which appears to be an effective deterrent to alcohol drinking. Thus, about 95% of Japanese alcoholics have the active rather than inactive form of ALDH2 (Hodgkinson *et al.*, 1991). So far, genes influencing susceptibility to alcoholism in Western populations have proved more elusive, although there have been suggestions of an association with a polymorphism at the dopamine D2 receptor gene (Uhl *et al.*, 1992).

(vi) Childhood disorders

Traditionally, child psychiatrists, unlike paediatricians, have shown little interest in genetic explanations of the disorders that they treat. However, over the past two decades, there has been a gradual but noticeable change. A seminal study was that on autism by Folstein and Rutter (1977) showing 36% concordance in monozygotic twins compared with zero concordance in dizygotic twins. When the definition of the phenotype was broadened to include other cognitive abnormalities, the concordance rates rose to 82% in monozygotic and 10% in dizygotic twins. Before the publication of this study, the predominant explanation of autism, which presents in the first 2 years of life with language delay, social deficits and eccentric stereotyped behaviour, was of disturbed family psychodynamics. It is now clear that although the risk of a full-blown autism in siblings is, at 3%, fairly small, this is nevertheless much elevated compared with the general population risk of around four per 10 000 (MacDonald *et al.*, 1989). A minority of children presenting autistic symptoms have an underlying single gene disorder, such as fragile X mental retardation, tuberous sclerosis or phenylketonuria, but the majority of cases are thought to have polygenic inheritance with a heritability in excess of 80%. Attempts to detect linked markers using affected sibling pair strategies are currently underway (Philippe *et al.*, 1999).

Another developmental disorder for which there is convincing twin evidence of substantial heritability is specific reading disorder or dyslexia. This is much commoner than autism, affecting 5% of children whose intelligence is otherwise normal. Part of the genetic variance in reading ability appears to be accounted for by QTLs on chromosome 6p (Cardon *et al.*, 1994) and on 1sq (Morris *et al.*, 2000).

Conduct disorder or persistent antisocial behaviour in childhood has a familial tendency, but this appears to be mainly accounted for by family environment (Thapar & McGuffin, 1996). A less common disorder with which, however, there is some phenotypic overlap is hyperkinetic disorder (also called attention deficit hyperactivity disorder in the American Psychiatric Association's nomenclature). Here, as the name implies, the characteristic features are abnormal levels of hyperactivity, restlessness and inattention. Recent twin studies suggest that the syndrome is highly heritable (Thapar *et al.*, 1995). The fact that symptoms improve on treatment with stimulants such as methylphenidate suggests the involvement of dopaminergic pathways in the brain. An association between attention deficit disorder and a polymorphism at the dopamine transporter gene has recently been reported (Thapar, 1998).

REFERENCES

American Psychiatric Association (1994) *Diagnostic and Statistical Manual of Psychiatric Disorders*, 4th edition, Washington, DC: American Psychiatric Association

Angst, J. (1966) *Zur Ätiologie und Nosologie Endogener Depressiver Psychosen*, Berlin and New York: Springer

Baron, M. (1997) Genetic linkage and bipolar affective disorder: progress and pitfalls. *Molecular Psychiatry* 2: 200–10

Bertelsen, A., Harvald, B. & Gauge, M. (1977) A Danish twin study of manic-depressive disorders. *British Journal of Psychiatry* 130: 330–51

Bohman, M., Sigvardsson, S. & Cloninger, C.R. (1981) Maternal inheritance of alcohol abuse: cross-fostering analysis of adopted women. *Archives of General Psychiatry* 38: 965–69

Brunner, H.G., Nelen, M.R., Van Zandvoort, P. *et al.* (1993) X-linked borderline mental retardation with prominent behavioural disturbance: phenotype, genetic localization and evidence for disturbed monoamine metabolism. *American Journal of Human Genetics* 52: 1032–39

Cadoret, R.J., Yates, W.R., Troughton, E., Woodworth, G. & Steward, M.A. (1995) Genetic–environmental interaction in the genesis of aggressivity and conduct disorders. *Archives of General Psychiatry* 52: 916–24

Cardno, A.I.G., Coid, B., MacDonald, A.M. *et al.* (1999) Heritability estimates for psychotic disorders: The Maudsley Twin Psychosis series. *Archives of General Psychiatry* 56: 162–68

Cardon, L.R., Smith, S.D., Fulker, D.W. *et al.* (1994) Quantitative trait locus for reading disability on chromosome 6. *Science* 265: 276–79

Cloninger, C.R. & Gottesman, I.I. (1987) Genetic and environmental factors in antisocial behavior disorders. In *The Causes of Crime: New Biological Approaches*, edited by S.A. Mednick, T.E. Moffitt & S.A. Stack, Cambridge and New York: Cambridge University Press

Crabb, J.C. & Belknap, J.K. (1992) Genetic approaches to drug dependence. *Trends in Pharmacological Sciences* 13: 212–19

Craddock, N.J. & Owen, M.J. (1996) Candidate gene association studies in psychiatric genetics: a SERTain future? *Molecular Psychiatry* 1: 434–36

Davies, J.L., Kawaguchi, Y., Bennett, S.T. *et al.* (1994) A genome-wide search for human type 1 diabetes susceptibility genes. *Nature* 371: 132–37

Edwards, J.H. (1960) The simulation of mendelism. *Acta Genetica* 10: 63–70

Falconer, D.S. (1965) The inheritance of liability to certain diseases, estimated from the incidence among relatives. *Annals of Human Genetics* 29: 51–76

Farmer, A.E., McGuffin, P., Gottesman, I.I. (1987) Twin concordance for DSM-III schizophrenia: scrutinizing the validity of the definition. *Archives of General Psychiatry* 44: 634–41

Folstein, S. & Rutter, M. (1977) Infantile autism: a genetic study of 21 twin pairs. *Journal of Child Psychology and Psychiatry* 18: 297–321

Goate, A.M., Chartier-Harlin, M.C., Mullan, M. *et al.* (1991) Segregation of a missense mutation in the amyloid precursor protein gene with familial Alzheimer's disease. *Nature* 349: 704–06

Gottesman, I.I. (1991) *Schizophrenia Genesis. Origins of Madness*, New York: Freeman

Gusella, J.F., Wexler, N.S., Conneally, P.M. & Naylor, S.L. (1983) A polymorphic DNA marker genetically linked to Huntington's disease. *Nature* 306: 234–38

Heath, A. (1993) What can we learn about the determinants of psychotherapy and substance abuse from studies of normal twins? In *Twins as a Tool of Behavioural Genetics*, edited by T.J. Bonehard & P. Propping, Chichester and New York: Wiley

Hebebrand, J.A. (1992) A critical appraisal of X-linked bipolar illness. Evidence for the assumed mode of inheritance is lacking. *British Journal of Psychiatry* 160: 7–11

Heston, A. (1966) Psychiatric disorders in foster home reared children of schizophrenic mothers. *British Journal of Psychiatry* 112: 819–25

Hodgkinson, S., Mullan, M. & Murray, R.M. (1991) The genetics of vulnerability to alcoholism. In *The New Genetics of Mental Illness*, edited by P. McGuffin & R. Murray, Oxford: Butterworth Heinemann

Holland, A.J., Hall, A. & Murray, R. (1984) Anorexia nervosa: evidence for a genetic basis. *Journal of Psychosomatic Research* 32: 561–71

Huntington's Disease Collaborative Group (1993) A novel gene containing a trinucleotide repeat that is expanded and unstable on Huntington's disease chromosomes. *Cell* 72: 971–83

Kendler, K.S., Heath, A.C., Martin, N.G. & Eaves, L.J. (1987) Symptoms of anxiety and symptoms of depression. *Archives of Surgery* 122: 451–57

Kendler, K.S., Neale, M.C., Kessler, R.C., Heath, A.C. & Eaves, L.J. (1993) The lifetime history of major depression in women: reliability of diagnosis and heritability. *Archives of General Psychiatry* 50: 863–70

Lesch, K.-P., Bengel, P. & Heils, A. (1997) Association of anxiety-related traits with a polymorphism in the serotonin transporter gene regulatory region. *Science* 274: 1527–31

Lyons, M.J., True, W.R., Eisen, S.A. *et al.* (1995) Differential heritability of adult and juvenile antisocial traits. *Archives of General Psychiatry* 52: 906–15

MacDonald, H., Rutter, M. & Rios, P. (1989) Cognitive and social abnormalities in the siblings of autistic and Down's syndrome probands. Paper given at the First World Congress on Psychiatric Genetics, Churchill College, Cambridge, 3–5 August

McGue, M. (1993) From proteins to cognitions: the behavioural genetics of alcoholism. In *Nature, Nurture and Psychology*, edited by R. Plomin & G.E. McClearn, Washington, DC: American Psychological Association

McGuffin, P. & Huckle, P. (1990) Simulation of Mendelism revisited: the recessive gene for attending Medical School. *American Journal of Human Genetics* 46: 994–99

McGuffin, P., Owen, M.J., O'Donovan, M.C., Thapar, A. & Gottesman, I.I., (1994) *Seminars in Psychiatric Genetics*, London: Gaskell

McGuffin, P., Owen, M.J. & Farmer, A.E. (1995) Genetic basis of schizophrenia. *Lancet* 346: 678–82

McGuffin, P., Katz, R., Watkin, S. & Rutherford, J. (1996) A hospital-based twin register of the heritability of DSM-IV unipolar depression. *Archives of General Psychiatry* 53: 129–36

Mednick, S.A., Gabrielli, J.F. & Hutchings, B. (1984) Genetic influences in criminal convictions: evidence from an adoption court. *Science* 224: 891–94

Morris, D.W., Robinson, L., Turic, D. *et al.* (2000) Family-based association mapping provides evidence for a gene for reading disability on chromosome 15q. *Human Molecular Genetics* 9(5): 843–48

Morton, N.E. (1982) *Outline of Genetic Epidemiology*, Basel and New York: Karger

Mullan, M., Houlden, H., Windelspecht, M. *et al.* (1992) A locus for familial early-onset Alzheimer's disease on the long arm of chromosome 14, proximal to the A1-antichymotrypsin gene. *Nature Genetics* 2: 340–42
Nurnberger, J.I., Blehar, M.C., Kauffman, C.A. *et al.* (1994) Diagnostic interview for genetic-studies – rationale, unique features and training. *Archives of General Psychiatry* 51: 849–59
Ogilvie, A.D., Battersby, S., Bubb, V.J. *et al.* (1996) Polymorphism in serotonin transporter gene associated with susceptibility to major depression. *Lancet* 347: 731–33
Onstad, S., Skre, I., Torgersen, S. *et al.* (1991) Twin concordance for DSM-III-R schizophrenia. *Acta Psychiatrica Scandinavica* 83: 395–401
Perris, C. (1966) A study of bipolar (manic depressive) and unipolar recurrent depressive psychoses. *Acta Psychiatrica et Neurologica Scandinavica* (Suppl. 42)
Philippe, A., Martinez, M., Guilloud-Bataille, M. *et al.* (1999) Genome-wide scan for autism sceptibility genes. Paris Autism Research International Sibpair Study. *Human Molecular Genetics* 8: 805–12
Plomin, R., Owen, M.J. & McGuffin, P. (1994) The genetic basis of complex human behaviours. *Science* 264: 1733–39
Reich, T., Clayton, P.J. & Winokur, G. (1969) Family history studies. V. The genetics of mania. *American Journal of Psychiatry* 125: 1358–69
Reich, T., James, J.W. & Morris, C.A. (1972) The use of multiple thresholds in determining the mode of transmission of semi-continuous traits. *Annals of Human Genetics* 36: 163–84
Riley, B.P. & McGuffin, P. (2000) Linkage and associated studies of schizophrenia. *American Journal of Medical Genetics* 97: 23–44
Rutherford, J., McGuffin, P., Katz, R.J. *et al.* (1993) Genetic influences on eating attitudes in a normal female twin population. *Psychological Medicine* 23: 425–36
Snell, R.G., MacMillan, J.C., Cheadle, J.P. *et al.* (1993) Relationship between trinucleotide repeat expansion and phenotypic variation in Huntington's disease. *Nature Genetics* 4: 394–97
Thapar, A. (1998) Attention deficit hyperactivity disorder: unravelling the molecular genetics. *Molecular Psychiatry* 3/5: 370–72
Thapar, A. & McGuffin, P. (1996) A twin study of antisocial and neurotic symptoms in childhood. *Psychological Medicine* 26: 1111–18
Thapar, A., Hervas, A. & McGuffin, P. (1995) Childhood hyperactivity scores are highly heritable and show sibling competition effects: twin study evidence. *Behavior Genetics* 25: 537–44
Tsuang, M.T. & Faraone, S.B. (1990) *The Genetics of Mood Disorders*, Baltimore, Maryland: Johns Hopkins University Press
Uhl, G.R., Persico, A.M. & Smith, S.S. (1992) Current excitement with D2 dopamine receptor genes alleles in substance abuse. *Archives of General Psychiatry* 49: 157–60
Williams, J., McGuffin, P., Nothen, M., Owen, M.J. & The EMASS Collaborative Group (1997) Meta-analysis of association between the 5-HT_{2a} receptor T102C polymorphism and schizophrenia. *Lancet* 349: 1221
Wing, J.K., Babor, T., Brughha, T. *et al.* (1990) SCAN: Schedules for Clinical Assessment in Neuropsychiatry. *Archives of General Psychiatry* 47: 589–93
Witkin, H.A., Mednick, S.A. & Schulsinger, F. (1976) Criminality in XYY and XXY men. *Science* 193: 547–55
World Health Organization (1993) *The ICD 10 Classification of Mental and Behavioural Disorders: Diagnostic Criteria for Research*, Geneva: World Health Organization

GLOSSARY

aetiology the cause of a disease
bipolar disorder major psychological disorder characterized by episodes of mania, depression or mixed mood
cosegregation where two genotypes are inherited together
dizygotic (DZ) developing from separately fertilized eggs, as in non-identical twins
monozygotic (MZ) developing from a single fertilized egg, as in identical twins
neuropathology study of the characteristics, causes and effects of neurological disease
pathogenesis source/cause of an illness or abnormal condition
trisomy/trisomic state where there is one chromosome additional to the normal diploid complement; in somatic nuclei one particular chromosome is represented three times, rather than twice
unipolar disorder major psychological disorder characterized by persistent depressive mood, loss of interest, self-reproach, suicidal ideas and appetite and sleep disturbances

See also **Trinucleotide repeat stability as a cause of human genetic disease** (p.396); **Genetics of animal and human prion diseases** (p.451); **Genetics of Huntington's disease** (p.459); **Twinning** (p.516)

Encyclopaedia of
GENETICS

Volume 2

Encyclopaedia of

GENETICS

Volume 2

Encyclopaedia of GENETICS

Volume 2

Editor

Eric C.R. Reeve

Assistant Editor

Isobel Black

PENTAGON PRESS

Encyclopedia of GENETICS / Eric C.R. Reeve (Editor)

ISBN 978-81-8274-241-2 (Set)

2010

Published with arrangement FITZROY DEARBORN PUBLISHERS, USA.

Published by

PENTAGON PRESS
206, Peacock Lane, Shahpur Jat, New Delhi-110049
Phones: 011-64706243, 26491568
Telefax: 011-26490600
email: rajan@pentagonpress.in
website: www.pentagonpress.in

Printed at
New Elegant Printers, New Delhi-110064

CONTENTS

GENETICS OF ANIMAL AND HUMAN PRION DISEASES

Nora Hunter
Institute for Animal Health, Neuropathogenesis Unit, University of Edinburgh, Edinburgh, UK

1. Introduction

Transmissible spongiform encephalopathies (TSEs) are slowly progressive, inevitably fatal, neurodegenerative disorders of mammals characterized by vacuolated brain neurones and the deposition of an abnormal form of a host protein, PrP (**Table 1**). Most TSEs have also been shown to be experimentally transmissible. There is a strong genetic component in the patterns of disease incidence of scrapie in sheep and of some forms of human TSE, and there is overwhelming evidence that the genetic component is the PrP gene. Although there are different strains of TSEs (Bruce & Dickinson, 1987), no infectious agent has been characterized. There are many hypotheses on the aetiology of the disease, partly because of its unusual physicochemical resistances (e.g. Brown, P. *et al.*, 1990), which have led to suggestions that the infectious agent may be made up mostly or completely of protein. A disease which is both the result of, and transmitted by, such a mutant protein has been variously described as a prion disease (Prusiner *et al.*, 1982), an infectious amyloidosis (Brown, P. *et al.*, 1991) or simply a genetic mutation (Parry, 1960; Brown, P. *et al.*, 1991; Collinge & Palmer, 1994). Others, citing evidence that scrapie carries its own genetic information, believe it to be a virus (Diringer, 1995; Manuelidis *et al.*, 1995) or a virino – mostly protein, but with a small additional information-carrying component (Somerville, 1991). Many years of effort have gone into the search for the DNA or RNA believed to form part of such structures, so far without success. What is universally accepted, however, is the central importance of prion protein (PrP) in the control of susceptibility to TSEs and in the resultant pathology. Depending on which hypothesis is supported, an abnormal isoform of PrP (PrP^{SC}) either is itself the infectious agent (prion) or forms a vital part of the infection process, perhaps as the coat of the virino. Despite what many textbooks say, the nature of the agent is not yet finally resolved and, indeed, a full explanation of the characteristics of TSEs presents problems whichever theory is being proposed.

Table 1 TSEs of animals and humans.

scrapie	sheep and goats
bovine spongiform encephalopathy (BSE)	cattle
feline spongiform encephalopathy (FSE)	cats
chronic wasting disease (CWD)	mule deer
transmissible mink encephalopathy (TME)	mink
Creutzfeldt-Jakob disease (CJD)	humans
Gerstmann-Sträussler-Scheinke syndrome (GSS)	humans
fatal familial insomnia (FFI)	humans
variant CJD (vCJD)	humans

2. PrP protein

PrP consists of approximately 250 amino acids (exact length depends on the species), is glycosylated at either one, or both, of two possible glycosylation sites, and is attached to the outside of the neuronal cell membrane by a GPI anchor (Hope, 1993). The normal form of PrP is denoted as PrP^{C} and, during TSE development, PrP^{C} changes its tertiary structure to form the disease-associated PrP^{SC} (**Table 2**) and scrapie-associated fibrils (SAF), both hallmarks of all TSEs. In addition, variant sequences or allotypes of the protein are associated with differences in the incubation period of experimental scrapie in laboratory mice (Carlson *et al.*, 1986) and in sheep (Goldmann *et al.*, 1991a) and with the incidence of natural disease in sheep and humans. Since the discovery of PrP^{C} and PrP^{SC}, it has been found that the aggregated form of the protein is so closely associated with infection, whether or not it is itself the infectious agent, that it can be used as a marker for infectivity.

3. Strains of TSEs

Many strains of scrapie have been described which have been passaged in laboratory mice and have precise incubation periods and brain region pathology in affected animals of different mouse lines (Dickinson & Meikle, 1971; Bruce *et al.*, 1991). These strain characteristics have been most recently used to great effect in the identification of bovine spongiform encephalopathy (BSE)-like infections of mammals other than cattle (Bruce *et al.*, 1994), including variant Creutzfeldt–Jakob disease (vCJD) in young humans (Bruce *et al.*, 1997). vCJD transmits very easily to mice, producing a BSE-like pattern of incubation periods and pathology and is clearly different from the normal forms of CJD (sporadic CJD), since sporadic CJD, including that occurring in farmers who had had cases of BSE in their cattle herds, has dramatically different transmission characteristics and does not transmit well to mice at all (Bruce *et al.*,

Table 2 Normal and abnormal PrP.

	PrP^{C}	PrP^{SC}
PK	sensitive	partially resistant
Detergent	soluble	insoluble
Molecular wt (–PK)	33–35 kDa	33–35 kDa
Molecular wt (+PK)	degraded	27–30 kDa
Location	cell surface GPI anchor	brain, CNS, lymph nodes, spleen, tonsil
Turnover	rapid	slow
Infectivity	does not copurify	copurifies

PK, proteinase K; GPI, glycophosphoinositol; CNS, central nervous system.

1997). In addition, there are also indications from sheep studies that different strains of natural scrapie may exist (Hunter *et al.*, 1997a). Strains of TSEs are explained, depending on the hypothesis adopted, in terms of the necessity for a scrapie-specific informational molecule in addition to PrP or as the result of distinct and transmissible tertiary structures of PrP^{SC}.

4. TSE genetics in rodent models

TSEs are usually studied in laboratory rodents where a great deal of work on genetics, particularly in transgenic lines, has been carried out. The earliest demonstration of genetic control of susceptibility came from the work of Alan Dickinson and colleagues (Dickinson *et al.*, 1968) who described the gene *Sinc*, now also known as *Prni*, as the major gene controlling the incubation period of scrapie in mice. The incubation period is the time between injection with TSE infectivity and the appearance of clinical signs or death with TSE-like brain pathology. The first clues to the product of the *Sinc/Prni* gene came from the finding of restriction fragment length polymorphism (RFLP) linkages of both genes to the mouse PrP protein-encoding gene *Prnp* (Carlson *et al.*, 1986; Hunter *et al.*, 1987). PrP allotypes differing at amino acids 108 and 189 were also linked to incubation time differences (Westaway *et al.*, 1987; Hunter *et al.*, 1992) in that mouse lines of distinct *Sinc/Prni* genotype also had either leucine and threonine at positions 108 and 189, respectively (allotype A: $L_{108}T_{189}$) or had phenylalanine and valine (allotype B: $F_{108}V_{189}$). The formal proof of identity of *Sinc/Prni* and *Prnp* has come recently with an elegant experiment involving targeted changes in the endogenous PrP gene (*Prnp*) in mice using embryonic stem cell technology. Mice which were of PrP A/A homozygous allotype ($LL_{108}TT_{189}$) were altered by gene-targeted mutagenesis such that amino acids 108 and 189 were changed to the PrP B/B allotype ($FF_{108}VV_{189}$). In all other respects, the mice remained genetically A/A. These two single amino acid changes altered the incubation time characteristics of the mouse lines dramatically (**Table 3**), such that an expected incubation period of the 301V TSE strain of 244 days in the parent A/A ($LL_{108}TT_{189}$) mice was reduced to 133 days in the transgenic A/A ($FF_{108}VV_{189}$) mice, almost the same as in the control B/B ($FF_{108}VV_{189}$) line (119 days) (Moore *et al.*, 1998). The inevitable conclusion is that PrP genotype controls incubation period.

Transgenic studies have shown that host range is also controlled by the PrP gene. In the first such experiment, the hamster PrP gene (HaPrP) was inserted into the mouse genome and changed the response of the mouse to a hamster-passaged strain of scrapie in a copy-number dependent way (Prusiner *et al.*, 1990). Transgenic mouse lines became susceptible to the hamster scrapie agent with incubation periods of 75–179 days, the length of which was related inversely to the level of HaPrP mRNA and protein. Non-transgenic mice were much more resistant to hamster scrapie, with only a few succumbing more than 400 days after inoculation. The strain of scrapie pathogen inoculated into the transgenic mice was replicated and reproduced, i.e. injection with the hamster scrapie agent resulted in high titres of the hamster scrapie agent, while a mouse-passaged scrapie strain produced high titres of the mouse scrapie agent. Pathology typical of hamster scrapie disease was found in the brain tissue from dying transgenic mice and included large deposits or plaques of PrP protein that reacted with HaPrP-specific antibodies. Such studies have given support to the idea that a "match" between incoming PrP^{SC} in the inoculum (from the individual which is the source of the infection) and the PrP^{C} in the infected individual produces the shortest incubation period or survival time. The "match" would in this case, therefore, be between the hamster PrP^{SC} in the inoculum and the hamster PrP^{C} being produced in the transgenic mouse. Subsequently, the incoming $HaPrP^{SC}$ would convert the transgenic mouse $HaPrP^{C}$ into more $HaPrP^{SC}$, differing only in conformation and in the acquisition of the ability to convert the "healthy" PrP isoform into the "diseased" isoform. The region of the PrP molecule believed to be involved in the PrP^{SC}/PrP^{C} interaction is at the amino terminus end of the protein.

Subsequently, many lines of transgenic mice expressing various forms of the PrP gene have been created; however, some of the most interesting animals are those in which the PrP gene has been ablated or rendered inactive. These PrP-null (or $PrP^{0/0}$) mice remain apparently healthy and breed normally and although they have more subtle abnormalities in phenotype – impaired electrophysiology of neurones (Collinge *et al.*, 1994; Manson *et al.*, 1995), lack of sleep continuity (Tobler *et al.*, 1997) and deficits in copper binding to their brain cell membranes (Brown, D.R. *et al.*, 1997) – they have not been able to reveal the normal function of the PrP^{C} protein. PrP-null mice are, however, resistant to scrapie (Bueler *et al.*, 1993) and the hemizygotes with only one functioning PrP allele have greatly extended incubation periods (Bueler *et al.*, 1994).

The crucial importance of expression of the PrP gene in order to allow disease to occur (Bueler *et al.*, 1993) has many parallels in conventional virology. For example, expression of the human poliovirus receptor in transgenic mice rendered the animals susceptible to poliovirus to which they are normally resistant (Ren *et al.*, 1990) and a mediator (herpes virus entry mediator, HVEM) of the entry of human herpes simplex virus into human cells allowed entry of the virus into previously resistant hamster and swine cells (Montgomery *et al.*, 1996). These similarities do not necessarily mean that PrP is a viral receptor, simply that perhaps a similar recognition and "fit" is required before infection is established. Indeed, the prion hypothesis itself involves the idea that PrP^{SC} in the inoculum would act as a seed, encouraging PrP^{C} in the new host to adopt

Table 3 301V incubation times.

Mouse line	PrP alleles	IP (days)
129/Ola	A/A (LT/LT)*	244
Vm/Dk	B/B (FV/FV)	119
TG: 129/Ola	A/A (FV/FV)	133

*108 189 allotype; IP, incubation period. Adapted from Moore *et al.*, 1998.

the PrP^{SC} conformation and to aggregate. This seeding activity would work most quickly in "matching" combinations of primary sequence of PrP proteins, giving shorter incubation periods. It may be that the rate of PrP to PrP^{SC} conversion (which can be persuaded to occur *in vitro*; Caughey *et al.*, 1995) is allele dependent (Raymond *et al.*, 1997) and that the more PrP^{SC} aggregating in the brain of an affected animal, the more likely that symptoms would occur. Indeed, PrP^{SC} is believed to be toxic to neurones (Hope *et al.*, 1996).

The resistance of PrP-null mice to scrapie has facilitated a series of experiments involving the grafting of tissue expressing wild-type PrP into the brains of PrP-null mice and attempting by various means to infect the grafted tissue. Stereotactic injection into the graft produces infection of the graft and heavy deposition of PrP^{SC}; however, the infection does not spread and the animal survives. Peripheral infection of the grafted PrP-null mice does not result in the infection being transported to the PrP-expressing graft, which remains healthy. The conclusion from these studies is that PrP expression is also required for the transport of infectivity throughout the body from the site of infection (Blattler *et al.*, 1997). Although spread of infectivity along peripheral nerves into the central nervous system (CNS) has been demonstrated (Kimberlin & Walker, 1980), other studies also indicate the involvement of cells of the lymphoreticular system (LRS). Severe combined immunodeficient (SCID) mice, deficient in T and B cells, resist peripheral infection with low doses of scrapie and BSE (Brown, K.L. *et al.*, 1997a). It is known, for example, that whole-body irradiation does not prevent scrapie development (Fraser & Farquhar, 1985) and, because of such studies, follicular dendritic cells (FDCs), which also stain positively with antibodies for PrP (McBride *et al.*, 1992), have been implicated as supporting scrapie replication in the periphery. SCID mice have only immature FDCs and reconstituting them with wild-type bone marrow causes the FDCs to mature and restores the scrapie susceptibility to the mice (Brown, K.L. *et al.*, 1997b). PrP-null mice reconstituted with wild-type bone marrow can sustain infection in peripheral tissues, but do not usually transport infectivity to the CNS (Blattler *et al.*, 1997).

In contrast to this work, recent studies with immunological knockout mice have implied that it is not the FDCs, but, in fact, B cells which are important in transport of infectivity in the early stages of the disease (Klein *et al.*, 1997); however, the two laboratories involved have used quite different strains of scrapie and are currently investigating the FDC/B cell controversy as another potential scrapie strain difference. Scrapie strains target different areas in the CNS, causing characteristic differences in the regions of the brain which become vacuolated (Bruce *et al.*, 1991); it is, therefore, not beyond the bounds of possibility that different strains might also target different cells in the LRS.

5. TSE genetics in sheep and goats

Studies of natural scrapie in sheep have confirmed the importance of three amino acid codons in the sheep PrP gene (136, 154 and 171) (Belt *et al.*, 1995; Clouscard *et al.*, 1995; Hunter *et al.*, 1996) originally shown to be associated with differing incubation periods following the experimental challenge of sheep with different sources of scrapie and BSE (Goldmann *et al.*, 1991a, 1994). A diagram of the sheep PrP gene structure (similar in all species) is shown in **Figure 1**. Genotypes herein are presented using the single letter amino acid code for each codon in turn with the codon number in subscript (V = valine, A = alanine, R = arginine, Q = glutamine). Alleles have one letter per codon (e.g. $V_{136}R_{154}Q_{171}$) and genotypes have two letters per codon (e.g. homozygote $VV_{136}RR_{154}QQ_{171}$ or heterozygote $VA_{136}RR_{154}QR_{171}$) indicating each allele which makes up the genotype. Although there are breed differences in PrP allele frequencies and in disease-associated alleles, some clear genetic rules have emerged. The most resistant genotype is $AA_{136}RR_{154}RR_{171}$. Out of hundreds of scrapie-affected sheep worldwide, only one animal of this genotype has been reported with scrapie – a Japanese Suffolk sheep (Ikeda *et al.*, 1995). This

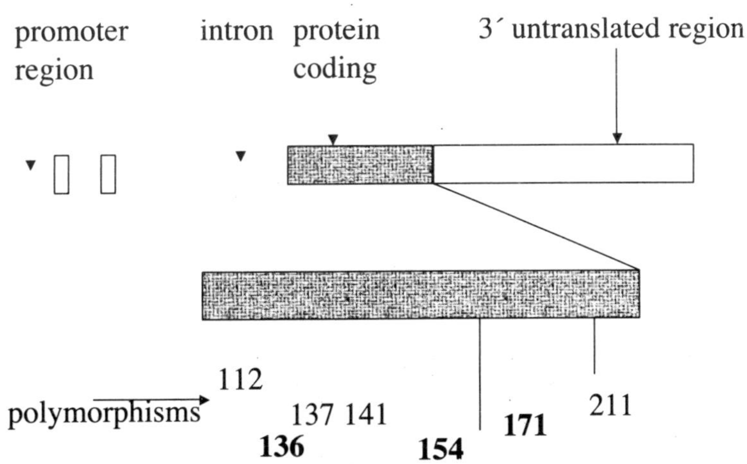

Figure 1 Sheep PrP gene structure. Codon 136 is either valine or alanine (V_{136} or A_{136}); codon 154 is arginine or histidine (R_{154} or H_{154}); and codon 171 is arginine, glutamine or histidine (R_{171}, Q_{171} or H_{171}).

genotype is also resistant to experimental challenge with both scrapie and BSE (Goldmann *et al.*, 1994). On the other hand, genotypes which encode QQ_{171} are more susceptible to scrapie. For example, in Suffolk sheep, the genotype $AA_{136}RR_{154}QQ_{171}$ is most susceptible, although not all animals of this genotype succumb to disease and it is quite a common genotype among healthy animals (Westaway *et al.*, 1994a; Hunter *et al.*, 1997b). The PrP genetic variation in Suffolk sheep is much less than that in some other breeds, the so-called "valine breeds". Breeds such as Cheviots, Swaledales and Shetlands encode PrP gene alleles with valine at codon 136 and sheep with the genotype $VV_{136}RR_{154}QQ_{171}$ appear to be exquisitely susceptible to scrapie (Hunter *et al.*, 1994a, 1996). $VV_{136}RR_{154}QQ_{171}$ is a rare genotype and, when it does occur, is almost always in scrapie-affected sheep and, as a result, it has been suggested that scrapie may simply be a genetic disease (Ridley & Baker, 1995). However, healthy animals of this genotype can live up to 8 years of age, well past the usual age-at-death from scrapie (2–4 years; Hunter *et al.*, 1996), and susceptible sheep genotypes are easily found in countries that are free of scrapie disease (Hunter *et al.*, 1997c). The genetic disease hypothesis seems less likely than an aetiology which involves host genetic control of susceptibility to an infecting agent.

In some "valine breed" sheep flocks affected by scrapie, there is a survival advantage if genotypes encode certain PrP alleles, such as $A_{136}H_{154}Q_{171}$ and $A_{136}R_{154}R_{171}$, so that, despite having a high-risk allele such as $V_{136}R_{154}Q_{171}$, animals are unlikely to develop scrapie if their genotypes are $VA_{136}HR_{154}QQ_{171}$ or $VA_{136}RR_{154}QR_{171}$ (Hunter *et al.*, 1996). This has not been found to be the case in all outbreaks, however.

There are some intriguing differences between flocks and between breeds. Why, for example, is the genotype which is most susceptible in Suffolks ($AA_{136}RR_{154}QQ_{171}$) usually found to be resistant to natural scrapie in Cheviots, a "valine breed"? In addition, why do some flocks of "valine breeds" show scrapie only in animals encoding the $V_{136}R_{154}Q_{171}$ allele (homozygotes and heterozygotes) whereas, in other flocks, despite the existence of $V_{136}R_{154}Q_{171}$-encoding sheep, scrapie appears to target only QQ_{171} genotypes and occurs in both $VA_{136}RR_{154}QQ_{171}$ and $AA_{136}RR_{154}QQ_{171}$ sheep (Clouscard *et al.*, 1995)? The answer could come from studies of experimental TSE in sheep where different sources of scrapie and BSE apparently target sheep according to their genotype at either codon 136 or codon 171. Following challenge by injection, the scrapie source SSBP/1 affects Cheviot sheep encoding the $V_{136}R_{154}Q_{171}$ allele, whereas the scrapie source CH1641 and BSE target sheep primarily according to codon 171 genotype, producing disease with the shortest incubation period in sheep which are QQ_{171} (Goldmann *et al.*, 1994).

Extending these findings to the naturally affected sheep, it is possible that there are also various types or strains of natural scrapie which target either particular sheep breeds and/or different PrP codons. The best way to investigate this at the moment is by the passage of natural scrapie sources into a panel of mouse strains where the characteristics of incubation periods, and the precise brain areas which become vacuolated, give distinct profiles or patterns (Bruce *et al.*, 1994). However, it is also possible that scrapie strains produce PrP^{SC} protein with distinct patterns on Western blots, and this possibility is currently under investigation as a strain-typing method in several laboratories throughout the world.

6. BSE in cattle

When BSE was found in cattle, the cattle PrP gene was searched for markers of resistance or susceptibility to disease similar to those which had been found in mice, sheep and humans (see below). The cattle gene, however, is remarkably invariant compared both with the sheep and with the human PrP genes. There has been one major polymorphism described – that of a difference in the numbers of an octapeptide repeat within the PrP protein (Goldmann *et al.*, 1991b). In humans, many variations in the octapeptide repeat number have been described, in a range of 4–14, some of which have clear linkage to the incidence of human TSE (Poulter *et al.*, 1992). In cattle, the most common number of octapeptide repeats is six and three genotypes have been described as 5:5, 6:6 and, with the heterozygote, 6:5. In a study which compared the frequencies of these genotypes in BSE-affected and healthy cattle, there was no difference in frequency between the two groups, with around 90% of animals being 6:6 and 10%, 6:5; the genotype 5:5 was rare (<1%) and was not found in the BSE-affected animals (Hunter *et al.*, 1994b). It is possible that the 5:5 genotype is associated with resistance, because all BSE cattle had at least one copy of the six-repeat allele, but this remains to be established by direct experimental challenge of either 5:5 cattle or transgenic mice expressing bovine PrP gene alleles.

Cattle appear to be unusual in not (so far) demonstrating a PrP-related link with TSE incidence. BSE challenge does show such differences in other species, including mice (Bruce *et al.*, 1994), sheep (Goldmann *et al.*, 1994) and goats (Goldmann *et al.*, 1996). It may be that all cattle would be susceptible to BSE if they received a high enough dose of infection or it may be that a polymorphism in the gene region controlling expression of the gene could provide such a link and, indeed, cattle family studies indicate that this is a possibility (Neibergs *et al.*, 1994).

7. Genetics of human TSEs

Each species in which genetic linkage has been demonstrated between PrP genotype and disease has been unique in precisely which amino acids are disease linked, and humans are no exception (**Table 4**). However, sporadic forms of CJD in humans are not linked to any mutations in the PrP gene and, instead, a codon 129 polymorphism (methionine/valine) is associated with differences in susceptibility to disease in that homozygous individuals (either MM_{129} or VV_{129}) are over-represented in CJD cases and heterozygosity seems to confer some protection (Palmer *et al.*, 1991). For example, at the time of writing, all reported vCJD cases are MM_{129} genotype (Ironside, 1999).

Other forms of TSEs in humans appear to be genetic

Table 4 Human PrP gene mutations.

CJD	No mutation + 129 Met/Val
fCJD	180, 200, 210, 219, 232 Octapeptide repeats (codons 51–91) (48 bp, 96 bp,120 bp,144 bp,168 bp, 192 bp, 216 bp)
GSS	102, 105, 117, 145, 198, 217
FFI	178 Asp/Asn 129 Met/Met

Diseases are named in full in Table 1 except fCJD which is familial CJD.

diseases, for example GSS (Gerstmann-Sträussler-Scheinker syndrome) which is linked to a codon 102 proline to leucine mutation (Hsiao *et al.*, 1989). This mutation, when introduced into the mouse PrP gene in transgenic mice (the equivalent mouse codon is 101) and expressed at extremely high levels, resulted in a spontaneous scrapie-like disease which, although no PrP^{SC} was detectable by standard methods, transmitted infection to hamsters and other transgenic mice and not to normal mice (Hsiao *et al.*, 1994). This experiment supports the "protein only" hypothesis, because apparently the only requirement for disease to develop is a single amino acid mutation. However, the interpretation of the results has been disputed on the grounds of the lack of PrP^{SC}, the odd transmission characteristics and the high levels of expression needed to see the effect – single copy transgenes do not make the mice sick (Chesebro, 1998). Mice genetically engineered to have only a single copy of the codon 101 proline to leucine change, show no spontaneous disease but instead show altered strain susceptibility patterns (Manson *et al.*, 1999) and it is known that overexpression of normal PrP genes can result in illness in transgenic mice (Westaway *et al.*, 1994b).

There are several other human PrP gene mutations associated with disease (Prusiner, 1998); one form of familial CJD, for example, is linked to an insert of 144 base pairs coding for six extra octapeptide repeats at codon 53 (Poulter *et al.*, 1992) and a codon 200 mutation (glutamic acid to lysine) is linked to CJD in Israeli Jews of Libyan origin, Slovaks in north-central Slovakia, a family in Chile and a German family in the US (Prusiner & Scott, 1997). One of the most intriguing human prion diseases is fatal familial insomnia, which is apparently linked to a PrP gene mutation at codon 178 (Gambetti *et al.*, 1992) but can be modified in phenotype by the codon 129 polymorphism (Medori & Tritschler, 1993).

Various forms of human PrP protein expressed by naturally occurring mutant genes have been studied *in vitro* in Chinese hamster ovary cells and have been found both to be abnormally processed (e.g. not appearing on the cell membrane) and to acquire characteristics of the disease-associated PrP^{SC} isoform. Wild-type PrP protein is processed normally and remains in the PrP^{C} isoform (Lehmann & Harris, 1996; Lehmann *et al.*, 1997). This suggests that mutations in the PrP gene may cause illness directly through loss of function of the PrP protein by misprocessing or that the mutant protein forms deposits and poisons the surrounding cells.

In attempts to develop transgenic mice which express human PrP genes and which might act as good bioassays for human TSE infectivity, it has been found that the presence of the endogenous mouse PrP gene has an inhibitory effect, apparently preventing human TSE infectivity from replicating in the transgenic mice. The crossing of the human PrP transgenic mice with PrP-null mice to produce animals which only express the human PrP gene was shown to "work" efficiently as the offspring of such crosses became fully susceptible to human TSE (Kaneko *et al.*, 1997). These results have been taken to imply that a second factor – Protein X – is required for successful infection and that the endogenous mouse Protein X has more affinity for the mouse PrP protein than the human PrP produced from the transgene. Once the mouse PrP gene has been bred out of the line, mouse Protein X will then interact with the human PrP protein and allow infection to proceed (Kaneko *et al.*, 1997). Protein X is thought to interact with the carboxy terminal region of the PrP protein molecule.

8. Conclusions

Amino acids throughout the PrP molecule have been shown to be associated with or linked to disease in different species. An understanding of the presumably common mechanisms involved may come from studies of the tertiary structure of both PrP^{C} and PrP^{SC} in these species, but, for the moment, it is clear that PrP genotype and expression are of crucial importance in TSE development and transmission. Transmission seems to work more efficiently if there is a match between the primary amino acid sequence of the PrP protein from the mammal which is the source of the infection and the mammal which becomes infected. This matching process may also depend on glycosylation of the molecules (Hill *et al.*, 1997). However, many still believe that, interesting and relevant as these ideas are, an additional factor other than PrP conformation is required for disease. For example, in transmission studies, BSE caused disease in C57BL/6 mice without, in around 55% of cases, producing the disease-related PrP^{SC} (Lasmezas *et al.*, 1997). The conclusion from this work was that, although PrP was of major importance, some further unidentified agent may be involved – perhaps Protein X of the prion hypothesis (Kaneko *et al.*, 1997) or the informational molecule of the virino hypothesis (Farquhar *et al.*, 1998).

Whatever the nature of the TSE agents, their modes of transmission are not yet understood. It is not known, for example, whether apparently resistant sheep or people are truly resistant and able to fight off infection or whether they merely have a longer-than-lifespan incubation period and harbour infectivity. If the latter idea is correct, these individuals could act as sources of infection (vectors) for others who are more genetically susceptible. The idea of hidden, yet still transmissible, infection is not without precedent – examples include chronic infections of neonates by rubella, inapparent polio infection (90% of infections produce no symptoms) and the slowly progressive measles infection which, rarely, can pass to the brain causing degeneration after an apparently healthy gap of several years (Dimmock & Primrose, 1994). If TSEs behave in a similar fashion, symptomless carriers could be the means by which susceptible individuals acquire infection.

REFERENCES

Belt, P.B.G.M., Muileman, I.H., Schreuder, B.E.C. *et al.* (1995) Identification of five allelic variants of the sheep PrP gene and their association with natural scrapie. *Journal of General Virology* 76: 509–17

Blattler, T., Brandner, S., Raeber, A.J. *et al.* (1997) PrP-expressing tissue required for transfer of scrapie infectivity from spleen to brain. *Nature* 389: 69–73

Brown, D.R., Qin, K., Herms, J.W. *et al.* (1997) The cellular prion protein binds copper *in vivo*. *Nature* 390: 684–87

Brown, K.L., Stewart, K., Bruce, M.E. & Fraser, H. (1997a) Severe combined immunodeficient (SCID) mice resist infection with bovine spongiform encephalopathy. *Journal of General Virology* 78: 2707–10

Brown, K.L., Stewart, K., Bruce, M.E. & Fraser, H. (1997b) Scrapie in immunodeficient mice. *Biochemical Society Transactions* 25: S173

Brown, P., Liberski, P., Wolff, A. & Gajdusek, D. (1990) Resistance of scrapie infectivity to steam autoclaving after formaldehyde fixation and limited survival after ashing at 360°C: practical and theoretical implications. *Journal of Infectious Diseases* 161: 467–72

Brown, P., Goldfarb, L. & Gajdusek, D. (1991) The new biology of spongiform encephalopathy: infectious amyloidoses with a genetic twist. *Lancet* 337: 1019–22

Bruce, M.E. & Dickinson, A.G. (1987) Biological evidence that scrapie agent has an independent genome. *Journal of General Virology* 68: 79–89

Bruce, M.E., McConnell, I., Fraser, H. & Dickinson, A.G. (1991) The disease characteristics of different strains of scrapie in Sinc congenic mouse lines: implications for the nature of the agent and host control of pathogenesis. *Journal of General Virology* 72: 595–603

Bruce, M., Chree, A., McConnell, I. *et al.* (1994) Transmission of bovine spongiform encephalopathy and scrapie to mice – strain variation and the species barrier. *Philosophical Transactions of the Royal Society of London Series B* 343: 405–11

Bruce, M.E., Will, R.G., Ironside, J.W. *et al.* (1997) Transmissions to mice indicate that 'new variant' CJD is caused by the BSE agent. *Nature* 389: 488–501

Bueler, H., Aguzzi, A., Sailer, A. *et al.* (1993) Mice devoid of PrP are resistant to scrapie. *Cell* 73: 1339–47

Bueler, H., Raeber, A., Sailer, A. *et al.* (1994) High prion and PrPsc levels but delayed-onset of disease in scrapie-inoculated mice heterozygous for a disrupted PrP gene. *Molecular Medicine* 1: 19–30

Carlson, G.A., Kingsbury, D.T., Goodman, P.A. *et al.* (1986) Linkage of prion protein and scrapie incubation time genes. *Cell* 46: 503–11

Caughey, B., Kocisko, D.A., Raymond, G.J. & Lansbury, P.T. (1995) Aggregates of scrapie-associated prion protein induce the cell-free conversion of protease-sensitive prion protein to the protease-resistant state. *Chemistry and Biology* 2: 807–17

Chesebro, B. (1998) BSE and prions: uncertainties about the agent. *Science* 279: 42–43

Clouscard, C., Beaudry, P., Elsen, J.M. *et al.* (1995) Different allelic effects of the codons 136 and 171 of the prion protein gene in sheep with natural scrapie. *Journal of General Virology* 76: 2097–101

Collinge, J. & Palmer, M.S. (1994) Human prion diseases. *Baillieres Clinical Neurology* 3: 241–55

Collinge, J., Whittington, M.A., Sidle, K.C.L. *et al.* (1994) Prion protein is necessary for normal synaptic function. *Nature* 370: 295–97

Dickinson, A.G. & Meikle, V.M. (1971) Host-genotype and agent effects in scrapie incubation: change in allelic interaction with different strains of agent. *Molecular and General Genetics* 112: 73–79

Dickinson, A.G., Meikle, V.M. & Fraser, H. (1968) Identification of a gene which controls the incubation period of some strains of scrapie agent in mice. *Journal of Comparative Pathology* 78: 293–99

Dimmock, S. & Primrose, N. (1994) *Introduction to Modern Virology*, 4th edition, Oxford and Boston: Blackwell Science

Diringer, H. (1995) Proposed link between transmissible spongiform encephalopathies of man and animals. *Lancet* 346: 1208–10

Farquhar, C., Somerville, R. & Bruce, M. (1998) Straining the prion hypothesis. *Nature* 391: 345–46

Fraser, H. & Farquhar, C.F. (1985) Are mitotically dormant, radiation insensitive cells involved in the peripheral pathogenesis of scrapie? *Neuropathology and Applied Neurobiology* 11: 71

Gambetti, P., Medori, R., Tritschler, H.J. *et al.* (1992) Fatal familial insomnia (FFI) – a prion disease with a mutation at codon 178 of the prion protein gene. *Journal of Neuropathology and Experimental Neurology* 51: 353

Goldmann, W., Hunter, N., Benson, G., Foster, J.D. & Hope, J. (1991a) Different scrapie-associated fibril proteins (PrP) are encoded by lines of sheep selected for different alleles of the Sip gene. *Journal of General Virology* 72: 2411–17

Goldmann, W., Hunter, N., Martin, T., Dawson, M. & Hope, J. (1991b) Different forms of the bovine PrP gene have five or six copies of a short, G-C-rich element within the protein coding exon. *Journal of General Virology* 72: 201–04

Goldmann, W., Hunter, N., Smith, G., Foster, J. & Hope, J. (1994) PrP genotype and agent effects in scrapie – change in allelic interaction with different isolates of agent in sheep, a natural host of scrapie. *Journal of General Virology* 75: 989–95

Goldmann, W., Martin, T., Foster, J. *et al.* (1996) Novel polymorphisms in the caprine PrP gene: a codon 142 mutation associated with scrapie incubation period. *Journal of General Virology* 77: 2885–91

Hill, A.F., Desbruslais, M., Joiner, S. *et al.* (1997) The same prion strain causes vCJD and BSE. *Nature* 389: 448–50

Hope, J. (1993) The biology and molecular biology of scrapie-like diseases. *Archives of Virology* 7: 201–14

Hope, J., Shearman, M., Baxter, H. *et al.* (1996) Cytotoxicity of prion protein peptide (PrP106–126) differs in mechanism from the cytotoxic activity of the Alzheimer's disease amyloid peptide, AB25–35. *Neurodegeneration* 5: 1–11

Hsiao, K., Baker, H.F., Crow, T.J. *et al.* (1989) Linkage of a prion protein missense variant to Gerstmann–Straussler syndrome. *Nature* 338: 342–45

Hsiao, K.K., Groth, D., Scott, M. *et al.* (1994) Serial transmission in rodents of neurodegeneration from transgenic mice expressing mutant prion protein. *Proceedings of the National Academy of Sciences USA* 91: 9126–30

Hunter, N., Hope, J., McConnell, I. & Dickinson, A.G. (1987)

Linkage of the scrapie-associated fibril protein (PrP) gene and Sinc using congenic mice and restriction fragment length polymorphism analysis. *Journal of General Virology* 68: 2711–16

Hunter, N., Dann, J.C., Bennett, A.D. *et al.* (1992) Are Sinc and the PrP gene congruent? Evidence from PrP gene analysis in Sinc congenic mice. *Journal of General Virology* 73: 2751–55

Hunter, N., Goldmann, W., Smith, G. & Hope, J. (1994a) The association of a codon 136 PrP gene variant with the occurrence of natural scrapie. *Archives of Virology* 137: 171–77

Hunter, N., Goldmann, W., Smith, G. & Hope, J. (1994b) Frequencies of PrP gene variants in healthy cattle and cattle with BSE in Scotland. *Veterinary Record* 135: 400–03

Hunter, N., Foster, J., Goldmann, W. *et al.* (1996) Natural scrapie in a closed flock of Cheviot sheep occurs only in specific PrP genotypes. *Archives of Virology* 141: 809–24

Hunter, N., Goldmann, W., Foster, J., Cairns, D. & Smith, G. (1997a) Natural scrapie and PrP genotype: case-control studies in British sheep. *Veterinary Record* 141: 137–40

Hunter, N., Moore, L., Hosie, B., Dingwall, W. & Greig, A. (1997b) Natural scrapie in a flock of Suffolk sheep in Scotland is associated with PrP genotype. *Veterinary Record* 140: 59–63

Hunter, N., Cairns, D., Foster, J. *et al.* (1997c) Is scrapie a genetic disease? Evidence from scrapie-free countries. *Nature* 386: 137

Ikeda, T., Horiuchi, M., Ishiguro, N. *et al.* (1995) Amino acid polymorphisms of PrP with reference to onset of scrapie in Suffolk and Corriedale sheep in Japan. *Journal of General Virology* 76: 2577–81

Ironside, J. (1999) Update on new variant Creutzfeldt-Jacob disease. *Neuropathology and Applied Neurobiology* 25: 33–34

Kaneko, K., Zulianello, L., Scott, M. *et al.* (1997) Evidence for protein X binding to a discontinuous epitope on the cellular prion protein during scrapie prion propagation. *Proceedings of the National Academy of Sciences USA* 94: 10069–74

Kimberlin, R.H. & Walker, C.A. (1980) Pathogenesis of mouse scrapie: evidence for neural spread of infection to the CNS. *Journal of General Virology* 51: 183–87

Klein, M.A., Frigg, R., Flechsig, E. *et al.* (1997) A crucial role for B cells in neuroinvasive scrapie. *Nature* 390: 687–90

Lasmezas, C., Deslys, J.-P., Robain, O. *et al.* (1997) Transmission of the BSE agent to mice in the absence of detectable abnormal prion protein. *Science* 275: 402–05

Lehmann, S. & Harris, D.A. (1996) Mutant and infectious prion proteins display common biochemical properties in cultured cells. *Journal of Biological Chemistry* 271: 1633–37

Lehmann, S., Daude, N. & Harris, D.A. (1997) A wild-type prion protein does not acquire properties of the scrapie isoform when co-expressed with a mutant prion protein in cultured cells. *Molecular Brain Research* 52: 139–45

Manson, J.C., Hope, J., Clarke, A. *et al.* (1995) PrP gene dosage and long term potentiation. *Neurodegeneration* 4: 113–15

Manson, J.C., Jamieson, E., Baybutt, H. *et al.* (1999) A single amino acid alteration (101L) introduced into murine PrP dramatically alters incubation time of transmissible spongiform encephalopathy. *EMBO Journal* 18: 6855–64

Manuelidis, L., Sklaviadis, T., Akowitz, A. & Fritch, W. (1995) Viral particles are required for infection in neurodegenerative Creutzfeldt–Jakob-disease. *Proceedings of the National Academy of Sciences USA* 92: 5124–28

McBride, P.A., Eikelenboom, P., Kraal, G., Fraser, H. & Bruce, M.E. (1992) PrP protein is associated with follicular dendritic cells of spleens and lymph nodes in uninfected and scrapie-infected mice. *Journal of Pathology* 168: 413–18

Medori, R. & Tritschler, H.J. (1993) Prion protein gene analysis in 3 kindreds with fatal familial insomnia (FFI) – codon-178 mutation and codon-129 polymorphism. *American Journal of Human Genetics* 53: 822–27

Montgomery, R., Morgyn, S., Lum, B. & Spear, P. (1996) Herpes simplex virus-1 entry into cells mediated by a novel member of the TNF/NGF receptor family. *Cell* 87: 427–36

Moore, R.C., Hope, J., McBride, P.A. *et al.* (1998) Mice with gene targetted prion protein alterations show that *Prnp*, *Sinc* and *Prni* are congruent. *Nature Genetics* 18: 118–25

Neibergs, H.L., Ryan, A.M., Womack, J.E., Spooner, R.L. & Williams, J.L. (1994) Polymorphism analysis of the prion gene in BSE-affected and unaffected cattle. *Animal Genetics* 25: 313–17

Palmer, M.S., Dryden, A.J., Hughes, J.T. & Collinge, J. (1991) Homozygous prion protein genotype predisposes to sporadic Creutzfeldt–Jakob disease. *Nature* 352: 340–42

Parry, H. (1960) Scrapie: a transmissible hereditary disease of sheep. *Nature* 185: 441–43

Poulter, M., Baker, H.F., Frith, C.D. *et al.* (1992) Inherited prion disease with 144 base pair gene insertion. 1. Genealogical and molecular studies. *Brain* 115: 675–85

Prusiner, S.B. (1998) Prions. *Proceedings of the National Academy of Sciences USA* 23: 13363–83

Prusiner, S.B., Groth, D.F., McKinley, M.P. *et al.* (1982) Scrapie agent is a novel proteinaceous infectious particle. *Federation Proceedings* 41: 695

Prusiner, S.B., Scott, M., Foster, D. *et al.* (1990) Transgenetic studies implicate interactions between homologous PrP isoforms in scrapie prion replication. *Cell* 63: 673–86

Prusiner, S.B. & Scott, M.R. (1997) Genetics of prions. *Annual Review of Genetics* 31: 139–75

Raymond, G.J., Hope, J., Kocisko, D. *et al.* (1997) Molecular assessment of the potential transmissibles of BSE and scrapie to humans. *Nature* 388: 285–87

Ren, R., Constantini, F., Gorgacz, E., Lee, J. & Racaniello, V. (1990) Transgenic mice expressing a human poliovirus receptor: a new model for poliomyelitis. *Cell* 63: 353–62

Ridley, R.M. & Baker, H.F. (1995) The myth of maternal transmission of spongiform encephalopathy. *British Medical Journal* 311: 1071–75

Somerville, R.A. (1991) The transmissible agent causing scrapie must contain more than protein. *Reviews in Medical Virology* 1: 131–39

Tobler, I., Gaus, S.E., Deboer, T. *et al.* (1997) Altered circadian activity rhythms and sleep in mice devoid of prion protein. *Journal of Neurosciences* 17: 1869–79

Westaway, D., Goodman, P.A., Mirenda, C.A. *et al.* (1987) Distinct prion proteins in short and long scrapie incubation period mice. *Cell* 51: 651–62

Westaway, D., Zuliani, V., Cooper, C.M. *et al.* (1994a) Homozygosity for prion protein alleles encoding glutamine-171 renders sheep susceptible to natural scrapie. *Genes and Development* 8: 959–69

Westaway, D., Dearmond, S., Cayetanocanlas, J. *et al.* (1994b) Degeneration of skeletal-muscle, peripheral-nerves, and the CNS in transgenic mice overexpressing wild-type prion proteins. *Neurology* 44: 260

GLOSSARY

aetiology (etiology) the cause of a disease
allotype antigenic determinant typifying allelic differences in light and heavy immunoglobulin chains
follicular dendritic cells non-mobile antigen presenting cells found in the B cell areas of lymph nodes and spleen. They present antigen to B cells
glycosylation process of adding sugars (oligosaccharides) to proteins to form glycoproteins
GPI anchor (glycosylphosphatidylinositol anchor) a linkage that attaches some membrane proteins to the lipid bilayer
isoform form of a protein with slightly different amino acid sequences, which may differ in activity, function, distribution, etc
Western blotting a electroblotting method of transferring proteins from a gel to a medium on which they can be further analysed by treatment with specific antibodies

See also **Fungal prions** (p.279)

GENETICS OF HUNTINGTON'S DISEASE: WHEN MORE IS LESS

Jamal Nasir
Division of Genomic Medicine, Royal Hallamshire Hospital, Sheffield, UK

1. Clinical

Huntington's disease (HD) is a progressive neurodegenerative disorder associated with involuntary movements (chorea), dementia and major psychiatric symptoms, which eventually lead to death within 15–20 years of onset (Hayden, 1981; Harper, 1996). The disease typically strikes during mid-life, but can manifest during childhood, in which case it tends to be more severe, with pronounced rigidity, seizures and dystonia, as well as a more rapid progression (Kremer *et al.*, 1992; Harper, 1996). These changes are accompanied by atrophy within the basal ganglia region of the brain, affecting both the caudate nucleus and the putamen, which together constitute the striatum. Within these affected regions, the loss of neurones is highly selective. The gamma amino butyric acid (GABA)-expressing, medium spiny neurones are the most vulnerable (Kremer *et al.*, 1992). To a lesser degree, there also appears to be loss of neurones within the cortex, but other regions of the brain, including the cerebellum do not appear to be affected. The chain of molecular and biochemical events leading to the disease has yet to be elucidated (Nasir *et al.*, 1996). However, recent findings derived from transgenic animal models are beginning to offer remarkable new insights into this incurable disease.

2. Genetics

HD is an autosomal dominant disorder with complete penetrance and a frequency of approximately one in 10 000 (Hayden, 1981; Harper, 1996). The cloning of the HD gene, using a positional cloning approach, led to the identification of a relatively large (10.5 kilobase [kb]) transcript which appears to be expressed in all tissues, including both affected and non-affected tissues (HDCRG, 1993). The gene encompasses a glutamine-encoding polymorphic CAG repeat within its first exon. The length of this repeat can vary from 8–36 in the general population, but repeat lengths beyond this range result in HD (Andrew *et al.*, 1993; Kremer *et al.*, 1994; Rubinsztein *et al.*, 1994, 1996; Brinkman *et al.*, 1997). Moreover, there is an inverse correlation between the age of onset and CAG repeat length, whereby longer repeats result in earlier onset (Andrew *et al.*, 1993). Sixty or more repeats generally result in juvenile onset HD and repeat lengths can go up to around 121 CAG (**Table 1**). This upper limit implies that longer repeats are either toxic to germ cells or lead to failure during embryogenesis. For any given repeat size, there can be considerable variation with respect to the age of onset, suggesting that other environmental and genetic components may be involved (Andrew *et al.*, 1993).

3. Polyglutamine disorders

Remarkably, seven other adult-onset neurodegenerative conditions are also associated with polyglutamine repeat expansions (**Table 1**) within their respective genes (Lunkes & Mandel, 1997; Mandel, 1997; Ross, 1997; Ross *et al.*, 1998). For each of these disorders, there is again a strong inverse correlation between length of CAG repeat and age of onset and, with the exception of spinocerebellar ataxia type 6 (SCA6), they display rather ubiquitous expression patterns. Yet, each disorder is associated with a unique and distinctive neurodegenerative pattern (Ross, 1995), indicating that the "context" of the polyglutamine tract (the gene) plays an important role in pathogenesis. For example, while the cerebellum is largely spared in HD, it is the principal site of pathology in SCA1.

With the exception of spinal and bulbar muscular atrophy (SBMA), an X-linked disorder associated with the

Table 1 Polyglutamine-induced neurodegenerative disorders. A list of genes associated with CAG repeat expansions leading to neurodegenerative disorders is given. The CAG repeat range associated with normal *vs.* affected individuals, as well as the major sites of neuropathology, are provided.

Gene	Product	Normal	Affected	Pathology
HD	huntington	9–35	36–121	Striatum, cerebral cortex
SCA1	ataxin-1	6–39	40–82	Cerebellar cortex (Purkinje cells), brainstem, spinocerebellar tracts
SCA2	ataxin-2	14–32	33–77	Cerebellum, brainstem nuclei, sustantia nigra
SCA3/MJD1	ataxin-3	13–42	61–84	Dentate and pontine nuclei, spinocerebellar tracts, globus pallidus, substantia nigra
SCA6	α 1a-calcium channel	4–18	21–30	Cerebellar cortex and mild brainstem atrophy
SCA7	ataxin-7	7–17	37–130	Photoreceptors and bipolar cells, cerebellar cortex, brainstem
DRPLA	atrophin	7–36	49–88	Globus pallidus, dentate nucleus, subthalamic nucleus, red nucleus
SBMA	androgen receptor	11–34	40–66	Motor neurons in spinal cord and brainstem, dorsal root ganglia

androgen receptor gene, all other polyglutamine disorders are autosomal dominant and mostly involve novel genes with no homologies to other sequences in the databases. The gene associated with SCA6 is a previously identified calcium channel α1-subunit gene, but this gene is an anomaly in many respects, suggesting that the underlying molecular mechanism associated with SCA6 may be different from other polyglutamine-related disorders. The anomalies include, first, a repeat size associated with the disease which is small (21–30 CAG), corresponding to a normal repeat length for all other polyglutamine disorders (see **Table 1**). Secondly, the polyglutamine tract is located at the C-terminus of the protein, whereas this domain is found in the N-terminus of all other proteins (barring ataxin-3) associated with neurodegenerative disorders. Finally, mutations elsewhere in this gene are involved with other disorders including familial hemiplegic migraine and episodic ataxia type-2 (Ophoff *et al.*, 1996). This has raised doubts as to whether the polyglutamine repeat expansion associated with SCA6 exerts a true "gain-of-function" effect (Zhuchenko *et al.*, 1997).

The pathogenic threshold for most polyglutamine disorders is strikingly similar, at around 35–40 CAG repeats (see **Table 1**), and it is widely agreed that the polyglutamine tract exerts a novel effect ("gain of function") which triggers the process culminating in neuronal cell death. This is supported by a large number of studies showing aberrant migration on polyacrylamide gels of proteins carrying large polyglutamine tracts compared with the same proteins carrying smaller polyglutamine tracts (Aronin *et al.*, 1995; Jou & Myers, 1995; Trottier *et al.*, 1995a), which presumably reflects the abnormal conformations of the former. Furthermore, a polyglutamine-specific antibody can discriminate between proteins carrying long or short polyglutamine tracts (Trottier *et al.*, 1995b), binding selectively to long polyglutamine tracts, but failing to detect wild-type proteins with normal length polyglutamine repeats. This property has been successfully exploited in cloning genes associated with polyglutamine expansion (Trottier *et al.*, 1995b), and provides further evidence that long polyglutamine tracts undergo abnormal and inappropriate conformations.

(i) Why polyglutamine?

It is quite remarkable that polyglutamine repeat expansion has occurred independently in eight different genes associated with neurodegenerative disorders. This raises the question of why we do not see similar expansions involving other coding sequences. It also suggests that polyglutamine tracts normally play an important biological function which has been maintained during evolution by selection. Indeed, there appears to be a good correlation between CAG repeat length and complexity of organisms; for the TATA-binding protein, we see a remarkable correlation between species complexity and glutamine repeat length, corresponding to 38, 26, 13 and 4 glutamines in human, orang-utan, mouse and *Xenopus*, respectively (Hashimoto *et al.*, 1992; Gostout *et al.*, 1993).

The murine HD gene encompasses seven glutamine repeats (Lin *et al.*, 1994), while the pufferfish, *Fugu rubripes*, has only four repeats (Baxendale *et al.*, 1995). Similarly, the rodent homologues of the genes associated with dentatorubral and pallidoluysian atrophy (DRPLA), SBMA and SCA1 harbour shorter repeats of 3, 4 and 2 CAGs, respectively (Banfi *et al.*, 1994, Oyake *et al.*, 1997; Chang *et al.*, 1988). Rubinsztein and colleagues have noticed a strong tendency for longer CAG repeats within the HD gene in humans compared with other primates, including chimpanzees, gorillas, orang-utans, baboons and macaques (Rubinsztein *et al.*, 1994). Based on these observations, they have proposed a model for mutational bias leading to directional evolution (Rubinsztein *et al.*, 1994).

Interestingly, a naturally occurring polyglutamine repeat expansion leading to a mutant phenotype has never been identified in mice. This could be accounted for by the fact that the repeats in mice are small to begin with and are usually interrupted with other sequences. For example, the CAG repeat associated with the murine HD gene harbours a CAA interruption (CAG CAG CAA CAG CAG CAG CAG) (Lin *et al.*, 1994). This presumably inhibits processes such as slippage and unequal exchange, which would normally be expected to lead to expansion of repeats. Similarly, for DRPLA, SBMA and SCA1 genes, the CAG repeats are interrupted in mice, which would account for the lack of polymorphisms (Chang *et al.*, 1988; Banfi *et al.*, 1994; Lin *et al.*, 1994; Oyake *et al.*, 1997). Presumably, the interruption was lost at some point in the primate lineage, allowing the expansion of CAG repeats. This might offer a distinct advantage to higher primates, perhaps contributing to processes such as language or higher cognitive functions. Furthermore, the ability to control gene regulation by adjusting the repeat length offers the potential to regulate or fine tune a gene in a very precise, intricate and deliberate manner. These "tuning knobs", it has been suggested, offer a further evolutionary advantage, in that it is not an all or nothing effect, as is the case for most random mutations, which result in either no functional change or which may be deleterious (King *et al.*, 1997).

(ii) β-sheets, polar zippers and protein–protein interactions

It is suggested that polyglutamine motifs play an important role in protein–protein interactions. Studies by Max Perutz and colleagues have demonstrated that polyglutamine repeats are able to form stable β-sheets, strongly held together by hydrogen bonds (Perutz *et al.*, 1994; Stott *et al.*, 1995; Perutz, 1996). They have suggested that polyglutamine repeats act as polar zippers which enables them to join protein molecules together and to self-associate and make oligomers. They also speculated that long glutamine repeats may form excessively high affinities for one another or acquire non-specific affinities.

Another proposed function for glutamine residues is their ability to form cross-links to other proteins with the help of transglutaminases (Green, 1993). It has been suggested that this process could become irreversible with excessive glutamines, leading to the formation of aggregates composed of the polyglutamine-encompassing protein and other proteins (Green, 1993). Recent studies showing that polyglutamine repeat expansion can lead to aggregate formation both *in vitro* and *in vivo* are discussed later; the *in vitro* data also demonstrate that such protein aggregates can form similar structures to scrapie prions and β-amyloid fibrils (Scherzinger *et al.*, 1997).

4. Interacting proteins

Any attempt towards elucidating the molecular mechanisms involved in HD must account for how a ubiquitously expressed protein leads to selective neuronal death. One possibility is that the HD protein, huntingtin, interacts with other proteins whose expression is largely confined to the brain. Using either the yeast two-hybrid technology (Fields & Song, 1989) or affinity columns, several interacting proteins have been identified including the Huntingtin Associated Protein (HAP-1), ubiquitin conjugating enzyme (hE2–25K), glyceraldehyde-3-phosphate dehydrogenase (GAPDH), calmodulin and the Huntingtin Interacting Protein (HIP-1) (Li *et al.*, 1995; Bao *et al.*, 1996; Burke *et al.*, 1996; Kalchman *et al.*, 1996, 1997; Wanker *et al.*, 1997). GAPDH also appears to interact with atrophin and ataxin-1, the proteins involved in DRPLA and SCA1, respectively (see Roses, 1996).

The size of CAG repeats associated with huntingtin appears to influence the strength of its interaction with at least some of these proteins. In the case of HAP-1, calmodulin and GAPDH, there is an increased affinity for huntingtin (**Table 2**) with increased CAG repeat length, whereas, with HIP-1, the extent of interaction is reduced with increasing CAG repeat size (Kalchman *et al.*, 1997). Increased CAG repeat size might have the effect of sequestering and/or allowing the toxic build-up of huntingtin, depending on whether there is excessive or insufficient binding. However, the loss of one copy of the HD gene in "knockout" mice does not manifest with HD (Nasir *et al.*, 1995), indicating that a simple dosage effect alone is not enough to cause disease. These gene-targeting experiments provide further evidence that HD is due to a "gain of function", arising from CAG repeat expansion (Nasir *et al.*, 1995), which might involve engaging huntingtin in novel interactions. Nevertheless, mouse models overexpressing HIP-1 and HAP-1, and with targeted disruptions within these genes, will still provide a valuable resource for further insights into the molecular mechanisms involved in HD.

While HAP-1 and HIP-1 appear to be predominantly expressed in the brain, the expression patterns of calmodulin, GAPDH and hE2–25K are more widespread (**Table 2**). In addition, the lack of any obvious CAG modulation in the case of hE2–25K (Kalchman *et al.*, 1996), plus our understanding of the roles of some of these proteins in metabolism (GAPDH) (Burke *et al.*, 1996) and ubiquitination (hE2–25K), would suggest that they do not play a primary role in neuronal cell death. Instead, they could be involved in downstream events, perhaps acting as scavengers in disassembling or clearing up protein complexes including huntingtin.

Based on the success so far in identifying interacting partners for huntingtin, finding the interactors of these interacting proteins in turn promises to reveal further clues to the pathogenesis of HD. It is likely to yield a more detailed picture of the chain of events leading to neuronal loss. For HAP-1, this philosophy has led to the identification of other proteins, including a novel protein which interacts with HAP-1 and which is also enriched in the brain (Colomer *et al.*, 1997).

5. Animal models for Huntington's disease

While *in vitro* studies may provide invaluable insights into the molecular basis for HD, ultimately *in vivo* studies must be undertaken in order to develop a true understanding of the molecular basis for HD and to test therapeutic strategies. Generating an animal model for HD has thus been a major goal ever since the cloning of the HD gene. A variety of transgenic strategies have been deployed in generating animal models for HD, with the common goal of integrating HD gene constructs carrying long CAG repeats into the mouse genome. By and large, long CAG repeats have been derived directly by PCR amplification of the region from patients.

Attempts at mimicking the disease with cDNA transgenic models have been unpredictable and disappointing due to the limitations of this approach. Finding a promoter which can express the protein at endogenous levels in appropriate regions of the brain at the required time in development has proved to be a major problem. Yeast artificial chromosome transgenics can overcome these limitations, faithfully recapitulating the native expression profile of huntingtin (Hodgson *et al.*, 1996), but even they can be subject to position and integration events. To avoid these obstacles, the

Table 2 Proteins which interact with the Huntington's disease (HD) protein. The table indicates whether or not the interactions are modulated by CAG repeat length associated with the HD gene. In some cases, there is increased affinity (+) for the HD protein, but in other instances there is decreased (–) affinity. Information on whether or not a given protein is selectively expressed in the brain and the region of the HD protein (amino acid positions) involved in the interaction is also included. For hE2–25K, only certain isoforms are selectively expressed in the brain. In some instances, the desired information is unavailable (?).

Protein	CAG modulation	Selective expression in brain	Region of interaction
HAP-1	✔ (+)	YES	1–230
hE2–25K	?	YES	1–540
GAPDH	✔ (+)	NO	PolyQ
Calmodulin	✔ (+)	NO	?
HIP-1	✔ (–)	YES	1–540

small CAG repeat motif within the murine HD gene has been replaced by longer repeats (CAG = 48) by homologous recombination in embryonic stem cells (White *et al.*, 1997). However, no phenotype was reported in the CAG = 48 mice at 6 months of age, so it will be of great interest to see when they develop a phenotype. In the quest to develop an animal model and to expedite the phenotype, there has been much emphasis on introducing longer CAG repeats into mice and on achieving high levels of expression.

(i) Exon 1 mice

The best characterized and most successful transgenic mouse model for HD was generated serendipitously as part of a study to investigate trinucleotide repeat instability, by the introduction of a small (1.9 kb) genomic fragment. This fragment encompasses exon 1 with 130 CAG repeats, and includes both the promoter region and intron 1 (Mangiarini *et al.*, 1996). Remarkably, it generated mice with severe neurological symptoms including tremor and seizures, leading to death by around 3 months of age, which provides strong evidence that the polyglutamine segment is necessary and sufficient to induce a phenotype. This notion is supported by a more recent study (Ordway *et al.*, 1997), where a polyglutamine domain was inserted into the HPRT gene to generate mice with a neurological phenotype and intranuclear inclusions, which are a major characteristic of the disease (see next section).

6. Neuronal intranuclear inclusions

While the "exon 1" mice generated by Mangiarini *et al.* (1996) appear to present with a progressive neurological phenotype with similarities to HD, initial observations failed to provide strong evidence for any of the neuropathological changes consistent with HD. However, further characterization has led to remarkable new insights, since detailed neuropathology now reveals an accumulation of nuclear inclusions in neurones normally affected in HD (Davies *et al.*, 1997).These amyloid-like aggregates of huntingtin were found in brains from the exon 1 transgenic mice (Davies *et al.*, 1997; Scherzinger *et al.*, 1997) and in postmortem tissues from HD brains (Difiglia *et al.*, 1997). The inclusions precede the neurological phenotype, are largely restricted to neurones which are lost in HD, occur more frequently with increasing CAG size, are not found in unaffected individuals and are ubiquinated (Davies *et al.*, 1997; Difiglia *et al.*, 1997). Furthermore, it appears that the accumulation of aggregates is greatly facilitated by cleavage of the protein. Using a series of HD constructs with varying CAG repeat sizes and truncations throughout the HD protein, Martindale and colleagues have been able to demonstrate that truncated versions of the protein with longer CAG repeat lengths form intracellular aggregates more readily (Martindale *et al.*, 1998). Nuclear inclusions have also been found in patients with SCA1 (Skinner *et al.*, 1997) and in patients with Machado–Joseph disease (SCA3) (Paulson *et al.*, 1997). Again, these inclusions appear to be ubiquitinated, predominantly present in affected regions of the brain and not seen in unaffected individuals (Paulson al., 1997; Skinner *et al.*, 1997).

These data provide a unifying theme for polyglutamine-induced neurodegenerative diseases (Ross, 1997). The specificity in each disease could be accounted for by interacting proteins, as discussed above. However, aggregate formation may simply reflect an excess of mutant protein in the nucleus rather than being the primary cause of neuronal loss.

7. Subcellular mechanisms

Huntingtin is normally distributed in the cytoplasm (Trottier *et al.*, 1995a), but in HD brains it is also found in the nucleus, as discussed above. How does this apparently large protein freely enter the nucleus? The clustering of DEVD caspase-3 cleavage sites close to the polyglutamine region suggested the HD protein may be a natural substrate for this protease. Caspase-3 is a key enzyme in initiation of apoptotic cell death (Nicholson *et al.*, 1995), which raised the possibility that cleavage of the N-terminal fragment of huntingtin, encompassing the CAG repeat, promotes its entry into the nucleus, resulting in a toxic effect. Goldberg *et al.* (1996) have now demonstrated that huntingtin is cleaved by caspase-3 *in vitro*. Furthermore, it appears that cleavage is enhanced with increasing CAG repeat sizes associated with huntingtin (Goldberg *et al.*, 1996), thereby allowing faster accumulation of the toxic protein within nuclear structures. This could account for the earlier onset associated with increased CAG repeat sizes. The notion of increased and inappropriate cell death in HD involving caspase-3 is supported by studies using TUNEL and other methods to demonstrate apoptosis in postmortem HD brains (Portera-Cailliau *et al.*, 1995).

A shift in the subcellular compartments following cleavage or alternate splicing has previously been noted for other proteins, including the sterol regulatory element binding protein 1 (SREPB-1) and the Wilms' tumor gene product (Gasic, 1994; Wang *et al.*, 1994; Englert *et al.*, 1995). Cleavage of SREBP-1 in the cytosol is followed by translocation of the 68 kilodalton N-terminal fragment into the nucleus, which triggers transcriptional activation of the genes for LDL receptor and HMG CoA synthase (Wang *et al.*, 1994). This event is exquisitely regulated by the amount of sterol in the cytosol: depletion of sterol promotes cleavage of SREBP 1, but cleavage is inhibited in the presence of sterol. Perhaps, in HD, the cleavage might also be regulated in a similar manner, where the presence of antiapoptotic factors, including bcl-2, could dictate whether or not huntingtin is cleaved.

The cleaved N-terminal fragment encompassing the polyglutamines could then potentially engage in aberrant interactions with other proteins in the cell. For example, it could mimic transcription factors and lead to the activation of other genes. Alternatively, it could bind to transcription factors and inhibit the transcription of certain genes. Polyglutamine repeats are a common feature of transcription factors and, in particular, are associated with genes involved in the development of the nervous system (Gerber *et al.*, 1994). Moreover, the polyglutamine domain alone is able to confer transcriptional activation (Gerber *et al.*, 1994).

8. Phenotypic assessment of animal models

In the near future, it is likely that a variety of transgenic animal models will emerge, each with a unique combination of CAG repeat size, expression levels, tissue distribution and phenotype, associated with the transgene. All of these animals are likely to be useful, but a major challenge will be phenotype assessment, paying particular regard to the well-documented contribution of strain differences on behaviour in mice (Gerlai, 1996; Lathe, 1996; Banbury Conference Report, 1997; Simpson *et al.*, 1997). Perhaps the major problem will be in identifying and quantifying behavioural and cognitive traits. This is a situation we are likely to encounter repeatedly in our pursuit of animal models for other neurological disorders.

A wide variety of tests for cognition, behaviour and motor functions are available (Nasir *et al.*, 1995). Many of the tests and paradigms have been adapted from studies on rats, which have enabled, for instance, sophisticated analyses of memory and learning. Many of the tasks allow determination of deficits in specific regions of the brain; for example, the water maze experiment is designed to test hippocampal function (Morris *et al.*, 1982). Furthermore, double or triple dissociation studies enable a variety of brain structures to be examined simultaneously, allowing one to discriminate between lesions in a variety of brain structures including the amygdala, hippocampus and the caudate (Amalric & Koob, 1987; Packard & McGaugh, 1992; McDonald & White, 1993; Baunez *et al.*, 1995). The use of these tests in combination with neuronal counts, and assessment of neurotransmitter levels involving both biochemical (high performance liquid chromatography, or HPLC) and molecular techniques (immunohistochemistry), should provide a thorough analysis of a neurological phenotype.

Recently, the SHIRPA protocol was proposed, providing a uniform battery of tests for assessment of neurological and behavioural phenotypes (Rogers *et al.*, 1997). This is likely to make it much easier to compare different transgenic models. Imaging techniques are also increasingly likely to be applied to the animal models.

9. Future

Clearly, the transgenic animal models will remain a central focus of future research, since they offer, for the first time, the opportunity to follow the progression of HD *in vivo* by studying the brain at fixed time points and, of course, to test therapeutic approaches. In principle, the existing animal models should be suitable for testing novel therapies. The initial aim will be to identify compounds which can slow down or halt the progression of the disease. This may require direct injections of compounds into the striatum, perhaps using liposomes, capsules or viral vectors for delivery (Breakefield, 1993). Alternatively, tissue transplantation may also become a feasible alternative.

We are beginning to get a much clearer picture of the chain of events leading to cell death, providing early opportunities for pharmaceutical intervention (**Figure 1**). In particular, careful scrutiny must be directed at (1) interfering with the cleavage or entry of the HD protein into the nucleus or (2) preventing its aggregation in the nucleus. Perhaps the

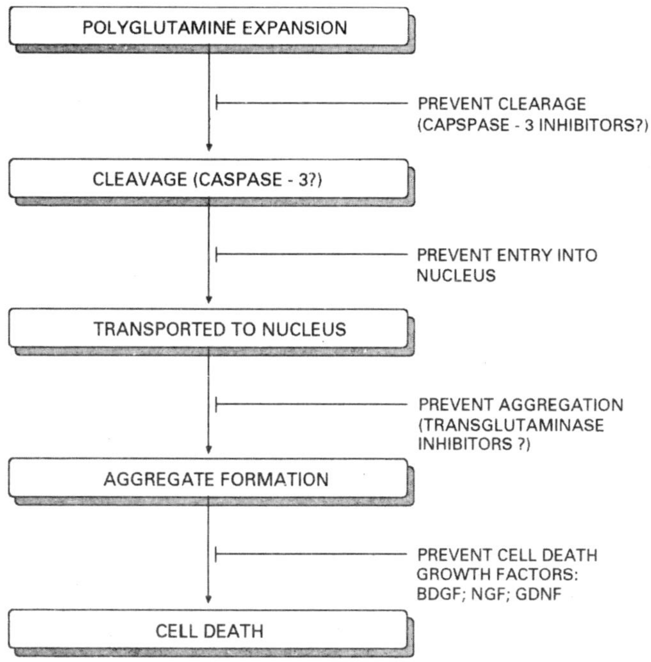

Figure 1 Sequence of events leading to the formation of neuronal intranuclear inclusions and routes for therapeutic intervention. The subcellular events associated with formation of huntingtin aggregates leading to cell death are outlined. Checkpoints for therapeutic intervention are also indicated.

initial compounds to test might include neuronal-specific growth factors, apoptosis inhibitors and transglutaminase inhibitors; indeed, a recent study has shown that transglutaminase inhibitors can suppress polyglutamine-induced apoptotic cell death and nuclear aggregate formation in cells expressing mutant atrophin, the protein associated with DRPLA (Igarashi *et al.*, 1998). There has also been considerable interest in developing specific inhibitors of apoptotic cell death – it has already been shown, for example, that synthetic caspase-3 inhibitors work efficiently in inhibiting cleavage of huntingtin (Nicholson *et al.*, 1995; Goldberg *et al.*, 1996). Finally, the ability to discuss treatments in tangible terms highlights the incredible progress that has been made in such a short space of time, which surely must be good news for patients.

REFERENCES

Amalric, M. & Koob, G.F. (1987) Depletion of dopamine in the caudate nucleus but not in the nucleus accumbens impairs reaction-time performance in rats. *Journal of Neuroscience* 7: 2129–34

Andrew, S.E., Goldberg, Y.P., Kremer, B. *et al.* (1993) The relationship between trinucleotide (CAG) repeat length and clinical features of Huntington's disease. *Nature Genetics* 4: 387–92

Aronin, N., Chase, K., Young, C. *et al.* (1995) CAG expansion affects the expression of mutant huntingtin in the Huntington's Disease brain. *Neuron* 15: 1193–201

Banbury Conference Report (1997) Genetic background in mice. *Neuron* 19: 755–59

Banfi, S., Servadio, A., Chung, M-Y. *et al.* (1994) Characterization of the gene causing type 1 spinocerebellar ataxia and identification of the murine homolog. *American Journal of Human Genetics* 55 (Suppl.): A17

Bao, J., Sharp, A.H., Wagster, M. *et al.* (1996) Expansion of polyglutamine repeat in huntingtin leads to abnormal protein interactions involving calmodulin. *Proceedings of the National Academy of Sciences USA* 93: 5037–42

Baunez, C., Nieoullon, A. & Amalric, M. (1995) In a rat model of Parkinsonism, lesions of the subthalamic nucleus reverse increases of reaction time but induce a dramatic premature responding deficit. *Journal of Neuroscience* 15: 6531–41

Baxendale, S., Abdulla, S., Elgar, G. *et al.* (1995) Comparative sequence analysis of the human and pufferfish Huntington's disease genes. *Nature Genetics* 10: 67–76

Breakefield, X.O. (1993) Gene delivery into the brain using virus vectors. *Nature Genetics* 3: 187–89

Brinkman, R.R., Mezei, M.M., Theilmann, J., Almqvist, E. & Hayden, M.R. (1997) The likelihood of being affected with Huntington disease by a particular age, for a specific CAG size. *American Journal of Human Genetics* 60: 1202–10

Burke, J.R., Enghild, J.J., Martin, M.E. *et al.* (1996) Huntingtin and DRPLA proteins selectively interact with the enzyme GAPDH. *Nature Medicine* 2: 347–50

Chang, C., Kokontis, J. & Liao, S. (1988) Structural analysis of complementary DNA and amino acid sequences of human and rat androgen receptors. *Proceedings of the National Academy of Sciences USA* 85: 7211–15

Colomer, V., Engelender, S., Sharp, A.H. *et al.* (1997) Huntingtin-associated protein 1 (HAP1) binds to a Trio-like polypeptide, with a rac1 guanine nucleotide exchange factor domain. *Human Molecular Genetics* 6: 1519–25

Davies, S.W., Turmaine, M., Cozens, B.A. *et al.* (1997) Formation of neuronal intranuclear inclusions underlies the neurological dysfunction in mice transgenic for the HD mutation. *Cell* 90: 537–48

Difiglia, M., Sapp, E., Chase, K.O. *et al.* (1997) Aggregation of Huntingtin in neuronal intranuclear inclusions and dystrophic neurites in brain. *Science* 277: 1990–93

Englert, C., Vidal, M., Maheshwaran, S. *et al.* (1995) Truncated WT1 mutants alter the subnuclear localization of the wild-type protein. *Proceedings of the National Academy of Sciences USA* 92: 11960–64

Fields, S. & Song, O. (1989) A novel genetic system to detect protein–protein interactions. *Nature* 340: 245–47

Gasic, G.P. (1994) Basic-helix–loop–helix transcription factor and sterol sensor in a single membrane-bound molecule. *Cell* 77: 17–19

Gerber, H-P., Seipel, K., Georgiev, O. *et al.* (1994) Transcriptional activation modulated by homopolymeric glutamine and proline stretches. *Science* 263: 808–11

Gerlai, R. (1996) Gene-targeting studies of mammalian behavior: is it the mutation or the background genotype? *Trends in Neuroscience* 19: 177–81

Goldberg, Y.P., Nicholson, D.W., Rasper, D.M. *et al.* (1996) Cleavage of huntingtin by apopain, a proapoptotic cysteine protease, is modulated by the polyglutamine tract. *Nature Genetics* 13: 442–49

Gostout, B., Liu, Q. & Sommer, S.S. (1993) "Cryptic" repeating triplets of purines and pyrimidines (cRRY(i)) are frequent and polymorphic: analysis of coding cRRY(i) in the preopiomelanocortin (POMC) and TATA-binding protein (TBP) genes. *American Journal of Human Genetics* 52: 1182–90

Green, H. (1993) Human genetic diseases due to codon reiteration: relationship to an evolutionary mechanism. *Cell* 74: 955–56

Harper, P.S. (ed.) (1996) *Huntington's Disease*, 2nd edition, London, Philadelphia: Saunders

Hashimoto, S., Fujita, H., Hasegawa, S., Roeder, R.G. & Horikoshi, M. (1992) Conserved structural motifs within the N-terminal domain of the TFIIDτ from *Xenopus*, mouse and human. *Nucleic Acids Research* 20: 3788

Hayden, M.R. (1981) *Huntington's Chorea*, Berlin and New York: Springer

Hodgson, J.G., Smith, D.J., McCutcheon, K. *et al.* (1996) Human huntingtin derived from YAC transgenes compensates for loss of murine huntingtin by rescue of the embryonic lethal phenotype. *Human Molecular Genetics* 5: 1875–85

Huntington's Disease Collaborative Research Group (HDCRG) (1993) A novel gene containing a trinucleotide repeat that is expanded and unstable on Huntington's disease chromosomes. *Cell* 72: 971–83

Igarashi, S., Koide, R., Shimohata, T. *et al.* (1998) Suppression of aggregate formation and apoptosis by transglutaminase inhibitors in cells expressing truncated DRPLA protein with an expanded polyglutamine stretch. *Nature Genetics* 18: 111–17

Jou, Y-S. & Myers, R.M. (1995) Evidence from antibody studies that the CAG repeat in the Huntington disease gene is expressed in the protein. *Human Molecular Genetics* 4: 465–69

Kalchman, M.A., Graham, R.K., Xia, G. *et al.* (1996) Huntingtin is ubiquitinated and interacts with a specific ubiquitin-conjugating enzyme. *Journal of Biological Chemistry* 271: 19385–94

Kalchman, M.A., Koide, B.H., McCutcheon, K. *et al.* (1997) HIP1, a human homologue of *S. cerevisiae Sla2p*, interacts with membrane-associated huntingtin in the brain. *Nature Genetics* 16: 44–53

King, D.G., Soller, M. & Kashi, Y. (1997) Evolutionary tuning knobs. *Endeavour* 21: 36–40

Kremer, B., Weber, B. & Hayden, M.R. (1992) New insights into the clinical features, pathogenesis and molecular genetics of Huntington's disease. *Brain Pathology* 2: 321–35

Kremer, B., Goldberg, P., Andrew, S.E. *et al.* (1994) A worldwide study of the Huntington's Disease mutation: the sensitivity and specificity of measuring CAG repeats. *New England Journal of Medicine* 330: 1401–06

Lathe, R. (1996) Mice, gene targeting and behaviour: more than just genetic background. *Trends in Neuroscience* 19: 183–86

Li, X-J., Li, S-H., Sharp, A.H. *et al.* (1995) A huntingtin-associated protein enriched in brain with implications for pathology. *Nature* 378: 398–402

Lin, B., Nasir, J., MacDonald, H. *et al.* (1994) Sequence of the murine Huntington disease gene: evidence for conservation, alternate splicing and polymorphism in a triplet (CCG) repeat. *Human Molecular Genetics* 3: 85–92

Lunkes, A. & Mandel, J-L. (1997) Polyglutamines, nuclear inclusions and neurodegeneration. *Nature Genetics* 3: 1201–02

Mandel, J-L. (1997) Breaking the rule of three. *Nature* 386: 767–69

Mangiarini, L., Sathasivam, K., Seller, M. *et al.* (1996) Exon 1 of the HD gene with an expanded CAG repeat is sufficient to cause a progressive neurological phenotype in transgenic mice. *Cell* 87: 493–506

Martindale, D., Hackam, A., Wieczorek, A. *et al.* (1998) Length of huntingtin and its polyglutamine tract influences localization and frequency of intracellular aggregates. *Nature Genetics* 18: 150–54

McDonald, R.J. & White, N.M (1993) A triple dissociation of memory systems: hippocampus, amygdala, and dorsal striatum. *Behavioural Neuroscience* 107: 3–22

Morris, R.G.M., Garrud, P., Rawlins, J.N.P. & O'Keefe, J. (1982) Place navigation impaired in rats with hippocampal lesions. *Nature* 297: 681–83

Nasir, J., Floresco, S.B., O'Kusky, J.R. *et al.* (1995) Targeted disruption of the Huntington's disease gene results in embryonic lethality and behavioral and morphological changes in heterozygotes. *Cell* 81: 811–23

Nasir, J., Goldberg, Y.P. & Hayden, M.R. (1996) Huntington disease: new insights into the relationship between CAG expansion and disease. *Human Molecular Genetics* 5: 1431–35

Nicholson, D.W., Ali, A., Thornberry, N.A. *et al.* (1995) Identification and inhibition of the ICE/CED-3 protease necessary for mammalian apoptosis. *Nature* 376: 37–43

Ophoff, R.A., Terwindt, G.M., Vergouwe, M.N. *et al.* (1996) Familial hemiplegic migraine and episodic ataxia type-2 are caused by mutations in the Ca^{2+} channel gene CACNL1A4. *Cell* 87: 543–52

Ordway, J.M., Tallaksen-Greene, S., Gutenkunst, C-A. *et al.* (1997) Ectopically expressed CAG repeats cause intranuclear inclusions and a progressive late onset neurological phenotype in the mouse. *Cell* 91: 753–63

Oyake, M., Onodera, O., Shiroishi, T. *et al.* (1997) Molecular cloning of murine homologue dentatorubral-pallidoluysian atrophy (DRPLA) cDNA: strong conservation of a polymorphic CAG repeat in the murine gene. *Genomics* 40: 205–07

Packard, M.G. & McGaugh, J.L. (1992) Double dissociation of fornix and caudate nucleus lesions on acquisition of two water maze tasks: further evidence for multiple memory systems. *Behavioural Neuroscience* 106: 439–46

Paulson, H.L., Perez, M.K., Trottier, Y. *et al.* (1997) Intranuclear inclusions of expanded polyglutamine protein in spinocerebellar ataxia type 3. *Neuron* 19: 333–44

Perutz, M.F. (1996) Glutamine repeats and inherited neurodegenerative diseases: molecular aspects. *Current Opinion in Structural Biology* 6: 848–58

Perutz, M.F., Johnson, T., Suzuki, M. & Finch, J.T. (1994) Glutamine repeats as polar zippers: their possible role in inherited neurodegenerative diseases. *Proceedings of the National Academy of Sciences USA* 91: 5355–58

Portera-Cailliau, C., Hedreen, J.C., Price, D.L. & Koliatsos, V.E. (1995) Evidence for apoptotic cell death in Huntington disease and excitotoxic animal models. *Journal of Neuroscience* 15: 3775–87

Rogers, D.C., Fisher, E.M.C., Brown, S.D.M. *et al.* (1997) Behavioral and functional analysis of mouse phenotype: SHIRPA, a proposed protocol for comprehensive phenotype assessment. *Mammalian Genome* 8: 711–13

Roses, A.D. (1996) From genes to mechanisms to therapies: lessons to be learned from neurological disorders. *Molecular Medicine* 2: 267–69

Ross, C.A. (1995) When more is less: pathogenesis of glutamine repeat neurodegenerative diseases. *Neuron* 15: 493–96

Ross, C.A. (1997) Intranuclear neuronal inclusions: a common pathogenic mechanism for glutamine-repeat neurodegenerative diseases? *Neuron* 19: 1147–50

Ross, C.A., Margolis, R.L., Becher, M.W. *et al.* (1998) Pathogenesis of polyglutamine neurodegenerative diseases: towards a unifying mechanism. In *Genetic Instabilities and Hereditary Neurological Diseases*, edited by R.D. Wells & S.T. Warren, San Diego: Academic Press

Rubinsztein, D.C., Amos, W., Leggo, J. *et al.* (1994) Mutational bias provides a model for the evolution of Huntington's disease and predicts a general increase in disease prevalence. *Nature Genetics* 7: 525–30

Rubinsztein, D.C., Leggo, J., Coles, R. *et al.* (1996) Phenotypic characterization of individuals with 30–40 CAG repeats in the Huntington Disease (HD) gene reveals HD cases with 36 repeats and apparently normal elderly individuals with 36–39 repeats. *American Journal of Human Genetics* 59: 16–22

Scherzinger, E., Lurz, R., Turmaine, M. *et al.* (1997) Huntingtin-encoded polyglutamine expansions form amyloid-like protein aggregates in vitro and in vivo. *Cell* 90: 549–58

Simpson, E.M., Linder, C.C., Sargent, E.E. *et al.* (1997) Genetic variation among 129 substrains and its importance for targeted mutagenesis in mice. *Nature Genetics* 16: 19–27

Skinner, P.J., Koshy, B.T., Cummings, C.J. *et al.* (1997) Ataxin-1 with an expanded glutamine tract alters nuclear matrix-associated structures. *Nature* 389: 971–74

Stott, K., Blackburn, J.M., Butler, P.J.G. & Perutz, M. (1995) Incorporation of glutamine repeats makes protein oligomerize: implications for neurodegenerative diseases. *Proceedings of the National Academy of Sciences USA* 92: 6509–13

Trottier, Y., Devys, D., Imbert, G. *et al.* (1995a) Cellular localization of the Huntington's disease protein and discrimination of the normal and mutant form. *Nature Genetics* 10: 104–10

Trottier, Y., Lutz, Y., Stevanin, G. *et al.* (1995b) Polyglutamine expansion as a pathological epitope in Huntington's disease and four dominant cerebellar ataxias. *Nature* 378: 403–06

Wang, X., Sato, R., Brown, M.S., Hua, X. & Goldstein, J.L. (1994) SREBP-1, a membrane-bound transcription factor released by sterol-regulated proteolysis. *Cell* 77: 53–62

Wanker, E.E., Rovira, C., Scherzinger, E. *et al.* (1997) HIP-1: a huntingtin interacting protein isolated by the yeast two-hybrid system. *Human Molecular Genetics* 6: 487–95

White, J.K., Auerbach, W., Duyao, M.P. *et al.* (1997) Huntingtin is required for neurogenesis and is not impaired by the Huntington's disease CAG expansion. *Nature Genetics* 17: 404–10

Zhuchenko, O., Bailey, J., Bonnen, P. *et al.* (1997) Autosomal dominant cerebellar ataxia (SCA6) associated with small polyglutamine expansions in the α1a-voltage-dependent calcium channel. *Nature Genetics* 15: 62–69

GLOSSARY

apoptosis active process of cell death requiring metabolic activity by the dying cell.

β-sheet secondary structure common in proteins comprising a folded polypeptide chain where the strands that lie parallel to each other are bound together by hydrogen bonding

C-terminus end of a polypeptide chain that contains a free COO-group, the last part of the chain synthesized

N-Terminus end of a polypeptide chain that contains a free amino acid group, the first part of the chain synthesized

oligomer a molecule comprised of only a few monomer units

slippage (strand-slippage) during replication, dissociation of a newly synthesized DNA strand from its template and reassociation at a new location

TATA [box] binding protein a DNA binding factor that recruits other proteins via protein–protein interactions

TATA box a tetranucleotide sequence found in the proximal upstream region of most polymerase II-transcribed genes. TBP binding to this sequence element is required to initiate transcription

TUNEL terminal deoxynucleotidyl transferase-mediated dUTP nick-end-labeling. A method for detecting apoptotic cells based on DNA strand breaks during apoptosis

See also **Trinucleotide repeat instability as a cause of human genetic disease** (p.396)

GENETICS OF DUCHENNE MUSCULAR DYSTROPHY

S.C. Brown and G. Dickson
Department of Biochemistry, Royal Holloway College, Egham, Surrey, UK

1. Pathology of Duchenne muscular dystrophy

Duchenne muscular dystrophy (DMD) and its milder form, Becker muscular dystrophy (BMD), are allelic X-linked muscle-wasting disorders in man. DMD affects one in 3500 males, one-third of whom are sporadic cases with no previous family history (Emery, 1993). BMD is less common and has an estimated incidence of approximately one in 18 500 births (Bushby *et al.*, 1991). DMD patients are clinically normal at birth, although serum levels of the muscle isoform of creatine kinase are elevated. Muscle degeneration nonetheless ensues and proximal muscle weakness leads to the loss of ambulation around 11 years of age. The regeneration of damaged fibres eventually fails to compensate for the recurrent phases of degeneration, and death due to respiratory or cardiac failure usually occurs by the third decade (Emery, 1993). By contrast, BMD presents a much more varied phenotype with some patients never losing the ability to walk. Both DMD and BMD are now known to be caused by mutations in the gene that codes for dystrophin (Koenig *et al.*, 1988). The identification and cloning of the gene responsible for DMD has led to major advances in diagnostic and genetic counselling for this disease, although no effective treatment is yet available.

2. Identification of the DMD gene

The locus for DMD was mapped to band Xp21 on the X chromosome by several different methods (Roberts, 1995). Cytogenetic analyses of affected females who possessed balanced X-autosome translocations revealed an apparent spread of breakpoints between Xp21.1 and Xp21.2, which was perhaps an indication of the large size of this locus. Linkage studies were first successful when a restriction fragment length polymorphism (RFLP) for the Xp21–Xp22.3 probe RC8 was found to be linked to DMD (Murray *et al.*, 1982) (see **Figure 1**). Further probes in Xp21 (L1.28, C7, 754) were subsequently isolated and shown to cosegregate with DMD and BMD (Dorkins *et al.*, 1985; Hofker *et al.*, 1985; Davies *et al.*, 1986). Additional evidence came from a study of a patient, BB, who presented with a complex syndrome of chronic granulomatous disease, McLeod erythrocyte phenotype, retinitis pigmentosa, mild mental retardation and DMD (Franke *et al.*, 1985). Cytogenetic

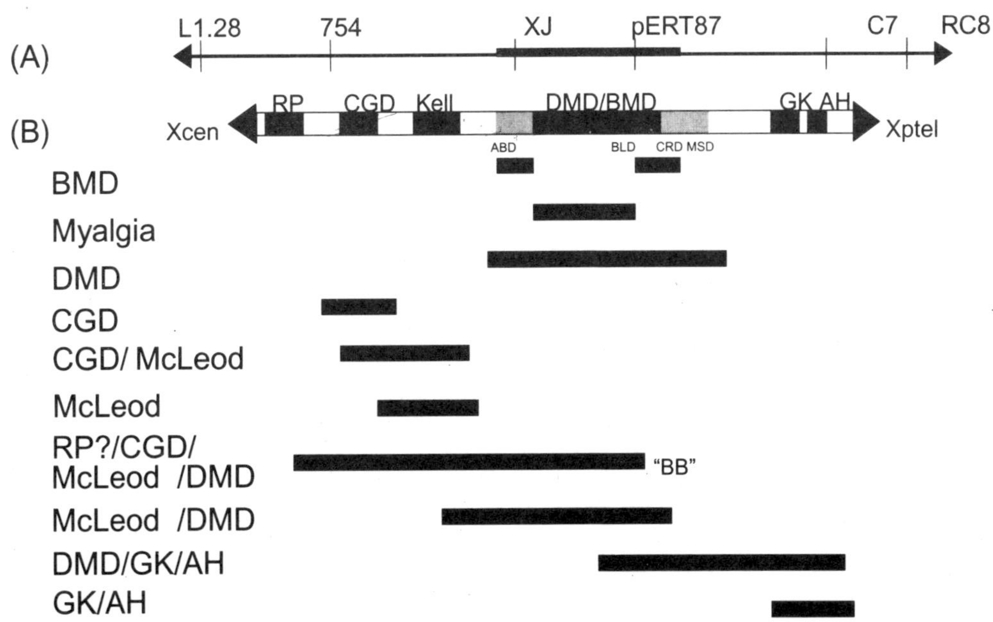

Figure 1 (A) Genetic map of the Xp12–p22 region showing the polymorphic marker loci initially used to localize the DMD/BMD gene to this region. (B) A map of the Xp21.1–p21.3 region indicating the genes that flank the dystrophin gene, and the complex inherited diseases that result from different DNA deletions within this region. The genes are RP (retinitis pigmentosa), CGD (chronic granulomatous disease), Kell (McLeod syndrome), DMD/BMD (dystrophin), GK (glycerol kinase deficiency) and AH (adrenal hypoplasia). The large X chromosome deletion of patient "BB" involved the retinitis pigmentosa, chronic granulomatous disease, Kell and dystrophin genes (RP/CGD/McLeod/DMD). From *Molecular Genetics of Human Inherited Disease*, edited by Duncan J. Shaw. © 1995 John Wiley & Sons Limited. Reproduced with permission.

examination showed that band Xp21 was shortened and light sub-band Xp21.2 absent. Screening of this patient's DNA with the aforementioned X-linked probes showed 754 to be deleted.

The isolation of DNA sequences that were located close to or within the DMD locus was achieved by two complementary strategies. One of these involved the hybridization of sheared DNA from patient BB with trace amounts of restriction enzyme-digested DNA from a 49,XXXXY lymphoblastoid cell line (Kunkel *et al.*, 1985). Shearing left BB's DNA with blunt ends, while digestion with *Mbo*I left DNA from the 4X-cell line with sticky ends. Since only DNA with sticky ends could be cloned into the chosen vector, this strategy could be utilized to enrich for sequences deleted in BB's DNA. Briefly, this was achieved by denaturing and extensively reassociating the DNA mixture using the technique of phenol enhancement (PERT). This strategy worked on the principle that the vast majority of the sequences from the 4X-cell line should anneal to BB's DNA, because they were present in excess. However, those sequences present in the 4X-cell line, but absent from BB's DNA as a consequence of the deletion, could only anneal to their complementary strands with sticky ends. These sequences were then selected by cloning into a suitable plasmid vector.

Seven clones were isolated in this way, one of which (pERT87) was thought to reside within or near to the DMD locus on the basis that it failed to hybridize in five of 57 unrelated DMD males (Monaco *et al.*, 1985). A larger survey of 1346 affected males found 6.5% to have deletions of pERT87 (Kunkel, 1986). A similar strategy was used by other workers to isolate another sequence called HIP25 (Smith *et al.*, 1987), which was shown to be deleted in DMD patients and located to the proximal (centromeric) side of pERT87. The second approach isolated sequences within the DMD locus based on the identification of a female with an X:21 translocation (Verellen-Dumoulin *et al.*, 1984). The site of translocation in the autosome was found to occur in the tandemly repeated genes encoding the 18S and 28S ribosomal RNAs (Worton *et al.*, 1984). This was exploited in 1986 by Ray and colleagues who used a probe from the 28S rRNA gene to identify the junctional fragment in clones derived from the region of the translocation site. The region spanning the translocation breakpoint (and which presumably contained at least part of the DMD locus) was then cloned. A sequence derived from the clone was found to detect an RFLP tightly linked to DMD. This probe (referred to as XJ probe) failed to hybridize with DNA from the genomes of a number of affected boys indicating that, in these patients, there was a deletion of the region complementary to the probe.

Bidirectional chromosomal walks from the 200 base pair (bp) pERT87 sequence and the 11 kilobase (kb) of X chromosome sequence of XJ1 expanded the loci identified by each clone. These walks resulted in the joining of the DXS164 (pERT87) and DXS206 (XJ1) loci to give approximately 380 kb of contiguous genomic DNA. Subclones of both loci which identified RFLPs were then used to analyse the DNA of DMD and BMD patients. This work showed a clear pattern of non-overlapping deletions extending in both directions at each locus. Since the identified loci failed to segregate with DMD/BMD mutations in approximately 5% of meioses, the implication was that the DMD/BMD locus was large. The subsequent isolation of other genomic probes and their physical localization showed the DMD locus to encompass approximately 2300 kb in length.

The isolation of the complete DMD cDNA was achieved by identifying transcribed regions at the DMD/BMD locus using unique sequence fragments of the DXS164 and DXS206 loci. Monaco and colleagues (1986) screened subclones of the DXS164 locus for cross-species DNA homology on the basis that evolutionary conserved sequences would be those most likely to be expressed. Using this methodology, a 16 kb transcript was identified in a Northern blot of foetal skeletal muscle. In 1987, Burghes and colleagues probed an adult muscle cDNA library with fragments of the DXS206 locus and identified a 16 kb transcript in adult muscle. The isolation of further cDNAs rapidly followed, such that cDNAs covering the entire DMD gene were cloned (Koenig *et al.*, 1987).

3. Mutations at the DMD locus

The genomic locus extends over 2400 kb and has one of the highest mutation rates so far reported (1×10^{-4} genes per generation). It consists of a minimum of 85 exons which have been well conserved throughout vertebrate evolution. Up to 64% of DMD cases are due to a gross deletion and 6% due to duplication within the gene. The vast majority of deletions cluster around two mutation hot spots, namely around exons 45–55 (1200 kb from the 5′ end of the gene) or close to the first 20 exons (500 kb from the 5′ end of the gene). Deletion size has no influence over whether the phenotype is DMD or BMD since deletions removing up to 50% of the dystrophin protein-coding sequences have been found to be associated with only mild forms of BMD, while other individuals with only small deletions display a DMD phenotype.

The underlying reason for this finding appears to depend on the influence of partial gene deletions on the open reading frame (ORF) of triplet codons in the mRNA transcript. Deletions causing a shift in the ORF (frameshift deletions) result in the DMD phenotype, presumably due either to metabolic instability of the mutant mRNAs or proteins, or most likely to functional inactivity of the latter. By contrast, those which maintain the ORF (in-frame deletions) give rise to the milder BMD, and are associated with the expression of mutant forms of dystrophin which, although internally deleted, exhibit variable but significant functional activity. Exceptions to this so-called "reading frame" hypothesis are thought to be due to variant alternative splicing patterns or the skipping of juxtaposed exons which effectively overcome the effect on the ORF of the original deletion mutation.

Gene duplications account for 5–10% of DMD cases and are probably the result of unequal sister-chromatid exchange (Emery, 1993); they also follow the reading frame rule. Point mutations causing a stop codon or amino acid substitution can also lead to severe disease.

4. Structure of the dystrophin gene

The 85 exons of the dystrophin gene span approximately 15 kb of genomic DNA with another 10–50 kb involved in promoter/enhancer functions (**Figure 2A**). It is consequently the largest known human gene. Approximately 98% of the 2.4 megabase gene is composed of introns and the exon length to intron length ratio is small (1:180). This large size appears to be conserved in mouse and chicken. There are at least 80 introns, the largest being 400 kb which is also the largest intron identified in a human gene. The largest exon is exon 79 (2.7 kb) which contains the entire 3′ untranslated region. The remaining exons are less than 200 bp.

At least seven different dystrophin promoters generate five different protein isoform size classes (**Figure 2A**). These promoters are named after their major, though not exclusive, sites of expression. The cortical (C), muscle (M) and Purkinje cell (P) encode full-length forms of dystrophin, which consist of unique first exons spliced to a shared set of 78 exons. Full-length dystrophin is confined to various sublocations within muscle and the central nervous system (CNS). The transcripts from these promoters are approximately 14 kb long and generate proteins of approximately 427 kilodalton (kDa). Several cases of X-linked cardiomyopathy (in which skeletal muscle involvement is minimal) are caused by dystrophin gene mutations; in some of these cases, the mutation inactivates the muscle promoter, which appears to be due to a compensation effect in skeletal but not cardiac muscle by the C and/or P promoters.

There are, in addition, at least four internal promoters with unique first exons that splice into exons 30, 45, 56 and 63. These are referred to, respectively, as the retinal (R), brain-3 (B_3), Schwann cell (S) and glial (G) promoters, and they give rise to proteins of 260, 140, 116 and 71 kDa (Dp260, Dp140, Dp116 and Dp71). Further diversity is generated by a range of alternative splicing events at the 3′ end of the gene and may reflect the functional adaptation of different isoforms in various locations (Ahn & Kunkel, 1993). Despite the widespread distribution of these isoforms, only striated muscle manifests a clear phenotypic consequence of dystrophin gene mutations. To date, no pathological abnormalities have been observed in the CNS of DMD patients. However, DMD is associated with variable degrees of selective cognitive impairment and a number of reports suggest that rearrangements in the second part of the gene, particularly those affecting Dp71, seem to be preferentially associated with such defects.

5. Evolution of the dystrophin gene

The dystrophin gene shares sequence homology with three other classes of genes, namely (1) utrophin, an autosomally encoded protein with an identical domain structure to dystrophin, (2) the spectrin gene family, including α-actinin, and (3) dystrobrevin, the mammalian orthologue of an 87 kDa postsynaptic protein originally identified in Torpedo electric organ. Since the 5′ and central portions of the dystrophin gene are homologous with members of the spectrin gene family, and the 3′ end is homologous with dystrobrevin, it has been suggested that dystrophin and utrophin may have evolved from the juxtaposition of an ancestral spectrin and 87 kDa-like gene to form a larger transcription unit.

The dystrophin/utrophin family seems to predate vertebrate radiation since dystrophins from such distantly related species as man and Torpedo still show greater similarity to each other than either does to human utrophin/DRP. Characterization of the C-terminal domains of dystrophin and utrophin from vertebrates and invertebrates shows that, while all the vertebrates examined express molecules with affinity to either dystrophin or utrophin branches, protochordates appear to have a single "protodystrophin" which is equally related to both vertebrate branches. This seems to suggest that a gene duplication event occurred in an amphioxus-like ancestor of the vertebrates, enabling one copy to fulfil a novel role. The rate of divergence of dystrophins is approximately one-quarter that of the utrophins and protodystrophins,

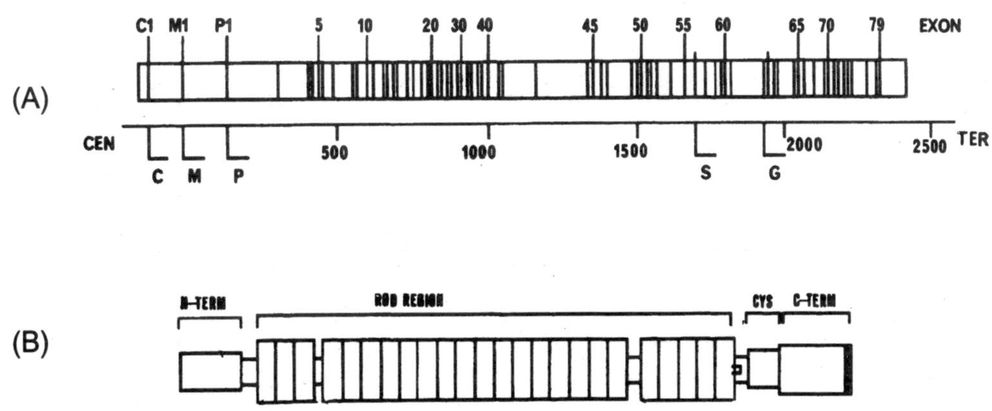

Figure 2 Transcript (A) and protein (B) maps of the dystrophin gene. The transcript map (A) shows the approximate location of the 79 exons and the position of the five different cell-type specific expression promoters, namely the C (cortical), M (muscle), P (Purkinje cell), S (Schwann cell) and G (glial). (B) The protein map identifies the four specific domains of the dystrophin gene product, namely the N-terminus (actin-binding domains), the rod-like region, the cysteine-rich domain (CYS) and the C-terminal domain (DAGC binding sites). From *Molecular Genetics of Human Inherited Disease*, edited by Duncan J. Shaw. © 1995 John Wiley & Sons Limited. Reproduced with permission.

suggesting that utrophin may fulfil an "ancestral role", while dystrophin has acquired specialized functions which further constrain its sequence (Roberts *et al.*, 1996).

6. Structure of dystrophin

Amino acid sequence analysis indicates that dystrophin is comprised of four contiguous domains (Koenig *et al.*, 1988) (**Figure 2B**). These are: (1) an N-terminal region (240 amino acids) which is similar to the conserved, actin-binding domain of α-actinin, spectrin and Dictyostelium actin-binding protein 120, and which binds to filamentous actin; (2) a large domain comprising 24 triple-helical sequences which resemble the repeats in spectrin and may be elastic; (3) a cysteine-rich region between residues 3080 and 3360 which shows significant homologies to the C-terminal domain of α-actinin in *Dictyostelium discoideum* and may contain two calcium-binding sites, although this is by no means proven; and (4) a sequence of 420 amino acids at the C-terminal end which appears to be unique apart from a similarity to the carboxyl terminus of utrophin encoded on chromosome 6 (see below). Thus, dystrophin appears to belong to the spectrin superfamily of proteins (Ahn & Kunkel, 1993). Based on the amino acid sequence similarity to spectrin, it was predicted that dystrophin would form a flexible rod-shaped molecule with a length of 125 nm (Koenig *et al.*, 1988). This has since been confirmed by rotary-shadowed electron microscopy of isolated dystrophin.

7. Dystrophin and its association with other proteins

Dystrophin locates to the cytoplasmic face of the normal adult muscle fibre membrane (Byers *et al.*, 1991), where it comprises approximately 5% of the membrane cytoskeleton. This association with the membrane is mediated by both the N- and C-terminal domains of dystrophin (Dunckley *et al.*, 1994). Patients with domain-specific "in-frame" deletions show that mutations in the putative actin-binding domain of the N terminus tend to be associated with a severe or intermediate BMD phenotype, whereas the absence of the cysteine-rich and proximal half of the carboxyl-terminal domain invariably leads to a severe DMD phenotype (Koenig *et al.*, 1989).

The cysteine-rich and C-terminal (CRCT) domains of dystrophin associate with the membrane via a large oligomeric complex of glycoproteins and proteins. This complex is referred to as the dystrophin-associated glycoprotein complex or DAGC (Ervasti & Campbell, 1991). The DAGC may be divided into: (1) the dystroglycan complex which is composed of α-dystroglycan (156 kDa) and β-dystroglycan (43 kDa); and (2) the sarcoglycan complex which is composed of α-sarcoglycan (also known as adhalin; 50 kDa), β-sarcoglycan and γ-sarcoglycan (35 kDa) (see **Figure 3**). α-Dystroglycan is a highly glycosylated peripheral-surface proteoglycan which is able to bind to the G domain of laminin, merosin and agrin in the extracellular matrix. β-Dystroglycan is a transmembrane protein, the extreme C terminus of which constitutes a unique binding site for the second half of hinge 4 and the cysteine-rich region of dystrophin. α- and β-Dystroglycan are tightly associated and form the principal linkage between the fibre cytoskeleton and the surrounding extracellular matrix. Dystroglycan is widely expressed in muscle and non-muscle tissues. The carbohydrate moieties rather than the protein sequence differ

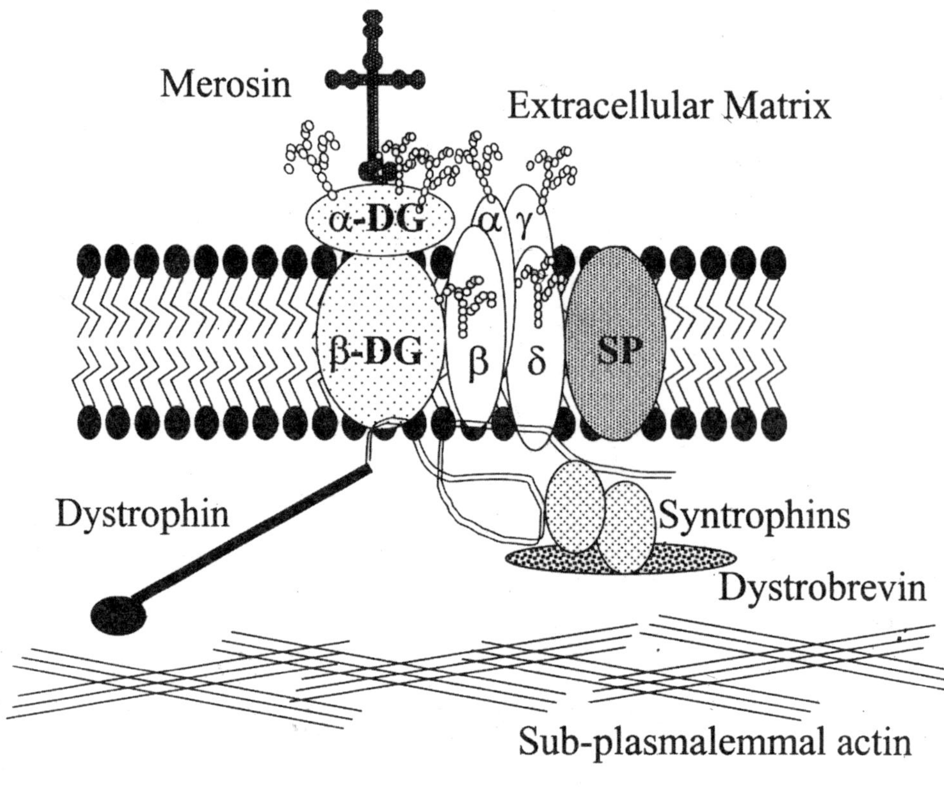

Figure 3 Diagrammatic representation of dystrophin and the dystrophin-associated glycoprotein complex (DAGC). The DAGC consists of α- and β-dystroglycan (α- and β-DG) and α-, β-, δ- and γ-sarcoglycan (α,β,γ,δ). α-dystroglycan binds to merosin in the extracellular matrix while dystrophin through its binding to β-dystroglycan links the entire complex to the actin associated cytoskeleton. The roles of syntrophin and dystrobrevin are as yet unclear but their association with the neuronal form of nitric oxide synthase is suggestive of a role in cell signalling.

between tissues, an example being cardiac muscle where α-dystroglycan is of a lower molecular weight compared with skeletal muscle. This variable glycosylation of the protein core has been suggested to modulate laminin binding.

Defects in either the α-, β-, γ- or δ sarcoglycan genes result in a group of autosomal recessive muscular dystrophies collectively referred to as the "sarcoglycanopathies" (Worton, 1995). When one gene is mutated all sarcoglycans are absent or severely reduced at the muscle fibre membrane. In addition, a specific deficiency of merosin, the laminin M chain, results in a classical non-Japanese form of congenital muscular dystrophy and dystrophia muscularis in mice (Sunada *et al.*, 1994).

8. Animal models of dystrophin deficiency

Several animal models of dystrophin deficiency have been described, namely the *mdx* mouse (Bulfield *et al.*, 1984), *xmd* dog (Cooper & Valentine, 1988) and the dystrophic cat (Gaschen *et al.*, 1992). The consequences of dystrophin deficiency in each of these species are very different, with the *mdx* mouse showing a markedly milder phenotype than that observed in patients with DMD. However, many skeletal muscles, and the diaphragm and intercostal muscles of older *mdx* mice, eventually exhibit a progressively severe and fibrotic myopathy similar to the human disease (Stedman *et al.*, 1991). In the *xmd* dog, the myopathy follows a similar course to DMD in man, with progressive muscular weakness, hypertrophy of defined muscle groups and premature death due to respiratory insufficiency. The cat model shows marked diaphragmatic and glossal hypertrophy, but, unlike the human disease and the *xmd* dog, there is no progressive loss of muscle fibres, fibrosis or weakness, despite the deficiency of dystrophin.

9. Future strategies for therapeutic intervention in DMD

As yet, gene therapy for DMD has not reached the stage of clinical trials and is currently restricted to somatic gene transfer experiments in the *mdx* mouse. Full-length mouse and human dystrophin cDNAs derived from the 14 kb major skeletal muscle transcript have been cloned and expressed in mammalian cells under the control of strong constitutive promoters generating functional recombinant proteins of the appropriate size (Dickson *et al.*, 1991; Lee *et al.*, 1991). These cDNAs are, however, too large for inclusion into conventional adenoviral or retroviral vectors and so their use is restricted to plasmid DNA injections or for inclusion in advanced vectors such as those based on herpes simplex virus or novel minimal-sequence adenovirus backbones.

Gene transfer into skeletal muscle via plasmid injection has to date proved to be of too low efficiency, with experiments in the *mdx* mouse resulting in only 1–2% of skeletal muscle fibres being transduced after a single injection into the *rectus femoris* (Acsadi *et al.*, 1991). While efficiency may be enhanced by inducing muscle regeneration prior to injection, it remains too low at present to be of any therapeutic value. In terms of future strategies, it is fortunate that some mild BMD phenotypes have been associated with very large in-frame deletions within the DMD locus. A 6.3 kb cDNA has been cloned from the skeletal muscle dystrophin mRNA of one such patient bearing a deletion of exons 17–48 and expressing a 220 kDa dystrophin protein which lacks 46% of the central rod domain (England *et al.*, 1990). This "mini-dystrophin" gene does not exceed the packaging limits of currently available adenoviral and retroviral vectors and has been used successfully to deliver the transgene to *mdx* skeletal muscle. Overall, viral-mediated gene transfer, based on retroviral and adenoviral vectors, has generally proved to be more effective than the direct injection of plasmid DNA. Retroviruses infect dividing cells with high efficiency, integrating their genome into the host DNA. Adenoviruses infect both dividing and non-dividing cells, but the DNA remains extrachromosomal and may, therefore, prove to be less stable in target cells. However, with both types of viral vector, the stability of infection and immune responses to viral proteins remain a major challenge to their routine clinical use.

One variant to classical gene transfer therapy is the reactivation of an embryonic or foetal isoform that might complement the deficiency of a mutated adult gene. The feasibility of such an approach is shown by the upregulation of foetal haemoglobin in sickle-cell anaemia patients. Given the homology between utrophin and dystrophin, utrophin would seem to be an obvious candidate. Utrophin is developmentally regulated in skeletal muscle, showing a maximum level of expression at 17–18 weeks' gestation which subsequently decreases as the expression of dystrophin increases (Clerk *et al.*, 1993). There is also some indication that utrophin binds to a complex antigenically identical to the DAGC. Moreover, recent work suggests that the overexpression of utrophin is able to compensate for the absence of dystrophin in mouse muscle (Campbell & Crosbie, 1996; Tinsley *et al.*, 1996). It now remains to be determined whether or not utrophin may be upregulated pharmacologically, since this would avoid many of the problems associated with many of the viral gene transfer strategies currently under development.

REFERENCES

Acsadi, G., Dickson, G., Love, D.R. *et al.* (1991) Human dystrophin expression in mdx mice after intramuscular injection of DNA constructs. *Nature* 352: 815–18

Ahn, A.H. & Kunkel, L.M. (1993) The structural and functional diversity of dystrophin. *Nature Genetics* 3: 283–91

Bulfield, G., Siller, W.G., Wight, P.A.L. & Moore, K.J. (1984) X chromosome-linked muscular dystrophy (mdx) in the mouse. *Proceedings of the National Academy of Sciences USA* 81: 1189–92

Burghes, A.H.M., Logan, C., Hu, X. *et al.* (1987) A cDNA clone from the Duchenne/Becker muscular dystrophy gene. *Nature* 328: 434–37

Bushby, K.M.D., Thambyayah, M. & Gardner-Medwin, D. (1991) Prevalence and incidence of Becker Muscular-Dystrophy. *Lancet* 337: 1022–24

Byers, T.J., Kunkel, L.M. & Watkins, S.C. (1991) The subcellular-distribution of dystrophin in mouse skeletal, cardiac, and smooth-muscle. *Journal of Cell Biology* 115: 411–21

Campbell, K.P. & Crosbie, R.H. (1996) Muscular-dystrophy – utrophin to the rescue. *Nature* 384: 308–09

Clerk, A., Morris, G.E., Dubowitz, V., Davies, K.E. & Sewry, C.A. (1993) Dystrophin-related protein, utrophin, in normal and dystrophic human fetal skeletal-muscle. *Histochemical Journal* 25: 554–61

Cooper, B.J. & Valentine, B.A. (1988) X-linked muscular-dystrophy in the dog. *Trends in Genetics* 4: 30

Davies, K.E., Ball, S.P., Dorkins, H.R. *et al.* (1986) Molecular analysis of human X-linked diseases. *Cold Spring Harbor Symposia on Quantitative Biology* 50: 337–43

Dickson, G., Love, D.R., Davies, K.E. *et al.* (1991) Human dystrophin gene-transfer – production and expression of a functional recombinant DNA-based gene. *Human Genetics* 88: 53–58

Dorkins, H., Junien, C., Mandel, J.L. *et al.* (1985) Segregation analysis of a marker localised Xp21.2–Xp21.3 in Duchenne and Becker muscular dystrophy families. *Human Genetics* 71: 103–07

Dunckley, M.G., Wells, K.E., Piper, T.A., Wells, D.J. & Dickson, G. (1994) Independent localization of dystrophin N-terminal and C-terminal regions to the sarcolemma of mdx mouse myofibers in-vivo. *Journal of Cell Science* 107: 1469–75

Emery, A.E.H. (1993) *Duchenne Muscular Dystrophy*, 2nd edition, Oxford and New York: Oxford University Press

England, S.B., Nicholson, L.V.B., Johnson, M.A. *et al.* (1990) Very mild muscular-dystrophy associated with the deletion of 46-percent of dystrophin. *Nature* 343: 180–82

Ervasti, J.M. & Campbell, K.P. (1991) Membrane organization of the dystrophin–glycoprotein complex. *Cell* 66: 1121–31

Franke, U., Ochs, H.D., de Martinville, B. *et al.* (1985) Minor Xp21 chromosome deletion in a male associated with expression of Duchenne muscular dystrophy, chronic granulomatous disease, retinitis pigmentosa, and McLeod syndrome. *American Journal of Human Genetics* 37: 250–67

Gaschen, F.P., Hoffman, E.P., Gorospe, J.R.M. *et al.* (1992) Dystrophin deficiency causes lethal muscle hypertrophy in cats. *Journal of the Neurological Sciences* 110: 149–59

Hofker, M.H., Wapenaar, M., Goor, N. *et al.* (1985) Isolation of probes detecting restriction fragment length polymorphisms from chromosome specific libraries: potential use for diagnosis of Duchenne muscular dystrophy. *Human Genetics* 70: 148–56

Koenig, M., Hoffman, E.P., Bertelson, C.J. *et al.* (1987) Complete cloning of the Duchenne muscular dystrophy (DMD) cDNA and preliminary genomic organisation of the DMD gene in normal and affected individuals. *Cell* 53: 219–28

Koenig, M., Monaco, A.P. & Kunkel, L.M. (1988) The complete sequence of dystrophin predicts a rod-shaped cytoskeletal protein. *Cell* 53: 219–28

Koenig, M., Beggs, A.H., Moyer, M. *et al.* (1989) The molecular basis for Duchenne versus Becker muscular dystrophy: correlation of severity with type of deletion. *American Journal of Human Genetics* 45: 498–506

Kunkel, L.M. *et al.* (1986) Analysis of deletions in DNA from patients with Becker and Duchenne muscular dystrophy. *Nature* 322: 73–77

Kunkel, L.M., Monaco, A.P., Middlesworth, W., Ochs, H.D. & Latt, S.A. (1985) Specific cloning of DNA fragments absent from the DNA of a male patient with an X chromosome deletion. *Proceedings of the National Academy of Sciences USA* 82: 4778–82.

Lee, C.C., Pearlman, J.A., Chamberlain, J.S. & Caskey, C.T. (1991) Expression of recombinant dystrophin and its localization to the cell-membrane. *Nature* 349: 334–36

Monaco, A.P., Bertelson, C.J., Middlesworth, W. *et al.* (1985) Detection of deletions spanning the Duchenne muscular dystrophy locus using a tightly linked DNA segment. *Nature* 316: 842–45

Monaco, A.P., Neve, R.L., Colletti-Feener, C. *et al.* (1986) Isolation of candidate cDNAs for portions of the Duchenne muscular dystrophy gene. *Nature* 323: 646–50

Murray, J.M., Davies, K.E., Harper, P.S. *et al.* (1982) Linkage relationship of a cloned DNA sequence on the short arm of the X chromosome to Duchenne Muscular Dystrophy. *Nature* 300: 69–71

Ray, P.N., Belfall, B., Duff, C. *et al.* (1986) Cloning of the breakpoint of an X:21 translocation associated with Duchenne muscular dystrophy. *Nature* 318: 672–75

Roberts, R.G. (1995) Dystrophin, its gene, and the dystrophinopathies. *Advances in Genetics* 33: 177–231

Roberts, R.G., Freeman, T.C., Kendall, E. *et al.* (1996) Characterization of drp2, a novel human dystrophin homolog. *Nature Genetics* 13: 223–26

Smith, T.J., Wilson, L., Kenwrick, S.J. *et al.* (1987) Isolation of a conserved sequence deleted in Duchenne muscular dystrophy patients. *Nucleic Acid Research* 15: 2167–74

Stedman, H.H., Sweeney, H.L., Shrager, J.B. *et al.* (1991) The mdx mouse diaphragm reproduces the degenerative changes of Duchenne muscular-dystrophy. *Nature* 352: 536–39

Sunada, Y., Bernier, S.M., Kozak, C.A., Yamada, Y. & Campbell, K.P. (1994) Deficiency of merosin in dystrophic dy mice and genetic-linkage of laminin m-chain gene to dy locus. *Journal of Biological Chemistry* 269: 13729–32

Tinsley, J.M., Potter, A.C., Phelps, S.R. *et al.* (1996) Amelioration of the dystrophic phenotype of mdx mice using a truncated utrophin transgene. *Nature* 384: 349–53

Verellen-Dumoulin, C., Freund, M., de Meyer, R. *et al.* (1984) Expression of an X-linked muscular dystrophy in a female due to translocation involving Xp21 and non-random inactivation of the normal X chromosome. *Human Genetics* 67: 115–19

Worton, R. (1995) Muscular-dystrophies – diseases of the dystrophin–glycoprotein complex. *Science* 270: 755–56

Worton, R.G., Duff, C., Sylvester, J.E., Schmickel, R.D. & Willard, H. (1984) Duchenne muscular dystrophy involving translocation of the DMD gene next to ribosomal RNA genes. *Science* 224: 1447–49

GLOSSARY

adenoviral relating to a group of viruses first isolated from cultures of adenoids. Cause of a variety of human respiratory infections

blunt end end of a double-stranded DNA molecule, cut at the same site on both strands without producing sticky ends

chromosomal walk method of mapping chromosomes from a collection of overlapping restriction fragments

enhancer DNA sequence that will increase the level of expression from a gene in manner independent of position and orientation

glycoprotein a protein linked to a sugar or polysaccharide which are components of receptor molecules on the outer surface of cells

glycosylation process of adding sugars (oligosaccharides) to proteins to form glycoproteins

hypertrophy increase in the size of a structure resulting from an increase in the size of the cells rather than the number of cells; overgrowth

lymphoblastoid cell line immature cells that develop into lymphocytes following an antigenic or mitogenic challenge

myopathy abnormal condition of skeletal muscle characterized by muscle weakness, wasting and histological changes within the muscle tissue

Northern blot technique in which RNA molecules, separated by electrophoresis in an agarose gel, are transferred to "blotting" to a nitrocellulose filter, then hybridized with radioactively labelled nucleic acid probes for identification and isolation of sequences of interest

oligomeric relating to a molecule comprised of only a few monomer units

orthologue genes related by common phylogenetic descent

postsynaptic protein highly conserved peripheral membrane protein believed to help attach acetylcholine receptors in the postsynaptic membrane

sticky end cut end of a short stretch of single-stranded DNA that will hybridize (stick) to a complementary strand or to sequences on other DNA fragments that were cut by the same restriction endonuclease

GENETICS OF OSTEOPOROSIS

Fiona E.A. McGuigan and Stuart H. Ralston
Bone Research Group, Department of Medicine and Therapeutics, University of Aberdeen, UK

1. Definition

Osteoporosis is a common disease characterized by a reduction in bone mass, microarchitectural deterioration of bone tissue and an increased risk of fracture. The importance of reduced bone mass in the pathogenesis of osteoporotic fractures led a World Health Organisation (WHO) working group to define osteoporosis on the basis of bone mineral density (BMD) measurements. Using the WHO definition, individuals who have BMD values more than 2.5 standard deviations (SD) below the mean in young healthy adults are classified as having osteoporosis; those with BMD values between −1 and −2.5 SD below normal are said to have osteopaenia and those with BMD values above −1.0 SD are said to have normal bone mass. Patients who have osteoporosis on the grounds of low BMD and who also suffer a fragility fracture are said to have severe osteoporosis. According to these definitions, approximately 50% of all postmenopausal women will be osteoporotic by the age of 80 years, since bone mass falls progressively with increasing age (World Health Organization, 1994).

2. Epidemiology

The clinical importance of osteoporosis lies in its association with fractures, which increase in incidence with age corresponding with age-related reductions in bone mass. These fractures are common, an important cause of morbidity and mortality and are expensive to treat. The most common sites for osteoporosis-related fractures are the hip, spine and distal forearm, but fractures of many other bones including the humerus, ribs and hand are also associated with reduced bone mass (Seeley *et al.*, 1991). The incidence of osteoporosis-related fractures increases dramatically above the age of 50 in both sexes, but women are affected more commonly than men because of an increased rate of bone loss after the menopause. This bone loss in postmenopausal women is due to an oestrogen deficiency and can be prevented by hormone replacement therapy. Although bone mass is strongly related to the risk of fracture, other factors also contribute. For example, individuals with an increased hip axis length are predisposed to hip fracture in the event of a fall, since the force transmitted to the femoral neck is dependent on this aspect of femoral neck geometry (Faulkner *et al.*, 1993). Most fractures in the elderly are precipitated by falls and these occur in about one-third of the population over the age of 65; it has been estimated that about 1% of such falls results in a fracture (Dennison *et al.*, 1998).

3. Secondary osteoporosis

Osteoporosis may occur as a complication of medical conditions such as thyrotoxicosis, Cushing's syndrome, hypogonadism, neoplasia, malabsorption and inflammatory diseases. The disease may also occur as the result of certain drug treatments such as corticosteroids, which reduce bone mass, and other drugs such as sedatives, which increase the likelihood of a fall. Osteoporosis may also be observed in monogenic disorders such as osteoporosis–pseudoglioma syndrome (OPS), Prader–Willi syndrome, homocystinuria and Gaucher's disease. Severe osteoporosis is observed in osteogenesis imperfecta (OI), a rare disease which is genetically heterogeneous, but which most commonly occurs as the result of mutations in the collagen Iα1 or collagen Iα2 genes. Since these diseases are dealt with in more detailed medical texts (see Online Mendelian Inheritance in Man; also Resnick, 1995; Rowe & Shapiro, 1998), this review will concentrate on the genetics of idiopathic osteoporosis.

4. Pathophysiology

Osteoporosis occurs when there is an imbalance between the amount of bone laid down during skeletal growth and that which is lost as the result of ageing. Bone mass increases linearly during childhood with a marked acceleration during puberty until a peak is reached between the ages of 25 and 35. Subsequently, there is a gradual reduction in bone mass in both sexes. Bone loss is accelerated in women after the menopause, as the result of oestrogen deficiency, which causes the "uncoupling" of bone resorption from bone formation, which occur continually throughout life in the normal skeleton. This process of bone remodelling is important in maintaining the mechanical integrity of the skeleton since it provides a mechanism by which old or damaged bone can be removed and replaced with new bone.

The cells responsible for bone resorption are termed osteoclasts. These are multinucleated cells from the monocyte–macrophage lineage, which resorb bone matrix by secreting acid and proteolytic enzymes. Bone formation is carried out by osteoblasts, which are derived from cells of the mesenchymal lineage. Some osteoblasts become buried in the newly formed bone matrix and undergo further differentiation to form osteocytes, which are thought to act as sensors of mechanical strain. The reduction in bone mass which occurs as the result of uncoupling between bone formation and resorption is accompanied by structural abnormalities of bone, such as trabecular thinning and perforation, which reduce the mechanical strength of bone.

5. Environmental and anthropomorphic determinants of bone mass

Many environmental factors contribute to the regulation of bone mass and to the pathogenesis of osteoporotic fractures. These include falls, low calorie and protein intake, low calcium and vitamin D intake, excessive caffeine and alcohol consumption, and reduced physical activity (Cummings *et al.*, 1985; Cooper *et al.*, 1992; Krall & Dawson-Hughes, 1993; Dawson-Hughes *et al.*, 1997; Etherington *et al.*, 1999). These environmental influences, while relevant, play relatively minor roles in determining bone mass, except under extreme conditions. Body weight is also strongly correlated with bone mass and, although this is subject to environmental influences, twin studies have shown that both height and weight are under strong genetic control (Arden & Spector, 1997) and it is probable that many of the genes which regulate body size are also important genetic determinants of bone mass.

6. Genetics and osteoporosis

Several lines of evidence indicate that genetic factors play an important role in the pathogenesis of osteoporosis. Bone mass is one of the most important determinants of osteoporotic fracture risk and it has been estimated from twin studies that up to 85% of the variance in BMD is genetically determined (Smith *et al.*, 1973; Pocock *et al.*, 1987; Christian *et al.*, 1989; Kelly *et al.*, 1991, 1993). Segregation studies in families also support a strong genetic contribution to the regulation of bone mass, with a pattern of inheritance that is most consistent with the effects of several genes, each with modest effects, rather than a few genes with large effects (Gueguen *et al.*, 1995). Other determinants of fracture risk also have a genetic component; family studies have shown that a maternal or grandmaternal history of hip fracture is associated with an increased risk of fracture independent of bone mass (Torgerson *et al.*, 1995). In keeping with this observation, several other traits that predispose to osteoporotic fracture – including hip axis length (Faulkner *et al.*, 1993), ultrasound properties of bone (Arden *et al.*, 1996) and biochemical markers of bone turnover (Kelly *et al.*, 1991; Tokita *et al.*, 1994) – have been found to have a heritable component in twin studies.

7. Molecular genetic studies in osteoporosis

Identifying the genes responsible for regulation of bone mass and other determinants of osteoporotic fracture risk is made difficult by the fact that the traits in question are determined by a complex interaction of genetic, metabolic and environmental factors. Moreover, current evidence suggests that, in normal subjects, the effect of individual genes is small and influenced by an interaction with environmental factors and by gene–gene interactions. A number of approaches are commonly used to dissect out the genetic contribution to complex diseases. These include (i) linkage studies, (ii) allele sharing studies and (iii) candidate gene studies.

(i) Linkage studies

Relatively few linkage studies have been performed in the field of osteoporosis because of the difficulties associated with obtaining multigenerational families who are suitable for analysis and because the mode of inheritance of the disease is unclear. A notable exception is the recent study by Johnson *et al.* which described a large family in which unusually high bone mass was inherited in a simple autosomal dominant fashion through several generations. A genome-wide search in this family resulted in the identification of a candidate locus for the high bone mass (HBM) trait on chromosome 11q12–13 (Johnson *et al.*, 1997). This locus is of special interest since it harbours the gene responsible for OPS, an autosomal recessive disease, which presents with ocular abnormalities and juvenile osteoporosis (Gong *et al.*, 1996). Together, these observations raise the possibility that both traits may be caused by allelic variation of the same gene. If this were the case, osteoporosis would presumably result from a loss-of-function (recessive) mutation and high bone mass from a gain-of-function (dominant) mutation. Alternatively, these diseases could result from mutations in two separate genes, both of which are involved in regulating bone metabolism. Fine mapping in the high bone mass family has narrowed the region of interest to 1–2 Mb, but the gene responsible has not yet been identified.

(ii) Allele sharing studies in sib-pairs

Non-parametric linkage analysis in sib-pairs has been successfully used in mapping the genes responsible for several complex diseases and has recently been applied to the study of osteoporosis. Koller *et al.* (1998) used this approach to determine if the HBM locus on chromosome 11q12–13 might also contribute to the genetic variation in BMD in a normal population. The study, which employed 835 premenopausal siblings of Caucasian and African–American origin, showed strong evidence of linkage to the region of interest with a peak lod score of 3.51 at the marker D11S987, close to the threshold of 3.6 which is generally accepted as definite evidence in favour of linkage. Although positive lod scores were observed in both groups, the effect was greater among Caucasian women, raising the possibility that the genes responsible for regulation of bone mass may differ in different ethnic groups.

Devoto *et al.* (1998) also used non-parametric linkage analysis in an extended sib-pair study to screen for chromosomal regions involved in the determination of spine and femoral neck bone density. A genome search identified three regions with lod scores above 2.5 on chromosomes 1p36, 2p23–24 and 4q32–34. Another locus on chromosome 11q, distinct from that identified by Johnson (Johnson *et al.*, 1997) and other workers (Koller *et al.*, 1998), had a lod score of 2.08, indicating that a broad area of chromosome 11 may be involved in bone mass regulation. A further genome-wide scan was performed in 153 Chinese sib-pairs by Nui *et al.* in relation to forearm BMD measurements (Nui *et al.*, 1999). The highest lod score was 2.15 for a locus on chromosome 2p21.1–24, which almost meets the criteria for suggestive linkage (Lander & Kruglyak, 1995) and overlaps with the region identified by Devoto as being linked to spine BMD (Devoto *et al.*, 1998). A further region showing possible linkage was identified on chromosome 13q21–24. These studies indicate that several genes in these areas of the genome contribute to the regulation of bone mass, although their identity remains unclear at present.

(iii) Candidate gene studies

This approach has been most widely used in studying the genetic basis of osteoporosis. Research so far has focused on candidate genes such as cytokines and growth factors (which regulate bone turnover), those that encode components of bone matrix and those that encode receptors for calciotropic hormones.

8. Vitamin D receptor

Vitamin D plays an important role in calcium homeostasis by regulating bone cell growth and differentiation, intestinal calcium absorption and parathyroid hormone secretion. The pleiotropic effects of this hormone on bone metabolism led Morrison and colleagues to study the relationship between common polymorphisms of the vitamin D receptor (VDR) gene and various aspects of bone metabolism (Morrison *et al.*, 1992). Morrison identified three common polymorphisms in the VDR gene situated between exons 8 and 9, which are recognized by the restriction enzymes *Bsm*I, *Apa*I and *Taq*I. These polymorphisms show strong linkage disequilibrium with each other, but do not cause a change in the coding sequence of VDR, since two are intronic and one is a conservative change in exon 9. Despite this, reporter gene constructs prepared from exons 8 and 9 of the VDR gene in different individuals have shown haplotype-specific effects in reporter assays. This suggests that the polymorphisms could act as a marker for other sequence variations in the 3′ untranslated region of VDR, which might affect RNA stability (Morrison *et al.*, 1992).

A strong association between the above polymorphisms and circulating levels of the osteoblast-specific protein osteocalcin has been reported in twin studies (Morrison *et al.*, 1992). Subsequent work showed associations with bone mass in twin studies and population studies (Morrison *et al.*, 1994). Other twin studies and population-based studies have been conflicting, however, with positive results in some populations (Eisman, 1995; Fleet *et al.*, 1995; Spector *et al.*, 1995; Viitanen *et al.*, 1996; Sainz *et al.*, 1997) and negative results (Garnero *et al.*, 1995; Peacock, 1995; Jorgensen *et al.*, 1996) in others. A possible explanation for the discrepancy between studies is that the effects of VDR genotype on bone loss and bone mass seem to be modified by calcium and vitamin D intake (Tarlow *et al.*, 1993; Krall *et al.*, 1995; Kiel *et al.*, 1997). For example, VDR genotype has been reported to have little effect on bone mass or bone loss in individuals with a high daily dietary calcium intake, whereas positive effects have been noted in the presence of reduced calcium intake (Dawson-Hughes *et al.*, 1995; Ferrari *et al.*, 1998a,b). VDR genotype has also been found to influence the response of bone mass to vitamin D supplementation (Graffmans *et al.*, 1997). A meta-analysis of 16 studies published in 1996 concluded that the effect of VDR on bone mass diminished with age, suggesting that the polymorphisms may act principally as determinants of peak bone mass (Cooper & Umbach, 1996). This is supported by the work of Riggs *et al.* (1995) and Sainz *et al.* (1997) who found a strong association between VDR gene in prepubertal girls and premenopausal women, which then diminished with increasing age.

A fourth polymorphism in VDR has more recently been described in exon 2, which is recognized by the restriction enzyme *Fok*I. This polymorphism creates an alternative translational start site, resulting in the production of two isoforms of the VDR protein, which differ in length by three amino acids. Evidence for functional differences between the isoforms has been conflicting; some groups have reported functional differences between the isoforms *in vitro* using certain assays, whereas other groups have found no differences using different assays. Clinical studies of the *Fok*I polymorphism have also been conflicting, with positive results in some studies and negative results in others (Gross *et al.*, 1996; Minamitani *et al.*, 1996; Arai *et al.*, 1997; Harris *et al.*, 1997).

9. Type I collagen

The genes encoding type I collagen (COLIA1 and COLIA2) are important candidates for the regulation of bone mass since collagen is the major protein of bone and mutations in the coding regions of the type 1 collagen genes give rise to the disease OI (Rowe, 1991). Working on the assumption that osteoporosis may result from subtle mutations in the collagen genes, Spotila *et al.* (1994) screened the coding regions of COLIA1 and COLIA2 for mutations in a series of families with premature osteoporosis. A disease-associated coding mutation was found in one family, which on clinical evaluation was thought to represent a mild form of OI. In a subsequent study, Grant and Ralston screened the regulatory regions of COLIA1 (promoter and intron 1) for mutations in a case-control study and a population-based study (Grant & Ralston, 1996). Three polymorphisms were found, all situated in the first intron of COLIA1, although two of these were rare (allele frequency <5%). Since osteoporosis is a common disease, further work concentrated on a relatively common G/T polymorphism (allele frequency 22%) which was situated at the first base of a binding site for the transcription factor Sp1. Grant found over-representation of the T allele (designated "s") in a case-control study of patients with osteoporotic vertebral fracture and found an association with bone mass in a population-based study.

In a subsequent population-based study of over 1800 individuals, Uitterlinden *et al.* (1998) also found over-representation of the "s" allele in patients with non-vertebral fractures and an association between the presence of the "s" allele and reduced bone mass. An interesting feature of this study was that the genotype-specific effect increased with increasing age, leading the authors to suggest that the polymorphism might act as a marker for bone loss. The polymorphism was also found to predict the presence of fracture, even after correcting for genotype-related differences in bone mass, suggesting that the polymorphism could predispose to fractures by affecting bone quality or skeletal geometry. Positive associations between the COLIA1 Sp1 polymorphism and bone mass have been found in other population-based studies (Garnero *et al.*, 1998) and in case-control studies in both men (Langhdal *et al.*, 1998) and women (Langhdal *et al.*, 1998; Roux *et al.*, 1998; Keen *et al.*, 1999).

Not all studies of the COLIA1 polymorphism have been positive; a small study in a Swedish population showed no association with bone mass or fracture (Liden *et al.*, 1998). Other studies have shown significant differences in population prevalence of the "s" allele. Although this polymorphism is relatively common in Caucasian populations, it is rare in Africans and so far has not been found in Chinese or Korean populations (Han *et al.*, 1999). These studies have led to the suggestion that population differences in frequency of the "s" allele may partly explain the ethnic differences in fracture risk (Bevan *et al.*, 1998). The mechanism by which the Sp1 polymorphism predisposes to osteoporosis remains to be fully defined, but preliminary data have shown evidence of allele-specific differences in binding of the Sp1 protein to the polymorphic recognition site, differences in allele-specific transcription and differences in collagen protein production in patients of different genotype (Dean *et al.*, 1998).

10. Transforming growth factor

Transforming growth factor beta (TGFβ) is abundant in bone matrix and is released during the process of osteoclastic bone resorption. Although there are three isoforms of TGFβ, most attention has focused on TGFβ-1 since this is abundant in bone. TGFβ regulates both osteoblast and osteoclast activity *in vitro* and it has been speculated that release of TGFβ during bone remodelling may act as a local mediator of osteoblast–osteoclast coupling (Pfeilschifter *et al.*, 1988); in addition, it has been implicated as a mediator of the skeletal effects of oestrogen (Pacifici *et al.*, 1991). A polymorphism in intron 4 of the TGFβ-1 gene has been identified in some osteoporotic individuals with very low BMD and fracture (Knudsen *et al.*, 1995), but the effect of this polymorphism on TGFβ function is unclear. Another polymorphism of the TGFβ-1 coding region, a T to C transition, has been described by Yamada *et al.* (1998), which results in a leucine to proline amino acid substitution. A significant association has been found between the T variant and low bone density in postmenopausal Japanese women. The polymorphism is also associated with circulating serum levels of TGFβ, indicating that it may influence the efficiency of protein secretion. This is consistent with the suggested role of TGFβ in bone remodelling, but the significance of these observations for the regulation of bone mass remains to be fully determined.

11. Oestrogen receptor alpha

Oestrogen is known to increase bone mass and inhibit bone turnover. The potential importance of oestrogen receptor alpha (ERα) as a candidate gene for osteoporosis stems from the clinical observation of severe osteoporosis in a man with a protein-coding mutation of the gene (Smith *et al.*, 1994). This prompted further studies which looked at the relationship between other polymorphisms at the ERα locus and bone mass. An initial study reported associations between a TA repeat polymorphism in the ER promoter and bone mass in a Japanese population (Sano *et al.*, 1995). Other studies looked at possible associations between restriction enzyme *Pvu*II and *Xba*I polymorphisms in the first intron of the gene and bone mass (Hosoi *et al.*, 1995; Kobayashi *et al.*, 1996; Mizunuma *et al.*, 1997). These polymorphisms are in strong linkage disequilibrium with each other and with the TA repeat polymorphism. Studies in Japanese populations have shown significant associations between the intronic polymorphisms and bone mass (Kobayashi *et al.*, 1996), whereas similar studies in a Korean population have yielded negative results (Han *et al.*, 1997). The mechanism by which these polymorphisms influence bone mass or ER function is unclear, since the intronic polymorphisms lie in an apparently non-functional area of the gene. Although the promoter polymorphism could theoretically influence gene transcription, this has not yet been investigated.

12. Interleukin-6

Interleukin-6 (IL-6) is produced by several cell types within the bone microenvironment and is known to act as a modulator of osteoclast differentiation and function (Roodman, 1992); data from some studies also indicate that IL-6 mediates some of the effects of oestrogen on bone (Ralston, 1994). An association has been described between an AT-rich minisatellite repeat in the 3′ region of the gene and bone density in both pre- and postmenopausal women from a Caucasian population (Murray *et al.*, 1997). These studies have not been repeated in other populations, however, and the mechanism by which this polymorphism might influence IL-6 function is unclear.

13. Apolipoprotein E

Apolipoprotein plays an important role in the transport of vitamin K from the intestine to target sites such as liver and bone. Its relevance to the pathogenesis of osteoporosis lies in the fact that vitamin K is an essential cofactor for carboxylation of osteocalcin, an osteoblast-specific protein. Single nucleotide polymorphisms have been described at two locations in the fourth exon of the gene (Kontula *et al.*, 1990) which define the E2, E3 and E4 alleles. These polymorphisms result in amino acid substitutions and have previously been shown to influence apolipoprotein E (APOE) function. Two studies have reported an association between APOE alleles and osteoporosis; in one study, the APOE4 allele was found to be associated with low bone mass (Shiraki *et al.*, 1997) and in another with osteoporotic fracture (Kohlmeier *et al.*, 1998). Since the APOE4 allele has also been associated with Alzheimer's disease, it might be expected to increase the risk of fracture by acting as a marker for impaired cognitive function, which could increase the susceptibility to falls.

14. Osteocalcin

Osteocalcin, also known as bone-gla protein, is the most abundant non-collagenous protein in bone, but its precise function is unknown. Some osteocalcin, which is synthesized by osteoblasts, is released into the circulation, providing a biochemical marker of osteoblast activity. A cytosine to thymine (C→T) polymorphism has been identified in the promoter region of the gene just upstream of exon 1. This region contains a number of important regulatory elements in addition to those necessary for osteocalcin gene expression; these include the region responsible for glucocorticoid repression and osteoblast-specific regulatory elements. This polymorphism has been shown to be associated with low bone mass in a Japanese population (Dohi *et al.*, 1998), although these observations remain to be confirmed in other populations. The mechanism by which osteocalcin polymorphisms could influence bone mass remain unclear, but gene knockout studies have shown that mice which are deficient in osteocalcin have increased bone mass, raising the possibility that osteocalcin may act as a negative regulator of osteoblast function (Ducy *et al.*, 1996).

15. Calcitonin receptor

Calcitonin is a hormone produced by the parafollicular cells of the thyroid gland. It exerts inhibitory effects on osteoclast function and stimulates urinary calcium excretion by interacting with its receptor, which is highly expressed on osteoclasts. Different isoforms of the calcitonin receptor result from gene splicing, and differential expression of these isoforms may explain the variable responsiveness to calcitonin in patients with high turnover metabolic bone diseases. The calcitonin receptor has seven potential transmembrane domains and a coding polymorphism has been identified in intracellular domain 4 (Nakamura *et al.*, 1997). The polymorphism, in the 1377th nucleotide, results in a proline-to-leucine amino acid substitution and has been found to be associated with bone mass in an Italian population (Masi *et al.*, 1998a,b). It is not yet clear how the polymorphism affects bone mass, but it has been speculated that the amino acid change could cause a conformational change in the receptor that might alter receptor signalling.

16. Parathyroid hormone

Parathyroid hormone (PTH) plays an important role in the regulation of calcium and bone metabolism via absorption of calcium from the kidneys and intestine. Extracellular calcium levels and vitamin D regulate its synthesis in the parathyroid glands and its principal target is the osteoclast. The age-related changes in these factors might modulate bone metabolism by altering PTH secretion. The gene is located on the short arm of chromosome 11, an area already identified in linkage studies to be associated with the regulation of bone mass. A polymorphism in intron 2 of the PTH gene has been described in a Japanese population, where it was found to be associated with reduced bone mass and increased bone turnover (Hosoi *et al.*, 1999). The mechanism by which this polymorphism influences bone mass is not yet clear, since it lies in a non-regulatory region of the gene.

17. Interleukin-1 receptor antagonist

The loss of ovarian function, observed at the menopause, is associated with increased production of various cytokines, including IL-1, which are responsible for the increase in bone turnover and subsequent bone loss observed during this period. Inhibition of the IL-1 receptor antagonist (IL-1ra) is one consequence of oestrogen deficiency and, in ovariectomized animals, it has been shown that osteoclast formation and bone loss is blocked following treatment with IL-1ra. Consequently, this gene is a possible candidate for the regulation of bone mass and postmenopausal bone loss. An 86 base pair variable number tandem repeat (VNTR) polymorphism has been identified in intron 2 of the gene (Tarlow *et al.*, 1993) and it has been found to be associated with differential rates of bone loss at the lumbar spine (Keen *et al.*, 1998). No effect of allelic variation was observed on absolute levels of BMD, suggesting that the effect is not mediated via peak bone mass. This polymorphism may be functionally significant since the sequence contains three potential transcription factor binding sites and VNTR repeats have been shown to influence gene expression at the insulin locus (Bennet *et al.*, 1995; Lucassen *et al.*, 1995). Further studies are necessary to determine if there is a relationship with BMD and fracture risk in other populations.

18. Animal studies

(i) Gene knockout studies

Gene knockout studies in experimental animals have provided an important insight into the role of specific candidate molecules in regulating bone cell differentiation and bone mass. Targeted inactivation of the *c-fos* gene results in failure of osteoclast formation with greatly increased bone density, giving a phenotype which resembles osteopetrosis (Grigoriadis *et al.*, 1994). Related studies have shown that the same bone phenotype is observed in animals with knockout of other molecules such as the haemopoietic transcription factor PU.1, the cytokine M-CSF, the proto-oncogene *c-src*, components of the signalling molecule NF-kB and the proteolytic enzyme cathepsin K (Soriano *et al.*, 1991; Johnell *et al.*, 1992; Grigoriadis *et al.*, 1994; Iotsova *et al.*, 1997; Tondravi *et al.*, 1997). Knockout of the transcription factor CBFA1 has revealed an essential role in osteoblast differentiation since mice with targeted inactivation of this gene have a skeleton that consists of cartilage and fibrous tissue (Rodan & Harada, 1997).

Although these studies have enormous relevance in understanding the molecular basis of bone cell differentiation and function, the phenotypes observed are not similar to osteoporosis. An exception is in the case of osteoprotegerin (OPG), which acts as an antagonist at the osteoprotegerin ligand receptor (also known as RANK). Knockout of OPG results in a phenotype very similar to that observed in osteoporosis, although it is currently unclear if polymorphisms of this gene occur in humans and, if they do, whether or not they are associated with the presence of osteoporosis.

(ii) Linkage studies

Linkage studies in laboratory animals have been successfully used to identify candidate loci for quantitative traits and complex diseases such as epilepsy. Based on the observation that different strains of mice have different levels of bone mass (Beamer *et al.*, 1996), several workers have begun to explore the possibility that quantitative trait loci for the regulation of bone mass may be identified by molecular genetic studies in recombinant inbred strains of mice. Klein *et al.* (1998) performed a genome search with 1522 polymorphic markers in 24 inbred strains of mice with varying BMD. Several loci were identified as showing evidence of linkage in these studies. It is of interest that some of these loci are homologous to those which have been found to regulate bone mass in humans or to contain candidate genes which have been implicated in the regulation of bone mass in humans (**Table 1**). In a further study, Shimizu *et al.* (1999)

Table 1 Candidate QTL sites for peak whole body BMD in mice.

Marker	Mouse chromosomal position	Human chromosomal position	Whole body BMD p-value	Candidate gene
Cfh	1:74	1q32	0.0093	?
IL-2ra	2:6	10p15–14	0.0071	IL-1ra
lapls2-4	2:86	20q11–12	0.0020	BMP2
D7mit234	7:44	11q13(?)/21(?)	0.0007	IGF1r, PTH, CT
D11mit14	11:57	17q11–12	0.0104	COLIA1
Ptprg	14:2	3p14	0.0007	BMP4, BMP2/4r
Atf4	15:43	11p13(?)/22q13(?)	0.0099	PDGFβ
Hmg1-rs7	16:19	3q27–28	0.0055	Ca^{2+} sens rec
D18Ncvs23	18:48	18q21(?)	0.0094	?
D19Ncvs21	19:53	10q23–26	0.0093	?

(Reproduced from Klein *et al*., 1998, with permission)

performed a whole genome search in F_2 intercrosses of two inbred strains of mice SAMP6 (low bone mass) and SAMP2 (high bone mass). Two loci were identified with significant linkage to peak bone mass, one on chromosome 13 (lod score 10.8) and the other on chromosome 11 (lod score 5.8). These loci are distinct from those identified by Klein *et al.* (1998), indicating that different genes may regulate bone mass in different strains of mouse.

19. Summary

Osteoporosis is a common disease characterized by low bone mass, microarchitectural deterioration of bone tissue and an increased risk of fracture. Osteoporosis is a complex disease that is subject to several environmental influences, but genetic factors play an important role in pathogenesis. Twin and family studies have shown that up to 85% of the variance in bone density between individuals is genetically determined and similar studies have shown that other predictors of fracture risk, such as ultrasound properties of bone, skeletal geometry and bone turnover, also receive a genetic contribution. A number of approaches have been used to try to identify the genes responsible for osteoporosis. Linkage studies have mapped a locus for autosomal dominant inheritance of high bone mass to chromosome 11q12, which also contains the gene for the autosomal recessively inherited OPS. Non-parametric linkage studies in healthy sib-pairs support the existence of a locus for regulation of bone mass on 11q12, whereas a genome search in osteoporotic sib-pairs identified several other potential loci for bone mass on chromosomes 1p36, 2p23–24 and 4q32–34. Both population-based and case-control studies have identified a number of polymorphisms in candidate genes such as cytokines, growth factors, hormone receptors and bone matrix components that have been associated with bone mass or osteoporotic fracture in specific populations. The most widely studied candidate genes are VDR and the collagen type I alpha 1 gene (COLIA1). Anonymous polymorphisms in the 3′ region of the VDR have been associated with bone mass in some populations and there is evidence to suggest that this association may be modified by calcium and vitamin D intake. A potentially functional polymorphism, which affects an Sp1 binding site in the COLIA1 gene, has been associated with bone mass in several Caucasian populations and has also been found to predict osteoporotic fractures independently of bone mass, suggesting that it may act as a genetic marker for bone quality. Although some advances have been made in recent years, much further work needs to be done to define the genes which regulate bone mass and bone quality, and to determine how they interact with each other and with environmental factors to cause osteoporotic fractures.

REFERENCES

Arai, H., Miyamoto, K.-I., Taketani, Y. *et al.* (1997) A vitamin D receptor gene polymorphism in the translation initiation codon: effect on protein activity and relation to bone mineral density in Japanese women. *Journal of Bone and Mineral Research* 12: 915–21

Arden, N.K. & Spector, T.D. (1997) Genetic influences on muscle strength, lean body mass, and bone mineral density: a twin study. *Journal of Bone and Mineral Research* 12: 2076–81

Arden, N.K., Baker, J.R., Hogg, C., Baan, K. & Spector, T.D. (1996) The heritability of bone mineral density, ultrasound of the calcaneus and hip axis length: a study of postmenopausal twins. *Journal of Bone and Mineral Research* 11: 530–34

Beamer, W.G., Donahue, L.R., Rosen, C.J. & Baylink, D.J. (1996) Genetic variability in adult bone density among inbred strains of mice. *Bone* 18: 397–403

Bennet, S.T., Lucassen, A.M., Gough, S.C.L. *et al.* (1995) Susceptibility to human type 1 diabetes at IDDM2 is determined by tandem repeat variation at the insulin gene minisatellite locus. *Nature Genetics* 9: 284–92

Bevan, S., Prentice, A., Dibba, B. *et al.* (1998) Polymorphism of the collagen type I alpha 1 gene and ethnic differences in hip-fracture rates. *New England Journal of Medicine* 339: 351–52

Christian, J.C., Yu, P.L., Slemenda, C.W. & Johnston, C.C. Jr (1989) Heritability of bone mass: a longitudinal study in ageing male twins. *American Journal of Human Genetics* 44: 429–33

Cooper, C., Atkinson, E.J., Wahner, H.W. *et al.* (1992) Is caffeine consumption a risk factor for osteoporosis? *Journal of Bone and Mineral Research* 7: 465–71

Cooper, G.S. & Umbach, D.M. (1996) Are vitamin D receptor polymorphisms associated with bone mineral density? A meta-analysis. *Journal of Bone and Mineral Research* 11: 1841–49

Cummings, S.R., Kelsey, J.L., Nevitt, M.C. & O'Dowd, K.J. (1985) Epidemiology of osteoporosis and osteoporotic fractures [Review]. *Epidemiologic Reviews* 7: 178–208

Dawson-Hughes, B., Harris, S.S. & Finneran, S. (1995) Calcium absorption on high and low calcium intakes in relation to VDR genotype. *Journal of Clinical Endocrinology and Metabolism* 80: 3657–61

Dawson-Hughes, B., Harris, S.S., Krall, E.A. & Dallal, G.E. (1997) Effect of calcium and vitamin D supplementation on bone density in men and women 65 years of age or older. *New England Journal of Medicine* 337: 670–76

Dean, V., Hobson, E., Aspden, R.M., Robins, S.P. & Ralston, S.H. (1998) Relationship between COLIA1 Sp1 alleles, gene transcription, collagen production and bone strength. *Bone* 25: S161

Dennison, E., Reynolds, R. & Cooper, C. (1998) Epidemiology of osteoporotic fractures. *CME Bulletin Endocrinology and Diabetes* 1: 40–42

Devoto, M., Shimoya, K., Caminis, J. *et al.* (1998) First-stage autosomal genome screen in extended pedigrees suggests genes predisposing to low bone mineral density on chromosomes 1p, 2p and 4q. *European Journal of Human Genetics* 6: 151–57

Dohi, Y., Iki, M., Ohgushi, H. *et al.* (1998) A novel polymorphism in the promoter region for the human osteocalcin gene: the possibility of a correlation with bone mineral density in postmenopausal Japanese women. *Journal of Bone and Mineral Research* 13: 1633–39

Ducy, P., Desbois, C., Boyce, B. *et al.* (1996) Increased bone formation in osteocalcin-deficient mice. *Nature* 382: 448–52

Eisman, J.A. (1995) Vitamin D receptor gene alleles and osteoporosis: an affirmative role. *Journal of Bone and Mineral Research* 10: 1289–93

Etherington, J., Harris, P.A., Nandra, D. *et al.* (1999) The effect of weight-bearing exercise on bone mineral density: a study of female ex-elite athletes and the general population. *Journal of Bone and Mineral Research* 11: 1333–38

Faulkner, K., Cummings, S.R., Black, D.M. *et al.* (1993) Simple measurement of femoral geometry predicts hip fracture: the study of osteoporotic fractures. *Journal of Bone and Mineral Research* 8: 1211–17

Ferrari, S., Rizzoli, R., Manen, D., Slosman, D. & Bonjour, J.-P. (1998a) Vitamin D receptor gene start codon polymorphisms (Fok I) and bone mineral density: interaction with age, dietary calcium and 3′ end region polymorphisms. *Journal of Bone and Mineral Research* 13: 925–30

Ferrari, S., Rizzoli, R., Slosman, D. & Bonjour, J.-P. (1998b) Do dietary calcium and age explain the controversy surrounding the relationship between bone mineral density and vitamin D receptor gene polymorphisms? *Journal of Bone and Mineral Research* 13: 363–70

Fleet, J.C., Harris, S.S., Wood, R.J. & Dawson-Hughes, B. (1995) The Bsm-I vitamin D receptor restriction length polymorphism (BB) predicts low bone density in premenopausal black and white women. *Journal of Bone and Mineral Research* 10: 985–90

Garnero, P., Borel, O., Sornay-Rendu, E. & Delmas, P.D. (1995) Vitamin D receptor gene polymorphisms do not predict bone turnover and bone mass in healthy premenopausal women. *Journal of Bone and Mineral Research* 10: 1283–88

Garnero, P., Borel, O., Grant, S.F.A., Ralston, S.H. & Delmas, P.D. (1998) Collagen IA1 Sp1 polymorphism, bone mass and bone turnover in healthy French premenopausal women: the OFELY study. *Journal of Bone and Mineral Research* 13: 813–17

Gong, Y., Vikkula, M., Boon, L. *et al.* (1996) Osteoporosis–pseudoglioma syndrome, a disorder affecting skeletal strength and vision, is assigned to chromosome region 11q12–13. *American Journal of Human Genetics* 59: 146–51

Graffmans, W.C., Lips, P., Ooms, M.E. *et al.* (1997) The effect of Vitamin D supplementation on the bone mineral density of the femoral neck is associated with Vitamin D receptor genotype. *Journal of Bone and Mineral Research* 12: 1241–45

Grant, S.F.A. & Ralston, S.H. (1996) Reduced bone density and osteoporosis associated with a polymorphic Sp1 binding site in the collagen type I alpha 1 gene. *Nature Genetics* 14: 203–05

Grigoriadis, A., Wang, Z.Q., Cecchini, M.G. *et al.* (1994) c-Fos: a key regulator of osteoclast–macrophage lineage determination and bone remodeling. *Science* 266: 443–48

Gross, C., Eccleshall, R., Malloy, P. *et al.* (1996) The presence of a polymorphism at the translational start site of the vitamin D receptor gene is associated with low bone mineral density in postmenopausal Mexican–American women. *Journal of Bone and Mineral Research* 11: 1850–55

Gueguen, R., Jouanny, P., Guillemin, F. *et al.* (1995) Segregation analysis and variance components analysis of bone mineral density in healthy families. *Journal of Bone and Mineral Research* 12: 2017–22

Han, K.O., Moon, I.G., Kang, Y.S. *et al.* (1997) Non association of estrogen receptor genotypes with bone mineral density and estrogen responsiveness to hormone replacement therapy in Korean postmenopausal women. *Journal of Clinical Endocrinology and Metabolism* 82: 991–95

Han, K.O., Moon, I.G., Hwang, C.S. *et al.* (1999) Lack of an intronic Sp1 binding-site polymorphism at the collagen type I alpha 1 gene in healthy Korean women. *Bone* 24: 135–37

Harris, S.S., Eccleshall, T.R., Gross, C., Dawson-Hughes, B. & Feldman, D. (1997) The vitamin D receptor start codon polymorphism (FokI) and bone mineral density in premenopausal American black and white women. *Journal of Bone and Mineral Research* 12: 1043–48

Hosoi, T., Inoue, S., Kobayashi, S. *et al.* (1995) Association of bone mineral density with polymorphisms of the estrogen receptor gene in post-menopausal women. *Journal of Bone and Mineral Research* 10 (Suppl. 1): s170

Hosoi, T., Miyao, M., Inoue, S. *et al.* (1999) Association study of parathyroid hormone gene polymorphism and bone mineral density in Japanese postmenopausal women. *Calcified Tissue International* 64: 205–08

Iotsova, V., Caamano, J., Loy, J. *et al.* (1997) Osteopetrosis in mice lacking NF-kappaB1 and NF-kappaB2. *Nature Medicine* 3: 1285–89

Johnell, O., Gullberg, B., Allander, E. & Kanis, J.A. (1992) The apparent incidence of hip fracture in Europe: a study of national register sources. MEDOS Study Group. *Osteoporosis International* 2: 298–302

Johnson, M.L., Gong, G., Kimberling, W. *et al.* (1997) Linkage of a gene causing high bone mass to human chromosome 11 (11q 12–13). *American Journal of Human Genetics* 60: 1326–32

Jorgensen, H.L., Scholler, J., Sand, J.C. *et al.* (1996) Relation of common allelic variation at vitamin D receptor locus to bone mineral density and postmenopausal bone loss: cross sectional and longitudinal population study. *British Medical Journal* 313: 586–90

Keen, R.W., Woodford-Richens, K.L., Lanchbury, J.S. & Spector,

T.D. (1998) Allelic variation at the interleukin-1 receptor antagonist gene is associated with early postmenopausal bone loss at the spine. *Journal of Bone and Mineral Research* 23: 367–71

Keen, R.W., Woodford-Richens, K.L., Grant, S.F.A. *et al.* (1999) Association of polymorphism at the type I collagen (COLIA1) locus with reduced bone mineral density, increased fracture risk, and increased collagen turnover. *Arthritis and Rheumatism* 42: 285–90

Kelly, P.J., Hopper, J.L., Macaskill, G.T. *et al.* (1991) Genetic factors in bone turnover. *Journal of Clinical Endocrinology and Metabolism* 72: 808–13

Kelly, P.J., Nguyen, T., Hopper, J. *et al.* (1993) Changes in axial bone density with age: a twin study. *Journal of Bone and Mineral Research* 8: 11–17

Kiel, D.P., Myers, R.H., Cupples, L.A. *et al.* (1997) The BsmI vitamin D receptor restriction fragment length polymorphism (bb) influences the effect of calcium intake on bone mineral density. *Journal of Bone and Mineral Research* 12: 1049–57

Klein, R.F., Mitchell, S.R., Phillips, T.J. & Belknap, J.K. (1998) Quantitative trait loci affecting peak bone mineral density. *Journal of Bone and Mineral Research* 13: 1648–56

Knudsen, J.Y., Langdahl, B.L., Jensen, H.K., Gregersen, N. & Eriksen, E.F. (1995) Mutations in the transforming growth factor beta gene are correlated to very low bone mass in osteoporosis. *Bone* 1: 84S (abstract)

Kobayashi, S., Inoue, S., Hosoi, T. *et al.* (1996) Association of bone mineral density with polymorphism of the estrogen receptor gene. *Journal of Bone and Mineral Research* 11: 306–11

Kohlmeier, M., Saupe, J., Schaefer, K. & Asmus, G. (1998) Bone fracture history and prospective bone fracture risk of haemodialysis patients are related to Apolipoprotein E genotype. *Calcified Tissue International* 62: 278–81

Koller, L., Rodriguez, L.A., Christian, J.C. *et al.* (1998) Linkage of a QTL contributing to normal variation in bone mineral density to chromosome 11q12–13. *Journal of Bone and Mineral Research* 13: 1903–08

Kontula, K., Aalto-Setala, K., Kuusi, T., Hamalainen, L. & Syvanen, A. (1990) Apolipoprotein E polymorphism determined by restriction enzyme analysis of DNA amplified by polymerase chain reaction: convenient alternative to phenotyping by isoelectric focussing. *Clinical Chemistry* 36: 2087–92

Krall, E.A. & Dawson-Hughes, B. (1993) Heritable and life-style determinants of bone mineral density. *Journal of Bone and Mineral Research* 8: 1–9

Krall, E.A., Parry, P., Lichter, J.B. & Dawson-Hughes, B. (1995) Vitamin D receptor alleles and rates of bone loss: influence of years since menopause and calcium intake. *Journal of Bone and Mineral Research* 10: 978–84

Lander, E.S. & Kruglyak, L. (1995) Genetic dissection of complex traits: guidelines for interpreting and reporting linkage results. *Nature Genetics* 11: 241–47

Langhdal, B.L., Ralston, S.H., Grant, S.F.A. & Erikson, E.F. (1998) An Sp1 binding site polymorphism in the ColIA1 gene predicts osteoporotic fractures in both men and women. *Journal of Bone and Mineral Research* 13: 1384–89

Liden, M., Wilen, B. & Melhus, H. (1998) Polymorphism at the Sp1 binding site in the collagen type IA1 gene does not predict bone mineral density in postmenopausal women in Sweden. *Calcified Tissue International* 63: 293–95

Lucassen, A.M., Screaton, G.R., Julier, C. *et al.* (1995) Regulation of insulin gene expression by the IDDM associated, insulin locus haplotype. *Human Molecular Genetics* 4: 501–06

Masi, L., Becherini, L., Colli, E. *et al.* (1998a) Polymorphisms of the calcitonin receptor gene are associated with bone mineral density in postmenopausal Italian women. *Biochemical and Biophysical Research Communications* 248: 190–95

Masi, L., Becherini, L., Gennari, L. *et al.* (1998b) Allelic variants of human calcitonin receptor: distribution and association with bone mass in postmenopausal Italian women. *Biochemical and Biophysical Research Communications* 245: 622–26

Minamitani, K., Takahashi, Y., Minagawa, M. *et al.* (1996) Exon 2 polymorphism in the human vitamin D receptor gene is a predictor of peak bone mineral density. *Journal of Bone and Mineral Research* 11 (Suppl. 1): s207

Mizunuma, H., Hosoi, T., Okano, H. *et al.* (1997) Estrogen receptor gene polymorphism and bone mineral density at the lumbar spine of pre- and postmenopausal women. *Bone* 21: 379–83

Morrison, N.A., Yeoman, R., Kelly, P.J. & Eisman, J.A. (1992) Contribution of trans-acting factor alleles to normal physiological variability: vitamin D receptor gene polymorphisms and circulating osteocalcin. *Proceedings of the National Academy of Sciences USA* 89: 6665–69

Morrison, N., Cheng Qi Jian, Tokita, A. *et al.* (1994) Prediction of bone density from vitamin D receptor alleles. *Nature* 367: 284–87

Murray, R., McGuigan, F.E.A., Grant, S., Reid, D. & Ralston, S. (1997) Polymorphisms of the interleukin-6 gene are associated with bone mineral density. *Journal of Bone and Mineral Research* 21: 89–92

Nakamura, M., Zhang, Z., Shan, L. *et al.* (1997) Allelic variants of human calcitonin receptor in the Japanese population. *Human Genetics* 99: 38–41

Nui, T., Chen, C., Cordell, H. *et al.* (1999) A genome-wide scan for loci linked to forearm bone mineral density. *Human Genetics* 104: 226–33

Pacifici, R., Brown, C., Puscheck, E. *et al.* (1991) Effect of surgical menopause and estrogen replacement on cytokine release from human blood mononuclear cells. *Proceedings of the National Academy of Sciences USA* 88: 5134–38

Peacock, M. (1995) Vitamin D receptor gene alleles and osteoporosis: a contrasting view. *Journal of Bone and Mineral Research* 10: 1294–97

Pfeilschifter, J., Seyedin, S.M. & Mundy, G.R. (1988) Transforming growth factor beta inhibits bone resorption in fetal rat bone cultures. *Journal of Clinical Investigation* 82: 680–85

Pocock, N.A., Eisman, J.A., Hopper, J.L. *et al.* (1987) Genetic determinants of bone mass in adults. A twin study. *Journal of Clinical Investigation* 80: 706–10

Ralston, S.H. (1994) Analysis of gene expression in human bone biopsies by PCR: evidence for enhanced cytokine expression in postmenopausal osteoporosis. *Journal of Bone and Mineral Research* 9: 883–90

Resnick, D. (1995) *Diagnosis of Bone and Joint Disorders*, 3rd edition, Philadelphia and London: W.B. Saunders

Riggs, B.L., Nguyen, T.V., Melton, L.J. *et al.* (1995) The contribution of vitamin D receptor gene alleles to the determination of bone mineral density in normal and osteoporotic women. *Journal of Bone and Mineral Research* 10: 991–96

Rodan, G. & Harada, S.I. (1997) The missing bone. *Cell* 89: 677–80

Roodman, G.D. (1992) Interleukin-6: an osteotropic factor? *Journal of Bone and Mineral Research* 7: 475–77

Roux, C., Dougados, M., Abel, L., Mercier, G. & Lucotte, G. (1998) Association of a polymorphism in the CollαI gene with osteoporosis in French women. *Arthritis and Rheumatism* 4: 187–88

Rowe, D.W. (1991) Osteogenesis imperfecta. In *Bone and Mineral Research*, edited by J.N.M. Heersche & J.A. Kanis, Amsterdam: Elsevier

Rowe, D.W. & Shapiro, J.R. (1998) Osteogenesis imperfecta. In *Metabolic Bone Disease*, edited by L.I. Avioli & S.M. Krane, 3rd edition, San Diegos Academic Press

Sainz, J., Van Tornout, J.M., Loro, M.L. *et al.* (1997) Vitamin D receptor gene polymorphisms and bone density in pre-pubertal American girls of Mexican descent. *New England Journal of Medicine* 337: 77–82

Sano, M., Inoue, S., Hosoi, T. *et al.* (1995) Association of estrogen receptor dinucleotide repeat polymorphism with osteoporosis. *Biochemical and Biophysical Research Communications* 217: 378–83

Seeley, D.G., Browner, W.S., Nevitt, M.C. *et al.* (1991) Which fractures are associated with low appendicular bone mass in elderly women? The Study of Osteoporotic Fractures Research Group. *Annals of Internal Medicine* 115: 837–42

Shimizu, M., Higuchi, K., Bennett, B. *et al.* (1999) Identification of peak bone mass QTL in a spontaneously osteoporotic mouse strain. *Mammalian Genome* 10: 81–87

Shiraki, M., Shiraki, Y., Aoki, C. *et al.* (1997) Association of bone mineral density with apolipoprotein E phenotype. *Journal of Bone and Mineral Research* 12: 1438–45

Smith, D.M., Nance, W.E., Kang, K.W., Christian, J.C. & Johnston, C.C. (1973) Genetic factors in determining bone mass. *Journal of Clinical Investigation* 52: 2800–08

Smith, E.P., Boyd, J., Frank, G.R. *et al.* (1994) Estrogen resistance caused by a mutation in the estrogen-receptor gene in a man. *New England Journal of Medicine* 331: 1056–61

Soriano, P., Montgomery, C., Geske, R. & Bradley, A. (1991) Targeted disruption of the c-src proto-oncogene leads to osteopetrosis in mice. *Cell* 64: 693–702

Spector, T.D., Keen, R.W., Arden, N.K. *et al.* (1995) Influence of vitamin D receptor genotype on bone mineral density in postmenopausal women: a twin study in Britain. *British Medical Journal* 310: 1357–60

Spotila, L.D., Colige, A., Sereda, L. *et al.* (1994) Mutation analysis of coding sequences for type I procollagen in individuals with low bone density. *Journal of Bone and Mineral Research* 9: 923–32

Tarlow, J.K., Blakemore, A.I., Lennard, A. *et al.* (1993) Polymorphism in human IL-1 receptor antagonist gene intron 2 is caused by variable numbers of an 86-bp tandem repeat. *Human Genetics* 91: 403–04

Tokita, A., Kelly, P.J., Nguyen, T.V. *et al.* (1994) Genetic influences on type I collagen synthesis and degradation: further evidence for genetic regulation of bone turnover. *Journal of Clinical Endocrinology and Metabolism* 78: 1461–66

Tondravi, M.M., McKercher, S.R., Anderson, K. *et al.* (1997) Osteopetrosis in mice lacking haematopoietic transcription factor PU.1. *Nature* 38: 81–84

Torgerson, D.J., Campbell, M.K. & Reid, D.M. (1995) Lifestyle, environmental and medical factors influencing peak bone mass in women. *British Journal of Rheumatology* 34: 620–24

Uitterlinden, A.G., Burger, H., Huang, Q. *et al.* (1998) Relation of alleles of the collagen type Ialpha1 gene to bone density and the risk of osteoporotic fractures in postmenopausal women. *New England Journal of Medicine* 338: 1016–21

Viitanen, A., Karkkainen, M., Laitinen, K. *et al.* (1996) Common polymorphism of the vitamin D receptor gene is associated with variation of peak bone mass in young Finns. *Calcified Tissue International* 59: 231–34

World Health Organization (1994) Assessment of fracture risk and its application to screening for postmenopausal osteoporosis. Report of a WHO Study Group. *World Health Organization Technical Report Series* 843: 1–129

Yamada, Y., Miyauchi, A., Goto, J. *et al.* (1998) Association of a polymorphism of the transforming growth factor-beta 1 gene with genetic susceptibility to osteoporosis in postmenopausal Japanese women. *Journal of Bone and Mineral Research* 13: 1569–76

GLOSSARY

bone mineral density (BMD) the estimated mineral content of bone as assessed by comparison of the photon absorption spectrum of the bone compared with that of a standard (containing usually calcium hydroxyapatite) measured either contemporaneously or at the beginning of the day. The resultant bone mineral content equivalent is divided by the area of bone detected to obtain the resultant BMD in g/cm^2

Cushing's syndrome metabolic condition resulting from the chronic and excessive production of cortisol by the adrenal cortex

cytokine any protein factor that is a product of a cell and which affects functions of other cells

hip axis length the distance from the greater trochanter through the femoral neck axis to the inner rim of the acetabulum

linkage disequilibrium the alliance of two linked alleles more often than would be expected by chance

non-parametric linkage analysis model-free method of testing for linkage analysis in affected sib-pairs when one of the markers is not inherited in a Mendelian fashion. It does not rely on the construction of a model. Even in the presence of confounding factors such as phenocopy, incomplete penetrance and genetic heterogeneity in the population, relatives who are affected should share an excess of alleles compared with unaffected individuals

osteoporosis–pseudoglioma syndrome (OPS) an autosomal recessive disease that is characterised by juvenile osteoporosis and congenital or juvenile-onset blindness

proteolytic enzyme enzyme that breaks down proteins and peptides into their constituent amino acids by hydrolysis of peptide bonds

thyrotoxicosis condition characterized by pronounced hyperthyroidism and commonly associated with an enlarged thyroid gland and abnormal protrusion of the eyeballs

CYSTIC FIBROSIS: THERAPY BASED ON A MOLECULAR UNDERSTANDING OF CFTR

David N. Sheppard[1], Katherine A. Lansdell[2] and Lynda S. Ostedgaard[3]
[1]*Department of Physiology, University of Bristol, Bristol, UK*
[2]*Department of Medical Sciences, University of Edinburgh, Western General Hospital, Edinburgh, UK*
[3]*Department of Internal Medicine, University of Iowa, Iowa City, USA*

1. Introduction

Cystic fibrosis (CF) is the most common lethal autosomal recessive disease in Caucasian populations, with an incidence of about one in 2500 births (Welsh *et al.*, 1995). The disease has a complex phenotype, which includes respiratory airway disease, pancreatic failure, meconium ileus (obstruction of the small intestine), male infertility and elevated levels of NaCl in sweat (Welsh *et al.*, 1995). The elevated concentration of NaCl in the sweat of CF patients is a benign abnormality, which remains the standard diagnostic test for CF. It is indicative of the underlying cellular defect in CF, a loss of Cl^- channel function in the apical membrane of epithelia (Quinton, 1990). In normal epithelia, apical Cl^- channels are activated by agonists which increase the intracellular concentration of cyclic 3′,5′ adenosine monophosphate (cAMP). However, epithelia in patients with CF fail to respond to cAMP agonists. The impermeability of the apical membrane to Cl^- in CF disrupts the transport of NaCl and water across epithelia and, hence, alters the quantity and composition of epithelial fluids (Quinton, 1990). Defective epithelial ion transport in CF disrupts the function of a variety of organs lined by epithelia. This explains the wide-ranging manifestations of the disease.

CF is an important medical problem of children and young adults, accounting for most cases of severe lung disease and exocrine pancreatic dysfunction in these patients. There are two major causes of debilitation and death in CF: first, chronic obstructive lung disease caused by thick mucous secretions that prevent normal mucociliary clearance; second, persistent bacterial infections, commonly with *Pseudomonas aeruginosa*, that result in bronchiectasis (the permanent dilation of one or more bronchi), respiratory failure and, eventually, death. Current therapies for CF include postural drainage with chest percussion, antibiotics and recombinant DNase to treat lung disease, and pancreatic enzyme replacement and proper nutrition to overcome the failure to thrive (Welsh *et al.*, 1995). Median survival in the US and Western Europe is about 30 years.

2. CFTR: a Cl^- channel with novel regulation

In 1989, the defective gene responsible for CF was identified by positional cloning (Riordan *et al.*, 1989). Its protein product was called the cystic fibrosis transmembrane conductance regulator (CFTR) because, at that time, it was not clear whether CFTR was itself the Cl^- channel or functioned only to regulate epithelial Cl^- channels (Riordan *et al.*, 1989). **Figure 1** shows that the predicted structure of CFTR contains five domains: two membrane-spanning domains (MSDs), each composed of six transmembrane segments that span the lipid bilayer; two nucleotide-binding domains (NBDs) predicted to interact with adenosine triphosphate (ATP); and an R domain that contains multiple consensus phosphorylation sites. The predicted structure of CFTR did not resemble that of known ion channels, but instead that of members of a family of transport proteins called ATP-binding cassette (ABC) transporters (Hyde *et al.*, 1990). Since ABC transporters utilize the energy of ATP hydrolysis to actively transport substrates across cell membranes, it was hypothesized that CFTR might pump a regulator of Cl^- channels into or out of epithelial cells.

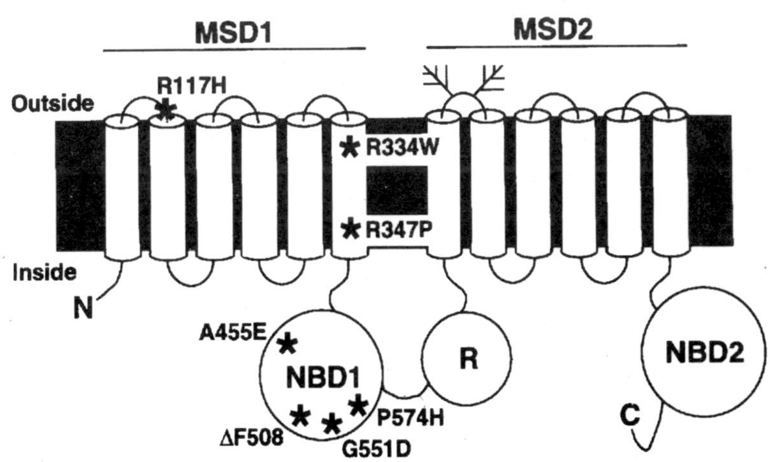

Figure 1 Domain structure of the cystic fibrosis transmembrane conductance regulator (CFTR) indicating the predicted location of some CF-associated mutations (asterisks). MSD, membrane-spanning domain; NBD, nucleotide-binding domain; R, R domain; Inside, intracellular; and Outside, extracellular. Glycosylation sites located in the fourth extracellular loop are indicated by branched structures. Modified from Welsh *et al.* (1995).

To investigate the function of CFTR, two strategies were adopted. First, wild-type CFTR and variants containing site-directed mutations were expressed in a variety of different cell types which do not normally contain cAMP-activated Cl^- channels and which have little or no endogenous CFTR. Second, recombinant CFTR was purified and reconstituted into planar lipid bilayers. These studies demonstrated that CFTR is a Cl^- channel with novel regulation (for review see Riordan, 1993; Gadsby & Nairn, 1994; Welsh *et al.*, 1995). They have also begun to provide insight into the relationship between the structure and the function of CFTR. The MSDs contribute to the formation of a pore through which Cl^- flows. The R domain is phosphorylated at multiple serine residues by the cAMP-dependent protein kinase (PKA). Once the R domain is phosphorylated, the opening and closing of the channel is regulated by cycles of ATP hydrolysis at the NBDs: ATP hydrolysis at NBD1 opens the channel, and ATP hydrolysis at NBD2 closes the channel. Finally, protein phosphatases dephosphorylate the R domain and return the channel to its quiescent state.

3. How do CF-associated mutations cause a loss of Cl^- channel function?

With the identification of the function of CFTR, the molecular mechanisms by which CF-associated mutations cause a loss of Cl^- channel function could be investigated. To date, over 900 disease-causing mutations have been identified in the CFTR gene (CF Genetic Analysis Consortium; http://www.genet.sickkids.on.ca/cftr/). They are located throughout the entire coding sequence and include deletions, missense and nonsense mutations. Most mutations are very rare; the exception is the deletion of a phenylalanine residue at position 508 of the CFTR sequence (termed ΔF508). This mutation, located in NBD1, accounts for about 70% of CF mutations worldwide. A few other mutations are found with a worldwide frequency of 1–5% and, in some ethnic groups, such as the Ashkenazi Jews, other mutations are even more common (Welsh *et al.*, 1995).

CF has a variable phenotype that is best described clinically by the level of pancreatic function. Most patients suffer pancreatic failure and are termed pancreatic insufficient (PI), but some retain significant pancreatic function and are termed pancreatic sufficient (PS). In general, PS patients tend to have less severe disease. Several studies have suggested a relationship between genotype, i.e. mutations, and clinical phenotype, i.e. PS *vs.* PI. A PI phenotype occurs in patients who have two "severe" mutations, for example ΔF508 on both chromosomes, whereas a PS phenotype occurs in patients who have either a "mild" mutation on each chromosome or one "mild" and one "severe" mutation (Kristidis *et al.*, 1992). CF mutations associated with a PS phenotype include: R117H, R334W, R347P, A455E and P574H (**Figure 1**) (Kristidis *et al.*, 1992).

When CF-associated mutants were expressed in heterologous cells, the amount of Cl^- current generated was reduced with wild-type CFTR > PS mutants > PI mutants (Drumm *et al.*, 1991). These results suggested a relationship between genotype, clinical phenotype and Cl^- channel function. To understand this relationship, researchers investigated the effect of CF-associated mutations on the biosynthesis and function of CFTR. They identified four general mechanisms: defective protein production, defective protein processing, defective channel regulation and defective channel conduction (Welsh & Smith, 1993). **Figure 2** shows a schematic representation of these mechanisms.

(i) Class I mutations: defective protein production

Some CF-associated mutations generate premature stop codon signals, such as deletions, insertions and nonsense mutations, which alter the open reading frame. CF patients bearing these mutations probably fail to produce full-length CFTR protein because either the mRNA or the protein is unstable and present at very low levels. Although there has been speculation that patients bearing nonsense mutations might have a milder clinical phenotype, current data suggest

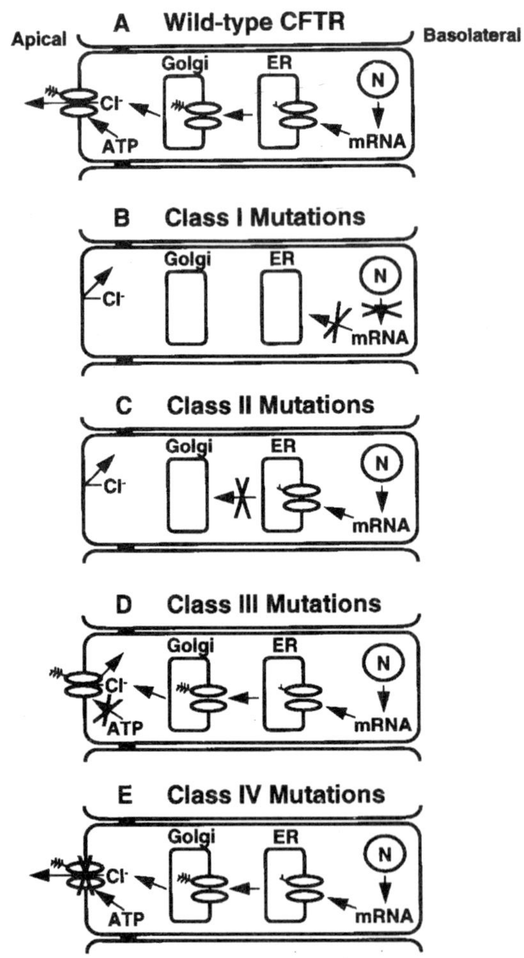

Figure 2 Models showing the synthesis and function of wild-type CFTR in an epithelial cell and the mechanism of dysfunction of CF-associated mutations. (A) Wild-type CFTR. (B) Class I mutations: defective protein production. (C) Class II mutations: defective protein processing. (D) Class III mutations: defective channel regulation. (E) Class IV mutations: defective channel conduction. N, nucleus; ER, endoplasmic reticulum. The immature, core-glycosylated and mature, fully glycosylated forms of CFTR are represented by the single- and multibranched structures, respectively.

that nonsense mutations are associated with severe disease (The Cystic Fibrosis Genotype–Phenotype Consortium, 1993). Thus, class I mutations cause a loss of Cl^- channel function because they are missing from the apical membrane of affected epithelia.

(ii) Class II mutations: defective protein processing

Many CF-associated mutations disrupt the processing and intracellular transport of CFTR. They have been identified in each of the functional domains of CFTR and include the most common mutation, ΔF508. The defective biosynthesis of these mutants can be monitored by assessing their state of glycosylation. **Figure 3** shows that, when electrophoresed on SDS-polyacrylamide gels, recombinant wild-type CFTR was present as two bands: a 135 kilodalton (kDa) band (band B), representing the immature or core-glycosylated protein, and a 170 kDa band (band C), representing the mature or fully glycosylated protein (Cheng *et al.*, 1990). In contrast, electrophoresis of recombinant ΔF508 resulted in only band B, the immature protein (**Figure 3**) (Cheng *et al.*, 1990). This result suggests that, in cells expressing recombinant CFTR, ΔF508 was retained in the endoplasmic reticulum (ER) and degraded. It was neither processed through the Golgi, where complex sugars are added, nor delivered to the plasma membrane (**Figure 2**). Consistent with this idea, studies using respiratory airway epithelia and sweat gland ducts expressing native CFTR have demonstrated that ΔF508 was mislocalized in CF epithelia (Welsh *et al.*, 1995). Thus, class II mutations also cause a loss of Cl^- channel function because they are missing from the apical membrane of affected epithelia.

Studies of the biosynthesis of CFTR suggest that the "quality control" machinery of the cell that determines the fate of ΔF508 is complex and involves multiple steps including ER retention, ATP-dependent conformational change and degradation by the ubiquitin–proteasome system.

ER retention. The ΔF508 mutant is retained in the ER by chaperones which prevent newly synthesized proteins from folding incorrectly during processing (Yang *et al.*, 1993; Pind *et al.*, 1994). Wild-type CFTR associates only transiently with chaperones before moving to the Golgi. However, ΔF508 is tightly bound by two chaperones, calnexin and hsp70, and retained in the ER (Yang *et al.*, 1993; Pind *et al.*, 1994).

ATP-dependent conformational change. The core-glycosylated form of wild-type CFTR undergoes an ATP-dependent conformational change to a stable form prior to exit from the ER (Lukacs *et al.*, 1994; Ward & Kopito, 1994). This process is very inefficient; only 20–25% of immature wild-type CFTR achieves the stable form that is competent to exit the ER. By contrast, ΔF508 rarely matures to the stable form that is processed to the plasma membrane (Lukacs *et al.*, 1994; Ward & Kopito, 1994).

Degradation by the ubiquitin–proteasome system. ΔF508 is degraded by the ubiquitin–proteasome system (Jensen *et al.*, 1995; Ward *et al.*, 1995). It is targeted for degradation by conjugation to multiple molecules of ubiquitin and then destroyed by the proteasome. Due to the prevalence of the ΔF508 mutation, class II mutations account for the molecular basis of most cases of CF.

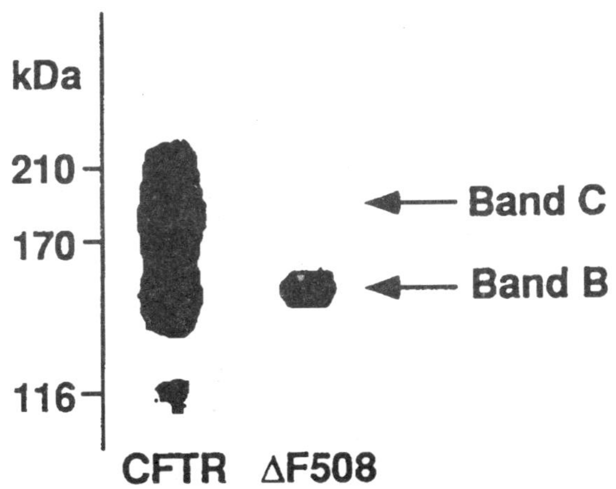

Figure 3 Immunoprecipitation of wild-type CFTR and the ΔF508 mutant. Immunoprecipitates were phosphorylated with PKA and [γ^{32}P]ATP and separated on a SDS-polyacrylamide gel. Arrows indicate the positions of band B (the immature, core-glycosylated form) and band C (the mature, fully glycosylated form). Recombinant wild-type CFTR and ΔF508 were transiently expressed in Fischer rat thyroid cells using the vaccinia virus-T7 polymerase hybrid expression system. Adapted by permission from Sheppard *et al. Nature* 362: 160–64 © 1993 Macmillan Magazines Ltd.

(iii) Class III mutations: defective channel regulation

Some CF-associated mutations are correctly processed and delivered to the apical membrane. However, they form Cl^- channels with altered single-channel properties and regulation. To learn how CF-associated mutations disrupt the Cl^- channel function of CFTR, researchers use the patch-clamp technique (Hamill *et al.*, 1981). Using the patch-clamp technique, the passage of Cl^- ions through individual CFTR Cl^- channels can be visualized as discrete current steps between a closed (non-conducting) and an open (conducting) state. Wild-type CFTR Cl^- channels are characterized by small single-channel conductance (8–10 pS), linear current-voltage relationship, selectivity for anions over cations, anion selectivity sequence of $Br^- \geq Cl^- > I^-$, time- and voltage-independent gating behaviour, and regulation by cAMP-dependent phosphorylation and ATP hydrolysis (Welsh *et al.*, 1995).

The NBDs hydrolyse ATP to control the gating behaviour of the channel. CF-associated mutations in the NBDs affect channel regulation in a number of different ways. First, some NBD mutants, such as G551D, had little or no function (Drumm *et al.*, 1991); second, others, such as G551S, G1244E, S1255P and G1349D, had a markedly reduced single-channel open probability (P_o; Anderson & Welsh, 1992); and, third, the activity of some, such as S1255P, was less potently stimulated by ATP (Anderson & Welsh, 1992). However, none of these CF-associated NBD mutations altered the pore properties of CFTR. Thus, class III mutations cause a loss of Cl^- channel function because they disrupt the regulation of CFTR Cl^- channels, with the result that there is little or no cAMP-activated Cl^- channel activity at the apical membrane.

The R domain is phosphorylated at multiple serine residues by PKA prior to channel activation. Although CF-associated mutations have been identified in the R domain, there appear to be fewer mutations than in other parts of the protein. Since the structural requirements of the R domain for channel activation are somewhat flexible, it was speculated that mutations in the R domain may be tolerated without compromising function. However, a recent study indicated that the R domain mutations D614G and I618T disrupt the regulation of CFTR Cl^- channels (Pasyk *et al.*, 1997).

CF-associated mutations in the intracellular loops (sequences that connect the transmembrane segments of the MSDs) also disrupt channel regulation. Most mutants in the intracellular loops have a reduced P_o compared with wild-type (Cotten *et al.*, 1996; Seibert *et al.*, 1996, 1997). In addition, ATP has been shown to stimulate less potently the activity of the mutant A1067T, and the response of the mutants F1052V and R1066L to pyrophosphate is decreased (Cotten *et al.*, 1996). Interestingly, one intracellular loop mutation, H949Y, and two NBD1 mutations, A455E and P574H, which are associated with a milder clinical phenotype, compensate for the effects of decreased processing by maximizing channel function (Champigny *et al.*, 1995; Sheppard *et al.*, 1995; Seibert *et al.*, 1996). The P_o of A455E is equivalent to that of wild-type, and the P_o of P574H and H949Y is even greater than that of wild-type (Champigny *et al.*, 1995; Sheppard *et al.*, 1995; Seibert *et al.*, 1996).

(iv) Class IV mutations: defective channel conduction

The MSDs, which contribute to the formation of the channel pore, are the site of a number of CF-associated mutations. Three mutations (R117H, R334W and R347P) associated with a milder clinical phenotype affect basic residues located within MSD1 (**Figure 1**). When expressed in heterologous epithelial cells, these MSD1 mild mutants were correctly processed and they generated cAMP-activated apical Cl^- currents, in contrast to the ΔF508 mutant (Sheppard *et al.*, 1993). However, the amount of apical Cl^- current was reduced with wild-type CFTR > R347P > R117H > R334W (Sheppard *et al.*, 1993). To determine why apical Cl^- current was reduced, we investigated the single-channel properties of the MSD1 mild mutants. Like wild-type CFTR, mutant Cl^- channels were regulated by phosphorylation and intracellular ATP. However, these mutants formed Cl^- channels with altered pore properties that impaired the rate of Cl^- flow through the channel, although they retained their Cl^- selectivity (**Figure 4**). Each of these mutant Cl^- channels had a reduced single-channel conductance (Sheppard *et al.*, 1993). In addition, the P_o of R117H was reduced to one-third that of wild-type CFTR (Sheppard *et al.*, 1993).

An explanation for why mutations at R347 decrease Cl^- flow through CFTR has been suggested by Tabcharani and colleagues (1993). They demonstrated that wild-type CFTR normally functions as a multi-ion pore, containing several Cl^- ions at once. However, the CF-associated mutation R347H reduced the single-channel conductance of CFTR and converted CFTR to a single-ion pore (Tabcharani *et al.*,

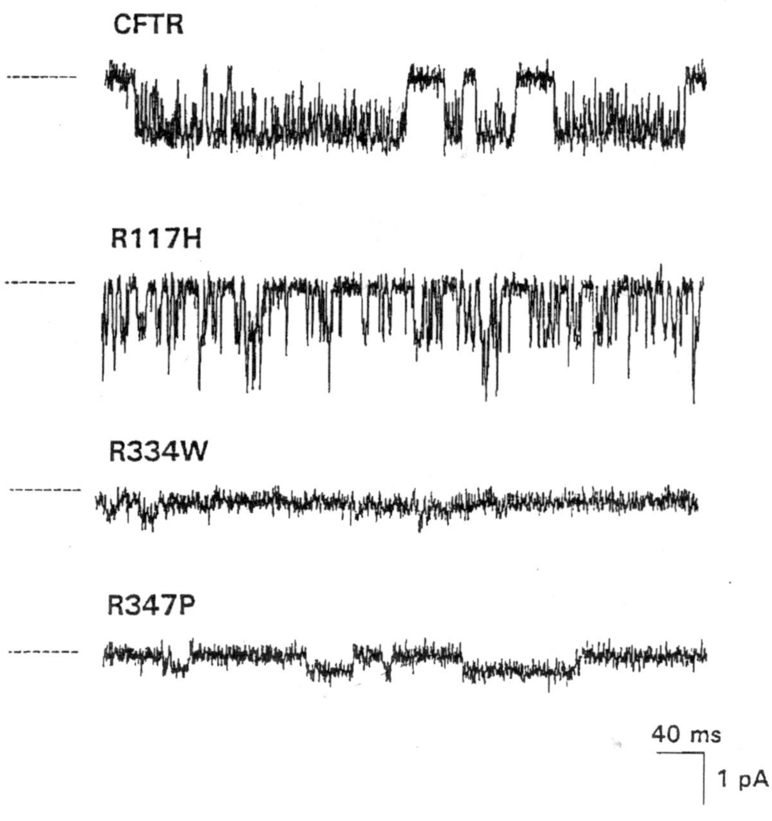

Figure 4 Mutations in MSD1 associated with a milder clinical phenotype have altered single-channel properties. Representative single-channel recordings are from excised inside-out membrane patches from HeLa cells expressing either wild-type CFTR, R117H, R334W or R347P. Membrane voltage was clamped at −90 mV and there was a large Cl^- concentration gradient across the membrane patch: external (pipette) $[Cl^-]$ = 10 mM; internal (bath) $[Cl^-]$ = 147 mM. All recordings were obtained in the presence of the catalytic subunit of PKA (75 nM) and ATP (1 mM) in the intracellular solution. Dashed lines indicate the closed-channel state, and downward deflections of the trace correspond to channel openings. The wild-type CFTR and R347P patches contained a single active channel and R117H two active channels, but the number of channels present in the R334W patch could not be determined. Reprinted by permission from Sheppard *et al.* *Nature* 362: 160–64 © 1993 Macmillan Magazines Ltd.

1993). This suggests that CF mutations at R347 probably cause CF because they abolish the multi-ion pore behaviour of CFTR Cl^- channels. Thus, class IV mutations cause a loss of Cl^- channel function because they disrupt the flow of Cl^- ions through the CFTR pore.

Some CF-associated mutations have been identified that disrupt the function of CFTR in more than one way. For example, although the major abnormality of the ΔF508 mutant is defective protein processing, it also disrupts channel regulation (Denning *et al.*, 1992). Similarly, the MSD1 mild mutant, P99L, forms a Cl^- channel with altered pore properties and is also misprocessed (Sheppard *et al.*, 1996). The data suggest that the mechanisms by which CF-associated mutations produce defective Cl^- channels are complex. They also suggest that it is not possible to predict the mechanism of dysfunction of CFTR based solely on the site of mutation.

4. Relationship between loss of CFTR function and lung disease

A dilemma for genotype–phenotype studies in CF has been the failure of lung disease to correlate with genotype and Cl^- channel function. The severity of lung disease can vary substantially, even among patients carrying the same mutation. Moreover, with the exception of A455E, CF mutations which are associated with a PS phenotype and which retain residual Cl^- channel function are not associated with milder lung disease (Gan *et al.*, 1995). Some variability may result from environmental factors and genetic background. However, a different explanation has been suggested by Clarke and colleagues. They studied the expression of a different apical membrane Cl^- channel, activated by an increase in intracellular Ca^{2+}, in a mouse model of CF. Their results suggest that expression of Ca^{2+}-activated Cl^- channels correlates with disease severity; epithelia expressing this Cl^- channel exhibited mild disease, whereas epithelia lacking this Cl^- channel exhibited severe disease (Clarke *et al.*, 1994). Consistent with these ideas, Rozmahel *et al.* (1996) identified a secondary genetic factor that modulates the severity of intestinal disease in CF mice by regulating the activity of Ca^{2+}-activated Cl^- channels. If a CF modifier gene also exists in humans, this may help to explain some of the variation in clinical phenotype observed.

In addition to its best-characterized function as a regulated Cl^- channel, CFTR has been suggested to have other roles in epithelial cells, including the regulation of other ion channels (Pasyk & Foskett, 1997; Stutts *et al.*, 1997). These additional functions of CFTR may help to explain some of the abnormalities in CF epithelia that have been difficult to account for solely by a defect in a Cl^- channel. For example, in CF airway epithelia, defective regulation of epithelial Na^+ channels results in enhanced Na^+ absorption, and dysfunction of CFTR-associated adenine nucleotide channels may explain altered glycoprotein processing (Pasyk & Foskett, 1997; Stutts *et al.*, 1997). Similarly, loss of CFTR-mediated clearance of *P. aeruginosa* from the lung may explain the increased susceptibility of CF patients to pulmonary infections with *P. aeruginosa* (Pier *et al.*, 1997). However, Smith *et al.* (1996) demonstrated that a loss of CFTR Cl^- channel function prevents the killing of bacteria by salt-sensitive antimicrobial factors which are present in airway surface fluid. The data suggest that chronic inflammation and infection occur when CFTR mutations raise the salt concentration of airway surface fluid, impairing the ability of CFTR to protect the airway epithelium from invading pathogens.

5. Implications for therapy

Knowledge of how CF-associated mutations disrupt the function of CFTR is leading to new approaches to CF therapy. Strategies to overcome the loss of CFTR function in CF include: replacing mutant CFTR with normal wild-type CFTR; altering the processing of mutant CFTR to facilitate its delivery to the apical membrane; increasing the activity of mutant Cl^- channels which are present in the apical membrane by altering channel regulation; and stimulating other apical membrane Cl^- channels regulated by different signalling pathways.

Therapies based on the delivery of wild-type CFTR to affected epithelial cells by either gene or protein transfer have the potential to correct all the defects in CFTR function. Although protein replacement therapy has been successfully used to treat haemoglobinopathies and Gaucher's disease, it is unlikely to form a treatment for CF because CFTR is a low abundance, large integral membrane protein that would probably be difficult to produce in the quantities required. Nevertheless, the feasibility of protein replacement therapy has been investigated. Recombinant CFTR has been purified, reconstituted in a functionally active form and delivered to the nasal epithelium of CF mice, where it transiently corrected defective epithelial ion transport (Ramjeesingh *et al.*, 1995).

At present, much emphasis is being placed on gene therapy as a treatment for CF. One reason is that somatic gene therapy has the potential to provide a definite treatment for pulmonary disease in CF. A variety of different vectors for gene delivery are being investigated, including viral-based vectors, DNA-liposome complexes and DNA-ligand complexes (for review see Crystal, 1995). Each vector has specific advantages and disadvantages. Adenovirus vectors efficiently transfer wild-type CFTR cDNAs to airway epithelial cells, but cause inflammatory responses in humans. In contrast, DNA-liposome vectors cause little or no inflammation in humans, but less efficiently transfer wild-type CFTR cDNAs to airway epithelial cells. Numerous phase I clinical trials have been initiated in the US and Europe. These trials have demonstrated that transfer of wild-type CFTR cDNAs to nasal epithelia transiently corrects defective epithelial Cl^- transport in CF patients.

In 1992, Denning and colleagues demonstrated that lowering the temperature at which cultured cells grow is sufficient to induce processing of the ΔF508 mutant to the cell membrane, where it has some functional activity. This observation suggests that other manoeuvres which allow ΔF508 to be delivered to the apical membrane may have therapeutic potential. In search of agents that rescue the folding defect of the ΔF508 mutant, Brown *et al.* (1996) demonstrated that high concentrations of the cellular osmolytes glycerol and trimethylamine N-oxide could overcome the processing defect of ΔF508. These results suggest that "chemical chaperones" may be of value in the treatment of CF. An alternative strategy is to enhance the expression of ΔF508 using agents that stimulate gene transcription, such as sodium 4-phenylbutyrate (4PBA), a drug that is approved for clinical use. When cells expressing the mutant ΔF508 were treated with 4PBA, ΔF508 was correctly processed and delivered to the plasma membrane (Zeitlin & Rubenstein, 1997). Based on these results, a clinical trial of oral 4PBA therapy in CF patients was undertaken. This trial demonstrated that 4PBA transiently corrects defective epithelial Cl^- transport in the nasal epithelia of CF patients (Zeitlin & Rubenstein, 1997). Pharmacological therapy for patients with premature stop mutations may also be feasible. When cells expressing the mutants R553X or G542X were treated with the aminoglycoside antibiotic G-418, full-length CFTR was synthesized and cAMP agonists stimulated Cl^- channel activity (Howard *et al.*, 1996). These results suggest that aminoglycoside antibiotics may restore CFTR function to patients with premature stop mutations.

Patients bearing CF mutations that are correctly processed and delivered to the apical membrane may be treated by altering channel regulation. The novel regulation of CFTR Cl^- channels by phosphorylation and ATP hydrolysis suggests a number of different strategies to increase the activity of mutant Cl^- channels. Several studies have demonstrated that mutant Cl^- channels can be stimulated by increasing the phosphorylation of CFTR by PKA. One approach is to prevent the breakdown of cAMP by phosphodiesterases. In airway epithelia, Kelley *et al.* (1997) identified inhibitors of class III phosphodiesterases, such as milrinone, as the phosphodiesterase inhibitors that most potently activate mutant CFTR Cl^- channels. A second approach is to inhibit the dephosphorylation of CFTR by protein phosphatases. Becq *et al.* (1994) showed that, in the absence of cAMP agonists, the alkaline phosphatase inhibitor (-)-bromotetramisole activates wild-type CFTR and the CF-associated mutants R117H and G551D. The effect on G551D was particularly striking because this mutant does not normally respond to cAMP agonists (Becq *et al.*, 1994). In airway and intestinal epithelia, CFTR is dephosphorylated by protein phosphatase 2C (PP2C) (Travis *et al.*, 1997). This suggests that inhibitors of PP2C will probably be of value as a therapeutic intervention in CF.

Once CFTR is phosphorylated, the opening and closing of CFTR Cl^- channels are regulated by cycles of ATP hydrolysis at the NBDs: ATP hydrolysis at NBD1 opens the channel, and ATP hydrolysis at NBD2 closes the channel (Gadsby & Nairn, 1994). Evidence supporting a requirement for ATP hydrolysis includes the ability of inorganic phosphate analogues and polyphosphates to "lock" CFTR Cl^- channels in the open configuration (Gadsby & Nairn, 1994; Carson *et al.*, 1995). Based on these results, Carson and colleagues (1995) investigated whether polyphosphates might activate mutant Cl^- channels. They demonstrated that pyrophosphate stimulates the activity of R117H, ΔF508 and G551S Cl^- channels, suggesting that polyphosphates may be of value in the treatment of CF.

Novel drugs which activate mutant CFTR Cl^- channels have been identified. One such compound is the adenosine receptor antagonist 8-cyclopentyl-1,3-dipropylxanthine (CPX) that stimulated Cl^- efflux from cells expressing the ΔF508 mutant (Eidelman *et al.*, 1992). A second compound is the substituted benzimidazolone NS004 which modulates the activity of several ion channels (Gribkoff *et al.*, 1994). Interestingly, studies using the patch-clamp technique suggested that NS004 may directly interact with CFTR to stimulate channel activity (Gribkoff *et al.*, 1994). A third agent is the tyrosine kinase inhibitor genistein (Hwang *et al.*, 1997). When applied alone, genistein was without effect, but, in combination with cAMP agonists, it potently stimulated the activity of ΔF508 Cl^- channels (Hwang *et al.*, 1997). How genistein activates CFTR Cl^- channels is presently unknown. It may inhibit a tyrosine kinase, inhibit a protein phosphatase or directly interact with CFTR. Thus, these studies suggest that it might be possible to develop specific, potent activators of mutant CFTR Cl^- channels present at the apical membrane of CF epithelia.

Cl^- channels with properties and regulation distinct from those of CFTR have been identified in the apical membrane of epithelia affected by CF. These include Ca^{2+}-activated Cl^- channels and volume-regulated Cl^- channels. The activation of these Cl^- channels in CF epithelia has been investigated as a way to bypass the defect in CFTR Cl^- channel function in CF. **Figure 5** shows that, when apical Na^+ channels are inhibited by amiloride, Ca^{2+}-activated Cl^- channels in normal and CF airway epithelia are activated by extracellular nucleotide triphosphates, such as ATP and UTP, which interact with purinergic receptors to stimulate an increase in intracellular Ca^{2+}. To investigate whether or not extracellular nucleotides might be of value in the treatment of CF, Knowles and colleagues conducted a clinical trial which demonstrated that ATP and UTP potently stimulate Cl^- secretion in nasal epithelia of CF patients (Knowles *et al.*, 1991).

Epithelial Cl^- secretion is regulated by both apical membrane Cl^- channels and basolateral membrane K^+ channels (**Figure 5**). This suggests that drugs which activate basolateral membrane K^+ channels might stimulate Cl^- secretion in CF epithelia. Consistent with this idea, Devor *et al.* (1996) demonstrated that, when apical Na^+ channels were inhibited with amiloride, the benzimidazolone, 1-ethyl-2-benzimidazolinone 1-EBIO stimulated Cl^- secretion in airway and intestinal epithelia by activating basolateral membrane K^+ channels. These results suggest that a combination of K^+ and Cl^- channel activators may be used to maximize the magnitude and duration of Cl^- secretion in CF epithelia.

6. Summary

In summary, the identification of the defective gene in CF was a landmark in molecular medicine. This achievement has made possible the exciting progress in CF research over the past few years, including the identification of the function of CFTR, an understanding of how CF mutations disrupt CFTR function and the design of new therapies for the disease. The knowledge gained from CF research may provide valuable lessons for the molecular understanding and treatment of other genetic disorders.

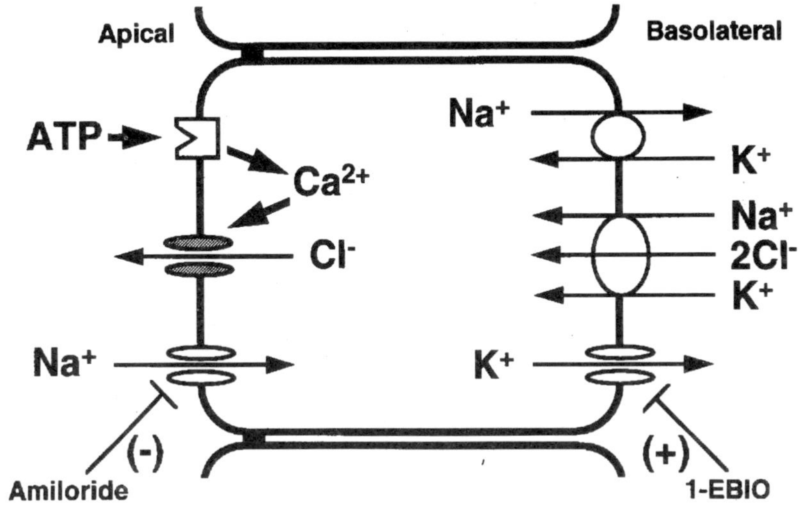

Figure 5 Bypassing the loss of CFTR Cl^- channel function by activation of Ca^{2+}-activated Cl^- channels. A schematic representation of the cellular mechanism of electrolyte transport in a CF airway epithelial cell is shown. The calcium-activated Cl^- channel is indicated by shading; all other channels, transporters, pumps and receptors are indicated by open symbols. Extracellular nucleotides, such as ATP, bind purinergic receptors located in the apical membrane to stimulate an increase in intracellular Ca^{2+}. 1-EBIO, 1-ethyl-2-benzimidazolinone – an activator of basolateral membrane K^+ channels.

Acknowledgements

We thank our laboratory colleagues for their advice and critical comments. The authors are supported by the BBSRC and CF Trust.

REFERENCES

Anderson, M.P. & Welsh, M.J. (1992) Regulation by ATP and ADP of CFTR chloride channels that contain mutant nucleotide-binding domains. *Science* 257: 1701–04

Becq, F., Jensen, T.J., Chang, X.B. *et al.* (1994) Phosphatase inhibitors activate normal and defective CFTR chloride channels. *Proceedings of the National Academy of Sciences USA* 91: 9160–64

Brown, C.R., Hong-Brown, L.Q., Biwersi, J., Verkman, A.S. & Welch, W.J. (1996) Chemical chaperones correct the mutant phenotype of the ΔF508 cystic fibrosis transmembrane conductance regulator protein. *Cell Stress Chaperones* 1: 117–25

Carson, M.R., Winter, M.C., Travis, S.M. & Welsh, M.J. (1995) Pyrophosphate stimulates wild-type and mutant cystic fibrosis transmembrane conductance regulator Cl^- channels. *Journal of Biological Chemistry* 270: 20466–72

Champigny, G., Imler, J.L., Puchelle, E. *et al.* (1995) A change in gating mode leading to increased intrinsic Cl^- channel activity compensates for defective processing in a cystic fibrosis mutant corresponding to a mild form of the disease. *EMBO Journal* 14: 2417–23

Cheng, S.H., Gregory, R.J., Marshall, J. *et al.* (1990) Defective intracellular transport and processing of CFTR is the molecular basis of most cystic fibrosis. *Cell* 63: 827–34

Clarke, L.L., Grubb, B.R., Yankaskas, J.R. *et al.* (1994) Relationship of a non-cystic fibrosis transmembrane conductance regulator-mediated chloride conductance to organ-level disease in *cftr*(−/−) mice. *Proceedings of the National Academy of Sciences USA* 91: 479–83

Cotten, J.F., Ostedgaard, L.S., Carson, M.R. & Welsh, M.J. (1996) Effect of cystic fibrosis-associated mutations in the fourth intracellular loop of cystic fibrosis transmembrane conductance regulator. *Journal of Biological Chemistry* 271: 21279–84

Crystal, R.G. (1995) The gene as the drug. *Nature Medicine* 1: 15–17

Denning, G.M., Amara, J.F., Anderson, M.P. *et al.* (1992) Processing of mutant cystic fibrosis transmembrane conductance regulator is temperature-sensitive. *Nature* 358: 761–64

Devor, D.C., Singh, A.K., Frizzell, R.A. & Bridges, R.J. (1996) Modulation of Cl^- secretion by benzimidazolones. I. Direct activation of a Ca^{2+}-dependent K^+ channel. *American Journal of Physiology* 271: L775–84

Drumm, M.L., Wilkinson, D.J., Smit, L.S. *et al.* (1991) Chloride conductance expressed by ΔF508 and other mutant CFTRs in *Xenopus* oocytes. *Science* 254: 1797–99

Eidelman, O., Guay-Broder, C., van Galen, P.J. *et al.* (1992) A_1 adenosine-receptor antagonists activate chloride efflux from cystic fibrosis cells. *Proceedings of the National Academy of Sciences USA* 89: 5562–66

Gadsby, D.C. & Nairn, A.C. (1994) Regulation of CFTR channel gating. *Trends in Biochemical Sciences* 19: 513–18

Gan, K.-H., Veeze, H.J., van den Ouweland, A.M. *et al.* (1995) A cystic fibrosis mutation associated with mild lung disease. *New England Journal of Medicine* 333: 95–99

Gribkoff, V.K., Champigny, G., Barbry, P. *et al.* (1994) The substituted benzimidazolone NS004 is an opener of the cystic fibrosis chloride channel. *Journal of Biological Chemistrv* 269: 10983–986

Hamill, O.P., Marty, A., Neher, E., Sakmann, B. & Sigworth, F.J. (1981) Improved patch-clamp techniques for high-resolution current recording from cells and cell-free membrane patches. *Pflügers Archives* 391: 85–100

Howard, M., Frizzell, R.A. & Bedwell, D.M. (1996) Aminoglycoside antibiotics restore CFTR function by overcoming premature stop mutations. *Nature Medicine* 2: 467–69

Hwang, T.-C., Wang, F., Yang, I.C.-H. & Reenstra, W.W. (1997) Genistein potentiates wild-type and ΔF508-CFTR channel activity. *American Journal of Physiology* 273: C988–98

Hyde, S.C., Emsley, P., Hartshorn, M.J. *et al.* (1990) Structural model of ATP-binding proteins associated with cystic fibrosis, multidrug resistance and bacterial transport. *Nature* 346: 362–65

Jensen, T.J., Loo, M.A., Pind, S. *et al.* (1995) Multiple proteolytic systems, including the proteasome, contribute to CFTR processing. *Cell* 83: 129–35

Kelley, T.J., Thomas, K., Milgram, L.J.H. & Drumm, M.L.

(1997) *In vivo* activation of the cystic fibrosis transmembrane conductance regulator mutant ΔF508 in murine nasal epithelium. *Proceedings of the National Academy of Sciences USA* 94: 2604–08

Knowles, M.R., Clarke, L.L. & Boucher, R.C. (1991) Activation by extracellular nucleotides of chloride secretion in the airway epithelia of patients with cystic fibrosis. *New England Journal of Medicine* 325: 533–38

Kristidis, P., Bozon, D., Corey, M. *et al.* (1992) Genetic determination of exocrine pancreatic function in cystic fibrosis. *American Journal of Human Genetics* 50: 1178–84

Lukacs, G.L., Mohamed, A., Kartner, N. *et al.* (1994) Conformational maturation of CFTR but not its mutant counterpart (ΔF508) occurs in the endoplasmic reticulum and requires ATP. *EMBO Journal* 13: 6076–86

Pasyk, E.A. & Foskett, J.K. (1997) Cystic fibrosis transmembrane conductance regulator-associated ATP and adenosine 3′-phosphate 5′-phosphosulfate channels in endoplasmic reticulum and plasma membranes. *Journal of Biological Chemistry* 272: 7746–51

Pasyk, E.A. *et al.* (1997) Single channel properties of disease-causing mutations within the R domain of CFTR. *Pediatric Pulmonology* (Suppl.) 14: 215

Pier, G.B., Grout, M. & Zaidi, T.S. (1997) Cystic fibrosis transmembrane conductance regulator is an epithelial cell receptor for clearance of *Pseudomonas aeruginosa* from the lung. *Proceedings of the National Academy of Sciences USA* 94: 12088–93

Pind, S., Riordan, J.R. & Williams, D.B. (1994) Participation of the endoplasmic reticulum chaperone calnexin (p88, IP90) in the biogenesis of the cystic fibrosis transmembrane conductance regulator. *Journal of Biological Chemistry* 269: 12784–88

Quinton, P.M. (1990) Cystic fibrosis: a disease in electrolyte transport. *FASEB Journal* 4: 2709–17

Ramjeesingh, M. *et al.* (1995) Treatment of the nasal epithelium of CF mice with liposomes containing purified CFTR protein. *Pediatric Pulmonology* (Suppl.) 12: S10.7

Riordan, J.R. (1993) The cystic fibrosis transmembrane conductance regulator. *Annual Review of Physiology* 55: 609–30

Riordan, J.R., Rommens, J.M., Kerem, B. *et al.* (1989) Identification of the cystic fibrosis gene: cloning and characterization of complementary DNA. *Science* 245: 1066–73

Rozmahel, R., Wilschanski, M., Matin, A. *et al.* (1996) Modulation of disease severity in cystic fibrosis transmembrane conductance regulator deficient mice by a secondary genetic factor. *Nature Genetics* 12: 280–87

Seibert, F.S., Linsdell, P., Loo, T.W. *et al.* (1996) Cytoplasmic loop three of cystic fibrosis transmembrane conductance regulator contributes to regulation of chloride channel activity. *Journal of Biological Chemistry* 271: 27493–99

Seibert, F.S., Jia, Y., Mathews, C.J. *et al.* (1997) Disease-associated mutations in cytoplasmic loops 1 and 2 of cystic fibrosis transmembrane conductance regulator impede processing or opening of the channel. *Biochemistry* 36: 11966–74

Sheppard, D.N., Rich, D.P., Ostedgaard, L.S. *et al.* (1993) Mutations in CFTR associated with mild disease form Cl^- channels with altered pore properties. *Nature* 362: 160–64

Sheppard, D.N., Ostedgaard, L.S., Winter, M.C. & Welsh, M.J. (1995) Mechanism of dysfunction of two nucleotide binding domain mutations in cystic fibrosis transmembrane conductance regulator that are associated with pancreatic sufficiency. *EMBO Journal* 14: 876–83

Sheppard, D.N., Travis, S.M., Ishihara, H. & Welsh, M.J. (1996) Contribution of proline residues in the membrane-spanning domains of cystic fibrosis transmembrane conductance regulator to chloride channel function. *Journal of Biological Chemistry* 271: 14995–15001

Smith, J.J., Travis, S.M., Greenberg, E.P. & Welsh, M.J. (1996) Cystic fibrosis airway epithelia fail to kill bacteria because of abnormal airway surface fluid. *Cell* 85: 229–36

Stutts, M.J., Rossier, B.C. & Boucher, R.C. (1997) Cystic fibrosis transmembrane conductance regulator inverts protein kinase A-mediated regulation of epithelial sodium channel single channel kinetics. *Journal of Biological Chemistry* 272: 14037–40

Tabcharani, J.A., Rommens, J.M., Hou, Y.X. *et al.* (1993) Multi-ion pore behaviour in the CFTR chloride channel. *Nature* 366: 79–82

The Cystic Fibrosis Genotype–Phenotype Consortium (1993) Correlation between genotype and phenotype in patients with cystic fibrosis. *New England Journal of Medicine* 329: 1308–13

Travis, S.M., Berger, H.A. & Welsh, M.J. (1997) Protein phosphatase 2C dephosphorylates and inactivates cystic fibrosis transmembrane conductance regulator. *Proceedings of the National Academy of Sciences USA* 94: 11055–60

Ward, C.L. & Kopito, R.R. (1994) Intracellular turnover of cystic fibrosis transmembrane conductance regulator: inefficient processing and rapid degradation of wild-type and mutant proteins. *Journal of Biological Chemistry* 269: 25710–18

Ward, C.L., Omura, S. & Kopito, R.R. (1995) Degradation of CFTR by the ubiquitin–proteasome pathway. *Cell* 83: 121–27

Welsh, M.J. & Smith, A.E. (1993) Molecular mechanisms of CFTR chloride channel dysfunction in cystic fibrosis. *Cell* 73: 1251–54

Welsh, M.J., Tsui, L.-C., Boat, T.F. & Beaudet, A.L. (1995) Cystic fibrosis. In *The Metabolic and Molecular Basis of Inherited Disease*, 7th edition, edited by C.R. Scriver, A.L. Beaudet, W.S. Sly and D. Valle, New York: McGraw-Hill

Yang, Y., Janich, S., Cohn, J.A. & Wilson, J.M. (1993) The common variant of cystic fibrosis transmembrane conductance regulator is recognized by hsp70 and degraded in a pre-Golgi nonlysosomal compartment. *Proceedings of the National Academy of Sciences USA* 90: 9480–84

Zeitlin, P.L. & Rubenstein, R.C. (1997) Phenylbutyrate therapy for cystic fibrosis. *Pediatric Pulmonology* (Suppl.) 14: 132

GLOSSARY

agonist an extracellular signal (e.g. hormone) that binds to a cell membrane receptor to trigger a cellular response

apical membrane cell membrane of epithelial cells that faces either the outside of an organ or the lumen of ducts

basolateral membrane cell membrane of epithelial cells that faces the inside of an organ

cAMP (cyclic 3′,5′ adenosine monophosphate) this universal "second messenger" is generated from adenosine triphosphate (ATP) by adenylate cyclase. It acts as a signalling molecule to trigger reactions that produce a cellular response following the activation of cell membrane receptors by extracellular signalling molecules

chloride (Cl^-) channel a pore that allows chloride ions to cross cell membranes, moving down a favourable electrochemical gradient

DNase (deoxyribonuclease) an enzyme that cleaves DNA into shorter oligonucleotides or degrades it totally into its constituent deoxyribonucleotides

epithelia a sheet of polarised cells that form a barrier between different organs of the body

exocrine [gland] a gland that secretes its products via fluid-filled ducts

gating behaviour a characteristic property of ion channels that describes the pattern of transitions between the closed (non-conducting) conformation and the open (conducting) conformation

heterologous exogenous DNA or protein from either the same or a different species that is expressed in cells for functional studies

lipid bilayer (phospholipid bilayer) the major constituent of the membranes that envelop cells and organelles within cells

meconium ileus in the newborn, obstruction of the small intestine at/near the ileocaecal valve by impacted meconium (thick, sticky, greenish/black material that forms the first stools)

multi-ion pore an ion channel that contains more than one ion at a time

nonsense mutation a mutation generating one of the termination codons which results in premature termination of polypeptide synthesis during translation (*see also* stop (termination) codon)

patch-clamp a powerful and versatile method to measure electrical currents flowing across cell membranes through ion channels. Currents flowing through either microscopic "single-channels" or macroscopic "whole-cells" can be recorded using different configurations of the technique

phospholipid bilayer (*see* lipid bilayer)

SDS-polyacrylamide gel electrophoresis a method to investigate proteins: a protein complex is first solubilized in the detergent sodium dodecyl sulphate (SDS) and then subjected to electrophoresis in a polyacrylamide gel to separate individual proteins according to their size and electrical charge

selectivity the ability of an ion channel to discriminate between different ions allowing some types of ions to flow through the channel and others not

single-channel conductance a measure of the electrical current that flows through an open ion channel in response to a given electrochemical driving force

single-channel open probability (P_o) a measure of the fraction of time that an ion channel is open

site-directed mutagenesis an *in vitro* technique in which a change is made at a specific site in a DNA molecule, which is then reintroduced into a cell

somatic gene therapy improvement/alleviation of the symptoms of a genetic disease by replacement or supplementation of affected somatic (non-sex) cells with genetically corrected cells, or by introducing correct copies of the gene directly into affected cells

stop mutation mutation in stop codon that acts as a signal for the termination of protein synthesis (*see* stop (termination) codon in Main Glossary)

HEREDITARY HEARING LOSS

Vasiliki Kalatzis and Christine Petit
Unité de Génétique des Déficits Sensoriels, Institut Pasteur, Paris, France

1. Structure and function of the ear

The adult mammalian ear is made up of three distinct parts, the external, middle and inner ear, which function as one anatomical unit responsible for both hearing and control of equilibrium (**Figure 1**). The external ear is the sound-collecting funnel. It comprises the auricle and the external auditory canal which transfer sound to the tympanic membrane. The middle ear transmits the sound from the tympanic membrane to the inner ear and consists of a chain of three ossicles, the malleus, incus and stapes, which pick up the vibrations received by the tympanic membrane and transmit them to the oval window of the inner ear. The inner ear is composed of an elaborate system of endolymph-filled, epithelially lined chambers and canals constituting the so-called membranous labyrinth. The membranous labyrinth lies, bathed in perilymph, within the temporal bone cavities (the bony labyrinth). The sound-processing portion of the membranous labyrinth is the snail-shaped cochlea. Within the cochlea lies the cochlear duct which contains the auditory transduction sensory apparatus, the organ of Corti. The remaining portion of the membranous labyrinth is the vestibular complex, composed of the saccule, utricle and three semicircular canals, which responds to linear and angular acceleration, hence aiding balance.

The sensory epithelia of the inner ear – i.e. the organ of Corti (**Figure 2**; see also **Plate 10** in the colour plate section), the maculae of the utricle and saccule, and the cristae ampullae of the semicircular canals – consists of a highly organized array of supporting and sensory hair cells. Each sensory hair cell possesses a distinct bundle of actin-filled stiff microvilli, called stereocilia, at their apical surface. All three types of inner ear neuroepithelium are covered by an acellular gelatinous membrane, the tectorial membrane over the organ of Corti, the otoconial membranes over the maculae and the thick, dome-shaped cupula over the cristae. Sound

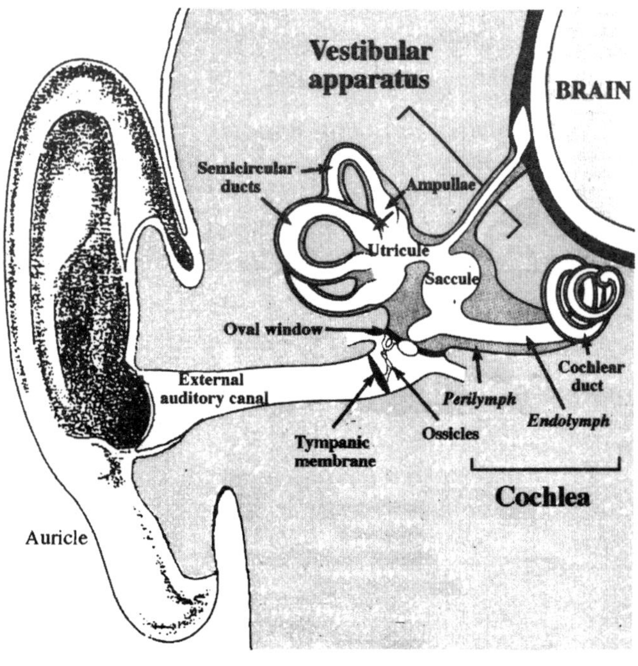

Figure 1 Schematic representation of the mammalian inner ear. Modified and reprinted with permission from Petit (1996).

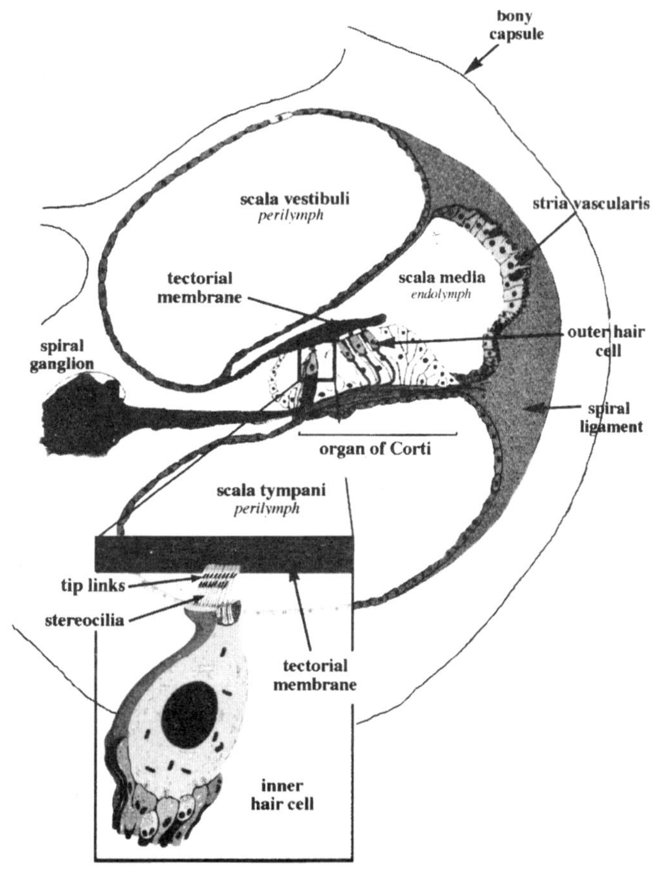

Figure 2 (see also **Plate 10**) Cross-section of the cochlea. The organ of Corti is situated on the floor of the endolymph-filled cochlear duct, and is made up of an array of sensory and supporting cells and the overlying tectorial membrane. Each sensory cell is capped by a stereociliary bundle (inset) which is deflected by shearing of the tectorial membrane. The stria vascularis on the lateral wall of the cochlear duct is responsible for the unique ionic composition of the endolymph. The cochlear duct is surrounded above and below by perilymph-filled spaces. Modified and reprinted from Petit (1996).

transfer or head movements cause a displacement of these acellular membranes relative to the neuroepithelia in the cochlea and vestibule, respectively. This displacement leads to the deflection of the sensory hair cell stereociliary bundles which, in turn, opens up the mechanoelectrical transduction channels located at the tip of the stereocilia (**Figure 2**; see also Plate 10 in the colour plate section). The gating spring of these channels has been proposed to be the tip link, a filamentous connection attaching the tip of a stereocilium to the nearest taller stereocilium. The consequent influx of potassium through the mechanotransduction channels alters the membrane potential, which affects the rate of release of a synaptic transmitter from the hair cell. On neurotransmitter release, an afferent nerve fibre at the base of the hair cell transmits to the brain a pattern of action potentials encoding the characteristics of the stimulus: intensity, frequency and time course (Hudspeth, 1989).

2. General aspects of hereditary hearing loss

Hearing loss can appear at any age and with any degree of severity. The overall impact of hearing impairment is greatly influenced by the age of onset as, if present in early childhood, it also alters speech acquisition and, thereby, cognitive and psychosocial development. Hearing impairment is classified, first, according to whether it is associated with other symptoms (syndromic) or whether it is the sole defect (non-syndromic or isolated). Secondly, it is classified according to the type of defect, i.e. conductive, sensorineural or a mix of the two; conductive hearing loss refers to external and/or middle ear defects and sensorineural hearing loss refers to a transmission anomaly of the sound signal from the inner ear to the cortical auditory centres of the brain. Most cases of sensorineural hearing loss are due to cochlear defects. Thirdly, hearing impairment is classified according to the degree of severity as being mild, moderate, moderate severe, severe or profound, determined by the degree of hearing loss, from 27–40, 41–55, 56–70, 71–90 and superior to 90 decibels hearing level (dB HL), respectively, for the better-hearing ear. Finally, the remaining criterion is the age of onset.

About one in 1000 children is affected by severe or profound deafness at birth or during early childhood; that is, during the prelingual period. In developed countries, it has been estimated that about 60% of prelingual or early childhood deafness cases with no obvious environmental origin are due to genetic defects. This proportion is continuously increasing as public health improves, leading to a decrease in the incidence of hearing loss due to infections. A further one in 1000 children becomes deaf before adulthood. These forms of deafness are usually less severe and progressive, and the proportion of such cases with a genetic basis is not well documented. In addition, 0.3% and 2.3% of the population manifest a hearing loss greater than 65 dB HL between the ages of 30 and 50 years and between 60 and 70 years, respectively. Among these late-onset forms, otosclerosis (i.e. conductive deafness due to fixation of the stapes footplate to the oval window) accounts for a large proportion of them. To date, the work aimed at assessing the genetic causes of deafness has been concentrated on the prelingual or young adult-onset forms of hearing loss. Late-onset forms are generally considered to result from a combination of genetic and environmental causes and, hence, this has dampened the enthusiasm for analysing the genetic basis of these forms.

3. Syndromic hearing loss

Among the prelingual forms of deafness, 30% of cases are syndromic, i.e. the hearing loss is associated with a variety of anomalies (such as eye, musculoskeletal, renal, nervous and pigmentary disorders). Several hundred such syndromes have been described (Gorlin *et al.*, 1995). Syndromic hearing loss can be conductive, sensorineural or mixed with different modes of transmission (autosomal dominant or recessive, X-linked or maternally inherited due to a mitochondrial mutation). At least some of these syndromes can be classified as early developmental defects, evidenced by the association of facial anomalies and/or dysmorphogenesis of the middle and inner ear as detected by computerized tomography. The elucidation of the molecular basis of the syndromic deafness forms provides an entry point for studying the embryogenesis of the ear which cannot be approached via the classical invertebrate genetic models, *Drosophila melanogaster* and *Caenorhabditis elegans*, that lack an auditory organ. Nevertheless, key families of signalling molecules, which have been identified as controlling a variety of early embryonic events in Drosophila, are also implicated in the formation of the inner ear. The elucidation of the function of such molecules is more easily addressed in this animal model.

Along this line is the cloning of the human gene *EYA1* responsible for an ear developmental defect, Branchio–Oto–Renal syndrome (Abdelhak *et al.*, 1997). *EYA1* is homologous to the Drosophila gene *eyes absent* which plays a role in the development of the compound eye (Bonini *et al.*, 1993). It has been postulated that the corresponding gene product could act to prevent the activation of apoptosis and/or promote cell differentiation. *EYA1* belongs to a gene family composed of at least two other members (Abdelhak *et al.*, 1997; Xu *et al.*, 1997), whereas *eyes absent* seems to be a singleton in the Drosophila genome; this situation is reminiscent of other genes such as *Hedgehog* (Zardoya *et al.*, 1996). Such duplications of an ancestral gene during evolution could permit the diversification of its regulation and/or a slight alteration of its function.

Numerous zebrafish (Whitfield *et al.*, 1996; Malicki *et al.*, 1997) and mouse mutants (Steel, 1995) with inner ear morphological defects, either associated or not with other anomalies, have been described. Once the corresponding genes have been characterized, these animal models will also contribute to our understanding of the development of this sensory system. Moreover, they should also contribute to the understanding of the pathophysiological processes of some forms of syndromic (and even isolated) deafness in humans. For example, the orthologues of the *pax3* and *mitf* genes defective in *splotch* and *microphthalmia* mouse mutants were identified as being responsible for the auditory pigmentary diseases, Waardenburg syndromes type I and type II, respectively. Subsequently, *pax3* was identified as having a role in the migration of the precursor inner ear melanoblasts from the neural crest (Tremblay *et al.*, 1995) and *mitf* was implicated in their survival (Motohashi *et al.*, 1994; Opdecamp *et al.*, 1997). Interestingly, this upstream role for *pax3* has also been shown at the molecular level, with recent work demonstrating that the expression of human *MITF* is under the control of *PAX3* (Watanabe *et al.*, 1997). Studies of knockout mice can also highlight candidate genes for syndromic and isolated forms of deafness in humans; examples are given in **Table 1** and include transcription factors, neurotrophins, growth factors, receptor molecules and ionic channels.

To date, causative genes for about 80 forms of syndromic deafness have been mapped to human chromosomes and approximately half of these genes have been cloned. The genes responsible for syndromic deafness encode diverse molecules including extracellular matrix components, enzymes, transcription factors, cytoskeletal components and membrane components (see **Table 2**). For some of these genes, the cells whose impaired function accounts for the phenotype have already been identified: the cartilaginous otic capsule and connective tissue components of the

Table 1 Genes with a role in the ear identified by knockout studies.

Category	Mutant	Phenotype	References
Transcription factors	*Hox1.6*	Defects in the formation of the outer, middle and inner ear	Chisaka *et al.*, 1992
	Krox20	Abnormalities of the vestibular ganglion	Swiatek & Gridley, 1993
	Pax2	Agenesis of the cochlea and spiral ganglion	Torres *et al.*, 1996
	Brn3.1	Loss of sensory hair cells	Erkman *et al.*, 1996
Neurotrophins (and their receptors)	*BDNF* (and *trkB*)	Pronounced loss of select populations of vestibular neurons	Enfors *et al.*, 1995; Schimmang *et al.*, 1995
	NT3 (and *trkC*)	Pronounced loss of select populations of cochlear neurons	Enfors *et al.*, 1995; Schimmang *et al.*, 1995
Growth factors	*FGF3*	Absence of the endolymphatic duct, as well as malformations of the cochlear duct and semicircular canals	Mansour *et al.*, 1993
	TGFb2	Dysdifferentiation of cochlear duct components	Sanford *et al.*, 1997
Receptors	Thyroid hormone (β receptor)	Absence of auditory function with no morphological abnormalities	Forrest *et al.*, 1996
	Retinoic acid (αγ receptor)	Absence of the stapes and an uncharacterized early developmental inner ear defect	Lohnes *et al.*, 1994
Ionic channels	*Isk* (that associates with the K^+ channel KvLQT1)	Collapsed vestibular and cochlear duct walls	Vetter *et al.*, 1996

Table 2 Molecules encoded by genes underlying syndromic forms of deafness. See Petit *et al.* (2001) for a more comprehensive list.

Category	Gene	Encoded molecule	Syndrome
• Extracellular matrix components	*COL4A3, -A4,*	Type IV (α3, α4,) collagen	Alport -autosomal recessive
	COL4A5, -A6	Type IV (α5, α6) collagen	Alport -X-linked
	COL2A1	Type II (α1) collagen	Stickler
	KAL	Anosmin-1	X-linked Kallmann's
	NDP	Norrin	Norrie
• Enzymes	*IDUA*	α-L-iduronidase	Hurler
	IDS	iduronate-2-sulfatase	Hunter
	ERCC3	Helicase	Cockayne's
• Transcription factors	*PAX3*	PAX3	Waardenburg type 1/3
	MITF	MITF	Waardenburg type 2
	SOX10	SOX10	Waardenburg-Hirschsprung
	EYA1	EYA1	Branchio-Oto-Renal
	SALLI	SALLI	Townes-Brocks
• Cytoskeletal components	*NF2*	Merlin	Neurofibromatosis type II
	MYO7A	Myosin VIIA	Usher type IB
• Membrane components			
–Two molecules forming a functional ionic channel	*KvLQT1*	KvLQT1	Jervell and Lange-Nielsen
	KCNE1/IsK	minK/IsK	Jervell and Lange-Nielsen
–Receptors plus their ligands	*FGFR2*	Fibroblast growth factor receptor 2	Crouzon
	EDNRB	Endothelin-B receptor	Waardenburg-Hirschsprung
	EDN3	Endothelin 3	Waardenburg-Hirschsprung
–A putative sulphate transporter	*PDS*	Pendrin	Pendred

membranous cochlea for type II (a1) collagen (Khetarpal *et al.*, 1994); the sensory neuroepithelia and associated ganglia of the inner ear, as well as the periotic mesenchyme for EYA1 (Abdelhak *et al.*, 1997); the middle ear ossicles and bony capsule of the inner ear for FGFR2 (Orr-Urtreger *et al.*, 1993); Schwann cells, glial cells and neurones for NF2 (Huynh *et al.*, 1996); sensory hair cells for myosin VIIA (Hasson *et al.*, 1995; El-Amraoui *et al.*, 1996); and the stria vascularis of the cochlear duct for KvLQT1 (Neyroud *et al.*, 1997) and its associated molecule, IsK (Sakagami *et al.*, 1991; Schulze-Bahr *et al.*, 1997).

4. Non-syndromic hearing loss

Among the prelingual forms of non-syndromic hearing loss, autosomal recessive forms (collectively referred to as DFNB; DFN for non-syndromic deafness and B for recessive) account for about 85% of the cases, autosomal dominant forms (DFNA; A for autosomal) for 15% and X-linked forms (DFN) for 1–3%. Maternally inherited hearing loss due to a mitochondrial mutation, combined or not with an autosomal recessive mutation, has also been described. The autosomal recessive forms of hearing loss usually are the most severe and account for almost all congenital profound deafness; they are almost exclusively sensorineural due to cochlear defects. The number of *DFNB* genes has been estimated to be 30–100. The postlingual forms manifesting before adulthood have not benefited from an extensive analysis of their mode of inheritance. However, as more pedigrees are described, these forms seem to be mainly sensorineural, being transmitted in an autosomal dominant mode, or are maternally inherited due to mitochondrial mutations.

While the mapping of genes responsible for syndromic deafness does not encounter particular difficulties, the mapping of autosomal genes responsible for isolated deafness is hindered by a combination of (i) the extreme genetic heterogeneity of this disorder, (ii) the absence of clinical or paraclinical criteria allowing differentiation of the inner ear defects caused by the various genes and (iii) the high frequency of unions between deaf people, at least in developed countries. During recent years, these obstacles have been circumvented, with three approaches having been used to map the genes responsible for the autosomal recessive forms (Petit, 1996). The first approach consists of analysing large consanguineous affected families living in geographically isolated regions for several generations. In such families, the probability of more than one defective *DFNB* gene segregating is minimized, and genetic linkage can be detected by lodscore analysis. Subsequently, the gene interval is refined to the regions defined as homozygous using polymorphic markers. The second approach is to analyse a population isolated from immigration in which segregation of the defect is indicative of the involvement of a single gene; the gene is then mapped by a direct search for homozygous regions presenting in addition with an allele frequency disequilibrium. Finally, several *DFNB* genes have been mapped in small consanguineous families by a direct search of homozygous regions. For the autosomal dominant forms, several genes have also been mapped, often by analysing large geographically isolated families. To date, 15 *DFNB*, 14 *DFNA* and four *DFN* loci have been mapped to human chromosomes.

Cloning the genes responsible for isolated deafness in man, which has represented an important challenge in the past, is now in the process of becoming easier due to a series of technical improvements. With regards to positional cloning, the extreme genetic heterogeneity of isolated deafness represents an obstacle. Indeed, the definition of a chromosomal interval for a deafness gene necessitates linkage analysis of sufficiently large individual families (which are rare) in order to obtain a significant lodscore with the locus of interest. This problem has prompted the positional cloning of causative genes in mouse mutants as a stepping stone to cloning the orthologous human deafness genes. Comparisons of human and murine genetic maps indicate that, for half of the identified *DFNA* and *DFNB* loci, a deaf mouse mutant may involve the orthologous gene. The current widespread use of large-scale DNA sequencing will largely facilitate the direct search for human deafness genes, as was the recent situation for *DFNA1* (Lynch *et al.*, 1997).

With regards to cloning via the candidate gene approach, the paucity of molecular data concerning the development of the inner ear and auditory function constitutes a serious obstacle. Causative genes for isolated forms of deafness could be genes expressed in several tissues, but whose defect is only deleterious for auditory function. The corresponding transcripts could be isolated by exploiting their anticipated sequence homology deduced from the expected function of their encoded proteins. Included in this category are certain cytoskeletal proteins and, indeed, a striking feature of the cochlea is its highly ordered architecture, which relies on the remarkable structural and spatial organization of the sensory hair cells themselves (Tilney *et al.*, 1992). Candidate genes for isolated forms of deafness may also be genes whose expression is restricted to the inner ear. Due to the development of certain cDNA substraction techniques, genes preferentially (Robertson *et al.*, 1994) or specifically (Cohen-Salmon *et al.*, 1997) expressed within the inner ear can now be isolated. The genes encoding specific inner ear ionic channels and associated molecules, as well as components of the tip link and tectorial membrane (Killick *et al.*, 1995; Cohen-Salmon *et al.*, 1997; Legan *et al.*, 1997), would certainly represent good candidate deafness genes.

The past difficulties in gene cloning explain the small number of causative genes identified to date. A total of only four genes for isolated forms of deafness have been identified: *POU3F4*, responsible for a progressive, mixed form of hearing loss DFN3; *diaphanous*, responsible for the progressive, sensorineural form of hearing loss DFNA1; and *Cx26* and *MYO7A*, responsible for the mainly prelingual and severe sensorineural forms DFNB1 and DFNB2, respectively. The cloning of three of these genes benefited from certain favourable situations: a chromosomal deletion for DFN3 (de Yok *et al.*, 1995), a good candidate gene for DFNB1 (Kelsell *et al.*, 1997) and a mouse model for DFNB2 (Gibson *et al.*, 1995; Weil *et al.*, 1995). These already identified genes seem to encode a variety of molecules with roles in different physiological pathways. *POU3F4* belongs to a family of homeobox genes encoding transcription factors sharing the conserved POU DNA-binding domain. The rat homologue of *POU3F4* is expressed in the otic vesicle (the inner ear primordium) (Le Moine & Young, 1992), but the role that this gene plays in inner ear morphogenesis is, as yet, unknown. *Cx26* encodes the connexin 26 monomers, which hexamerize to form intercellular channels, connexons, that are permeable to certain small molecules and ions. Connexin 26, which is present in a number of inner ear cells, excluding the sensory hair cells (Kikuchi *et al.*, 1995), has been proposed to be involved in potassium recycling, which is essential for the auditory transduction process. *MYO7A* encodes the unconventional myosin, myosin VIIA, which is a motor protein that moves along the actin filaments and is expected to be tethered to certain membranous compartments. Its expression in the inner ear has been found to be restricted to the sensory hair cells (**Figure 3**; see also **Plate 11** in the colour plate section), thus indicating that deafness results from a primary sensory cell defect (Hasson *et al.*, 1995; El-Amraoui *et al.*, 1996). Unconventional myosins have been proposed to play a crucial role in the mechanotransduction process by moving up and down the upper termination of the tip link, thus continuously controlling its tension. However, the subcellular localization of myosin VIIA argues more for a role in the formation and maintenance of the structural integrity of the stereociliary hair bundle (El-Amraoui *et al.*, 1996; Hasson *et al.*, 1997). The expression pattern of the human *diaphanous* gene in the inner ear has not yet been studied. However, the diaphanous protein in both Drosophila and mouse is known to interact with proteins involved in actin polymerization (Lynch *et al.*, 1997). Finally, two mutations in two different mitochondrial genes (12S rRNA and tRNA-ser) have been reported as being responsible for sensorineural deafness. The deafness in patients carrying the first of these mutations (mutation 1555G) is enhanced by treatment with aminoglycosides (Prezant *et al.*, 1992). It is not yet understood why these mitochondrial mutations give rise to defects restricted to the inner ear.

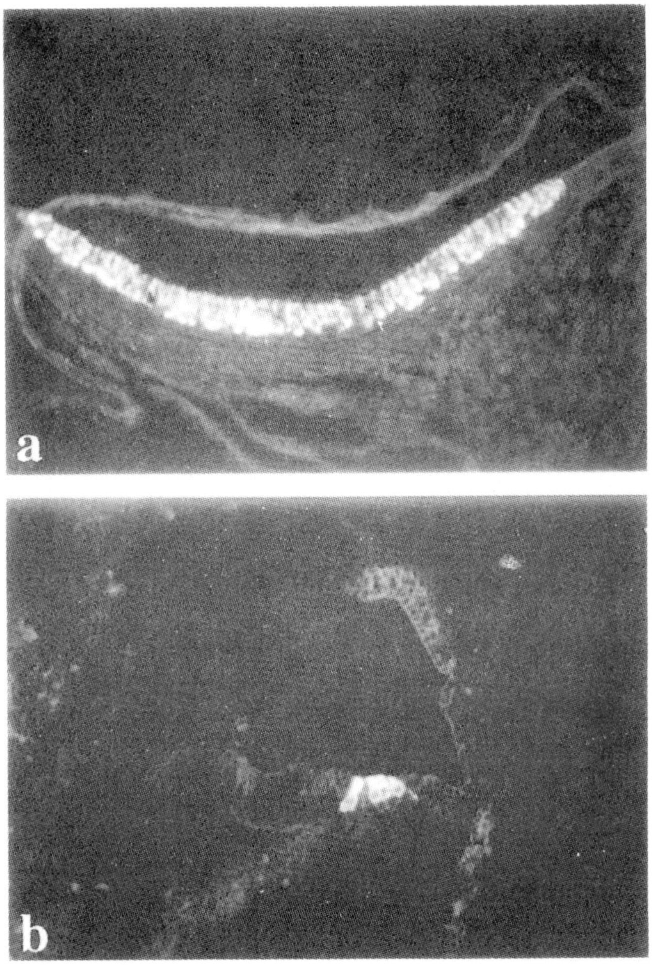

Figure 3 (see also **Plate 11**) Myosin VIIA in the sensory hair cells of the mouse inner ear. Immunohistofluorescence using a polyclonal antibody to myosin VIIA results in a signal in the sensory hair cells of (a) the utricle of the vestibular apparatus (at postnatal day 8) and (b) the cochlea (at postnatal day 2).

5. One gene, several forms of deafness

The genes responsible for several recessive and dominant forms of non-syndromic deafness in humans have been assigned to chromosomal intervals which largely overlap, i.e. both *DFNB1* and *DFNA3* to 13q11, *DFNB2* and *DFNA11* to 11q13.5, and *DFNB5* and *DFNA9* to 14q12, thus suggesting that different mutations within the same gene may underlie both forms. Such a proposal has already been proven for DFNB1/DFNA3, both of which are due to mutations in connexin 26 (Kelsell *et al.*, 1997). Whereas DFNB1 seems to be due to loss-of-function mutations, DFNA3 may be due to dominant negative mutations, with the mutated monomers likely impairing the function of the connexon which they form. Similarly, different allelic variants of the myosin VIIA gene have been shown to underlie a syndromic form of deafness, the Usher 1B syndrome (which associates deafness and retinitis pigmentosa) (Weil *et al.*, 1995), and two isolated forms of deafness, DFNB2 (Liu *et al.*, 1997a; Weil *et al.*, 1997) and DFNA11 (Liu *et al.*, 1997b). The myosin VIIA protein has been shown to dimerize (Weil *et al.*, 1997) and it is likely that DFNA11 results from a dominant negative mutation involving the dimerization process. The discovery that syndromic and isolated forms of deafness (autosomal recessive and dominant) arise from mutations within the same gene suggests that the predicted number of deafness genes may have been overestimated.

6. Clinical implications

Progress in the identification of genes for human hereditary hearing loss is expected to have medical impacts. Hereditary non-syndromic deafness is currently mainly defined as being either a sensorineural, conductive or mixed defect. The identification of the genes responsible for these hearing impairments should stimulate the search for clinical signs specific to the corresponding forms, and particular attention should be paid to vestibular dysfunction. Indeed, almost all the 25 identified mouse mutants exhibiting non-syndromic deafness have been identified on the basis of their abnormal behaviour (circling and head shaking, reflecting a defective sense of balance). The association of these two sensory defects is not surprising since the auditory and vestibular sensory epithelia are closely related, both embryologically and functionally. This observation has stimulated the search for vestibular dysfunction in deaf children. A delay in the age of walking (which can represent an *a minima* manifestation of vestibular dysfunction) has been detected in a number of these children. Moreover, accurate vestibular testing has shown that about half of the children with hearing impairment have some degree of vestibular dysfunction. Comprehensive cochlear and vestibular investigations should divide the vast collection of sensorineural deficiencies into subgroups and/or entities which are nosologically distinct.

Molecular diagnosis should prove helpful for genetic counselling. Indeed, a majority of the families attending genetic counselling clinics are composed of normal hearing parents with a single deaf child wanting to know the risk of recurrence of the defect. In most cases, given the important contribution of environmental causes to prelingual deafness (accounting for about a third of the cases), it is practically impossible to determine whether the deafness even has a genetic origin. In such a situation, only the detection of mutations can provide answers. However, due to the high genetic heterogeneity of isolated deafness, molecular diagnosis has been anticipated to be a difficult task.

Recent results, however, have led to a reassessment of this view. They have brought to light the fact that mutations in connexin 26 underlie approximately half of the cases of prelingual isolated deafness within the Caucasian population (Denoyelle *et al.*, 1997). This gene contains a single coding and a single non-coding exon, thus facilitating a search for mutations within the gene. Moreover, one specific mutation, 30delG, is responsible for the majority of mutant connexin 26 alleles. Thus, due to frequent involvement of this gene and the predominance of this particular mutation, it should be possible to determine whether the deafness is due to a genetic origin in numerous families with a single affected child. This will introduce profound changes into this area of medicine. However, the use of molecular diagnosis at a prenatal stage, in the absence of therapeutic possibilities, raises substantial ethical questions which need to be urgently addressed.

Molecular diagnosis should also prove helpful in late childhood-onset hearing loss, where a presymptomatic diagnosis might guide professional orientation. Due to its frequency and its possible iatrogenic consequences, the detection of the mitochondrial mutation in the 12S rRNA is of particular importance as it would lead to the prevention of the use of aminoglycoside treatment.

Presently, the main treatment for hereditary hearing loss is prosthetics. The continual identification of deafness genes would lead to significant progress in the understanding of the development and function of the ear at both molecular and cellular levels. This should guide the search for alternative treatments for auditory impairments due to either genetic or environmental causes.

Acknowledgements

The authors thank J.-P. Hardelin and E. Verpy for critical reading of the manuscript, and A. El-Amraoui for helpful suggestions and the kind gift of Figure 3.

REFERENCES

Abdelhak, S., Kalatzis, V., Heilig, R. *et al.* (1997) A human homologue of the Drosophila *eyes absent* gene underlies Branchio-Oto-Renal (BOR) syndrome and identifies a novel gene family. *Nature Genetics* 15: 157–64

Bonini, N.M., Leiserson, W.M. & Benzer, S. (1993) The *eyes absent* gene: genetic control of cell survival and differentiation in the developing Drosophila eye. *Cell* 72: 379–95

Chisaka, O., Musci, T.S. & Capecchi, M.R. (1992) Developmental defects of the ear, cranial nerves and hindbrain resulting from targeted disruption of the mouse homeobox gene *Hox-1.6*. *Nature* 355: 516–20

Cohen-Salmon, M., El-Amraoui, A., Leibovici, M. & Petit, C. (1997) Otogelin: a glycoprotein specific to the acellular membranes of the inner ear. *Proceedings of the National Academy of Sciences USA* 94: 14450–55

de Yok, Y.J.M., van der Maarel, S.M., Bitner-Glindzicz, M. *et al.* (1995) Association between X-linked mixed deafness and mutations in the POU domain gene POU3F4. *Science* 267: 685–88

Denoyelle, F., Weil, D., Maw, M.A. *et al.* (1997) Prelingual deafness: high prevalence of a 30delG mutation in the connexin 26 gene. *Human Molecular Genetics* 6: 2173–77

El-Amraoui, A., Sahly, I., Picard, S. *et al.* (1996) Human Usher AB/mouse *shaker-1*: the retinal phenotype discrepancy explained by the presence/absence of myosin VIIA in the photoreceptor cells. *Human Molecular Genetics* 5: 1171–78

Enfors, P., Van De Water, T., Loring, J. & Jaenisch, R. (1995) Complementary roles of BDNF and NT-3 in vestibular and auditory development. *Neuron* 14: 1153–64

Erkman, L., McEvilly, R.J., Luo, L *et al.* (1996) Role of transcription factors Brn-3.1 and Brn-3.2 in auditory and visual system development. *Nature* 381: 603–06

Forrest, D., Haneburth, E., Smeyne, R.J. *et al.* (1996) Recessive resistance to thyroid hormone in mice lacking thyroid hormone receptor b: evidence for tissue-specific modulation of receptor function. *EMBO Journal* 15: 3006–15

Gibson, F., Walsh, J., Mburu, P. *et al.* (1995) A type VII myosin encoded by the mouse deafness gene *Shaker-1*. *Nature* 374: 62–64

Gorlin, R.J., Toriello, H.V. & Cohen, M.M. (1995) *Hereditary Hearing Loss and Its Syndromes*, Oxford and New York: Oxford University Press

Hasson, T., Heintzelman, M.B., Santos-Sacchi, J., Corey, D.P. & Mooseker, M.S. (1995) Expression in cochlea and retina of myosin VIIA, the gene product defective in Usher syndrome type 1B. *Proceedings of the National Academy of Sciences USA* 92: 9815–19

Hasson, T., Gillespie, P.G., Garcia, J.A. *et al.* (1997) Unconventional myosins in inner-ear sensory epithelia. *Journal of Cell Biology* 137: 1287–307

Hudspeth, A.J. (1989) How the ear's works work. *Nature* 341: 397–404

Huynh, D.P., Tran, T.M.D., Nechiporuk, T. & Pulst, S.M. (1996) Expression of neurofibromatosis 2 transcript and gene product during mouse fetal development. *Cell Growth Differentiation* 7: 1551–61

Kelsell, D.P., Dunlop, J., Stevens, H.P. *et al.* (1997) Connexin 26 mutations in hereditary non-syndromic sensorineural deafness. *Nature* 387: 80–83

Khetarpal, U., Robertson, N.G., Yoo, T.J. & Morton, C.C. (1994) Expression and localization of COL2A1 mRNA and type II collagen in human fetal cochlea. *Hearing Research* 79: 59–73

Kikuchi, T., Kimura, R.S., Paul, D.L. & Adams, J.C. (1995) Gap junctions in the rat cochlea: immunohistochemical and ultrastructural analysis. *Anatomy and Embryology* 191: 101–18

Killick, R., Legan, P.K., Malenczak, C. & Richardson, G.P. (1995) Molecular cloning of chick b-tectorin, an extracellular matrix molecule of the inner ear. *Journal of Cell Biology* 129: 535–47

Le Moine, C. & Young, W.S. III. (1992) RHS2, a POU domain-containing gene, and its expression in developing and adult rat. *Proceedings of the National Academy of Sciences USA* 89: 3285–89

Legan, P.K., Rau, A., Keen, J.N. & Richardson, G.P. (1997) The mouse tectorins: modular matrix proteins of the inner ear homologous to components of the sperm–egg adhesion system. *Journal of Biological Chemistry* 272: 8791–801

Liu, X.-Z., Walsh, J., Mburu, P. *et al.* (1997a) Mutations in the myosin VIIA gene cause non-syndromic recessive deafness. *Nature Genetics* 16: 188–90

Liu, X.-Z., Walsh, J., Tamagawa, Y. *et al.* (1997b) Autosomal dominant non-syndromic deafness caused by a mutation in the myosin VIIA gene. *Nature Genetics* 17: 268–69

Lohnes, D., Mark, M., Mendelsohn, C. *et al.* (1994) Function of the retinoic acid receptors (RARs) during development. (I) Craniofacial and skeletal abnormalities in RAR double mutants. *Development* 120: 2723–48

Lynch, E.D., Lee, M.K., Morrow, J.E. *et al.* (1997) Nonsyndromic deafness DFNA1 associated with mutation of a human homolog of the *Drosophila* gene *diaphanous*. *Science* 278: 1315–18

Malicki, J., Schier, A.F., Solnica-Krezel, L. *et al.* (1997) Mutations affecting development of the zebrafish ear. *Development* 123: 275–83

Mansour, S.L., Goddard, J.M. & Capecchi, M.R. (1993) Mice homozygous for a targeted disruption of the proto-oncogene *int-2* have developmental defects in the tail and inner ear. *Development* 117: 13–28

Motohashi, H., Hozawa, K., Oshima, T., Takeuchi, T. & Takasaka, T. (1994) Dysgenesis of melanocytes and cochlear dysfunction in mutant microphthalmia (*mi*) mice. *Hearing Research* 80: 10–20

Neyroud, N., Tesson, F., Denjoy, I. *et al.* (1997) A novel mutation in the potassium channel gene KVLQT1 causes the Jervell and Lange–Nielsen cardioauditory syndrome. *Nature Genetics* 15: 186–89

Opdecamp, K., Nakayama, A., Nguyen, M.-T.T. *et al.* (1997) Melanocyte development in vivo and in neural crest cell cultures: crucial dependence on the Mitf basic-helix–loop–helix–zipper transcription factor. *Development* 124: 2377–86

Orr-Urtreger, A., Bedford, M.T., Burakova, T. *et al.* (1993) Developmental localization of the splicing alternatives of fibroblast growth factor receptor-2 (FGFR2). *Developmental Biology* 158: 475–86

Petit, C. (1996) Genes responsible for human hereditary deafness: symphony of a thousand. *Nature Genetics* 14: 385–91

Petit, C., Levilliers, J., Marlin, S. & Hardelin, J.-P. (2001) Hereditary hearing loss. In *The Metabolic and Molecular Basis of Inherited Disease*, edited by C.R. Scriver *et al.*, Montreal: McGraw Hill (in press)

Prezant, R.T., Shohat, M., Jaber, L., Pressman, S. & Fischel-Ghodsian, N. (1992) Biochemical characterization of a pedigree with mitochondrially inherited deafness. *American Journal of Medical Genetics* 44: 465–72

Robertson, N.G., Khetarpal, U., Gutierrez-Espeleta, G.A., Bieber, F.R. & Morton, C.C. (1994) Isolation of novel and known genes from a human fetal cochlear cDNA library using subtractive hybridisation and differential screening. *Genomics* 23: 42–50

Sakagami, M., Fukazawa, K., Matsunaga, T. *et al.* (1991) Cellular localization of rat IsK protein in the stria vascularis by immunohistochemical observation. *Hearing Research* 56: 168–72

Sanford, L.P., Ormsby, I., Gittenberger-de Groot, A.C. *et al.* (1997) TGFb2 knockout mice have multiple developmental defects that are non-overlapping with other TGFb knockout phenotypes. *Development* 124: 2659–70

Schimmang, T., Minichiello, L., Vazquez, E. *et al.* (1995) Developing inner ear sensory neurons require TrkB and TrkC receptors for innervation of their peripheral targets. *Development* 121: 3381–91

Schulze-Bahr, E., Wang, Q., Wedekind, H. *et al.* (1997) *KCNE1* mutations cause Jervell and Lange–Nielsen syndrome. *Nature Genetics* 17: 267–68

Steel, K. (1995) Inherited hearing defects in mice. *Annual Review of Genetics* 29: 675–701

Swiatek, P.J. & Gridley, T. (1993) Perinatal lethality and defects in hindbrain development in mice homozygous for a targeted mutation of the zinc finger gene *Krox20*. *Genes and Development* 7: 2071–84

Tilney, L.G., Tilney, M.S. & De Rosier, D.J. (1992) Actin filaments, stereocilia and hair cells: how cells count and measure. *Annual Review of Cell Biology* 8: 257–74

Torres, M., Gomez-Pardo, E. & Gruss, P. (1996) *Pax2* contributes to inner ear patterning and optic nerve trajectory. *Development* 122: 3381–91

Tremblay, P., Kessel, M. & Gruss, P. (1995) A transgenic neuroanatomical marker identifies cranial neural crest deficiencies associated with the *Pax3* mutant *Splotch*. *Developmental Biology* 171: 317–29

Vetter, D.E., Mann, J.R., Wangemann, P. *et al.* (1996) Inner ear defects induced by null mutation of the *isk* gene. *Neuron* 17: 1251–64

Watanabe, A., Takeda, K., Ploplis, B. & Tachibana, M. (1997) Epistatic relationship of genes for Waardenburg syndrome type 1 and type 2: *MITF* is regulated by *PAX3*. *American Journal of Human Genetics* 61 (Suppl.): A323

Weil, D., Blanchard, S., Kaplan, J. *et al.* (1995) Defective myosin VIIA gene responsible for Usher syndrome 1B. *Nature* 374: 60–61

Weil, D., Küssel, P., Blanchard, S. *et al.* (1997) The autosomal recessive isolated deafness, DFNB2, and the Usher 1B syndrome are allelic defects of the myosin-VIIA gene. *Nature Genetics* 16: 191–93

Whitfield, T.T., Granato, M., van Eeden, F.J.M. *et al.* (1996) Mutations affecting development of the zebrafish inner ear and lateral line. *Development* 123: 241–54

Xu, P.-X., Woo, I., Her, H., Beier, D.R. & Maas, R.L. (1997) Mouse *Eya* homologues of the Drosophila *eyes absent* gene require *Pax6* for expression in lens and nasal placode. *Development* 124: 219–31

Zardoya, R., Abouheif, E. & Meyer, A. (1996) Evolutionary analyses of hedgehog and *Hoxd-10* genes in fish species closely related to the zebrafish. *Proceedings of the National Academy of Sciences USA* 93: 13036–41

GLOSSARY

apoptosis active process of cell death requiring metabolic activity by the dying cell

linkage analysis the process of determining the degree of co-segregation of markers and/or phenotypic traits across generations

neuroepithelium the epithelium housing the sensory hair cells which transduce the mechanical sound stimulation into an electrical response

orthologue genes related by common phylogenetic descent

GENETIC EYE DISORDERS

Alan Wright
MRC Human Genetics Unit, Western General Hospital, Edinburgh, UK

1. Introduction

Vision is a highly specialized sense which requires unique light-sensitive photoreceptors and complex neural processing, both in the retina and in the brain, one-third of which is concerned with vision. Light-focusing structures within the eye have been shaped over millions of years of evolution for the requirements of each species; the differing visual demands of nocturnal mammals and deep water fish, for example, are reflected in their visual apparatus. A visual defect is almost certainly a selective disadvantage for almost all species including, until relatively recent times, our own. Relaxation of that selection in humans, and in domesticated and laboratory animals, has made it possible to explore the extent of genetic variation in visual function in vertebrates and invertebrates. The accessibility of the eye and ease of visual testing has made it a model system for discovering the effects of mutation on a complex organ. The visual system is unusual in that a significant proportion of the mutants which have contributed to understanding the structure and function of the eye have been discovered in humans rather than simpler organisms. Studies on signal transduction pathways in the compound eye of the invertebrate fruit fly, *Drosophila melanogaster*, have, however, not only pioneered our knowledge of widely used signal transduction pathways, such as Ras/MAP kinase, but have also provided a wealth of insights into specific developmental and transduction mechanisms, many of which are also relevant to the more complex vertebrate eye.

The major structural components of the visual apparatus in a typical vertebrate eye are shown in **Figure 1** (**Plate 12** in the colour plate section). The eyes of different species show an extraordinary diversity of external appearances. It is less clear that the individual developmental and phototransduction pathways are so diverse. One of the unique components of all eyes is the photoreceptor, which, in most vertebrates, consists of a small number (usually 3–5) of highly specialized neuronal receptors expressing different visual pigments. The phototransduction process in all animal species is initiated by the highly unstable *11-cis* retinal isomer of vitamin A_1 or A_2 aldehyde covalently coupled to the protein opsin. This is referred to as a visual pigment because *11-cis* retinal absorbs light, which changes its colour from reddish to yellow as the retinoid is isomerized to the more stable *all-trans* isomer (photo-bleaching), the major form in non-ocular tissues. A cascade of reactions follows, which can amplify the signal generated by a few photons of light by as much as six orders of magnitude, altering the state of polarization of the cell and conveying a neural signal to the second-order neurones (**Figure 2**; **Plate 13** in the colour plate section). A complex series of

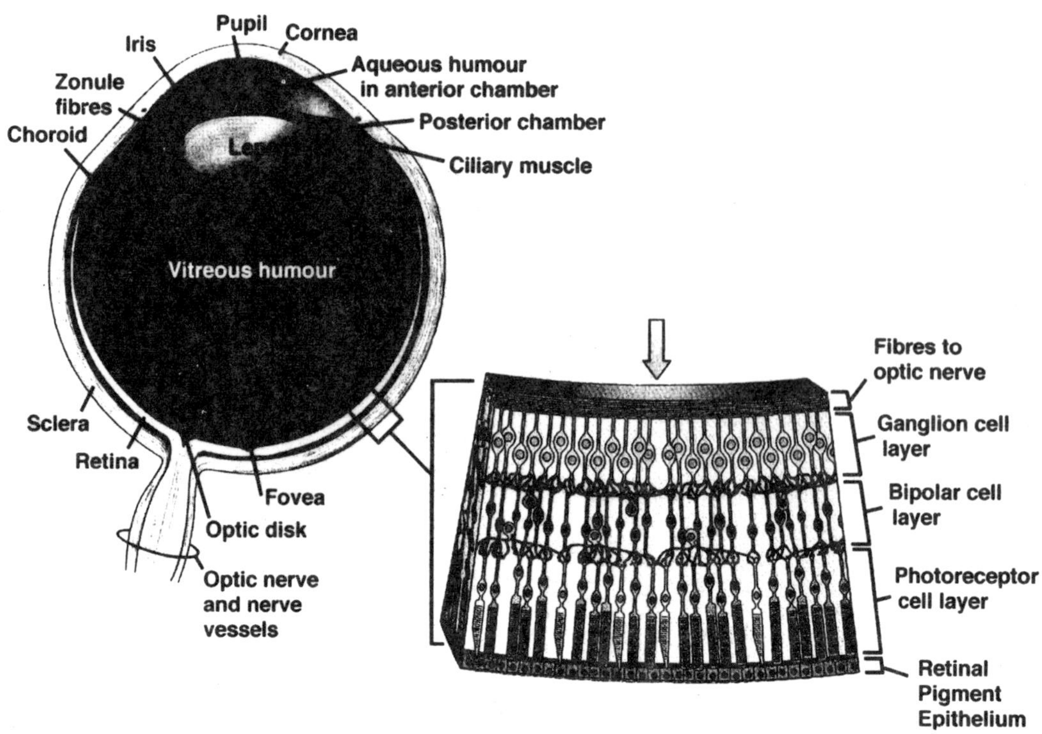

Figure 1 (see also **Plate 12**) Anatomy of a vertebrate (human) eye.

horizontal and vertical processing and filtering steps generate an axon potential in the retinal ganglion cells, whose fibres leave the eye as the optic nerve. The first genetic defects affecting vision were found in components of the phototransduction machinery within the photoreceptor (**Figure 2**; **Plate 13** in the colour plate section).

2. Phototransduction defects

The observation that photoreceptors contain a number of proteins that are uniquely expressed in the eye led to a search for mutations in eye-specific genes, the majority being within the phototransduction pathway. The first ocular genetic defect to be proposed was that of colour blindness by John Dalton in 1794, and it was in colourblind individuals that the first mutations causing a genetic eye disorder were found almost 200 years later (see below). Shortly afterwards, mutations were found both in the Drosophila and human rhodopsin genes (**Figure 2**, see also **Plate 13** in the colour plate section) in progressive forms of retinal degeneration. A large human kindred segregating for an autosomal dominant form of progressive retinal degeneration, called retinitis pigmentosa (RP), was found to show genetic linkage to markers on the long arm of human chromosome 3 (Dryja *et al.*, 1990). The chromosome 3 site was known to harbour the rhodopsin gene and mutations were soon found to account for about 10% of all RP patients. Over 100 different mutations have been found in this gene, predominantly in patients with RP, but also in some with a congenital form of stationary night blindness (CSNB).

RP affects 1–2 in 5000 of the general population in most ethnic groups and is now known to be a highly heterogeneous group of disorders in which there is a primary or secondary dysfunction of rod photoreceptors. This, in turn,

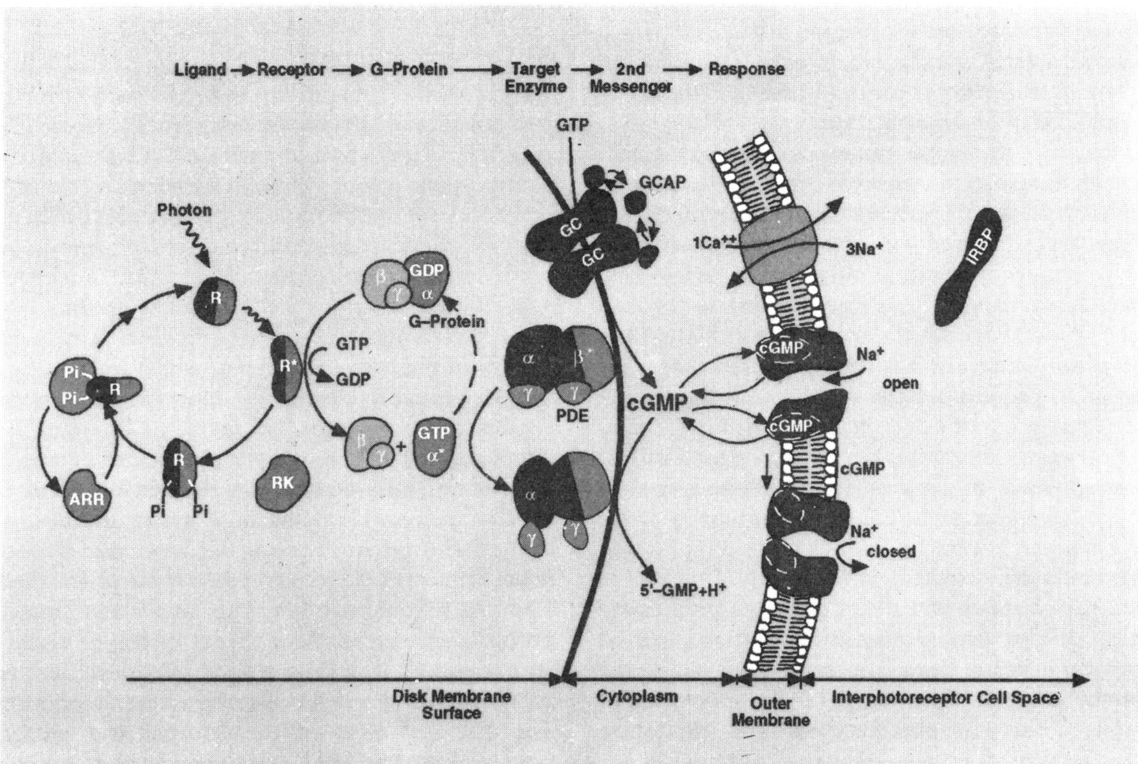

Figure 2 (see also **Plate 13**) Vertebrate phototransduction cascade. In the photoreceptor outer segment, light activates rhodopsin (R→R*) which promotes the exchange of GDP for GTP on the α subunit of the heterotrimeric G protein transducin with dissociation of the β and γ subunits. The activated transducin αGTP subunit stimulates cGMP phosphodiesterase (PDE) activity which hydrolyses cGMP to 5′-GMP, with closure of the cyclic nucleotide gated channels and hyperpolarization of the cell membrane. The $Na^{+}Ca^{2+}/K^{+}$ exchanger pumps calcium ions out of the cell, lowering the intracellular calcium concentration, which helps to restore intracellular cGMP levels by means of guanylate cyclase-activating protein (GCAP) and retinal guanylate cyclase (GC). In the dark, the cyclic nucleotide gated channels are open and sodium and calcium ions enter the cell, depolarizating the cell membrane.

Rhodopsin is deactivated by rhodopsin kinase (RK)-mediated phosphorylation of carboxyl terminal serine residues which promote binding of arrestin (ARR). The final deactivation step is the removal of phosphate (P_i) from phosphorylated residues by a protein phosphatase. Inter-photoreceptor retinoid-binding protein (IRBP) is concerned with the shuttling of retinoids between RPE and retina.

Proteins that can give rise to retinal degeneration when mutated are shown in red (human) or green (transgenic mice). Proteins shown in light blue can give rise to a stationary retinal disorder. Modified from Falk & Applebury (1988).

leads to a progressive loss of vision in dim light (night blindness) and "tunnel vision" – a constriction of the field of vision. Vision at ambient light levels is primarily due to the less numerous cone photoreceptors, which initiate the sensation of colour and the fine discrimination required for reading and many everyday tasks. In the human retina, rod photoreceptors ("rods") outnumber cone photoreceptors ("cones") by about 20:1, but provide low sensitivity and resolution in daylight or even in a well-lit room. In dim light, on the other hand, rods are capable of increasing their light sensitivity by about 75 000-fold and a neural signal is detectable in man following as few as five to seven photoisomerization reactions in *11-cis* retinal. A decline in the number of functional rods leads to a reduced ability to see in dim light, but it is more difficult to explain the dramatic reduction in visual field (tunnel vision) experienced by RP patients, since this is predominantly a cone-mediated response. It is now thought to be due to a secondary dysfunction and loss of cones in the mid-peripheral visual field where the rod density is highest, but the reasons for this are unclear. In the late stages of the disease, vision can be lost even within the central macular region of the retina, which contains the highest density of cones and which is the site of greatest visual acuity or discrimination.

The reasons why rhodopsin mutations cause a late-onset (second to fifth decade) and progressive retinal degeneration are not fully established. There are examples of rhodopsin alleles which are associated with the non-progressive disorder CSNB, caused by missense mutations at amino acid residues which are thought to affect coupling to the heterotrimeric G protein transducin (see **Figure 2**; **Plate 13** in the colour plate section). It has been proposed that some CSNB mutations produce a form of rhodopsin that is in a low but continuous (constitutive) state of activation, desensitizing the downstream effector pathway to signals of low intensity, which leads to night blindness. While it is clear that some rhodopsin mutations cause a constitutively active protein *in vitro*, the *in vivo* situation is more complex and this mechanism is still disputed.

The majority of rhodopsin alleles causing a progressive degeneration is also due to missense mutations, but at residues which, in many cases, appear to compromise the folding of the protein. There are 4×10^7 molecules of rhodopsin in a human photoreceptor and rhodopsin accounts for 70% of all protein in the rod "outer segment" (ROS), a stack of disc-like membranes packed with visual pigment and suitably oriented for catching light (**Figure 1**; Plate 12 in the colour plate section). These discs are turned over at a very high rate as a result of phagocytosis by the adjacent retinal pigment epithelium (RPE). The 500–800 discs of the human ROS are entirely replaced every 10 days. An abnormally folded rhodopsin molecule, even in the heterozygous state, may compromise the integrity of disc membranes and the uptake or degradation of discs by the RPE. Additionally, the endoplasmic reticulum (ER) in the rod "inner segment", towards which misfolded proteins are returned and degraded, may be compromised, the sheer abundance of misfolded rhodopsin molecules perhaps overwhelming the capacity of the ER to refold or degrade the proteins via the energy-requiring proteasome pathway.

In *Drosophila melanogaster*, rhodopsin (*NinaE*) mutations were also discovered in retinal degeneration mutants in the late 1980s, the majority of which showed recessive inheritance, suggesting a different disease mechanism (Palk, 1995). More recently, a number of dominant rhodopsin mutants have also been identified, some of which are at equivalent residues to those causing human retinal degeneration, illustrating a striking conservation of structure and function across more than 500 million years of evolution. Most human mutations are probably semidominant, so the difference between these and the recessive Drosophila mutants may only reflect the greater opportunity for identifying subtle phenotypic effects in a long-lived mammal. There are a small number of human rhodopsin alleles which are truly recessive, most of which are null alleles, suggesting that 50% of wild-type rhodopsin is sufficient for normal function, provided that the other allele does not produce an abnormally folded protein.

3. Colour blindness

As early as the 16th century, scientists such as Francis Bacon had noted that "all colours disappear in the dark", because dim light is inefficient at exciting cone photoreceptors, the source of our colour vision. In addition to light of sufficient intensity, at least two retinal pigments with different absorption spectra are necessary for colour perception and three are required to distinguish colours in the wavelength range 400–700 nm, which is the "visible" spectrum.

In vertebrates, the visual pigments are located in a variety of photoreceptor types, which in humans and Old World primates consist of blue or short wavelength-sensitive cones, green or medium wavelength-sensitive cones and red or long wavelength-sensitive cones. In 1794, the chemist John Dalton was the first to report the existence of colour blindness in a group of men, including himself and his brother, on the basis that they could see fewer spectral colours generated by a prism than others. This was referred to a few years later when Thomas Young proposed the trichromatic theory of colour vision in 1807 and suggested that Dalton's red blindness must be due to an absence or paralysis of certain fibres within the retina. The X-linked basis of the disorder was recognized by George Wilson in 1911, but the molecular details were not established until after the cloning of the red, green and blue pigment genes by Nathans and coworkers in 1986. In man, colour blindness is due to a group of over ten mostly rare conditions among which are some of the commonest genetic disorders in man.

Colour blindness provides the first example of a genetic disorder affecting a human behavioural trait and may prove to be a paradigm for other behavioural disorders. Red–green colour blindness affects about 8% of males and includes dichromats, who have only two of the three pigments, and anomalous trichromats, who have anomalous pigments that affect the ability to discriminate colours. The two forms of dichromat either lack red gene function (protanopia) or green gene function (deuteranopia), usually as a result of the deletion of the red or green opsin genes. These genes exist as a tandem array on the X chromosome (band Xq28). Deletion of one or other gene results from unequal crossing-over between normal pigment gene arrays, which contain a single red pigment gene and two, three or more green pigment genes, and which are 98% identical in sequence. Protanopia and deuteranopia each affect about 1% of Caucasian males.

Anomalous colour vision is due to the presence of hybrid red–green pigment genes, which cause mild red–green confusion and, again, result from unequal crossing-over between gene arrays. Anomalous trichromats may have protanomaly (~1% of Caucasians), who have a 5′ red–green 3′ fusion gene with an abnormal green-shifted absorption spectrum and a normal green gene. An additional 4–5% of Caucasians have deuteranomaly, which usually results from an abnormal red-shifted 5′ green–red 3′ fusion gene and a normal red gene. An amino acid polymorphism is also present in most populations, in which serine or alanine is present at position 180 of the red pigment, which alters the absorbance maximum from 530 nm (Ala^{180}) to 560 nm (Ser^{180}); individuals with Ser^{180} are, therefore, more sensitive to red light than others. Position 180 occurs within the fourth membrane-spanning domain, but influences the retinal chromophore lying within a hydrophobic pocket in opsin. Defective blue cone sensitivity (tritanopia) is an uncommon autosomal dominant trait which results from missense mutations within the blue pigment gene on chromosome 7 (band 7q31.3–q32).

4. Structural defects in photoreceptors

Mutations in at least two structural proteins that are necessary for maintaining the specialized outer segments of photoreceptors are found in patients with RP-type retinal degeneration. The first was in the peripherin/RDS (*RDS*) gene in the *retinal degeneration slow* (*rds*) mouse mutant. The RDS protein is found in the rims of rod and cone photoreceptor discs and appears to play a role in maintaining the flattened shape of the disc. Initially, human *RDS* mutations were identified in patients with autosomal dominant forms of RP, but the picture soon emerged of a protein in which structural variation could produce extraordinary clinical diversity. *RDS* mutations have been found in several different retinal degenerative disorders, predominantly affecting central vision (cone dystrophy, cone–rod dystrophy, Stargardt-type disease and other central receptor dystrophies), but also peripheral vision (e.g. RP, retinitis punctata albescens) and mixed central and peripheral visual defects (e.g. RPE pattern dystrophies).

The variety of phenotypes associated with mutations in the *RDS* gene is a paradigm for understanding the diverse effects of mutation. The phenotype depends, first, on which domain of the protein is altered; some domains may be necessary for critical homophilic interactions across the disc lumen, while others interact with a related protein in rods and cones (ROM-1, see below). Secondly, the nature of the mutations determines the phenotype. Some amino acid substitutions have no phenotypic effect – there are some parts of the protein that are highly variable and, yet, these substitutions produce no detectable abnormality. Others, however, cause very subtle phenotypic effects, which in a long-lived mammal may be evident comparatively late in life. Factors such as the presence or absence of functional protein are also important; in the case of RDS, half the amount of normal protein, in the absence of mutant protein, is insufficient for normal function (haplo-insufficiency). Finally, RDS provides one of the first examples of digenic inheritance in humans. RDS normally interacts with a structurally related protein called ROM-1 in the rod photoreceptor disc rim. The absence of ROM-1 due to a null allele has no phenotypic effect unless it is accompanied by a mutant RDS protein. Only individuals inheriting a mutant copy of both the RDS and ROM-1 genes are found to show signs of retinal degeneration (RP). Those inheriting either mutant gene individually are quite normal (Kajiwara, Berson & Dryja, 1994).

5. Deafness and retinal degeneration

Usher syndrome is a disorder in which a congenital neurosensory hearing loss is accompanied by progressive retinal degeneration of RP type. It affects about one in 20 000 of the human population and accounts for about one-half of those with both hearing and visual impairment. There is, again, genetic diversity within this group and mapping studies show that there are at least nine causative genes (types IA-F, IIA-B, III); one of these genes has been isolated, the myosin VIIA gene in type IB Usher syndrome.

Myosin VIIA is one of the family of atypical myosins which, in man, is expressed in the vestibulo-cochlear apparatus cochlea and the retina. The gene was first isolated by a positional cloning project on the *shaker* mutant mouse, which shows deafness and vestibular abnormalities, but no visual defect. Type IB Usher syndrome had been mapped by linkage analysis to chromosomal region 11q13, which is syntenic with the location of *shaker* in the mouse, so that the isolation of the *shaker* presented a strong candidate for a similar, if not identical, human disorder. Mutations were subsequently found in human myosin VIIA in type I Usher patients. The presence of blindness in Usher syndrome and not in the *shaker* mouse is explained by the fact that myosin VIIA is not expressed in the mouse photoreceptor, but that, in humans, it is a component of the connecting cilium between photoreceptor inner and outer segments. Loss-of-function mutations appear to affect either the structural integrity of the photoreceptor, the movement of molecules such as rhodopsin between inner and outer segments or both.

6. Defects in other phototransduction proteins

Once it was appreciated that retina-specific genes associated with the phototransduction process could give rise to retinal degeneration, many other phototransduction defects came to light in patients with either progressive or stationary retinal disorders. In most cases, these defects were rare, affecting only a few per cent of patients at most. The strategy was highly successful, however, and, by screening a large pool of several hundred patients with a variety of retinal disorders, a small number could generally be found with mutations in almost any phototransduction gene tested. Why the retina should be so sensitive to the degenerative process remains unclear, since the proximate consequences of mutation in different transduction components seem to differ quite widely.

Stationary forms of night blindness are associated with mutations in the α-subunit of the heterotrimeric G protein, transducin, in the catalytic β-subunit of cGMP phosphodiesterase (PDE) and in two proteins concerned with the "switch-off" mechanism for rhodopsin activation (arrestin, rhodopsin kinase). Degenerative retinal disorders are associated with mutations in both α- and β-subunits of PDE, in the α-subunit of the cGMP gated ion channel, and in retinal guanylyl cyclase (retGC-1) and its activating protein (GCAP-1) (**Figure 2**; **Plate 13** in the colour plate section).

Some of these mutant genes were first discovered in laboratory mouse models of retinal degeneration. The β-subunit of cGMP phosphodiesterase was identified as the gene responsible for the *retinal degeneration* (*rd*) mouse and peripherin/RDS for the *retinal degeneration slow* (*rds*) mouse. For those in which human or animal mutations could not be found, genetic "knockout" mice have been created using embryonic stem cells in which the gene has been experimentally disrupted. Many of these animals have proved highly informative for probing the molecular details of normal and abnormal vision. Some produced unexpected results, such as the loss of PDE activity caused by a knock-out of the γ-subunit of cGMP phosphodiesterase, previously thought to be inhibitory. Genetic knockout mice defective for a growing number of retinal proteins have been made, including rhodopsin, arrestin, Norrin, Rep-1, rhodopsin kinase, interstitial retinol-binding protein, recoverin and the ABCR transporter.

7. Retinal pigment epithelium function and retinal degeneration

Initially, it was thought that most defects leading to retinal degeneration are associated with the photoreceptor itself. The first indication that the RPE could be the primary site of disease came from a laboratory rat strain carrying the *RCS* mutation. Chimaeric animals carrying distinguishable clonal patches of wild-type and mutant cells were found to show defective phagocytosis of rod outer segments only where the mutant RPE cells were underlying the photoreceptors. The genetic basis of the RCS defect remains to be identified. Recently, mutations have been found in two proteins expressed in RPE cells which are concerned with the shuttling of retinoids between photoreceptors and RPE (the visual cycle). In vertebrates, retinoids are cycled between photoreceptors and RPE in order to regenerate *11-cis* retinal after it is bleached by light. A series of tightly regulated intercellular transport, cellular uptake and enzymatic conversion steps are necessary to achieve this. Mutations in the cellular retinaldehyde-binding protein and RPE-65 protein were among the first non-photoreceptor proteins shown to cause retinal disease – a childhood-onset severe retinal dystrophy and a severe congenital form of RP (Leber's congenital amaurosis).

8. Disorders of metabolite and vitamin transport

Abetalipoproteinaemia is an autosomal recessive condition associated with a retinal degeneration of RP type, symptoms of fat malabsorption, abnormal red blood cell membranes (acanthocytosis) and peripheral neuropathy. The primary defect is in a microsomal triglyceride transport protein, which is concerned with the assembly of lipids into apolipoprotein B, compromising the transport of fat-soluble vitamins such as vitamin A to the eye.

The superfamily of ATP-binding cassette (ABC) transporters are integral membrane proteins concerned with the transport of a variety of ions, fatty acids, drugs and metabolites. It includes several clinically important proteins such as the cystic fibrosis transmembrane regulator and the multidrug resistance P-glycoprotein. The ABCR transporter gene was recently identified as being specifically expressed in photoreceptors. Its precise function is not yet known, but it is emerging as an important cause of retinal disease. Mutations in the ABCR gene were initially shown to be the cause of a severe, autosomal recessive form of central retinal dystrophy called Stargardt disease, which results in the loss of central vision in early adult life with preservation of the peripheral visual field. Sufferers rarely complain of night vision problems, so this was, until recently, regarded as a disorder of cone photoreceptors. However, this has turned out not to be the case. The primary site of expression of the ABCR gene product, a 220 kDa "rim protein", is at the rims or margins of rod and cones photoreceptor discs (Weng *et al.*, 1999). Loss-of-function mutations were found in both copies of the gene in Stargardt disease. Further complexity followed when it was shown that two ABCR null alleles result in a more severe disease with both central and peripheral retinal degeneration resembling a mixed cone–rod type of RP. It now appears that the presence of even a small amount of residual ABCR activity in at least one allele results in a disorder more or less confined to central photoreceptors (cone–rod dystrophy, Stargardt disease, fundus flavimaculatus). An attractive, but unproven, hypothesis is that the ABCR protein transports retinoids across rod disc membranes, but it is unclear why central vision and acuity are often severely compromised when the gene is expressed in rod and cone photoreceptors (Weng *et al.*, 1999).

9. Invertebrate phototransduction mutants

The phototransduction cascade in invertebrates differs in a number of important respects from that of vertebrates. However, there are important differences even within vertebrate and invertebrate phyla, emphasizing that eyes are often highly specialized for particular functions. For example, the lizard parietal eye uses the cGMP second messenger system of other vertebrates, but the photoreceptor depolarizes in response to light, while the normal vertebrate response is a hyperpolarization. The normal invertebrate response to light is a depolarization, but the second messenger is inositol 1,4,5-trisphosphate (IP_3) instead of cGMP. The hyperpolarizing photoreceptor of molluscs is an exception; it utilizes a light-induced increase in cGMP as a signal for increased potassium channel conductance and hyperpolarization of the cell.

Over 70 visual mutants have been isolated in the fruit fly *Drosophila melanogaster*; about a dozen of these genes have been isolated, each providing important insights into disordered vision. In addition to the rhodopsin (*NinaE*) mutants discussed above, a number of different phototransduction proteins have been shown to be mutated in flies with abnormal electroretinograms. These include *norpA* mutants which have a defect in phospholipase C (PLC), consistent with the view that this enzyme is central to the phototransduction cascade in most invertebrates. Light-activated rhodopsin stimulates an eye-specific heterotrimeric G protein which, in turn, activates PLC, at the equivalent position to PDE in vertebrates. PLC catalyses the hydrolysis of phosphatidylinositol 4,5-bisphosphate (PIP_2) to diacyl glycerol (DAG) and IP_3. Inositol trisphosphate binds to a membrane IP_3 receptor and releases calcium from intracellular stores. This, in turn, leads to the opening of both a calcium-permeable channel and a cation channel in the plasma membrane and to depolarization of the membrane. The entry of calcium ions replenishes intracellular stores, while DAG activates an eye-specific protein kinase C, which may be concerned with adaptation of the response to differing levels of light. Mutations have been found in most of these genes, including eye protein kinase C (*inaC*), one of the plasma membrane calcium channels (*trp*), two subunits of the eye-specific heterotrimeric G protein (*dgq*, *dbe*) and two proteins concerned with the rhodopsin switch-off process, arrestin (*arr2*) and protein phosphatase (*rdgC*). A variety of other genes associated with abnormalities in photoreceptor function have also been identified, which are discussed in detail by Pak (1995). Many of these genes have helped to identify orthologues in the vertebrate eye which are the focus of considerable research interest.

10. Extracellular matrix modelling and the eye

The extracellular matrix is an important site of disease in the eye. The commonest cause of blind registration in the western world is a condition called age-related macular degeneration (AMD). With advancing age, changes accumulate in the elasticity and permeability of the basement membrane underlying the retina and RPE (Bruch's membrane). Retinoid and other lipid or protein breakdown products are deposited both within the RPE, leading to circumscribed areas of atrophy, and between the RPE and Bruch's membrane. In the most severe form of AMD, new blood vessels invade Bruch's membrane from the choroid, often leading to haemorrhagic destruction of the central macular photoreceptors, with profound loss of visual acuity. Some degree of AMD is found in about 5% of western populations over the age of 70.

AMD is a complex and heterogeneous disorder and while genetic factors are probably significant, they are hard to pin down. One approach favoured by geneticists for understanding complex disorders is to study Mendelian conditions showing similar features. One such disorder is Sorsby's fundus dystrophy, a rare autosomal dominant condition associated with subretinal deposits, formation of new blood vessels (neovascularization) and often bleeding into the macula. The gene for Sorsby's fundus dystrophy was mapped by genetic linkage to chromosome 22, which led to the identification of the causal gene as the tissue inhibitor of metalloproteinases-3 (*TIMP3*). The precise role of TIMP3 is not known, but it is likely to be concerned with the modelling and turnover of the extracellular matrix and constituents of Bruch's membrane. It remains to be seen whether this information can be exploited therapeutically.

Another group of eye disorders associated with abnormal deposits in the extracellular matrix are the hereditary corneal dystrophies. These are autosomal dominant conditions which present with recurrent bouts of painful corneal erosions and lead to progressive loss of vision due to opacification of the cornea. In many cases, corneal transplantation may become necessary at a late stage. In some corneal dystrophies, granular deposits accumulate in the cornea with age and progressively obscure vision; others show deposition of typical amyloid fibrils in the corneal stroma and subepithelial layers. Four clinically distinguishable forms of autosomal dominant corneal dystrophy were first mapped to a common region of chromosome 5 (5q31), suggesting the involvement of a candidate gene, *βig-h3*, coding for a widely expressed structural protein called kerato-epithelin. The four disorders, granular dystrophy (Groenouw type I), lattice type I dystrophy, Reis-Bucklers syndrome and Avellino dystrophy, were later all found to be associated with missense mutations in the *βig-h3* gene. Kerato-epithelin is a secreted protein containing a carboxyl terminal RGD motif found in many extracellular matrix proteins, suggesting integrin-binding and cell adhesion functions. In the eye, it is expressed in corneal epithelium and stromal keratocytes. The presence of abnormal fibrillar amyloid deposits in two of these disorders was associated with different missense mutations affecting the same codon, suggesting a critical role of the protein structure. These conditions fall into the class of amyloidoses, which share the common property of forming insoluble extracellular protein aggregates.

The cytoskeletal intermediate filaments of corneal epithelium are composed of cornea-specific keratins, K3 and K12. Mutations in these genes have also been found in an autosomal dominant form of corneal dystrophy called Meesmann's corneal dystrophy. Although no other genes have yet been identified, the heterogeneity of human corneal dystrophies is significant, since six other loci have been mapped to specific chromosomal sites (1p, 10q, 12q, 16q, 17p and 20p).

11. Positional cloning of novel genes causing eye disorders

The most successful strategies for identifying human eye disease genes have been either to screen large numbers of patients for mutations in selected "candidate" genes or to map the disease genes within families by genetic linkage, followed by analysis of local "positional" candidate genes. However, in the degenerative disorders of the retina, currently identified genes only account for about 20% of patients. Most of the more than 40 genetically mapped loci remain to be isolated and do not map to sites containing known candidate genes. There are major technical problems associated with the isolation of these genes solely on the basis of their chromosomal location (positional cloning) and, to date, only two have been found in this way, compared with some 15 using the candidate or positional candidate gene approaches. The two positionally cloned genes both cause retinal degeneration and provide examples of completely novel genes and unpredicted disease mechanisms.

12. Choroideraemia and REP-1

The first example of a positionally cloned gene came from the isolation of the X-linked choroideraemia (CHM) gene. The cause of this rare degenerative disorder of the vascular choroid and retina was found to be a mutated Rab escort protein-1 (*REP-1*) gene. *REP-1* is associated with the enzyme Rab geranylgeranyl transferase (RabGGT) which attaches isoprenoid lipids to members of the large Rab family of small GTP-binding proteins, facilitating their association with particular membranes. The *REP-1* gene is widely expressed, so it was not clear why the only abnormality should be a retinal degeneration. The function of REP-1 appears to be concerned with escorting unprenylated members of the Rab family to RabGGT and, following their isoprenylation, facilitating their entry into membranes, which is essential for normal function. One of the RabGGT substrates, Rab27, which is strongly expressed in choroid and RPE, was serendipitously discovered to be unprenylated in CHM patients. While Rab27 is a good substrate for REP-1, a closely related Rab escort protein, REP-2, does not interact with it and cannot compensate for the lack of REP-1. This suggests that the lack of functional redundancy of REP proteins for a particular Rab protein within the eye makes the eye vulnerable to the loss of REP-1. Almost all CHM mutations result from loss-of-function mutations in the *REP-1* gene. The precise function of Rab27 remains to be elucidated.

13. X-linked RP and RPGR

The second positionally cloned gene, retinitis pigmentosa GTPase regulator (*RPGR*), was found to be mutated in patients with an X-linked form of RP. Its function remains to be elucidated, but it is again widely expressed and sequence analysis suggests that it also regulates a small GTP-binding protein, perhaps functioning as a guanine nucleotide exchange factor. These are proteins that activate small GTP-binding proteins by promoting the exchange of GDP for GTP in their nucleotide-binding sites. Small GTP-binding proteins form an extremely diverse family of proteins concerned with signal transduction, cytoskeletal organization and membrane trafficking. It appears, therefore, that retinal dysfunction can arise from mutations in widely expressed genes with fundamental cellular roles in addition to those that are eye specific, such as phototransduction genes. It may turn out that the former represent the majority – the eye has a number of highly specialized structures and functions, but many of these may require novel combinations of widely expressed rather than eye-specific proteins.

14. Metabolic disorders associated with inherited eye disorder

A wide variety of disorders are associated with ocular abnormalities resulting from metabolic defects. A few examples will be given to illustrate the potential for developing therapeutic or preventative measures following elucidation of the biochemical defect. Galactosaemia is an autosomal recessive disorder affecting one of three different enzymes concerned with the metabolism of the sugar galactose. The resultant accumulation of toxic products such as galactitol within the lens leads to swelling and damage to the lens. The resulting cataracts can be prevented by the early application of a diet low in galactose, although other aspects of the disorder do not respond as well.

Many other rare metabolic disorders are associated with the build-up of toxic products within the eye, such as the long-chain fatty acid, phytanic acid, in Refsum disease, where there is a deficiency of the peroxisomal enzyme phytanoyl-CoA hydroxylase. Institution of a diet low in phytanic acid in infancy can prevent neurological deterioration, but only appears to arrest the retinal degeneration. In another metabolic disorder, gyrate atrophy of the retina, there is a degeneration of the choroid and retina associated with raised levels of plasma ornithine, resulting from a defect in the liver enzyme ornithine aminotransferase. Reduction of plasma ornithine can be achieved with an arginine-restricted diet which has been shown to slow progression of the retinal disorder.

15. Ocular tumours

Human eye disorders provide a number of paradigms for disease mechanisms. An example is the discovery of the first "tumour suppressor" gene in patients with the hereditary ocular tumour, retinoblastoma. This is the commonest intraocular tumour of childhood and presents as a tumour of the developing retina in the first few years of life; it frequently necessitates the removal of the eye. In about 40% of patients, the disease is hereditary, since a mutated copy of the *RB1* gene is either passed down from an affected parent or, more commonly, appears for the first time as a new mutation. The disease can, therefore, behave clinically as a typical autosomal dominant disorder. The idea known as the two-hit hypothesis that was initially proposed by Knudson (Knudson, 1971) was later confirmed with the isolation of the *RB1* gene, which showed that both gene copies must be lost or inactivated in order to initiate the growth of a tumour. This can occur by inactivation or loss of the second, normal gene copy within the developing retina as a result of deletion, point mutation, gene conversion, sister chromatid exchange or interchromosomal recombination. The commonest form of retinoblastoma is not heritable and results from two independent somatic mutations within a rare cell which is then subject to powerful cellular selection as a result of unrestrained cellular proliferation. Non-heritable forms of retinoblastoma always have a single ocular tumour, whereas patients with heritable retinoblastoma either have a single or multiple tumours (most commonly bilateral), some of which are highly malignant and affect non-ocular tissues (e.g. sarcomas).

The discovery of inactivating somatic mutations in the normal *RB1* allele provided the first of many examples of tumour suppressor genes, which are required to prevent malignant transformation. Several tumour suppressor genes, including *RB1*, affect critical checkpoints within the cell cycle. The *RB1* product, pRB, is a nuclear protein that regulates cell cycle progression by binding to members of the E2F transcription factor family, which, in turn, regulates several cell cycle progression genes. The pRB protein is one of a group of negative growth regulators known as pocket proteins, which include the p107 and p130 proteins. One of their other functions is to bind and regulate replication "licensing factor" proteins such as MCM7, which prevent the replication of DNA. pRB, therefore, plays a central and complex role in restraining cell growth and proliferation.

16. Mitochondrial disorders and the eye

Another human disease paradigm came from the discovery that mutations within the mitochondrial genome can cause human disease. The first example was a condition affecting the function of the optic nerve called Leber's hereditary optic neuropathy (LHON). This condition characteristically causes a relatively sudden loss of central vision in one eye followed, within a few months, by the other eye in young adults. It was found to be due to pathogenic mutations in mitochondrially encoded subunits of respiratory chain complexes I–IV. Several mutations have been shown to be specifically associated with LHON, the commonest of which results in the substitution of a histidine for an arginine residue in the ND4 subunit of complex I (11778 mutation), which is present in about 50% of patients. This discovery led to the identification of many other pathogenic mitochondrial DNA mutations in a variety of other disorders, mainly affecting brain, eye or muscle. Several puzzling features of LHON remain to be explained. For example, why does a maternally inherited mitochondrial mutation, which is passed on to all the offspring of a carrier mother, predominantly affect males (e.g. 11778 mutation)? And, why does LHON present clinically as a relatively sudden and often catastrophic event after 20–30 years of normal vision? There is a well-established decline in mitochondrial function with age and it may be that the presence of a relatively mild missense mutation alters the threshold at which the retinal ganglion cells, whose axons form the optic nerve, show impairment or loss of function.

17. Hereditary cataracts and lens defects

The ocular lens has an unusual structure which, in vertebrates, is derived from an invagination of the surface ectoderm. This forms a vesicle which becomes progressively filled with elongated lens fibre cells. These cells are arrayed in a highly compact fashion and are densely packed with crystallins, the major soluble proteins of the vertebrate lens. The function of the lens is to focus light onto the underlying retina and the maintenance of its transparency is essential for clear vision. The unique spatial arrangement and biophysical properties of crystallins are thought to underlie the transparency and refractive properties of the lens.

Cataract refers to a loss of transparency within the lens. It can arise from many causes, including mutations within the crystallin genes. One of the more common causes of blindness in children in the western world is congenital cataract, accounting for 10–38% of cases, about one-half of which have a heritable cause. Crystallin mutations were first identified in mouse models of cataract, including a mutation in one of the seven β-crystallin genes in the *Philly* mouse and a γ-crystallin gene mutation in the *Elo* mouse, which has a small eye (microphthalmia) secondary to abnormal lens development. Mutations have since been found in BA1, BB2, γC and γD crystallins in human autosomal dominant congenital cataract (ADCC). Mutations in genes for membranous proteins with critical roles in maintaining the gap junctions of lens fibres, such as lens connexins, have now been identified in human and mouse cataracts. The major intrinsic protein (MIP) of the lens fibre cell membrane is mutated in the *cataract Fraser* (*CatFr*) and *lens opacity* (*Lop*) mouse mutants. In addition to these identified mutations, nine loci for human ADCC have been mapped, several to known crystallin gene clusters (e.g. βB1–3/A4, βA1/A3, γA-F). In the mouse, 14 loci associated with autosomal dominant cataract have been mapped, of which three have been identified (*Pax6*, *Pax2*, *Mip*). Finally, a wide variety of inherited metabolic defects result in the loss of lens transparency.

18. Connective tissue disorders and the eye

One of the more common connective tissue disorders in man is Marfan syndrome, an autosomal dominant disorder in which there are abnormalities of the eyes, heart, skeleton, major blood vessels, lung and skin. Genetic defects have been identified in the fibrillin gene (*FBN1*), whose product is the principal component of a class of connective tissue microfibrils found in almost all extracellular matrices. The major ocular manifestations of fibrillin mutations are dislocation of the crystalline lens from its central location, myopia, retinal detachment and presenile cataract. Patients have reduced fibrillin content and aberrant fibrillin-containing fibres in their lens capsules where fibrillin serves as a scaffold for elastin deposition or as a link between elastin fibres and other extracellular components.

Mutations in another gene with an important role in connective tissue, collagen, occur in patients with an autosomal disorder called Stickler syndrome. This rather variable condition is associated with a combination of degenerative bone and joint changes, deafness and severe myopia, with retinal and vitreous degeneration often leading to detachment of the retina and severe loss of vision. About two-thirds of patients have loss-of-function mutations in the type II procollagen (COL2A1) gene, but some have mutations in the gene for type XI collagen (COL11A1), a minor fibrillar collagen that associates with type II collagen fibrils.

19. Developmental anomalies of the eye

The development of the eye has been the subject of intense investigation, both in vertebrates and invertebrates. A large number of genes with homology to transcription factors, homeotic genes, signalling and growth factors have been found to be concerned with eye development. In the invertebrate eye of *D. melanogaster*, ectopic expression of key developmental genes such as *Pax6*, *dachshund* and *eyes absent* appears to switch on master programmes controlling eye development and can result in the ectopic expression of normal eye structures (see article on "The role of Pax-6 homologues in the development of eye and other tissues during the embryogenesis of vertebrates and invertebrates"). These types of regulatory gene appear to activate networks of cell-specific genes coordinately. Their vertebrate homologues are beginning to be isolated and initial work on *Pax6*, *Prox1*, *Notch*, *mash1*, *Rax*, *Chx10* and *Brn3* transcription factor genes suggests that these may operate in a similar manner in the vertebrate eye. Loss of activity of the *Pax6* gene causes the *eyeless* phenotype in Drosophila, and anophthalmia (absent eyes) in humans with homozygous *PAX6* mutations. Heterozygous *PAX6* mutations cause cataracts or anterior chamber defects in humans (see below). A targeted loss-of-function mutation in the mouse *Rax* homeobox gene causes complete failure of eye development.

Microphthalmia is associated with mutations in the *Chx10* gene in the *ocular retardation* (*or*) mouse and in the *Mitf^mi6^* gene in the *microphthalmia* mouse. *Mitf^mi6^* codes for a basic helix–loop–helix leucine zipper transcription factor concerned with the differentiation of neural crest cells and ocular and skin melanocytes, in particular. Mutations in the human *MITF* orthologue cause Waardenburg syndrome type 2A, a disorder of melanocyte differentiation associated with hearing loss and mild pigmentary disturbances in the skin and eye. Finally, mutation in the paired-domain homeobox gene *Pax-2* causes developmental anomalies of the optic nerve (coloboma), lens (cataract) and kidneys in man (renal–coloboma syndrome) and in mouse (*1Neu*).

Recently, mutations have been found in an important transcription factor which appears to switch on a battery of phototransduction genes, including rhodopsin, in the later stages of photoreceptor differentiation. The cone–rod homeobox or *Crx* gene produces a paired homeodomain protein which is specifically expressed in developing and adult photoreceptors. Heterozygous mutations were first identified in patients suffering from cone–rod dystrophy, a progressive autosomal dominant disorder of cone and, to a lesser extent, rod photoreceptors, which had previously been mapped to human chromosome 19, suggesting the *Crx* gene as a candidate. *De novo* loss-of-function mutations were also found in patients with a severe congenital form of retinal degeneration categorized as Leber's congenital amaurosis. These results show that both missense and nonsense mutations in this gene can give rise to congenital or adult onset of retinal degeneration and suggest both developmental and maintenance roles for the CRX protein. The apparent predilection of the disease for the cone-rich macular region may relate to the complexity of the postnatal development of this region. This may be one of the first examples of disease associated with the less well-characterized secondary role of some transcription factors in maintaining the integrity of mature cells, in addition to ensuring their correct development.

20. Anterior segment dysgenesis and glaucoma

The development of the anterior segment of the eye – cornea, anterior chamber and associated drainage structures, iris, lens and ciliary body – also depends on the concerted action of transcription factors. Mutations in such genes have been shown to cause complex anomalies of the anterior segment, leading in many cases to structural anomalies of the iris, cataracts or problems with aqueous fluid drainage. Glaucoma is one of the most serious consequences of these defects. This is a disorder of optic nerve function which is often accompanied by elevated intraocular pressure, which is at least partly responsible for the optic neuropathy and consequent reduction in the field of vision. In humans with heterozygous mutations in the paired-domain transcription factor PAX6, the commonest phenotype is an absence of the iris (aniridia), often leading to secondary glaucoma, and a reduction or absence of the region in the central retina specialized for high-resolution vision (fovea). In others, there may be more complex anterior segment defects or associated cataracts.

Mutations in another human homeobox gene, *PITX2*, are the cause of Rieger syndrome, in which there are anomalies of the anterior chamber, secondary glaucoma due to obstruction of the aqueous outflow mechanism, and dental and umbilical anomalies. Mutations within the same gene can cause an autosomal dominant form of iris hypoplasia. The closely related human *PITX3* gene is associated with disordered anterior chamber development and congenital cataract.

Glaucoma is a very common disorder in man. The most common type is primary open-angle glaucoma (POAG), where there is no obstructive component to the drainage of aqueous fluid. POAG affects 1–2% of the general population over the age of 40 and, until recently, the causes were poorly understood. A major advance occurred when the gene for a rare autosomal dominant juvenile form of open-angle glaucoma was mapped to human chromosome 1 and later identified as the myocilin gene. Mutations in myocilin were then found in the more common adult variety of the disease. Heterozygous mutations were found in 4% of POAG patients with a family history of the disorder and in 3% of unselected POAG patients. The myocilin protein is known to be expressed in the trabecular meshwork, the site of aqueous drainage, and contains an evolutionarily conserved carboxyl terminal domain with homology to olfactomedin, an extracellular matrix glycoprotein, which is the commonest site of mutation. Myocilin is overexpressed in response to glucocorticoids and may contribute to the steroid-induced rise in intraocular pressure seen in some glaucoma patients.

The congenital form of glaucoma (buphthalmos) is an autosomal recessive disorder which affects about one in 10 000 people and is thought to result from a developmental defect in the anterior segment of the eye, with a resultant defect in aqueous outflow. Two loci have been mapped in human congenital glaucoma families and at one of these loci a gene has been identified as coding for the mixed function mono-oxygenase cytochrome P4501B1 (*CYP1B1*). It is proposed that loss-of-function alleles in this gene lead to the abnormal metabolism of an as-yet unidentified oxygenated signalling molecule required for normal anterior segment development.

FURTHER READING

Chang, H.C., Karim, F.D., O'Neill, E.M. *et al.* (1994) Ras signal transduction pathway in Drosophila eye development. *Cold Spring Harbor Symposium in Quantitative Biology* 59: 147–53

Dryja, T.P., McGee, T.L., Reichel, E. *et al.* (1990) A point mutation in the rhodopsin gene in one form of retinitis pigmentosa. *Nature* 343: 364–66

Falk, J.D. & Applebury, M.L. (1988) The molecular genetics of photoreceptor cells. *Progress in Retina Research* 7: 89–112

Freund, C.L., Gregory-Evans, C.Y., Furukawa, T. *et al.* (1997) Cone–rod dystrophy due to mutations in a novel photoreceptor-specific homeobox gene (*CRX*) essential for maintenance of the photoreceptor. *Cell* 91: 543–53

Gregory-Evans, K. & Bhattacharya, S.S. (1998) Genetic blindness: current concepts in the pathogenesis of human outer retinal dystrophies. *Trends in Genetics* 14: 103–08

Halder, G., Callaerts, P. & Gehring, W.J. (1995) Induction of ectopic eyes by targeted expression of the eyeless gene in *Drosophila*. *Science* 267: 1788–92

Hansen, M.F. & Cavenee, W.K. (1988) Retinoblastoma and the progression of tumor genetics. *Trends in Genetics* 4: 125–28

Kajiwara, K., Berson, E.L. & Dryja, T.P. (1994) Digenic retinitus pigmentosa due to mutations at the unlinked peripherin/RDS and ROM1 loci. *Science* 264: 1604–08

Knudson, A.G. (1971) Mutation and cancer: statistical study of retinoblastoma. *Proceedings of the National Academy of Sciences USA* 68: 820–23

Munier, F.L., Korvatska, E., Djemai, A. *et al.* (1997) Keratoepithelin mutations in four 5q31-linked corneal dystrophies. *Nature Genetics* 15: 247–51

Nathans, J., Thomas, D. & Hogness, D.S. (1986) Molecular genetics of human color vision: the genes encoding blue, green, and red pigments. *Science* 232: 193–202

Pak, W.L. (1995) *Drosophila* in vision research. *Investigative Ophthalmology and Visual Science* 36: 2340–57

Stone, E.M., Fingert, J.H., Alward, W.L.M. *et al.* (1997) Identification of a gene that causes primary open angle glaucoma. *Science* 275: 668–70

Wallace, D.C., Singh, G., Lott, M.T. *et al.* (1988) Mitochondrial DNA mutation associated with Leber's Hereditary Optic Neuropathy. *Science* 242: 1427–30

Weng, J., Mata, N.L., Azarian, S.M. *et al.* (1999) Insights into the function of rim protein in photoreceptors and etiology of Stargardt's disease from the phenotype in *abcr* knockout mice. *Cell* 98: 13–23

Wright, A.F. & Jay, B. (eds) (1994) *Molecular Genetics of Inherited Eye Disorders*, Chur, Switzerland: Harwood Academic

GLOSSARY

ABCR transporter (ATP binding cassette receptor) a member of the superfamily of ABC transporters mediating the transport of a variety of small molecules across cell membranes

arrestin inhibitory protein that blocks and terminates the signalling of tyrosine-phosphorylated receptors. Competes with transducin for light-activated rhodopsin, thereby inhibiting response to light

axon potential cytoplasmic process that carries electrical impulses away from the cell body of a neurone

cGMP (cyclic guanosine monophosphate) molecule formed from guanosine monophosphate by the enzyme guanylate cyclase; acts as a second messenger in some cellular reactions

chromophore part of a visibly coloured molecule responsible for absorbing light energy over a range of wavelengths, thereby giving rise to the perception of light

cytochrome enzyme pigmented as a result of its haem prosthetic groups

ectopic situated in an usual place, displaced from its usual location

endoplasmic reticulum convoluted internal membrane in eukaryotic cells, continuous with the outer nuclear membrane and enclosing a continuous internal space

homeotic relating to mutations that transform part of the body into another part

homophilic interactions binding of cell adhesion molecules to the same molecules within or between cells

isoprenylation addition of prenyl groups to a protein following translation in order to promote membrane association

leucine zipper structural motif found in many proteins concerned with gene regulation; thought to be involved in protein–protein interactions

MAP (mitogen-activated protein) kinase a serine–threonine kinase is activated when quiescent cells are treated with mitogens. Lying at the end of a multicomponent signal transduction pathway it phosphorylates transcription factors, enabling them to stimulate gene expression

neural crest cells embryonic cells that separate from the neural plate during formation of the neural tube and migrate to give several different lines of adult cells

null allele mutant allele lacking a functional gene product

orthologue genes related by common phylogenetic descent

phagocytosis a five-step process by which particular cells engulf and destroy micro-organisms and cell debris: (1) invagination; (2) engulfment; (3) formation of phagocyte vacuole; (4) digestion of phagocytosed material; (5) release of digested products

photoreceptor a nerve cell receptive to light stimuli

phototransduction reception and interpretation by a photoreceptor cell of a signal in the form of light

recoverin calcium-binding retinal protein

retinoid metabolite of retinol (vitamin A) with a role in embryonic development

RGD motif domain found in fibronectin and related proteins, recognized by integrins

rhodopsin light-sensitive protein found in photoreceptor cells; member of opsin protein family

signal transduction pathway cascade of processes for converting an extracellular signal (e.g. from a hormone or neurotransmitter) into an intracellular response (*see* MAP kinase)

somatic mutation mutation in body cell(s) as opposed to cells of the germline

transducin a GTP-binding protein involved in transducing the signal from activated rhodopsin in rod and cone cells, thus producing an electrical signal for transmission to the brain

unprenylated lacking a carboxy-terminal prenyl isoprenoid group in a protein

See also **The role of *Pax-6* homologues in the development of the eye and other tissues** (p.207)

GENETICS OF BALDNESS

Giuseppe Novelli
Department of Biopathology and Diagnostic Imaging, Tor Vergata University of Rome, Rome, Italy

1. Introduction

Hair loss is a common human trait. It is estimated that in the US about 60 million people (two-thirds of them men) show this phenotype. Is this trait really genetic? Although we know that partial and reversible hair loss is observed in some infections, it is undoubted that this trait can occur more often among genetic relations of probands than among the general population. A positive family history is elicited in 10–20% of patients, and a 6% risk to first-degree relatives has been estimated.

Hairs are important and characteristic appendages of the outer skin of homeotherm organisms, such as mammals and birds. However, hairs in lower animals are thread-like processes of the outer surface and differ from mammalian hairs in their characteristically finer structure and in their mode of development. The mammalian hair is a typical structure deriving from epidermal cells specially differentiated and arranged (Jones & Steinert, 1996). Initially, they appear in the embryo as solid plug-shaped processes of the epidermis which penetrate into the underlying leather-skin (*corium*), as do the sebaceous and sweat glands (**Figure 1A**). The simple plug consists originally of the ordinary epidermal cells, within which a firmer central cellular mass of conical shape soon forms (**Figure 1B**). This increases considerably in length, detaches itself from the surrounding cellular mass, the "root-sheath", and finally makes its way to the outside, appearing above the outer surface as a hair stem. This process is controlled by nutritional and regulatory factors (Jones & Steinert, 1996). The initiation of hair follicle formation is mediated by the expression in the epidermis of different proteins, including Wnt10b, Shh and BMP2/BMP4 (St-Jacques *et al.*, 1998). These proteins play diverse developmental roles and, in combination, are able to induce morphogenesis and differentiation. Human hair follicles normally exist in three phases of growth known as anagen (active phase), catagen (regressive phase) and telogen (resting phase). During normal hair growth, approximately 70–90% are in the anagen growth phase and may remain so for up to seven years. The length of the phase cycles is regulated by a precise action of apoptotic phenomena which involves different caspase activities (anagen phase) and transforming growth factor-β pathways (catagen phase) (Soma *et al.*, 1998).

In the human embryo, the first hairs make their appearance at the end of the fifth or in the beginning of the sixth month. During the three months preceding full-term birth, the human embryo is usually covered by a thick coating of delicate woolly hairs. This embryonic wool covering (lanugo) is often lost during the last weeks of embryonic life and, soon after birth, it is replaced by a thinner permanent hair covering. This woolly covering is a direct inheritance from our long-haired ancestors. Sexual selection and climatic conditions have caused the loss of the human original coat, with the exception of certain parts, in consequence of adaptation. Various pathological conditions, such as infectious diseases, hormonal disturbances and hereditary disorders, are frequently the cause of lack of scalp hair and lack of body hair.

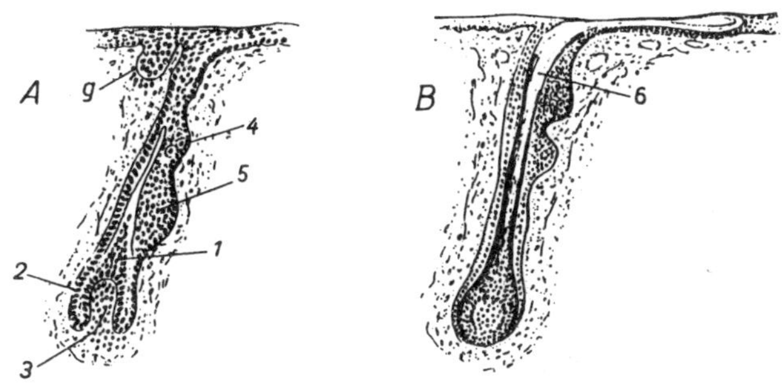

Figure 1 Schematic representation of the human hair development. (A) Hair follicle in early stage of development (g) and in a more advanced stage on the right. Inner root sheath (1–2), dermal papilla (3), sebaceous gland (4), erectile muscle (5). (B) Hair shaft channel (6).

2. Genetic hair disorders

A large number of genetic conditions affect hair growth, structure and distribution, either as isolated forms of diseases or as part of more generalized syndromes (**Table 1**) (Sybert, 1997). Unusual hair patterns are common in pleiotropic syndromes, associated with mental retardation and/or dysmorphic features due to chromosome abnormalities, and suggest the involvement of numerous genes active in hair growth and differentiation. Little is understood about the genetic control of body hair distribution in humans, but hair abnormality is a major component of well over 100 clearly delineated Mendelian syndromes and undoubtedly many more remain to be recognized (see Online Mendelian Inheritance in Man [OMIM]). Their classification is arbitrarily based on their occurrence in isolated abnormalities, in which the hair defect is the outstanding and invariable feature, or in syndromic forms, in which hair abnormalities appear to be inconstant, difficult to appreciate or less prominent than other cutaneous defects.

3. Male pattern baldness (MPB)

A gradual thinning and ultimate loss of hair on the vertex of the scalp with scanty hair in the parietal and occipital regions is common in males. By the age of 70, only one-third

Table 1 Principal genetic hair disorders in man.

Disorder	Inheritance	Gene	Hair phenotype	OMIM*
Alopecia universalis	AR	HAIRLESS	No hair in hair follicles on skin biopsy	203655
Congenital alopecia	XL	?	Hair follicles are absent, attenuated or plugged	300042
Björnstad syndrome	AR	?	Sparse, curly scalp hair	262000
Alopecia areata	AD	YMX1?	Hair loss area	104000
T-cell immunodeficiency, congenital alopecia and nail dystrophy	AR	WHN	Congenital absence of hair and severe immunodeficiency	601705
Atrichia with papular lesions	AR	HAIRLESS	Hair follicles cystic; hair that falls out and is not replaced	209500
Monilethrix	AD; AR	Keratin-II (hHb6/hHb1)	Beaded and/or spindle hair shaft	158000; 252200**
Pili torti	AD; AR; XL in Menkes' disease	?; ATP7A (XL)	Twisted (spiral) hair shaft	261900; 309400
Trichorrhexis invaginata	AR	SPINK5	Bamboo hair, hair shaft invagination, twisting and flattening	
Trichorrhexis nodosa syndrome	AR	SPINK5	Brittle hair with grayish nodules on the hair shaft	275550
Woolly hair	AD; AR	?	Coiled and matted hair	194300; 278150
Trichothiodystrophy	AR; XL also reported	TFIIH (AR) ?	Hair shaft show alternating light and dark zones	242170
Uncombable hair syndrome	AD	?	Spun glass hair, triangular appearance in hair shaft, unmanageable hair	191480
Ectodermal dysplasia anhidrotic (EDA)	XL	EDA1	Hypotrichosis, alopecia	305100
Pachyonychia congenita (Jadassohn–Lewandowsky type)	AD	KRT16	Hair anomalies, alopecia	167200
Pachyonychia congenita (Jackson–Lawler type)	AD	KRT17?	Hair anomalies, alopecia	167210

AD, autosomal dominant inheritance; AR, autosomal recessive inheritance; XL, X-linked inheritance
*Mendelian Inheritance in Man (OMIM)
**, not confirmed

of human males have no evidence of significant balding (Rubin, 1997). Although this trait is often termed a "polygenic" condition, the transmission through many successive generations, as in the descendants of the US president John Adams, suggests the action of a single major gene. In fact, the father-to-son transmission (**Figure 2**) in the Adams family rules out X-linked inheritance in favour of autosomal inheritance, but the exact mode or modes of transmission are still not clear. It has been suggested that premature male baldness (or PMB; OMIM *109200) is transmitted by an autosomal gene which acts dominantly in males and recessively in females; females transmit the trait when heterozygous, but are bald only if homozygous. In affected females, the hairs generally do not completely miniaturize and the density of follicles may decrease. MPB may occur in heterozygous females after exposure to exogeneous or endogeneous androgens, suggesting that the excess of androgen production is directly correlated with male baldness. MPB is considered a sex-limited trait similar to other rare genetic diseases such as hydrometrocolpos (OMIM *236700), a condition characterized by the accumulation of fluid in the uterus and vagina, which occurs in women homozygous for an autosomal recessive gene.

The PMB gene frequency appears to be highest in Caucasians, lower in Africans and lowest in Amerindians, Asians, Inuits and Yupiks. It is possible that cases described in the native Alaskan Yupik Eskimos are a consequence of the high incidence of congenital adrenal hyperplasia due to mutation within the CYP17 gene (cytochrome P450 enzymes). In fact, Carey *et al.* (1993) have described families in which the polycystic ovarian syndrome (PCO; OMIM *184700) and premature MPB segregated as an autosomal dominant phenotype. The same author has identified a single base change in the 5′ promoter region of the CYP17 gene, which creates an additional SP1-type (CCACC box) promoter site that is thought to cause increased expression of the gene. SP1 is a general transcription factor which is able to activate the expression of a large number of genes. There was a significant association between the presence of this base change and the affected state for consecutively identified Caucasian women with PCO as compared either with consecutively matched controls or with a random population. Within the 14 families, members with PCO or MPB had a significant association with the occurrence of at least one of the variant alleles compared with their normal relatives. The base change did not cosegregate, however,

Figure 2 Pattern baldness in the Adams family. (A) John Adams (1735–1826) at 65, second US president; (B) John Quincy Adams (1767–1848) at 52, sixth US president; (C) Charles Francis Adams (1807–1886) at 41, diplomat; (D) Henry Adams (1838–1918) at 45, historian.

with the affected phenotype within the families showing association, demonstrating that the mutation in CYP17 is not the cause of PCO/MPB. Although variation represented by the so-called A2 allele of the CYP17 gene appeared to be a significant factor modifying the expression of PCO/MPB in families in which the phenotype was demonstrated to segregate as a single gene disorder, it was excluded as the primary genetic defect.

A strong association between MPB and benign prostatic hyperplasia (BPH) has been recently demonstrated, but the nature of this association is not well understood. In fact, no significant differences have been found in the human 5-α-reductase gene, which is involved in BPH, by genotyping subjects with MPB. A significant difference was found in the CAG repeat numbers of the androgen receptor gene, suggesting that the androgen receptor may affect androgen-mediated gene expression in hair follicles (Sawaya & Shalita, 1998). CAG repeats code for polyglutamine stretches and, when their number increases, they cause degenerative diseases, such as Huntington's or Kennedy diseases. The androgen receptor CAG repeat is the cause of Kennedy disease (a spinal and bulbar type of muscular atrophy).

4. Alopecias

There are several forms of hereditary baldness (known as alopecias from the Greek word "αλωπηξ" meaning *fox*), transmitted as autosomal dominant, autosomal recessive or sex-linked traits. However, the majority of families with complete or partial congenital absence of hair (congenital alopecia) have been reported to follow an autosomal recessive mode of inheritance (alopecia universalis [AU]; OMIM 203655). Affected persons show complete absence of hair development over the entire scalp, eyebrows and eyelashes; skin biopsy from the scalp shows hair follicles without hair.

Linkage studies performed in a large inbred Pakistani family have allowed the localization and the successive isolation of the first human gene mutated in alopecia (**Figure 3**) (Ahmad *et al.*, 1998). The researchers first mapped the disease locus on the human chromosome 8p12 in a region where they also found the human homologue of the murine *hairless* gene (*hr*). This gene is the source of several allelic mutations in mice which cause complete baldness due to the shedding of apparently normal first pelage hairs. The *hr* gene is expressed in hair follicles, epidermis and brain, and codes for a transcription factor belonging to the zinc finger family, directly regulated by the thyroid hormone. Alterations in the expression of this gene in the mouse are responsible not only for the *hairless* phenotype, but also for immunological and reproductive abnormalities, sensitivity to ultraviolet radiation and susceptibility to skin tumours and lymphomas.

It is suggested that the absence of the hairless protein in mice and humans activates a premature and abnormal catagen due to an anomalous apoptosis and dysregulation of cell adhesion. The human and rodent hr amino acid sequences share approximately 85% identity, highlighting the high degree of conservation among different mammalian species (Panteleyev *et al.*, 1998). The human gene is organized into 19 exons encoding a polypeptide of 1189 amino acids. All the patients of the initial Pakistani pedigree investigated showed a homozygous A→G transition which results in a missense mutation converting threonine to alanine at amino acid residue 1022 of the human hairless protein. This finding clearly demonstrates that AU is a consequence of a single gene defect rather than an autoimmune disorder.

Other mutations have been described since this first report, providing convincing evidence that the human *hairless* gene is the causative gene for the autosomal recessive universal congenital alopecia. Interestingly, some of these mutations have been found in patients affected by atrichia with papular lesions (APL; OMIM* 209500). This demonstrates that AU and APL are a single clinical entity. However,

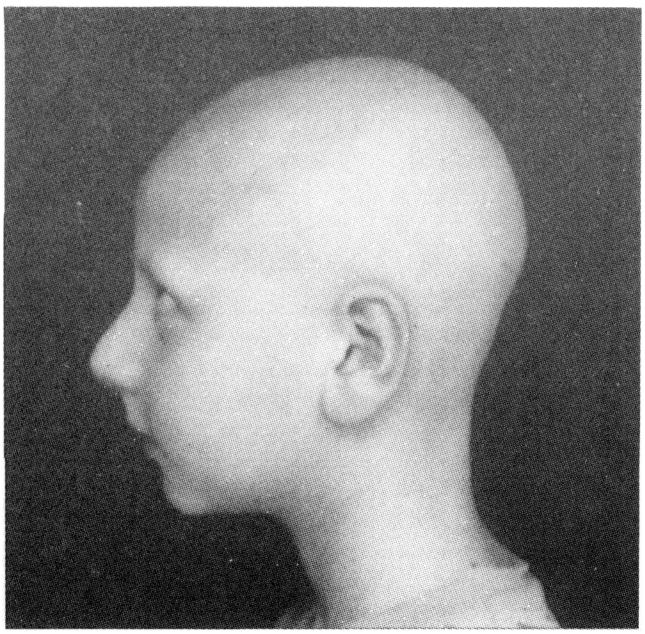

Figure 3 Phenotype of a classical patient with alopecia. With thanks to Dr G. Fabrizi (Catholic University of Rome) for the gift of this illustration.

evidence for phenotypic heterogeneity related to *hairless* mutations has been recently documented (Sprecher *et al.*, 1999), suggesting that mutations in selected regions of the gene may have different effects on the development of the hair follicle and may generate distinct disease phenotypes. The identification of mutations in the *hr* gene in humans and mice with congenital AU and APL underscores a crucial role of the *hr* gene product in the regulation of hair growth.

To date, only one other genetic defect associated with heritable forms of alopecia has been described. This results from a mutation in the human *WHN* gene, homologous to the mouse *winged–helix–nude* gene, which encodes a member of the forkhead/winged-helix transcription factor family with restricted expression in thymus and skin (Frank *et al.*, 1999). A nonsense mutation (R255X) has been recently found in two affected sisters characterized by a complete absence of scalp hair, eyebrows and eyelashes, and with dystrophic nails and absence of the thymus.

Unlike these two rare alopecias, the genetic defect of the most common form of alopecia, alopecia areata (AA; OMIM 104000), still remains unclear (McDonagh & Messenger, 1996). Different authors have suggested that this disorder has an autoimmune mechanism and, in fact, an association between this disease and human leucocyte antigens (HLA) of classes I and II has been documented. In addition to HLA, an association between the gene encoding the Km1 allotype of the immunoglobulin kappa light chain determinant and AA has also been observed. However, the existence of a susceptibility locus within the HLA region on chromosome 6 is also supported by the association with polymorphisms of the tumour necrosis factor-α.

5. Isolated hair shaft abnormalities

This group of diseases includes dermatological disorders that consist of solitary defects of hair growth and/or structure and those in which the hair abnormality is one of the major specific phenotypes. The use of molecular biology techniques and the development of transgenic mice have demonstrated that mutations in genes coding for proteins active in the keratinization of hair are responsible for these defects. To date, 11 human keratin genes have been associated with genetic defects (Jones & Steinert, 1996). These genes encode two keratin families – acidic (or type I) and basic (or type II) – which, as a rule, are coordinately synthesized in pairs so that at least one member of each family is expressed in each epithelial cell. For example, keratins 1 and 10 are specific for the epidermis, keratins 3 and 12 for cornea, keratins 6 and 16 for hyperproliferative conditions of the epidermis, and so on. Mutations of type I keratin genes are usually associated with epidermolysis, ichthyosis and other autosomal dominant epidermic disorders (Korge *et al.*, 1998). Members of keratin type II (hHb1, hHb3 and hHb6) are sequentially expressed in the cortex of the hair shaft and are, therefore, considered as candidates for diseases involving hair shaft anomalies. This has been confirmed by the discovery of mutations in the hHb6 keratin gene in patients affected by monilethrix ("necklace hair") (OMIM *158000), a rare autosomal dominant disease characterized by localized regions of hair loss to total lack of scalp hair. The degree of alopecia is variable from patient to patient and from time to time in the same individual; perifollicular hyperkeratosis is a consistent feature. Microscopically, the hair is beaded, which is the result of a periodic narrowing of the shaft with nodes separated by about 0.7 mm. Clinical expression of monilethrix is variable; in mild cases, dystrophic hair may be confined to the occiput, but more severely affected individuals have near total alopecia. In some cases, alopecia persists throughout life; in others, regrowth of apparently normal hair may occur in adolescence or, temporarily, in pregnancy. The first two monilethrix mutations described in two unrelated families were located in the highly conserved helix termination motif (HTM) of hHb6 and both changed a glutamic acid residue at codon 413. Interestingly, a subsequent mutation described in a Canadian family involved a substitution of the parallel glutamic acid codon E413 in an another keratin II gene, the hHb1 (Korge *et al.*, 1998).

These findings demonstrate definitively that mutations of keratin type II genes are associated with monilethrix. Mutations of hHb6 segregate as an autosomal dominant trait; however, severely affected homozygous individuals have been described. This suggests that hHb6 mutations may have a codominant effect which, in part, explains the variable expression. The position 413 of hHb6 is a hotspot for mutations in monilethrix. In fact, other families of different geographical origin showed the substitution of the glutamic acid at this position. Genetic heterogeneity of monilethrix has been also documented in a family not segregating with chromosome 12q13 where the keratin type II gene cluster maps. It is, therefore, possible that mutation in at least one other gene results in a similar phenotype.

Mutations of type I keratin genes K16 and K17 have been found in other hair shaft anomalies, including pachyonychia congenita type 1 (OMIM *167200) and type 2 (OMIM *167210). Pachyonychia congenita is a group of autosomal dominant disorders whose most prominent phenotype is hypertrophic nail dystrophy (see OMIM). Alopecia is present in about 7% of cases and the presence of cysts, which in some patients may represent essentially the steatocystoma phenotype, may be one of the better distinguishing features of type 2.

6. Conclusion

The genetics of hair loss is only just beginning, and the first evidence indicates that this trait is complex with autosomal dominant, codominant, autosomal recessive, X-linked, sex-influenced and polygenic inheritance. In this article, the numerous other types of disorder in which hair abnormality is part of a larger group of ectodermal defects (i.e. ectodermal dysplasia) were not included; the aetiology of these disorders is poorly understood. The identification of genes involved in human hair loss may, therefore, not only clarify the biological basis of this trait, but also contribute to the understanding of a wide spectrum of disorders which involve hair, dental, nail and sweating abnormalities. This will be a goal of medical genetics for the future.

REFERENCES

Ahmad, W., Ul Haque, M.F., Brancolini, V. *et al.* (1998) Alopecia universalis associated with a mutation in the human hairless gene. *Science* 279: 720–24

Carey, A.H., Chan, K.L., Short, F. *et al.* (1993) Evidence for a single gene effect causing polycystic ovaries and male pattern baldness. *Clinical Endocrinology* 38: 653–58

Frank, J., Pignata, C., Panteleyev, A.A. *et al.* (1999) Exposing the human nude phenotype. *Nature* 398: 473–74

Jones, L.N. & Steinert, P.M. (1996) Hair keratinization in health and disease. *Dermatologic Clinics* 14: 633–49

Korge, B.P., Healy, E., Munro, S.C. *et al.* (1998) A mutational hotspot in the 2B domain of human hair basic keratin (hHb6) in monilethrix patients. *Journal of Investigative Dermatology* 111: 896–99

McDonagh, A.J. & Messenger, A.G. (1996) The pathogenesis of alopecia areata. *Dermatologic Clinics* 14: 661–70

Online Mendelian Inheritance in Man (OMIM) http://www.ncbi.nlm.nih.gov/Omim

Panteleyev, A.A., Paus, R., Ahmad, W., Sudberg, J.P. & Christiano, A.M. (1998) Molecular and functional aspects of the hairless (hr) gene in laboratory rodents and humans. *Experimental Dermatology* 7: 249–67

Rubin, M. (1997) Androgenetic alopecia. Battling a losing proposition. *Postgraduate Medicine* 102: 129–31

Sawaya, M.E. & Shalita, A.R. (1998) Androgen receptor polymorphisms (CG repeat lengths) in androgenetic alopecia, hirsuitism, and acne. *Journal of Cutaneous Medicine and Surgery* 3: 9–15

Soma, T., Ogo, M., Suzuki, J., Takabashi, T. & Hibino, T. (1998) Analysis of apoptotic cell death in human hair follicle in vivo and in vitro. *Journal of Investigative Dermatology* 111: 948–54

Sprecher, E., Bergman, R., Szargel, R., Friedman-Birnbaum, R. & Cohen, N. (1999) Identification of a genetic defect in the hairless gene in atrichia with papular lesions: evidence for phenotypic heterogeneity among inherited atrichias. *American Journal of Human Genetics* 64: 1323–29

St-Jacques, B., Dassule, H.R., Karanova, I. *et al.* (1998) Sonic hedgehog signaling is essential for hair development. *Current Biology* 8: 1058–68

Sybert, V.P. (1997) *Genetic Skin Disorders*, Oxford and New York: Oxford University Press

GLOSSARY

allotype antigenic determinant typifying allelic differences in light and heavy immunoglobulin chains

apoptotic relating to cell death

caspase protease involved in apoptosis

codominant genes in which both alleles of a pair are fully expressed in the heterozygous state

cytochrome P450 specialized cytochrome (an enzyme pigmented as a result of its haem prosthetic groups) of the electron transport chain in adrenal mitochondria and liver microsomes, involved in hydroxylating reactions

etiology (aetiology) the cause of a disease

TWINNING

Judith G. Hall
Departments of Pediatrics and Medical Genetics, Faculty of Medicine, University of British Columbia, Canada

1. Introduction

Twins have always fascinated human beings. They are a relatively rare occurrence among human births, but multiple births are relatively common among other animals where polyovulation (release of multiple eggs) occurs regularly. The study of twins has led to important insights in genetics regarding the mechanisms of disease and early developmental processes. Approximately one in 40 individuals is a twin and, therefore, one in 80 births is a twin birth; probably many additional conceptions of twins occur in humans, but twin pregnancies are at increased risk for miscarriage and stillbirth (Hall & Lopez-Rangel, 1997).

Generally speaking, twins can be divided into two types, the first being "dizygotic" twins in which there is ovulation of two different eggs which are then fertilized by two different sperm. This produces two siblings who are genetically the same as any other two siblings, but who share a uterus during their pregnancy. This type of twin is also called fraternal. The second type of twin is "monozygotic". Such twins represent the fertilization of one egg by one sperm with the subsequent development of two separate individuals. This type of twin has been called identical, because it had been thought that they were genetically identical. However, it has become clear that many monozygotic twins are genetically discordant in some way and almost always there is a difference in the placental vascular flow to the two monozygotic twins. Consequently, it can be expected that monozygotic twin pairs will have some genetic differences between them (Machin, 1996). There are a number of subtypes of twins which will be mentioned later, but in general it is important when talking about twins to determine whether zygosity has been established and on what basis. In the future, the description of placental blood flow will also be important in establishing disease risk and degree of intrauterine environmental discordance.

2. Twins in studies of heredity

Twin research can be traced to Sir Francis Galton who, in the late 1800s, suggested that twins could be used to determine the importance of inheritance in determining various traits. Through his study of twins, he concluded that some sets who were very similar were derived from a single fertilized egg whereas those who had many differences originated from the fertilization of two different eggs. He reasoned that both types of twins shared the same environment *in utero*, but that only monozygotic twins shared all the same genes and, therefore, the differences between monozygotic twins were caused by environmental factors, while differences between dizygotic twins could be on the basis of either environmental or genetic factors. These notions were based on the concept that monozygotic twins have identical genetic information, which, for the most part is true, although many exceptions are known (Hall & Lopez-Rangel, 1997).

In the early 1900s, Sir Ronald Fisher and Carl Holtzinger formalized Galton's concept into statistical formulae for calculating the heritability of traits. This was particularly useful at that time because there was a strong controversy concerning the role of nature *vs.* nurture (genetic *vs.* environmental) influences. In the middle part of the 20th century, a number of groups began to establish registries of twins and institutes for the study of twins. Luigi Gedda in Italy led the way, but many other groups (e.g. in Denmark, Minnesota, Australia and Virginia) have developed very large twin registries and have done population-based studies which have contributed enormously to understanding the role of heredity. These groups have used comparisons of monozygotic and dizygotic twins to study complex traits such as behaviour, ageing and heart disease (Hall & Lopez-Rangel, 1997). The article "Genetics of psychiatric disorders" describes the use of twins in studies of common behavioural traits and disease.

3. Determining zygosity

Dizygotic twins, who result from the fertilization of two different ova by two different sperm, usually have separate placentas. However, recently it has become clear that dizygotic twins can also share placental vascular supply and even, on very rare occasions, placental membranes. Monozygotic twins result from one ovum fertilized by one sperm that divides during development in such a way as to form two embryos. In the past, the way of differentiating monozygotic twins and dizygotic twins was their gender and their appearance. If they were like sexed and looked alike, they were said to be monozygotic. Through the study of placentas and their membranes, it was recognized that 70% of monozygotic twins shared their placentas in that they have monochorionic and diamniotic membranes (Benirschke & Kim, 1973). A very small number of twins share both their amnion and chorion; one-third of monozygotic twins who come to birth have separate placentas. Although it is not always reliable, placentation is a means of establishing zygosity in >70% of monozygotic twins. When there is a single monochorionic diamniotic placenta, the twins can be considered monozygotic.

However, other methods were recognized to be useful in determining zygosity including blood types, blood protein polymorphorisms, HLA typing and dermatoglyphic studies including the documentation of finger and palm prints. Most recently, it has become clear that DNA testing using DNA polymorphisms and microsatellites is the most accurate way to determine zygosity. It should be noted, however, that many twins, both dizygotic (8%) and monozygotic (75%), have vascular connections *in utero* and thus will have cells from the other twin in their blood. Thus, if one is to do absolutely reliable DNA zygosity determination, it may be necessary to use a tissue other than blood, such as fibroblasts, for comparison between the twins. This is particularly appropriate in situations of apparently discordant twins who appear to have a single placenta. A unique situation does exist in which a male and female twin pair may actually be monozygotic. This usually occurs when the female has Turner syndrome, having lost a Y chromosome to become 45X which gives her a female phenotype, and the male has a normal complement of 46XY chromosomes.

4. Vanishing twin

Ultrasound studies done early in pregnancy show that at least 10% of twin pregnancies are either lost as a miscarriage or reduced to a singleton (Jeanty *et al.*, 1981). The earlier such studies are done, the higher the percentage of twin pregnancies that are observed to have loss of one twin. The number of twins among spontaneous abortions is three times more frequent than that seen in live births. The number of twins at delivery must be considerably less than the number of twins at conceptions, and only about 30% of women with twin pregnancies which have been diagnosed before 10 weeks of gestation actually give birth to twins. This suggests at least a 70% loss or reduction rate among twin conceptions. Twin pregnancies resulting from *in vitro* fertilization also demonstrate spontaneous reduction to singletons. Thus, a proper estimate of the rate of twin conceptions and "vanishing" twins is difficult to establish in humans, but such occurrences are definitely common.

5. Dizygotic twins

Multiple ovulation – that is, the release of more than one ovum from the ovary(ies) – is necessary for dizygotic twinning to occur. This tendency seems to run in families and is more common in certain ethnic groups, such as among African blacks, but less common in others – dizygotic twinning is very rare in Japan, for example. Dizygotic twinning is associated with higher than average levels of follicle-stimulating hormone and luteinizing hormone in the mothers delivering dizygotic twins (i.e. higher than the levels found in mothers who deliver singletons).

Dizygotic twinning occurs more frequently in certain families. The female members of such families are thought to have inherited a gene predisposing to multiple ovulation related to the higher gonadotrophin levels. Fertility drugs artificially duplicate this genetic tendency. It has been estimated that one in 15 individuals in the Australian and North American populations inherits this trait. In the course of evolution, there would have been a selective advantage for the increased fertility produced by such a gene; however, that would have to be balanced by the increased risk to the mother of a pregnancy with two foetuses (Meulemans *et al.*, 1996).

Certain herds of cattle have been developed where monozygotic twinning is relatively common, but, in general, the only animal to consistently produce monozygotic twins is the armadillo which may have as many as 16 monozygotic embryos. All other animals with multiple births usually represent multiple ovulation.

Dizygotic twins produced by the fertilization of two ova may be the result of "superfecundation". Superfecundation occurs when two different ova are fertilized by two different sperm in more than one act of coitus, either during the same ovulatory cycle or in a subsequent cycle. There are a few documented cases of dizygotic twins in which one is Caucasian and one is black. A subtype of superfecundation is "superfetation" which occurs when a second fertilized ovum implants in the uterus already containing a pregnancy of at least 1 month. Superfetation has been suggested in cases of dizygotic twins with marked discordance for birth size, but has also been documented by timing of conception in some cases. A third type of dizygotic twin is "polar body twinning", which is thought to arise from the simultaneous fertilization of the meiotic products of a single oocyte (i.e. the fertilization of the oocyte pronucleus and a polar body by two different sperm). This has been documented to occur by DNA molecular studies in very rare situations.

By definition, dizygotic twins would be expected to have two separate placentas, each with its own membranes. However, there are some dizygotic twin exceptions where the placentas are fused and have vascular connections; very rarely, there can even be one chorion and two amnions. When there are vascular connections between dizygotic twins, these twins become chimeras, containing in their haemopoietic system (and thus possibly among their macrophages) cells from both twins, even though they are genetically distinct. This situation allows each twin to be immunologically tolerant to cells from their genetically distinct twin.

The incidence of dizygotic twinning varies from country to country. Until DNA studies were developed, the rate of dizygotic twinning was estimated on the basis of the Weinberg method, which is based on a rough mathematical estimation using the relative number of like-sex and unlike-sex twins. The assumption is that unlike-sex twins are all dizygotic and there would be the same number of like-sex and unlike-sex dizygotic twins: thus, the excess of like-sex twins will all be monozygotic. Using this method, the rate of dizygotic twinning varies in a range of 2–7 per 1000 births. The highest rate of dizygotic twinning is seen in the black African population, where as many as 50 per 1000 twins are seen. The dizygotic twin rate varies with maternal age, parity, height, weight and gonadotrophin level, with an increased number of dizygotic twins seen with increased maternal age and high parity, and with tall and heavy mothers.

Dizygotic twinning is frequently seen using the new reproductive technologies. Gonadotrophins are given to cause ovulation and, because of the high doses, polyovulation leading to dizygotic twins is common. When *in vitro* fertilization is performed, frequently two or three genetically distinct embryos are implanted in the uterus. Thus, it would be expected that there will be an increased number of multiple births of genetically distinct individuals among pregnancies where reproductive technologies have been used.

It may be worth noting that when mothers have had supplementation with multivitamins including folic acid starting prior to becoming pregnant, there appears to be an increased occurrence of both dizygotic and monozygotic twin pregnancies coming to term, implying that adequate folic acid intake by mother leads to increased live births and maintenance of twin pregnancies (Czeizel *et al.*, 1994).

6. Monozygotic twins

Monozygotic twins are the result of the fertilization of one ovum by one sperm which then divides into two embryos. There are many theories about why this occurs during the course of development, but the cause of monozygotic twinning is as yet unknown. Monozygotic twins are thought to have the same genetic constitution and, for this reason, have, in the past, been called identical twins. However, it has become quite clear using the new molecular genetic techniques that many monozygotic twins are not completely identical. In any large multicellular organism, it would be expected that somatic mutations would occur. In addition, monozygotic twins frequently have different intrauterine vascular supplies leading to differential growth. Finally, at the time of separation into two distinct cell masses, there may be unequal contributions either of genetic material or of cytoplasmic material (including the mitochondria) such that the two embryos do not really carry the same genetic and epigenetic material. Interestingly, at birth, monozygotic twins are more discordant for height and weight than dizygotic twins.

It has been hypothesized (because experimental work on human embryos cannot be done) that monozygotic twinning occurs during four different phases (Benirschke & Kim, 1973).

1. Up until 3 days postconception, it is anticipated that if the conception breaks into two masses (from the 2-cell to the 32-cell stage), a dichorionic, diamniotic placenta would be produced. This is because the placental membranes do not begin to form until after day 3 postconception. Slightly more than 25% of monozygotic twin pregnancies which come to birth have dichorionic, diamniotic placentas.
2. It is hypothesized that if the split into the two embryos occurs between 4 and 6 days of development, the chorion will have already begun to form. Thus, if the inner cell mass separates into two embryos at that time, there would be a single chorion and two amnions. Some 70–75% of monozygotic twin pregnancies surviving to birth represent this situation. The two embryos/foetuses will always have a shared blood supply in this situation. If there is a compromise in blood supply, it can lead to a vascular accident and a subsequent congenital anomaly in one or both of the twins. There can also be shunting of blood from one twin to the other giving the twin-to-twin transfusion syndrome (Lopriore *et al.*, 1995).
3. If the monozygotic twinning event occurs after 7 days, but prior to 13 days, it is thought that a monoamniotic, monochorionic placenta would occur. Less than 1% of monozygotic twin pregnancies coming to term represent this situation. Since both twins would be in a single amniotic cavity, there would be great risk of tangling umbilical cords and having vascular compromise, with subsequent death of one or both twins.
4. After 13 days, in humans, it is anticipated that there will be a trilayered inner cell mass giving rise to gastrulation. Thus, a twin arising after 13 days would be expected to be a conjoined (Siamese) twin, a foetus-in-foetu or a teratoma (an undifferentiated mass of tissue from multiple cell type origin).

The estimates of timing of days of gestation are dependent on the normal sequence of events which has been worked out for human pregnancies. However, it may be that monozygotic twin pregnancies have a slightly different timescale and that these developmental events may occur in a different time-frame or sequence in the presence of a monozygotic twinning event. It is important to remember that monozygotic twins live off the cytoplasm of the one ovum (even though it has split into two embryos) until the time of implantation which normally takes place between 6 and 9 days postconception. It becomes relatively obvious that the egg's folic acid content (as well as that of other nutrients) is important for normal cell division, normal gene control and normal embryonic development until implantation.

There is an increased occurrence of congenital anomalies associated with monozygotic twin conceptions and births (Schinzel *et al.*, 1979), and there are some anomalies that are unique to the monozygotic twinning process itself. True malformations occur with increased frequency among monozygotic twins, which is thought to be related to incomplete development of the embryo. Disruptions leading to congenital anomalies occur more frequently than in singletons and are thought to be primarily related to vascular insufficiency and compromise. Deformations occur more frequently and are thought to be related to the presence of two embryos/foetuses in the space that is usually taken up by one. The rate of congenital anomalies among liveborn monozygotic twins is 2–3 times that of singletons. There are also congenital anomalies that are unique to the monozygotic twinning process, such as: conjoined twins (Siamese twins), thought to be due to a defect of incomplete embryogenesis; acardiac twins, thought to be due to abnormal vascular connections; foetus papyraceous occurring secondary to the death of a co-twin; and disorders such as foetus-in-foetu and teratomas which are thought to represent embryologically late attempts at twinning.

The causes of monozygotic twinning are not yet known. Monozygotic twinning is extremely rare in animals, but could be expected to occur if the zona pellucida was deficient. It is known that monozygotic twinning occurs at an increased rate following *in vitro* fertilization, particularly in cases where the zona pellucida has been disrupted. It makes sense then that, at the 2-, 4-, 6- or 8-cell stage, the cells of an embryo could become detached if they are not held together by the zona pellucida and each has the potential to develop a full embryo. There are some rare families with familial monozygotic twinning and these may be related to zona pellucida defects (Harvey *et al.*, 1977). The studies of folic acid supplementation (Czeizel *et al.*, 1994) suggest that periconceptual maternal folic acid level intake leads to a higher occurrence than expected of the birth of monozygotic twins (without any congenital anomalies). Thus, folic acid deficiency in the mother may play a role in the monozygotic twinning process itself. In animal studies, late fertilization, heat, X-rays and carcinogenic drugs can produce monozygotic twinning, suggesting that cell injury may predispose to monozygotic twinning. However, there is no evidence to that effect in human studies.

There are many types of monozygotic twins. These include:

1. a foetus-in-foetu which appears to be a small parasitic dead twin attached to a normal twin or placenta, and which is often confused with a tumour and may be localized on the superior mesenteric vessels;
2. a foetus papyraceous – a monozygotic twin which is a mummified dead foetus usually attached to the placenta and presenting at birth with a normal, viable twin;
3. an acardia twin with an absent or rudimentary non-functioning heart whose circulation has been sustained by the normal twin (acardia twins are associated with a higher rate of chromosome abnormalities);
4. conjoined twins – incomplete twins resulting from an abnormality in the twinning process, derived from a single zygote and always of the same sex, but having a marked predominance of females (the incidence is in the range of one in 50 000 to one in 100 000 live births;
5. mirror-image twins (see below);
6. discordant monozygotic twins; and
7. apparently concordant monozygotic twins.

(i) Mirror-image twins

Approximately 10% of liveborn monozygotic twins have mirror-image feature (Sperber *et al.*, 1994). Most do not have *situs inversus*, but rather have minor asymmetries (mirror images) of facial features. It is thought that these are late-forming monozygotic twins which have had a problem readjusting the body axis plan. Interestingly, if special forms of communication occur between monozygotic twins, it seems to be among mirror-image twins.

7. Sex ratio in twins

The sex ratio of monozygotic twins demonstrates a greater proportion of females than might be expected. Dizygotic twins and singletons have a slight excess of males whereas all monozygotic twins have a slight excess of females; conjoined twins are almost 80% female, as are sacral teratomas. This suggests that twinning is somewhat different in males and females, and that female conceptions may be a higher risk of a monozygotic twinning process. Interestingly, female embryos are somewhat delayed in their development as compared with male embryos (Tan *et al.*, 1993). Female twins also have to go through X inactivation, which is thought to occur at the time of tissue differentiation. If, by chance, a patch of cells entirely inactivates the maternal X and another patch inactivates the paternal X, it is conceivable that they would be sufficiently discordant to establish completely different embryos. When discordant X inactivation has been examined among normal female monozygotic twins, a small percentage (5–10%) have marked discordance between their X inactivation (Goodship *et al.*, 1996).

8. Discordance in monozygotic twins

The aetiology(ies) of monozygotic twinning is not clear. However, a very high occurrence of discordance between monozygotic twinning has now been observed. Discordance for chromosomal abnormalities, single gene disorders, uniparental disomy, X inactivation, imprinting, mitochondrial inheritance and several other forms of discordance have been observed (Hall & Lopez-Rangel, 1997). As a result of this, the hypothesis has been put forward that all monozygotic twins are discordant in some way, and that the arising of discordance during early development actually leads to the formation of two separate embryos. Thus, it is hypothesized that if a mutation or "change" in genetic information occurs early enough in development, it may actually lead to monozygotic twinning.

9. Population rates of twinning

Historically, the rate of dizygotic twinning was associated with advanced maternal age, large family size and rural living, while monozygotic twinning was thought to occur at a constant rate around the world. However, recently, with more careful determination of zygosity, it has become clear that the spontaneous rate of dizygotic twinning is decreasing (Bryan, 1994), while fertility drugs and *in vitro* fertilization are responsible for the increasing occurrence of dizygotic twins, which is observed for all dizygotic twins. Interestingly, like-sex dizygotic twins and those with the same HLA type are more likely to come to term and be born than would have been expected by chance alone. So, there may be selective forces at work in dizygotic pregnancies. Until recently, monozygotic twinning has been thought to be constant throughout the world. However, recent secular studies show an increased number of monozygotic twins coming to term. This observation suggests that a change in the environment (e.g. an increased mutation rate or better maternal nutrition) may be influencing monozygotic twinning rates. It should be pointed out that many monozygotic twins do not survive embryogenesis and/or vascular accidents. Thus, what is observed at birth may not really reflect the complete range of developmental and reproductive processes.

10. Practical aspects of twins

Besides the fascination with the twinning process, there are several practical aspects of twinning. The first is that twin pregnancies are much more prone to medical complications during the pregnancy. The mother is much more likely to have toxaemia, hypertension and difficulty at the time of delivery (Baldwin, 1994). Each embryo/foetus is more likely to have congenital anomalies, twin–twin transfusion, prematurity, metabolic complications and growth retardation at the time of birth than singletons. As a result of the high occurrence rate of congenital anomalies among twins, careful prenatal evaluation and evaluation at birth are indicated to rule out a (several) congenital anomaly(ies) (Hall, 1996a).

Over their lifetime, monozygotic twins are more likely to have or develop the same disease processes. Numerous studies of complex disorders such as hypertension, diabetes and cardiovascular disorders demonstrate a major heritability for those disorders and monozygotic twins mostly share the same genes. Since 75% of monozygotic twins and 8% of dizygotic twins have shared vascular supplies *in utero*, they are or have become immune tolerant and thus will be able to accept organ transplantation from their twin. Kidney, bowel and liver transplantations between monozygotic twins are possible without immunosuppression (Falik-Borenstein *et al.*, 1994).

11. Summary

In summary, there are two types of twins and it is essential to determine, at birth if possible, what type of twin pregnancy has occurred. The study of twins has led to an understanding of many embryological and developmental processes. The study of twins after birth has led to insight about complex disorders such as behaviour and adult degenerative diseases (Hall, 1996b). Twin pregnancies are more likely than singletons to have a chromosome abnormality and prenatal diagnosis, therefore, is indicated, with careful monitoring of the pregnancy.

REFERENCES

Baldwin, V.J. (1994) *Pathology of Multiple Pregnancy*, New York: Springer

Benirschke, K. & Kim, K. (1973) Multiple pregnancy. *New England Journal of Medicine* 288: 1276–84, 1329–36

Bryan, E. (1994) Trends in twinning rates. *Lancet* 343: 1151–52

Czeizel, A., Metneki, J. & Dudas, I. (1994) Higher rate of multiple births after periconceptual vitamin supplementation. *Lancet* 330: 23–24

Falik-Borenstein, T.C., Korenberg, J.R. & Schreck, R.R. (1994) Confined placental chimerism: prenatal and postnatal cytogenetic and molecular analysis, and pregnancy outcome. *American Journal of Medical Genetics* 50: 51–56

Goodship, J., Carter, J. & Burn, J. (1996) X-inactivation patterns in monozygotic and dizygotic female twins. *American Journal of Medical Genetics* 61: 205–08

Hall, J.G. (1996a) Twinning: mechanisms and genetic implications. *Current Opinion in Genetic Development* 6: 343–47

Hall J.G. (1996b) Twins and twinning. *American Journal of Medical Genetics* 61: 202–04

Hall, J.G. & Lopez-Rangel, E. (1997) Twins and twinning. In *Emery and Rimoin's Principles and Practice of Medical Genetics*, 3rd edition, edited by D.L. Rimoin, J.M. Connor & R.E. Pyeritz, London and New York: Churchill Livingstone

Harvey, M.A.S., Huntley, R.M.C. & Smith, D.W. (1977) Familial monozygotic twinning. *Journal of Pediatrics* 90: 246–50

Jeanty, P., Rodesch, F. & Verhoogen, C. (1981) The vanishing twin. *Ultrasonics* 2: 25–30

Lopriore, E., Vandenbussche, F.P.H.A., Tiersma, S.M.E., De Beaufort, A.J. & Leeuw, P.J. (1995) Twin-to-twin transfusion syndrome. *Journal of Pediatrics* 127: 675–80

Machin, G.A. (1996) Some causes of genotypic and phenotypic discordance in monozygotic twin pairs. *American Journal of Medical Genetics* 61: 216–28

Meulemans, W.J., Lewis, C.M., Boosma, D.I. *et al.* (1996) Genetic modelling of dizygotic twinning in pedigrees of spontaneous dizygotic twins. *American Journal of Medical Genetics* 61: 258–63

Schinzel, A.G.L., Smith, D.W. & Miller, J.R. (1979) Monozygotic twinning and structural defects. *Journal of Pediatrics* 95: 921–30

Sperber, G.H., Machin, G.A. & Bamborth, F.J. (1994) Mirror-image dental fusion and discordance in monozygotic twins. *American Journal of Medical Genetics* 51: 41–45

Tan, S.S., Williams, E.A. & Tam, P.P.L. (1993) X-chromosome inactivation occurs at different times in different tissues of the post-implantation mouse embryo. *Nature Genetics* 3: 170–74

GLOSSARY

etiology (aetiology) the cause of a disease

haemopoietic system system of development of blood cells from stem cells

macrophage large, phagocytic, mononuclear white blood cell of the reticuloendothelial system that ingests invading microorganisms and eliminates cellular debris

X inactivation process of inactivation of one of the X chromosomes in females

See also **Genetics of psychiatric disorders** (p.443)

DNA VACCINATION

R.A. Strugnell and D.R. Drew
Department of Microbiology and Immunology, University of Melbourne, Australia

1. Introduction

DNA immunization, or the use of eukaryotic expression constructs to generate immunogens *in vivo*, captured the imagination of all vaccinologists when it was first described in the early 1990s. The technique is relatively simple, does not require significant investment in equipment or intellectual capital, and can be applied to the prevention of viral, bacterial, protozoan or metazoan diseases (Donnelly *et al.*, 1997). Equally, the technique can be used to generate laboratory reagents, e.g. specific antisera, or to screen rapidly a large number of different antigens for their capacity to induce protective immune responses (Ulmer & Liu, 1996).

2. How DNA vaccines are designed

The most basic DNA vaccine comprises a bacterial plasmid, ideally of high copy number, into which a strong eukaryotic promoter and polyadenylation signal are inserted (**Figure 1**). DNA encoding the antigen gene is inserted between these two features in the correct orientation for transcription to occur from the promoter. Typically, the immediate early promoter from cytomegalovirus (CMV) is used in DNA vaccine plasmids for its powerful and sustained transcriptional activity. The CMV promoter is combined with a termination/polyadenylation signal derived from either the SV40 virus or the bovine growth hormone gene to generate a functional expression cassette. Refinements which have been added to this basic construct include the insertion of an artificial intron immediately upstream of the AUG start codon of the immunogen, and sequence mutagenesis to a Kozak optimal translation initiation sequence around the start codon. The improvements afforded by such modifications are often modest unless the gene is from a bacterium, for example, where translation initiation signals are inherently different. Where the antigen gene does not carry a signal peptide, the coding sequence for a signal peptide from, for example, the human tissue plasminogen activator protein can be fused in-frame to the start of the antigen gene to facilitate secretion from transfected cells. This modification often has the most significant impact on vaccine immunogenicity (Gurunthan *et al.*, 2000).

DNA for use in DNA vaccination is usually prepared in standard *Escherichia coli* strains which are used in recombinant DNA experiments. There has been little published work on strain optimization, a potentially important parameter because of the effects of DNA methylation on vaccine efficacy. Bacterial methylation of CpG motifs is apparently recognized as a "danger signal" by cell surface receptors on professional antigen presenting cells (APCs), the activation of which triggers APC maturation, migration and cytokine production. An increase or reduction in the number or spacing of CpG motifs can significantly affect vaccine immunogenicity (Cohen *et al.*, 1998; Gurunthan *et al.*, 2000). Vaccine plasmid DNA is prepared by standard purification techniques such as caesium chloride density gradient centrifugation or ionic/hydrophobic interaction chromatography. Contaminants such as lipopolysaccharides are carefully removed and efforts are made to retain the plasmids in a supercoiled form which appears to increase transfection and durability *in vivo*.

3. DNA vaccination in use

DNA vaccines are easily administered through intradermal or intramuscular injection and, in experimental murine systems, generally doses of 50–100 μg are resuspended in physiological saline. DNA vaccines can also be administered using a specialist device known as a "gene gun" (see article

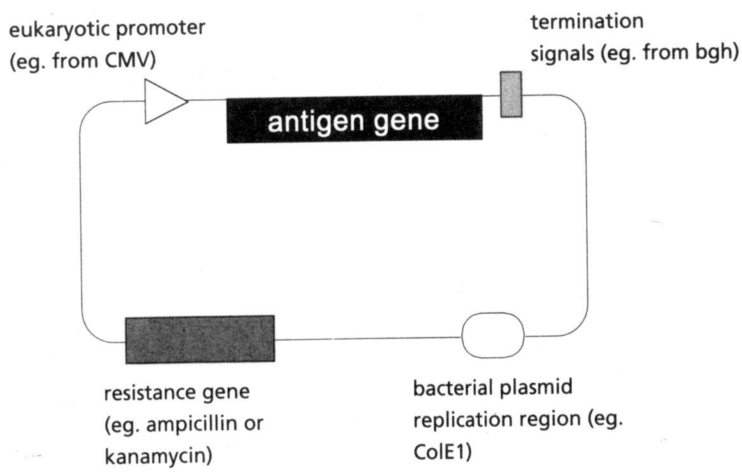

Figure 1 Schematic representation of a DNA vaccine expression plasmid.

on "Techniques in molecular genetics") or, more simply, using a Panjet apparatus. These devices use biolistic or hydrostatic pressure to push DNA-coated gold spheres, or DNA in suspension, through the highly keratinized epidermis. Gene gun delivery of DNA-coated gold particles is very efficient and can generate potent responses in mice with less than 2 μg of DNA (Cohen *et al.*, 1998). DNA immunization relies on transfection of cells *in vivo* followed by antigen expression. However, even with improvements such as the gene gun, transfection efficiency is very low, with most of the vaccine DNA failing to enter cells. In fact, histological studies of vaccinated muscle often reveal that as few as 2–3 muscle bundles are transfected.

The immunological basis of DNA vaccination is only partially resolved. Following intramuscular injection of DNA, the majority of cells transfected are myocytes which are not capable of presenting antigen directly to the immune system, as they do not express significant levels of MHC class II proteins and lack the co-stimulatory molecules necessary to activate naive T cell differentiation. The absence of essential APC functional activities in myocytes suggests that, while muscle cells may be a source of vaccine antigen, it will be other cells, most probably dendritic cells, which drive the early and essential differentiation of T cells within lymph nodes (Gurunthan *et al.*, 2000). Dendritic cells are specialized APCs which strongly stimulate T cells following exposure to minute amounts of antigen. It is clear that physical trauma associated with inoculation of the vaccine can result in the rapid migration of transfected migratory cells, including dendritic cells, to sites of immune induction – typically, the lymph nodes. This suggests that at least some of the immunogenicity of DNA vaccines is due to either the direct transfection of dendritic cells or the "cross-priming" activity of these cells (Gurunthan *et al.*, 2000). Dendritic cells are known to capture and actively present antigen from cells which undergo apoptosis – it is possible that the trauma associated with vaccination, combined with the local proinflammatory cytokine production caused by the abundance of methylated CpG sequences, could lead to apoptosis. Alternatively, dendritic cells may scavenge antigen released from viable transfected cells, and this may explain why DNA vaccines expressing a secreted form of an antigen are more immunogenic than DNA vaccines expressing a non-secreted form of the same antigen.

4. Conclusion

In conclusion, DNA immunization represents a powerful new approach for the design and administration of vaccines and immunotherapies. There are several advantages to the use of DNA vaccines: they eliminate the need to purify antigen, are relatively simple to produce, have the potential to be manufactured using generic processes and do not require the coadministration of an adjuvant. Also, as plasmid DNA is relatively stable, it can be freeze-dried and distributed without a cold-chain. However, the key attraction of DNA vaccines is that they are capable of inducing a full spectrum of cell-mediated and antibody responses against defined antigens, without the risks associated with live attenuated vaccine vectors. While this area remains a nascent field of immunology, it is already apparent that by altering the routes of vaccine administration and the design of antigen genes, and through the coadministration of specific cytokine genes, it is possible to modulate both the magnitude and orientation of subsequent immune responses (Cohen *et al.*, 1998; Gurunthan *et al.*, 2000). Hence, as the mechanisms underpinning the generation of immune responses to DNA vaccines are identified, future vaccine researchers may be able to tailor the immune responses generated by DNA vaccines specifically to the requirements of individual diseases.

REFERENCES

Cohen, A.D., Boyer, J.D. & Weiner, D.B. (1998) Modulating the immune response to genetic immunization. *FASEB Journal* 12: 1611–26

Donnelly, J.J., Ulmer, J.B., Shiver, J.W. & Liu, M.A. (1997) DNA vaccines. *Annual Reviews in Immunology* 15: 617–48

Gurunthan, S., Klinman, D.M. & Seder, R.A. (2000) DNA vaccines: immunology, application and optimization. *Annual Reviews in Immunology* 18: 927–74

Ulmer, J.B. & Liu, M.A. (1996) ELI's coming: expression library immunization and vaccine antigen discovery. *Trends in Microbiology* 4: 169–71

GLOSSARY

immunogenicity ability of an antigen or vaccine to stimulate an immune response

methylation [DNA methylation] process by which methyl groups are added to certain nucleotides in genomic DNA. This affects gene expression, as methylated DNA is not easily transcribed

polyadenylation addition of a poly(A) tail to eukaryotic messenger RNA precursors in the nucleus

transfect insertion of viral DNA into a cell

See also **Techniques in molecular genetics** (p.22)

US GENE THERAPY IN CRISIS

Doris Teichler Zallen
Center for Interdisciplinary Studies, Virginia Polytechnic Institute and State University, Blacksburg, Virginia, USA

1. Introduction

Gene therapy is the term most often used to describe the introduction of new genetic material into humans. It is now widely recognized that this term may be misleading because it implies that real medical benefit (in the form of a therapy or a treatment) is possible. The current trend is to use "gene transfer" as a more accurate term and one less likely to give rise to unfounded expectations by research subjects.

Human gene therapy experiments are currently being actively pursued around the world. Results, however, have been meager. Introducing genes into target tissue and getting them to work has turned out to be harder than anticipated (an assessment of the status of the field can be found in a report by S. Orkin and A. Motulsky at http://www.nih.gov/news/panelrep.html). Despite the lack of efficacy, gene therapy researchers had come to believe that, at least, the procedures were safe (Marshall, 1995). However, the death in 1999 of a research subject, Jesse Gelsinger, as a direct result of a gene transfer has challenged that view. Brought to wide attention by newspaper headlines such as "Death leads to concerns for future of gene therapy" (Weiss & Nelson, 1999) and "Some had said research moved too quickly" (Wade, 1999), the field of gene therapy is now under intense scrutiny from scientists, the government and the public.

2. Forging the infrastructure

The present policy framework for gene therapy research in the US can be traced to events that occurred in 1973. In that year, the new-found ability of scientists to recombine DNA between species raised fears that these new techniques might result in the creation of microbes with dangerous properties. The scientific community took decisive – and unprecedented – action, with researchers declaring a voluntary moratorium. An international group of scientists conferred to set up procedures that would guide the continuation of recombinant DNA research. Human gene therapy was not high on anyone's agenda at that time.

When the voluntary moratorium was lifted in 1976, a new oversight body had been created. Within the US National Institutes of Health (NIH), there was now a national-level body: the Recombinant DNA Advisory Committee (RAC). The RAC was established to formulate standards for recombinant DNA research and to oversee its progress. Local biosafety committees were set up in institutions where recombinant DNA experiments were underway to work in conjunction with the RAC. There was also a new partner in decision making: the general public. Alerted by extensive media coverage, various citizen groups began to be formed to add their voices to the discussions that were taking place (Wright, 1994). In 1978, the RAC expanded to include "public members" – typically lawyers, ethicists, political scientists and consumer advocates – as a means of incorporating a wider range of views into its deliberations.

Recombinant DNA research then proceeded in a fashion so dramatic that what many call a biological "revolution" was set in motion. As experience was gained, much of the oversight responsibility was delegated to local biosafety committees or, in the case of agricultural organisms, was transferred to the Department of Agriculture and the Environmental Protection Agency.

3. Anticipating human genetic manipulation

In 1980, the possibility of applying genetic engineering procedures to humans catapulted to the centre of attention. The US Supreme Court decision allowing patent protection for a living organism, a genetically engineered "oil-eating" bacterium (Diamond v. Chakrabarty, 1980), provided powerful financial inducements for private companies to expand their research efforts. Also in that year, a California researcher made a premature attempt to alter human genetic makeup in order to treat the genetic disorder thalassaemia. (This was the so-called Cline affair, in which an investigator proceeded to do studies overseas in the absence of approval from his home institution; National Institutes of Health, 1981.)

Worried that scientists were embarking on "a new era of fundamental danger triggered by the rapid growth of genetic engineering", three leaders of the Protestant, Jewish and Catholic religions requested that the then-president Jimmy Carter act swiftly to "provide a way for representatives of a broad spectrum of our society to consider these matters and advise the government on its necessary role" (Randall *et al.*, 1982). The President's Commission, a body impanelled (in 1980–1983) to study ethical issues in medicine and research, responded in a report *Splicing Life*. That report strongly endorsed the acceptability of research intended to treat, and perhaps cure, genetic diseases through genetic manipulation of body tissues (somatic cell gene therapy). It was decidedly more cautious about genetic manipulations that could alter gametes and be transferred to future generations (germline gene therapy) or that might be used to "improve" otherwise healthy individuals (enhancement gene therapy).

The Commission took seriously what it termed the "Frankenstein factor": the public's deep-rooted concerns that human genetic manipulation could lead to harmful outcomes for individual research subjects or pose a threat to widely held societal values. It recommended that human genetic engineering activities should proceed only under the watchful scrutiny of a body with both scientific and public representatives. The process of deliberation was to be open; secrecy was to be avoided. To underscore the latter point, the report cited a particularly prophetic comment made by George E. Brown, Jr, a member of Congress with expertise in science policy: "If ... industry follows the path that appears easier initially, the cloistered avoidance of other forces of society, it will pay a penalty years hence should some event force a public inquiry" (Brown, 1981).

4. Developing a multilevel oversight mechanism for gene therapy

The RAC responded to the *Splicing Life* report by establishing a Working Group on Human Gene Therapy (later renamed the Human Gene Therapy Subcommittee). The group's recommendations built on the pre-existing policy framework.

The RAC was to be central in the oversight process. Gene therapy protocols were to be considered on a case-by-case basis, just as the RAC had done for almost a decade for other forms of recombinant DNA research. Protocols would be reviewed first by local-level biosafety committees and by institutional review boards. Already required by federal law (45CFR 46), institutional review boards would render judgments on whether human subjects could be used in the research. The protocols would then proceed for national-level review by the Human Gene Therapy Subcommittee and, finally, by the RAC itself. The presence of public members on both the Subcommittee and the RAC would ensure that broader ethical and social issues, not just scientific feasibility, would be considered. All meetings would be held in public to permit input from interested individuals or members of citizen and consumer groups. The RAC would make its recommendations to the Director of NIH, whose official approval was required before the research could proceed. Thus, US policy regarding human gene manipulation would emerge, over time, from the decisions rendered by the RAC operating in full view of the general public.

In reality, RAC jurisdiction only extended to research conducted at institutions receiving public funding for recombinant DNA research through the NIH. However, the expectation was that private companies engaged in human gene therapy research would bring their protocols before the RAC voluntarily, just as they had done for other types of recombinant DNA research. Companies had seen it as an advantage to gain RAC endorsement of their research.

Along a parallel pathway, the Food and Drug Administration (FDA) would also have a central role in the approval of gene therapy experiments. The FDA has statutory authority over any drug (including, in this case, any system for DNA delivery) intended for human use. In contrast to the RAC, however, the FDA focuses on safety and efficacy, not ethics. To protect the proprietary interests of the sponsors of research, the FDA is compelled by law to carry out its work in private and keep all its data confidential.

The standards to be used by the Human Gene Therapy Subcommittee and the RAC in making decisions had to be developed. The issues were complex and the uncertainties profound. What were the real risks – and how safe was safe enough? Could patient privacy be protected? How would the selection process be managed?

To help address this ethical dimension, the Subcommittee added a new element. It produced a document, known as *The Points to Consider in the Design of Human Gene Therapy Protocols*, that was incorporated into the NIH Guidelines (Office of Recombinant DNA Activities, 1997). (*The Points to Consider* were accepted in September 1985 and amended in 1989 following consideration of the first approved protocol. The section on the consent process and the consent document was rewritten and expanded in 1994.)

The Points to Consider contain seven areas of questions for investigators to answer that help to determine whether investigators had met the three ethical criteria underpinning all human subjects research:

- beneficence, which refers to an overall balance of benefit over harm;
- respect for persons, which requires that adequate and accurate information be given to each prospective subject and that the decision to participate is uncoerced; and
- justice, which is seen as the fairness of selection criteria for enrollment in the study (Walters & Palmer, 1997).

By 1985, the RAC was ready to render judgements in a controversial area of research.

In 1988, the very first experiment to alter human genes wended its way through the approval process. This experiment was not an attempt at treatment. It was a study to insert a neomycin-resistance gene into lymphocytes of individuals with advanced melanoma in order to mark the cells and determine their distribution and duration in the body. The first attempt to use genetic manipulation to correct an inherited genetic disorder was approved in 1990 (Blaese, 1990). This was a study to transform lymphocytes from children afflicted with a form of severe combined immunodeficiency disease (SCID-ADA) by introducing the gene for adenosine deaminase. Since then, the pace of research has picked up remarkably. According to current estimates, about 350 experiments involving gene transfer are now underway in the US. Similar research in progress in the UK and other countries brings the worldwide total to over 400.

The protocols that have been approved employ different types of vectors and seek to alter human cells, either *in vivo* or *ex vivo*, in a variety of ways (**Table 1**). Initially, the expectations had been that most protocols would be concerned

Table 1 US human gene transfer protocols[a].

Type of study		Number of protocols
Therapy related		320
Single gene disorders	43	
Cancer	218	
Infectious diseases (HIV)	30	
Other diseases/disorders	29	
Marking		35
Non-therapeutic		2
Total		357

[a] US National Institutes of Health, Office of Biotechnology Activities; protocols approved as of December 1999

with single gene disorders. In reality, the emphasis has been on experiments designed to introduce new genetic information into cells as a way to treat major health problems such as cancer and AIDS. This emphasis reflects the increasing involvement of private industry as collaborators in research efforts. Such collaboration is nearly inevitable because most university scientists lack the resources to bring their research along from the laboratory bench through clinical trials while meeting stringent FDA standards. However, industry's quest for a profitable return on its investment leads it to prefer gene therapy applications aimed at widely prevalent disorders.

5. Reducing RAC oversight

Almost as soon as it started, calls came to modify the oversight process. These efforts first aimed to eliminate the Human Gene Therapy Subcommittee so that protocols would proceed directly from the local-level committees to the RAC. In November 1991, the Subcommittee responded to the growing pressure by voting itself out of existence.

Following this, a crescendo of criticism about the RAC itself began to be heard. Often, criticism took the form of complaints from biotech companies about possible delays resulting from the RAC's quarterly meeting schedule. It was also likely that they were concerned about the sometimes harsh critiques of scientific merit or of the proposed informed consent process which were expressed by RAC members at the public meetings. These critiques were often in sharp contrast to the enthusiastic reports of disease-curing possibilities simultaneously provided by biotech companies to the media.

The NIH also had its complaints. The media hype surrounding any protocol approval was sufficient to drive up the net worth of a company and make it a prime candidate for a lucrative buyout. The NIH found itself used to promote the financial self-interest of private companies. It was an awkward and untenable position for the nation's premier medical research agency.

In 1995, NIH Director Harold Varmus, perhaps disillusioned by this link between NIH and industry or by the low yield of much of the research, decided to examine the role of the RAC itself. He convened an Ad Hoc Review Committee, chaired by scientist Inder Verma. The Committee's Executive Summary can be found at: http://www4.od.nih.gov/oba/adhoc-re.htm.

The Verma committee ultimately recommended that the RAC continue its work, calling it a "credible forum for airing a wide range of public concerns". It endorsed the streamlining procedures, already in place, that ceased the RAC practice of case-by-case review of all protocols. The RAC was to consider only protocols with novel features (e.g. those using new vectors or targeting new diseases) or which moved the field into problematic areas (e.g. such as *in utero* interventions or germline modifications). Additionally, the Verma committee urged that the RAC "continue to be provided with the data needed for monitoring clinical gene transfer protocols".

It was under this system of consolidated review that the ornithine transcarbamylase deficiency (OTC) protocol, into which Jesse Gelsinger was enrolled, was approved by the RAC in December 1995. It had come before the RAC because it possessed a number of novel features: the disease targeted, the adenovirus vector used, the delivery directly into the liver and the use of relatively healthy individuals as research subjects. RAC reviewers raised a number of concerns and made recommendations to alter the mode of delivery and to improve the consent form.

In 1996, the RAC faced an unexpected challenge to its existence, one that came from the NIH itself. Contrary to the recommendation of his Ad Hoc Committee, NIH Director Varmus floated a plan to disband the RAC and replace it with a six- to ten-member committee operating in-house, but out of public view. This new committee would look at the broader policy issues related to gene therapy, but neither it nor the NIH Director would retain any role in the evaluation of protocols. Through occasional conferences and from the database, the NIH and the public would be informed of areas of promise as well as of problems.

The Biotechnology Industry Organization and individual biotech firms offered strong support to the plan to dismantle the RAC. However, a surge of opposition to the proposal, especially from consumer groups, forced a compromise. In the end, the RAC was retained, but it was reduced in size to 15 members and, while it could discuss them, it could no longer vote to approve or block any protocols; nor would the NIH Director play any part in the approval process. The only decision-making authority resided with the FDA. It is this system of oversight that governs US gene therapy research to the present day.

6. The current crisis

What the death of Jesse Gelsinger has shown, beyond the still-to-be-determined details of the reasons for the overwhelming response by his immune system to the adenoviral vector, is that the present oversight system is seriously flawed. At a hearing held on 2 February 2000 by the Subcommittee on Public Health, US Senate Health and Education Committee, it become clear that the marginalization of the RAC has produced a number of unfortunate outcomes.

Gaps in communication with the FDA left the RAC with a diminished ability to know what was really going on or what actions, if any, had been taken on its recommendations. For instance, the RAC was never informed that the FDA had authorized a change in the mode of administration of the adenoviral vector in the OTC protocol.

A pattern developed of failure to report serious adverse events to the NIH. The researchers responsible for the OTC protocol did inform the NIH, and thereby the public, as soon as the death occurred. However, many investigators and sponsors have either failed to report serious adverse events to the RAC or have requested that such reports be considered "proprietary" and, thus, shielded from public view. According to an NIH audit of adenoviral protocols following Mr Gelsinger's death, only 5% (39/691) of observed serious adverse events, including deaths, were ever reported to the NIH (Varmus, Harold, letter to Waxman,

Henry, US House of Representatives, 21 December 1999; Adams, 2000). Investigators maintained that the unreported events were disease related, unconnected to the gene transfer. All of them had been reported to the FDA.

The quality of the consent process has deteriorated. Once the consent process and associated consent documents were no longer subject to public review, some investigators reverted to the use of consent documents that the RAC had previously found to be inadequate. (One example of this can be found in RAC minutes at: http://www4.od.nih.gov/oba/dec96.htm; Item IX, Data Management.) Even in the OTC clinical trial, investigators could not account for the deletion of information on animal deaths from the consent documents.

There is a lot at stake. The NIH and the RAC itself have formed separate working groups to determine what appropriate procedures and standards there should be. Three areas are of prime concern, as listed below.

(i) Ensuring public participation

The lesson is clear from the growing public resistance to genetically modified foods: failure to guide gene therapy properly can lead to wide public disillusionment and even opposition. Decisions made behind closed doors create a public distrust that ultimately can obstruct entire lines of work. Unless there is public confidence in the way that those in authority make decisions regarding gene therapy research, there is the real risk that this area, despite its promise, can be similarly derailed.

The argument has been often raised that gene therapy research should not be subjected to a stiffer standard of oversight than other areas of frontier research. This view overlooks the fact that genetics has an unfortunate history. In the US, as elsewhere, people have been punished for their genes under the banner of eugenics movements (Paul, 1995). Moreover, possible *in utero*, germline and enhancement applications raise ethical and social dilemmas and require that whatever decisions are made be open to public view.

The RAC is still widely respected. It would appear essential that the RAC – with its well-established links to the public – be returned to a central role in the oversight process with its ability to veto protocols restored. The Internet can be a valuable tool in communicating the status of gene therapy research to public audiences.

(ii) Reporting procedures for adverse events

Labelling adverse events as proprietary information might make sense from a business standpoint, but it carries with it unacceptable consequences Quite simply, it puts at risk those generous human beings who are putting their bodies on the line on behalf of science. Knowing what has happened to others in the same or similar research is vital information that prospective subjects are entitled to have. To deny prospective subjects this information makes a mockery of achieving informed consent.

Furthermore, without full and accurate information on the status of clinical trials, it is impossible for the RAC or any other similar public group to make sound decisions or to identify potential dangers. Part of the purpose of any clinical trial is to look for patterns of unexpected outcomes and to explain them. What seems at first to be simply a disease-related event can, in conjunction with other similar events, provide enough data to reveal that certain treatments or practices may be harmful. It is hard to justify keeping such matters a secret when human lives are at stake.

Some researchers have offered the view that a publicly accessible database could compromise confidentiality. The current NIH Guidelines have anticipated this concern. Investigators are required to have appropriate procedures in place to protect the privacy of research subjects. Additionally, prospective subjects must be alerted to the fact that they may be sought out by the media. Serious adverse events are to be communicated with any identifying data removed.

These efforts seem to have worked. When individuals in gene therapy studies have become known, it has been by their own choice. Ashanti DeSilva, a subject in the first SCID-ADA experiment, has agreed to be featured in numerous articles. One day after the RAC discussed gene therapy for limb girdle muscular dystrophy in September 1999, a photograph of the first subject treated was a front page story (Friend, 1999). If investigators are not loathe to share their work when it can bring credit to them and their sponsors, it would appear hard to justify any refusal to share data when possible harms have occurred. The FDA and NIH are now working to clarify the reporting requirement so that adverse events are transmitted to both agencies. The safeguards already in place should be sufficient to provide the desired privacy protection.

(iii) Minimizing real or perceived conflicts of interest

The significant collaboration between private companies and university researchers engaged in gene therapy – and the various types of financial ties that have resulted – has raised substantial concern about possible conflicts of interest (DiStefano *et al.*, 2000). Such conflicts could arise:

- in the clinic when a prospective subject is invited to enter a gene therapy trial by an investigator with a financial interest in the outcome;
- when review boards, charged with looking out for the well-being of the research subjects, shift their focus to favour, instead, the well-being of the research mission of their institutions; and
- in the federal oversight process itself (the RAC is located administratively within the National Institutes of Health which has a strong commitment to human gene therapy research).

It is important to make certain that gene therapy research is untainted by any hint of conflicts of interest. This may mean adopting measures such as (1) involving outside individuals (with no financial or other links to the research programme) as resources for potential subjects during the informed-consent process – something Paul Gelsinger, Jesse's father, called for at the 2 February 2000 hearing; (2) enacting procedures that permit local and federal decision-making entities to be carefully scrutinized for their freedom from such conflicts; and (3) considering other administrative locales for the RAC.

This might well be a defining moment for human gene

therapy research not only in the US but also around the world. A credible partnership between the public and scientists is needed, not only to deal with the current challenges of the field but also to allow it to face those of the future when gene therapy experiments move into uncharted and contentious areas. Other countries may wish to consider the US experience as they evaluate or design their own systems of gene therapy oversight.

Acknowledgements

Drs Vivien Kane, LeRoy Walters and Jennifer Zallen generously gave assistance and advice in the preparation of this article. This article is reproduced from *Trends in Genetics*, June 2000, "US Gene therapy in crisis" by Doris Zallen, with permission from Elsevier Science.

REFERENCES

45 CFR (Code of Federal Regulations) Part 46 *Protection of Research Subjects*, Washington, DC: US Government Printing Office

Adams, C. (2000) NIH not told of bad events in gene trials. *Wall Street Journal* 10 January, section B, p.4

Blaese, R.M. (1990) Treatment of severe combined immune deficiency (SCID) due to adenosine deaminase (ADA) deficiency with autologous lymphocytes transduced with the human ADA gene. Office of Recombinant DNA Activities, protocol 9007–002. Bethesda, Maryland: US Department of Health and Human Services

Brown, G.E. Jr (1981) The policymaking challenge of the bioengineering industry. *Recombinant DNA Bulletin* 4: 121–23

Diamond v. Chakrabarty (1980) *U.S. Reports* 447: 303–22

Distefano, J.N., Collins, H. & Vedantam, S. (2000) Penn reviewing gene institute's ties to company. *Philadelphia Inquirer* 27 February, p.A1

Friend, T. (1999) Patient gets first muscular dystrophy gene therapy. *USA Today* 3 September, p.1

Marshall, E. (1995) Gene therapy's growing pains. *Science* 269: 1050–55

National Institutes of Health (1981) Report of the NIH Ad Hoc Committee on the UCLA Report concerning certain research activities of Dr Martin J. Cline

Office of Recombinant DNA Activities (1997) *Appendix M: The Points to Consider in the Design and Submission of Protocols for the Transfer of Recombinant DNA Molecules into One or More Human Subjects*, Bethesda, Maryland: US Department of Health and Human Services

Paul, D. (1995) *Controlling Human Heredity: 1865 to the Present*, Atlantic Highlands, New Jersey: Humanities Press

Randall, C., Mandelbaum, B. & Kelly, T. (1982) Letter from Three General Secretaries. In *Splicing Life: a Report on the Social and Ethical Issues of Genetic Engineering with Human Beings*, Washington DC: US Government Printing Office, pp.95–96

Wade, N. (1999) Death leads to concerns for future of gene therapy. *The New York Times* 30 September, p.22

Walters, L. & Palmer, J.G. (1997) *The Ethics of Human Gene Therapy*, Oxford and New York: Oxford University Press

Weiss, R. & Nelson, D. (1999) Patient, 18, dies at Penn during gene-therapy test. *Philadelphia Inquirer* 29 September, p.A1

Wright, S. (1994) *Molecular Politics: Developing American and British Regulatory Policy for Genetic Engineering, 1972–1982*, Chicago: University of Chicago Press

I HUMAN NON-CLINICAL GENETICS

Introduction to Human Non-Clinical Genetics

Most future research into human non-clinical genetics will result from knowledge of the complete human genome sequence. Ethical and legal aspects of the use of knowledge of the human genome will play an important role, as articles at the end of this section describe. The section begins however with human chromosome arrangements and cytogenetics, moves on to discuss possible mutations caused by the atomic bombs, and concludes with various articles on behavioural genetics. Research into human genetics is constantly advancing, and the following short summaries indicate some of the current results.

1. Mutations in human genes

Mutations provide the raw material for evolutionary progress and are also a concern when detected in the human genome as the cause of inherited diseases. A.S. Kondrashov (1995) introduced a novel problem with his article entitled "Contamination of the genomes by very slightly deleterious mutations. Why have we not died a hundred times over?" J.F. Crow (1997) then discussed whether the high human spontaneous mutation rate is actually a health risk. Further light is thrown on the human mutation problem by J.V. Neel's encyclopedia article below on the genetic effects of the atom bombs, which were, in fact, less than many people had expected.

Eyre-Walker and Keightley (1999) provided the first estimates of the deleterious mutation rates in humans, chimpanzees and gorillas, filling an important gap in our knowledge, though not without certain reservations they point out. They estimate that an average of 4.2 amino-acid-altering mutations per diploid per generation have occurred in the human lineage since humans separated from chimpanzees, and at least 38% of these were lost due to natural selection. They conclude that more than 1.6 new deleterious mutations occurred per diploid per generation, so the deleterious mutation rate specifically in protein-coding sequences is close to the upper limit tolerable by a species with as low a reproduction rate as humans.

2. Human genome sequence

The first human chromosome to be sequenced was chromosome 22 (Dunham *et al.*, 1999). In May 2000 *Nature* (Hattori *et al.*, 2000) reported the sequencing of chromosome 21. This is the smallest of all the human chromosomes, representing around 1–1.5% of the human genome, and became famous in 1959 when it was shown that the presence of an extra chromosome 21 was the cause of Down's syndrome. Its sequence has revealed 127 known genes, 98 predicted genes and 59 pseudogenes, all embedded in the 33.8 Mb of genomic DNA of this chromosome.

The two analyses support the conclusion that chromosome 22 is gene-rich while chromosome 21 is gene-poor. The two chromosomes contain 770 genes and represent about 2% of the human genome. The publication of the working draft of the full human genome sequence in 2001 is not a substitute for finished sequence: gaps will be closed and annotated information is continually being added.

3. Estimating the number of genes in the human genome

Three articles in the June 2000 issue of *Nature Genetics*, shortly after the draft human genome sequence was announced, provided widely varying estimates of human gene number. Two of these papers supported a number of approximately 30 000 genes, while the third shows evidence for 120 000. However, publication of the working draft sequence in February 2001 gives a much narrower range of estimates: 30–40 000 genes from the publicly funded project (International Human Genome Sequencing Consortium, 2001) and 26–38 000 genes from the private Celera project (Venter *et al.*, 2001).These estimates, though still far from exact, are just two to three times as large as the *c.*13 000 genes estimated for the fruit fly Drosophila, making it difficult to understand how the complexity of humans can be characterized by such a relative paucity of genes.

4. Recent research on Down's syndrome

Down's syndrome, while known to be a trisomy of chromosome 21, has a very variable phenotype, which could result from variations of both the environment during early growth and the genetic background, in addition to the 1.5–3-fold increase in levels of transcript and protein from the trisomy. This results in a constellation of more than 80 clinical traits, a subset of which is present in any given individual, though the subset differs widely between individuals. The Drosophila gene *minibrain* (*mnb*) has a close homologue in human and mouse, the homologue being within the critical region of the human chromosome 21. A recent review of the different theories to explain the basis of Down's syndrome is given by Reeves *et al.* (2001), and discusses recent research including the very variable Down syndrome, and the problems in accepting either theory as a satisfactory explanation.

REFERENCES

Crow, J.F. (1997) The high spontaneous mutation rate: is it a health risk? *Proceedings of the National Academy of Sciences USA*, 94: 8380–86

Dunham, I. *et al.* (1999) The DNA sequence of chromosome 22. *Nature* 402: 489–95

Eyre-Walker, A. & Keightley, P.D. (1999) High genomic deleterious mutation rates in hominoids. *Nature* 397: 344–47

International Human Genome Sequencing Consortium (2001) Initial sequencing and analysis of the human genome. *Nature* 409: 860–921

Hattori, M. *et al.* (2000) The DNA sequence of human chromosome 21. *Nature* 405: 311–19

Kondrashov, A.S. (1995) Contamination of the genomes by very slightly deleterious mutations. Why have we not died a hundred times over? *Journal of Theoretical Biology* 175: 483–594

Ouellette, F. (2001) Users must help to keep public databases correct. *Nature* 409: 452

Reeves, R.H., Baxter, L.L. & Richtsmeier, J.T. (2001) Too much of a good thing: mechanisms of gene action in Down syndrome. *Trends in Genetics* 17: 83–88

Venter, J.C. *et al.* (2001) The sequence of the human genome. *Science* 291: 1304–51

ANEUPLOIDY

Jeff Bond
Institute of Cell and Molecular Biology, University of Edinburgh, Edinburgh, UK

1. Introduction

Aneuploidy was originally defined as a condition in which a cell or an organism contains a chromosome number that is not a simple multiple of the haploid set (Täckholm, 1922). Most commonly, there will be one extra chromosome or one will be missing. Later, the definition was extended to include cells or organisms that have a duplication or deletion of a chromosome segment. This segmental aneuploidy can be important in determining which parts of particular chromosomes are significant determinants of genetically determined conditions – for example, Down's syndrome (see Section 3(ii)). Aneuploidy arises in the most part from an error at cell division. If one of the chromosomes fails to separate at cell division then the result will be two daughter cells each with an aneuploid chromosome number; one daughter will contain one chromosome more than normal and one will have one fewer. This non-disjunction event can occur at either mitosis (giving rise to a clone of aneuploid somatic cells) or at meiosis (giving rise to aneuploid gametes). It is the consequences of the latter with which this article is primarily concerned.

2. The frequency of human aneuploidy

Whole chromosome aneuploidy is an important feature of human genetics largely because it occurs so frequently. It is thought that about 25% of all conceptions might be aneuploid, although both higher and lower estimates exist in the literature (Bond & Chandley, 1983; Hassold *et al.*, 1996; Warburton & Kinney, 1996). It has proved difficult to estimate the frequency of aneuploidy accurately, because so many aneuploid conceptions abort spontaneously, many of them before the pregnancy is clinically recognized. About 0.3% of liveborn children are chromosomally abnormal and surveys of stillbirths show that 4% have an abnormal number of chromosomes. Among clinically recognized pregnancies, a large proportion (about 25%) of spontaneous miscarriages are chromosomally abnormal and many of these possess an extra chromosome, i.e. they are trisomic. There is no doubt that aneuploidy is responsible for a large percentage of spontaneous abortions. In most cases, trisomy is incompatible with a viable existence and spontaneous miscarriages are a "goldmine" in the study of aneuploidy, yielding much valuable cytogenetic material for the investigation of this phenomenon. Although the frequency of aneuploidy is not accurately known, it is clear that in humans it is exceptionally high (Bond & Chandley, 1983). No other species investigated to date has anything like the frequency recorded for our species, although there is a need to investigate other non-human primates to see whether our high frequency is peculiar to us or is a feature of primates in general. It is evident that the frequency in well-studied experimental organisms is much lower. For example, in the mouse, aneuploidy occurs with about one-tenth of the human frequency (Chandley & Speed, 1979; Brook *et al.*, 1984).

Originally, the frequency of aneuploidy was estimated from cytogenetic studies on spontaneous miscarriages and estimates of the real underlying frequency were attempted by extrapolating from these measurements. Techniques have now been developed to measure the frequency of aneuploidy directly in the gametes. The first reliable technique involved fusing human sperm with Chinese hamster oocytes after which the number of human chromosomes in the sperm could be visualized (Rudak *et al.*, 1978; Martin *et al.*, 1991). However, the number of sperm analysed was limited by the technical difficulty of the procedure and, lately, greater progress has been made by using fluorescent molecular probes to analyse the chromosomes indirectly. Sperm with an abnormal chromosome number can be detected because they possess an extra fluorescent spot corresponding to one extra chromosome. Fluorescent probes, which are specific for each chromosome, have been developed and so the frequency of extra chromosomes can be estimated separately for different chromosomes (Jacobs, 1992; Spriggs *et al.*, 1996). Although originally, and long before the development of the modern probes, the reliability of techniques which measured aneuploidy directly in the gametes was questioned, the modern molecular methods are very specific for each chromosome and, consequently, the reliability of the data is not in doubt (Martin & Rademaker, 1995).

The frequency of sperm with an abnormal chromosome number ranges, in different surveys, from 2% to 6% and this range agrees very well with estimates obtained from the Chinese hamster technique. Data obtained using fluorescent probes have revealed that the frequency of aneuploidy for chromosome 21 and for the sex chromosomes is higher than for other chromosomes (Spriggs *et al.*, 1996). Several groups have reported this effect for the sex chromosomes and it might be that the X and Y chromosomes are more susceptible to error leading to aneuploidy in male meiosis.

The *in vitro* fertilization programme also provides some data on the frequency of chromosome abnormality before implantation. Examination of the chromosomes of oocytes recovered during *in vitro* fertilization procedures reveals that about 20% are aneuploid, with about 8–9% having an extra chromosome (Eichenlaub-Ritter, 1996). A note of caution is needed when interpreting these data, however, because the donors of the eggs for the programme are not necessarily a representative sample of women in general. For example, they tend to be older than average and this may well be significant given the maternal age effect (see Section 5). Martin *et al.* (1991) compared aneuploid frequency found in sperm and oocytes and showed that, for individual chromosomes, chromosome 21 appears to be prone to aneuploidy in the gametes of both sexes.

3. Liveborn aneuploid conditions

Although the majority of aneuploid conceptions abort spontaneously, some are compatible with a viable existence and are born live. Of these, aneuploidy for the chromosomes associated with sex determination is more frequent than for the other chromosomes (the autosomes). Overall, about two individuals in 1000 are born with an abnormal number of sex chromosomes.

(i) Sex chromosome aneuploids

Turner's syndrome. Individuals who have one chromosome fewer than the normal number are said to be monosomic. In humans, only monosomy for the X chromosome is viable, leading to the development of Turner's syndrome (Ford *et al.*, 1959). The majority of X monosomy (written XO) conceptions miscarry and only about 1% survive to birth. Of newborn females, about one in 5000 are XO. They exhibit characteristic phenotypic features – short stature, rudimentary sexual development with no functional ovaries, a webbed neck and congenital heart defects. Although every somatic cell in normal females has one X chromosome inactivated, the abnormal development of females who have only one X chromosome indicates that the possession of two functional X chromosomes is necessary, at some point, for the development of full female characteristics.

An interesting feature of XO development in humans is the contrast with the equivalent condition in the mouse. XO mice do not show the high level of prenatal loss and develop more normally. It has been suggested that this difference results from a difference in X inactivation pattern between mice and humans. For normal development in humans it is necessary for an individual (both male and female) to have two functional copies of particular genes. At least one of these genes is thought to reside on the part of the X and Y chromosome which have common homology (the pseudoautosomal region). Recent molecular analysis has identified, at least tentatively, a gene in this region which is responsible for the short stature characteristic of Turner's syndrome (Rao *et al.*, 1997).

Another report (Skuse *et al.*, 1997) has analysed the behavioural characteristics of Turner's syndrome by analysing, separately, the behaviour of Turner's females whose X chromosome was paternally derived and those whose X came from the mother (see article on "Genomic imprinting"). It was claimed that there were significant differences between the two sets, which could be accounted for by postulating a gene on the X chromosome which was inactivated on the maternal chromosome and which influenced behaviour.

Klinefelter's syndrome. An individual with an XXY chromosome constitution develops Klinefelter's syndrome. Those affected develop as males but develop female-like secondary sexual characteristics. In 1959, Jacobs and Strong showed that the underlying basis of this syndrome was an XXY chromosome constitution. Most affected individuals are sterile and develop juvenile internal genitalia, about 50% show some degree of breast development and they are taller than the average XY male. About two in every 1000 newborn boys are XXY.

XYY. Males with an extra Y chromosome made prominent headlines when it was claimed that the presence of this extra chromosome was correlated with antisocial, aggressive behaviour. The possibility that the extra Y chromosome adversely affected behaviour was raised through the discovery that seven out of 197 males in a maximum-security prison were XYY (Jacobs *et al.*, 1965, 1968). Subsequent work tended to confirm that there was a high incidence of men with an XYY karyotype in mental penal institutions, but surveys also revealed that XYY individuals were much more common than had been previously thought – about one male in 1000 is of XYY karyotype. It follows that the majority of XYY men are law-abiding individuals without a history of mental illness. The question of possible disturbed behaviour of some XYY men remains unresolved and, until there are long-term studies of XYY men, the possibility that there is an influence of the Y chromosome on behaviour will remain an unanswered question.

At one time, there was a proposal to carry a survey of newborn males with the intention of identifying XYY individuals at birth and then studying their subsequent development. In the US, this proposal led to widespread opposition and was finally abandoned; among other things, the risk of self-fulfilling prophecy was deemed to be too great. It is interesting to note that attempts to use the presence of an extra Y chromosome as a mitigating circumstance in courts of law have been made, but have not been accepted. It is also worth making another point about the labelling of particular genetic conditions as predisposing factors towards a specific pattern of behaviour. It is well established that XYYs tend to be taller than average and it would not, therefore, be at all surprising if they were found with increased incidence within any body of men selected for height, such as the police force. Had the discovery of XYY karyotype been made within this body of men then, no doubt, it would be associated with characteristics of virtue such as the desire to uphold law and order and to serve the public. As far as the author knows, no survey of incidence within the police force of any country has, however, been carried out.

An important issue, of some concern to XYY men, is the frequency with which their sperm are disomic for the Y chromosome. There is no evidence that the sons of XYY men are themselves more likely to be XYY or XXY. This might be expected if XYY fathers had an elevated frequency of abnormal sperm. A study of semen samples from XYY men revealed a normal ratio of X- and Y-bearing sperm (Han *et al.*, 1994; Martini *et al.*, 1996). It looks as though the extra Y chromosome is eliminated in the early stages of spermatogenesis so that the meiotic cells giving rise to the sperm are diploid.

(ii) Autosomal trisomics

Only three autosomal trisomic conditions are compatible with a viable existence and two of these do not survive for long after birth.

Trisomy 21 – Down's syndrome. This is the best known

of the autosomal trisomic conditions. The discovery in 1959 that the syndrome of abnormalities was associated with the presence of an extra chromosome was the first documented case of a human chromosome abnormality (Lejeune *et al.*, 1959). Trisomy 21 leads to Down's syndrome and affected individuals have a characteristic appearance with a distinctive fold of the skin called the epicanthic fold and a protruding tongue, which causes the mouth to remain partially open. Growth, behaviour and mental development are all retarded; many cases exhibit defects of the heart. Modern health care has improved survival quite dramatically, although life expectancy is still shortened. There is an increased likelihood of leukaemia and older Down's syndrome individuals are very likely to exhibit characteristics of Alzheimer's disease.

Trisomy 21 is the most common autosomal trisomy and occurs in about one in 1000 live births. About 96% of cases in England and Wales have a free chromosome 21 and arise *de novo*, but 4% are translocation cases in which chromosome 21 material has been moved to another chromosome, often chromosome 14. In the case of translocation trisomy 21, the condition can be passed on from one generation to another and does not arise from a *de novo* non-disjunction event. Translocation Down's cases have proved very valuable in identifying the regions of chromosome 21 that are important in bringing about the typical features of trisomy 21. The short arm of chromosome 21 does not contribute to the phenotype of trisomy 21; the critical region turns out to be on the long arm of chromosome 21. The classical features of Down's syndrome including the characteristic appearance, the heart defects and some of the mental retardation are caused by three doses of genes in the Down Syndrome Critical Region between *D21S17* and *ETS2* (see Antonarakis, 1998, for a review). This is a region of about 4 megabases towards the tip of the long arm. However, other regions of this chromosome arm are not irrelevant to the syndrome and some genes in the more proximal region of the arm contribute to the mental retardation.

Information of some importance is promised, too, from the study of mouse aneuploidy. Normal mice have 20 pairs of chromosomes. Mice that are trisomic for any autosome are severely affected. Most are lethal early in embryonic development and only trisomy 19 survives for any significant period after birth. Trisomy 16 in the mouse is of interest because this chromosome shares a great deal of genetic material with the human chromosome 21. Mice that are trisomic for chromosome 16 survive, at best, for only a few hours after birth and these are not, therefore, a very good animal model for the study of Down's syndrome. Two mouse strains have been developed that are partially trisomic for chromosome 16 and these promise to be useful. Ts65Dn mice are aneuploid ($2n = 41$) and have an extra chromosome consisting of the centromere of chromosome 17 and a portion of chromosome 16 (Davisson *et al.*, 1993). Another strain, Ts1Cje (Sago *et al.*, 1998), has a normal chromosome number but one of the chromosomes is a chromosome translocation in which a piece of chromosome 16 has been moved into chromosome 12. The mice are, therefore, partial triploids for a part of chromosome 16. Both these strains have learning and behavioural deficiencies (Reeves *et al.*, 1995). The significance of this for the study and identification of the genes responsible for Down's syndrome in humans is obvious.

Trisomy 13 – Patau's syndrome. Trisomy 13, resulting in gross abnormalities of the mouth, eyes, limb extremities, heart and the brain, was described by Patau *et al.* in 1960. Life expectancy of affected individuals is measured in months, with half dying within the first month. Trisomy 13 is rare and occurs in one in 20 000 live births.

Trisomy 18 – Edwards' syndrome. Infants with trisomy 18 are severely affected and live for only a few months. Characteristic features are clenched fists and "rocker-bottom" feet with protruding heels. The extra chromosome responsible was detected by Edwards *et al.* in 1960. Severe heart defects are the usual cause of death. Trisomy 18 occurs in one in 8000 live births.

4. The origin of aneuploidy

The most obvious possible cause of aneuploidy following meiosis is non-disjunction, when homologous chromosomes fail to separate. Sturtevant and Beadle (1939) introduced the term "non-conjunction", to distinguish aneuploidy (arising from a failure of homologous chromosomes to pair) from non-disjunction in the strict sense of failure to separate. This distinction has not gained widespread acceptance, possibly because, in practice, it is often difficult to distinguish the two. Failure of chromosomes to separate can arise either at the first or the second division of meiosis and can be distinguished by the use of genetic markers situated at the centromere. In addition, however, precocious division of a centromere during meiosis I can result in misdivision and aneuploidy. Polani and Jagiello (1976) called this process "pre-division" while Hansmann and El-Nahass (1979) used the term "pre-segregation". More recently, Angell and her co-workers (Angell, 1991; Angell *et al.*, 1994) observed meiosis II oocytes with 23 chromosomes and an extra chromatid. These could have arisen through premature division of the centromere and led the authors to suggest that this might be an important origin of aneuploid gametes. In a fungal test system, in which it is possible to distinguish different origins of aneuploidy, Fulton and Bond (1983) concluded that about 10% of aneuploid products arose by premature centromere division. The relevant figure for human meiosis has still to be established unambiguously, however, because there is some increase in the frequency of the phenomenon while oocytes are cultured in the laboratory.

(i) Tracking down the parental origin of the aneuploidy

The extra chromosome present in aneuploid individuals can arise from either the mother or father and, assuming that the extra chromosome arises at meiosis during the formation of the gametes, there can be an error at either the first or second meiotic division (**Figure 1**). The development of molecular cloning techniques and the identification of DNA polymorphisms have allowed the origin of the extra chromosome to be traced (Zaragosa *et al.*, 1994; Fisher *et al.*, 1995; Eggermann *et al.*, 1996).

As far as human aneuploidy is concerned, the conclusions are not straightforward. For most aneuploid conditions, the error occurs at the first division of the maternal meiosis. Thus, the extra chromosome in trisomies 13, 14, 15, 21 and 22 comes from the mother about 90–95% of the time; trisomy 16 is the most extreme, with almost 100% resulting from maternal meiotic error. In contrast, the extra chromosome in XXY individuals is donated by the father in about 50% of cases. Obviously, XYY males all derive the extra Y from the father.

There is also some variation in the stage of meiosis at which the error occurs. As shown in **Figure 1**, if non-disjunction of a particular chromosome takes place at the first division of meiosis, this can be distinguished from a similar error at the second division by the use of genetic markers located near the centromere of the chromosome concerned. Any heterozygous marker will continue to be so following a first division meiotic error, but will become homozygous if an error has occurred at the second division. In trisomy 16 individuals, the error seems to occur almost invariably at the first division of meiosis, while, in trisomy 21, a mistake at the second meiotic division is responsible about 30% of the time. In the case of trisomy 18, on the other hand, 68% of investigated cases arose from an error at the second meiotic division. No explanation for these differences has been convincingly advocated, but one obvious conclusion to be drawn is that the underlying causes of aneuploidy vary and that there is unlikely to be one all-embracing explanation for the high frequency (Hassold *et al.*, 1996; Nicolaidis & Petersen, 1998; Ballesta *et al.*, 1999).

Molecular analysis of the X chromosome in individuals with Turner's syndrome reveals that in the majority of cases the paternal chromosome is missing (Sanger *et al.*, 1977; Hassold *et al.*, 1992). However, the explanation in this case

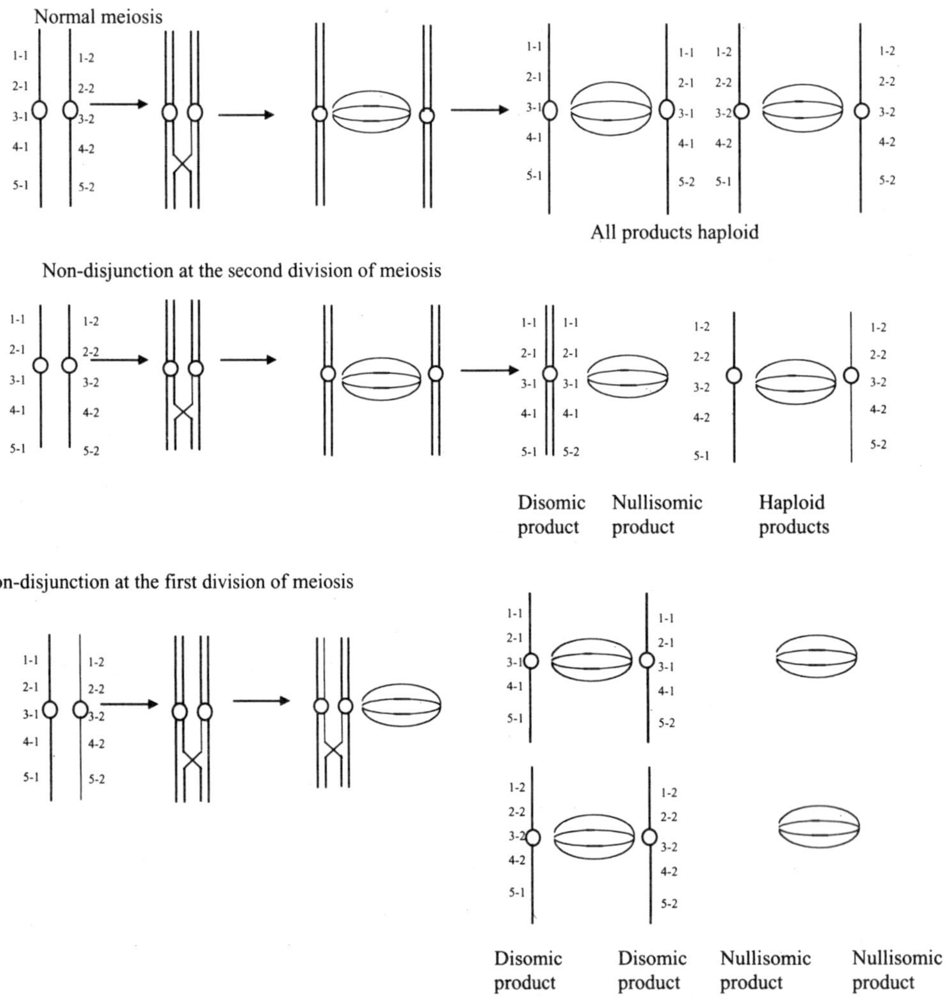

Figure 1 The use of genetic markers to trace the origin on aneuploid meiotic products. A chromosome pair is illustrated as marked at five loci along its length. One of the markers 3–1, and its allele 3–2, is located at the centromere. A single crossover event is illustrated between markers 4 and 5. If non-disjunction occurs at the second division of meiosis, the disomic product is homozygous for one of the centromere-located alleles. Non-disjunction at the first division of meiosis results in a disomic product which is heterozygous for the centromere marker. The precocious division of a centromere at the first division of meiosis can also result in a disomic product, but this is not illustrated because it results in heterozygosity of the centromere marker and is indistinguishable from non-disjunction at the first division of meiosis using genetic markers.

is likely to be different from the other aneuploid conditions. There is evidence to suggest that the majority of Turner's cases arise not from meiotic errors but from postfertilization mitotic loss of one of the sex chromosomes from a chromosomally normal conception. It is possible, but not proven, that the excess of Turner's syndrome individuals with a maternal, but no paternal, X chromosome comes about through the loss of the Y chromosome from a conception that would, without the lost chromosome, have resulted in a male.

(ii) Tracking the molecular defects at meiosis

The presence of molecular markers along the relevant chromosome has allowed some analysis of the meiotic process giving rise to aneuploid gametes. In particular, the process of crossing-over, which recombines the two homologous chromosomes, has been compared in normal meiosis and in those meioses giving rise to aneuploidy. The results are intriguing. Recombination is clearly abnormal in the latter case, but the abnormalities are not simple; again, this probably reflects the fact that there is more than one cellular defect giving rise to aneuploidy. It has been suspected for a long time that successful meiosis depends on the presence of chiasmata along each bivalent. One simple idea would be that aneuploid gametes arise from those meioses in which there is a failure of chromosomes to pair, in which case recombination by crossing-over would be completely absent. This turns out not to happen in most cases, although it does seem to happen occasionally. In the case of trisomy 21, which is the most extensively studied to date, there is some reduction in the frequency of crossing-over, with 40% of cases having no detectable crossing-over (nullichiasmate meiosis). However, an altered distribution of crossing-over has also been observed when aneuploid-generating meioses are compared with normal ones. This leads to the conclusion that the position of chiasmata along the chromosome is also important in the successful error-free completion of meiosis.

The importance of the correct positioning of chiasmata is also inferred for chromosome 16 because trisomy 16 individuals have a marked reduction in chiasmata near the centromere of the chromosome. For trisomy 18, on the other hand, there is also an increased frequency of nullichiasmate meioses but no indication that there is any alteration to the distribution of chiasmata when they do occur. For trisomy 15, an estimated 21% of meioses leading to aneuploidy after first division failure are nullichiasmate. To further complicate this question, XXY arising as a consequence of error in female meiosis shows both an increase in meioses with no chiasmata and an increase in meioses with a larger number. This leads to the conclusion that failure to segregate properly can arise from either no chiasmata or from entanglement of chromosomes when there are several (Sherman *et al.*, 1991; MacDonald *et al.*, 1994; Hassold *et al.*, 1995; Nicolaidis & Petersen, 1998; Robinson *et al.*, 1998; Savage *et al.*, 1998).

5. The maternal age effect

It is a very old observation that a Down's syndrome child was likely to be the youngest in the family. It is now very well established that older mothers are at much higher risk of carrying a foetus with aneuploidy. Rather surprisingly, however, the underlying cause of this maternal age effect is not known. There have been many theories put forward to explain it, but none has gained universal acceptance. Most centre on the very different meiotic cycle of males and females. In the female, meiosis starts before birth, but each meiosis is halted during prophase of the first meiotic division at the dictyate stage and re-starts at ovulation when the rest of the meiotic cycle is rapidly completed. It is an obvious speculation that the delay in completing meiosis is the factor responsible for the age effect. It is much less obvious which aspect(s) of meiosis becomes defective as a woman gets older.

Several things are known and have to be accommodated in any explanation of why older mothers are at greater risk. The maternal age effect is not confined to trisomy 21; most, if not all, human trisomies show an age-related increase. In the case of trisomy 21 and trisomy 18, both first and second divisions of meiosis appear to be affected by increasing maternal age. On the other hand, in sex chromosome aneuploidy and in trisomy 16, the increased frequency in older mothers is confined to aneuploidy arising at the first division of meiosis. The altered recombination found in the chromosomes of aneuploid individuals noted above is most pronounced in older mothers, at least for sex chromosome aneuploidy and for trisomy 16. For chromosome 21 of older women, a decrease in the number of crossover events in the telomeric region has been recorded, but this reduction was less marked in the centromeric region (Tanzi *et al.*, 1992; Sherman *et al.*, 1994; Hassold *et al.*, 1995).

The development of molecular techniques, which identify the parent and stage of meiosis responsible for the aneuploidy, has eliminated some ideas, but others remain as viable explanations. For example, the idea that the uterine environment in older women is more "forgiving" of aneuploidy and consequently fewer chromosomally abnormal conceptions miscarry in older mothers – the "relaxed selection" model – is discounted by the observation that aneuploidy of paternal origin is not increased in frequency. One of the most intensively studied ideas is the "production-line" hypothesis (Henderson & Edwards, 1968). Meiosis starts in the female foetus before birth, but is completed only at ovulation. According to the production-line hypothesis, oocytes which are the first to enter meiosis in the female foetus are the first to be ovulated more than a decade later and oocytes which are the last to commence meiosis are the last to be ovulated some 50 years later. It was originally suggested that a decrease in the frequency of recombination accompanied this production line and resulted in increased meiotic errors. There is now evidence that recombination is altered in older women. However, the observation that other parts of the meiotic cycle also become defective in older women has led to a modification of this hypothesis. It has been observed that oocytes in the prenatal ovary show a high frequency of pairing abnormalities, but most of these oocytes appear to degenerate before birth. This leads to the idea that there is a "surveillance" mechanism that eliminates aberrant oocytes in the female foetus however they arise and that this becomes less efficient over time (or the oocytes entering meiosis last are surveyed for a much shorter time). It is very likely that oocytes in older women have an increased frequency of aneuploidy from a variety of disturbances of meiosis (see Eichenlaub-Ritter, 1996, for an extensive review).

6. The paternal age effect

An obvious question, given the pronounced maternal age effect discussed above, is whether aneuploidy also increases with paternal age. Some groups have consistently claimed the existence of a paternal effect (notably Stene & Stene, 1977, 1981) while others regard this as controversial or non-existent (Carothers, 1988; Hook *et al.*, 1990). Although epidemiological studies have failed to detect any paternal age effect, it is possible that a small effect might exist which is swamped by the much larger and more obvious maternal effect. Direct observation on sperm that compares aneuploidy in men of different ages might resolve this question and, indeed, some results do exist in the literature. Some have reported more aneuploid sperm in older men compared with younger ones (Wyrobek *et al.*, 1994), but there is a fairly large inter-individual variation which makes a definitive conclusion difficult. It is clear, however, that any paternal age effect is small compared with the effect of advancing maternal age.

7. Uniparental disomy

A predictable outcome of the high frequency of chromosomal error in human meiosis is that the probability of simultaneous but complementary error occurring in both the meiosis giving rise to the sperm and to the egg is not vanishingly rare. Should a gamete lacking a particular chromosome and a gamete with two copies of that same chromosome fuse, the result will be a zygote with the correct number of chromosomes, but with one parent contributing both copies and the other none. This process has been called uniparental disomy (UPD) (Engel, 1980). The first example was reported by Spence *et al.* (1988) when they investigated a female with cystic fibrosis and short stature. Analysis of DNA markers on chromosome 7 failed to reveal any paternal-specific alleles and the conclusion was reached that the female in question had inherited two copies of the maternal chromosome 7 and none from the father. Since this initial report, several other instances have come to light.

UPD can lead to genetic disease if it occurs for a chromosome for which a contribution from one particular parent is important (see article on "Genomic imprinting"). Regions of chromosome 15 have been identified as important in this respect (Knoll *et al.*, 1989; Nicholls *et al.*, 1989), with UPD leading to Prader–Willi syndrome when both chromosomes are maternal (affected children have neonatal hypotonia, short stature, obesity and some mental retardation) and Angelman syndrome when both are paternal (affected children have severe motor and intellectual retardation, facies with a large mandible, wide mouth and protruding tongue, and an ataxic gait; the jerky movements and inappropriate outbursts of laughter give rise to the alternative name of "happy puppet" syndrome). It is estimated that about 25% of cases of Prader–Willi syndrome are caused by UPD.

The genetic disease results from genetic imprinting where some chromosome regions are silenced. In the case of Prader–Willi syndrome, a maternal gene (or genes) is not expressed due to imprinting and the only source of functional gene product comes from the paternal chromosome. If the father's contribution is missing (25% of the time through the occurrence of UPD; the remainder by mutation of the paternal gene), the genetic disease will result. The observation that Prader–Willi syndrome arising from UPD has no more severe phenotype than that arising from deletion suggests that the imprinting effect on chromosome 15 is confined to one small area of the chromosome. Conversely, in Angelman syndrome, expression of a maternally derived gene is important and the disease can arise from UPD resulting in the absence of the critical maternal gene.

REFERENCES

Angell, R.R. (1991) Predivision in human oocytes at meiosis I: a mechanism for trisomy formation in man. *Human Genetics* 86: 383–87

Angell, R.R., Xian, J., Keith, J., Ledger, W. & Baird, D.T. (1994) First meiotic division abnormalities in human oocytes: mechanism of trisomy formation. *Cytogenetics and Cell Genetics* 65: 194–202

Antonarakis, S.E. (1998) 10 years of *Genomics*, chromosome 21 and Down syndrome. *Genomics* 51: 1–16

Ballesta, F., Queralt, R., Gomez, D. *et al.* (1999) Parental origin and meiotic stage of non-disjunction in 139 cases of trisomy 21. *Annales de Genetique* 42: 11–15

Bond, D.J. & Chandley, A.C. (1983) *Aneuploidy*, Oxford and New York: Oxford University Press

Brook, J.D., Gosden, R.G. & Chandley, A.C. (1984) Maternal age and aneuploid embryos: evidence from the mouse that biological and not chronological age is the important influence. *Human Genetics* 66: 41–45

Carothers, A.D. (1988) Controversy concerning paternal age effect in 47,+21 Down's syndrome. *Human Genetics* 78: 384–85

Chandley, A.C. & Speed, R.M. (1979) Testing for nondisjunction in the mouse. *Environmental Health Perspectives* 31: 123–29

Davisson, M.T., Schmidt, C., Reeves, R.H. *et al.* (1993) Segmental trisomy as a mouse model for Down syndrome. *Progress in Clinical Biology Research* 384: 117–33

Edwards, J.H., Harnden, D.G., Cameron, A.H. *et al.* (1960) A new trisomic syndrome. *Lancet* 1: 787–90

Eggermann, T., Nöthe, M.M., Eiben, B. *et al.* (1996) Trisomy of human chromosome 18: molecular studies on parental origin and cell stage of nondisjunction. *Human Genetics* 97: 218–23

Eichenlaub-Ritter, U. (1996) Parental age-related aneuploidy in human germ cells and offspring: a story of past and present. *Environmental and Molecular Mutagenesis* 28: 211–36

Engel, E. (1980) A new genetic concept: uniparental disomy and its potential effect, isodisomy. *American Journal of Medical Genetics* 6: 137–43

Fisher, J.M., Harvey, J.F., Morton, N.E. & Jacobs, P.A. (1995) Trisomy 18: studies of the parent and cell division origin and

the effect of aberrant recombination on nondisjunction. *American Journal of Human Genetics* 56: 669–75
Ford, C.E., Miller, O.J., Polani, P.E., Almeida, J.C. de & Briggs, J.H. (1959) A sex chromosome anomaly in a case of gonadal dysgenesis (Turner's syndrome) *Lancet* 1: 711–13
Fulton, A.M. & Bond, D.J. (1983) An investigation into the origins of meiotic aneuploidy using ascus analysis. *Genetical Research* 41: 165–79
Han, T.L., Ford, J.H., Flaherty, S.P., Webb, G.C. & Matthews, C.D. (1994) A fluorescent *in situ* hybridization analysis of the chromosome constitution of ejaculated sperm in a 47,XYY male. *Clinical Genetics* 45: 67–70
Hansmann, I. & El-Nahass, E. (1979) Incidence of nondisjunction in mouse oocytes. *Cytogenetics and Cell Genetics* 24: 115–21
Hassold, T., Pettay, D., Robinson, A. & Uchida, I. (1992) Molecular studies of parental origin and mosaicism in 45,X conceptuses. *Human Genetics* 89: 647–52
Hassold, T., Merrill, M., Adkins, K., Freeman, S. & Sherman, S. (1995) Recombination and maternal-age dependent non-disjunction: molecular studies of trisomy 16. *American Journal of Human Genetics* 57: 867–74
Hassold, T., Abruzzo, M., Adkins, K. *et al.* (1996) Human aneuploidy: incidence, origin and etiology. *Environmental and Molecular Mutagenesis* 28: 167–75
Henderson, S.A. & Edwards, R.G. (1968) Chiasma frequency and maternal age in mammals. *Nature* 218: 22–28
Hook, E.B., Cross, P.K. & Regal, R.R. (1990) Factual, statistical and logical issues in the search for a paternal age effect for Down syndrome. *Human Genetics* 85: 387–88
Jacobs, P.A. (1992) The chromosome complement of human gametes. *Oxford Reviews in Reproductive Biology* 14: 48–72
Jacobs, P.A. & Strong, J.A. (1959) A case of human intersexuality having a possible XXY sex determining mechanism. *Nature* 183: 302–03
Jacobs, P.A., Brittain, R.P. & McClermont, W.F. (1965) Aggressive behaviour, mental subnormality and the XYY male. *Nature* 208: 1351
Jacobs, P.A., Price, W.H. & Court Brown, W.M. (1968) Chromosome studies on men in a maximum security hospital. *Annals of Human Genetics* 31: 339–47
Knoll, J.H.M., Nicholls, R.D., Magenis, R.E. *et al.* (1989) Angelman and Prader–Willi syndrome share a common chromosome 15 deletion but differ in parental origin of the deletion. *American Journal of Medical Genetics* 32: 285–90
Lejeune, J., Gautier, M. & Turpin, M.R. (1959) Étude des chromosomes somatiques de neuf enfants Mongoliens. *Comptes Rendus de l'Academie des Sciences* 248: 1721–22
MacDonald, M., Hassold, T., Harvey, J. *et al.* (1994) The origin of 47,XXY and 47,XXX aneuploidy: heterogeneous mechanisms and the role of aberrant recombination. *Human Molecular Genetics* 3: 1365–71
Martin, R.H. & Rademaker, A. (1995) Reliability of aneuploidy estimates in human sperm: results of fluorescence *in situ* hybridization studies using two different scoring criteria. *Molecular Reproduction and Development* 42: 89–93
Martin, R.H., Ko, E. & Rademaker, A. (1991) Distribution of aneuploidy in human gametes: comparison between human sperm and oocytes. *American Journal of Medical Genetics* 39: 321–31
Martini, E., Geraedts, J.P.M., Liebaers, I. *et al.* (1996) Constitution of semen samples from XYY and XXY males as analysed by *in situ* hybridization. *Human Reproduction* 11: 1638–43
Nicholls, R.D., Knoll, R.A., Butler, M.G., Karam, S. & Lalande, M. (1989) Genetic imprinting suggested by maternal heterodisomy in nondeletion Prader–Willi syndrome. *Nature* 342: 281–85
Nicolaidis, P. & Petersen, M.B. (1998) Origin and mechanisms of non-disjunction in human autosomal trisomies. *Human Reproduction* 13: 313–19
Patau, K.A., Smith, D.W., Therman, E.M., Inhorn, S.L. & Wagner, H.P. (1960) Multiple congenital abnormality caused by an extra autosome. *Lancet* 1: 790–93
Polani, P.E. & Jagiello, G.M. (1976) Chiasmata, meiotic univalents and age in relationship to aneuploid imbalance in mice. *Cytogenetics and Cell Genetics* 16: 505–29
Rao, E., Weiss, B., Fukami, M. *et al.* (1997) Pseudoautosomal deletions encompassing a novel homeobox gene cause growth failure in idiosympathic short stature and Turner syndrome. *Nature Genetics* 16: 54–63
Reeves, R.H., Irving, N.G., Moran, T.H. *et al.* (1995) A mouse model for Down syndrome exhibits learning and behaviour deficits. *Nature Genetics* 11: 177–84
Robinson, W.P., Kuchinka, B.D., Bernasconi, F. *et al.* (1998) Maternal meiosis I non-disjunction of chromosome 15: dependence of the maternal age effect on level of recombination. *Human Molecular Genetics* 7: 1011–19
Rudak, E., Jacobs, P.A. & Yanagamachi, R. (1978) Direct analysis of the chromosome constitution of human spermatozoa. *Nature* 274: 911–13
Sago, H., Carlson, E.J., Smith, D.J. *et al.* (1998) Ts1Cje, a partial trisomy 16 mouse model for Down syndrome, exhibits learning and behavioral abnormalities. *Proceedings of the National Academy of Sciences USA* 95: 6256–61
Sanger, R., Tippett, P., Gavin, J., Teesdale, P. & Daniels, G.L. (1977) Xg groups and sex chromosome abnormalities in people of northern European ancestry: an addendum. *Journal of Medical Genetics* 14: 210–13
Savage, A.R., Petersen, M.B., Pettay, D. *et al.* (1998) Elucidating the mechanism of paternal non-disjunction of chromosome 21 in humans. *Human Molecular Genetics* 7: 1221–27
Sherman, S.L., Takaesu, N., Freeman, S.B. *et al.* (1991) Trisomy 21: association between reduced recombination and non-disjunction. *American Journal of Human Genetics* 49: 608–20
Sherman, S.L., Petersen, M.B., Freeman, S.B. *et al.* (1994) Non disjunction of chromosome in maternal meiosis I: evidence for a maternal-age dependent mechanism involving reduced recombination. *Human Molecular Genetics* 3: 1529–35
Skuse, D.H., James, R.S., Bishop, D.V.M. *et al.* (1997) Evidence from Turner's syndrome of an imprinted X-linked locus affecting cognitive function. *Nature* 387: 705–08
Spence, J.E., Perciaccante, R.G., Greig, G.M. *et al.* (1988) Uniparental disomy as a mechanism for human genetic disease. *American Journal of Human Genetics* 42: 217–26
Spriggs, E.L., Rademaker, A.W. & Martin, R.H. (1996) Aneuploidy in human sperm: the use of multicolor FISH to test various theories of nondisjunction. *American Journal of Human Genetics* 58: 356–62
Stene, J. & Stene, E. (1977) Statistical methods for detecting a

moderate paternal age effect on incidence of disorder when a maternal one is present. *Annals of Human Genetics* 40: 343–53

Stene, J. & Stene, E. (1981) Paternal age and Down's syndrome. *Human Genetics* 59: 119–24

Sturtevant, A.H. & Beadle, G.W. (1939) *An Introduction to Genetics*, Philadelphia and London: Sanders

Täckholm, G. (1922) Zytologische Studien über die Gattung Rosa. *Acta Horti Bergiani* 7: 97–381

Tanzi, R.E., Watkins, P.C., Stewart, G.D. *et al.* (1992) A genetic linkage map of human chromosome 21: analysis of recombination as a function of sex and age. *American Journal of Human Genetics* 50: 551–58

Warburton, D.A. & Kinney, A. (1996) Chromosomal differences in susceptibility to meiotic aneuploidy. *Environmental and Molecular Mutagenesis* 28: 237–47

Wyrobek, A.J., Lowe, X., Holland, N.T. & Robbins, W. (1994) Paternal-age effects on sperm aneuploidy investigated in mice and humans by three-chromosome fluorescence *in situ* hybridization. *American Journal of Human Genetics* 55 (Suppl.): 698

Zaragosa, M.V., Jacobs, P.A., James, R.S. *et al.* (1994) Non-disjunction of human acrocentric chromosomes: studies of 432 trisomic fetuses and liveborns. *Human Genetics* 94: 411–17

GLOSSARY

bivalents structures composed of four double-stranded DNA molecules generated in meiosis from two homologous chromosomes that have replicated and paired. Each bivalent is usually held together by chiasmata

dictyate stage the stage during diplotene when meiosis is arrested, characteristic of female meiosis in mammals. In humans, meiosis in the female starts at about the fourth month of gestation; after arrest in the dictyate stage, meiosis resumes at ovulation

disomic an individual or cell with an abnormal number of chromosomes in which one particular chromosome is present twice

genetic imprinting parent-specific expression (or repression) of genes or chromosomes in offspring

non-disjunction failure of separation of the two chromosomes of a pair during production of the egg or sperm at meiosis

nullichiasmate meiosis a meiosis in which there are no chiasmata. May refer to one pair of homologous chromosomes in which chiasmata formation fails to occur

pseudoautosomal region small region in the mammalian X chromosome, homologous with a region at the tip of the short arm of the Y chromsome; involved in pairing of the sex chromosomes at meiosis

trisomic state where there is one chromosome additional to the normal diploid complement; in somatic nuclei one particular chromosome is represented three times, rather than twice

See also **Genomic imprinting** (p.757)

CHROMOSOME BANDING AND THE LONGITUDINAL DIFFERENTIATION OF CHROMOSOMES

Adrian T. Sumner
North Berwick, East Lothian, UK

1. Introduction

In the past, with the then-standard methods of preparation, the chromosomes of most organisms appeared as uniform objects, with usually only the centromeric constriction forming any sort of differentiation. This had always made the identification of individual chromosome pairs difficult, particularly in those organisms with large numbers of chromosomes having similar morphology. Therefore, it caused a revolution in cytogenetics when, at the end of the 1960s and in the early 1970s, chromosome banding methods were discovered that produced patterns of bands on each chromosome that permitted their unequivocal identification, and, moreover, showed that chromosomes were not the uniform objects that they had previously been shown to be, but were divided up into segments having different properties.

Chromosome banding has been defined as "a lengthwise variation in staining properties along a chromosome . . . normally independent of any immediately obvious structural variation" (Sumner, 1990). As such, chromosome banding excludes the patterns seen on polytene chromosomes of Drosophila, for example, which have an underlying visible structural component. In fact, the first observations of what could be called chromosome banding were made at the end of the 19th century, when the differential staining of kinetochores and of nucleolus organizer regions (NORs) was reported (see Sumner, 1990).

Following the description of heterochromatin by Heitz (1928), a number of methods were described for its differential staining (see Sumner, 1990 for more details), but such methods were generally devised for chromosomes of particular species prepared in specific ways and were not generally applicable. By contrast, the chromosome banding methods that were introduced from 1968 onwards were of general application, and could be applied to chromosomes of a wide variety of species with no more than slight modifications. Following the introduction of Q-banding by Caspersson and his colleagues in 1968, Pardue and Gall (1970) inadvertently produced differential staining of heterochromatin in their pioneering *in situ* hybridization studies, leading directly to C-banding, and, in 1971, G-banding was discovered by several authors (e.g. Seabright, 1971; Sumner *et al.*, 1971). R-banding was also introduced in 1971 (Dutrillaux & Lejeune, 1971). Over the next few years, many other banding techniques, too numerous to mention individually, were introduced, many of them using fluorochromes. It is, however, worth mentioning the reintroduction of silver staining for NORs in 1975 (Howell *et al.*, 1975), the invention of methods to show chromosome replication (Dutrillaux *et al.*, 1973; Perry & Wolff, 1974) and the use of autoimmune sera to label kinetochores immunocytochemically (Moroi *et al.*, 1980).

In the remainder of this article, the classification of chromosome bands will be discussed, and methods of chromosome banding described briefly. This will be followed by an outline of the applications of banding in various fields, and, finally, the functional significance of chromosome bands will be considered.

2. Classification of chromosome bands

Four classes of bands can be recognized, and these are listed, together with the banding techniques that produce them, in **Table 1**. Heterochromatic bands correspond to classically defined constitutive heterochromatin – that is, regions of chromosomes that normally remain condensed throughout interphase – and they are largely concentrated as blocks around centromeres, and sometimes terminally or occasionally interstitially on chromosomes. Facultative heterochromatin, such as the inactive X chromosome in female mammals, is not stained distinctively by heterochromatic banding methods (nor by any other banding methods except replication banding). Euchromatic bands form a pattern of

Table 1 Classification of chromosome bands.

Class	Principal methods	Other methods
Heterochromatic	C-banding	G-11 banding; Q-banding; N-banding; Distamycin/DAPI; Immunolabelling for 5-methylcytosine; Restriction enzyme banding
Euchromatic	G-banding; Q-banding; R-banding	T-banding; Various fluorochromes; Replication banding; Restriction enzyme banding
Nucleolus organizer regions (NORs)	Ag-NOR staining	N-banding
Kinetochores	Immunolabelling with autoimmune CREST serum	C_d-banding; Silver staining; *In situ* hybridization with alphoid satellite sequences

alternating positively and negatively stained (or fluorescent) bands throughout the length of the chromosomes, and it is these patterns that are of inestimable value for identifying chromosomes. NORs are the segments of chromosomes that contain the genes for ribosomal RNA and which give rise to the interphase nucleoli. Finally, kinetochores are the special structures at the centromeres by which the chromosomes are attached to the spindle microtubules at mitosis and meiosis.

3. Banding techniques

Full practical details of banding techniques are given in various excellent laboratory manuals (Barch, 1991; Rooney & Czepulkowski, 1992; Gosden, 1994) and only outlines of the principal methods will be given here.

Q-banding (**Figure 1**), the first practical modern banding technique, is carried out simply by staining the chromosome preparation with a solution of quinacrine (originally quinacrine mustard was used; Caspersson *et al.*, 1968). In many plants and some animals, certain heterochromatic segments show bright fluorescence, but, in addition, mammalian chromosomes show a pattern of euchromatic bands throughout the length of the chromosomes. Various other fluorochromes, such as Hoechst 33258 and DAPI, produce similar patterns, although the euchromatic banding is less clear. Other compounds, for example, chromomycin, produce a complementary banding pattern, and the quality of the banding can be enhanced by a combination of DAPI and chromomycin, the different patterns being visualized simply by changing filters on the fluorescence microscope. However, a combination of the non-fluorescent DNA-binding compound distamycin with DAPI suppresses euchromatic banding and highlights only certain blocks of heterochromatin in a limited range of organisms.

Fluorescent banding is very attractive and is simple to produce, but, on the other hand, the fluorescence is not permanent, it fades quickly when illuminated and it needs a special (and expensive!) microscope to view it. G-banding, on the other hand, produces permanent preparations that can be viewed with an ordinary microscope. The original ASG (Acetic-Saline-Giemsa) method (Sumner *et al.*, 1971) involved the incubation of chromosome preparations in warm (60°C) saline, followed by Giemsa staining (**Figure 2**). In Seabright's method (1971), which has now become more popular, a brief trypsin digestion is used, followed by Giemsa staining. G-banding is the method of choice for identifying chromosomes of humans and other higher vertebrates, and is in routine use in cytogenetics laboratories throughout the world. If chromosome preparations are incubated in an appropriate buffer at a higher temperature (around 90°C) before Giemsa or acridine orange staining, a complementary pattern known as R-banding is produced (**Figure 3**); this method has the advantage that the terminal regions of the chromosomes are strongly stained, whereas it

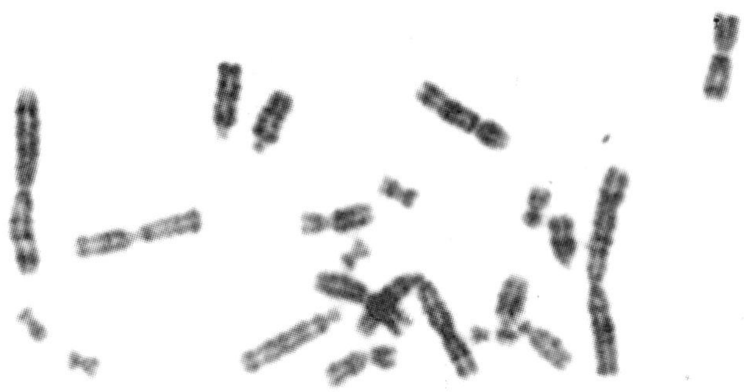

Figure 2 G-banded CHO (Chinese hamster ovary) chromosomes.

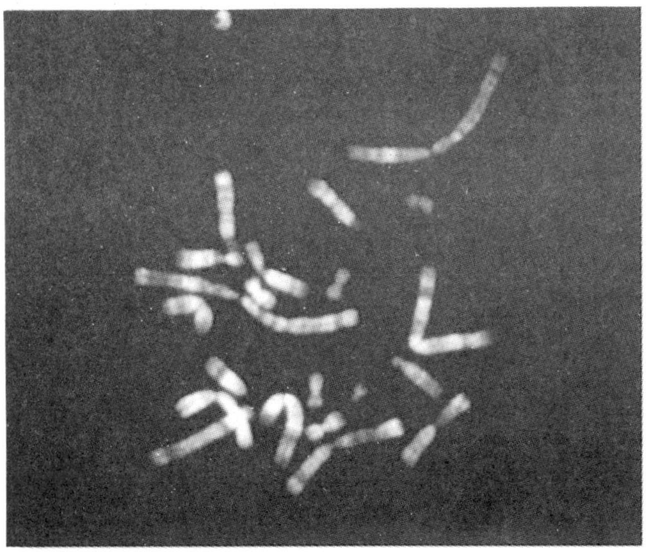

Figure 1 Q-banded CHO (Chinese hamster ovary) chromosomes.

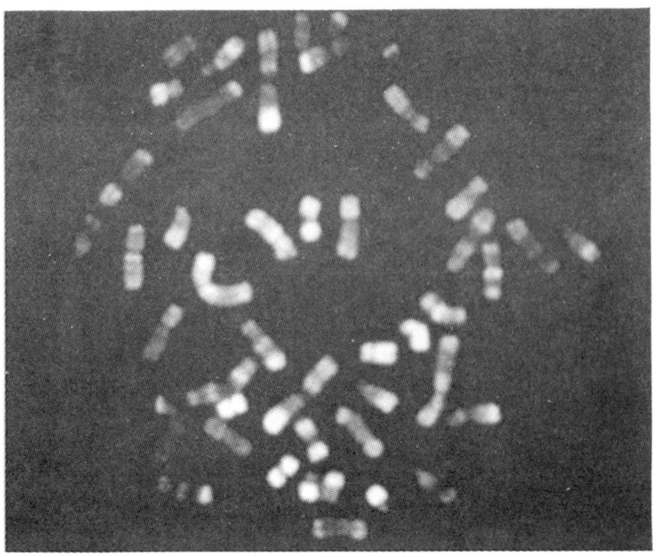

Figure 3 Human metaphase chromosomes, R-banded with acridine orange. Note that the ends of most chromosomes show bright fluorescence.

is often difficult to discern exactly where the ends of chromosomes are after G- or Q-banding.

When the first *in situ* hybridization experiments were done on mouse chromosomes (Pardue & Gall, 1970), it was noticed that the centromeric heterochromatin was more strongly stained with Giemsa. The original C-banding techniques were simply *in situ* hybridization procedures without the labelled complementary RNA, and the currently standard BSG (Barium hydroxide-Saline-Giemsa) method (Sumner, 1972) does not differ in principle. Chromosome preparations are first treated with dilute acid, then with moderately strong alkali, followed by a warm saline incubation prior to Giemsa staining (**Figure 4**). Various other techniques can be used to stain constitutive heterochromatin, but these normally only stain specific subsets of the blocks coloured by C-banding.

If cells are grown in the presence of bromodeoxyuridine (BrdU), this substance becomes incorporated into newly replicated DNA. If the BrdU is present only during the early part of the S phase, then its pattern of incorporation can be used to mark early replicating DNA; conversely, if the BrdU is present during the later part of the S phase, late replicating DNA is labelled. Incorporation of BrdU is commonly detected using the FPG (Fluorescence Plus Giemsa) technique (Perry & Wolff, 1974), which involves photolysis of the DNA that contains the BrdU, followed by Giemsa staining. Nowadays, however, immunolabelling of the incorporated BrdU is often preferred (**Figure 5**).

Another method for inducing bands on chromosomes involves digesting chromosome preparations with restriction endonucleases, followed by Giemsa staining. The pattern produced depends on the enzyme used, with some producing G-bands, while others induce C-bands or a subset of them. There is no particular advantage in using restriction endonucleases to obtain G-bands – in fact, it is a rather expensive way of doing it – but it is claimed that greater discrimination of different types of C-bands can be obtained using these enzymes. It is, however, difficult to relate the pattern produced to the known specificity of the enzymes and the DNA composition of particular chromosome segments.

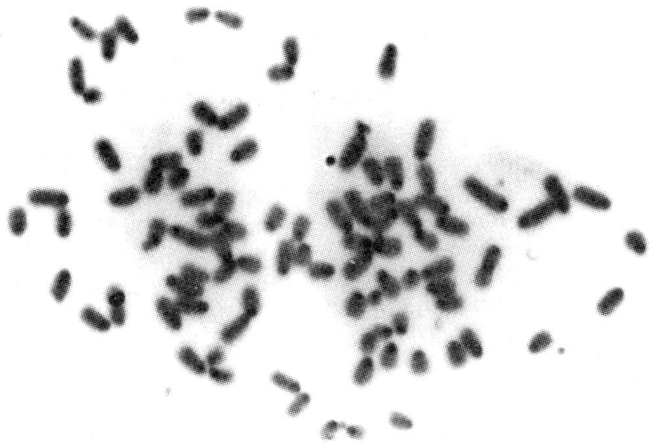

Figure 4 C-banded mouse chromosomes from a cell line. Most of the chromosomes are telocentric, and show a dark C-band at the centromeric end.

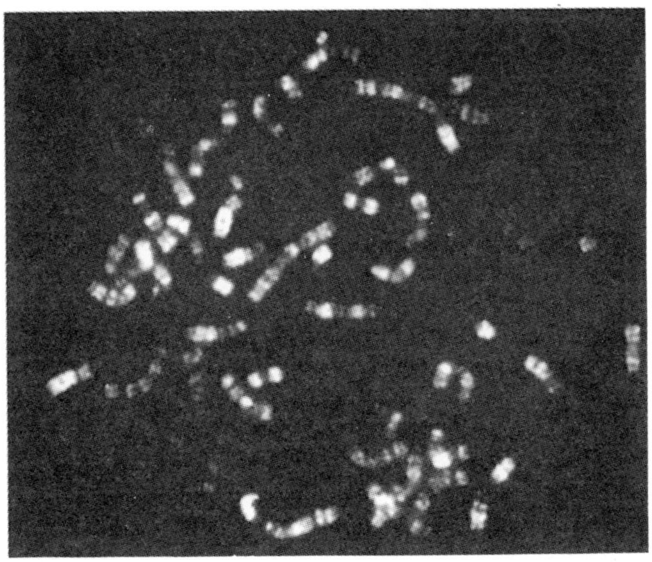

Figure 5 Human chromosomes labelled with BrdU, which has been detected by immunofluorescence, to show replication patterns.

Two other methods demonstrate more restricted regions of chromosomes. Ag-NOR (silver) staining for NORs labels those NORs that have been actively synthesizing rRNA, rather than every site of ribosomal genes. Thus, not every NOR is labelled; in humans, there are five pairs of chromosomes that bear NORs, but, of these ten sites, only six or seven are usually labelled with silver. Kinetochore labelling is now normally done by immunocytochemical means; autoimmune sera from patients with the CREST (calcinosis, Raynaud's phenomenon, esophageal dysmotility, sclerodactyly and telangiectasia) variant of scleroderma contain antibodies that react with kinetochores (Tan, 1989). This method differs from all other banding methods in that it cannot be applied to chromosomes fixed with alcohol–acetic acid; normally, cytocentrifuge preparations are used, with very mild fixation. If it is necessary to demonstrate kinetochores on chromosomes fixed with acetic acid, the C_d-banding method (Eiberg, 1974) is available.

4. Applications of chromosome banding

The greatest value of chromosome banding is as a means of identifying individual chromosomes. For this purpose, euchromatic banding techniques (G-banding in particular) are ideal; however, apart from higher vertebrates, there are few organisms in which euchromatic bands have been demonstrated. However, even in those organisms that lack euchromatic bands, the distinctive patterns of C-banding and other types of heterochromatic bands can generally be used to distinguish between chromosomes. In humans, G-banding is used to identify chromosome abnormalities and rearrangements in genetic diseases and cancers (Heim & Mitelman, 1995; Mitelman *et al.*, 1998).

Chromosome banding is also valuable in evolutionary

studies (Sumner, 1990, chapter 5), allowing the identification of the various chromosome rearrangements that have occurred in the course of speciation. Even in organisms in which it is not possible to induce euchromatic bands, C-banding can be used to some extent, to identify, for example, changes in the size or distribution of blocks of heterochromatin between closely related species or between races. Even in rodents, where good G-banding can be obtained, differences in C-banding patterns can be found between closely related species; for example, some species have wholly heterochromatic short arms, while their close relations have chromosomes that lack short arms.

Polymorphisms of heterochromatic bands have become a study in their own right, as well as being useful tools for distinguishing between paternal and maternal homologues. Although there is no evidence for any phenotypic effects in humans, it is intriguing that in maize there is a correlation between the rate of growth and quantity of heterochromatin (Rayburn *et al.*, 1985), suggesting that, at least in some cases, correlations might be found between the quantity of heterochromatin and various phenotypic features.

Ag-NOR staining can be used not only to identify the location of nucleolus organizers on chromosomes, but also to assess their activity. In a species with multiple NORs, such as man (who has five pairs), only a proportion are stainable with silver, while, in hybrids, it often happens that only the NORs from one parent are active (e.g. Miller *et al.*, 1976). Silver staining of interphase nuclei also has prognostic value in various cancers (Crocker, 1990). CREST labelling of kinetochores is an important tool in identifying centromeres and in distinguishing active and inactive centromeres in dicentric chromosomes, and has become invaluable in understanding centromeric organization.

5. Functional and structural significance of chromosome bands

Heterochromatin is widely believed to be functionless ("junk DNA"), a view supported by the lack of obvious phenotypic effect of C-band polymorphisms in many cases (see above), and by its content (in most cases) of highly repetitive DNA sequences that could certainly not code for proteins, and which, it seems to many people, could have no other conceivable function. Such views are almost certainly incorrect. In Drosophila, a number of genes, as well as some non-genic functions, have been localized to heterochromatin (Gatti & Pimpinelli, 1992), and it could well be that, when the heterochromatin of other organisms has been examined in the same amount of detail, it will be found that these too have various functions in their heterochromatin. In addition, it has been suggested that centromeric heterochromatin has an essential role in holding sister chromatids together until the end of metaphase and of ensuring their controlled separation at the beginning of anaphase (Sumner, 1991); the highly repeated DNAs in heterochromatin might well act as a substrate for the action of the enzyme topoisomerase II, which is required for chromatid separation.

Many differences have now been found between positive and negative euchromatic bands, and these are summarized in **Table 2** (for more detailed discussion of individual points, see Sumner, 1994). In fact, although there are many ways in which positive G-bands differ from negative G-bands, most of them are related to the fact that the former have relatively few genes, while the latter are much richer in genes (Craig & Bickmore, 1993); roughly speaking, 80% of the genes are in the negative G-bands, although these bands only constitute about half of the genome. The highest concentrations of genes turn out to be in the T-bands, a subset of R-bands (negative G-bands) that are found largely (though not exclusively) at the ends of chromosomes.

The reasons for the division of chromosomes into gene-rich and gene-poor segments is not at all clear, but it nevertheless seems probable that it is universal. Although differentiation of chromosomes into regions differing in DNA base composition, which can be demonstrated *in situ* as banding with base-specific fluorochromes, appears to be largely, if not exclusively, confined to higher vertebrates, early and late replicating segments and pachytene chromomeres seem to be virtually universal among eukaryotes. In higher vertebrates, early replicating segments and the interchromomeric regions correspond to the gene-rich regions, although it has yet to be shown directly that this is true of lower organisms. Nevertheless, it has now been shown that

Table 2 Characteristics of euchromatic bands in mammalian chromosomes.

Positive G-bands	Negative G-bands
Positive Q-bands	Negative Q-bands
Negative R-bands	Positive R-bands
Pachytene chromomeres	Interchromomeric regions
Late replicating DNA	Early replicating DNA
Early condensation	Late condensation
A + T-rich DNA	G + C-rich DNA
Low concentration of genes	High concentration of genes
DNase insensitive	DNase hypersensitive
Low level of histone acetylation	High level of histone acetylation
High level of histone H1 subtypes	Low level of histone H1 subtypes
HMG†-I protein present	HMG†-I protein absent
Rich in LINEs (long intermediate repetitive DNA sequences)	Rich in SINEs (short intermediate repetitive DNA sequences)
Low level of chromosome breakage	High level of chromosome breakage
Little recombination	Meiotic pairing and recombination

† High Mobility Group

DNase-hypersensitive regions, which are regions that contain potentially active genes, and the pattern of which corresponds to positive R-bands in mammals, are also non-uniformly distributed in lower organisms (de la Torre *et al.*, 1996). Chromosome banding, therefore, is not simply an invaluable method of identifying chromosomes, but has also become very important in drawing our attention to the longitudinal differentiation of chromosomes and in stimulating research on fundamental aspects of their organization.

REFERENCES

Barch, M.J. (ed.) (1991) *The ACT Cytogenetics Laboratory Manual*, 2nd edition, New York: Raven Press

Caspersson, T., Farber, S., Foley, G.E. *et al.* (1968) Chemical differentiation along metaphase chromosomes. *Experimental Cell Research* 49: 219–22

Craig, J.M. & Bickmore, W.A. (1993) Chromosome bands: flavours to savour. *BioEssays* 15: 349–54

Crocker, J. (1990) Nucleolar organiser regions. *Current Topics in Pathology* 82: 91–149

de la Torre, J., Herrero, P., Garcia de la Vega, C., Sumner, A.T. & Gosálvez, J. (1996) Patterns of DNase I sensitivity in the chromosomes of the grasshopper *Chorthippus parallelus* (Orthoptera). *Chromosome Research* 4: 56–60

Dutrillaux, B. & Lejeune, J. (1971) Sur une nouvelle technique d'analyse du caryotype humain. *Comptes Rendus de l'Academie des Sciences Paris D* 272: 2638–40

Dutrillaux, B., Laurent, C., Couturier, J. & Lejeune, J. (1973) Coloration des chromosomes humains par l'acridine orange après traitement par le 5 bromodeoxyuridine. *Comptes Rendus de l'Academie des Sciences Paris D* 276: 3179–81

Eiberg, H. (1974) New selective Giemsa technique for human chromosomes, C_d staining. *Nature* 248: 55

Gatti, M. & Pimpinelli, S. (1992) Functional elements in *Drosophila melanogaster* heterochromatin. *Annual Review of Genetics* 26: 239–75

Gosden, J.R. (ed.) (1994) *Chromosome Analysis Protocols*, Totowa, New Jersey: Humana Press

Heim, S. & Mitelman, F. (1995) *Cancer Cytogenetics*, 2nd edition, New York: Wiley

Heitz, E. (1928) Das Heterochromatin der Moose. I. *Jahrbuch Wissenschaftliche Botanik* 69: 762–818

Howell, W.M., Denton, T.E. & Diamond, J.R. (1975) Differential staining of the satellite regions of human acrocentric chromosomes. *Experientia* 31: 260–62

Miller, D.A., Dev, V.G., Tantravahi, R. & Miler, O.J. (1976) Suppression of human nucleolus organizer activity in mouse–human somatic hybrid cells. *Experimental Cell Research* 101: 235–43

Mitelman, F., Johansson, B. & Mertens, F. (1998) *Catalog of Chromosomal Aberrations in Cancer 98*, Version 1 (CD-ROM), New York: Wiley

Moroi, Y., Peebles, C., Fritzler, M.J., Steigerwald, J. & Tan, E.M. (1980) Autoantibody to centromere (kinetochore) in scleroderma sera. *Proceedings of the National Academy of Sciences USA* 77: 1627–31

Pardue, M.L. & Gall, J.G. (1970) Chromosomal localization of mouse satellite DNA. *Science* 168: 1356–58

Perry, P. & Wolff, S. (1974) New Giemsa method for the differential staining of sister chromatids. *Nature* 251: 156–58

Rayburn, A.L., Price, H.J., Smith, J.D. & Gold, J.R. (1985) C-band heterochromatin and DNA content in *Zea mays*. *American Journal of Botany* 72: 1610–17

Rooney, D.E. & Czepulkowski, B.H. (1992) *Human Cytogenetics: A Practical Approach*, 2nd edition, Oxford and New York: IRL Press

Seabright, M. (1971) A rapid banding technique for human chromosomes. *Lancet* 2: 971–72

Sumner, A.T. (1972) A simple technique for demonstrating centromeric heterochromatin. *Experimental Cell Research* 75: 304–06

Sumner, A.T. (1990) *Chromosome Banding*, London and Boston: Unwin Hyman

Sumner, A.T. (1991) Scanning electron microscopy of mammalian chromosomes from prophase to telophase. *Chromosoma* 100: 410–18

Sumner, A.T. (1994) Functional aspects of the longitudinal differentiation of chromosomes. *European Journal of Histochemistry* 38: 91–109

Sumner, A.T., Evans, H.J. & Buckland, R.A. (1971) New technique for distinguishing between human chromosomes. *Nature New Biology* 232: 31–32

Tan, E.M. (1989) Antinuclear antibodies: diagnostic markers for autoimmune diseases and probes for cell biology. *Advances in Immunology* 44: 93–151

GLOSSARY

anaphase stage of cell division when chromosomes leave the equatorial plate and migrate to opposite poles of the spindle

deoxyribonuclease (DNase) enzyme that cleaves DNA into shorter oligonucleotides or degrades it totally into its constituent deoxyribonucleotides

fluorochromes fluorescent dyes, which have the advantage, in general, when compared with dyes that merely absorb light, of much greater sensitivity. Valuable for *in situ* hybridization and for immunocytochemistry, where very small amounts of material have to be detected. Also very important for staining chromosomes: fluorochromes such as acridine orange, ethidium, and propidium bind to DNA without any base specificity, produce uniformly stained chromosomes, and are useful as counterstains (e.g. for immunocytochemical labelling of chromosomes). Others show base-specific binding to DNA, or base-specific enhancement of fluorescence, and produce banding patterns on chromosomes

immunocytochemistry methods for staining cells by using antibodies against the appropriate antigen

interphase period between one mitosis and the next, when no distinct chromosomes are visible and the chromatin is contained in a nucleus

kinetochore structure within the centromere of a chromosome to which spindle microtubules affix during meiosis or mitosis

pachytene stage during prophase of meiosis when homologous chromosomes are associated as bivalents

polytene chromosomes giant chromosomes produced by repeated replication of chromosomes without chromosome separation or nuclear division (endoreduplication)

restriction endonuclease enzyme which cuts DNA at a

specific base sequence, producing fragments of DNA of different lengths

topoisomerase action change in the degree of supercoiling in DNA by enzymic cutting of one or both strands; type I topoisomerase cuts only one strand; type II topoisomerase cuts both strands

See also **Human cytogenetics** (p.545)

HUMAN CYTOGENETICS

Adrian T. Sumner
North Berwick, East Lothian, UK

1. History of human cytogenetics

The practical study of human chromosomes began in 1956. Although human chromosomes were probably first seen in the late 19th century, and it became accepted (erroneously) in the first half of the 20th century that the human diploid chromosome number was 48, it was not until 1956 that it was finally established that the human chromosome complement consisted of 46 chromosomes (Tjio & Levan, 1956; see Hsu (1979) and Ferguson-Smith (1993) for very readable accounts of the various stages in the development of human cytogenetics). In addition, it was about this time that various technical advances were made that turned human cytogenetics into a practical study.

(i) Technical advances

During the 1950s, there were several innovations that made the routine study of human chromosomes feasible. Prior to this, chromosomes were only observed in the rare, naturally occurring mitotic cells by squashing or sectioning, and their scarcity, combined with the extensive overlapping of chromosomes, made accurate counting virtually impossible. The first of the essential technical innovations was the use of hypotonic treatment to swell the cells and thereby produce much clearer chromosome spreads with few overlapping chromosomes, while the introduction of colchicine treatment allowed the accumulation of large numbers of metaphase cells. A third important technique was the use of cultured cells, which had the advantage over solid tissues of allowing greater spreading of chromosomes.

These techniques enabled Tjio and Levan (1956), working with embryonic cells, to establish that the diploid human chromosome number is 46, which was confirmed very shortly afterwards by Ford and Hamerton (1956), who found only 23 bivalents in human spermatocytes. Subsequent technical developments in this period were the discovery that blood lymphocytes, probably the most easily accessible source of human cells, could be stimulated to divide by phytohaemagglutinin (Nowell, 1960), and the air-drying method of fixing and spreading chromosomes. These two methods have been the basis of human cytogenetics ever since.

(ii) Human chromosomal syndromes

Even before all these techniques were available, many laboratories started investigating people with congenital abnormalities: the first discovery was trisomy in Down's syndrome (Lejeune *et al.*, 1959), but several others soon followed (**Table 1**). Investigation of human spontaneous abortions showed that a substantial proportion of the high level of human foetal wastage was due to chromosomal abnormalities (Boué & Boué, 1966; Carr, 1967). It had long been believed that chromosome abnormalities were involved in cancers, and this belief was reinforced by the discovery of the Philadelphia chromosome (Ph[1]) in patients with chronic myelogenous leukaemia (Nowell & Hungerford, 1960); however, this appeared to be an isolated case, as no other specific chromosome abnormalities could be detected in cancers with the techniques then available.

2. Chromosome banding

By the late 1960s, human cytogenetics had become an established field, although no major discoveries had been made for a few years. Although a number of obvious chromosomal syndromes had been reported, it seemed quite likely that a considerable number of more subtle chromosomal alterations might occur. The problem was that it was impossible, with very few exceptions, to identify individual chromosome pairs. Although attempts were made, using autoradiographic labelling, to detect replication patterns and to identify chromosomes by their DNA content, these techniques were still not sufficiently precise. However, human cytogenetics (as well as the cytogenetics of all other eukaryotes) was revolutionized at the beginning of the 1970s by the discovery of chromosome banding. By using a variety of comparatively simple staining techniques, characteristic patterns could be induced on each pair of chromosomes. The techniques involved, and their applications and significance, are described in detail in Sumner (1990) and are summarized in the article on "Chromosome banding and the longitudinal differentiation of chromosomes".

The principal banding methods are Q-, G-, R- and C-banding (Quinacrine-, Giemsa-, Reverse- and Constitutive heterochromatin-banding); the first three produce detailed patterns of bands along each chromosome, so that each homologous pair can be identified unambiguously (**Figure 1**). The Q- and G-banding patterns are essentially the same, except that the former is fluorescent; R-banding produces a complementary pattern and has the advantage that the ends of chromosomes are strongly stained, whereas, with Q- and G-banding, the exact position of the chromosome ends is often difficult to discern. However, R-banding

Table 1 Syndromes associated with chromosome abnormalities identified in the prebanding era.

Syndrome	Chromosome constitution
Down's syndrome	Trisomy 21
Patau's syndrome	Trisomy 13
Edward's syndrome	Trisomy 18
Turner's syndrome	Monosomy X (XO)
Klinefelter's syndrome	47, XXY
Cri-du-chat	del 5p

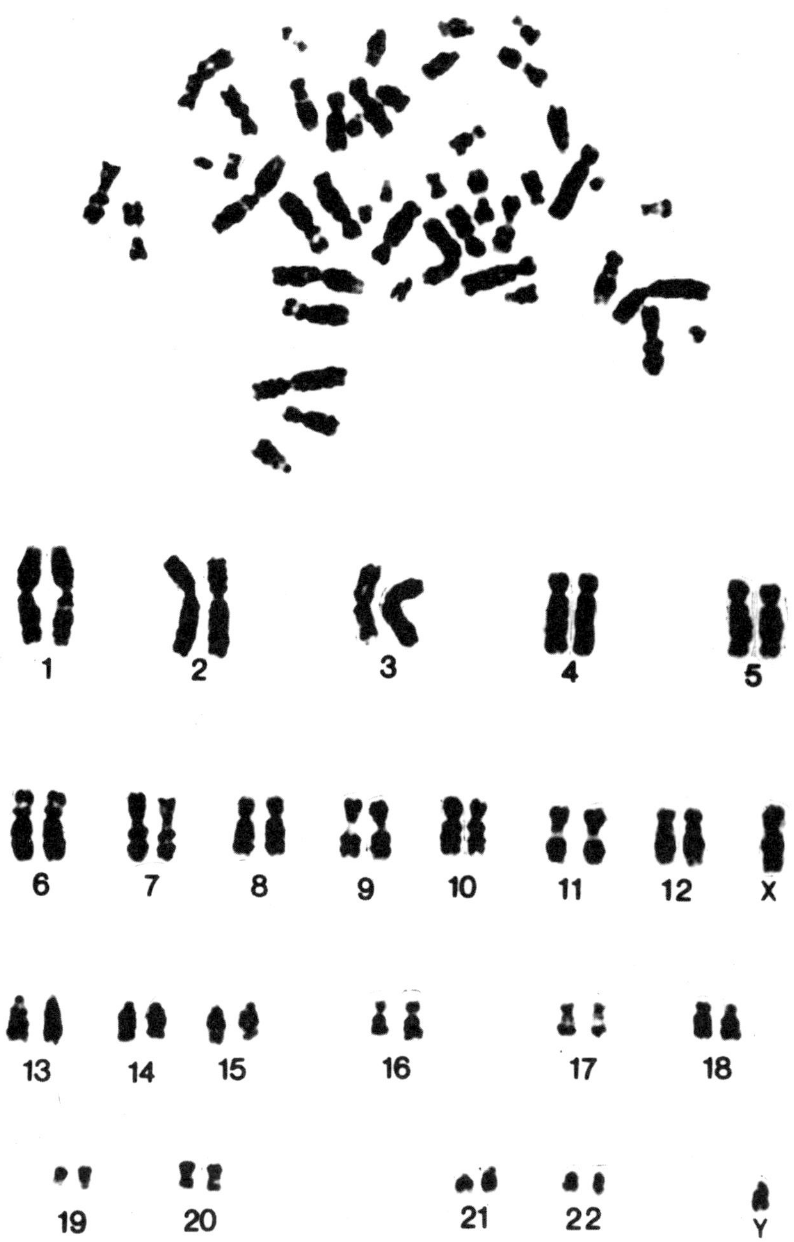

Figure 1 G-banded human metaphase chromosomes (top) with karyotype (below). Each pair of chromosomes has a distinctive banding pattern, which is identical in the two homologues. Reprinted by permission from *Nature New Biology* 232: 31–32 © 1971 Macmillan Magazines Limited.

has not normally been used as the principal method of banding, except in France.

The introduction of chromosome banding not only permitted the unambiguous identification of each chromosome, both in normal and abnormal karyotypes, but also the position of chromosome breaks, whether spontaneous or induced by radiation or chemicals, or occurring in numerous cancers. The precision of these methods was greatly increased with the introduction of high-resolution banding. By harvesting synchronized cells at a specific time-point in prophase or treating the cells with various substances that inhibit chromosome condensation, it is possible to induce chromosomes showing much more detailed banding patterns – up to 1250 bands per haploid chromosome set, compared with about 350 bands for a standard metaphase chromosome set. In practice, it is more usual to work with extended chromosomes showing 750–850 bands. It should be noted that bands include both positively and negatively stained bands – there are no "interbands" – and, indeed, a band that is positive by one method (e.g. G-banding) would be negative by another (i.e. R-banding).

(i) Classification of bands

The classification of human chromosome bands began with the Paris Conference in 1971, and subsequent reports have only refined this original classification; the latest version appeared in 1995 (ISCN, 1995). The numbering of human chromosomes, and the designation of short and long arms as "p" and "q", respectively, had already been decided at the Denver Conference in 1960. Chromosome bands were numbered outwards from the centromeres to the ends of the chromosomes. Major bands ("landmarks") were given a number to define the main regions of chromosomes and, within these, a second figure was used to define particular bands. Thus, a particular band on the short arm of chromosome 1 would be designated 1p36 (see **Figure 2**). With the discovery of high-resolution banding, it was a simple matter to label the subbands by extra figures after a decimal point (e.g. 1p36.12; **Figure 2**). The reports on the International System for Human Cytogenetic Nomenclature (ISCN) also give instructions on how to describe deletions, translocations and other chromosome alterations in relation to the bands (ISCN, 1995).

(ii) Applications of banding

Cytogenetic analysis has now become routine in individuals with the clinical features of certain genetic diseases (e.g. Down's syndrome, trisomy 21; DiGeorge syndrome, deletion of chromosome segment 22q11), for spontaneous abortions (at least 20% of which have chromosomal abnormalities), for abnormal sexual development and infertility (e.g. Klinefelter's syndrome, XXY males, Turner's syndrome, XO females) and for certain types of mental retardation (Wolstenholme & Burn, 1992).

The use of chromosome banding has helped to define a number of previously undetected chromosomal syndromes; in particular, the use of high-resolution banding has allowed the detection of microdeletions in a number of genetic diseases (**Table 2**). As mentioned above, chromosome alterations have long been thought to be important in cancers and, indeed, many show bizarre, highly altered karyotypes that seem to bear little relationship to the normal human diploid chromosome complement. However, with the exception of the Philadelphia chromosome, no consistent chromosome alteration had been identified in any cancer in the prebanding era. The use of banding has shown that, in fact, many cancers have specific chromosome abnormalities associated with them, although other, apparently less

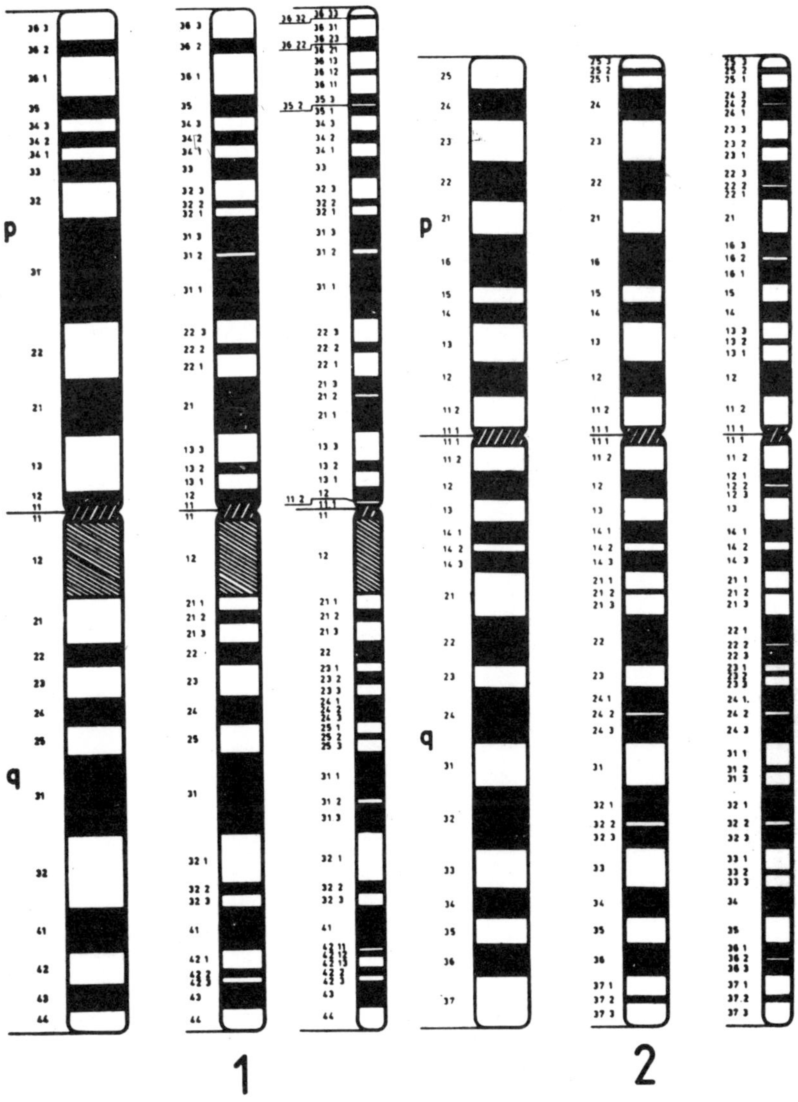

Figure 2 The standard human G-banded karyotype of human chromosomes 1 and 2. Left: pattern of a haploid karyotype with approximately 400 bands; middle, approximately 550 bands; right, high-resolution banding with approximately 850 bands. Reproduced with permission from ISCN (1995).

Table 2 Some examples of microdeletions and human diseases.

Syndrome	Deletion
Wilms' tumour	11p13 or 11p15
Prader–Willi syndrome Angelman syndrome	15p11–13
Langer–Giedion syndrome	8q23–24
DiGeorge syndrome	22q11

Table 3 Some examples of specific chromosome alterations in cancers.

Cancer	Chromosome change
Chronic myelogenous leukaemia	9;22 translocation
Acute myeloblastic leukaemia	8;21 translocation
Acute promyelocytic leukaemia	15;17 translocation
Childhood B cell acute lymphoblastic leukaemia	12;21 translocation
Burkitt's lymphoma	8;14 translocation
Ewing's sarcoma	11;22 translocation

specific, abnormalities usually develop as the cancer progresses (Heim & Mitelman, 1995; Mitelman *et al*., 1998). As a result of the relative ease of obtaining and preparing the material, the best information is available for leukaemias and proportionately little has yet been obtained for solid tumours; some of these alterations are listed in **Table 3**. It is of considerable interest that the Philadelphia chromosome, originally thought to be a deleted chromosome 22, turned out, after the application of banding, to be a translocation between chromosomes 9 and 22. It has been found, in many cases, that translocations in leukaemias move an oncogene to a position where it is under abnormal control, by, for example, an immunoglobulin gene. In other cases, cancer is associated with the presence of a microdeletion, as, for example, in some cases of Wilms' tumour and retinoblastoma, in which development of the cancer is associated with the loss of a tumour suppressor gene (e.g. Feinberg, 1994).

The use of chromosome banding has also allowed the accurate measurement of DNA content and centromeric index in chromosomes that have been unequivocally identified (**Table 4**); these are fundamental parameters in defining the human genome. Chromosome banding is not usually very successful on meiotic chromosomes (except for C-banding, see below), but Q-banding of adequate quality can sometimes be obtained and this allows the counting of the number of chiasmata in specific human chromosomes (**Table 4**).

3. Constitutive heterochromatin

C-banding differs from Q-, G- and R-banding in that it stains selectively only specific, limited regions of human chromosomes, the constitutive heterochromatin or C-bands (**Figure 3**). Each chromosome has a C-band at the centromere; chromosomes 1, 9 and 16 have particularly large paracentromeric C-bands, while the Y chromosome has a large C-band occupying most of the long arm (as well as

Table 4 DNA content, centromeric index and chiasma frequency of human chromosomes.

Chromosome	DNA content[1]	Centromeric index[2]	Number of chiasmata[3]
1	8.56	0.485	3.88
2	8.38	0.387	3.52
3	6.98	0.462	3.22
4	6.62	0.272	2.82
5	6.36	0.267	2.80
6	5.94	0.355	2.72
7	5.52	0.377	2.57
8	5.04	0.315	2.35
9	4.76	0.347	2.37
10	4.70	0.303	2.48
11	4.68	0.395	2.24
12	4.62	0.274	2.49
13	3.76	0.130	1.90
14	3.56	0.146	1.94
15	3.46	0.158	1.99
16	3.22	0.401	2.04
17	2.94	0.309	2.04
18	2.80	0.234	1.86
19	2.18	0.437	1.97
20	2.30	0.418	1.95
21	1.64	0.224	1.06
22	1.80	0.224	1.13
X	5.28	0.383	–
Y	1.94	0.223	–

[1] DNA contents of chromosomes expressed as percentages of the total haploid autosome DNA content (Mayall *et al*., 1984).
[2] DNA-based centromeric indices, i.e. the short arm as a fraction of the whole chromosome (Mayall *et al*., 1984).
[3] Mean number of chiasmata of Q-banded male bivalents at diakinesis (Laurie & Hultén, 1985); total number of chiasmata is about 50. Note that the XY pair has a single obligatory chiasma.

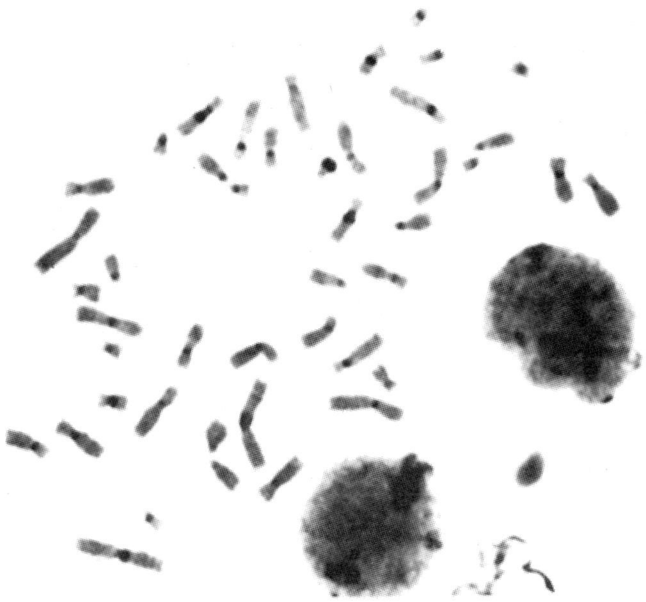

Figure 3 A C-banded human metaphase. Large blocks of heterochromatin are present on chromosomes 1, 9, 16 and Y.

having a centromeric C-band). The acrocentric chromosomes of group D (13, 14 and 15) and group G (21 and 22) have centromeric heterochromatin and also a block on the satellites at the end of the short arms; in between are the nucleolus organizer regions (NORs, which can be stained with silver, **Figure 4**). Most, if not all, blocks of human heterochromatin are polymorphic, varying not only in size, but also in their reaction to other banding methods such as Giemsa-11 and distamycin/DAPI (see Sumner, 1990, Chapter 14), and their position (pericentric inversions). These variations are particularly clear in the large blocks of heterochromatin on chromosomes 1, 9, 16 and Y (**Figure 5**). Certain blocks of heterochromatin, as well as showing variations in their size and position, also show variation in the brightness of their quinacrine fluorescence. In spite of extensive study, there is no evidence that extreme polymorphisms of heterochromatin are associated with any significant phenotypic effect (Bobrow, 1985; Hsu *et al.*, 1987).

(i) Applications

C-banding is not used for identifying human chromosomes, since, with few exceptions, the C-bands do not show sufficiently distinctive patterns. It is, nevertheless, of some help in identifying meiotic chromosomes, on which, as pointed out above, it is difficult to produce satisfactory Q- or G-banding patterns routinely. However, in general, the application of C-banding and other methods for staining constitutive heterochromatin depends on the polymorphism of this material. Since the size of blocks of heterochromatin, or their degree of quinacrine fluorescence, can differ between homologues, they provide a means of distinguishing between them and, by comparing the sizes of the heterochromatic blocks in an individual with those in the parents, the maternal and paternal chromosomes can be identified in many cases. In this way, it is often possible to determine the origin of the extra chromosome in trisomies or of the extra sets of chromosomes in triploids and tetraploids (Hassold & Jacobs, 1984). Although polymorphic DNA markers have supplemented chromosomal polymorphisms in these applications (and have completely supplanted them for paternity testing), the two methods are, in fact, complementary, as the heterochromatic polymorphisms provide markers at the centromeres, where DNA polymorphisms are less satisfactory. Polymorphisms are also useful in bone marrow transplantation and prenatal

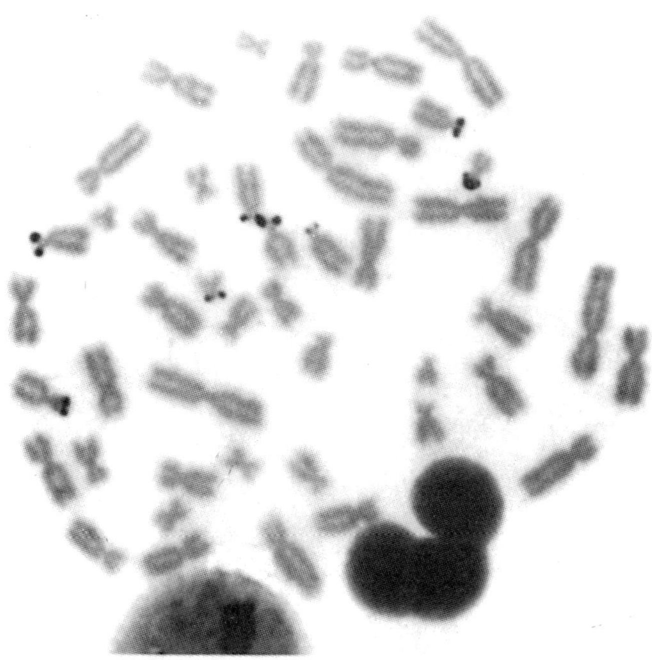

Figure 4 Human nucleolus organizer regions stained with silver (Ag-NOR staining).

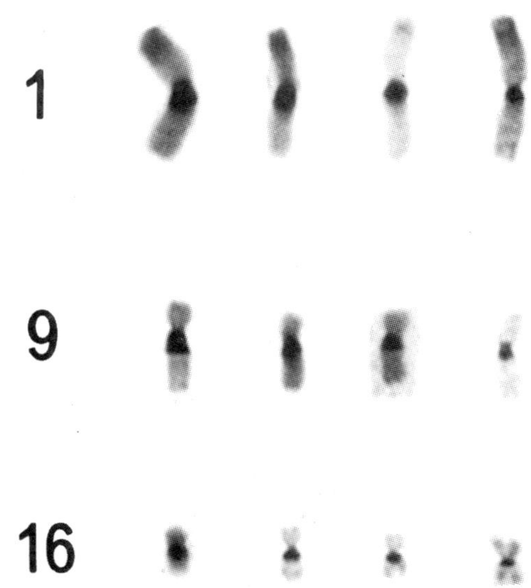

Figure 5 Heterochromatic polymorphisms of human chromosomes 1, 9 and 16. For each chromosome, the size of the C-band decreases from left to right.

diagnosis, where they can be used to identify the origin of individual cells – donor or recipient in the case of transplants, and mother or foetus in the case of prenatal diagnosis.

4. *In situ* hybridization

At the same time that banding techniques were discovered, *in situ* hybridization was invented; in fact, C-banding methods are derived directly from those for *in situ* hybridization. *In situ* hybridization has been of great importance in elucidating several aspects of the organization of human chromosomes. The earlier work on human chromosomes showed that the blocks of heterochromatin of human chromosomes, particularly those on chromosomes 1, 9, 16 and Y, contained large quantities of the so-called "classical" satellite DNA fractions. More recent refinements of the technique, especially fluorescence *in situ* hybridization (FISH), have permitted the localization of less highly repeated sequences. It would not be practicable to list all the applications, but several are worth mentioning. The localization of alphoid satellite DNA at the centromeres (not merely around the centromeric regions of chromosomes, as was found for the classical satellites) has not only provided a valuable marker for centromeres, but has also helped to provide insights into the organization and functioning of centromeres.

A second important application of FISH to human chromosomes is in the field of gene mapping, where FISH has become an important procedure for locating the sites of genes on chromosomes; high-resolution mapping is possible because *in situ* hybridization of the gene can be combined with an appropriate banding method. Either a fluorescent banding procedure, such as one using DAPI or chromomycin, can be employed or it is possible to produce a pattern of bands using *in situ* hybridization, since the distributions of LINEs (long intermediate repetitive DNA sequences) and SINEs (short intermediate repetitive DNA sequences) correspond to those of positive and negative G-bands, respectively (Korenberg & Rykowski, 1988).

Finally, although G-banding is the basic method of chromosome identification in routine cytogenetics laboratories, and indeed is likely to remain so for the foreseeable future, "chromosome painting" is becoming a serious rival to G-banding for certain applications. For chromosome painting, *in situ* hybridization is carried out using only sequences specific to a particular chromosome. In this way, it is possible to identify a particular chromosome, not only at metaphase, but also in interphase, the latter being particularly valuable for non-dividing cells or for those that only divide slowly, or in cases where the dividing cells may not be typical of the tissue as a whole. Thus, chromosome painting can be used to determine whether spermatozoa are X- or Y-bearing, or to determine the sex of early embryos or to detect trisomy in them. In early studies, only a single homologous pair of chromosomes was labelled, with the disadvantage that only pre-selected chromosomes could be studied. However, by combining fluorochromes of different colours, it became possible to label several different chromosomes simultaneously, until the technical *tour de force* was achieved of labelling every human chromosome pair in a different colour (Schröck *et al.*, 1996; Speicher *et al.*, 1996). This provides an immediate method of identifying and counting each type of chromosome, without having to analyse the banding patterns, and draws immediate attention to any translocations, however subtle. On the other hand, there is no means of identifying breakpoints or deletions, so that banding is still required for these purposes, and chromosome painting for all 24 types of human chromosome is technically far more demanding than routine chromosome banding. However, for interphase chromosome identification, which is undoubtedly becoming an important technique, there is no alternative to chromosome painting.

5. Concluding remarks

Human cytogenetics is now some 50 years old, and the discoveries in that period have been remarkable. There seems to be no doubt that more exciting discoveries will be made in the future. Like so many fields of scientific endeavour, human cytogenetics has been propelled by a series of technical advances: first, the methods for producing good metaphase spreads; secondly, the introduction of chromosome banding; and, now, the development of *in situ* hybridization and, in particular, chromosome painting. The first two advances produced explosions of new data, and there is no reason to believe that the latest techniques will be any less productive in advancing our understanding of the human genome.

REFERENCES

Bobrow, M. (1985) Heterochromatic chromosome variation and reproductive failure. *Experimental and Clinical Immunogenetics* 2: 97–105

Boué, J.G. & Boué, A. (1966) Les aberrations chromosomiques dans les avortements spontantés humains. *Comptes Rendus de l'Academie des Sciences Paris* 263: 2054–58

Carr, D.H. (1967) Chromosome anomalies as a cause of spontaneous abortion. *American Journal of Obstetrics and Gynecology* 97: 283–93

Feinberg, A.P. (1994) A developmental context for multiple genetic alterations in Wilms' tumor. *Journal of Cell Science* Suppl. 18: 7–12

Ferguson-Smith, M.A. (1993) From chromosome number to chromosome map: the contribution of human cytogenetics to genome mapping. In *Chromosomes Today*, volume 11, edited by A.T. Sumner & A.C. Chandley, London and New York: Chapman and Hall

Ford, C.E. & Hamerton, J.H. (1956) The chromosomes of man. *Nature* 178: 1020–23

Hassold, T. & Jacobs, P.A. (1984) Trisomy in man. *Annual Review of Genetics* 18: 69–97

Heim, S. & Mitelman, F. (1995) *Cancer Cytogenetics*, 2nd edition, New York: Wiley

Hsu, L.Y.F., Benn, P.A., Tannenbaum, H.L., Perlis, T.E. & Carlson, A.D. (1987) Chromosomal polymorphisms of 1, 9, 16 and Y in 4 major ethnic groups: a large prenatal study. *American Journal of Medical Genetics* 26: 95–101

Hsu, T.C. (1979) *Human and Mammalian Cytogenetics: An Historical Perspective*, Heidelberg and New York: Springer

ISCN (1995) *An International System for Human Cytogenetic Nomenclature*, edited by F. Mitelman, Basel: Karger

Korenberg, J.R. & Rykowski, M.C. (1988) Human genome organisation: Alu, Lines, and the molecular structure of metaphase chromosome bands. *Cell* 53: 391–400

Laurie, D.A. & Hultén, M.A. (1985) Further studies on bivalent chiasma frequency in human males with normal karyotypes. *Annals of Human Genetics* 49: 189–201

Lejeune, J., Gautier, M. & Turpin, R. (1959) Etude des chromosomes somatiques de neuf enfants mongoliens. *Comptes Rendus de l'Academie des Sciences Paris* 248: 1721–22

Mayall, B.H., Carrano, A.V., Moore, D.H. *et al.* (1984) The DNA-based human karyotype. *Cytometry* 5: 376–85

Mitelman, F., Johansson, B. & Mertens, F. (1998) *Catalog of Chromosomal Aberrations in Cancer 98*, Version 1 (CD-ROM), New York: Wiley

Nowell, P.C. & Hungerford, D.A. (1960) A minute chromosome in human chronic granulocytic leukemia. *Science* 132: 1497

Nowell, P.C. (1960) Phytohemagglutinin: an initiator of mitosis in cultures of normal human leukocytes. *Cancer Research* 20: 462–66

Schröck, E., du Manoir, S., Veldman, T. *et al.* (1996) Multicolor spectral karyotyping of human chromosomes. *Science* 273: 494–97

Speicher, M.R., Ballard, S.G. & Ward, D.C. (1996) Karyotyping human chromosomes by combinatorial multi-fluor FISH. *Nature Genetics* 12: 368–75

Sumner, A.T. (1990) *Chromosome Banding*, London and Boston: Unwin Hyman

Tjio, J.H. & Levan, A. (1956) The chromosome number of man. *Hereditas* 42: 1–6

Wolstenholme, J. & Burn, J. (1992) In *Human Cytogenetics: A Practical Approach*, vol. 1: *Constitutional Analysis*, 2nd edition, edited by D.E. Rooney & B.H. Czepulkowski, Oxford and New York: IRL Press

GLOSSARY

chromomycin one of the chromomycinone antibiotics which, together with the closely related mithramycin and olivomycin, are valuable fluorochromes for obtaining chromosome banding. These substances show a preference for G+C-rich DNA, and produce a fluorescent R-banding pattern on chromosomes. In combination with fluorochromes that show a preference for A+T-rich DNA (e.g. Hoechst 33258) they are also useful for two-parameter flow cytometry

chromosome painting the labelling of specific whole chromosomes by *in situ* hybridization. Hybridization probes are usually made by selecting a particular chromosome by flow sorting, and using the DNA from such chromosomes, after removing repeated sequences that are common to other chromosomes. Chromosome painting can be used to identify specific chromosomes not only in normal metaphases, but also in interphase nuclei, in chromosomal rearrangements, and to study chromosomal changes during evolution

CREST labelling immunocytochemical labelling of the kinetochores of chromosomes with autoimmune serum from patients with the CREST variant of scleroderma. (The acronym indicates the symptoms of this syndrome: Calcinosis, Raynaud's phenomenon, oEsophageal dysmotility, Sclerodactyly, and Telangiectasia.) Different CREST sera contain antibodies against a variety of proteins (CENPs, or CENtromeric Proteins) associated with the centromere or kinetochore. In general, successful CREST labelling is only obtained on the chromosomes of higher vertebrates, particularly mammals

fluorochromes fluorescent dyes, which have the advantage, in general, when compared with dyes that merely absorb light, of much greater sensitivity. Valuable for *in situ* hybridization and for immunocytochemistry, where very small amounts of material have to be detected. Also very important for staining chromosomes: fluorochromes such as acridine orange, ethidium, and propidium bind to DNA without any base specificity, produce uniformly stained chromosomes, and are useful as counterstains (e.g. for immunocytochemical labelling of chromosomes). Others show base-specific binding to DNA, or base-specific enhancement of fluorescence, and produce banding patterns on chromosomes

myelogenous relating to cells produced in the bone marrow or to the tissue from which such cells originate

phytohaemagglutinin (PHA) a haemagglutinin (a lectin that coagulates red blood cells) derived from the kidney bean, Phaseolus, which stimulates lymphocytes to divide

quinacrine a fluorochrome used to produce Q-banding of chromosomes. It was originally introduced as an antimalarial drug under the names Atebrine or Mepacrine, but became one of the first substances used to produce banding patterns on chromosomes, which it does because its fluorescence is enhanced when it binds to A+T-rich DNA

trisomy/trisomic state where there is one chromosome additional to the normal diploid complement; in somatic nuclei one particular chromosome is represented three times, rather than twice

See also **The *Wilms' tumour 1* gene *WT1*** (p.387); **Chromosome banding and the longitudinal differentiation of chromosomes** (p.539)

GENETIC EFFECTS OF THE ATOMIC BOMBS

James V. Neel

Department of Human Genetics, University of Michigan Medical School, Ann Arbor, USA

1. Background

The fact that exposure to ionizing radiation increases the mutation rate was first unequivocally demonstrated by H.J. Muller in 1927, using the fruit fly, *Drosophila melanogaster*, as his experimental organism. Over the next 18 years, Muller's original observations were greatly extended, not only with respect to Drosophila, but also by employing a variety of other experimental organisms. Thus, when the atomic bombs were detonated over Hiroshima and Nagasaki in 1945, the scientific community immediately raised the question of the genetic (and other) implications of these exposures, and speculation was rampant.

The great concern over the potential delayed effects of exposures to the atomic bombs eventually led, through a chain of intermediate events, to the president of the United States, then Harry Truman, assigning to the US National Academy of Sciences the task of organizing appropriate follow-up studies. The National Academy is a quasi-governmental organization, a traditional and prestigious intermediary between the government and its citizenry with respect to scientific issues. The joint Japanese–American operation set up in Japan in 1947 was first known as the Atomic Bomb Casualty Commission (ABCC), but, in 1975, reorganized into the Radiation Effects Research Foundation (RERF).

The organization and conduct of these studies presented extraordinary difficulties. Up to this time, the study of radiation effects had been based on carefully designed laboratory studies where the radiation exposure could be precisely controlled and the experimental organism – animal or plant – bred according to protocols specially designed to reveal the genetic damage created by the experiment. In obvious contrast, in the Japanese situation, the radiation exposure of survivors was unknown, but obviously quite variable, ranging from zero to the maximum whole-body radiation compatible with survival, and the population, of course, could not be subject to manipulation and was also quite mobile. Furthermore, the devastation and economic necessities of post-war Japan certainly were not the ideal setting for a detailed and meticulous scientific undertaking.

There were, on the other hand, certain aspects of the Japanese culture that were more favourable to a proper study than might have obtained for a similar study in many other parts of the world. The registration of births and deaths was quite complete. The proposed follow-up was a government-sanctioned study, and the Japanese respect for authority led to a high level of cooperation. In addition, in the immediate post-war years, the wartime rice ration system was still in place, by virtue of which pregnant women who registered the fact after completion of the fifth lunar month of pregnancy received access to special rations. Registration was almost 100%. The genetic study that was undertaken could ascertain these women at the time they applied for the special rations; this resulted in the necessary *prospective* complete coverage of the outcome of over 99% of all pregnancies in Hiroshima and Nagasaki.

2. The study

Initially, attention centred on attempting to conduct a complete physical examination of every infant born in the two cities. The initial battery of observations on each newborn included occurrence of major congenital defect/sentinel phenotype, stillbirth, survival of liveborn children through the neonatal period, sex of child and birth weight. (A sentinel phenotype is any one of some 15 easily diagnosed syndromes that may be recognized at birth or relatively soon thereafter and which result from a dominant mutation.) There was a further clinical examination of a subsample of these children at the age of 9 months. In 1953, this major clinical programme was discontinued, but births in the two cities were, as they were registered for civil purposes, screened for parental radiation history and, where indicated, added to a growing cohort for future study. By 1984, there were very few births in the two cities to exposed parents, and the study cohort was closed out, with 31 150 children in the cohort of children one or both parents of whom had been within 2 km of the hypocentre of the bombings (the point on the earth directly beneath the detonation), the so-called proximally exposed. A suitably matched control cohort that had been accumulating over the years, of 41 066 children, was also closed.

In 1967, Dr A.A. Awa and associates initiated major cytogenetic studies of a subset of this cohort. In the 1970s, Dr T. Furusho and Dr M. Otake analysed from school records the physical development of a subset of these children who were in middle and senior high school. In 1972, a search for mutational damage in a battery of serum proteins and erythrocyte enzymes was launched utilizing this cohort, a study which came under the direction of Dr C. Satoh. Finally, the children in these cohorts were followed for survival and malignancy, the studies on malignancy utilizing the newly established Cancer Registries in Hiroshima and Nagasaki, with Dr H. Kato playing a major role in those studies. It must be obvious that a study of this magnitude is the work of many more hands and minds that those just mentioned. On the US side, Dr W.J. Schull, with whom the author has been associated in these studies since 1949, played an especially prominent role. Altogether, there were perhaps 100 professionals involved in these studies over the years. This has been the largest and the longest-running study in genetic epidemiology ever undertaken.

(i) Estimation of radiation exposures

Estimating the precise amount of ionizing radiation that the bombs delivered to the germ cells of survivors remains a troublesome issue right down to the present time. There is general agreement that those beyond 2 km from the hypocentre (the so-called distally exposed) received negligible amounts of radiation and their children can serve as a comparison population. For those within 2 km of the hypocentre, the exposures were extremely variable. Humans are relatively radiation sensitive; it is estimated that about 450 roentgens (r) of whole-body gamma radiation delivered to the surface will kill half of an exposed population in 60 days. Of this surface dose, only about half will reach the gonads. However, most persons receiving the higher doses of radiation also sustained traumatic or thermal injury, decreasing their chance of survival. Thus, few people who received more than about 300 r to the gonads survived. It is estimated that the vast majority of persons within a 1 km radius of the hypocentre perished (Los Alamos Scientific Laboratory, 1950). The fact that only approximately 31 150 children were ever born to the survivors within 2 km of the hypocenter attests to how heavy the mortality (which cannot be exactly estimated) must have been within this area. The uncertainties in estimating radiation exposures for those within 2 km of the hypocentre who survived the exposure (the proximally exposed) are primarily of two types: first, as regards the spectrum of radiation delivered by the bombs and, secondly, as regards the shielding from radiation afforded to people who were within houses, large buildings or industrial plants at the time of the bombings (ATB).

Spectrum of radiation. With respect to the amount and kind of radiation exposure, the bombs delivered two types of radiation: gamma and neutron. The estimation of the magnitude of the latter component, especially in Hiroshima, has been particularly difficult. Since there have been about twice as many children born to proximally exposed survivors in Hiroshima than in Nagasaki, this is a non-trivial issue. To compound the problem, as measured in arbitrary units of ionization, the effectiveness of neutron radiation in producing mutation is greater than that of gamma radiation. To derive a single-dose figure for an exposed person, one must estimate the relative biological effectiveness (RBE) of the neutron component. For the Japanese studies, we have, based on animal experiments, employed an RBE of 20, but this conversion factor must be regarded as an approximation. The total exposure is obtained by adding both the gamma and the adjusted neutron contributions, to obtain a number termed "roentgen equivalents man" (rem). (Recently, there has been a move to express exposure in sieverts [Sv], 1 Sv being 100 rem.) The initial exposure estimate is for the surface of the body. In a genetic study, the dose of concern is to the gonads of the survivors. The high water content of human tissues attenuates the surface dose, the amount of attenuation depending on an individual's exact position ATB. It is often difficult to recreate this position for the purposes of dosage estimation.

Shielding. With respect to the shielding that protected people ATB, a substantial fraction of persons within 2 km of the hypocentre and in the open ATB were killed by the blast and/or thermal effects of these weapons, as noted earlier. The survivors almost always had been saved by some type of shielding, most commonly the tile roofs of the typical Japanese home, but sometimes the shielding of a concrete building or factory. For all the proximally exposed survivors, the ABCC attempted a meticulous reconstruction of their position and shielding ATB and, then, on the basis of the estimates of the air dose at the distance of these persons from the hypocentre, assigned an individual gonad dose.

Since the children under study reflect the results of mutations occurring in both parents, an analysis of the findings requires an estimate of the total gonadal dose of both parents. For the analyses to be presented, dosages are estimated from a very complex algorithm (DS86) developed by the staff of ABCC (Roesch, 1987). This algorithm requires an attempt to recreate the exact position of each proximally exposed person in the study ATB and all the shielding from the bomb that might have attenuated the radiation. At the present time, the best estimate of that combined average parental gonadal dose for the 31 150 children born to proximally exposed parents is approximately 40–50 rem (0.4–0.5 Sv). By the standards of the experimental radiation geneticist, this is a *small* dose. The fact that the gonadal exposures would be relatively small, and any effects probably correspondingly small, was recognized at the outset of the study (although no exact estimate was available). This required that unusual efforts had to be made to prevent bias of any sort from creeping into the study.

(ii) The findings and their statistical analysis

For the analysis of the data, summarized in a book published in 1991 (Neel & Schull, 1991), the observations on the frequency of stillbirths, congenital defects/sentinel phenotypes and early death (first week) were combined into a single entity, termed an "untoward pregnancy outcome" (UPO), since these outcomes are so interrelated and cannot be treated as separate events. The analysis consisted in estimating the regression of frequency of indicator on combined parental exposure for UPOs and four other of the indicators. The four other indicators were death of liveborn infants up to an average expected age of 26.2 years, cancer up to an average expected age of 19 years, frequency of a particular cytogenetic abnormality (sex-chromosome aneuploidy) and mutations altering the electrophoretic mobility or activity of a battery of some 30 enzymes/proteins. The results of the analyses are summarized in **Table 1**. None of these regression terms achieved statistical significance. The fact that several of the regressions were (insignificantly) negative presumably reflects sampling errors at these low exposures. There were two other indicators (sex ratio and physical development) that did not lend themselves to a regression-type analysis, but neither of these by other analytical techniques exhibited any evidence for a radiation effect. Inasmuch as all these observations were made on the same two cohorts of children, but involved different and independent aspects of a radiation effect, these regressions can be combined additively to obtain a single regression term, and this slightly positive term, of 0.00375, is also presented in Table 1. In other words, the combined frequency of those

Table 1 Summary of regression of various indicators on parental radiation exposure and of impact of spontaneous mutation on indicator.

Trait	Regression/combined parental dose in Sv	Contribution of spontaneous mutation
UPO[1]	+0.00264	
F1 mortality	+0.00076	0.0033–0.0053
Protein mutations	−0.00001	
Sex-chromosome aneuploids	+0.00044	0.0030
F1 cancer	−0.00008	0.00002–0.00005
	0.00375	0.00632–0.00835

[1] untoward pregnancy outcome

five indicators changes by 3.75 per thousand outcomes per parental Sv equivalent of radiation.

(iii) The results in perspective

The question of how best to summarize these data and evaluate their significance has been somewhat daunting and controversial. To begin with, the fact that a statistically significant difference between the exposed and their controls with respect to these indicators was not observed most emphatically does not prove the absence of a genetic effect of the bombs. In experimental organisms, radiation doses at the lowest levels to which studies have been carried out reveal genetic effects, and it seems almost beyond doubt that some mutations affecting the various indicators were induced in the two cities. However, the increase was not great enough to create a statistically significant difference between corresponding end-points in the children of exposed and control couples. A very conservative person might wish to stop the analysis at this point.

On the other hand, it is possible, albeit at the risk of simply manipulating chance differences between the two sets of children, rather than a meaningful difference based on the exposure histories of their parents, to take the data at face value and calculate a so-called "doubling dose" of ionizing radiation for humans. The doubling dose is the exposure of a population to ionizing radiation that will produce the same amount of genetic damage as occurs spontaneously each generation. It can be expressed either as per haploid gamete or per diploid zygote (i.e. person); the studies in Japan yielded a zygotic estimate, whereas most of the experimental studies result in gametic estimates. The doubling dose is a convenient concept, but the many assumptions and practical difficulties in actually deriving a doubling dose were well enumerated by Muller (Muller, 1959). The situation has not changed materially in the ensuing 40+ years. Ideally, the concept embraces the whole spectrum of mutational morbidity and mortality, from mutations involving entire chromosomes to single nucleotide substitutions, thus requiring the study and integration of a wide range of genetic damage. In addition, for the Japanese data, this calculation required specifying the contribution of spontaneous mutation in the preceding generation to the various indicators, including congenital defect and early death; that estimate totalled to somewhere between 0.0063 and 0.0084 (i.e. 6.3–8.4 births per thousand) under the conditions in Japan. The estimate of the amount of ionizing radiation necessary to double the normal burden of mutation with respect to these traits – the zygotic doubling dose – is obtained simply by dividing the background rate by the increase per Sv. We have estimated upper and lower limits for the background rate and the use of these limits leads to an estimate of the zygotic doubling dose of between 0.0063/0.00375 = 1.7 Sv and 0.0084/0.00375 = 2.2 Sv, which we can round off to 2.0 Sv. In an imperfect world, the doubling dose concept supplies a perspective, if blurred, that is difficult to obtain by any other approach.

This estimate carries a wide, but, for several reasons, essentially indeterminate error. We believe that, as befits the situation, the assumptions in reaching this estimate have been very conservative. This estimate may be biased downward by the somewhat lower socioeconomic status of the proximally exposed parents than that of the distally exposed population in the decade following the bombing. For instance, if only 50% of the small increase in UPOs among the children born to survivors of the bombing were socioeconomic in origin, a reflection of more difficult living conditions for this group, the estimate of the doubling dose of acutely administered radiation would become 4.0 Sv. It needs to be emphasized that this is a zygotic rather than gametic doubling dose. The calculations revealed that the doubling dose was unlikely to be less than 1.0 Sv, but in the absence of statistical significance an upper bound could not be assigned to the estimate. To be specific, the data do not exclude estimates of the zygotic doubling dose of acute radiation as high as 3 or 4 or even 5 Sv.

Most of the radiation human populations receive is in small dribbles, as in industrial exposures, or even more or less continuously, as from cosmic radiation or radon. In the mouse, at the higher experimental doses employed, chronic radiation is genetically only about one-third as effective in producing mutations as acutely delivered radiation, such as was involved in the Japanese exposures (Russell *et al.*, 1958). For the gonadal doses experienced in Japan, the reduction factor should almost certainly be less than in the mouse; a factor of 2 seems reasonable. The human zygotic doubling dose for chronic radiation thus becomes in the region of 4 Sv. For those for whom these radiation units are unfamiliar, some perspective to the numbers being used in this presentation is provided by the following. The average US citizen is receiving about 0.004 Sv a year from all sources of radiation in the environment, but especially from radon. This annual exposure is about 1/1000 of a doubling dose.

Otherwise stated, it would require some 1000 years to accumulate a doubling dose of radiation in our industrialized society – and there is a long-running debate as to whether, at these very low doses of radiation, the body's complex and efficient DNA repair mechanisms might be able to heal almost all the potential genetic damage caused by the radiation. For comprehensive recent summaries, the reader interested in repair may wish to consult Friedberg *et al.* (1995), and Vos (1995). In an additional effort to provide perspective, it should be pointed out that, in the decade following the atomic bombings, no less a scientific figure than the geneticist J.B.S. Haldane could speculate that the doubling dose of radiation for humans could be as low as 0.05 Sv; from this the reader can readily grasp the contribution brought to this issue by the studies in Japan.

(iv) Dispelling the uncertainties in these estimates

Throughout this treatment, the potential for uncertainties in the estimate of the doubling dose has been emphasized. The uncertainties range from difficulties in specifying the exact radiation spectrum of the bombs and errors in assigning gonadal exposures (at the one extreme), to problems in estimating the contribution of spontaneous mutation each generation to some of the end-points, especially the UPOs, at the other extreme. It is important to recognize, however, that these uncertainties at present are more apt to result in an *overestimate* of genetic risk than an *underestimate*.

Some of these uncertainties can almost surely be dispelled in the future. The question of the neutron component in the spectrum of radiation released by the bomb will probably be clarified in due time. An improving knowledge of the genetic basis for UPOs should sharpen up the estimate of the contribution of mutation in the preceding generation to this indicator and allow a recalculation of the significance of the findings regarding UPOs. Finally, the collection of data on the survival of the children in the two cohorts continues, as does the collection of data on cancer in these two groups.

New technologies may also be brought to bear on the issue. For instance, it is now possible to digest with a battery of enzymes suitably prepared DNA from a human (or other plant or animal) source, radiolabel it with an appropriate isotope, spread the labeled DNA fragments out in two dimensions by the technique of electrophoresis and visualize the position of the various fragments by techniques sensitive to the radiation emitted by the isotopes. The fragments can then be scored for various types of genetic variation. A mutation would be a DNA finding in a child not present in either parent. Radiation is especially prone to produce deletion-type mutations in chromosomes, ranging from a few nucleotides to segments several megabases in length, and this new technique is quite efficient in the detection of these deletion-type mutations. The RERF staff has in recent years, using blood samples, been preparing "immortalized" human cell lines in family constellations of mother, father and as many children as possible that can serve as the basis for such molecular studies for the foreseeable future, and pilot studies employing this technology are even now underway. A brief discussion of the present status of these techniques will be found in Neel (1995).

Unfortunately, despite the amazing genetic developments at the molecular level in recent years, there remain substantial difficulties in extrapolating from a change at the DNA level to a phenotypic effect. There are relatively extensive areas in DNA whose function is as yet unknown. Fortunately, the techniques of molecular genetics make it possible to recover and sequence any mutant fragments or to identify deletions precisely. However, even with a high level of automation, these new DNA technologies are demanding, expensive and laborious, and, after all possible studies, there may still be some uncertainty about the phenotypic impact of an alteration in DNA – and it is the phenotypic impact that concerns the public. The findings from these new approaches will not replace, but rather supplement, the results of the earlier approaches. For this reason, it will be important to continue the ongoing study on the survival of these children and their propensity to cancer (and possibly other diseases) for the foreseeable future.

3. The results of parallel studies on mice

Extensive studies of the genetic effects of ionizing radiation on mice have been carried out since World War II, on the thesis that the results from a mammal such as the mouse would be a better guide to human risks than the results from Drosophila, and, indeed, while the results of the human studies have been slowly accumulating, the results of the work with mice have supplied the principal guideline for human risks. The very extensive studies of W.L. and L.B. Russell at Oak Ridge, in the US, have been especially influential. The experiments with mice have largely employed gamma radiation, expressed as roentgens or, more recently, gray (Gy; 1 Gy = 100 r). Various individuals, national and international committees, manipulating these and a limited amount of other data, have suggested that for mice the *gametic* doubling dose for acute radiation was in the region of 0.40–0.80 Gy (Lüning & Searle, 1971; United Nations Scientific Committee on the Effects of Atomic Radiation, 1988; Committee on the Biological Effects of Ionizing Radiations, 1990; United Nations Scientific Committee on the Effects of Atomic Radiation, 1993). Based on a 3-fold reduction of mutation yield when radiation is administered chronically, doubling dose values in the range of 1.0–2.5 Gy have been suggested. In a reappraisal of all the data from male mice, we have suggested a higher gametic doubling dose for acute radiation than that usually employed, in the region of 1.35 Gy (Neel & Lewis, 1990) or approximately 4.0 Gy when the radiation is delivered chronically. (For present purposes, we can treat 1 Gy and 1 Sv as equivalent in the genetic damage they create.) These results are based on experiments with male mice. The radiation of female mice produces in the first litters following the treatment about the same frequency of mutations as results from male radiation, but the yield then declines very sharply in the later litters of these females (Russell, 1965).

How to combine the results to yield a *zygotic* doubling dose, such as is necessary to compare with the human data, has presented a quandary. The most *conservative* course would be simply to double the committee-sanctioned figures based largely on the Russell data on males, resulting in figures of 0.8–1.6 Gy for exposure to acute and 2.4–7.2 Gy for exposure to chronic radiation, but this procedure almost surely results in an underestimate of the doubling dose, especially since other mouse genetic systems using other genetic indicators of radiation damage do not show as high a sensitivity as the Russell system. Recently, it has become apparent that certain types of mutation occurring in the mouse experiments were not properly factored into the doubling dose estimates (Russell & Russell, 1996) and that the true value of the doubling dose estimate from mice could be substantially higher than the value employed over the past several decades, as suggested above (Neel, 1998). It will, undoubtedly, come as a surprise to many readers that there is still this uncertainty about the genetic risks of radiation, but, in many ways, the uncertainty reflects increasing knowledge of the complexity of the genome and the subtlety of its repair mechanisms. Perhaps the most judicious statement at this juncture is that both the doubling dose estimates for humans and mice carry many uncertainties, but are not in conflict with each other.

4. Epilogue

Ionizing radiation intrudes into the lives of all of us, and governments go to some pains to regulate these exposures. The horror of the atomic bomb blasts can never be undone, but, in the aftermath, studies such as those in Japan can help provide important guidance to responsible governmental agencies regarding the regulation of exposure to ionizing radiation. At the time of writing, it appears that the genetic risks of human exposures to ionizing radiation are not as great as feared in the immediate post-World War II years, or even up to 10 or 20 years ago.

From the scientific standpoint, the (unfortunate) circumstances in Japan probably resulted in a better opportunity to understand all the effects on humans of whole-body exposure to ionizing radiation than will ever occur again. Thus, it is a matter of some importance to the unfortunates exposed to the bomb blasts and their children, and to society as a whole, to learn as much as possible from the Japanese experience, so as to deal better with the questions that continue to arise regarding radiation exposures.

REFERENCES

Committee on the Biological Effects of Ionizing Radiations (1990) *Health Effects of Exposure to Low Levels of Ionizing Radiation (BEIR V)*, Washington, DC: National Academy Press; 2nd edition edited by W.R. Hendee and F.M. Richards, Bristol and Philadelphia: Institute of Physics Publishing, 1996

Friedberg, E.C., Walker, G.C. & Siede, W. (1995) *DNA Repair and Mutagenesis*, Washington, DC: ASM Press

Los Alamos Scientific Laboratory (1950) *The Effects of Atomic Weapons*, Washington, DC: US Government Printing Office

Lüning, K.G. & Searle, A.G. (1971) Estimates of the genetic risks from ionizing radiation. *Mutation Research* 12: 291–304

Muller, H.J. (1959) Advances in radiation mutagenesis through studies on Drosophila. In *Progress in Nuclear Energy Series VI*, volume 2, edited by J.C. Bugher, New York: Pergamon Press

Neel, J.V. (1995) Invited editorial: new approaches to evaluating the genetic effects of the atomic bombs. *American Journal of Human Genetics* 57: 1263–66

Neel, J.V. (1998) A reappraisal of studies concerning the genetic effects of the radiation of humans, mice, and Drosophila. *Environmental and Molecular Mutagenesis* 31: 4–10

Neel, J.V. & Lewis, S.E. (1990) The comparative radiation genetics of humans and mice. *Annual Review of Genetics* 24: 327–62

Neel, J.V. & Schull, W.J. (1991) *The Children of Atomic Bomb Survivors: A Genetic Study*, Washington, DC: National Academy Press

Roesch, W.C. (ed.) (1987) *US–Japan Joint Reassessment of Atomic Bomb Radiation Dosimetry in Hiroshima and Nagasaki: DS 86 (Dosimetry System 1986): Final Report*, Hiroshima: Radiation Effects Research Foundation

Russell, L.B. & Russell, W.L. (1996) Spontaneous mutations recovered as mosaics in the mouse specific-locus test. *Proceedings of the National Academy of Sciences USA* 93: 13072–77

Russell, W.L. (1965) Effect of the interval between irradiation and conception on mutation frequency in female mice. *Proceedings of the National Academy of Sciences USA* 54: 1552–57

Russell, W.L., Russell, L.B. & Kelly, E.M. (1958) Radiation dose rate and mutation frequency. *Science* 128: 1546–50

United Nations Scientific Committee on the Effects of Atomic Radiation (1988) *Sources, Effects and Risks of Ionizing Radiation Report to the General Assembly, with annexes*, New York: United Nations

United Nations Scientific Committee on the Effects of Atomic Radiation (1993) *Sources and Effects of Ionizing Radiation* (UNSCEAR 1993 Report), New York: United Nations

Vos, J.-M.H. (1995) *DNA Repair Mechanisms: Impact on Human Diseases and Cancer*, Austin, Texas: Landes

GLOSSARY

epidemiology study of the determinants of disease in populations

erythrocyte mature red blood cell

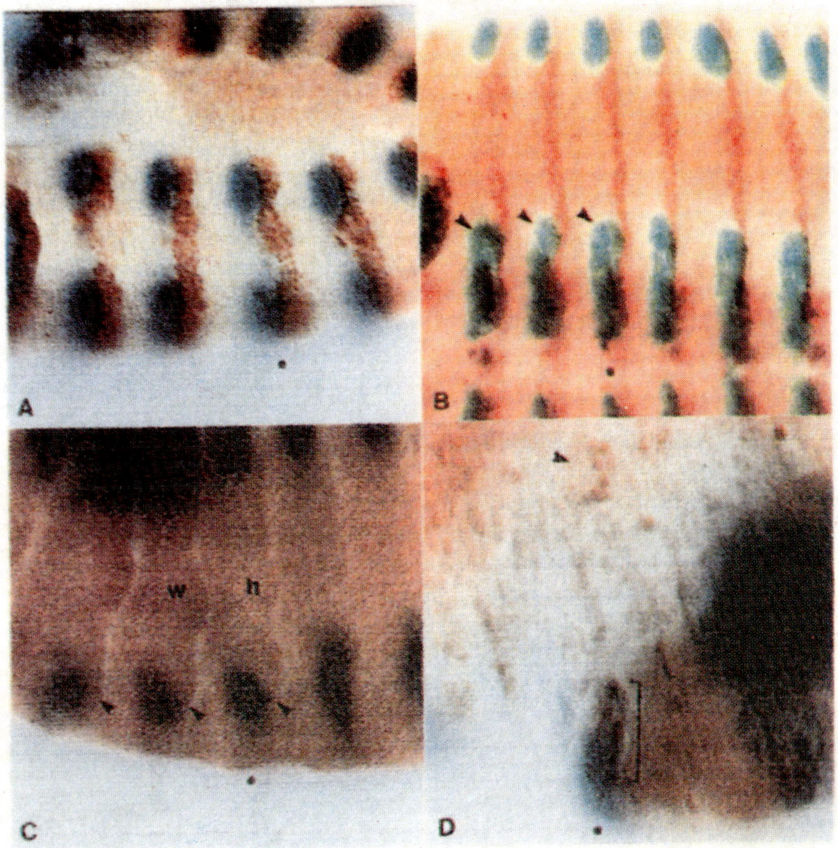

Plate 1 Genetic dissection of wing patterning in *Drosophila melanogaster*: Origin of imaginal disc primordia within segments in the embryo. A reporter gene activity (blue) indicates cells expressing *wingless* and an antibody (brown) identifies the *engrailed* gene product at two stages: (A) at 7 hours of development and (B) at 11 hours. Discs arise at this interface (arrowheads). (C) *wingless* expression is indicated by blue–black staining. While the three leg disc primordia all show staining (arrowheads), the primordia of the dorsal derivatives, the wing (w) and haltere (h) discs, do not stain. (D) shows the embryo after invagination of the leg (bracket) and the haltere discs (arrowhead). The haltere disc has *engrailed* but not *wingless* activity in some of its cells. In each micrograph (A–D), the black spot identifies the third thoracic segment. (Reprinted with permission from *Science* 259: 484–89. © 1993 American Association for the Advancement of Science.)

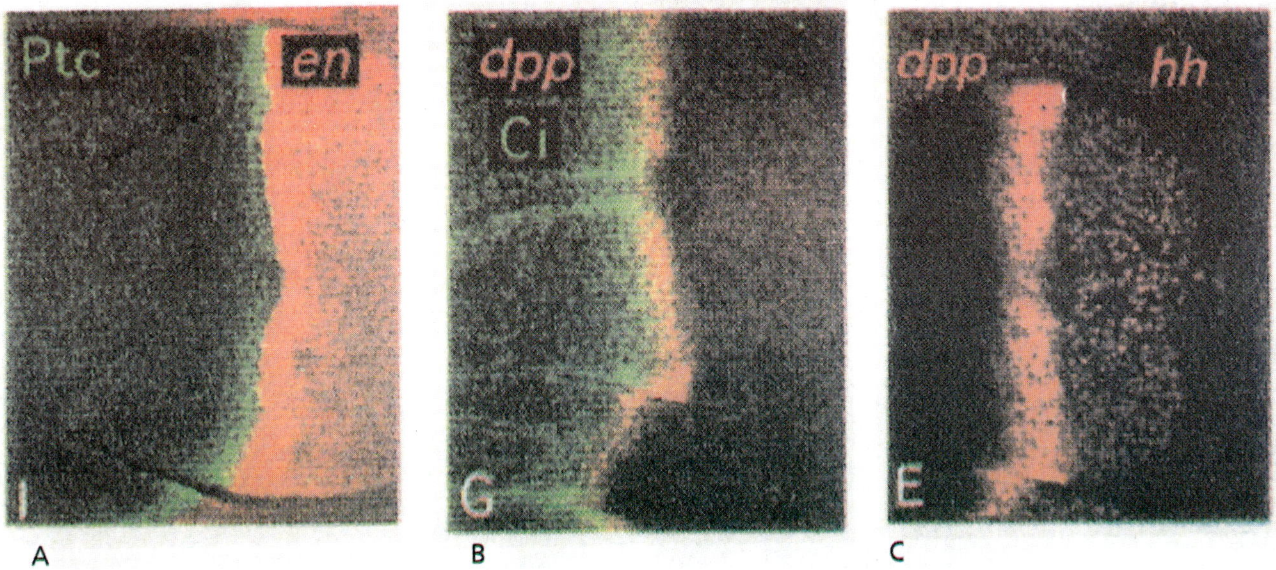

Plate 2 Genetic dissection of wing patterning in *Drosophila melanogaster*: Gene expression at the antero-posterior (A–P) border. (A) All posterior compartment cells express Engrailed (red), while anterior cells show a graded expression of Patched (green) which is most intense at the A–P border. (B) All A cells express the gene product Ci (green), but only those A cells immediately adjacent to the A–P border express the gene *dpp* (red, but indicated as yellow where red and green are simultaneously present). (C) Posterior cells express the gene *hedgehog* (purple), whose gene product triggers *dpp* expression (red) only in immediately adjacent A cells. (© Cell Press, reproduced with permission from Chen & Struhl, 1996.)

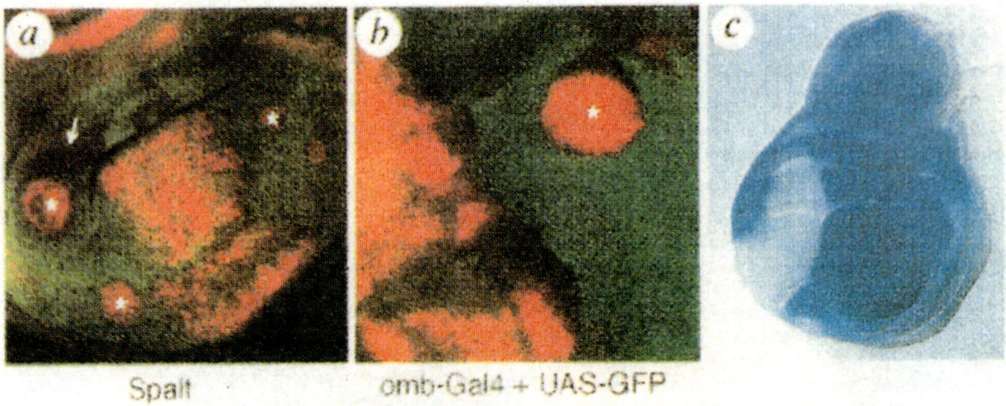

Plate 3 Genetic dissection of wing patterning in *Drosophila melanogaster*: The effect on target genes of activating the Dpp receptor in a clone of cells. (a) The normal expression pattern of *spalt*, a target gene regulated by Dpp signalling, is shown in red (or as orange in cells simultaneously expressing the two indicators, red and green). Clones of cells in which *thickveins* (*tkv*) is activated (not-green marks the clones) autonomously express *spalt* (red), but with no expression outside the border of the clone (no orange halo). (b) *tkv-activated* clones (not-green) autonomously express *optomotor blind* (red), another target gene regulated by Dpp signalling, but this response is also limited to the clone (no orange halo). (c) *tkv-activated* clones (not-blue) overgrow normal cells in the disc (blue). (Reprinted by permission from Lecuit *et al.*, *Nature* 381: 387–93 © 1996 Macmillan Magazines Ltd.)

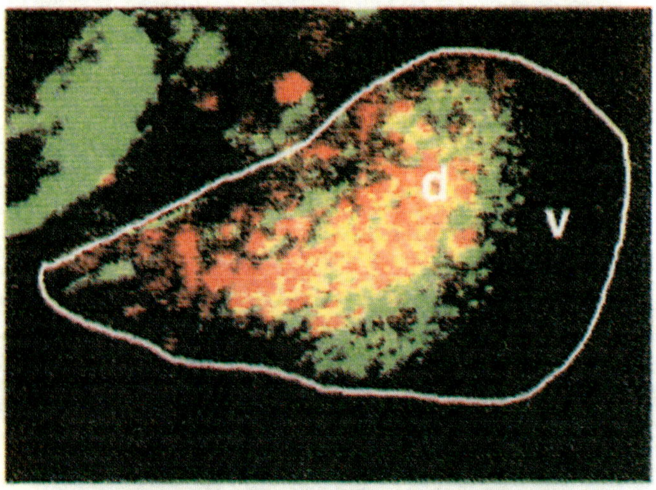

Plate 4 Genetic dissection of wing patterning in *Drosophila melanogaster*: Gene activities at the dorso-ventral (D–V) border of the wing blade. In this wing disc, at the moult between the second and third larval instars, the gene *apterous* (red) is expressed only in D cells. The *Serrate* gene, encoding a signalling ligand, appears in ventral cells (green), but also overlaps the expression domain of *apterous* (yellow, indicating the combination of green plus red) and consequently leads to the creation of wing margin structures at the D–V interface only. (Reprinted from Couso *et al.*, © 1995, with permission from Elsevier Science.)

Plate 5 Genetic dissection of wing patterning in *Drosophila melanogaster*: A compartment-specific signal inducing *cut* expression at the wing margin. The attempt to induce the gene *cut* in cells along the length of the A–P border (by deliberately upsetting its activation by genetic means in these cells) results in *cut* expression in D cells (white arrowhead) at the A–P border, but not in V cells along the A–P border (the line indicated by * *). The white arrow identifies the D–V border. (Reproduced with permission from Doherty *et al.*, 1996.)

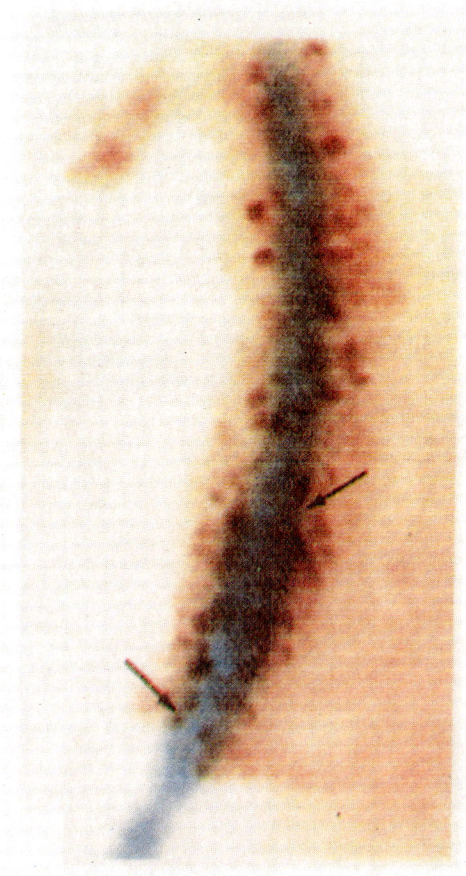

Plate 6 Genetic dissection of wing patterning in *Drosophila melanogaster*: Emergence of spatial order and cell fates at the D–V border of the wing. The *cut* gene is expressed at the D–V border (blue) and results in the reduction of *achaete* expression (purple) within this central zone so that parallel rows of cells only on either side of the Cut stripe (black arrows) become competent to form sense organs. (Reprinted from Couso *et al*., © 1994, with permission from Company of Biologists Ltd.)

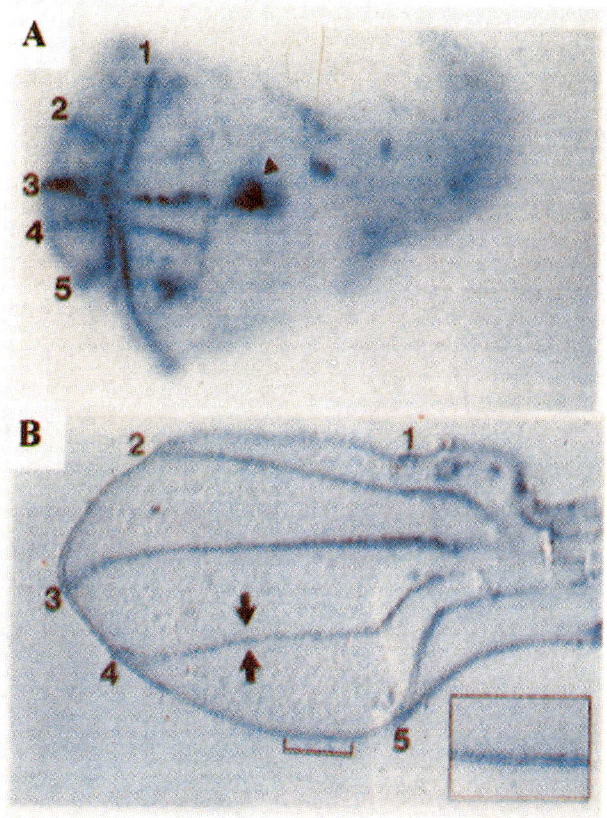

Plate 7 Genetic dissection of wing patterning in *Drosophila melanogaster*: Expression of the *veinlet* gene prefigures the formation of wing veins. (A) In the wing pouch during the third larval instar, stripes of *veinlet* expression (blue) indicate the positions where the veins (numbered, see also **Figure 1**) will appear. Vein 1 runs along the anterior part of the D–V margin. (B) At a later stage, 30 hours after pupation, the D and V wing surfaces have become apposed. The prospective veins are still identified by *veinlet* expression (blue), but the cells on both sides of each vein (dark arrows) have a different appearance even though they do not stain. (Reproduced with permission from Sturtevant *et al*., 1993.)

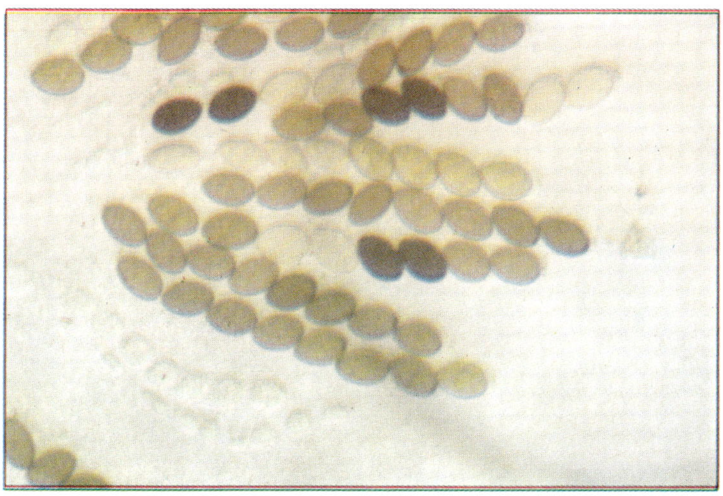

Plate 8 Fungal genetics: The photograph shows asci resulting from crossing buff and yellow spore colour mutant strains of the fungus *Sordaria brevicollis*. Some of the asci show the result of recombination between the genes; such an event results in a tetratype ascus. Three tetratypes are present, each containing a pair of black, white yellow and buff spores. In the other four asci no recombination has occurred. These contain four buff and four yellow spores.

Plate 9 Pharmaceutical proteins from milk of transgenic animals: Tracy – PPL Therapeutics' transgenic ewe – produces 30 g/l of human α-1-antitrypsin in her milk.

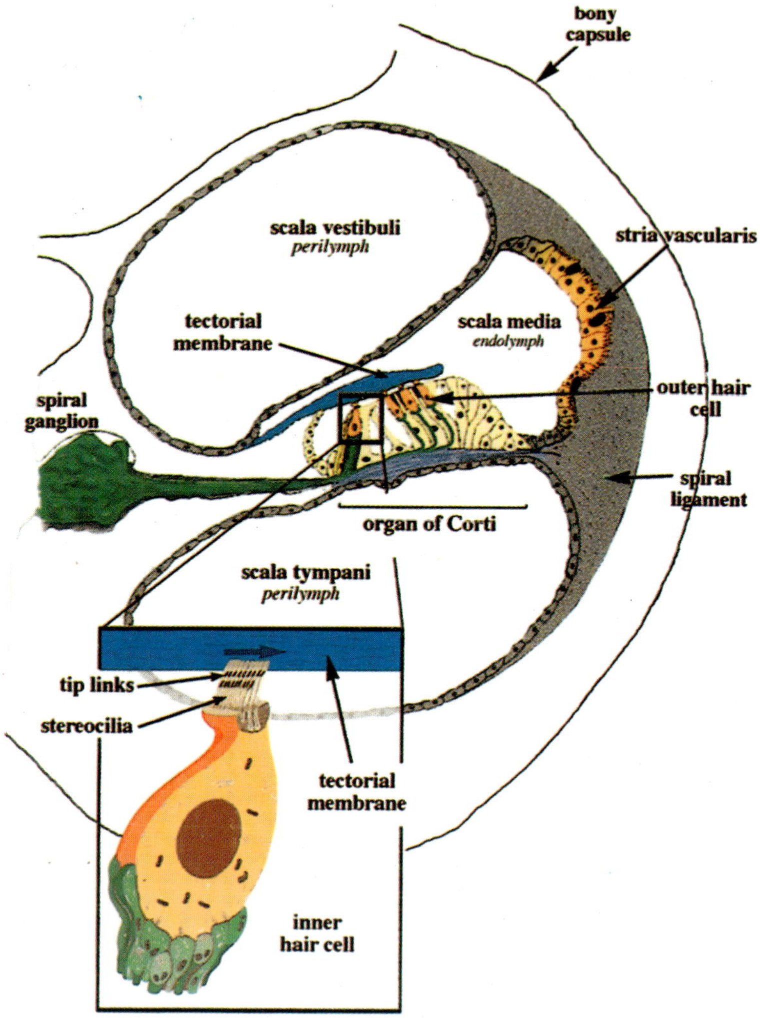

Plate 10 Hereditary hearing loss: Cross-section of the cochlea. The organ of Corti is situated on the floor of the endolymph-filled cochlear duct, and is made up of an array of sensory and supporting cells and the overlying tectorial membrane. Each sensory cell is capped by a stereociliary bundle (inset) which is deflected by shearing of the tectorial membrane. The stria vascularis on the lateral wall of the cochlear duct is responsible for the unique ionic composition of the endolymph. The cochlear duct is surrounded above and below by perilymph-filled spaces. Modified and reprinted from Petit (1996).

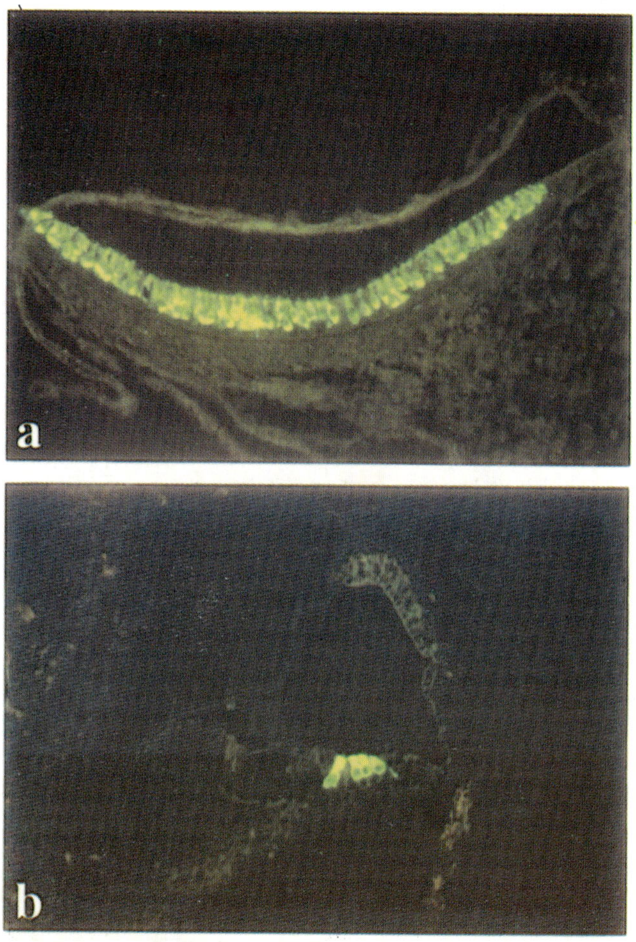

Plate 11 Hereditary hearing loss: Myosin VIIA in the sensory hair cells of the mouse inner ear. Immunohistofluorescence using a polyclonal antibody to myosin VIIA results in a signal in the sensory hair cells of (a) the utricle of the vestibular apparatus (at postnatal day 8) and (b) the cochlea (at postnatal day 2).

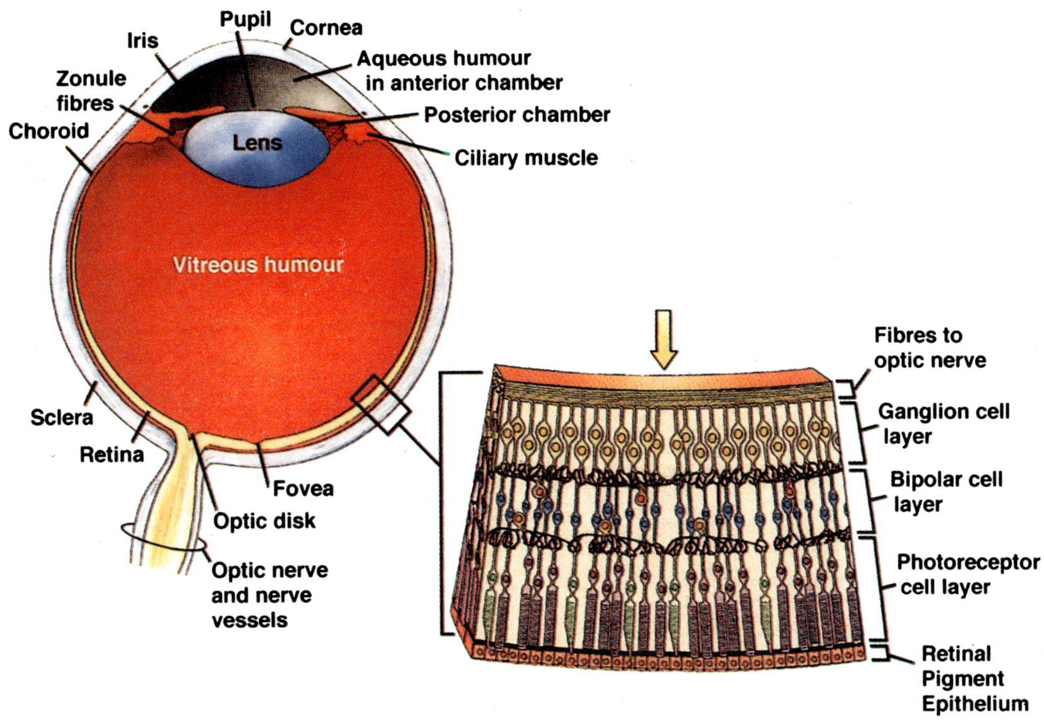

Plate 12 Genetic eye disorders: Anatomy of a vertebrate (human) eye.

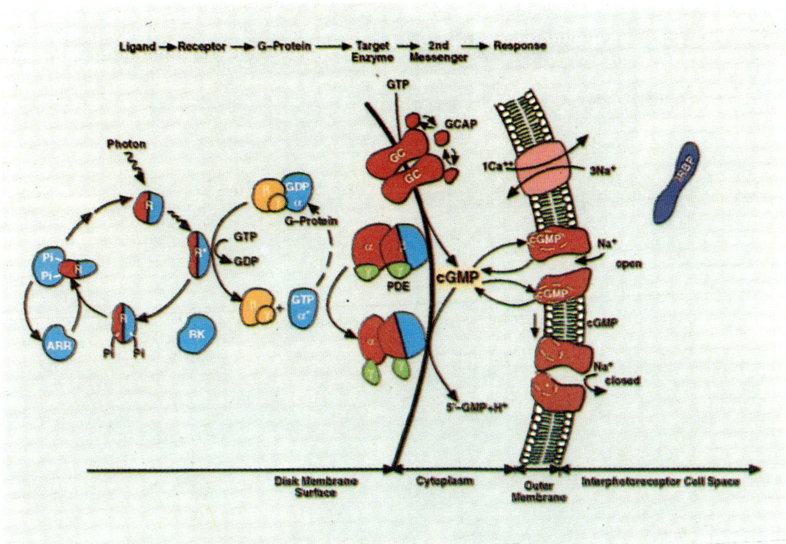

Plate 13 Genetic eye disorders: Vertebrate phototransduction cascade. In the photoreceptor outer segment, light activates rhodopsin (R→R*) which promotes the exchange of GDP for GTP on the α subunit of the heterotrimeric G protein transducin with dissociation of the β and γ subunits. The activated transducin αGTP subunit stimulates cGMP phosphodiesterase (PDE) activity which hydrolyses cGMP to 5′-GMP, with closure of the cyclic nucleotide gated channels and hyperpolarization of the cell membrane. The Na^+Ca^{2+}/K^+ exchanger pumps calcium ions out of the cell, lowering the intracellular calcium concentration, which helps to restore intracellular cGMP levels by means of guanylate cyclase-activating protein (GCAP) and retinal guanylate cyclase (GC). In the dark, the cyclic nucleotide gated channels are open and sodium and calcium ions enter the cell, depolarizating the cell membrane.

Rhodopsin is deactivated by rhodopsin kinase (RK)-mediated phosphorylation of carboxyl terminal serine residues which promote binding of arrestin (ARR). The final deactivation step is the removal of phosphate (P_i) from phosphorylated residues by a protein phosphatase. Inter-photoreceptor retinoid-binding protein (IRBP) is concerned with the shuttling of retinoids between RPE and retina.

Proteins that can give rise to retinal degeneration when mutated are shown in red (human) or green (transgenic mice). Proteins shown in light blue can give rise to a stationary retinal disorder. Modified from Falk and Applebury (1988).

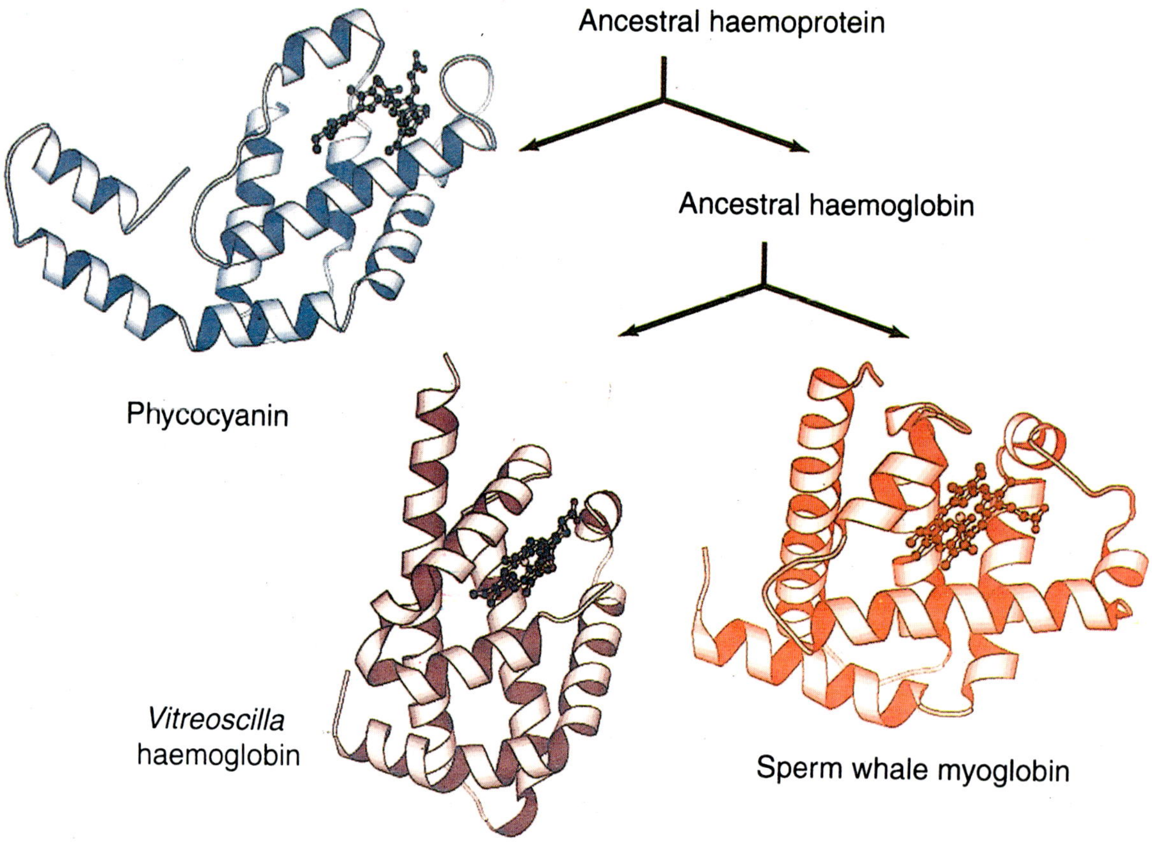

Plate 14 Evolution of the haemoglobins: Ribbon diagrams of the 3-dimensional structures of sperm whale myoglobin, Vitreoscilla haemoglobin and C-phycocyanin from Mastigocladus, arranged in the deduced evolutionary relationships. The ribbons diagrams were generated by Dr Greg Farber using the atomic coordinates deposited in the Protein DataBank.

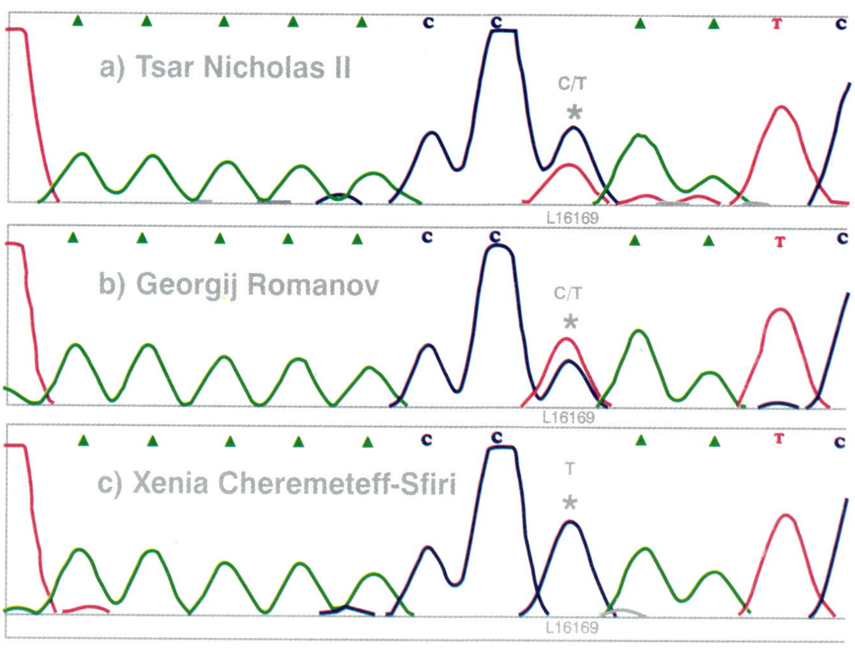

Plate 15 Forensic DNA analysis of the last Russian royal family: Automated DNA sequence electropherograms comparing mtDNA sequences at position 16169. (a) sequence from putative Tsar Nicholas II, showing heteroplasmy with cytosine predominating thymine; (b) sequence from bones of Grand Duke Georgij Romanov, showing heteroplasmy with thymine predominating cytosine; (c) sequence from Countess Xenia Cheremeteff-Sfiri, homoplasmic for thymine.

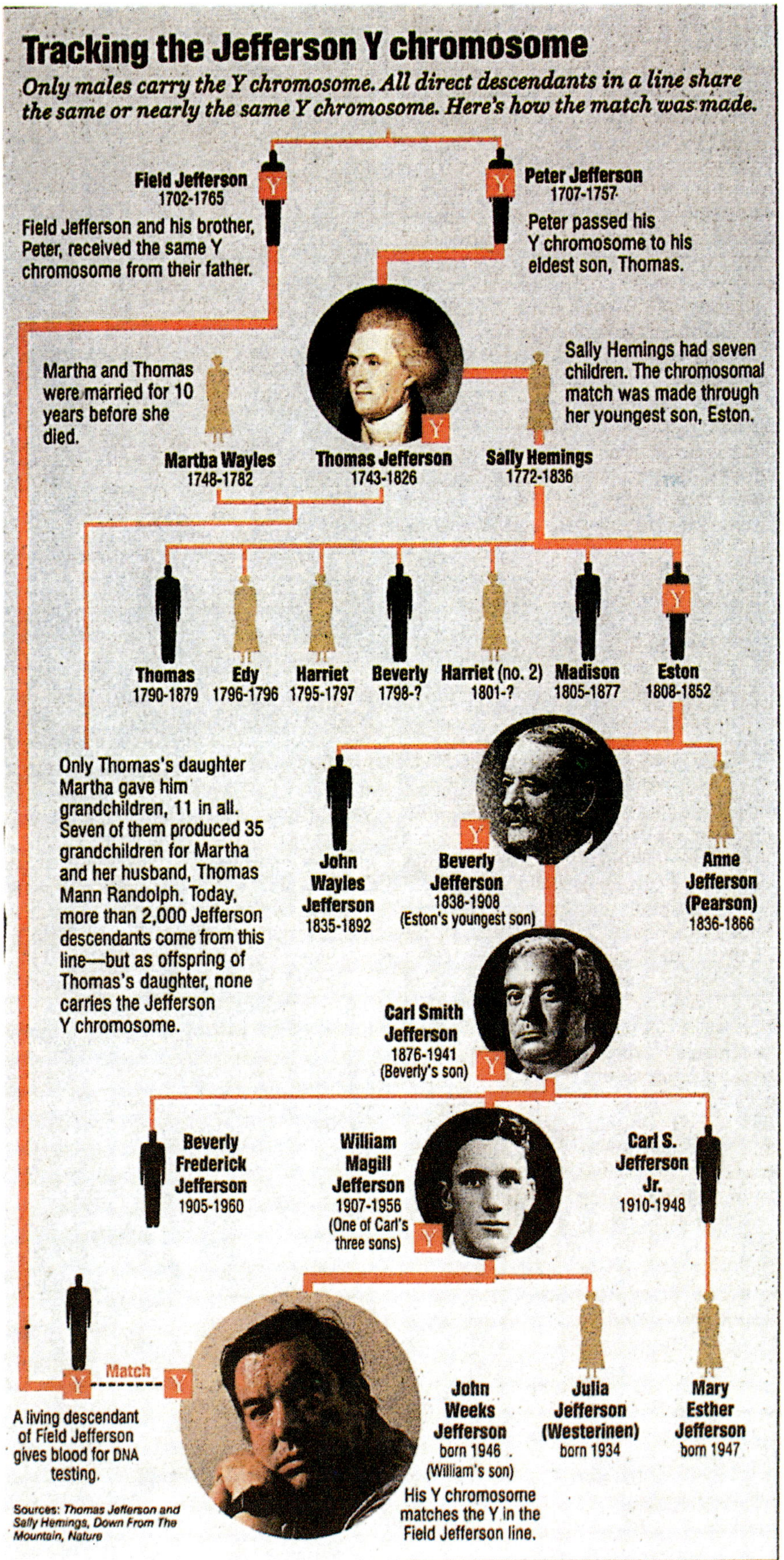

Plate 16 Y-chromosomal DNA analysis and the Jefferson–Hemings controversy: Tracking the Jefferson Y chromosome. A simplified version of the family tree, showing the transmission of the matching Y chromosomes. (Reproduced from USNEWS, 1998.)

Plate 17 Apples – the genetics of resistance to diseases and insect pests: Apple seedlings segregating for the major gene Pl_w, showing resistance and susceptibility to mildew (*Podosphaera leucotricha*).

Plate 18 Apples – the genetics of resistance to diseases and insect pests: Saturn – a variety from HRI–East Malling's resistance breeding programme. Saturn contains the V_f gene for resistance to scab (*Venturia inaequalis*) and has polygenic resistance to mildew (*Podosphaera leucotricha*).

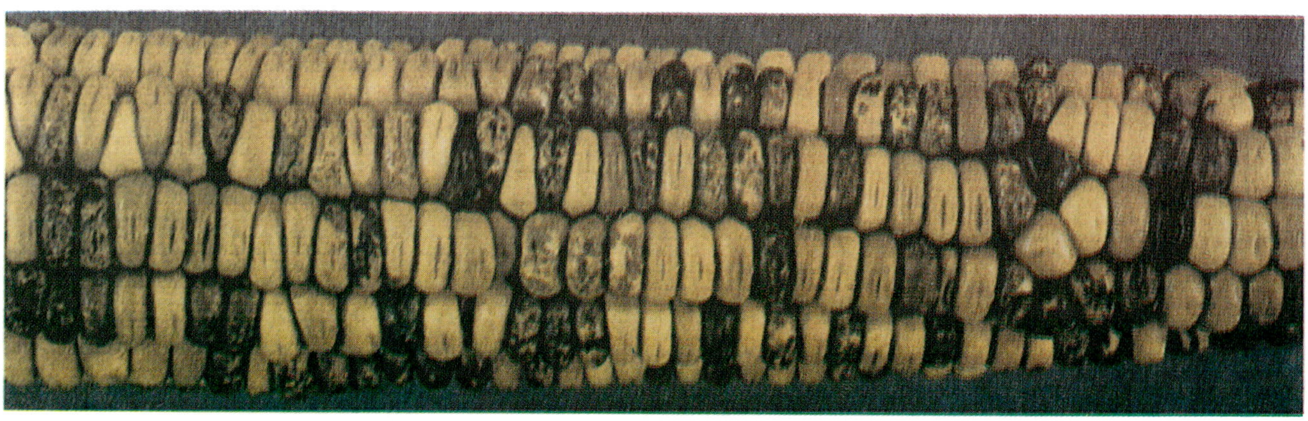

Plate 19 Barbara McClintock and transposable genetic sequences in maize (*Zea mays*): Maize cob showing 1:1 segregation of presence/absence of *Ac*, with all kernels carrying a *bronze-mutable* (bz^m) allele in heterozygous combination with stable recessive *bz.* (Reproduced with permission from Neuffer *et al.,* 1997.)

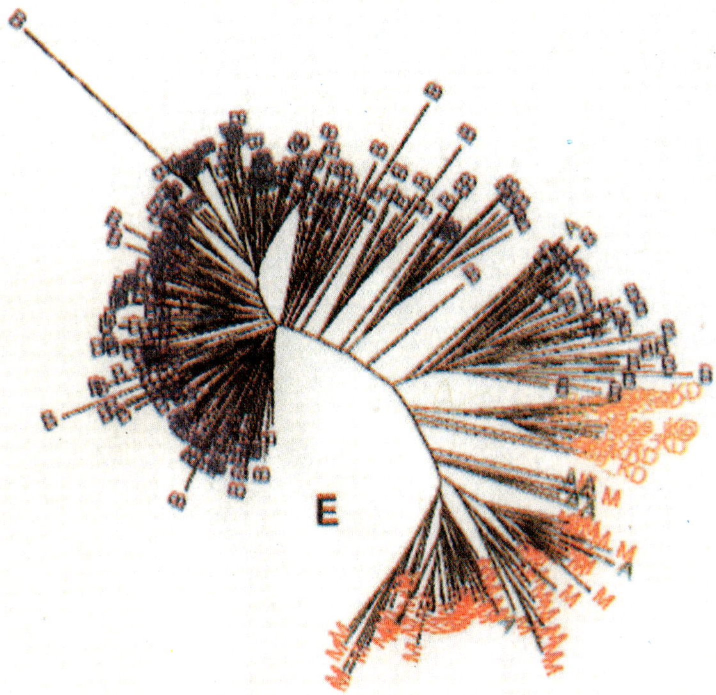

Plate 20 Pinpointing einkorn wheat domestication by DNA fingerprinting: Unrooted tree with all fingerprinted lines: red, cultivated einkorns; green, *T. aegilopoides*; orange, *T. boeoticum* from the Karacadag; blue/purple, remaining *T. boeoticum*.

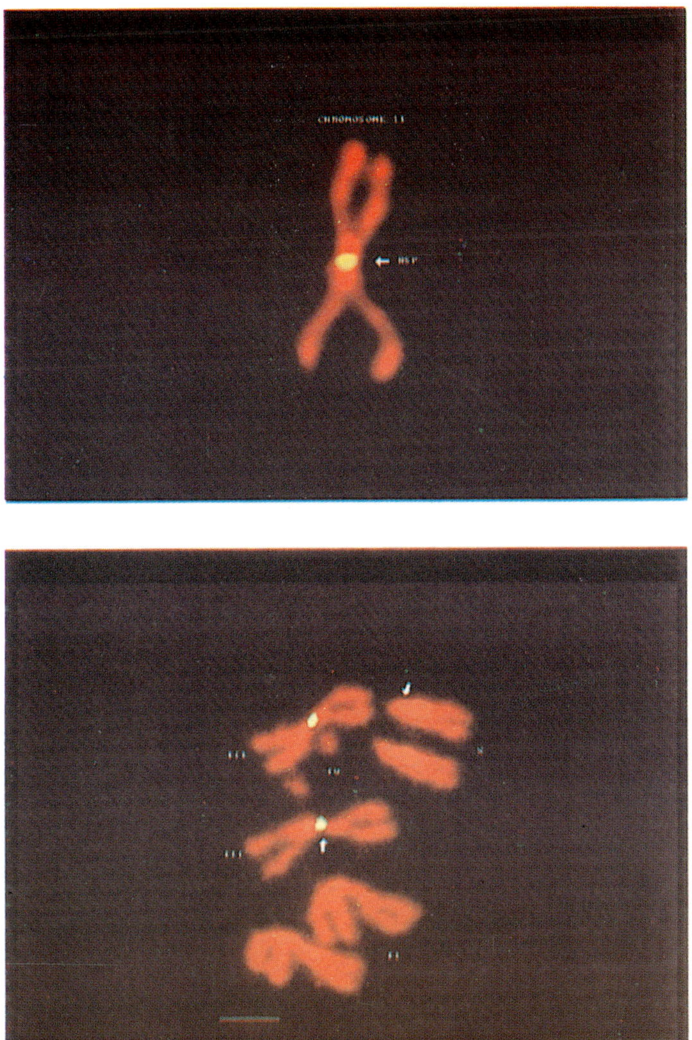

A

B

Plate 21 Centromeres and kinetochores: from yeast to man: (A) Fluorescence *in situ* hybridization (FISH) of chromosome II of *Drosophila melanogaster* with a probe for the RSP repeated element. This repeated element is found only in chromosome II. The DNA has been stained with propidium iodide (red) and the probe with fluorescein (green). The overlay of the two images gives a yellow colour (Paula Coelho, unpublished results). (B) FISH of a metaphase spread from a female *D. melanogaster* neuroblast. The different chromosomes are labelled as: chromosome II (II), chromosome III (III), chromosome IV (IV) and the X chromosome (X). The preparation was hybridized with the dodecasatellite repeat specific for chromosome III. The primary constriction of chromosome III and the X chromosome are shown with arrows. The DNA has been stained with propidium iodide (red) and the dodecasatellite probe with fluorescein (green). The overlay of the two images gives a yellow colour (Paula Coelho, unpublished results, used with permission). Bar is 5 μm.

Each cell in the **Oxford Grid** represents a comparison of two chromosomes, one from each of the selected species. The number of homologies appears inside each colored cell and the color indicates a range in the number: Grey (1), Blue (2-10), Green (11-25), Orange (26-50), Yellow (50+). **Click on a colored cell to retrieve homology details.** A note about printing an Oxford grid...

Click on a mouse chromosome (*blue numbers next to grid frame*) to retrieve a **comparative map** showing all homologies between the selected mouse chromosome and the comparison species displayed on the grid.

Total Homologies: 4059
Total mapped in both species: 2941

mouse, laboratory (columns); human (rows)

human \ mouse	1	2	3	4	5	6	7	8	9	10	11	12	13	14	15	16	17	18	19	X	Y	XY	UN	MT	
1	91	1	81	92	2			5		1	1	1	6					1					74		1
2	70	43			6	27					8	16			1		9	1					43		2
3	1		19	1		25		1	52					10		30	1						45		3
4			24		69	4		20				1	1						1				21		4
5	1										43		52		11		3	26					31		5
6	5			6	1		1		7	29			18				105						27		6
7		1			59	57	1				12	11	5				1						24		7
8	6		7	9				23						16	30	3				1			19		8
9		41		46									8						10				20		9
10		11				3	11			12			1	16				3	44				18		10
11	1	31		1			65		45		1								42				51		11
12		1	1		18	63	1			40					44								38		12
13	2		2		8			12						19				1					12		13
14	1		1		1		1	1	1	1		44		24									23		14
15		23					32		32		1		1										11		15
16		1					20	49			8				1	14	15						22		16
17			1			1		1			185	2		1									41		17
18	3				2											1	4	29					6		18
19						1	91	23	9	24	1	1					15						45		19
20		77																					11		20
21										18				1		26	6						12		21
22	1				6	2		3		8	9				33	27							11		22
X											2			1					1	157			18		X
Y																					6				Y
XY																									XY
UN	10	16	11	13	10	2	12	9	7	13	14	4	8	2	8	5	11	2	5	3			300		UN
MT																								13	MT
	1	2	3	4	5	6	7	8	9	10	11	12	13	14	15	16	17	18	19	X	Y	XY	UN	MT	

mouse, laboratory

http://www.informatics.jax.org/searches/oxfordgrid.cgi

Plate 22 Comparative mapping: tracking gene homologues among mammals: Oxford grid, providing a genome-wide overview of comparative mapping data.

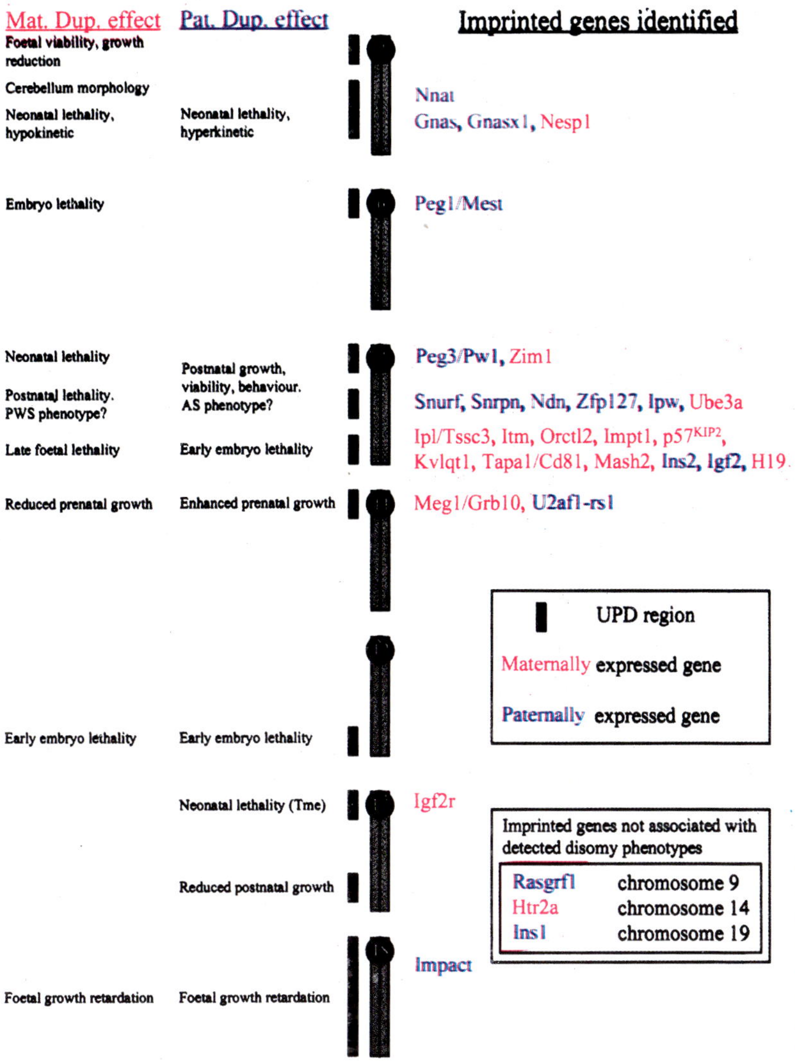

Plate 23 Genomic imprinting: Uniparental disomy (UPD) phenotypes in mice and the imprinted genes associated with them. The seven chromosomes with known UPD phenotypes are displayed in the centre of the diagram (black circles indicate the centromeric ends of the chromosomes), together with the approximate locations of the UPDs (green rectangles).

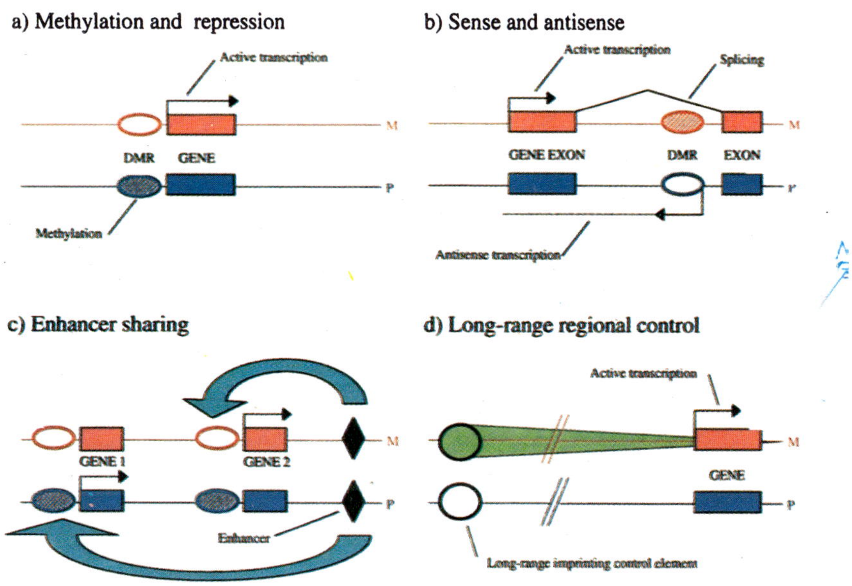

Plate 24 Genomic imprinting: The four principal mechanisms by which imprinted gene expression is regulated. Genes (or their constitutive exons) are shown as red (maternally derived) or blue (paternally derived) rectangles. Differentially methylated regions (DMRs) are represented as ovals which are either open (unmethylated) or hatched (methylated). Transcription is denoted by arrows. See text for details of each mechanism. M, maternal; P, paternal.

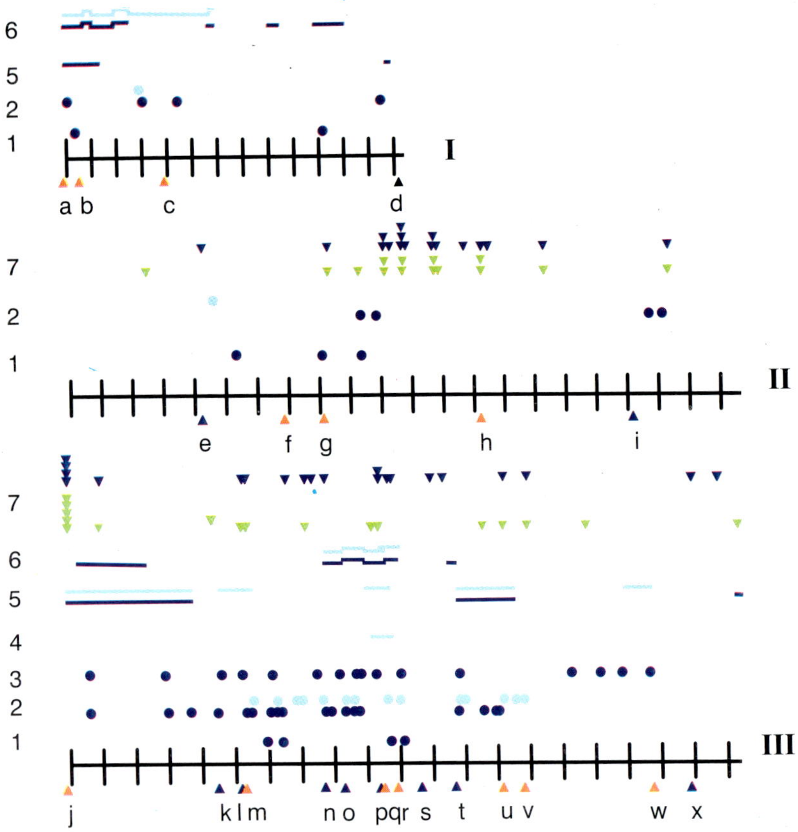

Plate 25 Genetics of quantitative characters: Map positions of QTLs and candidate loci affecting bristle number, in *Drosophila melanogaster*. (Adapted from Mackay, 1996. © 1996, reprinted by permission of Wiley-Liss, Inc., John Wiley & Sons, Inc.) Horizontal lines labelled (I), (II) and (III) at the right, depict chromosomes 1, 2 and 3, and short vertical strokes crossing these lines indicate 5 cM intervals along them. Lettered triangles point to the locations of candidate bristle number QTLs, with red triangles showing neurogenic loci necessary for normal sensory organ development. The gene names, abbreviations and map positions (cM) of the loci are as follows (Lindsley & Zimm, 1992): a, *achaete-scute* complex *(ASC)*, 1–0.0; b, *Notch (N)*, 1–3.0; c, *cut (ct)* 1–20.0; d, *bobbed (bb)* 1–66.0; e, *Sternopleural (Sp)*, 2–22.0; f, *numb (numb)*, 2–35.0; g, *daughterless (da)*, 2–41.3; h, *scabrous (sca)*, 2–66.7; i , *smooth (sm)*, 2–91.5; j, *extramachrochaetae (emc)*, 3–0.0; k, *que mao (qm)*, 3–23.0 (Lai *et al*., 1998); l, *abdominal (abd)*, 3–26.5; m, *hairy (h)*, 3–27.0; n, *polychaetoid (pyd)*, 3–39.0; o, *Bearded (Brd)*, 3–42.0; p, *malformed abdomen (mab)*, 3–47.5; q, *atonal (atonal)*, 3–48.0; r, *neuralized (neu)*, 3–50.0; s, *Tuft (Tft)*, 3–53.6; t, *abdominal A, abdominal B (abdA, abdB)*, 3–58.8; u, *Delta (Dl)*, 3–66.2; v, *Hairless (H)*, 3–69.5; w, *Enhancer-of-split (E(spl))*, 3–89.1; x, *brief (bf)*, 3–95.0.

Closed circles above the chromosomes are QTL map positions determined by interval mapping using visible morphological markers; dark blue shows sternopleural bristle number and light blue abdominal bristle number QTLs. Numbers on the leftmost column: line 1 gives the map positions of Gibson & Thoday (1962), Wolstenholme & Thoday (1963), Thoday *et al*. (1964) and Spickett *et al*. (1966); line 2 gives map positions from Davies (1971); line 3 shows map positions from Shrimpton & Robertson (1988) and line 4 indicates the interval with largest effect detected by Breese & Mather (1957). The solid lines refer to interval positions of QTLs detected using *roo* transposable element markers; progeny testing was not used to pinpoint the QTLs within the intervals. Again, dark blue refers to sternopleural bristle number and light blue to abdominal bristle number. Line 5 is data from Long *et al*. (1995) and Line 6 is from Gurganus *et al*. (1999). The green (sternopleural bristle number) and mauve (abdominal bristle number) triangles on line 7 point to cytogenetic map positions (converted to cM) of single *P* element inserts with quantitative effects on bristle number (Lyman *et al*. 1996). The mauve triangles on line 7 were yellow in Mackay's original figure but were not found to reprint sharply enough.

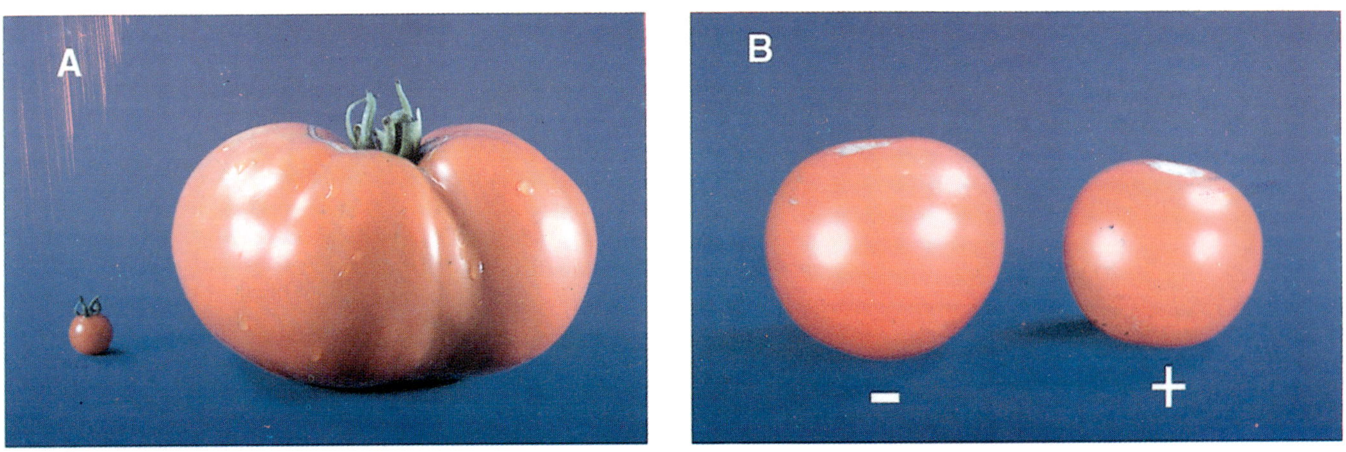

Plate 26 Genetics of quantitative characters: (A) Fruit size extremes in the genus Lycopersicon. On the left is a fruit from the wild tomato species *L. pimpinellifolium* that, like all other wild tomato species, bears very small fruit. On the right is a fruit from *L. esculentum* cv Giant Red, bred to produce extremely large tomatoes. (B) Phenotypic effect of the *fw2.2* transgene in the cultivar Mogeor. Fruit are from R1 progeny of fw107 segregating for the presence (+) or absence (–) of cos50 containing the small-fruit allele. (Reprinted from Frary *et al*., 2000, with permission. © 2000 American Association for the Advancement of Science.).

ABSOLUTE PITCH

Peter K. Gregersen

Division of Biology and Human Genetics, North Shore University Hospital, Manhasset, New York, USA

Absolute pitch (AP), also known as "perfect" pitch, is the relatively rare cognitive ability to identify the pitch of a musical note or ambient sound without the use of a reference pitch. In many individuals, this trait arises spontaneously in early childhood and it is likely that inheritance plays a significant role in its development. Individuals with AP who have some knowledge of music will instantaneously associate sounds, for example, a passing car horn, with the pitch-naming conventions of the western 12-tone scale. Despite the fact that professional musicians spend a lifetime working within these conventions, only a minority of musicians possess AP ability. In contrast, all competent musicians possess the ability to hear "relationships" between two or more pitches, independent of their absolute frequency. This ability is termed relative pitch.

Pitch is objectively defined by the fundamental frequency, in vibrations per second, emanating from a sound source, such as a vibrating string on a violin or piano (Rasch & Plomp, 1982). In reality, strings usually vibrate at multiple frequencies, with wavelengths that correspond to the entire length of the string (the fundamental frequency) as well as fractional string lengths ("overtone" frequencies). One can generate pure fundamental tones electronically, but even here the resonant characteristics of speakers may produce overtones. Nevertheless, a casual listener usually hears the fundamental frequency as "the" pitch of the sound source. The particular mixture and timing of the overtones establish the tone quality, or timbre, differences between different instruments (Risset & Wessel, 1982).

The human ear is able to apprehend pitch frequencies in a range of ~20–5000 Hz (vibrations per second), which is approximately the range of a piano keyboard. The cochlear basement membrane of the inner ear is structured so that it vibrates at a particular resonant frequency at each point along its length, within this range of frequencies. These membrane locations are, in turn, neuronally linked to particular locations in the temporal cortex (so-called "tonotopic" organization). Thus, there is an exquisitely precise link between pitch perception and brain organization in all individuals, independent of musical ability. This emphasizes that both relative and absolute pitch ability involve higher cortical processes, and are not a reflection of a special kind of ear.

AP is an appealing cognitive phenotype for scientific study because it is relatively easy to develop objective tests for it. A test generally involves presenting subjects with a series of randomly generated tones in rapid succession, and asking them to identify the pitch name, either verbally or by pressing a keyboard or other apparatus (Miyazaki, 1988). The vast majority of individuals who report AP ability perform extremely well on these tests, often with 100% accuracy. Nevertheless, there is some variation in ability among subjects with AP (Baharloo *et al.*, 1998, Miyazaki, 1988) and accuracy may even vary over time in one individual (Wynn, 1993). In contrast, individuals without AP will generally achieve a level of accuracy which approximates chance (1/12 possible pitch names in the western scale).

Until recently, the role of genetic vs. environmental factors in the development of AP was largely a matter of anecdote and conjecture (Ward & Burns, 1982). There is clearly an association between AP and musical exposure in early childhood, with AP subjects reporting musical training beginning on average 2 years earlier than in musicians without AP (Baharloo *et al.*, 1998; Gregersen *et al.*, 1999). This association has provoked the hypothesis that AP development requires an exposure to music at a "critical period" of brain development. However, it may also be that the presence of AP ability in a child contributes to the early onset of musical activities, through the spontaneous expression of interest in music by the child. While the development of AP in adults has been claimed to be possible, usually with heroic training efforts, there are very few documented cases of this, and the phenotype is not complete (Ward & Burns, 1982).

The argument for genetics stems from the observation that AP occasionally runs in families (Profita & Bidder, 1988), and AP ability often appears without any specific training in early childhood. Recently, several groups have attempted to develop more objective evidence for familial aggregation (Baharloo *et al.*, 1998; Gregersen, 1998; Gregersen *et al.*, 1999). The prevalence of AP is markedly elevated in the first-degree relatives of probands with AP; 15–25% of their siblings also have AP and approximately 5% of their parents also report AP. In order to interpret these data, one must compare these rates to the background population prevalence of AP. Accurate estimates for the general population are not available, in part because ascertainment of this ability requires some musical education. However, it is apparent that the AP rate is highly variable in musically educated populations, from less than 1% to as high as 35% in some music conservatories (Baharloo *et al.*, 1998; Gregersen *et al.*, 1999). Overall, the data suggest that, even controlling for a musical family environment, AP is approximately ten times more likely to occur if a sibling possesses AP ability, thus indicating a substantial genetic component (Baharloo *et al.*, 2000).

A balanced view of the "nature/nurture" question for AP would posit that there is an underlying genetic susceptibility for developing the higher cortical neural networks involved in AP, but that the probability of these connections forming is enhanced by early childhood exposure to a musical environment. The fact that cortical processing of pitch information is different in AP subjects has recently been elegantly demonstrated using functional brain imaging (positron emission tomography, Zatorre *et al.*, 1998). Anatomical differences in the temporal lobe have also been reported in AP subjects (Schlaug *et al.*, 1995). Of course, it is difficult to know whether these differences precede, or are a result of developing, AP ability.

An extremely interesting and relatively unexplored area concerns the possibility that other unusual cognitive abilities associate with AP or that the genetics of some conditions overlap with AP. Among these other abilities and conditions are unusual mathematical or memory abilities, as well as synesthesia. Synesthesia refers to a spontaneous perceptual overlap between two sensory modalities, such as sound and colour (Baron-Cohen *et al.*, 1996; Yoon, 1997). Strong associations, including true synesthesia between pitch and colour or shape, are reported by a small percentage of subjects with AP. Synesthesia was clearly present in a number of famous musicians with AP, including Sibelius and Scriabin (Profita & Bidder, 1988). Other areas of potential genetic overlap with AP include neurological disorders such as Williams' syndrome and autism, although there have been no well-controlled studies of this issue (Gregersen, 1998).

It is commonly asked whether the presence of AP enhances musical ability or musicianship. Certainly, some types of ear training and music dictation tasks are trivial for AP possessors, whereas these skills require extensive training in musicians who rely on relative pitch. However, there is no evidence that musical expressiveness or understanding is necessarily related to AP ability. Indeed, since the essence of musical communication involves the relationships between notes, the possession of relative pitch is generally much more relevant to musical meaning. While no objective study has been done of this issue, anecdotal reports from ear training specialists suggest that, in some cases, AP subjects may overlook pitch relationships, presumably because instantaneous absolute perception is more natural or much easier for them. On the other hand, for some musicians, AP may provide another realm of musical meaning, with certain keys having particular colours or emotional associations.

A major challenge for modern genetics is to achieve understanding of complex, multifactorial disorders and phenotypes. In this context, AP is valuable for study because the phenotype can be relatively cleanly defined, and yet the underlying causes are clearly diverse, with genetic, environmental, developmental and stochastic elements likely to be contributing to its appearance. While a dominant mode of inheritance of AP has been postulated, it is likely that the genetic predisposition is complex and heterogeneous. The future mapping and identification of these genes will provide the tools to assess more accurately the role of early childhood environment on brain development, and provide a window on the molecular mechanisms involved in this process.

REFERENCES

Baharloo, S., Johnston, P.A., Service, S.K., Gitschier, J. & Freimer, N.B. (1998) Absolute pitch: an approach for identifying genetic and non-genetic components. *American Journal of Human Genetics* 62: 224–31

Baharloo, S., Service, S.K., Risch, N., Gitschier, J. & Freimer, N.B. (2000) Familial aggregation of absolute pitch. *American Journal of Human Genetics* 67: 755–58

Baron-Cohen, S., Burt, L., Smith-Laittan, F., Harrison, J. & Bolton, P. (1996) Synaesthesia: prevalence and familiality. *Perception* 25: 1073–79

Gregersen, P.K. (1998) Instant recognition: the genetics of pitch perception. *American Journal of Human Genetics* 62: 221–23

Gregersen, P.K., Kowalsky, E., Kohn, N. & Marvin, E.W. (1999) Absolute pitch: prevalence, ethnic variation, and estimation of the genetic component. *American Journal of Human Genetics* 65(3): 911–12

Miyazaki, K. (1988) Musical pitch identification by absolute pitch possessors. *Perception and Psychophysics* 44: 501–12

Profita, J. & Bidder, G.T. (1988) Perfect pitch. *American Journal of Medical Genetics* 29: 763–71

Rasch, R.A. & Plomp, R. (1982) The perception of musical tones. In *The Psychology of Music*, edited by D. Deutsch, New York: Academic Press; 2nd edition 1999

Risset, J.-C. & Wessel, D.L. (1982) Exploration of timbre by analysis and synthesis. In *The Psychology of Music*, edited by D. Deutsch, New York: Academic Press; 2nd edition 1999

Schlaug, G., Jancke, L., Yanxiong, H. & Steinmetz, H. (1995) In vivo evidence of structural brain asymmetry in musicians. *Science* 267: 699–701

Ward, W.A.D. & Burns, E.M. (1982) Absolute pitch. In *The Psychology of Music*, edited by D. Deutsch, New York: Academic Press; 2nd edition 1999

Wynn, V.T. (1993) Accuracy and consistency of absolute pitch. *Perception* 22: 113–21

Yoon, C.K. (1997) Synesthesia: the taste of music, the sound of color. *Journal of NIH Research* 9: 25–27

Zatorre, R.J., Perry, D.W., Beckett, C.A., Westbury, C.F. & Evans, A.C. (1998) Functional anatomy of musical processing in listeners with absolute pitch and relative pitch. *Proceedings of the National Academy of Sciences USA* 95: 3172–77

THE TESTIS: THE WITNESS OF THE MATING SYSTEM, THE SITE OF MUTATION AND THE ENGINE OF DESIRE

R.V. Short
Department of Obstetrics and Gynaecology, Royal Women's Hospital, University of Melbourne, Australia

1. Introduction

There is now abundant evidence in a wide range of mammalian and non-mammalian species to show that the relative size of the testis, and the morphology of the spermatozoa, are infallible predictors of the mating system. Species with the largest testis: body weight ratios and the best (most uniform in appearance) spermatozoa have a multimale or promiscuous mating system in which sperm competition operates. Judged by these criteria, men were not designed to be promiscuous.

There is increasing evidence in humans to show that most spontaneous mutations of the germline occur in the testis. Since these provide the variability on which natural selection can operate, the testis holds the key to evolution. Genes on the Y chromosome which control male fertility are particularly prone to mutations, perhaps because of the mutagenic metabolites produced by the metabolically active testis. Testicular descent into a scrotum, and cooling by countercurrent heat exchange between the spermatic artery and vein, may have evolved as a way of holding the mutation rate in check. The hormones secreted by the testis, which control libido and aggression, ensure that these male mutations are disseminated as widely as possible throughout the population.

2. The testis: the witness of the mating system

The Latin word "testis" literally means "witness" and it gets its name from the ancient Roman custom of holding the testicles when taking an oath. Recent studies of testicular size relative to body weight have shown that the weight of the testis also bears witness to the mating system, and that this relationship holds true across a wide range of species.

When I first stumbled across this phenomenon in 1977, as a result of an opportunity to examine the genitalia of a tranquillized adult male gorilla, chimpanzee and orang-utan at Bristol Zoo in England, I was struck by the enormous contrast between the minute (18 g) testis of the giant silverback gorilla (200 kg) and the vast (60 g) testis of the small (47 kg) chimpanzee. Knowing that the polygamous male gorilla in the wild may easily go for up to a year without having an opportunity to copulate with one of his oestrous females, whereas the promiscuous male chimpanzee copulates almost daily with any oestrous female in the troop (the reason, no doubt, why the chimpanzee was given the generic name Pan), I concluded that these marked differences in relative testicular size must be a reflection of the different copulatory frequencies (Short, 1977). However, this was only part of the explanation and, at that time, I was unaware of the theoretical work of Parker on the concept of sperm competition in insects (Parker, 1970). Putting these ideas together, however, it became apparent that there must be two reasons for the increased testicular size in chimpanzees: not only does the high copulatory frequency demand a high rate of sperm production, but also, since many males could copulate with one female when she was in oestrus, sperm competition would be intense and the male who deposited the most spermatozoa in the female's reproductive tract would be the most likely to sire the offspring, further increasing the selection pressure for increased sperm production. Since the rate of sperm production per gram of testicular tissue is relatively constant within species, the only way of increasing sperm production is by increasing the volume of seminiferous tubular tissue and, hence, testicular size (Short, 1979).

When we put this idea to the test across a wide range of primates, it was obvious that all those with promiscuous, multimale mating systems, like the chimpanzee, had much larger testes:body weight ratios than species that were polygamous, like the gorilla, or monogamous, like the gibbons (Harcourt *et al.*, 1981). The relatively small size of the human testis clearly indicates that we are not promiscuous by nature (see **Figure 1**). Kenagy and Trombulak (1986) then extended the analysis to all mammals, and found that this relationship still held true (see **Figures 2** and **3**); recent evidence shows that it is also true of birds (Møller, 1988a), fish (G.A. Parker, personal communication) and insects (Gage, 1994).

We then examined the quality of the ejaculated spermatozoa in man and the great apes. Human and gorilla spermatozoa show a high degree of pleiomorphism in head size and shape, and many abnormal forms are present, including diploid spermatozoa. This is in marked contrast to the chimpanzee, whose spermatozoa are remarkably uniform (Seuanez *et al.*, 1977). The most plausible explanation for these rather striking morphological differences is that they are a result of intense sperm selection in species with multimale mating systems (Møller, 1988b), again highlighting the fact that humans are not by nature promiscuous.

In contrast to the testis, whose size is such a reliable indicator of the mating system, the ovary provides no such information; the ovary:body weight ratios in humans, chimpanzees and gorillas are almost identical (Short, 1984). However, there may be a correlation between the ovulation rate in women (which is not necessarily revealed by ovarian weight) and testicular size, since Oriental women are known to have the world's lowest incidence of dizygotic twinning and Oriental men have testes that are half the size of those

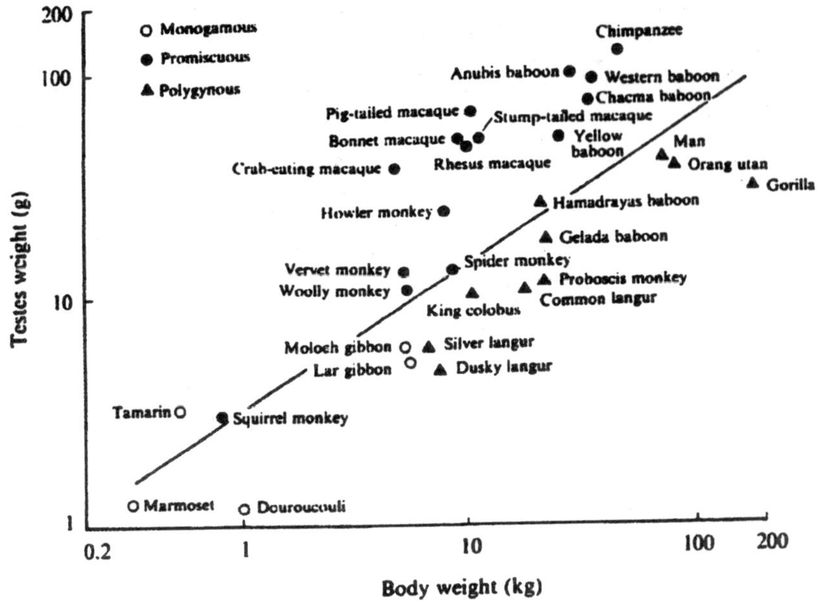

Figure 1 Testes size:body weight ratios in primates with a variety of different mating systems. (Adapted from Harcourt *et al.*, 1981.)

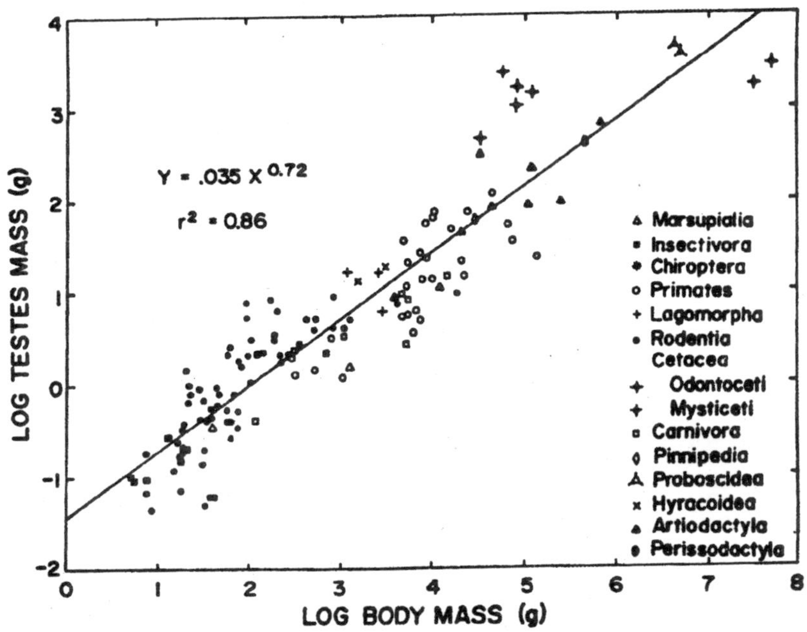

Figure 2 Testes size:body weight ratios in all mammals. Note that the dolphins and porpoises, which are toothed whales or Odontocetes, have particularly high testes:body weight ratios, presumably a reflection of their extremely high copulatory frequencies and promiscuous mating system. (Reproduced with permission from Kenagy & Trombulak, 1986.)

of Caucasians (Short, 1984). Rather than postulating that Oriental and Caucasian men have completely different patterns of sexual behaviour, for which there is no evidence, it seems more likely that dizygotic twinning has been selected against in Oriental women over the centuries and that testis size and ovulation rate may be genetically linked in some way, as they are known to be in animals (Short, 1984).

3. The testis: the site of mutation

In 1947, the English geneticist J.B.S. Haldane addressed the key question of how frequently mutations occur in the germ cells, and whether there is any difference in the mutation rate between the male and the female (Haldane, 1947). By studying the spontaneous mutation rate of the human X-linked gene for haemophilia, he concluded that it was up to 10 times higher in males than in females. This has since become known as the male-driven molecular evolution hypothesis, and there is growing evidence to support it. For example, Montandon and colleagues have been able to make a direct estimate of the human haemophilia B mutation rate by analysing pedigree data from Sweden, and they concluded that the mutation occurred 11 times more commonly in the male germline (Montandon *et al.*, 1992). Somewhat similar figures were provided in 1993 by Ketterling *et al.*, who confirmed a predominance of male germline mutations in the haemophilia B gene, with single-base substitutions being at least 11 times more common in men, whereas

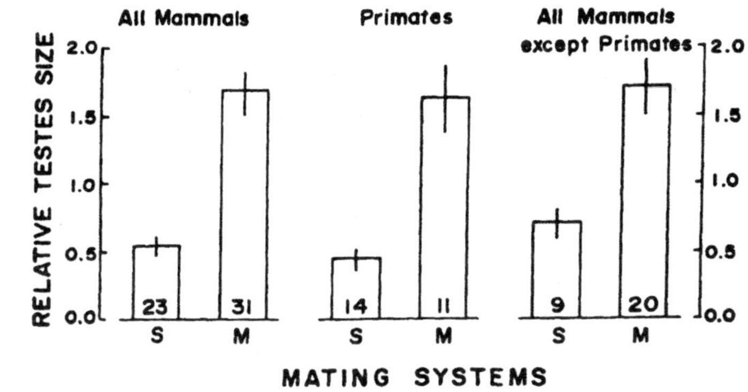

Figure 3 Relative sizes of mammals by mating system. S, single male, monogamous or polygamous; M, multimale, promiscuous or polyandrous; Vertical lines indicate ± standard error of the mean. (Reproduced with permission from Kenagy & Trombulak, 1986.)

deletions (presumably resulting from copying or other errors during cell division) showed no sex ratio bias.

With the recent identification, mapping and sequencing of the sex-determining gene on the mammalian Y chromosome, SRY, it has been possible to compare the amino acid sequences of the coded protein both within and between species. To the initial surprise of molecular biologists, SRY showed great interspecific variability, especially in the N and C terminal regions that flank the central 79 amino acid homeobox region, which is the DNA-binding domain and the presumed functional region of the gene. Although the homeobox region itself is much more conserved across species, it also shows some interspecific variability (Tucker & Lundrigan, 1995). Taken together, these studies suggest that the mutation rate of Y-linked genes is significantly higher than the mutation rate for X-linked or autosomal genes, and this effect is most pronounced in species with long generation times (Tucker & Lundrigan, 1995). This high mutation rate of genes on the Y chromosome may explain why the only genes to survive appear to be those essential for the transmission of life, concerned either with testicular formation or the control of spermatogenesis. The large amount of repeated DNA sequences on the Y chromosome may represent dead or dying genes, not essential for survival, which have mutated into oblivion (Graves, 1995).

There are several reasons why the mutation rate of Y-linked genes should be higher than those on the X chromosome. Apart from the small pseudoautosomal region, the Y chromosome does not recombine with the X during male meiosis, so there is no possibility for recombinant "repair" of DNA on the Y chromosome; all of its genes are clonally transmitted from father to son. X-linked genes, on the other hand, can be "repaired" by crossing-over with the homologous X chromosome during female meiosis, and only one of the four meiotic products of a female primordial germ cell gives rise to a functional oogonium, the others being discarded as polar bodies.

However, the most important reason of all for the high mutation rate of Y-linked genes relates to the male-driven molecular evolution hypothesis. The testis would seem to be a "hotspot" for germ cell mutations, and, as far as the germline is concerned, the Y chromosome is particularly vulnerable, since it never escapes from the testis. In contrast to this, the X chromosome spends much of its evolutionary life "at rest" in the ovary, in the arrested dictyate stage of meiosis, where it may be much less susceptible to mutation. Most germline mutations are thought to be due to endogenous factors rather than to mutagens in the external environment, and oxygen-free radicals or tissue metabolites are probably more mutagenic than copying errors during DNA replication (Sommer, 1992). Since the male germ cells are so metabolically active – the human testis produces 176×10^6 spermatozoa per day – whereas female germ cells have ceased all mitotic divisions and entered the arrested dictyate stage of meiosis before birth, it is perhaps not surprising that the male germline should be more mutation prone than the female.

There are some interesting consequences of the high mutation rate of Y-linked genes. This may be the explanation for Haldane's law, which states that, in interspecific hybrids, it is the heterogametic sex that is likely to be absent, rare or sterile. The best-known mammalian interspecific hybrid is the mule, a cross between a jack donkey and a mare. The female-skewed sex ratio of mules, and the sterility of male mules, could be explained by basic differences in the Y-linked testis-determining and spermatogenesis genes of the horse and donkey (Short, 1997). Male hybrid sterility as a result of Y-linked gene mutation could be an important mechanism for preventing the introgression of species when, for example, the geographical barriers that have allowed their independent evolution are removed.

In humans, we have achieved remarkable success in diagnosing and treating female infertility, whereas a high proportion of cases of male infertility are of unknown aetiology and untreatable. Could these be a result of spontaneous mutations of Y-linked fertility-regulating genes? Will the technique of (in which a single spermatozoon is microinjected into the cytoplasm of the egg) for the treatment of male infertility inadvertently propagate such deleterious mutations? One obvious way of reducing the production of mutagenic metabolites in the testis would be by lowering its metabolic rate, and this can be achieved by cooling. As long ago as 1957, three Swedish scientists suggested that testicular cooling might be a way of keeping the human mutation rate at a low level (Ehrenberg *et al.*, 1957). The descent of the testis into the scrotum, and the development of a countercurrent heat exchange mechanism between the spermatic artery and vein, enables testicular temperatures to be maintained at 4–7°C below core body temperature in most mammals (Setchell, 1982). Even in species like dolphins, where there is no scrotum and the testis is retained in the abdominal fat, it is still cooled by venous blood draining the dorsal fin and tail flukes (Pabst *et al.*, 1995). It is interesting that dolphins have the largest testes:body weight rations of any mammal (Kenagy & Trombulak, 1986; see **Figure 3**), and this is probably related to the fact that the males have an extremely high copulatory frequency, apparently using sex for pleasure as well as for procreation, and being promiscuous into the bargain.

One puzzling exception to the mammalian rule of testicular descent is the elephant, the intra-abdominal location of its testes being first noted by Aristotle. I have dissected a number of male African elephants and found that the spermatic artery ran a short course straight from the dorsal aorta into the testis, lying adjacent to the kidneys, and likewise the spermatic vein ran straight into the posterior vena cava. There was no vestige of a pampiniform plexus vein complex and, hence, no possibility for testicular cooling below the core body temperature of about 37°C (Short *et al.*, 1967). Another peculiarity of elephants is that they continue to grow in height throughout most of their life, the males faster than the females, and recent behavioural observations in Amboseli, Kenya, by Dr Joyce Poole have shown that it is the very biggest and hence by definition the oldest males in their 50s that sire most of the offspring (Poole, 1994). Male elephants come into musth once a year when there is a 5-fold increase in their testosterone levels and they become extremely aggressive. The oldest bulls in Amboseli come into musth at a time when most of the females are coming into oestrus, and the combination of their large body size, their enormous tusks and their aggressive behaviour ensures their reproductive success. Male elephants have the longest generation times of any mammal. This might disadvantage them in evolutionary terms when compared with species with much shorter generation times, such as the rodents. Perhaps by increasing their testicular temperatures they can increase the germline mutation rate per generation and, hence, compensate for this apparent handicap.

Another exception to the rule of testicular cooling by scrotal descent is the situation in all birds. Not only do they maintain a higher body temperature than mammals, but the testes are always intra-abdominal, with no evident cooling mechanism. In birds, it is the male that is the homogametic sex (ZZ), and the female is heterogametic (ZW), so they are spared the problem of exposing the heterogametic sex chromosome to the mutagenic environment of a testis, which may reduce the need for testicular cooling. In any case, perhaps, birds are more able to cope with a high mutation rate because of their oviparity. Since most mutations are likely to be deleterious, then laying an infertile egg in a clutch or having a chick die in the nest is not a great waste of reproductive effort.

However, to return to the mammalian argument, there seems no doubt that it is the male germline that is most susceptible to mutation, and this is likely to be due to something peculiar to the testis. The high metabolic activity of the seminiferous epithelium may be responsible for generating mutagenic metabolites, and testicular cooling by countercurrent heat exchange and/or testicular descent into a scrotum may be one way of keeping the mutation rate in check. It is interesting that testes that are retained in the abdomen are well known to be particularly susceptible to neoplasia and one of the commonest types of testicular tumour is the seminoma, involving the germline.

Since the testis appears to be the principal site of germline mutations, it deserves a new respect in our eyes, since it holds the key to evolution: natural selection depends for its success on the genetic variability generated by the testis.

4. The testis: the engine of desire

Scientific views about the factors responsible for the development of different mating systems have been in a constant state of flux, depending on their social acceptability at the time. Thus, when the late Lord Zuckerman wrote his pioneering book on the social life of monkeys and apes in 1932, he was able to conclude that: "The nucleus of the societies of monkeys and apes is the family party, consisting of an overlord and his harem, held together presumably by the interest of the male in his females and by their interest in their young" (Zuckerman, 1932). Then, there was a growing realization that it was the female that had the greatest energy investment in reproduction and, hence, she must be the limiting resource. This led to a complete re-evaluation of the genesis of primate mating systems. Where food was abundant, the female could afford to have a male permanently in attendance and monogamy was, therefore, the preferred strategy, as is seen in marmosets and gibbons. Where food was more restricted, it paid several females to share a male between them, resulting in polygyny, gorilla style. If food was really scarce, requiring a large area to meet an individual's demands, it paid females to share a number of males, who could be responsible for group defence of a large territory in return for shared sexual favours from all the females – hence the promiscuity of chimpanzees (Wrangham, 1979). Human reproductive strategies are often a compromise; although at any one time the man and woman may be pair-bonded in a mutually monogamous relationship, these relationships may not withstand the test of time, with serial monogamy as a result. Since the menopause restricts a woman's fertile life, whereas men are usually fertile throughout adult life, and since men, on remarriage, almost invariably choose a younger wife, the net result is that more men will have children by more than one wife than women will have children fathered by more than one husband. In other words, serial monogamy effectively results in a polygynous mating system. No doubt this accounts for the fact that men are significantly taller and heavier than women, and that women generally prefer taller men as husbands and *vice versa*. As first pointed out by Charles Darwin, in monogamous mating systems the sexes are of similar size.

Thus, there are many factors that may determine the mating system, but the basic "passion between the sexes", to use Malthus's description, ultimately depends on the hormones secreted by the gonads in foetal, neonatal and adult life. The whole reproductive strategy of the female seems aimed at preserving the status quo. It is the female who transmits all the mitochondrial DNA from generation to generation, and the sleeping ovary, with its oocytes presumably protected from mutation in the arrested dictyate stage, does its best to hand on unmutated mitochondrial DNA from generation to generation (Short, 1994). The hormones secreted by the ovary reflect this conservatism; women are more concerned with attracting a mate and keeping him than aggressively competing for the favours of several men at one time.

The male's reproductive strategy is very different. Since most male mammals show little parental investment in the rearing of their young, and do not to lactate (although endocrinologically it would not be difficult for them to do so), they have used the testicular hormone testosterone to imprint the developing brain, suppressing female characteristics and preparing it for a life of intermale competition and aggression in the pursuit of females. The mating system that eventuates is inevitably a compromise between the very different reproductive agendas of the two sexes and, hence, the eternal tension between the sexes. However, at the end of the day, it is the hormones secreted by the testis, that engine of desire, which ensure that the mutations in the nuclear DNA, originating in the testis, will be as widely disseminated throughout the female population as possible.

REFERENCES

Ehrenberg, L., von Ehrenstein, G. & Hedgran, A. (1957) Gonad temperature and spontaneous mutation-rate in man. *Nature* 180: 1433–34

Gage, M.J.G. (1994) Association between body size, mating pattern, testis size and sperm lengths across butterflies. *Proceedings of the Royal Society of London Series B* 258: 247–54

Graves, J.A.M. (1995) The evolution of mammalian sex chromosomes and the origin of sex determining genes. *Philosophical Transactions of the Royal Society Series B* 350: 305–11

Haldane, J.B.S. (1947) The mutation rate in the gene for haemophilia, and its segregation ratios in males and females. *Annals of Eugenics* 13: 262–71

Harcourt, A.H., Harvey, P.H., Larson, S.G. & Short, R.V. (1981) Testis weight, body weight and breeding system in primates. *Nature* 293: 55–57

Kenagy, G.J. & Trombulak, C. (1986) Size and function of mammalian testes in relation to body size. *Journal of Mammalogy* 67: 1–22

Ketterling, R.P., Vielhaber, E., Bottema, C.D.K. *et al.* (1993) Germ-line origins of mutation in families with Hemophilia B: the sex ratio varies with the type of mutation. *American Journal of Human Genetics* 52: 152–66

Møller, A.P. (1988a) Testis size, ejaculate quality and sperm competition in birds. *Biological Journal of the Linnean Society* 33: 273–83

Møller, A.P. (1988b) Ejaculate quality, testis size and sperm competition in primates. *Journal of Human Evolution* 17: 489–502

Montandon, A.J., Green, P.M., Bentley, D.R. *et al.* (1992) Direct estimate of the haemophilia B (factor IX deficiency) mutation rate and of the ratio of the sex-specific mutation rates in Sweden. *Human Genetics* 89: 319–22

Pabst, D.A., Rommel, S.A., McLellan, W.A.,Williams, T.M. & Rowles, T.K. (1995) Thermoregulation of the intra-abdominal testes of the bottlenose dolphin (*Tursiops truncatus*) during exercise. *Journal of Experimental Biology* 198: 221–26

Parker, G.A. (1970) Sperm competition and its evolutionary consequences in the insects. *Biological Reviews* 96: 281–94

Poole, J.H. (1994) Sex differences in the behaviour of African elephants. In *The Differences Between the Sexes*, edited by R.V. Short & E. Balaban, Cambridge: Cambridge University Press

Setchell, B.P. (1982) Spermatogenesis and spermatozoa. In *Reproduction in Mammals I. Germ Cells and Fertilization*, edited by C.R. Austin & R.V. Short, Cambridge: Cambridge University Press

Seuanez, H.N., Carothers, A.O., Martin, D.E. & Short, R.V. (1977) Morphological abnormalities in the shape of spermatozoa of man and the great apes. *Nature* 270: 345–47

Short, R.V. (1977) Sexual selection and the descent of man. In: *Reproduction and Evolution*, edited by J.H. Calaby & C.H. Tyndale-Biscoe, Canberra: Australian Academy of Science

Short, R.V. (1979) Sexual selection and its component parts, somatic and genital selection, as illustrated by man and the great apes. *Advances in the Study of Behaviour* 9: 131–58

Short, R.V. (1984) Testis size, ovulation rate and breast cancer. In *One Medicine*, edited by O.A. Ryder & M.L. Byrd, Berlin: Springer-Verlag

Short, R.V. (1994) Why sex? In *The Differences Between the Sexes*, edited by R.V. Short & E. Balaban, Cambridge: Cambridge University Press

Short, R.V. (1997) An introduction to mammalian interspecific hybrids. *Journal of Heredity* 88: 355–57

Short, R.V., Mann, T. & Hay, M.F. (1967) Male reproductive organs of the African elephant *Loxodonta africana*. *Journal of Reproductive Fertility* 13: 517–36

Sommer, S.S. (1992) Assessing the underlying pattern of human germline mutations: lessons from the factor IX gene. *FASEB Journal* 6: 2767–74

Tucker, P.K. & Lundrigan, B.L. (1995) The nature of gene evolution on the mammalian Y chromosome: lessons from Sry. *Philosophical Transactions of the Royal Society Series B* 350: 221–27

Wrangham, R.W. (1979) On the evolution of ape social systems. *Social Science Information* 18: 335–68

Zuckerman, S. (1932) *The Social Life of Monkeys and Apes*, London: Kegan Paul

GLOSSARY

aetiology the cause of a disease

dictyate stage the stage during diplotene when meiosis is arrested, characteristic of female meiosis in mammals. In humans, meiosis in females starts at about the fourth month of gestation; after arrest in the dictyate stage, meiosis resumes at ovulation

dizygotic (DZ) developing from separately fertilized eggs, as in non-identical twins

neoplasia abnormal development of cells; may be benign or malignant

oogonium diploid precursor to female germ cells

oviparity ability to lay eggs

pleiomorphism variability in shape

polar body smaller of the products of meiotic division of human oocytes

seminoma malignant tumour of the testis

GENETICS OF HUMAN BEHAVIOUR

Thalia C. Eley and Robert Plomin
Social, Genetic and Developmental Psychiatry Research Centre, Institute of Psychiatry, King's College, London, UK

1. Introduction

Behavioural genetics is the genetic analysis of traits largely in the domains of psychology and psychiatry, including quantitative dimensions, such as personality and cognitive abilities, and qualitative disorders, such as psychopathology and cognitive disabilities. This brief overview of behavioural genetics will focus on cognitive abilities and disabilities rather than on personality and psychopathology (for these aspects, see article on "Genetics of psychiatric disorders").

Behavioural genetics includes both quantitative genetics and molecular genetics (Plomin *et al.*, 1997a). Quantitative genetic research assesses the "bottom line" impact of genetics on complex traits using inbred strain and selection designs to study non-human behaviour and twin and adoption designs to investigate human behaviour. These are quasi-experimental designs that take the first step in genetic research by asking the extent to which behavioural differences among individuals are due to the genetic differences among them. Quantitative genetic research can go beyond this first step of estimating heritability by investigating the relationships among traits, the developmental course of genetic influence and the interface between nature and nurture. Such research can guide molecular genetic research towards the most heritable components and constellations of behaviour in order to identify specific genes.

Quantitative genetics and molecular genetics both have their origins early in the 20th century following the rediscovery of Mendel's laws of heredity. Early quantitative geneticists focused on complex, quantitatively distributed traits such as stature and their origins in naturally occurring variation caused by multiple genes and multiple environmental factors. The origins of molecular genetics lie in single-gene traits and on experimental approaches such as mutational analysis that changed genes in order to study their effects. This tradition continues today in research on knockouts and other gene targeting techniques that create novel genetic variation rather than studying naturally occurring variation. These two worlds of genetics diverged throughout the century until the 1990s, when they began to come together in the study of complex quantitative traits now that it was possible to harness the power of molecular genetics to identify specific genes for such traits, called quantitative trait loci (QTLs; Plomin *et al.*, 1994; see article on "Quantitative trait loci: statistical methods for mapping their positions").

Although there are thousands of single-gene traits, most of them quite rare, there is increasing recognition that complex traits, which include most common medical disorders as well as behavioural traits, are not due to single "sledgehammer" genes necessary and sufficient to cause a disorder. Single-gene disorders that drastically disrupt development are likely to be just the most easily noticed tip of the iceberg of genetic influence on complex traits. It seems reasonable to expect that complex traits like cognitive abilities and disabilities (defined later) are influenced by many genes that nudge development up as well as down across the population rather than producing major effects for a few individuals. Complex traits are often referred to as multifactorial because not only are they influenced by multiple genes but also by multiple environmental factors. Such traits are not called complex because they are difficult to define or to measure – these traits can be measured as reliably as traits of simpler origin.

Complex traits are also increasingly called quantitative traits. If genetic influence on a trait is due to multiple genes, the trait is likely to be distributed as a quantitative dimension rather than as a qualitative disorder, despite the tendency in psychiatry and clinical psychology to diagnose individuals at the extreme of a dimension as having a disorder. This recognition has led to an interest in QTLs and in designs that can identify such multiple-gene influences in complex systems. This QTL perspective suggests that there may be no genes for disorders, just genes for dimensions. For example, a QTL associated with reading disability may not be a gene specific to the disorder, but rather a gene that contributes to individual differences throughout the normal range of variation in reading ability. In other words, the same genetic factors may link the normal and the abnormal, which implies that the genetic origins of the extreme ends of dimensions of variation may be quantitatively, not qualitatively, different from the rest of the distribution. This appears to be the case for reading disability and mild mental retardation (Plomin *et al.*, 1997a). This new perspective has important implications for identifying genes for complex traits such as cognitive abilities and disabilities (Plomin, 1997).

2. Quantitative genetics

Quantitative genetic research consistently points to an important role of genetics for behaviour both in animal models and in the human species (Plomin *et al.*, 1997a). Although biologists tend to be receptive towards the notion of genetic influence, until the 1960s the behavioural sciences were dominated by environmentalism which assumes that we are what we learn. In the 1970s, and especially the 1980s, a dramatic shift occurred in the behavioural sciences towards the more balanced contemporary view that recognizes genetic as well as environmental influences (Plomin & McClearn, 1993). In this overview, we will focus on the domain of behaviour that has been the most contentious:

cognitive abilities and disabilities (Bouchard, 1993; Plomin & Petrill, 1997). Another well-studied area of behavioural genetics is personality, although much of the research in this area relies on self-report questionnaires (Loehlin, 1992). It is interesting that research in this area shows ubiquitous genetic influence – indeed, there is no self-report personality dimension that reliably shows no genetic influence. Moreover, in recent years, specific genes have begun to be reported for self-report personality data (Plomin *et al.*, 1997a). More recently, personality researchers have begun to use more objective measures such as observations or ratings by friends. Although it is too early for clear conclusions from this research, objective measures appear to yield more varied genetic results (Plomin *et al.*, 1997a). As a result of the high level of misinterpretation in this area, we begin with a brief discussion of what genetic influence does and does not mean in relation to behaviour.

The word "genetics" in behavioural genetics is used very narrowly to refer to inter-individual DNA differences that are inherited from generation to generation. It does not refer to the vast majority of DNA that is the same for all of us or to the many DNA events that are not inherited, such as mutations in DNA in cells other than the sex cells. Furthermore, genetic research describes *what is*, the genetic and environmental origins of differences among individuals in a particular population at a particular time. For example, the relative impact of genetic and environmental factors might be different in different cultures or in families with dissimilar economic circumstances. Moreover, genetic research does not predict *what could be*, nor does it prescribe *what should be*. Evidence for genetic influence (*what is*) does not imply that differences among individuals are immutable – novel environmental factors could make a difference (*what could be*). Conversely, expert training that can produce impressive gains in certain skills (*what could be*) does not contradict evidence for genetic influence in the population with its normal variations in experiences (*what is*). No research tells us *what should be* because this is a matter of goals and values.

We focus on *individual differences* in cognitive abilities and disabilities rather than *group differences*, such as average differences between sexes, races or classes, for three reasons. First, the vast majority of genetic research addresses the aetiology of individual differences, not average group differences. The second reason is related to the first: genetic research is much more capable of investigating the causes of individual differences than average differences between groups. Thirdly, it is critically important to recognize that the causes of individual differences are not necessarily related to the causes of average differences between groups.

(i) Animal models

Although the focus of this article is on human cognitive abilities and disabilities, research on animal models is warranted for two reasons. First, such research provides more powerful methods than human research for quantitative genetic and molecular genetic designs. Secondly, research on learning and memory continues to be at the cutting edge of research using animal models, which makes this research directly relevant to the genetics of human cognitive abilities and disabilities. Quantitative genetic research on animal models primarily consists of inbred strain and selection studies. Inbred strains are created by mating brothers with sisters for at least 20 generations. This intensive inbreeding makes each animal within the inbred strain virtually a genetic clone of all other members of the strain. Four frequently studied inbred strains of mice are shown in **Figure 1.** Since inbred strains differ genetically from each other, genetically influenced traits will show average differences between inbred strains reared in the same laboratory environment. Differences within strains are due to environmental influences.

Comparisons between inbred strains of mice have shown large differences between strains in most aspects of learning. For example, **Figure 2A** shows the differing performances of three inbred strains of mice on a maze-learning task (Bovet *et al.*, 1969). **Figure 2B** shows similar results for a very different type of learning, called active avoidance learning, in which mice learn to avoid a shock by moving from one compartment of a cage to another whenever a light is flashed on. Inbred strain studies have also indicated genetic effects on passive avoidance learning, escape learning, lever pressing for reward, reversal learning, discrimination

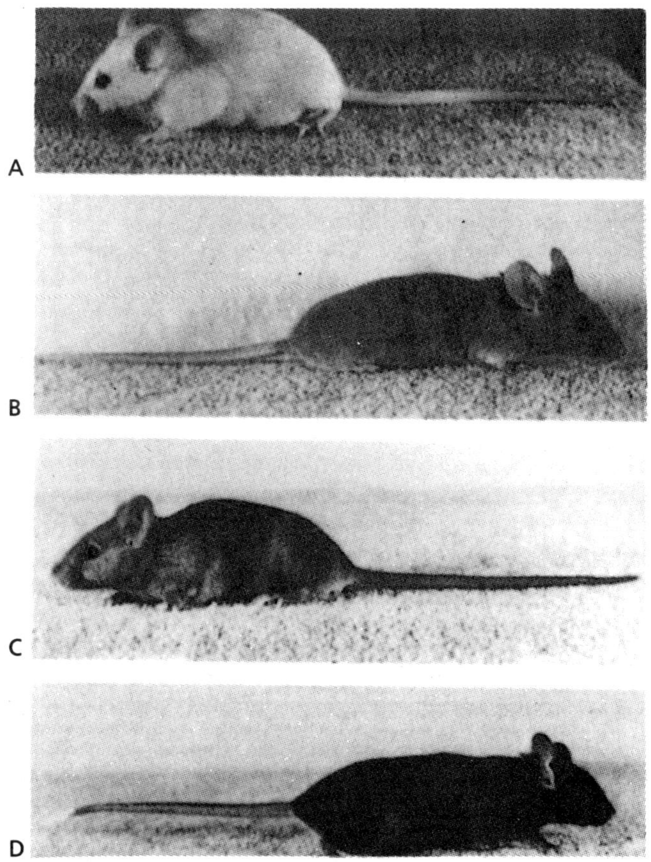

Figure 1 Four common inbred strains of mice: (A) BALB/c; (B) DBA/2; (C) C2H/2; and (D) C57BL/6. (Reproduced with permission from Plomin *et al.*, 1997a and from W.H. Freeman., Inc.)

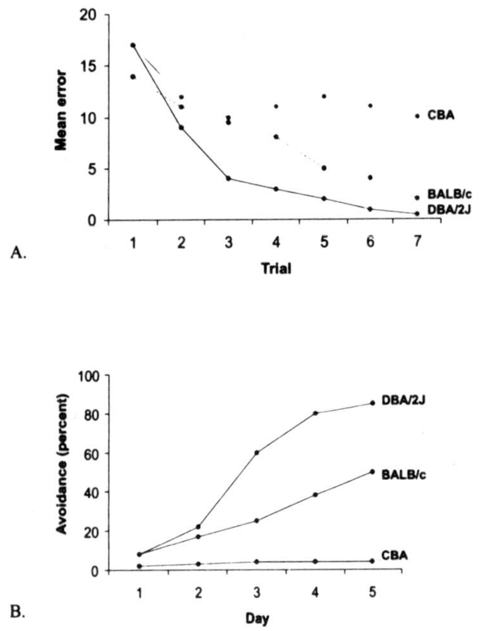

Figure 2 (A) Maze-learning errors (Lashley III maze) for three inbred strains of mice. (B) Avoidance learning for three inbred strains of mice. (Reprinted with permission from Bovet *et al.*, 1969. © 1969 by the American Association for the Advancement of Science.)

learning and heart rate conditioning (Bovet, 1977). Although inbred strain studies now tend to be overshadowed by more sophisticated genetic analyses, inbred strains still provide a simple and highly efficient test for the presence of genetic influence and for investigating interactions between genotype and environment.

The most impressive evidence for genetic influence on behaviour comes from selection studies. As animal breeders have known for centuries, if a behavioural trait is heritable, the trait can be bred. A familiar example of selection is the variety of breeds of dogs, which illustrates dramatically the extent of within-species genetic effects on behaviour. Most dog breeding has involved working dogs bred specifically for behaviour. For example, sheepdogs herd, retrievers retrieve, trackers track and pointers point, all with minimal training. Breeds also differ strikingly in their responsiveness to training for different tasks (Scott & Fuller, 1965).

One of the earliest laboratory selection studies, begun in 1924 by E.C. Tolman, involved rats selected for learning a maze in order to find food. Substantial response to selection was achieved for "maze-bright" rats (few errors) and "maze-dull" rats (many errors), as shown in **Figure 3** (McClearn, 1962). There was practically no overlap between the maze-bright and maze-dull lines; that is, nearly all rats in the maze-bright line were able to learn to run through a maze with fewer errors than any of the rats in the maze-dull line. The difference between the bright and dull lines did not increase after the first seven generations, possibly because brothers and sisters were often mated in this study. Such inbreeding greatly reduces the amount of genetic variability within selected lines, which inhibits progress in a

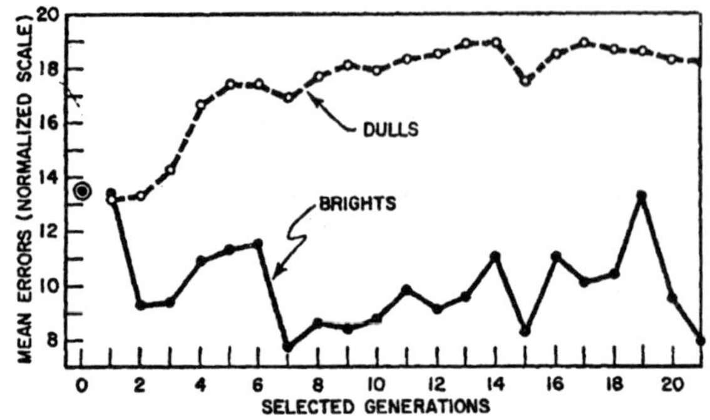

Figure 3 The results of Tryon's selective breeding for maze brightness and maze dullness in rats. (Reprinted from McClearn, 1962 with permission from McGraw-Hill. Inc.)

selection study. Selection studies of behaviour that avoid inbreeding typically find that the difference between high and low lines steadily increases each generation, which strongly suggests that many genes contribute to variation in behaviour.

(ii) Human studies

Genetic research on human behaviour begins with family studies that assess the *sine qua non* of genetic influence, familial resemblance. However, behavioural traits can run in families for reasons of nature or nurture. Twin and adoption designs are used by quantitative geneticists to disentangle the relative roles of genetic and environmental influences (Plomin *et al.*, 1997a). Twin studies capitalize on the experiment of nature in which about a third of twins – monozygotic or identical twins – are derived from the same fertilized egg (zygote) and are thus genetically identical, and other twins – dizygotic or non-identical (fraternal) twins – develop from separately fertilized eggs and are, therefore, only 50% similar genetically, as are other siblings (see **Figure 4**). If genetic factors are important for a trait, monozygotic twins must be more similar than dizygotic twins. Half of dizygotic pairs are same-sex twins, which provide a better comparison group for monozygotic twins, who are always same-sex pairs.

The other major quasi-experimental design of quantitative genetics is the adoption design. Adoption creates pairs of genetically related individuals who are adopted apart and do not share family environment, such as birth parents and their adopted-away offspring. Adoption also produces family members who share family environment but are not genetically related, such as adoptive parents and their adopted children. During the past two decades, behavioural geneticists began to use designs that combine the family, twin and adoption methods in order to bring more power to bear on these analyses. For example, the adoption design is combined with the twin design in studies of monozygotic and dizygotic twins adopted apart and twins reared together.

The earliest and still best-studied trait in behavioural genetics is general cognitive ability. A hierarchical model of cognitive ability is widely accepted in which cognitive abilities are organized hierarchically from specific tests to broad factors, called specific cognitive abilities, to general

Figure 4 Twinning is an experiment of nature that produces identical twins (above) and non-identical or fraternal twins (below). DNA markers can be used to test whether twins are identical or fraternal, although, for most twin pairs, it is easy to tell because identical twins are usually much more similar physically than fraternal twins. (From *Behavioral Genetics* 3rd edition by Plomin, DeFries, McClearn, Rutter © 1997 by W.H. Freeman and Company. Used with permission.)

cognitive ability (Carroll, 1993). Examples of specific cognitive abilities include verbal ability, spatial ability, memory and speed of processing. Such specific cognitive abilities intercorrelate modestly, and general cognitive ability ("g" or intelligence) is an index of what such diverse tests have in common. Intelligence tests, often called IQ tests, assess diverse cognitive abilities and yield total scores that are reasonable indices of general cognitive ability (Brody, 1992). Few genetic studies have as yet used information-processing reaction time measures (Petrill *et al.*, 1995) or neuroscience measures that directly assess brain function (Vernon, 1993).

The first twin study (Merriman, 1924) and the first adoption study (Theis, 1924) investigated general cognitive ability and, since then, twin and adoption studies have consistently found evidence for a strong genetic influence. **Figure 5** summarizes family, twin and adoption data on general cognitive ability. First-degree relatives (parents and offspring and siblings) living together correlate moderately. The adoption data in **Figure 5** suggest that about half of this resemblance is due to genetic factors. The twin method supports this conclusion. Monozygotic twins are nearly as similar as the same person tested twice (test–retest reliability correlations are generally about 0.85). Monozygotic twins reared apart are almost as similar as monozygotic twins who grow up together and both are significantly more similar than dizygotic twins reared together. Maximum-likelihood, model-fitting analyses (Neale & Cardon, 1992; Plomin *et al.*, 1997a) that simultaneously analyse all the family, adoption and twin data yield heritability estimates of about 50%, meaning that about half of the total variance can be accounted for by genetic differences among individuals (Loehlin, 1989; Chipuer *et al.*, 1990). Corrected for unreliability of measurement, heritability estimates would be higher. Regardless of the precise estimate of heritability, the point is that genetic influence on general cognitive ability is not only statistically significant, it is also substantial, although considerably lower than the 80% heritability typically estimated for height (Plomin, 1986). In the behavioural sciences, it is rare for one influence to account for 5% of the variance. For example, despite the interest in birth order, it accounts for less than 1% of the variance of IQ scores. Accounting for 50% of the variance of a trait as complex as general cognitive ability is unprecedented in the behavioural sciences.

Quantitative genetic analyses can go beyond this rudimentary first step of simply demonstrating genetic influence. One example is developmental genetic analysis. The summary of research on general cognitive ability in **Figure 5** is based primarily on data for children and adolescents because children still in their families are easier to study. An interesting developmental finding is that genetic factors become increasingly important for general cognitive ability throughout the life span (Plomin, 1986; McCartney *et al.*, 1990; McGue *et al.*, 1993; Plomin *et al.*, 1997a). For example, an ongoing longitudinal adoption study called the Colorado Adoption Project provides parent–offspring correlations for general cognitive ability from infancy through adolescence (Plomin *et al.*, 1997b). Correlations for biological mothers and their adopted-away children increase steadily from <0.20 in infancy, to about 0.20 in middle childhood and to >0.30 in adolescence. Although these children were adopted away from their mothers at birth, their pattern of resemblance to their biological mothers is just the same as for "control" children reared by their biological parents. In contrast, parent–offpsring correlations for adoptive parents and their adopted children consistently hover around zero. These results suggest not only that genetic factors become increasingly important from infancy to adolescence, but also that family environment shared by parents and offspring does not contribute importantly to parent–offspring resemblance for general cognitive ability.

Twin studies also support the hypothesis that genetic influences increase by age. **Figure 6** shows that the difference between identical and fraternal twins increases slightly

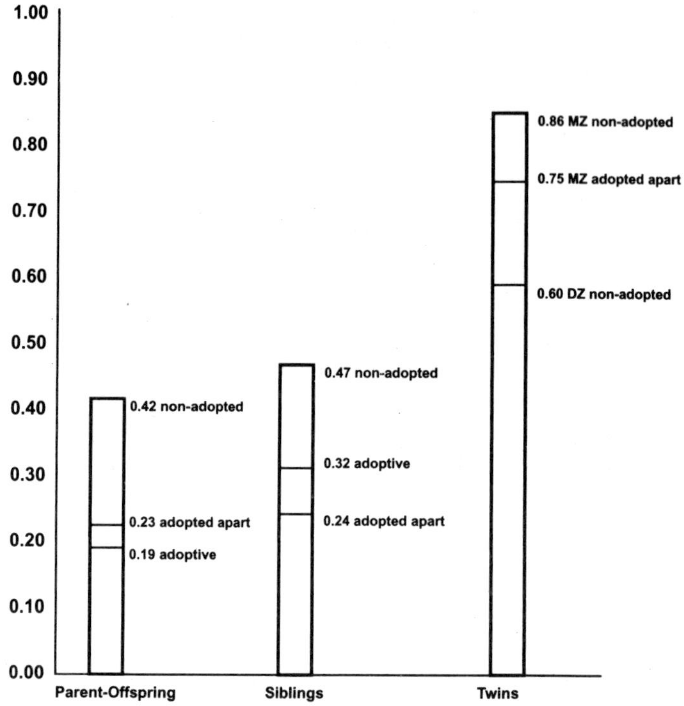

Figure 5 Average family, adoption and twin correlations for IQ scores. "Non-adopted" refers to genetically related family members living together; "adopted apart" refers to genetically related family members adopted apart; and "adoptive" refers to genetically unrelated family members living together. The numbers of non-adopted, adopted-apart, and adoptive parent and offspring pairs are 8433, 720 and 1397, respectively. For siblings, the numbers of pairs are 26,473, 203 and 714, respectively. The numbers of twin pairs are 4672 for non-adopted monozygotic, 158 for adopted-apart monozygotic and 5533 for non-adopted dizygotic. (Based on reviews by Bouchard & McGue (1981), as amended by Loehlin (1989). Data for adopted-apart monozygotic twins includes Bouchard *et al.* (1990) and Pedersen *et al.* (1992).)

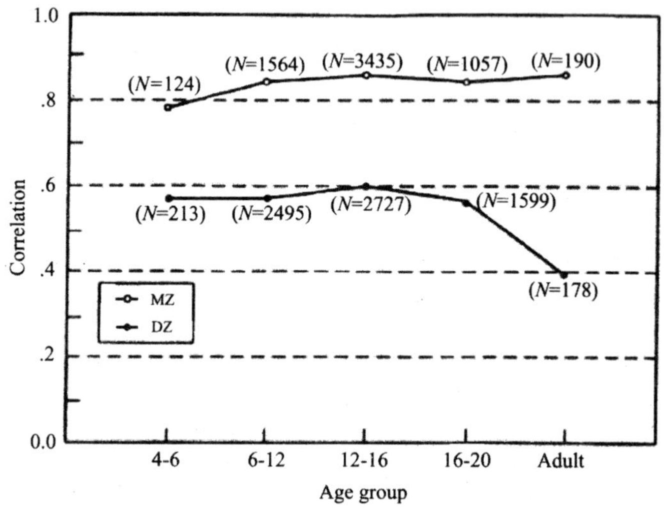

Figure 6 The difference between monozygotic and dizygotic twin correlations for general cognitive ability (*g*) increases during adolescence and adulthood, a trend suggesting increasing genetic influence. (From McGue *et al.* (1993) © 1993 by the American Psychological Association. Reprinted with permission.)

from early to middle childhood and then increases dramatically in adulthood (McGue *et al.*, 1993). Until recently, there have been few studies of older adults. However, one study of Swedish twins at the average age of 60 years reported that the genetic factor accounted for 80% of the variance in general cognitive ability (Pedersen *et al.*, 1992). More recently, the first twin study of Swedish twins aged 80 years and older also reported substantial heritability for general cognitive ability (McClearn *et al.*, 1997).

A second way in which quantitative genetics can go beyond heritability estimates is multivariate genetic analysis, which focuses on the covariance among traits rather than the variance of one trait, an analysis which is described elsewhere (Plomin *et al.*, 1997a). An example in the cognitive domain involves specific cognitive abilities. These abilities show almost as much genetic influence as general cognitive ability, although verbal and spatial abilities appear to be more heritable than other abilities such as processing speed and memory (Plomin, 1988). A surprising finding from multivariate genetic research is that specific cognitive abilities are not specific in relation to their genetic origins. In other words, most of the genetic variance on specific cognitive abilities covaries with genetic variance on general cognitive ability; once genetic effects on general cognitive ability are controlled, there is little genetic variance left that is unique to measures of specific cognitive abilities. This implies that genes that affect one specific cognitive ability are not specific to that ability, but rather affect most cognitive abilities. Similarly, tests of school achievement, which also show substantial genetic influence, correlate moderately with tests of general cognitive ability. Again, multivariate genetic research suggests that this overlap is largely genetic in origin, which carries the interesting implication that the differences between ability and achievement (overachievement and underachievement) are largely due to non-genetic factors (Plomin *et al.*, 1997a).

A third example of going beyond heritability involves the environment. Quantitative genetic research has made two of the most important discoveries about environmental influences relevant to cognitive development. First, although shared family environmental influence accounts for a substantial portion of variance in IQ scores in childhood, this diminishes to negligible importance after adolescence (Plomin *et al.*, 1994). This means that, in the long run, the salient environmental influences on cognitive development are those that are *not* shared by children growing up in the same family (this could include, for example, early hospitalization of one child). Environmental theories of cognitive development have not as yet taken on board the far-reaching ramifications of this discovery.

The second discovery has been called the *nature of nurture* (Plomin & Bergeman, 1991). Genetic research has shown that environmental influences thought to be importantly related to cognitive development, such as parenting and the cognitive stimulation of the family environment, relate to cognitive development for reasons of nature rather than nurture (Plomin, 1994). For example, when measures of parenting style were included in a genetically sensitive design like an adoption study, these measures were found to reflect genetic difference among children in that parents respond more similarly to adopted siblings who are genetically related to each other than to genetically unrelated siblings (adoptive siblings in which two genetically unrelated children are adopted into the same family) (Braungart *et al.*, 1992). Moreover, associations between such environmental measures and cognitive development are mediated in part by genetic factors. For example, the correlation between such measures of parenting and children's cognitive development is significantly greater in the usual family where parents and offspring are genetically related than in adoptive families where parents and offspring are genetically unrelated (Plomin *et al.*, 1985). Although space does not permit a full explanation of these findings, they suggest that research on the effects of family environment needs to be embedded in genetic designs that can disentangle nurture from nature (Rowe, 1994). More generally, research on the developmental interface between nature and nurture needs to consider a more active role for children in selecting and creating their own environments (Scarr, 1992). This is the more profound meaning of the discovery of genetic influence on measures of the environment: genes contribute to experience itself (Plomin *et al.*, 1985).

3. Molecular genetics

Although quantitative genetic research has concentrated on cognitive abilities more than any other behaviours, molecular genetic research on cognitive abilities has only just begun (Plomin *et al.*, 1995). In contrast, cognitive disabilities have been the area of greatest progress in relation to the molecular genetics of behaviour, but have scarcely been touched by quantitative geneticists. For example, there has never been a twin or adoption study of mental retardation (see article on "Genetics of mental retardation").

Attempts to use molecular genetics to identify genes for complex traits such as cognitive abilities and disabilities is difficult because complex traits fall between the realms of the two major molecular genetic strategies: linkage and association. Linkage is systematic, but not powerful, and association is powerful, but not systematic (Risch & Merikangas, 1996; Plomin, 1997). The problem is that, if complex quantitative traits are as influenced by multiple genes as by multiple environmental factors, the magnitude of QTL effects for such traits is likely to be small in the population. One direction that QTL research has taken is to use the experimental power of animal models of genetics.

(i) Animal models

The substantial similarity between the genomes of mouse and man means that candidate QTLs for human behaviour can be nominated by mouse QTL research (Crabbe *et al.*, 1994). Mouse models will be even more valuable when it comes to characterizing QTLs neurodevelopmentally. Although mouse QTL research on learning has not yet been reported, knockout genes have been shown to affect learning in mice. The first study of this type found a deficit in spatial learning for mice with a knockout version of a gene (α-*CaMKII*) for a protein enriched in the hippocampus and thought to be involved in long-term memory (Silva *et al.*, 1992). Several other knockout genes have subsequently been shown to affect learning and memory in mice (Wehner *et al.*, 1996). Recently, a new type of selective knockout in mice was identified that affects α-CaMKII expression only in certain cells in the hippocampus and also greatly impairs spatial learning and memory (McHugh *et al.*, 1996)

Some of the most sophisticated work has involved fruit flies. For example, a gene (cAMP-responsive element-binding protein or *CREB*) regulates other genes responsible for synthesizing proteins involved in long-term memory (Yin *et al.*, 1995). *CREB* was altered to produce large amounts of repressors or large amounts of activators, with the addition of a promoter that turned on the gene only when the flies were exposed to heat. At room temperature, the two types of flies learned normally. However, when the temperature was raised, the high-repressors had difficulty in learning, whereas the high-activators learned much more quickly. It is not yet known whether these genes affect human learning and memory or whether DNA variations in these genes are responsible for genetic effects on individual differences in cognitive abilities.

(ii) Human studies

Much more is known about single-gene causes of mental retardation than any other behavioural disorder (see article on "Genetics of mental retardation"). In the 1940s, phenylketonuria, a recessive gene on chromosome 12, caused severe mental retardation in about one in 10 000 individuals or about 1% of the institutionalized mentally retarded population. Understanding how the recessive form of the gene failed to metabolize phenylalanine led to the preventive intervention of diets low in phenylalanine. In the late 1950s, the single most important cause of mental retardation, Down's syndrome, was discovered to be due to an extra copy of one of the smallest chromosomes (chromosome 21). The second most important cause of mental retardation was identified in 1991 as a gene on the X chromosome called fragile X (Verkerk *et al.*, 1991). More than 100 rare single-gene disorders include mental retardation among their symptoms, such as Duchenne's muscular dystrophy, Lesch–Nyhan syndrome, neurofibromatosis type 1 and tuberous sclerosis (Wahlström, 1990).

Although single-gene causes of mental retardation have devastating effects on affected individuals, they are quite rare and account for only a small portion of mental retardation in the population. Most cognitive disabilities are likely to be complex quantitative traits influenced by multiple genes (QTLs). The first QTL linkage was reported in 1994 between markers on chromosome 6 and reading disability (Cardon *et al.*, 1994) and has recently been replicated (Grigorenko *et al.*, 1997). Although linkage is systematic, it can only detect genes of relatively large effect size, accounting for about 10% of the variance, even using new QTL linkage approaches (Fulker & Cardon, 1994).

Since linkage analysis cannot be made much more powerful, at least with realistic samples, QTL research in the human species has relied on association analysis using DNA markers in or near "candidate" genes. Such DNA markers may be close enough to a functional polymorphism in a gene to be in linkage disequilibrium for scores of generations, assuming low mutation rates at both loci. Association analysis is much more direct when the DNA marker itself is the functional polymorphism that affects the trait. The best example to date is the association, replicated in scores of studies, between allele 4 of the apolipoprotein-E gene on chromosome 19 and late-onset Alzheimer's disease, which affects as many as 15% of individuals over 80 years of age (Corder *et al.*, 1993). The association accounts for about 15% of the variance in liability to the disorder, although the allele is neither necessary nor sufficient to develop the disorder (Owen *et al.*, 1994). This allele has also been reported to be associated with cognitive decline in a normal sample in later life (Feskens *et al.*, 1994). Other ongoing research is attempting to identify QTLs for cognitive abilities (Plomin, 1997).

4. Conclusion

The simple finding that genetics contributes importantly to the development of individual differences in cognitive abilities and disabilities carries far-reaching implications. For example, theories and research on cognitive development and education need to take on board the fact that children differ in their ability to learn in part for genetic reasons. As a more specific example, educationalists need to recognize that tests of school achievement show genetic influence due to the genetic overlap between tests of achievement and general cognitive ability. As another example, developmentalists' research on environmental effects on cognitive development needs to be embedded in genetically sensitive designs in order to disentangle nature and nurture.

However, finding that heredity contributes to cognitive abilities and disabilities *does not* imply that intelligence is hard-wired in the genes. As discussed earlier, genetic research investigates *what is*, not *what could be* and certainly not *what should be*. The latter issue moves beyond science to policies which are based on values. These findings do not upset the foundations of democracy. When the signers of the American Declaration of Independence stated that all men are created equal, they did not mean that everyone is created the same. What they meant was that all people should be equal in political and legal rights; the fundamental reason for democracy is that people should be treated equally despite their differences.

One of the most exciting developments in research into behaviour is the ability to identify genes in multiple-gene systems, called quantitative trait loci (QTLs). Identifying even a few replicable QTLs associated with cognitive abilities and disabilities will revolutionize genetic research, as has already happened in gerontology with the finding of a QTL (the apolipoprotein E gene) – the only known risk factor for late-onset Alzheimer's disease. Identified QTLs will make it possible to use measured genotypes rather than indirect inferences about heritable influence based on familial resemblance in order to investigate development, multivariate relationships, links between normal dimensions and abnormal disorders, and gene–environment interplay (Plomin & Rutter, 1998). They will also provide discrete windows through which to view neurobiological pathways between genes and development.

REFERENCES

Bouchard, T.J. Jr (1993) The genetic architecture of human intelligence. In *Biological Approaches to the Study of Human Intelligence*, edited by P.A. Vernon, Norwood, New Jersey: Ablex

Bouchard, T.J. Jr & McGue, M. (1981) Familial studies of intelligence: a review. *Science* 212: 1055–59

Bouchard, T.J. Jr, Lykken, D.T., McGue, M., Segal, N.L. & Tellegen, A. (1990) Sources of human psychological differences: The Minnesota Study of Twins Reared Apart. *Science* 250: 223–50

Bovet, D. (1977) Strain differences in learning in the mouse. In *Genetics, Environment and Intelligence*, edited by A. Oliverio, Amsterdam and New York: North Holland

Bovet, D., Bovet-Nitti, F. & Oliverio, A. (1969) Genetic aspects of learning and memory in mice. *Science* 163: 139–49

Braungart, J.M., Fulker, D.W. & Plomin, R. (1992) Genetic influence of the home environment during infancy: a sibling adoption study of the HOME. *Developmental Psychology* 28: 1048–55

Brody, N. (1992) *Intelligence*, 2nd edition, San Diego: Academic Press

Cardon, L.R., Smith, S.D., Fulker, D.W. *et al.* (1994) Quantitative trait locus for reading disability on chromosome 6. *Science* 266: 276–79

Carroll, J.B. (1993) *Human Cognitive Abilities: A Survey of Factor-Analytic Studies*, Cambridge and New York: Cambridge University Press

Chipuer, H.M., Rovine, M.J. & Plomin, R. (1990) LISREL modeling: genetic and environmental influences on IQ revisited. *Intelligence* 14: 11–29

Corder, E.H., Saunders, A.M., Strittmatter, W.J. *et al.* (1993)

Gene dose of apolipoprotein E type 4 allele and the risk of Alzheimer's disease in late onset families. *Science* 261: 921–23
Crabbe, J.C., Belknap, J.K. & Buck, K.J. (1994) Genetic animal models of alcohol and drug abuse. *Science* 264: 1715–23
Feskens, E.J.M., Havekes, L.M., Kalmijn, S. *et al.* (1994) Apolipoprotein e4 allele and cognitive decline in elderly men. *British Medical Journal* 309: 1202–06
Fulker, D.W. & Cardon, L.R. (1994) A sib-pair approach to interval mapping of quantitative trait loci. *American Journal of Human Genetics* 54: 1092–103
Grigorenko, E.L, Wood, F.B., Meyer, M.S. *et al.* (1997) Susceptibility loci for distinct components of dyslexia on chromosomes 6 and 15. *American Journal of Human Genetics* 60: 27–39
Loehlin, J.C. (1989) Partitioning environmental and genetic contributions to behavioral development. *American Psychologist* 44: 1285–92
Loehlin, J.C. (1992) *Genes and Environment in Personality Development*, Newbury Park, California: Sage
McCartney, K., Harris, M.J. & Bernieri, F. (1990) Growing up and growing apart: a developmental meta-analysis of twin studies. *Psychological Bulletin* 107: 226–37
McClearn, G.E. (1962) The inheritance of behavior. In *Psychology in the Making: Histories of Selected Research Problems*, edited by L.J. Postman, New York: Knopf
McClearn, G.E, Johansson, B., Berg, S. *et al.* (1997) Substantial genetic influence on cognitive abilities in twins 80 or more years old. *Science* 276: 1560–63
McGue, M., Bouchard, T.J. Jr, Iacono, W.G. & Lykken, D.T. (1993) Behavioral genetics of cognitive ability: a life-span perspective. In *Nature, Nurture and Psychology*, edited by R. Plomin & G.E. McClearn, Washington, DC: American Psychological Association
McHugh, T.J., Blum, K.I., Tsien, J.Z., Tonegawa, S. & Wilson, M.A. (1996) Impaired hippocampal representation of space in CA1-specific NMDAR1 knockout mice. *Cell* 87: 1339–49
Merriman, C. (1924) The intellectual resemblance of twins. *Psychological Monographs* 33: 1–58
Neale, M.C. & Cardon, L.R. (1992) *Methodology for Genetic Studies of Twins and Families*, Dordrecht and Boston: Kluwer
Owen, M.J., Liddell, M.B. & McGuffin, P. (1994) Alzheimer's disease: an association with apolipoprotein e4 may help unlock the puzzle. *British Medical Journal* 308: 672–73
Pedersen, N.L., Plomin, R., Nesselroade, J.R. & McClearn, G.E. (1992) A quantitative genetic analysis of cognitive abilities during the second half of the life span. *Psychological Science* 3: 346–53
Petrill, S.A., Thompson, L.A. & Detterman, D.K. (1995) The genetic and environmental variance underlying elementary cognitive tasks. *Behavior Genetics* 25: 199–209
Plomin, R. (1986) *Development, Genetics, and Psychology*, Hillsdale, New Jersey: Erlbaum
Plomin, R. (1988) The nature and nurture of cognitive abilities. In *Advances in the Psychology of Human Intelligence*, edited by R.J. Sternberg, Hillsdale, New Jersey: Erlbaum
Plomin, R. (1994) *Genetics and Experience: The Developmental Interplay Between Nature and Nurture*, Thousand Oaks, California: Sage
Plomin, R. (1997) Identifying genes for cognitive abilities. In *Intelligence: Heredity and Environment*, edited by R.J. Sternberg & E.L. Grigorenko, Cambridge and New York: Cambridge University Press
Plomin, R. & Bergeman, C.S. (1991) The nature of nurture: genetic influence on "environmental" measures. *Behavior and Brain Sciences* 14: 373–427 (With Open Peer Commentary)
Plomin, R. & McClearn, G.E. (eds) (1993) *Nature, Nurture, and Psychology*, Washington, DC: American Psychological Association
Plomin, R. & Petrill, S.A. (1997) Genetics and intelligence: what's new? *Intelligence* 24: 53–77
Plomin, R. & Rutter, M. (1998) Child development, molecular genetics, and what to do with genes once they are found? *Child Development* 69: 1223–42
Plomin, R., Loehlin, J.C. & DeFries, J.C. (1985) Genetic and environmental components of "environmental" influences. *Developmental Psychology* 21: 391–402
Plomin, R., Owen, M.J. & McGuffin, P. (1994) The genetic basis of complex human behaviors. *Science* 264: 1733–39
Plomin, R., McClearn, G.E., Smith, D.L. *et al.* (1995) Allelic associations between 100 DNA markers and high versus low IQ. *Intelligence* 21: 31–48
Plomin, R., DeFries, J.C., McClearn, G.E. & Rutter, M. (1997a) *Behavioral Genetics*, 3rd edition, New York: Freeman
Plomin, R., Fulker, D.W., Corley, R. & DeFries, J.C. (1997b) Nature, nurture and cognitive development from 1 to 16 years: a parent–offspring adoption study. *Psychological Science* 8: 442–47
Risch, N. & Merikangas, K. (1996) The future of genetic studies of complex human diseases. *Science* 273: 1516–17
Rowe, D.C. (1994) *The Limits of Family Influence: Genes, Experience, and Behavior*, New York: Guilford Press
Scarr, S. (1992) Developmental theories for the 1990s: development and individual differences. *Child Development* 63: 1–19
Scott, J.P. & Fuller, J.L. (1965) *Genetics and the Social Behavior of the Dog*, Chicago: University of Chicago Press
Silva, A.J., Paylor, R., Wehner, J.M. & Tonegawa, S. (1992) Impaired spatial learning in α-calcium-calmodulin kinase mutant mice. *Science* 257: 206–11
Theis, S.V.S. (1924) *How Foster Children Turn Out*, New York: State Charities Aid Association
Tolman, E.C. (1924) The inheritance of maze learning ability in rats. *Journal of Comparative Psychology* 4: 1–18
Verkerk, A.J., Piretti, M., Sutcliffe, J.S. *et al.* (1991) Identification of a gene (FMR-1) containing a CGG repeat coincident with a breakpoint cluster region exhibiting length variation in fragile X syndrome. *Cell* 65: 904–14
Vernon, P.A. (ed.) (1993) *Biological Approaches to the Study of Human Intelligence*, Norwood, New Jersey: Ablex
Wahlström, J. (1990) Gene map of mental retardation. *Journal of Mental Deficiency Research* 34: 11–27
Wehner, J.M., Bowers, B.J. & Paylor, R. (1996) The use of null mutant mice to study complex learning and memory processes. *Behavior Genetics* 26: 301–12
Yin, J.C.P., Vecchio, M.D., Zhou, H. & Tully, T. (1995) CRAB as a memory modulator: induced expression of a dCREB2 activator isoform enhances long-term memory in *Drosophila*. *Cell* 8: 107–15

GLOSSARY

dizygotic (DZ) developing from separately fertilized eggs, as in non-identical twins

etiology (aetiology) the cause of a disease

monozygotic (MZ) developing from a single fertilized egg, as in identical twins.

See also **Genetics of mental retardation** (p.437); **Genetics of psychiatric disorders** (p.443); **Quantitative trait loci** (p.904)

HANDEDNESS: A LOOK AT LATERALITY

Ira B. Perelle[1] and Lee Ehrman[2]
[1]*Mercy College, Dobbs Ferry, New York, USA*
[2]*SUNY at Purchase, Purchase, New York, USA*

1. Introduction

Behavioural asymmetry in general, and handedness or pawedness in particular, has been the subject of considerable interest and research for many years. Human handedness appears to have been a favourite concern of investigators since Sir Thomas Browne's work in 1646 (*Pseudodoxia Epidemica* IV:v Of the right and left Hand), and non-human pawedness has been studied in chimpanzees, monkeys, cats, rats, mice, parrots, macaws and fruit flies (Cole, 1955; Collins, 1968, 1977; Perelle *et al.*, 1978; McGrew & Marchant, 1992). Although many non-human animals exhibit a paw, claw or wing preference, none appears to show a consistent lateralization within species as do humans with handedness. There is no controversy about the existence of human handedness: most humans are righthanded; some are lefthanded. The controversy surrounding handedness involves, primarily, the aetiology of handedness, and, because the behavioural definition will ordain the results, the classification of human handedness (left or right) by hand use for certain behaviours.

2. Handedness proportions

Most current research reports the proportion of lefthanders as approximately 10%. Our recent research among 12 000 subjects from 32 countries showed an overall lefthanded proportion of 9.5% – with lower and upper limits of 2.5% in Mexico and 12.8% in Canada (see Perelle & Ehrman, 1994, and references therein). In the past, the proportion of humans who are lefthanded has been reported to be as low as 1% and as high as 29.3% (a review of these estimates can be found in Hardyk & Petrinovich, 1977). Anthropological studies date differences in handedness back at least 1.4 million years (Babcock, 1993), with the earliest written record of the proportion of lefthanders in the Old Testament *Book of Judges*. There, the description of the Benjamite army includes that of a highly accurate slingshot unit of 700 lefthanders among the 27 000 regulars. This low proportion of lefthanders (2.6%) is probably the original source of many of the early (*c.*1800) estimates of lefthandedness. The biblical account, however, does not tell us whether the special unit included all lefthanders in the army, whether all lefthanded slingshooters were included or how the determination of lefthandedness was made. This brings us to one of the serious controversies in the study of human handedness – how to determine the handedness of an individual.

3. Handedness determination

This problem is serious for several reasons. Most research into human handedness is designed to answer questions about the aetiology of handedness, or about some difference between left- and righthanders. In order to examine the two groups, there must be some way to differentiate between the two groups and, in the half century that active research has been taking place, very few investigators have used the same differentiating protocol. Many researchers employ a questionnaire to determine the number of behaviours for which the left hand or the right hand is used. One of the first such questionnaires used was the Edinburgh Handedness Inventory (Oldfield, 1971); it became the model for several other questionnaires, all of which were used to make a handedness determination. In practice, methods of scoring handedness with questionnaires followed two patterns. First, if the subject performed all behaviours with the right hand, the subject was considered righthanded. If the subject performed any behaviours with the left hand, the subject was considered non-righthanded, the extent of which was determined by the number of behaviours performed with the left hand (Rife, 1994). The second method was to provide a numerical score on a 0–10 scale or a −5 to +5 scale. In both cases, the number of righthanded behaviours determined the score and the extremes of the scale indicated totally right or totally lefthanded behaviours (Provins & Cunliffe, 1972; Bradshaw & Nettleton, 1983).

To understand the problem with this method of handedness determination, it pays to consider why righthanders are righthanded in the first place. The most important behaviour in the human repertoire is certainly verbal behaviour. It is so important that we have evolved specialized sites in the brain that function exclusively to process verbal information. The area we use to process writing, gesturing and similar verbal output is Broca's area, located in the left hemisphere of the brain (Geschwind, 1975). It is well known that the control of both fine and coarse muscles takes place in the contralateral hemisphere (Geschwind, 1975; Corballis & Morgan, 1978). For maximum efficiency, therefore, verbal behaviour, i.e. writing, should be controlled by the same brain hemisphere as contains the verbal processing area. Righthanded people write with their right hand because they process verbal material in their left hemisphere. To write with their left hand, they would have to send the verbal message across the corpus callosum to the right hemisphere for left hand control, a relatively inefficient process at best.

4. Righthanders

Since righthanders use their right hand for writing, they tend to use it for other behaviours requiring fine motor control and, in many instances, for behaviours requiring gross motor control such as batting a ball or using a wrench. However, we must point out that many righthanders can and do learn to use their left hand for non-verbal behaviours and use it with equal dexterity as they do their right hand, as do lefthanders with their right hand (Perelle *et al.*, 1981). Consider, for example, musicians: pianists, clarinettists and others must use both hands equally well. Typists and word-processor users must do so also, and most professional mechanics use whichever hand is more appropriate for the job. None of these individuals, if they are righthanded, uses their left hand for writing because writing is the one behaviour that belongs exclusively to the right hand in righthanders and to the left hand in lefthanders. It is the single behaviour that shows a clear and consistent dichotomy.

The critical hand, therefore, is the writing hand and our research has shown that most individuals classify their handedness on the basis of the hand used for writing (Perelle & Ehrman, 1994). Whether one eats with a fork in the right hand or left hand depends on cultural norms, not on a unique area of the brain; other non-verbal behaviours depend on learning. For most people, only writing and similar verbal behaviours are programmed for the right hand, by way of the left brain hemisphere, and this should be the only behaviour used to determine the handedness of an individual.

5. Lefthanders

This brings us to the second controversy: why are there lefthanders at all? Lefthanders, after all, face an inherent disadvantage in life: most tools (utensils, stringed instruments, telephones, sports equipment) are designed for right-hand use, and lefthanders must either learn to use the tool with the right hand or learn to use it backwards with the left hand. To gain an understanding of this, we suggest that righthanders try to use a pair of scissors with their left hand. To answer the question, why are there lefthanders, we must consider all the research results that show how lefthanders differ from righthanders. Sodium amytal testing, indicating the active part of the brain during various tasks, shows that some lefthanders use their right brain hemisphere for verbal processing, some use their left hemisphere and some use both hemispheres (Sperry, 1971). A higher proportion of lefthanders are associated with birth trauma than are righthanders (Bakan *et al.*, 1973), and lefthanders are more likely to have suffered pre- or postnatal insult to the left hemisphere than righthanders. Lefthanders are more likely to be part of a monozygotic twin pair (Torgerson, 1950), and to have poor handwriting. Lefthanders are over-represented among the mentally retarded (McBurney & Dunn, 1976; Gregory & Paul, 1980), and lefthanders are over-represented among the gifted (Ehrman & Perelle, 1983). From all these results, it is apparent that there cannot be one single aetiology of lefthandedness. We believe there must be at least two, and probably three, aetiologies to account for the wide variation in the behaviour and test results of people who write with their left hand.

6. Aetiology of lefthandedness

(i) Pathological lefthandedness

The least controversial aetiology of lefthandedness is that of the pathological lefthander. Lefthanders comprise almost 20% of the mentally retarded population and 28% of the severely and profoundly mentally retarded population. We believe these individuals are both retarded and lefthanded for the same reason: a massive insult to their left hemisphere as a result of a prenatal or postnatal event. It is also possible that a nutritional insult resulted in left hemisphere aberration. If the verbal processing area in the left hemisphere was damaged even partially early in life, the right hemisphere would assume verbal processing functions, along with other hemispheric functions. This would account for the lefthanders who process verbal material in their right hemisphere and, depending on the severity of the brain damage, would also account for the higher proportion of lefthanders found in the retarded population. There is no genetic component to this type of lefthandedness.

(ii) Natural lefthandedness

The second type of lefthander is the natural or genetic lefthander. This individual seems to function normally, but is the reverse of a normal righthander; both brain hemispheres are intact, yet verbal processing takes place in the right hemisphere (Healy *et al.*, 1981; Noonan & Axelrod, 1981). Since many of these lefthanders are part of a monozygotic twin pair, the *situs inversus* phenomenon (mirror-image twinning) is probably involved (see article on Twinning). Concomitant with discordant handedness in monozygotic twins it has been found that there is a high probability that the lefthanded twin will be first born (Christian *et al.*, 1979). It has been suggested that when a lefthander is born as a singleton the conception was originally a monozygotic twin set, but one twin, the righthanded one, died and was reabsorbed *in utero*. This is plausible, although highly speculative. It has been suggested that there is a genetic component to monozygotic twinning and, if this is true, it could then provide a genetic basis for some lefthandedness (Torgerson, 1950; Harvey *et al.*, 1977; Lichtenstein *et al.*, 1996). As yet, no single gene or gene combination that determines left or righthandedness has been identified.

(iii) Learned lefthandedness

The third type of lefthander is the learned lefthander. This lefthander writes with the left hand, but has relatively poor handwriting (Ramos, 1968) and shows dual hemispheric activation during verbal processing. Since preverbal children are not lateralized for hand use (Roberts, 1949; Cernacek & Jagr, 1969), these lefthanders may have initially chanced to manipulate successfully some toy with their left hand and continued to use their left hand for toy manipulation. When eventually given a pencil or crayon, because of past reinforcement, they employ their left hand and continue to use their left hand when they write. This, of course, is quite inefficient neurologically, as described above, and, because of the additional processing time required, may be the reason quite a few lefthanders stutter when they are young and have notoriously poor handwriting. We believe that eventually these lefthanders develop verbal processing function in their right hemisphere too and that these individuals become the lefthanders who show dual hemispheric activation during verbal processing.

7. Switching handedness

The least frequently investigated aspect of handedness is "switching", the detrimental practice of forcing young left-handers to use their right hand. Switching usually takes place in school, when a child is first learning to write, but it also happens at home, when parents believe it is wrong or even evil to be lefthanded. We have found that countries such as Nigeria, that have the lowest proportion of left-handers, also have the highest proportion of reported switching (Perelle & Ehrman, 1994). This custom, fortunately obsolete in most developed countries, skewed the results of many studies and caused many children to be classified as learning disabled. The most obvious case of the former is the Halpern and Coren paper (1988) ostensibly showing that righthanders live longer than lefthanders. There is no question that this is not true. Our data have shown that the proportion of subjects reporting switching increases as the age of the subject increases. The proportion of lefthanders in each age cohort is smaller in older probands because older adults were much more likely to have been switched when they were children than were younger adults. There are, of course, many countries where left-hand use is a strict taboo, particularly in middle eastern countries, and data from these countries should not be merged with data from developed countries if valid results are desired. Analyses of several studies of twin laterality show the drawbacks of forming conclusions from populations in which social pressure against lefthandedness has artificially inflated the proportion of righthanders (Carlier *et al.*, 1996).

The second serious problem with switching is that it forces children into an unnatural behaviour. In our work, we have found that several students, classified as learning disabled, were switched lefthanders. After counselling them to start writing with their left hand, the so-called learning disabilities disappeared. We strongly discourage parents from attempting to switch their lefthanded children, and we strongly encourage them to prohibit any teachers from doing so.

8. Laterality in non-human species

As to laterality (pawedness, wingedness) in other species, there are about as many investigators who claim to have found some evidence for it as there are those who refute such claims. Collins (1985), who has been breeding mice for laterality for many generations, has found that he can select for mice that will be highly lateralized, but not for a particular side; that is, a highly lateralized mouse pair will produce highly lateralized pups, but there will be both left and right lateralized pups in each litter. Diamond and McGrew (1994) seem to have found that captive cotton-top tamarin use their left hand for visually guided behaviours and use their right hand for manipulative behaviours. However, Marchant and McGrew (1996) found that chimpanzees of Gombe National Park exhibited no lateral preference for any behaviours at a population level, although McGrew and Marchant (1999) did find strong lateral preferences for food processing behaviours at an individual level.

The only instance where lateralized behaviour seems consistent in a non-human animal is in songbirds. Birds that depend on song for their livelihood, that is to attract and keep mates and to defend territories, use their left hemisphere to process singing (Nottebohm & Nottebohm, 1976). If their left hemisphere is destroyed, they will lose their ability to sing. We find it interesting that both songbirds and (most) humans process "verbal" material in their left hemispheres. What are the evolutionary implications of this relationship? Clearly, the genetic models of Rife (1994) and Stern (1973), though still of historic interest, must be discarded, and we can offer nothing completely satisfactory in their place (for attempts, using Drosophila wing, leg and bodily movements, see Perelle & Ehrman, 1983).

9. Conclusions

What are the "real world" implications of the results of our handedness studies? The most important, we believe, is that lefthanders should not be combined into a single category. Pathological lefthanders need tutorial assistance, particularly in their early years. They must reorganize their brains to process verbal operations in their right hemisphere, having lost the use of the area that has been evolutionarily selected for such operations. This is not a simple task, and is the reason that some lefthanders do poorly on tests of verbal ability. The normal and learned lefthanders grow up in a world that is biased against their natural tool use and must learn their own way. This creates a sense of individualism in many lefthanders and may be at least part of the reason why so many lefthanders turn out to be leaders in their fields. A list of distinguished lefthanders would be much too long to include here, but would include Alexander the Great, Napoleon Bonaparte, Queen Victoria, Queen Elizabeth the Queen Mother, Queen Elizabeth, Princes Charles and William, several American Presidents including Reagan, Bush and Clinton, Lewis Carroll, H.G. Wells, Michelangelo, Picasso and a large number of professional athletes. An imposing list can be obtained at the World Wide Web site: Famous Left-Handers (Holder, 1996).

REFERENCES

Babcock, L.R. (1993) The right and the sinister. *Natural History* 102: 32–39

Bakan, P., Dibb, G. & Reid, P. (1973) Handedness and birth stress. *Neuropsychologia* 11: 363–66

Bradshaw, J.L. & Nettleton, N.C. (1983) *Human Cerebral Asymmetry*, Englewood Cliffs, New Jersey: Prentice Hall

Carlier, M., Spitz, E., Vacher-Lavenu, M.C. *et al.* (1996) Manual performance and laterality in twins of known chorion type. *Behavior Genetics* 26: 409–18

Cernacek, J. & Jagr, J. (1969) Laterality of the hand and foot during childhood. *Annales de Medico-Psychologiques* 1: 778

Christian, J.C., Hunter, D.S., Evans, M.M. & Standeford, F.M. (1977) Association of handedness and birth order in monozygotic twins. *Acta Genetica Medicae et Gemellologiae* 28(1): 67–68

Cole, J. (1955) Paw preference related to hand preference in animals and man. *Journal of Comparative and Physiological Psychology* 48: 137–40

Collins, R.L. (1968) On the inheritance of handedness. I: Laterality in inbred mice. *Journal of Heredity* 59: 9–12

Collins, R.L. (1977) Origins of the sense of asymmetry: mendelian and non-mendelian models of inheritance. *Annals of the New York Academy of Science* 299: 283–305

Collins, R.L. (1985) On the inheritance of direction and degree of laterality. In *Cerebral Lateralization in Non-human Species*, edited by S.D. Glick, New York: Academic Press

Corballis, M.C. & Morgan, M.J. (1978) On the biological basis of human laterality. *Behavioral and Brain Sciences* 2: 261–336

Diamond, A.C. & McGrew, W.C. (1994) True handedness in the cotton-top tamarin. *Primates* 35: 69–77

Ehrman, L. & Perelle, I.B. (1983) Laterality. *Mensa Research Journal* 16: 3–32

Geschwind, N. (1975) The apraxias: neural mechanisms of disorders of learned movements. *American Scientist* 63: 188–95

Gregory, P. & Paul, J. (1980) The effects of handedness and writing posture on neuropsychological test results. *Neuropsychologia* 18: 231–85

Halpern, D.F. & Coren, S. (1988) Do right-handers live longer? *Nature* 333: 213

Hardyk, C. & Petrinovich, L.F. (1977) Left-handedness. *Psychological Bulletin* 84: 385–404

Harvey, M.A., Huntley, R.M.C. & Smith, D.W. (1977) Familial monozygotic twinning. *Journal of Pediatrics* 90: 24608.

Healy, J., Rosen, J. & Gerstman, L. (1981) *Effects of Sex, Handedness, and Family Sinistrality on the Pattern and Degree of Visual Field Asymmetries*, New York: Eastern Psychological Association

Holder, M.K. (1996) Famous left handers http://www.indiana.edu/~primate/left.html

Lichtenstein, P., Olausson, P.O. & Bengt Kallen, A.J. (1996) Twin births to mothers who are twins: a registry-based study. *British Medical Journal* 312: 879–81.

Marchant, L.F. & McGrew, W.C. (1996) Laterality of limb function in wild chimpanzees of Gombe National Park: comprehensive study of spontaneous activities. *Journal of Human Evolution* 30: 427–43

McBurney, A. & Dunn, H.G. (1976) Handedness, footedness, eyedness: a prospective study with special reference to the development of speech and language skills. In *The Neuropsychology of Learning Disorders*, edited by R.M. Knights & D.J. Bakker, Baltimore: University Park Press

McGrew, W.C. & Marchant, L.F. (1992) Chimpanzees, tools, and termites: hand preference or handedness? *Current Anthropology* 33: 114–19

McGrew, W.C. & Marchant, L.F. (1999) Manual laterality in anvil use: wild chimpanzees cracking *Strychos* fruits. *Laterality* 4: 79–87

Noonan, M. & Axelrod, S. (1981) *Earedness: Its Measurement and Its Relationship to Other Lateral Preferences*. New York: Eastern Psychological Association

Nottebohm, F. & Nottebohm, M.E. (1976) Left hypoglossal dominance in the control of canary and white-crowned sparrow song. *Journal of Comparative Psychology* 108: 171–92

Oldfield, R.C. (1971) The assessment and analysis of handedness: the Edinburgh Inventory. *Neuropsychologia* 9: 97–113

Perelle, I.B. & Ehrman, L. (1983) The development of laterality. *Behavioral Science* 28: 284–97

Perelle, I.B. & Ehrman, L. (1994) An international study of human handedness: the data. *Behavior Genetics* 24: 217–27

Perelle, I.B., Saretsky, T. & Ehrman, L. (1978) Lateral consistency in *Drosophila*. *Animal Behaviour* 27: 622–23

Perelle, I.B., Ehrman, L. & Manowitz, W. (1981) Human handedness: the influence of learning. *Perceptual and Motor Skills* 53: 967–77

Provins, K.A. & Cunliffe, P. (1972) The reliability of some motor performance tests of handedness. *Neuropsychologia* 10: 199–206

Ramos, R. (1968) Left-handed writing. *Academic Therapy* 4: 47–48

Rife, D.C. (1994) Handedness, with special reference to twins. *Genetics* 25: 178–86

Roberts, W.W. (1949) The interpretation of some disorders of speech. *Journal of Mental Science* 95: 567–88

Sperry, R.W. (1971) How a developing brain gets itself properly wired for adaptive function. In *The Biopsychology of Development*, edited by E. Tobach, L.R. Aronson & E. Shaw, New York: Academic Press

Stern, C. (1973) *Principles of Human Genetics*, 3rd edition, San Francisco: Freeman

Torgerson, J. (1950) *Situs inversus*, asymmetry, and twinning. *American Journal of Human Genetics* 2: 361–70

GLOSSARY

monozygotic (MZ) developing from a single fertilized egg, as in identical twins

proband affected individual from whom a pattern of inheritance of a familial disorder is traced

See also **Twinning (p.516)**

GENETICS OF AGEING

Daniel E. L. Promislow[1] and Gordon J. Lithgow[2]
[1]Department of Genetics, University of Georgia, Athens, Georgia, USA
[2]Biological Gerontology Group, School of Biological Sciences University of Manchester, Manchester, UK

1. Introduction

The shrew-like marsupial Antechinus, the naked mole rat and the common house mouse are all roughly the same size. In the marsupial, males exhibit a sudden and catastrophic die-off at 11 months of age, within a few days of the end of the week-long breeding season. By contrast, in the naked mole rat, males may live for 20 years or more. As for the mouse, while many laboratory strains typically have a mean lifespan on the order of months and a maximum lifespan of 1 year, some strains have been shown to survive almost 5 years (Finch, 1990).

Why do some species show such rapid rates of ageing, while others have only a gradual decline in survival and fertility? Biologists have developed both evolutionary and molecular theories to explain why we see the age-related decline in both fertility and survival that we call ageing. A decade ago, a review article identified no fewer than 300 such theories (Medvedev, 1990), each of which claimed to provide a fundamental explanation for this undesirable but inevitable phenomenon. Since then, the number of theories has continued to increase.

To develop a comprehensive understanding of the genetics of ageing, we need to integrate molecular and evolutionary genetic approaches. Until recently, however, molecular geneticists and evolutionary biologists have pursued quite distinct lines of inquiry towards an understanding of ageing. On the one hand, many molecular biologists believe that a literal pot of gold awaits the scientist who identifies a single gene or process as the fundamental cause of ageing. Putative candidates suggested so far include genes involved in DNA repair, free radical scavenging, telomere shortening, the heat shock response and more (Finch, 1990). At the same time, evolutionary biologists have focused primarily on testing evolutionary genetic theories of ageing, which argue that there will not be a single "magic bullet" as the fundamental cause of ageing. In this article, we aim to present a balanced view of both molecular and evolutionary genetic perspectives on senescence, and to suggest ways to reconcile the "magic bullet" approach of molecular biology with the evolutionary gerontologists' assumption that a multitude of genes will eventually be found to affect ageing.

The field of ageing research is at a vibrant stage, with many hotly contested areas of research. In the first part of this article, we will address the evidence for evolutionary theories of why we age. The second part of our article will address possible molecular genetic mechanisms of how we age. By looking back to prior work with a critical eye, we hope to give the reader a sense of the many promising, and sometimes controversial, avenues still open for future inquiry.

For readers who wish to pursue issues raised here in greater detail, there are several excellent books that cover the field (e.g. Finch, 1990; Rose, 1991; Austad, 1997; Wachter & Finch, 1997; Kirkwood, 2000).

2. Evolutionary theories of ageing

Raymond Pearl and his colleagues laid the groundwork in the 1920s for the genetics of ageing. They found a significant heritable component to patterns of ageing in the fruit fly, *Drosophila melanogaster*. Subsequent work has found similar genetic effects in many other species, including mice, flour and bean beetles, and nematodes. Having established that allelic variants exist that can decrease lifespan, we need to understand why these alleles exist. If death reduces fitness, why has natural selection not removed these alleles from the population? Compelling evolutionary answers to this question were put forward by two scientists in the 1950s – Sir Peter Medawar and George C. Williams.

The evolutionary answer to the problem of ageing lies in the fact that the strength of selection declines with age. R.A. Fisher and J.B.S. Haldane were among the first to point out that the strength of selection on genes expressed at late ages will be much lower than on genes expressed early in life. Sir Peter Medawar (1952) developed this line of reasoning to create an explicit argument for the evolution of ageing. Medawar pointed out that a deleterious, inherited mutation that decreases some component of fitness (i.e. survival rate, fecundity rate) at an early age will be strongly selected against. A mutation whose effect is confined to a much later age will not be as strongly selected against. An individual carrying a late-acting deleterious mutation will likely have already passed on this mutation to its offspring before selection can "see" the mutation's phenotypic effects. From one generation to the next, Medawar argued, these late-acting mutations will accumulate at a much faster rate than early-acting mutations and lead, in turn, to the inevitable decline in fitness components that we call senescence.

George Williams (1957) extended Medawar's argument, arguing that these late-acting deleterious genes may actually experience favourable selection if they have early-acting, beneficial pleiotropic effects. As yet, there is no single gene that has been shown definitively to increase an early-age fitness trait at the expense of a late-age fitness trait, although Albin (1994) and others have suggested that the gene associated with Huntington's disease may also be associated with increased fertility early in life. Later in this article, we will examine other experimental evidence for this antagonistic pleiotropy model.

A third evolutionary model for ageing has been put forward by Tom Kirkwood. Beginning in the late 1970s,

Kirkwood argued that ageing evolved due to a trade-off between investment in somatic maintenance and repair on the one hand, and investment in reproduction on the other (Kirkwood, 1977). Consider a mutation that prevents an animal from investing in reproductive output. Resources previously devoted to reproduction will now be available for investment in somatic maintenance and repair, and the animal may be able to ward off ageing. However, with no reproduction, it would have zero fitness, and the mutation could not spread. On the other hand, a mutation that causes an organism to invest only in reproduction may also cause it to die very quickly. The optimal investment strategy, Kirkwood suggests, is shaped by natural selection to be in an intermediate ground. This genetically determined strategy leads to sufficient reproduction, but inevitable somatic decline and senescence. Kirkwood suggests a variety of novel predictions stemming from his model; however, from a genetic perspective, we consider the model to be a special case of Williams's antagonistic pleiotropy theory, and thus we will focus on the first two of these three evolutionary models.

These models – Medawar's "mutation accumulation" and Williams's "antagonistic pleiotropy" hypothesis in particular – have been the central focus of modern evolutionary genetic studies of ageing. Yet, which of these models is more likely to explain the origin and maintenance of senescence is the subject of much debate (Partridge & Barton, 1993). A host of studies has substantiated both mutation accumulation and antagonistic pleiotropy models, suggesting that both may play a role in the ageing process. These results come from careful study of a limited number of species, including yeast (*Saccharomyces cerevisiae*), nematodes (*Caenorhabditis elegans*), mice (*Mus musculus*), flour beetles (Tribolium spp.), bean beetles (*Callosobruchus maculatus*) and fruit flies (*D. melanogaster*), with most of the work focusing on Drosophila.

(i) Antagonistic pleiotropy

In Williams's original paper on antagonistic pleiotropy, he derived nine distinct predictions from his model. The final prediction was that "selection for increased longevity should result in decreased vigour in youth" (Williams, 1957, p.410). In particular, if ageing were caused by antagonistic pleiotropy, then selection for long lifespan should reduce fecundity at early ages. The first explicit test of Williams's idea did not appear until over two decades after his original paper. To determine if antagonistic pleiotropy was responsible for variation in lifespan in Drosophila, Rose and Charlesworth (1980) bred selectively from older and older fruit flies. After only 12 generations, they found that flies bred from relatively old females did, indeed, have extended lifespan and reduced early-age fecundity.

Although many subsequent studies have found additional support for Williams's prediction, not all have supported the hypothesized trade-off between survival and fecundity. In a recent review on the genetics of ageing, Curtsinger *et al.* (1995) suggest a variety of alternatives to antagonistic pleiotropy to explain the genetic trade-offs observed by Rose and others. For example, in selecting for long lifespan, most studies have also selected unconsciously for high, late-age fecundity and low, early fecundity. Furthermore, the sign of genetic correlations for life history traits may not be a good indicator of actual life history trade-offs. The definitive test of antagonistic pleiotropy may well come from quantitative trait locus (QTL) mapping, which integrates evolutionary and molecular genetic approaches. We will come back to the use of QTL mapping in ageing studies later in this article.

(ii) Mutation accumulation

It is important to stress that Medawar's evolutionary theory of mutation accumulation is distinct from the "somatic mutation accumulation" hypothesis developed a few years later by Szilard (1959) and Orgel (1963). The somatic mutation theory argues that ageing arises due to the accumulation of novel mutations within the lifetime of an individual. Older individuals carry more mutations and, thus, have higher mortality rates. Whereas somatic mutations cannot be passed on to offspring, the mutations discussed in Medawar's evolutionary mutation accumulation theory are passed on through the germline. Somatic mutations accumulate in an individual's lifetime; the germline mutations that Medawar describes accumulate over tens or hundreds of generations. In the latter case, individuals of all ages have the same number of mutations on average, but the effects of these mutations are limited to later ages.

According to Medawar, late-acting mutations accumulate from one generation to the next because the strength of selection to get rid of them is weaker at later ages. W.D. Hamilton (1966) showed in mathematical terms exactly how and why the strength of selection changes as a function of age. In particular, he was interested in determining the effect of changes in demographic parameters at different ages on population fitness, which he defined as r, or the intrinsic rate of increase in the population. To do this calculation, one needs to determine the partial derivative of r with respect to age-specific survival or fecundity. The result from this calculation is shown in **Figure 1A**. The important point of this model is that a decrease in survival or fecundity early in life will decrease fitness to a much greater extent than a similar decrease in survival or fecundity late in life. As a consequence, mutations that affect demographic parameters early in life will be removed by selection. Mutations of similar magnitude whose effects are confined to late ages are much less likely to be removed from the population by selection, and so will accumulate from generation to generation. Thus, older individuals will express far more deleterious mutations than younger individuals and will, therefore, have lower rates of survival and fecundity.

Charlesworth later developed a quantitative genetic model of mutation accumulation (Charlesworth, 1990). His model provided a testable prediction of the mutation accumulation theory of senescence. According to the model, variation among genotypes in survival rate and fecundity should be influenced primarily by the environment at early ages and by genes at late ages. In quantitative genetic terms, the additive genetic variance (V_A) for fitness traits should increase with age (**Figure 1B**). (Additive genetic variance gives us an estimate of how much of the variation that we

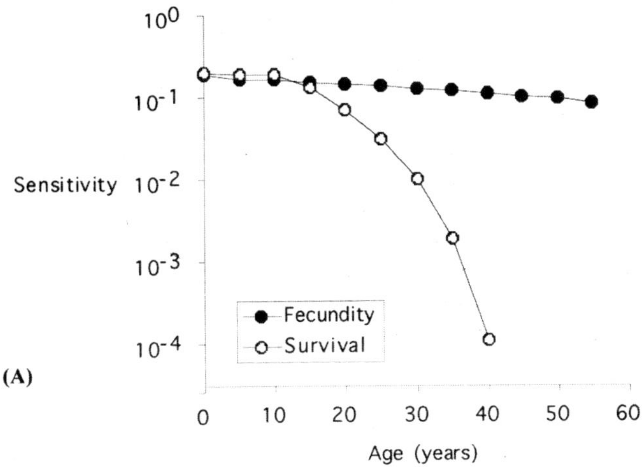

(A)

(B)

Figure 1 (A) Sensitivity of population fitness to a small perturbation in fecundity (filled circles) or survival (open circles) at a specific age. The figure illustrates that an increase or decrease in survival early in life has a much greater impact on population fitness than a similar change to survival late in life (from Caswell, 1989). This decline in sensitivity leads to the prediction, shown in (B), that additive genetic variance, V_A, for age-specific fecundity and survival should increase with age (Charlesworth, 1990). (Figure 1A is reproduced from *World Population* by N. Keyfitz and W. Flieger, 1968, with permission from University of Chicago Press.)

observe for some trait within a population is due to genetic effects, as opposed to environmental effects, and how quickly that trait can respond to selection.) If genetic variance for fitness increases with age, then we would not expect to see substantial differences for fecundity or mortality rates among different genotypes early in life. Late in life, however, we would expect to see different genotypes exhibit different mortality rates.

Until a few years ago, almost all evolutionary studies of ageing focused on tests of antagonistic pleiotropy as a cause of ageing. However, a few studies have tested the prediction that genetic variance for rates of mortality or fecundity should increase with age. Those studies that have tested this prediction have obtained mixed results. Studies of genetic variance for fecundity have found no change with age, a steady increase with age or an initial decline followed by a gradual increase with age (see Tatar *et al.*, 1996). Only recently have scientists begun to test the mutation accumulation model directly, by asking how genetic variance for mortality rates changes with age. In the first study of this kind, Hughes and Charlesworth (1994) measured mortality rates among 16 different genotypes of the fruit fly, *D. melanogaster*. They found that, at early ages (between 5 and 7 weeks), there were no consistent differences among genotypes in rates of ageing. In contrast, between 9 and 11 weeks of age, genetic variance was dramatically higher – there were clearly "short-lived" and "long-lived" genotypes, consistent with Medawar's model. Promislow and colleagues also used fruit flies to conduct a study similar to Hughes and Charlesworth's experiment, but with substantially larger sample sizes. They found that, while genetic variance for mortality initially increased with age, at late ages it declined, contrary to theoretical expectation (Promislow *et al.*, 1996). At present, the gap between theoretical prediction and empirical observation has yet to be reconciled.

Ideally, we would like to have theoretical models that make distinct predictions for mutation accumulation *vs.* antagonistic pleiotropy. Charlesworth's work suggested that mutation accumulation should lead to an increase in genetic variance for fitness at late ages. Recently, Charlesworth and Hughes (1996) pointed out that, at least under certain circumstances, the same prediction holds for genes with antagonistic pleiotropic effects. Thus, the two theories for the evolution of senescence cannot be distinguished on the basis of this prediction alone. However, Charlesworth and Hughes have developed a mathematical model that provides a prediction that is unique to mutation accumulation. They argue that for mutation accumulation, but not for antagonistic pleiotropy, the deleterious effects of inbreeding should be greater on fitness traits at late ages than at early ages. If we assume that most weakly deleterious mutations are rare and partially or fully recessive, in an outbred population they will have little or no effect. With inbreeding, recessive mutations will often be homozygous, with deleterious consequences. Due to the age-related decline in the force of selection, more mutations with late-age effects will be segregating than mutations with early-age effects. Thus, according to Charlesworth and Hughes' theoretical model, if we inbreed a population, the decline in fitness due to inbreeding in subsequent inbred generations should be greater at late ages than at early ages. The authors present data that support this prediction, suggesting that at least part of the genetic basis of ageing is due to the accumulation of deleterious mutations with late-age effects.

Finally, Service *et al.* (1988) carried out reverse selection experiments on Michael Rose's long-lived selected lines. They reasoned that, if the selection response were due to segregating antagonistic pleiotropic alleles, they should be able to reverse the selection process. Alternatively, if the initial response were due to the purging of accumulated deleterious mutations, no reverse selection response would appear. They found that the response in some traits (early-age fecundity, starvation resistance) was consistent with the presence of antagonistic pleiotropic loci, whereas the response in other traits (ethanol tolerance, desiccation resistance) was consistent with a mutation accumulation mechanism.

(iii) Antagonistic pleiotropy or mutation accumulation?

Most evolutionary biologists who study ageing agree that the fundamental evolutionary cause of ageing is the declining force of natural selection with age (Rose, 1991). It is just how this shapes the underlying genetic basis of the ageing process that is still a matter of great debate. There is more experimental evidence for antagonistic pleiotropy than for mutation accumulation, but this is primarily a reflection of where researchers have concentrated efforts in the past, rather than the relative merit of the two theories. In fact, one can find evidence both for and against each of these two theories. This does not mean that the theories are incorrect; rather, we would argue, they are simply incomplete. We now need to refine our mathematical models of both processes and develop much more demographically sophisticated experimental tests of these theoretical models. We also need to expand the scope of our studies to reach beyond the flies, beetles and worms that have provided the central focus of most work on ageing. Some of the vertebrate species that have been used to study development, such as the zebra fish, may provide especially powerful tools to understand the genetics of ageing in humans.

(iv) Senescence and its correlates

Any complex trait such as mortality rate is likely to be influenced not only by genetic factors, but also by environmental factors, and will also be genetically correlated with a suite of other traits. Nutrition is known to be a key environmental variable in the ageing process in laboratory organisms, and may play a critical role in the interaction between genes, reproduction and ageing (Tatar & Carey, 1995). In a series of elegant studies on Drosophila, Partridge and Harvey (1985) have shown that reproduction can dramatically decrease longevity. In addition, genes that influence patterns of ageing may also influence other components of the life history strategy, such as early developmental traits (Chippindale *et al.*, 1994), with long-lived genotypes exhibiting slower developmental rates. To develop a comprehensive understanding of the genetic basis of ageing, we will need to incorporate the effects of all of these interrelated factors.

3. The future of evolutionary gerontology

Evolutionary gerontology is a dynamic and lively field, with many questions still waiting to be addressed. In the following section, we address a few of those areas that we think are going to be particularly fruitful avenues of inquiry in the next few years.

(i) Age specificity of mutations

Evolutionary theories of ageing are based on the assumption that there exist mutations with age-specific effects on fitness. However, at present, researchers are still debating the extent of the effects of novel mutations (e.g. Peck & EyreWalker, 1997) and are even less sure about their age specificity. Recently, several laboratories have carried out experiments with Drosophila in order to determine the effect of novel mutations on the ageing process.

In the first such study, Clark and Guadalupe (1995) used P-element mutagenesis to determine the effect on survival of knocking out single random genes in the fruit fly. P elements are transposable DNA elements that can be inserted artificially into a single gene, thereby knocking out any expression in that gene. Whole populations of flies can then be raised that have the same, single gene knocked out. P-element mutagenesis has been used to study the effect of gene expression on many developmental, behavioural and life history traits. Clark and Guadalupe used this approach to compare survival rates in over 50 lines, each having a single gene knocked out.

If a gene has age-specific effects on mortality rate (or any fitness trait), then, when we knock the gene out by P-element insertion, we should see mortality rates change at some ages but not at other ages. If, for example, the loss of function of a gene usually has effects only at early ages, then we should find a high degree of variation among different knockout lines for early-age mortality. Mortality at late ages, on the other hand, should be relatively similar among lines. Clark and Guadalupe's lines differ substantially in mortality rates at very early ages and at very late ages (Promislow & Tatar, 1998). At intermediate ages, there is little variation among lines with respect to mortality rates. These data suggest that the phenotypic effects of most genes on mortality are seen either early or late in life.

Of course, natural mutations are likely to have much subtler effects on gene expression than a complete knockout. In the first study to look at the effect of natural mutations on ageing, David Houle and his colleagues found that the effects of mutations were strongly correlated across ages in fruit flies (Houle *et al.*, 1994). Thus, a mutation that increased mortality rate early in life would also tend to increase it late in life. Recall that the mutation accumulation theory relies on the presence of mutations with effects that are confined to late ages; such age-specific mutations should not give rise to positive correlations across ages. Thus, Houle argued that the positive correlations he observed mitigated against mutation accumulation playing much of a role in the ageing process. A subsequent study by Pletcher and colleagues (1998) also analysed lines of fruit flies that had been set up so as to accumulate natural mutations in the absence of selection. Although they found mutations had effects on mortality that were positively correlated across ages, these correlations declined with the time spanned between age classes. More interestingly, they found significant effects of novel mutations on early-age mortality rates, but no effect after approximately 30 days of age, suggesting that there are mutations with early-age effects, but not with late-age effects.

How are we to reconcile these diverse results? First, it is not surprising that the results from the P-element mutagenesis assay differed from the results from the lines that had been allowed to accumulate novel mutations. The former lines have a single mutation with a strong effect, whereas the latter lines are likely to have much larger numbers of mutations but with weaker deleterious effects. This does not, however, explain why Pletcher *et al.* (1998) observed a different pattern of correlation between ages than Houle *et al.* (1994). One possibility is simply that, with significantly larger sample sizes, Pletcher *et al.* were able to analyse mortality rates at relatively older ages than Houle *et al.* – had Houle *et al.* looked at these older flies, they too might have seen a decline in the correlated effects of novel mutations. Yet, why might Pletcher *et al.* have only seen mutations with early-age effects? Pletcher and colleagues' "mutation accumulation" lines, reared over multiple generations with weak or no selection, had already been in the laboratory for roughly 400 generations. Given the standard protocol under which flies are reared in the laboratory, there is little or no selection to remove mutations, with effects confined to late ages (Promislow & Tatar, 1998). Thus, Pletcher *et al.* initiated a mutation accumulation experiment using flies that may have already been virtually saturated with mutations with late-age effects, but which had relatively few mutations with effects confined to early ages. New mutations with late-age effects may have not been visible in the background of an already high late-age-specific mutation load.

Theories of senescence are based on understanding the equilibrium between mutation, which adds genetic variation to a population, and selection, which removes this variation. We understand a great deal about how selection shapes existing genetic variation, but very little about the other side of the equation – how mutation adds variation. As is evident from the above discussion, this is one of the most critical areas in need of resolution. Further studies are underway in several laboratories to determine the age specificity and magnitude of both P-element induced and natural mutations.

(ii) Demography

Senescence is defined as the age-related decline in fitness components due to physiological deterioration (Rose, 1991), but how should we measure senescence? Mean or maximum lifespan tells us how long an organism lives, but not whether it ages. To determine the rate of ageing, we generally measure the degree to which mortality rates increase with age. Mortality rate estimates are prone to sampling error, so substantial sample sizes (on the order of hundreds, if not thousands, of individuals) are needed in order to have any degree of accuracy.

It is relatively straightforward to measure mortality rates in the laboratory. We start with a cohort of same-aged, newly emerged adults (in holometabolous insects, age 0 is defined as the day at which the adult ecloses from its pupal case). Each individual is monitored until it dies. Once the entire cohort has died, we use individual ages at death to determine mortality rates for all ages. First, we define survival rate:

$$P_x = N_x/N_{(x-1)} \qquad (1)$$

as the number of individuals at age x (N_x) divided by the number of individuals that were alive at age $x-1$, N_{x-1}. The instantaneous mortality rate is given by:

$$\mu_x = -\ln(P_x) \qquad (2)$$

which has a lower limit of 0 and no upper limit. In laboratory-maintained organisms, as well as in human populations, μ_x increases exponentially with age (or linearly on a logarithmic scale – see **Figure 2**). This commonly observed pattern is described by the Gompertz equation:

$$\mu_x = Ae^{Bx} \qquad (3)$$

where A is the initial, age-independent mortality rate and e^B is the rate at which mortality rate increases during each time interval. A related measure, the mortality rate doubling time, is simply the time it takes for mortality rates to double and is equal to $\ln(2)/B$. In humans, this value is approximately 8 years. Powerful maximum likelihood methods now exist for estimating mortality rates, in general, and the Gompertz equation parameters in particular. However, these statistical methods lie beyond the scope of this article.

Until just a few years ago, the standard paradigm for mortality patterns was that mortality rates follow the Gompertz trajectory, increasing exponentially, and indefinitely, with age. This assumption was usually based on mortality estimates from small cohorts – generally fewer than 200 individuals. With few individuals to start with, very few members of the cohort are still alive at very old ages, so that mortality estimates become quite inaccurate. In the hypothetical human cohort in **Figure 2A**, if we start with 100 individuals, we can expect only ten of those individuals to surpass age 95. Obviously, it will be impossible to obtain accurate estimates of mortality rates for people older than this. However, if we start with 1000 individuals, we can expect to have more than 90 individuals aged 95 or older, though only four people, on average, will make it past age 100.

To alleviate the problem of inaccurate estimates due to diminishing sample size, experimental demographers have now carried out studies on such a large scale that many individuals are still alive at very late ages. The two landmark studies were based on mortality analysis of two unrelated species of fly – *D. melanogaster* and the Mediterranean fruit fly, *Ceratitis capitata*.

Curtsinger's group measured age-specific mortality in over 5700 male *D. melanogaster* (Curtsinger *et al.*, 1992), while Carey pushed the boundary even further, estimating age-specific mortality in over 1 million Mediterranean fruit flies (Carey *et al.*, 1992). Both studies started with known-age, newly eclosed flies, with dead flies being removed each day from the containers in which the flies were housed. One of Carey's studies included an analysis of mortality rates in flies kept individually in cups, to avoid the problem of changing density in the population cages as flies died. Both research teams found that the increase in mortality rate slowed at late ages. In the case of Drosophila, the curve became relatively flat, while in the medfly there was actually a dramatic decline in age-specific mortality among the oldest old (**Figure 3**). The most commonly accepted explanation for this levelling-off phenomenon is that it reflects underlying heterogeneity within a cohort (Vaupel & Yashin, 1985). It is unclear whether there is a genetic basis to this pattern. Promislow *et al.* (1996) found that some genotypes

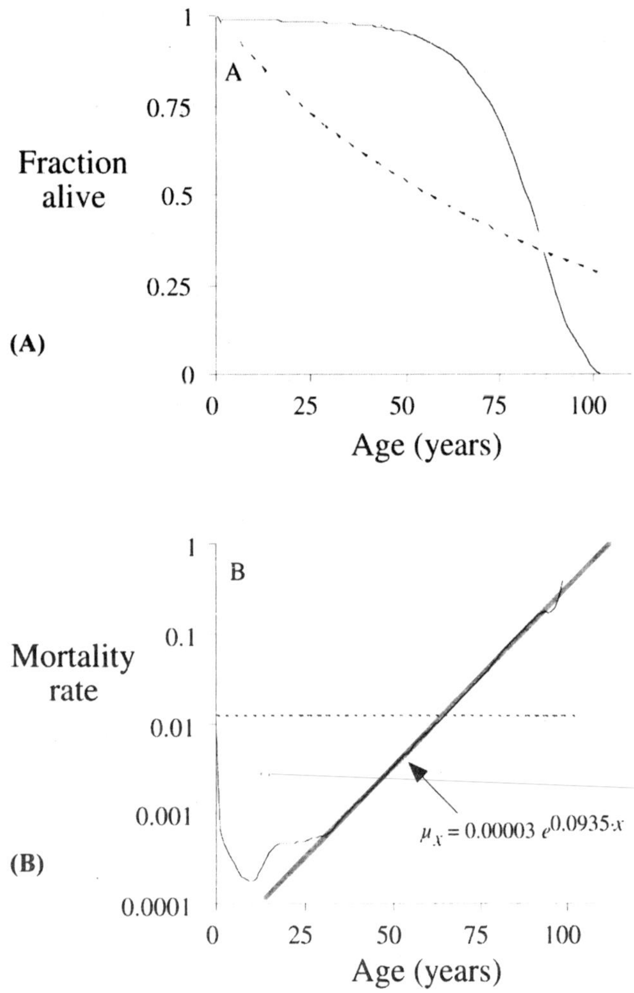

Figure 2 (A) Percentage surviving for a cohort of Canadian women in 1980 (solid line), with a life expectancy of approximately 80 years. The dashed line shows a hypothetical cohort with life expectancy of 80 years, but with no age-specific increase in mortality rate. This difference is illustrated more clearly in figure (B), which shows the age-specific instantaneous mortality rate, on a log scale, for the actual cohort (solid line) and the hypothetical cohort, with an age-independent mortality rate of 1.25% per year. The thick grey line in (B) shows the line for the Gompertz equation, assuming no levelling off in mortality late in life (see text). (Data provided by Statistics Canada.)

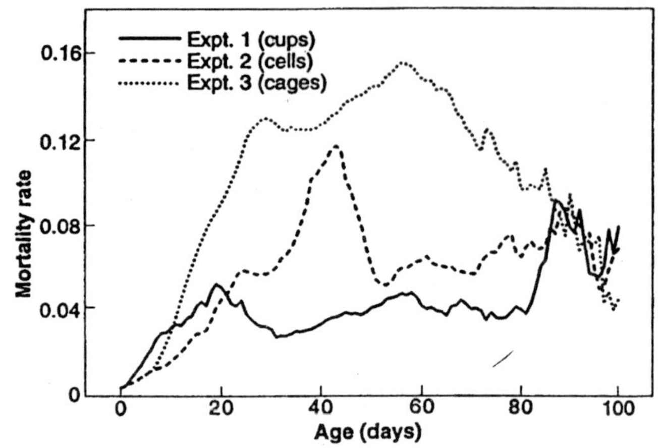

Figure 3 Log mortality *vs.* age for the medfly, *Ceratitis capitata*. Mortality rates increase exponentially over much of the lifespan (linear on the logarithmic scale shown here), but then appear to decelerate at late ages. The same pattern of deceleration has been found in many species, including humans. (Reprinted with permission from Carey *et al.*, *Science* 258: 457–61. © 1992 American Association for the Advancement of Science.)

of flies exhibited a greater degree of levelling off than others, but this does not necessarily imply that there are "genes for" levelling off. Future studies may help us to determine whether mortality levelling off is due in any part to age-specific patterns of gene expression.

This novel pattern is important not only because it overturns the commonly accepted wisdom regarding mortality patterns, but also because of its implications for the existence of a limit to lifespan. Human lifespan has increased dramatically over the past century. Some believe it will continue to increase indefinitely, whereas others believe that there are intrinsic limits to maximum lifespan. If mortality rates continue to increase exponentially throughout the lifespan, this will set a virtual limit of maximum lifespan for statistical reasons. However, consider Jeanne Calment, who died in 1998 at the age of 122 – the oldest living human being in the 20th century. If mortality rates shown in **Figure 2B** were to increase indefinitely at the same rate that they increase between the ages of 60 and 80 years, then the probability of a single individual surviving to age 122 would be approximately 10^{-12} (one in 1 trillion). Given the current global population size of approximately 10^{10}, we would expect to observe one person living to age 122 every 10 000 years. Many individuals alive today are approaching Calment's record age, suggesting that mortality rates in humans must level off at some point (Kannisto *et al.*, 1994).

Clearly, accurate estimates of ageing depend on large-scale demographic analyses. The relevance of such analyses extends even beyond the importance of levelling off in late-age mortality rates: large sample sizes are not only important for determining the extent of levelling off, but also for determining the type and magnitude of effects of single genes on ageing (Promislow & Tatar, 1998).

(iii) Genetics and human demography

Jeanne Calment was the oldest well-documented person on record. Do those of us with long-lived grandparents have a higher probability of surviving to a ripe old age ourselves? In general, the correlation of lifespan between parents and offspring has been relatively weak – between 0.01 and 0.15. In contrast, correlations between siblings have been found to be as high as 0.35 (Christensen & Vaupel, 1996). This could be due either to the shared environment in which offspring are reared or to non-additive genetics effects. In recent years, twin studies have provided estimates of the genetic effects for a variety of traits, ranging from morphological and physiological traits to intelligence and social behaviour. Twin studies of lifespan suggest that about one-quarter of the variation in lifespan is due to genetic effects. The rest of the variation is attributable to environmental influences throughout the lifespan.

Although 25% or more of the variation in lifespan may be due to genetic factors, there are likely to be many single factors contributing to the overall genetic effect. To date, the best single gene predictor of lifespan is apolipoprotein E, which explains only 1% of the total variation in lifespan (see below). In addition, this and other factors may interact in complex ways with the environment. For example, one might imagine a gene that extends lifespan by reducing cholesterol, but which only confers an observable advantage in individuals with a high-cholesterol diet. Later in this article, we will discuss other candidate longevity loci in humans.

(iv) Comparative approaches

Most of what we know about the evolutionary genetics of ageing comes from studies of *D. melanogaster* and *C. elegans*. Although the short lifespans of these species make them tractable as experimental organisms, they may tell us little about natural variation in ageing in more distantly related species. To develop a comprehensive understanding of ageing, we need not only to develop other experimental model animal systems, but also to generalize our findings to large numbers of species. With data from many species, we can use comparative evolutionary approaches to test existing theory. For example, by comparing data from a dozen different species of Drosophila, Promislow (1995) found support for George Williams's prediction of a trade-off between early-age fecundity and lifespan, as has been observed within species. However, many of the predictions outlined in Williams's paper over 40 years ago have yet to be tested. Comparative approaches may provide the ideal means of testing these hypotheses.

Comparative approaches can also be used to extend our understanding of ageing from animals to plants (Roach, 1993). Many plant species live well beyond the maximum known lifespan of any animal species. As discussed earlier, senescence is thought to arise because the strength of selection against deleterious mutations decreases with age. Recent mathematical models of clonal organisms (such as the long-lived creosote bush, *Larrea tridentata*) show that the ability of selection to weed out deleterious mutations can actually increase with age under some circumstances (Gardner & Mangel, 1997). This could lead; in turn, to the reverse of ageing, with older individuals having progressively lower mortality than younger individuals.

(v) Quantitative trait loci

Until very recently, evolutionary genetic studies of ageing relied on experimental methods that have been in use since the early part of the 20th century. Evolutionary geneticists have typically assumed that most quantitative traits, such as body height or heart rate or lifespan, are determined by a large number of genes, each with small effect. Yet, while classical approaches allowed us to determine which chromosomes accounted for variation in ageing, we could not dream of identifying specific genes. We now have the ability, through QTL mapping, not only to obtain minimum estimates of the number and magnitude of loci that influence rates of ageing, but also to locate those loci with substantial accuracy on specific chromosomes, and, ultimately, to pin down the role of specific genes.

We are now seeing a growing number of QTL analyses of ageing (Shook *et al.*, 1996; Nuzhdin *et al.*, 1997; Pasyukova *et al.*, 2000). In an analysis of lifespan in *C. elegans*, Shook *et al.* identified QTLs on three different chromosomes which, when combined, explained 90% of the genetic variance for lifespan. Nuzhdin *et al.* compared lines of *D. melanogaster* and identified five loci that showed significant effects on lifespan. The loci identified by Nuzhdin *et al.* map to candidate genes that influence a variety of other traits, including fertility, cellular ageing and stress resistance. Interestingly, some of these loci only affected rates of ageing in one sex. This sex-specific effect could explain how genetic variation for a fitness trait such as ageing could be maintained in the population.

The future of evolutionary genetic studies of ageing will likely see a close integration of population-level, demographic approaches and molecular genetic approaches. In the next section, we discuss the current state of our knowledge of the molecular genetics of ageing.

4. Molecular genetics of ageing

Since lifespan is likely to be influenced by a large number of genes, reductionist genetic and reverse genetic experimental approaches are perhaps of limited utility in ageing research. Nevertheless, a spate of genetic manipulations that alter lifespan have reminded us of the power of these experimental

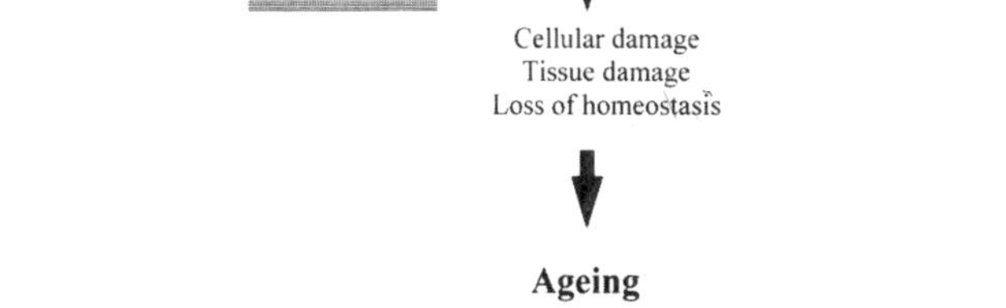

Figure 4 Molecular mechanisms of ageing. The diagram illustrates a model of ageing in which the rate of accumulation of damaged macromolecules determines the rate of ageing. The thick arrows represent ageing mechanisms and the thin arrows represent anti-ageing processes. Studies in Drosophila and *C. elegans* implicate metabolic rate, stress response protein levels and reactive oxygen species (ROS) detoxification rates in determining lifespan (see text). In this model, degradative processes, such as the removal of damaged proteins by the proteosome and the removal of damaged cells by apoptosis, slow the rate of ageing.

approaches. Although no single ageing mechanism is as yet fully understood, many independent components of mortality rate-determining processes are being identified by methods that have previously been successfully employed in solving complex problems such as development. Debate continues as to what really constitutes genetic control or genetic influence on rates of ageing. Mutations or laboratory genetic manipulations that shorten lifespan are unlikely to point the way to a pre-existing mechanism that is intrinsically involved in determining ageing rate. Rather, most major mutations simply lead to deleterious processes that increase mortality at all ages. We shall, therefore, emphasize those rarer genetic changes that increase lifespan.

Our focus in the following section is on specific genes and molecular mechanisms of ageing. However, we would stress that, generally, genetic manipulations shown to alter ageing rates also change other aspects of the life history, an observation which is consistent with the evolutionary models of ageing discussed earlier.

(i) Systems to study the molecular genetics of ageing

Three experimental approaches have been successful to different degrees in identifying ageing genes: the first approach is to identify genes that are differentially expressed in young *vs.* old organisms and subsequently to test their effects on ageing; the second is simply to make hypothesis-driven choices of candidate genes likely to influence the process and test their effects ("reverse genetics"); and the third is to introduce genetic mutations randomly and screen for those mutations that alter ageing rates ("forward genetics").

These approaches have been applied to experimental systems, including the most popular eukaryotic genetic systems – the yeast *S. cerevisiae*, the nematode *C. elegans* and the fruit fly *D. melanogaster*. These organisms have the advantage of having short lifespans and can be maintained readily in large populations.

(ii) The ageing of a single-celled organism

S. cerevisiae divides by asymmetric budding, which produces a mother cell and a smaller daughter cell. The mother cell divides again, but there is a limit to the number of divisions and it is this number which defines the yeast cell's lifespan. Following the production of the final daughter cell, the yeast mother cell becomes granulated and death follows. Yeast ageing may be the consequence of the accumulation of a "senescence factor". Evidence for the accumulation of such a factor comes from experiments (Jazwinski, 1996a) which show that:

- daughter cells from old mothers have a decreased lifespan in comparison with daughters from young mothers;
- like old mother cells, daughter cells from old mothers exhibit a slowed cell cycle time; and
- cells which are the products of cell fusions between young and old cells show the characteristic mortality rates of old cells.

The first genes demonstrated to have a role in defining the budding yeast's lifespan were identified by the generation of transformant strains that maintained additional copies of candidate genes. In this way, yeast *RAS* genes have been implicated in lifespan control (Jazwinski *et al.*, 1993). Yeast has two *RAS* genes, either one of which is sufficient for growth. Overexpression of *RAS2* results in a 30% increase in yeast lifespan, whereas deletion of *RAS1* is associated with lifespan extension (Sun *et al.*, 1994). It is not known how RAS proteins bring about alterations in lifespan, but it will undoubtedly involve their effects on the expression of other genes. The general strategy of screening for differentially expressed genes during ageing led to the identification of *LAG1* (Egilmez *et al.*, 1989) in yeast. Mutation of this potential membrane-spanning protein gene causes a 50% increase in mean and maximum lifespan.

The classical mutation approach then led to the discovery of four additional ageing genes, *UTH1*, *UTH3*, *UTH4* and *SIR4*. Mutations in these genes extend yeast replicative lifespan and were identified by looking for a "surrogate" phenotype – namely, starvation resistance. *SIR4* was previously characterized as a transcriptional regulator of silent mating-type locus. *SIR4*'s role as an ageing gene may be partly explained by the observation that silencing at the mating-type locus is diminished with age, causing sterility. It follows that a similar parting of derepression may occur across the genome. It is not known how the *UTH* mutations relate to the accumulation of a "senescence factor"; however, more recent research has uncovered a potential candidate for this factor – ribosomal DNA circles. Mutation of a gene called *SGS1* leads to the accumulation of circular rDNA fragments and significantly shortens lifespan. The *SGS1* gene is a homologue of the human gene associated with Werner's syndrome (WRN), which is characterized by premature ageing, and both genes are thought to be RecQ-like DNA helicases. It is proposed that *SGS1* plays a role in recombination in the rDNA region and that the accumulation of rDNA circles could contribute to the ageing phenotype (Sinclair & Guarente, 1997).

(iii) Multicellular microbial systems

The filamentous fungi *Podospora anserina* and *Neurospora crassa* are useful systems for studying aspects of ageing. In these systems, senescence is defined by limited vegetative growth and appears as a limitation of mycelial growth which is due, in both species, to a maternally inherited factor (reviewed in Jazwinski, 1996b). In Podospora, senescence is associated with the production of a circular DNA molecule (α-senDNA) which originates from the cytochrome oxidase subunit I gene in the mitochondrial genome. Single-gene mutations in the nuclear genes *gr* (*greisea*), *i* or *viv* result in the repression of the production of α-senDNA (Osiewacz & Nuber, 1996; Rossignol & Silar, 1996). Combinations of these mutations can result in indefinite mycelial propagation. The *gr* gene has been cloned and encodes a protein homologous to the budding yeast *ACE1* transcription factor (Rossignol & Silar, 1996) which is known to regulate the superoxide dismutase (*SOD1*) and metallothionein genes.

Deleted forms of mtDNA are also known to accumulate during ageing in invertebrates and mammals. However, the relationship between this and the accumulation of species like α-senDNA is unknown. While differences in the mtDNA structure between the fungi and more complex eukaryotes restrict direct comparison, these ageing mechanisms in filamentous fungi may prove to be important generally.

(iv) Invertebrate systems for ageing studies

The nematode *C. elegans* is a popular experimental system for the genetic dissection of complex processes such as development and programmed cell death. This organism has also become an important system for studying the molecular biology of ageing, with the identification of single-gene mutations that extend lifespan (Age mutations) (Lithgow, 1996).

C. elegans is a small (1.2 mm), free-living, soil-dwelling roundworm. It exists primarily as a self-fertilizing hermaphrodite, thus lacking inbreeding depression for life history traits (Johnson & Hutchinson, 1993). The hermaphrodite lives for approximately 20 days at 20°C and the male (which arises at low frequency) lives for approximately 18 days (Johnson & Hutchinson, 1993). *C. elegans* has a 3-day life cycle with four larval stages (L1–L4) before the final moult into the reproducing adult. Poor nutritional conditions or overcrowding lead L1 larvae to develop into an L3 alternative larval stage called the dauer (enduring) larva. The dauer larva is non-feeding, non-reproducing, stress resistant and lives 4–8 times longer than the adult. This long-lived life cycle stage attracted many early researchers to the nematode as a system to study ageing.

With few exceptions (Brown-Borg *et al.*, 1996), all known mutations that confer an increase in metazoan lifespan have been identified from studies in this species (Friedman & Johnson, 1988a,b; Van Voorhies, 1992; Ishii *et al.*, 1994; Wong *et al.*, 1995; Lakowski & Hekimi, 1996; Kimura *et al.*, 1997). Age mutations lead to increases in lifespan due to a decrease in the acceleration of mortality rate with age. The first Age locus identified was defined by the *age-1(hx546)* mutation (Johnson, 1990). This mutation confers an increase of 65% on mean lifespan and 110% on the maximum lifespan and decreases the acceleration in age-specific mortality rate, but it also displays normal development and near-normal fertility.

There is evidence of multiple genetic pathways defining lifespan, and one of these pathways partially overlaps with the genetic pathway that determines dauer formation during development (Riddle & Albert, 1997). This was first uncovered with the discovery that some conditional mutations which lead to constitutive dauer formation (Daf-c phenotype) also conferred extended lifespan on adult hermaphrodites (Kenyon *et al.*, 1993; Larsen *et al.*, 1995). Later, conditions were found where *age-1(hx546)* also leads to dauer formation, placing it too in the dauer developmental pathway (Malone *et al.*, 1996). Five genes in the dauer formation pathway are known to influence lifespan; mutations of *age-1*, *daf-2* or *daf-28* extended lifespan by 70%, 100% and 30%, respectively, and lifespan extension by these mutations requires a wild-type *daf-16* gene (Kenyon *et al.*, 1993; Dorman *et al.*, 1995; Malone *et al.*, 1996). Thus, the *daf-16* gene acts downstream in the dauer formation pathway and, indeed, mutation of *daf-16* results in the formation of a defective dauer (Daf-d phenotype). Mutation of another Daf-c gene, *daf-12*, in a *daf-2* mutant background results in a lifespan extension of 300% (Larsen *et al.*, 1995).

The identification of the protein products of the *C. elegans* Age genes is one the most important developments in ageing research in recent years. The *age-1* gene has been cloned and encodes a homologue of mammalian phosphatidylinositol-3-kinase catalytic subunit (PI3K) (Morris *et al.*, 1996). The *daf-2* gene (Kenyon *et al.*, 1993) encodes a protein which is 35% identical to the human insulin receptor and 34% identical to the insulin growth factor 1 receptor (Kimura *et al.*, 1997); the *daf-16* gene encodes a protein which is similar to the family of human fork head transcription factors (Ogg *et al.*, 1997). DAF-2, AGE-1 and DAF-16 proteins constitute an insulin-like signalling pathway. The existence in worms of this pathway may have implications for our understanding of the relationship between sensing nutrition levels, physiological status and the determination of ageing rates.

It is now important to determine whether the knowledge of insulin signalling in mammals can be integrated into our understanding of nematode ageing. The nature of the ligand for the *daf-2*-encoded receptor is unknown, but it is suggested that the receptor may interact directly with the AGE-1-encoded PI3K. By analogy with mammals, this interaction may regulate glucose uptake and fat and glycogen anabolism (Kimura *et al.*, 1997). However, we should be cautious and not assume that mammalian orthologues of the nematode PI3K will influence ageing rates.

Genetic evidence exists for a second separate lifespan pathway in the nematode, containing the genes *clk-1*, *clk-2*, *clk-3* and *gro-1* (Lakowski & Hekimi, 1996). The *clk* mutations (for "abnormal function of biological clocks") slow down the cell cycle and also embryonic and postembryonic development. In addition, these mutations also affect the timing of food pumping, defecation and swimming. The *clk* pathway does not appear to influence the formation of dauer larvae. *clk-1* mutant strains have a mean lifespan over 50% longer than wild-type and a maximum lifespan which is 70% longer. The gene is a functional homologue of a yeast gene (COQ7/CAT5) that affects the synthesis of the electron carrier, ubiquinone (Lakowski & Hekimi, 1996; Marbois & Clarke, 1996; Ewbank *et al.*, 1997). Loss of this carrier in yeast leads to a respiratory deficiency. Why would such a respiratory defect cause a lifespan increase in *C. elegans*? If we assume that the dauer formation pathway described previously also acts to alter adult metabolism, then a general relationship between metabolic rate (or metabolic state) and ageing rates emerges.

There is certainly nothing new in this notion. The production of reactive oxygen species (ROS or "free radicals") during oxidative metabolism has long been seen as a possible cause of age-associated functional decline in mammalian species. The free radical theory and its associated modifications (Sohal, 1993) form a well-developed mechanistic theory of ageing and the rationale for much current experimentation. The modern theory states that the decline in function and increase in mortality rate associated with ageing is due to the metabolic production of ROS and the subsequent reaction of ROS with cellular components causing irreparable damage. Consequently, the Age phenotype in *C. elegans* could simply be a manifestation of a reduced rate of ROS production by metabolic alteration. However, metabolic changes are not the complete picture – the repair of molecular damage may also play a role.

In the course of analysing Age mutants of *C. elegans*, researchers have uncovered a dramatic relationship between ageing and stress resistance. Almost every example of lifespan prolongation by either environmental or genetic means is also associated with elevated levels of response to stress and stress resistance (Martin *et al.*, 1996). All of the Age alleles that have been tested confer resistance to thermal stress, oxidative stress and ultraviolet radiation (Larsen, 1993; Lithgow *et al.*, 1995; Murakami & Johnson, 1996). The relationship between stress resistance and lifespan is such that Age mutations can be identified simply by selecting for mutations which confer thermotolerance (Walker *et al.*, 1998).

D. melanogaster has long been an important organism for ageing studies. While it has been ideally suited to the study of polygenic determination of history, it has been of less use in identifying major genetic effects such as those described for *C. elegans*. Nevertheless, Drosophila transgenic technology is somewhat more advanced than that available for other invertebrates and is consequently a good system to test the effects of small numbers of candidate genes. In one such case, biologists created transgenic flies with augmented antioxidant defences to test the free radical theory of ageing. In support of this theory, most transgenic lines containing additional copies of the Cu/Zn SOD and catalase genes were also long lived (Orr & Sohal, 1994) and had elevated metabolic rates (Sohal *et al.*, 1996). This work remains to be confirmed in other laboratories, and a number of recent studies have reported possible confounding factors in the use of transgenic lines (Kaiser *et al.*, 1997). Nevertheless, this line of investigation continues to produce interesting results; a recent study (Parkes *et al.*, 1998) unexpectedly demonstrated that expression of the human Cu/Zn SOD gene (*SOD1*) from a motor neurone-specific transcriptional promoter is sufficient to extend Drosophila lifespan by up to 40%.

Transgenic approaches have also been used to assess the role of heat shock protein 70 (HSP-70) in determining lifespan. Tatar *et al.* (1997) found that transgenic lines with elevated levels of the HSP-70 had lower rates of mortality during normal ageing. A number of different lines of experimentation are consistent with a relationship between HSPs and ageing. For example, *hsp* genes are induced late in the lifespan of Drosophila in the absence of heat shock (Wheeler *et al.*, 1996), which is consistent with the presence of conformationally altered protein in aged animals. In addition, a much earlier series of experiments, published in the late 1950s, showed increases in Drosophila lifespan as a consequence of heat shock treatments (Maynard Smith, 1958a,b). Recent experiments confirm these findings in Drosophila and extend them to *C. elegans* (Lithgow *et al.*, 1995); exposure of synchronous populations of Drosophila and *C. elegans* to transient, non-lethal heat shocks induces both lifespan extension and thermotolerance (often called acquired thermotolerance). Acquired thermotolerance is the result of the expression of HSPs and, consequently, the HSPs may be responsible for lifespan extension as well (Lithgow *et al.*, 1995).

(v) Genetic influences on mammalian ageing

Very little is known about the key genetic factors that determine mammalian ageing rates and species-specific lifespan. This is clearly due to the practical difficulties of maintaining large mammalian populations for the length of time required for such studies. Genetic studies are also confounded by inbreeding depression. Consequently, the most significant research on rodent ageing centres on the lifespan-prolonging effects of caloric restriction and is outside the scope of this article. However, the work on caloric restriction and ageing may be related to the previously described genetic studies of stress resistance and ageing – reduction of caloric intake is known to prolong rodent lifespan by anything up to 30% and is associated with an enhanced response to stress (Heydari *et al.*, 1993).

Human studies have concentrated on the genetics of specific age-related diseases such as Alzheimer's disease (AD). The genetics of familial AD is very well developed (Finch & Tanzi, 1997), but the relationship of the disease to basic ageing processes is unknown. Here, we shall focus on research of more direct relevance to the underlying ageing processes.

(vi) Cellular senescence and telomeres

When mammalian cells are grown *in vitro*, they display a finite number of population doublings before the cell population irreversibly ceases to grow. This is sometimes termed replicative senescence or the "Hayflick limit", after the original observation by Leonard Hayflick (1965). Senescent cells are metabolically active, morphologically distinct and display specific patterns of gene expression and other biochemical markers. The notion that this phenomenon could contribute to organismal ageing arose from the observation of a direct correlation between the age of a donor animal and the proliferative potential of cells explanted into *in vitro* culture. The additional observation that cells from donors with premature ageing syndromes exhibit a reduced proliferative capacity supports the view that senescent cells may contribute to tissue degeneration with age (reviewed in Campisi *et al.*, 1996). These observations alone are not enough to implicate cellular senescence in organismal ageing. However, cellular senescence remains important in the context of ageing, because of the more recent observation of senescent cells *in vivo* (Dimri *et al.*, 1995) and the discovery in a number of laboratories that senescent cells upregulate the production of extracellular metalloproteinase enzymes. Consequently, even a low frequency of senescent cells may have significant effects on the integrity of a tissue due to its effect on the extracellular matrix (reviewed in Campisi *et al.*, 1996). The significance of senescent cells in organismal ageing can only be properly gauged when the factors controlling the division potential of cells are manipulated. This has recently become possible.

The expression of a single gene, the human telomerase catalytic subunit gene (hTRT), is sufficient to extend the proliferative capacity of human fibroblasts in culture (Bodnar *et al.*, 1998). This finding follows from a train of research which relates the integrity of the extreme ends of the chromosomes (telomeres) to cellular senescence (e.g. Allsopp *et al.*, 1992). Telomeric repeats (in humans, this repeat is C_3TA_2) are lost during normal cell division in somatic cells due to a failure of the normal DNA replicative machinery to replicate the chromosome ends. This results in a progressive decline in the length of the telomere repeat region during cell division. In germ cells, some stem cells and most transformed cells, telomere length is maintained by the activity of the ribonucleoprotein enzyme telomerase, which synthesizes the telomere repeats using the RNA component as a template. It has been proposed that telomeres reach a threshold length in somatic cells, at which point the cells do not re-enter the cell cycle, but rather enter the senescent state (Harley *et al.*, 1992). Consistent with this hypothesis is the finding that, when hTRT cDNA is expressed in normal hTRT cells, the number of population doublings the population undergoes is increased and the cells fail to exhibit senescence biochemical markers.

Mice lines have been produced with a germline deletion of the telomerase RNA component (mTR) and these have proven highly informative. The laboratory mouse, *M. musculus*, has unusually long telomeres (Kipling & Cooke, 1990), but the mTR -/- mice lacking telomerase activity exhibited telomere shortening at a rate of approximately 5 kilobases per generation (Blasco *et al.*, 1997). The mTR -/- mice had defective spermatogenesis and increased apoptosis in the testis; other highly proliferative cells such as the haematopoietic cells, bone marrow and spleen cells were also affected. Clearly, telomerase activity is essential for cell survival and organismal replication. However, there is little indication that telomerase activity is instrumental in determining organismal ageing rate or species-specific lifespan.

Two types of experiment will be critical in determining the influence of cellular senescence on organismal ageing rates. First, we need to examine animals in which telomerase is activated in somatic tissue, thus preventing the accumulation of senescent cells. Secondly, researchers should introduce senescent cells into the tissues of young animals and examine the consequences in terms of ageing.

(vii) Progeroid syndromes

Human progeroid diseases are rare, heritable conditions and are associated with short lifespan and premature ageing-like characteristics (reviewed in Finch, 1990).

Hutchinson–Guilford syndrome is a childhood disease characterized by balding, wrinkling of the skin, subcutaneous fat loss and intense atherosclerosis resulting in cardiovascular-associated death. Fibroblast cultures show rapid replicative senescence, but the disease differs markedly from normal ageing; for example, there is no evidence of accelerated brain ageing. The disorder is thought to be the result of a rare autosomal dominant condition.

Werner's syndrome (WS) is an adult progeria and is characterized by intense atherosclerosis, heart valve calcification, skeletal muscle atrophy, premature greying and hypogonadism (Rossignol & Silar, 1996). In addition, carriers of this autosomal recessive mutation also exhibit neoplasms of connective tissues. Like Hutchinson–Guilford syndrome, WS exhibits many characteristics that are not indicative of accelerated ageing. WS is due to the mutation of a gene (WRN) encoding a member of the RecQ family DNA helicases (Gray *et al.*, 1997). This helicase may function in transcription, DNA replication, recombination or repair. The ways in which mutations lead to the complex pathologies of the WS remain to be elucidated in full.

(viii) Human longevity loci

Recent advances in the tools for rapid mapping of quantitative traits using sequence-tagged sites has led to attempts to uncover polymorphic loci associated with extreme human longevity. A pioneering study by Schachter *et al.* (1994) examined the allelic frequencies of the apolipoprotein E (ApoE) gene and the angiotensin-converting enzyme (ACE) gene in a population of 338 centenarians. ApoE and ACE are important determinants of coronary heart disease (CHD) risk, and the ε4 allele of APOE is known to promote premature atherosclerosis while the ε3 allele is associated with hyperlipidaemia. The study found a significant association of the ε2 allele with longevity and, perhaps surprisingly, also found an association between an ACE CHD-risk allele and longevity.

While such studies may be important in determining differential individual survival, it is unclear whether they will reveal genetic factors which determine ageing rates in the greater population.

(ix) The future of the molecular biology of ageing

There is an increasing optimism that current experimental protocols will produce a detailed understanding of the molecular processes defining organismal lifespan. We have seen that the isolation of mutations that confer extended lifespan in yeast and *C. elegans* is straightforward, and we are confident that other species will soon yield to this approach. When ageing genes are identified by mutation, the discovery of the underlying physiological processes of ageing is not a significant hurdle. While a wealth of detail will soon emerge on the genetic and metabolic pathways that govern the rates of ageing in animals, evolutionary theory tells us that the genetic basis of ageing will necessarily be complex and that differences in mechanistic detail will exist among species.

Some lessons have already been learned; it appears, for instance, that even for very simple genetic alterations, extension in lifespan is usually associated with changes in other traits. It also appears that lifespan extension is inherently related to stress resistance. The major challenge for the molecular biologist will be to trace the connections between traits at the metabolic and molecular level.

5. Conclusion

Until now, the genetics of ageing has been a disjointed field, with little communication between evolutionary and molecular biologists. By contrast, studies of development provide an excellent example of how researchers can integrate evolutionary and molecular genetic approaches to generate a comprehensive understanding of a biological phenomenon (e.g. Brakefield *et al.*, 1996). We are now on the cusp of a new synthesis of ageing. Eventually, it is our hope that we will see strong links forged between the molecular and organismal biology, mechanistic and evolutionary explanations, and theoretical, empirical and comparative approaches in the study of senescence. In this article, we have highlighted a few of the current trends in this dynamic field. Although we have discussed some of the most prevalent approaches currently being used to resolve the problem of ageing, just where the final answers will come from remains to be seen.

REFERENCES

Albin, R.L. (1994) Antagonistic pleiotropy, mutation accumulation, and human genetic disease. In *Genetics and Evolution of Aging*, edited by M.R. Rose & C.E. Finch, Dordrecht and Boston: Kluwer

Allsopp, R.C., Vaziri, H., Patterson, C. *et al.* (1992) Telomere length predicts replicative capacity of human fibroblasts. *Proceedings of the National Academy of Sciences USA* 89: 10114–18

Austad, S.N. (1997) *Why We Age: What Science is Discovering about the Body's Journey Through Life*, New York: Wiley

Blasco, M.A., Lee, H.W., Hande, M.P. *et al.* (1997) Telomere shortening and tumor formation by mouse cells lacking telomerase RNA. *Cell* 91: 25–34

Bodnar, A.G., Ouellette, M., Frolkis, M. *et al.* (1998) Extension of life-span by introduction of telomerase into normal human cells. *Science* 279: 349–52

Brakefield, P.M., Gates, J., Keys, D. *et al.* (1996) Development, plasticity and evolution of butterfly eyespot patterns. *Nature* 384: 236–42

Brown-Borg, H.M., Borg, K.E., Meliska, C.J. & Bartke, A. (1996) Dwarf mice and the ageing process. *Nature* 384: 33

Campisi, J., Dimri, G. & Hara, E. (1996) Control of replicative senescence. In *Handbook of the Biology of Aging*, 4th edition, edited by E.L. Schneider & J.W. Rowe, San Diego: Academic Press

Carey, J.R., Liedo, P., Orozco, D. & Vaupel, J.W. (1992) Slowing of mortality rates at older ages in large medfly cohorts. *Science* 258: 457–61

Caswell, H. (1989) *Matrix Population Models: Construction, Analysis and Interpretation*, Sunderland, Massachusetts: Sinauer

Charlesworth, B. (1990) Optimization models, quantitative genetics, and mutation. *Evolution* 44: 520–38

Charlesworth, B. & Hughes, K.A. (1996) Age-specific inbreeding depression and components of genetic variance in relation to the evolution of senescence. *Proceedings of the National Academy of Sciences USA* 93: 6140–45

Chippindale, A.K., Hoang, D.T., Service, P.M. & Rose, M.R. (1994) The evolution of development in *Drosophila melanogaster* selected for postponed senescence. *Evolution* 48: 1880–99

Christensen, K. & Vaupel, J.W. (1996) Determinants of longevity: genetic, environmental and medical factors. *Journal of Internal Medicine* 240: 333–41

Clark, A.G. & Guadalupe, R.N. (1995) Probing the evolution of senescence in *Drosophila melanogaster* with P-element tagging. *Genetica* 96: 225–34

Curtsinger, J.W., Fukui, H.H., Townsend, D.R. & Vaupel, J.W. (1992) Demography of genotypes: failure of the limited life-span paradigm in *Drosophila melanogaster*. *Science* 258: 461–63

Curtsinger, J.W., Fukui, H.H., Khazaeli, A.A. *et al.* (1995) Genetic variation and aging. *Annual Review of Genetics* 29: 553–75

Dimri, G.P., Lee, X., Basile, G. *et al.* (1995) A biomarker that identifies senescent human cells in culture and in aging skin *in vivo*. *Proceedings of the National Academy of Sciences USA* 92: 9363–67

Dorman, J.B., Albinder, B., Shroyer, T. & Kenyon, C. (1995) The age-1 and daf-2 genes function in a common pathway to control the lifespan of *Caenorhabditis elegans*. *Genetics* 141: 1399–406

Egilmez, N.K., Chen, J.B. & Jazwinski, S.M. (1989) Specific alterations in transcript prevalence during the yeast lifespan. *Journal of Biological Chemistry* 264: 14312–17

Ewbank, J.J., Barnes, T.M., Lakowski, B. *et al.* (1997) Structural and functional conservation of the *Caenorhabditis elegans* limiting gene clk-1. *Science* 275: 980–83

Finch, C.E. (1990) *Longevity, Senescence and the Genome*, Chicago: University of Chicago Press

Finch, C.E. & Tanzi, R.E. (1997) Genetics of aging. *Science* 278: 407–11

Friedman, D.B. & Johnson, T.E. (1988a) Three mutants that extend both mean and maximum lifespan of the nematode, *Caenorhabditis elegans*, define the age-1 gene. *Journal of Gerontology and Biological Science* 43: B102–09

Friedman, D.B. & Johnson, T.E. (1988b) A mutation in the age-1 gene in *Caenorhabditis elegans* lengthens life and reduces hermaphrodite fertility. *Genetics* 118: 75–86

Gardner, S.N. & Mangel, M. (1997) When can a clonal organism escape senescence? *American Naturalist* 150: 462–90

Gray, M.D., Shen, J.C., Kamath-Loeb, A.S. *et al.* (1997) The Werner syndrome protein is a DNA helicase. *Nature Genetics* 17: 100–03

Hamilton, W.D. (1966) The moulding of senescence by natural selection. *Journal of Theoretical Biology* 12: 12–43

Harley, C.B., Vaziri, H., Counter, C.M. & Allsopp, R.C. (1992) The telomere hypothesis of cellular aging. *Experimental Gerontology* 27: 375–82

Hayflick, L. (1965) The limited *in vitro* lifetime of human diploid cell strains. *Experimental Cell Research* 37: 614–36

Heydari, A.R., Wu, B., Takahashi, R., Strong, R. & Richardson, A. (1993) Expression of heat shock protein 70 is altered by age and diet at the level of transcription. *Molecular and Cellular Biology* 13: 2909–18

Houle, D., Hughes, K.A., Hoffmaster, D.K. *et al.* (1994) The effects of spontaneous mutation on quantitative traits. I. Variances and covariances of life history traits. *Genetics* 138: 773–85

Hughes, K.A. & Charlesworth, B. (1994) A genetic analysis of senescence in Drosophila. *Nature* 367: 64–66

Ishii, N., Suzuki, N., Hartman, P.S. & Suzuki, K. (1994) The effects of temperature on the longevity of a radiation-sensitive mutant rad-8 of the nematode I. *Journal of Gerontology and Biological Science* 49: B117–20

Jazwinski, S.M. (1996a) Longevity, genes, and aging. *Science* 273: 54–59

Jazwinski, S.M. (1996b) Longevity-assurance genes and mitochondrial DNA alterations: yeast and filamentous fungi. In *Handbook of the Biology of Aging*, 4th edition, edited by E.L. Schneider & J.W. Rowe, San Diego: Academic Press

Jazwinski, S.M., Chen, J.B. & Sun, J. (1993) A single gene change can extend yeast lifespan: the role of Ras in cellular senescence. *Advances in Experimental and Medical Biology* 330: 45–53

Johnson, T.E. (1990) Increased life-span of age-1 mutants in *Caenorhabditis elegans* and lower Gompertz rate of aging. *Science* 249: 908–12

Johnson, T.E. & Hutchinson, E.W. (1993) Absence of strong heterosis for lifespan and other life history traits in *Caenorhabditis elegans*. *Genetics* 134: 465–74

Kaiser, M., Gasser, M., Ackermann, R. & Stearns, S.C. (1997) P-element inserts in transgenic flies: a cautionary tale. *Heredity* 78: 1–11

Kannisto, V., Lauritsen, J. & Vaupel, J.W. (1994) Reductions in mortality at advanced ages: several decades of evidence from 27 countries. *Population and Development Review* 20: 793–810

Kenyon, C., Chang, J., Gensch, E., Rudner, A. & Tabtiang, R. (1993) A *C. elegans* mutant that lives twice as long as wild type. *Nature* 366: 461–64

Kimura, K.D., Tissenbaum, H.A., Liu, Y. & Ruvkun, G. (1997) daf-2: an insulin receptor-like gene that regulates longevity and diapause in *Caenorhabditis elegans*. *Science* 277: 942–46

Kipling, D. & Cooke, H.J. (1990) Hypervariable ultra-long telomeres in mice. *Nature* 347: 400–02

Kirkwood, T.B.L. (1977) Evolution and ageing. *Nature* 270: 301–04

Kirkwood, T.B.L. (2000) *Time of Our Lives: The Science of Human Aging*, Oxford and New York: Oxford University Press

Lakowski, B. & Hekimi, S. (1996) Determination of life-span in *Caenorhabditis elegans* by four clock genes. *Science* 272: 1010–13

Larsen, P.L. (1993) Aging and resistance to oxidative damage in *Caenorhabditis elegans*. *Proceedings of the National Academy of Sciences USA* 90: 8905–09

Larsen, P.L., Albert, P.S. & Riddle, D.L. (1995) Genes that regulate development and longevity in *Caenorhabditis elegans*. *Genetics* 139: 1567–83

Lithgow, G.J. (1996) Invertebrate gerontology: the age mutations of *Caenorhabditis elegans*. *BioEssays* 18: 809–15

Lithgow, G.J., White, T.M., Melov, S. & Johnson, T.E. (1995) Thermotolerance and extended lifespan conferred by single-gene mutations and induced by thermal stress. *Proceedings of the National Academy of Sciences USA* 92: 7540–44

Malone, E.A., Inoue, T. & Thomas, J.H. (1996) Genetic analysis of the roles of daf-28 and age-1 in regulating *Caenorhabditis elegans* dauer formation. *Genetics* 143: 1193–205

Marbois, B.N. & Clarke, C.F. (1996) The COQ7 gene encodes a protein in *Saccharomyces cerevisiae* necessary for ubiquinone biosynthesis. *Journal of Biological Chemistry* 271: 2995–3004

Martin, G.M., Austad, S.N. & Johnson, T.E. (1996) Genetic analysis of ageing: role of oxidative damage and environmental stresses. *Nature Genetics* 13: 25–34

Maynard Smith, J. (1958a) The effects of temperature and of egg-laying on the longevity of *Drosophila subobscura*. *Journal of Experimental Biology* 35: 832–43

Maynard Smith, J. (1958b) Prolongation of the life of *Drosophila subobscura* by brief exposure of adults to a high temperature. *Nature* 181: 496–97

Medawar, P.B. (1952) *An Unsolved Problem in Biology*, London: Lewis

Medvedev, Z.A. (1990) An attempt at a rational classification of theories of ageing. *Biological Reviews* 65: 375–98

Morris, J.Z., Tissenbaum, H.A. & Ruvkun, G. (1996) A phosphatidylinositol-3-OH kinase family member regulating longevity and diapause in *Caenorhabditis elegans*. *Nature* 382: 536–39

Murakami, S. & Johnson, T.E. (1996) A genetic pathway conferring life extension and resistance to UV stress in *Caenorhabditis elegans*. *Genetics* 143: 1207–18

Nuzhdin, S.V., Pasyukova, E.G., Dilda, C.L., Zeng, Z.-B. & Mackay, T.F.C. (1997) Sex-specific quantitative trait loci affecting longevity in *Drosophila melanogaster*. *Proceedings of the National Academy of Sciences USA* 94: 9734–39

Ogg, S., Paradis, S., Gottlieb, S. *et al.* (1997) The Fork head transcription factor DAF-16 transduces insulin-like metabolic and longevity signals in *C. elegans*. *Nature* 389: 994–99

Orgel, L.E. (1963) The maintenance of accuracy of protein synthesis and its relevance to aging. *Proceedings of the National Academy of Sciences USA* 49: 512–17

Orr, W.C. & Sohal, R.S. (1994) Extension of life-span by overexpression of superoxide dismutase and catalase in *Drosophila melanogaster*. *Science* 263: 1128–30

Osiewacz, H.D. & Nuber, U. (1996) GRISEA, a putative copper-activated transcription factor from *Podospora anserina* involved in differentiation and senescence. *Molecular and General Genetics* 252: 115–24

Parkes, T.L., Elia, A.J., Dickinson, D. *et al.* (1998) Extension of Drosophila lifespan by overexpression of human SOD1 in motor neurons. *Nature Genetics* 19: 171–74

Partridge, L. & Barton, N.H. (1993) Optimality, mutation and the evolution of aging. *Nature* 362: 305–11

Partridge, L. & Harvey, P. (1985) Costs of reproduction. *Nature* 316: 20

Pasyukova, E.G., Vieira, C. & Mackay, T.F. (2000) Deficiency mapping of quantitative trait loci affecting longevity in *Drosophila melanogaster*. *Genetics* 156: 1129–46

Peck, J.R. & EyreWalker, A. (1997) Evolutionary genetics: the muddle about mutations. *Nature* 387: 135–36

Pletcher, S.D., Houle, D. & Curtsinger, J.W. (1998) Age-specific properties of spontaneous mutations affecting mortality in *Drosophila melanogaster*. *Genetics* 148: 287–303

Promislow, D.E.L. (1995) New perspectives on comparative tests of antagonistic pleiotropy using Drosophila. *Evolution* 49: 394–97

Promislow, D.E.L. & Tatar, M. (1998) Mutation and senescence: where genetics and demography meet. *Genetica* 102/103: 299–314

Promislow, D.E.L., Tatar, M., Khazaeli, A. & Curtsinger, J.W. (1996) Age-specific patterns of genetic variance in *Drosophila melanogaster*. I. Mortality. *Genetics* 143: 839–48

Riddle, D.L. & Albert, P.S. (1997) Genetic and environmental regulation of dauer larva development. In C. elegans *II*, edited by D.L. Riddle, T. Blumenthal, B.J. Meyer & J.R. Priess, Plainview, New York: Cold Spring Harbor Laboratory Press

Roach, D. (1993) Evolutionary senescence in plants. *Genetica* 91: 53–64

Rose, M.R. (1991) *Evolutionary Biology of Aging*, Oxford and New York: Oxford University Press

Rose, M.R. & Charlesworth, B. (1980) A test of evolutionary theories of senescence. *Nature* 287: 141–42

Rossignol, M. & Silar, P. (1996) Genes that control longevity in *Podospora anserina*. *Mechanisms of Ageing and Development* 90: 183–93

Schachter, F., Faure-Delanef, L., Guenot, F. *et al.* (1994) Genetic associations with human longevity at the APOE and ACE loci. *Nature Genetics* 6: 29–32

Service, P.M., Hutchinson, E.W. & Rose, M.R. (1988) Multiple genetic mechanisms for the evolution of senescence in *Drosophila melanogaster*. *Evolution* 42: 708–16

Shook, D.R., Brooks, A. & Johnson, T.E. (1996) Mapping quantitative trait loci affecting life-history traits in the nematode *Caenorhabditis elegans*. *Genetics* 142: 801–17

Sinclair, D.A. & Guarente, L. (1997) Extrachromosomal rDNA circles – a cause of aging in yeast. *Cell* 91: 1033–42

Sohal, R.S. (1993) The free radical hypothesis of aging: an appraisal of the current status. *Aging and Clinical Experimental Research* 5: 3–17

Sohal, R.S., Agarwal, A., Agarwal, S. & Orr, W.C. (1996) Simultaneous overexpression of copper- and zinc-containing superoxide dismutase and catalase retards age-related oxidative damage and increases metabolic potential in *Drosophila melanogaster*. *Journal of Biological Chemistry* 270: 15671–74

Sun, J., Kale, S.P., Childress, A.M., Pinswasdi, C. & Jazwinski, S.M. (1994) Divergent roles of RAS1 and RAS2 in yeast longevity. *Journal of Biological Chemistry* 269: 18638–45

Szilard, L. (1959) On the nature of the aging process. *Proceedings of the National Academy of Sciences USA* 45: 30–45

Tatar, M. & Carey, J.R. (1995) Nutrition mediates reproductive trade-offs with age-specific mortality in the beetle *Callosobruchus maculatus*. *Ecology* 76: 2066–73

Tatar, M., Promislow, D.E.L., Khazaeli, A. & Curtsinger, J. (1996) Age-specific patterns of genetic variance in *Drosophila melanogaster*. II. Fecundity and its genetic correlation with age-specific mortality. *Genetics* 143: 849–58

Tatar, M., Khazaeli, A. & Curtsinger, J.W. (1997) Chaperoning extended life. *Nature* 390: 30

Van Voorhies, W.A. (1992) Production of sperm reduces nematode lifespan. *Nature* 360: 456–58

Vaupel, J.W. & Yashin, A.I. (1985) Heterogeneity's ruses: some surprising effects of selection on population dynamics. *American Statistician* 39: 176–95

Wachter, K.W. & Finch, C.E. (1997) *Between Zeus and the Salmon: The Biodemography of Longevity*, Washington, DC: National Academy Press

Walker, G.A., Walker, D.W. & Lithgow, G.J. (1998) Genes that determine both thermotolerance and rate of aging in *Caenorhabditis elegans*. *Annals of the New York Academy of Sciences* 851: 444–49

Wheeler, J.C., Bieschke, E.T. & Tower, J. (1996) Muscle-specific expression of Drosophila hsp70 in response to aging and oxidative stress. *Proceedings of the National Academy of Sciences USA* 92: 10408–12

Williams, G.C. (1957) Pleiotropy, natural selection, and the evolution of senescence. *Evolution* 11: 398–411

Wong, A., Boutis, P. & Hekimi, S. (1995) Mutations in the clk-1 gene of *Caenorhabditis elegans* affect developmental and behavioral timing. *Genetics* 139: 1247–59

GLOSSARY

actuarial ageing rate the rate at which instantaneous mortality rates increase. Sometimes defined as the parameter *B* in the Gompertz equation

additive genetic variance the genetic variance that accounts for the genetic component of similarity between parents and offspring, and which is due to the additive effect of genes on a trait

age mutation a mutation which increases mean and maximum lifespan due to a slowed acceleration of mortality with age

ageing for a population of animals, an increase in the rate of mortality with increasing chronological age

antagonistic pleiotropy theory of ageing, first proposed by George C. Williams. Williams argued that ageing arises because deleterious mutations with late-age effects are favoured by natural selection due to their beneficial (antagonistic pleiotropic) effects early in life

dauer the dauer larva of *Caenorhabditis elegans*. Non-feeding, non-reproducing, stress resistant, diapause stage that lives four to eight times longer than the adult and is formed during starvation or overcrowding

fitness trait a biological trait that has a strong and direct influence on the fitness of its bearer. This includes life history traits, such as age-specific fertility and mortality rates, age at maturity, and offspring size and number

genetic variance the amount of total phenotypic variance in a trait that is due to genetic factors (*see also* additive genetic variance)

Gompertz equation this equation fits an exponential increase to the instantaneous mortality rate ($\mu_x = Ae^{Bx}$, where μ_x is the instantaneous mortality rate at age x)

holometabolous insects that undergo a complete metamorphosis, with morphologically distinct larval and adult stages

instantaneous mortality rate (μ_x) defined as the probability of dying between age x and age x + Dx. In practice, usually measured as $\mu_x = -\ln(P_x)$, where P_x is the proportion of individuals that survive from age x to age x + Dx

longevity assurance gene (LAG) a gene which when overexpressed increases that organism's mean and maximum lifespan

mortality rate (*see* instantaneous mortality rate)

GENETIC ASPECTS OF BIOETHICS

Sandy Raeburn
Centre for Medical Genetics, City Hospital, Nottingham, UK

1. Introductory ideas

Genetic advances have occurred with such speed that geneticists themselves, and the public spurred on by the media, have become anxious lest scientific knowledge outstrips wisdom in developing or restraining some applications (Collins, 1999; Morton, 1999). The enthusiastic use of genetic ideas 100 years ago led to the development of eugenics; this led on, in the 1920s and onwards, to what is now universally regarded as a gross misuse of genetics. People with genetic conditions were sterilized against their wishes; so too were people with significant learning defects, who were also denied life-saving treatments. The holocaust demonstrated the dangers of racial discrimination based on genetics.

During the 1950s, Victor McKusick in the US and Cedric Carter in the UK began to use genetic principles to answer the questions about disease recurrence risks being posed by families. This led to the development of genetic counselling services, and, because of the eugenic legacy, these services emphasized individual patient autonomy, following genetic counselling, not group control. The 1960s, 1970s and 1980s led to the establishment of genetic centres in the UK, mostly in association with university medical schools (see Harper, 1998).

The principle of genetic counselling is that information is given non-directively. The geneticist or counsellor lists the options available from which the individual or couple can choose. The structure and process of genetic counselling as provided in the UK and many EC, Australasian and N American countries is in **Table 1**.

2. The importance of individual autonomy

Individual autonomy is the most important ethical principle in genetics. However, it must be accompanied by accurate information so that the individual can make an informed choice. A well-informed individual can then put the genetic information in the context of his or her family situation. There will be limits to individual autonomy set by the law of the country; an additional factor influencing an individual's autonomy will be the availability of appropriate services.

Table 1 The sequence of clinical genetic activities – duties of the clinical genetics team.

a)	referral;
b)	diagnosis and confirmatory investigations;
c)	preparation of available options
d)	offer of counselling and if wished relevant genetic testing
e)	disclosure of all results (positive or negative)
f)	accuracy of information

3. The unhealthy legacy of eugenics

Francis Galton correctly deduced that many human characteristics are genetically determined. Among these characteristics, he included "exceptional ability" or "man's natural abilities". He discussed this in detail in his monograph *Hereditary Genius* (1869). He (see article on "Francis Galton"), and those who followed him, created the eugenic movement, believing that it was imperative to plan future families in such a way that the intelligence of the population increased. No serious consideration was given to the need for any community to have a healthy mixture of people with differing skills; neither was consideration given to the destructive social effects of blind adherence to the eugenic idea, nor was there any investigation of whether eugenic policies could ever achieve their aims.

4. Dangers of neoeugenics

As genetic studies continue to add new knowledge about human heredity, it is likely that there will be further evidence to show that IQ (which is difficult to define, but probably a relevant measurement of a mixture of human characteristics) is truly determined by hereditary factors. In some specific (but very rare) families, it may be possible to identify genes that play a major causative part. The danger in the new millennium will be that this information, obtained from (and only relevant in) specific families, will be misapplied generally. This could lead to eugenic policies just as unethical and ineffective as were the policies of 60 years ago.

5. Genetic determinism

Many people believe that the presence of a specific mutation virtually guarantees the development of a particular phenotype. This is very rarely so. In practice, phenotype is influenced not only by genes of major effect, but also by a wide range of other genes acting in association with environmental factors. Genetic determinism, i.e. the belief or assumption that a particular genotype leads inevitably to a particular phenotype, is almost always misleading. In practice, for the purpose of understanding (and teaching a biological topic), it is helpful to break down knowledge into individual subsets of information, some being genetic. However, any genetic data need to be interpreted in the holistic context, particularly where the prediction of future phenotypes is involved. A practical consequence of this is that the interpretation of the same genotype varies from person to person.

6. Genetic discrimination

There are many different situations in which discrimination may take place because of an individual's genetic constitution (Launis, 1999). Clear and obviously unethical areas would be discrimination based on ethnic subgroups or on sexual differences. In many countries, these types of discrimination are reduced or prevented by appropriate legislation. Unfair discrimination could take place in selecting people for jobs or in preparing insurance contracts. Thus, knowing of a person's genetic test result, an ill-informed employer or insurer may overinterpret that information; the unfairness is due to an excessive faith in genetic determinism (see above). The ways of preventing such discrimination must include essential discussion with the geneticists who gather accurate information about the range of phenotypes likely if an individual has a particular genotype.

7. Insurance

In the UK, the Association of British Insurers has published a Genetics Code of Practice (1997) that sets out the responsibilities of insurers and other individuals who might use genetic information for underwriting. In addition, this Code sets out rules for insurers, some of which are in **Table 2**. Similar Codes of Practice could reduce or prevent unfair discrimination in other non-medical areas, such as employment. There continue to be concerns from geneticists and social scientists about this issue, especially in countries without any National Health Service (Harper, 1995; Lapham *et al.*, 1996; Hodge, 1998; Low *et al.*, 1998).

Table 2 Genetic Code of Practice.

1. No one will be forced to have a genetic test to obtain insurance.
2. Genetic test results will only be used in underwriting when the reliability and relevance to the insurance product has been established.
3. Negative genetic tests will negate the adverse effects of a family history of the condition. conversely, a genetic test showing an increase of risk does not necessarily justify an increase in premium.
4. The insurance company will appoint a nominated underwriter for genetic issues and the company medical officer will be involved , consulting a genetic specialist if necessary.
5. Insurers will not "cherry pick" i.e. use genetic tests which exclude disease to select "preferred lives" for lower insurance premiums.
6. There must also be a confidentiality policy in place, ensuring security of medical records and any other sensitive information.
7. Compliance with the Code will be monitored and each year the Chief Executive will certify to the ABI that there has been compliance
8. A robust appeals procedure is available to anyone who feels that the Code of Practice has not been followed. This is not at the individual applicant's expense.

8. Areas of particular ethical difficulty

(i) Presymptomatic testing for late-onset disorders

The difficulty here stems from the likely uncertainty about future risks and disease susceptibilities. Prospective studies are needed to add to the information already gathered in genetic centres where presymptomatic testing has been studied. However, in a small proportion of families (e.g. with special subtypes of breast cancer or in Huntington's disease), certain quite accurate predictions are possible.

One area of agreement is that children should not be tested for late-onset disorders (Clarke, 1997). This is because such testing would infringe the child's right to autonomy; only if there is a clear opportunity for effective preventive treatment should such tests be considered in children.

(ii) Prenatal diagnosis

As a result of the relative ease of terminating pregnancy (which in no way implies that a couple's decision to terminate, or the after-effects of such action, are in any way simple or non-traumatic), there is a great danger that genetic determinism will influence the decisions of couples who have prenatal diagnosis. Thus, although a number of genetic conditions can be accurately identified prenatally (usually by genetic studies on chorionic villus cells to check the genotype of the embryo), other genetic findings may be revealed opportunistically. This "unwanted" information might influence a couple's decision about proceeding (or not) with the pregnancy. Many of the unexpected findings might show conditions which are not serious, but which, without being understood, cause concern to both obstetrician and patient. Thus, genetic discrimination can operate unwittingly.

9. Ethics and the new genetics

It is likely that genetics will play a major part in our future understanding of disease, and in disease prevention. Therefore, it is very important that applications of genetics are soundly based on the strongest ethical principles. Doctors and scientists should not be directive, but should identify the available options from which a couple can choose. These professionals must realize too that even quite disadvantaged people may well be capable of making wise decisions within their own context and circumstances. By attention to any individual's right to make choices, and by moving well away from any centralized decision making about genetic futures, it should be possible to avoid the gross excesses and unforgivable mistakes in the earlier part of the 20th century by the supporters of eugenics.

10. Controls over sensitive issues

There has been no general agreement about the way in which individual countries tackle the sensitive issues of ethical issues created by the new genetics. However, there have been four workshops, each leading to reports and recommendations, which were run by the Professional and Public Policy Committee of the European Society of Human Genetics. These covered genetic screening, genetics and insurance/employment, DNA storage and genetic services. These can be accessed via the ESHG website (http://www.eshg.org/). Whether a country invokes legislation, or relies on voluntary codes of practice, depends on the nation's core provision for health or social services.

REFERENCES

ABI (Association of British Insurers) (1997) *ABI Genetics Code of Practice*, London: Association of British Insurers; revised edition 1999

Clarke, A.J. (1997) The genetic testing of children. In *Genetics Society and Clinical Practice*, edited by P.S. Harper & A.J. Clarke, Oxford: BIOS Scientific

Collins, F.S. (1999) Shattuck Lecture: Medical and Societal Consequences of the Human Genome Project. *New England Journal of Medicine* 341: 28–37

Galton, F. (1869) *Hereditary Genius: An Inquiry into its Laws and Consequences*, London: Macmillan and New York: Appleton, 1870; reprinted London: Friedmann and New York: St Martin's Press, 1978

Harper, P.S. (1995) Science and Technology Committee's report on genetics. *British Medical Journal* 311: 275–76

Harper, P.S. (1998) *Practical Genetic Counselling*, 5th edition, Oxford: Butterworth Heinemann

Hodge, J.G. (1998) Privacy and antidiscrimination issues: genetics legislation in the United States. *Community Genetics* 1: 169–74

Lapham, E.V., Kozma, C. & Weiss, J.O. (1996) Genetic discrimination: perspectives of consumers. *Science* 274: 621–24

Launis, V. (1999) Genetic discrimination. In *Insurance and Genetics*, edited by T. McGleenan, U. Wiesing & F. Ewald, Oxford: BIOS Scientific

Low, L., King, F. & Wilkie, T. (1998) Genetic discrimination in life insurance. Empirical evidence from a cross-sectional survey of genetic support groups in the United Kingdom. *British Medical Journal* 317: 1632–35

Morton, N.E. (1999) Genetic aspects of population policy. *Clinical Genetics* 56: 105–09

FURTHER READING

Advisory Committee on Genetic Testing (1998) *Genetic Testing for Late Onset Disorders*, London: Department of Health

Alper, J.S. & Natowicz, M.R. (1993a) Genetic testing and insurance. *British Medical Journal* 307: 1506–07

Alper, J.S. & Natowicz, M.R. (1993b) Genetic discrimination and the public entities and public accommodations Titles of the Americans with Disabilities Act. *American Journal of Human Genetics* 53: 26–32

American Society for Human Genetics Ad Hoc Committee on Insurance Issues in Genetic Testing (1995) Background statement. Genetic testing and insurance. *American Journal of Human Genetics* 56: 327–31

Bankowski, Z. (1991) Genetics, ethics and human values. *Bulletin of the World Health Organization* 69: 487–92

Cardno, A.G., Marshall, E.J., Coid, B. *et al.* (1999) Heritability estimates for psychotic disorders. *Archives of General Psychiatry* 56: 162–68

Department of Trade and Industry, Office of Science and Technology, Department of Health (1998) Government Response to the Human Genetics Advisory Commission's Report on The Implications of Genetic Testing for Insurance. London: HMSO

Disability Discrimination Act (1996). London: HMSO

El-Hashemite, N. (1997) The Islamic view in genetic preventive procedures. *Lancet* 350: 223

Harper, P.S. (1997) Genetic testing, life insurance, and adverse selection. *Philosophical Transactions of the Royal Society of London, Series B* 352: 1063–66

Harris, J. (1998) *Clones, Genes, and Immortality: Ethics and the Genetic Revolution*, Oxford and New York: Oxford University Press

Holtzman, N.A. & Shapiro, D. (1998) The new genetics. Genetic testing and public enquiry. *British Medical Journal* 316: 852–56

House of Commons Science and Technology Committee, Human Genetics: The Science and its Consequences, London: HMSO (1994–95 Session, Updated January 1996)

Hudson, K.L., Rothenberg, K.H., Andrews, L.B. *et al.* (1995) Genetic discrimination and health insurance: an urgent need for reform. *Science* 270: 391–93

Human Genetics Advisory Commission (1997) *The Implications of Genetic Testing for Insurance*, London: HGAC

Human Genetics Advisory Commission (1999) *The Implications of Genetic Testing for Employment*, London: HGAC

Kitcher, P. (1996) *The Lives to Come: The Genetic Revolution and Human Possibilities*, London: Allen Lane; New York: Simon and Schuster

Macdonald, A.S. (1997) How will improved forecasts of individual lifetimes affect underwriting? *British Actuarial Journal* 3: 1009–25

New England Regional Genetics Group (1996) Optimizing genetics services in a social, ethical, and policy context. *The Genetic Resource* 10: 2

Nuffield Council on Bioethics (1993) *Genetic Screening: Ethical Issues*, London: Nuffield Council on Bioethics

Olick, R.S. (1998) Physician's duty to warn third parties about the risk of genetic diseases (Letters to the Editor). *Pediatrics* 102: 855

O'Neill, O. (1998) Insurance and genetics: the current state of play. In *Human Genetics and the Law: Regulating a Revolution*, edited by R. Brownsword, W.R. Cornish & M. Llewelyn, Oxford: Hart

Ostrer, H., Allen, W., Crandall, L.A. *et al.* (1993) Insurance and genetic testing: where are we now? *American Journal of Human Genetics* 52: 565–77

Post, S.G., Whitehouse, P.J., Binstock, R.H. *et al.* (1997) The clinical introduction of genetic testing for Alzheimer's disease. *Journal of the American Medical Association* 277: 832–36

Raeburn, J.A. (1997) Using genetic information: critical issues in today's society. In *The Patient's Network*, Special Supplement to *Human Genomics* 21–22

Raeburn, J.A. (1999) Clinical aspects of genetics and insurance. The development and roles of Codes of Practice. In *Genetics and Insurance*, edited by T. McGleenan, U. Wiesing & F. Ewald, Oxford: BIOS Scientific

Reilly, P.R., Boshar, M.F. & Holtzman, S.H. (1997) Ethical issues in genetic research: disclosure and informed consent. *Nature Genetics* 15: 16–20

Smith, C. (1998) Huntington's chorea. A mathematical model for life insurance. Swiss Re Life & Health. Zurich: Swiss Reinsurance

The Council for Responsible Genetics (1997) Position statement on genetic privacy. A discussion on DNA data banking: see www.gene-watch.org

Wilcke, J.T.R. (1998) Late onset genetic disease: where ignorance is bliss, is it folly to inform relatives? *British Medical Journal* 317: 744–47

See also **Law, ethics and genetics** (p.594)

LAW, ETHICS AND GENETICS

Graeme T. Laurie
Faculty of Law, University of Edinburgh, Edinburgh, UK

1. Introduction

While many have argued that the pursuit of scientific knowledge is a morally neutral exercise *in se*, few can deny that the potential for exploitation which arises from such work invokes a moral imperative for all of us to reflect on the legitimacy, or otherwise, of any application of such knowledge. The promised benefits of genetic research are considerable and the "value" which genetic knowledge is perceived to have is appreciated by a number of different bodies, groups and individuals. Yet, the potential for conflicting claims to this knowledge is also considerable and, accordingly, the prospect of increased availability of genetic knowledge poses a number of dilemmas for the discipline of medical ethics. In turn, this becomes a problem for the law, which must respond with practical solutions and remedies to such dilemmas. In this section, we shall consider some of the ethical and legal rejoinders which have emerged as a result of the so-called "new genetics". We must begin, however, by asking what, if anything, is new about the new genetics.

Much of western ethical discourse over the last two centuries has placed reliance on the belief that individuals are worthy of respect as "moral choosers" (Benn, 1988). This is the view that the *autonomous* individual who has capacity to direct his or her life without undue interference from third parties deserves to have such life choices respected. In medical ethics, this view has been enshrined in the doctrine of respect for patient autonomy, an approach which is mirrored in the laws of most Western states. This requires that patient consent be obtained and that patient refusal be respected in relation to all medical interventions, in order to prevent the conduct of health-care professionals (HCPs) from coming under the scrutiny of the courts.

This view of the world is not, however, without its critics. Such an atomistic approach fails to appreciate the interconnectedness that each of us invariably has with other human beings around us, and it can often ignore the interaction that we inevitably encounter with our social and cultural institutions. "No man is an island", as John Donne wrote, and the *double involvement* of individuals and social institutions cannot be denied (Giddens, 1987); among these social institutions, we must include the family. Nonetheless, the net effect of much of medical ethics and law to date has been to pay considerable deference to the notion of patient autonomy. Limits have been imposed only when patient choice will lead to third party harm. The paradigm that has become established has been one of rights and duties; notably, the rights of the patient and the duties of HCPs in respect of their patients. Rarely have notions of patient *duties* to others been mooted or accepted (see Sommerville & English, 1999).

Yet, it is precisely in this respect that genetic information calls us to challenge as never before these assumptions about the appropriateness of such legal and ethical paradigms. In particular, the facility of genetic information about "X" also to reveal knowledge about the genetic constitution of the blood relatives of X, raises a number of issues about the correctness of leaving matters of control and access to that information solely in the hands of one individual. While X can claim that inappropriate use of, or access to, a genetic test result can have adverse consequences for him in his private life, precisely the same reasons can be given by his relatives because, in essence, the information is also theirs, in that it is also about them and can be used to the same adverse effect. One significant difference between a proband and his relatives, however, is that the proband will have made a conscious choice to acquire the information. The same might not be true of a blood relative. Yet, once such information exists questions of security, access and control arise. The variance of claims and the potential for conflict are significant. For example, relatives might claim that they have a "right" to know this information because of the reasons advanced above, yet the proband might not wish to share the information in question. Can he be required to do so? Alternatively, should it be incumbent on a HCP in possession of this information to reveal it to relatives? By corollary, a relative might not wish to know that genetic disease is prevalent within his family, while the proband might consider that he is under an obligation to disclose the risk to other family members. Might, then, a relative claim a "right" *not* to know in such circumstances? The role of ethics in resolving such dilemmas is of paramount importance. And yet, as will be argued below, it is not clear that existing ethical constructs, such as the principle of respect for autonomy, can alone meet the challenge of the new genetics.

A further problem posed by the increased availability of genetic knowledge concerns its perceived predictive value. In the familial context, genetic information can serve to facilitate a number of life choices, both for those who have been tested and their relatives. These include choices of diet, exercise, ingestion of drugs or alcohol, treatment decisions about whether or not to accept therapies or cures, and, of course, reproductive choices. It is frequently argued that through informed choice based on prediction of future harm, the harm in question can be avoided or significantly diminished. We shall return to this presently. Outside the family context, the predictive value of genetic information has raised the interest of employers, insurers and even the state because genetic test results can reveal a likelihood of future ill health in persons who are currently asymptomatic. To insure or employ such persons might not be in the longer-term financial interests of such parties, and so those parties claim an interest in knowing genetic information that reveals an increased risk. The counter to this, of course, is that such predictive power will be misunderstood and subject to abuse, leading to discrimination in the provision of insurance or the exclusion of persons from employment or certain state services. Once again, the role for ethics is to find a way through such competing claims. The role of the law, in turn, is to respond with ethically acceptable and socially viable mechanisms which ensure that valuable interests are not ignored or jeopardized and that, where possible, an acceptable balance is struck between the various claims to know and control genetic information.

Finally, an important factor to bear in mind when discussing the ethical and legal dimensions of genetics is the reality that we currently face a therapeutic shortfall in this field. Thus, while it is possible to offer tests for a high number of the most common diseases and a respectable number of rarer ones, little if anything can currently be done in most cases to alleviate or eliminate genetic disease. This knowledge is important because it calls on us to reflect on our reasons and motives for seeking genetic information *ab initio*. For example, why would a pregnant woman seek an antenatal test for Huntington's disease when there is nothing that can be done about a positive result, save to terminate the pregnancy? How permissible is it, in ethical terms, to offer such testing or, indeed, to permit abortion on such grounds? Indeed, genetic gender testing can now be used to facilitate choices which have nothing to do with disease, for example, the "choice" of the sex of one's baby. But how legitimate is this as a practice, and how valid a basis is the "wrong sex" of a baby as a reason for seeking an abortion? Similarly, can it be justifiable for parents to demand that a child be tested for an incurable and untreatable genetic condition before the child's health has begun to suffer? Is it necessarily in the child's best interests to undergo such testing? On what basis, if any, can such actions be justified?

2. Finding one's way in the ethical maze

A starting point in seeking ethically appropriate answers to such questions is to begin by appreciating that knowledge brings with it responsibility. Often we imagine that knowledge is always and necessarily a good thing and that with knowledge comes the obligation on those who receive it to act responsibly. And so, some may argue, it is irresponsible to test for Huntington's disease prenatally without contemplating the termination of the pregnancy if a positive result is returned. To do otherwise would be to burden the future child with the certain knowledge of his/her own appalling demise. By the same token, it might be argued that a parent is entitled to know whether a child will develop disease in order to prepare for the onset of disease (even if this will not be for many decades to come). Yet, one must recognize that the terms "knowledge" and "responsibility" are both value laden. It is not possible to look objectively at a given scenario and to deduce through reason alone that A or B is responsible for a particular event or outcome. A or B may have *caused* the event or outcome in question, but this is not to say that they are to be held responsible for it, nor, indeed, that they were irresponsible for allowing it to happen. "Responsibility" is a label that we attach to a set of circumstances to reflect our moral or intuitive response to those circumstances. Frequently, this can become enshrined in law, with the effect that A or B will be held to be legally responsible for a particular outcome and so will be punished or otherwise held to account for any loss which has been suffered. However, the very act of taking such a step – that is, of attaching the label of responsibility or irresponsibility – is in itself a moral judgement on the conduct of others, and such judgements are procured through the language that we use and can reflect values towards which it can predispose us. "Knowledge" that a child will have "disease", that s/he will be "disabled", that s/he will "suffer" and that s/he will be a "victim" who is "burdened" by a genetic condition presents a very bleak picture. Knowledge and fact are not necessarily the same thing, and we cannot escape the fact that knowledge and disability are socially constructed. Thus, when we talk of "genetic knowledge", we must ask, "whose knowledge"? Medical knowledge of genetic disease runs the risk of offering only a two-dimensional image of life with disease. Such knowledge can offer insight into the disease and its processes, but can say nothing of the experience of living with disease. For this type of knowledge, we must engage with those who have direct experience of the disease itself. As Gillon has said, "*. . . from their own point of view, having come into existence,* many people, even in very poor circumstances are glad that they have come into existence" (Gillon, 1998). We must guard against attaching labels of responsibility and irresponsibility based on narrow views about what genetic knowledge teaches us (Newell, 1999).

A particular concern must be to ensure that individuals are always seen as more than the sum of their genetic parts. To do otherwise might lead to what has been called geneticism. Wolf has posited, for example, that such a reductionist approach will ". . . subdivide communities by their genetic characteristics, and promote the idea that genetic differences are real, biological, and neutral grounds for different treatment" (Wolf, 1995). In turn, this will lead to discrimination and stigmatization based on misinformed and ill-guided uses of genetic information. Such potential for harmful uses of genetic information, axiomatically, must also be guarded against.

3. Resolving conflicting claims to genetic information: family issues

So, on what basis can genetic information be used responsibly and legitimately and how can we prevent illegitimate uses of the information? One way to address this question is to consider the nature of the range of interests that are at stake. We have already considered, for example, the personal interests of probands and their relatives in ensuring control over the use and access of their genetic data. The claims of such persons are grounded in strong moral and ethical terms, since many of them relate to aspects of what it means to be a human being in a society such as ours. Thus, claims that relate to choices about how information is used, and whether it should be known by others, are in reality demands for respect for individual autonomy, which is a widely recognized ethical norm and a well-protected legal right. However, an appeal solely to autonomy is not necessarily determinative of the question of who should retain control and access over genetic information in the familial context. Why, for example, should the choice of a relative to know familial information trump the choice of the proband not to disclose it? Even if it is argued that access to familial genetic information can facilitate healthier life choices or more informed reproductive choices, it is not clear that a conflict can be resolved solely by an appeal to autonomy (Ngwena & Chadwick, 1993). Other ethical concepts such as non-maleficence (avoid doing harm) and beneficence (seek to do good) can be invoked – in that it might be harmful to relatives not to reveal the information and beneficial to receive it – but it is in this respect that a further quality of genetic information must be taken into account. While it cannot be denied that more access to information can lead to more informed life choices, it should not thereby be assumed that "to know" is necessarily and always a good thing.

Consider a recent international study which has shown that the suicide rate among persons given a positive genetic test result for Huntington's disease was 10 times higher than the US average (Almqvist *et al.*, 1999). Knowledge can be helpful when there is something that can be done with the information. Thus, if a cure or a therapy is available, disclosure can clearly achieve a good end. However, in cases where individuals are presently asymptomatic and have no knowledge that they might be affected by disease, and when there is a significant therapeutic shortfall, one must reflect on the reasons why information is given. To argue that this promotes preparedness for the onset of disease presupposes that people are able to prepare adequately for the harm to come. Moreover, it ignores the possibility of *causing* psychological harm by burdening people with information that forces them into a period of self-reflection and self-reassessment that they would not otherwise have experienced. What is known can never then be "un-known", and knowledge of genetic disease can lead directly to what the Danish Council of Ethics has called "morbidification": the notion of "falling victim" to some inescapable "fate" through knowledge about risk of disease (Danish Council of Ethics, 1993). This can affect the way people feel about themselves, as well as the way they treat their children and the way they view future progeny, all of whom might also be afflicted by disease. Indeed, there is some evidence to suggest that there is a tendency for people who are given knowledge of risk of disease to behave as though disease is already present, especially when that knowledge is about a child. In 1974, the Swedish government abandoned a nationwide screening programme of newborns for alpha$_1$-antitrypsin deficiency because follow-up studies showed that more than half of the families with affected children suffered adverse psychological consequences, some of which continued for 5–7 years (Thelin *et al.*, 1985; McNeil *et al.*, 1988). Many asymptomatic children were "victimized", in the sense that they were treated by their parents as though already ill (Sveger *et al.*, 1999). Similarly, Hoffmann and Wulfsberg (1995) note that cystic fibrosis screening programmes in the US, which commenced as early as 1968, were abandoned because "many people think (even in cases where there is a familial risk for the disease) that early detection has no value and may, in fact, cause the family significant psychological distress prior to the time when the individual might become symptomatic". For these reasons, the authors assert that the US has not instituted a programme of screening newborns for Duchenne muscular dystrophy (Hoffmann & Wulfsberg, 1995). The conclusion to draw, as has been argued above, is that one should not assume that it is always in a person's best interests to know information about their genetic constitution. In certain circumstances, there may be a stronger interest *not* to know (Chadwick *et al.*, 1997).

4. Challenging existing medico-legal paradigms

Of interest to ethicists and lawyers is the question of how one might protect such an interest in not knowing. It is in this respect that many of the existing ethico-legal paradigms are called into question. The principle of respect for self-determination (autonomy) is of limited utility because autonomy is concerned with exercising choices based on information provided. In the context of protecting an interest in not knowing, one cannot approach an individual to facilitate their choice *not* to know, because to attempt to do so would defeat the interest with which one is concerned. For, in order to choose meaningfully one must have information about the choice to be made, but if I approach a person and ask "would you like to know if Tay Sachs disease is present in your family and whether or not you are at risk?", I alert them to the fact that there must be something to know (which, indeed, there must be, for otherwise I would not have made the approach at all). Thereby, I run the risk of causing the psychological harm that I am trying to avoid.

In like manner, the doctrine of confidentiality cannot protect the interest in not knowing. Confidentiality protects an individual's interest in controlling personal information about himself, and it is the duty of all HCPs (and arguably all persons who come into contact with such information) to keep it confidential. Yet, this duty relates to disclosure to third parties outside the relationship of confidence. It is simply nonsensical to suggest that an HCP breaches a duty of confidentiality owed to a person by giving that same person information about himself. Yet, this is the concern that we have with the interest not to know.

These inadequacies lead to a search for other means by which we can recognize and protect the interest in not knowing. This author has argued elsewhere that one means to do so is to recognize that the interest in question is part of broader privacy concerns which we have about ourselves and our place in the wider community (Laurie, 1996, 1999, 2001). Privacy is concerned with ensuring a state of separateness for ourselves from others. This can either be physical separateness (isolation) or separateness of intimate aspects of ourselves and our lives which, if disclosed, could cause harm or result in a net reduction of control over our lives. Two of those intimate aspects of our lives concern personal information (informational privacy) and psychological separateness (spatial privacy). The first of these is normally protected by confidentiality, which, as we have seen, ensures that personal information is not imparted to third parties without our consent. In contrast, spatial privacy protects our inner sense of "self" and allows us to define for ourselves "who we are". It is arguable, therefore, that to impose burdensome knowledge on a person who has not themselves solicited it could be an invasion of spatial privacy for the reasons advanced above: it forces self-reflection and fundamental self-reassessment leading to a series of life choices which would otherwise not have been considered.

None of this should lead the reader to conclude that an absolute right to spatial privacy is proposed. Rather, it is submitted that the nature of genetic information and the uses to which it may be put require an appreciation of a more subtle approach to the dissemination of information to persons who are not aware that it exists. One may have very good reasons for approaching relatives with information about familial genetic disease; namely, that a cure or therapy is available and, in such cases, few could deny that relatives should be approached. However, when this is not the case, considerably more caution should be exercised. The following factors should, in particular, be considered:

- the availability of a cure or therapy;
- the severity of the condition and the likelihood of onset;
- the nature of the likely affliction (i.e. carrier status or actual disease);
- the nature of any testing that would be required;
- the question of how the individual might be thought to react if exposed to unwarranted information; and
- the availability of evidence of what the individual would/would not want.

The importance of recognizing an interest in not knowing has already been appreciated by a number of bodies (Danish Council of Ethics, 1993; Advisory Committee on Genetic Testing, 1997; National Human Genome Research Institute, 1997; Advisory Committee on Genetic Testing, 1998), and it has been incorporated into a number of international legal instruments (Council of Europe, 1997; UNESCO, 1997). This is most commonly seen in the context of children; for example, the US Genetic Privacy Act – which was drafted in 1995 as part of the ELSI (Ethical, Legal and Social Issues) division of the Human Genome Project – is a piece of federal legislation designed for possible adoption by individual states. The Act provides that an individually identifiable DNA sample source shall not be taken from a minor under 16 years to detect any genetic condition which, in reasonable medical judgement, does not produce signs or symptoms of disease before the age of 16 unless an effective intervention is available to delay onset or ameliorate the severity of the disease, and the said intervention must be made before the age of 16 to be effective and written authorization is received from the minor's representative. This is an express recognition of the spatial privacy interests of children in this context (Mason & McCall-Smith, 1999).

5. Measuring legal responses to date

To date, the law's response to the regulation of the flow of genetic information between family members has been somewhat muted. The Genetic Privacy Act has not been adopted in the US, although a few states have modelled laws on its content. In the UK, no specific legislative initiative has been introduced, although a new body – The Human Genetics Commission – was established in 1999 with the remit to examine and advise on social, ethical and legal issues arising from genetic research (Human Genetics Commission, 2000). Now, in many respects, this legal *inaction* can be seen as propitious. There are few things more irksome to a lawyer than a law passed as a knee-jerk reaction to social developments, and there are few laws more useless for our society and our communities. However, as an alternative to hastily passed legislation, creeping developments of the common law – judge-made law – are often witnessed, whereby principles and rules are extended and applied to new situations and, as a result, new precedents are created. This process relies almost exclusively on the judges themselves and the willingness with which they are prepared to exercise a degree of activism in their interpretation of existing law. It is a process that is driven by policy considerations. What are the needs of the individual who is bringing the case (the plaintiff)? What are the implications for him of winning his case? What are the implications for the defendant if the plaintiff should win, or indeed, for society as a whole? And, just as importantly, what are the implications for the law? Will it be so twisted out of shape as to become unrecognizable, unprincipled or just plain unfair?

It is in the context of such judicial activism that the law has been used in the US to influence familial relationships in respect of genetic information and how it is handled within the familial milieu. The common law action that has been extended into the field of genetics is that of negligence. This action is concerned with excessively careless conduct. Compensation is payable if, as a result of such conduct, a person is harmed. The negligence action requires that three elements be proven by the person bringing an action in court:

1. that the defendant owed the plaintiff a duty of care;
2. that the defendant breached his duty of care by failing to take enough care in the particular circumstances; and
3. that, as a direct result of the defendant's breach of care, the plaintiff was harmed.

All three elements must be proven before the plaintiff will win his case. If this is possible, then damages are payable to him as compensation for the loss. Thus, for example, a supermarket owes a duty of care to all of its customers while on its premises. Should they be harmed as a result of the supermarket's failure to institute and maintain adequate safety procedures, then a negligence claim could arise against the supermarket if someone is directly harmed as a result of a failure in that system or a failure to have a system (e.g. slipping on a wet floor when no indication had been given that there was a hazard and when no one had been allocated the responsibility of ensuring that such hazards were identified, that customers were alerted to the danger and that it was dealt with promptly and effectively).

In these circumstances, the law imposes a duty of care from above. It is a duty to take *reasonable care* in light of all of the facts and circumstances of the case. A relationship of care is effectively deemed to exist between the parties, on one of whom it is incumbent to exercise at all times an acceptable standard of care towards the other party. In the context of genetics, it is a tripartite relationship with which the courts have been concerned. Consider this scenario:

Health-care professional ——— Proband (A)
(positive genetic result)
**/ **
A^1 A^2
(relatives)

If the HCP discovers that the proband has a particular genetic condition, then, depending on the condition in question, he might also know or at least strongly suspect that certain family members of the proband could also have the condition. It is also possible, however, that the HCP might know or suspect that the relatives have no knowledge that they are at risk. Late-onset dominant disorders such as Huntington's disease do not normally manifest themselves until middle age. Also, recessive diseases raise the prospect of asymptomatic carriers, who once again might not know that they have a genetic defect. While their own health is not necessarily at risk, these persons could pass on the recessive gene to their progeny and, if they do so with another carrier, there is a 25% chance in each case that any future child will be affected by disease. The following diagram represents the possible scenario:

Health-care professional ——— Proband (A)
(positive genetic result)
**/ **
A^1 A^2 —————— B
(relatives) :
:
(progeny)

Thus, the knowledge that A has a genetic disease can have implications for the lives of A^1, A^2 and B, and the knowledge that they are at risk, if they are made privy to it, might lead those persons to live differently and/or to make different life choices. For example, this information might help them to seek therapy, if it is available, or to prepare for the onset of disease or, indeed, to decide whether or not to have children. Hence, should the HCP disclose A's test results to these persons? Should he do so even if A objects? And, most importantly from the legal perspective, should the law require him to disclose?

The beginnings of a trend to extend a legal duty of care to relatives of persons diagnosed with genetic disease can be discerned in a number of American states. Thus, in the decision of *Pate v Threlkel* (661 So. 2d. 278 (1995)), the Florida Supreme Court specifically addressed the question: "Does a physician owe a duty of care to the children of a patient to warn the patient of the genetically transferable nature of the condition for which the physician is treating the patient?" The case concerned a woman who was diagnosed with thyroid carcinoma 3 years after her mother had been diagnosed with, and treated for, the same condition. In upholding the claim of the daughter against her mother's HCP, the court concluded that:

> . . . when the prevailing standard of care creates a duty that is obviously for the benefit of certain identified third parties and the physician knows of the existence of those third parties, then the physician's duty runs to those third parties . . . a patient's children fall within the zone of foreseeable risk.

The Court stressed, however, that the duty did not require that relatives be approached directly by the physician: "the duty will be satisfied by warning the patient". The Court was of the opinion that to require more would compromise the doctor's obligation of confidentiality to his primary patient and would place too burdensome a task on his/her shoulders to seek out and inform relatives.

However, in a subsequent decision of the Superior Court of New Jersey, this concern was no hurdle to the imposition of a more stringent duty of care. In *Safer v Estate of Pack* (677 A.2d. 1188 (N.J. 1996)), the daughter of a man who had died from multiple polyposis, and who in turn developed the condition at 36 years of age, brought an action against the estate of the doctor who had treated her father some 30 years previously. In upholding her cause of action, the court did not follow the Florida court's limitation of the duty and ". . . declin[ed] to hold . . . that, in all circumstances, the duty to warn will be satisfied by informing the patient", although it continued, "[i]t may be necessary, at some stage, to resolve a conflict between the physician's broader duty to warn and his fidelity to an expressed preference of the patient that nothing be said to family members about the details of the disease".

Here, the court seems to prefer an HCP's duty of care to third parties to his duty of confidentiality to his patients. Such extensions of the law must be very carefully considered before they are adopted by any other courts in any other legal systems. A number of key factors must be considered:

- How great a burden would be placed on the shoulders of HCPs to discharge such a duty?
- Who should be contacted and how?
- How should an HCP navigate a course between Scylla and Charybdis? Should one risk breaching confidentiality to discharge a duty to warn, or should one respect confidentiality and risk a negligence action?

While the law of confidentiality permits disclosure to third parties to avoid harm to them, the possible detrimental or fatal effect which such disclosures would have on the primary physician–patient relationship is none the less very real. Moreover, considerable public interests are also at stake. If doctors come to be seen as generally unreliable in their obligations of confidence, this could undermine the public's faith in the profession as a whole. The consequentialist effects of this could be very far-reaching. Those in need of care and treatment might not receive it and, in cases where their conditions pose a threat to others, e.g. if they are suffering from infectious or contagious diseases, then there would be a real and continuing threat to other members of the community if these persons remained untreated. It is not in the public interest generally to undermine physician–patient confidentiality.

An additional challenge relates to the premise on which these decisions have been taken. In *Safer v Pack*, the court justified imposing a duty on the HCP by saying that "... [t]he individual or group at risk is easily identified, and substantial future harm may be averted or minimized by a timely and effective warning". However, this is questionable on a number of grounds. As more work is done on the Human Genome Project, it becomes increasingly clear that most "genetic" diseases do not fall into the straightforward categories of dominant or recessive disorders, but rather form part of the burgeoning category of multifactorial diseases, for which the causes are numerous and complex. How are relative risks to be calculated in light of this knowledge? At the present time, it is also clear that we can detect many more diseases in which genetics has an operative role than we can treat or cure. Should disclosure take place even if no treatment or cure is available? As a strict matter of negligence law, arguably the answer is that there would be no imperative to disclose because no subsequent court action would ever be successful. To be successful, it must be shown that the failure to act "caused" the harm in question or at least contributed materially to that harm. Thus, if a cure or treatment is available and the HCP fails to disclose this, the individual in question is deprived of the opportunity to avail him or herself of the therapy or cure. However, if nothing can be done, the harm that is likely to arise, namely disease, would arise *even if* disclosure had been made. In this way, it cannot be said that a failure to warn in any way caused the harm. And yet, of course, much depends on the nature of the harm with which we are concerned. It is often argued that, even in the absence of effective medical intervention, genetic knowledge can bring benefits such as preparedness for the onset of disease or more informed reproductive choices. Should being deprived of these opportunities be a basis for a claim in negligence?

Before we can answer this question, we must answer a more fundamental one: should the courts rely unquestioningly on an assumption that non-disclosure is necessarily a legal harm? As has been argued above, the interest in *not* knowing can be very important. Disclosures can themselves "cause" harm rather than alleviate it, and it is not necessarily to respect persons that we offer them potentially burdensome information with a view to "facilitating" their choices. Indeed, to do so is arguably to disrespect the person's private or personal sphere. Certainly, the interest in not knowing will not be respected by imposing a blanket duty on HCPs to make disclosures without first considering the consequences for those to whom disclosure will be made. To refuse to extend tort law in this way would thereby assist in protecting the interest in not knowing. Such a non-activist stance is, however, solely within the domain of the courts. The protection of individual interests is left to the whims of the judiciary. Moreover, such an approach does nothing *actively* to protect from infringement the interest in not knowing. No right to compensation would be available. For that, it would be necessary for the law to deem an unauthorized disclosure a cause of action leading to the payment of damages for infringement of privacy. No such legal rights are currently in force. What is required, therefore, is specific action along legislative lines.

6. Resolving conflicting claims to genetic information: employers and insurers

In law, "rights" are invariably bound together with "remedics" and this leads us to ask, against whom might a person have a legal remedy if their privacy has been invaded by unwarranted disclosures of genetic information? It may be imprudent to embody in law a right to privacy that is actionable against relatives themselves, for the nature of family dynamics is such that it may be impossible to stem the tidal flow of information between family members. Furthermore, the prospect of intrafamilial legal suits is objectionable on a number of policy grounds. Thus, one should ask whether due account of the interest in not knowing might best be taken elsewhere. An example might be within an "ethic of care"; that is, when deciding how best to handle information and to respond to the needs and interests of the parties involved, for example, through appropriately tailored counselling services (Wertz & Fletcher, 1991). Nonetheless, the legal recognition of an actionable "right not to know" could be of significance in protecting individual interests against other claims to know information and, most notably, against the claims of employers and insurers.

It has already been noted that employers and insurers have a considerable financial interest in knowing genetic information, yet this form of interest must be set against the range of personal interests which individuals and their relatives have in the same data. The legitimacy of the former may then be called into question as being of significantly less weight. Genetic information is of interest to employers and insurers in two respects. First, if information is already available about a person's genetic constitution this can assist employers and insurers to take decisions as to the ability of a person to do a job of work or their "insurability" in terms of risk. Secondly, if employers and insurers are able to require that prospective employees/insured persons take genetic tests then, it is argued, they can better plan for the future in terms of deciding whether or not to enter contracts of employment or insurance with such persons. The fear, of course, is that unrestricted access to genetic information will lead to people being excluded from employment and insurance cover and stigmatized as members of a "genetic underclass". There has been much legislative activity in the US that has been driven by just such a fear. Sixteen states had introduced laws regulating the privacy of genetic information by March 1998 and some 150 bills had been proposed in state legislatures (Wertz, 1998). While such frantic legislative activity reveals the depth of concern which surrounds genetic information, the extent to which the Acts which make it to the statute books provide adequate protection for the range of interests at stake remains to be seen.

In the UK, matters have been regulated in a much more piecemeal fashion (Nuffield Council on Bioethics, 1993), with insurers and employers exercising self-restraint and issuing Codes of Practice (e.g. Association of British Insurers,

1997; Human Genetics Advisory Commission, 1997; Association of British Insurers, 1999; Human Genetics Advisory Commission, 1999). Nonetheless, there remains considerable public apprehension about the motives of these groups (Human Genetics Advisory Commission, 1997; Office of Science and Technology, 1999) and the prospect of direct legislative intervention to protect individual interests continues to loom large. In 1999, the Genetics and Insurance Committee was established in the UK to consider the acceptability of permitting insurers access to certain genetic test results. As of May 2000, seven such tests were remitted for consideration. In October 2000 the GALC approved the use of pre-existing test results for Huntington's disease in the context of life insurance. This was so only after the committee received clear and compelling evidence of the actuarial significance of such data. This cautious and *ad hoc* approcah will prevail in the UK until, or unless, it is found wanting. No legislative intervention is currently proposed. In some countries, such as France, the Netherlands and Australia, moratoria exist on the requesting of genetic tests and/or the results of prior tests by insurers, often voluntarily imposed by the industries themselves. More rarely, specific legislation has been passed in some countries to protect the security of genetic data from requests by insurers or employers. Examples include Austria and Belgium (Genetics & Insurance Forum, 2000). However, specific legal protection of individual genetic privacy interests remains sparse.

In addressing the question of protection against claims by parties outside the family relationship, the concept of spatial privacy has a role to play. To require that someone take a genetic test is, arguably, to invade their spatial privacy as it requires them to know information about themselves that they would not otherwise have known. The legitimacy of this, when it is to further purely financial interests, is highly questionable. For this reason, and others, it is therefore widely argued that the permissibility of access to genetic information for employers and insurers should be severely limited. Thus: (a) employers should have access only to information which is currently available and which has a direct bearing on the person's present ability to do the job of work; (b) screening is only acceptable when the working environment poses a high risk to susceptible individuals and no other means are available to reduce risk to health (and, in particular, screening must always be voluntary); and (c) insurers should have access only to known risks to health, through, for example, family history or pre-existing test results. This is by virtue of the legal nature of the contract of insurance, for all forms of insurance are *uberrimae fidei* – contracts of the utmost good faith – and require prospective insured persons to disclose all known facts which might have a material bearing on the decision of the insurer to offer insurance cover. This does not extend, however, to facts that become known after insurance has been granted nor, importantly, does it permit requests to be made that prospective insured persons undertake genetic testing solely for the purposes of calculating actuarial risk. While strong ethical argument can be made to support limited access to genetic information by insurers or employers, legal measures to give force to such argument remain elusive.

7. Public interests

The range of private interests which individuals have in protecting their genetic information is supplemented by the significant *public interest* that exists in this protection generally; that is, it is broadly in the public interest that individual interests be protected. Against this, however, can stand other public interests that might call for a reduction in the degree of protection accorded to individuals. This depends, of course, on the respective weights of each public interest. One such (competing) public interest is the detection and prosecution of crime. It is trite to point out that the advent of direct DNA analysis has revolutionized criminal justice (National Research Council, 1992, 1996). Space does not permit a broad-ranging account of its use, but it is suffice to say that its arrival has also led to an increase in concern about the taking of samples, their security and the creation of DNA databases from them, all of which potentially impinge to a significant degree on the civil liberties of citizens. In this regard, we see a fundamental tension between two public interests and, as is so often the case, the resolution of this is a compromise consisting of a balancing exercise. This exercise has as its aim, on the one hand, that DNA samples can be collected, stored and used in the effective detection and prosecution of crime, and, on the other, that adequate safeguards are in place to ensure that unauthorized uses of samples are not permitted, that the operation of databases is carried out within strictly limited parameters, that cross-referencing of data is kept to a minimum, and that "innocent" persons are excluded as soon as possible from such databases and that their samples are destroyed. Surprisingly, similar measures are employed across a range of jurisdictions to achieve this sort of balance (Chalmers, 1998). Compromise and balance is the order of the day when two such important public interests are at stake.

8. Protecting human interests and encouraging research

Before the benefits of the genetic revolution can be realized, much time and effort will require to be expended and much painstaking research conducted. Yet, the promise of substantial "goods" from genetic knowledge does not foreclose the debate on the appropriateness of seeking those goods in the first place, nor, indeed, the debate on the *means* by which such ends are to be reached (generally, see Burley, 1999). In the final section of this entry, we will consider both of these issues. First, *how* do we ensure that ethically acceptable research is conducted in the pursuit of genetic knowledge? Secondly, *what*, if any, limits should be placed on the pursuit of that knowledge? That is, is there any kind of knowledge that we should *not* seek to discover or use?

(i) Conducting ethical genetic research

How can we ensure that, once the decision is taken that certain research goals are to be pursued, the research is conducted in the most ethically acceptable and legally defensible manner possible?

No specific instruments – international or otherwise – have introduced new and unique provisions concerning the regulation of research in the context of genetics. Arguably, this is so because there is no need to cater specifically for the issues surrounding such research. The work to be scrutinized can be subjected to ethical and legal regulation within the rubric of existing instruments, principles, guidelines and laws. Moreover, the principles, guidelines and rules in question have been brought together in a number of instruments that display a high degree of homogeneity in the standards that they set for research.

Thus, for example, the *Declaration of Helsinki* (revised 1996) embodies recommendations guiding physicians in all forms of biomedical research involving human subjects (Helsinki Declaration, 1996). The Council of Europe *Convention for the Protection of Human Rights and Dignity of the Human Being with regard to the Application of Biology and Medicine: Convention on Human Rights and Biomedicine* seeks to "protect the dignity and identity of all human beings and guarantee everyone, without discrimination, respect for their integrity and other rights and fundamental freedoms with regard to the application of biology and medicine". Moreover, "[t]he interests and welfare of the human being shall prevail over the sole interest of society or science" (Council of Europe, 1997). The UNESCO *Universal Declaration on the Human Genome and Human Rights* recognizes "that research on the human genome and the resulting applications open up vast prospects for progress in improving the health of individuals and of humankind as a whole", while emphasizing "that such research should fully respect human dignity, freedom and human rights, as well as the prohibition of all forms of discrimination based on genetic characteristics" (UNESCO, 1997).

Common to all of these instruments is an approach that lays down certain key guiding principles in respect of research. These are:

- respect for research subjects at all times (people should never be treated as a means to an end);
- thorough risk/benefit analysis of the proposed research;
- free, full and informed consent must be obtained from all research subjects;
- privacy and confidentiality of research data must be protected;
- proper approval for each research project must be sought from a suitably constituted body, such as a research ethics committee, to ensure not only that the research will be carried out in an ethically acceptable manner, but also that the research has a solid scientific basis; research that does not reveal scientifically and statistically relevant data is unethical *in se*; and
- transparency is crucial – the public should not be kept uninformed about the research that is to be conducted, and the opportunity for full and frank discussions about the work should be made available.

It is unfortunate, however, that such internationally agreed standards embody, in the main, aspirational rights only as far as research subjects are concerned. Nonetheless, these principles and guidelines act as a beacon to guide the way forward through the ethical mire. In order to have teeth, they require to be incorporated into the domestic laws of nations which seek to conduct the research in question. The extent to which this has happened varies widely across legal systems.

This is not to say, however, that new laws are necessarily required in any particular legal system, but it does mean that each nation must reflect critically on its current position. To offer the example of the UK, no specific laws exist which are designed to regulate scientific research, and certainly not genetic research in particular. Rather, the common law approach has ensured that, for the most part, research is conducted within ethically appropriate limits. Indeed, it is arguable that one of the reasons why the UK has such a dearth of specific rules regulating research is because the research in question has until very recently been conducted out of the public arena and, so, beyond the scrutinizing power of public bodies. This is not an experience unique to the UK. Yet, this is a culture that must change very rapidly if genetic researchers are to retain public confidence in their work. This is likely to be true all the more so in the future as increasingly controversial possibilities emerge from the realm of genetic science.

(ii) Should certain matters remain unknown?

This brings us to the final section of this entry, and it is appropriate that we end as we began. We began this discussion with an assertion that scientific research is not a morally neutral exercise. More should be said on this proposition. It is imperative that we accept that scientific research cannot be conducted as though it were in isolation from the society on which it will impact nor can it be treated as having no consequences beyond the generation of new knowledge. We are each morally responsible in our own way for the future course of our society. Moreover, we owe duties of care to future generations who will inherit that society. The concerns and issues which surround this matter can be summed up in one simple question: "should we do things simply because we can?" Two examples offer a basis on which to explore this dilemma.

Sex selection. Advances in genetic technologies do not simply herald new opportunities to ameliorate health or avoid disease; the facilitation of a wider range of life choices is also possible. However, when those life choices impact on other persons, including future persons, their legitimacy can be called into question. The prospect of selecting the sex of one's child is a paradigm example of the ethical tensions that can arise. The determination of the sex of a foetus *in utero* has been possible for many years and, in circumstances where sex is associated with increased risk of disease (e.g. in the context of X-linked disorders such as haemophilia or Duchenne muscular dystrophy), then abortion of a foetus on the grounds of sex is widely accepted in both ethical and legal terms. In this case, the relationship between sex and disease remains. How permissible is it, however, to terminate a pregnancy on the grounds of sex alone?

The moral opprobrium with which many people view abortion tends to increase when abortion choices are seen to be taken "on demand". And yet, while some sex-based abortions might be based solely on the preference of a woman for a child of one sex rather than the other, it must also be appreciated that in a multicultural, pluralistic society the sex of a child, even a healthy child, can have adverse consequences for the life of a woman. As Mason has said: "... consider the woman of several ethnic minorities who is expected to produce a male child but is pregnant with her third female; there can be no doubt that both her physical and mental health are at risk ..." (Mason, 1998). Such circumstances justify abortion in the legal systems of most Western states. Yet, even where no such risk is likely, advocates of choice see no great harm in selecting the sex of one's future child. That choice can come either in deciding whether or not to continue with a pregnancy, or at the preimplantation stage when embryos have been created *in vitro* through artificial fertilization. For some, sex selection at this stage is more ethically acceptable, for it does not entail the destruction of a foetus by abortion. It does, of course, entail the preference and implantation of embryos of one sex over the other, and the consequent rejection and destruction of those that remain. Yet, as Mary Warnock, the lead architect of the ethical and legal framework for the regulation of reproductive services in the UK, has argued of sex selection:

> ... [i]f a couple has three sons and wanted a daughter, or three daughters and wanted a son, what ill consequences would follow? Not everyone would be able to afford such luxury of self-indulgence, it is true. But it would not constitute by any means the worst kind of inequality, even in medicine.
>
> (Warnock, 1992)

Certainly, in Western states, sex selection is not at all likely to upset the gender balance of the wider community; not only is there no clear preference for one sex over another, but there is no real evidence that sufficient numbers of potential parents would avail themselves of the choice to make a difference (Lancet, 1993). And, if preimplantation determination becomes the primary means by which sex selection occurs, this will be limited to the very small number of persons who are able to afford artificial reproductive services or who can claim them under a national health system. However, this does not preclude further restrictions on choice being imposed by those who deliver or authorize such services. While the measure of morality is never a matter to be determined by majoritarianism alone, evidence of a significant degree of public concern about certain practices can scarcely be ignored by law and policy makers, even when such views are open to sound ethical challenge. Thus, when the UK's Human Fertilisation and Embryology Authority (HFEA) carried out a public consultation in 1993 into the question of sex selection, its future policy was largely dictated by the result that 67% of those who responded were opposed to sex selection of children for social reasons. The current policy of the HFEA is accordingly to ban sex selection in clinics which hold an Authority Licence to treat childlessness (HFEA, 1998, paragraph 7.20). Fertility treatment cannot legally be provided in the UK without such a licence. HFEA reasoned that sex selection for social reasons is unacceptable because: (a) it accepts unequal values placed on the sexes, and would reinforce gender stereotyping to the disadvantage of women; (b) children may be seen as commodities; (c) there may be adverse consequences for a family trying to select the sex of one child whether this is successful or unsuccessful; and (d) it is an inappropriate use of medical resources (Deech, 1998). On the international plane, such sentiment is reflected in Article 14 of the Council of Europe Convention, which states that: "... [t]he use of techniques of medically assisted procreation shall not be allowed for the purposes of choosing a future child's sex, except where serious hereditary sex-related disease is to be avoided" (Council of Europe, 1997). As of December 2000, only seven countries had ratified this Convention (Denmark, Georgia, Greece, Saint-Marin, Slovakia, Slovenia and Spain).

Matters are different, however, elsewhere in the world where cultural and religious beliefs lead to a distinct preference for male children over female children. This is particularly prevalent in India and China where, despite the existence of laws which outlaw abortions and practices which prefer male children to females, sex selection in favour of males continues to flourish (Kusum, 1995). These practices give rise to a genuine ethical dilemma. On the one hand, there are the interests of women who face stigmatization, ostracism and, in some cases, death, if they are not permitted to abort or otherwise avoid the birth of a female child. On the other hand, the gender balances of parts of these countries have become dangerously unstable, signalling serious economic and social problems ahead. For example, in India the ratio of females to males dropped from 935:1000 in 1981 to 927:1000 in 1991, and, in the northern states of Bihar and Rajasthan, the figure is as low as 600:1000, which is also the lowest in the world. As a result, the Indian Medical Association and the Medical Council of India have urged HCPs to desist in the provision of sex selection tests and sex-preference abortions. The licences of errant practitioners will be revoked by their professional bodies (Mudur, 1999). While the increased prevalence of such practices has been facilitated primarily by ultrasonic technology, the increased availability of genetic testing in the future will only serve to exacerbate the problem. Thus, we have a classic example of new technology and new knowledge fuelling a serious ethical and social problem. While this fact in no way leads to the conclusion that genetic research should be tempered, it does throw into stark relief the realization that not all uses of new technologies result in net benefit for the community or its individuals. This, in turn, must lead us to question whether such types of knowledge should be made available and, if so, how that availability can be justified in light of the potentially adverse outcomes which might accompany its exploitation.

Reproductive and therapeutic cloning. The considerable disquiet that surrounded the birth of the first "cloned adult vertebrate" (Wilmut *et al.*, 1997) has intensified with the prospect that cloning technology may, one day, be applied to humans. A number of groups around the world have admitted that they are conducting research to these ends, and in June 1999 American Cell Technology claimed that they had created the first cloned human embryo in November 1998 using a cell from a man's leg and a cow's egg. The initial legal response to the prospect of human reproductive cloning has, however, been swift and almost universally condemnatory. The Council of Europe adopted a Protocol to its Convention on Human Rights and Biomedicine prohibiting the cloning of human beings (Council of Europe, 1998) and the UNESCO Declaration on the Human Genome and Human Rights specifically disallows cloning as contrary to human dignity (UNESCO, 1997). The UK Government sought to confirm its position that any work which was designed to produce cloned human beings was unethical and illegal (House of Commons, 1997) and in the US President Clinton asked the National Bioethics Advisory Commission (NBAC) to report, which it duly did in June 1997, concluding that the risks of research into human cloning involving clinical trials were too great and that legislation should be passed to prohibit research into cloning "complete people" (National Bioethics Advisory Commission, 1997). At the state level, 19 states had proposed 22 bills by the end of 1997 and, in that year, California passed a law imposing a 5-year moratorium on the cloning of an entire human being.

Since then, the debate has moved on. Therapeutic cloning is now thought to be possible, whereby individual body parts can be grown as "replacements" for those that may be diseased or become damaged. This requires the use of human embryonic stem cells. A joint report in the UK by the Human Genetics Advisory Commission and the Human Fertilisation and Embryology Authority recommended in December 1998 that the cloning of early-stage human embryos for research into the creation of transplant parts or for the treatment of disease be approved (Joint Report, 1998). The response of the UK Government was to commission yet another body of inquiry. It reported in April 2000 (Nuffield Council on Bioethics, 2000) and its recommendations endorsed in large part those of the US National Bioethics Advisory Commission which reported on the same matter to President Clinton in 1999 (National Bioethics Advisory Commission, 1999). The general tenor of the recommendations of both bodies supports research that uses human embryos for the purposes of developing tissues to treat diseases from derived embryonic stem cells.

Finally, in December 2000, British MPs voted to amend the existing law to permit the production of embryos for stem cell research. Thus, we see that in a relatively short space of time significant developments have occurred and, arguably, important attitudinal shifts have taken place (for debate see Burley & Harris, 1999; McMahan, 1999; Savulescu, 1999). Whether these developments serve as an indication of "moral progress" or "regress" is a matter of conjecture. Human beings, by our very nature, are fickle beasts. Morality and ethics compound this trait because they are, by their nature, malleable disciplines that are open to subjectivism (see Macklin, 1999). Thus, what is immoral and unethical today might be seen very differently in a number of years' time; it may even be only a matter of months. For the law, the problem is how to respond to such shifts. Specific laws which immediately outlaw certain practices are usually ineffective: more often than not, they are drafted in haste and so contain lacunae, or they serve only to drive practices underground and so beyond the reach of any regulation. This is clearly not in the interests of either the public or the scientific community.

Perhaps, then, what can at best be achieved is a broadly defined and flexible system of regulation which is open and responsive to scientific developments and to public opinion. It should also be honest enough to accept that, being what we are, we shall find it very hard to resist the fruit of forbidden knowledge. Thus, while moratoria are a defensibly cautious beginning to the process of responding to certain scientific advances, we must accept that their enduring power will quickly diminish and must soon be replaced by more effective means of addressing the issues surrounding "dangerous knowledge". Yet, of course, in the final analysis it is not the knowledge in and of itself that is dangerous, but rather the uses to which it is put. However, this truism does not strip the knowledge of its moral significance. Indeed, it requires us to assume more fully the responsibility that we all have towards each other and towards future generations when decisions must be taken about genetic information. The main concern should not be to suppress human knowledge, for that is anti-intellectual and can so often be driven by ill-informed and partisan agendas. Rather, our first responsibility must be to inform *ourselves* of the content and the limits of new knowledge in a mature, reasoned and logical fashion. Thereafter, we should embrace the task of engaging in the debate about the most appropriate uses of such knowledge in order to arrive at the most sensitive, balanced and socially valuable outcome of which we, and thereby our society, are capable.

REFERENCES

Advisory Committee on Genetic Testing (1997) *Code of Practice for Genetic Testing Offered Commercially Direct to the Public*, London: Department of Health

Advisory Committee on Genetic Testing (1998) *Report on Genetic Testing for Late Onset Disorders*, London: Department of Health

Almqvist, E.W., Bloch, M., Brinkman, R., Craufurd, D. & Hayden, M.R. (1999) A worldwide assessment of the frequency of suicide, suicide attempts, or psychiatric hospitalization after predictive testing for Huntington disease. *American Journal of Human Genetics* 64: 1293

Association of British Insurers (1997) *Policy Statement on Life Insurance and Genetics*, London: ABI (February 1997)

Association of British Insurers (1999) *Genetic Testing: ABI Code of Practice*, London: ABI (August 1999)

Benn, S.L. (1988) *A Theory of Freedom*, Cambridge and New York: Cambridge University Press

Burley, J. (ed.) (1999) *The Genetic Revolution and Human Rights*, Oxford and New York: Oxford University Press

Burley, J. & Harris, J. (1999) Human cloning and child welfare. *Journal of Medical Ethics* 25: 108–13

Chadwick, R., Levitt, M. & Shickle, D. (eds) (1997) *The Right to Know and the Right Not to Know*, Aldershot: Averbury

Chalmers, D. (1998) *Genetic Testing and the Criminal Law: A General Comparative Report for the International Academy of Comparative Law*. (July 1998) International Academy of Comparative Law

Council of Europe (1997) *Convention for the Protection of Human Rights and Dignity of the Human Being with regard to the Application of Biology and Biomedicine: Convention on Human Rights and Medicine*, Oviedo, April 1997, Article 10(2). Strasbourg: Council of Europe

Council of Europe (1998) *Additional Protocol to the Convention for the Protection of Human Rights and Dignity of the Human Being with regard to the Application of Biology and Medicine, on the Prohibition of Cloning Human Beings*, Paris, January 1998. Strasbourg: Council of Europe

Danish Council of Ethics (1993) *Ethics and Mapping of the Human Genome*, Copenhagen: Danish Council of Ethics

Deech, R. (1998) Family law and genetics. In *Law and Human Genetics: Regulating a Revolution*, edited by R. Brownsword, W.R. Cornish & M. Llewelyn, Oxford: Hart

Genetics & Insurance Forum (2000) http: //www.geneticsinsuranceforum.org.uk/InternDev/menu.asp

Giddens, A. (1987) *Sociology: A Brief But Critical Introduction*, 2nd edition, London: Macmillan, and San Diego: Harcourt Brace

Gillon, R. (1998) Wrongful life claims. *Journal of Medical Ethics* 24: 363–64

Helsinki Declaration (1996) *Recommendations Guiding Physicians in Biomedical Research involving Human Subjects* (revised, 1996). Reproduced in J.K. Mason & R.A.A. McCall-Smith (1999) *Law and Medical Ethics*, 5th edition, London: Butterworths (Appendix F)

Hoffmann, D.E. & Wulfsberg, E.A. (1995) Testing children for genetic predispositions: is it in their best interest? *Journal of Law, Medicine and Ethics* 23: 331–44

House of Commons Official Report (1997) Parliamentary Debates (Hansard), 26 June 1997, column 615ff.

Human Fertilisation and Embryology Authority (1998) *Code of Practice*, London: Department of Health

Human Genetics Advisory Commission (1997) *The Implications of Genetic Testing for Insurance*, London: Office of Science and Technology (December 1997)

Human Genetics Advisory Commission (1999) *The Implications of Genetic Testing for Employment*, London: Office of Science and Technology (July 1999)

Human Genetics Commission (2000) http: //www.hgc.gov.uk/

Joint Report (1998) Human Genetics Advisory Commission and Human Fertilisation and Embryology Authority: *Cloning Issues in Reproduction, Science and Medicine*, London: Department of Trade and Industry (December 1998)

Kusum, K. (1995) Sex selection. In *Ethical Aspects of Human Reproduction*, edited by C. Sureau & F. Shenfield, Paris: John Libbey Eurotext

Lancet (1993) Jack or Jill? (Editorial) *The Lancet* 341: 727

Laurie, G.T. (1996) The most personal information of all: an appraisal of genetic privacy in the shadow of the Human Genome Project. *International Journal of Law, Policy and the Family* 10: 74

Laurie, G.T. (1999) In defence of ignorance: genetic information and the right not to know. *European Journal of Health Law* 6(2): 119–32

Laurie, G.T. (2001) *Legal and Ethical Aspects of Genetic Privacy*, Cambridge: Cambridge University Press

Macklin, R. (1999) *Against Relativism: Cultural Diversity and the Search for Ethical Universals in Medicine*, New York: Oxford University Press

Mason, J.K. (1998) *Medico-Legal Aspects of Reproduction and Parenthood*, 2nd edition, Aldershot: Ashgate

Mason, J.K. & McCall-Smith, R.A.A. (eds) (1999) *Law and Medical Ethics*, 5th edition, London: Butterworths

McMahan, J. (1999) Cloning, killing and identity. *Journal of Medical Ethics* 25: 77–86

McNeil, T.F., Sveger, T. & Thelin, T. (1988) Psychological effects of screening for somatic risk: the Swedish alpha$_1$-antitrypsin experience. *Thorax* 43: 505–07

Mudur, G. (1999) Indian medical authorities act on antenatal sex selection. *British Medical Journal* 319: 401

National Bioethics Advisory Commission (1997) *Cloning Human Beings*, Washington, DC: National Bioethics Advisory Commission (June 1997); see http: //bioethics.gov/pubs/cloning1/cloning.pdf

National Bioethics Advisory Commission (1999) *Research Involving Human Biological Materials: Ethical Issues and Policy Guidance*, Rockville, Maryland: National Bioethics Advisory Commission (August 1999)

National Human Genome Research Institute (1997) *Promoting Safe and Effective Genetic Testing in the United States: Final Report of the Task Force on Genetic Testing*, Baltimore: Johns Hopkins University Press (September 1997)

National Research Council (1992) *DNA Technology in Forensic Science*, Washington, DC: National Academy Press

National Research Council (1996) *The Evaluation of Forensic DNA Evidence*, Washington, DC: National Academy Press

Newell, C. (1999) The social nature of disability, disease and genetics: a response to Gillam, Persson, Holtug, Draper and Chadwick. *Journal of Medical Ethics* 25: 172

Ngwena, C. & Chadwick, R. (1993) Genetic diagnostic information and the duty of confidentiality: ethics and law. *Medical Law International* 1: 73

Nuffield Council on Bioethics (1993) *Genetic Screening: Ethical Issues*, London: Nuffield Council on Bioethics

Nuffield Council on Bioethics (2000) *Stem Cell Therapy: The Ethical Issues*, London: Nuffield Council on Bioethics

Office of Science and Technology (1999) *The Advisory and Regulatory Framework for Biotechnology: Report from the Government's Review*, London: Department of Trade and Industry (May 1999)

Savulescu, J. (1999) Should we clone human beings? Cloning as a source of tissue for transplantation. *Journal of Medical Ethics* 25: 87–95

Sommerville, A. & English, V. (1999) Genetic privacy: orthodoxy or oxymoron? *Journal of Medical Ethics* 25: 144–50

Sveger, T., Thelin, T. & McNeil, T. (1999) Neonatal $\alpha 1$-antitrypsin screening: parents' views and reactions 20 years after the identification of the deficiency state. *Acta Paediatrica* 88: 315

Thelin, T., McNeil, T.F., Aspergren-Jansson, E. & Sveger, T. (1985) Psychological consequences of neonatal screening for alpha$_1$-antitrypsin deficiency (ATD). Parental reactions to the first news of their infants' deficiency. *Acta Paediatrica Scandinavica* 74: 787–93

UNESCO (1997) *Universal Declaration on the Human Genome and Human Rights*, Article 5c. Paris: United Nations Educational, Scientific and Cultural Organization

Warnock, M. (1992) Ethical challenges in embryo manipulation. *British Medical Journal* 304: 1045

Wertz, D. (1998) Genetic privacy/nondiscrimination legislation in Congress. *The Gene Letter*: (March issue)

Wertz, D.C. & Fletcher, J.C. (1991) Privacy and disclosure in medical genetics in an ethic of care. *Bioethics* 5: 212

Wilmut, I., Schnieke, A.E., McWhir, J., Kind, A.J. & Campbell, K.H. (1997) Viable offspring from fetal and adult mammalian cells. *Nature* 385: 810–13

Wolf, S.M. (1995) Beyond "genetic discrimination": towards the broader harm of geneticism. *Journal of Law, Medicine and Ethics* 23: 345

CONTACT ADDRESSES

Advisory Committee on Genetic Testing
Department of Health
Wellington House, Room 401
133–155 Waterloo Road
London SE1 8UG
UK

Association of British Insurers
51 Gresham Street
London EC2V 7HQ
UK

Council of Europe Publishing
Council of Europe
F-67075 Strasbourg
France

Danish Council of Ethics
Ravnsborggade 2–4
2200 Kobenhavn N
Denmark

Human Fertilisation and Embryology Authority
Paxton House
30 Artillery Lane
London E1 7LS
UK

Human Genetics Advisory Commission
Office of Science and Technology
Albany House
94–98 Petty France
London SW1H 9ST
UK

Human Genetics Commission
Department of Health
Area 652C, Skipton House
80 London Road
London
SE1 6LH

National Bioethics Advisory Commission
6100 Executive Boulevard, Suite 5B01
Rockville, MD 20892–7508
USA

National Human Genome Research Institute
National Institutes of Health
Bethesda, MD 20892
USA

National Research Council
Office of News and Public Information
The National Academies
2101 Constitution Ave. N.W.
Washington, DC 20418
USA

Nuffield Council on Bioethics
28 Bedford Square
London WC1B 3EG
UK

United Nations Educational, Scientific and Cultural Organization (UNESCO)
Documentation and Reference Centre
7 Place de Fontenoy
75352 Paris 07 SP
France

See also **Genetic aspects of bioethics** (p.591); **Human Genome Project** (p.606)

HUMAN GENOME PROJECT

Gillian Lindsey
London, UK

1. Introduction

The Human Genome Project (HGP) is an international research programme to determine the complete nucleotide sequence of the estimated 3 billion DNA subunits (base pairs [bp]) in the human genome. The project first constructed maps of the genome to localize each of the estimated 50–100 000 genes, and determined the order of genes or other markers and the spacing between them on each chromosome. As part of the HGP, parallel analyses have been carried out on the genomes of several model organisms such as *Escherichia coli* to help develop sequencing technology and computational support, and to help interpret human gene function.

The sequencing of the human genome has captured the public imagination and media interest, with comparisons to the boldness and technological achievement of the lunar landings. On 26 June 2000, the HGP public consortium announced that it had assembled a working draft of the sequence of the human genome. In a parallel announcement, Celera Genomics, a privately funded genome project, announced that they had completed a "first assembly" of a complete genome sequence. Following the final years of often acrimonious competition between the two projects and accelerating progress (more than 60% of the sequence was produced in the penultimate 6 months), the high-profile joint announcement generated a fanfare of public and political praise. The two groups have since separately published their results (International Human Genome Sequencing Consortium, 2001; Venter *et al.*, 2001).

The draft sequence, covering 3.2 billion bp and 97% of the human genome, is, however, incomplete. The remaining gaps will be closed and, with the improved accuracy to be developed over the next 3 years, a complete, high-quality human DNA reference sequence should be achieved by 2003. Even when the genome has been fully sequenced, it will not provide an instant "genetic blueprint for a human being" or an "instruction book for human life". Identifying all the genes, and determining the sequences of the 3 billion bases that make up human DNA, will not provide an instant revolution in medical diagnosis and treatment. However, knowing how these genes, gene products and even the non-coding DNA work together will lead in future decades to better diagnosis of diseases, detection of genetic predispositions to disease, rational drug design and gene therapies.

2. Why sequence the human genome?

The HGP was conceived in the mid-1980s and was widely discussed within the scientific community and public press. The importance of genes in a range of human diseases (such as cystic fibrosis and haemophilia) which appeared to be inherited with Mendelian characteristics was recognized, and various inherited disorders were already known to map to specific chromosomes (e.g. colour blindness and haemophilia were assigned to the X chromosome in the 1930s). With the development in the 1970s of powerful new technologies for replicating, manipulating and analysing DNA fragments, advances had led to an explosion of knowledge about the location of human genes. Meanwhile, with a method developed by Fred Sanger and his colleagues in 1977 to determine the order of the bases, or the sequence, in any piece of DNA from any organism, it became theoretically possible to determine the complete DNA sequence of any organism.

A genome map and sequence listing the precise order of each A, G, C and T of every base of human DNA would provide a benchmark for detecting the mutagenic effects of radiation and toxins, and for tracing the causative gene or mutation for diseases. It was envisaged that, in addition to genetic diagnosis in pregnancy and prediction for diseases in mid and later life, this might lead to new treatments, possibly even to correction of the gene defect itself.

3. History, the main players and data access

(i) Early plans

In March 1986, the US Department of Energy (DoE) initiated US government discussions on plans to sequence the human genome, and was supported in 1988 by the publication of a report from the US National Research Council (National Research Council, 1988). The initial planning process culminated in 1990 with a National Institute of Health (NIH)–DoE joint research plan (US Department of Health and Human Services & US Department of Energy, 1990) covering mapping and sequencing for the first 5 years of what was projected to be a 15-year project. The Human Genome Organization (HUGO), established in 1989 to coordinate international collaboration, took on responsibility for the single chromosome workshops and for fostering the exchange of data and technologies. The Wellcome Trust in the UK joined the public funding of genome projects in 1992 with projects including the sequencing of *Caenorhabditis elegans*. By this time, the HGP was an international endeavour, the international Human Genome Sequencing Consortium ultimately including scientists at 16 nationally funded institutions in France, Germany, Japan, China, Britain and the US. (In the 18 months leading to the announcement of the working draft, the European Bioinformatics Institute and the National Center for Biotechnology Information at NIH also played a key role in providing computational support and analysis.)

The original objective – to sequence the human genome – was expanded in subsequent 5-year plans (Collins &

Galas, 1993; Collins *et al.*, 1998) to include the study of genetic variation and functional analysis of the genome.

(ii) Progress

In September 1994, a complete genetic linkage map depicting the relative chromosomal locations of DNA markers on the human genome was published by French and US researchers, a year ahead of schedule (Murray *et al.*, 1994). Of the 5800 markers (including 427 genes), the average gap was about 0.7 centiMorgans.

The linkage map paved the way for a high resolution physical map. In December 1995, another collaboration of scientists from the US and France published a physical map of 94% of the human genome, containing 15 000 sequence tagged sites (STSs) with an average spacing of 199 kilobases (kb), localizing over 3000 genes (Hudson *et al.*, 1995). Increasingly comprehensive maps utilizing radiation hybrid (RH) mapping (Schuler *et al.*, 1996, 16 000 genes; Deloukas *et al.*, 1998, 30 000 genes) provided a framework and focus for accelerated sequencing efforts by highlighting gene-rich regions of chromosomes.

Access to the vast amount of data generated by the laboratories involved in the sequencing was facilitated by the Internet, but freedom of access became an issue in the later stages of the HGP. The first International Strategy Meeting on Human Genome Sequencing, organized by the Wellcome Trust, was held in Bermuda in February 1996. The "Bermuda statement" that sequence information should be "freely available and in the public domain in order to encourage research and development and to maximize its benefit to society" was endorsed unanimously by all participants (Bentley, 1996).

The complete sequence of a single-celled eukaryote was a major goal of the HGP, and was achieved in April 1996 for the yeast *Saccharomyces cerevisiae*. When human genes are compared with those of yeast, their function in humans (a more complex eukaryotic organism) can be deduced through experiments with yeast, which can be easily mutated and bred in the laboratory to follow through many generations in a relatively short time. It is estimated that about 40% of yeast genes may be homologous in structure or even function to humans.

(iii) Private funding accelerates progress

A significant development came in May 1998, when only 3% of the genome had been sequenced. Geneticist Craig Venter, who had left NIH in 1992 to form an independent laboratory called "The Institute of Genomic Research" (TIGR), set up a new biotechnology company, Celera Genomics, to sequence the human genome privately within 3 years – 4 years earlier than the HGP target and for a tenth of the public project's US$3 billion budget. Celera was funded by Perkin-Elmer, a manufacturer of automated DNA sequencing machines. Venter and TIGR scientists had previously developed a new strategy for DNA sequencing known as the whole-genome shotgun approach that had been successful in sequencing the first complete genome in history (that of *Haemophilus influenzae*; Fleischmann *et al.*, 1995). Many predicted that the whole-genome shotgun method, successful for small bacterial genomes, would present significant reassembly problems when sequencing the human genome. The methods used by Celera and the publicly funded project are described in section 4 below.

Following the announcement of the rival private genome project, the public project plan was ambitiously revised in a new 5-year plan in October 1998 (Collins *et al.*, 1998): new technologies would advance the target completion date for a "working draft" to 2003. Francis Collins, the director of the National Human Genome Research Institute (NHGRI), said to news reporters "This is action, not reaction", although the accelerated timetable was clearly stimulated by Celera's announcement (see news item in *Nature* 395: 207, 1998).

Separate from the difference in methodologies was a difference over commercial use and data access, since Celera would not follow the Bermuda declaration on immediate release, and planned to create a database for which subscriptions would be sold. A proposed collaboration between Celera and DoE scientists (an exchange of sequence information on the three chromosomes of the genome being sequenced by the DoE) was blocked in December 1998 over this issue. The NHGRI and the UK's Wellcome Trust – the two largest participants in the international genome project – felt that the agreement was inappropriate as it give DoE scientists access to data that had not been released immediately into the public domain (see news item in *Nature* 397: 93, 1999).

Just 6 months after the public genome project had cut its target completion date by 2 years, the NHGRI announced another jump: to complete the working draft by Spring 2000 (see news, *Nature* 398: 177, 1999). Funding for three academic sequencing centres in the US, and the Sanger Centre in Cambridge, funded by the Wellcome Trust, was brought forward, but increased sequencing capacity resulting from economies of scale would also play a part in meeting the new goal. The accelerated pace was marked in November 1999 (see news, *Nature* 402: 331, 1999), when the one-billionth base pair was sequenced – approximately one-third of the way towards the full sequence of about 3 billion nucleotides – and by the publication of the first complete chromosome in December 1999 (see section 5 below).

Celera did not begin sequencing the human genome until September 1999, when, together with an international public project, they completed the *Drosophila melanogaster* genome, another model organism (Adams *et al.*, 2000), and claimed that this was a vindication of their whole-genome shotgun approach. Following the failure of the 1998 public–private collaboration talks, there were further widely publicized discussions, but, although Celera could constantly make cross-checks with the public consortium discoveries which were made publicly available on the Internet (and updated every 24 hours), there was no reciprocal pooling of resources or data. Towards the joint announcement of the working draft in June 2000, relationships did thaw and there were plans to hold a joint meeting of the two research groups once the two separate sequences had been published. Following publication in *Science* (Venter *et al.*, 2001) Celera's entire sequence data will be freely available to publicly funded scientists, who will be allowed to download up to 1Mb of data. Longer downloads, and access by commercial users will require signing a material transfer agreement stating that they will not commercialize their results or redistribute the sequence. In addition, Celera's annotated database and bioinformatics tools will be sold to subscribers.

4. Methods of sequencing

The genome from the public consortium HGP is a single consensus sequence derived from a large number of anonymous donors. However, Celera has sequenced the genomes of five different donors who identified their ethnic background as Hispanic, Asian, Caucasian and African–American. In each case, DNA was sequenced from blood (female) or sperm (male) samples.

The public and private projects use similar automation and sequencing technology, but different sequencing approaches. In each case, the genomic DNA is broken into fragments, and the fragments cloned using yeast or bacterial artificial chromosomes or P1-derived artificial chromosomes (YACs, BACs or PACs; see article on "Techniques in molecular genetics") or, in Celera's case, plasmids. Each fragment from a large library of clones is sequenced and automatically analysed on sequencing machines by gel electrophoresis.

The HGP used a "clone-by-clone" approach to sequence the clones of known position, 500 bases at a time. The Human Genome Sequencing Consortium project first broke the genome into large fragments (typically 100–300 kb long) and used the clones to create a physical map, based on overlapping clones which represented all the human chromosomes. The Consortium has several BAC libraries (some derived from one person and others from panels of several to many individuals) each of around 300 000 clones, with about 12- to 25-fold overlap. Different laboratories sequenced different chromosomes, using different libraries. A subset of the mapped clones with minimum overlaps (a "tile path") was selected for random shotgun sequencing to provide the assembled unfinished sequence of each chromosome. DNA from each BAC or PAC was broken up randomly into short fragments (1–2 kb) which were subcloned. The sequences of all the subclones of a single BAC or PAC are analysed together. Many clones contain the same STSs: these overlaps allow the ordering and linking of the STS landmarks to form a "contig", and a consensus sequence is obtained.

The Celera project, using the "whole-genome shotgun" approach, shatters the genome into much smaller fragments of unknown origin, sequences the clones simultaneously, then uses one of the world's largest supercomputers to assemble the pieces. The shotgun strategy for the human genome (Venter *et al.*, 1998), though far quicker than the clone-by-clone technique, has potential problems with incorrect sequence assemblies due to the presence of repeated DNA. This is a problem because the human genome contains many such repeat sequences both due to its size (the region being sequenced is 25 times larger than the Drosophila equivalent) and because the repeats it contains are about tenfold more frequent and much more complex. If a whole-genome shotgun is performed in isolation, these repeats result in an ambiguous sequence. However, as with the *D. melanogaster* genome sequenced by Celera, information from separate mapping data from the public project allows the shotgun assembly to be checked. In *D. melanogaster*, map information was obtained by public research groups by sequencing both ends of BAC clones with large inserts (10 kb and 150 kb). In the human genome, Celera's paired clone end sequences are a key tool for "map-as-you-go". Incorporating mapping data from the public project will also increase the depth of coverage and confirm the assembly.

5. Current status

The working draft (International Human Genome Sequencing Consortium, 2001; Venter *et al.*, 2001) was never intended as a substitute for a complete or finished sequence. The draft sequence is of lower accuracy than the finished sequence will be and it is not continuous – to "finish" the sequence, gaps must be closed and the 7-fold coverage increased to 10 (the "gold standard") to ensure only a single error in 10 000 bases, or 99.99% accuracy.

This has already been achieved for chromosomes 22 (Dunham *et al.*, 1999) and 21 (Hattori *et al.*, 2000) which are accurate to 99.99%. Work on the completion of the other chromosomes is proceeding in parallel, with the responsibility for finishing each chromosome being undertaken by one consortium centre, although on most chromosomes the work is shared between multiple centres.

The HGP map of the human genome at the end of 2000 comprises about 35 000 clones. The public consortium has cloned 97% of the human genome: as of January 2001, 93% was available on the Internet as a working draft (in three public databases: Genbank, EMBL and DDBJ), of which 30% was finished reference sequence. The sequence contains 95% of all known genes and also 97% of all genes that are implicated in human disease.

Gene identification methods (sequence comparisons, gene predictions, cDNA libraries) suggest there are perhaps 30–35 000 genes. It is important to remember that in the definitions of the HGP, one gene can produce several (on average 1–7) proteins.

6. Postsequencing

There are ongoing efforts to finish the sequence and annotate it with information such as dispersed repeat locations, and homology to other proteins and predicted genes (from identified protein-coding regions). Annotation brings data release and curation (actively gathering information, correcting and updating) issues. The Ensembl annotation project run by groups at the Sanger Centre and the European Bioinformatics Institute is addressing these issues through the creation of a new systematic database of genome functional annotation (http://ensembl.ebi.ac.uk/), which is freely available.

Beyond annotation, the evolving collection of a vast amount of DNA sequence information has been the catalyst for the development of technologies to exploit its use further.

There is a new "postgenomic era" of "functional genomics". This is not simply a case of assigning functions to each gene in the catalogue, but also understanding the organization and control of genetic pathways, and the function of non-coding regions and repeats. The methods include expression profiles by the use of DNA-array technology, protein–protein interactions, computational approaches and the response to loss of function by mutation. Review articles by Eisenberg *et al.* (2000), Lockhart and Winzeler (2000) and Pandey and Mann (2000) describe these approaches to functional genomics. The functional characterization of some genes that have defied such analyses may be enabled by a new tool in chemical genetics: Bishop *et al.* (2000) describe chemical switches that can selectively disable particular genes and their protein products, and so the researchers can directly investigate their precise function. Finally, researchers are using computers to try to predict the 3-dimensional structure of proteins from DNA sequence. Structural genomics compares the gene sequences of proteins of known structure to the sequences of unsolved proteins (see article on "The role of bioinformatics in the postsequencing analysis of genomic data"). Other approaches to identifying proteins are still in early stages of development; however, identifying the properties and functions of every protein expressed in an organism, its "proteome", is seen as the next step beyond the human genome project (Pandey & Mann, 2000). Celera has announced its intention to launch a private sector "Human Proteome Project", and there are an increasing number of biotechnology companies devoted solely to proteomics.

In addition to listing genes and their functions, several groups including the public HGP and a consortium of pharmaceutical companies aim to produce a map of the most common "single nucleotide polymorphisms" (SNPs; see Altshuler *et al.*, 2000). These DNA sequence variations occur when a single nucleotide (A, T, C or G) in the genome sequence is altered. Many SNPs have no effect on cell function, but specific combinations of SNPs probably cause most of the common genetic disorders, and can cause variation in how humans respond to bacteria, viruses, toxins, chemicals, and drugs and other therapies. In addition to the medical implications (see section 7), SNP maps are expected to identify thousands of additional markers along the genome, thus adding to the much larger genome map being generated by researchers in the HGP.

7. Prospects for human medicine

Many genes that lead to disease have been isolated, including genes for early-onset breast and ovarian cancer, cystic fibrosis, Huntington's disease, hereditary colon cancer and the most common form of skin cancer. There are sequences on chromosome 22 thought to be involved in schizophrenia, and further database "mining" of the human genome sequence will allow researchers to search for genes implicated in other conditions. The complete sequence of the human major histocompatibility complex (MHC Sequencing Consortium, 1999) provides an invaluable target particularly for immunological research. More diseases are associated with this region than with any other locus in the human genome, and, although many of the 224 identified gene loci have unknown functions, 40% of them are estimated to have some immune system function.

In diseases such as cancer, heart disease and diabetes, where multiple genes interplay with environmental factors, knowledge of the entire genome will have an enormous impact on our understanding of, for example, the environmental regulation of gene expression and the genetic contribution to the intermediate phenotypes linking genes and disease, and thus the biology of the disorder, as in atherosclerotic disease (Kaprio, 2000). Knowledge of genetic susceptibility in these complex disorders will lead to the development of preventive strategies – environmental or behavioural interventions – and counselling directed at those individuals at high genetic risk.

Genetic testing as a diagnostic tool in pregnancy or before symptoms appear has been developed for many of these conditions, although the ethics of such testing is frequently debated (see below, and also article on "Genetic aspects of bioethics"). The prospect of manipulating genes leads clinical medicine away from diagnostics and towards therapy such as the development of highly targeted drugs that compensate for defects (see article on "Genetics of Duchenne muscular dystrophy"). Another approach is to correct or replace the altered gene through gene therapy in somatic (body) cells, or by modifying a prospective parent's gametes or an embryo. Most human clinical trials of gene therapy are only in the research stages, with high costs, regulations and problems with the viral vectors used to transport the gene having brought the method under intense scrutiny (see article on "US gene therapy in crisis"). Germline therapy, which is not actively being investigated for humans, can produce genetic modifications that are passed on to future generations and thus eliminate inheritance of monogenetic disease. However, a recent American Association for the Advancement of Science working group report (AAAS, 2000) recommends that no research that could cause inheritable modifications should proceed in humans until it is shown to be safe and reliable with no unintended genetic changes; social, ethical and theological considerations have been widely discussed and an oversight body is in place to monitor and regulate research and development.

SNP coding – knowledge of the presence or absence of a particular SNP in the genetic sequence of a particular individual – is an advanced diagnostic tool, but could also eventually lead to customized drugs. Pharmacogenomics (the study of the genetic basis of drug response) could be a powerful tool in drug design for specific subpopulations, and allow direct prescription in order to amplify drug response or avoid adverse side-effects. However, the feasibility of using SNPs to discover useful pharmacogenomic markers is uncertain (McCarthy & Hilfiker, 2000).

8. Ethics

Consideration of the ethical, legal and social implications of the HGP was implicit in the original DoE/NIH plan. While the HGP produced a composite sequence that was not specific to an individual, clinical genetic testing is on the increase and human genome diversity studies are beginning to profile individual genotypes. Public concerns have focused on privacy, potential misuse and commercialization of the information resulting from the HGP. Separate articles in this volume on "Genetic aspects of bioethics", and "Law, ethics and genetics" describe general issues in the use of genetics. Issues specific to the human genome and genetic variation in the human population include:

(i) Privacy and fairness in the use of genetic information

Who should have access to our personal genetic information? If information is disclosed, how can we prevent discrimination by health insurers, employers, courts and schools? There should be informed consent in genetic research – participants should know beforehand how the samples will be used – but should family members as well as the individual be informed about the results of a genetics study?

(ii) Testing when there is no cure or treatment

Knowledge of the single gene responsible for disease does not guarantee a cure or treatment. Should the test be offered when there is no cure, and what will result from knowledge of a fatal disease before symptoms appear? Although a predictive test for families at high risk from Huntington's disease has been available for years, only a minority of these individuals has decided to be tested. Should parents have the right to have their children tested for adult-onset diseases?

(iii) Psychological impact due to an individual's genetic differences

Knowledge of "defective" genes may lead to anxiety about disease, relationship breakdown and social stigmatization. Anxiety may be increased if there is doubt over the reliability or accuracy of tests, or if the information is incomplete or indeterminate (such as a 25% increase in the risk of cancer).

(iv) Reproductive issues

Genetic information is already used in reproductive decision making, by the testing of prospective parents and foetal or *in vitro* embryo testing for single-gene defects such as cystic fibrosis. At present, selection is purely on disease grounds rather than cosmetic ones. However, somatic or germline therapy that could enhance normal human characteristics (such as eye or hair colour, height, intelligence, though this can't yet be done) has led to concerns of eugenics and "designer babies". Researchers have created artificial human chromosomes (Harrington *et al.*, 1997) and, in principle, we have the knowledge to create a baby with any of the genes a geneticist – or prospective parents – desired.

(v) Rich vs. poor

Screening for susceptibility to diseases is likely to benefit few people in the developing world; even in rich countries, those deprived of health insurance or employment may suffer financially while those who can afford customized drugs or gene therapy will receive better disease treatment. It has even been postulated that a genetic "aristocracy" could arise if those wealthy enough to afford genetic "enrichment" develop a hereditary class that within a few centuries could become a distinct biological class.

(vi) Commercialization of products

How are the "products" of the HGP to be commercialized? Who has property rights, can there be patents and trade secrets, and will there be public access to data? For example, a private company was granted a license by the Icelandic government in 2000 to use medical records (including genetic data) to produce a genomic map of the Icelandic people, and this company has exclusive rights to the commercial exploitation of the data for 12 years. The participants (citizens are assumed to agree to participate unless they opt out) will not share the profits from products derived from genes discovered in the data.

9. Genes and patents

The principle of patenting an invention, the right to prevent others from exploiting that invention commercially, is seen by the biotechnology industry as the only way that investment in research and development and clinical trials can be protected, and recovered through licensing. However there is controversy over whether materials from the human body, as products of nature, can be classified as inventions, and ethical debate on whether patenting human genes violates a human right to protection of the body from ownership by an individual, institution, or corporation. (For patenting of agricultural gene products, there are environmental concerns in addition to these ethic and moral debates.)

Discovery of a gene sequence itself is not sufficient to grant a patent. US, Japanese and European patent law differ, but all routinely allow exclusive rights to be granted for genetic material extracted and modified or concentrated in the laboratory, provided the gene or gene product can be shown to be useful, novel and non-obvious. However a European Union directive (98/44/EC) on the legal protection of biotechnological inventions has to date been adopted only by Britain, Denmark, Germany, Finland and the Republic of Ireland, while the Netherlands is challenging the Directive through the European Court of Justice, and France has national patent laws prohibiting patenting of genes.

The scope and number of patent claims, particularly provisional applications, have raised concerns: often the gene is not fully characterized, and its function not known. For example, US-based Human Genome Sciences was issued a patent in February 2000 for a gene sequence that could lead to an AIDS vaccine or a new class of AIDS drugs, even though the firm knew nothing of the gene's role in AIDS when it sought the patent. Critics argue that if gene patents are granted to cover all possible functions of a gene sequence, development of drugs and therapies could be blocked or prices unfairly controlled if one company controls access to use of the gene in research. New guidelines have been released (January 2001) by the US Patent and Trademark Office clarifying when patents can be granted on gene sequences of predicted function (i.e. the sequence is homologous to a separate sequence of known function).

Of the 40 000[-]120 000 genes that make up the human genome, some 1000 are already patented. Thousands more applications await approval, and although the hunt for "profitable" genes leading to new drugs, tests or treatments is led by drug companies and private firms, universities and medical charities have also been granted gene patents. The interest from the financial markets in products from the HGP was illustrated by the response to a statement from President Clinton and British Prime Minister Tony Blair in March 2000 on the need for raw sequence data to be freely available to all researchers. Seen as a possible move to restrict patents on individual genes, the statement caused the NASDAQ biotech index to fall almost 13% in one day. The commercialization of human gene products also raises ethical issues (see section 8 *(vi)* above). In 1997, UNESCO, in its Universal Declaration on the Human Genome and Human Rights, declared that the "human genome in its natural state shall not give rise to financial gains". The HUGO Ethics Committee has called for the common heritage of mankind to be recognized in genetic research, and recommended that the benefits be shared among humanity; for example, that profit-making entities dedicate a percentage (e.g. 1[-]3%) of their annual net profit to healthcare infrastructure and/or to humanitarian efforts (HUGO, 2000).

10. Conclusions

The concerns over ethical and commercial misuse of sequence data are the greatest detractors from the prospects of the HGP. Data from the HGP will take many years to be analysed, but, although the medical benefits may be decades away, early concerns over the value against cost have been replaced by enthusiasm for one of the great success stories in modern biology.

The working draft sequence has cost about $360 million to produce; the Wellcome Trust in the UK and national funding in each of the countries that have genome projects have contributed to the initial US investment. The government projects did not predict the large investment of private funds by pharmaceutical and biotechnology companies.

The complete sequences of yeast and prokaryotes have had a huge impact on the HGP, both through the development of tools and in comparative genomics.

It was thought that new methods of sequencing would be developed, but the methodology developed by Fred Sanger is still in use. However, the cost of sequencing has dropped from over US$2 per "finished" base to less than 10 cents, mainly due to a new generation of automated capillary DNA sequencing machines. Informatics (databases, data analysis tools and management software) has been an integral part of the HGP's unexpected rate of progress. Celera's sequencing and computing facility processed more than 70 terabytes of data in under a year, described as some of the most complex computations in the history of supercomputing. These developments in tools and computational power have had benefits for the concurrent non-human genome projects, and will have applications in projects to map (and ultimately improve) the genomes of economically important farm animals and crops.

The Internet has allowed the broad and rapid dissemination of sequence data, and a front end for sophisticated analytical software such as the BLAST search, a rapid comparison of a search sequence to a database of known sequences, and the Baylor College of Medicine's *Gene Finder* program (see article on "The role of bioinformatics in the postsequencing analysis of genomic data"). The Internet will also provide the essential infrastructure for new tools to allow physically distant researchers to edit and annotate sequences, run analyses, share results and respond to other laboratories' interpretations.

REFERENCES

AAAS (2000) *Human Inheritable Genetic Modifications: Assessing Scientific, Ethical, Religious and Policy Issues*, http://www.aaas.org/spp/dspp/sfrl/germline/main.htm

Adams, M.D., Celniker, S.E., Holt, R.A. *et al.* (2000) The genome sequence of *Drosophila melangaster*. *Science* 287: 2185–95

Altshuler, D., Pollara, V.J., Cowles, C.R. *et al.* (2000) An SNP map of the human genome generated by reduced representation shotgun sequencing. *Nature* 407: 513–16

Bentley, D.R. (1996) Genomic sequence data should be released immediately and freely in the public domain. *Science* 274: 533–34

Bishop, A.C., Ubersax, J.A., Petsch, D.T. *et al.* (2000) A chemical switch for inhibitor-sensitive alleles of any protein kinase. *Nature* 407: 395–401

Collins, F. & Galas, D. (1993) A new five-year plan for the US human genome program. *Science* 262: 43–46

Collins, F.S., Patrinos, A., Jordan, E. *et al.* (1998) New goals for the US Human Genome Project: 1998–2003. *Science* 282: 682–89

Deloukas, P. *et al.* (1998) A physical map of 30,000 human genes. *Science* 282: 744–46

Dunham, I. *et al.* (1999) The DNA sequence of chromosome 22. *Nature* 402: 489–95

Eisenberg, D., Marcotte, E.M., Xenarious, I. & Yeates, T.O. (2000) Protein function in the post-genomic era. *Nature* 405: 823–26

Fleischmann, R.D. *et al.* (1995) Whole-genome random sequencing and assembly of *Haemophilus influenzae* Rd. *Science* 269: 496–512

Harrington, J.J. *et al.* (1997) Formation of de novo centromeres and construction of first-generation artificial chromosomes. *Nature Genetics* 15: 345

Hattori, M. *et al.* (2000) The DNA sequence of human chromosome 21. *Nature* 405: 311–19

Hudson, T.J., Stein, L.D. & Gerety, S.S. *et al.* (1995) An STS-based map of the human genome. *Science* 270: 1945–54

HUGO Ethics Committee (2000) Statement on benefit-sharing. http://ash.gene.ucl.ac.uk/hugo/benefit.html

International Human Genome Sequencing Consortium (2001) Initial sequencing and analysis of the human genome. *Nature* 409: 860–921

Kaprio, J. (2000) Genetic epidemiology. *British Medical Journal* 320: 1257–59

Lockhart, D.J. & Winzeler, E.W. (2000) Genomics, gene expression and DNA arrays. *Nature* 405: 827–36

McCarthy, J.J. & Hilfiker, R. (2000) The use of single-nucleotide polymorphism maps in pharmacogenomics. *Nature Biotechnology* 18(5): 505–08

MHC Sequencing Consortium (1999) Complete sequence and gene map of a human major histocompatibility complex. *Nature* 401: 921–23

Murray, J.C., Buuetow, K.H., Weber, J.L. *et al.* (1994) A comprehensive human linkage map with centiMorgan density. *Science* 265: 2049–54

National Research Council (1988) *Mapping and Sequencing the Human Genome*, Washington, DC: National Academy Press

Pandey, A. & Mann, M. (2000) Proteomics to study genes and genomes. *Nature* 405: 837–46

Schuler, G.D., Boguski, M.S., Stewart, E.A. *et al.* (1996) A gene map of the human genome. *Science* 274: 540–46

US Department of Health and Human Services & US Department of Energy (1990) *Understanding Our Genetic Inheritance. The US Human Genome Project: The First Five Years*, Springfield, Virginia: National Technical Information Service (NIH Publication No. 90–1590)

Venter, C., Adams, M.D., Sutton, G.G. *et al.* (1998) Shotgun sequencing of the human genome. *Science* 280: 1540–42

Venter, J.C. *et al.* (2001) The sequence of the human genome. *Science* 291: 1304–51

FURTHER READING

Science yearly Genome Issues, usually published in October

Cantor, C.R. (1999) *The Science and Technology behind the Human Genome Project*, New York: Wiley

Davies, K. (2001) *The Sequence: Inside the Race for the Human Genome*, London: Weidenfeld and Nicolson

Dear, P.H. (ed.) (1997) *Genome Mapping: A Practical Approach*, Oxford and New York: Oxford University Press

Knoppers, B.M. (1999) Status, sale and patenting of human genetic material. *Nature Genetics* 22: 23–25

Ridley, M. (1999) *Genome: The Autobiography of a Species in 23 Chapters*, New York: HarperCollins, and London: Fourth Estate

Sloan, P.R. (2000) *Controlling our Destinies: Historical, Philosophical, Ethical, and Theological Perspectives on the Human Genome Project*, Notre Dame, Indiana: University of Notre Dame Press

GLOSSARY

contig a continuous sequence of DNA assembled from overlapping cloned DNA fragments

DNA sequence relative order of base pairs in a fragment of DNA, a gene, a chromosome or an entire genome. DNA molecules are made up of thousands of four different kinds of bases (abbreviated A, C, G and T). The sequence of the genome or of large segments of bases is unique for each individual

genetic linkage map map of the relative position and distance between genes, with distances measured in centiMorgans. The map is based on careful analyses of how often two markers on one chromosome are passed together from parent to child

genome all the genetic material of an individual or species, including chromosomes, plastids and prophages; size is generally given as the total number of base pairs. The HGP is for nuclear DNA – the mtDNA sequence of humans is already known

genomic library an unordered collection of clones made from a set of randomly generated overlapping DNA fragments. The library represents the entire genome of an organism

physical map map of gene locations and the distance between them (measured in base pairs). The highest resolution map would be the complete nucleotide sequence of all the chromosomes in a genome

radiation hybrid (RH) mapping one of several methods for ordering markers along a chromosome, and estimating the physical distances between them. Human cell lines are irradiated (which fragments the DNA) and fused with rodent cells to make hybrids. Assessing the frequency of marker sites remaining together in the same fragment can establish the order and distance between the markers

sequence tagged site (STS) any site that has a single occurrence in a chromosome or genome and is identified by a known unique DNA sequence. STSs are useful for forming genetic maps

shotgun method sequencing method which involves randomly sequencing tiny cloned pieces of the genome, with no foreknowledge of where on a chromosome the piece originally came from. This can be contrasted with "directed" strategies, in which pieces of DNA from adjacent stretches of a chromosome are sequenced. Directed strategies eliminate the need for complex reassembly techniques

single nucleotide polymorphisms (SNPs) the most common type of genetic variation: an alteration of a single base in a DNA molecule

See also **Techniques in molecular genetics** (p.22); **Genetics of Duchenne muscular dystrophy** (p.467); **US gene therapy in crisis** (p.523); **Genetic aspects of bioethics** (p.591); **Law, ethics and genetics** (p.594); **The role of bioinformatics in the postsequencing analysis of genomic data** (p.893)

J PLANT GENETICS

Introduction to Plant Genetics

Several articles in this section describe analysis of the evolution of plant species, and the use, since plants were first domesticated, of selective breeding to improve yield or resistance to disease, drought or flood. Advanced genetic engineering technologies are increasingly allowing desirable nutritional or agricultural traits to be transferred or manipulated in plant genomes, as is described in the articles below on improving rice and maize, and in two articles on the agricultural impact of transgenic crop plants.

1. Sequencing the genome of crop plants

The introduction to section M on Model Organisms describes the completion in 2000 of the sequencing of the first plant genome, *Arabidopsis thaliana* (*Nature* 408: 791–826). The *Nature* article examines current opinions in the plant sequencing community as to the need and feasibility of mounting major sequencing efforts on the most important crop plants, starting with rice. While a private company announced in January 2001 that it had sequenced the rice genome, the unpublished results will only be available for a fee, and the publicly funded rice genome project will not be complete until the end of 2004. The high reliance on maize as a staple diet and livestock food, particularly in North and South America, has led to a number of maize mapping projects, and, boosted by technology developments in other genome sequencing projects, it seems likely that a major maize sequencing project will be launched very soon. As pointed out in the "Cereal chromosome evolution and pairing" article below, wheat, rye, barley, maize, sorghum and millet all seem to have a similar genetic layout to rice; their genes are in much the same order but the larger genomes have more junk DNA between the genes, which adds great difficulty to the sequencing process.

2. Genetically modified (GM) foods

In reviewing two new books on GM crops in *Nature* (*Pandora's Picnic Basket: The Potential and Hazards of Genetically Modified Foods*, by Alan McHughen, 2000, and *Food's Frontier, The Next Green Revolution*, by Richard Manning, 2000), Dick Taverne pointed out that transgenic crops are likely to play a vital part in feeding the extra two billion people expected to be born from 2000 to 2025, but only if they are applied in a way that fits in with local communities and cultures and does not lead to a greater degree of monoculture. Since both books favour the application of transgenic crops to feeding areas of the world where food or food containing all the essential components is in very short supply, the question remains as to whether this can be achieved without convincing the populations of Britain, Europe and the US that they should accept such GM foods as they are offered, and this may be very difficult.

SEPARATION OF THE SEXES AND SEX CHROMOSOMES IN PLANTS

D. Charlesworth
Institute of Cell, Animal and Population Biology, University of Edinburgh, Edinburgh, UK

1. Introduction

Most flowering plants function as both male and female (sometimes termed "cosexual"), but their sex systems include considerable diversity (see **Table 1**). Among cosexual species, some are hermaphroditic, having "perfect" flowers with both male and female parts, while others are monoecious, having male and female flowers. In contrast to animals, less than 4% of species are dioecious (with separate sexes). Among dioecious species, some, but not all, have chromosomal sex determination. Dioecy or systems involving individual plants with separate sexes are also known among bryophytes and gymnosperms (Givnish, 1980; Mishler, 1988). The taxonomic distribution of separate sexes and chromosomal sex determination systems in angiosperms strongly suggests recent evolutionary origins and replicated independent events (e.g. Westergaard 1958), in contrast to the ancient origins of mammalian and insect sex chromosomes. A recent origin of separation of the sexes in plants is also suggested by the observation that flowers of dioecious plants are often not fully unisexual, but frequently have distinct opposite sex organs: anthers in the case of females, and ovary rudiments in the case of males (Darwin, 1877).

Plants, therefore, offer opportunities to study the most interesting early stages of the evolution of sex separation and sex chromosome evolution. The only other comparable case where evolution of sex chromosomes is open to study is the situation when autosomal genomic segments are translocated onto sex chromosomes, generating neo-sex chromosome systems. This has happened a number of times in the genus Drosophila (Patterson & Stone, 1952) and it has been discovered that genes on the neo-Y chromosomes undergo genetic degeneration and become inactivated, and dosage compensation then evolves (Bone & Kuroda, 1996). The repeatability of these evolutionary events can be examined in plants too, with the added interest that sex chromosomes in dioecious plants from different angiosperm families are probably entirely independent evolutionary replicates.

Table 1 Plant and animal sex systems.

Plant term	Definition of plant term	Occurrence in plants and examples	Animal cases and term when different
Sexually monomorphic			
Hermaphrodite	Flowers have both male and female functions	90% of flowering plants (e.g. roses) (rare in animals)	Many slugs, snails, some fish
Monoecious	Separate sex flowers on the same individuals	5% of flowering plants, often those with catkins (e.g. hazel), and many gymnosperms (e.g. pines)	No corresponding animal system
Sexually polymorphic			
Dioecious	Separate sex individuals (male and female plants)	5% of flowering plants (e.g. holly), and some gymnosperms (e.g. yew, cycads)	Many animals
Gynodioecious	Individuals either female or hermaphrodite	e.g. ribwort plantain (*Plantago lanceolata*), bladder campion (*Silene vulgaris*)	Rare
Androdioecious	Individuals either male or hermaphrodite	Very rare	Rare

2. Distribution and evolution of dioecy in flowering plants (Figure 1)

Although it is infrequent in plants, dioecy is widely distributed. Perhaps as many as half of all flowering plant families have some dioecious members (Charlesworth, 1985; Renner & Ricklefs, 1995), but dioecy is often present in only a minority of genera in these families (reviewed in Charlesworth, 1985), again suggesting relatively recent evolution. For example, in Silene, species are primarily gynodioecious (having both female and hermaphroditic individuals; see **Table 1**) or hermaphroditic. However, phylogenetic analysis based on the DNA sequence of a part of the ribosomal RNA genes suggests that dioecy has evolved independently at least twice within different sections of the genus (Desfeux *et al.*, 1996; Zhang *et al.*, 1998). Transitions in the opposite direction, from dioecy back to hermaphroditism, have also occurred. These can be detected as isolated cosexual populations or species within otherwise dioecious taxonomic groups (e.g. Rohwer & Kubitzki, 1984; Rieseberg *et al.*, 1992; Pannell, 1997).

Why dioecy should evolve from hermaphroditism is still controversial. Models for the evolution of dioecy (Charnov *et al.*, 1976; Charlesworth & Charlesworth, 1978; Seger & Eckhart, 1996) involve two major types of selective factor.

(i) *Avoidance of inbreeding*. This can be favoured when the starting state is a self-compatible cosexual population that is partially self-fertilizing, and in which there is inbreeding depression.
(ii) *Selection on the allocation of reproductive resources*. If there is a trade-off between male and female sex functions, females might be able to produce more seeds than hermaphrodites, due to resources saved by not making pollen.

Some models focus on the second of these two selective forces, assuming that avoidance of inbreeding is unimportant and that the ancestral cosexual population was self-incompatible and random mating (Charnov *et al.*, 1976; Seger & Eckhart, 1996). This approach is very useful in suggesting ecological circumstances that can destabilize cosexuality and might thus be expected to induce the evolution of separate sexes. However, although sex determination systems can exist with both sexes under the control of a single locus (some genotypes being male while others are female; see, for example, Dellaporta & Urrea, 1994), a population with males and females cannot evolve in a single mutational step from an initial hermaphroditic state. Dioecy must have evolved by two or more steps, involving at least one male-sterility gene and one female-sterility gene invading the initial cosexual population (Lewis, 1942; Charlesworth & Charlesworth, 1978). An understanding of the evolution of dioecy from cosexuality must, therefore, include consideration of the genetics of the evolutionary pathway. Reversion from dioecy to unisexuality, however, can occur in a single mutational step if dioecy is the derived state (see below).

Theoretical models for the genetic details of the evolution of dioecy therefore assume a stepwise pathway, usually first establishing unisexual females in the population alongside the hermaphrodites (such a situation is known as gynodioecy). The view that females evolved first, and males afterwards, is suggested by the theoretical finding that females can establish under less implausible conditions than males. This view is supported by the empirical observation that dioecy and gynodioecy are often found in the same genera (Maurice *et al.*, 1993).

Females gain both the benefits listed above, and it is simple to work out the amount of inbreeding depression required for females to have higher fitness than cosexes with

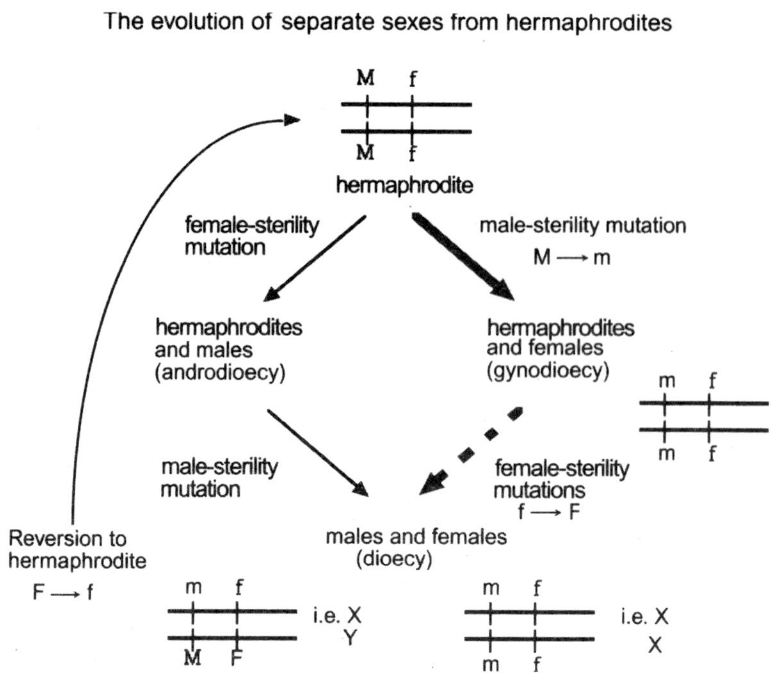

Figure 1 The pathways possible for the evolution of dioecy from hermaphroditism, showing the successive replacement of alleles of at least two genes, *M/m* and *F/f*. The initial hermaphrodite would have the genotype *M/M f/f* (male-fertile and female-fertile), and the final state would include females with the genotype *m/m f/f* (recessive male-sterile and female-fertile) and males with the genotype *M/m f/F* (male-fertile and female-sterile, where the male-determining genotype *MF* is dominant).

a given rate of self-fertilization, if their female fertility is increased by any given proportion of the fertility of the initial cosexual phenotype (Lloyd, 1975; Charlesworth & Charlesworth, 1978). The reason that males can invade cosexual populations less readily than females is that self-fertilization makes it less likely that males will be able to invade a cosexual population whereas, for females, it makes it more likely (if there is inbreeding depression and thus an advantage to avoiding inbreeding). This is because some fraction of ovules are self-fertilized, so that investing a higher proportion of reproductive resources in pollen generates a less than proportionate increase in number of offspring sired, i.e. it is difficult to increase fitness through male function unless pollen output is greatly increased (Lloyd, 1975; Charlesworth & Charlesworth, 1978).

After invasion of a cosexual population by females, and establishment of a gynodioecious population, there would be selection on the cosexual morph to become more male biased than before, since the availability of the females' ovules should lead to greater returns on investment in pollen output. The conditions for fully or partially female-sterile phenotypes to invade gynodioecious populations are, therefore, less stringent than those in the absence of females, and phenotypes with decreased female function are therefore favoured by selection once females are present, under conditions that would not permit their spread in the absence of females.

However, even if a suitable mutation arises in the population, further steps in the evolution of dioecy may be slow. Modifiers that make the cosex morph more male in function will probably also reduce female fertility. Unless these modifier genes can act in cosexes alone (and so be without effect on females), this produces counter-selection that will hinder the spread of such factors, as shown in **Table 2**. This also tends to prevent gynodioecious populations responding to selection and moving towards full dioecy (with purely male and female phenotypes). The result of this is that the invasion condition for a modifier producing a more male form differs according to the linkage between the initial male-sterility locus and the modifier locus. Linked modifiers can sometimes invade even though unlinked modifiers would fail to do so (Charlesworth & Charlesworth, 1978). If modifier alleles do invade, there is also selection for tighter linkage between the male-sterility locus and modifier loci in order to reduce the frequency of progeny carrying both male-sterility and female-sterility alleles. This will probably lead to a cluster of linked loci in a particular chromosomal region, in which recombination will be suppressed, and this may start the evolution of a sex chromosome system (see Charlesworth, 1991).

Table 2 Sex phenotypes of genotypes at two loci in the evolution of dioecy from gynodioecy. The table shows how the second gene (the dominant $f \rightarrow F$ mutation) would be advantageous in hermaphrodites but disadvantageous in females.

Genotype at modifier locus	Genotype at male-sterility locus *M/m* or *M/M*	*m/m*
f/f	Hermaphrodite ↓	Female ↓
F/f	Male	Neuter

3. Distribution and evolution of plant sex chromosomes

In plants, many dioecious species have no detectable chromosome heteromorphism between the sexes, but, in those that do, male heterogamety is the rule, with only a few exceptions (Westergaard, 1958; Bull, 1983). To understand the evolution of sex chromosomes, it is necessary to determine how the non-recombining differential segment of the Y chromosome evolved its distinctive properties. The theory for the initial stages of the evolution of sex chromosomes just outlined predicts that the X and the Y evolved from a pair of homologous ancestors, each carrying active loci. Once sex-determining genes evolved, crossing-over between the pair was reduced, to prevent recombination between different sex determination loci. This, in turn, led to the slow evolutionary loss of function of the alleles on the chromosome that is heterozygous in males (assuming male heterogamety) and, finally, to dosage compensation – the process whereby gene expression is equalized in males and females, despite females having two functional gene copies and males only one (Lucchesi, 1978; Charlesworth, 1996). Neither of the final two stages is known to have been reached in angiosperms. In species with sex chromosomes, it is not known with certainty if the Y chromosomes have degenerated genetically in a similar fashion to the degeneration of mammalian and insect Y chromosomes. However, heterochromatinization of Y chromosomes does seem to occur in some Rumex species (Réjon *et al.*, 1993) and YY homozygotes are non-viable in several, though not all, dioecious plants (Westergaard, 1958; Ye *et al.*, 1990).

The best studied plant sex chromosome system is in the white campion, *Silene latifolia*, also known as *S. alba* and *Melandrium album* (Westergaard, 1958; Nigtevecht, 1966a,b). *S. latifolia* is estimated to have diverged from its most recent, non-dioecious ancestor between 8 and 24 million years ago (Desfeux & Lejeune, 1996). This diploid species has morphologically distinct X and Y sex chromosomes that are, respectively, 85% and 170% larger than the mean of the 11 pairs of morphologically similar autosomes (Westergaard, 1958). That the Y is larger than the X is common in dioecious angiosperms (Bull, 1983); the reason for this is not yet understood, but it is likely that much of the extra DNA is repetitive (Clark *et al.*, 1993). The *S. latifolia* Y chromosome differs from ancient mammalian Y chromosomes by being primarily euchromatic (Vyskot *et al.*, 1993), but it is similar in having a pseudoautosomal region that pairs with the X chromosome (Westergaard, 1958; Zhang *et al.*, 1998) and the overall structure of the sex chromosomes shows remarkable similarities to that of mammalian ones (see Charlesworth, 1991). The *S. latifolia* Y chromosome is probably deficient in essential genes, since androgenic haploid plants (having a Y but no X chromosome) are non-viable (Ye *et al.*, 1990).

The details of sex determination have yet to be worked

out in flowering plants (Dellaporta & Urrea, 1994), although more is known for ferns (Banks, 1997). Of the few angiosperm species investigated, most have male-determining Y chromosomes, but, within the single genus Rumex, some species have this sex determination system, while others have an X-autosome balance system (Smith, 1969). The extremely interesting sex chromosomes of bryophytes have been little studied (reviewed in Bull, 1983). In *S. latifolia*, Y-linked RAPD (random amplified polymorphic DNA) markers have been developed and used to characterize hermaphroditic mutants caused by deletions of Y-chromosomal regions. Study of such deletions has revealed that at least three active loci appear to be present on the Y chromosome: a female suppressor locus and loci responsible for anther maturation and early stamen development (Westergaard, 1958; Donnison *et al.*, 1996; Farbos *et al.*, 1999; Lardon *et al.*, 1999). If plant Y chromosomes carry large numbers of expressed loci, it may prove difficult to isolate the loci responsible for these effects.

Based on the evolutionary hypothesis outlined above, pairs of active (X) and degenerate (Y) loci should exist in the non-pairing region of the sex chromosomes, due to the common autosomal ancestry of these two chromosomal segments (Charlesworth, 1996), and such pairs have been found in human and mouse (Graves, 1995). Plant sex chromosomes differ from animal ones in that there has been less time for degeneration and loss of Y-chromosomal loci. In addition, Y chromosomes probably harbour genes that are actively expressed in the haploid male gametophytes during pollination (Bull, 1983), whereas, in animals, there is no gene expression in male gametes. Experiments to examine expression in pollen led to estimates, based on the large numbers of genes tested, that about 70% of genes expressed in the diploid phase of the life cycle also express cDNAs in pollen (Stinson *et al.*, 1987). Some Y-linked genes in plants may, therefore, be under natural selection pressure to preserve function, even if they are located in the differential segment of the Y. A pair of homologous X- and Y-linked genes with very similar sequences has recently been discovered in *S. latifolia* and, although they are completely sex linked (i.e. the Y-linked gene is in the non-recombining portion of the Y), the alleles on both sex chromosomes are expressed (Delichére *et al.*, 1999).

4. Conclusion

It is not yet known if there is a dosage compensation system to ensure that X-linked loci are expressed at similar levels in the two sexes of plants. Before this can be determined, X-chromosomal loci that are expressed in females must be found. To date, few genetic map data are available from dioecious plants, and the markers used have not been derived from expressed loci. Only one plant X-linked-expressed single-copy gene has been studied, in *S. latifolia* (Guttman & Charlesworth, 1998). As this is expressed only in males, it cannot be used to test for dosage compensation. It does, however, appear to have a degenerated Y-linked homologue, which suggests that dosage compensation, at least at the single locus level, is not a precursor of genetic degeneration.

REFERENCES

Banks, J.A. (1997) The TRANSFORMER genes of the fern *Ceratopteris* simultaneously promote meristem and archegonia development and repress antheridia development in the developing gametophyte. *Genetics* 147: 1885–97

Bone, J.R. & Kuroda, M.I. (1996) Dosage compensation regulatory proteins and the evolution of sex-chromosomes in *Drosophila. Genetics* 144: 705–13

Bull, J.J. (ed.) (1983) *Evolution of Sex Determining Mechanisms*, Menlo Park, California: Benjamin Cummings

Charlesworth, B. (1991) The evolution of sex chromosomes. *Science* 251: 1030–33

Charlesworth, B. (1996) The evolution of chromosomal sex determination and dosage compensation. *Current Biology* 6: 149–62

Charlesworth, B. & Charlesworth, D. (1978) A model for the evolution of dioecy and gynodioecy. *American Naturalist* 112: 975–97

Charlesworth, D. (1985) Distribution of dioecy and self-incompatibility in angiosperms. In *Evolution: Essays in Honour of John Maynard Smith*, edited by P.J. Greenwood, P.H. Harvey & M. Slatkin, Cambridge and New York: Cambridge University Press

Charnov, E.L., Smith, J.M. & Bull, J.J. (1976) Why be an hermaphrodite? *Nature* 263: 125–26

Clark, M.S., Parker, J.S. & Ainsworth, C.C. (1993) Repeated DNA and heterochromatin structure in *Rumex acetosa*. *Heredity* 70: 527–36

Darwin, C.R. (1877) *The Different Forms of Flowers on Plants of the Same Species*, London: John Murray; revised 1880

Delichére, C., Veuskens, J., Hernould, M. *et al.* (1999) SlY1, the first active gene cloned from a plant Y chromosome, encodes a WD-repeat protein. *EMBO Journal* 18: 4169–79

Dellaporta, S.L. & Urrea, A.C. (1994) The sex determination process in maize. *Science* 266: 1501–05

Desfeux, C. & Lejeune, B. (1996) Systematics of euromediterranean *Silene* (Caryophyllaceae): evidence from a phylogenetic analysis using its sequences. *Comptes Rendus de l'Academie des Sciences Serie III – Sciences de la Vie – Life Sciences* 319: 351–58

Desfeux, C., Maurice, S., Henry, J.P., Lejeune, B. & Gouyon, P.H. (1996) Evolution of reproductive systems in the genus *Silene*. *Proceedings of the Royal Society of London, Series B* 263: 409–14

Donnison, I.S., Siroky, J., Vyskot, B., Saedler, H. & Grant, S.R. (1996) Isolation of Y chromosome-specific sequences from *Silene latifolia* and mapping of male sex determining genes using representational difference analysis. *Genetics* 144: 1893–901

Farbos, I., Veuskens, J., Vyskot, B. *et al.* (1999) Sexual dimorphism in white campion: complex deletion on the Y chromosome results in a floral asexual type. *Genetics* 151: 1187–96

Givnish, T.J. (1980) Ecological constraints on the evolution of breeding systems in seed plants: dioecy and dispersal in gymnosperms. *Evolution* 34: 959–72

Graves, J.A.M. (1995) The origin and function of the mammalian Y chromosome and Y-borne genes – an evolving understanding. *BioEssays* 17: 311–21

Guttman, D.S. & Charlesworth, D. (1998) An X-linked gene has a degenerate Y-linked homologue in the dioecious plant *Silene latifolia*. *Nature* 393: 263–66

Lardon, A., Georgiev, S., Aghmir, A., Merrer, G.L. & Negrutiu, I. (1999) Sexual dimorphism in white campion: complex control of carpel number is revealed by Y chromosome deletions. *Genetics* 151: 1173–85

Lewis, D. (1942) The evolution of sex in flowering plants. *Biological Reviews* 17: 46–67

Lloyd, D.G. (1975) The maintenance of gynodioecy and androdioecy in angiosperms. *Genetica* 45: 325–39

Lucchesi, J.C. (1978) Gene dosage compensation and the evolution of sex chromosomes. *Science* 202: 711–16

Maurice, S., Couvet, D., Charlesworth, D. & Gouyon, P.-H. (1993) The evolution of gender in hermaphrodites of gynodioecious populations: a case in which the successful gamete method fails. *Proceedings of the Royal Society of London, Series B* 251: 253–61

Mishler, B.D. (1988) Reproductive ecology of bryophytes. In *Plant Reproductive Ecology: Patterns and Strategies*, edited by J.Lovett Doust &. L.Lovett Doust, Oxford and New York: Oxford University Press

Nigtevecht, G. van (1966a) Genetic studies in dioecious *Melandrium*. I. Sex-linked and sex-influenced inheritançe in *Melandrium album* and *Melandrium dioicum*. *Genetica* 37: 281–306

Nigtevecht, G. van (1966b) Genetic studies in dioecious *Melandrium*. II. Sex determination in *Melandrium album* and *Melandrium dioicum*. *Genetica* 37: 307–44

Pannell, J. (1997) Widespread functional androdioecy in *Mercurialis annua* L. (Euphorbiaceae). *Biological Journal of the Linnean Society* 61: 95–116

Patterson, J.T. & Stone, W.S. (eds) (1952) *Evolution in the Genus Drosophila*, New York: Macmillan

Réjon, C.R., Jamilena, M., Ramos, M.G., Parker, J.S. & Rejon, M.R. (1993) Cytogenetic and molecular analysis of the multiple sex-chromosome system of *Rumex acetosa*. *Heredity* 72: 209–15

Renner, S.S. & Ricklefs, R.E. (1995) Dioecy and its correlates in the flowering plants. *American Journal of Botany* 82: 596–606

Rieseberg, L.H., Hanson, M.A. & Philbrick, C.T. (1992) Androdioecy is derived from dioecy in Datiscaceae: evidence from restriction site mapping of PCR amplified chloroplast DNA. *Systematic Botany* 17: 324–36

Rohwer, J. & Kubitzki, K. (1984) *Salix martiana*, a regularly hermaphrodite willow. *Plant Systematics and Evolution* 144: 99–101

Seger, J. & Eckhart, V.M. (1996) Evolution of sexual systems and sex allocation in annual plants when growth and reproduction overlap. *Proceedings of the Royal Society of London, Series B* 263: 833–41

Smith, B.W. (1969) Evolution of sex-determining mechanisms in *Rumex*. *Chromosomes Today* 2: 172–82

Stinson, J.R., Eisenberg, A.J., Willing, R.P. *et al.* (1987) Genes expressed in the male gametophyte of flowering plants and their isolation. *Plant Physiology* 83: 442–47

Vyskot, B., Araya, A., Veuskens, J., Negrutiu, I. & Mouras, A. (1993) DNA methylation of sex chromosomes in a dioecious plant, *Melandrium album*. *Molecular and General Genetics* 239: 219–24

Westergaard, M. (1958) The mechanism of sex determination in dioecious plants. *Advances in Genetics* 9: 217–81

Ye, D., Installé, P., Ciúperescu, C. *et al.* (1990) Sex determination in the dioecious *Melandrium*. I. First lessons from androgenic haploids. *Sexual Plant Reproduction* 3: 179–86

Zhang, Y.H., DiStilio, V.S., Rehman, F. *et al.* (1998) Y chromosome specific markers and the evolution of dioecy in the genus Silene. *Genome* 41: 141–47

GLOSSARY

euchromatic relating to euchromatin, those regions of chromosomes, generally forming the greater part of the chromosome, that decondense at the end of telophase, and that contain most of the genes

gametophyte haploid phase in the alternation of generations in plants; gametes are formed by the gametophytes

heterogamety having different chromosome constitutions (karyotypes) in the two sexes, due to different sex chromosome morphology (in humans, female XX and male XY)

pseudoautosomal region a region of the Y chromosome that is homologous with an X chromosome region, involved in pairing of the sex chromosomes at meiosis

RAPD (random amplified polymorphic DNA) DNA sequence variation detected by the polymerase chain reaction (PCR) using random primer sequences

PLANT SELF-INCOMPATIBILITY

D. Charlesworth
Institute of Cell, Animal and Population Biology, University of Edinburgh, Edinburgh, UK

1. Introduction

In plants with self-incompatibility (SI) systems, their own pollen is recognized and rejected, and such plants can thus avoid self-fertilization. Darwin (1876) first appreciated self-incompatibility as a widespread feature of angiosperms. In about 25–27 flowering plant families, the incompatibility types are associated with different placement of the anthers and stigmas (heterostyly), and plants fall into just two or three types (Darwin, 1877; Barrett, 1992). Most species of self-incompatible plants, however, are homomorphic, with no morphological differences betraying the incompatibility types of their flowers. These "homomorphic" systems have multiallelic polymorphisms, controlling large numbers of different incompatibility "types". Self-incompatibility has long been a problem of interest to geneticists and population geneticists, as well as to developmental biologists, but the limits of classical genetic resolution were reached many years ago. In recent years, new progress has been made by molecular genetic approaches.

2. Homomorphic self-incompatibility polymorphisms

(i) Nature and maintenance of multiallelic polymorphisms

The chief genetic interest of the homomorphic self-incompatibility systems is the nature and maintenance of the multiallelic polymorphisms. This polymorphism is similar to that seen in fungal incompatibility systems (May & Matzke, 1995) and mammalian major histocompatibility complex (MHC) systems (e.g. Hughes & Yeager, 1998; Takahata & Satta, 1998), although there are also important differences. Plants with this kind of system are known in at least 40 (and suspected in as many as 100) different angiosperm families, but not in gymnosperms (Weller *et al.*, 1995). Although incompatibility systems operate in such a way that a plant rejects its own pollen, but accepts that of most other individuals, these are not, strictly speaking, self-recognition systems. Specificity is genetically controlled, although the strength of incompatibility can be affected by modifier loci (see Ai *et al.*, 1991; Nasrallah *et al.*, 1994), and, unlike immune systems, is unaffected by prior exposure to given pollen types. Homomorphic incompatibility systems are classified into two major types – gametophytic and sporophytic (see **Figure 1**) – based on the genetic control of pollen incompatibility reactions.

Gametophytic inheritance. In many angiosperm families, the polymorphism in pollen incompatibility types is controlled by a gametophytically expressed locus, i.e. the pollen expresses the allele carried in its own haploid genotype and plants reject pollen carrying either of their own alleles. Gametophytic inheritance with a single incompatibility (*S*) locus is known in a variety of flowering plant families. These include the Scrophulariaceae, including self-incompatible species of Antirrhinum (snapdragons), and this is the family in which the genetics of self-incompatibility was first elucidated (in a Verbascum species, by East & Mangelsdorf, 1925). Gametophytic self-incompatibility is also known in some Evening Primrose species (Onagraceae; Emerson, 1939), in the poppy family (Papaveraceae; Lawrence, 1975), Solanaceae (with Nicotiana, Petunia, Lycopersicon and Solanum species; reviewed by Kao & McCubbin, 1996), and Rosaceae and several other flowering plant families (reviewed in Weller *et al.*, 1995). Systems with two or more loci are known from several families, including grasses (Lundqvist, 1954), but have been less studied, no doubt because of the difficult genetic work necessary (Lundqvist *et al.*, 1981).

Sporophytic inheritance. Single-locus "sporophytic" inheritance (pollen incompatibility types determined by the genotypes of the diploid plants producing the pollen) occurs

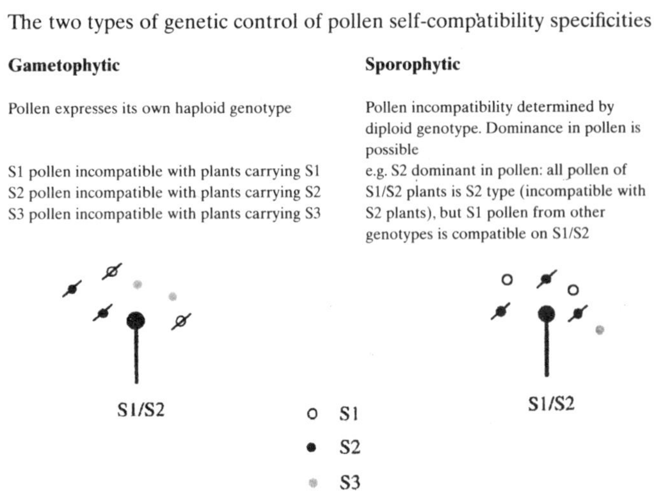

Figure 1 Homomorphic (non-heterostyled) self-incompatibility and its genetic basis in single-locus gametophytic and sporophytic systems. With either genetic control, all pollen of any genotype is incompatible, so that self-fertilization is prevented. With gametophytic self-incompatibility, half-compatibility is possible (e.g. if a pollen donor of genotype S1/S3 pollinates genotype S1/S2, and only the S3 pollen is compatible), but reciprocal differences in compatibility do not occur. With gametophytic self-incompatibility, reciprocal differences in compatibility are possible (e.g. if in pollen the dominance hierarchy is S3 > S2 >S1, then S1/S2 plants produce pollen of type S2, incompatible with S2/S3, assuming codominance in female reproduction, but S2/S3 plants produce S3-type pollen, compatible with S1/S2).

in homomorphic incompatibility systems in Brassicaceae, Asteraceae and several other flowering plant families (Kowyama *et al.*, 1980; Goodwillie, 1997). Dominance/recessivity of alleles is common in both pollen and pistil in these SI systems (Bateman, 1952; Crowe, 1954; Stevens & Kay, 1988).

In both gametophytic and sporophytic types of SI system, the *S*-loci have spectacular polymorphism in terms of *S*-allele numbers within populations (e.g. Emerson, 1939; Sampson, 1967; Lane & Lawrence, 1993). In terms of allele numbers, the *S*-loci are among the most highly polymorphic loci known, rivalling the polymorphism of the mammalian MHC loci (Zangenberg *et al.*, 1995) or fungal incompatibility loci (Raper *et al.*, 1958; Metzenberg, 1990). Different plant populations sometimes have strongly overlapping sets of *S*-alleles (O'Donnell & Lawrence, 1984; Stevens & Kay, 1988; Lawrence *et al.*, 1993; O'Donnell *et al.*, 1993), but low allelic overlap was found in *Brassica campestris* (Nou *et al.*, 1993).

(ii) Molecular studies of the loci controlling pistil incompatibility proteins

Molecular genetic studies have recently opened the way for more detailed evolutionary and population studies of the self-incompatibility polymorphism. In several self-incompatible plants, alleles have now been cloned that segregate with the incompatibility types of plants in families and encode sequences of cosegregating pistil proteins (Anderson *et al.*, 1986; Nasrallah & Nasrallah, 1986; Hayman & Richter, 1992; Walker *et al.*, 1996).

In the Solanaceae, Rosaceae and Scrophulariaceae, the S-proteins are related to RNases (e.g. McClure *et al.*, 1989; Huang *et al.*, 1994; Murfett *et al.*, 1994; Broothaerts *et al.*, 1995; Sassa *et al.*, 1996; Xue *et al.*, 1996), although it is not clear whether this similarity is due to independent evolution or to common origin (Sassa *et al.*, 1996; Richman *et al.*, 1997). Quite different gene products are involved in *Papaver rhoeas* (Franklin-Tong & Franklin, 1993; Walker *et al.*, 1996; Kurup *et al.*, 1998).

The system in the Brassicaceae, the only sporophytic system yet characterized at the molecular level, is unique in that two linked loci, *SLG* and *SRK*, both encoding stigmatically expressed loci, play essential roles in incompatibility. The protein product of the *SLG* locus is probably not involved in recognition. The closely linked *SRK* locus contains an *S*-domain, homologous in sequence to the *SLG* locus, and also an apparent transmembrane domain, and a domain with homology to members of a serine/threonine kinase gene family (Stein *et al.*, 1991; Goring & Rothstein, 1996). The kinase is therefore surmised to be an extracellular receptor that probably mediates the recognition function (Stein *et al.*, 1991).

At present, we do not know what parts of the incompatibility molecules are important for recognition, nor how many amino acid differences in the protein are required for there to be a difference in incompatibility type. Studies of allele sequence diversity may provide some information. Even just the first few alleles sequenced immediately revealed astonishing divergence between alleles for different incompatibility types, in both Solanaceae (Anderson *et al.*, 1989; Kheyr-Pour *et al.*, 1990) and Brassicaceae (Nasrallah *et al.*, 1987; Takayama *et al.*, 1987). Both silent differences, and multiple amino acid differences, were found between *S* alleles. The polymorphism is found throughout the *S*-locus sequences, but is unevenly distributed. In genes of both gametophytic and sporophytic systems, "hypervariable" regions are seen (Sims, 1993). High diversity has been confirmed using relatively conserved regions to design primers for PCR-based analysis of allelic diversity, both in large sets of Brassica alleles of known incompatibility type (Kusaba *et al.*, 1997) and sets of alleles of unknown incompatibility type from natural populations of species in the Solanaceae (Richman *et al.*, 1996). In measuring the polymorphism by diversity per nucleotide site (Nei, 1987), differences found between Brassica alleles are overall rather less than those found in the natural populations of Solanaceae, but more extensive than differences at the MHC loci (see references in Hughes & Yeager, 1997; Charlesworth & Awadalla, 1998). The only similar system known is fungal incompatibility (May & Matzke, 1995).

To ask what amount of change in an allele is necessary to change its specificity and produce a new one, we cannot simply compare allele sequences to test whether single amino acid differences cause the incompatibility type differences between alleles, because we have no way of knowing which differences are just neutral differences that accumulate over time (and have no effect on specificity) and which are the functionally important ones. Recently, transgenic experiments have succeeded in changing pistil reactions. Gain of function (making a plant transgenic for an allele that it originally does not have) has not yet proven possible in self-incompatible plants of the Brassica family (Toriyama *et al.*, 1991), but this has been achieved in the Solanum type of self-incompatibility. In both *Petunia inflata* (Lee *et al.*, 1994) and Nicotiana (Murfett *et al.*, 1994), transgenic plants that had extra DNA bands that hybridized with *S*-locus probes, showing that they had acquired new *S*-locus sequences, rejected pollen from plants with the introduced allelic type. Studies of chimaeric constructs between different *Nicotiana alata* alleles are starting to produce evidence on whether allelic differences are controlled by single or multiple amino acid differences, but the results of tests using different alleles have yielded different results (Matton *et al.*, 1997; Zurek *et al.*, 1997) so there are, as yet, no reliable answers to this question. In *P. rhoeas*, *in vitro* mutagenesis experiments have identified single amino acid residues at which a change can abolish a given incompatibility specificity, but it has not yet proved possible to change one specificity into another (Katsuyuki *et al.*, 1998).

(iii) Evolution and maintenance of self-incompatibility

Unlike the origin of self-incompatibility, which is still quite obscure (Charlesworth & Charlesworth, 1979b; Uyenoyama, 1988a,b, 1989; Holsinger & Steinbachs, 1997), the maintenance of variability at the self-incompatibility loci is well understood (Wright, 1939, 1964). Rare alleles have a fertility advantage because pollen carrying such alleles will not be rejected by incompatibility reactions of recipient plants. In a large enough population, this "frequency-dependent" selection favours each new incompatibility-type allele until equilibrium is reached with equal allele frequencies. The equilibrium number of alleles will thus be high (Wright, 1939, 1964), depending on population sizes (since alleles may be lost by chance in small populations) and mutation rates to new specificities (since frequent mutation tends to increase allele numbers). The fertility advantage of low frequency also means that alleles lost by genetic drift are soon restored if there is gene flow from other populations. Thus, alleles should be maintained in species for long evolutionary times (Wright, 1939, 1964; Vekemans & Slatkin, 1994; Schierup, 1998).

An important result of sequencing studies has been to show that self-incompatibility loci have evolved from independent origins several times in flowering plants and to identify in each case the gene family in which the *S*-locus belongs. These findings support the inferences from the taxonomic distribution of self-incompatibility (Charlesworth, 1985). While self-incompatibility may well be ancient in several angiosperm lineages, the commonly held view that it is ancestral to the angiosperms as a whole (Whitehouse, 1950) is therefore implausible (Weller *et al.*, 1995).

It is almost certain that self-incompatibility has broken down many times, i.e. reversion to self-compatibility and inbreeding readily occurs (Stebbins, 1957). This is not surprising, as these systems are complex, with many steps necessary for SI to function, and so mutational losses of function are easy to imagine. Several kinds of such mutation are known (Stebbins, 1957; Lloyd, 1965; Brauner & Gottlieb, 1987; Reinartz & Les, 1994), sometimes apparently due to changes at the *S*-locus (Dana & Ascher, 1986a,b). Self-incompatibility systems are, therefore, in an evolutionarily dynamic state. With the recent possibility of phylogenetic analysis using sequences of loci, it should be possible in the near future to estimate the time since self-compatible taxa arose and the number of times self-compatibility has arisen in genera that have self-incompatible species (see Schoen *et al.*, 1996).

(iv) The search for the pollen loci

Several other very interesting questions are, as yet, unanswered, including how self-incompatible pollen is recognized. An important part of this genetic question is whether the pollen of the same plant also expresses the recognition sequences expressed by the female parts of flowers (pistils). On the single-gene model, matching of two identical sequences encoded by the *S*-locus produces an incompatibility reaction (like-with-like models). Incompatible pollen grains, or the pollen tubes that have grown out from them, are then blocked from further growth, or sometimes killed, even when non-matching sequences are also present (e.g. when S_x pollen arrives on a S_xS_y stigma). Alternatively, matching need not involve identical sequences. Pollen and pistil may express recognition sequences that are encoded by different loci (signal and receptor, or lock and key, models). Although two-gene models present difficulties, mating type control in fungi, involving loci with multiple different specificities, has been shown to involve dimerization between the products of two different genes. In these systems, the numbers of different allelic types that are alternatives at truly orthologous loci are limited, and the total number of specificities is increased by the existence of more than one copy of the loci involved (reviewed by Casselton, 1998). It is possible that a similar situation exists in flowering plants. In either model, it is quite possible that the pollen and pistil sequences do not interact directly. There may well be one or more other loci mediating recognition and triggering the consequences of the recognition reaction. The subsequent steps in the incompatibility reaction, once the recognition reaction has happened, are as yet unknown, but it is well established that both linked and unlinked modifiers exist that can abolish incompatibility (e.g. Ai *et al.*, 1991; Nasrallah *et al.*, 1994).

A promising approach to discovering the pollen locus might be to generate mutations and find out what actually causes specificity changes. In a classic series of beautiful experiments, however, Lewis found that mutations to new specificities were undetectably rare, a finding that has held up to subsequent testing (see Lewis, 1960). This suggests that mutations from existing *S*-alleles to new functional ones involve some extremely rare type of change. Perhaps this is evidence that more than a single amino acid change is required or that perhaps **simultaneous changes in two genes are necessary, one changing the pollen reaction and one to change the receptor in stigmas**. Lewis did, however, find a series of allelic mutations with the extremely interesting phenotype that self-incompatibility was lost, in some cases only in pollen ("pollen-part mutations") and in others only in the pistil. Such mutations have also been found by others (Dana & Ascher, 1986a,b). The recovery of these mutations shows that there is no insuperable technical problem with experiments attempting to detect mutations to new specificities, and, furthermore, that self-incompatibility must differ in some manner between pollen and pistils. This might be because different loci exist for the pollen and pistil functions (Lewis, 1963; Hodgkin *et al.*, 1988; Sims, 1993; Dodds *et al.*, 1996), or, alternatively, there may be one locus with different domains for the two kinds of function or with different control of the expression of the genes. Either of these possibilities is plausible, given the different nature of the two types of tissue. It is intriguing that similar results (rarity of mutations that change specificity, but occurrence of loss-of-function mutations) have also been obtained in mutagenesis studies of fungal systems (Metzenberg, 1990).

Pollen-part mutations often involve duplications of the *S*-locus, often on centric chromosomal fragments (Brewbaker & Natarajan, 1960; see also review by Nettancourt, 1977). This finding also explains the frequent compatibility of polyploid plants in species and genera that have SI systems, as these have diploid pollen (Lewis, 1963). These findings show that pollen functions must, therefore, be controlled by a locus that is either identical to, or else tightly linked to, the pistil gene. This makes sense, because, under the two-gene lock and key hypothesis, the pollen and pistil incompatibility type loci must recombine only rarely, since recombinants would have one pollen type, and a different pistil type, and would be self-compatible. The two-gene view, however, encounters difficulties when one considers the appearance of new alleles. In a population with several established incompatibility alleles, mutation to a new specificity for pollen, say, would produce a self-compatible plant, not a complete new *S*-allele, because its pistil type would be unchanged (Lewis, 1960). We therefore have to invoke the additional assumption of a mechanism to ensure that changes in one sequence are accompanied by appropriate changes at the other, to preserve incompatibility in the newly arisen type. This would require a double mutation, yet one which must preserve the mechanism of incompatibility.

A direct test of whether the pollen and pistil genes could be one and the same would be the detection of pollen expression of the locus cloned from pistils. This, however, has been difficult, perhaps because the pollen and pistil loci really are different. Expression at low levels has, however, been detected in the anthers of Brassica (Dzelzkalns *et al.*, 1993) and in Nicotiana pollen (Dodds *et al.*, 1993), so the conclusion is not yet certain. Another approach to finding the specificity locus for pollen involves genetic engineering. If specificities can be changed by transgenic methods, this should allow a test of whether the pollen type changes whenever the pistil type is changed. This experiment is not, however, as straightforward as it sounds, because, when pollen of species with gametophytic incompatibility systems carries two alleles (instead of the normal haploid one), incompatibility is frequently lost (see previous paragraph). Unless the alleles were carefully chosen to be ones that do not have such an effect with one another, this approach will probably not be helpful in testing ideas about the pollen locus.

The complementary type of experiment, to achieve loss of a particular *S* specificity, would seem more promising, because one can then ask whether pollen expressing the transgene loses its incompatibility type along with such loss in the pistil. If the pollen type is controlled by the same locus as the pistil type, this should occur. Using a transgene for antisense S_3 RNA, complete agreement was again demonstrated in *P. inflata* between the allelic types of mRNA in pistils, the protein variants produced and the types of pollen rejected, although some transgenic plants unexpectedly had lowered expression of the native *S*-allele as well (Lee *et al.*, 1994); this may be due to the phenomenon of "co-suppression" (Jorgensen, 1990). In this experiment, however, the pollen incompatibility was expressed normally. This does not prove that the pollen specificity is controlled by a different locus from the pistil type, because the upstream sequence that was introduced with the transgene might not have included all sequences necessary for expression in the pollen. In Brassica, control of pollen expression differs from that for the pistil, although the pollen sequences are all in the immediate upstream region from the *S*-locus (Dzelzkalns *et al.*, 1993).

In view of these technical problems, the most promising approaches to discovering the pollen gene would seem to be molecular study of pollen-part mutations and physical mapping of the *S*-locus region to try to find candidate pollen incompatibility genes. Assuming that pollen-part mutations are in the same gene where pollen specificities are controlled, isolation and cloning of such mutant alleles should tell us whether the locus is the same or different from the pistil locus that is already known. It will also be important to know whether these mutations are loss- or gain-of-function mutations. However, the alternative approach is currently yielding some very interesting results. A candidate for the pollen gene involved in self-incompatibility of Brassica has recently been identified and the gene in question (SCR) has been shown to be capable of changing the pollen specificity in transgenic plants (Schopfer *et al.*, 1999; Suzuki *et al.*, 1999). Although the SCR sequence is known, its role in the self-incompatibility mechanism is as yet completely unknown.

(v) The role of RNase and the mechanism of self-incompatibility

The fact that in some systems RNase is involved in self-incompatibility (see above) suggests that, for these species, destruction of RNA in incompatible tubes is part of the mechanism. It is important to stress that it is a separate question, as yet unanswered, whether the recognition function, with the specificity for incompatibility type, has anything to do with the RNase function. The possibility remains that RNase is essential for incompatibility functions that occur after the recognition stage and is mediated by a different domain of the protein. It is highly unlikely that the pistil RNase proteins recognize the allelic types of pollen RNA and destroy the type corresponding to the same incompatibility allele (Dodds *et al.*, 1996). More likely, pollen and pistil proteins interact, and pollen RNA is destroyed if this combination is incompatible.

What is the nature of such a combination? Isolated pistil S-RNase is active and can be detected by staining after separation by gel electrophoresis (e.g. Huang *et al.*, 1994). It is not known how digestion of pistil cells' RNA is prevented, although possibly S-RNase does not become active until it is secreted into the intercellular matrix of the conducting tissue (it is also necessary to assume that it cannot enter pistil cells). If pollen also contains S-RNases, however, these are intracellular and their activity must presumably be blocked in some manner. One possibility is that matching of a pistil with a pollen protein, perhaps encoded by the same gene, removes such blocks and leads to RNase activity of a protein involving both pollen and pistil S-gene products. In either the one- or the two-gene version of this type of model, the three known types of self-compatible mutation (Lewis, 1960) could be losses of function of the following kinds:

- mutations that lack RNase function (in single-gene models, these could affect both male and female incompatibility functions);
- pollen-part mutations (where either the pollen protein is not correctly deployed or the blocking of the pollen component fails to be relieved); and
- pistil-part mutations (which either cannot enter pollen tubes, or cannot activate their RNase).

Other mutations affecting the pollen function, such as loss of the ability to block its RNase activity, would be lethal, and would therefore not be expected to be recovered.

In the two-gene (lock and key) versions of the above type of model, RNase activity results when corresponding pollen and pistil proteins interact (say, RNaseS1 + pollenS1), and mutants with loss of both pollen and pistil incompatibility would presumably have to have deletions of both genes (or, at least, loss of expression of both). Various other types of two-gene model are also possible. One recent model (Dodds *et al.*, 1996) proposes that each pollen type actively blocks pistil RNase entry into pollen tubes (or causes a lasting block or reduction in the activity of pistil RNase) of all different pistil allelic products other than that of plants of the same incompatibility type (e.g. pollenS1 blocks or degrades RNaseSi, where i is any allele type except 1). However, it is then difficult to understand the occurrence of pollen-part mutations (loss of function in the pollen would lead to entry or activity of RNase, and hence lethality). If, however, the phenotype of pollen-part mutations results from duplication of the pollen *S*-locus (see above), rather than sequence changes in the locus, this objection is irrelevant. Indeed, although this model seems unlikely *a priori*, it is attractive in that it can neatly account for the effect of duplication, which is known to require two different *S*-alleles to produce self-compatibility (Lewis, 1960; Chawla *et al.*, 1997). On the above model, self-compatibility arises because pollen with the constitution pollenS1/pollenS2 prevents activity of both pistil RNaseS1 and RNaseS2. As already mentioned, characterization of pollen-part mutations is probably the only way to advance our understanding of the mechanism of S-RNase action, but it seems unlikely that all such mutants are duplications. For example, Lewis's results showing rescue of inactive alleles in diploid pollen are not explained by this model, but seem to be a case of interallelic complementation and to require physical interaction between alleles with different incompatibility types, one of which has lost a function necessary for incompatibility, even though its specificity function is intact (reviewed in Lewis, 1960).

3. Heterostyly

Heterostyly is a completely evolutionarily independent type of self-incompatibility system. Heterostyled plants have only two or three different incompatibility types, associated with different floral types ("distyly" and "tristyly"). The types have complementary placement of the floral reproductive organs. For instance, in distylous primroses (see **Figure 2a**),

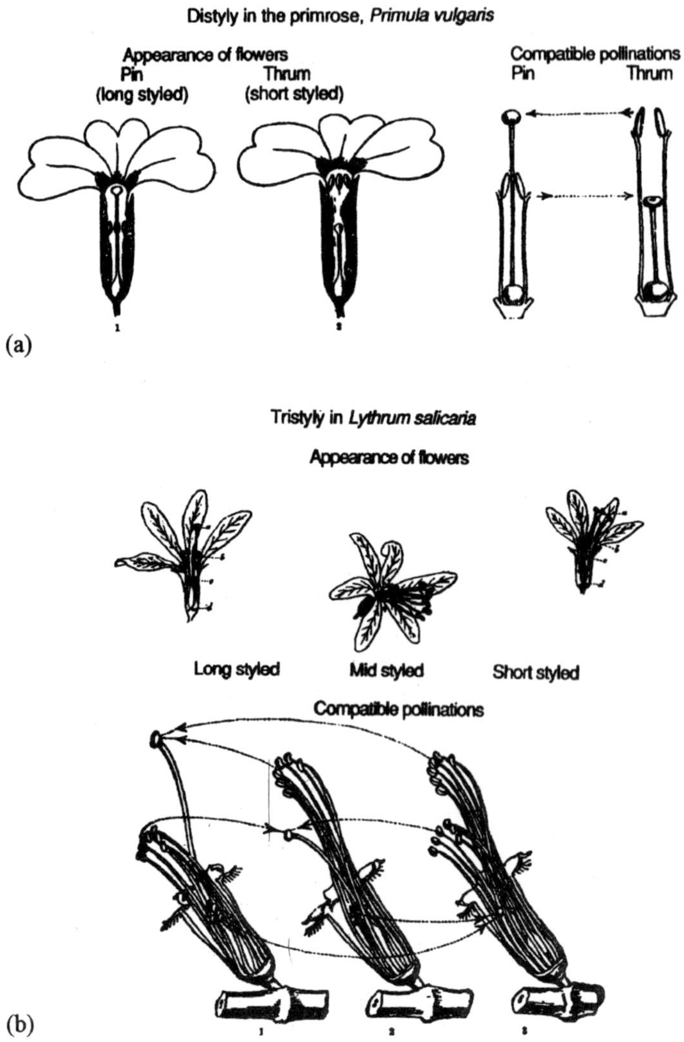

Figure 2 Heterostyly (from Darwin, 1877).

one "morph" (called pin) has long stigmas and anthers placed low in the corolla tube, while the other morph (thrum) has the stigma low down and the anthers at the mouth of the corolla. The morphs are controlled by two alleles at one (distyly) or two loci (tristyly), each with two alleles (Barrett, 1992). The genes are sporophytic in their control of pollen types, with dominance of certain alleles over others. The genetic control leads to populations consisting of equal frequencies of the two or three different morphs. In distyly, for instance, pin is the homozygote ss, while thrum plants are heterozygous Ss. Thus, all matings are backcrosses.

(i) Distyly supergenes, and breakdown of distyly

Distyly is known in plants from about 20 different angiosperm families (Ganders, 1979), including two crops, buckwheat and carambola (star fruit). In all species studied, the two morphs of distylous plants differ in several characteristics. The stigma and anther positions and incompatibility types have already been mentioned, but there are also frequently differences in pollen grain size and surface sculpturing and in stigma papilla morphology (Ganders, 1979). It is hard to believe that all of these are under the control of a single locus with two alleles, so it is likely that these alleles are really sets of tightly linked loci. For instance, s may be gpa (GPA in the S plants), where *g* controls the pistil characteristics, *p* controls the pollen characteristics and *a* controls the other anther characteristics.

There is at present no direct molecular evidence for this "supergene" theory, as these loci have not yet been cloned, but there is some genetic evidence from the occurrence of aberrant, putatively recombinant types. One of these has the L type of pistil, but the S type of anthers and pollen. This "homostyle" type has the thrum pollen type together with the pin pistil type (Crosby, 1949) and is thus self-compatible. It is controlled by a third "allele", S′, that segregates from s and S just as they do from one another. It therefore seems likely that this new allele is a recombinant product between S and s, having the alleles *gPA* (Haldane, 1933). In further work, all the eight possible types that could be generated from recombination between the three hypothesized loci have been found in different Primulas (Ernst, 1936). Unlike the case of homomorphic incompatibility, recombinant combinations with the pollen reaction of one incompatibility type and the pistil reaction of the other are probably responsible for the loss of self-incompatibility in distylous plants. Many Primula species are homostyled, presumably in many cases secondarily.

Recent identification of marker loci linked to the *S*-locus gives some promise that mapping can identify the region where these genes are located and perhaps test the hypothesis just given. There is still, however, much work to be done before the genes can be fully characterized (Athanasiou & Shore, 1997).

(ii) Tristyly

Tristyly is similar to distyly, but is even more complex as there are three forms of flowers and of plants (see **Figure 2b**) – L, M (mid) and S styled. Each morph has two whorls of anthers, at two different levels, and pollen succeeds in fertilization only if it comes from the same level as the stigma

Table 1 Compatibility in tristylous plants.

Morph	Stigma position	Anther positions	Pollen compatible with stigma position
L	long	m	mid
		s	short
M	mid	l	long
		s	short
S	short	m	mid
		l	long

(e.g. l pollen, from the M or S morph, to L stigmas; see **Table 1**). Thus, this system has the astonishing property that each morph produces pollen of two different incompatibility types. For example, L produces pollen that will fertilize M and S pistils. As with distyly, these reactions were first discovered by Darwin (1877).

Like distyly, tristyly has probably evolved several times in the history of the angiosperms, as it is found in a small number of unrelated families (see Barrett, 1992), but the details of how either system evolved are far from clear (see Charlesworth, 1979; Charlesworth & Charlesworth, 1979c; Lloyd & Webb, 1992). Just like homomorphic incompatibility systems, heterostyly has broken down many times (Crosby, 1949; Charlesworth & Charlesworth, 1979a; Barrett, 1992).

REFERENCES

Ai, Y., Kron, E. & Kao, T. (1991) S-alleles are retained and expressed in a self-compatible cultivar of *Petunia hybrida*. *Molecular and General Genetics* 230: 353–58

Anderson, M.A., Cornish, E.C., Mau, S.-L. *et al.* (1986) Cloning of cDNA for a stylar glycoprotein associated with expression of self-incompatibility in *Nicotiana alata*. *Nature* 321: 38–44

Anderson, M.A., McFadden, G.I., Bernatzky, R. *et al.* (1989) Sequence variability of three alleles of the self-incompatibility gene of *Nicotiana alata*. *The Plant Cell* 1: 483–91

Athanasiou, A. & Shore, J.S. (1997) Morph-specific proteins in pollen and styles of distylous Turnera (Turneraceae). *Genetics* 146: 669–79

Barrett, S.C.H. (ed.) (1992) Heterostylous genetic polymorphisms: model systems for evolutionary analysis. In *Evolution and Function of Heterostyly*, Berlin and New York: Springer

Bateman, A.J. (1952) Self-incompatibility systems in angiosperms. 1. Theory. *Heredity* 6: 285–310

Brauner, S. & Gottlieb, L.D. (1987) A self-compatible plant of *Stephanomeria exigua* subsp. *coronaria* (Asteraceae) and its relevance to the origin of its self-pollinating derivative *S. malheurensis*. *Systematic Botany* 12: 299–304

Brewbaker, J.L. & Natarajan, A.T. (1960) Centric fragments and pollen part mutation of incompatibility alleles in Petunia. *Genetics* 45: 699–704

Broothaerts, W., Janssens, G.A., Proost, P. & Broekaert, W.F. (1995) cDNA cloning and molecular analysis of two self-incompatibility alleles from apple. *Plant Molecular Biology* 27: 499–511

Casselton, L.A. (1998) Molecular genetics of mating recognition in Basidiomycete fungi. *Microbiology and Molecular Biology Reviews* 62: 55–70

Charlesworth, D. (1979) The evolution and breakdown of tristyly. *Evolution* 33: 486–98
Charlesworth, D. (1985) Distribution of dioecy and self-incompatibility in angiosperms. In *Evolution: Essays in Honour of John Maynard Smith*, edited by P.J. Greenwood, P.H. Harvey & M. Slatkin, Cambridge and New York: Cambridge University Press
Charlesworth, D. & Awadalla, P. (1998) The molecular population genetics of flowering plant self-incompatibility polymorphisms. *Heredity* 81: 1–9
Charlesworth, B. & Charlesworth, D. (1979a) The maintenance and breakdown of distyly. *American Naturalist* 114: 499–513
Charlesworth, D. & Charlesworth, B. (1979b) The evolution and breakdown of S-allele systems. *Heredity* 43: 41–55
Charlesworth, D. & Charlesworth, B. (1979c) A model for the evolution of distyly. *American Naturalist* 114: 467–98
Chawla, B., Bernatzky, R., Liang, W. & Marcotrigiano, M. (1997) Breakdown of self-incompatibility in tetraploid *Lycopersicon peruvianum*: inheritance and expression of *S*-related proteins. *Theoretical and Applied Genetics* 95: 992–96
Crosby, J.L. (1949) Selection of an unfavourable gene-complex. *Evolution* 3: 212–30
Crowe, L.H. (1954) Incompatibility in *Cosmos bipinnatus*. *Heredity* 8: 1–11
Dana, M.N. & Ascher, P.D. (1986a) Sexually localized expression of pseudo-self compatibility (PSC) in *Petunia* × *hybrid* Hort. 1. Pollen inactivation. *Theoretical and Applied Genetics* 71: 573–77
Dana, M.N. & Ascher, P.D. (1986b) Sexually localized expression of pseudo-self compatibility (PSC) in *Petunia* × *hybrid* Hort. 2. Stylar inactivation. *Theoretical and Applied Genetics* 71: 578–84
Darwin, C.R. (1876) *The Effects of Cross and Self Fertilisation in the Vegetable Kingdom*, London: John Murray; revised 1878, 1891
Darwin, C.R. (1877) *The Different Forms of Flowers on Plants of the Same Species*, London: John Murray; revised 1880
Dodds, P.N., Bönig, I., Du, H. *et al.* (1993) S-RNase gene of *Nicotiana alata* is expressed in developing pollen. *Plant Cell* 5: 1771–82
Dodds, P.N., Clarke, A.E. & Newbigin, E. (1996) A molecular perspective on pollination in flowering plants. *Cell* 85: 141–44
Dzelzkalns, V.A., Thorsness, M.K., Dwyer, K.G. *et al.* (1993) Distinct *cis*-acting elements direct pistil-specific and pollen-specific activity of the Brassica *S* locus glycoprotein gene promoter. *Plant Cell* 5: 855–63
East, E.M. & Mangelsdorf, A.J. (1925) A new interpretation of the hereditary behaviour of self-incompatible plants. *Proceedings of the National Academy of Sciences, Washington* 11: 166–71
Emerson, S. (1939) A preliminary survey of the *Oenothera organensis* population. *Genetics* 24: 524–37
Ernst, A. (1936) Weitere untersuchungen zur Phänanalyse zum Fertilitätsproblem und zur Genetik heterostyler Primeln. II. *Primula hortensis*. *Archiv der Julius-Klaus Stiftung für Vererbungsforchung, Sozial-Anthropoligischer u Rassenhygiene* 11: 1–280
Franklin-Tong, V.E. & Franklin, F.C.H. (1993) Gametophytic self-incompatibility: contrasting mechanisms for *Nicotiana* and in *Papaver*. *Trends in Cell Biology* 3: 340–45
Ganders, F.R. (1979) The biology of heterostyly. *New Zealand Journal of Botany* 17: 607–35
Goodwillie, C. (1997) The genetic control of self-incompatibility in *Linanthus parviflorus* (Polemoniaceae). *Heredity* 79: 424–32
Goring, D.F. & Rothstein, S.J. (1996) S-locus receptor kinase genes and self-incompatibility in *Brassica napus* ssp. *oleifera*. In *Signal Transduction in Plant Growth and Development*, edited by D.P.S. Verma, New York: Springer
Haldane, J.B.S. (1933) Two new allelomorphs for heterostyly in *Primula*. *American Naturalist* 67: 559–60
Hayman, D.L. & Richter, J. (1992) Mutations affecting self-incompatibility in *Phalaris coerulescens* Desf. (Poaceae). *Heredity* 68: 495–503
Hodgkin, T., Lyon, G.D. & Dickinson, H.G. (1988) Recognition in flowering plants: a comparison of the brassica self-incompatibility system and plant pathogen interactions. *New Phytologist* 110: 557–69
Holsinger, K.E. & Steinbachs, J.E. (1997) Mating systems and evolution in flowering plants. In *Evolution and Diversification of Land Plants*, edited by K. Iwatsuki & P. Raven, New York: Springer
Huang, S., Lee, H.-S., Karunanandaa, B. & Kao, T.-H. (1994) Ribonuclease activity of *Petunia inflata* S proteins is essential for rejection of self-pollen. *The Plant Cell* 6: 1021–28
Hughes, A. & Yeager, M. (1997) Molecular evolution of the vertebrate immune system. *BioEssays* 19: 777–86
Hughes, A. & Yeager, M. (1998) Natural selection and the evolutionary history of the major histocompatibility complex loci. *Frontiers in Bioscience* 3: 510–16
Jorgensen, R. (1990) Altered gene expression in plants due to *trans* interactions between homologous genes. *Trends in Biotechnology* 8: 340–44
Kao, T.-H. & McCubbin, A.G. (1996) How flowering plants discriminate between self and non-self pollen to prevent inbreeding. *Proceedings of the National Academy of Sciences USA* 93: 12059–65
Katsuyuki, K., Jordan, N.D., Conner, A. *et al.* (1998) Identification of residues in a hydrophilic loop of the *Papaver rhoeas* S protein that play a crucial role in recognition of incompatible pollen. *The Plant Cell* 10: 1723–32
Kheyr-Pour, A., Bintrim, S.B., Ioerger, T.R. *et al.* (1990) Sequence diversity of pistil S-proteins associated with gametophytic self-incompatibility in *Nicotiana alata*. *Sexual Plant Reproduction* 3: 88–97
Kowyama, Y., Shimano, N. & Kawasi, T. (1980) Genetic analysis of incompatibility in the diploid species *Ipomoea* closely related to the sweet potato. *Theoretical and Applied Genetics* 58: 149–55
Kurup, S., Ride, J.P., Jordan, N. *et al.* (1998) Identification and cloning of related self-incompatibility *S*-genes in *Papaver rhoeas* and *Papaver nudicaule*. *Sexual Plant Reproduction* 11: 192–98
Kusaba, M., Nishio, T., Satta, Y., Hinata, K. & Ockendon, D. (1997) Striking similarity in inter- and intra-specific comparisons of class I SLG alleles from *Brassica oleracea* and *Brassica campestris*: implications for the evolution and recognition mechanism. *Proceedings of the National Academy of Sciences USA* 94: 7673–78
Lane, M.D. & Lawrence, M.J. (1993) The population genetics of

the self-incompatibility polymorphism in *Papaver rhoeas*. VII. The number of S-alleles in the species. *Heredity* 71: 596–602

Lawrence, M.J. (1975) The genetics of self-incompatibility in *Papaver rhoeas*. *Proceedings of the Royal Society of London, Series B* 188: 275–85

Lawrence, M.J., Lane, M.D., O'Donnell, S. & Franklin-Tong, V.E. (1993) The population genetics of the self-incompatibility polymorphism in *Papaver rhoeas*. V. Cross-classification of the S-alleles from three natural populations. *Heredity* 71: 581–90

Lee, H.-S., Huang, S. & Kao, T.-H. (1994) S proteins control rejection of incompatible pollen in *Petunia inflata*. *Nature* 367: 560–63

Lewis, D. (1960) Genetic control of specificity and activity of the S antigen in plants. *Proceedings of the Royal Society of London, Series B* 151: 468–77

Lewis, D. (1963) A protein dimer hypothesis on incompatibility. In *Genetics Today*, edited by S.J. Geerts. Proceedings of the XI International Congress, The Hague, Netherlands

Lloyd, D.G. (1965) Evolution of self-compatibility and racial differentiation in *Leavenworthia* (Cruciferae). *Contributions from the Gray Herbarium of Harvard University* 195: 3–134

Lloyd, D.G. & Webb, C.J. (1992) The evolution of heterostyly. In *Evolution and Function of Heterostyly*, edited by S.C.H. Barrett, Berlin and New York: Springer

Lundqvist, A. (1954) Studies on self-incompatibility in rye, *Secale cereale* L. *Hereditas* 40: 278–94

Lundqvist, A., Østerbye, U. & Larsen, K. (1981) The backcross analysis of complex gametophytic complementary *S* gene systems in diploid angiosperms. *Hereditas* 95: 253–58

Matton, D.P., Maes, O., Laublin, G. *et al.* (1997) Hypervariable domains of self-incompatibility RNases mediate allele-specific pollen recognition. *Plant Cell* 9: 1757–66

May, G. & Matzke, E. (1995) Recombination and variation at the a mating-type of coprinus-cinereus. *Molecular Biology and Evolution* 12: 794–802

McClure, B.A., Haring, V., Ebert, P.R. *et al.* (1989) Style self-incompatibility gene products of *Nicotiana alata* are ribonucleases. *Nature* 342: 955–57

Metzenberg, R.L. (1990) The role of similarity and difference in fungal mating. *Genetics* 125: 457–62

Murfett, J., Atherton, T.L., Mou, B., Gasser, C.S. & McClure, B.A. (1994) S-RNase expressed in transgenic *Nicotiana* causes S-allele-specific pollen rejection. *Nature* 367: 563–66

Nasrallah, M.E. & Nasrallah, J.B. (1986) Molecular biology of self-incompatibility in plants. *Trends in Genetics* 2: 239–44

Nasrallah, J.B., Kao, T.-H., Chen, C.-H., Goldberg, M. & Nasrallah, M.E. (1987) Amino-acid sequence of glycoproteins encoded by three alleles of the S locus of *Brassica oleracea*. *Nature* 326: 617–19

Nasrallah, J.B., Rundle, S.J. & Nasrallah, M.E. (1994) Genetic evidence for the requirement of the *Brassica S*-locus receptor kinase in the self-incompatibility response. *The Plant Journal* 5: 373–84

Nei, M. (1987) *Molecular Evolutionary Genetics*, New York: Columbia University Press

Nettancourt, D.D. (1977) *Incompatibility in Angiosperms*, Berlin and New York: Springer

Nou, I.S., Watanabe, M., Isogai, A. & Hinata, K. (1993) Comparison of S-alleles and S-glycoproteins between two populations of *Brassica campestris* in Turkey and Japan. *Sexual Plant Reproduction* 6: 79–86

O'Donnell, S. & Lawrence, M.J. (1984) The population genetics of the self-incompatibility polymorphism of *Papaver rhoeas*. IV. The estimation of numbers of alleles in a population. *Heredity* 53: 495–507

O'Donnell, S., Lane, M.D. & Lawrence, M.J. (1993) The population genetics of the self-incompatibility polymorphism in *Papaver rhoeas*. VI. Estimation of the overlap between the allelic complements of a pair of populations. *Heredity* 71: 591–95

Raper, J.R., Krongelb, G.S. & Baxter, M.G. (1958) The number and distribution of incompatibility factors in *Schizopyllum*. *American Naturalist* 92: 221–32

Reinartz, J.A. & Les, D.H. (1994) Bottleneck-induced dissolution of self-incompatibility and breeding system consequences in *Aster furcatus* (Asteraceae). *American Journal of Botany* 81: 446–55

Richman, A.D., Uyenoyama, M.K. & Kohn, J.R. (1996) Allelic diversity and gene genealogy at the self-incompatibility locus in the Solanaceae. *Science* 273: 1212–16

Richman, A.J., Broothaerts, W. & Kohn, J.R. (1997) Self-incompatibility RNases from three plant families: homology or convergence? *American Journal of Botany* 84: 912–17

Sampson, D.R. (1967) Frequency and distribution of self-incompatibility alleles in *Raphanus raphanistrum*. *Genetics* 56: 241–51

Sassa, H., Nishio, T., Kowyama, Y. *et al.* (1996) Self-incompatibility (*S*) alleles of the Rosaceae encode members of a distinct class of the T_2/S ribonuclease superfamily. *Molecular and General Genetics* 250: 547–57

Schierup, M.H. (1998) The number of self-incompatibility alleles in a finite, subdivided population. *Genetics* 149: 1153–62

Schoen, D.J., Morgan, M.T. & Bataillon, T. (1996) How does self-pollination evolve? Inferences from floral ecology and molecular genetic variation. *Philosophical Transactions of the Royal Society of London, Series B* 351: 1281–90

Schopfer, C.R., Nasrallah, M.E. & Nasrallah, J.B. (1999) The male determinant of self-incompatibility in Brassica. *Science* 286: 1697–700

Sims, T.M. (1993) Genetic regulation of self-incompatibility. *Critical Reviews in Plant Science* 12: 129–67

Stebbins, G.L. (1957) Self fertilization and population variation in the higher plants. *American Naturalist* 91: 337–54

Stein, J., Howlett, B., Boyes, D.C., Nasrallah, M.E. & Nasrallah, J.B. (1991) Molecular cloning of a putative receptor protein kinase gene encoded at the self-incompatibility locus of *Brassica oleracea*. *Proceedings of the National Academy of Sciences USA* 88: 8816–20

Stevens, J.P. & Kay, Q.O.N. (1988) The number, dominance relationships and frequencies of self-incompatibility alleles in a natural population of *Sinapis arvensis* L. in South Wales. *Heredity* 62: 199–205

Suzuki, G., Kai, N., Hirose, T. *et al.* (1999) Genomic organization of the S locus: identification and characterization of genes in SLG/SRK region of S9 haplotype of *Brassica campestris* (syn. *rapa*). *Genetics* 153: 391–400

Takahata, N. & Satta, Y. (1998) Footprints of intragenic recombination at *HLA* loci. *Immunogenetics* 47: 430–41

Takayama, S., Isogai, A., Tsukamoto, C. *et al.* (1987) Sequences of S-glycoproteins, products of the *Brassica campestris* self-incompatibility locus. *Nature* 326: 102–05

Toriyama, K., Stein, J.C., Nasrallah, M.E. & Nasrallah, J.B. (1991) Transformation of *Brassica oleracea* with an S-locus gene from *B. campestris* changes the incompatibility phenotype. *Theoretical and Applied Genetics* 81: 769–76

Uyenoyama, M.K. (1988a) On the evolution of genetic incompatibility systems. II. Initial increase of strong gametophytic self-incompatibility under partial selfing and half-sib mating. *American Naturalist* 131: 700–22

Uyenoyama, M.K. (1988b) On the evolution of genetic incompatibility systems: incompatibility as a mechanism for regulating outcrossing distance. In *The Evolution of Sex: An Examination of Current Ideas*, edited by R.E. Michod & B.R. Levin, Sunderland, Massachusetts: Sinauer

Uyenoyama, M.K. (1989) On the evolution of genetic incompatibility systems. V. Origin of sporophytic self-incompatibility in response to overdominance in viability. *Theoretical and Population Biology* 36: 339–65

Vekemans, X. & Slatkin, M. (1994) Gene and allelic genealogies at a gametophytic self-incompatibility locus. *Genetics* 137: 1157–65

Walker, E.A., Ride, J.P., Kurup, S. *et al.* (1996) Molecular analysis of two functional homologues of the S3 allele of the *Papaver rhoeas* self-incompatibility gene isolated from different populations. *Plant Molecular Biology* 30: 983–94

Weller, S.G., Donoghue, M.J. & Charlesworth, D. (1995) The evolution of self-incompatibility in the flowering plants: a phylogenetic approach. In *Experimental and Molecular Approaches to Plant Biosystematics*, edited by P.C. Hoch & A.G. Stephenson, St. Louis: Missouri Botanic Garden

Whitehouse, H.L.K. (1950) Multiple-allelomorph incompatibility of pollen and style in the evolution of the angiosperms. *Annals of Botany, N.S.* 14: 198–216

Wright, S. (1939) The distribution of self-sterility alleles in populations. *Genetics* 24: 538–52

Wright, S. (1964) The distribution of self-sterility alleles in populations. *Evolution* 18: 609–19

Xue, Y., Carpenter, R., Dickinson, H.G. & Coen, E.S. (1996) Origin of allelic diversity in Antirrhinum S locus RNases. *The Plant Cell* 8: 805–14

Zangenberg, G., Huang, M.-M., Arnheim, N. & Erlich, H. (1995) New HLA-DPB1 alleles generated by interallelic gene conversion detected by analysis of sperm. *Nature Genetics* 10: 407–14

Zurek, D.M., Mou, B., Beecher, B. & McClure, B.A. (1997) Exchanging sequence domains between S-RNase from *Nicotiana alata* disrupts pollen recognition. *The Plant Journal* 11: 797–808

GLOSSARY

angiosperm (Angiospermae) class of flowering plants that bear seeds in closed fruits

antisense RNA RNA complementary to the normal RNA transcript of a gene; able to block expression by hybridizing to the RNA transcript and preventing its translation

digestion (by restriction enzymes) use of enzymes that cleave DNA to produce fragments smaller than the original DNA

gel electrophoresis (*see* pulsed field gel electrophoresis in Main Glossary)

gymnosperm group of woody plants, including the conifers, whose seeds are borne on the surface of the sporophylls

modifier loci term used to refer to loci that affect the expression of a phenotype caused by another locus; for example restorers of male fertility in plants with cytoplasmic male sterility factors

pistil the female organs of a flower, comprising the stigma, style and ovary

RNase (ribonuclease) enzyme that cleaves ribonucleic acid (RNA)

CEREAL CHROMOSOME EVOLUTION AND PAIRING

Graham Moore
John Innes Centre, Norwich Research Park, Norwich, UK

1. Introduction

Extensive comparative mapping of the cereal genomes has allowed us to describe them as a single genetic system. Such a framework provides us with an opportunity to relate chromosome evolution to chromosome pairing in cereals. This article reviews these developments.

For much of the past 50 years, individual cereals (maize, wheat, rice, sorghum and the millets) have been studied largely in isolation. Plant cytogeneticists undertaking studies using wide crosses between them had noted that pairing could occur between the two sets of parental chromosomes. These observations suggested there is some conservation in structure of the paired parental chromosomes. In fact, the basis of alien introgression in species such as hexaploid wheat is dependent on the assumption that gene order is conserved between wheat and its wild species. The advent of the molecular markers (restriction fragment length polymorphisms, RFLPs) in the 1980s enabled the actual gene order along the chromosomes of these species to be revealed.

Moreover, locating related molecular markers on the genetic maps of species from which hybrids cannot be produced enabled the comparisons of gene order to be extended beyond species studied by classical pairing methods. It has been the analysis of these comparisons that has provided a framework by which all data generated in the last 50 years can be collated. My colleagues and I have proposed that the cereal genomes are described by a series of conserved units based on the linkage of genes found within their genomes and that the rice genome is pivotal for the analysis of other cereal genomes. Subsequent publications have reinforced this approach. Such a framework has yet to be established for dicot genomes.

The field of comparative mapping has spawned a number of terms such as synteny, conserved linkage, conserved synteny, homology segment, colinearity and microsynteny (reviewed in Nadeau & Sankoff, 1998). Many of these terms, which were coined by mammalian geneticists, have often been used in a different context by plant geneticists and, therefore, will not be used in this article unless defined. However, for both mammals and plants, the basic observation is the same: genes that are found linked on a chromosome in one species are also found linked on a chromosome in another related species. This observation does not apply to any repetitive sequences that could occur between genes. Clearly, during evolution, rearrangements can occur which either disrupt the gene order but maintain the linkage of the genes, or disrupt both the gene order and the linkage of genes. The important issue is how much disruption has occurred and whether linkage of genes can still be observed between distantly related species.

During evolution, genome size can vary as a result of expansion of the repetitive sequences between genes, duplication of chromosome segments (both repetitive sequences and genes) within a genome or duplication of the whole genome. Most mammalian genomes are of a similar size; however, the genomes of the diploids rice and barley differ in size >10-fold, with rice possessing one of the smallest cereal genomes. More than 50% of angiosperms are polyploid as the result of either multiplication of a basic set of chromosomes (autopolyploidy) or combination of unrelated genomes (allopolyploidy). Pairing between chromosomes possessing similar gene orders needs to be restricted in polyploids in order to allow proper segregation. This can be achieved by two approaches. First, the parental chromosomes are already structurally distinct or become distinct by rearranging themselves rapidly on hybridization. Secondly, there is a modification to the process of chromosome assortment and pairing. Chromosome evolution and the organization of polyploids are likely to be a reflection of the pairing control used. Genome evolution has evolved around whole genome duplication followed by diploidizing events (Aparicio, 1998).

2. Chromosome evolution

The closest relatives of the cereals (Gramineae), the rushes and sedges (Juncaceae and Cypericeae), have holocentric/polycentric chromosomes possessing many sites along their chromosome length for microtubule attachment. Such chromosomes naturally fragment to create smaller viable chromosomes (Malheiros-Gards & Garde, 1950). In hybrids between parental lines carrying the original larger chromosomes and these small fragmentated chromosomes, the large and small chromosomes align (Nordenskiold, 1961), suggesting that species are very adept at being able to assort and correctly pair their chromosomes.

The ancestors of rice, maize/sorghum and barley/wheat speciated some 60 million years ago. Their genomes are now markedly different in size and are represented by different numbers of chromosomes. Initial comparative mapping of rice, maize and wheat genomes suggested that genes on the rice genetic map could be divided into sets and that the genetic maps of wheat and maize could be described by the same sets of genes (rice linkage segments) (Moore *et al.*, 1995a). It was apparent that this analysis could be extended to include the sorghum, foxtail millet and sugarcane genomes (Moore *et al.*, 1995b). Thus, a series of sets of genes could be used to describe most of the major cereal genomes.

This analysis revealed the tetraploid structure of the maize genome for the first time and the diploid nature of the sorghum genome. The ten chromosomes of maize could be divided into two parental sets of five chromosomes. These two sets of five maize chromosomes possess different arrangements of the linkage segments and, moreover, one set contains the largest and the other the smallest chromosomes. This implied that one of the parents on hybridization possessed larger chromosomes than the other.

Since this analysis, many comparative mapping studies have been undertaken (Van Deynze *et al.*, 1995a,b; Dufour, 1996; Dufour *et al.*, 1996; Saghai Maroof *et al.*, 1996; Van Deynze *et al.*, 1996; Dufour *et al.*, 1997; Devos *et al.*, 1998; Wilson *et al.*, 1999; Devos *et al.*, 2000; also reviewed in Moore, 1995). In essence, none of these studies changes the basic framework initially proposed. However, the more detailed analyses do indicate that, for example, there have been inversions of regions in the maize genome (Dufour, 1996; Wilson *et al.*, 1999) and that some of the linkage segments can be subdivided further (Wilson *et al.*, 1999).

At the gross level, rice linkage segments can be used to reveal the general gene content of the regions of different cereal chromosomes and provide a framework for analysing genome evolution. Individual genes contained within these segments can undergo amplification and deletion during evolution. This is clearly observed when comparing the location of disease resistance genes in the different cereal genomes (Leister *et al.*, 1998). Despite their flanking markers showing extensive conservation of gene order, many resistance genes have been either amplified or deleted. Analyses of the gene order over regions of several megabases in a number of cereal genomes reveal not only conservation of gene order at this level, but also disruption (Bennetzen *et al.*, 1995; Dunford *et al.*, 1995; Kilian *et al.*, 1995; Chen *et al.*, 1997). The level of conservation of gene order depends on which regions are chosen.

The collation of all the information generated from studies on the different cereals requires a simple method for display. A limited number of combinations of linkage segments are found in the genomes of the various cereals so far studied. A generalized genome structure using the rice linkage segments can be formed which, when cleaved at a different number of rice linkage segment junctions, produces combinations of segments that describe the chromosome structures found in the two sets of maize chromosomes, the wheat and barley chromosomes, and the sorghum and millet chromosomes (Moore *et al.*, 1995a,b).

The breakage and fusion of rice linkage segments is reminiscent of the evolution of rush and sedge holocentric chromosomes (Moore *et al.*, 1997). Comparison of the location of centromeric sites in the rice, wheat and maize genomes indicates a conservation of the location of sites within their genomes. The borders of the linkage segments are defined by centromeric and telomeric sites (Moore *et al.*, 1997a; Wilson *et al.*, 1999). In fact, telomeric heterochromatin in maize, rye, wheat and Bromus (a grass genus) can, under certain conditions, function as neocentromeres (reviewed in Moore *et al.*, 1997a).

The reason that the rice genome reveals the basic structure of the other genomes is that it has the greatest number of centromere (12) and telomere (24) sites in the species studied. The eventual number of linkage segments identified from comparing cereal and grass genomes is likely to reflect the maximum number of chromosomes possessed by a diploid species in such comparisons. The conservation of the location of centromeric sites in the cereal genomes implies that these sites might possess conserved sequences.

A family of conserved sequences (CCS1) located at the centromeres of all rice, maize, wheat, barley and rye chromosomes has been identified (Aragon-Alcaide *et al.*, 1996). Certain members of this family also detect the telomeric heterochromatin (Aragon-Alcaide *et al.*, 1996). In fact, CCS1 is not the only sequence described as being part of the cereal centromeres. Using *in situ* hybridization, Jiang and colleagues have described a probe that detected the centromeres of cereals (Jiang *et al.*, 1996). The stringency of this *in situ* hybridization suggests that this sequence has diverged more significantly than CCS1 in the genomes of these species. Later studies have indicated that both sequences are part of the same unit – a retrotransposon (Ananiev *et al.*, 1998; Miller *et al.*, 1998; Presting *et al.*, 1998). This family of retrotransposons is specifically located in cereal centromeres. Satellite sequences have been described at the cereal centromeres (Dong *et al.*, 1998) and it is, therefore, likely that retrotransposons inserting into such sequences would be rapidly deleted and so this family of retrotransposons could well be highly active.

Breakage and fusion at the centromeric and telomeric sites (either active or ancient) during chromosome evolution has resulted in a loss of marker synteny (Moore *et al.*, 1997a; Wilson *et al.*, 1999). Rice linkage segments in one maize parental chromosome set are often inverted with respect to those in sorghum and to the other parental chromosome set (Wilson *et al.*, 1999). Many of the breakpoints for these inversions involve functional centromeres or sites that align with centromeres in other cereal genomes (Wilson *et al.*, 1999). Even "ancient" centromere sites may still have a role in genome evolution. The two parental chromosome sets of maize had major structural differences either prior to hybridization or as a result of hybridization. In common with maize, the genetic mapping of the polyploids, oilseed rape (*Brassica napus*) (Parkin *et al.*, 1995) and cotton (*Gossypium barbadense*) (Reinisch *et al.*, 1994), demonstrates that homology between their parental chromosomes is restricted mainly to chromosome segments.

Comparative mapping of the Brassica and Arabidopsis genomes reveals that the Brassica genome has also evolved through breakage and fusion at centromere and telomere sites (Lagercrantz & Lydiate, 1996; Lagercrantz, 1998). There is no reason to suppose that the process of chromosome pairing in maize, brassica and cotton will be different. Despite the two parental sets of maize chromosomes apparently being distinct (a difference in chromosome size), it is also apparent that, on hybridization, chromosome pairing and recombination occurred between the parental sets. Each maize chromosome now represents a mixture of genes derived from both parents (Gaut & Doebley, 1997). Genetic mapping of wheat reveals that the gene order and recombination interval is similar between the homoeologous chromosomes (Devos & Gale, 1992).

In contrast to cotton, maize and oilseed rape, there were no major rearrangements of three wheat genomes after hybridization. Similarly, the genetic map of tetraploid rice reveals no extensive rearrangements of the two parental chromosomes (Huang & Kochert, 1994). This suggests that both hexaploid/tetraploid wheats and tetraploid rice possess mechanisms for discriminating between parental chromosomes.

3. Chromosome pairing

Comparative cereal genome studies reveal that centromeres and telomeres have been important for cereal genome evolution. Are they important for chromosome pairing? Homologous chromosomes start in a premeiotic nucleus randomly organized with respect to each other, but end up during the pachytene stage of meiotic prophase in close association. The process of bringing the chromosomes together (chromosome pairing) involves reciprocal recognition, coalignment and synapsis.

Recent developments in imaging have enabled this process to be studied 3-dimensionally in humans, maize and hexaploid wheat, providing a comparison with studies of *Saccharomyces cerevisiae* (budding yeast) and *Schizosaccharomyces pombe* (fission yeast). Essentially, in both the yeasts, the centromeres are clustered at one pole (spindle pole body) and the telomeres are spread around the other pole of the nucleus (Chikashige *et al.*, 1997; Jin *et al.*, 1998). The chromosomes are arranged in a "Rabl" configuration. In both budding and fission yeast during meiosis, the centromeres decluster from the spindle pole body as the telomeres cluster to the spindle pole body (Jin *et al.*, 1998; Trelles-Sticken *et al.*, 1999). This implies a major reorganization of chromosomes at this stage. While the telomeres are clustered, the rest of the chromosomes intimately align. There is some discussion within the yeast meiosis field as to whether premeiotic association of homologues occurs in budding yeast (Loidl *et al.*, 1994; Weiner & Kleckner, 1994). Initial studies had reported between 30–50% association of homologues premeiotically. However, further analysis has questioned such a high level of association (Loidl, 1998). Unfortunately, budding yeast does not lend itself to 3-dimensional studies using confocal microscopy and, therefore, the studies are based on squashed preparations.

Although the telomeres are dispersed around the premeiotic nucleus of humans and maize, as they are in fission and budding yeasts, the centromeres are not clustered at the opposite pole, but dispersed (Dawe *et al.*, 1994; Scherthan *et al.*, 1996; Bass *et al.*, 1997; Scherthan *et al.*, 1998). There is also no premeiotic association of homologues. However, during the leptotene and zygotene transition stages of meiotic prophase, human and maize telomeres cluster to form a bouquet (Bass *et al.*, 1997; Scherthan *et al.*, 1998). This is suggested to be the first stage of homologue pairing. The homologues then intimately associate along the rest of their length while the telomeres are still clustered.

In hexaploid wheat, barley and rye, the chromosomes are organized differently from those of human and maize, possessing a Rabl configuration with the centromeres grouped at one pole of the nucleus and the telomeres dispersed around the other pole (Abranches *et al.*, 1998). In hexaploid wheat during floral development, the centromeres associate in pairs (Aragon-Alcaide *et al.*, 1997a, 1998; Martinez-Perez *et al.*, 1999). These pairs are initially non-homologous in nature, but later become homologous in the meiocyte and tapetal precursors. Just prior to meiotic prophase, the homologues associate along their length, but the telomeres do not pair. At the onset of meiotic prophase, the telomere bouquet is formed (Martinez-Perez *et al.*, 1999). By the time the telomeres cluster, the chromosomes have replicated and both sister chromatids and homologues are often visibly separated along their length. However, the homologues remain associated by their centromeres and telomeres. The sister chromatids and then the homologues associate. Chromosomes, therefore, are intimately associated while in a Rabl configuration in contrast to those in fission and budding yeast, humans and maize. Moreover, it has been shown very recently that polyploidization itself induces early centromere association during floral development. Thus, polyploid relatives of wheat also exhibit centromere association during floral development. This contrasts with the diploid progenitors of these polyploids which associate both their centromeres and telomeres during meiotic prophase (Martinez-Perez *et al.*, 2000).

Hexaploid wheat ($2n = 42$) is composed of three ancestral genomes or three sets of seven chromosomes. Although each chromosome of hexaploid wheat has the potential to pair with either its homologue or the four homoeologues from the other two genomes, chromosome pairing is largely restricted to homologues. Thus, hexaploid wheat behaves as a diploid with 21 bivalents observed at metaphase 1. The major locus (*Ph1*) controlling this behaviour is located on chromosome 5B (Riley & Chapman, 1958; Sears, 1976). Deletion of this locus results in pairing between homoeologous chromosomes (Sears, 1976). Analysis of homologue pairing in a mutant carrying such a deletion reveals that, although centromeres associate in pairs premeiotically, these pairs do not later become homologous during floral development. The telomere cluster occurs later in the mutant when the centromeres have dispersed from the nuclear pole, at a similar stage to the occurrence of the telomere bouquet in humans and maize (Aragon-Alcaide *et al.*, 1997a; Martinez-Perez *et al.*, 1999). The chromosomes have lost the Rabl configuration when intimate homologue association occurs in the mutant. Thus, the presence of the *Ph1* locus results in the early association of homologues via centromeres during floral development and of telomere regions at the onset of meiotic prophase. In fact, the *Ph1* locus affects the structure of centromeres (Aragon-Alcaide *et al.*, 1997b). Studies have also shown that the locus eliminates recombination between homoeologous chromosomes or segments (Dubcousky *et al.*, 1995; Luo *et al.*, 1996). As yet, it is unclear whether all these effects are controlled by a single gene or by two or more genes affecting different processes. The characterization of the locus will reveal which is correct (Foote *et al.*, 1997; Roberts *et al.*, 1999).

In conclusion, it is clear that centromeres and telomeres have been important in influencing cereal chromosome evolution and are important for the pairing of cereal chromosomes.

Acknowledgements

The author wishes to thank the following for their work in this area: L. Aragon-Alcaide, A. Bevan, C. Dalgliesh, T. Foote, E. Martinez-Perez, T. Miller, S. Reader, M. Roberts, P. Shaw and J. Snape.

REFERENCES

Abranches, R., Beven, A.F., Aragon-Alcaide, L. & Shaw, P. (1998) Transcription sites are not correlated with chromosome domains in wheat nuclei. *Journal of Cellular Biology* 143: 5–12

Ananiev, E.V., Phillips, R.L. & Rines, H.W. (1998) Chromosome-specific molecular organisation of maize (*Zea mays* L.) centromeric regions. *Proceedings of the National Academy of Sciences USA* 95: 13073–78

Aparicio, S. (1998) Exploding vertebrate genomes. *Nature Genetics* 18: 301–03

Aragon-Alcaide, L., Miller, T.E., Schwarzacher, T., Reader, S.M. & Moore, G. (1996) A cereal centromeric sequence. *Chromosoma* 105: 261–68

Aragon-Alcaide, L., Reader, S.M., Shaw, P.J. *et al.* (1997a) Association of homologous chromosomes during floral development. *Current Biology* 7: 905–08

Aragon-Alcaide, L., Reader, S., Miller, T. & Moore, G. (1997b) Centromeric behaviour in wheat with high and low homoeologous chromosomal pairing. *Chromosoma* 106: 327–33

Aragon-Alcaide, L., Bevan, A., Moore, G. & Shaw, P. (1998) The use of vibratome sections of cereal spikelets to study anther development and meiosis. *Plant Journal* 14: 503–08

Bass, M.W., Marshall, W.F., Sedat, J.W., Agard, D.A. & Cande, W.Z. (1997) Telomere clusters de novo before the initiation of synapsis, a three dimensional spatial analysis of the telomere positions before and during meiotic prophase. *Journal of Cell Biology* 137: 5–18

Bennetzen, J.L., SanMiguel, P., Liu, C.-N. *et al.* (1995) Microcolinearity and segmental duplication in the evolution of grass nuclear genomes. In *Unifying Plant Genomes*, edited by J. Heslop-Harrison, Cambridge: Company of Biologists

Chen, M., SanMiguel, P., De Oliveira, A.C. *et al.* (1997) Microcolinearity in sh2-homologous regions of maize, rice and sorghum genomes. *Proceedings of the National Academy of Sciences USA* 94: 3431–35

Chikashige, Y., Ding, D.Q., Imai, Y. *et al.* (1997) Meiotic nuclear reorganisation: switching the position of centromeres and telomeres in the fission yeast *Schizosaccharomyces pombe*. *EMBO Journal* 16: 193–202

Dawe, R.K., Sedat, J.W., Agard, D.A. & Cande, W.Z. (1994) Meiotic chromosome pairing in maize is associated with a novel chromatin organisation. *Cell* 76: 901–12

Devos, K.M. & Gale, M.D. (1992) The genetic maps of wheat and their potential in plant breeding. *Outlook on Agriculture* 22: 93–99

Devos, K.M., Wang, Z.M., Beales, J., Sasaki, T. & Gale, M.D. (1998) Comparative genetic maps of foxtail millet (*Setaria italica*) and rice (*Oryza sativa*). *Theoretical and Applied Genetics* 96: 63–68

Devos, K.M., Pittaway, T.S., Reynolds, A. & Gale, M.D. (2000) Comparative mapping reveals a complex relationship between the pearl millet genome and those of foxtail millet and rice. *Theoretical and Applied Genetics* 100(2): 184–89

Dong, F., Miller, J.T., Jackson, S.A. *et al.* (1998) Rice centromeric regions consist of complex DNA. *Proceedings of the National Academy of Sciences USA* 95: 8135–40

Dubcousky, J., Luo, M.C. & Dvorak, J. (1995) Differentiation between homoeologous chromosomes 1A of wheat and 1A (M) of *Triticum monococcum* and its recognition by the wheat *Ph1* locus. *Proceedings of the National Academy of Sciences USA* 92: 6645–47

Dufour, P. (1996) Cartographie moleculaire du genome du sorgho (*Sorghum bicolor* L. Moench): application en selection varietale: cartographie compare chez les andropogonees. PhD Thesis, Université de Paris Sud, France

Dufour, P., Grivet, L., D'Hont, A. *et al.* (1996) Comparative genetic mapping between duplicated segments on maize chromosomes 3 and 8 and homoeologous regions in sorghum and sugarcane. *Theoretical and Applied Genetics* 92: 1024–30

Dufour, P., Due, M., Grivet, L. *et al.* (1997) Construction of a composite sorghum map and comparative mapping with sugarcane, a related complex polyploid. *Theoretical and Applied Genetics* 94: 409–18

Dunford, R.P., Kurata, N., Laurie, D.A. *et al.* (1995) Conservation of fine-scale DNA marker order in the genome of rice and the Triticeae. *Nucleic Acids Research* 23: 2724–28

Foote, T., Roberts, M., Kurata, N., Sasaki, T. & Moore, G. (1997) Detailed comparative mapping of cereal chromosome regions corresponding to the *Ph1* locus in wheat. *Genetics* 147: 801–07

Gaut, B.S. & Doebley, J.F. (1997) DNA sequence evidence for the segmental allotetraploid origin of maize. *Proceedings of the National Academy of Sciences USA* 94: 6809–14

Huang, H.C. & Kochert, G. (1994) Comparative RFLP mapping of allotetraploid wild rice species (*Oryza latifolia*) and cultivated rice (*Oryza sativa*). *Plant Molecular Biology* 25: 633–48

Jiang, J., Nasuda, S., Dong, F. *et al.* (1996) A conserved repetitive DNA element located in the centromeres of cereal chromosomes. *Proceedings of the National Academy of Sciences USA* 93: 14210–13

Jin, Q.U., Trelles-Sticken, E., Scherthan, H. & Loidl, J. (1998) Yeast nuclei display prominent centromere clustering that is reduced in non-dividing cells and in meiotic prophase. *Journal of Cellular Biology* 141: 21–29

Kilian, A., Kudma, D.A., Kleinhofs, A. *et al.* (1995) Rice–barley synteny and its application to saturation mapping of barley Rpg1 region. *Nucleic Acids Research* 23: 2729–33

Lagercrantz, U. (1998) Comparative mapping between *Arabidopsis thaliana* and *Brassica nigra* indicates that Brassica genomes have evolved through extensive genome replication accompanied by chromosome fusions and frequent rearrangements. *Genetics* 150: 1217–28

Lagercrantz, U. & Lydiate, D.J. (1996) Comparative genome mapping in the Brassica. *Genetics* 144: 1903–10

Leister, D., Kurth, J., Laurie, D.A. *et al.* (1998) Rapid reorganisation of resistance gene homologues in cereal genomes. *Proceedings of the National Academy of Sciences USA* 95: 370–75

Loidl, L. (1998) Non-random interphase chromosome arrangement in bakers yeast and its impact on meiotic chromosome pairing. Meiosis and Recombination. The Society for Experimental Biology Annual Meeting, York

Loidl, J., Klein, F. & Scherthan, H. (1994) Homologous pairing is reduced but not abolished in asynatic mutants in yeast. *Journal of Cellular Biology* 125: 1191–200

Luo, M.C., Dubcousky, J. & Dvorak, J. (1996) Recognition of homeology by wheat *Ph1* locus. *Genetics* 144: 1195–203

Malheiros-Gards, N. & Garde, A. (1950) Fragmentation as a possible evolutionary process in the genus Luzula DC. *Genetica Iberica* 2: 257–62

Martinez-Perez, E., Shaw, P., Reader, S. *et al.* (1999) Homologous chromosome pairing in wheat. *Journal of Cell Science* 112: 1761–69

Martinez-Perez, E., Shaw, P.J. & Moore, G. (2000) Polyploidy induces centromere association. *Journal of Cell Biology* 148: 233–38

Miller, J.T., Dong, F., Jackson, S.A., Song, J. & Jiang, J. (1998) Retrotransposon-related DNA sequences in the centromeres of grass chromosomes. *Genetics* 150: 1615–23

Moore, G. (1995) Cereal genome evolution: pastoral pursuits with lego genomes. *Current Opinion in Genetics and Development* 5: 717–24

Moore, G., Foote, T., Helentjaris, T. *et al.* (1995a) Was there a single ancestral cereal chromosome? *Trends in Genetics* 11: 81–82

Moore, G., Devos, K.M., Wang, Z. & Gale, M.D. (1995b) Cereal genome evolution, grasses, line up and form a circle. *Current Biology* 5: 737–39

Moore, G., Roberts, M., Aragon-Alcaide, L. & Foote, T. (1997a) Centromeric sites and cereal chromosome evolution. *Chromosoma* 35: 17–23

Moore, G., Aragon-Alcaide, L., Roberts, M. *et al.* (1997b) Are rice chromosomes components of a holocentric chromosome ancestor? *Plant Molecular Biology* 35: 17–23

Nadeau, J.H. & Sankoff, D. (1998) Counting on comparative maps. *Trends in Genetics* 14: 495–99

Nordenskiold, H. (1961) Tetrad analysis and course of meiosis in three hybrids of *Luzula campestris. Hereditas* 47: 203–38

Parkin, I.A.P., Sharpe, A.G., Keith, D.J. & Lydiate, D.J. (1995) Identification of the A and C genomes of amphidiploid *Brassica napus* (Oilseed rape). *Genome* 38: 1122–31

Presting, G.G., Malysheva, L., Fuchs, J. & Schubert, I. (1998) A Ty3/Gypsy retrotransposon-like sequence localises to the centromeric regions of cereal chromosomes. *Plant Journal* 16: 721–28

Reinisch, A.J., Dong, J., Brubaker, C.L. *et al.* (1994) A detailed RFLP map of cotton, *Gossypium hirsutum* × *Gossypium barbadense*: chromosome organisation and evolution in a disomic polyploid genome. *Genetics* 138: 829–47

Riley, R. & Chapman, V. (1958) Genetic control of the cytologically diploid behaviour of hexaploid wheat. *Nature* 182: 713–15

Roberts, M.A., Reader, S.M., Dalgliesh, C. *et al.* (1999) Induction and characterisation of Ph1 wheat mutants. *Genetics* 153: 1909–18

Saghai Maroof, M.A., Yang, G.P., Biyashev, R.M., Maughan, P.J. & Zhang, Q. (1996) Analysis of barley and rice genomes by comparative RFLP linkage mapping. *Theoretical and Applied Genetics* 92: 541–51

Scherthan, H., Weich, S., Schweigler, H. *et al.* (1996) Centromere and telomere movements during meiotic prophase of mouse and man are associated with the onset of chromosome pairing. *Journal of Cellular Biology* 134: 1109–25

Scherthan, H., Eils, R., Trelles-Sticken, E. *et al.* (1998) Aspects of three-dimensional chromosome reorganisation during the onset of human male meiotic prophase. *Journal of Cell Science* 111: 2337–51

Sears, E.R. (1976) Genetic control of chromosome pairing in wheat. *Annual Reviews of Genetics* 10: 31–51

Trelles-Sticken, E., Loidl, J. & Scherthan, H. (1999) Bouquet formation in budding yeast: initiation of recombination is not required for meiotic telomere clustering. *Journal of Cell Science* 112: 651–58

Van Deynze, A.E., Dubcousky, J., Gill, K.S. *et al.* (1995a) Molecular-genetic maps for group 1 chromosomes of Triticeae species and their relation to chromosomes in rice and oat. *Genome* 38: 45–59

Van Deynze, A.E., Nelson, J.C., Yglesias, E.S. *et al.* (1995b) Comparative mapping in grasses:wheat relationships. *Molecular and General Genetics* 248: 744–54

Van Deynze, A.E., Nelson, J.C., O'Donoughue, L.S. *et al.* (1996) Comparative mapping in grass:oats relationships. *Molecular and General Genetics* 249: 349–56

Weiner, B. & Kleckner, N. (1994) Chromosome pairing via multiple interactions before and during meiosis in yeast. *Cell* 77: 977–91

Wilson, W.A., Harrington, S.E., Woodman, W.L. *et al.* (1999) Can we infer the genome structure of the progenitor maize through comparative analysis of rice, maize and the domesticated panicoids? *Genetics* 153: 453–73

GLOSSARY

dicot genome genome of dicotyledenous plants, i.e. having two seed leaves (cotyledons)

holocentric chromosome chromosome possessing a diffuse rather than a discrete centromere; when fragmented each part of the chromosome acts at mitosis as though it possesses a centromere

homoeologous partly homologous; applies to sets of chromosomes in a polyploid species similar (but not identical) to other chromosome sets in the genome

polycentric chromosome chromosome possessing several centromeres

synteny when genetic loci lie on the same chromosome in different species

premeiotic nucleus an interphase nucleus prior to meiotic DNA replication

spindle pole body found at the poles of the mitotic spindles of many fungi, this structure acts as a centromere in organising the spindle microtubules

zygotene stage of prophase of meiosis at which homologous chromosomes pair

PINPOINTING EINKORN WHEAT DOMESTICATION BY DNA FINGERPRINTING

Manfred Heun[1], Basilio Borghi[2] and Francesco Salamini[3]

[1]*Institut Biologi & Nàturforvaltning, Norges Landbrukshøgskole, AAs-NLH, Norway*
[2]*Istituto Agrario Di San Michele all'Adige, Italy*
[3]*Max-Planck-Institut Züchtungsforschung, Cologne, Germany*

1. Establishment of agriculture

The emergence of agriculture is the central event of the Neolithic revolution. Around 10 000 years ago, the population size of hunter–gatherer societies reached saturation. The availability of food resources – wild animals and plants – was the major factor limiting further significant population increase (Cavalli-Sforza, 1996). This limitation, combined with the change in vegetation resulting from the warming trend that ended the last glacial epoch, stimulated the search for increased food supplies. The solution adopted was to domesticate wild plants and animals. How often this idea has arisen and how it was realized in different parts of the world can only be a matter of speculation, but agriculture was so successful in providing large and stable food supplies that a significant increase in population size resulted. This, in turn, stimulated human migration and agriculture spread from its sites of origin.

The Fertile Crescent is the area where today's European agriculture began. This large geographical area extends from Khuzistan in south-west Iran, through the Zagros and Taurus Mountains of Iran, Iraq and Turkey, into Anatolia and south as far as Palestine. At several archaeological sites in this region, excavation levels containing only remains of wild plants are followed, with transitions, by horizons containing only remains derived from domesticated plants. From the early sites in south-west Asia, agriculture spread to Europe at a rate of about 1 km per year (Cavalli-Sforza, 1996). Wild relatives of the "founder" crops – chickpea, lentil, pea, barley, Emmer and einkorn wheats, bitter vetch (Zohary, 1996) – are still found today in the Fertile Crescent. The study of the genetic relationships between cultivated varieties and their wild relatives contributes to the clarification of important aspects of plant domestication. For example, by comparing cultivated lines with wild relatives collected in defined areas, we have been able to pinpoint precisely the place of origin of einkorn wheat within the Fertile Crescent, a puzzle which archaeology has been unable to solve.

2. A favourable crop species

Einkorn wheats are diploid species ($2n = 2x = 14$) belonging to the family Poaceae and carrying the A genome (see also article on "Cereal chromosome evolution and pairing"). This genome has played a role in the evolution of the polyploid species bread wheat (A, B and D genomes, $2n = 6x = 42$) and durum wheat (A and B genomes, $2n = 2x = 28$). *Triticum urartu* is assumed to be the source of the A genome of the cultivated wheats mentioned above, but is reproductively isolated from other einkorn wheats and, thus, will not be discussed further here. We focus, instead, on the domestication process by considering two other einkorn wheats: *T. monococcum* ssp. *monococcum* (*T. monococcum*) and *T. monococcum* ssp. *boeoticum* (*T. boeoticum*). These are the scientific names of the domesticated and the wild einkorn wheat, respectively. *T. monococcum* ssp. *aegilopoides* (*T. aegilopoides*) is a further einkorn wheat that is fully fertile in crosses with both of the other subspecies. This taxon occurs wild mainly in the Balkans and is of interest because it shows domestication traits similar to those of *T. monococcum*.

The domesticated einkorn wheat is a forgotten crop, seldom grown in modern agricultural settings. Einkorn wheat has one seed per spikelet, a character that limits its yield potential. In fact, the importance of einkorn has steadily decreased since the Bronze Age. However, although playing only a marginal role, einkorn wheat does continue to be cultivated – traits like good cold tolerance and high protein and carotene contents may be the underlying reasons. Today, einkorn still grows wild in the Fertile Crescent and in the Balkans, while cultivated strains are all landraces to which modern plant breeding methods have not been applied. Einkorn wheat appears ideally suited to pinpointing the site(s) of its first domestication, based on DNA fingerprinting of wild and cultivated genotypes.

3. A representative collection of einkorn lines

Ten gene banks worldwide (for details see Heun *et al.*, 1997) provided einkorn wheat samples for the study. The 1362 samples were grown in fields in Italy and Germany to verify their taxonomic assignment and to evaluate morphological differences between domesticated and wild lines. The collection sites for about 900 samples were also provided by the seed suppliers. For the Fertile Crescent, as well as for most of the samples from Turkey, we have considered only lines for which the collection site was known to within about 5 km. Outside the primary habitat of wild einkorn, most lines are frequently known only by their country of origin. Moreover, since agriculture led to the spread of cultivated types, the consideration of their sites of collection could be misleading. The geographical distribution of the *T. boeoticum* and *T. aegilopoides* lines in our collection is in agreement with the distribution of wild einkorn published by Harlan and Zohary (1966). In their Figure 3, the primary habitats of *T. boeoticum* are shown to include the

Taurus–Zagros region from south-eastern Turkey through north-eastern Iraq into western Iran (i.e. the eastern half of the Fertile Crescent). *T. aegilopoides* grows wild mainly in the Balkans and Western Anatolia, where it occupies marginal habitats. In Central Anatolia and Transcaucasia, the two wild einkorns occur in marginal habitats together with cultivated einkorns (Harlan & Zohary, 1966). West of the Balkans, only cultivated einkorns occur.

The *T. boeoticum* samples collected in the Fertile Crescent were divided into nine geographical groups (A, B, C, D, E, G, H, I and L). All *T. aegilopoides* samples were included in the "Aegi" group and the cultivated einkorn in the "Mono" group. To test for the monophyletic origin of the cultivated types, this last group was also separated into four subgroups based on their geographical origin (Central Europe, the Balkans, Mediterranean countries and Turkey). **Figure 1** shows the sampling sites of the 338 einkorns used for DNA fingerprinting. To reduce our collection to this number, samples were chosen at random within the above mentioned 11 groups. Phenotypic scores of the 338 lines are presented below. It should be pointed out, however, that these are based on analyses carried out only after all lines collected had been grown for several years both in Italy and Germany.

4. Domestication traits

In **Table 1**, phenotypic traits showing significant differences between wild and domesticated einkorns are summarized. *T. boeoticum* has a brittle rachis (trait 6), very pronounced hairs on leaves (traits 1–3) and ear rachis (trait 4) and a two-seeded spikelet (trait 8, which correlates with trait 7). *T. monococcum* has a less brittle rachis, few and short hairs on leaves and rachis and single-seeded spikelets. On the basis of these traits, *T. aegilopoides* appears to be closely related to the cultivated einkorn. For this reason, and considering that *T. aegilopoides* grows in wild habitats, it was assumed in the past that *T. monococcum* derived from *T. aegilopoides*. Also, a trait like the number of spikelets per spike (**Table 2**, trait 4) – which is lower in *T. boeoticum* than in either of the other two taxa – indicates that *T. monococcum* and *T. aegilopoides* are very similar. However, when species comparisons are based on seed weight (**Table 2**, trait

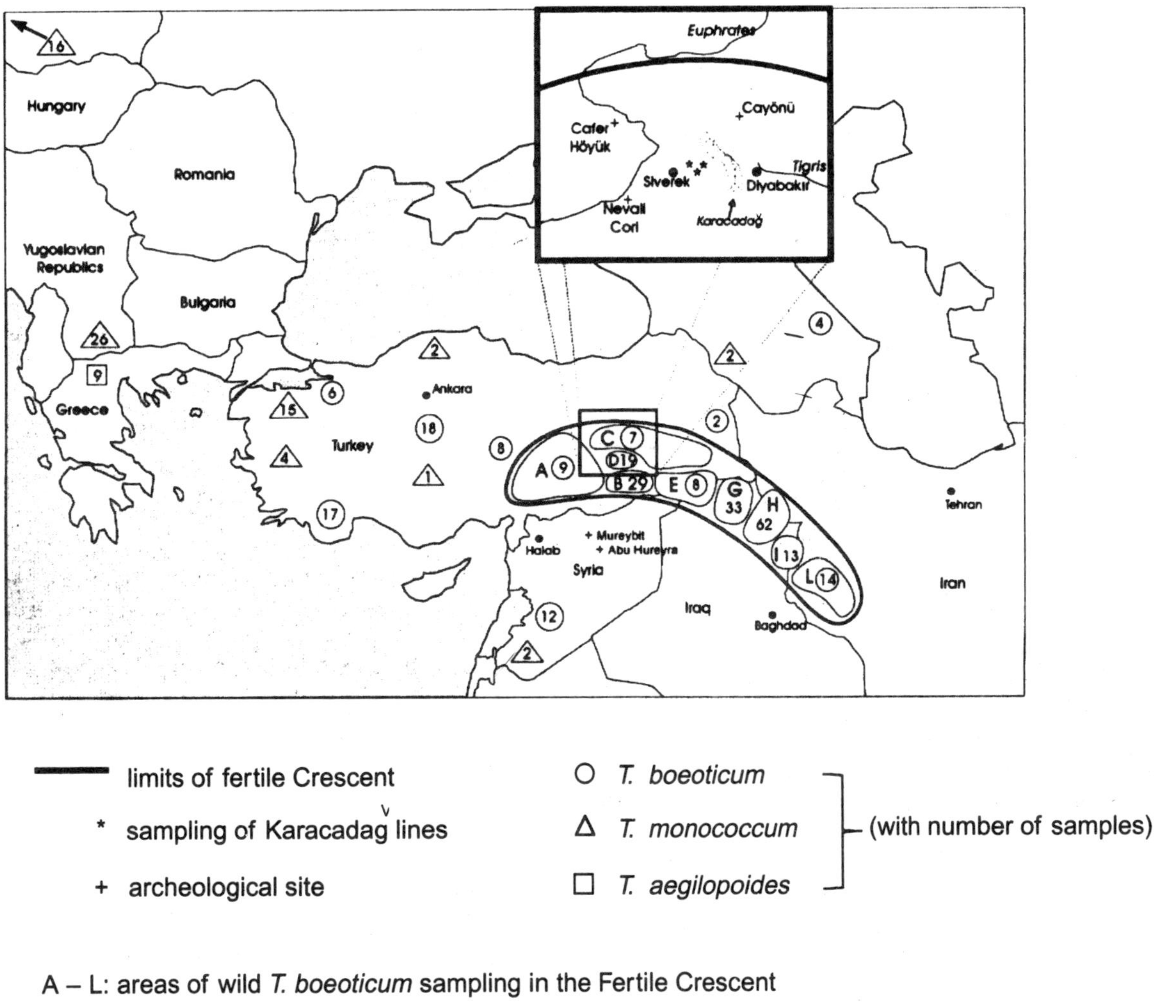

Figure 1 Sampling sites of 338 einkorn wheats. Insert: the Karacadag region. For the area of the Fertile Crescent, where einkorn occurs in primary habitats, nine groups were formed (for details see Heun *et al.*, 1997).

Table 1 Mean values (± standard error) of eight morphological traits of the three Einkorn wheats *T. boeoticum*, *T. monococcum* and *T. aegilopoides* and for 11 *T. boeoticum* originating from the Karacadag. Traits 1–4, 6b and 8 were recorded in 1994 in Köln (K) and 5, 6a and 7 in 1993 and 1994 in San Angelo (SA).

Species	*T. boeoticum*	*T. monococcum*	*T. aegilopoides*	*T. boeoticum Karacadag*
Number of lines	250	68	9	11
HLU[1] (K)	2.60 ± 0.02	0	0.42 ± 0.37	1.75 ± 0.08
HLL[2] (K)	1.75 ± 0.03	0	0.42 ± 0.30	1.33 ± 0.46
H1[3] (K)	5.03 ± 1.09	2.12 ± 0.18	3.00 ± 0.80	4.17 ± 0.26
HR[4] (K)	1.11 ± 0.03	0.27 ± 0.02	0.86 ± 0.13	0.88 ± 0.13
SD[5] (SA)	4.00 ± 0.02	6.80 ± 0.89	5.40 ± 0.60	4.00 ± 0
R[6a] (SA)	1.60 ± 0.05	0.03 ± 0.03	0	2.00 ± 0
R[6b] (K)	1.90 ± 0.02	0.30 ± 0.05	1.14 ± 0.29	2.00 ± 0
DA[7] (SA)	2.50 ± 0.05	1.00 ± 0.04	1.10 ± 0.14	2.33 ± 0.14
SS[8] (K)	0.2 ± 0.03	1.8 ± 1.21	0.7 ± 1.25	0

[1] HLU, hair on upper leaf (0 = absent; 1 = short; 2 = long; 3 = very long)
[2] HLL, hair on the lower leaf (0 = absent; 1 = short; 2 = long; 3 = very long)
[3] H1, hairs on the first 1 cm of the leaf border (1 = absent; 3 = weak presence; 5 = intermediate; 7 = dense; 9 = very dense)
[4] HR, hair length on the ear rachis (mm)
[5] SD, spike density (3 = very sparse; 5 = sparse; 7 = dense; 9 = very dense)
[6] R, rachis (0 = tough; 1 = brittle; 2 = very brittle)
[7] DA, awns per spikelet (0 = awnless; 1 = 1 awn well developed; 2 = 1 awn developed and a short one on the second floret; 3 = 2 awns equally well developed)
[8] SS, spikelet with 2 seeds (0 = 75–100%; 1 = 55–70%; 2 = 25–50%; 3 = 0–20%)

Table 2 Mean values (± standard error) of five agronomic traits recorded for *T. boeoticum*, *T. monococcum* and *T. aegilopoides* and for 11 *T. boeoticum* originating from the Karacadag.

Species	*T. boeoticum*	*T. monococcum*	*T. aegilopoides*	*T. boeoticum Karacadag*
Number of lines	250	68	9	11
GH[1] (SAII)	4.8 ± 0.11	7.00 ± 0.20	4.4 ± 1.00	5.3 ± 0.41
GH[1] (K)	2.5 ± 0.08	6.1 ± 0.19	3.8 ± 0.37	3.7 ± 0.13
HD[2] (SAI)	56.3 ± 0.18	64.7 ± 0.46	67.1 ± 2.54	55.8 ± 0.38
HD[2] (K)	67.3 ± 0.57	72.2 ± 0.50	73.0 ± 0.50	63.2 ± 0.11
PL[3] (SAI)	50.3 ± 0.54	39.0 ± 0.63	45.7 ± 2.38	47.4 ± 1.47
NS[4] (SAII)	21.1 ± 0.21	30.9 ± 0.57	31.6 ± 1.69	20.8 ± 0.82
SW[5] (SAII)	21.1 ± 0.24	30.2 ± 0.74	22.9 ± 0.71	21.5 ± 0.86
SW[5] (K)	20.3 ± 0.22	32.3 ± 4.80	25.0 ± 1.15	19.8 ± 1.19

[1] GH, growth habit (1 = prostrate; 9 = erect)
[2] HD, heading date (days from 1 April)
[3] PL, last internode length (in cm excluding the spike)
[4] NS, number of spikelets/spike
[5] SW, seed weight (mg)
SAI, characters measured in 1993 and 1994 in San Angelo, Italy; SAII, characters measured in 1994 in San Angelo, Italy; K, characters measured in 1994 in Köln, Germany.

5) or growth habit (**Table 2**, trait 1), *T. monococcum* appears unique, whereas *T. boeoticum* and *T. aegilopoides* cannot be differentiated. Flaksberger (1925) assumes that grouping beyond the subspecies level can be based on traits like seed and glume colour. In our case, however, such traits proved to be taxonomically irrelevant when the grouping of genotypes beyond the subspecies level was attempted. One exception was the case of four samples of free-threshing einkorn (*T. sinskajae*; Filatenko & Kurkiev, 1975), which were identical in several respects and which fell into the *T. monococcum* group. They have softer glumes, allowing the seed to be released naked during threshing. One gene with pleiotropic effects is responsible for the trait; *T. sinskajae* was, therefore, considered as a *T. monococcum* form. The phenotypic description of the 1362 lines present in the collection, undertaken in different years under different climatic conditions, allowed us reliably to distinguish *T. boeoticum* from *T. monococcum* and from *T. aegilopoides*. The differences between *T. monococcum* and *T. aegilopoides* are so small that they are quite difficult to distinguish on a single-line basis. Only when one compares groups of lines, does the difference become significant. We concluded that, beyond the subspecies level, any attempt to separate groups of genotypes based on morphological traits would not be worthwhile.

5. DNA fingerprinting

DNA-based information, which shows higher polymorphism rates than morphological characters, was needed to reveal the genetic relationships within and between the three subspecies. Amplified fragment length polymorphism (AFLP) markers were generated by a recently described high-volume DNA technique (Vos *et al.*, 1995). AFLP fragments are inherited in a dominant fashion and their polymorphisms cover the whole genome, as has been shown for barley (Becker *et al.*, 1995) which carries A genome homoeologous chromosomes. The AFLP procedure is, however, not technically simple and involves multiple reactions and careful interpretation of the X-ray films, so that various types of controls are needed. A total of 288 stable and reliably readable AFLPs were scored for presence *vs.* absence in the 338 lines analysed. Different genetic distance measures (e.g. NEI72 and NEI-UB) were used to construct several phylogenetic trees based on neighbour-joining and restricted maximum likelihood estimation methods. Almost identical topologies were detected by all methods employed. Finally, a consensus tree based on ten different tree-building procedures was obtained (for details see Heun *et al.*, 1997), which summarized all phylogenetic data.

6. Wild ancestors of cultivated einkorn

Figure 2A shows that the nine geographical groups of *T. boeoticum* collected in the Fertile Crescent can be distinguished genetically. Group D – originating from the Karacadag mountains, south-east Turkey – is the group most distant from all the others. Adding two further groups, the cultivated einkorns (Mono) and the wild einkorns from the Balkans (Aegi) to these nine groups, generates the results in **Figure 2B**. Cultivated einkorn appears closely related to *T. aegilopoides*. The Karacadag lines of group D link Mono and Aegi with the remaining eight groups. This result is a major achievement, since for the first time cultivated einkorns can be traced back to a locally restricted group of wild einkorns showing all the characteristics of a wild species, whereas the lines that grow wild in the Balkans show clear signs of domestication (see above). It is concluded that both *T. monococcum* and *T. aegilopoides* are derived from group D wheats. **Figure 2C** clearly demonstrates the monophyletic origin of the cultivated einkorn and strongly suggests that *T. aegilopoides* is a derivative of the cultivated forms. Group D is again positioned between *T. monococcum* and all other *T. boeoticum* forms. The second result that emerges is that all group D lines were collected from a relatively small area on the slopes of the Karacadag mountains. A gradient ranging from high to very high relationships within the 19 representatives of group D is evident (**Figure 2F**). This indicates that a more precise delineation of the site of origin of the domesticated types might be possible in the future. In **Figure 2D** and **2E** (see also **Plate 20** in colour plate section), the role of group D as the intermediate between other wild groups of *T. boeoticum* and the two subspecies that show domestication traits is confirmed.

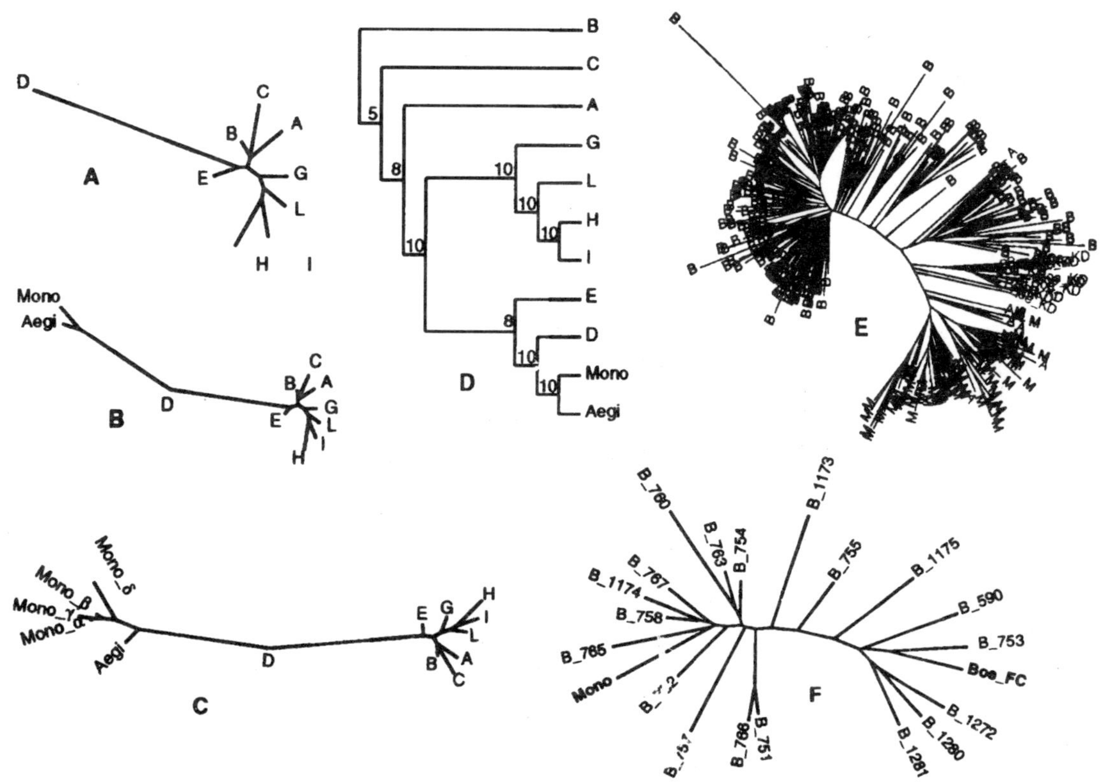

Figure 2 (see also **Plate 20** in colour plate section) (A), (B) and (C) Unrooted trees with the nine *T. boeoticum* groups alone, with the same nine groups plus *T. monococcum* (Mono) and *T. aegilopoides* (Aegi) and the result when splitting up the Mono group into four distinct subgroups. (D) Consensus tree summarizing the results with the nine *T. boeoticum* groups and the groups Mono and Aegi. (E) Unrooted tree with all fingerprinted lines: red, cultivated einkorns; green, *T. aegilopoides*; orange, *T. boeoticum* from the Karacadag; blue/purple, remaining *T. boeoticum*. (F) Unrooted tree for the 19 Karacadag lines aligned to one consensus genotype of the remaining *T. boeoticum* and one consensus genotype of the cultivated einkorn. For details on the tree building procedures etc., see Heun *et al.* (1997).

7. Connecting DNA studies with archaeology

The localization of the origin of cultivated einkorn to the Karacadag mountains stimulates questions concerning the human community which achieved this domestication: are there neighbouring human settlements with signs of early einkorn cultivation? Cafer Höyük, Nevali Cori and Cayönü are located in the vicinity of these mountains. They are among the oldest settlements where palaeontologists have found wild and domesticated einkorn seeds in different horizons. In Table 2 of Nesbitt and Samuel (1996), all archaeological data relevant to the origin of agriculture are summarized. From these, it becomes evident that farming of einkorn began between 7800 and 7500BC in the settlements cited. In the case of other excavated sites, such as Abu Hureyra and Mureybit in Northern Syria, wild seeds of *T. boeoticum* seem to have been collected elsewhere and transported there (van Zeist & Casparie, 1968). At the excavated sites in the Jordan valley mentioned by Jones *et al.* (1998), no remains of einkorn have been found (Heun *et al.*, 1998), emphasizing the importance of the northern Fertile Crescent in einkorn domestication (Nesbitt, 1998). In summary, the wild ancestors of cultivated einkorn are found in the Karacadag mountains. This, and the archaeological evidence from neighbouring excavations, imply that einkorn domestication was initiated there about 9500 years ago. The genetic data also indicate that the domestication event was monophyletic and that the cultivated lines differentiated, but only to a limited extent, during the spread of agriculture to western Europe. *T. aegilopoides* is most probably a feral form of the cultivated types that reached the Balkans as a result of the spread of agriculture.

8. Perspectives

DNA markers are powerful tools for assessing the relationships within species that show a limited range of phenotypic variation. Finding the ancestors of a cultivated crop that was separated from its wild counterparts 9500 years ago is only possible when the number of genes underlying the large phenotypic differences resulting from domestication is small or limited. If polygenic changes (such as mutations) in the genome were necessary, tightly linked DNA markers could have been changed, and due to hitchiking, the analysis would have become complicated. This could have made it impossible to find wild lines highly related to cultivated types. One must conclude that only a limited number of genes must be altered during the domestication process (e.g. see Doebley, 1992, for maize). Nevertheless, all the relevant einkorn traits now need to be mapped and the results must be placed within the context of a synteny-based analysis of the family Poaceae (Paterson *et al.*, 1995). The evolution of the Poaceae from a common ancestral genome has confirmed the phenomenon that the same genes in different species have undergone similar changes during domestication.

A different perspective concerns the sampling of genetic diversity to be stored in gene banks. From our data, it is evident that all cultivated einkorns are the result of one domestication event; when searching for additional genetic variation, breeders should now be more interested in using lines from wild stands different from the Karacadag lines. Crucial questions here are whether a limitation of genetic variability that may restrict future breeding progress exists, and whether particular genetic characteristics allowed the domestication of only the Karacadag lines. Present limitations to the genetic progress of this crop cannot be discussed, since modern breeding techniques have not yet been applied to this forgotten species. This introduces a third perspective: modern einkorn breeding programmes can now be carried out to enlarge today's crop diversity and to take advantage of the unique characteristics of this species. In this respect, the finding of good baking quality in some lines of einkorns, while on the one hand unexpected (Borghi *et al.*, 1996), on the other hand justifies the investment of resources in einkorn breeding. Finally, the study of the origin of other Fertile Crescent crops, if possible, should answer whether the Neolithic revolution in this part of the world had a common origin or whether the primary Mesopotamian crops were domesticated independently.

Acknowledgements

This article is based on Heun *et al.* (1997). The persons acknowledged therein provided seeds, data on collection sites and help.

REFERENCES

Becker, J., Kuiper, M., Vos, P., Salamini, F. & Heun, M. (1995) Combined mapping of AFLP and RFLP markers in barley. *Molecular and General Genetics* 249: 65–73

Borghi, B., Castagna, R., Corbellini, M., Heun, M. & Salamini, F. (1996) Breadmaking quality of Einkorn wheat (*T. monococcum* ssp. *monococcum*). *Cereal Chemistry* 73: 208–14

Cavalli-Sforza, L.L. (1996) The spread of agriculture and nomadic pastoralism: insights from genetics, linguistics and archaeology. In *The Origins and Spread of Agriculture and Nomadic Pastoralism in Eurasia*, edited by D.R. Harris, Washington, DC: Smithsonian Institution Press

Doebley, J. (1992) Mapping the genes that made maize. *Trends in Genetics* 8: 302–07

Filatenko, A.A. & Kurkiev, U.K. (1975) A new species: *Triticum sinskajae* A. *Genetike i Selektsii* 54: 239–41

Flaksberger, C.A. (1925) Pshenitsa odnozernyahki (wheat-einkorn). *Bulletin of Applied Botany and Plant Breeding (Trudy po Prikladnoi Botanike i Selektsii)* 15: 207–27

Harlan, J.R. & Zohary, D. (1966) Distribution of wild wheats and barley. *Science* 153: 1074–80

Heun, M., Schäfer-Pregl, R., Klawan, D. *et al.* (1997) Site of einkorn wheat domestication identified by DNA fingerprinting. *Science* 278: 1312–14

Heun, M., Borghi, B. & Salamini, F. (1998) Wheat domestication; response. *Science* 279: 303–04

Jones, M.K., Allaby, R.G. & Brown, T.A. (1998) Wheat domestication. *Science* 279: 302–03

Nesbitt, M. (1998) Where was einkorn wheat domesticated? *Trends in Plant Science* 3: 82–83

Nesbitt, M. & Samuel, D. (1996) From staple to extinction? The archaeology and history of hulled wheats. In *Hulled Wheats:*

Proceedings of the First International Workshop on Hulled Wheats, edited by S. Padulosi, K. Hammer & J. Heller, Rome: IPGRI Press

Paterson, A.H., Lin, Y.R., Li, Z. *et al.* (1995) Convergent domestication of cereal crops by independent mutations at corresponding loci. *Science* 269: 1714–18

van Zeist, W. & Casparie, W.A. (1968) Wild einkorn wheat and barley from Tell Mureybit in northern Syria. *Acta Botanica Neerlandica* 17: 44–53

Vos, P., Hogers, R., Bleeker, M. *et al.* (1995) AFLP: a new technique for DNA fingerprinting. *Nucleic Acids Research* 23: 4407–14

Zohary, D. (1996) The mode of domestication of the founder crops of Southwest Asian agriculture. In *The Origins and Spread of Agriculture and Nomadic Pastoralism in Eurasia*, edited by D.R. Harris, Washington, DC: Smithsonian Institution Press, and London: UCL Press

Note added in proof.

Lev-Yadun *et al.* report in *Science* 288: 1602–03 (2000) on the overlapping of the geographical distribution of the seven Neolithic founder crops in the Fertile Crescent. From their figure entitled "East of Eden" it becomes evident that the pinpointing of einkorn wheat described here is exactly in the centre of that overlapping.

GLOSSARY

amplified fragment length polymorphism (AFLP) a molecular marker generated by amplifying restriction fragments with PCR

rachis main axis of a compound leaf or flower cluster

spikelet the smallest unit of florets or flowers in the grass flower cluster, consisting of outer and inner bracts called glumes, and one or more florets

synteny conserved gene order in different species

See also **Cereal chromosome evolution and pairing** (p.629); **Origin and evolution of modern maize** (p.647); **Origin of the apple** (p.674)

BARBARA McCLINTOCK AND TRANSPOSABLE GENETIC SEQUENCES IN MAIZE (*ZEA MAYS*)

John R.S. Fincham
Institute of Cell and Molecular Biology, University of Edinburgh, Edinburgh, UK

1. Introduction

Barbara McClintock was the discoverer of moveable genetic elements – "jumping genes" as they have sometimes been called. Her first published paper on the subject appeared in the *Proceedings of the National Academy of Sciences* in 1950 (McClintock, 1950), and she contributed a further description of these phenomena to the *Cold Spring Harbor Symposium* of 1951 (McClintock, 1951). These papers presented in very compressed form the results of several years' very intensive work at the Cold Spring Harbor Laboratory, and their dense style combined with the novelty of their content made them difficult for the genetics community to assimilate. Yet, McClintock enjoyed too high a reputation as a maize cytologist and geneticist for her claims to be readily discounted – among the contributions that she had already made was her proof (with Barbara Creighton) that genetic crossovers did indeed involve physical exchanges between homologous chromosomes. However, it took another 30 years, and studies by others at the DNA level, for McClintock's observations on transposable elements to be brought into the mainstream of genetic thinking.

2. The activator–dissociation (*Ac–Ds*) systems

Most of the maize genes studied by McClintock affected the endosperm of the seed, which reveals the constitution of the potential seedling within, while the seeds (kernels) are still on the cob. Of particular importance for her initial observations were the marker genes *C* and *Bz* (*bronze*) necessary for normal pigmentation, and *Wx* (*waxy*) and *Sh* (*shrunken*), necessary for normal starch development, all located on the short arm of chromosome 9.

The starting point for her work on transposable elements was her discovery of a genetic element, which she called *Dissociation* (*Ds*), initially located on that chromosome arm close to the centromere, which was the site of frequent chromosome breaks. The consequence was spots and sectors in the endosperm showing the effects of loss of chromosome fragments containing dominant alleles of the genes *C*, *Sh*, *Bz* and *Wx* (**Figure 1**). It should be noted that, where this article refers to variegation due to the gain or loss of function of a particular gene, the activity or inactivity of that gene was visible in the phenotype because it was in heterozygous combination with a stable recessive allele.

The second key observation was that the chromosome breaks due to *Ds* occurred only in the presence, somewhere else in the maize genome, of another element which McClintock called *Activator* (*Ac*) (**Figure 2**) (see also **Plate 19** in the colour plate section). The most extraordinary finding, however, was that in certain plants the loci of the breaks seemed to have been shifted to different positions, still on the same chromosome arm, but further from the centromere, so that a shorter chromosome fragment bearing fewer marker genes was lost. The location of *Ds* appeared to be stable in the absence of *Ac*, which thus seemed to activate both of the novel properties of *Ds*: its induction of breaks and its own movement.

The next step in McClintock's analysis was the demonstration that *Ac* was itself transposable. Its transposition was detected when it resulted in the inactivation of a gene at or closely adjacent to its new site of insertion, but with subsequent frequent reversion to gene activity as it made further moves. Transposition of *Ac* to the *Bz* locus, for example, resulted in a mutable allele $bz^{m\text{-}1}$ that, in heterozygous combination with stable recessive *bz*, gave palely pigmented ("bronze") kernels with fully pigmented spots and sectors. This is an example of a one-element system, with *Ac* acting autonomously, in contrast to the *Ds–Ac* two-element system, where the movement of *Ds* depended on *Ac*. In several cases, however, originally autonomous one-element systems were shown to give rise to two-element systems, with the *Ac* element losing its ability to transpose autonomously, but retaining its ability to move, with restoration of the spotting/sectoring pattern, in response to the presence of an autonomous *Ac* located elsewhere in the genome. These non-autonomous *Ac* derivatives were similar to *Ds* except that they mostly did not produce breaks on transposition. It was thus reasonable to suppose that *Ds* was also a defective derivative of *Ac*, and this conjecture was eventually confirmed by DNA analysis (see **Figure 4A**). The generic term *Ds* is applied to all non-autonomously transposing derivatives of *Ac*, whether or not they are chromosome breakers.

Although they sometimes transposed from one chromosome to another, *Ds* and *Ac* both moved more often within chromosomes, with a preference for short rather than long jumps. This may mean that the donor and recipient loci have to be in contact for transposition to occur.

3. Suppressor–mutator (*Spm–dSpm*) systems

As McClintock's analysis of *Ac*-controlled systems progressed, she began to find new mutable alleles that neither activated *Ds* nor depended on *Ac* (McClintock, 1956). These proved to be due to a second autonomously movable element that she called *Suppressor–mutator* (*Spm*) because of its dual effect on some mutable alleles (see below; see review by Fincham & Sastry, 1974). What was essentially the same element was discovered independently by P.A. Peterson of the University of Iowa, which he called

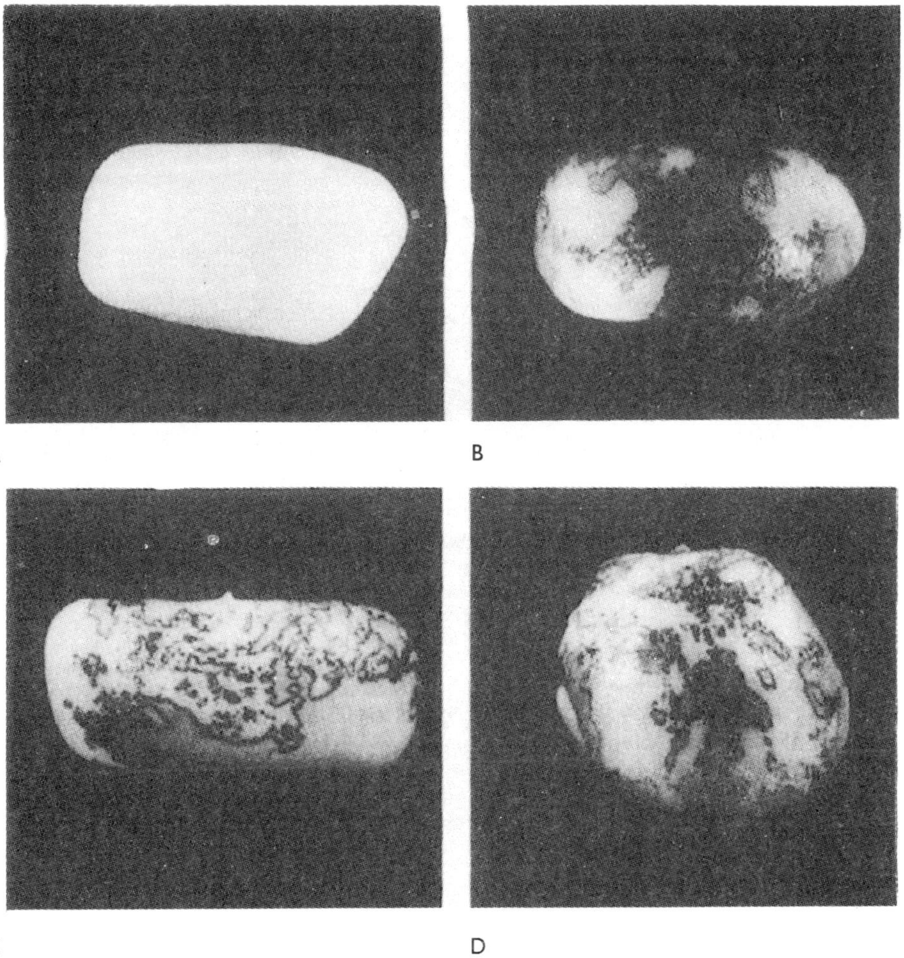

Figure 1 Variegation of maize kernels, due to chromosome breaks at *Ds* on the short arm of chromosome 9, and consequent loss of chromosome fragments. Spots and sectors of endosperm tissue acquire the ability to produce pigment when they lose a chromosome fragment containing *C-I*, a dominant–negative allele of the pigment gene *C*. These spots and sectors have also lost functional *Bz* and have, therefore, only pale pigment except at their rims, which are strongly pigmented because of diffusion of the *Bz* product across the sector boundaries. Kernels a, b, c and d have, respectively, zero, one, two and three copies of *Ac* in their endosperm genomes. *Ac* is necessary for *Ds* breakage, but additional copies delay and reduce the frequency of breakage events. (A, D reproduced from McClintock, 1951; B, C reproduced with permission from Neuffer *et al.*, 1997.)

Enhancer (*En*). Like *Ac*, *Spm* transposed autonomously, but gave rise to non-autonomous derivatives, now called *dSpm*, which underwent transposition only in response to an autonomous *Spm* located elsewhere. McClintock now had two families of moveable elements, each recognizing only its own signals; *Spm* did not activate *Ds* and nor did *Ac* activate *dSpm*. Several more families have now been recognized in maize and other plants, but this review will concentrate on those studied by McClintock.

Three different kinds of *dSpm*-controlled mutable alleles were distinguished (**Figure 3**). First, there were the *Spm*-suppressible alleles that gave the element its name. They retained some gene activity in the absence of *Spm*, but in its presence were active only by mutation. Thus, in the absence of *Spm*, a_1^{m-1} (A_1 is a gene essential for pigmentation) gave a moderate-to-high level of uniform pigmentation, but in its presence both suppression and mutation were seen; pigmentation was suppressed, except in the fully coloured spots where the *dSpm* element had been excised. Secondly, there was a rather rare class of *Spm*-dependent alleles, exemplified by a_1^{m-2}, which were inactive in the absence of *Spm*, but stimulated to a relatively high level of activity in its presence. Thus, a_1^{m-2} gave white kernels by itself, but, with the help of *Spm*, fully coloured mutational spots against a medium-coloured background. Finally, there was a class of mutable null alleles, such as wx^{m-8}, which were inactive, except by mutation, whether *Spm* was present or not.

4. Changes of state of transposable elements

Another strange property of *Ac* and *Spm*, even more difficult to understand because less predictable than their ability to transpose, was their fairly frequent switching between

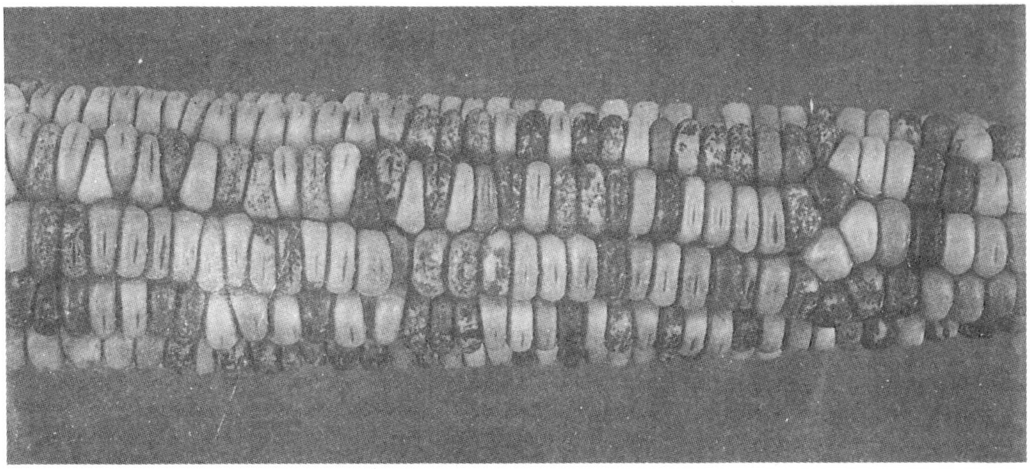

Figure 2 (see also **Plate 19** in the colour plate section) A maize cob showing 1:1 segregation of presence/absence of *Ac*, with all kernels carrying a *bronze-mutable* (bz^m) allele in heterozygous combination with stable recessive *bz*. (Reproduced with permission from Neuffer *et al.*, 1997.)

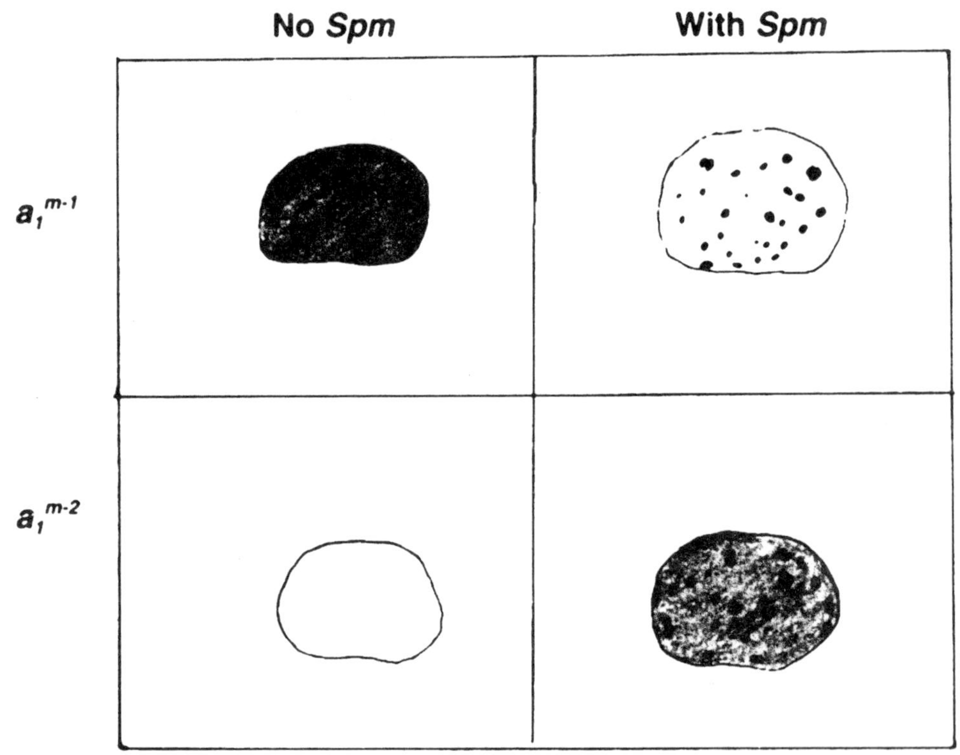

Figure 3 The effect of the *Spm* element on two *dSpm*-containing mutable alleles, one (a_1^{m-1}) *Spm* suppressible and the other (a_1^{m-2}) *Spm* dependent. (Maize kernels drawn approximately five times real size, after the photographs of Fedoroff, 1989b.)

greater and lesser levels of effectiveness. The effect of *Ac* is dependent on its dosage, which, since endosperm is a triploid tissue with two chromosome sets from the maternal parent and one from the pollen parent, can vary between one and three copies with respect to any one *Ac* location. It turned out that *Ac* was less effective if present in two copies and much less effective again if in triplicate (**Figure 1**). The progeny of certain crosses, which should have contained only one *Ac* copy, included kernels with a markedly reduced frequency of *Ds*-induced breaks, as if the single *Ac* had acquired the potency of more than one copy. This apparently enhanced strength of *Ac* was heritable, but with fairly frequent reversion to the weaker state. In other plants, the transposition-inducing activity of *Ac*, monitored through its effect on a *Ds* derivative at the *Wx* locus, appeared to have been lost in some kernels. However, the continued presence of the element was shown by the occurrence of endosperm sectors with restoration of the *Wx*-spotting pattern, which was revealed by application of iodine which stains black the starch product of the *Wx* gene.

Reversible changes of state were also characteristic of the *Spm* element. It appeared that this element could exist in a number of alternative heritable but unstable states, some showing virtually complete and others only partial loss of transposition activity, and most showing occasional reversion to the fully active condition. It was shown that the suppressor activity of *Spm* was generally retained when the mutator activity was lost. McClintock's observations of the shift of activity of *Spm* led her to other novel conclusions. One was that an inactive *Spm* element could be at least temporarily reactivated through residing in the same genome as an active one. Another, which was less convincingly documented, was that different *Spm* derivatives cycled between active and inactive states on a predictable time scale, some with a shorter and others with a longer cycle time. Not all weak states of *Spm* were reversible, however, and two kinds of stable *Spm-weak* are considered in the next section.

Another phenomenon, perhaps related to the changes of state just considered, is what McClintock termed "presetting". This was the ability of a normally non-autonomous *Spm* element to maintain some activity even after the fully competent *Spm* copy that formerly activated it had been segregated away in meiosis. Thus, a plant carrying the normally *Spm*-dependent a_1^{m-2}, together with a fully competent *Spm*, yielded – when crossed to a plant without *Spm* and with a stable recessive allele of a_1 – progeny with a_1^{m-2} either with or without *Spm*. The former class showed the expected fully coloured spots on a more lightly pigmented background, and the latter, surprisingly, some patches of pigmentation of low to medium intensity, apparently due to the recent association of a_1^{m-2} with the *Spm* that was now absent. This residual activity was not usually transmitted to the next plant generation, but, in one family, it was reported to persist, at low frequency, for a further two generations.

5. DNA analysis of McClintock's transposable elements

(i) Cloning of AC *and* Spm

In the 1970s and 1980s, the idea of transposable genetic elements gained more general acceptance with the discovery of such elements in bacteria, yeast and the fruit fly Drosophila. In these new examples, the genetic evidence was backed by analysis of DNA sequence. By this time, the basic tools of DNA analysis – Southern blotting, the mapping of restriction fragments, the use of DNA probes for specific sequences, and the cloning and sequencing of DNA fragments – were becoming increasingly available, and the time was ripe for molecular analysis of the McClintock phenomena. Nina Fedoroff and her colleagues at the Carnegie Institute of Washington, Baltimore, made a decisive advance in isolating the *Ac* and *Spm* elements as DNA sequences (Fedoroff, 1989a,b). The first step was the preparation, by reverse transcription of endosperm messenger RNA, of a cDNA probe for the *Waxy* gene. This was then used to identify a clone of a mutable *wx* allele known to harbour the *Ac* element, and to isolate from this clone the DNA corresponding to *Ac* itself. The *Ac* DNA was then used to probe for and clone an *Ac*-controlled mutable allele of *bronze* and, from this, the DNA of *Bz* without the *Ac* element was isolated. The *Bz* DNA was then used to probe for and capture the DNA of a different mutable *bz* allele known to harbour *Spm*.

The *Ac* and *Spm* sequences so isolated were 4.6 and 8.3 kilobases (kb), respectively, in length. Their structures are shown diagrammatically in **Figure 4**. Typical of DNA transposons, they are flanked by mutually inverted repeats of 11 base pairs (bp) for *Ac* and 13 bp for *Spm*. The outer ends of these repeats evidently provide the sites for the cutting and rejoining involved in the transposition process. It should be noted that the other main family of transposable elements, the retrotransposons of Drosophila and other organisms, which are not excised as DNA but transpose by transcription into RNA and then reverse transcription back to DNA, are typically flanked by direct, not inverted repeats. Like transposable sequences of all kinds, McClintock's elements are usually flanked by short direct repeats generated from the recipient chromosome, indicating that the gaps into which they are inserted are generated by staggered cuts in the DNA. When the element is transposed, these host-sequence duplications are left behind as a sign that the transposon has visited. Another feature which *Ac* and *Spm* have in common is that they each contain a single transcription unit extending along most of their length which, after the splicing-out of introns, encodes a single protein which, since it is essential for transposition, may be called a transposase (sometimes symbolized as Tpn). Mutable alleles have a transposable element, or one of its non-autonomous derivatives, integrated into the transcribed sequence of the gene concerned.

(ii) Properties of the Ac *and* Ac–Ds *systems*

The properties of the *Ac* and *Ac–Ds* systems can be explained quite simply. The insertion of one of the elements blocks transcription of the gene or disrupts its coding sequence, and excision, which depends on the *Ac* protein, restores the gene to activity. Excision and transposition result from the action of the transposase on the terminal inverted repeats. The *Ds* elements are all derived from *Ac* by internal deletions which remove part or all of the transposase coding sequence (**Figure 4A**), but leave the terminal inverted repeats intact. Some *Ds* isolates, in fact, contain almost nothing from *Ac* except the terminal repeats. Most *Ds* elements do not cause chromosome breaks on excision and the original *Ds*, which does have this property, has an exceptional double structure, with one *Ds* element inserted, in inverted orientation, into the middle of another one (**Figure 4A**, bottom line). Perhaps as a result of the presence of four terminal repeats, and the possibility that the

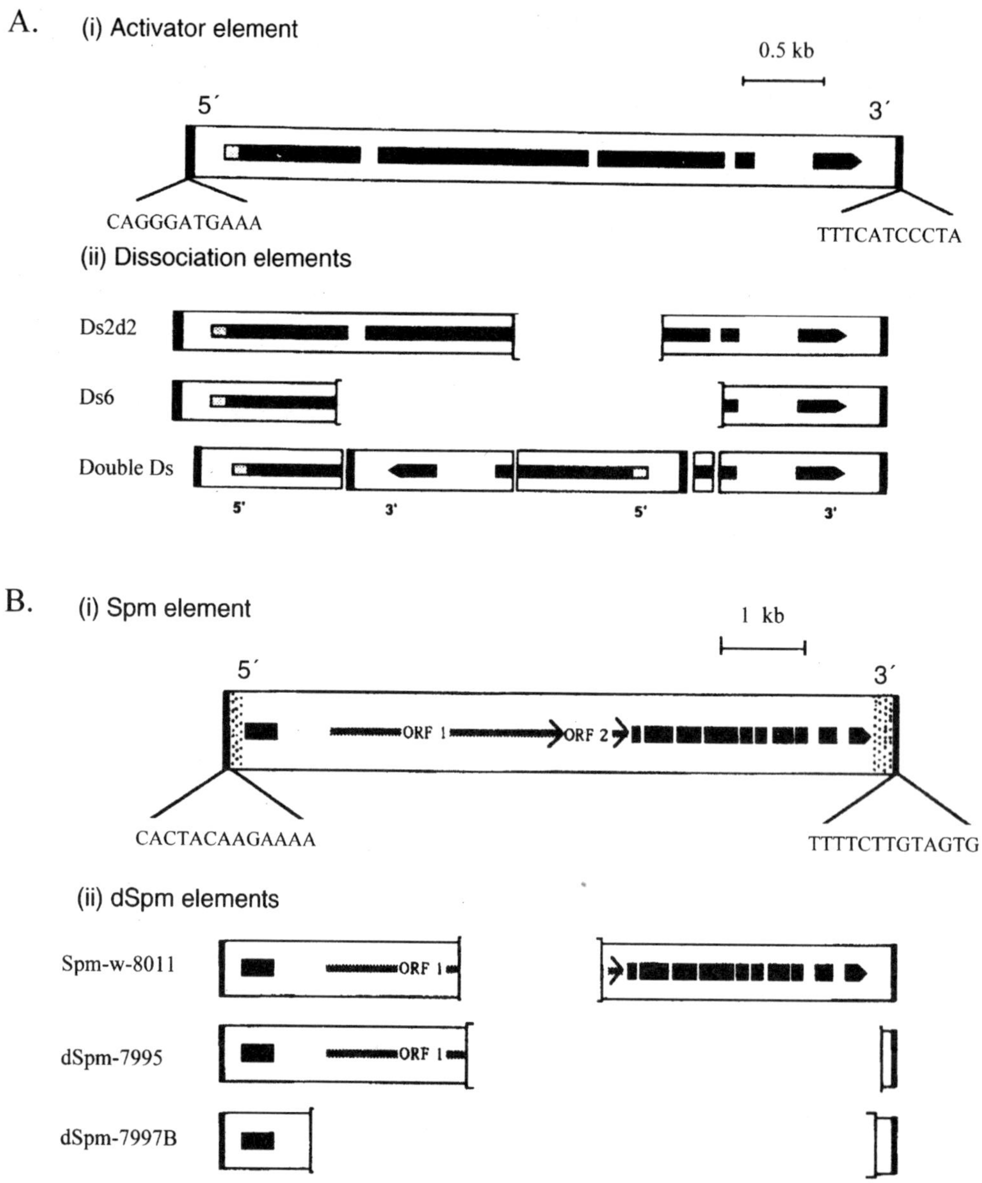

Figure 4 (A) Structure of *Ac* and some *Ds* derivatives. The transcribed sequence is shown as black bars (exons) interrupted by introns; the stippled box at the 5′ end contains a number of alternative transcription start-points. The thin terminal black boxes represent the 11 bp inverted repeats; the sequence of the single strand complementary to the transcribed sequence, and corresponding to the transcript, is shown. *Ds* elements have internal deletions and cannot transpose autonomously; the chromosome-breaking double-*Ds* has one internally deleted copy inserted, in inverse orientation, into the middle of another. (B) Structure of *Spm* and some internally deleted derivatives. Conventions as in (A) except that the stippling adjacent to each end represents numerous 12 bp repeats (not the same as the terminal repeats) to which the transposon protein binds. Note the two open reading frames (ORF1 and ORF2) in the long first intron. The 3′ terminal inverted repeat sequence, with TAG preceded by a thymine-rich tract (uracil-rich in the transcript), resembles a 3′ splice site for intron removal. The *Spm-weak* element (second from the top), with a deletion confined to the first intron, can still transpose autonomously, but with reduced efficiency. The two elements shown below cannot transpose autonomously. (Adapted from Fedoroff, 1989a.)

two cut ends may be in similar rather than mutually inverted orientation, the action of the transposes on the double *Ds* has a variety of complicated outcomes, some with cutting without the normal religation of cut ends. A feature of *Ac* not yet explained in detail by molecular analysis is the inhibitory effect on transposition and *Ds*-induced breakage of multiple copies of the element (**Figure 1**); apparently, the transposase protein acts as a repressor of the action of its own gene.

(iii) Structure of Spm

The structure of *Spm* has turned out to be rather more complicated (**Figure 4B**). Three structural features not shared by *Ac* are important: first, the presence, just inside the terminal repeats, of numerous 12 bp repeats in both orientations; secondly, an extremely large (4.4 kb) intron containing two contiguous open reading frames (ORFs); and, thirdly, the presence of a sequence at the extreme end of the 3′ inverted repeat of the transcript which resembles an acceptor site for the splicing-out of introns.

dSpm elements are all derived from *Spm* by internal deletions removing some of the transposase-encoding sequence. As might have been expected, deletions which remove parts of the long first intron do not abolish autonomous transposability, but they do substantially reduce the frequency of transposition and account for the stable *Spm-weak* which had been identified by McClintock (**Figure 4B**, second line). One possible explanation for this apparent function of the intronic ORFs is suggested by the demonstration that the primary *Spm* transcript can be spliced in a number of different ways, some of them resulting in the inclusion of parts of these ORFs in the mRNA and the consequent synthesis of modified transposase, perhaps with some enhanced function. A second kind of weakening of *Spm* function results from small deletions that remove some of the subterminal 12 bp repeats. It has been shown that the transposase protein binds to these sequences, rather than to the inverted terminal repeats as such, and their partial removal evidently reduces the efficiency of transposition.

(iv) Effects of dSpm elements

The different effects of *dSpm* elements on gene expression are mostly due to their different orientations and sites of insertion rather than to their own internal structures. The first question regarding the *Spm*-suppressible class of alleles is how they come to have any activity to suppress. Part of the answer is that all the suppressible alleles have the inserted element in inverted orientation relative to the direction of transcription, and so the *Spm* transcription terminator is without effect. The more surprising part is that the suppressible–mutable alleles can produce approximately normal mRNA by excising the intrusive element from the transcript as if it were an intron. This is possible because, if the 5′ end of the *Spm* transcript happens to be fairly close to the 5′ splicing sequence of an intron of the host gene, the normal intron-removal machinery can splice out the entire sequence between that splice junction and the fortuitous 3′ splicing TAG sequence at the 3′ terminus of the *Spm* transcript. In this way, the entire *Spm* sequence will be removed, together with the flanking intron. Some intronic sequence may be left behind, but, if this addition is fairly short and contains a whole number of codons, the protein product may still be functional. To explain the suppressor function, one has to suppose that the binding of the transposase to the subterminal *dSpm* sequence blocks transcription and, therefore, forestalls the rescue mechanism.

(v) Properties of Spm-null and Spm-dependent mutable alleles

The *Spm*-null and *Spm*-dependent mutable alleles have other explanations. At least some of the null class have been shown to have *dSpm* elements inserted in the same orientation, with respect to the direction of transcription, as the gene into which they are inserted. These alleles will be transcribed only as far as the transcription-termination sequence at the 3′ end of the inserted element and will only function when the element is excised. The *Spm*-dependent class is exemplified by the a_1^{m-2} allele described by McClintock: *dSpm* derivatives of this originally autonomous allele give no pigment in the endosperm and no mutation in the absence of *Spm*, but, with the help of *Spm*, give fully coloured spots against a moderately coloured background. It turns out to have a *dSpm* element inserted within the transcribed region of the A_1 gene and upstream of its initiation codon. Evidently, the insertion of the *dSpm* in the leader region of the A_1 gene blocks normal initiation of transcription, but, in the presence of transposase protein, the *dSpm* element can provide an alternative transcription start-point. This is one of several indications that the *Spm* transposase can activate transcription of its own gene.

(vi) Epigenetic effects

The many reversible changes that McClintock observed in her transposable elements can now be seen as examples of epigenetic effects – i.e. changes in the chromatin structure that have at least some heritability, sometimes extending through many cycles of cell division, but not based on any change in DNA sequence. The phenomenon of gene silencing, now observed to many different organisms, is usually accompanied by methylation of the DNA, predominantly of cytosine residues. Derivatives of *Ac* and *Spm* that are weakened without having any internal deletions have been shown to be heavily methylated, particularly in their transcription-initiation regions. McClintock's observation that a weak or inactive *Spm* may be revitalized by the presence of an active *Spm* has been confirmed, and the reactivation has been shown to involve loss of methylation. Among the other properties of the versatile transposase protein is its ability to demethylate the *Spm* promoter region; presumably this accounts also for its activation, mentioned above, of *Spm*-dependent alleles, which tend to become methylated in the absence of *Spm* (Schlappi *et al.*, 1994). More surprisingly, the transposase protein has been shown to be a repressor of transcription of its own gene when the gene is not methylated and, in this respect, may resemble the *Ac* transposase.

The "presetting" of *Spm*-dependent alleles, so that they remain active for a time in the absence of *Spm*, must also be an epigenetic effect. It is known that chromatin is maintained in a transcriptionally active state by a complex of proteins, and that this complex, once in place, may be to some extent self-renewing through cycles of cell division, even though all the conditions that permitted its initial establishment are no longer present.

6. The significance of the maize transposable elements

Barbara McClintock called *Ac* and *Spm* "controlling elements" and believed that they, or others like them, had essential roles for the plant. She liked the analogy between the *cis*-acting *Ds* and *trans*-acting *Ac* and the *cis/trans* control of bacterial operons, revealed by the seminal work of Monod and Jacob. However, modern opinion does not, on the whole, favour the view that transposable sequences of the apparently anarchic kind found in maize have any constructive role. They are regarded rather as molecular parasites. It has been shown that they often transpose from chromosome loci that have just replicated to others that are about to replicate, and so they have the capacity to multiply faster than their host genomes. Their relatively conspicuous presence in maize may be due to human selection, which may well have favoured the striking and decorative patterns to which they give rise. So, although one cannot rule out the possibility that transposable elements may sometimes have been recruited by natural selection to serve the interests of the host plant, one may doubt McClintock's view of their general functional significance.

This judgement does not, however, detract from her achievement. She was the main pioneer in two biological fields of very great importance: transposable sequences, previously considered impossible but now evidently ubiquitous, and epigenetic control of gene activity, then a virtual heresy but now one of the major themes of developmental genetics. The then bizarre observations that she presented to the largely incredulous biological world in 1950 showed these two unprecedented phenomena acting together. It is no wonder that it took so many decades for geneticists to come to terms with them.

REFERENCES

Fedoroff, N.V. (1989a) About maize transposable elements and development. *Cell* 56: 181–91

Fedoroff, N.V. (1989b) Maize transposable elements. In *Mobile DNA*, edited by D.E. Berg & M.M. Howe, Washington, DC: American Society for Microbiology

Fincham, J.R.S. & Sastry, G.K. (1974) Controlling elements in maize. *Annual Reviews of Genetics* 8: 15–50

McClintock, B. (1950) The origin and behavior of mutable loci in maize. *Proceedings of the National Academy of Sciences USA* 36: 344–55

McClintock, B. (1951) Chromosome organization and gene expression. *Cold Spring Harbor Symposia on Quantitative Biology* 16: 19–47

McClintock, B. (1956) Controlling elements and the gene. *Cold Spring Harbor Symposia on Quantitative Biology* 21: 197–216

Neuffer, M.G., Coe, E.H. & Wessler, S.R. (1997) *Mutants of Maize*, Cold Spring Harbor, New York: Cold Spring Harbor Laboratory Press

Schlappi, M., Raina, R. & Fedoroff, N. (1994) Epigenetic regulation of the maize *Spm* transposable element: novel activation of a methylated promoter by TpnA. *Cell* 77: 427–37

ORIGIN AND EVOLUTION OF MODERN MAIZE

Walton C. Galinat
University of Massachusetts, Amherst, USA

1. Introduction

Maize (*Zea mays* L.), developed by the domestication of teosinte, is humankind's greatest gift from the first peoples of Mesoamerica, and one of the earliest examples of plant breeding. Archaeological evidence places the time of maize domestication between 3000 and 8000BC. The origin of maize by the domestication of teosinte was until recently controversial (**Figure 1**). This article explains why and how the philosophical differences between those involved in the controversy developed. The questions of interpretation arise from the nature of teosinte and whether or not it is the wild ancestor of maize. Would the combined power of recombination and domestication under the direction of the earliest agriculturists be sufficient to bring about the transformation of teosinte into maize?

2. Nuclear status

What is the nature of these plants? Teosintes are wild grasses occurring naturally in the Mexican highlands, while no maize varieties exist in the wild. Both maize and most teosintes are monoecious diploids ($n = 10$) with one teosinte being a perennial tetraploid ($n = 20$). All races of maize are annuals while the perennial teosinte occurs in both diploid and tetraploid forms.

The genus Tripsacum, the second closest wild relative of maize, has also been involved in the controversy over the origin of both teosinte and maize. Tripsacum is also monoecious and it is known only as a perennial genus of about 16 species, native only to the New World with a wide distribution. It ranges in ploidy from a diploid ($n = 18$) to a hexaploid ($n = 54$). Like wheat and most other grasses, but unlike maize, it follows the evolutionary model of allopolyploidy (the addition of different genomes) and apomixis (asexual seed development). The C4 system of photosynthesis in maize is more efficient at producing food to eat and oxygen to inhale than the C3 system of wheat and other cereals.

3. Domestic development

Maize is the only cereal grass that was developed by a combination of domestication and recombination from meiosis I and II. During meiosis in the F_1 hybrid of diverse parents, the genes are recombined into joint interactions expressed as new phenotypes in the F_2 progeny of maize. Selection (natural or domestic) and evolution then starts with diploids like teosinte and maize. But polyploids such as the bread wheat have an evolutionary handicap that almost defeats recombination, which comes from the ancient evolution of sexual reproduction. With the wheat handicap of three double genomes, the same recessive mutant gene would have to be in all six homologous chromosomes to be expressed. Thus, evolution would slow down by being

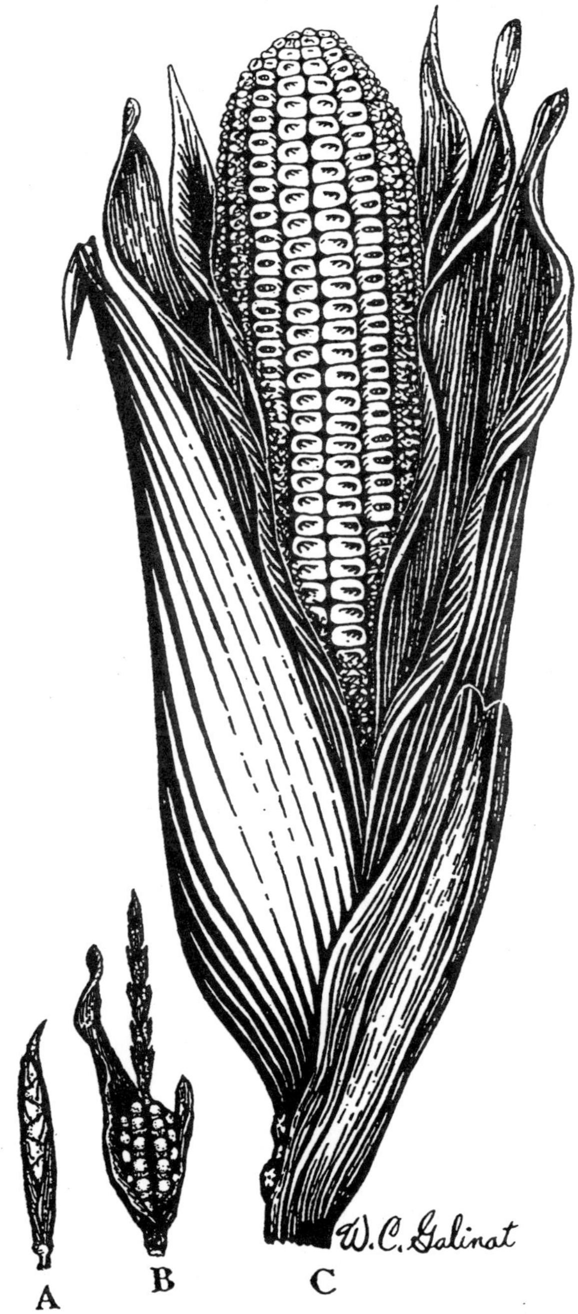

Figure 1 The origin of maize from teosinte, the greatest plant breeding accomplishment known to humankind. (A) Teosinte, the wild ancestor. (B) The oldest known archaeological ear of maize of c.5500 years ago (Long *et al.*, 1989). (C) The modern ear of maize (drawing by W.C. Galinat). Adapted from *Economic Botany* 49: 3–12, 1995, with permission from the New York Botanical Garden, Bronx, NY.

confined to dominant mutations, such as the dominant dwarfing gene transferred to most wheat by breeders. In maize, evolution is rapid and dramatic because maize is a cross-pollinated diploid of great diversity due to the expression of both dominant and recessive mutant genes. New mutations in maize and teosinte are constantly being generated by mutagenic factors originally derived from Tripsacum introgession into teosinte, according to research still underway by Dr Mary Eubanks (see below). Apparently, such alien introgession into teosinte produced the key diversity that was domesticated into the first maize. Figures 2, 3 and 4 illustrate the role of recombination in this process.

Nevertheless, commercial maize seed companies have artificially inbred maize, and then controlled the crossing between special male and female inbred lines so as to make the best F_1 hybrid to sell to the farmers. The farmer cannot profitably save some of his crop for seed to plant the next year, because the F_2 generation has a big drop in yield. In this, and many other respects, the survival of maize is tightly bound to that of our own.

The combined power of recombination and domestication that, together with genetic diversity, transformed teosinte into maize was anticipated in vivid terms by Darwin (1859, chapter 1: Variation under domestication) in pre-Mendelian times as: "that which enables the agriculturist, not only to modify the character of his flock but to change it altogether. It is the magician's wand by means of which he may summon into life whatever form and mold he pleases." In comments that could apply to the maize ear, he elaborates:

> Domestic races often have a somewhat monstrous character; by which I mean, that, although differing from each other, and from other species in the same genus, in several trifling respects, they often differ in an extreme degree in some one part, both when compared one with another, and more especially when compared with the species under nature to which they are nearest allied.

Starting with unconscious selection for traits such as grain size when harvesting, the first teosinte cultivators would later recognize that the progeny of superior plants resembled its parents, and deliberately select those seeds for breeding. When the farmer–breeders selected these more useful mutants of teosinte from wild populations and grew them together in isolated gardens, they also unconsciously introduced mutagenic factors. Eubanks (1995) has suggested that the mutagenic factors are small pieces of Tripsacum chromosome derived from hybrids between Tripsacum and teosinte. These factors would continue to generate additional diversity needed to increase the frequency of mutated genes in the planted population and transform teosinte into maize. The evolutionary development of teosinte in this way is described in section 6.

The differences in ear morphology between teosinte and maize were at first interpreted by the early taxonomists to be of major taxonomic significance. They placed teosinte in a separate genus and species as *Euchlaena mexicana* from that of maize (*Z. mays* L.). Eventually, the developing field of cytogenetics caught up as a tool to reveal taxonomic relationships. The cytological homology of maize and teosinte was reported by Beadle (1932a,b); they have the same chromosome number ($n = 10$), their F_1 hybrids are as fertile and their chromosomes pair as close at the pachytene stage, when the crossing-over of recombination occurs. To this, Emerson and Beadle (1932) added that, for the loci tested, maize–teosinte hybrids have about the same frequencies of crossing-over as in maize itself. Longley (1937, 1941) showed that the idiograms (meaning the genetic map) for the same teosintes are also identical to those of maize (excluding exceptional inversions and terminal knobs). From all of this cytogenetic information, Beadle (1939) concluded that teosinte was the wild ancestor of maize. He suggested that prehistoric man, after first utilizing teosinte for food, possibly as popped kernels that separate from the hard inedible fruitcase, eventually "selected the combination of the five or more major and the many minor genes and chromosome mutations that now distinguish cultivated maize from its presumed wild ancestor (teosinte)".

4. Wild maize hypothesis

Meanwhile, a very different hypothesis that might account for both these cytogenetic similarities and the large morphological differences between the female spikes of maize and teosinte was suggested (Mangelsdorf & Reeves, 1939). According to their "tripartite hypothesis", there was a wild maize with the key traits of modern maize; teosinte was not the wild progenitor, but just a maize–Tripsacum hybrid; and teosinte contributed to the diversity and heterosis of the evolving races of maize by serving as a bridge for these postulated segments of Tripsacum germ plasm. This was a time when the role of introgression in evolution, as suggested by Edgar Anderson (Anderson & Hubricht, 1938; Anderson & Erickson, 1941), carried academic appeal. The controversy started by Beadle was not immediately pursued because of his preoccupation with other research, and the wild maize hypothesis of Mangelsdorf and Reeves was generally accepted for the next 30 years. Mangelsdorf had several graduate students including Garrison Wilkes who prepared a monographic study of teosinte as his Ph.D. thesis. When this monograph and the general acceptance of the wild maize hypothesis, as reflected in a statement of Wilkes (1967), came to the attention of George Beadle, his earlier belief that teosinte had to be the direct wild progenitor of maize was sparked into immediate action. The activating statement of Wilkes was as follows: "The theories that teosinte is the primitive ancestor of maize, although they cannot be dismissed entirely, are on the whole crude attempts to explain the origin of maize."

It appears that Mangelsdorf's thinking as expressed in the tripartite hypothesis and that of his subsequent research at Harvard was unconsciously directed by his previous experience with wheat breeding involving domestication from large sudden changes by alien genome addition, as in the hexaploid bread wheats. In the case of maize and teosinte evolution, where the ploidy was held at the diploid level, Mangelsdorf had genome evolution relegated to introgressive hybridization from Tripsacum or teosinte introgression into the various races of maize. In 1974, Mangelsdorf postulated that at least six different geographical races of wild maize were still living in the modified form of primitive cultivated races (Mangelsdorf, 1974). As with small grain breeding, when there was selection for single-gene traits, such as a non-shattering rachis or insect and disease resistance, the domesticate and its original wild race remained almost identical. Accordingly, this concept of only a one or two gene difference between the wild and cultivated forms ruled teosinte out as the wild ancestor of maize because the teosinte transformation took at least four or five key genes. Mangelsdorf's key pivotal gene that would convert the wild podcorn races into domestic maize was at the compound tunicate locus near gl_3 on the long arm of chromosome 4 (Mangelsdorf, 1948; Mangelsdorf & Galinat, 1964). Mangelsdorf was fascinated by the tunicate locus. By a single gene change in maize, grain protected by long glumes developed in a fragile tassel capable of survival as a wild grass. The analogy to wheat evolution seemed clear to him.

5. Role of the tunicate locus

One of the few points that Mangelsdorf and Beadle seemed to agree on was that the tunicate locus converting the hard erect glumes of teosinte into the soft reflexed glumes in one step may have had a role in the origin of maize. However, there was a difference. Mangelsdorf (1948) would have the tunicate condition in his hypothetical wild maize and non-tunicate in the first domestic maize. In contrast, Beadle (1972) would have the teosinte wild ancestor be like present-day teosinte in being non-tunicate, while one of the selected mutant teosintes that may have been included in the teosinte garden of useful forms was a tunicate teosinte with a softer fruitcase and partly exposed kernels that would serve a temporary rôle in making teosinte threshable and amenable towards further domestication. In support of this suggestion, a weak tunicate allele still occurs in the primitive race, Chapalote.

6. Evolutionary development

(i) Molecular evidence

Mangelsdorf did not take into account the power of recombination in a teosinte garden of useful mutants selected from wild populations. He also did not appreciate that teosinte forms the culmination of an evolutionary sequence extending back into the Andropogonae (Galinat, 1956). If maize did not come from teosinte, it has no identified ancestors. All of the key traits that separate maize and teosinte are exactly those that people would select in attempting to make teosinte a more useful food plant.

Under the encouragement of George Beadle, who had now seriously returned to the problem of the origin of maize, Galinat (1970) showed that the cupule of the corn cob had a direct derivation from the teosinte fruitcase. In a comprehensive review of 91 relevant references, Galinat (1971) concluded that there was no compelling evidence against accepting teosinte as the wild ancestor of maize, and that the available evidence (including archaeological maize cobs, which have been interpreted by some authors as supporting the wild maize hypothesis) was at most circumstantial. This review (Galinat, 1971) revitalized research on the origin of maize in concluding that teosinte might be the wild ancestor of maize. More recent reviews (see Galinat, 1992a,b, 1994 and references within) continued to update the story with the newly developing evidence from the field of molecular technology spearheaded by John Doebley. That teosinte is the progenitor of corn is now generally accepted. For example, Doebley (1990) considers the molecular evidence for at least one origin of corn by domestication of teosinte (sp. *parviglumis*, race Balsas) as definitive (see also Beadle, 1972; Iltis, 1972, 1983; Kato, 1984; Doebley *et al.*, 1990).

In an extended project of intergenomic mapping of maize, teosinte and Tripsacum using standard marker genes of the time, Galinat (1973) attempted a reality check on the hybrid origin of teosinte. If teosinte was just a product of Tripsacum introgression into a wild maize, then it should be possible to identify phenotypically those key segments of Tripsacum germ plasm responsible for this conversion and then reconstruct teosinte out of a maize–Tripsacum hybrid. For example, a gene on Tripsacum chromosome 9, which has several genes in common with the short arm of maize chromosome 2, reduces the ranking from four to two, which would be a key trait of teosinte. Yet, a much more important key trait of teosinte that was first identified by Mangelsdorf and Reeves (1939), and later confirmed by many others, was a presumed Tripsacum segment closely linked to the *-Su-su* locus on the short arm of chromosome 4 and that was essential to development of the fruitcase (erect, indurated female spikelet glumes) and, according to their hybrid origin of teosinte hypothesis, originally came from Tripsacum. This was to be the undoing of the great tripartite hypothesis.

In cross-mapping maize and Tripsacum chromosomes, the Tripsacum chromosome marked by the *Su* locus which happened to be its 7th chromosome should have carried genes such as *Tga* (teosinte glume architecture), as described later, if this gene introgressed into maize chromosome 4 during an origin of teosinte. No such relationship between Tripsacum chromosome 7 and maize–teosinte chromosome 4 was observed (Galinat, 1973). Also, the gl_3 locus on the long arm of maize chromosome 4 was carried elsewhere on Tripsacum chromosome 13. The tripartite hypothesis based on the hybrid origin of teosinte by means of Tripsacum introgression is untenable. We may now return to understanding just how maize may have evolved under domestication from teosinte.

(ii) Maize domesticated from teosinte

The origin of maize from synergistic recombination within teosinte is one of the world's earliest and most important examples of plant breeding. The first Mesoamericans had to combine hunting and gathering with also being farmer–breeders in order to obtain adequate food. The teosinte gatherers must have saved the more useful variants found in wild populations and planted them together in isolated teosinte gardens. One accidental consequence of the random intercrossings in the gardens of more useful teosinte was the synergistic convergence by recombination between two parallel pathways each with a different way of making a four-rowed ear and each descended from the two-rowed wild-type ear of teosinte. One had four rows because it had two ranks of paired female spikelets (pd_m on chromosome 3). Teosinte ears (spikes) are normally paired and female spikelets single. The pd_m gene extends the pairing up into the spikelets of the uppermost single female spike within a cluster (fascicle) of paired spikes with solitary spikelets. In advanced levels of maize, the ears are usually solitary and the spikelets usually paired. The other type of four-rowed teosinte ear evolved along a different pathway of condensation (fatter meristem) in the uppermost spike, again uppermost within a cluster of typical teosinte two ranked–two rowed spikes. The uppermost spike had decussate yokes of solitary spikelets that were controlled by a recessive gene (rk_m, on chromosome 2). The author has described elsewhere (Galinat, 1995) what happened in the garden of useful forms (see **Figures 2** and **3**).

Figure 2 The origin of maize in the F_2 from a hybrid between two kinds of four-rowed teosinte shown in intact longitudinal view. Only the eight-rowed recombinant was selected for illustration here (see also **Figure 3** for F_2 types) (drawn especially for the *Encyclopedia of Genetics* by author W.C. Galinat).

7. Morphogenesis: female spikelet architecture, outer glume induration and rachilla elongation

The first observation of a factor on the short arm of chromosome 4 that would transform the female spikelet–rachis architecture and increase the glume induration of maize into that of teosinte was made by Mangelsdorf and Reeves (1939) in their hypothesis of an origin of teosinte by Tripsacum introgression into maize. The position as close to the *Su-su* locus was confirmed by Rogers (1950). Although a research associate of Mangelsdorf at the time, the author (Galinat, 1963) made a unique series of drawings reproduced here in **Figure 4** illustrating this transformation only in the *reverse* direction, starting with Tripsacum and teosinte and terminating in modern maize with intermediates reconstructed from experimentally known introgression. Apparently, the significance of the sequence concealed in the art work escaped Mangelsdorf at the time. Recently, a genetic symbol of *Tga* (teosinte glume architecture) was assigned to this chromosome 4S factor by Dorweiler *et al.* (1993). Although useful, the symbol is inaccurate because the change in architecture is the upward orientation of the whole spikelet – not just the outer glume – except for the degree of outer glume induration. Furthermore, the direction of the evolution is now accepted as being from teosinte to maize, so that the symbol of *msa* (maize spikelet architecture) was coined for the recessive mutant allele leading toward maize because the genetic effect of a recessive gene

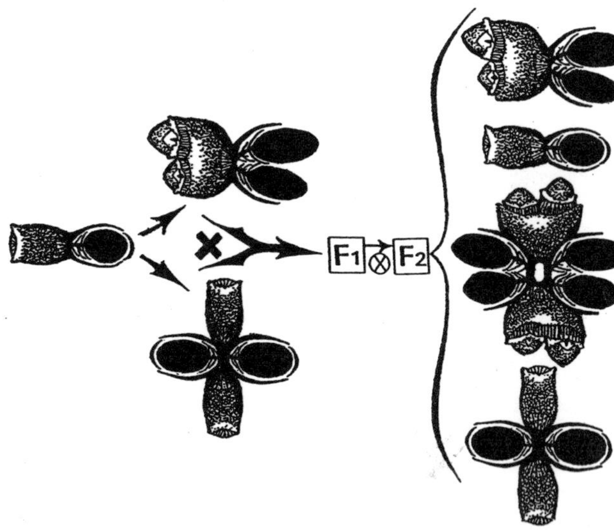

Figure 3 The origin of maize by recombination in the F_2 from a hybrid between two kinds of four-rowed teosinte. The fruitcases are shown with enclosed spikelets and kernels from two successive internode rachis segments represented here in horizontal views. The upper fruitcase in solid black is a cross-sectional view while the lower alternate fruitcase is intact in apical view. The F_2 segregation has a typical 9-3-3-1 ratio in which the frequency of the double recessive is the smallest (6.25%). In a teosinte background, this represents the origin rate of eight-rowed maize as an infrequent recombinant. (drawing by W.C. Galinat). Adapted from *Economic Botany* 49: 3–12, 1995, with permission from the New York Botanical Garden, Bronx, NY.

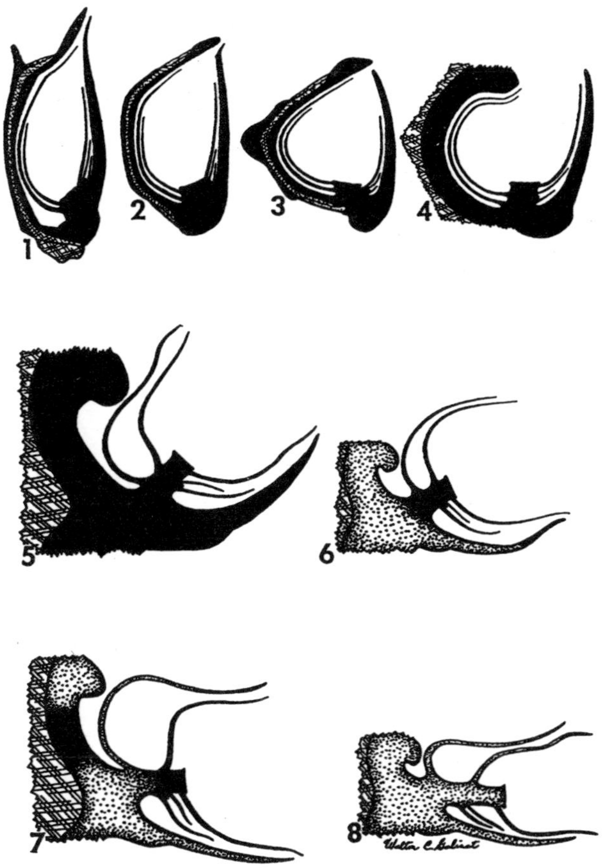

Figure 4 Transitional specimens in longitudinal view extending from Tripsacum (1), teosintes (Florida 2, Nobogame 3), a maize–teosinte hybrid (4), to their homologues in modern maize (8) created by experimental introgression of known teosinte segments into modern maize (5,6,7). The *Tga* gene is manifest in specimen 5 while its allele *msa* is manifest in specimen 8, as described in the text. Specimen 6 has a reduced rachilla; specimen 7 an elongate cupule increasing ear length. Solid black indicates teosinte effects. Drawing by W.C. Galinat from *Economic Botany* 17: 51–59, 1963, with permission from the New York Botanical Garden, Bronx, NY.

is the traditional basis for coining a new symbol rather than the original dominant wild-type represented here by teosinte.

Since all known races of extant maize carry the *msa* allele and all teosinte species the *Tga* allele, this is an important key trait for their separation. From the oldest known archaeological cobs from Tehuacan, Mexico, to modern cobs, the *msa* allele causes the female spikelet to be reflexed down to a right angle position from the rachis, the outer glume to have reduced induration and the rachilla elongate – all of which combine to expose the grain and make it threshable. Rather than by mutation of teosinte's *Tga* allele to maize's *msa* allele, similar phenotypic changes could have been made, in a more unstable manner, by the modifying action on *Tga* expression of a different gene, an intermediate tunicate allele down on the long arm of chromosome 4 (Beadle, 1972; Mangelsdorf & Galinat, 1964), but this course of evolution now seems to be ruled out in the origin of maize.

The condensation that had increased ranking from two to four continued to escalate in the uppermost spike where the growing point is largest (fattest). Not only was ranking increased to higher levels, but both the interspace above the cupule (backside of alternate cupule) and cupule length (internode cupule) were condensed shut. With reflexion of the spikelet (*msa*), the apex of the closed cupule was able to fuse onto the glume cushion of the spikelet now directly above. The rachis became rigid and non-shattering. The abscission layers of teosinte across the rind and pith became non-functional and were usually lost. The presence *vs.* absence of abscission layers was a major factor in the domestication of the other cereals, but, in the case of the origin of maize from teosinte, it is less important because of the substitution of husk leaves and condensation initially to serve the same purpose.

In any case, the evolutionary emergence of maize was soon to include a continuous non-disarticulating rachis. Where would such a non-wild, but highly domestic, trait come from? Eubanks (1995) has apparent evidence of teosinte–Tripsacum hybridization. She suggests that it may have been involved in the origin of maize by contributing to the diversity of ancestral teosinte. Teosinte and Tripsacum have morphologically and genetically different means of rachis disarticulation. The products of recombination between the two systems would include non-disarticulation, as in maize.

The garden of useful teosinte mutants also included other genes soon to be incorporated into the mother of maize as follows: tru_m (ears replace lateral tassels) or *pff* (precocious female florets) on chromosome 3L, and *msa* (maize spikelet architecture) on chromosome 4S.

Teosinte as the wild ancestor originally carried the dominant wild-type alleles in relation to the divergent recessives creating maize. As maize became the majority and more evolved type, its key traits reversed their dominance. Now the symbolic identity of the genes must be coded with a subscript, *t* for teosinte and *m* for maize, without definition of phenotype in order to retain their previous symbols coined by maize-centric geneticists. An alternative is to recoin synonyms that clearly recognize the teosinte alleles as wild-type (Galinat, 1996; unpublished) and avoid Mangelsdorfian implications that teosinte was derived from maize.

The genetic symbol has been used to indicate either progression from or regression to the wild-type. However, in the case of domestication, the cultivar is often perceived as the wild-type because it is better known and more obvious. The coining of different symbols for the same allele seen from the opposite perspectives of the wild *vs.* domestic backgrounds seems justified in the case of teosinte and maize because there has been a reversal of key trait dominance. In this case, the traditional use of upper and lower case to designate dominance does not work across backgrounds.

The symbol *tru* (tassel replaces upper ear in a lateral branch) has been coined and located by Sheridan (1988) on the long arm of chromosome 3. It is a good example of a regressive (throwback) mutant in maize that reveals the original genetic condition of teosinte. A progeny test from the F_1 of a cross between *tru tru* maize and a strain of teosinte fixed for a similar phenotype showed 100% expression of *tru* in all 235 plants grown, indicating allelism of the genes (Galinat, unpublished).

At first, the dominant allele to the *tru* mutant of maize was just a recessive mutant in a teosinte background. Then, it gradually evolved an antithetical (reversed) type of dominance in a maize background, as described with presumed Tripsacum introgression by Anderson and Erickson (1941), from disruptive selection between the breeding of maize as a better food plant and natural selection in teosinte for survival as a wild plant – as described by Doggett (1965). The selection is for modifier genes that influence the degree of dominance in relation to the plant's adaptability. When such selection is confined to the maize end of the maize–teosinte spectrum, it will tend to make the key maize alleles dominant and their teosinte alleles recessive and visa versa with teosinte in the wild.

Looking at the *tru* gene from the perspective of how the lateral branch of teosinte was reconstructed into a husk-enclosed ear of maize, we need to recoin a different symbol for the progressive allele leading toward maize in order to recognize symbolically dominance reversal and to avoid implication that teosinte was derived from maize rather than visa versa. The new symbol coined here is *pff*, standing for precocious female florets in the terminal inflorescence of lateral branches. The *pff* gene has a cascade of pleiotropic effects that replace both the lateral tassels of teosinte and the elongate branches that they terminate, usually by a single potentially large husk-enclosed ear borne in the axil of a leaf – now centrally located, although potentially for every leaf carried alternately along the main stalk.

The triggering of the cascade of changes within the lateral branches by the *pff* gene follows its precocious feminization of the florets together with increased fatness (less divergence) of the apical meristem. As the rachis becomes thickened and more vascularized, the ear's potential for largeness is fulfilled in combination with other genes such as *rk* and *pd* for many ranks of paired female spikelets in the uppermost (central) spike terminating the lateral branch and genes for increased kernel size. Meanwhile, the main stalk remains unchanged and is terminated by a male tassel such as had previously also terminated the lateral branches in the presence of the *tru* gene of teosinte. The direct cascade from *pff* feminization of the terminal inflorescences to lateral branches starts by increasing condensation (inhibiting internode elongation) of the internodes in the lateral branch; then, by suppression of usually all the axillary buds along the branch; and, as a result, by the permanent entrapment of the developing ear within the protective confinement of the leaves from along the branch which it terminates. Meanwhile, there is a great elongation of the styles (silks) as they reach upward toward freedom above the husk leaves where they become pollinated and pollen tubes start their growth downward with sperm nuclei to cause double fertilization and kernel development. After the ear and kernels mature, the farmer–breeder, like his ancestors for thousands of years before him, will husk out and shell some of the ears for immediate use as food with the rest stored on the cob for use in another day.

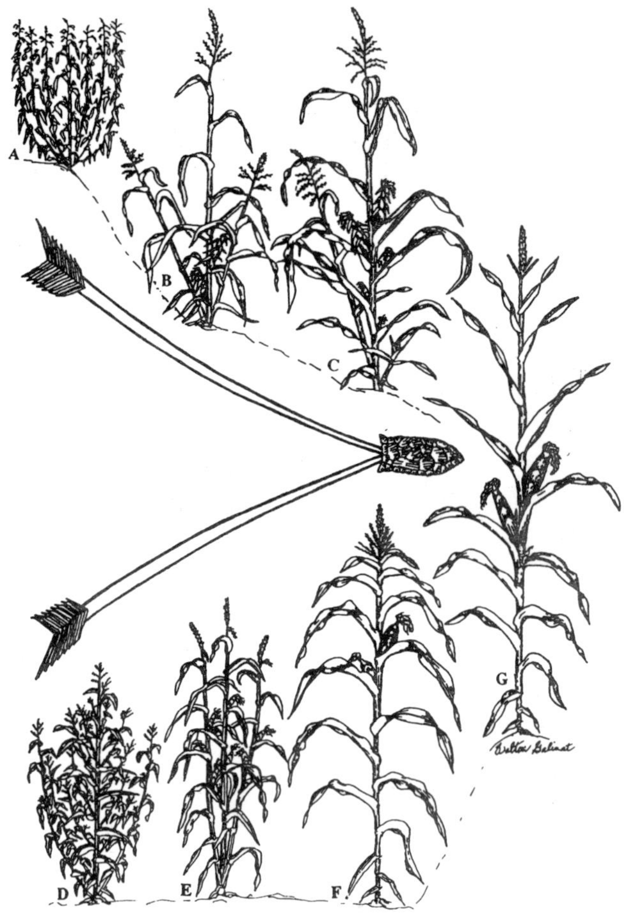

Figure 5 The evidence from plant habit shown here corresponds to that from cob morphology (Galinat, 1992a) in supporting the two origins hypothesis with parallel pathways tracing back to two different teosintes. The first and probably oldest starts with parviglumis teosinte and its sympatic maize NalTel, leading to the eight-rowed flint and flour races of maize shown here in the upper sequence (A, B, C). They are characterized by the plant habit of basal branching. The second pathway in the lower sequence (D, E, F) stems from Chalco teosinte and its sympatic maize, Palomero Toluqueño leading to the many-rowed, gourd-seed Southern Dents. They have the plant habit of lateral branching. Both pathways advance toward their intermediate hybrid type, the Cord Belt Dent (G). Drawing by W.C. Galinat from *Economic Botany* 49: 3–12, 1995, with permission from the New York Botanical Garden, Bronx, N.Y.

Then, as now, the best ears are always saved for seed to plant the following year and so complete the symbiotic relationship between the biological evolution of maize and the cultural evolution of the human farmer–breeders.

8. Conclusion

The immense diversity of maize in its hundreds of races and thousands of varieties greatly exceeds the diversity of any other diploid plant species in which humans have directed the evolution. The ability of farmer–breeders to observe and consciously select both dominant and recessive mutants of teosinte in their isolated gardens would eventually result in recombination between the two morphologically different types of mutant four-rowed teosinte that produced eight-rowed maize in the F_2 segregation. This origin process happened at least twice independently: once in the highland valley of Mexico starting with Chalco teosinte, and again in the lowland areas in southwestern Mexico with parviglumis teosinte. Evidence for this two origins hypothesis comes from identical patterns of chromosome knobs between sympatic races of maize and teosinte, although Kato (1984) suggests on this basis that there were many origins of maize. The two origins hypothesis fits more precisely with the high degree of heterosis between the Northern Flint plant habitat type of basal branching found in parviglumis teosinte, and the Southern Dent habit of elevated branching found in Chalco teosinte, both expressed in maize in the US and correlated to the degree of hybrid productivity when their pathways hybridize (see **Figure 5**). Successful hybrid corn breeding in the US depends on reconstructing the interpathway heterosis by crossing segregates representing the two ancestral lines. This history is documented by Wallace and Brown (1956). It is significant that these two authors were founder (Wallace) and president (Brown) of one of the most successful hybrid corn seed companies (Pioneer Hybrid Int.).

REFERENCES

Anderson, E. & Hubricht, L. (1938) Hybridization, in *Tradescantia*. III. The evidence for introgressive hybridization. *American Journal of Botany* 25: 396–402

Anderson E. & Erickson R.O. (1941) Antithetical dominance in North American maize. *Proceedings of the National Academy of Sciences USA* 27: 436–40

Beadle, G.W. (1932a) Studies of *Euchlaena* and its hybrids with *Zea*. I. Chromosome behavior in *Euchlaena mexicana* and its hybrids with *Zea mays*. *Zeitschrift für inductive Abstammungs und Vererbungslehre* 62: 291–304

Beadle, G.W. (1932b) The relation of crossing over to chromosome association *in Zea–Euchlaena* hybrids. *Genetics* 17: 481–501

Beadle, G.W. (1939) Teosinte and the origin of maize. *Journal of Heredity* 30: 245–47

Beadle, G.W. (1972) The mystery of maize. *Field Museum of Natural History Bulletin* 43(10): 2–11

Darwin, C. (1859) *On the Origin of Species by Means of Natural Selection, or the Preservation of Favoured Races in the Struggle for Life*, London: John Murray; many reprints

Doebley, J. (1990) Molecular evidence and the evolution of maize. *Economic Botany* 44 (Suppl. 3): 6–27

Doebley, J., Stec, A., Wendel, J. & Edwards, M. (1990) Genetic and morphological analysis of a maize–teosinte F_2 population: implications for the origin of maize. *Proceedings of the National Academy of Sciences USA* 87: 9888–92

Doggett, H. (1965) Disruptive selection in crop development. *Nature* 206: 279–80

Dorweiler, J., Stec, A., Kermicle J. & Doebley, J. (1993) Teosinte glume architecture 1. A genetic locus controlling a key step in maize evolution. *Science* 262: 233–35

Emerson, R.A. & Beadle, G.W. (1932) Studies of *Euchlaena* and its hybrids with *Zea*. II. Crossing-over between the chromosomes of *Euchlaena* and those of *Zea*. *Zeitschrift für inductive Abstammungs und Vererbungslehre* 62: 305–15

Eubanks, M. (1995) A cross between two maize relatives: *Tripsacum dactyloides* and *Zea diploperennis* (Poaceae). *Economic Botany* 49: 172–82

Galinat, W.C. (1956) Evolution leading to the formation of the *cupulate fruitcase* in the American Maydeae. *Bot. Mus. Leafl. Harvard University* 17: 217–39

Galinat, W.C. (1963) Form and function of plant structures in the American Maydeae and their significance for breeding. *Economic Botany* 17: 51–59

Galinat, W.C. (1970) The cupule and its role in the origin and evolution of maize. *Massachusetts Agricultural Experimental Station Bulletin* 585: 1–18

Galinat, W.C. (1971) The origin of maize. *Annual Review of Genetics* 5: 447–78

Galinat, W.C. (1973) Intergenomic mapping of maize, teosinte and *Tripsacum*. *Evolution* 27: 644–55

Galinat, W.C. (1992a) Evolution of corn. *Advances in Agronomy* 47: 203–31

Galinat, W.C. (1992b) Corn, Columbus, and culture. *Perspectives in Biology and Medicine* 36: 1–12

Galinat, W.C. (1994) Plant habit as evidence of a biphyletic domestication of teosinte in the origins of maize. In *Corn and Culture in the Prehistoric New World*, edited by S. Johannessen and C.A. Hastorf, Boulder, Colorado: Westview Press

Galinat, W.C. (1995) The origin of maize. Grain of humanity. *Economic Botany* 49: 3–12

Iltis, H.H. (1972) The taxonomy of *Zea Mays* L. (Gramineae). *Phytologia* 23: 248–49

Iltis, H.H. (1983) From teosinte to maize: the catastrophic sexual transmutation. *Science* 22: 886–94

Kato, Y.T.A. (1984) Chromosome morphology and the origin of maize and its races. *Evolutionary Biology* 17: 219–53

Long, A., Benz, B.F., Donahue, D.J. *et al.* (1989) First direct AMS dates on early maize from Tehucan, Mexico. *Radiocarbon* 31: 1035–40

Longley, A.E. (1937) Morphological characters of teosinte chromosomes. *Journal of Agricultural Research* 54: 835–62

Longley, A.E. (1941) Chromosome morphology in maize and its relatives. *Botanical Review* 7: 263–89

Mangelsdorf, P.C. (1948) The role of pod corn in the origin and evolution of maize. *Annals of Missouri Botanical Garden* 36: 377–406

Mangelsdorf, P.C. (1974) *Corn: Its Origin, Evolution and Improvement*, Cambridge, Massachusetts: Harvard University Press

Mangelsdorf, P.C. & Reeves, R.G. (1939) The origin of Indian corn and its relatives. *Texas Agricultural Experimental Station Bulletin* 574: 1–315

Mangelsdorf, P.C. & Galinat, W.C. (1964) The truncate locus in maize dissected and reconstituted. *Proceedings of the National Academy of Sciences USA* 51: 147–50

Rogers, J.S. (1950) The inheritance of inflorescence characters in maize–teosinte hybrids. *Genetics* 35: 541–58

Sheridan, W.F. (1988) Maize developmental genetics. *Annual Review of Genetics* 22: 353–85

Wallace, H.A. & Brown, W.L. (1956) *Corn and its Early Fathers*. East Lansing: Michigan State University Press

Wilkes, H.G. (1967) *Teosinte: The Closest Relative of Maize.* Thesis, Cambridge, Massachusets: Bussey Institute, Harvard University

GLOSSARY

abscission layer a group of cells at the base of a leaf which can separate from nearby cells so that the leaf dies and falls from the plant

apomixis in plants, a seed reproductive process without fertilization

axillary developing in the axil of a leaf, i.e. between the stalk and the stem from which it grew

cupule cup-shaped structure made of hardened fused bracts

meristem group of undifferentiated cells from which the growing point for a plant organ forms

monoecious having unisexual male and female flowers on the same plant

pachytene stage during prophase of meiosis when homologous chromosomes are associated as bivalents

rachis main axis of a compound leaf or flower cluster

rachilla secondary axis bearing the individual florets

spikelet the smallest unit of florets or flowers in the grass flower cluster, consisting of outer and inner bracts called glumes, and one or more florets with lemma subtending

tunicate covered with a papery outer skin or membrane

See also **Pinpointing einkorn wheat domestication by DNA fingerprinting** (p.634); **Barbara McClintock and transposable genetic sequences in maize** (p.640); **Maize: the long trail to QPM** (p.657)

LONG-TERM SELECTION FOR OIL AND PROTEIN IN MAIZE

J.W. Dudley
Department of Agronomy, University of Illinois, Urbana, USA

The longest continuous selection experiment in higher plants began at the University of Illinois in 1896 when C.G. Hopkins initiated selection for oil and protein concentration in maize grain. A cycle per year has been completed each year since 1896 except for 4 years during World War II. The 95th generation was completed in 1994. The initial objectives were to determine whether oil and protein concentration in maize grain could be altered by selection. As these objectives were met, the objectives changed to the development of corn with increased feeding or processing value. Later, the objectives changed again to determining the limits to selection and the implications of the results to selection theory and quantitative genetics.

The continued existence of this experiment would not have been possible without the persistence and vision of the scientists who have been responsible for the experiment including C G. Hopkins (1896–1900), L.H. Smith (1901–25), C.M. Woodworth (1921–51), E.R. Leng (1951–65) and, from 1965 to date, J.W. Dudley, R.J. Lambert and D.E. Alexander.

Hopkins initiated selection by determining the oil and protein percentages in the grain for each of 196 ears from the open-pollinated maize cultivar Burr's White. The 24 ears highest in protein percentage were the foundation stock for the Illinois High Protein (IHP) strain, while the 12 ears lowest in protein percentage founded the Illinois Low Protein (ILP) strain. Likewise, the 24 ears highest in oil percentage were the beginning of the Illinois High Oil (IHO) strain and the 12 ears lowest in oil were the beginning of the Illinois Low Oil (ILO) strain. Each year, the selected ears were inter-mated to form the foundation stock for the next generation. Essentially, a form of mass selection was used. Details of the selection experiment have been reported and key papers reprinted in Dudley (1974). Following 48 generations of selection, reverse selection was initiated in each of the original strains to determine whether or not genetic variance had been exhausted. Reverse selection gave rise to the Illinois Reverse High Protein (RHP), Reverse Low Protein (RLP), Reverse High Oil (RHO) and Reverse Low Oil (RLO) strains, and easily demonstrated the continued existence of genetic variability.

Response to selection, as measured by changes in population means, is shown in **Figures 1** and **2**. Limits to selection have been reached only in the ILO and ILP strains. Selection was stopped after 87 generations in ILO because oil percentage was so low it could no longer be measured accurately and the population was difficult to maintain. No progress has been made in ILP for at least 25 generations, suggesting a physiological limit has been reached for low protein – at about 4% protein. In the high strains, selection has continued to the point that IHP has approximately 32% protein and IHO has approximately 22% oil. IHP and IHO show no evidence of having reached an upper limit. Dudley (1977) suggested the results could be accounted for by segregation of a relatively large number of loci with low average frequencies of favourable alleles in the original population.

The number of effective factors segregating in the original population was estimated, using data from 90 generations

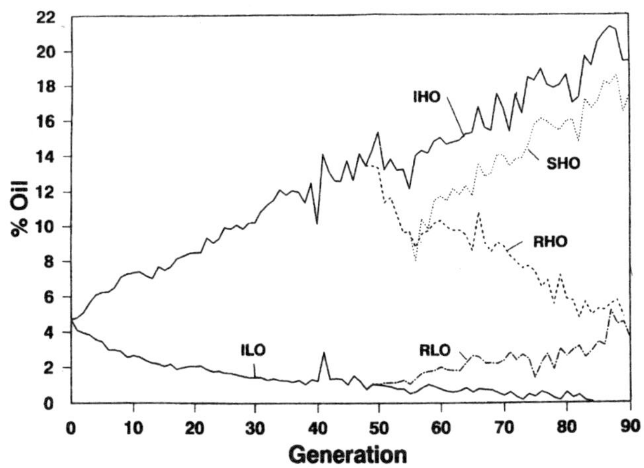

Figure 1 Mean oil percentage by cycle for strains selected for oil concentration in the grain. IHO, Illinois High Oil; RHO, Reverse High Oil; SHO, Switchback High Oil; ILO, Illinois Low Oil; RLO, Reverse Low Oil. Reproduced with permission from Dudley & Lambert (1992).

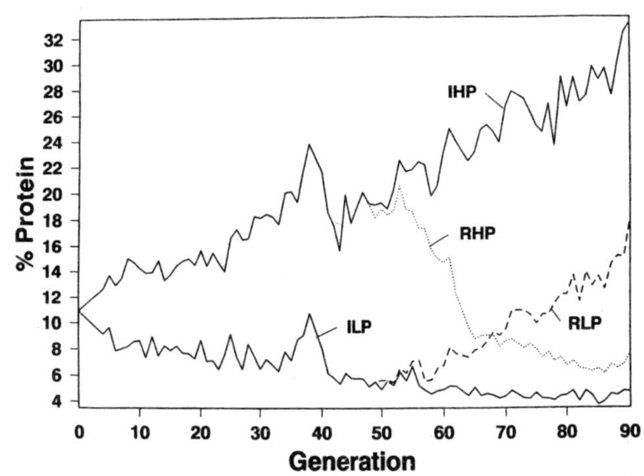

Figure 2 Mean protein percentage by cycle for strains selected for protein concentration in the grain. IHP, Illinois High Protein; RHP, Reverse High Protein; ILP, Illinois Low Protein; RLP, Reverse Low Protein. Reproduced with permission from Dudley & Lambert (1992).

(Dudley & Lambert, 1992), as 69 for oil and 173 for protein. Average frequencies of alleles for high oil and high protein in the original Burr's White were estimated as 0.22 and 0.32, respectively. Progress in IHO was approximately 20 and, in IHP, 26 additive genetic standard deviations. Dudley (1994) suggested that these results had major implications for breeding for yield. If the same amount of progress could be made for grain yield as has been made for oil in IHO and for protein in IHP, yields of approximately 700 bushels/acre (44 tons/hectare) could be achieved (the current yield record is 370 bushels/acre or 23.5 tons/hectare).

Correlated responses noted with selection include increased embryo size, oil percentage in the germ and germ percentage in the kernel in the IHO strain. Embryo size, oil percentage in the germ and germ percentage were reduced in the ILO strain. In addition, weight per 100 kernels in IHO was reduced relative to ILO. Starch percentage decreased in both IHP and IHO with selection (Dudley & Lambert, 1992). In fact, the IHP and ILP strains have the largest divergence in starch percentage of any two corn strains we have measured. Since the effective population size has been relatively small for such a large number of generations, the inbreeding coefficient is of the order of 90%. Thus, the strains appear highly uniform and are low yielding.

These strains are now being used as tools to help understand the genetic control of chemical composition in maize grain. Using molecular marker technology (restriction fragment length polymorphisms), chromosomal regions controlling protein, oil and starch percentages are being identified (Goldman *et al.*, 1993, 1994). In addition to basic research information, the ease of increasing oil percentage demonstrated in this experiment provided the impetus for a breeding effort aimed at making high oil corn feasible for commercial production.

As the centenary of the beginning of the experiment approaches, and passes, selection will continue with the aim of determining the upper limits to concentration of oil and protein in corn grain and of developing a greater understanding of the genetic control of chemical composition in the maize kernel.

REFERENCES

Dudley, J.W. (ed.) (1974) *Seventy Generations of Selection for Oil and Protein in Maize*, Madison, Wisconsin: Crop Science Society of America

Dudley, J.W. (1977) Seventy-six generations of selection for oil and protein percentage in maize. In *Proceedings of the International Conference on Quantitative Genetics*, edited by E. Pollak, O. Kempthorne & T.B. Bailey, Jr, Ames: Iowa State University Press

Dudley, J.W. (1994) Plant breeding: a vital part of improvement in crop yields, quality, and production efficiency. In *Historical Perspectives in Plant Science*, edited by K.J. Frey, Ames, Iowa: Iowa State University Press

Dudley, J.W. & Lambert, R.J. (1992) Ninety generations of selection for oil and protein in maize. *Maydica* 37: 1–7

Goldman, I.L., Rocheford, T.R. & Dudley, J.W. (1993) Quantitative trait loci affecting protein and starch concentration in the Illinois Long Term Selection maize strains. *Theoretical and Applied Genetics* 87: 217–24

Goldman, I.L., Rocheford, T.R. & Dudley, J.W. (1994) Molecular markers associated with maize kernel oil concentration in an Illinois high protein × Illinois low protein cross. *Crop Science* 34: 908–14

GLOSSARY

cultivar plant variety found only in cultivation

germ percentage percentage by weight of the corn kernel which is germ

reverse selection selection in the opposite direction from the original selection

See also **Maize: the long trail to QPM** (p.657)

MAIZE: THE LONG TRAIL TO QPM

Oliver E. Nelson
Laboratory of Genetics, University of Wisconsin-Madison, Madison, USA

1. Introduction

QPM is an abbreviation of "quality protein maize". This is the designation given to genotypes homozygous for the recessive mutant, opaque2 (*o2*), and incorporating the semi-dominant modifying genes that convert the soft, floury endosperms characteristic of most *o2* genotypes to vitreous endosperms similar to those of non-mutant maize. This phenotypic shift, together with selection for more productive synthetic varieties and hybrids, has made the QPMs promising candidates to replace non-mutant maizes on much of the world's acreage devoted to maize cultivation. Plant breeders have realized the potential envisioned for opaque2 maize ever since the initial reports of the increased lysine (Mertz *et al.*, 1964) and tryptophan (Nelson *et al.*, 1965) in *o2* endosperms. To save space, *o2*/*o2*/*o2* endosperms or *o2*/*o2* seeds will be described simply as *o2*. The full genotype will be entered for heterozygotes.

Maize is the cereal staple of diet for millions of people in Latin America and subSaharan Africa as well as in some areas of Asia. In addition, it is an important livestock food, particularly in North America. However, the collective proteins of maize seeds are deficient in two essential amino acids, lysine and tryptophan, which must be supplied from other sources for an adequate diet for humans or monogastric animals. This is especially important for the young and for pregnant females. The idea was prevalent in the 25 years following World War II that those segments of populations subsisting largely or wholly on this cereal staple were severely disadvantaged. This prompted efforts to ascertain whether there were variants within the major cereals that could redress to some extent these deficiencies in essential amino acids. In the 1970s, however, some influential nutritionists asserted that the so-called "protein gap" was a myth and that maize would provide adequate protein nutrition even for very young children and for pregnant females provided that sufficient maize was available to supply the need of the individual for calories. While this idea was prevalent, it was a marked disincentive to attempts to improve further the already greatly improved *o2* varieties or to induce their production by populations that could have benefited from their enhanced protein value. With more recent data (Young, 1992), showing that maize protein is indeed limiting for children and some adults in situations where little or no animal protein is available, the time is propitious for widespread use of the QPMs that have been available for over a decade.

The hope of detecting variants of maize that would alleviate this deficiency of lysine and tryptophan in the collective seed proteins prompted Edwin Mertz at Purdue in the US to establish a screening programme of some tropical varieties and of mutants that alter the seed phenotype. It was based on the hypothesis that, in a variant with an impaired ability to synthesize the alcohol-soluble storage proteins of the endosperm, the zeins, the nitrogen used ordinarily in zein synthesis might enhance the synthesis of non-zein proteins, which should have higher contents of these two amino acids. It had been known since early in the 20th century that the zeins, which constitute over 50% of endosperm protein, are devoid of lysine and tryptophan.

In the fall of 1962, after attending a lecture by Professor Mertz in which he outlined the hypothesis on which he was working, I suggested that the opaque and floury mutants were reasonable candidates to be deficient in zein. The basis was the knowledge that the long-term selection at the University of Illinois for maize with higher seed protein resulted in seeds with especially vitreous endosperms, and it had been shown that the increase in protein was due largely to increased zein. The selection for maize with lower protein content had resulted in seeds in which zein was reduced, and the endosperm lacked vitreous starch and was soft and floury. After providing samples of opaque1, opaque2, floury1 and floury2 for analysis by Mertz's laboratory, the answers were quickly available: both opaque2 and floury2 endosperms contained less zein and had significantly higher lysine and tryptophan contents than non-mutant maize. There was no difference in the amino acid composition of the embryo proteins, and the opaque1 and floury1 mutants were not different from ordinary maize.

These data showed that Mertz's hypothesis (that the type of change in protein synthesis, which would presumably give increased lysine, might exist) was productive, but I should note that his contribution to the development of QPMs went well beyond this heuristic idea. Beyond the tests on rats described below that validated the nutritional superiority of opaque2 maize as compared with common maize, he has been a constant contributor to the improvement of QPM maize even years after his formal retirement. From assisting scientists at the Centro Internacional de Mejoramiento de Maiz y Trigo (CIMMYT) in developing the biochemical tests that were vital to the success of their breeding programme to organizing meetings at which research results could be reported and then editing the proceedings of the meetings, he has greatly helped in the recognition of the potentiality of the QPMs.

After the realization that the kernels of these mutants had higher lysine and tryptophan, priority was placed on testing whether these increases would be biologically available and support better growth for experimental animals when maize supplied the only protein in the diet. Weanling rats fed on a diet containing 90% opaque2 maize plus corn oil, a salt mixture and a vitamin supplement gained weight more than three times as rapidly as rats on a similar diet in which a standard hybrid maize replaced the opaque2 maize. The rats on the opaque2 maize grew as rapidly as rats on a diet that had 20% soybean oil meal, with corn starch supplying the carbohydrate content (Mertz *et al.*, 1965). The rat trials prompted trials with swine (Cromwell *et al.*, 1967) in which young pigs fed opaque2 maize had a growth rate 3.5 times faster than those fed on common maize. It was clear that the increased lysine and tryptophan are available to support growth. The nutritive value of *o2* was tested before floury2 since it has somewhat higher lysine. Subsequent tests of floury2 showed it to be inferior to *o2* in nutritive value except for chicks where the higher methionine content of floury2 is useful. The floury2 mutant will not be further discussed.

The trials testing the nutritional value of *o2* maize for children were equally encouraging. Nitrogen balance experiments with Guatemalan children fed on *o2* maize demonstrated that the nutritive value of *o2* protein was nearly equivalent to milk protein when fed at protein levels of 1.8–2.0 g/kg per day (Bressani, 1966). Subsequently, it was shown in Columbia that children could recover from kwashiorkor (malnutrition due to severe protein deficiency) on diets in which part of the nitrogen in their diets came from *o2* protein and the remainder from non-essential nitrogen sources, such as glycine and diammonium citrate, to compensate for the protein content of the maize being too low for children in such a debilitated state (Pradilla *et al.*, 1973).

The realization that a mutant gene could condition this increase in the two most limiting amino acids in maize proteins stimulated considerable activity in two areas: one was the attempt to exploit the nutritional advantage of *o2* by deriving *o2* versions of elite hybrids and varieties; the second was the attempt to find variants in other cereals that would similarly increase the nutritive value of those cereals. Such mutants have been identified in barley (Munck *et al.*, 1970) and in sorghum (Singh & Axtell, 1973), but not in oats, rice or wheat. In both barley and sorghum, as in maize, a large fraction of the endosperm protein consists of alcohol-soluble prolamines. These proteins are reduced in the mutants with a concomitant increase in the water- and salt-soluble proteins. Since both oats and rice seeds have relatively small proportions (<10%) of prolamines in their endosperms, mutations reducing the prolamines would have little effect. Wheat is a different problem owing to its hexaploid genotype; a recessive mutant on one of the three homeologous chromosomes would have little effect.

2. The breeding efforts

Initial attempts to capitalize on the desirable attributes of *o2* were not encouraging Conversions of inbred parents of elite hybrids to *o2* and making *o2* versions of the hybrids showed that there was a yield penalty of at least 7% and more in some conversions. In addition, the *o2* versions dried slowly and, therefore, were susceptible to fungal infection. The soft endosperms often broke during machine harvesting and predisposed the seeds during storage to insect infestation. The negative aspects associated with *o2* hybrids discouraged almost all hybrid seed corn companies and university research programmes from substantial research. The same factors had a similarly inhibitory effect on national programmes, even in countries where maize is the cereal staple of human diets and the greater nutritive value of *o2* maize would be especially useful.

There were, however, suggestions of approaches that might rectify the negative aspects. A report from the University of Missouri (Paez *et al.*, 1969) indicated that in some outcrosses of *o2* to non-opaque lines advanced generations had some kernels with vitreous sectors in the endosperm, and that these kernels had as high lysine as the wholly opaque kernels. So, one might select in such materials for modifiers of the *o2* mutation in the hope of completing the transformation to vitreosity. A second suggestion was that other mutants could modify the phenotype of *o2* endosperms in double-mutant combinations. A third suggestion was that selection in non-opaque lines might raise the content of lysine to equal that observed in *o2* endosperms.

The most intensive exploration of these proposed paths to high lysine maize with hard endosperms has been carried out at the CIMMYT in Mexico in a programme supported by the United Nations Development Fund and in which Dr S.K. Vasal has provided continuity and leadership from its inception. The CIMMYT scientists satisfied themselves that selection in non-opaque varieties, while showing some initial progress, did not lead to lysine contents approaching those in *o2* maize. The modification of *o2* phenotypes by a second mutant was also found to be unlikely to achieve the desired goal.

However, preliminary investigations at CIMMYT indicated that continued selection in *o2* varieties for kernels with the most vitreous endosperms, combined with assays of the limiting amino acids in those kernels, had the potential of producing seeds with hard endosperms and nearly as high lysine as the original *o2* lines. The assays were essential to the selection process since not all kernels with modified or partially modified endosperms retained the desirable protein quality. As an extreme example, Pollaczek, Caenen and Rousset (Pollaczek *et al.*, 1972) identified two simply inherited recessive suppressors of *o2* that condition vitreous endosperms and a normal amino acid content.

The selection process demanded the close cooperation of a laboratory capable of making a large number of analyses for the critical amino acids, and this was provided by a laboratory headed by Evangelina Villegas. Analyses with an amino acid analyser are much too slow and provide more information than needed, so the laboratory adapted colorimetric methods to assay tryptophan and lysine.

With this collaboration, CIMMYT breeders launched a major effort to select modified endosperm *o2* families in tropical and subtropical germplasm by a programme of within- and between-ear selection continued for a few cycles. This programme resulted in a number of hard endosperm *o2* families which could be used as the sources for converting a wide range of tropical and subtropical races by a process of backcrossing with recurrent selection. During this procedure, emphasis was also placed on yield and disease resistance. By 1977–78, CIMMYT had developed hard endosperm QPM versions of most of their populations. Many of these QPM selections performed as well in yield trials as non-opaque progenitors. At this point, genetically similar populations were merged to reduce the volume of QPM material, with the merged populations being improved by further selection.

This continued until 1987 when population improvement was discontinued in favour of developing inbred lines that could be used either to produce hybrids or combined to form synthetic varieties. An excellent account of the CIMMYT programme to improve *o2* maize can be found in a review by Vasal (1994).

Many of these CIMMYT QPMs, or hybrids descended from them, are being grown around the world, and the approximate acreages of the preferred maize during the last growing season are given below for the participating countries. In South America, Bolivia grows 5000 hectares (ha) of Tuxpeno Opaco. Brazil, which has a very active breeding programme, grows 120 000 ha of BR 451 and 30 000 ha of BR473. Ecuador grows 5000 ha of INIAP-528, Peru 4000 ha of Opaco Huascaran and Venezuela has 12 000 ha of Funit-4 and 3000 ha of Funiprot. In Africa, Ghana is a major grower with 70 000 ha of Obtanpa. In Asia, China is seriously involved in the breeding and growing of QPM maize. In southern China, approximately 250 000 ha of Tuxpeno 102 are grown, while in northern China 70 000 ha of Lu Yu 13 are grown. Other hybrids are grown as well, but the acreages are not available (Vasal, personal communication). The Ukraine is also interested, but data concerning the hectares grown were not available.

Another successful breeding programme to correct the deficiencies inherent in the first *o2* hybrids was initiated by Gevers in South Africa (Gevers & Lake, 1992). This has produced productive modified endosperm *o2* hybrids for use in South Africa. The experience there has paralleled that at CIMMYT except that the necessary analyses were for lysine using a short column on an amino acid analyser. By 1977, this programme released its first white high-lysine hybrid, HL1. This hybrid was competitive in yield with the best non-opaque hybrids, but the endosperm was still too soft to be acceptable to dry millers who utilize most of the white corn to prepare products popular with the large maize-eating population. It has been supplanted by HL hybrids (e.g. HL8) with much harder endosperms and satisfactory yields. The first yellow endosperm, HL2, was released in 1982. Intended for animal feed, it is satisfactory in yield and endosperm hardness. In 1995, it was estimated that approximately 10 000 hectares were planted to HL hybrids (Gevers, personal communication).

In the US, the Crow's Hybrid Corn Company has been the seed company with the greatest commitment to the breeding of improved *o2* hybrids. Incorporating modifiers found in midwestern germ plasm, their breeders have produced hybrids which have excellent yields and correct the tendency of the original *o2* hybrids to dry slowly and to be more susceptible to fungal attacks and insect infestation. At the same time, they have tried to preserve an endosperm texture that is less hard than non-opaque maize since their researchers believe that such a texture contributes to improved digestibility by domestic animals (Briggs, 1994).

3. Pertinent research

It would be amiss not to note some of the research that has been informative concerning the prolamine fraction of maize storage proteins and how the significant changes in amino acid composition of *o2* endosperms occur. A number of laboratories have participated, but the most sustained investigations have been carried out by Brian Larkins, at Purdue and now at the University of Arizona.

The zeins are a heterogeneous group of multigene families that agree only in their solubility in alcohol and complete lack of lysine and tryptophan. The major components of the zeins are the a-zeins (Mr: 19 000 and 22 000), where Mr is the apparent molecular weight determined by the protein's migration rate relative to those of several standards of known molecular weight on a denaturing gel. There are also b-zein (Mr: 14 000), d-zein (Mr: 10 000) and the g-zeins (Mr: 12, 16 and 27 000), The a-zeins have low contents of cysteine and methionine, but the other zeins are all sulfur-rich proteins and require a reducing agent for solubilization. It is the a-zeins that are reduced most by the *o2* mutations. The *o2* locus was cloned first at the Brookhaven National Laboratory by transposon tagging and shown to be a transcription factor for the 22 000 a-zeins (Schmidt *et al.*, 1987).

Following the realization that modifiers of *o2* existed, which conditioned an approach to the desirable vitreous endosperm phenotype without greatly diminishing the lysine content, the Bergamo laboratory of Francesco Salamini demonstrated that the zeins extractable by alcohol with a reducing agent were increased in the modified endosperms (Gentinetta *et al.*, 1975). The Larkins laboratory (Or *et al.*, 1993) found that the 27 000 g-zeins are responsible for the increase, which is 2–3-fold higher than in unmodified *o2* endosperms. Of the two contiguous genes, A and B, present at the 27 000 g-zein locus in most genotypes, it is specifically the product of the A gene that is enhanced. Since the transcription of the g-zeins is reduced by about 50% in both *o2* and modified *o2* endosperms, the authors suggest that the modifying genes post-transcriptionally stabilize the A mRNA.

The same laboratory has recently reported two investigations which should be useful to corn breeders in predicting which inbreds would produce the highest lysine contents if converted to modified *o2* versions and then in converting those inbreds. The initial report (Mertz *et al.*, 1964) of the increase in lysine in *o2* endosperms noted that the albumin and globulin fractions of endosperm proteins are increased in *o2*. However, the contribution of specific proteins to the lysine increase was unknown. Habben and colleagues (1995) have shown that a lysine-rich protein, elongation factor-1a (EF-1a), is significantly increased in *o2* endosperms in most but not all inbred backgrounds. More significantly, there is a high positive correlation between EF-1a content and lysine content in *o2* endosperms ($r^2 = 0.91$) and in non-opaque endosperms ($r^2 = 0.89$). For comparison, the correlation between two other abundant endosperm proteins, elongation factor-2 and sucrose synthase, and lysine content is lower – $r^2 = 0.07$ and 0.57, respectively. The ELISA test for EF-1a developed here is a rapid method for estimating lysine content in breeding operations and would offer some predictability of the best candidates for conversion into modified *o2* lines. However, the EF-1a protein could not account for more than 10% of the lysine increase in *o2* endosperms, indicating that there must be more lysine-rich proteins that are increased in amount.

The second investigation (Lopes *et al.*, 1995) showed that the CIMMYT-modified *o2* investigated differs from a non-opaque control by two semidominant modifier genes, both of which are located on chromosome 7. The g-zein locus is close to the centromere on the short arm of 7. One modifier gene is linked closely to that locus. The second modifier locus maps to an area close to the telomere on the long arm of 7. The obvious utility of this information for breeding programmes is clear.

It is ironic that, as the CIMMYT programme began to produce some excellent QPM hybrids, the breeding programme was terminated in the early 1990s by the CIMMYT board, which was strongly influenced by a nutritionist member who believed that the protein gap is not real (Cohen, 1995). The good news is that the advantages of QPM maize are beginning to be realized and the QPMs are being used in a number of countries where they can make a substantial contribution to the health of the populace. In response to these encouraging signs, CIMMYT reinstated their breeding programme for QPMs in 1996 and started campaigns to promote their use in targeted countries in Latin America (Mexico, Guatemala, Bolivia and Brazil), in Asia (China and India) and in Africa (Ghana, Mozambique, Ethiopia and South Africa).

The most recent meeting to focus on QPMs was the International Symposium on Quality Protein Maize: Improvement and Use held in December 1994 in Sete Lagoas, Brazil, under the auspices of the governmental agencies, CNPMS and EMBRAPA. The proceedings, edited by E.T. Mertz and Brian Larkins, have been published by the Agricultural Communication Service of Purdue University, and a copy may be obtained by applying to Professor Bruce Hamaker, Food Science Department at that university.

Acknowledgements

I am indebted to Dr Vasal for supplying the publications from CIMMYT relevant to QPM and for the estimates of current production of QPM maize, to Dr Gevers for estimates of production of their HL hybrids, and to Dr Larkins and Dr Motto for reprints of their *o2* research.

REFERENCES

Bressani, R. (1966) Protein quality of opaque2 maize in children. In *Proceedings of the Conference on High Lysine Corn*, Washington, DC: Corn Industries Research Federation

Briggs, R.W. (1994) Thirty years of commercial development of opaque2 high lysine corn. In *Quality Protein Maize: 1964–1994: Proceedings of the International Symposium on Quality Protein Maize Improvement and Use, Sete Lagoas, MG, Brazil, 1–3 December 1994*, CNPMS/EMBRPA, edited by E.T. Mertz & B.A. Larkins, Purdue: Purdue University Press

Cohen, J. (1995) A Mexican-bred super maize. *Science* 267: 825

Cromwell, G.L., Pickett, R.A. & Beeson, W.M. (1967) Nutritional value of opaque2 corn for swine. *Journal of Animal Science* 26: 1325–31

Gentinetta, E., Maggiore, T., Salamini, F., Lorenzoni, C. & Pioli, F. (1975) Protein studies in 46 opaque2 strains with modified endosperm texture. *Maydica* 20: 145

Gevers, H.O. & Lake, J.K. (1992) Development of modified opaque2 maize in South Africa. In *Quality Protein Maize*, edited by E.T. Mertz, St. Paul, Minneapolis: American Association of Cereal Chemists

Habben, J.E., Moro, G.O., Hunter, B.G., Hamaker, B.R. & Larkins, B.A. (1995) Elongation factor 1alpha is highly correlated with the lysine content of maize endosperm. *Proceedings of the National Academy of Sciences USA* 92: 8640–44

Lopes, M.A., Takasaki, K., Boswick, D.E., Helentjaris, T. & Larkins, B.A. (1995) Identification of two opaque2 modifier loci in Quality Protein Maize. *Molecular and General Genetics* 247: 603–13

Mertz, E.T., Bates, L.S. & Nelson, O.E. (1964) Mutant gene that changes protein composition and increases lysine content of maize endosperm. *Science* 145: 279–80

Mertz, E.T., Veron, O.A., Bates, L.S. & Nelson, O.E. (1965) Growth of rats on opaque2 maize. *Science* 148; 1741–42

Munck, L., Karlsson, K.E., Hagberg, A. & Eggum, B.O. (1970) Gene for improved nutritional value in barley seed protein. *Science* 168: 985–87

Nelson, O.E., Mertz, E.T. & Bates, L.S. (1965) Second mutant gene affecting the amino acid pattern of maize endosperm proteins. *Science* 150: 1469–70

Or, E., Boyer, S.K. & Larkins, B.A. (1993) Opaque2 modifiers act posttranscriptionally and in a polar manner on g-zein gene expression in maize endosperm. *The Plant Cell* 5: 1599–609

Paez, A.V., Helm, J.L. & Zuber, M.S. (1969) Lysine content of opaque2 maize kernels having different phenotypes. *Crop Science* 9: 251–52

Pollaczek, M., Caenen, M. & Rousset, M. (1972) Mise en evidence d'un deuxieme gene suppresseur du gene opaque2 chez le mais. *Annales de Genetique* 15: 173–76

Pradilla, A., Frances, C.A. & Linares, F.A. (1973) Studies on protein quality of flint phenotypes of opaque-2 modified maize. *Archivo Latinoamericano de Nutricion* 23: 217–23

Schmidt, R.J., Burr, F.A. & Burr, B. (1987) Transposon tagging and molecular analysis of the maize regulatory locus opaque2. *Science* 238: 960–63

Singh, R. & Axtell, J.D. (1973) High lysine mutant gene (hl) that improves protein quality and biological value of grain sorghum. *Crop Science* 13: 535–39

Vasal, S.K. (1994) High quality protein corn. In *Speciality Corns*, edited by A.R. Hallauer, Boca Raton, Florida: CRC Press

Young, V.R. (1992) Protein and amino acid requirements in humans: metabolic basis and current recommendations. *Scandinavian Journal of Nutrition* 36: 47–56

GLOSSARY

enzyme-linked immunosorbent assay (ELISA) serological assay where antibodies used to identify a particular substance are labelled by linkage to an enzyme.

endosperm nutritional tissue surrounding embryo in most seeds. In cereals, it constitutes the major portion of the seed

transposon tagging if one has a clone of a transposable element, it is possible to isolate a clone of any gene in which the transposable element has inserted itself to produce a discernible effect

See also **Long-term selection for oil and protein in maize** (p.655)

PEA GENETICS

T.H.N. Ellis
John Innes Centre, Norwich, UK

1. Historical background

Pea (*Pisum sativum*) has been the subject of genetical interest at least since Knight's experiments reported in 1799 and, according to Olby (1985), it seems likely that it is from Knight that Mendel (1866) later took his justification for studying this organism. (Mendel's experiments with peas are fully described in the article on "Gregor Mendel"; see also Tschermak-Seysenegg, 1951). This early genetical interest was also matched by cytogenetical studies (Cannon, 1903; Richardson, 1929), and linkage was reported by Vilmorin and Bateson (1911) before Bateson accepted its existence. The early start to pea genetics was not followed through as incisively as was the case for other organisms studied at that time. Nevertheless, an enormous amount of genetic analysis has been undertaken, such as the early work on flowering time (Hoshino, 1915; see also Murfet & Reid, 1993; Weller *et al.*, 1997), while the range and depth of these studies is exemplified by the *Contributions to the Genetics of Pisum* from de Haan (1930). One of de Haan's mutants would likely have been the key to the development of a transposon tagging system for pea had it not been subsequently lost.

2. Linkage

By 1925, a linkage map for pea was established (Wellensiek, 1925) and the early pea genetics was dominated by Lamprecht who undertook a great deal of linkage analysis based on this early start. Unfortunately, linkage analysis and the assignment of chromosomal locations of genes using translocation stocks was pursued with premature vigour and some dispute. For example, Lamm (1956) concluded that *gp* and *r* lay on the same chromosome (the current interpretation), but Lamprecht disagreed and Lamprecht's assessment has persisted in more recent maps (Blixt, 1974; Ellis *et al.*, 1992). The two issues of linkage groups and the assignment of linkage groups to chromosomes became a progressively entangled morass of conflicts; however, several laboratories have now independently derived equivalent maps of the seven pea linkage groups (Gilpin *et al.*, 1997; Hall *et al.*, 1997; Ellis *et al.*, 1998; Laucou *et al.*, 1998; Weeden *et al.*, 1998). These linkage maps have, in turn, been related to the seven chromosomes of the haploid pea karyotype (Fuchs *et al.*, 1998).

3. Mutagenesis

Despite these problems, a great deal of mutation genetics was undertaken, as is reviewed extensively by Blixt (1972). As a result, there is a huge number of mutants described for pea. Reference stocks for alleles at about 300 mutant loci are held at the John Innes Pisum germplasm collection (http://www.jic.bbsrc.ac.uk/germplas/pisum) and these cover a very wide range of phenotypic characters from meiotic to agravitropic mutants. Recent mutagenesis experiments have generated mutants affecting symbiotic associations with Rhizobia and Mycorrhiza, for which about 40 loci have been identified (La Rue & Weeden, 1992; Sagan *et al.*, 1993, 1994), starch biosynthesis (Wang *et al.*, 1990; Smith *et al.*, 1995) and stem branching characters (Symons & Murfet, 1997), augmenting the extant mutants such as the suite of mutants altered in compound leaf development (Marx, 1987; Hofer & Ellis, 1998). Several of these genes have been cloned (Bhattacharyya *et al.*, 1990; Hofer *et al.*, 1997; Lester *et al.*, 1997; Martin *et al.*, 1997) showing that the large genome size ($c.5 \times 10^9$ base pairs) of this organism, while presenting many difficulties (and much interest), does not preclude molecular analysis. The development of transformation procedures for pea (e.g. Schroeder *et al.*, 1993; Bean *et al.*, 1997) opens the way to new types of analysis, but is not yet sufficiently efficient for large-scale T-DNA tagging experiments.

4. Future investigations

The pea mutants reviewed by Blixt (1972), together with those isolated more recently, represent a resource for molecular analysis as genes we can clone and as phenotypes we can understand at a molecular level. To a large degree, pea as a model species has been superseded by maize, rice, Antirrhinum and Arabidopsis, but several aspects of biology are still best studied in pea. Mutants not known in other species are exemplified by the set of leaf form mutants (Marx, 1987; Hofer & Ellis, 1998), while it is not yet clear whether the genes regulating the Rhizobium and mycorrhizal symbiosis in pea (Sagan *et al.*, 1993, 1994) correspond to those identified in other legumes. The alignment of the pea linkage map with those of other legumes (Weeden *et al.*, 1992; Simon & Muehlbauer, 1997) opens the possibility of exploiting "model" legumes (Handberg & Stougaard, 1992; Cook *et al.*, 1997) to further the genetics of pea.

REFERENCES

Bean, S.J., Gooding, P.S., Mullineaux, P.M. & Davies, D.R. (1997) A simple system for pea transformation. *Plant Cell Reports* 16: 513–19

Bhattacharyya, M.K., Smith, A.M., Ellis, T.H.N., Hedley, C. & Martin, C. (1990) The wrinkled-seed character of pea described by Mendel is caused by a transposon-like insertion in a gene encoding starch-branching enzyme. *Cell* 60: 115–22

Blixt, S. (1972) Mutation genetics in *Pisum. Agricultural and Horticultural Genetics* 30: 1–293

Blixt, S. (1974) The pea. In *Handbook of Genetics*, volume 2, edited by R.C. King, New York: Plenum Press

Cannon, W.A. (1903) Studies in plant hybrids: the spermatogenesis of hybrid peas. *Bulletin of the Torrey Botanical Club* 30: 519–43

Cook, D., VandenBosch, K., De Bruijn, F.J. & Huguet, T. (1997) Model legumes get the nod. *Plant Cell* 9: 275–81

Ellis, T.H.N., Turner, L., Hellens R.P. *et al.* (1992) Linkage maps in pea. *Genetics* 130: 649–63

Ellis, T.H.N., Poyser, S.J., Knox, M.R., Vershinin, A.V. & Ambrose, M.J. (1998) *Ty1-copia* class retrotransposon insertion site polymorphism for linkage and diversity analysis in pea. *Molecular and General Genetics* 260: 9–19

Fuchs, J., Kühne, M. & Schubert, I. (1998) Assignment of linkage groups to pea chromosomes after karyotyping and gene mapping by fluorescent in situ hybridization. *Chromosoma* 107: 272–76

Gilpin, B.J., McCallum, J.A., Frew, T.J. & Timmerman-Vaughan, G.M. (1997) A linkage map of the pea (*Pisum sativum* L.) genome containing cloned sequences of known function and expressed sequence tags (ESTs). *Theoretical and Applied Genetics* 95: 1289–99

de Haan, H. (1930) Contributions to the genetics of *Pisum. Genetica* 12: 321–439

Hall, K.J., Parker, J.S., Ellis, T.H.N. *et al.* (1997) The relationship between genetic and cytogenetic maps of pea. II. Physical maps of linkage mapping populations. *Genome* 40: 755–69

Handberg, K. & Stougaard, J. (1992) *Lotus japonicus*, an autogamous, diploid legume species for classical and molecular genetics. *The Plant Journal* 2: 487–96

Hofer, J.M.I. & Ellis, T.H.N. (1998) The genetic control of patterning in pea leaves. *Trends in Plant Science* 3: 439–44

Hofer, J., Turner, L., Hellens, R. *et al.* (1997) *Unifoliata* regulates leaf and flower morphogenesis in pea. *Current Biology* 7: 581–87

Hoshino, Y. (1915) On the inheritance of the flowering time in peas and rice. *Journal of the College of Agriculture (Sapporo)* 6: 229–88

Knight, T.A. (1799) Experiments on the fecundation of vegetables. *Philosophical Transactions of the Royal Society* 89: 504–06

La Rue, T. & Weeden, N.F. (1992) The symbiosis genes of pea. *Pisum Genetics* 24: 5–12

Lamm, R. (1956) Localization of *Gp* and *R* in *Pisum. Hereditas* 42: 520–21

Laucou, V., Haurogné, K., Ellis, N. & Rameau, C. (1998) Genetic mapping in pea. 1. RAPD-based genetic linkage map of *Pisum sativum. Theoretical and Applied Genetics* 97: 905–15

Lester, D., Ross, J.J., Davies, P.J. & Reid, J.B. (1997) Mendel's stem length gene (Le) encodes a gibberellin 3-hydroxylase. *The Plant Cell* 9: 1435–43

Martin, D.N., Probesting, W.M. & Hedden, P. (1997) Mendel's dwarfing gene: cDNAs from the *Le* alleles and function of the expressed proteins. *Proceedings of the National Academy of Sciences USA* 94: 8907–11

Marx, G.A. (1987) A suite of mutants that modify pattern formation in pea leaves. *Plant Molecular Biology Reports* 5: 311–35

Mendel, G. (1866) Versuche über Pflanzen-Hybriden. *Verhandlungen der Naturforschendern Vereins Brünn* 4: 3–47

Murfet, I.C. & Reid, J.B. (1993) Developmental mutants. In *Peas: Genetics, Molecular Biology and Biotechnology*, edited by R. Casey & D.R. Davies, Wallingford: CAB International

Olby, R. (1985) *Origins of Mendelism*, 2nd edition, Chicago: University of Chicago Press

Richardson, E. (1929) A chromosome ring in *Pisum. Nature* 124: 578

Sagan, M., Messsager, A. & Duc, G. (1993) Specificity of the Rhizobium–legume symbiosis obtained after mutagenesis in pea (*Pisum sativum* L.). *New Phytologist* 125: 757–61

Sagan, M., Huguet, T. & Duc, G. (1994) Phenotypic characterization of nodulation mutants of pea (*Pisum sativum* L.). *Plant Science* 100: 59–70

Schroeder, H.E., Schotz, A.H., Wardley-Richardson, T., Spencer, D. & Higgins, T.J.V. (1993) Transformation and regeneration of two cultivars of pea (*Pisum sativum* L.). *Plant Physiology* 101: 751–57

Simon, C.J. & Muehlbauer, F.J. (1997) Construction of a chickpea linkage map and its comparison with maps of pea and lentil. *Journal of Heredity* 88: 115–19

Smith, A.M., Denyer, K. & Martin, C. (1995) What controls the amount and structure of starch in storage organs? *Plant Physiology* 107: 673–77

Symons, G.M. & Murfet, I.C. (1997) Inheritance and allelism tests on six further branching mutants in pea. *Pisum Genetics* 29: 1–6

Tschermak-Seysenegg, E. von (1951): The rediscovery of Gregor Mendel's work. *Journal of Heredity* 42: 163–71

Vilmorin, P. de & Bateson, W. (1911) A case of gametic coupling in *Pisum. Philosophical Transactions of the Royal Society, Series B* 84: 9–11

Wang, T.L., Hadavizideh, A., Harwood, A. *et al.* (1990) An analysis of seed development in *Pisum sativum* XIII. The chemical induction of storage product mutants. *Plant Breeding* 105: 311–20

Weeden, N.F., Muehlbauer, F.J. & Ladizinsky, G. (1992) Extensive conservation of linkage relationships between pea and lentil genetic maps. *Journal of Heredity* 83: 123–29

Weeden, N.F., Ellis, T.H.N., Timmerman-Vaughan, G.M. *et al.* (1998) A consensus linkage map for *Pisum sativum. Pisum Genetics* 30: 1–4

Wellensiek, S.J. (1925) *Pisum*-crosses II. *Genetica* 7: 337–59

Weller, J.L., Reid, J.B., Taylor, S.A. & Murfet, I.C. (1997) The genetic control of flowering in pea. *Trends in Plant Science* 2: 412–18

GLOSSARY

agravitropic movement or growth against the direction of gravity, away from ground, as in plant shoots

T-DNA (transferred DNA) a segment of DNA inserted into the chromosomes of a host plant cell. In T-DNA tagging a large number of independent transgenic plants are generated by Agrobacterium–mediated transformation. Such plant populations will include gene disruptions and thus "tag" mutant genes in a similar way to transposons

See also **Gregor Mendel** (p.62)

SUPER RICE FOR INCREASING THE GENETIC YIELD POTENTIAL

Gurdev S. Khush
International Rice Research Institute, Manila, Philippines

1. Introduction

Major increases in rice production have occurred during the last 25 years because of the large-scale adoption of high-yielding semidwarf varieties and improved management practices. World rice production doubled in a 25-year period, from 256 million tons in 1966 to 520 million tons in 1990. During this period, rice production increased at a slightly higher rate than the population. However, the rate of increase of rice production is now lower (1.5% per year) than the rate of increase of population (1.8% per year). If this trend is not reversed, severe food shortage will occur in the next century. The 2000 world population of 6.1 billion is likely to reach 8.2 billion in 2025. The population of rice consumers is increasing at a faster rate than that of the rest of the world and the number of rice consumers will probably increase by 60% during the next 25 years. It is estimated that demand for rice will exceed production by the early part of the 21st century (Pinstrup-Anderson, 1994).

Major increases in the area planted to rice are unlikely. The area planted to rice has remained stable since 1980. Moreover, a diminution in the area planted to rice is likely because of the pressures on good rice land from urbanization and industrialization. The increased demand for rice will have to be met from less land, with less water, less labour and less pesticides. Therefore, rice varieties with higher yield potential and better management practices are needed to meet the goals of increased rice production. In its strategy document for 2000 and beyond, the International Rice Research Institute (IRRI) accorded highest priority to increasing the genetic yield potential of rice (IRRI, 1989).

The yield potential of current high-yielding rice varieties in the tropics is 10 tons/hectare (ha) during the dry season and 6.5 tons/ha during the wet season. Plant physiologists have suggested that the physical environment in the tropics is not a limiting factor for increasing rice yields. Maximum yield potential has been estimated to be 9.5 tons/ha during the wet season and 15.9 tons/ha during the dry season (Yoshida, 1981).

2. Modification of the rice plant in the 1960s

Prior to the so-called green revolution, rice varieties were tall and leafy with weak stems and produced a total biomass (leaves, stems and grain) of about 12 tons. When nitrogenous fertilizer was applied at rates exceeding 40 kg/ha, these traditional varieties tillered profusely, grew excessively tall, lodged early and the biomass could not be increased by fertilization. These varieties had a harvest index (ratio of dry grain weight to total dry matter or biomass) of 0.3. Thus, out of 12 tons biomass, about 30% were grain indicating a maximum yield of about 4 tons/ha. To increase the yield potential of tropical rice, it was necessary to improve the harvest index as well as total biomass or nitrogen responsiveness. This was accomplished by reducing the plant height through the incorporation of a recessive gene *Sd-1* for short stature from a Chinese variety Dee-geo-woo-gen (**Figure 1**) (Khush, 1995). The first semidwarf variety IR8, developed at IRRI, also had a combination of other

Figure 1 Plots of a conventional rice variety on the left and an improved high-yielding variety on the right.

desirable features such as profuse tillering, dark green and erect leaves and sturdy stems. It had a harvest index of 0.45 and did not lodge when high doses of nitrogenous fertilizer were applied. With proper fertilization, it could produce 18 tons biomass/ha and a grain yield of 8 tons/ha. Thus, the yield potential of tropical rice was doubled from 4 to 8 tons/ha through modification of plant type. Since the development of IR8 in 1966, yield potential of rice has been improved at the rate of about 1% per year. Thus, IR72 released in 1980 produces a biomass of about 20 tons/ha and has a harvest index of 0.5; it yields 10 tons/ha under best management.

3. New plant type for increased yield potential

As discussed above, yield is a function of total dry matter or biomass and the harvest index. Therefore, in order to increase the yield potential of tropical rice further, we have to increase either the total biomass production or the harvest index or both. We conceptualized a plant type to increase the biomass to about 23 tons and harvest index to 0.55. Such a plant should produce a grain yield of about 12.5 tons or an increase of 25% (**Table 1**) over the yield of existing high-yielding varieties.

The harvest index can be increased by increasing the proportion of energy that is stored in the grain or by increasing the sink size. The sink size can be increased by:

- larger numbers of spikelets per panicle or ear;
- greater partition of photosynthesis in spikelet formation;
- increased spikelet filling;
- slow leaf senescence;
- maintenance of healthy root system; and
- increased lodging resistance.

Biomass can be increased by genetic manipulation as well as by better management practices. Varietal characteristics for increasing the biomass include:

- establishment of desirable leaf canopy structure;
- rapid leaf area development;
- rapid nutrient uptake; and
- increased lodging resistance.

To achieve these goals, a new plant type was conceptualized with the following attributes (**Figure 2**) (Khush, 1994):

- lower tillering capacity;
- no unproductive tillers;
- 200–250 grains per panicle;
- 90–100 cm height;
- very sturdy stems;
- dark green thick and erect leaves;
- thicker and deeper roots;
- multiple disease and insect resistance; and
- acceptable grain quality.

Table 1 Yield and components of rice.

	Conventional Rice	IR8	IR72	New Plant Type
Total biomass (tons/ha)	12	18	19	23
Harvest index	0.3	0.45	0.5	0.55
Yield (tons/ha)	4.0	8.0	9.5	12.5

(i) Reduced tillering

Increases in the yield potential of other cereals such as maize and sorghum have resulted from increases in sink size (ear size). Selection and breeding for large sink size were accompanied by a decrease in tiller number. Modern maize and sorghum varieties have a single culm (stem), whereas primitive maize and sorghum had a large number of tillers with small cobs (or ears) (Khush, 1993). Teosinte, the ancestor of maize, has 20–25 tillers and small cobs with a few grains. The agriculturists who domesticated maize in the Americas continued to select maize with large cobs and this resulted in reduced tiller number. By the 15th century, when maize was introduced into Europe, it had only 4–5 tillers. Further selection resulted in uniculm plants with very large cobs. Single culm sorghums were bred in the post-Mendelian era.

By contrast, modern rice varieties produce 20–25 tillers under favourable growth conditions. Only 14–15 of these tillers produce panicles, which are small, and the rest remain unproductive. Unproductive tillers compete with productive tillers for solar energy and mineral nutrients – particularly nitrogen. Elimination of the unproductive tillers could direct more nutrients to grain production; furthermore, dense foliage that results from excess tiller production creates a humid microenvironment favourable to disease and insect build-up. Reduced tillering also facilitates synchronous flowering and maturity and more uniform panicle size. Genotypes with lower tiller number are also reported to produce a larger proportion of heavier grains (Padmaja Rao, 1987).

The number of grains produced per unit of land area primarily determines the grain yield in cereal crops grown in high-yield environments without stress (Takeda, 1984). The number of grains per unit land area can be increased by increasing the number of panicles or the number of grains per panicle. The modern high-yielding varieties have a much greater panicle number than the traditional rice varieties they replaced, but there is a limit to how much the number of panicles can be increased. Additional tillers become unproductive. They produce excessive vegetative growth and have a higher proportion of unfilled grains – hence the approach to increase the number of grains per panicle rather than the number of panicles per unit area.

(ii) Grain density and grain filling percentage

Larger grains tend to be chalky and thus have lower market value. Medium-sized grains with high density (higher specific gravity) are more desirable. High-density grains tend to occur on primary branches of the panicle, whereas the grains of the secondary branches have a lower density (Ahn, 1986). Low tillering varieties have more primary tillers per unit land area and, thus, they should produce a higher proportion of high-density grains and contribute to increased yield potential.

A variable proportion of grains in rice varieties remains unfilled. Higher grain-filling rates are essential to achieve

Figure 2 Sketches of different plant types of rice. Left: tall traditional plant type; centre: improved high-yielding, high-tillering plant type; right: proposed low-tillering ideotype of super rice.

maximum yield. Photosynthate (carbohydrate) production in leaves and stems, and its translocation and accumulation in the grains, are prerequisites for higher grain filling rates. For higher photosynthate production, thick dark-green leaves which senesce (die) slowly are desirable. Thicker stems have more vascular bundles, which provide a more effective system to transport photosynthates to the grains. Thicker stems also contribute to lodging resistance.

(iii) Leaf characteristics

Light is used more efficiently if the leaves are erect (Yoshida, 1976) and, therefore, an erect leaf angle is a desirable trait for achieving high yield – photosynthesis is greater when leaves are exposed to light on both sides. Droopy leaves are exposed to light only on one side and also raise the relative humidity and reduce the temperature since they lessen light penetration and air movement. Thicker leaves generally contain more nitrogen per unit leaf area and have a higher photosynthetic rate. Thick and green leaves may show slower leaf senescence.

(iv) Growth duration

The optimum growth duration for maximum rice yield in the tropics is thought to be 120 days from seed to seed. In transplanted rice, varieties of shorter growth duration usually give lower yields when planted at conventional spacings, since they do not produce sufficient vegetative growth for maximum yield (Yoshida, 1976). A growth duration of about 120 days allows the plant to utilize more soil, nitrogen and solar radiation, and results in higher yields. However, for adaptation to various cropping systems, varieties with a varying growth duration of 100–130 days are required.

(v) Plant height, stem thickness, biomass production

Short stature reduces the susceptibility of rice crop to lodging and leads to higher harvest index. A plant height of 90–100 cm is considered ideal for maximum yield. Increased biomass production is not difficult to achieve when the rice crop is grown in a high solar radiation environment, provided there is an ample supply of nitrogen (Akita, 1989). Without strong, thick culms, however, increased biomass results in lodging, mutual shading, increased disease incidence and lower yield (Vergara, 1988) – hence the importance of sturdy stems for lodging resistance in raising the biomass production.

(vi) Root system

Roots are the foundations of plants, yet they remain relatively unstudied compared with the rest of the plant. Rice varieties differ as much in the plant parts below the soil surface as in the parts above ground. For example, different cultivars are known to have different lengths, degrees of branching, volumes and thickness. Thicker and deeper roots provide better anchorage and lodging resistance and healthy roots are more efficient at supplying nutrients, particularly during the grain-filling period.

(vii) Disease and insect resistance

For the full expression of yield potential, genetic resistance to diseases and insects is essential. Resistance is even more important under tropical conditions. The major diseases of rice are blast, bacterial blight, sheath blight, grassy stunt and tungro viruses; the major insects are brown planthopper, green leafhopper and stem borers.

(viii) Grain quality

In the tropics and subtropics, consumers prefer long, slender and translucent grains with intermediate amylose content and intermediate gelatinization temperature. Rices with such characteristics are soft and moist when cooked and remain soft on cooling. Short-grained rices with low amylose content and low gelatinization temperature cook sticky and are preferred in temperate areas such as Japan, Korea and China.

4. Breeding for new plant type

Breeding work on the new plant type popularly known as “super rice” was started in 1989. About 2000 entries from the IRRI germplasm bank were grown to identify parents or donors for various traits. Parents for low tillering, large panicles, thick stems, a vigorous root system and thick dark-green leaves were identified. Hybridization was undertaken in the 1990 dry season. To begin with, these parents were crossed with a short-statured parent and breeding lines with short stature and the above-mentioned traits were selected. These lines were intercrossed and plant types with the proposed ideotype were selected. To date, over 2100 crosses have been made and more than 110 000 pedigree nursery rows have been grown. Numerous breeding lines with targeted traits have been selected. These lines have short stature, 6–10 tillers, dark green and erect leaves and large panicles with 200–250 grains per panicle. A plant of one such line and its panicle are shown in **Figure 3**. The yield potential of these new plant type lines is being evaluated in replicated yield trials under various management practices. On the basis of observational trials, the following characteristics of the new plant type (NPT) lines have been observed:

- The NPT lines produced 6–10 tillers compared with 25–27 tillers for modern high-yielding lines.
- The number of grains per panicle in the NPT lines was 2–3 times greater than IR72, the modern high-yielding variety.
- Some of the NPT lines had 20% more grains per square metre of land than IR72 and thus a 20% larger potential sink size.

Figure 3 Plants of super rice on the left and modern high-yielding variety on the right.

- Photosynthesis per unit area of single leaves in the new plant type lines was 10–15% higher than that of IR72 at the vegetative and reproductive phases. This advantage was mainly due to a higher concentration of leaf nitrogen in the NPT lines.
- The NPT lines had greener, thicker and more erect leaves than those of IR72. They had one or two more functional leaves at flowering compared with IR72.
- The flag leaves of NPT lines (the keys to photosynthesis during grain filling) appeared to function, photosynthetically, longer than those of IR72. This may result in a longer grain-filling period and contribute to increased yield potential.
- The NPT lines had thicker and sturdier stems, and much greater lodging resistance.
- The growth duration of NPT lines was similar (110–120 days) to that of IR72.

(i) Improvements still needed

Rice germplasm is classified into two broad groups – indica and japonica. Indica varieties are grown in the tropics and subtropics. Modern high-yielding varieties, grown in the tropics and subtropics, belong to the indica group and those grown in the temperate areas belong to the japonica group. Some japonica varieties are grown in the tropics and these are called tropical japonicas, compared with temperate japonicas grown in Japan, Korea and northern China. Improved indicas have long slender grains whereas most of the japonicas have short bold grains. For developing the NPT lines, tropical japonica parents were used because many of them have large panicles, some have low-tillering ability and others have thick stems and vigorous root systems. However, they do also have short bold grains and, consequently, the NPT lines also have short bold grains. Preference in the tropics is for long slender grains. Therefore, a few tropical japonica parents with long slender grains were identified and crossed with NPT lines. NPT lines with long slender grains are now being selected.

The majority of the NPT lines lack resistance to tropical diseases and insects as the parents used for developing these lines are susceptible. However, donors for blast and bacterial blight were identified within the tropical japonica germplasm and utilized in the hybridization programme. NPT lines with resistance to blast and bacterial blight have now been selected. Genes for resistance to brown planthopper, green leafhopper and tungro viruses are being incorporated from the indica germplasm through backcrossing.

5. Prospects for super rice

It is expected that, by 2003, NPT lines with all the desirable traits will become available. These lines will be shared with the national rice improvement programmes and evaluated under local conditions. Those found suitable will be multiplied and released for on-farm production. Thus, NPT varieties should be grown widely by 2005. When planted widely in the tropics and subtropics, they will help to feed 300 million more rice consumers.

NPT lines would not be adapted for growing in the temperate areas as they lack cold tolerance, which is one of the most important adaptability traits of temperate rices. However, NPT lines would be useful parents for increasing the yield potential of temperate rice. Collaboration has been established with the rice improvement programmes of Korea and Egypt to breed NPT varieties adapted to local conditions.

NPT varieties will also be useful for increasing the level of heterosis (yield advantage) of hybrid rice. Most of the hybrid rices grown in China and elsewhere are based on indica germplasm and show yield advantage of 10–15%. The level of heterosis depends on the genetic distance between the parents. Since the NPT lines are based on japonica germplasm, hybrids between them and the elite indica lines show a yield advantage of 20–25% (Khush & Aquino, 1994). Thus, the NPT rices will also be useful in increasing the yield potential of hybrid rice.

6. Summary

Yield potential of tropical rice was doubled through modification of plant type in the 1960s. Modern high-yielding varieties have a harvest index of 0.5 and total biomass of 20 tons/ha and, thus, can yield 10 tons/ha under best management. Old varieties, on the other hand, have a harvest index of 0.3 and produce a biomass of about 12 tons/ha. Thus, their yield potential is about 4 tons/ha. The new plant type lines (super rices) have a harvest index of 0.55 and produce a biomass of about 23 tons/ha. Thus, their yield potential is likely to be 12.5 tons/ha. Numerous NPT lines have been produced. Ongoing breeding work aims to improve the grain quality and enable the incorporation of genes for disease and insect resistance. NPT lines with all the desired traits should become available by 2003. They will be shared with national rice improvement programmes for widescale evaluation. It is expected that these lines will be grown widely in the tropics and subtropics by 2005 and will help to feed 300 million more rice consumers. NPT lines will also be useful for increasing the yield potential of hybrid rices.

REFERENCES

Ahn, J.K. (1986) Physiological factors affecting grain filling in rice. Ph.D. Thesis, University of the Philippines at Los Baños, Philippines

Akita, S. (1989) Improving yield potential in tropical rice. In *Progress in Irrigated Rice Research*, Manila, Philippines: International Rice Research Institute

IRRI (International Rice Research Institute) (1989) *IRRI Towards 2000 and Beyond*, Los Baños, Philippines: International Rice Research Institute

Khush, G.S. (1993) Varietal needs for different environments and breeding strategies. In *New Frontiers in Rice Research*, edited by K. Muralidharan & E.A. Siddiq, Hyderabad, India: Directorate of Rice Research

Khush, G.S. (1994) Increasing the genetic yield potential of rice: prospects and approaches. *International Rice Commission Newsletter* 43: 1–8

Khush, G.S. (1995) Modern varieties – their real contribution to food supply and equity. *GeoJournal* 35: 275–84

Khush, G.S. & Aquino, R.C. (1994) Breeding tropical japonicas for hybrid rice production. In *Hybrid Rice Technology: New Developments and Future Prospects*, edited by S.S. Virmani, Manila, Philippines: International Rice Research Institute

Padmaja Rao, S. (1987) High density grain among primary and secondary tillers of short- and long-duration rices. *International Rice Research Newsletter* 12: 12

Pinstrup-Anderson, P. (1994) World food trends and future food security. Food Policy Report. Washington, DC: International Food Policy Research Institute

Takeda, T. (1984) Physiological and ecological characteristics of high yielding varieties of lowland rice. In *Proceedings of the International Crop Science Symposium*, Fukuoka, Japan

Vergara, B.S. (1988) Raising the yield potential of rice. *Philippine Technical Journal* 13: 3–9

Yoshida, S. (1976) Physiological consequences of altering plant type and maturity. In *Proceedings of the International Rice Research Conference*, Los Baños, Philippines: International Rice Research Institute

Yoshida, S. (1981) *Fundamentals of Rice Crop Science*, Los Baños, Philippines: International Rice Research Institute

GLOSSARY

bacterial blight disease of rice caused by a bacterium
grassy stunt a disease of rice caused by a virus
ideotype architect of a plant to give specific appearance
lodged early crop fell over at early stage after flowering
panicle part of the plant which bears grains
rice blast a disease of rice caused by a fungus
spikelet a part of panicle which develops into a grain
tiller flowering stem of a grass
tillered early plants produced tillers early
tungro virus a disease of rice caused by two virus particles
vascular bundle strand of plant vascular tissue, composed of phloem and xylem

See also **Rice genetics** (p.668)

RICE GENETICS: ENGINEERING VITAMIN A

Eleanore T. Wurtzel
Lehman College and The Graduate Center of The City University of New York, USA

1. Introduction

Carotenoids, derived from plant food sources, are converted in humans to vitamin A and other important compounds needed for growth and development. Certain carotenoids have been distinguished in protection against cancer, in promoting immune responses, as antioxidants and as photoprotectants. Carotenoids are also useful as natural colouring agents in foods and cosmetics. Endosperms of food crops, such as maize and wheat, are low in provitamin A (1–10%) as compared with non-provitamin A carotenoids (Graham, 1997). Rice, an important food staple worldwide, accumulates no carotenoids in its endosperm and is, therefore, associated with vitamin A deficiency affecting 250 million children in developing countries (Underwood & Arthur, 1996). It has been estimated that improved vitamin A nutrition would eliminate approximately 1.3–2.5 million annual deaths (Humphrey *et al.*, 1992). The effects of vitamin A deficiency are manifested as xerophthalmia (visual impairment), blindness, increased mortality due to heightened severity of childhood diseases and greater risk of maternal transmission of viruses such as HIV (Semba *et al.*, 1994). Vitamin A intervention programmes have proven effective over the short term, but difficulties have been found in reaching children at the highest risk and at maintaining serum levels required to eliminate vitamin deficiency for the long term (Underwood & Arthur, 1996).

In the 1990s, molecular biology research led to the discovery of genes encoding enzymes required for carotenoid biosynthesis. Such an achievement provided a successful alternative to alleviating worldwide vitamin A deficiency through the use of genetic engineering to improve levels of provitamin A carotenoids in rice. This biotechnological achievement is a success both for genetics and for public health. This accomplishment can be similarly applied to improving the carotenoid content and composition of other important food staples such as wheat and maize.

2. What are carotenoids?

Carotenoids are a large class of yellow, red and orange pigments derived from isoprenoids. Carotenoids are synthesized by all photosynthetic organisms, as well as some bacteria and fungi. Animals do not have the ability to synthesize carotenoids, but must obtain these nutritionally important compounds through dietary sources, typically plants. In plants, the biosynthesis of carotenoids is essential for plant growth and development; carotenoids function as accessory pigments in photosynthesis, as photoprotectors preventing photooxidative damage and as precursors to the plant hormone, abscisic acid. The presence of carotenoids in endosperm tissue also adds nutritional value; in humans and animals, dietary carotenoids are essential precursors to vitamin A and to retinoid compounds needed in development (Lee *et al.*, 1981; Bendich & Olson, 1989). Other non-provitamin A carotenoids, such as lycopene, lutein, zeaxanthin and others, also play beneficial roles in human health (Giovannucci *et al.*, 1995; Kohlmeier *et al.*, 1997; Sommerburg *et al.*, 1998). Carotenogenic bacteria and other non-photosynthetic organisms synthesize carotenoids to provide protection in high-light, oxygen-containing environments.

3. The carotenoid biosynthetic pathway in bacteria and plants

(i) Location of the biosynthetic pathway

In plant cells, the biosynthetic pathway takes place in the plastid compartment of the cell and the enzymes of the pathway are encoded by genes located in the nucleus (Cunningham & Gantt, 1998).

(ii) Carotenoids are terpenoids derived from isopentenyl pyrophosphate

Carotenoids are derived from a five-carbon isoprenoid building block, isopentenyl pyrophosphate (IPP), which is common to all terpenoid compounds such as rubber, menthol and taxol. All plastids have the ability to manufacture these IPP precursors through a plastid-specific biosynthetic route that is also found in bacteria (Lichtenthaler, 1999). Four molecules of IPP are combined to produce the 20-carbon isoprenoid GGPP (geranylgeranyl pyrophosphate), the first precursor to carotenoids and to a variety of other isoprenoid-derived pathways, including gibberellins, the phytol chain of chlorophyll, prenylquinones, tocopherols and other natural products (Chappell, 1995).

(iii) From GGPP to the first carotenoid intermediate, phytoene

The biosynthesis of all carotenoids (**Figure 1**) begins with the combination of two molecules of GGPP to produce phytoene, the first compound specific to the carotenoid biosynthetic pathway (Cunningham & Gantt, 1998). This step is catalysed by the enzyme PSY (phytoene synthase).

(iv) From phytoene to lycopene

Phytoene, a colourless compound, undergoes the addition of double bonds resulting in the red-coloured carotenoid intermediate, lycopene, most notably seen in red tomato fruits. In higher plants and cyanobacteria, these steps are catalysed by two enzymes, PDS (phytoene desaturase) and ZDS (zeta-carotene desaturase), while in bacteria, such as *Erwinia uredovora*, only one enzyme, CRTI, is required to convert phytoene to lycopene.

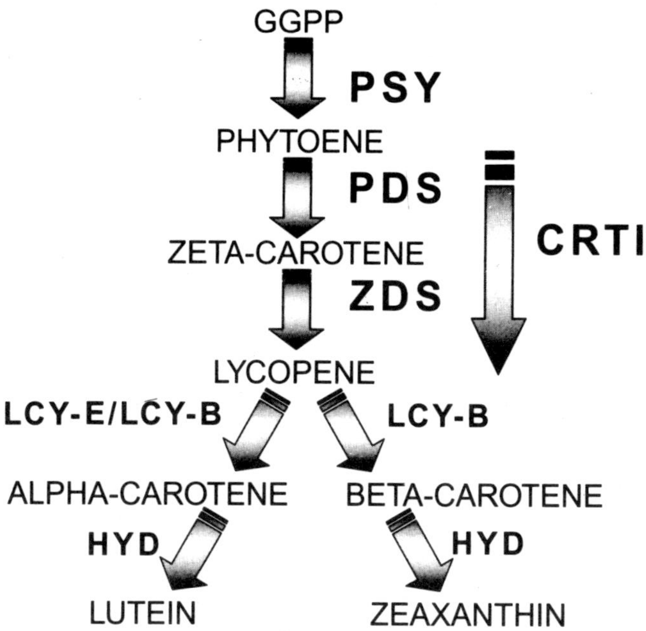

Figure 1 Carotenoid biosynthetic pathway. Enzymes are in bold and arrows point from enzyme substrate to product. Compounds: GGPP, geranylgeranyl pyrophosphate. Enzymes: PSY, phytoene synthase; PDS, phytoene desaturase (plant type); ZDS, zeta-carotene desaturase; CRTI, phytoene desaturase (bacterial type); LCY-B, lycopene beta cyclase; LCY-E, lycopene epsilon cyclase; and HYD, hydroxylase enzymes.

(v) From lycopene to provitamin A carotenes

Rings added by the enzyme LCY-B (lycopene beta cyclase) to both ends of the lycopene molecule result in the most active provitamin A carotenoid, beta-carotene, having two "beta" rings. Alternatively, LCY-E (lycopene epsilon cyclase), in combination with LCY-B, catalyses the biosynthesis of alpha-carotene, with one "epsilon" ring and one "beta" ring. In humans and animals, the central cleavage of beta-carotene results in two molecules of vitamin A (see **Figure 2**); cleavage of alpha-carotene results in only one molecule of vitamin A, which is derived from that half of alpha-carotene having the "beta" ring. As a result of the "epsilon" ring, alpha-carotene has only *half* the provitamin A activity compared with that of beta-carotene. Therefore, it is after lycopene formation that the pathway diverges, producing either more or less provitamin A active carotenoid, depending on the relative levels of the two cyclase enzymes LCY-E and LCY-B.

(vi) Conversion of provitamin A carotenes to non-provitamin A xanthophylls

After ring addition, both beta-carotene and alpha-carotene undergo addition of oxygen by HYD (hydroxylase) enzymes, giving rise to xanthophylls (oxygenated carotenoids) such as lutein (derived from alpha-carotene) or zeaxanthin (derived from beta-carotene). However, addition of oxygen further diminishes provitamin A activity. Other types of structural modifications give rise to the diversity of carotenoid compounds found in nature.

(vii) Summary of genes and enzymes needed for provitamin A carotenoid accumulation

Biosynthesis of the provitamin A carotenoid beta-carotene requires enzyme activity of PSY, PDS, ZDS and LCY-B (**Figure 1**). Alternatively, the bacterial enzyme CRTI takes the place of the two plant enzymes, PDS plus ZDS. HYD and LCY-E enzyme activities diminish the provitamin A value of carotenoids. In order to genetically engineer the pathway for provitamin A accumulation in any organism, the manipulation of the genes encoding the above enzymes is critical, as are genes encoding enzymes needed for accumulation of the carotenoid building blocks, IPP and GGPP.

4. Genetic engineering of carotenoid accumulation in rice and other plants

Survival of all photosynthetic organisms, including rice, depends on the production of carotenoids in green tissue. Therefore, one can assume that rice and other cereal crops all possess the genes needed for carotenoid biosynthesis, although these genes may not necessarily be expressed in seed tissue. From research on the pathway, it is understood that if pathway precursors are present, the accumulation of carotenoids such as beta-carotene is dependent on expression of the messenger RNAs encoding pathway enzymes. When efforts were first made to engineer the rice endosperm, it was determined that the plastid-containing endosperm lacked carotenoids, but did contain the carotenoid building block GGPP, making this tissue biochemically competent to biosynthesize carotenoids. To test this hypothesis, a gene encoding PSY from daffodil flowers was introduced into rice so that it was only expressed in endosperm tissue. As a result, phytoene accumulated in the rice endosperm (Burkhardt *et al.*, 1997). This indicated a potential for carotenoid accumulation in endosperm, but also that later pathway enzymes were still missing, since no compounds downstream of phytoene accumulated. Subsequently, the genes encoding daffodil PSY, LCY-B and a bacterial CRTI were introduced into rice and resulted in "golden rice" which accumulated beta-carotene (Ye *et al.*, 2000). However, it was surprising that some of the engineered rice also accumulated significant levels of xanthophylls, oxygenated carotenes that could only have been produced if the rice endosperm contained HYD enzyme activity. Since the HYD genes were not introduced as part of the genetic engineering effort, this result suggested that

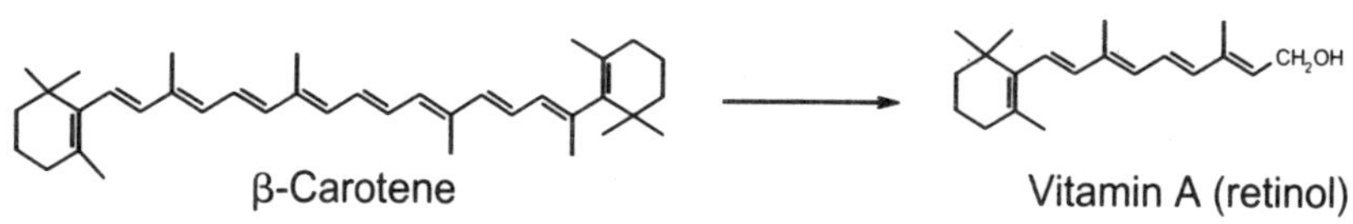

Figure 2 Conversion of betacarotene to vitamin A.

the rice endosperm already contained HYD enzyme activity.

In another study, where canola seeds were similarly engineered with the gene encoding PSY, a 50-fold increase in carotenoids was achieved (Shewmaker *et al.*, 1999). These transgenic (genetically engineered) canola seeds contained primarily beta-carotene and alpha-carotene. Compared with non-transgenic rice endosperm, which lacked carotenoid, non-transgenic canola seeds contained some carotenoids – mainly the xanthophyll lutein. The presence of xanthophylls indicated that non-transgenic canola seeds must have had HYD enzyme activity. However, this activity was insufficient to keep pace with a pathway engineered with an increased level of PSY enzyme, since the transgenic canola seeds accumulated mostly carotenes and little xanthophyll.

In contrast to the engineered rice endosperm where several pathway enzymes beyond PSY were required to confer carotenoid accumulation, increased carotenoid in canola seed was accomplished by introducing only one enzyme, PSY, suggesting that in non-transgenic canola seed PSY levels limit pathway throughput. This is not always the case, as was seen in the genetic engineering of a model bacterium where enhanced carotenoid accumulation was accomplished simply by increasing production of the carotenoid building block IPP (Matthews & Wurtzel, 2000).

5. Implications and future prospects

These first experiments to manipulate the carotenoid biosynthetic pathway are exciting and encouraging as they point to the enormous potential of introducing biochemical traits that will have a great impact on worldwide nutrition. From the rice and canola studies, it is clear that other important food crops, such as maize and wheat, will be good candidates for manipulation. The success in rice is the first example of a genetic engineering accomplishment in plants whereby a trait of nutritional value has been introduced and has potential for global impact. The pathway must be further introduced into native varieties, and other gene combinations incorporated to maximize pathway output for specific provitamin A carotenoids.

However, while the scientific problem appears to have been solved, there are still potential commercial problems. The technology used to create this rice involved as many as 32 companies holding 70 patents, each of which represents a barrier to further commercial development. Furthermore, the goal of the inventors of golden rice was to provide it at no cost to farmers in developing countries. In an effort to set a good example, Monsanto announced that they would grant patent licenses at no charge to developers of the "golden rice".

Another biotechnological success of engineering tobacco flowers to accumulate astaxanthin, a commercially valuable carotenoid (Mann *et al.*, 2000), points to a future of manipulating plants as biochemical factories instead of using traditional methods of organic synthesis, which sometimes provides impure mixtures. Plants genetically engineered for improved carotenoid composition will provide access to improved nutrition that is otherwise unavailable to much of the world's population.

REFERENCES

Bendich, A. & Olson, J. (1989) Biological actions of carotenoids. *FASEB Journal* 3: 1927–32

Burkhardt, P.K., Beyer, P., Wünn, J. *et al.* (1997) Transgenic rice (*Oryza sativa*) endosperm expressing daffodil (*Narcissus pseudonarcissus*) phytoene synthase accumulates phytoene, a key intermediate of provitamin A biosynthesis. *The Plant Journal* 11: 1071–78

Chappell, J. (1995) Biochemistry and molecular biology of the isoprenoid biosynthetic pathway in plants. *Annual Reviews of Plant Physiology and Plant Molecular Biology* 46: 521–47

Cunningham, F.X. & Gantt, E. (1998) Genes and enzymes of carotenoid biosynthesis in plants. *Annual Reviews of Plant Physiology and Plant Molecular Biology* 49: 557–83

Giovannucci, E., Ascherio, A., Rimm, E.B. *et al.* (1995) Intake of carotenoids and retinol in relation to risk of prostate cancer. *Journal of the National Cancer Institute* 87: 1767–76

Graham, R. (1997) Wheat: Research at Waite Agricultural Research Institute in Australia. *CGIAR Micronutrients Project* Update No. 2

Humphrey, J., West, K. Jr & Sommer, A. (1992) Vitamin A deficiency and attributable mortality among under-5-year-olds. *Bulletin of the World Health Organization* 72: 225–32

Kohlmeier, L., Kark, J.D., Gomez-Gracia, E. *et al.* (1997) Lycopene and myocardial infarction risk in the EURAMIC Study. *American Journal of Epidemiology* 146: 618–26

Lee, C., McCoon, P. & LeBowitz, J. (1981) Vitamin A value of sweet corn. *Journal of Agricultural Food Chemistry* 29: 1294–95

Lichtenthaler, H.K. (1999) The 1-deoxy-d-xylulose-5-phosphate pathway of isoprenoid biosynthesis in plants. *Annual Reviews of Plant Physiology and Plant Molecular Biology* 50: 47–65

Mann, V., Harker, M., Pecker, I. & Hirschberg, J. (2000) Metabolic engineering of astaxanthin production in tobacco flowers. *Nature Biotechnology* 18: 888–92

Matthews, P.D. & Wurtzel, E.T. (2000) Metabolic engineering of carotenoid accumulation in *Escherichia coli* by modulation of the isoprenoid precursor pool with expression of deoxyxylulose phosphate synthase. *Applied Microbiology and Biotechnology* 53: 396–400

Semba, R.D., Miotti, P.G. & Chiphangwi, J.D. *et al.* (1994) Maternal vitamin A deficiency and mother-to-child transmission of HIV-1. *Lancet* 343: 1593–97

Shewmaker, C.K., Sheehy, J.A., Daley, M., Colburn, S. & Ke, D.Y. (1999) Seed-specific overexpression of phytoene synthase: increase in carotenoids and other metabolic effects. *Plant Journal* 20: 401–12

Sommerburg, O., Keunen, J.E., Bird, A.C. & van Kuijk, F.J. (1998) Fruits and vegetables that are sources for lutein and zeaxanthin: the macular pigment in human eyes. *British Journal of Ophthalmology* 82: 907–10

Underwood, B.A. & Arthur, P. (1996) The contribution of vitamin A to public health. *FASEB Journal* 10: 1040–49

Ye, X., Al-Babili, S., Klöti, A., Zhang, J. *et al.* (2000) Engineering the provitamin A (beta-carotene) biosynthetic pathway into (carotenoid-free) rice endosperm. *Science* 287: 303–5

GLOSSARY

endosperm nutritional tissue surrounding the embryo in most seeds

biosynthetic pathway metabolic pathway in which small molecules are synthesized into larger organic molecules

isoprenoids compounds composed of one or more "isoprene units", a five-carbon saturated hydrocarbon

pathway engineering the application of recombinant DNA technology to alter or construct a sequence of enzymatically catalysed biochemical reactions that result from the activity of several gene products

See also **Super rice for increasing the genetic yield potential** (p.663)

WINEGRAPE ORIGINS REVEALED BY DNA MARKER ANALYSIS

Carole P. Meredith and John E. Bowers
Department of Viticulture and Enology, University of California, Davis, California, USA

Grapes are among the very oldest of fruit crops. Like many crops, grape offers interesting scientific challenges to plant geneticists. However, unlike most other crops, grape is also intensely interesting to the many people who enjoy its products and are intrigued by the history and complexity associated with it.

The wines with which most wine drinkers are familiar are produced from a relatively small number of classic European cultivars of *Vitis vinifera*. Most of these have existed for centuries and, since they are propagated vegetatively, they persist as genotypes that are largely unchanged since their conception. The origin of the famous wine varieties is a popular theme in the wine literature and is the subject of much speculation (Levadoux, 1956; Bouquet, 1982). Ancient texts have been scrutinized in an effort to match old names and vague descriptions of wine grapes cultivated long ago to the varieties cultivated today. The advent of molecular markers, particularly microsatellite markers, has provided a means to analyse objectively genetic relationships among wine varieties and, thereby, to identify the origins of some of them with a high degree of certainty.

As a wild species, *V. vinifera* was once widespread throughout southern Europe and western Asia (Zohary, 1996). Only remnant populations of truly wild *V. vinifera* remain today. The cultivated varieties are thought to have originated in several ways – initially, by the propagation of desirable wild individuals and, later, by the selection of individuals resulting from spontaneous crosses, both between cultivars and wild plants and also between cultivars. Some cultivars are the result of deliberate crosses made in the 19th and 20th centuries. While the original wild genotypes that may have contributed to the pedigrees of today's wine varieties are no doubt lost, relationships among cultivated genotypes still in existence can easily be deduced by the analysis of microsatellite loci.

One of the most highly esteemed winegrape varieties is Cabernet Sauvignon, long associated with the Bordeaux region of France where it has been grown for at least 300 years (Viala & Vermorel, 1901). Microsatellite analysis has now shown that Cabernet Sauvignon is, in fact, the progeny of two other wine varieties associated with Bordeaux, Cabernet franc and Sauvignon blanc (Bowers & Meredith, 1997). At each of 30 microsatellite loci analysed, Cabernet Sauvignon shares one allele with each of the parents (**Table 1**). By means of likelihood analysis, it was further shown that the probability that the alleles of Cabernet Sauvignon came from the two putative parents is 10^{14} times higher than the probability that they came from two random parents. Thus, it is not only possible that Cabernet franc and Sauvignon blanc are the parents, but it is also highly probable. Subsequent to our report, this parentage was further strengthened with data from 15 additional microsatellite loci (Sefc *et al.*, 1997).

Cabernet franc is not as well known as Cabernet Sauvignon, but it is a significant component of many Bordeaux wines and is also grown in the Loire Valley. Historical records suggest that it was being grown in Bordeaux several centuries before the first specific mention of Cabernet Sauvignon. Since the two cultivars are morphologically similar, a close relationship between the two has long been suspected. Sauvignon blanc, as its name indicates, is a white wine variety that is important in both Bordeaux and the Loire. As would be expected in the case of a red-fruited variety with a white-fruited parent, selfed seedlings of Cabernet Sauvignon segregate for fruit colour (Walker & Meredith, unpublished results). A close genetic relationship between

Table 1 Microsatellite alleles of Cabernet Sauvignon and its presumptive parents Cabernet franc and Sauvignon blanc.

Locus	Cabernet franc	Cabernet Sauvignon	Sauvignon blanc
VVMD5	226 240	232 240	228 232
VVMD6	205 211	211 212	205 212
VVMD7	239 263	239 239	239 257
VVMD8	147 157	143 157	143 143
VVMD14	228 235	222 228	222 232
VVMD16	168 168	168 168	168 168
VVMD17	212 221	221 222	220 222
VVMD21	249 258	249 258	243 249
VVMD23	178 178	178 178	178 178
VVMD24	210 210	210 219	219 219
VVMD25	243 259	243 253	245 253
VVMD26	249 249	249 251	251 251
VVMD27	181 189	175 189	175 189
VVMD28	231 239	237 239	237 239
VVMD31	206 216	206 210	210 216
VVMD32	241 259	241 241	241 257
VVMD34	240 240	240 248	240 248
VVMD36	254 254	254 264	264 295
VVS1	181 181	181 181	181 190
VVS2	139 147	139 151	133 151
VVS4	167 175	168 175	168 169
VVS16	284 284	284 284	284 284
VVS29	175 181	179 181	171 179
VHS43	210 213	210 213	210 213

Sauvignon blanc and Cabernet Sauvignon had not previously been suspected.

Cabernet Sauvignon already existed in the 17th century, before the earliest reports of deliberate plant hybridization (Sturtevant, 1965). Thus, it is likely that this cross was spontaneous. Mixtures of red and white varieties were common in medieval vineyards and pollen from one of the parents could easily have landed on a flower of the other parent. The direction of the cross is still unknown. It is clear, however, that all Cabernet Sauvignon vines existing today have descended from a single seedling – all Cabernet Sauvignon vines tested to date, including 11 registered clones, have exactly the same microsatellite genotype (Bowers & Meredith, unpublished results).

Several other winegrapes have now been shown to be the offspring of other varieties. Monbadon, a minor variety from southwestern France, is a cross between Folle blanche and Ugni blanc, found in the same part of France and used for brandy. (Ugni blanc is also widely grown in Italy under the name Trebbiano.) Durif, a variety reported to have been a seedling of the old French variety Peloursin selected in the late 1800s, has now been shown to be a cross between Peloursin and Syrah, the distinguished variety of the Rhone Valley (Meredith *et al.*, 1999). Some Durif is grown in California, where, interestingly, it has long been called Petite Sirah, even though no suggestion of a genetic relationship between it and Syrah has ever appeared in the literature.

Surprising genetic relationships have been found among the varieties associated with northeastern France. Pinot noir, a noble variety that is generally acknowledged to be ancient, and a variety called Gouais blanc, once widespread but considered so mediocre that it was eventually banned, are the direct parents of almost all of the varieties grown in northeastern France today, including Chardonnay, Gamay noir, Melon, Aligoté and Auxerrois as well as at least eleven others (Bowers *et al.* 1999). That Pinot noir and Gouais blanc have been such successful parents is a testament to heterosis. Gouais blanc is genetically distant from Pinot noir and is actually an eastern European grape. Historical writings suggest that it may have been brought to France by the 3rd-century Roman emperor Probus as a gift to the Gauls.

In providing answers to long-standing questions about the origin of the classic wine cultivars, the tools of molecular genetics have opened a window into history. For those who would preserve the mystique of wine, such revelations may not always be welcome, but for others this new genetic knowledge only enhances their appreciation of wine as an expression of geography and human history.

REFERENCES

Bouquet, A. (1982) Origine et évolution de l'encépagement français à travers les siècles. *Le progrès agricole et viticole* 99: 110–21

Bowers, J.E. & Meredith, C.P. (1997) The parentage of a classic wine grape, Cabernet Sauvignon. *Nature Genetics* 16: 84–87

Bowers, J.E., Boursiquot, J.-M. & This, P. *et al.* (1999) Historical genetics: the parentage of Chardonnay, Gamay and other wine grapes of northeastern France. *Science* 285: 1562–65

Levadoux, L. (1956) Les populations sauvages et cultivées de *Vitis vinifera* L. *Annales de l'Amélioration des Plantes* 6: 59–118

Meredith, C.P., Bowers, J.E. & Riaz, S. *et al.* (1999) The identity and parentage of the variety known in California as "Petite Sirah". *American Journal of Enology and Viticulture* 50: 236–42

Sefc, K.M., Steinkellner, H., Wagner, H.W., Glossl, J. & Regner, F. (1997) Application of microsatellite markers to parentage studies in grapevine. *Vitis* 36: 179–83

Sturtevant, A.H. (1965) *A History of Genetics*, New York: Harper and Row

Viala, P. & Vermorel, V. (1901) *Ampélographie, Tome II*, Paris: Masson

Zohary, D. (1996) The domestication of the grapevine *Vitis vinifera* L. in the Near East. In *The Origins and Ancient History of Wine*, edited by P.E. McGovern, S.J. Fleming & S.H. Katz, Amsterdam: Gordon and Breach

ORIGIN OF THE APPLE (*MALUS DOMESTICA* BORKH.)

B.E. Juniper, J. Robinson, S.A. Harris and R. Watkins*
Department of Plant Sciences, University of Oxford, Oxford, UK
**Andrewshayes Cottage, Axminster, Devon, UK*

1. Introduction

The apple, *Malus domestica*, is one of the most important fruit crops of the colder and temperate parts of the Old World, New World and Southern Hemisphere (Vavilov, 1951; Brown, 1975). Evidence for the collection of apples from the wild can be found in Neolithic (11 200 years ago) and Bronze Age (*c*.4500 years ago) sites throughout Europe (Hopf, 1973; Schweingruber, 1979). Evidence for apple cultivation as early as 1000BC has been found in Israel (Zohary & Hopf, 1994).

Little is known about the time and place of apple domestication except that, possibly in the Persian city-states and certainly in Graeco-Roman times, orchards (which would imply grafting) were well established. In addition, some 25–35 wild species are widely distributed throughout Europe, Asia and North America and, in most cases, are interfertile (Watkins, 1981, 1995). The diploid chromosome number of the domestic apple, as well as all the wild species of apple examined, is 34, with some commercial triploids, e.g. the famous cultivar Bramley's Seedling, and occasional tetraploids, but the latter are generally of no major commercial value. All commercial apples are more or less self-incompatible and, although the feature has not been studied in detail, all the wild apple species are assumed to be likewise. Apple seed requires isolation from the placental tissue and a 60-day cold-chill period before successful germination can be achieved, and apples (both domesticated and wild), along with most other Maloideae, cannot be grown from cuttings.

The word "crab", a Saxon word meaning bitter or sour, has no precise taxonomic value and, although widely current, will not be used in this article.

2. Natural distribution of the genus *Malus*

The genus *Malus* is found almost continuously throughout the temperate Old and New Worlds. However, the primary centre of *Malus* species diversity appears to be within the region which includes Asia Minor, the Caucasus, Central Asia, Himalayan India and Pakistan, and the western provinces of China (Watkins, 1981; Way *et al.*, 1990; Zhang *et al.*, 1993). Over the greater area of central Asia, Rehder (1940) and Phipps *et al.* (1990) identified 25 native species of *Malus* arranged in five sections. Janick *et al.* (1996) have, more precisely, suggested the area around Almaten in eastern Kazakhstan (Almaten means the father of apples) as the possible centre for the domestic apple.

There are 329 *Malus* names recorded while the total worldwide number of *Malus* species recognized varies between eight (Likhonos, 1974) and 78 (Ponomarenko, 1986) with Phipps *et al.* (1990) arguing for 55. Similarly, infrageneric classifications of the genus vary between authorities. In the absence of a comprehensive monograph for the genus, the large number of names is probably the result of the presumed level of hybridization within the genus (Dickson *et al.*, 1991a,b) and the large number of *Malus* cultivars, for a diverse range of uses, that have been planted and selected (Morgan & Richards, 1993). In the absence of detailed phylogenetic information, Watkins (1981, 1995) has recognized 34 species arranged into three geographical groups: a Chinese group, a European group and an American group.

The progenitors of the American *Malus* group, but oddly enough not those of the genus *Pyrus* (the pear), crossed the land bridge over the Bering Straits, almost certainly on more than one occasion (Dickson *et al.*, 1991a,b), apparently before the Americas were separated from the Old World. This Malus invasion may have originated from *M. kansuensis* or *M. yunnanensis* or their precursors which may, in their turn, have originated on the lower mountain slopes of what is now western China. Evidence for these events can be obtained from studies of species distribution. For example, *M. fusca* (a species restricted to the west coast of North America) is considered part of section Sorbomalus, and has affinities with species *M. kansuensis* and *M. yunnanensis* in China.

Some five or six species of *Malus* can presently be found scattered from the central to the eastern parts of North America. The most important of these in current apple breeding, both for crop and ornament, are probably *M. ioensis* and *M. coronaria* (the Garland Tree). However, *M. fusca*, found only along the west coast of North America, may also play some part.

3. Prehistory

It is probably no coincidence, in this evolutionary story, that one of the world's most ancient trade routes, the Old Silk Road (or Old Silk Roads since, for climatic and political reasons, there were diversions) crossed from western China through Almaten, Tashkent and Samarkand to the Caspian and the Black Sea, the near East and the Danube valley. Travellers on foot, and with camel and pack-horse trains, must have begun to traverse at least parts of this great route from near the beginning of the Neolithic period and it would have been well established by the Bronze Age. Recent discoveries of human mummies buried on the edge of the Taklakmakan Desert suggest that "Celtic", i.e. western races, were resident in what is now western China *c*.4000 years ago (Barber, 1999). Such evidence reinforces the idea that there were extensive human migrations along these ancient routes at a very early date.

Ruminants, such as deer and antelope native to the area,

wild horses and horses, donkeys and mules in human employ, will avidly eat all manner of top fruit. However, apple and pear seed, since these animals (and humans) possess no gizzard, will pass mostly undamaged through their alimentary canals. (Camels would also eat such fruit but masticate so fiercely that most seeds would be destroyed.) Thus, seeds of diverse species of *Malus*, but also related genera such as *Pyrus*, *Sorbus*, *Crataegus*, *Docynia* and *Cydonia*, between which genera hybrids are still possible, would have been scattered randomly, in a fertile medium, along the whole length of the Silk Roads. The possibilities for hybridization between previously geographically isolated species, and possibly genera, would have been legion (Watkins, 1995).

4. Immediate prehistory and the development of grafting

Despite the importance of apple crops, the origin of *M. domestica* is controversial (Vavilov, 1930, 1951; Knight, 1963; Watkins, 1981). Current thinking suggests that *M. sieversii*, which has a central Asian distribution and spreads somewhat towards the Chinese (eastern) end of the Old Silk Road, played a seminal role in the origin of the cultivated apple (Vavilov, 1930, 1951; Forsline, 1995; Hokanson *et al.*, 1997). Fruits of *M. sieversii*, in the wild, show a large range of colours, forms and flavours from almost inedible bitter specimens to fruits not very dissimilar to some modern cultivars (Vavilov, 1930, 1951; Forsline, 1995). Although the possibilities of random hybridization with both eastern and western species of wild apple remain, we have found no evidence for such hybridizations, not even with the northern European *M. sylvestris*.

Individual selected cultivars, playing a significant role in Old World horticulture, appeared following the introduction of grafting technologies. These predominantly maloid species crops are what Zohary and Hopf (1994) have called the second wave of cultivation. We shall probably never know who invented grafting, but to implicate the Persians or the Chinese of the Bronze Age does not seem unreasonable.

5. The historical period

The Greeks were fully familiar with the art of grafting and Theophrastus (*c.*370–285BC) writes about the maintenance of orchards using this technique (Hort, 1916, p. 129). The Roman horticulturalists knew almost everything about budding, grafting and rootstocks; there is, in fact, a detailed and comprehensible depiction, in mosaic, of the techniques of grafting (and other horticultural practices), dating from about the 2nd century AD at Saint-Romain-en-Gal in southern France (Lancha, 1981). Pliny (Plinius) writes of about 20 distinct varieties of apple known to him, and Columella writes *c.*50AD of importing selected cultivars from consuls stationed in the Middle East, Greece and Egypt. (Columella's belief that certain of his scions came from Pelusia in Egypt is almost certainly mistaken, since Egypt was far too hot for normal apple cultivation.)

The Romans certainly brought selected apple cultivars and the technique of grafting up through western Europe to Britain, but whether these isolates, or the techniques, survived through the Dark Ages is still debatable. It is probable that the cultivar we now recognize as Decio, still grown commercially in Italy, is similar to, if not virtually identical with, Roman cultivars. Cultivars were certainly brought back to Britain after the Norman Conquest, and among these early "colonists" may be some we recognize as being of great antiquity, e.g. Pearmain, Costard, Nonpareil, Court Pendu Plat and Leathercoat, still growing in specialist collections in this country today. Apple cultivars deriving originally, we believe, from the eastern end of the Old Silk Roads were, from the Middle Ages, cut off from all their parental origins. We think it now very unlikely that the native British apple *M. sylvestris* has produced any hybrids of commercial value even though apple growing spread widely in Britain after the Middle Ages. *M. sylvestris* might have contributed to local adaptations to local conditions, but otherwise any influence of *M. sylvestris* is likely to have been, if significant at all, on northern French and English cider apples.

Alien apple cultivars found, when they reached the British Isles, a fertile climate. They hybridized, principally among themselves we surmise, mutated to a degree and proliferated to the point whereby, at the close of the 19th century, virtually every town and village in central/southern England could claim a local apple, and the known list of cultivars probably exceeded 2500 (Morgan & Richards, 1993; Case, 1996). However, for a comparison, in what was the Soviet Union and where principally we think the apple originated, there are known to be in excess of 6000 cultivars (Ponomarenko, personal communication; Ponomarkenko, 1986).

6. The agricultural invasion of the North American continent

A European style of horticulture was established by the early colonists on the eastern seaboard of North America. From the end of the 16th century, European crops were pouring into the new colonies. Apples, probably in the form of apple pips in the first instance, and selected cultivars rather later, were soon established in the Tidewater region and elsewhere.

Grafting was rarely practised there in the 17th and 18th centuries, even into the early 19th century, and the settlers relied on the fecund germination of apple seed (Calhoun, 1995). The native Amerindians who, in some areas, practised a very sophisticated horticulture with certain crops do not seem to have exploited the potential of the native apples, such as *M. coronaria*, *M. ioensis* and *M. fusca*. English cultivars as such were not spectacularly successful in the American climate, but they did often set seed, where they were fertile and could find compatible partners. The seeds would have germinated more readily after the extreme continental winters. The "European" apples may have hybridized to some extent with the local native apples, but, although these hybrids may now have some importance in the raising of ornamental *Malus*, they do not appear to have entered into *M. domestica* lineages. Dickson *et al.* (1991b) have indicated that *M. domestica* rarely acts as a pollen source in hybridizing with native American *Malus*.

The genetic diversity that accumulated as a result of the

spectacular germination after the cold winters soon became considerably greater than that in their European counterparts (Watkins, 1981). The potential for cross-pollination was almost infinite and eastern North America became, literally, a vast experimental station, crudely screening seedling apple varieties (Calhoun, 1995). Itinerant horticulturalists, such as John "Johnny Appleseed" Chapman (1774–1825), spread these random populations across the whole continent, giving rise to a distinctive new cluster of North American cultivars, e.g. Jonathan, Wagener and Golden Delicious, successful in the more extreme climates of the US and Canada and, subsequently, in South Africa, Australia and the Mediterranean region. (Golden Delicious was not, as is commonly supposed, an act of deliberate hybridization, but was found as a chance seedling of unknown parentage in a hedgerow in West Virginia in about 1890.)

Despite the economic value of the genus, and the importance of systematics to genetic conservation efforts, little rigorous work has been undertaken to assess the relationships between the domesticated apple and other species. Investigations of genetic variation in the domesticated apple have focused on the generation of linkage maps (e.g. Conner *et al.*, 1997), the discrimination of cultivars (Mulcahy *et al.*, 1993; Gardiner *et al.*, 1996; Guilford *et al.*, 1997) and the identification of cultivar origins (Gardiner *et al.*, 1996). Fewer studies have been concerned with wild species, particularly species from the important Chinese group (Dickson *et al.*, 1991a,b; Lamboy *et al.*, 1996).

Apples, as with other Old World crops, have had a long history of human interaction, and have been modified, often unconsciously, through hybridization and selection (Schweingruber, 1979; Calhoun, 1995; Janick *et al.*, 1996). The problem, therefore, of trying to identify those species that have contributed to the apparently intricate hybrid origin of the domesticated apple is a complex one which requires detailed analysis of neutral characteristics and the assumption that all *Malus* species are extant. These problems have been faced in the study of the bean *Phaseolus* (Beccara Velásquez & Gepts, 1994), the peanut *Arachis* (Paik-Ro *et al.*, 1992), maize (Doebley, 1990) and hexaploid wheat *Triticum x aestivum* (Hsiao *et al.*, 1995). Detailed studies of DNA sequences from biparentally inherited nuclear DNA and uniparentally inherited chloroplast DNA potentially provide a means of assessing information about contributors to complex hybrid situations (Arnold, 1997).

So far, we are in the process of assessing whether or not sequence analysis of two regions, the *mat*K region of the chloroplast DNA (Johnson & Soltis, 1994; Steele & Vilgalys, 1994) and the internally transcribed spacers (ITS) of nuclear ribosomal DNA (Baldwin *et al.*, 1995; Campbell *et al.*, 1995), are adequate to distinguish Maloid taxa. Initial results from sequencing the ITS of the nuclear rDNA are promising. Although only a few taxa have been investigated so far, certain tentative hypotheses can be drawn from the results. There appears to be a "core" group of Malus taxa and a number of "satellite" groupings of taxa, e.g. the "American Group". Within the core group, the validity of previous classifications (e.g. Watkins, 1981; Phipps *et al.*, 1990; Huckins, 1992) and of Watkins' geographical groupings of taxa is still uncertain. The few cultivars of *M. domestica* analysed, Leathercoat and Esopus Spitzenburg, appear to be related to the *M. sieversii* accessions analysed. The relationships of these cultivars to other germplasm donors, such as *M. prunifolia* and *M. sylvestris*, have yet to be determined. It is hoped that further results in this ongoing investigation will help us to understand the relationships of the wild Malus species and to identify which wild species have contributed to the *M. domestica* gene pool.

REFERENCES

Arnold, M.L. (1997) *Natural Hybridization and Evolution*, Oxford and New York: Oxford University Press

Baldwin, B.G., Sanderson, M.J., Porter, J.M. *et al.* (1995) The ITS region of nuclear ribosomal DNA: a valuable source of evidence on angiosperm phylogeny. *Annals of the Missouri Botanical Garden* 82: 247–77

Barber, E.W. (1999) *The Mummies of Urumchi*, London: Macmillan, and New York: Norton

Beccara Velásquez, V.L. & Gepts, P. (1994) RFLP diversity of common bean (*Phaseolus vulgaris*) in its centres of origin. *Genome* 37: 256–63

Brown, A.G. (1975) Apples. In *Advances in Fruit Breeding*, edited by J. Janick & J.N. Moore, West Lafayette, Indiana: Purdue University Press

Calhoun, C.L. (1995) *Old Southern Apples*, Blacksburg, Virginia: McDonald and Woodward

Campbell, C.S., Donoghue, M.J., Baldwin, B.G. & Wojciechowski, M.F. (1995) Phylogenetic relationships in Maloideae (Rosaceae): evidence from sequences of the internal transcribed spacers of nuclear ribosomal DNA and its congruence with morphology. *American Journal of Botany* 82: 903–18

Case, H.J. (1996) European *Malus* Germplasm. Proceedings of a Workshop held 21–24 June 1995, Wye College, University of London

Conner, P.J., Brown, S.K. & Weeden, N.F. (1997) Randomly amplified polymorphic DNA-based genetic linkage maps of three apple cultivars. *Journal of the American Society for Horticultural Science* 122: 350–59

Dickson, E.E., Arumuganathan, K., Kresovich, S. & Doyle, J.J. (1991a) DNA contents of *Malus* section *Chloromeles* and other genera within the Rosaceae as revealed by flow cytometry. *American Journal of Botany* (Suppl.) 78: 180

Dickson, E.E., Kresovich, S. & Weeden, N.F. (1991b) Isozymes in North American *Malus* (Rosaceae): hybridisation and species differentiation. *Systematic Botany* 16: 363–75

Doebley, J. (1990) Molecular systematic of *Zea* (Gramineae). *Maydica* 35: 143–50

Forsline, P.L. (1995) Adding diversity to the national apple germplasm collection: collecting wild apple in Kazakhstan. *New York Fruit Quarterly* 3: 3–6

Gardiner, S.E., Bassett, H.C.M., Madie, C. & Noiton, D.A.M. (1996) Isozyme, randomly amplified polymorphic DNA (RAPD) and restriction-fragment length polymorphism (RFLP) markers used to deduce a putative parent for the 'Braeburn' apple. *Journal of the American Society for Horticultural Science* 121: 996–1001

Guilford, P., Prakash, S., Zhu, J.M. *et al.* (1997) Microsatellites in *Malus x domestica* (apple): abundance, polymorphism and cultivar identification. *Theoretical and Applied Genetics* 94: 249–54

Hokanson, S.C., McFerson, J.R., Forsline, P.L. *et al.* (1997) Collecting and managing wild *Malus* germplasm in its center of diversity. *HortScience* 32: 173–76

Hopf, M. (1973) Äpfel (*Malus communis* L.); Aprikose (*Prunus armeniaca* L.). In *Reallexikon der Germanischen Altertumskunde*, edited by H. Beck, H. Jankuhn, K. Ranke & R. Wenskus, Berlin: de Gruyter

Hort, A. (1916) *Theophrastus. Enquiry into Plants*, London: Heinemann

Hsiao, C., Chatterton, N.J., Asay, K.H. & Jensen, K.B. (1995) Phylogenetic relationships of the monogenomic species of the wheat tribe Triticeae (Poaceae) inferred from nuclear rDNA (internal transcribed spacer) sequences. *Genome* 38: 211–33

Huckins, C.A. (1992) *A Revision of the Sections of the Genus* Malus *Miller*. Ph.D. Thesis, Ithaca, New York: Cornell University

Janick J., Cummins, J.N., Brown, S.K. & Hemmatt, M. (1996) Apples. In *Fruit Breeding*, vol. 1: *Tree and Tropical Fruits*, edited by J. Janick & J.N. Moore, New York: Wiley

Johnson, L.A. & Soltis, D.E. (1994) *mat*K-sequences and phylogenetic reconstruction in Saxifragaceae sensu stricto. *Systematic Botany* 19: 143–56

Knight, R.L. (1963) Abstract bibliography of fruit breeding and genetics to 1960: *Malus* and *Pyrus*. *Technical Communications of the Commonwealth Bureau of Horticultural and Plantation Crops*

Lamboy, W.F., Yu, J., Forsline, P.L. & Weeden, N.F. (1996) Partitioning of allozyme diversity in wild populations of *Malus sieversii* L. and implications for germplasm collection. *Journal of the American Society for Horticultural Science* 121: 982–87

Lancha, J. (1981) Recueil général des mosaiques de la Gaule. Xe Supplément a Gallia II. Province de Narbonne 2. Vienne, Paris: Ouvrage publié avec le concours du Ministère de la Culture sous-Direction de l'Archeologie, Editions du Centre National de la Recherche Scientifique

Likhonos, A. (1974) A survey of the species in the genus *Malus* Mill. *Sbornik Nauchnykh Trudov po Prikladnoi Botanike, Genetike i Seleksii* 52: 16–34

Morgan, J. & Richards, A. (1993) *The Book of Apples*, London: Ebony Press

Mulcahy, D.L., Cresti, M., Sansavini, S. *et al.* (1993) The use of random amplified polymorphic DNA's to fingerprint apple genotypes. *Scientia Horticulturae* 54: 89–96

Paik-Ro, O.G., Smith, R.L. & Knauft, D.A. (1992) Restriction fragment length polymorphism evaluation of six peanut species within the *Arachis* section. *Theoretical and Applied Genetics* 84: 201–08

Phipps, J.B., Robertson, K.R., Smith, P.G. & Rohrer, J.R. (1990) A checklist of the subfamily Maloideae (Rosaceae). *Canadian Journal of Botany* 68: 2209–69

Ponomarenko, V.V. (1986) Obsor vidov roda *Malus* Mill. *Sbornik Nauchnykh Trudov po Prikladnoi Botanike, Genetike i Seleksii* 106: 16–27

Rehder, A. (1940) *Manual of Cultivated Trees and Shrubs Hardy in North America*, 2nd edition, New York: Macmillan

Schweingruber, F.H. (1979) Wildäpfel und prähistorische Äpfel. *Archaeo-Physika* 8: 283–94

Steele, K.P. & Vilgalys, R. (1994) Phylogenetic analyses of Polemoniaceae using nucleotide sequences of the plastid gene *mat*K. *Systematic Botany* 19: 126–42

Vavilov, N.I. (1930) Wild progenitors of the fruit trees of Turkistan and the Caucasus and the problem of the origin of fruit trees. International Horticultural Congress, London, 1930

Vavilov, N.I. (1951) *The Origin, Variation, Immunity and Breeding of Cultivated Plants*, Waltham, Massachusetts: Chronica Botanica

Watkins, R. (1981) Apples (genus *Malus*), pears (genus *Pyrus*), and plums, apricots, almonds, peaches, cherries (genus *Prunus*). In *The Oxford Encyclopaedia of Trees of the World*, edited by F.B. Hora, Oxford and New York: Oxford University Press

Watkins, R. (1995) Apple and pear. In *Evolution of Crop Plants*, 2nd edition, edited by J. Smartt & N.W. Simmonds, Harlow: Longman, and New York: Wiley

Way, R.D., Aldwinckle, H.S., Lamb, R.C. *et al.* (1990) Apples *(Malus)*. In *Genetic Resources of Temperate Fruit and Nut Crops*, edited by J.N. Moore & J.R. Ballington, Wageningen: International Society for Horticultural Science

Zhang, W., Zhang, J. & Hu, X. (1993) Distribution and diversity of *Malus* germplasm resources in Yunnan, China. *HortScience* 28: 978–80

Zohary, D. & Hopf, M. (1994) *Domestication of Plants in the Old World: The Origin and Spread of Cultivated Plants in West Asia, Europe and the Nile Valley*, Oxford: Clarendon Press, and New York: Oxford University Press

GLOSSARY

infrageneric within a genus

internal transcribed spacers stretches of DNA found between tandemly repeated copies of the ribosomal RNA genes. Once transcribed into RNA they are subsequently cut out leaving a separate RNA for each gene

APPLES: THE GENETICS OF FRUIT QUALITY

K.M. Evans and F.H. Alston
Horticulture Research International-East Malling, West Malling, UK

1. Introduction

Apple (*Malus pumila*) is an allopolyploid species. It is usually a functional diploid ($2n$ = 34), although a few important varieties are triploid ($3n$ = 52) and tetraploid ($4n$ = 68) sports occur. A relatively high genetic variance controls most fruit quality traits and, in most cases, the mean progeny expression of a trait is represented by the mid-parent value. Careful observation of a trait in prospective parents can therefore result in the identification of parental combinations which will enhance the mid-parent value of particular traits. In addition, a few important quality traits have been found to be controlled by single genes.

The combining of fruit quality traits with all the other characters that are necessary in a successful apple variety is hampered by the heterozygosity of apples. Another problem is their long juvenile phase; prior to cropping, apple seedlings pass through a juvenile phase of 4–5 years. The selection and establishment of new varieties is also affected since the non-productive phase of adult trees is related closely to the length of the juvenile phase.

Although direct assessment of fruit quality is not possible until fruiting occurs, attempts have been made to correlate subsequent fruit characters with vegetative characters in 1- and 2-year-old seedlings. It has been shown (Visser & Verhaegh, 1978a) that there are good prospects of preselecting for fruit acidity on the basis of the pH of seedling leaf sap among large populations. Later, an isoenzymic locus was found to be linked to the primary locus governing fruit skin colour, thus providing a further possibility for screening at the young seedling stage (Weeden *et al.*, 1994). Current research (Alston *et al.*, 1996) aims to relate key quality components and consumer preferences with molecular markers to facilitate reliable selection in the years preceding fruiting.

2. Fruit quality

Fruit quality in dessert apples may be considered in three main categories: appearance, eating quality and storage quality. Appearance includes skin colour, skin finish, size and shape. Some consumers associate appearance with eating quality as a result of their experience of varieties currently on sale. Green apples are expected to be sharp and crisp, yellow and red-blushed apples sweet and soft, and bicoloured apples crisp and aromatic. However, some of these preconceived impressions are being overcome as new varieties are introduced.

Eating quality covers all aspects of flavour, flesh colour and texture. A combination of acids, sugars and volatiles forms the basis of flavour, but the texture of the flesh has the most important effect on overall eating quality for today's consumer. Flesh colour, skin colour and skin finish can also have significant effects on consumers' appreciation of the fruit.

Consumers now expect to be able to purchase apples all year round, necessitating the development of variety-specific storage regimes. These rely on low temperatures and controlling the concentrations of oxygen and carbon dioxide in the atmosphere of the store. Storage quality and shelf life depend on a number of factors, including resistance of the fruit to shrivelling, bitter pit, low temperature breakdown, superficial scald, disease and control of ethene production.

All these components of fruit quality can be modified greatly by environmental variation during the growing season and in the fruit store.

3. Appearance

The presence of anthocyanin, the red pigment in apple skin, is determined by a single dominant gene *Rf* (Alston & Watkins, 1975). Additional genes are undoubtedly responsible for the intensity of this overcover pigment and its expression in stripes, a blush or a solid overcover. Investigations attributing further loci as modifiers of anthocyanin pigmentation in apple skin have produced inconclusive results so far (White & Lespinasse, 1986; Schmidt, 1988). However, more recently (Weeden *et al.*, 1994; Cheng *et al.*, 1996), the gene R_f was confirmed and was found to be linked closely to the isoenzyme locus *Idh-2* and to several RAPD (randomly amplified polymorphic DNA) markers. This presents the potential for preselection for skin colour among young seedlings prior to fruiting. There are a large number of sports, naturally occurring from the widely grown bicoloured commercial varieties, which show increased anthocyanin intensity or an increased distribution of the pigment. These sports, which arose presumably from mutations at specific loci, support the hypotheses that propose additional loci in the control of apple skin colour.

Fruits can be disfigured by patchy russetting and cracking, defects which can be accentuated in spring and early summer by low temperatures and some fungicide applications. This problem occurs with Cox's Orange Pippin, the most widely grown English apple, where russetting is said to be controlled by a single dominant gene, *Ru*, the effects of which can be modified by minor genes (Alston & Watkins, 1975). Some varieties, however, have a complete cover of russetting, without cracks, and are sought after for certain niche markets despite their tendency to shrivel severely in store. There is further evidence from some of these types of apple (in particular, bimodal segregations in a progeny derived from the fully russetted variety d'Arcy Spice and the occurrence of fully russetted sports from certain clear skinned varieties) that russetting is under simple genetic control. Some varieties, such as Idared, have been selected as good parents for skin finish as, in crosses with Cox, they can give up to 40% of seedlings with a clean skin. Reliable selection requires careful observation, repeated over a number of growing seasons, with due regard to climatic conditions.

Another deleterious character is greasy skin, the inheritance of which is determined by a single dominant gene, *Gr*.

Fruit size, an important aspect of appearance, is under polygenic control. In general, large fruit size is recessive and progeny means are below the mid-parent value (Brown, 1960). In most progenies, over half the seedlings carry fruit which is too small. Cox is a very poor parent, with 60–70% of derivatives being too small (less than 60 mm diameter); stringent selection is necessary to achieve good commercial fruit size.

Fruit shape in apples varies from very flat oblate to oblong and conical. The distribution frequency in progenies follows a normal distribution about the mean, with most seedlings falling between the parental values for this polygenically inherited character (Brown, 1960).

4. Eating quality

Malic acid, one of the most important flavour components, is controlled by a single dominant gene, *Ma* (Nybom, 1959). Most apple progenies show distinct segregations for acidity because most dessert varieties are heterozygous for *Ma*. The 25% of segregants with the homozygous genotype *ma/ma* have an unacceptably low acidity which produces a bland flavour; such seedlings can be detected easily and discarded in the field at fruiting, following testing of fruit slices with the indicator bromo-cresol green. A polygenic inheritance pattern is superimposed on the oligogenic system; it controls the distribution of acidity levels among the acidic seedlings. Where one of the parents has a high acidity, a large proportion of plants will be discarded as being too acid; in some progenies, this can amount to a further 30% (Visser & Verhaegh, 1978b). Fruit acidity can be estimated most conveniently for screening purposes by assessing the pH of the juice. Preselection for fruit acidity became a promising possibility after the establishment of a close correlation between the pH of leaf sap in 2-year-old seedlings and the pH of fruit juice in mature trees (Visser & Verhaegh, 1978a). More recently, the *Ma* locus has been located on the reference genetic linkage map of apple (Maliepaard *et al*., 1998).

Fructose, sucrose and glucose are the principal fruit sugars, the amounts of which are under polygenic control with progeny means approaching the mid-parent values (Brown & Harvey, 1971). Although acidity and sugar content are inherited independently, it is their balance that determines the acceptability of successful dessert apples.

By tasting derivatives of Cox, an acknowledged source of good flavour, seedlings can be classified into two distinct groups: those with an aroma similar to Cox and those without. This flavour component appeared to be determined by a single dominant gene (Alston *et al*., 1996). Two gas chromatographic peaks were shown to correlate well with good flavour in a Cox selfed progeny. These peaks were related later to hexylacetate and hexylbutyrate which have been identified as important flavour components in Golden Delicious and Cox. They should be regarded as important flavour impact compounds in the apple since they are not variety specific.

The colour of apple flesh appears to have an effect on organoleptic perception, with yellow or creamy flesh being associated with aromatic flavour. The yellow/cream flesh colour is dominant (Crane & Lawrence, 1933) and is a characteristic of the variety Cox; a number of its derivatives breed true for the character.

Although texture is the most important component of fruit quality, little is known of the genetics of textural components, in particular firmness and crispness, except that they are controlled polygenically. At present, penetrometers are used to assess firmness, but this provides no more than a crude representation of eating quality; however, the technique has been used in an examination of the breeding potential of some varieties (Alston, 1988). Non-destructive texture assessments, which rely on the vibrational behaviour of the fruit or its acoustic resonance, have also been developed. Tasting is necessary for the best judgement of crispness and juiciness of the apple flesh and to obtain some appreciation of skin toughness. Work is underway (Evans, 1999) to identify and quantify texture components physically and biochemically, with a view to devising efficient realistic selection techniques, including molecular markers that can be applied in breeding programmes.

5. Storage

Naturally occurring, late maturing apples are not necessarily suited to long-term commercial storage at low temperatures in controlled atmospheres. In particular, the high humidities in storage result in apples being more prone to low temperature breakdown and superficial scald. Fruit samples of new varieties have to be tested in conditions that take these factors into account. Parental varieties have been identified which show high resistance to these disorders, but only a few progenies have been examined (Alston, 1988) and there is no firm information on their genetics.

Some storage disorders are associated with low concentrations of calcium and phosphorus in the fruit. One very long-storing variety, Malling Kent, shows a high concentration of calcium in the flesh (Perring, 1978) and is relatively free from the related disorders, bitter pit and breakdown, during storage. Standard techniques of leaf analysis, which are used to predict storage potential in commercial fruit, show large between-progeny variation for calcium content, suggesting that the technique has potential as a preselection agent. Such a method has been used routinely in apple rootstock breeding to select rootstocks with improved calcium uptake (Kennedy *et al*., 1982).

A low propensity to produce ethene (ethylene) is a useful characteristic for long-storing apples as its absence naturally delays the ripening process. It was concluded (Stow *et al*., 1992) that ethene production is under genetic control. Later, it was suggested that the production of ethene in apple is controlled by two complementary genes, *E-1* and *E-2*, low ethene producers being homozygous at each locus (Batlle, 1993).

6. Breeding strategy

Combining important quality components with all the other necessary characters, such as yield and resistance to diseases, is hampered by several constraints including heterozygosity and the long juvenile phase of the crop; several generations are needed to combine all the required features. A further problem is likely to arise as the generations progress; the apple has a gametophytic incompatibility system controlled by a wide range of *S* alleles (Bošković & Tobutt, 1999). Recent breeding and selection has resulted in a parental breeding pool with a very narrow range of *S* alleles; consequently, some chosen combinations might well prove to be incompatible or semicompatible, producing either no seeds or a restricted number. These effects can be modified genetically or environmentally in some varieties (Petropoulou & Alston, 1998), but, while being sufficient to set a commercial crop, they are insufficient to increase seed production substantially to the levels achieved by fully compatible combinations.

An estimate of the size of progeny required for an effective breeding programme (Alston, 1987) was based on information from previous progenies, variety trials and estimates of the varying parental contributions for each character. Six quality components – skin finish, fruit colour, fruit size, acidity, storage and texture – were included, together with yield. By considering these characters and applying optimum selection values for each, it was concluded that even the best combinations might have only a 1% success rate. This implies that, with a requirement of ten selections for advanced orchard trials, a minimum progeny size of 1000 fruiting trees would be required. After making due allowance for trees that do not fruit within 4 years and heavy selection for disease and plant habit in the 2 years before fruiting, initial progenies should approach 10 000 per combination. More precise preselection procedures are being developed. This work is facilitated by the construction of a reference apple genetic linkage map developed in the European Apple Genome Mapping Project (Maliepaard *et al.*, 1998).

It is essential to relate selection for quality components to the requirements of consumers. Following consumer surveys, sensory profile maps have been constructed (Daillant-Spinnler *et al.*, 1996) that define the quality characteristics of apples most appreciated by consumers. Apple breeders use these when planning a crossing programme and prioritizing selection criteria. In addition, the profile maps are related to the results of trained sensory panels. After examining a series of selections from a progeny in this way, it is expected that it will be possible to identify the genes that determine key features of consumer preference and to relate them to molecular markers (Evans, 1999).

Genetic transformation related to quality has so far been confined to work on fruit ripening (James & Gittins, 1997), but further genetic manipulation affecting flavour components and key cell wall components related to texture can be expected in the near future.

REFERENCES

Alston, F.H. (1987) Strategy for apple and pear breeding. In *Improving Vegetatively Propagated Crops*, edited by A.J. Abbott & R.K. Atkin, London and San Diego: Academic Press

Alston, F.H. (1988) Breeding apples for long storage. *Acta Horticulturae* 224: 109–17

Alston, F.H. & Watkins, R. (1975) Apple breeding at East Malling. *Proceedings of the Eucarpia Symposium on Tree Fruit Breeding, Canterbury 1973*, pp. 14–29

Alston, F.H., Evans, K.M., King, G.J., MacFie, H.G.H. & Beyts, P.K. (1996) The potential for improving organoleptic quality in apples through marker assisted breeding related to consumer preference studies. In *Agri-Food Quality: An Interdisciplinary Approach*, edited by G.R. Fenwick, C. Hedley, R.L. Richards & S. Khokhar, Cambridge: Royal Society of Chemistry

Batlle, I. (1993) *Linkage of Isoenzymic Genes with Agronomic Characters in Apple*. Ph.D. Thesis, London: University of London

Bošković, R. & Tobutt, K.R. (1999) Correlation of stylar ribonuclease allozymes with incompatibility alleles in apple. *Euphytica* 107: 29–43

Brown, A.G. (1960) The inheritance of shape, size and season of ripening in progenies of the cultivated apple. *Euphytica* 9: 327–37

Brown, A.G. & Harvey, L.M. (1971) The nature and inheritance of sweetness and acidity in the cultivated apple. *Euphytica* 20: 68–80

Cheng, F.S., Weeden, N.F. & Brown, S.K. (1996) Identification of co-dominant RAPD markers tightly linked to fruit skin color in apple. *Theoretical and Applied Genetics* 93: 222–27

Crane, M.M. & Lawrence, W.G.C. (1933) Genetical studies in cultivated apples. *Journal of Genetics* 28: 265–96

Daillant-Spinnler, B., MacFie, H.J.H., Beyts, P. & Hedderley, D. (1996) The effect of peeling on the perceived sensory properties and major preference directions of 12 varieties from the Southern Hemisphere. *Food Quality and Preference* 7: 113–26

Evans, K.M. (1999) New approaches to apple scion breeding at HRI-East Malling. *Acta Horticulturae* 484: 171–74

James, D.J. & Gittins, J. (1997) Genetic transformation to reduce ethylene production. *Horticulture Research International. Annual Report 1996–97*, p. 81

Kennedy, A.G., Werts, J.M. & Watkins, R. (1982) An analysis of mineral uptake in apple rootstock seedlings. *Theoretical and Applied Genetics* 61: 141–44

Maliepaard, C., Alston, F.H., van Arkel, G. *et al.* (1998) Aligning male and female linkage maps of apple (*Malus pumila* Mill.) using multi-allelic markers. *Theoretical and Applied Genetics* 97: 60–73

Nybom, N. (1959) On the inheritance of acidity in cultivated apples. *Hereditas* 45: 332–50

Perring, M.A. (1978) Influence of orchard conditions on mineral composition of apples – effects of scion variety. *Report of East Malling Research Station for 1977*, p. 142

Petropoulou, S.P. & Alston, F.H. (1998) Selecting for improved pollination at low temperatures in apple. *Journal of Horticultural Science and Biotechnology* 73: 507–12

Schmidt, H. (1988) The inheritance of anthocyanin in apple fruit skin. *Acta Horticulturae* 224: 89–93

Stow, J.R., Alston, F.H., Hatfield, S.G.S. & Genge, J.M. (1992)

New apple selections with inherently low ethene production. *Acta Horticulturae* 326: 85–92

Visser, J. & Verhaegh, G.G. (1978a) Inheritance and selection of some fruit characters of apple. II. The relation between leaf and fruit pH as a basis for pre-selection. *Euphytica* 27: 761–65

Visser, J. & Verhaegh, J.J. (1978b) Inheritance and selection of some fruit characters of apple. I. Inheritance of low and high acidity. *Euphytica* 27: 753–60

Weeden, N.F., Hemmat, M., Lawson, L.M. *et al.* (1994) Development and application of molecular marker linkage maps in woody fruit crops. *Euphytica* 77: 71–75

White, A.G. & Lespinasse, Y. (1986) The inheritance of colour in apple (*Malus pumila* Mill.). *Agronomie* 6: 105–08

GLOSSARY

allopolyploid a polyploid containing genetically different sets of chromosomes, for example, sets from two or more different species

gametophytic incompatibility incompatibility by response to the pollen tube in the style of stigma

isoenzymic locus location of variants of enzymes that catalyse the same reaction

mid-parent value the mean of the values of a quantitative phenotype for two specific parents

oligogenic relating to qualitative characters controlled by a few major genes responsible which show normal Mendelian inheritance

polygenically controlled (or caused) by the small action/interaction of many genes

RAPD (random amplified polymorphic DNA) use of the polymerase chain reaction (PCR) with short oligonucleotide primers of random sequence to generate genetic markers

APPLES: THE GENETICS OF RESISTANCE TO DISEASES AND INSECT PESTS

F.H. Alston and K.M. Evans
Horticulture Research International-East Malling, West Malling, UK

1. Introduction

The apple (*Malus pumila*) is an outbreeding species. Varieties are heterozygous and they are usually propagated clonally, whereas new varieties are invariably selected from segregating F_1 progenies. It is usual for varieties to be grown on rootstocks and sometimes a third component, an interstock, is also included. Thus, trees are produced which can carry characters from four or six parental varieties and provide a means of combining characters from several widely differing parents after only one generation of breeding.

In apple, most traits are controlled by a relatively high genetic variance; thus, the selection of superior phenotypes and their subsequent random mating could be considered an effective way of increasing the numbers of favourable alleles in subsequent generations. However, a careful matching of parents according to genotype is preferable, since single dominant or recessive genes determine several important commercial characters. Suitable parental combinations are determined from the results of earlier crosses and variety trials, as well as on the basis of known genotype, thus combining the two approaches to genetic improvement. From the resulting progenies, breeders select individuals which are exceptional – usually those that exhibit a heterotic effect in the F_1. Clonal propagation of these élite selections permits the exploitation of the total genetic variance, including that due to epistasis and dominance, and thus largely offsets the slow pace of genetic gain in apple which is a consequence of the long juvenile phase.

Prior to cropping, apple seedlings pass through a juvenile period of up to eight years. This delays the selection of new varieties, although there are considerable environmental and genetic effects on the juvenile phase which can be shortened by various cultural practices and by intercrossing precocious parents. The long juvenile phase delays the authoritative delineation of specific genes and, consequently, many genes are based on observations in F_1 progenies alone, although the progressive introduction of some resistance genes through series of backcrosses has enabled a more definitive identification of those genes.

The delineation of genes controlling important economic characters and the development of early selection techniques for use in the early years of growth prior to fruiting are priority aims in apple breeding. Several lists of genes have been produced; the most recent lists some 145 loci, of which 29 determine resistance to pests and diseases (Alston *et al.*, 2000). In addition, at least 69 isoenzymes in 22 enzyme systems are determined by single genes, some of which are linked to important resistance genes. More recent work on the search for genetic markers as aids in early selection has led to the establishment of maps of the apple genome including various DNA markers as well as isoenzymes and resistance genes (Maliepaard *et al.*, 1998).

2. Diseases

The fungal foliar diseases mildew (*Podosphaera leucotricha*) and scab (*Venturia inaequalis*) account for the largest pesticide outlay in apple growing. Progressive infections from the shoot and stem disease canker (*Nectria galligena*) and the root and stem disease crown rot (Phytophthora spp.) can result in severe losses of trees in plantations. There is no satisfactory chemical control for these diseases. In addition, other foliar diseases, including cedar apple rust (*Gymnosporangium juniperi-virginianae*), apple blotch (*Phyllosticta solitaria*) and Glomerella leaf blotch, occur in some regions, as does the bacterial shoot disease fireblight (*Erwinia amylovora*). Finally, high losses during storage can occur from fruit-rotting diseases, notably Gloeosporium spp.

(i) Mildew

Most new varieties of apple carry a moderate level of resistance, achieved by stringent selection in the field during the second or third growing season after germination. This resistance, which is under polygenic control (Brown, 1959), is rarely sufficient to permit a complete relaxation of spray programmes; moreover, it is transmitted to only a small proportion of seedlings. Attempts to incorporate such resistance can drastically handicap progress in complex breeding programmes designed primarily to produce new varieties which will give high yields of good quality fruit. However, high levels of simply inherited resistance were found among Malus species (Knight & Alston, 1968) and transferred through four generations of crosses to the cultivated apple (Alston *et al.*, 1997). These genes are *Pl-1* from *M.* × *robusta* and *Pl-2* from *M.* × *zumi*. Other sources of high resistance under monogenic control are the ornamental crab White Angel (Gallot *et al.*, 1985), in which resistance is controlled by the gene *Pl-w* (Batlle & Alston, 1996; see **Figure 1** and also **Plate 17** in the colour plate section), the D series selections, *M. pumila* open pollinated (Visser & Verhaegh, 1977), where the resistance gene *Pl-d* has been identified in one selection (Batlle, 1993) and MIS derived from an unidentified Malus species (Korban & Dayton, 1983), which does not appear to be universally resistant.

The resistance observed from *M.* × *robusta* and *M.* × *zumi* was attributed later in each case to the effects of two genes modified by a polygenic background (Alston, 1977).

Figure 1 Apple seedlings segregating for the major gene *Pl-w*, showing resistance and susceptibility to mildew (*Podosphaera leucotricha*). See also **Plate 17**.

This situation was explained further after Batlle and Alston (1996) found that resistance in White Angel is controlled by complementary genes. The gene *Pr-Pl-w*, present also in many cultivated varieties, was found to be necessary for the expression of resistance determined by *Pl-w*. While this work was being developed, a similar relationship was observed for the genes *Pl-1*, *Pl-2* and *Pl-d* for which the complementary genes *Pr-Pl-1*, *Pr-Pl-2* and *Pr-Pl-d* appeared to be necessary.

The gene *Pl-w* is linked closely in White Angel to the isoenzymic gene *Lap-2* which could serve as a marker for mildew resistance (Batlle & Alston, 1996). DNA markers for *Pl-1* (Markussen *et al.*, 1995) and *Pl-2* (Dunemann *et al.*, 1998) have been identified and promise to improve the efficiency of screening for those genes and facilitate the selection of plants carrying both resistance genes.

The apomictic, male-sterile triploid ($3n = 51$) species *M. hupehensis*, which is highly resistant, produced mainly triploid apomictic progeny when crossed with the cultivated apple ($2n = 34$), although a few tetraploid ($4n = 68$) hybrids were also produced. The production and identification of resistant diploid hybrids in later generations proved difficult. However, it was facilitated by using alleles of the codominant isoenzymic gene *GOT-1* as ploidy indicators in the progenies of $4n \times 2n$ crosses (Batlle & Alston, 1994).

(ii) Scab

As with mildew, polygenic and monogenic types of resistance are available. The monogenic resistances provide the most complete resistance and are used more widely in breeding programmes. Even small amounts of sporulation on fruit surfaces result in the crop being unmarketable. The most durable type of resistance appears to be that determined by a combination of both types and many breeding programmes aim to achieve this.

Six genes, each providing strong resistance, have been identified (Dayton & Williams, 1968). They are: *Va*, from the variety Antonovka, *Vb* from *M. baccata*, *Vbj* from *M. baccata jackii*, *Vf* from *M. floribunda*, *Vr* from a Russian apple seedling and *Vm* from *M. micromalus*, which is susceptible to a virulent pathotype, Race 5, in Europe. The most widely used gene in breeding programmes is *Vf;* about 70 varieties have been introduced, although few have been planted extensively. Varieties carrying *Vf* have been observed free from scab at sites all over the world for at least 50 years (see **Figure 2** and also **Plate 18** in the colour plate section). Resistance based on *Vf* represents the most promising type of resistance, although the considerable variations in the incidence of disease within progenies can hinder efficient screening.

Another pathotype, Race 6, has been isolated which is virulent on several varieties carrying *Vf*, while the donor source *M. floribunda* 821 is resistant (Parisi & Lespinasse, 1996). Yet another pathotype has been identified (Roberts & Crute, 1994) to which *M. floribunda* 821 and some of the varieties derived from it are susceptible. In later studies, it was shown that *M. floribunda* 821 carries a gene *Vfn*, which gives resistance to Race 6, in addition to *Vf* (Gianfranceschi *et al.*, 1998). Despite these problems, *Vf*, in combination with polygenes, continues to provide the most promising basis for resistance to scab in most breeding programmes. Variations in genetic background and environmental variations sometimes interfere with the easy recognition of plants carrying *Vf*. Consequently, there has been an emphasis on searching for genetic markers. An isoenzymic marker has been identified (Manganaris *et al.*, 1994), the gene *Vf* being linked closely to the isoenzymic gene pgm-1. A number of DNA markers linked to *Vf* have also been located on the apple map (Koller *et al.*, 1994; Yang & Kruger, 1994; Tartarini, 1996). The effectiveness of selection using genetic markers compared with current glasshouse selection procedures has been confirmed over a range of environments (King *et al.*, 1998). An examination of the value of molecular markers in a breeding programme for resistance to scab showed that they can be used to distinguish plants homozygous for *Vf* in progenies from crosses between heterozygous parents. Moreover, such selections were also

Figure 2 Saturn – a variety from HRI–East Malling's resistance breeding programme. Saturn contains the *Vf* gene for resistance to scab (*Venturia inaequalis*) and has polygenic resistance to mildew (*Podosphaera leucotricha*). See also **Plate 18**.

shown to carry a higher level of resistance than the heterozygous selections (Benauf *et al.*, 1997).

(iii) Apple canker

While sources of strong resistance have been found among apple varieties and Malus species, and progeny studies have suggested suitable parents for resistance breeding (Kruger, 1983), there is no definite information on the genetics of resistance. It appears to be under polygenic control and transmitted to a small proportion of seedlings only.

(iv) Crown rot

Resistance to this disease appears to be under polygenic and monogenic control. A single dominant gene, *Pc*, controls resistance to *Phytophthora cactorum* in the variety Northern Spy (Alston, 1970). This gene is responsible for the moderate level of resistance exhibited by some of the rootstocks from the Malling–Merton series, which are derived from Northern Spy. Higher levels of resistance are inherited from *M.* × *sublobata* Novole, *M. angustifolia* and *M.* × *magdeburgensis* Hartwig. Rapid screening techniques to aid rootstock breeding have been developed. These are based on flooding 2–3-week-old seedlings, growing in peat and sand, with a zoospore suspension of pathogen. Effective selection of resistant plants is possible after only a further 2 weeks.

(v) Apple rust

A number of established varieties are resistant to this disease in North America. The resistance is controlled by two dominant genes, *Gy-a* and *Gy-b* (Aldwinckle *et al.*, 1977).

(vi) Apple blotch

Susceptibility to this disease is believed to be controlled by two dominant genes, *Ps-1* and *Ps-2* (Mowry & Dayton, 1964).

(vii) Glomerella *leaf blotch*

One variety, Golden Delicious, is known to be heterozygous for the dominant gene *Gb* which confers susceptibility to a strain of *Glomerella cingulata* (Janick *et al.*, 1996).

(viii) Fireblight

All commercial varieties are susceptible, but a few show only a low degree of susceptibility. Very high resistance has been found in Robusta 5 and *M. fusca*. A number of the varieties which carry the *Vf* scab resistance gene from *M. floribunda* 821 have a reasonable level of resistance. Experience has shown that a reasonable level of resistance results from a careful choice of parents in breeding programmes, but genes for resistance have not yet been identified. However, successful genetic transformation of Royal Gala and the rootstock M7 has been achieved using *Agrobacterium tumefaciens* to introduce the insect-derived lytic protein gene attacin E (Norelli *et al.*, 1998).

(ix) Storage diseases

An extensive survey which involved the inoculation of over 200 varieties prior to storage showed that a few were very resistant to *Gloeosporium perennans*. One of these, Cravert Rouge, transmitted a high level of resistance in a cross with the highly susceptible variety Cox's Orange Pippin, while derivatives of another, Jonathan, showed good resistance. No further information indicating the genetics of resistance to storage diseases in apple is available.

3. Insect pests

Most insecticide regimes control two or more pests; therefore, if resistance to one pest only is incorporated, it might not result in the removal of a particular insecticide from the spray programme. While avoidance of a range of spring pests, e.g. apple grass aphid (*Rhopalosiphum insertum*), apple sucker (*Psylla mali*), rosy apple aphid (*Dysaphis plantaginea*) and various caterpillars, could be achieved by selecting for late flowering, these benefits would be outweighed by the inconsistent cropping levels common in late flowering varieties. However, it is not essential to incorporate resistance to all pests, as some can be controlled by predators and other components of integrated control systems. Resistance has been found to most insect pests, but interest has concentrated on strong resistance that is under simple genetic control, for which seedling selection techniques are available. Consequently, there has been most interest in three aphid pests: woolly aphid (*Eriosoma lanigerum*), rosy apple aphid and rosy leaf curling aphid (*Dysaphis devecta*). Resistance to sawfly (*Hoplocampa testudinea*) and codling moth (*Cydia pomonella*) is also being investigated.

(i) Woolly aphid

As well as attacking the trunk and stems of apple trees, large colonies of woolly aphid, which are difficult to control chemically, can be established on the root systems in warmer regions. The 190-year-old variety Northern Spy has been known for many years to be very resistant and has been used as a parent in several breeding programmes. Most notable is the work at East Malling, UK, which led to the release of the Malling–Merton (MM) series of resistant rootstocks (Crane *et al.*, 1936). Today, three of these, MM106, MM111 and Merton 793, are used widely. Although there have been isolated reports of woolly aphid biotypes which can infest these rootstocks, this resistance, which is controlled by a single dominant gene *Er-1* (Knight *et al.*, 1962), is widely effective. It has enabled the establishment of successful orchards in areas of the world, such as New Zealand and South Africa, where the pest is particularly active. The resistance observed in the varieties Irish Peach and Winter Majetin is under polygenic control, while that observed in *M.* × *robusta*, which is controlled by a single gene, *Er-2*, does not appear to be effective in the roots. Screening is normally possible on seedlings from three months after germination; nymphs are transferred to the more basal nodes with a fine brush. It normally takes a further three months, including repeated inoculations, before resistance can be confirmed.

(ii) Rosy apple aphid

There is, at present, no means of controlling this pest other than by insecticides. This is now the most serious insect pest in UK apple orchards. Heavy infestations can have severe effects on the growth of both shoots and fruit; however, several Malus species are resistant to this aphid. The most useful source of resistance, *M.* × *robusta*, which is also the source of the mildew resistance gene *Pl-1*, carries a single gene for hypersensitivity, *Sm-h* (Alston & Briggs, 1970), which is very effective in unsprayed orchards. This gene has been transferred at East Malling to the cultivated apple through four backcrosses. There are also indications that modifier genes which affect the expression of *Sm-h* are present in many varieties. After the inoculation of 2-month-old seedlings with adult aphids, hypersensitive plants can be identified a week later.

(iii) Rosy leaf curling aphid

Colonization by this pest results in severe leaf curling and galling and thus leads to damage to apple crops. Several commercial varieties are resistant, including the main English variety, Cox's Orange Pippin. Three aphid biotypes, three distinct resistance genes (*Sd-1*, *Sd-2* and *Sd-3*) and a precursor gene *Pr-Sd*, without which the resistance genes are not expressed, are known (Alston & Briggs, 1977). Cox carries the gene *Sd-1* and is resistant to biotypes 1 and 2. Selection is carried out on 2-month-old seedlings, each inoculated with an adult aphid; typical leaf galling symptoms appear on susceptible plants after five days. This clear expression of symptoms and the distinct segregation of the gene *Sd-1* facilitated the mapping of the gene to a single locus on a chromosome of the resistant variety Fiesta (Cox × Idared) within 2 centiMorgan of three tightly linked restriction fragment length polymorphism (RFLP) markers (Roche *et al.*, 1997). This first report of molecular markers for aphid resistance in tree fruit crops shows the potential for the use of markers in resistance breeding programmes.

(iv) Sawfly

A high level of resistance to this pest of apple fruit is present in the *M.* × *zumi* derivative which is the source of the mildew resistance gene *Pl-2*. Crosses with the cultivated apple have yielded resistant and susceptible segregants.

(v) Codling moth

Various levels of resistance to this serious apple fruit pest have been reported. Whereas selection in established plantations appears to be the most effective, it is essentially long term and has to await the establishment of mature plantations, together with the build-up of populations of the pest large enough to ensure its dispersal in the orchard. While special attention to orchard hygiene might reduce pest populations to levels low enough to alleviate the need for a special breeding programme, in regions where pest populations are large this will not be sufficient. Genetic manipulation offers the best approach in these situations, using non-apple genes such as the Bt toxin (Dandekar *et al.*, 1992).

4. Resistance breeding strategy

The inputs of chemical pesticides are necessarily large in commercial apple orchards. More than 20 sprays are required each season to control pests and diseases in European orchards, as most established varieties are highly susceptible. Fungicides, aimed at scab and mildew, are responsible for most of the spray applications; therefore, these two diseases have received the most attention.

Although information on the segregation of the various resistances studied is limited, because of the relatively small populations involved and the limited numbers of generations on which the conclusions were based, recent studies using molecular markers have confirmed the presence and position of important major genes controlling resistance to scab, mildew and one species of aphid. These genes, most of which have been transferred to the cultivated apple from wild species through a series of backcrosses, have striking effects on the response of the apple to pathogens. For many years, all appeared to be durable. However, after long exposure in some locations, new pathogenic isolates arose.

The products of genetic manipulation programmes will also be subject to the build-up of virulent biotypes. Genetic manipulation presents the possibility of introducing resistance genes to commercially established varieties. In future, strategically managed manipulation programmes, involving the introduction of alternative resistance genes, might play an important part in the establishment of orchards not requiring treatments with chemical pesticides. Few suitable genes are known currently, but more will become available as useful genes are located in genomic maps. For the present, conventional transfer systems involve intercrossing and subsequent segregation and recombination of all the genes in the background, including all those controlling commercial quality. Currently, emphasis is on the production of durable resistance by the pyramiding of the various resistance genes and the combination of monogenic resistance and polygenic resistance in carefully managed crossing and selection programmes.

Nine research organizations, including Horticulture Research International–East Malling, are collaborating in Europe to develop durably resistant apple breeding material and to assess the risks of the appearance of new virulent pathotypes (Durel *et al.*, 1998). Additionally, numerous molecular markers will be developed to locate the genomic regions involved with polygenic resistance.

REFERENCES

Aldwinckle, H.S., Lamb, R.C. & Gustafson H.L. (1977) Nature and inheritance of resistance to *Gymnosporangium juniperi-virginianae* in apple cultivars. *Phytopathology* 67: 259–66

Alston, F.H. (1970) Resistance to collar rot *Phytophthora cactorum* (Leb and Cohn) Schroet in apple. *Report East Malling Research Station for 1969*, pp. 143–45

Alston, F.H. (1977) Practical aspects of breeding for mildew (*Podosphaera leucotricha*) resistance in apples. *Proceedings of the Eucarpia Fruit Section Symposium VII, Wageningen 1976*, pp. 4–17

Alston, F.H. & Briggs, J.B. (1970) Inheritance of hypersensitivity to rosy apple aphid *Dysaphis plantaginea* in apple. *Canadian Journal of Genetics and Cytology* 12: 257–58

Alston, F.H. & Briggs, J.B. (1977) Resistance genes in apple and biotypes of *Dysaphis devecta*. *Annals of Applied Biology* 87: 75–81

Alston, F.H., Kellerhals, M., Goerre, M. & Rapillard, C. (1997) Resistenzzüchtung bei Apfel und Birne: 10 Jahre Zusammenarbeit zwischen East Malling, Wädenswil und Fougères. *Schweizerische Zeitschrift für Obst- und Weinbau* 132: 668–70

Alston, F.H., Phillips, K.L. & Evans, K.M. (2000) A *Malus* gene list. *Acta Horticulturae* 538: 561–70

Batlle, I. (1993) *Linkage of Isoenzymic Genes with Agronomic Characters in Apple*. PhD Thesis, London: London University, p. 181

Batlle, I. & Alston, F.H. (1994) Isoenzyme aided selection in the transfer of mildew (*Podosphaera leucotricha*) resistance from *Malus hupehensis* to the cultivated apple. *Euphytica* 77: 11–14

Batlle, I. & Alston, F.H. (1996) Genes determining leucine aminopeptidase and mildew resistance from the ornamental apple 'White Angel'. *Theoretical and Applied Genetics* 93: 179–82

Benauf, G., Parisi, L. & Laurens, F. (1997) Inheritance of *Malus floribunda* clone 821 resistance to *Venturia inaequalis*. Proceedings of the 4th Workshop on Integrated Control of Pome Fruit Diseases, Croydon, 1996. *IOBC/WPRS Bulletin* 20: 1–7

Brown, A.G. (1959) The inheritance of mildew resistance in progenies of the cultivated apple. *Euphytica* 8: 81–88

Crane, M.B., Greenslade, R.M., Massee, R.M. & Tydeman, H.M. (1936) Studies on the resistance and immunity of apples to woolly aphids *Eriosoma lanigerum* (Hausm). *Journal of Pomology* 14: 137–63

Dandekar, A.M., McGranahan, G.H., Uratsu, S.L. *et al.* (1992) Engineering for apple and walnut resistance to codling moth. *Proceedings of the Brighton Crop Protection Conference – Pests and Diseases 1992*, pp. 741–47

Dayton, R.F. & Williams, E.B. (1968) Independent genes in *Malus* for resistance to *Venturia inaequalis*. *Proceedings of the American Society of Horticultural Science* 62: 334–40

Dunemann, F., Bracker, G., Markussen, I. & Roche, P.A. (1998) Identification of molecular markers for the major mildew resistance gene Pl_2 in apple. *Acta Horticulturae* 484: 411–16

Durel, C.E., Lespinasse, Y., Chevalier, M. *et al.* (1998) Genetic dissection of apple resistance regarding pathogen variability: co-ordination of European research programmes. *Acta Horticulturae* 484: 435–41

Gallot, J.C., Lamb, R.C. & Aldwinckle, H.S. (1985) Resistance to powdery mildew from some small fruited *Malus* cultivars. *HortScience* 20: 1085–87

Gianfranceschi, L., Seglias, N., Kellerhals, M. & Gessler, C. (1998) Molecular markers applied to apple breeding, analysis of oligogenic and single gene resistances. *Acta Horticulturae* 484: 417–28

Janick, J., Cummins, J.N., Brown, S.K. & Hemmat, M. (1996) Apples. In *Fruit Breeding: Tree and Tropical Fruits*, edited by J. Janick & J.N. Moore, New York: Wiley

King, G.J., Alston, F.H., Brown, L.M. *et al.* (1998) Multiple field and glasshouse assessments increase the reliability of linkage mapping of the V_f source of scab resistance in apple. *Theoretical and Applied Genetics* 96: 699–708

Knight, R.L. & Alston, F.H. (1968) Sources of field immunity to mildew (*Podosphaera leucotricha*) in apple. *Canadian Journal of Genetics and Cytology* 10: 294–98

Knight, R.L., Briggs, J.B., Massee, R.M. & Tydeman H.M. (1962) The inheritance of resistance to woolly aphid, *Eriosoma lanigerum* in the apple. *Journal of Horticultural Science* 37: 207–18

Koller, B., Gianfranceschi, L., Seglias, N., McDermott, G. & Gessler, C. (1994) DNA markers linked to *Malus floribunda* 821 scab resistance. *Plant Molecular Biology* 26: 597–602

Korban, S.S. & Dayton, R.F. (1983) Evaluation of *Malus* germplasm for resistance to powdery mildew. *HortScience* 18: 219–20

Kruger, J. (1983) Anfälligkeiten von Apfelsorten und Kreuzungsnach-kommerschaften für den Obstbaumkrebs nach natürlicher und künstlicher Infektion. *Erwerbsobstbau* 25: 114–16

Maliepaard, C., Alston, F.H., van Arkel, G. *et al.* (1998) Aligning male and female linkage maps of apple (*Malus pumila* Mill.) using multi-allelic markers. *Theoretical and Applied Genetics* 97: 60–73

Manganaris, A.G., Alston, F.H., Weeden, R.F. *et al.* (1994) Isoenzyme locus *Pgm-1* linked to a gene (V_f) for scab resistance in apple. *Journal of the American Society of Horticultural Science* 119: 1286–88

Markussen, J., Kruger, J., Schmidt, H. & Dunemann, F. (1995) Identification of PCR based markers linked to the powdery mildew resistance gene Pl_1 from *Malus robusta* in cultivated apple. *Plant Breeding* 114: 530–34

Mowry, J.B. & Dayton, D.F. (1964) Inheritance of susceptibility to apple blotch. *Journal of Heredity* 55: 129–32

Norelli, J.L., Mills, J.Z., Jensen, L.A., Momol, M.T. & Aldwinckle, H.S. (1998) Genetic engineering of apple for increased resistance to fireblight. *Acta Horticulturae* 484: 541–46

Parisi, L. & Lespinasse, Y. (1996) Pathogenicity of *Venturia inaequalis* strains of Race 6 on apple clones (*Malus* sp). *Plant Diseases* 80: 1179–83

Roberts, A.L. & Crute, I.R. (1994) Apple scab resistance from *Malus floribunda* 821 (V_f) is rendered ineffective by isolates of *Venturia inaequalis* from *Malus floribunda*. *Norwegian Journal of Agricultural Sciences* (Suppl.) 17: 403–06

Roche, P., Alston, F.H., Maliepaard, C. *et al.* (1997) RFLP and RAPD markers linked to the rosy leaf curling aphid resistance gene (Sd_1) in apple. *Theoretical and Applied Genetics* 94: 528–33

Tartarini, S. (1996) RAPD markers linked to the V_f gene for scab resistance in apple. *Theoretical and Applied Genetics* 92: 803–10

Visser, J. & Verhaegh, J.J. (1977) Review of tree fruit breeding carried out at the Institute for Horticultural Plant Breeding at Wageningen from 1951–76. *Proceedings of the Eucarpia Fruit Breeding Section Symposium VII, Wageningen 1976*, pp. 113–32

Yang, H. & Kruger, J. (1994) Identification of a RAPD marker linked to the V_f gene for scab resistance in apple. *Plant Breeding* 112: 323–29

GLOSSARY

apomixis/apomictic reproduction in which sexual organs or related structures take part but fertilization does not occur, so that the resulting seed is vegetatively produced

clonal propagation multiplication of genetically identical plants by vegetative (asexual) means, such as budding or grafting

heterotic hybrid vigour

isoenzymes variants of enzymes that catalyse the same reaction but as a result of differences in amino acid sequence can be distinguished by techniques such as electrophoresis or isoelectric focusing

pathotype a subdivision of a species distinguished by common characters of pathogenicity, particularly in relation to host range

polygenic control phenotypic characters that are under the control/caused by the actions of many genes

zoospore asexual reproductive cell in fungi

TREES: STILL CHALLENGING BREEDERS AT THE MILLENNIUM

Peter Kanowski
Department of Forestry, Australian National University, Canberra, Australia

1. Introduction

Trees are the defining feature of forests and woodlands, the most biologically diverse terrestrial ecosystems which occupy around a third of the world's ice-free land surface. These long-lived woody plants have been little domesticated relative to the crop plants or animals on which agricultural systems are based and for which genetic knowledge and manipulation are most advanced. Relatively few tree species – around 500 of the presumed 50 000 or more – have been subjected to any level of deliberate selection or breeding. The domestication histories of these species are quite contrasting, with contrasting implications for the domesticated populations and for breeding, as listed below.

- Those with the longest history of domestication, in the order of millennia, are species important in traditional land-use systems and are familiar to us from those landscapes (e.g. those of the genera Artocarpus, the jak- and bread-fruits of Asia; Faidherbia, the widespread and widespreading leguminous tree of African savannahs; Leucaena, the typically smaller legumes long-cultivated in Central America and Mexico; or Quercus, the oaks of Europe). The situation of these species, subject to many generations of informal selection and often to induced hybridization, parallels that of other extensively but long-domesticated crops, with an often imprecise distinction between natural and naturalized populations.
- Around 200 species have been subject to at least one cycle of breeding (*viz.* selection, mating and testing); a similar number has simply been included in tests. Those subject to the most intensive breeding efforts are the 60 or so species, principally of the genera Acacia, Eucalyptus, Picea, Pinus, Pseudotsuga, Populus, Larix and Tectona, which have been improved for industrial wood production (i.e. for solid or reconstituted wood and for pulp) over – at most – the past 50 years. Few of these improvement programmes have yet progressed to the third or subsequent generations; nevertheless, they provide the bulk of our experience and information about the genetic improvement of trees.
- Another 60 or so more taxonomically disparate species – among them some of the long-domesticated species – have become the subject of breeding for non-industrial objectives in the past few decades. Examples include species of the tropical and subtropical Acacia, Azadirachta, Calliandra, Calycophyllum, Casuarina, Dalbergia, Faidherbia, Gliricidia, Grevillia, Irvingia, Leucaena and Prosopis, and the temperate Acacia, Alnus and Salix. Breeding objectives and strategies for these species are typically more diverse than for those bred for industrial wood production.
- A fourth group of species, those of horticultural and arboricultural importance, is usually considered the realm of these disciplines rather than of tree breeders. The horticultural and arboricultural preference for use of only a few genotypes, and their propagation as cultivars, contrasts markedly with the emphasis on the maintenance of genetic diversity in breeding and production populations of trees domesticated for other purposes.

The overwhelming majority of tree species, though, have not been subject to either informal or to organized domestication. Although their gene pools have been altered, sometimes profoundly, by a wide range of human activities – in the forms of deforestation, exploitation, population fragmentation, demographic and habitat alterations, environmental modification and translocation – most tree species remain essentially in the genetically wild state. Despite the accelerating rate of forest loss and degradation globally – for example, half the world's croplands were forested a century ago, as were half the tropics' agricultural lands only a half century ago – the genetic resources of most tree species remain rich, a consequence of the longevity of individual trees, the extensive geographical distribution of many species and the high proportion of genes common across most populations in most species. Even for those species which have been somewhat domesticated, tree breeders are still in the enviable position of having relatively easy access to extensive gene pools which are highly diverse compared with those of other plant species.

2. Tree breeding in context

The size and importance of the agricultural economy, and the fundamental role of improved crops and animals in agricultural production systems, have fostered high levels of activity in crop and animal breeding. Tree breeding programmes remain modest by comparison, partly because products harvested from natural forests have been dominant in world markets and partly as a consequence of historical perceptions that trees played a relatively minor role in sustaining rural livelihoods. As economic and environmental forces shift the production of wood and other forest products dramatically, from natural sources to plantations and farm forests, and as our appreciation grows of the fundamental importance of trees in many land-use systems, so the emphasis accorded to genetic improvement of trees is increasing.

Principally because of the long time periods inherent in tree growing, the genetic improvement of trees has often appeared to be less commercially attractive than alternative investments; consequently, most tree breeding programmes began, and remain, in the public domain. Similarly, access to the genetic resources of tree species has seldom been restricted; more usually, access has been facilitated by strong cooperation, both internationally and between public and private sectors. However, tree breeding activities and tree genetic resources are becoming increasingly proprietary, mirroring more general trends in the declining role of the public sector and in changes to intellectual property regimes often – though not necessarily – associated with the application of biotechnologies. Since almost all tree species are at early stages of domestication, because the majority of wild genetic resources and many improved populations remain in the public domain, and because at least some of the benefits of tree growing are public rather than private goods, these changes pose particular challenges to tree breeders.

As perennial plants with predominantly outcrossing mating systems, trees present breeders with something of a paradox. Like the annual or short-lived plants familiar to crop breeders, trees produce many individual flowers and progeny, facilitating mating and testing; species are variously amenable to clonal propagation; and field testing is free of the complications of individual mobility and behaviour. However, trees also share many biological and genetic commonalities with animals: individuals are long lived, often with a marked juvenile phase and with long generation intervals; many tree products from the wild state are useful without improvement; most species are strongly outcrossing, necessitating breeding at the level of population improvement rather than as inbred lines; the large size of individuals, and the heterogeneity of both test and production sites, complicate field testing and the prediction of gains; and individual plant rather than plot mean performance is of primary interest to breeders. Thus, trees present the breeder with a distinct suite of advantages and constraints.

3. A brief history of tree breeding

Although foresters have been aware for centuries of the variation in performance associated with provenance, formal tree breeding began only in the 1940s in Scandinavia. This classical approach, soon adopted elsewhere – initially in the US, but soon thereafter in Australia, New Zealand and Southern Africa, where plantation forests were already important – followed the cycle of selection of phenotypically superior individuals, progeny testing of their (usually open-pollinated) offspring, the concurrent establishment of their ramets in seed orchards, the culling of those orchards as progeny test results became available and – often somewhat later – the incorporation of selections from a wider range of sources than had been accessed initially. Almost invariably, testing at the species and provenance-within-species levels revealed differences of major consequence to breeders. Selection criteria reflected the requirements of the wood-processing industries: principally, growth rate, stem and crown quality, and wood properties. Typical productivity improvements from the first generation of breeding, reflecting changes in both yield and quality, were substantial; for example, the gains realized by a number of programmes in utilizable stem volume, the initial focus for most industrial tree breeding, were 30–50% over unimproved population means. Gains realized reflected, variously, selection of well-adapted provenance(s), effective within-provenance selection and the release from inbreeding depression achieved by restricting selection to avoid sampling relatives in the wild.

During this first era of tree improvement, breeders were necessarily preoccupied with understanding the reproductive and propagation biology of candidate species, and with the mechanics of artificial pollination, field testing and assessment. It was generally not until the 1970s and 1980s that tree breeders began seriously to address the more strategic issues of clarifying breeding objectives and structuring programmes to optimize economic returns, of structuring breeding populations to realize both short- and long-term objectives, and of applying analytical tools to maximize the utility of the increasing wealth of information that their labours had generated. Around this time, tree breeders also began to work with non-industrial species, with their more diverse breeding objectives (e.g. fodder production from leaves and fruit, wood for fuel and poles rather than for industrial processing, compatibility with other components of farming systems or the capacity to help rehabilitate degraded sites).

4. Issues challenging tree breeders

A number of issues particularly challenge tree breeders at the beginning of the 21st century.

(i) Continuing to capitalize on interprovenance variation

The gains achieved through simple provenance selection in most tree breeding programmes have, almost invariably, been enormous and very cost effective. Selection on this simple basis, for both immediate gain and as the foundation for subsequent breeding, remains an imperative for new programmes, new environments or new breeding objectives. It also emphasizes the links between tree breeding programmes and genetic conservation strategies.

(ii) Developing robust, progressive and cost-effective breeding strategies

Breeders of industrial tree species, for which programmes are most advanced and commercial imperatives strongest, have necessarily taken the lead in defining breeding strategies which satisfy genetic, economic and operational imperatives. Experience demonstrates that such strategies are seldom simple, even for the simplest forest production systems of short-rotation trees grown as monocultures for industrial use. Their common features are the maintenance of a broad genetic base and the use of efficient selection, testing and multiplication procedures.

Breeders of non-industrial species are even more strongly challenged: by the diversity of species for which improvement would be advantageous, by the diversity of production systems in which these species are or could be used, by the variety of germplasm dissemination pathways among growers, by the relative poverty of basic information about and genetic resources of many of the species of interest and by the relative lack of financial resources available to support their improvement.

The increasing levels of investment required to support

advanced programmes and technologies, and the restricted public funds available to support improvement of non-industrial trees, will maintain the recent focus on the refinement of breeding strategies to guide resource allocations; tree breeders will continue to benefit from the corresponding experiences of their colleagues in plant and animal breeding.

(iii) Integrating biotechnologies and tree breeding programmes

Strategic decisions about investments in biotechnologies, and the integration of biotechnologies into breeding programmes, pose particular challenges to tree breeders. The long generation intervals, constraints to breeding and lack of genetic information for many tree species enhance the appeal of biotechnologies; conversely, the relative genetic complexity of trees, the relatively wild state of most gene pools and the consequent relative uncertainty of outcomes add to the risk of investments in tree biotechnologies. The proprietary nature of much biotechnological research also challenges the collaborative foundations and prospects of tree breeding programmes; its relatively high costs generally strain rather than supplement already inadequate budgets and the long lead times inherent in the development and application of many biotechnologies have led to the characterization of "biotechnological constipation" in the delivery of results. As in plant and animal breeding generally, biotechnologies offer enormous gains, but also carry substantial uncertainties.

Currently, the biotechnologies of most relevance to tree breeding are genomic mapping, molecular markers, transformation and micropropagation. The implementation of many biotechnologies are interdependent and most are dependent for delivery on successful clonal propagation. The optimal integration of these biotechnologies into tree breeding will be programme specific.

(iv) Clarifying breeding objectives and improving selection decisions

The economic and operational advantages of clearly defining breeding objectives are apparent from experience in plant and animal breeding. However, lack of economic and genetic information has limited the clarification of breeding objectives for trees; the increasing use of short-generation tree crops grown on short rotations (e.g. Acacia and Eucalyptus species) is helping to overcome these constraints and provide guidance as to how breeding objectives may be better defined for trees with longer rotations and more complex production systems and end uses.

Since even short crop rotations in forestry are long by agricultural standards – a minimum of 5 years, more typically 10–25 years and often greater than 40 years – indirect selection is ubiquitous in tree breeding programmes. Early selection has been described as the "Holy Grail" of tree improvement, and has been the focus of considerable research using the tools of both quantitative and molecular genetics. Current early selection decisions are based on the outcomes of physiological and quantitative studies; marker-assisted selection remains a tantalizing prospect and the subject of intensive research. The prospective availability of hypervariable codominant markers will greatly facilitate the mapping of multiple-allele quantitative trait loci, but we do not yet know whether economically important traits in trees are typically under the control of multiple alleles per locus at relatively few loci. Capitalizing on early selection will require concomitant promotion of early flowering in some species.

(v) Optimizing recombination and transformation

Much early work with industrial species focused on the completion of elaborate mating schemes, to both generate genetic information and exploit any heterotic effects. While molecular geneticists are now benefiting from the availability of some of the material produced from these schemes, the emphasis in operational improvement programmes has shifted to simplicity wherever possible, a wider range of complementary mating schemes to meet differing objectives and to differential schemes favouring known or presumed elite material.

Intra- and interspecies hybridization has yielded outstanding results in some genera, notably Eucalyptus, Leucaena, Pinus and Populus. Interspecific hybrids have been the basis of the most successful clonal plantation programmes, of which the Aracruz (Brazil) programme based on Eucalyptus is perhaps the most spectacular. Progress with genetic transformation – for example, of sterility, cold tolerance and herbicide resistance – has been promising, but little material has yet reached the field testing stage.

(vi) Improving testing and analysis

Much of the progress realized in tree improvement over the past decade has resulted from improved testing strategies and analytical methodologies. Field testing to even the minimum age for early selection – seldom less than 3 years for short-rotation trees, and more usually at least twice that for those grown on longer rotations – is expensive of time and resources; research to clarify the value of short-term or glasshouse tests continues, frequently in association with that seeking molecular markers to aid selection decisions. Contemporary statistical techniques, many derived from animal breeding, are better able to cope with the environmental heterogeneity inevitably associated with testing large numbers of trees over long periods of time and the recovery of the maximum information from diverse arrays of relatives across generations and testing sites. The costs of testing and the opportunity costs of not maximizing recovery of information will demand continuing improvements in testing and analysis.

(vii) Capturing gain through multiplication and deployment

The past decade has also seen radical changes in multiplication strategies, with traditional open-pollinated clonal orchards superseded by new forms of pollen and orchard plant management, and by advances in both low- and high-technology methods for the vegetative propagation of both industrial and non-industrial species. These advances greatly enhanced the flexibility of breeding strategies and their capacity to deliver gains. Consequently, an increasing proportion of trees established are clonal propagules; where propagation is linked to an advanced breeding programme, the cuttings are likely to originate from controlled crosses between elite parents. The rise of clonal forestry has been accompanied by research to clarify the adaptability or specificity of genotypes to typically heterogeneous planting environments, and by the development of regulations or practices to ensure that some level of genetic heterogeneity is maintained, at least at a landscape scale. The latter issue, in particular, will continue to be the subject of debate and research, especially in the case of industrial plantations.

5. Conclusion

Tree breeding is a young discipline with a tradition of cooperative research, now establishing an identity complementary to those of crop and animal breeding. The tradition of cooperation that has facilitated access to genetic resources, and which has permitted genetic progress that would have been unlikely were breeders working in isolation, is challenged by the changing roles of the public and private sectors, and by the increasing prevalence of proprietary research associated particularly with investment in biotechnologies. The gains delivered by tree breeding will depend as much on how breeders engage with these strategic issues as on the progress they make in science and its applications.

FURTHER READING

Eldridge, K.G., Davidson, J., Harwood, C. & van Wyk (1994) *Eucalypt Domestication and Breeding*, Oxford: Clarendon Press, and New York: Oxford University Press

Haines, R.J. (1994) Biotechnology in forest tree improvement. *FAO Forestry Paper* 118

Kanowski, P.J. (1993) Mini-review: forest genetics and tree breeding. *Plant Breeding Abstracts* 63: 717–26

Ledig, F.T. (1992) Human impacts on genetic diversity in forest ecosystems. *Oikos* 63: 87–108

Libby, W.J. (1992) Use of genetic variation for breeding forest trees. In *Plant Breeding in the 1990s*, edited by H.T. Stalker & J.P. Murphy, Wallingford: CAB International

Namkoong, G., Kang, H.-C. & Brouard, J.S. (1988) *Tree Breeding: Principles and Strategies*, New York: Springer

National Research Council, Committee on Managing Global Genetic Resources: Agricultural Imperatives (1991) *Forest Trees*, Washington, DC: National Academy Press

GLOSSARY

cultivar plant variety found only in cultivation

outcrossing outbreeding; mating of unrelated individuals

transformation any change in the properties of a cell stably inherited by its progeny

vegetative (or clonal) propagation asexual reproduction when the offspring develops from a group of cells of the parent rather than from an egg, e.g. budding, reproduction with bulbs, runners or cuttings

DETERMINING THE IMPACT AND CONSEQUENCES OF GENETICALLY MODIFIED CROPS

J.B. Sweet
National Institute of Agricultural Botany, Cambridge, UK

1. Introduction

Genetically modified (GM) crops will have impacts on agriculture and both the agricultural and natural environment. Analysing the consequences for the agricultural environment requires study of the characteristics of the GM crop and its hybridizing relatives, and study of the management systems involved in growing the GM crop and other crops grown in rotation with it. GM crops may also have impacts on uncultivated land and natural environments. Thus, risk assessments are concentrating on whether the genetically modified characteristics of a GM crop are likely to change the behaviour of the plants in their environments to the extent that ecological balances are altered.

2. Risk assessment in GM crops

GM crops have potential benefits for growers, processors and consumers. However, many will have impacts on agriculture and both the agricultural and natural environment while others will have food quality and safety implications. Pest-, disease- and herbicide-resistant crops will require different (often reduced) pesticide and herbicide inputs in order to exploit these novel traits. These modified management systems will themselves have an impact on agricultural environments. Thus, analysing the consequences for the agricultural environment requires study of the characteristics of the GM crop and its hybridizing relatives, together with study of the management systems involved in growing the GM crop and other crops grown in rotation with it. For example, GM herbicide-tolerant (GMHT) crops will be treated with different herbicides, with different activity spectra, at different crop development growth stages. Thus, the effect on botanical diversity in the GMHT crop and in subsequent crops will be a product of the GM crop and the herbicide treatment.

GM crops may also have impacts on uncultivated land and natural environments. These environments may be affected by the characteristics of crop and wild species induced by novel genetic constructs and their products. Thus, risk assessments concentrate on whether the genetically modified characteristics of a GM crop, and of similarly modified hybridizing relatives, are likely to change the behaviour of the plants in their environments to the extent that ecological balances are altered. Risk assessments study both the severity and extent of the hazard or damage as well as the likelihood and frequency at which the damage will occur. Hence, the definition of risk as:

$$\text{RISK (IMPACT)} = \text{FREQUENCY} \times \text{HAZARD}$$

3. Gene flow measurements

The impact of transgenes on particular wild species will depend on whether genes will introgress into wild populations and the rate of that introgression. Introgression of a transgene is a product of cross-pollination with transgenic relatives, survival of the hybrid and its hybridization with the wild population to the extent that the gene becomes established in a proportion of the population. If there is sexual incompatibility between species, the assumption is that no gene transfer will take place and that the risk is zero. However, in experiments, certain wild brassica and other crucifer species previously considered incompatible with *Brassica napus* (oilseed rape) have shown some ability to hybridize (Scheffler & Dale, 1994). Thus, there is the possibility that hybridization can occur, albeit at very low frequencies, and that transgenes have a route to introgression. There are numerous studies to determine the frequency of interspecific hybridizations. However, information is also required on whether the hybrids can survive and backcross with their wild parental species so that the gene can become established in the wild population. Studies by Scott and Wilkinson (1999) of *B. rapa* (*B. campestris*) populations growing outside fields of *B. napus* in England indicated that cross-pollination frequencies were low (0.4–1.5%) in 7% of *B. rapa* populations surveyed. The remaining 93% of populations contained no hybrid seed. In addition, they found that, on average, <2% of all seedlings survive, so that, unless the transgene conferred higher survival characteristics, establishment of GM *rapa* × *napus* hybrids would be very poor and introgression of the gene into *rapa* populations would be very slow. Hybridization frequencies appear to be much higher where *B. rapa* occurs as a weed in *B. napus* crops (Jorgensen *et al.*, 1998; Sweet & Norris, unpublished data). In the UK, however, there is no indication that *B. rapa* has modified its behaviour or is spreading as a weed in *B. napus* crops as a consequence of hybridization. The relationship between weedy and wild *B. rapa* is yet to be determined, but may be a significant route for introgression of a transgene from oilseed rape.

Chevre *et al.* (1998) have shown that hybridization and backcrossing with wild Raphanus and Sinapis can occur at low frequencies under particular conditions. Studies are underway in the UK to determine whether oilseed rape genes can be detected in Sinapis species (Dale & Moyes, personal communication). However, little is known of the ability and frequency of hybridization of weedy crucifers, e.g. Sinapis and Raphanus, with *B. napus* under natural conditions and the survival and reproductive characteristics of the interspecific hybrids. Thus, we appear to be far from determining whether natural hybridization frequencies are zero or close to zero for many related cruciferous species.

Hybridization between weedy crucifers and oilseed rape would tend to lead to introgression of advantageous adaptive characters into the weedy populations such as increased waxyness and reduced leaf hairs associated with tolerance to the herbicides currently used on oilseed rape. There are indications that this may have occurred in weedy *B. rapa* populations growing in oilseed rape crops in Denmark (Jorgensen *et al.*, 1998) and in England (Sweet & Norris, unpublished data). However, no morphological modifications associated with enhanced herbicide tolerance have been reported in populations of other weedy crucifers. Some Sinapis and Raphanus populations appear to have partial tolerance to some oilseed rape herbicides, although whether this is due to gene exchange, evolution or selection from an inherent natural variability is unknown.

There is also a need to study indirect gene flow through intermediary species that are compatible with both oilseed rape and with other wild crucifers. Thus, we need to determine whether transgenes are able to introgress from *B. napus* into *B. rapa* and, hence, to Sinapis or Raphanus species.

There have been and continue to be numerous studies of transgene flow through seed dispersal, pollination and hybridization within the same species. These studies are gradually being scaled up to levels that reflect field-scale releases and the results confirm those found with non-GM pollen. There are fewer reports of studies of the persistence and survival of introgressed genes in volunteers (self-sown plants appearing in the succeeding crop). Sweet and Shepperson (1998), and Norris *et al.* (1999) studied volunteer populations of winter and spring oilseed rape occurring at release sites in subsequent crops. There are also reports on volunteer GM rape from France (Messean, 1997), but none of the studies has reported increased numbers or fitness of herbicide-tolerant volunteers.

(i) Impact of plant species

Many studies have concentrated on measuring frequency phenomena such as gene flow and interspecific hybridization without necessarily considering the impact of the transgene when it has dispersed or introgressed into other populations or species. Frequency and hazard are dependent on the characteristics of both the crop that is modified and of the GM trait and so risk assessments require the measurement and study of the hazard or impact of both the crop/plant and the trait. In addition, the impact of the release of the GM plant will depend on the type and location of the environment into which it is being released. This means that risk assessments are not necessarily transferable from one site, area, region or country to another.

Plants vary in the degree to which they dominate or are invasive of certain environments and in their ability to disperse genes to different populations and species. Thus, they will have different environmental impacts when genetically modified and, for any particular country or region, plants can be classified as being of high, medium and low impact.

High impact plants. Plants in this group are hardy, perennial, competitive, open pollinating and prolific, have a wide range of relatives with which they hybridize and an ability to colonize a range of natural and seminatural habitats. Examples of such plants are perennial grasses (e.g. *Lolium perenne*, perennial ryegrass) and certain indigenous and introduced trees and shrubs which form a significant proportion of forests and woodlands (e.g. Populus spp., poplar). Modifications of these plants that affect their competitiveness and behaviour could have significant impacts on the ecology of a range of environments.

Medium impact plants. Plants in this group are open pollinating, hybridize with some wild relatives, are prolific and colonize a limited range of habitats. Examples of such plants are oilseed rape, oats, sugar beet and rice, all of which have closely related wild relatives with which they hybridize and an ability to colonize disturbed ground. These plants and their close relatives rarely form climax populations except in particular environments such as coastal areas or in disturbed ground.

Low impact plants. These are usually annual or biennial species, largely self-pollinating and with few hybridizing relatives that are poorly adapted to the area in which they are cultivated. In the UK, examples include maize and sunflower.

It is important to appreciate that the impact of plant species will depend on the environment into which they are being released. Maize and potato are considered low impact plants in England. However, in countries of Central and South America, where their centres of genetic diversity occur, along with many wild relatives, their impact would be considered very high.

(ii) Impact of transgenes

Transgenes, operating through their expression in plants, will have different impacts on environments. Since genes often operate uniquely, it is not easy to classify transgenes as having high or low impact. In addition, their impact is also dependent on the nature of the receiving environment.

High impact transgenes. In general, genetic modifications that improve the fitness of GM plants by increasing their reproduction, competitiveness, invasiveness and/or persistence will have the greatest environmental impact. Transformations that significantly increase plant productivity and overcome constraints and stresses, such as pests, diseases, drought, etc., will have the highest impact. Hence, very high yielding and vigorous GM plants with enhanced and broad-spectrum pest, disease and stress tolerance will have the greatest impact.

Many pest- and disease-resistance genes will have effects on non-target species, either directly through gene products that destroy or debilitate non-targets or indirectly by altering relationships between pests and beneficials. It is important that these non-target effects are thoroughly understood before commercialization progresses.

Low impact transgenes. These are genes that do not noticeably enhance the fitness of the modified plant or of other organisms; consequently, they have minimum ecological impact. Examples would be herbicide tolerance and genes that modify seed composition, e.g. high lauric acid genes in oilseed rape and high starch genes in potato. However, it is important to verify that these genes do not significantly increase seed tuber overwinter survival through enhanced frost resistance or alter the dormancy characteristics of oilseed rape seed so that it has enhanced soil survival characteristics.

4. Agricultural impact, consequences and monitoring

Genetic modification can have a range of impacts on agricultural systems and hence will require specific management. Genetic modification can alter the nature of crop volunteers in subsequent crops and the GM trait can disperse to other crops and weeds through cross-pollination and seed dispersal. Low impact genes such as those for herbicide tolerance, which have little impact on natural environments, become highly significant in the agricultural environment because of the changes in herbicides required for their management. These herbicides will differ in the effect they have on plant and other species diversity in cropped fields. These aspects are now the subject of several research projects at the National Institute of Agricultural Botany (Cambridge, UK) and other European institutes.

Deployment of high impact genes such as those for pest and disease resistance will result in reductions and changes in pesticide usage and thus offer opportunities to enhance diversity in cropped fields, especially if the transgene products are very specific to selected pest species. It is important, however, that the selection pressures they impose on pests and diseases do not encourage the development of virulent races of pests and pathogens that are resistant to the genes and require additional pesticide treatments.

There are considerable concerns throughout Europe about the environmental effects of releases and the indirect effects through agriculture of the commercialization of GM crops. Risk assessments conducted for regulatory purposes tend to concentrate on the direct effects of the GM crop and its relatives on the natural environment. It is now becoming apparent, however, that the agricultural consequences of the deployment and management of GM crops could also have significant impacts on the environment in regions where a very high proportion of the total land area is managed by man. Plans are being developed for monitoring the early years of the commercialization of each GM crop so that its impact on both agriculture and the environment can be evaluated in farm-scale releases.

Acknowledgements

The National Institute of Agricultural Botany acknowledges MAFF, DETR, Aventis, Monsanto and other biotechnology companies for their support of studies of the environmental impacts and consequences for agriculture of releases of GM crops.

REFERENCES

Chevre, A.-M., Eber, F., Baranger, A. *et al.* (1998) Risk assessment on crucifer species. *Acta Horticulturae* 459: 219–24

Jorgensen, R.B., Andersen, B. & Thure, P. (1998) Introgression of crop genes from oilseed rape (*Brassica napus*) to related wild species – an avenue for the escape of engineered genes. *Acta Horticulturae* 459: 211–17

Messean, A. (1997) Management of herbicide tolerant crops in Europe. *Proceedings of the 1997 Brighton Conference*, pp. 947–54

Norris, C.E., Simpson, E.C., Sweet, J.B. & Thomas, J.E. (1999) Monitoring weediness and persistence of genetically modified oilseed rape (*Brassica napus*) in the UK. In *Gene Flow and Agriculture: Relevance for Transgenic Crops*, edited by P. Lutman, Farnham, UK: British Crop Protection Council

Scheffler, J.A. & Dale, P.J. (1994) Opportunities for gene transfer from transgenic oilseed rape to related species. *Transgenic Research* 3: 263–78

Scott, S.E. & Wilkinson, M.J. (1999) Low probability of chloroplast movement from oilseed rape (Brassica napus) into wild Brassica rapa. *Nature Biotchnology* 17: 390–92

Sweet, J.B. & Shepperson, R. (1998) The impact of releases of genetically modified herbicide tolerant oilseed rape in the UK. *Acta Horticulturae* 459: 225–34

See also **Environmental risk assessment and transgenic plants** (p.695)

ENVIRONMENTAL RISK ASSESSMENT AND TRANSGENIC PLANTS

John Barrett
Department of Genetics, University of Cambridge, Cambridge, UK

1. Introduction

The techniques that eventually became the basis of "genetic engineering" were published in the early 1970s. In 1974, the scientists involved in the development of these methods called for a moratorium on their application, until the safety implications could be assessed (Berg *et al.*, 1974, 1975). Throughout the industrial world, regulations and advisory bodies were put in place to ensure that development of the technology was carried out under strict safety conditions, which varied according to the perceived risks of each and every experiment. Probably for the first time in history, a "hold" was put on the exploitation of a new technology while its safety was scrutinized. As experience increased, the regulations regarding the level of "containment" were relaxed for some categories of experiment, but not for others (e.g. those involving DNA from human pathogens).

Up to about 1980, most of the experiments involved microorganisms as hosts for the DNA, such as the bacterium *Escherichia coli* and the yeast *Saccharomyces cerevisiae*, and were carried out in the laboratory. By the early 1980s, techniques had been developed that allowed DNA to be introduced into plant cells (transformation) and whole "transgenic" plants to be regenerated from the tissue culture. It was quickly realized that one of the major applications of plant genetic engineering would be in the production of crops with novel characteristics unavailable in conventional plant breeding. If such an objective were to be realized, then transgenic plants would have to be moved from the laboratory and glasshouse, where they could be safely contained, into field tests where the control of the experimenter is reduced and real hazards to the environment could exist. In 1986, before any field trials were planned, a set of guidelines (Recombinant DNA Safety Considerations – otherwise known as the "Blue Book" – OECD, 1986) was published; it included protocols for assessing the risk involved in field trials of transgenic plants prior to carrying out the experiments. This publication introduced the idea of "case-by-case" and "step-by-step" assessment of potential risks. The 1986 Blue Book has been followed by other guidelines expanding on its foundations and these guidelines have now been turned into legal requirements for risk assessment in most of the industrial world, e.g. CEC Directive 90/220/EC.

2. Risk assessment

The first step in a risk assessment is to determine whether a chemical, a process or even a transgenic plant poses a hazard, i.e. whether it has the potential to cause harm. The next step is to determine the degree of exposure to the hazard and this leads directly to an assessment of risk, as:

$$\text{RISK} = \text{HAZARD} \times \text{EXPOSURE}$$

Although "risk" and "hazard" are used interchangeably in everyday speech, it is important to keep their different technical meanings clearly in mind when assessing risks.

3. Assessing the risk

In assessing the environmental risk of transgenic plants or crops, a clear description of the plant or crop, the transgenic construct and the environment into which it is to be introduced is needed first of all. Then, the potential effects on the environment need to be listed. The information required by the UK Environmental Protection Act – The Genetically Modified Organisms (Deliberate Release) Regulations 1992 – runs to 89 questions for an experimental release, plus a further nine questions for "Consent to Market" (see also CEC Directive 90/220/EC). The information requirements are very similar in all countries where such regulations have been introduced.

The potential environmental hazards of transgenic plants can be listed as follows:

- Can the genetic modification be transferred to wild relatives of the crop by cross-pollination?
- Does the modification increase any tendency to weediness of the crop species?
- Does the modification have any direct effect on the environment?
- What are the possibilities that a transgenic plant line might show genetic or phenotypic instability in the field?
- Where a vector is based on a pathogenic organism, could pathogenicity be regained in the transgenic line?
- Are there any concerns for human health, particularly for the safety of agricultural workers?
(Modified from OECD, 1993.)

4. Gene transfer to wild relatives

There is considerable variation in the degree to which crops will crossbreed with wild relatives. In the UK, this ranges from introduced crops like wheat, maize and sunflower which have no close wild relatives, through to crops, such as sugar beet, carrot and cabbage, which are cultivated forms of indigenous wild species. In between, there are crops that may have wild relatives with which they can interbreed or with which they can be forced to do so under experimental conditions (with or without special techniques such as embryo rescue), e.g. some cultivated brassicas and other members of the Cruciferae, and others which are polyploid descendants of crosses between different indigenous species (e.g. oil-seed rape, *Brassica napus*) (Raybould & Gray, 1993, 1994).

Table 1 Hazards of genetically modified crops cross-pollinating wild relatives.

Construct	Hazard
Herbicide resistance	Loss of control of wild weedy species
	Possibility of recombination with genes for resistance to other herbicides, giving multiple resistance
Pest and disease resistance	Wild species previously kept in check by naturally occurring pests and disease, now no longer controlled
Insecticide	Effects on non-target organisms, e.g. if expressed in pollen, pollinators affected

Among those crops which could, in principle, interbreed with wild relatives, the environmental hazard will depend on the sort of construct transferred. **Table 1** gives an indication of the types of hazard that could occur. Although these hazards can be readily identified, an assessment now has to be made on the exposure of wild relatives to the pollen from a transgenic crop. For cross-pollination to occur, both crop and wild species must be growing in the same area (the size of this area being determined to a large extent by how far pollen can travel); whether they are flowering at the same time and the relative areas or densities of both species are also important factors. Thus, a wild relative of a crop grown for its seed, fruit or flowers has a greater potential exposure to pollen from the crop than one related to a crop species which is normally harvested before flowering, such as carrot or cabbage. The breeding system, i.e. whether the species involved are inbreeding or outbreeding and whether they are wind- or insect-pollinated, also has an influence on the exposure of wild relatives.

Finally, when cross-pollination can occur, the viability/fertility of the hybrids must be considered, and it should be noted that even when viability or fertility is low, the descendants of a hybrid may still persist for some time, allowing backcrosses into the wild species to occur. This process could be accelerated, for example, by applications of a herbicide to which resistance has been transferred from the transgenic crop.

It can now clearly be seen that evaluating the environmental risk of gene transfer to wild relatives depends on an interplay between the taxonomy (and ecology) of the species involved, their breeding system, pollination biology and farming practice, and that risk assessment must be on a "case-by-case" basis.

5. Weediness

The simplest definition of a "weed" is simply that of a plant growing in the wrong place. A plant species that may be an important crop in one environment may be a weed in another: as in, for example, "carry-over" of one crop into the succeeding crop and the establishment of feral populations away from where the crop is cultivated. The risk to be assessed here is whether the construct will increase the tendency to "weediness". In cultivation, crops that tend to weediness are well known and standard husbandry is used to control them. However, some constructs may limit or prevent the use of standard methods; transgenic herbicide resistance, for example, may restrict the choice of sprays in a following crop. A further example of a potential risk is in a crop that is not normally winter-hardy, but has been engineered for frost resistance to extend the growing season or the areas in which the crop can be grown.

6. Effects on the environment

These effects can be split into two groups. The first is where the construct is designed to have an impact on the environment – insect resistance, for example. Obviously, the purpose here is to control pest populations, but increased control over conventional methods may affect predators of the pest species because the pest population has been suppressed or the predators pick up the transgenic toxin from its prey. Also, persistence of the gene product in crop residues may have an effect on post-harvest ecological communities that would not normally be subject to control measures.

The second group of risks is indirect. A transgenic crop may allow changes in crop husbandry that have an effect on the environment. For example, a transgenic herbicide-resistant crop may permit more effective weed control than in a conventional crop. This may then impact on the farmland ecological communities, such as the invertebrates that feed on the weeds and their predators, e.g. birds.

7. Genotypic and phenotypic instability

By and large, crop varieties are selected to be stable, particularly for purposes of registration for royalties on seed sales: for registration, new varieties must be distinct, uniform and stable. Trials are carried out in a range of environments to assess these characters. However, even with conventionally bred varieties, unusual environmental conditions (e.g. the weather) or cultivation under conditions not covered by the trial sites may expose unsuspected instability, such as an increase in the level of self-sterility in a normally self-pollinating crop species leading to cross-pollination from other varieties or even wild relatives. In general, this is a problem for the grower (e.g. unstable expression of a transgene reducing insect or herbicide resistance), but if the instability led to increased weediness, or changes in the expression of the transgene affected beneficial insects, for example, there is an environmental risk.

8. Vectors and pathogens

Among the methods used for transforming plants, there are some that use genetic material from naturally occurring plant pathogens (e.g. bacteria and viruses, such as the Ti plasmid from *Agrobacterium tumefaciens*). Although these vectors are "disarmed" before use and not capable of producing disease, there is a small risk that the pathogen itself may remain associated with the plant and be capable of causing disease. However, routine screening during the early stages of development should detect this.

A further set of potential risks arises when the transgene itself expresses protein (particularly a viral coat protein) derived from a pathogenic plant virus. If another pathogenic virus then infects the transgenic plant, the transgenic coat protein may be incorporated into the coat of the replicating second virus (transcapsidation) and may alter its host specificity. In this case, however, no genetic material is transferred. The second problem is recombination via the mRNA produced from the viral gene incorporated in the transgene. This will produce a recombinant virus, which may have altered pathogenicity, or host range, and may infect wild plants.

9. Worker safety

It is well known that some individuals develop skin reactions or allergies on exposure to some plant products, e.g. sap and pollen. Many of the genes (and their gene products) in transgenic crops occur naturally in their original hosts in the agricultural environment, but at low levels. Elevated levels of these gene products, and frequent exposure, may present a risk to agricultural workers. For example, a gene product expressed in pollen may change the allergenicity of the pollen or a gene product expressed throughout the plant may sensitize individuals who handle these plants frequently.

10. Determining the risk

As can be seen from the outline above, potential hazards can be identified on the basis of field and laboratory experience and searches of the literature. In some cases, the risk can even be assessed (at least in relative terms) as a paper exercise; for example, a crop will not cross-pollinate wild relatives if they are not in the region where the crop is being grown.

Where information is lacking, preliminary experiments can be done in the laboratory or glasshouse, such as crosses between crops and their wild relatives and testing for transcapsidation, for example. However, these sorts of experiment will always be limited in scope and, for the purposes of risk assessment, the level at which an experimental procedure would detect an effect should always be specified. Where greater experimental precision is required, field experiments could be carried out using non-transgenic varieties.

The case-by-case/step-by-step approach would then move on to small-scale field trials. Again, the same criteria for the detection thresholds of the experiment should be specified. Assuming no detrimental effects (within the sensitivity of the experimental design) are detected at each stage, the sizes of experiments can be increased to field- and farm-scale trials. It should be remembered, however, that the failure to detect an adverse effect by increasing the scale of trials only allows a statement that the level of risk of adverse effects is below a certain level, not that there is no risk. At some point in this process, the estimated risk may be at a level that is considered "acceptable". On the other hand, this must be balanced against the observation that the impact of crops is often dependent on the scale of cultivation. For example, cereal crops, such as barley and wheat, bred conventionally for resistance to economically important diseases may be outstanding performers in the breeders' plots and during registration trials, but, when grown on tens or hundreds of thousands of acres, rapidly select pathogen genotypes which are able to overcome the disease resistance of the new varieties – a process which may take only a few years. In the same way, extensive use of, say, transgenic insect or herbicide resistance may also select for extremely rare genotypes that are no longer affected by the control agent in the target organisms. Although this sort of genetic variation occurs naturally, its rarity makes it difficult to detect at any level of testing short of full commercial release.

REFERENCES

Berg, P., Baltimore, D., Boyer, H.W. *et al.* (1974) Potential biohazards of recombinant DNA molecules. *Science* 185: 303

Berg, P., Baltimore, D., Brenner, S., Roblin, R.O. III & Singer, M.F. (1975) Asilomar Conference on Recombinant DNA Molecules. *Science* 188: 991–94

Council of the European Communities (1990) Council Directive of 23 April 1990 on the deliberate release into the environment of genetically modified organisms (90/220/EC). *Official Journal of the European Communities* L117: 15–27

OECD (1986) *Recombinant DNA Safety Considerations: Safety Considerations for Industrial, Agricultural, and Environmental Applications of Organisms Derived By Recombinant DNA Techniques*, Paris: Organisation for Economic Co-operation and Development

OECD (1993) *Safety Considerations for Biotechnology: Scale-up of Crop Plants*, Paris: Organisation for Economic Co-operation and Development

Raybould, A.F. & Gray, A.J. (1993) Genetically modified crops and hybridisation with wild relatives: a UK perspective. *Journal of Applied Ecology* 30: 199–219

Raybould, A.F. & Gray, A.J. (1994) Will hybrids of genetically modified crops invade natural communities? *Trends in Ecology and Evolution* 9(3): 85–89

FURTHER READING

Fincham, J.R.S. & Ravetz, J.R. (1991) *Genetically Engineered Organisms: Benefits and Risks*, Milton Keynes: Open University Press, and Toronto: University of Toronto Press

Russo, E. & Cove, D. (1995) *Genetic Engineering: Dreams and Nightmares*, Oxford and New York: Freeman

See also **Determining the impact and consequences of genetically modified crops** (p.692)

K GENETICS OF CELL ORGANELLES, STRUCTURES AND FUNCTION

Introduction to Genetics of Cell Organelles, Structures and Function

The structure and function of subcellular structures, or organelles, and their regulation during the processes of cellular growth and differentiation are of fundamental importance in medical research: it is at the subcellular level that the origin and cause of most diseases can be explained. Clearly, DNA damage by defects in DNA replication or recombination can cause cell death if not repaired; furthermore, as described in the following articles, the centromere and/or kinetochore has been implicated in the safeguard mechanisms to prevent abnormal chromosome segregation, while aberrations in DNA methylation have been implicated in ageing and various diseases including cancer. Telomere erosion during cell division and leaked free radicals from mitochondria are also implicated in ageing.

Molecular genetic approaches to chromosome structure and the mechanisms of cell division, DNA replication and recombination have supplied geneticists with a "parts list" of individual cell components and in many cases the fine structure of these is known. The function of many of the genes encoded in the organelles remains to be discovered.

Although genetic inheritance is usually through both the maternal and paternal copies of genes from nuclear chromosomes, there are several exceptions. Mitochondria have their own genome, and though over an evolutionary time scale most mitochondrial genes have been integrated into the host's nuclear genome, several common genes remain in all mitochondria and an article below describes several disorders that have been attributed to the human mitochondrial genome. A further exception is when control of gene expression depends on whether the gene is derived from the male or female germline. As described in the articles below, DNA methylation is regarded as a strong candidate for the molecular basis of X-chromosome inactivation and of imprinting in general.

GENETIC RECOMBINATION

David R.F. Leach
Institute of Cell and Molecular Biology, University of Edinburgh, Edinburgh, UK

1. Introduction

Genetic recombination is the process whereby genetic material is reassorted or rearranged into different combinations. This can be accomplished either by the independent assortment of unlinked loci (on different chromosomes) or by physical exchange between chromosomes. Recombination via physical exchange can be subdivided into four logical classes: homologous, site-specific, transpositional and illegitimate recombination. However, several complex events utilize more than one of these classes of mechanism. In this article, each class is described in turn and then three examples of complex recombination reactions are given. References are given to a few historically important papers and to a selection of reviews and books. Other reviews and the primary literature can be accessed via these references.

2. Genetic recombination

(i) Homologous recombination

In homologous recombination, one DNA sequence recognizes another by comparison of the two arrays of nucleotides. Proteins are involved in the catalysis of recognition and exchange, but the specificity of the reaction is determined by the nucleotide sequence. In 1964, Robin Holliday proposed a model for homologous recombination which today still serves to illustrate the main features of the reaction (Holliday, 1964). This model is shown in **Figure 1**. Initially, two homologous substrate molecules exchange strands to form a Holliday junction adjacent to a region of heteroduplex DNA. This junction is able to migrate through the DNA (branch migration) to increase the extent of heteroduplex DNA and is then resolved by cleavage at points A or B to give either splice or patch recombinants. The formation of Holliday junctions and their migration and their resolution are now accepted features of homologous recombination; however, the reciprocal swapping of single strands, envisaged in the Holliday model, to initiate the reaction has not been observed. Instead, homologous recombination appears to be initiated by single-stranded DNA (ssDNA) on one of the chromosomes. This may have resulted from a double-strand break that is resected to reveal single-stranded ends or to a single-strand gap produced during DNA replication or repair. Since homologous recombination is initiated at regions of ssDNA that have been generated by degradation of the complementary strand, completion of recombination requires some DNA synthesis to replace the DNA that has been lost. This means that the mechanism of most, if not all, homologous recombination reactions involves the breaking and copying of DNA. This is called break–copy recombination in contrast to the pure break–join recombination of the Holliday model or the pure copying of the early copy-choice models that envisaged the formation of recombinants simply by starting to copy one chromosome and then swapping to copy another. (For a more detailed analysis of homologous recombination see Kucherlapati & Smith, 1988; Low, 1988; Petes *et al.*, 1991; Kowalczykowski *et al.*, 1994; Myers & Stahl, 1994; Camerini-Otero & Hseih, 1995; Lichten & Goldman, 1995; Shinohara & Ogawa, 1995; Leach, 1996; West, 1997.)

This reaction has been extensively investigated in the bacterium *Escherichia coli*. In 1965, John Clark and Ann Margulies isolated the first recombination-deficient mutants and called the gene they identified *recA* (Clark & Margulies, 1965). The RecA gene product has proved to be the central catalyst in the reaction. This protein forms a spiral filament around ssDNA which places the bases in a configuration

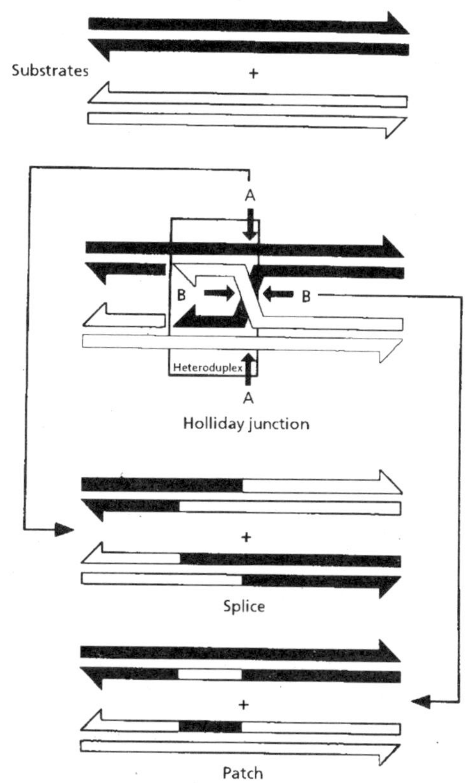

Figure 1 Homologous recombination is initiated by an exchange of DNA strands to form heteroduplex DNA adjacent to a Holliday junction. The Holliday junction is then resolved to recombinant products by cleavage and religation. Cleavage can either be at points A to generate splice products or at B to generate patch products.

where they can search for homologous double-stranded DNA (dsDNA). When homologous dsDNA is found, the strands are paired (synapsis) and then exchanged (strand-exchange). The consequence of synapsis and stand-exchange is the formation of a Holliday junction adjacent to heteroduplex DNA. Recombination can be initiated at sites of double-strand breaks or single-strand gaps. In the case of double-strand break repair, a nuclease is initially required to provide the ssDNA that RecA protein requires. In *E. coli*, this nuclease is the RecBCD enzyme which degrades DNA starting at a double-strand break until it reaches the sequence (5′-GCTGGTGG-3′) known as χ or *chi*. When a *chi* site is reached, the enzyme proceeds with reduced nucleolytic activity to provide a single strand that becomes coated with RecA protein. When RecBCD and RecA have done their work, the heteroduplex DNA adjacent to the Holliday junction can be increased by the action of the RuvABC branch migration complex. RuvA binds the junction and helps the assembly of two hexameric rings of RuvB circling two opposite arms of the Holliday junction. RuvC joins the complex and branch migration proceeds until the junction is cleaved by RuvC. Branch migration can proceed for several hundreds of nucleotides before cleavage occurs.

In eukaryotes, recombination can be observed between homologous chromosomes during meiosis. Meiosis is the process by which diploid organisms produce haploid germ cells. Initially, chromosomes are replicated, but sister chromatids remain attached at their centromeres. These replicated chromosomes then pair with their homologues to form bivalents composed of four dsDNA molecules. It is at this stage that they undergo recombination. Meiotic recombination is initiated at sites of double-strand breaks which, in *Saccharomyces cerevisiae*, have the Spo11 protein bound to the 5′ ends. In the presence of the Rad50/Mre11/Xrs2 complex, the Spo11 protein is removed and the strands to which it was bound are resected, leaving overhanging 3′ ends. The 3′ single strands then bind a complex of proteins that may include Rad51, Rad52, Rad55, Rad57 and Dmc1. This complex promotes synapsis and strand exchange with the uncleaved homologue. This results in the formation of two Holliday junctions and repair of the double-strand break – so-called double-strand break repair (Szostak *et al.*, 1983). As meiosis proceeds, the points at which crossing-over (the formation of splice recombinants) has occurred become visible as chiasmata. In the first meiotic division, the centromeres of homologous chromosomes are segregated into different cells. This is followed by a second division in which the sister chromatids are also segregated.

In certain fungi, the four products of meiosis are packaged as a tetrad of spores in an ascus. This is particularly valuable for study, since all the products of recombination can be identified and investigated. In some fungi, another round of cell division occurs after meiosis and so an octad of eight spores is deposited in an ascus. This allows the study of genetic markers that have remained heteroduplex in any of the four chromatids.

Homologous recombination can be used to map genes on chromosomes. Recombinant frequencies are measured by counting the number of recombinant products out of the total number of products. If these are spores, it may be sufficient to germinate them and observe the phenotypes of the haploid organisms. If the products are germ cells of an animal or plant, a cross will have to be carried out to reveal their genotypes. This is normally a test-cross to a homozygous recessive individual so that the genotypes of the germ cells are revealed in the phenotypes of the F_1 progeny. If markers are closely linked, recombination will rarely separate them, whereas totally unlinked markers will assort independently. Independent assortment of two markers will result in their recombination 50% of the time, so recombinant frequencies can vary between 0 and 50%.

If recombinant frequencies between three linked markers are studied, the frequency of double recombinants can be compared with the frequencies of single recombinants between the central marker and each of the outer markers. By simple probability, the frequency of double recombinants should equal the product of the two frequencies of single recombinants. However, this is rarely the case and recombination events appear to interfere with each other. If there are fewer double recombinants than predicted from single recombinant frequencies, there is interference (positive interference). On the other hand, if there are more double recombinants than predicted from single recombinant frequencies, there is an apparent stimulation of multiple events (negative interference). In eukaryotes, positive interference is generally observed except for very closely linked markers when interference becomes negative. The basis of positive interference is not well understood, but it is likely to involve the inhibition of crossing-over in the vicinity of an existing crossover. Negative interference is thought to come about via the processing of markers within heteroduplex DNA. Here, multiple recombinants can be generated via the processing of single recombination intermediates. The apparent stimulation of multiple events is really a reflection of the potential to generate multiple products from a single event. Interference can be measured by the ratio of the observed to the expected frequency of double recombinants. In a cross between three marked loci (a, b and c), the expected frequency of double recombinants is $R_{ab} \times R_{bc}$. Therefore, the ratio of observed to expected double recombinants (known as the coefficient of coincidence or S) is defined as:

$$S = R_{doubles} / R_{ab} \times R_{bc}$$

A measure of interference (I) is defined as:

$$I = 1 - S$$

For positive interference, $S < 1$ and I is positive. For negative interference, $S > 1$ and I is negative.

In a recombination event between two markers, one can ask whether both classes of recombinants are produced (i.e. does a+ × +b lead to the formation of both ab and ++ recombinants in a single event?). If, in one molecular event, both recombinants are made then recombination is said to be reciprocal; if not, it is non-reciprocal. Recombination between distant markers tends to be reciprocal whereas recombination between very closely linked markers tends to be non-reciprocal. The explanation is that, in recombination between distant markers, neither marker is involved in the recombination event itself. Recombination is generated by a crossover somewhere between the two markers. However, if the markers lie close to each other, there is a good chance that one or both markers is included in the region of heteroduplex DNA associated with the recombination event itself (i.e. the recombination event cannot be viewed as occurring at a point between the two markers). The final fate of the event will depend on events such as mismatch correction within the heteroduplex DNA which can easily lead to the formation of one recombinant without its reciprocal partner. Non-reciprocal recombination is often referred to as gene conversion.

Homologous recombination acts to "shuffle the pack" of existing alleles within a population. Therefore, it has an important role in accelerating evolution by natural selection. It allows the purification of desirable alleles from linked detrimental ones as well as bringing together beneficial alleles. The converse is also true; detrimental alleles can be brought together and beneficial alleles can be combined with detrimental ones. The important point is that recombination shuffles the alleles in a population and, therefore, provides natural selection with new combinations. The fittest individuals survive.

(ii) Site-specific recombination

In site-specific recombination, proteins recognize specific target sites and catalyse their juxtaposition and recombination. A protein, or a set of proteins, recognizes a particular DNA sequence and promotes its recombination with another copy of the same or related sequence. In this reaction, the site of recombination is a specific recognition sequence for the recombination machinery. Therefore, protein–DNA and protein–protein interactions form the basis of recognition. Site-specific recombination can occur only at the specific site for which a recombinase protein has evolved to function and the recombination reaction is normally restricted to an exchange within that site. An outline of site-specific recombination is shown in **Figure 2** (for further information see Cox, 1988; Hatfull & Grindley, 1988; Glasgow *et al.*, 1989; Landy, 1989; Stark *et al.*, 1989, 1992; Leach, 1996).

The prototype of this class of recombination reaction is the integration of bacteriophage lambda (λ) into the *E. coli* chromosome. This reaction is localized to a particular sequence on the bacteriophage DNA (*attP*) and a particular sequence on the host chromosome (*attB*). A bacteriophage-encoded protein integrase (Int) acts in conjunction with several accessory host proteins to mediate the integration of the circular bacteriophage genome into the host chromosome. Integration proceeds by two pairs of strand-transfer reactions – strand-transfers are the formation of new covalent connections between DNA strands. The first occurs at the left-hand side of the 7 base pair (bp) homologous overlap region of the *att* sites to form a four-way (Holliday) junction. This junction then rearranges and a second pair of strand-transfers to the right side of the overlap completes the reaction. The reaction proceeds via a covalent intermediate between the DNA and the protein. The 3′ phosphate of the

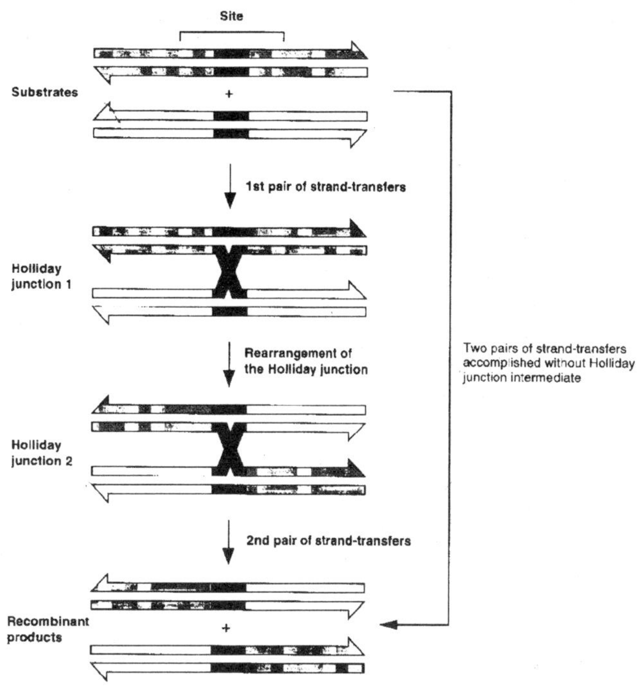

Figure 2 Site-specific recombination occurs only at specific sites. In the λ integration/excision reaction, a pair of strand-transfers at one side of the overlap region (shown in black) within the site generates a Holliday junction. The junction is rearranged and a second pair of strand-transfers leads to the formation of recombinant products. In the Tn*3* resolution reaction, both pairs of strand-transfers are accomplished without progressing through a Holliday junction intermediate.

cleaved DNA is joined to a tyrosine residue of integrase before strand-transfer is completed.

Site-specific recombination is used in biological systems for a variety of purposes. Bacteriophage λ integration is an example of its use to recombine two circular DNA molecules. The reverse of this reaction (excision) is also accomplished by the λ system. Biochemically, excision is not precisely the reverse of integration and requires an additional (phage-encoded) protein known as Xis. Reactions similar in outcome are the resolution of transposition-generated cointegrates by the transposon Tn*3*-encoded resolvase protein TnpR and the Xer system of *E. coli* proposed to resolve chromosomal dimers back to monomers. An alternative outcome of site-specific recombination is observed in natural systems where the recombining sites are inverted with respect to each other. In this situation, recombination results in inversion of the DNA between the two sites. In bacteriophage Mu, a region called G is located between two such inverted sites and the tail fibre gene is located partially within and partially outwith the G region. This means that the N-terminal part of the tail fibre is constant whereas there are two alternative C-terminal parts, the synthesis of which depends on the orientation of G. In this way, the bacteriophage makes two alternative classes of tail fibre allowing it to extend its host range. Another inversion system affects the flagella of *Salmonella typhimurium*. Here, a promoter is located within an invertable region and controls the expression of two alternative types of flagellum. The ability to switch between flagellar types helps the bacterium to escape its host's immune system.

(iii) Transpositional recombination

In transpositional recombination (transposition), two contiguous DNA sites are recognized by one or more proteins and the DNA between the sites is moved to a new location. This reaction involves the specific recognition of two DNA sequences by a protein (the transposase) and the interaction of this complex with a relatively non-specific target sequence. The specificity of the reaction is mediated by protein–DNA and protein–protein interactions. An outline of transposition is shown in **Figure 3**. (Further information can be obtained in Berg & Howe, 1989; Mizuuchi, 1992; Leach, 1996; Finnegan, 1997; Hallet & Sherratt, 1997; Labrador & Corces, 1997.)

Transposons can be subdivided into two types: the first transpose as DNA and the second transpose via an RNA intermediate. All bacterial transposons are of the first type and some eukaryotic transposons also transpose as DNA. These (DNA-only) elements have a characteristic organization. The ends of the transposons have at least two bases in inverted orientation plus longer subterminal regions (often in inverted orientation too) that bind transposase. Within an active copy of the transposon, there is at least one open reading frame that encodes a transposase which is the primary catalyst of transposition. This transposase mediates cleavage of the 3′ end of the element to expose a 3′-OH. This 3′-OH then attacks a phosphodiester bond at the target site in a reaction again catalysed by transposase. The reaction is a direct transesterification that leads to strand-transfer. There is no covalent intermediate between cleaved DNA and protein as there is in the site-specific recombination reactions described above. The cleavage at the 5′ end of the transposon is more variable in its location. For certain elements, there is no 5′ cleavage and this means that strand-transfer at both 3′ ends generates a structure with two replication forks (the Shapiro intermediate). DNA synthesis through the element generates two copies of the element. In the Tn*3* family of bacterial elements, directly repeated copies then recombine using an element-encoded site-specific recombination system. Other elements do cleave the 5′ ends at a variety of positions (relative to the end of the element) and DNA synthesis is used to restore an intact double-strand after strand-transfer at the 3′ end. In all cases, cleavage at the target site is staggered on the two strands. This means that, when strand-transfer and repair synthesis have been completed, the target site is duplicated. Target-site duplications vary in length between 2 and 12 bp depending on the stagger with which the target site is cleaved.

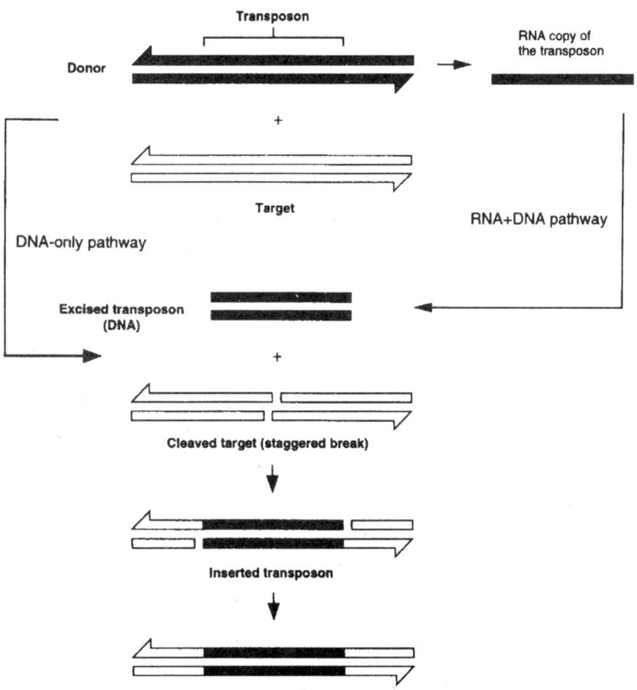

Figure 3 Transposition is accomplished by attack of a target sequence by the free ends of the transposon. The target is cleaved with a staggered break which results in the formation of a target-site duplication. In the DNA-only pathway, the transposon is either cleaved out (double-strand break, as shown above) or nicked at the 3′ end (single-strand break, not shown). In the RNA + DNA pathway, an RNA copy of the transposon is made which is then reverse transcribed into DNA. This DNA copy is then inserted at the cleaved target site.

The second type of transposable element is comprised of those that transpose via an RNA intermediate. This is a diverse group of elements. Some are clearly related to retroviruses, but have lost the ability to be encapsidated. These retrotransposons are bounded by long terminal repeats (LTRs) which are used during retrotransposition, as are the LTRs of retroviruses. Well-studied examples of these elements are *copia* in Drosophila and Ty1 in *S. cerevisiae*. Other elements do not have LTRs, but instead have a simple repeated sequence at one end (e.g. $(TAA)_n$); examples of these are the *I* factor of Drosophila and LINE elements of mammals. These elements encode a reverse transcriptase that allows an RNA copy of the element to be copied back into DNA during transposition. A third class of element, also believed to transpose via an RNA intermediate, has a highly repetitious structure consisting of two long terminal inverted repeats, each of which is composed of short direct repeats. Due to their overall structure, they have been called fold-back elements.

(iv) Illegitimate recombination

Illegitimate recombination is recombination between DNA sequences that share little or no homology. Nevertheless, microhomology in the range of 3–25 bp can play a role to direct the reassociation of DNA ends. Illegitimate recombination can be subdivided into end-joining and strand-slippage reactions. End-joining (**Figure 4**) is a relatively efficient reaction in mammalian cells where it accounts for the majority of double-strand break repairs and of integration events detected

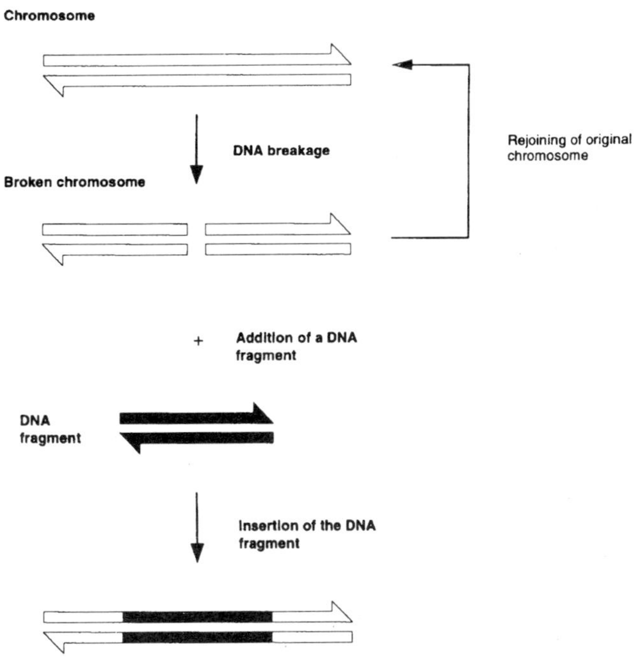

Figure 4 Illegitimate recombination (by end-joining) can occur if a DNA double-strand break occurs in a chromosome. The broken ends may rejoin to restore the parental DNA. Alternatively, in the presence of a DNA fragment, this may be joined via its ends to the broken ends of the chromosome. This reaction does not require homology between the two DNA molecules.

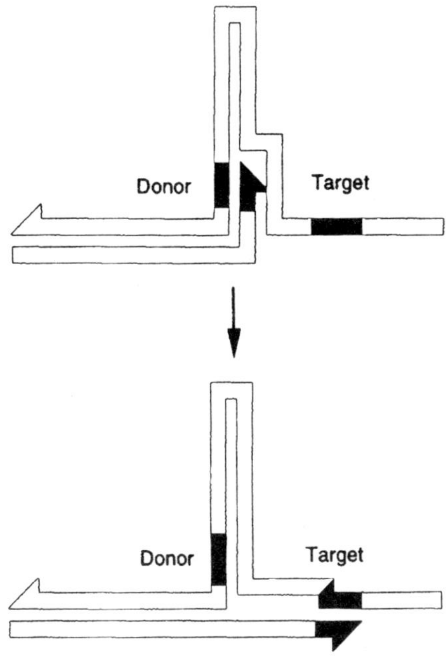

Figure 5 Illegitimate recombination (by strand-slippage) can occur during DNA replication if the newly synthesized strand dissociates from its template and reassociates in a new position. This reaction is stimulated by the formation of an intrastrand secondary structure (e.g. hairpin DNA). The short, directly repeated sequences used (black) are labelled "donor" and "target" since there is a polarity in this reaction consistent with the repeat within the folded structure acting as a donor and the repeat outside acting as a target.

after the introduction of DNA to cells. The reaction does not require any homology between the interacting ends, but it does normally use microhomology if this exists. DNA ends are juxtaposed, paired at microhomology if present, gaps filled in and nicks ligated. Random DNA sequences can be inserted between the recombining ends and it is thought that these originate from oligonucleotide sequences that happen to be in the vicinity of the ends at the time of recombination. Experiments in the yeast *S. cerevisiae* have revealed the requirement of a DNA repair complex for end-joining. This complex consists of the DNA end-binding protein Ku and the Rad50/Mre11/Xrs2 putative nuclease. In addition, there is a requirement for silencing factors Sir2, Sir3 and Sir4 that are known to generate heterochromatin-like structures (for further information see Allgood & Silhavy, 1988; Roth & Wilson, 1988; Ehrlich, 1989; Meuth, 1989; Ehrlich *et al.*, 1993; Leach, 1996; Tsukamoto & Ikeda, 1998).

Strand-slippage (**Figure 5**) occurs when a newly synthesized DNA strand dissociates from its template during replication and reassociates at a new location. When replication resumes, a rearrangement (usually deletion) results. Since the reassociation requires some base pairing, short direct repeats are generally observed at the end-points of strand-slippage events. In some events, no such repeats can be detected. Certain DNA sequences are known to favour strand-slippage events. In particular, long DNA inverted repeats (palindromes) stimulate strand-slippage by several orders of magnitude. This is because the inverted repeats can pair on the same strand of DNA to form a hairpin. Hairpin DNA is difficult to replicate and it brings into close proximity the newly replicated strand and its target for reassociation. Most strand-slippage events occur on the lagging-strand of the DNA replication fork and this may be because of the higher probability of ssDNA being found there.

Other simple repeated sequences are believed to be unstable due to strand-slippage. These include dinucleotide and trinucleotide repeats found in mammalian microsatellite sequences. These microsatellites are destabilized in mutants defective in mismatch repair, suggesting that intermediates in this strand-slippage reaction are subject to mismatch repair.

3. Complex reactions

(i) V(D)J recombination

V(D)J recombination generates diversity of immunoglobulins and T cell receptors. The variable regions of the immunoglobulin and T cell receptor loci are rearranged in a manner that combines site specificity with randomness. Initially, a D segment of the heavy chain locus is joined to a J segment and then a V segment becomes joined to DJ. A similar rearrangement then occurs at the light chain locus between V and J segments. The reaction is site specific in that a break is made adjacent to a DNA sequence known as a signal sequence. Signal sequences are composed of three parts: a 7 bp palindrome adjacent to the coding region, a spacer region of non-conserved nucleotides and a 9 bp A/T-rich region. There are two types of signal sequence distinguished by the length of their spacers of 12 or 23 bp. For a productive reaction to take place, there must be one signal sequence of each type present. The formation of the coding joint involves the formation of DNA hairpin structures on the coding flanks at the junctions with the heptamers of the signal sequences. The hairpin formation reaction is chemically the same as the direct transesterification reaction described in transposition. The hairpins are then cleaved to generate short inverted repeats and the opened hairpin ends are processed by exonucleolytic degradation and the addition of residues by terminal transferase. Finally, the ends are joined in a reaction that sometimes, but not always, uses DNA homology between the single-stranded ends. This is reminiscent of an end-joining reaction (for details see Engler & Storb, 1988; Lewis & Gellert, 1989; Lieber, 1992; Schatz *et al.*, 1992; Leach, 1996; Gellert, 1997).

(ii) Mating-type interconversion

Mating-type interconversion in *S. cerevisiae* combines homologous recombination with site-specificity and results in the movement of genetic information from one location to another (reminiscent of, but not mechanistically related to, transposition – see Strathern, 1988; Klar, 1989; Haber, 1992). Haploid yeast cells switch in a controlled way between the two mating types "a" and "α" which can then mate to produce a/α diploids. This switching involves four loci: MAT, the mating-type locus that can contain either a or α information and which defines the cell's mating type; HMR, that normally contains a silent (non-expressed) copy of "a"; HML, that normally contains a silent copy of a; and HO, an endonuclease that controls switching. Switching occurs by homologous recombination of the silent copies, HMR and HML, with MAT to alternate the information contained at the mating-type locus. The HO endonuclease recognizes the DNA at the MAT locus and cleaves it to generate a double-strand break. This break is then repaired in a controlled manner by the genetic information at the HM locus of the alternative mating type.

(iii) Tn10 transposition

Tn*10* transposition utilizes homologous recombination to ensure that the transposon is not lost from its original location, although transposition itself involves a cut-and-paste mechanism (cleavage at both 3′ and 5′ ends of the element). The transposition of Tn*10* is regulated by Dam methylation of GATC sequences in the transposable element. Expression of the transposase and its activity are optimal when the element is hemimethylated, a situation that only occurs directly after the element is replicated by its host. This means that when one copy of the transposon is excised to leave a double-strand break at the site of its original insertion, there is a second undisturbed copy of the element with which the broken chromosome can recombine (by homologous recombination) to restore the original chromosomal situation. In this way, the excised copy of the transposon can be inserted elsewhere while not being lost from the original location (for further details see Kleckner, 1990).

REFERENCES

Allgood, N.D. & Silhavy, T.J. (1988) Illegitimate recombination in bacteria. In *Genetic Recombination*, edited by R. Kulcherlapati & G.R. Smith

Berg, D.E. & Howe, M.M. (eds) (1989) *Mobile DNA*, Washington, DC: American Society for Microbiology

Camerini-Otero, D. & Hseih, P. (1995) Homologous recombination proteins in prokaryotes and eukaryotes. *Annual Review of Genetics* 29: 509–52

Clark, A.J. & Margulies, A.D. (1965) Isolation and characterisation of recombination-deficient mutants of *Escherichia coli* K12. *Proceedings of the National Academy of Sciences USA* 53: 451–59

Cox, M.M. (1988) FLP site-specific recombination system of *Saccharomyces cerevisiae*. In *Genetic Recombination*, edited by R. Kulcherlapati & G.R. Smith

Ehrlich, S.D. (1989) Illegitimate recombination in bacteria. In *Mobile DNA*, edited by D.E. Berg & M.M. Howe

Ehrlich, S.D., Bierne, H., d'Alençon, E. *et al.* (1993) Mechanisms of illegitimate recombination. *Gene* 135: 161–66

Engler, P. & Storb, U. (1988) Immunoglobulin gene rearrangement. In *Genetic Recombination*, edited by R. Kulcherlapati & G.R. Smith

Finnegan, D.J. (1997) Transposable elements: how non-LTR retrotransposons do it. *Current Biology* 7: R245–48

Gellert, M. (1997) Recent advances in understanding V(D)J recombination. *Advances in Immunology* 64: 39–64

Glasgow, A.C., Hughes, K.T. & Simon, M.I. (1989) Bacterial DNA inversion systems. In *Mobile DNA*, edited by D.E. Berg & M.M. Howe

Haber, J.E. (1992) Mating-type gene switching in *Saccharomyces cerevisiae*. *Trends in Genetics* 8: 446–52

Hallet, B. & Sherratt, D.J. (1997) Transposition and site-specific recombination; adapting DNA cut-and-paste mechanisms to a variety of genetic rearrangements. *FEMS Microbiology Reviews* 21: 157–78

Hatfull, G.F. & Grindley, N.D.F. (1988) Resolvases and DNA-invertases: a family of enzymes active in site-specific recombination. In *Genetic Recombination*, edited by R. Kulcherlapati & G.R. Smith

Holliday, R. (1964) A mechanism for gene conversion in fungi. *Genetical Research* 5: 282–304

Klar, A.J.S. (1989) The interconversion of yeast mating type: *Saccharomyces cerevisiae* and *Schizosaccharomyces pombe*. In *Mobile DNA*, edited by D.E. Berg & M.M. Howe

Kleckner, N. (1990) Regulating Tn*10* transposition. *Genetics* 124: 449–54

Kowalczykowski, S.C., Dixon, D.A., Eggleston, A.K., Lauder, S.D. & Rehrauer, W.M. (1994) Biochemistry of homologous recombination. *Microbiology Reviews* 58: 401–65

Kucherlapati, R. & Smith, G.R. (eds) (1988) *Genetic Recombination*, Washington, DC: American Society for Microbiology

Labrador, M. & Corces, V.G. (1997) Transposable element interactions: regulation of insertion and excision. *Annual Review of Genetics* 31: 381–404

Landy, A. (1989) Dynamic, structural, and regulatory aspects of λ site-specific recombination. *Annual Review of Biochemistry* 58: 913–49

Leach, D.R.F. (1996) *Genetic Recombination*, Oxford and Cambridge, Massachusetts: Blackwell Science

Lewis, S. & Gellert, M. (1989) The mechanism of antigen receptor gene assembly. *Cell* 59: 585–88

Lichten, M. & Goldman, A.S.H. (1995) Meiotic recombination hotspots. *Annual Review of Genetics* 29: 423–44

Lieber, M.R. (1992) The mechanism of V(D)J recombination: a balance of diversity, specificity, and stability. *Cell* 70: 873–76

Low, K.B. (ed.) (1988) *The Recombination of Genetic Material*, San Diego: Academic Press (a collection of review articles on homologous recombination)

Meuth, M. (1989) Illegitimate recombination in mammalian cells. In *Mobile DNA*, edited by D.E. Berg & M.M. Howe

Mizuuchi, K. (1992) Transpositional recombination: mechanistic insights from studies of Mu and other elements. *Annual Review of Biochemistry* 61: 1011–51

Myers, R.S. & Stahl, F.W. (1994) Chi and the RecBCD enzyme of *Escherichia coli*. *Annual Review of Genetics* 28: 49–70

Petes, T.D., Malone, R.E. & Symington, L.S. (1991) Recombination in yeast. In *The Molecular and Cellular Biology of Yeast Saccharomyces*, edited by J.R. Broach, J.R. Pringle & E.W. Jones, Cold Spring Harbor, New York: Cold Spring Harbor Laboratory Press

Roth, D. & Wilson, J. (1988) Illegitimate recombination in mammalian cells. In *Genetic Recombination*, edited by R. Kulcherlapati & G.R. Smith

Schatz, D.G., Oettinger, M.A. & Schlissel, M.S. (1992) V(D)J recombination; molecular biology and regulation. *Annual Review of Immunology* 10: 359–83

Shinohara, A. & Ogawa, T. (1995) Homologous recombination and the roles of double-strand breaks. *Trends in Biochemical Sciences* 20: 387–91

Stark, W.M., Boocock, M.R. & Sherratt, D.J. (1989) Site-specific recombination by Tn*3* resolvase. *Trends in Genetics* 5: 304–09

Stark, W.M., Boocock, M.R. & Sherratt, D.J. (1992) Catalysis by site-specific recombinases. *Trends in Genetics* 8: 432–39

Strathern, J.N. (1988) Control and execution of homothallic switching in *Saccharomyces cerevisiae*. In *Genetic Recombination*, edited by R. Kulcherlapati & G.R. Smith

Szostak, J.W., Orr-Weaver, T.L., Rothstein, R.J. & Stahl, F.W. (1983) The double-strand break repair model for conversion and crossing-over. *Cell* 33: 25–35

Tsukamoto, Y. & Ikeda, H. (1998) Double-strand break repair mediated by DNA end-joining. *Genes to Cells* 3: 135–44

West, S.C. (1997) Processing of recombination intermediates by the RuvABC proteins. *Annual Review of Genetics* 31: 213–44

GLOSSARY

ascus elongated spore case containing four or eight haploid sexual ascospores of ascomycete fungi

bivalents structures composed of four double-stranded DNA molecules generated in meiosis from two homologous chromosomes that have replicated and paired

branch migration ability of Holliday junctions to migrate through the DNA

break–copy recombination mechanism of recombination involving breaking and copying of DNA strands

break–join recombination mechanism of recombination involving only breaking and re-joining of DNA strands

coding joint (in V(D)J rearrangement) junction that lies within the coding sequence that is generated during rearrangement of immunoglobulin and T-cell receptor genes.

copy-choice recombination mechanism of recombination involving only copying of DNA strands

dinucleotide repeats stretches of repeated dinucleotides, often AC/GT, found throughout the genome. Useful as markers for gene mapping (*see also* trinucleotide repeats in Main Glossary)

double-strand break repair (DSBR) the repair of DNA which has been broken in both strands. This can either be accomplished by homologous recombination or by end-joining

end-joining reaction in mammalian somatic cells this reaction accounts for the majority of double-strand break repairs. DNA ends are juxtaposed, paired at microhomology if present, gaps filled in and nicks ligated. Random DNA sequences can be inserted between the recombining ends

fold back (FB) element transposable element with a highly repetitious structure consisting of two long terminal inverted repeats, each composed of short direct repeats

gene conversion a phenomenon in meiosis where alleles are segregated in a 3:1 not 2:2 ratio

hairpin DNA motif formed by base-pairing of two adjacent regions of a single polynucleotide chain

heteroduplex DNA DNA in which the two strands are different; either of different heritable origin or formed *in vitro* by annealing complementary strands containing some mismatches

Holliday junction four-way junction formed as an intermediate in recombination reactions. Proposed in 1964, formation, migration and resolution of Holliday junctions are now accepted features of homologous recombination

inverted repeat a palindromic nucleotide sequence

long terminal repeats (LTRs) identical DNA sequences, several hundred nucleotides long, found at either end of retrotransposons and proviral DNAs, formed by reverse transcription of retrotransposons or retroviral RNAs

mating-type interconversion combines homologous recombination with site-specificity, resulting in movement of genetic information conferring mating-type from one location to another

microsatellites arrays of variable length distributed widely throughout the genome composed of 2–5 base-pair sequences repeated many times over

mismatch correction DNA repair system that detects and replaces mismatched bases in newly replicated DNA

mismatch repair (*see* mismatch correction)

negative interference an apparent clustering of crossing-over events. The frequency of double, and multiple, recombinants is greater than that predicted from single recombinant frequencies

non-reciprocal a recombination event between two markers where only one recombinant is produced (recombination between very closely linked markers tends to be non-reciprocal)

octad eight haploid spores derived from eight homologous chromatids paired together during first meiotic prophase. More generally, any group of eight objects

overlap (in site-specific recombination) region of DNA sequence identity between cleavage sites present on the two partners undergoing site-specific recombination

patch recombinant a recombinant that retains a parental configuration of DNA flanking a sequence acquired by genetic exchange

positive interference an apparent reduction in multiple crossing-over events. The frequency of double, and multiple, recombinants is less than that predicted from single recombinant frequencies

recombinant frequency proportion of the number of recombinant products out of the total number of products

Shapiro intermediate intermediate generated during the transposition of a transposon where there is no 5′ cleavage, resulting in strand-transfer at both 3′ ends to generate a structure with two replication forks

signal sequence (in V(D)J rearrangement) sequence recognized by the Rag proteins that directs the position of immunoglobulin and T-cell receptor gene rearrangement

sister chromatid (in meiosis) one of the two chromatids making up a bivalent, both being semiconservative copies of the original chromatid

site-specific recombination recombination occurring between two specific short DNA sequences present in the same or in different molecules

splice recombinant a recombinant that shows crossing-over

strand-exchange process of exchange of DNA strands following synapsis

strand-slippage during replication, dissociation of a newly synthesized DNA strand from its template and reassociation at a new location

strand-transfer reactions formation of new covalent connections between DNA strands

synapsis specific pairing of homologous chromosomes allowing crossing-over to take place

target-site duplication duplication that occurs when the target of transposition is cleaved with a staggered break

tetrad four haploid spores derived from four homologous chromatids paired together during first meiotic prophase. More generally, any group of four objects

Tn*10* transposition via regulation by Dam methylation of GATC sequences in the transposable element, a type of transposition utilizing homologous recombination, ensuring that the transposon is not lost from its original location

transesterification (in DNA) chemical reaction where one phospho-ester bond is broken and another is formed

V(D)J recombination site-specific recombination that produces diversity of immunoglobulins and T-cell receptors, the variable regions of which are rearranged in a form combining site specificity with randomness

See also **Barbara McClintock and transposable genetic sequences in maize** (p.640)

TELOMERES

David Kipling

Department of Pathology, University of Wales College of Medicine, Cardiff, UK

Telomeres are the ends of linear chromosomes. Functionally, a telomere is whatever biological structure makes the natural end of a linear chromosome behave differently from a simple double-strand DNA break. That the natural ends of chromosomes behave differently was first reported by Muller when observing the effects of X irradiation on Drosophila cells (see Muller, 1962). Chromosome rearrangements such as terminal deletions which removed the natural end of the chromosome were almost never observed, whereas internal inversions and deletions which left the ends of the chromosomes in place were readily obtained. This suggested that the natural end of the chromosome has a special structure which is, in some way, required for chromosome stability. Muller termed this structure the telomere, from the Greek *telos*, meaning "end". A modern interpretation of this functional definition is that the ends of eukaryotic nuclear chromosomes have a special DNA sequence to which specific proteins are bound, and this DNA–protein complex is responsible for the many and various aspects of telomere function.

One function was determined from observations of the fate of the ends of broken chromosomes in maize. These were prone to fusion with other chromosomes, undergoing so-called breakage–fusion–bridge cycles. This behaviour was not seen for the ends of natural maize chromosomes, indicating that one telomere function is to stop the natural tendency of DNA ends to fuse in this fashion. Another role is to "hide" the end of the chromosome from cellular systems which monitor genome damage. An example of this is provided by the yeast *Saccharomyces cerevisiae*. Here, a single experimentally induced break in the genome caused by expression of the HO endonuclease results in cell cycle arrest, thus providing time for the genome to be repaired. The pathway which detects this single double-strand break in the yeast genome does so while, at the same time, ignoring 32 naturally occurring double-strand DNA (dsDNA) ends in the form of the natural chromosome termini. Thus, a telomere must somehow "mark" a natural chromosome end as being different from a double-strand break.

Another putative role for telomeres has been implicated during meiosis. Pairing of chromosomes during meiosis often initiates at or close to their termini. In many species, telomeres are attached to the meiotic nuclear envelope via electron-dense "attachment plaques". These associations may facilitate chromosome pairing by providing an initial, simple chromosome alignment – all ends together.

What is the molecular structure of a telomere? Although there are some exceptions, such as in Drosophila, the majority of eukaryotic nuclear chromosomes terminate in tandem repeat arrays of a simple short sequence. The exact sequence varies between species, but the majority are of the form T_nG_m or T_nAG_m. Examples include TTAGGG (vertebrates, slime moulds, Trypanosoma), TTGGGG (various ciliates), TTTAGGG (Arabidopsis) and TTTTAGGG (Chlamydomonas). The repeat sequence can sometimes be rather variable, such as the $(TG)_{1-6}TG_{2-3}$ repeats of *S. cerevisiae*, or much larger, as with the 23-mer repeats of Candida. Almost all repeat sequences have a strand bias whereby the G-rich strand runs 3′ to 5′ towards the end of the chromosome, and, in some cases, these G-rich repeats can form unusual secondary structures *in vitro*. This terminal repeat sequence is believed to be synthesized *de novo* by telomerase.

In addition to the terminal repeat sequences, many species have more complex repetitive sequence families located in the subtelomeric region. These sequences are often very species specific and it is not clear if they have any biological role. Indeed, analysis of telomeres that have formed *de novo*, such as by the healing of terminal deletions or following the introduction of cloned DNA, strongly suggests that all that is required for essential telomere function are the terminal repeat sequences.

Telomere-binding proteins have been identified in many species. Some of these proteins bind the end of the telomere, whereas others bind internal telomere repeat sequences. The proteins are diverse in their biochemical behaviour and often show little sequence homology between species. In the majority of cases, their role in telomere function is poorly understood. Probably the best understood role of telomeric DNA and telomere-associated proteins is in replicating the terminus of the chromosome. The ends of linear DNA molecules in both prokaryotes and eukaryotes face a particular problem. All known DNA polymerases require a 3′-OH group onto which to add the next base; they cannot initiate synthesis *de novo*. Lagging strand DNA synthesis at the DNA replication fork commences with the synthesis of a short RNA molecule by RNA primase. This then provides the necessary 3′-OH group for extension by DNA polymerase. However, this replication scheme causes a problem with the very last few nucleotides of a linear molecule (**Figure 1**). A short region (a "primer gap") is unavoidably left following removal of the RNA primer, and this region cannot be replicated by conventional DNA polymerases. This primer gap is at least as large as the size of an RNA primer and, although only a small amount of sequence is lost, this will occur every cell division.

Nature has devised a number of different strategies to overcome this "end-replication problem". One of these, exemplified by vaccinia virus, involves terminal hairpin structures which, in concert with virus-encoded endonucleases, allow complete replication of the termini. Another solution is to use protein-primed replication, as illustrated

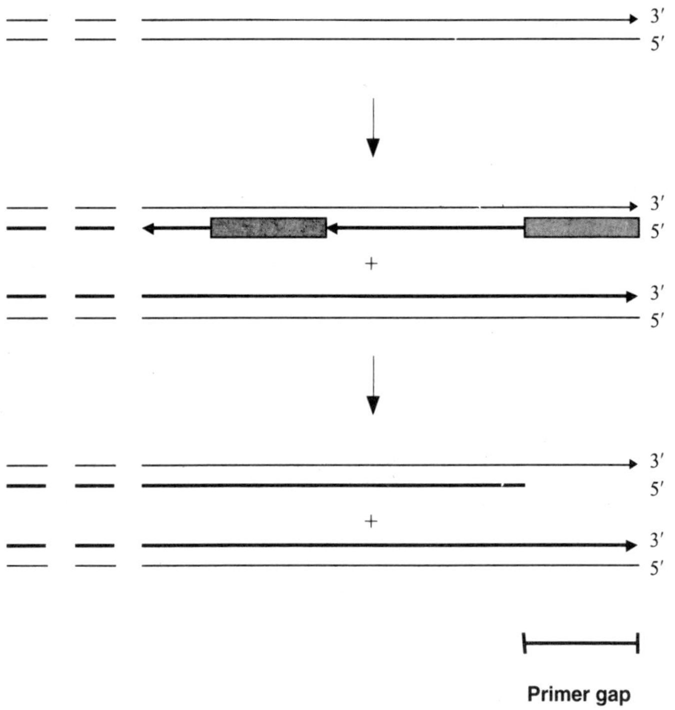

Figure 1 The end-replication problem. The very end of the chromosome is replicated by a fork moving from the left. Leading strand synthesis results in complete duplication of the upper strand. Lagging strand synthesis of the lower strand is initiated by RNA primers (hatched boxes). Their removal is followed by Okazaki fragment extension to fill the gaps. However, a region at the very end of the molecule cannot be synthesized, leaving a primer gap.

by adenovirus. Here, a virus-encoded terminal protein primes DNA synthesis, with the 3′-OH not being provided by an RNA or DNA molecule, but instead by the –OH group of a serine, threonine or tyrosine residue of this terminal protein. In other cases, terminal sequence loss is allowed to occur and is then compensated for by *de novo* addition of sequence to the end of the chromosome. For example, in Drosophila, retroposon-like elements called HeT-A and TART duplicatively transpose to the chromosome ends.

For the majority of eukaryotic nuclear telomeres, however, end-replication losses appear to be compensated for by *de novo* addition of telomere sequences by telomerase. Telomerase is a specialized DNA polymerase which can catalyse the *de novo* synthesis and addition of terminal repeat sequences onto the ends of chromosomes. The template for this synthesis is a telomerase-associated RNA molecule, which contains within it a region specifying the terminal repeat sequence. Telomerase was initially identified in ciliates, which are a rich biochemical source of telomerase because of the very large number of telomeres in their nuclei; ciliates have a specialized lifestyle involving genome fragmentation and *de novo* telomere addition and some species have over 10^7 telomeres per nucleus. The biochemistry of telomerase has been deduced primarily from *in vitro* studies of the ciliate holoenzyme. To date, telomerase activity has been identified in a wide range of species including ciliates, plants, vertebrates and yeast. The RNA templating component of telomerase (TERC) has been cloned and sequenced in many species. Telomerase activity can be reconstituted *in vitro* using TERC RNA together with an essential protein component (TERT). TERT genes have been cloned from ciliates, yeast, *Caenor habditis elegans*, mice and humans and contain reverse transcriptase-like motifs in their amino acid sequence.

One aspect of telomere biology that has been the subject of much recent study relates to the expression pattern of telomerase in normal and malignant human tissues. Telomerase is expressed at low or undetectable levels in most human cells, although there are exceptions (such as stem cells), whereas it is expressed at high levels in the germline and the majority of tumours. The absence of telomerase activity almost always correlates with the absence of TERT mRNA. This has raised intense interest in the potential utility of telomerase as a differential marker to improve the detection of malignant cells either as part of diagnosis or for screening purposes. Detection would focus either on enzyme activity, or expression of the TERT gene by either RNA *in situ* hybridization or immunocytochemical detection of the TERT protein. Telomerase inhibitors are also an exciting new class of potential anti-tumour agents, for which proof of principle has already been obtained by the genetic ablation of telomerase activity in cancer cells *in vitro*.

One of the most exciting discoveries in the telomere field in recent years has been the relationship between telomeres and cellular lifespan. Primary human cells have a finite division capacity *in vitro*, at the end of which they enter a viable though non-dividing state termed cellular (or replicative) senescence – the so-called Hayflick limit. This in-built limit to cellular lifespan may provide a potent barrier to the formation of that disease of uncontrolled and unlimited cell division that we think of as cancer. Until recently, however, the nature of the cell division "counter" was unknown. What was clear, however, was that telomere DNA length decreases with successive rounds of division for primary, telomerase-negative human cells when grown *in vitro*. Experimentally forcing expression of the human TERT gene in such cells is sufficient to restore telomerase activity and prevent further telomere erosion. Furthermore, such "telomerised" cells no longer undergo replicative senescence but continue to divide indefinitely. Data such as these have been obtained for many human cell types, and indicates that for some human cells replicative senescence is telomere-driven. The removal of cellular mortality following the expression of telomerase provides an elegant rationale to explain the observed expression of telomerase in human cancer.

However, cell division and telomere erosion also occurs during human life and has been suggested to contribute to human ageing. Although telomerase may at first seem a powerful route to intervening in the ageing process, such enthusiasm must be tempered by a recognition of the role of telomerase repression as a potential anti-cancer mechanism. More prosaically, perhaps, the ability to make effectively unlimited quantities of otherwise normal human cells using telomerase may have wide-ranging significance in the fields of drug discovery and tissue engineering.

Ironically, recent data indicate that Drosophila is the one eukaryote where true terminal deletions, uncapped by addition of new telomere sequences, can in fact be obtained (Kipling, 1995). The species where the telomere was first defined is the one where the original, classical definition of a telomere may not apply.

REFERENCES

Hahn, W.C., Counter, C.M., Lundberg, A.S. *et al.* (1999) Creation of human tumour cells with defined genetic elements. *Nature* 400: 464–68

Hahn, W.C., Stewart, S.A., Brooks, M.W. *et al.* (1999) Inhibition of telomerase limits the growth of human cancer cells. *Nature Medicine* 5: 1164–70

Greider, C.W. (1999) Telomerase activation – one step on the road to cancer? *Trends in Genetics* 15: 109–12

Kipling, D. (1995) *The Telomere*, Oxford and New York: Oxford University Press

Muller, H.J. (1962) *Studies in Genetics: The Selected Papers of H.J. Muller*, Bloomington, Indiana: Indiana University Press

Yeager, T.R. & Reddel, R.R. (1999) Constructing immortalized human cell lines. *Current Opinions in Biotechnology* 10: 465–69

GLOSSARY

endonuclease enzyme that splits the nucleic acid chain at internal sites (*see also* exonuclease in Main glossary)

holoenzyme complete enzyme complex comprising the protein portion (apoenzyme) and cofactor or coenzyme

telomerase (telomere terminal transferase) enzyme involved in forming the telomeres

INTRONS

Ian Dix and Jean D. Beggs
The Wellcome Trust Centre for Cell Biology, University of Edinburgh, Edinburgh, UK

1. Introduction

As the known genes of prokaryotes had uninterrupted coding sequences, it came as a surprise when, in the late 1970s, it was discovered that the coding sequences of many eukaryotic genes are interrupted by intervening DNA. Over the last 20 years, the mechanisms for the removal of these intervening sequences (introns) have been intensively researched and the possible reasons for their existence have been the subject of much debate. It was found that introns are normally removed from newly transcribed RNA and the regions present in the mature RNA (exons) are concomitantly joined in a process termed splicing. Although examples of introns have now been identified in a variety of RNA classes in all three divisions of life (eubacteria, archaebacteria and eukarya), introns represent a high percentage of the genome only in eukaryotes and, particularly, in the protein-coding genes of the higher eukaryotes.

There are a variety of intron types: group I, group II, group III, nuclear pre-mRNA, archaeal, eukaryotic nuclear tRNA, *HAC1* pre-mRNA and inteins. These can be considered to fall into four main categories according to the catalytic mechanism involved in their removal (**Figure 1**) and can be further grouped according to the *trans*- and *cis*-acting factors involved. Each of these types will be discussed below.

2. Mechanisms of splicing

(i) Group I introns

This class of intron has been found in chloroplasts and in the mitochondria of plants, fungi and other lower eukaryotes as well as in the nuclear rRNA genes of lower eukaryotes such as slime moulds. Cech (1990) provides a good review. They have also been found in the genomes of some eubacteria and bacteriophages. Some of these introns have the ability to self splice *in vitro* due to the autocatalytic activity of the RNA; however, *in vivo*, protein factors are involved which optimize and stabilize the intron structure, increasing the efficiency of intron removal. These introns are

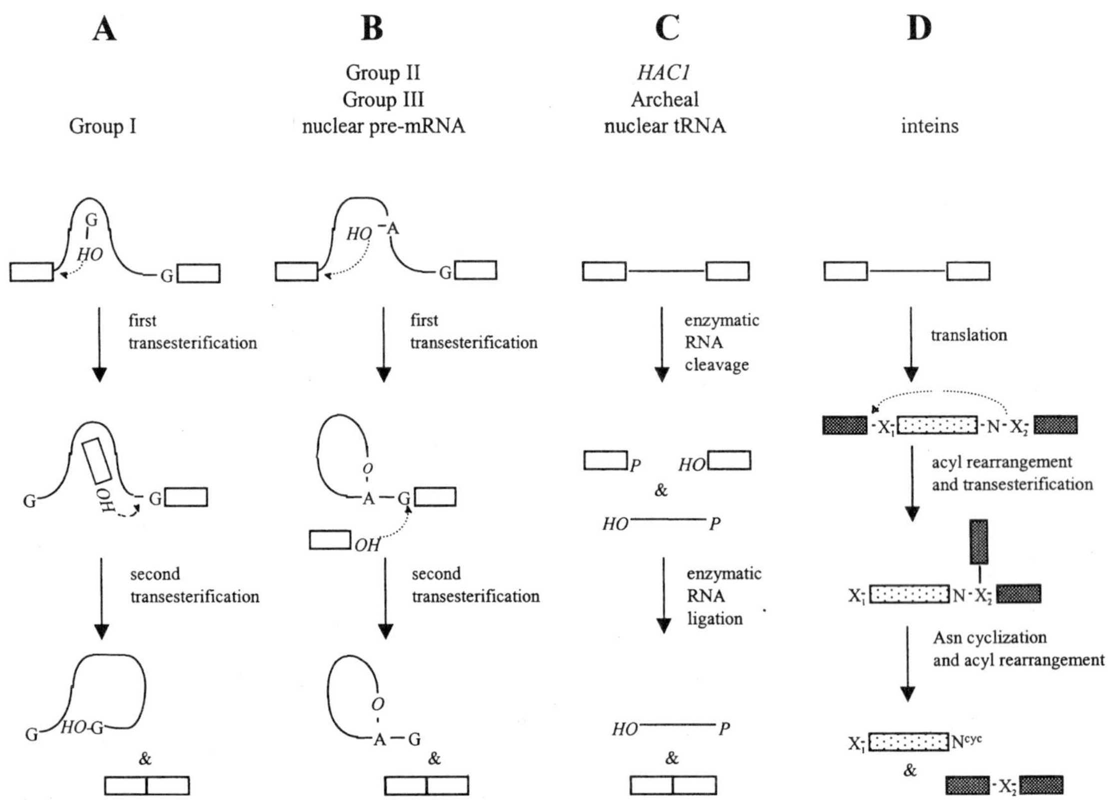

Figure 1 The four mechanisms of splicing (A, B, C and D). Exons are shown as unshaded boxes, introns as solid lines, translated exteins as dark shaded boxes and translated inteins as dotted boxes. G, guanosine; A, adenosine; *P*, phosphate group; *OH*, hydroxide group; N, Asn, asparagine; X, any amino acid.

variable in length, but all have a distinct secondary and tertiary structure of nine stem–loops (**Figure 2A**; P1–P9).

Group I intron splicing proceeds by two consecutive transesterification reactions as outlined below. First, the conformation of the intron brings a nucleophile (a guanosine) into close proximity with the 5′ exon–intron junction, causing the 3′ hydroxyl group of the guanosine to attack the phosphate at the 5′ splice site. This results in cleavage of the phosphodiester bond and addition of the guanosine to the 5′ end of the intron. The products of this reaction are the intron–exon 2, with the additional guanosine, and a separate exon 1 with a reactive hydroxyl group at its 3′ end. The 5′ splice site is located by virtue of its interaction with the internal guide sequence of the intron, being part of the P1 stem; there is little primary sequence conservation at the 5′ splice site, apart from a uridine residue preceding the cleaved phosphodiester bond. After this first transesterification reaction, the reactive hydroxyl at the 3′ end of exon 1 attacks the phosphate at the 3′ splice site. There is only limited sequence conservation at the 3′ splice site also, the last base of the intron always being guanosine. However, it is thought that an interaction between the 3′ end of the intron and a region between P7 and P9 is important for 3′ splice site identification. This second transesterification reaction results in the mature RNA and the excised intron. Some excised introns undergo cyclization, in which the 3′ end guanosine attacks a phosphate near the 5′ end of the intron producing a circular intron and a small oligonucleotide. This type of splicing only requires monovalent cation, divalent cation (Mg^{2+}) and a guanosine cofactor; no energy input is required.

(ii) Group II introns

Group II introns (Michel & Ferat, 1995) have only been identified in chloroplasts, fungal mitochondria and some eubacterial genomes. Like the unrelated group I introns, group II introns have a highly conserved secondary structure comprising six stem–loop domains (I–VI) radiating from a central axis (**Figure 2B**). The primary sequence is less well conserved, with only a few bases near the 5′ splice site and domain V being characteristic for the class.

As with group I introns, the mechanism of splicing involves two transesterification reactions; however, unlike group I introns, no guanosine cofactor is required. The 5′ splice site is identified by interactions between adjacent exon sequences (IBS1 and IBS2; **Figure 2B**) and domain I of the intron (EBS1 and EBS2). This complex tertiary RNA structure brings the 5′ splice site into close proximity with an adenosine that is bulged from the semiconserved domain VI of the intron. The 2′ hydroxyl group of this adenosine is the nucleophile which attacks the phosphate at the 5′ splice site. This results in the release of exon 1, while the guanosine at the 5′ end of the intron is covalently bound to the bulged adenosine through a 5′–2′ phosphodiester bond to form an intron lariat–exon 2 branched structure (**Figure 2B**). The 3′ splice site is usually 7 or 8 base pairs (bp) downstream of the bulged adenosine and is determined by limited base pairing of adjacent sequences with regions in the intron. The hydroxyl group at the 3′ end of exon 1 attacks the 3′ splice site in the second transesterification reaction to form the mature RNA and the excised intron lariat. As with group I introns, the efficiency of this autocatalytic process is increased by protein cofactors *in vivo*.

(iii) Group III introns

The most extensively studied group III introns are from the plastid genomes of *Euglena gracilis* and *Astasia longa*, both primitive eukaryotes of the kingdom Protista. Group III introns are usually short (91–119 bp) and contain domains similar to some of those found in group II introns. In particular, the 5′ splice site consensus for group II and group III introns is very similar, as is the 3′ splice site region where a stem–loop structure containing a bulged adenosine can be found in both (domain VI in group II). Some group III introns also have a domain akin to group II domain I that contains putative exon-binding sequences (EBS). These similarities, and the fact that the mechanism of splicing appears to be the same, with a lariat intermediate branched at the bulged adenosine, has led to the proposal that group III introns are group II introns that require some domains to be supplied *in trans*. This apparent requirement for *trans*-acting components is reminiscent of nuclear pre-mRNA splicing and may represent either convergent evolution or an intermediate in the evolution of nuclear pre-mRNA introns (Copertino & Hallick, 1993).

(iv) Nuclear pre-mRNA introns

Krämer (1996), Nilsen (1997) and Will & Lührman (1997) provide useful introductions to protein function in pre-mRNA splicing. In the most primitive eukaryotes, e.g. *Giardia*, nuclear pre-mRNA splicing is completely absent. There is an approximate correlation between the complexity of a eukaryote and both the number and size of the introns in its nuclear genome. In mammals, the average exon number for a gene is seven with the average pre-mRNA being about five times as long as the average mRNA. Introns in higher eukaryotes can be very large (up to 200 000 bp) while exons are usually small (50–400 bp), with some genes having more than 50 exons. This means that a single gene can cover hundreds of kilobases on the eukaryotic genome, raising the question of how these introns are accurately recognized and removed from such large sequences.

Comparing homologous genes that have highly conserved exons, intron sequences are not usually conserved. However, in the majority of pre-mRNA introns, there are four regions which show sequence conservation; the 5′ splice site, branchpoint, polypyrimidine tract and 3′ splice site (**Figure 2C**). Nuclear pre-mRNA splicing occurs by a mechanism similar to that of group II introns, raising the possibility that group II introns may be the progenitors of nuclear pre-mRNA splicing (Copertino & Hallick, 1993). Both are spliced in a two-step process involving sequential transesterification reactions in which a 2′ hydroxyl group from a bulged adenosine residue in the intron attacks the 5′ splice site phosphate, followed by an attack of the 3′ splice site phosphate by the 3′ end of exon 1. It is of note that, in nuclear pre-mRNA splicing, the reactive adenosine is bulged from an intermolecular helix (U2/intron; see below), while in group II splicing it is bulged from an intramolecular helix (intron), although the reactive adenosine is derived from the

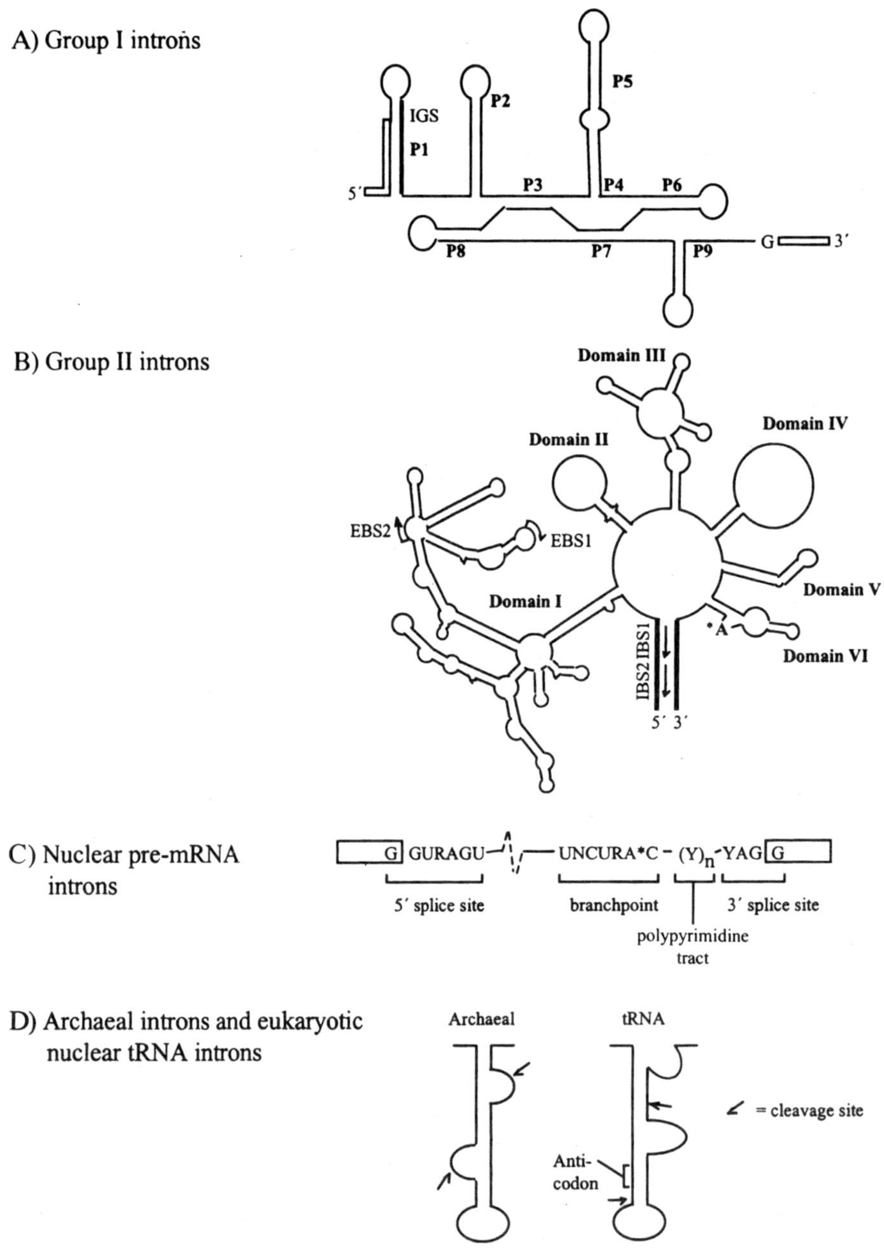

Figure 2 Intron structures. All boxed regions are exon sequences. (A) A typical secondary structure of group I introns as determined by sequence alignment. The conserved base pair elements (P1–P9) are indicated. (B) A typical secondary structure of group II introns. The six conserved domains (I–VI) are shown in bold. The EBS1, EBS2, IBS1 and IBS2 sequences and their orientation are indicated. The A* represents the bulged adenosine involved in the 2′, 5′ phosphodiester bond formation of the first transesterification reaction. (C) A typical mammalian nuclear pre-mRNA intron (major class). (D) The typical secondary structures of an archaeal intron and eukaryotic nuclear tRNA intron with cleavage sites indicated. No differentiation between exon and intron elements is made. G, guanosine; IGS, internal guide sequence involved in 5′ splice site selection; R, purine; Y, pyrimidine.

intron sequence in both cases. As with group II splicing, the end-products are the mature RNA and an intron–lariat containing a 2′–5′ phosphodiester branch. Both classes require divalent cations (Mg^{2+}) for splicing, but, unlike group II splicing, pre-mRNA splicing also requires energy input in the form of hydrolysis of both adenosine triphosphate and guanosine triphosphate.

While group II introns are highly conserved, with the reactive centre derived from the intron sequence, nuclear pre-mRNA introns are degenerate with the catalytic machinery being provided *in trans* by five small nuclear RNAs (snRNAs: U1, U2, U4, U5 and U6) and a large number of proteins. The snRNAs are packaged in four small nuclear ribonucleoprotein particles (snRNPs: U1, U2, U5 and U4/U6) which, along with a number of proteins, associate with the pre-mRNA in an ordered fashion to form a large

RNP complex termed the spliceosome. Studies of the spliceosome (Staley & Guthrie, 1998) have led to the theory that the snRNAs interact with each other and with the substrate pre-mRNA to form a catalytic centre for splicing in a similar way to the folding of a group II intron.

Figure 3 shows how the formation and dissociation cycle of the spliceosome can be divided into five stages, as defined by six RNP complexes which are distinguished by fractionation procedures: E (commitment complex), A (pre-spliceosome), B (assembled spliceosome), C (activated spliceosome), and D + I (splicing end-products). First, the U1 snRNP associates with the 5′ splice site of the pre-mRNA via base pairing of the U1 snRNA with conserved sequences, and a U2AF heterodimer interacts with the polypyrimidine tract near the 3′ splice site to form the commitment complex (E). It is thought that these two components interact with each other via proteins containing arginine–serine (Arg–Ser) repeats and RNA binding domains (SR proteins).

As the 3′ splice site is only weakly conserved, U2AF is recruited, via SR proteins, by the U1-bound 5′ splice site of the next intron downstream in a process known as exon definition. With terminal exons, it is believed that the 5′ cap and the 3′ polyadenylation machinery enhance splice-site recognition. It is thought that this process is essential to prevent exon skipping and may explain why exons in higher eukaryotes are usually short in comparison to introns. Subsequently, the U2 snRNP binds to the branchpoint (BP) region to form the A complex. This complex is promoted by SR proteins associating with the polypyrimidine tract and results in the U2 snRNA base pairing with the BP sequence and forming an imperfect RNA helix with a single unpaired adenosine. The B complex forms when the other three snRNAs join as a tri-snRNP (U5.U4/U6). On association with the spliceosome, the U5 snRNA interacts in a non-Watson–Crick fashion with the 5′ exon and is thought to anchor it in the spliceosome reaction centre. The U4 and U6 snRNAs contain extensive sequence complementarity and are initially base paired together. However, once in the spliceosome, the base pairing of the U4 and U6 snRNAs is disrupted; the U6 snRNA now base pairs with the U2 snRNA in a structure analogous to domain V of group II introns, and U6 also base pairs with the 5′ end of the intron, displacing the U1 snRNA. This brings the 5′ splice site into close proximity to the BP adenosine, so priming the spliceosome for the first transesterification reaction.

The cleavage of the 5(splice site results in the C complex. It is thought that the spliceosome is then reorganized enabling the second transesterification reaction. Exon I is retained in the spliceosome, being positioned in close proximity to the 3′ splice site, probably by interactions with the U5 snRNP. Once the two exons are joined, the spliceosome dissociates to form the D and I complexes which contain the mRNA and intron–lariat end-products, respectively. It is thought that the released snRNPs are recycled, while the intron is degraded by nucleases.

Recent work has demonstrated the existence of sequence

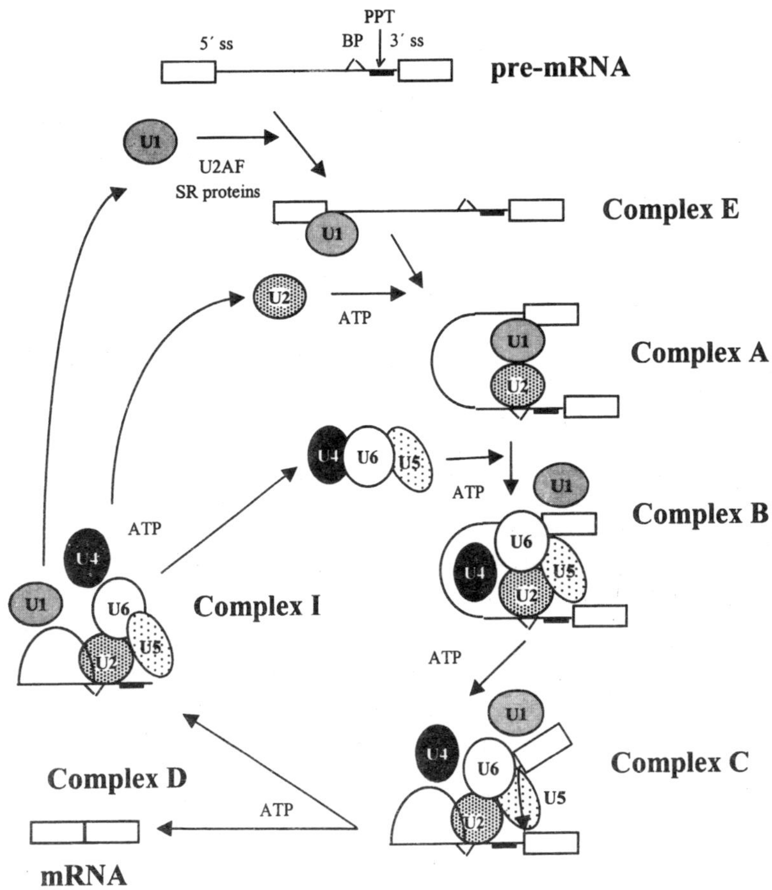

Figure 3 Schematic of the splicing cycle. Pre-mRNA is initially bound by U1 snRNP, U2AF and SR proteins to form the E complex. The addition of U2 snRNP leads to complex A formation which is, in turn, converted to complex B by the recruitment of the tri-snRNP. On complex B formation, both the U1 and U4 snRNP interactions with spliceosome components are destabilized. The completion of the first transesterification reaction results in C complex formation. After the second transesterification reaction, the spliceosome dissociates, culminating in complexes ID and I. The snRNPs present in complex I are recycled and the intron–lariat degraded. 3′ss, 3′ splice site; 5′ss, 5′ splice site; BP, branchpoint; PPT, polypyrimidine tract.

elements, other than the splice sites and BP, that affect splicing called splicing enhancers. Enhancers have been found in both exons (exonic splicing enhancers [ESEs]) and introns and are known to affect splice site usage in both constitutive and alternative splicing (see below). Little is known about intronic enhancers except that they appear to be target sites for cell-specific regulatory factors of alternative splicing (e.g. regulation of exon 5 in the cardiac troponin T gene) and appear to play no role in constitutive splicing. ESEs are better understood and are known to enhance splice site choice in both constitutive and alternative splicing, activating suboptimal splice sites. These purine or A/C rich elements are bound by SR proteins which are believed to recruit other splicing factors, thus enhancing nearby splice site use.

In the early 1990s, it was discovered that among nuclear pre-mRNA introns there is a rare subclass with different consensus sequences (Sharp & Burge, 1997; Tarn & Steitz, 1997). These occur at a frequency of about 1/5000 to 1/10 000 introns and were termed AT–AC introns as opposed to the canonical GT–AG introns, as the first examples had AT and AC instead of GT and AG at their 5′ and 3′ ends. These introns are spliced by a novel spliceosome (the U12 spliceosome as opposed to the canonical U2 spliceosome), containing the snRNPs U11, U12, U4atac/U6atac and U5. More recently, it has been demonstrated that the dinucleotides at the extremities of these introns can also be GT–AG and that they are more clearly distinguished from the major class of introns by other highly conserved intron sequences. To avoid ambiguity, the two classes of introns were renamed U2 type or U12 type as opposed to a nomenclature based on terminal sequences. It appears that the mechanism of splicing of these two spliceosomes is analogous.

(v) Archaeal, eukaryotic nuclear tRNA and HAC1 *introns* (Rosbash, 1996; Lykke-Andersen et al., 1997)

The introns found in archaeal rRNA and tRNA genes and those present in eukaryotic nuclear tRNA genes are thought to have a common evolutionary origin despite being in highly divergent life forms. This notion is based on the similarity of their locations in the RNAs and the mechanisms by which they are removed. Most archaeal tRNA introns and all eukaryotic tRNA introns are located one nucleotide 3′ to the anticodon, although the significance of this is unclear. Both types are cleaved by an endoribonuclease activity which generates intermediates with 2′–3′ cyclic phosphate and 5′-hydroxyl ends, followed by a distinct ligation reaction.

Comparison of the archaeal endoribonucleases with the heterotetrameric Sen complexes of eukaryotic tRNA processing has revealed sequence homologies which suggest a common mechanism in splicing. However, the splicing mechanisms seem to differ in the means of identification of the splice sites; archaeal intron boundaries are determined by a "bulge–helix–bulge" motif, whereas eukaryotic tRNA introns are recognized by a measuring mechanism which monitors the structure of the pre-tRNA (**Figure 2D**).

Recently, a third type of intron which falls into this category was discovered in *Saccharomyces cerevisiae*: the *HAC1* pre-mRNA intron. The *HAC1* gene is involved in the signalling pathway of the unfolded protein response which monitors the accumulation of unfolded proteins in the endoplasmic reticulum. There is some ambiguity over the precise location of the intron in the *HAC1* gene, with the 5′ splice site being CAG/CCGTGA (or possibly CA/GCCGTG) and the 3′ splice site being TCCG/A (or possibly TCC/GA). These sequences are predicted to form stem–loop structures which may be important in intron recognition. The intron is removed by a *HAC1*-specific endoribonuclease (Irelp) and the two exons are subsequently ligated by Rlglp (the nuclear tRNA ligase).

(vi) Inteins (Perler, 1998)

Inteins were first discovered in 1992 and are the exception among the introns described here in that they are intervening sequences which are not spliced out at the level of RNA, but are excised from the protein. They are removed in an autocatalyic process by multifunctional peptides which are encoded by these introns. Some of the multifunctional intein peptides are homologous to site-specific endonucleases, and they may promote the movement of the intein sequence into an allelic inteinless site. These endonucleases are similar to the endonucleases encoded by some group I and archaeal introns involved in intron homing (see section 4, Intron mobility). Inteins are widely distributed, being found in both prokaryotic and eukaryotic genomes. The catalytic mechanism by which inteins are removed from protein sequences is outwith the scope of this review, but details can be found in Perler (1998) and on the Internet at http://www.neb.com/neb/frame_tech.html.

3. *Trans*-splicing (Nilsen, 1993; Blumenthal, 1995)

Trans-splicing is the joining of distinct RNA molecules, each of which has associated partial intron sequences that are removed in the splicing process. In the case of group II *trans*-spliced introns, the secondary and tertiary RNA interactions between the 5(and 3(components of the split intron serve to anneal the two molecules to be joined. The intron may even be more fragmented. For example, the *Chlamydomonas reinhardrii* chloroplast gene *psa*A consists of three exons that are discontinuous on the chloroplast genome and are independently transcribed. The three transcripts are joined by two group II splicing events, with a fourth transcript, encoded by the chloroplast *tsc*A locus, required for splicing together exons 1 and 2. The *tsc*A RNA sequence resembles domains II and III of a group II intron, whereas these domains appear to be absent from the 5′ and 3′ components of intron 1. Thus, apparently, *tsc*A is a *trans*-acting component of this discontinuous intron and, as such, may be considered to play a role resembling that of snRNAs in nuclear pre-mRNA splicing (Goldschmidt-Clermont *et al.*, 1991).

Nuclear pre-mRNAs can also be *trans*-spliced, but this process does not involve extensive sequence complementarity between the RNAs to be joined. In trypanosomes, all nuclear mRNAs acquire a 39 bp "spliced leader" sequence derived from a longer precursor that contains a match to the 5(splice site consensus sequence of spliceosomal introns immediately downstream of the *trans*-spliced 39 bp exon. In nematode nuclei, many mRNAs acquire spliced leaders (SL1 and SL2) via *trans*-splicing. SL1 comes from the first 22 nucleotides of a small (~100 nucleotide) RNA that includes a 5(splice site sequence immediately downstream of the 22 nucleotide SL1 sequence. Both the trypanosome and the nematode spliced leader precursor RNAs associate with proteins to form snRNPs that appear to function instead of U1 snRNPs in *trans*-splicing spliceosomes. The mechanisms of *trans*- and *cis*-splicing are therefore very similar, except that the SL snRNP donates the 5(exon. In nematodes, a large number of mRNAs are produced as polycistronic transcripts. SL1 is *trans*-spliced to most 5(cistrons, whereas a second spliced leader RNA, SL2, is specifically *trans*-spliced to internal cistrons. Thus, in nematodes, *trans*-splicing functions to convert polycistronic primary transcripts into monocistronic mRNAs.

4. Intron mobility (Lambowitz & Belfort, 1993; Curcio & Belfort, 1996; Grivell, 1996; Belfort & Roberts, 1997)

Evidence of intron mobility has been found for group I, group II, archaeal introns and inteins, although each occurs by a different mechanism. Some group I, archaeal and inteins have been found to encode homing endonucleases containing the dodecapeptide LAGLIDADG motif. All of these endonucleases specifically cleave DNA at a defined target site, often present in intronless alleles of the cognate gene. A copy of the intron is then introduced into the recipient gene by a DNA recombination event. The situation in group II introns is more complicated, with some introns encoding a single protein which has DNA endonuclease, reverse transcriptase and RNA maturase activities. The sense strand of the specific target DNA is cleaved by the intron RNA and the antisense strand is cut by the endonuclease. How the intron is inserted is less well understood; however, the splicing activity of the intron does seem to be required for mobility. It is thought that excised intron RNA is covalently attached to the 3′ exon sense strand, followed by reverse transcription to produce the antisense strand, possibly using an unspliced intron as a template. This transposition is reminiscent of movement of non-LTR retrotransposons, suggesting the possibility that group II introns may be their progenitors.

5. Nuclear pre-mRNA intron evolution (Roger & Dolittle, 1993; Mattick, 1994; De Souza *et al.*, 1996)

Why nuclear pre-mRNA introns exist is still not clear. As the majority of introns do not encode proteins or functional RNA (a few do encode small nucleolar RNAs) and usually display a lack of sequence constraint between homologous genes, they appear to have no function *per se*. Also, it would appear that to process these apparently redundant pieces of genetic material is expensive for the cell, yet, in higher eukaryotes, introns comprise a large percentage of the nuclear genome.

An appreciation of why some genes contain introns should come from the knowledge of their origin. However, despite 20 years of research, there is still a division of opinion as to the origin of nuclear introns. Currently, two main hypotheses exist.

(i) The "intron early" hypothesis

This hypothesis postulates that introns were originally intervening sequences between ancient, independent exons which encoded distinct peptides. The activity of catalytic RNA in these early organisms joined these exons together, creating transcripts which encoded proteins composed of a number of distinct domains (the exon theory of genes, Gilbert, 1987). One possible reason why introns have not been lost during the course of evolution is that they may confer a selective advantage as DNA recombination points, facilitating the movement of exons around the genome ("exon shuffling"). This concept of exon shuffling proposes the rapid creation of novel genes by mixing exons, thereby conveying an evolutionary advantage to the organism.

The argument continues with the idea that initially all the splicing mechanisms were at the ribozyme level, such as in group I and group II splicing. However, as this primordial cell developed into present-day eukarya, eubacteria and archaebacteria, so the splicing evolved. In the eubacteria and archaebacteria, the genomes lost introns due to the pressures of rapid growth and genome replication, while in the eukarya a more flexible splicing mechanism developed – nuclear pre-mRNA splicing. It is considered that group II splicing could be the progenitor of nuclear pre-mRNA splicing due to the similarities between the two, and the observation of possible evolutionary intermediates (group III).

Based on this hypothesis, one would expect the exons and introns of genes to meet a number of criteria. First, it is predicted that there would be a correlation between exons and protein domains or modules; however, to date, there is no consensus of opinion as to whether this exists. Secondly, if the intron early hypothesis were true, one would expect to find introns in the same codon phase (i.e. preserving reading frame during exon shuffling). In the converse, if introns arose late in evolution (see below), then introns would be expected to be inserted randomly into the reading frame, with no phase correlation. When the phasing of over 13 000 exons and 11 000 introns was examined, a clear bias to phase 0 was found, which is consistent with the intron early hypothesis (reviewed in Long *et al.*, 1995).

Thirdly, this hypothesis implies that eubacteria, archaebacteria and simple eukarya are highly evolved in comparison to their progenitor, lacking intronic regions to achieve competitive growth rates. If this were the case, there should be examples of intron loss during evolution. Indeed, there are many examples of apparent intron loss with only a few examples of intron acquisition. For instance, there are two highly conserved rat insulin genes, which appear to have arisen by duplication of the original, two-intron gene, followed by loss of an intron from one of the genes. Also, *S. cerevisiae* has relatively few introns and it has been proposed that most of its introns have been eliminated through gene conversion by reverse transcript copies of spliced mRNAs (Fink, 1987).

Lastly, the lack of homology between introns and examples of conservation of intron position in ancient genes suggests that introns arose early in the evolution of life, but these observations do not exclude the possibility of late insertion of introns.

(ii) The "intron late" hypothesis

This hypothesis holds that introns were selfish elements which spread into eukaryotic genes after the origin of the nucleus. Again, this theory presumes the evolution of nuclear pre-mRNA introns from autocatalytic group II introns, but supposes that, instead of being originally present in the genes, the group II introns moved, possibly from organelle genomes, into the nuclear genome by retroposition. Fragmentation of one of these elements produced the genes for the snRNAs, thus providing *trans*-acting catalytic components and allowing the mutagenic drift of all other group II introns.

This hypothesis is based on the following premises. First, similarities between the mechanisms of group II and nuclear pre-mRNA splicing suggest that the latter evolved from the former; such similarities include related intermediates, end-products and RNA structures. The existence of group III introns could represent an intermediate stage in evolution. Secondly, there is the observation that at least some group II introns are mobile, an essential attribute if introns were to spread around the genome (see above). Thirdly, if introns migrated from organelles to the eukaryotic genome then group II introns should be present in eubacteria, the presumed ancestors of mitochondria and chloroplasts. It is now evident that group II introns are present in at least some eubacteria and they appear to be very ancient, probably having arisen before the endosymbiotic event that introduced eubacteria into eukaryotic cells. Fourthly, if nuclear pre-mRNA introns are derived from group II introns of organelles then it can be presumed that eukaryotes which diverged prior to organelle endosymbiosis lack introns. Although the phylogeny of protists is controversial, no introns have been reported in what are believed to be the earliest amitochondrial protists, Giardia and Vairimorpha, supporting the intron late hypothesis. However, such an observation is explained in the intron early theory by postulating the streamlining of genomes to maximize growth rates, as argued for prokaryotes.

The intron late hypothesis fails to address the observation that no group II intron has been identified in the nuclear genome of a eukaryote. If nuclear pre-mRNA introns spread through intron mobility, this must be due to the mobility of nuclear pre-mRNA introns and not group II introns. It is inconceivable that every nuclear intron is the product of the movement of a group II intron from an organelle to the nucleus and then subsequent convergent evolution of all the group II introns into nuclear pre-mRNA introns. Nuclear pre-mRNA mobility has yet to be demonstrated, although intron gain has been proposed for some genes – in particular, for the U6 snRNA genes of some fungi. It has been proposed that this occurred by a reverse splicing event that inserted an excised intron into the spliceosomal U6 snRNA followed by reverse transcription and gene conversion of the chromosomal sequence.

6. Alternative pre-mRNA splicing (Chabot, 1996; Manley & Tacke, 1996; Valcarcel & Green, 1996)

There are avid proponents of both arguments of intron evolution, with new evidence for each theory continuously being proposed. Whether introns are of ancient origin or a relatively recent addition to the genome, many introns have acquired functions which justify their retention. For example, higher eukaryotes have evolved mechanisms where they utilize nuclear pre-mRNA splicing to regulate gene expression in a process known as alternative splicing.

In vertebrates, it has been estimated that one in 20 genes is expressed by alternative pathways of RNA splicing. Alternatively spliced mRNAs are generated from the same pre-mRNA by utilizing different combinations of splice sites according to the environment of the RNA. Such regulation is common in developmental and tissue-specific gene expression. The potential of alternative splicing appears to be limitless, with examples of exon deletion, swapping, addition, extension, shortening, etc. being known. Such alternative splicing can have a wide range of effects on the final protein product, from splicing events which only affect the non-coding region of the mRNA, altering RNA stability or targeting, to domain swapping in the protein to produce protein isoforms, through to the production of completely unrelated proteins.

Some of the mechanisms by which splice site utilization can be regulated have only recently been elucidated. As the regions of conservation in functional splice sites of nuclear pre-mRNA introns are relatively short, there is a high probability that similar sequences occur in the pre-mRNA that are normally non-functional due to their suboptimal location and sequence. However, these "cryptic sites" may be activated in alternative modes of splicing. It was found that, with some RNAs, increasing the concentration of certain constitutive splicing factors, such as the SR protein ASF/SF2, alters the selection of the 5′ splice site in favour of a proximal (more 3′) site, possibly by actively recruiting U1 snRNPs. It is thought that, if an RNA has U1 snRNPs bound at multiple 5′ splice sites due to the action of SR proteins like ASF/SF2, the most proximal 5′ splice site is used by default. Some hnRNP proteins, such as hnRNP Al, antagonize the activity of SR proteins by favouring distal 5′ splice site selection (**Figure 4**).

Increased concentrations of SR proteins have also been shown to promote the use of more proximal 3′ splice sites, possibly by binding to splicing enhancers (ESEs; see above) located in the adjacent exon, and thereby recruiting U2AF and the U2 snRNP. In addition to the concentration-dependent effects of constitutive SR proteins on splice site choice, tissue-specific regulators have been identified. Tra and Tra-2 are female-specific Drosophila SR proteins involved in sex determination (MacDougall *et al.*, 1995). These two SR proteins activate weak ESEs in doublesex (*dsx*) pre-mRNA by binding to them and recruiting constitutive SR proteins, thereby promoting a female-specific splicing pattern and the production of female-specific proteins.

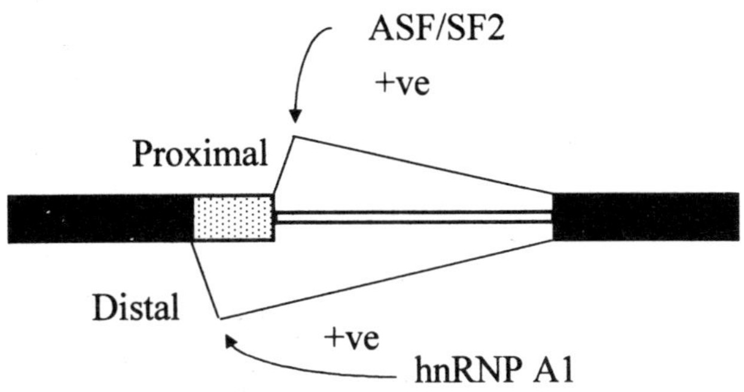

Figure 4 An example of splice site regulation by *trans*-acting factors. The protein ASF/SF2 is believed to recruit U1 snRNP to the 5′ splice sites, with the most proximal splice site preferentially used. The binding of hnRNP A1 protein, between the two potential 5′ splice sites, shifts the selection to the distal splice site, thus altering the mRNA produced. Exons, thick shaded boxes; intron, thin unshaded box.

Examples of negative regulation of splicing also exist where the use of certain splice sites is suppressed. Suppression can be due to *cis*-acting factors such as RNA secondary structure or to *trans*-acting factors. The viral REV protein is a *trans*-acting negative regulator that binds to the RRE (Rev response element) in the HIV pre-mRNA, preventing spliceosome formation and promoting the export of unspliced pre-mRNA. In addition, exonic negative elements have recently been found which suppress splicing, but the mechanism of action has yet to be determined.

Thus, there are many ways to regulate alternative splicing and numerous factors may be involved. Evidently, this is a combinatorial process, involving combinations of different sets of *cis*-elements (enhancers and splice sites) and *trans*-acting factors (constitutive factors such as SR proteins, hnRNPs, etc.); and cell-specific factors, e.g. Tra and Tra-2). Recent work has demonstrated that splice site choice is just one level of control in a complex series, with *trans*-acting factors themselves being subject to regulated expression such as control of transcription and/or alternative splicing and modulation of their activity through phosphorylation. Thus, cascades of regulatory events determine the alternative splicing patterns observed in tissue-specific, sex-specific and developmental gene expression.

7. Introns and disease (Treisman *et al.*, 1983; Cooper & Mattox, 1997; Mattaj, 1998)

It has been estimated that splicing defects account for ~15% of all human genetic diseases caused by point mutations. Many examples have been identified in β-thalassaemias, where the two-intron β-globin pre-mRNA is aberrantly spliced. In some thalassaemias, the first nucleotide of the first intron is mutated, rendering the 5′ splice site non-functional. This causes the activation of cryptic 5′ splice sites, altering the size of the intron. In other β-thalassaemias, mutations in the second intron create a novel 5′ splice site, which activates an upstream cryptic 3′ site that results in the production of a novel exon within the second intron. These mutations lead to non-functional β-globin chains which cause the thalassaemia phenotype.

As the conserved splice site and branchpoint sequences of introns are short, it is of no surprise that the majority of mutations which affect splicing lie within the larger enhancer elements. In the last few years, a number of unrelated diseases have been shown to be caused by alterations in pre-mRNA splicing as a result of ESE mutations. Conversely, mutations in the SR proteins which recognize these elements have also been associated with disease. A well-studied example is the cell-surface-adhesion molecule CD44 which is subject to extensive alternative splicing. It has been demonstrated that the production of some aberrantly spliced CD44 isoforms can result in the metastasis of some normally non-metastasizing cell lines. This perturbation in splicing has been correlated with the expression of certain SR proteins, suggesting a link between cell metastasis and changes in the components of the splicing machinery. In the future, it may be possible to treat such diseases by using gene therapy to alter ESEs or the *trans*-acting factors involved in splice site determination.

Other diseases such as spinal muscular atrophy have been linked to assembly defects of constitutive pre-mRNA splicing complexes. The gene linked with this neurone degenerative disease was identified in 1995 as SMN (survival motor neurone). Subsequent analysis of the SMN protein revealed its association with snRNPs and implicated it in the snRNP assembly process. Why SMN mutations result in a disease phenotype that is manifest only in motor neurones is unclear, but this may reflect a greater need for SMN in motor neurones where snRNPs have been shown to have a different composition compared with other tissues.

8. Conclusion

The discovery of introns in 1977 opened up a whole new area of molecular biology. Not only are the coding regions of genes interrupted, but these coding regions may be organized in unexpected ways in the mature transcript and, in some organisms, distinct transcripts may be joined by *trans*-splicing events. The existence of open reading frames or small RNA coding sequences in introns – that is, genes within genes – explains early anomalies observed in genetic studies of fungal mitochondria, for example, in which complementation groups were intermingled. Regulated splicing added a new and complex layer of control to gene expression in eukaryotes and provided a previously unimaginable mechanism for producing multiple, related proteins from the same genetic locus. The discovery of RNA catalysis transformed our thinking about enzymology, formed the basis of an entirely new kind of biotechnology and fuelled the evolutionary debates. Perhaps the most salutary message for the biologist has been: always remain open-minded and respect the unexpected. The RNA world probably has more surprises in store for us.

REFERENCES

Belfort, M. & Roberts, J. (1997) Homing endonucleases: keeping the house in order. *Nucleic Acids Research* 25: 3379–88

Blumenthal, T. (1995) Trans-splicing and polycistronic transcription in *Caenorhabditis elegans*. *Trends in Genetics* 11: 132–35

Cech, T. (1990) Self-splicing of group I introns. *Annual Review of Biochemistry* 59: 543–68

Chabot, B. (1996) Directing alternative splicing: cast and scenarios. *Trends in Genetics* 12: 472–78

Cooper, T.A. & Mattox, W. (1997) The regulation of splice-site selection, and its role in human disease. *American Journal of Human Genetics* 61: 259–66

Copertino, D.W. & Hallick, R.B. (1993) Group II and group III introns of twintrons: potential relationships and pre-mRNA introns. *Trends in Biochemical Sciences* 18: 467–71

Curcio, M.J. & Belfort, M. (1996) Retrohoming: cDNA-mediated mobility of group II introns requires a catalytic RNA. *Cell* 84: 9–12

De Souza, S.J., Long, M. & Gilbert, W. (1996) Introns and gene evolution. *Genes to Cells* 1: 493–505

Fink, G.R. (1987) Pseudogenes in yeast? *Cell* 49: 5–6

Gilbert, W. (1987) The exon theory of genes. *Cold Spring Harbor Symposium on Quantitative Biology* 52: 901–05

Goldschmidt-Clermont, M., Choquet, Y., Girard-Bascov, J. *et al.* (1991) A small chloroplast RNA may be required for trans-splicing in *Chlamydomonas reinhardtii*. *Cell* 65: 135–43

Grivell, L.A. (1996) Transposition: mobile introns get into line. *Current Biology* 6: 48–51

Krämer, A. (1996) The structure and function of proteins involved in mammalian pre-mRNA splicing. *Annual Review of Biochemistry* 65: 367–409

Lambowitz, A.M. & Belfort, M. (1993) Introns as mobile genetic elements. *Annual Review of Biochemistry* 62: 587–622

Long, M., De Souza, S.J. & Gilbert, W. (1995) Evolution of the exon-intron structure of eukaryotic genes. *Current Opinion in Genetics and Development* 5: 774–78

Lykke-Andersen, J., Aagaard, C., Semionenkov, M. & Garrett, R.A. (1997) Archaeal introns: splicing, intracellular mobility and evolution. *Trends in Biochemical Sciences* 22: 326–31

MacDougall, C., Harbison, D. & Bownes, M. (1995) The developmental consequences of alternative splicing in sex determination and differentiation in Drosophila. *Developmental Biology* 172: 353–76

Manley, J.L. & Tacke, R. (1996) SR proteins and splicing control. *Genes and Development* 10: 1569–79

Mattaj, I.W. (1998) Ribonucleoprotein assembly: clues from spinal muscular atrophy. *Current Biology* 8: R93–95

Mattick, J.S. (1994) Introns: evolution and function. *Current Opinion in Genetics and Development* 4: 823–31

Michel, F. & Ferat, J.-L. (1995) Structure and activities of group II introns. *Annual Review of Biochemistry* 64: 453–61

Nilsen, T.W. (1993) Trans-splicing of nematode pre-messenger RNA. *Annual Review of Microbiology* 47: 413–40

Nilsen, T.W. (1997) RNA–RNA interactions in nuclear pre-mRNA splicing. In *RNA Structure and Function*, edited by R. Simons & M. Grunberg-Mangano, Cold Spring Harbor, New York: Cold Spring Harbor Laboratory Press

Perler, F.B. (1998) Protein splicing of inteins and hedgehog autoproteolysis: structure, function and evolution. *Cell* 92: 1–4

Roger, A.J. & Dolittle, W.F. (1993) Why introns-in pieces? *Nature* 364: 289–90

Rosbash, M. (1996) Mixed mechanisms in yeast pre-mRNA splicing? *Cell* 87: 357–59

Sharp, A. & Burge, B. (1997) Classification of introns: U2-type or U12-type. *Cell* 91: 875–79

Staley, J.P. & Guthrie, C. (1998) Mechanical devices of the spliceosome: motors, clocks, springs and things. *Cell* 92: 315–26

Tarn, W-Y. & Steitz, J.A. (1997) Pre-mRNA splicing: the discovery of a new spliceosome doubles the challenge. *Trends in Biochemical Sciences* 22: 132–37

Treisman, R., Orkin, S.H. & Maniatis, T. (1983) Specific transcription and RNA splicing defects in five cloned beta-thalassaemia genes. *Nature* 302: 591–96

Valcarcel, J. & Green, R.G. (1996) The SR protein family: pleiotropic functions in pre-mRNA splicing. *Trends in Biochemical Sciences* 21: 296–301

Will, C.L. & Lührman, R. (1997) Protein functions in pre-mRNA splicing. *Current Opinion in Cell Biology* 9: 320–28

GLOSSARY

5′ or 3′ splice sites exon/intron junctions; 5′ and 3′ refer to the respective ends of introns

alternative splicing splicing an RNA in different ways, using alternative splice sites or introns

branch point (BP) position in a group II or nuclear pre-mRNA intron where a 2′–5′ phosphodiester bond is formed with the most 5′ nucleotide of the intron in the first step of splicing, creating a branched structure

exon binding sequence(s) (EBS) group II intron sequences that base pair with complementary IBS sequences in a flanking exon

exon splicing enhancer (ESE) a region in an exon that enhances the splicing efficiency of a neighbouring intron

hnRNP heterogeneous nuclear ribonucleoprotein particle. Complex of high molecular weight nuclear RNA (hnRNA or pre-mRNA) and a specific set of (hnRNP) proteins

intron binding sequence(s) (IBS) exonic sequences that base pair with complementary EBS sequences in a neighbouring group II intron

intein protein intron. An internal region of a precursor protein that is spliced out and therefore absent from the mature protein. As opposed to extein which is a region that is retained in the mature protein

polypyrimidine tract a pyrimidine-rich region between the branch point and the 3′ splice site of a nuclear pre-mRNA intron

precursor of mRNA (pre-mRNA) primary transcript of a protein-encoding gene

small nuclear RNA (snRNA) five snRNAs are involved in splicing the majority of nuclear pre-mRNAs U1, U2, U4, U5 and U6

small nuclear ribonucleoprotein particle (snRNP) discrete complex of an individual snRNA and a specific set of (snRNP) proteins

spliceosome large RNA-protein complex that catalyses nuclear pre-mRNA splicing. Formed by the assembly of five snRNAs and approximately 100 proteins on the pre-mRNA

SR protein member of a family of serine–arginine (SR)-rich proteins involved in nuclear pre-mRNA splicing that contain a region rich in arginine–serine dipeptides, and often one or more RNA-binding domains

U2AF U2 snRNP auxiliary factor, member of the SR protein family

CHLOROPLASTS AND MITOCHONDRIA

Nicholas W. Gillham
DCMB Group, Duke University, Durham, USA

1. Introduction[1]

Chloroplasts and mitochondria complement one another bioenergetically. Chloroplasts derive energy from light that is employed for splitting water and the production of molecular oxygen. The electrons produced from the splitting of water are used via the photosynthetic electron transport chain to drive photosynthetic phosphorylation. Ultimately, molecular CO_2 is reduced by the protons and electrons derived from water and is converted into carbohydrates by the soluble enzymes of the chloroplast stroma. The mitochondrion, in contrast, catalyses the aerobic oxidation of reduced carbon compounds via soluble enzymes of the tricarboxylic acid cycle found in its matrix. The electrons produced by the oxidation of reduced carbon compounds flow via the respiratory electron transport chain and drive oxidative phosphorylation. The electrons and protons derived from the oxidation of reduced carbon compounds convert molecular oxygen to water, and CO_2 is released as an oxidation product of the tricarboxylic acid cycle. In summary, the chloroplast reduces CO_2 and splits water with the release of O_2, while the mitochondrion oxidizes reduced carbon compounds with the formation of CO_2 and water. However, chloroplasts and mitochondria are not simply energy-generating and utilizing systems. A vast array of other metabolic processes goes on within their confines as well, which are just as much key to the health and well-being of the cell as electron transport and energy generation.

2. Endosymbiotic origin of chloroplasts and mitochondria

Until the early 1980s, there was a long-standing controversy as to whether chloroplasts and mitochondria evolved from free-living eubacteria (the endosymbiont hypothesis) or were derived by compartmentalization of genetic material within the cell. This controversy has now been settled decisively in favour of the endosymbiont hypothesis. Molecular phylogenies of conserved sequences of ribosomal RNA molecules from these organelles indicate that the chloroplast derives from a cyanobacterial endosymbiont, while the mitochondrion originated from a member of the α subdivision of the purple bacteria (Proteobacteria). Whether each organelle resulted from a single endosymbiotic event (a monophyletic origin) or whether they arose more than once via prokaryote–eukaryote associations in different eukaryotic lineages (a polyphyletic origin) is a question that has not been settled.

Land plant chloroplasts are surrounded by a double membrane envelope, but the number of bounding membranes increases to three in the dinoflagellate and euglenophyte algae and to four in the cryptophyta, chrysophyta, prymnesiophyta and phaeophyta. Such "complex" chloroplasts are thought to derive via successive endosymbiotic events.

Direct evidence for secondary endosymbiosis comes from the cryptophyte algae which possess a nucleomorph, the name used for the presumably vestigial nucleus of the first host, together with ribosomes and rRNA whose phylogeny is distinct from both plastid and cryptophyte cytoplasmic rRNAs.

3. Structure and function

(i) Chloroplasts

The chloroplast envelope encloses a soluble phase (stroma) and the photosynthetic membranes (thylakoids) which are stacked into grana connected by stroma lamellae (**Figure 1**). The envelope membranes have at least three major functions:

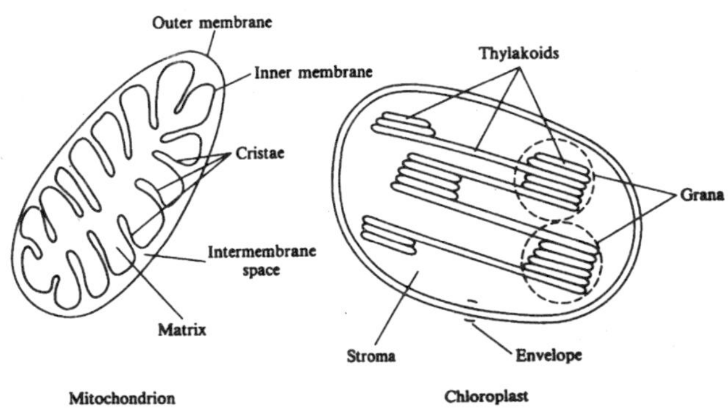

Figure 1 Diagram comparing a typical mitochondrion and a land plant chloroplast. The mitochondrion is surrounded by an outer membrane. Separating the outer and inner membranes is the intermembrane space containing soluble enzymes. Localized in the inner membrane, which is infolded into cristae, is the respiratory electron transport chain. Within the inner membrane is the mitochondrial matrix containing many soluble enzymes. The chloroplast is surrounded by a double membrane envelope. Within the envelope is the fluid phase or stroma, which, like the mitochondrial matrix, contains many soluble enzymes. The thylakoid membranes are the sites of photosynthetic electron transport and are not connected to the envelope. Thylakoids are differentiated into areas where adjacent membranes become tightly appressed into stacks called grana, which are connected by unstacked stromal thylakoids. (From *Organelle Genes and Genomes* by Nicholas W. Gillham, © 1994 by Oxford University Press, Inc. Used by permission of Oxford University Press, Inc.)

[1] With a few 1998 exceptions, only reviews or original papers of particular importance published from 1994–1997 and not cited in the book *Organelle Genes and Genomes* by Gillham (1994) are referenced in this article.

1. the inner envelope membrane contains specific translocators and thereby regulates metabolite transport between the cytoplasm and the plastid stroma;
2. the envelope is also involved in the biosynthesis of specific plastid components including glycerolipids, pigments and prenylquinones; and
3. the envelope membranes also function in transport of plastid proteins encoded in the nucleus.

Most envelope proteins are coded in the nuclear genome, but a few have been reported to be plastid gene products.

The stroma contains the chloroplast protein-synthesizing system and multiple copies of the chloroplast genome. The stroma is home to a remarkable array of biosynthetic pathways including the soluble enzymes required for CO_2 fixation and the enzymes involved in fatty acid synthesis and synthesis of aromatic amino acids. Chlorophyll biosynthesis from 5-aminolevulinate also takes place in the chloroplast (von Wettstein *et al.*, 1995). Certain enzymes and enzyme pathways include nuclear-encoded chloroplast and cytoplasmic isozymes (e.g. aromatic amino acid biosynthesis). In land plants, virtually all stromal polypeptides are nuclear gene products, but the large subunit of the CO_2-fixing enzyme ribulose-1,5-bisphosphate carboxylase oxygenase (RUBISCO-LSU) is a notable exception, being encoded by the chloroplast genome. However, a growing list of other stromal polypeptides is proving to be chloroplast encoded in different algal groups. The rRNAs and about a third of the chloroplast ribosomal proteins are encoded in the chloroplast genome, with the rest of the ribosomal proteins being nuclear gene products. Similarly, chloroplast transfer RNAs are chloroplast gene products, but the aminoacyl tRNA synthetases which charge them with their cognate amino acids are specified by nuclear genes.

The three complexes of the photosynthetic electron transport chain (**Figure 2**), photosystems I (PSI) and II (PSII), and the cytochrome b_6/f complex, are embedded in the thylakoid membrane within which is a soluble inner space referred to as the thylakoid lumen. Each complex is a mosaic of chloroplast- and nuclear-encoded components (Pakrasi, 1995; Nicholson *et al.*, 1996; Seidler, 1996). PSII initiates photosynthetic electron transport, with the absorption of light energy by chlorophyll *a/b*-protein complexes that serve as light-harvesting antennae. The light energy absorbed by the antenna chlorophylls is transferred to a pair of P_{680} reaction centre chlorophyll *a* molecules. An excited P_{680} molecule loses an electron to pheophytin; this electron then reduces the primary quinone acceptor, Q_A, oxidizing pheophytin, followed by reduction of the secondary electron acceptor, Q_B, and reoxidation of Q_A. The electron then proceeds to plastoquinone and the cytochrome b_6/f complex. Electron transport on the oxidizing side of PSII involves the donation of an electron to P_{680}^+ from Z, a tyrosine residue (Tyr_{161}) on the D1 reaction centre protein, and electron donation from the ligand Mn to Z. The stepwise accumulation of four positive charges on the oxidizing side of PSII is sufficient for the oxidation of two H_2O molecules and the release of four electrons, four protons and molecular O_2. Three hydrophilic extrinsic proteins, present in equimolar amounts, form the O_2 evolving complex of PSII. ATP formation (non-cyclic

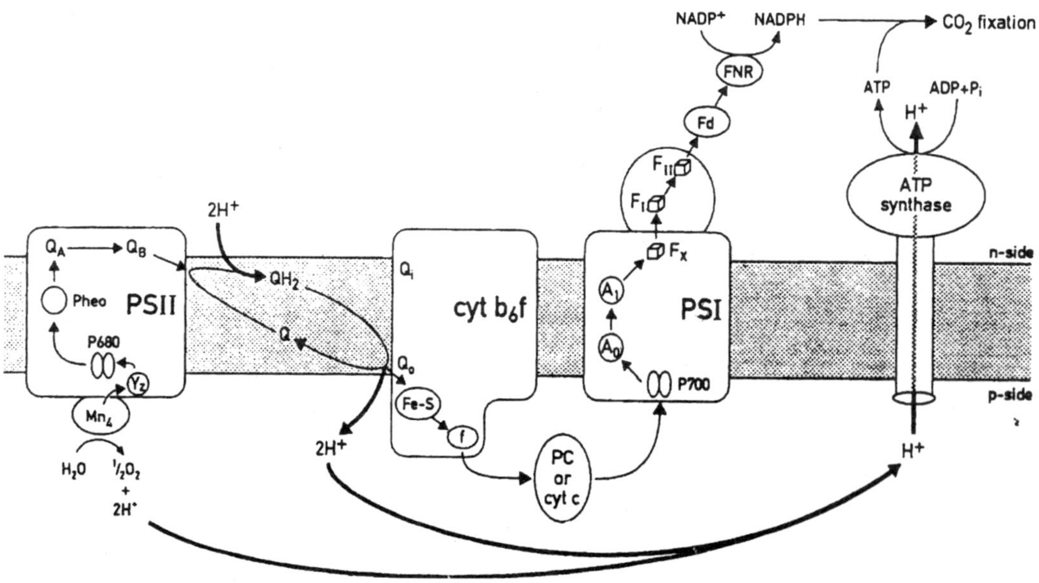

Figure 2 The classic Z scheme for photosynthetic electron transport. The cofactors involved in electron transport are shown: Mn_4, the Mn-cluster involved in O_2 evolution; Y_z, tyrosine serving as immediate donor to P_{680}; P_{680}, the primary donor of PSII (chlorophyll dimer); Pheo, primary electron acceptor of PSII (pheophytin); Q_A, secondary electron acceptor of PSII (quinone); Q_B, terminal electron acceptor of PSII (quinone); FeS, the Rieske iron-sulphur centre of Cyt *b6f*, f, Cyt *f*; PL, plastocyanin; P_{700}, primary electron donor of PSI (chlorophyll dimer); A_0, primary electron acceptor of PSI (chlorophyll); A_1, secondary electron acceptor of PSI (phylloquinone); F_x, tertiary electron acceptor of PSI (iron-sulphur centre); F_I and F_{II}, terminal electron acceptors of PSI (iron-sulphur centres) Fd, ferredoxin; FNR, ferredoxin: $NADP^+$ oxidoreductase. From Redding & Peltier (1998), with kind permission from Kluwer Academic Publishers.

photophosphorylation) is catalysed by the CF_1 ATP synthase. The PSII reaction centre includes the D1, D2 and cytochrome b_{559} proteins plus two chlorophyll *a*-binding proteins, CP47 and CP43. All of these proteins are encoded by chloroplast genes. Nuclear genes specify the three proteins of the oxygen-evolving complex, the chlorophyll *a/b* (CAB) proteins of the antenna and several other PSII components. The cytochrome b_6/f complex receives electrons from plastoquinone, the mobile carrier which accepts electrons from PSII. Three of the four proteins of this complex, including cytochromes b_6 and *f*, are encoded by chloroplast genes. Plastocyanin, which transfers electrons from the cytochrome b_6/f complex to photosystem I, is encoded by a nuclear gene. PSI catalyses the coupled photo-oxidation of plastocyanin and the reduction of ferredoxin. The reaction centre contains two proteins that bind a second special form of chlorophyll *a*, P_{700}. These proteins, plus nine others, form the PSI core complex. The PSI reaction centre proteins plus four other PSI core proteins are known to be chloroplast gene products, while two others derive from nuclear genes. Other peripheral proteins associated with PSI are nuclear gene products, as are the CAB proteins of the antenna complex. Electrons are transferred from PSI via ferredoxin: $NADP^+$ oxidoreductase to $NADP^+$, a two-electron carrier, and NADPH serves as the substrate for CO_2 reduction. Electrons may also flow from PSI back into the electron transport chain and catalyse ATP formation (cyclic photophosphorylation).

(ii) Mitochondria

Mitochondria are bounded by a phospholipid-rich outer membrane that also possesses many enzymatic activities, including those involved in phospholipid synthesis (**Figure 1**). The outer membrane contains the receptor complexes required for protein import which, in yeast, are made up of integral membrane proteins divided into two subcomplexes (Lithgow *et al.*, 1995). Between the outer and inner membranes is the intermembrane space, to which very few enzyme activities have been attributed unambiguously. The mitochondrial inner membrane is a highly specialized structure containing the respiratory electron transport chain components and the F_1 ATP synthase that couples oxidative phosphorylation with ATP formation. The inner membrane is highly convoluted into folds named cristae.

The four complexes involved in respiratory electron transport (**Figure 3**) are a mosaic of gene products encoded in the nucleus and the mitochondrion. NADH-ubiquinone reductase (complex I) catalyses the oxidation of NADH (nicotinamide adenine dinucleotide) by ubiquinone (coenzyme Q). This complex is the entry point for electrons travelling from NADH into the mitochondrial electron transport chain and is linked to the first of the three sites coupled with oxidative phosphorylation. Complex I contains upward of 40 polypeptides in beef heart and at least 30 in the fungus Neurospora. It is, however, much simpler in the yeast *Saccharomyces cerevisiae*, which also lacks the first phosphorylation site in electron transfer normally associated with complex I. Seven subunits of complex I are encoded in most mitochondria, but the number varies between zero and eight depending on taxonomic group. Succinate-ubiquinone oxidoreductase (complex II) catalyses the reduction of ubiquinone by succinate. This complex consists of four subunits, none of which was thought to be encoded in the mitochondrial genome. Recently, however, three of the subunits were found to be mitochondrially encoded in two phylogenetically distant eukaryotes, *Porphyra purpurea* (a red alga) and *Reclinomonas americana* (a heterotrophic zooflagellate) (Burger *et al.*, 1996). Ubiquinol-cytochrome c reductase (complex III) catalyses the oxidation of reduced coenzyme Q and possesses the second of three sites for oxidative phosphorylation. The complex contains 8–11 subunits, depending on species, of which the five largest are highly conserved. They include cytochromes *b* and c_1 plus two core proteins and a Rieske iron sulfur protein. So far, without exception, cytochrome *b* has proved to be a mitochondrial gene product, with all of the remaining subunits of the complex normally being encoded by nuclear genes. Cytochrome *c* mediates electron transfer from complex III to complex IV; the isozymes of this water-soluble protein are encoded by nuclear genes. Cytochrome *c* oxidase (complex IV) catalyses the oxidation of reduced cytochrome *c* by molecular O_2. The redox centre of both prokaryotic and eukaryotic aa_3-type cytochrome *c* oxidases consists of two haem moieties and two copper atoms. Bacterial aa_3-type cytochrome *c* oxidases possess three subunits that are homologous with the generally mitochondrially encoded subunits I, II and III. Six nuclear-encoded subunits of complex IV are common to yeast and mammals and four additional nuclear-encoded subunits are associated with mammalian complex IV.

The mitochondrial inner membrane surrounds the matrix

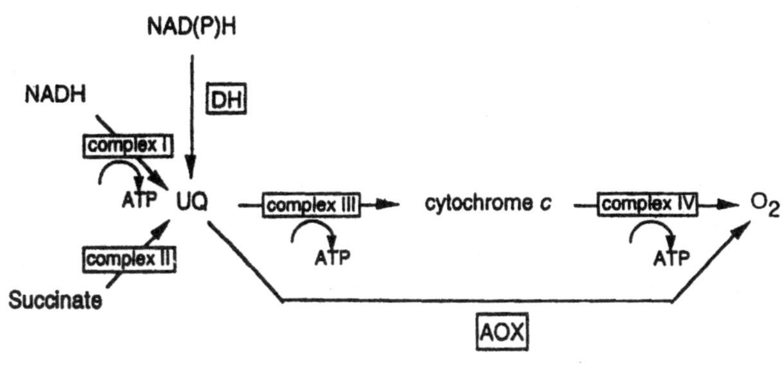

Figure 3 The electron transfer chain of the green alga Chlamydomonas and higher plants. Complex I: rotenone-sensitive NADH: ubiquinone oxidoreductase; complex II: succinate: ubiquinone oxidoreductase; complex III: ubiquinone: cytochrome *c* oxidorreductase; complex IV: cytochrome *c* oxidase; AOX, alternative oxidase (found in plants, but not most other organisms); DH, rotenone-insensitive NAD(P)H dehydogenase. The sites of proton translocation leading to ATP synthesis are also indicated. From Remacle & Matagne (1998).

where most of the soluble enzymes of the mitochondrion are found; among the most important of these are the enzymes of the tricarboxylic acid cycle and fatty acid oxidation, and the pyruvate dehydrogenase complex. No matrix protein has proven to be a mitochondrial gene product in any organism. The matrix also houses the mitochondrial protein-synthesizing system and multiple copies of the mitochondrial genome. The mitochondrial rRNAs are always encoded in the mitochondrial genome, but, depending on taxonomic group (see below), the mitochondrial genome may encode anywhere from a complete set of tRNAs (e.g. most animal groups) to none at all (trypanosomes). Similarly, while animal mitochondrial genomes encode no mitochondrial ribosomal proteins, the yeast mitochondrial genome encodes a single ribosomal protein, angiosperm mitochondrial genomes specify six ribosomal protein genes and the liverwort mitochondrial genome encodes 16 ribosomal protein genes.

(iii) ATP synthases

The chloroplast and mitochondrial ATP synthases, localized on the thylakoid membranes and cristae, respectively, are composed of both organelle- and nuclear-encoded proteins. The coupling factors of both the chloroplast (CF_1) and mitochondrial (F_1) ATP synthases are complexes of five subunits. Three subunits of CF_1 are normally encoded in the chloroplast. For mitochondrial F_1, the α subunit has been identified as a mitochondrial gene product in a few organisms. These soluble complexes are bound to the integral proteins of the CF_o and F_o complexes that contain the transmembrane channels through which proton transport is coupled to ATP synthesis. Three subunits of the CF_o complex are normally chloroplast encoded, while one is a nuclear gene product. In the case of F_o, one subunit usually is a mitochondrial gene product, the second a nuclear gene product, while the third may be encoded in either genome depending on taxonomic group.

(iv) Import

Most nuclear-encoded proteins destined for chloroplasts and mitochondria have N-terminal cleavable presequences, usually 20–80 amino acids in length, that are required for their import (Schnell, 1995; Baker *et al.*, 1996; Cline & Henry, 1996; Schatz, 1996; Schatz & Dobberstein, 1996). Although there is much evidence for post-translational import of these proteins, cotranslational import plays a role as well. Presequences have two roles: the first is to route the protein to the correct organelle and the second is to sort the protein to the proper compartment in the organelle. The imported proteins are directed to the proper compartment within the organelle via other molecular routing or sorting signals specified within by the amino acid sequence of the polypeptide. A multisubunit complex is required to transport precursor polypeptides across the two mitochondrial membranes and this involves more than 20 different proteins (Kubrich *et al.*, 1995; Lithgow *et al.*, 1995). The chloroplast import machinery is also elaborate (Schnell, 1995; Cline & Henry, 1996). Molecular chaperones in the cytoplasm are involved in aiding precursors to assume a transport-competent form, while intraorganellar chaperones are required for the folding and assembly of precursors (Hendrick & Hartl, 1995). Once inside the organelle, presequences are processed by specific proteases. While precursor import into chloroplasts and mitochondria is an ATP-dependent process, precursor import into mitochondria requires, in addition, a membrane potential.

4. Organelle genome structure and content

Mitochondrial genomes are naked DNA duplexes, usually circular, but occasionally linear as in certain groups of Cnidaria (e.g. Hydrazoa), the green alga *Chlamydomonas reinhardtii* and the yeast Hansenula. They vary *c.*400-fold in size (see Table 3.1, Gillham, 1994) from the small mitochondrial genomes of animals (*c.*14–18 kilobases[kb]) to the giant mitochondrial genomes of angiosperms (*c.*208–2400 kb). This size variation can be accounted for by four factors: repeated sequences, gene and intron content, incorporation of foreign DNA sequences and length mutations. Apart from a short region around the origin of replication, which can vary in size due to repeated sequences, animal mitochondrial genomes are similar in gene content. Their genes are tightly packed, lack introns and are separated by tRNA genes. A complete set of tRNAs and the rRNA molecules of the large and small mitochondrial ribosomal subunits are encoded in animal mitochondrial genomes (**Figure 4**). Many protein-coding genes end with incomplete termination codons, with T or TA following the last sense codon. The termination codon TAA is completed by polyadenylation of the mRNA. Only in the sea anemone *Metridium senile* (a cnidarian), found at the base of the metazoan phylogeny, is the mitochondrial genome radically different. This DNA molecule encodes only two tRNAs and the genes specifying cytochrome oxidase subunit I (*cox1*) and NADH dehydrogenase subunit 5 (*ndh5*). Both protein-encoding genes contain group I introns and the *ndh5* intron includes the genes for subunits 1 and 3 of NADH dehydrogenase. So, unlike other animals, mitochondrial gene expression in the sea anemone requires tRNA import and intron splicing (see article on "Introns").

The circular mitochondrial genomes of angiosperms are the largest known, with a size range of 208–2400 kb (Wolstenholme & Fauron, 1995). However, this increase in size is not reflected in a proportionate increase in gene content. The genes so far discovered encode the rRNAs, including a 5S species found only in plant mitochondrial ribosomes and in the mitochondrial ribosomes of several protists and the red alga *Chondrus crispus* (Gray *et al.*, 1998), the three largest subunits of cytochrome oxidase, cytochrome *b*, several ATPase and NADH subunits, a few mitochondrial ribosomal proteins and a number of tRNAs. The tRNAs encoded by the mitochondrial genomes of flowering plants, some of which are derived from the chloroplast, are insufficient to support mitochondrial protein synthesis so the difference is made up with tRNAs specified by nuclear genes and imported (Dietrich *et al.*, 1996). Flowering plant mitochondrial genomes also recombine to yield: (1) isomers following intramolecular recombination between inverted repeats and (2) subgenomic circles produced by recombination between direct repeats.

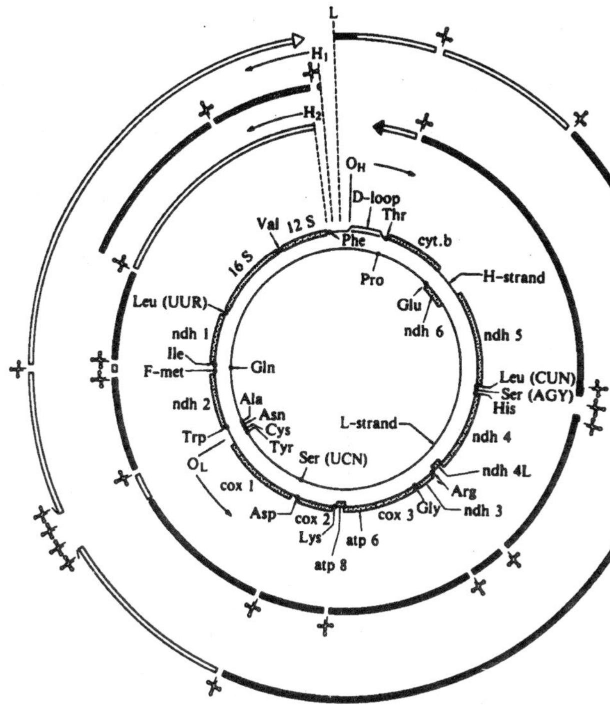

Figure 4 Genetic and transcription maps of the human mitochondrial genome. The two inner circles show the positions of the two rRNA genes (12S and 16S), 14 tRNAs (black circles) and 12 reading frames transcribed from the heavy (H)-strand, and the positions of eight tRNA genes a reading frame transcribed from the light (L)-strand. In the middle portion of the diagram, curved black bars represent the identified functional RNA species other than tRNAs resulting from processing of the two polycistronic primary transcripts of the H-strand starting at H_1 (rDNA transcription unit) and H_2 (total H-strand transcription unit). Dark bars in the outer circle represent the identified RNA species resulting from processing of the primary transcript of the L-strand. The white bars represent unstable presumably non-functional byproducts. Gene symbols *cox1*, *cox2* and *cox3* encode subunits I, II and III of cytochrome oxidase, respectively; *cytb* encodes apocytochrome *b*; *ndh1*, *ndh2*, *ndh3*, *ndh4*, *ndh4L*, *ndh5* and *ndh6* encode subunits of the NADH dehydrogenase; *atp6* and *atp8* encode subunits of the F_0 portion of the ATP synthetase; O_H, O_L are the origins of H- and L-strand synthesis, respectively; tRNA genes are Ala (alanine), Arg (arginine), Asn (asparagine), Asp (aspartic acid), Cys (cysteine), F-met (formylmethionine), Gln (glutamine), Glu (glutamic acid), Gly (glycine), His (histidine), Ile (isoleucine), Leu (leucine), Lys (lysine), Phe (phenylalanine), Pro (proline), Ser (serine), Thr (threonine), Trp (tryptophan), Tyr (tyrosine) and Val (valine). (Modified after Attardi & Schatz, 1988 from Gillham, 1994, with permission from Oxford University Press.)

Cytoplasmic male sterility in flowering plants results from the formation of novel, chimeric mitochondrial genes through recombination. Thus, in the *cms-T* cytoplasmic male sterile strain of maize, a new mitochondrial gene, *T-urf13*, is present which encodes a unique 13 kilodalton polypeptide that is responsible not only for male sterility, but also for increased sensitivity to Southern corn leaf blight caused by the fungus *Bipolaris maydis* race T (Rhoads *et al.*, 1995). *T-urf-13* has a chimeric structure and consists of 5′ flanking sequences with significant similarity to the 5′ flanking region of the *atp6* gene. The coding region of the gene consists of 88 codons with homology to an untranscribed 3′ flanking region of the 26S rRNA gene (*rrn26*) and 18 codons with homology to the coding region of *rrn26*.

Other plant mitochondrial genomes are quite different in gene content and structure from flowering plants. At one extreme is the completely sequenced 187 kb circular mitochondrial genome of liverwort which, with 94 putative genes, is genetically richer than almost any other known mitochondrial genome. At the other extreme is the minimal 15.8 kb, linear mitochondrial genome of the green alga *Chlamydomonas reinhardtii*, also completely sequenced (Boynton & Gillham, 1996), which encodes only the largest subunit of cytochrome oxidase, cytochrome *b*, five subunits of the NADH dehydrogenase complex, three tRNAs and the mitochondrial rRNAs. The genes encoding the rRNAs in this alga are segmented and the order of individual segments for the two subunits is scrambled. In contrast to Chlamydomonas, the sequenced mitochondrial genomes of the chlorophyte algae *Nephroselmis olivacea* (45 kb) and *Prototheca wickerhamii* (55 kb) are much larger and similar in gene content to liverwort (Gray *et al.*, 1998).

Fungal mitochondrial genomes are usually circular and vary in size among the ascomycetes alone from 17.3 kb (*Schizosaccharomyces pombe*) to 115 kb (*Cochliobolus heterostrophus*) (Wolstenholme & Fauron, 1995). It is not only size, however, but also gene organization and structure that vary greatly. Respiratory gene content is similar to that of other mitochondrial genomes, except that yeasts like *Saccharomyces cerevisiae* and *Schizosaccharomyces pombe* lack the NADH dehydrogenase complex and, therefore, the mitochondrial genes encoding certain components of this complex. Most fungal mitochondrial genomes encode no ribosomal proteins or only single unique proteins, as in the cases of *S. cerevisiae* and *Neurospora crassa*. The mitochondrial genome of the early diverging fungus *Allomyces macrogynus* has also been sequenced completely and its gene content is similar to that of more highly evolved fungi (Gray *et al.*, 1998).

Twenty-three protistan mitochondrial genomes have been completely sequenced and they show considerable variation in gene content (Gray *et al.*, 1998). For example, the 42 kb mitochondrial genome of *Acanthamoeba castellanii* possesses an array of protein-encoding genes very similar to that of liverwort and Prototheca. They specify several novel respiratory genes and a virtually identical array of ribosomal protein genes including a contiguous set organized like the *str S10 spc* and α ribosomal protein operons of *Escherichia coli*, with the same genes absent in each case and, presumably, transferred to the nucleus. The

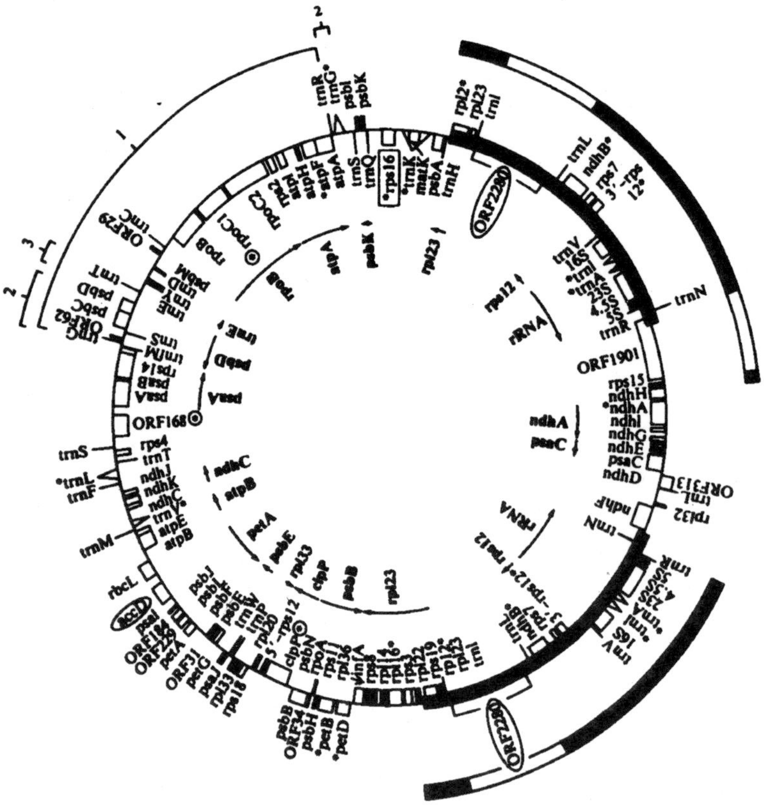

Figure 5 Map of the 156 kb chloroplast genome of tobacco (*Nicotiana tabacum*). Arrows on the inner circle indicate sets of genes thought to constitute operons. In general, the operons are named according to the first transcribed gene. The *rbcL* operon consists of a single gene in higher plants and chlorophytes, but two genes (*rbcL* and *rbcS*) in other algae. Circled gene names indicate genes present in tobacco and liverwort (Marchantia), but absent in rice. The solid box around *rps16* calls attention to the fact that this gene is also present in rice, but absent from liverwort. Asterisks denote genes that have the same intron(s) in all three sequenced genomes; circled asterisks denote genes that are split in only two of the three genomes. Thickened parts of the circle represent the 25.3 kb inverted repeat of tobacco. Thick lines outside the tobacco inverted repeat represent the extent of the inverted repeat in rice (20.8 kb), with regions deleted from rice indicated as open boxes. Numbered brackets outside the circle indicate regions that have been moved by three overlapping inversions in rice relative to tobacco. Gene nomenclature: rRNA genes are indicated by 16S, 23S, 4.5S and 5S; tRNA genes by *trn* followed by the one-letter amino acid code; RNA polymerase genes by *rpo*, followed by a subunit-specific letter; genes for the 50S and 30S ribosomal proteins by *rpl* and *rps*, respectively, followed by a number corresponding to the cognate *E. coli* protein; genes for components of the thylakoid membrane complexes ATP synthase, photosystem I, photosystem II and cytochrome b_6/f complex by *atp*, *psa*, *psb* and *pet*, respectively, followed by a subunit-specific letter; genes for NADH dehydrogenase subunits by *ndh*, followed by a letter (the corresponding mitochondrial genes are followed by a number, e.g. *ndhA* equals *ndh1*); gene for the large subunit of RUBISCO, by *rbcL*; and open reading frames (ORFs) conserved in at least two of the three sequenced genomes compared here are indicated by ORF, followed by their lengths in tobacco in codons. (Reproduced from Palmer (1991), with permission from Academic Press.)

linear 40 kb mitochondrial genome of Paramecium contains most of the usual respiratory chain genes plus several ribosomal protein genes. The early diverging freshwater protozoan *Reclinomonas americana* contains the greatest number of genes of any mitochondrial genome so far sequenced, resembling a miniature eubacterium in this regard (Lang *et al.*, 1997). The most unusual mitochondrial genomes reside in the single giant mitochondrion of trypanosomes, the kinetoplast (Shapiro & Englund, 1995). Kinetoplast DNA is comprised of networks of 5–10 000 small (0.5–2.8 kb), catenated minicircles interspersed with 20–50 larger (20–39 kb) maxicircles. The maxicircles are the true mitochondrial genomes, while the minicircles encode guide RNAs that are required for the process of RNA editing by which maxicircle mRNAs are rendered translatable (see below).

Although land plant chloroplast genomes vary from 50–220 kb in size (see Table 3.6, Gillham, 1994), most are fairly homogeneous ranging from 130–160 kb and containing two large inverted repeats (usually 10–30 kb) that divide them into large and small single copy regions (**Figure 5**). Certain legumes have lost the inverted repeat and only a remnant of this structure is found in black pine (Wakasugi *et al.*, 1994). The smallest of these chloroplast genomes belongs to colourless parasitic plants like Epifagus (71 kb) and the largest to plants like the geranium (*Pelargonium hortorum*, 220 kb) where almost half of the chloroplast genome has been duplicated as part of the inverted repeat. Completely sequenced land plant chloroplast genomes include four angiosperms (Epifagus, tobacco, maize and rice), a gymnosperm (black pine) and liverwort (Wakasugi *et al.*, 1994; Maier *et al.*, 1995). Land plant chloroplast genomes contain about 110–120 genes, the majority of which are located in the large unique sequence regions. The known products of most land plant chloroplast genes function either to promote chloroplast gene expression (four rRNAs, 30–31 tRNAs, 20 ribosomal proteins, four RNA polymerase subunits) or photosynthesis (28 thylakoid proteins), plus one soluble protein, RUBISCO-LSU. Eleven homologues of mitochondrial NADH dehydrogenase have been identified in angiosperm chloroplast DNA and liverwort, but they are missing or not intact in the black pine chloroplast genome (Wakasugi *et al.*, 1994).

Algal chloroplast genomes, although circular in structure and frequently containing inverted repeats, are much more variable in gene content. The plastid genomes of *Euglena gracilis*, the red alga *Porphyra purpurea* (Reith & Munholland, 1995), the phycobiliprotein-containing glaucocystophyte alga *Cyanophora paradoxa* (Stirewalt *et al.*, 1995) and the diatom *Odontella sinensis* (Kowallik *et al.*, 1995) have been sequenced completely. These algal chloroplast genomes are richer in genes (*c.*149 in Euglena to about 250 in Porphyra) than those of land plants. While most of these genes encode components of the photosynthetic apparatus and chloroplast translational apparatus like their counterparts in land plants, reading frames for proteins involved in biosynthesis, secretion and chaperone function in prokaryotes have also been detected. Parasitic protozoa of the phylum Apicomplexa, which includes the malaria parasite Plasmodium as well as other genera with species that cause animal or human disease (e.g. toxoplasma), are known to possess two sets of extrachromosomal DNA (McFadden *et al.*, 1996). One of these is a *c.*6–7 kb linear mitochondrial genome encoding only cytochrome *b*, two subunits of cytochrome oxidase and the mitochondrial rRNA genes (Gray *et al.*, 1998). The second DNA molecule appears to be a 35 kb residual plastid genome which localizes to a small ovoid organelle whose function is unknown.

Chloroplast genes are generally located in close proximity, with spacers only a few hundred base pairs in length. Many chloroplast genes, particularly in land plants, are clustered into operons (Sugita & Sugiura, 1996) and are cotranscribed into polycistronic mRNAs that are subsequently processed. One such operon, found in all chloroplast genomes studied to date, contains a subset of the genes encoding ribosomal proteins present in the *str*, *S10*, *spc* and α operons of *E. coli* (**Figure 6**). Chloroplast genes frequently contain group I or group II introns whose mRNAs undergo *cis*-splicing. Certain chloroplast genes, such as those encoding ribosomal protein S12 in angiosperms and the psaA protein in *Chlamydomonas reinhardtii*, are fragmented and their mRNAs are *trans*-spliced.

5. Mechanisms of organelle gene expression

The most thoroughly characterized RNA polymerases and promoters required for chloroplast and mitochondrial DNA (mtDNA) transcription differ markedly. The yeast mitochondrial RNA polymerase, encoded by a nuclear gene, is a single subunit enzyme (145–150 kDa) similar to the RNA polymerase of phages T3 and T7. In contrast, the predominant chloroplast RNA polymerase is much more complex and the different subunits have strong amino acid sequence similarity to the comparable enzyme from *E. coli*. Furthermore, the major subunits of this enzyme are encoded in the chloroplast genome. Interestingly, the mitochondrial genome of the ancestral protist *Reclinomonas americana* also encodes the major subunits of a eubacterial-type RNA polymerase (Lang *et al.*, 1997). To add to this complexity, there is evidence that a second, nuclear-encoded RNA polymerase is involved in chloroplast DNA transcription. A candidate single-subunit enzyme of the T7 type has been reported from spinach (Iratni *et al.*, 1994). Furthermore, deletion of a chloroplast gene encoding one of the subunits of the *E. coli*-like enzyme in tobacco reveals a low level of transcription of several chloroplast genes which recognize promoters distinct from those targeted by the *E. coli*-like enzyme (Hajdukiewicz *et al.*, 1997). Specificity factors (e.g. *MTF1* in yeast mitochondria and *E. coli*-like σ factors in

Figure 6 Conservation of ribosomal protein gene clusters in chloroplast genomes, showing retention of some (but not all) genes of the closely adjacent *str*, S10, *spc* and α operons of *E. coli*, in *Cyanophora paradoxa*, *Porphyra purpurea*, *Euglena gracilis*, *Chlamydomonas reinhardtii*, *Marchantia polymorpha*, *Epifagus virginiana* and *Nicotiana tabacum*. Shaded boxes indicate genes that have been lost from the corresponding operon, but have been identified elsewhere in a given plastid genome. For example, *rps7* and *tufA* are present in the *C. reinhardtii* chloroplast genome, but have become separated from *rps12*. In *C. paradoxa* and *P. purpurea*, the operon begins with the *rpl3* gene (▲) and ends with the *rps12*, *rps7*, *tufA* and *rps10* genes. The *rps12* gene in land plants is split and the 3′ portion of the gene remains proximal to *rps7*. (Reprinted from Harris *et al.*, 1994 with permission from American Society for Microbiology.)

chloroplasts) are also required for proper transcriptional initiation in these organelles.

In mammals, transcription of each strand of the mitochondrial genome is initiated from a single major promoter in the D-loop region where DNA replication also starts (**Figure 4**). Following initiation of heavy or H-strand transcription, elongation begins, but transcripts frequently terminate at a tridecamer sequence distal to the rRNA genes. When transcription termination fails to occur, a second major H-strand transcript is synthesized which is complementary to the entire H-strand. The light or L-strand is also copied over its entire length. Genes are asymmetrically distributed between the strands, with the H-strand being far richer in coding sequences than the L-strand (**Figure 4**). The principal mechanism by which primary transcripts of protein-encoding genes are processed from the H-strand transcript involves excision of tRNA transcripts lying between these sequences, following which the termination codons (U or UA) of the protein-coding transcripts are completed by polyadenylation. In *S. cerevisiae*, all mitochondrial genes are transcribed from one strand, except for a short transcript of a tRNA gene synthesized from its complement, and most are transcribed as parts of *c*.20 different multicistronic units with transcription initiating within a nine nucleotide consensus sequence. Endonucleolytic processing at a conserved dodecamer sequence downstream of protein-coding genes has been implicated in mRNA maturation. In Neurospora, but not yeast, tRNAs seem to act as important transcript processing signals, as in animals.

Intron processing has been extensively studied in *Saccharomyces cerevisiae* mitochondria where the genes encoding cytochrome *b* and cytochrome oxidase subunit I may have several to many group I and II introns, depending on the strain (Belfort & Perlman, 1995). Splicing requires specific *trans*-acting factors (maturases), some of which are specified by nuclear genes, while others are intron encoded and are in reading frame with the upstream exon. One of the latter maturases, encoded by the fourth intron of the cytochrome *b* gene (bI4 maturase), is required not only for the splicing of this group I intron, but also for splicing the first intron of the cytochrome oxidase subunit I gene. The latter intron also contains a coding sequence in frame with the first exon, specifying a protein closely related to the maturase required for the splicing of the fourth intron of the cytochrome *b* gene. This maturase can be activated by mutation and will splice both introns just like the bI4 maturase. The large rRNA gene in many yeast strains also contains a mobile group I intron (see section 8: Mobile elements). In Neurospora, splicing of group I introns from mitochondrial mRNAs is carried out by a bifunctional, nuclear-encoded protein that also acts as the mitochondrial tyrosyl tRNA synthetase.

Transcription of land plant mitochondrial genomes is less well characterized. However, since gene clustering is not a feature of these genomes, most mRNAs would be expected to be monocistronic. While the predicted simple transcripts are found in the case of *Brassica campestris*, more complex transcription patterns are observed in maize, reflecting length variability in the 3′ and 5′ untranslated sequences. Overall, plant mitochondrial gene promoters exhibit less conservation than seen in yeast and there may be a number of separate initiation sites for a single gene.

In the case of chloroplast genomes, sequences with striking similarity to the −10 and −35 promoter sequences of prokaryotes recognized by eubacterial RNA polymerases can be identified upstream of many genes, but functional analysis has only been carried out in a few cases. Thus, in Chlamydomonas, the −35 sequence was shown to be essential *in vivo* in one case (16S rRNA), but not another (*atpB*). In addition, a 10 nucleotide consensus sequence has been identified which is recognized by the nuclear-encoded plastid RNA polymerase (Hajdukiewicz *et al.*, 1997). Many chloroplast mRNAs are synthesized as polycistronic transcripts that are later processed, but experiments with maize indicate that both processed and unprocessed transcripts are found on polysomes and can be translated just so long as intron removal has occurred. Putative stem–loop structures that act as RNA processing and stabilizing elements are found in the 3′ untranslated regions of plastid mRNAs and many tRNA and rRNA transcripts.

Translation in chloroplasts has numerous features in common with prokaryotes (Harris *et al.*, 1994). Many chloroplast mRNAs have Shine–Dalgarno sequences upstream of the AUG initiator codon and the 16S rRNA of the 70S chloroplast ribosome has an anti-Shine–Dalgarno sequence near the 3′ end of the molecule. However, the spacing of the Shine–Dalgarno sequences upstream of the initiator codon is much more variable in chloroplast mRNAs than in *E. coli*. Experiments with *Chlamydomonas reinhardtii* show that the Shine–Dalgarno sequence is not required for translation initiation in chloroplasts (Fargo *et al.*, 1998). Homologues of two of three factors involved in initiation in prokaryotes (IF-2, IF-3) have been characterized in Euglena, as have all three elongation factors (Harris *et al.*, 1994). A reading frame specifying IF-1 has been identified in the chloroplast genomes of several land plants, but release factors remain to be reported. All three termination codons are used in chloroplast mRNAs, with UAA being the most abundant (70% in land plant genes).

Chloroplast ribosomes possess many similarities to prokaryotic ribosomes, but there are also a few differences. As in prokaryotes, the small subunit includes a 16S rRNA molecule while the large subunit contains 23S and 5S rRNA. In flowering plants, as well as certain other land plants (e.g. Marchantia), the 3′ end of the 23S rRNA is split off as a 4.5S rRNA species, while in *Chlamydomonas reinhardtii* the 5′ end of the molecule is cleaved to yield two species of approximately 7S and 3S. In *C. eugametos*, there is a further cleavage in the 23S molecule to yield two large species. Of the 57 ribosomal proteins that constitute the *E. coli* ribosome, the chloroplast equivalents of 53 have been identified by sequencing chloroplast or nuclear genes from one or more organisms (Harris *et al.*, 1994). A few novel chloroplast ribosomal proteins have also been discovered. With a few exceptions, chloroplast genomes of land plants encode the same set of ribosomal proteins, while marked variations occur in certain algal groups. In most chloroplast genomes, the genes encoding many of these ribosomal proteins are arranged in a large cluster which resembles the combined

E. coli str, *S10*, *spc* and α operons, except that certain genes are missing and relocated to the nucleus (**Figure 6**). Chloroplast ribosomes are sensitive to the same antibiotics (e.g. chloramphenicol, erythromycin, spectinomycin, streptomycin) that affect eubacterial ribosomes.

The factors involved in translation initiation and elongation have been identified in bovine mitochondria (Schwartzbach *et al.*, 1996). A single initiation factor, IF-2_{mt}, and three elongation factors, EF-Tu_{mt}, EF-Ts_{mt} and EF-G_{mt}, have been characterized to date. The *MRF1* nuclear gene of yeast encodes the mitochondrial release factor mRF-1 and mutations in this gene cause defects in translation termination.

Mitochondrial ribosomes, proteins and rRNAs are much more variable in size and secondary structure conservation than their counterparts in chloroplasts and their organization is highly variable. For example, vertebrate mitochondrial rRNAs are small, have short variable regions, are separated by a single tRNA gene and are cotranscribed. In yeast and Neurospora, mitochondrial rRNAs are larger, the genes encoding them are well separated within the genome and they are transcribed separately. With the exception of land plants, several protists and the red alga *Chondrus crispus* (Gray *et al.*, 1998), mitochondrial ribosomes lack a 5S rRNA species. Mammalian mitochondrial ribosomes have many more proteins (85 *vs.* 57) than bacterial ribosomes, but their rRNAs are much smaller and their proteins are all nuclear gene products. Yeast mitochondrial ribosomes are similarly protein rich, containing *c.*80 proteins (Mason *et al.*, 1996). Unlike those in animals, the mitochondrial genomes of yeast and Neurospora each encode a single protein of the small subunit of the mitochondrial ribosome. This number increases in the mitochondrial genomes of land plants and certain protists and algae, and the proteins encoded by these genes have clear sequence similarity to prokaryotic ribosomal proteins. In several cases (see section 4: Organelle genome structure and content), clusters of genes encoding ribosomal proteins are found in which the genes are ordered in the same way as they are in the chloroplast genome (Lang *et al.*, 1997). Mitochondrial ribosomes are sensitive to certain antibacterial antibiotics, including chloramphenicol and erythromycin.

With the exception of the colourless, parasitic flowering plant Epifagus, chloroplast genomes specify sufficient tRNAs to carry out protein synthesis. This is also true of animal and fungal mitochondrial genomes (except cnidarians), but not the mitochondrial genomes of land plants, certain algae (Chlamydomonas) and protists (e.g. Acanthamoeba, Paramecium, Tetrahymena and Trypanosomes) where tRNA import must be invoked for organellar protein synthesis to take place (Gray *et al.*, 1998). Land plant mitochondrial tRNAs fall into three groups; one set must be imported from the cytosol, but the other two are encoded by the mitochondrial genome. The "native" set of tRNAs is unique to plant mitochondria, but the "chloroplast-like" tRNAs were apparently derived by gene transfer from the plastid. Genes encoding organelle tRNAs do not specify the 3′ CCA terminal sequence and this must be added post-transcriptionally. Chloroplast tRNAs are similar in structure and sequence to prokaryotic tRNAs, while mitochondrial tRNAs, particularly those of animals, are more variable. For example, in nematodes, the TΨC arm and variable loop are replaced by a single, simple loop of between 6 and 12 nucleotides. All chloroplast and mitochondrial aminoacyl tRNA synthetases are specified by nuclear genes, except in the case of the chloroplast genome of the red alga *Porphyra purpurea* where two open reading frames (ORFs) coding for putative aminoacyl tRNA synthetases have been reported (Reith & Munholland, 1995).

The three-letter universal code is employed in translating mRNAs in chloroplasts and plant mitochondria, but in animal and fungal mitochondria, with 22 and 25–27 tRNAs, respectively, a modified two-letter code is used. Except in land plants, UGA is not read as a termination codon in mitochondria, but specifies tryptophan instead. In most animal mitochondria, AUA codes for methionine. Two distinct modes of RNA editing complicate organelle gene expression further (Benne, 1996; Maier *et al.*, 1996; Simpson & Emeson, 1996; Sollner-Webb, 1996). Modification–substitution of nucleotides in mRNAs, usually C→U editing, occurs in flowering plant mitochondria and chloroplasts. Editing of mismatches in the first three positions at the 5′ ends of the acceptor stems of most mitochondrial tRNA transcripts occurs in Acanthamoeba (Burger *et al.*, 1996; Gray *et al.*, 1998). Post-transcriptional insertion and deletion of U nucleotides takes place in trypanosome mitochondria under the direction of guide RNAs encoded by the different classes of mtDNA minicircles. A C-insertional editing system has been discovered in the mitochondria of the slime mold Physarum that affects both mRNA and rRNA.

6. Regulation of expression of organelle genomes

Since organelle gene products participate with their counterparts encoded in the nucleus in constructing the electron transport membranes and protein synthetic apparatus of chloroplasts and mitochondria, coordination of gene expression between organelle and nucleus is of paramount importance. The synthesis and accumulation of many chloroplast-encoded mRNAs occur as a function of light and developmental stages; processing and stabilization of these mRNAs operate as post-transcriptional controls (Rochaix, 1996; Sugita & Sugiura, 1996; Goldschmidt-Clermont, 1998). In particular, the 3′ ends of chloroplast transcripts, whether monocistronic or parts of polycistronic units, contain inverted repeat sequences capable of forming stem–loop structures which play an important role in transcript stabilization and 3′ end formation. In human mitochondria, differential regulation of transcript synthesis and turnover play significant roles in mitochondrial gene expression. The high steady state level of rRNA transcripts in human mitochondria reflects not only an increased rate of synthesis, but also the existence of a specific termination mechanism for these transcripts. In yeast, with its *c.*20 mitochondrial transcription units, promoter strength seems to be an important controlling factor, but transcripts overall increase in cells derepressed with respect to mitochondrial function.

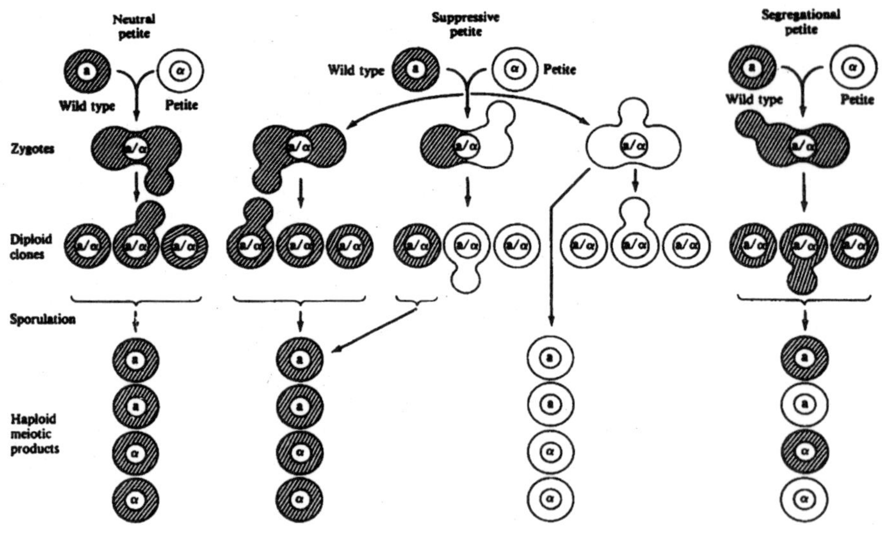

Figure 7 Comparison of inheritance patterns of neutral and suppressive mitochondrial (vegetative) petites of *S. cerevisiae* with nuclear-encoded segregational petites in crosses to wild-type. Neutral petites produce only wild-type diploid progeny in such crosses that segregate 4:0 wild-type:petite progeny following sporulation. Zygotes issuing from a cross between a suppressive petite and wild-type may yield wild-type progeny, petite progeny or a mixture, with the ratio depending on the level of suppressivity of the mutation. Petite vegetative diploids will not sporulate, but zygotes destined to yield petite progeny will do so, and the segregate 0:4 wild-type:petite progeny. Phenotypically wild-type progeny result from crosses between recessive segregational petites and wild-type. On sporulation of these diploids, the petite phenotype is found to segregate 2:2 wild-type:petite progeny in a typical Mendelian fashion. (From Gillham, 1994. © 1994 by Oxford University Press, Inc. Used by permission.)

Although transcriptional and post-transcriptional control mechanisms are important, translational control predominates in chloroplasts and mitochondria (Fox, 1996; Rochaix, 1996; Goldschmidt-Clermont, 1998). These control mechanisms have been particularly well studied in yeast and Chlamydomonas (Fox, 1996; Hauser *et al.*, 1998). These studies show that the stability and/or translation of chloroplast and mitochondrial transcripts are controlled in a gene-specific manner by nuclear-encoded proteins that interact with the 5′ untranslated regions of these organellar mRNAs. The expression of nuclear genes specifying polypeptides that complex with organelle-encoded proteins is largely under transcriptional control, but whether this is true of the *trans*-acting regulatory factors involved in translational regulation of organelle mRNAs is unknown. In yeast, oxygen signalling, via haem and carbon source and positive (*HAP*) and negative (*ROX*) transcriptional regulators, determines the expression of nuclear genes specifying proteins involved in respiratory chain function (e.g. cytochrome *c*). Similarly, exposure of dark-grown land plants to light signals chloroplast development via several different light reactions: red light-induced photoconversion of protochlorophyllide to chlorophyllide; conversion of phytochrome from its inactive form to its active form by red light; and blue/ultraviolet light receptors. This signal transduction pathway is being worked out in Arabidopsis using two classes of nuclear gene mutants whose phenotypes either: (1) partially resemble those of dark-grown wild-type seedlings when exposed to light or (2) develop in darkness as if exposed to light (Chory *et al.*, 1995). Sequences involved in light regulation have been identified upstream of several groups of nuclear genes involved in chloroplast biogenesis (e.g. the gene families encoding the small subunit of RUBISCO and the chlorophyll *a/b* binding proteins).

Changes in light intensity activate one or more molecular signalling systems in chloroplasts that recognize changes in the redox state of the photosynthetic electron transport chain (Allen *et al.*, 1995). By sensing the increased level of reduction of the plastoquinone pool, one signalling system in Dunaliella (a close relative of Chlamydomonas) appears to operate through a kinase–phosphatase system to down-regulate transcription of the nuclear genes encoding the chlorophyll *a/b* binding proteins when the cells are shifted from low to high light (Escoubas *et al.*, 1995), thus reducing the size of the light-harvesting antenna. In Chlamydomonas, a second system has been reported (Danon & Mayfield, 1994; Kim & Mayfield, 1997) that regulates expression of the D1 reaction centre protein encoded by the *psa* gene in response to light. Binding of the nuclear-encoded RB47 (47 kDa) protein to the 5′ untranslated region of the *psbA* mRNA upregulates translation of this rapidly turning over protein. Light stimulates ferredoxin reduction which in turn reduces RB47, thereby facilitating its binding to the *psbA* mRNA leader. ADP levels also influence binding via another protein RB60 (60 kDa). High levels of ADP (as attained *in vivo* in dark-grown plants) correlate with phosphorylation of RB60 via a serine/threonine phosphotransferase associated with the mRNA binding complex that transfers the β-phosphate of ADP to RB60. In the light, ADP levels and RB60 phosphorylation are reduced which promotes binding of the complex and accelerated D1 synthesis. Hence, chlorophyll antenna size is reduced and D1 synthesis increased in response to the light shift. Similarly, in yeast with dysfunctional mitochondria, a signalling system has been discovered that operates via the products of two nuclear genes that transcriptionally activate the nuclear gene encoding the peroxisomal glyoxalate cycle enzyme citrate synthetase 2 (Chelstowska & Butow, 1995). Hence, citrate

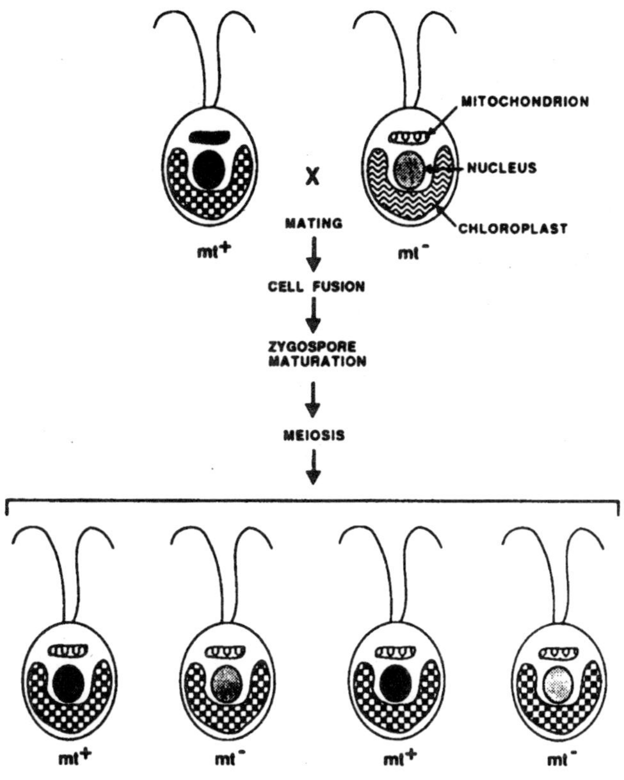

Figure 8 The three basic modes of gene transmission in sexual zygotes of *Chlamydomonas reinhardtii*. Chloroplast genes are normally transmitted by the mt^+ parent (>90–95% of the zygotes) to all four meiotic products, while the mt^- parent transmits the mitochondrial genome uniparentally (>99% of the zygotes). Nuclear genes segregate 2:2 reflecting the contributions of both parents. (From *Organelle Genes and Genomes* by Nicholas W. Gillham, © 1994 by Oxford University Press, Inc. Used by permission of Oxford University Press, Inc.)

synthetase 2 may be an important control point for cross-feeding from the glyoxalate cycle to mitochondria.

7. Genetics and gene manipulation

Chloroplasts and mitochondria and/or their genomes exhibit one of two transmission patterns: uniparental–maternal or biparental. Uniparental–maternal transmission is the most common (Birky, 1995). Organelles and organelle genomes transmitted biparentally subsequently segregate somatically. Thus, the heteroplasmic state, where two genotypically distinct organelle genomes are maintained in the same cell, is normally transitory. Yeast and Chlamydomonas, respectively, are the two most useful systems for studying the genetics of mitochondria and chloroplasts. Mitochondrial genes in yeast exhibit biparental transmission and undergo recombination when haploid cells of opposite mating type form diploids (**Figure 7**). Since yeast cells can be grown on a fermentable carbon source such as glucose, mitochondrial mutations can be isolated that block respiration, but which are lethal only when the cells are grown on a carbon source that must be respired, such as glycerol. Partial or complete deletions of the mitochondrial genome, called vegetative petite mutations, result from the loss of mitochondrial genes specifying components of the respiratory electron transport chain and/or mitochondrial protein synthesizing apparatus. Mitochondrial point mutations to antibiotic resistance affect the rRNA genes and certain components of the respiratory chain or cause loss of respiratory competence by blocking the function or synthesis of mitochondrially encoded components of the respiratory electron transport chain or ATP synthase.

Chloroplast and mitochondrial genes in *Chlamydomonas reinhardtii*, the most commonly used species of this genus, are transmitted uniparentally in most zygotes by opposite mating types with the mating type plus (mt^+) parent normally transmitting chloroplast genes, while the mating type minus (mt^-) parent transmits its mitochondrial genes (**Figure 8**). Chloroplast genes are rarely transmitted from the mt^- parent, but the frequency of transmission can be greatly increased if the mt^+ parent is pretreated with certain agents (e.g. ultraviolet light) prior to mating. Many chloroplast point mutations exist in the rRNA genes which confer resistance to antibiotics such as erythromycin, streptomycin and spectinomycin. Similarly, mutations to streptomycin resistance and dependence have been isolated, which cause alterations in the chloroplast-encoded gene specifying ribosomal protein S12. Since the alga can use acetate as well as CO_2 as a sole carbon source, mutations resulting in defects in chloroplast- or nuclear-encoded components of photosynthesis are viable when maintained on acetate. Chloroplast deletion mutations with a non-photosynthetic phenotype have also been characterized.

Both chloroplasts and mitochondria are polyploid genetic systems and intermolecular recombination is extensive within these organelles, but is detectable between genomes in different organelles in only a few systems such as yeast mitochondria and Chlamydomonas chloroplasts where organelle fusion occurs. Despite the high organelle genomic copy number in these two organisms, segregation of genetic markers is very rapid so there appear to be fewer segregating units than there are DNA molecules in these organelles. Random drift may be involved in this process as well as intracellular selection. The best population genetic model describing recombination of chloroplast genomes in Chlamydomonas and mitochondrial genomes in yeast is the phage analogy model which likens an organelle "cross" to a phage cross in terms of such parameters as input and output of genomes, upper limit of observable recombination (25% because of recombination between genetically identical genomes), multiple rounds of pairing and recombination, and random segregation. Genetic mapping experiments have been carried out both with yeast mitochondria and Chlamydomonas chloroplasts. In both cases, map distances can be determined only over very short segments of the organelle genome (yeast, *c.*1 kb; Chlamydomonas, *c.*12–14 kb). Limited genetic analysis has been performed on a variety of other organelle genetic systems.

Biolistic transformation of the chloroplast genomes of Chlamydomonas (Boynton & Gillham, 1994) and tobacco as well as the mitochondrial genome of yeast (Butow *et al.*, 1996) with DNA-coated microprojectiles has made gene manipulation possible with these organelle genomes. Introduced genes replace resident genes by homologous recombination and homoplasmic transformants are easily obtained using appropriate selective markers. Transformation methodology has revolutionized the study of organelle genome structure and function since techniques like site-directed *in vitro* mutagenesis and reporter genes can now be used routinely to study the function of different organelle genes.

In the case of human mitochondria, numerous mutations have now been characterized which result in specific diseases (e.g. Leber hereditary optic neuropathy, mitochondrial encephalomyopathy) and their telltale signature is uniparental transmission by the mother (Ozawa, 1995; Wallace *et al.*, 1995). Individuals inheriting these diseases are often relatively normal in early life with symptoms developing in childhood, adult life or old age, depending on the severity of the disease produced by the particular mitochondrial mutation. These unique characteristics of human mtDNA diseases may result because of the marked dependence of the target tissues (e.g. brain, heart and skeletal muscle) and organs on mitochondrial bioenergetics, together with the cumulative defect caused by the original mutation combined with the age-related accumulation of new mitochondrial mutations in postmitotic tissues. Techniques are now available for eliminating mtDNA entirely in tissue culture cell lines (King & Attardi, 1996a) and for introducing mitochondria containing mutated or wild-type mtDNA into these lines (King & Attardi, 1996b). This has simplified the study of the cellular effects of specific mitochondrial mutations. Mitochondrial mutations have also been characterized and their genetics studied in a number of other organisms, especially the fission yeast *Schizosaccharomyces pombe* and the filamentous fungi *Neurospora crassa* and *Aspergillus nidulans*.

8. Mobile elements

Chloroplasts and mitochondria may also contain mobile elements associated with their genomes. In yeast and Chlamydomonas, these include mobile group I introns that encode specific endonucleases required for their transposition (Belfort & Perlman, 1995; see article on "Introns"). For example, ω specifies an endonuclease I SceI that recognizes an 18 bp sequence in the large rRNA gene and catalyses a staggered 4 bp cut at this position into which the intron can insert. In addition, the yeast mitochondrial genome also possesses mobile G+C-rich clusters. Mobile group II introns encode reverse transcriptases (RTs) and insert site specifically into intronless alleles (Michel & Ferat, 1995; Grivell, 1996). In the case of the aI2 intron of yeast, intron homing occurs by reverse transcription at a double-strand break in the recipient DNA made by a site-specific endonuclease (Zimmerly *et al.*, 1995). The DNA endonuclease that initiates homing requires both the aI2 RT protein and aI2 RNA. Excision and amplification of a group II intron from the *cox1* gene in the mitochondrial genome of the filamentous fungus *Podospora anserina* plays a major role in the ultimately lethal condition called senescence. Mobile group I introns have also been reported in the chloroplast genomes of *Chlamydomonas reinhardtii* and *C. eugametos* and the mitochondrial genome of *C. smithii*.

Circular and linear mitochondrial plasmids have been characterized in the mitochondria of filamentous fungi and flowering plants. The circular Mauriceville and Varkud plasmids of Neurospora contain nucleotide sequence motifs characteristic of group I introns and encode RTs which catalyse their replication (Griffiths, 1995). The linear kalilo and maranhar plasmids of this fungus are killer elements whose integration into the mitochondrial genome leads to senescence and, ultimately, death of the mycelium. Circular and linear plasmids have also been reported from the mitochondria of maize. They have terminal inverted repeats, the 5′-termini of which are covalently linked to a threonine, serine or tyrosine residue of an attached protein. These elements have the ability to integrate into the mitochondrial genome. The S1 and S2 plasmids are found in the mitochondria of cytoplasmic male sterile maize having the cmsS cytoplasm. RNA plasmids have been reported in several plant species.

9. Conclusion

While chloroplasts and mitochondria were derived originally from free-living prokaryotes, they are now integral parts of eukaryotic cells retaining only vestiges of their original genomes. Yet the genes encoded in these organelles are vital to their function as are the ones they have shed into the nucleus over the millenia. The mechanisms that coordinate the coexpression of these genes are only now being unravelled. During the long period of coevolution that has taken place between chloroplasts and mitochondria and their eukaryotic hosts, unique features have developed in the way in which their genes are transcribed and translated, providing interesting contrasts to the more completely characterized gene expression systems of prokaryotes and eukaryotes. With the advent of organelle transformation, it has now become possible not only to manipulate their genes *in vitro* and reintroduce them into recipient organisms to study their functions more precisely, but to introduce and express foreign genes in these organelles.

REFERENCES

Allen, J.F., Alexciev, K. & Hakansson, G. (1995) Regulation by redox signalling. *Current Biology* 5: 869–72

Attardi, G. & Schatz G. (1988) Biogenesis of mitochondria. *Annual Review of Cell Biology* 4: 289–333

Baker, A., Kaplan, C.P. & Pool, M.R. (1996) Protein targeting and translocation: a comparative survey. *Biological Reviews of the Cambridge Philosophical Society* 71: 637–702

Belfort, M. & Perlman, P.S. (1995) Mechanisms of intron mobility. *Journal of Biological Chemistry* 270: 30237–40

Benne, R. (1996) RNA editing: how a message is changed. *Current Opinion in Genetical Development* 6: 221–31

Birky, C.W. Jr (1995) Uniparental inheritance of mitochondrial and chloroplast genes: mechanisms and evolution. *Proceedings of the National Academy of Sciences USA* 92: 11331–38

Boynton, J.E. & Gillham, N.W. (1996) Genetics and transformation of mitochondria in the green alga *Chlamydomonas reinhardtii*. *Methods in Enzymology* 264 (Part B): 279–96

Burger, G., Lang, B.F., Reith, M. & Gray, M.W. (1996) Genes

encoding the same three subunits of respiratory complex II are present in the mitochondrial DNA of two phylogenetically distant eukaryotes. *Proceedings of the National Academy of Sciences USA* 93: 2328–32

Butow, R.A., Henke, R.M., Moran, J.V., Belcher, S.M. & Perlman, P.S. (1996) Transformation of *Saccharomyces cerevisiae* mitochondria using the biolistic gun. *Methods in Enzymology* 264 (Part B): 265–78

Chelstowska, A. & Butow, R.A. (1995) RTG genes in yeast that function in communication between mitochondria and the nucleus are also required for expression of genes encoding peroxisomal proteins. *Journal of Biological Chemistry* 270: 18141–46

Chory, J., Cook, R.K., Dixon, R. *et al.* (1995) Signal-transduction pathways controlling light-regulated development in Arabidopsis. *Philosophical Transactions of the Royal Society of London, Series B* 350: 59–65

Cline, K. & Henry, R. (1996) Import and routing of nucleus-encoded chloroplast proteins. *Annual Review of Cellular and Developmental Biology* 12: 1–26

Danon, A. & Mayfield, S.P. (1994) Light-regulated translation of chloroplast messenger RNAs through redox potential. *Science* 266: 1717–19

Dietrich, A., Small, I., Cosset, A., Weil, J.H. & Marechal-Drouard, L. (1996) Editing and import: strategies for providing plant mitochondria with a complete set of functional transfer RNAs. *Biochimie* 78: 518–29

Escoubas, J.M., Lomas, M., LaRoche, J. & Falkowski, P.G. (1995) Light intensity regulation of *cab* gene transcription is signaled by the redox state of the plastoquinone pool. *Proceedings of the National Academy of Sciences USA* 92: 10237–41

Fargo, D.C., Zhang, M., Gillham, N.W. & Boynton, J.E. (1998) Shine–Dalgarno-like sequences are not required for translation of chloroplast mRNAs in *Chlamydomonas reinhardtii* or in *Escherichia coli. Molecular and General Genetics* 257: 271–82

Fox, T.D. (1996) Translational control of endogenous and recoded nuclear genes in yeast mitochondria: regulation and membrane targeting. *Experientia* 52: 1130–35

Gillham, N.W. (1994) *Organelle Genes and Genomes*, Oxford and New York: Oxford University Press

Goldschmidt-Clermont, M. (1998) Coordination of nuclear and chloroplast gene expression in plant cells. *International Review of Cytology* 177: 115–80

Gray, M., Lang, B.F., Cedergren, R. *et al.* (1998) Genome structure and gene content in protist mitochondrial DNAs. *Nucleic Acids Research* 26: 865–78

Griffiths, A.J. (1995) Natural plasmids of filamentous fungi. *Microbiology Reviews* 59: 673–85

Grivell, L.A. (1996) Transposition: mobile introns get into line. *Current Biology* 6: 48–51

Hajdukiewicz, P.T., Allison, L.A. & Maliga, P. (1997) The two RNA polymerases encoded by the nuclear and plastid compartments transcribe distinct groups of genes in tobacco plastids. *EMBO Journal* 16: 4041–48

Harris, E.H., Boynton, J.E. & Gillham, N.W. (1994) Chloroplast ribosomes and protein synthesis. *Microbiological Reviews* 58: 700–54

Hauser, C.R., Gillham, N.W. & Boynton, J.E. (1998) Regulation of chloroplast translation. In *The Molecular Biology of Chloroplasts and Mitochondria in Chlamydomonas*, edited by J.-D. Rochaix, M. Goldschmidt-Clermont & S. Merchant, Dordrecht and Boston: Kluwer

Hendrick, J.P. & Hartl, F.U. (1995) The role of molecular chaperones in protein folding. *FASEB Journal* 9: 1559–69

Iratni, R., Baeza, L., Andreeva, A., Mache, R. & Lerbs-Mache, S. (1994) Regulation of rDNA transcription in chloroplasts: promoter exclusion by constitutive expression. *Genes and Development* 8: 2928–38

Kim, J. & Mayfield, S.P. (1997) Protein disulfide isomerase as a regulator of chloroplast translational activation. *Science* 278: 1954–57

King, M.P. & Attardi, G. (1996a) Isolation of human cell lines lacking mitochondrial DNA. *Methods in Enzymology* 264 (Part B): 304–13

King, M.P. & Attardi, G. (1996b) Mitochondria-mediated transformation of human ρ^0 cells. *Methods in Enzymology* 264 (Part B): 313–34

Kowallik, K.V., Stoebe, B., Schaffran, I., Kroth-Pancic, P. & Freier, U. (1995) The chloroplast genome of a chlorophyll *a+c*-containing alga, *Odontella sinensis. Plant Molecular Biology Reporter* 13: 336–42

Kubrich, M., Dietmeier, K. & Pfanner, N. (1995) Genetic and biochemical dissection of the mitochondrial protein-import machinery. *Current Genetics* 27: 393–403

Lang, B.F., Burger, G., O'Kelly, C.J. *et al.* (1997) An ancestral mitochondrial DNA resembling a eubacterial genome in miniature. *Nature* 387: 493–97

Lithgow, T., Glick, B.S. & Schatz, G. (1995) The protein import receptor of mitochondria. *Trends in Biochemical Sciences* 20: 98–101

Maier, R.M., Neckermann, K., Igloi, G.L. & Kossel, H. (1995) Complete sequence of the maize chloroplast genome: gene content, hotspots of divergence and fine tuning of genetic information by transcript editing. *Journal of Molecular Biology* 251: 614–28

Maier, R.M., Zeltz., P., Kossel, H. *et al.* (1996) RNA editing in plant mitochondria and chloroplasts. *Plant Molecular Biology* 32: 343–65

Mason, T.L., Pan, C., Sanchirico, M.E. & Sirum-Connolly, K. (1996) Molecular genetics of the peptidyl transferase center and the unusual Var1 protein in yeast mitochondrial ribosomes. *Experientia* 52: 1148–57

McFadden, G.I., Reith, M.E., Munholland, J. & Lang-Unnasch, N. (1996) Plastid in human parasites. *Nature* 381: 482

Michel, F. & Ferat, J.L. (1995) Structure and activities of group I introns. *Annual Review of Biochemistry* 64: 435–61

Nicholson, W.V., Ford, R.C. & Holzenburg, A. (1996) A current assessment of photosystem II structure. *Bioscience Reports* 16: 159–87

Ozawa, T. (1995) Mechanism of somatic mitochondrial DNA mutations associated with age and diseases. *Biochimica et Biophysica Acta* 1271: 177–89

Pakrasi, H.B. (1995) Genetic analysis of the form and function of photosystem I and photosystem II. *Annual Review of Genetics* 29: 755–76

Palmer, J.D. (1991) Plastid chromosomes: structure and evolution. In *The Molecular Biology of Plastids*, edited by L. Bogorad and I.K. Vasil, San Diego: Academic Press

Redding, K. & Peltier, G. (1998) Reexamining the validity of the Z-scheme: is photosystem I required for oxygenic photosynthesis in Chlamydomonas? In *The Molecular Biology of Chloroplasts and Mitochondria in Chlamydomonas*, edited by J.-D. Rochaix, M. Goldschmidt-Clermont and S. Merchant, Dordrecht and Boston: Kluwer

Reith, M. & Munholland, J. (1995) Complete nucleotide sequence of the *Porphyra purpurea* chloroplast genome. *Plant Molecular Biology Reporter* 13: 333–35

Remacle, C. & Matagne, R.F. (1998) Mitochondrial genetics. In *The Molecular Biology of Chloroplasts and Mitochondria in Chlamydomonas*, edited by J.-D. Rochaix, M. Goldschmidt-Clermont and S. Merchant, Dordrecht and Boston: Kluwer

Rhoads, D.M., Levings, C.S. III & Siedow, J. N. (1995) URF13, a ligand-gated, pore-forming receptor for T-toxin in the inner membrane of cms-T mitochondria. *Journal of Biometrics and Biomembranes* 4: 437–45

Rochaix, J.D. (1996) Post-transcriptional regulation of chloroplast gene expression in *Chlamydomonas reinhardtii. Plant Molecular Biology* 32: 327–41

Schatz, G. (1996) The protein import system of mitochondria. *Journal of Biological Chemistry* 271: 31763–66

Schatz, G. & Dobberstein, B. (1996) Common principles of protein translocation across membranes. *Science* 271: 1519–526

Schnell, D.J. (1995) Shedding light on the chloroplast protein import machinery. *Cell* 83: 521–24

Schwartzbach, C.J., Farwell, M., Liao, H.X. & Spremulli, L.L. (1996) Bovine mitochondrial initiation and elongation factors. *Methods in Enzymology* 264: 248–61

Seidler, A. (1996) The extrinsic polypeptides of Photosystem II. *Biochimica et Biophysica Acta* 1277: 35–60

Shapiro, T.A. & Englund, P.T. (1995) The structure and replication of kinetoplast DNA. *Annual Review of Microbiology* 49: 117–43

Simpson, L. & Emeson, R.B. (1996) RNA editing. *Annual Review of Neuroscience* 19: 27–52

Sollner-Webb, B. (1996) Trypanosome RNA editing: resolved. *Science* 273: 1182–83

Stirewalt, V.L., Michalowski, C.B., Löffelhardt, W., Bohnert, H.J. & Bryant, D.A. (1995) Nucleotide sequence of the cyanelle genome from *Cyanophora paradoxa. Plant Molecular Biology Reporter* 13: 327–32

Sugita, M. & Sugiura, M. (1996) Regulation of gene expression in chloroplasts of higher plants. *Plant Molecular Biology* 32: 315–26

von Wettstein, D., Gough, G. & Kannangara, C.G. (1995) Chlorophyll biosynthesis. *Plant Cell* 7: 1039–57

Wakasugi, T., Tsudzuki, J., Ito, S. *et al.* (1994) Loss of all ndh genes as determined by sequencing the entire chloroplast genome of the black pine *Pinus thunbergii. Proceedings of the National Academy of Sciences USA* 91: 9794–98

Wallace, D.C., Shoffner, J.M., Trounce, I. *et al.* (1995) Mitochondrial DNA mutations in human degenerative diseases and aging. *Biochimica et Biophysica Acta* 1271: 141–51

Wolstenholme, D.R. & Fauron, C.M.R. (1995) Mitochondrial genome organization. In *The Molecular Biology of Plant Mitochondria*, edited by C.S. Levings III & I.K. Vasil, Dordrecht and Boston: Kluwer

Zimmerly, S., Guo, H., Perlman, P.S. & Lambowitz, A.M. (1995) Group II intron mobility occurs by target DNA-primed reverse transcription. *Cell* 82: 545–54

GLOSSARY

biolistic a method of transformation by which DNA-coated metallic particles are launched by bombardment onto cells or tissues, from plants or animals

***cis*-splicing** splicing together of exon sequences in which exons and introns are part of the same contiguous gene sequence and mRNA

cryptophyte phylum of mainly free-living but some parasitic unicellular protists containing both photosynthetic and non-photosynthetic types

cytochrome enzyme pigmented as a result of its haem prosthetic groups

elongation factor peptidyl transferance components of ribosomes that catalyse formation of the acyl bond between the incoming amino acid residue and the peptide chain

endonuclease enzyme that splits the nucleic acid chain at internal sites

eubacteria a major subdivision of the prokaryotes (including all except the archaebacteria); eubacteria lack a nucleus and other cellular organelles

isomers alternative stereochemical form of molecules that contain the same atoms

isozyme enzymes that exist in multiple molecular forms which can be detected biochemically

monocistronic mRNA coding for only one polypeptide-coding sequence

oxidative phosphorylation phosphorylation of ATP coupled to the respiratory chain

oxidoreductase oxidase that uses molecular oxygen as the electron acceptor

phosphatase any of a large group of enzymes that acts as a catalyst in the hydrolysis and synthesis of organic phosphate esters

polyadenylation addition of a poly(A) tail to eukaryotic messenger RNA precursors in the nucleus

polycistronic mRNA that contains more than one polypeptide-coding sequence

polysome (polyribosome) group of ribosomes associated with the same molecule of mRNA

protease term applied to endopeptidases that have broad specificity and which will split most proteins into small fragments

reading frame starting point on DNA or messenger RNA from which the base sequence is read off in triplet codons

redox centre a centre where reduction and oxidation occur in electron transport so that transfer of an electron from one component to the next results in oxidation of the first component and reduction of the second

respiratory electron transport chain intracellular oxidation of substrates, together with production of ATP and oxidized coenzymes, in aerobic conditions

ribosome a heterodimeric multisubunit enzyme composed of ribonucleoprotein and protein subunits; interacts with aminoacylated tRNAs and mRNAs and translates protein-coding sequences from mRNA

RNA polymerase an enzyme that is capable of synthesizing complementary RNA from a DNA template: normally producing messenger RNA from genes coding for proteins, or ribosomal and transfer RNA from their respective genes

Shine–Dalgarno sequence in bacterial messenger RNA, a polypurine sequence found approximately seven nucleotides in front of the initiation codon, UAG

trans-splicing splicing together of exon sequences in which exons and introns are not part of the same contiguous gene sequence, but are separated spatially on the genome so that their mRNAs must come together and interact for splicing to occur

See also **Introns** (p.712)

CELL CYCLE GENES

Murdoch Mitchison
Institute of Cell, Animal and Population Biology, University of Edinburgh, Edinburgh, UK

Cell cycle genes are those which affect the cell cycle and, although they have been found in a wide variety of eukaryotic cells, they have been most extensively studied in yeasts. The pioneer here was L.H. Hartwell working with the budding yeast *Saccharomyces cerevisiae*. In the late 1960s, he isolated and characterized a number of recessive temperature-sensitive conditional mutants which grew normally at the permissive temperature, but which were blocked at the restrictive temperature at particular stages of the cell cycle (e.g. nuclear division). These stages were mostly identified by morphology, but DNA measurements were also involved. He christened these genes *cdc* ("cell division cycle"). In all the earlier work, the restrictive temperature was higher than the permissive one, but a small number of *cdc* genes were later isolated as cold sensitives.

In an influential review of 1974, Hartwell and his colleagues set out the "circuitry" of the cell cycle at that time (**Figure 1**) (Hartwell *et al.*, 1974). Each of the events of the cycle depended on the functioning of a number of *cdc* genes and, at the restrictive temperature, the cells were blocked

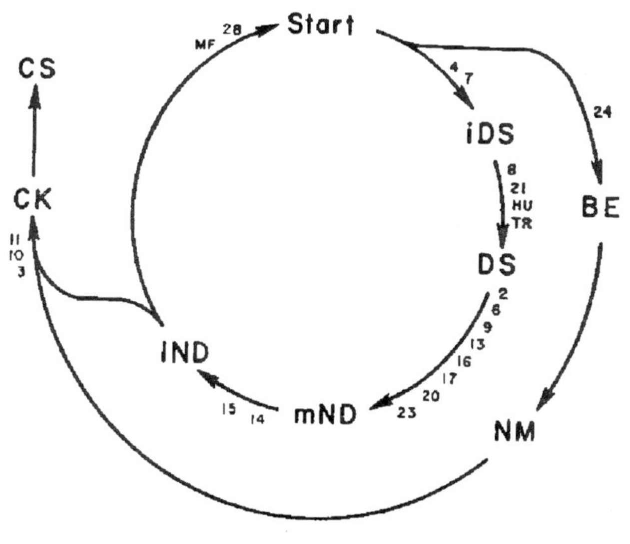

Figure 1 The circuitry of the yeast cell cycle. Events connected by an arrow are proposed to be related such that the distal event is dependent for its occurrence on the prior completion of the proximal event. iDS, initiation of DNA synthesis; BE, bud emergence; DS, DNA synthesis; NM, nuclear migration; mND, medial nuclear division; lND, late nuclear division; CK, cytokinesis; CS, cell separation. Numbers refer to *cdc* genes that are required for progress from one event to the next. HU and TR refer to the DNA synthesis inhibitors hydroxyurea and trenimom. MF refers to the mating factor α factor. (Reproduced with permission from Hartwell *et al.*, 1974. © 1974 by the American Association for the Advancement of Science.

before the event in mutants of these genes. The point where the gene had completed its function was called the "execution point" which, in most cases, was shortly before the event in question. Cells after this point, when raised to the restrictive temperature, would complete the cycle and only be blocked in the next cell cycle. There are two other points of importance in **Figure 1**. The first is the identification of what was christened "Start". This is a point of commitment early in the cell cycle. After "Start", the cell is committed to the normal vegetative cell cycle, whereas before it the cell can follow alternative pathways of development such as meiosis and sporulation. The second important point is the "parallel pathway" revealed by *CDC24*. At the restrictive temperature, cells of this mutant will complete nuclear division without the normal bud emergence, but will then stop; bud emergence is necessary for cytokinesis. This concept of pathways which diverge and then converge was developed and amplified later in more complex circuitry diagrams (Pringle & Hartwell, 1982).

Following in Hartwell's footsteps, Paul Nurse and his colleagues (Nurse *et al.*, 1976) isolated *cdc* genes in the fission yeast *Schizosaccharomyces pombe* and their circuitry diagram was broadly similar to that in **Figure l**, although different in detail. Their initial screen was somewhat different from that in budding yeast, since they were looking for long cells at the restrictive temperature. Blocked *cdc* mutants continue to grow in both yeasts, but they are more conspicuous in fission yeast which only grows in one dimension – length.

Work continued in the yeasts with the isolation of further *cdc* mutants. Other mutants were found in genes which were not *cdc*, but which affected the cell cycle in other ways. This inevitably is a more diverse collection of kinases and phosphatases, and genes affecting feedback controls, size at division, the cytoskeleton, mating and meiosis. One important and early example in fission yeast are the *wee* mutants which reduce the cell size at division, but which have little effect on growth rate or viability. The valuable appendix in Murray and Hunt (1993) lists a total of 54 *cdc* genes in budding yeast together with another 229 genes which affect the cell cycle. For fission yeast (also in this appendix), the comparable figures are 25 and 91, respectively.

The isolation of cell cycle mutants in *Escherichia coli* (Hirota *et al.*, 1968) started at about the same time as in budding yeast, although many of the earlier mutants were ones that affected the cell cycle rather than being the equivalent of *cdc* mutants. Work continued on them in the same way as in the yeasts. However, recent reviews by Donachie (1992, 1993) suggest that the number of genes obligatory for passage through the cell cycle is considerably lower than in eukaryotes such as yeast.

Genetic analyses of the cell cycle were not restricted to

yeasts and bacteria. Simchen (1978) reviewed them in *Aspergillus*, *Chlamydomonas* and *Tetrahymena*, and they are listed for *Aspergillus* in the appendix in Murray and Hunt (1993). Starting in the 1970s, cell cycle mutants were also isolated in various mammalian cell lines (Baserga, 1985). Their genetic status is less clear, especially when they appear in diploid strains. Another biological material which was exploited from the mid-1980s was the early embryo of Drosophila with its classical genetic background. Glover (1989) lists about 70 genes that play a role in mitosis, some of which have homologues in yeast.

As well as the isolation of mutants, the earlier genetic analyses involved the ordering of gene effects and dependencies – for example, by finding the terminal phenotypes of double mutants and by "reciprocal shifts". Revertants and suppressors were also analysed with profit. Again, *cdc* mutants proved useful tools in biochemical and physiological measurements. A shift-up in temperature produces a population arrested at one stage in the cycle, e.g. with unreplicated DNA, whose properties can be analysed. Shifting-down again after a period of time produces a culture with a large yield of synchronized cells which can also be analysed. Similar results can be obtained with chemical inhibitors, but the range of block points is less than with *cdc* mutants.

However, the major advances came in the 1980s, as in many other branches of genetics, with the development of the new techniques of molecular biology or molecular genetics. Genes could be identified from "libraries", cloned and sequenced. As the databases of gene sequences expanded, it became increasingly easy to find homologues which might define the gene function. Genes could be disrupted and site-specific mutations became possible. In some cells, foreign genes could be introduced to test for functional complementation and also integrated into the genome. With hybrid vectors, genes could be expressed in bacteria, so enabling an antibody to be raised without an initial chemical purification of the protein. Perhaps the major limitation was that the substrate for an enzyme *in vivo* was difficult to identify. The results of these techniques are illustrated in the gene lists in Murray and Hunt (1993).

Genetic analysis has proved a powerful tool in understanding the cell cycle and adding to biochemical and biophysical techniques. One striking example is in the regulatory network that precedes mitosis (reviewed in Glover, 1989; Murray & Hunt, 1993). In the mid-1980s, "maturation promoting factor" (MPF) was shown to induce mitosis in early amphibian embryos. Cyclin was also discovered as a protein in these cells which had the unusual property of building up in interphase and then breaking down rapidly at mitosis. Meanwhile, work on fission yeast had shown that a key player in mitosis is the protein kinase $p34^{cdc2}$, the product of $cdc2^{+}$ (whose homologue in budding yeast is *CDC28*). It is activated post-translationally by dephosphorylation of a tyrosine residue.

The separate work on embryos and on fission yeast came together in the late 1980s with the biochemical purification of MPF. It seemed likely from molecular weights that MPF was a complex of cyclin B and p34, but definitive proof came from recognition by antibodies. The discovery of these two novel key proteins, their association in a complex and its activation by changes in phosphorylation has been rightly regarded a major success of molecular genetics and biochemistry in understanding the preparations for mitosis. Since mitosis is nearly universal in eukaryotes, it is gratifying, but not altogether surprising, that these proteins have also been shown to be widely conserved.

The result of 35 years' work on cell cycle genes is a great increase in our knowledge of the genetic elements. The "parts list" is large, and is still increasing. This is essential for an understanding of cell cycle control, but the limitation is that it is not easy to know how the gene products are acting in the complex regulatory systems of the living cell. For example, the substrate on which the p34 kinase acts is not yet known nor what triggers its rise in activity before mitosis in growing cells. Also, the function of many of the *cdc* genes remains to be clarified. These are problems for the future.

REFERENCES

Baserga, R. (1985) *The Biology of Cell Reproduction*, Cambridge, Massachusetts: Harvard University Press

Donachie, W.D. (1992) What is the minimum number of dedicated functions required for a basic cell cycle? *Current Opinion in Genetic Development* 2: 792–98

Donachie, W.D. (1993) The cell cycle of *Escherichia coli*. *Annual Review of Microbiology* 47: 199–230

Glover, D.M. (1989) Mitosis in Drosophila. *Journal of Cell Science* 92: 137–46

Hartwell, L.H., Culotti, J., Pringle, J.R. & Reid, B.J. (1974) Genetic control of the cell cycle in yeast. *Science* 183: 46–51

Hirota, Y., Ryter, A. & Jacob, F. (1968) Thermosensitive mutants affected in the processes of DNA synthesis and cellular division in *Escherichia coli*. *Cold Spring Harbor Symposia on Quantitative Biology* 33: 677–93

Murray, A. & Hunt, T. (1993) *The Cell Cycle: An Introduction*, New York: Freeman

Nurse, P., Thuriaux, P. & Nasmyth, K. (1976) Genetic control of the cell division cycle in fission yeast. *Molecular and General Genetics* 146: 167–78

Pringle, J.R. & Hartwell, L.H. (1982) The *Saccharomyces cerevisiae* cell cycle. In *The Molecular Biology of the Yeast Saccharomyces: Life Cycle and Inheritance*, edited by J.N. Strathern, E.W. Jones & J.R. Broach, Cold Spring Harbor, New York: Cold Spring Harbor Laboratory Press

Simchen, G. (1978) Cell cycle mutants. *Annual Review of Genetics* 12: 161–91

GLOSSARY

cytokinesis process in which the cytoplasm of a cell is divided once mitosis is complete

hybrid vector a DNA molecule that can transfer DNA sequence from one organism to another

interphase period between one mitosis and the next, when no distinct chromosomes are visible and the chromatin is contained in a nucleus

phosphatase any of a large group of enzymes that acts as a catalyst in the hydrolysis and synthesis of organic phosphate esters

revertant organism or cell in which a reversion has occurred

suppressor [mutation] one that cancels out the effects of a primary mutation at a different locus

CENTROMERES AND KINETOCHORES: FROM YEAST TO MAN

Claudio E. Sunkel
Instituto de Biologia Molecular e Celular, Universidade do Porto, Porto, Portugal

1. Introduction

The term centromere was first used to describe the most prominent landmark of condensed mitotic chromosomes, namely the primary constriction (see Plate 21 in the colour plate section). This site was called the centromere since, in most chromosomes, it was invariably located in the middle between the ends of the two chromosome arms. Since then, the term centromere has been expanded to describe the primary constriction of all mitotic chromosomes even when it is not located in a central position. Years of research have implicated the centromere in multiple roles during mitosis, the process which results in the equal distribution of genetic material during cell division (Rattner, 1991; Earnshaw & Pluta, 1994; Yanagida, 1995). It is the site for the formation of the kinetochore, a multiprotein complex located at the surface of chromosomes that binds spindle microtubules and regulates chromosome movement in mitosis. It is also the final site of sister chromatid pairing (Miyazaki & Orr-Weaver, 1994) before segregation takes place and, therefore, must contain the necessary molecular machinery to allow ordered sister chromatid separation during the initiation of anaphase. Finally, recent work has implicated the centromere and/or kinetochore in the safeguard mechanisms that the cell uses to ensure the proper orientation of chromosomes at metaphase in order to prevent abnormal chromosome segregation.

Although it appears that the function of centromeres has been conserved throughout evolution, centromeres in different cell types show great structural variability. This has led to their classification into two different types (Pluta *et al.*, 1995). Diffused centromeres are present in the holocentric chromosomes of many arthropods, plants and worms. In these chromosomes, spindle microtubules attach along the entire length of the chromatids and the nature of their centromeric DNA (CEN-DNA) remains poorly understood. Localized centromeres are present in all other eukaryotes and are characterized by their discrete localization, providing a single region of attachment for spindle microtubules. Localized centromeres have been further subdivided into point and regional centromeres. Most of the molecular understanding of centromere structure and function has been obtained from localized centromeres and they will form the basis for this review.

2. Centromeric DNA

(i) Yeasts

The best characterized point centromere is that of the budding yeast *Saccharomyces cerevisiae* (**Figure 1**). Point centromeres have been well characterized because they are very small (160–220 base pairs [bp]) allowing for their *in vitro* manipulation and the development of functional assays (Hegemann & Fleig, 1993). The CEN-DNA of all 16 chromosomes of the budding yeast has now been cloned and extensive molecular and genetic analyses have shown that it can be divided into three distinct domains called "centromere determining elements": CDEI (8 bp), CDEII (78–86 bp) and CDEIII (26 bp). Of the three, CDEIII is absolutely essential for centromeric activity defined as the ability to confer stable segregation properties to either circular or linear minichromosomes. A single point mutation of the C14 nucleotide localized at the centre of CDEIII renders the entire CEN-DNA inactive. However, mutations or deletions of either CDEI or CDEII only cause a reduction in the stability of minichromosomes and do not abolish centromere activity. Several lines of evidence suggest that the entire CEN-DNA is organized differently from the rest of the chromatin in that a 200–250 bp core containing all three CEN-DNA elements is mostly resistant to digestion and does not show the normal nucleosome spacing present in other chromosomal regions (Jiang & Carbon, 1993). The idea that CEN-DNA might form distinct and specific higher order chromatin structure built around a modified nucleosome core has, over recent years, gained substantial experimental support.

The centromere of the fission yeast *Schizosaccharomyces pombe* is to date the best studied regional centromere (Clarke, 1990; Clarke *et al.*, 1993). CEN-DNA from all three chromosomes has been cloned and sequenced, and spans some 40–100 kilobases (kb). Surprisingly, centromeres of the fission yeast are variable genetic elements, varying both between different chromosomes and between the same chromosome among different strains, while the general organization is retained (Steiner *et al.*, 1993). CEN-DNA is characterized by a small central core element (cen1, cen2 or cen3 corresponding to each chromosome) surrounded by an array of repeats of variable size. These repeats have been classified into the K, L or B classes (Chikashige *et al.*, 1989). Extensive analysis of CEN-DNA has shown that the central core contains several functionally redundant domains capable of participating in centromere function. Any one of these domains placed next to a 2.1 kb KpnI–KpnI restriction fragment present in all K-type repeats is both necessary and sufficient to display substantial centromere activity (Baum *et al.*, 1994; Marschall & Clarke, 1995). As with the CEN-DNA in the budding yeast, CEN-DNA of the fission yeast also shows an atypical chromatin organization which is dependent on the presence of the 2.1 kb fragment from the K-type repeats. The larger size of the *S. pombe* centromere, as well as the presence of repeated DNA sequences, suggests that these centromeres are more representative of higher eukaryote centromeres.

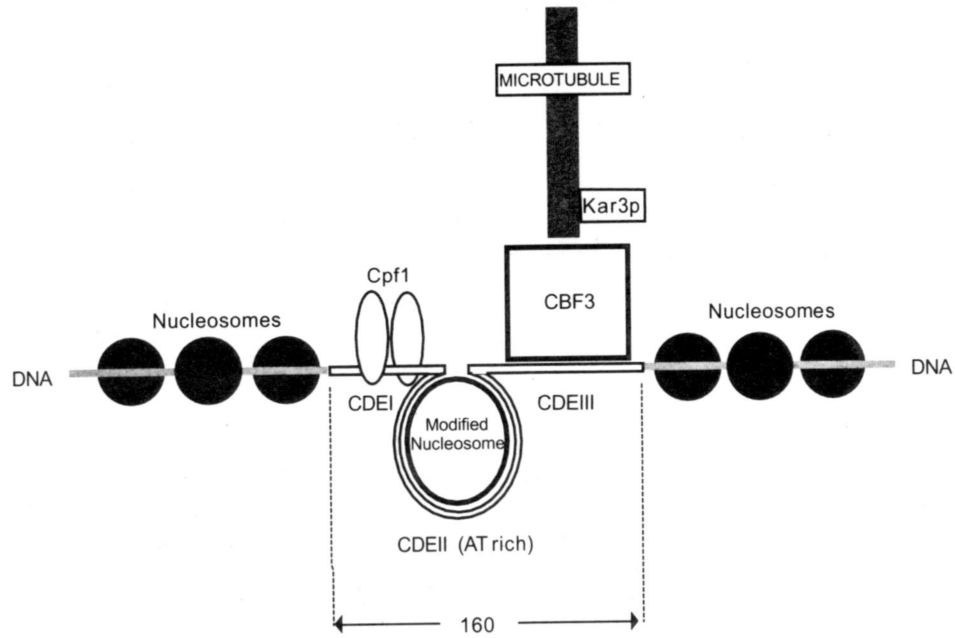

Figure 1 Molecular organization of the centromere/kinetochore of *Saccharomyces cerevisiae*. The diagram shows a model proposed for the organization of centromeric DNA (see Pluta *et al*., 1995). At either side of the three centromere elements (CDEI, CDEII and CDEIII), there is a regular nucleosome spacing for chromosomal DNA. CDEI is known to associate with the protein Cpf1p and a modified nucleosome, thought to contain the protein Cse4p, might bind the CDEII element. The third element, CDEIII, has been shown to interact with a protein complex known as CBF3 (see Table 1). The model also suggests that a single microtubule binds the CBF3 complex, although the factor involved in this association is not known. Finally, a kinesin-like motor protein (Kar3p) has been shown to be responsible for microtubule movement *in vitro*.

(ii) Drosophila melanogaster

The centromeres of most higher eukaryotes are embedded within large regions of heterochromatin which, due to its sequence organization, has remained poorly studied until recently (Cook & Karpen, 1994). This type of DNA is characterized by the presence of repeated sequences of variable length from a few bp to a few thousand bp that appear to be condensed (highly compacted) throughout the cell cycle and to have a low gene density. Although heterochromatin was always thought to be of little use for the cell, it is now thought to play an important structural role associated with the different functions of the centromere.

Recent studies have shed some light on the molecular organization of CEN-DNA of *D. melanogaster* (Murphy & Karpen, 1995), which is a very important model organism. Small chromosome derivatives that show stable mitotic and meiotic transmission have been produced by gamma radiation. This is the case with Dp(1;f)1187, a 1.3 Megabase (Mb) minichromosome marked with the visible recessive markers *yellow*$^+$ and *rosy*$^+$ (**Figure 2**). Since these two marker genes have been cloned, they can be used as probes to map restriction enzyme sites deep within heterochromatin. The results of this work indicate that within heterochromatin there are at least three large blocks of complex DNA (Tahiti, Moorea and Bora-Bora) surrounded by various types of satellite repeat and many copies of different transposable elements (Le *et al*., 1995). Smaller derivatives of this minichromosome were produced by irradiation and the stability correlated with the different islands of complex DNA was still present. The results were clear, indicating that for normal transmission some 220 kb of Bora-Bora flanked by another 200 kb of repetitive DNA are both necessary and sufficient for centromere activity (Murphy & Karpen, 1995). While the results of these experiments provide an entry point to analyse CEN-DNA function–structure relationships, one must be aware that Dp(1;f)1187 is itself the result of a rearranged X chromosome and we do not know the sequence organization of the wild-type X chromosome.

The general organization and nature of Drosophila heterochromatin has been studied for a number of years. Results from a number of different laboratories have shown that the arrangement of satellite repeats (Lohe *et al*., 1993), as well as that of transposable elements (Carmena & Gonzalez, 1995; Pimpinelli *et al*., 1995), is different between the different chromosomes of Drosophila; there are, however, some chromosome-specific satellite repeats (*Bari-I*, *Porto-I*, 359 and *dodecasatellite*) and there are others present within the heterochromatin of all chromosomes (Sunkel & Coelho, 1995). A similar pattern is also observed for transposable element repeats. This large variability in centromeric structure and organization parallels that observed in the fission yeast and, as will be discussed later, is a characteristic of CEN-DNA of all higher eukaryotes studied so far. Finally, it is important to remember that, without a functional test on which to rely for the assessment of CEN-DNA activity in Drosophila, we cannot yet obtain firm conclusions about the centromere sequence requirements in this species.

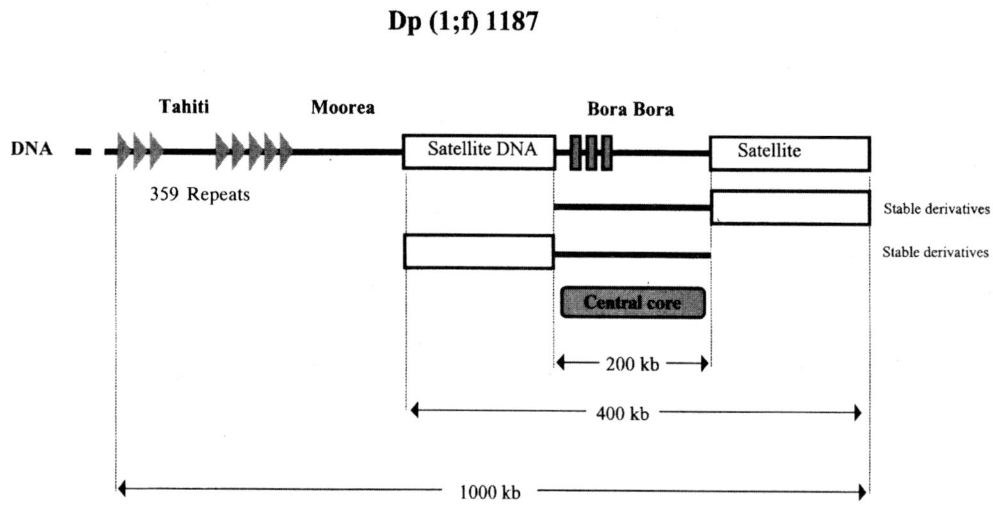

Figure 2 Molecular organization of the centromere region of the minichromosome Dp(1;f)1187 of *Drosophila melanogaster*. The diagram represents the heterochromatin of the minichromosome, in which the islands of complex DNA (Tahiti, Moorea and Bora-Bora) are shown as a thick black line and repeated DNA is shown as arrowheads (satellite repeat 359, also known as 1.688), white boxes (satellite 1.672) or grey bars (an unknown satellite repeat). Below, the two smallest stable derivatives that defined the central core (boxed) are shown.

(iii) Mammals

The molecular characterization of CEN-DNA from mammalian chromosomes has been the subject of intensive research over the last 10 years or so, and it progressed surprisingly fast in the case of human centromeres (Willard, 1990; Tyler-Smith & Willard, 1993). Although heterochromatic sequences comprise some 10% of the human genome, they appear to be of rather discrete classes. Of these, only the alphoid satellite repeats (α-satellite) have been shown to be present at the primary constriction of all human chromosomes. Alphoid DNA consists of a large family of repeats (170 bp) that show divergence within and between chromosomes. This divergence allows for unequivocal identification of chromosome-specific α-satellite repeats, a very useful tool for fluorescence *in situ* hybridization studies of human chromosomal aberrations. Although there is a good correlation between the presence of α-satellite DNA and centromere function, results published in recent years have shown that this correlation does not always hold true (Huxley, 1997). It is possible to have CEN-DNA activity without detectable alphoid DNA; also, abnormal chromosomes have been detected that contain alphoid sequences that do not induce the formation of a functional centromere. Due to the absence of functional tests for centromeric DNA in mammalian cells, it has not yet been possible to determine if alphoid DNA is indeed the human CEN-DNA (see later discussion).

The centromeric-associated DNA of at least two other mammals has been studied in some detail. In *Mus musculus*, two types of satellite repeat have been identified (minor and major satellites); the minor satellite repeats have been co-localized with the primary constriction and, to a large extent, correlated with centromere activity. Minor satellite repeats (120 bp monomers) appear to be organized in tandem arrays (reaching an average of 300 kb) that are predominantly uninterrupted by non-satellite sequences (Kipling *et al.*, 1991). As shown for the human alphoid family of repeats, significant intra- and interchromosome divergence has occurred among minor satellite repeats (Kipling *et al.*, 1994).

The centromere repeats of the Asian mouse, *Mus caroli*, have also been recently cloned and characterized. The results suggest that in this species there are two types of repeat present at or close to the primary constriction – a 60 bp repeat and a 79 bp repeat. It is thought that the 79 bp repeat might be related to CEN-DNA function, although the evidence at the moment is at best indirect (Kipling *et al.*, 1995).

Although overall DNA sequence has not been conserved among the centromeric-associated DNA of the various mammals so far studied, a short consensus motif (PyTTCGttggAaPuCGGGa) called the CENP-B box (Masumoto *et al.*, 1993) was identified within most α-satellite repeats. Furthermore, the CENP-B box is also present in the 79 bp repeat of *M. caroli*, the minor satellite repeat of primates and probably in centromeric repeats of other mammals. This motif was first identified as the DNA binding site for the major centromere antigen CENP-B using serum from patients suffering from a variety of autoimmune disorders (Earnshaw *et al.*, 1987; Masumoto *et al.*, 1989). CENP-B is a conserved centromere-binding protein present throughout the primary constriction (Earnshaw & Tomkiel, 1992). It is, however, significant that not all α-satellite repeats contain CENP-B boxes and, more importantly, that the α-satellite repeats present on the human Y chromosome do not have any, raising doubts about the importance of this DNA sequence for centromere activity. Nevertheless, some recent experiments aimed at testing the role of α-satellite repeats on centromere activity have provided surprising results. Large fragments of alphoid satellite DNA have been introduced into various cell types including Chinese hamster ovary cells, African green monkey cells, HT1080 human cells and mouse LA-9 cells (Heartlein *et al.*, 1988; Haaf *et al.*, 1992; Larin *et al.*, 1994; Taylor *et al.*, 1996). When these sequences are incorporated into chromosomes, a number of phenomena are observed, including: chromosome instability, formation of a constriction, binding of centromere proteins, sister chromosome pairing during anaphase, lagging chromosomes and unstable extrachromosomal elements. All of these observations suggested that the integration of alphoid satellite repeats could be associated with partial centromere formation. However, in most cases, it was not possible to determine whether the integrated alphoid DNA was rearranged and, more importantly, no *de novo* stable centromere was detected. Thus, the results of these experiments remained inconclusive.

However, more recently, the introduction of alphoid DNA plus telomeric DNA and carrier DNA into HT1080 human cells has led to the production of minichromosomes (Harrington *et al.*, 1997). These 6–10 Mb minichromosomes have a functional centromere since they segregate normally and bind centromere antigens that are known to be essential for centromere function during mitosis. While this is the first time *de novo* centromere function has been obtained, detailed analysis of some of these minichromosomes has shown that they may contain alphoid sequences from other chromosomes, suggesting that significant genome rearrangement takes place during the formation of the minichromosomes. Furthermore, the role of the carrier DNA continues to raise questions about the significance of alphoid satellite DNA in centromere formation.

3. What have we learned?

The first and perhaps most important conclusion from the molecular characterization of centromeric-associated DNA from all these species is that there has been very little sequence conservation. Indeed, detailed analyses of sequences in *S. cerevisiae*, *S. pombe*, humans and some rodents clearly show that even centromere DNA from different chromosomes within a species may vary significantly. Furthermore, studies in the fission yeast have revealed that centromeres can even vary between the same chromosome among different strains. Although, at the level of DNA sequence, there is little conservation among centromeric DNA of different species, it is possible to draw some parallels between the general organization of centromeres of some distantly related species. For example, functional centromeres of *S. pombe* and Drosophila can be obtained once sequences of the central core (*cen* locus in *S. pombe* or the Bora-Bora island in the Drosophila Dp(1;f)1187 minichromosome) are located next to a fragment of repetitive DNA. In both cases, it appears that the general organization of the sequences rather than any specific sequence is responsible for the organization of a functional centromere. Also, in both cases, the ability of the central core to provide the basis for centromere organization appears to be located in redundant sites. These results are difficult to reconcile within the very deterministic view of CEN-DNA obtained after analysing the point centromeres of the budding yeast. Indeed, a view that is gaining considerable support suggests that centromeres correspond to a higher order chromatin structure rather than to a specific sequence. A different chromatin structure from the regular nucleosome spacing has been observed in all centromeres so far studied and, in *S. pombe*, the transition from inactive to active centromeres has been observed experimentally (Steiner & Clarke, 1994). Furthermore, it has recently been shown in a human rearranged chromosome that a fully functional centromere can form on a sequence that is not normally associated with a centromere (du Sart *et al.*, 1997). It has been argued that the sequence requirements for the maintenance of centromere activity are not the same as those required for *de novo* formation (Huxley, 1997) so that, although infrequently, if a centromere forms on a rearranged chromosome on sequences not normally associated with the centromere, it will be stably maintained. Conversely, the introduction of centromeric-associated sequences on an ectopic site, like the case of alphoid satellite transfection experiments, will not necessarily be associated with the formation of a functional centromere. The nature of the proposed epigenetic mechanism that determines centromere formation is unknown; however, it is clearly very powerful since dicentric chromosomes (rearranged chromosomes that contain two centromeres) in which both centromeres appear active are extremely rare.

4. Proteins of the centromere and kinetochores

During the process of chromosome segregation in mitosis or meiosis, spindle microtubules attach to the condensed chromosomes via a specialized structure, the kinetochore, which is localized at the centromere. The kinetochore of most higher eukaryotes is a complex protein structure which, at the ultrastructural level, appears as a trilaminar disk that consists of an outer plate (layer) of condensed material approximately 40 nm thick and an inner plate (layer) of similar thickness which is closely adjacent to the centromeric chromatin. These two electron-dense layers are separated by a more translucent layer of about 30 nm. When mitotic cells are treated with drugs that cause microtubule depolymerization, the morphology of the outer layer is altered and a corona of fibrillar material is clearly visible. The molecular characterization of kinetochore components has advanced rapidly over the last decade. A surprising outcome of this work has indicated very clearly that, while centromeric-associated DNA appears to have undergone rapid divergence, the protein components of the kinetochore have been remarkably conserved from yeast to man.

(i) Yeasts

Although the chromosomes of the budding yeast never really condense and a proper kinetochore structure has never been seen, the identification of the conserved centromere-determining elements made it possible to isolate many proteins that are involved in kinetochore function (Plate 21A in the colour plate section, and **Table 1**). There is a single CDEI-binding protein, Cpf1p, which is not essential, but, in its absence, cells undergo a 10-fold increase in the frequency of chromosome loss. There have been no proteins identified that bind CDEII specifically. However, a histone H3 variant, Cse4p, was identified that is essential for kinetochore function and might participate together with other histones in the formation of a specialized nucleosome around which CDEII is organized (Pimpinelli *et al.*, 1995). Cse4p shows significant amino acid conservation with the human centromere antigen CENP-A (see later). In contrast, a whole complex of proteins termed CBF3 was shown to bind CDEIII specifically and to be essential for kinetochore function (Lechner & Carbon, 1991). CBF3 is composed of at least four different polypeptides (p110/Cbf3A, p64/Cbf3B, p58/Cbf3C and p23/SKP1). The detailed function of each of the components is not known, but p64 contains zinc finger motifs and has been suggested to be the DNA binding subunit. Associated with this complex is Kar3p, a kinesin-related protein that appears to be responsible for centromere movement on microtubules *in vitro* (Middleton & Carbon, 1994). Finally, the Mif2 has been isolated and shown to be required for chromosome segregation and spindle assembly and to interact genetically with p39, p64 and p110. More significantly, Mif2p shows some sequence similarity with CENP-C, a human kinetochore protein (Meluh & Koshland, 1995).

To date, only one centromere-associated protein has been identified in the fission yeast *S. pombe*. Abp1p ("autonomously replicating sequence-binding protein 1") was shown to bind the cen2 core DNA of chromosome II and both overexpression and inactivation lead to mitotic chromosome instability. Furthermore, sequence analysis of the Abp1 protein shows significant homology with the major mammalian centromere antigen CENP-B (Halverson *et al.*, 1997).

(ii) Mammalian centromere/kinetochore components

Many proteins have now been identified that localize to the mammalian kinetochore, although little is known about their function (**Table 1**). The first group of centromere antigens was identified using human sera from patients suffering from scleroderma spectrum diseases which are characterized by developing antibodies against various cellular components. A proportion of these sera contained anti-centromere antibodies and were shown to stain centromeres and kinetochores of several species including primates and rodents. These observations provided the first evidence that centromere proteins might be conserved. The first three antigens to be identified were designated CENP-A, CENP-B and CENP-C ("centromere protein") (Earnshaw & Tomkiel, 1992; Saitoh *et al.*, 1992). The human and bovine genes coding for CENP-A have now been cloned and the proteins shown to be a centromere-specific histone H3-like protein (Cooke *et al.*, 1990). The human, mouse and bovine CENP-B genes have also been cloned and shown to be highly conserved and homologues are likely to be present in many other mammals (Sullivan & Glass, 1991). CENP-B was localized by immunoelectron microscopy to the central domain of the centromere and, therefore, it is not considered a kinetochore protein. It binds the α-satellite DNA by a direct interaction with the 17 bp CENP-B box motif; however, its role in centromere function remains elusive. Expression of truncated versions of the protein or injection of specific antibodies into tissue culture cells have failed to established any essential role. Furthermore, this protein has never been localized to the centromere of the Y chromosome, although it has been localized to the inactive centromere of a dicentric chromosome.

Table I Proteins associated to the centromere/kinetochore.

Genes	Proteins	Protein motif	Homologues
S.cerevisiae			
CBF1	Cbf1p	Helix-loop-helix	
(CBF3 complex)			
NCD10	Cbf3A	Nucleotide-binding	
CBF3B	Cbf3B	DNA-binding	
CTF1	Cbf3C	Acidic serine-rich	CENP-B
KAR3	Kar3p	Kinesin	
CSE4	Cse4p	Histone H3	CENP-A
MIF2	Mif2p		CENP-C
S.pombe			
Abp1	Abp1p		CENP-B
Mammals			
	CENP-A	Histone H3	Cse4p
	CENP-B	Acidic serine-rich	Cbf3C/Abp1p
	CENP-C	Basic	Mif2p
	CENP-D	GTP-binding	
	CENP-E	Kinesin	
	MCAK	Kinesin	

CENP-C is a highly conserved protein, localized mainly in the inner kinetochore layer. It is present in the kinetochores of all human chromosomes and only localized to the active centromere of dicentric chromosomes. CENP-C is thought to be required for the maintenance of a functional kinetochore that is able to interact with spindle microtubules (Tomkiel *et al.*, 1994). One protein has been identified that binds to the outer layer and fibrous corona of the kinetochore. CENP-E shows significant sequence similarity to cytoplasmic dynein (Yen *et al.*, 1992), a type of protein known to bind microtubules and that is able to act as a molecular motor, transporting cargo from the plus end of microtubules (near the chromosomes) to the minus end (near the spindle poles) (Lombillo *et al.*, 1995).

(iii) The kinetochore checkpoint

It has recently become clear that the kinetochore participates in a safeguard mechanism used by the cell to ensure that all chromosomes are stably attached and correctly oriented to the spindle microtubules before sister chromatid separation begins during the transition from metaphase to anaphase (Gorbsky, 1995). This mechanism involves phosphorylated antigens identified by the monoclonal antibody 3F3/2. As the cell progresses through the early stages of mitosis, 3F3/2 phosphorylated antigens accumulate at the middle kinetochore layer, reaching a maximum at prometaphase. As a result of the stable interaction of spindle microtubules with a kinetochore, the 3F3/2 antigens of that kinetochore become progressively dephosphorylated. Once all chromosomes have become bi-oriented (each kinetochore from a sister chromatid pair bound to microtubules from the opposite spindle pole), the 3F3/2 phosphoepitopes are dephosphorylated and the cell can proceed into anaphase. It has been clearly demonstrated that, both in mammalian and insect cells, the presence of a lost chromosome outside of the metaphase plate which shows 3F3/2 staining is sufficient to cause a significant delay in metaphase–anaphase progression (Campbell & Gorbsky, 1995; Nicklas *et al.*, 1995). It is thought that a mechanism localized at the paired kinetochores senses tension applied by the spindle microtubules from the opposite poles. This causes the dephosphorylation of the 3F3/2 epitopes, thus removing a signal that causes the cell to delay the initiation of anaphase until all chromosomes are properly attached to the mitotic spindle. Once this process is completed, then the mechanism involved in sister chromatid separation can proceed.

The centromere is also involved in the cohesion of sister chromatids up to the initiation of anaphase. Although the nature of the molecules that maintain sister chromatid cohesion is not known, some aspects of the mechanism involved in their release are beginning to be understood. It is now clear that ubiquitin-dependent proteolysis plays an important part during this process (Holloway *et al.*, 1993). This has led to the idea that sister chromatid cohesion, up to the metaphase–anaphase transition, is achieved by a glue-like protein that binds the two chromatids at the inner centromere region. Once the cell is ready to proceed into anaphase and complete division, this glue is specifically degraded, and sister chromatids separate and migrate towards opposite poles of the mitotic spindle (Bickel & Orr-Weaver, 1996).

(iv) Kinetochore conservation

As opposed to the wide divergence of centromeric DNA sequence and organization, it has now become clear that at least some kinetochore components have been conserved during evolution. CENP-B, CENP-A and CENP-C homologues have now been found in organisms as diverse as yeast and mammals suggesting that the basic mechanism required to make the interface between microtubules and chromosomes is conserved. This poses an important question: how can a conserved protein structure be organized and built in close association with very different DNA sequences? We do not yet have any clear answers to this question, but the idea that centromeres might rely on higher order chromatin organization goes some way in the right direction. The other mechanism in which centromere/kinetochores also participate is the checkpoint control for accurate chromosome segregation. It has been shown that 3F3/2 mitotic phosphoepitopes and other proteins involved in this process are present in many distantly related species. Again, it would appear that control mechanisms required to achieve a high degree of genome stability developed during evolution have been conserved.

5. Future trends

One of the major aims of the work on centromeres has been the construction of functional artificial chromosomes. In the case of *S. cerevisiae* and *S. pombe*, yeast artificial chromosomes have proved to be efficient vectors for many different applications, although most of them are restricted to techniques of cloning and genome analysis. Mammalian artificial chromosomes, on the other hand, will have a much wider range of applications, including human gene therapy. One clear advantage for the use of artificial chromosomes is that large DNA fragments would be used which could accommodate the relatively large size of human genes, including all the necessary sequences that control the tissue-specific and temporal-specific expression of genes – sequences which, for most human genes, have not yet been fully characterized. Full details of the use of artificial chromosomes are given in the article "Techniques in molecular genetics".

Acknowledgements

I would like to thank Baldev Vig and Eric Reeve, as well as all the members of my laboratory for their comments on the early versions of the manuscript. I am grateful to Paula Coelho for allowing the use of Plate 21 in the colour section.

REFERENCES

Baum, M., Ngan, V.K. & Clarke, L. (1994) The centromeric K-type repeat and the central core are together sufficient to establish a functional *Schizosaccharomyces pombe* centromere. *Molecular Biology of the Cell* 5: 747–61

Bickel, S.E. & Orr-Weaver, T.L. (1996) Holding chromatids together to ensure they go their separate ways. *BioEssays* 18: 293–300

Campbell, M.S. & Gorbsky, G.J. (1995) Microinjection of mitotic cells with the 3F3/2 anti-phosphoepitope antibody delays the onset of anaphase. *Journal of Cell Biology* 129: 1195–204

Carmena, M. & Gonzalez, C. (1995) Transposable elements map in a conserved pattern of distribution extending from beta-heterochromatin to centromeres in *Drosophila*. *Chromosoma* 103: 676–84

Chikashige, Y., Kinoshita, N., Nakaseko, Y. *et al.* (1989) Composite motifs and repeat symmetry in *S. pombe* centromeres: direct analysis by integration of NotI restriction sites. *Cell* 57: 739–51

Clarke, L. (1990) Centromeres in budding and fission yeasts. *Trends in Genetics* 6: 150–54

Clarke, L., Baum, M., Marschall, L.G., Ngan, V.K. & Staeiner, N.C. (1993) The structure and function of *Schizosaccharomyces pombe* centromeres. *Cold Spring Harbor Laboratory Symposium on Quantitative Biology* 58: 687–95

Cook, K.R. & Karpen, G.H. (1994) A rosy future for heterochromatin. *Proceedings of the National Academy of Sciences* 91: 5219–21

Cooke, C.A., Bernat, R.L. & Earnshaw, W.C. (1990) CENP-B: a major human centromere protein located beneath the kinetochore. *Journal of Cell Biology* 110: 1475–88

du Sart, D., Cancilla, M.R., Earle, E. *et al.* (1997) A functional neo-centromere formed through activation of a latent human centromere and consisting of non-alpha-satellite DNA. *Nature Genetics* 16: 144–53

Earnshaw, W.C. & Pluta, A.F. (1994) Mitosis. *BioEssays* 16: 636–43

Earnshaw, W.C. & Tomkiel, J.E. (1992) Centromere and kinetochore structure. *Current Opinion in Cell Biology* 4: 86–93

Earnshaw, W.C., Sullivan, K.F., Machlin, P.S. *et al.* (1987) Molecular cloning of cDNA for CENP-B, the major human centromere autoantigen. *Journal of Cell Biology* 104: 817–29

Gorbsky, G. (1995) Kinetochores, microtubules and the metaphase checkpoint. *Trends in Cell Biology* 5: 143–48

Haaf, T., Warburton, P.E. & Willard, H.F. (1992) Integration of human α-satellite DNA into simian chromosomes: centromere protein binding and disruption of normal chromosome segregation. *Cell* 70: 681–96

Halverson, D., Baum, M., Stryker, J., Carbon, J. & Clarke, L. (1997) A centromere DNA-binding protein from fission yeast affects chromosome segregation and has homology to human CENP-B. *Journal of Cell Biology* 136: 487–500

Harrington, J.J., Van Bokkelen, G., Mays, R.W., Gustashaw, K. & Willard, H.F. (1997) Formation of de novo centromeres and construction of first-generation human artificial chromosomes. *Nature Genetics* 15: 345–55

Heartlein, M.W., Knoll, J.H. & Latt, S.A. (1988) Chromosome instability associated with human alphoid DNA transfected into the Chinese hamster genome. *Molecular and Cellular Biology* 8: 3611–18

Hegemann, J.H. & Fleig, U.N. (1993) The centromere of budding yeast. *BioEssays* 15: 451–60

Holloway, S.L., Glotzer, M., King, R.W. & Murray, A.W. (1993) Anaphase is initiated by proteolysis rather than by inactivation of maturation promoting factor. *Cell* 73: 1393–402

Huxley, C. (1997) Mammalian artificial chromosomes and chromosome transgenics. *Trends in Genetics* 13: 345–47

Jiang, W. & Carbon, J. (1993) Molecular analysis of the budding yeast centromere/kinetochore. *Cold Spring Harbor Laboratory Symposium on Quantitative Biology* 58: 669–76

Kipling, D., Ackford, H.E., Taylor, B.A. & Cooke, H.J. (1991) Mouse minor satellite DNA genetically maps to the centromere and is physically linked to the proximal telomere. *Genomics* 11: 235–41

Kipling, D., Wilson, H.E., Mitchell, A.R., Taylor, B.A. & Cooke, H.J. (1994) Mouse centromere mapping using oligonucleotide probes that detect variants of the minor satellite. *Chromosoma* 103: 46–55

Kipling, D., Mitchell, A.R., Masumoto, H. *et al.* (1995) CENP-B binds a novel centromeric sequence in the Asian mouse *Mus caroli*. *Molecular and Cellular Biology* 15: 4009–20

Larin, Z., Fricker, M.D. & Tyler-Smith, C. (1994) *De novo* formation of several features of a centromere following introduction of a Y alphoid YAC into mammalian cells. *Human Molecular Genetics* 3: 689–95

Le, M., Duricka, D. & Karpen, G.H. (1995) Islands of complex DNA are widespread in *Drosophila* centric heterochromatin. *Genetics* 141: 283–303

Lechner, J. & Carbon, J. (1991) A 240 kD multisubunit protein complex (CBF3) is a major component of the budding yeast centromere. *Cell* 64: 717–27

Lohe, A.R., Hiliker, A.J. & Roberts, P. (1993) Mapping simple repeated DNA sequence in heterochromatin of *Drosophila melanogaster*. *Genetics* 134: 1149–74

Lombillo, V.A., Nislow, C., Yen, T.J., Gelfand, V.I. & McIntosh, J.R. (1995) Antibodies to the kinesin motor domain and CENP-E inhibit microtubule depolymerization-dependent motion of chromosomes *in vitro*. *Journal of Cell Biology* 128: 107–15

Marschall, L. & Clarke, L. (1995) A novel cis-acting centromeric DNA element affects *S. pombe* centromeric chromatin structure at a distance. *Journal of Cell Biology* 128: 749–60

Masumoto, H., Masukata, H., Muro, Y., Nozaki, N. & Okazaki, T. (1989) A human centromere antigen (CENP-B) interacts with a short specific sequence in alphoid DNA, a human centromeric satellite. *Journal of Cell Biology* 109: 1963–73

Masumoto, H., Yoda, K., Ikeno, M. *et al.* (1993) Properties of CENP-B and its target sequence in a satellite DNA. In *Chromosome Segregation and Aneuploidy*, edited by B.K. Vig, Berlin and New York: Springer

Meluh, P.B. & Koshland, D. (1995) Evidence that the Mif2 gene of *Saccharomyces cerevisiae* encodes a centromere protein with homology to the mammalian centromere protein, CENP-C. *Molecular Biology of the Cell* 7: 793–807

Middleton, K. & Carbon, J. (1994) KAR3 kinesin is a minus-end-directed motor that functions with centromere binding proteins (CBF3) on an in vitro yeast kinetochore. *Proceedings of the National Academy of Sciences USA* 91: 7212–16

Miyazaki, W.Y. & Orr-Weaver, T.L. (1994) Sister-chromatid cohesion in mitosis and meiosis. *Annual Review of Genetics* 28: 167–87

Murphy, T.D. & Karpen, G.H. (1995) Localization of centromere function in a *Drosophila* minichromosome. *Cell* 82: 599–609

Nicklas, R.B., Ward, S.C. & Gorbsky, G.J. (1995) Kinetochore chemistry is sensitive to tension and may link mitotic forces to a cell cycle checkpoint. *Journal of Cell Biology* 130: 929–39

Pimpinelli, S., Berloco, M., Fanti, L. *et al.* (1995) Transposable elements are stable structural components of *Drosophila melanogaster* heterochromatin. *Proceedings of the National Academy of Sciences USA* 92: 3804–08

Pluta, A.F., Mackay, A.M., Ainsztein, A.M., Goldberg, I.G. & Earnshaw, W.C. (1995) The centromere: hub of chromosomal activities. *Science* 270: 1591–94

Rattner, J.B. (1991) The structure of the mammalian centromere. *BioEssays* 13: 51–56

Saitoh, H., Tomkiel, J., Cooke, C.A. *et al.* (1992) CENP-C, an autoantigen in scleroderma, is a component of the human inner kinetochore plate. *Cell* 70: 115–25

Steiner, N.C. & Clarke, L. (1994) A novel epigenetic effect can alter centromere function in fission yeast. *Cell* 79: 865–74

Steiner, N.C., Hahnenberger, K.M. & Clarke, L. (1993) Centromeres of the fission yeast *Schizosaccharomyces pombe* are highly variable genetic loci. *Molecular and Cellular Biology* 13: 4578–87

Sullivan, K.F. & Glass, C.A. (1991) CENP-B is a highly conserved mammalian centromere protein with homology to the helix-loop-helix family of proteins. *Chromosoma* 100: 360–70

Sunkel, C.E. & Coelho, P.A. (1995) The elusive centromere: sequence divergence and functional conservation. *Current Opinion in Genetical Development* 5: 756–67

Taylor, S.S., Larin, Z. & Tyler-Smith, C. (1996) Analysis of extrachromosomal structures containing human centromeric alphoid satellite DNA sequences in mouse cells. *Chromosoma* 105: 70–81

Tomkiel, J., Cooke, C.A., Saitoh, H., Bernat, R.L. & Earnshaw, W.C. (1994) CENP-C is required for maintaining proper kinetochore size and for a timely transition to anaphase. *Journal of Cell Biology* 125: 531–45

Tyler-Smith, C. & Willard, H.F. (1993) Mammalian chromosome structure. *Current Opinion in Genetics and Development* 3: 390–97

Willard, H.F. (1990) Centromeres of mammalian chromosomes. *Trends in Genetics* 6: 410–16

Yanagida, M. (1995) Frontier questions about sister chromatid separation. *BioEssays* 17: 519–26

Yen, T.J., Li, G., Schaar, B.T., Szilak, I. & Cleveland, D.W. (1992) CENP-E is a putative kinetochore motor that accumulates just before mitosis. *Nature* 359: 536–39

GLOSSARY

anaphase stage of cell division when the two sister chromatids of each chromosome separate and migrate to opposite poles of the cell

fluorescence *in situ* hybridization (FISH) the labelling of specific regions of chromosomes by *in situ* hybridization. Hybridization probes are usually made of DNA sequences (cDNA or genomic DNA). This allows the localization of a particular DNA sequence along the length of the chromosome. In conjunction with specific DNA dyes that stain specific landmarks of chromosomes it is possible to determine the relative position of a particular DNA sequence (gene)

histone any one of a set of simple basic proteins, rich in arginine and lysine, bound to DNA in eukaryotic chromosomes to form nucleosomes

kinetochore protein structure within the centromere of a chromosome to which spindle microtubules affix during meiosis or mitosis

minichromosomes small double-stranded DNA molecules that contain centromeric DNA and are capable of segregating by attaching to the mitotic spindle during mitosis. Linear minichromosomes must also have telomeric DNA sequences

monoclonal antibody antibody produced by a single clone of B cells, thereby consisting of a population of identical antibody molecules specific for a single antigenic determinant

phosphoepitope short amino acid sequence within a protein that contains an amino acid which can be modified by attaching phosphate groups and that can be recognized in its phosphorylated form by a specific antibody

satellite repeat short DNA sequence that can be repeated many times on a region of the genome. There are many different families of satellite repeats that are classified according to their DNA sequence. Most satellites are located close to the centromere and at the telomeres

telomeric relating to the ends of chromosomes

See also **Techniques in molecular genetics** (p.22)

DNA METHYLATION IN ANIMALS AND PLANTS

Susan Tweedie
Institute of Cell and Molecular Biology, University of Edinburgh, Edinburgh, UK

1. 5-Methylcytosine – a fifth base in DNA

DNA is composed of four different bases: adenine (A), cytosine (C), guanine (G) and thymine (T). In some species, these bases can be modified after DNA synthesis to produce variations on their basic structure. The most common type of modification seen in animal and plant DNA involves the addition of a methyl group to the 5 position of a cytosine residue resulting in 5-methylcytosine (m^5C), essentially forming a fifth base in DNA (**Figure 1**). Although this alteration changes the structure of the base, the methyl group lies in the major groove of the DNA helix in such a way that base pairing is not affected. Hence, m^5C stills pairs with a G residue on the opposite strand and the coding capacity of the DNA is not affected; methylated genes encode the same proteins as unmethylated genes.

So, what are the functional consequences of DNA methylation? This question has been subject to debate for many years. It is clear that the proportion of methylated cytosines varies extensively between species, and some animals, such as flies and worms, exist effectively without a trace of methylation in their genomes. These points may argue against a generally important function. However, it is now clear that methylation is essential for mammalian development at least and that it plays a similarly key role in plants. DNA methylation is not neutral; while it does not affect the coding capacity of a DNA sequence, it can affect the likelihood of whether that same sequence is active or inactive. Hence, two apparently identical DNA sequences may behave differently if one is methylated; methylated DNA sequences can be silenced or repressed such that they do not produce RNA. This effect is termed "epigenetic" rather than "genetic", because the primary DNA sequence is unaffected. It seems likely that the role of methylation in animals and plants is a result of this transcriptional silencing, but it is still unclear what the targets of such a control system are. This article discusses the characteristics of methylation, its possible functions and likely targets. The mechanism of repression by methylation and the significance of the substantial differences in DNA methylation patterns seen between species are also considered.

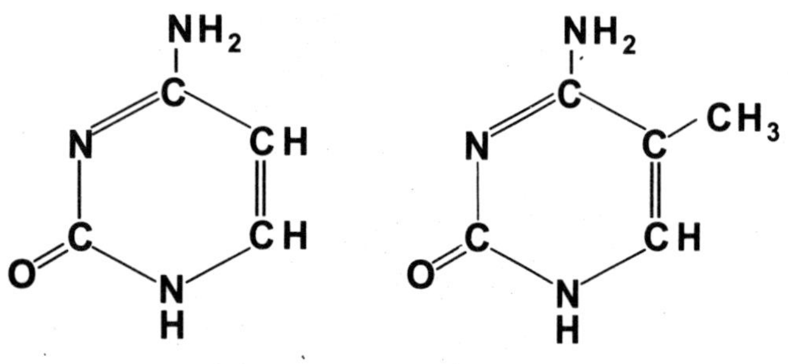

Cytosine 5- Methylcytosine

Figure 1 Structure of cytosine and 5-methylcytosine.

2. The ancestral function of DNA methylation

It should be stressed that 5-methylcytosine is not confined to eukaryotes, but is found in all biological kingdoms. In bacteria, cytosine may also be modified at the 4 position and adenine is also methylated. However, there is no evidence for methylation causing transcriptional silencing in bacteria. Instead, the proposed function is that of a neutral DNA marking system used in genome defence. Each bacterium methylates C or A bases at specific 4–8 base pair (bp) target sequences within its own genome. Invading DNA, such as that of viruses, which lacks the modification can then be distinguished as foreign and targeted for destruction by bacterially encoded restriction enzymes that cleave only unmodified DNA at the same 4–8 bp target sequence (Wilson & Murray, 1991). Only the modification function of this so called restriction–modification system appears to have been retained in eukaryotes.

3. Addition of m^5C is controlled by DNA methyltransferases

The methyl groups are added to cytosines, after they have been incorporated into DNA, by a DNA methyltransferase (MTase) enzyme that is targeted to replication foci (Leonhardt *et al.*, 1992). The fundamental structure of these enzymes is very similar in all species from bacteria to human. Each has ten conserved domains that are responsible for the catalytic reaction and a variable region that determines which DNA sequence is methylated (Posfai *et al.*, 1989). In animals, DNA methylation is principally found in cytosine residues that directly precede guanine residues (i.e. in the sequence CpG where p represents a phosphate group). Since C pairs with G, this sequence also reads CpG on the opposite strand of DNA and, consequently, this sequence is symmetrical in double-stranded DNA. Where present, methyl groups are found on cytosines on both strands within a site. This symmetry of methylation turns out to have important consequences for the inheritance of methylation patterns. After DNA is replicated during cell division, the newly synthesized DNA contains a methylated and a non-methylated strand and is described as hemimethylated. Given the choice of non-methylated or hemimethylated DNA, most eukaryotic methyltransferases prefer to methylate hemimethylated DNA by adding a methyl group to the cytosine on the unmethylated strand. This means that the same sites remain methylated throughout multiple rounds of cell divisions. Clearly, any associated transcriptional silencing would also be maintained at the same sequences within a clonal cell population. In contrast, bacterial MTase enzymes do not show the same preference for hemimethylated DNA, but prefer to act on unmethylated DNA, thus displaying a *de novo* methyltransferase activity rather than a maintenance activity.

Plant genomes are characteristically more extensively methylated than animal genomes; up to 30% of cytosines are methylated in flowering plants whereas 8% is the highest level seen in animals. This is partly due to the fact that DNA methylation is found in both CpG and CpNpG sequences (Gruenbaum *et al.*, 1981). CpNpG sequences are also symmetrical in nature and methylation at these sites can be maintained in the same way as CpG methylation. Two MTases purified from peas show distinct preferences for methylating CpG or CpNpG sites (Pradhan & Adams, 1995). Since each type of methylation is controlled independently, there is the possibility that they are also functionally distinct.

Although the vast majority of methylation in plants and animals is found at symmetrical sites, there are increasingly frequent reports of non-symmetrically methylated sites in both plants (Meyer *et al.*, 1994; Oakeley & Jost, 1996; Jacobson & Meyerowitz, 1997) and animals (Woodcock *et al.*, 1997). Identification of these minor sites reflects the growing use of bisulfite sequencing to assess the methylation status of every C within a stretch of genomic DNA (Frommer *et al.*, 1992). When genomic DNA is treated with bisulfite, all the unmodified cytosines are converted to thymine, but m^5C remains unchanged. Following this treatment, the DNA can be amplified and sequenced; any remaining Cs in the sequence must have been methylated in the original DNA. The technique has to be well controlled since incomplete chemical treatment would give a falsely high impression of the level of methylation. Such studies have also yielded evidence of low level CpNpG methylation in animals (Clark *et al.*, 1995). The discovery of non-symmetrically methylated sites raises the interesting questions of how this type of methylation is set up and maintained and whether it has a separate function.

4. DNA methylation required for normal development

The cloning of a mammalian methyltransferase (Dnmt1) provided an opportunity to test the functional significance of DNA methylation (Bestor *et al.*, 1988; Pradhan *et al.*, 1997). In a series of experiments, the enzyme activity was reduced or abolished from mice by targeted disruption of the gene encoding it (Li *et al.*, 1992; Lei *et al.*, 1996). The results have been unambiguous; homozygous Dnmt1 "knockout" mice die around mid-gestation. This indicates that, in mice at least, methylation is vital for the successful completion of development. Even though levels of methylation were dramatically reduced in these mice, there were still traces of m^5C which is consistent with there being more than one mammalian MTase. Subsequent to these experiments, three further MTases (Dnmt 2, Dnmt 3α and Dnmt 3β) have been identified in mice (Okano *et al.*, 1998a,b). The function of these enzymes is not yet clear, but they are obviously not capable of compensating for the loss of Dnmt1.

In analogous experiments in the thale cress *Arabidopsis thaliana*, plants with dramatically reduced levels of methylation were produced using antisense MTase constructs (Finnegan *et al.*, 1996; Ronemus *et al.*, 1996). The phenotype of these plants is not straightforward; they show a range of abnormalities affecting size, leaf, flower and root development (reviewed by Richards, 1997). Also, the phenotypes become more severe after several rounds of self-pollination even though there is no evidence of decreasing methylation in successive generations. Curiously, although overall methylation decreased, there was evidence that some sequences in these plants were hypermethylated relative to the wild-type (Jacobson & Meyerowitz, 1997). In addition, mutant plants with reduced levels of methylation were identified by screening programmes (Kakutani *et al.*, 1996). Some mutations identified by this method (ddm mutations) do not lie within MTase genes, but appear to affect the control of methylation. They also show a number of morphological changes, although these were only obvious after several generations of self-pollination. Whatever the targets, methylation clearly plays a role in plant development.

5. DNA methylation represses transcription

A substantial body of evidence suggests that the primary role of DNA methylation in eukaryotes is transcriptional repression. DNA sequences can be introduced into cultured cells and tested for transcriptional activity by assessing whether or not they produce RNA. In such experiments, transcription is frequently reduced or abolished when the test DNA is methylated, whereas unmethylated DNA remains active (e.g. Stein *et al.*, 1982; Vardimon *et al.*, 1982). Some investigators have also noticed that there is a delay in silencing from methylated templates introduced into cells (Buschhausen *et al.*, 1985, 1987; Kass *et al.*, 1997). This seems to be associated with the time it takes the naked DNA to become incorporated into a chromatin structure. Clearly, either the proteins that associate with DNA or folding into a higher order structure can be an important feature of silencing.

There are two related parameters that affect methylation-dependent silencing: density of methylation and promoter strength (Boyes & Bird, 1992; Hsieh, 1994). Promoters are regions of DNA that lie at the start of genes and contain binding sites for specific proteins called transcription factors. These factors determine whether the gene is switched on or off in any given cell. Not all promoters are equivalent; strong promoters allow much higher levels of RNA to be produced from a gene than weak promoters. Low density methylation (around 1 per 100 bases) is sufficient to inhibit transcription from a weak promoter, but not from a strong promoter. However, if the density of methylation is sufficiently high (around 1 per 10 bases), then even a strong promoter will be silenced (**Figure 2**).

Methylated sites must be close to the promoter to influence silencing, but the absolute positions of m^5C are less important than the overall density. With this in mind, it is

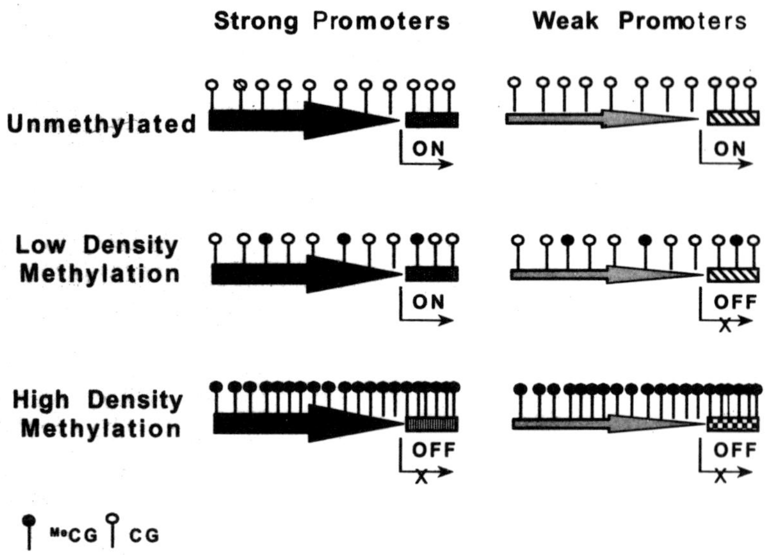

Figure 2 Methylation density and promoter strength combine to affect gene activity. Arrows denote promoter regions upstream of genes (shaded boxes); strong promoters are shown as thick black arrows and weak promoters as narrow grey arrows. Methylated cytosines are shown as filled lollipops and unmethylated cytosines as open lollipops. Low-density methylation can only silence weak promoters whereas high-density methylation silences even strong promoters. Low-density methylation of around one methyl-CpG per 100 bp is typical of vertebrate genomes.

interesting to note that bisulfite sequencing of the mouse adenosylphosphoribosyl transferase (*aprt*) gene from different cells of the same animals showed different patterns (Macleod *et al.*, 1994). Thus, although there is a general tendency for sites to remain methylated by the maintenance activity of Mtase, there is no rigid pattern for all cells in the same organism. In mice, some of this flexibility in pattern probably stems from the fact that methylation does not remain throughout all stages of development. Analysis of DNA from mouse embryos shows that there is a wave of demethylation in the early stages of development prior to implantation (Monk *et al.*, 1987; Kafri *et al.*, 1992, 1993). This is followed by extensive remethylation such that the pattern is re-established by the gastrulation stage (Jahner *et al.*, 1982; Monk *et al.*, 1987; Kafri *et al.*, 1992). This indicates that MTases must also act on unmethylated DNA templates and may explain why there are multiple mammalian MTases. Slight variations in establishing the *de novo* methylation pattern in each cell of the developing mouse could account for cell-to-cell methylation differences seen in adult mice. It is possible that such a dramatic developmental modulation of methylation is peculiar to mammals. Certainly, no such global changes were observed during sea urchin development (Pollock *et al.*, 1978).

(i) How does DNA methylation repress transcription?

There are a number of different mechanisms by which methylation could repress transcription. Methylation changes the structure of the base and this could directly prevent binding of transcription factors. There are some examples of such direct interference (Watt & Molloy, 1988; Inamdar *et al.*, 1991), but, in general, transcription factors such as the ubiquitous Sp1 are indifferent to the methylation status of the CpG sequences in their binding sites (Harrington *et al.*, 1988; Holler *et al.*, 1988). Given the tolerance of many transcription factors to methylation, and the fact that only a subset of factor binding sites contain CpG, it seems likely that repression by this action is restricted to a few cases.

The alternative mechanism proposes that methylation has an indirect effect. This view was supported by the finding that cells contain proteins which can bind specifically to methylated DNA (for review see Tate & Bird, 1993). These proteins fall into two classes: those that bind to a specific methylated sequence and those that bind to methylated CpG sequences in any wider context. Clearly, the first class of proteins could be important in silencing a specific subset of sequences, but the lack of sequence dependence shown by the second class implies that they have the potential to suppress multiple sites.

One such "non-specific" methyl-CpG-binding protein identified in mammalian cells is MeCP2. MeCP2 is an abundant chromosomal protein that has the ability to bind to a single methylated CpG pair (Lewis *et al.*, 1992; Meehan *et al.*, 1992). Studies using the cloned gene for MeCP2 revealed two key features: an 80 amino acid methyl-CpG-binding domain that is responsible for targeting the protein to methylated DNA (Nan *et al.*, 1993, 1996) and a repression domain that causes genes binding this protein to be silenced (Nan *et al.*, 1997). MeCP2 has also been shown to interact with proteins that are components of a histone deacetylase complex (Nan *et al.*, 1998). Histones are chromosomal proteins that bind to DNA at regular intervals and help to package DNA into a more compact form. Histone proteins can be modified by acetylation in a process analogous to methylation that is also correlated with gene expression. Histones with acetylated tails are found in open chromatin conformations where DNA is less compacted and more accessible to the proteins involved in transcription. Conversely, histones in more densely packed transcriptionally silent DNA are frequently deacetylated (for review see Struhl, 1998). The interaction between MeCP2 and histone deacetylase seems to indicate that the two activities work together to force chromatin into an inactive conformation. Although MeCP2 can bind a single CpG, there is evidence that the silencing activity is still dependent on methylation density and it seems likely that there are interactions between molecules of MeCP2 (Nan *et al.*, 1997). Like

MTase, MeCP2 is vital for mouse development (Tate *et al.*, 1996).

MeCP2 is not the only non-specific CpG-binding protein in mouse cells. Methyl-CpG-binding domains similar to that in MeCP2 have been found in a family of other proteins that share little other homology (Cross *et al.*, 1997; Hendrich & Bird, 1998). At least one of these Mbd ("methyl-CpG-binding domain") proteins is a component of MeCP1 (Cross *et al.*, 1997). MeCP1 is a large protein complex that requires a cluster of around ten methylated sites for efficient binding and, like MeCP2, can repress transcription (Meehan *et al.*, 1989; Boyes & Bird, 1991). Clearly, these proteins share a similar mechanism for binding to methylated DNA, but it is not yet known whether they also share a similar repression mechanism. Since MeCP1 binding is restricted to clusters of methylated sites, it will have a reduced number of potential targets, compared with MeCP2, in genomes with sparse methylation.

Most of the characterization of methyl-CpG-binding proteins has come from work in rodents, but there are homologues of MeCP2 in all the vertebrates that have been analysed including humans (D'Esposito *et al.*, 1996), chicken (Meehan *et al.*, 1992; Weitzel *et al.*, 1997) and the African clawed toad Xenopus (Jones *et al.*, 1998). No such activity could be detected in extracts from a Drosophila cell line, in keeping with the observation that these flies do not contain methylated DNA. The extent to which these proteins are conserved through evolution is not yet known.

Assays on nuclear extracts from plants indicate that they also contain methyl-CpG-binding proteins, but these await further characterization before it is known if there is any similarity to those of animals (Ehrlich, 1993).

6. What does methylation silence?

It is striking that target sequences for both plant and animal MTases are so simple. CpG and CpNpG sequences occur frequently throughout the genome. This gives the potential for methylation to be at almost any location in the genome. However, not all potential targets are modified and both the proportion of CpGs that become methylated and the distribution pattern vary enormously between different species. Clearly, there are factors other than sequence involved in determining what sites get methylated or remain unmethylated. From what is known of the action of methyl-CpG-binding proteins, DNA methylation must be carefully targeted if unwanted silencing is to be avoided. If repression is the main function of methylation, then what are the likely targets? Developmental gene regulation, tissue-specific gene control and the silencing of non-coding and foreign DNA elements are all good candidates for employing methylation. In considering these options, it is important to bear in mind where methylation is found in the genome.

7. How is DNA methylation distributed?

It is possible to measure the level of methylation in DNA by breaking the DNA down into its single bases and quantifying the proportions of m^5C in the mix. However, this does not give any information about how methylation was distributed throughout the DNA. The most detailed analysis comes from using bisulfite sequencing (Frommer *et al.*, 1992) to pinpoint the position of every methylated base in a given sequence, but this is not practical for analysing the genome as a whole. An easy way to assess the distribution of DNA methylation on a gross level makes use of methylation-sensitive restriction enzymes isolated from bacteria (Waalwijk & Flavell, 1978; Singer *et al.*, 1979a). The restriction enzyme *Hpa*II recognizes the sequence CCGG, but will only cleave that site if the internal C is unmethylated (**Figure 3**). Therefore, DNA that is heavily methylated at CGs within the sequence CCGG will be cut much less frequently by *Hpa*II than DNA that is unmethylated and genomic digest products will be large.

*Hpa*II digests of DNA from a wide range of animals suggest that methylation is not randomly distributed, but conforms to a pattern and that the vast majority of animal species share the same distribution pattern (**Figure 4A**) (Bird & Taggart, 1980). In each case, the digest has a very high molecular weight component and a smear of much smaller fragments. This indicates that these genomes have large stretches of DNA of >10 kilobases (kb) where every *Hpa*II

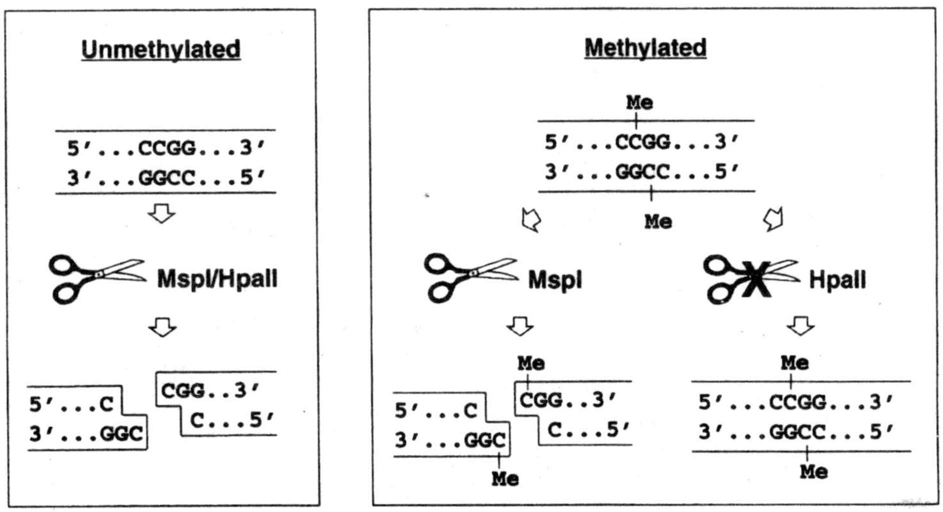

Figure 3 The methylation-sensitive restriction enzyme *Hpa*II is a useful tool for distinguishing methylated and unmethylated DNA. *Hpa*II will not cut DNA within the sequence CCGG if the internal cytosine is methylated, but will cut the same unmethylated sequence. *Msp*I is a different restriction enzyme that cuts the same site regardless of the methylation status of the internal cytosine.

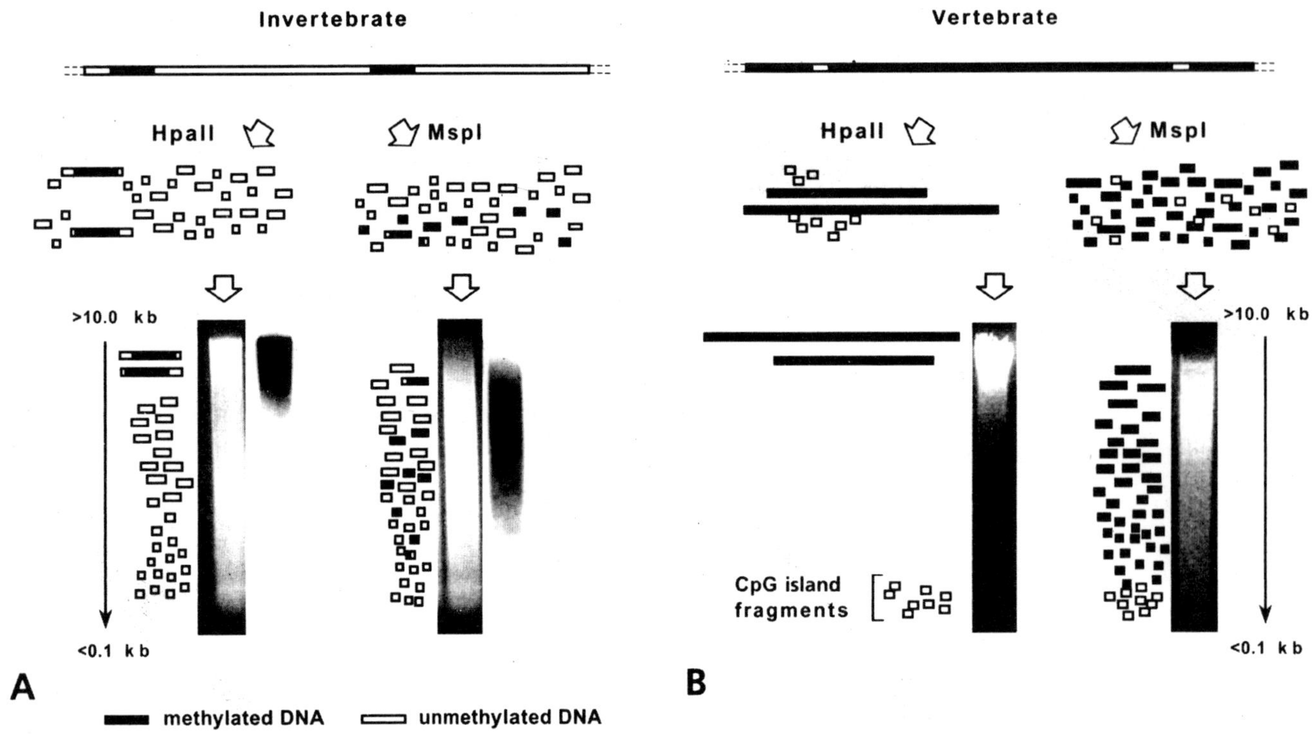

Figure 4 (A) Methylation in invertebrates (i.e. in most animals) is confined to small compartments of the genome. The figure shows a typical section of an invertebrate genome made up of alternating stretches of methylated and unmethylated DNA. These different compartments can be demonstrated by restriction analysis. Digestion with the methylation-sensitive enzyme *Hpa*II cuts the genomic DNA extensively to give numerous small fragments that can be separated according to size on an agarose gel and visualized as a smear by staining with ethidium bromide. The extensive digestion shows that the bulk of DNA is free of methylation. This is confirmed by incubating a blot of this DNA with an antibody against m^{5}C; none of the small fragments interacts with the antibody. The antibody shows that the methylated DNA is confined to very large DNA fragments that remain at the top of the gel. The large size of these fragments indicates that all CpGs in long stretches of DNA are methylated. The intensity of the large fragments is reduced when the DNA is digested with *Msp*I since this enzyme can cleave methylated DNA with the corresponding pattern of fragments reacting to the antibody changes. The DNA shown is from amphioxus. (B) Methylation patterns in vertebrate genomes are unusual. The figure shows a typical section of a vertebrate genome. The vast majority of DNA contains methylated cytosines (shown in black). The only regions that are free of methylation (shown in white) are short sequences of around 1 kb that are frequently found at the start of genes. These are the CpG islands that make up around 2% of the genome and are characterized by a high frequency of CpGs compared with the rest of the DNA. The widespread methylation in vertebrate genomes is clearly demonstrated by digestion with the methylation-sensitive enzyme *Hpa*II; very little digestion is observed and the DNA remains in very large fragments that look similar to untreated genomic DNA. This is very different from the pattern obtained with invertebrate DNA. Small fragments derived from the CpG islands are not visible by eye, but can be detected by radioactively labelling the ends of the fragments. Lack of digestion by *Hpa*II is due to methylation rather than lack of sites, since *Msp*I can cut the DNA to a smear of smaller fragments. The DNA shown here is from mouse.

site is methylated. These long stretches are not simply devoid of *Hpa*II sites, since they are cut down to small fragments by *Msp*I, an isoschizomer which also recognizes CCGG but is indifferent to the methylation status of the internal C. Since the remainder of the DNA is cut into very small pieces, it must be essentially free of methylation; this is confirmed by the finding that an antibody against m^{5}C reacts with DNA from the high molecular weight fraction but not the smaller digestion products (Tweedie *et al.*, 1997). Such analysis allows us to visualize these genomes as alternating compartments of methylated and unmethylated DNA, with the two types of DNA remaining separate. The ratio of the unmethylated to methylated DNA varies between species, but the majority of the genome is always unmethylated. This compartmentalized type of methylation pattern is seen in a wide range of species and is typical of the majority of species studied, including all methylated invertebrates and plants such as Arabidopsis. Vertebrate DNA has an unusual methylation pattern which will be discussed later.

8. What sequences reside in the methylated and unmethylated compartments?

Analysis of plant DNA suggests that there is a correlation between the size of the genome and the proportion of methylated DNA. The Arabidopsis genome is relatively small for a plant genome and only 5% of cytosines are methylated. Most of the m^{5}C in Arabidopsis has been found within repeated sequences, including the centromeric repeat, while single copy sequences are frequently unmethylated (Leutwiler *et al.*, 1984; Pruitt & Meyerowitz, 1986). In general, there is no direct relationship between genome size and gene content. Although the rice and wheat genomes vary hugely in size, comparative analysis of these genomes suggests they have similar numbers of genes and that the excess DNA is derived from transposable elements and non-coding repetitive sequences (Gale & Devos, 1998). Transposable elements in other plant species, such as maize, tobacco and the snapdragon Antirrhinum, are also found to be methylated (Martin *et al.*, 1989; Schlappi *et al.*, 1994). There is much less information on the methylation status of coding sequences in plants, but many are reported to be unmethylated.

The correlation between repeats and methylation is not unique to plants. The slime mold *Physarum polycephalum* has around 20% of its genome methylated and this is almost entirely made up of multiple copies of the single transposable element Tp1 (Rothnie *et al.*, 1991). In the sea urchin, a similar pattern of demarcation was originally thought to exist since several repeated sequences were found to be methylated while a selection of genes were completely free from methylation (Bird *et al.*, 1979).

9. A role for DNA methylation in genome defence?

The frequent finding that transposons, retroviruses and satellite sequences are methylated suggests that DNA methylation plays a role in silencing such sequences (Bird, 1986; Bestor, 1990). Mobile elements pose a substantial threat to the genome by moving from site to site and interrupting vital sequences. Since they encode the proteins required for their own movement, this is prevented by transcriptional repression. There is evidence that methylation does inhibit transposition of Spm and Ac elements in maize (Dennis & Brettell, 1990; Schlappi *et al.*, 1994) and the Tam3 transposon in Antirrhinum (Martin *et al.*, 1989). The role of genome defence is analogous to the proposed ancestral function of DNA methylation which allows the bacterium to protect its own DNA while destroying foreign viral DNA that could pose a threat to its genome.

If silencing mobile elements is the main function of DNA methylation, then it is interesting to consider the cases of animals that do not have methylation. Drosophila and the free-living worm *Caenorhabditis elegans* both lack DNA methylation (Rae & Steele, 1979; Simpson *et al.*, 1986), but have the usual spectrum of transposable elements in their genomes. If rampant transposition poses such a major threat of genome scrambling, then how do these animals survive? It is argued that Drosophila does in fact have a particularly difficult time in the absence of methylation since 50% of mutations found in these animals result from transposon insertions (Yoder *et al.*, 1997). Since bacteria contain methyltransferase enzymes that are similar in structure to the mammalian MTase, it seems likely that the common ancestor of invertebrates contained methylation. This is backed up by the finding that some unmethylated species contain vestigial remnants of a methylation system. For example, the yeast *Schizosaccharomyces pombe* contains a non-functional MTase, Pmt1, that has been inactivated by an amino acid insertion (Wilkinson *et al.*, 1995; Pinarbasi *et al.*, 1996). Do the fairly frequent cases of methylation loss in the course of evolution suggest that it is useful rather than vital to most species?

If DNA methylation does play a role in genome defence, it is clear that it does not always do a very efficient job. Detailed analysis of the sea squirt genome revealed that many of the transposable elements were not within methylated regions of the genome (Simmen *et al.*, 1999). These were clearly mobile elements since they appeared at different locations in the genomes of individual animals. Even though this animal has methylation, it does not use it effectively to stop movement of these elements around the genome.

10. Gene regulation by methylation in invertebrates?

While the methylated compartment was thought to be biased towards repetitive sequences, the unmethylated region of the genome was proposed to be rich in genes. In fact, it turns out that genes are not strictly limited to unmethylated regions of invertebrate genomes. Extensive analysis of coding sequences from sea urchins, sea squirts and amphioxus, all of which have fractionally methylated genomes, reveals that a substantial fraction of genes lie in the methylated compartment (Tweedie *et al.*, 1997). This opens up the possibility of gene regulation by methylation. If methylation is involved in the tissue-specific control of genes, then we may expect to find that a gene becomes methylated in a non-expressing cell type. However, there are no documented examples of this in invertebrates to date. In fact, several genes with a tissue-restricted expression pattern are found to be unmethylated in all cell types, suggesting that silencing can happen in the absence of methylation. Equally, we may have predicted that housekeeping genes, which are expressed in all cell types, would remain free from methylation. This is clearly not the case, as some housekeeping genes are located in the methylated fractions of the amphioxus and sea urchin genomes (Tweedie *et al.*, 1997). Housekeeping genes encode the essential components vital to survival of all cells types, so they must be expressed in spite of their methylation status.

Where it occurs, methylation in invertebrates is present at high density and transfection experiments have shown that this can silence even strong promoters (Boyes & Bird, 1992). It raises the question of how these methylated genes avoid complete silencing. Detailed analysis of the location of m^5C in relation to invertebrate promoters and comparison of the nature of the methyl-CpG-binding proteins in these animals is required.

Methylation has also been suggested to play a role in developmental gene expression. Demethylation of sea urchin embryos with 5-azacytidine leads to a developmental arrest (Chen *et al.*, 1993). The prediction is that the effects result from misexpression of genes which would normally be silenced by methylation. Unfortunately, this type of treatment is likely to affect multiple pathways and it is difficult to implicate DNA methylation alone. However, the types of pleiotropic effects of MTase reduction seen in Arabidopsis are consistent with methylation influencing a number of developmentally important genes (Richards, 1997).

While the global patterns of methylation in plants and invertebrates are similar, it would be naive to assume that the role of methylation is the same. There are few clues from the distribution of sequences between the methylated and unmethylated regions of the genomes since both genes and repeated elements are found in both regions. There is clearly a possibility that methylation influences gene expression in invertebrates, but this has still to be proven. Evidence that methylation is as important to the survival of invertebrates as it is to mice awaits a system that can specifically remove methylation from these animals.

11. The DNA methylation pattern in vertebrates is unusual

While the importance of methylation still hangs in the balance for the vast majority of species, it is not in any doubt for the vertebrates. Mice with reduced methylation die very early in development.

Overall levels of m^5C in vertebrates are quite high; around 4% of CpGs are methylated in humans. Yet, where is DNA methylation located in vertebrates? If we carry out a survey of vertebrate genomes using methylation-sensitive restriction enzymes, we find that the DNA is hardly touched by them (**Figure 4B**) (e.g. Singer *et al.*, 1979b). This indicates that methylation in these animals is spread over the entire genome. In fact, the proportion of unmethylated DNA represents only around 2% of the genome (Cooper *et al.*, 1983). This is in stark contrast to the majority of animals where well over half the genome is methylation free.

This change in patterns is striking and, from an evolutionary point of view, this seems a fairly surprising development. First, how could such a situation evolve when the likely consequence is transcriptional silencing of essential genes? Secondly, m^5C is a mutational hotspot which is a distinct disadvantage when it comes to maintaining intact gene sequences (Coulondre *et al.*, 1978). This mutational effect results from the fact that methylated cytosine frequently becomes deaminated to a thymine residue resulting in mismatched T–G base pairs. Although the cell devotes substantial efforts to repairing similar types of DNA damage, this particular error is repaired inefficiently. Consequently, on replication, the C can be lost altogether and the frequencies of CpG dinucleotides in methylated DNA become markedly reduced (Bird, 1980). Based on C + G content, human DNA would be expected to have a CpG every 25 bases, but, in reality, they are found only every 100 bp. Accordingly, although around 70% of CpGs are methylated in the human genome, the density of methylated sites is relatively low. This has important implications for the function of methylation in bulk genomic DNA. Studies on the methyl-CpG-binding proteins and transfection experiments suggest that this density of methylation will only repress weak promoters and so the overall effect on transcription may be mild.

12. Tissue-specific regulation by methylation in vertebrates

Low-density methylation may still regulate gene expression. Tissue-specific regulation results from the correct combination of activating transcription factors arriving together at a promoter sequence. The strength of a promoter is partly defined by the cell type it is in. A globin promoter is considered strong in a blood cell because the correct combination of transcription factors is present, but it would be considered weak if tested in, say, a skin cell. Since low-density methylation affects weak but not strong promoters, it could help to silence tissue-specific genes in cells where they should not be expressed (Bird, 1995; Bird & Tweedie, 1995). This effect could be achieved purely on the basis of methylation density without any need to adjust the pattern of methylation to suit the cell type or even to pay particular attention to which sites are methylated.

13. CpG islands – the unmethylated fraction of vertebrate DNA

The 2% of the vertebrate genome that remains free from methylation has different sequence characteristics from bulk genomic DNA. This DNA represents the CpG island fraction, so-called because it contains a high frequency of CpGs that is in line with the frequency expected from base composition (Bird *et al.*, 1985). It seems likely that there has been no suppression of CpG in this DNA, because it is always free from methylation. CpG islands are on average 1 kb long and are associated with the promoters of ~50% of mammalian genes (Larsen *et al.*, 1992; Antequera & Bird, 1993). Since these are the only regions of the vertebrate genome that are free from methylation, it seems plausible that methylation is the default state of these genomes and that a mechanism exists by which islands are protected. CpG islands replicate early in S-phase and have been mapped close to multiple origins of replication in hamster cells (Delgado *et al.*, 1998). Since the MTase acts in conjunction with the replication machinery (Leonhardt *et al.*, 1992), the earliest sequence to be synthesized may somehow miss out on the MTase activity. It is clear that transcription also plays a part in defining islands: the Sp1 sites in the mouse and hamster *aprt* genes are important for establishing methylation-free islands (Brandeis *et al.*, 1994; Macleod *et al.*, 1994). Therefore, it is likely that CpG islands represent the genomic "footprints" left by the combined processes of transcription and DNA replication (Delgado *et al.*, 1998).

14. Gene silencing by CpG island methylation

Whatever the origin of CpG islands, it is clearly important that they do remain free of methylation since the high density of CpGs is likely to silence even the strongest promoter. This is a recognized phenomenon in tissue culture cells where non-essential island-containing genes have been silenced by methylation (Antequera *et al.*, 1990; Jones *et al.*, 1990). However, it seems that there are at least two cases in normal mammalian development where high-density methylation could serve a purpose. X inactivation in females involves the silencing of one X chromosome to achieve equal gene doses in males and females (for review see Heard *et al.*, 1997). This is achieved in part through the action of *XIST* mRNA; *XIST* is the only gene expressed from the inactive X (Brown *et al.*, 1992). However, there is good evidence that methylation helps to maintain the inactive state by methylating CpG islands on the inactive X and by methylation of the *XIST* island on the active X (Panning & Jaenisch, 1996).

The second role is in mammalian imprinting. Imprinted genes are expressed from only one allele and this is determined by the parental origin of the allele. While the function of imprinting is still unresolved, it is clear that there are differences in the methylation patterns between the paternal and maternal alleles for these genes and that this is important in the phenomenon (Jaenisch, 1997).

Aberrant methylation of island sequences may also contribute to the development of cancers (Bird, 1996). Cancer can arise when both copies of a tumour-suppressor gene are inactivated. Frequently, this is by loss of one copy of the gene by mutation, but a number of tumour-suppressor genes have CpG islands and analysis of tumour DNA suggests that there are cases where one or both copies of a tumour-suppressor gene are shut down by accidental methylation of the CpG island (Jones & Laird, 1999).

15. Genome-wide methylation: a vertebrate innovation or an unfortunate error?

Genome-wide methylation is seen in all vertebrates, including lampreys and hagfish which are the most primitive living vertebrates. The switch from partial to complete genome methylation seems to be linked to the emergence of the vertebrates in evolution since the chordates amphioxus, the most vertebrate-like invertebrate, and sea squirt both have fractionally methylated genomes (Tweedie *et al.*, 1997). The invertebrate–vertebrate boundary is also associated with an increase in gene number. Invertebrates as diverse as worm, slimemold, fly and the chordate sea squirt have been estimated to have similar gene numbers (in the range 12 000–17 000), whereas the vertebrate estimate based on mouse, human and pufferfish lies in the range of 50 000–100 000. Perhaps these two observations are connected. The challenge is to explain why, in the face of the transcriptional repression and mutation that it causes, methylation has spread out from discrete compartments to cover the entire genome. The consequences of mutation are not trivial; as much as a third of human diseases results from C to T changes that can be ascribed to methylation (Cooper & Krawczak, 1990). There must have been a significant selective advantage that overcame this detrimental effect.

One theory suggests that the advantage was better control of gene expression (Bird, 1995; Bird & Tweedie, 1995). Animals are made up of a range of different cell types, and biological complexity has increased in the course of evolution, giving rise to increased numbers of ever more specialized cell types, a process that goes hand in hand with new genes arising. One problem associated with increasing gene number is that every new gene has to be controlled; not only do they have to be activated in the correct spatial and temporal manner, but they have to be silenced in cells where they are not required. In some respects, repression may be considered a bigger problem than activation since, in any mammalian cell, there will be many more genes switched off than turned on at any given time. The need for tight control is clear when you consider that most transcripts are normally present in the cell at relatively low (around 5–10) copy numbers (Hastie & Bishop, 1976). In theory, only a few inappropriately expressed transcripts could represent a significant problem in maintaining cell type integrity. It is unlikely that any biological system is completely watertight and genes will occasionally be misexpressed. This is clearly true as sensitive PCR assays can detect any transcript in any cell type if enough cells are examined (Chelly *et al.*, 1989). It also follows that, as gene number increases, then the number of inappropriate transcripts per cell will rise unless the silencing mechanism can be improved. Thus, for a given level of control, there is likely to be a maximum number of genes that can be controlled without the misexpressed genes causing the cell problems. It is suggested that this limit is around 15 000 in invertebrates and that vertebrate gene number was capable of rising to as high as 100 000 when global transcriptional repression was introduced and gene control improved. Low-level methylation, as described above, has the ideal characteristics to reduce transcriptional noise. In this respect, is it possible that the inefficient repair of T–G mismatches has actually been an advantage since it has allowed methylation density to drop to a point where the effects are advantageous rather than detrimental? In addition to misexpression of genes, DNA methylation could silence weak cryptic promoters that arise by chance in the non-coding bulk of the genome.

16. Genome defence in vertebrates

An alternative view suggests that the sole function for DNA methylation is to silence elements that threaten the genome (Yoder *et al.*, 1997). Such elements in vertebrates are frequently associated with methylation, just as they are in invertebrates. Mouse satellite sequences that contain CpGs are so abundant that they account for most of the m^5C in the mouse genome. Sequences such as Alu repeats crop up in all parts of the genome including introns, untranslated sequences and even embedded in coding sequences. If this is the sole purpose of methylation, then why do genes ever become methylated? The argument is that methylation was originally confined to such sequences, but has since spread out into adjacent coding regions, blanketing the entire genome in methylation and creating the illusion that genes are deliberate targets. There is evidence for such spreading of methylation, but it seems odd that this has not also happened in the invertebrates where transposable elements appear to be just as prevalent.

It seems likely that methylation does play a role in silencing transposon, retroviral and satellite sequences in somatic cells; Alu, L1 and retroviral elements are all heavily methylated in such cells. However, by some peculiar quirk, such sequences are frequently undermethylated and expressed in germ cells – the very cells where they are most likely to cause lasting damage (Bird, 1997). Hence, it is not clear how effective methylation is at controlling transposition. It is clear that the genomes of most animals are peppered with elements so that, at some stage, this control has been quite poor.

It seems likely that there is not just one role for DNA methylation, but that this general repression mechanism has been co-opted for multiple uses which may vary from species to species.

REFERENCES

Antequera, F. & Bird, A. (1993) Number of CpG islands and genes in human and mouse. *Proceedings of the National Academy of Sciences USA* 90: 11995–99

Antequera, F., Boyes, J. & Bird, A. (1990) High levels of de novo methylation and altered chromatin structure at CpG islands in cell lines. *Cell* 62: 503–14

Bestor, T.H. (1990) DNA methylation: evolution of a bacterial immune function into a regulator of gene expression and genome structure in higher eukaryotes. *Philosophical Transactions of the Royal Society of London, Series B* 326: 179–87

Bestor, T., Laudano, A., Mattaliano, R. & Ingram, V. (1988) Cloning and sequencing of a cDNA encoding DNA methyltransferase of mouse cells. *Journal of Molecular Biology* 203: 971–83

Bird, A.P. (1980) DNA methylation and the frequency of CpG in animal DNA. *Nucleic Acids Research* 8: 1499–594

Bird, A.P. (1986) CpG-rich islands and the function of DNA methylation. *Nature* 321: 209–13

Bird, A.P. (1995) Gene number, noise reduction and biological complexity. *Trends in Genetics* 11: 94–100

Bird, A.P. (1996) The relationship of DNA methylation to cancer. *Cancer Surveys* 28: 87–101

Bird, A. (1997) Does DNA methylation control transposition of selfish elements in the germline? *Trends in Genetics* 13: 469–70

Bird, A. & Taggart, M.H. (1980) Variable patterns of total DNA and rDNA methylation in animals. *Nucleic Acids Research* 8: 1485–97

Bird, A. & Tweedie, S. (1995) Transcriptional noise and the evolution of gene number. *Philosophical Transactions of the Royal Society of London, Series B* 349: 249–53

Bird, A.P., Taggart, M.H. & Smith, B.A. (1979) Methylated and unmethylated DNA compartments in the Sea Urchin genome. *Cell* 17: 889–901

Bird, A., Taggart, M., Frommer, M., Miller, O.J. & Macleod, D. (1985) A fraction of the mouse genome that is derived from islands of nonmethylated, CpG-rich DNA. *Cell* 40: 91–99

Boyes, J. & Bird, A. (1991) DNA methylation inhibits transcription indirectly via a methyl-CpG binding protein. *Cell* 64: 1123–34

Boyes, J. & Bird, A. (1992) Repression of genes by DNA methylation depends on CpG density and promoter strength: evidence for involvement of a methyl-CpG binding protein. *EMBO Journal* 11: 327–33

Brandeis, M., Frank, D., Keshet, I. *et al.* (1994) Sp1 elements protect a CpG island from de novo methylation. *Nature* 371: 435–38

Brown, C.J., Hendrich, B.D., Rupert, J.L. *et al.* (1992) The human XIST gene: analysis of a 17 kb inactive X-specific RNA that contains conserved repeats and is highly localized within the nucleus. *Cell* 71: 527–42

Buschhausen, G., Graessmann, M. & Graessmann, A. (1985) Inhibition of herpes simplex thymidine kinase gene expression by DNA methylation is an indirect effect. *Nucleic Acids Research* 13: 5503–13

Buschhausen, G., Wittig, B., Graessmann, M. & Graessmann, A. (1987) Chromatin structure is required to block transcription of the methylated herpes simplex virus thymidine kinase gene. *Proceedings of the National Academy of Sciences USA* 84: 1177–81

Chelly, J., Concordet, J.P., Kaplan, J.C. & Kahn, A. (1989) Illegitimate transcription: transcription of any gene in any cell type. *Proceedings of the National Academy of Sciences USA* 86: 2617–21

Chen, J., Maxson, R. & Jones, P.A. (1993) Direct induction of DNA hypermethylation in sea urchin embryos by microinjection of 5-methyl dCTP stimulates early histone gene expression and leads to developmental arrest. *Developmental Biology* 155: 75–86

Clark, S.J., Harrison, J. & Frommer, M. (1995) CpNpG methylation in mammalian cells. *Nature Genetics* 10: 20–27

Cooper, D.N. & Krawczak, M. (1990) The mutational spectrum of single base-pair substitutions causing human genetic disease: patterns and predictions. *Human Genetics* 85: 55–74

Cooper, D.N., Taggart, M.H. & Bird, A. (1983) Unmethylated domains in vertebrate DNA. *Nucleic Acids Research* 11: 647–58

Coulondre, C., Miller, J.H., Farabaugh, P.J. & Gilbert, W. (1978) Molecular basis of base substitution hotspots in *Escherichia coli*. *Nature* 274: 775–80

Cross, S.H., Meehan, R.R., Nan, X. & Bird, A. (1997) A component of the transcriptional repressor MeCP1 shares a motif with DNA methyltransferase and HRX proteins. *Nature Genetics* 16: 256–59

D'Esposito, M., Quaderi, N.A., Ciccodicola, A. *et al.* (1996) Isolation, physical mapping, and northern analysis of the X-linked human gene encoding methyl CpG-binding protein, MECP2. *Mammalian Genome* 7: 533–35

Delgado, S., Gomez, M., Bird, A. & Antequera, F. (1998) Initiation of DNA replication at CpG islands in mammalian chromosomes. *EMBO Journal* 17: 2426–35

Dennis, E.S. & Brettell, R.I. (1990) DNA methylation of maize transposable elements is correlated with activity. *Philosophical Transactions of the Royal Society of London, Series B* 326: 217–29

Ehrlich, K.C. (1993) Characterization of DBPm, a plant protein that binds to DNA containing 5-methylcytosine. *Biochimica et Biophysica Acta* 1172: 108–16

Finnegan, E.J., Peacock, W.J. & Dennis, E.S. (1996) Reduced DNA methylation in *Arabidopsis thaliana* results in abnormal plant development. *Proceedings of the National Academy of Sciences USA* 93: 8449–54

Frommer, M., McDonald, L.E., Millar D.S. *et al.* (1992) A genomic sequencing protocol that yields a positive display of 5-methylcytosine residues in individual DNA strands. *Proceedings of the National Academy of Sciences USA* 89: 1827–31

Gale, M.D. & Devos, K.M. (1998) Comparative genetics in grasses. *Proceedings of the National Academy of Sciences USA* 95: 1971–74

Gruenbaum, Y., Naven-Many, T., Cedar, H. & Razin, A. (1981) Sequence specificity of methylation in higher plant DNA. *Nature* 292: 860–62

Harrington, M.A., Jones, P.A., Imagawa, M. & Karin, M. (1988) Cytosine methylation does not affect binding of transcription factor Sp1. *Proceedings of the National Academy of Sciences USA* 85: 2066–70

Hastie, N.D. & Bishop, J.O. (1976) The expression of three

abundance classes of messenger RNA in mouse tissues. *Cell* 9: 761–74

Heard, E., Clerc, P. & Avner, P. (1997) X-chromosome inactivation in mammals. *Annual Review of Genetics* 31: 571–610

Hendrich, B.H. & Bird, A. (1998) Identification and characterisation of a family of mammalian methyl-CpG binding proteins. *Molecular and Cellular Biology* 18: 6538–47

Holler, M., Westin, G., Jiricny, J. & Schaffner, W. (1988) Sp1 transcription factor binds DNA and activates transcription even when the binding site is CpG methylated. *Genes and Development* 2: 1127–35

Hsieh, C.-L. (1994) Dependence of transcriptional repression on CpG methylation density. *Molecular and Cellular Biology* 14: 5487–94

Inamdar, N.M., Ehrlich, K.C. & Ehrlich, M. (1991) CpG methylation inhibits binding of several sequence-specific DNA-binding proteins from pea, wheat, soybean and cauliflower. *Plant Molecular Biology* 17: 111–23

Jacobson, S.E. & Meyerowitz, E.M. (1997) Hypermethylated SUPERMAN epigenetic alleles in *Arabidopsis*. *Science* 277: 1100–03

Jaenisch, R. (1997) DNA methylation and imprinting: why bother? *Trends in Genetics* 13: 323–29

Jahner, D., Stuhlmann, H., Stewart, C.L. *et al.* (1982) De novo methylation and expression of retroviral genomes during mouse embryogenesis. *Nature* 298: 623–28

Jones, P.A., & Laird, P.W. (1999) Cancer epigenetics comes of age. *Nature Genetics* 21: 163–67

Jones, P.A., Wolkowicz, M.J., Rideout, W.M. *et al.* (1990) De novo methylation of the MyoD1 CpG island during the establishment of immortal cell lines. *Proceedings of the National Academy of Sciences USA* 87: 6117–21

Jones, P.L. Veenstra, G.J., Wade, P.A. *et al.* (1998) Methylated DNA and MeCP2 recruit histone deacetylase to repress transcription. *Nature Genetics* 19: 187–91

Kafri, T., Ariel, M., Brandeis, M. *et al.* (1992) Developmental pattern of gene-specific DNA methylation in the mouse embryo and germ line. *Genes and Development* 6: 705–14

Kafri, T., Gao, X. & Razin, A. (1993) Mechanistic aspects of genome-wide demethylation in the preimplantation mouse embryo. *Proceedings of the National Academy of Sciences USA* 90: 10558–62

Kakutani, T., Jeddeloh, J.A., Flowers, S.K., Munakata, K. & Richards, E.J. (1996) Developmental abnormalities and epimutations associated with DNA hypomethylation mutations. *Proceedings of the National Academy of Sciences USA* 93: 12406–11

Kass, S.U., Landsberger, N. & Wolffe, A. (1997) DNA methylation directs a time-dependent repression of transcription initiation. *Current Biology* 7: 157–65

Larsen, F., Gunderson, G., Lopez, R. & Prydz, H. (1992) CpG islands as gene markers in the human genome. *Genomics* 13: 1095–107

Lei, H., Okano, M., Juttermann, R. *et al.* (1996) De novo DNA cytosine methyltransferase in mouse embryonic stem cells. *Development* 122: 3195–205

Leonhardt, H., Page, A.W., Weier, H.U. & Bestor, T.H. (1992) A targeting sequence directs DNA methyltransferase to sites of DNA replication in mammalian nuclei. *Cell* 71: 865–73

Leutwiler, L.S., Hough-Evans, B.R. & Meyerowitz, E.M. (1984) The DNA of *Arabidopsis thaliana*. *Molecular and General Genetics* 194: 15–23

Lewis, J.D., Meehan, R.R., Henzel, W.J. *et al.* (1992) Purification, sequence and cellular localisation of a novel chromosomal protein that binds to methylated DNA. *Cell* 69: 905–14

Li, E., Bestor, T.H. & Jaenisch, R. (1992) Targeted mutation of the DNA methyltransferase gene results in embryonic lethality. *Cell* 69: 915–26

Macleod, D., Charlton, J., Mullins, J. & Bird, A.P. (1994) Sp1 sites in the mouse aprt gene promoter are required to prevent methylation of the CpG island. *Genes and Development* 8: 2282–92

Martin, C., Prescott, A., Lister, C. & MacKay, S. (1989) Activity of the transposon Tam3 in *Antirrhinum* and tobacco: possible role of DNA methylation. *EMBO Journal* 8: 997–1004

Meehan, R.R., Lewis, J.D., McKay, S., Kleiner, E.L. & Bird, A. (1989) Identification of a mammalian protein that binds specifically to DNA containing methylated CpGs. *Cell* 58: 499–507

Meehan, R.R., Lewis, J.D. & Bird, A. (1992) Characterization of MeCP2, a vertebrate DNA binding protein with affinity for methylated DNA. *Nucleic Acids Research* 20: 5085–92

Meyer, P., Niedenhof, I. & ten Lohuis, M. (1994) Evidence for cytosine methylation of non-symmetrical sequences in transgenic *Petunia hybrida*. *EMBO Journal* 13: 2084–88

Monk, M., Boubelik, M. & Lehnert, S. (1987) Temporal and regional changes in DNA methylation in the embryonic, extraembryonic and germ cell lineages during mouse embryo development. *Development* 99: 371–82

Nan, X., Meehan, R.R. & Bird, A. (1993) Dissection of the methyl-CpG binding domain from the chromosomal protein MeCP2. *Nucleic Acids Research* 21: 4886–92

Nan, X., Tate. P., Li, E. & Bird, A. (1996) DNA methylation specifies chromosomal localization of MeCP2. *Molecular and Cellular Biology* 16: 414–21

Nan, X., Campoy, J. & Bird, A. (1997) MeCP2 is a transcriptional repressor with abundant binding sites in genomic chromatin. *Cell* 88: 471–81

Nan, X., Ng, H.H., Johnson, C.A. *et al.* (1998) Transcriptional repression by the methyl-CpG binding protein MeCP2 involves a histone deacetylase complex. *Nature* 393: 386–89

Oakeley, E.J. & Jost, J.-P. (1996) Non-symmetrical cytosine methylation in tobacco pollen DNA. *Plant Molecular Biology* 31: 927–30

Okano, M., Xie, S. & Li, E. (1998a) Dnmt2 is not required for de novo and maintenance methylation of viral DNA in embryonic stem cells. *Nucleic Acids Research* 26: 2536–40

Okano, M., Xie, S. & Li, E. (1998b) Cloning and characterization of a family of novel mammalian DNA (cytosine-5) methyltransferases. *Nature Genetics* 19: 219–20

Panning, B. & Jaenisch, R. (1996) DNA hypomethylation can activate Xist expression and silence X-linked genes. *Genes and Development* 10: 1991–2002

Pinarbasi, E., Elliott, J. & Hornby, D. (1996) Activation of a yeast pseudo DNA methyltransferase by deletion of a single amino acid. *Journal of Molecular Biology* 257: 804–13

Pollock, J.M., Swihard, M. & Taylor, J.H. (1978) Methylation of DNA in early development: 5-methylcytosine content of DNA in sea urchin sperm and embryos. *Nucleic Acids Research* 5: 4855–64

Posfai, J., Bhagwat, A.S., Posfai G. & Roberts R.J. (1989) Predictive motifs derived from cytosine methyltransferases. *Nucleic Acids Research* 17: 2421–35

Pradhan, S. & Adams, R.L.P. (1995) Distinct CG and CNG DNA methyltransferases in *Pisum sativum*. *Plant Journal* 7: 471–81

Pradhan, S., Talbot, D., Sha, M. *et al.* (1997) Baculovirus-mediated expression and characterization of the full-length murine DNA methyltransferase. *Nucleic Acids Research* 25: 4666–73

Pruitt, R.E. & Meyerowitz, E.M. (1986) Characterization of the genome of *Arabidopsis thaliana*. *Journal of Molecular Biology* 187: 169–83

Rae, P.M.M. & Steele, R.E. (1979) Absence of cytosine methylation at CCGG and GCGC sites in the rDNA coding regions and intervening sequences of Drosophila and the rDNA of other higher insects. *Nucleic Acids Research* 6: 2987–95

Richards, E.J. (1997) DNA methylation and plant development. *Trends in Genetics* 13: 319–23

Ronemus, M.J., Galbiati, M., Ticknor, C., Chen, J. & Dellaporta, S.L. (1996) Demethylation-induced developmental pleiotropy in *Arabidopsis*. *Science* 273: 654–57

Rothnie, H.M., McCurrach, K.J., Glover, L.A. & Hardman, N. (1991) Retrotransposon-like nature of Tp1 elements: implications for the organisation of highly repetitive, hypermethylated DNA in the genome of *Physarum polycephalum*. *Nucleic Acids Research* 19: 279–86

Schlappi, M., Raina, R. & Fedoroff, N. (1994) Epigenetic regulation of the maize Spm transposable element: novel activation of a methylated promoter by TnpA. *Cell* 77: 427–37

Simmen, M.W., Leitgeb, S., Charlton, J. *et al.* (1999) Non-methylated transposable elements and methylated genes in a chordate genome. *Science* 283: 1164–67

Simpson, V.J., Johnson, T.E. & Hammen, R.F. (1986) *Caenorhabditis elegans* DNA does not contain 5-methylcytosine at any time during development or aging. *Nucleic Acids Research* 14: 6711–19

Singer, J., Roberts-Ems, J. & Riggs, A.D. (1979a) Methylation of mouse liver DNA by means of restriction enzymes HpaII and MspI. *Science* 203: 1019–21

Singer, J., Roberts-Ems, J., Luthardt, F.W. & Riggs, A.D. (1979b) Methylation of DNA in mouse early embryos, teratocarcinoma cells and adult tissues of mouse and rabbit. *Nucleic Acids Research* 7: 2369–85

Stein, R., Razin, A. & Cedar, H. (1982) In vitro methylation of the hamster adenine phosphorybosyl transferase gene inhibits its expression in mouse L cells. *Proceedings of the National Academy of Sciences USA* 79: 3418–22

Struhl, K. (1998) Histone acetylation and transcriptional regulatory mechanisms. *Genes and Development* 12: 599–606

Tate, P.H. & Bird, A. (1993) Effects of DNA methylation on DNA-binding proteins and gene expression. *Current Biology* 3: 226–31

Tate, P., Skarnes, W. & Bird, A. (1996) The methyl-CpG binding protein MeCP2 is essential for embryonic development in the mouse. *Nature Genetics* 12: 205–08

Tweedie, S., Charlton, J., Clark, V. & Bird, A. (1997) Methylation of genomes and genes at the invertebrate–vertebrate boundary. *Molecular and Cellular Biology* 17: 1469–75

Vardimon, L., Kressmann, A., Cedar, H., Maechler, M. & Doerfler, W. (1982) Expression of a cloned adenovirus gene is inhibited by in vitro methylation. *Proceedings of the National Academy of Sciences USA* 79: 1073–77

Waalwijk, C. & Flavell, R.A. (1978) MspI, an isoschizomer of HpaII which cleaves both unmethylated and methylated HpaII sites. *Nucleic Acids Research* 5: 3231–36

Watt, F. & Molloy, P.L. (1988) Cytosine methylation prevents binding to DNA of a HeLa cell transcription factor required for optimal expression of the adenovirus late promoter. *Genes and Development* 2: 1136–43

Weitzel, J.M., Buhrmester, H. & Stratling, W.H. (1997) Chicken MAR-binding protein ARBP is homologous to rat methyl-CpG-binding protein MeCP2. *Molecular and Cellular Biology* 17: 5656–66

Wilkinson, C.R.M., Bartlett, R., Nurse, P. & Bird, A.P. (1995) The fission yeast gene pmt1+ encodes a DNA methyltransferase homologue. *Nucleic Acids Research* 23: 203–10

Wilson, G.G. & Murray, N.E. (1991) Restriction and modification systems. *Annual Review of Genetics* 25: 585–627

Woodcock, D.M., Lawler, C.B., Linsenmeyer, M.E., Doherty, J.P. & Warren, W.D. (1997) Asymmetric methylation in the hypermethylated CpG promoter region of the human L1 retrotransposon. *Journal of Biological Chemistry* 272: 7810–16

Yoder, J.A., Walsh, C. & Bestor, T.H. (1997) Cytosine methylation and the ecology of intragenomic parasites. *Trends in Genetics* 13: 335–40

GLOSSARY

Ac element activator transposable element found in maize

chromatin nuclear complex composed of DNA and associated proteins such as histones

imprinting parent-specific expression (or repression) of genes or chromosomes in offspring

isoschizomers restriction enzymes that recognize the same DNA sequence

satellite sequences generally restricted to the centromeric region of eukaryotic chromosomes, these highly repeated short DNA sequences are not transcribed and have no known function

Spm element suppressor–mutator transposable element found in maize

See also **Genomic imprinting** (p.757); **X-chromosome inactivation** (p.780)

GENOMIC IMPRINTING

Ben Pickard[1] and Wolf Reik[2]
[1]*Department of Psychiatry, Medical Genetics Section, Molecular Medicine Centre, University of Edinburgh, UK*
[2]*Laboratory of Developmental Genetics and Imprinting, Babraham Institute, Babraham, Cambridge, UK*

1. Definition

Genomic, genetic or gametic imprinting is a genetic mechanism by which some genes in the genome are expressed or repressed depending on whether they have been maternally or paternally inherited. Some imprinted genes are maternally expressed, others are paternally expressed. These expression differences are brought about by *epigenetic* marks (the imprints), such as DNA methylation, which are introduced in the parental germ cells and maintained in the offspring.

2. Development and genetics

(i) Consequences of genomic imprinting

Mendel's second law of genetic inheritance states that the coding potential of each parental set of chromosomes is equivalent. Experimentally manipulated embryos consisting of two maternal chromosome sets (known as *gynogenetic/parthenogenetic* embryos) or two paternal sets (*androgenetic* embryos) would, therefore, be viable and develop normally. This is indeed the case for some groups of animals and plants. However, in mammals, both situations result in lethality at an early embryonic stage. This proves that the maternal and paternal copies of chromosomes cannot be functionally equivalent, even if they are identical genetically (i.e. have the same DNA sequence); a copy of each chromosome from both parents is required in embryonic cells for successful development. It is the imprinting of certain genes that gives rise to these effects. As a result of imprinting, a parthenogenetic embryo will transcribe maternally expressed imprinted genes (at twice the normal level, as both alleles will be operational), but not paternally expressed genes, which are silenced. The reverse is true for androgenetic embryos.

Since these expression deficits and excesses manifest themselves as embryonic failure, imprinted genes may play important roles in developmental processes. As different sets of imprinted genes are affected in androgenetic and parthenogenetic embryos, there are differences in the precise developmental problems that arise from these two situations. Androgenetic embryos, although having well-developed extraembryonic tissues (e.g. trophoblast/placenta), do not develop past the 6–8 somite stage. In contrast, parthenogenetic embryos can reach the 25 somite stage, but have poor extraembryonic development (Surani *et al.*, 1990). Therefore, the differences between these two types of embryo seem to reveal a division of labour between maternally and paternally expressed imprinted genes.

(ii) Uniparental disomies and the imprinting map

In order to locate imprinted genes and to study their effects in relative isolation from each other, uniparental disomies (UPDs) have been used extensively (Cattanach & Beechey, 1990). Heterozygous crosses using reciprocal translocations and Robertsonian fusions give rise to duplications of chromosomes or chromosome regions from one parent, with corresponding absence of those from the other parent. By examining mice containing various UPDs for abnormalities in development and behaviour, it was possible to compile a chromosomal map indicating which regions of each chromosome contain genes which are expressed from just one parental allele. **Figure 1** (and **Plate 23** in the colour plate section) shows the locations which have been shown to contain imprinted genes as revealed by UPDs, together with those imprinted genes that have so far been chromosomally mapped.

It is immediately obvious from Figure 1 that most imprinted genes are chromosomally clustered. This raises important questions, particularly about regional mechanisms of imprinting (see Section 3, Mechanisms, below). It can also make difficult the interpretation of UPD phenotypes in relation to the individual imprinted genes that are known to map to that chromosomal region. For example, the growth deficiency of mouse embryos with maternal disomy of distal chromosome 7 can be explained by the absence of insulin-like growth factor 2 (*Igf2*) expression, but their *in utero* lethality cannot (and must be due to deficiency or excess of another imprinted gene in that cluster).

Generally, when new imprinted genes have been isolated, they map to known areas of imprinted gene action as defined by UPDs. *Rasgrf1* is one of several apparent exceptions as it maps to a region that has not previously revealed imprinting effects. However, the phenotypes of *Rasgrf1*, revealed by homologous recombination, are rather subtle (loss of long-term memory and postnatal growth effects). More sensitive testing procedures (especially for behavioural phenotypes) may be needed to identify certain imprinted regions. Imprinting is generally conserved in the human genome (with only a few exceptions), and the associations of naturally occurring human UPDs with specific phenotypes has again allowed the establishment of an imprinting map which reveals syntenic organization of imprinted segments and genes.

(iii) Categorizing imprinted gene action

It has been estimated that a minimum of 100 human genes (approximately 0.1% of the total number) may be subject to

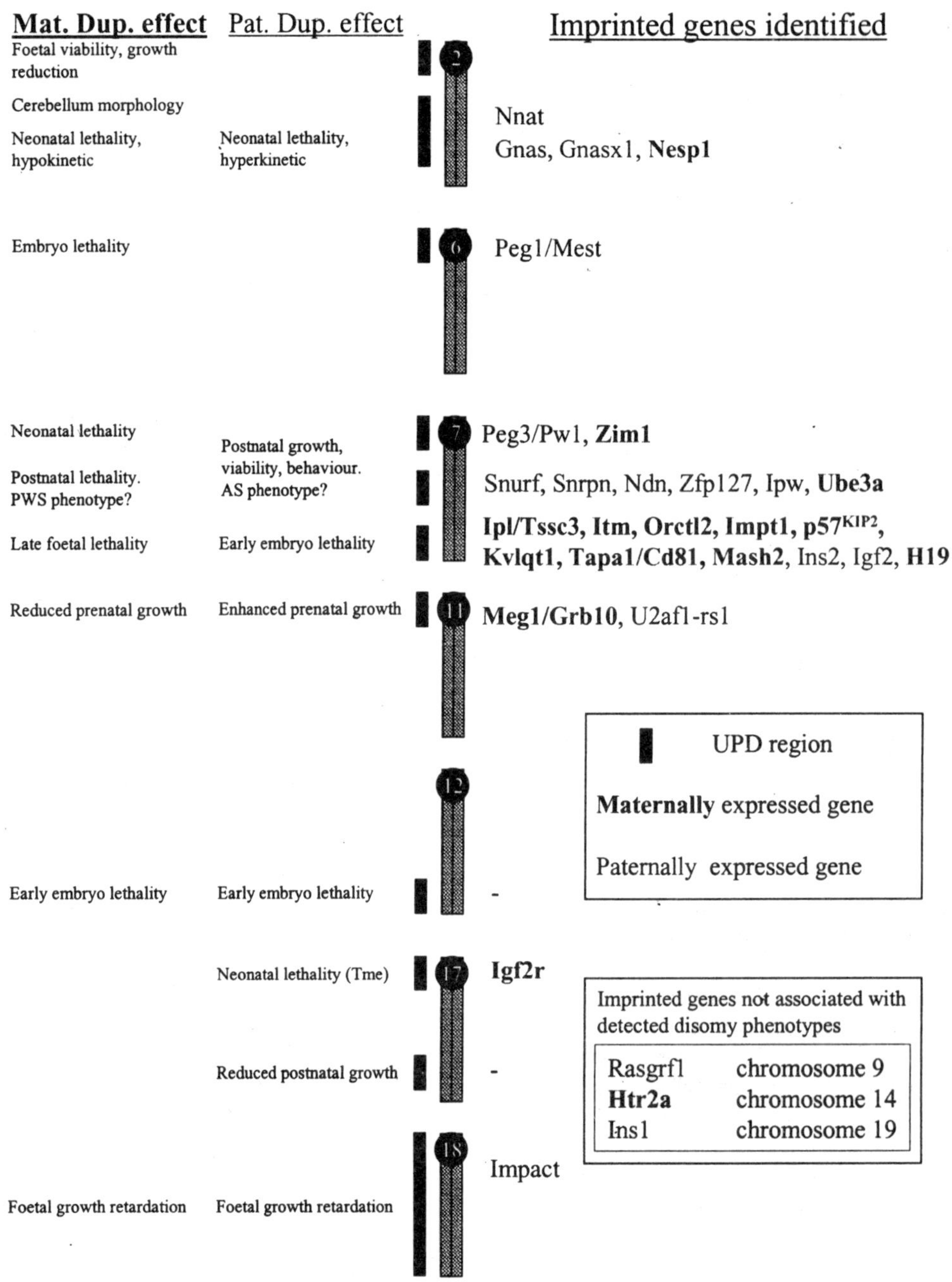

Figure 1 Uniparental disomy (UPD) phenotypes in mice and the imprinted genes associated with them. The seven chromosomes with known UPD phenotypes are displayed in the centre of the diagram (circles indicate the centromeric ends of the chromosomes), together with the approximate locations of the UPDs (rectangles). (See also **Plate 23** in colour plate section)

imprinting; currently, the number of identified imprinted genes is 31. In addition, a number of diseases/phenotypes have been studied which show the characteristic patterns of inheritance indicative of the involvement of imprinting (see Section 5, Imprinting and disease, below). The Medical Research Council Mammalian Genetics Unit at Harwell, UK, maintains a Web page detailing all published imprinted genes and chromosomal regions (http://www.mgu.har.mrc.ac.uk/).

Functional studies of imprinted genes show that it is possible to categorize them loosely into two groups: those which play a role in growth and intrauterine development and those which affect behaviour. Another emerging pattern matches the parent-of-origin of the expressed allele of an imprinted gene with the direction of its effect. Even though there are exceptions to the general rule, it appears that maternally expressed genes generally inhibit embryonic growth whereas paternally expressed genes promote it. This may have an evolutionary significance and is further discussed in Section 6, Evolution (below). Paternally expressed genes may preferentially contribute to the development of muscle and other mesodermal tissues, and may be expressed in the hypothalamus, a brain region involved in aggression and kinetic activity. Maternally expressed genes, on the other hand, tend to be expressed in the epidermis and other ectodermal tissues and may be expressed in the neocortex, a brain region involved in higher order processing (Allen *et al.*, 1995). The link between imprinted genes and higher order cognitive processing is intriguing, but may be quite complex. This is exemplified by the effect of parental origin of the X chromosome in Turner's syndrome females who have just one X chromosome. Those inheriting it from their mother seem to have impaired social skills compared with those with a paternally derived X chromosome implying that, in this case, a putative gene involved in higher cognitive function is paternally expressed (Skuse *et al.*, 1997).

3. Mechanisms

Imprinted genes must have molecular properties which enable them to "remember" from which parent they are derived. This imprint has to be reset during gametogenesis such that, for example, a father ensures that his maternally inherited chromosome complement is rewritten as being paternal in origin (i.e. as his) before being passed on to his progeny. At present, it seems that CpG methylation is a major mechanism of imprinting.

Methylation of DNA occurs on the 5′ position of cytosine residues present in the dinucleotide CpG. Since the sequence on the opposite strand in the double helix is also CpG, this can also become methylated. Methylation is catalysed by enzymes termed DNA methyltransferases, which can be divided into two groups: maintenance methyltransferases which prevent the dilution of the methylation signal (during DNA replication, for example) by fully methylating hemimethylated (only one strand methylated) CpG sites, and *de novo* methyltransferases which recognize and methylate completely unmethylated CpGs. Of the four mammalian DNA methyltransferases (encoded by *Dnmt* genes) characterized to date, Dnmt1 appears to be predominantly a maintenance enzyme whereas Dnmt3α and Dnmt3β seem to be *de novo* methyltransferases. In addition, demethylation during development can either arise from an absence of maintenance methylation (a passive process) or be due to the presence of demethylase activity.

The role of DNA methylation in imprinting was formally established by deletion of the mouse *Dnmt1* maintenance methyltransferase gene by targeted disruption ("knockout"). This resulted in embryos with a large (but not complete) decrease in global methylation levels, leading to postgastrulation lethality (Li *et al.*, 1992). The expression of imprinted genes became deregulated, with biallelic expression of some genes and repression of others. This establishes that DNA methylation is vital for imprinting, but does not further define whether it is involved in the initiation of imprinting in germ cells in addition to its maintenance in somatic cells.

The vast majority of imprinted genes studied have been found to be differentially methylated between the parental chromosomes in somatic tissues. Within each imprinted gene, there are often one or more discrete regions which show very clear differences in methylation between the two parental alleles; these are termed "differentially methylated regions" or DMRs (Feil *et al.*, 1994). In a number of DMRs (core DMRs), methylation differences are introduced in parental germ cells and maintained throughout development; in others, methylation differences are only established during differentiation. The core DMRs are thought to be important for imprinting control (Olek & Walter, 1997). The simplest mechanism for the imprinting of a gene might be that the future inactive copy (or a part thereof, the DMR) becomes methylated in one germline and remains so after fertilization and in all somatic tissues. This is indeed observed. However, more complex mechanisms are also found, particularly when linked imprinted genes in a cluster are coordinately regulated. The different mechanisms found so far are discussed below and illustrated in **Figure 2** (see also **Plate 24** in the colour plate section).

(i) Methylation and repression

In this category (containing *H19*, *Peg1*, *U2af1-rs1* and other genes), DNA methylation is introduced into the DMR of the future inactive copy in one germline. In *H19*, for example, methylation is introduced into the DMR in spermatogonia and is retained after fertilization, with methylation apparently spreading into the promoter and gene body postfertilization. The maternal copy, in contrast, is not methylated in the egg and remains unmethylated after fertilization. When the enhancers for the gene become active (immediately after implantation), the maternal copy is expressed, but the paternal one is silent. The DMR has an important part to play in the control of imprinting; when it is deleted, the paternal copy of *H19* is no longer methylated and is expressed.

(ii) Sense and antisense

A different type of regulation is observed for the *Igf2r* gene, which encodes a receptor for the Igf2 ligand. This gene has a differentially methylated region in an intron which is methylated on the expressed maternal allele. This is contrary to expectation because the methylation of an allele is usually associated with its silencing. Targeted deletion of this region in mice results in loss of repression of the paternal allele (Wutz *et al.*, 1997). The most likely explanation comes from the observation that the DMR acts as a promoter or enhancer for an antisense transcript (transcription in the opposite direction to that producing translatable mRNA). Hence, it may be that the paternal antisense expression (initiating from the unmethylated DMR) prevents sense expression on the same allele, which could occur by a variety of mechanisms.

Increasing numbers of untranslated antisense transcripts associated with imprinted genes are being discovered. For example, the maternally expressed Angelman syndrome gene, *UBE3A*, has a paternally expressed antisense transcript associated with it. The *Igf2* gene has several antisense transcripts, but, unlike *Igf2r* and *UBE3A*, they are expressed from the same allele as the sense transcript. Linked sense and antisense systems of imprinted genes could be one reason for gene clustering.

(iii) Enhancer sharing

As described above, imprinting of the *H19* gene is assumed to be regulated by a closely linked DMR. However, 100

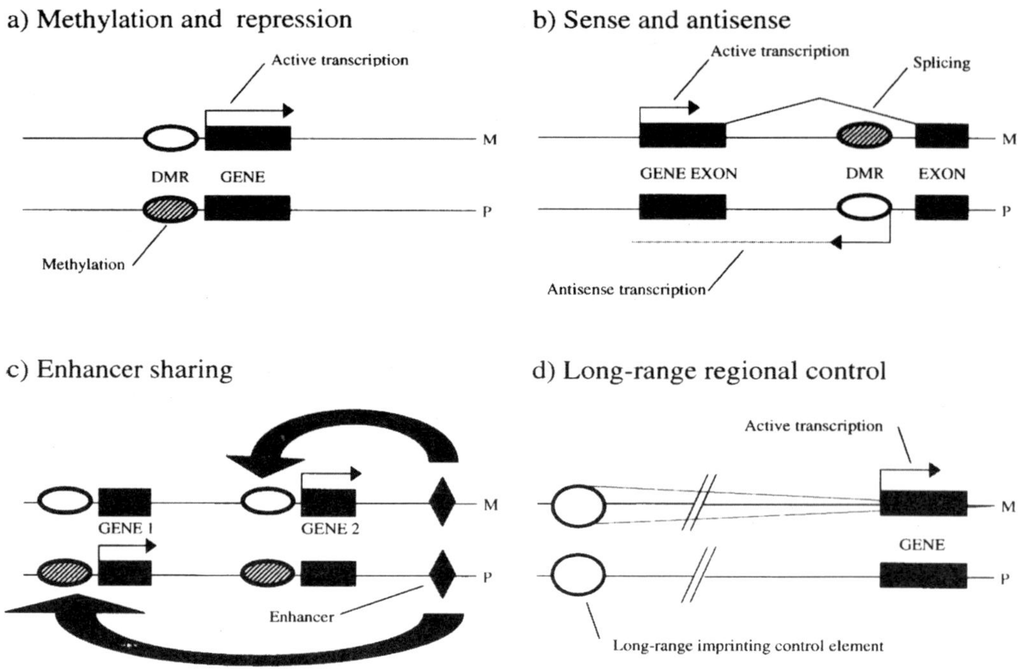

Figure 2 The four principal mechanisms by which imprinted gene expression is regulated. Genes (or their constitutive exons) are shown as rectangles (M for maternally derived, P for paternally derived). Differentially methylated regions (DMRs) are represented as ovals which are either open (unmethylated) or hatched (methylated). Transcription is denoted by arrows. See text for details of each mechanism. (see also **Plate 24** in colour plate section)

kilobases (kb) upstream of *H19* is the paternally expressed *Igf2* gene whose imprinting status is dependent, at least in part, on *H19* and its control sequences. An enhancer located 3′ to *H19* (which is active in the endodermal subset of tissues where *H19* is expressed) acts on *H19* on the maternal allele, but drives *Igf2* transcription on the paternal allele, on which *H19* is methylated. Whether the promoters of *Igf2* and *H19* actively compete for the enhancer (*H19*, when methylated, allows *Igf2* access) or whether part of the *H19* DMR acts as an "insulator" or "boundary" element that, when methylated, prevents access of the *H19* enhancer to the *Igf2* promoter is not yet clear (Jones *et al.*, 1998 and references within). Other systems of enhancer sharing and boundary elements between imprinted genes may well exist and could also contribute to the observed gene clustering.

(iv) Long-range regional control

Although *H19* imprinting and expression interact with the linked *Igf2* and *Ins2* genes, other imprinted genes in the same cluster, such as *p57*^{Kip2}*/Cdkn1c*, are apparently not affected by this type of control. Whether other longer-range regional controls exist in this particular cluster is not clear, but there are regional epigenetic features other than DNA methylation that seem to encompass the whole cluster. These include differences in the timing of DNA replication between the homologues (alleles) and differences in meiotic recombination frequencies between the two germlines.

There is genetic evidence for long-range imprinting control from another imprinting cluster on human chromosome 15q11–13. Imprinting mechanisms associated with the Prader–Willi and Angelman syndromes (see Section 5, Imprinting and disease, below) can result in altered epigenotypes (e.g. a maternal chromosome with a paternal pattern of methylation and expression) extending over several hundred kb. Mutations (such as deletions) have been found to be associated with a paternal-to-maternal switch in the promoter region of the *SNRPN* gene. The opposite switch is associated with mutations just upstream of this promoter where low-level untranslated transcripts are found. It has been proposed that these mutations make the whole region unable to switch imprints in the parental germline (see Section 4, The life cycle of the imprint, below), but whether they actually act in the germline or after fertilization has so far not been established; that these effects are due to long-range enhancer sharing and sense–antisense interactions as described is also not excluded.

4. The life cycle of the imprint

(i) From the germline to the embryo

The previous sections have described how imprinting (via methylation) affects gene expression and, therefore, development. However, this does not explain the origin of parental differences in imprinting. Imprinting takes place in the two germlines, and it seems that the methylation patterns are completely erased at an early time-point (day 12.5) in the cells that give rise to spermatocytes and oocytes. Each germline then puts in place the methylation patterns that will dictate the monoallelic expression of the imprinted genes within the genome; this probably happens during oocyte growth in females and in spermatogonia in males. After fertilization, transcriptional regulation complexes are able to distinguish the differently methylated parental alleles and initiate competent transcription from just one of them.

It must not be forgotten that, as well as controlling imprinted gene expression, methylation also has a general effect on non-imprinted gene transcription by the inactivation of tissue-specific genes outside of their expression domains. As a rule, sperm DNA is more methylated than egg

DNA. In preimplantation embryos, there is a wave of global demethylation which is followed by remethylation during gastrulation. However, this is the global picture of methylation; imprinted genes seem relatively impervious to these shifting methylation levels and undergo modification of the gametic methylation imprint according to their individual criteria and timetables. Often, despite the presence of an initial methylation imprint, there may be biallelic expression from imprinted genes in early embryonic stages or in particular tissues. This can be explained either by the immaturity of the methylation imprint or the failure of the associated regulatory complexes to read it correctly. In addition, a subset of the imprinted genes are only monoallelically expressed during early embryonic development (they are biallelic at later stages) which reinforces the role of imprinting in developmental processes and explains why imprinting defects are often manifest at early time-points.

Perhaps the clearest evidence for the initiation of the imprint in gamete-producing cells comes from the study of genes which become imprinted after the introduction of foreign sequences. *Fused* is a classical mutation affecting the development of the neural tube (Zeng *et al*., 1997). This mutation is present as several distinct alleles which vary in penetrance and expressivity, although they all seem to act through "gain of function". Interestingly, some alleles exhibit a marked reduction in potency when transmitted through the female germline: they show a form of genomic imprinting because their parental origin determines their activity. The *Fused* mutation has been mapped to a gene, *Axin*, which is a component of the signal transduction pathway responsible for setting up the body axis. The mutant alleles which possess imprinted activity all seem to have Intracisternal-A Particle (IAP) provirus insertions in the gene region. It may be the case that these proviral sequences are recognized and modified in female gamete-producing cells by the endogenous genomic imprinting machinery such that the mutated allele becomes transcriptionally silenced. This would result in expression from only the non-mutated paternally derived allele, permitting near normal development. If this is the case, the analysis of IAP sequences should permit insights into the imprinting process which may extend to naturally imprinted genes. Indeed, DMRs in imprinted genes and IAPs may share some structural features, such as direct repeats.

(ii) The cellular machinery of imprinting

Once the methylation component of the imprint is established, it is necessary for the cellular machinery to recognize that only one of the alleles is competent for transcription. Methylation might prevent the binding of transcription factors to promoters, thus rendering them inactive. Alternatively, methylated CpGs may recruit certain factors resulting in the formation of protein complexes which may promote (or more often repress) transcription. It has been shown for some imprinted genes that the protein complexes (forming chromatin) are preferentially associated with one of the two alleles.

Methylcytosine-binding protein 2 (MeCP2) has been cloned and shown to bind with high affinity to singly methylated stretches of DNA. Targeted disruption of this X-linked gene resulted in early embryonic lethality depending on the percentage contribution of "knocked out" cells to experimentally derived chimaeras (Tate *et al*., 1996). However, in humans, mutations in the corresponding gene, *MECP2*, are associated with some cases of Rett syndrome, a neurodevelopmental disorder affecting female children (Amir *et al*., 1999). Another methylcytosine-binding protein, MeCP1, is comprised of multiple subunits, one of which has been identified as PCM1. This subunit acts to bind methylated DNA where it has been shown to inhibit local transcription. Interestingly, recent work has shown that MeCP2 interacts with the histone deacetylase complex (reviewed in Ng & Bird, 1999). This protein complex removes acetyl moieties from DNA-bound histones resulting in silencing of the associated DNA sequences. Trichostatin A (an inhibitor of histone deacetylase activity) has been shown to perturb the imprinting of certain imprinted genes. Thus, the action of methylcytosine-binding proteins may be just the first step in the assembly of large chromatin complexes which affect the transcription of DNA over a wide distance – another possible explanation for the existence of clusters of imprinted genes.

5. Imprinting and disease

Imprinting diseases can arise through the direct mutation of individual imprinted genes themselves. This is exactly the same way in which heritable diseases occur in all genes, the only difference being that only the expressed allele has to be mutated in an imprinted gene before the effects are felt (i.e. mutations are always dominant in nature; Clarke, 1990).

However, there is a second class of mutations in imprinted genes – disruption of parental-specific expression. This is known as "loss of imprinting" (LOI) and underlies most of the classical imprinted diseases as well as being a contributory factor in several types of tumour such as Wilms' tumour. It is hard to locate the changes in sequence responsible for such diseases because the mutation can often lie at considerable distance from the affected gene(s) within the imprinted cluster. One possibility is that, within each cluster, there exists a control region which specifies that the local environment is subject to imprinting (although it does not define the parental direction as genes within a single cluster can be imprinted in different directions). A prediction from such a model would be that disruption of this region would result in the deregulation of all the members of the cluster. Another class of events that may lead to LOI are "epimutations" – changes in DNA methylation or imprinting without any sequence mutations.

The presence of such a region appears to be suggested by certain mutations in the *IGF2–H19* imprinting cluster on human chromosome 11p15.5 which give rise to Beckwith–Wiedemann syndrome (BWS; Reik & Maher, 1997). While the majority of mutations or epimutations in BWS seem to affect the individual genes within the cluster (e.g. *H19*, *IGF2* and *CDKN1C/p57*^{KIP2}), there are a number of patients whose pathology seems to arise from chromosomal translocation events within the *KCNA9/KvLQT* gene region at the centre of the cluster. Although disruption of a gene in this way might be expected to be sufficient to cause the BWS symptoms, it seems as if the *KCNA9* gene disruption and the effect the translocations have on the regulation of imprinting in the cluster are separable effects. Therefore, it has been postulated that there is a local imprinting control centre within the *KCNA9* gene region.

In another example of regional control, deletions on human chromosome 15q11–13 result in two different diseases: Prader–Willi syndrome (PWS) and Angelman syndrome (AS). This is the site of an imprinted gene cluster; one half of the cluster (affected in AS) contains at least one maternally expressed gene whereas the other half (affected in PWS) contains at least seven paternally expressed genes. Again, sporadic forms (affecting the individual imprinted genes) and familial forms (caused by deletions which result in LOI of the genes) of these diseases are observed (Nicholls *et al.*, 1998). The deletions reside in the 5′ region of a paternally expressed gene, *SNRPN*. One part of this 5′ region contains sequences which, when deleted, prevent the paternal imprint from being switched to a maternal imprint (e.g. if a mother passes her father's mutated chromosome to her children, 50% of her children will have AS). Another part of this region, slightly closer to the *SNRPN* gene, causes PWS when deleted. In this case, the switch from the maternal imprint to the paternal imprint is prevented (e.g. if a father passes on his mother's mutated chromosome to his children). These observations suggest that the 5′ end of *SNRPN* is acting to set the imprinting status for all genes present in the cluster.

6. Evolution

If, as suggested by the preceding sections, imprinting plays such a vital role in growth and development, why do many organisms thrive without its involvement? Mammals and angiosperms both possess *bona fide* imprinting. Marsupials only display a form of imprinting in that they inactivate solely the paternal X chromosome in female offspring (this also occurs in the extraembryonic tissues of eutherians) in contrast to the more usual random inactivation. Perhaps this tells us that imprinting has evolved several times over, as and when it is required.

The imprinting of certain genes has to provide a selective advantage for the initial members of a species that acquire it if the trait is to spread to encompass all members of the population. Evolutionary biologists have long sought an explanation for the selective advantages of (mammalian) monoallelic expression. A number of theories have been proposed which have been extensively reviewed elsewhere (see Moore & Haig, 1991; Hurst & McVean, 1997 and references within). One of these theories, the "conflict theory", is supported by many observations from experimental studies. The theory proposes that there is a shortfall of nutrients in the intrauterine environment in mammals and that the "selfish" genes of mother and offspring have different priorities as a result. It is not in the best (genetic) interests of a mother to invest all of her nutritional resources in one litter at the expense of subsequent litters. However, if an individual embryo actively extracted more of the mother's resources than its siblings, then it might stand a better chance of survival and the opportunity to pass on its genes.

This competition for resources might be especially pronounced in species where multiple matings with different males precede pregnancy (polyandry); here, it would be in the best interests of a particular father's "selfish" alleles to ensure the preferential survival of their host embryo over the other embryos also present in the uterus (which might be the result of matings with other males). In contrast, the mother has an equal genetic investment in each of her embryos and attempts to even out embryonic demands within a litter and between litters. The maternally expressed *H19* growth suppressor and the paternally expressed *Igf2* growth factor seem to provide a mechanistic example for this hypothesis, as it can be imagined that there would be a competition between embryonic nutrition demand and maternal supply based on the relative activities of the two genes. *H19* deletion mutations and additional transgenic copies of *Igf2* both result in foetal overgrowth (greater embryonic nutrition). Although counter-intuitive at first hearing, this overgrowth is also seen with Igf2 receptor (*Igf2r*, maternally expressed) mutations, albeit coupled with postnatal lethality. This can be explained by the fact that the Igf2r protein does not act to transduce the growth factor signal, but, in fact, sequesters serum Igf2; a failure to carry out this function in the mother results in greater levels of available growth factor and subsequent foetal overgrowth. This balancing act is confirmed by the fact that *Igf2–Igf2r* or *Igf2–H19* double mutants appear to develop relatively normally. In similar competitive systems, an evolutionary "arms race" is often observed such that one component gains the upper hand through an advantageous mutation until a mutation in the other component is selected for and the balance is restored temporarily. The *H19–Igf2* genes seem to exhibit only background level mutation rates which perhaps argues against this hypothesis or maybe suggests that the consequences of unbalanced expression (in either direction) outweigh selective advantages at this particular stage of evolution.

One set of experiments argues for the conflict hypothesis in the context of *H19–Igf2* (Rogers & Dawson, 1970). Interspecific crosses between *Peromyscus polionotus* (old field mouse) and *P. maniculatus* (deer mouse) produce very differently sized offspring depending on the direction of the cross. Large offspring are produced when the father is from the polygamous species (*P. maniculatus*), whereas small offspring are produced when the father is from the monogamous species (*P. polionotus*). In the first case, the embryos are "aggressive" in the appropriation of nutrients from the mother. In the second case, the mother unnecessarily restricts nutrient supply to the embryos. Recently, molecular investigation of these crosses has shown that *H19–Igf2* imprinting is perturbed in the hybrid offspring (Vrana *et al.*, 1998). This matching of the conflict theory with reproductive strategy may also explain why some humans express the IGF2 receptor (*IGF2R*) biallelically (even though some allelic methylation differences are still evident) – such an arms race holds no advantage for a primarily monogamous species.

REFERENCES

Allen, N.D., Logan, K., Lally, G. *et al.* (1995) Distribution of parthenogenetic cells in the mouse brain and their influence on brain development and behavior. *Proceedings of the National Academy of Sciences USA* 92: 10782–86

Amir, R.E., van den Veyver, I.B., Wan, M. *et al.* (1999) Rett syndrome is caused by mutations in X-linked *MECP2*, encoding methyl-CpG-binding protein 2. *Nature Genetics* 23: 185–88

Cattanach, B.M. & Beechey, C.V. (1990) Autosomal and X chromosomal imprinting. *Development* (Suppl.) 63–72

Clarke, A. (1990) Genetic imprinting in clinical genetics. *Development* (Suppl.) 131–39

Feil, R., Walter, J., Allen, N.D. & Reik, W. (1994) Developmental control of allelic methylation in the imprinted Igf2 and H19 genes. *Development* 120: 2933–43

Hurst, L.D. & McVean, G.T. (1997) Growth effects of uniparental disomies and the conflict theory of genomic imprinting. *Trends in Genetics* 13: 436–43

Jones, B.K., Levorse, J.M. & Tilghman, S.M. (1998) Igf2 imprinting does not require its own DNA methylation or H19 RNA. *Genes and Development* 12: 2200–07

Li, E., Bestor, T.H. & Jaenisch, R. (1992) Targeted mutation of the DNA methyltransferase gene results in embryonic lethality. *Cell* 69: 915–26

Moore, T. & Haig, D. (1991) Genomic imprinting in mammalian development: a parental tug-of-war. *Trends in Genetics* 7: 45–49

Ng, H.-H. & Bird, A. (1999) DNA methylation and chromatin modification. *Current Opinion in Genetics and Development* 9: 158–63

Nicholls, R.D., Saitoh, S. & Horsthemke, B. (1998) Imprinting in Prader–Willi and Angelman syndromes. *Trends in Genetics* 14: 194–200

Olek, A. & Walter, J. (1997) The pre-implantation ontogeny of the H19 methylation imprint. *Nature Genetics* 17: 275–76

Reik, W. & Maher, E.R. (1997) Imprinting in clusters: lessons from Beckwith–Wiedemann syndrome. *Trends in Genetics* 13: 330–34

Rogers, J.F. & Dawson, W.D. (1970) Foetal and placental size in a Peromyscus species cross. *Journal of Reproduction and Fertility* 21: 255–62

Skuse, D.H., James, R.S., Bishop, D.V. *et al.* (1997) Evidence from Turner's syndrome of an imprinted X-linked locus affecting cognitive function. *Nature* 387: 705–08

Surani, M.A., Kothary, R., Allen, N.D. *et al.* (1990) Genomic imprinting and development in the mouse. *Development* (Suppl.) 89–98

Tate, P., Skarnes, W. & Bird, A. (1996) The methyl-CpG binding protein MeCP2 is essential for embryonic development in the mouse. *Nature Genetics* 12: 205–08

Vrana, P.B., Guan, X.J., Ingram, R.S. & Tilghman, S.M. (1998) Genomic imprinting is disrupted in interspecific Peromyscus hybrids. *Nature Genetics* 20: 362–65

Wutz, A., Smrzka, O.W., Schweifer, N. *et al.* (1997) Imprinted expression of the Igf2r gene depends on an intronic CpG island. *Nature* 389: 745–49

Zeng, L., Fagotto, F., Zhang, T. *et al.* (1997) The mouse Fused locus encodes Axin, an inhibitor of the Wnt signalling pathway that regulates embryonic axis formation. *Cell* 90: 181–92

FURTHER READING

Bartolomei, M.S. & Tilghman, S.M. (1997) Genomic imprinting in mammals. *Annual Review of Genetics* 31: 493–525

Constancia, M., Pickard, B., Kelsey, G. & Reik, W. (1998) Imprinting mechanisms. *Genome Research* 8: 881–900

Reik, W. & Walter, J. (1998) Imprinting mechanisms in mammals. *Current Opinions in Genetics and Development* 8: 154–64

GLOSSARY

androgenetic androgenetic embryos have all their genetic information from their father: there is no maternal contribution

antisense the protein-encoding information in a gene can only be obtained by transcription in one direction along the gene. This is described as the sense direction. Antisense transcription (reading the gene in the reverse direction) doesn't usually produce a translatable product but may act to regulate that gene's expression

biallelic expression (*see* expression)

bisulphite sequencing a method allowing the position of every methylated CpG to be mapped in a defined stretch of sequence

CpG island a region rich in CpG dinucleotides. It is usually located at the 5′ end of an expressed housekeeping gene where it is found in the unmethylated state

differentially methylated region (DMR) a region of a gene which shows very different methylation states between the two parental alleles. DMRs are assumed to play an important role in the control of imprinted monoallelic gene expression

expression (monoallelic, biallelic) a term used to described the activity of a gene. An expressed gene is being actively transcribed to produce mRNA. This mRNA is translated to produce a functional protein. Most genes are expressed from both alleles (biallelic expression). Imprinted genes, in contrast, are only expressed from one parental allele (monoallelic expression)

gastrulation (*see* preimplantation)

gynogenetic/parthenogenetic gynogenetic/parthenogenetic embryos have all their genetic information from their mother. There is no paternal contribution

implantation (*see* preimplantation)

monoallelic expression (*see* expression)

parthenogenetic (*see* gynogenetic)

reciprocal translocation the reciprocal exchange of genetic material between two chromosomes

Robertsonian fusion (or centric fusion) the fusion of two acrocentric chromosomes to form a single chromosome with one centromere

targeted deletion/knock out an important technique used to examine the function of a gene by deleting it from the genome. The resulting mice are studied for phenotypes/deficits (*see also* embryonic stem cells)

See also **The *Wilms' tumour 1* gene *WT1*** (p.387); **DNA methylation in animals and plants** (p.746)

GENETIC CONTROL OF IMMUNOLOGICAL RESPONSES

David V. Serreze
The Jackson Laboratory, Bar Harbor, Maine, USA

1. Introduction

The function of the immune system is to defend an organism against infection by pathogenic microorganisms. It achieves the diversity necessary to combat a plethora of microorganisms through an enormously complex network of molecular and cellular events regulated by some of the most exotic control mechanisms known to exist in all of biology. Such complexity will limit this article to a broad, and by no means complete, overview of the genetic elements controlling immunological responses. For those who wish a more comprehensive survey of immunogenetics, the author recommends an outstanding textbook by Janeway and Travers (1997). For the sake of brevity, this article will cite review articles and textbooks, rather than primary publications.

Most attention will focus on the elegant genetic control of adaptive or acquired immunity. Adaptive immunity is so named because it is acquired during the lifetime of an organism as a set of specific responses elicited against particular stimulating agents known as antigens. These responses are distinguished from those of innate immunity. Innate immunity comprises inborn non-specific processes that protect against infection such as macrophage-mediated phagocytosis of microorganisms. Adaptive and innate immunity do not operate in a mutually exclusive fashion. Indeed, as noted throughout this article, these two arms of the immune system are usually triggered through interactive and coordinated processes. It should also be noted that while this article will mostly focus on knowledge gained from murine systems about the genetic control of immunity, such responses are regulated in a virtually identical fashion in humans.

2. Genetic control of immunoglobulin production

The discovery of antibodies, or immunoglobulin (Ig) molecules, had its roots in the work of Edward Jenner who found late in the 18th century that inoculation of humans with cowpox (i.e. vaccinia virus) resulted in protection against often fatal smallpox infections. In the 1890s, Emil von Behring and Shibasaburo Kitasato found that the serum in individuals vaccinated against particular pathogens contained substances, which they deemed antibodies, that mediated disease protection by specifically binding to the relevant microorganism. Subsequent work by Karl Landsteiner demonstrated that antibodies could be elicited against a virtually unlimited set of antigenic molecules. Proteins were found to be the most efficient antigenic molecules for eliciting antibody responses, but Landsteiner found that antibodies could also be raised against small organic molecules such as arsonates and nitrophenyls provided they were coupled to a protein carrier. Similarly, while antibodies can also be made to carbohydrates and nucleic acids, they generally must be bound to a protein carrier in order to elicit such responses. Nevertheless, the finding that the potential antibody repertoire is virtually unlimited, and hence vastly exceeds the total number of genes within a mammalian genome (~10^5), indicated that each specific antibody molecule is not encoded by a single gene that had been selected by evolution to confer resistance to a particular pathogen.

The first insights into the enigma of how a relatively limited set of genes can generate a virtually unlimited spectrum of antibody specificities were gleaned when the structure of these molecules was elucidated (reviewed in Reth, 1994). As shown in **Figure 1**, antibodies are comprised of four polypeptides that include two identical heavy (H) and two identical light (L) chains. The H chains are linked to each other and to an L chain by disulfide bonds. Antibodies are produced by B lymphocytes, both as cell membrane-bound and secreted proteins. An antibody molecule can utilize one of two types of L chain designated as kappa (κ) or lambda (λ). No functional differences have been found between antibodies utilizing κ or λ L chains. However, it is interesting to note that the ratio of antibodies utilizing κ or λ L chains varies between different species, being, in mice, 20:1. As discussed below, the antigenic specificity of a given antibody molecule results from amino acid sequence variation in approximately the first 110 residues comprising the paired NH2-terminal domains of both the H and L chains which is hence referred to as the variable (V) region. The

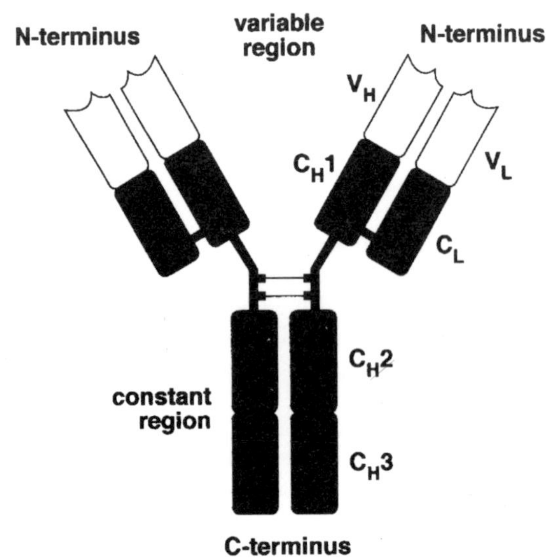

Figure 1 Structure of a representative Ig molecule. Open and shaded areas, respectively, depict the variable (V) and constant (C) regions of the heavy and light chains.

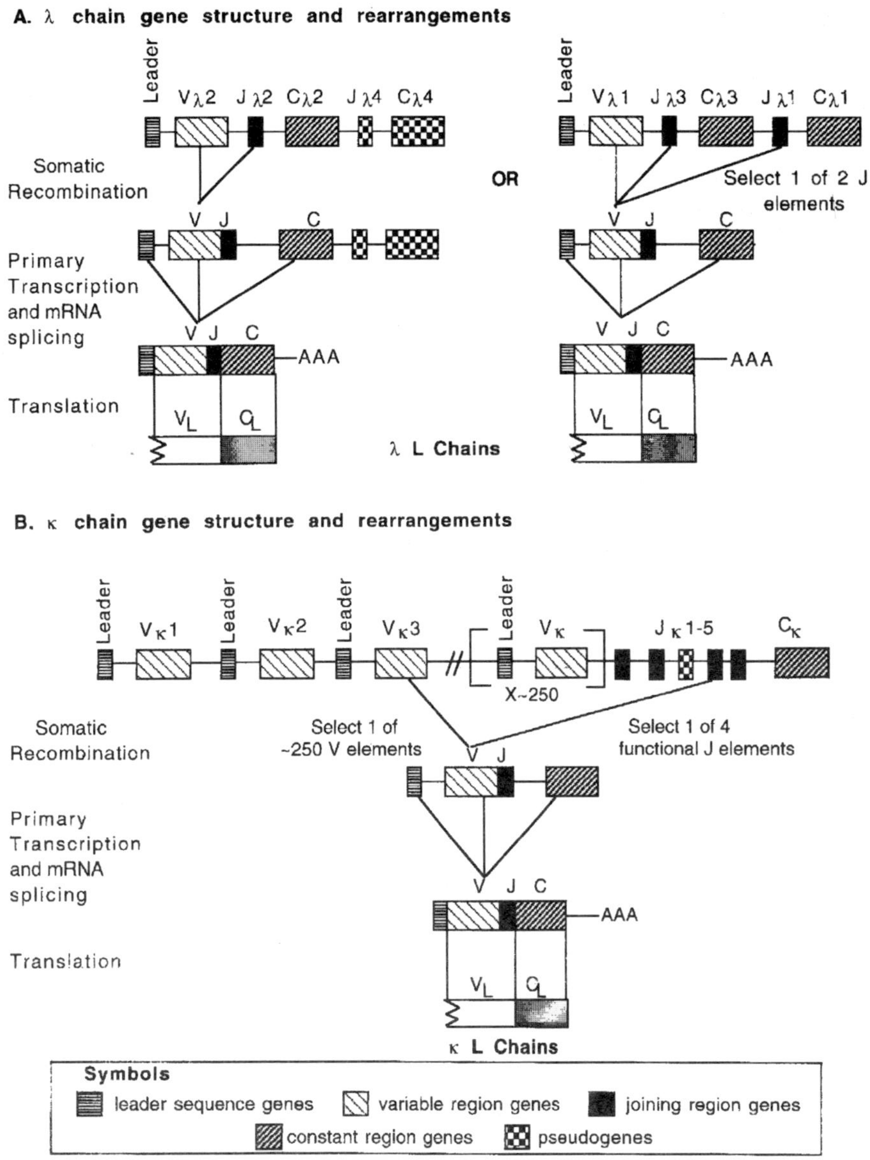

Figure 2 Ig L chain gene organization and rearrangement patterns.

COOH-terminus of each L chain does not vary with antigenic specificity and thus is termed the constant-light (C_L) region. There are five main classes of H chain referred to as isotypes which determine the functional characteristics of an antibody molecule. The five major antibody isotypes are designated as IgM, IgD, IgG, IgE and IgA, with their heavy chain components designated by the corresponding Greek letters (μ, δ, γ, ε, α). In addition, murine IgG antibodies are further divided into four H chain subtypes termed γ1, γ2a, γ2b and γ3. The COOH-terminus of the H chain for all antibodies within a given isotype does not vary and hence is referred to as the constant (C) region. The functional characteristics of a given Ig molecule, as defined by its ability to interact with other components of the immune system, are conferred by the C-region of the particular H chain the Ig molecule has utilized.

Cloning of the genes that encode the different domains of Ig molecules revealed that a repertoire of up to 10^{11} different antibody specificities can be generated from a relatively limited number of genes by DNA rearrangement, and then further amplified by somatic mutation of the rearranged sequences (reviewed in Schatz *et al.*, 1992; Fanning *et al.*, 1996). While the complete array is not yet known, two important enzymes which mediate these DNA rearrangements are encoded by genes termed RAG-1 and RAG-2 ("recombinase activator genes"). Another important enzyme in this process is encoded by the *Prkdc* (protein kinase, DNA-activated, catalytic polypeptide) gene. Initial insight that the antibody repertoire is generated by a DNA recombinational mechanism was provided by restriction fragment length variant analyses showing that the DNA encoding various Ig subcomponents is in a different configuration in antibody-producing B lymphocytes than in other cell types. As shown in **Figure 2**, the V region of Ig L chains is encoded by two separate segments of DNA that have been spliced together in differing patterns. These are referred to as V gene segments that encode approximately the first 95 amino acids of the L chain and the joining (J) gene segments which encode the next 12–14 residues. The germline sequences for the λ L chain on mouse chromosome 16 consist of two consecutive loci, each containing one V gene segment followed by one or two functional J and C gene segments (**Figure 2A**). Germline sequences for the κ L chain on mouse chromosome 6 consist of ~250 V gene segments followed by four functional J gene segments and a single C region gene segment (**Figure 2B**). A coding sequence for the V region of a particular L chain is generated by the splicing and rejoining of germline DNA elements for any V and J gene segment. During RNA splicing, the recombinant VJ element is subsequently joined to the downstream C gene segment and an upstream sequence encoding a leader peptide which partially contributes to the migration of mature Ig proteins to the cell surface.

Germline sequences for the subcomponents of Ig H chains are located on chromosome 12 in mice. As shown in **Figure 3**, the coding sequences for the V region of Ig H chains are generated by recombination of three gene segments, rather than two as observed for the L chain. Ig H chains can utilize 200–1000 V gene segments and four functional J gene segments which, unlike their L chain counterparts, are interspersed by 15 diversity (D) gene segments. The coding sequence for the V region of a particular Ig H chain is generated by an initial fusing of any D and J gene segment, followed by their subsequent splicing to any V gene segment. As with the L chain, a recombinant VDJ element encoding the V region of an Ig H chain is joined during RNA splicing to a downstream C gene element and an upstream leader peptide sequence which also contributes to the cell surface migration of mature Ig proteins.

Two additional factors greatly enhance diversity in the V region of Ig H and L chains. The first of these is junctional diversity, which results from the insertion of additional untemplated nucleotides to the double-stranded ends of V–J (L chains) or V–D–J (H chains) gene elements as they undergo recombination. As a result of variability in the original points of DNA fragmentation and the subsequent insertion of additional nucleotides to the double-stranded ends, the exact junctions between any two V–J, V–D or D–J elements will differ each time they are utilized in a rearrangement event, a process known as imprecise joining. Together, the processes of junctional diversity and imprecise joining elicit shifts in the reading frame of DNA comprising V–J, V–D or D–J junctions. These frameshifts further enhance the spectrum of Ig molecules which can be produced. The great flexibility in the way that germline V, D and J gene elements can fragment and rejoin, coupled with the combinatorial diversity resulting from the subsequent pairing of H and L chains produced by independent gene rearrangement events, generates an Ig repertoire capable of recognizing on the order of 10^{11} antigenic specificities.

While the total repertoire of antibodies an individual can produce is extremely large, the production of these molecules is regulated in such a way that any individual B lymphocyte only generates Ig molecules with a single antigenic specificity. This stems from the fact that once an Ig H or L chain gene rearrangement has occurred in a differentiating B lymphocyte, a process termed allelic exclusion then inhibits subsequent rearrangement of Ig gene elements remaining in the germline configuration on the homologous chromosome. As discussed later in this article, the

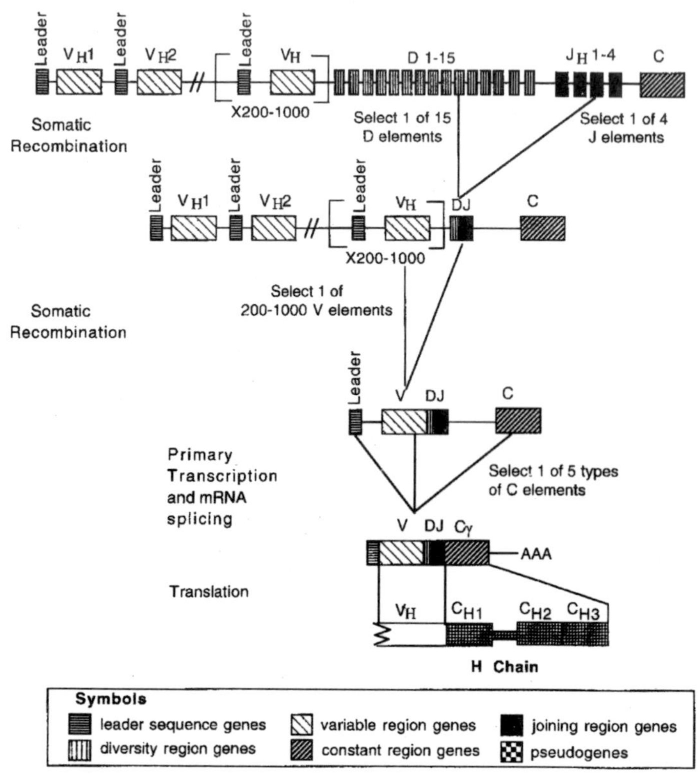

Figure 3 Ig H chain gene organization and rearrangement patterns.

generation by allelic exclusion of individual B lymphocytes, each producing Ig molecules with a single antigenic specificity, is a critical component in the normal prevention of autoimmune responses.

3. Further diversification of antibody specificity by somatic hypermutation

All the mechanisms described above for generating antibody diversity result from the rearrangement in differentiating B lymphocytes of gene segments that encode various portions of mature Ig molecules. However, an additional mechanism generates further diversity in the V region of Ig molecules after they have been generated by gene rearrangement events. This mechanism is referred to as somatic hypermutation and entails the generation of point mutations at a very high frequency in rearranged gene segments encoding the V region domain of Ig H and L chains (reviewed in Wagner & Neuberger, 1996). Interestingly, the gene sequences encoding the C region domain of Ig molecules are not affected by this process. Through mechanisms that are not yet understood, somatic hypermutation occurs when an Ig molecule expressed on the cell surface of a B lymphocyte encounters and responds to its cognate antigen. The resulting point mutations elicit amino acid changes in the portion of the Ig V region that makes contact with the antigen in question. This process results in the preferential outgrowth of any B lymphocytes which, as a result of somatic hypermutation, produce Ig molecules with a higher antigenic affinity than those produced by the original gene rearrangement events. In this way, the presence or absence of particular antigens can help determine the shape of an individual's Ig repertoire.

4. Mechanistic basis of isotype switching and its control of antibody function

As described earlier, Ig H chains consist of a highly variable V region domain at the NH2-terminus, encoded by rearranged VDJ gene elements, which contributes to antigenic specificity, and an isotype-specific non-variable COOH domain encoded by a C region gene segment. The functions that a given Ig molecule can exert are controlled by the isotype-specific C region of the H chain. Some of these isotype-specific Ig functions are summarized in **Table 1**.

One critical function of antibodies is the ability to bind and activate a series of serum complement proteins. Activated complement contributes to the destruction of pathogens, both directly and indirectly, through an ability to mediate the recruitment and activation of phagocytic cells such as macrophages and neutrophils. Complement proteins are activated by binding to the C region domain in the H chain of IgG or IgM molecules which have complexed with antigen. Another important function of antibodies is to enhance the ability of phagocytic cells such as macrophages to engulf antigens that they cannot take up efficiently by pinocytosis. This process is mediated by a series of proteins expressed on the surface of phagocytic cells known as Fc receptors. These receptors can bind the C region domain of certain IgG molecules that are complexed to antigen. Antigen–antibody complexes bound to Fc receptors are then internalized by the phagocytic cell. As discussed below, antigens taken up by macrophages through such an Fc receptor-mediated pathway can be processed subsequently in a way that activates specific populations of T lymphocytes, which comprise the other major arm of the adaptive immune response. In this way, the antibody–Fc receptor–T lymphocyte axis represents a classic example by which innate and adaptive immune responses interact.

Table 1 Summary of isotype-specific Ig functions.

Ig isotype	Representative immunological functions
M	complement activation
D	unknown
G	complement activation, placental transfer, binding to macrophage Fc receptors
E	High affinity binding to mast cells and basophils to elicit histamine/serotonin release
A	mucosal membrane immunity

Another way by which antibodies serve as links between adaptive and innate immune responses is by controlling the release of histamine and other vasodilators from mast cells and basophils. The release of such vasodilators allows leucocytes exerting various immunological functions to leave the bloodstream and enter tissues producing such factors. When complexed with their cognate antigen, antibodies of the IgE isotype bind to mast cells and basophils and trigger their production of vasodilators. However, through their ability to mediate the production of vasodilators, an inappropriate IgE response against certain antigens can trigger allergic reactions that sometimes lead to death through an oedema-mediated circulatory collapse. It is also important that antibodies be present in mucosal membranes that are often the portal of entry for pathogenic agents. Antibodies of the IgA isotype provide this function.

A process termed isotype switching allows for the generation of Ig molecules sharing the same antigen-specific V region domain, but differing in the C region of the H chain (reviewed in Stavnezer, 1996). In this way, isotype switching can vary the functional consequences of generating an antibody response to a given antigen. As discussed later in this article, isotype switching is not a random process, but is controlled by various cytokine proteins secreted by T lymphocytes. In the presence of the appropriate cytokines, isotype switching occurs when B lymphocytes are activated by binding of antigen to the initial cell membrane-bound Ig molecules they have produced. All B lymphocytes initially produce a μ H chain and, therefore, make Ig molecules of the IgM isotype. As shown in **Figure 4**, this occurs because the Cμ gene segment lies closest to the upstream rearranged VDJ element that encodes the V region domain of the H chain. A repetitive stretch of DNA known as the switch region resides in the intron between the rearranged VDJ gene segment and the Cμ element. This switch region provides a splicing signal to fuse the rearranged VDJ element to the start of the Cμ gene segment and hence allow production of a complete IgM mRNA transcript. There are similar switch regions 5′ to each of the other C region gene segments. Each switch region is given the same designation as

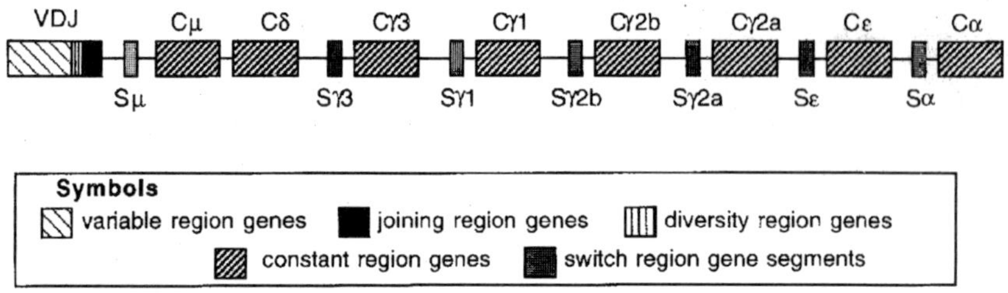

Figure 4 Organization of constant region genes used to generate Ig H chains in mice.

the corresponding C region gene, and hence are termed Sμ, Sγ3, Sγ1, Sγ2b, Sγ2a, Sε and Sα. While antibodies of all antigenic specificities are originally generated as an IgM isotype, they can class-switch to any other isotype after encountering antigen. This takes place in antigen-activated B lymphocytes through a recombination event between the Sμ and any other downstream switch unit, with the subsequent deletion of DNA containing intervening C region gene segments. The only exception to this process is the production of IgD molecules that can be coexpressed with IgM molecules on the plasma membrane of mature naive B lymphocytes which have not yet been activated by an encounter with their cognate antigen. This occurs because the μ and δ C region gene segments lie adjacent to each other with no intervening switch region element. As a result, by utilizing a differential mRNA splicing process, naive B lymphocytes can produce either or both IgM and IgD molecules of the same antigenic specificity. The function of IgD is not known, but the level of IgM and IgD expression appears to be a factor in determining whether developing B lymphocytes producing autoreactive antibodies are eliminated (see below).

5. The dual role of Ig molecules as cell-surface receptors and secreted immunological effectors

Antibodies were originally identified as secreted immunological effector molecules found in plasma following immunization with a particular antigen. However, it was subsequently found that individual B lymphocytes express a single antigen-specific clonotype of Ig molecules as plasma membrane-bound cell-surface proteins. These plasma membrane-bound Ig molecules act as cell-surface receptors which, on binding their cognate antigen, mediate the functional activation of B lymphocytes (reviewed in Cambier, 1992; DeFranco, 1995). As described above, one functional consequence of B lymphocyte activation is Ig isotype switching. Another outcome of B lymphocyte activation is the production of secreted as well as plasma membrane-bound Ig molecules. The production of plasma membrane-bound *vs.* secreted isoforms of each Ig isotype is regulated by generating different mRNA splice variants from the 3′ end of the C gene segment used to construct a given H chain.

Plasma membrane-bound Ig molecules cannot directly transmit B lymphocyte-activating signals following antigen ligation. In order to trigger intracellular activating signals, Ig molecules must be associated on the plasma membrane with two other proteins termed Igα and Igβ. Ig molecules also fail to be transported to the cell surface in the absence of Igα and Igβ proteins. The cytoplasmic domains of Igα and Igβ are associated with several different enzymes in the protein tyrosine kinase (PTK) family. Binding of antigens can induce an aggregation of plasma membrane-bound Ig molecules that initiates a signal through the Igα and Igβ proteins which then activates these PTK enzymes to add phosphate groups to other proteins and trigger their functional activity. One of the most important proteins activated by this PTK-dependent pathway is phospholipase C-γ. Activated phospholipase C-γ cleaves membrane-associated phosphatidyl inositol into inositol triphosphate, which releases calcium ions from intracellular stores, and diacylglycerol which mediates further protein phosphorylation reactions. Collectively, these mediators trigger a cascade of intracellular reactions that induces changes in gene expression patterns which result in B lymphocyte proliferation and functional activation, including antibody secretion.

Binding of only certain types of antigen can independently induce the aggregation of plasma membrane-bound Ig molecules required to initiate signalling pathways leading to B lymphocyte activation. These include antigens that have many identical Ig-binding sites (referred to as epitopes) such as bacterial cell wall polysaccharides. However, most antigens that bind to plasma membrane-bound Ig molecules do not have the capacity to trigger B lymphocyte activation pathways independently. B lymphocyte responses to these antigens also require signals from growth factors secreted by another specialized population of cells mediating adaptive immune responses called helper T (Th) lymphocytes.

6. Mechanistic basis for antigen recognition by T lymphocytes

As described above, antibodies produced by B lymphocytes recognize antigens consisting of conformationally dependent epitopes contained within structurally intact molecules which are usually proteins. In contrast, T lymphocytes, which represent the other major cell type mediating adaptive immune responses, recognize antigens consisting of linear peptide fragments derived from whole proteins that have been degraded in specialized processing pathways. Furthermore, T lymphocytes only recognize peptides that are bound to one of a specialized set of carrier molecules encoded by genes within the major histocompatibility complex (MHC). The ability of a given T lymphocyte to recognize a particular peptide/MHC carrier complex is imparted by clonally distributed T cell receptor (TCR) molecules expressed on their cell surface (reviewed in Gascoigne, 1995; Gilfillan *et al.*, 1995; Hilyard & Strominger, 1995). Similar to the case for Ig molecules, the total potential repertoire of TCR specificities ($\sim 10^{13}$) far exceeds the total number of genes in a mammalian genome ($\sim 10^5$). This occurs because, like Ig diversity, a virtually unlimited spectrum of TCR molecules is generated through the splicing and rejoining of germline DNA sequences from a relatively limited number of genes. Many of the same enzymes controlling the splicing and rejoining of Ig gene elements, such as RAG-1, RAG-2 and PRKDC, also contribute to TCR gene rearrangements. It is for this reason that defects in the *Prkdc* gene result in the severe combined immunodeficiency (SCID) syndrome which is manifest by a failure to generate both B and T lymphocytes (Blunt *et al.*, 1996). Most T lymphocytes (>95%) utilize a TCR comprised of an α and β chain heterodimer linked to each other through a disulfide bond (hence designated, TCR αβ T cells). A minor population of T lymphocytes (<5%), whose function is not yet clear, utilize a TCR comprised of a γ and δ chain heterodimer (hence designated, TCR γδ T cells) (Kaufmann, 1996). Individual T lymphocytes of both classes are derived from precursor cells of bone marrow origin that subsequently differentiate and acquire a distinct TCR clonotype within the thymus. However, given that TCR αβ T cells comprise the major class of T lymphocytes whose function is best understood, this article will focus on this subset.

The genes encoding various components of TCR αβ molecules were originally identified from cDNA clones that could be isolated from T but not B lymphocytes. Analyses of the amino acid sequences encoded by these cDNA clones revealed that both TCR αβ and β chains are comprised of a variable region at the NH2 terminus, with homology to the Ig V domain, and a constant region at the COOH terminus that shares homology with the Ig C domain. The COOH region of TCR α and β chains also contains a transmembrane and cytoplasmic tail domain that allows these heterodimeric molecules to anchor at the cell surface.

Given the high degree of homology between their gene products, it is not surprising that the genetic elements encoding various components of TCR α and β chains are similar in structure and arrangement to those encoding Ig molecules. Like the L chain of Ig molecules, the V domain of TCR α chains is encoded by fused V and J gene segments. As with the H chain of Ig molecules, the V domain of TCR β chains is encoded by fused V, D and J gene segments. As shown in **Figure 5**, the germline sequence for the murine TCR β chain resides on chromosome 6 and consists of approximately 50 V region gene segments and two consecutive downstream clusters, each consisting of 1 D, 6–7 J and 1 C region gene segment. The germline sequence for the TCR α chain on mouse chromosome 14 lacks D region segments, but contains 70–80 V region gene segments, ~60 J region segments and 1 C region segment. Interestingly, the TCR α chain gene complex is interrupted between the V and J segments by the germline sequences of the TCR δ chain gene complex. As a result of this organizational pattern, the rearrangements of

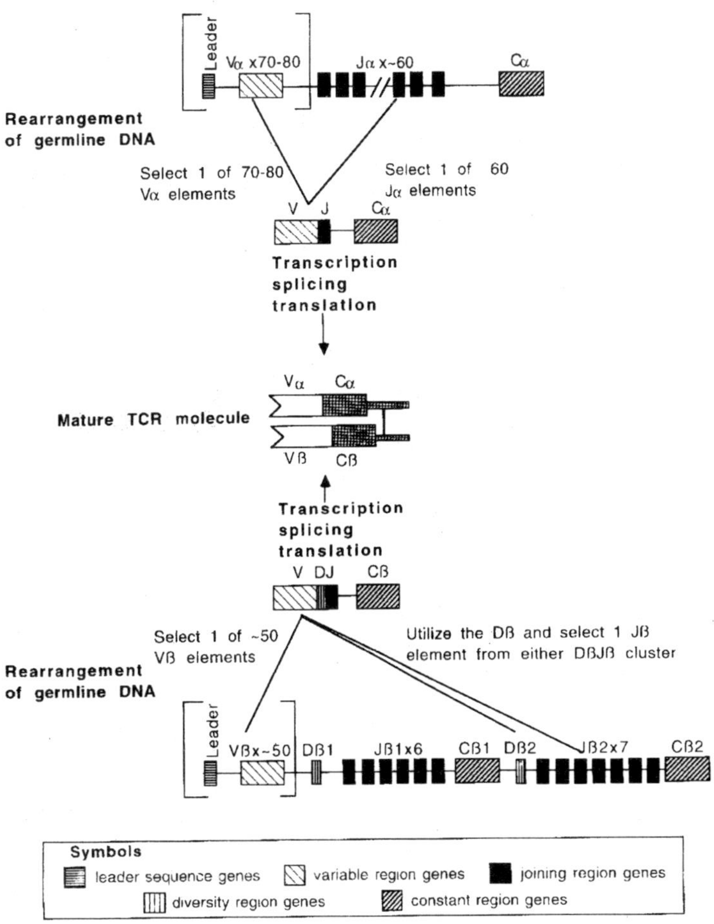

Figure 5 TCR α and β chain gene organization and rearrangement patterns.

TCR α or δ chain gene sequences are mutually exclusive processes in any individual T cell. This is a key factor in determining whether an individual T cell enters the TCR $\alpha\beta$ or $\gamma\delta$ lineage. The particular germline segments which are incorporated into any given VDJ (β chain) or VJ (α chain) gene rearrangement generate structural variability at the NH2-terminus of both resulting molecules, which, on pairing, contribute to the antigenic specificity of the TCR. As with Ig molecules, heterogeneity among rearranged TCR α and β chains is further enhanced by the processes of junctional diversity and imprecise joining during the splicing of V–J or V–D–J gene sequences. These V–J linkage sequences (which incorporate a D segment in the TCR β chain) are termed CDR3 regions and are also major contributors to the antigenic specificity of the rearranged TCR molecule. Unlike Ig molecules, the antigenic specificity of mature TCR molecules is not further altered by somatic hypermutation.

The C region domain at the COOH-terminus of rearranged TCR $\alpha\beta$ molecules non-covalently associates at the cell surface with a set of polypeptides termed CD3 γ, δ and ε which are highly homologous to the Igα and Igβ proteins associated with plasma membrane-bound Ig molecules on B lymphocytes. Antigen binding to the rearranged TCR $\alpha\beta$ molecules induces the CD3 complex to transmit signals that activate PTK-mediated second messenger pathways which trigger appropriate T cell effector functions. Once a TCR gene rearrangement has occurred in a differentiating T cell, subsequent rearrangement of TCR α and β sequences remaining in the germline configuration is suppressed through the mechanism of allelic exclusion, although this process is less complete for α than β chain sequences. Due to the allelic exclusion process, each individual T cell only produces TCR molecules with a single antigenic specificity. Mature TCR $\alpha\beta$ complexes non-covalently associate with one of two accessory molecules termed CD4 or CD8. As described below, CD4 or CD8 molecules stabilize TCR binding to one of the two types of antigenic peptide/MHC complexes that can be generated.

7. Generation of antigenic peptide/MHC complexes recognized by TCR molecules

Individual TCR $\alpha\beta$ molecules recognize a 3-dimensional antigenic structure formed by the binding of a single peptide to a specific type of MHC carrier molecule encoded by genes on chromosome 17 in mice and 6q in humans. MHC genes are provided the designation of *H2* in mice and *HLA* in humans, and the two primary types of MHC gene products are termed class I and class II. These MHC class I and class II molecules are encoded by the most allelically variable set of genes known to exist in biology (reviewed in Klein *et al.*, 1993; Parham *et al.*, 1995). Indeed, well over 100 MHC class I and class II allelic variants have been identified in both humans and mice. The chromosomal arrangement of some of the more important genes within the murine MHC are shown in **Figure 6**. While a similar linked set of MHC genes exists in humans, the chromosomal organization pattern differs from that of mice. The particular collection of alleles characterizing the linked set of MHC genes on an individual chromosome is referred to as an MHC haplotype. The presence of multiple MHC genes with extensive allelic variability appears to be an evolutionary adaptation to ensure that, at the species level, no pathogen can completely escape immune surveillance.

In mice, the most widely expressed and polymorphic MHC class I molecules are encoded by the H2K, D and L genes. These MHC class I gene products are expressed on the surface of virtually all cells and mediate the presentation to T cells of peptides derived from intracellular proteins, such as those of replicating viruses. As shown in **Figure 7**, this process requires contributions from other intra-MHC-encoded gene products which degrade cytosolic proteins into peptide fragments and then transport them into the rough endoplasmic reticulum (RER), where they can be bound to newly synthesized MHC class I molecules (reviewed in Hansen & Lee, 1997). Cytosolic protein degradation is mediated by a large protease complex known as the proteasome which is made up of multiple subunits. On incorporation of the LMP2 and LMP7 subunits, both encoded by genes within the MHC, the proteolytic activity of the proteasome is converted to an antigen-processing function. Peptide fragments generated by this process are transported into the RER by a heterodimeric molecule encoded by the intra-MHC genes *TAP1* and *TAP2* ("transporters associated with antigen processing"). MHC class I molecules are comprised of three extracellular domains termed α1, α2 and α3 as well as a short transmembrane domain. Within the RER, peptides 8–10 amino acids in length can be bound in a groove formed by folding between the allelically variable α1 and α2 domains located at the NH2 terminus of a given MHC class I molecule. The COOH terminus of MHC class I molecules is constant among all allelic variants and contains the α3 and transmembrane domains. However, the non-variant α3 domain must dimerize with the non-MHC-encoded protein β2-microglobulin (β2m), in order for the allelically variable α1 and α2 helices to fold properly and bind peptides.

The peptide-binding groove of each MHC class I allelic variant has a different shape and hence binds a different set of peptides. The set of peptides binding to a given MHC class I gene product is dictated by sequence similarity at 2–3 amino acid anchor residues (reviewed in Young *et al.*, 1995). These anchor motifs vary between peptides binding to different MHC class I molecules. Only those MHC class I molecules that have dimerized with β2m and bound peptide

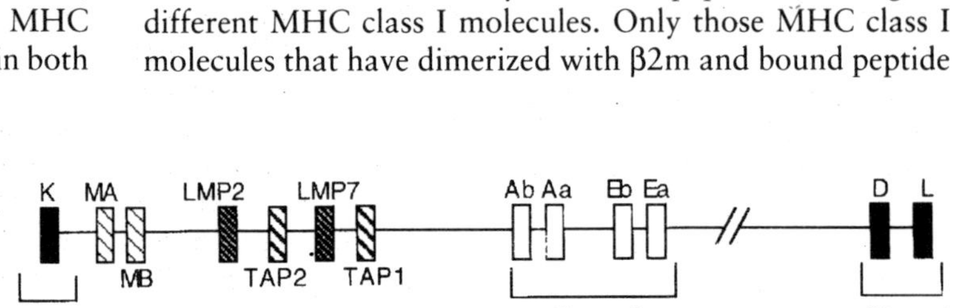

Figure 6 Organization of some important genes within the murine MHC.

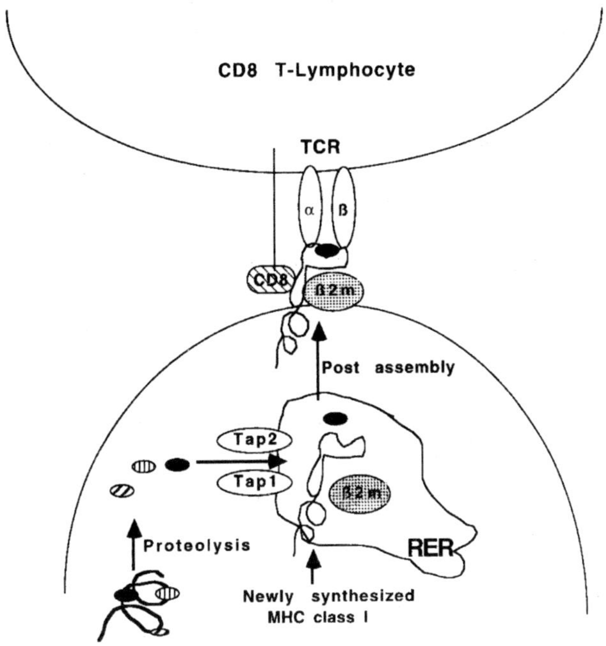

Figure 7 MHC class I-mediated antigen presentation to CD8+ T cells.

are then transported to, and expressed in a stable fashion on, the cell surface. Peptide/MHC class I complexes expressed on the cell surface provide a 3-dimensional antigenic structure that can be recognized by an appropriate TCR αβ complex associated with a CD8 molecule (Garcia *et al.*, 1996). The CD8 accessory molecule stabilizes this TCR αβ recognition process by binding to the non-polymorphic α3 domain of the MHC class I molecule. T cells expressing CD8 exert cytotoxic functions that are mediated through several different mechanisms. This leads to the destruction of cells bearing peptide/MHC class I complexes that are recognized by the TCR of a CD8+ T lymphocyte. In this way, cells harbouring intracellular pathogens can be eliminated and thus limit the spread of infection.

In addition to the highly polymorphic K, D and L gene products expressed on virtually all cells, many other MHC class I genes (>50 in mice) encode relatively non-polymorphic variants which are expressed in a β2m-dependent fashion in a relatively limited number of cell types. These are known as MHC class Ib genes and, in mice, are located distally to the H2D locus. The function of these MHC class Ib genes is mostly unknown. However, in mice, it is known that one of these class Ib gene products, H2-M3, can present peptides with N-formylated NH2 termini. This is of interest because prokaryotes initiate protein synthesis with N-formylmethionine and cells infected with cytosolic bacteria can be killed by CD8+ T cells which recognize N-formylated bacterial peptides bound to H2-M3. Other genes encoding β2m-dependent class Ib-like molecules map outside the MHC and include the CD1 locus on mouse chromosome 3. CD1 molecules may also play a role in generating antibacterial immune responses since these class Ib molecules bind the mycobacterial membrane components mycolic acid and lipoarabinomannan.

In contrast to MHC class I molecules that present peptides of intracellular origin, the function of MHC class II molecules is to present peptides derived from extracellular proteins. Mice produce two types of MHC class II molecules termed I-A and I-E that are each heterodimers comprised of paired α and β chains encoded by the genes shown in Figure 6. The NH2 termini of MHC class II α and β chains are each characterized by an allelically variable α1 domain that, on pairing, forms a groove into which peptides generated by the processes described below can be bound. The COOH regions of MHC class II α and β chains are characterized by a constant α2 and transmembrane domain shared by all alleles. MHC class II molecules bind peptides that are at least 13 and generally no more than 20 amino acids in length. As with class I molecules, each MHC class II allelic variant binds a different spectrum of peptides through interactions with particular arrays of amino acid anchor residues.

While MHC class I gene products are present on virtually all cell types, MHC class II expression is generally restricted to a specialized set of bone marrow-derived cells that can take up and degrade extracellular agents such as circulating pathogens (reviewed in Germain *et al.*, 1996). These are collectively referred to as antigen-presenting cells (APC) and include B lymphocytes, macrophages and dendritic cells. The various types of APC can utilize several different mechanisms to take up extracellular agents, and all types of APC share an ability to internalize extracellular agents by pinocytosis. As described earlier, APC can also utilize an Fc receptor-mediated pathway to internalize extracellular antigens that are bound to antibodies. In addition, B lymphocytes also have a unique means to take up extracellular antigens. This results from the fact that B lymphocytes internalize plasma membrane-bound Ig molecules that have captured antigen. Due to this highly efficient antibody-mediated capture mechanism, B lymphocytes have a much greater capacity than other types of APC to generate MHC class II-bound peptides from low abundance extracellular antigens. However, as shown in **Figure 8**, regardless of the mechanism by which they were initially internalized, extracellular agents taken up by APC are sequestered into membrane-bound vesicles known as endosomes. The internal compartment of endosomes is acidic in nature, resulting in the degradation of proteins from internalized extracellular agents into peptide fragments that can be bound to MHC class II molecules.

Like all proteins, MHC class II molecules are originally synthesized within the RER. However, following their synthesis, MHC class II molecules must then be transported to the endosomes in order to bind antigenic peptides. The transport of MHC class II molecules from the RER to endosomal compartments and their subsequent binding of peptides is mediated by gene products encoded both within and outwith the MHC. Newly synthesized MHC class II molecules associate with the invariant chain (Ii) protein encoded by a non-MHC gene on mouse chromosome 18. The Ii chain serves two functions, one of which is to act as a chaperone protein to direct migration of MHC class II molecules

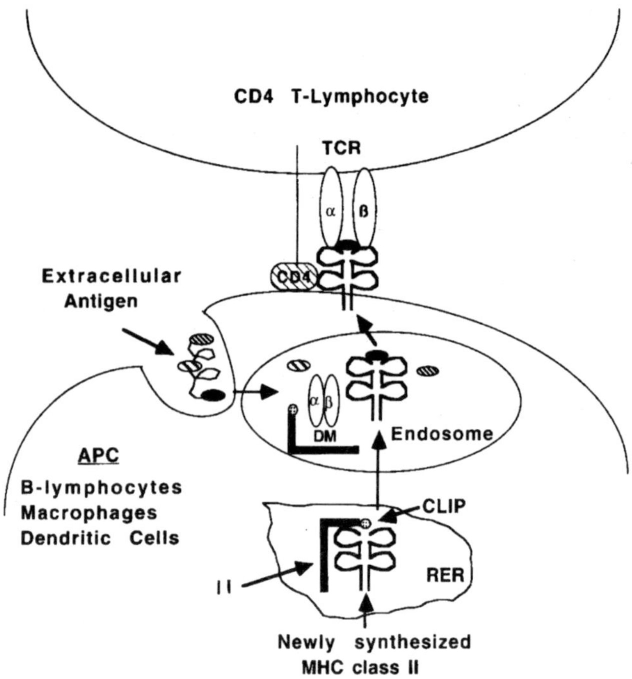

Figure 8 MHC class II-mediated antigen presentation to CD4⁺ T cells.

through the Golgi and to the endosomes. The Ii chain also contains a region known as the CLIP fragment which occupies the peptide-binding groove of MHC class II molecules being transported to endosomal compartments. Association with the CLIP fragment prevents MHC class II molecules from binding intracellular peptides while being transported to the endosomes. Within the endosomes, the MHC class II–Ii chain complex associates with heterodimeric H2-DM molecules which are comprised of paired α and β chains encoded by the intra-MHC *MA* and *MB* genes depicted in Figure 6. Association with H2-DM catalyses the removal of the Ii chain including the CLIP fragment from intraendosomal MHC class II molecules allowing them to now bind peptides. MHC class II molecules that have bound peptide are then transported to the plasma membrane surface of the APC. Peptide/MHC class II complexes expressed on the cell surface provide a 3-dimensional antigenic structure that can be recognized by an appropriate TCR αβ complex associated with a CD4 molecule. The CD4 accessory molecule stabilizes this TCR αβ recognition process by binding to the non-allelically variable COOH domain of the MHC class II molecule.

TCR ligation to a peptide/MHC class II antigenic complex only provides one of at least two signals required to activate the proliferation and effector functions of $CD4^+$ T cells. A second co-stimulatory signal required for T cell activation is most commonly provided through a cell surface receptor known as CD28 upon its interaction with the B7–1 or B7–2 ligand molecules expressed as plasma membrane-bound proteins on APC (reviewed in Lenschow *et al.*, 1996). The importance of the CD28/B7 receptor–ligand system is illustrated by the fact that in the absence of the co-stimulatory functions it provides, TCR ligation alone results in T cell anergy rather than activation. If the proper signals have been provided through the TCR and CD28/B7 co-stimulatory system, activated $CD4^+$ T cells secrete a series of communication molecules known as cytokines that aid in triggering and sustaining other immunological functions. It is for this reason that CD4 T cells are referred to as helper T (Th) cells.

The array of known cytokines and their modes of action are far too extensive for a detailed description in this article. For readers desiring a comprehensive overview of the known cytokines, their receptors and functions, the author recommends an excellent textbook by Callard and Gearing (1994). In general, the types of cytokines produced by $CD4^+$ Th cells can be divided into two major classes (reviewed in Paul & Seder, 1994) – the Th1 and Th2 cytokines. Types of Th1 cytokines include interleukin (IL)-2 and gamma-interferon. The predominant Th2 cytokines include IL-4 and IL-10. Th1 cytokines amplify cell-mediated immune responses such as cytotoxic $CD8^+$ T cell functions and macrophage phagocytic activity. The enhancement of non-specific macrophage phagocytic activity by Th1 cytokines released from $CD4^+$ T cells proliferating in response to a specific antigenic stimulus represents an excellent example of cooperation between the adaptive and innate arms of the immune response. Enhancement of cytotoxic $CD8^+$ T cell responses by Th1 cytokines secreted from $CD4^+$ T cells plays an important role in the control of viral infections. This results from a cooperative interaction where antigenic peptides, generated in APC from circulating extracellular virions, trigger $CD4^+$ T cells to secrete Th1 cytokines which then amplify the responses of cytotoxic $CD8^+$ T cells that recognize virally derived peptides within already infected cells. Th2 cytokines provide help for the activation of antibody secretion by B lymphocytes. Antibody production is particularly enhanced through a special form of help termed cognate recognition which results from the efficient ability of B lymphocytes to serve as APC for extracellular proteins that have been captured by plasma membrane-bound Ig molecules and then internalized. As a result, B lymphocytes can trigger $CD4^+$ helper T cell responses to MHC class II-bound peptides derived from the same extracellular proteins recognized by the specific Ig molecules they produce. Thus, their dual role as APC enables B lymphocytes to activate directly the source of helper cytokines they require to commence antibody secretion. However, it should be noted that, following ligation of their antigen-specific receptors (Ig or TCR molecules), helper-dependent immunological effectors such $CD8^+$ T cells and B lymphocytes can be activated in the presence of the appropriate cytokines regardless of the antigenic specificity of the $CD4^+$ T cell response producing them.

8. MHC restriction and allogeneic cross-reactivity of T cell responses

The term "MHC restriction" is applied to the fact that a given peptide must be bound to a specific MHC allelic variant in order to form the 3-dimensional antigenic structure recognized by a particular TCR αβ complex. Rolf Zinkernagel and Peter Doherty discovered the process of MHC restriction in the 1970s (Zinkernagel & Doherty, 1974). They found that virally infected mice generated cytotoxic T cells which recognized cells infected with the same virus, but not uninfected cells or those infected with a different virus. Thus, the cytotoxic T cells were virus specific. However, the most important outcome of their work was that the cytotoxic T cells only recognized virally infected cells expressing the same MHC molecules as the originally inoculated mice. This demonstrated that an antigen must be bound to one of an individual's own MHC allelic variants in order to elicit a T cell response. Work by multiple investigators subsequently demonstrated that the 3-dimensional antigenic structure recognized by a particular TCR jointly results from conformational variations imposed on peptides once they bind to different MHC molecules, as well as by conformational changes in the MHC molecules themselves when binding different peptides. X-ray crystallography studies have shown that MHC restriction results from an appropriately rearranged TCR αβ complex interacting with contact points on both the bound peptide as well as the MHC molecule itself (Garcia *et al.*, 1996).

The discovery that MHC restriction results from TCR interactions with sites on both the bound peptide and the MHC molecule itself provided an explanation for the previously mysterious observation that potent T cell responses were generated against non-self or allogeneic MHC molecules during the course of graft rejection. Such responses had long puzzled immunologists since it seemed unlikely that T cells had evolved an ability to recognize transplanted allogeneic tissues. Adding to this mystery were results from limiting dilution analyses demonstrating that 1–10% of all T cells in an individual were capable of responding to allogeneic MHC molecules. After the discovery of MHC restriction, it became apparent that allogeneic recognition results from cross-reactive responses by T cells that are normally specific for a variety of foreign peptides bound to "self" MHC molecules. Supporting this concept is the fact that monoclonal T cell lines selected in culture for their ability to respond to a given foreign peptide bound to a "self" MHC molecule can often be stimulated by cells expressing particular allogeneic MHC molecules. Based on these findings, it is currently thought that allogeneic reactivity results when a "non-self" MHC molecule and its bound peptide have a similar 3-dimensional structure as a foreign peptide bound to a "self" MHC molecule, so that both antigenic complexes can be recognized by the same TCR.

9. Selection of the T and B lymphocyte repertoires

Within a given species, there is relatively little allelic variability among germline TCR or Ig gene sequences. Thus, all species members have the theoretical capacity to generate the same wide spectrum of T and B lymphocyte clonotypes. Each species member must have a system for retaining from this virtually identical pool those T and B lymphocyte clonotypes that are individually beneficial in nature while eliminating those potentially capable of mediating autoimmune responses. This process is termed clonal selection.

(i) T cell repertoire selection

T cell precursors of bone marrow origin migrate to the thymus where they undergo clonal selection during their course of differentiation (reviewed in von Boehmer & Kisielow, 1993; Ashton-Rickardt & Tonegawa, 1994). As shown in **Figure 9**, the earliest T cell precursors express neither CD4 nor CD8 accessory molecules and, hence, are called double-negative (DN) cells. A small portion of these DN cells initiate expression of TCR γδ complexes. DN cells that have not initiated TCR γδ expression rapidly proliferate and progress to the double-positive (DP) stage of differentiation defined by coexpression of both CD4 and CD8 accessory molecules. TCR αβ expression initiates in DP cells, which represent the predominant cell type (>80%) within the thymus (Sebzda *et al.*, 1999). This occurs so that the proper CD8 or CD4 accessory molecule is available to any TCR αβ complex that, through its rearrangement pattern, functions in a MHC class I- or class II-restricted manner, respectively. DP T cell precursors that express TCR αβ molecules then undergo the clonal selection processes are described below. The few (<5%) DP cells that survive the clonal selection process differentiate into CD4 or CD8 single-positive (SP) cells by downregulating expression of the accessory molecule which is inappropriate for the particular rearranged TCR αβ complex they express. These CD4 and CD8 SP T cells are then exported from the thymus into the bloodstream and, subsequently, into peripheral lymphoid organs such as the spleen and lymph nodes. They are then available to exert their immunological effector functions if they should ever encounter an antigen to which they react.

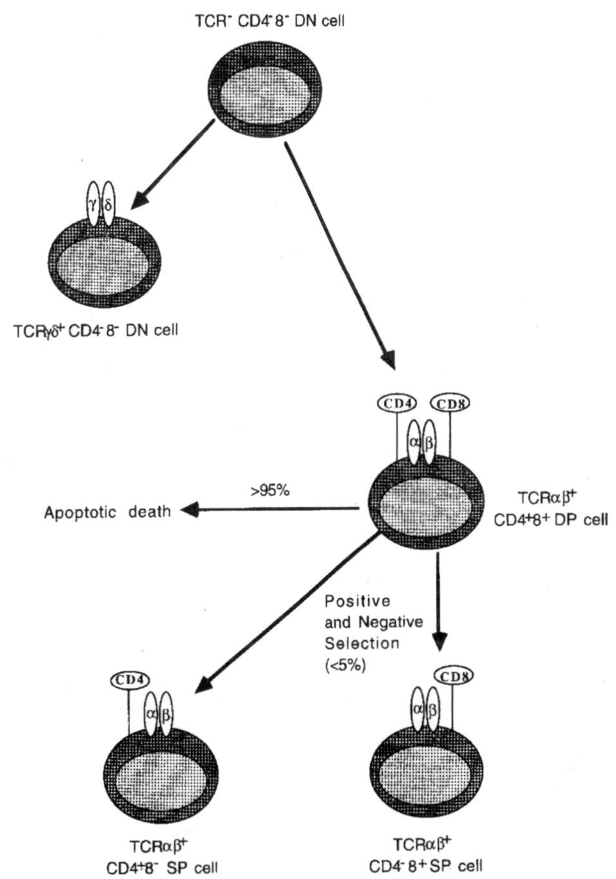

Figure 9 Intrathymic T cell differentiation pattern.

On acquisition of TCR expression, developing T cells undergo two processes, termed positive and negative selection, that determine which clonotypes are allowed to seed the blood and peripheral lymphoid organs. Positive selection assures the survival of only those T cells that express TCR complexes which recognize peptides bound to self MHC molecules. However, some of the T cells that survive the positive selection process could mediate deleterious autoimmune responses if their TCR happens to recognize a peptide derived from a normal endogenous protein bound to a self MHC molecule. Negative selection mediates the destruction of such autoreactive T cells through an apoptotic programmed cell death process.

Positive and negative selection are both mediated through TCR recognition of peptides bound to self MHC molecules. Any T cell differentiating within the thymus that fails to produce a TCR capable of recognizing a peptide–self MHC complex dies through an apoptotic default mechanism. This results in the death of most thymic T cell precursors. The peptide/MHC complexes controlling positive and negative selection within the thymus are expressed on bone marrow-derived macrophages and dendritic cells as well as non-haematopoietically derived thymic epithelial cells (TEC). TEC represent the only cell type other than marrow-derived APC that normally express MHC class II molecules in a constitutive fashion. Within the thymus, haematopoietically derived APC (macrophages and dendritic cells) and TEC can mediate both positive as well as negative selection. However, positive selection is primarily mediated by TEC, and negative selection by haematopoietically derived APC. One of the major questions in immunology is how signalling through the same TCR can promote the diametrically opposite results of either T cell survival or death through positive or negative selection. Recent studies (Sebzda *et al.*, 1999) indicate that this paradox can be explained through a "strength-of-signal" model.

Several different factors contribute to the overall strength of the activation signals any T cell may receive. The most important of these is the avidity of the interaction between the TCR and a peptide/MHC complex. Avidity is defined as the overall product of the inherent affinity of a TCR for a peptide/MHC complex and the number of such antigenic complexes available. As discussed earlier, provided that prerequisite TCR recognition of a peptide/MHC complex has occurred, T cell activation is also partially controlled by signals transmitted through CD28 receptors upon their interaction with B7 ligand molecules expressed on APC. Cytokines produced by APC, particularly IL-1, IL-6 and IL-12, also provide T cell activation signals. Quantitative and qualitative variations in any or all of these factors determine the extent of the activation signals any T cell may receive. T cells undergo different fates dependent on the threshold of activation signals they receive; those T cell precursors differentiating within the thymus that produce a TCR which fails to recognize any available peptide/MHC complex, and hence receives no activation signals, die through an apoptotic default mechanism. A portion of the intrathymic T cell precursors will produce a rearranged TCR that can undergo low avidity interactions with some self peptide/MHC complexes, but which also has, by chance, a high affinity for a particular foreign peptide bound to a self MHC molecule. Such low avidity interactions with self peptide/MHC complexes appear to provide a minimal threshold of TCR-mediated signalling required for positive selection and the prevention of T cell death.

A recent study has demonstrated that the minimal level of TCR-mediated signalling required for positive selection and T cell survival is less than that required to trigger an immunological effector response through the same TCR (Girao *et al.*, 1997). Thus, it is unlikely that the effector functions of T cells, which were positively selected through a low avidity interaction of their TCR with a self peptide/MHC complex, will be triggered if they should encounter the same antigenic complex after leaving the thymus. Other T cell precursors differentiating within the thymus will produce a TCR that can undergo high avidity interactions with self peptide/MHC complexes. When TCR-mediated signalling exceeds a certain threshold, T cells are destroyed through an activation-induced cell death (AICD) pathway that results in apoptosis. Such an AICD pathway mediates the negative selection of T cells expressing a TCR that can undergo high avidity interactions with self peptide/MHC complexes and, hence, could elicit potentially dangerous autoimmune responses if they were allowed to exit the thymus.

Only those T cells expressing a TCR specific for ubiquitously expressed self antigens can be deleted through negative selection in the thymus. These antigens include peptides derived from proteins expressed in all cell types, as well as soluble proteins that enter the thymus through the blood supply. However, there are other self proteins, such as those characterizing some endocrine cells, that will not be present in the thymus. Thus, any T cells that are potentially autoreactive against such non-ubiquitously expressed self antigens cannot be negatively selected within the thymus. However, such autoreactive T cells can also be deleted by an AICD pathway outside of the thymus when TCR-mediated signalling elicited by an encounter with their cognate antigen exceeds a critical threshold level (reviewed in Abbas, 1996). The quantity of any self protein outside of the thymus will likely be sufficient to trigger a potentially autoreactive T cell to the activation threshold required to induce its deletion. Thus, autoreactive T cells are much more likely than those that recognize foreign antigens to achieve a TCR-mediated activation threshold which triggers their extrathymic destruction. However, even T cells reactive against foreign proteins can undergo extrathymic deletion when a sufficient level of TCR-mediated signalling is achieved by artificially introducing high doses of antigen. This can actually be clinically advantageous since it allows for the elimination of T cells contributing to deleterious allergic responses.

(ii) Selection of the B lymphocyte repertoire

B lymphocyte precursors undergo their Ig gene rearrangements within the bone marrow. Unlike TCR complexes, rearranged Ig molecules produced during the course of B lymphocyte differentiation do not recognize antigens in an MHC-restricted fashion. Thus, developing B lymphocytes do not have to be positively selected for MHC restriction, but those expressing potentially autoreactive Ig molecules must still be negatively selected (reviewed in Goodnow *et al.*, 1995).

Immature B lymphocytes within the bone marrow initially only express plasma membrane-bound Ig molecules of the IgM isotype. On emigrating from the bone marrow, mature B lymphocytes commence alternate splicing of Ig H chain mRNA transcripts to produce both IgM and IgD molecules with an identical antigenic specificity. Any immature B lymphocyte within the bone marrow that expresses rearranged IgM molecules which aggregate following antigenic recognition of a highly abundant plasma membrane-bound protein, such as a self MHC molecule, are triggered to die via apoptosis (**Figure 10A**). This distinguishes immature from mature B lymphocytes that become functionally activated following antigen-induced aggregation of their plasma membrane-bound Ig molecules. Some immature B lymphocytes within the bone marrow will express plasma membrane-bound IgM molecules that recognize soluble self proteins (**Figure 10B**). Immature B lymphocytes expressing IgM molecules that bind a soluble antigen within the bone marrow are not killed, but become permanently inactivated; this inactive state is referred to as anergy. Developing B lymphocytes that have been anergized express greatly reduced levels of plasma membrane-bound IgM molecules. Such anergized B lymphocytes can emigrate from the bone marrow and commence to coexpress normal levels of IgD molecules on the cell surface along with the few remaining IgM molecules. However, these B lymphocytes do not become activated following antigen binding to their plasma membrane-bound IgD molecules. Furthermore, B lymphocytes with such imbalances in IgM and IgD expression are very short lived outside of the bone marrow since they cannot properly home to areas of lymphoid tissue that produce growth factors necessary for their survival. As a result of these negative selection processes, only immature B lymphocytes that are not activated by antigen during their early development emigrate from the bone marrow, express both IgM and IgD molecules at normal levels and successfully seed the peripheral lymphoid organs (**Figure 10C**). The repertoire of B lymphocytes that has been purged of autoreactive clonotypes by negative selection and seeded the peripheral lymphoid organs is then available to exert immunological effector functions if they should ever encounter their cognate antigen.

10. Functional consequences of defects in genes controlling immunological functions

Thus far, this article has discussed the normal genetic control of immunological responses. However, functional defects in genes regulating immunological responses can have devastating effects. These can include either immune deficiency diseases, which render an individual susceptible to overwhelming infections by pathogenic agents, or autoimmune diseases where an inappropriate immune response destroys cell types mediating physiological responses critical to life.

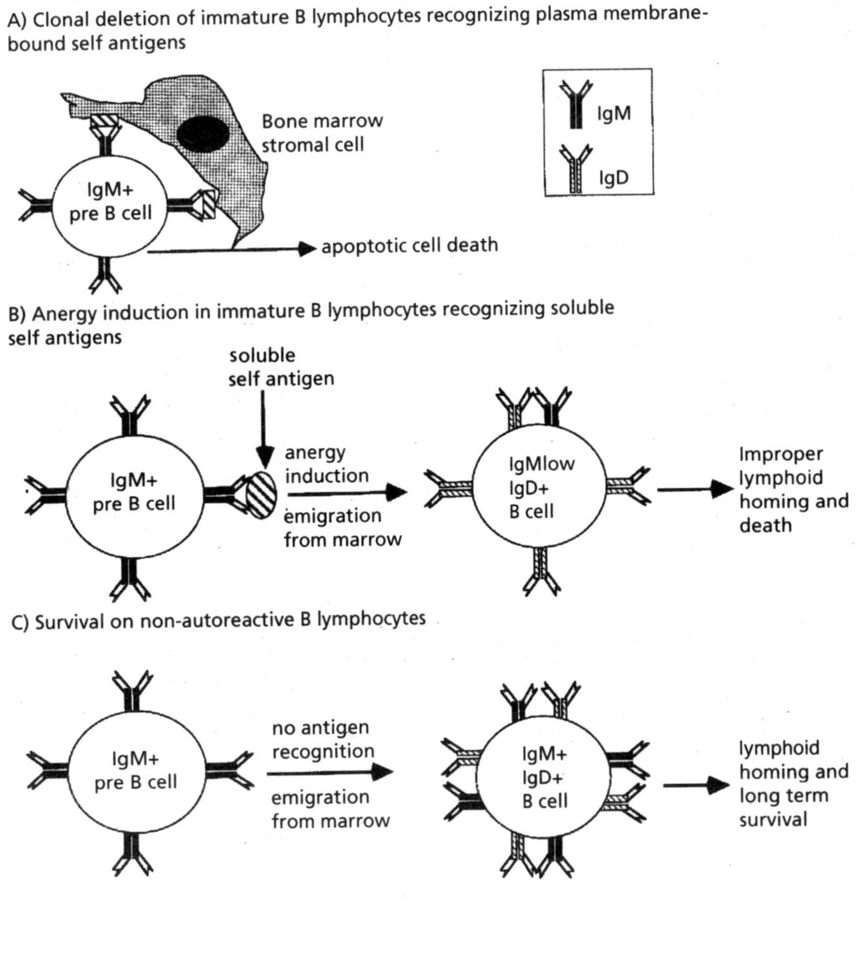

Figure 10 Selection of the B lymphocyte repertoire.

Most genetic defects causing immunodeficiency diseases are recessive in nature and include several X chromosome-linked syndromes which are usually only observed in males (reviewed in Hammarstrom *et al.*, 1993; Conley, 1995). Examples of some inherited immunodeficiency syndromes are outlined in **Table 2**. Among the more serious of these diseases is the SCID syndrome which is characterized by an inability to produce either T or B lymphocytes. Several different genetic defects can result in SCID syndromes. As described earlier, SCID syndromes can result from defects in the genes encoding DNA repair enzymes such as PRKDC, which mediates the rejoining of spliced TCR and Ig gene segments. SCID syndromes can also result from defects in the genes encoding the enzymes adenosine deaminase (ADA) and purine nucleotide phosphorylase (PNP). The loss of either ADA or PNP results in an overaccumulation of nucleotide metabolites that are toxic to T and B lymphocytes. Another severe immunodeficiency syndrome results from defects in the class II transactivator (*CIITA*) gene that leads to a loss of MHC class II expression and, hence, a failure to select positively $CD4^+$ T cells. Individuals have also been identified who produce no $CD8^+$ T cells due to a *Tap* gene defect which blocks expression of MHC class I molecules. Other gene defects have been identified that either prevent the development of B lymphocytes or their ability to initiate isotype switching away from IgM. Some individuals are also characterized by defects in complement genes that result in an impaired ability of antibodies to clear various pathogenic agents. Genetic defects have also been identified that impair important innate immune responses such as phagocytic activity, which, as discussed earlier, is also an essential step in triggering some aspects of adaptive immunity.

Defects in the genetic control elements normally preventing the development of T and B lymphocyte clonotypes that recognize self antigens can lead to a number of different autoimmune diseases (reviewed in Vyse & Todd, 1996). Examples of some autoimmune diseases are outlined in **Table 3**. In some of these diseases, antibodies represent the terminal mediators of pathogenic autoimmune damage. However, the B lymphocytes producing autoantibodies in these syndromes require T cell help in order to become functionally activated. In other autoimmune diseases, T cells themselves mediate pathogenic tissue destruction. Given the direct or indirect role of T cells in autoimmune diseases, it is not surprising that the development of many of these syndromes is most strongly associated with the presence of certain MHC allelic variants. In all likelihood, this is partially due to the fact that some MHC allelic variants will have a greater capacity than others to bind certain peptides derived from self proteins and present them to T cells. However, this mechanism does not address the issue of why T cells that recognize a self peptide/MHC complex are not negatively selected either intrathymically or extrathymically in the periphery of individuals with autoimmune diseases.

Insights into this issue have been gained through studies analysing the mechanistic basis for the development of insulin-dependent diabetes mellitus (IDDM), one of the most serious of all autoimmune diseases (reviewed in Serreze & Leiter, 1994; Serreze, 1998). These studies have demonstrated that, while certain MHC alleles provide the primary component of susceptibility, autoimmune IDDM development also requires contributions from a large number of non-MHC genes. Interestingly, it appears that some of these MHC and non-MHC diabetes susceptibility genes interact in a way that reduces the ability of APC to stimulate autoreactive T cells to an activation threshold that would normally induce their deletion, but not below the signalling threshold required to trigger the positive selection or immunological effector responses of such T cells. Since polygenic control seems to be a common feature of most other autoimmune diseases, it is possible that a similar mechanism also underlies their development.

11. Concluding remarks

The goal of this article was to provide an overview of the genetic elements controlling both the development and function of various components of the immune system. Hopefully, readers have been left with an appreciation of the

Table 2 Examples of some genetically caused immunodeficiency syndromes.

Immunodeficiency syndrome	Causative defect	Characteristic immune defect
Severe combined immunodeficiency (SCID)	Several including a defect in the *Prkdc* DNA repair gene, as well as ADA or PNP deficiencies	No T or B lymphocytes
MHC class I deficiency	*Tap* gene mutations	No $CD8^+$ T cells
MHC class II deficiency	At least five, including a defect in the class II transactivator (*CIITA*) gene	No $CD4^+$ T cells
X-linked agammaglobulinaemia	*Btk* tyrosine kinase gene	No B lymphocytes
X-linked hyper-IgM syndrome	Defective CD40 ligand (*Cd40*) gene	No isotype switching from IgM
Phagocyte deficiencies	Multiple gene defects	Loss of phagocytic activity
Complement deficiencies	Multiple gene defects	Loss of complement activity

Table 3 Examples of autoimmune diseases.

Autoimmune disease	Targeted antigen	Immunological effector mechanism	Disease consequences
Autoimmune haemolytic anaemia	Rh blood group I antigen	antibodies	red blood cell destruction, anaemia
Autoimmune thrombocytopenia purpura	platelet integrin GpIIb:IIIa	antibodies	abnormal bleeding
Myasthenia gravis	acetylcholine receptors	antibodies	progressive weakness
Systemic lupus erythematosus	DNA, histones, ribonuclear proteins	renal deposition of antibody–autoantigen complexes	glomerulonephritis, kidney failure
Insulin dependent diabetes mellitus	pancreatic β cells	T cells	loss of insulin production, hyperglycaemia
Rheumatoid arthritis	synovial joints	T cells	joint destruction
Multiple sclerosis	myelin basic protein, proteolipid protein	T cells	central nervous system degeneration

complicated, but elegant, genetic control mechanisms that normally enable an immune response to be generated against a virtually unlimited spectrum of foreign pathogenic invaders, but not against the body's own tissues. The broad overview approach taken in this article necessitated the omission of discussion of some important immunological aspects. These include, but by no means are limited to, the basis for allergic responses and immunological memory, and how immune cells home to lymphoid organs and sites of infection. The area of immunological memory is of special importance since it explains the basis by which permanent resistance can be induced to some infectious diseases through vaccinations. However, any areas of immunology not covered in this article will have been discussed in one or more of the excellent reviews cited.

Acknowledgements

Research in the author's laboratory is supported by National Institutes of Health grants DK46266, DK51090, AI41469 and DK27722, as well by grants from the Juvenile Diabetes Foundation International and Cancer Center Support (CORE) CA34196.

REFERENCES

Abbas, A.K. (1996) Die and let live: eliminating dangerous lymphocytes. *Cell* 84: 655–57

Ashton-Rickardt, P.G. & Tonegawa, S. (1994) A differential-avidity model for T-cell selection. *Immunology Today* 15: 362–66

Blunt, T., Gell, D., Fox, M. *et al.* (1996) Identification of a nonsense mutation in the carboxyl-terminal region of DNA-dependent protein kinase catalytic subunit in the *scid* mouse. *Proceedings of the National Academy of Sciences USA* 93: 10285–90

Callard, R.E. & Gearing, A.J.H. (1994) *The Cytokine FactsBook*, London and San Diego: Academic Press

Cambier, J.C. (1992) Signal transduction by T- and B-cell antigen receptors: converging structures and concepts. *Current Opinion in Immunology* 4: 257–64

Conley, M.E. (1995) Primary immunodeficiencies: a flurry of new genes. *Immunology Today* 16: 313–15

DeFranco, A.L. (1995) Transmembrane signalling by antigen receptors of B and T lymphocytes. *Current Opinion in Cell Biology* 7: 163–75

Fanning, L.J., Conner, A.M. & Wu, G.E. (1996) Development of the immunoglobulin repertoire. *Clinical Immunology and Immunopathology* 79: 1–14

Garcia, K.C., Degano, M., Stanfield, R.L. *et al.* (1996) An a(T cell receptor structure at 2.5 A and its orientation in the TCR-MHC complex. *Science* 274: 209–19

Gascoigne, N.R.J. (1995) Genomic organization of T cell receptor genes in the mouse. In *T cell receptors*, edited by J.I. Bell, M.J. Owen and E. Simpson, Oxford and New York: Oxford University Press

Germain, R.N., Castellino, F., Han, R.E. *et al.* (1996) Processing and presentation of endocytically acquired protein antigens by MHC class I and class II molecules. *Immunological Reviews* 151: 5–30

Gilfillan, S., Benoist, C. & Mathis, D. (1995) Emergence of the T cell receptor repertoire. In *T cell receptors*, edited by J.I. Bell, M.J. Owen and E. Simpson, Oxford and New York: Oxford University Press

Girao, C., Hu, Q., Sun, J. & Ashton-Rickardt, P.G. (1997) Limits to the differential avidity model of T cell selection in the thymus. *Journal of Immunology* 159: 4205–11

Goodnow, C.C., Cyster, J.G., Hartley, S.B. *et al.* (1995) Self-tolerance checkpoints in B lymphocyte development. *Advances in Immunology* 59: 279–368

Hammarstrom, L., Gillner, M. & Smith, C.I. (1993) Molecular basis for human immunodeficiencies. *Current Opinion in Immunology* 5: 579–84

Hansen, T.H. & Lee, D.R. (1997) Mechanism of class I assembly with β2-microglobulin and loading with peptide. *Advances in Immunology* 64: 105–37

Hilyard, K.L. & Strominger, J.L. (1995) The immune recognition unit: the TCR-peptide-MHC complex. In *T cell receptors*, edited by J.I. Bell, M.J. Owen and E. Simpson, Oxford and New York: Oxford University Press

Janeway, C.A. & Travers, P.A. (1997) *Immunobiology: the Immune System in Health and Disease*, 3rd edition, London: Current Biology and New York: Garland

Kaufmann, S.H.E. (1996) γδ and other unconventional T lymphocytes: what do they see and what do they do? *Proceedings of the National Academy of Sciences USA* 93: 2272–79

Klein, J., Satta, Y., O'hUigin, C. & Takahata, N. (1993) The molecular descent of the major histocompatibility complex. *Annual Review of Immunology* 11: 269–95

Lenschow, D.J., Walunas, T.L. & Bluestone, J.A. (1996) CD28/B7 system of T cell costimulation. *Annual Review of Immunology* 14: 233–58

Parham, P., Adams, E.J. & Arnett, K.L. (1995) The origins of HLA-A,B,C polymorphism. *Immunological Reviews* 143: 141–80

Paul, W.E. & Seder, R.S. (1994) Lymphocyte responses and cytokines. *Cell* 76: 241–51

Reth, M. (1994) B cell antigen receptors. *Current Opinion in Immunology* 6: 3–8

Schatz, D.G., Oettinger, M.A. & Schlissel, M.S. (1992) V(D)J recombination-molecular biology and regulation. *Annual Review of Immunology* 10: 359–83

Sebzda, E.S., Mariathasan, T., Ohteki, R. *et al.* (1999) Selection of the T cell repertoire. *Annual Review of Immunology* 17: 829–74

Serreze, D.V. (1998) The identity and ontogenic origins of autoreactive T lymphocytes in NOD mice. In *NOD Mice and Related Strains: Research Applications in Diabetes, AIDS, Cancer, and Other Diseases*, edited by E.H. Leiter and M.A. Atkinson, Austin, Texas: Landes Bioscience

Serreze, D.V. & Leiter, E.H. (1994) Genetic and pathogenic basis of autoimmune diabetes in NOD mice. *Current Opinion in Immunology* 6: 900–06

Stavnezer, J. (1996) Antibody class switching. *Advances in Immunology* 61: 79–146

von Boehmer, H. & Kisielow, P. (1993) Lymphocyte lineage commitment: instruction versus selection. *Cell* 73: 207–08

Vyse, T.J. & Todd, J.A. (1996) Genetic analysis of autoimmune disease. *Cell* 85: 311–18

Wagner, S.D. & Neuberger, M.S. (1996) Somatic hypermutation of immunoglobulin genes. *Annual Review of Immunology* 14: 441–57

Young, A.C.M., Nathenson, S.G. & Sacchettini, J.C. (1995) Structural studies of class I major histocompatibility complex proteins: insights into antigen presentation. *FASEB Journal* 9: 26–36

Zinkernagel, R.M. & Doherty, P.C. (1974) Restriction of *in vitro* T cell-mediated cytotoxicity in lymphocytic choriomeningitis within a syngeneic or semiallogeneic system. *Nature* 248: 701–02

GLOSSARY

allelic exclusion process where a locus (or loci) on one of a pair of homologous chromosomes is (are) inactivated so that individual cells express only one allelic form of the product of the locus

antibody a immunoglobulin protein produced by plasma cells in response to an antigenic stimulus and that reacts specifically with this antigen

apoptosis active process of cell death requiring metabolic activity by the dying cell

B-lymphocyte (*see* lymphocyte)

basophil type of while blood cell (granulocyte) that releases histamine and serotonin in specific immune reactions

CD (cluster of differentiation) antigen cell-surface antigen on white blood cells detected by specific sets of monoclonal antibodies; CD1 is mainly expressed in cortical thymocytes, CD3 in thymocytes, CD4 in T-helper/inducer cells and CD8 in T-cytotoxic/suppressor cells

chaperone proteins cytoplasmic proteins that bind to unfolded polypeptides to ensure correct folding or transport. They do not form part of the finished product

clonotype T or B lymphocyte respectively expressing a single TCR or Ig molecule

complement proteins of the complement system, a group of blood proteins of the globulin class involved in the lysis of foreign cells

cytokine any protein factor that is a product of a cell and which affects functions of other cells

cytotoxic attacking or destroying cells

dendritic cell subtype of antigen presenting cell

endosome a large membrane-bound structure that is the result after several coated vesicles fuse together following endocytosis. The structure contains the combined contents of all of the former vesicles

gamma interferon secreted by activated T lymphocytes during an immune response, this small protein is an important growth factor for lymphocytes and other immune system cells

Golgi (apparatus, body, complex, vesicles) intracellular stack of flattened membrane-bound sacs, involved in directing membrane lipid and proteins and secretory proteins to their precise cellular destinations

haematopoiesis normal formation and development of blood cells in the bone marrow

interleukin a variety of substances produced mainly by leucocytes that function during inflammatory responses

isotype switching switch of antigenic determinant on immunoglobulin molecules that occurs when the immune response progresses, e.g. IgM to IgG

L chain (light chain) the light chains of immunoglobulins are of two types: kappa (κ) and lambda (λ)

lymphocyte white blood cells derived from lymphoid stem cells, consisting of two types: B lymphocytes which (when activated) produce antibodies; T lymphocytes (comprising helper, suppressor, and cytotoxic T cells) involved in cell-mediated immunity and for stimulating B lymphocytes

macrophage large, phagocytic, mononuclear white blood cell of the reticuloendothelial system that ingests invading microorganisms and eliminates cellular debris

mast cell cell with large basophilic granules containing heparin, serotonin, bradykinin and histamine, which are released from the cell in response to injury or infection

monoclonal T-cell line line of T cells with a single clonal origin

neutrophil white blood cell, the granules of which stain only with neutral stains.

phagocyte a cell that is capable of phagocytosis

phagocytosis a five-step process by which particular cells engulf and destroy micro-organisms and cell debris: (1) invagination; (2) engulfment; (3) formation of phagocyte vacuole; (4) digestion of phagocytosed material; (5) release of digested microbial products

pinocytosis uptake of fluid-filled vesicles into a cell by the process of endocytosis

polypeptide organic compound of three or more amino acids

restriction fragment length variant analysis (RFLV) method to determine if distance between specific restriction enzyme digestion sites differs between test samples

rough endoplasmic reticulum (RER) eukaryotic organelle that is the site of membrane synthesis; binder of ribosomes

SCID (severe combined immunodeficiency) syndrome significant deficiency/complete absence of B and T cells, resulting in lack of immunity; may be X-linked recessive (affecting only males) or autosomal recessive (affecting both sexes)

somatic hypermutation higher than normal rate of mutation in body cell(s) as opposed to cells of the germline

somatic mutation mutation in body cell(s) as opposed to cells of the germline

T-lymphocyte (*see* lymphocyte)

vaccinia virus poxvirus injected as protection against development of smallpox

See also **Genetic recombination** (p.701)

X-CHROMOSOME INACTIVATION

Mary F. Lyon
MRC Mammalian Genetics Unit, Harwell, Didcot, UK

1. Introduction

All the somatic cells of female mammals carry two X chromosomes. However, only one of these is transcriptionally active and produces mRNA, whereas the other, although it replicates normally, is transcriptionally silent. This phenomenon is known as X-chromosome inactivation (XCI). The effect of XCI is that the chromosomally XX cells of females and XY cells of males both have a single dosage of X-linked gene products. Thus, this is a mechanism of dosage compensation for X-linked genes. Some animals in other groups with an XX:XY sex-determining system also show dosage compensation, including the fly Drosophila and the worm *Caenorhabditis elegans*, but the mechanisms of achieving it are different. XCI is found only in placental mammals and marsupials and in a rudimentary form in monotremes (see reviews in Gartler *et al.*, 1992; Migeon, 1994; Lyon, 1996).

In a young mammalian embryo, both X chromosomes are active and the onset of XCI occurs early in development, at the late blastocyst to egg cylinder stage in mice. Once XCI has occurred, each X chromosome retains its active or inactive state throughout all further mitotic divisions in the life of the animal. The result is that the adult female has large clumps of cells with the same X chromosome active. If the animal is heterozygous for an X-linked gene producing a visible effect, such as an effect on coat colour, this results in a variegated pattern. The best known example of this is the tortoiseshell cat, in which the pattern is due to the animal being heterozygous for an X-linked coat colour gene, one allele giving orange colour and the other black or tabby. For genes affecting other cell types, the presence of two cell populations in heterozygotes can be detected using appropriate means, including histochemical or immunological methods.

Rarely, individuals are found with additional or missing X chromosomes. In chromosomally XO humans or animals, the single X chromosome is active and, in XXY, XXX or XXXX individuals, all X chromosomes except one become inactive. Since they have a virtually normal dosage of X-linked gene products, this explains how individuals with abnormal numbers of X chromosomes can survive, whereas those with additional or missing autosomes die or are severely malformed. Thus, XCI should be considered as a means of ensuring that a single X chromosome remains active, rather than that one becomes inactive. Rare triploid individuals can be found in human and mouse, and, in contrast to diploids, these may have either one or two X chromosomes active, whereas tetraploids have two active X chromosomes. Thus, there is thought to be a counting mechanism, maintaining one X chromosome active per two autosome sets.

2. Properties of the inactive X chromosome

The inactive X chromosome (Xi) has various distinctive properties. It replicates its DNA later in the S phase of mitosis, and this can be made out by appropriate staining with H^3 thymidine or BrdU. During interphase, it is condensed to form a small dark-staining body, the sex chromatin body, which lies against the nuclear membrane. Thus, in its late replication and condensation, the inactive X exhibits the properties of heterochromatin. A further feature in which the inactive X resembles heterochromatin is in its lack of acetylated histone 4 (Jeppesen & Turner, 1993). This form of histone is known to be associated with transcriptionally active chromatin and is absent from heterochromatin. A differential pattern of DNA methylation is a further property of the Xi of eutherian mammals. The CpG islands in the 5′ promoter regions of X-linked genes are heavily methylated on the Xi, in contrast to their unmethylated state on the active X chromosome (Xa) and the autosomes. This differential methylation of CpG islands does not occur in marsupials (Gartler *et al.*, 1992; Cooper *et al.*, 1993).

3. Ohno's Law

Ohno (1967) pointed out that if, during evolution, an X-linked gene was translocated to an autosome, or vice versa, the correct balance of X-linked and autosomal gene products would be disturbed and, hence, such translocations would be eliminated by natural selection. He put forward the law that a gene X linked in one mammalian species would be X linked in all, and this has become known as Ohno's Law. The law is very widely obeyed, and has been of great value in X-chromosomal studies. It applies to marsupials as well as eutherian mammals, but, in marsupials, only the genes on the long arm of the present day human X chromosome are X linked, and those on the short arm are autosomal. This has led Graves to postulate that the short arm genes have been transferred to the eutherian X chromosome from autosomes during evolution, presumably while XCI itself was evolving (Graves & Watson, 1991).

4. X chromosome activity in germ cells

Single X chromosome activity in mammals is a property of somatic cells, and is stable once it has occurred. In germ cells, the situation is different (**Figure 1**). In female germ cells, XCI occurs in early development, as in somatic cells, but at the onset of meiosis the Xi is reactivated, so that in mature oocytes both X chromosomes are active. In male germ cells, the reverse situation occurs, and the single X chromosome becomes inactive early in spermatogenesis. The differential X chromosome activity is thought to be important for the survival of germ cells. Males with more than one X chromosome are sterile, with death of germ cells at the spermatogonial stage; females with a missing X chromosome, i.e. XO, are sterile in the human and of reduced fertility in the mouse, in both cases with excessive germ cell death. One possible explanation for this germ cell death is that abnormal numbers of X chromosomes lead to abnormalities of meiotic pairing, and that pairing failure is detrimental to germ cells (McKee & Handel, 1993). However, there may well also be harmful effects of incorrect dosage of X-linked gene products, as germ cell death in XX and XXY males occurs before the onset of meiosis.

5. Genomic imprinting in XCI

As previously mentioned, in eutherian mammals either of the two X chromosomes in any cell may become inactive at the onset of XCI in the embryo, the choice between the maternally derived (Xm) and paternally derived (Xp) X chromosomes being random. However, in marsupials, the situation is different, with the Xp becoming inactive in all cells (Cooper *et al.*, 1993). This is an example of genomic imprinting, a phenomenon in which homologous genes or chromosomes in a cell behave differently according to their parental origin, and which is seen for some autosomal genes of mammals as well as in the X chromosome. Imprinting of the X chromosome is also seen in the cells of some extraembryonic membranes of mice and rats, where, as in marsupials, the Xp is preferentially inactivated in all cells. Preferential Xp inactivation may also occur in human extraembryonic membranes, but this point is controversial.

6. Stability of XCI

In eutherians, XCI in somatic cells is remarkably stable. In the mouse, reactivation of the X-linked gene ornithine transcarbamylase (*Otc*) occurs at a low level throughout life, as also does reactivation of the autosomal coat colour gene for tyrosinase (or albino locus) when transferred to the X chromosome by a translocation. However, for other X-linked genes, no reactivation has been detected and in humans reactivation has not been found for any gene. Many experimental attempts to induce reactivation have been made (Gartler & Goldman, 1994), but success has only been obtained in somatic cell hybrids, particularly by treating such cells with demethylating agents. Some types of embryonic cells provide an exception to the great stability of XCI in eutherians. Fusion of adult cells with embryonal cells of another species has in some cases induced reactivation in the adult cell. Reactivation has also been induced by treating an embryonal cell line with a demethylating agent. In adult cells, such a procedure is only effective in species hybrid cell lines, suggesting that the embryonal cells concerned lacked some stabilizing mechanism present in adult cells.

By contrast, in marsupials, XCI is much less stable and reactivation can occur spontaneously either *in vivo* in certain tissues or in cultured cells. There is variation in stability of XCI among species and among different X-linked genes within a species (Cooper *et al.*, 1993).

7. Escape from XCI

XCI is a mechanism for dosage compensation. X-linked genes in the segment of the X chromosome that pairs with the Y chromosome (the pseudoautosomal region) have homologues on the Y chromosome and do not require dosage compensation. Hence, these genes might be expected to escape from XCI, and this is, indeed, found to be the case. In the human, various pseudoautosomal genes are known which do not undergo XCI. However, in addition, there are genes in other regions of the human X chromosome, notably in the short arm and proximal long arm, that also escape XCI (Disteche, 1995). By contrast, in the mouse, the corresponding genes undergo XCI normally; in this species, the few genes known to escape XCI either partly or wholly, including *Smcx* (Carrel *et al.*, 1996; Sheardown *et al.*, 1996) and *Sts*, have homologues on the Y chromosome.

The difference between human and mouse in escape from XCI may explain the difference between the two species in the effects of abnormal numbers of X chromosomes. In the human, the majority of XO females die as foetuses and the survivors are malformed, whereas in the mouse XO females are phenotypically normal. The abnormality of human XOs may result from a shortage of gene products from those genes that escape XCI. Similarly, in the human, the presence of extra X chromosomes results in malformations and this may be due to excess gene products of the same genes. Genes that escape XCI are interspersed on the human X chromosome with those that undergo XCI, indicating that single genes or small groups of genes can respond individually to the signals that invoke inactivation.

8. Applications of XCI

The presence of two cell populations in heterozygous females as a result of XCI has been valuable in studies of the patterns of growth of cell populations during development and of the origin of tumours. Since the onset of XCI occurs early in development, the sizes, shapes and spatial distribution of patches indicate the manner in which cells have migrated, mingled and clonally expanded during development (reviewed in Lyon, 1988). By histochemical studies of mice heterozygous for deficiency of glucose-6-phosphate dehydrogenase, Thomas and Williams (1988) showed that intestinal crypts have a monocellular origin whereas villi are polyclonal. Similarly, thyroid acini are polyclonal, but thyroid nodules monoclonal.

Although XCI typically affects either Xm or Xp at random, in some cases non-random XCI is seen and frequently this is due to cell selection against one or other cell population. In particular, this occurs in females heterozygous for some chromosome aberrations affecting the X chromosome. In heterozygotes for translocations between the X chromosome and an autosome, XCI spreads from the X chromosome into the attached autosomal material. In addition, only one of the two segments into which the X chromosome is broken undergoes XCI. Thus, in cells in which the translocated X chromosome undergoes XCI there is imbalance both of autosomal and of X-linked gene products. In most X-autosome translocations in humans and in a single such translocation, T16H, in the mouse, all cells have the normal X chromosome inactive. In the mouse, this has been shown to be due to random XCI followed by selection against the unbalanced cells, and it is assumed that this is so in humans also. Similarly, in heterozygotes for X chromosome deletions, all cells have the deleted X chromosome inactive, presumably resulting from selection against cells of the other type.

For some genes, cell selection may take place only in certain tissues of heterozygotes rather than in all tissues. In human heterozygotes for deficiency of hypoxanthine phosphoribosyl transferase, there is random X chromosome activity in fibroblasts, but, in blood, all cells have the normal allele of the gene active. In heterozygotes for Wiskott–Aldrich syndrome, B and T lymphocytes all have the normal allele active, whereas a typical mixture of cells with one or other X chromosome active (Greer *et al.*, 1989) is found in other cell types. The cells showing non-random XCI are those in which the abnormal gene is expressed and, hence, this type of apparently non-random XCI can provide information as to the cell types in which a gene acts. In addition, detection of non-random XCI either by X-linked gene products or by differential methylation can reveal which females in a family are heterozygous for an X-linked disease gene.

9. Mechanism of XCI

In consideration of its mechanism, XCI is usually divided under three headings. There is, first, the initiation of XCI in the early embryo. This is followed by spreading of the inactivation along the chromosome and this, in turn, by maintenance of the inactive state. Thus, the three phases to be elucidated are initiation, spreading and maintenance.

(i) Initiation of XCI

A key feature of the initiation of XCI in the early embryo is that it requires the action of the X-inactivation centre (XIC). The evidence for this came from the study of mouse X–autosome translocations. In heterozygotes for such translocations, only one of the segments into which the X chromosome is broken undergoes inactivation. The interpretation of this result is that there is an XIC located on the X chromosome and that only genes in physical continuity with it can undergo inactivation. Similar evidence indicates the presence of an XIC on the human X chromosome also. Comparison of the behaviour of different translocations and deletions has indicated the position of the XIC in both species. From this region, a gene termed *XIST/Xist* (X inactive-specific transcript) was cloned which, in both species, was transcribed from the inactive but not the active X chromosome. From a combination of its position and its unique pattern of transcription, the *Xist* gene was a strong candidate for a role in the XIC (reviewed in Lyon, 1996).

For *Xist* to have a role in XCI, it must be expressed in the embryo before the time of initiation of XCI. Kay *et al.* (1993, 1994) found that expression could first be seen at the 4-cell stage, well before initiation, and persisted throughout further development. Using embryos with distinctively marked X chromosomes, they showed that at early stages only the paternal allele of *Xist* was expressed, with expression of the maternal allele not being seen until the late blastocyst stage. Thus, both the time and the imprinted nature of its first expression are appropriate for *Xist* to have a causal role in XCI (**Figure 1**).

Kay *et al.* (1993) also studied *Xist* expression in embryonic stem cells (ES cells). Previously, Martin *et al.* (1978) had shown that when XX teratocarcinoma stem cells are maintained in an undifferentiated, pluripotent, state both X chromosomes are active. However, when differentiation is allowed, XCI occurs; ES cells behave similarly (Rastan & Robertson, 1985). Thus, ES cells provide a valuable system for study of the initiation of XCI. Kay *et al.* (1993) found that in undifferentiated ES cells only very slight expression of *Xist* was present and that, when differentiation occurred, *Xist* expression increased markedly. Thus, in ES cells

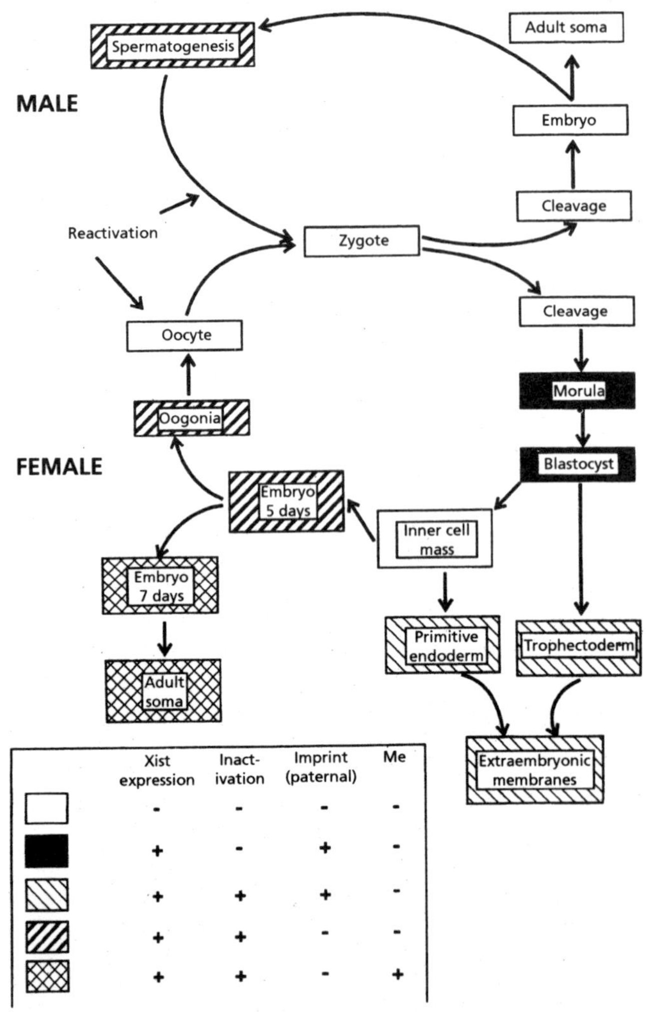

Figure 1 Cycle of X chromosome activity in the mouse, showing inactivation in the female embryo, followed by reactivation in the germ cells, and inactivation in the male germ cell, followed by reactivation in the new zygote. Also shown is the relation of expression, imprinting and methylation (Me) of the *Xist* gene to the cycle of inactivation and reactivation. (Reprinted with permission from Lyon, 1993.)

also, *Xist* expression was suitable for it to have a role in XCI.

Various authors have studied *Xist* expression in germ cells. In the male, *Xist* expression can be seen only in the testis and, specifically, in the male germ cells. In the female, *Xist* mRNA is present in the ovary, as in other tissues, but *Xist* expression disappears from oocytes at around the time of entry into meiosis, when reactivation of the Xi occurs. Thus, *Xist* expression may be involved in the changes of X chromosome activity that take place in germ cells (reviewed in Lyon, 1996).

In order to investigate the role of *Xist* further, the effects of a gene knockout and of generating animals transgenic for *Xist* have been studied. The aim was to find whether these procedures affected the functions of the XIC in the counting of X chromosomes, choice (including imprinting) of an X chromosome to remain active and spreading of inactivation into nearby genes. Penny *et al.* (1996) generated a deletion of 7 kilobases (kb) of the first exon of *Xist* by homologous recombination in XX ES cells. The X chromosomes in these cells were distinguishable by the alleles of X-linked genes they carried, one being from strain PGK and the other from strain 129, with the deletion generated in the 129 X chromosome. When these ES cells were allowed to differentiate, XCI occurred, as recognized by a late-replicating X chromosome. In order to find which X chromosome was inactivated, Penny *et al.* (1996) studied expression of the X-linked genes *Pgk1* and *Rps4x*. In all cells with an Xi, it was the normal PGK X chromosome that was inactive and the 129 X chromosome with the deletion remained active in all cells; some cells had two active X chromosomes. The interpretation was that the counting mechanism still recognized the deleted X chromosome, and the cell then behaved as other cells with two X chromosomes and XCI occurred. The authors suggested that the mechanism of choice of X chromosome for XCI was also still present, and that either X chromosome was being selected for XCI in different cells. If the deleted X chromosome was selected to remain active, the other X chromosome underwent XCI normally. If the normal X chromosome was selected to remain active, then the deleted X chromosome could not undergo XCI, and the cell then had two Xa. The authors also studied the effect of introducing ES cells bearing the knockout into chimaeric embryos generated by aggregating ES cells with normal 8-cell embryos. Again, XCI occurred and again only the normal PGK X chromosome underwent XCI. However, no cells were found with two Xa, in line with other work showing that cells with two Xa are rapidly eliminated from embryos by cell selection. Thus, this work provides good evidence for a role of the *Xist* gene in the function of the XIC in initiating the process of XCI, but leaves unanswered the question as to whether other sequences in the *Xist* gene, or other loci, are involved in the counting of X chromosomes and the choice of X chromosome to undergo XCI.

Further information has come from the work of Lee *et al.* (1996) who made chromosomally XY ES cells transgenic for a yeast artificial chromosome (YAC) carrying the *Xist* gene. The YAC was 450 kb in size and thus carried 100–200 kb of sequences either side of the *Xist* gene itself. Three lines were obtained, with the YAC inserted into three different autosomes. To detect expression of *Xist* when the ES cells underwent differentiation, the authors carried out *in situ* hybridization of RNA to the chromosome, *Xist* mRNA having the property of remaining attached to the chromosome. In the transgenic lines, either the inserted or the endogenous *Xist* gene could be expressed. The three lines differed in the numbers of copies of the transgene inserted, namely 20–30, 6–8 and 3–4 in the individual lines. The results were consistent with each *Xist* copy being recognized by the counting mechanism, since many cells showed *Xist* expression of both the transgenic and the endogenous *Xist*, and very few cells expressed the endogenous *Xist* only. Cells in which the endogenous *Xist* was expressed would be expected to die as a result of XCI and lack of X-linked gene products. The authors found that cell death was indeed increased in the transgenic cells during differentiation of the ES cells, when *Xist* expression was being upregulated.

To study the behaviour of the transgene in postnatal animals, chimaeric embryos were made by aggregating the ES cells with normal 8-cell embryos. In fibroblasts from the resulting chimaeric mice, *Xist* expression was indeed present, but it was entirely from the transgene and the endogenous *Xist* was not expressed. This is again consistent with selection against those cells in which the counting mechanism had chosen the endogenous *Xist* for expression. In order to detect whether expression of the transgenic *Xist* was inducing inactivation of nearby genes, a marker gene, *lacZ*, had been introduced into the YAC construct. In undifferentiated ES cells, where *Xist* was only very minimally expressed, the *lacZ* gene was active, but in transgenic postnatal fibroblasts, where the transgenic *Xist* was expressed in all cells, there was no *lacZ* expression. This is consistent with the transgenic *Xist* being able to inactivate nearby genes in *cis*. Thus, these results indicate that the 450 kb YAC construct was able to carry out the functions of the XIC in counting and choice of X chromosome for inactivation, and also in inactivation of nearby genes. The results of the gene knockout of *Xist* indicate that a functional *Xist* RNA is required for the inactivation function of the XIC, but not for counting and choice of X chromosome. Whether the counting and choice mechanisms involve other parts of the *Xist* gene itself, or of genes in the 100–200 kb flanking regions present in the 450 kb YAC, remains to be elucidated.

(ii) Function of the Xist gene

The exact functions of the *Xist* gene in counting, choice of X chromosome and in the spreading of inactivation are still largely unknown. An early step in the elucidation of these functions was the study of the *Xist* gene product. In both human and mouse, the gene has no substantial open reading frame, but produces a large mRNA, 17 kb in size in human (Brown *et al.*, 1992) and 15 kb in mouse (Brockdorff *et al.*, 1992), which does not migrate to the cytoplasm, but is retained in the nucleus. Clemson *et al.* (1996) showed that the *Xist* mRNA coats the Xi. Thus, this coating of the Xi may be in some way involved in spreading of the inactivation.

Further support for the possible role of coating of the X chromosome in spreading comes from work on *Xist* expression in very early embryos and undifferentiated ES cells. Kay *et al.* (1993) showed that *Xist* was expressed from the 4-cell stage onwards, which is a surprising result since XCI does not occur until the late blastocyst stage. However, Latham and Rhambhatla (1995) measured *Xist* expression quantitatively and found the level very low, only 100–200 copies per embryo, during early stages. It increased as development proceeded, rising to ~500 copies per embryo at the late blastocyst stage, but this was still much lower than the 1000–2000 copies per cell found by Buzin *et al.* (1994) in embryos after XCI had occurred. At early stages, only the Xp allele of *Xist* is expressed.

Latham and Rhambhatla studied the expression of X-linked genes in early normal embryos and in androgenotes and gynogenotes. Androgenotes are those embryos in which the maternal and paternal pronuclei of the 1-cell zygote have been replaced with two paternal pronuclei, whereas gynogenotes are those with two maternal pronuclei. The *Pgk1* gene, which lies close to *Xist*, was very weakly expressed in androgenotes whereas other genes more distant from *Xist* were expressed more strongly. They suggested that the low levels of *Xist* mRNA found in early stages might affect only nearby genes (**Figure 2**) (Latham & Rhambhatla, 1995; Latham, 1996). Latham suggested that only when the counting mechanism comes into play does long-range spreading of XCI occur. Lee *et al.* (1996) reported that the low levels of expression of *Xist* seen in undifferentiated ES cells resulted not in coating of the X chromosome, but in pinpoints of RNA lying over the *Xist* locus. After differentiation, when XCI occurs, the pinpoint expression was extinguished in normal male cells, and, in female cells, expression spread to coat the chromosome in the presumed Xi (**Figure 2**). Thus, the low levels in early embryos probably also result in pinpoints of RNA, and full coating with *Xist* RNA may be needed for spreading of XCI. Brockdorff *et al.* (1992) suggested that either the *Xist* RNA itself might be involved in bringing about a change in state of the chromatin of the Xi or that the presence of *Xist* RNA might affect the binding of other factors affecting the chromatin state.

It is of interest that the expression of *Xist* in male germ cells, although present, is at a very low level (Kay *et al.*, 1993). In view of the apparent ineffectiveness of low levels of *Xist* expression in early embryos, the role of the low expression in male germ cells remains unclear.

(iii) Imprinting and choice of X chromosome for inactivation

Information concerning imprinting and choice of X chromosome for inactivation has come from studies on experimentally produced androgenetic, gynogenetic and parthenogenetic mouse embryos (parthenogenotes are produced by activation of unfertilized eggs). In androgenotes, all chromosomes are of paternal origin, while in gynogenotes and parthenogenotes all are of maternal origin. Kay *et al.* (1994) found that, in androgenotes, *Xist* was expressed from the 4-cell stage onwards. Androgenotes can be chromosomally of three types: XpXp, XpY and YY. The YY

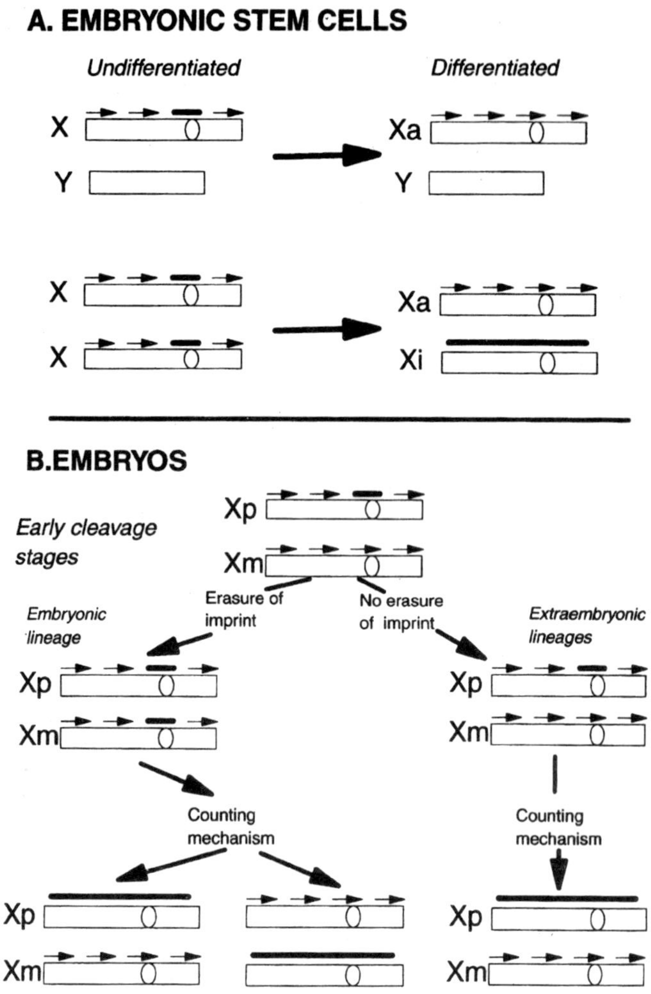

Figure 2 Expression of the *Xist* gene, coating of the X chromosome with *Xist* mRNA and the effect of this coating on transcription at various stages. The position of the XIC and *Xist* locus is denoted by an ellipse, *Xist* RNA by a heavy line and transcription from the X chromosome by arrows. (A) Relation of *Xist* mRNA to the X chromosome in embryonic stem (ES) cells, according to the work of Lee *et al.* (1996). In undifferentiated ES cells, there is pinpoint expression of *Xist* mRNA from X chromosomes of both male and female cells. After differentiation, the pinpoint expression is extinguished in the male and in one X chromosome of females, and the other female X chromosome is coated with *Xist* RNA. (B) Possible sequence of *Xist* mRNA production and effects on transcription in embryos, according to the combined results of Lee *et al.* (1996), Latham and Rhambhatla (1995) and Latham (1996). In cleavage stages, there is very low *Xist* expression from the paternal *Xist* allele, possibly giving pinpoint coating, and inhibition of transcription of nearby genes only. In the embryonic lineage, the imprint is lost and the maternal *Xist* allele becomes expressed. When the counting mechanism operates, *Xist* expression is strongly enhanced, the X chromosome is coated and there is chromosome-wide inhibition of transcription.

embryos are expected to die early because of their lack of X-linked gene products. The results of Kay *et al.* (1994) indicated that all other embryos, both XX and XY, were expressing *Xist*, showing that the counting mechanism was not functional at early stages. By contrast, in gynogenotes and parthenogenotes, expression of *Xist* was not seen until the morula to blastocyst stage. Latham and Rhambhatla (1995) also studied *Xist* expression in androgenotes and gynogenotes and did find some expression of *Xist* in 8-cell gynogenotes, but much less than in androgenotes. They also found that expression of *Xist* in androgenotes persisted to the blastocyst stage, in contrast to the findings of Kay *et al.* (1994). Thus, these results indicate that the paternal *Xist* allele bears an imprint permitting early expression, whereas the imprint on the maternal allele inhibits expression. The counting mechanism is not functional at early stages, but appears later, whereas the imprints are apparently lost over time, permitting expression of the maternal allele (**Figure 2**).

The difference between XCI in trophectoderm and primitive endoderm, where Xp is preferentially inactivated, and in the embryo proper, where XCI is random, is thought to be one of timing, with XCI occurring earlier in trophectoderm and primitive endoderm. Mouse embryos with two copies of Xm, such as XmXmXp, die with abnormalities of the extraembryonic membranes, which are thought to result because the imprint on Xm prevents it undergoing XCI in these lineages (Tada *et al.*, 1993). In the embryonic lineage, the imprint is thought to be erased before XCI.

To investigate the nature of the imprint on the *Xist* gene, studies have been made of methylation of various sites in the gene, since methylation of cytosine bases in DNA is regarded as a strong candidate for the molecular basis of imprinting in general. Norris *et al.* (1994) found that sites in the promoter and 5′ region of the body of the gene were fully methylated in somatic tissues of the male, whereas these sites were methylated only on the Xi in females. Thus, the methylated state of the gene was correlated with a lack of expression. In the testis, specifically in the germ cells, the *Xist* gene was demethylated and this demethylation occurred around the time of entry into meiosis and of inactivation of the X chromosome in the germ cells. This demethylation persisted to the mature sperm stage. Ariel *et al.* (1995) and Zuccotti and Monk (1995) compared methylation at 5′ sites in the gene in sperm and in oocytes and found some sites which were methylated in the oocyte, and which remained so in the early embryo, but were unmethylated in sperm. In addition, Beard *et al.* (1995) bred embryos deficient in DNA methyltransferase activity and found that demethylation of *Xist* resulted in its expression in XY male embryos. Thus, differential methylation of 5′ sites in *Xist* is a good candidate for the molecular basis of imprinting of the X chromosome.

A further factor involved in the choice of X chromosome for inactivation is the X chromosome controlling element locus or *Xce*. Alleles at this locus affect the probability of the X chromosome on which they lie becoming inactive, with Xce^c being "strongest" in its tendency to remain active and Xce^a "weakest". The *Xce* locus maps to the XIC and is regarded as being part of it. Some controversy exists on whether it is another manifestation of the *Xist* gene or a separate locus. The expression of *Xist* varies in mice with different *Xce* alleles, being inversely related to the strength of the allele. Moreover, sequence differences in *Xist* have been found in different *Xce* alleles. On the other hand, mapping experiments have revealed an apparent crossover between *Xce* and *Xist* (reviewed in Lyon, 1996).

(iv) Possible molecular bases of spreading and maintenance of XCI

Although the spreading and maintenance of XCI are regarded as separate parts of the process, evidence suggests that they may share some molecular mechanisms and, hence, they are more conveniently considered together.

The best evidence of a role of *Xist* in the spreading of XCI is the coating of the Xi with *Xist* RNA (Clemson *et al.*, 1996). However, what molecular effect this produces is not clear. The condensation of the Xi indicates a different state of the chromatin fibre, but it is not known how this is brought about. Riggs (1990) suggested that enzymes, similar to bacterial type I restriction enzymes, bind to the DNA and draw DNA towards themselves by a reeling-in process. This reeling-in, he suggested, would cause the DNA to be thrown into loops, with a protein scaffold at the bases of these loops. These sites would be the scaffold- or matrix-associated regions. This is consistent with the known structure of the Xi which has the shape of a metaphase chromosome, with the arms formed of loops surrounded by less dense loops (Gartler *et al.*, 1992).

Methylation of cytosines in DNA has been suggested as a mechanism of spreading of XCI, with the idea that a methyltransferase was a suitable enzyme to pass along the chromosome for the required long distances. However, although differential methylation of CpG islands in 5′ regions of housekeeping genes is clearly present in the Xi of eutherians, it is now regarded as a means of stabilizing, rather than spreading, XCI. The reasons for this are that XCI occurs without methylation of CpG islands in marsupials, in the germ cells of eutherians and probably also in the extraembryonic membranes of mice. In addition, in development, differential methylation occurs after XCI. The types of XCI in which methylation of CpG islands does not occur are those in which XCI is less stable than in somatic cells of eutherians and this is further regarded as evidence that methylation is involved in the stabilization of XCI (reviewed in Lyon, 1996). Asynchronous, usually late, replication of DNA is seen in all types of XCI, is found early in the onset of XCI and could be an underlying fundamental feature. Riggs and Pfeifer (1992) suggested that it could form a feedback loop for stabilizing inactivation if late replication prevents transcription and if transcription is required for early replication. Evidence that late replication and differential promoter methylation are both involved in the maintenance of XCI was obtained by Hansen *et al.* (1996) who studied the replication timing of individual genes and the susceptibility of X chromosomes carrying them to reactivation by the demethylating agent 5-azacytidine (5aC). Reactivation with 5aC was accompanied by earlier replication of the loci concerned; somatic cell hybrids tended to have earlier gene replication times than normal cells and reactivation occurred more readily in such hybrids, and, among different cell hybrids, more readily in those with earlier replication times. The authors suggested a two-stage model in which large chromosomal domains were placed into an inactive state by late replication and transcription was blocked locally by promoter methylation.

Whether *Xist* has any role in the maintenance of XCI is not clear. Hansen *et al.* (1996) found that *XIST* was still expressed in 5aC-treated cells with reactivated genes, indicating that *XIST* expression is not sufficient to maintain inactivation. Conversely, in two systems in which the *XIST* locus was deleted from X chromosomes after they had undergone XCI, reactivation did not occur (Brown & Willard, 1994; Rack *et al.*, 1994). Thus, *XIST* expression is not required to maintain XCI, and the role of the continued expression of *XIST/Xist* in adult somatic cells is not clear.

10. Evolution of XCI

Cooper (1971) suggested that the imprinted type of XCI seen in marsupials and the extraembryonic tissues of mice and rats was the primitive type and that the random type of XCI seen in eutherians evolved later. This suggestion is regarded as very plausible. Inactivation of the X chromosome of male germ cells may have occurred first – inactivation of the sex chromosomes in male meiosis is seen in some other animal groups with XX:XY sex determination systems and may be required to protect unsaturated pairing sites in the differentiating X and Y chromosomes. Some imprint may then have been carried over into the new zygote and provided a starting point for XCI. If XCI is of the imprinted type, a counting mechanism is not needed, since males are always XmY, and thus would have an active X chromosome, and females are XmXp and would have one active (Xm) and one inactive (Xp) X chromosome. In eutherians, it is known from many aneuploids and polyploids that a counting mechanism exists. However, in marsupials, very few aneuploids are known and it is possible that there is no counting mechanism. A selective advantage leading to the evolution of a counting mechanism and random inactivation may be that deleterious mutations are not expressed in the heterozygote due to the presence of two cell populations with mutant and normal genes active. The random type of XCI in eutherians is also more stable, providing better dosage compensation. However, it is not clear whether this is connected with the counting mechanism or whether the differential promoter methylation which confers greater stability is an entirely separate evolutionary development.

Discovery of the *XIST/Xist* gene has led to considerable advances in understanding of XCI and its mechanism. A great deal remains to be discovered, and one may hope for further strides in knowledge in the near future.

REFERENCES

Ariel, M., Robinson, E., McCarrey, J.R. & Cedar, H. (1995) Gamete-specific methylation correlates with imprinting of the murine *Xist* gene. *Nature Genetics* 9: 312–15

Beard, C., Li, E. & Jaenisch, R. (1995) Loss of methylation activates *Xist* in somatic but not in embryonic cells. *Genes and Development* 9: 2325–34

Brockdorff, N., Ashworth, A., Kay, G.F. *et al.* (1992) The product of the mouse *Xist* gene is a 15 kb inactive X-specific transcript containing no conserved ORF and located in the nucleus. *Cell* 71: 515–26

Brown, C.J. & Willard, H.F. (1994) The human X-inactivation centre is not required for maintenance of X-chromosome inactivation. *Nature* 368: 154–56

Brown, C.J., Hendrich, B.D., Rupert, J.L. *et al.* (1992) The human *XIST* gene: analysis of a 17 kb inactive X-specific RNA that contains conserved repeats and is highly localised within the nucleus. *Cell* 71: 527–42

Buzin, C.H., Mann, J.R. & Singer-Sam, J. (1994) Quantitative RT-PCR assays show *Xist* RNA levels are low in mouse female adult tissue, embryos and embryoid bodies. *Development* 120: 3529–36

Carrel, L., Hunt, P.A. & Willard, H.F. (1996) Tissue and lineage-specific variation in inactive X chromosome expression of the murine *Smcx* gene. *Human Molecular Genetics* 5: 1351–56

Clemson, C.M., McNeil, J.A., Willard, H. & Lawrence, J.B. (1996) *XIST* RNA paints the inactive X chromosome at interphase: evidence for a novel RNA involved in nuclear/chromosome structure. *Journal of Cell Biology* 132: 259–75

Cooper, D.W. (1971) A directed genetic change model for X-chromosome inactivation in eutherian mammals. *Nature* 231: 292–94

Cooper, D.W., Johnston, P.G., Watson, J.M. & Graves, J.A.M. (1993) X-inactivation in marsupials and monotremes. *Seminars in Developmental Biology* 4: 117–28

Disteche, C.M. (1995) Escape from X inactivation in human and mouse. *Trends in Genetics* 11: 17–22

Gartler, S.M. & Goldman M.A. (1994) Reactivation of inactive X-linked genes. *Developmental Genetics* 15: 504–14

Gartler, S.M., Dyer, K.A. & Goldman, M.A. (1992) Mammalian X-chromosome inactivation. In *Molecular Genetic Medicine*, edited by T. Friedmann, San Diego: Academic Press

Graves, J.A.M. & Watson, J.M. (1991) Mammalian sex chromosomes: evolution of organization and function. *Chromosoma* 101: 61–68

Greer, W.L., Kwong, P.C., Peacocke, M. *et al.* (1989) X-chromosome inactivation in the Wiskott–Aldrich syndrome: a marker for detection of the carrier state and identification of cell lineages expressing the gene defect. *Genomics* 4: 60–67

Hansen, R.S., Canfield, T.K., Fjeld, A.D. & Gartler, S.M. (1996) Role of late replication timing in the silencing of X-linked genes. *Human Molecular Genetics* 5: 1345–53

Jeppesen, P. & Turner, B.M. (1993) The inactive X chromosome in female mammals is distinguished by a lack of histone H4 acetylation, a cytogenetic marker for gene expression. *Cell* 74: 281–89

Kay, G.F., Penny, G.D., Patel, D. *et al.* (1993) Expression of *Xist* during mouse development suggests a role in the initiation of X chromosome inactivation. *Cell* 72: 171–82

Kay, G.F., Barton, S.C., Surani, M.A. & Rastan, S. (1994) Imprinting and X chromosome counting mechanisms determine *Xist* expression in early mouse development. *Cell* 77: 639–50

Latham, K. (1996) X chromosome imprinting and inactivation in the early mammalian embryo. *Trends in Genetics* 12: 134–38

Latham, K.E. & Rhambhatla, L. (1995) Expression of X-linked genes in androgenetic, gynogenetic, and normal mouse preimplantation embryos. *Developmental Genetics* 17: 212–22

Lee, J.T., Strauss, W.M., Dausman, J.A. & Jaenisch, R. (1996) A 450 kb transgene displays properties of the mammalian X-inactivation center. *Cell* 86: 83–94

Lyon, M.F. (1988) Clones and X-chromosomes. *Journal of Pathology* 155: 97–99

Lyon, M.F. (1993) Epigenetic inheritance in mammals. *Trends in Genetics* 9: 123–28

Lyon, M.F. (1996) Molecular genetics of X-chromosome inactivation. *Advances in Genome Biology* 4: 119–51

Martin, G.R., Epstein, C.J., Travis, B. *et al.* (1978) X-chromosome inactivation during differentiation of female teratocarcinoma stem cells *in vitro*. *Nature* 271: 329–33

McKee, B.D. & Handel, M.A. (1993) Sex chromosomes, recombination, and chromatin conformation. *Chromosoma* 102: 71–80

Migeon, B.R. (1994) X-chromosome inactivation: molecular mechanisms and genetic consequences. *Trends in Genetics* 10: 230–35

Norris, D.P., Patel, D., Kay, G.F. *et al.* (1994) Evidence that random and imprinted *Xist* expression is controlled by preemptive methylation. *Cell* 77: 41–51

Ohno, S. (1967) *Sex Chromosomes and Sex-Linked Genes*, Berlin and New York: Springer

Penny, G.D., Kay, G.F., Sheardown, S.A., Rastan, S. & Brockdorff, N. (1996) Requirement for *Xist* in X chromosome inactivation. *Nature* 379: 131–37

Rack, K.A., Chelly, J., Gibbons, R.J. *et al.* (1994) Absence of the XIST gene from late-replicating isodicentric X chromosomes in leukaemia. *Human Molecular Genetics* 3: 1053–59

Rastan, S. & Robertson, E.J. (1985) X-chromosome deletions in embryo-derived (EK) cell lines associated with lack of X-chromosome inactivation. *Journal of Embryology and Experimental Morphology* 90: 379–88

Riggs, A.D. (1990) DNA methylation and late replication probably aid cell memory, and type I DNA reeling could aid chromosome folding and enhancer function. *Philosophical Transactions of the Royal Society of London, Series B* 326: 1285–97

Riggs, A.D. & Pfeifer, G.P. (1992) X-chromosome inactivation and cell memory. *Trends in Genetics* 8: 169–74

Sheardown, S., Norris, D., Fisher, A. & Brockdorff, N. (1996) The mouse *Smcx* gene exhibits developmental and tissue specific variation in degree of escape from X inactivation. *Human Molecular Genetics* 5: 1355–60

Tada, T., Takagi, N. & Adler, I.-D. (1993) Parental imprinting on the mouse X chromosome: effects on the early development of XO, XXY and XXX embryos. *Genetical Research* 62: 139–48

Thomas, G.E. & Williams, E.D. (1988) The demonstration of tissue clonality by X-linked enzyme histochemistry. *Journal of Pathology* 155: 101–08

Zuccotti, M. & Monk, M. (1995) Methylation of the mouse *Xist* gene in sperm and eggs correlates with imprinted *Xist* expression and paternal X-inactivation. *Nature Genetics* 9: 316–20

GLOSSARY

androgenesis/androgenetic androgenetic embryos have all their genetic information from their father. There is no maternal contribution

azacytidine (5-azacytidine; 5ac) nucleotide analogue incorporated into DNA in place of cytosine. It cannot be methylated and thus acts as a demethylating agent (*see* methylation, demethylating agent)

bromodeoxyuridine (BrdU, BUdR) a deoxynucleoside of 5-bromouracil, used particularly as a marker for DNA synthesis because of its distinctive staining pattern

CpG island (i) a region rich in CpG dinucleotides. It is usually located at the 5′ end of an expressed housekeeping gene where it is found in the unmethylated state; (ii) a segment of genomic DNA, usually several hundred base pairs long, that has a higher than expected occurrence of the dinucleotide CpG. In many organisms, the dinucleotide CpG is under-represented throughout most of the genome, but this effect is not seen in CpG islands. Most characterized CpG islands are involved in gene regulation, including serving as promoters and enhancers

demethylating agents chemical agents which when incorporated into DNA cannot be methylated and thus lead to loss of methylation (*see* methylation)

fibroblast flat, elongated cell of connective tissue that secretes fibrillar procollagen, fibronectin and collagenase

gynogenetic/parthenogenetic gynogenetic/parthenogenetic embryos have all their genetic information from their mother. There is no paternal contribution

histone 4 the histones comprise a set of simple, highly conserved basic proteins, with a high arginine and lysine content, found in the nuclei of all eukaryotic cells

interphase period between one mitosis and the next, when no distinct chromosomes are visible and the chromatin is contained in a nucleus

matrix-associated region (MAR) another term for scaffold-associated region (*see* scaffold-associated region)

methylation [DNA methylation] process by which methyl groups are added to certain nucleotides in genomic DNA. This affects gene expression, as methylated DNA is not easily transcribed

morula mammalian embryo immediately before the blastocyst stage (8–16 cells in the mouse)

parthenogenetic (*see* gynogenetic)

pluripotent [cells] cells that are able to develop into various different types of cell

polyclonal tissue or structure originating from a number of founder cells

pronuclei haploid nuclei resulting from meiosis. The female pronucleus in animals is the nucleus of the ovum prior to fusion with the male pronucleus; the male pronucleus is the sperm nucleus after it has entered the ovum but before fusion with the female pronucleus

scaffold-associated region (SAR) sequences in DNA that bind to the protein scaffold which gives the chromosome its structure

trophectoderm also known as trophoblast, the trophectoderm comprises the tissue layer that forms the blastocyst wall

trophoblast (*see* trophectoderm)

X-chromosome inactivation (XCI) inactivation in female somatic cells of all but one copy of the X chromosome

X-inactivation centre (XIC) region on the X chromosome responsible for one inactivation of the two X chromosomes in female mammals

X-inactive specific transcript (*XIST/Xist*) [gene] gene responsible for X-chromosome inactivation

See also **DNA methylation in animals and plants** (p.746); **Genomic imprinting** (p.757)

L DNA-BASED GENETIC ANALYSIS AND BIOTECHNOLOGY

Introduction to DNA-based Genetic Analysis and Biotechnology

The basic DNA manipulation techniques (as described in section A) of cutting, splicing, amplification by PCR and cloning have led to a revolution in recombinant DNA technology and the ability to insert genes at specific sites, or to "knock out" or reduce the expression levels of genes, as described in the article on "Gene targeting". The development of DNA sequencing has led to new analysis techniques, both for genetic screening, and as described here for DNA fingerprinting and forensic DNA analysis – even from skeletons or from DNA millions of years old, as described in the article "DNA from museum specimens".

In the postgenomic era, the enormous mass of data from completed genome sequences in public databases provides a powerful "bioinformatics" opportunity, provided the sequence databases are well curated and annotated. Sequence similarity can be used to determine gene homology between species, and an article below describes the use of comparative mapping between mammalian species genomes. Current approaches to bioinformatics and functional genomics are described in another article. A recent *Nature* insight provides a forward-looking perspective on functional genomics and proteomics.

DNA microarrays or gene chips are adding another level to biotechnology, which is well described in a recent *Scientific American* article. DNA chips it is claimed, in handheld housings can sense the on/off state of up to 400 000 genes in a tissue sample, but scientists fear they may soon drown in the data, and numerous patent applications are likely to hold up progess.

REFERENCES

Nature Insight on Functional Genomics, *Nature* (2000) 405: 819–65
Scientific American (Februrary 2001): 23–24

GENE TARGETING

Elizabeth A. Lovejoy and Alan R. Clarke
Department of Veterinary Pathology, University of Edinburgh, Edinburgh, UK

1. Introduction

For some years, it has been possible to create transgenic rodents which carry additional DNA sequences introduced through pronuclear injection. These have been very successfully used to address the effects of overexpression of any chosen gene. However, this technology suffers three principal limitations: first, there is no control over where the foreign DNA integrates into the host genome; second, there is no control over the number of copies of the introduced gene; and, third, this technology is limited to the addition of DNA, and cannot therefore be used to look at loss of gene function. These limitations stimulated efforts to create a more controllable methodology, which finally arose out of the dual technologies of embryonic stem (ES) cell culture and gene targeting. In combination, these approaches now allow the production of murine strains which bear precise genetic alterations, including both the addition and subtraction of sequences. This approach has been termed "gene targeting".

2. Transgenic mice

The first reported transgenic mice were produced by injecting purified viral DNA into the blastocoel cavity of mouse blastocysts. Viral DNA sequences could be detected in the tissues of resulting animals indicating the successful integration of viral genomes into host chromosomes (Jaenisch & Mintz, 1974). Today, transgenic mice are most commonly produced by microinjection of DNA directly into the pronuclei of fertilized mouse eggs (Gordon *et al.*, 1980).

Since their development, transgenic strains of mice have proved to be valuable tools in the study of the genetics and pathogenesis of many human diseases. Phenotypic alterations which occur as a direct consequence of transgene expression can provide evidence to implicate specific genes in a given disease process. Where transgene expression can be shown to model human disease, such mice could be used as test systems to determine the efficacy of treatment regimens, and crossing onto different backgrounds may reveal extragenic modifiers of the phenotype.

However, the transgenic approach has limitations. For example, both the copy number and position of transgene integration cannot be predicted, yet both of these parameters can exert significant influence over phenotype (Palmiter & Brinster, 1986). Also, because of the random nature of integration, there is a significant risk of insertional mutagenesis. If the transgene integrates within a coding sequence and abolishes expression, the transgenic phenotype will be complicated by additional effects due to loss of expression of the endogenous gene. Perhaps the most significant drawback to transgenic models generated through pronuclear injection is the difficulty in creating null alleles, since exogenous genetic material can only be added to the genome.

A major advance in our ability to create mouse models of human disease was the development of gene-targeting technology which facilitates the introduction of specific mutations into the mouse germline.

3. ES cells

In 1981, two laboratories reported the derivation of non-transformed, pluripotential early ES cells (Evans & Kaufman, 1981; Martin, 1981). ES cells were derived from the inner cell mass (ICM) of 3.5 days postcoitum (dpc) blastocysts and were initially maintained *in vitro* by co-culture with feeder layers. Extensive ES cell differentiation was observed when the cells were cultured in the absence of feeder cells, implying that the maintenance of the undifferentiated stem cell phenotype is an active process. The need for co-culture with feeder layers was bypassed with the discovery that undifferentiated ES cells could be maintained in medium conditioned by the Buffalo rat liver cell line BRL. The macromolecule subsequently identified as a potent inhibitor of ES cell differentiation was differentiation inhibiting activity (DIA) (Williams *et al.*, 1988), identical to the previously described D-factor or leukaemia inhibitory factor (LIF). The culturing of ES cells in the presence of DIA/LIF and medium additionally supplemented with β-mercaptoethanol and serum is sufficient to maintain an undifferentiated pluripotential population. However, even with these culture conditions, a background of spontaneous differentiation is observed.

4. ES cell-mediated transgenesis

ES cells can be reintroduced into blastocysts to generate chimaeric progeny and their normal chromosome complement allows reproducible germline colonization (Bradley *et al.*, 1984). The pluripotential nature of ES cells made them an ideal system with which to attempt transgenic manipulation of the mouse genome, but, in order to achieve this goal, the twin technologies of genetic manipulation *in vitro* and ES cell germline transmission needed to be combined. Despite initial problems, in 1986 the first paper was published which reported the successful generation of transgenic mice via the ES cell route (Gossler *et al.*, 1986). This early work focused on the addition of genetic material to ES cells either by transfection or retroviral infection, but the logical next step was the manipulation of specific endogenous genes. This was achieved by two independent research groups working on the same gene, *hypoxanthine guanine phosphoribosyl transferase* (*Hprt*) (Hooper *et al.*, 1987; Kuehn *et al.*, 1987). The *Hprt* gene was ideally suited to this approach for several reasons. In both humans and mice, the gene is X linked;

thus, male cells are hemizygous, *Hprt* null cells can be separated from their wild-type counterparts on the basis of altered drug sensitivity and the generation of *Hprt* null mice offered the tantalizing possibility of a mouse model of the human disease, Lesch–Nyhan syndrome, which is caused by *Hprt* deficiency (Nyhan, 1973). Although both groups were successful in introducing mutations into the *Hprt* gene and the subsequent generation of mutant mice, no phenotypic alterations were observed in these animals. Further work suggested that differences in purine metabolism between mice and humans were to blame for the absence of a phenotype in the *Hprt* null animals. Nevertheless, this work was a crucial breakthrough that simultaneously demonstrated both the potential of this technology and the limitations of mouse models of human disease.

The generation of *Hprt* null mice was not achieved by gene targeting and, instead, relied on random mutagenesis, either chemical or insertional. While these approaches demonstrated the feasibility of ES cell genetic manipulation *in vitro* and subsequent return to host blastocysts, they were unsuitable for the specific modification of non-selectable genes. It was with this aim in mind that gene targeting was developed.

5. Homologous recombination

Homologous recombination (HR) describes the process of DNA strand breakage and reunion which occurs between chromosomes within regions of sequence homology. This process occurs in prokaryotic organisms and in eukaryotic organisms during meiosis, mitosis and DNA repair (Bollag *et al.*, 1989).

HR in mammalian cells can occur between introduced exogenous DNA molecules (extrachromosomal recombination), between homologous sequences within a chromosome (intrachromosomal recombination), between homologous sequences on sister chromatids (interchromosomal recombination) and, finally, between an introduced piece of DNA and a chromosomal sequence (gene targeting). All of these types of recombination are thought to proceed by similar, but poorly defined, molecular mechanisms within cells.

Two major models have been proposed to explain HR. Both postulate that HR is initiated by the generation of a single strand of DNA (ssDNA). The free end of the ssDNA molecule subsequently invades a homologous double-stranded DNA (dsDNA) duplex producing a heteroduplex structure. As the donor ssDNA molecule pairs with the recipient dsDNA sequence, a loop of uncut recipient DNA (D-loop) is displaced from the heteroduplex. The two models differ in the mechanism by which this structure is resolved. In the Meselson–Radding model of HR, the D-loop is degraded creating a free end of ssDNA in the recipient molecule (Meselson & Radding, 1975). The free ssDNA end in the recipient is then able to ligate to the remaining free end in the donor molecule forming an exchange structure called a Holliday junction (Holliday, 1964) (**Figure 1**). The double-strand break-repair model of HR also envisages the generation of a D-loop, but, rather than this sequence being degraded, it acts as a template for repair synthesis of the donor strand (Orr-Weaver *et al.*, 1981). After repair synthesis is complete, ligation of free ssDNA ends results in the formation of two Holliday junctions (**Figure 2**). In both of these models, the junction(s) between the two homologues are able to migrate along the DNA by continued strand transfer within the crossover(s). Depending on how the Holliday junctions are resolved, several outcomes are possible: the reciprocal exchange of genetic material (crossover), a non-reciprocal exchange (gene conversion), a gene conversion with an associated crossover or, finally, a non-conservative recombination event (**Figure 3**).

6. Gene targeting

The term gene targeting is used to describe the introduction of defined modifications into specific chromosomal loci via HR with exogenous DNA sequences (termed targeting vectors). The first endogenous gene to be modified by HR with a targeting vector was the human β-globin gene (Smithies *et al.*, 1985) and the targeted modification of different loci within other non-ES cell lines has since been reported by other groups (e.g. Adair *et al.*, 1989).

For the same reasons that the endogenous *Hprt* gene was the target of early mutagenesis strategies, it was also used for the development and optimization of gene targeting protocols in ES cells. Subsequently, two groups reported the disruption of the gene via HR in ES cells (Thomas & Capecchi, 1987; Doetschman *et al.*, 1988). The application of gene targeting to non-selectable genes required the development of novel targeting strategies and this work is summarized in the following sections.

7. Gene targeting strategies

(i) Insertion-type (O-type) targeting strategies

Insertion-type targeting strategies (**Figure 4**) have been developed to allow the introduction of subtle mutations into target genes by a two-step approach termed either "hit and run" (Hasty *et al.*, 1991) or "in and out" (Valancius & Smithies, 1991). In the first stage of this process, O-type targeting vectors, linearized within the region of homology, integrate into the genome via a single crossover event within the region of homology. This insertion event creates a duplication of the region of homology within the targeted genomic locus, but, importantly, specific mutations can be introduced into the targeting vector. In the second stage of this process, reversion, the duplication of homology is resolved via intrachromosomal HR. Two outcomes are possible: the first is that all integrated vector-derived sequences are lost and the locus is restored to wild-type; the second is that vector-derived homologous sequences, including any engineered mutations, are retained and a region of chromosomal sequence is lost.

Interestingly, significantly different targeting and reversion frequencies have been reported by groups using identical strategies on different loci. For example, when hit-and-run targeting was used to create duplications of homology at both the *Hprt* and *Hox-2.6* loci, the frequency of reversion was 1000 times greater at the *Hox-2.6* locus. Hence, while this work demonstrated the feasibility of insertion-type

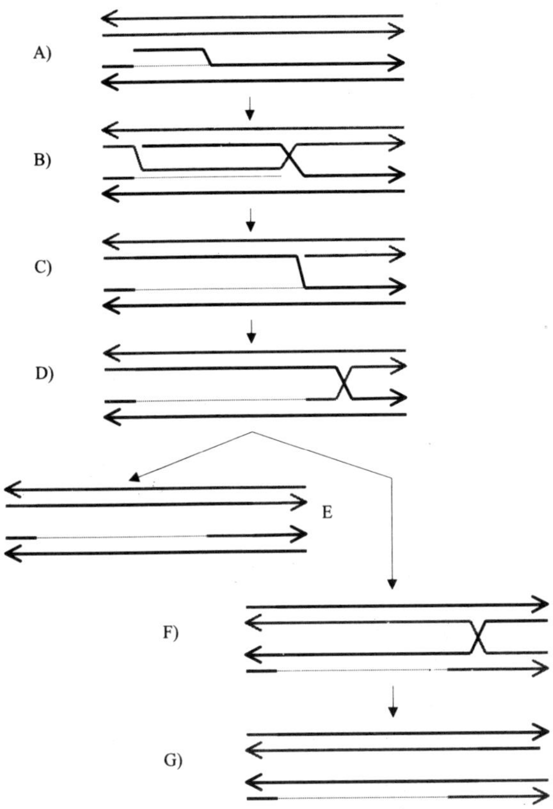

Figure 1 The Meselson–Radding model of homologous recombination. (A) Recombination is initiated by the formation of a single-strand nick in one DNA molecule. (B) At the 3′ end of the nick, site repair synthesis begins, leading to the displacement of the single strand of DNA which then invades the other DNA duplex. This invasion displaces a D-loop in the other homologue. (C) The D-loop is degraded and an asymmetric heteroduplex is created. (D) Ligation produces a Holliday junction that is able to migrate along the DNA molecules. (E) The Holliday junction can be resolved in three ways (E–G): first, by simply cutting the crossed strands giving the recombination products shown (E). Such products are the result of gene conversion without crossover. (F) In the second alternative, the Holliday junction undergoes isomerization and, thirdly (G), strand cleavage generating crossover products, also with segments that have undergone gene conversion.

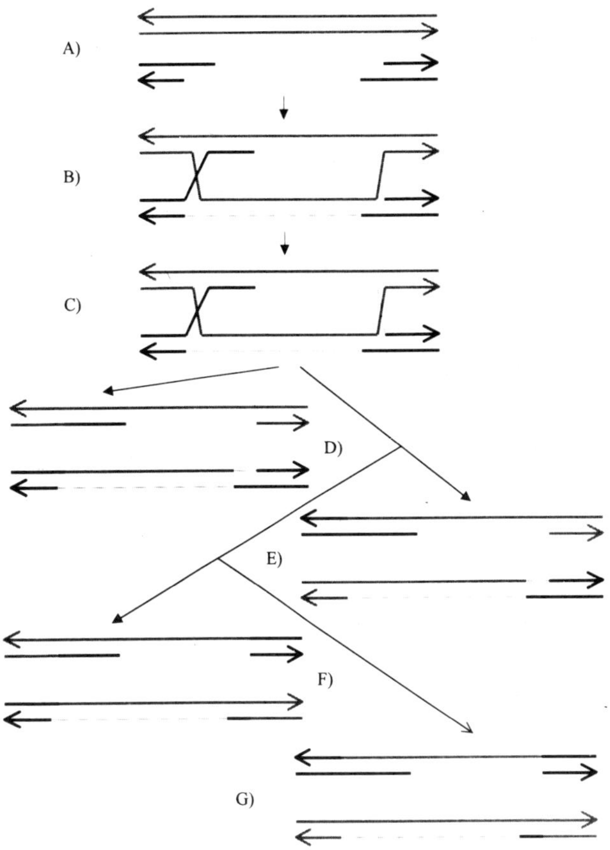

Figure 2 The double-strand break model of homologous recombination. (A) Recombination is initiated by the formation of a double-strand break in one DNA molecule. (B) One single-strand DNA molecule from the double-strand break site then invades the duplex of the other homologue, leading to the displacement of a D-loop which is enlarged by repair synthesis. To complete the repair of the double-strand break, repair synthesis is initiated from the second 3′ end at the original double-strand break site. (C) Ligation then generates two Holliday junctions, both of which can move via branch migration. The two Holliday junctions can be resolved in four different ways (D–G). (D) Neither junction undergoes isomerization and the crossed strands within the junctions are cut. (E–G) If the left-hand Holliday junction undergoes isomerization, the recombination products shown in (E) are the outcome; if the right-hand Holliday junction undergoes isomerization, the recombination products shown in (F) are the outcome; and if both the left- and right-hand Holliday junctions undergo isomerization, the recombination products shown in (G) are the outcome.

targeting strategies, it also highlighted the locus-dependent nature of both targeting and reversion frequencies.

(ii) Replacement-type (Ω-type) targeting strategies

A typical replacement targeting vector consists of two regions of DNA homologous to the target locus which are interrupted by a positive selection marker, e.g. *neomycin phosphotransferase* (*neo*) or *puromycin N-acetyltransferase* (*pac*). This type of vector is linearized outwith the regions of homology as the aim is to insert only the selectable marker, not the whole vector, into the chromosome. The selectable marker can either replace part of the genomic sequence and so introduce a deletion, or insert into the genomic sequence thereby disrupting the target gene. Thus, both outcomes are designed to disrupt expression of the endogenous gene (**Figure 5**). This simple replacement strategy can be used successfully to create null alleles, but it cannot be used to introduce subtle non-selectable mutations. To address this difficulty, the basic replacement strategy has been further developed in the "double replacement" (Stacey *et al.*, 1994) or "tag and exchange" methods (**Figure 6**) (Askew *et al.*, 1993).

Both of these methods involve two separate rounds of gene targeting with two separate vectors, the outcome of

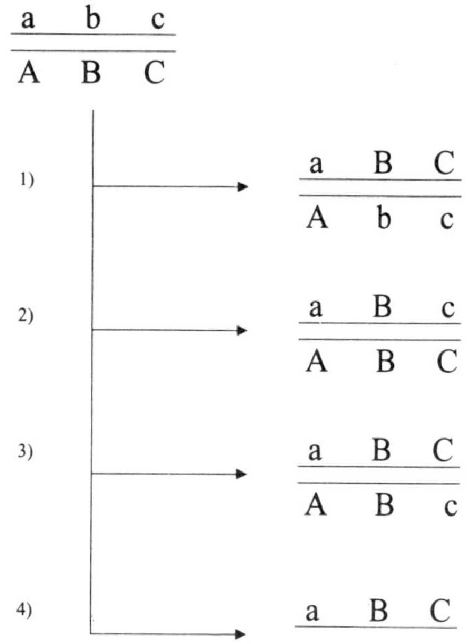

Figure 3 Types of recombination event. If the two DNA duplexes shown at the top of this figure undergo recombination, there are four types of outcome (one thick black line represents a double-stranded DNA duplex). (1) Reciprocal exchange when there is a straightforward exchange of genetic material between the DNA duplexes. (2) Gene conversion occurs after a non-reciprocal exchange of genetic material whereby the genetic information is copied from one duplex to another. (3) Gene conversion with an associated crossover. (4) Non-conservative recombination. As the name implies, when DNA duplexes undergo this type of recombination event, DNA is lost from opposite ends of the molecules and the remaining DNA is sufficient only to reconstitute one duplex.

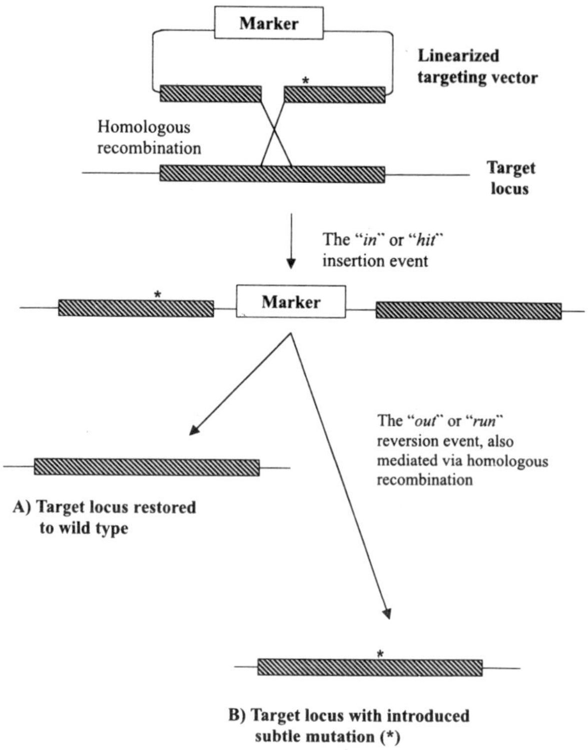

Figure 4 An insertion or "in and out" gene targeting strategy. Insertion-type targeting vectors are linearized within the region of homology (hatched boxes). When such vectors are introduced into cells, a single crossover event, mediated by homologous recombination between the chromosomal target sequence and the region of the homology, results in the integration of the complete vector into the target locus. This is referred to as the "in" or "hit" event and generates a duplication of homology within the target locus. The duplication of homology can be resolved in two ways. (A) Homologous recombination results in the loss of all targeting vector-derived sequences from the target locus, which is consequently restored to wild-type. (B) Homologous recombination results in the excision of some targeting vector-derived sequences, but also some chromosomally derived sequences from the target locus. Crucially, the original region of homology present within the targeting vector is retained within the target locus and if this had been previously engineered to contain subtle mutations (*), these will be introduced in the target locus. Both ways of resolving the duplication of homology are referred to as either the "out" or "run" event.

which is the introduction of subtle non-selectable mutations into a target gene. The first step uses a normal Ω-type vector with a positive selectable marker. The resultant clones are screened and any that have correctly undergone the first round of homologous recombination are electroporated with a second vector. This vector replaces the previously introduced selectable marker in the target locus with a sequence containing a non-selectable mutation via a second round of HR. The second recombination event is selected for using negative selection so that only clones which have lost the selective marker and gained the introduced mutation will survive.

Another variation on this theme is the "plug and socket" (**Figure 7**) (Detloff *et al.*, 1994). In this system, the first round targeting vector introduces one functional and one crippled positive selection marker into the target locus via HR. The second round vector carries DNA sequences which undergo HR with the integrated first vector, resulting in the deletion of the functional positive selection gene, simultaneous reconstitution of the crippled marker and the introduction of any predetermined mutations into the target locus. Such clones can then be selected for on the basis of expression of the crippled marker gene. This system has the advantage that expression of the crippled marker gene is strictly dependent on HR, unlike the double replacement or tag and exchange strategies in which loss of the negative selectable gene can occur by HR-independent mechanisms. Nevertheless, these methods do have the advantage of not leaving any marker sequence at the target locus after completion of the second round of targeting, whereas, after the plug and socket strategy is complete, the reconstituted crippled marker gene is retained.

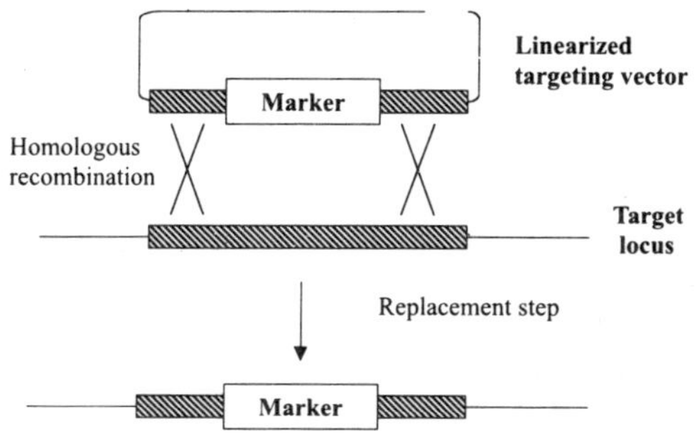

Figure 5 A replacement gene targeting strategy. Replacement-type targeting vectors are linearized outwith the region of homology (hatched boxes). When such vectors are introduced into cells, a double crossover event mediated by homologous recombination between the chromosomal target sequence and the vector-derived region of the homology results in the replacement of part of the target locus with the marker cassette and surrounding vector-derived homologous sequence. This type of replacement targeting strategy is commonly used to insert selectable marker cassettes within the coding sequence of a target gene generating a null allele.

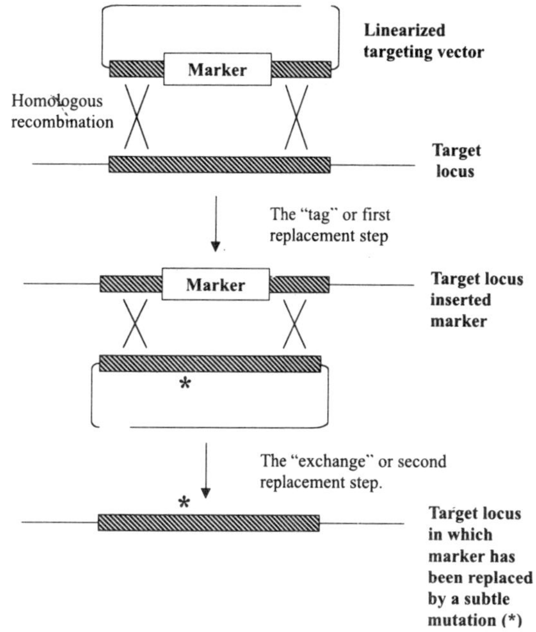

Figure 6 Double replacement and tag and exchange gene targeting. Tag and exchange gene targeting strategies are mediated by two rounds of replacement gene targeting (see also Figure 5). In the "tag" or first replacement step, a standard linearized replacement targeting vector is introduced into cells. The vector is linearized outwith the region of homology (hatched boxes) and a double crossover event, mediated by homologous recombination between the chromosomal target sequence and the vector-derived region of the homology, results in the replacement of part of the target locus with the marker cassette and surrounding vector-derived homologous sequence. In the second replacement step, or "exchange", another targeting vector is introduced into cells which are known to have undergone the first replacement step. A double crossover event mediated by homologous recombination then exchanges the chromosomally integrated marker gene for sequences derived from the second targeting vector. Importantly, if the second replacement vector was previously engineered to contain a subtle, non-selectable mutation, this will be introduced into the target locus.

8. Introducing exogenous DNA into ES cells and identifying recombinants

In order for recombination to occur between an exogenous DNA molecule and a chromosomal target, it is first necessary to introduce the DNA into ES cells. As discussed above, electroporation has become the method of choice. However, because of the relatively low transfection efficiencies, only a minority of cells will carry the introduced DNA (transformation frequencies of 10^{-5}–10^{-2} have been reported; reviewed in Mansour, 1990). Therefore, it is necessary to select the successfully transfected cells, usually by the acquisition of a novel drug resistance, via the introduction of appropriate expression cassettes included in the targeting construct. The strategies used to select transfected cells can be grouped into three types: positive selection, positive/negative selection (PNS) and the use of promoterless selectable genes. The relative advantages and limitations of all of these approaches are discussed below.

(i) Positive selection

All cells that acquire a positive selectable drug resistance gene should survive exposure to the corresponding drug at an appropriate concentration. Examples of this type of system include the bacterial *neomycin phosphotransferase* gene (*neo*), which confers resistance to the neomycin analogue G418, and the *hygromycin phosphotransferase* (*hyg*) and *puromycin N-acetyltransferase* (*pac*) genes which confer resistance to hygromycin B and puromycin, respectively.

The electroporation of a targeting vector into a population of cells will result in the uptake and integration of that DNA molecule into the genome of a proportion of those cells. If one of the above positive selection markers was included in the introduced targeting vector, then the exposure of the cell population to the corresponding drug will select for cells which have acquired the novel drug resistance. However, the introduced DNA may be incorporated into the cell genome in two ways: first, and most frequently, via random integration or, secondly, via HR. This simple positive selection strategy is unable to distinguish between these two alternatives and has been superseded by the more sophisticated strategies discussed below.

(ii) Positive/negative selection

PNS was devised in an attempt to enrich the selected cell population for those in which HR-mediated, rather than random, integration events have occurred.

The PNS strategy relies on the observations that random integration proceeds through and preserves the ends of incoming exogenous DNA vectors, whereas HR does not require that the ends of the vector be homologous with the target sequence. Therefore, a PNS replacement vector carries an independently expressed positive selectable marker and a

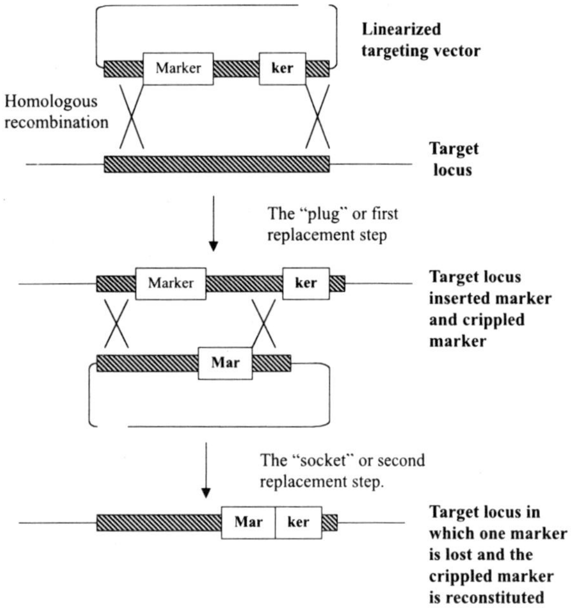

Figure 7 Plug and socket gene targeting. Plug and socket gene targeting strategies are mediated by two rounds of replacement gene targeting in a manner analogous to tag and exchange targeting. The only difference between the two types of strategy is the selectable marker used. In the first replacement step or "plug", a replacement targeting vector, linearized outwith the region of homology, is introduced into cells. A double crossover event, mediated by homologous recombination between the chromosomal target sequence and the vector-derived region of the homology, results in the replacement of part of the target locus with vector-derived sequences. The vector-derived sequences introduced into the target locus include a functional positive selectable marker and a crippled positive selectable marker. It should be noted that a single crossover event and subsequent branch migration of the Holliday junction could equally well mediate this type of targeting event. In the second replacement step, or "socket", another targeting vector is introduced into cells that are known to have undergone the first replacement step. A double crossover event mediated by homologous recombination then exchanges the chromosomally integrated marker gene for sequences derived from the second targeting vector. This reaction simultaneously deletes the first positive selectable marker from the target locus and reconstitutes the second positive selectable marker. Hence, unlike tag and exchange, the outcome of this type of strategy is the introduction of a marker cassette into the target locus rather than a subtle non-selectable mutation.

negative selectable marker that is placed within the plasmid backbone, outwith the region of homology used for HR. Random integration of the PNS vector will preserve the negative selectable marker and this can be used for counter-selection with an appropriate drug. During HR, however, the negative selectable gene is lost as it is located distal to the region of homology between the vector and target and so is excluded from the crossover event. The most frequently used negative drug selection marker is the *Herpes simplex* virus *thymidine kinase* gene (*HSV-tk*), although other genes such as a diphtheria toxin A (DT-A) have been used successfully. The expression of *HSV-tk* renders cells sensitive to the nucleoside analogues FIAU and gancyclovir, whereas the DT-A gene encodes an intrinsically toxic protein.

PNS has been used to target both ES cell-expressed and non-expressed genes and has been reported to yield enrichments (from 10-fold to 2000-fold) of HR relative to random integration events (Mansour *et al.*, 1988; Thomas & Capecchi, 1990).

(iii) Double replacement/tag and exchange targeting strategies

The second instance when the presence of a negative selection marker is required is during "double replacement" or "tag and exchange" targeting strategies. Both of these methods use two separate rounds of targeting to introduce a subtle non-selectable mutation into the locus of interest. To achieve this, a drug resistance marker is needed which can be both positively selected for after the first round of targeting and negatively selected against after the second round. Two such selectable markers are the *Hprt* gene and *neo/HSV-tk* expression cassette.

Expression of the *Hprt* gene can be selected for by culturing cells in HAT medium. Importantly, selection for the expression of an exogenously introduced *Hprt* gene is dependent on the host cell lacking any endogenous *Hprt* enzyme activity. Such ES cell lines are now available – such as the HM-1 ES cell line, for example (Magin *et al.*, 1992). HAT-medium selection relies on the role of the *Hprt* enzyme within the cellular purine metabolism pathway. Cells have two pathways to synthesize mononucleotides: the first is a salvage pathway in which *Hprt* operates and the second is *de novo* synthesis independent of *Hprt* status. HAT medium supplies hypoxanthine as a substrate for the salvage pathway while aminopterin is used to block the *de novo* synthesis route. The final component of HAT medium is thymidine; this is needed by cells to circumvent the aminopterin-dependent inhibition of the thymidylate synthase reaction. The outcome of this selection is that all cells will be unable to synthesize mononucleotides *de novo* due to aminopterin, but only those expressing *Hprt* will be able to take advantage of the supplied hypoxanthine via the *Hprt*-dependent salvage pathway.

To select against cells which express *Hprt*, the drug 6-thioguanine is used. This guanine analogue is a substrate for the *Hprt* enzyme and undergoes a phosphoribosylation reaction to generate 6-thioguanine monophosphate (6-thioGMP). High levels of 6-thioGMP within cells inhibit the biosynthesis of guanine nucleotides and this inhibition eventually results in the death of all cells that express *Hprt*.

The *neo/HSV-tk* expression cassette has also been used successfully in double replacement targeting strategies. The HR-mediated integration of the first round targeting vector is selected for using G418. The HR-mediated exchange reaction at the target locus with the incoming second round vector is predicted to result in the loss of the *neo/HSV-tk* expression cassette, and loss of the *HSV-tk* gene can be selected for using FIAU or gancyclovir.

(iv) Crippled selectable marker genes

An alternative strategy for the enrichment of HR events is to use a selectable marker gene in the construct lacking either a functional promoter (Doetschman *et al.*, 1988) or polyadenylation sequence (Schwartzberg *et al.*, 1989) or requiring the trapping of an enhancer element (Jasin & Berg, 1988). In this type of strategy, the expression of the positive selection marker is HR dependent as it is unlikely that random integration will juxtapose appropriate activating sequences adjacent to the crippled selection cassette.

High targeting efficiencies have been reported with the use of promoterless or otherwise crippled selectable genes. For example, the targeted disruption of an endogenous gene, *Hprt*, using selection with a promoterless *neo* gene has been demonstrated in ES cells with a targeting frequency of 67%. Interestingly, this targeting frequency was higher than that reported by Thomas and Capecchi, despite the fact that these authors targeted the same locus and used an insertion targeting vector containing a larger region of genomic homology (Thomas & Capecchi, 1987). The use of an enhancerless *xanthine guanine phophoribosyltransferase* (*gpt*) gene has been reported to yield an estimated 100-fold enrichment for targeted integrations without direct selection at the target locus (Jasin & Berg, 1988). Homologous integrations accounted for 44% of the *gpt*$^+$ cells. As well as using HR recombination to insert crippled drug-resistance genes into target loci, the technique has also been used to insert histochemically detectable reporter genes. For example, a promoterless *βgeo* gene, a fusion of the *neo* and *lacZ* genes, was targeted into the *Oct-4* locus (Mountford *et al.*, 1994). This strategy not only resulted in a very high targeting efficiency (70–80%), but also placed *βgeo* expression under the control of the endogenous *Oct-4* promoter allowing direct visualization of promoter activity *in vitro* and *in vivo* (Mountford *et al.*, 1994).

The major consideration about the use of promoterless or crippled positive selection marker genes is the expression levels of the target gene within ES cells. A promoterless marker gene can only be used for gene targeting if the target gene is actively expressed in ES cells. The level of target gene expression must also be considered; for example, if the target gene is more strongly expressed in differentiated derivatives of the ES cells, the selection procedure may enrich for these at the expense of the pluripotent population (reviewed in Hooper, 1992).

9. Inducible gene targeting

The advent of gene targeting has permitted the generation of numerous mouse strains that carry specifically engineered modifications in a gene of choice – knockout mice. The technique has become an established methodology of reverse genetics, i.e. the determination of gene function by the ablation or modification of target gene expression *in vitro* or *in vivo*.

While standard gene targeting strategies have furthered our understanding of the function of many genes, this approach is limited. Among the key problems is the phenomenon of embryonic lethality where target gene function is critical for the successful completion of the developmental programme, making it impossible to study loss of gene function in adult, differentiated tissues. Although embryonic lethality cannot be overcome with standard gene targeting, novel inducible gene targeting strategies now under development should allow conditional target gene expression and provide significant insights into gene function.

The aim of inducible gene targeting is to allow the experimenter to determine both the time and location at which gene function is ablated or restored. For example, genes can be switched off in the adult mouse, so circumventing the phenotype of embryonic lethality. The conditional control of gene expression in a tissue-specific manner will also permit a dissection of the often complex phenotypes of knockout animals, so increasing our understanding of the pleiotropic nature of gene function. For example, the ability to knockout genes in specific somatic tissues will provide a model system in which to study human somatically acquired genetic diseases such as cancer. This type of analysis is extremely difficult with standard gene targeting as both tumour suppressor genes and proto-oncogenes commonly have vital developmental roles, the absence of which can be incompatible with survival in the mutant mice. Finally, inducible gene targeting should also provide an *in vivo* model system for the study of gene therapy. Just as gene function can be specifically ablated in the adult mouse, expression could also be restored. This type of study should provide valuable data about both the feasibility and efficacy of gene therapy in humans.

The main technologies used to achieve inducible gene targeting are the recombinase-based Cre/loxP (Abremski *et al.*, 1983) and FLP/frt (Broach & Hicks, 1980) systems isolated from bacteriophage P1 and *Saccharomyces cerevisiae*, respectively, and are reviewed below.

10. The Flp/FRT system: background

The Flp gene of the 2 μm circle plasmid of *S. cerevisiae* encodes a conservative site-specific DNA recombinase protein that is involved in the amplification of that plasmid. Flp mediates a recombination event between two 48 base pair (bp) FRT sites. Each FRT site includes two inverted 13 bp symmetry elements (a and b) surrounding an 8 bp core region and a third, 13 bp symmetry, element (c) present in the direct orientation (**Figure 8**). When two Flp molecules are bound to symmetry elements a and b, a bend of approximately 60° is introduced into the DNA; binding of a third protein to element c increases the DNA bend to >144°. Intermolecular protein–protein interactions between the individual Flp molecules bring the two partner FRT sites together to form a synaptic complex. Site-specific cleavage results from a nucleophilic attack by Tyr343 on the phosphodiester backbone of DNA within the core of the FRT site. The Flp-mediated DNA bending is thought to be important in facilitating DNA strand separation and exchange in the core region.

The Flp recombinase system has been shown to work efficiently in a number of experimental systems including the mouse. For example, several Flp-transgenic mouse strains have been developed. These include TgN(hACTB::Flp) in which Flp expression is driven from the global human

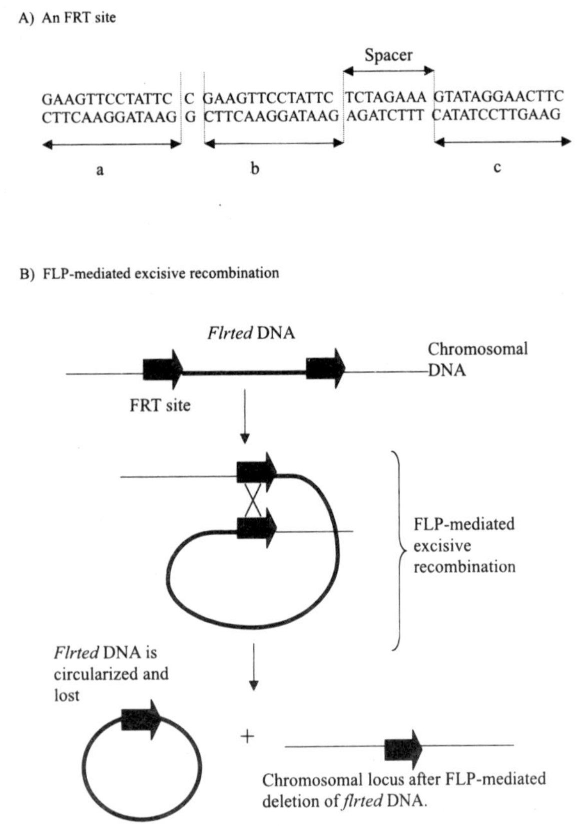

Figure 8 The FLP recombinase system.

β-actin promoter (Dymecki, 1996), Wnt1::Flp, in which Flp expression is directed to the developing nervous system (Dymecki & Tomasiewicz, 1998), and rPOMC-Flp in which sequences from the rat pro-opiomelanocortin promoter were used to direct Flp expression to the pituitary gland (Vooijs *et al.*, 1998) Furthermore, an inducible Flp recombinase has also been developed, allowing temporal as well as spatial control of gene expression (Logie & Stewart, 1995).

The Flp system has also been used in inducible gene targeting strategies. For example, the insertion of two FRT sites flanking exon 19 of the mouse retinoblastoma gene allowed its inducible inactivation *in vivo* and demonstrated the tumour-suppressor function of this gene (Vooijs *et al.*, 1998). Flp has also been used to manipulate FRT-containing transgene arrays in zygotes (Ludwig *et al.*, 1996) and to engineer large-scale chromosome manipulations *in vitro* (Aladjem *et al.*, 1997).

11. The Cre/loxP system: background

Cre recombinase (*c*auses *re*combination) is an enzyme produced by bacteriophage P1. Within the bacteriophage, Cre has two roles: first, to circularize the viral genome should this fail to be carried out by host cell recombination machinery; and, secondly, to maintain correct unit copy segregation of the prophage by resolving any dimeric molecules into monomers prior to cell division. Cre does not act randomly throughout the bacteriophage genome, but will only stimulate recombination at specific sequences encoded within loxP sites (*lo*cus of crossover [*x*]). A loxP site consists of 34 bp of DNA arranged in three components: two 13 bp inverted repeats separated by an 8 bp non-symmetrical spacer region (**Figure 9**). DNA footprinting studies have shown that one molecule of Cre binds to one 13 bp repeat, so that each complete loxP site is bound by two molecules of recombinase. Once bound to the loxP site, Cre catalyses recombination in a sequential manner with double-strand exchange occurring after two rounds of single-strand cleavage and religation.

Using this mechanism, Cre catalyses two types of recombination event depending on the relative orientation of the two loxP sites (**Figure 9**). As the two inverted repeats within a loxP site are palindromic, it is the non-symmetrical 8 bp spacer that confers directionality. When the spacer regions of two loxP sites are directly repeated, any intervening DNA (described as floxed) will be excised, whereas inversion will occur when the loxP sites are in opposite orientations.

The exciting potential of the Cre/loxP system became apparent when it was shown that the enzyme will catalyse recombination in the absence of ATP, cofactors or topoisomerase activity. Most importantly, the system will also

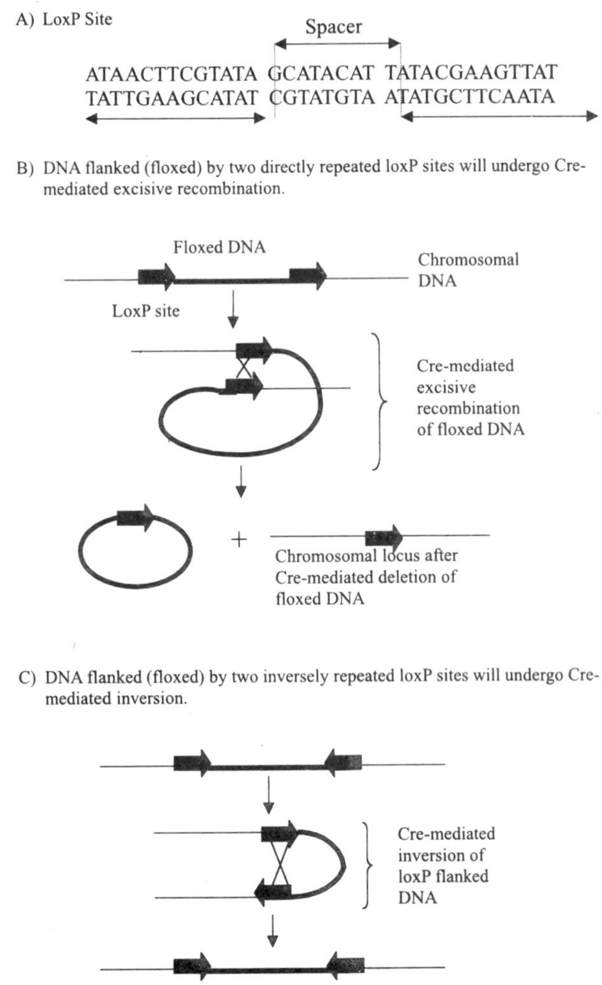

Figure 9 The Cre/loxP system.

function in eukaryotic cells (Sauer, 1987). These observations have subsequently been extended *in vivo* using transgenic mice and it was this work which first demonstrated the potential application of this system to conditional gene targeting (Lakso *et al.*, 1992; Orban *et al.*, 1992).

12. Applications of the Cre/loxP and Flp/FRT systems: inducible gene targeting

To achieve conditional gene expression using Cre/loxP technology, there are two basic requirements. First is the ability to regulate the expression of Cre recombinase *in vitro* or *in vivo*. Numerous such methods have been developed including tissue-specific or inducible promoters, Cre and Flp fusion proteins and viral delivery. The second requirement to achieve conditional gene expression is the introduction of loxP or FRT sites into the gene of interest. As already mentioned, Cre recombinase is able to catalyse more than one type of recombination reaction and, therefore, the technology can be adapted in a multitude of ways depending on the desired outcome of recombination – either target gene activation or inactivation.

13. Site-specific insertion

The presence of a single loxP or FRT site within a chromosome will act as a target for insertional recombination with any introduced exogenous DNA constructs that also contain a single loxP site. The target loxP site can be introduced via gene targeting into a specific location or positioned at random within the genome by illegitimate recombination. Whichever method is used, the site of the target loxP site remains fixed such that any subsequent insertional recombination events will occur at this predetermined locus. This strategy of site-specific integration has been successfully demonstrated using loxP-*tk* (Sauer & Henderson, 1990) and loxP-*neo* constructs, as well as a loxP-*lacZ* marker gene used to analyse the scale of position effect (the inactivation of a gene in some cells resulting from its chromosomal position) on marker gene expression (Fukushige & Sauer, 1992).

14. Site-specific excisional recombination of exogenous DNA sequences

The ability of Cre to mediate excisive recombination of exogenous DNA sequences has been used by many groups to achieve conditional gene expression, all of whom used the same basic strategy. The target gene is placed under the control of a given promoter, but the insertion of a floxed inactivating cassette between the transgene promoter and downstream coding sequence confers Cre-mediated control of expression. In the absence of Cre, transgene expression is blocked by the presence of the inactivating cassette. After the expression of Cre, the floxed cassette is lost and transgene expression activated.

15. Site-specific excisional recombination of endogenous DNA sequences

This type of strategy has been applied to achieve inducible inactivation of gene expression – temporal-specific or tissue-specific knockouts *in vitro* and *in vivo*, for example. Whatever the target gene, the basic approach is the same: the introduction of two directly repeated loxP sites into the gene of interest and then regulated expression of Cre recombinase.

To date, recombination-dependent gene inactivation has been demonstrated *in vivo* by several groups, but varying degrees of success have been reported in obtaining mutant mice with relevant phenotypes. The mice reported so far range from having no observable phenotype to one of embryonic lethality. For example, the DNA polymerase β gene has been conditionally inactivated in T cells (Gu *et al.*, 1994) and also in a temporally specific manner (Kuhn *et al.*, 1995), but both of these conditional knockouts had no obvious phenotype. At the other end of the scale, mice homozygous for a floxed allele of the mitochondrial transcription factor A (TFAM) gene die prior to day E10.5 (Larsson *et al.*, 1998). All of this work serves to highlight the fact that while Cre/loxP technology can offer a way to circumvent the problem of embryonic lethality, both the target gene and the Cre recombinase delivery strategy must be chosen carefully. For example, in the TFAM experiments, Cre expression was driven from the β-actin promoter, so directing expression to the preimplantation embryo. As a result, the floxed animals underwent recombination at an early stage in development, making them little different from constitutively null animals which could have been generated using standard gene targeting techniques.

Despite these problems, other groups have been more fortunate in both their choice of target gene and Cre delivery strategy. For example, Tarutani and colleagues used an inducible gene targeting strategy to construct a conditional allele of the murine X-linked *Phosphatidylinositolglycan class A* (*Pig-A*) gene (Tarutani *et al.*, 1997). Somatically acquired mutation(s) in the *PIG-A* gene in humans are responsible for paroxysmal nocturnal haemoglobinuria and *Pig-A* null ES cells are incapable of making a significant contribution to chimaeras, thereby precluding the generation of a *Pig-A* knockout mouse. To create a conditional allele of the *Pig-A* gene, two directly repeated loxP sites were inserted on either side of exon 6. The floxed mice were then mated with a second transgenic strain that expressed Cre recombinase under the control of the skin-specific keratin 5 promoter (Tarutani *et al.*, 1997). Cre-mediated inactivation of the *Pig-A* gene in hemizygous floxed male animals resulted in death 1–3 days after birth; however, lethality was not observed in the floxed female mice. Both sexes displayed altered skin characteristics relative to wild-type, including scaly and wrinkly skin with a reduced lipid component. The application of this inducible gene targeting strategy to the *Pig-A* gene provided insights into gene function that were unobtainable with standard gene targeting.

A variety of other genes have also been analysed with this approach. LoxP sites were targeted into the *N*-methyl-*D*-aspartate receptor 1 gene and, using forebrain-specific expression of Cre, the function of this receptor was ablated specifically in one region of the hippocampus (Tsien *et al.*, 1996). The Cre/loxP system also permitted the study of the developmentally important B cell antigen receptor (BCR) complex (Lam *et al.*, 1997). Complete ablation of the BCR would lead to a total block in B cell production during early development precluding an analysis of its function in mature B cells. Using the Cre/loxP system, mutant mice were generated in which BCR function could be specifically abolished in mature B cells, providing the first direct evidence for the role of BCR expression and the persistence of mature B cells in the peripheral immune system.

This technology has been taken one step further to generate an allelic series of mutations at one given locus (Meyers *et al.*, 1998). Using one round of standard gene targeting, two FRT sites (for the Flp recombinase), two loxP sites and a *neo* cassette were inserted into the *Fgf8* gene generating an *Fgf8*^{neo} allele. When these mice were crossed to a strain which expressed Cre from the human β-actin promoter, excisive recombination resulted in the generation of a null allele, FgF8$^{\Delta 2,3n}$. Mice with the null allele displayed a defect in gastrulation and, hence, embryonic lethality. This deletion phenotype was more severe than that displayed by mice that carried the *Fgf8*^{neo} insertion allele – these animals showed perinatal lethality. This observation led to the classification of the *Fgf8*^{neo} insertion allele as hypomorphic. The third allele that can be derived from the parental *Fgf8*^{neo} was generated by crossing the mice to a second transgenic strain which expressed Flp. Flp-mediated excisive recombination leads to the loss of the neo cassette from the *Fgf8*^{neo} allele; two intronic loxP sites are retained, but this allele, *Fgf8*^{flox}, is functionally equivalent to wild-type. Finally, the *Fgf8*^{flox} allele can be inactivated by subsequent expression of Cre giving the fourth possible allele in this allelic series.

Previously, gene targeting technologies have been directed at the creation of null alleles whereas the paper by Meyers *et al.* (1998) is the first to report the construction of an allelic series of mutations, all of which can be derived from one targeted progenitor mouse strain. This type of approach not only increases the productivity of gene targeting, but also permits a more complete analysis of gene function. The ability to use Flp and Cre recombinases in parallel also increases the complexity of genetic modifications possible.

Another advantage of the Cre/loxP-based inducible gene targeting relative to standard targeting techniques is the ability to analyse specific splice variants. For example, the α6 integrin gene encodes two mRNAs, α6A and α6B, generated by alternative splicing. Conditional gene targeting was used to delete the α6A-specific exon and a mild phenotype of altered cell migration resulted, demonstrating that the two splice variants are not functionally equivalent. The same exon-specific knockout approach has also been used successfully to investigate the function of the β1D integrin splice variant.

16. Chromosomal manipulations using Cre/loxP

The applications of the Cre/loxP system are not limited to the manipulation of specific transgenic loci; the technology can also be applied to chromosome engineering. The feasibility of using Cre/loxP technology to engineer large-scale chromosomal manipulations into the eukaryotic genome was first demonstrated in plants (Qin *et al.*, 1994); however, this technology was rapidly applied to the mammalian genome.

For example, the human chromosome translocation t(6;9), associated with a specific subtype of acute myeloid leukaemia, was modelled in ES cells using the Cre/loxP system. Two loxP sites of identical orientation were introduced by standard gene targeting techniques, one into each target chromosome. Transient expression of Cre recombinase in doubly targeted ES cell clones resulted in a site-specific reciprocal exchange of chromosome arms beyond the position of the loxP sites, generating a balanced translocation between non-homologous chromosomes (van Deursen *et al.*, 1995). An identical approach was also used to engineer successfully a t(12;15) translocation commonly found in mouse plasmacytomas in ES cells (Smith *et al.*, 1995). As well as translocations, the Cre/loxP system has also been used to introduce deletions into the genomes of ES cells; for example, deletions ranging from 90 kilobases to 3–4 centiMorgan have been successfully generated on mouse chromosome 11 (Ramirez-Solis *et al.*, 1995).

17. Conclusion

Prior to the development of gene targeting, it was possible to create addition transgenics by pronuclear injection. This approach has generated a wealth of data which has begun to reveal the true *in vivo* roles of many genes. However, it remained limited technically for a number of reasons. The advent of gene targeting has permitted another great leap forward in our understanding of gene function, by allowing precise genetic manipulation and also by permitting the analysis of loss of gene function. The basic principle of gene targeting has spawned many different approaches, discussed above, which have continually upgraded and refined the power of this technology. The most recent of these advances has been the application of Cre-lox and Flp-FRT technology to permit inducible and tissue-specific gene targeting strategies. The application of these methodologies heralds yet another revolution in our capability to understand gene function.

REFERENCES

Abremski, K., Hoess, R.H. & Sternberg, N. (1983) Studies on the properties of P1 site-specific recombination: evidence for topologically unlinked products following recombination. *Cell* 32: 1301–11

Adair, G.M., Nairn, R.S., Wilson, J.H. *et al.* (1989) Targeted homologous recombination at the endogenous adenine phosphoribosyltransferase locus in Chinese Hamster cells. *Proceedings of the National Academy of Sciences USA* 86: 4574–79

Aladjem, M.I., Brody, L.L., O'Gorman, S. & Wahl, G.M. (1997) Positive selection of FLP-mediated unequal sister chromatid exchange products in mammalian cells. *Molecular and Cellular Biology* 17: 857–61

Askew, G.R., Doetschman, T. & Lingrel, J.B. (1993) Site-directed point mutations in embryonic stem cells: a gene targeting tag and exchange strategy. *Molecular and Cellular Biology* 13: 4115–24

Bollag, R.J., Waldman, A.S. & Liskay, R.M. (1989) Homologous recombination in mammalian cells. *Annual Review of Genetics* 23: 199–225

Bradley, A., Evans, M., Kaufman, M.H. & Robertson, E. (1984)

Formation of germ-line chimaeras from embryo-derived teratocarcinoma cell lines. *Nature* 309: 255–56

Broach, J.R. & Hicks, J.B. (1980) Replication and recombination functions associated with yeast plasmid 2μM circle. *Cell* 21: 501–08

Detloff, P.J., Lewis, J., John, S.W.M. *et al.* (1994) Deletion and replacement of the mouse adult β-globin genes by a "plug and socket" repeated targeting strategy. *Molecular and Cellular Biology* 14: 6939–43

Doetschman, T., Maeda, N. & Smithies, O. (1988) Targeted mutation of the Hprt gene in mouse embryonic stem cells. *Proceedings of the National Academy of Sciences USA* 85: 8583–87

Dymecki, S.M. (1996) Flp recombinase promotes site-specific DNA recombination in embryonic stem cells and transgenic mice. *Proceedings of the National Academy of Sciences USA* 93: 6191–96

Dymecki, S.M. & Tomasiewicz, H. (1998) Using Flp-recombinase to characterise expansion of Wnt1-expressing neural progenitors in the mouse. *Developmental Biology* 201: 57–65

Evans, M.J. & Kaufman, M.J. (1981) Establishment in culture of pluripotential cells from mouse embryos. *Nature* 292: 154–56

Fukushige, S. & Sauer, B. (1992) Genomic targeting with a positive-selection lox integration vector allows highly reproducible gene expression in mammalian cells. *Proceedings of the National Academy of Sciences USA* 89: 7905–09

Gordon, J.W., Scangos, G.A., Plotkin, D.J., Barbosa, J.A. & Ruddle, F.H. (1980) Genetic transformation of mouse embryos by microinjection of purified DNA. *Proceedings of the National Academy of Sciences USA* 77: 7380–84

Gossler, A., Doetschman, T., Korn, R., Serfling, E. & Kemler, R. (1986) Transgenesis by means of blastocyst-derived embryonic stem cells lines. *Proceedings of the National Academy of Sciences USA* 83: 9060–65

Gu, H., Marth, J.D., Orban, P.C., Mossmann, H. & Rajewsky, K. (1994) Deletion of a DNA polymerase β gene segment in T cells using cell type specific gene targeting. *Science* 265: 103–06

Hasty, P., Ramirez-Solis, R., Krumlauf, R. & Bradley, A. (1991) Introduction of a subtle mutation into the Hox-2.6 locus in embryonic stem cells. *Nature* 350: 243–46

Holliday, R. (1964) A mechanism for gene conversion in fungi. *Genetical Research* 5: 282–304

Hooper, M.I., Hardy, K., Handyside, A., Hunter, S. & Monk, M. (1987) HPRT-deficient (Lesch–Nyhan) mouse embryos derived from germ-line colonisation by cultured cells. *Nature* 326: 292–94

Hooper, M.L. (1992) Embryonal stem cells; introducing planned changes into the animal germline. In *Modern Genetics*, edited by H.J. Evans, Switzerland Harwood Academic

Jaenisch, R. & Mintz, B. (1974) Simian virus 40 sequences in DNA of healthy adult mice derived from preimplantation blastocysts injected with viral DNA. *Proceedings of the National Academy of Sciences USA* 71: 1250–54

Jasin, M. & Berg, P. (1988) Homologous integration in mammalian cells without target gene selection. *Genes and Development* 2: 1353–63

Kuehn, M., Bradley, A., Robertson, E.J. & Evans, M.J. (1987) A potential animal model for Lesch–Nyhan syndrome through the introduction of HPRT mutations in mice. *Nature* 326: 295–98

Kuhn, R., Schwenk, F., Aguet, M. & Rajewsky, K. (1995) Inducible gene targeting in mice. *Science* 269: 1427–29

Lakso, M., Sauer, B., Mosinger, B. Jr *et al.* (1992) Targeted oncogene activation by site-specific recombination in transgenic mice. *Proceedings of the National Academy of Sciences USA* 89: 6232–36

Lam, K.-P., Kuhn, R. & Rajewsky, K. (1997) In vivo ablation of surface immunoglobulin on mature B cells by inducible gene targeting results in rapid cell death. *Cell* 90: 1073–83

Larsson, N.-G., Wang, J., Wilhelmsson, H. *et al.* (1998) Mitochondrial transcription factor A is necessary for mtDNA maintenance and embryogenesis in mice. *Nature Genetics* 18: 231–36

Logie, C. & Stewart, A.F. (1995) Ligand-regulated site-specific recombination. *Proceedings of the National Academy of Sciences USA* 92: 5940–44

Ludwig, D.L., Stringer, J.R., Wight, D.C., Doetschman, H.C. & Duffy, J.J. (1996) FLP-mediated site-specific recombination in micro-injected murine zygotes. *Transgenic Research* 5: 385–95

Magin, T.M., McWhir, J. & Melton, D.W. (1992) A new mouse embryonic stem cell line with good germ-line contribution and gene targeting frequency. *Nucleic Acids Research* 20: 3795–96

Mansour, S.L. (1990) Gene targeting in murine embryonic stem cells: introduction of specific alterations into the mammalian genome. *Genetic Analysis and Technical Applications* 7: 219–27

Mansour, S.L., Thomas, K.R. & Capecchi, M.R. (1988) Disruption of the proto-oncogene int-2 in mouse embryo derived stem cells: a general strategy for targeting mutations to non-selectable genes. *Nature* 336: 348–52

Martin, G.R. (1981) Isolation of a pluripotent cell line from early mouse embryos cultured in medium conditioned by teratocarcinoma stem cells. *Proceedings of the National Academy of Sciences USA* 78: 7634–36

Meselson, M.S. & Radding, C.M. (1975) A general model for genetic recombination. *Proceedings of the National Academy of Sciences USA* 72: 358–61

Meyers, E.N., Lewandoski, M. & Martin, G.R. (1998) An *Fgf8* mutant allelic series generated by Cre- and Flp-mediated recombination. *Nature Genetics* 18: 136–41

Mountford, P.S., Zevnik, B., Duwel, A. *et al.* (1994) Dicistronic targeting constructs: reporters and modifierss of mammalian gene expression. *Proceedings of the National Academy of Sciences USA* 91: 4303–07

Nyhan, W.L. (1973) The Lesch–Nyhan syndrome. *Annual Review of Medicine* 24: 41–60

Orban, P.C., Chui, D. & Marth, J.D. (1992) Tissue- and site-specific DNA recombination in transgenic mice. *Proceedings of the National Academy of Sciences USA* 89: 6861–65

Orr-Weaver, T., Szostak, J.W. & Rothstein, R.J. (1981) Yeast transformation: a model system for the study of recombination. *Proceedings of the National Academy of Sciences USA* 78: 6354–58

Palmiter, R.D. & Brinster, R.L. (1986) Germline transformation of mice. *Annual Review of Genetics* 20: 465–99

Qin, M., Bayley, C., Stockton, T. & Ow, D.W. (1994) Cre recombinase mediated site specific recombination between plant chromosomes. *Proceedings of the National Academy of Sciences USA* 91: 1706–10

Ramirez-Solis, R., Liu, P. & Bradley, A. (1995) Chromosome engineering in mice. *Nature* 378: 720–24

Sauer, B. (1987) Functional expression of the cre-lox site-specific recombination system in the yeast *Saccharomyces cerevisiae. Molecular and Cellular Biology* 7: 2087–96

Sauer, B. & Henderson, N. (1990) Targeted insertion of exogenous DNA into the eukaryotic genome by the cre recombinase. *New Biology* 2: 441–49

Schwartzberg, P.L., Goff, S.P. & Robertson, E.J. (1989) Germ-line transmission of a c-abl mutation produced by targeted gene disruption in ES cells. *Science* 246: 799–803

Smith, A.J.H., De Sousa, M.A., Kwabi-Addo, B. *et al.* (1995) A site directed chromosomal translocation induced in embryonic stem cells by cre-loxP recombination. *Nature Genetics* 9: 376–85

Smithies, O., Gregg, R.G., Boggs, S.S., Karalewski, M.A. & Kucherlapati, R.S. (1985) Insertion of DNA sequences into the human chromosomal β-globin locus by homologous recombination. *Nature* 317: 230–34

Stacey, A., Schnieke, A., McWhir, J. *et al.* (1994) Use of double-replacement gene targeting to replace the murine alpha-lactalbumin gene with its human couterpart in embryonic stem cells and mice. *Molecular and Cellular Biology* 14: 1009–16

Tarutani, M., Itami, S., Okabe, M. *et al.* (1997) Tissue-specific knockout of the mouse Pig-a gene reveals important roles for the GPI-anchored proteins in skin development. *Proceedings of the National Academy of Sciences USA* 94: 7400–05

Thomas, K.R. & Capecchi, M.R. (1987) Site-directed mutagenesis by gene targeting in mouse-embryo derived stem cells. *Cell* 51: 503–12

Thomas, K.R. & Capecchi, M.R. (1990) Targeted disruption of the murine int-1 proto-oncogene resulting in severe abnormalities in midbrain and cerebellar development. *Nature* 346: 847–50

Tsien, J.Z., Huerta, P.T. & Tonegawa, S. (1996) The essential role of hippocampal CA1 NMDA receptor-dependent synaptic plasticity in spatial memory. *Cell* 87: 1327–38

Valancius, V. & Smithies, O. (1991) Testing an "in-out" targeting procedure for making subtle genomic modifications in mouse embryonic stem cells. *Molecular and Cellular Biology* 11: 1402–08

van Deursen, J., Fornerod, M., van Rees, B. & Grosveld, G. (1995) Cre mediated site specific translocation between non-homologous mouse chromosomes. *Proceedings of the National Academy of Sciences USA* 92: 7376–80

Vooijs, M., van der Valk, M., te Reile, H. & Berns, A. (1998) FLP-mediated tissue-specific inactivation of the retinoblastoma tumour suppressor gene in the mouse. *Oncogene* 17: 1–12

Williams, R.L., Hilton, D.J., Pease, S. *et al.* (1988) Myeloid leukaemia inhibitory factor maintains the developmental potential of embryonic stem cells. *Nature* 336: 684–88

GLOSSARY

blastocoel cavity hollow, fluid-filled cavity inside the blastula

conditional allele relating to a modified allele from which gene expression can be regulated by the experimenter

DNA footprinting technique for identifying the recognition site of DNA-binding proteins

electroporation application of a brief electric pulse to cell membranes to make them temporarily permeable, allowing entry of large or hydrophilic molecules

expression cassette piece of DNA which contains all sequences necessary for gene expression, usually refers to selectable marker genes

hypomorphic relating to a mutant allele that behaves in a similar fashion to the normal gene but has a weaker effect

inner cell mass (ICM) thickening of the epiblast and primitive endoderm layers of the mammalian blastocyst

LacZ enzyme that hydrolyses lactose to galactose and glucose

plasmacytoma malignancy that develops within plasma cells, often characterized by cells that have undergone specific chromosomal rearrangements

pronuclear injection microinjection of genetic material directly into the pronuclei of fertilized mouse eggs

pronucleus haploid nucleus of unfertilized egg or of sperm

transfection introduction of DNA into a eukaryotic cell either transiently or permanently after its subsequent integration into the DNA of the recipient cell

Ω-type vector alternative name for a replacement type targeting vector

DNA FINGERPRINTING

John Brookfield
Institute of Genetics, University of Nottingham, Queen's Medical Centre, UK

Approximately one in every 1000 of the base pairs (bp) in human DNA is heterozygous in any individual. With three billion bp of DNA in our haploid chromosome set, there will be millions of differences between the DNAs of any two people. With the exception of identical sibs, all humans who have ever lived have been genetically unique. Thus, in principle, if one screened enough DNA sequences, one could, with certainty, identify any individual from their DNA and could do so using DNA from any tissue of the body. This theoretical possibility was not, however, made practical until the development in 1985, by Alec Jeffreys and his colleagues at the University of Leicester, England, of DNA fingerprinting. Jeffreys produced DNA probes that yield individual-specific patterns, identifying hypervariable regions in the chromosomes. While investigating, with Polly Weller, the structure of the human myoglobin gene, Jeffreys noticed, in an intron, a sequence of four tandem repeats of a 33 bp unit (Weller *et al.*, 1984). This sequence was found to be able to hybridize to and thereby detect other repetitive sequences in the DNA. Cloning and sequencing such "minisatellite" sequences showed that they are at scattered positions (loci) in the chromosomes and that all show tandem repetition of simple motifs. They share a core sequence which is similar but not identical at the different loci (Jeffreys *et al.*, 1985b). Probes consisting of tandem repeats of slightly different versions of this core sequence were produced, and used to probe human DNAs. The patterns discovered differed very greatly between individuals (Jeffreys *et al.*, 1985c), with each person's pattern resembling a supermarket bar code, with a collection of 20–30 bands in differing relative positions. The differences seen result from variation in the numbers of tandem repeats in the minisatellite arrays (hence the acronym VNTRs), and each band represents an allele at one of the minisatellite loci. The two most commonly used probes, 33.15 and 33.6, which differ slightly in sequence, detected different sets of bands, which segregation experiments showed to be almost entirely independent of each other. **Figure 1** shows the procedures involved in DNA fingerprinting, and the article "Chromosome banding and the longitudinal differentiation of chromosomes" describes the techniques used to detect these patterns.

The proportion of bands shared by unrelated individuals (often called the bandsharing, x, and estimated by its mean value, S) is typically around 20%, although it is less for the

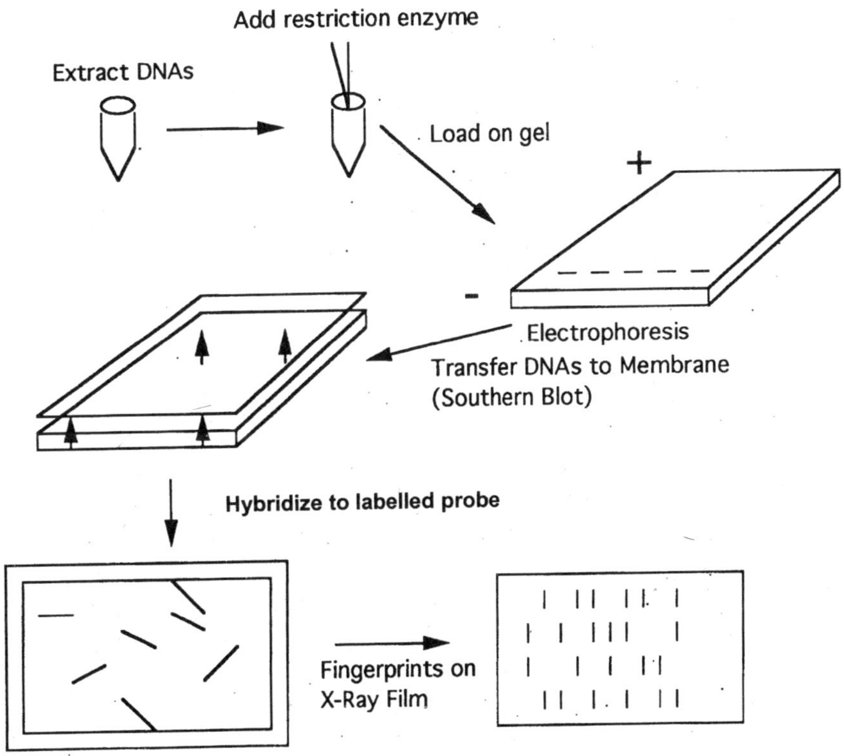

Figure 1 DNA fingerprints are produced by starting with the purification of the DNAs of the scene-of-crime tissue and a blood sample of the suspect. The DNAs are digested with a restriction enzyme, which separates them into approximately 10 million different fragments, of differing sizes; some of these fragments will contain the minisatellites. The DNAs are separated electrophoretically according to size, and then transferred from the gel to a membrane, using a technique devised by Edwin Southern and named after him. The millions of different fragments bound to the membrane have been made single stranded, and can hybridize (base pair) with a "probe" DNA with a complementary sequence. A ^{32}P-labelled single-stranded probe bearing the minisatellite core sequence is washed over the membrane, and this binds to the minisatellite sequences in the human DNAs. On an X-ray film, the β-particles from the labelled probe produce black bands, creating the DNA fingerprint. Variations in the number of repeat units in the minisatellite arrays create differences in the length of the restriction fragments that contain them. These, therefore, run at different speeds through the gel, and end up at different positions on the membrane.

larger bands. Since different bands in the pattern are derived from loci at diverse locations in the chromosomes, it was assumed that their occurrences are independent. This assumption allowed the calculation that the probability that an individual would have a particular fingerprint is x^n, where n is the number of bands seen. When the probabilities associated with the patterns detected by the two different probes, 33.15 and 33.6, were combined, the resulting probability was so low that no two individuals in the world would be expected to show the same overall pattern. For this reason, these were called "DNA fingerprints". The individual specificity comes with the caveat that they are genetically determined, and thus identical in identical sibs.

Why are minisatellites hypervariable? The answer is simply that their mutation rate is high. Population genetics theory predicts that the level of variability for neutral sequences such as these is determined solely by the balance between the production of variation by mutation and its loss by genetic drift. Mutations change the number of repeats in the arrays and show the very high average rate across loci of around 0.004, although this varies between loci and, indeed, between alleles within loci.

The significance of individual specificity was immediately apparent. If a criminal leaves a blood or semen stain at the scene of a crime, the police can match the DNA fingerprint derived from this with that of a suspect and be confident that the suspect left the scene-of-crime DNA. Furthermore, while minisatellite loci show high mutation rates, enough bands are faithfully inherited for paternity testing to be possible. A major industry has developed using DNA fingerprints in paternity tests. Typically, there is a mother, her child and one or more putative fathers. Maternal-specific bands are identified in the child's fingerprint; these are those shared with the mother. The remaining bands in the child, apart from perhaps one or two arising from mutation, must have come from the child's father. A true father will show all of these paternal-specific bands. Usually, the male tested shows either all of the paternal-specific bands, or only a minority of them. He can, therefore, be identified either as being the father or not. (Occasional cases in which the man shows an intermediate number of bands are probably explained by his being the brother of the true father, who will share many of the father's bands.)

The first use of DNA fingerprinting in human identification was a modified paternity test. A boy had legally entered the UK from Ghana, along with his mother and three sisters (Jeffreys *et al.*, 1985a). Immigration officials doubted that he was truly the woman's son. Jeffreys and his colleagues showed that every band in the boy's fingerprint could be identified either in the mother or in one or more of the three sisters. The boy was indeed her son and he shared the father of the three girls.

Soon after the development of Jeffreys' probes, they were used successfully in a wide range of other animals. Ornithologists were particularly interested in their use to establish the parentage of nestlings, looking for evidence for extra-pair paternity. The first papers demonstrating their usefulness in showing relationships in wild birds appeared in 1987 from the groups of David Parkin at the University of Nottingham (Wetton *et al.*, 1987) and Terry Burke in Leicester (Burke & Bruford, 1987). The technique has since been applied to vast numbers of other species, of almost all animal groups. It has also transpired that many other tandemly repeated DNA sequences can also give highly variable patterns when used to probe genomic DNA.

While it is DNA fingerprinting, in which bands from many loci are simultaneously visualized, that has dominated studies of non-human organisms, the forensic use of hypervariable DNAs has continued to develop technically. For various reasons, in order to compare DNA fingerprints, the DNAs need to be run on the same gel, which is not always possible. While an initial suspect's blood or semen may be run on the same gel as the scene-of-crime DNA, they may not match and further suspects must be identified. The DNA from the scene-of-crime having all been used, it might be impossible to compare these individuals' fingerprints directly with that of the crime sample. Also, to produce a DNA fingerprint requires DNA of high quality, which is not often possible with a blood or semen stain from the crime scene.

For these reasons, almost all forensic uses of minisatellites now use single-locus profiling (in contrast to the multi-locus profiling that is DNA fingerprinting). In single-locus profiling, the membrane bearing the DNAs is probed under conditions that visualize a single minisatellite locus. There will usually be two bands, showing heterozygosity, but occasionally a single-banded homozygote is seen. The probability of a match is calculated from the frequency in the population of the alleles showing the mobilities seen in the profile. Thus, the interpretation of the strength of the evidence uses databases derived from population samples, showing the relative frequencies of alleles of varying sizes at each of the loci used. Such databases now exist for various ethnic groups, particularly in the US, and show small but significant differences between these groups. The forensic evidence derived from just two bands is clearly much less than that from the 20–30 bands of a DNA fingerprint. This difference is reduced, however, by sequentially hybridizing three or four probes to the same membrane, revealing a total of six to eight bands. Calculations of the probability that two individuals will match at all loci usually give values between one in a million and one in a thousand million. Unlike DNA fingerprinting, the technique is not individual specific, and is simply called "DNA profiling" for this reason. This abandonment of individual specificity might seem foolhardy, in that the scene-of-crime DNA can no longer be said with certainty to be that of the suspect. In reality, such certainty is illusory anyway, since the probability of laboratory error (or deception), in which the suspect's DNA is substituted for that of the scene-of-crime sample, is higher than that of a chance match with an innocent suspect and will remain so however small this latter probability is made.

All agree that a DNA match between a suspect and the scene-of-crime sample is powerful evidence of guilt, but there have been major controversies as to exactly how powerful. An expert witness may quote a probability of matching that is one in many millions, yet no survey has been carried out of sufficient size to show that this is indeed the frequency of matches. Instead, these low probabilities must be calculated using assumptions. Normally, the "product rule" is used. Here, the probability of a match at multiple loci is calculated as the product of the probabilities of matching at each of the individual loci. It is assumed, in other words, that the probabilities of matching at the individual loci are independent. This assumption has been strongly challenged (Lewontin & Hartl, 1991). One obvious cause of an association, or linkage disequilibrium, between alleles at different loci is the subdivision of the population into subpopulations, differing in the frequencies of alleles at all the loci. If two individuals match at the first locus, this increases the probability that they come from the same subpopulation. If they do, then they are more likely to match at other loci also. The product rule will underestimate the probability of a chance match, which will act to the detriment of the defendant in a trial. While most of the evidence suggests that the inaccuracies in the product rule will be very small for most populations, it is possible that for some groups, such as native Americans, its inaccuracy may be considerable. In view of the perceived need to avoid inaccurate evidence being presented that is prejudicial to the defendant, it is becoming increasingly common for the probabilities of chance matches presented in court to be conservative estimates. This means that they are calculated to represent the upper bounds of the true values, thereby ensuring that they err in favour of the defence.

Also important in the calculation of the probability that an innocent man will match a crime sample is to remember that DNA profiles are inherited characteristics and will be similar in close relatives. For example, the probability that a brother of the truly guilty man matches the scene-of-crime DNA will normally be many orders of magnitude higher than the corresponding probability for an unrelated man.

Courts in the US, and also in the UK, use an adversarial system, and it is common for both the defence and the prosecution to call their own expert witnesses, who give the court different views about the strength of the DNA evidence. This happens when they take differing views of the degree of population substructure or of the appropriate degree of conservativeness. The end result is often confusion and misunderstanding of the evidence by the jury, which may lead to acquittal.

The use of DNA in forensic science continues to advance technically, and the focus has been on improving the probability of obtaining useful DNA profiles from the crime scene. While DNA can be remarkably stable under some conditions, particularly when dry, in others it rapidly degrades and there is, consequently, very little chance of DNA of sufficient quality and quantity remaining for single-locus minisatellite profiling. The solution to the degradation problem is the incorporation of PCR, which can detect minuscule amounts of surviving DNA. For PCR to work on badly degraded DNA, the length of DNA between the primers must be short, which rules out the amplification of the long fragments of DNA, of many kilobases, that contain minisatellite arrays. The alternative is to use microsatellites, which are often, in the forensic context, called short tandem repeats. These are again repetitive sequences, but now the length of the repetitive unit in the DNA may be only 2, 3 or 4 bp. Such microsatellites are very common in human DNAs, and may arise from slippage of the DNA polymerase during replication. The mutation rate to new alleles with different numbers of repeats is less at microsatellite loci than with minisatellites, but still very much greater than the mutation rate by base substitution. The result is that, while the level of variation in length in populations at microsatellite loci is less than for minisatellites, it is still impressively high, and even a few loci can give a probability of a chance match that is one in a hundred thousand. To amplify microsatellites, PCR primers are designed that bind to nonrepetitive DNAs surrounding the microsatellite array. The PCR yields short DNA fragments that differ between the homologous chromosomes in the number of repeat units. These can be separated on gels that can distinguish between changes in DNA mass caused by as little as a single base addition. A further use of PCR in forensics is the amplification of variable regions of mitochondrial DNAs. These can be sequenced and their high levels of variability allow them to be informative in identification.

A case in which PCR was essential was the identification of the remains of the family of Tsar Nicolas II from a collection of human bones recovered in Ekaterinberg, Russia (Gill *et al.*, 1994; see article on "Forensic DNA analysis of the last Russian royal family"). The bones, which represented the remains of nine people, had been buried since 1918, and only a PCR-based technique had any possibility of success. Microsatellites were amplified and showed that the three children and two of the adults were members of the same family. That this was the family of the Tsar was demonstrated by the mother and the children having a rather rare variant of the mitochondrial DNA which was known (through the testing of her living relatives) to have been possessed by the Tsarina.

Databases including microsatellite information from very large numbers of people are being produced and stored. This is furthest advanced in the UK. Large databases will revolutionize detection in criminal cases involving DNA evidence. A DNA profile from a crime scene can be compared with all those in the database. A huge improvement in the efficiency of detection will result, relative to the collection of new sets of profiles of suspects in each individual case. Care will be needed in the interpretation of matches that arise in such database screens. In previous casework, the probability that an innocent suspect will match the DNA from the scene of a crime has been so small that any DNA match has been regarded as overwhelming evidence that both DNAs came from the same individual. However, in these cases, suspects were initially identified through other evidence linking them with the crime, and only subsequently tested with DNA. When large DNA databases are systematically screened, the situation is very different. Since hundreds of thousands of stored DNA profiles are examined, it is quite likely that a matching individual from the database is not, in fact, the source of the scene-of-crime DNA. Indeed, multiple matches may be found in the database. In this context, a DNA match is no longer overwhelming evidence of guilt. The problem is soluble, however. A matching individual identified through the database can be subsequently compared with the scene-of-crime DNA with another set of marker loci. Only if the individual truly is the source of the scene-of-crime DNA will he/she also match with these loci; evidence other than DNA may also link this individual with the crime.

REFERENCES

Burke, T. & Bruford, M.W. (1987) DNA fingerprinting in birds. *Nature* 327: 149–52

Gill, P., Ivanov, P.L., Kimpton, C. *et al.* (1994) Identification of the remains of the Romanov family by DNA analysis. *Nature Genetics* 6: 130–35

Jeffreys, A.J., Brookfield, J.F.Y. & Semeonoff, R. (1985a) Positive identification of an immigration test-case using human DNA fingerprints. *Nature* 317: 818–19

Jeffreys, A.J., Wilson, V. & Thein, S.L. (1985b) Hypervariable "minisatellite" regions in human DNA. *Nature* 314: 67–73

Jeffreys, A.J., Wilson, V. & Thein, S.L. (1985c) Individual-specific 'fingerprints' of human DNA. *Nature* 316: 76–79

Lewontin, R.C. & Hartl, D.L. (1991) Population genetics in forensic DNA typing. *Science* 254: 1745–50

Weller, P., Jeffreys, A.J., Wilson, V. & Blanchetot, A. (1984) Organisation of the human myoglobin gene. *EMBO Journal* 3: 439–46

Wetton, J.H., Carter, R.E., Parkin, D.T. & Walters, D. (1987) Demographic study of a wild house sparrow population by DNA fingerprinting. *Nature* 327: 147–49

GLOSSARY

hypervariable region segment of a chromosome characterized by multiple alleles at a single locus, especially alleles whose variation is due to variable numbers of tandem repeats

minisatellite regions of DNA, comprising a very short nucleotide sequence, tandemly repeated. Even shorter sequence repeats are termed microsatellites

tandem repeat multiple copies of a DNA sequence at a defined locus. Repeat units can be short nucleotide sequences or entire sets of genes

See also **Chromosome banding and the longitudinal differentiation of chromosomes** (p.539); **Y-chromosomal DNA analysis and the Jefferson–Hemings controversy** (p.821); **Forensic DNA analysis of the last Russian royal family** (p.827)

COMPARATIVE MAPPING: TRACKING GENE HOMOLOGUES AMONG MAMMALS

J.T. Eppig
The Jackson Laboratory, Bar Harbor, USA

1. Introduction

The usefulness of comparing similarities of gene structure and location among species' genomes has long been recognized. Mammalian species genomes are highly related in content and organization due to their recent evolutionary divergence and rapid diaspora (Bush *et al.*, 1977).

Early observations included those of Haldane who noted that the phenotypes of two known Mendelian traits, pink-eyed dilution and albino, were genetically linked in mouse, deer mouse and rat, and possibly could represent gene equivalencies (Haldane, 1927). While phenotypic similarity can provide such hints, the formal test for these gene homologues had to wait until later in the century when the albino (Kwon *et al.*, 1988) and pink-eyed dilution (Gardner *et al.*, 1992) genes were cloned and the sequences of the genes could be compared.

Comparative maps provide a graphical overview of genomic regions that have been evolutionarily conserved as contiguous segments. Such maps can be used to identify candidate genes for hereditary diseases, to facilitate mapping in other species and to examine divergence of chromosomal organization during speciation. Detailed physical maps and sequence analyses are also providing new data showing small rearrangements and duplications that may provide new insights into the process of chromosomal evolution and genetic regulation.

2. Describing how genes are similar

The term homology has been loosely used to mean homology, orthology or paralogy, the key differences among the terms being the rigour of evidence that two genes in different species are derived from (and are functionally equivalent to) one another or whether similarity of gene sequence or function may be attributed to gene duplication and subsequent divergence (i.e. gene family members) within or among species.

(i) Homologous genes

Gene homology among species is defined as the similarity among two or more genes in two or more species where those genes derived from a common ancestor. Thus, closely related members of gene families may be thought of as homologous; examples between mouse and human would include "esterases" or "P450 cytochromes" (Nelson *et al.*, 1996). These gene families contain many members, yet a one-to-one correspondence between them is not possible due to their highly overlapping specificities.

(ii) Orthologous genes

Orthology has a more stringent definition. Orthologous genes arose from a common ancestor and may have a one-to-one functional correspondence between two species. Examples among mouse, rat and human would include "hypoxanthine phosphoribosyl transferase" and "tyrosinase".

(iii) Paralogous genes

Paralogy refers to genes within a species that are homologous. Most of these have resulted from gene duplication events, either pre- or postspeciation. Evidence of duplication is easily observed in the mammalian genome; for example, there are two genes for "glucose-6-phosphate isomerase" and four "Hox" homeobox clusters.

3. Describing homologous gene map locations

(i) Homology segment

A homology segment is the most basic of comparative mapping statements between two species. A homology segment identifies a gene in species x mapped to location *a*, with its homologue in species y mapped to location *b*. Thus, a homology segment provides an expectation that an unknown extent of chromosomal region has been conserved between the two species (**Figure 1A**).

(ii) Conserved segment

A conserved segment refers to two or more genes mapped on the same chromosome, where the homologous genes also map together in the other species. Thus, genes x and y mapping to chromosome a in species 1 and their homologues, genes x' and y', mapping to chromosome b in species 2 constitute a conserved segment. The order of genes within the chromosomes of the two species need not be the same (**Figure 1B**).

(iii) Conserved linkage segment

Conserved linkage represents the most evolutionarily similar chromosome arrangement between species. In a conserved linkage segment, both the gene content and the gene order on the chromosome of the species being compared are the same (**Figure 1C**).

4. Criteria for establishing gene homology

There are many ways to describe homology among species' genomes, and different criteria and statistical stringencies can be used. In this respect, it is important that the information consumer understands how particular data analyses compare genes among species to assert homology. Two reviews of human and rodent orthologues analysed using large sequence data sets of coding regions (Makalowski *et al.*, 1996, Makalowski & Boguski, 1998) present an entrée into the exploration of genome similarities using computational comparisons. Files listing the GenBank identifiers

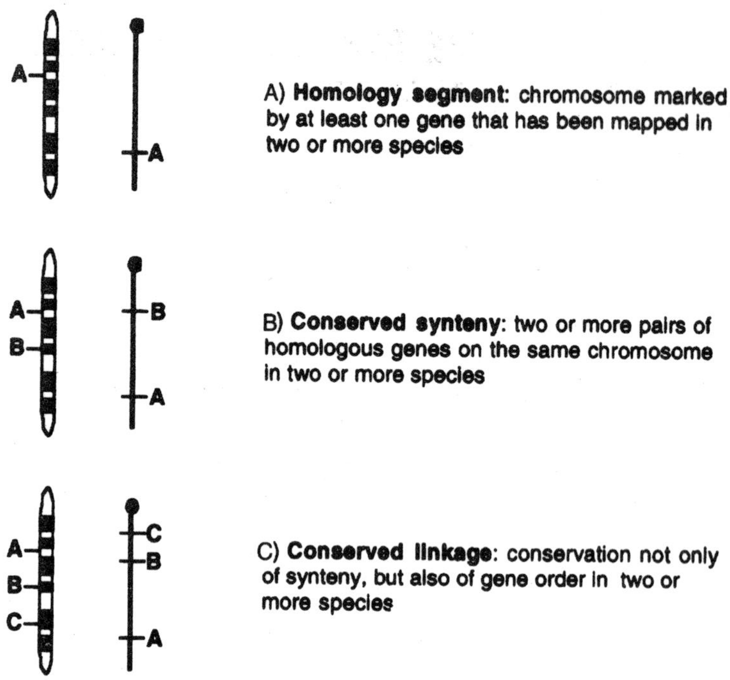

Figure 1 Definitions of homology segments.

used in these analyses can be found in the appendices to the full-text papers at http://www.ncbi.nlm.nih.gov/Makalowski/PNAS and http://www.ncbi.nlm.nih.gov/Makalowski/mus-art. Other efforts are being made to compare non-coding regions of mouse and human (e.g. Jareborg, Birney & Durbin, 1999; http://www.sanger.ac.uk/Software/Alfresco/mmhs.shtml).

The Mouse Genome Database (MGD) uses sequence similarities as one criterion for asserting homology among species' genes. In curating additional homology assertions from publications, the criteria recommended by the Human Genome Organization (HUGO) (Andersson *et al.*, 1996) are followed. These include:

- Formation of functional heteropolymers
- Immunological cross-reaction
- Similar response to specific inhibitors
- Similar subcellular location
- Similar substrate specificity
- Similar subunit structure
- Coincident expression
- Nucleotide sequence comparison
- Amino acid sequence comparison
- Cross-hybridization to same molecular probe
- Conserved map location
- Functional complementation.

The criteria associated with each species–species gene homology assertion in MGD are displayed along with the comparative data (**Figure 2A**). This allows users to evaluate the strength of homology evidence and supplement evidence that one might obtain by sequence similarity searches alone. Some criteria provide stronger evidence than do others; for example, functional complementation is stronger evidence than nucleotide sequence similarity. Multiple criteria for specific homologies generally imply stronger evidence. Conserved map location as a criterion must be cautiously evaluated as evidence, particularly when used alone, because this can be a circular argument for the existence of homology.

5. Pitfalls to establishment of homologies

Orthologous relationships among genes imply that there is a one-to-one correspondence between the genes in the respective species and that they arose from a common ancestor. There are a number of circumstances for which these relationships are difficult to establish. These include inter-species comparison of gene family similarities, differences in gene numbers in different species for functional proteins, convergent evolution resulting in different evolutionary pathways producing genes with the same function and the relative reliability of sequence similarity. For example, in the large family of olfactory receptors, very high sequence similarity and overlapping function make it impossible to know the precise relationships among all family members in different species.

In the realm of establishing homologies among genes or gene families, a number of considerations are in order. Duplications and/or deletions of gene family members and their functional specificities have been amply demonstrated. For example, the mouse gene *Uox* (urate oxidase) has no human counterpart, and presumably represents a genomic deletion that occurred since the divergence of the human and mouse lineages (Wu *et al.*, 1992). Similarly, the mouse CD1 family includes a d1 and a d2 isoform due to a duplication in the rodent lineage relative to human; but, the human equivalent isoform *a*, *b* and *c* genes have been deleted in mouse (Blumberg *et al.*, 1995). In addition, primates, artiodactyls and rats have a single active alpha-1-protease inhibitor, compared with up to five copies in mouse and guinea pig (Goodwin *et al.*, 1996). Evolution can result in similar molecular functions being performed by genes with distinct evolutionary histories, as has been shown for the human and hedgehog apolipoprotein-alpha genes (Lawn *et al.*, 1997).

Expression pattern differences among orthologues can lead to phenotypic consequences and confound the ability to use functional evidence to support orthologous relationships. For example, the *Myo7a* gene (myosin VIIa) (Steel *et al.*, 1997) is differentially expressed during embryogenesis in human and mouse, which may factor into the differences between the human Usher 1B syndrome and mouse shaker-1 phenotype caused by mutations in that gene, although the precise orthologous mutations have not yet been isolated. Imprinting provides another opportunity for differences in phenotypic manifestations and gene expression. One example of such a difference is the *Igf2r* (insulin-like growth factor 2 receptor) gene which is paternally imprinted in mouse, but not imprinted in human (Kalscheuer *et al.*, 1993).

6. Status of mammalian comparative mapping

The genomes of human and mouse contain more identified genes than any other mammalian species. Over 10 000 genes have been characterized sufficiently to be named in each of these species, which represents about 10% of the anticipated gene number. **Table 1** summarizes the number of homologous genes in the comparative maps of human or mouse with other relatively well-mapped mammals.

These homologous genes are distributed along the chromosomes of mammalian species in a non-random fashion, occurring in conserved syntenies and conserved linkage groups. For mouse and human, whose genomes are the most thoroughly mapped, the number of conserved segments represented by these 4000+ homologues has been estimated at approximately 200 (Nadeau & Sankoff, 1998). The size of a conserved segment can be very large; the largest in mouse (excluding the X chromosome) represents approximately 60% of chromosome 11, measures 48 centiMorgans (cM) and is conserved with much of human chromosome 17.

It should be noted that, at this genetic map resolution, many minor genomic rearrangements, local duplications and repeats will be invisible initially. As mapping resolution increases, or physical mapping or sequence level analysis becomes possible, more of the large conserved segments will be found that have small changes within them (e.g. small transpositions that place homologues from other chromosomes within the segment or small inversions that disrupt conserved regions). The evidence for this is already apparent with physical comparisons of certain regions such as proximal mouse chromosome 7 and human chromosome 19 (Stubbs *et al.*, 1996), where it has been demonstrated that micro-rearrangements and zinc finger gene expansions are present in the conserved regions of human relative to mouse. Sequence analysis of a region of human chromosome 12q13, representing a conserved segment of mouse chromosome 6, has revealed the same gene content and order, although two human pseudogenes are absent in mouse (Ansari-Lari *et al.*, 1998). Similarly, it has been known for some time that, although the major histocompatibility regions of human and mouse are generally conserved, the order of homologous genes within the complexes is arranged differently (Amadou, 1999).

7. Visualizing homologous regions

A variety of visual displays are available for users to obtain genomic views of segments of conservation among major mammalian species (**Table 2**). A common paradigm of comparative maps is that homologous genes or conserved segments are displayed relative to the chromosome map backbone of a reference species (**Figure 3**). The reference species must be a species that has significant mapping information available for the comparative map to be meaningfully constructed. Another useful display is the Oxford Grid, which provides an overview of the genome-wide linkage conservation between species. In this type of display, one species' chromosomes appear on one axis and the second species' chromosomes appear on the alternate axis. Cells in the grid are coloured when homologues have been mapped to the respective species' chromosomes represented by that

Table 1 Status of species genetic maps and comparative maps with mouse and human.

Species	Number of named genes[a]	Number with orthologs[b] in mouse	human	WWW address for species database
human	10 925	4059	–	http://www.ncbi.nlm.nih.gov/LocusLink/index.html
mouse	10 676	–	4059	http://www.informatics.jax.org
rat	2298[c]	1597	1484	http://ratmap.gen.gu.se
cattle	869[d]	566	573	http://138.102.88.140/cgi-bin/bovmap/intro.pl
sheep	403[e]	189	199	http://www.ri.bbsrc.ac.uk/sheep/ark/
pig	713[f]	244	249	http://www.ri.bbsrc.ac.uk/pigmap/pigbase/pigbase.html
cat	128[g]	65	70	http://www.ri.bbsrc.ac.uk/cat/ark

[a] Number of coding genes characterized sufficiently to be given names in that species. Data are from the specific species' databases as of 5 Jan. 2000.
[b] Data from Mouse Genome Database (MGD) (http://www.informatics.jax.org) 5 Jan. 2000.
[c] Calculated value (1643 genes from RatMap + 655 additional genes identified from MGD homology data)
[d] Calculated value (554 genes from Bovmap + 315 additional genes identified from MGD homology data)
[e] Calculated value (350 genes from ArkDB/sheep + 53 additional genes identified from MGD homology data)
[f] Calculated value (613 genes from ArkDB/pig + 100 additional genes identified from MGD homology data)
[g] Calculated value (115 genes from ArkDB/cat + 13 additional genes identified from MGD homology data)

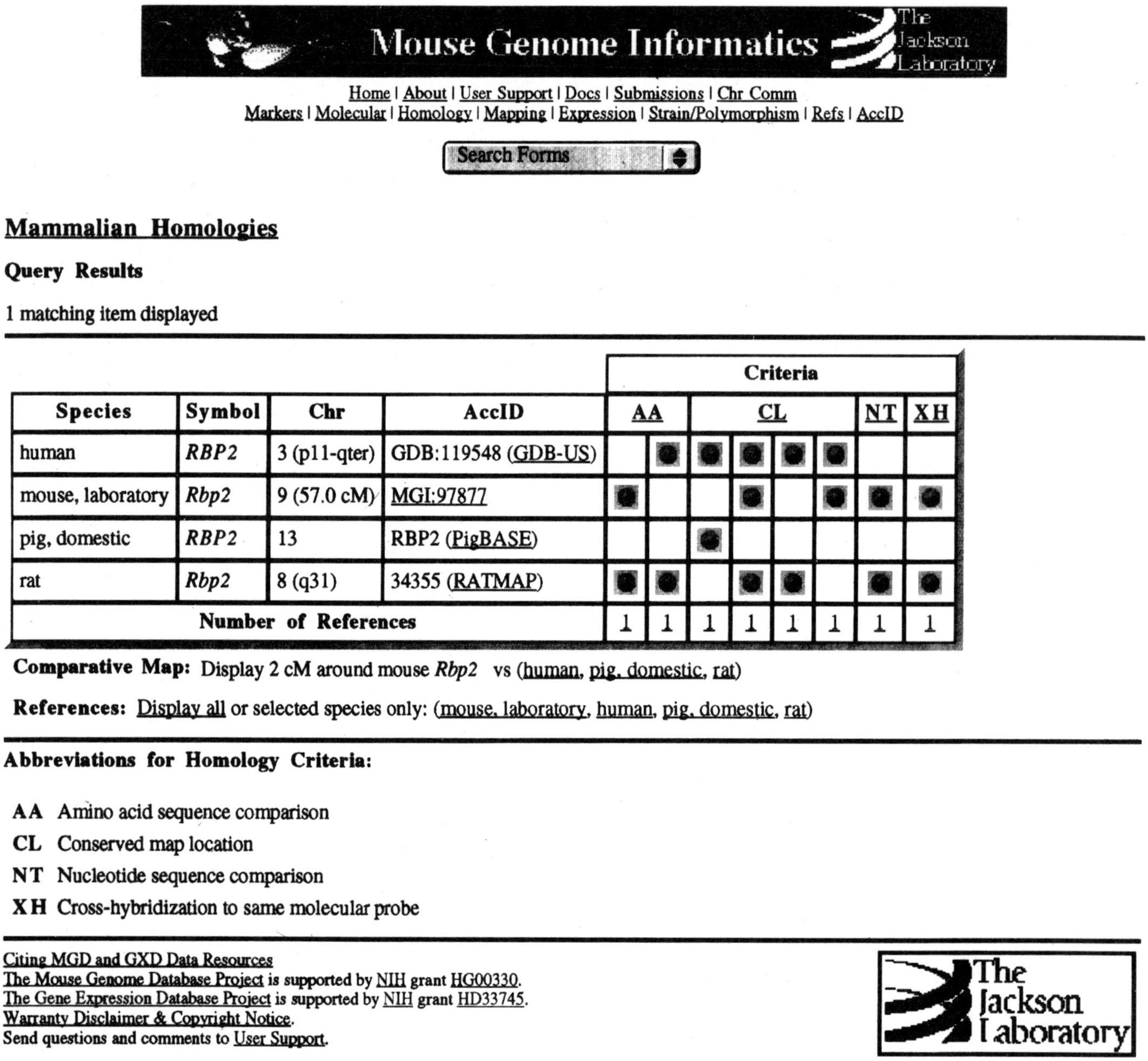

Mouse Genome Informatics — The Jackson Laboratory

Home | About | User Support | Docs | Submissions | Chr Comm
Markers | Molecular | Homology | Mapping | Expression | Strain/Polymorphism | Refs | AccID

Search Forms

Mammalian Homologies

Query Results

1 matching item displayed

				Criteria							
Species	**Symbol**	**Chr**	**AccID**	AA		CL				NT	XH
human	*RBP2*	3 (p11-qter)	GDB:119548 (GDB-US)		●	●	●	●	●		
mouse, laboratory	*Rbp2*	9 (57.0 cM)	MGI:97877	●			●		●	●	●
pig, domestic	*RBP2*	13	RBP2 (PigBASE)			●					
rat	*Rbp2*	8 (q31)	34355 (RATMAP)	●	●		●	●		●	●
Number of References				1	1	1	1	1	1	1	1

Comparative Map: Display 2 cM around mouse *Rbp2* vs (human, pig, domestic, rat)

References: Display all or selected species only: (mouse, laboratory, human, pig, domestic, rat)

Abbreviations for Homology Criteria:

AA Amino acid sequence comparison
CL Conserved map location
NT Nucleotide sequence comparison
XH Cross-hybridization to same molecular probe

Citing MGD and GXD Data Resources
The Mouse Genome Database Project is supported by NIH grant HG00330.
The Gene Expression Database Project is supported by NIH grant HD33745.
Warranty Disclaimer & Copyright Notice.
Send questions and comments to User Support.

The Jackson Laboratory

A

Figure 2 (A) Comparative mapping data from MGD showing homologues for retinol binding protein 2, cellular: human *RBP2*, mouse *Rbp2*, pig *RBP2* and rat *Rbp2*. The chromosomal location of these genes in the respective species is provided, as is a hypertext link to the gene detail information in the relevant species database. On the right of the table are listed the criteria for establishing these homology assertions with a key beneath the table. For this example, the genes from human, mouse, pig and rat were evaluated for amino acid sequence similarity and conserved map location, and mouse and rat were evaluated for nucleotide sequence comparison and cross-hybridization to the same molecular probe. The references for these data are found via a hypertext link beneath the table. Also available beneath the table is a link to a blow-up region of the genome surrounding the mouse *Rbp2*. (B) (see opposite) The comparative map surrounding (±2 cM) mouse *Rbp2*, obtained by linking from the display in (A) and choosing the comparative map with human. This map shows all mouse loci in this 4 cM region, and displays the homologues from human where they are known. Similar maps showing pig or rat homologues could have been chosen. This map view allows easy access to genes and markers located in a region of homology, but which have not been identified in the comparison species. Users find this helpful for map predictions and as an opportunity to view potential candidate genetic markers in a region of known homology.

Mouse Genome Informatics

Home | About | User Support | Docs | Submissions | Chr Comm
Markers | Molecular | Homology | Mapping | Expression | Strain/Polymorphism | Refs | AccID

Search Forms

MGD: Mouse Chromosome 9 Linkage Map with human homologies

Click on a marker symbol to retrieve a detailed marker record from MGD. (*Requires client-side imagemap support*)

- Page 0: 55.00-59.00 cM

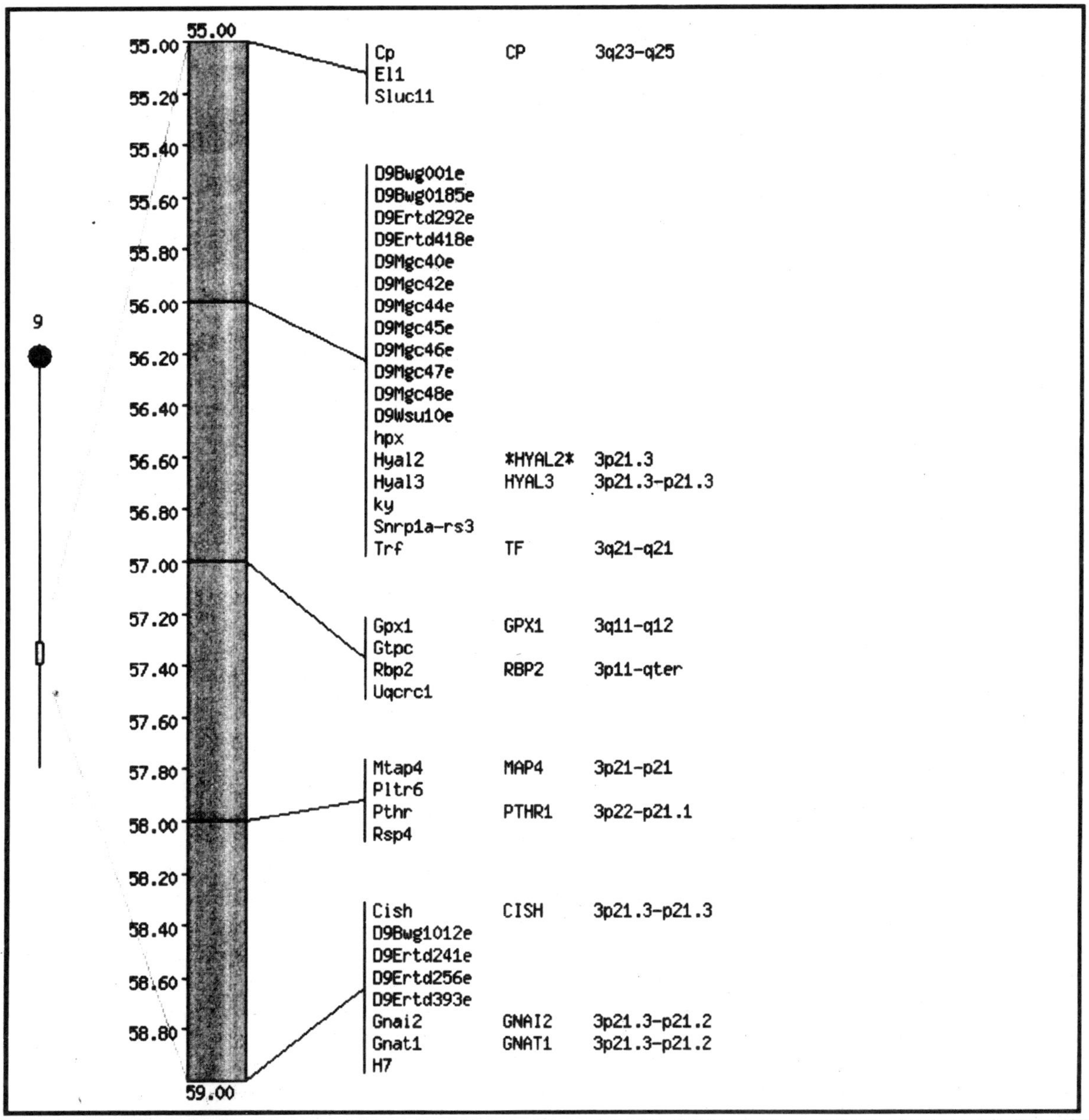

Citing MGD and GXD Data Resources
The Mouse Genome Database Project is supported by NIH grant HG00330.
The Gene Expression Database Project is supported by NIH grant HD33745.

Send questions and comments to User Support.

The Jackson Laboratory

B

Table 2 Internet access to human–mouse comparative mapping data.

Source	Data	Map	WWW address	Content synopsis
Mouse Genome Database (MGD) The Jackson Laboratory Bar Harbor, ME, USA	•	•	http://www.informatics.jax.org/menus/homology_menu.shtml	Searchable comparative mapping data for major mammalian species; mouse–human and mouse–rat full data reports generated daily; Oxford grid displays; Genetic maps that include homologies. As of Jan. 2000 4059 homologous loci for human/mouse were included. Links to MGD, GDB, PigMap, SheepBase, RatMap, and references. Daily data updates ensure current content.
Comparative Mapping Diagrammatic Database Interface, LaTrobe University, Australia		•	http://www.latrobe.edu.au/www/genetics/compmapddi.html	Human map with comparative display of many mammals (primates, monkeys, cat, dog, sheep, cattle, mouse, rat, horse, marsupials, etc). Created in 1996, from the reference: Wakefield, M.J. & Graves, J.A.M. Comparative maps of vertebrates. *Mammalian Genome* 7: 715–16.
The Comparative Animal Genome Database (TCAGDB), Roslin Institute, Edinburgh, UK	•		http://www.ri.bbsrc.ac.uk/cgi-bin/tcagdb/cbr.pl	Searchable repository of putative homologies among loci in various animal species (chicken, cattle, human, mouse, pig, sheep). As of Jan. 2000 contained 260 loci for human/mouse. Links to MGD, GDB, ArkDB (pig, sheep, cattle, chicken).
National Center for Biotechnology Information (NCBI) Bethesda, MD, USA		•	http://www3.ncbi.nlm.nih.gov/Homology/	Human/mouse comparative mapping based on published reference: *Genomics* (1996) 33: 337–51. 1793 homologous loci for human/mouse are represented. Links to MGD, OMIM.

GDB, Human Genome Database; MGD, Mouse Genome Database; OMIM, Online Mendelian Inheritance in Man.

cell, and the number in the cell represents the number of homologues in that chromosome-by-chromosome pair (**Figure 4**; see also **Plate 22**).

8. Homology in the MGD

Comparative mapping data can be searched and viewed in various ways through the MGD (http://www.informatics.jax.org/menus/homology_menu.shtml). The detailed comparative mapping data include information on homologous genes among the mouse, human and rat, as well as major primates, model and agricultural mammals. By specifying parameters in the search form, users can ask questions such as "What genes on human chromosome 1p13 have known homologues in mouse and pig?" or "Using all mammalian species represented in MGD, what ATPase genes have at least one identified homologue in another mammalian species and what are those genes/species associations?" The tabular display returned from such a search includes the species, genes, map locations and the evidence and citation for the homology assertion based on HUGO criteria (**Figure 2A**) (Andersson *et al.*, 1996). In addition, links to each species database records for specific genes are provided. Another link takes the user to a comparative map for a 2 cM region around the gene using the mouse backbone map. In particular, since all mouse loci are displayed for this 4 cM interval, with any homologues from the selected species, this comparative mini-map provides a focused view for those searching for candidate genes (**Figure 2B**).

Additionally, specific database reports, generated daily, are available for the complete listing of all mouse/human and mouse/rat homologues, sorted and formatted in various ways. These pre-generated database reports are provided to speed access to these highly sought after (and largest) mammalian comparative data sets. Also available as a report are mouse/Drosophila homologies from a joint curation effort of MGD and Flybase.

Oxford grid displays provide a graphical overview of mammalian homologies (**Figure 4**). Since these displays are only visually useful where the species being compared have a significant number of identified and mapped genes, Oxford grids are available for any pairs of the species mouse, human, rat, sheep, cattle, pig and cat. Each cell in the grid provides information on the number of homologous genes between chromosome m from species x and chromosome n from species y. The cells so represented are hypertext links to the underlying data. Links are also provided from the mouse chromosome designation to the whole chromosome comparative map between mouse and the species chosen.

Finally, comparative maps that can be user-customized are a powerful tool for viewing mammalian homologies in MGD. A map-building utility allows users to view a comparative map on the web or to produce a postscript file for publication-quality printing. Parameter selection includes specifying a particular chromosomal region, displaying homologues from a particular species, indicating whether to limit the display to a particular class of genes (e.g. homeoboxes) and the option of adding *de novo* markers to the map (**Figure 3**).

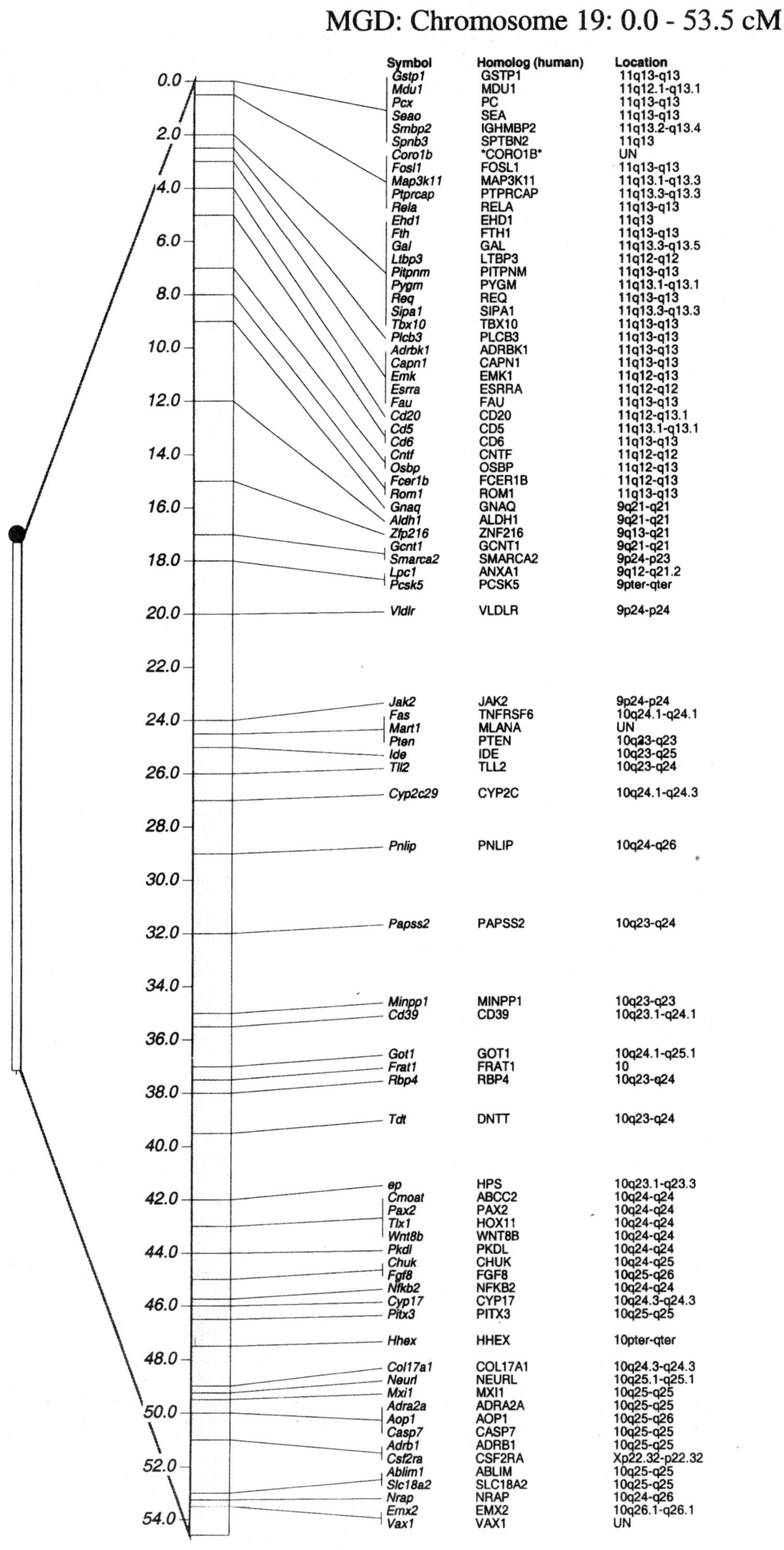

Figure 3 Comparative map showing mouse chromosome 19 and homologous human genes and their mapped locations. This version of the comparative map was created by customizing the map query form (http://www.informatics.jax.org/searches/linkmap_form.shtml) as follows: select "Postscript' format, display mouse 'chromosome 19", "do not include DNA segments or syntenic markers", "show homologues from human" and "show only those markers with homologues". The mouse chromosome 19 figure shows a stick-figure sketch of the chromosome extent and a cM-scaled blow-up of the chromosome to the immediate right. Each mouse gene is connected to its cM position. To the right of each mouse gene symbol is the corresponding human gene symbol and the location of the homologue in humans. For example, the mouse locus *Papss2* at cM 32.0 has a homologue in humans, PAPSS2, which is located on human chromosome 10q23–q24.

Mouse Genome Informatics The Jackson Laboratory

Home | About | User Support | Docs | Submissions | Chr Comm
Markers | Molecular | Homology | Mapping | Expression | Strain/Polymorphism | Refs | AccID

Search Forms

Each cell in the **Oxford Grid** represents a comparison of two chromosomes, one from each of the selected species. The number of homologies appears inside each colored cell and the color indicates a range in the number: Grey (1), Blue (2-10), Green (11-25), Orange (26-50), Yellow (50+). **Click on a colored cell to retrieve homology details.** A note about printing an Oxford grid...

Click on a mouse chromosome (*blue numbers next to grid frame*) to retrieve a **comparative map** showing all homologies between the selected mouse chromosome and the comparison species displayed on the grid.

Total Homologies: 4059
Total mapped in both species: 2941

mouse, laboratory (columns); human (rows)

human \ mouse	1	2	3	4	5	6	7	8	9	10	11	12	13	14	15	16	17	18	19	X	Y	XY	UN	MT
1	91	1	81	92	2			5		1	1	1	6					1					74	
2	70	43			6	27					8	16			1		9	1					43	
3	1		19	1		25		1	52					10		30	1						45	
4			24		69	4		20				1	1						1				21	
5	1										43		52		11		3	26					31	
6	5			6	1		1		7	29			18				105						27	
7		1			59	57	1				12	11	5				1						24	
8	6		7	9				23						16	30	3				1			19	
9		41		46									8						10				20	
10		11				3	11			12			1	16				3	44				18	
11	1	31		1			65		45		1								42				51	
12		1	1		18	63	1			40					44								38	
13	2		2		8			12						19				1					12	
14	1		1		1		1	1	1	1		44		24									23	
15		23					32		32		1		1										11	
16		1					20	49			8				1	14	15						22	
17			1			1		1			185	2		1									41	
18	3				2											1	4	29					6	
19						1	91	23	9	24	1	1					15						45	
20		77																					11	
21										18				1		26	6						12	
22	1				6	2		3		8	9				33	27							11	
X											2			1					1	157			18	
Y																					6			
XY																								
UN	10	16	11	13	10	2	12	9	7	13	14	4	8	2	8	5	11	2	5	3			330	
MT																								13

http://www.informatics.jax.org/searches/oxfordgrid.cgi

9. MGD support

User support staff are dedicated to assisting users with getting the most out of MGD, tutoring new users, addressing specific biological queries and expediting new data submissions. For information or assistance, contact User Support at mgi-help@informatics.jax.org.

10. The future of MGD homology data: the genomic era

As rapid sequencing of the human and mouse genomes is certain, the question of the role of curated homology data based on experimental (rather than computational) evidence arises. In particular, it is anticipated that sequence similarity and long-range genomic sequence alignments will quickly bring new perspectives and change our views of linkage conservation. Sequence analysis will support orthologue determination and assist with cross-species annotation. The ultimate connection between genomic and biological information will rely both on well-assembled sequence and on careful annotation and examination of data. By analysing long sequence alignments, we will learn much more about evolutionary input into mammalian genomes. Some surprises are undoubtedly ahead as the details of conserved segments are revealed, for which the importance of maintaining a current representation of comparative information will be critical.

11. Summary

Homology and orthology data continue to be examined in the laboratory (and reported in the literature) under very stringent conditions. Likewise, the availability of sequence data for genes allows both nucleotide and amino acid residue comparisons to be made. The combined genomic and biological approaches are already greatly enriching our understanding of the mammalian genome and the cross-species comparisons will continue to provide information transfer and insights into mammalian evolution.

Acknowledgements

The author gratefully acknowledges the Mouse Genome Database staff, particularly P.L. Grant, D.J. Reed, J.J. Merriam, S.F. Rockwood, C.G. Wray, for careful curation of mammalian homology data and the support of co-PIs J.A. Blake, M.T. Davisson, J.A. Kadin and J.E. Richardson. Supported by NIH grant HG00330 (to JTE).

Figure 4 Oxford grid (see also **Plate 22** in colour plate section). The Oxford grid provides a genome-wide overview of comparative mapping data. To generate this Oxford grid, "human" and "mouse" were chosen as the comparison species in the query form (http://www.informatics.jax.org/searches/oxfordgrid_form.shtml). The vertical grid axis shows the chromosome designations for human and the horizontal axis shows the chromosome designations for mouse. Where a homology between the two species exists, the cell corresponding to the intersection of the chromosomal locations of the homologues is coloured; the numbers in the cell correspond to the number of homologous genes represented. XY designates the pseudoautosomal region; UN indicates unmapped genes.

REFERENCES

Amadou, C. (1999) Evolution of the Mhc class I region: the framework hypothesis. *Immunogenetics* 49: 362–67

Andersson, L., Archibald, A., Ashburner, M. *et al.* (1996) Comparative genome organization of vertebrates. The First International Workshop on Comparative Genome Organization. *Mammalian Genome* 7: 717–34

Ansari-Lari, M.A., Oeltjen, J.C., Schwartz, S. *et al.* (1998) Comparative sequence analysis of a gene-rich cluster at human chromosome 12p13 and its syntenic region in mouse chromosome 6. *Genome Research* 8: 29–40

Blumberg, R.S., Gerdes, D., Chott, A., Porcelli, S.A. & Balk, S.P. (1995) Structure and function of the CD1 family of MHC-like cell surface proteins. *Immunological Reviews* 147: 5–29

Bush, G.L., Case, S.M., Wilson, A.C. & Patton, J.L. (1977) Rapid speciation and chromosomal evolution in mammals. *Proceedings of the National Academy of Sciences USA* 74: 3942–46

Gardner, J.M., Nakatsu, Y., Gpndo, Y. *et al.* (1992) The mouse pink-eyed dilution gene: association with human Prader–Willi and Angelman syndromes. *Science* 257: 1121–24

Goodwin, R.L., Baumann, H. & Berger, F.G. (1996) Patterns of divergence during evolution of alpha 1-proteinase inhibitors in mammals. *Molecular Biology and Evolution* 13: 346–58

Haldane, J.B.S. (1927) The comparative genetics of colour in rodents and carnivora. *Biological Reviews of the Cambridge Philosophical Society* 2: 199–212

Jareborg, N., Birney, E. & Durbin, R. (1999) Comparative analysis of non-coding regions of 77 orthologous mouse and human gene pairs. *Genome Research* 9: 815–24

Kalscheuer, V.M., Mariman, E.C., Schepens, M.T., Rehder, H. & Ropers, H.H. (1993) The insulin-like growth factor type-2 receptor gene is imprinted in the mouse but not in humans. *Nature Genetics* 5: 74–78

Kwon, B.S., Wakulchik, M., Haq, A.K., Halaban, R. & Kestler, D. (1988) Sequence analysis of mouse tyrosinase cDNA and the effect of melanotropin on its gene expression. *Biochemical and Biophysical Research Communications* 153: 1301–9

Lawn, R.M., Schwartz, K. & Patthy, L. (1997) Convergent evolution of apolipoprotein(a) in primates and hedgehog. *Proceedings of the National Academy of Sciences USA* 94: 11992–97

Makalowski, W. & Boguski, M.S. (1998) Evolutionary parameters of the transcribed mammalian genome: an analysis of 2,820 orthologous rodent and human sequences. *Proceedings of the National Academy of Sciences USA* 95: 9407–12

Makalowski, W., Zhang, J. & Boguski, M.S. (1996) Comparative analysis of 1,196 orthologous mouse and human full-length mRNA and protein sequences. *Genome Research* 6: 846–57

Nadeau, J.H. & Sankoff, D. (1998) Counting on comparative maps. *Trends in Genetics* 14: 495–501

Nelson, D.R., Koymans, L., Kamataki, T. *et al.* (1996) P450 superfamily: update on new sequences, gene mapping, accession numbers and nomenclature. *Pharmacogenetics* 6: 1–42

Steel, K.P., Mburu, P., Gibson, F. *et al.* (1997) Unravelling the genetics of deafness. *Annals of Otology, Rhinology and Laryngology* 168 (Suppl.): 59–62

Stubbs, L., Carver, E.A., Shannon, M.E. *et al.* (1996) Detailed comparative map of human chromosome 19q and related regions of the mouse genome. *Genomics* 35: 499–508

Wu, X.W., Muzny, D.M., Lee, C.C. & Caskey, C.T. (1992) Two independent mutational events in the loss of urate oxidase during hominoid evolution. *Journal of Molecular Evolution* 34: 78–84

GLOSSARY

apolipoprotein alpha apolipoprotein that covalently attaches to a LDL particle (of cholesterol, phospholipid, and apolipoprotein β-100), to make lipoprotein alpha, an important risk factor in heart disease

esterase a hydrolytic enzyme that attacks an ester (a compound formed by condensation of an acid with an alcohol), releasing an alcohol or thiol and acid

heteropolymer a protein or carbohydrate macromolecule composed of different types of subunits

isoform form of a protein with slightly different amino acid sequences, which may differ in activity, function, distribution, etc.

P450 cytochrome specialized cytochrome (an enzyme pigmented as a result of its haem prosthetic groups) of the electron transport chain in adrenal mitochondria and liver microsomes, involved in hydroxylating reactions

See also **The role of bioinformatics in the postsequencing phase of genomics** (p.833); **The pufferfish as a model organism for comparative vertebrate genome analysis** (p.849)

DNA FROM MUSEUM SPECIMENS

Jeremy J. Austin
Department of Palaeontology and Department of Zoology, The Natural History Museum, London, UK

Natural history museum collections contain a large and incredibly diverse representation of the earth's biodiversity, both past and present. The total number of specimens held in natural history collections around the world has been estimated at around 1.5 billion and they range from microscopic, single-celled algae and protists to entire skeletons of blue whales. Specimens from the distant past are represented as fossils (mineralized hard and soft tissues, amber-preserved animals and plants) or archaeological material (human, animal and plant remains), ranging in age from hundreds to many millions of years. Specimens of more recent origin represent extant or recently extinct species collected during the last 300 years. These are preserved either dry as whole organisms (e.g. insects) or as parts of organisms (e.g. many herbarium specimens, animal skins and skeletons), or wet in one of several types of liquid fixative or preservative. Natural history museum collections are clearly a very important resource, as records of biological diversity and as the primary source of material for the study of taxonomy, evolutionary biology, comparative anatomy and natural history. The value of museum specimens is derived from the preservation of their morphological features – the size, shape, number, distribution and/or colour of both internal and external structures. However, it has become clear over the last 10 years that fragments of DNA – the genetic material – can survive intact in tissues long after an organism dies, both in specimens collected during historical times and, under special circumstances, in archaeological material and sub-fossils thousands and tens of thousands of years old. Advances in molecular biology, particularly the development of PCR, have allowed this previously inaccessible genetic information to be obtained from museum specimens. In addition to their morphological value, museum collections are now being seen as valuable genetic resources.

DNA is a labile biomolecule. Spontaneous oxidation and hydrolysis, and endogenous and microbial enzymatic attack, rapidly degrade DNA after an organism dies. Fortuitously, many methods used to preserve the gross morphological features of museum specimens also act to retard the posthumous degradation of DNA. Several natural phenomena, such as peat bogs, dry caves and the Arctic permafrost, provide protection for DNA in archaeological and fossil materials. Preservation of DNA in museum specimens is primarily provided by protection from water through rapid dehydration in air or an organic preservative (e.g. ethanol), or through freezing; protection from oxygen and light are also contributing factors. DNA in museum specimens is not perfectly preserved and it has three general properties that distinguish it from DNA in fresh specimens. First, DNA in museum specimens is degraded into small fragments, generally <1000 base pairs (bp) and often <200 bp in length (Pääbo *et al.*, 1989). In contrast, fresh material contains DNA molecules many tens of thousands of bp long. There are no general rules about the size of intact DNA fragments that may be present in any given specimen, but extrinsic factors such as the method and conditions of preservation, and intrinsic factors, such as tissue composition, secondary compounds and the size of the organism, appear to be important. Second, DNA in museum specimens is chemically modified with an abundance of lesions such as baseless sites, oxidized pyrimidines and crosslinking between DNA strands and to surrounding proteins (Pääbo *et al.*, 1989; Lindahl, 1993). These chemical modifications may prevent or dramatically slow the PCR amplification of DNA or result in errors in the amplified DNA. Thirdly, the extent of this degradation means that the number of intact and largely undamaged endogenous DNA fragments in museum specimens is very low; generally, >99% of the specimen's DNA has been fragmented into such small pieces or so heavily chemically modified that it is of little value in terms of its genetic information.

There are two general problems that make the extraction and analysis of DNA from museum specimens technically challenging. First, extraction of DNA from museum specimens often results in the coextraction of organic and inorganic compounds which are strong inhibitors of subsequent enzymatic manipulations, including PCR. These inhibitory compounds may derive from posthumous modifications to endogenous organic material, or may originate from the environment (e.g. humic acids in the case of many archaeological specimens) or from the preservatives and fixatives used to preserve the museum specimen itself. A variety of techniques has been developed to extract DNA free of these inhibitory compounds, to remove them from DNA once it is extracted or to reduce their inhibitory effects in the PCR. The second and greater problem involves contamination with extraneous DNA. The sensitivity and power of the PCR makes it possible to amplify preferentially a specific DNA fragment from a very small number of intact DNA molecules in the presence of a vast excess of damaged molecules. However, this capacity also introduces the real threat that contaminating DNA molecules will be amplified. Contaminating DNA of recent origin is more likely to outcompete endogenous DNA during PCR because it is less damaged and often present in higher copy number. Specimens can become contaminated with DNA following posthumous invasion by bacteria and fungi, handling by humans and via cross-contamination from related samples. Much more insidious is carryover contamination with DNA from previous PCR experiments. Each successful PCR generates millions of copies of a specific DNA fragment which can be efficiently and preferentially amplified in subsequent PCR experiments. Contamination with human, bacterial or fungal DNA can often be circumvented by the judicious design of PCR primers specific to the group of organisms under study. In general, however, careful choice and handling of specimens and tissue samples, rigorous attention to protocols and the use of proper controls are necessary as precautions against contamination and to achieve believable results in any study using DNA from museum specimens. The precautions taken to reduce the possibilities for contamination are similar to those adopted by laboratories working on viruses and forensic samples. The problem of contamination requires that DNA putatively derived from museum specimens be rigorously authenticated – a process that can be expensive and time-consuming. Several criteria of authenticity have been published (Austin *et al.*, 1997; Cooper & Poiner, 2000), the most important being physically isolating work on museum specimens from that on fresh specimens; carrying out control DNA extractions and PCR amplifications to demonstrate that reagents used in these two procedures are free from contaminating DNA; and reproducibility of results – multiple and separate extractions and PCR amplifications from different tissues from the same individual should yield identical and unambiguous DNA sequences. Where possible, results should be verified by a completely independent laboratory.

The amount of genetic information retrievable from a single specimen is constrained by two factors. First, most studies involving DNA from museum specimens have been restricted to genes that occur in high copy number, such as nuclear ribosomal RNA genes and mitochondrial genes. Most molecular systematic studies have targeted similar gene sequences, but in the case of museum specimens the use of high copy number genes improves the chances of finding sufficient numbers of undamaged DNA fragments, per unit of material, to be retrievable by PCR. DNA sequences from most low- and single-copy genes may prove to be unobtainable from many museum specimens because of the degraded nature of their DNAs. Secondly, extraction of DNA from museum specimens is necessarily a destructive process. This places limits on the amount of material (and, therefore, the amount of genetic information) that can be removed from any one specimen without damaging excessively its morphological features. Some material in collections, such as type specimens or those that are historically important, rare or small, can never be subjected to destructive sampling and is, therefore, unavailable for genetic analysis.

Despite these problems, museum specimens have proved to be of great value in many molecular genetic studies. Museum collections contain many specimens of extant taxa from which it is impossible to obtain fresh material because there is legal protection of the species or habitat, the habitat is inaccessible, known collection localities have been destroyed or the cost of obtaining fresh material is prohibitive. Museum specimens are the only sources of material for studies involving extinct species. The Tasmanian wolf or thylacine (*Thylacinus cynocephalus*) is a large carnivorous marsupial that may or may not be a primitive remnant of the Australian marsupial stock. Establishing the evolutionary relationships between the Tasmanian wolf and other marsupials based on morphological characteristics has been problematic because it possesses both primitive and derived features. The thylacine is thought to have become extinct at the beginning of this century and is now only represented by museum skin and skeleton collections. Thomas *et al.* (1989) and Krajewski *et al.* (1992) extracted DNA from small pieces of skin taken from museum specimens of the Tasmanian wolf. The short DNA sequences they retrieved from the mitochondrial 12S rRNA and cytochrome *b* genes allowed the phylogenetic relationships between the marsupial wolf and other carnivorous Australian and South American marsupials to be resolved. The thylacine is clearly related to other Australian marsupials, but lacks a close relative among extant taxa. Thus, it appears that the Tasmanian wolf is a primitive species that diverged from other Australian marsupials very early in their evolutionary history.

DNA has also been used to test the authenticity of museum specimens. Steller's sea cow (*Hydrodamalis gigas*), a sirenian related to manatees and dugongs, became extinct in the late 1700s due to overhunting. All that remains of this species are bone and skin samples in various museums. Persson *et al.* (1994) extracted and amplified mtDNA from two of the three known skin samples and the short DNA sequences that they obtained were almost identical to those of the right whale, but considerably different from those of the closely related dugong. These results supported previous suspicions that the skins ascribed to Steller's sea cow were in fact derived from right whales.

Where collections are sufficiently extensive over space and time, museum specimens can provide data to examine the population genetic history of a species. Such information has implications for conservation biology and may help in making informed management decisions about remaining populations. A good example involves the Laysan duck (*Anas laysanensis*) an endangered species known only from the small and remote island of Laysan in the Hawaiian Islands. Bones of small ducks have been found on the main Hawaiian islands but it was not known whether these came from the Laysan duck, the koloa (*Anas wyvilliana*) another Hawaiian endemic, or a previously unknown, extinct species. Cooper *et al.* (1996) extracted and sequenced small pieces of mitochondrial DNA from the unidentified bones and showed that the sequences match those from the Laysan duck, indicating that the species was once much more widespread than at present. This study has important conservation implications, justifying the reintroduction of the Laysan duck to islands that were once part of its natural range.

Museum specimens can provide an important temporal component to the study of the origins and spread of disease organisms. One example of this was the search for the Lyme disease bacterium, *Borrelia burgdorferi*, in museum specimens of the deer tick, *Ixodes dammini*, collected up to 50 years before the disease was clinically diagnosed (Persing *et al.*, 1990). The use of Borrelia-specific PCR primers allowed DNA extracted from the ticks to be screened for the presence of the bacterial DNA. The Lyme disease bacterium was detected in a number of ticks collected from two locations in the US in the 1940s, indicating that it was present in suitable tick vectors many years before the disease became formally recognized. Archives of pathology specimens may also prove to be valuable historical records of the appearance of human diseases and parasites.

Molecular data from archaeological collections of human, plant and animal remains can provide valuable information on the origins, development and movement of humans, their pathogens and their domesticated plants and animals. The history of human colonization of islands in the Pacific has been and remains controversial, especially for the more remote islands such as Easter Island. Hagelberg *et al.* (1994) extracted DNA and amplified segments of the mitochondrial genome from a number of prehistoric human bones taken from Easter Island to investigate the genetic affinities of the original settlers. All the bone samples exhibited genetic markers typical of modern-day Pacific Islanders and south-east Asians, several of which have not been detected in native American populations. These results provide genetic evidence for a Polynesian origin of the original Easter Islanders, which is contrary to a previously proposed South American origin.

Extraction of DNA from much older material – tens of thousands and even millions of years old – is more controversial, technically demanding and requires rigorous authentication. Authenticated DNA sequences have been extracted from a range of Pleistocene age remains including woolly mammoths, moas, ground sloths, brown bears and Neanderthal man. These sequences have provided new insights into the origins, relationships and population genetics of these extinct species. However, for every specimen that does yield DNA there are at least as many, if not many more, that do not (Poinar *et al.*, 1996; Höss *et al.*, 1996). The value of any short DNA sequences that may be obtained from such specimens must be weighed against the expense and time required to retrieve and authenticate such sequences. Claims of ancient DNA retrieval from millions of year old palaeontological material have been made for Miocene plant fossils, Cretaceous dinosaur bone and Oligocene insects and bacteria preserved in amber. These claims created the exciting possibility that palaeontological specimens – fossils from the geological past – could be included in molecular genetic analyses and provide insights into the process of evolution over geological time scales.

However rigorous attempts to reproduce the work have failed and careful reanalysis of the data obtained have shown that the sequences are derived from experimental artefacts or contamination. Therefore it would seem that even under the best, naturally occurring conditions DNA has a lifespan limited to the recent past. Currently, the oldest authenticated ancient DNA comes from 50 000–100 000 year old mammoths from the Siberian permafrost.

The study of DNA from museum specimens has a clear role to play in many areas of evolutionary biology, medicine and conservation biology. Museum specimens provide a unique source of genetic information from extinct species and populations with which to examine changes in the population genetics of endangered species and the evolutionary relationships of extinct and extant organisms. Museum collections also represent historical genetic records that document the origins and spread of diseases as well as human, animal and plant populations. The benefits of museum specimens in molecular genetics must, however, be tempered against the destructive nature of sampling tissues. The morphological value of many specimens precludes them from sampling for genetic analysis, and limitations on the amount of tissue that can be taken from any one specimen will ultimately limit the amount of genetic information that can be obtained.

REFERENCES

Austin, J.J., Smith, A.B. & Thomas, R.H. (1997) Palaeontology in a molecular world: the search for authentic ancient DNA. *Trends in Ecology and Evolution* 12: 303–06

Cooper, A., Rhymer, J. & James, H.F. *et al.* (1996) Ancient DNA and island endemics. *Nature* 381: 484

Cooper, A. & Poinar, H.N. (2000) Ancient DNA: do it right or not at all. *Nature* 289: 1139

Hagelberg, E., Quevedo, S., Turbon, D. & Clegg, J.B. (1994) DNA from ancient Easter Islanders. *Nature* 369: 25–26

Höss, M., Dilling, A., Currant, A. & Pääbo, S. (1996) Molecular phylogeny of the extinct ground sloth *Mylodon darwinii*. *Proceedings of the National Academy of Sciences USA* 93: 181–85

Krajewski, C., Driskell, A.C., Baverstock, P.R. & Braun, M.J. (1992) Phylogenetic relationships of the thylacine (Mammalia: Thylacinidae) among dasyuroid marsupials: evidence from cytochrome *b* DNA sequences. *Proceedings of the Royal Society of London, Series B* 250: 19–27

Lindahl, T. (1993) Instability and decay of the primary structure of DNA. *Nature* 362: 709–15

Pääbo, S. (1989) Ancient DNA: extraction, characterization, molecular cloning, and enzymatic amplification. *Proceedings of the National Academy of Sciences USA* 86: 1939–43

Persing, D.H., Telford III, S.R., Rys, P.N. *et al.* (1990) Detection of *Borrelia burgdorferi* DNA in museum specimens of *Ixodes dammini* ticks. *Science* 249: 1420–23

Persson, P., Arnason, U. & Schlieman, H. (1994) DNA from remains ascribed to Steller's sea cow. *Ancient DNA Newsletter* 2: 16–17

Poinar, H.N., Höss, M., Bada, J.L., & Pääbo, S. (1996) Amino acid racemization and the preservation of ancient DNA. *Science* 272: 864–66

Thomas, R.H., Schaffner, W., Wilson, A.C. & Paabo, S. (1989) DNA phylogeny of the extinct marsupial wolf. *Nature* 340: 465–67

FURTHER READING

Austin, J.J., Smith, A.B. & Thomas, R.H. (1997) Palaeontology in a molecular world: the search for authentic ancient DNA. *Trends in Ecology and Evolution* 12: 303–06

Pääbo, S. (1993) Ancient DNA. *Scientific American* November: 60–66

Wayne, R.K., Leonard, J.A. & Cooper, A. (1999) Full of sound and fury: the recent history of ancient DNA. *Annual Review of Ecology and Systematics* 30: 457–77

GLOSSARY

dugong a marine mammal found in Asian seas and coasts (sometimes called a sea cow)

Lyme disease bacterium tickborne spirochaete, *Borrelia burgdorferi*. Disease is spread by two species of deer tick – *Ixodes dammini* and *I. pacificus*

right whale rich in whalebone and captured easily, this large-headed whale is a member of the family Balaenidae

See also **Forensic DNA analysis of the last Russian royal family** (p.827); **Conservation genetics** (p.910)

Y-CHROMOSOMAL DNA ANALYSIS AND THE JEFFERSON–HEMINGS CONTROVERSY

Eugene A. Foster[1] and Chris Tyler-Smith[2]

[1]*Charlottesville, Virginia, USA*

[2]*Department of Biochemistry, University of Oxford, Oxford, UK*

Thomas Jefferson (1743–1826), the third president of the United States of America and author of the Declaration of Independence, is a complex figure who has fascinated people for 200 years (**Figure 1**). One subject that has attracted attention is his attitude to slavery, and, in particular, the question of whether he was the father of the children of his slave Sally Hemings. Historians continue to debate the matter, and this article describes some recent DNA studies that have added to the evidence available.

1. The historical background

During Thomas Jefferson's first term as president, he was accused by the journalist James Callender, writing in *The Richmond Recorder* (1802), of being the father of Sally Hemings's son Tom, who allegedly bore a striking physical resemblance to Jefferson. Tom is said to have been born in 1790 shortly after Jefferson and his family members, including Sally Hemings, returned from France where Jefferson had been minister since 1787. Present-day members of the large African–American Woodson family believe that they are descendants of Tom. The Woodson family tradition, independently transmitted in three branches of the family, states that when Tom's presence at Monticello (Jefferson's home) became a political embarrassment, he was sent to live with a John Woodson, said to be a relative of Jefferson's mother. He then took, or was given, the name Thomas Woodson, married, moved to Ohio and eventually had 11 children, several of whom gave rise to the present family of about 1400 people.

The story of a sexual relationship between Jefferson and Sally Hemings was revived in 1873 when an Ohio newspaper, *The Pike County Republican* (Life among the lowly, 1873), published an extensive interview with Madison Hemings, one of Sally Hemings's later children. In it he said that his mother told him that Thomas Jefferson was the father of all her children and that his mother's first child, conceived in France, died in infancy. He also said that his mother had not wanted to return from France, where she was free, but finally agreed to do so when Jefferson promised to free her children when they reached 21 years of age. Eston Hemings, Sally's last child, was said to have resembled Jefferson in more than facial features. He was over 6 feet tall, had red hair and played the violin very well. He took the name Eston Hemings Jefferson and moved into white society in Madison, Wisconsin in 1852. His descendants (**Figure 1**) believe that they are descended from Thomas Jefferson.

Based on Madison Hemings's statement that Sally Hemings's first child died in infancy, as well as the lack of documentation of his existence at Monticello, some historians believe that Thomas Woodson, ancestor of the present-day Woodsons, was not the child of either Thomas Jefferson or Sally Hemings. At least one historian believes that Tom was simply a fiction invented by Callender and never really existed (Miller, 1977).

Many of Thomas Jefferson's recognized descendants believe that all of Sally Hemings's children except Tom were fathered by either Samuel or Peter Carr, sons of Martha Jefferson, Thomas Jefferson's sister. Present-day members of the Carr family strongly question this explanation for the Hemings children's light colour and alleged striking resemblance to Jefferson.

The controversy has been discussed at great length by historians, journalists and others, but the paucity of objective evidence has made a wide range of opinions seem plausible to various commentators (Brodie, 1974; Miller, 1977; Ellis, 1997; Gordon-Reed, 1997). We believed that additional evidence would be useful, and set out to determine whether genetic data could contribute to the subject.

2. The special properties of Y-chromosomal DNA markers

DNA fingerprinting and profiling techniques are now used routinely by forensic laboratories to investigate genetic relationships (see article on "DNA fingerprinting"). Could they be used to investigate the ancestry of present-day descendants of Sally Hemings's children? Conventional techniques use autosomal markers, which are ideal for testing whether two independent DNA samples come from the same individual, since they should all match in two samples from the same person, but are different in different people. They are also well suited to determining close relationships such as paternity since a child carries half of the father's markers. However, they are less useful for determining distant relationships. For example, H21 (see **Figure 2**), a descendant of Sally Hemings's last son Eston, would carry either none of Jefferson's autosomal markers (if Jefferson was not Eston's father) or 1 in 32 (if Jefferson was Eston's father), assuming that there are no other complications to the descent. Living descendants of Jefferson would similarly carry only a small proportion of his autosomal markers. Random individuals share some markers by chance, so it would be extremely difficult to detect the small excess of marker sharing due to such a distant relationship. Conventional techniques would, therefore, not be very useful in this case.

Fortunately, two regions of the genome have different properties and can be used to investigate more distant

Figure 1 President Thomas Jefferson and descendants of Sally Hemings. (A) President Thomas Jefferson (1743–1826); (B) Beverly Jefferson (1838–1908), son of Eston Hemings Jefferson and grandson of Sally Hemings; (C) Carl Smith Jefferson (1876–1941), Beverly's son; (D) William Magill Jefferson (1907–1956), Carl's son.

relationships. Mitochondrial DNA (mtDNA) is passed by the mother to all her children, but is probably never transmitted by the father. Consequently, it does not recombine and can be used to trace maternal lineages. In a similar way, the Y chromosome, which determines male sex, is present only in males and is passed on to half of their children, who are therefore sons. Apart from its tips, it does not recombine and thus the original set of markers it carries remains the same; throughout this article, we will use "Y chromosome" to refer to the major non-recombining part of the Y chromosome. Jefferson's children would not carry his mtDNA, but his sons would carry his Y chromosome. Thus, if male-line descendants or relatives of Jefferson, the Carrs and of Sally Hemings's children are available, Y-chromosomal DNA analysis can be used to compare their Y chromosomes. The extent to which this can help to determine whether Jefferson was the father of Sally Hemings's children is discussed below.

More than 99.9% of Y-chromosomal DNA sequences are the same in unrelated males. In recent years, several laboratories have searched for the <0.1% of sequences that differ, and now a wide variety of polymorphic Y-chromosomal DNA markers has been found. These markers are of several different types, with different properties. Base substitutions where, for example, a C is replaced by a T, and insertions occur rarely (perhaps about 1 per 10^7 generations at any one site) and are, therefore, stable. These "biallelic markers" are scored as 0 (ancestral) or 1 (derived) and those that are used

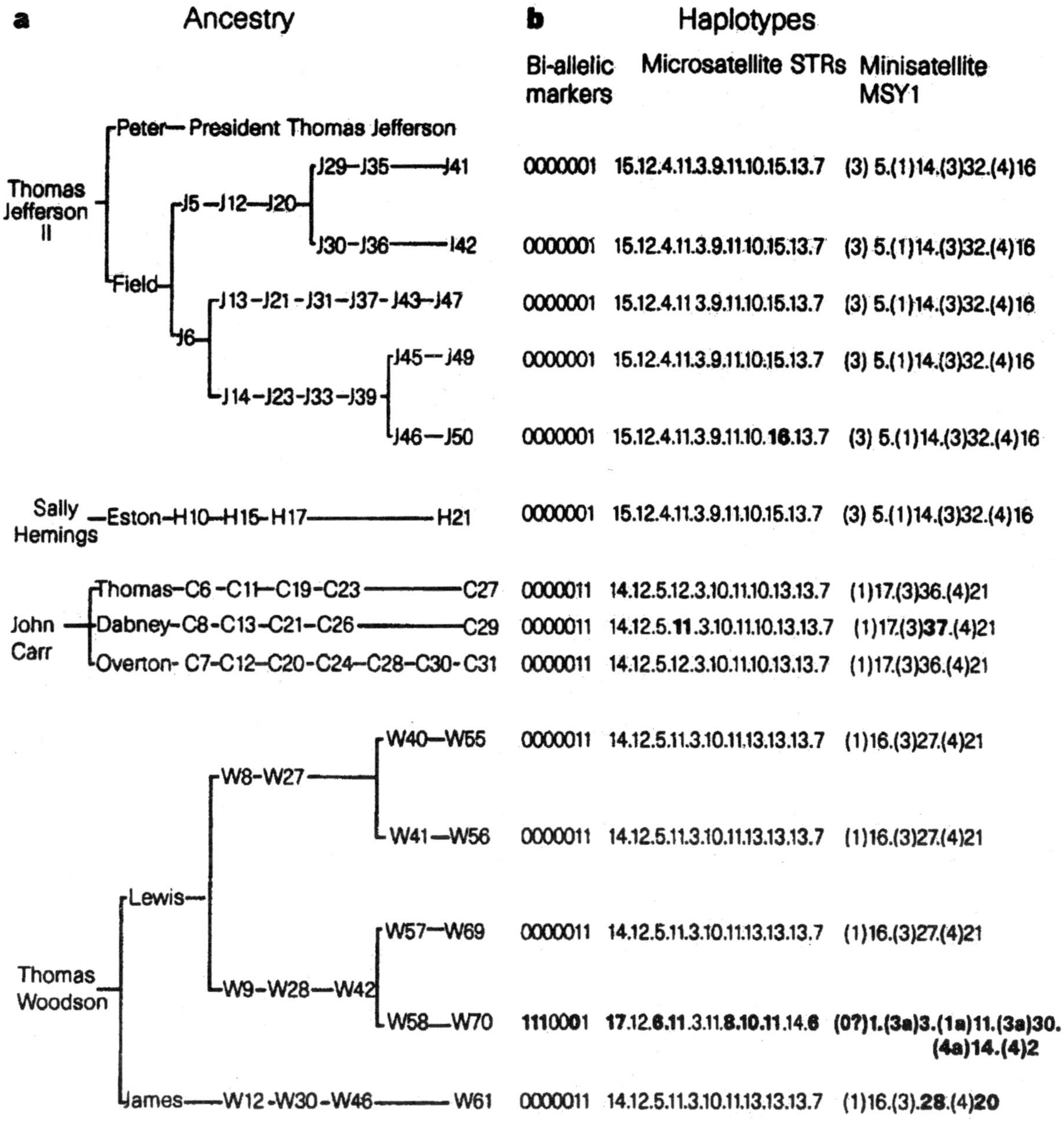

Figure 2 Y-chromosomal DNA haplotypes in the Jefferson, Hemings, Carr and Woodson families. Part a (Ancestry) shows simplified family trees, including only relevant males. Historical figures are named; more recent figures are identified by codes such as J41. Part b (Haplotypes) shows the Y-chromosomal DNA haplotype of each individual in the right-hand column of part a. The left-hand side of b gives the codes for seven biallelic markers. The ancestral state is scored as 0, the derived state as 1. The markers used were YAP, *SRY*-8299, sY81, LLY22g/*Hin*dIII, Tat, 92R7 and *SRY*-1532; sources of the markers are given by Foster *et al*. (1998). The central part of b gives the codes for 11 microsatellite loci (also known as short tandem repeats, STRs). Each number is the number of copies of the microsatellite repeat present at a locus. The loci were *DYS19*, *DYS388*, *DYS389A*, *DYS389B*, *DYS389C*, *DYS389D*, *DYS390*, *DYS391*, *DYS392*, *DYS393* and *DXYS156Y*. The right-hand side of b gives the code for the minisatellite MSY1. The number within the brackets indicates the sequence variant type of the minisatellite repeat units (1, 3 or 4); the number after it is the number of copies of these units. Further descriptions of these markers are given in the text. The complete Y-chromosomal haplotype for each individual thus consists of seven biallelic markers, 11 microsatellites and the minisatellite. Haplotypes from different members of the same family are usually similar to one another; deviations from the usual pattern for that family are shown in bold type. (Reprinted by permission from *Nature* (Foster *et al*., 396: 27–28) © 1998 Macmillan Magazines Ltd.)

are present in quite large subsets of the population. Thus, chromosomes that share a biallelic marker or set of biallelic markers may be closely or distantly related, but chromosomes that differ at these loci are not closely related. These markers also provide some information about the geographical origin of Y chromosomes.

Microsatellites are tandemly arranged repeats of a short (usually 2–5 base pairs [bp]) unit. One individual might have, for example, 14 copies of the repeat, and another 15 copies; these would be scored as 14 and 15, respectively. Microsatellite mutation rates are high, at about 1 per 500 generations per locus (Heyer *et al.*, 1997). Consequently, they are much more variable than biallelic markers and a set of 11 or so microsatellites can be specific to a family. However, the mutation rate is so high that occasional changes (usually of single units) can be detected even within a family, and this possibility must be taken into account when family relationships are being assessed. Minisatellites are repeats of longer units, 25 bp in the only Y-specific minisatellite known, MSY1 (Jobling *et al.*, 1998). Both the sequence of the units and their number can vary, and these are described by the MSY1 code. For example, the code (3)5.(1)14.(3)32.(4)16 refers to 5 repeats of sequence type 3 units, followed by 14 repeats of type 1 units, 32 repeats of type 3 units and finally 16 repeats of type 4 units. MSY1 mutation rates are even higher than those of microsatellites, about 1 per 10–50 generations for the whole locus, so again MSY1 can be family specific, but the possibility of mutation has to be taken into account. Any combination of markers on the same chromosome is called a "haplotype".

3. The analyses

Thus, we set out with the help of Winifred Bennett, Herbert Barger, Corinne Carr Nettleton, Byron Woodson and Judith Justus to trace male-line descendants of Jefferson, Thomas Woodson, Eston Hemings and the Carrs. Jefferson had no male line of descent because his only son died in infancy, but five male-line relatives, descendants of his paternal uncle Field Jefferson, were available. Five descendants of Thomas Woodson and one of Eston Hemings were enlisted. Single descendants of each of the Carr brothers – Thomas, Dabney (father of Peter and Samuel) and Overton – also participated. After informing each participant of the purpose and possible consequences of the study, and obtaining formal written permission, a sample of venous blood was obtained and assigned a randomly selected code number. After extraction of DNA from the blood samples in a laboratory at the University of Virginia in Charlottesville, aliquots of each sample, each bearing the same code number, were tested with 7 biallelic Y markers (in Oxford), 11 microsatellites (Leiden) and the minisatellite MSY1 (Leicester). When testing was complete, the sample codes were broken and the Y marker results compared with the genealogies (Foster *et al.*, 1998; see **Figure 2** for a reproduction of the data presented in the original paper).

The five Jefferson family members share the same haplotype with the biallelic markers and MSY1. Four of them are identical at all of the microsatellite loci and the fifth, J50, differs by a single repeat at one locus. This difference is likely to represent a mutation and the consensus haplotype can be considered as the Jefferson family haplotype. Similarly, the three Carr family members have closely related haplotypes, with one (C29) showing a single microsatellite unit difference at one locus and a single MSY1 unit difference, again probably due to mutations. The consensus haplotype can therefore be considered as the Carr family haplotype. It differs substantially from the Jefferson family haplotype, even showing a biallelic marker difference, and thus the two families can be readily distinguished. It is, therefore, possible to ask how the Hemings descendants and the Woodsons compare with these two haplotypes.

Four of the five Woodsons have closely related haplotypes, with one of the four (W61) showing two MSY1 unit differences, probably representing mutations. The fifth Woodson (W70) has a very different haplotype, which is unlikely to have arisen by mutation and indicates illegitimacy or adoption in that line of descent. The consensus Woodson family haplotype differs substantially from the Jefferson family haplotype, including one biallelic marker difference. It is closer to the Carr family haplotype, but still differs from this by at least three microsatellite steps and ten MSY1 steps. The single Eston Hemings descendant (H21) has an identical haplotype to the consensus Jefferson family haplotype. The work is summarized in a simple form in **Figure 3**. (*Editor's note*: The list of Sally Hemings's children shown in **Figure 3** appears to have been based on obsolete information. According to the historians at Monticello, Sally Hemings did **not** have a child named Edy. The list they believe to be correct is: Thomas 1790–1879, Harriet (1795–1797), Beverly (1798–1822+), Unnamed daughter (1799–1800), Harriet (1801–1822+), Madison (1805–1877) and Eston (1808–1856).)

4. Conclusions

What conclusions can be drawn from these findings? We will consider the Woodsons first. Similar haplotypes are found in three descendants of Lewis Woodson, one of Thomas Woodson's sons and one descendant of James Woodson, another of Thomas Woodson's sons. It is therefore likely that this represents the haplotype of Thomas Woodson himself. Alternative explanations are possible, but seem less likely. For example, the same man, or male-line relatives, could have illegitimately fathered both of the two sons, or they could both have been adopted from the same family. If Thomas Woodson did have the Woodson family haplotype and Jefferson had the consensus Jefferson family haplotype, Jefferson was not the father of Thomas Woodson (see **Table 1**). Similar considerations exclude the Carrs. Nevertheless, the biallelic markers, together with the MSY1 code, suggest a European origin for this haplotype.

Since Eston Hemings's descendant carries the Jefferson family haplotype, the Carrs are excluded as Eston's father and there are two questions to consider. First, did this chromosome come from a Jefferson family member or an unrelated male? Secondly, if it came from a Jefferson, which one? The first question can be addressed using genetic data. The Jefferson family haplotype has never been observed outside this study. It is therefore not common, and upper

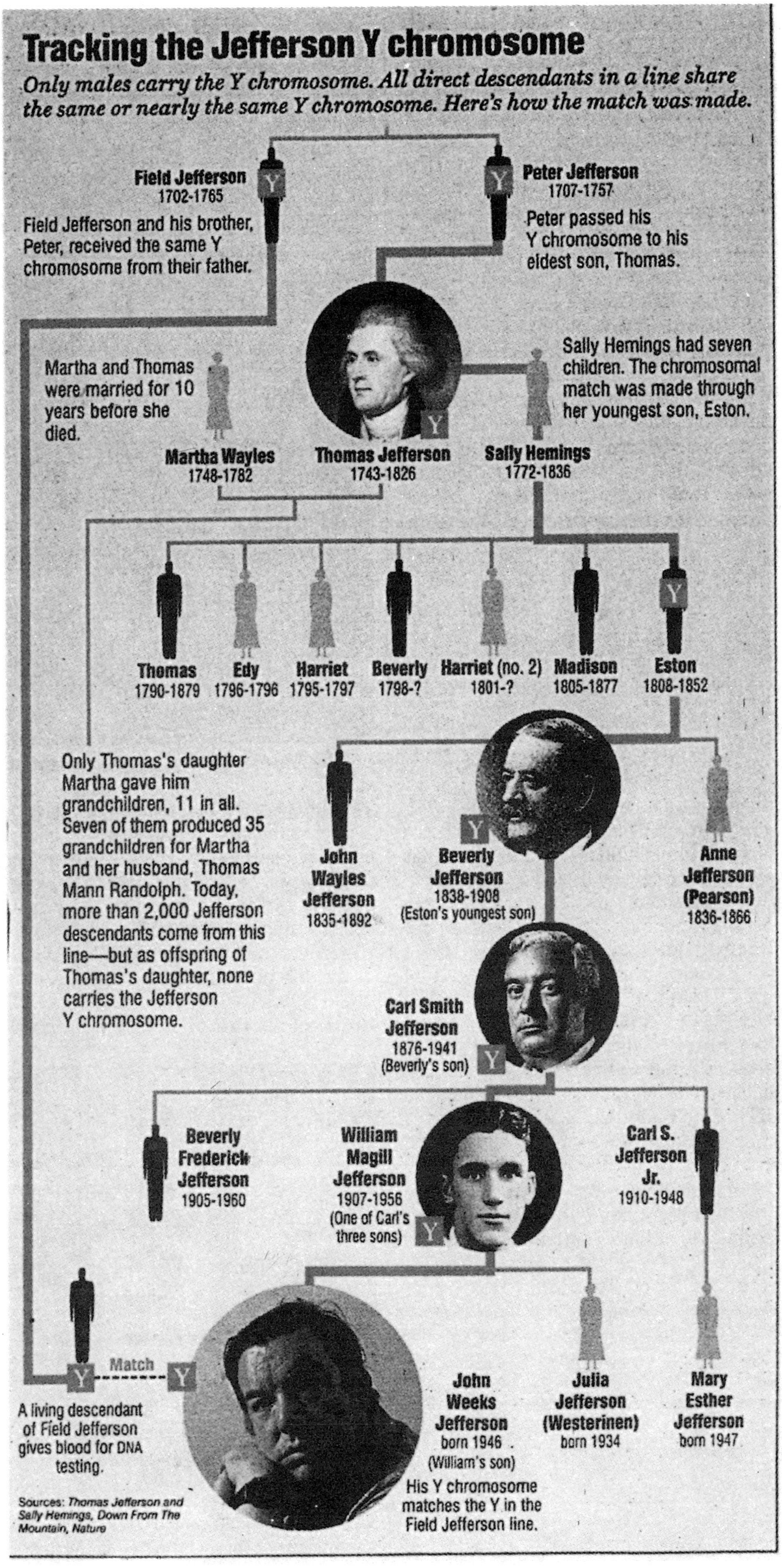

Figure 3 Tracking the Jefferson Y chromosome. A simplified version of the family tree, showing the transmission of the matching Y chromosomes (see also **Plate 16** in the colour plate section). (Reproduced from USNEWS, 1998.)

Table 1 Conclusions, based on both genetic and historical data.

Thomas Jefferson was the ancestor of Thomas Woodson's living descendants	**very unlikely**
Thomas Jefferson was Thomas Woodson's father	**very unlikely**
Thomas Jefferson was the father of a child of Sally Hemings conceived in France	**not ruled out**
Thomas Woodson was the child Sally Hemings conceived in France	**very unlikely**
Thomas Jefferson was Eston Hemings' father	**very likely**
a Jefferson other than Thomas was Eston Hemings' father	**unlikely**
a Carr was Eston Hemings' father	**very unlikely**

limits to the estimates of its frequency depend on the number of males tested – about 1200 have been tested with the microsatellites and 690 with MSY1. It should be noted that different sets of males were tested with these markers, and the combination microsatellite–MSY1 haplotype will be even rarer than the separate microsatellite and MSY1 haplotypes. Nevertheless, using an estimate of 1 in 1200 for the frequency of the Jefferson haplotype in the population, Donnelly (in Foster *et al.*, 1998) calculated the ratio of the likelihood of obtaining the data above under the hypothesis that Jefferson was the father of Eston Hemings, compared with the likelihood of obtaining the same data under the hypothesis that an unrelated male was the father. This likelihood ratio was about 100:1. This is a very conservative estimate and the true ratio may be much higher. It therefore seems unlikely that an unrelated male contributed this Y chromosome.

If, then, the chromosome came from a Jefferson, who was it? It must be appreciated that the Y-chromosomal genetic analysis, which was useful here precisely because all male-line family members share the same Y-chromosomal haplotype (apart from mutational changes, as noted), in the same way cannot distinguish between the different family members (Abbey, 1999; Davis, 1999; Foster *et al.*, 1999). Historical evidence is needed for this.

The historical evidence seems to point most strongly to Thomas Jefferson for several reasons. It is well documented that Thomas Jefferson was present at Monticello when Eston Hemings was conceived and, indeed, was available to father all of Sally Hemings's known children (Gordon-Reed, 1997). Although Jefferson's younger brother Randolph and his sons lived at, or relatively close to, Monticello at various times, possibly when Eston Hemings was conceived, there is no documentation that any of them were actually there at the time. It should also be noted that Jefferson's grandchildren, Thomas Jefferson Randolph and Ellen Randolph Coolidge, implicated the Carr brothers, rather than Randolph or his sons (Gordon-Reed, 1997).

Many historians also point to the fact that Sally Hemings's children and a few of her close relatives were the only slaves that Jefferson freed as evidence for Jefferson's paternity of the children. Beverly and Harriet Hemings, Sally's two older children, were helped to "escape" or "walk away" from Monticello and make their way to Washington. Jamey Hemings, Sally Hemings's nephew, was also allowed to escape. In his Will of 1826, in addition to Madison and Eston Hemings, Jefferson freed John Hemings, Burwell Colbert and Joe Fossett. The latter three were relatives of Sally Hemings, and also had been important craftsmen at Monticello for many years. The significance of these emancipations is discussed by Gordon-Reed (1997).

Thus, we conclude that the genetic evidence shows that a Jefferson is very likely to have been the father of at least one of Sally Hemings's children, and the historical evidence makes Thomas Jefferson the most probable family member.

REFERENCES

Abbey, D.M. (1999) The Thomas Jefferson paternity case. *Nature* 397: 32

Brodie, F.M. (1974) *Thomas Jefferson: An Intimate History*, New York: Norton and London: Eyre Methuen

Callender, J.T. (1802) Richmond Recorder, September 1, 1802, p. 22, quoted in A. Gordon-Reed (1997), *Thomas Jefferson and Sally Hemings: An American Controversy*

Davis, G. (1999) The Thomas Jefferson paternity case. *Nature* 397: 32

Ellis, J.J. (1997) *American Sphinx: The Character of Thomas Jefferson*, New York: Knopf

Foster, E.A., Jobling, M.A., Taylor, P.G. *et al.* (1998) Jefferson fathered slave's last child. *Nature* 396: 27–28

Foster, E.A., Jobling, M.A., Taylor, P.G. *et al.* (1999) The Thomas Jefferson paternity case. *Nature* 397: 32

Gordon-Reed, A. (1997) *Thomas Jefferson and Sally Hemings: An American Controversy*, Charlottesville: University Press of Virginia

Heyer, E., Puymirat, J., Dieltjes, P., Bakker, E. & de Knijff, P. (1997) Estimation of Y chromosome specific microsatellite mutation frequencies using deep rooting pedigrees. *Human Molecular Genetics* 6: 799–803

Jobling, M.A., Bouzekri, N. & Taylor, P.G. (1998) Hypervariable digital DNA codes for human paternal lineages: MVR-PCR at the Y-specific minisatellite, MSY1 *DYF155S1*. *Human Molecular Genetics* 7: 643–53

Life Among the Lowly, No 1, *Pike County (Ohio) Republican*, March 13, 1873 quoted in A. Gordon-Reed (1997), *Thomas Jefferson and Sally Hemings: An American Controversy*

Miller, J.C. (1977) *The Wolf by the Ears: Thomas Jefferson and Slavery*, New York: Free Press

See also **DNA fingerprinting** (p.803); **Forensic DNA analysis of the last Russian royal family** (p.827)

FORENSIC DNA ANALYSIS OF THE LAST RUSSIAN ROYAL FAMILY

Thomas J. Parsons, Suzanne M. Barritt and Mark J. Wadhams
Armed Forces DNA Identification Laboratory, Office of the Armed Forces Medical Examiner, Armed Forces Institute of Pathology, Rockville, Maryland, USA

1. Introduction

The case of Tsar Nicholas Romanov II and his family has been called the greatest forensic mystery of this century. The fate of the last Russian royals has now been made clear through a multidisciplinary forensic investigation, in which state-of-the-art DNA testing provided the final verdict. The case highlights the strengths of two DNA identification methods, targeting nuclear and mitochondrial loci, that are based on amplification of sample DNA through polymerase chain reaction (PCR). As a side note, the case also provides an interesting picture of various reactions to scientific surety, when it pertains to questions involving sentiment and political importance.

After the Bolshevik revolution in 1917, Tsar Nicholas II, the Empress Alexandra and their children Olga, Tatiana, Marie, Anastasia and Alexi (**Figure 1**) were held under arrest in the Impatiev House in Yekaterinburg, Siberia. In July 1918, it was announced that Nicholas had been shot, but there was no declaration regarding the fate of the family and no bodies were released. In 1919, a monarchist investigator determined that a slaughter occurred at the Impatiev House, and concluded that all family members had been executed and the bodies burned. However, disinformation released by the Leninist government in the ensuing years fuelled eager public speculation, spawning a great variety of reports and rumours that some or all of the Romanovs survived.

A breakthrough occurred in 1978, when two amateur investigators obtained a copy of a report that Alexander Yurovsky, the chief executioner of the royal family, filed with the Soviet government. In it, he detailed a brutal massacre. The family, their personal physician and three servants were awakened in the night and ordered into a basement room, where they were arranged, ostensibly for a photograph. Instead of a photographer, a firing squad entered and brought the group down in a barrage of bullets. Adding to the horror was the amazing circumstance that jewellery, sewn for safe keeping into the corsets of the women, served as body armour and deflected bullets. The execution was completed using rifle butts and bayonets. Yurovsky described an ensuing series of bumbling manoeuvres to hide the bodies from advancing White Russian forces. This culminated in a hastily dug grave in a remote swampy area, where all but two of the bodies were buried. According to Yurovsky, the two other bodies, those of the royal heir Alexi and a female (whose identity was uncertain), were burned nearby.

Yurovsky's report also gave the investigators clues regarding the burial site, which they found and clandestinely unearthed in 1979. Realizing the magnitude of their find, and fearing reprisal from the Soviet government, they reburied the remains and said nothing (a detailed account of the discovery and subsequent investigations can be found in Massie, 1995). After the disintegration of the Soviet Union,

Figure 1 The Romanov royal family photographed in 1914 (listed from left to right): Olga, Marie, Nicholas, Alexandra, Anastasia, Alexi and Tatiana. © Austrian Archives/CORBIS.

the location of the grave was reported and an official investigation was undertaken in 1991. Nine skeletons were uncovered, with evidence of brutally violent deaths. An international team of forensic experts conducted a detailed, multidisciplinary examination of the grave and skeletons, and concluded that the remains were unquestionably those of the royal family and entourage. The evidence was consistent with Yurovsky's account: two skeletons were not recovered, those of Alexi and one of the daughters. Forensic anthropologists remain divided whether the missing Grand Duchess is Anastasia or Marie.

2. DNA testing, round one: British Forensic Science Service

An investigation of this importance required that all available approaches be brought to bear. The decision was made to attempt recovery of DNA information from the skeletal remains, in order to test independently the conclusions reached by the forensic anthropologists. It was arranged for Dr Pavel Ivanov, a leading Russian expert on DNA forensics, to bring the samples to the laboratory of the British Forensic Science Service (FSS), headed by Dr Peter Gill. The age and condition of the bones suggested that any surviving DNA would likely be degraded into small fragments and, possibly, that very little DNA could be recovered. This requires techniques and safeguards similar to those employed in the ancient DNA field (for overviews see, for example, Hummel & Herrmann, 1994; Handt *et al.*, 1996; Parsons & Weedn, 1997, and the article on "DNA from museum specimens"). The approach is based on the use of PCR, which permits particular DNA fragments to be exponentially amplified to high abundance, from an original sample that may contain as little as a single copy of the target fragment.

As a result of the great sensitivity of PCR, it is highly prone to contamination by DNA that is not from the original source. Great care must be taken that specimens are properly handled and cleaned, and that no foreign DNA is introduced at any point in the analysis. The forensic and ancient DNA fields have established sets of control experiments and other guidelines for ensuring authenticity of results (Hummel & Herrmann, 1994; Handt *et al.*, 1996; Parsons & Weedn, 1997), the proper implementation of which requires extensive scientific experience and facilities dedicated to such work. The FSS laboratory is one of several state-of-the-art laboratories worldwide that have been designed specifically for contamination-free forensic DNA analysis.

The PCR-based testing targeted both nuclear and mitochondrial loci. These elements have quite different characteristics, and serve different forensic purposes. The Romanov case is an example where the strengths of nuclear and mitochondrial DNA testing complemented each other to produce an extremely informative genetic portrait – notwithstanding a problem which arose that required a second round of DNA testing for final resolution.

Mitochondrial DNA (mtDNA) is a small (16 569 base pair [bp]) circular genome that is found in cytoplasmic mitochondria and is present at ≥1000 copies in most cells. This abundance imparts a correspondingly higher likelihood of PCR recovery of mtDNA sequences compared with nuclear genes, which are present in only two copies per cell. mtDNA, therefore, is a common target for ancient DNA studies or forensic cases where DNA is highly degraded (Holland *et al.*, 1993, 1995). Additionally, because sperm contribute essentially no mtDNA to the fertilized egg, mtDNA is maternally inherited. Thus, barring mutation, an exact sequence match is expected between even distant maternal relatives. Finally, the non-coding mtDNA control region (sometimes called the D-loop) is highly variable in human populations. By DNA fingerprinting standards, the identification power of mtDNA is moderate: on average, there is approximately a 1% chance that two random individuals will match in mtDNA sequence (Piercy *et al.*, 1993). However, this is sufficient to be quite useful in many forensic cases, especially those where reference samples are available only from maternal relatives.

Nuclear DNA testing in the Romanov case was performed by single-tube PCR amplification of a battery of five single-locus short tandem repeat (STR) loci: HUMTHO1, HUMVWA31, HUMF13A1, HUMFES/FEPS and HUMACTPB2. STR typing is emerging as the prominent method in DNA fingerprinting because, as it combines the extreme sensitivity of PCR with a high statistical power of discrimination between individuals, combinations of STR loci can provide profiles that are statistically unique in the world (Hammond *et al.*, 1994). Among nuclear DNA loci, STRs are targets of choice in cases involving degraded DNA, because the size of the target template is small (in this case, with a maximum possible size of 310 bp). Since individuals have two copies of each nuclear locus, one deriving from each parent, STRs are useful for determining paternal as well as maternal relationships. However, the biparental inheritance causes the shuffling of nuclear alleles between generations and does not permit useful identity comparisons with distant relatives. In addition to STRs, the nuclear DNA testing in the Romanov case included analysis of a fragment of the amelogenin gene. This gene has different-sized alleles on the X and Y chromosomes, permitting PCR-based sex determination.

DNA was extracted from powdered bone using standard techniques, with controls and modifications designed to minimize any chance of contamination (Gill *et al.*, 1994). Previously, forensic STR testing had not been performed on bone more than 15 years old, and the prospects for success with bones buried for three-quarters of a century were uncertain. Only small quantities of human DNA were recovered (50–100 picograms per gram of bone), but these did permit amplification of STR and amelogenin loci, albeit at the limit of detectability. PCR products were visualized and typed on polyacrylamide gels using automated fluorescence detection (Kimpton *et al.*, 1993). The results provided independent confirmation of the expected sex and family relationships of the individuals: STR profiles of the presumed Romanov children were appropriate combinations of alleles present in the profiles of the presumed Tsar and Tsarina, and the recovered children were all females. However, the nuclear DNA typing could only show relatedness among the family group. It could not show which of the daughters were present (this would require some sort of reference sample from the daughters themselves), nor could it provide independent evidence that the family was in fact the Romanovs. For the latter, mtDNA analysis would be required.

Living maternal relatives of both the Tsar and Tsarina provided reference blood samples for mtDNA sequence analysis. Tsarina Alexandra is maternally related to Prince Philip, Duke of Edinburgh (**Figure 2A**). The two mtDNA hypervariable regions were amplified and the mtDNA sequence was determined from Prince Philip, the presumed Tsarina and the presumed children. The sequences all matched exactly and did not match any other sequences in a database of 307 random Caucasian sequences, providing a likelihood ratio of more than 300:1 favouring the identification. Likewise, mtDNA hypervariable sequences from the presumed Tsar were compared with those of two known maternal relatives of the Tsar (**Figure 2B**). In this case, the living relatives matched each other, but both differed from the presumed Tsar at a single base position. When examined closely, the sequence of the presumed Tsar appeared to have a mixture of bases at position 16169, with C predominating over T (the relatives had only T at position 16169). In theory, such a mixture could be due to a contaminating sequence in the DNA extract, but there was no evidence of contamination at other base positions in the control experiments or in the nuclear DNA testing from the same extracts.

The team concluded that they had encountered heteroplasmy – a genuine mixture of mtDNA sequence types within an individual. However, such heteroplasmy had been seen very rarely, and calculations of the significance of the DNA evidence had to take into account the occurrence of a mutation within the Tsar's lineage that caused the heteroplasmy. In the end, the team estimated only a 98.5% probability that the remains were the Romanovs (based solely on the mtDNA evidence). This level of certainty proved insufficient to withstand the partisan debate and public pressure that followed. The Russian authorities decided to postpone the final verdict until additional information could be obtained relating to the occurrence of heteroplasmy in the mtDNA lineage of the Tsar.

3. DNA testing, round two: Armed Forces Institute of Pathology

Heteroplasmic mtDNA mixtures can be passed between generations (Parsons *et al.*, 1997), so it seemed likely that the Tsar received the heteroplasmy from his mother. However, a bottleneck mechanism in mtDNA transmission generally causes mixtures to segregate to one type or the other over a small number of generations. The best recourse, then, for independent confirmation of heteroplasmy within the Tsar's lineage was to analyse a much closer maternal relative. After much negotiation, the remains of the Tsar's younger brother Georgij, who died of tuberculosis in 1899, were exhumed from Sts Peter and Paul Cathedral in St Petersburg. In 1995, Dr Ivanov brought a femur and tibia from Georgij Romanov, along with additional bone samples from the presumed Tsar, to the US Armed Forces DNA Identification Laboratory (AFDIL) at the Armed Forces Institute of Pathology (AFIP). Like the FSS, AFDIL is a high-technology facility designed specifically for mtDNA testing. The track record of AFDIL is based on extensive experience in identifying degraded skeletal remains of US military service personnel missing from previous military conflicts, such as the Vietnam and Korean wars.

The samples from Georgij Romanov were analysed first, followed by independent analysis of bones from the putative Tsar to confirm results of the previous testing. The mtDNA sequence from the presumed Tsar Nicholas II matched the sequence obtained previously, including the heteroplasmic mixture at position 16169. The sequence from Georgij Romanov matched the sequence from the putative Tsar: both shared six differences from the standard mtDNA sequence (Anderson *et al.*, 1981) and, importantly, Georgij Romanov also had a heteroplasmic mixture at position 16169 (Ivanov *et al.*, 1996). This established heteroplasmy in a known close relative of the Tsar and eliminated the discrepancy in the mtDNA testing. Interestingly, the ratio of C:T at position 16169 differed between the Romanov siblings: the Tsar had approximately 30% T and 70% C, while Georgij Romanov had 60% T and 40% C (see **Plate 15**). This indicates that the Tsar's mother was also heteroplasmic

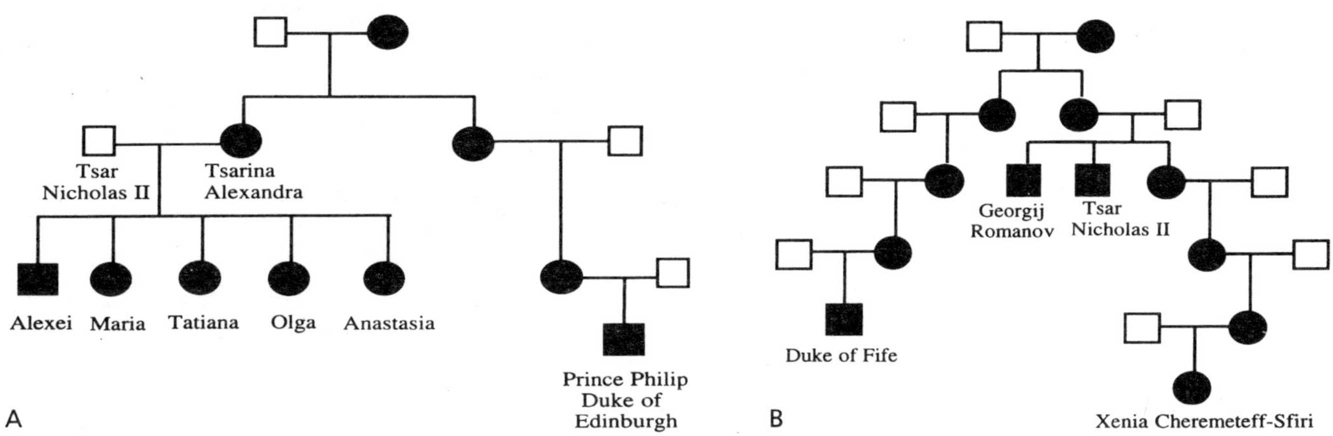

Figure 2 (A) Maternal lineage of Tsarina Alexandra, showing relationship to HRH Prince Philip who provided a reference blood sample. (B) Maternal lineage of Tsar Nicholas II, indicating individuals whose mtDNA sequences were determined.

and passed different subsets of her mtDNA pool to her sons, in accordance with the bottleneck theory of mtDNA transmission.

In addition to resolving what had been viewed by some as a problem in the mtDNA analysis, determination of heteroplasmy in Georgij Romanov actually provided greater statistical surety regarding the significance of the match. Not only did the sequence of Georgij Romanov and the presumed Tsar match throughout the entire sequence (matching only one other individual in a random database of 307 Caucasians), but they shared a genetic anomaly. Using a rough but conservative estimate of the frequency of heteroplasmy, it was possible to estimate that the total mtDNA data (from both the Tsar and Tsarina) favoured the authenticity of the remains by a likelihood ratio of greater than one hundred million to one. Given that anthropological and circumstantial evidence was also considered conclusive that the skeletons are those of the Romanovs, no reasonable scientific doubt now exists regarding the identification.

4. Anna Anderson: the Romanov claimant

Since the disappearance of the Romanovs, there have been numerous individuals claiming to be one of the lost children – or, as years go by, grandchildren – of Nicholas II and Alexandra. Interestingly, even the most preposterous claimants seem able to attract at least some ardent supporters. The most famous claimant was Anna Anderson, who surpassed her lesser rivals and convinced a great many people, including members of the Romanov lineage, of her claim to be the youngest Romanov daughter, Anastasia (**Figure 3**).

The first record of the woman who came to be known as Anna Anderson is from 1920, when she was hospitalized after jumping off a bridge into the Landwehr Canal in Berlin. While hospitalized, she refused to reveal her identity and was soon transferred to a mental institution. After several months, she encountered an article about the Romanovs and informed a nurse that she was the missing Anastasia. This initiated a life-long controversy, and a complex story of interviews, investigations and court hearings, that left many firmly convinced that she was Anastasia and others equally adamant that she was a clever and/or deluded impostor (Kurth 1983; Massie, 1995). There was, in fact, an alternative hypothesis regarding Anna Anderson's identity. A woman named Doris Wingender reported that she recognized Anderson as her former roommate, Franzisca Schanzkowska. Schanzkowska was born in 1896 and raised in a poor Polish farming family. In 1916, she worked in a munitions factory where she injured herself and caused the death of a co-worker when she inadvertently released a grenade. Schanzkowska was subsequently plagued with bouts of mental illness, migrating in and out of asylums to recuperate from her trauma, and was eventually taken in by the Wingenders. In 1920, Franzisca Schanzkowska disappeared without a trace, at around the same time that Anna Anderson emerged as Anastasia. Anna Anderson died in 1984 in Charlottesville, Virginia, where she had resided for over 15 years (Kurth, 1983; Massie, 1995).

Anna Anderson's claim seemed to be bolstered in 1991 when the Romanov grave was shown to be missing one of the daughters, possibly Anastasia. Further, the DNA testing of the Romanovs suggested a means for finally establishing her true identity. Although Anderson's body had been cremated, an alternative source for DNA testing was identified; in 1979, portions of Anderson's small bowel were resected, and pathology specimens were preserved in formalin and embedded in paraffin wax for storage at the hospital. Supporters of Anna Anderson's claim arranged for both the FSS and AFDIL to collect sections from the five paraffin blocks at the hospital for DNA testing. Demonstrating that the samples were correctly ascribed to Anna Anderson, microscopic examination of the tested samples matched them to

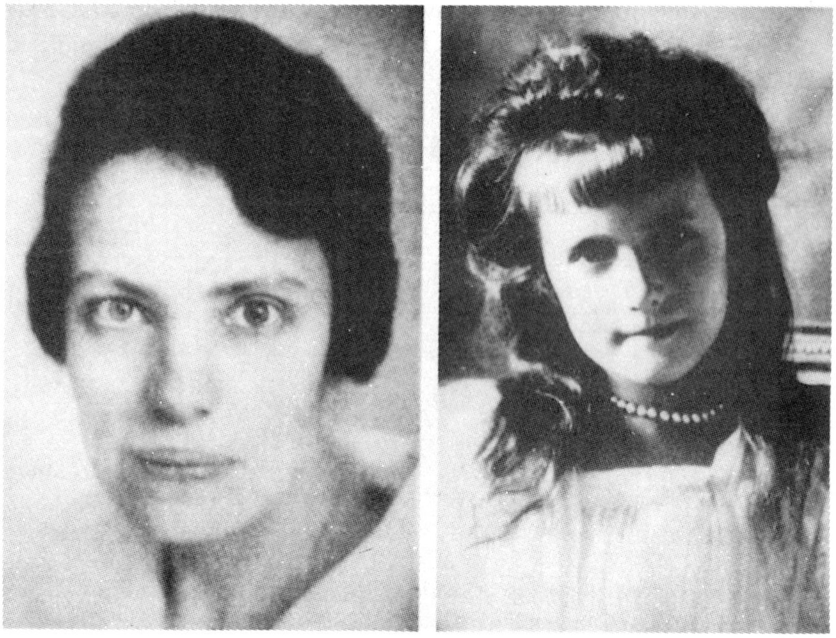

Figure 3 Pictures of Anna Anderson as a young woman (left), and Anastasia (right). © Rykoff Collection/CORBIS.

slides that were made from her samples at the time of surgery and stored in a separate coded archive. Additionally, another source of DNA was discovered in 1992 by a used bookshop owner in North Carolina. Anna Anderson's husband had sold his library to the book store and among the boxes of books was an envelope labelled "Anastasia's hair" containing hair presumably recovered from her hairbrush. These hair specimens were sent to Dr Mark Stoneking of Pennsylvania State University (PSU) in 1994 for mtDNA analysis (Massie, 1995; Gill *et al.*, 1995).

mtDNA sequencing was independently performed on the tissue and hair specimens. Results obtained from all three laboratories (FSS, AFDIL and PSU) gave sequences that matched exactly (Gill *et al.*, 1995). The sequence obtained has not been seen in any other individual in a large forensic database (currently including 1143 sequences, 530 of which are from Caucasians), so the match between the hair and the tissue samples strongly corroborates the presumption that they were from the same individual – namely, Anna Anderson. When the sequence was compared with the mtDNA sequence obtained from HRH Prince Philip, the great nephew of Tsarina Alexandra, there were differences at six nucleotide positions. Thus, Anna Anderson could not be of the same maternal line as Tsarina Alexandra. Additionally, the FSS performed STR analysis on the tissue sections and compared the STR profile with those from the remains of Tsar Nicholas II and Tsarina Alexandra. Four out of the five loci were inconsistent with a parental relationship, again invalidating Anna Anderson's claim (Gill *et al.*, 1995).

The FSS then performed mtDNA sequence analysis of a blood sample from Carl Maucher, a maternal relative of the woman, Franzisca Schanzkowska, who many believed Anna Anderson to be. In this case, there was an exact match. On the basis of the rarity of their mtDNA sequence, the chance that Anna Anderson and Carl Maucher would match if they were not maternally related is less than one in 500. This supports the hypothesis that Anna Anderson and Franzisca Schanzkowska were one and the same (Massie, 1995; Gill *et al.*, 1995).

5. Epilogue

DNA testing has positively identified the skeletons of the Romanov royal family. Replicate and independent testing in several state-of-the-art laboratories effectively rules out considerations of contamination, laboratory error or artefact – as do the results themselves. Spurious results or random contamination will generally produce uninterpretable data or will exclude rather than include the hypothesis of identification. Nevertheless, some (outside of the scientific community) have not accepted the findings. DNA testing has the ability to place the stamp of scientific surety on conclusions that, in many instances, people do not want to believe themselves or have others believe. In such cases, responses seem to be of two general classes: (1) unspecified and groundless allegations of evidence mishandling, investigator incompetence or inappropriateness of methods, and (2) great conspiracy theories involving agents with seemingly infinite knowledge, abilities and resources who have engineered biological samples to mislead investigators. The second response has been put forward by supporters of Anna Anderson unwilling to accept that she could have been of other than aristocratic origin. Both arguments have been employed by interested parties regarding the remains of the Romanovs, and have played a role in delaying a final burial (the ultimate fate of the remains is uncertain at the time of this writing). DNA identity testing continues to expand in use, and it will be interesting to follow the evolution of society's response to this powerful tool for revealing the truth.

The opinions and assertions expressed herein are solely those of the authors and are not to be construed as official or as the views of the US Department of Defense or the US Department of the Army.

REFERENCES

Anderson, S., Bankier, A.T., Barrell, B.G. *et al.* (1981) Sequence and organization of the human mitochondrial genome. *Nature* 290: 457–65

Gill, P., Ivanov, P.L., Kimpton C.P. *et al.* (1994) Identification of remains of the Romanov family by DNA analysis. *Nature Genetics* 6: 130–35

Gill, P., Kimpton, C., Aliston-Greiner, R. *et al.* (1995) Establishing the identity of Anna Anderson Manahan. *Nature Genetics* 9: 9–10

Hammond, H.A., Jin, L., Zhong, Y., Caskey, C.T. & Chakraborty, R. (1994) Evaluation of 13 short tandem repeat loci for use in personal identification applications. *American Journal of Human Genetics* 55: 175–89

Handt, O., Krings, M., Ward, R.H. & Paabo, S. (1996) The retrieval of ancient human DNA sequences. *American Journal of Human Genetics* 59: 368–76

Holland, M.M., Fisher D.L., Mitchell, L.G. *et al.* (1993) Mitochondrial DNA sequence analysis of human skeletal remains: identification of remains from the Vietnam War. *Journal of Forensic Science* 38: 542–53

Holland, M.M., Fisher, D.L., Roby, R.K *et al.* (1995) Mitochondrial DNA sequence analysis of human remains. *Crime Laboratory Digest* 22: 3–8

Hummel, S. & Herrmann, B. (1994) General aspects of sample preparation. In *Ancient DNA: Recovery and Analysis of Genetic Material from Paleontological, Archaeological, Museum, Medical, and Forensic Specimens*, edited by B. Herrmann & S. Hummel, New York and London: Springer

Ivanov, P.L., Wadhams, M.J., Roby, R.K. *et al.* (1996) Mitochondrial DNA sequence heteroplasmy in the Grand Duke of Russia Georgij Romanov establishes the authenticity of the remains of Tsar Nicholas II. *Nature Genetics* 12: 417–20

Kimpton, C.P., Gill, P., Walton, A. *et al.* (1993) Automated DNA profiling employing "multiplex" amplification of short tandem repeat loci. *PCR Methods and Applications* 13: 13–22

Kurth, P. (1983) *Anastasia: The Riddle of Anna Anderson*, Boston: Little Brown, and London: Cape

Massie, R.K. (1995) *The Romanovs: The Final Chapter*, New York: Random House, and London: Cape

Parsons, T.J. &. Weedn, V. (1997) Preservation and recovery of DNA in postmortem specimens and trace samples. In *Forensic Taphonomy*, edited by W.D. Haglund & M.H. Sorg, Boca Raton, Florida: CRC Press

Parsons, T.J., Muniec, D.S., Sullivan, K. *et al.* (1997) A high observed substitution rate in the human mitochondrial DNA control region. *Nature Genetics* 15: 363–67

Piercy, R., Sullivan, K.M., Benson, N. & Gill, P. (1993) The application of mitochondrial DNA typing to the study of white Caucasian identification. *International Journal of Legal Medicine* 106: 85–90

See also **DNA from museum specimens** (p.817)

THE ROLE OF BIOINFORMATICS IN THE POSTSEQUENCING PHASE OF GENOMICS

David Jones
Department of Biological Sciences, Brunel University, Middlesex, UK

1. Introduction

A common misconception is that the end of a genome project is marked by the completion of the entire genome sequence. In fact, the complete sequencing of a genome marks not the end of the genome project, but the start of the most important phase of the project, namely, the postsequencing phase. In this phase of the project, the tasks move from routine sequencing to the important problem of interpreting the huge amount of raw biological information that has been acquired. Generally speaking, postsequencing analysis does not wait until the genome has been completely sequenced, but runs in parallel with the sequencing. This can help to identify sequencing errors quite early on in the overall project.

Although the postsequencing phase of the human genome project is still some years away at the time of writing (2003 being the latest projected date for the completion of the finished human sequence although a working draft was announced in 2000), for a number of simpler organisms (11 different bacteria, yeast and the nematode worm, the postsequencing phase of their respective genome projects has already arrived.

In this article, the role of bioinformatics (the analysis and storage of biological data by computers) in the postsequencing phase of a genome project will be reviewed. Other technologies beyond bioinformatics are also exploited in postsequencing analysis, including mass spectrometry (Roepstorff, 1998), gene expression mapping (Wodicka *et al.*, 1997) and crystallography (Kim, 1998). Of these technologies, however, bioinformatics is by far the most heavily used and potentially most useful aspect of postsequencing analysis.

Genome bioinformatics divides roughly into two main areas of research: first, the detection of open reading frames (ORFs), the sections of the genome sequence that potentially code for expressed proteins; and, secondly, the identification of the function of each of these possible genes.

2. ORF finding

Given a long stretch of contiguous DNA sequence, the first problem faced is to locate the ORFs, and also to identify possible regulatory signals that might control the transcription of these genes. For prokaryotic sequences, this problem is greatly simplified due to the absence of introns in the genome and, for these genomes, various methods have been developed (Borodovsky *et al.*, 1994; Robison *et al.*, 1994) to detect ORFs automatically with very high reliability (rates as high as 95% are claimed by some authors). In addition to these pattern-matching approaches, another powerful means for identifying prokaryotic ORFs is to translate the genome in all six reading frames and to compare the resulting protein sequences with the sequences found in the available protein sequence data banks. Any predicted protein sequence that has high similarity to a known protein sequence relates to a correct ORF in the new genome.

For the genomes of higher organisms, however, the identification of ORFs is complicated by the fact that large fractions of the genes in these genomes are interrupted by sections of non-coding DNA (introns) which are spliced out during transcription. Unfortunately, there are no obvious sequence features that mark the start and end-points of introns, and so very sophisticated algorithms have had to be developed which attempt to recognize the entire structure of the genes. These methods employ various computing techniques, including neural networks (Farber *et al.*, 1992), linear discriminant functions (Solovyev *et al.*, 1995) and computational linguistics (Dong & Searls, 1994).

Not surprisingly, given the fact that identifying genes in the presence of introns is so much harder, the reliability is that much lower. Recent tests (Burset & Guigo, 1996) have shown that the sensitivity of gene-finding software for vertebrate genes can vary in a range of 15–60%, which means that as much as 40% of exons will not be correctly located by such methods even at best. Even worse is the fact that between 10% and 24% of predicted exons will not correspond to any true exons at all.

In practice, laboratories making use of gene-finding software will not rely on a single program. Most commonly, the results from three or four different methods will be considered by a genome annotation specialist when trying to locate genes in genome sequence data. Even with the assistance of these programs, gene location in human DNA is particularly difficult and labour intensive. Given that the reliable identification of genes is a necessary first step in genome analysis, this bottleneck is critical and is a problem that urgently needs a solution before the vast amount of human sequence data is in hand.

As soon as a probable ORF has been located in a genome, the next question usually addressed by means of bioinformatics is the assignment of a putative function to the encoded protein. Ideally, the function of a protein should be ascertained experimentally, either by applying a systematic set of biochemical assays to the expressed and purified protein or by knocking out the particular gene and observing changes to the organism. For the yeast genome, in particular, this work is already in progress. A European consortium of biochemistry laboratories, called EUROFAN, is currently attempting to determine experimentally the function of novel protein coding genes in yeast by gene disruption (Dujon, 1998).

Given that the discovery of gene function by experimental means is a time-consuming and labour-intensive process, it is not surprising that the most widely used approach to elucidating the function of a newly sequenced gene is computational genome analysis. In fact, on searching the published literature in biochemistry, it is found that the most commonly used "technique" in molecular biology is not any standard laboratory technique, such as gel filtration or blotting, but is in fact computational sequence comparison.

Sequence analysis in genome projects is somewhat different from that generally carried out during the analysis of a single protein. Typically, a researcher analysing a single gene already has some idea of the function of the gene in question. The main questions in this case are how the protein relates to other proteins in the sequence databank, what its mechanism might be and perhaps what its secondary or tertiary structure might be. These analyses are performed mostly to plan further experiments, such as site-directed mutagenesis. In computational genome analysis, however, things are somewhat different. For the most part, nothing at all will be already known about each ORF. Concerns about the mechanism and structure of the encoded protein are really secondary when the main challenge is to work out the actual function of the ORF.

3. Homology

Given that there is a virtually infinite number of different possible primary structures, it is reasonable to suppose that it is very unlikely that two proteins with "similar" amino acid sequences will have evolved independently. When such a similarity is detected, therefore, it is taken to indicate that the two proteins must be related and thus share a common evolutionary ancestor. Such related proteins are termed homologous proteins. The term homology is frequently misused, however. If, and only if, two proteins are evolutionarily related (i.e. stem from a common ancestor), may they be called homologous. Similarity and homology are two very different terms and must not be confused. People in the sequence analysis field sometimes loosely talk about "degrees of homology", but this is not the correct usage of the word; either two proteins are homologous, or they are not.

(i) Detecting homology

Over evolutionary timespans, genes are subject to mutation events. In other words, "errors" begin to creep into the nucleotide sequence and, thus, the encoded protein primary structure also becomes slowly altered, generally by one amino acid at a time (although more drastic single modifications can also occur). Not only do point mutations (the substitution of one amino acid for another) occur, a protein sequence may lose some of its amino acids (deletion mutation) or have amino acids inserted (insertion mutation). The fact that two sequences which are to be compared may be of different lengths, and that there is a need to allow for deletions and insertions, makes the optimal alignment (that alignment which gives the closest match, i.e. the smallest number of differences) of the sequences a difficult task. Despite these difficulties, a number of very powerful methods are now in regular use that can solve this problem. Optimal alignment of two sequences is the cornerstone of bioinformatics. As already mentioned, the power of sequence alignment stems from the empirical finding that, if two biological sequences are sufficiently similar, almost invariably they will be descended from a common ancestor.

(ii) Measuring sequence similarity

Although it is not correct to ask how homologous two proteins are, it is quite reasonable to ask how similar they are. Indeed, it is only by considering the degree of similarity that exists between two proteins that any inference of homology can be made.

The simplest measure of similarity is of "sequence identity". Two proteins that have a certain number of amino acids in common when aligned are said to be identical to that degree. So, for example, if they have 43 amino acid residues out of a total of 144 in common they would be said to have a sequence identity of 29.9%. Percentage sequence identity is a very widely used and generally understood measure of overall sequence similarity, but it is a very crude measurement of the degree of similarity between two sequences. Often a number of residues will be replaced by ones with similar physicochemical properties. Such mutations may be termed conservative and one may define various scoring schemes to quantify how similar the two sequences are, taking into account conservative mutations. Such scores will be measures of similarity rather than strict identity.

(iii) Measures of amino acid similarity

When comparing two protein sequences, some method must be used to score the similarity between one residue and another; amino acid substitution matrices contain such values. Most scoring schemes represent the 400 pairs of scores as a 20×20 matrix of similarities where identical amino acids and those with similar properties (e.g. isoleucine/leucine) give higher scores compared with those which have very different properties (e.g. isoleucine/asparagine). Many scoring schemes have been devised, but by far the most widely used scoring matrices are those similar to the ubiquitous "Dayhoff matrix". Dayhoff and co-workers (1978) examined alignments of very similar protein sequences and counted how many times each amino acid was replaced by every other. The result of this type of analysis is a matrix of mutation probabilities. These matrices are frequently called Dayhoff mutation data matrix, or per cent accepted mutation.

4. Methods for sequence alignment

Alignment methods aim to find the minimum number of mutations, insertions and deletions required to convert one sequence into another (the principle of maximum parsimony, or simpler solutions being preferred over more complex ones). Two classes of sequence alignment method are typically distinguished – local and global. In global alignment, the two sequences are assumed to be similar over their entire lengths; an example of this kind of similarity is between myoglobin and haemoglobin. Local alignment is appropriate when the sequences show isolated regions of similarity (e.g. modular proteins that comprise quite distinct functional domains). Local alignment methods have, in fact, been used to demonstrate the modular nature of recently evolved proteins (Altschul & Gish, 1996).

Although maximum parsimony is a useful concept, it does not necessarily guarantee that the most parsimonious alignment corresponds to the biologically correct alignment. For highly similar sequences, it can be assumed that the computed alignment does correspond to the correct biological alignment, but, for remotely related sequences, the "correct" alignment is often not the one which gives the highest score. If structural information is available, then this can be used to improve the accuracy of the alignment, otherwise some human intelligence is needed to adjust the calculated alignment manually. Manual editing of alignments is a very black art indeed!

5. Searching biological sequence data banks

No matter what the goal of a sequence analysis project is, the universal first step in beginning sequence analysis of a newly characterized protein is to perform a data bank search. There are now a number of very widely used methods for comparing a single sequence against a data bank of sequences. The main distinguishing features for these methods are speed and sensitivity. Speed is an obvious requirement for a good data bank search method. Given the large volume of sequences that must be analysed for genome annotation, a fast search method is clearly desirable. Sensitivity (the ability of the method to detect very weak similarities) is clearly another desirable feature for a search program. The more sensitive the method, the more relationships can be discovered between genome sequences and already annotated sequences. Unfortunately, speed and sensitivity are somewhat contrary requirements. Generally speaking, the fastest methods tend to be the least sensitive.

(i) Basic local alignment search tool (BLAST)

BLAST (Altschul *et al.*, 1990) is currently the most popular search method mainly due to its very high speed and reliable statistical properties. The method works by detecting ungapped regions of statistically significant similarity between protein sequences. Recent developments of the BLAST algorithm (Altschul *et al.*, 1997) now allow gaps to be handled, and this has substantially increased its sensitivity.

(ii) FastA

FastA (Pearson, 1990) is a modern development of the original tuple method developed by Wilbur and Lipman (1983). It was until recently the most popular data bank search method as it offered a reasonable compromise between speed and sensitivity. FastA tends to be a little more sensitive than BLAST (even the newer gapped version), but the statistics applied by FastA are empirically based as opposed to the rigorous mathematical treatment that underpins BLAST.

(iii) Dynamic programming methods

Data bank search methods that use the Smith–Waterman algorithm (Smith & Waterman, 1981) exhibit the highest sensitivity of all. Ideally, dynamic programming would be used for all data bank scanning, but, unfortunately, it is very slow and, thus, until recently, has not often been used. However, the problem of low speed has been overcome recently by the development of specialized computer circuitry for performing dynamic programming sequence comparisons (Fagin *et al.*, 1993).

(iv) Position-specific iteration BLAST (PSI-BLAST)

PSI-BLAST is the newest data bank search method, and is a further development of BLAST (Altschul *et al.*, 1990, 1997). Like newer versions of BLAST, PSI-BLAST is now capable of generating gapped alignments. However, of even more interest are the new position-specific iteration features of PSI-BLAST. Here, rather than searching a data bank of sequences with a single sequence and then finishing, the program builds a sequence profile (i.e. a multiple sequence alignment) based on the initial search results and uses this profile to search the data bank again. If the profile pulls any more sequences out of the data bank, then these new sequences are added to the profile and the revised profile used to search for more sequences. This procedure can be terminated after a fixed number of iterations or can be allowed to continue until no more sequences are detected (according to preset convergence criteria). Although these additional iterations greatly slow down the overall search speed, the increase in sensitivity is more than worthwhile. PSI-BLAST is by far the most sensitive sequence data bank searching method available at the time of writing.

6. Identifying protein function by homology

The first, and most important, step in identifying the function of a newly characterized protein sequence is to compare the sequence with a data bank of known sequences, as already described. The main problem with this lies in the interpretation of the results that can be categorized as follows (in increasing order of usefulness):

1. The new sequence is not significantly similar to any sequence in the data bank.
2. The new sequence matches entries in the data bank, but the function of the data bank sequences themselves is unknown.
3. The new sequence matches a data bank entry with known function, but the statistical significance of the match is only borderline.
4. The new sequence closely matches a data bank entry with known function.

Sequences that fall into category 1 are known as "sequence orphans". There is really little more that can be done to characterize these proteins further without resorting to experimental methods, although it might be worth attempting to use a more sensitive sequence comparison method (see later). Category 2 is also a dead end in terms of functional assignment, although it might be possible to identify possible functionally important residues by observing which amino acids are highly conserved. Functional assignment is only really possible for matches which fall into categories 3 or 4, with category 4, of course, being the ideal case – either the new protein itself or a very close homologue is found in the sequence data bank, and this data bank entry has a known function. In this case, the function of the new protein can simply be assumed to be identical to that of the data bank entry. Category 3 matches provide some insight into the function of the new sequence, but sufficient sequence divergence has obviously occurred for the function of the new protein to be quite different from the data bank entry. In reality, of course, the results obtained will tend to fall somewhere between category 3 and category 4, the degree of confidence in function assignment being directly proportional to the degree of sequence similarity between the new sequence and the data bank entry.

(i) Functional motifs

As the degree of homology between two sequences drops below 30%, it becomes difficult to locate the biologically optimal pairwise alignment between them. The 30% identity cut-off very roughly marks the outermost boundary of the so-called "twilight zone". The twilight zone is simply defined as the region of homology where an optimal alignment between random sequences is found to be no worse or better than the alignment between the trial protein sequences. As already seen, the problem is simply due to the extreme variability of protein sequences even where the structure and function may be seen to be highly conserved. Of course, where two proteins have a common function, it might reasonably be expected that certain key residues are conserved across the evolutionary tree, but how can these key residues be located amid the extreme evolutionary "noise"? The number of key residues may well be very small in comparison to the lengths of the sequences, but methods have been developed to enable such residues to be rapidly located (Bork & Gibson, 1996). These conserved "fingerprint" residues are commonly called "sequence motifs". Sequence motifs can often identify very remote relationships between sequences, far more remote relationships than can be detected by simple pairwise sequence alignment. Of even more interest is the fact that libraries of sequence motifs have now been constructed, of which the most well-known examples are the PROSITE motif library (Bairoch *et al.*, 1997) and the PRINTS library (Attwood *et al.*, 1998). Scanning a new protein sequence against one of these sequence motif libraries can provide significant clues as to its function.

(ii) Identifying the function of yeast ORFs

To illustrate how many of the proteins in a genome can be assigned a function, we will consider the example of the yeast genome. The yeast genome sequence was completed in 1996 and over 6000 ORFs have been predicted. Functional information about yeast genes is surprisingly complete in that about 30% of yeast genes have an experimentally determined function (Dujon, 1996). However, by homology detection with proteins of known function from other organisms, it is possible to extrapolate function for an additional 35% of the ORFs.

(iii) Sources of error in function assignment by homology

Given the apparent simplicity in assigning function by homology, it is not surprising that efforts are underway to automate genome analysis at different levels. The best known of these systems is known as GeneQuiz (Andrade *et al.*, 1997). Although automation of the labour-intensive process of sequence annotation is certainly desirable, it is important to bear in mind the sources of error that complicate the annotation process (Bork & Bairoch, 1996). Although by no means a complete list, the main sources of error are as follows.

Incorrect annotation in protein databases. The fact that a protein in a sequence data bank is described as having a particular function is no guarantee that that protein really has that function. Often the functions of many proteins in the data banks have also been determined by homology to other proteins, and so the annotation errors can multiply rapidly. This kind of annotation error can propagate through a whole family of proteins if the error in the original data bank entry is copied over and over again.

Low sequence complexity. Many protein sequences include regions of highly repetitive sequence patterns. Often these regions relate to coiled-coil structures or flexible domain linkers. Due to the low complexity of these regions, they can match regions of a target sequence with quite a high sequence similarity score (Wootton & Federhen, 1996), and so it might be falsely concluded that the proteins are homologous. These errors can be avoided quite easily by removing regions of low complexity from both the data bank sequences and the target sequence before the data bank search is performed.

Multidomain organization of proteins. Domains in proteins may be considered to be connected units that are independent in terms of their structure, function and folding behaviour. Examination of the database of known amino acid sequences has revealed that certain sequences are repeated many times throughout a number of different proteins. The segmental nature of these proteins indicates that the different domains have had different origins during the evolution of the genome. The domain structure of proteins adds an extra layer of complexity to gene function assignment. For example, in some bacteria, the enzymes involved in tryptophan synthesis are all encoded in different genes, whereas *Escherichia coli* has two bifunctional polypeptides, each the result of the fusion of two genes. A good example of annotation error attributable to domain organization has been highlighted by Eugene Koonin (personal communication). ORF MG262 in *Mycoplasma genitalium* has been repeatedly annotated as relating to DNA polymerase I by automated sequence analysis methods. While this ORF has very high sequence similarity to *E. coli* DNA polymerase I, the region of similarity only spans the N-terminal domain of this protein and, therefore, does not include the polymerase portion (Klenow fragment). Without the necessary Klenow fragment, MG262 cannot possibly encode a protein with DNA polymerase I activity. This kind of error is typical for entirely automatic sequence annotation, but future developments of sequence analysis tools and better data bank curation will hopefully allow this kind of problem to be circumvented.

Non-orthologous gene displacement. In some cases, two proteins may share a significant degree of sequence similarity and yet the function is not conserved. Generally, this happens due to mutations in the functionally important residues resulting in a change of substrate specificity. In many protein families, enzymes and binding proteins with entirely different specificities may be as similar to each other as proteins with the same specificity. Global sequence similarity cannot, therefore, be relied on to assign the function of a new protein with 100% confidence. Sequence motifs can help avoid this kind of error by highlighting whether functionally important residues are conserved between the proteins being compared.

7. Structural genomics

Perhaps the most exciting proposals being made in post-sequencing genome analysis are in the new field of structural genomics. This is essentially the structural equivalent of whole genome sequencing, i.e. rather than attempting to sequence the entire genome, the challenge, in this case, is to determine experimentally the 3-dimensional structure of every protein encoded in the genome (Kim, 1998). Why is the 3-dimensional structure of a protein such a valuable piece of information? The most direct answer to this is that, given the 3-dimensional structure of a protein, it is possible to gain insight into the mechanistic aspects of protein function. For example, by analysing the 3-dimensional structure of an enzyme both with and without a substrate analogue, it is possible to map accurately which amino acid residues form part of the active site and, subsequently, to begin to unravel the mechanism of the enzyme's functioning. It is also possible to plan mutation experiments from a known 3-dimensional structure in order to alter artificially the substrate specificity and thus re-engineer the protein for novel applications.

For genome analysis, however, perhaps the most important immediate benefit from having a 3-dimensional structure for each protein is that it is possible to infer extremely distant evolutionary relationships between proteins. It is now well known that the 3-dimensional structure of proteins is much more highly conserved than the sequence. Even where the amino acid sequences of two related proteins have diverged to such a great extent that sequence analysis methods cannot detect the similarity, the folded conformations of the 3-dimensional structures can still show a remarkable degree of similarity. Therefore, by comparing proteins at a structural level, rather than by sequence analysis, it is possible to look back much further in evolutionary time than might otherwise be possible.

Although attempts are now underway to determine experimentally the structures of a large fraction of the proteins encoded by small bacterial genomes, it will still be a long time before a similar level of structural information becomes available for larger genomes. Despite great technical advances, methods for structure determination are still slow, and it can take anywhere from 1 month to several years to solve the structure of a single protein. The complete sequencing of a number of microbial genomes has particularly highlighted the gap between the number of known protein sequences and the number of experimentally determined protein structures. There is hope that methods for protein structure prediction might offer an alternative means for narrowing this gap and, over recent years, a class of protein structure prediction method known as "fold recognition" has become particularly popular. Fold recognition generally refers to methods that take a protein sequence and attempt to find a structure (from a data bank of known structures) which is most likely to resemble the native conformation of the sequence, based on learned rules of protein structure. Blind testing of these methods has shown that they can be very effective (Lemer *et al.*, 1995); it is, therefore, not surprising that they are now being widely applied to genome analysis. By means of these techniques, around 40–46% of the ORFs in the small bacterial genome *M. genitalium* are currently believed to resemble (at least in part) an already known protein structure.

8. The future

In perhaps 10 or 20 years, every protein encoded in the genomes of many small organisms will be thoroughly characterized, both in terms of function and structure. This will not be the end of postsequencing analysis, however. At this point, it will be possible to start piecing together the complex network of molecular interactions which are fundamental to life itself. Understanding the life processes of a small bacterium will hopefully shed much light on the molecular biology of higher organisms, including *Homo sapiens*, and the prospects for medical advances based on such an all-encompassing information resource must be very bright indeed.

REFERENCES

Altschul, S.F. & Gish, W. (1996) Local alignment statistics. *Methods in Enzymology* 266: 460–80

Altschul, S.F., Gish, W., Miller, W., Myers, E.W. & Lipman, D.J. (1990) Basic local alignment search tool. *Journal of Molecular Biology* 214: 403–10

Altschul, S.F., Madden, T.L., Schaffer, A.A. *et al.* (1997) Gapped BLAST and PSI-BLAST: a new generation of protein database search programs. *Nucleic Acids Research* 25: 3389–402

Andrade, M., Casari, G., de Daruvar, A. *et al.* (1997) Sequence analysis of the *Methanococcus jannaschii* genome and the prediction of protein function. *Computer Applications in the Biosciences* 13: 481–83

Attwood, T.K., Beck, M.E., Flower, D.R., Scordis, P. & Selley, J.N. (1998) The PRINTS protein fingerprint database in its fifth year. *Nucleic Acids Research* 26: 304–08

Bairoch, A., Bucher, P. & Hofmann, K. (1997) The PROSITE database, its status in 1997. *Nucleic Acid Research* 25: 217–21

Bork, P. & Bairoch, A. (1996) Go hunting in sequence databases but watch out for the traps. *Trends in Genetics* 12: 425–27

Bork, P. & Gibson, T.J. (1996) Applying motif and profile searches. *Methods in Enzymology* 266: 162–84

Borodovsky, M., Rudd, K.E. & Koonin, E.V. (1994) Intrinsic and extrinsic approaches for detecting genes in a bacterial genome. *Nucleic Acids Research* 22: 4756–67

Burset, M. & Guigo, R. (1996) Evaluation of gene structure prediction programs. *Genomics* 34: 353–67

Dayhoff, M.O., Schwartz, R.M. & Orcutt, B.C. (1978) In *Atlas of Protein Sequence and Structure*, volume 5 (suppl. 3), Washington, DC: National Biomedical Research Foundation

Dong, S. & Searls, D.B. (1994) Gene structure prediction by linguistic methods. *Genomics* 23: 540–51

Dujon, B. (1996) The yeast genome project: what did we learn? *Trends in Genetics* 12: 263–70

Dujon, B. (1998) European Functional Analysis Network (EUROFAN) and the functional analysis of the *Saccharomyces cerevisiae* genome. *Electrophoresis* 19: 617–24

Fagin, B., Watt, J.G. & Gross, R. (1993) Special-purpose processor for gene sequence-analysis. *Computer Applications in the Biosciences* 9: 221–26

Farber, R., Lapedes, A. & Sirotkin, K. (1992) Determination of eukaryotic protein coding regions using neural networks and information theory. *Journal of Molecular Biology* 26: 471–79

Kim, S.H. (1998) Shining a light on structural genomics. *Nature Structural Biology* 5 (Suppl.): 643–45

Lemer, C.M.R., Rooman, M.J. & Wodak, S.J. (1995) Protein structure prediction by threading methods: evaluation of current techniques. *Proteins* 23: 337–55

Pearson, W.R. (1990) Rapid and sensitive sequence comparison with FastP and FastA. *Methods in Enzymology* 183: 63–98

Robison, K., Gilbert, W. & Church, G.M. (1994) Large scale bacterial gene discovery by similarity search. *Nature Genetics* 7: 205–14

Roepstorff, P. (1998) Protein sequencing or genome sequencing. Where does mass spectrometry fit into the picture? *Journal of Protein Chemistry* 17: 542–43

Smith, T.F. & Waterman, M.S. (1981) Identification of common molecular subsequences. *Journal of Molecular Biology* 147: 195–97

Solovyev, V.V., Salamov, A.A. & Lawrence, C.B. (1995) Identification of human gene structure using linear discriminant functions and dynamic programming. In *Proceedings of the Third International Conference on Intelligent Systems for Molecular Biology*, edited by C. Rawling, D. Clark, R. Altman *et al.*, Cambridge, UK: AAAI Press

Wilbur, W.J. & Lipman, D.J. (1983) Rapid similarity searches of nucleic-acid and protein data banks. *Proceedings of the National Academy of Sciences USA* 80: 726–30

Wodicka, L., Dong, H., Mittmann, M., Ho, M.H. & Lockhart D.J. (1997) Genome-wide expression monitoring in *Saccharomyces cerevisiae*. *Nature Biotechnology* 15: 1359–67

Wootton, J.C. & Federhen, S. (1996) Analysis of compositionally biased regions in sequence databases. *Methods in Enzymology* 266: 554–71

GLOSSARY

DNA polymerase removes the RNA primers and replaces them with DNA until it reaches the start of the previously made DNA segment

Klenow fragment fragment of DNA polymerase I from *E. coli* containing both the polymerase and $3' \rightarrow 5'$ exonuclease, but lacking $5' \rightarrow 3'$ exonuclease activity

site-directed mutagenesis an *in vitro* technique in which a change is made at a specific site in a DNA molecule, which is then reintroduced into a cell

See also ***Escherichia coli:* genome, genetic exchange, genetic analysis** (p.138); **Human Genome Project** (p.606); **Arabidopsis genetics and genome analysis** (p.855)

GENETIC INFORMATION ONLINE: A BRIEF INTRODUCTION TO FLYBASE AND OTHER ORGANISMAL DATABASES

Kathleen A. Matthews and Kevin R. Cook
Department of Biology, Indiana University, Bloomington, Indiana, USA

1. Introduction

The rampant pace of genetic research over the last three decades has produced a wealth of information about the role of genes in many areas of biology, including development, physiology, behaviour, ageing and disease. Much of this information has been organized into electronic databases available online. Although most of these databases are directed at specialists, many of them contain information of interest to a wide variety of scientists and to the motivated pilgrim. Databases that focus on the structure of genes and gene products – the nucleotide sequence of DNA and RNA and the amino acid sequence of proteins – typically include data from all species. Although more integration can be expected in the future, at present, information about the consequences of gene action – the normal roles of genes and their products and the effects of gene mutations – are largely in single-organism databases.

The most extensive genetic information is known for model organisms. Model organisms have attributes such as short life cycle, large numbers of offspring and adaptability, among others, that make them particularly well suited to the experimental study of certain aspects of biology. The primary genetic models are the mouse, the pufferfish, the fruit fly Drosophila, the worm *Caenorhabditis elegans*, the plants maize (corn) and Arabidopsis, the yeast *Saccharomyces cerevisiae* and the bacterium *Escherichia coli* (the Drosophila, yeast and C. *elegans* are the only organisms to be completely sequenced to date; see article on "The pufferfish as a model organism for comparative vertebrate genome analysis"). Genes of many other organisms of agricultural, human health and other economic importance are also studied and much of this information is accessible online.

2. FlyBase

The most intensively studied model organism is *D. melanogaster*. Over 10 000 of the approximately 14 000 Drosophila genes predicted from the complete genome sequence were already known from experimental studies, including well over 60 000 mutant alleles. Information on the genetics and molecular biology of Drosophila, drawn from the scientific literature and from large-scale genome projects, are brought together in FlyBase. Links to nucleic acid and protein databases are provided, and links to other organismal databases, including those for humans, are provided when corresponding genes are known. Special purpose datasets such as T.B. Brody's developmental pathways database The Inter*active* Fly, which synthesizes and summarizes information in ways that are particularly useful to the non-specialist, are also made available through FlyBase.

(i) Reaching and using FlyBase

The primary FlyBase server can be reached at http://flybase.bio.indiana.edu/ and is best used with a recent release web browser. International mirrors are available in the UK, France, Israel, Japan and Australia (select ***FlyBase Mirrors*** from the sidebar of the homepage for links to these servers). Details of the structure and use of FlyBase are documented in the Reference Manual, which is available from the sidebar (as ***FlyBase Reference***) of the homepage. In particular, the structure and format options of Gene Reports are explained in Section C.2. of the Reference Manual.

Links to Medline, the database of the National Library of Medicine, exist for many references cited in FlyBase. Selecting these links from FlyBase Reference Reports displays the relevant PubMed report, which includes an abstract of the reference, links to related articles and options for obtaining the full document.

FlyBase, as with most online databases, can change rapidly (a strength of the network approach that can, nevertheless, be disconcerting) to take advantage of new technology, to improve or replace previous offerings, or to add new features. Your session may differ in the details of names and locations of searches, files and folders from those described here, but the basic approaches to querying the data will apply.

(ii) Finding genetic information on FlyBase

Genetic information can be retrieved from FlyBase in a variety of ways. For the non-drosophilist, the most common starting points are:

- gene product (e.g. insulin, actin, neuropeptide);
- biological process in which a gene product participates (e.g. mitosis, muscle formation, eye development); or
- structure, tissue or organ in which a gene product is active (e.g. mitochondria, heart, antenna).

In some cases, both search and browse options are available through FlyBase. The following examples illustrate useful approaches given these different starting points. Server items are italicized, and hyperlinks are indicated in bold type.

Searching for genes by product characteristics. When possible, FlyBase categorizes genes using the standardized terms for molecular function, biological process and cellular component developed by the Gene Ontology Consortium (http://www.geneontology.org/). For many purposes of the non-drosophilist, the gene product description itself or these functional categorizations will be the best route to identification of Drosophila genes of interest. Characterizations include: enzyme activities and structural roles, such as "alcohol dehydrogenase" and "myosin"; physiological roles such as "insulin" and "Ca^{2+}-transporting protein"; more general properties of gene products such as "antibacterial", "drug-resistance" or "hormone receptor"; and associative properties such as "Lupus erythematosus-associated La/SS-B autoantigen" and "retinoblastoma-family protein".

To search for Drosophila genes that are known to encode, for example, collagen or a collagen-like or collagen-related protein, the simplest approach uses the *Search* option on the homepage. Type "*collagen*" into the input field with the *Genes* section selected and click on *Search Everything* to submit the search. Alternatively, use the *Search Genes* tool, available from the homepage or from the *Genes* data section, which supports field-specific searches within the Gene records. Select *Product Function*, *Product Process* or *Cellular Location* as the field name and enter the term to be matched in the input field and submit the query. The asterisks (*) are wild cards. By including these in your search, gene records that include the search word, but do not match it exactly, such as "procollagen", will be found. Select a hyperlinked gene symbol from the list of results to view Gene Reports individually, or use the *Batch download* option at the bottom of the Search Results page to view all, or any given subset, of the records at once.

Browsing for genes by gene product characteristics. Select *Browse Function Location, Process, Structure* from the *Genes* section of the homepage for access to alphabetized lists of terms applicable to the gene product in one of four domains: function, cellular location, process and structure. Select, for example, the letter I in the *D. melanogaster* section of *Function of gene product* to see a scrollable list of gene product descriptions (followed by the term's Gene Ontology Identifier and Enzyme Commission numbers if the gene product is an enzyme) and gene symbol links. Select the gene symbol(s) to display the relevant Gene Report. For example, the following entries are found for insulin and insulin-related proteins:

insulin; GO:0016088:
  IRP
insulin receptor; GO:0005009
  InR
insulin receptor ligand; GO:0005158
  CG6736
  CG14167
  CG14173
  IRP
  Ppi
insulin-like growth factor receptor binding protein; GO: 0005067
  chico
insulysin; GO:0004231; EC:3.4.24.56:
  ESTS:176C8S
  Ide

Genes can also be identified based on cellular location of the gene product, the biological process in which the gene product is involved, or structural domains within the gene product using the *Function, Location, Process, & Structure of Gene Product* page. For example, from the *Cellular location* section,

mitotic chromosome; GO:0005708:
  stg

from the *Process* section,
male courtship behaviour; GO:0008049:
  crl
  fru
  Ubc47D

and from the *Structure* section,
immunoglobulin-superfamily:
  kek1
  kek2
  Lac

Alternative approaches to gene information based on biological process. For many developmental processes, and a few others (memory and immune response at this writing), The Inter*active* Fly provides a useful summary of the pathway as it is understood in Drosophila, along with linked lists of Drosophila genes known, or thought, to be involved in the pathway. These summaries are excellent starting points for those wanting an overview of a process.

Select ***Interactive Fly*** from the side bar of the FlyBase homepage to display The Inter*active* Fly (IF) homepage. Choose a category, such as *Tissue and Organ Development*, and then select a subcategory, for example, ***Morphogenesis and organogenesis in the embryo***. A list of tissues and structures is displayed, with links to summaries and gene lists. Select *Malpighian tubules* (the Drosophila version of a kidney), for example, and a list of genes involved in various aspects of Malpighian tubule formation and function is given, with links from each gene name to the IF's *Biological Overview* of that gene, followed by a summary of the developmental origins, structure and physiology of the tubules. FlyBase Gene Reports can be displayed by selecting the FlyBase gene identifier from the header of the IF gene overview page.

Finding genes based on Drosophila anatomy. The *Anatomy & Images* section of FlyBase contains tools that allow the identification of genes known to be expressed in a given tissue, organ or other anatomical structure. The *Body Part Viewer* contains a set of anatomical images labeled with anatomy terms. Those terms are linked to Term Reports, which include a linked list of genes that are known to be expressed in that structure. The *Terms* hierarchy allows direct access to Term Reports via the anatomy terms. The *Interacive* Fly, as described above, also provides access to information on a subset o Drosophila genes based on their activity in specific tissues or organs.

Finding genes with known human homologues. If a human homologue of a Drosophila gene has been reported, the Gene Report will contain a *Homo sapiens* subheading with the human gene name and links to GDB or OMIM (databases for human genetic information) under the heading *Non-Drosophila homologue*. To find all Drosophila genes with identified human homologues, use *Search* option on the homepage with *Genes* selected ask for "homo sapiens" and use the *Search Everything* option to submit the query.

Finding genes by the gene symbol or gene name. Drosophila genes have names that are typically descriptive of some aspect of the phenotype of animals mutant for that gene or of the gene product. Gene symbols – short strings of characters with no spaces – are abbreviations of the full name. For example, *dnc* is the symbol for *dunce*, a gene involved in learning and memory, and *Pka-C1* is the symbol for *cAMP-dependent protein kinase 1*. Use the *Search* option on the homepage with *Genes* selected, enter a gene symbol, full name or a portion of the name flanked by asterisks, in the input field and use the *Search Symbol/Names* option to submit the query.

3. A few other organismal databases accessible to the non-specialist

(i) Mouse (MGD, http://www.informatics.jax.org/)

To retrieve information from the Mouse Genome Database, select *Genetic and Phenotypic Data* from the *Searches, Data and Reports* section of the homepage, and then select the *Genes, Markers and Phenotypes* search (or *Combined Mouse/Human Phenotypes*). The most useful fields for the non-specialist to search will usually be *Name* or *Phenotype*. For example, searching for the term "insulin" in the *Name* field identifies a set of gene records that include *insulin I, II* and *III*, *insulin receptor*, and *insulin dependent diabetes susceptibility 1*, among others. Searching for "insulin" in the *Phenotype* field identifies an additional set of genes that has been found to affect insulin levels when mutant, including the gene *pancreas defect*, for example.

(ii) Human (OMIM, http://www3.ncbi.nlm.nih.gov/Omim/searchomim.html)

The Online Mendelian Inheritance in Man database provides syntheses of published information on human genes and genetic disorders. Type a search term – "ageing", for example – in the input field and submit the search. A linked list of records that include your search term will be returned.

(iii) Other animals (OMIA, http://www.angis.su.oz.au/Databases/BIRX/omia/)

The Online Mendelian Inheritance in Animals database provides literature references for genetic traits and disorders in a wide range of domesticated, farm and wild animals, but does not provide an OMIM-style synopsis. Select a trait or disorder from the list of options, such as *inherited bleeding disorders*, or make a selection from the species list to display references for only that species. At the time of writing, links to Medline are not included in OMIA. Use the search option at http://www.ncbi.nlm.nih.gov/PubMed/ to see abstracts and related information for references in the Medline database.

(iv) Corn (MaizeDB, http://www.agron.missouri.edu/cgi-bin/sybgw_mdb/mdb3/Phenotype/query

Phenotypes of corn variants can be queried without specialized knowledge of corn genetics by using the pull-down menu of traits provided on this MaizeDB page. For example, searching for the trait *starch content* returns two result categories, *low starch content* and *high starch content*. The *high starch content* report includes a link to the *starch content* trait report and to the starch-associated anatomical structures *aleurone*, *endosperm* and *kernel*, which in turn link to descriptions of the structure and a list of references relevant to the trait and structure.

4. Links to other databases and resources

The Web pages at the following URLs provide well-maintained sets of links to a wide variety of genetic databases and genetics-related resources.

- http://www.nhgri.nih.gov/ – Links are provided to a variety of genome-related databases and resources on this page are maintained by the National Human Genome Research Institute.
- http://www.nalusda.gov/ – Browse and query interfaces and links to a wide variety of plant and animal databases are available at this site maintained by the National Agricultural Library.
- http://www.gene.com/ae/index.html – Science news, discussion and resources for science teachers can be found on this site maintained by Genentech, Inc.

See also **The pufferfish as as model organism for comparative vertebrate genome analysis** (p.849)

M MODEL ORGANISMS

Introduction to Model Organisms

The term "model organism" has been used rather widely, because if the term sticks it is likely to bring in both research interest and funds. The complete genome sequence of *Arabidopsis thaliana*, the tiny but much investigated weed, was published in December 2000: a notable achievement as it is the first plant genome completed. Its small genome of about 120 Mb contrasts with the maize and wheat genomes of 1500 and 16 000 Mb respectively, neither of which will be sequenced in the near future. However the rice genome is only about four times as large as that of Arabidopsis and in January 2001 was sequenced by a private company (although the publicly funded project will not be complete until the end of 2004). New methods will doubtless speed up the process, but meanwhile we can look for increasingly rapid developments in the application of Arabidopsis knowledge to other fields of plant genetics. *Nature* in its special issue covering the Arabidopsis sequence for December 14 2000 reminds us that "more than 250 000 species of flowering plants decorate the world", and there are already arguments to be heard as to whether the complete sequence of Arabidopsis will be of much help in the improvement of our main crop plants. The *Nature* report (p.793) provides a useful list of web links and there is much additional information about Arabidopsis in further articles in the same issue.

Of the two other model organisms included in this section, sequencing of the nematode worm *C. elegans* is essentially complete, and was published in 1998 in *Science* (282: 2012–18). The pufferfish genome should be ready in draft form by April 2001, and through comparative analysis is predicted to have a dramatic effect on annotation of the human genome.

CAENORHABDITIS ELEGANS

Mark Viney
School of Biological Sciences, University of Bristol, Bristol, UK

1. Introduction

Caenorhabditis elegans is a small soil-dwelling nematode, about 1 mm in length, found in most temperate regions. It was chosen as a model organism by Sydney Brenner in 1963 (first publication, Brenner, 1974) to study the genetics of the control mechanisms in cellular development and since then has become the metazoan whose anatomy, genetics and development is most thoroughly understood.

There are two sexes – self-fertilizing hermaphrodites and males (**Figure 1**). Most adult worms are hermaphrodites, with males occurring spontaneously at a rate of <0.5%. However, the progeny of a mating between males and hermaphrodites are 50% of each sex. The hermaphrodite lays embryonated eggs which hatch to give the first larval stage (L1), which then moults through three further larval stages (L2–L4), with the L4 finally moulting into adult worms (**Figure 2**). The generation time is about 3 days at 25°C and adult hermaphrodites live for about 10–15 days. There is also an alternative L3 morph called the dauer larva, which is morphologically, behaviourally and physiologically distinct from the "normal" L3. It develops under conditions of crowding and limited food supply and, biologically, it is the resistant, dispersive stage of the life cycle.

The hermaphrodite is able to undergo self-fertilization because it makes both sperm and eggs. During gonadogenesis, the germ cells of the hermaphrodite undergo meiosis and produce about 150 mature sperm which are stored in the spermatheca. Following this, the meiosis of the germline generates only oocytes. Therefore, a hermaphrodite has a fixed sperm supply which limits the brood size that can be generated by self-fertilization. If a hermaphrodite mates with a male worm, the ova are preferentially fertilized by the male's sperm rather than the hermaphrodite's sperm.

Worms can be maintained on agar plates seeded with *Escherichia coli* and kilogram quantities can be grown in liquid culture. The number of somatic nuclei is fixed at an early stage in development; the adult hermaphrodite has 959 somatic nuclei and the adult male 1031. This, combined with whole-body transparency, has made it possible to determine the entire cell lineage from zygote to adult. The whole anatomy has been described at the electron microscope level which has enabled a complete wiring diagram of cell–cell contacts to be made. The haploid genome size is $c.10 \times 10^7$ base pairs arranged in six linkage groups. There are aprroximately 19 000 genes.

2. Sex determination

The primary sex-determining mechanism is X chromosome dosage. Hermaphrodites have five autosome pairs and one X chromosome pair; the male also has five autosome pairs, but only one X chromosome. To better understand the role of the X chromosome, worms were constructed with different numbers of autosomes and X chromosomes, thereby varying the X:autosome ratios, and the sex of the worms assessed. For example, tetraploid animals with four sets of five autosomes and two X chromosomes were male. This demonstrates that the presence of two X chromosomes does not cause female development, but rather that the ratio of X:autosomes is the critical determinant of sex. The X:autosome ratio is the primary signal that initiates a cascade of other events that result in the different sexual morphs. This method of sex determination has analogies with that of *Drosophila spp*. Males normally occur in the progeny of self-fertilizing hermaphrodites at a rate of <0.5%. This occurs by the spontaneous loss of an X chromosome during meiosis. Mutants have been created

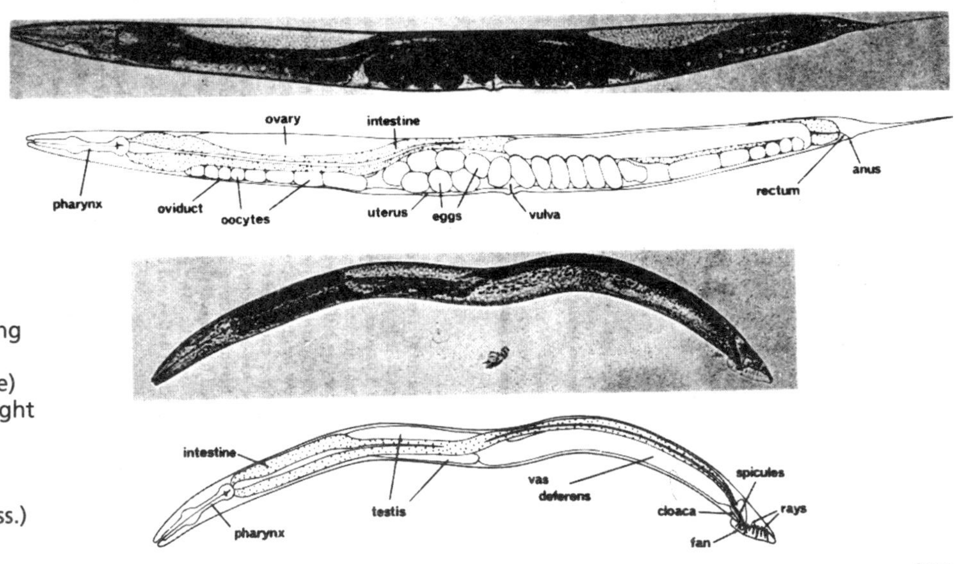

Figure 1 Photomicrographs showing the major anatomical features of *C. elegans* adult hermaphrodite (above) and male (below). Lateral views: bright field illumination. (Reprinted from Sulston and Horvitz (1977) *Developmental Biology* 56: 110–56 with permission from Academic Press.)

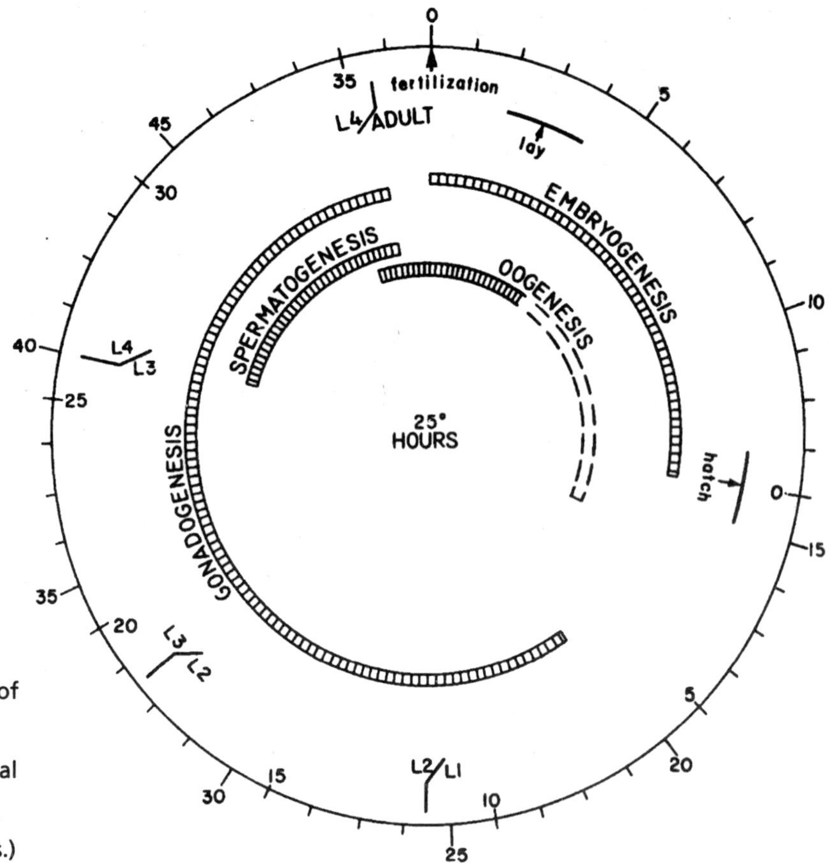

Figure 2 Diagrammatic representation of the *C. elegans* life cycle, showing durations of developmental stages. Numbers on the outside of the large circle indicate hours after fertilization at 25°C; numbers on the inside indicate approximate hours after hatching. The four larval moults are indicated by radial lines. (Reprinted from Wood *et al.* (1980) *Developmental Biology* 74: 446–69 with permission from Academic Press.)

which generate a higher proportion of males by increased non-disjunction at meiosis. For further information, see Meyer (1997).

3. Genetic methods

The genetic analysis of *C. elegans* has largely been by the analysis of mutations that have been artificially induced either with chemical mutagens, such as EMS (ethylmethanesulphonate), or with ionizing radiation. The life cycle of *C. elegans* is ideally suited to genetic analysis because it is a self-fertilizing hermaphrodite. Most mutations are loss-of-function mutations and are genetically recessive; thus, the progeny of mutated worms will be phenotypically wild-type. Such worms can be allowed to self-fertilize thereby making the mutation homozygous in one quarter of the progeny. The progeny of this self-fertilization will, therefore, display the mutant phenotype. In this way, thousands of genomes can be rapidly screened either visually or by a passive screen such as resistance to a drug.

To construct multiple mutants and to transfer alleles between different worm lines male worms are used. This is possible because in a hermaphrodite/male cross the oocytes are preferentially fertilized by the male's sperm. Hodgkin (1999) gives further information.

4. Mutant phenotypes

Most of the mutations that have been identified are those that in some way affect the morphology or behaviour of the worms on agar plates. Thus, there are "dumpy" (Dpy) worms that are shorter and fatter than wild-type, worms that move in an "uncoordinated" (Unc) way or are unable to move at all, "squat" (Sqt) worms, worms that are defective in egg laying (Egl) and those that roll (Rol) in a corkscrew fashion as they move. Phenotypes detected by other criteria include those with abnormal cell lineages (Lin), worms that are abnormal in their mechanosensory abilities (Mec), such as insensitivity to tactile stimulation, and those with abnormal dauer larva formation (Daf).

(i) An example

Using these methods, a full genetic analysis has been made of many mutant phenotypes. This has been followed by a molecular analysis that has given an understanding of the molecular basis of the observed phenotypes. One example concerns collagen genes. Over 100 collagen genes are known to be involved in the construction of the worm's cuticle alone, where they provide the structural support of the cuticle. Mutations in these genes cause a number of phenotypes, including Dpy, Rol and Sqt, and the molecular basis of many of these mutations is now understood. The structural roles played by collagen molecules are due to a region that has an amino acid repeat unit that helically stacks on itself, while other regions are involved in the molecular processing of the molecule. Molecular and genetic analyses of these mutants have shown that mutations in the structural regions, leading to defective arrangement of the protein, have the most severe phenotypes. Null alleles of the same loci actually give a less severe phenotype, presumably because the cuticle is less disrupted by the absence of a molecule than by the presence of one that is structurally aberrant. Mutations in the regions involved in the molecular processing of the molecule have very variable phenotypes that depend on the exact position of the mutation. This is because some sites are absolutely necessary for the correct processing of the molecule, and thus cannot tolerate alteration, whereas other sites within this region can be mutated without visible effect.

(ii) Temperature-sensitive and suppressor mutations

A number of mutants are temperature sensitive. This means that the mutant phenotype is only seen at restrictive temperatures; at permissive temperatures, the worms are phenotypically wild-type. Temperature-sensitive mutations that have a highly deleterious effect on the worms can be propagated by rearing them at permissive temperatures. Temperature-sensitive mutations can also be used to determine when a gene product is required, by switching worms between different temperatures at various points during development and assessing the result.

Mutant worms can be further mutagenized and screened for a second mutation at a different site that suppresses, at least partially, the first mutant phenotype. Such mutations are suppressor mutations. This suppression can occur through a mutation in the same gene as the first mutation (intragenic suppressor) or by a mutation in a different gene or regulatory region that in some way affects the expression of the first gene. These methods can be used to understand the interplay of the locus of interest and its regulatory mechanisms.

5. Epistatic analysis

In addition to the genetic analysis of simple phenotypes, complex developmental pathways can be analysed. This is done by epistatic analysis. If there is a developmental pathway in which a number of gene products interact sequentially, the relative positions of those gene products in the developmental pathway can be ordered by epistasis. This order can be established by constructing double mutants and assessing their phenotype in relation to the phenotype of the relevant single mutants. Thus, if mutant gene x has phenotype X and mutant gene y has phenotype Y, and the double mutant xy has phenotype X, gene x is epistatic to gene y. The developmental pathway that leads to the development of the dauer larva has been extensively analysed in this way. The loci responsible for sensory processing of the environmental conditions that initiate dauer development are found early in the pathway; those loci that are responsible for the morphogenesis specific to the dauer larva morph are found later in the pathway. A similar analysis has been made for the process by which sex determination is implemented following the establishment of the X:autosome ratio and for the process that results in the formation of the vulva (the ventral genital opening) in hermaphrodite worms and for the pathway by which developing dauer "choose" between alternative dauer development and "normal", non-dauer development. Both the pathways which induce the formation of the vulva and which control dauer formation use molecular pathways and processes that are used throughout the animal kingdom for various developmental and signal transduction processes. For a more detailed discussion of these processes, see Greenwald (1997) and Riddle & Albert (1997).

6. Reverse genetics

To attempt to determine the phenotype of a mutation in a gene that is only known molecularly, a number of reverse genetic methods can be used. Transposable elements are known in *C. elegans* and two of these (Tc1 and Tc3) undergo transposition in the germline. Strains of *C. elegans* which have a high number of transposons in their genome can be screened molecularly to identify worms in which a transposon is inserted into the gene of interest. The transposon can then be induced to undergo excision and in so doing generate a mutation in the locus of interest, thereby revealing the phenotype of the locus. Chemical mutagens will also induce random deletions throughout the genome. In a process analogous to the transposon analysis, chemically induced deletions of a gene of interest can be molecularly identified and, thereby, used to reveal the phenotype of the locus. A very recent technique has been found that generates a phenocopy of a null mutation of a gene. In this process, double-stranded RNA that matches the coding sequence of the gene of interest is injected into the gonad of worms. By a process that is not yet understood, this treatment can induce a phenocopy that is equivalent to a null mutation in that locus.

7. Transformation

C. elegans can be transformed by injecting DNA into the germline of adult hermaphrodites. A proportion of the progeny of transformed worms will contain the introduced DNA. The transforming DNA is usually coinjected with a marker gene whose expression gives a visible phenotype (most commonly, a Rol phenotype) that enables successfully transformed worms to be identified visually. Transformed DNA is usually maintained as extrachromosomal arrays with various degrees of stability, such that some of the progeny of a transformed hermaphrodite sometimes do not contain the transforming DNA. This technique is commonly used to rescue worms with a mutant phenotype with cloned DNA that is a likely candidate for the locus of interest. It can also be used to screen large cloned fragments of the genome (cosmids) for their ability to rescue mutant phenotypes, thereby beginning the process of identifying the physical part of the genome responsible for a phenotype.

8. Trans-splicing

In *C. elegans*, about 75% of mRNA molecules have a 22 nucleotide spliced leader (SL1) sequence added to the 5′ end of the molecule, a process known as trans-splicing. This phenomenon has been described only in the protozoan *Trypanosoma spp*., where all mRNAs are trans-spliced, and in nematodes, where only a proportion of the molecules are trans-spliced. In nematodes, genes that undergo trans-splicing are those that have an intron-like sequence in the 5′ untranslated region of the gene. Some genes have been found to be transcribed in a polycistronic unit. During the processing of such molecules, the first gene is trans-spliced to the SL1 sequence and the downstream genes are trans-spliced to the second spliced leader sequence (SL2). The function of trans-splicing is not known and there does not appear to be any similarity of function that separates those genes that are trans-spliced from those that are not. For further information see Blumenthal & Steward (1997).

9. The genetic and physical maps

The rapid progress of the *C. elegans* research field has been enhanced by the construction of a genetic map of the genome by the coordinated efforts of all workers in the field, the usefulness of which has been greatly enhanced by the use of a single terminology. Newly identified mutants are placed on the map by analysing their recombination in relation to other loci.

This map has been augmented by the construction of a physical map in which the entire genome has been cloned into cosmids that have been placed in order such that the entire genome is represented by overlapping cosmids. These cosmids are readily available in an organized array to the whole community of *C. elegans* workers so that newly cloned fragments of DNA can be unambiguously assigned to a cosmid. Alignment of these two maps has been responsible for the rapid rate of progress in the *C. elegans* field.

A logical extension of these activities is the sequencing of the entire genome of *C. elegans*. This was done by an international consortium based in St Louis, Washington, US and Cambridge, UK. This project was completed in 1998 and represents the first entire genome sequence of any metazoan (Hodgkin *et al*., 1998). The raw sequence data are analysed to detect putative genes and all this information is placed in a publicly available interactive database (ACEDB) that combines the genetic map, physical map and DNA sequence. The complete genome sequence of *C. elegans* predicted the existence of approximately 19 000 genes, which is greater than the predicted approximately 14 000 genes for the insect *Drosophila melanogaster*. Comparison of known genes from other organisms with the *C. elegans* genome information shows that *c*.75% of 5000 human genes, *c*.50% of 6000 yeast genes and *c*.25% of 4000 *E. coli* genes have matches in *C. elegans*. Many of the *C. elegans* genes identified by the genome project have no known function. To try and discover their function a systematic reverse genetic approach is being used. In this approach, each gene is being individually deleted and the consequent effect of this on the worm phenotype is being analysed to try and discover the function of that gene. The completion of the genome project is a highly significant achievement; the challenge for the future is understand how the worm's genome is coordinated into a fully functioning animal.

REFERENCES

Blumenthal, T. & Steward, K. (1997) RNA processing and gene structure. In *C. elegans II*, edited by D.L. Riddle, T. Blumenthal, B.J. Meyer & J.R. Priess

Brenner, S. (1974) The genetics of *Caenorhabditis elegans*. *Genetics* 77: 71–94

Greenwald, I. (1997) Development of the vulva. In *C. elegans II*, edited by D.L. Riddle, T. Blumenthal, B.J. Meyer & J.R. Priess

Hodgkin, J. *et al*. (1998) Genome sequence of the nematode *C. elegans*: a platform for investigating biology. *Science* 282: 2012–18

Hodgkin, J. (1999) Conventional genetics. In *C. elegans: A Practical Approach*, edited by I.A. Hope, Oxford and New York: Oxford University Press

Meyer, B.J. (1997) Sex determination and X chromosome dosage compensation. In *C. elegans II*, edited by D.L. Riddle, T. Blumenthal, B.J. Meyer & J.R. Priess

Riddle, D.L. & Albert, P.S. (1997) Genetic and environmental regulation of dauer larva development. In *C. elegans II*, edited by D.L. Riddle, T. Blumenthal, B.J. Meyer & J.R. Priess

FURTHER READING

Two major books have been produced by the *C. elegans* community. The 1988 book provides a comprehensive introduction to all aspects of *C. elegans* biology, while the 1997 book covers all aspects of *C. elegans* biology in considerable depth.

Riddle, D.L., Blumenthal, T., Meyer, B.J. & Priess, J.R. (eds) (1997) *C. elegans II*, Cold Spring Harbor, New York: Cold Spring Harbor Laboratory Press

Wood, W.B. (ed.) (1988) *The Nematode* Caenorhabditis elegans, Cold Spring Harbor, New York: Cold Spring Harbor Laboratory Press

For a practical introduction to working with *C. elegans*, see
Hope, I.A. (ed) (1999) *C. elegans: A Practical Approach*, Oxford and New York: Oxford University Press

See also the *C. elegans* community web sites:
http://elegans.swmed.edu or a UK mirror of this at
http://c.elegans.leeds.ac.uk/index.html

For *C. elegans* genome information, see
http://www.sanger.ac.uk and/or
http://genome.wustl.edu/gsc/index.html

GLOSSARY

epistatic describes a gene or character whose effect overrides that of another gene with which it is not allelic; analogous to dominant applied to genes at different loci. More generally, epistasis exists when the effect of two or more non-allelic genes in combination is not the sum of their separate effects

gonadogenesis development of the reproductive system

phenocopy phenotype resulting from environmental factors which simulates a genetically determined change

polycistronic messenger RNA that contains more than one polypeptide coding sequence

THE PUFFERFISH AS A MODEL ORGANISM FOR COMPARATIVE VERTEBRATE GENOME ANALYSIS

Greg Elgar
UK HGMP Resource Centre, Wellcome Trust Genome Campus, Hinxton, UK

1. Introduction

In the early 1990s, the Human Genome Project was in its infancy and, while some researchers developed methods in order to accommodate the huge sequencing task, others were looking at ways in which all the data could be interpreted. The majority of these strategies centred on cDNA technology, a route through which the coding potential of a genome can be accessed while reducing the sequencing effort. In mammalian genomes, it is likely that the coding sequences amount to less than 5% of the genomic DNA. Sydney Brenner, based at the MRC Molecular Genetics Unit in Cambridge, also realized that there was more to genes and gene function than the coding sequence, and considered ways in which other genomic elements, as well as the coding sequence, might be interpreted and understood.

Although there was good evidence that the events which gave rise to the first vertebrates involved a massive increase in gene number, Brenner proposed that the variation within the vertebrate lineage relied not on the introduction and creation of new genes, but by variation in the intricate pathways of gene regulation. He argued that all vertebrates have a similar gene repertoire and, therefore, comparing any two vertebrate genomes would be useful in deciphering which DNA sequences were of value and which were not.

He sought to find a good model for this kind of study and came up with a simple set of criteria. First, the genome size must be small, to reduce the amount of effort in genomic sequencing. Secondly, the genes contained within the model genome must be similar enough to human genes to be identifiable both through computer analysis and by hybridization. Thirdly, the model must have a similar gene repertoire to the human genome. Finally, it would be advantageous if the divergence time between the model and the human genome is large enough that non-essential sequences have diverged essentially to randomness. This last criterion is of particular importance in comparative genomics because it allows us to assume that where we do find sequence conservation between species, it is meaningful and we can consider that the sequence serves some common function in both organisms. Given the amount of "junk" DNA in mammals, this is a particularly powerful way of sifting through the data.

The most straightforward of these criteria, genome size, was also the easiest to select for, as this has been established for many vertebrates. Among the vertebrate class, the smallest genome sizes belong to teleost fish, with a mean size of around 1000 megabases (Mb). Ralph Hinegardner determined the size of over 200 teleost fish genomes in the late 1960s, and found that the smallest genome sizes of all belong to the Tetraodontoid fish, consisting of the pufferfishes (Hinegardner, 1968). Haploid cells of two species of puffer, *Spheroides maculatus* and *Tetraodon fluviatilis*, were found to contain just 0.4–0.5 picograms of DNA, which equates to a genome size of 400 Mb.There are about 300 species of pufferfish, of the Order Tetraodontiformes, which also includes the triggerfish, cowfish, filefish and sunfish. Pufferfish have the ability to inflate with air or water, as well as being able to eject the water in "coughs". Many species of pufferfish contain a potent neurotoxin, tetrodotoxin, which kills many people each year – pufferfish are considered an edible delicacy, particularly in Japan. They are found in marine, brackish and freshwater environments, and have powerful teeth and jaws that they use to break open their favourite prey, sea urchins and shellfish.

Molecular analysis of another pufferfish, *Arothron maculatus*, using density gradient centrifugation and reassociation kinetics, not only confirmed the small size of tetraodontoid genomes but also determined that very little of the DNA was repetitive (Pizon *et al.*, 1984). With a genome 7.5 times smaller than that of man (3000 Mb), can the pufferfish really contain a similar number of genes to a human being? By sequencing random fragments from the genome of the pufferfish *Fugu rubripes* (a commercially farmed species from Japan) and comparing the data generated against coding sequences from other mammals, Brenner *et al.* (1993) showed that the proportion of coding sequence was 7.5 times higher than in the human genome, so confirming that the coding content of the two genomes was approximately the same, but that the Fugu genome was 7.5 times more compact (see **Figure 1**). In addition, data presented by Brenner *et al.* showed that intron sizes were very small (the majority between 80 and 120 base pairs [bp]) in Fugu compared with those of mammalian genes and that there was very little highly repetitive sequence.

2. Genome structure

We now know quite a lot about the Fugu genome, particularly from work carried out at the UK HGMP Resource Centre at Hinxton, where a project to sequence over 50 000 genomic DNA fragments from the Fugu genome has allowed more detailed scrutiny of the genome as a whole (Elgar *et al.*, 1999) (http://fugu.hgmp.mrc.ac.uk). The 50 000 sequences are derived from just over 1000 Fugu cosmid clones, each cosmid having an average insert size of 40 kilobases (kb). The approach was designed to allow the analysis of many different areas (represented by each cosmid) of the genome, thereby generating an overall picture. Each cosmid was broken into small random fragments, about 50 of which

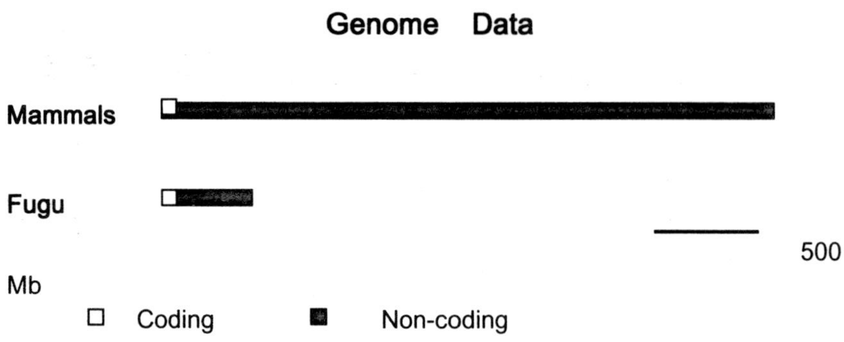

	Mammals	Fugu
Genome size	3000 Mb	400 Mb
Gene number	60—70000	60—70000
Average coding length	1.2 kb	1.2 kb
Total coding sequence	72 Mb	72 Mb
Average length of gene	40—50 kb	6—7 kb
Percentage coding	2.4	18

Figure 1 Comparison of genome size and coding sequence content between the genomes of mammals and the pufferfish, *Fugu rubripes*. While the genome size of mammals is nearly eight times larger than Fugu, the amount of coding sequence is very similar. This results in nearly 20% of the Fugu genome being coding, compared with less than 3% for mammals.

were sequenced from one end. The sequences were analysed in order to look at the gene content of each cosmid, and these data were compared with the gene order in mammalian (usually human) genomes, so generating comparative mapping data. In addition, the sequences from all the cosmids were pooled in order to carry out more global analyses of the Fugu genome. The cosmids show remarkably uniform G + C%, suggesting that there are either no G + C "rich" or "poor" regions of the genome, or that they are not well defined.

In summary, the Fugu genome is relatively uniform and highly gene dense, with a gene every 6–7 kb. From *in silico* approaches using open reading frame prediction packages, it appears that the genome is made up of nearly 20% coding sequence. As shown in **Table 1**, Fugu coding sequence is enriched in G/C in the third codon position, and, because of the high coding sequence content of the genome, this results in a genome that is G + C rich (47.7%) for a vertebrate (compare with figures of 42% for mammals).

Short tandem repeats, known as microsatellites, which have a repeat unit length of six or fewer bases, are frequent (in particular, $CA_{(n)}$ repeats) (Edwards *et al.*, 1998) and comprise about 1.3% of the genome. Minisatellites and dispersed repeats (such as short interspersed repeat sequences, SINEs, transposons and long interspersed repeat sequences, LINEs) are relatively rare, comprising another few per cent, and may cluster to some extent throughout the genome. In some cases, these repeats in the Fugu genome have been studied in depth (Poulter & Butler, 1998; Poulter *et al.*, 1999).

Another class of "repeat", which is a particular feature of mammalian genomes, is the pseudogene. This is a non-functional copy of a gene, rendered inactive by deletions or insertions or by base changes that introduce a stop codon in the sequence or a change that results in loss of function for that gene. Most pseudogenes are retrotranscribed – that is, they have been transcribed back into the genome from an mRNA – and, as such, they are usually intronless. A good example of a region rich in pseudogenes is the major histocompatibility complex class I region in man, where over 40 of the 50 genes at the centromeric end are pseudogenes (MHC Sequencing Consortium, 1999). Fugu, on the other hand, appears to possess very few pseudogenes, which, as discussed later, facilitates gene hunting with the Fugu genome.

Mammalian genomes have a 5 times lower than expected CpG dinucleotide level. This CpG suppression is due to the fact that many CpG dinucleotides are methylated on the cytosine, which then is liable to deamination to thymine. While the Fugu genome shows some CpG suppression (60% of expected value), it is nowhere near as extensive as it is in mammals (Elgar *et al.*, 1999). In mammalian genomes, CpG "islands" are markers for the 5′ ends of genes (in particular, of "housekeeping" genes), but, in the Fugu genome, the distribution of the CpG dinucleotide is far more random and uniform and, therefore, not indicative of the 5′ ends of genes.

3. Gene sequence and structure

Having laid the foundation for further investigation, interest turned from random sequence fragments to the sequencing of whole genes from the Fugu genome. Early data had suggested that Fugu and mammalian gene orthologues shared the same intron/exon organization and this meant that, by sequencing compact Fugu genes at the genomic level, gene structure could be predicted for the much larger human genes.

Table 1 Codon usage for *Fugu rubripes* calculated from 101 670 codons taken from 176 genes. The protein coding sequences were used to calculate codon usage bias. Codon frequencies and the relative synonymous codon usage (RSCU) values are quoted. The RSCU is calculated as follows; the frequency of codon/(the frequency of amino-acid × the number of synonyms for that amino acid). So for leucine this value could be as low as 0 and as high as 6. It allows the comparison of codon biases in six-, four- and twofold synonymous codons. Random codon usage generates a RSCU of 1 for all codons.

Phe	UUU	1527	0.80	Ser	UCU	1387	0.97	Tyr	UAU	813	0.59	Cys	UGU	1078	0.88
	UUC	2299	1.20		UCC	2209	1.54		UAC	1932	1.41		UGC	1375	1.12
Leu	UUA	455	0.28		UCA	963	0.67	TER	UAA	59	1.01	TER	UGA	81	1.38
	UUG	1042	0.64		UCG	789	0.55		UAG	36	0.61	Trp	UGG	1234	1.00
	CUU	987	0.60	Pro	CCU	1353	0.94	His	CAU	795	0.58	Arg	CGU	616	0.66
	CUC	2188	1.34		CCC	1999	1.39		CAC	1930	1.42		CGC	1168	1.24
	CUA	492	0.30		CCA	1399	0.97	Gln	CAA	986	0.42		CGA	544	0.58
	CUG	4658	2.85		CCG	1013	0.70		CAG	3655	1.58		CGG	813	0.87
Ile	AUU	1151	0.72	Thr	ACU	1179	0.81	Asn	AAU	1065	0.56	Ser	AGU	963	0.67
	AUC	3037	1.90		ACC	2296	1.57		AAC	2765	1.44		AGC	2312	1.61
	AUA	604	0.38		ACA	1256	0.86	Lys	AAA	2218	0.81	Arg	AGA	1108	1.18
Met	AUG	2500	1.00		ACG	1124	0.77		AAG	3235	1.19		AGG	1384	1.47
Val	GUU	1109	0.66	Ala	GCU	1748	1.00	Asp	GAU	1743	0.68	Gly	GGU	1237	0.78
	GUC	2023	1.20		GCC	2861	1.64		GAC	3363	1.32		GGC	2064	1.31
	GUA	482	0.29		GCA	1299	0.75	Glu	GAA	1934	0.60		GGA	1637	1.04
	GUG	3122	1.85		GCG	1054	0.61		GAG	4542	1.40		GGG	1380	0.87

Among the first Fugu genes to be sequenced were those for glucose-6-phosphate dehydrogenase (G6PD) (Mason *et al.*, 1995), p55 (Elgar *et al.*, 1995), Cctd (Yoda *et al.*, 1995), Huntington's disease (HD; Baxendale *et al.*, 1995) and tuberous sclerosis 2 (TSC2; Maheshwar *et al.*, 1996). All of these genes proved to be much smaller than their mammalian counterparts. The last two of these are large multiexon genes; the Fugu HD gene has 67 exons (identical to the human HD gene) and the Fugu TSC2 gene has 42 exons, one more than the human TSC2 gene. While the position of all splice sites is conserved, the additional intron in the Fugu TSC2 gene splits what is a particularly large exon in the human gene (488 bp, which is more than twice the size of any other coding exon within the gene). Indeed, in the few exceptions that show variation in intron/exon organization between Fugu and mammals, in the coding regions at any rate, there are additional introns in the Fugu genes, something of a paradox when comparative genome size is taken into consideration. An extreme example of this is the polycystic kidney disease 1 (PKD1) gene which has 54 exons in Fugu compared with only 46 in man (Sandford *et al.*, 1997). In all cases, the human splice sites are conserved, but the four largest human exons are interrupted by additional introns in the Fugu gene. By way of size comparison, the Fugu HD gene is just 23 kb in length, whereas the human gene is approximately 170 kb and this equates to a 7.3-fold reduction in size, much in line with the genome ratio. This is not consistent, however; the Fugu C9 gene is less than 3 kb in length, compared with 90 kb in man (a 30-fold difference; Yeo *et al.*, 1997) and the Fugu utrophin gene is less than 90 kb in length compared with approximately 1000 kb in man (11-fold difference). In other cases, particularly where the human gene is quite small, there is little difference in size and, on the odd occasion, the Fugu gene is slightly the larger of the two (Macrae & Brenner, 1995; Coutelle *et al.*, 1998).

While gene size varies greatly, the coding sequences of Fugu and mammalian genes are very similar in length and, as a general rule, code for very similar protein products. Regions of particularly high similarity are likely to be of functional significance to the protein, whereas regions of lower identity may be less critical and so subjected to less selection pressure. Therefore, comparisons between amino acid sequences from Fugu and mammalian genes may help to pinpoint important functional domains in uncharacterized proteins (Auf der Maur *et al.*, 1999; Bycroft *et al.*, 1999; Cottage *et al.*, 1999).

4. Gene regulation

As a result of the concerted effort to sequence coding sequences from genomes, particularly of the human genome, very little is known about sequences which control or affect the way in which genes are regulated. These regulatory sequences are very hard to identify from non-coding sequence for a number of reasons. They may be very short, we do not necessarily know where in the DNA to look for them, they appear to have no secondary code (like the genetic code for proteins) to help us identify them and they are not expressed, so we cannot use cDNA technology to help us. The fact that they appear to follow no rules whatsoever means that, for the moment at least, we cannot even use computers to predict them! Instead, we have to work from raw genomic DNA and one of the few ways of identifying these sequence stretches is by virtue of the fact that they may well be conserved between species, while the non-essential DNA surrounding them will not.

Comparative sequence analysis between two or more organisms allows us to look for quite small regions of conserved sequence. Since the evolutionary distance between mammals and Fugu is so vast (400 million years in both directions from a common ancestor), even very small regions of similarity, in a similar contextual position, suggest functional significance. A number of workers have looked for regulatory sequence conservation between Fugu and mammals, either in order to identify specific elements (Marshall *et al.*, 1994; Aparicio *et al.*, 1995; Schuddekopf *et al.*, 1996; Venkatesh *et al.*, 1997; Kleinjan *et al.*, 1998; Rowitch *et al.*, 1998; Wentworth *et al.*, 1999) or as part of large-scale genomic comparisons (Miles *et al.*, 1998). All these studies have had some measure of success and the use of cross-species genomic comparisons will become one of the key tools for identifying regulatory sequences in the human genome.

5. Gene families

It is still difficult to tell exactly how similar the gene contents of the Fugu and human genomes are, although, because all vertebrates appear to share a common ancestor, gene repertoire is likely to be similar. Indeed, despite morphological differences (and some minor physiological ones – lungs instead of gills, for instance), we all share the same developmental patterns and the same basic biochemical pathways. As we get a better overall idea of how vertebrates have evolved, it is clear that, in some cases, organisms have lost or gained genes or sets of genes. This is probably due to local or regional duplications and, in rarer cases, whole genome duplications. Fish, in particular, it seems are prone to this sort of event, as there are a number of polyploid fish (such as salmon, for instance) and – more contentiously – some species that appear to have undergone an additional round of whole genome duplication (after the divergence of the tetrapod lineage) and have subsequently reverted to a diploid state (the zebra fish, *Brachydanio rerio*, for instance). However, these fish have large genomes; the zebra fish genome is 1700 Mb, four times that of Fugu and almost twice the mean size for teleost fish (Hinegardner, 1968).

One approach to the question of gene diversity in vertebrates is to look at gene families from different species. This has been applied to the Fugu genome in a number of cases (Macrae & Brenner, 1995; Sarwal *et al.*, 1996; Venkatesh *et al.*, 1996; Yamaguchi *et al.*, 1996; Aparicio *et al.*, 1997; Koh *et al.*, 1997; Naito *et al.*, 1998; Cottage *et al.*, 1999). The Fugu genome, uncluttered with repetitive DNA and small in overall size, is ideal for this purpose. Furthermore, there is little evidence of the retrotranscribed pseudogenes that are so abundant in the genomes of mammals and this facilitates the identification of gene family members from Fugu genomic DNA by means of degenerate PCR amplification or from genomic libraries by hybridization. While there is variation between different species in gene number within families, it appears, in general, that the Fugu genome contains a very similar repertoire of genes to mammals, even within gene families, although it is often difficult to determine whether genes are orthologues or simply homologous – orthologues are derived from a single gene in the nearest common ancestor, whereas homologous genes are derived from separate ancestral genes. Phylogenetic analysis suggests that pathways may be more complex than at first sight. There are two cannabinoid receptors in both the Fugu and human genomes. However, it appears that both the Fugu genes are more closely related phylogenetically to one of the human genes and, therefore, one human gene has two orthologues in Fugu, whereas the other gene may have none (Yamaguchi *et al.*, 1996). Until gene families have been exhaustively searched for and characterized from both species, it is difficult to draw confident conclusions, but while gene family analysis from the Fugu genome is relatively straightforward, this is not the case with mammalian genomes.

6. Conservation of synteny and gene order

As vertebrate genomes have evolved, they have constantly rearranged, shuffling genes around the chromosomes. It is not clear whether this movement is random or whether there are selection pressures, particularly among certain groups of functionally related genes, that regulate this movement. Certainly, there is surprising conservation of synteny between distantly related genomes. There is now evidence that mammalian genomes have very large regions of synteny in common with chicken and even zebra fish at the gross chromosomal level (Postelthwait *et al.*, 1998; Gates *et al.*, 1999). From work on the pufferfish genome, however, it appears that local gene order is often rearranged, even when overall synteny between chromosomes is conserved.

A number of researchers have looked at close-range synteny between the pufferfish and mammals and the picture that emerges is a varied one. Miles *et al.* (1998) found absolute conservation of gene order and orientation across the WAGR (Wilms' tumour, aniridia, genital abnormalities and mental retardation) region on human chromosome 11p13 from the genes WT1 to PAX6. Conservation of gene order has also been found with regions of human chromosome 14q24 (Trower *et al.*, 1996), 16p13.3 (Sandford *et al.*, 1996), 2q32–35 (Schofield *et al.*, 1997) and Xp22 (Brunner *et al.*, 1999). Gellner and Brenner (1999) found 15 genes in 148 kb of Fugu genomic DNA; eight of these genes are known to map to human chromosome 12q13, whereas the other genes have either not been mapped or do not as yet have recognized orthologues in the human genome. Others have found conserved synteny, but with considerable rearrangement of gene order, with regions of human chromosome 9q34 (Gilley & Fried, 1999) and 17q11 (Kehrer-Sawatzki *et al.*, 1999). The evidence suggests that intrachromosomal gene movement is more common, or better "tolerated", than interchromosomal movement. What is not clear is whether there is a selection pressure for this or whether the mechanics of genome fluidity are simply more prone to one kind of movement than another. The construction of comparative gene maps from a wide range of vertebrates will allow us to address these questions.

7. Future perspectives

Progress is now so rapid on the Human Genome Project that what was considered costly and difficult last year is simple and routine this year; next year, of course, there will be new challenges and new techniques will have been developed. However, while large-scale sequencing is now essentially a factory industry, interpretation of all the resulting data will require many different approaches and some of these are not readily amenable to high throughput technology. Biological systems and pathways will have to be dissected individually, using data from many sources; comparative analyses, at the DNA and protein levels, will contribute significantly to these process.

On a wider biological theme, comparative analysis at the DNA level between different organisms provides an exquisitely powerful evolutionary tool. Subtle differences between species can be identified and their molecular basis scrutinized. The mechanisms by which chromosomes and whole genomes change and move can be deciphered. This not only allows us to follow the path of our past evolution, but, perhaps frighteningly, may allow us to model and predict our evolutionary future.

REFERENCES

Aparicio, S., Morrison, A., Gould, A. *et al.* (1995) Detecting conserved regulatory elements with the model genome of the Japanese puffer fish, *Fugu rubripes*. *Proceedings of the National Academy of Sciences USA* 92: 1684–88

Aparicio, S., Hawker, K., Cottage, A. *et al.* (1997) Organization of the *Fugu rubripes* Hox clusters: evidence for continuing evolution of vertebrate Hox complexes. *Nature Genetics* 16: 79–83

Auf der Maur, A., Belser, T., Elgar, G., Georgiev, O. & Schaffner, W. (1999) Characterization of the transcription factor MTF-1 from the Japanese pufferfish (*Fugu rubripes*) reveals evolutionary conservation of heavy metal stress response. *Biological Chemistry* 380: 175–85

Baxendale, S., Abdulla, S., Elgar, G. *et al.* (1995) Comparative sequence analysis of the human and pufferfish Huntington's disease genes. *Nature Genetics* 10: 67–76

Brenner, S., Elgar, G., Sandford, R. *et al.* (1993) Characterization of the pufferfish (*Fugu*) genome as a compact model vertebrate genome. *Nature* 366: 265–68

Brunner, B., Todt, T., Lenzner, S. *et al.* (1999) Genomic structure and comparative analysis of nine *Fugu* genes: conservation of synteny with human chromosome Xp22.2–p22.1. *Genome Research* 9: 437–48

Bycroft, M., Bateman, A., Clarke, J. *et al.* (1999) The structure of a PKD domain from polycystin-1: implications for polycystic kidney disease. *EMBO Journal* 18: 297–305

Cottage, A., Clark, M., Hawker, K. *et al.* (1999) Three receptor genes for plasminogen related growth factors in the genome of the puffer fish *Fugu rubripes*. *FEBS Letters* 443: 370–74

Coutelle, O., Nyakatura, G., Taudien, S. *et al.* (1998) The neural cell adhesion molecule L1: genomic organization and differential splicing is conserved between man and the pufferfish *Fugu*, *Gene* 208: 7–15

Edwards, Y.J., Elgar, G., Clark, M.S. & Bishop, M.J. (1998) The identification and characterization of microsatellites in the compact genome of the Japanese pufferfish, *Fugu rubripes*: perspectives in functional and comparative genomic analyses. *Journal of Molecular Biology* 278: 843–54

Elgar, G., Rattray, F., Greystrong, J. & Brenner, S. (1995) Genomic structure and nucleotide sequence of the p55 gene of the puffer fish *Fugu rubripes*. *Genomics* 27: 442–46

Elgar, G., Clark, M.S., Meek, S. *et al.* (1999) Generation and analysis of 25 Mb of genomic DNA from the pufferfish *Fugu rubripes* by sequence scanning. *Genome Research* 9: 960–71

Gates, M.A., Kim, L., Egan, E.S. *et al.* (1999) A genetic linkage map for zebrafish: comparative analysis and localization of genes and expressed sequences. *Genome Research* 9: 334–47

Gellner, K. & Brenner, S. (1999) Analysis of 148 kb of genomic DNA around the wnt1 locus of *Fugu rubripes*. *Genome Research* 9: 251–58

Gilley, J. & Fried, M. (1999) Extensive gene order differences within regions of conserved synteny between the *Fugu* and human genomes: implications for chromosomal evolution and the cloning of disease genes. *Human Molecular Genetics* 8: 1313–20

Hinegardner, R. (1968) Evolution of cellular DNA content in teleost fishes. *American Naturalist* 102: 517–23

Kehrer-Sawatzki, H., Maier, C., Moschgath, E., Elgar, G. & Krone, W. (1999) Characterization of three genes, AKAP84, BAW and WSB1, located 3′ to the neurofibromatosis type 1 locus in *Fugu rubripes*. *Gene* 235: 1–11

Kleinjan, D.A., Dekker, S., Guy, J.A. & Grosveld, F.G. (1998) Cloning and sequencing of the CRABP-I locus from chicken and pufferfish: analysis of the promoter regions in transgenic mice. *Transgenic Research* 7: 85–94

Koh, C.G., Oon, S.H. & Brenner, S. (1997) Serine/threonine phosphatases of the pufferfish, *Fugu rubripes*. *Gene* 198: 223–28

Macrae, A.D. & Brenner, S. (1995). Analysis of the dopamine receptor family in the compact genome of the puffer fish *Fugu rubripes*. *Genomics* 25: 436–46

Maheshwar, M.M., Sandford, R., Nellist, M. *et al.* (1996) Comparative analysis and genomic structure of the tuberous sclerosis 2 (TSC2) gene in human and pufferfish. *Human Molecular Genetics* 5: 131–37

Marshall, H., Studer, M., Pepperl, H. *et al.* (1994) A conserved retinoic acid response element required for early expression of the homeobox gene Hoxb-1. *Nature* 370: 567–71

Mason, P.J., Stevens, D.J., Luzzatto, L., Brenner, S. & Aparicio, S. (1995) Genomic structure and sequence of the *Fugu rubripes* glucose-6-phosphate dehydrogenase gene (G6PD). *Genomics* 26: 587–91

MHC Sequencing Consortium (1999) Complete sequence and gene map of a human major histocompatibility complex. *Nature* 401: 921–23

Miles, C., Elgar, G., Coles, E. *et al.* (1998) Complete sequencing of the *Fugu* WAGR region from WT1 to PAX6: dramatic compaction and conservation of synteny with human chromosome 11p13. *Proceedings of the National Academy of Sciences USA* 95: 13068–72

Naito, T., Saito, Y., Yamamoto, J. *et al.* (1998) Putative pheromone receptors related to the Ca^{2+}-sensing receptor in *Fugu*. *Proceedings of the National Academy of Sciences USA* 95: 5178–81

Pizon, V., Cuny, G. & Bernardi, G. (1984) Nucleotide sequence organization in the very small genome of a tetraodontid fish, *Arothron diadematus*. *European Journal of Biochemistry* 140: 25–30

Postelthwait, J.H., Yan, Y.L., Gates, M.A. *et al.* (1998) Vertebrate genome evolution and the zebrafish gene map. *Nature Genetics* 18: 345–49

Poulter, R. & Butler, M. (1998) A retrotransposon family from the pufferfish (*Fugu*) *Fugu rubripes*. *Gene* 215: 241–49

Poulter, R., Butler, M. & Ormandy, J. (1999) A LINE element from the pufferfish (*Fugu*) *Fugu rubripes* which shows similarity to the CR1 family of non-LTR retrotransposons. *Gene* 227: 169–79

Rowitch, D.H., Echelard, Y., Danielian, P.S. *et al.* (1998) Identification of an evolutionarily conserved 110 base-pair cis-acting regulatory sequence that governs Wnt-1 expression in the murine neural plate. *Development* 125: 2735–46

Sandford, R., Sgotto, B., Burn, T. & Brenner, S. (1996) The tuberin (TSC2), autosomal dominant polycystic kidney disease (PKD1), and somatostatin type V receptor (SSTR5) genes form a synteny group in the *Fugu* genome. *Genomics* 38: 84–86

Sandford, R., Sgotto, B., Aparicio, S. *et al.* (1997) Comparative analysis of the polycystic kidney disease 1 (PKD1) gene reveals an integral membrane glycoprotein with multiple evolutionary conserved domains. *Human Molecular Genetics* 6: 1483–89

Sarwal, M.M., Sontag, J.M., Hoang, L., Brenner, S. & Wilkie, T.M. (1996) G protein alpha subunit multigene family in the Japanese puffer fish *Fugu rubripes*: PCR from a compact vertebrate genome. *Genome Research* 6: 1207–15

Schofield, J.P., Elgar, G., Greystrong, J. *et al.* (1997) Regions of human chromosome 2 (2q32-q35) and mouse chromosome 1 show synteny with the pufferfish genome (*Fugu rubripes*). *Genomics* 45: 158–67

Schuddekopf, K., Schorpp, M. & Boehm, T. (1996) The whn transcription factor encoded by the nude locus contains an evolutionarily conserved and functionally indispensable activation domain. *Proceedings of the National Academy of Sciences USA* 93: 9661–64

Trower, M.K., Orton, S.M., Purvis, I.J. *et al.* (1996) Conservation of synteny between the genome of the pufferfish (*Fugu rubripes*) and the region on human chromosome 14 (14q24.3) associated with familial Alzheimer disease. *Proceedings of the National Academy of Sciences USA* 93: 1366–69

Venkatesh, B., Tay, B.H., Elgar, G. & Brenner, S. (1996) Isolation, characterization and evolution of nine pufferfish (*Fugu rubripes*) actin genes. *Journal of Molecular Biology* 259: 655–65

Venkatesh, B., Si-Hoe, S.L., Murphy, D. & Brenner, S. (1997) Transgenic rats reveal functional conservation of regulatory controls between the *Fugu* isotocin and rat oxytocin genes. *Proceedings of the National Academy of Sciences USA* 94: 12462–66

Wentworth, J.M., Schoenfeld, V., Meek, S. *et al.* (1999) Isolation and characterisation of the retinoic acid receptor-alpha gene in the Japanese pufferfish, *F. rubripes*. *Gene* 236: 315–23

Yamaguchi, F., Macrae, A.D. & Brenner, S. (1996) Molecular cloning of two cannabinoid type 1-like receptor genes from the puffer fish *Fugu rubripes*. *Genomics* 35: 603–05

Yeo, G.S., Elgar, G., Sandford, R. & Brenner, S. (1997) Cloning and sequencing of complement component C9 and its linkage to DOC-2 in the pufferfish *Fugu rubripes*. *Gene* 200: 203–11

Yoda, T., Morita, T., Kawatsu, K. *et al.* (1995) Cloning and sequencing of the chaperonin-encoding Cctd gene from *Fugu rubripes rubripes*. *Gene* 166: 249–53

GLOSSARY

CpG dinucleotide nucleotide composed of cytosine and guanine, linked together by a 3′,5′ phosphodiester bond

LINEs (long interspersed elements) group of interspersed repetitive retrotransposon-like sequences found in mammalian genomes

methylation addition of a methyl group to a chemical compound or macromolecule, e.g. protein or DNA

order taxonomic group of related organisms ranking between family and class.

orthologue genes related by common phylogenetic descent

SINEs (short interspersed elements) group of repeated sequences in the mammalian genome

splice sites exon/intron junctions

synteny when genetic loci lie on the same chromosome in different species

ARABIDOPSIS GENETICS AND GENOME ANALYSIS

Randy Scholl and Michael Tilley
Arabidopsis Biological Resource Center, Ohio State University, Columbus, USA

1. Introduction

The genetics of Arabidopsis, a small annual weed which belongs to the Brassicaceae family, has been studied since the early 20th century (Rédei, 1975). The first reported mutation study was conducted by Reinholz (1947), and physiological research on the plant was initiated by Laibach (1943). Biochemical mutants in the thiamine pathway were isolated by Langridge (1955) and Rédei (1965, 1975). Subsequently, numerous mutants have been induced in Arabidopsis by chemical, radiation and, more recently, insertional mutagenesis (Bouchez *et al.*, 1993; Feldmann *et al.*, 1994) and used for biological investigations and genetic mapping (Koornneef *et al.*, 1983). Molecular research was initiated in the 1980s, and the discovery that the genome was small (Leutwiler *et al.*, 1984) added impetus to the growing use of the plant for genetic and molecular research. Subsequently, many genes have been cloned, the physical map has been essentially completed, many new and interesting mutants have been isolated and the genome sequencing project was initiated (Kaiser, 1996). The main characteristics that favour Arabidopsis as a research subject are:

- rapid growth cycle (8–12 weeks);
- large seed set (thousands per plant);
- small plant size, particularly the facility to grow massive numbers of seedlings on plates to impose chemical and/or physical environmental selection;
- ease of mutagenesis;
- relatively small haploid genome size; and
- low repetitive DNA content (Rédei, 1992).

Since the discovery of its small genome size, research on Arabidopsis has grown exponentially, with a concomitant development of new analytical tools. It is intriguing that the coding and regulatory base sequences of plant genes are highly conserved so that Arabidopsis can be utilized as a genetic model and direct development tool for biotechnology. By contrast, the complexity of plant genomes varies substantially so that the small overall haploid size – 100 million base pairs (bp) – and content of repetitive DNA (<20%) place Arabidopsis at the lowest end of the spectrum for both measurements in flowering plants.

2. Genetic analysis

Mutagenesis can be conducted successfully by several means, and diverse types of informative mutants have been isolated. A simple gene nomenclature has been adopted (Meinke & Koornneef, 1997): Arabidopsis gene symbols are written in italics and consist of three letters, with mutant being lowercase and wild-type uppercase. Arabic numbers are used to designate loci of a common group and alleles of a locus. The locus number immediately follows the gene symbol and the allele number follows the gene symbol/number, being preceded by a dash (for example, *cys-3A*).

(i) Mutagenesis

Both physical and chemical mutagens have been utilized in Arabidopsis. Radiation sources such as X-rays, gamma rays, fast and thermal neutrons and ultraviolet radiation have been applied. Chemicals employed include ethylmethane sulfonate, nitrosoguanidine and nitrosourea (Feldmann *et al.*, 1994). Ionizing radiation tends to generate chromosomal deletions and rearrangements, and chemicals produce point mutations.

(ii) Mutants

Point mutations may affect individual genes through complete or partial loss of function or alteration of function. Small deletions or insertions usually result in loss of function, but may, in specific cases, reduce gene function. A large public collection of mutants, obtained by treatment with chemicals and radiation, has been accumulated (Anderson & Mulligan, 1992; see also current Arabidopsis Biological Resource Center (ABRC) Web Catalogue of the collections at http://aims.cse.msu.edu/aims/). Many of these mutants were identified by visible effects on plant/organ appearance; other classes include nutritional mutants, altered environmental response, altered interaction with organisms (including pathogens) and resistance to chemical agents.

The set of thiamine-requiring mutants collected by Rédei (1965, 1975) is well known as the largest set of auxotrophic mutants in plants. Most of these occur at two loci, with a few mutations at three additional loci also having been characterized. Other auxotrophic mutants have been difficult to isolate. Mutants affecting morphology have been exploited to facilitate analysis of plant development. Among these, the flower mutants are the best known, but many other morphological effects have been characterized. Mutants for pathogen resistance and variation in environmental response such as salt tolerance also exist. The identification of the *leafy* mutation (gene symbol *lfy*), its cloning and subsequent expression analysis represents a prominent example of the potential applications for developmental mutants (Weigel *et al.*, 1992). In the initial *leafy* mutant, Arabidopsis flowers (and subsequent fruits) are replaced on the inflorescence by stems and leaves. Since the mutant defect is the loss of flowers, the wild-type allele functions to stimulate flowering (Hempel *et al.*, 1997). Hence, this gene may be useful in streamlining tree breeding through acceleration of the generation time, among other applications.

(iii) Chromosomal variants

Chromosomal variants are difficult to characterize cytogenetically because the Arabidopsis chromosomes are very small. Nevertheless, the complete aneuploid series has been developed in two different genetic backgrounds (Koornneef, 1994). Telotrisomics have been isolated and utilized for mapping; polyploid strains have also been characterized (Rédei, 1964). Tetraploid strains ($2n = 10$) are generally fertile, with stable genomes.

(iv) Molecular variants

DNA polymorphisms identified in Arabidopsis have been utilized to facilitate genetic mapping and genome analysis/exploration. Restriction fragment length polymorphisms (RFLPs) were initially analysed by the comparison among wild strains (ecotypes) of DNA hybridization patterns of genomic DNA treated with restriction enzymes (Chang *et al.*, 1988; Kuiper, 1997). Various PCR-based procedures have since been utilized. These substantially streamline the typing of lines for the involved markers and make it much more feasible to find new polymorphisms in a region of interest.

3. Recombination mapping

The earliest mapping studies in Arabidopsis were conducted by Rédei and Hirono (1964), who identified linkage groups using mutant loci. Koornneef published the first map defining linkage groups corresponding to the five Arabidopsis chromosomes (Koornneef *et al.*, 1983).

The following approaches to mapping include defining types of mapping populations and acceleration of genetic mapping. The current status of genetic mapping is also addressed.

(i) Types of mapping populations

Genetic mapping of mutant loci has been conducted utilizing populations resulting from crosses between contrasting homozygous lines. Both visible markers and DNA polymorphisms have been mapped – in some cases, in integrated experiments. Due to the difficulty of performing crosses, F_2 or recombinant inbred populations rather than backcrosses are utilized (e.g. Chang *et al.*, 1988; Kuiper, 1997). To generate recombinant inbred lines (RILs), random members of F_2 populations are selfed, as separate lines, to homozygosity (Burr & Burr, 1991). Specifically, each F_2 plant is self-fertilized, and a single resulting F_3 progeny plant of each F_2 plant is again self-fertilized to produce F_4 progeny plants. In each subsequent generation, single progeny plants of the parental generation are self-fertilized to advance the population. This is continued to approximately the F_8 generation, which results in plants that are essentially homozygous. At this point, each single-seed descent line is bulk self-fertilized to obtain a large stock of seeds. These RILs can be shared among laboratories to allow combined mapping of many DNA markers in the same population. While the overall information content per individual may be less for RILs than for some other types of mapping population (Liu *et al.*, 1996), this method is very efficient for analysing linkage orders in short chromosomal regions. Several sets of RILs are publicly available (e.g. Reiter *et al.*, 1992; Lister & Dean, 1993; see ABRC Web site, http://aims.cse.msu.edu/aims/).

Tetrad analysis, made possible by the isolation of two mutants which results in the fusion of the walls of the four pollen grains from a single meiotic event, has recently been applied to Arabidopsis (Copenhaver *et al.*, 1998). This method allows the mapping of centromeres and direct determination of crossover numbers per pollen mother cell, and substantial progress has been made in pinpointing Arabidopsis centromeric locations utilizing this population. A backcross population was developed using these mutants as background. The four attached pollen grains of a tetrad of the F_1 were each used to pollinate single, homozygous parental plants and the resulting seeds were grown to be part of the mapping population.

(ii) Approaches to accelerate genetic mapping

Exploring DNA variation by hybridization of filters (e.g. Chang *et al.*, 1988; Nam *et al.*, 1989) is effective but labour intensive and requires large quantities of DNA. Consequently, a number of techniques have been developed to facilitate genetic recombination mapping – particularly the visualization of polymorphisms utilizing PCR. These techniques include the use of:

- cleaved amplified polymorphisms (CAPs);
- simple sequence length polymorphisms (SSLPs);
- random amplified polymorphic DNAs (RAPDs);
- amplified fragment length polymorphisms (AFLPs);
- acceleration of population screening via bulked segregant analysis (BSA) and chromosome landing; and
- the Arabidopsis physical map information.

The *CAP* principle involves amplification of a specific chromosomal region using an opposing pair of location-specific primers (Konieczny & Ausubel, 1993). The amplified band is then treated with a restriction enzyme and the digestion product separated by standard electrophoresis. If individuals differ for the presence of a restriction site within the amplified region, a difference between the banding patterns will result. This technique is reliable, but the sequence of the targeted region must be known and a restriction polymorphism available for its use.

SSLPs (Bell & Ecker, 1994) take advantage of repetitive sequences that are scattered throughout the genome and which typically have differing numbers of short repeats among strains (ecotypes). Unique sequences are located which flank both sides of such a region, and primers from the flanking regions are used for PCR. Electrophoresis of the PCR products from different strains typically yields different band sizes.

RAPD analysis utilizes PCR for the visualization of polymorphism and does not require prior knowledge of the local DNA sequence (Reiter *et al.*, 1992). Short (typically 9- or 10-nucleotide) single primers are used for PCR and subsequent electrophoresis. Such primer sequences are frequent enough in the genome that a number of bands will be amplified from opposing, identical primer-binding sites. Polymorphism for the presence of these bands exists, based on divergence of sequence in the target binding sites of the primers. Inheritance of RAPDs is dominant/recessive as opposed to codominant, which is a disadvantage for analysis in F_2, but not RIL populations. Identification and incorporation of new RAPD loci into a mapping project can be extremely quick.

AFLP is an alternative PCR-based approach that is becoming increasingly popular because, like RAPDs, it allows rapid identification of random new polymorphisms (Vos *et al.*, 1995). Presence of an AFLP band is genetically dominant. It may be possible, however, to distinguish heterozygotes from homozygotes for band presence by quantifying the band intensity (Kuiper, 1997). The process is as follows (see **Figure 1**):

1. genomic DNA is double-digested with a frequent-cutting enzyme (e.g. 4-cutter such as *Mse*I) and a rarer cutting enzyme (e.g. a 6-cutter such as *Eco*RI)
2. *Eco*RI and *Mse*I adaptors, which are *c.*20 bases in length and consist of non-Arabidopsis sequence, are ligated to the restricted genomic DNA
3. a preliminary non-specific PCR amplification is conducted with one primer homologous to the *Eco*RI adaptor and a second primer homologous to the *Mse*I primer, but with a single arbitrary selective base added at the 3′ end
4. a second amplification is conducted with 2–3 arbitrary bases added at the 3′ end of both restriction sequences (one of the primers is radioactively end-labelled for later visualization of the amplified fragments
5. the amplified products are separated on a sequencing gel at 55°C. PCR amplification produces 30–80 bands on the gel, with approximately one-third being polymorphic among typical Arabidopsis ecotypes. In this way, numbers of polymorphic sites can be visualized simultaneously. The repeatability of banding patterns is very high so that AFLP represents a robust approach to polymorphism analysis.

Chromosome landing. Determining the location of genetic markers that are close to a particular locus can be a daunting task, and yet this is vital if the mapping is being used in conjunction with attempts to clone the locus. The success of localized mapping relies on the identification of recombinants carrying crossovers near the target locus and the mapping of these relative to the target and its neighbouring loci. An approach (Tanksley *et al.*, 1995) that reduces the effort associated with this task is conducted as follows (see **Figure 2**):

1. produce a number of segregants in F_2 or other mapping populations;
2. map the target locus by standard methods relative to a number of standard mapping sites with the goal of identifying flanking sites that are as close as possible to the target gene;
3. organize the segregant lines into classes relative to recombination between these sites and the target (i.e. recombination between left-flanking locus and the target but not the right-flanking locus as one bulked

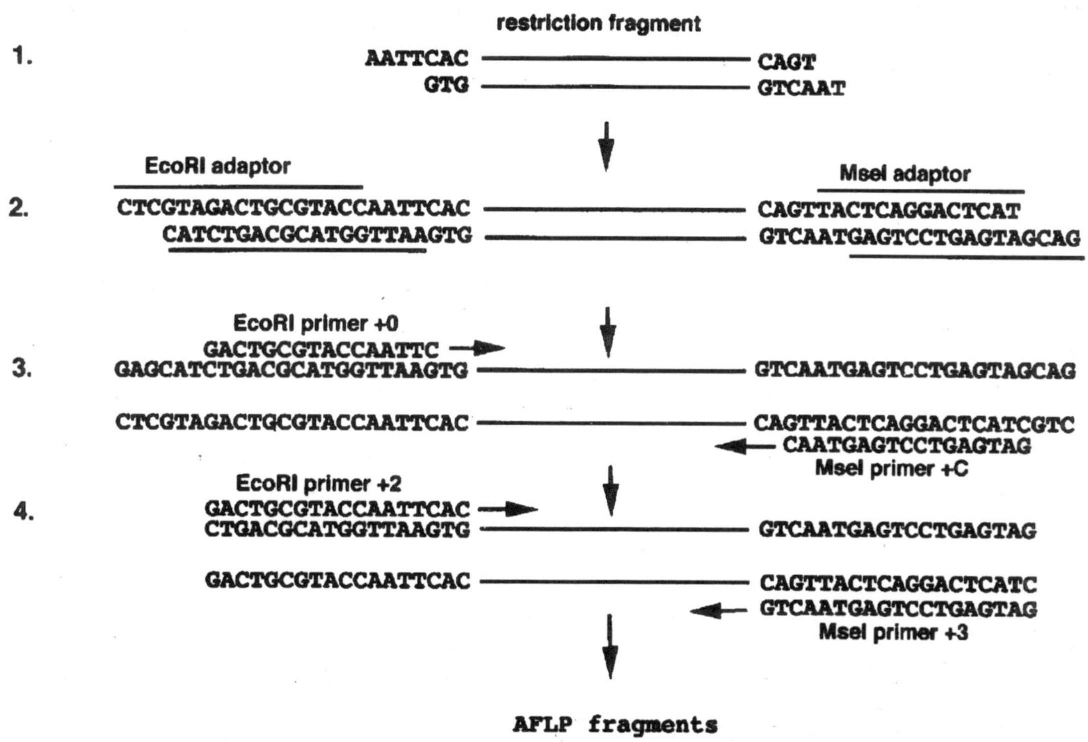

Figure 1 Schematic representation of the AFLP protocol. (1) *Eco*RI–*Mse*I restriction fragment. (2) Same fragment after ligation of the *Eco*RI and *Mse*I adapters (boxed), creating the template DNA for AFLP amplification. (3) Target sites of the *Eco*RI and *Mse*I preamplification primers on the template DNA; the arrows indicate the direction of DNA synthesis. (4) Target sites of AFLP primers on the preamplified DNA.

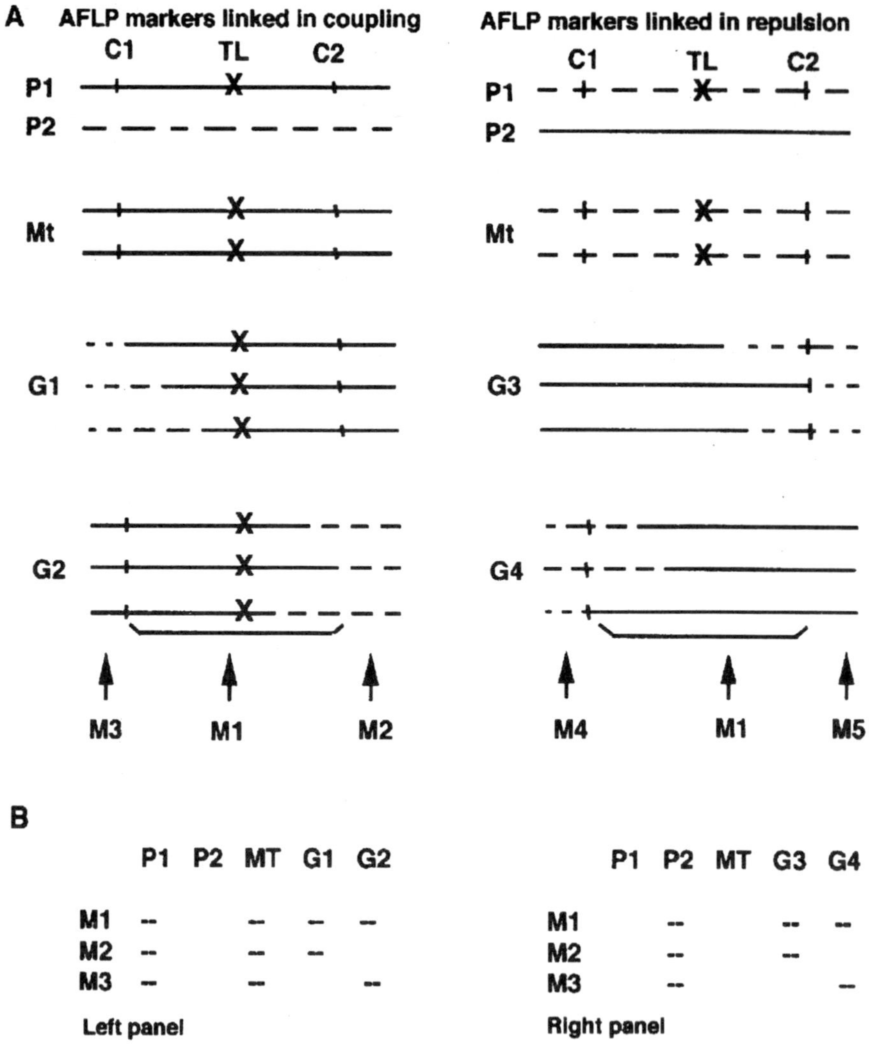

Figure 2 (A) The left and right panels show the identification of AFLP markers linked in coupling and repulsion, respectively. G_1 and G_2 each show three possible genotypes of recombination events between C1 and TL and C2 and TL, respectively, in homozygous F_2 plants. G_3 and G_4 represent possible genotypes of F_2 recombinants that lost the mutation, but are homozygous for C1 (G_4) or C2 (G_3), respectively. P1, mutant parental genotype; P2, non-mutant parental genotype; MT, homozygous mutant; TL, target locus; C1 and C2, flanking markers linked in coupling to the TL. (B) Marker M1 shows band pattern that identifies AFLP markers closely linked to the TL (left panel, in coupling; right panel, in repulsion). Markers M2 and M3 are absent in G_2 or G_3, respectively, and, therefore, situated further away from the locus than marker M1. Markers M4 and M5 are absent in G_4 and G_3, respectively, and thus not necessarily tightly linked to the locus.

class, and the reverse of recombination for a second bulked class);

4. classify the bulked DNAs for type relative to new random markers (AFLP or RAPD can be used efficiently to develop new markers); and
5. when markers very close to the locus are found, they can be used as probes to isolate large-insert clones such as bacterial artificial chromosomes (BACs) or yeast artificial chromosomes (YACs). If the mapped loci are sufficiently close to the target locus, the hybridizing clone will likely include the target gene.

While this approach has been shown to be feasible for plants with large genomes which have not been extensively physically mapped, its power is amplified for Arabidopsis with its small genome and essentially complete physical map. In Arabidopsis, fine recombination mapping by chromosome landing-based approaches should ensure that a clone can be quickly identified for subsequent analysis and cloning. The detailed contig assemblies, especially of BAC clones that are rapidly being sequenced, should greatly accelerate and simplify the efforts to clone genes based on phenotype.

BSA (Michelmore *et al.*, 1991) is a simpler technique in which the segregants of a particular phenotype are combined into a single group for which the PCR markers are typed. Hence, a bulk of such DNA will give a uniform type for closely linked markers and mixed type for distantly or non-linked markers. It represents a good strategy for finding markers that are within a few centiMorgan (cM) of a locus.

(iii) Status of genetic mapping

The status of genetic maps has been addressed by Koornneef (1994). The morphological marker map contained 284 loci. RFLP maps have been constructed in several laboratories. The map of Chang *et al.* (1988) contained 126 mapped clones, the map of Nam *et al.* (1989) located 94 cosmids and other clones in relation to a set of 17 previously mapped clones and the map of Reiter *et al.* (1992) contained 60 RFLP clones and 252 RAPDs. Both the Chang and Nam studies utilized F_2 populations derived from pairs of contrasting ecotypes; the Reiter map was based on RILs derived from a 9-mutant marker crossed to a wild-type of contrasting background. The more recent map of Lister and Dean (1993; see also http://www.nasc.ac.uk) encompasses >600 markers, including standard RFLPs, CAPs and SSLPs on a set of 300 RILs.

The above maps represent more than one site per cM. Many more visible marker loci exist from other data – as evidenced by the mapping information for the >500 markers on the Meinke laboratory Web page (http://mutant.lse.okstate.edu/). More than 1000 named genes are currently recorded on this site, most of which are being studied genetically. Many molecular markers in additions to those reported by Lister and Dean (1993) have been mapped so that the overall mapping information, if combined, would have a resolution approaching 0.1–0.2 cM, or roughly 20–40 kilobase (kb) molecular resolution. Hence, there is probably at least one and likely multiple mapped markers for each BAC clone comprising the Arabidopsis physical map. Clearly, the combined density of mapped markers in Arabidopsis may often make it unnecessary to conduct searches for new closely linked markers for a gene targeted for cloning.

4. Gene discovery by forward genetics

Although numerous markers for diverse phenotypic effects have been described in Arabidopsis, cloning the genes simply from their phenotype has represented a daunting task. The earlier cases of positional cloning in Arabidopsis involved complicated chromosomal walks, sorting through candidate loci and ultimate verification by complementation. Currently, while the basic premises of positional cloning remain the same, a number of improvements has been made so that each step of the process has been simplified and accelerated. In addition to map-based methods, alternatives exist including T-DNA and transposon tagging (see parts (ii) and (iii) in this section). Furthermore, direct analysis of gene function can be conducted using gene replacement and RNA-based methods such as antisense and gene silencing.

(i) Map-based approaches

The relatively small percentage of repetitive DNA and the favourable relationship between the meiotic recombination rate and the physical map (i.e. 1% recombination equals approximately 180 kb) makes this approach feasible in Arabidopsis. A number of important advances have recently enhanced this situation, as follows.

Large-insert libraries have lent substantial power to positional cloning. YAC libraries (Guzman & Ecker, 1988; Hauge *et al.*, 1990; Ward & Jen, 1990; Grill & Somerville, 1991; Creusot *et al.*, 1995) and then P1-derived artificial chromosomes (Liu *et al.*, 1995a) and BAC libraries (Choi *et al.*, 1995; use http://www.mpimp-golm.mpg.de/101/mpi_mp_map/bac.html at T. Altmann Web site, Max Planck Institut, Germany) have become widely available. The sizes of the insertions of these clones translate nearly to cM spans, and these clones, therefore, greatly reduce the number necessary to span a region. The ease with which BACs and P1s can be manipulated has also been extremely useful.

The simplification of polymorphism visualization, as described above, has also been an aid to positional cloning. PCR-based probes greatly reduce the effort associated with the map phase. Methods not requiring sequence knowledge (i.e. RAPDs and AFLPs) are extremely useful in this regard. While these are vital for exploration of species for which molecular literature is lacking, they provide powerful adjuncts to the other tools of Arabidopsis.

With the physical map of Arabidopsis essentially complete, researchers can choose BAC or YAC clones for any region of the genetic map. As the publicly available BACs are sequenced, their value to cloning efforts continues to increase. The identification of close flanking DNA polymorphisms is critical to the process of positional cloning. As described in the genetic mapping section, the relatively dense Arabidopsis map allows a reasonable chance of at least one existing marker being found close to a candidate locus. Positional information provided by close flanking markers should place the target gene on a BAC or YAC from the public collections and set the stage for analysis of a single clone to locate the correct gene.

Identification of potential coding regions should be substantially accelerated by the existence of public expressed sequence tag (EST) sequences (Höfte *et al.*, 1993; Newman *et al.*, 1994) and corresponding clones. Likewise, the final confirmation that a candidate gene is responsible for a phenotype, which is most often done by transformation of the mutant with a vector containing the wild-type allele of the candidate gene, can be conducted efficiently. The task of showing that a gene/location is responsible for the phenotypic effect has been made easier by the recent development of transformable artificial chromosomes (TACs) which allow BAC-sized DNAs to be transformed into plants for complementation analysis (Kaneko *et al.*, 1999). In fact, if genome-saturating numbers of such clones are available, it should be possible to clone genes by simply complementing the phenotype of mutants with these clones and then dissecting the positive clone(s) to isolate the target gene.

(ii) T-DNA populations

An alternative approach for cloning genes with interesting phenotypes is insertional mutagenesis with a characterized DNA so that the mutant becomes automatically tagged (Feldmann *et al.*, 1989). This approach has been popular in Arabidopsis due to the uncertainties of positional methods. While gene tagging in Arabidopsis can be achieved by either T-DNA transformation or transposon activity, the former has developed more quickly in Arabidopsis because native transposons for such experiments could not be located and transposon systems from maize had to be tailored and introduced. Hence, a number of populations of random T-DNA transformants have been created and screened intensively.

In T-DNA tagging, numbers of independently transformed plants are used to establish lines. These lines are examined for phenotypic variations. When an interesting variation is identified, cosegregation analyses between the phenotype and markers associated with the T-DNA (antibiotic resistance introduced with the T-DNA or molecular probes specific to the T-DNA) are conducted (Bouchez *et al.*, 1993). If the cosegregation results are positive, the T-DNA is used to rescue flanking genomic DNA from the T-DNA mutant line. This can be done in one of several ways, most commonly using PCR-based procedures and plasmid rescue. Thermal asymmetric interlaced PCR (TAIL PCR) (e.g. Liu *et al.*, 1995b) consists of repeated amplifications with nested, T-DNA-specific primers 3′ to, but inside, the T-DNA paired with a degenerate arbitrary primer to obtain the flanking chromosomal regions. The resulting amplification product can often be cloned directly and then characterized. Alternatively, plasmid rescue can be conducted when the initial T-DNA vector includes a bacterial origin of replication and selective marker. In this case, the DNA can be cut with an enzyme, which releases the plasmid region plus flanking genomic DNA from the majority of the T-DNA, bacteria can be transformed with the resulting DNA and transformed strains identified and analysed. In principle, the identification of the responsible gene is straightforward.

It should be noted that about half of the mutants produced through T-DNA transformation turn out not to be associated with identifiable T-DNA, and hence cannot be cloned via tagging. Nevertheless, this system has been the most popular avenue for tagging in Arabidopsis.

The usefulness of T-DNA-transformed mutants has been enhanced by using insertions which can be expressed under the control of the regulatory sequences which might flank the mutant gene in the transformed plants. These fall into several classes, as described below. All these systems involve expression of a coding sequence, such as the one for beta-glucuronidase (GUS), which can be visualized using coloured enzyme reaction products. The most popular of these are enhancer traps (e.g. Bancroft *et al.*, 1992), which involve GUS linked to a minimal promoter and which will drive transcription only in conjunction with a neighbouring enhancer. The second classes are promoter trap and gene trap constructs (Sundaresan *et al.*, 1995); these are promoterless genes that are activated if inserted adjacent to an active plant promoter. Some of these constructs also have intron splice junctions at their 5′ end so that insertion into an intron may result in synthesis of a GUS hybrid protein that will have activity – hence the term "gene trap" (Sundaresan *et al.*, 1995). The third type of expression construction is the "activation" tag approach (Hayashi *et al.*, 1992). In this case, a multiple enhancer construct is included in the T-DNA so that it may upregulate genes near which it inserts. All of these types of expression system enhance the usefulness of the T-DNA transformants, so that specific and/or novel expression patterns resulting from fusion of the insertion construction to genomic plant sequences can be identified.

As a result of the recognized usefulness of random T-DNA lines, numbers of laboratories have generated substantially sized populations. Some of these have been made publicly available through the ABRC and Nottingham, UK stock centres. Approximately 40 000 T-DNA lines are currently available in this fashion, and it is expected that the numbers will increase rapidly. It is estimated that >80 000 lines, which should carry 120 000 insertions based on the typical number of 1.5 per transformed line, would be sufficient to ensure a high probability of finding any unique locus tagged.

(iii) Transposon systems

Transposon tagging systems aimed at gene cloning have been developed in other species, including Drosophila and maize. The maize systems have proven transportable to Arabidopsis, so that transposon tagging using the Ac/Ds transposons, as well as Spm, has been utilized to build transposon tagging lines and populations (Bancroft *et al.*, 1992; Fedoroff & Smith, 1993; Aarts *et al.*, 1995; Osborne *et al.*, 1995).

The Ac and Ds transposons of maize have been engineered for two types of insertional tagging strategies – local and genome wide. The former relies on the fact that approximately 70% of transpositions result in insertions at targets within 20 cM of the donor site (on the same chromosome). Hence, different lines have been generated (using Agrobacterium transformation) in which the Ds element that will transpose is transformed into different, random locations which are then mapped. Hence, one of these Ds lines can be chosen which will transpose into a particular chromosomal region. The second useful aspect of Ds/Ac transposition systems is that the transpositions can be induced for a single generation and further transposition prohibited in subsequent generations. This is achieved by the utilization of lines having Ac in one strain and Ds in the other. The Ac transposon contains active transposase, but not the transposon ends, so that it promotes transposition but does not itself jump. Ds, conversely, is a transposon lacking transposase, so that it jumps only when induced by Ac. Hence, crossing an Ac and a Ds line results in Ds transposition activity in the F_1, and Ds transposed lines lacking Ac can be isolated by normal segregation in the F_2 generation. This type of system provides a very powerful tool for stably tagging genes, if a Ds line exists for which the element is near a mapped target site.

Genome-wide tagging using Ac/Ds was developed by Sundaresan and colleagues (Sundaresan *et al.*, 1995). The crossing scheme between Ac and Ds parents is also employed in the same fashion as for local Ac/Ds tagging. For genome-wide tagging, selective markers are available so that transpositions down the same chromosome can be selected against in favour of jumps to other chromosomes including the homologue. Among jumps to different chromosomes, the destination site is random; hence, genome-wide, random insertions are the result. Gene- and enhancer-trap construction are also utilized in transposon tagging systems and have been shown to be very effective in identifying interesting expression patterns.

5. Functional genomic analysis

(i) Reverse genetic PCR exploration of T-DNA populations

It is possible with Arabidopsis to utilize populations with random insertions, such as T-DNAs or transposons, to locate mutations for genes that have been cloned but not previously investigated genetically. Procedures involving exploration of genetics using cloned molecular sequences or gene products as a starting point are referred to as "reverse genetics". PCR can be utilized to identify insertions in a gene of interest as follows:

- primers are designed for both borders of the insertion construction;

- forward and reverse primers are also designed for different locations of the target gene;
- PCR amplifications are performed for combinations of target gene primers and T-DNA insertion primer; and
- if an insertion has occurred for a gene in a particular line, a PCR product will be observed when the insertion primer and the target primer are opposed in the PCR amplification.

Large populations of insertion candidate lines can be assayed simultaneously since the sensitivity of PCR allows the detection of a positive product even when the DNA of the critical line is diluted by non-positive DNA from many other lines with insertions at other locations.

Large populations of T-DNA lines have been developed and DNA isolated and successfully utilized to identify loss-of-function insertional mutants for a number of genes of interest (McKinney *et al.*, 1995; Krysan *et al.*, 1996). DNAs and lines for substantial T-DNA populations are available from ABRC so that screening of large pools can be followed by the screening of smaller subpools and then sorting of a few lines to isolate and characterize the disrupted gene. A service to which investigators can submit primers for screening of a 65 000-member T-DNA population has been established at the University of Wisconsin as part of the National Science Foundation's Plant Genome Project (http://www.nsf.gov).

The number of mutations that have been located by reverse genetics is increasing rapidly. The probabilities of finding insertions by PCR are similar to those for direct phenotypic screening. A population of 80 000 lines containing an average of 1.5 insertions per line should have a probability in the 0.95 range of carrying at least one insertion in any particular target gene. However, the likelihood of an insertion in a smaller than average sized gene (e.g. <1 kb, where the average is assumed to be approximately 2 kb) is much lower, so that the availability of as many as 2–300 000 insertions may prove useful.

(ii) PCR analysis with transposon systems

Transposons have the same potential for utilization in reverse genetics as T-DNA transformants. In particular, mapped Ds lines can be crossed to Ac lines to create saturating populations of insertions within 10–20 cM of the original location of the Ds element. Hence, for insertions not found in large random populations, local saturating mutagenesis is feasible using these resources. The PCR screening procedure for identification of the targeted insertion event can be conducted using a strategy similar to that employed for T-DNA populations, as outlined above.

(iii) Direct gene replacement/inactivation

The most direct approach for inactivating or altering expression of genes would be through gene replacement via homologous recombination. However, in the past, this technology has not been available in plants. Recently, however, a potentially useful gene replacement technique was described for Arabidopsis (Kempin *et al.*, 1997). Standard Agrobacterium transformation of the target strain is conducted and transformants identified by marker genes in the transforming T-DNA. Two novel aspects of the transformation are: 1) the T-DNA is small and includes an altered, target gene-specific insertion (e.g. disabled clone if the aim is gene disruption), so that, if the transformation site is the target locus, a functional replacement will result; and 2) transformations are created in large numbers (>1000) and then the insertion in target region is assayed by PCR with primers flanking the gene so that insertions at that location can be identified efficiently.

(iv) RNA-based approaches

Antisense RNA approaches to studying gene expression have been available for some time in Arabidopsis (e.g. Zhang *et al.*, 1992). Gene silencing and cosuppression (Brusslan *et al.*, 1993; Boerjan *et al.*, 1994) also occur with many expression transformation constructs, so, in conjunction with existing efficient transformation approaches, these represent viable alternatives for achieving over- or underexpression of Arabidopsis genes. The transformations necessary to integrate these types of construct into plants can be achieved with reasonable effort, although transient expression systems, which would greatly reduce the time required for completion of analysis, may not always work. Furthermore, for both antisense suppression and gene silencing, the degree of expression cannot always be controlled/predicted.

(v) Large-scale analysis of gene expression using DNA "chip" and microarray technology

A main goal of biology has been to understand how an organism responds to changes in its environment and controls development. With the advent of molecular biology and biochemistry, it has become possible to look at these phenomena at the level of gene regulation. However, until recently, researchers had been limited by the fact that they could look at only a few genes at a time and, thus, they were not able to grasp the global genetic effects of environmental stimuli or changes in expression effects for genes of interest. Recent advances allow scientists to study simultaneously the transcriptional regulation of many genes, even the entire genome, of a species (Schena *et al.*, 1995; Schena, 1996; Schena *et al.*, 1996; Shalon *et al.*, 1996; DeRisi *et al.*, 1997). This technology is called DNA microarray or chip technology.

DNA microarray and DNA chip analyses are similar technologies in which high density arrays of DNA sequences representing large numbers of genes are hybridized to probes such as cDNA populations derived from plant RNA so that comparative quantitation of messages comprising the RNA population(s) can be achieved. Chip technology uses oligonucleotides that are photolithographically synthesized at a very high density on the surface of a small synthetic wafer or "chip". To use chip technology, the researchers must know the sequence of each gene to be compared; also, it is very expensive to design and manufacture the chips (Southern, 1996).

Microarray technology differs in that larger individual DNAs, isolated from a source such as a standard cDNA library, are arrayed onto silylated glass slides by a robot (Shalon *et al.*, 1996). A 5′ amino modification of the cDNAs is made and used to bind the clones. It should be noted that microarrays do not require preexisting knowledge of sequences. The Arabidopsis public genomic projects currently being initiated are utilizing microarrays rather than chips for their gene expression analyses. To assay gene expression using a microarray, the slide with its array is simultaneously hybridized with two probes of interest (e.g. cDNAs generated from plants subjected to some form of stress *vs.* a control cDNA sample). Generation of the cDNA probes is as follows. A mixture of dNTPs containing Cy3- or Cy5-dUTP, respectively, is used as substrate, so that the resulting cDNAs will fluoresce. Following the hybridization step, a laser is used to stimulate the fluorescent tags on the DNAs of the probes so that the degree of hybridization to any location in the microarray can be visualized. The total fluorescence is proportional to the amount of probe bound to the microarray and is linear over a wide range. Also, the relative fluorescence of the respective tags can be ascertained to evaluate the proportion of signal being generated by the two probes employed. This shows visually as a colour gradient that is created by the proportion of the two fluorescent wavelengths (Schena, 1996; Shalon *et al.*, 1996). Therefore, the relative hybridization of the two probe sources to each individual spot on the microarray can be determined. Scanners have been developed specifically for the task of reading the large number of points that form an array. These data can then be compared with other results through computer analyses. In addition to the response of normal plants to stimuli, different developmental stages, varying temporal conditions and the effects of loss- or change-of-function mutant *vs.* wild-type can be compared.

Since the sequencing of the Arabidopsis genome is now complete (The Arabidopsis Genome Initiative, 2000) and a large set of cDNA and other clones are available, a series of functional genomics projects have been established which will conduct large-scale technologies for global expression and forward- and reverse-genetic initiatives. These are funded by a programme of the US National Science Foundation called the "Plant Genome Project" (PGP, see http://www.nsf.gov for a complete list of awardees). The data will be deposited in public databases and mutants etc. derived from the projects will be made publicly available. In one of these projects – a collaboration between the University of Arizona, Purdue University and others – microarrays will be used to study the glycophytes *Arabidopsis thaliana* and *Oryza sativa* and the halophytes *Dunaliella salina* and *Mesembryanthemum crystallinum*. The purpose behind the experiments is to compare the differences in gene expression between glycophytes and halophytes using microarray analysis (glycophytes being plants requiring more than 0.5% NaCl solution in the substratum, while halophytes are saline tolerant). Also, large numbers of mutations affecting the responses to abiotic stress will be isolated.

Another project – at Michigan State University, the University of Wisconsin, Yale University and Stanford University – will generate cDNA microarrays for public use. Hence, global effects resulting from mutations or growth conditions already being intensively studied from other perspectives will be ascertained. This project has several additional goals. One is to discover the global expression pattern of all available Arabidopsis genes in various tissues under different conditions. To achieve this, plants will be grown and RNA isolated under a standardized range of environmental conditions, known and patented as the "gauntlet". cDNAs are made from the mRNA isolated from treated plants. These cDNAs will be hybridized to microarrays, and the global expression pattern for each treatment can then be discovered. A second goal is to identify plant-specific genes (through comparative analysis of ESTs and concentration on those unique to plants) and the effects of loss-of-function mutations of these genes on plant phenotypes (as assessed by the gauntlet) and the effects of these mutations on expression of members of the microarray.

In addition to the above approaches, a service facility to aid researchers in identifying T-DNA inserts in genes of interest in Arabidopsis is being established by this group as described above. Initially, PCR of DNA isolated from large T-DNA populations (as outlined in section 5, part (i) above) will be used to scan rapidly for knockout mutations, and then the line of interest can be isolated. Ultimately, however, microarrays may be used to perform these screens. For example, the plant-specific loss-of-function mutants mentioned above could be identified by amplification of genomic DNAs flanking the T-DNA insertions using TAIL PCR (Liu *et al.*, 1995b), followed by hybridization of the PCR product to a known set of plant-specific DNAs in a microarray.

6. Physical genome exploration

The sequencing of the 100 million bp Arabidopsis genome, initiated in 1996, had an original target date of 2007, which was revised to 2004 and then to the year 2000 for the essential completion of the determination, annotation and publication of the completed DNA sequence.

(i) Physical mapping

The physical map of Arabidopsis is being completed rapidly and efficiently, aided by advanced techniques for contig assembly. This has been a key factor in the forward revisions of the target date for completing the genomic sequencing.

Methods and clones utilized for contig construction. Construction of a physical map initially involves the development of sets of clones that are contiguous to one another on a chromosome region (i.e. a contig). When various contigs can be linked into a continuous overlapping set covering the entire chromosome, its physical map is complete. The Arabidopsis physical map is currently being very rapidly covered in this fashion, and these clones are anchored to the genetic map (see Arabidopsis Genome Initiative (AGI) Web sites for current contig sets, http://pgec-genome.pw.usda.gov/agi.html). The initial Arabidopsis physical maps were constructed from cosmid libraries. Contigs covering the majority of the genome were constructed via comparative restriction fragment fingerprinting of clones, BAC-end sequencing and hybridization (Grill & Somerville, 1991; Bevan *et al.*, 1998). Subsequently, YAC, BAC and P1 libraries have become available, all of which have average clone sizes greater than 100 000 bp. These combined libraries represent a genome coverage greater than 20-fold, and it has been possible essentially to complete the physical map. Clones of these libraries are available at the ABRC.

In conjunction with the large-insert libraries, the use of AFLP, end-sequencing for establishing and evaluating overlap, and BAC fingerprinting from restriction patterns (Bevan *et al.*, 1999; Marra *et al.*, 1999), the physical map has essentially been completed. Much of this information can be viewed on the various Arabidopsis Web sequencing sites.

(ii) The expressed sequence tag (EST) collection

Systematic sequencing in Arabidopsis began in 1993, when the sequencing of short end sequences (*c*.300 bp) of random cDNAs was initiated at Michigan State University (Newman *et al.*, 1994) and the CNRS, France (Höfte *et al.*, 1993). The sequences of these clones are registered in GenBank, and the clones are maintained at ABRC. More than 25 000 clones were sequenced, with 12 000 unique sequences resulting so far from the Michigan State University project, as estimated from the actual end sequences from the project. It is currently estimated that there are >20 000 transcribed genes in Arabidopsis. The Michigan State University group is currently sequencing cDNAs from modified and specialized libraries with the expectation that up to 20 000 unique cDNAs can be identified.

The Arabidopsis EST collection represents one of the largest publicly accessible cDNA collections. The published rice ESTs number approximately 20 000, which means that the degree of gene coverage is similar to that of Arabidopsis.

(iii) The genome sequencing project and results

The decision to initiate an international, publicly funded Arabidopsis genome sequencing project aimed at determining the entire sequence of all five chromosomes was finalized in 1996 (Kaiser, 1996). Subsequently, three groups in the US were funded to conduct genome sequencing, and groups in the UK, Europe and Japan also began work around the same time. The groups (collectively, AGI) cooperate in that each is responsible for the final physical mapping and sequencing of clones in a specific chromosomal region. The same BAC libraries are utilized by all of the sequencing groups, and sequences of BACs are reported to GenBank promptly after their complete sequences have been determined.

It is projected that essentially all of the genome sequence will be determined and available on the public databases by 2001. Characterization and annotation of the sequence information are proceeding apace. Gene-calling software is utilized to identify potential coding regions of the new sequence (Bevan & Murphy, 1999). It is anticipated that the existing EST collection will be useful in identifying numbers of coding regions that might be missed by analytical software. The Arabidopsis genome project is already having a direct effect on Arabidopsis research, and it is expected that immediate benefits will be realized for other flowering plants due to the close sequence similarities of corresponding genes within this group.

The Arabidopsis genomic sequencing strategy relies on the accurate placement of BACs, with each group accumulating sequences of BACs associated with defined chromosomal locations. Shotgun sequencing of random plasmid isolated from each BAC results in multiple, independent, bidirectional sequences for each BAC. This approach ensures high accuracy – the error rate is less than 1/10 000. The resulting BAC sequences are compared with fingerprint data developed by restriction analysis, which can locate any major anomalies for a BAC sequence. Hence, the overall chromosomal sequence is reliable and multiply checked prior to final assembly.

Sufficient genomic sequence is now available so that a clear picture of its composition can be formed. Some basic aspects of the sequence will be highlighted here. The telomeric sequences are known (Richards & Ausubel, 1998) and the centromeres have been localized by tetrad analysis (Copenhaver *et al.*, 1998) and are being characterized. The telomeres consist of eukaryotic consensus tandem repeat sequences (Richards *et al.*, 1992); the centromeres likely contain a 180 bp element with other types of repeat elements occurring adjacently, but apparently not involved in centromere function (Round *et al.*, 1997; Franz *et al.*, 1998; Copenhaver *et al.*, 1999). The genomic tools now exist to dissect the function of plant centromeres, which will be conducted as part of the Plant Genome Project.

The approximately 730 rDNA genes are located in two clusters at the ends of the short arms of chromosomes 2 and 4 – the locations of the nucleolus organizer regions. In both cases, the rDNA repeats are immediately adjacent to the telomeric repeat regions.

From the analysis of the complete chromosome 4 and partial chromosome 5 sequences (Bevan *et al.*, 1999), exons plus introns comprise slightly greater than one-half of the genome, and the mean gene size is about 2 kb. Hence, *cis* regulatory sequences occupy a substantial portion of the remainder of the genome, so that non-functioning DNA (including repetitive sequences) can constitute little more than 10% of the genome. The transposon-generated repeats found in Arabidopsis are of the long terminal repeat (LTR) and non-LTR retroelement classes, which occur from single elements to clusters of three elements; these comprise a very small proportion of the genome. This situation is very different from corn, in which complex interspersions of various retroelement sequences comprise one-half of the genome and occur in many large clusters. Hence, the accumulating genome sequence confirms the early reports from lambda phage sampling (Leutwiler *et al.*, 1984) which characterized the Arabidopsis genome as consisting largely of coding and regulatory sequences with repetitive sequences being very low in abundance. It is also interesting to note that, while evidence of the above types indicates that transposon activity has occurred in Arabidopsis, highly active native transposons are absent from the genetic sources so far examined.

The best current estimate of the number of genes in Arabidopsis is 21 000. The GC content of exons (~44%, based on a segment of chromosome 4) is distinctly higher than that of introns (~33%) and intergenic DNA (32%). The mean number of introns for genes on chromosome 4 is 5.6, with a range of 0–29 (Bevan *et al.*, 1999) and has been estimated at 4.8 for a sample of the entire genome (Deustch & Long, 1999). The introns are typically 25–2984 bp in size with average size of 240 bp (Deustch & Long, 1999). There is approximately one gene every 4.8 kb, as assessed for part of chromosome 4 (Bevan *et al.*, 1999). About 55% of the putative genes have similarity to known genes, and about 53% match Arabidopsis ESTs; about 53% of the genes can be assigned cellular roles. From the EST and AGI efforts, large numbers of genes can be identified as being unique to plants and not found in animals or bacteria. Elucidation of the effects and functions of these genes will form a major focus of the functional genomics studies of the PGP-funded Arabidopsis projects. Codon bias exists for Arabidopsis proteins and has been analysed (Chiapello *et al.*, 1999); interestingly, codon usage differs for nuclear genes encoding mitochondrial-targeted protein compared with the same protein to be utilized in the cytoplasm (Chiapello *et al.*, 1999).

Arabidopsis proteins are typically encoded by families of genes. For some genes, the families may be smaller in Arabidopsis than for other plant species, although generalizations cannot be made. Particularly large gene families exist for a number of proteins (e.g. actin) and large gene families are commonplace for transcription factors (e.g. the myb factors). There are estimated to be 350 cytochrome P450 genes in Arabidopsis, although these are very diverse in sequence and probable functions in metabolism (Nelson, 1999). Recently, a large gene family consisting of at least 100 members was discovered through analysis of the genomic sequence data (Ride *et al.*, 1999). These genes have homology to the stigmatic self-incompatibility genes of *Papaver rhoeas*, and have been termed "S-protein homologues". These genes are unique to plants and two members have been shown to be most highly expressed in flowers, although no function has yet been assigned to any member of the family.

An interesting feature that has emerged from the Arabidopsis genomic sequencing is the clustering of protein-encoding genes having similar function (Bevan *et al.*, 1999). In one chromosome segment, 20% of the genes occurred in this arrangement and the same DNA strand is often the coding strand. The most notable such example is the leucine-rich repeat family, which likely confers disease resistance. The functional redundancy provided by families of genes poses a barrier to the initial elucidation of gene function by genetic approaches, but the application of postsequencing functional genomics to these systems should successfully reveal their role in metabolism and development, including any complex interactions.

Conserved arrangements of similar genes on corresponding chromosome segments (synteny) has been shown to occur in the grasses, including maize and rice (Gale & Devos, 1998). This will likely aid in the advancement of functional analysis of genes within this group. The degree of synteny that may occur between these monocots and dicots is still uncertain. A high degree of conservation would enhance the usefulness of Arabidopsis as a model for diverse angiosperms. Paterson *et al.* (1996) identified clusters of apparently conserved genes between Arabidopsis and sorghum. Conversely, similar synteny between Arabidopsis and rice has not been observed (Gale & Devos, 1998). The analyses of both the structure and function of the genes of diverse plant species will help illuminate this and other important issues in plant genomics. The Web site, http://www.arabidopsis.org/, is linked to all of the Arabidopsis sequencing and genomic initiatives and maintains all Arabidopsis sequences.

REFERENCES

Aarts, M.G., Corzaan, P., Stiekema, W.J. & Pereira, A. (1995) A two element enhancer inhibitor transposon system in *Arabidopsis thaliana. Molecular and General Genetics* 247: 555–64

Anderson, M. & Mulligan, B. (1992) Arabidopsis mutant collection. In *Methods in Arabidopsis Research*, edited by C. Koncz, N.H. Chua & J. Schell, River Edge, New Jersey: World Scientific

Arabidopsis Genome Initiative (2000) Analysis of the genome sequence of the flowering plant *Arabidopsis thaliana. Nature* 408: 796–815

Bancroft, I., Bhatt, A., Sjodin, C. *et al.* (1992) Development of a tagging system in *Arabidopsis thaliana. Molecular and General Genetics* 233: 449–61

Bell, C.J. & Ecker, J.R. (1994) Assignment of 30 microsatellite loci to the linkage map of Arabidopsis. *Genomics* 19: 137–44

Bevan, M. & Murphy, G. (1999) The small, the large and the wild: the value of comparison in plant genomics. *Trends in Genetics* 15: 211–14

Bevan, M., Bancroft, I., Bent, E. *et al.* (1998) Analysis of 1.9 Mb of contiguous sequence from chromosome 4 of *Arabidopsis thaliana. Nature* 391: 485–88

Bevan, M., Bancroft, I., Hans Werner, M., Martienssen, R. & McCombie, R. (1999) Clearing a path through the jungle: progress in Arabidopsis genomics. *BioEssays* 21: 110–20

Boerjan, W., Bauw, G., Van Montagu, M. & Inze, D. (1994) Distinct phenotypes generated by overexpression and suppression of S-adenosyl-L-methionine synthetase reveal developmental patterns of gene silencing in tobacco. *Plant Cell* 6: 1401–14

Bouchez, D., Camilleri, C. & Caboche, M. (1993) A binary vector based on Basta resistance in planta transformation of *Arabidopsis thaliana. Comptes Rendus de l'Académie des Sciences Série III Sciences de la Vie* 316: 1188–93

Brusslan, J.A., Karlin, N.G., Huang, L. & Tobin, E.M. (1993) An Arabidopsis mutant with a reduced level of cab140 RNA is a result of cosuppression. *Plant Cell* 5: 667–77

Burr, B. & Burr, F.A. (1991) Recombinant inbreds for molecular mapping in maize. Theoretical and practical considerations. *Trends in Genetics* 7: 55–60

Chang, C., Bowman, J.L., deJohn, A.W., Lander, E.S. & Meyerowitz, E.M. (1988) Restriction fragment length polymorphism linkage map for *Arabidopsis thaliana. Proceedings of the National Academy of Sciences USA* 85: 6856–60

Chiapello, H., Ollivier, E., Landes-Devauchelle, C., Nitschke, P. & Risler, J. (1999) Codon usage as a tool to predict the cellular location of eukaryotic ribosomal proteins and aminoacyl-tRNA synthetases. *Nucleic Acids Research* 27: 2848–51

Choi, S., Creelman, R.A., Mullet, J.E. & Wing, R.A. (1995) Construction and characterization of a bacterial artificial chromosome library of *Arabidopsis thaliana. Weeds World* 2: 17–20

Copenhaver, G.P., Browne, W.E. & Preuss, D. (1998) Assaying genome-wide recombination and centromere functions with Arabidopsis tetrads. *Proceedings of the National Academy of Sciences USA* 95: 247–52

Copenhaver, G., Nickel, T., Takashi, K. *et al.* (1999) Genetic definition and sequence analysis of *Arabidopsis* centromeres. *Science* 286: 2468–74

Creusot, F., Fouilloux, E., Dron, M. *et al.* (1995) The CIC YAC library: a large insert YAC library for genome mapping in *Arabidopsis thaliana. Plant Journal* 8: 763–70

DeRisi, J., Iyer, V. & Brown, P. (1997) Exploring the metabolic and genetic control of gene expression on a genomic scale. *Science* 278: 680–86

Deustch, M. & Long, M. (1999) Intron–exon structures of eukaryotic model organisms. *Nucleic Acids Research* 27(15): 3219–28

Fedoroff, N.V. & Smith, D.L. (1993) A versatile system for detecting transposition in Arabidopsis. *Plant Journal* 3: 273–89

Feldmann, K.A., Marks, M.D., Christianson, M.L. & Quatrano, R.S. (1989) A dwarf mutant of Arabidopsis generated by T DNA insertion mutagenesis. *Science* 243: 1351–54

Feldmann, K.A., Malmberg, R.L. & Dean, C. (1994) Mutagenesis in Arabidopsis. In *Arabidopsis*, edited by E.M. Meyerowitz & C.R. Somerville, Plain View, New York: Cold Spring Harbor Laboratory Press

Franz, P., Armstrong, S., Alonso-Blanco, C. *et al.* (1998) Cytogenetics for the model system *Arabidopsis thaliana. Plant Journal* 13: 867–76

Gale, M. & Devos, K. (1998) Plant comparative genetics after 10 years. *Science* 282: 656–58

Grill, E. & Somerville, C. (1991) Construction and characterization of a yeast artificial chromosome library of Arabidopsis which is suitable for chromosome walking. *Molecular and General Genetics* 226: 484–90

Guzman, P. & Ecker, J. (1988) Development of large DNA methods for plants: molecular cloning of large segments of Arabidopsis and carrot DNA into yeast. *Nucleic Acids Research* 16: 11091–105

Hauge, B., Giraudat, J., Hanley, S. *et al.* (1990) Physical mapping of the Arabidopsis genome and its applications. *Journal of Cellular Biochemistry* 244 (Suppl.): 259

Hayashi, H., Czaja, I., Lubenow, H., Schell, J. & Walden, R. (1992) Activation of a plant gene by T-DNA tagging: auxin-independent growth in vitro. *Science* 258: 1350–53

Hempel, F.D., Weigel, D., Mandel, M.A. *et al.* (1997) Floral determination and expression of floral regulatory genes in Arabidopsis. *Development* 124: 3845–53

Höfte, H., Desprez, T., Amselem, J. *et al.* (1993) An inventory of 1152 expressed sequence tags obtained by partial sequencing of cDNAs from *Arabidopsis thaliana. Plant Journal* 4: 1051–61

Kaiser, J. (1996) Plant genetics: first global sequencing effort begins. *Science* 274: 30

Kaneko, T., Katoh T., Sato S. *et al.* (1999) Structural analysis of *Arabidopsis thaliana* chromosome 5. IX. Sequence features of the regions of 1,011,550 bp covered by seventeen P1 and TAC clones. *DNA Research* 6: 183–95

Kempin, S.A., Liljegren, L.M., Block, L.M., Rounsley, S.D. & Yanofsky, M.F. (1997) Targeted disruption in Arabidopsis. *Nature* 389: 802–03

Konieczny, A. & Ausubel, F.M. (1993) A procedure for mapping Arabidopsis mutations using co-dominant ecotype-specific PCR-based markers. *Plant Journal* 4: 403–10

Koornneef, M. (1994) Arabidopsis genetics. In *Arabidopsis*, edited by E.M. Meyerowitz & C.R. Somerville, Cold Spring Harbor, New York: Cold Spring Harbor Laboratory Press

Koornneef, M., van Eden, J., Hanhart, C.J. *et al.* (1983) Linkage map of *Arabidopsis thaliana. Journal of Heredity* 74: 265–72

Krysan, P.J., Young, J.C., Tax, F. & Sussman, M.R. (1996) Identification of transferred DNA insertions within Arabidopsis gene involved in signal transduction and ion transport. *Proceedings of the National Academy of Sciences USA* 93: 8145–50

Kuiper, M.T.R. (1997) Building a high density genetic map using the AFLP[tm] technology. Arabidopsis protocols. *Methods in Molecular Biology* 82: 157–71

Laibach, F. (1943) *Arabidopsis thaliana* (L.) Heynh. als Object fur genetische und entiwichlungsphysiologische Untersuchungen. *Botanisch Archiv* 44: 439–55

Langridge, J. (1955) Biochemical mutations in the crucifer *Arabidopsis thaliana* (L.) Heynh. *Nature* 176: 260–61

Leutwiler, L.S., Hough Evans, B.R. & Meyerowitz, E.M. (1984) The DNA of *Arabidopsis thaliana. Molecular and General Genetics* 194: 15–23

Lister, C. & Dean, C. (1993) Recombinant inbred lines for mapping RFLP and phenotypic markers in *Arabidopsis thaliana. Plant Journal* 4: 745–50

Liu, Y.G., Mitsukawa, N., Vazquez Tello, A. & Whittier, R.F. (1995a) Generation of a high-quality P1 library of Arabidopsis suitable for chromosome walking. *Plant Journal* 7: 351–58

Liu, Y.G., Mitsukawa, N., Oosumi, T. & Whittier, R.F. (1995b) Efficient isolation and mapping of *Arabidopsis thaliana* T-DNA insert junctions by thermal asymmetric interlaced PCR. *Plant Journal* 8: 457–63

Liu, S.C., Kowalski, S.P., Lan, T.H., Feldmann, K.A. & Paterson, A.H. (1996) Genome-wide high-resolution mapping by recurrent intermating using *Arabidopsis thaliana* as a model. *Genetics* 142: 247–58

Marra, M., Kucaba, T., Sekhon, M. *et al.* (1999) A map for sequence analysis of the *Arabidopsis thaliana* genome. *Nature Genetics* 22: 265–70

McKinney, E.C., Ali, N. Traut, A. *et al.* (1995) Sequence-based identification of T-DNA insertion mutations in Arabidopsis: actin mutants act2–1 and act4–1. *Plant Journal* 8: 613–22

Meinke, D. & Koornneef, M. (1997) Community standards for Arabidopsis genetics. *Plant Journal* 12: 247–53

Michelmore, R.W., Paran, I. & Kesseli, R.V. (1991) Identification of markers linked to disease-resistance genes by bulked segregant analysis: a rapid method to detect markers in specific genomic regions by using segregating populations. *Proceedings of the National Academy of Sciences USA* 88: 9828–32

Nam, H.-G., Giraudat, J., den Boer, B. *et al.* (1989) Restriction fragment length polymorphism linkage map of *Arabidopsis thaliana. Plant Cell* 1: 699–705

Nelson, D. (1999) Cytochrome P450 and the individuality of species. *Archives of Biochemistry and Biophysics* 369: 1–10

Newman, T., de Bruijin, F.J., Green, P. *et al.* (1994) Genes galore: a summary of methods for accessing results from large scale partial sequencing of anonymous Arabidopsis cDNA clones. *Plant Physiology* 106: 1241–55

Osborne, B.I., Wirtz, U. & Baker, B. (1995) A system for insertional mutagenesis and chromosomal rearrangement using the Ds transposon. *Plant Journal* 7: 687–701

Paterson, A.H., Lan, T.H., Reischmann, K.P. *et al.* (1996) Towards a unified genetic map of higher plants, transcending the monocot–dicot divergence. *Nature Genetics* 14: 380–82

Rédei, G.P. (1964) Crossing experiments with polyploids. *Arabidopsis Information Service* 1: 13

Rédei, G.P. (1965) Genetic blocks in the thiamine synthesis of the angiosperm Arabidopsis. *American Journal of Botany* 52: 834–41

Rédei, G.P. (1975) *Arabidopsis thaliana* as a genetic tool. *Annual Review of Genetics* 9: 111–27

Rédei, G.P. (1992) A heuristic glance at the past of Arabidopsis genetics. In *Methods in Arabidopsis Research*, edited by C. Koncz, N.H. Chua & J. Schell, River Edge, New Jersey: World Scientific

Rédei, G.P. & Hirono, Y. (1964) Linkage studies. *Arabidopsis Information Service* 1: 9–10

Reinholz, E. (1947) X-ray mutations in *Arabidopsis thaliana* (L.) Heynh. and their significance for plant breeding and the theory of evolution. *Field Information Agency, Technical (FIAT) Report* 1006: 1–70

Reiter, R.S., Williams, J.G.K., Feldmann, K.A. *et al.* (1992) Global and local genome mapping in *Arabidopsis thaliana* by using recombinant inbred lines and random amplified polymorphic DNAs. *Proceedings of the National Academy of Sciences USA* 89: 1477–81

Richards, E. & Ausubel, F. (1998) Isolation of a higher eukaryotic telomere from *Arabidopsis thaliana. Cell* 53: 127–36

Richards, E., Chao, S., Vongs, A. & Yang, J. (1992) Characterization of *Arabidopsis thaliana* telomeres isolated in yeast. *Nucleic Acids Research* 20: 4039–46

Ride, J., Davies, E., Franklin, F. & Marshall, D. (1999) Analysis of Arabidopsis genome sequence reveals a large new gene family in plants. *Plant Molecular Biology* 39: 927–32

Round, E., Floweres, S. & Richards, E. (1997) *Arabidopsis thaliana* centromere regions: genetic map positions and repetitive DNA structure. *Genome Research* 7: 1045–54

Schena, M. (1996) Genome analysis with gene expression microarrays. *BioEssays* 18: 427–31

Schena, M., Shalon, D., Davis, R. & Brown, P. (1995) Quantitative monitoring of gene expression patterns with a complementary DNA microarray. *Science* 270: 467–70

Schena, M., Shalon, D., Heller, R. *et al.* (1996) Parallel human genome analysis, microarray-based expression monitoring of 1000 genes. *Proceedings of the National Academy of Sciences USA* 93: 10614–19

Shalon, D., Smith, S. & Brown, P. (1996) A DNA microarray system for analyzing complex DNA samples using two-color fluorescent probe hybridization. *Genome Research* 6: 639–45

Southern, E. (1996) DNA chips, analyzing sequence by hybridization to oligonucleotides on a large scale. *Trends in Genetics* 12: 110–15

Sundaresan, V., Springer, P., Volpe, T. *et al.* (1995) Patterns of gene action in plant development revealed by enhancer trap and gene trap transposable elements. *Genes and Development* 9: 1797–810

Tanksley, S.D., Ganal, M.W. & Martin, G.B. (1995) Chromosome landing: a paradigm for map based gene cloning in plants with large genomes. *Trends in Genetics* 11: 63–68

Vos, P., Hogers, R., Bleeker, M. *et al.* (1995) AFLP: a new technique for DNA fingerprinting. *Nucleic Acids Research* 23: 4407–14

Ward, E.R. & Jen, G.C. (1990) Isolation of single-copy-sequence clones from a yeast artificial chromosome library of randomly-sheared *Arabidopsis thaliana* DNA. *Plant Molecular Biology* 14: 561–68

Weigel, D., Alvarez, J., Smyth, D.R., Yanofsky, M.F. & Meyerowitz, E.M. (1992) *LEAFY* controls floral meristem identity in Arabidopsis. *Cell* 69: 843–59

Zhang, H., Scheirer, D.C., Fowle, W.H. & Goodman, H.M. (1992) Expression of antisense or sense RNA of an ankyrin repeat-containing gene blocks chloroplast differentiation in Arabidopsis. *Plant Cell* 4: 1575–88

GLOSSARY

antisense the protein-encoding information in a gene can only be obtained by transcription in one direction along the gene. This is described as the sense direction. Antisense transcription (reading the gene in the reverse direction) does not usually produce a translatable product but may act to regulate that gene's expression by forming an RNA–RNA duplex with the natural sense mRNA of the gene, thereby preventing its translation

auxotrophic any organism with a nutritional requirement for a specific substance

contig large contiguous DNA sequence formed from assembly of overlapping shorter sequences

cytochrome P450 specialized cytochrome (an enzyme pigmented as a result of its haem prosthetic groups) of the electron transport chain in adrenal mitochondria and liver microsomes, involved in hydroxylating reactions

dNTPs mixture or set of deoxynucloetide triphosphates, as would be utilized as substrates for DNA synthesis

ecotype adapted genetically to a specific habitat; subspecies within a true species

expressed sequence tag (EST) DNA sequence developed by sequencing an end of a random cDNA clone from a library of interest

gene silencing RNA-mediated reduction in expression of a gene at the transcriptional or post-transcriptional stage

insertional mutagenesis *in vitro* insertion of a portion of DNA into a gene which results in a mutation in that gene

linkage group the group of genes that are located on the same chromosome

oligonucleotide short chain of nucleotides joined by phosphodiester bonds

primer short RNA or DNA that must be present on a DNA template before DNA polymerase can commence elongation of a new DNA chain

restriction fragment fingerprinting characterization of DNA fragments produced by restriction digestion of DNAs, used as a unique identifier to differentiate among clones, etc

shotgun sequencing sequencing method which involves randomly sequencing tiny cloned pieces of the genome, with no foreknowledge of where on a chromosome the piece originally came from. This can be contrasted with "directed" strategies, in which pieces of DNA from adjacent stretches of a chromosome are sequenced. Directed strategies eliminate the need for complex reassembly techniques

splice junction junction between an intron and an exon in a primary transcript from a eukaryotic "split gene", the positions from which introns are excised from the transcript and flanking exons rejoined, and the immediate nucleotide sequences in this area

T-DNA for Agrobacterium, the DNA of its Ti plasmid which is actually transferred to a plant and inserted in the plant's chromosome as a result of infection. Insertion in the plant genome occurs at random locations

telotrisomics a chromosome for which one of the two arms has been duplicated and the other lost, creating an extra copy of the genes associated with the duplicated region

tetrad analysis genetic evaluation of all four products of single meioses, so that genetic events, such as crossing-over, can be analysed in detail

transposon tagging creation of inserted transposable DNA elements into genes so that the phenotype resulting from the interruption of gene function can be correlated with the presence of the transposon. The known transposon DNA sequence is then utilized to recover bordering chromosomal DNA and eventually a clone of the affected gene

N POPULATION GENETICS AND EVOLUTIONARY STUDIES

Introduction to Population Genetics and Evolutionary Studies

Genetic variability, facilitated by the evolution of sex, is nevertheless static within a population unless there are forces that change the overall genetic composition of the population. Mutations, natural selection and gene flow from the migration of populations are now accepted as major evolutionary processes that can change allele frequencies in populations over time. The development of the neutral theory provided another mechanism whereby selectively neutral mutations, in addition to the advantageous ones, can contribute to evolution through random drift. Selfish genes are another likely source of the spread of new alleles in a population, though it is yet to be determined how important they may be.

With the advent of molecular techniques, mitochondrial analysis has contributed to population genomics, as described in the "Mitochondrial Eve" article below on the migration and evolution of the human population "Out of Africa". Like mitochondrial DNA, the Y chromosome is inherited from only one parent, and can thus be used to reconstruct human evolutionary history through patrilineal descent. Until recently, very few polymorphisms had been found, but these rare differences have been utilized in several recent studies, for example on tracing the European Y chromosomes to Paleolithic and Neolithic ancestors from two separate waves of migration (Semino *et al.*, 2000), and confirming the "Out of Africa" origin of modern humans (Underhill *et al.*, 2000). Intriguingly, the Y chromosome analysis suggests that our most recent common paternal ancestor ("Y-chromosome Adam") would have been about 84 000 years younger than our maternal one (Mitochondrial Eve).

Where gene flow between populations is restricted by biological incompatibility, as described in the article on "Reproductive isolating mechanisms", the two populations can become divided into two separate species. Recent debate on isolated locales (refugia; for example islands, caves, glacial isolation in the Pleistocene) suggests this can also act as an evolutionary mechanism resulting in the development of new species (Willis & Whittaker, 2000).

Evolution of morphological traits with continuous variation is likely to have been by the accumulation of mutations at many locations. Analysis of quantitative traits thus presents a problem to population geneticists, since multiple trait loci must be mapped, in addition to separating out environmental effects. The recent cloning of one of the genes (*fw2.2*) responsible for the size difference between the small-fruited wild tomatoes and their huge cultivated counterparts (Frary *et al.*, 2000; see Plate 26) is a significant achievement in quantitative genetics, but one which took a decade to achieve. With the increasing amount of high-quality genetic marker maps from a number of genome projects, mapping of specific QTLs should become much faster.

Finally, any reduction in genetic diversity leads to fears of extinction, and thus conservation genetics is an important part of monitoring and maintenance of biodiversity.

REFERENCES

Frary, A., Nesbitt, T.C., Frary, A. *et al.* (2000) *fw2.2:* a quantitative trait locus key to the evolution of tomato fruit size. *Science* 289: 85–88

Semino, O., Passarino, G., Oefner, P.J., Lin, A.A. & Arbuzova, S. *et al.* (2000) The genetic legacy of Paleolithic *Homo sapiens sapiens* in extant Europeans: a Y chromosome perspective. *Science* 290: 1155–59

Underhill, P.A., Shen, P., Lin, A.A., Jin, L. & Passarino, G. *et al.* (2000) Y chromosome sequence variation and the history of human populations. *Nature Genetics* 26(3): 358–61

Willis, K.J. & Whittaker, R.J. (2000) The refugial debate. *Science* 281: 1406–07

MOTOO KIMURA AND THE NEUTRAL THEORY

James F. Crow
Genetics Department, University of Wisconsin, Madison, USA

1. Introduction

"Neutral theory" is short for the theory that the great majority of evolutionary changes at the molecular level are due to mutations whose effects on fitness are so minute that their dynamics are determined mainly by mutation and random genetic drift. In addition, the theory holds that molecular polymorphisms are mainly neutral and represent a phase of molecular evolution.

The theory first appeared in 1968 in a paper by Kimura which was soon followed by the same idea written by King and Jukes (1969) under the provocative title "Non-Darwinian evolution". The theory was immediately criticized by several evolutionists on the grounds that any mutational change can hardly be so nearly neutral as to be uninfluenced by selection over long evolutionary periods, when extremely weak selective forces have time to act. In the years after 1968, the theory was debated at great length and sometimes with bitter fervor. The arguments became increasingly quantitative and experimental, yet the answer to the controversy has remained elusive.

It is convenient to classify the theory into weak and strong forms. The weak form holds that nucleotide changes that do not affect coding or regulatory functions – in higher eukaryotes, the bulk of the DNA – follow the neutral paradigm. The strong form says that neutrality also applies to amino acid substitutions and polymorphisms.

2. Origins of the neutral theory

In the early days of evolutionary study, a number of authors (including Darwin himself) mentioned the possibility that some traits are of no consequence for natural selection and are subject to the vicissitudes of chance. The first person to treat random fluctuations quantitatively as part of a general theory of evolution was Sewall Wright (1988). Wright's "shifting balance" theory of evolution invokes random drift as a way of moving gene frequencies past the barrier of an unstable selective equilibrium to a higher average fitness. For Wright's theory to apply, the fitness differences among alleles must be of the order of the reciprocal of the effective population number, the population being the local population. In Kimura's theory, the selective differences must be less than the reciprocal of the global population size – usually a much smaller difference. Thus, Wright's and Kimura's theories, while both invoking random drift, have little else in common. Wright was interested in genes that interact to increase fitness; the neutral theory (in its original and simplest form) involves evolutionary changes that have no effect on fitness.

The earliest approach to a theory of evolution driven by mutation and random drift came, independently from two microbial geneticists, Freese (1962) and Sueoka (1962). Both had been impressed by the great variability in nucleotide frequencies among different bacterial species, despite much greater similarity in the amino acid content. To explain this discrepancy, they assumed that many nucleotide differences are selectively neutral so that their frequencies are determined by mutation rates. At the same time, selective constraints keep the amino acid content within narrower limits. It is remarkable that these papers were written before the genetic code was found to be redundant, but neither paper had any influence on the evolution community.

In the years shortly before Kimura's 1968 paper, a number of geneticists suggested that many molecular polymorphisms might be neutral, but no one developed the idea or argued convincingly for it. Jukes (1991) noted that many enzymes function normally when various amino acids are replaced, and argued from this that the changes are neutral. Evolutionists were not convinced, however. They thought that changes too small to be detected by any physiological test could still have small fitness effects that would be important in a long evolutionary period.

With all these antecedents, why is the theory routinely attributed to Kimura, sometimes with a nod to King and Jukes? It is like the relation of Darwin and Wallace. Wallace is sometimes mentioned along with Darwin, but Darwin gets most of the credit. The reason is that Darwin, in addition to formulating the theory, spent the rest of his life documenting and arguing the case – to an extent regarded by some as overkill. Kimura, like Darwin, spent the rest of his life arguing and gathering evidence for the theory. King and Jukes, like Wallace, had an equal part in the discovery, but were much less active in promoting it. Just as it is customary to give the major credit for the theory of natural selection to Darwin, it is customary to give Kimura the major credit for the neutral theory.

3. Evidence for the neutral theory

A great virtue of the neutral theory, and perhaps its greatest strength, is the predictions it makes, both qualitative and quantitative. The simplest quantitative prediction gives the rate of evolutionary substitution in terms of the mutation rate. Consider a population of N diploid individuals in which neutral mutations arise at a rate μ per gene per generation. Therefore, in a given generation, there will be $2N\mu$ new mutant genes. These will persist in the population for varying lengths of time, because of random extinctions and because some individuals happen to leave more descendants than others. If we wait a very long time, one of the genes will have replaced all the others. The probability that this lucky gene happens to be the new mutant, and not one of the $2N-1$ other alleles, is simply $1/(2N)$. Therefore, putting these together, the probability of a successful mutant gene arising in a single generation is $2N\mu(1/2N) = \mu$. Thus, successful mutations arise at a rate μ per generation; in other words, if we view the process over a very long time, the rate of molecular substitutions at a locus is equal to the mutation rate at that locus.

How long a time is required for a lucky mutation to become the prevailing type? The answer, thanks to Kimura, is $4N_e$, where N_e is the global effective population number (roughly the number of reproducing adults, or somewhat less if there are larger than random fertility differences or unequal sex ratios). Therefore, if this theory is to be applied, it should involve a time period that is long relative to four times the population number. Fortunately, the times involved are hundreds of millions of years.

Mutation rates among vertebrates in the evolutionary past are not known well enough to provide a very accurate quantitative test of the theory, although there is no major discrepancy. More important, however, is the fact that, although there should be substantial differences in the action of selection at different times and places, mutation rates are thought to be much more stable. Hence, one of the earliest and strongest arguments for the neutral theory is the approximate constancy of amino acid substitution rates in corresponding proteins in different lineages. A striking example emphasized by Kimura in 1983 is that the number of amino acid differences between two kinds of haemoglobin (α and β) in fish and mammals is no greater than the number of differences between the two kinds in the fish. The selective forces in land and water vertebrates should be very different, yet haemoglobins in fish and man are no more different than those that had been in the same cell since the original duplication. Comparisons of other proteins also show approximate constancy. The rates are not absolutely constant, nor even within the variability expected with a Poisson distribution, but Kimura argued that they are much more nearly constant than would be expected from a selection-driven process. It is certainly clear that morphological and molecular evolution are marching to different drummers. The rates of morphological evolution differ greatly in different phylogenies and show the effects of population size, habitat and life style; molecular evolution is largely independent of these.

Some of the most striking of Kimura's arguments are not quantitative but qualitative. He noted that the parts of protein molecules that are constrained by being important to the function of that protein evolve the most slowly. From this, he asserted that the weaker the selection, the faster the evolution rate, with neutral mutations being the fastest. Evidence supporting this came from examination of the rates of substitution at different codon positions in proteins; as predicted, silent substitutions evolved most rapidly. Similarly, Kimura predicted that the evolution rates for pseudogenes and introns would be very high, and they are. Noting the absence of correlation between molecular and morphological evolution rates, he predicted that living fossils (such as Coelocanths and the opossum), would be as polymorphic for molecular markers as their more rapidly evolving relatives.

In the 20 years following Kimura's 1968 paper, a large number of tests were developed for both evolutionary rates and molecular polymorphisms. The rough shape of the distribution of neutral polymorphisms with multiple alleles is well understood, as is that of alleles maintained by balancing selection. The distribution of neutral allele frequencies is J-shaped, with a high peak close to zero and a very long tail (i.e. most of the mutants are present only once or a few times, while a very small number are common). In contrast, the distribution of alleles maintained by balancing selection is unimodal. Some data fit one theory better, other data fit the other. Although a great deal of inventive mathematics was developed and careful observations were made, these did not really resolve the question. This would be expected if some loci were neutral and others maintained by selection. In recent years, clear examples of balancing selection have been found, especially at loci concerned with immunological processes such as the major histocompatibilty complex (Takahata, 1990).

A difficulty in the theory is that observed molecular substitutions are proportional to the time measured in years, whereas the mutation rate is more nearly constant per generation. One suggestion is that some of the "neutral" substitutions are really slightly deleterious, but follow neutral kinetics in small populations. Since small populations are characteristic of large mammals with long life cycles, this can help support the neutral theory. Also, newer data show a slowing up of the evolution rate, measured in years, in primates, but, although these data help the theory, the issue is not resolved.

Many other arguments in favour of the neutral theory are summarized in Kimura's book, written in 1983. The arguments are laid out systematically, fully and forcefully. His more recent thoughts in the subject are given in a series of papers published in 1994.

4. Alternative theories

What are the alternatives to the neutral theory? For polymorphism, balancing selection is one clear alternative. The balancing selection hypotheses are usually one of two classes, a fitness advantage of rare alleles or a fitness advantage of heterozygotes. An example is the histocompatibility loci, where the study of the age of haplotypes shows enormous persistence times – clearly, the result of 10 million years or more of balancing selection. However, as mentioned above, only a small fraction of loci in the genome have been demonstrated to be under the control of balancing selection.

The most fully developed general alternative to neutrality is fluctuating selection coefficients. A gene that is sometimes favourable and sometimes unfavourable can be close to neutrality on the average, and such a locus may be very difficult to distinguish from a neutral one. The theory of variable selection coefficients has been worked out most thoroughly by Gillespie (1991). Gillespie notes that, while the variability of evolution rates for silent DNA changes is consistent with neutrality, the variability for amino acid substitutions is too great to fit comfortably into the neutral paradigm. He also points out cases, immunoglobulins for example, where evolution is faster than the supposed neutral maximum and, therefore, the result of positive selection. He argues for temporal and spatial fluctuations in selection intensity as an evolutionary model. The mathematics of Gillespie's treatment are formidable, but they are set forth along with a full discussion in his book published in 1991.

5. The present status of the neutral theory

It is well established that some loci are selected; others are clearly consistent with the neutral paradigm. So, the question is no longer an all-or-none difference; neither neutrality nor selection is complete. The answer must be quantitative, and a case-by-case analysis may be needed. By now, there has been a steady accumulation of loci at which selection is shown to operate. Yet, these are a very small fraction of all loci, so the large bulk of molecular evolution could still be neutral.

The neutral theory has three salutary properties: (l) it is simple; (2) it leads to predictions, both qualitative and quantitative; and (3) it is heuristic, having led to theoretical advances and careful empirical observations. The theory has increasingly played the role of a null hypothesis from which departures can be assessed. The theory is also widely assumed to be nearly enough correct to be used for many purposes, such as phylogenetic reconstruction, assessment of evolutionary time scales and measurement of population subdivision (for example, as a background for the study of behavioural evolution).

Kimura has never denied the central role played by natural selection in the evolution of form and function. He also realized that many mutations are deleterious. His interest has been in changes at the molecular level. His view was that only a very small fraction of what goes on at this level is adaptive, the majority of changes being neutral. One reaction to the theory is that it may be correct, but uninteresting. Evolutionists concerned with adaptive changes have little interest in non-adaptive molecular noise.

Recent years have seen a tempering of the neutralist–selectionist debate. There seems to be a developing consensus that the weak form of the theory is mainly correct; the strong form remains uncertain. The answer to this debate may have to await the study of individual cases, and it will be interesting to see whether or not the theory will continue to be discussed in the absence of Kimura's tireless advocacy.

6. Kimura's other work

Prior to the neutral theory, which brought Kimura's name to the large community of evolutionists, he was already well known to population geneticists and was becoming recognized as the successor to the great triumvirate, Fisher, Haldane and Wright (see article "Darwin and Mendel united: the contributions of Fisher, Haldane and Wright up to 1932"). While still a graduate student, Kimura had done ground-breaking work in population genetics theory. He extended and made much more rigorous Wright's work on fluctuating selection coefficients. He worked out the complete process of random drift in a finite population from an arbitrary starting point, first for two alleles, then for multiple alleles. The great Russian mathematician Kolmogorov had developed two stochastic theories for diffusion-like processes. The first, or forward, equation had been widely used by physicists and was adopted by Sewall Wright. Kimura was the first to exploit the second, or backward, equation and used it to solve problems such as the time required for fixation of a mutant gene, the time required for its loss and the number of individuals affected by the gene during its meandering path in the population.

This is a small portion of the mountain of theoretical work that Kimura did, mainly in population genetics, but also in several other areas. He made important advances in the genetics of quantitative characters, inbreeding theory, intergroup selection, sexual reproduction, evolution of linkage and population structure. Fortunately, his greatest papers have been reprinted in a single volume (Kimura, 1994). This includes very useful commentaries by Takahata, which give the background for each paper and describe later developments stimulated by the work.

Remarkably, the theory that Kimura developed early in his career, particularly that using the Kolmogorov backward equation, turned out to be made to order for the neutral theory. Kimura had a privilege that few scientists enjoy, that of having his earlier work – done for its own sake – turn out to be just what was needed for a major new subject during his lifetime. His work, to use an evolutionary metaphor, was preadapted for the neutral theory.

7. Kimura's life

Motoo Kimura was born in Okazaki, Japan, on 13 November 1924. He showed an early interest in botany and spent many happy hours with a microscope, which his father had provided. Kimura was unusually adept at mathematics and his teacher encouraged him to go into this field, but he could see no useful connection between mathematics and biology. He studied chromosome morphology in high school and, in 1944, was admitted to Kyoto Imperial University. As a student of botany, he was exempt from military service until graduation. The war ended during his first year, and he escaped military service. Yet life was hard, both during and after the war. Living conditions were difficult and there was never enough food.

While at Kyoto University, Kimura discovered the work of the mathematical geneticists and, at last, realized that his love and aptitude for mathematics could be applied to important biological problems. He was especially interested in the work of Sewall Wright. He also learned about the forward and backward equations of Kolmogorov. In 1949, he joined the research staff of the National Institute of Genetics in Mishima, Japan. There, he continued his reading and research in population genetics. The beginnings of some of his greatest work are to be found in the earliest *Annual Reports of the Genetics Institute*. He hoped to study abroad, and the chance came through the help of American workers in the Atomic Bomb Casualty Commission in Hiroshima. Kimura had wanted to study with Sewall Wright in Chicago, but Wright was expecting to retire soon and was not taking more students; he recommended that Kimura study with J.L. Lush at Iowa State University.

Kimura moved to Iowa in 1963, but was not satisfied; he wished to study stochastic processes, and this was not emphasized there. After a year, he transferred to the University of Wisconsin and became my student. A few months later, Sewall Wright also moved to Wisconsin and, at last, Kimura had the chance to study with his long-time hero. He received the Ph.D. degree from the University of Wisconsin in 1956.

Kimura returned to his position at the Genetics Institute in Mishima, and remained there for the rest of his life. After returning to Japan, he continued to be highly productive, and was increasingly recognized in the small community of mathematical population geneticists. Then, with the neutral theory in 1968, he became both famous and controversial. He spent the rest of his life in vigorous defence of the theory. He eventually contracted amyotrophic lateral sclerosis and died on 13 November 1994, on his 70th birthday.

REFERENCES

Freese, E. (1962) On the evolution of base composition of DNA. *Journal of Theoretical Biology* 3: 82–101

Gillespie, J.H. (1991) *The Causes of Molecular Evolution*, Oxford and New York: Oxford University Press

Jukes, T.H. (1991) Early development of the neutral theory. *Perspectives in Biology and Medicine* 34: 473–85

Kimura, M. (1968) Evolutionary rate at the molecular level. *Nature* 217: 624–26

Kimura, M. (1983) *The Neutral Theory of Molecular Evolution*, Cambridge and New York: Cambridge University Press

Kimura, M. (1994) *Population Genetics, Molecular Evolution, and the Neutral Theory: Selected Papers*, edited with introductory essays by N. Takahata, Chicago: University of Chicago Press

King, J.L. & Jukes, T.H. (1969) Non-Darwinian evolution. *Science* 164: 788–98

Sueoka, N. (1962) On the genetic basis of variation and heterogeneity of DNA base composition. *Proceedings of the National Academy of Sciences USA* 48: 582–92

Takahata, N. (1990) Allelic genealogy and MHC polymorphisms. In *Population Biology of Genes and Molecules*, edited by N. Takahata and J.F. Crow, Tokyo: Baifukan

Wright, S. (1988) Surfaces of selective value revisited. *American Naturalist* 131: 115–23

GLOSSARY

molecular polymorphism the existence of two or more alleles of a gene in a population. Usually the definition requires that the frequency of the most common allele does not exceed 99%

Poisson distribution a frequency distribution that is produced when the probability of an individual event is very low but the total number of opportunities is large enough that a finite number of events occur. The distribution has the property that the variance equals the mean

See also **Darwin and Mendel united** (p.77)

SELFISH GENES

Laurence Hurst
Centre for Mathematical Biology, Department of Biology and Biochemistry, University of Bath, UK

1. Introduction

The use of the term "selfish gene" has had a turbulent and often confusing history (see **Figure 1**). In its original meaning, as coined by Dawkins (1976), the term was intended to apply to any allele that spread in a population under the force of selection (more appropriately, we should talk of selfish alleles rather than selfish genes). Dawkins argues that alleles which spread because of selection on them may be considered (in anthropomorphic terms) to be "looking after their own interests", above those of their allelic competitors, and so, in this sense, may be thought of as being "selfish".

Any selectively neutral or slightly deleterious allele that might have risen to appreciable frequencies by drift, rather than been favoured by selection, or those neutral or deleterious alleles that invade because they are tightly linked to selectively advantageous alleles (i.e. those that hitchhike) would, in Dawkins' understanding, be the only classes of allele not described as selfish.

Despite its inappropriate nature, the point Dawkins was attempting to make concerns more the choice of the word "gene" than the word "selfish". This usage of the term gene highlights the fact that selection does not necessarily act to maximize an individual's fitness, but rather that the fundamental currency of evolution is genic fitness. To put it another way, the only question worth asking about the evolution of organismic features (e.g. the giraffe's neck, the eye, sex, etc.) is not "is the feature good for the species", nor "is it good for the individual", but, rather, "is it good for the gene", meaning "will the new allele spread in the population?" The understanding that this is the correct question is often accredited to W.D. Hamilton (1964) and G.C. Williams (1966), although several antecedents exist (for example, R.A. Fisher and J.B.S. Haldane both made small but related comments). Dawkins' book and the coining of the term selfish gene can be understood as a means to popularize this insight.

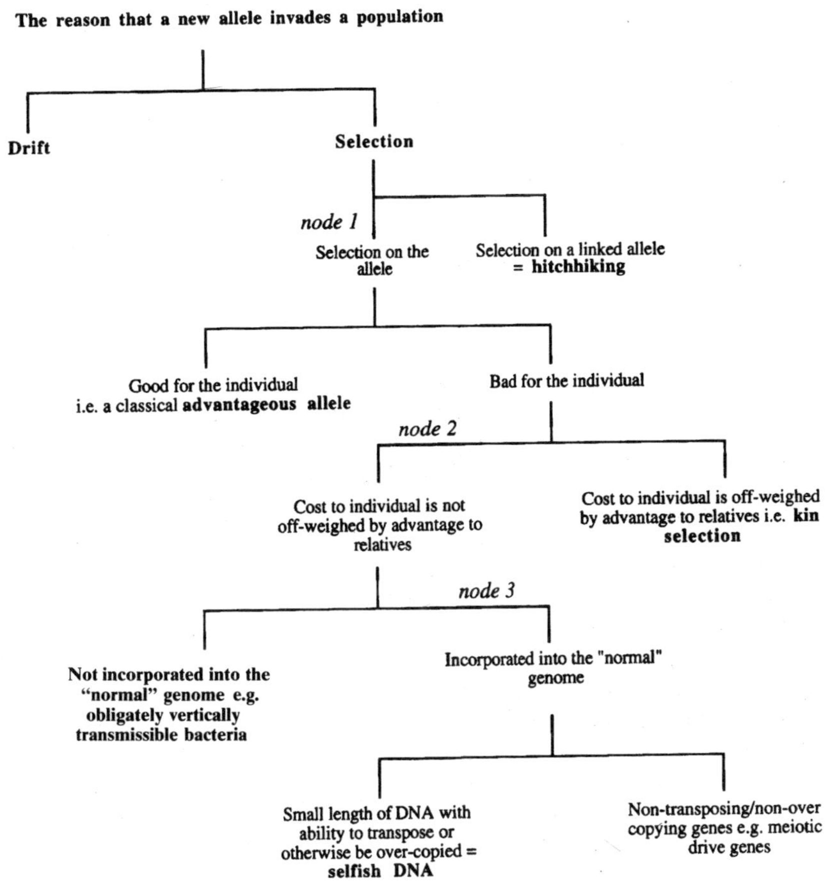

Figure 1 A key to the various definitions of selfish gene, selfish genetic element, etc. Up to node 2, the definitions depend on characterizing the reason for the spread of a mutant allele in a population. After node 2, they are dependent on genomic location and means of action. If the spread is not due to drift or hitchhiking (i.e. all categories below node 1), then the allele is a selfish gene in Dawkins' classification. However, more commonly, this term is restricted either for those alleles below node 2 or, in alternative usages, below node 3. All alleles below node 2 can correctly be referred to as either selfish genetic elements or as self-promoting elements. Often, people use selfish genes or ultraselfish genes to apply to everything below node 2. Others prefer that these terms should be restricted to instances where the allele is somehow incorporated into the host genome (i.e. below node 3).

2. Genic fitness *vs.* individual fitness

This insight can at first appear both trivial and facile. It is, however, an understanding that has greatly clarified and structured evolutionary thinking. Why then is genic fitness considered a more important currency than individual or group fitness? When discussing issues such as the evolution of the giraffe's neck, the distinctions that Dawkins makes seem to be merely semantic tinkering with the individual-based selectionist argument. Dawkins would argue that the allele for a longer neck would be a selfish gene that looks after its own interests by promoting the fitness of its "vehicle" (i.e. the bearer of the gene). Dawkins might then insist that selection favoured the allele rather than the individual. However, why should one ascribe the gene as that which is uniquely selfish, when it appears to be as much in the individual giraffe's interest as it is in a gene's interest to have a longer neck (assuming long necks are advantageous)? The answer is straightforward: there exist instances where alleles can spread deterministically (the trait is good for the gene), but where that trait does not benefit their bearers. Hence, the fundamental currency for evolution must be genic, not individual, fitness.

Consider, for example, the case of the exploding termites. In some termite species, there exist genes that, in certain circumstances, cause individuals to explode, so spreading glue all around. Consider when this allele was rare in the population. The death of the exploder is clearly not beneficial to the individual concerned. The reason such an allele can spread when rare, however, is that the termites do this only when invading ants are around. By spreading glue, the ants are prevented from invading the termite nest. Importantly, the termites in the nest have the very same allele. Hence, although there may be one less copy of the allele in the population (one individual just died), there could have been a much greater reduction. Indeed, all competing termite colonies without this rare allele would leave fewer copies of the competing allele under the same circumstance.

3. Altruism and Hamilton's Rule

More generally, as W.D. Hamilton (1964) has shown, so long as the reduction in the number of copies of an allele associated with an altruistic act (i.e. the cost, c) is less than the benefit (b) to individuals, scaled to the probability that the prospering individuals have the same allele (the relatedness, r), then the allele will spread, i.e. if $br > c$ then the altruistic trait will spread. This condition is known as Hamilton's Rule and the phenomenon is termed "kin selection". Similar examples of kin selection have been used extensively by Dawkins and others to demonstrate that genic fitness, and not individual fitness, is the critical currency.

The school of thought that grew up around the insight that it is important to ask about the invasion of alleles, can be contrasted with the views of group selectionists, such as Wynne-Edwards (1962). When asking about the evolution of altruism, Wynne-Edwards would postulate that organisms can act altruistically for the good of the species or the group (regardless of relatedness). Until examined critically, this can look like a reasonable argument – altruistic groups are probably more successful than non-altruistic ones, so selection would favour the former! So what is wrong with the argument? The problem stems from the fact that Wynne-Edwards' argument is one about the fitness of groups, not of individuals/genes within groups. We can ask what would happen to such a group of altruists if a cheat were to arrive – one that had its back scratched but did not reciprocate. An allele conferring such cheating behaviour will easily spread (it gains benefit at no cost). For this reason, groups of altruists are typically unstable. Which force then dominates: selection favouring groups of altruists over non-altruistic groups, or selection favouring cheats in groups? As the rate at which intergroup selection can occur is much lower than interindividual/gene selection, the instability of altruistic groups is considered the more important force. Hence, the success or failure of groups cannot be considered as an explanation for the behaviour of individuals within the group. The Wynne-Edwards class of argument is no longer taken seriously.

4. The "selfish gene"

The original usage of the term selfish gene, while perhaps heuristically helpful in understanding and clarifying the currency and operation of evolution, has been criticized on a number of counts. It is helpful to outline these so as to clarify the usage of the term and its limitations.

1. First, it is argued that the term appears to be ascribing motivation to genes. How can a few chemicals (bases of nucleic acid) be selfish when they do not have motivation as we do? This is a misunderstanding of the term and correctly it should simply be read as a statement that a certain trait had spread, or could spread, because of selection. Similarly, Dawkins and others frequently refer to the "interests of genes". To say that a gene has an interest in making the giraffe's neck longer is not to say that the gene has considered the issue and decided that making the neck longer would be a stimulating enterprise. It is a way of saying that if an allele were to arrive that could make the giraffe's neck longer, selection would favour that allele.
2. Secondly, it has been argued that Dawkins misuses the term "gene" as genes are, in the molecular geneticists' understanding, strips of DNA coding for protein. In Dawkins' meaning, the gene is the information that is transmitted that ensures that a particular trait is heritable (see also Williams, 1992). (Note: to be heritable, there must be polymorphism at the locus concerned and hence, to repeat myself, Dawkins should not have referred to genes but to alleles.) This may be a strip of DNA but need not be – it could, for example, be heritable epigenetic marks such as a methylation pattern or many alleles at many loci. Clearly, the two meanings intersect at some points, but they should not be confused as they are logically distinct. Which usage should have precedence is a questionable issue. Genes were first known about because they were heritable information, not because they were known to be DNA. Mendel's factors were describable only in terms of heritable information. Which idea has precedence is largely irrelevant, however, as both can be used with clarity in the appropriate context.

3. Thirdly, and perhaps most importantly, the term can be criticized for providing an overly simple view of the operation of selection. Dawkins and the mathematical evolutionary geneticists of the same tradition typically discuss the evolution of a trait as though it were coded for by one allele at a single locus. Hence, this approach ignores the complexities of quantitative genetics. At other times (as in "phenotypic modelling", often employed by behavioural ecologists), the genetics is ignored altogether and the evolution of the trait is mathematically modelled as if the organism were asexual (hence, there are none of the complexities of random assortment of genes, of heterozygosity, etc.). It is ironic that the school of thought that was popularized in *The Selfish Gene* (Dawkins, 1976) is the one that tends to ignore genes! While such simplifying assumptions make the mathematical analysis of the evolution of the particular traits more tractable, something must be being lost as many heritable traits (e.g. height, weight, etc.) are the results of interactions between numerous alleles at numerous independently segregating loci. The selfish gene view, and phenotypic modelling in particular, has hence been criticized, for example by Richard Lewontin (1984), as inevitably providing misleading results. Accordingly, there exist a few theoretical results in major evolutionary problems that require quantitative genetical models (see, for example, Charlesworth, 1993).

5. Phenotypic modelling

More significantly, however, recent theoretical work suggests that the simplifications employed by phenotypic modellers are defensible as they allow simple and accurate ways to work out the trajectory of long-term evolution (for review see Marrow *et al.*, 1996). Hammerstein (1996), for example, could show that in a population genetical framework one can characterize a phenotypic position that is uninvadable by new alleles. Importantly, this position, he could show, was also that defined by phenotypic modellers, despite the fact that they ignore the genetics. He summarizes the position by suggesting that evolution can work like a streetcar – it goes from temporary stop to temporary stop until it reaches its final destination. What he and colleagues have established is that population genetical and phenotypic modellers can agree on what the final stop might be. A corollary of this is that the selfish gene approach and phenotypic modelling have little relevance to understanding the details of micropopulation genetics.

The generality of the result that the final stop found by phenotypic modellers is consistent with population genetics assumes that genes are inherited in a Mendelian fashion. Some genes, however, break the conventional rules of inheritance and it is to these that the second, and presently more common, usage of the term selfish gene applies (**Figure 1**). This is of more restricted application than Dawkins' use and applies to what have otherwise been called selfish genetic elements (Werren *et al.*, 1988), genomic parasites, self-promoting elements (Hurst *et al.*, 1996) and, perhaps most confusingly, ultraselfish genes (Crow, 1988). There is, however, some confusion over the usages of the above terms, with some insisting that selfish genes are a subclass of selfish genetic elements (see **Figure 1**). In the broadest sense, these are all genes that subvert the host's genetic system to their own ends and act as parasites manipulating the individual. The activity of these genes is deleterious to the bearer of the gene, but does not have the cost to the individual offweighed against benefits to relatives. When they spread in a population, they thus create the context for the spread of other genes, acting in the same individual, which try to suppress the selfish element (much as an immune system attempts to suppress parasites).

6. Segregation distorters

Consider, for example, the segregation distorter (SD) system in *Drosophila melanogaster*. This is a gene that acts in males to kill sperm that do not contain that gene; a male heterozygous for the gene will have half of its sperm killed, which is likely to lead to a reduction in the fertility of the male. Importantly, however, so long as the male sires more than half the progeny that he would otherwise have sired, the gene can invade. To see this, imagine that the male would have sired ten offspring. If the distorting allele were not effective then half of the progeny would have inherited one allele (i.e. five offspring) and the other half would have inherited the competing allele. Let us assume then that the distorting allele is functioning and kills all of the sperm bearing the wild-type allele. This being so, all ten eggs in the female may potentially be fertilized exclusively by sperm bearing the distorting allele. So long as the male's fertility is not down over 50%, the remaining sperm will, by definition, fertilize between five and ten eggs. This being so, the distorting allele is doing better than it would were it not distorting, and hence may spread in the population when initially rare. As it becomes more common, the fate of the allele becomes dependent on the fitness of homozygotes for the distorting alleles (Lyttle, 1991). SD is one of a general class of genes known as meiotic drive genes or segregation distorters (Crow, 1988).

When segregation distorters are on sex chromosomes, not only may fertility be reduced, but the sex ratio of the progeny will be heavily skewed. As a consequence, it is easier to detect sex-linked drive than it is to find autosomal drive. A survey of nine species of fruit fly has found that X-linked drive may be remarkably common: five of the nine species observed had X-linked drive in polymorphism (Jaenike, 1996). It has also recently been shown that all populations of *D. simulans* have an X-linked meiotic drive gene, although in most its presence is hard to detect because of the existence of suppressor alleles.

Many of the distorters (SD and *t* haplotypes included) segregate as though they were one locus, but they actually comprise two very tightly linked loci. These two may be understood as loci coding for a toxin and an antitoxin. The spread of the gene pairs is dependent on restricted access to the antitoxin. In the case of SD, selection will favour such gene pairs only if they can remain together. These two-locus distorters are hence often associated with regions of low recombination (usually they are near centromeres) and are often contained in inversions (Lyttle, 1991).

Segregation analysis has demonstrated the presence of genes that may well be two-locus distorters, but that act in mothers to kill progeny not containing the same gene (Beeman *et al.*, 1992; Jaenike, 1996). The putative model for these systems is that the toxin locus operates in mothers and places toxin in the progeny. Only those progeny with the antitoxin locus will survive. Invasion of such a gene pair is trivial if the death of the offspring without the antitoxin provides increased nutrient for those with the allele (again assuming the two loci can be kept very tightly linked). In at least one instance, the locus is known to be close to the centromere (Hurst, 1993a) as expected of a locus under selection to remain tightly linked. From a human point of view, this class of distorters may be especially interesting as there exists a claim that many such elements exist in humans (Haig, 1996); selfish elements are otherwise remarkably rare in humans. However, in the absence of segregation data, the suggestion must be considered speculative.

7. Cytoplasmic factors

Not all selfish genes are on the nuclear chromosomes. There exist, for example, cytoplasmic factors in isopod crustaceans that force male individuals to become female. The fact that the factors are exclusively cytoplasmically (i.e. maternally) transmitted is critical to their spread. Cytoplasmic factors are not transmitted by males and, hence, a male is effectively a dead end for them (this is not true for autosomal genes that are transmitted equally by males and females). By converting a male into a female, the cytoplasmic factor is thus guaranteeing its own transmission to the next generation at a higher rate than non-feminizing cytotypes.

If we start with a population without such factors and one comes in at low frequency, it would increase in frequency and eventually it is possible that all cytoplasms will end up with the factor. Were this the case, the population could go extinct due to an absence of males. It is, however, important to realize that, although these cytoplasmic factors decrease the fitness of their host individuals and can send the population extinct, the factors can nonetheless spread in the population, possibly to fixation.

Not all selfish cytoplasmic factors convert males into females (for a review of cytoplasmic sex ratio distorters see Hurst, 1993b). In some wasps, the factors convert sexual females into asexual ones producing female progeny only (Stouthamer *et al.*, 1990). Other cytoplasmic factors kill males (Hurst & Majerus, 1993). In plants, there are many reports of mitochondrial genes that sterilize male tissue but that do not affect female tissue (Saumitou-Laprade *et al.*, 1994).

8. Selfish elements

An important aspect of selfish elements is that, regardless of the ancestry, they receive no significant levels of horizontal transfer between individuals in a population. While then it is quite correct to think of them as genomic parasites (they do harm to their host), they are different from conventional parasites as these do regularly undergo horizontal transmission (e.g. HIV, the influenza virus, bacteria responsible for food poisoning and meningitis, tapeworms, etc.). These boundaries can be blurred, however. While some examples of "mobile DNA", alias "selfish DNA", may be considered a subclass of selfish element (e.g. self-splicing introns, transposable elements and "homing" elements – see Figure 1), they are closely related to horizontally transmitting retroviruses.

Selfish elements are of interest for a variety of reasons. First, alongside examples of kin selection, their spread in a population is clear evidence that genic fitness really is the fundamental currency of evolution. Secondly, and more importantly, the spread in a population of a selfish element creates the conditions necessary for the invasion of a suppressor of the selfish gene acting within the same individual. In this regard, they may be important generators of evolutionary change even in an unchanging environment (cf. classical host–parasite coevolution). It also became apparent that some of what is going on in the workings of the cell need not be for the best of the cell or organism concerned and may represent a struggle between selfish elements and their suppressors.

Meiotic drive genes, for example, often have their effects opposed by other unlinked genes in the same genome. It is quite possible for the meiotic drive gene and the suppressor to evolve in an escalating arms race: the drive gene invades but is suppressed, a duplication of the drive gene results in stronger drive and invades again, thus providing the conditions for the invasion of a duplication of the suppressor, and so on. This is most probably the ancestry of a multicopy meiotic drive gene on the X chromosome of *D. melanogaster* that is suppressed during spermatogenesis by a multicopy gene on the Y chromosome (Hurst, 1996a). The apparent wastage and redundancy of this pair of genes is baffling until understood as a meiotic "battle" between opposing interests. This difference of interest between the selfish element and the rest of the genome is indicative of a "conflict of interests" between the two parties. There is then said to be "intragenomic conflict" (alias genetic conflict) (Hurst *et al.*, 1996).

A possible, albeit contentious, example of such a balance between selfish element and suppressor active in humans (and placental mammals as a whole) is witnessed in the activity of genomically imprinted genes in mammals and angiosperm plants (see article on "Genomic imprinting"). The antagonism between insulin-like growth factor II and its receptor has been interpreted as an antagonism between selfish paternally derived genes demanding more from a mother than she is prepared to give (Moore & Haig, 1991; Haig, 2000).

9. Suppressors and new genetic systems

Of more interest to evolutionary theorists are instances in which suppressors do more than simply stop the action of the selfish gene. The spread of these more important modifiers can result in the evolution of new genetic systems. Such novel suppressors have been shown to be central to the evolution of novel sex determination systems in crustaceans affected with feminizing factors and in lemmings affected with meiotic drive. They have been conjectured to be involved in the evolution of various curious sex-determining systems in scale insects (Bull, 1979). They have also been hypothesized to play a role in some of the major transitions in evolution (Maynard Smith & Szathmáry, 1995; Hurst *et al.*, 1996), including the evolution of gender (Hoekstra, 1987; Hastings, 1992; Hurst & Hamilton, 1992; Hurst, 1995), meiosis (Haig & Grafen, 1991; Haig, 1993; Reed & Hurst, 1996; but see Hurst & Randerson, 2000) and multicellularity (Buss, 1987). A role in speciation has also often been suggested (reviewed in Hurst *et al.*, 1996). For some of these hypotheses, there has been no attempt to test them and they must be considered speculative. In other instances, comparative tests have been performed.

(i) Evolution of mating types

Consider, for example, the problem of the evolution of sexes (Hurst & Hamilton, 1992). That unicellular organisms which have equal-sized gametes are typically divided into two mating types is paradoxical. The paradox stems from the fact that when a species has only two mating types, mate finding is at its most difficult – on average, a given gamete will meet a potential mate only half of the time, if it encounters other gametes at random. However, if a gamete were to appear that had a new mating type that could allow it to mate with all other gametes, except those with the same mating type, then such an allele would easily invade (it can mate with everybody). At equilibrium, the population would also be better off – any given gamete can mate with two-thirds of the population (whereas previously it could only mate with half). So why do so many isogamous protists have only two mating types? The problem has been explained by arguing that two mating type alleles evolved in response to selfish organelle genomes (Hoekstra, 1987; Hurst & Hamilton, 1992; Hutson & Law, 1993). Without uniparental inheritance of organelle genomes, fast replicating deleterious organelle genomes can easily spread (these would be selfish genomes) and their spread can indeed lead to the invasion of genes that stop inheritance from one of the two parents. Assortative mating alleles can then invade if closely linked (Hutson & Law, 1993) to the genes controlling organelles and two mating types will be the result. More than two alleles at this locus will reintroduce the problem of selfish cytoplasmic genomes and provide the conditions for a return to two mating types (Hurst, 1996b).

The theory can allow the prediction of which species should have none, two, few and many mating types (Hurst & Hamilton, 1992; Hurst, 1995), as well as predicting the structure and functions of mating type loci (Hurst, 1995). The hypothesis correctly predicts that the mating-type alleles define both who can and cannot be a potential mate, and also which partner should and should not transmit a particular class of cytoplasmic genes. In addition, it explains why fusion between cells that are closely related need not be associated with mating types and need not have controls over cytoplasmic gene inheritance, and why organisms in which sex involves the full fusion of cells that lack cytoplasmic organelles do not have mating types (a few archaebacteria are like this). Importantly, it correctly predicts that organisms in which sex does not involve the full fusion of cells, just the exchange of nuclear DNA (as in most ciliates and mushrooms), are not usually restricted to two mating types.

It would be remarkable if selfish genes were not important in some evolutionary processes, both major and minor. How common they are and how generally important have yet to be determined. What does seem to be clear is that they are not simply minor curiosities (as once assumed), but potentially key elements in the evolution of genetical systems.

REFERENCES

Beeman, R.W., Friesen, K.S. & Denell, R.E. (1992) Maternal-effect selfish genes in flour beetles. *Science* 256: 89–92

Bull, J.J. (1979) An advantage for the evolution of male haploidy and systems with similar genetic transmission. *Heredity* 43: 361–81

Buss, L. (1987) *The Evolution of Individuality*, Princeton, New Jersey: Princeton University Press

Charlesworth, B. (1993) Directional selection and the evolution of sex and recombination. *Genetical Research* 61: 205–24

Crow, J.F. (1988) Anecdotal, historical and critical commentaries on genetics: the ultraselfish gene. *Genetics* 118: 389–91

Dawkins, R. (1976) *The Selfish Gene*, Oxford and New York: Oxford University Press

Haig, D. (1993) Alternatives to meiosis: the unusual genetics of red algae, Microsporidia, and others. *Journal of Theoretical Biology* 163: 15–31

Haig, D. (1996) Gestational drive and the green-bearded placenta. *Proceedings of the National Academy of Sciences USA* 93: 6547–51

Haig, D. & Grafen, A. (1991) Genetic scrambling as a defence against meiotic drive. *Journal of Theoretical Biology* 153: 531–58

Haig, D. (2000) The kinship theory of genomic imprinting. *Annual Reviews of Ecology* Suppl. 31: 9–32

Hamilton, W.D. (1964) The genetical evolution of social behaviour I and II. *Journal of Theoretical Biology* 7: 1–16, 17–52

Hammerstein, P. (1996) Darwinian adaptation, population-genetics and the streetcar theory of evolution. *Journal of Mathematical Biology* 34: 511–32

Hastings, I.M. (1992) Population genetic-aspects of deleterious cytoplasmic genomes and their effect on the evolution of sexual reproduction. *Genetical Research* 59: 215–25

Hoekstra, R.F. (1987) The evolution of sexes. In *The Evolution of Sex and Its Consequences*, edited by S.C. Stearns, Basel and Boston: Birkhäuser

Hurst, G.D.D. & Majerus, M.E.N. (1993) Why do maternally inherited microorganisms kill males? *Heredity* 71: 81–95

Hurst, L.D. (1993a) *scat+* is a selfish gene analogous to *Medea* of *Tribolium castaneum*. *Cell* 75: 407–08

Hurst, L.D. (1993b) The incidences, mechanisms and evolution of cytoplasmic sex ratio distorters in animals. *Biological Review* 68: 121–93

Hurst, L.D. (1995) Selfish genetic elements and their role in evolution: the evolution of sex and some of what that entails. *Philosophical Transactions of the Royal Society, Series B* 349: 321–32

Hurst, L.D. (1996a) Further evidence consistent with *Stellate*'s involvement in meiotic drive. *Genetics* 142: 641–43

Hurst, L.D. (1996b) Why are there only 2 sexes? *Proceedings of the Royal Society of London, Series B* 263: 415–22

Hurst, L.D. & Hamilton, W.D. (1992) Cytoplasmic fusion and the nature of sexes. *Proceedings of the Royal Society of London, Series B* 247: 189–94

Hurst, L.D., Atlan, A. & Bengtsson, B.O. (1996) Genetic conflicts. *Quarterly Review of Biology* 71: 317–64

Hurst, L.D. & Randerson, J.P. (2000) Transitions in the evolution of meiosis. *Journal of Evolutionary Biology* 13: 466–79

Hutson, V. & Law, R. (1993) Four steps to two sexes. *Proceedings of the Royal Society of London, Series B* 253: 43–51

Jaenike, J. (1996) Sex-ratio meiotic drive in the *Drosophila quinaria* group. *American Naturalist* 148: 237–54

Lewontin R.A. (1984) The structure of evolutionary genetics. In *Conceptual Issues in Evolutionary Biology: An Anthology*, edited by E. Sober, Cambridge, Massachusetts: MIT Press; 2nd edition 1994

Lyttle, T.W. (1991) Segregation distorters. *Annual Review of Genetics* 25: 511–57

Marrow, P., Johnstone, R.A. & Hurst, L.D. (1996) Riding the evolutionary streetcar – where population-genetics and game-theory meet. *Trends in Ecology and Evolution* 11: 445–46

Maynard Smith, J. & Szathmáry, E. (1995) *The Major Transitions in Evolution*, Oxford and New York: Freeman

Moore, T. & Haig, D. (1991) Genomic imprinting in mammalian development: a parental tug-of-war. *Trends in Genetics* 7: 45–49

Reed, J.N. & Hurst, L.D. (1996) Dynamic analysis of the evolution of a novel genetic system – the evolution of ciliate meiosis. *Journal of Theoretical Biology* 178: 355–68

Saumitou-Laprade, P., Cuguen, J. & Vernet, P. (1994) Cytoplasmic male sterility in plants: molecular evidence and the nucleocytoplasmic conflict. *Trends in Ecology and Evolution* 9: 431–35

Stouthamer, R., Luck, R.F. & Hamilton, W.D. (1990) Antibiotics cause parthenogenetic trichogramma (hymenoptera, trichogrammatidae) to revert to sex. *Proceedings of the National Academy of Sciences USA* 87: 2424–27

Werren, J.H., Nur, U. & Wu, C.-I. (1988) Selfish genetic elements. *Trends in Ecology and Evolution* 3: 297–302

Williams G.C. (1966) *Adaptation and Natural Selection: A Critique of some Current Evolutionary Thought*, Princeton, New Jersey: Princeton University Press

Williams G.C. (1992) *Natural Selection: Domains, Levels and Challenges*, Oxford and New York: Oxford University Press

Wynne-Edwards, V.C. (1962) *Animal Dispersion in Relation to Social Behaviour*, Edinburgh: Oliver and Boyd

GLOSSARY

cytotype just as individuals may differ as regards the particular allelic composition of the nuclear alleles (i.e. those on the chromosomes), so too they can differ as regards their cytoplasmic constituents. An individual's genotype refers to the allelic composition of nuclear alleles. The cytotype is the comparable term for the cytoplasmic genes/genomes. Additionally the term cytotype can refer to presence or absence of some maternally heritable factors such as bacteria, as well as variants of mitochondrial and chloroplast genomes

horizontal transfer transfer of genes from one evolutionarily unrelated organism to another

methylation [DNA methylation] process by which methyl groups are added to certain nucleotides in genomic DNA. This affects gene expression, as methylated DNA is not easily transcribed

suppressor [mutation] one that cancels out the effects of a primary mutation at a different locus

See also **Genomic imprinting** (p.757)

EVOLUTION OF SEX

Szilvia Kövér and Eörs Szathmáry
Department of Plant Taxonomy and Ecology, Eötvös University, Budapest

1. The modes of reproduction

A genetic system is defined by the way in which the genomes of the offspring are created. Basically, there are two solutions: sexual and asexual. In the former, both parents contribute to the genome of the offspring, while, in the latter, the genome is inherited from only one parent. The most curious concomitant of sex is that it is only half as efficient in propagating genes as asexuality; this is the so-called "two-fold cost of sex", discussed in detail later. Looking around, one can realize that despite this cost the majority of multicellular plants and animals practise sex, although asexual clones (species) can be found in all major taxa except gymnosperms, birds and mammals. This indicates that sex must have some great advantage because most species chose it in spite of the fact that the cheaper possibility of being asexual seems widely attainable. After examining the basic types of reproduction, we will come back to discuss what this advantage might be.

An enormous diversity of modes of reproduction (genetic systems) exists in nature (Bell, 1982). Sex (amphimixis) in eukaryotes can be defined as the regular alternation of meiosis and outcrossing (union of two gametes of different origin). Generally, sex is tied to reproduction, which is why the term sexual reproduction is widespread; however, in some protists sex and reproduction are separated (e.g. the conjugation between two ciliates results in the exchange of genes, but not in proliferation; the latter is solved through simple cell divisions). The essence of sex is recombination: the maternally and paternally inherited genes are rearranged and mixed through the independent assortment of chromosomes and through molecular recombination in meiosis. In the course of outcrossing, every two alleles present in the population of gametes have the chance to meet. In consequence, every individual of a sexual population has a unique genotype.

In the case of asexual reproduction, there is no possibility of mixing genes descending from different individuals. Several different types of asexual reproduction are collectively called parthenogenesis, which indicates that the offspring develop from an unfertilized egg; the term's literal meaning is virgin birth. One can further classify parthenogenetic systems according to the presence of meiosis. Apomixis is ameiotic and the mitotically produced offspring have the same genotype as their parent, barring new mutations. Many apomictic "species" can be found among dandelions (Taraxacum). In selfing and automixis, meiosis does occur; however, the diploid state is then restored by the union of two gametes of the same parent in selfing, and by the union of the gametic pronuclei in automixis. These two processes are similar in that homozygosity is increased in the offspring; they are different since, in the case of selfing, there is the possibility of occasional outcrossing – in contrast, automixis does not allow outcrossing as no distinct gametes are formed.

The other type of asexual reproduction is vegetative reproduction, when the offspring develop from a group of somatic cells without meiosis so that they are genetically identical with their parent, as in apomixis. It is common among plants that proliferate with rhizomes, runners and bulbs, and some animals like Cnidaria with buds.

2. The consequences of sex

First let us have a closer look at the so-called "two-fold cost of sex". One can understand it most easily from a gene's eye view. Assume, for the time being, that an asexual and a sexual female have equal numbers of offspring and, for the sake of simplicity, that asexuality is caused by a single allele (**Figure 1**). In this case, the "asexual" allele will certainly be present in every offspring while the rival "sexual" allele has only a 50% chance of getting into each offspring. Thus, the "asexual" allele has a two-fold advantage, which means that it doubles in frequency in each generation when it is rare and outcompetes its sexual rival if the latter cannot add a two-fold advantage to its carrier. Note that the above assumption is not valid if there is paternal care. The sexual female can raise more offspring than the asexual one – this is why the cost of sex is less than two-fold; if the sexual pair has two-fold brood then there is no cost of sex.

The above reasoning is valid only for the anisogamous kind of sex when the gametes are different in size and the little sperm contributes only its genetic material to the

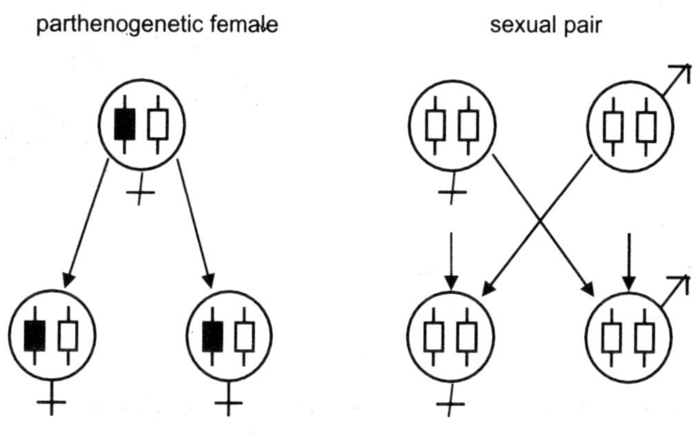

Figure 1 The cost of sex (after Maynard Smith, 1988a). The black rectangle represents the allele causing parthenogenesis and the white rectangle represents the allele for amphimixis. The former spreads with two-fold efficiency if the parthenogenetic and the amphimictic females have equal numbers of offspring.

zygote, but no nutrients. If the male does not care for the offspring, the sexual female has a two-fold disadvantage because she puts in all the necessary nutritive materials, but only half of her genes into the egg. If she were parthenogenetic, she could put all her genes into a parthenogenetic egg along with the same amount of nutrients.

In the case of isogamous sex, however, which is widespread among protists and algae, the two gametes are of equal size. As both parents contribute equally to the zygote, there is no cost of sex. Thus, the two-fold cost of sex means that anisogamous sex is only half as efficient a mode of propagation as the various kinds of asexual reproduction.

The second consequence of sex is that it maintains a moderately high degree of heterozygosity according to the Hardy–Weinberg rule. Some forms of asexual reproduction, namely automixis and selfing, result in a large decrease in heterozygosity and cause inbreeding depression. The main cause of the fitness reduction is that the majority of deleterious alleles are recessive and, thus, their detrimental effect manifests itself only in homozygotes. It is often said that the masking ability of sex is greater than that of selfing or automixis, because, in a sexual population, the deleterious recessives are mainly in the heterozygous state. Geneticists have different opinions on the importance of heterosis, when the heterozygote has superior fitness than either homozygote. The lack of heterosis can be another cause of inbreeding depression. Clearly, inbreeding depression can restrain sexual populations from switching to automixis or selfing, but not from apomixis. As apomixis is ameiotic (i.e. it produces eggs exclusively through mitoses), it does not entail segregation. Consequently, it conserves the highest level of heterozygosity (has the highest masking ability). This means that apomicts harbour more deleterious recessives in mutation–selection equilibrium than sexuals, and switching from sex to apomixis causes no inbreeding depression. In fact, it is advantageous. Thus, we need another explanation for the advantage of sex – one that can compensate for the two-fold cost.

The majority of evolutionary geneticists agree that the most important property of sex is that it generates genetic variability (alternative opinions are briefly discussed below). For every member of a sexually reproducing population, meiosis and syngamy make a new combination from the alleles of the previous generation. In contrast, asexual populations are genotypically depauperate; usually, one can find fewer than ten clones, i.e. less than ten different genotypes in a local population (**Figure 2**). It can happen, as in the case of an endomitotic lizard, *Cnemidophorus uniparens*, that the local populations are monoclonal. Since different local populations can contain different clones, sometimes the variation shown by the whole assemblage of clones is comparable with that of a sexual species, as in the case of *C. tesselatus*.

Studying the differences in the cost, heterozygosity and genetic variability of different reproductive systems in **Table 1**, one can conclude that the rarity of automixis in nature can well be explained by inbreeding depression, and the predominance of sex can be explained by the advantages of variability. The problem is that it is not so easy to show how variability can confer a two-fold advantage on sex.

3. Hypotheses on the advantage of sex

The difficulty of understanding the evolutionary role of sex is shown by the existence of more than 20 theories on the advantage of sex and by the lack of agreement as to their validity. However, the majority of hypotheses agree that the beneficial effect of recombination is indirect: it has no direct effect on fitness, but it creates new combinations of alleles

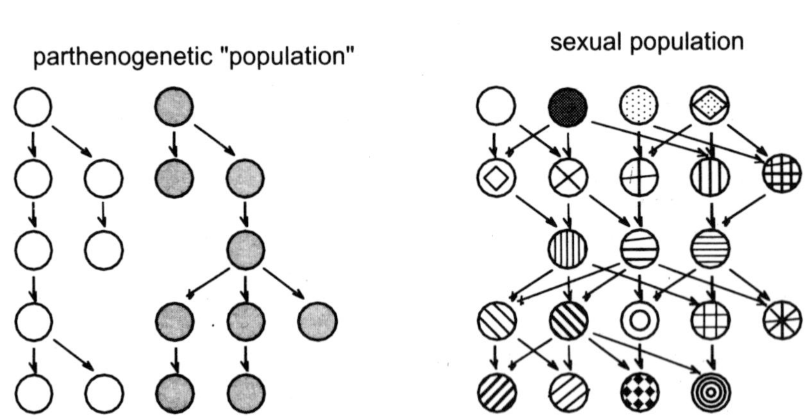

Figure 2 Genetic variability is greater in sexual than in asexual populations.

Table 1 The consequences of genetic systems (after Bell, 1982).

	Genetic variability	Heterozygosity (masking ability)	Costliness
Apomixis	minimal	maximal	minimal
Selfing, automixis	intermediate	low	low
Sex	maximal	high	maximal

at other loci. These hypotheses are often called "hitchhiking" hypotheses, because, according to them, a gene causing recombination increases in frequency when associated with favourable gene combinations that are directly selected. In contrast, the so-called "immediate benefit" hypotheses state that the variability caused by recombination is neutral or even detrimental and that the function of meiotic recombination is to repair double-strand DNA breaks or to decrease the deleterious mutation rate through biased gene conversion. There are theoretical problems with the immediate benefit hypotheses (Kondrashov, 1993; Hurst & Peck, 1996). Moreover, the data do not seem to support them; for example, it has been shown that meiotic recombination is initiated by enzymatically induced double-strand breaks, so the function of recombination cannot be the repair of spontaneous breaks (Cao *et al.*, 1990). Instead, we investigate in detail the hypotheses which state that the function of recombination is the generation of new allele combinations (i.e. genetic variability).

At first sight, the following simple explanation might seem convincing. If there is one lethal mutation on each homologous chromosome but at different loci, recombination can result in a perfect chromosome and a doubly mutant one (**Figure 3a**). Alternatively, one can imagine two chromosomes, with a beneficial mutation on each, and the function of recombination is then to create a superior double mutant (**Figure 3b**). The problem with this argument is that if the population is in linkage equilibrium then the above processes are exactly as frequent as their opposites; recombination creates a favourable gene combination as many times as it breaks it up and it has no *net* beneficial effect. Linkage equilibrium means that the alleles are not associated with each other; rather, they are independently distributed in the individual genotypes and recombination is useless. In contrast, in coupling linkage disequilibrium, the similar alleles at different loci tend to appear together and, in repulsion linkage disequilibrium, the alleles with opposite phenotypic effect are associated. (For the exact definition of linkage equilibrium, see glossary.) Thus, in linkage disequilibrium, recombination really has something to do: it breaks up the association between the alleles and creates novel genotypes that were previously missing or rare. It follows that, if we want to give an advantage to recombination, we must determine why the population is in linkage disequilibrium and why it is advantageous to destroy the association.

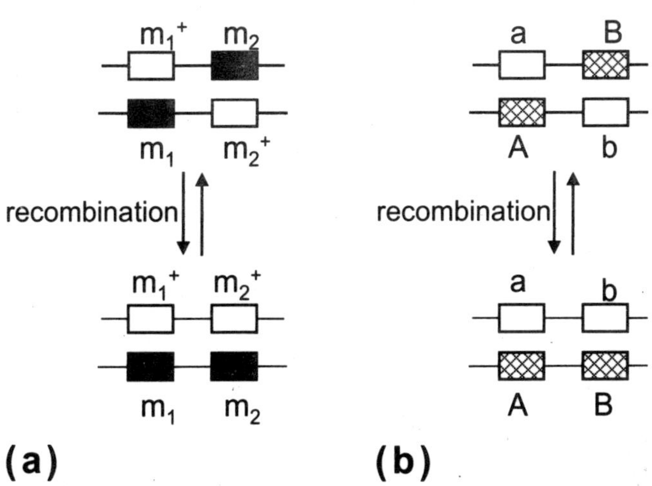

Figure 3 In linkage equilibrium, recombination creates and destroys favourable gene combinations at equal rates. (a) m_1 and m_2 are lethal mutations; (b) *A* and *B* are favourable mutations.

Let us consider the first question: "Why is a population in linkage disequilibrium?" The cause of the over-representation of some genotypes and the lack or rarity of others can be genetic drift (in stochastic hypotheses) or epistatic selection (in deterministic hypotheses). For the second question, as to why is it advantageous to destroy the association between alleles, we again have two possible answers. If the environment changes, then previously selected and, hence, over-represented genotypes become suboptimal (environmental hypotheses). Even in a constant environment, deleterious mutations can render over-represented genotypes suboptimal (mutational hypotheses). With these answers, we have arrived at the usual 2 × 2 classification of hypotheses shown in **Table 2** (Maynard Smith, 1988a; Kondrashov, 1993).

Before we consider the most important hypotheses of these four types, we must note that the destruction of associations by recombination is not always advantageous. The breaking up of coadapted gene complexes is deleterious and elicits a recombination load; coadapted genes may be closely linked to prevent recombination as is the case, for example, in the case of the genes for mimicry. (Here we will give only the most important references about the hypotheses; for more, see the references in Kondrashov, 1993 and Hurst & Peck, 1996.)

(i) Environmental–stochastic hypotheses

We begin with environmental–stochastic hypotheses, which consider the advantage of sex in a finite population and in a varying environment. Suppose that three mutant alleles, *A*, *B* and *C*, at different loci, which were previously deleterious, become advantageous due to an environmental change. Most probably, they occur in different individuals and as they begin to spread simultaneously in an asexual population they compete with each other and only one can go to fixation at one time. **Figure 4** shows that mutation *B* can spread only if it recurs after the fixation of *A*, and *C* can spread only after the fixation of *B*. In contrast, in a sexual population, recombination can combine these mutants in the same genotype, so that the sexual population does not have to wait for their reappearance. In other words, an asexual population can incorporate beneficial mutations only one by one, whereas in a sexual population more than one beneficial mutation can spread simultaneously; that is why the sexual population is capable of faster adaptive evolution. This old idea is often called the Fisher–Muller theory (Fisher, 1930; Muller, 1932; Maynard Smith, 1978).

It is worth investigating how stochasticity generates repulsion association between rare alleles (Felsenstein, 1988). Let the frequency of the rare beneficial alleles, *A* and *B*, be 10^{-3}, then the frequency of their co-occurrence (i.e. of

Table 2 Classification of hypotheses on the advantage of sex (after Maynard Smith, 1988a).

	Cause of excess of suboptimal genotypes	
Cause of linkage disequilibrium	**Environmental change**	**Deleterious mutations**
Genetic drift	*Fisher–Muller theory,"A ruby in the rubbish"*: sex accelerates adaptive evolution	*Muller's ratchet*: sex prevents the accumulation of deleterious mutations
Epistatic selection	*Directional selection* is more effective in a sexual population *Coevolutionary hypothesis*: sex helps to escape from parasites	*Mutational deterministic hypothesis*: sex decreases the mutational load

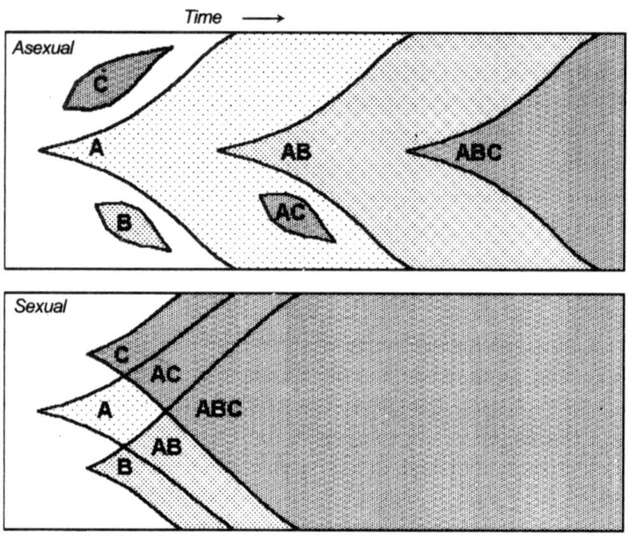

Figure 4 Adaptive evolution is faster in a sexual population (after Muller, 1932).

the *AB* genotype) would be 10^{-6} in an infinite population (assuming no epistatic selection). In the real world, it means that in a not too small population of size 10^4 (on the average) there will be in 99 generations out of 100 no individual with genotype *AB*, and there will be one in generation 100. It means that, although on the average there is no association between *A* and *B*, in 99 generations there is repulsion association and in one there is a strong coupling association. Thus, the population spends most of its time with repulsion association and in most of the cases recombination is advantageous.

The Fisher–Muller theory requires that more than one beneficial mutation must spread at the same time in the population. However, there is a more general explanation for faster adaptive evolution. If we take deleterious mutations into account as well, it is enough to consider the spread of only one beneficial mutation (Peck, 1994). Both in sexual and asexual populations, individuals may have more or fewer deleterious mutations, but the fate of a beneficial mutation appearing in a genotype with many deleterious mutations has different chances. In the asexual population, it remains stuck to the bad fellows and disappears together with them because in an asexual population only the descendants of the fittest individuals will be present in the distant future. In the sexual population, recombination can liberate the beneficial mutation from its original genetic environment; in Peck's words, it picks out "the ruby from the rubbish".

(ii) Mutational–stochastic hypothesis

In the mutational–stochastic category, we will discuss a famous hypothesis called Muller's ratchet (Muller, 1964). Muller's basic idea was that in a "finite asexual population there is a tendency for deleterious mutations to accumulate". Let us group the individuals of such a population into mutant classes: the zeroth mutant class contains no deleterious mutation, the first contains one deleterious mutation, the second class contains two, and so on. If n_0, the number of the mutation-free individuals is low, then it can happen that they have no descendants because of chance, despite being the fittest. In that case, there is no way to reconstitute the mutation-free genotype other than by back-mutation; this was shown, however, to be too infrequent to prevent the shift of the distribution of genotypes in the population. In Muller's phrase, the ratchet has clicked round one notch. The process can go further and the best class, now containing one mutation, can be lost again (**Figure 5**). Of course, such a process cannot take place in a sexual population because recombination can recreate the mutation-free genotype from two genotypes containing mutations at different loci.

The relevance of Muller's ratchet depends on n_0, the number of individuals in the mutation-free class. If $n_0 < 10$, the best class will be lost quickly and if $n_0 > 100$ the ratchet clicks slowly. n_0 and the speed of the ratchet depend on the following factors: *N*, the population size; *s*, the selective disadvantage per mutation; *U*, the genomic deleterious mutation rate; and the type of selection – namely, if it acts independently or not on different loci. Naturally, decreasing *N* and increasing *U* decreases n_0, and hence accelerates the ratchet.

If selection acts independently on the loci then each new mutation lowers the fitness by the same proportion; the fitness of individuals with 1, 2, 3 . . . mutations is $1-s$, $(1-s)^2$, $(1-s)^3$ . . . and, hence, this type is called multiplicative selection. For the multiplicative case, Haigh (1978) calculated that the expected value of n_0 is:

$$n_0 = N_e^{-U/s}.$$

Haigh thought that, as n_0 decreases with decreasing *s*, the mildly deleterious mutations are the most dangerous for an

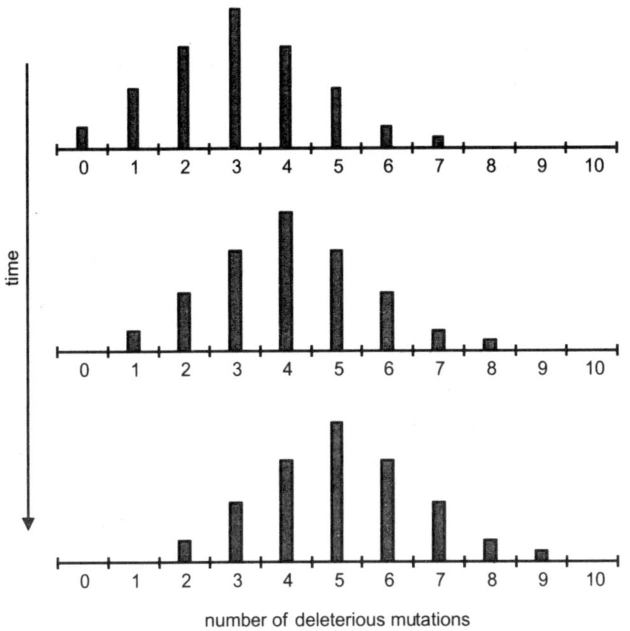

Figure 5 Muller's ratchet. The top diagram shows the initial distribution in an asexual population of individuals with 0, 1, 2, . . . deleterious mutations. The lower two diagrams show the same population after the ratchet has clicked round one and two notches, respectively (after Maynard Smith, 1988a).

asexual clone because for them selection is weak and cannot maintain the best genotype against drift. In contrast, strongly deleterious mutations are weeded out quickly and cannot accumulate; if only lethal mutations appeared then only the mutation-free individuals would survive and $n_0 = N$ would hold.

Recently, Lynch *et al.* (1993) argued that, as mutations accumulate in the clone, sooner or later its absolute fitness decreases below one, so that the population size begins to decrease. As the importance of drift compared with selection increases with decreasing population size, it accelerates the further accumulation of mutations. Lynch *et al.* called this process the mutational meltdown. As the accumulation of mildly deleterious mutations decreases mean fitness only moderately, they are not the most dangerous for the clone; it is rather the moderately deleterious ones (with s = 0.01–0.1) that lead to the fastest mutational meltdown. The reason is that the latter cause a greater decline in fitness at each turn of the ratchet and, consequently, decrease the population size more and more. However, beyond a critical value of about $s^* = 0.1$, the ratchet turns very slowly, as mentioned above. Selfing is less prone to accumulate mutations than apomixis for similar reasons: as self-fertilization brings the deleterious recessive alleles into homozygous state, selection can effectively weed them out. Thus, a selfing (or automictic) clone, which was founded by a relatively mutation-free individual, can survive inbreeding depression and goes to extinction more slowly than an apomictic one because of Muller's ratchet.

From the above arguments, it seems that Muller's ratchet gives a large advantage to sexuals over apomicts. However, we assumed multiplicative fitness and there is some evidence (see later), that the fitness function is actually synergistic (epistatic). This means that each new mutation decreases the fitness more and more (**Figure 6**), because the mutant alleles strengthen each other's deleterious effects. At the extreme case of truncation selection – in which having 1, 2, . . . k mutations does not reduce the fitness of the individual, but if it has $k + 1$ it dies – several authors claim (Charlesworth *et al.*, 1993; Lynch *et al.*, 1993; Kondrashov, 1994a) that sufficiently strong "synergistic selection halts Muller's ratchet" because, as the ratchet advances, the successive steps cause more and more serious decrease in fitness and the frequency of the best existing class is greater and greater.

(iii) Mutational deterministic hypothesis

The third category consists of one well-defined idea: the mutational deterministic hypothesis of Kondrashov (1988). In contrast to Muller's ratchet, it works only in the case of synergistic epistasis. The role it assigns to recombination is the increase of the variance of deleterious mutations per individual and, in this manner, the increase of the efficacy of selection and the decrease of the mutational load. We will not go into the mathematical details of the model here, but the beneficial effect of recombination can be understood intuitively. The first important point is that synergistic selection causes negative linkage disequilibrium (i.e. the excess of genotypes with moderate numbers of mutations and the depletion of those with many or few mutations). In the extreme case of truncation selection (**Figure 7**), all members of an asexual population have exactly k mutations after selection. Suppose, as seems probable (Kondrashov, 1988; Drake *et al.*, 1998), that the mutation rate per genome is U = 1. Then, in an asexual population, the majority of individuals who die in the course of selection have only one more mutation than the surviving ones; as Muller argued: "selection eliminates only one deleterious allele per selective death". In contrast, in the sexual population, recombination restores the genetic variance that was reduced by synergistic selection and there is a big difference in the number of mutations between individuals who die and who survive. As

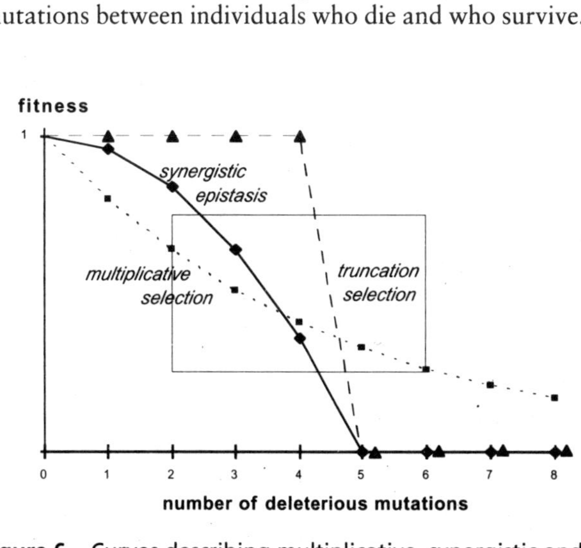

Figure 6 Curves describing multiplicative, synergistic and truncation selection.

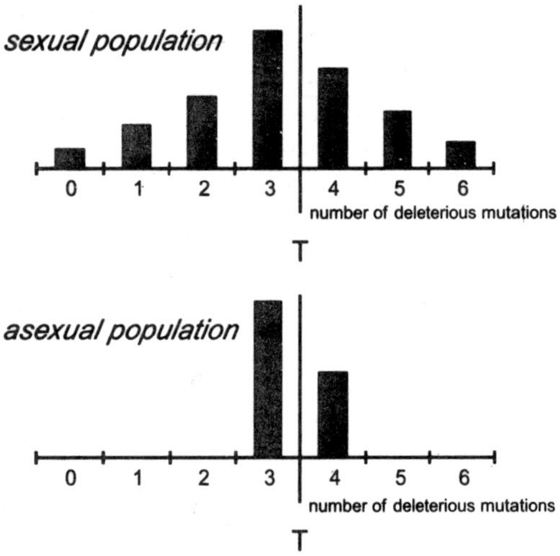

Figure 7 In a sexual population, recombination increases the variance of the number of deleterious mutations and, hence, it increases the efficiency of selection.

more than one mutation is eliminated by one genetic death, fewer individuals must die to eliminate the same number of mutations from the population. This means that the efficacy of selection is greater and the mutation load is lower in a sexual population.

As chance does not play any role here, the mutational deterministic hypothesis works in large populations. The conditions on which its validity depends are: first, the genomic deleterious mutation rate must be greater than about one and, secondly, the deleterious mutations must act synergistically. It turns out, however, that, at least in mammals, the mutation rate is higher in males than in females. If we take this into account, then being sexual is worthwhile for a female only in the case of rather strong truncation selection.

The models are not only deterministic by definition, but apparently require a great deal of determinism (i.e. a lack of stochasticity). It is generally known that stochastic effects can break rules set up in a deterministic context. Although it is true that, under the conditions of the hypothesis, the mean equilibrium load of the sexual population is lower than that of the asexual clone, this does not apply to transients (while the clone is not yet in mutation–selection equilibrium) and to stochastic effects. An asexual clone can successfully be founded by an individual whose genome is relatively uncontaminated by deleterious mutations. By the time the asexual mutation–selection balance sets in, the sexual population may go extinct. This was found in computer simulation of 10 000 individuals, unless the mutation rate was on the order of 2.0 per genome per generation (Howard, 1994).

(iv) Environmental–deterministic hypotheses

In the environmental–deterministic category of hypotheses, the advantage given to sex, as Haldane said, is that a sexual population is "more elastic" because it can respond to selection by rearranging its genes into "new combinations suited to the new environment". More concretely, it can respond to directional selection better or it can escape from parasites. We will consider these possibilities in turn.

Directional selection. First suppose that an environmental change causes directional selection through many generations or, alternatively, that environmental fluctuations result in a fluctuating selection. Then there is a load, called the lag load, because the population always lags behind the optimum. Many authors (Maynard Smith, 1988b; Crow, 1992; Charlesworth, 1993) have suggested that recombination is advantageous because it increases the response to selection through increasing the variability of the population.

Why then is genetic variability decreased? Let us use the usual "+" and "−" symbols for the alternative alleles affecting the quantitative trait. Generally, directional selection causes repulsion linkage disequilibrium – an excess of medium genotypes (e.g. + − + − and − − + +) and a lack of extreme genotypes (e.g. + + + + and − − − −). By destroying repulsion associations, recombination increases the genetic variability of the population. Note that the greatest response to directional selection could be achieved theoretically in a population with extreme coupling association (consisting exclusively of + + + + and − − − −), but recombination cannot make such a population. It only pushes the population towards linkage equilibrium, meaning increased variation compared with repulsion linkage disequilibrium. In a stable environment, selection is stabilizing and the increase of variability by recombination is harmful. Thus, the validity of the hypothesis stating that the advantage of recombination is a better responsiveness to selection depends on the relative amount of directional and stabilizing selection in nature. The fact that domesticated mammals do have a high frequency of recombination (Burt & Bell, 1987) can well be explained by the better responsiveness hypothesis because they have been under strong artificial directional selection.

Escape from parasites. The second version, when a sexual population can respond better to environmental changes, considers the selective pressures originating from the changes of the biotic environment. It is often referred to as the Red Queen hypothesis. The metaphor is from the tale *Through the Looking Glass*, and its message is that you must keep running if you want to remain at the same place. In the coevolutionary arms race of hosts and parasites, the hosts must escape from their parasites by changing their genotypes (Hamilton *et al.*, 1990; Dybdahl & Lively, 1998; Otto & Michalakis, 1998). If not, the parasites adapt to them more precisely and become more efficient because the parasites have a shorter generation time than their hosts. The host genotype that was rare during the past few generations is the most advantageous because parasites evolve to attack mainly the common ones. In technical terms, the interaction between hosts and parasites results in time-lagged frequency-dependent selection acting on the hosts, under which recombination is advantageous because it can produce new rare host genotypes from the existing alleles. The fact that the immune-system genes evolve faster than the other parts of the genome supports the Red Queen hypothesis; the correlation between the higher occurrence of sex and parasites in the tropics is consistent with it as well, although this correlation can also be explained with the mutational deterministic hypothesis because higher temperatures can cause higher mutation rates. Recently, it has been suggested that the presence of parasites makes an asexual clone more prone to accumulate mutations by Muller's ratchet because parasites cause the decrease of the frequency of every host genotype from time to time (Howard & Lively, 1994, 1998). This study calls attention to the fact that it is relatively easy to find a strong advantage of sex if several proposed mechanisms act together (similar to the model of the "ruby in the rubbish" discussed above).

Another interesting point is that dispersal (migration from subpopulation to subpopulation) could function as an alternative to sex. If the host disperses (e.g. as a dormant egg) when it is resistant to heat and drought, but its parasites are not resistant to these factors, then without sex it can enjoy the advantage of rarity in the new subpopulation where the parasites are not coadapted to it (Ladle *et al.*, 1993; Judson, 1997).

4. Discussion of the hypotheses

After having seen how many different types of advantages sex can have, a logical question is raised as to which ones are really effective in nature. A pluralistic point of view might be taken, based on the analogy that in physics a realistic situation is always understood with the vectorial summation of all acting forces, be they mechanistic, eletromagnetic or something else in nature (cf. Bell, 1982; West *et al.*, 1999; Kövér & Szathmáry, 1999).

Each hypothesis can explain why a sexual population has a higher average fitness than an asexual clone, if they are reproductively isolated. Williams (1975) called attention to the fact that, despite this group selection argument being valid for many species, some are cyclically parthenogenetic like *Daphnia pulex*, which sexually produces dormant embryos after many rounds of apomixis. If sex did not have any short-term individual advantage, then the sexual part would disappear from the life cycle. A short-term advantage can be conferred on sex by mutational–deterministic or environmental–deterministic (better responsiveness to directional selection, Red Queen) hypotheses. Unfortunately, it is difficult to decide which one to accept; we need more experimental results on the mutation rate, the interaction between deleterious mutations (if the interaction is synergistic), and the genetical and ecological background of host–parasite interactions. Muller's ratchet and the Fisher–Muller theory explain well the relatively short lifespan of asexual clones, since they accumulate deleterious and beneficial mutations faster and slower, respectively, than their sexual competitors. As these disadvantages appear only in the long run (in the course of 1000–10 000 generations), Muller's ratchet and the Fisher–Muller theory cannot give a short-term advantage to sex, and thus we cannot accept them as the main explanation. Some asexual taxa (such as the bdelloid rotifers) seem to be exceptionally ancient (Judson & Normark, 1996); if we knew the secret of their longevity, we could understand the role of sex better; for example, if they tend to disperse more than other asexual clones, then this supports the Red Queen theory (Judson, 1997). The vesicular–arbuscular mycorrhizae (a symbiotic association between microscopic soil fungi and roots) have been around for 440 million years (Remy *et al.*, 1994); in this mutualism, the fungal partner (Glomales, Zygomycotina) is strictly asexual. This would indicate that one abandons sex if conditions are fine; this would favour the Red Queen hypothesis.

There has been a fortunate increase in the number of experimental approaches in recent years. We know of two experiments showing that recombination increases the rate of adaptive evolution. Phage T4 populations evolve proflavine resistance faster if their rate of recombination is higher (Malmberg, 1977) and *Drosophila melanogaster* responds more slowly to artificial selection on bristle number if recombination is suppressed (McPhee & Robertson, 1970). Recent experiments on yeast (Zeyl & Bell, 1997) and on Chlamydomonas (de Visser *et al.*, 1996) support the mutational–deterministic hypothesis. Sex increases the mean fitness of yeast populations in an environment to which they were adapted. The Chlamydomonas case is particularly important since it provides an ingenious new method to test for mutational synergy (Zeyl & Bell, 1997). The effect of sex on mean log fitness of the offspring of a cross depends on the mode of mutation interaction and the relative difference in mutational contamination of the two parents (**Figure 8**). These findings are, at least partly, in accord with the metabolic control–theoretical analysis, showing that harmful mutations affecting enzymes of a biochemical pathway should show synergistic epistasis if: (1) selection is for optimal flux or concentration and the mutations affect the same or different enzymes; or if (2) selection is for maximal flux and the mutations affect the same enzyme (Szathmáry, 1993). As to Muller's ratchet, a clear experimental case has been demonstrated for RNA viruses (Chao, 1990).

Finally, there is a general problem with the hypotheses described above. They suggest that the optimal strategy is to have a small fraction of offspring produced sexually and the majority asexually. Thus, obligate sex seems to be superior to facultative sex only in the case of extreme assumptions (e.g. very high mutation rate, truncation selection). We must understand why "a little bit of sex is not as good as a lot" (Green & Noakes, 1995).

5. The origin of sex and sexes

The origin of sex is a problem different from that of its maintenance. Selective forces favouring the origin of component processes of contemporary sex may be insufficient to preserve it. Since the enzymes involved in the molecular machinery of recombination also deal with repair of genetic damage, recombination *per se* may have appeared as a by-product of selection for repair in early prokaryotes. Here, we are interested in the origin of eukaryotic sex only; there is recombination in prokaryotes associated with transformation, but it is the regular alternation of meiosis and syngamy that we will now address. It seems that even for eukaryotic sex the need for repair may have played a decisive role at the time of its origin.

Bernstein *et al.* (1981) have argued that the repair of

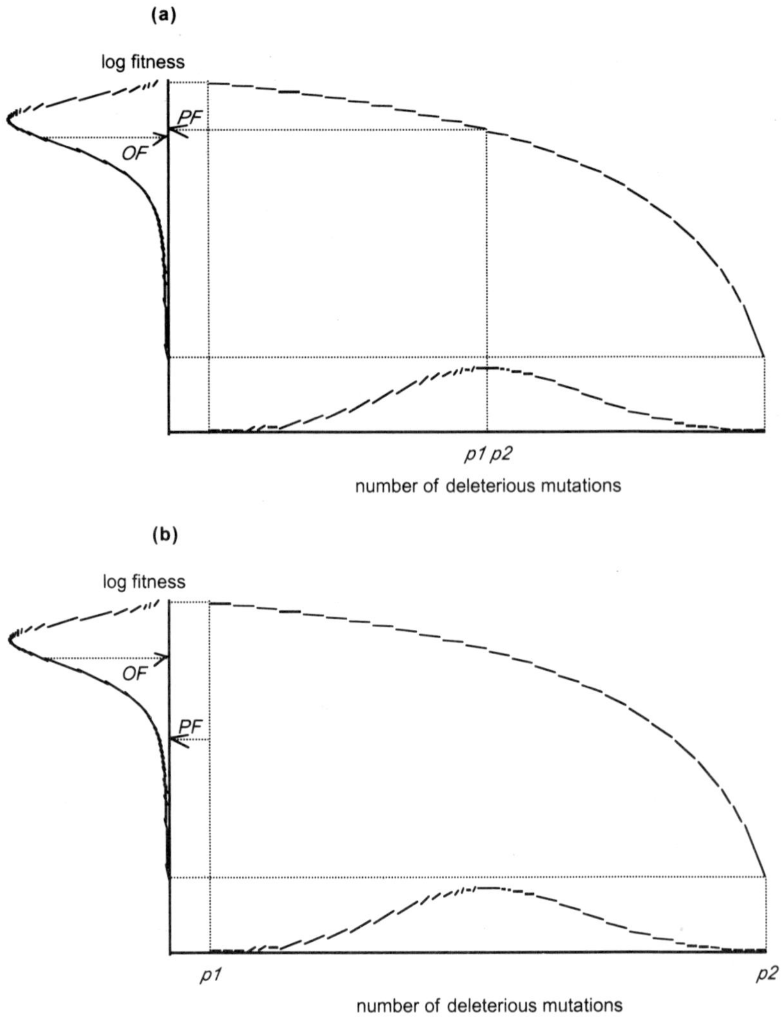

Figure 8 The effect of sex on mean log fitness of the offspring of a cross depends on the mode of mutation interaction and the relative difference in mutational load of the two parents. (a) If mutations act synergistically and both parents (*p1* and *p2*) have similar mutation loads, the effect of sex on mean log fitness is negative. (This might seem strange, but note that the parents in this experiment are not a population in mutation–selection balance.) Since equal numbers of offspring with higher and lower mutational loads than the mean of the parents are generated by sex, this negative effect is caused by the greater influence of offspring with many mutations. The bell-shaped curve on the horizontal axis represents the distribution of the number of mutations per individual among the offspring. (b) If the parents carry rather different numbers of deleterious mutations (*p1* has low, *p2* has high number of mutations), most offspring will obtain intermediate mutation numbers and, consequently, show higher mean log fitness. The reverse argument holds in the case of antagonistic epistasis, the opposite of synergistic epistasis (after de Visser *et al*., 1996). OF, offspring mean log fitness; PF, parental mean log fitness.

double-strand breaks (DSBs) in DNA may still be selecting for the maintenance of sex, whereas the need for complementation could, given repair, select for outcrossing. This idea seems generally untenable. For example, in the case of permanent diploids, repair of DSBs could be carried out while preserving a considerable degree of heterozygosity, provided repair were achieved through gene conversion (Szathmáry & Kövér, 1991).

The proposal (Bernstein *et al*., 1984) that the "need for repair may have been important originally" has a different merit. It is true that diploids are more resistant to damaging agents than haploids, and for non-phagotrophs it seems to hold (with exceptions) that haploids have an advantage in terms of growth rates (for review see Maynard Smith & Szathmáry, 1995). The question then is whether the fusion–reduction cycle originated before or after that of phagotrophy. There is some evidence that the former is true; there is such a cycle in the archaebacterium *Halobacterium volcanii* (Tchelet & Mevarech, 1994). Since archaebacteria are the sister group of eukaryotes, fusion–reduction may have been an ancestral trait in the earliest eukaryotes. If not, then the advantage of haploidy is less obvious, although Kondrashov (1994b) has suggested one advantage: an intermittent haploid state reduces the mutational load of a predominantly diploid, asexual cycle. (Note that such life cycles exist even today.) This confirms the earlier suggestion (e.g. Szathmáry *et al*., 1990) that an asexual alternation of haploid and diploid phases (achieved through endomitosis/endoreduplication and reduction) could have been an excellent preadaptation for sex. One can, for example, envisage an environment with an alternation of periods eliciting low and high levels of genetic damage, selecting for haploidy and diploidy, respectively. Later, endogenous reduplication could have been replaced by syngamy, facilitated by the advantage gained from complementation (for further details see Maynard Smith & Szathmáry, 1995).

Finally, we will briefly discuss the problem of the number of sexes and mating types. A crucial observation is that cytoplasmic fusion and the presence of organelles is coincident with two mating types, whereas lack of cytoplasmic fusion (e.g. ciliate conjugation) is coincident with many mating types (Hurst & Hamilton, 1992). This gives weight to the proposal that the existence of two sexes is due to selection for the efficient control of (potentially) harmful intracellular symbionts in the form of uniparental inheritance, which requires a difference in sexual roles controlled by the nucleus

(Hoekstra, 1987). It is remarkable that the mating-type locus in Chlamydomonas contains genes coordinating who mates with whom and those controlling organelle inheritance (Armbrust *et al.*, 1993).

Dynamic modelling (Hurst, 1996a) and further comparative evidence (Hurst, 1996b) confirm that this view is likely to be correct. For example, without the problem of intracellular genetic conflict, two mating types would be replaced by three, and so on, especially if there is a cost of finding a mate. If more than two mating types confer a disadvantage on matings, due to an inefficient control of organelle inheritance, two sexes are evolutionarily stable (Hurst, 1996a).

REFERENCES

Armbrust, E.V., Ferris, P.J. & Goodenough, U.W. (1993) A mating type-linked gene cluster expressed in *Chlamydomonas* zygotes participates in the uniparental inheritance of the chloroplast genome. *Cell* 74: 801–11

Bell, G. (1982) *The Masterpiece of Nature: The Evolution and Genetics of Sexuality*, Berkeley: University of California Press

Bernstein, H., Byers, G.S. & Michod, R.E. (1981) The evolution of sexual reproduction: the importance of DNA repair, complementation, and variation. *American Naturalist* 117: 537–49

Bernstein, H., Byerly, H., Hopf, F. & Michod, R. (1984) Origin of sex. *Journal of Theoretical Biology* 110: 323–51

Burt, A. & Bell, G. (1987) Mammalian chiasma frequencies as a test of two theories of recombination. *Nature* 326: 803–05

Cao, L., Alani, E. & Cleckner, N. (1990) A pathway for generation and processing of double-strand breaks during meiotic recombination in *S. cerevisiae*. *Cell* 61: 1089–101

Chao, L. (1990) Fitness of RNA virus decreased by Muller's ratchet. *Nature* 348: 454–55

Charlesworth, B. (1993) Directional selection and the evolution of sex and recombination. *Genetical Research* 61: 205–24

Charlesworth, D., Morgan, M.T. & Charlesworth, B. (1993) Mutation accumulation in finite outbreeding and inbreeding populations. *Genetical Research* 61: 39–56

Crow, J.F. (1992) An advantage of sexual reproduction in a rapidly changing environment. *Journal of Heredity* 83: 169–73

de Visser, J.A.G.M., Hoekstra, R.F. & Van den Ende, H. (1996) The effect of sex and deleterious mutations on fitness in *Chlamydomonas*. *Proceedings of the Royal Society of London, Series B* 263: 193–200

Drake, J.W., Charlesworth, B., Charlesworth, D. & Crow, J.F. (1998) Rates of spontaneous mutation. *Genetics* 148: 1667–86

Dybdahl, M.F. & Lively, C.M. (1998) Host-parasite coevolution: evidence for rare advantage and time-lagged selection in a natural population. *Evolution* 52: 1057–66

Felsenstein, J. (1988) Sex, and the evolution of recombination. In *The Evolution of Sex: An Examination of Current Ideas*, edited by R.E. Michod & B.R. Levin, Sunderland, Massachusetts: Sinauer

Fisher, R.A. (1930) *The Genetical Theory of Natural Selection*, Oxford: Clarendon Press

Green, R.F. & Noakes, D.L.G. (1995) Is a little bit of sex as good as a lot? *Journal of Theoretical Biology* 174: 87–96

Haigh, J. (1978) The accumulation of deleterious genes in a population – Muller's ratchet. *Theoretical and Population Biology* 14: 251–67

Hamilton, W.D., Axelrod, R. & Tanese, R. (1990) Sexual reproduction as an adaptation to resist parasites (a review). *Proceedings of the National Academy of Sciences USA* 87: 3566–73

Hoekstra, R. (1987) The evolution of sexes. In *The Evolution of Sex and Its Consequences*, edited by S.C. Stearns, Basel: Birkhauser

Howard, R.S. (1994) Selection against deleterious mutations and the maintenance of biparental sex. *Theoretical and Population Biology* 45: 313–23

Howard, R.S. & Lively, C.M. (1994) Parasitism, mutation accumulation and the evolution of sex. *Nature* 367: 554–57

Howard, R.S. & Lively, C.M. (1998) The maintenance of sex by parasitism and mutation accumulation under epistatic fitness functions. *Evolution* 52: 604–10

Hurst, L.D. (1996a) Why are there only two sexes? *Proceedings of the Royal Society of London, Series B* 263: 415–22

Hurst, L.D. (1996b) Selfish genetic elements and their role in evolution: the evolution of sex and some of what it entails. *Philosophical Transactions of the Royal Society of London, Series B* 349: 321–32

Hurst, L.D. & Hamilton, W.D. (1992) Cytoplasmic fusion and the nature of the sexes. *Proceedings of the Royal Society of London, Series B* 247: 189–94

Hurst, L.D. & Peck, J.R. (1996) Recent advances in understanding of the evolution and maintenance of sex. *Trends in Ecology and Evolution* 11: 46–62

Judson, O.P. (1997) A model of asexuality and clonal diversity: cloning the Red Queen. *Journal of Theoretical Biology* 186: 33–40

Judson, O.P. & Normark, B.B. (1996) Ancient asexual scandals. *Trends in Ecology and Evolution* 11: 41–46

Kondrashov, A.S. (1988) Deleterious mutations and evolution of sexual reproduction. *Nature* 336: 435–40

Kondrashov A.S. (1993) Classification of hypotheses on the advantage of amphimixis. *Journal of Heredity* 84: 372–87

Kondrashov, A.S. (1994a) Muller's ratchet under epistatic selection. *Genetics* 136: 1469–73

Kondrashov, A.S. (1994b) The asexual ploidy cycle and the origin of sex. *Nature* 370: 213–16

Kövér, S. & Szathmáry, E. (1999) Hybrid theories of sex. *Journal of Evolutionary Biology* 12: 1032–33

Ladle, R.J., Johnstone, R.A. & Judson, O.P. (1993) Coevolutionary dynamics of sex in a metapopulation: escaping the Red Queen. *Proceedings of the Royal Society of London, Series B* 253: 155–60

Lynch, M., Burger, R., Butcher, D. & Gabriel, W. (1993) The mutational meltdown in asexual populations. *Journal of Heredity* 84: 339–44

Malmberg, R.L. (1977) The evolution of epistasis and the advantage of recombination in populations of bacteriophage T4. *Genetics* 86: 607–21

Maynard Smith, J. (1978) *The Evolution of Sex*, Cambridge and New York: Cambridge University Press

Maynard Smith, J. (1988a) The evolution of recombination. In *The Evolution of Sex: An Examination of Current Ideas*, edited by R.E. Michod & B.R. Levin, Sunderland, Massachusetts: Sinauer

Maynard Smith, J. (1988b) Selection for recombination in a polygenic model – the mechanism. *Genetical Research* 51: 59–63

Maynard Smith, J. & Szathmáry, E. (1995) *The Major Transitions in Evolution*, Oxford and New York: Freeman

McPhee, C.P. & Robertson, A. (1970) The effect of suppressing crossing-over on the response to selection in *Drosophila melanogaster*. *Genetical Research* 16: 1–16

Muller, H.J. (1932) Some genetic aspects of sex. *American Naturalist* 66: 118–38

Muller, H.J. (1964) The relation of recombination to mutational advantage. *Mutation Research* 1: 2–9

Otto, S.P. & Michalakis, Y. (1998) The evolution of recombination in changing environments. *Trends in Ecology and Evolution* 13: 145–51

Peck, J.R. (1994) A ruby in the rubbish – beneficial mutations, deleterious mutations and the evolution of sex. *Genetics* 137: 597–606

Remy, W., Taylor, T.N., Hass, H. & Kerp, H. (1994) Four hundred-million-year-old vesicular arbuscular mycorrhizae. *Proceedings of the National Academy of Sciences USA* 91: 11841–43

Szathmáry, E. (1993) Do deleterious mutations act synergistically? Metabolic control theory provides a partial answer. *Genetics* 133: 127–32

Szathmáry, E. & Kövér, S. (1991) A theoretical test of the DNA repair hypothesis for maintenance of sex in eukaryotes. *Genetical Research* 58: 157–65

Szathmáry, E., Scheuring, I., Kotsis, M. & Gladkih, I. (1990) Sexuality of eukaryotic unicells: hyperbolic growth, coexistence of facultative parthenogens, and the repair hypothesis. In *Organizational Constraints on the Dynamics of Evolution*, edited by J. Maynard Smith & G. Vida, Manchester: Manchester University Press

Tchelet, R. & Mevarech, M. (1994) Interspecies transfer in halophilic Archaebacteria. *Systematic and Applied Microbiology* 16: 577–81

West, S.A., Lively, C.M. & Read, A.F. (1999) A pluralist approach to sex and recombination. *Journal of Evolutionary Biology* 12: 1003–12

Williams, G.C. (1975) *Sex and Evolution*, Princeton, New Jersey: Princeton University Press

Zeyl, C. & Bell, G. (1997) The advantage of sex in evolving yeast populations. *Nature* 388: 465–68

GLOSSARY

automixis self-fertilization in which the two nuclei that fuse to form a zygote are the products of the same meiosis during gametogenesis

endomitosis (i) process that occurs naturally in some differentiating cells or when induced by certain substances e.g. colchicine, in which the chromosomes divide without cell division giving double the original number of chromosomes in the cell; (ii) somatic polyploidization taking place within an intact nuclear envelope. No stages comparable with the natural mitotic cycle are observed but the DNA content increases in multiples of the haploid value. If the reduplicated chromosomes fail to separate and remain in register, polytene chromosomes are formed

linkage equilibrium for any two loci with two alleles: *A* and *a* on the first locus and *B* and *b* on the second one, the loci are in linkage quilibrium if the frequency of the gamete type AB equals the product of the frequencies of alleles A and B. Linkage equilibrium is a property of a population and not of an individual. If the presence of *A* on the first locus is not independent of whether *B* or *b* is present on the second locus, then there is linkage disequilibrium. In coupling linkage disequilibrium, *A* tends to occur with *B* and *a* with *b*, while, in repulsion linkage disequilibrium, the alleles with opposite effects are associated. In the first case, $p_{AB} > p_A p_B$ and, in the second, $p_{AB} < p_A p_B$. The departure from linkage equilibrium is measured by the difference between the expected and the observed frequency. Recombination cannot destroy linkage disequilibrium in one generation. The decrease of linkage disequilibrium depends on the frequency of recombination between the two loci

phagotrophy heterotrophic nutrition in which cells ingest solid food particles

stochastic (i) developing in accordance with a probabilistic model; random; (ii) a process in which there is an element of chance or randomness

synergistic acting together, often to produce an effect greater than the sum of the two agents acting separately

MITOCHONDRIAL "EVE"

John Brookfield
Institute of Genetics, University of Nottingham, Queens Medical Centre, Nottingham, UK

While we are understandably fascinated by our evolutionary origins, our knowledge is sparse – the fossil record is poor. We do now know that our relatives, the chimpanzee (*Pan troglodytes*) and the gorilla (*Gorilla gorilla*) are much closer to us than their morphology and behaviour led us to believe, and that Pan is indeed more closely related to *Homo sapiens* than it is to Gorilla. Increasingly, evolutionary inferences are drawn from molecular differences between man and apes and from the molecular variation between humans themselves. Studies of intraspecific variation have concentrated on mitochondrial DNAs.

Circles of DNA are found in mitochondria. Each mitochondrion has 5–10 identical circles and a cell may have thousands of mitochondria, depending on its size. In mammals, mtDNAs are small; humans, for example, have around 16 000 base pairs in their DNA circle. There are only 37 genes, most of which are the small genes producing transfer RNAs, but there are also two ribosomal RNA genes and 13 protein-coding genes. Their circular structure, and sequence similarities in the ribosomal RNA genes, reveal that mitochondria are descended from symbiotic bacteria. Since the symbiosis was established around a billion years ago, the DNAs have reduced in size, as genes have transferred from the mitochondrion to the nucleus. In mammals, mitochondria are almost exclusively inherited through the female line, and their DNAs do not recombine – although there is evidence that, in interspecific hybrids in mice, one molecule in every 10 000 comes from the father (Gyllensten *et al.*, 1991), and a recent, and highly controversial, report (Hagelberg *et al.*, 1999) gives suggestive evidence that there may be some low level of recombination in human DNAs. Despite the many mitochondria in the oocyte, which could, in principle, have different DNA sequences within them, there seems to be very little heteroplasmy – that is, there is almost no variation between mtDNAs within an individual. The mechanism for this rapid loss of variability is unknown.

An absence of recombination has an important consequence for the evolution of the mtDNA molecule. All copies of a non-recombining DNA molecule must trace back to a single ancestor. Thus, the mtDNAs of all the individuals in a population are connected by a single family tree (phylogeny), a tree that is exactly the same for each of the genes in the molecule. Within a species, the molecules are very similar and, due to shared descent, they largely retain the ancestral sequence. Variation results from mutations occurring at various positions in the tree connecting the DNAs, and this variation can be used to estimate the structure of the tree.

If all our mtDNAs are descended from a single molecule, then, obviously, the bearer of the ancestral mtDNA molecule must have been an ancestor of all humans alive on earth. Furthermore, she must have been female, and, since we have all received descendants of her mtDNA, we must all be derived from her by lineages in which all the individuals were also female. We can define "mitochondrial Eve" as the most recent female from whom all living humans are descended through purely female lines. We need to know when and where she lived, and what the branching pattern is in the family tree connecting her to people alive today.

MtDNAs are very easy to isolate and study. In addition, their rate of mutation is, in warm-blooded vertebrates, approximately ten times higher than that of DNAs in the nucleus. Therefore, in a given time period, mtDNAs diverge much more in their sequence than do nuclear DNAs and give much more information about phylogeny.

The first major study of human mtDNA variation was by Cann, Stoneking and Wilson (1987). They digested DNAs isolated from 147 individuals from a variety of human populations with a collection of 12 restriction enzymes, each recognizing a four-base sequence in the DNA. From this, 133 different restriction patterns were seen. The diversity between the DNAs from Africans was greater than that for the other groups studied (Caucasians, Asians, Australians and New Guineans). When the most parsimonious tree was rooted assuming a molecular clock, the first bifurcation separated a branch including only African sequences from a branch including sequences from Africa and from all other populations. The implication was that the mitochondrial Eve was African. The assumption of a constant evolutionary rate predicted a time for the root of the tree (i.e. for "Eve") as between 140 000 and 290 000 years ago. The work made a massive impact in the media throughout the world, largely traceable to the use of the unfortunate term, "Eve". It is trivial that humans must descend from a "mitochondrial Eve", yet some suggested that this was, in itself, a result of the study. Some even suggested that "Eve" was the only human female alive at the time! Probably hundreds of thousands of other females lived on earth at this time, and most of these were probably also ancestors of all living humans (although not now through purely female lines).

Eve is important because fossils have not resolved the origins of modern humans, *H. sapiens sapiens*. The earliest Homo outside Africa were *H. erectus*, fossils of which have been found in China from 750 000 years ago; even older fossils are known from Java. Other *H. erectus* populations lived in Africa at this time. European humans 300 000–100 000 years ago belonged to the subspecies *H. sapiens neanderthalensis*. Then, over the last 150 000 years, modern man, *H. sapiens sapiens*, appeared in Asia, Africa and Europe, before spreading to America, New Guinea and Australia. There are two different explanatory models.

1. The multiregional model postulates the independent development of modern man in each geographical area. It suggests that the morphological characters of ancient skulls show similarities to modern humans from the same geographical location, particularly in China, Europe and North Africa, indicating genetic continuity.
2. The alternative view is the "Out-of-Africa" model. In this model, modern humans developed in Africa about 200 000 years ago and spread rapidly throughout the world, completely replacing the indigenous populations.

These hypotheses represent extremes in a spectrum of possibilities, differing in the proportion of the genes in modern populations that are derived from the invading Africans. Eve rules out the most extreme multiregional model, which has been called the "candelabra" model, in which the populations in different regions have exchanged no genes in the last 500 000 years. However, the data are still consistent with less extreme multiregional models, which postulate continuous low-level gene flow between all populations throughout the last million years.

Two questions remain. Is it true that all mtDNAs are recently derived from an African source? If they are, does this mean that our other genes share this origin? Subsequent studies of mtDNA variation have concentrated on sequencing the very variable D loop region of the molecule and these studies confirm that the total variability is low (consistent with a time to common ancestry between 100 000 and 300 000 years ago). In addition, African samples are consistently more variable than are samples from outside Africa (Vigilant *et al.*, 1991).

One very direct piece of evidence that the mtDNAs of modern man have replaced those of earlier humans has come from the sequencing of DNAs derived from the bone of the upper arm of a Neanderthal man (Krings *et al.*, 1997; see "Neanderthals" article). The mtDNA which was amplified from this 50 000-year-old bone had a sequence which was distinct from any of the sequences in modern humans, and formed an outgroup to them. The mtDNA from the Neanderthaler had apparently split from the lineage leading to modern humans about 500 000 years ago, well before the time of common ancestry of modern humans. It seems that this mtDNA type has become extinct and that the Neanderthalers did not contribute any mtDNAs to modern man.

If mtDNAs have recently spread from Africa, does this mean that all our genes derived from these same Africans or does the multiregional model apply at other loci? Evolutionary biologists often represent the relationships between populations within species by family trees. It is common to see the indigenous populations of the world represented in this way. However, this may mislead us, since different individuals from the same population may have very different ancestries. Furthermore, with recombination, the trees for different genes will differ. There will not necessarily be any true tree linking the various populations. The finding that the mtDNAs have a recent African root does not mean that mankind as a whole does. We really do not know how likely it is that all or most of our genes share the same origins. We do know that modern humans coexisted with Neanderthalers around 100 000 years ago in what is now Israel, and recent work (Swisher *et al.*, 1996) implies that *H. erectus* persisted on Java until the time of the arrival of the modern human ancestors. So, the opportunity for some gene flow between earlier humans and our ancestors was present. Whether we should interpret the mtDNA evidence as ruling this out greatly depends on whether natural selection has acted to spread the African mtDNA.

Has selection operated on mtDNAs? Interestingly, and particularly for non-African sequences (Di Rienzo & Wilson, 1991), almost all the molecules from different people differ from each other by similar numbers of base substitutions, suggesting that they diverged in a short time period, perhaps 100 000 years ago. This is what one would expect from a continuing expansion in the population size, as would happen during the spread from an African source. However, the pattern is also consistent with selection. With 37 genes in the molecule, it seems reasonable that at least one favourable mutation has arisen in the sequence in the last 200 000 years, and that this has spread through the population as a result of its selective advantage. With no recombination, this would create shared ancestry throughout the whole molecule. If this has happened, the expansion of mtDNAs from an African source is entirely consistent with a multiregional model for the other genes.

The resolution must come from studies of the other, nuclear genes. β-globin alleles differ so greatly in sequence that the expected time to common ancestry of two sequences is nearer 1 million years ago than 200 000 years, and this seems to be typical of nuclear genes (Fullerton *et al.*, 1994). For human leucocyte antigen genes, times to the common ancestor are probably at least ten times older still, a result explained by diversifying selection, in which rare genetic types are generally favoured by natural selection, at this locus (Ayala, 1995). For all types of genetic loci, African populations tend to show the greatest genetic variation, as is seen, for example, with microsatellite sequences (Bowcock *et al.*, 1994), and this suggests that there has indeed been a spread of a subset of the African population to colonize the rest of the world.

One molecule which does seem to share recent common ancestry (within the last 200 000 years) is the non-recombining region of the Y-chromosome, although the data are insufficient to estimate the time precisely (Hammer, 1995). Again, in this long, non-recombining molecule, selection might have been important. The Y chromosome, in addition to being a nuclear DNA and thus having a lower rate of mutation than that of the mtDNA, also differs from the latter in that it is passed on solely by males, unlike the maternally inherited mitochrondrion. This means that the two molecules' patterns of variability might be different if the migration behaviours of males and females have been different in human prehistory. Indeed, studies of Y chromosomal variation in Africa (Seielstad *et al.*, 1998) reveal a higher degree of local differentiation in the Y-chromosome, as would be expected if females have migrated but males stayed close to their birthplaces. Recent studies of Y chromosomal variation (Shen *et al.*, 2000; Thomson *et al.*, 2000), however, seem to place the most recent common ancestor for the Y chromosome as existing around 50 000 years ago, a time considerably more recent than that estimated for the mitochondrial DNAs.

Agreement between different genetic loci, such as mtDNA and the Y-chromosome, in their estimated times to common ancestry suggests that their recent shared ancestry has not resulted from the action of natural selection, but rather from the action of random genetic drift in a small population. If so, then the times to common ancestry can tell us something about how large or small our ancestral population was. Most of the estimates of the size of the population which was ancestral to modern humans are remarkably small, around 10 000 people, but we still do not know how constant this population size was in the time prior to its recent explosive growth.

REFERENCES

Ayala, F.J. (1995) The myth of Eve: molecular biology and human origins. *Science* 270: 1930–36

Bowcock, A.M., Ruiz-Linares, A., Tomfohrde, J. *et al.* (1994) High resolution of human evolutionary trees with polymorphic microsatellites. *Nature* 368: 455–57

Cann, R., Stoneking, M. & Wilson, A.C. (1987) Mitochondrial DNA and human evolution. *Nature* 325: 31–36

Di Rienzo, A. & Wilson, A.C. (1991) Branching pattern in the evolutionary tree for human mitochondrial DNA. *Proceedings of the National Academy of Sciences USA* 88: 1597–601

Fullerton, S.M., Harding, R.M., Boyce, A.J. & Clegg, J.B. (1994) Molecular and population genetic analysis of allelic sequence diversity at the human β-globin locus. *Proceedings of the National Academy of Sciences USA* 91: 1805–09

Gyllensten, U., Wharton, D., Josefsson, A. & Wilson, A.C. (1991) Paternal inheritance of mitochondrial DNA in mice. *Nature* 352: 255–57

Hagelberg, E., Goldman, N., Lío, P. *et al.* (1999) Evidence for mitochondrial DNA recombination in a human population of island Melanesia. *Proceedings of the Royal Society, Series B* 266: 485–92

Hammer, M.F. (1995) A recent common ancestry for human Y chromosomes. *Nature* 378: 376–78

Krings, M., Stone, A., Schmitz, R.W. *et al.* (1997) Neandertal DNA sequences and the origin of modern humans. *Cell* 90: 19–30

Seielstad, M., Minch, E. & Cavalli-Sforza, L.L. (1998) Genetic evidence for a higher female migration rate in humans. *Nature Genetics* 20: 278–80

Shen, P., Wang, F., Underhill, P.A. *et al.* (2000) Population genetic implications from sequence variation in four Y chromosomal genes. *Proceedings of the National Academy of Sciences USA* 97: 7354–59

Swisher, C.C. III, Rink, W.J., Antón, S.C. *et al.* (1996) Latest *Homo erectus* of Java: potential contemporaneity with *Homo sapiens* in Southeast Asia. *Science* 274: 1870–74

Thomson, R., Pritchard, J.K., Shen, P., Oefner, P.J. & Feldman, M.W. (2000) Recent common ancestry of human Y chromosomes: evidence from DNA sequence data. *Proceedings of the National Academy of Sciences USA* 97: 7360–65

Vigilant, L., Stoneking, M., Harpending, H., Hawkes, K. & Wilson, A.C. (1991) African populations and the evolution of human mitochondrial DNA. *Science* 253: 1503–07

GLOSSARY

leukocyte antigen a glycoprotein encoded by the major histocopatibility complex, expressed on the surface of all cells but first discovered on the leukocytes (white blood cells). These antigens are recognised as "self" or "non-self" by antibodies

See also **Neanderthals (p.58)**

GENETICS OF QUANTITATIVE CHARACTERS

Eric C.R. Reeve
Institute of Cell, Animal and Population Biology, University of Edinburgh, Edinburgh, UK

1. Introduction

Since the rediscovery of Mendel's laws in 1900 geneticists have tended to concentrate on character variations that can be tied to identifiable genes. These may be followed from generation to generation, leading to the chromosomal location, mutant forms, DNA sequence and function of each gene. Breeders wanting to improve their livestock, however, rarely find such genes useful and instead have to work with so-called quantitative characters such as milk yield, meat quality, growth rate or egg production, which are controlled by probably many genes interacting with environmental effects that cannot be eliminated.

Selection of these characters has been very successful in the past but left the breeders unable to explain the genetic basis of their success or to decide how best to make further progress. Large-scale genetic studies on livestock to identify the genes responsible were obviously impractical for reasons of time and expense, and breeders and other geneticists turned to "model" laboratory animals for information on how various quantitative characters behave under different types of selection and on the genes responsible for the result. The view has also grown that evolutionary changes have been largely due to the effects of natural selection on such quantitative characters.

An obvious model animal for livestock was the laboratory mouse, on which studies of the effects of selection for increase in body weight were started by H.D. Goodale in 1930 (Goodale, 1938) and were followed by many other such experiments with various designs (see article on "Body weight limits in mice"). A laboratory animal of a very different kind is the fruit fly *Drosophila melanogaster*, on which studies of quantitative characters began in Moscow in the 1930s, and have been continuing in many laboratories until the present day. It should be noted that this small insect has made an enormous contribution to our knowledge of genetics ever since it was taken up by Thomas Hunt Morgan, who discovered the first mutation recognized in *D. melanogaster* in 1910 (see article on "*Drosophila melanogaster:* the fruit fly").

2. Studies of bristle characters in *D. melanogaster*

(i) Early research by Karp

M.L. Karp, working in Moscow, first proposed certain patches of chaetae or bristles on the adult integument (the outer covering of the insect, comprising epidermis and cuticle) of Drosophila as suitable for genetic study and gave some results in a short article entitled "The distribution of mutant genes affecting the number of sternital bristles in chromosome 3 of *Drosophila melanogaster*" (Karp, 1935). This article consists of the following paragraph and is of particular interest in proposing a theoretical basis for his observations, which influenced later studies.

> In chromosome 3 of *D. melanogaster* the presence of at least six mutant genes affecting the number of sternital bristles, independently of the possible effect of the gene markers, has been shown. These genes possess a considerable power of action, approximately 5–15 per cent of the manifestation of the character. Being opposite in tendency and alternately located they are more or less balanced, not only along the whole length of the chromosome but within its small regions as well. In the chromosome, causing the reduction of 5–6 bristles on 2 sternites of the abdomen, were detected genes which determine conjointly the reduction of 18–21 bristles on the same 2 sternites, and on the other hand there were found genes which together intensify the character by 12–20 bristles. Hence the genic balance of the chromosome examined offers the possibility of a considerable change as to the extent of the manifestation of the character.

The sternital bristles are those on the abdominal plates (sternites) shown in **Figure 1**. These plates lie on abdominal segments 2–7 in females and 2–5 in males, and the two sternites Karp refers to are likely to have been 4 and 5 in both sexes. To obtain the bristle counts he gives for different regions of chromosome 3, Karp must have identified these regions using visible mutations of marker genes having known locations on the chromosome, so that each region could be transferred to a separate line for testing (plenty of genes with suitable recessive mutations were available as markers by 1930).

We have been unable to trace any later paper by Karp on this research, but he published an article on "Heterosis and inbreeding" in 1940, in Russian with an English summary (Karp, 1940), which would have been submitted to the Russian journal by 1939 or earlier. This summary describes a complex study of *D. melanogaster* involving inbred strains with crossover suppressors in the three major chromosomes, and refers to totals of 125 000 flies of 86 types and 500 000 eggs having been examined. These details suggest that he had several assistants in Moscow between 1930 and 1939, before World War II started; so the absence of details about his study of bristle genes is surprising. He did not attend the International Congress of Genetics held at Edinburgh in 1939, nor did he send an Abstract of a paper to it (as many Russian geneticists did, not all of whom actually managed to attend the Congress), so he may have lost his life or position, like many Russian geneticists, before he could publish further.

D. melanogaster has three major chromosomes (the X,Y pair that determine sex and autosomal chromosome pairs 2 and 3, with in addition the very small 4th pair; pair 3 is the

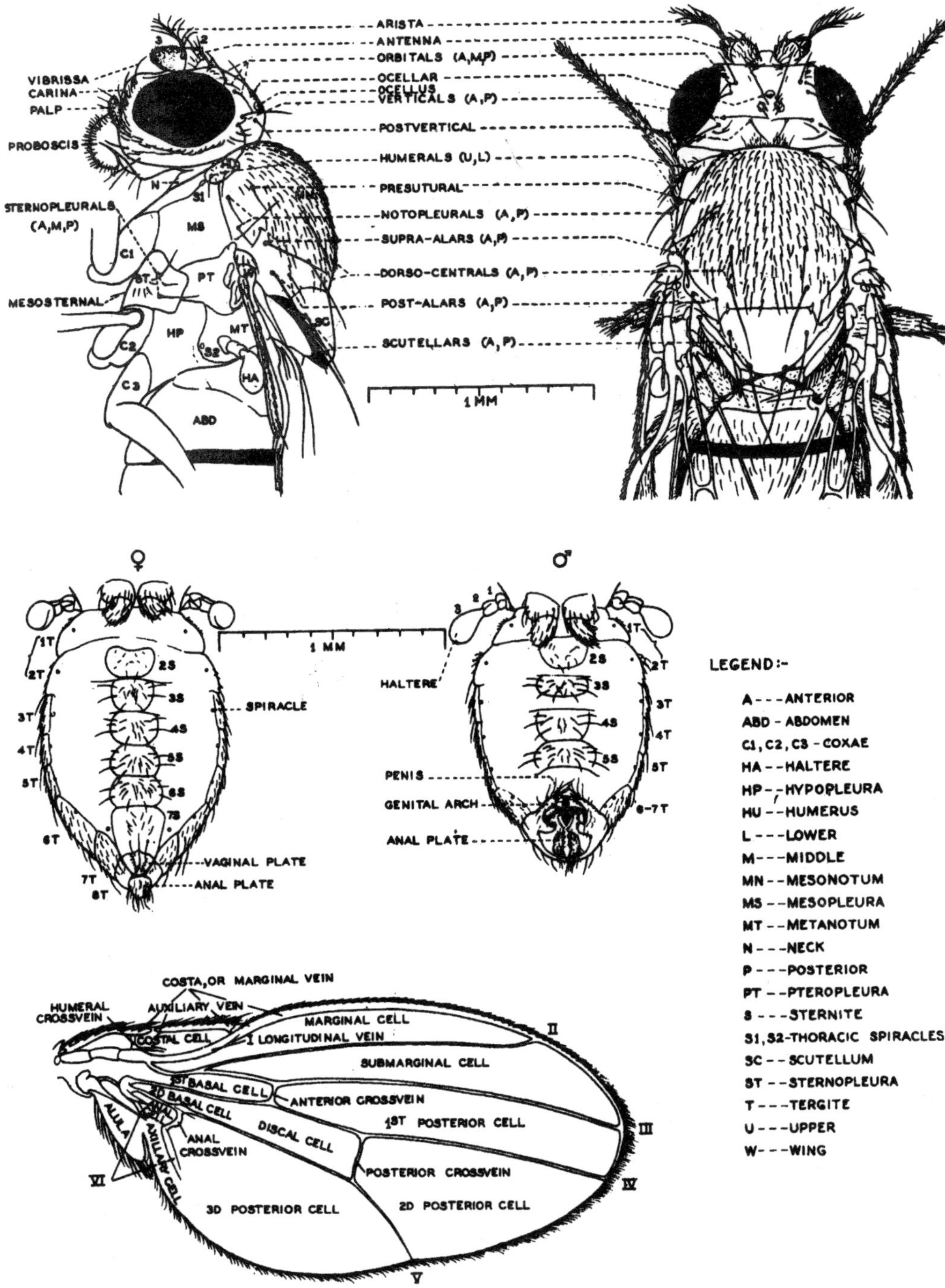

Figure 1 Drawings of the fruit fly *D. melanogaster* showing the external surface details of each sex and the labels attached to them. This worksheet was sent by Calvin Bridges to those who received *Drosophila Information Service* (*DIS*) during the 1930s. (Bridges was a leading research worker and keeper of the ever-increasing variety of mutants and special stocks in the famous "Flyroom" at Columbia University from about 1915 until his death.) The abdominal sternites with their patches of bristles are shown as 3S–5S on the male abdomen and 3S–6S on the female abdomen. The left sternopleural bristle patch is indicated on the upper left drawing.

largest). Karp claimed to have found at least six genes distributed along chromosome 3 with alternately plus and minus effects on the number of bristles, and he described this as a "genic balance" along the chromosome which allowed for considerable changes in bristle number.

This idea was taken up and developed enthusiastically by K. Mather (1941, 1942 and later papers), and both the title of his 1942 paper ("The balance of polygenic combinations") and his long promotion of the idea that sternite bristle numbers were determined by a series of genes of small effect in each chromosome having alternating plus and minus effects on the character strongly suggest that he took over this whole new field of study from Karp's short *Drosophila Information Service* article quoted above. *DIS* was essential reading for every geneticist studying or interested in Drosophila, so the absence of a reference to Karp's article in any of Mather's papers is surprising and regrettable, particularly since it has robbed Karp of recognition as the founder of an important branch of quantitative genetics; so I am glad to be able to put his name back in the history of genetics as an innovator. This may lead to more details of his life and research being rediscovered.

(ii) Counts on the abdominal sternite bristles

Figure 1 shows that four sternites in females (Nos. 3S–6S in **Figure 1**) and three in males (Nos. 3S–5S **Figure 1**) have similar bristle counts, while the sternites on other segments are clearly modified by sex differences. So an obvious question is whether there is independent genetic variation affecting the counts on each sternite. Reeve and Robertson (1954) took up this question, which has become of more general interest as the genetics of quantitative characters has gained in importance, and their results are summarized below. **Table 1** gives the average counts for four wild stocks of *D. melanogaster* and shows only small differences between stocks and between sternites; but females average 20% more bristles than males for sternites 3–5, probably because females are larger than males.

Progeny tests on two of the wild stocks and inbred lines derived from them were used to estimate the contributions of genetic and environmental factors to the individual variations in bristle number on sternites 3–5 for Renfrew males and 3–6 for both Renfrew and Crianlarich females. For this purpose pairs of flies chosen at random were mated to give separate families, and bristle counts were made on 10 progeny of each sex per family. From these Reeve and Robertson calculate, for each sternite, the variance between families from the average counts of each family, the variance within families as the average of the variances within each family, and the corresponding covariances between each pair of sternites (i.e. 3 & 4, 4 & 5, and 3 & 5 for males and in addition 3 & 6, 4 & 6, and 5 & 6 for females).

Let $\sigma^2\{G\}$ and $\sigma^2\{E\}$ be the genetic and environmental parts of the phenotypic variance of bristle number for one sternite, while the covariances between two sternites are indicated by cov{GG′} and cov{EE′}. Then if N individuals are counted per family for n families, the mean squares and mean products have the following expected values, where

Table 1 Mean sternite bristle numbers in wild stocks.

Stock	Males (segment)			Females (segment)			
	(3)	(4)	(5)	(3)	(4)	(5)	(6)
Crianlarich	19.8	20.2	20.5	24.4	25.2	24.8	22.5
Renfrew	18.4	19.4	19.7	23.8	23.9	24.3	22.5
Nettlebed	18.0	18.2	18.7	22.0	22.2	22.5	20.9
Edinburgh	21.9	22.0	23.2	24.0	24.9	24.8	22.9
Average*	19.5	20.0	20.5	23.5	24.0	24.1	22.2

* Counts are based on 50 flies of each sex and stock.

(B) and (W) stand for Between and Within Families in the progeny test:

Mean squares = variances
(B) $\frac{1}{2}\sigma^2\{G\} + \sigma^2\{E\} + \frac{1}{2} N \sigma^2\{G\}$
(W) $\frac{1}{2}\sigma^2\{G\} + \sigma^2\{E\}$
Whence $\sigma^2\{G\} = 2(B - W)/N$ and $\sigma^2\{E\} = [(N + 1)W - B]/N$

Mean products = covariances
(B) $\frac{1}{2}$ cov{GG′} + cov{EE′} + $\frac{1}{2}N$ cov{GG′}
(W) $\frac{1}{2}$ cov{GG′} + cov{EE′}
Whence cov{GG′} = $2(B - W)/N$ and cov{EE′} = $[(N + 1)W - B]/N$

The factor ½ appears in these equations because the flies in each family are full sibs. If the number of flies scored per family varies, and N is the mean sample size, n_r is the number counted in family r and F is the number of families, then the mean family size is

$$N = [1 / (F - 1)][\Sigma n_r^2 - (\Sigma n_r^2 / \Sigma n_r)]$$

For (W), each sample variance is estimated as the sum of squares of deviations of the individual counts from their mean, divided by the number of "degrees of freedom", which is $(N - 1)$ because one degree of freedom is taken up by using the sample mean; then the sample variances are added together and divided by the number of families. For (B), the variance is computed from the squared mean counts of the different families, in the same way. The covariances are calculated using products of the counts on two sternites instead of the squares of the counts on one sternite. These computations lead to the estimates of variance and covariance given in **Table 2** for the wild stocks of Renfrew males and females and Crianlarich females.

This table shows the type of result obtained when *D. melanogaster* is grown on rich food without crowding. The degrees of freedom for the B lines are each one less than the number of families in the progeny test, and for the W lines they are the total numbers of flies less the total number of families in each test. We can now estimate the genetic correlations between the variations in two different sternites, which will tell us how far different genes affect each. These genetic correlations are calculated from the B – W lines in Table 2 as cov(3 × 4) ÷ √(var3 × var4), i.e. 7.2 ÷ √(7.7 × 7.1) = 0.97 for Renfrew males, and so on. The genetic correlations between bristle numbers on the different sternites, estimated from this analysis, are given in **Table 3**. All the correlations are close to unity, with no reduction in correlation even between sternites 3 and 6 in females compared with two adjacent sternites, as **Table 4** shows. This suggests that there is little or no genetic variation affecting the bristle counts on individual sternites alone, so genes affecting bristle number must have a rather uniform effect on a large area of the adult ventral integument.

In contrast to these results, Reeve and Robertson (1954) found that non-genetic variations in bristle number on the same sternites, though substantial, were completely

Table 2 Mean squares and products for numbers of sternite bristles.

Source of variance	Degrees of freedom	Segment variances (3)	(4)	(5)	(6)	Segment covariances (3 × 4)	(4 × 5)	(5 × 6)	(3 × 5)	(4 × 6)	(3 × 6)
Renfrew males											
Between families (B)	45	10.8	10.5	12.9	–	8.0	9.5	–	8.7	–	–
Within families (W)	389	3.1	3.4	3.8	–	0.8	0.8	–	0.8	–	–
B – W		7.7	7.1	9.1		7.2	8.7		7.9		
Renfrew females											
Between families (B)	45	18.0	17.6	16.5	14.2	12.1	12.5	12.1	13.6	11.8	11.4
Within families (W)	410	3.8	5.3	4.3	4.0	1.0	1.2	0.9	0.8	1.0	0.6
B – W		14.2	12.3	12.2	10.2	11.1	11.3	11.2	12.8	10.8	10.8
Crianlarich females											
Between families (B)	41	12.8	14.2	10.8	13.3	10.9	8.6	8.8	9.4	9.5	9.8
Within families (W)	270	4.2	4.0	4.3	4.3	1.5	1.3	1.7	1.4	1.5	1.5
B – W		8.6	10.2	6.5	9.0	9.4	7.3	7.1	8.0	8.0	8.3

Table 3 Genetic correlations for number of sternite bristles.

	(3 × 4)	(4 × 5)	(5 × 6)	(3 × 5)	(4 × 6)	(3 × 6)
Renfrew males	0.97	1.08	–	0.94	–	–
Renfrew females	0.84	0.92	1.00	0.97	0.96	0.90
Crianlarich females	1.01	0.90	0.94	1.07	0.84	0.94

Table 4 Average genetic correlations.

Stock	Number of segments intervening between sternites correlated: 0	1	2	Overall average
Renfrew males	1.02	0.94	–	0.99
Renfrew females	0.92	0.96	0.90	0.93
Crianlarich females	0.95	0.95	0.94	0.95

uncorrelated with each other unless variable food conditions caused marked variation in body size and therefore in sternite area.

(iii) Selection for abdominal sternite bristle counts

Selection studies on this character have been made in a number of laboratories, and a particularly interesting example is shown in **Figure 2**, adapted from Clayton and Robertson (1957). This gives the frequency distributions of the sum of counts on the 4th and 5th sternites in females after 35 generations of selection for high and low count, respectively, together with that of the unselected wild stock. The wild stock had a mean count of 40 bristles and a small range between extremes; the high line had reached a mean of 87 bristles with the enormous range of 62 to 120, while the low line ranged from 0 to 30 bristles, with a mean of 12.

The much increased variability in both high and low lines suggests that 35 generations of selection either increased the number of segregating genes that affected the bristle counts, or more probably picked out genes of major effect, but Alan Robertson was particularly concerned with the relevance of such "model" experiments to the practical problems of improving livestock, including the calculation of the expected selection limits, and a detailed analysis of the genes causing this increased variation was not then feasible.

More recently an interesting hypothesis has been put forward by Richard Frankham (1988) whose team selected an isogenic (i.e. pure) line of *D. melanogaster* for increased and decreased numbers of abdominal sternite bristles during more than 100 generations, with controls at the same time. They found that the two lines selected for low bristle number produced bursts of selection response in females but not in males, one line around generation 20 and the other around generation 35. Similar sex-limited changes also occurred in reproductive fitness, emergence time and phenotype (Frankham *et al.*, 1978).

All the sex-limited changes in the two lines were shown to be due to "mutations" occurring at the *bobbed* (*bb*) locus on the X but not the Y chromosome. This locus codes for rRNA and is a multigene family of 200 to 250 tandemly repeated copies of the gene; it is also the only locus known to be present on both the X and Y chromosomes of *D. melanogaster*.

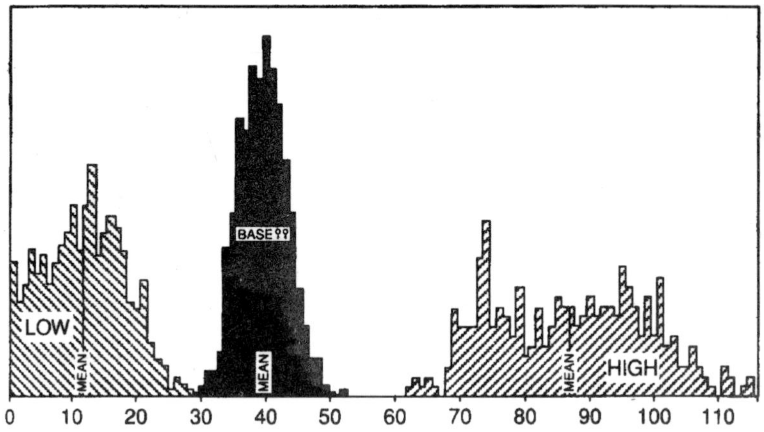

Figure 2 Frequency distributions of abdominal bristle number in *D. melanogaster* females, in the base population and in the most extreme high and low lines after 35 and 34 generations of selection from a wild strain. (Adapted from Clayton & Robertson, 1957.)

Further, *bb* mutations are known to be partial deletions in the number of copies of the *rRNA* genes rather than single locus base change mutations, which would have been lethal. Frankham suggests that the sex-limited response in his selected lines is due to unequal crossing over between the *bobbed* regions of the X and Y chromosomes, leading to reduced numbers of ribosomes in the (XX) females but not in the (XY) males. This would reduce the rate of protein synthesis in female development, and affect bristle number since a burst of protein synthesis is needed to push out a bristle (Coen & Dover, 1983; Gillings *et al.*, 1987). Frankham (1988) suggests that the lines of *D. melanogaster* selected for low abdominial bristle number by Clayton and Robertson (1957) might have shown the same "*bb* mutational" effect, which would explain the large range in the low selected line of Figure 2.

Figure 3 shows a long-term experiment by Yoo (1980) in which six replicate lines from a wild stock of *D. melanogaster* were selected for increased abdominal bristle number. The large base population was homozygous for the genes *scute* and *yellow body*, which made bristle counting and recognition of the lines easier. Fifty flies out of 250 of each sex scored per generation were selected for increased bristle number (on the 4th sternite in males and the 5th in females), and selection continued for 85–90 generations followed by recording of the counts without further selection for another 30 generations. The six lines all increased 3 to 4 times fairly steadily for 70–80 generations, but progress then virtually ceased. However, when selection was relaxed the count decreased to varying extents in all of them. This suggests that many genes affecting bristle number were segregating in the starting population, some of which could not be fixed because of their deleterious effects when homozygous.

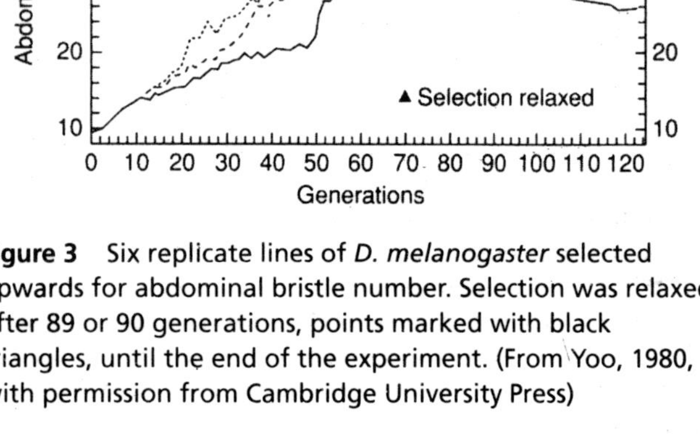

Figure 3 Six replicate lines of *D. melanogaster* selected upwards for abdominal bristle number. Selection was relaxed after 89 or 90 generations, points marked with black triangles, until the end of the experiment. (From Yoo, 1980, with permission from Cambridge University Press)

(iv) Mutation rate in inbred lines

Another aspect of great interest is the rate of mutation in such genes. A highly inbred line was maintained without selection for 25 generations to allow mutations affecting bristle number to accumulate, and then six replicate lines taken from it were selected for change in abdominal bristle number – three for increase and three for decrease. A parallel set of lines from the same origin were selected for a second character, number of sternopleural bristles. These form a patch on the sternopleural plate on each side of the integument between the bases of the second and third legs (**Figure 1**), and the sum of the bristle numbers on the two sides has been used in many selection studies. Further details about this character will be given below.

All twelve replicate lines were selected for increase or decrease in one of the two bristle characters for a total of 178 generations, which makes this probably the largest selection experiment ever undertaken (Mackay *et al.*, 1994; Mackay, 1995). Simple inspection (see **Figure 4**) suggests that abdominal bristle number, which was counted on the 5th and 6th abdominal sternite, respectively, in males and females, increased very slowly, by about 20% over the whole period, so that mutations to increase their number are likely to have been rare and of small effect (top graph). Low selection for the same character, however, produced a rapid decline from 15 to about 5 bristles over the first 40 generations, and the three replicate lines remained variable but settled down at different levels. This suggests the possibility that much of the reduction in abdominal bristle number was

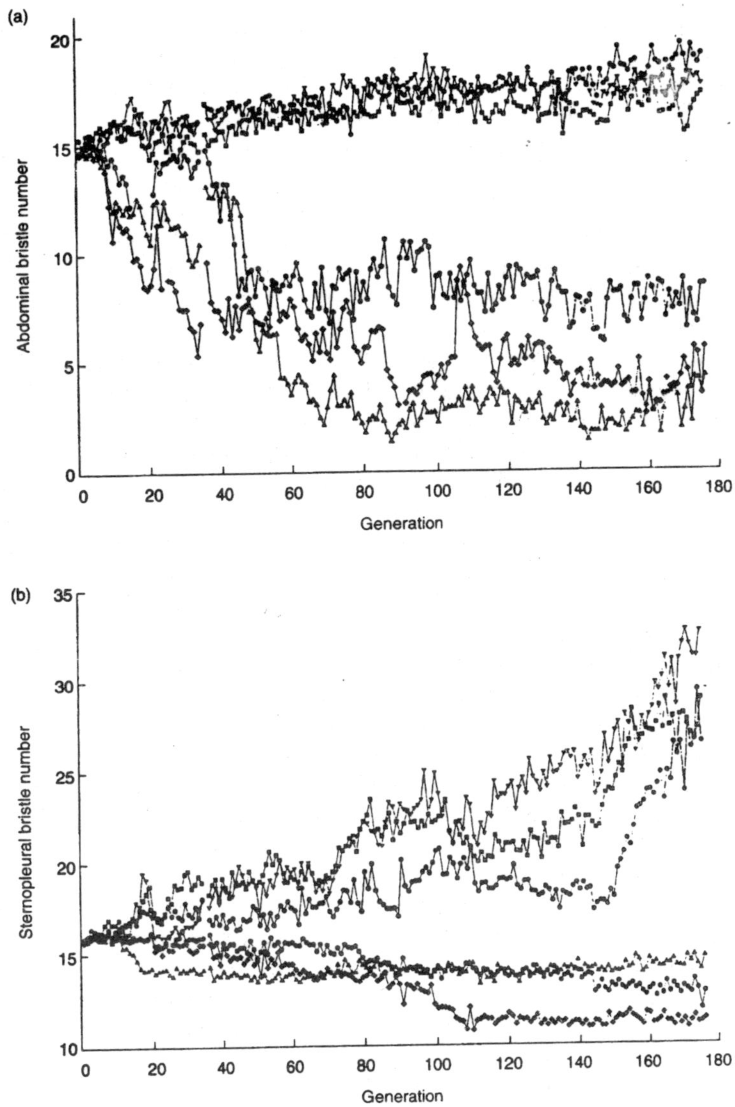

Figure 4 Response from new mutations to 178 generations of artificial selection for bristle number from an inbred base population of *Drosophila melanogaster*. There were three replicated selection lines for (a) high abdominal and (b) high sternopleural bristle number (●, ▼, ■), maintained by selecting the ten males and females with the highest number of bristles each generation. Similarly there were three replicated selection lines for low bristle number (▲, ◆, hexagon) for each trait, maintained by picking the ten lowest-scoring individuals of each sex every generation, Gaps indicate generations where the lines were not selected. (Reprinted from Mackay, 1995, with permission from Elsevier Science.)

again due to unequal crossing over between the *bobbed* regions in the X and Y chromosomes, as proposed by Frankham *et al.* (1978).

The lower two graphs in **Figure 4** show the effect of selection during 178 generations to increase and decrease the number of sternopleural bristles, starting in the same replicate inbred lines as used for the sternite bristle selection. The sternopleurals show almost a mirror image of the responses by the sternite bristles in their reaction to the long period of selection, up-selection giving a variable response with a maximum doubling effect while one of the low lines showed a marked decline, probably due to a single mutation, occurring between generations 90 and 100.

(v) Sternopleural bristle counts

As **Figure 1** shows, the sternopleural bristle patches normally consist of three large bristles or macrochaetae (labeled A, M, and P in the figure) located dorsally and about 10 small bristles (microchaetae) below them. The three large bristles may well be controlled by genes other than those controlling the small bristles, but I have not come across any selection experiments on the sternopleural bristles which reported whether increase in total number affected the macrochaetae as well as the microchaetae.

Another interesting question concerns the level of asymmetry between the two sides of the fly as measured by the absolute value of the difference, |L–R|. Mather (1953), Thoday (1958) and Reeve (1960) found this index could be both increased and mildly decreased by selection in wild stocks of *D. melanogaster*, but it remains to be tested whether other bilateral characters respond in the same way.

3. Search for quantitative trait loci (QTLs) in Drosophila

(i) QTLs affecting bristle number characters in Drosophila

D. melanogaster turned out to be a particularly happy discovery by Thomas Hunt Morgan as a subject for genetic research, since it is not only small, easy, rapid and cheap to breed, has three large and one very small chromosome, and normally only allows recombination to occur in females, but also presents its nuclear DNA to the investigator, in squashed larval salivary glands as giant polytene chromosomes marked by a total of just over 5000 bands along their cytological map. Several different transposable elements (TEs) – short DNA sequences which can move from one location to another in the Drosophila nuclear genome, leaving a copy behind to act as a visible positional marker by *in situ* hybridization with the polytene chromosomes – make it easy to identify a large number of marked locations throughout the genome. Finally, recombination tests of such marked lines in crosses with lines selected for high and low numbers of sternite or sternopleural bristles provide a unique insight into the genes which have an appreciable effect on these two classical quantitative characters.

These transposons and their full understanding have been acquired fairly recently (reviewed by Arkhipova *et al.*, 1995; and articles in Sherratt, 1995), and TEs have been given interesting titles such as *copia, roo, gypsy, P elements, I elements, 297* and *412*, all with recognizably different molecular characteristics. During the last decade Trudy F.C. Mackay, having moved from the limited facilities at the then Institute of Animal Genetics of the University of Edinburgh to North Carolina State Genetics Department when the experiment I shown in **Figure 4** had reached selection generation 34, collected a dedicated team to identify genes with effects on the sternite and sternopleural bristle characters.

Mackay has published two valuable reviews: "The genetic basis of quantitative variation: numbers of sensory bristles of *Drosophila melanogaster* as a model system" (Mackay, 1995) and "The nature of quantitative genetic variation revisited" (Mackay, 1996). The latter review gives a figure in colour to show the approximate positions of the 24 genes so far identified as candidate loci for one or both of the two characters. This figure has been adapted with the author's permission and is shown as **Plate 25** in the colour plate section with the different colours giving details of how the gene locations were estimated. It needs to be borne in mind that the locations estimated for these genes (especially those very provisionally located in earlier experiments) are by no means precise. Nevertheless the map represents a remarkable success which will no doubt become much more refined in the next few years. Details of the 24 genes, labeled a to x in **Plate 25** purely for identification purposes, are given in **Table 5** and, with some additional information regarding use of colour and recent references, in the legend to the plate.

One of the most interesting of these genes is *Delta*, in which a restriction enzyme survey of a 57 kb gene region was undertaken. The complete transcriptional unit of *Delta* occupied 22 kb of this region, within which 53 polymorphic markers were found in a study of 55 naturally occurring chromosomes. This analysis identified two common sites that independently affected bristle number: one, in the second intron of *Delta*, affected only sternopleural bristle number, in both sexes, and the other site, in the fifth intron, affected only abdominal bristle number and only in females (Long *et al.*, 1998; Lyman & Mackay, 1998). The authors suggest that, under an additive genetic model, the polymorphism in the second intron may account for 12% of the total genetic variation due to third chromosomes in sternopleural bristle number and the site in the fifth intron may account for 6% of the total variation in female abdominal bristle number due to third chromosomes. It is of particular interest that statistical removal of the effect of either of these sites did not alter the effect of the other, although the two sites are only separated by about 10 kb.

Table 5 Details of genes listed in Mackay (1996).

a, *achaete-scute* complex (ASC), 1–0.0
b, *Notch (N)*, 1–3.0
c, *cut (ct)*, 1–20.0
d, *bobbed (bb)*, 1–66.0
e, *Sternopleural (Sp)*, 2–22.0
f, *numb (numb)*, 2–35.0
g, *daughterless (da)*, 2–41.3
h, *scabrous (sca)*, 2–66.7
i, *smooth (sm)*, 2–91.7
j, *extramacrochaetae (emc)*, 3–0.0
k, *que mao (qm)*, 3–23.0
l, *abdominal (abd)*, 3–26.5
m, *hairy (h)*, 3–27.0
n, *polychaetoid (pyd)*, 3–39.0
o, *Bearded (Brd)*, 3–42.0
p, *malformed abdomen (mab)*, 3–47;5
q, *atonal (atonal)*, 3–48.0
r, *neuralized (neu)*, 3–50.0
s, *Tuft (Tft)*, 3–53.6
t, *Abdominal A, Abdominal B (AbdA, Abd B)*, 3–58.8
u, *Delta (Dl)*, 3–66.2
v, *Hairless (H)*, 3–69.5
w, *Enhancer-of-split (E(spl))*, 3–89.1
x, *brief (bf)*, 3–95.0

The product of the *Delta* gene is the ligand and that of the *Notch* gene is the receptor in the well-characterized *Notch* signal transduction pathway (Artavanis-Tsakonas *et al.*, 1995), and bristle position and spacing is believed to be the result of "lateral inhibition" whereby cell-cell interactions cause feedback loops between cells expressing *Notch* and *Delta*, so that some cells assume a neuronal (bristle) fate and others a non-neuronal (i.e. non-bristle) fate (Hettzler & Simpson, 1991).

(ii) QTLs when the complete gene sequence is available

It used to be rather widely assumed that quantitative characters were generally under the control of a large number of genes having very small and mainly additive effects, and it remains to be discovered whether, now that the complete gene sequence of *D. melanogaster* has been decoded, there remain many, few or even no further loci affecting either of the two bristle characters to be identified. This is an important question since it has bearing on evolutionary problems, and there are plenty of other species of Drosophila whose bristle genes could be compared with those identified in *D. melanogaster.*

(iii) Other quantitative characters in Drosophila

Forbes W. Robertson and I made a collaborative study of genetic and environmental variation in wild strains of *D. melanogaster* for thorax length and wing length, which are highly correlated characters that show some degree of independent variation when one or the other is selected for change. Progress was limited in our experiments, mainly no doubt because the labour of maintaining constant growth conditions and taking accurate measurements stretched our limited resources and restricted the size of our experiments. Selection on a larger scale might well lead to much more striking changes, since there is considerable variation in size as well as life style between different species of Drosophila. Much of this research was summarized by F.W. Robertson (1955), and he later went on to study other interesting quantitative characters in *D. melanogaster* such as ovary size and ovariole number, the relation of egg production rate to body size, cell size in relation to body size, the effect of changing the relative size of the body parts, and the ecological genetics of growth (Robertson, 1957a,b, 1959, 1962, 1966). These add up to a considerable number of related quantitative characters for all of which Robertson found selection produced changes. They all differ strikingly from the two bristle number characters on which so much attention has been focused, and the greater difficulty in studying them would be lessened by applying the new techniques of molecular genetics and isolation of QTLs with the help of TEs.

4. QTLs in other organisms: the tomato

Most fruit-bearing species of food plants are believed to have been domesticated in neolithic times, when early humans were changing from a hunter–gatherer to a settled farming existence (Zohary & Hopf, 1999), so little is known about their origin. But analysis of crosses between modern tomato cultivars and their wild relatives (i.e. between large-fruited and very small-fruited species) has generally given a continuous range in both fruit size and shape, which strongly suggests that progress occurred slowly by the accumulation of mutations at many locations (MacArthur & Butler, 1938; Banerjee & Kalloo, 1989). Thus identifying these genes and finding what they do in order to put them to further use has proved very difficult; and none of the genes controlling either fruit size or shape had, until very recently, been cloned or its exact genetic location settled.

Tomato was one of the first food-plant species in which QTL analysis using molecular markers was undertaken (Tanksley *et al.*, 1982; Paterson *et al.*, 1988); and in their review at the beginning of 1999, Grandillo *et al.* state that 15 molecular mapping studies involving 50 traits have been reported during the previous 10 years, with fruit size and shape being examined in many of them (see also Pillen *et al.*, 1996). A large amount of complex data are presented in the Grandillo *et al.* review, of which I can only note some major features. But first the reader should be reminded of some anatomical aspects that Grandillo *et al.* point out. Fruits can be defined as matured carpels with or without accessory structures and/or seeds, and tomato fruits are fleshy fruits classified as berries because the thick pericarp encloses many seeds, and when the ovary develops into a fruit the ovary wall becomes the pericarp. The shape of tomato fruit, particularly among cultivars, can also vary significantly, ranging from round to pear-shaped to plum, ovate, oblate or stuffer.

Fruit-size QTLs are designated *fw* (for fruit weight or size) followed by the number 1–12 to indicate the chromosome, and ".1, .2 etc." for more than one QTL of the same type on a single chromosome. A large number of *fw*'s have been proposed as possibles from the many tests so far conducted, and comparisons of these have suggested that 28 may be considered provisionally as conserved. Six of these – *fw1.1, fw2.1, fw2.2, fw3.1, fw3.2,* and *fw11.3* – are considered major QTLs that account for most of the variation in tomato evolution.

Fruit shape QTLs are designated *fs1, etc.* Six of these – *fs2.1, fs2.2, fs7.1, fs8.1, fs10.1* and *fs10.2* – show a large effect in more than one independent study. Of these, the two major QTLs *fs2.1* and *fs8.1* are discussed further in work very recently published or still in preparation.

Cloning of the first fw QTL. This is *fw2.2*, cloned recently by the Tanksley group and described by Frary *et al.* (2000). The authors point out that the wild progenitor of the domesticated tomato (*Lycopersicon esculentum*) probably had fruit of less than 1 cm in diameter and only a few grams in weight. Such fruit was small enough to be dispersed by small rodents or birds but large enough to contain hundreds of seeds. Modern tomatoes can weigh a thousand grams and be 15 cm or more in diameter, and the difference between the wild and domesticated forms is controlled by "quantitative" genes or QTLs. One of these, *fw2.2*, changes fruit size by up to 30%, and appears to have been responsible for a key transition during domestication, since all wild Lycopersicon species tested have small-fruit alleles at this locus, in contrast to the large-size alleles in the modern tomato cultivars.

Frary *et al.* describe the complex process of narrowing down the locus in *fw2.2* responsible for this major fruit size increase to a sequence which contains two introns and encodes a 163-amino acid polypeptide of ~22 kilodaltons (kD). This gene, named by the authors *ORFX*, has yet to be cloned into wild strains of the tomato to provide final proof, but the present arguments are compelling and the work represents a remarkable achievement for the new technique of molecular genetics after many years of slow progress. A colour photograph in **Plate 26** contrasts the wild and cultivated forms of the tomato.

Both the review by Grandillo *et al.* (1999) and the report on cloning and sequencing *OFRX* contain much information likely to interest the reader which I have not been able to include, and we can assume that further analysis of the *fw* and *fs* QTLs will soon be published, with the effect of moving tomato genes further up the scale of QTL analysis.

5. Temporary conclusions

The recent interest in QTLs affecting other characters and other organisms, including mice and human diseases, has of course been encouraged by the unexpected difficulties in solving major gene therapy problems. Drosophila has spread the word that there are plenty of QTLs out there waiting to be trapped by the expert and persistent fisherperson, and the tomato story summarized above shows that we don't need patches of bristles to help us run them down. The next obvious question is whether knowledge of the complete gene sequence of the organism under study will make it possible to identify the rest of the QTLs, of lesser effect, and to find out whether they can interact to give us better fruit or larger fruit flies, or to enable us to reproduce any first stages of the evolutionary changes seen by comparing species. A recent rough trawl through the genes in Flybase suggested to me the possibility that there may be plenty more bristle genes in the database waiting to be added to the list of 24 in Mackay (1996).

REFERENCES

Arkhipova, I.R., Lyubomirskaya, N.V. & Ilyin, Y.V. (1995) *Drosophila Retrotransposons*, Austin, Texas: R.G. Landes

Artavanis-Tsakonas, S., Matsuno, K. & Fortini, M.E. (1995) *Notch* signalling. *Science* 268: 225–32

Banerjee, M.K. & Kalloo (1989) The inheritance of earliness and fruit weight in crosses between cultivated tomatoes and two wild species of *Lycopersicon*. *Plant Breeding* 102: 148–52

Breeze, E.L. & Mather, K. (1957) The organization of polygenic activity within a chromosome in Drosophila. I. Hair characters. *Heredity* 11: 373–95

Clayton, G.A. & Robertson, A. (1957) An experimental check on quantitative genetical theory II. The effects of long term selection. *Journal of Genetics* 55: 152–70

Coen, E.S. & Dover, G.A. (1983) Unequal exchanges and the coevolution of X and Y chromosomal rDNA arrays in *Drosophila melanogaster. Cell* 33: 840–55

Davies, R.W. (1971) The genetic relationship of two quantitative characters in *Drosophila melanogaster*. II. Location of the effects. *Genetics* 69: 363–75

Frankham, R. (1988) Exchanges in the *rRNA* multigene family as a source of genetic variation. In *Proceedings of the 2nd International Conference on Quantitative Genetics,* edited by B.S. Weir, E.J. Eisen, M.M. Goodman & G. Namkoong, Sunderland, Massachusetts: Sinauer

Frankham, R., Briscoe, D.A. & Nurthen, R.J. (1978) Unequal crossing over at the *rRNA* locus as a source of quantitative genetic variation. *Nature* 272: 80–81

Frary, A., Nesbitt, T.C., Frary, A. *et al.* (2000) *fw2.2:* a quantitative trait locus key to the evolution of tomato fruit size. *Science* 289: 85–88

Gibson, J.B. & Thoday, J.M. (1962) Effects of disruptive selection. VI. Analysis of a second chromosome polymorphism. *Heredity* 17: 1–26

Gillings, M.R., Frankham, R., Spiers, J. & Whalley, M. (1987) X-Y exchange and coevolution of X and Y rDNA arrays in *Drosophila melanogaster. Genetics* 116: 241–51

Goodale, H.D. (1938) A study of the inheritance of body weight in the albino mouse by selection. *Journal of Heredity* 29: 101–12

Grandillo, S., Ku, H.M. & Tanksley, S.D. (1999) Identifying the loci responsible for natural variation in fruit size and shape in tomato. *Theoretical and Applied Genetics* 99: 978–87

Gurganus, M.C., Nuzhdin, S.V., Leips, J.W. & Mackay, T.F.C. (1999) High resolution mapping of quantitative trait loci for sternopleural bristle number in *Drosophila melanogaster. Genetics* 152: 1585–1604

Hettzler, P. & Simpson, P. (1991) The choice of cell fate in the epidermis of Drosophila. *Cell* 64: 1083–92

Jan, Y.N. & Jan, L.Y. (1994) Genetic control of cell fate specification in Drosophila peripheral nervous system. *Annual Reviews of Genetics* 28: 373–93

Karp, M.L. (1935) The distribution of mutant genes affecting the number of sternital bristles in chromosome 3 of *Drosophila melanogaster. Drosophila Information Service No. 4*

Karp, M.L. (1940) Inbreeding and heterosis (in Russian with English summary). *Bulletin de L'Académie des Sciences de L'URSS, Classe des Sciences Biologiques*, 219–51

Lai, C., McMahon, R., Young, C., Mackay, T.F.C. & Langley, C.H. (1998) *que mao,* a *Drosophila* bristle locus, encodes geranylgeranyl pyrophosphate synthase. *Genetics* 149: 1051–61

Lindsley, D.L. & Zimm, G.G. (1992) *The Genome of Drosophila melanogaster*, San Diego: Academic Press

Long, A.D., Lyman, R.F., Langley, C.H. & Mackay, T.F.C. (1998) Two sites in the *Delta* gene region contribute to naturally occurring variation in bristle number in *Drosophila melanogaster*. *Genetics* 149: 999–1017

Long, A.D., Mullaney, S.L., Reid, L.A. *et al.* (1995) High resolution mapping of genetic factors affecting abdominal bristle number in *Drosophila melanogaster*. *Genetics* 139: 1273–91

Lyman, R.F. & Mackay, T.F.C. (1998) Candidate quantitative trait loci and naturally occurring phenotypic variation for bristle number in *Drosophila melanogaster:* the *Delta-Hairless* gene region. *Genetics* 149: 993–98

Lyman, R.F., Lawrence, F., Nuzhdin, S.V. & Mackay, T.F.C. (1996) Effects of single *P* element insertions on bristle number and viability in *Drosophila melanogaster. Genetics* 143: 277–92

MacArthur, J.W. & Butler, L. (1938) Size inheritance and geometric growth processes in the tomato fruit. *Genetics* 23: 253–68

Mackay, T.F.C. (1995) The genetic basis of quantitative variation: numbers of sensory bristles of *Drosophila melanogaster* as a model system. *Trends in Genetics* 11: 464–70

Mackay, T.F.C. (1996) The nature of quantitative genetic variation revisited: lessons from Drosophila bristles. *BioEssays* 18: 113–21

Mackay, T.F.C., Fry J.D., Lyman R.F. & Nuzhdin, H.G. (1994) Polygenic mutation in *Drosophila melanogaster:* estimates from response to selection of inbred strains. *Genetics* 136: 837–51

Mather, K. (1941) Variation and selection of polygenic characters. *Journal of Genetics* 41: 159–93

Mather, K. (1942) The balance of polygenic combinations. *Journal of Genetics* 43: 309–36

Mather, K. (1953) Genetical control of stability in development. *Heredity* 7: 27–336

Nuzhdin, S.V., Pasyukova, E.G. & Mackay, T.F.C. (1997) Accumulation of transposable elements in laboratory lines of *Drosophila melanogaster. Genetica* 100: 167–75

Paterson, A.H., Lander, E.S., Hewitt, J.D. *et al.* (1988) Resolution of quantitative traits into Mendelian factors by using a complete linkage map of restriction fragment length polymorphisms. *Nature* 335: 721–26

Pillen, K., Pineda, O., Lewis, C.B. & Tanksley, S.D. (1996) Status of genome mapping tools in the taxon Solanaceae. In *Genome*

Mapping in Plants, edited by A.H. Paterson, Austin, Texas: R.G. Landes, and London: Academic Press

Reeve, E.C.R. & Robertson, F.W. (1954) Studies of quantitative inheritance VI. Sternite chaeta number in Drosophila: a metameric quantitative character. *Zeitschrift fuer Vererbungslehre, Bd* 86: 269–88

Reeve, E.C.R. (1960) Some genetic tests on asymmetry of sternopleural chaeta number in *Drosophila melanogaster. Genetical Research* 1: 151–72

Robertson, F.W. (1955) Selection response and the properties of genetic variation. *Cold Spring Harbor Symposia on Quantitative Biology* XX: 166–77

Robertson, F.W. (1957a) Studies in quantitative inheritance X. Genetic variation of ovary size in Drosophila. *Journal of Genetics* 55: 410–27

Robertson, F.W. (1957b) Studies in quantitative inheritance XI. Genetical and environmental correlation between body size and egg production in *Drosophila melanogaster. Journal of Genetics* 55: 428–43

Robertson, F.W. (1959) Studies in quantitative inheritance XII. Cell size and number in relation to genetical and environmental variation of body size in Drosophila. *Genetics* 44: 869–96

Robertson, F.W. (1962) Changing the relative size of the body parts of Drosophila by selection. *Genetical Research* 3: 169–80

Robertson, F.W. (1966) The ecological genetics of growth in Drosophila 8. Adaptation to a new diet. *Genetical Research* 8: 165–79

Sherratt, D.J. (ed) (1995) *Mobile Genetic Elements*, Oxford: IRL Press

Shrimpton, A.E. & Robertson, A. (1988) The isolation of polygenic factors controlling bristle score in *Drosophila melanogaster*. II. Distribution of third chromosome bristle effects within chromosome sections. *Genetics* 118: 445–59

Spickett, S.G. & Thoday, J.M. (1966) Regular responses to selection. 3. Interactions between located polygenes. *Genetical Research* 7: 98–121

Tanksley, S.D., Medina-Filho, H. & Rick, C.M. (1982) Use of naturally occurring enzyme variation to detect and map genes controlling quantitative traits in an interspecific backcross of tomato. *Heredity* 49: 11–25

Thoday, J.M.(1958) Homeostasis in a selection experiment. *Heredity* 12: 401–15

Thoday, J.M., Gibson, J.B. & Spickett, S.G. (1964) Regular responses to selection. 2. Recombination and accelerated response. *Genetical Research* 5: 1–19

Wolstenholme, D.R. & Thoday, J.M. (1963) Effects of disruptive selection. VII. A third chromosome polymorphism. *Heredity* 10: 413–31

Yoo, B.H. (1980) Long term selection for a quantitative character in large replicate populations of *Drosophila melanogaster.* I: Responses to selection, II: Lethals and visible mutants with large effects. *Genetical Research* 35: 1–17, 19–31

Zohary, D. & Hopf, M. (1999) *Domestication of Plants in the Old World: the Origin and Spread of Cultivated Plants in West Asia, Europe and the Nile Valley*, 3rd edition, Oxford: Clarendon Press

FURTHER READING

Cordell, H.J., Todd, J.A. & Lathrop, G.M. (1998) Mapping multiple linked quantitative trait loci in non-obese diabetic mice using a stepwise regression strategy. *Genetical Research* 71: 51–64

Falconer, D.S. & Mackay, T.F.C. (1996) *Introduction to Quantitative Genetics*, 4th edition, Harlow: Longman

Frankel, W.N. (1995) Taking stock of complex trait genetics in mice. *Trends in Genetics* 11: 471–77

Haley, C.S. (1995) Livestock QTLs – bringing home the bacon. *Trends in Genetics* 11: 488–92

Kohler, R.E. (1994) *Lords of the Fly: Drosophila Genetics and the Experimental Life*, Chicago: Chicago University Press

Lynch, M. & Walsh, B. (1998) *Genetics and Analysis of Quantitative Traits*, Sunderland, Massachusetts: Sinauer

McCouch, S.R. & Doerge, R.W. (1995) QTL mapping in rice. *Trends in Genetics* 11: 482–87

Nuzhdin, S.V., Keightley, P.D., Pasyukova, E.G. & Morozova, E.A. (1998) Mapping quantitative trait loci affecting sternopleural bristle number in *Drosophila melanogaster* using changes of marker allele frequencies in divergently selected lines. *Genetical Research* 72: 79–91

Phillips, P.C. (1999) From complex traits to complex alleles. *Trends in Genetics* 15: 6–8

Stuber, C.W. (1995) Mapping and manipulating quantitative traits in maize. *Trends in Genetics* 11: 477–81

Wang, R.-L., Stec, A., Hey, J., Lukens, L & Doebley, J. (1999) The limits of selection during maize domestication. *Nature* 398: 236–39

See also ***Drosophila melanogaster*** (p.157); **Body weight limits in mice** (p.337); **Quantitative trait loci** (p.904)

QUANTITATIVE TRAIT LOCI: STATISTICAL METHODS FOR MAPPING THEIR POSITIONS

Zhao-Bang Zeng
Program in Statistical Genetics, Department of Statistics, North Carolina State University, Raleigh, USA

1. Introduction

Often traits in plants and animals are quantitative in nature, influenced by many genes. For a long time, it has been an important aim in genetics and breeding to identify those individual genes contributing significantly to the variation of traits within and between populations or species. When the locus (or loci) that affects the phenotypes is unknown, researchers usually have to resort to indirect approaches, such as using correlations between relatives, responses to selection and hybridization or controlled crosses, in order to gain information about the genotype–phenotype relationship. Although insightful in many ways, these unmeasured genotype approaches have a number of limitations, offering little insight into the genetic architecture underlying the phenotypes. The genetic architecture of quantitative traits can be very complex and involves multiple loci, interaction among alleles (dominance) and loci (epistasis), linkage or linkage disequilibrium, pleiotropy, multiple level (random and fixed) environmental effects and genotype by environment interaction (see "Genetics of Quantitative Characters").

When a number of polymorphic genetic markers are available, studies of the genotype–phenotype relationship can take a very different approach. These markers can be used to construct a genetic linkage map for a genome. With such a complete linkage map, the probability of an individual at any genomic position being in different possible genotypic states can be inferred. This inference, together with quantitative genetic models, allows us to map segregating genes onto linkage maps and, at the same time, to estimate the genetic architecture of the quantitative trait variation in a population. This map-based strategy has revolutionized quantitative genetics. For the first time, we have access to information about the genetic architecture of complex quantitative traits.

Data for mapping quantitative trait loci (QTL) consist of genetic types of a number of polymorphic genetic markers and quantitative trait values for a number of individuals. Marker data are categorical and can be classified in different categories and recorded in digital form, such as 1 or 0 for denoting the presence or absence of a particular molecular band at a particular marker, or the two marker genotypes (homozygote and heterozygote) for a backcross population from two inbred lines. Based on segregation analysis, these markers can be ordered in linkage groups or linearly on chromosomes to represent a genetic linkage map. Quantitative trait data are usually continuous, such as body weight, but can also be discrete, such as litter size. While marker data contain information about the segregation of a genome in a population, quantitative trait data contain information about the variation of traits of interest in the same population. The statistical task of mapping QTL is to relate quantitative trait variation to genetic marker variation in terms of the number, positions, effects, interactions and pleiotropy of genes that affect the quantitative traits of interest.

Traditional experimental designs for locating QTL start with two parental lines differing both in trait values and in the marker variants they carry. Suppose two pure-breeding lines, P_1 and P_2, have marker genotypes *MN/MN* and *mn/mn* for two markers. Crossing these lines produces F_1 offspring that is doubly heterozygous. It is denoted as *MN/mn*, where the slash separates the contributions from the two parents. Each F_1 individual can produce four possible gametes or marker allele combinations for transmission to the next generation. The proportions of these four gametes *MN*, *Mn*, *mN* and *mn* are $(1 - r_{MN})/2$, $r_{MN}/2$, $r_{MN}/2$ and $(1 - r_{MN})/2$, respectively, where r_{MN} is the recombination frequency between the two markers. This segregation of gametes can be observed, for example, from backcross populations B_1 and B_2 ($B_1 = F_1 \times P_1$ and $B_2 = F_1 \times P_2$) and also F_2 populations ($F_2 = F_1 \times F_1$). If a number of genetic markers and quantitative traits are observed in these populations, mapping can be performed to locate QTL.

2. One marker analysis

The simplest method of associating markers with quantitative trait variation is to test for trait value differences between different marker groups of individuals for a particular marker. For example, let $\tilde{\mu}_{M/M}$ and $\tilde{\mu}_{M/m}$ be the observed trait means of the groups of individuals with marker genotypes *M/M* and *M/m* for a marker in a backcross population; we can test for significance between means $\tilde{\mu}_{M/M}$ and $\tilde{\mu}_{M/m}$ using the usual t test with the statistic

$$t = \frac{\tilde{\mu}_{M/M} - \tilde{\mu}_{M/m}}{\sqrt{s^2\left(\frac{1}{n_{M/M}} + \frac{1}{n_{M/m}}\right)}}$$

where s^2 is the pooled sampling variance, and $n_{M/M}$ and $n_{M/m}$ are the corresponding sample sizes in each marker class. The hypotheses to be tested can be H_0: $\tilde{\mu}_{M/M} = \tilde{\mu}_{M/m}$ and H_1: $\tilde{\mu}_{M/M} \neq \tilde{\mu}_{M/m}$.

To understand the relevance of this test to QTL mapping, we need to know what exactly is tested in genetic terms. Suppose that there are q QTL contributing to the genetic variation in a backcross population from two inbred lines. Ignoring epistasis, the expected difference between $\tilde{\mu}_{M/M}$ and $\tilde{\mu}_{M/m}$ is

$$\varepsilon\left(\tilde{\mu}_{M/M} - \tilde{\mu}_{M/m}\right) = \sum_{i=1}^{q}(1 - 2r_i)\,a_i$$

where ε denotes expectation, a_i is the effect of the ith QTL expressed as a difference in effects between the recurrent parent homozygote and the heterozygote, and r_i is the recombination frequency between the marker and the ith QTL. Essentially, this means that we test a composite parameter that constitutes gene effects and recombination frequencies for (potentially) a number of genes. Of course, many QTL may not be linked to the marker, and thus have 0.5 recombination frequency. The above hypotheses are then equivalent to H_0: all $r_i = 0.5$ and H_1: at least one $r_i < 0.5$, because a_i is usually non-zero by experimental design. If $\tilde{\mu}_{M/M}$ and $\tilde{\mu}_{M/m}$ are found to be significantly different, we conclude that the marker is linked to one or possibly more QTL. This analysis, however, cannot determine whether a significant marker effect is due to one or multiple QTL and whether the effect is due to distantly linked QTL with large effects or closely linked QTL with small effects. With a dense linkage map, the second problem can be alleviated.

3. Interval mapping

Since single marker analysis cannot separate r and a in test and estimation, even when there is only one QTL on a chromosome, Lander and Botstein (1989) proposed a maximum likelihood method that uses a pair of adjacent markers to test the effect of a genomic position within a chromosomal interval bracketed by two adjacent markers. This is an attempt to disentangle r and a in analysis, and this method is called interval mapping. Specifically, for a backcross population, they proposed the following linear model to test for a QTL located on an interval between two adjacent markers

$$y_i = \mu + b^* x_j^* + e_i \quad \text{for } j = 1, 2 \cdots, n$$

where y_j is a quantitative trait value of the jth individual, μ is the mean of the model, x_j^* is an indicator variable, taking a value 1 or 0 for the two possible QTL genotypes with probability depending on the genotypes of markers and the genomic position being tested, b^* is the effect of the putative QTL, e_j is a residual variable (usually assumed to be normally distributed with mean zero and variance σ^2) and n is the sample size. Since x_j^* is usually unobserved for a particular genomic position and can take different values, this is, statistically, a mixture model. The likelihood function of the model is

$$L\left(\mu, b^*, \sigma^2\right) =$$
$$\prod_{j=1}^{n}\left[p_{1j}\,\phi\left(y_i : \mu + b^*, \sigma^2\right) + p_{0j}\,\phi\left(y_i : \mu, \sigma^2\right)\right]$$

where $\phi(y_j$: $\mu + b^*$, $\sigma^2)$ is a normal density function of y_i with mean $\mu + b^*$ and variance σ^2, and p_{kj} is the probability of $x_j^* = k$ given marker data and the testing position of the putative QTL.

The test statistic can be constructed using a likelihood ratio (LR)

$$LR = -2\ln\frac{\hat{\hat{\mu}},\ b^* = 0,\ \hat{\hat{\sigma}}^2}{L\left(\hat{\mu},\ \hat{b}^*,\ \hat{\sigma}^2\right)}$$

to compare the null hypothesis H_0: $b^* = 0$ with the alternative hypothesis H_1: $b^* \neq 0$, assuming that the putative QTL is located at the point of consideration, where $\hat{\mu}$, $\hat{b}^*$ and $\hat{\sigma}^2$ are the maximum likelihood estimates of μ, b^* and σ^2 under H_1, and $\hat{\hat{\mu}}$ and $\hat{\hat{\sigma}}^2$ are the estimates of μ, σ^2 under H_0 with b^* constrained to zero.

In human linkage analysis, the likelihood ratio test statistic, however, has traditionally been expressed in terms of LOD (for log odds) score

$$\text{LOD} = -\log_{10}\frac{L\left(\hat{\hat{\mu}},\ b^* = 0,\ \hat{\hat{\sigma}}^2\right)}{L\left(\hat{\mu},\ \hat{b}^*,\ \hat{\sigma}^2\right)}$$

Extending this tradition, many QTL mapping analyses also use LOD score as a test statistic. There is a one-to-one correspondence between LR and LOD, and LR can be translated into LOD as LOD = $\frac{1}{2}(\log_{10}e)$LR = 0.217LR.

This test can be performed at any genomic position covered by markers and thus the method involves a systematic strategy of searching for QTL. If the likelihood ratio test statistic at a genomic region exceeds a predefined critical threshold, a QTL is estimated at the position of the maximum test statistic. The estimates of locations and effects of QTL are asymptotically unbiased statistically with this maximum likelihood approach if there is only one QTL on a chromosome.

It is important to determine an appropriate critical threshold for a test statistic above which a QTL can be claimed with a certain confidence. The determination of the critical threshold is based on the distribution of a test statistic under the null hypothesis. This distribution for LR at a given position is generally asymptotically chi-square, with a degree of freedom that is equal to the number of parameters under the test. However, because the test is usually performed in the whole genome, there is a multiple testing problem, and the distribution of the maximum LR or LOD score over the whole genome under the null hypothesis becomes very complicated. Theoretical and numerical analyses have indicated that the 5% significance threshold over a whole genome is generally between 2 and 3.5 on LOD score, depending on the size of genome, density of markers, sample size and genetic model (Lander & Botstein, 1989; Lander & Schork, 1994). Alternatively, the relevant threshold for a given data set can be estimated numerically from the data by using a permutation test (Church & Doerge, 1994).

The model of interval mapping is relatively simple in terms of genetics. As a result of this, it has a critical problem that, if there are two or more QTL on a chromosome, the test statistic at a genomic position will be affected by all those linked QTL. Therefore, the estimated positions and effects of "QTL" identified by this method can be biased.

Moreover, some genomic regions that do not contain QTL can still show a significant peak on LOD score if there are multiple QTL in the nearby regions. This is the so-called "ghost" gene phenomenon. This defect is similar to the defect in single marker analysis that is discussed above.

4. Composite interval mapping

Ideally, when we test a marker interval for a QTL, we would like our test statistic to be independent of the effects of possible QTL located in other regions of the chromosome. If such a test can be constructed, we can break down the effects of linked QTL by statistical means to avoid the confounding effects of multiple linked QTL in the search for each individual QTL. In other words, we can test independently each interval for the presence of a QTL. Such a test can be constructed by using a combination of interval mapping and multiple regression (Zeng, 1993).

In a multiple regression analysis of a trait on multiple markers or other explanatory variables, each regression coefficient becomes a partial regression coefficient conditional on other variables fitted in the model. Largely because of the linear structure of genes on chromosomes, a partial regression coefficient of a trait on a marker, or a testing position of interest, possesses a very important property in that the coefficient is expected to depend only on those QTL within an interval that is bracketed by two fitted flanking markers. The flanking markers are fitted in the model as cofactors to block the effects of other possibly linked QTL to the test. This treatment makes the partial regression coefficient independent of QTL effects on other linked or unlinked intervals, and is the basis of composite interval mapping (Zeng, 1994). The linear independence, however, depends on the assumption of no crossing-over interference and no epistasis. Interference and epistasis introduce non-linearity into the model.

Specifically, to test for a QTL on an interval between two adjacent markers, we can extend the interval mapping model to

$$y_j = \mu + b^* x_j^* + \sum_k b_k x_{jk} + e_j \quad \text{for } j = 1, 2, \ldots, n$$

where x_{jk} is an indicator variable referring to the genotype of marker k that is selected to control the genetic background, b_k is the partial regression coefficient associated with marker k and b^* now is also a partial regression coefficient associated with the putative QTL. In this case, the likelihood function becomes

$$L\left(b^*, \mathbf{b}, \sigma^2\right) = \prod_{j=1}^{n}\left[p_{1j}\,\phi\left(y_j : \mathbf{x}_j \mathbf{b} + b^*, \sigma^2\right) + p_{0j}\,\phi\left(y_j : \mathbf{x}_j \mathbf{b}, \sigma^2\right)\right]$$

A likelihood ratio test statistic can also be constructed to compare the hypotheses H_0: $b^* = 0$ with H_1: $b^* \neq 0$. However, since b^* is a partial regression coefficient, the null hypothesis is a composite hypothesis, conditional on other partial regression coefficients in the model. Thus, the method is called composite interval mapping. Many statistical issues of composite interval mapping are discussed in Zeng (1994).

As with Lander and Botstein's (1989) interval mapping, this test can be performed at any position in a genome covered by markers. Thus, it also gives a systematic strategy to search for QTL in a genome. The main advantage of composite interval mapping, as compared with interval mapping, is the ability to separate effects and locations of multiple linked QTL in mapping. This is shown in **Figure 1** as an example.

(i) Example of composite interval mapping analysis

Figure 1 summarizes the analyses of mapping body weight loci on mouse chromosome X from a backcross population (Dragani *et al.*, 1995). The test statistic (LOD score) of the interval mapping and composite interval mapping analyses is plotted against the linkage map location of the chromosome referenced by 14 microsatellite markers. The value of LOD score at each map position indicates the strength of evidence for a QTL at the position. If the LOD score at a

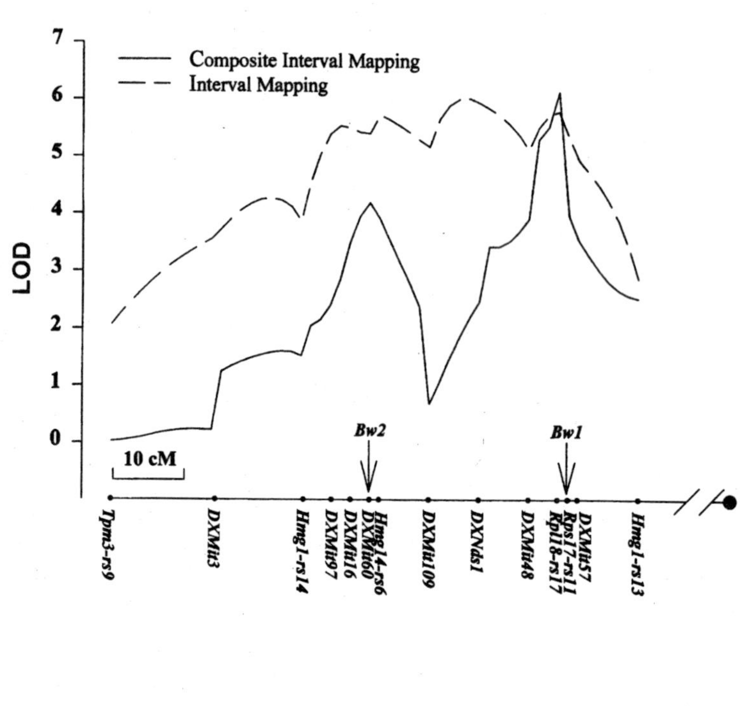

Figure 1 Genetic mapping of body weight loci on chromosome X in a mouse backcross population. LOD (log odds) score curves of a composite interval mapping analysis (solid curve) and an interval mapping analysis (dashed curve) are shown on a map containing 14 molecular markers. By the interval mapping analysis, it seems that most of the chromosome shows significant effects on body weight. The composite interval mapping analysis strongly indicates that there are two body weight loci on chromosome X segregating in the population. The arrows indicate the estimated positions of *Bw1* and *Bw2*. (Adapted from Dragani *et al.*, 1995.)

genomic region exceeds a predetermined threshold, one or more QTL are indicated in that region.

For the interval mapping, the threshold is 3.3 for the experimental design (Lander & Schork, 1994). By this criterion, the LOD score in the larger part of chromosome X is above the threshold, and shows significant peaks in several marker intervals. However, as pointed out above, not all significant peaks could be interpreted as QTL because of linkage effects, the "ghost" gene phenomenon and statistical sampling effects. Although the analysis strongly supports the existence of segregating QTL on chromosome X, it is not clear from the interval mapping analysis how many QTL are on the chromosome and where they are located.

The LOD score of the composite interval mapping analysis shows two distinct major peaks. This suggests that there are at least two body weight QTL on chromosome X in the mouse genome: one named *Bw1* is mapped near marker *Rp18-rs11* and the other (*Bw2*) near *DXMIT60*. The two QTL together explain 25% of the phenotypic variance in the mapping population. In this case, the composite interval mapping analysis achieved a much better resolution in mapping QTL.

5. Cloning a QTL

A critical question usually asked by geneticists about QTL mapping analyses is how reliable these analyses can be or how we know that a region mapped by these analyses for a QTL does contain a gene we want. One way to prove the latter is to obtain a clone of a small DNA segment that is shown to have a significant effect on the trait, indicating that the segment contains the gene. In an experiment of mapping fruit weight QTL in tomato, Alpert *et al.* (1995) crossed the domesticated tomato *Lycopersicon esculentum* with a near-isogenic introgression line (IL2–5) that contains a chromosomal segment on chromosome 2 derived from the small green-fruited wild tomato species *L. pennellii*. An F_2 population (called Pn population) of 540 individuals was generated from the cross and subjected to the analysis on the phenotype of fruit weight and a number of restriction fragment length polymorphism markers on chromosome 2. From the analysis, a major QTL named *fw2.2* was mapped between markers *TG91* and *TG167* (**Figure 2**). A separate experiment from a cross between *L. esculentum* and the small red-fruited wild tomato species *L. pimpinellifolium* (called Pp population) also mapped a fruit weight QTL to

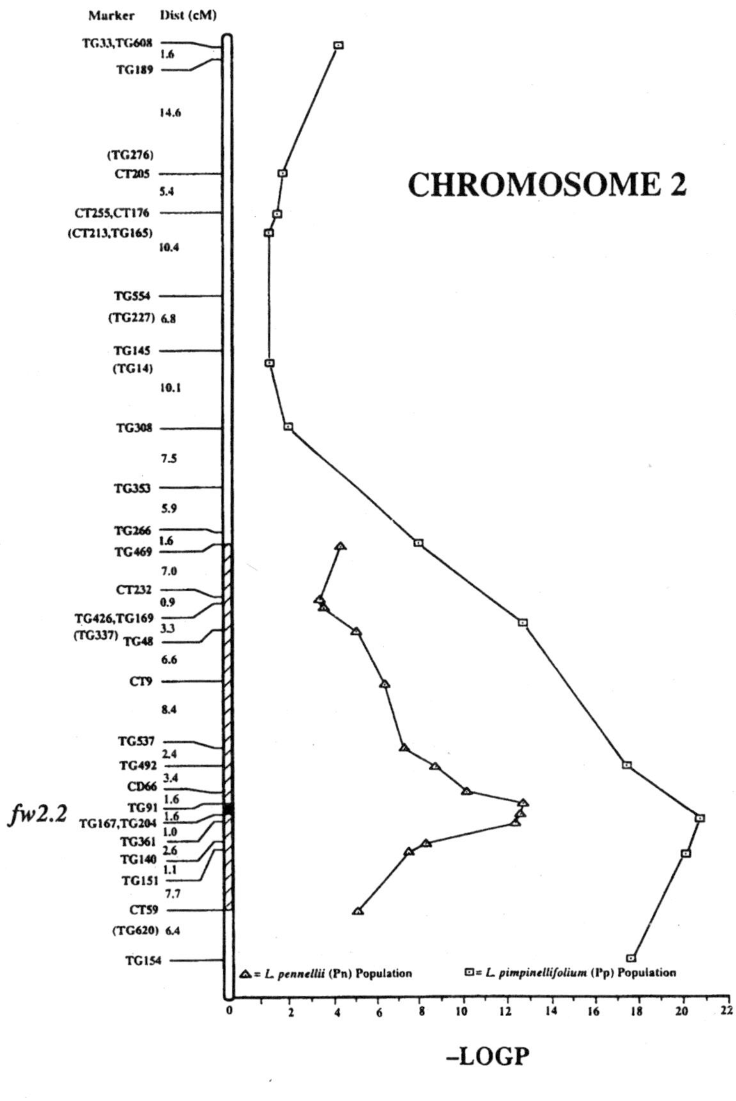

Figure 2 The association between reduced fruit weight and tomato chromosome 2 markers in the Pn population (triangles) and Pp population (squares). The F statistic was used to determine the corresponding P value for each marker that was converted to −logP for simplicity. The hatched box corresponds to the *L. pennellii* region of chromosome 2 contained in the introgression line (IL2–5). The black box indicates the likely position of *fw2.2*. (Reproduced with permission from Alpert *et al.*, © Springer Verlag 1995.)

the same region. **Figure 2** shows the single marker analysis in terms of the P value of the test statistic (the significance probability at which the null hypothesis would be rejected for the given observation) for the two experiments. Since the marker coverage is relatively dense and the genetic basis of fruit weight differences between the crosses on chromosome 2 appears to be relatively simple for the experiments, the single marker analysis is informative and a single QTL *fw2.2* is well supported.

To obtain the clone of *fw2.2*, Alpert and Tanksley (1996) raised a further 3472 F_2 plants, nearly isogenic for the region spanning the *fw2.2* locus, from the cross of *L. esculentum* and the *L. pennellii* introgression line and added a few more markers in the analysis. From the analysis, they were able to obtain a high-resolution map of the region and narrowed the placement of *fw2.2* to a region less than 150 kilobases and 0.13 ± 0.03 centiMorgans between molecular markers *TG91* and *HSF24*. A physical contig composed of six yeast artificial chromosomes (YACs) and encompassing the *fw2.2* region was isolated; *fw2.2* was included in two overlapping YACs. This marked the first time that a QTL was mapped with such a high precision and delimited to a segment of cloned DNA. Further DNA sequence analysis of the cloned segment (Frary *et al.*, 2000) identified *fw2.2* to a single gene, *ORFX* that is expressed early in floral development and controls carpet cell number. It was suggested that the effects of *fw2.2* alleles on fruit size are most likely due to changes in gene regulation rather than in the sequence of the encoded protein.

6. Summary

Mapping QTL is not restricted to backcross and F_2 populations of inbred lines. Mapping methods can be extended and applied to other crosses of populations or species or to segregating populations. For species such as human, the mapping of QTL has to be made with current segregating populations. No matter what population is analysed, the general idea of QTL mapping analysis is based on the inference of genotypes and an appropriate model that relates a trait to the genotypes or combinations of genotypes at a number of genomic positions. However, for mapping QTL with segregating populations, statistical analyses become much more complicated due to a number of limiting factors in data, such as small family size, unknown linkage phases between markers and QTL, and complicated family structures. Many statistical methods for mapping QTL from segregating populations have been developed. These include, for example, the sib-pair methods (Kruglyak & Lander, 1995), identity-by-descent mapping (Dupuis *et al.*, 1995) and some Bayesian methods (Satagopan *et al.*, 1996; Uimari *et al.*, 1996) that incorporate Markov chain Monte Carlo algorithms. More studies are needed to generalize these and other methods to make them applicable to the wide variety of populations and experimental designs, data structures and genetic models. Texts by Falconer and Mackay (1996) and Lynch and Walsh (1997) are useful sources for more general discussion on the genetic basis of QTL, and genetic and statistical analyses for mapping QTL.

REFERENCES

Alpert, K.B. & Tanksley, S.D. (1996) High-resolution mapping and isolation of a yeast artificial chromosome contig containing *fw2.2*: a major fruit weight quantitative trait locus in tomato. *Proceedings of the National Academy of Sciences USA* 93: 15503–07

Alpert, K.B., Grandillo, S. & Tanksley, S.D. (1995) *fw2.2*: a major QTL controlling fruit weight is common to both red- and green-fruited tomato species. *Theoretical and Applied Genetics* 91: 994–1000

Church, G.A. & Doerge, R.W. (1994) Empirical threshold values for quantitative trait mapping. *Genetics* 138: 963–71

Dragani, T.A., Zeng, Z.-B., Canzian, F. *et al.* (1995) Molecular mapping of body weight loci on mouse chromosome X. *Mammalian Genome* 6: 778–81

Dupuis, J., Brown, P.O. & Siegmund, D. (1995) Statistical methods for linkage analysis of complex traits from high-resolution maps of identity by descent. *Genetics* 140: 834–56

Falconer, D.S. & Mackay T.F.C. (1996) *Introduction to Quantitative Genetics*, 4th edition, Harlow, Essex: Longman

Frary, A., Nesbitt, T.C., Grandillo, S. *et al.* (2000) A quantitative trait locus key to the evolution of tomato fruit size. *Science* 289: 85–88

Kruglyak, L. & Lander, E.S. (1995) Complete multipoint sib-pair analysis of qualitative and quantitative traits. *American Journal of Human Genetics* 57: 439–54

Lander, E.S. & Botstein, D. (1989) Mapping Mendelian factors underlying quantitative traits using RFLP linkage maps. *Genetics* 121: 185–99

Lander, E.S. & Schork, N.J. (1994) Genetic dissection of complex traits. *Science* 265: 2037–48

Lynch, M. & Walsh, B. (1997) *Genetics and Analysis of Quantitative Traits*, Sunderland, Massachusetts: Sinauer Associates

Satagopan, J.M., Yandell, B.S., Newton, M.A. & Osborn, T.C. (1996) A Bayesian approach to detect quantitative trait loci using Markov chain Monte Carlo. *Genetics* 144: 805–16

Uimari, P., Thaller, G. & Hoeschele, I. (1996) The use of multiple markers in a Bayesian method for mapping quantitative trait loci. *Genetics* 143: 1831–42

Zeng, Z.-B. (1993) Theoretical basis of separation of multiple linked gene effects on mapping quantitative trait loci. *Proceedings of the National Academy of Sciences USA* 90: 10972–76

Zeng, Z.-B. (1994) Precision mapping of quantitative trait loci. *Genetics* 136: 1457–68

GLOSSARY

contig a continuous sequence of DNA assembled from overlapping cloned DNA fragments

dominance where a trait is expressed in heterozygotes, i.e. where the phenotype of the recessive gene is masked by the phenotype of the dominant gene, resulting in a phenotype that is the same as that manifest by homozygotes

genotype by environment interaction the genetic make-up of an individual as influenced by the environment

inbred line breeding of any animal or plant strain that leads to homozygosity, e.g. backcrossing of offspring with parents

interference the predilection of one crossover in meiosis to suppress further crossing over within the same chromosomal domain

isogenic homozygous
linkage disequilibrium the alliance of two linked alleles more often than would be expected by chance
polymorphic genetic markers DNA sequence polymorphism detected by a variety of molecular techniques, such as restriction fragment length polymorphism (RFLP), microsatellite, single nucleotide polymorphism (SNP)

See also **Genetics of quantitative characters** (p.894)

CONSERVATION GENETICS

Richard Frankham
Key Centre for Biodiversity and Bioresources, Department of Biological Sciences, Macquarie University, Australia

1. Introduction

Conservation genetics is the genetics of rarity and extinction. This discipline has arisen in response to the depletion of the biological diversity of the planet through habitat loss, overexploitation, pollution and the impact of introduced species (World Conservation Monitoring Centre, 1992). An unknown but large number of species are already extinct, while many others have reduced population sizes that put them at risk of extinction. Many species now require human intervention to ensure their survival. The scale of the problem is enormous: 56% of mammals and 53% of birds are categorized as threatened, and over 50% of all vertebrate taxa are threatened (IUCN, 1996).

There are four justifications for maintaining biodiversity: the economic value of biological resources (food, drugs, etc.), ecosystem services (nutrient recycling, pest control, etc.), aesthetics and the rights of living organisms to exist. The World Conservation Union (IUCN) recognizes the need to conserve biodiversity at three levels: genetic diversity, species diversity and ecosystem diversity. Genetics is directly involved in the first two of these. Habitat loss, introduced species, overexploitation and pollution typically reduce species to population sizes where they are susceptible to chance effects, whether environmental fluctuations (stochasticity), catastrophes, demographic stochasticity or genetic effects (inbreeding depression, loss of genetic variation and accumulation of deleterious mutations).

2. Genetics in conservation biology

Modern conservation biology arose in the late 1970s and 1980s as an applied field, drawing on ecology, genetics and wildlife and resource biology. Sir Otto Frankel, an Austrian-born Australian, was largely responsible for the recognition of genetic factors in conservation biology. He both pointed out the need to preserve the evolutionary potential of species, and strongly influenced Michael Soulé who was the dominating influence in the development of conservation biology during the 1980s.

There are seven major genetic issues in conservation biology:

1. the deleterious effects of inbreeding (inbreeding depression);
2. loss of genetic variation in small populations;
3. accumulation and loss of deleterious mutations;
4. genetic adaptation to captivity and its effect on the success of reintroduction into the wild;
5. fragmentation of populations and reduced migration;
6. outbreeding depression; and
7. taxonomic uncertainties.

Endangered species are unsuitable for evaluating these issues (apart from taxonomic uncertainties) as they are typically slow breeders, expensive to maintain, present in low numbers and too valuable to risk in experiments. Consequently, most of the information on these issues has come from studies using laboratory species (such as Drosophila, Tribolium, butterflies, fish and mice) or from combined analyses done on small data sets from many species (meta-analyses).

(i) Inbreeding, genetic variation and extinction

The presumption underlying the application of genetics within conservation biology is that inbreeding and loss of genetic variation increase the risk of extinction.

Inbreeding depression in wild animals and plants. Inbreeding depresses reproductive fitness in naturally outbreeding populations of domestic plants and animals (inbreeding depression). This occurs primarily because inbreeding increases homozygosity and exposes rare deleterious mutations. Nevertheless, there was scepticism that wildlife suffered from inbreeding depression. Katherine Ralls and Jon Ballou of the National Zoo in Washington, DC, provided compelling evidence that captive populations of mammals suffer from inbreeding depression (Ballou & Lacy, 1995). Offspring resulting from brother–sister matings have on average 33% lower juvenile survival than non-inbred offspring, with there being no clear differences between mammalian orders. However, the full effects of inbreeding depression are much greater than this; for a range of animal species, the effects on total reproductive fitness are about three times greater than for those on any single component.

The argument about inbreeding depression has now shifted to wild populations. Since inbreeding depression is typically more severe in harsher environments, it is expected to be more severe in the wild than in captivity. Inbreeding depression in wild or semi-wild environments has been reported in several species of fish, snails, snakes, lions, shrews, white-footed mice, golden lion tamarins, two species of birds and several species of outbreeding plants (Frankham, 1995). For example, Madsen *et al.* (1996) showed that a small isolated population of snakes in Sweden exhibited inbreeding depression – it had lower litter size, a higher proportion of abnormal offspring and less genetic variation than a large population. In addition, the introduction of a male from the large population reduced the frequency of inviable offspring. Several studies in birds have failed to find inbreeding depression, but that can be attributed to incorrect paternities (Frankham, 1995). The occurrence of inbreeding depression in cheetahs is controversial and has been the subject of much debate (Frankham, 1995); this controversy stems mainly from the lack of non-inbred cheetahs to use as controls. Overall, however, there is clear and unequivocal evidence for inbreeding depression in wild populations of animals and plants.

Inbreeding and extinction. Inbreeding increases the risk of extinction in deliberately inbred laboratory and domestic animals. Typically, they show a threshold relationship, with low initial extinction, but sharply increased extinction rates beginning at intermediate levels of inbreeding (Frankham, 1995).

Inbreeding in wildlife usually occurs as a consequence of the cumulative effects of finite population size, often over hundreds of generations, as described by equation 1.

$$F \sim 1 - e^{-t/2Ne} \qquad (1)$$

where F is the inbreeding coefficient, N_e the effective population size and t the number of generations. Plants, snakes, fish and Drosophila have all showed inbreeding depression due to finite population sizes (Loeschcke *et al.*, 1994; Frankham, 1995; Madsen *et al.*, 1996). For example, 25% of captive populations of *D. melanogaster* with population sizes of 67 went extinct over 210 generations.

In the wild, demographic and environmental fluctuations (stochasticity) and catastrophes contribute to extinction. It has been claimed that these factors cause extinctions in wildlife before the effects of inbreeding become evident. However, inbreeding and loss of genetic variation influence the responses of populations to these factors. Extinctions can be incorrectly attributed to "non-genetic" factors, when interactions between genetic and non-genetic factors have actually led to extinctions. Birth and death rates are susceptible to inbreeding depression, and loss of genetic variation decreases the ability of wild populations to survive climatic extremes, pollutants, diseases, pests and parasites.

Claims that demographic and environmental stochasticity are likely to cause extinctions in wild populations before inbreeding and loss of genetic variation become important have been refuted by recent theoretical studies. For example, Mills and Smouse (1994) developed a computer simulation model with all these factors operating, and showed that inbreeding is likely to contribute to population decline, especially for species with low reproductive rates.

(ii) Loss of genetic variation

Since genetic variation is required for populations to evolve in response to environmental change, its maintenance is a major issue in conservation genetics. Genetic variation is lost over time in finite populations due to random sampling (genetic drift) in an exponential decay process according to equation 2.

$$H_t/H_0 \sim e^{-t/2Ne} \qquad (2)$$

where H_t is heterozygosity after t generations, H_0 initial heterozygosity, N_e the effective population size and t the number of generations. In random mating populations, loss of genetic variation and inbreeding are intimately related; the proportion of the genetic variation that is lost is equal to the inbreeding coefficient. The predictions of equation 2 have been verified for allozyme variation in experimental populations of Drosophila. Furthermore, large populations of wild animals and plants typically have more genetic variation than small populations.

The genetic consequences of finite population size depend on the effective population size, rather than the census size. The effective size depends not only on the number of sexually mature adults, but on the variation in family size, inequalities in sex ratio and on fluctuation in numbers over generations (Frankham, 1995). The ratio of effective population size to actual size is required to estimate the genetically effective size of wildlife populations from their actual sizes. Estimates of the ratio that include the effects of all relevant factors (unequal sex-ratio, variance in family sizes and fluctuations in population size), based on data from over 100 species of animals and plants, averaged 10–11%, much lower than generally assumed. Consequently, wild animal and plant populations are genetically about an order of magnitude smaller than their actual sizes (Frankham, 1995).

Low genetic diversity, endangerment and extinction in wildlife. A majority of endangered species have low levels of genetic variation compared with related non-endangered species (Frankham, 1995). Whether inbreeding has caused endangerment or low population size associated with endangerment has reduced genetic variation is unclear. Even in the latter case, inbreeding depression will exacerbate proneness to extinction. Consequently, genetic concerns appear to be of significance in the majority of endangered species and populations whose genetic variation has been measured.

Clear evidence for the role of inbreeding and loss of genetic variation in decline of a wild population was provided by Vrijenhoek and colleagues (Vrijenhoek, 1994). They showed that a genetically variable sexual species of fish numerically dominated a related parthenogenetic clonal species until a drought eliminated their habitat. When the populations were subsequently re-established, the sexual species possessed reduced genetic variation from a founding event and was consistently less abundant than the genetically constant parthenogenetic species. The sexual species re-established its numerical dominance following the deliberate addition of genetic variation via the replacement of 30 sexual individuals by ones from elsewhere. Further, low genetic diversity was associated with increased parasite loads in these fish populations. Suggestive evidence for the role of inbreeding and loss of genetic variation in the decline and extinction of wild populations exists for Florida panthers, Puerto Rican parrots, Isle Royale grey wolves, inbreeding colonial spiders, native mice, heath hens, bighorn sheep, middle spotted woodpeckers and island species (Frankham, 1995).

Island populations are much more prone to extinction than mainland populations (World Conservation Monitoring Centre, 1992). This is expected on both genetic and ecological grounds. Populations on islands lose genetic variation and become more inbred due to bottlenecks at foundation (with as few as one or two individuals) and through subsequent low average population sizes. As predicted, island populations generally are found to have less genetic variation than mainland populations. Further, many island populations are inbred to levels where captive populations show an elevated risk of extinction from inbreeding. While the susceptibility of island populations to extinction has been interpreted as being due to non-genetic causes, both genetic and non-genetic factors, and their interactions, are probably involved.

Genetic management of endangered species. The objectives of genetic management of endangered species are to retain reproductive fitness and evolutionary potential. The major issues are the number of founding individuals, the subsequent effective population size, the degree of population fragmentation, the time frame and the adverse effects of genetic adaptation to captive conditions.

Population and quantitative genetic principles have been introduced into the management of threatened species since the early to mid-1980s, especially in captivity. The Conservation Breeding Specialist Group of IUCN has played a major global role in this. The principle guiding the management of endangered species in captivity is to retain 90% of the original genetic variation for 200 or 100 years, the presumption being that human populations will stabilize and decrease over this time so that wild habitat is available for reintroductions. Procedures that maximize effective population size and the generation interval are recommended; these include equalization of family sizes, equalizing the sex ratio of breeding individuals and avoiding fluctuations in population size. The required size to retain 90% of genetic variation, obtained from equation 2, is strongly dependent on the generation interval. For example, a large mammal with a generation interval of 20 years requires an effective population size of 24 to retain 90% of its genetic variation for 100 years, while an insect with five generations per year requires a size of almost 2400 per generation.

The generation interval can be maximized by cryopreservation or by breeding from older animals or plants. Embryo and semen freezing technology is being used in the conservation of domestic animals, but is not available for most wildlife. For the majority of non-domesticated animal species, breeding from older animals is the only current means for extending the generation interval and that is difficult to achieve.

Many captive populations of wildlife have been founded when few individuals are left. For example, the US captive population of Speke's gazelle was founded from one male and three females in 1969–72 and has steadily increased since then. At foundation, it was presumed to be the only existing population of the species, but subsequently another captive population has been found in the Middle East. Founders of captive populations often contribute unequally, such that the rate of inbreeding and the loss of genetic variation is increased. Consequently, it has been recommended that such populations be managed to equalize founder representation, and this was done with Speke's gazelle.

Ballou and Lacy (1995) predicted that minimizing kinship is the optimum means for managing small pedigreed populations with unequal founder contributions to maximize the retention of heterozygosity. This procedure combines the benefits of equalizing family sizes and adjusting founder representation. Experimental results in Drosophila have verified its predicted benefits, and minimizing kinship is being widely used in the captive management of endangered animals.

Alleviating inbreeding depression: immigration. The obvious way to overcome inbreeding depression and loss of genetic variation in small populations is to introduce immigrants. There is ample evidence that this improves reproductive fitness, though occasionally it results in outbreeding depression (see (vi) below). Immigration is often not an option for endangered species, as they frequently exist only as a single population.

(iii) Accumulation and loss of deleterious mutations

Deleterious mutations arise by mutation and are removed by natural selection. In small populations, some become fixed and reduce reproductive fitness. Lande (1995) and Lynch and colleagues (1995) have predicted that the accumulation of new, mildly deleterious mutations may be a more important cause of extinction than demographic stochasticity, and of similar importance to environmental stochasticity in populations up to effective sizes of 1000. However, the only experimental study of this threat failed to find evidence for it in Drosophila populations of 25–500 maintained for 45–50 generations (Gilligan *et al.*, 1997).

Rare recessive deleterious mutations are exposed by inbreeding and so can be more effectively removed from inbred than outbred populations through natural selection. Such purging has been documented in many species of animals and plants. However, recent theoretical and experimental studies indicate that, while purging may ameliorate inbreeding depression, it does not eliminate it (Frankham, 1995).

(iv) Genetic deterioration in captivity

Captive breeding is being used as a means for saving many endangered species from extinction, especially mammals, birds, reptiles and plants. Reintroduction into the wild is usually considered the desired end-point. Three adverse genetic changes occur in captivity: inbreeding depression, loss of genetic variation and genetic adaptation to the captive environment. Selection for tameness and other adaptations to the captive environment are likely to jeopardize reintroduction success. Genetic adaptation to captivity has been documented in a wide range of species, and it is generally disadvantageous on return to the natural environment. Considerable difficulty has been encountered in the reintroduction of endangered vertebrates into the wild. Genetic adaptation to captivity is one of many possible reasons for this, but there is no critical evidence to separate this from other possible causes.

Genetic adaptation to captivity is minimized when the number of generations, and amount of selection in captivity, are minimized and when generation length and the proportion of immigrants are maximized. Equalizing family sizes reduces genetic adaptation by approximately 50%, since it eliminates selection among families. However, severe genetic deterioration still occurred in captive populations of

D. melanogaster maintained for 50 generations with equal family sizes under benign conditions. Genetic adaptation to captivity appears to be a major problem in the captive breeding of endangered species, but this is not widely recognized at present.

How large? How large do populations have to be to (1) avoid inbreeding depression and (2) retain their evolutionary potential? It is widely asserted that an effective population size of 50 is sufficient to avoid inbreeding depression in the short term. However, Latter and Mulley (1995) found inbreeding depression in long-term populations with effective sizes of approximately 50.

Franklin (1980) suggested that an effective size of 500 should be sufficient for an indefinite retention of evolutionary potential due to the balance between chance loss of quantitative genetic variation in finite populations and gain through mutation. However, Lande (1995) argued that an effective size of 5000 is required to retain evolutionary potential since mutations are predominantly deleterious. Preliminary data from *D. melanogaster* suggest that an effective size in the hundreds, rather than thousands, is required.

While effective population sizes of 500 are recommended, they are rarely attainable in captivity due to a shortage of captive breeding space. In the wild, this translates into a population size of about 5000, a figure that cannot be attained for large herbivores and carnivores even in the largest existing reserves throughout the world.

(v) Consequences of population fragmentation

Wild populations of many species are being fragmented by human actions (World Conservation Monitoring Centre, 1992). This results in small, partially isolated populations that are more susceptible to inbreeding and loss of genetic variation. These have been equated to island populations that have a much higher risk of extinction than larger mainland populations. Extinction rates in isolated populations have been shown to be related to population sizes in bighorn sheep (**Figure 1**), in arboreal mammals in the western US and in mammals in US national parks. However, the contribution of genetic *vs.* ecological factors to these extinctions is unclear.

Deliberate movement of individuals among isolated populations is an obvious solution to alleviate their genetic problems. This is being done with the Florida panther, which exists as a small inbred population in and near the Everglades (O'Brien, 1994). Individuals from the nearest subspecies in Texas are being introduced in an attempt to overcome the presumed effects of inbreeding. However, such immigration is a highly controversial issue in wildlife circles due to concerns about outbreeding depression and the spread of disease. Many conservation geneticists consider that current attitudes to the exchange of individuals between populations are unnecessarily restrictive.

(vi) Outbreeding depression

Reproductive fitness may be reduced following hybridization of populations within species, either in the F_1 or in later generations. Such outbreeding depression is associated with local adaptations (coadapted gene complexes) and low mobility, and may be common in plants. However, larger and more mobile animals show less local adaptation, so it is expected to be rarer in them. While evidence is limited, outbreeding depression seems to be uncommon in animals and is clearly much less important than inbreeding depression (Frankham, 1995).

(vii) Taxonomic uncertainties

Since crosses among different species or subspecies may result in sterility, it is critical that the taxonomic status of

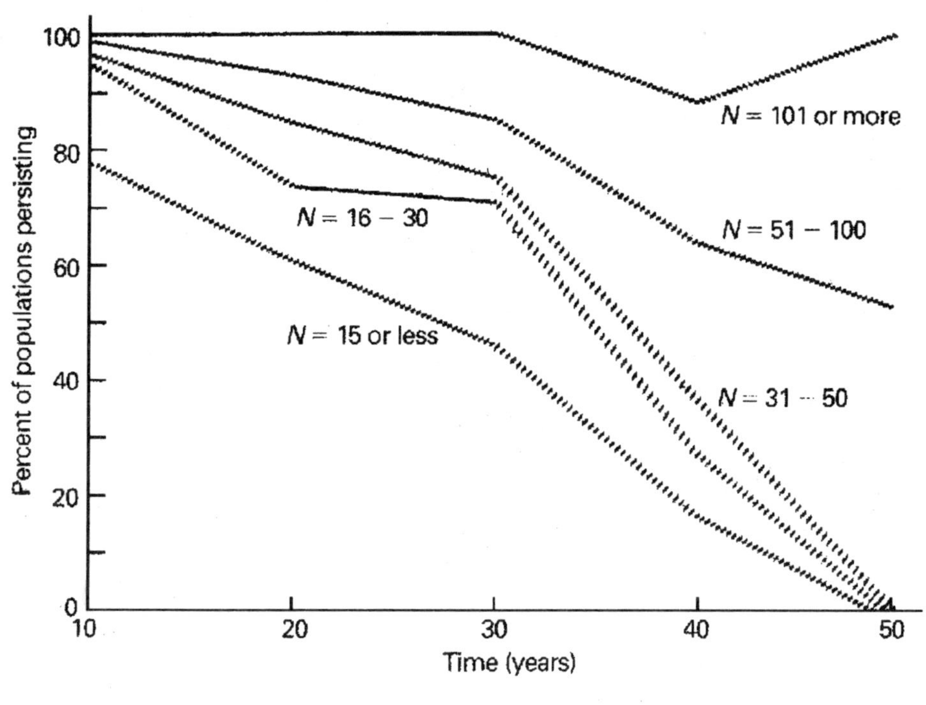

Figure 1 The relationship between the size of populations of bighorn sheep (*N*) and the percentage of populations that persisted over time (reproduced from Primack, R.B. 1993. *Essentials of Conservation Biology*, Sinauer Associates; data of Berger, 1990). Larger populations show greater persistence than smaller ones.

managed populations be clarified. Genetic markers provide valuable means for resolving cases of ambiguous taxonomic status. Chromosomes, allozymes, sequences of mitochondrial or nuclear DNA, restriction fragment length polymorphisms (RFLPs), DNA fingerprints and microsatellites have all been used to resolve ambiguities (O'Brien, 1994). Several captive populations have been shown to be mixtures of subspecies or species, including orang-utans and lions. Przewalski's horse from central Asia traces to only 13 founders, one of which is a domestic horse (a separate species with different chromosome number). Much effort has been made to minimize the contributions of the domestic horse, but its genetic contribution cannot be entirely eliminated.

Crosses between species occur in wild animals and plants, and cause much concern about the genetic "swamping" of rare species by related common species. The red wolf is a controversial case. It contains genes from coyotes and grey wolves and there is argument as to whether it is a distinct species or simply a hybrid population that does not justify conservation (O'Brien, 1994).

Use of genetic markers in population management. Genetic markers have been used to aid in the management of threatened species. They have been used to identify genetically distinct populations for augmentation of small inbred populations (to alleviate inbreeding and increase genetic variation), to identify the best populations for reintroductions and to infer the relationship among founders in populations of endangered species, so that their genetic management can be optimized. Genetic markers have been used to determine paternity in studies of the basic biology of endangered species. Molecular genetic markers have also been used to detect illegal hunting of whales. The use of genetic distances has been advocated for setting conservation priorities according to taxonomic distinctness. PCR provides a non-destructive means for genotyping endangered species. Animals can be monitored from hairs, feathers, museum specimens or excrement, birds can be sexed and diet can even be determined. Andrea Taylor and colleagues (Taylor *et al.*, 1994) used such technology to study the endangered northern hairy-nosed wombat, which exists as a remnant population of approximately 70 individuals in northern Australia. As these burrowing marsupials are difficult to capture, PCR was used to genotype individuals for microsatellite loci based on hair samples. Studies of the extant population, of museum skins from an extinct population in southern Australia and from its closest relative (the southern hairy-nosed wombat) demonstrated lowered genetic variation in the endangered wombats and established that the extinct southern population was of the same species, thus providing a potential site for a second population to be re-established. Molecular genetic data have also been used to identify and sex individuals, and to infer relatedness of mates and dispersal patterns.

Evidence concerning the reality of genetic concerns in conservation biology has been greatly strengthened in recent years (Frankham, 1995). Further, many of the concepts underlying the genetic management of endangered species have been resolved using experiments in laboratory animals and meta-analyses of data from wildlife.

REFERENCES

Ballou, J.D. & Lacy, R.C. (1995) Identifying genetically important individuals for management of genetic variation in pedigreed populations. In *Population Management for Survival and Recovery: Analytical Methods and Strategies in Small Population Conservation*, edited by J.D. Ballou, M. Gilpin & T.J. Foose, New York: Columbia University Press

Berger, J. (1990) Persistence of different-sized populations: an empirical assessment of rapid extinctions in bighorn sheep. *Conservation Biology* 4: 91–98

Frankham, R. (1995) Conservation genetics. *Annual Review of Genetics* 29: 305–27

Franklin, I.R. (1980) Evolutionary change in small populations. In *Conservation Biology: An Evolutionary–Ecological Perspective*, edited by M.E. Soulé & B.A. Wilcox, Sunderland, Massachusetts: Sinauer Associates

Gilligan, D.M., Woodworth, L.M., Montgomery, M.E., Briscoe, D.A. & Frankham, R. (1997) Is mutation accumulation a threat to the survival of endangered species? *Conservation Biology* 11: 1235–41

IUCN (1996) *Red List of Threatened Animals*, edited by J. Baillie & B. Groombridge, Gland, Switzerland: IUCN

Lande, R. (1995) Mutation and conservation. *Conservation Biology* 9: 782–91

Latter, B.D.H. & Mulley, J.C. (1995) Genetic adaptation to captivity and inbreeding depression in small laboratory populations of *Drosophila melanogaster*. *Genetics* 139: 255–66

Loeschcke, V., Tomiuk, J. & Jain, S.K. (1994) *Conservation Genetics*, Basel and Boston: Birkhäuser

Lynch, M., Conery, J. & Burger, R. (1995) Mutational meltdowns in sexual populations. *Evolution* 49: 1067–80

Madsen, T., Stille, B. & Shine, R. (1996) Inbreeding depression in an isolated population of adders *Vipera berus*. *Biological Conservation* 75: 113–18

Mills, L.S. & Smouse, P.E. (1994) Demographic consequences of inbreeding in remnant populations. *American Naturalist* 144: 412–31

O'Brien, S.J. (1994) Genetic and phylogenetic analyses of endangered species. *Annual Review of Genetics* 28: 467–89

Taylor, A.C., Sherwin, W.B. & Wayne, R.K. (1994) Genetic variation of microsatellite loci in a bottlenecked species: the northern hairy-nosed wombat *Lasiorhinus krefftii*. *Molecular Ecology* 3: 277–90

Vrijenhoek, R.C. (1994) Genetic diversity and fitness in small populations. In *Conservation Genetics*, edited by V. Loeschcke, J. Tomiuk & S.K. Jain, Basel and Boston: Birkhäuser

World Conservation Monitoring Centre (1992) *Global Biodiversity: Status of the Earth's Living Resources: A Report*, London and New York: Chapman and Hall

See also **DNA from museum specimens (p.817)**

REPRODUCTIVE ISOLATING MECHANISMS

Lee Ehrman
SUNY College at Purchase, Purchase, USA

1. Introduction

One may represent the processes of race or population, subspecies or semispecies divergence – and perhaps subsequent speciation – by the following sequences of events:

1. Expansion of the distribution area occupied by a species, leading to the formation of allopatric (geographical) populations more or less isolated by distance from one another, or, alternatively, secular changes taking place in the environments of the different parts of the geographic area that the species inhabits may make its populations face novel conditions.
2. Genetic divergence of the populations leading to formation of allopatric races. This divergence occurs through natural selection and random genetic drift, making the populations different in the incidence of gene alleles, chromosomal variations or in both their gene and chromosomal pools. The genetic differences may be adaptive to the environmental conditions in which the populations live.
3. Integration of the diverging corporate genotypes of the races or populations. As a result of the epistatic interactions between the constituent genes in different populations – or simply because of the accumulation of genetic differences that make the populations, specialized to exploit different environments – mixed or hybrid genotypes become disadvantageous. Gene exchange between the populations produces mostly or only genotypes that cause inferior fitness in their carriers.
4. Natural selection favours situations that minimize or eliminate the gene exchange between the divergent populations. These selection processes involve the appearance of a group of several reproductive isolating mechanisms, or, rarely, a single one. Very often none of these isolating barriers alone is sufficient to make the isolation complete, but they do so in combination.
5. Most of the processes of population divergence described here, and of the formation of isolating mechanisms, take place while the populations are allopatric or only slightly overlapping in their geographic distributions. Reproductively isolated species may continue to be more or less strictly allopatric, but perhaps more usually these distribution regions will overlap, making them sympatric in more or less extensive territories. It should be noted that although sympatric coexistence does not necessarily require the reproductive isolation to be complete, it must be robust enough to make the gene exchange between the populations involved not exceed the rate at which diffused genes can be eliminated by natural selection.

2. Genetically closed systems

Species of sexually reproducing organisms are genetically closed systems. They are closed systems because they do not exchange genes or they do so sufficiently rarely that the species differences are not swamped. Races are, on the contrary, genetically open systems. They do exchange genes by peripheral gene flow, unless they are isolated by extrinsic causes such as spatial separation. The biological meaning of the closure of a genetic system is simple, but important – it is evolutionary independence. Consider these four species – man, chimpanzee, gorilla and orang-utan. No mutation and no gene combination arising in any one of them, no matter how favourable, can benefit any of the others. It cannot do so for the simple reason that no gene can be transferred from the gene pool of one species to that of another. On the contrary, races composing a species are not independent in their evolution: a favourable genetic change arising in one race is, at least potentially, capable of becoming a genetic characteristic of the species as a whole.

Species are genetically closed systems because the gene exchange between them is impeded or prevented by reproductive isolating mechanisms. The term "isolating mechanism" was proposed by Dobzhansky in 1937 (third edition, Dobzhansky, 1951) as a common name for all genetically conditioned barriers to gene exchange between sexually reproducing populations. According to Mayr (1963), isolating mechanisms are "... perhaps the most important set of attributes a species has". It is a remarkable fact that isolating mechanisms are physiologically and ecologically a most heterogeneous collection of phenomena. It is another remarkable fact that the isolating mechanisms that maintain the genetic separateness of species are quite different, not only in different groups of organisms, but also between different pairs of species in the same genus.

3. Premating and postmating barriers

Several classifications of the reproductive isolating mechanisms (RIMs) have been proposed. That of Mayr is a simple and convenient one: the two major groups of RIMs are the premating barriers that prevent the formation of hybrid zygotes and the postmating barriers that impede the survival or reproduction of these zygotes. For examples and extended discussion see Ehrman & Maxson (2001).

Three of the premating barriers are:

- potential mates do not meet – seasonal and habitat isolation;
- potential mates meet, but do not mate – ethological, psychological or sexual isolation (the phenomenon observed is usually that the mutual attraction between conspecific females and males is greater than the attraction between males and females of different species); and

- copulation attempted, but no transfer of gametes takes place – mechanical isolation.

Four of the postmating barriers are:

- gametes transferred, but no fertilization and hence no zygote formation takes place – gamete mortality;
- death of the zygotes – hybrid inviability;
- the F_1 zygotes are viable, but partly or completely sterile – hybrid sterility; and
- the F_1 hybrids are fertile, but the fitness of the F_2 or back-cross hybrid is reduced – hybrid breakdown.

4. Sterility

Sterility – the most irrevocable of all the RIMs and recognized even by Aristotle – implies the partial elimination or total suppression of reproduction. Hybrid sterility may further be defined as the inhibition or suppression of the reproductive capacity of F_1 or later generation hybrids between genetically different strains or populations, usually belonging to different species. The term should not be used, as it is occasionally, to refer to the inability of representatives of different strains or species to produce hybrid offspring (which may be due to several different mechanisms; see Dobzhansky, 1951, 1970).

Hybrid sterility may, however, result from a number of different causes; it is consequently a common name for an assemblage of different phenomena. We can distinguish three or four kinds of hybrid sterility, including genic, chromosomal and cytoplasmic sterility.

(i) Genic sterility

Genic sterility is the outcome of physiological derangements in the developmental processes that would normally lead to the formation of functional gametes. A hybrid gene complement may be discordant; aberrant developmental processes may lower the somatic vigour of the hybrid organism (hybrid inviability) or may specifically disturb the reproductive processes and thus make the hybrids sterile.

(ii) Chromosomal sterility

Chromosomal sterility is a product of structural differences between chromosomes, differences which interfere with chromosome pairing and disjunction at meiosis. These structural differences are usually or always the result of rearrangements of the genes in the chromosomes.

(iii) Hybrid sterility

Hybrid sterility due to intersexuality of the hybrids is really a special case of genic sterility, the developmental processes interfered with being those concerned with the normal divergence of female and male sexes.

One may distinguish between sterility due to the disturbance of the haploid or of the diploid phase of the life cycle (haplontic or diplontic sterility). The distinction is descriptively simpler than that between genic and chromosomal sterility, but is probably less meaningful for analytical purposes. Hybrid sterility in invertebrates, and probably in most or in all animals, is diplontic. According to Dobzhansky (1951, 1970), the behaviour of allopolyploidal hybrids furnishes one method of distinguishing between genic and chromosomal sterility. If the sterility of a hybrid is due to a mechanical non-correspondence of gene arrangements in the chromosomes, a doubling of the chromosomal complement may furnish to each chromosome an exact pairing partner and, in this way, make gametogenesis normal. Chromosomal sterility of a diploid hybrid often contrasts with fertility of an allotetraploid obtained from the diploid, most often in plants. Doubling the chromosome complement does not alleviate genic sterility. Thus, it is convenient to distinguish four classes of hybrid sterility: (1) genic; (2) that due to intersexuality of the hybrids; (3) chromosomal; and (4) cytoplasmic, due to discordance between the chromosomal gene complement and the cytoplasm, the properties of the latter being due either to autonomous self-reproducing particles or to formative influences of the chromosomal genes having been present in the egg cell before meiosis. It is these autonomously self-reproducing particles, inherited microorganisms, that concern current researchers (Kidwell, 1983; Schwemmler, 1983; Huettel, 1986; Hoffmann *et al.*, 1990; Rousset & Raymond, 1991; Prout, 1994, Ehrman *et al.*, 1995). They represent a newly known, bravura evolutionary excursion.

Members of a species routinely mate with each other rather than with members of other species. Prior to copulation, information exchange between prospective mates is transmitted via several types of signal – acoustic, visual, chemical and tactile – which may be substrate or airborne. Courtship behaviour is species-specific and is a major component of the ethological isolation that leads to reproductive isolation between closely related species (Mayr, 1963; Spiess, 1987). This is desirable in that ill-adapted sterile hybrids are then not routinely produced. Such production would be eminently wasteful and no species could tolerate the continuous production of relatively defective hybrids without facing extinction.

REFERENCES

Dobzhansky, Th. (1951) *Genetics and the Origin of Species*, 3rd edition, New York: Columbia University Press (1st edition 1937)

Dobzhansky, Th. (1970) *Genetics of the Evolutionary Process*, New York: Columbia University Press

Ehrman, L., Perelle, I. & Factor, J. (1995) Endosymbiotic infectivity in *Drosophila paulistorum*. In *The Continuing Influence of Theodosius Dobzhansky*, edited by L. Levine, New York: Columbia University Press

Ehrman, L. & Maxson, S. (2001) *The Genetics of Behavior*, Oxford and New York: Oxford University Press

Hoffmann, A., Turelli, M. & Harshman, L. (1990) Factors affecting the distribution of cytoplasmic incompatibility in *Drosophila simulans*. *Genetics* 126: 933–48

Huettel, M. (ed.) (1986) *Evolutionary Genetics of Invertebrate Behavior: Progress and Prospects*, New York: Plenum Press

Kidwell, G.M. (1983) Intraspecific hybrid sterility. In *The Genetics and Biology of Drosophila*, vol. 3c, edited by M. Ashburner, H.L. Carson & J.N. Thompson, London and New York: Academic Press

Mayr, E. (1963) *Animal Species and Evolution*, Cambridge, Massachusetts: Belknap Press

Prout, T. (1994) Some evolutionary possibilities for a microbe that causes incompatibility in its host. *Evolution* 48: 909–11

Rousset, R. & Raymond, M. (1991) Cytoplasmic incompatibility in insects: why sterilize females? *Trends in Ecology and Evolution* 6: 54–57

Schwemmler, W. (1983) Analysis of possible gene transfer between an insect host and its bacteria-like endocytobionts. *International Review of Cytology* 14 (Suppl.): 247–66

Spiess, E.B. (1987) Discrimination among prospective mates in *Drosophila*. In *Kin Recognition in Animals*, edited by D.J.C. Fletcher & C.D. Michener, Chichester and New York: Wiley

GLOSSARY

adaptation the process of adjustment to an environment, through inheritance of a phenotype that gives a selection advantage

allele animals exist in the state of diploidy, that is, they possess two copies of every chromosome except the sex chromosomes. Therefore, every gene is present in two copies, or alleles. One allele is inherited from the father and the other from the mother

allogeneic unlike genetically, when applied to individuals of the same species

allopatric having separate and mutually exclusive areas of geographical distribution

allopolyploidy combining unrelated genomes **OR** state where contributing genomes are dissimilar; common in plants but not in animals

amino acid a subunit of protein. The sequence of amino acids in a protein is determined by the sequence of nucleotide bases in the DNA of a corresponding gene

aneuploidy any chromosome number that is not an exact multiple of the haploid number; usually refers to an extra or missing copy of a chromosome

antigen a foreign substance that, on introduction into a vertebrate animal, stimulates the production of homologous antibodies

artificial selection selection of specific forms by deliberate imposition of environmental pressures, e.g. in animal breeding, whole plant or *in vitro* culture

autopolyploidy multiplication of a basic set of chromosomes

autosomal relating to a gene carried on an autosome, i.e. any chromosome other than a sex chromosome

autosome any chromosome other than the sex chromosomes

auxotrophic any organism with a nutritional requirement for a specific substance

backcross cross between the heterozygous first filial (F_1) generation and the homozygous recessive parent, allowing different genotypes present in the F_1 to be distinguished

bacterial artificial chromosome (BAC) vector used to clone large DNA fragments (100–300 kb insert size; average, 150 kb) from another species into *E. coli* cells

bases the building blocks of DNA and RNA: adenine (A), cytosine (C), guanine (G), thymine (T), and in RNA only, uracil (U)

base pairs the pairs of complementary bases that form the structure of DNA: adenine (A) pairs with thymine (T), cytosine (C) pairs with guanine (G)

biallelic expression *see* expression

blastocyst mammalian embryo between the morula stage and gastrulation, comprising an outer sphere of trophectoderm cells surrounding an inner cell mass (epiblast and primitive endoderm layers) and a fluid-filled cavity (typically 32–128 cells)

carrier an individual who is heterozygous for a mutant allele but does not show the phenotype because of a dominant allele

cDNA (complementary DNA) DNA complementary to messenger RNA

cell line line of cells that can be propagated in culture indefinitely

centiMorgan the metric used to denote linkage distance. A centiMorgan (cM) is the distance between two genes that will undergo recombination with a frequency of 1%. Named after Thomas Hunt Morgan, who first conceptualized linkage while working with Drosophila

centromere point where two chromatids of a chromosome are joined; region of the chromosome attached to the spindle during cell division

chiasma (pl. chiasmata) junction point(s) between non-sister chromatids seen at the diplotene stage of meiosis, the consequence of a crossing-over event between maternal and paternally derived chromatids

chromatid one of the two side-by-side subunits of a prophase or metaphase chromosome, which become the daughter chromosomes when the chromosome splits at metaphase

chromatin nuclear complex composed of DNA and associated proteins such as histones

chromosome a self-replicating structure carrying the genetic material (genes and other DNA, sometimes with associated protein and RNA) in viruses and cells

cloning (i) isolating a gene or DNA fragment and using a vector molecule to produce multiple, identical copies for further study (ii) producing many genetically identical copies of an entire cell or organism

codominance the situation in which two different alleles are both expressed equally in the phenotype, i.e. neither allele is dominant or recessive

codon triplet of bases that codes for one amino acid or chain termination (*see also* stop (termination) codon)
consensus sequence in related DNA, RNA or protein sequences, the sequence that mirrors the most common choice of base or amino acid at each position
continuous variation variation between individuals in a population where changes are slight and grade into each other
cosmid cloning vector that consists of a bacterial plasmid into which the *cos* sequences of phage lambda have been inserted
crossing-over process of exchange of genetic material between homologous chromosomes during meiosis
cytogenetics the cytological (relating to the development or formation of cells) approach to genetics, mainly involving microscopic studies on chromosomes
cytoplasmic inheritance inheritance of genes on an extranuclear basis (e.g. via chloroplasts or mitochondria); usually inherited from the maternal parent only
deletion a mutation involving the loss of a DNA segment from a chromosome
differentiation in cells, the development of cells with specialized structure and function from unspecialized precursor cells; in an embryo, the increasing specialization of different parts in the development of a multicellular organism from an undifferentiated fertilized egg
dimorphism/dimorphic existing in two distinct forms
diploid possessing two copies of every chromosome except the sex chromosomes
directional selection selection that removes individuals from one end of a phenotypic distribution and results in a shift in the population mean (*see also* stabilizing selection)
DNA (deoxyribonucleic acid) the long chain of nucleotides that encodes genetic information and encodes the information for proteins. Consists of base pairs of nucleotides (adenine [A], guanine [G], cytosine [C], and thymine [T]) linked together. Most DNA is double-stranded
DNA marker *see* marker
DNA fingerprint the characteristic banding pattern produced when DNA is segmented using enzymes and sorted by gel electrophoresis. Due to a simple repeating unit scattered throughout the human genome, a DNA fingerprint is the same for every cell of a person, but highly variable between individuals
drift (genetic drift) change of allele frequency due to chance. Drift can be more important than selection in small populations or when the allele considered is nearly neutral
dominant term describing the relationship of one allele to a second at the same locus when an animal heterozygous for these alleles expresses the same phenotype as an animal homozygous for the first allele. The second allele of the pair is considered recessive
ectoderm outer of the three embryonic germ layers. The inner and middle layers are known as endoderm and mesoderm respectively
electrophoresis separation of molecules based on their mobility in an electric field (*see also* pulsed field gel electrophoresis)
embryonic stem (ES) cells pluripotent cell lines derived from the epiblast layer of the blastocyst. Capable of colonising all foetal tissues, including the germline, in chimaeras but usually not the primitive endoderm or trophectoderm derivatives. Previously used term "EK cells" is now redundant
endoderm inner of the three embryonic germ layers. The outer and middle layers are known as ectoderm and mesoderm respectively
enhancer a DNA sequence that will increase the level of expression from a gene in manner independent of position and orientation
enucleate to remove the nucleus from a cell; without a nucleus
epigenetic the sequence of the DNA encodes all the genetic information heritable from a parent. However, it appears that this is not the only route through which heritable information can travel. These other mechanisms are described as being epigenetic. The most familiar example of an epigenetic mechanism is DNA methylation
epistasis/epistatic describes a gene or character whose effect overrides that of another gene with which it is not allelic; analogous to dominant applied to genes at different loci. More generally, epistasis exists when the effect of two or more non-allelic genes in combination is not the sum of their separate effects
euchromatin regions of chromosomes, generally forming the greater part of the chromosome, that decondense at the end of telophase, and that contain most or all of the genes
eugenics controlled human breeding based on presumed desirable and undesirable genotypes
eukaryotes/eukaryotic multicellular (higher) organisms (e.g. fungi, yeasts, plants, animals), with true nuclei, separated from the cytoplasm by a two-membrane nuclear envelope
exons regions that flank introns in a gene or RNA and are retained in the mature RNA following the removal of introns
expression (monoallelic, biallelic) a term used to described the activity of a gene. An expressed gene is being actively transcribed to produce mRNA. This mRNA is translated to produce a functional protein. Most genes are expressed from both alleles (biallelic expression). Imprinted genes, in contrast, are only expressed from one parental allele (monoallelic expression)
F_1 hybrid in the first filial generation, the result of crossing of two dissimilar parents
founder effect describes genetic disorders common in isolated populations where all members are descended from a small number of ancestors, one or more of whom may have had the disorder
frequency of recombination for linked genes, which recombine less than 50% of the time, the frequency of recombination is a useful measure of the distance separating the two genes
gamete (micro- and macro-) germ cell containing a haploid number of chromosomes
gastrulation (i) in insects, gastrulation initiates the formation of the germ layers, i.e. the ectoderm, mesoderm

and endoderm; (ii) in mammals, gastrulation is the third stage in the development of the early embryo (preimplantation, implantation, gastrulation) and is the process where the body plan is set up

gene a region of a chromosome with a specific DNA sequence that encodes peptide or RNA; the unit of heredity physically transmitted from generation to generation resulting in inheritance of particular identifiable traits

gene conversion a meiotic process resulting in a genetic change of one allele to another

gene expression *see* expression

gene pool all genes present in a population

genera (sing. genus) groups of closely related species, classified into families

genetic code relationship between the sequence of bases in DNA (read as sets of three) and the order of amino acids of a polypeptide chain

genotype the genetic make-up of an individual

germ cells cells which produce the gametes (oocyte or spermatozoon). Genetic or epigenetic changes in these cells have the capability of being transmitted to offspring

germline line of cells from which gametes (ova and sperm) are produced

germplasm collection of varieties from which new improved varieties can be developed

Gram-negative bacteria bacteria with thin peptidoglycan walls bounded by an outer membrane containing endotoxin (lipopolysaccharide)

Gram-positive bacteria bacteria with thick cell walls containing teichoic and lipoteichoic acid complexed to the peptidoglycan

haploid possessing one set of unpaired chromosomes. Gametes are haploid

haplotype set of alleles on one of a pair of homologous chromosomes

hemizygote/hemizygous a gene present as only one copy in a diploid organism, cell or nucleus. For example, X-linked genes in a male mammal

heritability proportion of variation in a trait that is explained by additive genetic effects

heritable capacity for being transmitted from one generation to the next

heterochromatin regions of chromosomes that do not decondense at the end of the telophase, but remain condensed in the interphase nucleus. Often contains highly repeated DNA sequences and few, if any, genes

heterologous derived from tissues or DNA of a different species

heteromorphic having different forms at different times; in chromosomes can refer to chromosome pairs of different size

heterosis case where fitness traits are greater in heterozygote than either homozygote

heterozygote/homozygote (heterozygous/homozygous) a mutation can be present on one of the two alleles (it is said to be heterozygous/a heterozygote for the mutation) or on both alleles (homozygous/a homozygote for the mutation)

homeobox conserved DNA sequence originally detected by DNA–DNA hybridization in many of the genes that give rise to homeotic mutants and segmentation mutants in Drosophila

homeotic relating to mutations that transform part of the body into another part

homologue structure of similar evolutionary and developmental origin to another structure, but serving different functions

homologous recombination genetic recombination involving exchange of homologous loci

homozygote/heterozygote (homozygous/ heterozygous) a mutation can be present on one of the two alleles (it is said to be heterozygous/a heterozygote for the mutation) or on both alleles (homozygous/a homozygote for the mutation)

hybrid a heterozygote; an offspring from any cross involving parents of differing genotypes

hybridization (DNA hybridization) technique to determine similarity of two DNA strands whereby single-stranded nucleic acids are allowed to interact so that complexes (or hybrids) are formed by molecules with sufficiently similar complementary sequences

implantation *see* preimplantation

integration incorporation of the genetic material of a virus into the host genome

interphase period between one mitosis and the next, when no distinct chromosomes are visible and the chromatin is contained in a nucleus

introns internal regions of a gene or RNA that are spliced out of the primary transcript and therefore absent from the mature RNA

inversion chromosomal mutation where part of the chromosome is inverted

karyotype number, size and shape of chromosomes in a somatic cell; photomicrograph of chromosomes arranged in a standard manner

kinase (abbreviation of phosphokinase) any enzyme catalysing transfer of phosphate from ATP to an acceptor

linkage where two genes are located close together on the same chromosome

linkage group the group of genes that are located on the same chromosome

linkage map map of the sites of gene loci on a chromosome

locus (pl. loci) the position of a gene on a chromosome

mapping (genetic mapping) the production of a linear map of the relative position of genes along a chromosome

marker [genetic] any chromosome feature of known location whose inheritance can be easily detected by molecular techniques or phenotype. Genetic markers are used in mapping and in linkage analysis: genes closely linked to the marker will generally be inherited with it

mating type even in the case of isogamy in most species the gametes belong to two mating types: a "+" type gamete can fuse only with a "–" and vice versa

meiosis the process by which diploid organisms produce haploid germ cells

meiotic drive mechanism operating in heterozygotes during meiosis that results in a disproportionate representation in the gametes of one member of a chromosome pair

mesoderm middle of the three cell layers of the developing embryo. The inner and outer layers are known as endoderm and ectoderm respectively

messenger RNA (mRNA) found in all cells, this single-stranded RNA molecule acts as a template for protein synthesis by specifying the amino acid sequence of one or more polypeptide chains

metaphase the second phase of mitosis, or of one of the divisions of meiosis, when the chromosomes align on the equator of the cell (metaphase plate)

microsatellites arrays of DNA of variable length distributed evenly throughout the genome and composed of 2–5 base-pair sequences repeated many times over

missense mutation mutation that results in a change in an amino acid-specifying codon

mitochondrial DNA (mtDNA) the DNA contained within the mitochondria (small cytoplasmic organelles) of a cell, separate from the DNA in the nucleus. In mammals, mtDNA makes up less than 1% of the total cellular DNA, but in plants the amount is variable. mtDNA is maternally inherited

mitosis process of cell division that occurs in somatic cells of eukaryotes

molecular clock the theory that molecules evolve at an approximately constant rate. Any differences between a form of a molecule in two species can be used to infer the time elapsed since they shared a common ancestor

monoallelic expression *see* expression

mutation change in the DNA sequence of an organism

natural selection process whereby one genotype in a population survives and leaves more offspring than another genotype as a result of adaptation to the environment

neo-Darwinism contemporary version of Darwin's theory of evolution that encompasses the principles of genetics while emphasizing natural selection as a principal motivating force of evolution

neutral theory theory, first proposed by Motoo Kimura in 1968, that the majority of evolutionary changes at the molecular level are due to mutations whose effects on fitness are so minute that their dynamics are determined mainly by mutation and random genetic drift

non-reciprocal a recombination event between two markers where only one recombinant is produced (recombination between very closely linked markers tends to be non-reciprocal)

nonsense mutation mutation that results in the introduction of a termination codon and premature termination of polypeptide synthesis during translation (*see also* stop (termination) codon)

nucleic acid e.g. deoxyribonucleic acid (DNA) and ribonucleic acid (RNA); macromolecules formed from long chains of nucleotide subunits

nucleotide the structural unit of DNA or RNA molecules consisting of a nitrogenous base, a phosphate molecule, and a sugar molecule

oncogene genetic entity promoting tumour development

oocyte a stage in the development of the female gamete during which the progenitor primordial germ cell develops into an egg

open reading frame (ORF) stretch of DNA containing a signal for the start of translation, followed by an adequate number of amino acid-encoding triplets in the correct register to form a protein, followed by a signal for termination of translation

organelles membrane-bound intracellular structures having specialized functions

oxidation addition of oxygen and loss of hydrogen/electrons from a compound

palindrome a double-stranded nucleic acid (nucleotide) sequence that reads exactly the same from left to right as from right to left

pathogenic capable of causing disease

pathological relating to characteristic symptoms/signs of a disease

penetrance proportion of individuals with a particular genotype who show the associated phenotype

phenotype visible or measurable characteristics (physical, biochemical, physiological) of an individual resulting from interaction of environment and genotype

phyla (sing. phylum) in classification, a primary grouping composed of animals constructed on a similar general plan, considered to be evolutionarily related

phylogenetic tree diagram setting out the genealogy of a species

phylogeny tracing of evolutionary relationships by comparing DNA and protein sequences from different organisms

plasmid small, circular or linear DNA that replicates independently of the chromosome in bacteria and yeasts

plastid cellular organelle containing pigment, e.g. chloroplast in plants

pleiotropic [gene] a gene with multiple effects

ploidy number of chromosomes or DNA molecules in a cell or organelle (*see also* allopolyploidy, aneuploidy, autopolyploidy, diploid, haploid, polyploid, tetraploid, triploid)

point mutation mutation involving a change at a single base pair in DNA

polygenic controlled (or caused) by the action/interaction of many genes

polymerase chain reaction (PCR) method used to synthesize DNA starting from minute quantities of material

polymorphism in a population, the existence of two or more alleles of a gene with a frequency of at least 1%

polyploid nucleus, cell or organism having more than two haploid sets of chromosomes

population genetics study of the application of genetic principles to groups of interbreeding individuals (a population) as a whole

positional cloning isolation of a gene that commences from information about its location on a chromosome

preimplantation, implantation, gastrulation three stages in the development of the early embryo (in order of occurrence) where methylation changes have been observed. In preimplantation development there is cell division without cell growth. Implantation is the attachment of the embryo to the uterine wall allowing the placental connection to be made. Gastrulation is the process where the body plan is set up

prion (proteinaceous infectious particle) an abnormal protein, containing no nucleic acid, believed to be the agent responsible for spongiform encephalopathies, including scrapie and bovine spongiform encephalopathy (BSE; mad cow disease)
prokaryotes unicellular (lower) organisms (bacteria and cyanobacteria) with small, simple cells lacking membrane-bound organelles; DNA is usually of circular structure
promoter start site of transcription
prophase first stage of cell division where chromosomes first become visible
protein a large molecule composed of one or more chains of amino acids in a specific order; the order is determined by the base sequence of nucleotides in the gene coding for the protein
pseudogene non-functional DNA sequences very similar to the sequences of known genes
pulsed field gel electrophoresis molecular genetic (electrophoretic) technique particularly suited for separating large pieces of DNA
quantitative genetics study of the inheritance of characteristics that show continuous variation in a population. The differences may be determined by different, independently acting genes that each has only a small effect on a specific character, e.g. height or weight
quantitative trait locus (QTL) an area of a chromosome in which lies a gene believed to influence a quantitative trait
R loop single stranded loop section of DNA formed by the correlation of a section of ssRNA with the other strand of the DNA, whereby one DNA strand is displaced as the loop
recessive allele or mutation expressed phenotypically only when present in the homozygous state
reciprocal a recombination event between two markers where both recombinants are produced (recombination between distant markers tends to be reciprocal)
recombination through intermolecular exchange, the creation of chromosomes combining genetic information from different sources, usually two genomes of a given species
reduction addition of hydrogen/electrons from a compound
restriction fragment length polymorphism (RFLP) polymorphism due to the presence or absence of a particular restriction site
restriction enzyme bacterial enzyme that cuts DNA at specific sites
retrotransposon a transposable element with a transpositional mechanism requiring reverse transcriptase
retrovirus virus with a single-stranded RNA genome (Class VI)
reverse genetics method of ascertaining a gene's function by sequencing it, mutating it, and then attempting to identify the nature of the change in the phenotype
reverse transcriptase RNA-directed DNA polymerase
reverse transcription synthesis of DNA on an RNA template, catalysed by the enzyme reverse transcriptase
ribosomal RNA (rRNA) most abundant RNA species in cells, in eukaryotes it is synthesized in the nucleolus from rRNA genes tandemly repeated in the chromosomes
ribosome a heterodimeric multisubunit enzyme composed of ribonucleoprotein and protein subunits; interacts with aminoacylated tRNAs and mRNAs and translates protein coding sequences from mRNA
RNA (ribonucleic acid) a chain of repeating nucleotides similar to DNA but having ribose sugar rather than deoxyribose sugar and uracil rather than thymine as one of the bases. The genetic information encoded in DNA is translated by RNA into specific proteins
RNA polymerase an enzyme that is capable of synthesizing complementary RNA from a DNA template: normally producing messenger RNA from genes coding for proteins, or ribosomal and transfer RNA from their respective genes
segregate/segregation act of separation of parental homologous chromosomes at meiosis, subsequent separation of alleles of a gene at a locus and their transmission to different gametes
senescence advancing age; ageing process that leads to death
sex chromosomes the 23rd chromosome pair in a human karyotype that determines the sex of an individual. Females inherit an X from each parent, males get an X from the mother and a Y from the father
sex-linked genes carried on the sex chromosomes
sibs (siblings) brothers or sisters
somatic describes body cell(s) as opposed to cells of the germline
S-phase phase of the cell cycle when DNA replication takes place
speciation a process by which one species evolves into a different species or splits over time into two or more new descendant species
splicing removal of intervening sequences (introns) and joining of exons in RNA, i.e. introns are spliced out, exons are spliced together
stabilizing selection selection that removes individuals from both ends of a phenotypic distribution and thus maintains the same distribution mean (*see also* directional selection)
stop (termination) codon the three codons UAA (ochre), UAG (amber) and UGA (opal) do not code for an amino acid but act as signals for the termination of protein synthesis. They are not represented by any tRNA and termination is catalysed by protein release factors
taxonomy analysis of an organism's characteristics for the purpose of classification
telomere the tip of a chromosome
termination codon *see* stop codon
tetraploid nucleus, cell or organism possessing four sets of chromosomes
transcription transmission of genetic information from the DNA in the chromosomes to messenger RNA
transcription factor protein required for recognition by RNA polymerases of specific stimulatory sequences in eukaryotic genes
transfer RNA (tRNA) small RNA found in all cells.

Transports amino acids to the ribosomes for protein synthesis; each tRNA is specific for one amino acid only

transgene a gene which has been artificially introduced into the genome of an organism

translation process that occurs in the ribosome whereby genetic information from messenger RNA is translated into protein synthesis

translocation chromosomal rearrangement such that one chromosome or part of a chromosome is attached to another chromosome

transposable element *see* transposon

transposase protein specified by a transposon and responsible for its transposition

transposition movement from one location to another, especially movement of a DNA sequence (*see also* transposon)

transposon mobile genetic element with the ability to replicate and insert a copy at a new location in the genome

trinucleotide repeat repetitive part of a genome consisting of directly repeated copies of a three base-pair sequence (e.g. CAG/CTG)

triplet code three consecutive bases in DNA or RNA that encode an amino acid (*see* genetic code)

triploid nucleus, cell or organism having three haploid sets of chromosomes

vector common term for a plasmid that can be used to transfer DNA sequences from one organism to another

virus an infectious particle consisting of nucleic acid (DNA or RNA) covered by protein, and which is dependent on a host cell for replication

wild-type organism with the normal (unaltered) form of a gene

X-linked relates to any gene carried on the X chromosome

yeast artificial chromosome (YAC) large fragments of DNA maintained in yeast by fitting them with components (a centromere and telomeres) of a yeast chromosome

zinc finger hairpin fold of amino acid chain held together by a zinc atom, characteristic of numerous proteins that bind to DNA and function as transcriptional regulators

zygote fertilized ovum

APPENDIX: USEFUL WEB ADDRESSES

Web addresses are given in many of the individual articles as references for particular documents or projects. This appendix, though not comprehensive, provides a more general list of resources, per section of the Encyclopedia.

Web users may also find the News Feature "Souped-up search engines" in *Nature* 11 May 2000 helpful. Web sites and the data and other information they offer evidently multiply with time, and may have a short life or present unreliable information.

Section B: The origins of genetics

Mendel web
http://www.netspace.org/MendelWeb/homepage.html
As well as the origins of classical genetics and the history and literature of science, this site explores elementary plant science and introductory data analysis.

Evolution and population genetics
http://www.hoflink.com/~house/evolution.html
Includes information on the origin of life.

Section C: Genetics of bacteria and viruses

Bacterial genetics
http://www.medmicro.mds.qmw.ac.uk/underground/bactnat/bactgen.html
A good introduction to bacterial genetics.

http://www.bact.wisc.edu/MicrotextBook/BactGenetics/TOC.html
Another university course site containing information on bacterial genetics.

http://www.biol.unt.edu/~farinha/medbac99/genetics.html
Excellent site covering all aspects of bacterial genetics.

Magpie
http://www.mcs.anl.gov/home/gaasterl/genomes.html
The reference site for genome sequencing projects. As of February 2000 there are links to 20 completed bacterial genomes and 45 genomes in progress.

Bacterial genome projects
http://www.pasteur.fr/recherche/unites/tcruzi/minoprio/genomics/bacteria.htm
Jump station to current bacterial genome sequencing projects.

Prokaryotic genome databases
http://www.hgmp.mrc.ac.uk/GenomeWeb/prokaryote-gen-db.html
Another jump station to bacterial genome sequencing projects.

All the virology on the WWW
http://www.Tulane.EDU:80/~dmsander/garryfavweb.html
All you need to find information on viruses.

Institute for Molecular Virology
http://www.bocklabs.wisc.edu/Welcome.html
University of Wisconsin site containing links to good virology resources.

Viruses
http://www.uky.edu/~rebeat1/virus.html
University of Kentucky site containing links to viral genetics pages.

The Big Picture Book of Viruses
http://www.virology.net/Big_Virology/BVFamilyGenome.html
An excellent resource for virologists.

Virus genome projects
http://www.pasteur.fr/recherche/unites/tcruzi/minoprio/genomics/virus.htm
Jump station to all viral genome projects.

Section D: Genetics of Drosophila and other insects

Flybase
http://flybase.bio.indiana.edu:82/
The only site needed for information on Drosophila genes and current research.

Flyview
http://pbio07.uni-muenster.de/
Comprehensive image database for Drosophila.

Mutant fruit flies
http://www.exploratorium.edu/exhibits/mutant_flies/mutant_flies.html
An excellent site for those unfamiliar with fruit flies, who would like to know the basics.

Section E: Genetics of eukaryotic micro-organisms and their organelles

Structure and function of organelles
http://esg-www.mit.edu:8001/esgbio/cb/org/organelles.html
Excellent site full of information and links on the structure and function of organelles.

Organelle genome resources
http://www.ncbi.nlm.nih.gov/PMGifs/Genomes/organelles.html
This site currently contains 181 reference sequences of eukaryotic organelles and information on organelle genomes is updated regularly.

Fungal genetics stock centre
http://www.hgmp.mrc.ac.uk/research/fgsc/main.html
Jump station to all fungal collections, cloned genes and libraries as well as methods, recipes and hints and online bibliographies.

Candida albicans information
http://alces.med.umn.edu/Candida.html
Links to information on sequencing of *Candida albicans* as well as methods for Candida genetics. Also links to other yeast and mycology sites.

The Fungal Web
http://helios.bto.ed.ac.uk/bto/microbes/fungalwe.htm#crest
Edinburgh University site containing an excellent range of information on fungi.

Section F: Mouse genetics

The Jackson Laboratory
http://www.jax.org/resources/documents/
List of mutant resources, backcrossing, cytogenetics, model resources and much more.

The mouse atlas and gene expression database
http://genex.hgu.mrc.ac.uk/
New site containing the mouse atlas page of development and also a gene expression database.

Mouse genetic map information
http://www-genome.wi.mit.edu/genome_data/mouse/mouse_index.html
Contains a comprehensive list of links along with "clickable maps" for all of the mouse chromosomes.

Section G: Genetics of other mammals and birds

Online Mendelian Inheritance in Animals (OMIA)
http://angis.su.oz.au/Databases/BIRX/omia/
A database of genes that have been documented in a wide range of animal species other than those for which databases already exist.

Genetics of budgerigars
http://www.budgerigars.co.uk/genetics/index.html
All the information you could need on the genetics of budgerigars.

Gerbil genetics
http://www.rodent.demon.co.uk/gerbils/genetics.htm
Up to date information on gerbil mutations as well as links to other rodent genetics pages.

Roslin Institute
http://www.roslin.ac.uk
Sheep, poultry and pig genetics, in addition to the cloning and nuclear transfer research made famous by Dolly the sheep.

Feline information page
http://www.best.com/~sirlou/cat.shtml
Good introduction to cat genetics containing information on feline evolution and breeding.

Dog genetics
http://dmoz.org/Recreation/Pets/Dogs/Genetics/
Jump station to sites on all aspects of dog genetics from breeding to genetic disorders.

Canine genetics resources
http://www.workingdogs.com/genetics.htm
Another good jump station to pages on dog genetics including links to the dog genome projects.

Horse genetics
http://www.vgl.ucdavis.edu/~lvmillon/
Links to all you need to know about horse genetics from coat colour to karyotyping.

Section H: Human clinical genetics

Genetic professional societies
http://www.kumc.edu/gec/prof/soclist.html
List of all councils, boards or societies involved in human medical and clinical genetics, grouped by region.

Rare genetic diseases in children
http://mcrcr2.med.nyu.edu/murphp01/homenew.htm
Jump station to information on genetic disorders in children, and a resource directory.

UK Genetic Interest Group
http://www.gig.org.uk
Information on a range of genetic disorders and national contact groups.

Genetic Alliance
http://www.geneticalliance.org/
The American equivalent to the UK Genetic Interest Group

British Society for Human Genetics
http://www.bshg.org.uk/
Contains links to current research projects of human genetic conditions, to information for patients and also anyone interested in a career in research.

European Directory of DNA Laboratories
http://www.eddnal.com/scripts/request.idc?
Directory of laboratories that deal with DNA testing for genetic disorders.

Online Mendelian Inheritance in Man
http://www.ncbi.nlm.nih.gov/Omim/
An online catalogue of human genes and genetic disorders.

Webliography
http://www.faseb.org/genetics/webliog.htm
Excellent set of links for clinical geneticists.

Gene cards
http://bioinformatics.weizmann.ac.il/cards/
Database of human genes, their products and involvement in disease.

Gene clinics
http://www.geneclinics.org/
Links to gene-specific information, genetic laboratories, clinics and educational material.

The Genetics Forum
http://www.geneticsforum.org.uk
Forum for all aspects of human genetical research.

Genetics and Insurance Committee
http://www.doh.gov.uk/genetics/gaic.htm
UK Department of Health site discussing use of genetic testing in insurance risk assessment.

Human Genetics Advisory Commission
http://www.dti.gov.uk/hgac/papers.htm
UK department of trade and industry site discussing implications of genetic testing on employment and insurance and also implications of cloning technologies.

Section J: Plant genetics

Plant Breeding Laboratory
http://www.spg.wau.nl/pv/nindex.htm
Wageningen laboratory web site that includes free downloads of genetics and breeding-related software.

Cold Spring Harbor Laboratory
http://www.cshl.org/
Excellent resource not only for plant genetics but also for anything you need to know about DNA from the basics to current and future techniques.

Corn Breeding and Genetics Home Page
http://www.plant.uoguelph.ca/research/corn_breeding/
Site maintained at the University of Guelph containing links for all of your corn questions.

Bionet
http://www.bionet.nsc.ru/ICIG/plant/
A very comprehensive site covering all aspects of wheat genetics from heterosis, mutagenesis and cytology to genetic engineering and developmental genetics of plants.

Section K: Genetics of cell organelles, structures and function

Molecular genetics jump station
http://www.horizonpress.com/gateway/genetics.html
The only page needed for links to every aspect of molecular genetics.

Functional genomics jump station
http://genomics.co.uk/
Another excellent site for links to information and tools for functional genomics.

The Bio-Web
http://cellbiol.com/
Jump station to all the tools you need for molecular biology.

Lab Velocity
http://www.labvelocity.com
A further site containing links to literature, protocols, and other molecular genetics tools.

Section L: DNA-based genetic analysis and biotechnology

DNA from the beginning
http://vector.cshl.org/dnaftb/
Online textbook covering genetic inheritance, DNA analysis and much more. Also from the Cold Spring Harbor Laboratory.

Homeobox page
http://copan.bioz.unibas.ch/homeo.html
Anything you want to find out about homeobox genes may be found from this site.

Online protocols
http://www.protocol-online.net/
Online protocols and discussion forums for molecular and cellular biology.

Current Protocols
http://www.wiley.com/cp/
Publisher-run site containing links to protocols in cell and molecular biology, cytometry, immunology, pharmacology, human genetics and more.

The DNA Diagnostic Institute
http://www.tddi.org/althomepage.htm
A good site for information on forensic techniques and the application of molecular techniques to monitor the welfare of both humans and wildlife.

Section M: Model organisms

Wormbase
http://www.wormbase.org
Brings together information on *Caenorhabditis elegans*, the leading model oranism in developmental biology.

Fugu Genome Project
http://jgi.doe.gov/tempweb/programs/fugu.htm
Information about the International Fugu Sequencing Consortium, at pages hosted by the DOE Joint Genome Institute.

The Arabidopsis Information Resource
http://www.arabidopsis.org/home.html
A comprehensive resource, including a searchable database, news and useful links.

The Biology Project
http://www.biology.arizona.edu/developmental_bio/developmental_bio.html
University of Arizona site containing articles and tutorials on brain development and links to "the virtual embryo" and "the visible embryo". Also links to the developmental biology page at Loyola, Chicago, and the "Frog2" page of amphibian development.

Zygote
http://zygote.swarthmore.edu/
A virtual library of developmental biology

Centre for Developmental Genetics
http://www.shef.ac.uk/~biomsc/research/links.html
Based at the University of Sheffield this site contains good links to their own developmental genetics laboratories as well as links to journal sites and other molecular biology pages.

Section N: Population genetics and evolutionary studies

Human Genome Project
http://www.ornl.gov/TechResources/Human_Genome/home.html
Links to comparative and functional genomics pages, and to the primary Human Genome Project sequencing sites.

PopGen
http://cc.oulu.fi/~jaspi/popgen/popgen.htm
Simulation programme designed to describe population genetic events.

Evolution on the internet
http://acad.uwsuper.edu/morden/WebClass/305web/Eweb.htm
An excellent site covering all aspects of evolution.

Nearctica
http://www.nearctica.com/evolve/evolve.htm
A comprehensive site containing an introduction to evolution as well as information on the principles of evolution and evolutionary theory.

Molecular evolution and population genetics
http://www-hto.usc.edu/papers/abstracts/lists/molecularEvolution.html
A useful site for publications on molecular evolution and population genetics.

Population genetics and evolution
http://library.adelaide.edu.au/guide/sci/Biochemistry/popevol.html
Adelaide University site containing biolinks, phylogenetics resources, glossaries and much more.

Evolution and population genetics
http://www.hoflink.com/~house/evolution.html
Another comprehensive site containing information on the basic evolution, taxonomy, fossil records and hyperlinks to much more.

GENERAL

Genetics WWW resources
http://www.gen.cam.ac.uk/Library/geneticswww.html
Comprehensive list covering links to all aspects of genetics.

Genetics
http://www.biology.about.com/cs/genetics/
Containing links to over 700 sites, this web page is all that is needed for information on genetics.

Statistical genetics websites
http://www.rdg.ac.uk/~sns99kla/links.html
Links to websites of interest to statistical geneticists.

BioNetbook
http://www.pasteur.fr/recherche/BNB/bnb-en.html
An excellent search tool for software, courses analysis tools, or any information on genetics, biochemistry or molecular biology of humans, bacteria, zebrafish and much more. Maintained by the Institut Pasteur.

The Biologist's Assistant
http://biocfarm.unibo.it/~bassist/
Another excellent site containing links to biological science databases, newsgroups, universities and research centres as well as sites for developmental biology, genetics, evolution, ecology, science history and statistical resources.

INDEX

When page numbers are shown in **bold** this indicates that these are subjects of individual essays. The plate section is also indexed. Common terms not indexed may be found in the Glossary.